Family						
Ether	**Amine**	**Aldehyde**	**Ketone**	**Carboxylic Acid**	**Ester**	**Amide**
CH_3OCH_3	CH_3NH_2	$\overset{\displaystyle O}{\underset{\displaystyle }{CH_3CH}}$	$\overset{\displaystyle O}{\underset{\displaystyle }{CH_3CCH_3}}$	$\overset{\displaystyle O}{\underset{\displaystyle }{CH_3COH}}$	$\overset{\displaystyle O}{\underset{\displaystyle }{CH_3COCH_3}}$	$\overset{\displaystyle O}{\underset{\displaystyle }{CH_3CNH_2}}$
Methoxy-methane	Methan-amine	Ethanal	Propanone	Ethanoic Acid	Methyl ethanoate	Ethanamide
Dimethyl ether	Methyl-amine	Acetal-dehyde	Acetone	Acetic acid	Methyl acetate	Acetamide
ROR	RNH_2 R_2NH R_3N	$\overset{\displaystyle O}{\underset{\displaystyle }{RCH}}$	$\overset{\displaystyle O}{\underset{\displaystyle }{RCR}}$	$\overset{\displaystyle O}{\underset{\displaystyle }{RCOH}}$	$\overset{\displaystyle O}{\underset{\displaystyle }{RCOR}}$	$\overset{\displaystyle O}{\underset{\displaystyle }{RCNH_2}}$ $\overset{\displaystyle O}{\underset{\displaystyle }{RCNHR}}$ $\overset{\displaystyle O}{\underset{\displaystyle }{RCNR_2}}$
$-\overset{\mid}{\underset{\mid}{C}}-O-\overset{\mid}{\underset{\mid}{C}}-$	$-\overset{\mid}{\underset{\mid}{C}}-\overset{\mid}{N}-$	$\overset{\displaystyle O}{\underset{\displaystyle }{-\overset{\mid}{\underset{\mid}{C}}-H}}$	$-\overset{\mid}{\underset{\mid}{C}}-\overset{\displaystyle O}{\underset{\displaystyle }{\overset{\mid}{C}}}-\overset{\mid}{\underset{\mid}{C}}-$	$\overset{\displaystyle O}{\underset{\displaystyle }{-\overset{\mid}{\underset{\mid}{C}}-OH}}$	$\overset{\displaystyle O}{\underset{\displaystyle }{-\overset{\mid}{\underset{\mid}{C}}-O-\overset{\mid}{\underset{\mid}{C}}-}}$	$\overset{\displaystyle O}{\underset{\displaystyle }{-\overset{\mid}{\underset{\mid}{C}}-\overset{\mid}{N}-}}$

ORGANIC CHEMISTRY

ACQUISITIONS EDITOR   Nedah Rose
SENIOR MARKETING MANAGER   Catherine Faduska
SENIOR PRODUCTION EDITOR   Suzanne Magida
TEXT DESIGNER   Madelyn Lesure
COVER DESIGNER   Carolyn Joseph
COVER ART   Joseph W. Lauher
MANUFACTURING MANAGER   Mark Cirillo
PHOTO RESEARCHER   Lisa Passmore
ILLUSTRATION EDITOR   Sigmund Malinowski

This book was set in 10.5/12 Times Roman by Progressive Information Technologies and printed and bound by Von Hoffmann Press. The cover was printed by Phoenix color corp.

Electronic Illustrations were provided by Fine Line; new spectra courtesy of Craig Fryhle.

Library of Congress Cataloging in Publication Data:

Solomons, T. W. Graham.
    Organic chemistry / T. W. Graham Solomons. — 6th ed.
        p. cm.
    Includes index.
    ISBN 0-471-01342-0 (cloth : alk. paper)
    1. Chemistry, Organic.   I. Title.
QD253.S65   1996   95-16672
547—dc20   CIP

Printed in the United States of America

10 9 8 7 6 5 4 3 2 1

SIXTH EDITION

ORGANIC CHEMISTRY

T. W. GRAHAM SOLOMONS

University of South Florida

JOHN WILEY & SONS, INC.

New York   Chichester   Brisbane   Toronto   Singapore

FOR JUDITH

ABOUT THE AUTHOR

T. W. GRAHAM SOLOMONS did his undergraduate work at The Citadel and received his doctorate in organic chemistry in 1959 from Duke University where he worked with C. K. Bradsher. Following this he was a Sloan Foundation Postdoctoral Fellow at the University of Rochester where he worked with V. Boekelheide. In 1960 he became a charter member of the faculty of the University of South Florida and became Professor of Chemistry in 1973. In 1992 he was made Professor *Emeritus.* In 1994 he was a visiting professor with the Faculté des Sciences Pharmaceutiques et Biologiques, Université René Descartes (Paris V). He is a member of Sigma Xi, Phi Lambda Upsilon, and Sigma Pi Sigma. He has received research grants from the Research Corporation and the American Chemical Society Petroleum Research Fund. For several years he was director of an NSF-sponsored Undergraduate Research Participation Program at USF. His research interests have been the areas of heterocyclic chemistry and unusual aromatic compounds. He has published papers in the *Journal of the American Chemical Society,* the *Journal of Organic Chemistry,* and the *Journal of Heterocyclic Chemistry.* He has received several awards for distinguished teaching. His organic chemistry textbooks have been widely used for 20 years and have been translated into Japanese, Chinese, Korean, Malaysian, Arabic, Portuguese, Spanish, and Italian.

He and his wife Judith have a daughter who is a geophysicist and two younger sons.

TO THE STUDENT

A Study Guide for the textbook is available through your college bookstore under the title **Study Guide to accompany *ORGANIC CHEMISTRY,* Sixth Edition** by **T. W. Graham Solomons.** The Study Guide can help you with course material by acting as a tutorial, review, and study aid. If the Study Guide is not in stock, ask the bookstore manager to order a copy for you.

About The Cover

The molecule on the cover is called dopamine. Although this is a relatively simple organic molecule, it is a major neurotransmitter in the brain. Dopamine plays a pivotal role in the regulation and control of movement, motivation, and cognition. It also is closely linked to reward, reinforcement, and addiction. Abnormalities in brain dopamine are associated with many neurological and psychiatric disorders including Parkinson's disease, schizophrenia, and substance abuse. This close association between dopamine and neurological and psychiatric diseases and with substance abuse make the dopamine system an important topic in neurosciences research and an important target for drug development.

PREFACE

This new edition has the same goal as its predecessors: *to provide students with the most current and effective text possible for studying organic chemistry, an important and fascinating subject.*

As with previous editions, I have been guided in my revision by many suggestions from students and colleagues. My aim, as always, has been to provide a book that will emphasize the biological, medical, and environmental applications of organic chemistry that are so effective in stimulating the interest of students. I also want to provide a text that will bring real functional group chemistry into the first term of the course, that will — through a series of well-placed special topics — provide instructors with greater flexibility in designing their course, and that will provide opportunities for inquisitive students to delve deeper into subjects that are often passed over lightly.

I also believe it is vitally important that we provide our students with the means *to master the fundamentals of the subject.* It is important that we teach them organic chemistry itself, not just what organic chemistry is about. For this reason, I have placed even more emphasis on detailed explanations of mechanisms, structure, and theory in this new edition. I have also provided many new opportunities for students to apply what they have learned by problem solving.

Organization

Organic chemists know that two aspects of their field, more than any others, make their subject comprehensible to students: the fact that the properties of organic compounds can be explained on the basis of their functional groups, and that the reactions of organic compounds can be explained in terms of their mechanisms. The basic organization of this text combines the best aspects of the traditional *functional group approach* with the *reaction mechanisms approach.* The primary organization is by functional groups. In most instances, the mechanisms that unify the underlying chemistry are presented in the context of a chapter devoted to a particular functional group. For example, nucleophilic substitutions and eliminations are introduced in the context of alkyl halides, electrophilic additions in a chapter on alkenes, nucleophilic additions in a chapter on aldehydes and ketones, and so on. I do not follow the traditional method of describing radical chemistry in the chapter on alkanes, however. Because ionic reactions are fundamentally simpler, I introduce them first beginning with acid–base chemistry and simple nucleophilic substitution and elimination reactions. Because I want to give radical chemistry a broader scope, I have a separate chapter (Chapter 9) on radical chemistry.

One other aspect of the organization of my book that sets it apart is an introduction to the mechanisms of organic reactions (Chapter 3), which is set in the context of acid–base chemistry. There are several reasons for doing this. Acid–base chemistry is so fundamental it finds its way into almost every chapter that follows. When looked at in the broadest sense of Lewis acids and bases, most organic reactions are acid–base reactions. Acid–base reactions, moreover, are relatively simple, and they are reactions that students are familiar with from their general chemistry course. They also lend themselves to an introduction of several important topics that students need to know about early in the course: (1) the curved arrow notation for illustrating mechanisms; (2) free-energy changes and their relationship to equilibrium constants; (3) enthalpy and entropy changes and how they affect reactions under equilibrium control; (4) inductive and resonance effects; and (5) solvent effects.

New Features in this Edition

The reception given the fifth edition of *Organic Chemistry* made it the most widely used edition so far. Since its publication four years ago I have received helpful comments from users of that edition. In addition, Wiley has obtained many extensive and thoughtful reviews to guide me in the preparation of this new edition. With this aid I believe I have improved my text substantially. There are many changes, but the most important ones are:

- An increased emphasis on reaction mechanisms
- An earlier presentation of organic reactions
- More than 350 new problems
- Spectroscopy now comes earlier and the chapter has been extensively revised to include new 300 MHz FT NMR spectra
- New and updated material emphasizes biological applications
- Carbonyl chemistry is now covered in consecutive chapters
- A new Special Topic on Two-Dimensional NMR Techniques
- The chapter on radical reactions has been moved to a later position
- A glossary has been added

Reaction Mechanisms

I now discuss mechanisms earlier (in Chapter 3), and have revised all of the representative mechanisms in the book to give them more prominence and more explanation. All of these important mechanisms are given a special presentation called "**A Mechanism for the Reaction**." This new format includes much more step-by-step detail, with careful attention to the use of curved arrows, and, in most instances, with accompanying explanations of each step of the mechanisms.

Early Presentation of Organic Reactions

Responding to several users who thought that my book should have more material on reactions earlier, I have revised and expanded Chapter 3 in order to accomplish this. This chapter, which was very well received in the fifth edition, still has all of the acid–base chemistry it had then. Now, however, the chapter goes further. After an introduction to the different types of organic reactions and a discussion of the meaning and importance of mechanisms, the chapter gives many more examples of organic acid–base reactions. It then uses these examples and the principles developed in the chapter to present an introductory mechanism. This example, the reaction of *tert*-butyl alcohol with concentrated HCl, is, I believe, especially appropriate as a leadoff mechanism, because each of the three steps of the mechanism is a simple acid–base reaction. This simple example reinforces the central point of the chapter: the importance of acid–base chemistry in organic chemistry.

New Problems

The sixth edition has more than 350 new problems in it. Because many problems have multiple parts, these problems provide as many as 1000 new exercises for students to work. As with earlier editions, these new problems range from drill problems to challenging problems that require considerable thought and that require the students to draw on a variety of concepts learned in the chapter at hand and on material learned earlier.

Spectroscopy

The placement of the chapter on spectroscopy was an item that produced split recommendations from the reviewers. After much thought, I have decided to move the chapter forward (Chapter 13) so that it comes before the two chapters on aromatic compounds. This avoids the awkward splitting of these two chapters by a chapter on spectroscopy that caused some concern in the fifth edition.

In order to allow spectroscopy to be brought forward, I have expanded the discussion of aromatic compounds in Chapter 2 so that the students will be prepared to examine spectra of compounds having aromatic groups.

The chapter on spectroscopy has been written in a way *that will allow instructors considerable latitude in when they cover* it in their courses. Some instructors may want to cover spectroscopy very early, as early as after Chapter 4. This should not present any problems, because all of the structural ideas needed to deal with spectroscopy will have been presented in the first four chapters. Other instructors may want to delay coverage of the chapter on spectroscopy until the second semester, and still others, according to the reviews, prefer to allocate coverage of spectroscopy to the lab course. All of these approaches will now be possible.

The chapter on spectroscopy has been revised extensively in order to present new FT methods of ^{13}C spectroscopy. Almost all of the ^1H spectra have been replaced with new state-of-the-art 300 MHz FT spectra as well. These new ^1H spectra have the advantage of much greater clarity because signal separation is much greater. To avoid the problem of compression of signals associated with spectra run at high frequencies, all important signals are shown in an expansion above the main spectrum. All of the new ^1H spectra have superimposed integral curves.

The new technique of generating DEPT (Distortionless Enhanced Polarization Transfer) spectra on FT NMR instruments is also described briefly, and data from DEPT spectra are included in most ^{13}C NMR spectra. The old technique of including multiplicities from proton-off-resonance-decoupled spectra has been abandoned because this technique is now little used in practice.

New sections on spectroscopy have been added to several chapters later in the book.

New Material and Biological Applications

As noted earlier, this new edition of *Organic Chemistry* incorporates some important new material: new methods for adding HX to alkenes and alkynes (based on the work of Paul J. Kropp at the University of North Carolina) are described in Chapter 8, the use of magnesium monoperoxyphthalate as a safer method for epoxidation is discussed in Chapter 10, and a new subsection on fullerenes is given in Chapter 14.

I am always looking for new ways to show the biological, medical, and industrial applications of organic chemistry. For example, there is now a new subsection on enantioselective synthesis in Chapter 5 as well as two other new sections in the same chapter: one discussing the biological importance of chirality and another describing chiral drugs. Chapter 7 contains new material on alkenes in industry and nature, and a subsection on hydrogenation in the food industry. Chapter 20 contains new material on neurotransmitters, and a subsection in Chapter 24 describes the p53 protein said to be ''the guardian of the genome'' and designated by the journal *Science* as ''molecule of the year'' in 1993.

Carbonyl Chemistry

Responding to suggestions from reviewers, I have rearranged the chapters on carbonyl chemistry so as to present them in four consecutive chapters (16–19) and two special topics (H and I). The chapter on amines now comes after the chapter on β-dicarbonyl compounds instead of before it. This will give students a more coherent view of this important chemistry.

Special Topic on Two-Dimensional NMR Techniques

This new special topic, contributed by Craig Fryhle of Pacific Lutheran University, follows the chapter on spectroscopy. It provides a clear and brief introduction to the important COSY and HETCOR techniques of 2-D NMR. As with other special topics, this one will allow students and instructors, who choose to do so, an oppportunity to go beyond the material usually covered in a first-year course.

Radical Reactions

Several reviewers thought that the introduction of radical reactions should be postponed so as to allow students to consolidate their understanding of ionic reactions thoroughly before launching into a new type of reaction. Persuaded by this idea (one that I had used in the 4th Edition), I have moved the chapter on Radical Reactions to a later position (it is now Chapter 9, not Chapter 7). The one slight problem this presents is that it does not allow a discussion of the mechanism of anti-Markovnikov addition of HBr to alkenes when it is first presented in Chapter 8. While this change may have this disadvantage, I believe the advantages that will accrue from it will be more important.

Glossary

A glossary of more than 350 important terms is given in the appendix. This is new to this edition.

Study Guide

A *Study Guide for Organic Chemistry, 6th Edition* contains explained solutions to all of the problems in this text.

The Study Guide also contains a quiz for each chapter that students can use in preparing for the quizzes their instructor gives. In addition it has sections describing the calculation of molecular formulas and a set of molecular model exercises.

Instructor's Supplements

To aid instructors, a Test Bank is available, containing more than 1600 questions. The Test Bank is available in both a printed and a computerized (Macintosh and IBM) version. Also available is a set of color overhead transparencies and ChemGraphics, designed by Darrell Woodman of the University of Washington. This helps to explain chemical concepts involving complex spatial relationships and three-dimensional structures and dynamic processes. The units are available on a CD-ROM for both IBM and Macintosh platforms.

T. W. GRAHAM SOLOMONS

ACKNOWLEDGMENTS

I am especially grateful to the following people who provided the reviews that helped me prepare this new edition of *Organic Chemistry.*

George Bandik
University of Pittsburgh

Kenneth Berlin
Oklahoma State University

Bruce Branchaud
University of Oregon

Ed Brusch
Tufts University

Sidney Cohen
Buffalo State College

Randolph Coleman
College of William & Mary

Eric Edstrom
Utah State University

Craig Fryhle
Pacific Lutheran University

Brad Glorvigen
University of St. Thomas

Frank Guziec
New Mexico State University

Kenneth Hartman
Geneva College

Michael Hearn
Wellesley College

John Jewett
University of Vermont

A. William Johnson
University of North Dakota

Paul Langford
David Lipscomb University

Patricia Lutz
Wagner College

Paul Papadopoulos
University of New Mexico

Michael Richmond
University of North Texas

Christine Russell
College of DuPage

Warren Sherman
Chicago State University

Jean Stanley
Wellesley College

Frank Switzer
Xavier University

Arthur Watterson
U. Massachusetts-Lowell

Mark Welker
Wake Forest University

Darrell Woodman
University of Washington

I am also grateful to the many people who have provided the reviews that have guided me in preparing earlier editions of my textbooks:

Winfield M. Baldwin
University of Georgia

David Ball
California State University, Chico

Paul A. Barks
North Hennepin State Junior College

Ronald Baumgarten
University of Illinois at Chicago

Harold Bell
Virginia Polytechnic Institute
and State University

Newell S. Bowman
The University of Tennessee

Wayne Brouillette
University of Alabama

Edward M. Burgess
Georgia Institute of Technology

Robert Carlson
University of Minnesota

George Clemans
Bowling Green State University

William D. Closson
State University of New York at Albany

Brian Coppola
University of Michigan

Phillip Crews
University of California, Santa Cruz

James Damewood
University of Delaware

O. C. Dermer
Oklahoma State University

Phillip DeShong
University of Maryland

Trudy Dickneider
University of Scranton

Paul Dowd
University of Pittsburgh

Robert C. Duty
Illinois State University

Stuart Fenton
University of Minnesota

Gideon Fraenkel
The Ohio State University

Jeremiah P. Freeman
Notre Dame University

M. K. Gleicher
Oregon State University

Roy Gratz
Mary Washington College

Wayne Guida
Eckerd College

Philip L. Hall
Virginia Polytechnic Institute
and State University

Lee Harris
University of Arizona

John Helling
University of Florida

William H. Hersh
University of California, Los Angeles

Jerry A. Hirsch
Seton Hall University

John Hogg
Texas A & M University

John Holum
Augsburg College

Robert G. Johnson
Xavier University

Stanley N. Johnson
Orange Coast College

John F. Keana
University of Oregon

David H. Kenny
Michigan Technological University

Robert C. Kerber
State University of New York
at Stony Brook

Karl R. Kopecky
The University of Alberta

Paul J. Kropp
University of North Carolina
at Chapel Hill

Michael Kzell
Orange Coast College

John A. Landgrebe
University of Kansas

Allan K. Lazarus
Trenton State College

Philip W. LeQuesne
Northeastern University

Robert Levine
University of Pittsburgh

Samuel G. Levine
North Carolina State University

John Mangravite
West Chester University

Jerry March
Adelphi University

John L. Meisenheimer
Eastern Kentucky University

Gerado Molina
Universidad de Puerto Rico

Everett Nienhouse
Ferris State College

John Otto Olson
Camrose Lutheran College

William A. Pryor
Louisiana State University

Thomas R. Riggs
University of Michigan

Frank Robinson
University of Victoria, British Columbia

Stephen Rodemeyer
California State University, Fresno

Yousry Sayed
University of North Carolina
at Wilmington

Jonathan Sessler
University of Texas at Austin

John Sevenair
Xavier University of Louisiana

Doug Smith
University of Toledo

Ronald Starkey
University of Wisconsin — Green Bay

James G. Traynham
Louisiana State University

Daniel Trifan
Fairleigh Dickinson University

James Van Verth
Canisius College

Darrell Watson
GMI Engineering and Management
Institute

Desmond M. S. Wheeler
University of Nebraska

James K. Whitesell
The University of Texas at Austin

Joseph Wolinski
Purdue University

Darrell J. Woodman
University of Washington

Herman E. Zieger
Brooklyn College

I thank my colleagues at the University of South Florida for the many helpful suggestions that they have offered over the years. In this regard, I think especially of: Raymond N. Castle, Jack E. Fernandez, George R. Jurch, Leon Mandell, George R. Newkome, Terence C. Owen, George R. Wenzinger, and Robert D. Whitaker.

Craig Fryhle of Pacific Lutheran University has been invaluable. I cannot thank him enough. He not only contributed the fine special topic on 2-D NMR, he provided a detailed and especially helpful review, he provided all of the new NMR and IR spectra contained in this edition, and he corrected the page proofs. He was also a constant source of advice.

I am also much indebted to Joseph Lauher of the State University of New York at Stony Brook for providing the splendid computer graphics used on the cover of the book and to open all of the chapters and special topics. I also thank Bill Fowler of SUNY Stony Brook for all of the help he has given me, and I thank Joanna Fowler of the Brookhaven National Laboratories, for suggesting dopamine as a cover molecule and for contributing the caption for it.

I also thank Bob Johnson of Xavier University for all of the help he has given me once again. He contributed new problems, he read the galley proofs, and he prepared the new test bank, which will accompany this edition.

I am very grateful to Henri-Philippe Husson of the Faculté des Sciences Pharmaceutiques et Biologiques at the Université René Descartes (Paris V) for his kind invitation to visit with his faculty. This visit gave me an opportunity to learn much more about chiral synthesis. I also thank him and his colleagues for the generous hospitality they provided, and for the opportunity to use their fine library.

I am much indebted to many people at Wiley for their help, especially Nedah Rose, Chemistry Editor; Suzanne Magida, Senior Production Editor; Madelyn Lesure, Design Director; Sigmund Malinowski, Illustration Editor; and Joan Kalkut, Associate Editor.

And finally, I thank my son Allen for generating on the computer many of the new reaction mechanisms included in this edition. And, once again I thank my wife, Judith Taylor Solomons, for her encouragement, for her editing, and for her proofreading.

T. W. GRAHAM SOLOMONS

CONTENTS

CHAPTER 3

AN INTRODUCTION TO ORGANIC REACTIONS: ACIDS AND BASES

CHAPTER 4

ALKANES AND CYCLOALKANES. CONFORMATIONS OF MOLECULES

CHAPTER 5
STEREOCHEMISTRY. CHIRAL MOLECULES

CHAPTER 6
IONIC REACTIONS — NUCLEOPHILIC SUBSTITUTION AND ELIMINATION REACTIONS OF ALKYL HALIDES

CHAPTER 7
ALKENES AND ALKYNES I. PROPERTIES AND SYNTHESIS

CHAPTER 8
ALKENES AND ALKYNES II. ADDITION REACTIONS

CHAPTER 9
RADICAL REACTIONS

CHAPTER 10
ALCOHOLS AND ETHERS

CHAPTER 11

ALCOHOLS FROM CARBONYL COMPOUNDS, OXIDATION–REDUCTION AND ORGANOMETALLIC COMPOUNDS

CHAPTER 12

CONJUGATED UNSATURATED SYSTEMS

CHAPTER 13

SPECTROSCOPIC METHODS OF STRUCTURE DETERMINATION

CHAPTER 17
ALDEHYDES AND KETONES II. ALDOL REACTIONS

CHAPTER 18
CARBOXYLIC ACIDS AND THEIR DERIVATIVES. NUCLEOPHILIC SUBSTITUTION AT THE ACYL CARBON

CHAPTER 19

SYNTHESIS AND REACTIONS OF β-DICARBONYL COMPOUNDS: MORE CHEMISTRY OF ENOLATE IONS

CHAPTER 20

AMINES

SPECIAL TOPIC J   **REACTIONS AND SYNTHESIS OF HETEROCYCLIC AMINES**

CHAPTER 21

PHENOLS AND ARYL HALIDES: NUCLEOPHILIC AROMATIC SUBSTITUTION

CHAPTER 2 4
AMINO ACIDS AND PROTEINS

CHAPTER 2 5
NUCLEIC ACIDS AND PROTEIN SYNTHESIS

CHAPTER 1

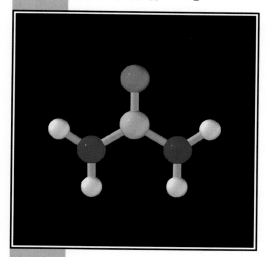

Urea
(Section 1.2A)

CARBON COMPOUNDS AND CHEMICAL BONDS

1.1 INTRODUCTION

Organic chemistry is *the chemistry of the compounds of carbon.* The compounds of carbon constitute the central chemicals of all living things on this planet. Carbon compounds include deoxyribonucleic acids (DNAs), the giant molecules that contain the genetic information for all living species. Carbon compounds make up the proteins of our blood, muscle, and skin. They make up the enzymes that catalyze the reactions that occur in our bodies. Together with oxygen in the air we breathe, carbon compounds in our diets furnish the energy that sustains life.

Considerable evidence indicates that several billion years ago most of the carbon atoms on the earth existed in the form of the gas CH_4, called methane. This simple organic compound, along with water, ammonia, and hydrogen, were the main components of the primordial atmosphere. It has been shown experimentally that when electrical discharges and other forms of highly energetic radiation pass through this kind of atmosphere, many of these simple compounds become fragmented into highly reactive pieces. These pieces combine into more complex compounds. Compounds called amino acids, formaldehyde, hydrogen cyanide, purines, and pyrimidines can form in this way. It is thought that these, and other compounds produced in the primordial atmosphere in the same way, were carried by rain into the sea until the sea became a vast storehouse containing all of the compounds necessary for the emergence of life. Amino acids apparently reacted with each other to form the first proteins. Molecules of formaldehyde reacted with each other to become sugars, and some of these sugars, together with inorganic phosphates, combined with purines and pyrimidines to become simple molecules of ribonucleic acids (RNAs) and DNA. Molecules of RNA, because they can carry genetic information and can act as enzymes, were apparently instrumental in the emergence of the first primitive self-replicating systems. From these first systems, in a manner far from understood, through the long process of natural selection came all the living things on Earth today.

Not only are we composed largely of organic compounds, not only are we derived from and nourished by them, *we also live in an Age of Organic Chemistry.* The clothing we wear, whether a natural substance such as wool or cotton or a synthetic such as nylon or a polyester, is made up of carbon compounds. Many of the materials that go into the houses that shelter us are organic. The gasoline that propels our automobiles, the rubber of their tires, and the plastic of their interiors are all organic. Most of the medicines that help us cure diseases and relieve suffering are organic. Organic pesticides help us eliminate many of the agents that spread diseases in both plants and animals.

Organic chemicals are also factors in some of our most serious problems. Many of the organic chemicals introduced into the environment have had consequences far beyond those originally intended. A number of insecticides, widely used for many years, have now been banned because they harm many species other than insects and they pose a danger to humans. Organic compounds called polychlorobiphenyls (PCBs) are responsible for pollution of the Hudson River that may take years to reverse. Organic compounds used as propellants for aerosols have been banned because they threatened to destroy the ozone layer of the outer atmosphere, a layer that protects us from extremely harmful radiation.

Thus for good or bad, organic chemistry is associated with nearly every aspect of our lives. We would be wise to understand it as best we can.

1.2 THE DEVELOPMENT OF ORGANIC CHEMISTRY AS A SCIENCE

Humans have used organic compounds and their reactions for thousands of years. Their first deliberate experience with an organic reaction probably dates from their discovery of fire. The ancient Egyptians used organic compounds (indigo and alizarin) to dye cloth. The famous "royal purple" used by the Phoenicians was also an organic substance, obtained from mollusks. The fermentation of grapes to produce ethyl alcohol and the acidic qualities of "soured wine" are both described in the Bible and were probably known earlier.

As a science, organic chemistry is less than 200 years old. Most historians of science date its origin to the early part of the nineteenth century, a time in which an erroneous belief was dispelled.

1.2A Vitalism

During the 1780s scientists began to distinguish between **organic compounds** and **inorganic compounds.** Organic compounds were defined as compounds that could be obtained from *living organisms.* Inorganic compounds were those that came from *nonliving sources.* Along with this distinction, a belief called "vitalism" grew. According to this idea, the intervention of a "vital force" was necessary for the synthesis of an organic compound. Such synthesis, chemists held then, could take place only in living organisms. It could not take place in the flasks of a chemistry laboratory.

Between 1828 and 1850 a number of compounds that were clearly "organic" were synthesized from sources that were clearly "inorganic." The first of these syntheses was accomplished by Friedrich Wöhler in 1828. Wöhler found that the organic compound urea (a constituent of urine) could be made by evaporating an aqueous solution containing the inorganic compound ammonium cyanate.

$$
NH_4{}^+NCO^- \xrightarrow{\text{heat}} H_2N-\overset{\overset{\displaystyle O}{\|}}{C}-NH_2
$$

Ammonium cyanate   **Urea**

Although "vitalism" disappeared slowly from scientific circles after Wöhler's synthesis, its passing made possible the flowering of the science of organic chemistry that has occurred since 1850.

Despite the demise of vitalism in science, the word "organic" is still used today by some people to mean "coming from living organisms" as in the terms "organic vitamins" and "organic fertilizers." The commonly used term "organic food" means that the food was grown without the use of synthetic fertilizers and pesticides. An "organic vitamin" means to these people that the vitamin was isolated from a natural source and not synthesized by a chemist. While there are sound arguments to be made against using food contaminated with certain pesticides, while there may be environmental benefits to be obtained from organic farming, and while "natural" vitamins may contain beneficial substances not present in synthetic vitamins, it is impossible to argue that pure "natural" vitamin C, for example, is healthier than pure "synthetic" vitamin C, since the two substances are identical in all respects. In science today, the study of compounds from living organisms is called natural product chemistry.

1.2B Empirical and Molecular Formulas

In the eighteenth and nineteenth centuries extremely important advances were made in the development of qualitative and quantitative methods for analyzing organic substances. In 1784 Antoine Lavoisier first showed that organic compounds were composed primarily of carbon, hydrogen, and oxygen. Between 1811 and 1831, *quantitative* methods for determining the composition of organic compounds were developed by Justus Liebig, J. J. Berzelius, and J. B. A. Dumas.

A great confusion was dispelled in 1860 when Stanislao Cannizzaro showed that the earlier hypothesis of Amedeo Avogadro (1811) could be used to distinguish between **empirical** and **molecular formulas.** As a result, many molecules that had appeared earlier to have the same formula were seen to be composed of different numbers of atoms. For example, ethylene, cyclopentane, and cyclohexane all have the same empirical formula: CH_2. However, they have molecular formulas of C_2H_4, C_5H_{10}, and C_6H_{12}, respectively. Appendix A of the Study Guide that accompanies this book contains a review of how empirical and molecular formulas are determined and calculated.

1.3 THE STRUCTURAL THEORY OF ORGANIC CHEMISTRY

Between 1858 and 1861, August Kekulé, Archibald Scott Couper, and Alexander M. Butlerov, working independently, laid the basis for one of the most fundamental theories in chemistry: **the structural theory.** Two central premises are fundamental:

1. The atoms of the elements in organic compounds can form a fixed number of bonds. The measure of this ability is called **valence.** Carbon is *tetravalent;* that is, carbon atoms form four bonds. Oxygen is *divalent;* and hydrogen and (usually) the halogens are *monovalent.*

$$-\overset{\displaystyle |}{\underset{\displaystyle |}{C}}- \qquad -O- \qquad H- \quad Cl-$$

| **Carbon atoms** | **Oxygen atoms** | **Hydrogen and halogen** |
| **are tetravalent** | **are divalent** | **atoms are monovalent** |

2. A carbon atom can use one or more of its valences to form bonds to other carbon atoms.

Carbon–carbon bonds

$$-\overset{\displaystyle |}{\underset{\displaystyle |}{C}}-\overset{\displaystyle |}{\underset{\displaystyle |}{C}}- \qquad \overset{\displaystyle \diagdown}{\underset{\displaystyle \diagup}{C}}=\overset{\displaystyle \diagup}{\underset{\displaystyle \diagdown}{C}} \qquad -C\equiv C-$$

| **Single bond** | **Double bond** | **Triple bond** |

In his original publication Couper represented these bonds by lines much in the same way that most of the formulas in this book are drawn. In his textbook (published in 1861), Kekulé gave the science of organic chemistry its modern definition: *a study of the compounds of carbon.*

1.3A Isomers: The Importance of Structural Formulas

The structural theory allowed early organic chemists to begin to solve a fundamental problem that plagued them: the problem of **isomerism.** These chemists frequently found examples of **different compounds that have the same molecular formula.** Such compounds are called **isomers.**

Let us consider an example. There are two compounds with the molecular formula C_2H_6O that are clearly different because they have different properties (see Table 1.1). These compounds, therefore, are classified as being isomers of one another; they are said to be **isomeric.** Notice that these two isomers have different boiling points, and, because of this, one isomer, called *dimethyl ether,* is a gas at room temperature; the other isomer, called *ethyl alcohol,* is a liquid. The two isomers also have different melting points, and they show a striking difference in their reactivity towards sodium. Dimethyl ether *does not react* with sodium; ethyl alcohol *does react,* and the reaction produces hydrogen gas.

Because the molecular formula (C_2H_6O) for these two compounds is the same, it

TABLE 1.1 Properties of ethyl alcohol and dimethyl ether

	ETHYL ALCOHOL C_2H_6O	DIMETHYL ETHER C_2H_6O
Boiling point (°C)	78.5	−24.9
Melting point (°C)	−117.3	−138
Reaction with sodium	Displaces hydrogen	No reaction

gives us no basis for understanding the differences between them. The structural theory remedies this situation, however. It does so by giving us different **structures** (Fig. 1.1), and different **structural formulas** for the two compounds.

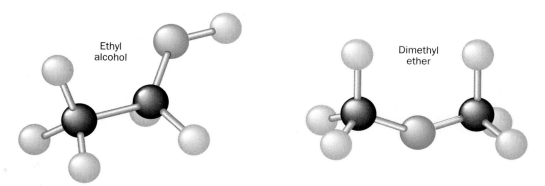

FIGURE 1.1   Ball-and-stick models show the different structures of ethyl alcohol and dimethyl ether.

One glance at the structural formulas for these two compounds reveals their difference. The two compounds differ in their **connectivity:** The atoms of ethyl alcohol are connected in a way that is different from those of dimethyl ether. In ethyl alcohol there is a C—C—O linkage; in dimethyl ether the linkage is C—O—C. Ethyl alcohol has a hydrogen atom attached to oxygen; in dimethyl ether all of the hydrogen atoms are attached to carbon. It is the hydrogen atom covalently bonded to oxygen in ethyl alcohol that is displaced when this alcohol reacts with sodium:

This is just the way water reacts with sodium:

$$2 \text{ H—O—H} + 2 \text{ Na} \longrightarrow 2 \text{ H—O}^- \text{Na}^+ + \text{H}_2$$

Hydrogen atoms that are covalently bonded to carbon are normally unreactive toward sodium. As a result, none of the hydrogen atoms in dimethyl ether is displaced by sodium.

The hydrogen atom attached to oxygen also accounts for the fact that ethyl alcohol is a liquid at room temperature. As we shall see in Section 2.16, this hydrogen atom allows molecules of ethyl alcohol to form hydrogen bonds to each other and gives ethyl alcohol a boiling point much higher than that of dimethyl ether.

Ethyl alcohol and dimethyl ether are examples of what are now called **constitutional isomers.*** *Constitutional isomers are different compounds that have the same molecular formula, but differ in their connectivity, that is, in the sequence in which their atoms are bonded together.* Constitutional isomers usually have different physical properties (e.g., melting point, boiling point, and density) and different chemical properties. The differences, however, may not always be as large as those between ethyl alcohol and dimethyl ether.

1.3B The Tetrahedral Shape of Methane

In 1874, the structural formulas originated by Kekulé, Couper, and Butlerov were expanded into three dimensions by the independent work of J. H. van't Hoff and J. A. Le Bel. van't Hoff and Le Bel proposed that the four bonds of the carbon atom in methane, for example, are arranged in such a way that they would point toward the corners of a regular tetrahedron, the carbon atom being placed at its center (Fig. 1.2). The necessity for knowing the arrangement of the atoms in space, taken together with an understanding of the order in which they are connected, is central to an understanding of organic chemistry, and we shall have much more to say about this later, in Chapters 4 and 5.

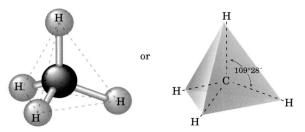

FIGURE 1.2   The tetrahedral structure of methane.

1.4   CHEMICAL BONDS: THE OCTET RULE

The first explanations of the nature of chemical bonds were advanced by G. N. Lewis (of the University of California, Berkeley) and W. Kössel (of the University of Munich) in 1916. Two major types of chemical bonds were proposed.

1.   The **ionic** (or **electrovalent**) bond, formed by the transfer of one or more electrons from one atom to another to create ions.

2.   The **covalent** bond, a bond that results when atoms share electrons.

The central idea in their work on bonding is that atoms without the electronic configuration of a noble gas generally react to produce such a configuration.

The concepts and explanations that arise from the original propositions of Lewis and Kössel are satisfactory for explanations of many of the problems we deal with in organic chemistry today. For this reason we shall review these two types of bonds in more modern terms.

---

*An older term for isomers of this type was **structural isomers.** The International Union of Pure and Applied Chemistry (IUPAC) now recommends that use of the term ''structural'' when applied to isomers of this type be abandoned.

1.4A Ionic Bonds

Atoms may gain or lose electrons and form charged particles called *ions*. An ionic bond is an attractive force between oppositely charged ions. One source of such ions is a reaction between atoms of widely differing electronegativities (Table 1.2). ***Electronegativity measures the ability of an atom to attract electrons.*** Notice in Table 1.2 that electronegativity increases as we go across a horizontal row of the periodic table from left to right:

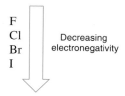

Li Be B C N O F

Increasing electronegativity

and that it decreases as we go down a vertical column:

F
Cl
Br
I

Decreasing electronegativity

An example of the formation of an ionic bond is the reaction of lithium and fluorine atoms.

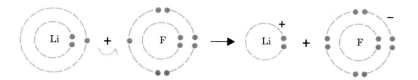

Lithium, a typical metal, has a very low electronegativity; fluorine, a nonmetal, is the most electronegative element of all. The loss of an electron (a negatively charged species) by the lithium atom leaves a lithium cation (Li^+); the gain of an electron by the fluorine atom gives a fluoride anion (F^-). Why do these ions form? In terms of the Lewis–Kössel theory both atoms achieve the electronic configuration of a noble gas by becoming ions. The lithium cation with two electrons in its valence shell is like an atom of the nobel gas helium, and the fluoride anion with eight electrons in its valence

TABLE 1.2 Electronegativities of some of the elements

			H			
			2.1			
Li	**Be**	**B**	**C**	**N**	**O**	**F**
1.0	1.5	2.0	2.5	3.0	3.5	4.0
Na	**Mg**	**Al**	**Si**	**P**	**S**	**Cl**
0.9	1.2	1.5	1.8	2.1	2.5	3.0
K						**Br**
0.8						2.8

shell is like an atom of the noble gas neon. Moreover, crystalline lithium fluoride forms from the individual lithium and fluoride ions. In this process negative fluoride ions become surrounded by positive lithium ions, and positive lithium ions by negative fluoride ions. In this crystalline state, the ions have substantially lower energies than the atoms from which they have been formed. Lithium and fluorine are thus ''stabilized'' when they react to form crystalline lithium fluoride.

We represent the formula for lithium fluoride as LiF, because this is the simplest formula for this ionic compound.

Ionic substances, because of their strong internal electrostatic forces, are usually very high melting solids, often having melting points above 1000°C. In polar solvents, such as water, the ions are solvated (see Section 2.16E), and such solutions usually conduct an electric current.

Ionic compounds form only when atoms of very different electronegativities transfer electrons to become ions.

1.4B Covalent Bonds

When two or more atoms of the same or similar electronegativities react, a complete transfer of electrons does not occur. In these instances the atoms achieve noble gas configurations by *sharing electrons. Covalent* bonds form between the atoms, and the products are called *molecules.* Molecules may be represented by electron-dot formulas or, more conveniently, by dash formulas, where each dash represents a pair of electrons shared by two atoms. Some examples are shown here. These formulas

$$H_2 \qquad H\cdot + \cdot H \longrightarrow H\colon H \quad \text{or} \quad H-H$$

$$Cl_2 \qquad \colon \ddot{C}l\cdot + \cdot \ddot{C}l\colon \longrightarrow \colon \ddot{C}l\colon \ddot{C}l\colon \quad \text{or} \quad \colon \ddot{C}l - \ddot{C}l\colon$$

$$CH_4 \qquad \cdot \dot{C}\cdot + 4\,H\cdot \longrightarrow H\colon \overset{..}{\underset{..}{C}}\colon H \quad \text{or} \quad H - \overset{\overset{\textstyle H}{|}}{\underset{\underset{\textstyle H}{|}}{C}} - H$$

are often called **Lewis structures;** in writing them we show only the electrons of the valence shell.

In certain cases, multiple covalent bonds are formed; for example,

$$N_2 \qquad \colon N\colon\colon N\colon \quad \text{or} \quad \colon N \equiv N\colon$$

and ions themselves may contain covalent bonds.

$$\overset{+}{NH_4} \qquad H\colon \overset{\overset{\textstyle H}{..}}{\underset{\underset{\textstyle H}{..}}{N}}\overset{+}{\colon} H \quad \text{or} \quad H - \overset{\overset{\textstyle H}{|}}{\underset{\underset{\textstyle H}{|}}{N}}\overset{+}{-} H$$

1.5 WRITING LEWIS STRUCTURES

When we write Lewis structures (electron-dot formulas) we assemble the molecule or ion from the constituent atoms showing only the valence electrons (i.e., the electrons of the outermost shell). By having the atoms share or transfer electrons, we try to give each atom the electronic configuration of the noble gas in the same horizontal row of the periodic table. For example, we give hydrogen atoms two electrons because by doing so we give them the structure of helium. We give carbon, nitrogen, oxygen, and fluorine atoms eight electrons because by doing this we give them the electronic configuration of neon. The number of valence electrons of an atom can be obtained from the periodic table because it is equal to the group number of the atom. Carbon, for example, is in Group **IVA** and it has four valence electrons; fluorine, in Group **VIIA** has seven; hydrogen in Group **IA**, has one.

SAMPLE PROBLEM

Write the Lewis structure of CH_3F.

ANSWER:

1. We find the total number of valence electrons of all the atoms:

$$4 + 3(1) + 7 = 14$$
$$\uparrow \qquad \uparrow \qquad \uparrow$$
$$C \quad H_3 \quad F$$

2. We use pairs of electrons to form bonds between all atoms that are bonded to each other. We represent these bonding pairs with lines. In our example this requires four pairs of electrons (8 of our 14 valence electrons).

$$\begin{array}{c} H \\ | \\ H-C-F \\ | \\ H \end{array}$$

3. We then add the remaining electrons in pairs so as to give each hydrogen 2 electrons (a duet) and every other atom 8 electrons (an octet). In our example, we assign the remaining 6 valence electrons to the fluorine atom in three nonbonding pairs.

$$\begin{array}{c} H \\ | \\ H-C-\overset{..}{\underset{..}{F}}: \\ | \\ H \end{array}$$

If the structure is an ion, we add or subtract electrons to give it the proper charge.

SAMPLE PROBLEM

Write the Lewis structure for the chlorate ion (ClO_3^-).

ANSWER:

1. We find the total number of valence electrons of all the atoms including the extra electron needed to give the ion a negative charge:

$$7 + 3(6) + 1 = 26$$

Cl O_3 e^-

2. We use three pairs of electrons to form bonds between the chlorine atom and the three oxygen atoms:

$$O$$
$$|$$
$$O—Cl—O$$

3. We then add the remaining 20 electrons in pairs so as to give each atom an octet.

$$\left[\begin{array}{c} \text{:}\ddot{O}\text{:} \\ | \\ \text{:}\ddot{O}—\overset{..}{Cl}—\ddot{O}\text{:} \end{array} \right]^{-}$$

If necessary, we use multiple bonds to give atoms the noble gas configuration. The carbonate ion ($CO_3{}^{2-}$) illustrates this.

$$\left[\begin{array}{c} \text{:}\ddot{O} \\ \| \\ C \\ \overset{..}{\underset{..}{O}} \qquad \overset{..}{\underset{..}{O}} \end{array} \right]^{2-}$$

The organic molecules ethene (C_2H_4) and ethyne (C_2H_2) have a double and triple bond, respectively.

$$\begin{array}{cc} H & H \\ \diagdown & \diagup \\ & C{=}C \\ \diagup & \diagdown \\ H & H \end{array} \quad \text{and} \quad H—C{\equiv}C—H$$

1.6 EXCEPTIONS TO THE OCTET RULE

Atoms share electrons, not just to obtain the configuration of an inert gas, but because sharing electrons produces increased electron density between the positive nuclei. The resulting attractive forces of nuclei for electrons is the "glue" that holds the atoms together (cf. Section 1.11). Elements of the second period of the periodic table can have a maximum of four bonds (i.e., have eight electrons around them) because these elements have only $2s$ and $2p$ orbitals available for bonding and a total of eight electrons fills these orbitals (Section 1.11). The octet rule, therefore, only applies to these elements, and even here, as we shall see in compounds of beryllium and boron, fewer than eight electrons are possible. Elements of the third period, and beyond, have d orbitals that can be used for bonding. These elements can accommodate more than eight electrons in their valence shell and therefore can form more than four covalent bonds. Examples are compounds such as PCl_5 and SF_6.

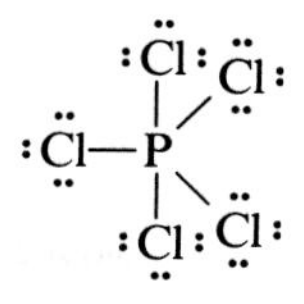

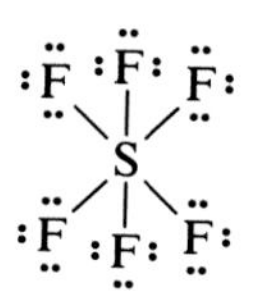

$$:\overset{..}{\underset{}{C}}l: \overset{..}{\underset{}{C}}l:$$

(Lewis structures of PCl₅ and SF₆ shown)

SAMPLE PROBLEM

Write a Lewis structure for the sulfate ion ($SO_4{}^{2-}$).

ANSWER:

1. We find the total number of valence electrons including the extra 2 electrons needed to give the ion a negative charge:

    $$6 + 4(6) + 2 = 32$$

    ↑   ↑   ↑

    S O_4 $2e^-$

2. We use pairs to form bonds between the sulfur atom and the four oxygen atoms:

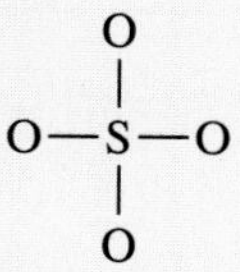

    ```
        O
        |
    O—S—O
        |
        O
    ```

3. We add the remaining 24 electrons as unshared pairs on oxygen atoms and as double bonds between the sulfur atom and two oxygen atoms. This gives each oxygen 8 electrons and the sulfur atom 12.

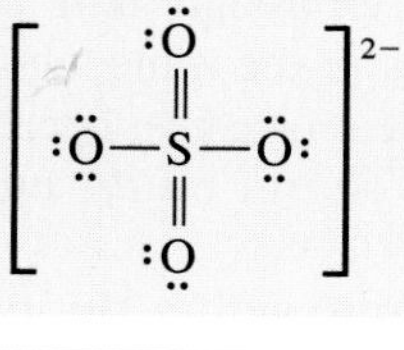

    (Lewis structure of sulfate ion with 2− charge shown)

Some highly reactive molecules or ions have atoms with fewer than eight electrons in their outer shell. An example is boron trifluoride (BF_3). In the BF_3 molecule the central boron atom has only six electrons around it.

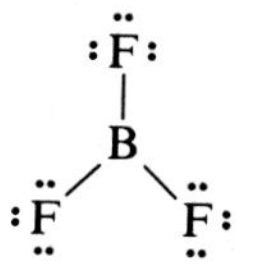

(Lewis structure of BF_3 shown)

Finally, one point needs to be stressed: Before we can write some Lewis structures, *we must know how the atoms are connected to each other.* Consider nitric acid, for example. Even though the formula for nitric acid is often written HNO_3, the hydrogen is actually connected to an oxygen, not to the nitrogen. The structure is $HONO_2$ and not HNO_3. Thus the correct Lewis structure is

This knowledge comes ultimately from experiments. If you have forgotten the structures of some of the common inorganic molecules and ions (such as those listed in Problem 1.1), this may be a good time for a review of the relevant portions of your general chemistry text.

PROBLEM 1.1

Write Lewis structures for each of the following:

(a) HF   (d) HNO_2 (g) H_3PO_4

(b) F_2 (e) H_2SO_3 (h) H_2CO_3

(c) CH_3F (f) BH_4^- (i) HCN

1.7 FORMAL CHARGE

When we write Lewis structures, it is often convenient to assign unit positive or negative charges, called **formal charges,** to certain atoms in the molecule or ion. This is nothing more than a bookkeeping method for electrical charges, because *the arithmetic sum of all of the formal charges equals the total charge on the molecule or ion.*

We calculate formal charges on individual atoms **by subtracting the number of valence electrons assigned to an atom in its bonded state from the number of valence electrons it has as a neutral free atom.** (Recall that the number of valence electrons in a neutral free atom is equal to its **group number** on the periodic table.)

We assign valence electrons to atoms in the bonded state by apportioning them. **We divide shared electrons equally between the atoms that share them and we assign unshared pairs to the atom that possesses them.**

Consider first the ammonium ion, an ion that has no unshared pairs. We divide all of the valence electrons equally between the atoms that share them. Each hydrogen is assigned *one electron* (e^-) and we subtract this from *one* (the number of valence electrons in a neutral hydrogen atom) to give a formal charge of 0 for each hydrogen atom. The nitrogen atom is assigned *four electrons.* We subtract this from *five* (the number of valence electrons in a neutral nitrogen atom) to give a formal charge of $+1$. In effect, we say that because the nitrogen atom in the ammonium ion lacks one electron when compared to a neutral nitrogen atom (in which the number of protons and electrons are equal) it has a formal charge of $+1$.*

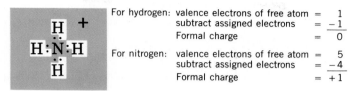

For hydrogen:
valence electrons of free atom = 1
subtract assigned electrons = -1
Formal charge = 0

For nitrogen:
valence electrons of free atom = 5
subtract assigned electrons = -4
Formal charge = $+1$

Charge on ion = (4)(0) + 1 = +1

Let us next consider the nitrate ion (NO_3^-), an ion that has oxygen atoms with unshared electron pairs. Here we find that the nitrogen atom has a formal charge of $+1$, that two oxygen atoms have formal charges of -1, and that one oxygen has a formal charge equal to 0.

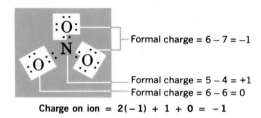

Formal charge = 6 − 7 = −1
Formal charge = 5 − 4 = +1
Formal charge = 6 − 6 = 0

Charge on ion = 2(−1) + 1 + 0 = −1

Molecules, of course, have no net electrical charge. Molecules, by definition, are neutral. Therefore, the sum of the formal charges on each atom making up a molecule must be zero. Consider the following examples:

Ammonia

H—N̈—H or

Formal charge = 5 − 5 = 0
Formal charge = 1 − 1 = 0

Charge on molecule = 0 + 3(0) = 0

Water

H—Ö—H or

Formal charge = 6 − 6 = 0
Formal charge = 1 − 1 = 0

Charge on molecule = 0 + 2(0) = 0

* An alternative method for calculating formal charge is to use the equation:

$$F = Z - S/2 - U$$

where F is the formal charge, Z is the group number, S equals the number of shared electrons, and U is the number of unshared electrons.

PROBLEM 1.2

Write a Lewis structure for each of the following negative ions, and assign the formal negative charge to the correct atom.

(a) NO_2^- (d) HSO_4^-

(b) NH_2^- (e) HCO_3^-

(c) CN^- (f) HC_2^-

1.7A Summary of Formal Charges

With this background it should now be clear that each time an oxygen atom of the type —Ö: appears in a molecule or ion it will have a formal charge of −1, and that each time an oxygen atom of the type =Ö: or —Ö— appears it will have a formal charge of 0. Similarly, —N— will be +1, and —N̈— will be zero. These common structures are summarized in Table 1.3.

TABLE 1.3 A summary of formal charges

GROUP	FORMAL CHARGE OF +1	FORMAL CHARGE OF 0	FORMAL CHARGE OF −1
3		—B—	—B⁼
4	—Ċ⁺— =Ċ⁺ ≡Ċ⁺	—C̈— =C̈— ≡C̈—	—C̈⁼ =C:⁻ ≡C:⁻
5	—N⁺⁼ =N⁺— ≡N⁺—	—N̈— =N̈— ≡N:	—N̈⁼ =N:⁻
6	—Ö⁺⁼ =Ö⁺—	—Ö— =Ö:	—Ö:⁻
7	—Ẍ⁺⁼	—Ẍ: (X = F, Cl, Br, or I) :Ẍ:⁻	

PROBLEM 1.3

Assign the proper formal charge to the colored atom in each of the following structures:

$$
\text{(a) }
\begin{array}{ccc}
& H & H \\
& | & | \\
H- & C- & C \\
& | & | \\
& H & H
\end{array}
\qquad
\text{(d) }
\begin{array}{c}
:\ddot{O}: \\
| \\
H-C-H \\
| \\
H
\end{array}
$$

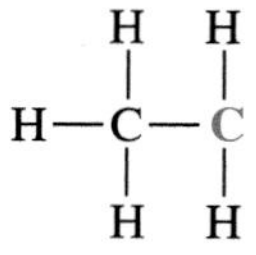

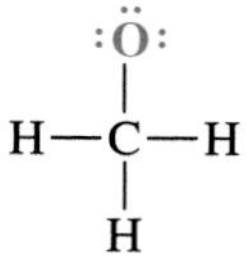

(b) 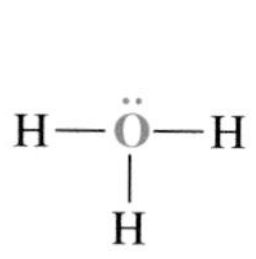 $H - \ddot{O} - H$
$\qquad | $
$\qquad H$

(e) 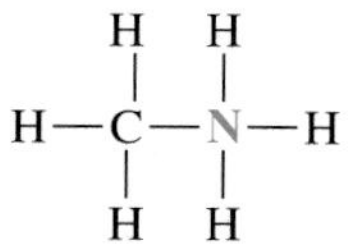 $H - \overset{\displaystyle H \quad H}{\underset{\displaystyle H \quad H}{C - N}} - H$

(c) $H - \overset{\displaystyle :\ddot{O}}{C} - \ddot{O}:$

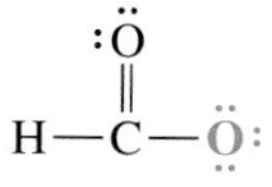

(f) 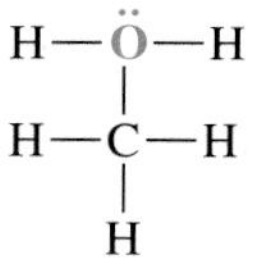 $H - \overset{\displaystyle H - \ddot{O} - H}{\underset{\displaystyle H}{C}} - H$

1.8 RESONANCE

One problem with Lewis structures is that they impose an artificial **location** on the electrons. As a result, more than one *equivalent* Lewis structure can be written for many molecules and ions. Consider, for example, the carbonate ion ($CO_3{}^{2-}$). We can write three *different* but *equivalent* structures, **1–3.**

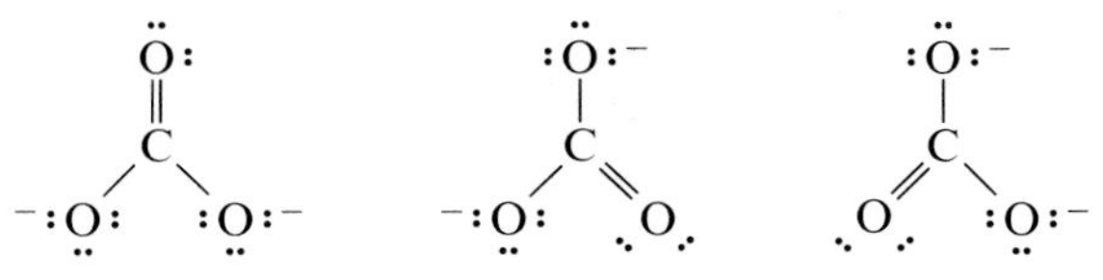

Notice two important features of these structures. First, each atom has the noble gas configuration. Second, *and this is especially important,* we can convert one structure into any other by *changing only the positions of the electrons.* We do not need to change the relative positions of the atomic nuclei. For example, if we move the electron pairs in the manner indicated by the curved arrows* in structure **1**, we change structure **1** into structure **2**:

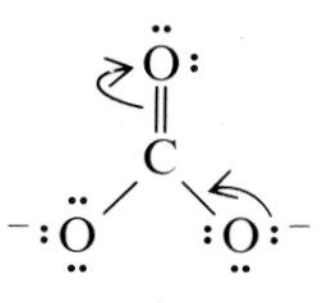

becomes

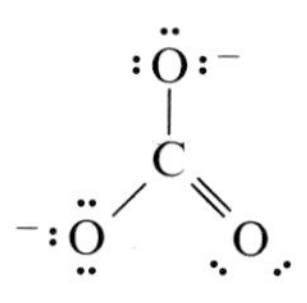

In a similar way we can change structure **2** into structure **3**:

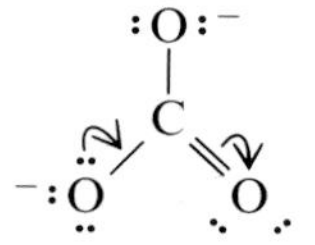

becomes

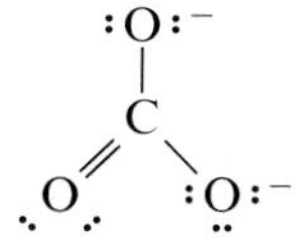

* The use of curved arrows is described in more detail later (Section 3.4). We should point out now that the curved arrows show movement of pairs of electrons, not atoms, and that the tail of the arrow begins at the current position of the electron pair, while the head of the arrow shows the new location of the electron pair.

Structures **1–3**, although not identical, *are equivalent.* None of them, however, fits important data about the carbonate ion.

X-Ray studies have shown that carbon–oxygen double bonds are shorter than single bonds. The same kind of study of the carbonate ion shows, however, that all of its carbon–oxygen bonds *are of equal length.* One is not shorter than the others as would be expected from the representations **1**, **2**, or **3**. Clearly none of the three structures agrees with this evidence. In each structure, **1–3**, one carbon–oxygen bond is a double bond and the other two are single bonds. None of the structures, therefore, is correct. How, then, should we represent the carbonate ion?

One way is through a theory called **resonance theory.** This theory states that whenever a molecule or ion can be represented by two or more Lewis structures *that differ only in the positions of the electrons,* two things will be true:

1. None of these structures, which we call **resonance structures** or **resonance contributors,** will be a correct representation for the molecule. None will be in complete accord with the physical or chemical properties of the substance.
2. The actual molecule or ion will be better represented by a *hybrid of these structures.*

   Resonance structures, then, are not structures for the actual molecule or ion; they exist only in theory. As such they can never be isolated. No single contributor adequately represents the molecule or ion. In resonance theory we view the carbonate ion, which is, of course, a real entity, as having a structure that is a **hybrid** of these three **hypothetical** resonance structures.

What would a hybrid of structures **1–3** be like? Look at the structures and look especially at a particular carbon–oxygen bond, say, the one at the top. This carbon–oxygen bond is a double bond in one structure (**1**) and a single bond in the other two (**2** and **3**). The actual carbon–oxygen bond, since it is a hybrid, must be something in between a double bond and a single bond. Because the carbon–oxygen bond is a single bond in two of the structures and a double bond in only one it must be more like a single bond than a double bond. It must be like a one- and one-third bond. We could call it a partial double bond. And, of course, what we have just said about any one carbon–oxygen bond will be equally true of the other two. Thus all of the carbon–oxygen bonds of the carbonate ion are partial double bonds, and *all are equivalent.* All of them *should be* the same length, and this is exactly what experiments tell us. They are all 1.28 Å long, a distance which is intermediate between that of a carbon–oxygen single bond (1.43 Å) and that of a carbon–oxygen double bond (1.20 Å).

One other important point: By convention, when we draw resonance structures, we connect them by double-headed arrows to indicate clearly that they are hypothetical, not real. For the carbonate ion we write them this way:

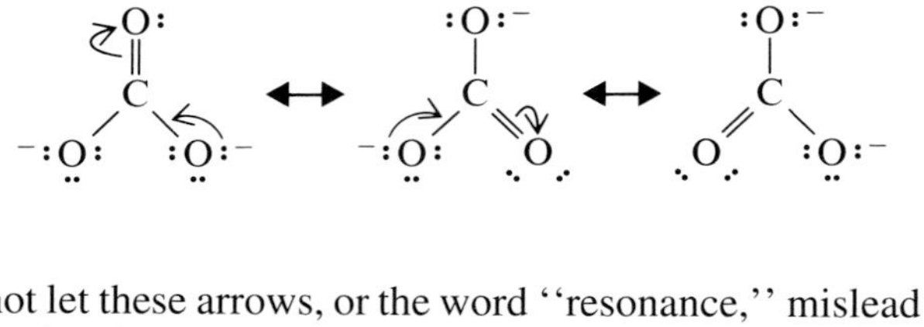

We should not let these arrows, or the word ''resonance,'' mislead us into thinking that the carbonate ion fluctuates between one structure and another. These structures exist only on paper; therefore, the carbonate ion cannot fluctuate among them. It is also important to distinguish between resonance and **an equilibrium.** In an equilibrium between two, or more, species, it is quite correct to think of different structures and moving (or fluctuating) atoms, *but not in the case of resonance* (as in the carbonate ion). Here the atoms do not move, and the "structures" exist only on paper. An equilibrium is indicated by ⇌ and resonance by ↔.

How can we write the structure of the carbonate ion in a way that will indicate its actual structure? We may do two things: we may write all of the resonance structures as we have just done and let the reader mentally fashion the hybrid or we may write a non-Lewis structure that attempts to represent the hybrid. For the carbonate ion we might do the following:

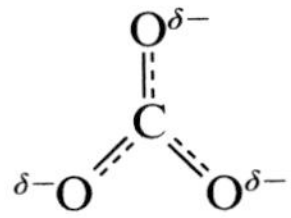

The bonds are indicated by a combination of a solid line and a dashed line. This is to indicate that the bonds are something in between a single bond and a double bond. As a rule, we use a solid line whenever a bond appears in all structures, and a dashed line when a bond exists in one or more but not all. We also place a $\delta -$ (read partial minus) beside each oxygen to indicate that something less than a full negative charge resides on each oxygen atom. (In this instance each oxygen atom has two-thirds of a full negative charge.)

SAMPLE PROBLEM

The following is one way of writing the structure of the nitrate ion.

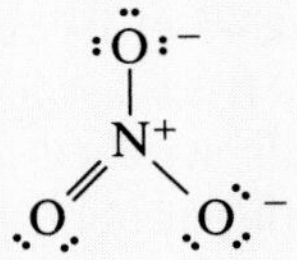

However, considerable physical evidence indicates that all three nitrogen–oxygen bonds are equivalent and that they have the same length, a bond distance between that expected for a nitrogen–oxygen single bond and a nitrogen–oxygen double bond. Explain this in terms of resonance theory.

ANSWER:
We recognize that if we move the electron pairs in the following way, we can write three *different* but *equivalent* structures for the nitrate ion:

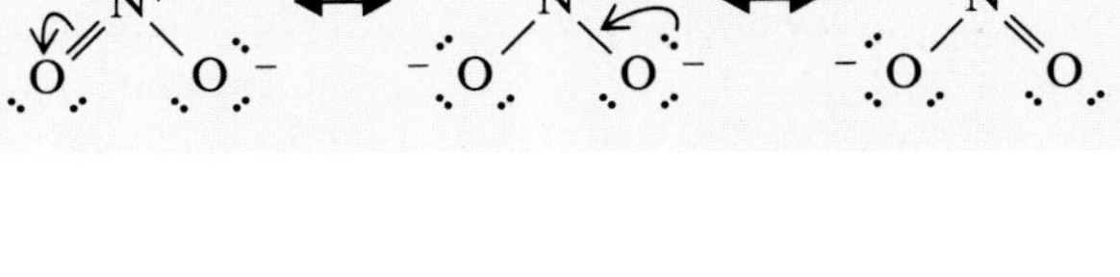

Since these structures differ from one another *only in the positions of their electrons,* they are *resonance structures* or *resonance contributors.* As such, no single structure taken alone will adequately represent the nitrate ion. The actual molecule will be best represented by a *hybrid of these three structures.* We might write this hybrid in the following way to indicate that all of the bonds are equivalent and that they are more than single bonds and less than double bonds. We also indicate that each oxygen atom bears an equal partial negative charge. This charge distribution corresponds to what we find experimentally.

$$\begin{array}{c} O^{\delta-} \\ \| \\ N^+ \\ {}_{\delta-}O \diagdown \diagdown O^{\delta-} \end{array}$$

Hybrid structure for the nitrate ion

PROBLEM 1.4

(a) Write two resonance structures for the formate ion HCO_2^-. (The hydrogen and oxygen atoms are bonded to the carbon.) (b) Explain what these structures predict for the carbon–oxygen bond lengths of the formate ion, and (c) for the electrical charge on the oxygen atoms.

1.9 ENERGY CHANGES

Since we will be talking frequently about the energies of chemical systems, perhaps we should pause here for a brief review. *Energy* is defined as the capacity to do work. The two fundamental types of energy are **kinetic energy** and **potential energy.**

Kinetic energy is the energy an object has because of its motion; it equals one half the object's mass multiplied by the square of its velocity (i.e., $\frac{1}{2} mv^2$).

Potential energy is stored energy. It exists only when an attractive or repulsive force exists between objects. Two balls attached to each other by a spring can have their potential energy increased when the spring is stretched or compressed (Fig. 1.3). If the spring is stretched, an attractive force will exist between the balls. If it is compressed, a repulsive force will exist. In either instance releasing the balls will cause the potential energy (stored energy) of the balls to be converted into kinetic energy (energy of motion).

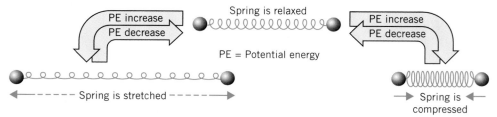

FIGURE 1.3 Potential energy (PE) exists between objects that either attract or repel each other. When the spring is either stretched or compressed, the PE of the two balls increases. (Adapted with permission from J. E. Brady and G. E. Humiston, *General Chemistry: Principles and Structure,* 1st ed., Wiley, New York, p. 18.)

Chemical energy is a form of potential energy. It exists because attractive and repulsive electrical forces exist between different pieces of the molecules. Nuclei attract electrons, nuclei repel each other, and electrons repel each other.

It is usually impractical (and often impossible) to describe the *absolute* amount of potential energy contained by a substance. Thus we usually think in terms of their *relative potential energies.* We say that one system has *more* or *less* potential energy than another.

Another term that chemists frequently use in this context is the term **stability** or **relative stability.** *The relative stability of a system is inversely related to its relative potential energy.* The *more* potential energy an object has, the *less stable* it is. Consider, as an example, the relative potential energy and the relative stability of snow when it lies high on a mountainside and when it lies serenely in the valley below. Because of the attractive force of gravity, the snow high on the mountain *has greater potential energy and is much less stable* than the snow in the valley. This greater potential energy of the snow on the mountainside can become converted to the enormous kinetic energy of an avalanche. By contrast, the snow in the valley, with its lower potential energy and with its greater stability, is incapable of releasing such energy.

1.9A Potential Energy and Covalent Bonds

Atoms and molecules possess potential energy—often called chemical energy—that can be released as heat when they react. Because heat is associated with molecular motion, this release of heat results from a change from potential energy to kinetic energy.

From the standpoint of covalent bonds, the state of greatest potential energy is the state of free atoms, the state in which the atoms are not bonded to each other at all. This is true because the formation of a chemical bond is always accompanied by the lowering of the potential energy of the atoms (cf. Fig. 1.9). Consider as an example the formation of hydrogen molecules from hydrogen atoms:

$$\text{H} \cdot + \text{H} \cdot \longrightarrow \text{H—H} \quad \Delta H° = -104 \text{ kcal mol}^{-1} \, (-435 \text{ kJ mol}^{-1})\text{*}$$

The potential energy of the atoms decreases by 104 kcal mol^{-1} as the covalent bonds form. This potential energy change is illustrated graphically in Fig. 1.4.

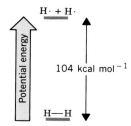

FIGURE 1.4 The relative potential energies of hydrogen atoms and hydrogen molecules.

---

*A kilocalorie of energy (1000 cal) is the amount of energy in the form of heat required to raise by 1°C the temperature of 1 kg (1000 g) of water at 15°C. The unit of energy in SI units is the joule, J, and 1 cal = 4.184 J. (Thus 1 kcal = 4.184 kJ.)

A convenient way to represent the relative potential energies of molecules is in terms of their relative **enthalpies** or **heat contents,** *H.* (*Enthalpy* comes from *en* + Gk: *thalpein* to heat.) The difference in relative enthalpies of reactants and products in a chemical change is called the enthalpy change and is symbolized by $\Delta H°$. [The Δ (delta) in front of a quantity usually means the difference, or change, in the quantity. The superscript ° indicates that the measurement is made under standard conditions.]

By convention, the sign of $\Delta H°$ for **exothermic** reactions (those evolving heat) is negative. **Endothermic** reactions (those that absorb heat) have a positive $\Delta H°$. The heat of reaction, $\Delta H°$, measures the change in enthalpy of the atoms of the reactants as they are converted to products. For an exothermic reaction the atoms have a smaller enthalpy as products than they do as reactants. For endothermic reactions, the reverse is true.

1.10 QUANTUM MECHANICS

In 1926 a new theory of atomic and molecular structure was advanced independently and almost simultaneously by three men: Erwin Schrödinger, Werner Heisenberg, and Paul Dirac. This theory, called **wave mechanics** by Schrödinger or **quantum mechanics** by Heisenberg, has become the basis from which we derive our modern understanding of bonding in molecules.

The formulation of quantum mechanics that Schrödinger advanced is the form that is most often used by chemists. In Schrödinger's publication the motion of the electrons is described in terms that take into account the wave nature of the electron.* Schrödinger developed a way to convert the mathematical expression for the total energy of the system consisting of one proton and one electron—the hydrogen atom —into another expression called a **wave equation.** This equation is then solved to yield not one but a series of solutions called **wave functions.**

Wave functions are most often denoted by the Greek letter psi (ψ), and each wave function (ψ function) corresponds to a different state for the electron. Corresponding to each state, and calculable from the wave equation for the state, is a particular energy.

Each state is a sublevel where one or two electrons can reside. *The solutions to the wave equation for a hydrogen atom can also be used* (with appropriate modifications) *to give sublevels for the electrons of higher elements.*

A wave equation is simply a tool for calculating two important properties: These are the energy associated with the state and the relative probability of an electron residing at particular places in the sublevel (Section 1.11). When the value of a wave equation is calculated for a particular point in space relative to the nucleus, the result may be a positive number or a negative number (or zero). These signs are sometimes called **phase signs.** They are characteristic of all equations that describe waves. We do not need to go into the mathematics of waves here, but a simple analogy will help us understand the nature of these phase signs.

Imagine a wave moving across a lake. As it moves along, the wave has crests and troughs; that is, it has regions where the wave rises above the average level of the lake or falls below it (Fig. 1.5). Now, if an equation were to be written for this wave, the wave function (ψ) would be plus (+) in regions where the wave is above the average level of the lake (i.e., in crests) and it would be minus (−) in regions where the wave is

---

*The idea that the electron has the properties of a wave as well as those of a particle was proposed by Louis de Broglie in 1923.

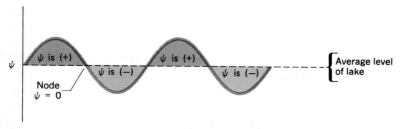

FIGURE 1.5 A wave moving across a lake is viewed along a slice through the lake. For this wave the wave function, ψ, is plus $(+)$ in crests and minus $(-)$ in troughs. At the average level of the lake it is zero; these places are called nodes.

below the average level (i.e., in troughs). The relative magnitude of ψ (called the amplitude) will be related to the distance the wave rises above or falls below the average level of the lake. At the places where the wave is exactly at the average level of the lake, the wave function will be zero. Such a place is called a **node.**

One other characteristic of waves is their ability to reinforce each other or to interfere with one another. Imagine two waves approaching each other as they move across a lake. If the waves meet so that a crest meets a crest, that is, so that *waves of the same phase sign meet each other,* the waves **reinforce** each other, they add together, and the resulting wave is larger than either individual wave. On the other hand, if a crest meets a trough, that is, if waves of opposite sign meet, the waves **interfere** with each other, they subtract from each other, and the resulting wave is smaller than either individual wave. (If the two waves of opposite sign meet in precisely the right way, complete cancellation can occur.)

The wave functions that describe the motion of an electron in an atom or molecule are, of course, different from the equations that describe waves moving across lakes. And when dealing with the electron we should be careful not to take analogies like this too far. Electron wave functions, however, are like the equations that describe water waves in that they have phase signs and nodes, and *they undergo reinforcement and interference.*

1.11 ATOMIC ORBITALS

For a short time after Schrödinger's proposal in 1926, a precise physical interpretation for the electron wave function eluded early practitioners of quantum mechanics. It remained for Max Born, a few months later, to point out that the square of ψ *could* be given a precise physical meaning. According to Born, ψ^2 for a particular location (x,y,z) expresses the **probability** of finding an electron at that particular location in space. If ψ^2 is large in a unit volume of space, the probability of finding an electron in that volume is great—we say that the **electron probability density** is large. Conversely if ψ^2 for some other unit volume of space is small, the probability of finding an electron there is low.* Plots of ψ^2 in three dimensions generate the shapes of the familiar s, p, and d atomic orbitals, which we use as our models for atomic structure.

The f orbitals are practically never used in organic chemistry, and we shall not concern ourselves with them in this book. The d orbitals will be discussed briefly later

*Integration of ψ^2 over all space must equal 1; that is, the probability of finding an electron somewhere in all of space is 100%.

when we discuss compounds in which *d* orbital interactions are important. The *s* and *p* orbitals are, by far, the most important in the formation of organic molecules and, at this point, we shall limit our discussion to them.

An orbital is a region of space where the probability of finding an electron is large. The shapes of *s* and *p* orbitals are shown in Fig. 1.6. There is a finite, but very small, probability of finding an electron at greater distances from the nucleus. The volumes that we typically use to illustrate an orbital are those volumes that would contain the electron 90–95% of the time.

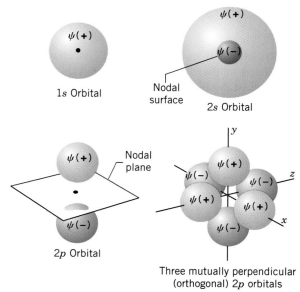

FIGURE 1.6   The shapes of some *s* and *p* orbitals.

Both the 1*s* and 2*s* orbitals are spheres (as are all higher *s* orbitals) (Fig. 1.6). The 2*s* orbital contains a nodal surface, that is, an area where $\psi = 0$. In the inner portion of the 2*s* orbital, ψ_{2s} is negative.

The 2*p* orbitals have the shape of two almost-touching spheres. The phase sign of the 2*p* wave function, ψ_{2p}, is positive in one lobe (or sphere) and negative in the other. A nodal plane separates the two lobes of a *p* orbital, and the three *p* orbitals are arranged in space so that their axes are mutually perpendicular.

You should not associate the sign of the wave function with anything having to do with electrical charge. As we said earlier the $(+)$ and $(-)$ signs associated with ψ are simply the arithmetic signs of the wave function in that region of space. The $(+)$ and $(-)$ signs **do not** imply a greater or lesser probability of finding an electron either. The probability of finding an electron is ψ^2, and ψ^2 is always positive. (Squaring a negative number always makes it positive.) Thus the probability of finding the electron in the $(-)$ lobe of a *p* orbital is the same as that of the $(+)$ lobe. The significance of the $(+)$ and $(-)$ signs will become clear later when we see how atomic orbitals combine to form molecular orbitals and when we see how covalent bonds are formed.

There is a relationship between the number of nodes of an orbital and its energy: *The greater the number of nodes, the greater the energy.* We can see an example here; the 2*s* and 2*p* orbitals have one node each and they have greater energy than a 1*s* orbital, which has no nodes.

The relative energies of the lower energy orbitals are as follows. Electrons in $1s$ orbitals have the lowest energy because they are closest to the positive nucleus. Electrons in $2s$ orbitals are next lowest in energy. Electrons of $2p$ orbitals have equal but still higher energy. (Orbitals of equal energy are said to be **degenerate orbitals**.)

We can use these relative energies to arrive at the electronic configuration of any atom in the first two rows of the periodic table. We need only follow a few simple rules.

1. **The aufbau principle:** Orbitals are filled so that those of lowest energy are filled first. (*Aufbau* is German for "building up.")

2. **The Pauli exclusion principle:** A maximum of two electrons may be placed in each orbital *but only when the spins of the electrons are paired.* An electron spins about its own axis. For reasons that we cannot develop here, an electron is permitted only one or another of only two possible spin orientations. We usually show these orientations by arrows, either ↑ or ↓. Thus two spin-paired electrons would be designated ↑↓. Unpaired electrons, which are not permitted in the same orbital, are designated ↑ ↑ (or ↓ ↓).

3. **Hund's rule:** When we come to orbitals of equal energy (degenerate orbitals) such as the three p orbitals, we add one electron to each *with their spins unpaired* until each of the degenerate orbitals contains one electron. (This allows the electrons, which repel each other, to be farther apart.) Then we begin adding a second electron to each degenerate orbital so that the spins are paired.

If we apply these rules to some of the second-row elements of the periodic table, we get the results shown in Fig. 1.7.

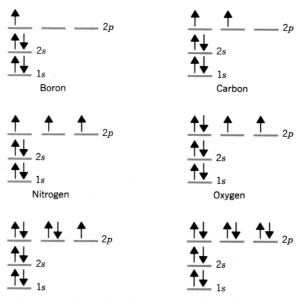

FIGURE 1.7 The electron configuration of some second-row elements.

1.12 MOLECULAR ORBITALS

For the organic chemist the greatest utility of atomic orbitals is in using them as models for understanding how atoms combine to form molecules. We shall have much more to say about this subject in subsequent chapters, for, as we have already said, covalent bonds are central to the study of organic chemistry. First, however, we shall concern ourselves with a very simple case: the covalent bond that is formed when two hydrogen atoms combine to form a hydrogen molecule. We shall see that the description of the formation of the H—H bond is the same as, or at least very similar to, the description of bonds in more complex molecules.

Let us begin by examining what happens to the total energy of two hydrogen atoms with electrons of opposite spins when they are brought closer and closer together. This can best be shown with the curve shown in Fig. 1.8.

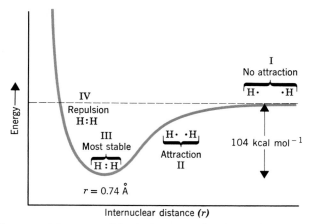

FIGURE 1.8 The potential energy of the hydrogen molecule as a function of internuclear distance.

When the atoms of hydrogen are relatively far apart (**I**) their total energy is simply that of two isolated hydrogen atoms. As the hydrogen atoms move closer together (**II**), each nucleus increasingly attracts the other's electron. This attraction more than compensates for the repulsive force between the two nuclei (or the two electrons), and the result of this attraction *is to lower the energy of the total system.* When the two nuclei are 0.74 Å apart (**III**), the most stable (lowest energy) state is obtained. This distance, 0.74 Å, corresponds to the *bond length* for the hydrogen molecule. If the nuclei are moved closer together (**IV**) the repulsion of the two positively charged nuclei predominates, and the energy of the system rises.

There is one serious problem with this model for bond formation. We have assumed that the electrons are essentially motionless and that as the nuclei come together they will be stationary in the region between the two nuclei. Electrons do not behave that way. Electrons move about, and according to the **Heisenberg uncertainty principle,** we cannot know simultaneously the position and momentum of an electron. That is, we cannot pin the electrons down as precisely as our explanation suggests.

We avoid this problem when we use a model based on quantum mechanics and *orbitals,* because now we describe the electron in terms of probabilities (ψ^2) of finding it at particular places. By treating the electron in this way we do not violate the uncertainty principle, because we do not talk about where the electron is precisely. We talk instead about where the *electron probability density* is large or small.

Thus an orbital explanation for what happens when two hydrogen atoms combine to form a hydrogen molecule is the following: As the hydrogen atoms approach each other, their 1s orbitals (ψ_{1s}) begin to overlap. As the atoms move closer together, orbital overlap increases until the **atomic orbitals** (AOs) combine to become **molecular orbitals (MOs).** The molecular orbitals that are formed encompass both nuclei and, in them, the electrons can move about both nuclei. They are not restricted to the vicinity of one nucleus or the other as they were in the separate atomic orbitals. Molecular orbitals, like atomic orbitals, *may contain a maximum of two spin-paired electrons.*

When atomic orbitals combine to form molecular orbitals, ***the number of molecular orbitals that result always equals the number of atomic orbitals that combine.*** Thus in the formation of a hydrogen molecule the *two* atomic orbitals combine to produce *two* molecular orbitals. Two orbitals result because the mathematical properties of wave functions permit them to be combined by either *addition* or *subtraction.* That is, they can combine either *in* or *out of* phase. What are the natures of these new molecular orbitals?

One molecular orbital, called the **bonding molecular orbital** (ψ_{molec}) contains both electrons in the lowest energy state, or *ground* state, of a hydrogen molecule. It is formed when the atomic orbitals combine in the way shown in Fig. 1.9. Here atomic orbitals combine by *addition,* and this means that atomic *orbitals of the same phase sign overlap.* Such overlap leads to *reinforcement* of the wave function in the region between the two nuclei. Reinforcement of the wave function not only means that the value of ψ is larger between the two nuclei, it means that ψ^2 is larger as well. Moreover, since ψ^2 expresses the probability of finding an electron in this region of space, we can now understand how orbital overlap of this kind leads to bonding. It does so by increasing the electron probability density in exactly the right place — in the region of space between the nuclei. When the electron density is large here, the attractive force of the nuclei for the electrons more than offsets the repulsive force acting between the two nuclei (and between the two electrons). This extra attractive force is, of course, the "glue" that holds the atoms together.

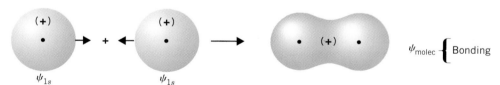

FIGURE 1.9   The overlapping of two hydrogen $1s$ atomic orbitals to form a bonding molecular orbital.

The second molecular orbital, called the **antibonding molecular orbital** (ψ^*_{molec}) contains no electrons in the ground state of the molecule. It is formed by subtraction in the way shown in Fig. 1.10. [Subtraction means that the phase sign of one orbital has been changed from $(+)$ to $(-)$.] Here, because *orbitals of opposite phase overlap,* the wave functions *interfere* with each other in the region between the two nuclei and a node is produced. At the node $\psi = 0$, and on either side of the node ψ is small. This means that in the region between the nuclei ψ^2 is also small. Thus if electrons were to occupy the antibonding orbital, the electrons would avoid the region between the nuclei. There would be only a small attractive force of the nuclei for the electrons. Repulsive forces (between the two nuclei and between the two electrons) would be greater than the attractive forces. Having electrons in the antibonding orbital would not tend to hold the atoms together; it would tend to make them fly apart.

What we have just described has its counterpart in a mathematical treatment called the LCAO (linear combination of atomic orbitals) method. In the LCAO treatment,

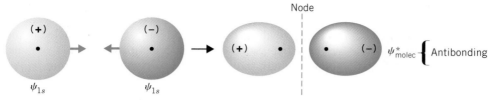

FIGURE 1.10 The overlapping of two hydrogen $1s$ atomic orbitals to form an antibonding molecular orbital.

wave functions for the atomic orbitals are combined in a linear fashion (by addition or subtraction) in order to obtain new wave functions for the molecular orbitals.

Molecular orbitals, like atomic orbitals, correspond to particular energy states for an electron. Calculations show that the relative energy of an electron in the bonding molecular orbital of the hydrogen molecule is substantially less than its energy in a ψ_{1s} atomic orbital. These calculations also show that the energy of an electron in the antibonding molecular orbital is substantially greater than its energy in a ψ_{1s} atomic orbital.

An energy diagram for the molecular orbitals of the hydrogen molecule is shown in Fig. 1.11. Notice that electrons are placed in molecular orbitals in the same way that they were in atomic orbitals. Two electrons (with their spins opposed) occupy the bonding molecular orbital, where their total energy is less than in the separate atomic orbitals. This is, as we have said, the *lowest electronic energy state* or *ground state* of the hydrogen molecule. (An electron may occupy the antibonding orbital in what is called an *excited state* for the molecule. This state forms when the molecule in the ground state absorbs a photon of light of proper energy.)

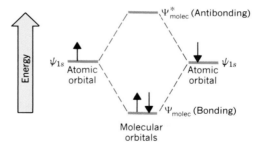

FIGURE 1.11 Energy diagram for the hydrogen molecule. Combination of two atomic orbitals. ψ_{1s} gives two molecular orbitals, ψ_{molec} and ψ^*_{molec}. The energy of ψ_{molec} is lower than that of the separate atomic orbitals, and in the lowest electronic energy state of molecular hydrogen it contains both electrons.

1.13 THE STRUCTURE OF METHANE: sp^3 HYBRIDIZATION

The s and p orbitals used in the quantum mechanical description of the carbon atom, given in Section 1.11, were based on calculations for hydrogen atoms. These simple s and p orbitals do not, when taken alone, provide a satisfactory model for the *tetravalent-tetrahedral* carbon of methane (see Problem 1.5). However, a satisfactory model of methane's structure that is based on quantum mechanics *can* be obtained through an approach called **orbital hybridization.** Orbital hybridization, in its simplest terms, is nothing more than a mathematical approach that involves the combining of individual wave functions for s and p orbitals to obtain wave functions for new orbitals. The new orbitals have, *in varying proportions,* the properties of the original orbitals taken separately. These new orbitals are called **hybrid atomic orbitals.**

According to quantum mechanics the electronic configuration of a carbon atom in its lowest energy state—called the *ground state*—is that given here.

$$C \; \uparrow\downarrow \quad \uparrow\downarrow \quad \uparrow \quad \uparrow \quad \underline{\hphantom{xx}}$$

$$1s \quad 2s \quad 2p \quad 2p \quad 2p$$

Ground state of a carbon atom

The valence electrons of a carbon atom (those used in bonding) are those of the *outer level,* that is, the $2s$ and $2p$ electrons.

Hybrid atomic orbitals that account for methane's structure can be obtained by combining the wave functions of the $2s$ orbital of carbon with those of the three $2p$ orbitals. The mathematical procedure for hybridization can be approximated by the illustration that is shown in Fig. 1.12.

In this model, four orbitals are mixed—or hybridized—and four new hybrid orbitals are obtained. The hybrid orbitals are called sp^3 orbitals to indicate that they have one part the character of an s orbital and three parts the character of a p orbital. The mathematical treatment of orbital hybridization also shows that *the four sp^3 orbit-*

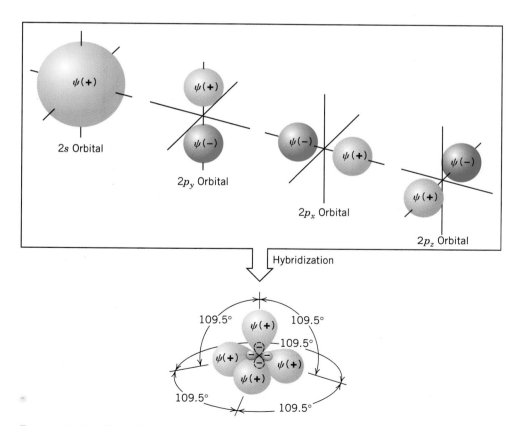

FIGURE 1.12 Hybridization of atomic orbitals of a carbon atom to produce sp^3-hybrid orbitals.

als should be oriented at angles of 109.5° with respect to each other. This is precisely the spatial orientation of the four hydrogen atoms of methane.

If, in our imagination, we visualize the formation of methane from an sp^3-hybridized carbon atom and four hydrogen atoms, the process might be like that shown in Fig. 1.13. For simplicity we show only the formation of the *bonding molecular orbital* for each carbon–hydrogen bond. We see that an sp^3-hybridized carbon gives a *tetrahedral structure for methane, and one with four equivalent C—H bonds.*

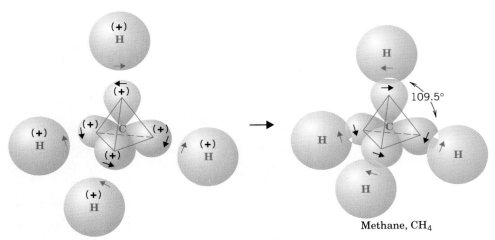

Methane, CH_4

FIGURE 1.13   The formation of methane from an sp^3-hybridized carbon atom. In orbital hybridization we combine orbitals, *not* electrons. The electrons can then be replaced in the hybrid orbitals as necessary for bond formation, but always in accordance with the Pauli principle of no more than two electrons (with opposite spin) in each orbital. In this illustration we have placed one electron in each of the hybrid carbon orbitals. In this illustration, too, we have shown only the bonding molecular orbital of each C—H bond because these are the orbitals that contain the electrons in the lowest energy state of the molecule.

PROBLEM 1.5

(a) Consider a carbon atom in its ground state. Would such an atom offer a satisfactory model for the carbon of methane? If not, why not? (*Hint:* Consider whether or not a ground state carbon atom could be tetravalent, and consider the bond angles that would result if it were to combine with hydrogen atoms.)
(b) What about a carbon atom in the excited state:

C   ⇅   ↑   ↑   ↑   ↑

$1s$ $2s$ $2p_x$ $2p_y$ $2p_z$
Excited state of a carbon atom

Would such an atom offer a satisfactory model for the carbon of methane? If not, why not?

In addition to accounting properly for the shape of methane, the orbital hybridization model also explains the very strong bonds that are formed between carbon and hydrogen. To see how this is so, consider the shape of the individual sp^3 orbital shown in Fig. 1.14. Because the sp^3 orbital has the character of a p orbital, the positive lobe of the sp^3 orbital is large and is extended quite far into space.

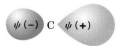

FIGURE 1.14   The shape of an sp^3 orbital.

It is the positive lobe of the sp^3 orbital that overlaps with the positive $1s$ orbital of hydrogen to form the bonding molecular orbital of a carbon–hydrogen bond (Fig. 1.15). Because the positive lobe of the sp^3 orbital is large and is extended into space, the overlap between it and the $1s$ orbital of hydrogen is also large, and the resulting carbon–hydrogen bond is quite strong.

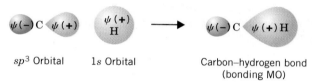

sp^3 Orbital $1s$ Orbital   Carbon–hydrogen bond
                                                (bonding MO)

FIGURE 1.15   Formation of a C—H bond.

The bond formed from the overlap of an sp^3 orbital and a $1s$ orbital is an example of a **sigma bond** (Fig. 1.16). The term *sigma bond* is a general term applied to those bonds in which orbital overlap gives a bond that is *circularly symmetrical in cross section when viewed along the bond axis*. **All purely single bonds are sigma bonds.**

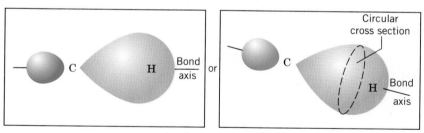

A sigma (σ) bond

FIGURE 1.16   A sigma (σ) bond.

From this point on we shall often show only the bonding molecular orbitals because they are the ones that contain the electrons when the molecule is in its lowest energy state. Consideration of antibonding orbitals is important when a molecule absorbs light and in explaining certain reactions. We shall point out these instances later.

1.14 THE STRUCTURE OF BORANE: *sp²* HYBRIDIZATION

Borane (BH_3), a molecule that can be detected only at low pressures, has a triangular (trigonal planar) shape with three equivalent boron–hydrogen bonds. In its ground state the boron atom has the following electronic configuration. Only one orbital contains a single electron that might be used to overlap with an s orbital containing the unpaired electron in a hydrogen atom.

B ⇅ ⇅ ↑ — —
  $1s$ $2s$ $2p$ $2p$ $2p$

Boron atom ground state

Triangular structure of BH$_3$

Clearly, the s and p orbitals of the ground state will not furnish a satisfactory model for the trivalent and triangularly bonded boron of BH$_3$.

PROBLEM 1.6

(a) What valence would you expect a boron atom in its ground state to have?

(b) Consider an excited state of boron in which one $2s$ electron is promoted to a vacant $2p$ orbital. Show how this state of boron also fails to account for the structure of BH$_3$.

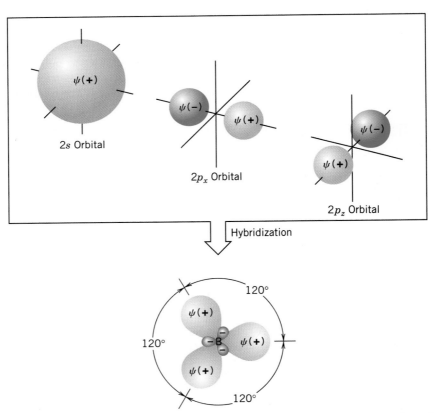

FIGURE 1.17   A representation of the mathematical procedure for the hybridization of one $2s$ orbital and two $2p$ orbitals of boron to produce three sp^2-hybrid orbitals.

Once again we use a model based on the mathematical process of orbital hybridization. Here, however, we combine the $2s$ orbital with only two of the $2p$ orbitals. Mixing three orbitals as shown in Fig. 1.17 gives three equivalent hybrid orbitals and these orbitals are sp^2 orbitals. They have one part the character of an s orbital and two parts the character of a p orbital. Calculations show that these orbitals are pointed toward the corners of an equilateral triangle with angles of 120° between their axes. These orbitals, then, are just what we need to account for the trivalent, trigonal planar boron atom of borane.

By placing one of the valence electrons in each of the three sp^2 orbitals and allowing these orbitals to overlap with an s orbital containing one electron from each of three hydrogen atoms, we obtain the structure shown in Fig. 1.18. Notice that the boron atom still has a vacant p orbital, the one that we did not hybridize.

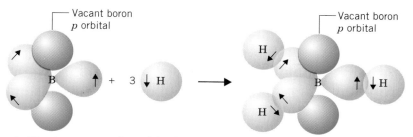

FIGURE 1.18   A representation of the formation of the bonding MOs of borane from an sp^2-hybridized boron atom and three hydrogen atoms.

We shall see in Section 2.4 that sp^2 hybridization offers a satisfactory model for carbon atoms that form double bonds.

1.15   THE STRUCTURE OF BERYLLIUM HYDRIDE: *sp* HYBRIDIZATION

Beryllium hydride (BeH_2) is a linear molecule; the bond angle is 180°.

$$H-Be-H \quad \overset{180°}{\frown}$$

In its ground state the beryllium atom has the following electronic configuration:

$$Be \quad \uparrow\downarrow \quad \uparrow\downarrow \quad \underline{} \ \underline{} \ \underline{}$$

$$\quad\quad 1s \quad 2s \quad 2p \ \ 2p \ \ 2p$$

In order to account for the structure of BeH_2 we again need a model based on orbital hybridization. Here (Fig. 1.19) we hybridize one s orbital with one p orbital and obtain two sp orbitals. Calculations show that these sp orbitals are oriented at an angle of 180°. The two p orbitals that were not mixed are vacant. Beryllium can use these hybrid orbitals to form bonds to two hydrogen atoms in the way shown in Fig. 1.20.

We shall see in Section 2.5 that sp hybridization offers a satisfactory model for carbon atoms that form triple bonds.

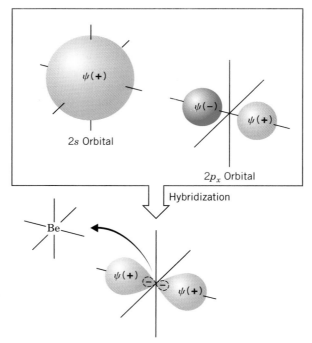

FIGURE 1.19 A representation of the mathematical procedure for the hybridization of one 2s orbital and one 2p orbital of beryllium to produce two sp-hybrid orbitals.

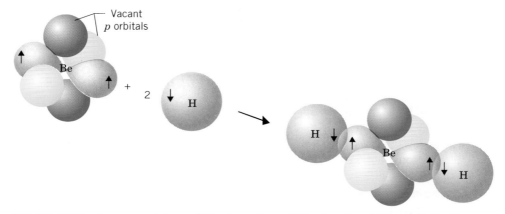

FIGURE 1.20 A representation of the formation of the bonding MOs of BeH$_2$ from an sp-hybridized beryllium atom and two hydrogen atoms.

1.16 A SUMMARY OF IMPORTANT CONCEPTS THAT COME FROM QUANTUM MECHANICS

1. An **atomic orbital (AO)** corresponds to a region of space about the nucleus of a single atom where there is a high probability of finding an electron. Atomic orbitals called s orbitals are spherical; those called p orbitals are like two almost-tangent spheres. Orbitals can hold a maximum of two electrons when their spins are paired. Orbitals are described by a wave function, ψ, and each orbital has a characteristic energy. The phase signs associated with an orbital may be $(+)$ or $(-)$.

2. When atomic orbitals overlap, they combine to form **molecular orbitals (MOs).** Molecular orbitals correspond to regions of space encompassing two (or more) nuclei where electrons are to be found. Like atomic orbitals, molecular orbitals can hold up to two electrons if their spins are paired.

3. When atomic orbitals with the same phase sign interact they combine to form a **bonding molecular orbital:**

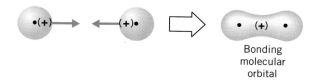

Bonding molecular orbital

The electron probability density of a bonding molecular orbital is large in the region of space between the two nuclei where the negative electrons hold the positive nuclei together.

4. An **antibonding molecular orbital** forms when orbitals of opposite phase sign overlap:

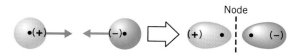

An antibonding orbital has higher energy than a bonding orbital. The electron probability density of the region between the nuclei is small and it contains a **node**—a region where $\psi = 0$. Thus, having electrons in an antibonding orbital does not help hold the nuclei together. The internuclear repulsions tend to make them fly apart.

5. The **energy of electrons** in a bonding molecular orbital is less than the energy of the electrons in their separate atomic orbitals. The energy of electrons in an antibonding orbital is greater than that of electrons in their separate atomic orbitals.

6. The **number of molecular orbitals** always equals the number of atomic orbitals from which they are formed. Combining two atomic orbitals will always yield two molecular orbitals—one bonding and one antibonding.

7. **Hybrid atomic orbitals** are obtained by mixing (hybridizing) the wave functions for orbitals of a different type (i.e., s and p orbitals) but from the same atom.

8. Hybridizing three p orbitals with one s orbital yields four sp^3 **orbitals.** Atoms that are sp^3 hybridized direct the axes of their four sp^3 orbitals toward the corners of a tetrahedron. The carbon of methane is sp^3 hybridized and **tetrahedral.**

9. Hybridizing two p orbitals with one s orbital yields three sp^2 **orbitals.** Atoms that are sp^2 hybridized point the axes of three sp^2 orbitals toward the corners of an equilateral triangle. The boron atom in BF_3 is sp^2 hybridized and **trigonal planar.**

10. Hybridizing one p orbital with one s orbital yields two sp **orbitals.** Atoms that are sp hybridized orient the axes of their two sp orbitals in opposite directions (at an

angle of 180°). The beryllium atom of BeH_2 is sp hybridized and BeH_2 is a **linear** molecule.

11. **A sigma bond** (a type of single bond) is one in which the electron density has circular symmetry when viewed along the bond axis. In general, the skeletons of organic molecules are constructed of atoms linked by sigma bonds.

1.17   MOLECULAR GEOMETRY: THE VALENCE SHELL ELECTRON-PAIR REPULSION (VSEPR) MODEL

We have been discussing the geometry of molecules on the basis of theories that arise from quantum mechanics. It is possible, however, to predict the arrangement of atoms in molecules and ions on the basis of a theory called the **valence shell electron-pair repulsion (VSEPR) theory.** Consider the following examples found in Sections 1.17A–F.

We apply VSEPR theory in the following way:

1. We consider molecules (or ions) in which the central atom is covalently bonded to two or more atoms or groups.

2. We consider all of the valence electron pairs of the central atom—both those that are shared in covalent bonds, called **bonding pairs,** and those that are unshared, called **nonbonding pairs** or **unshared pairs.**

3. Because electron pairs repel each other, the electron pairs of the valence shell tend to stay as far apart as possible. The repulsion between nonbonding pairs is generally greater than that between bonding pairs.

4. We arrive at the geometry of the molecule by considering all of the electron pairs, bonding and nonbonding, but we describe the shape of the molecule or ion by referring to the positions of the nuclei (or atoms) and not by the positions of the electron pairs.

Consider the following examples.

1.17A Methane

The valence shell of methane contains four pairs of bonding electrons. Only a tetrahedral orientation will allow four pairs of electrons to have the maximum possible separation (Fig. 1.21). Any other orientation, for example, a square planar arrangement, places the electron pairs closer together.

Thus, in the case of methane, the VSEPR model accommodates what we have known since the proposal of van't Hoff and Le Bel (Section 1.3B): The molecule of methane has a tetrahedral shape.

PROBLEM 1.7 _____

Part of the reasoning that led van't Hoff and Le Bel to propose a tetrahedral shape for molecules of methane was based on the number of isomers possible for substituted methanes. For example, only one compound with the formula CH_2Cl_2 has ever been found (i.e., there is no isomeric form). Consider both a square planar structure and a tetrahedral structure for CH_2Cl_2 and explain how this observation supports a tetrahedral structure.

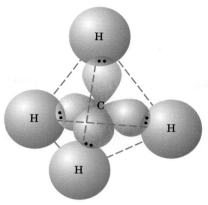

FIGURE 1.21   A tetrahedral shape for methane allows the maximum separation of the four bonding electron pairs.

The bond angles for any atom that has a regular tetrahedral structure are 109.5°. A representation of these angles in methane is shown in Fig. 1.22.

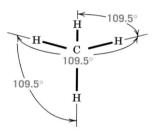

FIGURE 1.22   The bond angles of methane are 109.5°.

1.17B Ammonia

The geometry of a molecule of ammonia is a **trigonal pyramid.** The bond angles in a molecule of ammonia are 107°, a value very close to the tetrahedral angle (109.5°). We can write a general tetrahedral structure for the electron pairs of ammonia by placing the nonbonding pair at one corner (Fig. 1.23). A *tetrahedral arrangement* of the electron pairs explains the *trigonal pyramidal* arrangement of the four atoms. The bond angles are 107° (not 109.5°) because the nonbonding pair occupies more space than the bonding pairs.

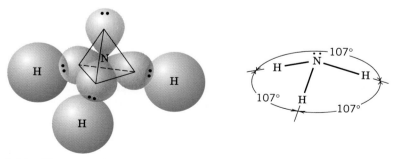

FIGURE 1.23   The tetrahedral arrangement of the electron pairs of an ammonia molecule that results when the nonbonding electron pair is considered to occupy one corner. This arrangement of electron pairs explains the trigonal pyramidal shape of the NH_3 molecule.

1.17C Water

A molecule of water has an **angular** or **bent geometry.** The H—O—H bond angle in a molecule of water is 105°, an angle that is also quite close to the 109.5° bond angles of methane.

We can write a general tetrahedral structure for the electron pairs of a molecule of water *if we place the two nonbonding electron pairs at corners of the tetrahedron.* Such a structure is shown in Fig. 1.24. A *tetrahedral arrangement* of the electron pairs accounts for the *angular arrangement* of the three atoms. The bond angle is less than 109.5° because the nonbonding pairs are effectively ''larger'' than the bonding pairs and, therefore, the structure is not perfectly tetrahedral.

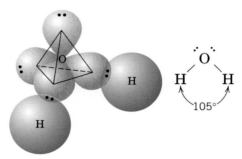

FIGURE 1.24 An approximately tetrahedral arrangement of the electron pairs of a molecule of water that results when the pairs of nonbonding electrons are considered to occupy corners. This arrangement accounts for the angular shape of the H_2O molecule.

1.17D Boron Trifluoride

Boron, a Group **IIIA** element, has only three outer level electrons. In the compound boron trifluoride (BF_3) these three electrons are shared with three fluorine atoms. As a result, the boron atom in BF_3 has only six electrons (three bonding pairs) around it. Maximum separation of three bonding pairs occurs when they occupy the corners of an equilateral triangle. Consequently, in the boron trifluoride molecule the three fluorine atoms lie in a plane at the corners of an equilateral triangle (Fig. 1.25). Boron trifluoride is said to have a *trigonal planar structure.* The bond angles are 120°.

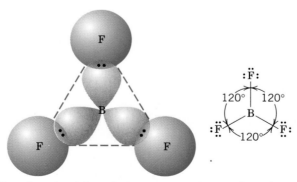

FIGURE 1.25 The triangular (trigonal planar) shape of boron fluoride maximally separates the three bonding pairs.

1.17E Beryllium Hydride

The central beryllium atom of BeH_2 has only two electron pairs around it; both electron pairs are bonding pairs. These two pairs are maximally separated when they are on opposite sides of the central atom as shown in the following structures. This arrangement of the electron pairs accounts for the linear geometry of the BeH_2 molecule and its bond angle of 180°.

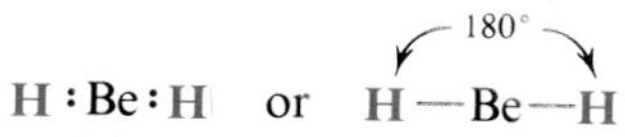

$$H : Be : H \quad \text{or} \quad \overset{\overset{\displaystyle 180°}{\curvearrowright}}{H - Be - H}$$

Linear geometry of BeH$_2$

PROBLEM 1.8

Use VSEPR theory and predict the geometry of each of the following molecules and ions:

(a) BH_4^- (c) NH_4^+ (e) BH_3 (g) SiF_4

(b) BeF_2 (d) H_2S (f) CF_4 (h) $: CCl_3^-$

1.17F Carbon Dioxide

The VSEPR method can also be used to predict the shapes of molecules containing multiple bonds if we assume that *all of the electrons of a multiple bond act as though they were a single unit,* and, therefore, are located in the region of space between the two atoms joined by a multiple bond.

This principle can be illustrated with the structure of a molecule of carbon dioxide (CO_2). The central carbon atom of carbon dioxide is bonded to each oxygen atom by a double bond. Carbon dioxide is known to have a linear shape; the bond angle is 180°.

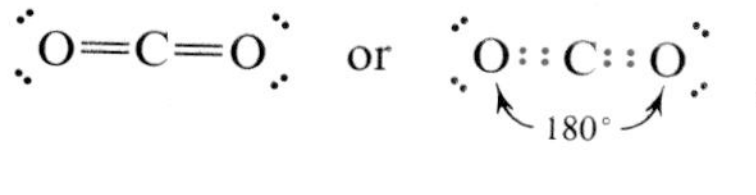

$$\overset{\cdot\cdot}{O} = C = \overset{\cdot\cdot}{O} \quad \text{or} \quad \overset{\cdot\cdot}{O} :: C :: \overset{\cdot\cdot}{O}$$

The four electrons of each double bond act as a single unit and are maximally separated from each other

Such a structure is consistent with a maximum separation of the two groups of four bonding electrons. (The nonbonding pairs associated with the oxygen atoms have no effect on the shape.)

PROBLEM 1.9

Predict the bond angles of   (a) $H_2C = CH_2$,   (b) $HC \equiv CH$,   (c) $HC \equiv N$.

The shapes of several simple molecules and ions as predicted by VSEPR theory are shown in Table 1.4. In this table we have also included the hybridization state of the central atom.

TABLE 1.4   *Shapes of molecules and ions from VSEPR theory*

NUMBER OF ELECTRON PAIRS			HYBRIDIZATION STATE OF CENTRAL ATOM	SHAPE OF MOLECULE OR ION[a]	EXAMPLES
Bonding	Nonbonding	Total			
2	0	2	sp	Linear	BeH_2
3	0	3	sp^2	Trigonal planar	BF_3, CH_3^+
4	0	4	sp^3	Tetrahedral	CH_4, NH_4^+
3	1	4	$\sim sp^3$	Trigonal pyramidal	NH_3, CH_3^-
2	2	4	$\sim sp^3$	Angular	H_2O

[a] Excluding nonbonding pairs.

1.18 POLAR COVALENT BONDS

When two atoms of different electronegativities form a covalent bond, the electrons are not shared equally between them. The atom with greater electronegativity draws the electron pair closer to it, and a **polar covalent bond** results. (One definition of *electronegativity* is *the ability of an element to attract electrons that it is sharing in a covalent bond.*) An example of such a polar covalent bond is the one in hydrogen chloride. The chlorine atom, with its greater electronegativity, pulls the bonding electrons closer to it. This makes the hydrogen atom somewhat electron deficient and gives it a *partial* positive charge ($\delta +$). The chlorine atom becomes somewhat electron rich and bears a *partial* negative charge ($\delta -$).

$$\overset{\delta +}{H} \quad : \overset{\delta -}{\ddot{\underset{\cdot\cdot}{Cl}}} :$$

Because the hydrogen chloride molecule has a partially positive end and a partially negative end, it is a dipole, and it has a **dipole moment.**

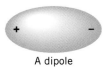

A dipole

The dipole moment is a physical property that can be measured experimentally. It is defined as the product of the magnitude of the charge in electrostatic units (esu) and the distance that separates them in centimeters (cm).

$$\text{Dipole moment} = \text{charge (in esu)} \times \text{distance (in cm)}$$

$$\mu = e \times d$$

The charges are typically on the order of 10^{-10} esu and the distances are on the order of 10^{-8} cm. Dipole moments, therefore, are typically on the order of 10^{-18} esu cm. For convenience, this unit, 1×10^{-18} esu cm, is defined as one **debye** and is abbreviated D. (The unit is named after Peter J. W. Debye, a chemist born in the Netherlands, who taught at Cornell University from 1936–1966. Debye won the Nobel prize in 1936.)

The direction of polarity of a polar bond can be symbolized by a vector quantity $\longmapsto$. The crossed end of the arrow is the positive end and the arrow head is the negative end.

$$(\text{positive end}) \longmapsto (\text{negative end})$$

In HCl, for example, we would indicate the direction of the dipole moment in the following way:

$$\text{H}\!-\!\text{Cl}$$
$$\longmapsto$$

If necessary, the length of the arrow can be used to indicate the magnitude of the dipole moment. Dipole moments, as we shall see in Section 1.19, are very useful quantities in accounting for physical properties of compounds.

PROBLEM 1.10 _____

Give the direction of the dipole moment (if any) for each of the following molecules:

(a) HF ,   (b) IBr,   (c) Br_2,   (d) F_2

1.19 POLAR AND NONPOLAR MOLECULES

In the discussion of dipole moments in Section 1.18, our attention was restricted to simple diatomic molecules. Any *diatomic* molecule in which the two atoms are *different* (and thus have different electronegativities) will, of necessity, have a dipole moment. If we examine Table 1.5, however, we find that a number of molecules (e.g.,

TABLE 1.5   Dipole moments of some simple molecules

FORMULA	μ (D)	FORMULA	μ (D)
H_2	0	CH_4	0
Cl_2	0	CH_3Cl	1.87
HF	1.91	CH_2Cl_2	1.55
HCl	1.08	$CHCl_3$	1.02
HBr	0.80	CCl_4	0
HI	0.42	NH_3	1.47
BF_3	0	NF_3	0.24
CO_2	0	H_2O	1.85

CCl_4, CO_2) consist of more than two atoms, have *polar* bonds, *but have no dipole moment.* Now that we have an understanding of the shapes of molecules we can understand how this can occur.

Consider a molecule of carbon tetrachloride (CCl_4). Because the electronegativity of chlorine is greater than that of carbon, each of the carbon–chlorine bonds in CCl_4 is polar. Each chlorine atom has a partial negative charge, and the carbon atom is considerably positive. Because a molecule of carbon tetrachloride is tetrahedral (Fig. 1.26), however, *the center of positive charge and the center of negative charge coincide, and the molecule has no net dipole moment.*

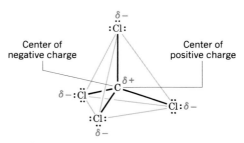

FIGURE 1.26 Charge distribution in carbon tetrachloride.

This result can be illustrated in a slightly different way: If we use arrows (+→) to represent the direction of polarity of each bond, we get the arrangement of bond moments shown in Fig. 1.27. Since the bond moments are vectors of equal magnitude arranged tetrahedrally, their effects cancel. Their vector sum is zero. The molecule has *no net dipole moment.*

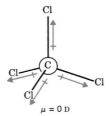

FIGURE 1.27 A tetrahedral orientation of equal bond moments causes their effects to cancel.

The chloromethane molecule (CH_3Cl) has a net dipole moment of 1.87 D. Since carbon and hydrogen have electronegativities (Table 1.2) that are nearly the same, the contribution of three C—H bonds to the net dipole is negligible. The electronegativity difference between carbon and chlorine is large, however, and this highly polar C—Cl bond accounts for most of the dipole moment of CH_3Cl (Fig. 1.28).

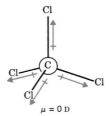

$\mu = 1.87$ D

FIGURE 1.28 The dipole moment of chloromethane arises mainly from the highly polar carbon–chlorine bond.

PROBLEM 1.11

Boron trifluoride (BF_3) has no dipole moment ($\mu = 0$). Explain how this observation confirms the geometry of BF_3 predicted by VSEPR theory.

PROBLEM 1.12

Tetrachloroethene ($CCl_2{=}CCl_2$) does not have a dipole moment. Explain this fact on the basis of the shape of $CCl_2{=}CCl_2$.

PROBLEM 1.13

Sulfur dioxide (SO_2) has a dipole moment ($\mu = 1.63$ D); on the other hand, carbon dioxide (CO_2) has no dipole moment ($\mu = 0$). What do these facts indicate about the geometries of the two molecules?

Unshared pairs of electrons make large contributions to the dipole moments of water and ammonia. Because an unshared pair has no atom attached to it to partially neutralize its negative charge, an unshared electron pair contributes a large moment directed away from the central atom (Fig. 1.29). (The O—H and N—H moments are also appreciable.)

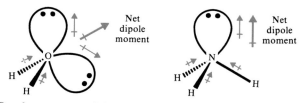

FIGURE 1.29   Bond moments and the resulting dipole moment of water and ammonia.

PROBLEM 1.14

Using a three-dimensional formula, show the direction of the dipole moment of CH_3OH.

PROBLEM 1.15

Trichloromethane ($CHCl_3$, also called *chloroform*) has a larger dipole moment than $CFCl_3$. Use three-dimensional structures and bond moments to explain this fact.

1.20 REPRESENTATION OF STRUCTURAL FORMULAS

Organic chemists use a variety of ways to write structural formulas. The most common types of representations are shown in Fig. 1.30. The **dot structure** shows all of the valence electrons, but writing it is tedious and time consuming. The other representations are more convenient and are, therefore, more often used.

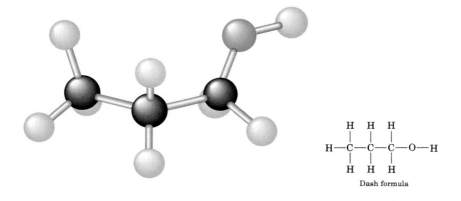

H—C—C—C—O—H
Dash formula

CH$_3$CH$_2$CH$_2$OH
Condensed formula

FIGURE 1.30   Structural formulas for propyl alcohol.

In fact, we often omit unshared pairs when we write formulas unless there is a reason to include them. For example,

$$H:\overset{H}{\underset{H}{C}}:\overset{..}{\underset{..}{O}}:\overset{H}{\underset{H}{C}}:H \quad = \quad H-\overset{H}{\underset{H}{C}}-O-\overset{H}{\underset{H}{C}}-H \quad = \quad CH_3OCH_3$$

Dot structure   **Dash formula**   **Condensed formula**

1.20A Dash Structural Formulas

If we look at the model for propyl alcohol given in Fig. 1.30 and compare it with the formulas given there, we find that the chain of atoms is straight in all the formulas. In the model, which corresponds more accurately to the actual shape of the molecule, the chain of atoms is not at all straight. Also of importance is this: *Atoms joined by single bonds can rotate relatively freely with respect to one another.* (We discuss this point further in Section 2.2B.) This relatively free rotation means that the chain of atoms in propyl alcohol can assume a variety of arrangements like those that follow:

It also means that all of the dash structures that follow are *equivalent* and all represent propyl alcohol. (Notice that in these formulas we represent the bond angles as being 90° not 109.5°. This convention is followed simply for convenience in printing.)

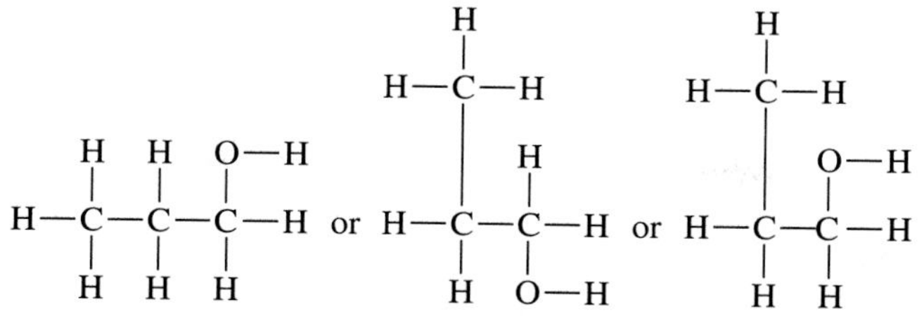

Equivalent dash formulas for propyl alcohol

Structural formulas such as these indicate the way in which the atoms are attached to each other and *are not* representations of the actual shapes of the molecule. They show what is called the **connectivity** of the atoms. *Constitutional isomers (Section 1.3A) have different connectivity, and, therefore, must have different structural formulas.*

Consider the compound called isopropyl alcohol, whose formula we might write in a variety of ways:

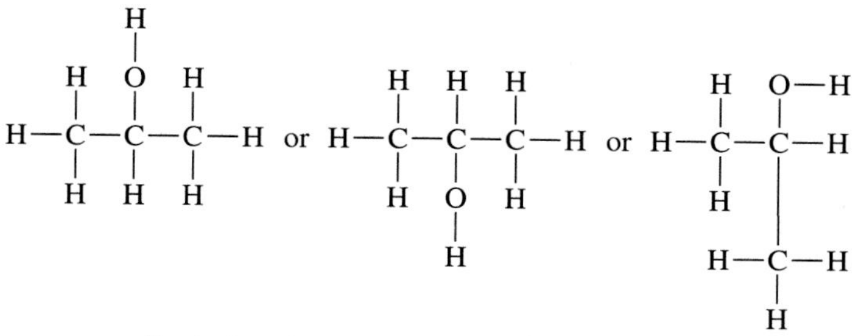

Equivalent dash formulas for isopropyl alcohol

Isopropyl alcohol is a constitutional isomer (Section 1.3A) of propyl alcohol because its atoms are connected in a different order and both compounds have the same molecular formula, C_3H_8O. In isopropyl alcohol the OH group is attached to the central carbon; in propyl alcohol it is attached to an end carbon.

One other point: In problems you will often be asked to write structural formulas for all the isomers with a given molecular formula. Do not make the error of writing several equivalent formulas, like those that we have just shown, mistaking them for different constitutional isomers.

PROBLEM 1.16 _____

There are actually three constitutional isomers with the molecular formula C_3H_8O. We have seen two of them in propyl alcohol and isopropyl alcohol. Write a dash formula for the third isomer.

1.20B Condensed Structural Formulas

Condensed structural formulas are easier to write than dash formulas and, when we become familiar with them, they will impart all the information that is contained in the dash structure. In condensed formulas all of the hydrogen atoms that are attached to a particular carbon are usually written immediately after that carbon. In fully condensed formulas, all of the atoms that are attached to the carbon are usually written immediately after that carbon. For example,

$$H-\overset{\overset{\displaystyle H}{|}}{\underset{\underset{\displaystyle H}{|}}{C}}-\overset{\overset{\displaystyle H}{|}}{\underset{\underset{\displaystyle Cl}{|}}{C}}-\overset{\overset{\displaystyle H}{|}}{\underset{\underset{\displaystyle H}{|}}{C}}-\overset{\overset{\displaystyle H}{|}}{\underset{\underset{\displaystyle H}{|}}{C}}-H$$

$$CH_3CHCH_2CH_3 \quad \text{or} \quad CH_3CHClCH_2CH_3$$
$$\underset{Cl}{|}$$

Dash formula   **Condensed formulas**

The condensed formula for isopropyl alcohol can be written in four different ways:

$$H-\overset{\overset{\displaystyle H}{|}}{\underset{\underset{\displaystyle H}{|}}{C}}-\overset{\overset{\displaystyle H}{|}}{\underset{\underset{\displaystyle O}{|}}{C}}-\overset{\overset{\displaystyle H}{|}}{\underset{\underset{\displaystyle H}{|}}{C}}-H$$
$$\underset{H}{|}$$

$$CH_3CHCH_3 \qquad CH_3CH(OH)CH_3$$
$$\underset{OH}{|}$$

$$CH_3CHOHCH_3 \quad \text{or} \quad (CH_3)_2CHOH$$

Dash formula   **Condensed formulas**

SAMPLE PROBLEM

Write a condensed structural formula for the compound that follows:

$$H-\overset{\overset{\displaystyle H}{|}}{\underset{\underset{\displaystyle H}{|}}{C}}-\overset{\overset{\displaystyle H}{|}}{\underset{\underset{\displaystyle\overset{\displaystyle |}{C}}{|}}{C}}-\overset{\overset{\displaystyle H}{|}}{\underset{\underset{\displaystyle H}{|}}{C}}-\overset{\overset{\displaystyle H}{|}}{\underset{\underset{\displaystyle H}{|}}{C}}-H$$

$$H-\overset{\overset{\displaystyle H}{|}}{\underset{\underset{\displaystyle H}{|}}{C}}-H$$

ANSWER:

$$CH_3CHCH_2CH_3 \text{ or } CH_3CH(CH_3)CH_2CH_3 \text{ or } (CH_3)_2CHCH_2CH_3$$
$$\underset{CH_3}{|}$$

$$\text{or } CH_3CH_2CH(CH_3)_2 \text{ or } CH_3CH_2CHCH_3$$
$$\underset{CH_3}{|}$$

1.20C Cyclic Molecules

Organic compounds not only have their carbon atoms arranged in chains, they can also have them arranged in rings. The compound called cyclopropane has its carbon atoms arranged in a three-membered ring.

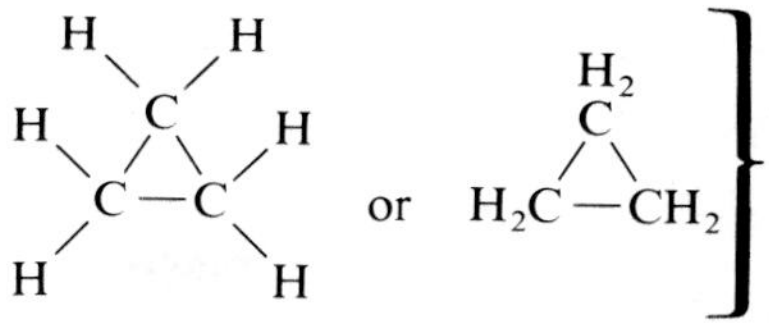

Formulas for cyclopropane

1.20D Bond-Line Formulas

More and more organic chemists are using a very simplified formula called a **bond-line formula** to represent structural formulas. The bond-line representation is the quickest of all to write because it shows only the carbon skeleton. The number of hydrogen atoms necessary to fulfill the carbon atoms' valences are assumed to be present, but we do not write them in. Other atoms (e.g., O, Cl, N) *are* written in. Each intersection of two or more lines and the end of a line represents a carbon atom unless some other atom is written in

$CH_3CHClCH_2CH_3 =$

$CH_3CH(CH_3)CH_2CH_3 =$

$(CH_3)_2NCH_2CH_3 =$

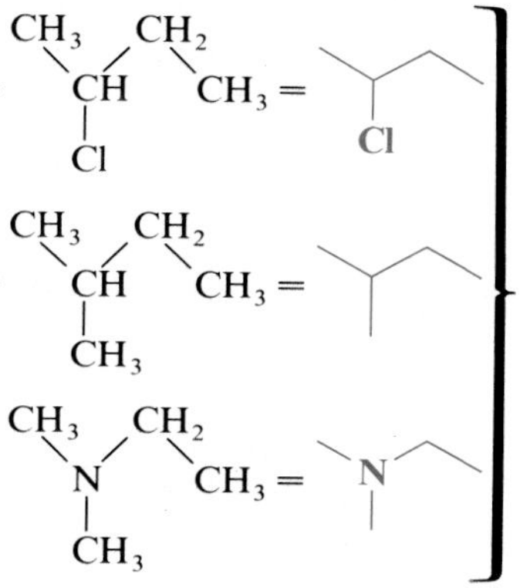

Bond-line formulas

Bond-line formulas are often used for cyclic compounds:

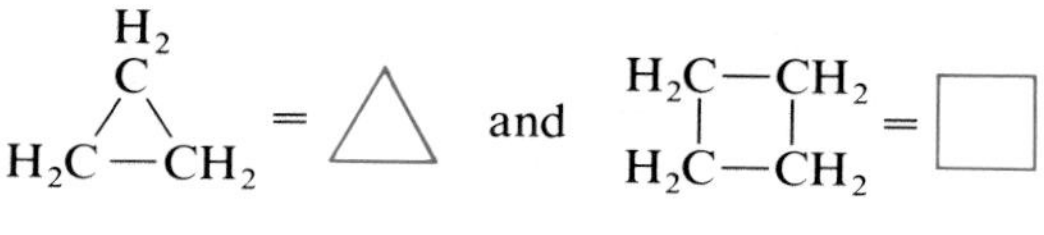

Multiple bonds are also indicated in bond-line formulas. For example,

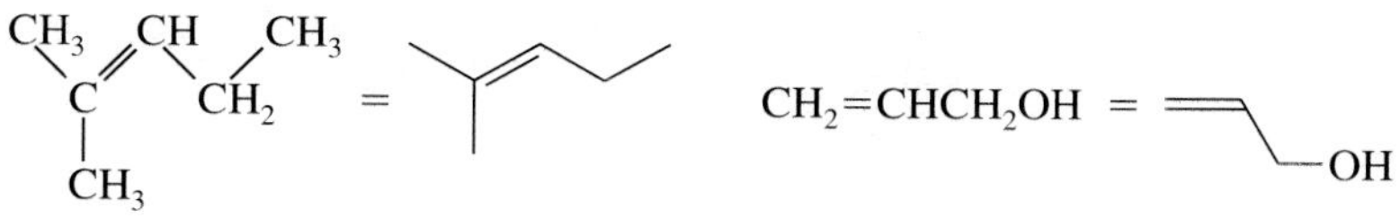

SAMPLE PROBLEM

Write the bond-line formula for $CH_3\underset{\displaystyle CH_3}{CH}CH_2CH_2CH_2OH$.

ANSWER:

First, we outline the carbon skeleton, including the OH group as follows:

$$
\begin{array}{c}
\text{CH}_3 \quad \text{CH}_2 \quad \text{CH}_2 \\
\qquad \text{CH} \quad \text{CH}_2 \quad \text{OH} \\
\qquad \quad \text{CH}_3
\end{array}
\quad = \quad
\begin{array}{c}
\text{C} \quad \text{C} \quad \text{C} \\
\text{C} \quad \text{C} \quad \text{OH} \\
\text{C}
\end{array}
$$

Thus, the bond-line formula is

OH

PROBLEM 1.17 —————————————————————

Outline the carbon skeleton of the following condensed structural formulas and then write each as a bond-line formula.

(a) $(CH_3)_2CHCH_2CH_3$ (f) $CH_2{=}C(CH_2CH_3)_2$

(b) $(CH_3)_2CHCH_2CH_2OH$

(c) $(CH_3)_2C{=}CHCH_2CH_3$

(d) $CH_3CH_2CH_2CH_2CH_3$ $\quad\quad\quad\;\; \overset{\displaystyle O}{\overset{\|}{}}$

(e) $CH_3CH_2CH(OH)CH_2CH_3$ (g) $CH_3CCH_2CH_2CH_2CH_3$

(h) $CH_3CHClCH_2CH(CH_3)_2$

PROBLEM 1.18 —————————————————————

Which molecules in Problem 1.17 form sets of constitutional isomers?

PROBLEM 1.19 —————————————————————

Write dash formulas for each of the following bond-line formulas:

OH

(a)   Cl   (b)   (c)

O

1.20E Three-Dimensional Formulas

None of the formulas that we have described so far conveys any information about how the atoms of a molecule are arranged in space. There are several types of representations that do this. The type of formula that we shall use is shown in Fig. 1.31. In his representation, bonds that project upward out of the plane of the paper are indicated by a wedge (◄), those that lie behind the plane are indicated with a dashed wedge (·····), and those bonds that lie in the plane of the page are indicated by a line (—). Generally we only use three-dimensional formulas when it is necessary to convey information about the shape of the molecule.

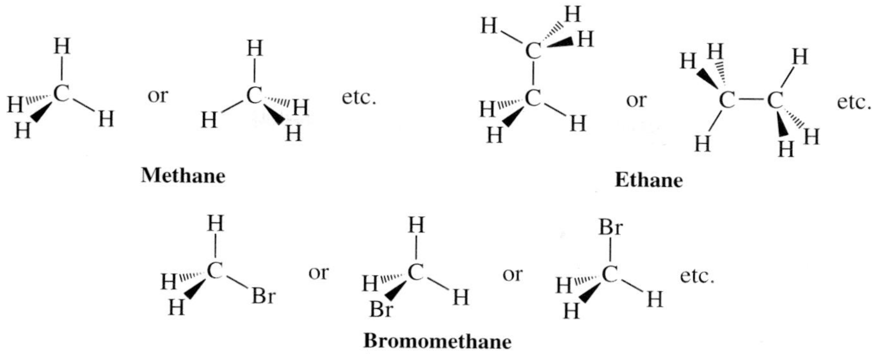

FIGURE 1.31   Three-dimensional formulas using wedge–dashed wedge–line formulas.

PROBLEM 1.20

Write three-dimensional (wedge–dashed wedge–line) representations for each of the following: (a) CH_3Cl, (b) CH_2Cl_2, (c) CH_2BrCl, (d) CH_3CH_2Cl.

ADDITIONAL PROBLEMS

1.21   Which of the following ions possess the electron configuration of a noble gas?
(a) Na^+ (c) F^+ (e) Ca^{2+} (g) O^{2-}
(b) Cl^- (d) H^- (f) S^{2-} (h) Br^+

1.22   Write Lewis structures for each of the following:
(a) $SOCl_2$ (b) $POCl_3$ (c) PCl_5 (d) $HONO_2$ (HNO_3)

⨯1.23   Give the formal charge (if one exists) on each atom of the following:

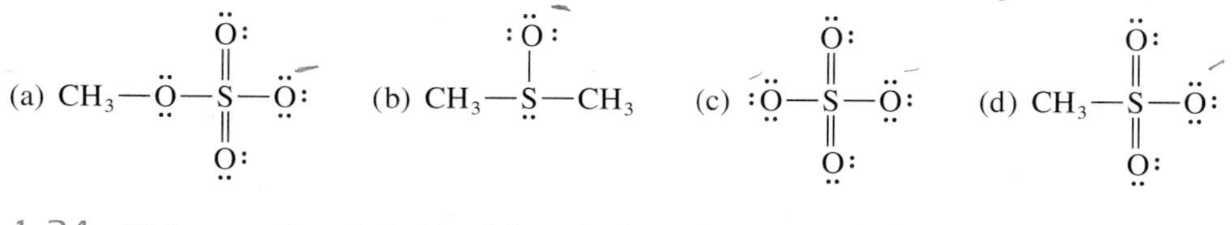

1.24   Write a condensed structural formula for each compound given here.

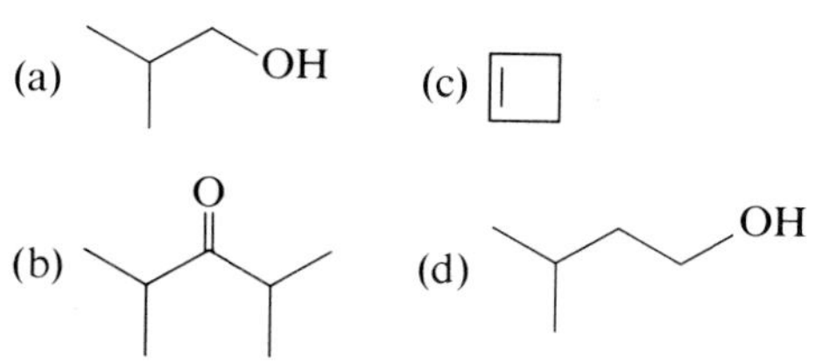

1.25   What is the molecular formula for each of the compounds given in Problem 1.24?

1.26   Consider each pair of structural formulas that follow and state whether the two formulas represent the same compound, whether they represent different compounds that are constitutional isomers of each other, or whether they represent different compounds that are not isomeric.

(a) 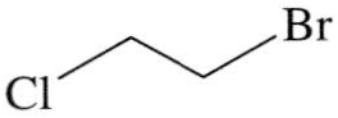 and 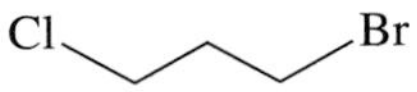

(b) 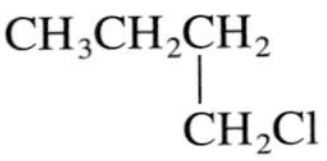 and $ClCH_2CH(CH_3)_2$

(c) 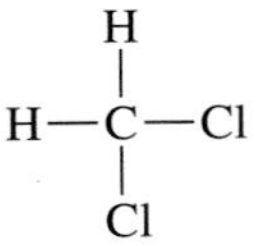 and 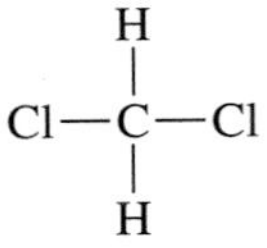

(d) 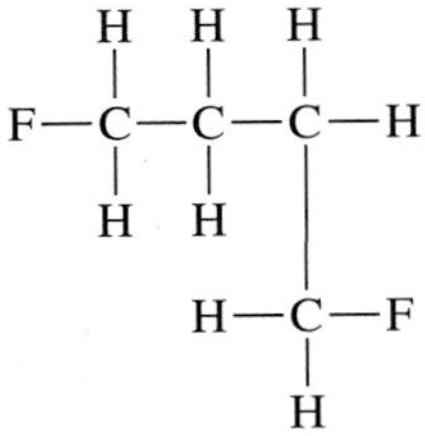 and $CH_2FCH_2CH_2CH_2F$

(e) 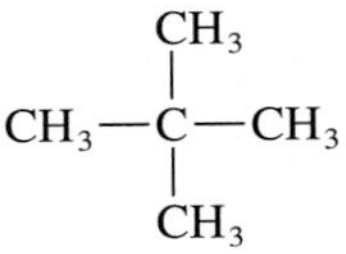 and $(CH_3)_3C{-}CH_3$

(f) $CH_2{=}CHCH_2CH_3$ and 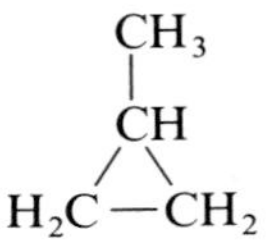

(g) $CH_3OCH_2CH_3$ and 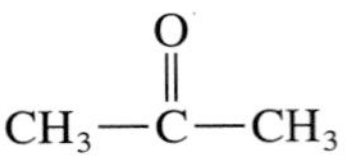

(h) 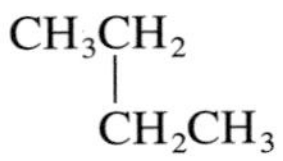 and $CH_3CH_2CH_2CH_3$

(i) $CH_3OCH_2CH_3$ and 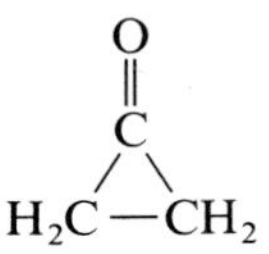

(j) $CH_2ClCHClCH_3$ and $CH_3CHClCH_2Cl$

(k) $CH_3CH_2CHClCH_2Cl$ and 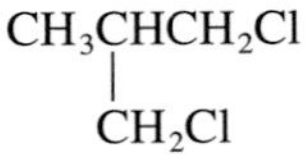

(l) 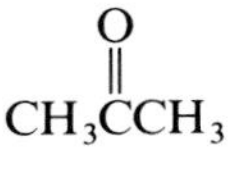 and 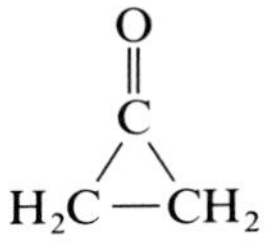

(m) 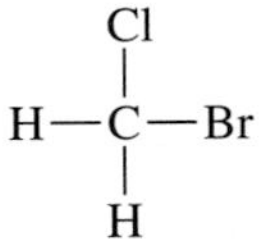 and 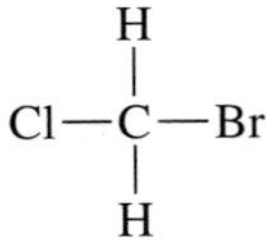

(n) $CH_3-\underset{\underset{H}{|}}{\overset{\overset{CH_3}{|}}{C}}-H$ and $CH_3-\underset{\underset{H}{|}}{\overset{\overset{H}{|}}{C}}-CH_3$

(o) [structure] and [structure]

(p) [structure] and [structure]

1.27 Write a three-dimensional formula for each of the following molecules. If the molecule has a net dipole moment, indicate its direction with an arrow, $\longmapsto$. If the molecule has no net dipole moment, you should so state. (You may ignore the small polarity of C—H bonds in working this and similar problems.)

(a) CH_3F (c) CHF_3 (e) CH_2FCl (g) BeF_2 (i) CH_3OH

(b) CH_2F_2 (d) CF_4 (f) BCl_3 (h) CH_3OCH_3 (j) CH_2O

1.28 Rewrite each of the following using bond-line formulas:

(a) $CH_3CH_2CH_2\overset{\overset{O}{\|}}{C}CH_3$

(b) $CH_3\underset{\underset{CH_3}{|}}{C}HCH_2CH_2\underset{\underset{CH_3}{|}}{C}HCH_2CH_3$

(c) $(CH_3)_3CCH_2CH_2CH_2OH$

(d) $CH_3CH_2\underset{\underset{CH_3}{|}}{C}HCH_2\overset{\overset{O}{\|}}{C}OH$

(e) $CH_2{=}CHCH_2CH_2CH{=}CHCH_3$

(f) [cyclohexanone structure with HC, C, CH₂, HC, CH₂, C, H₂]

1.29 Write a dash formula for each of the following showing any unshared electron pairs:

(a)

(b)

(c) $(CH_3)_2NCH_2CH_3$

(d) [tetrahydrofuran structure]

*1.30 Write structural formulas of your choice for all of the constitutional isomers with the molecular formula C_4H_8.

1.31 Write structural formulas for at least three constitutional isomers with the molecular formula CH_3NO_2. (In answering this question you should assign a formal charge to any atom that bears one.)

1.32   Chloromethane (CH_3Cl) has a larger dipole moment ($\mu = 1.87$ D) than fluorometh-ane (CH_3F) ($\mu = 1.81$ D), even though fluorine is more electronegative than chlorine. Explain.

1.33   Cyanic acid ($H\!-\!O\!-\!C\!\equiv\!N$) and isocyanic acid ($H\!-\!N\!=\!C\!=\!O$) differ in the positions of their electrons but their structures do not represent resonance structures.   (a) Explain. (b) Loss of a proton from cyanic acid yields the same anion as that obtained by loss of a proton from isocyanic acid. Explain.

1.34   Consider a chemical species (either a molecule or an ion) in which a carbon atom forms three single bonds to three hydrogen atoms, and in which the carbon atom possesses no other electrons.   (a) What formal charge would the carbon atom have?   (b) What total charge would the species have?   (c) What shape would you expect this species to have?   (d) What would you expect the hybridization state of the carbon atom to be?

1.35   Consider a chemical species like the one in the previous problem in which a carbon atom forms three single bonds to three hydrogen atoms, but in which the carbon atom possesses an unshared electron pair.   (a) What formal charge would the carbon atom have?   (b) What total charge would the species have?   (c) What shape would you expect this species to have? (d) What would you expect the hybridization state of the carbon atom to be?

1.36   Consider another chemical species like the ones in the previous problems in which a carbon atom forms three single bonds to three hydrogen atoms, but in which the carbon atom possesses a single unpaired electron.   (a) What formal charge would the carbon atom have? (b) What total charge would the species have?   (c) Given that the shape of this species is trigonal planar, what would you expect the hybridization state of the carbon atom to be?

1.37   Ozone (O_3) is found in the upper atmosphere where it absorbs highly energetic ultra-violet (UV) radiation and thereby provides the surface of the earth with a protective screen (cf. Section 9.11C). One possible resonance structure for ozone is the following:

$$:\ddot{O}\diagdown_{}^{}\overset{\ddot{O}}{\diagup}\ddot{O}:$$

(a) Assign any necessary formal charges to the atoms in this structure. (b) Write another equivalent resonance structure for ozone. (c) What do these resonance structures predict about the relative lengths of the two oxygen-oxygen bonds of ozone?   (d) The structure above, and the one you have written, assumes an angular shape for the ozone molecule. Is this shape consistent with VSEPR theory? Explain your answer.   (e) What experimentally measurable quantity would allow you to prove that the ozone molecule is angular and not linear? Explain.

1.38   Write resonance structures for the azide ion, N_3^-. Explain how these resonance structures account for the fact that both bonds of the azide ion have the same length.

1.39   Write structural formulas of the type indicated: (a) Bond-line formulas for seven constitutional isomers with the formula $C_4H_{10}O$.   (b) Condensed structural formulas for two constitutional isomers with the formula C_2H_7N.   (c) Condensed structural formulas for four constitutional isomers with the formula C_3H_9N.   (d) Bond-line formulas for three constitutional isomers with the formula C_5H_{12}.

1.40   Consider each of the following molecules in turn: (a) dimethyl ether, $(CH_3)_2O$, (b) trimethylamine, $(CH_3)_3N$,   (c) trimethylboron, $(CH_3)_3B$, and   (d) dimethylberyllium, $(CH_3)_2Be$. Describe the hybridization state of the central atom (i.e., O, N, B, or Be) of each molecule, tell what bond angles you would expect at the central atom, and state whether or not the molecule would have a dipole moment.

1.41   Analyze the statement: For a molecule to be polar, the presence of polar bonds is necessary, but it is not a sufficient requirement.

1.42   In Chapter 15 we will learn how the nitronium ion, NO_2^+, forms when concentrated nitric and sulfuric acid are mixed.   (a) Write a Lewis structure for the nitronium ion.   (b) What geometry does VSEPR theory predict for the NO_2^+ ion?   (c) Give a species that has the same number of electrons as NO_2^+.

CHAPTER 2

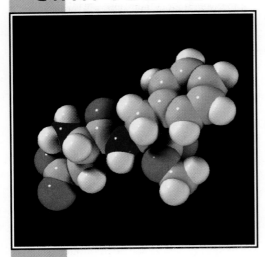

Aspartame
(Problem 2.15)

REPRESENTATIVE CARBON COMPOUNDS

2.1 CARBON–CARBON COVALENT BONDS

Carbon's ability to form strong covalent bonds to other carbon atoms is the single property of the carbon atom that—more than any other—accounts for the very existence of a field of study called organic chemistry. It is this property too that accounts in part for carbon being the element around which most of the molecules of living organisms are constructed. Carbon's ability to form as many as four strong bonds to other carbon atoms and to form strong bonds to hydrogen, oxygen, sulfur, and nitrogen atoms as well, provides the necessary versatility of structure that makes possible the vast number of different molecules required for complex living organisms.

2.2 METHANE AND ETHANE: REPRESENTATIVE ALKANES

Methane (CH_4) and ethane (C_2H_6) are two members of a broad family of organic compounds called **hydrocarbons.** Hydrocarbons, as the name implies, are compounds whose molecules contain only carbon and hydrogen atoms. Methane and ethane also belong to a subgroup of hydrocarbons known as **alkanes** whose members do not have multiple bonds between carbon atoms. *Hydrocarbons whose molecules have a carbon–carbon double bond* are called *alkenes,* and *those with a carbon–carbon triple bond* are called *alkynes.* Hydrocarbons that contain a special ring that we shall introduce in Section 2.7 and study in Chapter 14 are called aromatic hydrocarbons.

Generally speaking, compounds such as the alkanes, whose molecules contain only single bonds, are referred to as **saturated compounds** because these compounds contain the maximum number of hydrogen atoms that the carbon compound can possess. Compounds with multiple bonds, such as alkenes, alkynes, and aromatic hydrocarbons are called **unsaturated compounds,** because they possess fewer than the maximum number of hydrogen atoms, and are capable of reacting with hydrogen under the proper conditions. We shall have more to say about this in Chapter 7.

2.2A Sources of Methane

Methane was one major component of the early atmosphere of this planet. Methane is still found in the atmosphere of Earth, but no longer in appreciable amounts. It is, however, a major component of the atmosphere of Jupiter, Saturn, Uranus, and Neptune. Recently, methane has also been detected in interstellar space—far from the Earth (10^{16} km) in a celestial body that emits radio waves in the constellation Orion.

On Earth, methane is the major component of natural gas, along with ethane and other low molecular weight alkanes. The United States is currently using its large reserves of natural gas at a very high rate. Because the components of natural gas are important in industry, efforts are being made to develop coal-gasification processes to provide alternative sources.

Some living organisms produce methane from carbon dioxide and hydrogen.

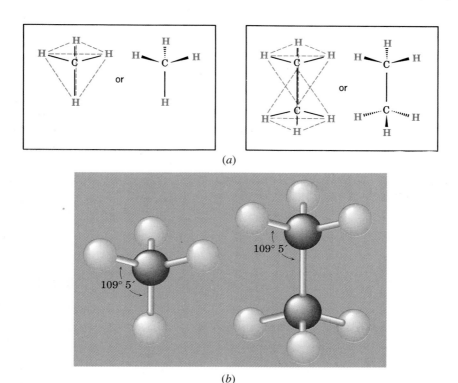

(a)

(b)

FIGURE 2.1 (a) Two ways of representing the structures of methane and ethane that show the tetrahedral arrangements of the atoms around carbon. (b) Ball-and-stick models of methane and ethane.

These very primitive creatures called *methanogens* may be the Earth's oldest organisms, and they may represent a separate form of evolutionary development. Methanogens can survive only in an anaerobic (i.e., oxygen-free) environment. They have been found in ocean trenches, in mud, in sewage, and in cows' stomachs.

2.2B The Structure of Ethane

The bond angles at the carbon atoms of ethane, and of all alkanes, are also tetrahedral like those in methane. In the case of ethane (Fig. 2.1), each carbon atom is at one corner of the other carbon atom's tetrahedron; hydrogen atoms are situated at the other three corners.

A satisfactory model for ethane (and for other alkanes as well) can be provided by sp^3-hybridized carbon atoms (Section 1.13). Figure 2.2 shows how we might imagine the bonding molecular orbitals of an ethane molecule being constructed from two sp^3-hybridized carbon atoms and six hydrogen atoms.

The carbon–carbon bond of ethane is a *sigma bond* (Section 1.13), formed by two overlapping sp^3 orbitals. (The carbon–hydrogen bonds are also sigma bonds. They are formed from overlapping carbon sp^3 orbitals and hydrogen s orbitals.)

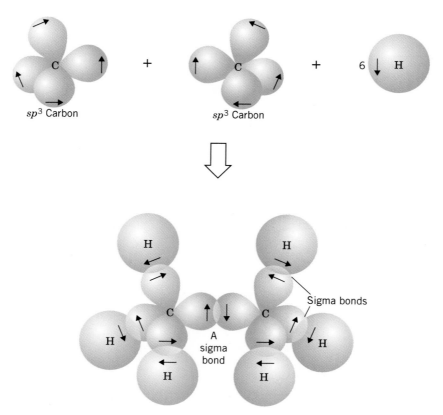

FIGURE 2.2 The formation of the bonding molecular orbitals of ethane from two sp^3-hybridized carbon atoms and six hydrogen atoms. All of the bonds are sigma bonds. (Antibonding sigma molecular orbitals—called σ^* orbitals—are formed in each instance as well, but for simplicity these are not shown.)

Because a sigma bond (i.e., any nonmultiple bond) has circular symmetry along the bond axis, *rotation of groups joined by a single bond does not usually require a large amount of energy.* Consequently, groups joined by single bonds rotate relatively freely with respect to one another. (We discuss this point further in Section 4.6.)

2.3 ALKENES: COMPOUNDS CONTAINING THE CARBON– CARBON DOUBLE BOND; ETHENE AND PROPENE

The carbon atoms of virtually all of the molecules that we have considered so far have used their four valence electrons to form four single covalent bonds to four other atoms. We find, however, that many important organic compounds exist in which carbon atoms share more than two electrons with another atom. In molecules of these compounds some bonds that are formed are multiple covalent bonds. When two carbon atoms share two pairs of electrons, for example, the result is a carbon–carbon double bond.

$$\ddot{C}::\ddot{C} \quad \text{or} \quad \diagdown \!\! C\!\!=\!\!C \diagup$$

Hydrocarbons whose molecules contain a carbon–carbon double bond are called **alkenes.** Ethene (C_2H_4) and propene (C_3H_6) are both alkenes. (Ethene is also called ethylene, and propene is sometimes called propylene.)

$$\begin{matrix} H & & H \\ \diagdown & & \diagup \\ & C\!\!=\!\!C & \\ \diagup & & \diagdown \\ H & & H \end{matrix} \qquad \begin{matrix} H & & H \\ \diagdown & & \diagup \\ & C\!\!=\!\!C & \\ \diagup & & \diagdown \\ H_3C & & H \end{matrix}$$

Ethene   **Propene**

In ethene the only carbon–carbon bond is a double bond. Propene has one carbon–carbon single bond and one carbon–carbon double bond.

The spatial arrangement of the atoms of alkenes is different from that of alkanes. The six atoms of ethene are coplanar, and the arrangement of atoms around each carbon atom is triangular (Fig. 2.3). In Section 2.4 we shall see how the structure of ethene can be explained on the basis of the same kind of orbital hybridization, sp^2, that we learned about for BH_3 (Section 1.14).

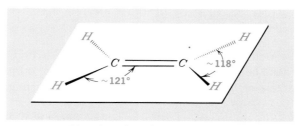

FIGURE 2.3 The structure and bond angles of ethene.

2.4 ORBITAL HYBRIDIZATION AND THE STRUCTURE OF ALKENES

A satisfactory model for the carbon–carbon double bond can be based on sp^2-hybridized carbon atoms.*

The mathematical mixing of orbitals that furnish the sp^2 orbitals for our model can be visualized in the way shown in Fig. 2.4. The $2s$ orbital is mathematically mixed (or hybridized) with two of the $2p$ orbitals. (The hybridization procedure applies only to the orbitals, not to the electrons.) One $2p$ orbital is left unhybridized. One electron is then placed in each of the sp^2-hybrid orbitals and one electron remains in the $2p$ orbital.

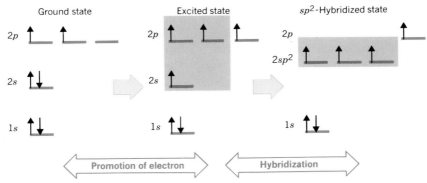

FIGURE 2.4 A process for obtaining sp^2-hybridized carbon atoms.

The three sp^2 orbitals that result from hybridization are directed toward the corners of a regular triangle (with angles of 120° between them). The carbon p orbital that is not hybridized is perpendicular to the plane of the triangle formed by the hybrid sp^2 orbitals (Fig. 2.5).

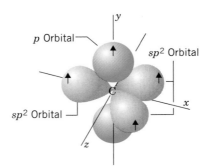

FIGURE 2.5 An sp^2-hybridized carbon atom.

In our model for ethene (Fig. 2.6) we see that two sp^2-hybridized carbon atoms form a sigma (σ) bond between them by the overlap of one sp^2 orbital from each. The remaining sp^2 orbitals of the carbon atoms form σ bonds to four hydrogen atoms

---

*An alternative model for the carbon–carbon bond is discussed in an article by W. E. Palke, *J. Am. Chem. Soc.*, **1986**, *108*, 6543–6544.

through overlap with the $1s$ orbitals of the hydrogen atoms. These five bonds account for 10 of the 12 bonding electrons of ethene, and they are called the **σ-bond framework.** The bond angles that we would predict on the basis of sp^2-hybridized carbon atoms (120° all around) are quite close to the bond angles that are actually found (Fig. 2.3).

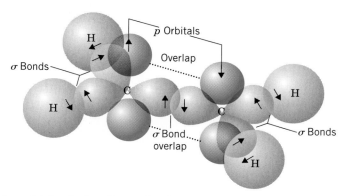

FIGURE 2.6 A model for the bonding molecular orbitals of ethene formed from two sp^2-hybridized carbon atoms and four hydrogen atoms.

The remaining two bonding electrons in our model are located in the p orbitals of each carbon atom. We can better visualize how these p orbitals interact with each other if we replace the σ bonds by lines. This is shown in Fig. 2.7. We see that the parallel p orbitals *overlap above and below the plane of the σ framework.* This sideways overlap of the p orbitals results in a new type of covalent bond, known as a **pi (π) bond.** Note the difference in shape of the bonding molecular orbital of a π bond as contrasted to that of a σ bond. A σ bond has cylindrical symmetry about a line connecting the two bonded nuclei. A π bond has a nodal plane passing through the two bonded nuclei.

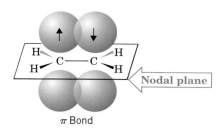

FIGURE 2.7 The overlapping p orbitals of ethene to make a π bond.

According to molecular orbital theory, both bonding and antibonding π molecular orbitals are formed when p orbitals interact in this way to form a π bond. The bonding π orbital (Fig. 2.8) results when p-orbital lobes of like signs overlap; the antibonding π orbital is formed when p-orbital lobes of opposite signs overlap.

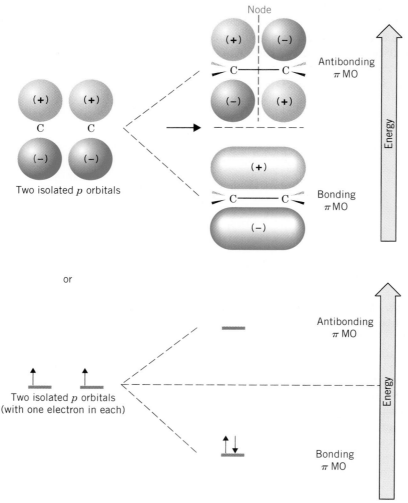

FIGURE 2.8 How two *p* orbitals combine to form two π (pi) molecular orbitals. The bonding MO is of lower energy. The higher energy antibonding MO contains an additional node. (Both orbitals have a node in the plane containing the C and H atoms.)

The bonding π orbital is the lower energy orbital and contains both π electrons (with opposite spins) in the ground state of the molecule. The region of greatest probability of finding the electrons in the bonding π orbital is a region generally situated above and below the plane of the σ-bond framework between the two carbon atoms. The antibonding π^* orbital is of higher energy, and it is not occupied by electrons when the molecule is in the ground state. It can become occupied, however, if the molecule absorbs light of the right frequency and an electron is promoted from the lower energy level to the higher one. The antibonding π^* orbital has a nodal plane between the two carbon atoms.

To summarize: In our model based on orbital hybridization, the carbon−carbon double bond is viewed as consisting of two different kinds of bonds, *a σ bond and a π bond*. The σ bond results from two overlapping sp^2 orbitals end-to-end and is symmetrical about an axis linking the two carbon atoms. The π bond results from a sideways overlap of two p orbitals; it has a nodal plane like a p orbital. In the ground state the electrons of the π bond are located between the two carbon atoms but generally above and below the plane of the σ-bond framework.

Electrons of the π bond have greater energy than electrons of the σ bond. The relative energies of the σ and π molecular orbitals (with the electrons in the ground state) are shown in the following figure. (The σ^* orbital is the antibonding sigma orbital.)

2.4A Restricted Rotation and the Double Bond

The $\sigma-\pi$ model for the carbon–carbon double bond also accounts for an important property of the double bond: *There is a large barrier to free rotation associated with groups joined by a double bond.* Maximum overlap between the p orbitals of a π bond occurs when the axes of the p orbitals are exactly parallel. Rotating one carbon of the double bond 90° (Fig. 2.9) breaks the π bond, for then the axes of the p orbitals are perpendicular and there is no net overlap between them. Estimates based on thermochemical calculations indicate that the strength of the π bond is 63 kcal mol^{-1}. This, then, is the barrier to rotation of the double bond. It is markedly higher than the rotational barrier of groups joined by carbon–carbon single bonds (3–6 kcal mol^{-1}). While groups joined by single bonds rotate relatively freely at room temperature, those joined by double bonds do not.

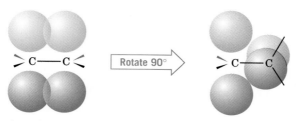

FIGURE 2.9 Rotation of a carbon atom of a double bond through an angle of 90° results in the breaking of the π bond.

2.4B Cis–Trans Isomerism

Restricted rotation of groups joined by a double bond causes a new type of isomerism that we illustrate with the two dichloroethenes written in the following structures.

cis-1,2-Dichloroethene   *trans*-1,2-Dichloroethene

These two compounds are isomers; they are different compounds that have the same molecular formula. We can tell that they are different compounds by trying to super-pose a model of one on a model of the other. We find that it cannot be done. By superpose we mean that we place one model on the other *so that all parts of each coincide.*

We indicate that they are different isomers by attaching the prefixes cis or trans to their names (cis, Latin: on this side; trans, Latin: across). *cis*-1,2-Dichloroethene and *trans*-1,2-dichloroethene are not constitutional isomers because the connectivity of the atoms is the same in each. The two compounds *differ only in the arrangement of their atoms in space.* Isomers of this kind are classified formally as **stereoisomers,** but often they are called simply cis–trans isomers. (We shall study stereoisomerism in detail in Chapters 4 and 5.)

The structural requirements for cis–trans isomerism will become clear if we con-sider a few additional examples. 1,1-Dichloroethene and 1,1,2-trichloroethene do not show this type of isomerism.

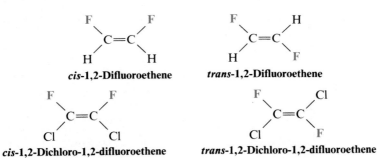

1,2-Difluoroethene and 1,2-dichloro-1,2-difluoroethene do exist as cis–trans isomers. Notice that we designate the isomer with two identical groups on the same side as being cis.

Clearly, then, *cis–trans isomerism of this type is not possible if one carbon atom of the double bond bears two identical groups.*

PROBLEM 2.1

Which of the following alkenes can exist as cis–trans isomers? Write their structures.

(a) CH_2=$CHCH_2CH_3$ (c) CH_2=$C(CH_3)_2$

(b) CH_3CH=$CHCH_3$ (d) CH_3CH_2CH=$CHCl$

Cis–trans isomers have different physical properties. They have different melting points and boiling points, and often cis–trans isomers differ markedly in the magni-

TABLE 2.1   *Physical properties of cis–trans isomers*

COMPOUND	MELTING POINT (°C)	BOILING POINT (°C)	DIPOLE MOMENT (D)
cis-1,2-Dichloroethene	− 80	60	1.90
trans-1,2-Dichloroethene	− 50	48	0
cis-1,2-Dibromoethene	− 53	112.5	1.35
trans-1,2-Dibromoethene	− 6	108	0

tude of their dipole moments. Table 2.1 summarizes some of the physical properties of two pairs of cis–trans isomers.

PROBLEM 2.2 _____

Indicate the direction of the important bond moments in each of the following compounds (neglect C—H bonds). You should also give the direction of the net dipole moment for the molecule. If there is no net dipole moment state that $\mu = 0$.

(a) *cis*-CHF=CHF   (c) CH_2=CF_2
(b) *trans*-CHF=CHF   (d) CF_2=CF_2

PROBLEM 2.3 _____

Write structural formulas for all of the alkenes with the formula (a) $C_2H_2Br_2$, and (b) with the formula $C_2Br_2Cl_2$. In each instance designate compounds that are cis–trans isomers of each other.

2.5 ALKYNES: COMPOUNDS CONTAINING THE CARBON–CARBON TRIPLE BOND; ETHYNE (ACETYLENE) AND PROPYNE

Hydrocarbons in which two carbon atoms share three pairs of electrons between them, and are thus bonded by a triple bond, are called **alkynes.** The two simplest alkynes are ethyne and propyne.

$$H-C\equiv C-H \qquad CH_3-C\equiv C-H$$

Ethyne   **Propyne**
(acetylene)   **(C_3H_4)**
(C_2H_2)

Ethyne, a compound that is also called acetylene, consists of linear molecules. The H—C≡C bond angles of ethyne molecules are 180°.

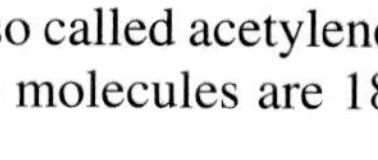

$$H-C\equiv C-H$$
180°   180°

2.6 ORBITAL HYBRIDIZATION AND THE STRUCTURE OF ALKYNES

We can account for the structure of ethyne on the basis of orbital hybridization as we did for ethane and ethene. In our model for ethane (Section 2.2B) we saw that the carbon orbitals are sp^3 hybridized, and in our model for ethene (Section 2.4) we saw that they are sp^2 hybridized. In our model for ethyne we shall see that the carbon atoms are *sp hybridized* and resemble the hybrid orbitals of BeH_2 (Section 1.15).

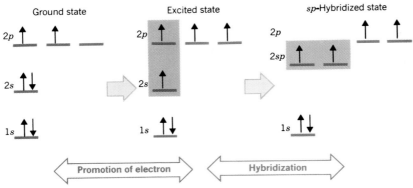

FIGURE 2.10   A process for obtaining *sp*-hybridized carbon atoms.

The mathematical process for obtaining the *sp*-hybrid orbitals of ethyne can be visualized in the following way (Fig. 2.10). The $2s$ orbital and one $2p$ orbital of carbon are hybridized to form two *sp* orbitals. The remaining two $2p$ orbitals are not hybridized. Calculations show that the *sp*-hybrid orbitals have their large positive lobes oriented at an angle of 180° with respect to each other. The $2p$ orbitals that were not hybridized are perpendicular to the axis that passes through the center of the two *sp* orbitals (Fig. 2.11).

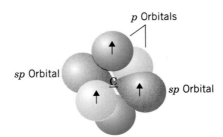

FIGURE 2.11   An *sp*-hybridized carbon atom.

We envision the bonding molecular orbitals of ethyne being formed in the following way (Fig. 2.12). Two carbon atoms overlap *sp* orbitals to form a sigma bond between them (this is one bond of the triple bond). The remaining two *sp* orbitals at each carbon atom overlap with *s* orbitals from hydrogen atoms to produce two sigma C—H bonds. The two *p* orbitals on each carbon atom also overlap side to side to form two π bonds. These are the other two bonds of the triple bond. If we replace the σ bonds of this illustration with lines, it is easier to see how the *p* orbitals overlap. Thus we see that the carbon–carbon triple bond consists of two π bonds and one σ bond.

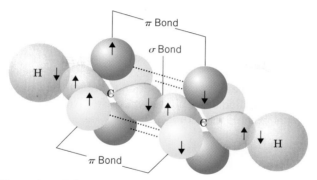

FIGURE 2.12 Formation of the bonding molecular orbitals of ethyne from two *sp*-hybridized carbon atoms and two hydrogen atoms. (Antibonding orbitals are formed as well but these have been omitted for simplicity.)

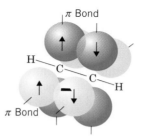

The two π bonds create a cylinder of π electrons around the central sigma bond; as a result there is no restriction of rotation of groups joined by a triple bond.

2.6A Bond Lengths of Ethyne, Ethene, and Ethane

The carbon–carbon triple bond is shorter than the carbon–carbon double bond, and the carbon–carbon double bond is shorter than the carbon–carbon single bond. The carbon–hydrogen bonds of ethyne are also shorter than those of ethene, and the carbon–hydrogen bonds of ethene are shorter than those of ethane. This illustrates a general principle: *The shortest C—H bonds are associated with those carbon orbitals with the greatest s character.* The *sp* orbitals of ethyne—50% *s* (and 50% *p*) in character—form the shortest C—H bonds. The *sp*³ orbitals of ethane—25% *s* (and 75% *p*) in character—form the longest C—H bonds. The differences in bond lengths and bond angles of ethyne, ethene, and ethane are summarized in Fig. 2.13.

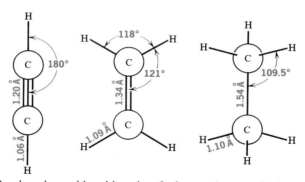

FIGURE 2.13 Bond angles and bond lengths of ethyne, ethene, and ethane.

2.7 BENZENE: A REPRESENTATIVE AROMATIC HYDROCARBON

In Chapter 14 we shall study in detail a group of unsaturated cyclic hydrocarbons known as **aromatic compounds.** The compound known as **benzene** is the prototypical aromatic compound. Benzene can be written as a six-membered ring with alternating single and double bonds, called a Kekulé structure after August Kekulé (Section 1.3), who first conceived of this representation.

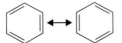
                                        or

Kekulé structure   **Bond line representation**
for benzene   **of Kekulé structure**

Even though the Kekulé structure is frequently used for benzene compounds, there is much evidence that this representation is inadequate and incorrect. For example, if benzene had alternating single and double bonds as the Kekulé structure indicates, we would expect the lengths of the carbon–carbon bonds around the ring to be alternately longer and shorter, as we typically find with carbon–carbon single and double bonds (Fig. 2.3). In fact, the carbon–carbon bonds of benzene are all the same length (1.39 Å), a value in between that of a carbon–carbon single bond and a carbon–carbon double bond. There are two ways of dealing with this problem: with resonance theory or with molecular orbital theory.

If we use resonance theory, we visualize benzene as being represented by either of two equivalent Kekulé structures:

Two contributing Kekulé structures   **A representation of the**
for benzene   **resonance hybrid**

Based on the principles of resonance theory (Section 1.8) we recognize that benzene cannot be represented adequately by either structure, but that, instead, *it should be visualized as a hybrid of the two structures.* We represent this hybrid by a hexagon with a circle in the middle. Resonance theory, therefore, solves the problem we encountered in understanding how all of the carbon–carbon bonds are the same length. According to resonance theory, the bonds are not alternating single and double bonds, they are a resonance hybrid of the two: Any bond that is a single bond in the first contributor is a double bond in the second, and vice versa. Therefore, we should expect *all of the carbon–carbon bonds to be the same,* to be one- and one-half bonds, and to have a bond length in between that of a single bond and a double bond. That is, of course, what we actually find.

In the molecular orbital explanation, which we shall describe in much more depth in Chapter 14, we begin by recognizing that the carbon atoms of the benzene ring are sp^2 hybridized. Therefore, each carbon has a p orbital that has one lobe above the plane of the ring and one lobe below.

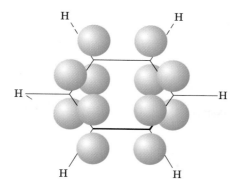

Although the depiction of *p* orbitals in our illustration does not show this, each lobe of each *p* orbital, above and below the ring, overlaps with the lobe of a *p* orbital on each atom on either side of it. This kind of overlap of *p* orbitals leads to a set of bonding molecular orbitals that encompass all of the carbon atoms of the ring. Therefore, the six electrons associated with these *p* orbitals (one from each) are **delocalized** about all six carbon atoms of the ring. This delocalization of electrons explains how all the carbon–carbon bonds are equivalent and have the same length. In Section 14.11, when we study nuclear magnetic resonance spectroscopy, we shall present convincing physical evidence for this delocalization of the electrons.

When the benzene ring is attached to some other group of atoms in a molecule, it is called a **phenyl group** and it is represented in several ways:

or   or C_6H_5—   or   Ph—   or ϕ—

Ways of representing a phenyl group

The combination of a phenyl group and a —CH_2— group is called a **benzyl group.**

—CH_2—   or   —CH_2—   or $C_6H_5CH_2$—

Ways of representing a benzyl group

2.8 FUNCTIONAL GROUPS

One great advantage of the structural theory is that it enables us to classify the vast number of organic compounds into a relatively small number of families based on their structures. (The end papers inside the front cover of this text give the most important of these families.) The molecules of compounds in a particular family are characterized by the presence of a certain arrangement of atoms called a **functional group.**

A functional group is the part of a molecule where most of its chemical reactions occur. It is the part that effectively determines the compound's chemical properties (and many of its physical properties as well). The functional group of an alkene, for example, is its carbon–carbon double bond. When we study the reactions of alkenes in greater detail in Chapter 8, we shall find that most of the chemical reactions of alkenes are the chemical reactions of the carbon–carbon double bond.

The functional group of an alkyne is its carbon–carbon triple bond. Alkanes do not have a functional group. Their molecules have carbon–carbon single bonds and carbon–hydrogen bonds, but these bonds are present in molecules of almost all organic molecules, and C—C and C—H bonds are, in general, much less reactive than common functional groups.

2.8A Alkyl Groups and the Symbol R

Alkyl groups are the groups that we identify for purposes of naming compounds. They are groups that would be obtained by removing a hydrogen atom from an alkane:

Alkane	*Alkyl Group*	*Abbreviation*
CH_4	CH_3—	Me—
Methane	**Methyl group**	
CH_3CH_3	CH_3CH_2— or C_2H_5—	Et—
Ethane	**Ethyl group**	
$CH_3CH_2CH_3$	$CH_3CH_2CH_2$—	Pr—
Propane	**Propyl group**	
$CH_3CH_2CH_3$	CH_3CHCH_3 or $CH_3\overset{\displaystyle CH_3}{\underset{\displaystyle \|}{CH}}$—	*i*-Pr—
Propane	**Isopropyl group**	

While only one alkyl group can be derived from methane and ethane (the **methyl** and **ethyl** groups, respectively), two groups can be derived from propane. Removal of a hydrogen from one of the end carbon atoms gives a group that is called the **propyl** group; removal of a hydrogen from the middle carbon atom gives a group that is called the **isopropyl** group. The names and structures of these groups are used so frequently in organic chemistry that you should learn them now.

We can simplify much of our future discussion if, at this point, we introduce a symbol that is widely used in designating general structures of organic molecules: The symbol R. *R is used as a general symbol to represent any alkyl group.* For example, R might be a methyl group, an ethyl group, a propyl group, or an isopropyl group.

$$\left. \begin{array}{ll} CH_3- & \text{Methyl} \\ CH_3CH_2- & \text{Ethyl} \\ CH_3CH_2CH_2- & \text{Propyl} \\ CH_3CHCH_3 & \text{Isopropyl} \end{array} \right\} \begin{array}{l} \textbf{All of} \\ \textbf{these} \\ \textbf{can be} \\ \textbf{designated} \end{array} \textbf{by R}$$

Thus, the general formula for an alkane is R—H.

2.9 ALKYL HALIDES OR HALOALKANES

Alkyl halides are compounds in which a halogen atom (fluorine, chlorine, bromine, or iodine) replaces a hydrogen atom of an alkane. For example, CH_3Cl and CH_3CH_2Br are alkyl halides. Alkyl halides are also called **haloalkanes.**

Alkyl halides are classified as being primary (1°), secondary (2°), or tertiary (3°).* *This classification is based on the carbon atom to which the halogen is directly attached.* If the carbon *atom* that bears the halogen is attached to only one other carbon, the carbon atom is said to be a **primary carbon atom** and the alkyl halide is classified as a **primary alkyl halide.** If the carbon that bears the halogen is itself attached to two other carbon atoms, then the carbon is a **secondary carbon** and the alkyl halide is a **secondary alkyl halide.** If the carbon that bears the halogen is attached to three other carbon atoms, then the carbon is a **tertiary carbon** and the alkyl halide is a **tertiary alkyl halide.** Examples of primary, secondary, and tertiary alkyl halides are the following:

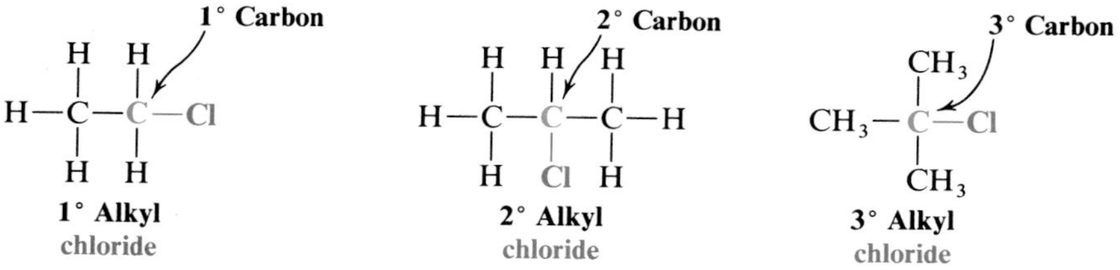

PROBLEM 2.4

Write structural formulas   (a) for two constitutionally isomeric primary alkyl bromides with the formula C_4H_9Br,   (b) for a secondary alkyl bromide, and (c) for a tertiary alkyl bromide with the same formula.

PROBLEM 2.5

Although we shall discuss the naming of organic compounds later when we discuss the individual families in detail, one method of naming alkyl halides is so straightforward that it is worth describing here. We simply name the alkyl group attached to the halogen and add the word *fluoride, chloride, bromide, or iodide.* Write formulas for   (a) ethyl fluoride and   (b) isopropyl chloride. What are names for   (c) $CH_3CH_2CH_2Br$,   (d) CH_3CHFCH_3, and   (e) C_6H_5I ?

2.10 ALCOHOLS

Methyl alcohol (more systematically called methanol) has the structural formula CH_3OH and is the simplest member of a family of organic compounds known as **alcohols.** The characteristic functional group of this family is the hydroxyl (OH) group attached to a sp^3-hybridized carbon atom. Another example of an alcohol is ethyl alcohol, CH_3CH_2OH (also called ethanol).

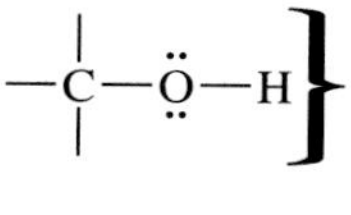

This is the functional group of an alcohol

*Although we use the symbols 1°, 2°, 3°, we do not *say* first degree, second degree, and third degree; we say *primary, secondary,* and *tertiary.*

Alcohols may be viewed in two ways structurally: (1) as hydroxy derivatives of alkanes, and (2) as alkyl derivatives of water. Ethyl alcohol, for example, can be seen as an ethane molecule in which one hydrogen has been replaced by a hydroxyl group, or as a water molecule in which one hydrogen has been replaced by an ethyl group.

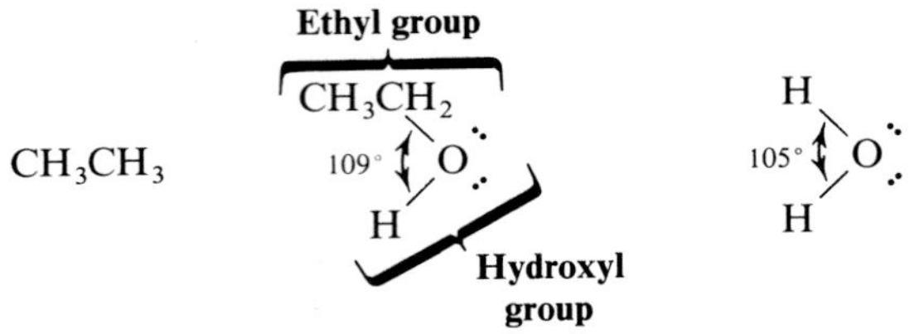

CH_3CH_3

Ethane

Ethyl group

CH_3CH_2

Hydroxyl group

Ethyl alcohol
(ethanol)

Water

As with alkyl halides, alcohols are classified into three groups: primary (1°), secondary (2°), or tertiary (3°) alcohols. *This classification is also based on the degree of substitution of the carbon to which the hydroxyl group is directly attached.* If the carbon has only one other carbon attached to it, the carbon is said to be a **primary carbon** and the alcohol is a **primary alcohol.**

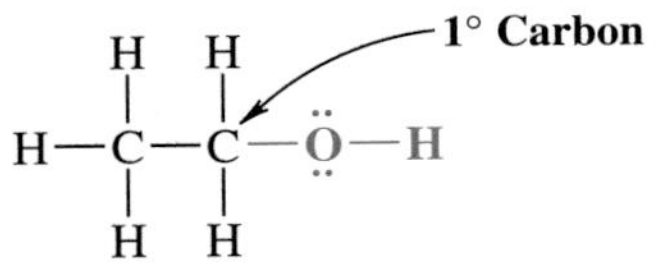

1° Carbon

Ethyl alcohol
(a 1° alcohol)

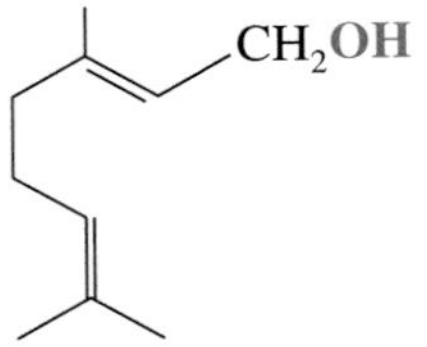

CH_2OH

Geraniol
(a 1° alcohol with
the odor of roses)

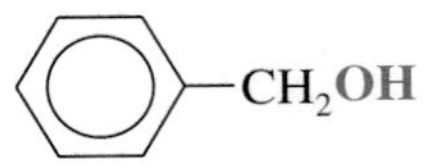

CH_2OH

Benzyl alcohol)
(a 1° alcohol)

If the carbon atom that bears the hydroxyl group also has two other carbon atoms attached to it, this carbon is called a secondary carbon, and the alcohol is a secondary alcohol, and so on.

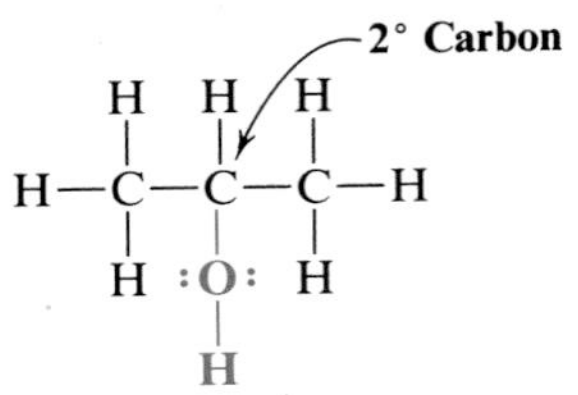

2° Carbon

(a 2° alcohol)

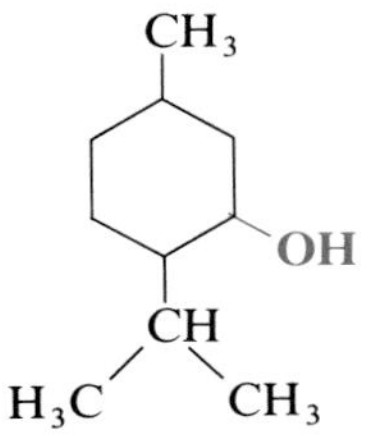

CH_3

OH

CH

H_3C CH_3

Menthol
(a 2° alcohol found
in peppermint oil)

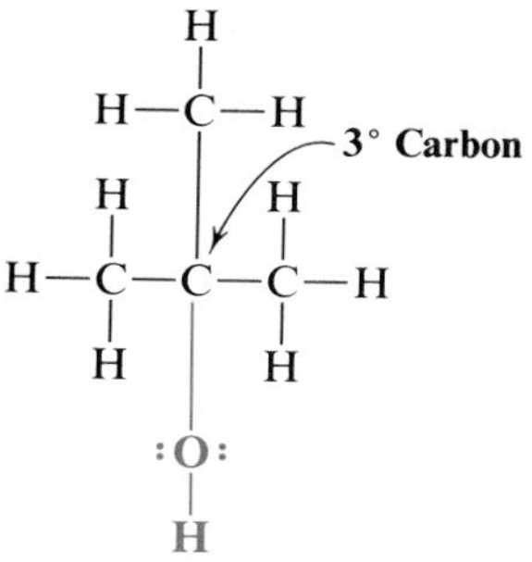

(a 3° alcohol)

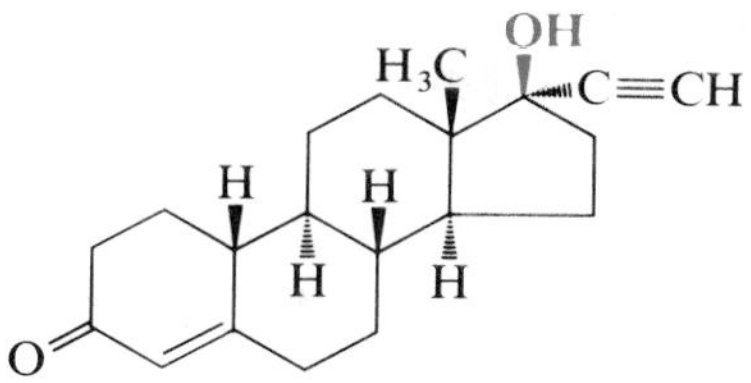

Norethindrone
(an oral contraceptive that contains a 3° alcohol
group, as well as a ketone group and carbon–
carbon double and triple bonds)

PROBLEM 2.6

Write structural formulas for (a) two primary alcohols, (b) a secondary alcohol, and (c) a tertiary alcohol all having the molecular formula $C_4H_{10}O$.

PROBLEM 2.7

One way of naming alcohols is to name the alkyl group that is attached to the —OH and add the word *alcohol*. Write the structures of (a) propyl alcohol and (b) isopropyl alcohol.

2.11 ETHERS

Ethers have the general formula R—O—R or R—O—R' where R' may be an alkyl group different from R. They can be thought of as derivatives of water in which both hydrogen atoms have been replaced by alkyl groups. The bond angle at the oxygen atom of an ether is only slightly larger than that of water.

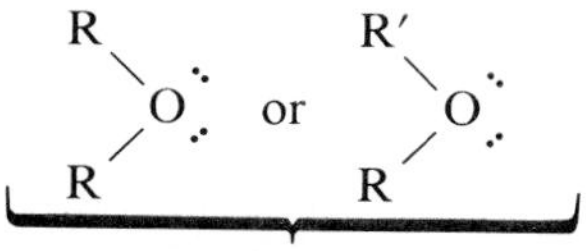

General formula for an ether

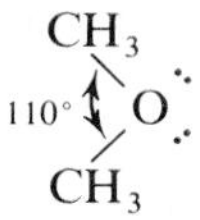

Dimethyl ether
(a typical ether)

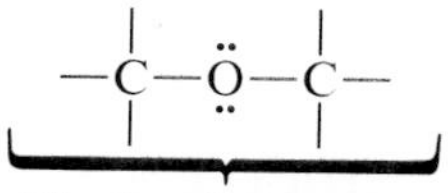

The functional group
of an ether

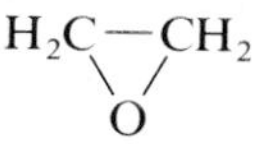

**Ethylene
oxide**

**Tetrahydrofuran
(THF)**

Two cyclic ethers

PROBLEM 2.8

One way of naming ethers is to name the two alkyl groups attached to the oxygen atom in alphabetical order and add the word *ether*. If the two alkyl groups are the same, we use the prefix *di-*, for example, as in *dimethyl ether*. Write structural formulas for (a) diethyl ether, (b) ethyl propyl ether, and (c) ethyl isopropyl ether. (d) What name would you give to $CH_3OCH_2CH_2CH_3$, (e) to $(CH_3)_2CHOCH(CH_3)_2$, and (f) to $CH_3OC_6H_5$?

2.12 AMINES

Just as alcohols and ethers may be considered as organic derivatives of water, amines may be considered as organic derivatives of ammonia.

$$H-\overset{..}{\underset{|}{N}}-H \qquad R-\overset{..}{\underset{|}{N}}-H \qquad C_6H_5CH_2\underset{|}{CH}CH_3 \qquad H_2NCH_2CH_2CH_2CH_2NH_2$$
$$\quad H \qquad\qquad H \qquad\qquad\quad NH_2$$

Ammonia   **Amphetamine**   **Putrescine**
    (an amine)   **(a dangerous stimulant)**   **(found in decaying meat)**

Amines are classified as primary, secondary, or tertiary amines. **This classification is based on** *the number of organic groups that are attached to the nitrogen atom:*

$$R-\overset{..}{\underset{|}{N}}-H \qquad R-\overset{..}{\underset{|}{N}}-H \qquad R-\overset{..}{\underset{|}{N}}-R''$$
$$\quad H \qquad\qquad\quad R' \qquad\qquad\quad R'$$

A primary (1°)   **A secondary (2°)**   **A tertiary (3°)**
  amine   **amine**   **amine**

Notice that this is quite different from the way alcohols and alkyl halides are classified. Isopropylamine, for example, is a primary amine even though its $-NH_2$ group is attached to a secondary carbon atom. It is a primary amine because only one organic group is attached to the nitrogen atom.

$$H-\overset{\overset{\displaystyle H}{|}}{\underset{\underset{\displaystyle H}{|}}{C}}-\overset{\overset{\displaystyle H}{|}}{\underset{\underset{\displaystyle :NH_2}{|}}{C}}-\overset{\overset{\displaystyle H}{|}}{\underset{\underset{\displaystyle H}{|}}{C}}-H$$

Isopropylamine
(a 1° amine)   **(a cyclic 2° amine)**

PROBLEM 2.9

One way of naming amines is to name in alphabetical order the alkyl groups attached to the nitrogen atom, using the prefixes *di-* and *tri-* if the groups are the same. An example is *isopropylamine* for $(CH_3)_2CHNH_2$. Write formulas for (a) propylamine, (b) trimethylamine, and (c) ethylisopropylmethylamine. What are names for (d) $CH_3CH_2CH_2NHCH(CH_3)_2$, (e) $(CH_3CH_2CH_2)_3N$, (f) $C_6H_5NHCH_3$, and (g) $C_6H_5N(CH_3)_2$?

PROBLEM 2.10

Which amines in Problem 2.9 are   (a) primary amines,   (b) secondary amines, and   (c) tertiary amines?

Amines are like ammonia (Section 1.17B) in having a trigonal pyramidal shape. The C—N—C bond angles of trimethylamine are 108.7°, a value very close to the H—C—H bond angles of methane. Thus, for all practical purposes, the nitrogen atom of an amine can be considered to be sp^3 hybridized. This means that the unshared electron pair occupies an sp^3 orbital, and so it is considerably extended into space. This is important because, as we shall see, the unshared electron pair is involved in almost all of the reactions of amines.

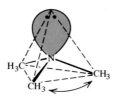

Bond angle = 108.7 °

PROBLEM 2.11

Amines are like ammonia in being weak bases. They do this by using their unshared electron pair to accept a proton.   (a) Show the reaction that would take place between trimethylamine and HCl.   (b) What hybridization state would you expect for the nitrogen atom in the product of this reaction?

2.13 ALDEHYDES AND KETONES

Aldehydes and ketones both contain the **carbonyl group**—a group in which a carbon atom has a double bond to oxygen.

$$\ce{>C=\ddot{O}:}$$

The carbonyl group

The carbonyl group in aldehydes is bonded to at least one *hydrogen atom,* and in ketones it is bonded to *two carbon atoms.* Using R, we can designate the general formula for an aldehyde as

$$\ce{R-\overset{\ddot{O}:}{\overset{\|}{C}}-H} \qquad \text{R may also be H}$$

and the general formula for a ketone as

$$\ce{R-\overset{\ddot{O}:}{\overset{\|}{C}}-R} \quad \text{or} \quad \ce{R-\overset{\ddot{O}:}{\overset{\|}{C}}-R'}$$

(where R′ is an alkyl group different from R).

Some examples of aldehydes and ketones are

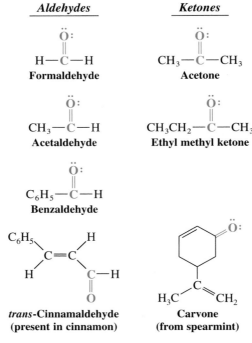

| Aldehydes | Ketones |

Formaldehyde

Acetaldehyde

Benzaldehyde

***trans*-Cinnamaldehyde**
(present in cinnamon)

Acetone

Ethyl methyl ketone

Carvone
(from spearmint)

Aldehydes and ketones have a trigonal planar arrangement of groups around the carbonyl carbon atom. The carbon atom is sp^2 hybridized. In formaldehyde, for example, the bond angles are as follows:

$$H \xrightarrow{121°}$$
$$118° \; C{=}O$$
$$H \; 121°$$

2.14 CARBOXYLIC ACIDS, AMIDES, AND ESTERS

2.14A Carboxylic Acids

Carboxylic acids have the general formula $R{-}\overset{\overset{\textstyle O}{\|}}{C}{-}O{-}H$. The functional group, $-\overset{\overset{\textstyle O}{\|}}{C}{-}O{-}H$, is called the **carboxyl group** (**carb**onyl + hydro**xyl**). (Colloquially, carboxylic acids are often just called "organic acids.")

$$R{-}C\overset{\ddot{\text{O}}:}{\underset{\ddot{\text{O}}{-}H}{\Big\langle}} \qquad \text{or} \qquad RCO_2H \qquad \text{or} \qquad RCOOH$$

A carboxylic acid

$$-C\overset{\ddot{\text{O}}:}{\underset{\ddot{\text{O}}{-}H}{\Big\langle}} \qquad \text{or} \qquad -CO_2H \qquad \text{or} \qquad -COOH$$

The carboxyl group

Esters can be made from an acid and an alcohol through the loss of a molecule of water. For example:

$$CH_3-C \begin{matrix} O \\ \\ OH \end{matrix} \quad + \; HOCH_2CH_3 \xrightarrow{H^+} \; CH_3-C \begin{matrix} O \\ \\ OCH_2CH_3 \end{matrix} \quad + \; H_2O$$

Acetic acid   **Ethyl alcohol**   **Ethyl acetate**

2.15 SUMMARY OF IMPORTANT FAMILIES OF ORGANIC COMPOUNDS

A summary of the important families of organic compounds is given in Table 2.2. You should learn to identify these common functional groups as they appear in other more complicated molecules.

2.16 PHYSICAL PROPERTIES AND MOLECULAR STRUCTURE

So far, we have said little about one of the most obvious characteristics of organic compounds, that is, *their physical state or phase.* Whether a particular substance is a solid, or a liquid, or a gas would certainly be one of the first observations that we would note in any experimental work. The temperatures at which transitions occur between phases, that is, melting points (mp) and boiling points (bp), are also among the more easily measured physical properties. Melting points and boiling points are also useful in identifying and isolating organic compounds.

Suppose, for example, we have just carried out the synthesis of an organic compound that is known to be a liquid at room temperature and 1-atm pressure. If we know the boiling point of our desired product, and the boiling points of byproducts and solvents that may be present in the reaction mixture, we can decide whether or not simple distillation will be a feasible method for isolating our product.

In another instance our product might be a solid. In this case, in order to isolate the substance by crystallization, we need to know its melting point and its solubility in different solvents.

The physical constants of known organic substances are easily found in handbooks and other reference books.* Table 2.3 lists the melting and boiling points of some of the compounds that we have discussed in this chapter.

Often in the course of research, however, the product of a synthesis is a new compound—one that has never been described before. In these instances, success in isolating the new compound depends on making reasonably accurate estimates of its melting point, boiling point, and solubilities. Estimations of these macroscopic physical properties are based on the most likely structure of the substance and on the forces that act between molecules and ions. The temperatures at which phase changes occur are an indication of the strength of these intermolecular forces.

2.16A Ion–Ion Forces

The **melting point** of a substance is the temperature at which an equilibrium exists between the well-ordered crystalline state and the more random liquid state. If the

*Two useful handbooks are *Handbook of Chemistry,* N. A. Lange, Ed., McGraw-Hill, New York and *CRC Handbook of Chemistry and Physics,* CRC, Boca Raton, FL.

TABLE 2.3 *Physical properties of representative compounds*

COMPOUND	STRUCTURE	mp (°C)	bp (°C) (1 atm)
Methane	CH_4	− 182.6	− 162
Ethane	CH_3CH_3	− 183	− 88.2
Ethene	$CH_2{=}CH_2$	− 169	− 102
Ethyne	$HC{\equiv}CH$	− 82	− 84 subl[a]
Chloromethane	CH_3Cl	− 97	− 23.7
Chloroethane	CH_3CH_2Cl	− 138.7	13.1
Ethyl alcohol	CH_3CH_2OH	− 115	78.5
Acetaldehyde	CH_3CHO	− 121	20
Acetic acid	CH_3CO_2H	16.6	118
Sodium acetate	CH_3CO_2Na	324	dec[a]
Ethylamine	$CH_3CH_2NH_2$	− 80	17
Diethyl ether	$(CH_3CH_2)_2O$	− 116	34.6
Ethyl acetate	$CH_3CO_2CH_2CH_3$	− 84	77

[a] In this table dec = decompose and subl = sublimes.

substance is an ionic compound, such as sodium acetate (Table 2.3), the forces that hold the ions together in the crystalline state are the strong electrostatic lattice forces that act between the positive and negative ions in the orderly crystalline structure. In Fig. 2.14 each sodium ion is surrounded by negatively charged acetate ions, and each acetate ion is surrounded by positive sodium ions. A large amount of thermal energy is required to break up the orderly structure of the crystal into the disorderly open structure of a liquid. As a result, the temperature at which sodium acetate melts is quite high, 324°C. The *boiling points* of ionic compounds are higher still, so high that most ionic organic compounds decompose before they boil. Sodium acetate shows this behavior.

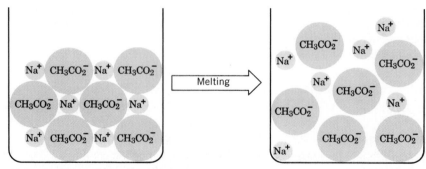

FIGURE 2.14 The melting of sodium acetate.

2.16B Dipole–Dipole Forces

Most organic molecules are not fully ionic but have instead a *permanent dipole moment* resulting from a nonuniform distribution of the bonding electrons (Section 1.19). Acetone and acetaldehyde are examples of molecules with permanent dipoles because the carbonyl group that they contain is highly polarized. In these compounds, the attractive forces between molecules are much easier to visualize. In the liquid or solid state, dipole–dipole attractions cause the molecules to orient themselves so that the positive end of one molecule is directed toward the negative end of another (Fig. 2.15).

FIGURE 2.15 Dipole–dipole interactions between acetone molecules.

2.16C Hydrogen Bonds

Very strong dipole–dipole attractions occur between hydrogen atoms bonded to small, strongly electronegative atoms (O, N, or F) and nonbonding electron pairs on other such electronegative atoms (Fig. 2.16). This type of intermolecular force is called a **hydrogen bond.** The hydrogen bond (bond dissociation energy about $1-9$ kcal mol^{-1}) is weaker than an ordinary covalent bond, but is much stronger than the dipole–dipole interactions that occur in acetone.

$$\overset{\delta-}{:Z}\!\!-\!\!\overset{\delta+}{H}\cdots\cdots\overset{\delta-}{:Z}\!\!-\!\!\overset{\delta+}{H}$$

FIGURE 2.16 The hydrogen bond. Z is a strongly electronegative element, usually oxygen, nitrogen, or fluorine.

Hydrogen bonding accounts for the fact that ethyl alcohol has a much higher boiling point ($+78.5°C$) than dimethyl ether ($-24.9°C$) even though the two compounds have the same molecular weight. Molecules of ethyl alcohol, because they have a hydrogen atom covalently bonded to an oxygen atom, can form strong hydrogen bonds to each other.

$$CH_3CH_2 \!\!\overset{\delta-}{\underset{}{\cdot\cdot\cdot}}\!\!O\!\!\overset{\delta+}{-}\!\!H \cdots \overset{\delta-}{:O}\!\!\overset{\delta+}{\underset{}{H}}\!\!\diagdown CH_2CH_3$$

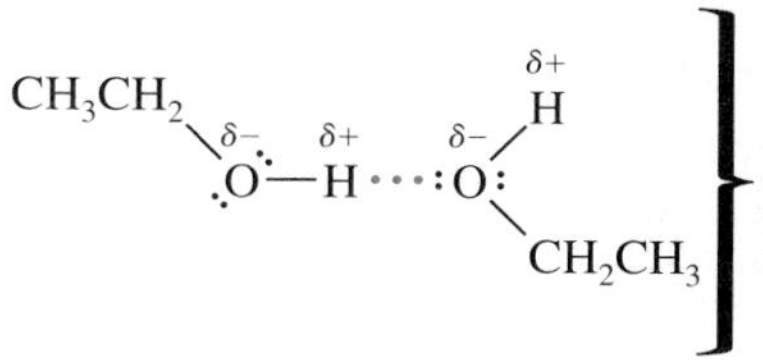

The dotted bond is a hydrogen bond. Strong hydrogen bonding is limited to molecules having a hydrogen atom attached to an O, N, or F atom

Molecules of dimethyl ether, because they lack a hydrogen atom attached to a strongly electronegative atom, cannot form strong hydrogen bonds to each other. In dimethyl ether the intermolecular forces are weaker dipole–dipole interactions.

PROBLEM 2.12

The compounds in each part below have the same (or similar) molecular weights. Which compound in each part would you expect to have the higher boiling point? Explain your answers.   (a) $CH_3CH_2CH_2CH_2OH$ or $CH_3CH_2OCH_2CH_3$, (b) $(CH_3)_3N$ or $CH_3CH_2NHCH_3$, (c) $CH_3CH_2CH_2CH_2OH$ or $HOCH_2CH_2CH_2OH$.

A factor (in addition to polarity and hydrogen bonding) that affects the *melting point* of many organic compounds is the compactness and rigidity of their individual molecules. Molecules that are symmetrical generally have abnormally high melting points. *tert*-Butyl alcohol, for example, has a much higher melting point than the other isomeric alcohols shown here.

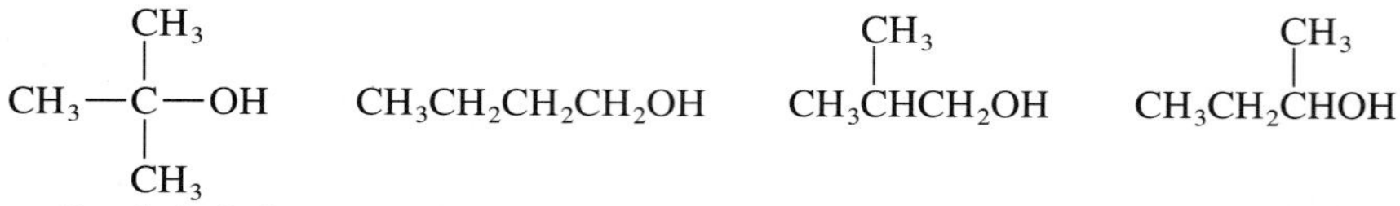

$$CH_3-\overset{\overset{\displaystyle CH_3}{|}}{\underset{\underset{\displaystyle CH_3}{|}}{C}}-OH \qquad CH_3CH_2CH_2CH_2OH \qquad CH_3\overset{\overset{\displaystyle CH_3}{|}}{CH}CH_2OH \qquad CH_3CH_2\overset{\overset{\displaystyle CH_3}{|}}{CH}OH$$

tert-Butyl alcohol	Butyl alcohol	Isobutyl alcohol	*sec*-Butyl alcohol
mp, 25°C	mp, −90°C	mp, −108°C	mp, −114°C

PROBLEM 2.13

Which compound would you expect to have the higher melting point, propane or cyclopropane? Explain your answer.

2.16D van der Waals Forces

If we consider a substance like methane where the particles are nonpolar molecules, we find that the melting point and boiling point are very low: −182.6°C and −162°C, respectively. Instead of asking, ''Why does methane melt and boil at low temperatures?'' a more appropriate question might be ''Why does methane, a nonionic, nonpolar substance, become a liquid or a solid at all?'' The answer to this question can be given in terms of attractive intermolecular forces called **van der Waals forces** (or **London forces** or **dispersion forces**).

An accurate account of the nature of van der Waals forces requires the use of quantum mechanics. We can, however, visualize the origin of these forces in the following way. The average distribution of charge in a nonpolar molecule (such as methane) over a period of time is uniform. At any given instant, however, *because electrons move,* the electrons and therefore the charge may not be uniformly distributed. Electrons may, in one instant, be slightly accumulated on one part of the molecule and, as a consequence, *a small temporary dipole will occur* (Fig. 2.17). This temporary dipole in one molecule can induce opposite (attractive) dipoles in surrounding molecules. It does this because the negative (or positive) charge in a portion of one molecule will distort the electron cloud of an adjacent portion of another molecule causing an opposite charge to develop there. These temporary dipoles change constantly, but the net result of their existence is to produce attractive forces between nonpolar molecules, and thus make possible the existence of their liquid and solid states.

FIGURE 2.17 Temporary dipoles and induced dipoles in nonpolar molecules resulting from a nonuniform distribution of electrons at a given instant.

One important factor that determines the magnitude of van der Waals forces is the relative **polarizability** of the electrons of the atoms involved. By polarizability we mean *the ability of the electrons to respond to a changing electric field.* Relative polarizability depends on how loosely or tightly the electrons are held. In the halogen family, for example, polarizability increases in the order F < Cl < Br < I. Fluorine atoms show a very low polarizability because their electrons are very tightly held; they are close to the nucleus. Iodine atoms are large and hence are more easily polarized. Their valence electrons are far from the nucleus. Atoms with unshared pairs are generally more polarizable than those with only bonding pairs. Thus a halogen substituent is more polarizable than an alkyl group of comparable size. Table 2.4 gives the relative magnitude of van der Waals forces and dipole–dipole interactions for several simple compounds. Notice that except for the molecules where strong hydrogen bonds are possible, van der Waals are far more important than dipole–dipole interactions.

The *boiling point* of a liquid is the temperature at which the vapor pressure of the liquid equals the pressure of the atmosphere above it. For this reason, the boiling points of liquids are *pressure dependent,* and boiling points are always reported as occurring at a particular pressure, as 1 atm (or at 760 torr), for example. A substance that boils at 150°C at 1-atm pressure will boil at a substantially lower temperature if the pressure is reduced to, for example, 0.01 torr (a pressure easily obtained with a vacuum pump). The normal boiling point given for a liquid is its boiling point at 1 atm.

In passing from a liquid to a gaseous state the individual molecules (or ions) of the substance must separate considerably. Because of this, we can understand why ionic organic compounds often decompose before they boil. The thermal energy required to completely separate (volatilize) the ions is so great that chemical reactions (decompositions) occur first.

Nonpolar compounds, where the intermolecular forces are very weak, usually boil at low temperatures even at 1-atm pressure. This is not always true, however, because of other factors that we have not yet mentioned: the effects of molecular weight and

TABLE 2.4 *Attractive energies in simple molecular solids*

MOLECULE	DIPOLE MOMENT (D)	ATTRACTIVE ENERGIES (kcal mol^{-1})		MELTING POINT (°C)	BOILING POINT (°C)
		DIPOLE– DIPOLE	VAN DER WAALS		
H_2O	1.85	8.7[a]	2.1	0	100
NH_3	1.47	3.3[a]	3.5	− 78	− 33
HCl	1.08	0.8[a]	4.0	− 115	− 85
HBr	0.80	0.2	5.2	− 88	− 67
HI	0.42	0.006	6.7	− 51	− 35

[a] These dipole–dipole attractions are called hydrogen bonds.

molecular size. Heavier molecules require greater thermal energy in order to acquire velocities sufficiently great to escape the liquid surface, and because their surface areas are usually much greater, intermolecular van der Waals attractions are also much larger. These factors explain why nonpolar ethane (bp, $-88.2°C$) boils higher than methane (bp, $-162°C$) at a pressure of 1 atm. It also explains why, at 1 atm, the even heavier and larger nonpolar molecule decane ($C_{10}H_{22}$) boils at $+174°C$.

Fluorocarbons (compounds containing only carbon and fluorine) have extraordinarily low boiling points when compared to hydrocarbons of the same molecular weight. The fluorocarbon C_5F_{12}, for example, has a slightly lower boiling point than pentane (C_5H_{12}) even though it has a far higher molecular weight. The important factor in explaining this behavior is the very low polarizability of fluorine atoms that we mentioned earlier, resulting in very small van der Waals forces. The fluorocarbon polymer called *Teflon* [$\{CF_2CF_2\}_n$, see Section 9.10] has self-lubricating properties, which are exploited in making "nonstick" frying pans and lightweight bearings.

2.16E Solubilities

Intermolecular forces are of primary importance in explaining the **solubilities** of substances. Dissolution of a solid in a liquid is, in many respects, like the melting of a solid. The orderly crystal structure of the solid is destroyed, and the result is the formation of the more disorderly arrangement of the molecules (or ions) in solution. In the process of dissolving, too, the molecules or ions must be separated from each other, and energy must be supplied for both changes. The energy required to overcome lattice energies and intermolecular or interionic attractions comes from the formation of new attractive forces between solute and solvent.

Consider the dissolution of an ionic substance as an example. Here both the lattice energy and interionic attractions are large. We find that water and only a few other very polar solvents are capable of dissolving ionic compounds. These solvents dissolve ionic compounds by **hydrating** or **solvating** the ions (Fig. 2.18).

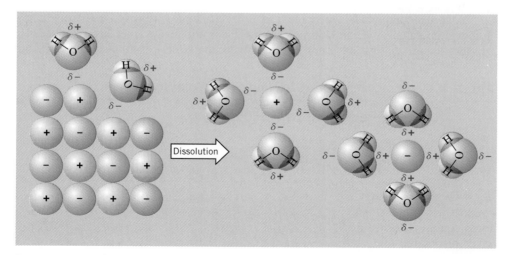

FIGURE 2.18 The dissolution of an ionic solid in water, showing the hydration of positive and negative ions by the very polar water molecules. The ions become surrounded by water molecules in all three dimensions, not just the two shown here.

Water molecules, by virtue of their great polarity, as well as their very small compact shape, can very effectively surround the individual ions as they are freed from the crystal surface. Positive ions are surrounded by water molecules with the negative end of the water dipole pointed toward the positive ion; negative ions are solvated in exactly the opposite way. Because water is highly polar, and because water is capable of forming strong hydrogen bonds, the *dipole–ion* attractive forces are also large. The energy supplied by the formation of these forces is great enough to overcome both the lattice energy and interionic attractions of the crystal.

A rule of thumb for predicting solubilities is that "like dissolves like." Polar and ionic compounds tend to dissolve in polar solvents. Polar liquids are generally miscible with each other. Nonpolar solids are usually soluble in nonpolar solvents. On the other hand, nonpolar solids are insoluble in polar solvents. Nonpolar liquids are usually mutually miscible, but nonpolar liquids and polar liquids "like oil and water" do not mix.

Methanol and water are miscible in all proportions; so too are mixtures of ethanol and water, and mixtures of both propyl alcohols and water. In these cases the alkyl groups of the alcohols are relatively small, and the molecules therefore resemble water more than they do an alkane. Another factor in understanding their solubility is that the molecules are capable of forming strong hydrogen bonds to each other.

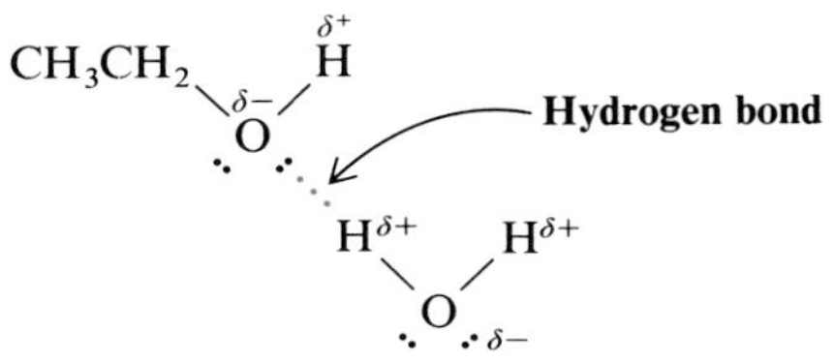

$$CH_3CH_2 \underset{\ddot{\cdot}}{\overset{\delta-}{O}} \cdots \overset{\delta+}{H} \qquad \text{Hydrogen bond}$$

$$H^{\delta+} \quad H^{\delta+}$$

$$\underset{\ddot{\cdot} \cdot \delta-}{O}$$

If the carbon chain of an alcohol is long, however, we find that the alcohol is much less soluble in water. Decyl alcohol (see following structure) with a chain of 10 carbon atoms is only very slightly soluble in water. Decyl alcohol resembles an alkane more than it does water. The long carbon chain of decyl alcohol is said to be **hydrophobic** (*hydro,* water; *phobic,* fearing or avoiding — "water avoiding"). Only the OH group, a rather small part of the molecule, is **hydrophilic** (*philic,* loving or seeking — "water seeking"). (On the other hand, decyl alcohol is quite soluble in nonpolar solvents.)

Hydrophobic portion   **Hydrophilic group**

$$\overbrace{CH_3CH_2CH_2CH_2CH_2CH_2CH_2CH_2CH_2CH_2}OH \longleftarrow$$

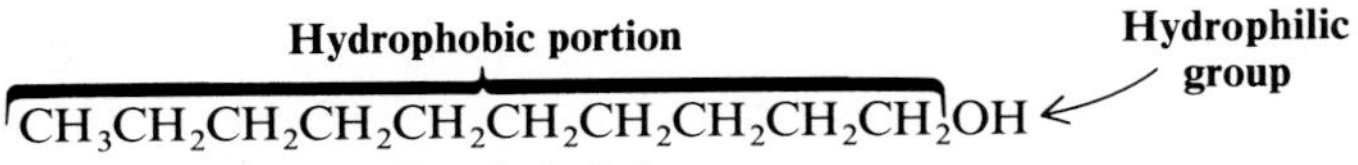

Decyl alcohol

An explanation for why nonpolar groups such as long alkane chains avoid an aqueous environment, for the so-called **hydrophobic effect**, is complex. The most important factor seems to involve an **unfavorable entropy change** in the water. Entropy changes (Section 3.8) have to do with changes from a relatively ordered state to a more disordered one, or the reverse. Changes from order to disorder are favorable, whereas changes from disorder to order are unfavorable. For a nonpolar hydrocarbon chain to be accommodated by water, the water molecules have to form a more ordered structure around the chain, and for this, the entropy change is unfavorable.

2.16F Guidelines for Water Solubility

Organic chemists usually define a compound as water soluble if at least 3 g of the organic compound dissolves in 100 mL of water. We find that for compounds containing one hydrophilic group—and thus capable of forming strong hydrogen bonds—the following approximate guidelines hold: Compounds with one to three carbon atoms are water soluble, compounds with four or five carbon atoms are borderline, and compounds with six carbon atoms or more are insoluble.

When a compound contains more than one hydrophilic group these guidelines do not apply. Polysaccharides (Chapter 22), proteins (Chapter 24), and nucleic acids (Chapter 25) all contain thousands of carbon atoms *and all are water soluble*. They dissolve in water because they also contain thousands of hydrophilic groups.

2.16G Intermolecular Forces in Biochemistry

Later, after we have had a chance to examine in detail the properties of the molecules that make up living organisms, we shall see how intermolecular forces are extremely important in the functioning of cells. Hydrogen bond formation, the hydration of polar groups, and the tendency of nonpolar groups to avoid a polar environment all cause complex protein molecules to fold in precise ways—ways that allow them to function as biological catalysts of incredible efficiency. The same factors allow molecules of

TABLE 2.5 Attractive electric forces

ELECTRIC FORCE	RELATIVE STRENGTH	TYPE	EXAMPLE
Cation–anion (in a crystal)	Very strong	⊕ ⊖	Lithium fluoride crystal lattice
Covalent bonds	Strong ($36-125$ kcal mol^{-1})	Shared electron pairs	H—H (104 kcal mol^{-1}) CH$_3$—CH$_3$ (88 kcal mol^{-1}) I—I (36 kcal mol^{-1})
Ion–dipole	Moderate		Na$^+$ in water (see Fig. 2.18)
Dipole–dipole (including hydrogen bonds)	Moderate to weak ($1-9$ kcal mol^{-1})	$-\overset{\delta-}{Z}:\cdots\overset{\delta+}{H}-$ and	and
van der Waals	Variable	Transient dipole	Interactions between methane molecules

hemoglobin to assume the shape needed to transport oxygen. They allow proteins and molecules called glycosphingolipids to function as cell membranes. Hydrogen bonding alone gives molecules of certain carbohydrates a globular shape that makes them highly efficient food reserves in animals. It gives molecules of other carbohydrates a rigid linear shape that makes them perfectly suited to be structural components in plants.

2.17 SUMMARY OF ATTRACTIVE ELECTRIC FORCES

The attractive forces occurring between molecules and ions that we have studied so far are summarized in Table 2.5.

ADDITIONAL PROBLEMS

2.14   Classify each of the following compounds as an alkane, alkene, alkyne, alcohol, aldehyde, amine, and so forth.

(a) [structure]   (b) — ≡   (c) [structure with OH]   (d) [structure with O and H]

(e) [structure with OH]   **(obtained from oil of cloves)**

(f) $CH_3(CH_2)_7$ $(CH_2)_{12}CH_3$ [structure] H   H   **(sex attractant of common house fly)**

2.15   Identify all of the functional groups in each of the following compounds:

(a) [structure]   **Vitamin D₃**

HO

(b) HO—C—H₂C [structure] N   H   CH₂   CH   O   CH₃   **Aspartame**
     H₂N   H   C
                                    O

(c) [structure] —CH₂—CH—NH₂   **Amphetamine**
                      CH₃

(d) 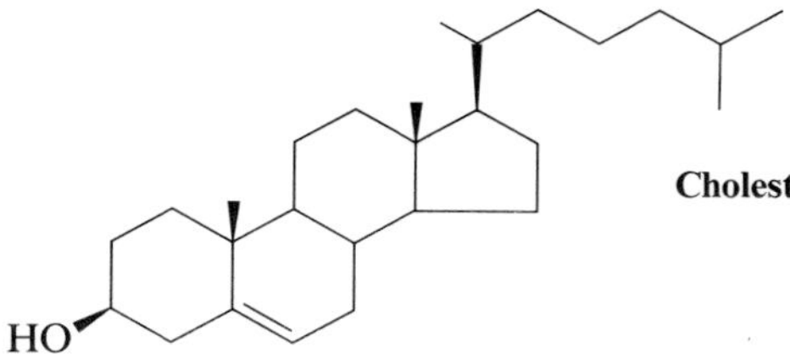   Cholesterol

HO

(e) 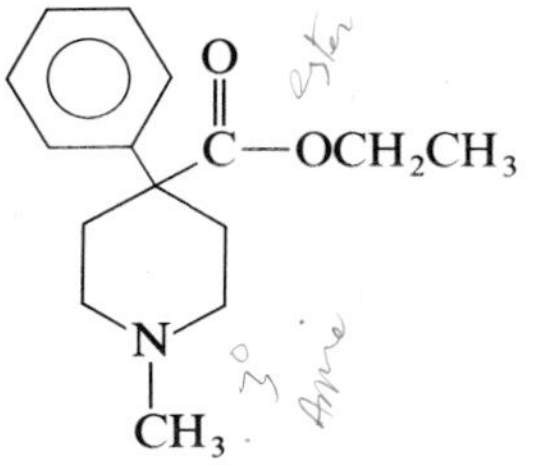   Demerol

$C-OCH_2CH_3$

N
|
CH_3

(f)

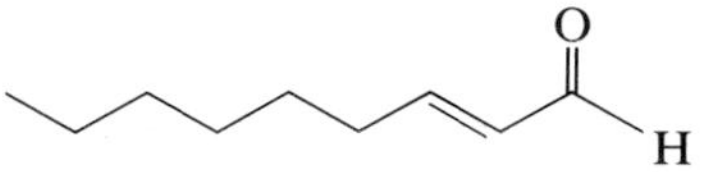

A cockroach repellent found
in cucumbers

(g)

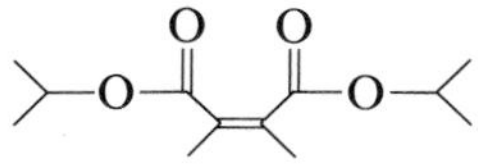

A synthetic cockroach repellent

2.16   There are four alkyl bromides with the formula C_4H_9Br. Write their structural formulas and classify each as to whether it is a primary, secondary, or tertiary alkyl bromide.

2.17   There are seven isomeric compounds with the formula $C_4H_{10}O$. Write their structures and classify each compound according to its functional group.

2.18   Write structural formulas for four compounds with the formula C_3H_6O and classify each according to its functional group.

2.19   Classify the following alcohols as primary, secondary, or tertiary:

(a) $(CH_3)_3CCH_2OH$ (d)

(b) $CH_3CH(OH)CH(CH_3)_2$

(c) $(CH_3)_2C(OH)CH_2CH_3$ (e) 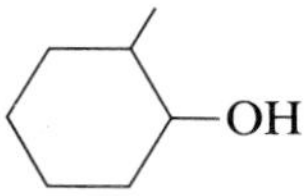 $-OH$

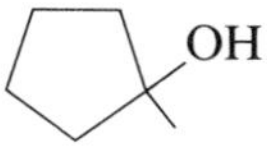

2.20   Classify the following amines as primary, secondary, or tertiary:

(a) $CH_3NHCH(CH_3)_2$ (e) HN

(b) $CH_3CH_2CH(CH_3)CH_2NH_2$

(c) $(CH_3CH_2)_3N$

(d) $(C_6H_5)_2CHCH_2NHCH_3$ (f)

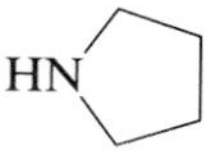

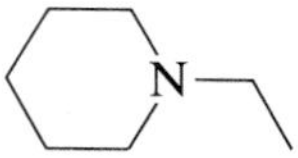

2.21   Write structural formulas for each of the following:

(a)   Three ethers with the formula $C_4H_{10}O$.

(b)   Three primary alcohols with the formula C_4H_8O.

(c)   A secondary alcohol with the formula C_3H_6O.

(d)   A tertiary alcohol with the formula C_4H_8O.

(e)   Two esters with the formula $C_3H_6O_2$.

(f)   Four primary alkyl halides with the formula $C_5H_{11}Br$.

(g)   Three secondary alkyl halides with the formula $C_5H_{11}Br$.

(h)   A tertiary alkyl halide with the formula $C_5H_{11}Br$.

(i)   Three aldehydes with the formula $C_5H_{10}O$.

(j)   Three ketones with the formula $C_5H_{10}O$.

(k)   Two primary amines with the formula $C_3H_{11}N$.

(l)   A secondary amine with the formula $C_3H_{11}N$.

(m)   A tertiary amine with the formula $C_3H_{11}N$.

(n)   Two amides with the formula C_2H_5NO.

2.22   Which compound in each of the following pairs would have the higher boiling point? Explain your answers.

(a) $CH_3CH_2CH_2OH$ or $CH_3CH_2OCH_3$

(b) $CH_3CH_2CH_2OH$ or $HOCH_2CH_2OH$

(c) 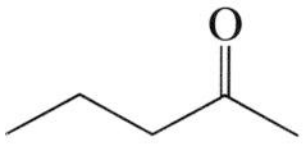   or

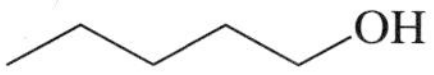

(d) 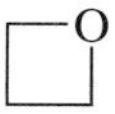   or

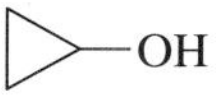

(e)    or

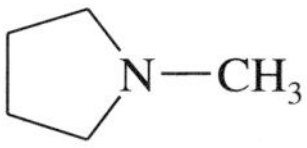

(f) 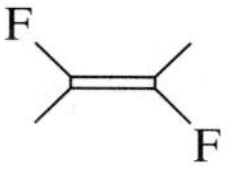   or

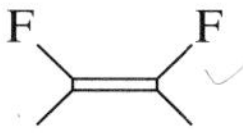

(g) 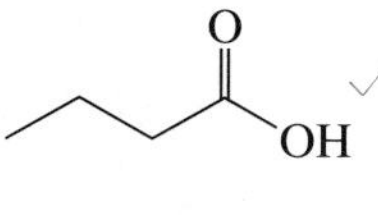   or

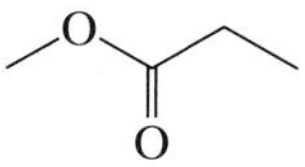

(h)   Hexane $CH_3(CH_2)_4CH_3$ or nonane $CH_3(CH_2)_6CH_3$

(i) 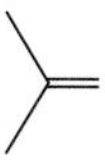   or

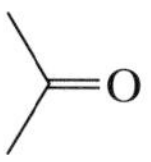

2.23   There are four amides with the formula C_3H_7NO.   (a) Write their structures.   (b) One of these amides has a melting and boiling point that is substantially lower than that of the other three. Which amide is this? Explain your answer.

2.24   Cyclic compounds of the general type shown here are called lactones. What functional group does a lactone contain?

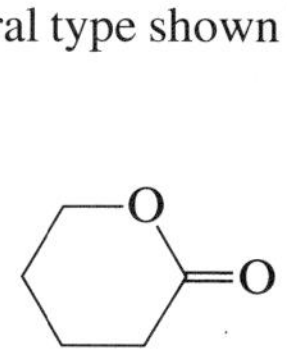

2.25   Hydrogen fluoride has a dipole moment of 1.82 D; its boiling point is 19.34°C. Ethyl fluoride (CH_3CH_2F) has an almost identical dipole moment and has a larger molecular weight, yet its boiling point is -37.7°C. Explain.

2.26   Which of the following solvents should be capable of dissolving ionic compounds? (a) liquid SO_2;   (b) liquid NH_3;   (c) benzene;   (d) CCl_4

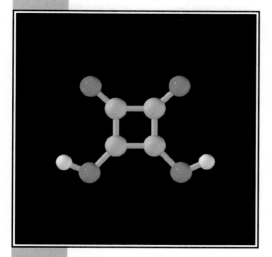

CHAPTER 3

Squaric acid
(Problem 3.37)

AN INTRODUCTION TO ORGANIC REACTIONS: ACIDS AND BASES

3.1 REACTIONS AND THEIR MECHANISMS

Virtually ? organic reactions fall into one of four categories: They are either *substitutions, additions, eliminations, or rearrangements.*

Substitutions are the characteristic reactions of saturated compounds such as alkanes, alkyl halides, and aromatic compounds. In a substitution, *one group replaces another.* For example, methyl chloride reacts with sodium hydroxide to produce methyl alcohol and sodium chloride:

$$\text{H}_3\text{C}-\text{Cl} + \text{Na}^+\text{OH}^- \xrightarrow{\text{H}_2\text{O}} \text{H}_3\text{C}-\text{OH} + \text{Na}^+\text{Cl}^-$$

A substitution reaction

In this reaction a hydroxide ion from sodium hydroxide replaces the chlorine of methyl chloride. We shall study this reaction in detail in Chapter 6.

Additions are characteristic of compounds with multiple bonds. Ethene, for example, reacts with bromine by an addition.

$$\underset{\displaystyle \text{H}}{\overset{\displaystyle \text{H} \quad \text{H}}{\text{H}-\text{C}=\text{C}-\text{H}}} + \text{Br}-\text{Br} \xrightarrow{\text{CCl}_4} \underset{\displaystyle \text{Br} \quad \text{Br}}{\overset{\displaystyle \text{H} \quad \text{H}}{\text{H}-\text{C}-\text{C}-\text{H}}}$$

An addition reaction

In an addition *all parts of the adding reagent appear in the product; two molecules become one.*

Eliminations are the opposite of additions. *In an elimination one molecule loses the elements of another small molecule.* Elimination reactions give us a method for preparing compounds with double and triple bonds. In Chapter 7, for example, we shall study an important elimination, called *dehydrohalogenation,* a reaction that is used to prepare alkenes. In dehydrohalogenation, as the word suggests, the elements of a hydrogen halide are eliminated. An alkyl halide becomes an alkene:

$$H\!-\!\underset{\underset{\displaystyle H}{|}}{\overset{\overset{\displaystyle H}{|}}{C}}\!-\!\underset{\underset{\displaystyle Br}{|}}{\overset{\overset{\displaystyle H}{|}}{C}}\!-\!H \xrightarrow[\text{(− HBr)}]{\text{KOH}} H\!-\!\overset{\overset{\displaystyle H}{|}}{C}\!=\!\overset{\overset{\displaystyle H}{|}}{C}\!-\!H$$

An elimination reaction

In a **rearrangement** *a molecule undergoes a reorganization of its constituent parts.* For example, heating the following alkene with a strong acid causes the formation of another isomeric alkene.

$$H_3C\!-\!\underset{\underset{\displaystyle CH_3}{|}}{\overset{\overset{\displaystyle CH_3}{|}}{C}}\!-\!\!-\!\overset{\overset{\displaystyle H}{|}}{C}\!=\!\overset{\overset{\displaystyle H}{|}}{C}\!-\!H \xrightarrow{\text{H}^+} H_3C\!-\!\overset{\overset{\displaystyle CH_3}{|}}{C}\!=\!\!=\!\overset{\overset{\displaystyle CH_3}{|}}{C}\!-\!CH_3$$

A rearrangement

In this rearrangement not only has the position of the double bond and a hydrogen atom changed, but a methyl group has moved from one carbon to another.

In the following sections we shall begin to learn some of the principles that explain how these kinds of reactions take place.

3.1A Mechanisms of Reactions

To the uninitiated, a chemical reaction must seem like an act of magic. A chemist puts one or two reagents into a flask, warms them for a period of time, and then takes from the flask one or more completely different compounds. It is, until we understand the details of the reaction, like a magician who puts apples and oranges in a hat, shakes it, and then pulls out rabbits and parakeets.

One of our goals in this course will be, in fact, to try to understand how this chemical magic takes place. We will want to be able to explain *how the products of the reaction are formed.* This explanation will take the form of **a mechanism for the reaction—a description of the events that take place on a molecular level as reactants become products.** If, as is often the case, the reaction takes place in more than one step, we will want to know what chemical species, called **intermediates,** intervene between each step along the way.

By postulating a mechanism, we may take some of the magic out of the reaction, but we will put rationality in its place. Any mechanism we propose must be consistent with what we know about the reaction, and with what we know about the reactivity of organic compounds generally. In later chapters we shall see how we can glean evidence for or against a given mechanism from studies of reaction rates, from isolating intermediates, and from spectroscopy. We cannot actually see the molecular events because molecules are too small, but from solid evidence and from good chemical intuition, we can propose reasonable mechanisms. If at some later time an experiment gives results that contradict our proposed mechanism, then we change it, because in the final analysis, our mechanism must be consistent with all our experimental observations.

One of the most important things about approaching organic chemistry mechanistically is this: It helps us organize what otherwise might be an overwhelmingly complex body of knowledge into a form that makes it understandable. There are millions of organic compounds now known, and there are millions of reactions that these compounds undergo. If we had to learn them all by rote memorization, then we would soon give up. But, we don't have to do this. In the same way that functional groups help us organize compounds in a comprehensible way, mechanisms help us organize reactions. Fortunately, too, there is a relatively small number of basic mechanisms.

3.1B Homolysis and Heterolysis of Covalent Bonds

Reactions of organic compounds always involve the making and breaking of covalent bonds. A covalent bond may break in two fundamentally different ways. The bond may break so that one fragment takes away both electrons of the bond, leaving the other fragment with an empty orbital. This kind of cleavage, called **heterolysis** (Gr: *hetero-,* different, + *lysis,* loosening or cleavage) produces charged fragments or **ions.** The bond is said to have broken *heterolytically.*

$$A : B \longrightarrow \underbrace{A^+ \ + \ :B^-}_{\text{Ions}} \quad \textbf{Heterolytic bond cleavage}$$

The other possibility is that the bond breaks so that each fragment takes away one of the electrons of the bond. This process, called **homolysis** (Gr: *homo-,* the same, + *lysis*), produces fragments with unpaired electrons called **radicals.** *

$$A : B \longrightarrow \underbrace{A \cdot \ + \ \cdot B}_{\text{Radicals}} \quad \textbf{Homolytic bond cleavage}$$

We shall postpone further discussions of reactions involving radicals and homolytic bond cleavage until we reach Chapter 9. At this point we shall focus our attention on reactions involving ions and heterolytic bond cleavage.

Heterolysis of a bond normally requires that the bond be polarized.

$$^{\delta+}A : B^{\delta-} \longrightarrow A^+ \ + \ :B^-$$

Polarization of a bond usually results from differing electronegativities (Section 1.18) of the atoms joined by the bond. The greater the difference in electronegativity,

---

*Notice in these two illustrations we have used curved arrows to show the movement of electrons. We will have more to say about this convention in Section 3.4, but for the moment, notice that we use a double-barbed curved arrow to show the movement of a pair of electrons, and a single-barbed curved arrow to show the movement of a single electron.

the greater the polarization. In the instance just given, atom B is more electronegative than A.

Even with a highly polarized bond, heterolysis rarely occurs without assistance. The reason: *Heterolysis requires separation of oppositely charged ions.* Because oppositely charged ions attract each other, their separation requires considerable energy. Often, heterolysis is assisted by a molecule with an unshared pair that can form a bond to one of the atoms:

$$C: + \,^{\delta+}A : B^{\delta-} \longrightarrow \overset{+}{C}:A \; + \; :B^-$$

or

$$C: + \,^{\delta+}A - B^{\delta-} \longrightarrow \overset{+}{C} - A \; + \; :B^-$$

Formation of the new bond furnishes some of the energy required for the heterolysis.

3.2 ACID–BASE REACTIONS

We begin our study of chemical reactions by examining some of the basic principles of acid–base chemistry, starting with inorganic compounds. There are several reasons for doing this: Many of the reactions that occur in organic chemistry are either acid–base reactions themselves, or they involve an acid–base reaction at some stage. An acid–base reaction is also a simple fundamental reaction that will enable you to see how chemists use curved arrows to represent mechanisms of reactions, and how they depict the processes of bond breaking and bond making that occur as molecules react. Acid–base reactions also allow us to examine important ideas about the relationship between the structures of molecules and their reactivity, and to see how certain thermodynamic parameters can be used to predict how much of the product will be formed when a reaction reaches equilibrium. Acid–base reactions also provide an illustration for the important role solvents play in chemical reactions. They even give us a brief introduction to organic synthesis. Finally, acid–base chemistry is something that you will find familiar because of your studies in general chemistry. We begin, therefore, with a brief review.

3.2A The Brønsted–Lowry Definition of Acids and Bases

According to the Brønsted–Lowry theory, an acid is a substance that can donate (or lose) a proton, and a base is a substance that can accept (or remove) a proton. Let us consider, as an example of this concept, the reaction that occurs when gaseous hydrogen chloride dissolves in water:

$$H-\overset{..}{O}: \; + \; H-\overset{..}{\underset{..}{Cl}}: \longrightarrow H-\overset{..}{\overset{+}{O}}-H \; + \; :\overset{..}{\underset{..}{Cl}}:^-$$

Base	**Acid**	**Conjugate**	**Conjugate**
(**proton**	(**proton**	**acid**	**base**
acceptor)	**donor**)	**of H₂O**	**of HCl**

Hydrogen chloride, a very strong acid, transfers its proton to water. Water acts as a base and accepts the proton. The products that result from this reaction are a hydronium ion (H_3O^+) and a chloride ion (Cl^-).

The molecule or ion that forms when an acid loses its proton is called **the conjugate base** of that acid. The chloride ion, therefore, is the conjugate base of HCl. The molecule or ion that forms when a base accepts a proton is called **the conjugate acid** of that base. The hydronium ion, therefore, is the conjugate acid of water.

Other strong acids that completely transfer a proton when dissolved in water are hydrogen iodide, hydrogen bromide, and sulfuric acid.*

$$HI + H_2O \longrightarrow H_3O^+ + I^-$$

$$HBr + H_2O \longrightarrow H_3O^+ + Br^-$$

$$H_2SO_4 + H_2O \longrightarrow H_3O^+ + HSO_4^-$$

$$HSO_4^- + H_2O \rightleftharpoons H_3O^+ + SO_4^{2-}$$

Because sulfuric acid has two protons that it can transfer to a base, it is called a diprotic (or dibasic) acid. The proton transfer is stepwise; the first proton transfer occurs completely, the second only to the extent of $\sim 10\%$.

Hydronium ions and hydroxide ions are the strongest acids and bases that can exist in aqueous solution in significant amounts. When sodium hydroxide (a crystalline compound consisting of sodium ions and hydroxide ions) dissolves in water, the result is a solution containing solvated sodium ions and solvated hydroxide ions.

$$Na^+OH^-_{(solid)} \longrightarrow Na_{(aq)}^+ + OH_{(aq)}^-$$

Sodium ions (and other similar cations) become solvated when water molecules donate unshared electron pairs to their vacant orbitals. Hydroxide ions (and other anions with unshared electron pairs) become solvated when water molecules form hydrogen bonds to them.

Solvated sodium ion   **Solvated hydroxide ion**

When an aqueous solution of sodium hydroxide is mixed with an aqueous solution of hydrogen chloride (hydrochloric acid), the reaction that occurs is between hydronium and hydroxide ions. The sodium and chloride ions are called **spectator ions** because they play no part in the acid–base reaction.

---

*The extent to which an acid transfers protons to a base, such as water, is a measure of its strength as an acid. Acid strength is therefore a measure of the percentage of ionization and not of concentration.

Total Ionic Reaction

$$H-\overset{+}{\underset{H}{\overset{..}{O}}}-H + :\overset{..}{\underset{..}{Cl}}:^- + Na^+ \quad {}^-:\overset{..}{\underset{..}{O}}-H \longrightarrow 2\ H-\overset{..}{\underset{H}{O}}: + Na^+ + :\overset{..}{\underset{..}{Cl}}:^-$$

Spectator ions

Net Reaction

$$H-\overset{+}{\underset{H}{\overset{..}{O}}}-H + {}^-:\overset{..}{\underset{..}{O}}-H \longrightarrow 2\ H-\overset{..}{\underset{H}{O}}:$$

What we have just said about hydrochloric acid and aqueous sodium hydroxide is true when solutions of all aqueous strong acids and bases are mixed. The net ionic reaction is simply:

$$H_3O^+ + OH^- \longrightarrow 2\ H_2O$$

3.2B The Lewis Definition of Acids and Bases

Acid–base theory was broadened considerably by G. N. Lewis in 1923. Striking at what he called "the cult of the proton," Lewis proposed that acids be defined as **electron-pair acceptors** and bases be defined as **electron-pair donors.** In the Lewis theory, the proton is not the only acid; many other species are acids as well. Aluminum chloride and boron trifluoride, for example, react with ammonia in the same way that a proton does. Using curved arrows to show the donation of the electron pair of ammonia (the Lewis base), we have the following examples:

$$H^+ + :NH_3 \longrightarrow H-\overset{+}{N}H_3$$

Lewis acid   **Lewis base**
(electron-pair   **(electron-pair**
acceptor)   **donor)**

$$Cl-\overset{\overset{\textstyle Cl}{|}}{\underset{\underset{\textstyle Cl}{|}}{Al}} + :NH_3 \longrightarrow Cl-\overset{\overset{\textstyle Cl}{|}}{\underset{\underset{\textstyle Cl}{|}}{Al}}\overset{+}{=}NH_3$$

Lewis acid   **Lewis base**
(electron-pair   **(electron-pair**
acceptor)   **donor)**

$$F-\overset{\overset{\textstyle F}{|}}{\underset{\underset{\textstyle F}{|}}{B}} + :NH_3 \longrightarrow F-\overset{\overset{\textstyle F}{|}}{\underset{\underset{\textstyle F}{|}}{B}}\overset{+}{=}NH_3$$

Lewis acid   **Lewis base**
(electron-pair   **(electron-pair**
acceptor)   **donor)**

In these examples, aluminum chloride and boron trifluoride accept the electron pair of ammonia just as a proton does, by using it to form a covalent bond to the nitrogen atom. They do this because the central aluminum and boron atoms have only a sextet of electrons and are, therefore, electron deficient. When they accept the electron pair, aluminum chloride and boron trifluoride are, in the Lewis definition, *acting as acids.*

Bases are much the same in the Lewis theory and the Brønsted–Lowry theory, because in the Brønsted–Lowry theory a base must donate a pair of electrons in order to accept a proton.

The Lewis theory, by virtue of its broader definition of acids, allows acid–base theory to include all of the Brønsted–Lowry reactions and, as we shall see, a great many others.

Any *electron-deficient atom* can act as a Lewis acid. Many compounds containing Group **IIIA** elements such as boron and aluminum are Lewis acids because Group **IIIA** atoms have only a sextet of electrons in their outer shell. Many other compounds that have atoms with vacant orbitals also act as Lewis acids. Zinc and iron(III) halides (ferric halides) are frequently used as Lewis acids in organic reactions. Two examples that we shall study later are the following:

$$R-\overset{..}{\underset{..}{O}}-H \;+\; ZnCl_2 \;\longrightarrow\; R-\overset{\overset{+}{\underset{|}{O}}}{\underset{H}{}}\!\!-\overset{-}{Z}nCl_2$$

Lewis base (electron-pair donor)	**Lewis acid** (electron-pair acceptor)

$$:\!\overset{..}{\underset{..}{Br}}-\overset{..}{\underset{..}{Br}}\!: \;+\; FeBr_3 \;\longrightarrow\; :\!\overset{..}{\underset{..}{Br}}-\overset{+}{\underset{..}{Br}}\!\!-\overset{-}{F}eBr_3$$

Lewis base (electron-pair donor)	**Lewis acid** (electron-pair acceptor)

PROBLEM 3.1

Write equations showing the Lewis acid–base reaction that takes place when:

(a) Methyl alcohol reacts with BF_3.

(b) Methyl chloride reacts with $AlCl_3$.

(c) Dimethyl ether reacts with BF_3.

PROBLEM 3.2

Which of the following are potential Lewis acids and which are potential Lewis bases?

(a) $CH_3CH_2-\overset{..}{N}-CH_3$
    $\quad\quad\quad\quad\quad |$
    $\quad\quad\quad\quad\quad CH_3$

(b) $H_3C-\overset{\overset{CH_3}{|}}{\underset{\underset{CH_3}{|}}{C}}{}^+$

(c) $(C_6H_5)_3P:$

(d) $:\overset{..}{\underset{..}{Br}}:^-$

(e) $(CH_3)_3B$

(f) $H:^-$

3.3 HETEROLYSIS OF BONDS TO CARBON: CARBOCATIONS AND CARBANIONS

Heterolysis of a bond to a carbon atom can lead to either of two ions: either to an ion with a positive charge on the carbon atom, called a **carbocation***—or to an ion, with a negatively charged carbon atom, called a **carbanion.**

$$
-\overset{|}{\underset{|}{C}} : Z^{\delta-} \xrightarrow{\text{heterolysis}} \quad -\overset{|}{\underset{|}{C}}{}^{+} \quad + \quad :Z^{-}
$$

Carbocation

$$
-\overset{|}{\underset{|}{C}} : Z^{\delta+} \xrightarrow{\text{heterolysis}} \quad -\overset{|}{\underset{|}{C}} :^{-} \quad + \quad Z^{+}
$$

Carbanion

Carbocations are electron deficient. They have only six electrons in their valence shell, and because of this carbocations are Lewis acids. In this way they are like BF_3 and $AlCl_3$. Most carbocations are also short-lived and highly reactive. They occur as intermediates in some organic reactions. Carbocations react rapidly with Lewis bases—with molecules or ions that can donate the electron pair that they need to achieve a stable octet of electrons (i.e., the electronic configuration of a noble gas).

$$
-\overset{|}{\underset{|}{C}}{}^{+} \quad + \quad :B^{-} \quad \longrightarrow \quad -\overset{|}{\underset{|}{C}} : B
$$

| **Carbocation** | **Anion** |
| (a Lewis acid) | (a Lewis base) |

$$
-\overset{|}{\underset{|}{C}}{}^{+} \quad + \quad :\overset{..}{\underset{|}{O}}-H \quad \longrightarrow \quad -\overset{|}{\underset{|}{C}} : \overset{..}{\underset{|}{O}}{}^{+}-H
$$

| **Carbocation** | **Water** |
| (a Lewis acid) | (a Lewis base) |

Because carbocations are electron-seeking reagents, chemists call them electrophiles. *Electrophiles are reagents, which in their reactions, seek the extra electrons that will give them a stable valence shell of electrons.* All Lewis acids, including protons, are electrophiles. By accepting an electron pair, a proton achieves the valence shell configuration of helium; carbocations achieve the valence shell configuration of neon.

Carbanions are Lewis bases. In their reactions they seek a proton or some other positive center to which they can donate their electron pair and thereby neutralize their negative charge. *Reagents, like carbanions, that seek a proton or some other positive center are called nucleophiles* (because the nucleus is the positive part of an atom).

---

* Some chemists refer to carbocations as **carbonium ions.** This older term, however, no longer has wide usage; therefore, we shall use the term *carbocation.* Its meaning is clear and distinct.

$$-\overset{|}{\underset{|}{C}}{:}^{-} \;+\; \overset{\delta+}{H}{\underset{}{-}}A^{\delta-} \longrightarrow -\overset{|}{\underset{|}{C}}{:}H \;+\; {:}A^-$$

Carbanion   **Lewis acid**

$$-\overset{|}{\underset{|}{C}}{:}^{-} \;+\; -\overset{|}{\underset{|}{C}}{\underset{}{-}}L^{\delta-} \longrightarrow -\overset{|}{\underset{|}{C}}{:}\overset{|}{\underset{|}{C}}- \;+\; {:}L^-$$

Carbanion   **Lewis acid**

3.4 THE USE OF CURVED ARROWS IN ILLUSTRATING REACTIONS

In the preceding sections we have shown the movement of an electron pair with a curved arrow. This type of notation is commonly used by organic chemists to show *the direction of electron flow in a reaction. The curved arrow does not show the movement of atoms,* however. The atoms are assumed to follow the flow of electrons. Consider as an example the reaction of hydrogen chloride with water.

A MECHANISM FOR THE REACTION

REACTION:

$$H_2O + HCl \longrightarrow H_3O^+ + Cl^-$$

MECHANISM:

$$H-\overset{..}{O}{:} + H\overset{\delta+}{\underset{}{-}}\overset{..}{\underset{..}{Cl}}{:}^{\delta-} \longrightarrow H-\overset{..}{\underset{|}{O}}{\overset{+}{-}}H + {:}\overset{..}{\underset{..}{Cl}}{:}^-$$
$$\underset{H}{} \qquad\qquad\qquad \underset{H}{}$$

A water molecule uses one of its electron pairs to form a bond to a proton of HCl. The bond between the hydrogen and chlorine breaks, with the electron pair going to the chlorine atom.

This leads to the formation of a hydronium ion and a chloride ion.

Typically, a curved arrow begins with a covalent bond or unshared electron pair (a site of higher electron density) and points toward a site of electron deficiency. We see here that as the water molecule collides with a hydrogen chloride molecule, it uses one of its unshared electron pairs (shown in blue) to form a bond to the proton of HCl. This bond forms because the negatively charged electrons of the oxygen atom are attracted to the positively charged proton. As the bond between the oxygen and the proton forms, the hydrogen–chlorine bond of HCl breaks, and the chlorine of HCl departs with the electron pair that formerly bonded it to the proton. (If this did not happen, the proton would end up forming two covalent bonds, which, of course, a proton cannot do.) We, therefore, use a curved arrow to show the bond cleavage as well. By pointing to the chlorine, the arrow indicates that the electron pair leaves with the chloride ion.

The following acid–base reactions give other examples of the use of the curved-arrow notation:

$$H\overset{+}{\underset{\overset{|}{H}}{\ddot{O}}}H + {:}\ddot{O}{-}H \longrightarrow H{-}\underset{\overset{|}{H}}{\ddot{O}}{:} + H{-}\ddot{O}{-}H$$

Acid   Base

$$CH_3{-}\overset{\displaystyle :\ddot{O}}{\overset{\|}{C}}{-}\ddot{O}{-}H + {:}\ddot{O}{-}H \;\rightleftharpoons\; CH_3{-}\overset{\displaystyle :\ddot{O}}{\overset{\|}{C}}{-}\ddot{O}{:}^{-} + H{-}\overset{+}{\underset{\overset{|}{H}}{\ddot{O}}}H$$

Acid   Base

$$CH_3{-}\overset{\displaystyle :\ddot{O}}{\overset{\|}{C}}{-}\ddot{O}{-}H + {:}\ddot{O}{-}H \longrightarrow CH_3{-}\overset{\displaystyle :\ddot{O}}{\overset{\|}{C}}{-}\ddot{O}{:}^{-} + H{-}\ddot{O}{-}H$$

Acid   Base

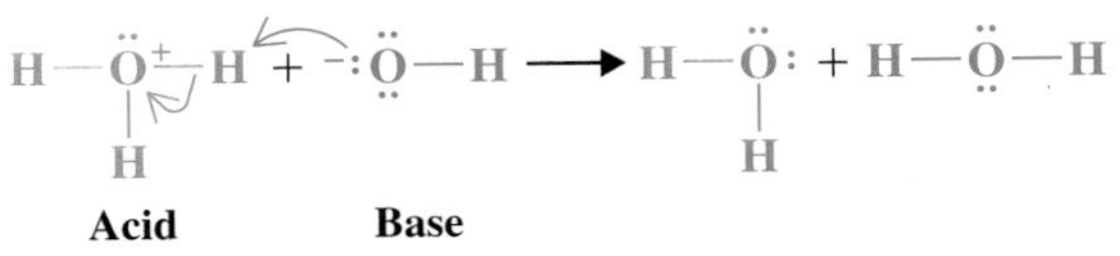

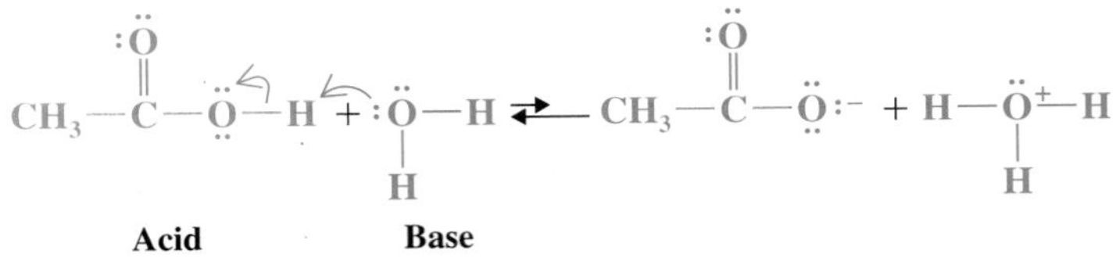

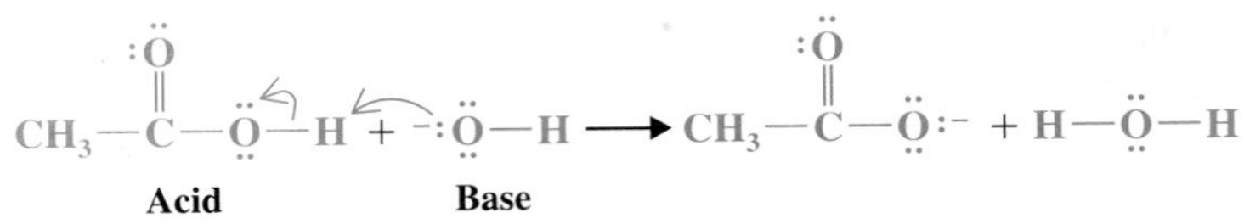

PROBLEM 3.3

Use the curved arrow notation to write the reaction that would take place between dimethylamine and boron trifluoride. Identify the Lewis acid and Lewis base and assign appropriate formal charges.

3.5 THE STRENGTH OF ACIDS AND BASES: K_a AND pK_a

In contrast to the strong acids, such as HCl and H_2SO_4, acetic acid is a much weaker acid. When acetic acid dissolves in water, the following reaction does not proceed to completion.

$$CH_3{-}\overset{O}{\overset{\|}{C}}{-}OH + H_2O \rightleftharpoons CH_3{-}\overset{O}{\overset{\|}{C}}{-}O^- + H_3O^+$$

Experiments show that in a $0.1M$ solution of acetic acid at 25°C only about 1% of the acetic acid molecules ionize by transferring their protons to water.

3.5A The Acidity Constant, K_a

Because the reaction that occurs in an aqueous solution of acetic acid is an equilibrium, we can describe it with an expression for the equilibrium constant.

$$K_{eq} = \frac{[H_3O^+]\,[CH_3CO_2{}^-]}{[CH_3CO_2H]\,[H_2O]}$$

For dilute aqueous solutions, the concentration of water is essentially constant ($\sim 55.5M$), so we can rewrite the expression for the equilibrium constant in terms of a new constant (K_a) called **the acidity constant.**

$$K_a = K_{eq} [H_2O] = \frac{[H_3O^+] [CH_3CO_2^-]}{[CH_3CO_2H]}$$

At 25°C, the acidity constant for acetic acid is 1.76×10^{-5}.

We can write similar expressions for any weak acid dissolved in water. Using a generalized hypothetical acid (HA) the reaction in water is

$$HA + H_2O \rightleftharpoons H_3O^+ + A^-$$

and the expression for the acidity constant is

$$K_a = \frac{[H_3O^+] [A^-]}{[HA]}$$

Because the concentrations of the products of the reaction are written in the numerator and the concentration of the undissociated acid in the denominator, **a large value of K_a means the acid is a strong acid, and a small value of K_a means the acid is a weak acid.** If the K_a is greater than 10, the acid will be, for all practical purposes, completely dissociated in water.

PROBLEM 3.4

Formic acid (HCO_2H) has a $K_a = 1.78 \times 10^{-4}$.   (a) What are the molar concentrations of the hydronium ion and formate ion (HCO_2^-) in a 0.1M aqueous solution of formic acid?   (b) What percentage of the formic acid is ionized?

3.5B Acidity and pK_a

Chemists usually express the acidity constant, K_a, as its negative logarithm, pK_a.

$$pK_a = -\log K_a$$

This is analogous to expressing the hydronium ion concentration as pH.

$$pH = -\log[H_3O^+]$$

For acetic acid the pK_a is 4.75:

$$pK_a = -\log(1.76 \times 10^{-5}) = -(-4.75) = 4.75$$

Notice that there is an inverse relationship between the magnitude of the pK_a and the strength of the acid. **The larger the value of the pK_a, the weaker is the acid.** For example, acetic acid with a $pK_a = 4.75$ is a weaker acid than trifluoroacetic acid with a $pK_a = 0$ ($K_a = 1$). Hydrochloric acid with a $pK_a = -7$ ($K_a = 10^7$) is a far stronger acid than trifluoroacetic acid. (It is understood that a positive pK_a is larger than a negative pK_a.)

$$CH_3CO_2H \ < \ CF_3CO_2H \ < \qquad HCl$$

$$pK_a = 4.75 \qquad pK_a = 0 \qquad pK_a = -7$$

Weak acid   **Very strong acid**

Increasing acid strength

Table 3.1 lists pK_a values for a selection of acids relative to water as the base. The values in the middle of the table are the most accurate because they can be measured in aqueous solution. Special methods must be used to estimate the pK_a values for the very strong acids at the top of the table and for the very weak acids at the bottom.* The pK_a values for these very strong and weak acids are therefore approximate. All of the acids that we shall consider in this book will have strengths in between that of ethane (an extremely weak acid) and that of $HSbF_6$ (an acid that is so strong that it is called a "superacid"). As you examine Table 3.1 take care not to lose sight of the vast range of acidities that it represents.

TABLE 3.1   Relative strength of selected acids and their conjugate bases

	ACID	APPROXIMATE pK_a	CONJUGATE BASE	
Strongest Acid	$HSbF_6$	> -12	SbF_6^-	Weakest Base
	HI	-10	I^-	
	H_2SO_4	-9	HSO_4^-	
	HBr	-9	Br^-	
	HCl	-7	Cl^-	
	$C_6H_5SO_3H$	-6.5	$C_6H_5SO_3^-$	
	H_3O^+	-1.74	H_2O	
	HNO_3	-1.4	NO_3^-	
	CF_3CO_2H	0.18	$CF_3CO_2^-$	
	HF	3.2	F^-	
	CH_3CO_2H	4.75	$CH_3CO_2^-$	
	NH_4^+	9.2	NH_3	
	C_6H_5OH	9.9	$C_6H_5O^-$	
	$CH_3NH_3^+$	10.6	CH_3NH_2	
	H_2O	15.7	OH^-	
	CH_3CH_2OH	16	$CH_3CH_2O^-$	
	$(CH_3)_3COH$	18	$(CH_3)_3CO^-$	
	$HC\equiv CH$	25	$HC\equiv C^-$	
	H_2	35	H^-	
	NH_3	38	NH_2^-	
	$CH_2=CH_2$	44	$CH_2=CH^-$	
Weakest Acid	CH_3CH_3	50	$CH_3CH_2^-$	Strongest Base

Increasing acid strength (left, upward arrow)   *Increasing base strength* (right, downward arrow)

* Acids that are stronger than a hydronium ion and bases that are stronger than a hydroxide ion react completely with water (see Sections 3.1A and 3.8). Therefore, it is not possible to measure acidity constants for these acids in water. Other solvents and special techniques are used, but we do not have the space to describe these methods here.

PROBLEM 3.5

(a) An acid (HA) has a $K_a = 10^{-7}$. What is its pK_a? (b) Another acid (HB) has a $K_a = 5$; what is its pK_a? (c) Which is the stronger acid?

Water, itself, is a very weak acid and undergoes self-ionization even in the absence of acids and bases.

$$H-\overset{..}{\underset{H}{O}}: + H-\overset{..}{\underset{H}{O}}: \rightleftharpoons H-\overset{+}{\underset{H}{\overset{..}{O}}}-H + {}^-:\overset{..}{\underset{..}{O}}-H$$

In pure water at 25°C, the concentrations of hydronium and hydroxide ions are equal to $10^{-7}M$. Since the concentration of water in pure water is $55.5M$, we can calculate the K_a for water.

$$K_a = \frac{[H_3O^+][OH^-]}{[H_2O]} \qquad K_a = \frac{(10^{-7})(10^{-7})}{(55.5)} = 1.8 \times 10^{-16} \qquad pK_a = 15.7$$

PROBLEM 3.6

Show calculations proving that the pK_a of the hydronium ion (H_3O^+) is -1.74 as given in Table 3.1.

3.5C Predicting the Strength of Bases

In our discussion so far we have dealt only with the strengths of acids. Arising as a natural corollary to this is a principle that allows us to estimate the strengths of bases. Simply stated, the principle is this: **The stronger the acid, the weaker will be its conjugate base.**

We can, therefore, **relate the strength of a base to the pK_a of its conjugate acid. The larger the pK_a of the conjugate acid, the stronger is the base.** Consider the following as examples:

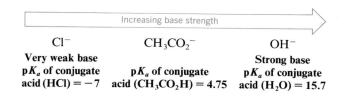

We see that the hydroxide ion is the strongest base of these three bases because its conjugate acid, water, is the weakest acid. (We know that water is the weakest acid because it has the largest pK_a.)

Amines are like ammonia in that they are weak bases. Dissolving ammonia in water brings about the following equilibrium.

$$\overset{..}{N}H_3 \ + \ H-\overset{..}{\underset{..}{O}}-H \rightleftharpoons H-\overset{\overset{\displaystyle H}{|}}{\underset{\underset{\displaystyle H}{|}}{N}}{}^{\pm}-H \ + \ {}^{-}\!:\overset{..}{\underset{..}{O}}-H$$

Base	Acid	Conjugate acid	Conjugate base
		$pK_a = 9.2$	

Dissolving methylamine in water causes the establishment of a similar equilibrium.

$$CH_3\overset{..}{N}H_2 \ + \ H-\overset{..}{\underset{..}{O}}-H \rightleftharpoons CH_3-\overset{\overset{\displaystyle H}{|}}{\underset{\underset{\displaystyle H}{|}}{N}}{}^{+}-H \ + \ {}^{-}\!:\overset{..}{\underset{..}{O}}-H$$

Base	Acid	Conjugate acid	Conjugate base
		$pK_a = 10.6$	

Again we can relate the basicity of these substances to the strength of their conjugate acids. The conjugate acid of ammonia is the ammonium ion, $NH_4{}^+$. The pK_a of the ammonium ion is 9.2. The conjugate acid of methylamine is the $CH_3NH_3{}^+$ ion. This ion, called the methylaminium ion, has a $pK_a = 10.6$. Since the conjugate acid of methylamine is a weaker acid than the conjugate acid of ammonia, we can conclude that methylamine is a stronger base than ammonia.

PROBLEM 3.7

The pK_a of the anilinium ion ($C_6H_5\overset{+}{N}H_3$) is equal to 4.6. On the basis of this fact, decide whether aniline ($C_6H_5NH_2$) is a stronger or weaker base than methylamine.

3.6 PREDICTING THE OUTCOME OF ACID–BASE REACTIONS

Table 3.1 gives the approximate pK_a values for a range of representative compounds. While you may not be expected to memorize all of the pK_a values in Table 3.1 now, it is a good idea to begin to learn the general order of acidity and basicity for some of the common acids and bases. The examples given in Table 3.1 are representative of their class or functional group. For example, acetic acid has a $pK_a = 4.75$, and carboxylic acids generally have pK_a values near this value (in the range $pK_a = 3-5$). Ethyl alcohol is given as an example of an alcohol, and alcohols generally have pK_a values near that of ethyl alcohol (in the range $pK_a = 15-18$), and so on. (There are exceptions, of course, and we shall learn what these exceptions are as we go on.)

By learning the relative scale of acidity of common acids now, you will be able to predict whether or not an acid–base reaction will occur as written. The general principle to apply is this: **Acid–base reactions always favor the formation of the weaker acid and the weaker base.** The reason for this is that the outcome of an acid–base reaction is determined by the position of an equilibrium. Acid–base reactions, are said, therefore, to be **under equilibrium control,** and reactions under equilibrium control always favor the formation of the most stable (lowest potential energy) species. The weaker acid and weaker base are more stable (lower in potential energy) than the stronger acid and stronger base.

Using this principle, we can predict that a carboxylic acid (RCO_2H) will react with aqueous NaOH in the following way because the reaction will lead to the formation of the weaker acid (H_2O) and weaker base (RCO_2^-).

$$R-\overset{\overset{\ddot{O}}{\|}}{C}-\ddot{O}-H + Na^+ \,{}^-\!:\!\ddot{O}-H \longrightarrow R-\overset{\overset{:\ddot{O}}{\|}}{C}-\ddot{O}:^-Na^+ + H-\ddot{O}-H$$

Stronger acid	**Stronger base**	**Weaker base**	**Weaker acid**
$pK_a = 3{-}5$			$pK_a = 15.7$

Because there is a large difference in the value of the pK_a of the two acids, the position of equilibrium will greatly favor the formation of the products. In instances like these we commonly show the reaction with a one-way arrow even though the reaction is an equilibrium.

Although acetic acid and other carboxylic acids containing fewer than five carbon atoms are soluble in water, many other carboxylic acids of higher molecular weight are not appreciably soluble in water. Because of their acidity, however, *water-insoluble carboxylic acids dissolve in aqueous sodium hydroxide;* they do so by reacting to form water-soluble sodium salts.

Insoluble in water   **Soluble in water**

We can also predict that an amine will react with aqueous hydrochloric acid in the following way:

Stronger base	**Stronger acid**	**Weaker acid**	**Weaker base**
	$pK_a = 1.74$	$pK_a = 9{-}10$	

While methylamine and most amines of low molecular weight are very soluble in water, amines with higher molecular weights, such as aniline ($C_6H_5NH_2$), have limited water solubility. However, these water-insoluble amines dissolve readily in hydrochloric acid because the acid–base reactions convert them to soluble salts.

Water insoluble   **Water-soluble salt**

3.7 THE RELATIONSHIP BETWEEN STRUCTURE AND ACIDITY

The strength of an acid depends on the extent to which a proton can be separated from it and transferred to a base. Removing the proton involves breaking a bond to the proton, and it involves making the conjugate base more electrically negative. *When we compare compounds in a vertical column of the periodic table, the strength of the bond to the proton is the dominating effect.* The acidities of the hydrogen halides furnish an example:

Acidity increases →

H—F	H—Cl	H—Br	H—I
$pK_a = 3.2$	$pK_a = -7$	$pK_a = -9$	$pK_a = -10$

Acidity increases as we descend a vertical column: H—F is the weakest acid and H—I is the strongest. The important factor is the strength of the H—X bond, the *stronger* the bond the *weaker* the acid. The H—F bond is by far the strongest and the H—I bond is the weakest.

   Because HI, HBr, and HCl are such strong acids, *their conjugate bases (I^-, Br^-, Cl^-) are all very weak bases.* The fluoride ion is considerably more basic. Overall the basicity of the halide ions increases in the following way:

← Basicity increases

F^- Cl^- Br^- I^-

   We see the same trend of acidities and basicities in other vertical columns of the periodic table. Consider, for example, the column headed by oxygen:

Acidity increases →

H_2O H_2S H_2Se

and

← Basicity increases

OH^- SH^- SeH^-

Here the strongest bond is the O—H bond and H_2O is the weakest acid; the weakest bond is the Se—H bond and H_2Se is the strongest acid.

   When we compare compounds in the same horizontal row of the periodic table, bond strengths are roughly the same and *the dominant factor becomes the electronegativity of the atom bonded to the hydrogen.* The electronegativity of this atom affects acidity in two related ways. It affects the polarity of the bond to the proton and it affects the relative stability of the anion (conjugate base) that forms when the proton is lost. Let us compare two hypothetical acids, H—A and H—B.

$\overset{\delta+}{H}\!-\!\overset{\delta-}{A}$ and $\overset{\delta+}{H}\!-\!\overset{\delta-}{B}$

Let us assume that A is more electronegative than B. The greater electronegativity of A will cause atom A to be more negative than atom B, and the hydrogen (proton) of

H—A will be more positive than that of H—B. The proton of H—A, consequently, will be held less strongly, and it will separate and be transferred to a base more readily. The greater electronegativity of A will also mean that atom A will acquire a negative charge more readily than B, and that the A$^-$ anion will be more stable than the B$^-$ anion. H—A, therefore, will be the stronger acid.

We can see an example of this effect when we compare the acidities of the compounds CH_4, NH_3, H_2O, and HF. These compounds are all hydrides of first-row elements, and electronegativity increases across a row of the periodic table from left to right (see Table 1.2).

Electronegativity increases →

C   N   O   F

Because fluorine is the most electronegative, the bond in H—F is most polarized, and the proton in H—F is the most positive. Therefore, H—F loses a proton most readily and is the most acidic:

Acidity increases →

$$\overset{\delta-\delta+}{H_3C—H} \qquad \overset{\delta-\delta+}{H_2N—H} \qquad \overset{\delta-\delta+}{HO—H} \qquad \overset{\delta-\delta+}{F—H}$$
$$pK_a = 48 \qquad\qquad pK_a = 38 \qquad\qquad pK_a = 15.7 \qquad pK_a = 3.2$$

Because H—F is the strongest acid, its conjugate base, the fluoride ion (F$^-$), will be the weakest base. Fluorine is the most electronegative atom and it accommodates the negative charge most readily.

← Basicity increases

$$CH_3^- \qquad H_2N^- \qquad HO^- \qquad F^-$$

The methanide ion (CH_3^-) is the least stable anion of the four, because carbon being the least electronegative element is least able to accept the negative charge. The methanide ion, therefore, is the strongest base. [The methanide ion, a **carbanion,** and the amide ion (NH_2^-) are exceedingly strong bases because they are the conjugate bases of extremely weak acids. We shall discuss some uses of these powerful bases in Section 3.13.]

3.7A The Effect of Hybridization

The protons of ethyne are more acidic than those of ethene, which in turn, are more acidic than those of ethane.

$$H—C≡C—H$$

Ethyne
$pK_a = 25$

Ethene
$pK_a = 44$

Ethane
$pK_a = 50$

We can explain this order of acidities on the basis of the hybridization state of carbon in each compound. Electrons of 2s orbitals have lower energy than those of 2p orbitals because *electrons in 2s orbitals tend, on the average, to be much closer to the nucleus than electrons in 2p orbitals.* (Consider the shapes of the orbitals: 2s orbitals are spherical and centered on the nucleus; 2p orbitals have lobes on either side of the nucleus and are extended into space.) With hybrid orbitals, therefore, **having more s character means that the electrons of the anion will, on the average, be lower in energy, and the anion will be more stable.** The sp orbitals of the C—H bonds of ethyne have 50% s character (because they arise from the combination of one s orbital and one p orbital), those of the sp^2 orbitals of ethene have 33.3% s character, while those of the sp^3 orbitals of ethane have only 25% s character. This means, in effect, that the sp carbon atoms of ethyne act as if they were the most electronegative when compared to the sp^2 carbon atoms of ethene, and the sp^3 carbon atoms of ethane. (Remember: Electronegativity measures an atom's ability to hold bonding electrons close to its nucleus, and having electrons closer to the nucleus makes them more stable.)

Now we can see how the order of relative acidities of ethyne, ethene, and ethane parallels the effective electronegativity of the carbon atom in each compound:

Relative Acidity of the Hydrocarbons

$$HC \equiv CH > H_2C = CH_2 > H_3C — CH_3$$

Being the most electronegative, the sp-hybridized carbon atom of ethyne polarizes its C—H bonds to the greatest extent, causing its hydrogens to be most positive. Therefore, ethyne donates a proton to a base more readily. And in the same way, the ethynide ion is the weakest base because the more electronegative carbon of ethyne is best able to stabilize the negative charge.

Relative Basicity of the Carbanions

$$H_3C — CH_2 : ^- > H_2C = CH : ^- > HC \equiv C : ^-$$

Notice that the explanation given here is the same as that given to account for the relative acidities of HF, H_2O, NH_3, and CH_4.

3.7B Inductive Effects

The carbon–carbon bond of ethane is completely nonpolar because at each end of the bond there are two equivalent methyl groups.

$$CH_3 — CH_3$$
Ethane
The C—C bond is nonpolar

This is not the case with ethyl fluoride, however.

$$\overset{\delta+}{CH_3} \!\!\rightarrow\!\! \overset{\delta+}{CH_2} \!\!\rightarrow\!\! \overset{\delta-}{F}$$
$$\quad 2 \qquad\quad 1$$

One end of the bond, the one nearer the fluorine atom is more positive than the other. This polarization of the carbon–carbon bond results from an intrinsic electron-attracting ability of the fluorine (because of its electronegativity) that is transmitted *through space* and *through the bonds of the molecule.* Chemists call this kind of effect an **inductive effect.** The inductive effect here is **electron attracting** (or **electron with-drawing**), but we shall see later, inductive effects can also be **electron releasing.** *Inductive effects weaken steadily as the distance from the substituent increases.* In this instance, the positive charge that the fluorine imparts to C-**1** is greater than that im-parted to C-**2** because the fluorine is closer to C-**1**.

Transmission of the effect through bonds results from the polarization of one bond causing polarization of an adjacent bond. In ethyl fluoride, the C—F bond is polarized (C-**1** is made positive because the highly electronegative fluorine atom draws in its direction the electrons it is sharing with C-**1**). The C—C bond becomes polarized, too, because the positively charged C-**1** pulls in its direction the electrons that it is sharing with C-**2**. The positive charge on C-**2**, however, is smaller than that on C-**1**. (Even the hydrogen atoms of C-**2** are made slightly positive because the C—H bonds are slightly polarized by the positive charge on C-**2**.)

Transmission of the effect through space is known to be the more important mode of transmission; it results from simple electrostatic effects. In this instance, because the positive end of the C—F dipole is closer to C-**2**, it attracts the electrons around C-**2** and makes C-**2** positive, and so on.

3.8 THE RELATIONSHIP BETWEEN THE EQUILIBRIUM CONSTANT AND THE STANDARD FREE-ENERGY CHANGE, $\Delta G°$

An important relationship exists between the equilibrium constant and the standard free-energy change* ($\Delta G°$) that accompanies the reaction.

$$\Delta G° = -2.303 \, RT \log K_{eq}$$

R is the gas constant and equals 1.987 cal K^{-1} mol^{-1}; T is the absolute temperature in Kelvins (K).

It is easy to show with this equation that **a negative value of $\Delta G°$ is associated with reactions that favor the formation of products when equilibrium is reached,** and for which the equilibrium constant is greater than 1. Reactions with a $\Delta G°$ more negative than about -3 kcal mol^{-1} are said *to go to completion,* meaning that almost all ($>99\%$) of the reactants are converted to products when equilibrium is reached. Conversely, **a positive value of $\Delta G°$ is associated with reactions for which the formation of products at equilibrium is unfavorable** and for which the equilibrium constant is less than 1. Inasmuch as K_a is an equilibrium constant, it is related to $\Delta G°$ in the same way.

The free-energy change ($\Delta G°$) has two components, the enthalpy change ($\Delta H°$)

---

* By standard free-energy change ($\Delta G°$) we mean that the products and reactants are taken as being in their standard states (1 atm of pressure for a gas, and 1M for a solution). The free-energy change is often called the **Gibbs free-energy change,** to honor the contributions to thermodynamics of J. Willard Gibbs, a professor of mathematical physics at Yale University from 1871 until the turn of the century. Gibbs ranks as one of the greatest scientists produced by the United States.

and the entropy change ($\Delta S°$). The relationship between these three thermodynamic quantities is

$$\Delta G° = \Delta H° - T\Delta S°$$

We have seen (Section 1.9) that $\Delta H°$ is associated with changes in bonding that occur in a reaction. If, collectively, stronger bonds are formed in the products than existed in the starting materials, then $\Delta H°$ will be negative (i.e., the reaction is *exothermic*). If the reverse is true, then $\Delta H°$ will be positive (the reaction is *endothermic*). A negative value for $\Delta H°$, therefore, will contribute to making $\Delta G°$ negative, and will, consequently favor the formation of products. For the ionization of an acid, the less positive or more negative the value of $\Delta H°$ the stronger the acid will be.

Entropy changes have to do with *changes in the relative order of a system.* **The more random a system is, the greater is its entropy.** Therefore, a positive entropy change $(+ \Delta S°)$ is always associated with a change from a more ordered system to a less ordered one. A negative entropy change $(- \Delta S°)$ accompanies the reverse process. In the equation $\Delta G° = \Delta H° - T\Delta S°$, the entropy change (multiplied by T) is preceded by a negative sign; this means that *a positive entropy change (from order to disorder) makes a negative contribution to $\Delta G°$ and is energetically favorable for the formation of products.*

For many reactions in which the number of molecules of products equals the number of molecules of reactants (e. g., when two molecules react to produce two molecules) the entropy change will be small. This means that except at high temperatures (where the term $T\Delta S°$ becomes large even if $\Delta S°$ is small) a large value of $\Delta H°$ will determine whether or not the formation of products will be favored. If $\Delta H°$ is large and negative (if the reaction is exothermic), then the reaction will favor the formation of products at equilibrium. If $\Delta H°$ is positive (if the reaction is endothermic) then the formation of products will be unfavorable.

PROBLEM 3.8

State whether you would expect the entropy change, $\Delta S°$, to be positive, negative, or approximately zero for each of the following reactions. (Assume the reactions take place in the gas phase.)

(a) $A + B \longrightarrow C$ (c) $A \longrightarrow B + C$

(b) $A + B \longrightarrow C + D$

PROBLEM 3.9

(a) What is the value of $\Delta G°$ for a reaction where the $K_{eq} = 1$?   (b) Where the $K_{eq} = 10$? (The change in $\Delta G°$ required to produce a tenfold increase in the equilibrium constant is a useful term to remember.)   (c) Assuming that the entropy change for this reaction is negligible (or zero), what change in $\Delta H°$ is required to produce a tenfold increase in the equilibrium constant?

3.9 THE ACIDITY OF CARBOXYLIC ACIDS

Both carboxylic acids and alcohols are weak acids. However, *most carboxylic acids are much more acidic than the corresponding alcohols.* Unsubstituted carboxylic acids have pK_a's in the range of 3 to 5; alcohols have pK_as in the range of 15 to 18. Consider as examples two compounds of about the same molecular proportions, but with very different acidities: acetic acid and ethanol.

$$CH_3-\overset{\overset{\displaystyle O}{\|}}{C}-OH \qquad\qquad CH_3-CH_2-OH$$

Acetic acid   **Ethanol**

$pK_a = 4.75$ $pK_a = 16$

$\Delta G^\circ = 6.5 \text{ kcal mol}^{-1}$ $\Delta G^\circ = 21.7 \text{ kcal mol}^{-1}$

From the pK_a for acetic acid ($pK_a = 4.75$) one can calculate (Section 3.8) that the free energy change for the ionization of acetic acid is positive and equal to 6.5 kcal mol^{-1}. For ethanol ($pK_a = 16$), the free energy change for the ionization is much larger and is equal to 21.7 kcal mol^{-1}. These values (see Fig. 3.1) show that whereas both compounds are weak acids, ethanol is a much weaker acid than acetic acid.

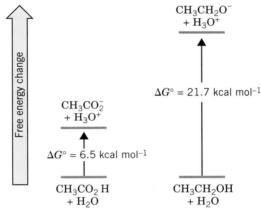

FIGURE 3.1 A diagram comparing the free-energy changes that accompany ionization of acetic acid and ethanol. Ethanol has a larger positive free-energy change and is a weaker acid because its ionization is more unfavorable.

How do we explain the greater acidity of carboxylic acids relative to alcohols? Two explanations have been proposed, one based on effects arising from resonance theory (and called, therefore, **resonance effects**) and another explanation based on **inductive effects** (Section 3.7B). Although both effects make a contribution to the greater acidity of carboxylic acids, the question as to which effect is more important, as we shall see, has not been resolved.

3.9A An Explanation Based on Resonance Effects

For many years, the greater acidity of carboxylic acids was attributed primarily to **resonance stabilization of the carboxylate ion.** This explanation invokes a principle of resonance theory (Section 1.8) that states that molecules and ions *are stabilized by resonance* especially *when the molecule or ion can be represented by two or more* **equivalent** *resonance structures (i.e., by resonance structures of equal stability).*

Two resonance structures can be written for a carboxylic acid and two for its anion (Fig. 3.2), and it can be assumed that *resonance stabilization of the anion is greater*

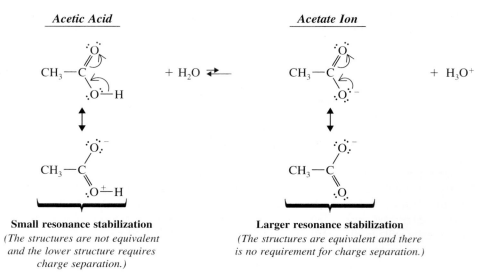

Acetic Acid   *Acetate Ion*

Small resonance stabilization
(The structures are not equivalent and the lower structure requires charge separation.)

Larger resonance stabilization
(The structures are equivalent and there is no requirement for charge separation.)

FIGURE 3.2 Two resonance structures that can be written for acetic acid and two that can be written for the acetate ion. According to a resonance explanation of the greater acidity of acetic acid, the equivalent resonance structures for the acetate ion provide it greater resonance stabilization and reduce the positive free-energy change for the ionization.

because the resonance structures for the anion are equivalent and because no separation of opposite charges occurs in them. [Separation of opposite charges requires energy, and resonance structures that have separated charges are assumed to be less important in providing stability than those that do not (Section 12.5).] The greater stabilization of the anion of a carboxylic acid (relative to the acid itself) lowers the free energy of the anion and thereby decreases the positive free energy change required for the ionization. Remember: *Any factor that makes the free energy change for the ionization of an acid less positive (or more negative) makes the acid stronger* (Section 3.8).

No stabilizing resonance structures are possible for an alcohol or its anion:

$$CH_3—CH_2—\ddot{O}—H + H_2O \rightleftharpoons CH_3—CH_2—\ddot{O}:^- + H_3O^+$$

No resonance stabilization   **No resonance stabilization**

The positive free energy change for the ionization of an alcohol is not reduced by resonance stabilization, and this explains why the free energy change is much larger than that for a carboxylic acid. An alcohol, consequently, is much less acidic than a carboxylic acid.

3.9B An Explanation Based on Inductive Effects

In 1986 an alternative explanation of the greater acidity of carboxylic acids was put forward, stating that resonance stabilization of the carboxylic acid anion is a relatively unimportant factor in explaining the acidity of carboxylic acids. [Resonance, however, does explain other properties of acid derivatives (see Problem 3.10) and this has not been questioned.] According to this newer explanation, the most important factor in accounting for the acidity of carboxylic acids is **the inductive effect of the carbonyl** group of the carboxylic acid.* To understand this explanation based on inductive effects, let us consider the same two compounds.

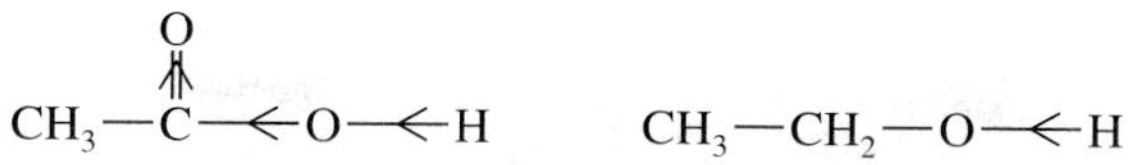

$$
\begin{array}{cc}
\underset{\begin{array}{c}\textbf{Acetic acid}\\ \textbf{(stronger acid)}\end{array}}{CH_3-\overset{\overset{\displaystyle O}{\|}}{C}\!\leftarrow\!O\!\leftarrow\!H} &
\underset{\begin{array}{c}\textbf{Ethanol}\\ \textbf{(weaker acid)}\end{array}}{CH_3-CH_2-O\!\leftarrow\!H}
\end{array}
$$

In both compounds the O—H bond is highly polarized by the greater electronegativity of the oxygen atom. The key to the much greater acidity of acetic acid, in this explanation, is the powerful electron-attracting inductive effect of its carbonyl group (C=O group) when compared with the CH_2 group in the corresponding position of ethanol. The carbonyl group is highly polarized; the carbon of the carbonyl group bears a large positive charge *because the second resonance structure below appears to be the more important contributor to the overall resonance hybrid.* In other words, the carbonyl group should show a strong resemblance to the second structure below.

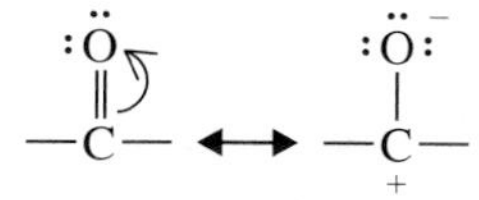

$$
\underset{-C-}{\overset{:\ddot{O}:}{\|}} \longleftrightarrow \underset{\underset{+}{-C-}}{\overset{:\ddot{O}:^{-}}{|}}
$$

Resonance structures for the carbonyl group

Because the carbon atom of the carbonyl group of acetic acid bears a large positive charge, it adds its electron-attracting inductive effect to that of the oxygen atom of the hydroxyl group attached to it; *these combined effects make the hydroxyl proton much more positive than the proton of the alcohol.* This greater positive charge on the proton of the carboxylic acid explains why the proton separates more readily. This electron-attracting inductive effect of the carbonyl group also stabilizes the acetate ion that forms from acetic acid, and therefore the acetate ion is a weaker base than the ethoxide ion.

$$
\begin{array}{cc}
\underset{\textbf{Weaker base}}{CH_3-\overset{\overset{\displaystyle O^{\delta-}}{\|}}{C}\!\leftarrow\!O^{\delta-}} &
\underset{\textbf{Stronger base}}{CH_3-CH_2\!\leftarrow\!O^{-}}
\end{array}
$$

The carbon of the CH_2 group of ethanol, by contrast, has a negligible effect on the alcohol or on its anion.

G. W. Wheland, who many years ago formulated the resonance explanation for the greater acidity of carboxylic acids, recognized that the inductive effect of the carbonyl group could be important and saw that deciding which effect, resonance or inductive, was more important would be difficult. Today this difficulty persists. A vigorous and interesting debate on the relative importance of the two effects is still being carried on.†

* See M. R. Siggel and T. D. Thomas, *J. Am. Chem. Soc.* **1986,** *108,* 4360–4362, and M. R. F. Siggel, A. R. Streitwieser, Jr., and T. D. Thomas, *J. Am. Chem. Soc.* **1988,** *110,* 8022–8028.

† The explanation based on the greater importance of inductive effects has been challenged recently. Those who may be interested in pursuing this debate further should consult the following article and the references given in it: F. G. Bordwell and A. V. Satish, *J. Am. Chem. Soc.* **1994,** *116,* 8885–8889.

PROBLEM 3.10

Resonance theory may or may not provide the best explanation for the acidity of a carboxylic acid; however, it does provide an appropriate explanation for two related facts: The carbon–oxygen bond distances in the acetate ion are the same, and the oxygens of the acetate ion bear equal negative charges. Provide the resonance explanation for these facts.

3.9C Inductive Effects of Other Groups

The acid-strengthening effect of other electron-attracting groups (other than the carbonyl group) can be shown by comparing the acidities of acetic acid and chloroacetic acid:

$$CH_3-\overset{\overset{\textstyle O}{\|}}{C}\text{---O---H} \qquad\qquad Cl\text{---}CH_2\text{---}\overset{\overset{\textstyle O}{\|}}{C}\text{---O---H}$$

$$pK_a = 4.75 \qquad\qquad\qquad pK_a = 2.86$$

The greater acidity of chloroacetic acid can be attributed, in part, to the extra electron-attracting inductive effect of the electronegative chlorine atom. By adding its inductive effect to that of the carbonyl group and the oxygen, it makes the hydroxyl proton of chloroacetic acid even more positive than that of acetic acid. It also stabilizes the chloroacetate ion that is formed when the proton is lost *by dispersing its negative charge.*

$$Cl\text{---}CH_2\text{---}\overset{\overset{\textstyle O}{\|}}{C}\text{---O---H} + H_2O \rightleftarrows {}^{\delta-}Cl\text{---}CH_2\text{---}\overset{\overset{\textstyle O^{\delta-}}{\|}}{C}\text{---O}^{\delta-} + H_3O^+$$

The negative charge is more spread out in the chloroacetate ion because it resides partially on the chlorine atom. Dispersal of charge always makes a species more stable, and, as we have seen now in several instances, **any factor that stabilizes the conjugate base of an acid increases the strength of the acid.** (In Section 3.10, we shall see that entropy changes in the solvent are also important in explaining the increased acidity of chloroacetic acid.)

PROBLEM 3.11

Which would you expect to be the stronger acid? Explain your reasoning in each instance.

(a) CH_2ClCO_2H or $CHCl_2CO_2H$

(b) CCl_3CO_2H or $CHCl_2CO_2H$

(c) CH_2FCO_2H or CH_2BrCO_2H

(d) CH_2FCO_2H or $CH_2FCH_2CO_2H$

3.10 THE EFFECT OF THE SOLVENT ON ACIDITY

In the absence of a solvent (i.e., in the gas phase), most acids are far weaker than they are in solution. In the gas phase, for example, acetic acid is estimated to have a pK_a of about 130 (a K_a of $\sim 10^{-130}$)! The reason for this: When an acetic acid molecule donates a proton to a water molecule in the gas phase, the ions that are formed are oppositely charged particles and these particles must become separated.

$$CH_3-\overset{\overset{\displaystyle O}{\|}}{C}-OH + H_2O \rightleftharpoons CH_3-\overset{\overset{\displaystyle O}{\|}}{C}-O^- + H_3O^+$$

In the absence of a solvent, separation is difficult. In solution, solvent molecules surround the ions, insulating them from one another, stabilizing them, and making it far easier to separate them than in the gas phase.

In a solvent such as water, called a **protic solvent,** solvation by hydrogen bonding is important (Section 2.16C). **A protic solvent is one that has a hydrogen atom attached to a strongly electronegative element such as oxygen or nitrogen.** Molecules of a protic solvent, therefore, can form hydrogen bonds to the unshared electron pairs of oxygen (or nitrogen) atoms of an acid and its conjugate base, but they may not stabilize both equally.

Consider, for example, the ionization of acetic acid in aqueous solution. Water molecules solvate both the undissociated acid (CH_3CO_2H) and its anion ($CH_3CO_2^-$) by forming hydrogen bonds to them (Section 3.1A). However, hydrogen bonding to $CH_3CO_2^-$ is much stronger than to CH_3CO_2H because the water molecules are more attracted by the negative charge. This differential solvation, moreover, has important consequences for the entropy change that accompanies the ionization. Solvation of any species decreases the entropy of the solvent because the solvent molecules become much more ordered as they surround molecules of the solute. Because solvation of $CH_3CO_2^-$ is stronger, the solvent molecules become more orderly around it. The entropy change ($\Delta S°$) for the ionization of acetic acid, therefore, is negative. This means that the $-T\Delta S°$ term in the equation $\Delta G° = \Delta H° - T\Delta S°$, makes an acid-weakening positive contribution to $\Delta G°$. In fact, as Table 3.2 shows, the $-T\Delta S°$ term contributes more to $\Delta G°$ than $\Delta H°$ does, and accounts for the fact that the free-energy change for the ionization of acetic acid is positive (unfavorable).

We saw in Section 3.5B that chloroacetic acid is a stronger acid than acetic acid, and we attributed this increased acidity to the presence of the electron-withdrawing chlorine atom. Table 3.2 shows us that both $\Delta H°$ and $-T\Delta S°$ are more favorable for the ionization of chloroacetic acid ($\Delta H°$ is more negative by 1.0 kcal mol^{-1}, and $-T\Delta S°$ is less positive by 1.6 kcal mol^{-1}). The larger contribution is clearly in the entropy term. Apparently, by stabilizing the chloroacetate anion, the chlorine atom makes the chloroacetate ion less prone to cause an ordering of the solvent because it requires less stabilization through solvation.

TABLE 3.2 Thermodynamic values for the dissociation of acetic and chloroacetic acids in H_2O at 25°C[a]

ACID	pK_a	$\Delta G°$ (kcal mol^{-1})	$\Delta H°$ (kcal mol^{-1})	$-T\Delta S°$ (kcal mol^{-1})
CH_3CO_2H	4.75	$+6.5$	-0.1	$+6.6$
$ClCH_2CO_2H$	2.86	$+3.9$	-1.1	$+5.0$

[a] Table adapted from J. March, *Advanced Organic Chemistry,* 3rd ed., Wiley, New York, 1985 p. 236.

3.11 ORGANIC COMPOUNDS AS BASES

If an organic compound contains an atom with an unshared electron pair, it is a potential base. We saw in Section 3.5C that compounds with an unshared electron pair on a nitrogen atom (i.e., amines) act as bases. Let us now consider several examples in which organic compounds having an unshared electron pair on an oxygen atom act in the same way.

Dissolving gaseous HCl in methanol brings about an acid–base reaction much like the one that occurs with water (Section 3.2A).

$$
\underset{\substack{\text{Methanol}}}{H_3C\!-\!\overset{..}{\underset{|}{\overset{|}{O}}}\!:\ +\ H\!-\!\overset{..}{\underset{..}{Cl}}:} \longrightarrow \underset{\substack{\text{Methyloxonium ion}\\ \text{(a protonated alcohol)}}}{H_3C\!-\!\overset{..}{\underset{|}{\overset{\pm}{O}}}\!-\!H} \ +\ :\!\overset{..}{\underset{..}{Cl}}:^-
$$

The conjugate acid of the alcohol is often called a **protonated alcohol,** although more formally it is called an **alkyloxonium ion**.

Alcohols, in general, undergo this same reaction when they are treated with solutions of strong acids such as HCl, HBr, HI, or H_2SO_4.

$$
\underset{\substack{\text{Alcohol}}}{R\!-\!\overset{..}{\underset{|}{\overset{|}{O}}}\!:\ +\ }\underset{\substack{\text{Strong}\\ \text{acid}}}{H\!-\!A} \longrightarrow \underset{\substack{\text{Alkyloxonium ion}}}{R\!-\!\overset{..}{\underset{|}{\overset{\pm}{O}}}\!-\!H} \ +\ \underset{\substack{\text{Weak}\\ \text{base}}}{:\!A^-}
$$

So, too do ethers:

$$
\underset{\substack{\text{Ether}}}{R\!-\!\overset{..}{\underset{R}{\overset{|}{O}}}\!:\ +\ }\underset{\substack{\text{Strong}\\ \text{acid}}}{H\!-\!A} \longrightarrow \underset{\substack{\text{Dialkyloxonium}\\ \text{ion}}}{R\!-\!\overset{..}{\underset{R}{\overset{\pm}{O}}}\!-\!H} \ +\ \underset{\substack{\text{Weak}\\ \text{base}}}{:\!A^-}
$$

Compounds containing a carbonyl group also act as bases in the presence of a strong acid.

$$
\underset{\substack{\text{Ketone}}}{\overset{H_3C}{\underset{H_3C}{>}}C\!=\!\overset{..}{O}\!:\ +\ }\underset{\substack{\text{Strong}\\ \text{acid}}}{H\!-\!A} \rightleftharpoons \underset{\substack{\text{Protonated}\\ \text{ketone}}}{\overset{H_3C}{\underset{H_3C}{>}}C\!=\!\overset{..}{\overset{\pm}{O}}\!-\!H}\ +\ \underset{\substack{\text{Weak}\\ \text{base}}}{:\!A^-}
$$

Proton transfer reactions like these are often the first step in many reactions that alcohols, ethers, aldehydes, ketones, esters, amides, and carboxylic acids undergo.

An atom with an unshared electron pair is not the only locus that confers basicity on an organic compound. The π bond of an alkene can have the same effect. Later we shall study many reactions in which alkenes react with a strong acid by accepting a proton in the following way.

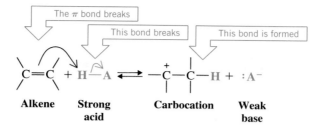

Alkene   **Strong acid**   **Carbocation**   **Weak base**

In this reaction the electron pair of the π bond of the alkene is used to form a bond between one carbon of the alkene and the proton donated by the strong acid. Notice that two bonds are broken in this process: the π bond of the double bond and the bond between the proton of the acid and its conjugate base. One new bond is formed: a bond between a carbon of the alkene and the proton. This process leaves the other carbon of the alkene trivalent, electron deficient, and with a formal positive charge. A compound containing a carbon of this type is called a **carbocation** (Section 3.3).

PROBLEM 3.12

It is a general rule that any organic compound containing oxygen, nitrogen, or a multiple bond will dissolve in concentrated sulfuric acid. Explain the basis of this rule.

3.12 A MECHANISM FOR AN ORGANIC REACTION

In Chapter 6 we shall begin our study of mechanisms of organic reactions in earnest. Let us consider now one mechanism as an example; one that allows us to apply some of the chemistry we have learned in this chapter, and one that, at the same time, will reinforce what we have learned about how curved arrows are used to illustrate mechanisms.

Dissolving *tert*-butyl alcohol in concentrated (conc.) aqueous hydrochloric acid soon results in the formation of *tert*-butyl chloride. The reaction is a substitution reaction:

$$H_3C-\underset{\underset{CH_3}{|}}{\overset{\overset{CH_3}{|}}{C}}-OH \; + \; H-\overset{+}{\underset{|}{\overset{..}{O}}}-H \; + \; :\overset{..}{\underset{..}{Cl}}:^- \; \xrightarrow{H_2O} \; H_3C-\underset{\underset{CH_3}{|}}{\overset{\overset{CH_3}{|}}{C}}-Cl \; + \; 2\,H_2O$$

tert-**Butyl alcohol**   **Conc. HCl**   *tert*-**Butyl chloride**
(soluble in H_2O)   (insoluble in H_2O)

That a reaction has taken place is obvious when one actually does the experiment. *tert*-Butyl alcohol is soluble in the aqueous medium; however, *tert*-butyl chloride is not, and consequently it separates from the aqueous phase as another layer in the flask. It is easy to remove this nonaqueous layer, purify it by distillation, and thus obtain the *tert*-butyl chloride.

Considerable evidence, described later, indicates that the reaction occurs in the following way.

A MECHANISM FOR THE REACTION

Step 1

tert-Butyl alcohol acts as a base
and accepts a proton from the
hydronium ion

***tert*-Butyloxonium ion**

The product is a protonated alcohol
and water (the conjugate acid and base).

Step 2

Carbocation

The bond between the carbon and oxygen of the *tert*-butyl
oxonium ion breaks heterolytically, leading to the
formation of a carbocation and a molecule of water.

Step 3

***tert*-Butyl chloride**

The carbocation, acting as a Lewis acid,
accepts an electron pair from a chloride
ion to become the product.

Notice that all of these steps involve acid–base reactions. Step 1 is a straightforward Brønsted acid–base reaction in which the hydronium ion transfers a proton to the alcohol. Step 2 is the reverse of a Lewis acid–base reaction. In it, the carbon–oxygen bond of the protonated alcohol breaks heterolytically as a water molecule departs with the electrons of the bond. This happens, in part, because the alcohol is protonated. The presence of a formal positive charge on the oxygen of the protonated alcohol weakens the carbon–oxygen bond by drawing the electrons in the direction of the positive oxygen. Step 3 is a Lewis acid–base reaction, in which the carbocation (a Lewis acid) reacts with a chloride ion (a Lewis base) to form the product.

> A question might arise: Why doesn't the carbocation react with a molecule of water (also a Lewis base) instead of a chloride ion? After all, there are many water molecules around, since water is the solvent. The answer is that this step does occur, but it is simply the reverse of Step 2. That is to say, not all of the carbocations that form go on directly to become product. Some react with water to become protonated alcohols again. However, these will dissociate again to become carbocations (even if, before they do, they lose a proton to become the alcohol again). Eventually, however, most of them are converted to

the product because, under the conditions of the reaction, the equilibrium of the last step lies far to the right (and drives the reaction to completion).

3.13 ACIDS AND BASES IN NONAQUEOUS SOLUTIONS

If you were to add sodium amide ($NaNH_2$) to water in an attempt to carry out a reaction using the very powerful base, the amide ion (NH_2^-) in aqueous solution, the following reaction would take place immediately.

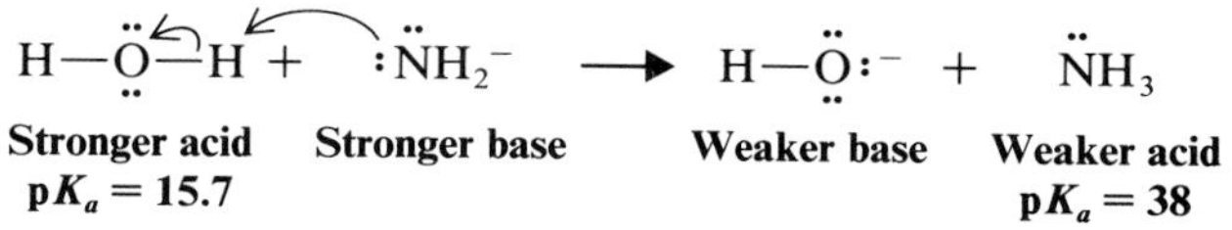

$$H-\overset{..}{\underset{..}{O}}-H + \ :\overset{..}{N}H_2^- \longrightarrow H-\overset{..}{\underset{..}{O}}:^- + \ \overset{..}{N}H_3$$

Stronger acid	Stronger base	Weaker base	Weaker acid
pK_a = 15.7			pK_a = 38

The amide ion would react with water to produce a solution containing hydroxide ions (a much weaker base) and ammonia. This example illustrates what is called **the leveling effect** of the solvent. The solvent here, water, converts any base stronger than a hydroxide ion to a hydroxide ion by donating a proton to it. Therefore, *it is not possible to use a base stronger than hydroxide ion in aqueous solution.*

We can use bases stronger than hydroxide ion, however, by using solvents that are weaker acids than water. We can use amide ion (e.g., $NaNH_2$) in a solvent such as hexane, diethyl ether, or in liquid NH_3 (the liquified gas, b.p. $-33°C$, not the aqueous solution that you may have used in your general chemistry laboratory). All of these solvents are very weak acids, and therefore they will not convert the amide ion to a weaker base by donating a proton to it.

We can, for example, convert ethyne to its conjugate base, a carbanion, by treating it with sodium amide in liquid ammonia.

$$H-C\equiv C-H + \ :\overset{..}{N}H_2^- \xrightarrow[NH_3]{liquid} H-C\equiv C:^- + \ :NH_3$$

Stronger acid	Stronger base	Weaker base	Weaker acid
pK_a = 25	(from NaNH$_2$)		pK_a = 38

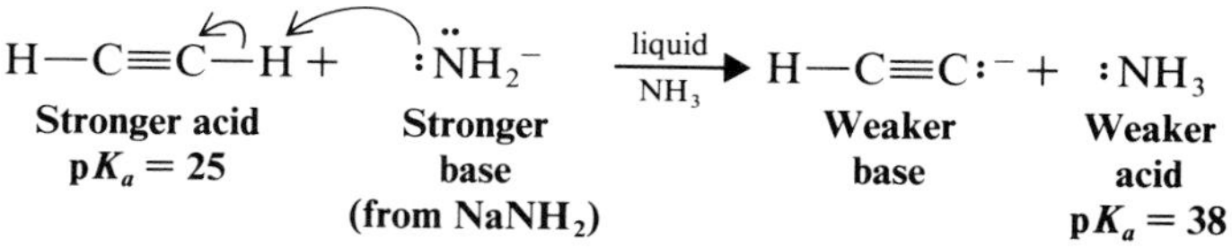

Most alkynes with a proton attached to a triply bonded carbon (called **terminal alkynes**) have pK_a values of about 25; therefore, all react with sodium amide in liquid ammonia in the same way that ethyne does. The general reaction is

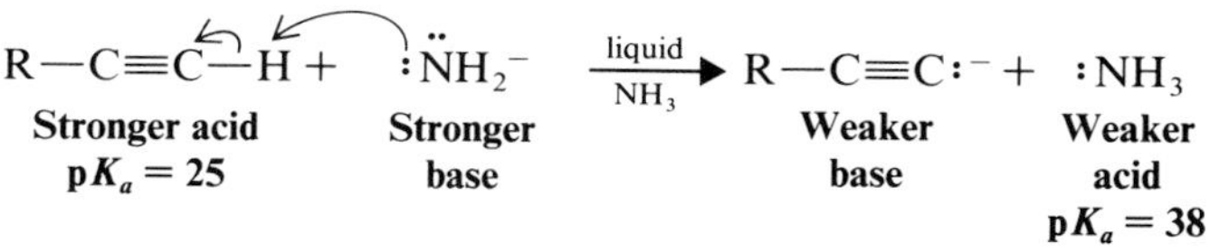

$$R-C\equiv C-H + \ :\overset{..}{N}H_2^- \xrightarrow[NH_3]{liquid} R-C\equiv C:^- + \ :NH_3$$

Stronger acid	Stronger base	Weaker base	Weaker acid
pK_a = 25			pK_a = 38

Alcohols are often used as solvents for organic reactions because being somewhat less polar than water, they dissolve less polar organic compounds. Using alcohols as solvents also offers the advantage of using RO^- ions (called **alkoxide ions**) as bases. Alkoxide ions are somewhat stronger bases than hydroxide ions because alcohols are weaker acids than water. For example, we can create a solution of sodium ethoxide (CH_3CH_2ONa) in ethyl alcohol by adding sodium hydride (NaH) to ethyl alcohol. We use a large excess of ethyl alcohol because we want it to be the solvent. Being a very strong base, the hydride ion reacts readily with ethyl alcohol:

$$\text{CH}_3\text{CH}_2\overset{..}{\underset{..}{\text{O}}}\text{—H} + \quad :\text{H}^- \quad \xrightarrow{\text{ethyl alcohol}} \quad \text{CH}_3\text{CH}_2\overset{..}{\underset{..}{\text{O}}}:^- + \quad \text{H}_2$$

Stronger acid	Stronger	Weaker	Weaker
$pK_a = 16$	base	base	acid
	(from NaH)		$pK_a = 35$

The *tert*-butoxide ion, $(\text{CH}_3)_3\text{CO}^-$, in *tert*-butyl alcohol, $(\text{CH}_3)_3\text{COH}$, is a stronger base than the ethoxide ion in ethyl alcohol, and it can be prepared in a similar way.

$$(\text{CH}_3)_3\text{C}\overset{..}{\underset{..}{\text{O}}}\text{—H} + \quad :\text{H}^- \quad \xrightarrow[\text{alcohol}]{\text{tert-butyl}} \quad (\text{CH}_3)_3\text{C}\overset{..}{\underset{..}{\text{O}}}:^- + \quad \text{H}_2$$

Stronger acid	Stronger	Weaker	Weaker
$pK_a = 18$	base	base	acid
	(from NaH)		$pK_a = 35$

Although the carbon–lithium bond of an alkyllithium (RLi) has covalent character, it is polarized so as to make the carbon negative.

$$\overset{\delta-}{\text{R}} \longleftarrow \overset{\delta+}{\text{Li}}$$

Alkyllithiums react as though they contained alkanide ($\text{R}:^-$) ions, and being the conjugate bases of alkanes, alkanide ions are the strongest bases that we shall encounter. Ethyllithium ($\text{CH}_3\text{CH}_2\text{Li}$), for example, acts as though it contains an ethanide ($\text{CH}_3\text{CH}_2:^-$) carbanion. It reacts with ethyne in the following way:

$$\text{H—C}\equiv\text{C—H} + \quad ^-:\text{CH}_2\text{CH}_3 \quad \xrightarrow{\text{hexane}} \quad \text{H—C}\equiv\text{C}:^- + \text{CH}_3\text{CH}_3$$

Stronger acid	Stronger	Weaker	Weaker
$pK_a = 25$	base	base	acid
	(from $\text{CH}_3\text{CH}_2\text{Li}$)		$pK_a = 50$

Alkyllithiums can be easily prepared by allowing an alkyl bromide to react with lithium metal in an ether solvent (such as diethyl ether).

General Reaction

$$\text{RBr} + 2\,\text{Li} \xrightarrow{\substack{\text{diethyl}\\\text{ether}}} \underset{\textbf{Alkyllithium}}{\text{RLi}} + \text{LiBr}$$

Specific Example

$$\text{CH}_3\text{CH}_2\text{Br} + 2\,\text{Li} \xrightarrow{\substack{\text{diethyl}\\\text{ether}}} \underset{\textbf{Ethyllithium}}{\text{CH}_3\text{CH}_2\text{Li}} + \text{LiBr}$$

In this reaction lithium acts as a reducing agent; each lithium atom donates an electron to the alkyl bromide producing the alkyllithium and lithium bromide.

$$2\,\text{Li}\cdot + \text{R—}\overset{..}{\underset{..}{\text{Br}}}: \longrightarrow \text{R}:^-\text{Li}^+ + \text{Li}^+ + :\overset{..}{\underset{..}{\text{Br}}}:^-$$

PROBLEM 3.13 _____

Write equations for the acid–base reaction that would occur when each of the following compounds or solutions are mixed. In each case label the stronger acid and stronger base, and the weaker acid and weaker base. (If no appreciable acid–base reaction would occur, you should indicate this.)

(a) NaH is added to CH_3OH

(b) $NaNH_2$ is added to CH_3CH_2OH

(c) Gaseous NH_3 is added to ethyllithium in hexane

(d) NH_4Cl is added to sodium amide in liquid ammonia

(e) $(CH_3)_3CONa$ is added to H_2O

(f) NaOH is added to $(CH_3)_3COH$

3.14 ACID–BASE REACTIONS AND THE SYNTHESIS OF DEUTERIUM- AND TRITIUM-LABELED COMPOUNDS

Chemists often use compounds in which deuterium or tritium atoms have replaced one or more hydrogen atoms of the compound as a method of ''labeling'' or identifying particular hydrogen atoms. Deuterium (^2H) and tritium (^3H) are isotopes of hydrogen with masses of 2 and 3 atomic mass units (amu), respectively.

For most chemical purposes, deuterium and tritium atoms in a molecule behave in much the same way that ordinary hydrogen atoms behave. The extra mass and additional neutrons associated with a deuterium or tritium atom often makes its position in a molecule easy to locate by certain spectroscopic methods that we shall study later. Tritium is also radioactive, which makes it very easy to locate. (The extra mass associated with these labeled atoms may also cause compounds containing deuterium or tritium atoms to react more slowly than compounds with ordinary hydrogen atoms. This effect, called an ''isotope effect,'' has been useful in studying the mechanisms of many reactions.)

One way to introduce a deuterium or tritium atom into a specific location in a molecule is through the acid–base reaction that takes place when a very strong base is treated with D_2O or T_2O. For example, treating a solution containing $(CH_3)_2CHLi$ (isopropyllithium) with D_2O results in the formation of propane labeled with deuterium at the central atom:

$$
\begin{array}{cc}
CH_3 & CH_3 \\
| & | \\
CH_3CH:{}^-Li^+ \ + \ D_2O \ \xrightarrow{\text{hexane}} \ CH_3CH{-}D \ + \ OD^- \\
\textbf{Isopropyl-} & \textbf{2-Deuterio-} \\
\textbf{lithium} & \textbf{propane} \\
\textit{(stronger} & \textit{(stronger} & \textit{(weaker} & \textit{(weaker} \\
\textit{base)} & \textit{acid)} & \textit{acid)} & \textit{base)}
\end{array}
$$

SAMPLE PROBLEM _____

Assuming you have available propyne, a solution of sodium amide in liquid ammonia, and T_2O, show how you would prepare the tritium-labeled compound $(CH_3C{\equiv}CT)$.

ANSWER:

First add the propyne to the sodium amide in liquid ammonia. The following acid–base reaction will take place:

$$CH_3C\equiv CH + NH_2^- \xrightarrow{\text{liq. ammonia}} CH_3C\equiv C:^- + NH_3$$

| Stronger | Stronger | | Weaker | Weaker |
| acid | base | | base | acid |

Then adding T_2O (a much stronger acid than NH_3) to the solution will produce $CH_3C\equiv CT$.

$$CH_3C\equiv C:^- + T_2O \xrightarrow{\text{liq. ammonia}} CH_3C\equiv CT + OT^-$$

| Stronger | Stronger | | Weaker | Weaker |
| base | acid | | acid | base |

PROBLEM 3.14

Complete the following acid–base reactions:

(a) $HC\equiv CH + NaH \xrightarrow{\text{hexane}}$

(b) The solution obtained in (a) + $D_2O \longrightarrow$

(c) $CH_3CH_2Li + D_2O \xrightarrow{\text{hexane}}$

(d) $CH_3CH_2OH + NaH \xrightarrow{\text{hexane}}$

(e) The solution obtained in (d) + $T_2O \longrightarrow$

(f) $CH_3CH_2CH_2Li + D_2O \xrightarrow{\text{hexane}}$

3.15 SOME IMPORTANT TERMS AND CONCEPTS

A *Brønsted–Lowry acid* is a substance that can donate a proton; a *Brønsted–Lowry base* is a substance that can accept a proton.

A *Lewis acid* is an electron-pair acceptor; a *Lewis base* is an electron-pair donor.

A *mechanism for a reaction* is a step-by-step description of how a reaction takes place.

Curved arrows (⤳) are used to show the direction of electron flow when mechanisms are written. The arrow begins with a site of higher electron density and points toward a site of electron deficiency.

The *strength of an acid* can be expressed by its acidity constant, K_a,

$$K_a = \frac{[H_3O^+][A^-]}{[HA]}$$

or by its pK_a,

$$pK_a = -\log K_a$$

The *larger* the value of the K_a, or the *smaller* the value of the pK_a, the *stronger* is the acid.

The *strength of a base* is inversely related to the strength of its conjugate acid; the *weaker* the conjugate acid, the *stronger* is the base. Therefore the larger the pK_a of the conjugate acid, the stronger is the base.

The *outcome of acid–base reactions* can be predicted on the basis of the principle that acid–base reactions proceed toward equilibrium *so as to favor the formation of the weaker acid and the weaker base.*

An inductive effect reflects the ability of a substituent to attract or release electrons because of its electronegativity. The effect, transmitted through space and, less effectively, through bonds, weakens steadily as the distance from the substituent increases.

Dispersal of electrical charge always makes a chemical entity more stable.

The *relationship between K_{eq} and the standard free-energy change ($\Delta G°$)* is as follows:

$$\Delta G° = -2.303\ RT \log K_{eq}$$

A negative value of $\Delta G°$ is associated with reactions that favor the formation of products when equilibrium is reached.

The relationship between $\Delta G°$, and the *enthalpy change ($\Delta H°$)* and the *entropy change ($\Delta S°$)* is as follows:

$$\Delta G° = \Delta H° - T\Delta S°$$

A *negative enthalpy change* (associated with an exothermic reaction) and a *positive entropy change* (associated with the products being less ordered than the reactants) favors the formation of products when equilibrium is reached.

A *protic solvent* is one that has a hydrogen atom attached to a strongly electronegative atom (i.e., to an oxygen, nitrogen, or fluorine atom).

ADDITIONAL PROBLEMS

3.15 What is the conjugate base of each of the following acids?
(a) NH_3 (d) $HC \equiv CH$
(b) H_2O (e) CH_3OH
(c) H_2 (f) H_3O^+

3.16 List the bases you gave as answers to Problem 3.15 in order of decreasing basicity.

3.17   What is the conjugate acid of each of the following bases?

(a) HSO_4^- (d) NH_2^-

(b) H_2O (e) $CH_3CH_2^-$

(c) CH_3NH_2 (f) $CH_3CO_2^-$

3.18   List the acids you gave as answers to Problem 3.17 in order of decreasing acidity.

3.19   Designate the Lewis acid and Lewis base in each of the following reactions:

(a) 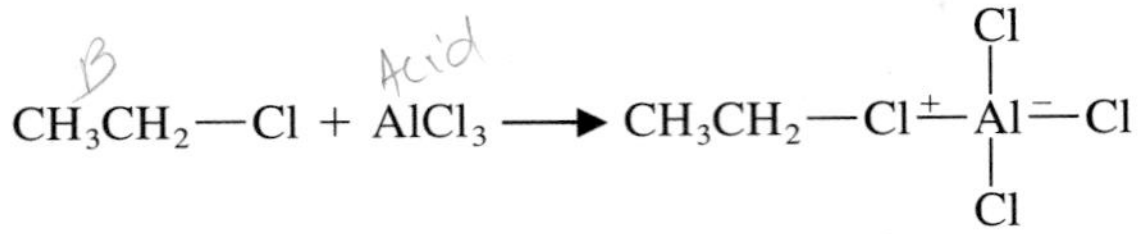 CH_3CH_2—Cl + $AlCl_3$ ⟶ CH_3CH_2—Cl$\overset{+}{\underset{-}{}}Al=$Cl (with Cl, Cl on Al)

(b) 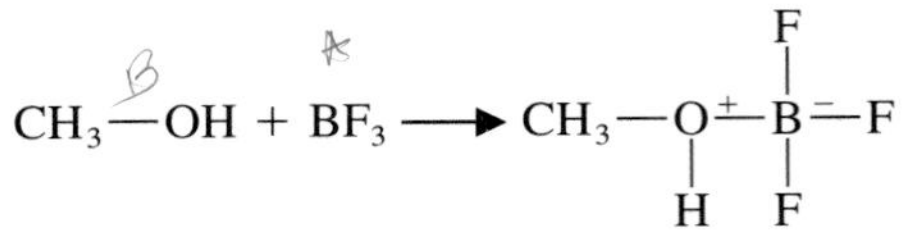 CH_3—OH + BF_3 ⟶ CH_3—O$\overset{+}{\underset{-}{}}B=$F (with H, F substituents)

(c) 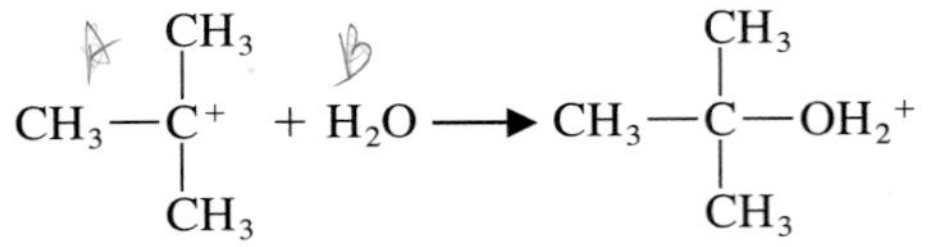 CH_3—C$^+$ (with two CH_3) + H_2O ⟶ CH_3—C—OH_2^+ (with two CH_3)

3.20   Rewrite each of the following reactions using curved arrows and show all nonbonding electron pairs.

(a) CH_3OH + HI ⟶ $CH_3OH_2^+$ + I^-

(b) CH_3NH_2 + HCl ⟶ $CH_3NH_3^+$ + Cl^-

(c) 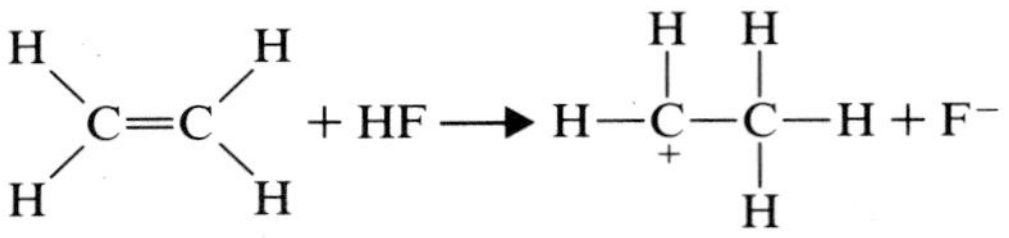 $H_2C=CH_2$ + HF ⟶ H—C—C—H + F^- (ethyl cation shown)

3.21   When methyl alcohol is treated with NaH, the product is $CH_3O^-Na^+$ (and H_2) and not Na^+ $^-CH_2OH$ (and H_2). Explain why this is so.

3.22   What reaction will take place if ethyl alcohol is added to a solution of $HC\equiv C:^-Na^+$ in liquid ammonia?

3.23   (a) The K_a of formic acid (HCO_2H) is 1.77×10^{-4}. What is the pK_a?   (b) What is the K_a of an acid whose $pK_a = 13$?

3.24   Acid HA has a $pK_a = 20$; acid HB has a $pK_a = 10$.   (a) Which is the stronger acid? (b) Will an acid–base reaction with an equilibrium lying to the right take place if Na^+A^- is added to HB? Explain your answer.

3.25   Write an equation, using the curved-arrow notation, for the acid–base reaction that will take place when each of the following are mixed. If no appreciable acid–base reaction takes place, because the equilibrium is unfavorable, you should so indicate.

(a) Aqueous NaOH and $CH_3CH_2CO_2H$

(b) Aqueous NaOH and $C_6H_5SO_3H$

(c) CH_3CH_2ONa in ethyl alcohol and ethyne

(d) CH_3CH_2Li in hexane and ethyne

(e) CH_3CH_2Li in hexane and ethyl alcohol

3.26   Show how you would synthesize each of the following, starting with an alkyl bromide and using any other needed reagents.

(a) $CH_3CH_2CH_2D$

(b) CH_3CHDCH_3

(c)

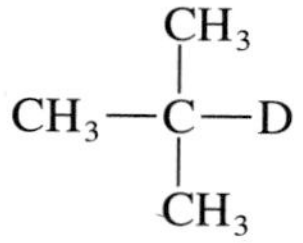

$$CH_3-\underset{\underset{CH_3}{|}}{\overset{\overset{CH_3}{|}}{C}}-D$$

3.27   Starting with appropriate unlabeled organic compounds show syntheses of each of the following:

(a) $CH_3-C\equiv C-T$ (b) $CH_3-\underset{\underset{CH_3}{|}}{CH}-O-D$ (c) $CH_3CH_2CH_2OD$

3.28   (a) Arrange the following compounds in order of decreasing acidity and explain your answer: $CH_3CH_2NH_2$, CH_3CH_2OH, and $CH_3CH_2CH_3$.   (b) Arrange the conjugate bases of the acids given in part (a) in order of increasing basicity and explain this answer.

3.29   Arrange the following compounds in order of decreasing acidity:

(a) $CH_3CH=CH_2$, $CH_3CH_2CH_3$, $CH_3C\equiv CH$

(b) $CH_3CH_2CH_2OH$, $CH_3CH_2CO_2H$, $CH_3CHClCO_2H$

(c) CH_3CH_2OH, $CH_3CH_2OH_2^+$, CH_3OCH_3

3.30   Arrange the following ions in order of increasing basicity:

(a) CH_3NH_2, $CH_3NH_3^+$, CH_3NH^-

(b) CH_3O^-, CH_3NH^-, $CH_3CH_2^-$

(c) $CH_3CH=CH^-$, $CH_3CH_2CH_2^-$, $CH_3C\equiv C^-$

3.31   Whereas H_3PO_4 is a triprotic acid, H_3PO_3 is a diprotic acid. Draw structures for these two acids that account for this difference in behavior.

3.32   Supply the curved arrows necessary for the following reactions.

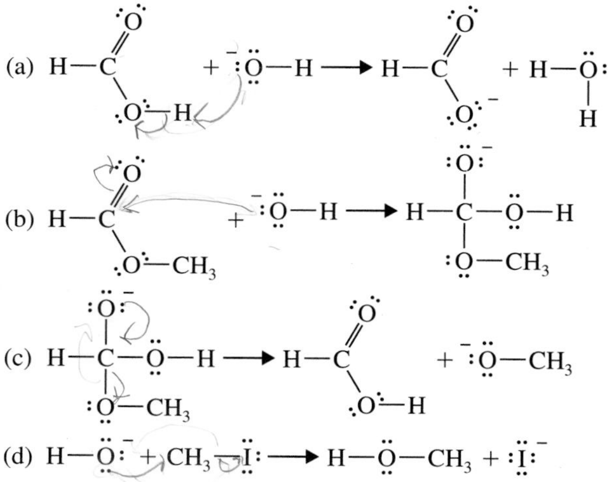

(e)

$$H-\overset{..}{\underset{..}{O}}: + H-CH_2-\overset{\overset{\displaystyle CH_3}{|}}{\underset{\underset{\displaystyle CH_3}{|}}{C}}-\overset{..}{\underset{..}{Cl}}: \longrightarrow CH_2=C\overset{\diagup CH_3}{\diagdown CH_3} \quad + :\overset{..}{\underset{..}{Cl}}:^-$$

3.33 Glycine is an amino acid that can be obtained from most proteins. In solution, glycine exists in equilibrium between the two forms:

$$H_2NCH_2CO_2H \rightleftharpoons H_3\overset{+}{N}CH_2CO_2^-$$

(a) Consult Table 3.1 and state which form is favored at equilibrium. (b) A handbook gives the melting point of glycine as 262°C (with decomposition). Which structure better represents glycine?

3.34 Malonic acid, $HO_2CCH_2CO_2H$, is a diprotic acid. The pK_a for the loss of the first proton is 2.83; the pK_a for the loss of the second proton is 5.69. (a) Explain why malonic acid is a stronger acid than acetic acid ($pK_a = 4.75$). (b) Explain why the anion, $-O_2CCH_2CO_2H$, is so much less acidic than malonic acid itself.

3.35 The free energy change, $\Delta G°$, for the ionization of acid HA is 5 kcal mol^{-1}; for acid HB it is -5 kcal mol^{-1}. Which is the stronger acid?

3.36 At 25°C the enthalpy change, $\Delta H°$, for the ionization of trichloroacetic acid is $+1.5$ kcal mol^{-1} and the entropy change, $\Delta S°$, is $+0.0020$ kcal mol^{-1} K^{-1}. What is the pK_a of trichloroacetic acid?

3.37 The compound below has (for obvious reasons) been given the trivial name **squaric acid.** Squaric acid is a diprotic acid, with both protons being more acidic than acetic acid. In the dianion obtained after the loss of both protons, all of the carbon–carbon bonds are the same length as well as all of the carbon–oxygen bonds. Provide a resonance explanation for these observations.

Squaric acid

CHAPTER 4

Dodecahedrane
(Section 4.13)

ALKANES AND CYCLOALKANES: CONFORMATIONS OF MOLECULES

4.1 INTRODUCTION TO ALKANES AND CYCLOALKANES

We noted earlier that the family of organic compounds called hydrocarbons can be divided into several groups based on the type of bond that exists between the individual carbon atoms. Those hydrocarbons in which all of the carbon–carbon bonds are single bonds are called *alkanes;* those hydrocarbons that contain a carbon–carbon double bond are called *alkenes;* and those with a carbon–carbon triple bond are called *alkynes.*

Cycloalkanes are alkanes in which all or some of the carbon atoms are arranged in a ring. Alkanes have the general formula C_nH_{2n+2}; cycloalkanes containing a single ring have two fewer hydrogen atoms and thus have the general formula C_nH_{2n}.

Alkanes and cycloalkanes are so similar that many of their properties can be considered side by side. Some differences remain, however, and certain structural features arise from the rings of cycloalkanes that are more conveniently studied separately. We shall point out the chemical and physical similarities of alkanes and cycloalkanes as we go along.

4.1A Sources of Alkanes: Petroleum

The primary source of alkanes is petroleum. Petroleum is a complex mixture of organic compounds, most of which are alkanes and aromatic hydrocarbons (cf. Chapter 14). It also contains small amounts of oxygen-, nitrogen-, and sulfur-containing compounds.

TABLE 4.1 Typical fractions obtained by distillation of petroleum

BOILING RANGE OF FRACTION (°C)	NUMBER OF CARBON ATOMS PER MOLECULE	USE
Below 20	C_1—C_4	Natural gas, bottled gas, petrochemicals
20–60	C_5—C_6	Petroleum ether, solvents
60–100	C_6—C_7	Ligroin, solvents
40–200	C_5—C_{10}	Gasoline (straight-run gasoline)
175–325	C_{12}—C_{18}	Kerosene and jet fuel
250–400	C_{12} and higher	Gas oil, fuel oil, and diesel oil
Nonvolatile liquids	C_{20} and higher	Refined mineral oil, lubricating oil, grease
Nonvolatile solids	C_{20} and higher	Paraffin wax, asphalt, and tar

Adapted with permission from John R. Holum, *Elements of General and Biological Chemistry,* 8th ed., Wiley, New York, 1991.

4.1B Petroleum Refining

The first step in refining petroleum is distillation; the object here is to separate the petroleum into fractions based on the volatility of its components. Complete separation into fractions containing individual compounds is economically impractical and virtually impossible technically. More than 500 different compounds are contained in the petroleum distillates boiling below 200°C and many have almost the same boiling points. Thus the fractions taken contain mixtures of alkanes of similar boiling points (cf. Table 4.1). Mixtures of alkanes, fortunately, are perfectly suitable for uses as fuels, solvents, and lubricants, the primary uses of petroleum.

4.1C Cracking

The demand for gasoline is much greater than that supplied by the gasoline fraction of petroleum. Important processes in the petroleum industry, therefore, are concerned with converting hydrocarbons from other fractions into gasoline. When a mixture of alkanes from the gas oil (C_{12} and higher) fraction is heated at very high temperatures (~ 500°C) in the presence of a variety of catalysts, the molecules break apart and rearrange to smaller, more highly branched alkanes containing 5–10 carbon atoms (see Table 4.1). This process is called **catalytic cracking.** Cracking can also be done in the absence of a catalyst—called **thermal cracking**—but in this process the products tend to have unbranched chains, and alkanes with unbranched chains have a very low "octane rating."

The highly branched compound 2,2,4-trimethylpentane (called "isooctane" in the petroleum industry) burns very smoothly (without knocking) in internal combustion engines and is used as one of the standards by which the octane rating of

$$CH_3-\underset{\underset{\displaystyle CH_3}{|}}{\overset{\overset{\displaystyle CH_3}{|}}{C}}-CH_2-\underset{\overset{\displaystyle CH_3}{|}}{CH}-CH_3$$

**2,2,4-trimethylpentane
("isooctane")**

gasolines is established. According to this scale 2,2,4-trimethylpentane has an octane rating of 100. Heptane, $CH_3(CH_2)_5CH_3$, a compound that produces much knocking when it is burned in an internal combustion engine, is given an octane rating of 0. Mixtures of 2,2,4-trimethylpentane and heptane are used as standards for octane ratings between 0 and 100. A gasoline, for example, that has the same characteristics in an engine as a mixture of 87% 2,2,4-trimethylpentane and 13% heptane would be rated as 87-octane gasoline.

4.2 SHAPES OF ALKANES

A general tetrahedral orientation of groups—and thus sp^3 hybridization—is the rule for the carbon atoms of all alkanes and cycloalkanes. We can represent the shapes of alkanes as shown in Fig. 4.1.

Butane and pentane are examples of alkanes that are sometimes called "straight-chain" alkanes. One glance at their three-dimensional models shows that because of

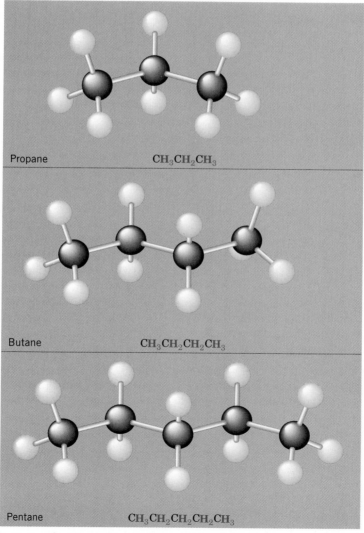

Propane $CH_3CH_2CH_3$

Butane $CH_3CH_2CH_2CH_3$

Pentane $CH_3CH_2CH_2CH_2CH_3$

FIGURE 4.1 Ball-and-stick models for three simple alkanes.

the tetrahedral carbon atoms their chains are zigzagged and not at all straight. Indeed, the structures that we have depicted in Fig. 4.1 are the straightest possible arrangements of the chains, for rotations about the carbon–carbon single bonds produce arrangements that are even less straight. The better description is **unbranched.** This means that each carbon atom within the chain is bonded to no more than two other carbon atoms and that unbranched alkanes contain only primary and secondary

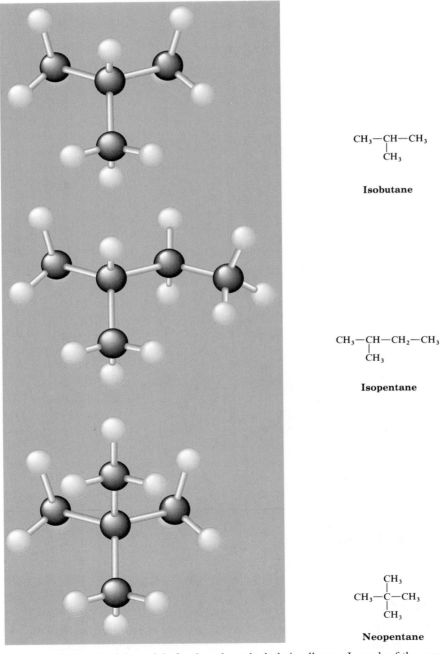

$$CH_3-CH-CH_3$$
$$|$$
$$CH_3$$

Isobutane

$$CH_3-CH-CH_2-CH_3$$
$$|$$
$$CH_3$$

Isopentane

$$CH_3$$
$$|$$
$$CH_3-C-CH_3$$
$$|$$
$$CH_3$$

Neopentane

FIGURE 4.2 Ball-and-stick models for three branched-chain alkanes. In each of the compounds one carbon atom is attached to more than two other carbon atoms.

TABLE 4.2   *Physical constants of the butane, pentane, and hexane isomers*

MOLECULAR FORMULA	STRUCTURAL FORMULA	mp (°C)	bp (°C)[a] (1 atm)	DENSITY[b] (g mL^{-1})	INDEX OF REFRACTION[c] (n_D 20°C)
C_4H_{10}	$CH_3CH_2CH_2CH_3$	− 138.3	− 0.5	0.6012^0	1.3543
C_4H_{10}	CH_3CHCH_3 　　$\vert$ 　　CH_3	− 159	− 12	0.603^0	
C_5H_{12}	$CH_3CH_2CH_2CH_2CH_3$	− 129.72	36	0.6262^{20}	1.3579
C_5H_{12}	$CH_3CHCH_2CH_3$ 　　$\vert$ 　　CH_3	− 160	27.9	0.6197^{20}	1.3537
C_5H_{12}	CH_3 　　　$\vert$ CH_3-C-CH_3 　　　$\vert$ 　　　CH_3	− 20	9.5	0.61350^{20}	1.3476
C_6H_{14}	$CH_3CH_2CH_2CH_2CH_2CH_3$	− 95	68	0.65937^{20}	1.3748
C_6H_{14}	$CH_3CHCH_2CH_2CH_3$ 　　$\vert$ 　　CH_3	− 153.67	60.3	0.6532^{20}	1.3714
C_6H_{14}	$CH_3CH_2CHCH_2CH_3$ 　　　　$\vert$ 　　　　CH_3	− 118	63.3	0.6643^{20}	1.3765
C_6H_{14}	$CH_3CH-CHCH_3$ 　　$\vert$　　$\vert$ 　　CH_3　CH_3	− 128.8	58	0.6616^{20}	1.3750
C_6H_{14}	CH_3 　　　$\vert$ $CH_3-C-CH_2CH_3$ 　　　$\vert$ 　　　CH_3	− 98	49.7	0.6492^{20}	1.3688

[a] Unless otherwise indicated, all boiling points are at 1 atm or 760 torr.

[b] The superscript indicates the temperature at which the density was measured.

[c] The index of refraction is a measure of the ability of the alkane to bend (refract) light rays. The values reported are for light of the D line of the sodium spectrum (n_D).

carbon atoms. (Unbranched alkanes used to be called ''normal'' alkanes or *n*-alkanes, but this designation is archaic and should not be used now.)

Isobutane, isopentane, and neopentane (Fig. 4.2) are examples of branched-chain alkanes. In neopentane the central carbon atom is bonded to four carbon atoms.

Butane and isobutane have the same molecular formula: C_4H_{10}. The two compounds have their atoms connected in a different order and are, therefore, *constitutional isomers.* Pentane, isopentane, and neopentane are also constitutional isomers. They, too, have the same molecular formula (C_5H_{12}) but have different structures.

PROBLEM 4.1

Write condensed structural formulas for all of the constitutional isomers with the molecular formula C_7H_{16}. (There is a total of nine constitutional isomers; see Table 4.3.)

TABLE 4.3   Number of alkane isomers

MOLECULAR FORMULA	POSSIBLE NUMBER OF CONSTITUTIONAL ISOMERS
C_4H_{10}	2
C_5H_{12}	3
C_6H_{14}	5
C_7H_{16}	9
C_8H_{18}	18
C_9H_{20}	35
$C_{10}H_{22}$	75
$C_{15}H_{32}$	4,347
$C_{20}H_{42}$	366,319
$C_{30}H_{62}$	4,111,846,763
$C_{40}H_{82}$	62,481,801,147,341

Constitutional isomers, as stated earlier, have different physical properties. The differences may not always be large, but constitutional isomers are always found to have different melting points, boiling points, densities, indexes of refraction, and so forth. Table 4.2 gives some of the physical properties of the C_4H_{10}, C_5H_{12}, and C_6H_{14} isomers.

As Table 4.3 shows, the number of constitutional isomers that is possible increases dramatically as the number of carbon atoms in the alkane increases.

The large numbers in Table 4.3 are based on calculations that must be done with a computer. Similar calculations, which take into account stereoisomers (Chapter 5) as well as constitutional isomers, indicate that an alkane with the formula $C_{167}H_{336}$ would, in theory, have more possible isomers than there are particles in the observed universe!

4.3   IUPAC NOMENCLATURE OF ALKANES, ALKYL HALIDES, AND ALCOHOLS

The development of a formal system for naming organic compounds did not come about until near the end of the nineteenth century. Before that time many organic compounds had already been discovered. The names given these compounds sometimes reflected a source of the compound. Acetic acid, for example, can be obtained from vinegar; it takes its name from the Latin word for vinegar, *acetum*. Formic acid can be obtained from some ants; it is named from the Latin word for ants, *formicae*. Ethanol (or ethyl alcohol) was at one time called grain alcohol because it was obtained by the fermentation of grains.

These older names for organic compounds are now called "common" or "trivial" names. Many of these names are still widely used by chemists and biochemists and in commerce. (Many are even written into laws.) For this reason it is still necessary to learn the common names for some of the common compounds. We shall point out these common names as we go along, and we shall use them occasionally. Most of the time, however, the names that we shall use will be those called IUPAC names.

The formal system of nomenclature used today is one proposed by the International Union of Pure and Applied Chemistry (IUPAC). This system was first developed in 1892 and has been revised at irregular intervals to keep it up to date. Underlying the IUPAC system of nomenclature for organic compounds is a fundamental principle: *Each different compound should have a different name.* Thus, through a systematic set of rules, the IUPAC system provides different names for the more than 7 million known organic compounds, and names can be devised for any one of millions of other compounds yet to be synthesized. In addition, the IUPAC system is simple enough to allow any chemist familiar with the rules (or with the rules at hand) to write the name for any compound that might be encountered. In the same way, one is also able to derive the structure of a given compound from its IUPAC name.

The IUPAC system for naming alkanes is not difficult to learn, and the principles involved are used in naming compounds in other families as well. For these reasons we begin our study of the IUPAC system with the rules for naming alkanes, and then study the rules for alkyl halides and alcohols.

The names for several of the unbranched alkanes are listed in Table 4.4. The ending for all of the names of alkanes is *-ane*. The stems of the names of most of the alkanes (above C_4) are of Greek and Latin origin. Learning the stems is like learning to count in organic chemistry. Thus, one, two, three, four, five, becomes meth-, eth-, prop-, but-, pent-.

TABLE 4.4   *The unbranched alkanes*

NAME	NUMBER OF CARBON ATOMS	STRUCTURE	NAME	NUMBER OF CARBON ATOMS	STRUCTURE
Methane	1	CH_4	Heptadecane	17	$CH_3(CH_2)_{15}CH_3$
Ethane	2	CH_3CH_3	Octadecane	18	$CH_3(CH_2)_{16}CH_3$
Propane	3	$CH_3CH_2CH_3$	Nonadecane	19	$CH_3(CH_2)_{17}CH_3$
Butane	4	$CH_3(CH_2)_2CH_3$	Eicosane	20	$CH_3(CH_2)_{18}CH_3$
Pentane	5	$CH_3(CH_2)_3CH_3$	Heneicosane	21	$CH_3(CH_2)_{19}CH_3$
Hexane	6	$CH_3(CH_2)_4CH_3$	Docosane	22	$CH_3(CH_2)_{20}CH_3$
Heptane	7	$CH_3(CH_2)_5CH_3$	Tricosane	23	$CH_3(CH_2)_{21}CH_3$
Octane	8	$CH_3(CH_2)_6CH_3$	Triacontane	30	$CH_3(CH_2)_{28}CH_3$
Nonane	9	$CH_3(CH_2)_7CH_3$	Hentriacontane	31	$CH_3(CH_2)_{29}CH_3$
Decane	10	$CH_3(CH_2)_8CH_3$	Tetracontane	40	$CH_3(CH_2)_{38}CH_3$
Undecane	11	$CH_3(CH_2)_9CH_3$	Pentacontane	50	$CH_3(CH_2)_{48}CH_3$
Dodecane	12	$CH_3(CH_2)_{10}CH_3$	Hexacontane	60	$CH_3(CH_2)_{58}CH_3$
Tridecane	13	$CH_3(CH_2)_{11}CH_3$	Heptacontane	70	$CH_3(CH_2)_{68}CH_3$
Tetradecane	14	$CH_3(CH_2)_{12}CH_3$	Octacontane	80	$CH_3(CH_2)_{78}CH_3$
Pentadecane	15	$CH_3(CH_2)_{13}CH_3$	Nonacontane	90	$CH_3(CH_2)_{88}CH_3$
Hexadecane	16	$CH_3(CH_2)_{14}CH_3$	Hectane	100	$CH_3(CH_2)_{98}CH_3$

4.3A Nomenclature of Unbranched Alkyl Groups

If we remove one hydrogen atom from an alkane, we obtain what is called **an alkyl group.** These alkyl groups have names that end in **-yl**. When the alkane is **unbranched,** and the hydrogen atom that is removed is a **terminal** hydrogen atom, the names are straightforward:

Alkane		*Alkyl Group*	*Abbreviation*
CH_3—H **Methane**	becomes	CH_3— **Methyl**	Me—
CH_3CH_2—H **Ethane**	becomes	CH_3CH_2— **Ethyl**	Et—
$CH_3CH_2CH_2$—H **Propane**	becomes	$CH_3CH_2CH_2$— **Propyl**	Pr—
$CH_3CH_2CH_2CH_2$—H **Butane**	becomes	$CH_3CH_2CH_2CH_2$— **Butyl**	Bu—

4.3B Nomenclature of Branched-Chain Alkanes

Branched-chain alkanes are named according to the following rules:

1. **Locate the longest continuous chain of carbon atoms; this chain determines the parent name for the alkane.** We designate the following compound, for example, as a *hexane* because the longest continuous chain contains six carbon atoms.

$$CH_3CH_2CH_2CH_2CHCH_3$$
$$|$$
$$CH_3$$

The longest continuous chain may not always be obvious from the way the formula is written. Notice, for example, that the following alkane is designated as a *heptane* because the longest chain contains seven carbon atoms.

$$CH_3CH_2CH_2CH_2CH—CH_3$$
$$|$$
$$CH_2$$
$$|$$
$$CH_3$$

2. **Number the longest chain beginning with the end of the chain nearer the substituent.** Applying this rule, we number the two alkanes that we illustrated previously in the following way.

$$\overset{6}{C}H_3\overset{5}{C}H_2\overset{4}{C}H_2\overset{3}{C}H_2\overset{2}{C}H\overset{1}{C}H_3$$
$$|$$
$$CH_3$$

Substituent ⟶

Substituent ⟶

$$\overset{7}{C}H_3\overset{6}{C}H_2\overset{5}{C}H_2\overset{4}{C}H_2\overset{3}{C}HCH_3$$
$$|$$
$$^2CH_2$$
$$|$$
$$^1CH_3$$

3. **Use the numbers obtained by application of rule 2 to designate the location of the substituent group.** The parent name is placed last, and the substituent group, preceded by the number designating its location on the chain, is placed first. Numbers are separated from words by a hyphen. Our two examples are 2-methylhexane and 3-methylheptane, respectively.

$$\overset{6}{C}H_3\overset{5}{C}H_2\overset{4}{C}H_2\overset{3}{C}H_2\overset{2}{C}HCH_3$$
$$\underset{CH_3}{|}$$

$$\overset{7}{C}H_3\overset{6}{C}H_2\overset{5}{C}H_2\overset{4}{C}H_2\overset{3}{C}HCH_3$$
$$\underset{^2CH_2}{|}$$
$$\underset{^1CH_3}{|}$$

2-Methylhexane   **3-Methylheptane**

4. **When two or more substituents are present, give each substituent a number corresponding to its location on the longest chain.** For example, we designate the following compound as 4-ethyl-2-methylhexane.

$$CH_3CH-CH_2-CHCH_2CH_3$$
$$\underset{CH_3}{|}\qquad\quad\underset{CH_2}{|}$$
$$\underset{CH_3}{|}$$

4-Ethyl-2-methylhexane

The substituent groups should be listed *alphabetically* (i.e., ethyl before methyl).* In deciding on alphabetical order disregard multiplying prefixes such as ''di'' and ''tri.''

5. **When two substituents are present on the same carbon atom, use that number twice.**

$$\overset{CH_3}{\underset{|}{}}$$
$$CH_3CH_2-C-CH_2CH_2CH_3$$
$$\underset{CH_2}{|}$$
$$\underset{CH_3}{|}$$

3-Ethyl-3-methylhexane

6. **When two or more substituents are identical, indicate this by the use of the prefixes di-, tri-, tetra-,** and so on. Then make certain that each and every substituent has a number. Commas are used to separate numbers from each other.

$$CH_3CH--CHCH_3$$
$$\underset{CH_3}{|}\quad\underset{CH_3}{|}$$
2,3-Dimethylbutane

$$\overset{CH_3}{\underset{|}{}}$$
$$CH_3CHCHCHCH_3$$
$$\underset{CH_3}{|}\quad\underset{CH_3}{|}$$
2,3,4-Trimethylpentane

$$\overset{CH_3}{\underset{|}{}}\ \overset{CH_3}{\underset{|}{}}$$
$$CH_3CCH_2CCH_3$$
$$\underset{CH_3}{|}\quad\underset{CH_3}{|}$$
2,2,4-Tetramethylpentane

Application of these six rules allows us to name most of the alkanes that we shall encounter. Two other rules, however, may be required occasionally.

*Some handbooks also list the groups in order of increasing size or complexity (i.e., methyl before ethyl). Alphabetical listing, however, is now by far the most widely used system.

7. **When two chains of equal length compete for selection as the parent chain, choose the chain with the greater number of substituents.**

$$\overset{7}{CH_3}\overset{6}{CH_2}-\overset{5}{CH}-\overset{4}{CH}-\overset{3}{CH}-\overset{2}{CH}-\overset{1}{CH_3}$$

$$\qquad\qquad CH_3\quad CH_2\quad CH_3\quad CH_3$$

$$\qquad\qquad\qquad CH_2$$

$$\qquad\qquad\qquad CH_3$$

2,3,5-Trimethyl-**4-**propylheptane
(four substituents)
(*not* 4-*sec*-butyl-2,3-dimethylheptane)
(three substituents)

8. **When branching first occurs at an equal distance from either end of the longest chain, choose the name that gives the lower number at the first point of difference.**

$$\overset{6}{CH_3}-\overset{5}{CH}-\overset{4}{CH_2}-\overset{3}{CH}-\overset{2}{CH}-\overset{1}{CH_3}$$

$$\qquad CH_3\qquad\qquad CH_3\quad CH_3$$

2,3,5-Trimethyl**hexane**
(*not* **2,4,5-**trimethylhexane)

4.3C Nomenclature of Branched Alkyl Groups

In Section 4.3A you learned the names for the unbranched alkyl groups such as methyl, ethyl, propyl, butyl, and so on, groups derived by removing a terminal hydrogen from an alkane. For alkanes with more than two carbon atoms, more than one derived group is possible. Two groups can be derived from propane, for example; the **propyl group** is derived by removal of a terminal hydrogen, and the **1-methylethyl** or **isopropyl group** is derived by removal of a hydrogen from the central carbon:

Three-Carbon Groups

$$CH_3CH_2CH_2-$$
Propyl group

$$CH_3CH_2CH_3-$$
Propane

$$CH_3-CH-$$
$$\qquad\;\; CH_3$$
1-Methylethyl or isopropyl group

1-Methylethyl is the systematic name for this group; isopropyl is a common name. Systematic nomenclature for alkyl groups is similar to that for branched-chain alkanes, with the provision that *numbering always begins at the point where the group is attached to the main chain.* There are four C$_4$ groups. Two are derived from butane and two are derived from isobutane.*

---

* Isobutane is a common name for 2-methylpropane that is approved by the IUPAC.

Four-Carbon Groups

CH$_3$CH$_2$CH$_2$CH$_3$ —
Butane

→ CH$_3$CH$_2$CH$_2$CH$_2$ —
Butyl group

$$\overset{\displaystyle H}{\underset{\displaystyle CH_3}{\text{CH}_3\text{CH}_2\text{C}}}-$$

1-Methylpropyl or *sec*-butyl group

$$\underset{\displaystyle CH_3}{CH_3CHCH_3}-$$
Isobutane

$$\underset{\displaystyle CH_3}{CH_3CHCH_2}-$$
2-Methylpropyl or isobutyl group

$$\overset{\displaystyle CH_3}{\underset{\displaystyle CH_3}{CH_3C}}-$$
1,1-Dimethylethyl or *tert*-butyl group

The following examples show how the names of these groups are employed.

$$CH_3CH_2CH_2CHCH_2CH_2CH_3$$
$$\underset{\displaystyle CH_3}{CH_3-CH}$$

4-(1-Methylethyl)heptane or 4-isopropylheptane

$$CH_3CH_2CH_2CHCH_2CH_2CH_2CH_3$$
$$\underset{\displaystyle CH_3}{CH_3-C-CH_3}$$

4-(1,1-Dimethylethyl)octane or 4-*tert*-butyloctane

The common names, **isopropyl, isobutyl, *sec*-butyl,** and ***tert*-butyl** are approved by the IUPAC for the unsubstituted groups, and they are still very frequently used. You should memorize these groups so well that you will recognize them any way that they are written. In deciding on alphabetical order for these groups you should disregard structure-defining prefixes that are written in italics and separated from the name by a hyphen. Thus *tert*-butyl precedes ethyl, but ethyl precedes isobutyl.

There is one five-carbon group with an IUPAC approved common name that you should also know: the 2,2-dimethylpropyl group, commonly called the **neopentyl group.**

$$\overset{\displaystyle CH_3}{\underset{\displaystyle CH_3}{CH_3-C-CH_2}}-$$

2,2-Dimethylpropyl or neopentyl group

PROBLEM 4.2

(a) In addition to the 2,2-dimethylpropyl (or neopentyl) group just given, there are seven other five-carbon groups. Write their structures and give each structure its systematic name.   (b) Provide IUPAC names for the nine C_7H_{16} isomers that you gave as answers to Problem 4.1.

4.3D Classification of Hydrogen Atoms

The hydrogen atoms of an alkane are classified on the basis of the carbon atom to which they are attached. A hydrogen atom attached to a primary carbon atom is a primary hydrogen atom, and so forth. The following compound, 2-methylbutane, has primary, secondary, and tertiary hydrogen atoms.

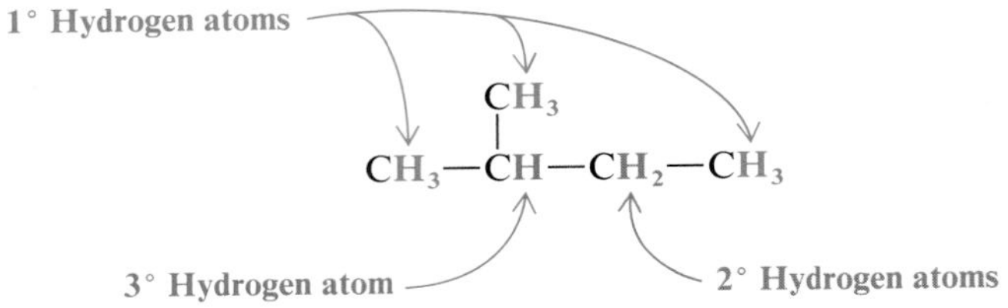

1° Hydrogen atoms

$$CH_3-\overset{\overset{\displaystyle CH_3}{|}}{CH}-CH_2-CH_3$$

3° Hydrogen atom   2° Hydrogen atoms

On the other hand, 2,2-dimethylpropane, a compound that is often called **neopentane,** has only primary hydrogen atoms.

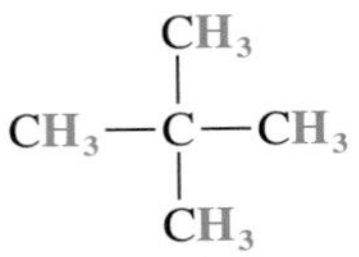

$$CH_3-\overset{\overset{\displaystyle CH_3}{|}}{\underset{\underset{\displaystyle CH_3}{|}}{C}}-CH_3$$

2,2-Dimethylpropane
(neopentane)

4.3E Nomenclature of Alkyl Halides

Alkanes bearing halogen substituents are named in the IUPAC substitutive system as haloalkanes:

CH_3CH_2Cl $CH_3CH_2CH_2F$ $CH_3CHBrCH_3$
Chloroethane   **1-Fluoropropane**   **2-Bromopropane**

When the parent chain has both a halo and an alkyl substituent attached to it, number the chain from the end nearer the first substituent, regardless of whether it is halo or alkyl. If two substituents are of equal distance from the end of the chain, then number the chain from the end nearer the substituent that has alphabetical precedence.

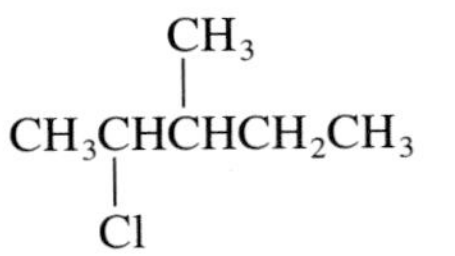

$$CH_3\overset{\overset{\displaystyle CH_3}{|}}{\underset{\underset{\displaystyle Cl}{|}}{CH}}CHCH_2CH_3$$

2-Chloro-3-methylpentane

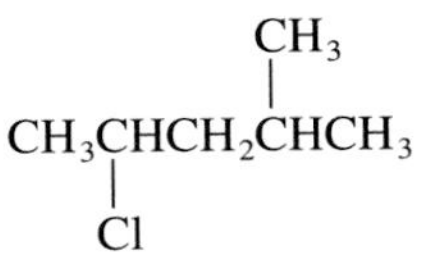

$$CH_3\underset{\underset{\displaystyle Cl}{|}}{CH}CH_2\overset{\overset{\displaystyle CH_3}{|}}{CH}CH_3$$

2-Chloro-4-methylpentane

Common names for many simple haloalkanes are still widely used, however. In this common nomenclature system, called *radicofunctional nomenclature,* haloalkanes are named as alkyl halides. (The following names are also accepted by the IUPAC.)

$$CH_3CH_2Cl \qquad \underset{\underset{Br}{|}}{CH_3CHCH_3} \qquad (CH_3)_3CBr \qquad \underset{\underset{CH_3}{|}}{CH_3CHCH_2Cl} \qquad \overset{\overset{CH_3}{|}}{\underset{\underset{CH_3}{|}}{CH_3CCH_2Br}}$$

Ethyl chloride	**Isopropyl bromide**	***tert*-Butyl bromide**	**Isobutyl chloride**	**Neopentyl bromide**

PROBLEM 4.3

Give IUPAC substitutive names for all of the isomers of (a) C_4H_9Cl and (b) $C_5H_{11}Br$.

4.3F Nomenclature of Alcohols

In what is called IUPAC **substitutive nomenclature** a name may have as many as four features: **locants, prefixes, parent compound,** and **one suffix.** Consider the following compound as an illustration without, for the moment, being concerned as to how the name arises.

$$CH_3CH_2\underset{\underset{CH_3}{|}}{CH}CH_2CH_2CH_2OH$$

4-Methyl-1-hexanol

locant   prefix   locant   parent   suffix

The *locant* **4-** tells that the substituent **methyl** group, named as a *prefix,* is attached to the *parent compound* at C4. The *parent compound* contains six carbon atoms and no multiple bonds, hence the parent name **hexane,** and it is an alcohol, therefore it has the *suffix* **-ol.** The locant **1-,** tells that C1 bears the hydroxyl group. **In general, numbering of the chain always begins at the end nearer the group named as a suffix.**

The following procedure should be followed in giving alcohols IUPAC substitutive names:

1. Select the longest continuous carbon chain *to which the hydroxyl is directly attached.* Change the name of the alkane corresponding to this chain by dropping the final *e* and adding the suffix *ol.*
2. Number the longest continuous carbon chain so as to give the carbon atom bearing the hydroxyl group the lower number. Indicate the position of the hydroxyl group by using this number as a locant; indicate the positions of other substituents (as prefixes) by using the numbers corresponding to their positions along the carbon chain as locants.

The following examples show how these rules are applied.

$$\overset{3}{C}H_3\overset{2}{C}H_2\overset{1}{C}H_2OH \qquad \overset{1}{C}H_3\overset{2}{C}HCH_2\overset{4}{C}H_3 \qquad \overset{5}{C}H_3\overset{4}{C}HCH_2\overset{2}{C}H_2\overset{1}{C}H_2OH$$

$$\underset{OH}{|} \qquad\qquad \underset{CH_3}{|}$$

1-Propanol   **2-Butanol**   **4-Methyl-1-pentanol**
                                                  (*not* 2-methyl-5-pentanol)

$$CH_3$$
$$\overset{3}{C}lCH_2\overset{2}{C}H_2\overset{1}{C}H_2OH \qquad\qquad \overset{1}{C}H_3\overset{2}{C}HCH_2\overset{4|}{C}\overset{5}{C}H_3$$

$$\underset{OH}{|} \quad \underset{CH_3}{|}$$

3-Chloro-1-propanol   **4,4-Dimethyl-2-pentanol**

PROBLEM 4.4

Give IUPAC substitutive names for all of the isomeric alcohols with the formulas   (a) $C_4H_{10}O$ and   (b) $C_5H_{12}O$.

Simple alcohols are often called by *common* radicofunctional names that are also approved by the IUPAC. We have seen several examples already (Section 2.10). In addition to *methyl alcohol, ethyl alcohol,* and *isopropyl alcohol,* there are several others including the following:

$$CH_3CH_2CH_2OH \qquad CH_3CH_2CH_2CH_2OH \qquad CH_3CH_2CHCH_3$$

$$\underset{OH}{|}$$

Propyl alcohol   **Butyl alcohol**   *sec*-**Butyl alcohol**

$$\overset{CH_3}{|} \qquad\qquad \overset{CH_3}{|} \qquad\qquad \overset{CH_3}{|}$$
$$CH_3COH \qquad\qquad CH_3CHCH_2OH \qquad CH_3CCH_2OH$$
$$\underset{CH_3}{|} \qquad\qquad\qquad\qquad\qquad\qquad \underset{CH_3}{|}$$

tert-**Butyl alcohol**   **Isobutyl alcohol**   **Neopentyl alcohol**

Alcohols containing two hydroxyl groups are commonly called glycols. In the IUPAC substitutive system they are named as **diols.**

$$CH_2\!-\!CH_2 \qquad CH_3CH\!-\!CH_2 \qquad CH_2CH_2CH_2$$
$$\underset{OH}{|}\quad\underset{OH}{|} \qquad\quad \underset{OH}{|}\quad\underset{OH}{|} \qquad\quad \underset{OH}{|}\qquad\underset{OH}{|}$$

| Common | Ethylene glycol | Propylene glycol | Trimethylene glycol |
| Substitutive | 1,2-Ethanediol | 1,2-Propanediol | 1,3-Propanediol |

4.4 NOMENCLATURE OF CYCLOALKANES

4.4A Monocyclic Compounds

Cycloalkanes with only one ring are named by attaching the prefix cyclo- to the names of the alkanes possessing the same number of carbon atoms. For example,

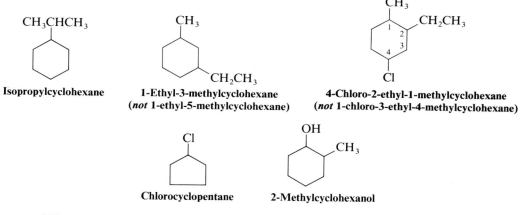

Cyclopropane   Cyclopentane

Naming substituted cycloalkanes is straightforward: We name them as *alkyl-cycloalkanes, halocycloalkanes, alkylcycloalkanols,* and so on. If only one substituent is present, it is not necessary to designate its position. When two substituents are present, we number the ring *beginning with the substituent* first in the alphabet, and number in the direction that gives the next substituent the lower number possible. When three or more substituents are present, we begin at the substituent that leads to the lowest set of locants.

CH$_3$CHCH$_3$

Isopropylcyclohexane

CH$_3$

CH$_2$CH$_3$

1-Ethyl-3-methylcyclohexane
(*not* 1-ethyl-5-methylcyclohexane)

CH$_3$

CH$_2$CH$_3$

Cl

4-Chloro-2-ethyl-1-methylcyclohexane
(*not* 1-chloro-3-ethyl-4-methylcyclohexane)

Cl

Chlorocyclopentane

OH

CH$_3$

2-Methylcyclohexanol

When a single ring system is attached to a single chain with a greater number of carbon atoms, or when more than one ring system is attached to a single chain, then it is appropriate to name the compounds as *cycloalkylalkanes.* For example:

—CH$_2$CH$_2$CH$_2$CH$_2$CH$_3$

1-Cyclobutylpentane

1,3-Dicyclohexylpropane

PROBLEM 4.5

Give names for the following substituted alkanes:

(a) (CH$_3$)$_3$C

(CH$_3$)$_2$CH

(c) CH$_3$(CH$_2$)$_2$CH$_2$—

(b) (CH$_3$)$_2$CHCH$_2$

H$_3$C

(d)

Cl

CH$_3$

CH$_3$

(e) [structure: cyclopentane ring with Cl and OH substituents]

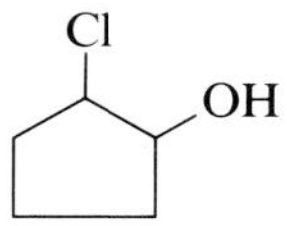

(f) [structure: cyclohexane ring with OH and C(CH₃)₃ substituents]

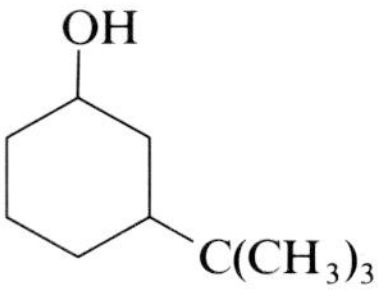

4.4B Bicyclic Compounds

We name compounds containing two fused or bridged rings as **bicycloalkanes** and we use the name of the alkane corresponding to the total number of carbon atoms in the rings as the parent name. The following compound, for example, contains seven carbon atoms and is, therefore, a bicycloheptane. The carbon atoms common to both rings are called bridgeheads, and each bond, or chain of atoms connecting the bridgehead atoms, is called a bridge.

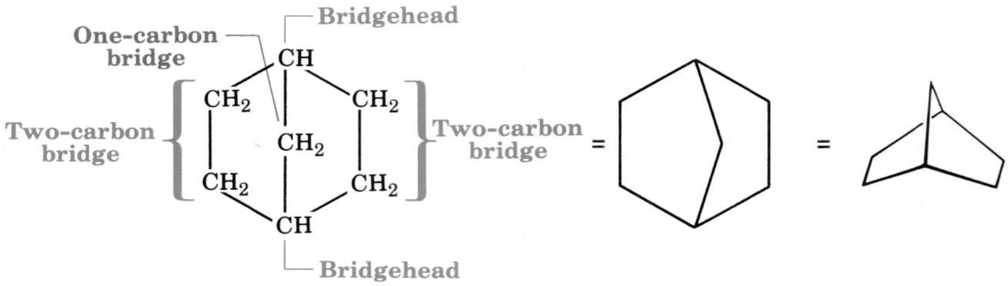

A bicycloheptane

Then we interpose in the name an expression in brackets that denotes the number of carbon atoms in each bridge (in order of decreasing length). For example,

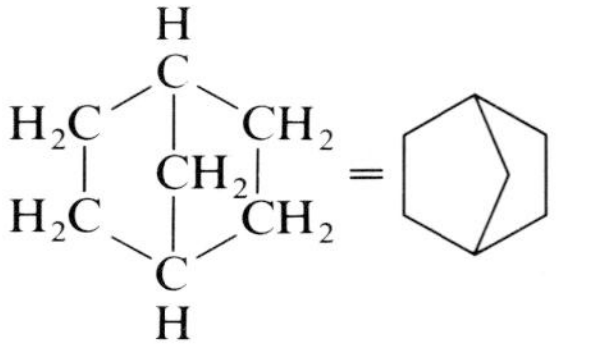

Bicyclo[2.2.1]heptane
(also called *norbornane*)

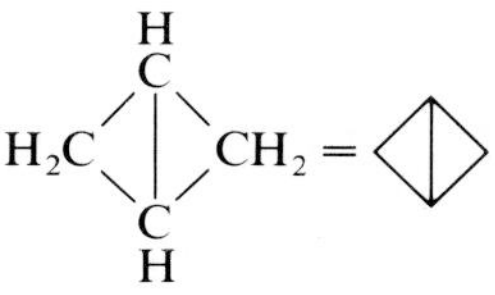

Bicyclo[1.1.0]butane

If substituents are present, we number the bridged ring system beginning at one bridgehead, proceeding first along the longest bridge to the other bridgehead, then along the next longest bridge back to the first bridgehead. The shortest bridge is numbered last.

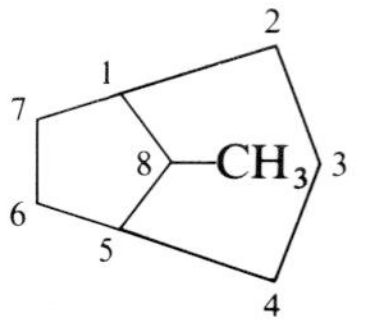

8-Methylbicyclo[3.2.1]octane

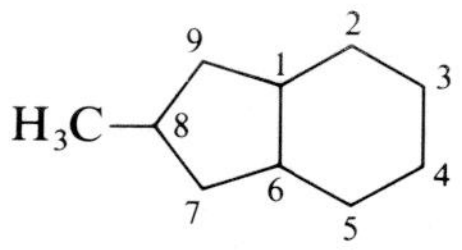

8-Methylbicyclo[4.3.0]nonane

PROBLEM 4.6

Give names for each of the following bicyclic alkanes:

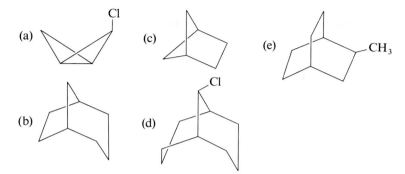

(f) Write the structure of a bicyclic compound that is an isomer of bicyclo-[2.2.0]hexane and give its name.

4.5 PHYSICAL PROPERTIES OF ALKANES AND CYCLOALKANES

If we examine the unbranched alkanes in Table 4.4 we notice that each alkane differs from the preceding one by one —CH$_2$— group. Butane, for example, is CH$_3$(CH$_2$)$_2$CH$_3$ and pentane is CH$_3$(CH$_2$)$_3$CH$_3$. A series of compounds like this, where each member differs from the next member by a constant unit, is called a **homologous series.** Members of a homologous series are called **homologs.**

At room temperature (25°C) and 1-atm pressure the first four members of the homologous series of unbranched alkanes (Table 4.5) are gases; the C$_5$–C$_{17}$ unbranched alkanes (pentane to heptadecane) are liquids; and the unbranched alkanes with 18 and more carbon atoms are solids.

Boiling Points   The boiling points of the unbranched alkanes show a regular increase with increasing molecular weight (Fig. 4.3). Branching of the alkane chain, however, lowers the boiling point. As examples consider the C$_6$H$_{14}$ isomers in Table 4.2. Hexane boils at 68°C and 2-methylpentane and 3-methylpentane, each having one branch, boil lower at 60.3 and 63.3°C, respectively. 2,3-Dimethylbutane and 2,2-dimethylbutane, each with two branches, boil lower still at 58 and 49.7°C, respectively.

Part of the explanation for these effects lies in the van der Waals forces that we studied in Section 2.16D. With unbranched alkanes, as molecular weight increases, so too does molecular size, and even more importantly molecular surface areas. With increasing surface area, the van der Waals forces between molecules increase; therefore, more energy (a higher temperature) is required to separate molecules from one another and produce boiling. Chain branching, on the other hand, makes a molecule more compact, reducing its surface area, and with it the strength of the van der Waals forces operating between it and adjacent molecules; this has the effect of lowering the boiling point.

Melting Points   The unbranched alkanes do not show the same smooth increase in melting points with increasing molecular weight (black line in Fig. 4.4) that they show

TABLE 4.5 *Physical constants of unbranched alkanes*

NUMBER OF CARBON ATOMS	NAME	bp (°C) (1 atm)	mp (°C)	DENSITY d^{20} (g mL^{-1})
1	Methane	−161.5	−182	
2	Ethane	−88.6	−183	
3	Propane	−42.1	−188	
4	Butane	−0.5	−138	
5	Pentane	36.1	−130	0.626
6	Hexane	68.7	−95	0.659
7	Heptane	98.4	−91	0.684
8	Octane	125.7	−57	0.703
9	Nonane	150.8	−54	0.718
10	Decane	174.1	−30	0.730
11	Undecane	195.9	−26	0.740
12	Dodecane	216.3	−10	0.749
13	Tridecane	235.4	−5.5	0.756
14	Tetradecane	253.5	6	0.763
15	Pentadecane	270.5	10	0.769
16	Hexadecane	287	18	0.773
17	Heptadecane	303	22	0.778
18	Octadecane	316.7	28	0.777
19	Nonadecane	330	32	0.777
20	Eicosane	343	36.8	0.789

in their boiling points. There is an alternation as one progresses from an unbranched alkane with an even number of carbon atoms to the next one with an odd number of carbon atoms. For example, propane (mp, −188°C) melts lower than ethane (mp, −183°C) and also lower than methane (mp, −182°C). Butane (mp, −138°C) melts 53°C higher than propane and only 5°C lower than pentane (mp, −130°C). If, however, the even- and odd-numbered alkanes are plotted on *separate* curves (white and red lines in Fig. 4.4), there *is* a smooth increase in melting point with increasing molecular weight.

X-Ray diffraction studies, which provide information about molecular structure, have revealed the reason for this apparent anomaly. Alkane chains with an even number of carbon atoms pack more closely in the crystalline state. As a result, attractive forces between individual chains are greater and melting points are higher.

The effect of chain branching on the melting points of alkanes is more difficult to predict. Generally, however, branching that produces highly symmetrical structures results in abnormally high melting points. The compound 2,2,3,3-tetramethylbutane, for example, melts at 100.7°C. Its boiling point is only six degrees higher, 106.3°C.

$$\begin{array}{ccc} & CH_3 & CH_3 \\ & | & | \\ CH_3\!-\!C & \!-\!-\!-\! & C\!-\!CH_3 \\ & | & | \\ & CH_3 & CH_3 \end{array}$$

2,2,3,3-Tetramethylbutane

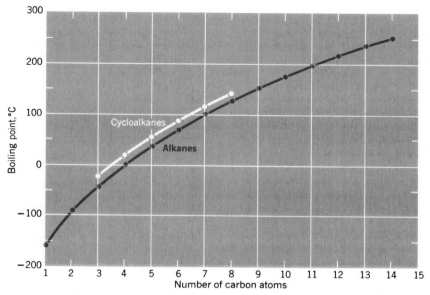

FIGURE 4.3 Boiling points of unbranched alkanes (in red) and cycloalkanes (in white).

Cycloalkanes also have much higher melting points than their open-chain counterparts (Table 4.6).

Density As a class, the alkanes and cycloalkanes are the least dense of all groups of organic compounds. All alkanes and cycloalkanes have densities considerably less than 1.00 g mL^{-1} (the density of water at 4°C). As a result, petroleum (a mixture of hydrocarbons rich in alkanes) floats on water.

Solubility Alkanes and cycloalkanes are almost totally insoluble in water because of their very low polarity and their inability to form hydrogen bonds. Liquid alkanes and cycloalkanes are soluble in one another, and they generally dissolve in solvents of low

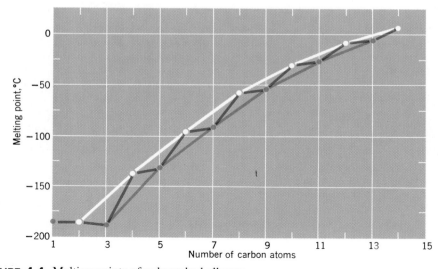

FIGURE 4.4 Melting points of unbranched alkanes.

TABLE 4.6 *Physical constants of cycloalkanes*

NUMBER OF CARBON ATOMS	NAME	bp (°C) (1 atm)	mp (°C)	DENSITY d^{20} (g mL^{-1})	REFRACTIVE INDEX (n_D^{20})
3	Cyclopropane	− 33	− 126.6		
4	Cyclobutane	13	− 90		1.4260
5	Cyclopentane	49	− 94	0.751	1.4064
6	Cyclohexane	81	6.5	0.779	1.4266
7	Cycloheptane	118.5	− 12	0.811	1.4449
8	Cyclooctane	149	13.5	0.834	

polarity. Good solvents for them are benzene, carbon tetrachloride, chloroform, and other hydrocarbons.

4.6 SIGMA BONDS AND BOND ROTATION

Groups bonded only by a sigma bond (i.e., by a single bond) can undergo rotation about that bond with respect to each other. The temporary molecular shapes that result from rotations of groups about single bonds are called **conformations** of the molecule. An analysis of the energy changes that a molecule undergoes as groups rotate about single bonds is called a **conformational analysis.***

Let us consider the ethane molecule as an example. Obviously an infinite number of different conformations could result from rotations of the CH_3 groups about the carbon–carbon bond. These different conformations, however, are not all of equal stability. The conformation (Fig. 4.5) in which the hydrogen atoms attached to each carbon atom are perfectly staggered when viewed from one end of the molecule along

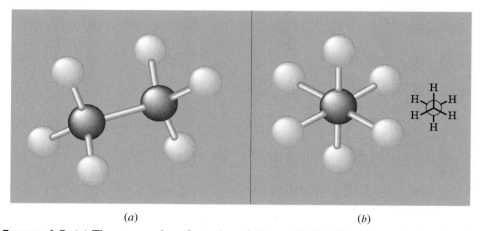

(*a*)   (*b*)

FIGURE 4.5 (*a*) The staggered conformation of ethane. (*b*) The Newman projection formula for the staggered conformation.

*Conformational analysis owes its modern origins largely to the work of O. Hassel of Norway and D. H. R. Barton of Great Britain. Hassel and Barton won the Nobel Prize in 1969, mainly for their contributions in this area. The idea that certain conformations of molecules are favored, however, originates from the work of van't Hoff.

the carbon–carbon bond axis is the *most stable* conformation (i.e., it is the conformation of *lowest potential energy*). This is easily explained in terms of repulsive interactions between bonding pairs of electrons. The staggered conformation allows the maximum possible separation of the electron pairs of the six carbon–hydrogen bonds and therefore it has the lowest energy.

In Fig. 4.5*b* we have drawn what is called a **Newman projection formula*** for ethane. In writing a Newman projection we imagine ourselves viewing the molecule from one end directly along the carbon–carbon bond axis. The bonds of the front carbon atom are represented as $\curlyvee$ and those of the back atom as $\curlyvee$.

Another kind of formula that is sometimes used to show conformations is the **sawhorse formula.**

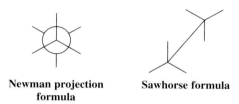

Newman projection
formula   Sawhorse formula

This formula depicts the conformation as seen in Fig. 4.5*a*. For clarity we have omitted the atoms at the ends of the bonds.

The least stable conformation of ethane is the **eclipsed conformation** (Fig. 4.6). When viewed from one end along the carbon–carbon bond axis, the hydrogen atoms attached to each carbon atom in the eclipsed conformation are in direct opposition to each other. This conformation requires the maximum repulsive interaction between the electrons of the six carbon–hydrogen bonds. It is, therefore, of highest energy and has the least stability.

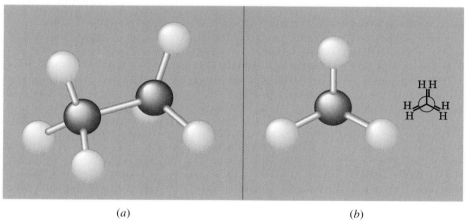

(*a*)   (*b*)

FIGURE 4.6 (*a*) The eclipsed conformation of ethane. (*b*) The Newman projection formula for the eclipsed conformation.

We represent this situation graphically by plotting the energy of an ethane molecule as a function of rotation about the carbon–carbon bond. The energy changes that occur are illustrated in Fig. 4.7.

*These formulas are named after their inventor Melvin S. Newman of The Ohio State University.

In ethane the difference in energy between the staggered and eclipsed conformations is 2.8 kcal mol^{-1} (12 kJ mol^{-1}). This small barrier to rotation is called the **torsional barrier** of the single bond. Unless the temperature is extremely low ($-250°C$) many ethane molecules (at any given moment) will have enough energy to surmount this barrier. Some molecules will wag back and forth with their atoms in staggered or nearly staggered conformations. The more energetic ones, however, will rotate through eclipsed conformations to other staggered conformations.

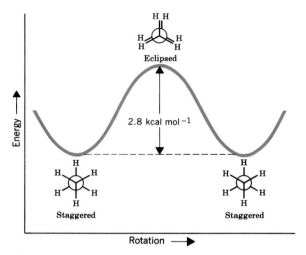

FIGURE 4.7 Potential energy changes that accompany rotation of groups about the carbon–carbon bond of ethane.

What does all this mean about ethane? We can answer this question in two different ways. If we consider a single molecule of ethane, we can say, for example, that it will spend most of its time in the lowest energy, staggered conformation, or in a conformation very close to being staggered. Many times every second, however, it will acquire enough energy through collisions with other molecules to surmount the torsional barrier and it will rotate through an eclipsed conformation. If we speak in terms of a large number of ethane molecules (a more realistic situation), we can say that at any given moment most of the molecules will be in staggered or nearly staggered conformations.

If we consider substituted ethanes (G is a group or atom other than hydrogen) such as GCH$_2$CH$_2$G, the barriers to rotation are somewhat larger but they are still far too small to allow isolation of the different staggered conformations or **conformers** (see following figure), even at temperatures considerably below room temperature.

**These conformers cannot be
isolated except at extremely
low temperatures**

4.7 CONFORMATIONAL ANALYSIS OF BUTANE

When we concern ourselves with *the three-dimensional aspects of molecular structure,* we are involved in the field of study called **stereochemistry.** We have already had some experience with stereochemistry because we began considering the shapes of molecules as early as Chapter 1. In this chapter, however, we begin our study in earnest as we take a detailed look at the *conformations* of alkanes and cycloalkanes. In many of the chapters that follow we shall see some of the consequences of this stereochemistry in the reactions that these molecules undergo. In Chapter 5, we shall see further basic principles of stereochemistry when we examine the properties of molecules that, because of their shape, are said to possess "handedness" or **chirality.** Let us begin, however, with a relatively simple molecule, butane, and study its conformations and their relative energies.

4.7A A Conformational Analysis of Butane

The study of the energy changes that occur in a molecule when groups rotate about single bonds is called *conformational analysis.* We saw the results of such a study for ethane in Section 4.6. Ethane has a slight barrier (2.8 kcal mol^{-1}) to free rotation about the carbon–carbon single bond. This barrier causes the potential energy of the ethane molecule to rise to a maximum when rotation brings the hydrogen atoms into an eclipsed conformation. This barrier to free rotation in ethane is due to the **torsional strain** of an eclipsed conformation of the molecule.

If we consider rotation about the C2—C3 bond of butane, torsional strain plays a part, too. There are, however, additional factors. To see what these are, we should look at the important conformations of butane **I–VI.**

I	**II**	**III**
An *anti* conformation	An eclipsed conformation	A *gauche* conformation
IV	**V**	**VI**
An eclipsed conformation	A *gauche* conformation	An eclipsed conformation

The *anti* **conformation (I)** does not have torsional strain because the groups are staggered and the methyl groups are far apart. Therefore, the *anti* conformation is the most stable. The methyl groups in the *gauche* **conformations** are close enough to each other that the van der Waals forces between them are *repulsive;* the electron clouds of the two groups are so close that they repel each other. This repulsion causes the *gauche* conformations to have approximately 0.9 kcal mol^{-1} (3.7 kJ mol^{-1}) more energy than the *anti* conformation.

The eclipsed conformations (**II**, **IV**, and **VI**) represent energy maxima in the potential energy diagram (Fig. 4.8). Eclipsed conformations **II** and **VI** not only have torsional strain, they have additional van der Waals repulsions arising from the eclipsed methyl groups and hydrogen atoms. Eclipsed conformation **IV** has the greatest energy of all because, in addition to torsional strain, there is the added large van der Waals repulsive force between the eclipsed methyl groups.

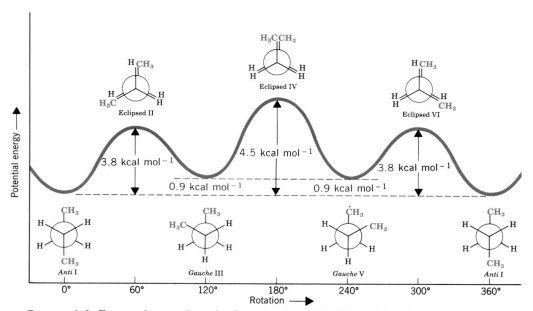

FIGURE 4.8 Energy changes that arise from rotation of the C2—C3 bond of butane.

Although the barriers to rotation in a butane molecule are larger than those of an ethane molecule (Section 4.6), they are still far too small to permit isolation of the *gauche* and *anti* conformations at normal temperatures. Only at extremely low temperatures would the molecules have insufficient energies to surmount these barriers.

We saw earlier that van der Waals forces can be *attractive*. Here, however, we find that they can also be *repulsive*. Whether van der Waals interactions lead to attraction or to repulsion depends on the distance that separates the two groups. As two nonpolar groups are brought closer and closer together, the first effect is one in which a momentarily unsymmetrical distribution of electrons in one group induces an opposite polarity in the other. The opposite charges induced in those portions of the two groups that are in closest proximity lead to attraction between them. This attraction increases to a maximum as the internuclear distance of the two groups decreases. The internuclear distance at which the attractive force is at a maximum is equal to the sum of what are called the *van der Waals radii* of the two groups. The van der Waals radius of a group is, in effect, a measure of its size. If the two groups are brought still closer—closer than the sum of their van der Waals radii—the interaction between them becomes repulsive. Their electron clouds begin to penetrate each other, and strong electron–electron interactions begin to occur.

PROBLEM 4.7 _____

Sketch a curve similar to that in Fig. 4.8 showing in general terms the energy changes that arise from rotation of the C2—C3 bond of 2-methylbutane. You need not concern yourself with the actual numerical values of the energy changes, but you should label all maxima and minima with the appropriate conformations.

4.8 THE RELATIVE STABILITIES OF CYCLOALKANES: RING STRAIN

Cycloalkanes do not all have the same relative stability. Data from heats of combustion (Section 4.8A) show that cyclohexane is the most stable cycloalkane and cyclopropane and cyclobutane are much less stable. The relative instability of cyclopropane and cyclobutane is a direct consequence of their cyclic structures and for this reason their molecules are said to possess **ring strain.** To see how this can be demonstrated experimentally, we need to examine the relative heats of combustion of cycloalkanes.

4.8A Heats of Combustion

The **heat of combustion** of a compound is the enthalpy change for the complete oxidation of the compound.

For a hydrocarbon complete oxidation means converting it to carbon dioxide and water. This can be accomplished experimentally, and the amount of heat evolved can be accurately measured in a device called a calorimeter. For methane, for example, the heat of combustion is -192 kcal mol^{-1} (-803 kJ mol^{-1}).

$$CH_4 + 2\,O_2 \longrightarrow CO_2 + 2\,H_2O \qquad \Delta H° = -192 \text{ kcal mol}^{-1} \quad (-803 \text{ kJ mol}^{-1})$$

For isomeric hydrocarbons, complete combustion of 1 mol of each will require the same amount of oxygen and will yield the same number of moles of carbon dioxide and water. We can, therefore, use heats of combustion to measure the relative stabilities of the isomers.

Consider, as an example, the combustion of butane and isobutane:

$$CH_3CH_2CH_2CH_3 + 6\tfrac{1}{2}\,O_2 \longrightarrow 4\,CO_2 + 5\,H_2O \qquad \Delta H° = -687.5 \text{ kcal mol}^{-1}$$
$$(C_4H_{10}) \qquad\qquad\qquad\qquad\qquad\qquad\qquad\qquad\qquad (-2877 \text{ kJ mol}^{-1})$$

$$CH_3CHCH_3 + 6\tfrac{1}{2}\,O_2 \longrightarrow 4\,CO_2 + 5\,H_2O \qquad \Delta H° = -685.5 \text{ kcal mol}^{-1}$$
$$\overset{|}{CH_3} \qquad\qquad\qquad\qquad\qquad\qquad\qquad\qquad (-2868 \text{ kJ mol}^{-1})$$
$$(C_4H_{10})$$

Since butane liberates more heat on combustion than isobutane, it must contain relatively more potential energy. Isobutane, therefore, must be _more stable._ Figure 4.9 illustrates this comparison.

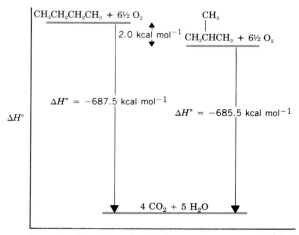

FIGURE 4.9 Heats of combustion show that isobutane is more stable than butane by 2.0 kcal mol^{-1} (8.4 kJ mol^{-1}).

4.8B Heats of Combustion of Cycloalkanes

The cycloalkanes constitute a *homologous series;* each member of the series differs from the one immediately preceding it by the constant amount of one —CH$_2$— group. Thus, the general equation for combustion of a cycloalkane can be formulated as follows:

$$(CH_2)_n + \tfrac{3}{2}\, n\, O_2 \longrightarrow n\, CO_2 + n\, H_2O + \text{heat}$$

Because the cycloalkanes are not isomeric, their heats of combustion cannot be compared directly. However, we can calculate the amount of heat evolved *per CH$_2$ group.* On this basis, the stabilities of the cycloalkanes become directly comparable. The results of such an investigation are given in Table 4.7.

Several observations emerge from a consideration of these results.

TABLE 4.7   Heats of combustion of cycloalkanes

CYCLOALKANE (CH$_2$)$_n$	n	HEAT OF COMBUSTION (kcal mol^{-1})	(kJ mol^{-1})	HEAT OF COMBUSTION PER CH$_2$ GROUP (kcal mol^{-1})	(kJ mol^{-1})
Cyclopropane	3	499.8	2091	166.6	697.5
Cyclobutane	4	655.9	2744	164.0	686.2
Cyclopentane	5	793.5	3220	158.7	664.0
Cyclohexane	6	944.5	3952	157.4	658.6
Cycloheptane	7	1108.2	4636.7	158.3	662.3
Cyclooctane	8	1269.2	5310.3	158.6	663.6
Cyclononane	9	1429.5	5981.0	158.8	664.4
Cyclodecane	10	1586.0	6635.8	158.6	663.6
Cyclopentadecane	15	2362.5	9984.7	157.5	659.0
Unbranched alkane				157.4	658.6

TABLE 4.8   *Ring strain of cycloalkanes*

CYCLOALKANE	RING STRAIN	
	(kcal mol^{-1})	(kJ mol^{-1})
Cyclopropane	27.6	115
Cyclobutane	26.3	110
Cyclopentane	6.5	27
Cyclohexane	0	0
Cycloheptane	6.4	27
Cyclooctane	10.0	42
Cyclononane	12.9	54
Cyclodecane	12.0	50
Cyclopentadecane	1.5	6

1. Cyclohexane has the lowest heat of combustion per CH_2 group (157.4 kcal mol^{-1}). This amount does not differ from that of unbranched alkanes, which, having no ring, can have no ring strain. We can assume, therefore, that cyclohexane has no ring strain and that it can serve as our standard for comparison with other cycloalkanes. We can calculate ring strain for the other cycloalkanes (Table 4.8) by multiplying 157.4 kcal mol^{-1} by *n* and then subtracting the result from the heat of combustion of the cycloalkane.

2. The combustion of cyclopropane evolves the greatest amount of heat per CH_2 group. Therefore, molecules of cyclopropane must have the greatest ring strain (27.6 kcal mol^{-1}, cf. Table 4.8). Since cyclopropane molecules evolve the greatest amount of heat energy per CH_2 group on combustion, they must contain the greatest amount of potential energy per CH_2 group. Thus what we call ring strain is a form of potential energy that the cyclic molecule contains. The more ring strain a molecule possesses, the more potential energy it has and the less stable it is compared to its ring homologs.

3. The combustion of cyclobutane evolves the second largest amount of heat per CH_2 group and, therefore, cyclobutane has the second largest amount of ring strain (26.3 kcal mol^{-1}).

4. While other cycloalkanes possess ring strain to varying degrees, the relative amounts are not large. Cyclopentane and cycloheptane have about the same modest amount of ring strain. Rings of 8, 9, and 10 members have slightly larger amounts of ring strain and then the amount falls off. A 15-membered ring has only a very slight amount of ring strain.

4.9   THE ORIGIN OF RING STRAIN IN CYCLOPROPANE AND CYCLOBUTANE: ANGLE STRAIN AND TORSIONAL STRAIN

The carbon atoms of alkanes are sp^3 hybridized. The normal tetrahedral bond angle of an sp^3-hybridized atom is 109.5°. In cyclopropane (a molecule with the shape of a regular triangle) the internal angles must be 60° and therefore they must depart from this ideal value by a very large amount—by 49.5°.

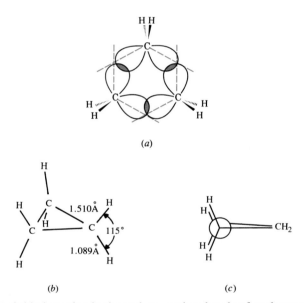

This compression of the internal bond angle causes what chemists call **angle strain.** Angle strain exists in a cyclopropane ring because the sp^3 orbitals of the carbon atoms cannot overlap as effectively (Fig. 4.10a) as they do in alkanes (where perfect end-on overlap is possible). The carbon–carbon bonds of cyclopropane are often described as being "bent." Orbital overlap is less effective. (The orbitals used for these bonds are not purely sp^3, they contain more p character). The carbon–carbon bonds of cyclopropane are weaker, and as a result, the molecule has greater potential energy.

FIGURE 4.10 (a) Orbital overlap in the carbon–carbon bonds of cyclopropane cannot occur perfectly end-on. This leads to weaker "bent" bonds and to angle strain. (b) Bond distances and angles in cyclopropane. (c) A Newman projection formula as viewed along one carbon–carbon bond shows the eclipsed hydrogens. (Viewing along either of the other two bonds would show the same picture.)

While angle strain accounts for most of the ring strain in cyclopropane, it does not account for it all. Because the ring is (of necessity) planar, the hydrogen atoms of the ring are all *eclipsed* (Fig. 4.10b and c), and the molecule has torsional strain as well.

Cyclobutane also has considerable angle strain. The internal angles are 88°—a departure of more than 21° from the normal tetrahedral bond angle. The cyclobutane ring is not planar but is slightly "folded" (Fig. 4.11a). If the cyclobutane ring were planar, the angle strain would be somewhat less (the internal angles would be 90° instead of 88°), but torsional strain would be considerably larger because all eight hydrogen atoms would be eclipsed. By folding or bending slightly the cyclobutane

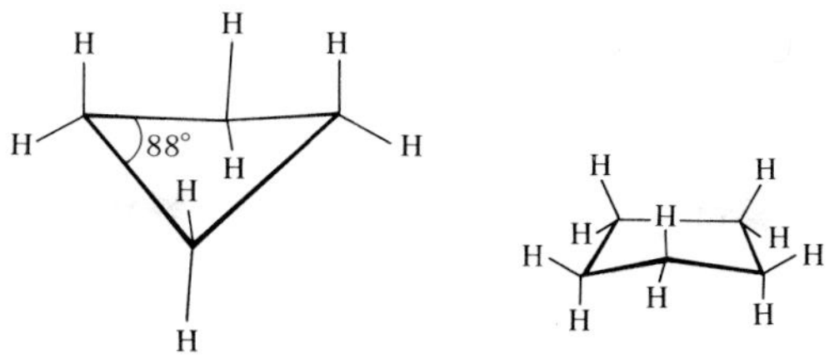

FIGURE 4.11 (a) The ''folded'' or ''bent'' conformation of cyclobutane. (b) The ''bent'' or ''envelope'' form of cyclopentane. In this structure the front carbon atom is bent upwards. In actuality, the molecule is flexible and shifts conformations constantly.

ring relieves more of its torsional strain than it gains in the slight increase in its angle strain.

4.9A Cyclopentane

The internal angles of a regular pentagon are 108°, a value very close to the normal tetrahedral bond angles of 109.5°. Therefore, if cyclopentane molecules were planar, they would have very little angle strain. Planarity, however, would introduce considerable torsional strain because all 10 hydrogen atoms would be eclipsed. Consequently, like cyclobutane, cyclopentane assumes a slightly bent conformation in which one or two of the atoms of the ring are out of the plane of the others (Fig. 4.11b). This relieves some of the torsional strain. Slight twisting of carbon–carbon bonds can occur with little change in energy and causes the out-of-plane atoms to move into plane and causes others to move out. Therefore, the molecule is flexible and shifts rapidly from one conformation to another. With little torsional strain and angle strain, cyclopentane is almost as stable as cyclohexane.

4.10 CONFORMATIONS OF CYCLOHEXANE

There is considerable evidence that the most stable conformation of the cyclohexane ring is the ''chair'' conformation illustrated in Fig. 4.12.* In this nonplanar structure the carbon–carbon bond angles are all 109.5° and are thereby free of angle strain. The chair conformation is free of torsional strain as well. When viewed along any carbon–carbon bond (viewing the structure from an end, Fig. 4.13), the atoms are seen to be perfectly staggered. Moreover, the hydrogen atoms at opposite corners of the cyclohexane ring are maximally separated.

By partial rotations about the carbon–carbon single bonds of the ring, the chair conformation can assume another shape called the ''boat'' conformation (Fig. 4.14). The boat conformation is like the chair conformation in that it is also free of angle strain.

The boat conformation, however, is not free of torsional strain. When a model of the boat conformation is viewed down carbon–carbon bond axes along either side (Fig. 4.15a), the hydrogen substituents at those carbon atoms are found to be eclipsed. Additionally, two of the hydrogen atoms on C1 and C4 are close enough to each other

---

*An understanding of this and subsequent discussions of conformational analysis can be aided immeasurably through the use of a molecular model.

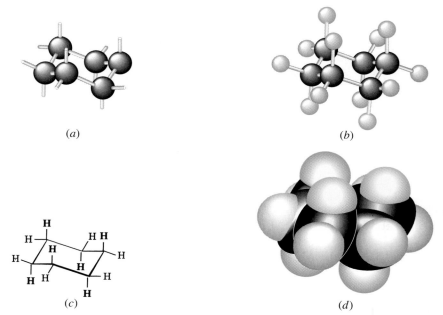

FIGURE 4.12 Representations of the chair conformation of cyclohexane: (*a*) carbon skeleton only; (*b*) carbon and hydrogen atoms; (*c*) line drawing; (*d*) space-filling model of cyclohexane. Notice that there are two types of hydrogen substituents—those that project up or down (shown in red) and those that lie generally in the plane of the ring (shown in black or gray). We shall discuss this further in Section 4.11.

to cause van der Waals repulsion (Fig. 4.15*b*). This latter effect has been called the "flagpole" interaction of the boat conformation. Torsional strain and flagpole interactions cause the boat conformation to have considerably higher energy than the chair conformation.

Although it is more stable, the chair conformation is much more rigid than the boat conformation. The boat conformation is quite flexible. By flexing to a new form—the twist conformation (Fig. 4.16)—the boat conformation can relieve some of its torsional strain and, at the same time, reduce the flagpole interactions. Thus, the twist conformation has a lower energy than the boat conformation. *The stability gained by flexing is insufficient, however, to cause the twist conformation of cyclohexane to be more stable than the chair conformation.* The chair conformation is

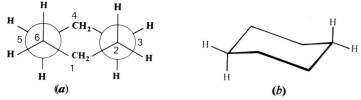

FIGURE 4.13 (*a*) A Newman projection of the chair conformation of cyclohexane. (Comparison with an actual molecular model will make this formulation clearer and will show that similar staggered arrangements are seen when other carbon–carbon bonds are chosen for sighting.) (*b*) Illustration of large separation between hydrogen atoms at opposite corners of the ring (designated C1 and C4) when the ring is in the chair conformation.

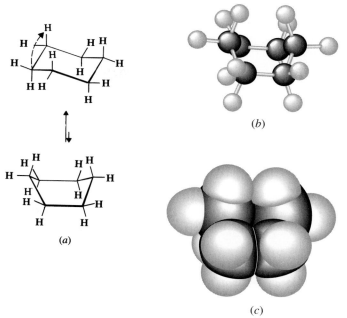

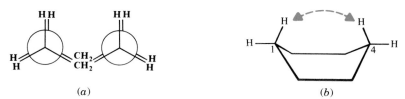

FIGURE 4.14 (*a*) The boat conformation of cyclohexane is formed by "flipping" one end of the chair form up (or down). This flip requires only rotations about carbon–carbon single bonds. (*b*) Ball-and-stick model of the boat conformation. (*c*) A space-filling model.

FIGURE 4.15 (*a*) Illustration of the eclipsed conformation of the boat conformation of cyclohexane. (*b*) Flagpole interaction of the C1 and C4 hydrogen atoms of the boat conformation.

estimated to be lower in energy than the twist conformation by approximately 5 kcal mol^{-1} (21 kJ mol^{-1}).

The energy barriers between the chair, boat, and twist conformations of cyclohexane are low enough (Fig. 4.17) to make their separation impossible at room temperature. At room temperature the thermal energies of the molecules are great enough to cause approximately 1 million interconversions to occur each second and, ***because of***

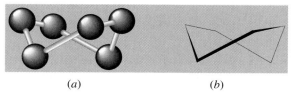

FIGURE 4.16 (*a*) Carbon skeleton and (*b*) line drawing of the twist conformation of cyclohexane.

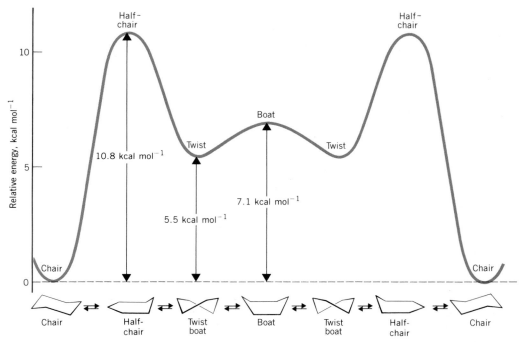

FIGURE 4.17 The relative energies of the various conformations of cyclohexane. The positions of maximum energy are conformations called half-chair conformations, in which the carbon atoms of one end of the ring have become coplanar.

its greater stability, more than 99% of the molecules are estimated to be in a chair conformation at any given moment.

4.10A Conformations of Higher Cycloalkanes

Cycloheptane, cyclooctane, and cyclononane and other higher cycloalkanes also exist in nonplanar conformations. The small instabilities of these higher cycloalkanes (Table 4.7) appear to be caused primarily by torsional strain and van der Waals repulsions between hydrogen atoms across rings, called *transannular strain*. The nonplanar conformations of these rings, however, are essentially free of angle strain.

X-ray crystallographic studies of cyclodecane reveal that the most stable conformation has carbon–carbon–carbon bond angles of 117°. This indicates some angle strain. The wide bond angles apparently allow the molecule to expand and thereby minimize unfavorable repulsions between hydrogen atoms across the ring.

There is very little free space in the center of a cycloalkane unless the ring is quite large. Calculations indicate that cyclooctadecane, for example, is the smallest ring through which a —$CH_2CH_2CH_2$— chain can be threaded. Molecules have been synthesized, however, that have large rings threaded on chains and that have large rings which are interlocked like links in a chain. These latter molecules are called **catenanes.**

$$(CH_2)_n \quad \bigcirc \quad (CH_2)_n$$

A catenane
($n \geqslant 18$)

In 1994 J. F. Stoddart and his coworkers at the University of Birmingham (England) achieved a remarkable synthesis of a catenane containing a linear array of five interlocked rings. Because the rings are interlocked in the same way as those of the olympic symbol, they named this compound **olympiadane**.

4.11 SUBSTITUTED CYCLOHEXANES: AXIAL AND EQUATORIAL HYDROGEN ATOMS

The six-membered ring is the most common ring found among nature's organic molecules. For this reason, we shall give it special attention. We have already seen that the chair conformation of cyclohexane is the most stable one and that it is the predominant conformation of the molecules in a sample of cyclohexane. With this fact in mind, we are in a position to undertake a limited analysis of the conformations of substituted cyclohexanes.

If we look carefully at the chair conformation of cyclohexane (Fig. 4.18), we can see that there are only two different kinds of hydrogen atoms. One hydrogen atom attached to each of the six carbon atoms lies in a plane generally defined by the ring of carbon atoms. These hydrogen atoms, by analogy with the equator of the earth, are called **equatorial** hydrogen atoms. Six other hydrogen atoms, one on each carbon, are oriented in a direction that is generally perpendicular to the average plane of the ring. These hydrogen atoms, again by analogy with the earth, are called **axial** hydrogen atoms. There are three axial hydrogen atoms on each face of the cyclohexane ring and their orientation (up or down) alternates from one carbon atom to the next.

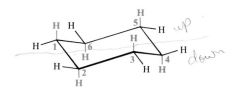

FIGURE 4.18 The chair conformation of cyclohexane. The axial hydrogen atoms are shown in color.

STUDY AID

You should now learn how to draw the important chair conformation. Notice (Fig. 4.19) the sets of parallel lines that constitute the bonds of the ring and the equatorial hydrogen atoms. Notice, too, that when drawn this way, the axial bonds are all vertical, and when the vertex of the ring points up, the axial bond is up; when the vertex is down, the axial bond is down.

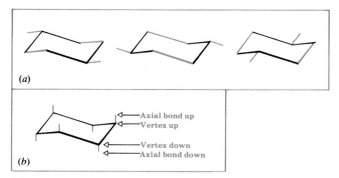

(a)

(b)

FIGURE 4.19 (a) Sets of parallel lines that constitute the ring and equatorial C—H bonds of the chair conformation. (b) The axial bonds are all vertical. When the vertex of the ring points up, the axial bond is up and vice versa.

We saw in Section 4.10 (and Fig. 4.17) that at room temperature, the cyclohexane ring rapidly flips back and forth between two *equivalent* chain conformations. An important thing to notice now is that **when the ring flips, all of the bonds that were axial become equatorial and vice versa:**

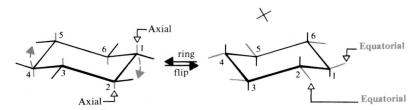

The question one might next ask is: What is the most stable conformation of a cyclohexane derivative *in which one hydrogen atom has been replaced by an alkyl*

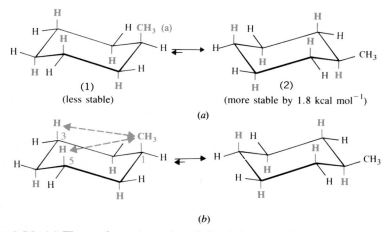

FIGURE 4.20 (a) The conformations of methylcyclohexane with the methyl group axial (1) and equatorial (2). (b) 1,3-Diaxial interactions between the two axial hydrogen atoms and the axial methyl group in the axial conformation of methylcyclohexane are shown with dashed arrows. Less crowding occurs in the equatorial conformation.

substituent? That is, what is the most stable conformation of a monosubstituted cyclohexane? We can answer this question by considering methylcyclohexane as an example.

Methylcyclohexane has two possible chair conformations (Fig. 4.20), and these are interconvertible through the partial rotations that constitute a ring flip. In one conformation (Fig. 4.20*a*) the methyl group occupies an *axial* position, and in the other the methyl group occupies an *equatorial* position. Studies indicate that the conformation with the methyl group equatorial is more stable than the conformation with the methyl group axial by about 1.8 kcal mol^{-1}. Thus, in the equilibrium mixture, the conformation with the methyl group in the equatorial position is the predominant one. Calculations show that it constitutes about 95% of the equilibrium mixture (Table 4.9).

The greater stability of methylcyclohexane with an equatorial methyl group can be understood through an inspection of the two forms as they are shown in Fig. 4.20*a* and *b*.

Studies done with scale models of the two conformations show that when the methyl group is axial, it is so close to the two axial hydrogen atoms on the same side of the molecule (attached to C3 and C5 atoms) that the van der Waals forces between them are repulsive. This type of steric strain, because it arises from an interaction between axial groups on carbon atoms 1 and 3 (or 5) is called a **1,3-diaxial interaction.** Similar studies with other substituents indicate that *there is generally less repulsive interaction when the groups are equatorial rather than axial.*

The strain caused by a 1,3-diaxial interaction in methylcyclohexane is the same as the strain caused by the close proximity of the hydrogen atoms of methyl groups in the *gauche* form of butane (Section 4.7A). Recall that the interaction in *gauche*-butane (called, for convenience, a *gauche interaction*) causes *gauche*-butane to be less stable than *anti*-butane by 0.9 kcal mol^{-1}. The following Newman projections will help you to see that the two steric interactions are the same. In the second projection we view axial methylcyclohexane along the C1—C2 bond and see that what we call a 1,3-diaxial interaction is simply a *gauche* interaction between the hydrogen atoms of the methyl group and the hydrogen atom at C3.

TABLE 4.9   Relationship between free-energy difference and isomer percentages for isomers at equilibrium at 25°C

FREE-ENERGY DIFFERENCE, $\Delta G°$ (kcal mol^{-1})	MORE STABLE ISOMER (%)	LESS STABLE ISOMER (%)
0	50	50
0.41	67	33
0.65	75	25
0.82	80	20
0.95	83	17
1.4	91	9
1.8	95	5
2.7	99	1
4.1	99.9	0.1
5.5	99.99	0.01

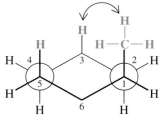

gauche-Butane
(0.9 kcal mol⁻¹ steric strain)

Axial methylcyclohexane
(two gauche interactions =
1.8 kcal mol⁻¹ steric strain)

Equatorial methylcyclohexane

Viewing methylcyclohexane along the C1—C6 bond (do this with a model) shows that it has a second identical *gauche* interaction between the hydrogen atoms of the methyl group and the hydrogen atom at C5. The methyl group of *axial*-methylcyclohexane, therefore, has two *gauche* interactions and, consequently, it has 2 × 0.9 = 1.8 kcal mol⁻¹ of strain. The methyl group of equatorial methylcyclohexane does not have a *gauche* interaction because it is *anti* to C3 and C5.

PROBLEM 4.8

Show by a calculation that a free-energy difference of 1.8 kcal mol⁻¹ between the axial and equatorial forms of methylcyclohexane at 25°C (with the equatorial form being more stable) does correlate with an equilibrium mixture in which the concentration of the equatorial form is approximately 95%.

In cyclohexane derivatives with larger alkyl substituents, the strain caused by 1,3-diaxial interactions is even more pronounced. The conformation of *tert*-butylcyclohexane with the *tert*-butyl group equatorial is estimated to be more than 5 kcal mol⁻¹ more stable than the axial form (Fig. 4.21). This large energy difference between the two conformations means that, at room temperature, 99.99% of the molecules of *tert*-butylcyclohexane have the *tert*-butyl group in the equatorial position. (The molecule is not conformationally ''locked,'' however; it still flips from one chair conformation to the other.)

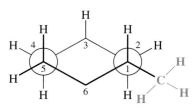

Equatorial *tert*-
butylcyclohexane

Axial *tert*-
butylcyclohexane

FIGURE 4.21 Diaxial interactions with the large *tert*-butyl group axial cause the conformation with the *tert*-butyl group equatorial to be the predominant one to the extent of 99.99%.

4.12 DISUBSTITUTED CYCLOALKANES: CIS–TRANS ISOMERISM

The presence of two substituents on the ring of a molecule of any cycloalkane allows for the possibility of cis–trans isomerism. We can see this most easily if we begin by examining cyclopentane derivatives because the cyclopentane ring is essentially planar. (At any given moment the ring of cyclopentane is, of course, slightly bent, but we know that the various bent conformations are rapidly interconverted. Over a period of time, the average conformation of the cyclopentane ring is planar.) Since the planar representation is much more convenient for an initial presentation of cis–trans isomerism in cycloalkanes, we shall use it here.

cis-1, 2-Dimethylcyclopentane
bp, 99.5 °C

trans-1, 2-Dimethylcyclopentane
bp, 91.9 °C

FIGURE 4.22 cis- and trans-1,2-Dimethylcyclopentanes.

Let us consider 1,2-dimethylcyclopentane as an example. We can write the structures shown in Fig. 4.22. In the first structure the methyl groups are on the same side of the ring, that is, they are cis. In the second structure the methyl groups are on opposite sides of the ring; they are trans.

The *cis*- and *trans*-1,2-dimethylcyclopentanes are stereoisomers: They differ from each other only in the arrangement of the atoms in space. The two forms cannot be interconverted without breaking carbon–carbon bonds. As a result, the cis and trans forms can be separated, placed in separate bottles, and kept indefinitely.

1,3-Dimethylcyclopentanes show cis–trans isomerism as well:

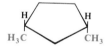

cis-1, 3-Dimethylcyclopentane   trans-1, 3-Dimethylcyclopentane

The physical properties of cis–trans isomers are different; they have different melting points, boiling points, and so on. Table 4.10 lists these physical constants of the dimethylcyclohexanes.

TABLE 4.10 Physical constants of *cis*- and *trans*-disubstituted cyclohexane derivatives

SUBSTITUENTS	ISOMER	mp (°C)	bp (°C)[a]
1,2-Dimethyl-	cis	− 50.1	130.04[760]
1,2-Dimethyl-	trans	− 89.4	123.7[760]
1,3-Dimethyl-	cis	− 75.6	120.1[760]
1,3-Dimethyl-	trans	− 90.1	123.5[760]
1,2-Dichloro-	cis	− 6	93.5[22]
1,2-Dichloro-	trans	− 7	74.7[16]

[a]The pressures (in units of torr) at which the boiling points were measured are given as superscripts.

PROBLEM 4.9

Write structures for the cis and trans isomers of (a) 1,2-dichlorocyclopropane and (b) 1,3-dibromocyclobutane.

The cyclohexane ring is, of course, not planar. A "time average" of the various interconverting chair conformations however, would be planar and, as with cyclopentane, this planar representation is convenient for introducing the topic of cis–trans isomerism of cyclohexane derivatives. The planar representations of the 1,2-, 1,3-, and 1,4-dimethylcyclohexane isomers follow:

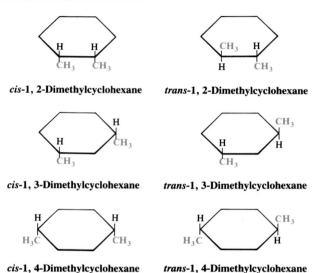

cis-1, 2-Dimethylcyclohexane   *trans*-1, 2-Dimethylcyclohexane

cis-1, 3-Dimethylcyclohexane   *trans*-1, 3-Dimethylcyclohexane

cis-1, 4-Dimethylcyclohexane   *trans*-1, 4-Dimethylcyclohexane

4.12A Cis–Trans Isomerism and Conformational Structures

If we consider the *actual* conformations of these isomers, the structures are somewhat more complex. Beginning with *trans*-1,4-dimethylcyclohexane, because it is easiest to visualize, we find there are two possible chair conformations (Fig. 4.23). In one conformation both methyl groups are axial; in the other both are equatorial. The diequatorial conformation is, as we would expect it to be, the more stable conformation, and it represents the structure of at least 99% of the molecules at equilibrium.

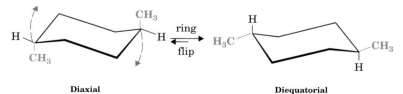

Diaxial   Diequatorial

FIGURE 4.23 The two chair conformations of *trans*-1,4-dimethylcyclohexane. (*Note:* All other C—H bonds have been omitted for clarity.)

That the diaxial form of *trans*-1,4-dimethylcyclohexane is a trans isomer is easy to see; the two methyl groups are clearly on opposite sides of the ring. The trans relationship of the methyl groups in the diequatorial form is not as obvious, however. The trans relationship of the methyl groups becomes more apparent if we imagine ourselves "flattening" the molecule by turning one end up and the other down.

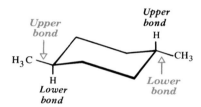

A second *and general* way to recognize a *trans*-disubstituted cyclohexane is to notice that one group is attached by the *upper* bond (of the two to its carbon) and one by the *lower* bond.

trans-1, 4-Dimethylcyclohexane

In a *cis*-disubstituted cyclohexane both groups are attached by an upper bond or both by a lower bond. For example,

cis-1, 4-Dimethylcyclohexane

cis-1,4-Dimethylcyclohexane actually exists in two *equivalent* chair conformations (Fig. 4.24). The cis relationship of the methyl groups, however, precludes the possibility of a structure with both groups in an equatorial position. One group is axial in either conformation.

Equatorial-axial   Axial-equatorial

FIGURE 4.24 Equivalent conformations of *cis*-1,4-dimethylcyclohexane.

SAMPLE PROBLEM

Consider each of the following conformational structures and tell whether each is cis or trans.

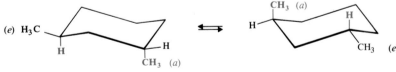

(a)   (b)   (c)

ANSWER:

(a) Each chlorine atom is attached by the upper bond at its carbon; therefore, both chlorine atoms are on the same side of the molecule and this is a cis isomer. This is *cis*-1,2-dichlorocyclohexane.   (b) Here both chlorine atoms are attached by a lower bond; therefore, in this example, too, both chlorine atoms are on the same side of the molecule and this, too, is a cis isomer. It is *cis*-1,3-dichlorocyclohexane.   (c) Here one chlorine atom is attached by a lower bond and one by an upper bond. The two chlorine atoms, therefore, are on opposite sides of the molecule, and this is a trans isomer. It is *trans*-1,2-dichlorocyclohexane.

PROBLEM 4.10

(a) Write structural formulas for the two chair conformations of *cis*-1-isopropyl-4-methylcyclohexane.   (b) Are these two conformations equivalent?   (c) If not, which would be more stable?   (d) Which would be the preferred conformation at equilibrium?

trans-1,3-Dimethylcyclohexane is like the *cis*-1,4-compound in that no chair conformation is possible with both methyl groups in the favored equatorial position. The following two conformations are of equal energy and are equally populated at equilibrium.

(e) H_3C ⟷ H

CH_3 (a)

H CH_3 (e)

H CH_3 (a)

trans-1, 3-Dimethylcyclohexane

If, however, we consider some other *trans*-1,3-disubstituted cyclohexane in which one alkyl group is larger than the other, the conformation of lower energy is the one having the larger group in the equatorial position. For example, the more stable conformation of *trans*-1-*tert*-butyl-3-methylcyclohexane, shown here, has the large *tert*-butyl group occupying the equatorial position.

CH_3

(e) H_3C—C

CH_3 H

H

CH_3 (a)

TABLE 4.11 Conformations of dimethyl-cyclohexanes

COMPOUND	CIS ISOMER			TRANS ISOMER		
1,2-Dimethyl-	*a,e*	or	*e,a*	**e,e**	or	*a,a*
1,3-Dimethyl-	**e,e**	or	*a,a*	*a,e*	or	*e,a*
1,4-Dimethyl-	*a,e*	or	*e,a*	**e,e**	or	*a,a*

PROBLEM 4.11

(a) Write the two conformations of *cis*-1,2-dimethylcyclohexane. (b) Would these two conformations have equal potential energy? (c) What about the two conformations of *cis*-1-*tert*-butyl-2-methylcyclohexane? (d) Would the two conformations of *trans*-1,2-dimethylcyclohexane have the same potential energy?

The different conformations of the dimethylcyclohexanes are summarized in Table 4.11. The more stable conformation, where one exists, is set in heavy type.

4.13   BICYCLIC AND POLYCYCLIC ALKANES

Many of the molecules that we encounter in our study of organic chemistry contain more than one ring (Section 4.4B). One of the most important bicyclic systems is bicyclo[4.4.0]decane, a compound that is usually called by its common name, *decalin*.

$$
\begin{array}{c}
\text{H}_2 \quad \text{H} \quad \text{H}_2 \\
\text{C} \qquad \text{C} \\
\text{H}_2\text{C}_9 \quad {}^{10}\text{C}_1 \quad {}_2 \quad {}_3\text{CH}_2 \\
\text{H}_2\text{C}_8 \quad {}_7 \quad \text{C}_6 \quad {}_5 \quad {}_4\text{CH}_2 \\
\text{C} \qquad \text{C} \\
\text{H}_2 \quad \text{H} \quad \text{H}_2
\end{array}
\qquad \text{or}
$$

Decalin (bicyclo[4.4.0]decane)
(carbon atoms 1 and 6 are bridgehead carbon atoms)

Decalin shows cis–trans isomerism:

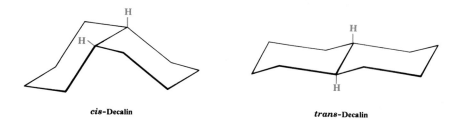

cis-Decalin   *trans*-Decalin

In *cis*-decalin the two hydrogen atoms attached to the bridgehead atoms lie on the same side of the ring; in *trans*-decalin they are on opposite sides. We often indicate this by writing their structures in the following way:

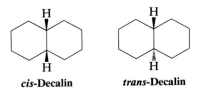

cis-**Decalin**   *trans*-**Decalin**

Simple rotations of groups about carbon–carbon bonds do not interconvert *cis*- and *trans*-decalins. In this respect they resemble the isomeric *cis*- and *trans*-disubstituted cyclohexanes. (We can, in fact, regard them as being *cis*- or *trans*-1,2-disubstituted cyclohexanes in which the 1,2-substituents are the two ends of a four-carbon bridge, that is, $-CH_2CH_2CH_2CH_2-$.)

The *cis*- and *trans*-decalins can be separated. *cis*-Decalin boils at 195°C (at 760 torr) and *trans*-decalin boils at 185.5°C (at 760 torr).

Adamantane (see the following figure) is a tricyclic system that contains a three-dimensional array of cyclohexane rings, all of which are in the chair form. Extending the structure of adamantane in three dimensions gives the structure of diamond. The great hardness of diamond results from the fact that the entire diamond crystal is actually one very large molecule—a molecule that is held together by millions of strong covalent bonds.*

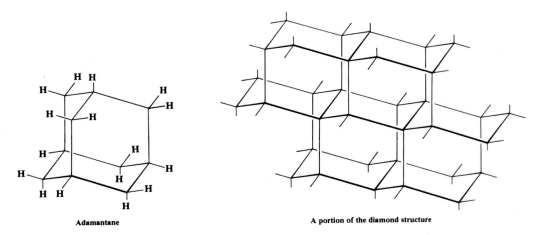

Adamantane   **A portion of the diamond structure**

One goal of research in recent years has been the synthesis of unusual, and sometimes highly strained, cyclic hydrocarbons. Among those that have been prepared are the compounds that follow:

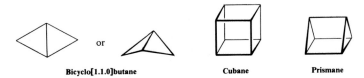

Bicyclo[1.1.0]butane   **Cubane**   **Prismane**

* There are other allotropic forms of carbon, including graphite, diamond, Wurzite carbon [with a structure related to Wurzite (ZnS)], and a new group of compounds called fullerenes; see Section 13.8C.

In 1982, Leo A. Paquette and his co-workers at The Ohio State University announced the successful synthesis of the ''complex, symmetric, and aesthetically appealing'' molecule called dodecahedrane.

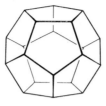

Dodecahedrane

4.14 CHEMICAL REACTIONS OF ALKANES

Alkanes, as a class, are characterized by a general inertness to many chemical reagents. Carbon–carbon and carbon–hydrogen bonds are quite strong; they do not break unless alkanes are heated to very high temperatures. Because carbon and hydrogen atoms have nearly the same electronegativity, the carbon–hydrogen bonds of alkanes are only slightly polarized. As a consequence, they are generally unaffected by most bases. Molecules of alkanes have no unshared electrons to offer sites for attack by acids. This low reactivity of alkanes toward many reagents accounts for the fact that alkanes were originally called *paraffins* (Latin: *parum affinis*, little affinity).

The term paraffin, however, is probably not an appropriate one. We all know that alkanes react vigorously with oxygen when an appropriate mixture is ignited. This combustion occurs in the cylinders of automobiles, and in oil furnaces, for example. When heated, alkanes also react with chlorine and bromine, and they react explosively with fluorine. We shall study these reactions in Chapter 9.

4.15 SYNTHESIS OF ALKANES AND CYCLOALKANES

Mixtures of alkanes as they are obtained from petroleum are suitable as fuels. However, in our laboratory work we often have the need for a pure sample of a particular alkane. For these purposes, the chemical preparation — or synthesis — of that particular alkane is often the most reliable way of obtaining it. The preparative method that we choose should be one that will lead to the desired product alone or, at least, to products that can be easily and effectively separated.

Several such methods are available, and three are outlined here. In subsequent chapters we shall encounter others.

4.15A Hydrogenation of Alkenes

Alkenes react with hydrogen in the presence of metal catalysts such as nickel and platinum to produce alkanes. The general reaction is one in which the atoms of the hydrogen molecule add to each atom of the carbon–carbon double bond of the alkene. This converts the alkene to an alkane.

General Reaction

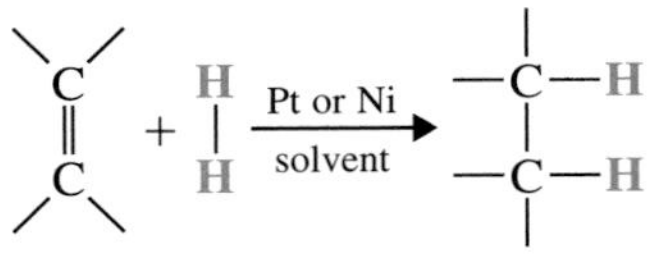

Alkene   **Alkane**

The reaction is usually carried out by dissolving the alkene in a solvent such as ethyl alcohol (C_2H_5OH), adding the metal catalyst, and then exposing the mixture to hydrogen gas under pressure in a special apparatus. (We shall discuss the mechanism of this reaction—called **hydrogenation**—in Chapter 7.)

Specific Examples

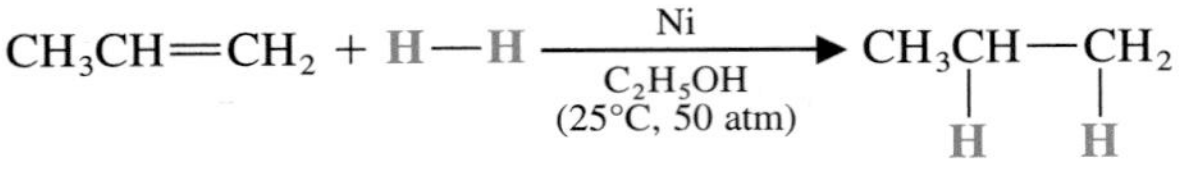

Propene   **Propane**

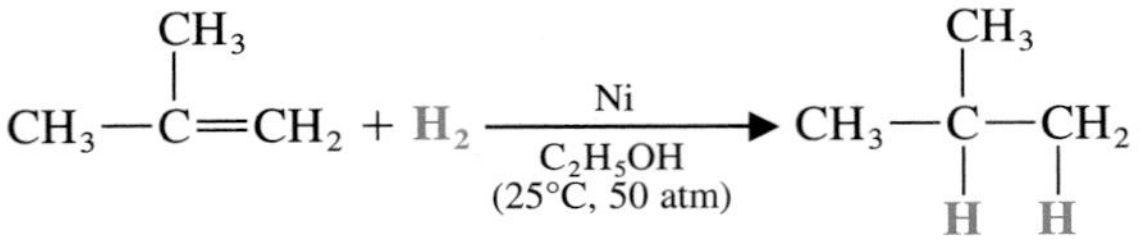

2-Methylpropene   **Isobutane**

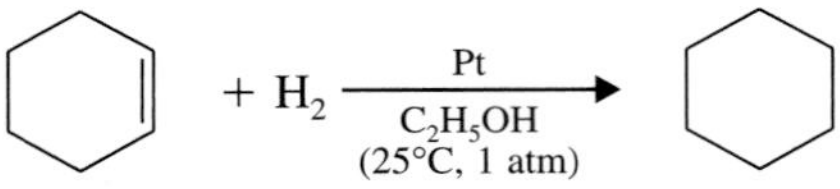

Cyclohexene   **Cyclohexane**

PROBLEM 4.12

Three different alkenes react with hydrogen in the presence of a platinum or nickel catalyst to yield 2-methylbutane. Show the reactions involved.

4.15B Reduction of Alkyl Halides

Most alkyl halides react with zinc and aqueous acid to produce an alkane. The general reaction is as follows:

General Reaction

$$2\ R\!-\!X + Zn + 2\ H^+ \longrightarrow 2\ R\!-\!H + ZnX_2$$

or* $R\!-\!X \xrightarrow[(-ZnX_2)]{Zn,\ H^+} R\!-\!H$

*This illustrates the way organic chemists often write abbreviated equations for chemical reactions. The organic reactant is shown on the left and the organic product on the right. The reagents necessary to bring about the transformation are written over (or under) the arrow. The equations are often left unbalanced and sometimes byproducts (in this case, ZnX_2) are either omitted or are placed under the arrow in parentheses

Specific Examples

$$2\ CH_3CH_2\underset{\overset{|}{Br}}{C}HCH_3 \xrightarrow[Zn]{H^+} 2\ CH_3CH_2\underset{\overset{|}{H}}{C}HCH_3 + ZnBr_2$$

sec-**Butyl bromide**   **Butane**
(2-bromobutane)

$$2\ CH_3\underset{\overset{|}{CH_3}}{C}HCH_2CH_2{-}Br \xrightarrow[Zn]{H^+} 2\ CH_3\underset{\overset{|}{CH_3}}{C}HCH_2CH_2{-}H + ZnBr_2$$

Isopentyl bromide   **Isopentane**
(1-bromo-3-methylbutane)   **(2-methylbutane)**

In these reactions zinc atoms transfer electrons to the carbon atom of the alkyl halide. Therefore, the reaction is a **reduction** of the alkyl halide. Zinc is a good reducing agent because it has two electrons in an orbital far from the nucleus, which are readily donated to an electron acceptor. The mechanism for the reaction is complex because the reaction takes place in a separate phase at or near the surface of the zinc metal. It is possible that an alkylzinc halide forms first and then reacts with the acid to produce the alkane:

$$Zn: \quad + \quad \overset{\delta+}{R}{-}\overset{\overset{\cdot\cdot}{}}{\underset{\overset{\cdot\cdot}{}}{X}}\overset{\delta-}{:} \longrightarrow \left[R:Zn^{2+}:\overset{\cdot\cdot}{\underset{\cdot\cdot}{X}}: \right] \xrightarrow{H^+} R{-}H + Zn^{2+} + :\overset{\cdot\cdot}{\underset{\cdot\cdot}{X}}:$$

Reducing   **Alkylzinc halide**   **Alkane**
agent

PROBLEM 4.13

Your goal is the synthesis of 2,3-dimethylbutane by treating an alkyl halide with zinc and aqueous acid. Show two methods (beginning with different alkyl halides) for doing this.

4.15C Lithium Dialkylcuprates: The Corey–Posner, Whitesides–House Synthesis

A highly versatile method for the synthesis of alkanes and other hydrocarbons from organic halides has been developed by E. J. Corey* (Harvard University) and G. H. Posner (The Johns Hopkins University) and by G. M. Whitesides (MIT) and H. O. House (Georgia Institute of Technology). The overall synthesis provides, for example, a way for coupling the alkyl groups of two alkyl halides to produce an alkane:

$$R{-}X + R'{-}X \xrightarrow[\substack{steps \\ (-2\,X)}]{several} R{-}R'$$

In order to accomplish this coupling, we must transform one alkyl halide into a lithium dialkylcuprate (R_2CuLi). This transformation requires two steps. First, the alkyl halide is treated with lithium metal in an ether solvent to convert the alkyl halide into an alkyllithium, RLi (Section 3.8).

* Corey was awarded the Nobel Prize for Chemistry in 1990 for finding new ways of synthesizing organic compounds, which, in the words of the Nobel committee ''have contributed to the high standards of living and health enjoyed . . . in the Western world.''

$$R-X + 2\,Li \xrightarrow[\text{ether}]{\text{diethyl}} RLi + LiX$$

$$\textbf{Alkyllithium}$$

Then the alkyllithium is treated with cuprous iodide (CuI). This converts it to the lithium dialkylcuprate.*

$$2\,RLi + CuI \longrightarrow R_2CuLi + LiI$$

Alkyllithium		**Lithium**
		dialkylcuprate

When the lithium dialkylcuprate is treated with the second alkyl halide ($R'-X$), coupling takes place between one alkyl group of the lithium dialkylcuprate and the alkyl group of the alkyl halide, $R'-X$.

$$R_2CuLi + R'-X \longrightarrow R-R' + RCu + LiX$$

Lithium	**Alkyl halide**	**Alkane**
dialkylcuprate		

For the last step to give a good yield of the alkane, the alkyl halide $R'-X$ must be either a methyl halide, a primary alkyl halide, or a secondary cycloalkyl halide. The alkyl groups of the lithium dialkylcuprate may be methyl, 1°, 2°, or 3°.† Moreover, the two alkyl groups being coupled need not be different.

The overall scheme for this alkane synthesis is given in Fig. 4.25.

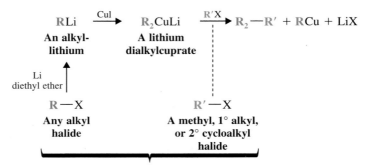

These are the organic starting materials. The $R-$ and $R'-$
groups need not be different.

FIGURE 4.25 A scheme outlining the synthesis of alkanes via the Corey–Posner, Whitesides–House method.

Consider the following two examples: the synthesis of hexane from methyl iodide and pentyl iodide, and the synthesis of nonane from butyl bromide and pentyl bromide:

$$CH_3-I \xrightarrow[\text{diethyl ether}]{\text{Li}} CH_3Li \xrightarrow{\text{CuI}} (CH_3)_2CuLi \xrightarrow{CH_3CH_2CH_2CH_2CH_2I}$$

$$CH_3-CH_2CH_2CH_2CH_2CH_3$$

Hexane

(98%)

---

* Special techniques, which we shall not discuss here, are required when R is tertiary. For an excellent review of these reactions see Gary H. Posner, ''Substitution Reactions Using Organocopper Reagents,'' *Organic Reactions,* Vol. 22, Wiley, New York, 1975.

† Lithium dialkylcuprates were first synthesized by Henry Gilman (of Iowa State University) and are often called **Gilman Reagents**.

$$CH_3CH_2CH_2CH_2Br \xrightarrow[\text{diethyl ether}]{Li} CH_3CH_2CH_2CH_2Li \xrightarrow{CuI}$$

$$(CH_3CH_2CH_2CH_2)_2CuLi \xrightarrow{CH_3CH_2CH_2CH_2CH_2Br} CH_3CH_2CH_2CH_2-CH_2CH_2CH_2CH_2CH_3$$

Nonane
(98%)

Lithium dialkylcuprates couple with other organic groups. Coupling reactions of lithium dimethylcuprate with two cycloalkyl halides are shown here.

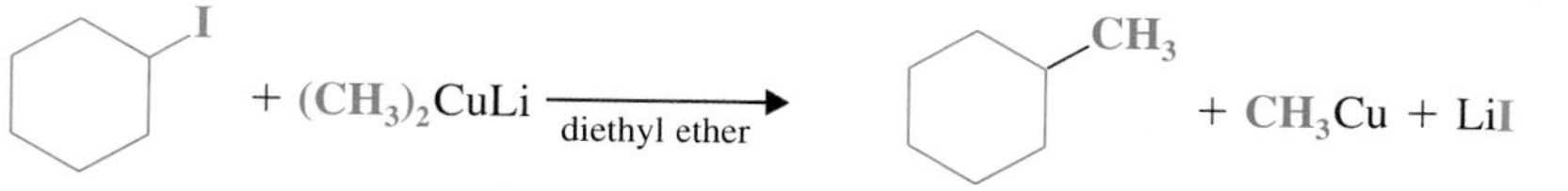

Methylcyclohexane
(75%)

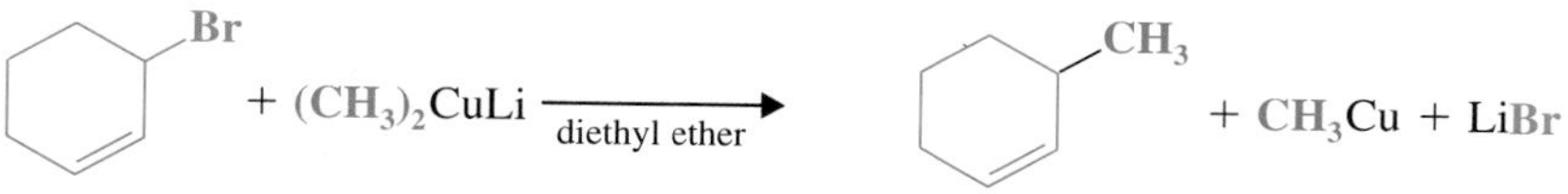

3-Methylcyclohexene
(75%)

Lithium dialkylcuprates also couple with phenyl halides. An example is the following synthesis of butyl benzene.

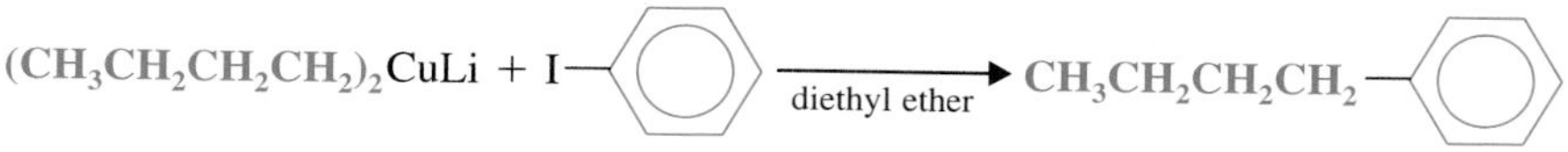

Butylbenzene
(75%)

The following scheme summarizes the coupling reactions of lithium dialkyl-cuprates.

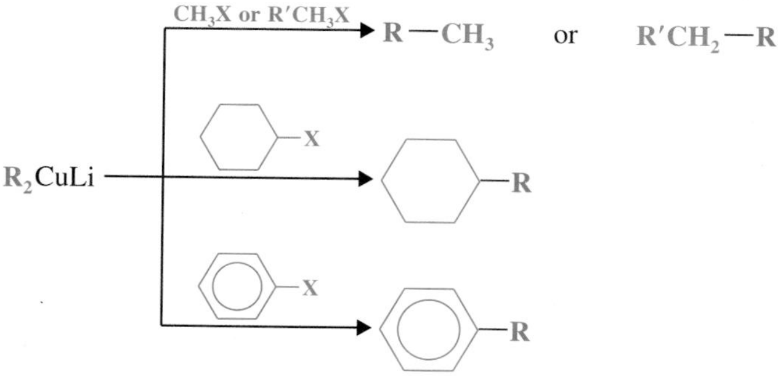

The mechanism of the Corey–Posner, Whitesides–House synthesis is beyond our scope but is of a type discussed in Special Topic M.

4.16 PLANNING ORGANIC SYNTHESES

Most of the more than 7 million organic compounds that are now known have come about because organic chemists have synthesized them. Only a small fraction of these compounds have been isolated from natural sources. Even then, in the isolation of a naturally occurring compound, the final proof of its structure is the synthesis of the compound by an unambiguous route from simpler molecules.

Syntheses are carried out for many reasons. We may need a particular compound to test some hypothesis about a reaction mechanism or about how a certain organism metabolizes the compound. In cases like these, we will need a particular compound, and we may need it with a "labeled" atom (e. g., deuterium or tritium) at a particular position.

The syntheses that you are asked to design in this text have a teaching purpose. At first, you will be asked to design syntheses of relatively simple compounds that are, in most instances, commercially available. Nevertheless, successfully planning these syntheses can offer an intellectual challenge and reward.

In planning syntheses we are required to think backward, to work our way backward from relatively complex molecules to simpler ones that can act as the precursor (or precursors if two molecules are combined) for our target molecule. We carry out what is called a **retrosynthetic analysis,** and we represent this reasoning process in the following way:

Target Molecule ⟹ Precursors

The open arrow is a symbol that relates the target molecule to its most immediate precursors; it signifies a **retro** or **backward** step.*

In most instances more than one step is required to bring about a synthesis. We then repeat the analytical process: The precursors become our target molecules, and we reason backward to another level of precursors, and so on, until we reach the level of the compounds that we have available as starting materials.

Target Molecule ⟹ 1st Precursor ⟹ 2nd Precursor ⟹ ⟹ Starting Compound

There will almost always be more than one way to carry out a synthesis. We usually face the sort of situation given in Fig. 4.26.

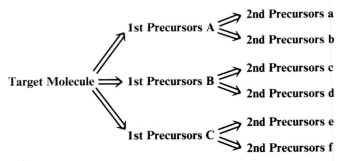

FIGURE 4.26 The retroanalytic process often discloses several routes from the target molecule back to varied precursors.

Often more than one route will give the desired product in a reasonable percentage yield. However, not all pathways will be available to us because of restrictions brought

---

* Although organic chemists have used this approach intuitively for many years, E. J. Corey originated the term *retrosynthetic analysis* and was the first person to state its principles formally. His studies, dating from the 1960s, have made the designing of organic syntheses systematic enough to be aided by computers. You may want to examine the book: E. J. Corey and Xue-Min Chen, *The Logic of Chemical Synthesis,* Wiley, New York, 1989.

about by the reactions that we wish to use. If a step, for example, involves coupling of a lithium dialkylcuprate with an alkyl halide, we should not try to couple the lithium dialkylcuprate with a 2° or 3° alkyl halide, because the yields for this type of coupling are generally very low (Section 4.15C).

Let us consider an example: the synthesis of 2-methylbutane. One way to disclose retrosynthetic pathways is to look for **disconnections** in the target molecule. We might, for example, imagine the following disconnection, which discloses two possible syntheses using lithium dialkylcuprates. Only the second synthesis is likely to give a good yield, however.

Analysis

$$CH_3 \xrightarrow{} CH-CH_2-CH_3 \quad \Rightarrow \quad (CH_3)_2CuLi + X-\underset{\underset{CH_3}{|}}{C}HCH_2CH_3$$

$$CH_3-X + LiCu\left(\underset{\underset{CH_3}{|}}{C}HCH_2CH_3\right)_2$$

Syntheses

$$(CH_3)_2CuLi + X-\underset{\underset{CH_3}{|}}{C}HCH_2CH_3 \xrightarrow[\text{ether}]{\text{diethyl}} CH_3\underset{\underset{CH_3}{|}}{C}HCH_2CH_3 \qquad \textbf{Poor yield because the alkyl halide is 2°}$$

$$CH_3-X + LiCu\left(\underset{\underset{CH_3}{|}}{C}HCH_2CH_3\right)_2 \xrightarrow[\text{ether}]{\text{diethyl}} CH_3\underset{\underset{CH_3}{|}}{C}HCH_2CH_3$$

Other disconnections and the syntheses that they reveal are the following:

Analysis

$$CH_3-\underset{\underset{CH_3}{|}}{C}H-CH_2-CH_3 \quad \Rightarrow \quad \left(CH_3\underset{\underset{CH_3}{|}}{C}H\right)_2CuLi + CH_3CH_2-X$$

$$CH_3\underset{\underset{CH_3}{|}}{C}H-X + (CH_3CH_2)_2CuLi$$

Syntheses

$$\left(CH_3\underset{\underset{CH_3}{|}}{C}H\right)_2CuLi + CH_3CH_2-X \xrightarrow[\text{ether}]{\text{diethyl}} CH_3\underset{\underset{CH_3}{|}}{C}HCH_2CH_3$$

$$CH_3\underset{\underset{CH_3}{|}}{C}H-X + (CH_3CH_2)_2CuLi \xrightarrow[\text{ether}]{\text{diethyl}} CH_3\underset{\underset{CH_3}{|}}{C}HCH_2CH_3 \qquad \textbf{Poor yield because the alkyl halide is 2°}$$

Analysis

$$CH_3-\underset{\underset{CH_3}{|}}{C}H-CH_2-CH_3 \quad \Rightarrow \quad \left(CH_3\underset{\underset{CH_3}{|}}{C}HCH_2\right)_2CuLi + CH_3-X \xrightarrow[\text{ether}]{\text{diethyl}} CH_3\underset{\underset{CH_3}{|}}{C}HCH_2CH_3$$

$$CH_3\underset{\underset{CH_3}{|}}{C}HCH_2-X + (CH_3)_2CuLi \xrightarrow[\text{ether}]{\text{diethyl}} CH_3\underset{\underset{CH_3}{|}}{C}HCH_2CH_3$$

Syntheses

$$\left(\underset{\underset{CH_3CHCH_2}{|}}{\overset{\overset{CH_3}{|}}{}} \right)_2 CuLi + CH_3-X \xrightarrow[\text{ether}]{\text{diethyl}} \underset{\underset{CH_3CHCH_2CH_3}{|}}{\overset{\overset{CH_3}{|}}{}}$$

$$\underset{\underset{CH_3CHCH_2}{|}}{\overset{\overset{CH_3}{|}}{}} -X + (CH_3)_2CuLi \xrightarrow[\text{ether}]{\text{diethyl}} \underset{\underset{CH_3CHCH_2CH_3}{|}}{\overset{\overset{CH_3}{|}}{}}$$

PROBLEM 4.14

Give a retrosynthetic analysis and then outline a Corey–Posner, Whitesides–House synthesis of each of the following alkanes starting with appropriate alkyl halides.

(a) Propane   (d) 2,7-Dimethyloctane   (f) Isopropylcyclopentane

(b) Butane   (e) Ethylcyclohexane   (g) 3-Methylcyclopentene

(c) 2-Methylbutane

PROBLEM 4.15

Using any other needed reagents outline methods showing how butane could be synthesized starting with

(a) Bromoethane   (c) A bromobutane

(b) A bromopropane   (d) A butene

4.17 PHEROMONES: COMMUNICATION BY MEANS OF CHEMICALS

Many animals, *especially insects,* communicate with other members of their species using a language based not on sounds or even visual signals, but on the odors of chemicals called **pheromones** that these animals release. For insects, this appears to be the chief method of communication. Although pheromones are secreted by insects in extremely small amounts, they can cause profound and varied biological effects. Insects use some pheromones in courtship as sex attractants. Others use pheromones as warning substances, and still others secrete chemicals called ''aggregation compounds'' to cause members of their species to congregate. Often these pheromones are relatively simple compounds, and some are hydrocarbons. For example, a species of cockroach uses undecane as an aggregation pheromone.

<table>
<tr><td style="text-align:center">$CH_3(CH_2)_9CH_3$</td><td style="text-align:center">$(CH_3)_2CH(CH_2)_{14}CH_3$</td></tr>
<tr><td style="text-align:center">**Undecane**</td><td style="text-align:center">**2-Methylheptadecane**</td></tr>
<tr><td style="text-align:center">(cockroach aggregation pheromone)</td><td style="text-align:center">(sex attractant of female tiger moth)</td></tr>
</table>

When a female tiger moth wants to mate, she secretes 2-methylheptadecane, a perfume that the male tiger moth apparently finds irresistable.

The sex attractant of the common housefly (*Musca domestica*) is a 23-carbon alkene with a cis double bond between atoms 9 and 10, called muscalure.

$$CH_3(CH_2)_7 \quad (CH_2)_{12}CH_3$$
$$\underset{H}{\diagup} C = C \underset{H}{\diagdown}$$

Muscalure
(sex attractant of common housefly)

Many insect sex attractants have been synthesized and are used to lure insects into traps as a means of insect control, a much more environmentally sensitive method than the use of insecticides.

PROBLEM 4.16

The final step of a commercial synthesis of muscalure is based on a reaction of the following alkyl halide.

$$CH_3(CH_2)_7 \quad (CH_2)_7CH_2Br$$
$$\underset{H}{\diagup} C = C \underset{H}{\diagdown}$$

(a) What is the structure of the other reagent, and (b) how would you synthesize it?

4.18 SOME IMPORTANT TERMS AND CONCEPTS

Alkanes are hydrocarbons with the general formula C_nH_{2n+2}. Molecules of alkanes have no rings (i.e., they are **acyclic**) and they have only single bonds between carbon atoms. Their carbon atoms are sp^3 hybridized.

Cycloalkanes are hydrocarbons whose molecules have their carbon atoms arranged into one or more rings. They have only single bonds between carbon atoms, and their carbon atoms are sp^3 hybridized. Cycloalkanes with only one ring have the general formula C_nH_{2n}.

Conformational analysis is a study of the energy changes that occur in a molecule when groups rotate about single bonds.

Torsional strain refers to a small barrier to free rotation about the carbon–carbon single bond that is associated with the eclipsed conformation. For ethane this barrier is 2.8 kcal mol^{-1} (11.7 kJ mol^{-1}).

van der Waals forces are weak forces that act between nonpolar molecules or between parts of the same molecule. Bringing two groups together first results in an *attractive* van der Waals force between them because a temporary unsymmetrical distribution of electrons in one group induces an opposite polarity in the other. When the groups are brought closer than their *van der Waals radii,* the force between them becomes repulsive because their electron clouds begin to interpenetrate each other. The methyl groups of the *gauche* form of butane, for example, are close enough for the van der Waals forces to be repulsive.

Ring strain Certain cycloalkanes have greater potential energy than open-chain compounds. This extra potential energy is called ring strain. The principal sources of ring strain are *torsional strain* and *angle strain.*

Angle strain is introduced into a molecule because some factor (e.g., ring size) causes the bond angles of its atoms to deviate from the normal bond angle. The normal bond angles of an sp^3 carbon are 109.5°, but in cyclopropane, for example, one pair of bonds at each carbon atom is constrained to a much smaller angle. This introduces considerable angle strain into the molecule, causing molecules of cyclopropane to have greater potential energy per CH_2 group than cycloalkanes with less (or no) angle strain.

Cis–trans isomerism is a type of stereoisomerism that occurs with certain alkenes (Section 2.4B) and with disubstituted cycloalkanes. Cis–trans isomers of 1,2-dimethylcyclopropane are shown here.

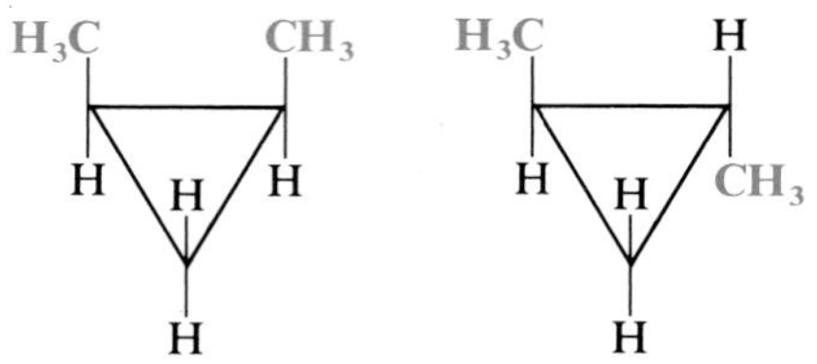

These two isomers can be separated and they have different physical properties. The two forms cannot be interconverted without breaking carbon–carbon bonds.

Conformations of molecules of cyclohexane The most stable conformation is a chair conformation. Twist conformations and boat conformations have greater potential energy. These conformations (chair, boat, and twist) can be interconverted by rotations of single bonds. In a sample of cyclohexane more than 99% of the molecules are in a chair conformation at any given moment. A group attached to a carbon atom of a molecule of cyclohexane in a chair conformation can assume either of two positions: *axial* or *equatorial,* and these are interconverted when the ring flips from one chair conformation to another. A group has more room when it is equatorial; thus most of the molecules of substituted cyclohexanes at any given moment will be in the chair conformation that has the largest group (or groups) equatorial.

ADDITIONAL PROBLEMS

4.17   Write a structural formula for each of the following compounds:

(a) 1,4-Dichloropentane

(b) *sec*-Butyl bromide

(c) 4-Isopropylheptane

(d) 2,2,3-Trimethylpentane

(e) 3-Ethyl-2-methylhexane

(f) 1,1-Dichlorocyclopentane

(g) *cis*-1,2-Dimethylcyclopropane

(h) *trans*-1,2-Dimethylcyclopropane

(i) 4-Methyl-2-pentanol

(j) *trans*-4-Isobutylcyclohexanol

(k) 1,4-Dicyclopropylhexane

(l) Neopentyl alcohol

(m) Bicyclo[2.2.2]octane

(n) Bicyclo[3.1.1]heptane

(o) Cyclopentylcyclopentane

4.18 Give systematic IUPAC names for each of the following:

(a) $CH_3CH_2C(CH_3)_2CH(CH_2CH_3)CH_3$

(b) $CH_3CH_2C(CH_3)_2CH_2OH$

(c) 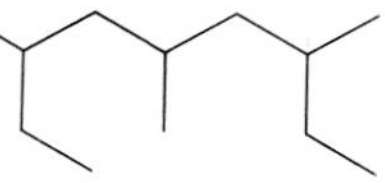

(f) 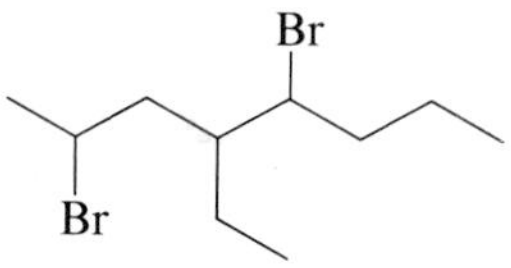

(d) 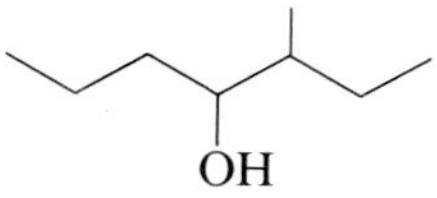

(g) 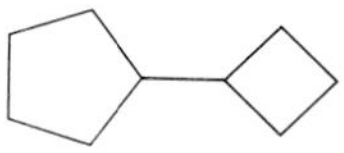

(e) 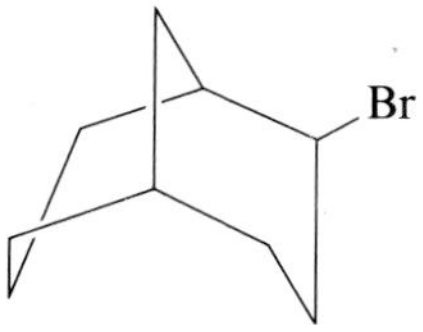

(h) 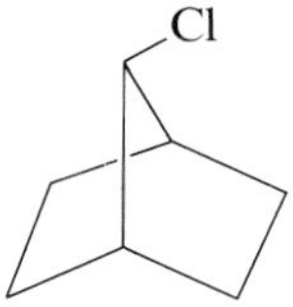

4.19 The name *sec*-butyl alcohol defines a specific structure but the name *sec*-pentyl alcohol is ambiguous. Explain.

4.20 Write the structure and give the IUPAC systematic name of an alkane or cycloalkane with the formula: (a) C_8H_{18} that has only primary hydrogen atoms, (b) C_6H_{12} that has only secondary hydrogen atoms, (c) C_6H_{12} that has only primary and secondary hydrogen atoms, and (d) C_8H_{14} that has 12 secondary and 2 tertiary hydrogen atoms.

4.21 Three different alkenes yield 2-methylbutane when they are hydrogenated in the presence of a metal catalyst. Give their structures and write equations for the reactions involved.

4.22 An alkane with the formula C_6H_{14} can be synthesized by treating (in separate reactions) five different alkyl chlorides ($C_6H_{13}Cl$) with zinc and aqueous acid. Give the structure of the alkane and the structures of the alkyl chlorides.

4.23 An alkane with the formula C_6H_{14} can be prepared by reduction (with Zn and H^+) of only two alkyl chlorides ($C_6H_{13}Cl$) and by the hydrogenation of only two alkenes (C_6H_{12}). Write the structure of this alkane, give its IUPAC name, and show the reactions.

4.24 Four different cycloalkenes will all yield methylcyclopentane when subjected to catalytic hydrogenation. What are their structures? Show the reactions.

4.25 The heats of combustion of three pentane (C_5H_{12}) isomers are $CH_3(CH_2)_3CH_3$, 845.2 kcal mol^{-1}; $CH_3CH(CH_3)CH_2CH_3$, 843.4 kcal mol^{-1}; and $(CH_3)_3CCH_3$, 840.0 kcal mol^{-1}. Which isomer is most stable? Construct a diagram such as that in Fig. 4.9 showing the relative potential energies of the three compounds.

4.26 Tell what is meant by a homologous series and illustrate your answer by writing a homologous series of alkyl halides.

4.27 Write the structures of two chair conformations of 1-*tert*-butyl-1-methylcyclohexane. Which conformation is more stable? Explain your answer.

4.28   Ignoring compounds with double bonds, write structural formulas and give names for all of the isomers with the formula C_5H_{10}.

4.29   Write structures for the following bicyclic alkanes.

(a) Bicyclo[1.1.0]butane   (c) 2-Chlorobicyclo[3.2.0]heptane

(b) Bicyclo[2.1.0]pentane   (d) 7-Methylbicyclo[2.2.1]heptane

4.30   Rank the following compounds in order   (a) of increasing heat of combustion, and (b) in order of increasing stability.

4.31   Sketch curves similar to the one given in Fig. 4.8 showing the energy changes that arise from rotation about the C2—C3 bond of (a) 2,3-dimethylbutane and (b) of 2,2,3,3-tetra-methylbutane. You need not concern yourself with actual numerical values of the energy changes, but you should label all maxima and minima with the appropriate conformations.

4.32   Without referring to tables, decide which member of each of the following pairs would have the higher boiling point. Explain your answers.

(a) Pentane or 2-methylbutane   (d) Butane or 1-propanol

(b) Heptane or pentane   (e) Butane or CH_3COCH_3

(c) Propane or 2-chloropropane

4.33   Which compound would you expect to be the more stable: *cis*-1,2-dimethylcyclopro-pane or *trans*-1,2-dimethylcyclopropane? Explain your answer.

4.34   Which member of each of the following pairs would you expect to have the larger heat of combustion? (a) *cis*- or *trans*-1,2-dimethylcyclohexane, (b) *cis*- or *trans*-1,3-dimethylcyclo-hexane, (c) *cis*- or *trans*-1,4-dimethylcyclohexane. Explain your answers.

4.35   Write the two chair conformations of each of the following and in each part designate which conformation would be the more stable. (a) *cis*-1-*tert*-butyl-3-methylcyclohexane, (b) *trans*-1-*tert*-butyl-3-methylcyclohexane, (c) *trans*-1-*tert*-butyl-4-methylcyclohexane, (d) *cis*-1-*tert*-butyl-4-methylcyclohexane.

4.36   *trans*-1,3-Dibromocyclobutane has a measurable dipole moment. Explain how this proves that the cyclobutane ring is not planar.

4.37   Write structural formulas for the missing reagent(s) in each of the following syntheses:

(a) $A + CH_3CH_2CH_2Br \longrightarrow$

(b) $B + (CH_3)_2CuLi \longrightarrow$

(c) $C + (CH_3CH_2)_2CuLi \longrightarrow$

(d) $(CH_3CH_2)_2CuLi + D \longrightarrow$

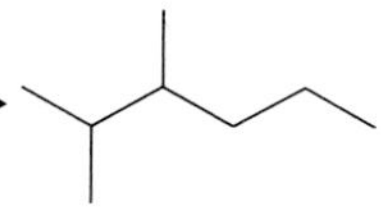

(e) $E \xrightarrow[H^+]{Zn}$ Neopentane

(f)

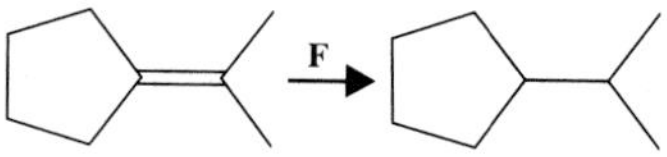

4.38   When 1,2-dimethylcyclohexene (below) is allowed to react with hydrogen in the presence of a platinum catalyst, the product of the reaction is a cycloalkane that has a melting point of $-50°C$ and a boiling point of $130°C$ (at 760 torr). (a) What is the structure of the product of this reaction? (b) Consult an appropriate table and tell which stereoisomer it is. (c) What does this experiment suggest about the mode of addition of hydrogen to the double bond?

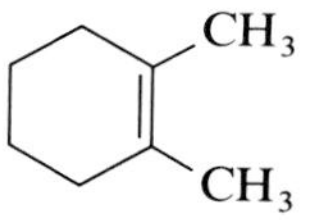

1,2-Dimethylcyclohexene

4.39   When cyclohexene is dissolved in an appropriate solvent and allowed to react with chlorine, the product of the reaction, $C_6H_{10}Cl_2$, has a melting point of $-7°C$ and a boiling point (at 16 torr) of $74°C$. (a) Which stereoisomer is this? (b) What does this experiment suggest about the mode of addition of chlorine to the double bond?

4.40   Consider the cis and trans isomers of 1,3-di-*tert*-butylcyclohexane. What unusual feature accounts for the fact that one of these isomers apparently exists in a twist boat conformation rather than a chair conformation?

CHAPTER 5

(+)-Carvone and (−)-Carvone
(Problem 5.14)

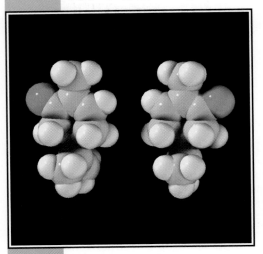

STEREOCHEMISTRY: CHIRAL MOLECULES

5.1 ISOMERISM: CONSTITUTIONAL ISOMERS AND STEREOISOMERS

Isomers are different compounds that have the same molecular formula. In our study of carbon compounds, thus far, most of our attention has been directed toward those isomers that we have called constitutional isomers.

Constitutional isomers are isomers that differ because their atoms are connected in a different order. They are said to have a different **connectivity.** Several examples of constitutional isomers are the following:

Molecular Formula	Constitutional Isomers	
C_4H_{10}	$CH_3CH_2CH_2CH_3$ and **Butane**	$\overset{\displaystyle CH_3}{\underset{}{\overset{\mid}{CH_3CHCH_3}}}$ **Isobutane**
C_3H_7Cl	$CH_3CH_2CH_2Cl$ and **1-Chloropropane**	CH_3CHCH_3 $\mid$ Cl **2-Chloropropane**
C_2H_6O	CH_3CH_2OH and **Ethanol**	CH_3OCH_3 **Dimethyl ether**

Stereoisomers are not constitutional isomers—they have their constituent atoms connected in the same sequence. *Stereoisomers differ only in arrangement of their atoms in space.* The cis and trans isomers of alkenes are stereoisomers (Section 2.4B); we can see that this is true if we examine the *cis*- and *trans*-1,2-dichloroethenes shown here.

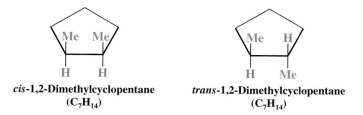

αCH=CHCl

cis-**1,2-Dichloroethene**
($C_2H_2Cl_2$)

Cl CH=CHCl

trans-**1,2-Dichloroethene**
($C_2H_2Cl_2$)

cis-1,2-Dichloroethene and *trans*-1,2-dichloroethene are isomers because both compounds have the same molecular formula ($C_2H_2Cl_2$) but they are different. They cannot be easily interconverted because of the large barrier to rotation of the carbon–carbon double bond. Stereoisomers are *not* constitutional isomers, because the order of connections of the atoms in both compounds is the same. Both compounds have two central carbon atoms joined by a double bond, and both compounds have one chlorine atom and one hydrogen atom attached to the two central atoms. The *cis*-1,2-dichloro-ethene and *trans*-1,2-dichloroethene isomers differ only in the arrangement of their atoms in space. In *cis*-1,2-dichloroethene the hydrogen atoms are on the same side of the molecule, and in *trans*-1,2-dichloroethene the hydrogen atoms are on opposite sides. Thus, *cis*-1,2-dichloroethene and *trans*-1,2-dichloroethene are stereoisomers (see Section 2.4B).

Stereoisomers can be subdivided into two general categories: **enantiomers** and **diastereomers.** Enantiomers are stereoisomers whose molecules *are nonsuperposable mirror images of each other.* Diastereomers are stereoisomers whose molecules *are not mirror images of each other.*

Molecules of *cis*-1,2-dichloroethene and *trans*-1,2-dichloroethene *are not* mirror images of each other. If one holds a model of *cis*-1,2-dichloroethene up to a mirror, the model that one sees in the mirror is not *trans*-1,2-dichloroethene. But *cis*-1,2-dichloro-ethene and *trans*-1,2-dichloroethene *are* stereoisomers and, since they are not related to each other as an object and its mirror image, they are diastereomers.

Cis and trans isomers of cycloalkanes furnish us with another example of stereo-isomers that are diastereomers of each other. Consider the following two compounds.

cis-**1,2-Dimethylcyclopentane**	*trans*-**1,2-Dimethylcyclopentane**
(C_7H_{14})	**(C_7H_{14})**

These two compounds are isomers of each other *because they are different compounds* that are *not* interconvertible, and *because they have the same molecular formula* (C_7H_{14}). They are not constitutional isomers because their atoms are joined in the same sequence. They are therefore *stereoisomers. They differ only in the arrange-ment of their atoms in space.* They are not enantiomers because their molecules are not mirror images of each other. They are therefore *diastereomers*. (In Section 5.12 we shall find that the *trans*-1,2-dimethylcyclopentane also has an enantiomer.)

Cis–trans isomers are not the only kind of diastereomers that we shall encounter. In Section 5.10 we shall study diastereomers that are not cis–trans isomers. The essential requirements that must be fulfilled for two compounds to be diastereomers of each other are that the two compounds be stereoisomers of each other, and that they not be mirror images of each other.

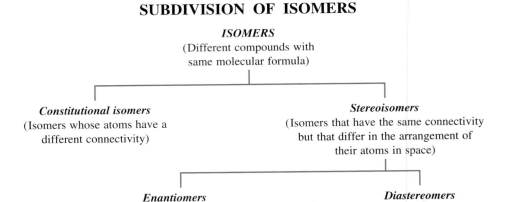

SUBDIVISION OF ISOMERS

ISOMERS
(Different compounds with same molecular formula)

Constitutional isomers
(Isomers whose atoms have a different connectivity)

Stereoisomers
(Isomers that have the same connectivity but that differ in the arrangement of their atoms in space)

Enantiomers
(Stereoisomers that are nonsuperposable mirror images of each other)

Diastereomers
(Stereoisomers that are not mirror images of each other)

5.2 ENANTIOMERS AND CHIRAL MOLECULES

Enantiomers occur only with those compounds whose molecules are ***chiral. A chiral molecule is defined as one that is not identical with its mirror image.*** The chiral molecule and its mirror image are enantiomers, and the relationship between the chiral molecule and its mirror image is defined as enantiomeric.

The word chiral comes from the Greek word *cheir,* meaning ''hand.'' Chiral objects (including molecules) are said to possess ''handedness.'' The term chiral is used to describe molecules of enantiomers because they are related to each other in the same way that a left hand is related to a right hand. When you view your left hand in a mirror, the mirror image of your left hand is a right hand (Fig. 5.1). Your left and right hands, moreover, are not identical and this can be shown by observing that they are not superposable* (Fig. 5.2).

Many familiar objects are chiral and the chirality of some of these objects is clear because we normally speak of them as having ''handedness.'' We speak, for example, of nuts and bolts as having right- or left-handed threads or of a propeller as having a right- or left-handed pitch. The chirality of many other objects is not obvious in this sense, but becomes obvious when we apply the test of nonsuperposability of the object and its mirror image.

Objects (and molecules) that *are* superposable on their mirror images are **achiral.** Most socks, for example, are achiral, whereas gloves are chiral.

Remember: To be *superposable* means that we can place one thing on top of the other *so that all parts of each coincide* (cf. Section 2.4B).

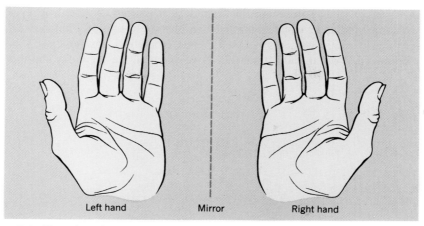

FIGURE 5.1 The mirror image of a left hand is a right hand.

PROBLEM 5.1

Classify each of the following objects as to whether it is chiral or achiral:

(a) A screwdriver   (d) A tennis shoe   (g) A car

(b) A baseball bat   (e) An ear   (h) A hammer

(c) A golf club   (f) A woodscrew   (i) A shoe

The chirality of molecules can be demonstrated with relatively simple compounds. Consider, for example, 2-butanol.

$$CH_3CHCH_2CH_3$$
$$|$$
$$OH$$

2-Butanol

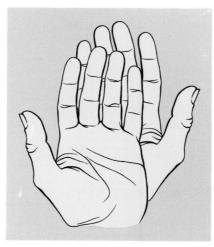

FIGURE 5.2 Left and right hands are not superposable.

Until now, we have presented the formula just written as though it represented only one compound and we have not mentioned that molecules of 2-butanol are chiral. Because they are, there are actually two different 2-butanols and these two 2-butanols are enantiomers. We can understand this if we examine the drawings and models in Fig. 5.3.

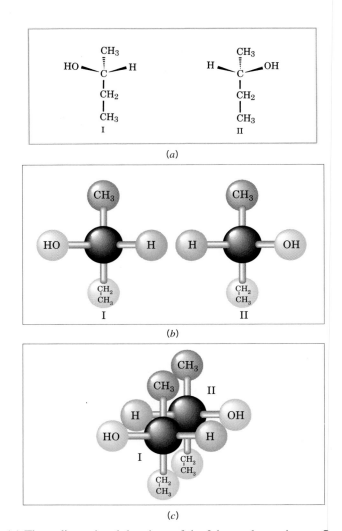

FIGURE 5.3 (*a*) Three-dimensional drawings of the 2-butanol enantiomers **I** and **II**. (*b*) Models of the 2-butanol enantiomers. (*c*) An unsuccessful attempt to superpose models of **I** and **II**.

If model **I** is held before a mirror, model **II** is seen in the mirror and vice versa. Models **I** and **II** are not superposable on each other; therefore they represent different, but isomeric, molecules. *Because models I and II are nonsuperposable mirror images of each other, the molecules that they represent are enantiomers.*

PROBLEM 5.2

If models are available, construct the 2-butanols represented in Fig. 5.3 and demonstrate for yourself that they are not mutually superposable. (a) Make similar models of 2-bromopropane. Are they superposable? (b) Is a molecule of 2-bromopropane chiral? (c) Would you expect to find enantiomeric forms of 2-bromopropane?

How do we know when to expect the possibility of enantiomers? One way (but not the only way) is to recognize that a pair of enantiomers is always possible for molecules that contain **one tetrahedral atom with four different groups attached to it.*** In 2-butanol (Fig. 5.4) this atom is C2. The four different groups that are attached to C2 are a hydroxyl group, a hydrogen atom, a methyl group, and an ethyl group.

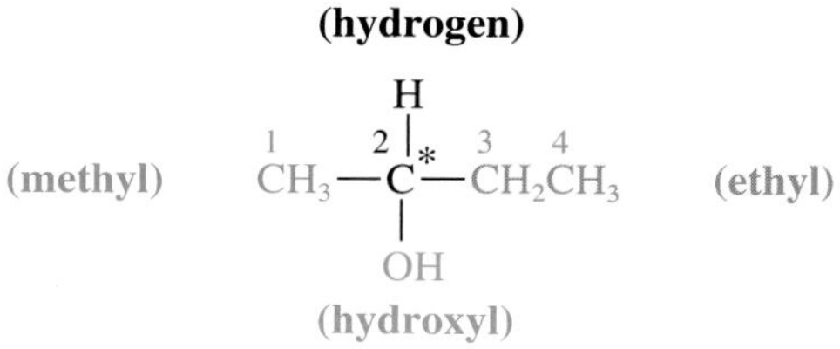

(hydrogen)

$$\text{(methyl)} \quad \overset{1}{\text{CH}_3}\!-\!\overset{2}{\underset{|}{\overset{|}{\text{C}}}}\!\!\overset{\text{H}}{\underset{*}{}}\!-\!\overset{3}{\text{CH}_2}\overset{4}{\text{CH}_3} \quad \text{(ethyl)}$$
$$\text{OH}$$
(hydroxyl)

FIGURE 5.4 The tetrahedral carbon atom of 2-butanol that bears four different groups. [By convention such atoms are often designated with an asterisk (*).]

An important property of enantiomers such as these is that *interchanging any two groups at the tetrahedral atom that bears four different groups converts one enantiomer into the other.* In Fig. 5.3b it is easy to see that interchanging the hydroxyl group and the hydrogen atom converts one enantiomer into the other. You should now convince yourself with models that interchanging any other two groups has the same result.

Because interchanging two groups at C2 converts one stereoisomer into another, C2 is an example of what is called a **stereocenter.** A **stereocenter** is defined as **an atom bearing groups of such nature that an interchange of any two groups will produce a stereoisomer.** Carbon-2 of 2-butanol is an example of a **tetrahedral stereocenter.** Not all stereocenters are tetrahedral, however. The carbon atoms of *cis-* and *trans-*1,2-dichloroethene (Section 5.2) are examples of *trigonal planar stereocenters* because an interchange of groups at either atom also produces a stereoisomer (a diastereomer). In general, when referring to organic compounds, the term stereocenter implies a tetrahedral stereocenter unless otherwise specified.

When we discuss interchanging groups like this, we must take care to notice that what we are describing is *something we do to a molecular model* or *something we do on paper.* An interchange of groups in a real molecule, if it can be done, requires

---

* We shall see later that enantiomers are also possible for molecules that contain more than one tetrahedral atom with four different groups attached to it, but some of these molecules (Section 5.11A) do not exist as enantiomers.

breaking covalent bonds, and this is something that requires a large input of energy. This means that enantiomers such as the 2-butanol enantiomers *do not interconvert* spontaneously.

At one time, *tetrahedral atoms* with four different groups were called *chiral atoms* or *asymmetric atoms*. Then, in 1984, K. Mislow (of Princeton University) and J. Siegel (now at the University of California, San Diego) pointed out that the use of such terms as this had represented a source of conceptual confusion in stereochemistry that had existed from the time of van't Hoff (Section 5.3A). Chirality is a geometric property that pervades and affects all parts of a chiral molecule. All of the atoms of 2-butanol, for example, are in a chiral environment and, therefore, all are said to be *chirotopic*. When we consider an atom such as C2 of 2-butanol in the way that we describe here, however, we are considering it as a *stereocenter* and, therefore, we should designate it as such, and not as a "chiral atom." Further consideration of these issues is beyond our scope here, but those interested may wish to read the original paper; cf. K. Mislow and J. Siegel, *J. Am. Chem. Soc.,* **1984,** *106,* 3319–3328.

Figure 5.5 demonstrates the validity of the generalization that enantiomeric com-

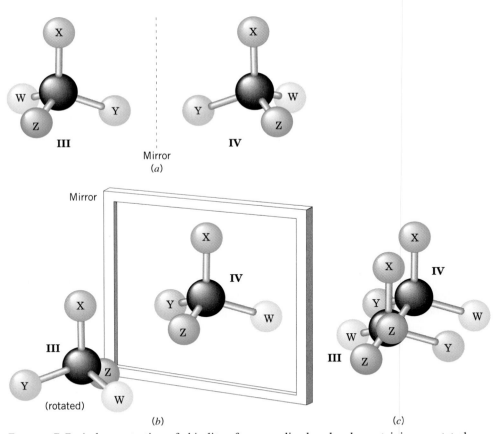

FIGURE 5.5 A demonstration of chirality of a generalized molecule containing one tetrahedral stereocenter. (*a*) The four different groups around the carbon atom in **III** and **IV** are arbitrary. (*b*) **III** is rotated and placed in front of a mirror. **III** and **IV** are found to be related as an object and its mirror image. (*c*) **III** and **IV** are not superposable; therefore, the molecules that they represent are chiral and are enantiomers.

pounds necessarily result whenever a molecule contains a single tetrahedral stereo-center.

PROBLEM 5.3

Demonstrate the validity of what we have represented in Fig. 5.3 by constructing models. Demonstrate for yourself that **III** and **IV** are related as an object and its mirror image *and that they are not superposable* (i.e., that **III** and **IV** are chiral molecules and are enantiomers). Now take **IV** and exchange the positions of any two groups. (a) What is the new relationship between the molecules? (b) Now take either model and exchange the positions of any two groups. What is the relationship between the molecules now?

If all of the tetrahedral atoms in a molecule have two or more groups attached that *are the same* the molecule does not have a stereocenter. The molecule is superposable on its mirror image and is **achiral.** An example of a molecule of this type is 2-propanol; carbon atoms 1 and 3 bear three identical hydrogen atoms and the central atom bears two identical methyl groups. If we write three-dimensional formulas for 2-propanol, we find (Fig. 5.6) that one structure can be superposed on its mirror image.

Thus, we would not predict the existence of enantiomeric forms of 2-propanol, and experimentally only one form of 2-propanol has ever been found.

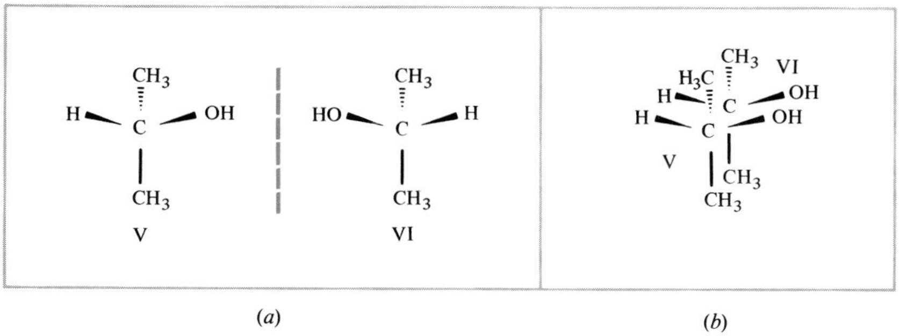

FIGURE 5.6 (*a*) 2-Propanol (**V**) and its mirror image (**VI**). (*b*) When either one is rotated, the two structures are superposable and so do not represent enantiomers. They represent two molecules of the same compound. 2-Propanol does not have a stereocenter.

PROBLEM 5.4

Some of the molecules listed here have stereocenters; some do not. Write three-dimensional formulas for the enantiomers of those molecules that do have stereocenters.

(a) 2-Propanol   (e) 2-Bromopentane

(b) 2-Methylbutane   (f) 3-Methylpentane

(c) 2-Chlorobutane   (g) 3-Methylhexane

(d) 2-Methyl-1-butanol   (h) 1-Chloro-2-methylbutane

5.3 THE BIOLOGICAL IMPORTANCE OF CHIRALITY

Chirality is a phenomenon that pervades the universe. The human body is structurally chiral, with the heart lying to the left of center, and the liver to the right. For evolutionary reasons, far from understood, most people are right handed. Helical seashells are chiral, and most spiral like a right-handed screw. Many plants show chirality in the way they wind around supporting structures. The honeysuckle, *Lonicera sempervirens,* winds as a left-handed helix; bindweed, *Convolvulus arvensi,* winds in a right-handed way. Most of the molecules that make up plants and animals are chiral, and usually only one form of the chiral molecule occurs in a given species. All but one of the 20 amino acids that make up naturally occurring proteins are chiral, and all of them are classified as being left handed. The molecules of natural sugars are almost all classified as being right handed, including the sugar that occurs in DNA.* DNA, itself, has a helical structure and all naturally occurring DNA turns to the right.

Chiral molecules can show their different handedness in many ways, including the way they affect human beings. One enantiomeric form of a compound called limonene (Section 23.3) is primarily responsible for the odor of oranges, the other enantiomer, for the odor of lemons.

Limonene

One enantiomer of a compound called carvone (Problem 5.14) is the essence of caraway, the other, the essence of spearmint.

Differences in chirality may have much more dramatic, indeed tragic, effects on humans. For several years before 1963 the drug thalidomide was used to alleviate the symptoms of morning sickness in pregnant women. In 1963 it was discovered that thalidomide was the cause of horrible birth defects in many children born subsequent to the use of the drug.

Thalidomide

Even later, evidence began to appear indicating that whereas one of the thalidomide enantiomers (the right-handed molecule) has the intended effect of curing morning sickness, the other enantiomer, which was also present in the drug (in equal amounts), may be the cause of the birth defects. The evidence regarding the effects of the two enantiomers is complicated by the fact that under physiological conditions, the two enantiomers are interconverted.

PROBLEM 5.5

Which atom is the stereocenter   (a) of limonene and   (b) of thalidomide?

---

*For an interesting article, see R. A. Hegstrum and D. K. Kondepudi, ''The Handedness of the Universe,'' *Sci. Am.,* **1990,** *262* (January), 98–105.

5.4 HISTORICAL ORIGIN OF STEREOCHEMISTRY

In 1877, Hermann Kolbe (of the University of Leipzig), one of the most eminent organic chemists of the time, wrote the following:

> Not long ago, I expressed the view that the lack of general education and of thorough training in chemistry was one of the causes of the deterioration of chemical research in Germany. . . . Will anyone to whom my worries seem exaggerated please read, if he can, a recent memoir by a Herr van't Hoff on 'The Arrangements of Atoms in Space,' a document crammed to the hilt with the outpourings of a childish fantasy. . . . This Dr. J. H. van't Hoff, employed by the Veterinary College at Utrecht, has, so it seems, no taste for accurate chemical research. He finds it more convenient to mount his Pegasus (evidently taken from the stables of the Veterinary College) and to announce how, on his bold flight to Mount Parnassus, he saw the atoms arranged in space.

Kolbe, nearing the end of his career, was reacting to a publication of a 22-year-old Dutch scientist. This publication had appeared two years earlier in September 1874, and in it, van't Hoff had argued that the spatial arrangement of four groups around a central carbon atom is tetrahedral. A young French scientist, J. A. Le Bel, had independently advanced the same idea in a publication in November 1874. Within 10 years after Kolbe's comments, however, abundant evidence had accumulated that substantiated the "childish fantasy" of van't Hoff. Later in his career (in 1901), and for other work, van't Hoff was named the first recipient of the Nobel Prize for chemistry.

Together, the publications of van't Hoff and Le Bel marked an important turn in a field of study that is concerned with the structures of molecules in three dimensions: *stereochemistry*. Stereochemistry, as we shall see in Section 5.15, had been founded earlier by Louis Pasteur.

It was reasoning based on many observations such as those we presented earlier in this chapter that led van't Hoff and Le Bel to the conclusion that the spatial orientation of groups around carbon atoms is tetrahedral when a carbon atom is bonded to four other atoms. The following information was available to van't Hoff and Le Bel.

1. Only one compound with the general formula CH_3X is ever found.
2. Only one compound with the formula CH_2X_2 or CH_2XY is ever found.
3. Two enantiomeric compounds with the formula CHXYZ are found.

By working Problem 5.6 you can see more about the reasoning of van't Hoff and Le Bel.

PROBLEM 5.6 _____

Show how a square planar structure for carbon compounds can be eliminated from consideration by considering CH_2Cl_2 and CH_2BrCl as examples of disubstituted methanes. (a) How many isomers would be possible in each instance if the carbon had a square planar structure? (b) How many isomers are possible in each instance if the carbon is tetrahedral? Consider CHBrClF as an example of a trisubstituted methane. (c) How many isomers would be possible if the carbon atom were square planar? (d) How many isomers are possible for CHBrClF if carbon is tetrahedral?

5.5   TESTS FOR CHIRALITY: PLANES OF SYMMETRY

The ultimate way to test for molecular chirality is to construct models of the molecule and its mirror image and then determine whether they are superposable. If the two models are superposable, the molecule that they represent is achiral. If the models are not superposable, then the molecules that they represent are chiral. We can apply this test with actual models, as we have just described, or we can apply it by drawing three-dimensional structures and attempting to superpose them in our minds.

There are other aids, however, that will assist us in recognizing chiral molecules. We have mentioned one already: **the presence of a *single* tetrahedral stereocenter.** The other aids are based on the absence in the molecule of certain symmetry elements. A molecule **will not be chiral,** for example, if it possesses **a plane of symmetry.**

A plane of symmetry (also called a **mirror plane**) is defined as *an imaginary plane that bisects a molecule in such a way that the two halves of the molecule are mirror images of each other.* The plane may pass through atoms, between atoms, or both. For example, 2-chloropropane has a plane of symmetry (Fig. 5.7a), whereas 2-chlorobutane does not (Fig. 5.7b). **All molecules with a plane of symmetry are achiral.**

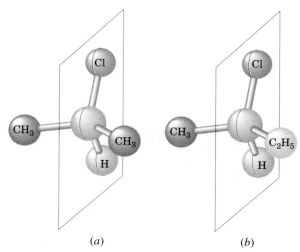

(a)   (b)

FIGURE 5.7 (a) 2-Chloropropane has a plane of symmetry and is achiral. (b) 2-Chlorobutane does not possess a plane of symmetry and is chiral.

PROBLEM 5.7

Which of the objects listed in Problem 5.1 possess a plane of symmetry and are, therefore, achiral?

PROBLEM 5.8

Write three-dimensional formulas and designate a plane of symmetry for all of the achiral molecules in Problem 5.4. (In order to be able to designate a plane of symmetry you may have to write the molecule in an appropriate conformation. This is permissible with all of these molecules because they have only single bonds and groups joined by single bonds are capable of essentially free rotation at room temperature. We discuss this matter further in Section 5.11.)

5.6 NOMENCLATURE OF ENANTIOMERS: THE *(R–S)* SYSTEM

The two enantiomers of 2-butanol are the following:

$$
\begin{array}{cc}
\text{CH}_3 & \text{CH}_3 \\
\text{HO} \diagdown \underset{|}{\text{C}} \diagup \text{H} \qquad\qquad & \text{H} \diagdown \underset{|}{\text{C}} \diagup \text{OH} \\
\text{CH}_2 & \text{CH}_2 \\
| & | \\
\text{CH}_3 & \text{CH}_3 \\
\mathbf{I} & \mathbf{II}
\end{array}
$$

If we name these two enantiomers using only the IUPAC system of nomenclature that we have learned so far, both enantiomers will have the same name: 2-butanol (or *sec*-butyl alcohol) (Section 4.3F). This is undesirable because *each compound must have its own distinct name.* Moreover, the name that is given a compound should allow a chemist who is familiar with the rules of nomenclature to write the structure of the compound from its name alone. Given the name 2-butanol, a chemist could write either structure **I** or structure **II**.

Three chemists, R. S. Cahn (England), C. K. Ingold (England), and V. Prelog (Switzerland), devised a system of nomenclature that, when added to the IUPAC system, solves both of these problems. This system, called the *(R–S)* system, or the Cahn–Ingold–Prelog system, is now widely used and is part of the IUPAC rules.

According to this system, one enantiomer of 2-butanol should be designated *(R)*-2-butanol and the other enantiomer should be designated *(S)*-2-butanol. [*(R)* and *(S)* are from the Latin words *rectus* and *sinister,* meaning right and left, respectively.]

The *(R)* and *(S)* designations are assigned on the basis of the following procedure. You should use models as you follow the steps.

1. Each of the four groups attached to the stereocenter is assigned a **priority** or **preference** *a*, *b*, *c*, or *d*. Priority is first assigned on the basis of the **atomic number** of the atom that is directly attached to the stereocenter. The group with the lowest atomic number is given the lowest priority, *d*; the group with next higher atomic number is given the next higher priority, *c*; and so on. (In the case of isotopes, the isotope of greatest atomic mass has highest priority.)

We can illustrate the application of this rule with the 2-butanol enantiomer, **I**.

$$
\begin{array}{c}
\text{CH}_3 \;\; (b \text{ or } c) \\
(a) \;\; \text{HO} \diagdown \underset{\text{C}}{\Vert} \diagup \text{H} \;\; (d) \\
| \\
\text{CH}_2 \;\; (b \text{ or } c) \\
| \\
\text{CH}_3 \\
\mathbf{I}
\end{array}
$$

Oxygen has the highest atomic number of the four atoms attached to the stereocenter and is assigned the highest priority, *a*. Hydrogen has the lowest atomic number and is assigned the lowest priority, *d*. A priority cannot be assigned for the methyl group and the ethyl group by this approach because the atom that is directly attached to the stereocenter is a carbon atom in both groups.

2. When a priority cannot be assigned on the basis of the atomic number of the atoms that are directly attached to the stereocenter, then the next set of atoms in the unassigned groups are examined. This process is continued until a decision can be made. *We assign a priority at the first point of difference.**

When we examine the methyl group of enantiomer **I**, we find that the next set of atoms consists of three hydrogen atoms (**H, H, H**). In the ethyl group of **I** the next set of atoms consists of one carbon atom and two hydrogen atoms (**C, H, H**). Carbon has a higher atomic number than hydrogen so we assign the ethyl group the higher priority, *b*, and the methyl group the lower priority, *c*.

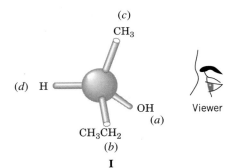

3. We now rotate the formula (or model) so that the group with lowest priority (*d*) is directed away from us.

I

Then we trace a path from *a* to *b* to *c*. If, as we do this, the direction of our finger (or pencil) is *clockwise,* the enantiomer is designated (*R*). If the direction is *counter-clockwise,* the enantiomer is designated (*S*). On this basis the 2-butanol enantiomer **I** is (*R*)-2-butanol.

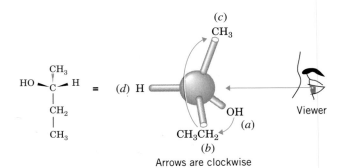

Arrows are clockwise

* The rules for a branched chain require that we follow the chain with the highest priority atoms.

PROBLEM 5.9 _____

Write the enantiomeric forms of bromochlorofluoromethane and assign each enantiomer its correct *(R)* or *(S)* designation.

PROBLEM 5.10 _____

Give *(R)* and *(S)* designations for each pair of enantiomers given as answers to Problem 5.4.

_____

The first three rules of the Cahn–Ingold–Prelog system allow us to make an *(R)* or *(S)* designation for most compounds containing single bonds. For compounds containing multiple bonds one other rule is necessary.

4. Groups containing double or triple bonds are assigned priorities as if both atoms were duplicated or triplicated, that is,

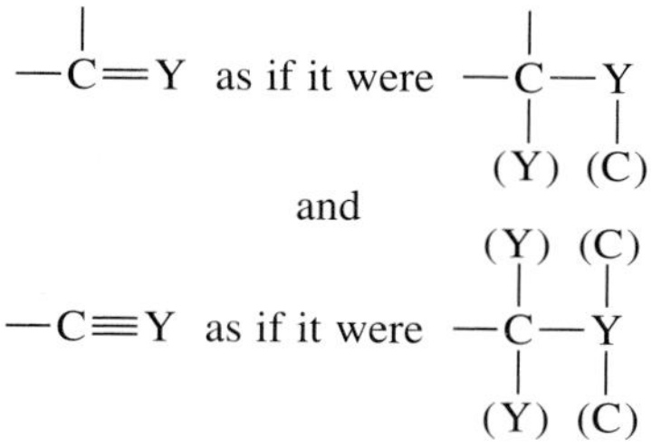

$$-\overset{|}{C}=Y \text{ as if it were } -\overset{|}{\underset{(Y)\ (C)}{C}}-Y$$

and

$$-\overset{|}{C}\equiv Y \text{ as if it were } -\overset{(Y)\ (C)}{\underset{(Y)\ (C)}{C}}-Y$$

where the symbols in parentheses are duplicate or triplicate representations of the atoms at the other end of the multiple bond.

Thus, the vinyl group, $-CH=CH_2$, is of higher priority than the isopropyl group, $-CH(CH_3)_2$.

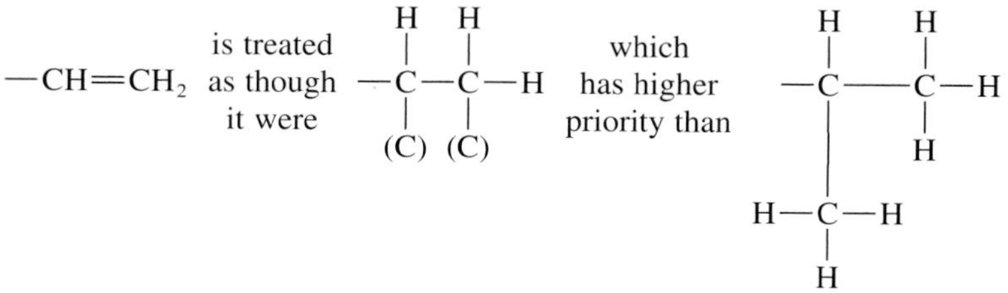

$-CH=CH_2$ is treated as though it were $-\underset{(C)\ (C)}{\overset{H\ \ H}{C-C}}-H$ which has higher priority than $-\overset{H}{C}\overset{H}{-C}-H$ with $H-C-H$ branch

because at the third set of atoms out, the vinyl group (see the following structure) is **C**, **H**, **H**, whereas the isopropyl group along either branch is **H**, **H**, **H**. (At the first and second set of atoms both groups are the same: **C**, then **C**, **C**, **H**.)

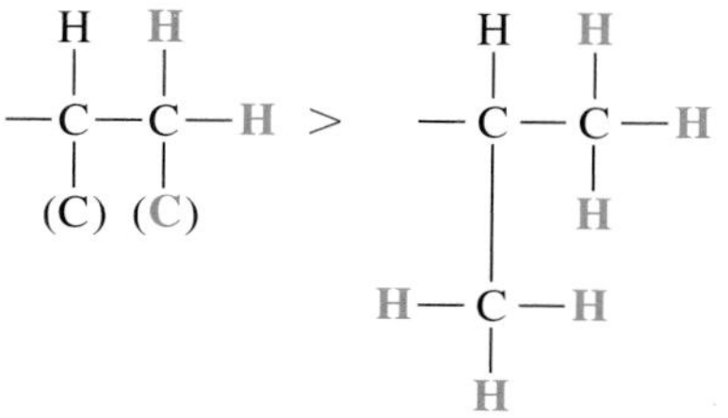

$$-\overset{H\ \ H}{\underset{(C)\ (C)}{C-C}}-H \quad > \quad -\overset{H\ \ H}{C-C}-H \text{ with } H-C-H$$

C, H, H   >   H, H, H
Vinyl group   Isopropyl group

Other rules exist for more complicated structures, but we shall not study them here.

PROBLEM 5.11

List the substituents in each of the following sets in order of priority, from highest to lowest:

(a) $-Cl, -OH, -SH, -H$

(b) $-CH_3, -CH_2Br, -CH_2Cl, -CH_2OH$

(c) $-H, -OH, -CHO, -CH_3$

(d) $-CH(CH_3)_2, -C(CH_3)_3, -H, -CH=CH_2$

(e) $-H, -N(CH_3)_2, -OCH_3, -CH_3$

PROBLEM 5.12

Assign (R) or (S) designations to each of the following compounds:

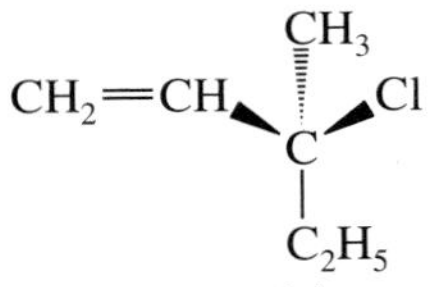

(a) 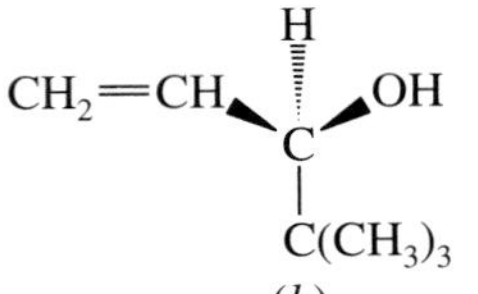 $\qquad$ (b) 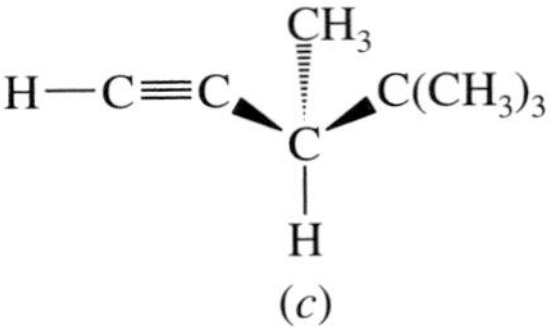 $\qquad$ (c)

SAMPLE PROBLEM

Consider the following pair of structures and tell whether they represent enantiomers or two molecules of the same compound in different orientations.

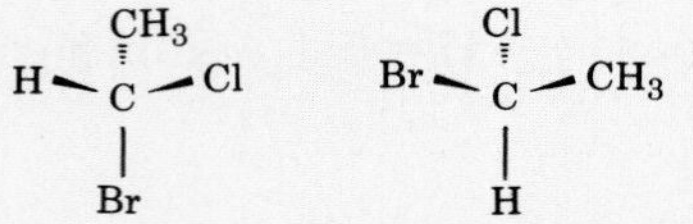

A $\qquad$ **B**

ANSWER:

One way to approach this kind of problem is to take one structure and in your mind, hold it by one group. Then rotate the other groups until at least one group is in the same place as it is in the other structure. (Until you can do this easily in your mind, practice with models.) By a series of rotations like this you will be able to convert the structure you are manipulating into one that is either identical with, or the mirror image of the other. For example, take **B**, hold it by the Cl atom and then rotate the other groups about the C*—Cl bond until the bromine is at the bottom (as it is in **A**). Then hold it by the Br and rotate the other groups about the C*—Br bond. This will make **B** identical with **A**.

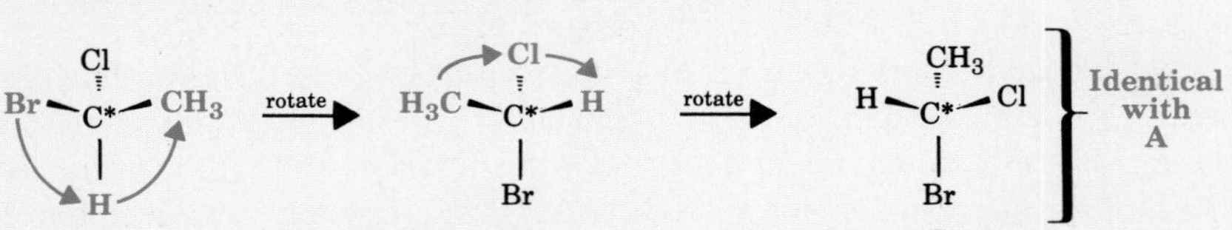

Another approach is to recognize that exchanging two groups at the stereocenter *inverts the configuration of* that carbon atom and converts a structure *with only one stereocenter* into its enantiomer; a second exchange recreates the original molecule. So we proceed this way, keeping track of how many exchanges are required to convert **B** into **A**. In this instance we find that two exchanges are required, and, again, we conclude that **A** and **B** are the same.

$$
Br\overset{Cl}{\underset{H}{\overset{|}{\underset{|}{C}}}}CH_3 \quad \xrightarrow[\text{Cl and CH}_3]{\text{exchange}} \quad Br\overset{CH_3}{\underset{H}{\overset{|}{\underset{|}{C}}}}Cl \quad \xrightarrow[\text{Br and H}]{\text{exchange}} \quad H\overset{CH_3}{\underset{Br}{\overset{|}{\underset{|}{C}}}}Cl
$$

$$
\quad\quad\quad\quad\quad\quad\quad\quad\quad\quad\quad\quad \mathbf{B} \quad\quad\quad\quad\quad\quad\quad\quad\quad\quad\quad\quad\quad\quad\quad\quad\quad\quad\quad\quad \mathbf{A}
$$

A useful check is to name each compound including its $(R-S)$ designation. If the names are the same, then the structures are the same. In this instance both structures are (R)-1-bromo-1-chloroethane.

PROBLEM 5.13

Tell whether the two structures in each pair represent enantiomers or two molecules of the same compound in different orientations.

(a) $Br\overset{H}{\underset{Cl}{\overset{|}{\underset{|}{C}}}}F$ and $Br\overset{H}{\underset{F}{\overset{|}{\underset{|}{C}}}}Cl$ *entiomer*

(b) $F\overset{H}{\underset{Cl}{\overset{|}{\underset{|}{C}}}}CH_3$ and $H\overset{F}{\underset{CH_3}{\overset{|}{\underset{|}{C}}}}Cl$ *Same.*

(c) $H\overset{CH_3}{\underset{\underset{CH_3}{|}}{\overset{|}{\underset{CH_2}{\overset{|}{C}}}}}OH$ and $HO\overset{H}{\underset{CH_3}{\overset{|}{\underset{|}{C}}}}CH_2{-}CH_3$ *enand.*

enantiomers are different compounds.

5.7 PROPERTIES OF ENANTIOMERS: OPTICAL ACTIVITY

The molecules of enantiomers are not superposable one on the other and, on this basis alone, we have concluded that enantiomers are different compounds. How are they different? Do enantiomers resemble constitutional isomers and diastereomers in having different melting and boiling points? The answer is *no*. Enantiomers have *identical* melting and boiling points. Do enantiomers have different indexes of refraction, different solubilities in common solvents, different infrared spectra, and different rates of reaction with ordinary reagents? The answer to each of these questions is also *no*.

Many of these properties (e.g., boiling points, melting points, and solubilities) are dependent on the magnitude of the intermolecular forces operating between the mole-

TABLE 5.1   *Physical properties of (R)- and (S)-2-butanol*

PHYSICAL PROPERTY	(R)-2-BUTANOL	(S)-2-BUTANOL
Boiling point (1 atm)	99.5°C	99.5°C
Density (g mL^{-1} at 20°C)	0.808	0.808
Index of refraction (20°C)	1.397	1.397

cules (Section 2.16), and for molecules that are mirror images of each other these forces will be identical.

We can see examples if we examine Table 5.1 where some of the physical properties of the 2-butanol enantiomers are listed.

Enantiomers show different behavior only when they interact with other chiral substances. Enantiomers show different rates of reaction toward other chiral molecules — that is, toward reagents that consist of a single enantiomer or an excess of a single enantiomer. Enantiomers also show different solubilities in solvents that consist of a single enantiomer or an excess of a single enantiomer.

One easily observable way in which enantiomers differ is in *their behavior toward plane-polarized light.* When a beam of plane-polarized light passes through an enantiomer, the plane of polarization **rotates.** Moreover, separate enantiomers rotate the plane of plane-polarized light equal amounts *but in opposite directions.* Because of their effect on plane-polarized light, separate enantiomers are said to be **optically active compounds.**

In order to understand this behavior of enantiomers we need to understand the nature of plane-polarized light. We also need to understand how an instrument called a **polarimeter** operates.

5.7A Plane-Polarized Light

Light is an electromagnetic phenomenon. A beam of light consists of two mutually perpendicular oscillating fields: an oscillating electric field and an oscillating magnetic field (Fig. 5.8).

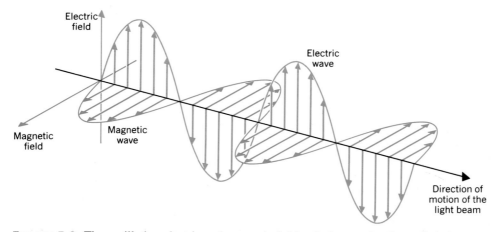

FIGURE 5.8 The oscillating electric and magnetic fields of a beam of ordinary light in one plane. The waves depicted here occur in all possible planes in ordinary light.

If we were to view a beam of ordinary light from one end, and if we could actually see the planes in which the electrical oscillations were occurring, we would find that oscillations of the electric field were occurring in all possible planes perpendicular to the direction of propagation (Fig. 5.9). (The same would be true of the magnetic field.)

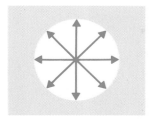

FIGURE 5.9 Oscillation of the electrical field of ordinary light occurs in all possible planes perpendicular to the direction of propagation.

When ordinary light is passed through a polarizer, the polarizer interacts with the electrical field so that the electrical field of the light that emerges from the polarizer (and the magnetic field perpendicular to it) is oscillating only in one plane. Such light is called plane-polarized light (Fig. 5.10).

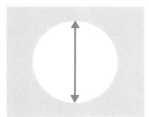

FIGURE 5.10 The plane of oscillation of the electrical field of plane-polarized light. In this example the plane of polarization is vertical.

The lenses of Polaroid sunglasses have this effect. You can demonstrate for yourself that this is true with two pairs of Polaroid sunglasses. If two lenses are placed one on top of the other so that the axes of polarization coincide, then light passes through both normally. Then if one lens is rotated 90° with respect to the other, no light passes through.

5.7B The Polarimeter

The device that is used for measuring the effect of plane-polarized light on optically active compounds is a polarimeter. A sketch of a polarimeter is shown in Fig. 5.11. The principal working parts of a polarimeter are (1) a light source (usually a sodium lamp), (2) a polarizer, (3) a tube for holding the optically active substance (or solution) in the light beam, (4) an analyzer, and (5) a scale for measuring the number of degrees that the plane of polarized light has been rotated.

The analyzer of a polarimeter (Fig. 5.11) is nothing more than another polarizer. If the tube of the polarimeter is empty, or if an optically *inactive* substance is present, the axes of the plane-polarized light and the analyzer will be exactly parallel when the instrument reads 0°, and the observer will detect the maximum amount of light passing through. If, by contrast, the tube contains an optically active substance, a solution of

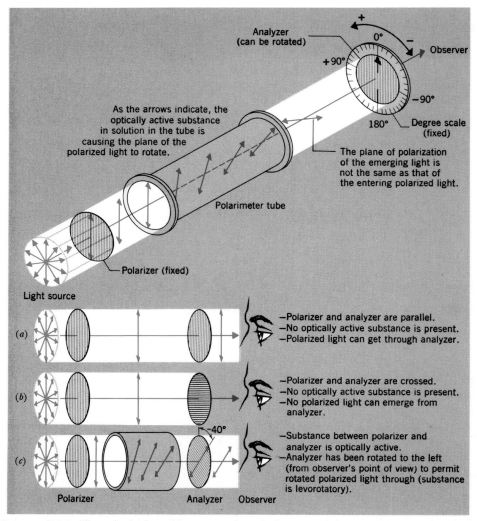

FIGURE 5.11 The principal working parts of a polarimeter and the measurement of optical rotation. (From John R. Holum, *Organic Chemistry: A Brief Course,* Wiley, New York, 1975, p. 316.)

one enantiomer, for example, the plane of polarization of the light will be rotated as it passes through the tube. In order to detect the maximum brightness of light the observer will have to rotate the axis of the analyzer in either a clockwise or counterclockwise direction. If the analyzer is rotated in a clockwise direction, the rotation, α (measured in degrees), is said to be positive (+). If the rotation is counterclockwise, the rotation is said to be negative (−). A substance that rotates plane-polarized light in the clockwise direction is also said to be **dextrorotatory,** and one that rotates plane-polarized light in a counterclockwise direction is said to be **levorotatory** (from the Latin: *dexter,* right and *laevus,* left).

5.7C Specific Rotation

The number of degrees that the plane of polarization is rotated as the light passes through a solution of an enantiomer depends on the number of chiral molecules that it encounters. This, of course, depends on the length of the tube and the concentration of the enantiomer. In order to place measured rotations on a standard basis, chemists calculate a quantity called the **specific rotation,** $[\alpha]$, by the following equation:

$$[\alpha] = \frac{\alpha}{c \cdot l}$$

where $[\alpha]$ = the specific rotation

α = the observed rotation

c = the concentration of the solution in grams per milliliter of solution (or density in g mL^{-1} for neat liquids)

l = the length of the tube in decimeters (1 dm = 10 cm)

The specific rotation also depends on the temperature and the wavelength of light that is employed. Specific rotations are reported so as to incorporate these quantities as well. A specific rotation might be given as follows:

$$[\alpha]_D^{25} = +3.12°$$

This means that, the D line of a sodium lamp (λ = 599.6 nm) was used for the light, that a temperature of 25°C was maintained, and that a sample containing 1.00 g mL^{-1} of the optically active substance, in a 1-dm tube, produced a rotation of 3.12° in a clockwise direction.*

The specific rotations of (R)-2-butanol and (S)-2-butanol are given here.

(R)-2-Butanol
$[\alpha]_D^{25} = -13.52°$

(S)-2-Butanol
$[\alpha]_D^{25} = +13.52°$

The direction of rotation of plane-polarized light is often incorporated into the names of optically active compounds. The following two sets of enantiomers show how this is done.

(R)-(+)-2-Methyl-1-butanol
$[\alpha]_D^{25} = +5.756°$

(S)-(−)-2-Methyl-1-butanol
$[\alpha]_D^{25} = +5.756°$

* The magnitude of rotation is dependent on the solvent when solutions are measured. This is the reason the solvent is specified when a rotation is reported in the chemical literature.

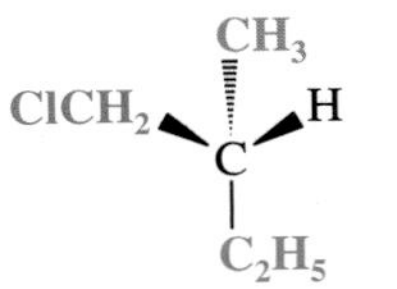

(R)-(−)-1-Chloro-2-methylbutane
$[\alpha]_D^{25} = -1.64°$

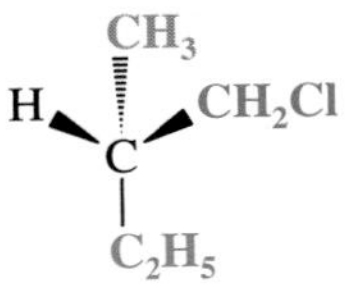

(S)-(+)-1-Chloro-2-methylbutane
$[\alpha]_D^{25} = +1.64°$

The previous compounds also illustrate an important principle: *No obvious correlation exists between the configurations of enantiomers and the direction [(+) or (−)] in which they rotate plane-polarized light.*

(R)-(+)-2-Methyl-1-butanol and (R)-(−)-1-chloro-2-methylbutane have the same *configuration,* that is, they have the same general arrangement of their atoms in space. They have, however, an opposite effect on the direction of rotation of the plane of plane-polarized light.

(R)-(+)-2-Methyl-1-butanol   Same configuration   (R)-(−)-1-Chloro-2-methylbutane

These same compounds also illustrate a second important principle: *No necessary correlation exists between the (R) and (S) designation and the direction of rotation of plane-polarized light.* (R)-2-Methyl-1-butanol is dextrorotatory (+), and (R)-1-chloro-2-methylbutane is levorotatory (−).

A method based on the measurement of optical rotation at many different wavelengths, called optical rotatory dispersion, has been used to correlate configurations of chiral molecules. A discussion of the technique of optical rotatory dispersion, however, is beyond the scope of this text.

PROBLEM 5.14

Shown below is the configuration of (+)-carvone. (+)-Carvone is the principal component of caraway seed oil and is responsible for its characteristic odor. (−)-Carvone, its enantiomer, is the main component of spearmint oil and gives it its characteristic odor. The fact that the carvone enantiomers do not smell the same suggests that the receptor sites in the nose for these compounds are chiral, and only the correct enantiomer will fit its particular site (just as a hand requires a glove of the correct chirality for a proper fit). Give the correct (R) and (S) designations for (+)- and (−)-carvone.

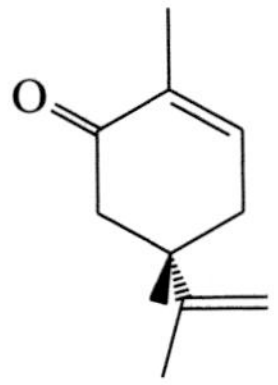

(+)-Carvone

5.8 THE ORIGIN OF OPTICAL ACTIVITY

It is not possible to give a complete, condensed account of the origin of the optical activity observed for separate enantiomers. An insight into the source of this phenomenon can be obtained, however, by comparing what occurs when a beam of plane-polarized light passes through a solution of *achiral* molecules with what occurs when a beam of plane-polarized light passes through a solution of *chiral* molecules.

Almost all *individual* molecules, whether chiral or achiral, are theoretically capable of producing a slight rotation of the plane of plane-polarized light. The direction and magnitude of the rotation produced by an individual molecule depends, in part, on its orientation at the precise moment that it encounters the beam. In a solution, of course, billions of molecules are in the path of the light beam and at any given moment these molecules are present in all possible orientations. If the beam of plane-polarized light passes through a solution of the achiral compound 2-propanol, for example, it should encounter at least two molecules in the exact orientations shown in Fig. 5.12. The effect of the first encounter might be to produce a very slight rotation of the plane of polarization to the right. Before the beam emerges from the solution, however, it should encounter at least one molecule of 2-propanol that is in exactly the mirror-image orientation of the first. The effect of this second encounter is to produce an equal and opposite rotation of the plane: a rotation that exactly cancels the first rotation. The beam, therefore, emerges with no net rotation.

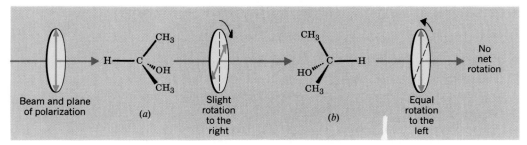

FIGURE 5.12 A beam of plane-polarized light encountering a molecule of 2-propanol (an achiral molecule) in orientation (*a*) and then a second molecule in the mirror-image orientation (*b*). The beam emerges from these two encounters with no net rotation of its plane of polarization.

What we have just described for the two encounters shown in Fig. 5.12 can be said of all possible encounters of the beam with molecules of 2-propanol. Because so many molecules are present, it is statistically certain that *for each encounter with a particular orientation there will be an encounter with a molecule that is in a mirror-image orientation.* The result of all of these encounters is such that all of the rotations produced by individual molecules are canceled and 2-propanol is found to be **optically inactive.**

What, then, is the situation when a beam of plane-polarized light passes through a solution of one enantiomer of a chiral compound? We can answer this question by considering what might occur when plane-polarized light passes through a solution of pure (*R*)-2-butanol. Figure 5.13 illustrates one possible encounter of a beam of plane-polarized light with a molecule of (*R*)-2-butanol.

When a beam of plane-polarized light passes through a solution of (*R*)-2-butanol, *no molecule is present that can ever be exactly oriented as a mirror image of any given*

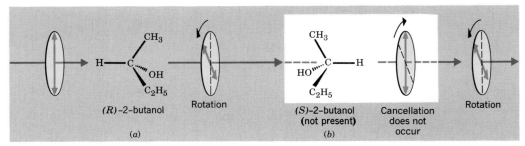

FIGURE 5.13 (*a*) A beam of plane-polarized light encounters a molecule of (*R*)-2-butanol (a chiral molecule) in a particular orientation. This encounter produces a slight rotation of the plane of polarization. (*b*) Exact cancellation of this rotation requires that a second molecule be oriented as an exact mirror image. This cancellation does not occur because the only molecule that could ever be oriented as an exact mirror image at the first encounter is a molecule of (*S*)-2-butanol, which is not present. As a result, a net rotation of the plane of polarization occurs.

orientation of an (*R*)-*2-butanol molecule.* The only molecules that could do this would be molecules of (*S*)-2-butanol, and they are not present. Exact cancellation of the rotations produced by all of the encounters of the beam with random orientations of (*R*)-2-butanol does not happen and, as a result, a net rotation of the plane of polarization is observed. (*R*)-2-Butanol is found to be *optically active.*

5.8A Racemic Forms

The net rotation of the plane of polarization that we observe for a solution consisting of molecules of (*R*)-2-butanol alone would not be observed if we passed the beam through a solution that contained equimolar amounts of (*R*)-2-butanol and (*S*)-2-butanol. In the latter instance, molecules of (*S*)-2-butanol would be present in a quantity equal to those of (*R*)-2-butanol and for every possible orientation of one enantiomer, a molecule of the other enantiomer would be in a mirror-image orientation. Exact cancellations of all rotations would occur, and the solution of the equimolar mixture of enantiomers would be *optically inactive.*

An equimolar mixture of two enantiomers is called a **racemic form** (either a **racemate** or a **racemic mixture**). A racemic form shows no rotation of plane-polarized light; as such, it is often designated as being (±). A racemic form of (*R*)-(−)-2-butanol and (*S*)-(+)-2-butanol might be indicated as

$$(\pm)\text{-2-Butanol} \quad \text{or as} \quad (\pm)\text{-CH}_3\text{CH}_2\text{CHOHCH}_3$$

5.8B Racemic Forms and Enantiomeric Excess

A sample of an optically active substance that consists of a single enantiomer is said to be **enantiomerically pure** or to have an **enantiomeric excess** of 100%. An enantiomerically pure sample of (*S*)-(+)-2-butanol shows a specific rotation of +13.52° ($[\alpha]_D^{25} = +13.52°$). On the other hand, a sample of (*S*)-(+)-2-butanol that contains less than an equimolar amount of (*R*)-(−)-2-butanol will show a specific rotation that is less than +13.52° but greater than 0°. Such a sample is said to have an *enantiomeric excess* less than 100%. The **enantiomeric excess (ee)** is defined as follows:

% Enantiomeric excess

$$= \frac{\text{moles of one enantiomer} - \text{moles of other enantiomer}}{\text{total moles of both enantiomers}} \times 100$$

The enantiomeric excess can be calculated from optical rotations:

$$\% \text{ Enantiomeric excess*} = \frac{\text{observed specific rotation}}{\text{specific rotation of the pure enantiomer}} \times 100$$

Let us suppose, for example, that a mixture of the 2-butanol enantiomers showed a specific rotation of $+6.76°$. We would then say that the enantiomeric excess of the (S)-$(+)$-2-butanol is 50%.

$$\text{Enantiomeric excess} = \frac{+6.76°}{+13.52°} \times 100 = 50\%$$

When we say that the enantiomeric excess of this mixture is 50% we mean that 50% of the mixture consists of the $(+)$ enantiomer (the excess) and the other 50% consists of the racemic form. Since for the 50% that is racemic, the optical rotations cancel one another out, only the 50% of the mixture that consists of the $(+)$ enantiomer contributes to the observed optical rotation. The observed rotation is, therefore, 50% (or half) of what it would have been if the mixture had consisted only of the $(+)$ enantiomer.

SAMPLE PROBLEM

What is the actual stereoisomeric composition of the mixture referred to above?

ANSWER:
Of the total mixture, 50% consists of the racemic form, which contains equal numbers of the two enantiomers. Therefore, half of this 50%, or 25%, is the $(-)$ enantiomer and 25% is the $(+)$ enantiomer. The other 50% of the mixture (the excess) is also the $(+)$ enantiomer. Consequently the mixture is 75% $(+)$ enantiomer and 25% $(-)$ enantiomer.

PROBLEM 5.15

A sample of 2-methyl-1-butanol (cf. Section 5.7C) has a specific rotation, $[\alpha]_D^{25}$, equal to $+1.151°$. (a) What is the % enantiomeric excess of the sample? (b) Which enantiomer is in excess, the (R) or the (S)?

---

* This calculation should be applied to a single enantiomer or to mixtures of enantiomers only. It should not be applied to mixtures in which some other compound is present.

5.9 THE SYNTHESIS OF ENANTIOMERS

5.9A Racemic Forms

Many times in the course of working in the organic laboratory a reaction carried out with reactants whose molecules are achiral results in the formation of products whose molecules are chiral. In the absence of any chiral influence (from the solvent or a catalyst), the outcome of such a reaction is the formation of a racemic form. The reason: The chiral molecules of the product are obtained as a 50 : 50 mixture of enantiomers.

An example is the synthesis of 2-butanol by the nickel-catalyzed hydrogenation of 2-butanone. In this reaction the hydrogen molecule adds across the carbon–oxygen double bond in much the same way that it adds to a carbon–carbon double bond (Section 4.15A).

$$CH_3CH_2CCH_3 \; + \; H{-}H \; \xrightarrow{Ni} \; (\pm) \; CH_3CH_2\overset{*}{C}HCH_3$$

$\overset{\|\|}{O}$		OH
2-Butanone	**Hydrogen**	**(±)-2-Butanol**
(achiral	**(achiral**	**[chiral molecules**
molecules)	**molecules)**	**but 50 : 50 mixture (R) and (S)]**

Molecules of neither reactant (2-butanone nor hydrogen) are chiral. The molecules of the product (2-butanol) are chiral. The product, however, is obtained as a racemic form because the two enantiomers, (R)-(−)-2-butanol and (S)-(+)-2-butanol, are obtained in equal amounts.

> This is not the result if reactions like this are carried out in the presence of a chiral influence such as an optically active solvent or, as we shall see below, an enzyme. The nickel catalyst used in this reaction does not exert a chiral influence.

Figure 5.14 shows why a racemic form of 2-butanol is obtained. Hydrogen, adsorbed on the surface of the nickel catalyst, adds with equal facility at either face of 2-butanone. Reaction at one face produces one enantiomer; reaction at the other face produces the other enantiomer, and the two reactions occur at the same rate.

5.9B Enantioselective Syntheses

If a reaction that leads to the formation of enantiomers produces a preponderance of one enantiomer over its mirror image, the reaction is said to be **enantioselective.** To be enantioselective, a chiral reagent, solvent, or catalyst must assert an influence on the course of the reaction.

In nature, where most reactions are enantioselective, the chiral influences come from protein molecules called **enzymes.** Enzymes are biological catalysts of extraordinary efficiency. They not only have the ability to cause reactions to take place much more rapidly than they would otherwise, they also have the ability to assert a dramatic *chiral influence* on a reaction. Enzymes do this because they, too, are chiral, and they possess an active site where the reactant molecules are bound, momentarily, while the reaction takes place. This active site is chiral, and only one enantiomer of a chiral reactant fits it properly and is able to undergo reaction.

Many enzymes have also found use in the organic chemistry laboratory, where organic chemists take advantage of their properties to bring about enantioselective reactions. One enzyme used frequently in this way is an enzyme called **lipase.** Lipase

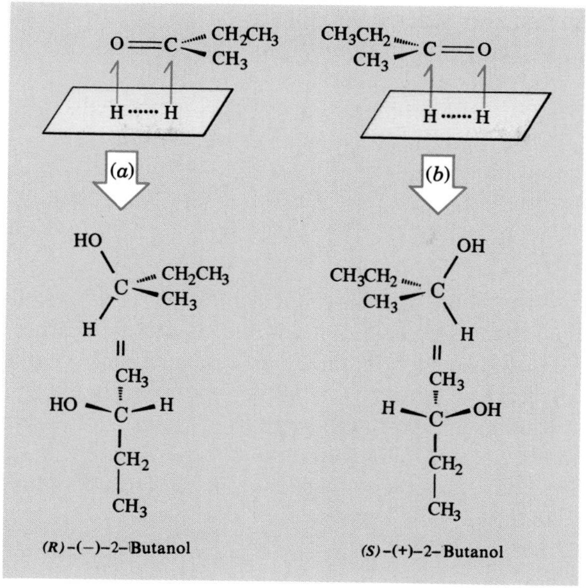

FIGURE 5.14 The reaction of 2-butanone with hydrogen in the presence of a nickel catalyst. The reaction rate by path (*a*) is equal to that by path (*b*). (*R*)-(−)-2-butanol and (*S*)-(+)-2-butanol are produced in equal amounts, as a racemate.

catalyzes a reaction, called **hydrolysis,** whereby esters (Section 2.14C) react with a molecule of water and are converted to a carboxylic acid and an alcohol. (This is the reverse of a reaction by which esters are synthesized.)

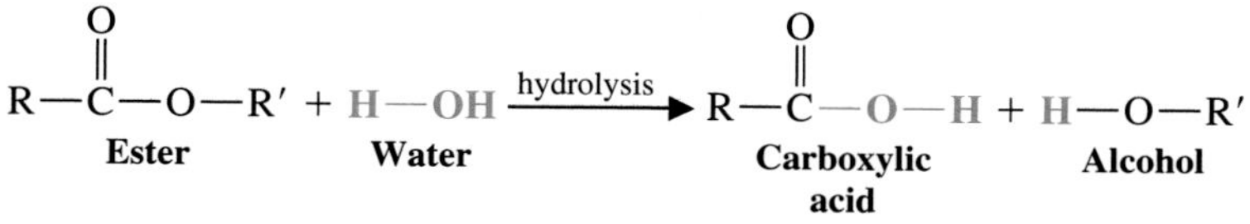

$$
\underset{\textbf{Ester}}{R-\overset{\overset{\displaystyle O}{\|}}{C}-O-R'} + \underset{\textbf{Water}}{H-OH} \xrightarrow{\text{hydrolysis}} \underset{\substack{\textbf{Carboxylic}\\\textbf{acid}}}{R-\overset{\overset{\displaystyle O}{\|}}{C}-O-H} + \underset{\textbf{Alcohol}}{H-O-R'}
$$

Hydrolysis, which means literally *cleavage (lysis) by water,* can be carried out in the laboratory in a variety of ways that do not involve the use of an enzyme. However, use of the enzyme lipase allows the hydrolysis to be used to prepare almost pure enantiomers. The following hydrolysis using lipase is a good example.

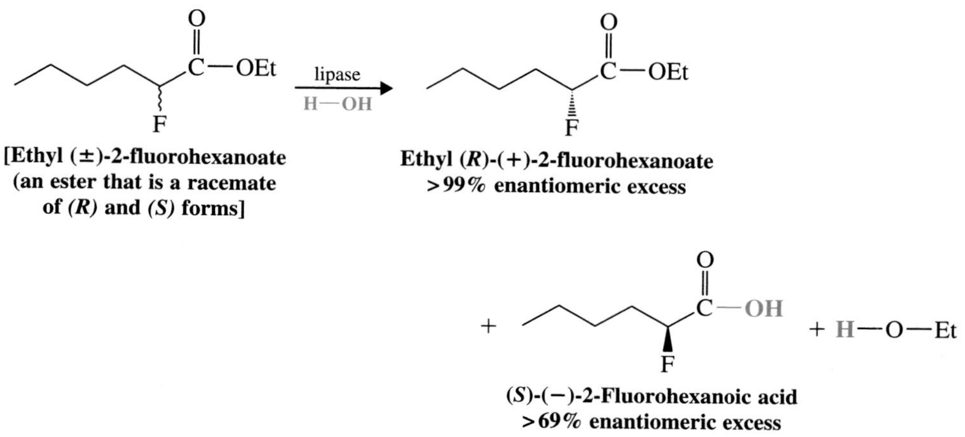

[Ethyl (±)-2-fluorohexanoate
(an ester that is a racemate
of (*R*) and (*S*) forms]

Ethyl (*R*)-(+)-2-fluorohexanoate
>99% enantiomeric excess

(*S*)-(−)-2-Fluorohexanoic acid
>69% enantiomeric excess

The (R) enantiomer of this ester does not fit the active site of the enzyme and is, therefore, unaffected. Only the (S) enantiomer of the ester fits the active site and undergoes hydrolysis. After the reaction is over, therefore, one can isolate the unchanged (R)-ester in 99% enantiomeric purity. This method also has the added benefit of producing the (S)-($-$) acid in 69% enantiomeric purity.

5.10 CHIRAL DRUGS

Of much recent interest to the pharmaceutical industry is the production and sale of "chiral drugs," that is, drugs that contain a single enantiomer, rather than a racemate.* In some instances a drug has been marketed as a racemate for years even though only one enantiomer is the active agent. Such is the case with the antiinflammatory agent **ibuprofen** (Advil, Motrin, Nupren). Only the (S) isomer is effective. The (R) isomer has no antiinflammatory action, and even though the (R) isomer is slowly converted to the (S) isomer in the body, a medicine based on the (S) isomer alone takes effect more quickly than the racemate.

$$H_3C—\underset{\underset{H}{|}}{\overset{\overset{CH_3}{|}}{C}}—CH_2—\overset{CH_3}{\underset{}{}}—CH—\underset{\underset{O}{\|}}{C}—OH$$

Ibuprofen

The antihypertensive drug, **methyldopa** (Aldomet) also owes its effect exclusively to the (S) isomer.

$$HO—\overset{}{\underset{HO}{}}—CH_2—\underset{\underset{NH_2}{|}}{\overset{\overset{CH_3}{|}}{C}}—CO_2H \qquad H_3C—\underset{SH\ \ NH_2}{\overset{CH_3}{|}}—CH—CO_2H$$

Methyldopa   **Penicillamine**

And, with penicillamine, the (S) isomer is a highly potent therapeutic agent for primary chronic arthritis; the (R) isomer has no therapeutic action, it is highly toxic.

PROBLEM 5.16

Write three-dimensional formulas for the (S) isomers of   (a) ibuprofen (b) methyldopa, and   (c) penicillamine.

There are many other examples of drugs like these, including drugs where the enantiomers have distinctly different effects. The preparation of enantiomerically pure

*See, for example, "Chiral drugs," *C & E News,* **1993,** *71*(39), 38–64.

drugs, therefore, is one factor that makes enantioselective synthesis (Section 5.9B) an active area of research today.

5.11 MOLECULES WITH MORE THAN ONE STEREOCENTER

So far all of the chiral molecules that we have considered have contained only one stereocenter. Many organic molecules, especially those important in biology, contain more than one stereocenter. Cholesterol (Section 23.4B), for example, contains eight stereocenters. (Can you locate them?) We can begin, however, with simpler molecules. Let us consider 2,3-dibromopentane shown here—a structure that has two stereocenters.

$$CH_3\overset{*}{C}H\overset{*}{C}HCH_2CH_3$$
$$| \quad |$$
$$Br \ Br$$

2,3-Dibromopentane

A useful rule gives the maximum number of stereoisomers: In compounds whose stereoisomerism is due to tetrahedral stereocenters, *the total number of stereoisomers will not exceed 2^n, where n is equal to the number of tetrahedral stereocenters.* For 2,3-dibromopentane we should not expect more than four stereoisomers ($2^2 = 4$).

Our next task is to write three-dimensional formulas for the stereoisomers of the compound. We begin by writing a three-dimensional formula for one stereoisomer and then by writing the formula for *its* mirror image.

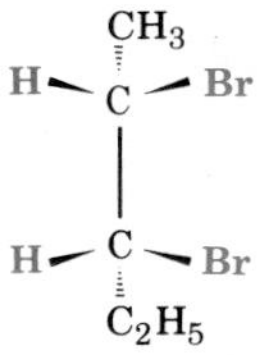

1 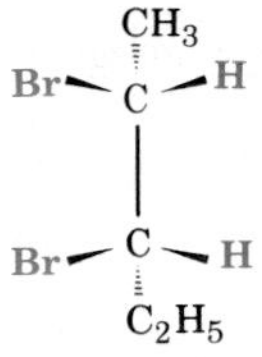   **2**

It is helpful to follow certain conventions when we write these three-dimensional formulas. For example, we usually write our structures in eclipsed conformations. When we do this we do not mean to imply that eclipsed conformations are the most stable ones—they most certainly are not. We write eclipsed conformations because, as we shall see later, they make it easy for us to recognize planes of symmetry when they are present. We also write the longest carbon chain in a generally vertical orientation on the page; this makes the structures that we write directly comparable. As we do these things, however, *we must remember that molecules can rotate in their entirety* and that *at normal temperatures rotations about all single bonds are also possible.* If rotations of the structure itself or rotations of groups joined by single bonds make one structure superposable with another, then *the structures do not represent different compounds;* instead, they represent different orientations or different conformations of two molecules of the same compound.

Since structures **1** and **2** are not superposable, they represent different compounds. Since structures **1** and **2** differ *only* in the arrangement of their atoms in space, they represent stereoisomers. Structures **1** and **2** are also mirror images of each other; thus **1** and **2** represent enantiomers.

Structures **1** and **2** are not the only possible structures, however. We find that we can write a structure **3** that is different from either **1** or **2**, and we can write a structure **4** that is a nonsuperposable mirror image of structure **3**.

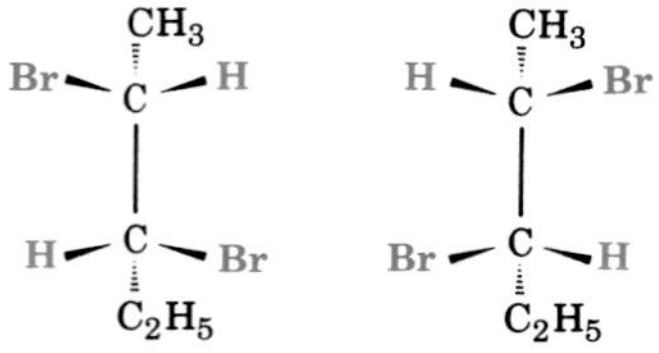

Structures **3** and **4** correspond to another pair of enantiomers. Structures **1–4** are all different, so there are, in total, four stereoisomers of 2,3-dibromopentane. At this point you should convince yourself that there are no other stereoisomers by writing other structural formulas. You will find that rotation of the single bonds (or of the entire structure) of any other arrangement of the atoms will cause the structure to become superposable with one of the structures that we have written here. Better yet, using different colored balls, make molecular models as you work this out.

The compounds represented by structures **1–4** are all optically active compounds. Any one of them, if placed separately in a polarimeter, would show optical activity.

The compounds represented by structures **1** and **2** are enantiomers. The compounds represented by structures **3** and **4** are also enantiomers. But what is the isomeric relation between the compounds represented by **1** and **3**?

We can answer this question by observing that **1** and **3** *are stereoisomers* and that they *are not mirror images of each other*. They are, therefore, *diastereomers*. **Diastereomers have different physical properties** — different melting points and boiling points, different solubilities, and so forth. In this respect these diastereomers are just like diastereomeric alkenes such as *cis-* and *trans-*2-butene.

PROBLEM 5.17

(a) If **3** and **4** are enantiomers, what are **1** and **4**?   (b) What are **2** and **3** and **2** and **4**?   (c) Would you expect **1** and **3** to have the same melting point? (d) The same boiling point?   (e) The same vapor pressure?   (f) Could you separate **1** and **3** using gas–liquid chromatography?

5.11A Meso Compounds

A structure with two stereocenters does not always have four possible stereoisomers. Sometimes there are only *three*. This happens because some molecules are *achiral even though they contain stereocenters.*

To understand this, let us write stereochemical formulas for 2,3-dibromobutane shown here.

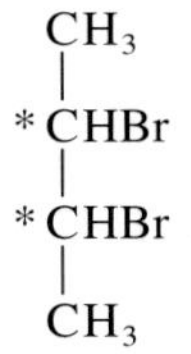

$$CH_3$$
$$|$$
$$*CHBr$$
$$|$$
$$*CHBr$$
$$|$$
$$CH_3$$

2,3-Dibromobutane

We begin in the same way as we did before. We write the formula for one stereoisomer and for its mirror image.

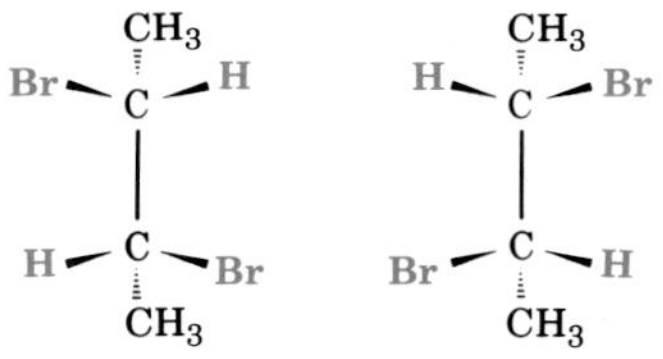

A   **B**

Structures **A** and **B** are nonsuperposable and represent a pair of enantiomers.

When we write structure **C** (see the following structure) and its mirror image **D**, however, the situation is different. *The two structures are superposable.* This means that **C** and **D** do not represent a pair of enantiomers. Formulas **C** and **D** represent two different orientations of the same compound.

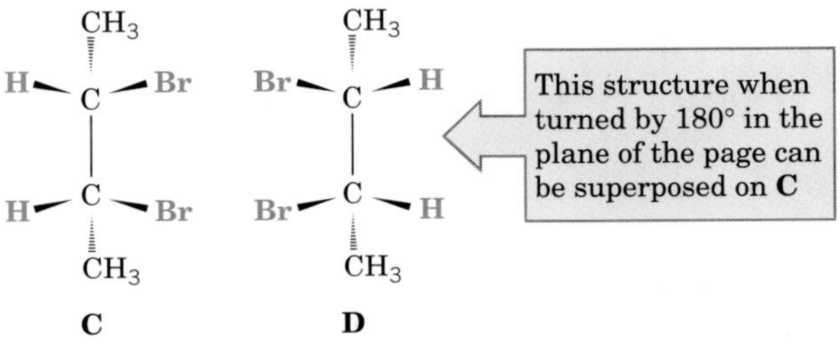

This structure when turned by 180° in the plane of the page can be superposed on **C**

C   **D**

The molecule represented by structure **C** (or **D**) is not chiral even though it contains tetrahedral atoms with four different attached groups. Such molecules are called *meso compounds.* Meso compounds, *because they are achiral,* are optically inactive.

The ultimate test for molecular chirality is to construct a model (or write the structure) of the molecule and then test whether or not the model (or structure) is superposable on its mirror image. If it is, the molecule is achiral: If it *is not,* the molecule is chiral.

We have already carried out this test with structure **C** and found that it is achiral. We can also demonstrate that **C** is achiral in another way. Figure 5.15 shows that structure **C** *has a plane of symmetry* (Section 5.5).

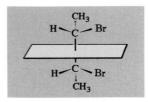

FIGURE 5.15 The plane of symmetry of *meso*-2,3-dibromobutane. This plane divides the molecule into halves that are mirror images of each other.

PROBLEM 5.18

Which of the following would be optically active?
(a) A pure sample of **A**   (c) A pure sample of **C**
(b) A pure sample of **B**   (d) An equimolar mixture of **A** and **B**

PROBLEM 5.19

The following are formulas for three compounds, written in noneclipsed conformations. In each instance tell which compound (**a**, **b**, or **c**) each formula represents.

(a)   (b)   (c)

PROBLEM 5.20

Write three-dimensional formulas for all of the stereoisomers of each of the following compounds.

(a) $CH_3CHClCHClCH_3$ (e) $CH_3CHBrCHFCH_3$
(b) $CH_3CHOHCH_2CHOHCH_3$ (f) In answers to parts (a)–(e) label
(c) $CH_2ClCHFCHFCH_2Cl$ pairs of enantiomers and label meso
(d) $CH_3CHOHCH_2CHClCH_3$ compounds.

5.11B Naming Compounds with More Than One Stereocenter

If a compound has more than one tetrahedral stereocenter, we analyze each center separately and decide whether it is (*R*) or (*S*). Then, using numbers, we tell which designation refers to which carbon atom.

Consider the stereoisomer **A** of 2,3-dibromobutane.

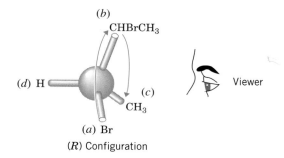

A

2,3-Dibromobutane

When this formula is rotated so that the group of lowest priority attached to C2 is directed away from the viewer it resembles the following.

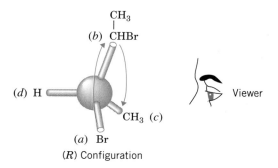

The order of progression from the group of highest priority to that of next highest priority (from —Br, to —CHBrCH₃, to —CH₃) is clockwise. So C2 has the (R) configuration.

When we repeat this procedure with C3 we find that C3 also has the (R) configuration.

Compound **A**, therefore, is (2R,3R)-2,3-dibromobutane.

PROBLEM 5.21

Give names that include (R) and (S) designations for compounds **B** and **C** in Section 5.11A.

PROBLEM 5.22

Give names that include (R) and (S) designations for your answers to Problem 5.20.

PROBLEM 5.23

Chloramphenicol (below) is a potent antibiotic, isolated from *Streptomyces venezuelae*, that is particularly effective against typhoid fever. It was the first naturally occurring substance shown to contain a nitro ($-NO_2$) group attached to an aromatic ring. Both stereocenters in chloramphenicol are known to have the (R) configuration. Identify the two stereocenters and write a three-dimensional formula for chloramphenicol.

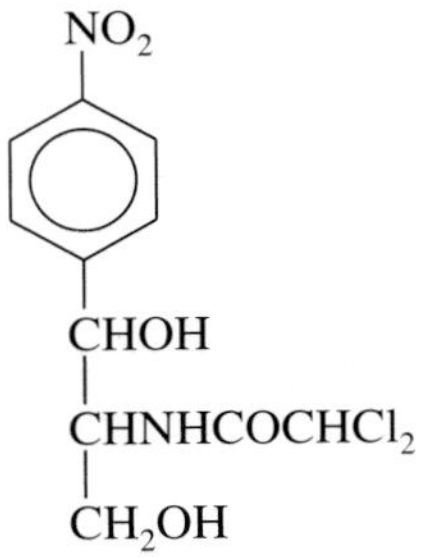

$$NO_2$$

CHOH

CHNHCOCHCl$_2$

CH$_2$OH

Chloramphenicol

5.12 FISCHER PROJECTION FORMULAS

In writing structures for chiral molecules so far, we have used only three-dimensional formulas, and we shall continue to do so until we study carbohydrates in Chapter 22. The reason: Three-dimensional formulas are unambiguous and can be manipulated on paper in any way that we wish, as long as we do not break bonds. Their use, moreover, teaches us to see molecules (in our mind's eye) in three dimensions, and this ability will serve us well.

Chemists sometimes represent structures for chiral molecules with *two-dimensional formulas* called **Fischer projection formulas.** These two-dimensional formulas are especially useful for compounds with several stereocenters because they save space and are easy to write. Their use, however, requires a rigid adherence to certain conventions. *Used carelessly, these projection formulas can easily lead to incorrect conclusions.*

The Fischer projection formula for (2R,3R)-2,3-dibromobutane is written as follows:

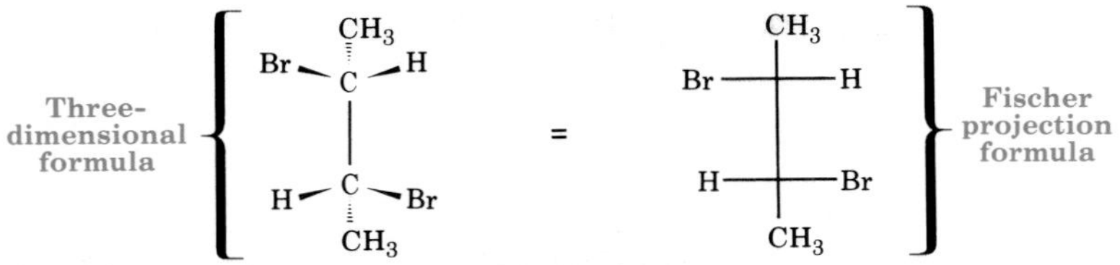

Three-dimensional formula

CH$_3$
Br—C—H
|
H—C—Br
CH$_3$

=

Fischer projection formula

CH$_3$
Br——H
H——Br
CH$_3$

A   **A**

By convention, Fischer projections are written with the main carbon chain extending from top to bottom and with all groups eclipsed. *Vertical lines represent bonds that project behind the plane of the paper (or that lie in it). Horizontal lines represent bonds that project out of the plane of the paper.* The intersection of vertical and horizontal lines represents a carbon atom, usually one that is a stereocenter.

In using Fischer projections to test the superposability for two structures, we are permitted to rotate them in the plane of the paper by 180° *but by no other angle.* We must always keep them in the plane of the paper, and *we are not allowed to flip them over.*

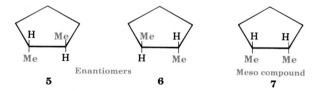

Your instructor will advise you about the use you are to make of Fischer projections.

5.13 STEREOISOMERISM OF CYCLIC COMPOUNDS

Because the cyclopentane ring is essentially planar, cyclopentane derivatives offer a convenient starting point for a discussion of the stereoisomerism of cyclic compounds. For example, 1,2-dimethylcyclopentane has two stereocenters and exists in three stereoisomeric forms **5**, **6**, and **7**.

The trans compound exists as a pair of enantiomers **5** and **6**. *cis*-1,2-Dimethylcyclopentane is a meso compound. It has a plane of symmetry that is perpendicular to the plane of the ring.

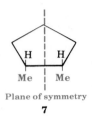

Plane of symmetry

7

PROBLEM 5.24

(a) Is the *trans*-1,2-dimethylcyclopentane (**5**) superposable on its mirror image (i.e., on compound **6**)?   (b) Is the *cis*-1,2-dimethylcyclopentane (**7**) superposable on its mirror image?   (c) Is the *cis*-1,2-dimethylcyclopentane a chiral molecule?   (d) Would *cis*-1,2-dimethylcyclopentane show optical activity? (e) What is the stereoisomeric relationship between **5** and **7**?   (f) Between **6** and **7**?

PROBLEM 5.25

Write structural formulas for all of the stereoisomers of 1,3-dimethylcyclopentane. Label pairs of enantiomers and meso compounds if they exist.

5.13A Cyclohexane Derivatives

1,4-Dimethylcyclohexanes   If we examine a formula of 1,4-dimethylcyclohexane we find that it does not contain any tetrahedral atoms with four different groups. However, we learned in Section 4.12 that 1,4-dimethylcyclohexane exists as cis–trans isomers. The cis and trans forms (Fig. 5.16) are *diastereomers*. Neither compound is

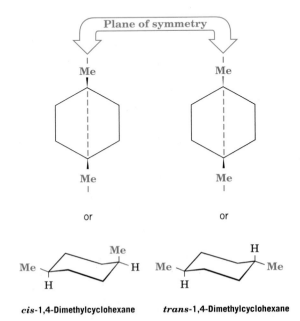

cis-1,4-Dimethylcyclohexane   *trans*-1,4-Dimethylcyclohexane

FIGURE 5.16 The cis and trans forms of 1,4-dimethylcyclohexane are diastereomers of each other. Both compounds are achiral.

chiral and, therefore, neither is optically active. Notice that both the cis and trans forms of 1,4-dimethylcyclohexane have a plane of symmetry.

1,3-Dimethylcyclohexanes A 1,3-dimethylcyclohexane has two stereocenters; we can, therefore, expect as many as four stereoisomers ($2^2 = 4$). In reality there are only three. *cis*-1,3-Dimethylcyclohexane has a plane of symmetry (Fig. 5.17) and is achiral. *trans*-1,3-Dimethylcyclohexane does not have a plane of symmetry and exists as a pair of enantiomers (Fig. 5.18). You may want to make models of the *trans*-1,3-dimethyl-cyclohexane enantiomers. Having done so, convince yourself that they cannot be superposed as they stand, and that they cannot be superposed after one enantiomer has undergone a ring flip.

FIGURE 5.17 *cis*-1,3-Dimethylcyclohexane has a plane of symmetry and is therefore achiral.

FIGURE 5.18 *trans*-1,3-Dimethylcyclohexane does not have a plane of symmetry and exists as a pair of enantiomers. The two structures (*a*) and (*b*) shown here are not superposable as they stand, and flipping the ring of either structure does not make it superposable on the other.

1,2-Dimethylcyclohexanes A 1,2-dimethylcyclohexane also has two stereo-centers and again we might expect as many as four stereoisomers. *There are four;* however, we find that we can can isolate only three stereoisomers. *trans*-1,2-Dimeth-ylcyclohexane (Fig. 5.19) exists as a pair of enantiomers. Its molecules do not have a plane of symmetry.

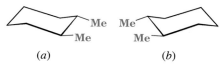

(*a*)   (*b*)

FIGURE 5.19 *trans*-1,2-Dimethylcyclohexane has no plane of symmetry and exists as a pair of enantiomers (*a*) and (*b*). [Notice that we have written the most stable conformations for (*a*) and (*b*). A ring flip of either (*a*) or (*b*) would cause both methyl groups to become axial.]

With *cis*-1,2-dimethylcyclohexane, the situation is somewhat more complex. If we consider the two conformational structures (*c*) and (*d*) shown in Fig. 5.20 we find

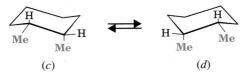

FIGURE 5.20 *cis*-1,2-Dimethylcyclohexane exists as two rapidly interconvertible chair conformations (*c*) and (*d*).

that these two mirror-image structures are not identical. Neither has a plane of symmetry and each is a chiral molecule, *but they are interconvertible by a ring flip*. (You should prove this to yourself with models.) Therefore, although the two structures represent enantiomers *they cannot be separated* because at temperatures even considerably below room temperature they interconvert rapidly. They simply represent *different conformations of the same compound*. In effect (*c*) and (*d*) comprise an interconverting racemic form. Structures (*c*) and (*d*) are not configurational stereoisomers; they are *conformational stereoisomers*. This means that at normal temperatures there are only three *isolable stereoisomers* of 1,2-dimethylcyclohexane.

PROBLEM 5.26

Write formulas for all of the isomers of each of the following. Designate pairs of enantiomers and achiral compounds where they exist.

(a) 1-Bromo-2-chlorocyclohexane   (c) 1-Bromo-4-chlorocyclohexane

(b) 1-Bromo-3-chlorocyclohexane

PROBLEM 5.27

Give the (*R*–*S*) designation for each compound given as an answer to Problem 5.26.

5.14 RELATING CONFIGURATIONS THROUGH REACTIONS IN WHICH NO BONDS TO THE STEREOCENTER ARE BROKEN

If a reaction takes place in a way so that no bonds to the stereocenter are broken, the product will of necessity have the same general configuration of groups around the stereocenter as the reactant. Such a reaction is said **to proceed with retention of configuration.** Consider as an example the reaction that takes place when (*S*)-(−)-2-methyl-1-butanol is heated with concentrated hydrochloric acid.

Same configuration

$$H\!-\!\underset{\underset{CH_3}{\overset{|}{\underset{|}{CH_2}}}}{\overset{CH_3}{\overset{|}{C}}}\!-\!CH_2\!-\!OH + H\!-\!Cl \quad \xrightarrow{\text{heat}} \quad H\!-\!\underset{\underset{CH_3}{\overset{|}{\underset{|}{CH_2}}}}{\overset{CH_3}{\overset{|}{C}}}\!-\!CH_2\!-\!Cl + H\!-\!OH$$

(*S*)-(−)-2-Methyl-1-butanol   (*S*)-(+)-1-Chloro-2-methylbutane
$[\alpha]_D^{25} = -5.756°$ $[\alpha]_D^{25} = +1.64°$

We do not need to know now exactly how this reaction takes place to see that the reaction must involve breaking of the CH_2—OH bond of the alcohol because the —OH group is replaced by a —Cl. There is no reason to assume that any other bonds are broken. All that we shall learn about reactions such as this tells us they are not. (We shall study how this reaction takes place in Section 10.14.) Since no bonds to the stereocenter are broken, the reaction must take place with retention of configuration, and the product of the reaction *must have the same configuration of groups around the stereocenter that the reactant had.* By saying that the two compounds have the same configuration we simply mean that comparable or identical groups in the two compounds occupy the same relative positions in space around the stereocenter. (In this instance the —CH_2OH group and the —CH_2Cl are comparable and they occupy the same relative position in both compounds; all the other groups are identical and they occupy the same positions.)

Notice that in this example while the $(R-S)$ designation *does not change* [both reactant and product are (S)] the direction of optical rotation *does change* [the reactant is $(-)$ and the product is $(+)$]. Neither occurrence is a necessity when a reaction proceeds with retention of configuration. In the next section we shall see examples of reactions in which configurations are retained and where the direction of optical rotation does not change. The following reaction is an example of a reaction that proceeds with retention of configuration but involves a change in $(R-S)$ designation.

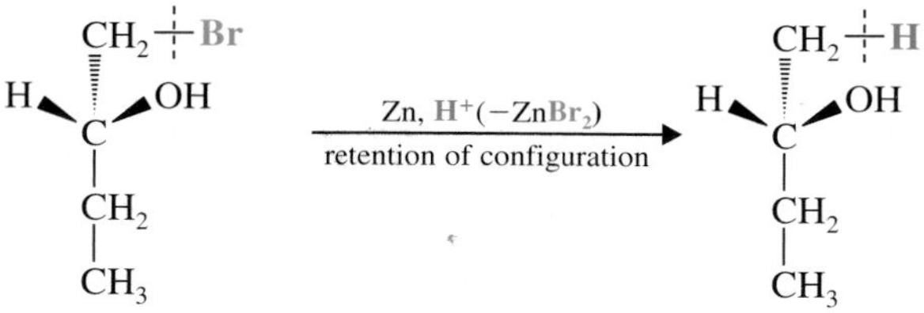

(R)-1-Bromo-2-butanol   (S)-2-Butanol

In this example the $(R-S)$ designation changes because the —CH_2Br group of the reactant (—CH_2Br has a higher priority than —CH_2CH_3) changes to a —CH_3 group in the product (—CH_3 has a lower priority than —CH_2CH_3).

5.14A Relative and Absolute Configurations

Reactions in which no bonds to the stereocenter are broken are useful in relating configurations of chiral molecules. That is, they allow us to demonstrate that certain compounds have the same **relative configuration.** In each of the examples that we have just cited, the products of the reactions have the same *relative configurations* as the reactants.

Before 1951 only relative configurations of chiral molecules were known. No one prior to that time had been able to demonstrate with certainty what the actual spatial arrangement of groups was in any chiral molecule. To say this another way, no one had been able to determine the **absolute configuration** of an optically active compound.

Configurations of chiral molecules were related to each other *through reactions of known stereochemistry.* Attempts were also made to relate all configurations to a single compound that had been chosen arbitrarily to be the standard. This standard compound was glyceraldehyde.

$$\underset{\text{Glyceraldehyde}}{\underset{\displaystyle CH_2OH}{\overset{\displaystyle *CHOH}{\overset{\displaystyle CH}{\overset{O}{\|}}}}}$$

Glyceraldehyde molecules have one tetrahedral stereocenter; therefore, glyceraldehyde exists as a pair of enantiomers.

$$
\begin{array}{ccc}
\textbf{(R)-Glyceraldehyde} & \text{and} & \textbf{(S)-Glyceraldehyde}
\end{array}
$$

In the older system for designating configurations (*R*)-glyceraldehyde was called D-glyceraldehyde and (*S*)-glyceraldehyde was called L-glyceraldehyde. This system of nomenclature is still widely used in biochemistry.

One glyceraldehyde enantiomer is dextrorotatory (+) and the other, of course, is levorotatory (−). Before 1951 no one could be sure, however, which configuration belonged to which enantiomer. Chemists decided arbitrarily to assign the (*R*) configuration to the (+)-enantiomer. Then configurations of other molecules were related to one glyceraldehyde enantiomer or the other through reactions of known stereochemistry.

For example, the configuration of (−)-lactic acid can be related to (+)-glyceraldehyde through the following sequence of reactions.

(+)-Glyceraldehyde → HgO (oxidation) → **(−)-Glyceric acid** ← HNO₂/H₂O → **(+)-Isoserine** → HNO₂/HBr →

(−)-3-Bromo-2-hydroxy-propanoic acid → Zn, H⁺ → **(−)-Lactic acid**

The stereochemistry of all of these reactions is known. Because bonds to the stereocenter (shown in red) are not broken in any of them, they all proceed with retention of configuration. If the assumption is made that the configuration of (+)-glyceraldehyde is as follows:

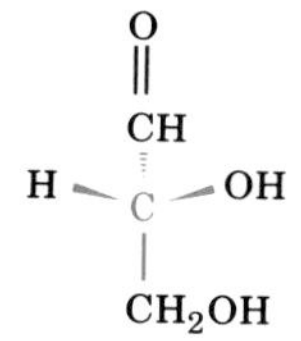

$$
\begin{array}{c}
O \\
\| \\
CH \\
H \diagdown \underset{C}{} \diagup OH \\
| \\
CH_2OH
\end{array}
$$

(R)-(+)-Glyceraldehyde

then the configuration of (−)-lactic acid is

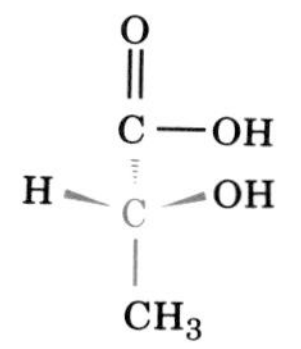

$$
\begin{array}{c}
O \\
\| \\
C-OH \\
H \diagdown \underset{C}{} \diagup OH \\
| \\
CH_3
\end{array}
$$

(R)-(−)-Lactic acid

PROBLEM 5.28

Write three-dimensional formulas for the starting compound, the product, and all of the intermediates in a synthesis similar to the one just given that relates the configuration of (−)-glyceraldehyde with (+)-lactic acid. Label each compound with its proper (R)–(S) and (+)–(−) designation.

The configuration of (−)-glyceraldehyde was also related through reactions of known stereochemistry to (+)-tartaric acid.

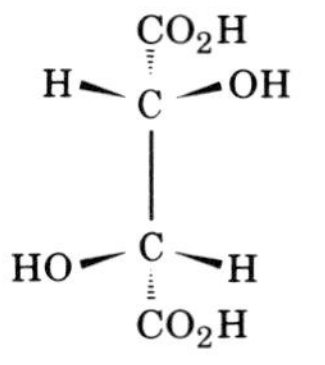

$$
\begin{array}{c}
CO_2H \\
H \diagdown \underset{C}{} \diagup OH \\
| \\
HO \diagup \underset{C}{} \diagdown H \\
CO_2H
\end{array}
$$

(+)-Tartaric acid

In 1951 J. M. Bijvoet, the director of the van't Hoff Laboratory of the University of Utrecht in Holland, using a special technique of X-ray diffraction, was able to show conclusively that (+)-tartaric acid had the absolute configuration shown above. This meant that the original arbitrary assignment of the configurations of (+)- and (−)-glyceraldehyde was also correct. It also meant that the configurations of all of the compounds that had been related to one glyceraldehyde enantiomer or the other were now known with certainty and were now **absolute configurations.**

PROBLEM 5.29 ───

How would you synthesize (*R*)-1-deuterio-2-methylbutane? [Hint: Consider one of the enantiomers of 1-chloro-2-methylbutane in Section 5.7C as a starting compound.]

5.15   SEPARATION OF ENANTIOMERS: RESOLUTION

So far we have left unanswered an important question about optically active compounds and racemic forms: How are enantiomers separated? Enantiomers have identical solubilities in ordinary solvents, and they have identical boiling points. Consequently, the conventional methods for separating organic compounds, such as crystallization and distillation, fail when applied to a racemic form.

It was, in fact, Louis Pasteur's separation of a racemic form of a salt of tartaric acid in 1848 that led to the discovery of the phenomenon called enantiomerism. Pasteur, consequently, is often considered to be the founder of the field of stereochemistry.

(+)-Tartaric acid is one of the byproducts of wine making (nature usually only synthesizes one enantiomer of a chiral molecule). Pasteur had obtained a sample of racemic tartaric acid from the owner of a chemical plant. In the course of his investigation Pasteur began examining the crystal structure of the sodium ammonium salt of racemic tartaric acid. He noticed that two types of crystals were present. One was identical with crystals of the sodium ammonium salt of (+)-tartaric acid that had been discovered earlier and had been shown to be dextrorotatory. Crystals of the other type were *non*superposable mirror images of the first kind. The two types of crystals were actually chiral. Using tweezers and a magnifying glass, Pasteur separated the two kinds of crystals, dissolved them in water, and placed the solutions in a polarimeter. The solution of crystals of the first type was dextrorotatory, and the crystals themselves proved to be identical with the sodium ammonium salt of (+)-tartaric acid that was already known. The solution of crystals of the second type was levorotatory; it rotated plane-polarized light in the opposite direction and by an equal amount. The crystals of the second type were of the sodium ammonium salt of (−)-tartaric acid. The chirality of the crystals themselves disappeared, of course, as the crystals dissolved into their solutions *but the optical activity* remained. Pasteur reasoned, therefore, that the molecules themselves must be chiral.

Pasteur's discovery of enantiomerism and his demonstration that the optical activity of the two forms of tartaric acid was a property of the molecules themselves led, in 1874, to the proposal of the tetrahedral structure of carbon by van't Hoff and Le Bel.

Unfortunately, few organic compounds give chiral crystals as do the (+)- and (−)-tartaric acid salts. Few organic compounds crystallize into separate crystals (containing separate enantiomers) that are visibly chiral like the crystals of the sodium ammonium salt of tartaric acid. Pasteur's method, therefore, is not one that is generally applicable.

The most useful procedure for separating enantiomers is based on allowing a racemic form to react with a single enantiomer of some other compound. This changes a *racemic form into a mixture of diastereomers;* and ***diastereomers, because they have different melting points, different boiling points, and different solubilities, can be separated by conventional means.*** We shall see how this is done in Section 20.3D. The separation of the enantiomers of a racemic form is called **resolution.**

5.16 COMPOUNDS WITH STEREOCENTERS OTHER THAN CARBON

Any tetrahedral atom with four different groups attached to it is a stereocenter. Listed here are general formulas of compounds whose molecules contain stereocenters other than carbon. Silicon and germanium are in the same group of the periodic table as carbon. They form tetrahedral compounds as carbon does. When four different groups are situated around the central atom in silicon, germanium, and nitrogen compounds, the molecules are chiral and the enantiomers can be separated.

$$R_4 \overset{R_1}{\underset{R_3}{\overset{|}{\underset{|}{\mathrm{Si}}}}} R_2 \qquad R_4 \overset{R_1}{\underset{R_3}{\overset{|}{\underset{|}{\mathrm{Ge}}}}} R_2 \qquad R_4 \overset{R_1}{\underset{R_3}{\overset{|}{\underset{|}{\mathrm{N}^+}}}} R_2 \quad X^-$$

5.17 CHIRAL MOLECULES THAT DO NOT POSSESS A TETRAHEDRAL ATOM WITH FOUR DIFFERENT GROUPS

A molecule is chiral if it is not superposable on its mirror image. The presence of a tetrahedral atom with four different groups is only one focus that can confer chirality on a molecule. Most of the molecules that we shall encounter do have such stereocenters. Many chiral molecules are known, however, that do not. An example is 1,3-dichloroallene.

Allenes are compounds whose molecules contain the following double bond sequence:

$$\overset{\diagdown}{\diagup}C{=}C{=}C\overset{\diagup}{\diagdown}$$

The planes of the π bonds of allenes are perpendicular to each other.

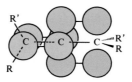

This geometry of the π bonds causes the groups attached to the end carbon atoms to lie in perpendicular planes and, because of this, allenes with different substituents on the end carbon atoms are chiral (Fig. 5.21). (Allenes do not show cis–trans isomerism.)

Mirror

FIGURE 5.21 Enantiomeric forms of 1,3-dichloroallene. These two molecules are nonsuperposable mirror images of each other and are therefore chiral. They do not possess a tetrahedral atom with four different groups, however.

5.18 SOME IMPORTANT TERMS AND CONCEPTS

Stereochemistry. Chemical studies that take into account the spatial aspects of molecules.

Isomers are different compounds that have the same molecular formula. All isomers fall into either of two groups: *constitutional* isomers or *stereoisomers.*

Constitutional isomers are isomers that have their atoms connected in a different order.

Stereoisomers have their atoms joined in the same order but differ in the way their atoms are arranged in space. Stereoisomers can be subdivided into two categories: *enantiomers* and *diastereomers.*

Enantiomers are stereoisomers that are related as an object and its mirror image. Enantiomers occur only with compounds whose molecules are chiral, that is, with molecules that are *not* identical with their mirror images. Separate enantiomers rotate the plane of plane-polarized light and are said to be *optically active.* They have equal but opposite specific rotation.

Diastereomers are stereoisomers that are not enantiomers, that is, they are stereoisomers that are not related as an object and its mirror image.

Chirality is equivalent to ''handedness.'' A chiral molecule is one that is not identical with its mirror image. An *achiral* molecule is one that can be superposed on its mirror image. Any tetrahedral atom that has four different attached groups is a **stereocenter.** A pair of enantiomers is possible for all molecules that contain a single tetrahedral stereocenter. For molecules with more than one stereocenter, the number of stereoisomers will not exceed 2^n where n is the number of stereocenters.

Plane of symmetry. A plane that bisects a molecule in such a way that the two halves of the molecule are mirror images of each other. Any molecule that has a plane of symmetry is achiral.

Configuration. The particular arrangement of atoms (or groups) in space that is characteristic of a given stereoisomer. The configuration at each tetrahedral stereocenter can be designated as (R) or (S) by using the rules given in Section 5.6.

Racemic form (racemate or racemic mixture). An equimolar mixture of enantiomers.

Meso compound. An achiral molecule that contains tetrahedral stereocenters. Since it is achiral it will be optically inactive.

Resolution. The separation of the enantiomers of a racemic form.

ADDITIONAL PROBLEMS

5.30   Which of the following are chiral and, therefore, capable of existing as enantiomers?

(a) 1,3-Dichlorobutane
(b) 1,2-Dibromopropane
(c) 1,5-Dichloropentane
(d) 3-Ethylpentane
(e) 2-Bromobicyclo[1.1.0]butane
(f) 2-Fluorobicyclo[2.2.2]octane
(g) 2-Chlorobicyclo[2.1.1]hexane
(h) 5-Chlorobicyclo[2.1.1]hexane

5.31 (a) How many carbon atoms does an alkane (not a cycloalkane) need before it is capable of existing in enantiomeric forms? (b) Give correct names for two sets of enantiomers with this minimum number of carbon atoms.

5.32 (a) Write the structure of 2,2-dichlorobicyclo[2.2.1]heptane. (b) How many stereocenters does it contain? (c) How many stereoisomers are predicted by the 2^n rule? (d) Only one pair of enantiomers is possible for 2,2-dichlorobicyclo[2.2.1]heptane. Explain.

5.33 Shown below are Newman projection formulas for (*R,R*)-, (*S,S*)-, and (*R,S*)-2,3-dichlorobutane. (a) Which is which? (b) Which formula is a meso compound?

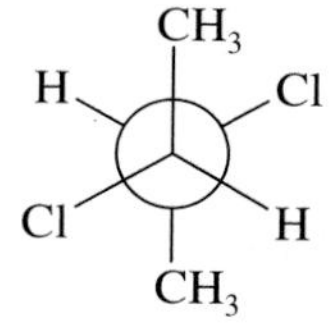

A

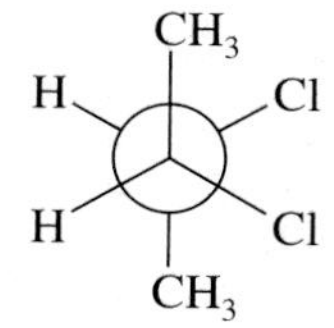

B

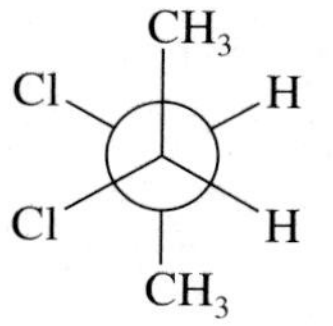

C

5.34 Write appropriate structural formulas for (a) a cyclic molecule that is a constitutional isomer of cyclohexane, (b) molecules with the formula C_6H_{12} that contain one ring and that are enantiomers of each other, (c) molecules with the formula C_6H_{12} that contain one ring and that are diastereomers of each other, (d) molecules with the formula C_6H_{12} that contain no ring and that are enantiomers of each other, and (e) molecules with the formula C_6H_{12} that contain no ring and that are diastereomers of each other.

5.35 Consider the following pairs of structures. Identify the relationship between them by describing them as representing enantiomers, diastereomers, constitutional isomers, or two molecules of the same compound.

(a) 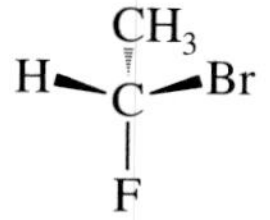 and 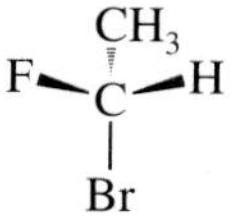

(b) 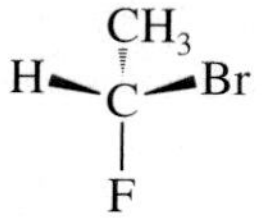 and 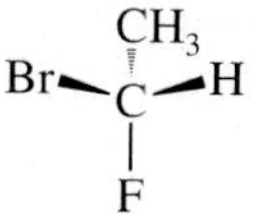

(c) 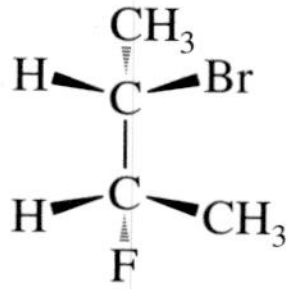 and 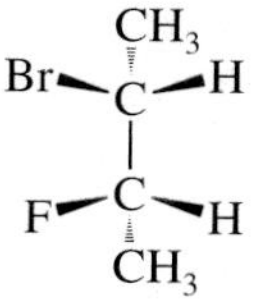

(d) 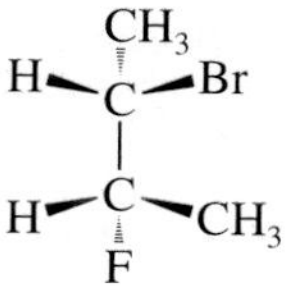 and 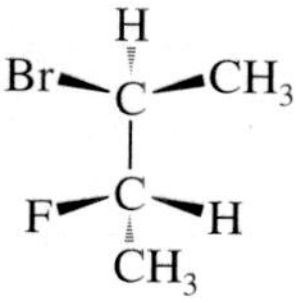

(e) 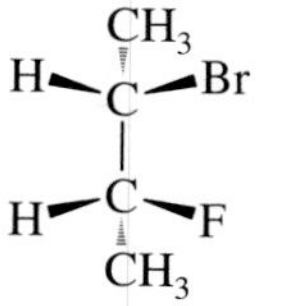 and 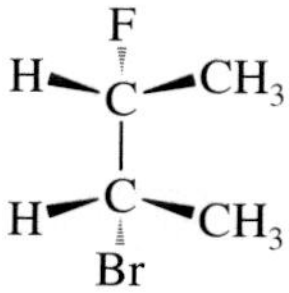

(f) 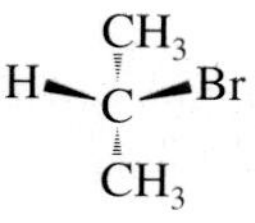 and 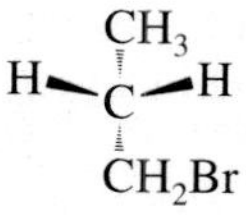

(g) 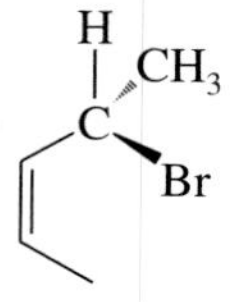 and 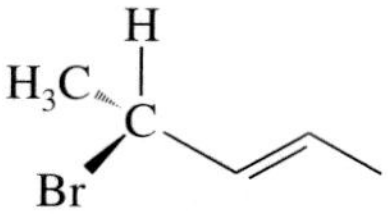

(h) 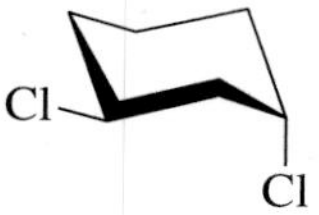 and 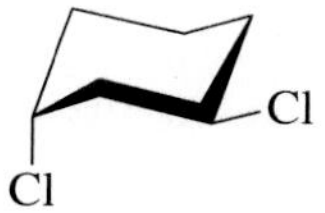

(i) 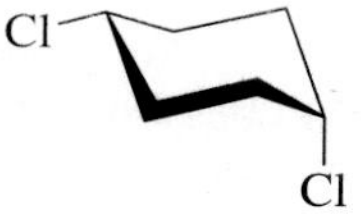 and 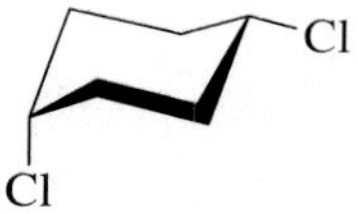

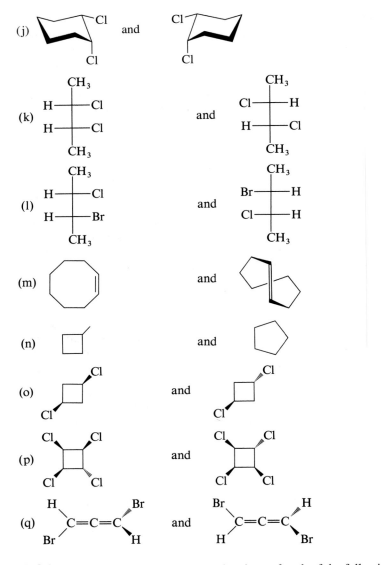

5.36   Discuss the anticipated stereochemistry of each of the following compounds.

(a) $ClCH=C=C=CHCl$ (c) $ClCH=C=C=CCl_2$

(b) $CH_2=C=C=CHCl$

✗ 5.37   There are four dimethylcyclopropane isomers.   (a) Write three-dimensional formulas for these isomers.   (b) Which of these isomers are chiral?   (c) If a mixture consisting of 1 mol of each of these isomers were subjected to simple gas–liquid chromatography, how many fractions would be obtained and which compounds would each fraction contain?   (d) How many of these fractions would be optically active?

5.38   (Use models to solve this problem.)   (a) Write a conformational structure for the most stable conformation of *trans*-1,2-diethylcyclohexane and write its mirror image.   (b) Are these two molecules superposable?   (c) Are they interconvertible through a ring ''flip''?   (d) Repeat the process in part (a) with *cis*-1,2-diethylcyclohexane.   (e) Are these structures superposable?   (f) Are they interconvertible?

5.39 (Use models to solve this problem.) (a) Write a conformational structure for the most stable conformation of *trans*-1,4-diethylcyclohexane and for its mirror image. (b) Are these structures superposable? (c) Do they represent enantiomers? (d) Does *trans*-1,4-diethylcyclohexane have a stereoisomer, and if so, what is it? (e) Is this stereoisomer chiral?

5.40 (Use models to solve this problem.) Write conformational structures for all of the stereoisomers of 1,3-diethylcyclohexane. Label pairs of enantiomers and meso compounds if they exist.

✳**5.41** Tartaric acid [HO₂CCH(OH)CH(OH)CO₂H] was an important compound in the history of stereochemistry. Two naturally occurring forms of tartaric acid are optically inactive. One form has a melting point of 206°C, the other a melting point of 140°C. The inactive tartaric acid with a melting point of 206°C can be separated into two optically active forms of tartaric acid with the same melting point (170°C). One optically active tartaric acid has $[\alpha]_D^{25} = +12°$, the other $[\alpha]_D^{25} = -12°$. All attempts to separate the other inactive tartaric acid (melting point 140°C) into optically active compounds fail. (a) Write the three-dimensional structure of the tartaric acid with melting point 140°C. (b) What are possible structures for the optically active tartaric acids with melting points of 170°C? (c) Can you be sure which tartaric acid in (b) has a positive rotation and which has a negative rotation? (d) What is the nature of the form of tartaric acid with a melting point of 206°C?

5.42 (a) Show the conversion of (S)-(+)-1-chloro-2-methylbutane (Section 5.6C) to a lithium dialkylcuprate through its reaction with lithium in diethyl ether followed by reaction with CuI. (b) Would you expect these reactions to proceed with retention of configuration or not? Explain. (c) What product would you expect to obtain if this lithium dialkylcuprate were allowed to react with (S)-(+)-1-chloro-2-methylbutane? (d) Would you expect the product from (c) to be optically active? (e) What product would you expect to obtain if the lithium dialkylcuprate from (a) were allowed to react with (R)-(−)-1-chloro-2-methylbutane? (f) Would you expect this product to be optically active? (g) How would you synthesize (3R,6R)-3,6-dimethyloctane?

CHAPTER 6

Transition State for an S_N2 reaction
(Section 6.8)

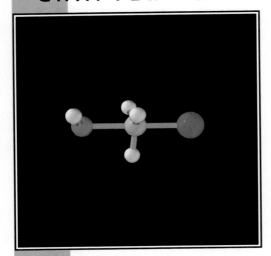

IONIC REACTIONS— NUCLEOPHILIC SUBSTITUTION AND ELIMINATION REACTIONS OF ALKYL HALIDES

6.1 INTRODUCTION

The halogen atom of an alkyl halide is attached to an sp^3-hybridized carbon. The arrangement of groups around the carbon atom, therefore, is generally tetrahedral. Because halogen atoms are more electronegative than carbon, the carbon–halogen bond of alkyl halides is *polarized;* the carbon atom bears a partial positive charge, the halogen atom a partial negative charge.

$$\overset{\delta+}{C} \longrightarrow \overset{\delta-}{X}$$

The size of the halogen atom increases as we go down the periodic table: fluorine atoms are the smallest and iodine atoms the largest. Consequently, the carbon–halogen bond length (Table 6.1) also increases as we go down the periodic table.

In the laboratory and in industry, alkyl halides are used as solvents for relatively nonpolar compounds, and they are used as the starting materials for the synthesis of many compounds. As we shall learn in this chapter, the halogen atom of an alkyl halide can be easily replaced by other groups, and the presence of a halogen atom on a carbon chain also affords us the possibility of introducing a multiple bond.

Compounds in which a halogen atom is bonded to an sp^2-hybridized carbon are

TABLE 6.1 Carbon–
halogen bond lengths

BOND	BOND LENGTH (Å)
CH_3—F	1.39
CH_3—Cl	1.78
CH_3—Br	1.93
CH_3—I	2.14

called **vinylic halides** or **phenyl halides.** The compound CH_2=CHCl has the common name **vinyl chloride** and the group, CH_2=CH—, is commonly called the **vinyl group.** *Vinylic halide,* therefore, is a general term that refers to a compound in which a halogen is attached to a carbon atom that is also forming a double bond to another carbon atom. *Phenyl halides* are compounds in which a halogen is attached to a benzene ring (Section 2.7). Phenyl halides belong to a larger group of compounds that we shall study later, called **aryl halides.**

A vinylic halide   **A phenyl halide or aryl halide**

Together with alkyl halides, these compounds comprise a larger group of compounds known simply as **organic halides** or **organohalogen compounds.** The chemistry of vinylic and aryl halides is, as we shall also learn later, quite different from alkyl halides, and it is on alkyl halides that we shall focus most of our attention in this chapter.

6.2 PHYSICAL PROPERTIES OF ORGANIC HALIDES

Most alkyl and aryl halides have very low solubilities in water, but as we might expect, they are miscible with each other and with other relatively nonpolar solvents. Dichloromethane (CH_2Cl_2, also called *methylene chloride*), trichloromethane ($CHCl_3$, also called *chloroform*), and tetrachloromethane (CCl_4, also called *carbon tetrachloride*) are often used as solvents for nonpolar and moderately polar compounds. Many chloroalkanes, including $CHCl_3$ and CCl_4, have a cumulative toxicity and are carcinogenic, however, and should therefore be used only in fume hoods and with great care.

Methyl iodide (bp, 42°C) is the only monohalomethane that is a liquid at room temperature and 1-atm pressure. Ethyl bromide (bp, 38°C) and ethyl iodide (bp, 72°C) are both liquids but ethyl chloride (bp, 13°C) is a gas. The propyl chlorides, bromides, and iodides are all liquids. In general, alkyl chlorides, bromides, and iodides are all liquids and tend to have boiling points near those of alkanes of similar molecular weights.

Polyfluoroalkanes, however, tend to have unusually low boiling points (Section 2.16D). Hexafluoroethane boils at −79°C, even though its molecular weight (MW = 138) is near that of decane (MW = 144; bp, 174°C).

Table 6.2 lists the physical properties of some common organic halides.

TABLE 6.2 Organic halides

GROUP	FLUORIDE		CHLORIDE		BROMIDE		IODIDE	
	bp (°C)	DENSITY (g mL^{-1})	bp (°C)	DENSITY (g mL^{-1})	bp (°C)	DENSITY (g mL^{-1})	bp (°C)	DENSITY (g mL^{-1})
Methyl	−78.4	0.84[−60]	−23.8	0.92[20]	3.6	1.73[0]	42.5	2.28[20]
Ethyl	−37.7	0.72[20]	13.1	0.91[15]	38.4	1.46[20]	72	1.95[20]
Propyl	−2.5	0.78[−3]	46.6	0.89[20]	70.8	1.35[20]	102	1.74[20]
Isopropyl	−9.4	0.72[20]	34	0.86[20]	59.4	1.31[20]	89.4	1.70[20]
Butyl	32	0.78[20]	78.4	0.89[20]	101	1.27[20]	130	1.61[20]
sec-Butyl			68	0.87[20]	91.2	1.26[20]	120	1.60[20]
Isobutyl			69	0.87[20]	91	1.26[20]	119	1.60[20]
tert-Butyl	12	0.75[12]	51	0.84[20]	73.3	1.22[20]	100 dec[a]	1.57[0]
Pentyl	62	0.79[20]	108.2	0.88[20]	129.6	1.22[20]	155[740]	1.52[20]
Neopentyl			84.4	0.87[20]	105	1.20[20]	127 dec[a]	1.53[13]
CH$_2$=CH—	−72	0.68[26]	−13.9	0.91[20]	16	1.52[14]	56	2.04[20]
CH$_2$=CHCH$_2$—	−3		45	0.94[20]	70	1.40[20]	102–103	1.84[22]
C$_6$H$_5$—	85	1.02[20]	132	1.10[20]	155	1.52[20]	189	1.82[20]
C$_6$H$_5$CH$_2$—	140	1.02[25]	179	1.10[25]	201	1.44[22]	93[10]	1.73[25]

[a] Decomposes is abbreviated dec.

6.3 NUCLEOPHILIC SUBSTITUTION REACTIONS

There are many reactions of the general type shown here.

$$\text{Nu}:^- \ + \ \text{R} - \ddot{\text{X}}: \longrightarrow \text{R} - \text{Nu} \ + \ :\ddot{\text{X}}:^-$$

| Nucleophile | Alkyl halide (substrate) | Product | Halide ion |

Following are some examples:

$$\text{H}\ddot{\text{O}}:^- + \text{CH}_3 - \ddot{\text{Cl}}: \longrightarrow \text{CH}_3 - \ddot{\text{O}}\text{H} + :\ddot{\text{Cl}}:^-$$

$$\text{CH}_3\ddot{\text{O}}:^- + \text{CH}_3\text{CH}_2 - \ddot{\text{Br}}: \longrightarrow \text{CH}_3\text{CH}_2 - \ddot{\text{O}}\text{CH}_3 + :\ddot{\text{Br}}:^-$$

$$:\ddot{\text{I}}:^- + \text{CH}_3\text{CH}_2\text{CH}_2 - \ddot{\text{Cl}}: \longrightarrow \text{CH}_3\text{CH}_2\text{CH}_2 - \ddot{\text{I}}: + :\ddot{\text{Cl}}:^-$$

In this type of reaction a ***nucleophile, a species with an unshared electron pair,*** reacts with an alkyl halide (called the **substrate**) by replacing the halogen substituent. A *substitution reaction* takes place and the halogen substituent, called the leaving group, departs as a halide ion. Because the substitution reaction is initiated by a nucleophile, it is called a **nucleophilic substitution reaction.**

In nucleophilic substitution reactions the carbon–halogen bond of the substrate undergoes *heterolysis,* and the unshared pair of the nucleophile is used to form a new bond to the carbon atom:

$$\underset{\text{Nucleophile}}{\text{Nu}:^-} + \ \text{R} \overset{\overset{\text{Leaving group}}{\frown}}{\underset{\substack{\text{Heterolysis} \\ \text{occurs} \\ \text{here}}}{:\ddot{\text{X}}:}} \longrightarrow \text{Nu}:\text{R} + :\ddot{\text{X}}:^-$$

One of the questions we shall want to address later in this chapter is, when does the carbon–halogen bond break? Does it break at the same time that the new bond between the nucleophile and the carbon forms?

$$\text{Nu}:^- + \text{R}:\ddot{\text{X}}: \longrightarrow \overset{\delta-}{\text{Nu}}\text{- -R- - }\overset{\delta-}{\ddot{\text{X}}}: \longrightarrow \text{Nu}:\text{R} + :\ddot{\text{X}}:^-$$

Or does the carbon–halogen bond break first?

$$\text{R}:\ddot{\text{X}}: \longrightarrow \text{R}^+ + :\ddot{\text{X}}:^-$$

then

$$\text{Nu}:^- + \text{R}^+ \longrightarrow \text{Nu}:\text{R}$$

We shall find that the answer depends primarily on the structure of the alkyl halide.

6.4   NUCLEOPHILES

A nucleophile is a reagent that seeks a positive center. (The word nucleophile comes from nucleus, the positive part of an atom plus *phile* from the Greek word *philein* meaning to love.) When a nucleophile reacts with an alkyl halide, the positive center that the nucleophile seeks is the carbon atom that bears the halogen atom. This carbon atom carries a partial positive charge because the electronegative halogen pulls the electrons of the carbon−halogen bond in its direction (see Section 1.19).

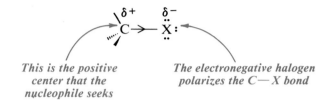

This is the positive center that the nucleophile seeks   *The electronegative halogen polarizes the C—X bond*

A nucleophile is any negative ion or any neutral molecule that has at least one unshared electron pair. For example, both hydroxide ions and water molecules can act as nucleophiles by reacting with alkyl halides to produce alcohols.

General Reaction

$$\text{H}-\overset{..}{\text{O}}{:}^{-} \; + \; \text{R}-\overset{..}{\overset{..}{\text{X}}}{:} \; \longrightarrow \; \text{R}-\overset{..}{\text{O}}-\text{H} \; + \; {:}\overset{..}{\overset{..}{\text{X}}}{:}^{-}$$

Nucleophile   Alkyl halide   Alcohol   Leaving group

General Reaction

$$\text{R}-\overset{..}{\overset{..}{\text{X}}}{:} \; + \; \text{H}-\overset{..}{\underset{|}{\text{O}}}{:} \; \longrightarrow \; \text{R}-\overset{..}{\underset{|}{\text{O}}}{\overset{+}{=}}\text{H} \; + \; {:}\overset{..}{\overset{..}{\text{X}}}{:}^{-}$$

H   H

Alkyl   Nucleophile   Alkyloxonium
halide   ion

$$\text{H}_2\text{O}\Big\updownarrow$$

$$\text{R}-\text{OH} + \text{H}_3\text{O}^+ + {:}\overset{..}{\overset{..}{\text{X}}}{:}^{-}$$

In this last reaction the first product is an alkyloxonium ion, $\text{R}-\overset{+}{\underset{|}{\text{O}}}-\text{H}$, which

H

then loses a proton to a water molecule to form an alcohol.

PROBLEM 6.1 ─────────────────────────────────────

Write the following as *net ionic reactions* and designate the nucleophile, substrate, and leaving group in each reaction.

(a) $\text{CH}_3\text{I} + \text{CH}_3\text{CH}_2\text{ONa} \longrightarrow \text{CH}_3\text{OCH}_2\text{CH}_3 + \text{NaI}$

(b) $\text{NaI} + \text{CH}_3\text{CH}_2\text{Br} \longrightarrow \text{CH}_3\text{CH}_2\text{I} + \text{NaBr}$

(c) $2\,\text{CH}_3\text{OH} + (\text{CH}_3)_3\text{CCl} \longrightarrow (\text{CH}_3)_3\text{COCH}_3 + \text{CH}_3\text{OH}_2{}^+ + \text{Cl}^-$

(d) $\text{CH}_3\text{CH}_2\text{CH}_2\text{Br} + \text{NaCN} \longrightarrow \text{CH}_3\text{CH}_2\text{CH}_2\text{CN} + \text{NaBr}$

(e) $\text{C}_6\text{H}_5\text{CH}_2\text{Br} + 2\,\text{NH}_3 \longrightarrow \text{C}_6\text{H}_5\text{CH}_2\text{NH}_2 + \text{NH}_4\text{Br}$

6.5 LEAVING GROUPS

Alkyl halides are not the only substances that can act as substrates in nucleophilic substitution reactions. We shall see later that other compounds can also react in the same way. To be reactive — that is, to be able to act as the substrate in a nucleophilic substitution reaction — a molecule must have a good **leaving group** (or **nucleofuge**). In alkyl halides the leaving group is the halogen substituent — it leaves as a halide ion. *To be a good leaving group the substituent must be able to leave as a relatively stable, weakly basic molecule or ion.* (We shall see why in Section 6.15E.) Because halide ions are relatively stable and are very weak bases, they are good leaving groups. Other groups can function as good leaving groups as well. We can write more general equations for nucleophilic substitution reactions using L to represent a leaving group.

$$\text{Nu}:^- + \text{R—L} \longrightarrow \text{R—Nu} + :\text{L}^-$$

or

$$\text{Nu}: + \text{R—L} \longrightarrow \text{R—Nu}^+ + :\text{L}^-$$

Specific Examples

$$\text{HÖ}:^- + \text{CH}_3\text{—Cl}: \longrightarrow \text{CH}_3\text{—ÖH} + :\text{Cl}:^-$$

$$\text{H}_3\text{N}: + \text{CH}_3\text{—Br}: \longrightarrow \text{CH}_3\text{—NH}_3^+ + :\text{Br}:^-$$

Later we shall also see reactions where the substrate bears a positive charge and a reaction like the following takes place:

$$\text{Nu}: + \text{R—L}^+ \longrightarrow \text{R—Nu}^+ + :\text{L}$$

Specific Example

$$\text{CH}_3\text{—Ö}: + \text{CH}_3\text{—Ö}^+\text{—H} \longrightarrow \text{CH}_3\text{—Ö}^+\text{—CH}_3 + :\text{Ö—H}$$
$$\quad\quad \overset{|}{\text{H}} \quad\quad\quad \overset{|}{\text{H}} \quad\quad\quad\quad\quad \overset{|}{\text{H}} \quad\quad\quad\quad \text{H}$$

Nucleophilic substitution reactions are more understandable and useful if we know something about their mechanisms. How does the nucleophile replace the leaving group? Does the reaction take place in one step, or is more than one step involved? If more than one step is involved, what kinds of intermediates are formed? Which steps are fast and which are slow? In order to answer these questions we need to know something about the rates of chemical reactions.

6.6 THERMODYNAMICS AND KINETICS OF CHEMICAL REACTIONS

With a chemical reaction we are usually concerned with two features: *the extent* to which it takes place and *the rate* of the reaction. By the extent of the reaction, we mean how completely the reactants will be converted to products when an equilibrium is established between them. By the rate of the reaction (also called the *kinetics* of the reaction) we mean how rapidly the reactants will be converted to products. In simpler terms then, with chemical reactions we are concerned with "how far" and "how fast."

We saw in Section 3.6 that the extent to which a reaction takes place can be expressed by an equilibrium constant, K_{eq}, and that the equilibrium constant is related to the change in the standard free energy for the reaction, $\Delta G°$:

$$\Delta G° = -2.303 \, RT \log K_{eq}$$

This relationship means that a large negative value of $\Delta G°$ ensures that the reaction *proceeds to completion*—that the reactants, for all practical purposes, will be completely converted to products when equilibrium is reached. If a reaction, on reaching equilibrium, is capable of producing several products, the major product will be the one for which the value of $\Delta G°$ is most favorable (most negative); that is, **reactions that reach equilibrium lead to the most stable product or products.** Such reactions are said to be under **thermodynamic control** or **equilibrium control.**

The value of $\Delta G°$, however, tells us nothing about how rapidly the reaction takes place—about how long it will take for equilibrium to be reached. The reaction of hydrogen with oxygen to produce water, for example,

$$2\,H_2 + O_2 \longrightarrow 2\,H_2O$$

has a very large negative free-energy change. However, in the absence of a catalyst or a flame, the reaction takes place so slowly as to be imperceptible.

For many other reactions (Fig. 6.1), pathways to several different products may have favorable free-energy changes, but equilibrium is never reached, and **the product that we actually obtain in greatest amount is the one that comes from the reaction that occurs most rapidly.** These reactions are said to be under **kinetic control** or **rate control.** We can see from this that a knowledge of the factors that determine reaction rates is of considerable importance as well.

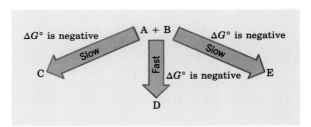

FIGURE 6.1 The hypothetical reaction of A and B to form different products, C, D, and E. All of the reactions have favorable free-energy changes. Compound D is the product that is mainly obtained, however, because it is produced by the reaction that proceeds most rapidly. Compounds E and C are not isolated in appreciable amounts because the reactions by which they are formed are slow.

6.7 KINETICS OF A NUCLEOPHILIC SUBSTITUTION REACTION: AN S$_N$2 REACTION

To understand how the rate of a reaction might be measured let us consider an actual example: the reaction that takes place between methyl chloride and hydroxide ion in aqueous solution.

$$CH_3—Cl + OH^- \xrightarrow[\text{H}_2\text{O}]{60°C} CH_3—OH + Cl^-$$

Although methyl chloride is not highly soluble in water, it is soluble enough to allow us to carry out our kinetic study. The presence of hydroxide ion in the aqueous solution can be assured by simply adding sodium hydroxide. We carry out the reaction at a specific temperature because reaction rates are known to be temperature dependent (Section 6.9).

The rate of the reaction can be determined experimentally by measuring the rate at which methyl chloride or hydroxide ion *disappears* from the solution, or the rate at which methanol or chloride ion *appears* in the solution. We can make any of these measurements by withdrawing a small sample from the reaction mixture soon after the reaction begins and analyzing it for the concentrations of CH_3Cl or OH^- and CH_3OH or Cl^-. We are interested in what are called *initial rates,* because as time passes the concentrations of the reactants change. Since we also know the initial concentrations of reactants (because we measured them when we made up the solution), it will be easy to calculate the rate at which the reactants are disappearing from the solution or the products are appearing in solution.

We perform several such experiments keeping the temperature the same, but varying the initial concentrations of the reactants. The results that we might get are shown in Table 6.3.

Notice that the experiments show that the rate depends on the concentration of methyl chloride *and* on the concentration of hydroxide ion. When we doubled the concentration of methyl chloride in experiment 2, the rate *doubled.* When we doubled the concentration of hydroxide ion in experiment 3, the rate *doubled.* When we doubled both concentrations in experiment 4 the rate increased by *four times.*

We can express these results as a proportionality,

$$\text{Rate} \propto [CH_3Cl] [OH^-]$$

and this proportionality can be expressed as an equation through the introduction of a proportionality constant (k) called the rate constant:

$$\text{Rate} = k[CH_3Cl] [OH^-]$$

For this reaction at this temperature we find that $k = 4.9 \times 10^{-4}$ L mol^{-1} s^{-1}. (Verify this for yourself by doing the calculation.)

TABLE 6.3   Rate study of reaction of CH_3Cl with OH^- at 60°C

EXPERIMENT NUMBER	INITIAL $[CH_3Cl]$	INITIAL $[OH^-]$	INITIAL RATE (mol L^{-1} s^{-1})
1	0.0010	1.0	4.9×10^{-7}
2	0.0020	1.0	9.8×10^{-7}
3	0.0010	2.0	9.8×10^{-7}
4	0.0020	2.0	19.6×10^{-7}

This reaction is said to be **second order overall.*** It is reasonable to conclude, therefore, that *for the reaction to take place a hydroxide ion and a methyl chloride molecule must collide.* We also say that the reaction is **bimolecular.** (By *bimolecular* we mean that two species are involved in the step whose rate is being measured.) We call this kind of reaction an **S$_N$2** reaction, meaning **Substitution, Nucleophilic, bimolecular.**

6.8   A MECHANISM FOR THE S$_N$2 REACTION

A modern mechanism for the S$_N$2 reaction—one based on ideas proposed by Edward D. Hughes and Sir Christopher Ingold in 1937[†]—is outlined below.

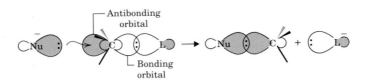

According to this mechanism, the nucleophile approaches the carbon bearing the leaving group from the **backside,** that is, from the side directly opposite the leaving group. The orbital that contains the electron pair of the nucleophile begins to overlap with an empty (antibonding) orbital of the carbon atom bearing the leaving group. As the reaction progresses the bond between the nucleophile and the carbon atom grows, and the bond between the carbon atom and the leaving group weakens. As this happens, the carbon atom has its configuration turned inside out, it becomes *inverted,*[‡] and the leaving group is pushed away. The formation of the bond between the nucleophile and the carbon atom provides most of the energy necessary to break the bond between the carbon atom and the leaving group. We can represent this mechanism with methyl chloride and hydroxide ion as shown at the top of page 233.

The Hughes–Ingold mechanism for the S$_N$2 reaction involves only one step. There are no intermediates. The reaction proceeds through the formation of an unstable arrangement of atoms called the **transition state.**

The transition state is a fleeting arrangement of the atoms in which the nucleophile and the leaving group are both partially bonded to the carbon atom undergoing attack.

---

* In general the overall order of a reaction is equal to the sum of the exponents *a* and *b* in the rate equation

$$\text{Rate} = k[A]^a [B]^b$$

If in some other reaction, for example, we found that the

$$\text{Rate} = k[A]^2 [B]$$

then we would say that the reaction is second order with respect to [A], first order with respect to [B], and third order overall.

† Ingold and Hughes of the University College, London, were pioneers in this field. Their work provided the foundation on which our modern understanding of nucleophilic substitution and elimination is built.

‡ Considerable evidence had appeared in the years prior to Hughes and Ingold's 1937 publication indicating that in reactions like this an inversion of configuration of the carbon bearing the leaving group takes place. The first observation of such an inversion was made by the Latvian chemist Paul Walden in 1896, and such inversions are called **Walden inversions** in his honor. We shall study this aspect of the S$_N$2 reaction further in Section 6.10.

A MECHANISM FOR THE S_N2 REACTION

REACTION:

$$HO^- + CH_3Cl \longrightarrow CH_3OH + Cl^-$$

MECHANISM:

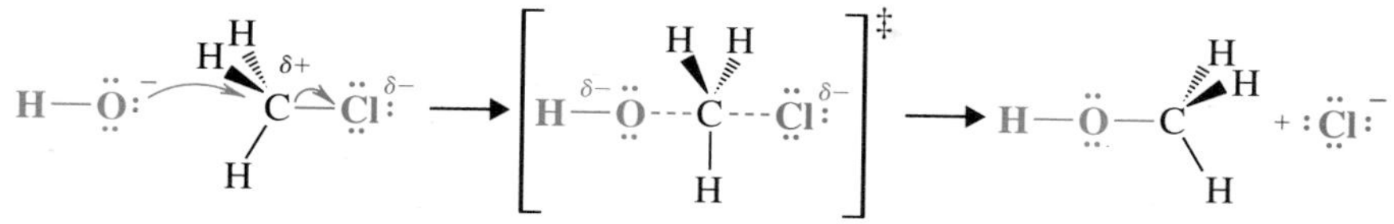

Transition state

| The hydroxide ion pushes a pair of electrons into the carbon from the backside. The chlorine begins to move away with the pair of electrons that have bonded it to the carbon. | In the transition state, a bond between oxygen and carbon is partially formed and the bond between carbon and chlorine is partially broken. The configuration of the carbon atom begins to invert. | Now the bond between the oxygen and carbon has formed and the chloride ion has departed. The configuration of the carbon has inverted. |

Because the transition state involves both the nucleophile (e.g., a hydroxide ion) and the substrate (e.g., a molecule of methyl chloride), this mechanism accounts for the second-order reaction kinetics that we observe. (Because bond formation and bond breaking occur in a single transition state, the S_N2 reaction is an example of what is called a concerted reaction.)

The transition state has an extremely brief existence. It lasts only as long as the time required for one molecular vibration, about 10^{-12} s. The energy and structure of the transition state are highly important aspects of any chemical reaction. We shall, therefore, examine this subject further in Section 6.9.

6.9 TRANSITION STATE THEORY: FREE-ENERGY DIAGRAMS

A reaction that proceeds with a negative free-energy change is said to be **exergonic;** one that proceeds with a positive free-energy change is said to be **endergonic.** The reaction between methyl chloride and hydroxide ion in aqueous solution is highly exergonic; at 60°C (333 K), $\Delta G° = -24$ kcal mol^{-1}. (The reaction is also exothermic, $\Delta H° = -18$ kcal mol^{-1}).

$$CH_3-Cl + OH^- \longrightarrow CH_3-OH + Cl^- \qquad \Delta G° = -24 \text{ kcal mol}^{-1}$$

The equilibrium constant for the reaction is extremely large:

$$\Delta G° = -2.303 \, RT \log K_{eq}$$

$$\log K_{eq} = \frac{-\Delta G°}{2.303 \, RT}$$

$$\log K_{eq} = \frac{-(-24 \text{ kcal mol}^{-1})}{2.303 \times 0.001987 \text{ kcal K}^{-1} \text{ mol}^{-1} \times 333 \text{ K}}$$

$$\log K_{eq} = 15.75$$

$$K_{eq} = 5.6 \times 10^{15}$$

An equilibrium constant as large as this means that the reaction goes to completion.

Because the free-energy change is negative, we can say that in energy terms the reaction goes **downhill.** The products of the reaction are at a lower level of free energy than the reactants.

However, considerable experimental evidence exists showing that **if covalent bonds are broken in a reaction, the reactants must go up an energy hill first,** before they can go downhill. This will be true even if the reaction is exergonic.

We can represent this graphically by plotting the free energy of the reacting particles against the reaction coordinate. Such a graph is given in Fig. 6.2. We have chosen as our example a generalized S_N2 reaction.

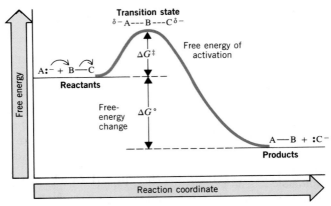

FIGURE 6.2 A free-energy diagram for a hypothetical S_N2 reaction that takes place with a negative $\Delta G°$.

The reaction coordinate is a quantity that measures the progress of the reaction. It represents the changes in bond orders and bond distances that must take place as the reactants are converted to products. In this instance the B—C distance could be used as the reaction coordinate because as the reaction progresses the B—C distance becomes longer.

In our illustration (Fig. 6.2), we can see that **an energy barrier** exists between the reactants and products. The height of this barrier (in kilocalories per mole) above the level of reactants is called **the free energy of activation, $\Delta G^{\ddagger}$.**

The top of the energy hill corresponds to the **transition state.** *The difference in free energy between the reactants and the transition state is the free energy of activation, $\Delta G^{\ddagger}$. The difference in free energy between the reactants and products is the free-energy change for the reaction, $\Delta G°$.* For our example, the free-energy level of the products is lower than that of the reactants. In terms of our analogy, we can say that the reactants in one energy valley must traverse an energy hill (the transition state) in order to reach the lower energy valley of the products.

If a reaction in which covalent bonds are broken proceeds with a positive free-energy change (Fig. 6.3), there will still be a free energy of activation. That is, if the products have greater free energy than reactants, the transition state will have a free energy even higher. ($\Delta G^{\ddagger}$ will be larger than $\Delta G°$.) In other words, in the **uphill** (endergonic) reaction an even larger energy hill lies between the reactants in one valley and the products in a higher one.

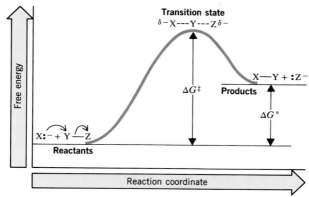

FIGURE 6.3 A free-energy diagram for a hypothetical reaction with a positive free-energy change.

Just as the overall free-energy change for a reaction contains enthalpy and entropy components (Section 3.6):

$$\Delta G° = \Delta H° - T\Delta S°$$

The free energy of activation has similar components:

$$\Delta G^{\ddagger} = \Delta H^{\ddagger} - T\Delta S^{\ddagger}$$

The enthalpy of activation ($\Delta H^{\ddagger}$) is the difference in bond energies between the reactants and the transition state. It is, in effect, the energy necessary to bring the reactants close together and to bring about the partial breaking of bonds that must happen in the transition state. Some of this energy may be furnished by the bonds that are partially formed. The entropy of activation ($\Delta S^{\ddagger}$) is the difference in entropy between the reactants and the transition state. Most reactions require the reactants to come together with a particular orientation. (Consider, e.g., the specific orientation required in the S_N2 reaction.) This requirement for a particular orientation means that the transition state must be more ordered than the reactants and that $\Delta S^{\ddagger}$ will be negative. The more highly ordered the transition state, the more negative $\Delta S^{\ddagger}$ will be. When a three-dimensional plot of free energy versus the reaction coordinate is made, the transition state is found to resemble a mountain pass or *col* (Fig. 6.4) rather than the top of an energy hill as we have shown in Figs. 6.2 and 6.3. (A plot such as that seen in Figs. 6.2 or 6.3 is simply a two-dimensional slice through the three-dimensional energy surface for the reaction.) That is, the reactants and products appear to be separated by an energy barrier resembling a mountain range. While an infinite number of possible routes lead from reactants to products, the transition state lies at the top of the route that requires the lowest energy climb. Whether or not the pass is a wide or narrow one depends on $\Delta S^{\ddagger}$. A wide pass means that there is a relatively large number of orientations of reactants that allow a reaction to take place. A narrow pass means just the opposite.

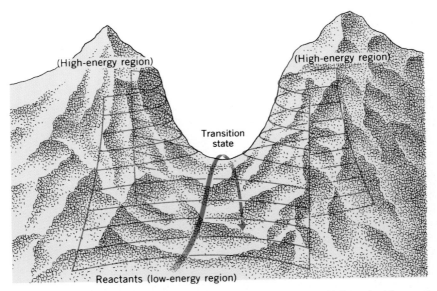

(High-energy region)   (High-energy region)

Transition
state

Reactants (low-energy region)

FIGURE 6.4 Mountain pass or col analogy for the transition state. (Adapted with permission from J. E. Leffler and E. Grunwald, *Rates and Equilibria of Organic Reactions,* Wiley, New York, 1963, p. 6.)

The existence of an activation energy ($\Delta G^{\ddagger}$) explains why most chemical reactions occur much more rapidly at higher temperatures. *For many reactions taking place near room temperature, a 10°C increase in temperature will cause the reaction rate to double.*

This dramatic increase in reaction rate results from a large increase in the number of collisions between reactants that together have sufficient energy to surmount the barrier at the higher temperature. The kinetic energies of molecules at a given temperature are not all the same. Figure 6.5 shows the distribution of energies brought to collisions at two temperatures (that do not differ greatly), labeled T_1 and T_2. Because of the way energies are distributed at different temperatures (as indicated by the shapes of the curves), increasing the temperature by only a small amount causes a large increase in the number of collisions with larger energies. In Fig. 6.5 we have designated a particular minimal free energy as being required to bring about a reaction between colliding molecules. The number of collisions having sufficient energy to allow reaction to take place at a given temperature is proportional to the area under the portion of the curve that represents free energies greater than or equal to $\Delta G^{\ddagger}$. At the lower temperature (T_1) this number is relatively small. At the higher temperature (T_2), however, the number of collisions that take place with enough energy to react is very much larger. Consequently, a modest temperature increase produces a large increase in the number of collisions with energy sufficient to lead to a reaction.

There is also an important relationship between the rate of a reaction and the magnitude of the free energy of activation. The relationship between the rate constant (k) and $\Delta G^{\ddagger}$ is an *exponential one.*

$$k = k_0 e^{-\Delta G^{\ddagger}/RT}$$

In this equation, e is 2.718, the base of natural logarithms, and k_0 is the absolute rate constant, which equals the rate at which all transition states proceed to products. At

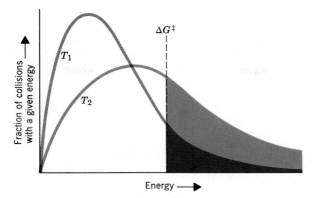

FIGURE 6.5 The distribution of energies at two different temperatures, T_1 and T_2 ($T_2 > T_1$). The number of collisions with energies greater than the free energy of activation is indicated by the appropriately shaded area under each curve.

25°C, $k_0 = 6.2 \times 10^{12}$ s^{-1}. Because of this exponential relationship, **a reaction with a lower free energy of activation will occur very much faster than a reaction with a higher one.**

Generally speaking, if a reaction has a $\Delta G^{\ddagger}$ less than 20 kcal mol^{-1}, it will take place readily at room temperature or below. If $\Delta G^{\ddagger}$ is greater than 20 kcal mol^{-1} heating will be required to cause the reaction to occur at a reasonable rate.

A free-energy diagram for the reaction of methyl chloride with hydroxide ion is shown in Fig. 6.6. At 60°C, $\Delta G^{\ddagger} = 24.5$ kcal mol^{-1}, which means that at this temperature, the reaction reaches completion in a matter of a few hours.

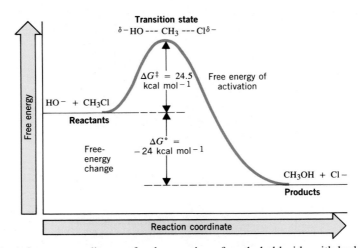

FIGURE 6.6 A free-energy diagram for the reaction of methyl chloride with hydroxide ion at 60°C.

6.10 THE STEREOCHEMISTRY OF S_N2 REACTIONS

As we learned earlier (Section 6.8), in an S_N2 reaction *the nucleophile attacks from the backside, that is, from the side directly opposite the leaving group.* This mode of attack (see following figure) causes **a change in the configuration** of the carbon atom that is the object of nucleophilic attack. (The configuration of an atom *is the particular arrangement of groups around that atom in space,* Section 5.6C) As the displacement takes place, the configuration of the carbon atom under attack **inverts** —it is turned inside out in much the same way that an umbrella is turned inside out, or inverts, when caught in a strong wind.

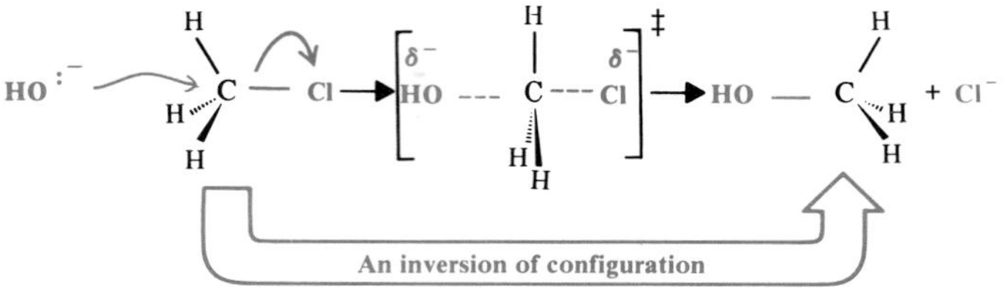

An inversion of configuration

With a molecule such as methyl chloride, however, there is no way to prove that attack by the nucleophile inverts the configuration of the carbon atom because one form of methyl chloride is identical to its inverted form. With a cyclic molecule such as *cis*-1-chloro-3-methylcyclopentane, however, we can observe the results of a *configuration inversion.* When *cis*-1-chloro-3-methylcyclopentane reacts with hydroxide ion in an S_N2 reaction the product is *trans*-3-methylcyclopentanol. *The hydroxide ion ends up being bonded on the opposite side of the ring from the chloride it replaces:*

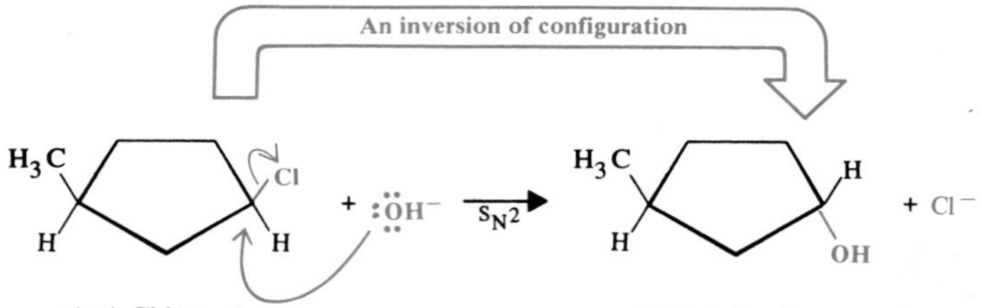

An inversion of configuration

cis-1-Chloro-3-methylcyclopentane

trans-3-Methylcyclopentanol

Presumably, the transition state for this reaction is like that shown here.

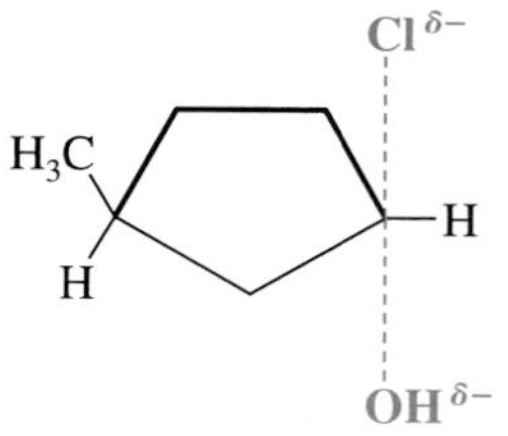

Leaving group departs from this side

Nucleophile attacks from this side

PROBLEM 6.2

Use conformational structures and show the nucleophilic substitution reaction that would take place when *trans*-1-bromo-4-*tert*-butylcyclohexane reacts with iodide ion. (Show the most stable conformation of the reactant and the product.)

We can also observe an inversion of configuration with an acyclic molecule *when the S_N2 reaction takes place at a stereocenter.* Here, too, we find that **S_N2 reactions always lead to inversion of configuration.**

A compound that contains one stereocenter, and therefore exists as a pair of enantiomers, is 2-bromooctane. These enantiomers have been obtained separately and are known to have the configurations and rotations shown here.

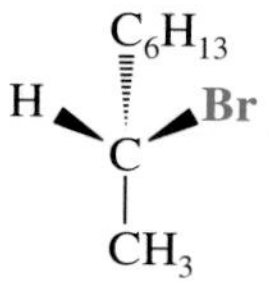

(R)-(−)-2-Bromooctane
$[\alpha]_D^{25} = -34.25°$

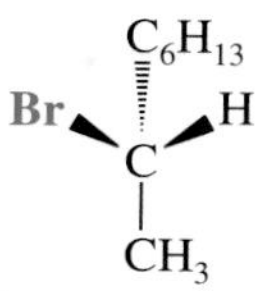

(S)-(+)-2-Bromooctane
$[\alpha]_D^{25} = +34.25°$

The alcohol 2-octanol is also chiral. The configurations and rotations of the 2-octanol enantiomers have also been determined:

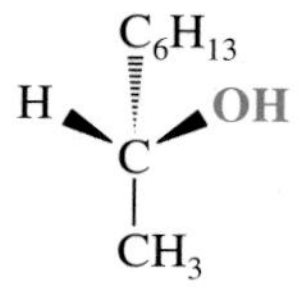

(R)-(−)-2-Octanol
$[\alpha]_D^{25} = -9.90°$

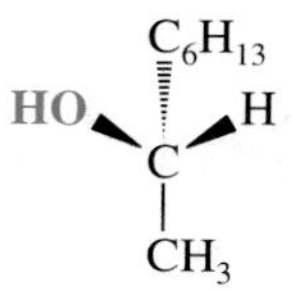

(S)-(+)-2-Octanol
$[\alpha]_D^{25} = +9.90°$

When (R)-(−)-2-bromooctane reacts with sodium hydroxide, the only substitution product that is obtained from the reaction is (S)-(+)-2-octanol. The following reaction is S_N2 and takes place with *complete inversion of configuration.*

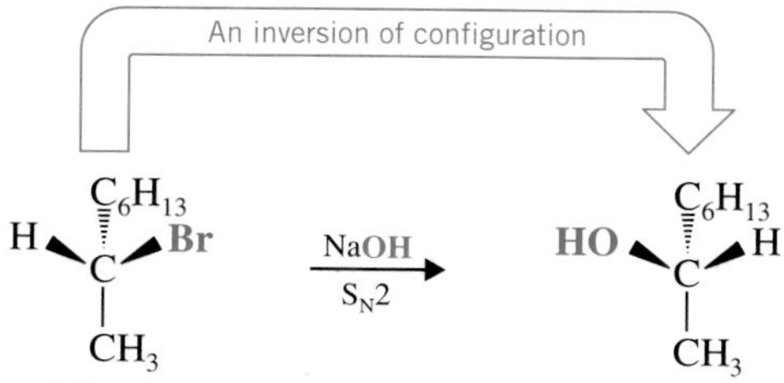

An inversion of configuration

(R)-(−)-2-Bromooctane
$[\alpha]_D^{25} = -34.25°$
Enantiomeric purity = 100%

(S)-(+)-2-Octanol
$[\alpha]_D^{25} = +9.90°$
Enantiomeric purity = 100%

PROBLEM 6.3

S_N2 reactions that involve breaking a bond to a stereocenter can be used to relate configurations of molecules because the *stereochemistry* of the reaction is known. (a) Illustrate how this is true by assigning configurations to the 2-chlorobutane enantiomers based on the following data. [The configuration of (−)-2-butanol is given in Section 5.6C.]

$$\text{(+)-2-Chlorobutane} \xrightarrow[S_N2]{OH^-} \text{(−)-2-Butanol}$$

$[\alpha]_D^{25} = +36.00°$ $[\alpha]_D^{25} = -13.52°$
Enantiomerically pure   **Enantiomerically pure**

(b) When optically pure (+)-2-chlorobutane is allowed to react with potassium iodide in acetone in an S_N2 reaction, the 2-iodobutane that is produced has a minus rotation. What is the configuration of (−)-2-iodobutane? Of (+)-2-iodobutane?

6.11 THE REACTION OF *TERT*-BUTYL CHLORIDE WITH HYDROXIDE ION: AN S_N1 REACTION

When *tert*-butyl chloride reacts with sodium hydroxide in a mixture of water and acetone, the kinetic results are quite different. The rate of formation of *tert*-butyl alcohol is dependent on the concentration of *tert*-butyl chloride, but it is *independent of the concentration of hydroxide ion*. Doubling the *tert*-butyl chloride concentration *doubles* the rate of the reaction, but changing the hydroxide ion concentration (within limits) has no appreciable effect. *tert*-Butyl chloride reacts by substitution at virtually the same rate in pure water (where the hydroxide ion is $10^{-7}M$) as it does in $0.05M$ aqueous sodium hydroxide (where the hydroxide ion concentration is 500,000 times larger). (We shall see in Section 6.12 that the important nucleophile in this reaction is a molecule of water.)

Thus the rate equation for this substitution reaction is first order with respect to *tert*-butyl chloride and *first order overall*.

$$(CH_3)_3C-Cl + OH^- \xrightarrow[H_2O]{acetone} (CH_3)_3C-OH + Cl^-$$

$$\text{Rate} \propto [(CH_3)_3CCl]$$

$$\text{Rate} = k[(CH_3)_3CCl]$$

We can conclude, therefore, that hydroxide ions do not participate in the transition state of the step that controls the rate of the reaction, and that only molecules of *tert*-butyl chloride are involved. This reaction is said to be **unimolecular.** We call this type of reaction an **S_N1** reaction (**Substitution, Nucleophilic, unimolecular).**

How can we explain an S_N1 reaction in terms of a mechanism? To do so we shall need to consider the possibility that the mechanism involves more than one step. But what kind of kinetic results should we expect from a multistep reaction? Let us consider this point further.

6.11A Multistep Reactions and the Rate-determining Step

If a reaction takes place in a series of steps, and if the first step is intrinsically slower than all the others, then the rate of the overall reaction will be essentially the same as the rate of this slow step. This slow step, consequently, is called the **rate-limiting step** or the **rate-determining step.**

Consider a multistep reaction such as the following:

Step 1   Reactant $\xrightarrow{\text{slow}}$ intermediate-1

Step 2   Intermediate-1 $\xrightarrow{\text{fast}}$ intermediate-2

Step 3   Intermediate-2 $\xrightarrow{\text{fast}}$ product

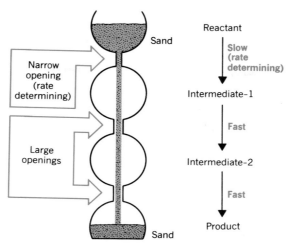

FIGURE 6.7 A modified hourglass that serves as an analogy for a multistep reaction. The overall rate is limited by the rate of the slow step.

When we say that the first step is intrinsically slow, we mean that the rate constant for step 1 is very much smaller than the rate constant for step 2 or for step 3:

Step 1 Rate = k_1 [Reactant]

Step 2 Rate = k_2 [Intermediate-1]

Step 3 Rate = k_3 [Intermediate-2]

$$k_1 \ll k_2 \quad \text{or} \quad k_3$$

When we say that steps 2 and 3 are *fast,* we mean that because their rate constants are larger, they could (in theory) take place rapidly if the concentrations of the two intermediates ever became high. In actuality, the concentrations of the intermediates are always very small because of the slowness of step 1, and steps 2 and 3 actually occur at the same rate as step 1.

An analogy may help clarify this. Imagine an hourglass modified in the way shown in Fig. 6.7. The opening between the top chamber and the one just below is considerably smaller than the other two. The overall rate at which sand falls from the top to the bottom of the hourglass is limited by the rate at which sand passes through this small orifice. This step, in the passage of sand, is analogous to the rate-determining step of the multistep reaction.

6.12 A MECHANISM FOR THE S$_N$1 REACTION

The mechanism for the reaction of *tert*-butyl chloride with water (Section 6.11) apparently involves three steps. Two distinct **intermediates** are formed. The first step is the slow step—it is the rate-determining step. In it a molecule of *tert*-butyl chloride ionizes and becomes a *tert*-butyl cation and a chloride ion. Carbocation formation in general takes place slowly because it is usually a highly endothermic process and is uphill in terms of free energy.

A MECHANISM FOR THE $S_N 1$ REACTION

REACTION:

$$(CH_3)_3CCl + 2 H_2O \longrightarrow (CH_3)_3COH + H_3O^+ + Cl^-$$

MECHANISM:

Step 1

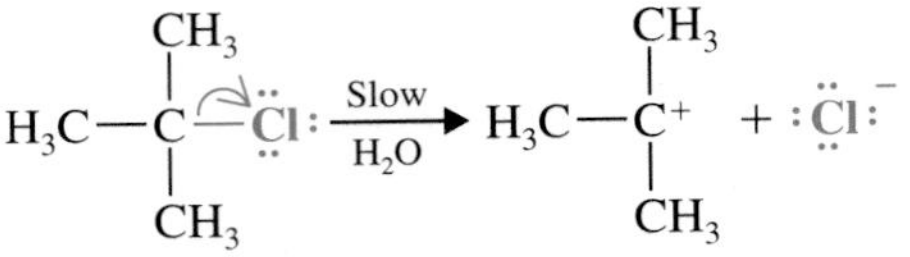

Aided by the polar solvent a chlorine departs with the electron pair that bonded it to the carbon.

This slow step produces the relatively stable 3° carbocation and a chloride ion. Although not shown here, the ions are solvated (and stabilized) by water molecules.

Step 2

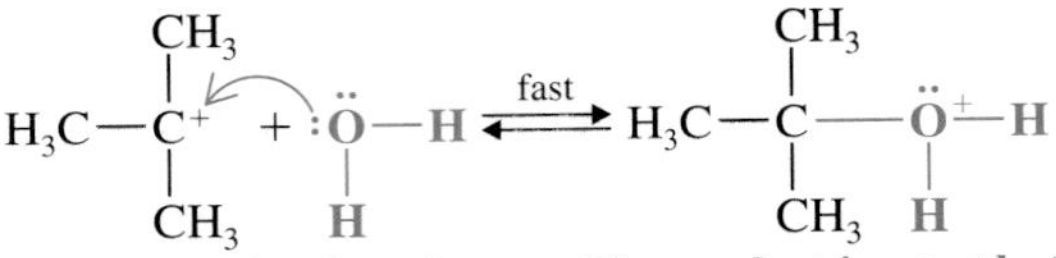

A water molecule acting as a Lewis base donates an electron pair to the carbocation (a Lewis acid). This gives the cationic carbon eight electrons.

The product is a *tert*-butyl oxonium ion (or protonated *tert*-butyl alcohol).

Step 3

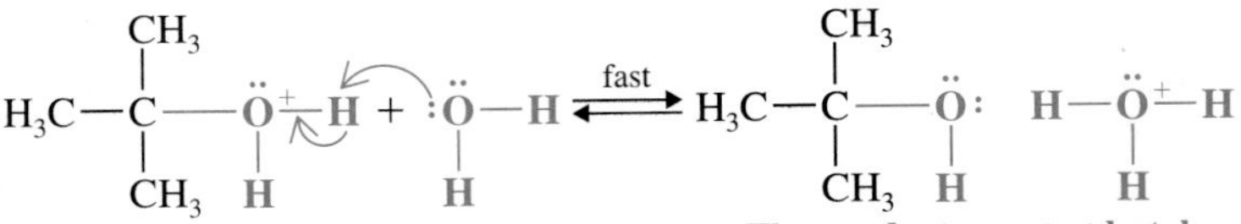

A water molecule acting as a Brønsted base accepts a proton from the *tert*-butyloxonium ion.

The products are *tert*-butyl alcohol and a hydronium ion.

In the second step the intermediate *tert*-butyl cation reacts rapidly with water to produce a *tert*-butyloxonium ion (another intermediate); this, in the third step, rapidly transfers a proton to a molecule of water producing *tert*-butyl alcohol.

The first step requires heterolytic cleavage of the carbon–chlorine bond. Because no other bonds are formed in this step, it should be highly endothermic and it should have a high free energy of activation. That it takes place at all is largely because of the ionizing ability of the solvent, water. Experiments indicate that in the gas phase (i.e., in the absence of a solvent), the free energy of activation is about 150 kcal mol^{-1}! In aqueous solution, however, the free energy of activation is much lower—about 20 kcal mol^{-1}. Water molecules surround and stabilize the cation and anion that are produced (cf. Section 2.16E).

Even though the *tert*-butyl cation produced in step 1 is stabilized by solvation, it is still a highly reactive species. Almost immediately after it is formed, it reacts with one of the surrounding water molecules to form the *tert*-butyloxonium ion, $(CH_3)_3COH_2^+$. (It may also occasionally react with a hydroxide ion, but water molecules are far more plentiful.)

A free-energy diagram for the S_N1 reaction of *tert*-butyl chloride and water is given in Fig. 6.8.

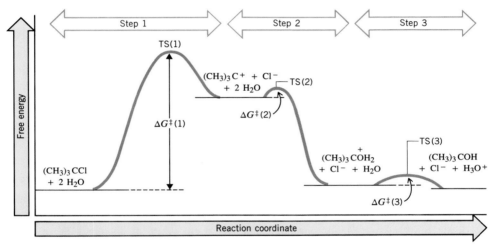

FIGURE 6.8 A free-energy diagram for the S_N1 reaction of *tert*-butyl chloride with water. The free energy of activation for the first step, $\Delta G^{\ddagger}(1)$, is much larger than $\Delta G^{\ddagger}(2)$ or $\Delta G^{\ddagger}(3)$. TS(1) represents transition state (1), and so on.

The important transition state for the S_N1 reaction is the transition state of the rate-determining step [TS(1)]. In it the carbon–chlorine bond of *tert*-butyl chloride is largely broken and ions are beginning to develop:

$$CH_3 - \overset{\overset{\displaystyle CH_3}{|}}{\underset{\underset{\displaystyle CH_3}{|}}{C}} \overset{\delta+}{\cdots} Cl^{\delta-}$$

The solvent (water) stabilizes these developing ions by solvation.

6.13 CARBOCATIONS

Beginning in the 1920s much evidence began to accumulate implicating simple alkyl cations as intermediates in a variety of ionic reactions. However, because alkyl cations are highly unstable and highly reactive they were, in all instances studied before 1962, very short-lived, transient species that could not be observed directly.* However, in 1962 George A. Olah (now at the University of Southern California) and his co-workers published the first of a series of papers describing experiments in which alkyl cations were prepared in an environment in which they were reasonably stable and in which they could be observed by a number of spectroscopic techniques.†

6.13A The Structure of Carbocations

Considerable experimental evidence indicates that the structure of carbocations is **trigonal planar** like that of BH_3 (Section 1.14). Just as the trigonal planar structure of BH_3 can be accounted for on the basis of sp^2 hybridization so, too (Fig. 6.9), can the trigonal planar structure of carbocations.

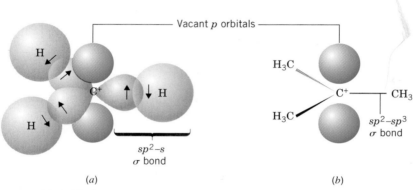

(a)   (b)

FIGURE 6.9 (a) Orbital structure of the methyl cation. The bonds are sigma bonds (σ) formed by overlap of the carbon atom's three sp^2 orbitals with the $1s$ orbitals of the hydrogen atoms. The p orbital is vacant. (b) A dashed line–wedge representation of the *tert*-butyl cation. The bonds between carbon atoms are formed by overlap of sp^3 orbitals of the methyl groups with sp^2 orbitals of the central carbon atom.

The central carbon atom in a carbocation is electron deficient; it has only six electrons in its outside energy level. In our model (Fig. 6.9) these six electrons are used to form sigma covalent bonds to hydrogen atoms (or to alkyl groups). The p orbital contains no electrons.

6.13B The Relative Stabilities of Carbocations

A large body of experimental evidence indicates that the relative stabilities of carbocations are related to the number of alkyl groups attached to the positively charged trivalent carbon atom. Tertiary carbocations are the most stable, and the methyl cation is the least stable. The overall order of stability is as follows:

* As we shall learn later, carbocations bearing aromatic groups can be much more stable; one of these had been studied as early as 1901.

† Olah was awarded the Nobel Prize for his work in 1994.

$$R-\overset{\displaystyle R}{\underset{\displaystyle R}{\overset{|}{\underset{|}{C}}}}{}^{+} > R-\overset{\displaystyle R}{\underset{\displaystyle H}{\overset{|}{\underset{|}{C}}}}{}^{+} > R-\overset{\displaystyle H}{\underset{\displaystyle H}{\overset{|}{\underset{|}{C}}}}{}^{+} > H-\overset{\displaystyle H}{\underset{\displaystyle H}{\overset{|}{\underset{|}{C}}}}{}^{+}$$

3°	>	2°	>	1°	>	Methyl
(most stable)						(least stable)

This order of stability of carbocations can be explained on the basis of a law of physics that states that *a charged system is stabilized when the charge is dispersed or delocalized.* Alkyl groups, when compared to hydrogen atoms, are **electron releasing.** This means that alkyl groups will shift electron density toward a positive charge. Through electron release, *alkyl groups* attached to the positive carbon atom of a carbocation **delocalize** the positive charge. In doing so, the attached alkyl groups assume part of the positive charge themselves and thus *stabilize* the carbocation. We can see how this occurs by inspecting Fig. 6.10.

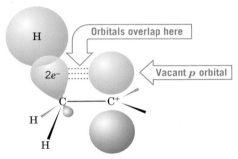

FIGURE 6.10 *How a methyl group helps stabilize the positive charge of a carbocation.* Electron density from one of the carbon–hydrogen sigma bonds of the methyl group flows into the vacant *p* orbital of the carbocation because the orbitals can partly overlap. Shifting electron density in this way makes the sp^2-hybridized carbon of the carbocation somewhat less positive and the hydrogens of the methyl group assume some of the positive charge. Delocalization (dispersal) of the charge in this way leads to greater stability. This interaction of a bond orbital with a *p* orbital is called **hyperconjugation.**

In the *tert*-butyl cation (see the following structure) three electron-releasing methyl groups surround the central carbon atom and assist in delocalizing the positive charge. In the isopropyl cation there are only two attached methyl groups that can serve to delocalize the charge. In the ethyl cation there is only one attached methyl group, and in the methyl cation there is none at all. As a result, *the delocalization of charge and the order of stability of the carbocations parallel the number of attached methyl groups.*

***tert*-Butyl cation (3°) (most stable)**   **Isopropyl cation (2°)**   **Ethyl cation (1°)**   **Methyl cation (least stable)**

The relative stability of carbocations is 3° > 2° > 1° > methyl.

6.14 THE STEREOCHEMISTRY OF S$_N$1 REACTIONS

Because the carbocation formed in the first step of an S$_N$1 reaction has a trigonal planar structure (Section 6.13A) when it reacts with a nucleophile, it may do so from either the frontside or the backside (see following illustration). With the *tert*-butyl cation this makes no difference because the same product is formed by either mode of attack.

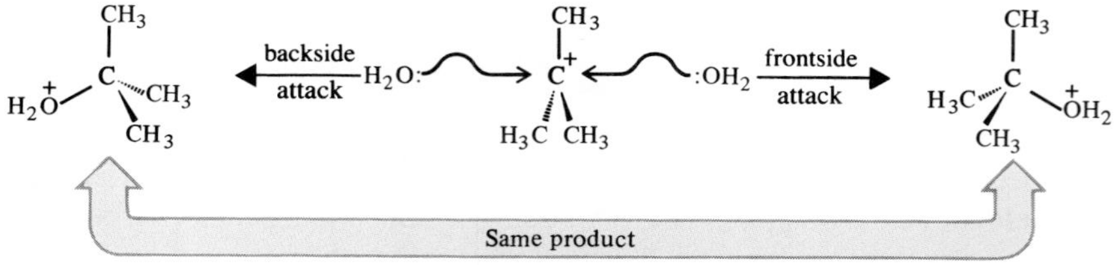

With some cations, however, different products arise from the two reaction possibilities. We shall study this point next.

6.14A Reactions that Involve Racemization

A reaction that transforms an optically active compound into a racemic form is said to proceed with **racemization.** If the original compound loses all of its optical activity in the course of the reaction, chemists describe the reaction as having taken place with *complete* racemization. If the original compound loses only part of its optical activity, as would be the case if an enantiomer were only partially converted to a racemic form, then chemists describe this as proceeding with *partial* racemization.

Racemization takes place *whenever the reaction causes chiral molecules to be converted to an achiral intermediate.*

Examples of this type of reaction are S$_N$1 reactions in which the leaving group departs from a stereocenter. These reactions almost always result in extensive and sometimes complete racemization. For example, heating optically active (*S*)-3-bromo-3-methylhexane with aqueous acetone results in the formation of 3-methyl-3-hexanol as a racemic form.

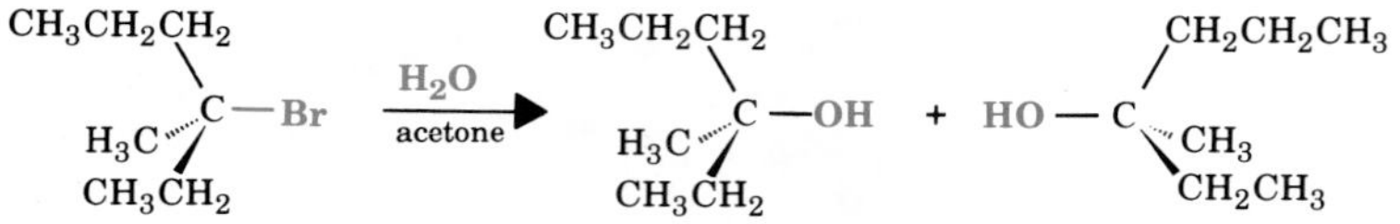

(S)-3-Bromo-3-
methylhexane
(optically active)

(S)-3-Methyl-
3-hexanol

(R)-3-Methyl-
3-hexanol

(optically inactive, a racemic form)

The reason: The S$_N$1 reaction proceeds through the formation of an intermediate carbocation (Fig. 6.11) and the carbocation, because of its trigonal planar configuration, *is achiral.* It reacts with water with equal rates from either side to form the enantiomers of 3-methyl-3-hexanol in equal amounts.

PROBLEM 6.4 _____

Keeping in mind that carbocations have a trigonal planar structure,   (a) write a

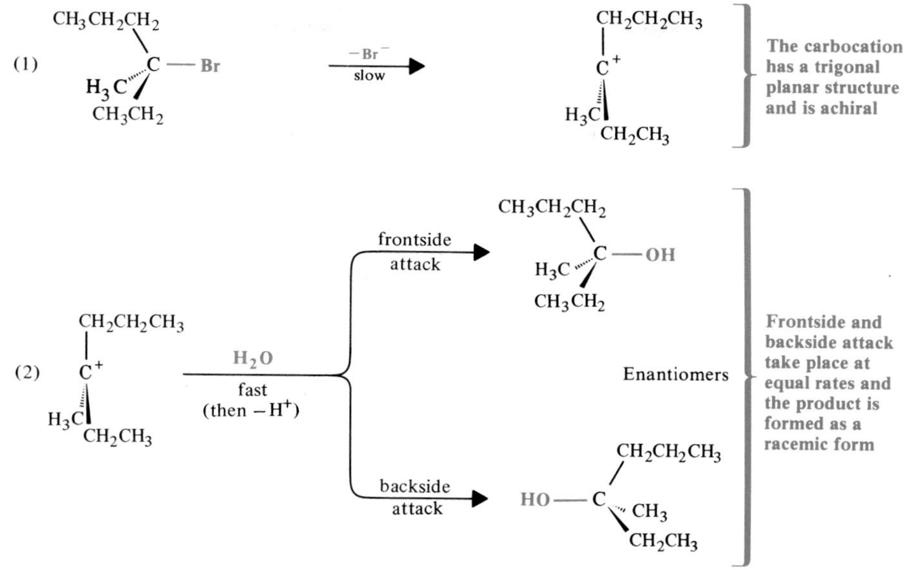

(1)

(2)

FIGURE 6.11 The S$_N$1 reaction of 3-bromo-3-methylhexane proceeds with racemization because the intermediate carbocation is achiral.

structure for the carbocation intermediate and   (b) write structures for the alcohol (or alcohols) that you would expect from the following reaction:

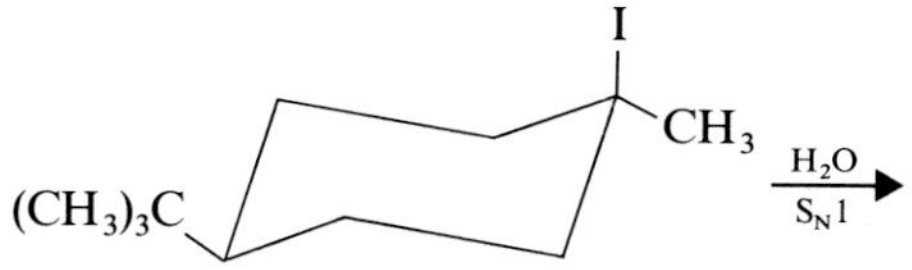

6.14B Solvolysis

The S$_N$1 reaction of an alkyl halide with water is an example of **solvolysis.** A solvolysis is a nucleophilic substitution in which *the nucleophile is a molecule of the solvent* (*solvent + lysis:* cleavage by the solvent). Since the solvent in this instance is water, we could also call the reaction a **hydrolysis.** If the reaction had taken place in methanol we would call the reaction a **methanolysis.**

Examples of Solvolysis

$$(CH_3)_3C—Br + H_2O \longrightarrow (CH_3)_3C—OH + HBr$$

$$(CH_3)_3C—Cl + CH_3OH \longrightarrow (CH_3)_3C—OCH_3 + HCl$$

$$(CH_3)_3C—Cl + H\overset{\overset{\displaystyle O}{\|}}{C}OH \longrightarrow (CH_3)_3C—O\overset{\overset{\displaystyle O}{\|}}{C}H + HCl$$

These reactions all involve the initial formation of a carbocation and the subsequent reaction of that cation with a molecule of the solvent. In the last example the solvent is formic acid (HCO$_2$H) and the following steps take place:

Step 1

$$(CH_3)_3C\overset{\frown}{-}Cl \xrightarrow{\text{slow}} (CH_3)_3C^+ + Cl^-$$

Step 2

$$(CH_3)_3C^+ + H\overset{..}{\underset{..}{O}}-\overset{\overset{O}{\|}}{CH} \xrightarrow{\text{fast}} (CH_3)_3C-\underset{\underset{H}{|}}{\overset{+}{O}}\overset{\overset{O}{\|}}{-CH}$$

Step 3

$$(CH_3)_3C-\underset{\underset{H}{|}}{\overset{+}{O}}\overset{\overset{O}{\|}}{-CH} \xrightarrow{\text{fast}} (CH_3)_3C-O-\overset{\overset{O}{\|}}{CH} + H^+$$

PROBLEM 6.5

What product(s) would you expect from the methanolysis of the cyclohexane derivative given as the reactant in Problem 6.4?

6.15 FACTORS AFFECTING THE RATES OF S_N1 AND S_N2 REACTIONS

Now that we have an understanding of the mechanisms of S_N2 and S_N1 reactions, our next task is to explain why methyl chloride reacts by an S_N2 mechanism and *tert*-butyl chloride by an S_N1 mechanism. We would also like to be able to predict which pathway—S_N1 or S_N2—would be followed by the reaction of any alkyl halide with any nucleophile under varying conditions.

The answer to this kind of question is to be found in the *relative rates of the reactions that occur.* If a given alkyl halide and nucleophile react *rapidly* by an S_N2 mechanism but *slowly* by an S_N1 mechanism under a given set of conditions, then an S_N2 pathway will be followed by most of the molecules. On the other hand, another alkyl halide and another nucleophile may react very slowly (or not at all) by an S_N2 pathway. If they react rapidly by an S_N1 mechanism, then the reactants will follow an S_N1 pathway.

Experiments have shown that a number of factors affect the relative rates of S_N1 and S_N2 reactions. The most important factors are

1. The structure of the substrate.
2. The concentration and reactivity of the nucleophile (for bimolecular reactions only).
3. The effect of the solvent.
4. The nature of the leaving group.

6.15A The Effect of the Structure of the Substrate

S_N2 Reactions   Simple alkyl halides show the following general order of reactivity in S_N2 reactions:

methyl > primary > secondary $\gg$ (tertiary)

Methyl halides react most rapidly and tertiary halides react so slowly as to be un-reactive by the S$_N$2 mechanism. Table 6.4 gives the relative rates of typical S$_N$2 reactions.

TABLE 6.4   Relative rates of reactions of alkyl halides in S$_N$2 reactions

SUBSTITUENT	COMPOUND	RELATIVE RATE
Methyl	CH_3—X	30
1°	CH_3CH_2—X	1
2°	$(CH_3)_2CHX$	0.02
Neopentyl	$(CH_3)_3CCH_2X$	0.00001
3°	$(CH_3)_3CX$	~0

Neopentyl halides, even though they are primary halides, are very unreactive.

$$CH_3-\underset{\underset{CH_3}{|}}{\overset{\overset{CH_3}{|}}{C}}-CH_2-X$$

A neopentyl halide

The important factor behind this order of reactivity is a **steric effect.** A steric effect is an effect on relative rates caused by the space-filling properties of those parts of a molecule attached at or near the reacting site. One kind of steric effect—the kind that is important here—is called **steric hindrance.** By this we mean that the spatial arrangement of the atoms or groups at or near the reacting site of a molecule hinders or retards a reaction.

For particles (molecules and ions) to react, their reactive centers must be able to come within bonding distance of each other. Although most molecules are reasonably flexible, very large and bulky groups can often hinder the formation of the required transition state. In some cases they can prevent its formation altogether.

An S$_N$2 reaction requires an approach by the nucleophile to a distance within bonding range of the carbon atom bearing the leaving group. Because of this, bulky substituents on *or near* that carbon atom have a dramatic inhibiting effect (Fig. 6.12). They cause the potential energy of the required transition state to be increased and, consequently, they increase the free energy of activation for the reaction. Of the simple alkyl halides, methyl halides react most rapidly in S$_N$2 reactions because only three small hydrogen atoms interfere with the approaching nucleophile. Neopentyl and ter-tiary halides are the least reactive because bulky groups present a strong hindrance to the approaching nucleophile. (Tertiary substrates, for all practical purposes, do not react by an S$_N$2 mechanism.)

S$_N$1 Reactions   *The primary factor that determines the reactivity of organic sub-strates in an S$_N$1 reaction is the relative stability of the carbocation that is formed.*

Except for those reactions that take place in strong acids, which we shall study later, the only organic compounds that undergo reaction by an S$_N$1 path at a reasonable rate are *those that are capable of forming relatively stable carbocations.* Of the simple

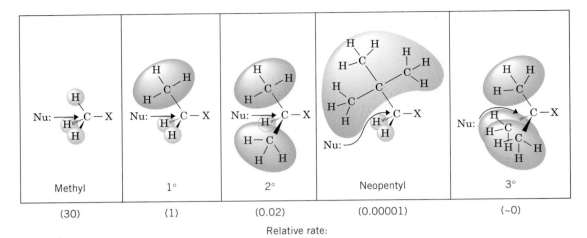

Relative rate:

FIGURE 6.12 Steric effects in the S_N2 reaction.

alkyl halides that we have studied so far, this means (for all practical purposes) that only tertiary halides react by an S_N1 mechanism. (Later we shall see that certain organic halides, called *allylic halides* and *benzylic halides*, can also react by an S_N1 mechanism because they can form relatively stable carbocations; cf. Section 13.11.)

Tertiary carbocations are stabilized because three alkyl groups release electrons to the positive carbon atom and thereby disperse its charge (see Section 6.13B).

Formation of a relatively stable carbocation is important in an S_N1 reaction because it means that the free energy of activation for the slow step for the reaction (i.e., $R—X \longrightarrow R^+ + X^-$) will be low enough for the overall reaction to take place at a reasonable rate. If you examine Fig. 6.8 again, you will see this step (step 1) is *uphill in terms of free energy* ($\Delta G°$ for this step is positive). It is also uphill in terms of enthalpy ($\Delta H°$ is also positive), and, therefore, this step is *endothermic*. According to a postulate made by G. S. Hammond (then at the California Institute of Technology) and J. E. Leffler (Florida State University) **the transition state for a step that is uphill in energy should show a strong resemblance to the product of that step.** Since the product of this step (actually an intermediate in the overall reaction) is a carbocation, any factor that stabilizes it—such as dispersal of the positive charge by electron-releasing groups—should also stabilize the transition state in which the positive charge is developing.

Step (1)

$$CH_3—\overset{\overset{\displaystyle CH_3}{|}}{\underset{\underset{\displaystyle CH_3}{|}}{C}}—Cl \xrightarrow{H_2O} \left[CH_3 \rightarrow \overset{\overset{\displaystyle CH_3}{|}}{\underset{\underset{\displaystyle CH_3}{|}}{C}}{}^{\delta+}\text{-}Cl^{\delta-} \right]^{\ddagger} \xrightarrow{H_2O} CH_3 \rightarrow \overset{\overset{\displaystyle CH_3}{|}}{\underset{\underset{\displaystyle CH_3}{|}}{C}}{}^{+} + Cl^-$$

Reactant · **Transition state** *resembles product of step because $\Delta G°$ is positive* · **Product of step** *stabilized by three electron-releasing groups*

For a methyl, primary, or secondary halide to react by an S_N1 mechanism it would have to ionize to form a methyl, primary, or secondary carbocation. These carbocations, however, are much higher in energy than a tertiary carbocation, and the transition states leading to these carbocations are even higher in energy. The activation energy for an S_N1 reaction of a simple methyl, primary or secondary halide, consequently, is so large (the reaction is so slow) that, for all practical purposes, an S_N1 reaction does not compete with the corresponding S_N2 reaction.

The **Hammond–Leffler postulate** is quite general and can be better understood through consideration of Fig. 6.13. One way that the postulate can be stated is to say that *the structure of a transition state resembles the stable species that is nearest it in free energy.* For example, in a highly **endergonic** step (blue curve) the transition state lies close to the products in free energy and we assume, therefore, that **it resembles the products of that step in structure.** Conversely, in a highly exergonic step (red curve) the transition state lies close to the reactants in free energy, and we assume **it resembles the reactants in structure** as well. The great value of the Hammond–Leffler postulate is that it gives us an intuitive way of visualizing those important, but fleeting, species that we call transition states. We shall make use of it in many future discussions.

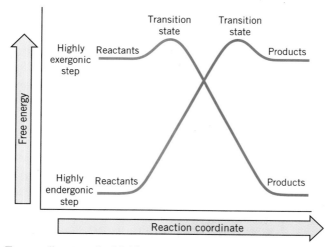

FIGURE 6.13 Energy diagrams for highly exergonic and highly endergonic steps of reactions. (Adapted from William A. Pryor, *Introduction to Free Radical Chemistry* © 1966. Reprinted by permission of Prentice–Hall, Inc., Englewood Cliffs, NJ, p. 53.)

PROBLEM 6.6

The relative rates of ethanolysis of four primary alkyl halides are as follows: CH_3CH_2Br, 1.0; $CH_3CH_2CH_2Br$, 0.28; $(CH_3)_2CHCH_2Br$, 0.030; $(CH_3)_3CCH_2Br$, 0.00000042.

(a) Are these reactions likely to be S_N1 or S_N2?

(b) Provide an explanation for the relative reactivities that are observed.

6.15B The Effect of the Concentration and Strength of the Nucleophile

Since the nucleophile does not participate in the rate-determining step of an S_N1 reaction, the rates of S_N1 reactions are unaffected by either the concentration or the identity of the nucleophile. The rates of S_N2 reactions, however, depend on *both* the concentration *and* the identity of the attacking nucleophile. We saw in Section 6.7 how increasing the concentration of the nucleophile increases the rate of an S_N2 reaction. We can now examine how the rate of an S_N2 reaction depends on the identity of the nucleophile.

We describe nucleophiles as being *good* or *poor*. When we do this we are really describing their relative reactivities in S_N2 reactions. A good nucleophile is one that reacts rapidly with a given substrate. A poor nucleophile is one that reacts slowly with the same substrate under the same reaction conditions.

The methoxide ion, for example, is a good nucleophile. It reacts relatively rapidly with methyl iodide to produce dimethyl ether.

$$CH_3O^- + CH_3I \xrightarrow{\text{rapid}} CH_3OCH_3 + I^-$$

Methanol, on the other hand, is a poor nucleophile. Under the same conditions it reacts very slowly with methyl iodide.

$$CH_3OH + CH_3I \xrightarrow{\text{very slow}} CH_3\overset{+}{\underset{|}{O}}CH_3 + I^-$$
$$H$$

The relative strengths of nucleophiles can be correlated with two structural features:

1. **A negatively charged nucleophile is always a more reactive nucleophile than its conjugate acid.** Thus HO^- is a better nucleophile than H_2O and RO^- is better than ROH.

2. **In a group of nucleophiles in which the nucleophilic atom is the same, nucleophilicities parallel basicities.** Oxygen compounds, for example, show the following order of reactivity:

$$RO^- > HO^- \gg RCO_2^- > ROH > H_2O$$

This is also their order of basicity. An alkoxide ion (RO^-) is a slightly stronger base than a hydroxide ion (HO^-), a hydroxide ion is a much stronger base than a carboxylate ion (RCO_2^-), and so on.

6.15C Solvent Effects on S_N2 Reactions: Polar Protic and Aprotic Solvents

The relative strengths of nucleophiles do not always parallel their basicities *when the nucleophilic atoms are not the same*. When we examine the relative nucleophilicity of compounds or ions within the same group of the periodic table, we find that *in hydroxylic solvents such as alcohols and water* the nucleophile with the larger nucleophilic atom is better. Thiols (R—SH) are stronger nucleophiles than alcohols (ROH); RS$^-$ ions are better than RO$^-$ ions; and the halide ions show the following order:

$$I^- > Br^- > Cl^- > F^-$$

This effect is related to the strength of the interactions between the nucleophile and its surrounding layer of solvent molecules. A molecule of a solvent such as water or an alcohol—called a **protic solvent** (Section 3.7)—*has a hydrogen atom attached to an atom of a strongly electronegative element (oxygen).* Molecules of protic solvents can, therefore, form hydrogen bonds to nucleophiles in the following way:

Molecules of the protic solvent, water, solvate a halide ion by forming hydrogen bonds to it

A small nucleophile, such as a fluoride ion, because its charge is more concentrated, is more strongly solvated than a larger one. Hydrogen bonds to a small atom are stronger than those to a large atom. For a nucleophile to react, it must shed some of its solvent molecules because it must closely approach the carbon bearing the leaving group. A large ion, because the hydrogen bonds between it and the solvent are weaker, can shed some of its solvent molecules more easily and so it will be more nucleophilic.

The greater reactivity of nucleophiles with large nucleophilic atoms is not entirely related to solvation. Larger atoms are more **polarizable** (their electron clouds are more easily distorted), therefore, a larger nucleophilic atom can donate a greater degree of electron density to the substrate than a smaller nucleophile whose electrons are more tightly held.

While nucleophilicity and basicity are related, they are not measured in the same way. Basicity, as expressed by pK_a, is measured *by the position of an equilibrium* involving an electron-pair donor (base), a proton, the conjugate acid, and the conjugate base. Nucleophilicity is measured *by relative rates of reaction,* by how rapidly an electron-pair donor reacts at an atom (usually carbon) bearing a leaving group. For example, the hydroxide ion (OH^-) is a stronger base than a cyanide ion (CN^-); at equilibrium it has the greater affinity for a proton (the pK_a of H_2O is ~ 16, while the pK_a of HCN is ~ 10). Nevertheless, cyanide ion is a stronger nucleophile; it reacts more rapidly with a carbon bearing a leaving group than a hydroxide ion.

The relative nucleophilicities of some common nucleophiles in protic solvents are as follows:

Relative Nucleophilicity in Protic Solvents

$$SH^- > CN^- > I^- > OH^- > N_3^- > Br^- > CH_3CO_2^- > Cl^- > F^- > H_2O$$

Polar Aprotic Solvents

Aprotic solvents are those solvents whose molecules do not have a hydrogen atom that is attached to an atom of a strongly electronegative element. Most aprotic solvents (benzene, the alkanes, etc.) are relatively nonpolar, and they do not dissolve most ionic compounds. (In Section 10.22 we shall see how they can be induced to do so, however.) In recent years a number of **polar aprotic solvents** have come into wide use by chemists; *they are especially useful in S_N2 reactions.* Several examples are the following:

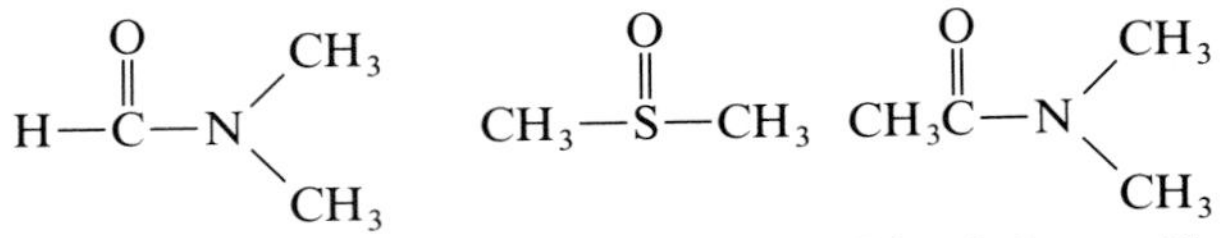

N,N-Dimethylformamide   Dimethyl sulfoxide   Dimethylacetamide
(DMF)   (DMSO)   (DMA)

All of these solvents (DMF, DMSO, and DMA) dissolve ionic compounds, and they solvate cations very well. They do so in the same way that protic solvents solvate cations: by orienting their negative ends around the cation and by donating unshared electron pairs to vacant orbitals of the cation:

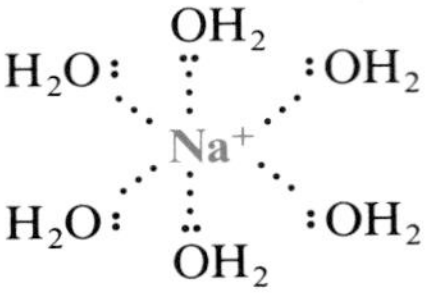

A sodium ion solvated
by molecules of the
protic solvent water

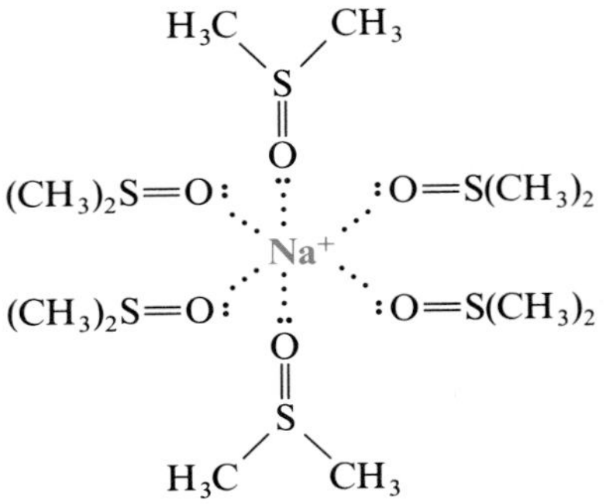

A sodium ion solvated by
molecules of the aprotic
solvent dimethyl sulfoxide

However, because they cannot form hydrogen bonds and because their positive centers are well shielded from any interaction with anions, *aprotic solvents do not solvate anions to any appreciable extent.* In these solvents anions are unencumbered by a layer of solvent molecules and they are therefore poorly stabilized by solvation. These ''naked'' anions are highly reactive both *as bases and nucleophiles.* In DMSO, for example, the relative order of reactivity of halide ions is the same as their relative basicity:

$$F^- > Cl^- > Br^- > I^-$$

This is the opposite of their strength as nucleophiles in alcohol or water solutions:

$$I^- > Br^- > Cl^- > F^-$$

The rates of S_N2 reactions generally are vastly increased when they are carried out in polar aprotic solvents. The increase in rate can be as large as a millionfold.

PROBLEM 6.7

Classify the following solvents as being protic or aprotic: formic acid, $HCOH$;

acetone, CH_3CCH_3; acetonitrile, $CH_3C \equiv N$; formamide, $HCNH_2$; sulfur dioxide, SO_2; ammonia, NH_3; trimethylamine, $N(CH_3)_3$; ethylene glycol, $HOCH_2CH_2OH$.

PROBLEM 6.8

Would you expect the reaction of propyl bromide with sodium cyanide (NaCN), that is,

$$CH_3CH_2CH_2Br + NaCN \longrightarrow CH_3CH_2CH_2CN + NaBr$$

to occur faster in DMF or in ethanol? Explain your answer.

PROBLEM 6.9

Which would you expect to be the stronger nucleophile in a protic solvent: (a) $CH_3CO_2^-$ or CH_3O^-? (b) H_2O or H_2S? (c) $(CH_3)_3P$ or $(CH_3)_3N$?

6.15D Solvent Effects on S_N1 Reactions. The Ionizing Ability of the Solvent

Because of its ability to solvate cations *and* anions so effectively, the use of a **polar protic solvent** will greatly increase the rate of ionization of an alkyl halide *in any S_N1 reaction.* It does this because solvation stabilizes the transition state leading to the intermediate carbocation and halide ion more than it does the reactants; thus the free energy of activation is lower. The transition state for this endothermic step is one in which separated charges are developing and thus it resembles the ions that are ultimately produced.

$$(CH_3)_3C-Cl \longrightarrow \left[(CH_3)_3\overset{\delta+}{C}\cdots\overset{\delta+}{Cl} \right]^{\ddagger} \longrightarrow (CH_3)_3C^+ + Cl^-$$

Reactant   **Transition state**   **Products**
                Separated charges are
                developing

A rough indication of a solvent's polarity is a quantity called the **dielectric constant.** The dielectric constant is a measure of the solvent's ability to insulate opposite charges from each other. Electrostatic attractions and repulsions between ions are smaller in solvents with higher dielectric constants. Table 6.5 gives the dielectric constants of some common solvents.

Water is the most effective solvent for promoting ionization, but most organic compounds do not dissolve appreciably in water. They usually dissolve, however, in alcohols, and quite often mixed solvents are used. Methanol–water and ethanol–water are common mixed solvents for nucleophilic substitution reactions.

PROBLEM 6.10

When *tert*-butyl bromide undergoes solvolysis in a mixture of methanol and water, the rate of solvolysis (measured by the rate at which bromide ions form in the mixture) *increases* when the percentage of water in the mixture is increased.

(a) Explain this occurrence.   (b) Provide an explanation for the observation that the rate of the S_N2 reaction of ethyl chloride with potassium iodide in methanol and water *decreases* when the percentage of water in the mixture is increased.

TABLE 6.5   *Dielectric constants of common solvents*

	SOLVENT	FORMULA	DIELECTRIC CONSTANT
	Water	H_2O	80
	Formic acid	$\overset{\displaystyle O}{\overset{\|}{H C O H}}$	59
	Dimethyl sulfoxide (DMSO)	$\overset{\displaystyle O}{\overset{\|}{CH_3 S CH_3}}$	49
Increasing solvent polarity	N,N-Dimethylformamide (DMF)	$\overset{\displaystyle O}{\overset{\|}{H C N(CH_3)_2}}$	37
	Acetonitrile	$CH_3C\equiv N$	36
	Methanol	CH_3OH	33
	Hexamethylphosphoric triamide (HMPT)	$[(CH_3)_2N]_3P{=}O$	30
	Ethanol	CH_3CH_2OH	24
	Acetone	$\overset{\displaystyle O}{\overset{\|}{CH_3 C CH_3}}$	21
	Acetic acid	$\overset{\displaystyle O}{\overset{\|}{CH_3 C OH}}$	6

6.15E The Nature of the Leaving Group

The best leaving groups are those that become the most stable ions after they depart. Since most leaving groups leave as a negative ion, the best leaving groups are those ions that stabilize a negative charge most effectively. Because weak bases do this best, the best leaving groups are weak bases. The reason that stabilization of the negative charge is important can be understood by considering the structure of the transition states. In either an S_N1 or S_N2 reaction the leaving group begins to acquire a negative charge as the transition state is reached.

S_N1 Reaction (rate-limiting step)

Transition state

S$_N$2 Reaction

$$\text{Nu:}^- \longrightarrow \overset{\diagdown}{\underset{\diagup}{C}} \longrightarrow X \quad\longrightarrow\quad \left[\overset{\delta^-}{\text{Nu}} -- \overset{\diagdown | \diagup}{\underset{|}{C}} -- \overset{\delta^-}{X}\right]^{\ddagger} \quad\longrightarrow\quad \text{Nu} \longrightarrow \overset{\diagup}{\underset{\diagdown}{C}} \;+\; X^-$$

Transition state

Stabilization of this developing negative charge by the leaving group stabilizes the transition state (lowers its potential energy); this lowers the energy of activation, and thereby increases the rate of the reaction. Of the halogens, an iodide ion is the best leaving group and a fluoride ion is the poorest:

$$I^- > Br^- > Cl^- \gg F^-$$

The order is the opposite of the basicity:

$$F^- \gg Cl^- > Br^- > I^-$$

Other weak bases that are good leaving groups that we shall study later are alkanesulfonate ions, alkyl sulfate ions, and the *p*-toluenesulfonate ion.

$$^-O-\overset{\overset{O}{\|}}{\underset{\underset{O}{\|}}{S}}-R \qquad\qquad ^-O-\overset{\overset{O}{\|}}{\underset{\underset{O}{\|}}{S}}-O-R \qquad\qquad ^-O-\overset{\overset{O}{\|}}{\underset{\underset{O}{\|}}{S}}-\bigcirc\!\!\!\!\!\!\!-CH_3$$

An alkanesulfonate ion   **An alkyl sulfate ion**   ***p*-Toluenesulfonate ion**

These anions are all the conjugate bases of very strong acids.

The trifluoromethanesulfonate ion ($CF_3SO_3^-$, commonly called the **triflate ion**) is one of the best leaving groups known to chemists. It is the anion of CF_3SO_3H, an exceedingly strong acid—one that is much stronger than sulfuric acid.

$$CF_3SO_3^-$$

Triflate ion
(a "super" leaving group)

Strongly basic ions rarely act as leaving groups. The hydroxide ion, for example, is a strong base and thus reactions like the following do not take place:

$$X^- \longrightarrow R \overset{\frown}{-} OH \;\cancel{\longrightarrow}\; R-X + OH^-$$

This reaction does not
take place because the
leaving group is a strongly
basic hydroxide ion

However, when an alcohol is dissolved in a strong acid it can react with a halide ion. Because the acid protonates the —OH group of the alcohol, the leaving group no longer needs to be a hydroxide ion; it is now a molecule of water—a much weaker base than a hydroxide ion.

$$X^- \rightarrow R\overset{\curvearrowleft}{\underset{|}{\overset{+}{O}H}} \longrightarrow R—X + H_2O$$

This reaction takes place because the leaving group is a weak base

PROBLEM 6.11

List the following compounds in order of decreasing reactivity toward CH_3O^- in an S_N2 reaction carried out in CH_3OH: CH_3F, CH_3Cl, CH_3Br, CH_3I, $CH_3OSO_2CF_3$, $^{14}CH_3OH$.

Very powerful bases such as hydride ions ($H:^-$) and alkanide ions ($R:^-$) never act as leaving groups. Therefore, reactions such as the following never take place:

$$Nu:^- + CH_3CH_2\overset{\curvearrowleft}{—H} \longrightarrow\!\!\!\!\times\ CH_3CH_2—Nu + H:^-$$
or
$$Nu:^- + CH_3\overset{\curvearrowleft}{—CH_3} \longrightarrow\!\!\!\!\times\ CH_3—Nu + CH_3:^-$$

These are not leaving groups

6.15F Summary: S_N1 Versus S_N2

Reactions of alkyl halides by an S_N1 mechanism are favored by the use of substrates that can form relatively stable carbocations, by the use of weak nucleophiles, and by the use of highly ionizing solvents. S_N1 mechanisms, therefore, are important in solvolysis reactions of tertiary halides, especially when the solvent is highly polar. In a solvolysis the nucleophile is weak because it is a neutral molecule (of the solvent) rather than an anion.

If we want to favor the reaction of an alkyl halide by an S_N2 mechanism, we should use a relatively unhindered alkyl halide, a strong nucleophile, a polar aprotic solvent, and a high concentration of the nucleophile. For substrates, the order of reactivity in S_N2 reactions is

$$CH_3—X > R—CH_2—X > R\overset{\overset{\textstyle R}{|}}{—}CH—X$$
$$\textbf{Methyl} > \quad \textbf{1°} \quad > \quad \textbf{2°}$$

Tertiary halides do not react by an S_N2 mechanism.

The effect of the leaving group is the same in both S_N1 and S_N2 reactions: alkyl iodides react fastest; fluorides react slowest. (Because alkyl fluorides react so slowly, they are seldom used in nucleophilic substitution reactions.)

$$R—I > R—Br > R—Cl \qquad S_N1 \ \text{ or } \ S_N2$$

6.16 ORGANIC SYNTHESIS: FUNCTIONAL GROUP TRANSFORMATIONS USING S$_N$2 REACTIONS

The process of making one compound from another is called **synthesis.** When, for one reason or another, we find ourselves in need of an organic compound that is not available in the stockroom, or perhaps even of one that has never been made before, the task of synthesis starts. We shall have the job of making the compound we need from other compounds that are available.

S$_N$2 reactions are highly useful in organic synthesis because they enable us to convert one functional group into another—a process that is called a **functional group transformation** or a **functional group interconversion.** With the S$_N$2 reactions shown in Fig. 6.14, the functional group of a methyl, primary, or secondary alkyl halide can be transformed into that of an alcohol, ether, thiol, thioether, nitrile, ester, and so on. (*Note:* The use of the prefix *thio* in a name means that a sulfur atom has replaced an oxygen atom in the compound.)

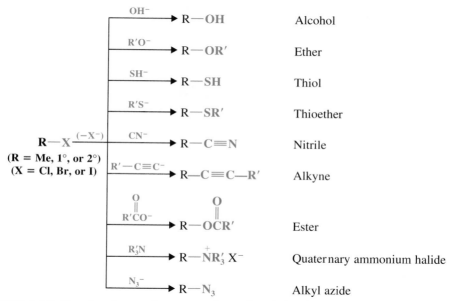

FIGURE 6.14 Functional group interconversions of methyl, primary, and secondary alkyl halides using S$_N$2 reactions.

Alkyl chlorides and bromides are also easily converted to alkyl iodides by nucleophilic substitution reactions.

$$R-Cl \atop R-Br \Big\} \xrightarrow{I^-} R-I \quad (+Cl^- \quad \text{or} \quad Br^-)$$

One other aspect of the S$_N$2 reaction that is of great importance in synthesis is its **stereochemistry** (Section 6.10). S$_N$2 reactions always occur **with inversion of configuration** at the atom that bears the leaving group. This means that when we use S$_N$2 reactions in syntheses we can be sure of the configuration of our product if we know the configuration of our reactant. For example, suppose we need a sample of the following nitrile with the (*S*) configuration.

$$:N \equiv C - C \overset{CH_3}{\underset{CH_2CH_3}{\overset{|}{\diagdown}}} H$$

(*S*)–2–Methylbutanenitrile

If we have available (*R*)-2-bromobutane, we can carry out the following synthesis:

$$:N\equiv C:^- \; + \; H\overset{CH_3}{\underset{CH_3CH_2}{\overset{|}{\diagdown}}}C - Br \xrightarrow[\text{(inversion)}]{S_N2} \; :N\equiv C - C\overset{CH_3}{\underset{CH_2CH_3}{\overset{|}{\diagdown}}}H \; + \; Br^-$$

(*R*)-2-Bromobutane   (*S*)-2-Methylbutanenitrile

PROBLEM 6.12

Starting with (*S*)-2-bromobutane, outline syntheses of each of the following compounds:

(a) (*R*)-CH$_3$CHCH$_2$CH$_3$
      |
      OCH$_2$CH$_3$

(b) (*R*)-CH$_3$CHCH$_2$CH$_3$
      |
      ·OCCH$_3$
       ‖
       O

(c) (*R*)-CH$_3$CHCH$_2$CH$_3$
      |
      SH

(d) (*R*)-CH$_3$CHCH$_2$CH$_3$
      |
      SCH$_3$

6.16A The Unreactivity of Vinylic and Phenyl Halides

As we learned in Section 6.1, compounds that have a halogen atom attached to one carbon atom of a double bond are called **vinylic halides;** those that have a halogen atom attached to a benzene ring are called **phenyl halides.**

$$\overset{\diagdown}{\diagup}C=C\overset{\diagup}{\underset{X}{\diagdown}}$$

A vinylic halide   Phenyl halide

Vinylic halides and phenyl halides are generally unreactive in S_N1 or S_N2 reactions. Vinylic and phenyl cations are highly unstable and do not form readily. This explains the unreactivity of vinylic and phenyl halides in S_N1 reactions. The carbon–halogen bond of a vinylic or phenyl halide is stronger than that of an alkyl halide (we shall see why later) and the electrons of the double bond or benzene ring repel the approach of a nucleophile from the backside. These factors explain the unreactivity of a vinylic or phenyl halide in an S_N2 reaction.

6.17   ELIMINATION REACTIONS OF ALKYL HALIDES

Another characteristic of alkyl halides is that they undergo elimination reactions. In an elimination reaction the fragments of some molecule (YZ) are removed (eliminated) from adjacent atoms of the reactant. This elimination leads to the introduction of a multiple bond:

$$-\overset{|}{\underset{Y}{C}}-\overset{|}{\underset{Z}{C}}- \xrightarrow[(-YZ)]{\text{elimination}} \underset{/}{\overset{\backslash}{C}}=\overset{/}{\underset{\backslash}{C}}$$

6.17A Dehydrohalogenation

A widely used method for synthesizing alkenes is the elimination of HX from adjacent atoms of an alkyl halide. Heating the alkyl halide with a strong base causes the reaction to take place. The following are two examples:

$$CH_3CHCH_3 \xrightarrow[C_2H_5OH, 55°C]{C_2H_5ONa} CH_2{=}CH{-}CH_3 + NaBr + C_2H_5OH$$
$$\underset{Br}{|} \qquad\qquad\qquad (79\%)$$

$$CH_3{-}\underset{\underset{CH_3}{|}}{\overset{\overset{CH_3}{|}}{C}}{-}Br \xrightarrow[C_2H_5OH, 25°C]{C_2H_5ONa} CH_3{-}\underset{}{\overset{\overset{CH_3}{|}}{C}}{=}CH_2 + NaBr + C_2H_5OH$$
$$(91\%)$$

Reactions like these are not limited to the elimination of hydrogen bromide. Chloroalkanes also undergo the elimination of hydrogen chloride, iodoalkanes undergo the elimination of hydrogen iodide and, in all cases, alkenes are produced. When the elements of a hydrogen halide are eliminated from a haloalkane in this way, the reaction is often called **dehydrohalogenation.**

$$-\overset{\overset{\displaystyle H}{|}}{\underset{|}{C}}{}^{\beta}\overset{|}{\underset{\underset{X}{|}}{C}}{}^{\alpha} + \;\; :B^- \;\; \longrightarrow \;\; \overset{\backslash}{\underset{/}{C}}{=}\overset{/}{\underset{\backslash}{C}} + H{:}B + {:}X^-$$

A base

Dehydrohalogenation

In these eliminations, as in S_N1 and S_N2 reactions, there is a leaving group and an attacking particle (the base) that possesses an electron pair.

Chemists often call the carbon atom that bears the substituent (e.g., the halogen atom in the previous reaction) the **alpha (α) carbon atom** and any carbon atom adjacent to it a **beta (β) carbon atom.** A hydrogen atom attached to the β carbon atom is called a **β hydrogen atom.** Since the hydrogen atom that is eliminated in dehydrohalogenation is from the β carbon atom, these reactions are often called **β eliminations.** They are also often referred to as **1,2 eliminations.**

We shall have more to say about dehydrohalogenation in Chapter 8, but we can examine several important aspects here.

6.17B Bases Used in Dehydrohalogenation

Various strong bases have been used for dehydrohalogenations. Potassium hydroxide dissolved in ethanol is a reagent sometimes used, but the sodium salts of alcohols often offer distinct advantages.

The sodium salt of an alcohol (a sodium alkoxide) can be prepared by treating an alcohol with sodium metal:

$$2\ R-\ddot{O}H\ +\ 2\ Na \longrightarrow 2\ R-\ddot{O}:^-Na^+\ +\ H_2$$

Alcohol   **Sodium alkoxide**

This reaction involves the displacement of hydrogen from the alcohol and is, therefore, an **oxidation–reduction reaction.** Sodium, an alkali metal, is a very powerful reducing agent and always displaces hydrogen atoms that are bonded to oxygen atoms. The vigorous (at times explosive) reaction of sodium with water is of the same type.

$$2\ H\ddot{O}H\ +\ 2\ Na \longrightarrow 2\ H\ddot{O}:^-Na^+\ +\ H_2$$

Sodium hydroxide

Sodium alkoxides can also be prepared by allowing an alcohol to react with sodium hydride (NaH). The hydride ion ($H:^-$) is a very strong base.

$$R-\ddot{O}H\ +\ Na^+:H^- \longrightarrow R-\ddot{O}:^-Na^+\ +\ H_2$$

Sodium (and potassium) alkoxides are usually prepared by using an excess of the alcohol, and the excess alcohol becomes the solvent for the reaction. Sodium ethoxide is frequently employed in this way.

$$2\ CH_3CH_2OH\ +\ 2\ Na \longrightarrow 2\ CH_3CH_2O^-Na^+\ +\ H_2$$

Ethanol (excess)   **Sodium ethoxide**

Potassium *tert*-butoxide is another highly effective dehydrohalogenating reagent.

$$2\ CH_3\overset{\displaystyle CH_3}{\underset{\displaystyle CH_3}{\overset{|}{\underset{|}{C}}}}-\ddot{O}H\ +\ 2\ K \longrightarrow 2\ CH_3\overset{\displaystyle CH_3}{\underset{\displaystyle CH_3}{\overset{|}{\underset{|}{C}}}}-\ddot{O}:^-K^+\ +\ H_2$$

tert-**Butyl alcohol (excess)**   **Potassium *tert*-butoxide**

6.17C Mechanisms of Dehydrohalogenations

Elimination reactions occur by a variety of mechanisms. With alkyl halides, two mechanisms are especially important because they are closely related to the S_N2 and S_N1 reactions that we have just studied. One mechanism is a bimolecular mechanism called the E2 reaction; the other is a unimolecular mechanism called the E1 reaction.

6.18 THE E2 REACTION

When isopropyl bromide is heated with sodium ethoxide in ethanol to form propene, the reaction rate depends on the concentration of isopropyl bromide and on the concentration of ethoxide ion. The rate equation is first order in each reactant and second order overall.

$$\text{Rate} \propto [CH_3CHBrCH_3]\,[C_2H_5O^-]$$

$$\text{Rate} = k[CH_3CHBrCH_3]\,[C_2H_5O^-]$$

From this we infer that the transition state for the rate-determining step must involve both the alkyl halide and the alkoxide ion. The reaction must be bimolecular. Considerable experimental evidence indicates that the reaction takes place in the following way:

A MECHANISM FOR THE E2 REACTION

REACTION:

$$C_2H_5O^- + CH_3CHBrCH_3 \longrightarrow CH_2{=}CHCH_3 + C_2H_5OH + Br^-$$

MECHANISM:

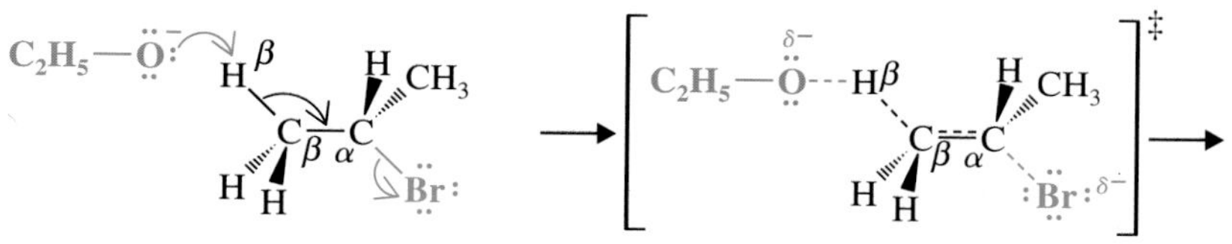

The basic ethoxide ion begins to remove a proton from the β carbon using its electron pair to form a bond to it. At the same time, the electron pair of the β C—H bond begins to move in to become the π bond of a double bond and the bromine begins to depart with the electrons that bonded it to the α carbon.

Transition state

Partial bonds now exist between the oxygen and the β hydrogen and between the α carbon and the bromine. The carbon–carbon bond is developing double bond character.

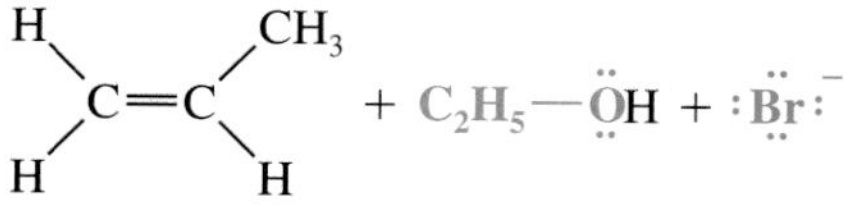

Now the double bond of the alkene is fully formed and the alkene has a trigonal planar geometry at each carbon atom. The other products are a molecule of ethanol and a bromide ion.

When we study the E2 reaction further in Section 7.12C, we shall find that the orientations of the hydrogen atom being removed and the leaving group are not arbitrary and that the orientation shown above is required.

6.19 THE E1 REACTION

Eliminations may take a different pathway from that given in Section 6.18. Treating *tert*-butyl chloride with 80% aqueous ethanol at 25°C, for example, gives *substitution products* in 83% yield and an elimination product (2-methylpropene) in 17% yield.

$$
\text{CH}_3\text{C}-\text{Cl} \xrightarrow[\substack{20\% \text{ H}_2\text{O} \\ 25°\text{C}}]{80\% \text{ C}_2\text{H}_5\text{OH}}
$$

$$
\xrightarrow{\text{S}_N1} \quad \text{CH}_3\text{C}-\text{OH} \;+\; \text{CH}_3\text{C}-\text{OCH}_2\text{CH}_3
$$

tert-Butyl chloride

tert-Butyl alcohol   *tert*-Butyl ethyl ether

(83%)

$$
\xrightarrow{\text{E1}} \quad \text{CH}_2\!\!=\!\!\text{C}\begin{smallmatrix}\text{CH}_3\\[2pt]\text{CH}_3\end{smallmatrix}
$$

2-Methylpropene
(17%)

The initial step for both reactions is the formation of a *tert*-butyl cation. This is also the rate-determining step for both reactions; thus both reactions are unimolecular.

$$
\text{CH}_3\text{C}-\overset{..}{\underset{..}{\text{Cl}}}\!: \xrightarrow{\text{slow}} \text{CH}_3\text{C}^+ \;+\; :\!\overset{..}{\underset{..}{\text{Cl}}}\!:^-
$$

(solvated)   (solvated)

Whether substitution or elimination takes place depends on the next step (the fast step). If a solvent molecule reacts as a nucleophile at the positive carbon atom of the *tert*-butyl cation, the product is *tert*-butyl alcohol or *tert*-butyl ethyl ether and the reaction is S_N1.

$$
\text{CH}_3\text{C}^+ \; \text{Sol}-\overset{..}{\underset{..}{\text{O}}}\text{H} \xrightarrow{\text{fast}} \text{CH}_3\text{C}-\overset{+}{\underset{..}{\text{O}}}: \rightleftharpoons \text{CH}_3\text{C}-\text{O}-\text{Sol} + \text{H}^+ \left.\right\}\; \substack{\text{S}_N1 \\ \text{reaction}}
$$

(Sol = H—   or   CH₃CH₂—)

If, however, a solvent molecule acts as a base and abstracts one of the β hydrogen atoms as a proton, the product is 2-methylpropene and the reaction is E1.

E1 reactions almost always accompany S_N1 reactions.

$$
\text{Sol}-\overset{..}{\underset{..}{\text{O}}}\!:\; \text{H}-\text{CH}_2-\text{C}^+ \xrightarrow{\text{fast}} \text{Sol}-\overset{+}{\underset{\text{H}}{\text{O}}}-\text{H} + \text{CH}_2\!\!=\!\!\text{C}\begin{smallmatrix}\text{CH}_3\\[2pt]\text{CH}_3\end{smallmatrix} \left.\right\}\; \substack{\text{E1} \\ \text{reaction}}
$$

2-Methylpropene

A MECHANISM FOR THE E1 REACTION

REACTION:

$$(CH_3)_3CCl + H_2O \longrightarrow CH_2{=}C(CH_3)_3 + H_3O^+ + Cl^-$$

MECHANISM:

Step 1

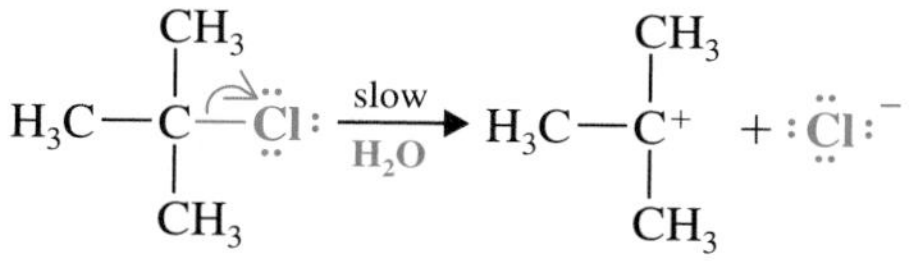

Aided by the polar solvent a chlorine departs with the electron pair that bonded it to the carbon.

This slow step produces the relatively stable 3° carbocation and a chloride ion. The ions are solvated (and stablized) by surrounding water molecules.

Step 2

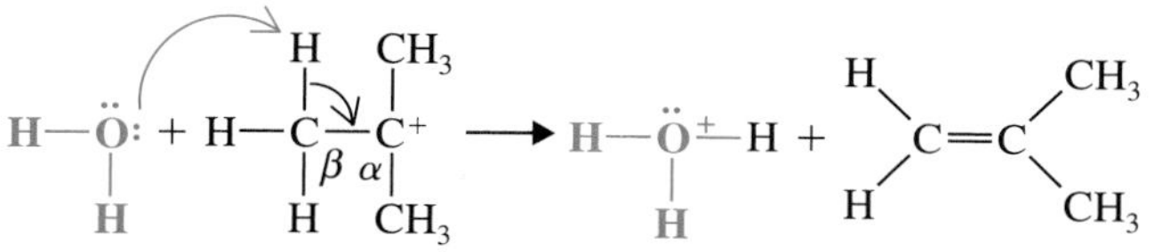

A molecule of water accepts a proton from the β carbon of the carbocation. An electron pair moves in to form a double bond between the α and β carbon atoms.

This step produces the alkene and a hydronium ion.

6.20 SUBSTITUTION VERSUS ELIMINATION

Because the reactive part of a nucleophile or a base is an unshared electron pair, all nucleophiles are potential bases and all bases are potential nucleophiles. It should not be surprising, then, that nucleophilic substitution reactions and elimination reactions often compete with each other.

6.20A S_N2 Versus E2

Since eliminations occur best by an E2 path when carried out with a high concentration of a strong base (and thus a high concentration of a strong nucleophile), substitution reactions by an S_N2 path often compete with the elimination reaction. When the nucleophile (base) attacks a β hydrogen atom, elimination occurs. When the nucleophile attacks the carbon atom bearing the leaving group, substitution results.

When the substrate is a primary halide and the base is ethoxide ion, substitution is highly favored.

$$CH_3CH_2O^-Na^+ + CH_3CH_2Br \xrightarrow[\substack{55°C \\ (-NaBr)}]{C_2H_5OH} CH_3CH_2OCH_2CH_3 + CH_2{=}CH_2$$

$$\underset{\substack{S_N2 \\ (90\%)}}{} \qquad \underset{\substack{E2 \\ (10\%)}}{}$$

With secondary halides, however, the elimination reaction is favored.

$$C_2H_5O^-Na^+ + \underset{\underset{Br}{|}}{CH_3CHCH_3} \xrightarrow[\substack{55°C \\ (-NaBr)}]{C_2H_5OH} \underset{\underset{\underset{C_2H_5}{|}}{O}}{CH_3CHCH_3} + CH_2{=}CHCH_3$$

$$\underset{\substack{S_N2 \\ (21\%)}}{} \qquad \underset{\substack{E2 \\ (79\%)}}{}$$

With tertiary halides an S_N2 reaction cannot take place, so the elimination reaction is highly favored, especially when the reaction is carried out at higher temperatures. Any substitution that occurs probably takes place through an S_N1 mechanism.

$$C_2H_5O^-Na^+ + \underset{\underset{Br}{|}}{\overset{\overset{CH_3}{|}}{CH_3CCH_3}} \xrightarrow[\substack{25°C \\ (-NaBr)}]{C_2H_5OH} \underset{\underset{\underset{C_2H_5}{|}}{O}}{\overset{\overset{CH_3}{|}}{CH_3CCH_3}} + \overset{\overset{CH_3}{|}}{CH_2{=}CCH_3}$$

$$\underset{\substack{S_N1 \\ (9\%)}}{} \qquad \underset{\substack{\textbf{Mainly E2} \\ (91\%)}}{}$$

$$C_2H_5O^-Na^+ + \underset{\underset{Br}{|}}{\overset{\overset{CH_3}{|}}{CH_3CCH_3}} \xrightarrow[\substack{55°C \\ (-NaBr)}]{C_2H_5OH} \overset{\overset{CH_3}{|}}{CH_2{=}CCH_3} + C_2H_5OH$$

$$\underset{\substack{\textbf{E2 + E1} \\ (100\%)}}{}$$

Increasing the temperature favors eliminations (E1 and E2) over substitutions. The reason: Eliminations have higher free energies of activation than substitutions because eliminations have a greater change in bonding (more bonds are broken and formed). By giving more molecules enough energy to surmount the energy barriers, increasing the temperature increases the rates of both substitutions and eliminations; however, because the energy barriers for eliminations are higher, the proportion of molecules able to cross them is significantly higher.

Increasing the reaction temperature is one way of favorably influencing an elimination reaction of an alkyl halide. Another way is to use a **strong sterically hindered base** such as the *tert*-butoxide ion. The bulky methyl groups of the *tert*-butoxide ion appear to inhibit its reacting by substitution, so elimination reactions take precedence. We can see an example of this effect in the following two reactions. The relatively unhindered methoxide ion reacts with octadecyl bromide primarily by *substitution;* the bulky *tert*-butoxide ion gives mainly *elimination.*

$$CH_3O^- + CH_3(CH_2)_{15}CH_2CH_2\!-\!Br \xrightarrow[65°C]{CH_3OH}$$

$$CH_3(CH_2)_{15}CH\!=\!CH_2 + CH_3(CH_2)_{15}CH_2CH_2OCH_3$$

E2	S_N2
(1%)	(99%)

$$CH_3\!-\!\underset{\underset{CH_3}{|}}{\overset{\overset{CH_3}{|}}{C}}\!-\!O^- + CH_3(CH_2)_{15}CH_2CH_2\!-\!Br \xrightarrow[40°C]{(CH_3)_3COH}$$

$$CH_3(CH_2)_{15}CH\!=\!CH_2 + CH_3(CH_2)_{15}CH_2CH_2\!-\!O\!-\!\underset{\underset{CH_3}{|}}{\overset{\overset{CH_3}{|}}{C}}\!-\!CH_3$$

E2	S_N2
(85%)	(15%)

Another factor that affects the relative rates of E2 and S_N2 reactions is the relative basicity and polarizability of the base/nucleophile. Use of a strong, slightly polarizable base such as amide ion (NH_2^-) or alkoxide ion (especially a hindered one) tends to increase the likelihood of elimination (E2). Use of a weakly basic ion such as a chloride ion (Cl^-) or an acetate ion ($CH_3CO_2^-$) or a weakly basic and highly polarizable one such as Br^-, I^-, or RS^- increases the likelihood of substitution (S_N2). Acetate ion, for example, reacts with isopropyl bromide almost exclusively by the S_N2 path:

$$CH_3\overset{\overset{O}{\|}}{C}\!-\!O^- + CH_3\overset{\overset{CH_3}{|}}{CH}\!-\!Br \xrightarrow[(\sim 100\%)]{S_N2} CH_3\overset{\overset{O}{\|}}{C}\!-\!O\!-\!\overset{\overset{CH_3}{|}}{C}HCH_3 + Br^-$$

The more strongly basic ethoxide ion (Section 6.17B) reacts with the same compound mainly by an E2 mechanism.

6.20B Tertiary Halides: S_N1 Versus E1

Because the E1 reaction and the S_N1 reaction proceed through the formation of a common intermediate, the two types respond in similar ways to factors affecting reactivities. E1 reactions are favored with substrates that can form stable carbocations (i.e., tertiary halides); they are also favored by the use of poor nucleophiles (weak bases) and they are generally favored by the use of polar solvents.

It is usually difficult to influence the relative partition between S_N1 and E1 products because the energy of activation for either reaction proceeding from the carbocation (loss of a proton or combination with a molecule of the solvent) is very small.

In most unimolecular reactions the S_N1 reaction is favored over the E1 reaction, especially at lower temperatures. *In general, however, substitution reactions of tertiary halides do not find wide use as synthetic methods. Such halides undergo eliminations much too easily.*

Increasing the temperature of the reaction favors reaction by the E1 mechanism at the expense of the S_N1 mechanism. *If the elimination product is desired, however, it is more convenient to add a strong base and force an E2 reaction to take place instead.*

6.21 OVERALL SUMMARY

The most important reaction pathways for the substitution and elimination reactions of simple alkyl halides can be summarized in the way shown in Table 6.6.

TABLE 6.6 Overall summary of S_N1, S_N2, E1, and E2 reactions

CH_3X	RCH_2X	$\overset{\displaystyle R}{\underset{\displaystyle \mid}{RCHX}}$	$R-\overset{\displaystyle R}{\underset{\displaystyle R}{\overset{\displaystyle \mid}{\underset{\displaystyle \mid}{C}}}}-X$
Methyl	1°	2°	3°
	Bimolecular reactions only		$S_N1/E1$ or E2
Gives S_N2 reactions	Gives mainly S_N2 except with a hindered strong base [e.g., $(CH_3)_3CO^-$] and then gives mainly E2	Gives mainly S_N2 with weak bases (e.g., I^-, CN^-, RCO_2^-) and mainly E2 with strong bases (e.g., RO^-)	No S_N2 reaction. In solvolysis gives $S_N1/E1$, and at lower temperatures S_N1 is favored. When a strong base (e.g., RO^-) is used, E2 predominates.

Let us examine several sample exercises that will illustrate how the information in Table 6.6 can be used.

SAMPLE PROBLEM

Give the product (or products) that you would expect to be formed in each of the following reactions. In each case give the mechanism (S_N1, S_N2, E1, or E2) by which the product is formed and predict the relative amount of each (i.e., would the product be the only product, the major product, or a minor product?).

(a) $CH_3CH_2CH_2Br + CH_3O^- \xrightarrow[CH_3OH]{50°C}$

(b) $CH_3CH_2CH_2Br + (CH_3)_3CO^- \xrightarrow[(CH_3)_3COH]{50°C}$

(c) $\underset{\underset{H}{\overset{CH_3}{\big|}}}{CH_3CH_2} C-Br + HS^- \xrightarrow[CH_3OH]{50°C}$

(d) $(CH_3CH_2)_3CBr + OH^- \xrightarrow[CH_3OH]{50°C}$

(e) $(CH_3CH_2)_3CBr \xrightarrow[CH_3OH]{25°C}$

ANSWER:

(a) The substrate is a 1° halide. The base/nucleophile is CH_3O^-, a strong base (but not a hindered one) and a good nucleophile. According to Table 6.6 we should expect an S_N2 reaction mainly, and the major product should be $CH_3CH_2CH_2OCH_3$. A minor product might be $CH_3CH=CH_2$ by an E2 pathway.

(b) Again the substrate is a 1° halide, but the base/nucleophile, $(CH_3)_3CO^-$, is a strong hindered base. We should expect, therefore, the major product to be $CH_3CH=CH_2$ by an E2 pathway, and a minor product to be $CH_3CH_2CH_2OC(CH_3)_3$ by an S_N2 pathway.

(c) The reactant is (S)-2-bromobutane, a 2° halide, and one in which the leaving group is attached to a stereocenter. The base/nucleophile is HS^-, a strong nucleophile, but a weak base. We should expect mainly an S_N2 reaction, causing an inversion of configuration at the stereocenter and producing the (R)-stereoisomer:

$$HS - \underset{\underset{H}{\big|}}{\overset{\overset{CH_3}{\big/}}{C}}\cdots CH_2CH_3$$

(d) The base/nucleophile is OH^-, a strong base and a strong nucleophile. However, the substrate is a 3° halide; therefore, we should not expect an S_N2 reaction. The major product should be $CH_3CH=C(CH_2CH_3)_2$ via an E2 reaction. At this higher temperature, and in the presence of a strong base, we should not expect an appreciable amount of the S_N1 product, $CH_3OC(CH_2CH_3)_3$.

(e) This is solvolysis; the only base/nucleophile is the solvent, CH_3OH, which is a weak base (therefore, no E2 reaction) and a poor nucleophile. The substrate is tertiary (therefore, no S_N2 reaction). At this lower temperature we should expect mainly an S_N1 pathway leading to $CH_3OC(CH_2CH_3)_3$. A minor product, by an E1 pathway would be $CH_3CH=C(CH_2CH_3)_2$.

6.22 SOME IMPORTANT TERMS AND CONCEPTS

Carbocation. A positive ion formed by heterolysis of a bond to a carbon atom as follows:

$$-\overset{\big|}{\underset{\big|}{C}}-X \longrightarrow -\overset{\big|}{\underset{\big|}{C}}{}^+ + :X^-$$

Carbocations show the relative stabilities:

$$3° > 2° > 1° > \text{methyl}$$

Nucleophile. A negative ion or a molecule that has at least one unshared pair of electrons. In a chemical reaction a nucleophile attacks a positive center of some other molecule or positive ion.

Nucleophilic substitution reaction. (Abbreviated as S_N reaction.) A substitution reaction brought about when a nucleophile reacts with a *substrate* that bears a *leaving group*.

S_N2 reaction. A nucleophilic substitution reaction for which the rate-determining step is *bimolecular* (i.e., the transition state involves two species). The reaction of methyl chloride with hydroxide ion is an S_N2 reaction. According to the Ingold mechanism it takes place in a *single step* as follows:

$$HO^- + CH_3{-}Cl \longrightarrow \left[\overset{\delta-}{HO}{-}{-}{-}CH_3{-}{-}{-}\overset{\delta-}{Cl} \right]^{\ddagger} \longrightarrow HO{-}CH_3 + Cl^-$$

Transition state

The order of reactivity of alkyl halides in S_N2 reactions is

$$\underset{\textbf{Methyl}}{CH_3{-}X} > \underset{\textbf{1°}}{RCH_2X} > \underset{\textbf{2°}}{R_2CHX}$$

S_N1 reaction. A nucleophilic substitution reaction for which the rate-determining step is *unimolecular*. The hydrolysis of *tert*-butyl chloride is an S_N1 reaction that takes place in three steps as follows. The rate-determining step is step 1.

Step 1 $(CH_3)_3CCl \xrightarrow{\text{slow}} (CH_3)_3C^+ + Cl^-$

Step 2 $(CH_3)_3C^+ + H_2O \xrightarrow{\text{fast}} (CH_3)_3C\overset{+}{O}H_2$

Step 3 $(CH_3)_3C\overset{+}{O}H_2 + H_2O \xrightarrow{\text{fast}} (CH_3)_3COH + H_3O^+$

S_N1 reactions are important with tertiary halides and with other substrates that can form relatively stable carbocations.

Solvolysis. A nucleophilic substitution reaction in which the nucleophile is a molecule of the solvent.

Steric effect. An effect on relative reaction rates caused by the space-filling properties of those parts of a molecule attached at or near the reacting site. *Steric hindrance* is an important effect in S_N2 reactions. It explains why methyl halides are most reactive and tertiary halides are least reactive.

Elimination reaction. A reaction in which the fragments of some molecule are eliminated from adjacent atoms of the reactant to give a multiple bond. Dehydrohalogenation is an elimination reaction in which HX is eliminated from an alkyl halide, leading to the formation of an alkene.

$$\underset{X}{\overset{H}{-C-C-}} + :B^- \longrightarrow \,{>}C{=}C{<} + H:B + :X^-$$

E1 reaction. A unimolecular elimination. The first step of an E1 reaction, formation of a carbocation, is the same as that of an S_N1 reaction, consequently E1 and S_N1 reactions compete with each other. E1 reactions are important when tertiary halides are subjected to solvolysis in polar solvents especially at higher temperatures. The steps in the E1 reaction of *tert*-butyl chloride are the following:

Step 1

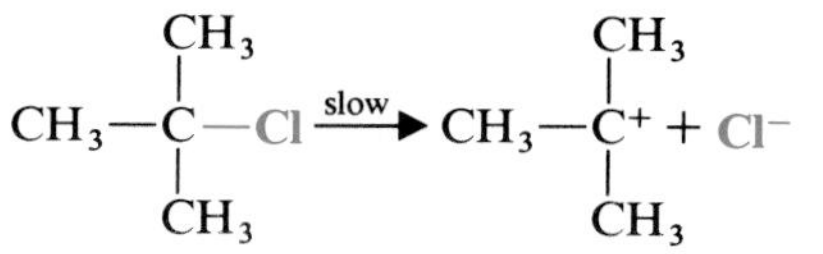

Step 2

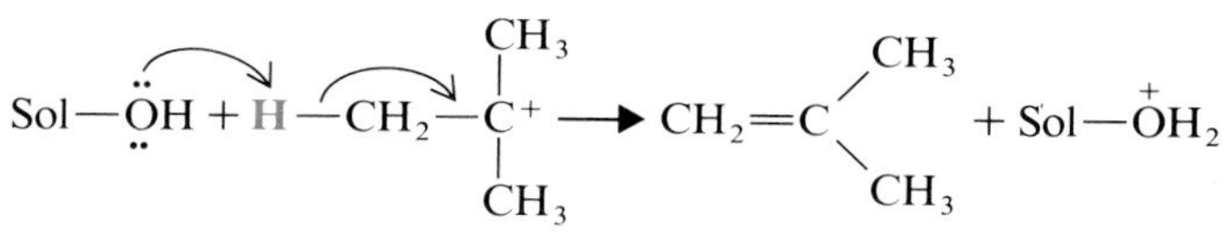

E2 reaction. A bimolecular elimination that often competes with S_N2 reactions. E2 reactions are favored by the use of a high concentration of a strong, bulky, and slightly polarizable base. The order of reactivity of alkyl halides toward E2 reactions is $3° \gg 2° > 1°$. The mechanism of the E2 reaction involves a single step:

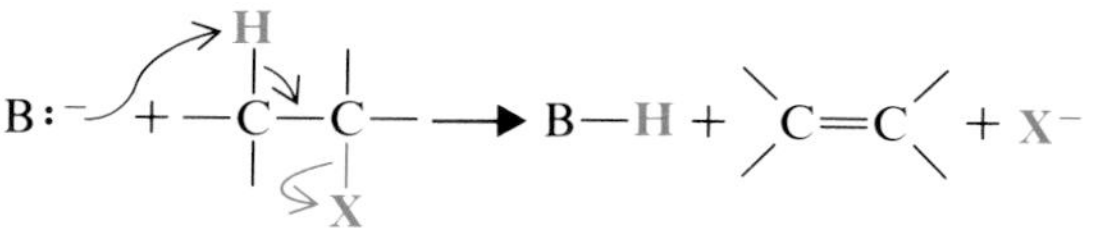

ADDITIONAL PROBLEMS

6.13 Show how you might use a nucleophilic substitution reaction of propyl bromide to synthesize each of the following compounds. (You may use any other compounds that are necessary.)

(a) $CH_3CH_2CH_2OH$

(b) $CH_3CH_2CH_2I$

(c) $CH_3CH_2OCH_2CH_2CH_3$

(d) $CH_3CH_2CH_2$—S—CH_3

(e) $CH_3\overset{\displaystyle O}{\overset{\|}{C}}OCH_2CH_2CH_3$

(f) $CH_3CH_2CH_2N_3$

(g) CH_3—$\overset{\displaystyle CH_3}{\underset{\displaystyle CH_3}{\overset{|}{\underset{|}{N}}}}{}^+$—$CH_2CH_2CH_3$ Br^-

(h) $CH_3CH_2CH_2CN$

(i) $CH_3CH_2CH_2SH$

6.14 Which alkyl halide would you expect to react more rapidly by an S_N2 mechanism? Explain your answer.

(a) $CH_3CH_2CH_2Br$ or $(CH_3)_2CHBr$

(b) $CH_3CH_2CH_2CH_2Cl$ or $CH_3CH_2CH_2CH_2I$

(c) $(CH_3)_2CHCH_2Cl$ or $CH_3CH_2CH_2CH_2Cl$

(d) $(CH_3)_2CHCH_2CH_2Cl$ or $CH_3CH_2CH(CH_3)CH_2Cl$

(e) C_6H_5Br or $CH_3CH_2CH_2CH_2CH_2CH_2Cl$

6.15 Which S_N2 reaction of each pair would you expect to take place more rapidly in a protic solvent?

(a) (1) $CH_3CH_2CH_2Cl + CH_3CH_2O^- \longrightarrow CH_3CH_2CH_2OCH_2CH_3 + Cl^-$
     or
  (2) $CH_3CH_2CH_2Cl + CH_3CH_2OH \longrightarrow CH_3CH_2CH_2OCH_2CH_3 + HCl$

(b) (1) $CH_3CH_2CH_2Cl + CH_3CH_2O^- \longrightarrow CH_3CH_2CH_2OCH_2CH_3 + Cl^-$
     or
  (2) $CH_3CH_2CH_2Cl + CH_3CH_2S^- \longrightarrow CH_3CH_2CH_2SCH_2CH_3 + Cl^-$

(c) (1) $CH_3CH_2CH_2Br + (C_6H_5)_3N \longrightarrow CH_3CH_2CH_2N(C_6H_5)_3^+ + Br^-$
     or
  (2) $CH_3CH_2CH_2Br + (C_6H_5)_3P \longrightarrow CH_3CH_2CH_2P(C_6H_5)_3^+ + Br^-$

(d) (1) $CH_3CH_2CH_2Br \, (1.0M) + CH_3O^- \, (1.0M) \longrightarrow CH_3CH_2CH_2OCH_3 + Br^-$
     or
  (2) $CH_3CH_2CH_2Br \, (1.0M) + CH_3O^- \, (2.0M) \longrightarrow CH_3CH_2CH_2OCH_3 + Br^-$

6.16 Which S_N1 reaction of each pair would you expect to take place more rapidly? Explain your answer.

(a) (1) $(CH_3)_3CCl + H_2O \longrightarrow (CH_3)_3COH + HCl$
     or
  (2) $(CH_3)_3CBr + H_2O \longrightarrow (CH_3)_3COH + HBr$

(b) (1) $(CH_3)_3CCl + H_2O \longrightarrow (CH_3)_3COH + HCl$
     or
  (2) $(CH_3)_3CCl + CH_3OH \longrightarrow (CH_3)_3COCH_3 + HCl$

(c) (1) $(CH_3)_3CCl \, (1.0M) + CH_3CH_2O^- \, (1.0M) \xrightarrow{EtOH} (CH_3)_3COCH_2CH_3 + Cl^-$
     or
  (2) $(CH_3)_3CCl \, (2.0M) + CH_3CH_2O^- \, (1.0M) \xrightarrow{EtOH} (CH_3)_3COCH_2CH_3 + Cl^-$

(d) (1) $(CH_3)_3CCl \, (1.0M) + CH_3CH_2O^- \, (1.0M) \xrightarrow{EtOH} (CH_3)_3COCH_2CH_3 + Cl^-$
     or
  (2) $(CH_3)_3CCl \, (1.0M) + CH_3CH_2O^- \, (2.0M) \xrightarrow{EtOH} (CH_3)_3COCH_2CH_3 + Cl^-$

(e) (1) $(CH_3)_3CCl + H_2O \longrightarrow (CH_3)_3COH + HCl$
     or
  (2) $C_6H_5Cl + H_2O \longrightarrow C_6H_5OH + HCl$

6.17 With methyl, ethyl, or cyclopentyl halides as your organic starting materials and using any needed solvents or inorganic reagents, outline syntheses of each of the following. More than one step may be necessary and you need not repeat steps carried out in earlier parts of this problem.

(a) CH_3I (d) CH_3CH_2OH (g) CH_3CN (j) $CH_3OCH_2CH_3$

(b) CH_3CH_2I (e) CH_3SH (h) CH_3CH_2CN (k) Cyclopentene

(c) CH_3OH (f) CH_3CH_2SH (i) CH_3OCH_3

6.18 Listed below are several hypothetical nucleophilic substitution reactions. None is synthetically useful because the product indicated is not formed at an appreciable rate. In each case provide an explanation for the failure of the reaction to take place as indicated.

(a) $CH_3CH_2CH_3 + OH^- \nrightarrow CH_3CH_2OH + CH_3^-$

(b) $CH_3CH_2CH_3 + OH^- \nrightarrow CH_3CH_2CH_2OH + H^-$

(c) $\square + OH^- \nrightarrow CH_3CH_2CH_2CH_2OH$

(d) $CH_3CH_2-\underset{\underset{CH_3}{|}}{\overset{\overset{CH_3}{|}}{C}}-Br + CN^- \nrightarrow CH_3CH_2-\underset{\underset{CH_3}{|}}{\overset{\overset{CH_3}{|}}{C}}-CN + Br^-$

(e) $NH_3 + CH_3OCH_3 \nrightarrow CH_3NH_2 + CH_3OH$

(f) $NH_3 + CH_3OH_2^+ \nrightarrow CH_3NH_3^+ + H_2O$

6.19   You have the task of preparing styrene ($C_6H_5CH{=}CH_2$) by dehydrohalogenation of either 1-bromo-2-phenylethane or 1-bromo-1-phenylethane using KOH in ethanol. Which halide would you choose as your starting material to give the best yield of the alkene? Explain your answer.

6.20   Your task is to prepare isopropyl methyl ether, $CH_3OCH(CH_3)_2$, by one of the following reactions. Which reaction would give the better yield? Explain your choice.

(1) $CH_3ONa + (CH_3)_2CHI \longrightarrow CH_3OCH(CH_3)_2$

(2) $(CH_3)_2CHONa + CH_3I \longrightarrow CH_3OCH(CH_3)_2$

6.21   Which product (or products) would you expect to obtain from each of the following reactions? In each part give the mechanism (S_N1, S_N2, E1, or E2) by which each product is formed and predict the relative amount of each product (i.e., would the product be the only product, the major product, a minor product, etc.?).

(a) $CH_3CH_2CH_2CH_2CH_2Br + CH_3CH_2O^- \xrightarrow[CH_3CH_2OH]{50°C}$

(b) $CH_3CH_2CH_2CH_2CH_2Br + (CH_3)_3CO^- \xrightarrow[(CH_3)_3COH]{50°C}$

(c) $(CH_3)_3CBr + CH_3O^- \xrightarrow[CH_3OH]{50°C}$

(d) $(CH_3)_3Br + (CH_3)_3CO^- \xrightarrow[(CH_3)_3COH]{50°C}$

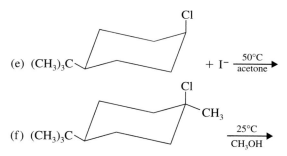

(e) $(CH_3)_3C{-} \quad + I^- \xrightarrow[acetone]{50°C}$

(f) $(CH_3)_3C{-} \xrightarrow[CH_3OH]{25°C}$

(g) 3-Chloropentane + $CH_3O^- \xrightarrow[CH_3OH]{50°C}$

(h) 3-Chloropentane + $CH_3CO_2^- \xrightarrow[CH_3CO_2H]{50°C}$

(i) $HO^- + (R)$-2-bromobutane $\xrightarrow{25°C}$

(j) (S)-3-bromo-3-methylhexane $\xrightarrow[CH_3OH]{25°C}$

(k) (S)-2-bromooctane $+ I^- \xrightarrow[CH_3OH]{50°C}$

6.22   Write conformational structures for the substitution products of the following deuterium-labeled compounds:

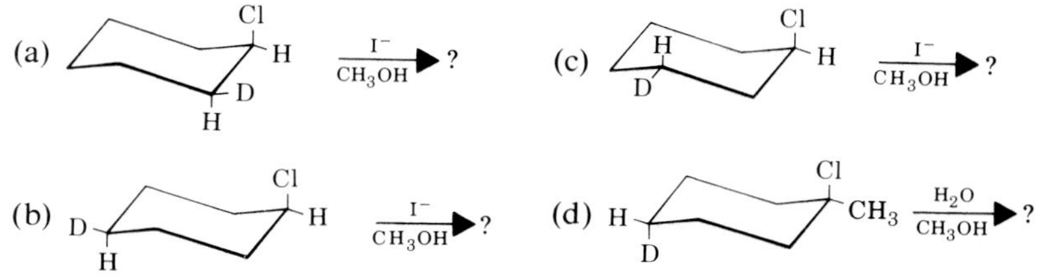

(a) [structure: cyclohexane with Cl, H, D, H substituents] $\xrightarrow[\text{CH}_3\text{OH}]{\text{I}^-}$?　　(c) [structure: cyclohexane with Cl, H, H, D substituents] $\xrightarrow[\text{CH}_3\text{OH}]{\text{I}^-}$?

(b) [structure: cyclohexane with Cl, H, D, H substituents] $\xrightarrow[\text{CH}_3\text{OH}]{\text{I}^-}$?　　(d) [structure: cyclohexane with Cl, CH₃, H, D substituents] $\xrightarrow[\text{CH}_3\text{OH}]{\text{H}_2\text{O}}$?

6.23　Although ethyl bromide and isobutyl bromide are both primary halides, ethyl bromide undergoes S_N2 reactions more than 10 times faster than isobutyl bromide does. When each compound is treated with a strong base/nucleophile ($CH_3CH_2O^-$), isobutyl bromide gives a greater yield of elimination products than substitution products, whereas with ethyl bromide this behavior is reversed. What factor accounts for these results?

6.24　Consider the reaction of I^- with CH_3CH_2Cl.　(a) Would you expect the reaction to be S_N1 or S_N2? The rate constant for the reaction at 60°C is 5×10^{-5} L mol^{-1} s^{-1}.　(b) What is the reaction rate if $[I^-] = 0.1$ mol L^{-1} and $[CH_3CH_2Cl] = 0.1$ mol L^{-1}?　(c) If $[I^-] = 0.1$ mol L^{-1} and $[CH_3CH_2Cl] = 0.2$ mol L^{-1}?　(d) If $[I^-] = 0.2$ mol L^{-1} and $[CH_3CH_2Cl] = 0.1$ mol L^{-1}?　(e) If $[I^-] = 0.2$ mol L^{-1} and $[CH_3CH_2Cl] = 0.2$ mol L^{-1}?

6.25　Which reagent in each pair listed here would be the more reactive nucleophile in a protic solvent?

(a) CH_3NH^-　or　CH_3NH_2

(b) CH_3O^-　or　$CH_3\overset{\overset{\displaystyle O}{\|}}{C}O^-$

(c) CH_3SH　or　CH_3OH

(d) $(C_6H_5)_3N$　or　$(C_6H_5)_3P$

(e) H_2O　or　H_3O^+

(f) NH_3　or　NH_4^+

(g) H_2S　or　HS^-

(h) $CH_3\overset{\overset{\displaystyle O}{\|}}{C}O^-$　or　OH^-

6.26　Write mechanisms that account for the products of the following reactions:

(a) 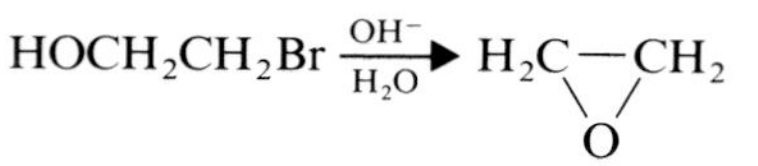 $HOCH_2CH_2Br \xrightarrow[\text{H}_2\text{O}]{\text{OH}^-} H_2C\underset{O}{\overset{\diagdown\diagup}{}}CH_2$

(b) 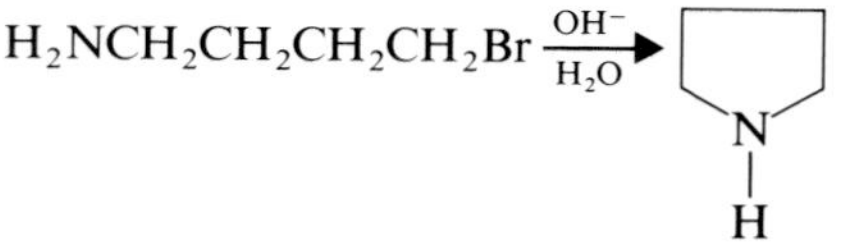 $H_2NCH_2CH_2CH_2CH_2Br \xrightarrow[\text{H}_2\text{O}]{\text{OH}^-}$ [structure: pyrrolidine ring with N–H]

6.27　Many S_N2 reactions of alkyl chlorides and alkyl bromides are catalyzed by the addition of sodium or potassium iodide. For example, the hydrolysis of methyl bromide takes place much faster in the presence of sodium iodide. Explain.

6.28　Explain the following observations: When *tert*-butyl bromide is treated with sodium methoxide in a mixture of methanol and water, the rate of formation of *tert*-butyl alcohol and *tert*-butyl methyl ether does not change appreciably as the concentration of sodium methoxide is increased. Increasing the concentration of sodium methoxide, however, causes a marked increase in the rate at which *tert*-butyl bromide disappears from the mixture.

6.29　(a) Consider the general problem of converting a tertiary alkyl halide to an alkene, for example, the conversion of *tert*-butyl chloride to 2-methylpropene. What experimental conditions would you choose to insure that elimination is favored over substitution? (b) Consider the opposite problem, that of carrying out a substitution reaction on a tertiary alkyl halide. Use as your example the conversion of *tert*-butyl chloride to *tert*-butyl ethyl ether. What experimental conditions would you employ to insure the highest possible yield of the ether?

6.30   1-Bromobicyclo[2.2.1]heptane is extremely unreactive in either S_N2 or S_N1 reactions. Provide explanations for this behavior.

6.31   When ethyl bromide reacts with potassium cyanide in methanol, the major product is CH_3CH_2CN. Some CH_3CH_2NC is formed as well, however. Write Lewis structures for the cyanide ion, for both products, and provide a mechanistic explanation of the course of the reaction.

6.32   Starting with an appropriate alkyl halide, and using any other needed reagents, outline syntheses of each of the following. When alternative possibilities exist for a synthesis you should be careful to choose the one that gives the better yield.

(a) Butyl *sec*-butyl ether

(b) $CH_3CH_2SC(CH_3)_3$

(c) Methyl neopentyl ether

(d) Methyl phenyl ether

(e) $C_6H_5CH_2CN$

(f) $CH_3CO_2CH_2C_6H_5$

(g) (S)-2-Pentanol

(h) (R)-2-Iodo-4-methylpentane

(i) $(CH_3)_3CCH{=}CH_2$

(j) *cis*-4-Isopropylcyclohexanol

(k) (R)-$CH_3CH(CN)CH_2CH_3$

(l) *trans*-1-Iodo-4-methylcyclohexane

6.33   Give structures for the products of each of the following reactions:

(a)

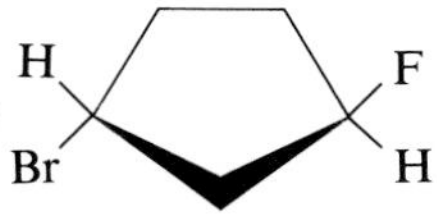

$+ \ NaI \ (1 \ mol) \xrightarrow[\text{acetone}]{} C_5H_8FI + NaBr$

(b) 1,4-Dichlorohexane (1 mol) + NaI (1 mol) $\xrightarrow[\text{acetone}]{} C_6H_{12}ClI + NaCl$

(c) 1,2-Dibromoethane (1 mol) + $NaSCH_2CH_2SNa \longrightarrow C_4H_8S_2 + 2\ NaBr$

(d) 4-Chloro-1-butanol + NaH $\xrightarrow[\text{Et}_2\text{O}]{(-H_2)} C_4H_8ClONa \xrightarrow[\text{Et}_2\text{O}]{\text{heat}} C_4H_8O + NaCl$

(e) Propyne + $NaNH_2 \xrightarrow[\text{liq. NH}_3]{(-NH_3)} C_3H_3Na \xrightarrow{CH_3I} C_4H_6 + NaI$

6.34   When *tert*-butyl bromide undergoes S_N1 hydrolysis, adding a "common ion" (e.g., NaBr) to the aqueous solution has no effect on the rate. On the other hand, when $(C_6H_5)_2CHBr$ undergoes S_N1 hydrolysis, adding NaBr retards the reaction. Given that the $(C_6H_5)_2CH^+$ cation is known to be much more stable than the $(CH_3)_3C^+$ cation (and we shall see why in Section 15.12B), provide an explanation for the different behavior of the two compounds.

6.35   When the alkyl bromides (listed here) were subjected to hydrolysis in a mixture of ethanol and water (80% C_2H_5OH/20% H_2O) at 55°C, the rates of the reaction showed the following order:

$$(CH_3)_3CBr > CH_3Br > CH_3CH_2Br > (CH_3)_2CHBr$$

Provide an explanation for this order of reactivity.

6.36 The reaction of 1° alkyl halides with nitrite salts produces both RNO_2 and RONO. Account for this behavior.

6.37 What would be the effect of increasing solvent polarity on the rate of each of the following nucleophilic substitution reactions?

(a) $Nu\colon + R—L \longrightarrow R—Nu^+ + \colon L^-$

(b) $R—L^+ \longrightarrow R^+ + \colon L$

6.38 Competition experiments are those in which two reactants at the same concentration (or one reactant with two reactive sites) compete for a reagent. Predict the major product resulting from each of the following competition experiments.

(a)
$$\text{Cl—CH}_2\text{—}\underset{\underset{\textstyle CH_3}{|}}{\overset{\overset{\textstyle CH_3}{|}}{C}}\text{—CH}_2\text{—CH}_2\text{—Cl} + I^- \xrightarrow[\text{DMF}]{}$$

(b)
$$\text{Cl—}\underset{\underset{\textstyle CH_3}{|}}{\overset{\overset{\textstyle CH_3}{|}}{C}}\text{—CH}_2\text{—CH}_2\text{—Cl} + H_2O \xrightarrow[\text{acetone}]{}$$

6.39 In contrast to S_N2 reactions, S_N1 reactions show relatively little nucleophile selectivity. That is, when more than one nucleophile is present in the reaction medium, S_N1 reactions show only a slight tendency to discriminate between weak nucleophiles and strong nucleophiles, whereas S_N2 reactions show a marked tendency to discriminate. (a) Provide an explanation for this behavior. (b) Show how your answer accounts for the fact that $CH_3CH_2CH_2CH_2Cl$ reacts with $0.01M$ NaCN in ethanol to yield primarily $CH_3CH_2CH_2CH_2CN$, whereas under the same conditions, $(CH_3)_3CCl$ reacts to give primarily $(CH_3)_3COCH_2CH_3$.

6.40 In the gas phase, the homolytic bond dissociation energy (Section 9.2A) for the carbon–chlorine bond of *tert*-butyl chloride is $+78.5$ kcal mol^{-1}; the ionization potential for a *tert*-butyl radical is $+171$ kcal mol^{-1}; and the electron affinity of chlorine is -79 kcal mol^{-1}. Using these data, calculate the enthalpy change for the gas phase ionization of *tert*-butyl chloride to a *tert*-butyl cation and a chloride ion (this is the heterolytic bond dissociation energy of the carbon–chlorine bond).

6.41 The reaction of chloroethane with water *in the gas phase* to produce ethanol and hydrogen chloride has a $\Delta H° = +6.36$ kcal mol^{-1} and a $\Delta S° = +1.15$ cal k^{-1} mol^{-1} at 25°C. (a) Which of these terms, if either, favors the reaction going to completion? (b) Calculate $\Delta G°$ for the reaction. What can you now say about whether the reaction will proceed to completion? (d) Calculate the equilibrium constant for the reaction. (e) In aqueous solution the equilibrium constant is very much larger than the one you just calculated. How can you account for this fact?

Nicotine

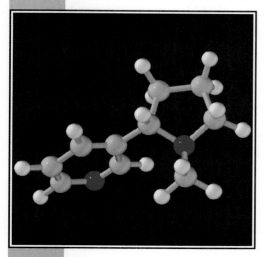

A BIOLOGICAL NUCLEOPHILIC SUBSTITUTION REACTION: BIOLOGICAL METHYLATION

The cells of living organisms synthesize many of the compounds they need from smaller molecules. Often these biosyntheses resemble the syntheses organic chemists carry out in their laboratories. Let us examine one example now.

Many reactions take place in the cells of plants and animals that involve the transfer of a methyl group from an amino acid called methionine to some other compound. That this transfer takes place can be demonstrated experimentally by feeding a plant or animal methionine containing a radioactive carbon atom (^{14}C) in its methyl group. Later, other compounds containing the "labeled" methyl group can be isolated from the organism. Some of the compounds that get their methyl groups from methionine are the following. The radioactively labeled carbon atom is shown in color.

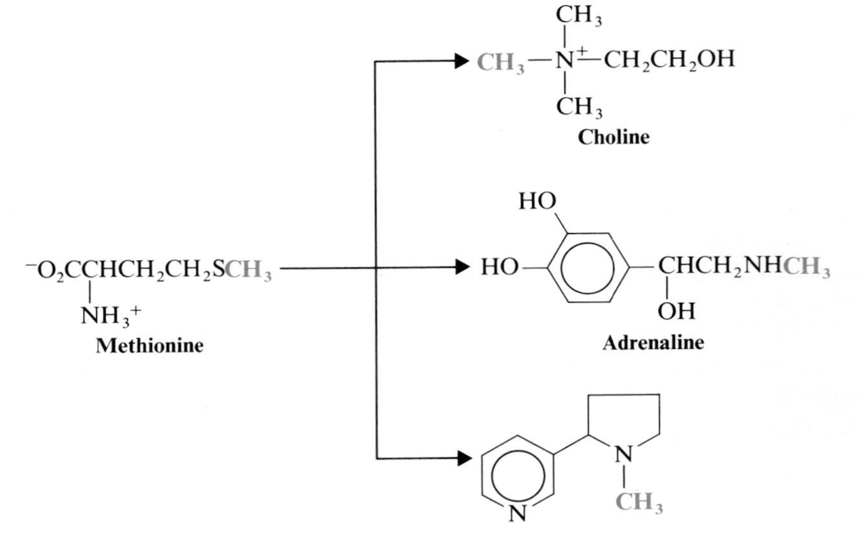

$$CH_3 - \overset{\overset{\displaystyle CH_3}{|}}{\underset{\underset{\displaystyle CH_3}{|}}{N^{+}}} - CH_2CH_2OH$$

Choline

$$^{-}O_2CCHCH_2CH_2SCH_3 \longrightarrow$$
$$\underset{NH_3^{+}}{|}$$

Methionine

Adrenaline

HO—⟨○⟩—CHCH$_2$NHCH$_3$ with HO above and OH below

Nicotine

Choline is important in the transmission of nerve impulses, adrenaline causes blood pressure to increase, and nicotine is the compound contained in tobacco that makes smoking tobacco addictive. (In large doses nicotine is poisonous.)

The transfer of the methyl group from methionine to these other compounds does not take place directly. The actual methylating agent is not methionine; it is *S*-adenosyl-methionine,* a compound that results when methionine reacts with adenosine triphosphate (ATP):

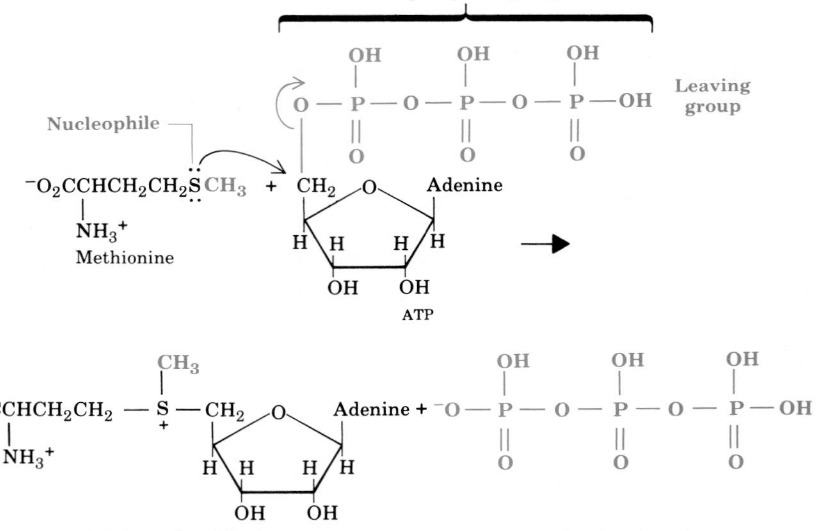

Triphosphate group

Nucleophile

$$^{-}O_2CCHCH_2CH_2\overset{\cdot\cdot}{\underset{\cdot\cdot}{S}}CH_3 \; + \; CH_2 \quad O$$
$$\underset{NH_3^{+}}{|}$$
Methionine

Leaving group

ATP

$$^{-}O_2CCHCH_2CH_2 - \overset{\overset{\displaystyle CH_3}{|}}{\underset{+}{S}} - CH_2 \quad O \quad \text{Adenine} \; + \; ^{-}O - \overset{OH}{\underset{\|}{P}} - O - \overset{OH}{\underset{\|}{P}} - O - \overset{OH}{\underset{\|}{P}} - OH$$
$$\underset{NH_3^{+}}{|}$$

S-Adenosylmethionine

Triphosphate ion

* The prefix *S* is a locant meaning "on the sulfur atom" and should not be confused with the (*S*) used to define absolute configuration. Another example of this kind of locant is *N* meaning "on the nitrogen atom."

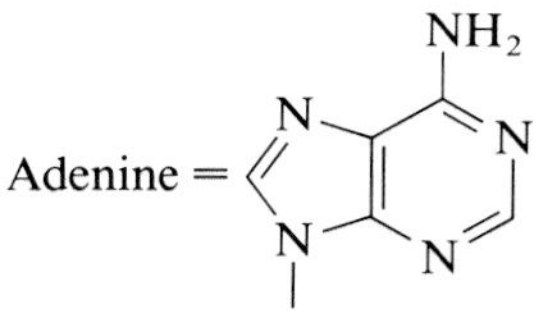

$$Adenine = $$

This reaction is a nucleophilic substitution reaction. The nucleophilic atom is the sulfur atom of methionine. The leaving group is the weakly basic triphosphate group of adenosine triphosphate. The product, S-adenosylmethionine, contains a methylsulfonium group, $CH_3 - \overset{|}{\underset{..}{S}} \overset{+}{-}$.

S-Adenosylmethionine then acts as the substrate for other nucleophilic substitution reactions. In the biosynthesis of choline, for example, it transfers its methyl group to a nucleophilic nitrogen atom of 2-(N,N-dimethylamino)ethanol:

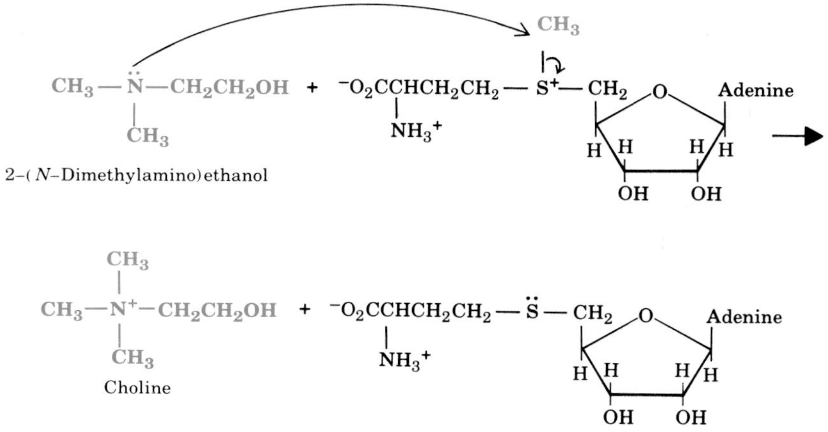

These reactions appear complicated only because the structures of the nucleophiles and substrates are complex. Yet conceptually they are simple and they illustrate many of the principles we have encountered in Chapter 6. In them we see how nature makes use of the high nucleophilicity of sulfur atoms. We also see how a weakly basic group (e.g., the triphosphate group of ATP) functions as a leaving group. In the reaction of 2-(N,N-dimethylamino)ethanol we see that the more basic $(CH_3)_2N$— group acts as the nucleophile rather than the less basic —OH group. And when a nucleophile attacks S-adenosylmethionine, we see that the attack takes place at the less hindered CH_3— group rather than at one of the more hindered —CH_2— groups.

PROBLEM A.1

(a) What is the leaving group when 2-(N,N-dimethylamino)ethanol reacts with S-adenosylmethionine?   (b) What would the leaving group have to be if methionine itself were to react with 2-(N,N-dimethylamino)ethanol?   (c) Of what special significance is this difference?

Malic acid isomers

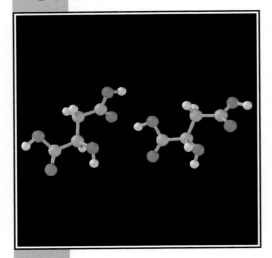

NEIGHBORING GROUP PARTICIPATION IN NUCLEOPHILIC SUBSTITUTION REACTIONS

Not all nucleophilic substitutions take place with racemization or with inversion of configuration. Some take place with overall *retention of configuration.*

One factor that can lead to retention of configuration in a nucleophilic substitution is a phenomenon known as *neighboring group participation.* Let us see how this operates by examining the stereochemisty of two reactions in which 2-bromopropanoic acid is converted to lactic acid.

$$CH_3CHCO_2H \longrightarrow CH_3CHCO_2H$$
$$\qquad | \qquad\qquad\qquad\qquad |$$
$$\quad Br \qquad\qquad\qquad\qquad OH$$

2-Bromopropanoic   **Lactic acid**
acid

When (*S*)-2-bromopropanoic acid is treated with concentrated sodium hydroxide, the reaction is *bimolecular* and it takes place with *inversion of configuration.* This, of course, is the normal stereochemical result for an S_N2 reaction.

$$HO^- + \underset{H}{\overset{CO_2^-}{\underset{CH_3}{C}}}-Br \longrightarrow HO\cdots\overset{CO_2^-}{\underset{H\ CH_3}{C}}\cdots Br \longrightarrow HO-\overset{CO_2^-}{\underset{CH_3}{C}}H + Br^-$$

(*S*)-**2-Bromopropanoate**   (*R*)-**Lactate ion**
ion

Inversion of configuration

However, when the same reaction is carried out with a low concentration of hydroxide ion in the presence of Ag_2O, it takes place with an overall *retention of configuration*. In this case, the mechanism for the reaction involves the participation of the carboxylate group. In step 1 (see following reaction) an oxygen of the carboxylate group attacks the stereocenter from the backside and displaces bromide ion. (Silver ion aids in this process in much the same way that protonation assists the ionization of an alcohol.) The configuration of the stereocenter inverts in step 1, and a cyclic ester called an α-lactone forms.

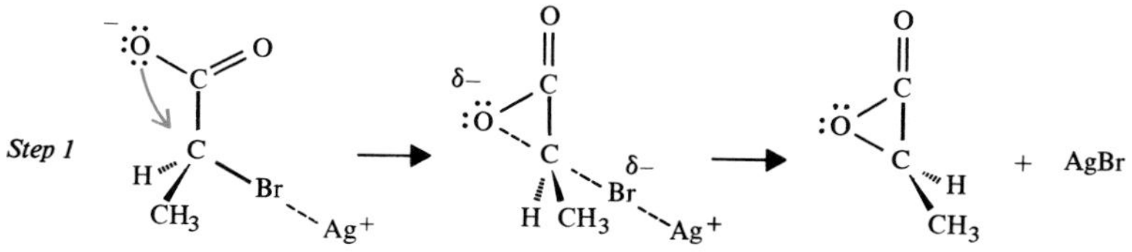

An α-lactone

The highly strained three-membered ring of the α-lactone opens when it is attacked by a water molecule in step 2. *This step also takes place with an inversion of configuration.*

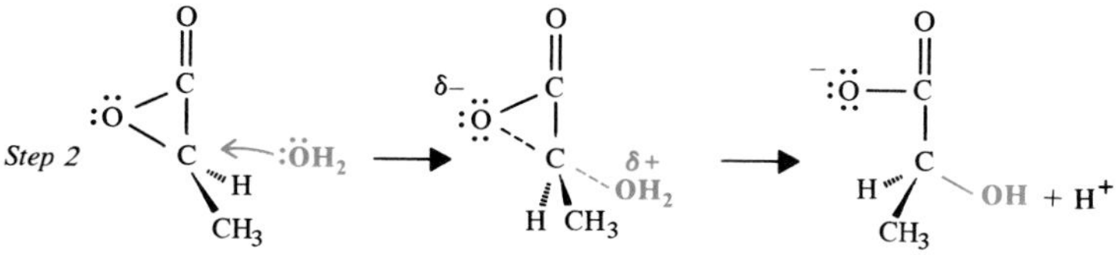

The net result of two inversions (in steps 1 and 2) is an overall *retention of configuration.*

PROBLEM B.1

The phenomenon of configuration inversion in a chemical reaction was discovered in 1896 by Paul von Walden (Section 6.8). Walden's proof of configuration inversion was based on the following cycle:

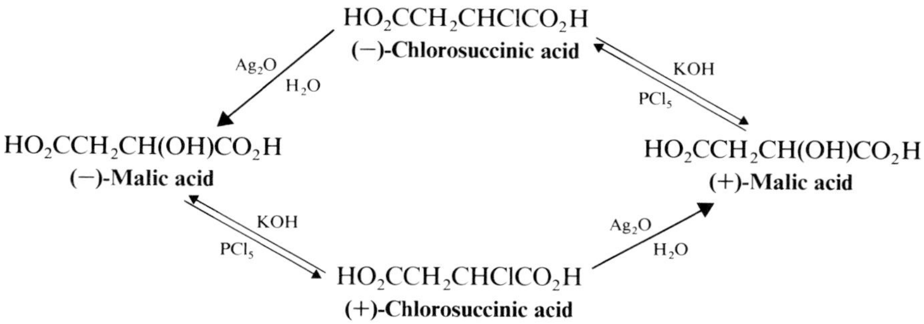

The Walden cycle

(a) Basing your answer on the preceding discussion, which reactions of the Walden cycle are likely to take place with overall inversion of configuration and which are likely to occur with overall retention of configuration? (b) Malic acid with a negative optical rotation is now known to have the (S) configuration. What are the configurations of the other compounds in the Walden cycle? (c) Walden also found that when (+)-malic acid is treated with thionyl chloride (rather than PCl$_5$), the product of the reaction is (+)-chlorosuccinic acid. How can you explain this result? (d) Assuming that the reaction of (−)-malic acid and thionyl chloride has the same stereochemistry, outline a Walden cycle based on the use of thionyl chloride instead of PCl$_5$.

Neighboring group participation can also lead to *cyclization reactions*. Compounds called epoxides, for example, can be prepared from 2-bromo alcohols by treating them with sodium hydroxide. This reaction involves the following steps:

An epoxide
(see Section 10.18)

CHAPTER 7

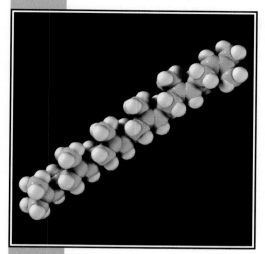

Squalene
(Problem 7.44)

ALKENES AND ALKYNES I.
PROPERTIES AND SYNTHESIS

7.1 INTRODUCTION

Alkenes are hydrocarbons whose molecules contain the carbon–carbon double bond. An old name for this family of compounds that is still often used is the name *olefins*. Ethene, the simplest olefin (alkene), was called olefiant gas (Latin: *oleum*, oil + *facere*, to make) because gaseous ethene (C_2H_4) reacts with chlorine to form $C_2H_4Cl_2$, a liquid (oil).

Hydrocarbons whose molecules contain the carbon–carbon triple bond are called alkynes. The common name for this family is *acetylenes*, after the first member, $HC{\equiv}CH$.

$$
\underset{\textbf{Ethene}}{\begin{array}{c} H \\ \diagdown \\ C \end{array} = \begin{array}{c} H \\ \diagup \\ C \\ \diagup \\ H \end{array}} \qquad
\underset{\textbf{Propene}}{\begin{array}{c} H \\ \diagdown \\ C \end{array} = \begin{array}{c} CH_3 \\ \diagup \\ C \\ \diagup \\ H \end{array}} \qquad
\underset{\textbf{Ethyne}}{H-C{\equiv}C-H}
$$

Ethene and propene are among the most important industrial chemicals produced in the United States. Each year, our chemical industry produces more than 30 billion pounds of ethene and about 15 billion pounds of propene. Ethene is used as a starting material for the synthesis of many industrial compounds, including ethanol, ethylene oxide (Section 10.8), ethanal (Section 16.2) and the polymer, polyethylene (Section 9.10). Propene is used in making the polymer, polypropylene (Section 9.10 and Special Topic C) and, in addition to other uses, propene is the starting material for a synthesis of acetone and cumene (Section 21.4).

Ethene also occurs in nature as a plant hormone. It is produced naturally by fruits such as tomatoes and bananas and is involved in the ripening process of these fruits. Much use is now made of ethene in the commercial fruit industry to bring about the ripening of tomatoes and bananas picked green, because the green fruits are less suceptible to damage during shipping.

There are many other naturally occuring alkenes. Two examples are the following:

β-Pinene
(a component of
turpentine)

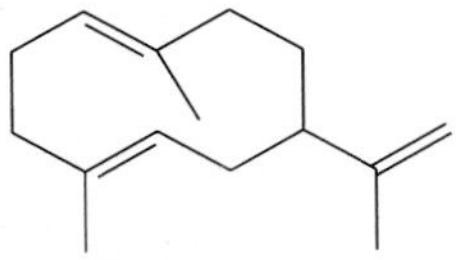

An aphid alarm pheromone

Naturally occuring alkynes are not as numerous as alkenes, but more than a thousand have been isolated. An example is capillin, an antifungal agent.

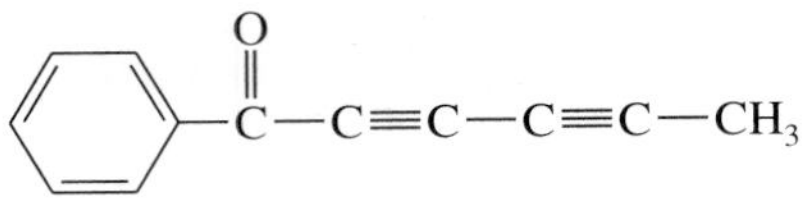

Capillin

7.2 NOMENCLATURE OF ALKENES AND CYCLOALKENES

Many older names for alkenes are still in common use. Propene is often called propylene, and 2-methylpropene frequently bears the name isobutylene.

$$CH_2{=}CH_2 \qquad CH_3CH{=}CH_2 \qquad CH_3-\overset{\displaystyle CH_3}{\underset{\displaystyle |}{C}}{=}CH_2$$

| IUPAC: Ethene | IUPAC: Propene | IUPAC: 2-Methylpropene |
| or ethylene* | or propylene* | Common: Isobutylene |

The IUPAC rules for naming alkenes are similar in many respects to those for naming alkanes:

1. **Determine the base name by selecting the longest chain that contains the double bond and change the ending of the name of the alkane of identical length from -ane to -ene.** Thus, if this longest chain contains five carbon atoms, the base name for the alkene is *pentene;* if it contains six carbon atoms, the base name is *hexene,* and so on.

2. **Number the chain so as to include both carbon atoms of the double bond, and begin numbering at the end of the chain nearer the double bond. Designate the location of the double bond by using the number of the first atom of the double bond as a prefix:**

*The IUPAC system also retains the names ethylene and propylene when no substituents are present.

$$\overset{1}{C}H_2\!\!=\!\!\overset{2}{C}H\overset{3}{C}H_2\overset{4}{C}H_3 \qquad CH_3CH\!\!=\!\!CHCH_2CH_2CH_3$$

1-Butene   **2-Hexene**
(*not* 3-butene)   (*not* 4-hexene)

3. **Indicate the locations of the substituent groups by the numbers of the carbon atoms to which they are attached.**

$$\underset{1}{CH_3}\underset{2}{\overset{CH_3}{C}}\!\!=\!\!\underset{3}{C}H\underset{4}{C}H_3 \qquad \underset{1}{CH_3}\underset{2}{\overset{CH_3}{C}}\!\!=\!\!\underset{3}{C}H\underset{4}{C}H_2\underset{5}{\overset{CH_3}{C}}H\underset{6}{C}H_3$$

2-Methyl-2-butene   **2,5-Dimethyl-2-hexene**
(*not* 3-methyl-2-butene)   (*not* 2,5-dimethyl-4-hexene)

$$\underset{1}{CH_3}\underset{2}{C}H\!\!=\!\!\underset{3}{C}H\underset{4}{C}H_2\underset{5}{\overset{CH_3}{\underset{CH_3}{C}}}\!\!-\!\!\underset{6}{C}H_3 \qquad \overset{4}{C}H_3\overset{3}{C}H\!\!=\!\!\overset{2}{C}H\overset{1}{C}H_2Cl$$

5,5-Dimethyl-2-hexene   **1-Chloro-2-butene**

4. **Number substituted cycloalkenes in the way that gives the carbon atoms of the double bond the 1 and 2 positions and that also gives the substituent groups the lower numbers at the first point of difference.** With substituted cycloalkenes it is not necessary to specify the position of the double bond since it will always begin with C1 and C2. The two examples listed here illustrate the application of these rules.

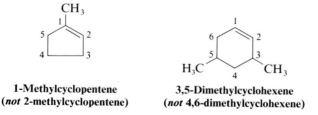

1-Methylcyclopentene   **3,5-Dimethylcyclohexene**
(*not* 2-methylcyclopentene)   (*not* 4,6-dimethylcyclohexene)

5. Name compounds containing a double bond and an alcohol group as alkenols (or cycloalkenols) and give the alcohol carbon the lower number.

$$\underset{5}{CH_3}\underset{4}{\overset{CH_3}{C}}\!\!=\!\!\underset{3}{C}H\underset{2}{\overset{}{\underset{OH}{C}}}H\underset{1}{C}H_3$$

4-Methyl-3-penten-2-ol   **2-Methyl-2-cyclohexen-1-ol**

6. Two frequently encountered alkenyl groups are the *vinyl group* and the *allyl group*.

$$CH_2\!\!=\!\!CH\!\!-\!\! \qquad CH_2\!\!=\!\!CHCH_2\!\!-\!\!$$

The vinyl group   **The allyl group**

The following examples illustrate how these names are employed:

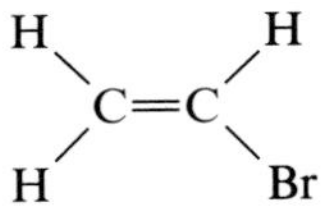

Bromoethene
or
vinyl bromide
(common)

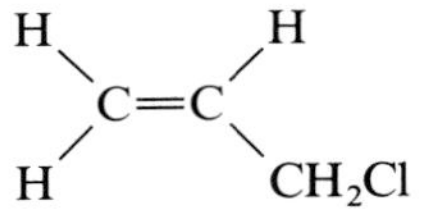

3-Chloropropene
or
allyl chloride
(common)

7. If two identical groups are on the same side of the double bond, the compound can be designated *cis*; if they are on opposite sides it can be designated *trans*.

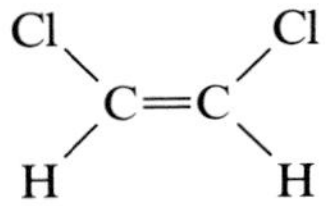

cis-1,2-Dichloroethene 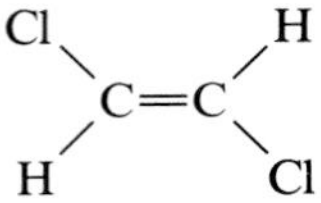   *trans*-1,2-Dichloroethene

In Section 7.2A we shall see another method for designating the geometry of the double bond.

PROBLEM 7.1

Give IUPAC names for the following alkenes:

(a)

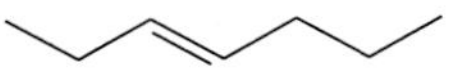

(b)

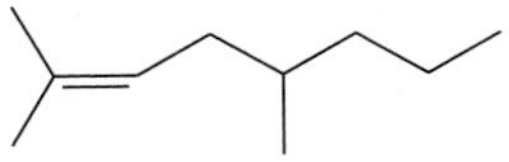

(c)

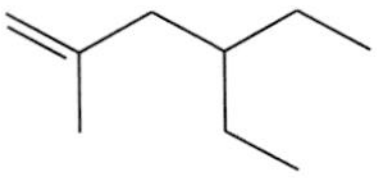

(d)

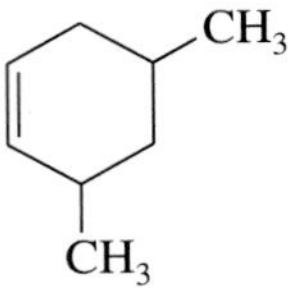

(e)

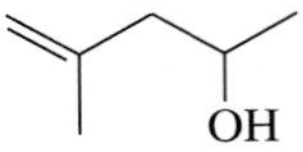

(f)

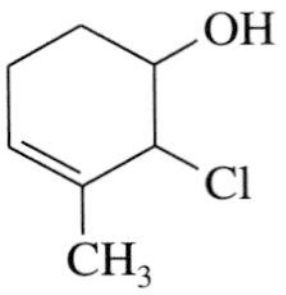

PROBLEM 7.2

Write structural formulas for:

(a) *cis*-3-Octene

(b) *trans*-2-Hexene

(c) 2,4-Dimethyl-2-pentene

(d) *trans*-1-Chloro-2-butene

(e) 4,5-Dibromo-1-pentene

(f) 1,3-Dimethylcyclohexene

(g) 3,4-Dimethylcyclopentene

(h) Vinylcyclopentane

(i) 1,2-Dichlorocyclohexene

(j) *trans*-(4R)-1,4-Dichloro-2-pentene

7.2A The (E)–(Z) System for Designating Alkene Diastereomers

The terms cis and trans, when used to designate the stereochemistry of alkene diastereomers, are unambiguous only when applied to disubstituted alkenes. If the alkene is trisubstituted or tetrasubstituted, the terms cis and trans are either ambiguous or do not apply at all. Consider the following alkene as an example.

$$Br\diagdown C=C\diagup Cl$$
$$H\diagup \diagdown F$$

A

It is impossible to decide whether **A** is cis or trans since no two groups are the same.

A newer system is based on the priorities of groups in the Cahn–Ingold–Prelog convention (Section 5.6). This system, called the (E)–(Z) system, applies to alkene diastereomers of all types. In the (E)–(Z) system, we examine the two groups attached to one carbon atom of the double bond and decide which has higher priority. Then we repeat that operation at the other carbon atom.

Higher priority → Cl   F
Higher priority → Br   H

(Z)-2-Bromo-1-chloro-1-fluoroethene

F   Cl ← Higher priority
Higher priority → Br   H

(E)-2-Bromo-1-chloro-1-fluoroethene

Cl > F

Br > H

We take the group of higher priority on one carbon atom and compare it with the group of higher priority on the other carbon atom. If the two groups of higher priority are on the same side of the double bond, the alkene is designated (Z) (from the German word *zusammen,* meaning together). If the two groups of higher priority are on opposite sides of the double bond, the alkene is designated (E) (from the German word *entgegen,* meaning opposite). The following examples illustrate this:

$CH_3 > H$

$$H_3C\diagdown C=C\diagup CH_3$$
$$H\diagup \diagdown H$$

(Z)-2-Butene
(cis-2-butene)

$$H_3C\diagdown C=C\diagup H$$
$$H\diagup \diagdown CH_3$$

(E)-2-Butene
(trans-2-butene)

$$Cl\diagdown C=C\diagup Cl$$
$$H\diagup \diagdown Br$$

(E)-1-Bromo-1,2-dichloroethene

$Cl > H$
$Br > Cl$

$$Cl\diagdown C=C\diagup Br$$
$$H\diagup \diagdown Cl$$

(Z)-1-Bromo-1,2-dichloroethene

PROBLEM 7.3

Using (E)–(Z) designation [and in parts (e) and (f) the (R–S) designation as well] give IUPAC names for each of the following:

(a)
$$Br\diagdown C=C\diagup H$$
$$Cl\diagup \diagdown CH_2CH_2CH_3$$

(b)
$$Cl\diagdown C=C\diagup Br$$
$$I\diagup \diagdown CH_2CH_3$$

(c)

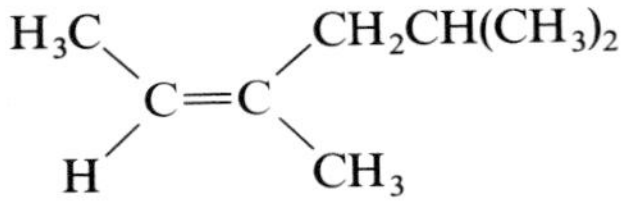

$$H_3C \diagdown C=C \diagup CH_2CH(CH_3)_2$$
$$H \diagup \diagdown CH_3$$

(d)

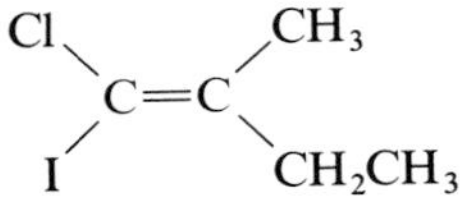

$$Cl \diagdown C=C \diagup CH_3$$
$$I \diagup \diagdown CH_2CH_3$$

(e)

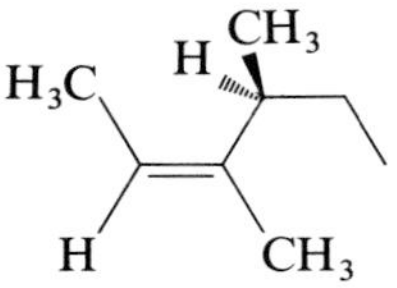

$$H_3C \diagdown C=C \diagup{\overset{H}{\underset{}{}} CH_3}$$
$$H \diagup \diagdown CH_3$$

(f)

$$\overset{Br}{\diagdown} C= \overset{Cl}{C}$$
$$H \diagdown \overset{}{H_3C} H$$

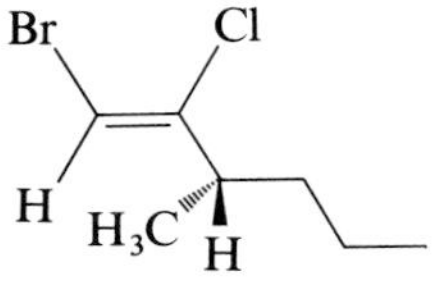

7.3 NOMENCLATURE OF ALKYNES

7.3A IUPAC Nomenclature

Alkynes are named in much the same way as alkenes. Unbranched alkynes, for example, are named by replacing the **-ane** of the name of the corresponding alkane with the ending **-yne.** The chain is numbered to give the carbon atoms of the triple bond the lower possible numbers. The lower number of the two carbon atoms of the triple bond is used to designate the location of the triple bond. The IUPAC names of three unbranched alkynes are shown here.

$$H-C\equiv C-H \qquad CH_3CH_2C\equiv CCH_3 \qquad H-C\equiv CCH_2CH=CH_2$$

Ethyne or
acetylene*
2-Pentyne
1-Penten-4-yne†

The locations of substituent groups of branched alkynes and substituted alkynes are also indicated with numbers. An —OH group has priority over the triple bond when numbering the chain of an alkynol.

$$\overset{3}{Cl}-\overset{2}{CH_2}\overset{1}{C}\equiv CH \qquad \overset{4}{CH_3}\overset{3}{C}\equiv \overset{2}{C}\overset{1}{CH_2}Cl \qquad H\overset{4}{C}\equiv \overset{3}{C}\overset{2}{CH_2}\overset{1}{CH_2}OH$$

3-Chloropropyne
1-Chloro-2-butyne
3-Butyn-1-ol

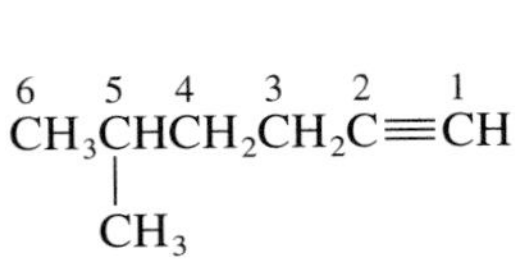

$$\overset{6}{CH_3}\overset{5}{CH}\overset{4}{CH_2}\overset{3}{CH_2}\overset{2}{C}\equiv \overset{1}{CH}$$
$$| $$
$$CH_3$$

5-Methyl-1-hexyne

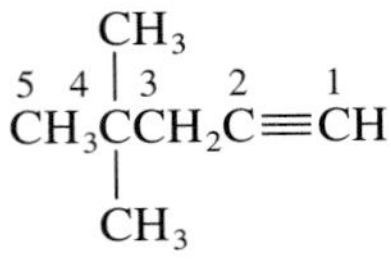

$$\overset{5}{CH_3}\overset{4}{C}\overset{3}{C}H_2\overset{2}{C}\equiv \overset{1}{CH}$$
with CH_3 above and CH_3 below C-4

4,4-Dimethyl-1-pentyne

$$\overset{1}{CH_3}\overset{2}{C}\overset{3}{CH_2}\overset{4}{C}\equiv \overset{5}{CH}$$
with OH above and CH_3 below C-2

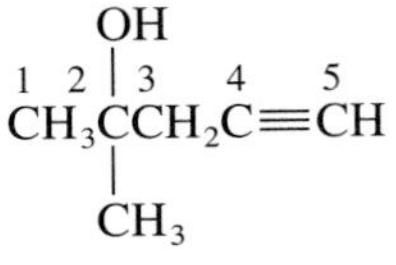

2-Methyl-4-pentyn-2-ol

PROBLEM 7.4

Give the structures and IUPAC names [including the $(R-S)$ designation] for enantiomeric alkynes with the formula C_6H_{10}.

*The name acetylene is retained by the IUPAC system for the compound $HC\equiv CH$ and is used frequently.
†Where there is a choice the double bond is given the lower number.

Monosubstituted acetylenes or 1-alkynes are called **terminal alkynes,** and the hydrogen attached to the carbon of the triple bond is called the acetylenic hydrogen.

$$R—C≡C—H \qquad \text{← Acetylenic hydrogen}$$

**A terminal
alkyne**

The anion obtained when the acetylenic hydrogen is removed is known as *an alkynide ion* or an acetylide ion. As we shall see in Section 7.20 these ions are useful in synthesis.

$$R—C≡C:^- \qquad CH_3C≡C:^-$$

**An alkynide ion
(an acetylide ion)**   **The propynide ion**

7.4 PHYSICAL PROPERTIES OF ALKENES AND ALKYNES

Alkenes and alkynes have physical properties similar to those of corresponding alkanes. The lower molecular weight alkenes and alkynes (Tables 7.1 and 7.2) are gases at room temperature. Being relatively nonpolar themselves, alkenes and alkynes dissolve in nonpolar solvents or in solvents of low polarity. Alkenes and alkynes are only *very slightly soluble* in water (with alkynes being slightly more soluble than alkenes). The densities of alkenes and alkynes are less than that of water.

7.5 HYDROGENATION OF ALKENES

Alkenes react with hydrogen in the presence of a variety of finely divided metal catalysts (cf. Section 4.15). The reaction that takes place is an **addition reaction;** one atom of hydrogen *adds* to each carbon atom of the double bond. Without a catalyst the

TABLE 7.1 *Physical constants of alkenes*

NAME	FORMULA	mp (°C)	bp (°C)	DENSITY d_4^{20} (g mL^{-1})
Ethene	$CH_2{=}CH_2$	-169	-104	0.384^a
Propene	$CH_3CH{=}CH_2$	-185	-47	0.514
1-Butene	$CH_3CH_2CH{=}CH_2$	-185	-6.3	0.595
(Z)-2-Butene	$CH_3CH{=}CHCH_3$ (cis)	-139	3.7	0.621
(E)-2-Butene	$CH_3CH{=}CHCH_3$ (trans)	-106	0.9	0.604
1-Pentene	$CH_3(CH_2)_2CH{=}CH_2$	-165	30	0.641
2-Methyl-1-butene	$CH_2{=}C(CH_3)CH_2CH_3$	-138	31	0.650
1-Hexene	$CH_3(CH_2)_3CH{=}CH_2$	-140	63	0.673
1-Heptene	$CH_3(CH_2)_4CH{=}CH_2$	-119	94	0.697

^{a} Density at $-10°C$.

TABLE 7.2   Physical constants of alkynes

NAME	FORMULA	mp (°C)	bp (°C)	DENSITY d_4^{20} (g mL^{-1})
Ethyne	HC≡CH	− 80.8	− 84.0$^{760}_{(sub)}$	
Propyne	$CH_3C≡CH$	− 101.51	− 23.2	
1-Butyne	$CH_3CH_2C≡CH$	− 125.7	8.1	
2-Butyne	$CH_3C≡CCH_3$	− 32.3	27	0.691
1-Pentyne	$CH_3(CH_2)_2C≡CH$	− 90	39.3	0.695
2-Pentyne	$CH_3CH_2C≡CCH_3$	− 101	55.5	0.714
1-Hexyne	$CH_3(CH_2)_3C≡CH$	− 132	71	0.715
2-Hexyne	$CH_3(CH_2)_2C≡CCH_3$	− 88	84	0.730
3-Hexyne	$CH_3CH_2C≡CCH_2CH_3$	− 101	81.8	0.724

reaction does not take place at an appreciable rate. (We shall see how the catalyst functions in Section 7.6.)

$$CH_2{=}CH_2 + H_2 \xrightarrow[\substack{\text{or Pt} \\ 25°C}]{\text{Ni, Pd}} CH_3{-}CH_3$$

$$CH_3CH{=}CH_2 + H_2 \xrightarrow[\substack{\text{or Pt} \\ 25°C}]{\text{Ni, Pd}} CH_3CH_2{-}CH_3$$

The product that results from the addition of hydrogen to an alkene is an alkane. Alkanes have only single bonds and contain the maximum number of hydrogen atoms that a hydrocarbon can possess. For this reason, alkanes are said to be **saturated compounds.** Alkenes, because they contain a double bond and possess fewer than the maximum number of hydrogen atoms, are capable of adding hydrogen and are said to be **unsaturated.** The process of adding hydrogen to an alkene is sometimes described as being one of **reduction.** Most often, however, the term used to describe the addition of hydrogen is **catalytic hydrogenation.**

7.5A Hydrogenation in the Food Industry

The food industry makes use of catalytic hydrogenation to convert liquid vegetable oils to semisolid fats in making margarine and solid cooking fats. Examine the labels of many prepared foods and you will find that they contain ''partially hydrogenated vegetable oils.'' There are several reasons why foods contain these oils, but one is that partially hydrogenated vegetable oils have a longer shelf life.

Fats and oils (Section 23.2) are glyceryl esters of carboxylic acids with long carbon chains, called ''fatty acids.'' Fatty acids are saturated (no double bonds), monounsaturated (one double bond), or polyunsaturated (more than one double bond). Oils typically contain a higher proportion of fatty acids with one or more double bonds than fats do. Hydrogenation of an oil converts some of its double bonds to single bonds, and this conversion has the effect of producing a fat with the consistency of margarine or a semisolid cooking fat.

Our bodies are incapable of making polyunsaturated fats and, therefore, they must be present in our diets in moderate amounts in order to maintain health. Saturated fats can be made in the cells of our body from other food sources, for example, from carbohydrates (i.e., from sugars and starches). For this reason saturated fats in our diet are not absolutely necessary and, indeed, too much saturated fat has been implicated in the development of cardiovascular disease.

One potential problem that arises from using catalytic hydrogenation to produce partially hydrogenated vegetable oils is that the catalysts used for hydrogenation cause isomerization of some of the double bonds of the fatty acids (some of those that do not absorb hydrogen). In all natural fats and oils, the double bonds of the fatty acids have the cis configuration. The catalysts used for hydrogenation convert some of these cis double bonds to the unnatural trans configuration, and there is evidence that these unnatural trans fatty acids may be harmful.

7.6 HYDROGENATION: THE FUNCTION OF THE CATALYST

Hydrogenation of an alkene is an exothermic reaction ($\Delta H° \cong -30$ kcal mol^{-1}).

$$R-CH=CH-R + H_2 \xrightarrow{\text{hydrogenation}} R-CH_2-CH_2-R + \text{heat}$$

Hydrogenation reactions usually have high free energies of activation. The reaction of an alkene with molecular hydrogen does not take place at room temperature in the absence of a catalyst, but often *does* take place at room temperature when a metal catalyst is added. The catalyst provides a new pathway for the reaction with a *lower free energy of activation* (Fig. 7.1).

The most commonly used catalysts for hydrogenation (finely divided platinum, nickel, palladium, rhodium, and ruthenium) apparently serve to adsorb hydrogen mol-

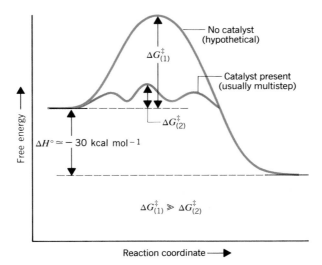

FIGURE 7.1 Free-energy diagram for the hydrogenation of an alkene in the presence of a catalyst and the hypothetical reaction in the absence of a catalyst. The free energy of activation for the uncatalyzed reaction [$\Delta G^{\ddagger}_{(1)}$] is very much larger than the largest free energy of activation for the catalyzed reaction [$\Delta G^{\ddagger}_{(2)}$].

ecules on their surfaces. This adsorption of hydrogen is essentially a chemical reaction; unpaired electrons on the surface of the metal *pair* with the electrons of hydrogen (Fig. 7.2*a*) and bind the hydrogen to the surface. The collision of an alkene with the surface bearing adsorbed hydrogen causes adsorption of the alkene as well (Fig. 7.2*b*). A stepwise transfer of hydrogen atoms takes place, and this produces an alkane before the organic molecule leaves the catalyst surface (Fig. 7.2*c* and *d*). As a consequence, *both hydrogen atoms usually add from the same side of the molecule.* This mode of addition is called a **syn** addition (Section 7.6A).

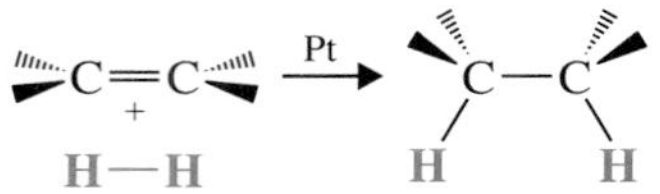

Catalytic hydrogenation is a syn addition

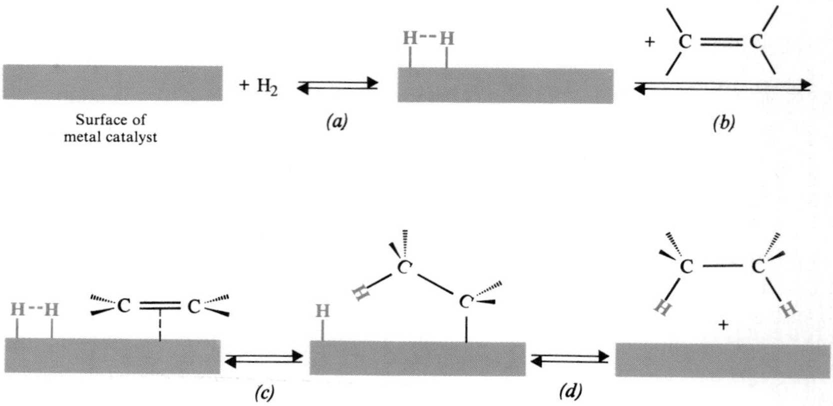

FIGURE 7.2 The mechanism for the hydrogenation of an alkene as catalyzed by finely divided platinum metal. Notice that both hydrogen atoms add from the same side of the double bond.

7.6A Syn and Anti Additions

An addition that places the parts of the adding reagent on the same side (or face) of the reactant is called a **syn addition.** We have just seen that the platinum-catalyzed addition of hydrogen (X = Y = H) is a syn addition.

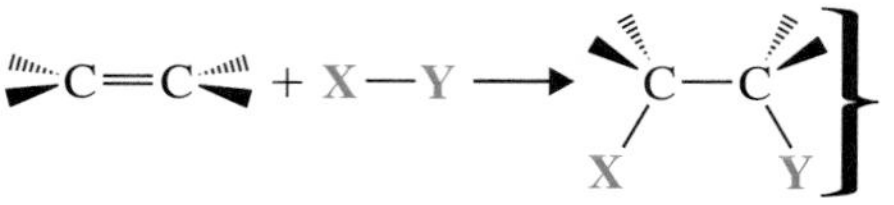

The opposite of a syn addition is an **anti addition.** An anti addition places the parts of the adding reagent on opposite faces of the reactant.

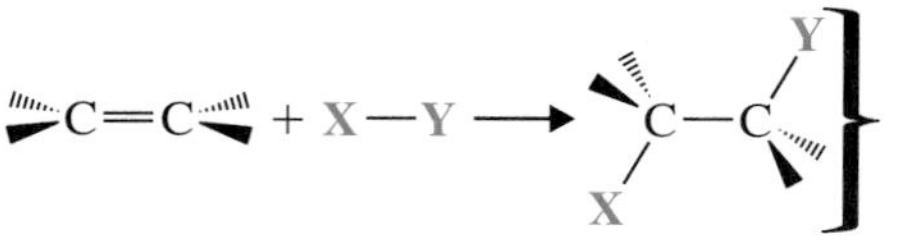

In Chapter 8 we shall study a number of important syn and anti additions.

7.7 HYDROGENATION OF ALKYNES

Depending on the conditions and the catalyst employed, one or two molar equivalents of hydrogen will add to a carbon–carbon triple bond. When a platinum catalyst is used, the alkyne generally reacts with two molar equivalents of hydrogen to give an alkane.

$$CH_3C{\equiv}CCH_3 \xrightarrow[H_2]{Pt} [CH_3CH{=}CHCH_3] \xrightarrow[H_2]{Pt} CH_3CH_2CH_2CH_3$$

However, hydrogenation of an alkyne to an alkene can be accomplished through the use of special catalysts or reagents. Moreover, these special methods allow the preparation of either (E) or (Z) alkenes from disubstituted alkynes.

7.7A Syn Addition of Hydrogen: Synthesis of *Cis*-Alkenes

A catalyst that permits hydrogenation of an alkyne to an alkene is the nickel boride compound called P-2 catalyst. This catalyst can be prepared by the reduction of nickel acetate with sodium borohydride.

$$Ni\left(\overset{\overset{O}{\|}}{OCCH_3}\right)_2 \xrightarrow[C_2H_5OH]{NaBH_4} Ni_2B \quad (P\text{-}2)$$

Hydrogenation of alkynes in the presence of P-2 catalyst causes **syn addition of hydrogen** to take place, and the alkene that is formed from an alkyne with an internal triple bond has the (Z) or cis configuration. The hydrogenation of 3-hexyne (see Section 7.7B) illustrates this method. The reaction takes place on the surface of the catalyst (Section 7.6) accounting for the syn addition.

$$CH_3CH_2C{\equiv}CCH_2CH_3 \xrightarrow[\text{(syn addition)}]{H_2/Ni_2B\ (P\text{-}2)}$$

3-Hexyne (Z)-**3-Hexene**
                                              (*cis*-**3-hexene**)
                                              (97%)

Other specially conditioned catalysts can be used to prepare *cis*-alkenes from disubstituted alkynes. Metallic palladium deposited on calcium carbonate can be used in this way after it has been conditioned with lead acetate and quinoline (Section 20.1B). This special catalyst is known as Lindlar's catalyst.

$$R{-}C{\equiv}C{-}R \xrightarrow[\substack{\text{quinoline} \\ \text{(syn addition)}}]{\substack{H_2,\ Pd/CaCO_3 \\ \text{(Lindlar's catalyst)}}}$$

7.7B Anti Addition of Hydrogen: Synthesis of *Trans*-Alkenes

An **anti addition** of hydrogen atoms to the triple bond occurs when alkynes are reduced with lithium or sodium metal in ammonia or ethylamine at low temperatures. This reaction called a **dissolving metal reduction** produces an (*E*) or *trans*-alkene.

$$CH_3(CH_2)_2 - C \equiv C - (CH_2)_2CH_3 \xrightarrow[\text{(2) } NH_4Cl]{\text{(1) Li, } C_2H_5NH_2, \ -78°C}$$

4-Octyne

(product structure)

(*E*)-**4-Octene**
(*trans*-**4-octene**)
(52%)

The mechanism for this reduction is shown in the following outline. It involves successive electron transfers from the lithium (or sodium) atom and proton transfers from the amine (or ammonia). In the first step, the lithium atom transfers an electron to the alkyne to produce an intermediate that bears a negative charge and has an unpaired electron, called a **radical anion.** In the second step, the amine transfers a proton to produce a **vinylic radical.** Then, transfer of another electron gives a **vinylic anion.** It is this step that determines the stereochemistry of the reaction. The *trans*- vinylic anion is formed preferentially because it is more stable; the bulky alkyl groups are farther apart. Protonation of the *trans*-vinylic anion leads to the *trans*-alkene.

A MECHANISM FOR THE REDUCTION

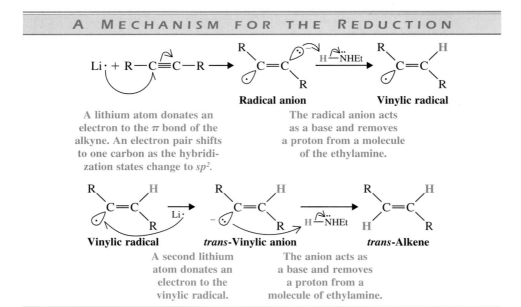

A lithium atom donates an electron to the π bond of the alkyne. An electron pair shifts to one carbon as the hybridization states change to sp^2.

The radical anion acts as a base and removes a proton from a molecule of the ethylamine.

Radical anion   **Vinylic radical**

Vinylic radical   ***trans*-Vinylic anion**   ***trans*-Alkene**

A second lithium atom donates an electron to the vinylic radical.

The anion acts as a base and removes a proton from a molecule of ethylamine.

7.8 MOLECULAR FORMULAS OF HYDROCARBONS: THE INDEX OF HYDROGEN DEFICIENCY

Alkenes whose molecules contain only one double bond have the general formula C_nH_{2n}. They are isomeric with cycloalkanes. For example, 1-hexene and cyclohexane have the same molecular formula (C_6H_{12}):

$$CH_2=CHCH_2CH_2CH_2CH_3$$

1-Hexene
(C_6H_{12})

Cyclohexane
(C_6H_{12})

Cyclohexane and 1-hexene are constitutional isomers.

Alkynes and alkenes with two double bonds (alkadienes) have the general formula C_nH_{2n-2}. Hydrocarbons with one triple bond and one double bond (alkenynes) and alkenes with three double bonds (alkatrienes) have the general formula C_nH_{2n-4}, and so forth.

$$CH_2=CH—CH=CH_2 \qquad CH_2=CH—CH=CH—CH=CH_2$$

1,3-Butadiene   **1,3,5-Hexatriene**
(C_4H_6)   **(C_6H_8)**

A chemist working with an unknown hydrocarbon can obtain considerable information about its structure from its molecular formula and its **index of hydrogen deficiency.** The index of hydrogen deficiency is defined as the number of *pairs* of hydrogen atoms that must be subtracted from the molecular formula of the corresponding alkane to give the molecular formula of the compound under consideration.*

For example, both cyclohexane and 1-hexene have an index of hydrogen deficiency equal to one (meaning one *pair* of hydrogen atoms). The corresponding alkane (i.e., the alkane with the same number of carbon atoms) is hexane.

C_6H_{14} = formula of corresponding alkane (hexane)
$\underline{C_6H_{12}}$ = formula of compound (1-hexene or cyclohexane)
H_2 = difference = 1 pair of hydrogen atoms

Index of hydrogen deficiency = 1

The index of hydrogen deficiency of acetylene or of 1,3-butadiene equals 2; the index of hydrogen deficiency of 1,3,5-hexatriene equals 3. (Do the calculations.)

Determining the number of rings present in a given compound is easily done experimentally. Molecules with double bonds and triple bonds add hydrogen readily at room temperature in the presence of a platinum catalyst. **Each double bond consumes one molar equivalent of hydrogen; each triple bond consumes two. Rings are not affected by hydrogenation at room temperature.** Hydrogenation, therefore, allows us to distinguish between rings on the one hand and double or triple bonds on the other. Consider as an example two compounds with the molecular formula C_6H_{12}: 1-hexene and cyclohexane. 1-Hexene reacts with one molar equivalent of hydrogen to yield hexane; under the same conditions cyclohexane does not react.

$$CH_2=CH(CH_2)_3CH_3 + H_2 \xrightarrow[25°C]{Pt} CH_3(CH_2)_4CH_3$$

$$\bigcirc + H_2 \xrightarrow[25°C]{Pt} \text{no reaction}$$

---

*Some organic chemists refer to the index of hydrogen deficiency as the "degree of unsaturation" or "the number of double-bond equivalencies."

Or consider another example. Cyclohexene and 1,3-hexadiene have the same molecular formula (C_6H_{10}). Both compounds react with hydrogen in the presence of a catalyst, but cyclohexene, because it has a ring and only one double bond, reacts with only one molar equivalent. 1,3-Hexadiene adds two molar equivalents.

Cyclohexene

$$CH_2=CHCH=CHCH_2CH_3 + 2\ H_2 \xrightarrow[25°C]{Pt} CH_3(CH_2)_4CH_3$$

1,3-Hexadiene

PROBLEM 7.5

(a) What is the index of hydrogen deficiency of 2-hexene?   (b) Of methylcyclopentane?   (c) Does the index of hydrogen deficiency reveal anything about the location of the double bond in the chain?   (d) About the size of the ring? (e) What is the index of hydrogen deficiency of 2-hexyne?   (f) In general terms, what structural possibilities exist for a compound with the molecular formula $C_{10}H_{16}$?

PROBLEM 7.6

Zingiberene, a fragrant compound isolated from ginger, has the molecular formula $C_{15}H_{24}$ and is known not to contain any triple bonds.   (a) What is the index of hydrogen deficiency of zingiberene?   (b) When zingiberene is subjected to catalytic hydrogenation using an excess of hydrogen, 1 mol of zingiberene absorbs 3 mol of hydrogen and produces a compound with the formula $C_{15}H_{30}$. How many double bonds does a molecule of zingiberene have?   (c) How many rings?

OPTIONAL MATERIAL

More on Calculating the Index of Hydrogen Deficiency (IHD)

Calculating the index of hydrogen deficiency (IHD) for compounds other than hydrocarbons is relatively easy.

For compounds containing halogen atoms we simply count the halogen atoms as though they were hydrogen atoms. Consider a compound with the formula $C_4H_6Cl_2$. To calculate the IHD, we change the two chlorine atoms to hydrogen atoms, considering the formula as though it were C_4H_8. This formula has two hydrogen atoms fewer than the formula for a saturated alkane (C_4H_{10}), and this tells us that the compound has an IHD = 1. It could, therefore, have either one ring or one double bond. [We can tell which it has from a hydrogenation experiment: If the compound adds one molar equivalent of hydrogen (H_2) on catalytic hydrogenation at room temperature, then it must have a double bond; if it does not add hydrogen, then it must have a ring.]

For compounds containing oxygen we simply ignore the oxygen atoms and calculate the IHD from the remainder of the formula. Consider as an example a compound with the formula C_4H_8O. For the purposes of our calculation we consider the compound to be simply C_4H_8 and we calculate an IHD = 1. Again, this means that the compound contains either a ring or a double bond. Some structural possibilities for this compound are shown next. Notice that the double bond may be present as a carbon–oxygen double bond.

$$CH_2{=}CHCH_2CH_2OH \qquad CH_3CH{=}CHCH_2OH \qquad CH_3CH_2\overset{\overset{\displaystyle O}{\|}}{C}CH_3$$

$$CH_3CH_2CH_2\overset{\overset{\displaystyle O}{\|}}{C}H \qquad \qquad \text{and so on}$$

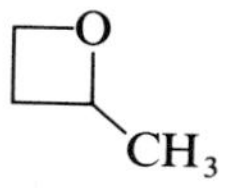

For compounds containing nitrogen atoms we subtract one hydrogen for each nitrogen atom, and then we ignore the nitrogen atoms. For example, we treat a compound with the formula C_4H_9N as though it were C_4H_8, and again we get an IHD = 1. Some structural possibilities are the following:

$$CH_2{=}CHCH_2CH_2NH_2 \qquad CH_3CH{=}CHCH_2NH_2 \qquad CH_3CH_2\overset{\overset{\displaystyle NH}{\|}}{C}CH_3$$

$$CH_3CH_2CH_2CH{=}NH \qquad \qquad \text{and so on}$$

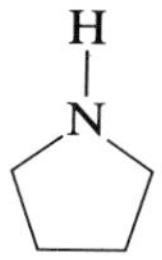

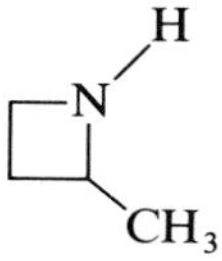

7.9 RELATIVE STABILITIES OF ALKENES

7.9A Heats of Hydrogenation

Hydrogenation also provides a way to measure the relative stabilities of certain alkenes. The reaction of an alkene with hydrogen is an exothermic reaction; the enthalpy change involved is called **the heat of hydrogenation.** Most alkenes have heats of hydrogenation near -30 kcal mol^{-1}. Individual alkenes, however, have heats of hydrogenation that may differ from this value by more than 2 kcal mol^{-1}.

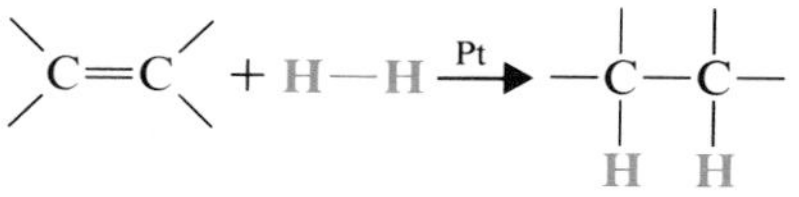

$$\overset{\diagdown}{\diagup}C{=}C\overset{\diagup}{\diagdown} \ + \ H{-}H \ \xrightarrow{\ \text{Pt}\ } \ \underset{\underset{\displaystyle H \ \ H}{|\ \ |}}{-\overset{|}{C}-\overset{|}{C}-} \qquad \Delta H° \approx -30 \text{ kcal mol}^{-1}$$

These differences permit the measurement of the relative stabilities of alkene isomers *when hydrogenation converts them to the same product.*

Consider, as examples, the three butene isomers that follow:

$$CH_3CH_2CH{=}CH_2 \ + \ H_2 \ \xrightarrow{\ \text{Pt}\ } \ CH_3CH_2CH_2CH_3 \qquad \Delta H° = -30.3 \text{ kcal mol}^{-1}$$

$$\underset{\substack{\textbf{1-Butene}\\ \textbf{(C}_4\textbf{H}_8\textbf{)}}}{} \qquad\qquad\qquad\qquad \underset{\textbf{Butane}}{}$$

$$H_3C, \quad CH_3$$
$$\diagdown C = C \diagup$$
$$H \diagup \diagdown H$$
$$+ H_2 \xrightarrow{\text{Pt}} CH_3CH_2CH_2CH_3 \qquad \Delta H° = -28.6 \text{ kcal mol}^{-1}$$

cis-2-Butene
(C_4H_8)

Butane

$$H_3C, \quad H$$
$$\diagdown C = C \diagup$$
$$H \diagup \diagdown CH_3$$
$$+ H_2 \xrightarrow{\text{Pt}} CH_3CH_2CH_2CH_3 \qquad \Delta H° = -27.6 \text{ kcal mol}^{-1}$$

trans-2-Butene
(C_4H_8)

Butane

In each reaction the product (butane) is the same. In each case, too, one of the reactants (hydrogen) is the same. A different amount of *heat* is evolved in each reaction, however, and these differences must be related to different relative stabilities (different heat contents) of the individual butenes. 1-Butene evolves the greatest amount of heat when hydrogenated, and *trans*-2-butene evolves the least. Therefore 1-butene must have the greatest energy (enthalpy) and be the least stable isomer. *trans*-2-Butene must have the lowest energy and be the most stable isomer. The energy (and stability) of *cis*-2-butene falls in between. The order of stabilities of the butenes is easier to see if we examine the energy diagram in Fig. 7.3.

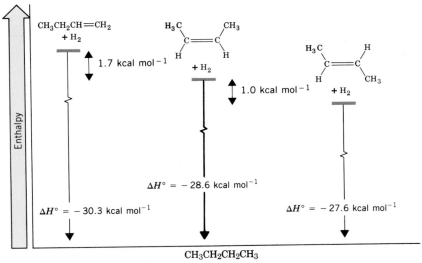

FIGURE 7.3 An energy diagram for the three butene isomers. The order of stability is *trans*-2-butene > *cis*-2-butene > 1-butene.

The greater stability of the *trans*-2-butene when compared to *cis*-2-butene illustrates a general pattern found in cis–trans alkene pairs. The 2-pentenes, for example, show the same stability relationship: **trans isomer > cis isomer.**

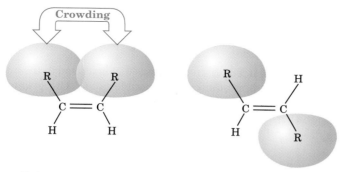

$$CH_3CH_2CH_2CH_2CH_3 \qquad \Delta H^\circ = -28.6 \text{ kcal mol}^{-1}$$

cis-2-Pentene   Pentane

$$CH_3CH_2CH_2CH_2CH_3 \qquad \Delta H^\circ = -27.6 \text{ kcal mol}^{-1}$$

trans-2-Pentene   Pentane

The greater enthalpy of cis isomers can be attributed to strain caused by the crowding of two alkyl groups on the same side of the double bond (Fig. 7.4).

FIGURE 7.4 *cis*- and *trans*-Alkene isomers. The less stable cis isomer has greater strain

7.9B Relative Stabilities from Heats of Combustion

When hydrogenation of isomeric alkenes does not yield the same alkane, *heats of combustion can be used to measure their relative stabilities.* For example, 2-methylpropene cannot be compared directly with the other butene isomers (1-butene, *cis*-, and *trans*-2-butene) because on hydrogenation 2-methylpropene yields isobutane, *not butane:*

$$CH_3C{=}CH_2 \ + H_2 \xrightarrow{\text{Pt}} CH_3CHCH_3$$

2-Methylpropene   Isobutane

Isobutane and butane do not have the same enthalpy so a direct comparison of heats of hydrogenation is not possible.

However, when 2-methylpropene is subjected to complete combustion, the products are the same as those produced by the other butene isomers. Each isomer consumes six molar equivalents of oxygen and *produces four molar equivalents of CO$_2$ and four molar equivalents of H$_2$O.* Comparison of the heats of combustion shows that 2-methylpropene is the most stable of the four isomers because it evolves the least heat.

$$CH_3CH_2CH{=}CH_2 + 6\,O_2 \longrightarrow 4\,CO_2 + 4\,H_2O \qquad \Delta H^\circ = -649.8 \text{ kcal mol}^{-1}$$

$$C{=}C \qquad + 6\,O_2 \longrightarrow 4\,CO_2 + 4\,H_2O \qquad \Delta H^\circ = -648.1 \text{ kcal mol}^{-1}$$

$$\underset{H}{\overset{H_3C}{>}}C=C\underset{CH_3}{\overset{H}{<}} + 6\ O_2 \longrightarrow 4\ CO_2 + 4\ H_2O \qquad \Delta H° = -647.1\ \text{kcal mol}^{-1}$$

$$\underset{}{\overset{CH_3}{\underset{|}{CH_3C}}}=CH_2 + 6\ O_2 \longrightarrow 4\ CO_2 + 4\ H_2O \qquad \Delta H° = -646.1\ \text{kcal mol}^{-1}$$

The heat evolved by each of the other three isomers, moreover, confirms the order of stability measured by heats of hydrogenation. Therefore, the stability of the butene isomers overall is

$$\underset{|}{\overset{CH_3}{CH_3C}}=CH_2 > trans\text{-}CH_3CH=CHCH_3 > cis\text{-}CH_3CH=CHCH_3 > CH_3CH_2CH=CH_2$$

7.9C Overall Relative Stabilities of Alkenes

Studies of numerous alkenes reveal a pattern of stabilities that is related to the number of alkyl groups attached to the carbon atoms of the double bond. **The greater the number of attached alkyl groups (i.e., the more highly substituted the carbon atoms of the double bond), the greater is the alkene's stability.** This order of stabilities can be given in general terms as follows:*

Relative Stabilities of Alkenes

| Tetrasubstituted | Trisubstituted | ← ———————— Disubstituted ———————— → | Monosubstituted | Unsubstituted |

PROBLEM 7.7

Heats of hydrogenation of three alkenes are as follows:

2-methyl-1-butene ($-28.5\ \text{kcal mol}^{-1}$)

3-methyl-1-butene ($-30.3\ \text{kcal mol}^{-1}$)

2-methyl-2-butene ($-26.9\ \text{kcal mol}^{-1}$)

(a) Write the structure of each alkene and classify it as to whether its doubly bonded atoms are monosubstituted, disubstituted, trisubstituted, and so on. (b) Write the product formed when each alkene is hydrogenated. (c) Can heats of hydrogenation be used to relate the relative stabilities of these three alkenes? (d) If so, what is the predicted order of stability? If not, why not? (e) What other alkene isomers are possible for these alkenes? Write their structures. (f) What data would be necessary to relate the stabilities of all of these isomers?

---

*This order of stabilities may seem contradictory when compared with the explanation given for the relative stabilities of cis and trans isomers. Although a detailed explanation of the trend given here is beyond our scope, the relative stabilities of substituted alkenes can be rationalized. Part of the explanation can be given in terms of the electron-releasing effect of alkyl groups (Section 6.13B), an effect that satisfies the electron-withdrawing properties of the sp^2-hybridized carbon atoms of the double bond.

PROBLEM 7.8

Predict the more stable alkene of each pair.   (a) 2-Methyl-2-pentene or 2,3-di-methyl-2-butene,   (b) *cis*-3-hexene or *trans*-3-hexene,   (c) 1-hexene or *cis*-3-hexene, and   (d) *trans*-2-hexene or 2-methyl-2-pentene.

PROBLEM 7.9

Reconsider the pairs of alkenes given in Problem 7.8. For which pairs could you use heats of hydrogenation to determine their relative stabilities? For which pairs would you be required to use heats of combustion?

7.10 CYCLOALKENES

The rings of cycloalkenes containing five carbon atoms or fewer exist only in the cis form (Fig. 7.5). The introduction of a trans double bond into rings this small would, if it were possible, introduce greater strain than the bonds of the ring atoms could accommodate. *trans*-Cyclohexene might resemble the structure shown in Fig. 7.6. There is evidence that it can be formed as a very reactive short-lived intermediate in some chemical reactions.

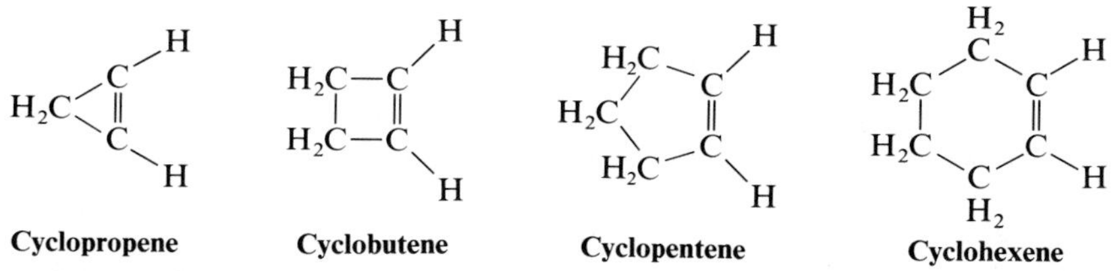

FIGURE 7.5 *cis*-Cycloalkenes.

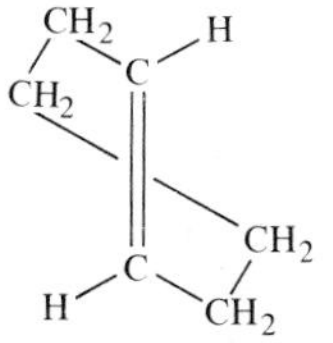

FIGURE 7.6 Hypothetical *trans*-cyclohexene. This molecule is apparently too highly strained to exist at room temperature.

trans-Cycloheptene has been observed with instruments called spectrometers, but it is a substance with a very short lifetime and has not been isolated.

trans-Cyclooctene (Fig. 7.7) has been isolated, however. Here the ring is large enough to accommodate the geometry required by a trans double bond and still be stable at room temperature. *trans*-Cyclooctene is chiral and exists as a pair of enantiomers.

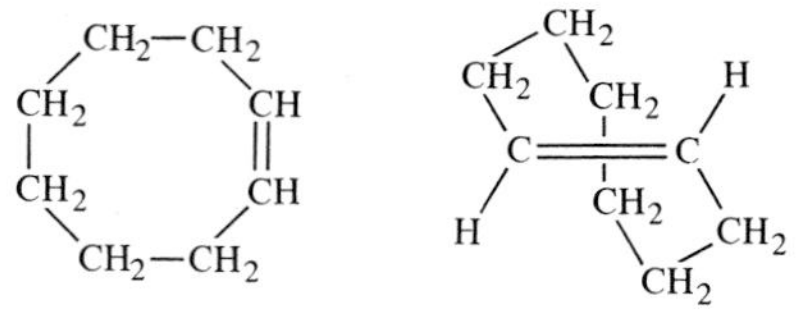

cis-Cyclooctene   *trans*-Cyclooctene

FIGURE 7.7 The cis and trans forms of cyclooctene.

7.11 SYNTHESIS OF ALKENES VIA ELIMINATION REACTIONS

Because an elimination reaction can introduce a double bond into a molecule, eliminations are widely used for synthesizing alkenes. In this chapter we shall study three methods based on eliminations. The following examples of each of these methods are given using a simple two-carbon starting reagent.

Dehydrohalogenation of Alkyl Halides (Sections 6.17, 6.18, and 7.12)

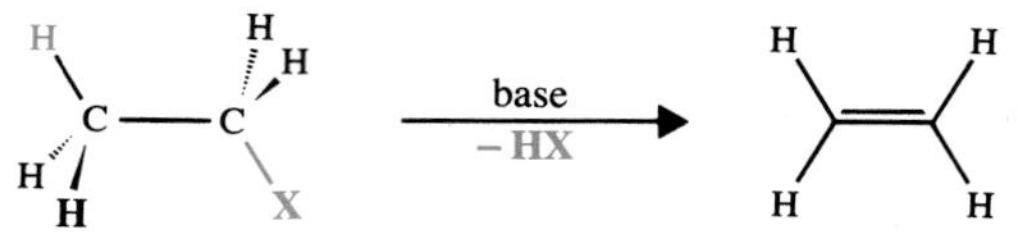

Dehydration of Alcohols (Sections 7.13–7.15)

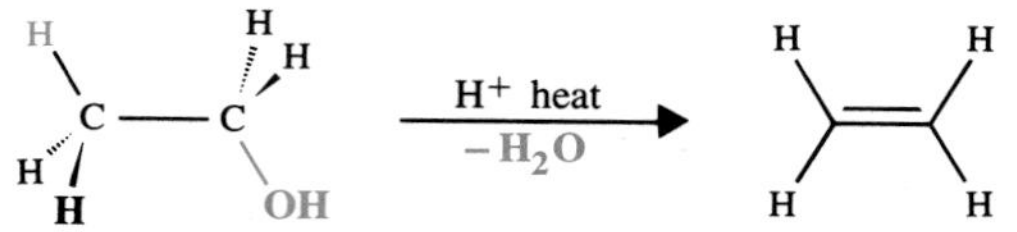

Debromination of vic-Dibromides (Section 7.16)

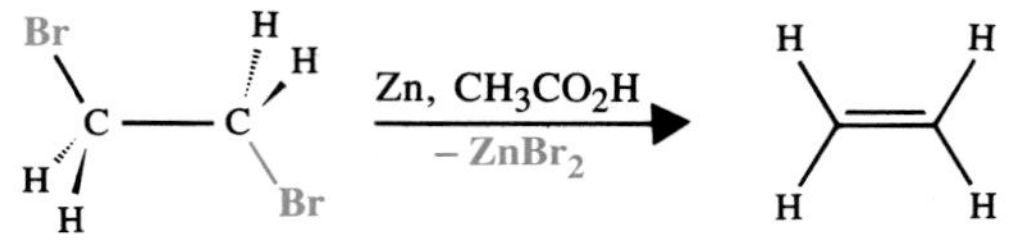

7.12 DEHYDROHALOGENATION OF ALKYL HALIDES

Synthesis of an alkene by dehydrohalogenation is almost always better achieved by an E2 reaction:

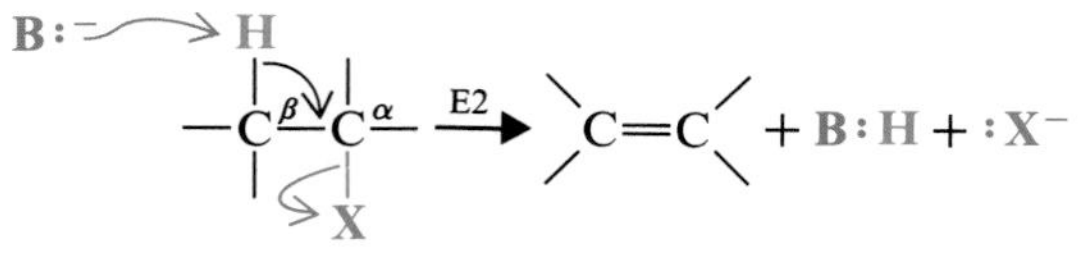

The reason for this choice is that dehydrohalogenation by an E1 mechanism is too variable. Too many competing events are possible, one being rearrangement of the carbon skeleton (Section 7.15). In order to bring about an E2 reaction, a secondary or tertiary alkyl halide is used if possible. (If the synthesis must begin with a primary halide, then a bulky base is used.) To avoid E1 conditions a high concentration of a strong, relatively nonpolarizable base, such as an alkoxide ion, is used and a relatively nonpolar solvent such as an alcohol. To favor elimination generally, a relatively high temperature is used. The typical reagents for dehydrohalogenation are sodium ethoxide in ethanol and potassium *tert*-butoxide in *tert*-butyl alcohol. Potassium hydroxide in ethanol is also used sometimes; in this reagent the reactive bases probably include the ethoxide ion formed by the following equilibrium.

$$OH^- + C_2H_5OH \rightleftharpoons H_2O + C_2H_5O^-$$

7.12A E2 Reactions: The Orientation of the Double Bond in the Product. Zaitsev's Rule

In earlier examples of dehydrohalogenations (Sections 6.17–6.19) only a single elimination product was possible. For example:

$$CH_3CHCH_3 \xrightarrow[\substack{C_2H_5OH \\ 55°C}]{C_2H_5O^-Na^+} CH_2{=}CHCH_3$$
$$\underset{Br}{|} \qquad\qquad\qquad \textbf{(79\%)}$$

$$\underset{\underset{Br}{|}}{\overset{\overset{CH_3}{|}}{CH_3CCH_3}} \xrightarrow[\substack{C_2H_5OH \\ 55°C}]{C_2H_5O^-Na^+} \underset{}{\overset{\overset{CH_3}{|}}{CH_2{=}C{-}CH_3}}$$
$$\textbf{(100\%)}$$

$$CH_3(CH_2)_{15}CH_2CH_2Br \xrightarrow[\substack{(CH_3)_3COH \\ 40°C}]{(CH_3)_3CO^-K^+} CH_3(CH_2)_{15}CH{=}CH_2$$
$$\textbf{(85\%)}$$

Dehydrohalogenation of most alkyl halides, however, yields more than one product. For example, dehydrohalogenation of 2-bromo-2-methylbutane can yield two products: 2-methyl-2-butene and 2-methyl-1-butene.

If we use a base such as ethoxide ion or hydroxide ion, the major product of the reaction will be **the more stable alkene.** The more stable alkene, as we know from Section 7.9, has the more highly substituted double bond.

$$CH_3CH_2O^- + CH_3CH_2\underset{\underset{Br}{|}}{\overset{\overset{CH_3}{|}}{C}}—CH_3 \xrightarrow[CH_3CH_2OH]{70°C} CH_3CH{=}C\overset{CH_3}{\underset{CH_3}{\diagdown}} + CH_3CH_2C\overset{CH_2}{\underset{CH_3}{\diagdown}}$$

2-Methyl-2-butene　　**2-Methyl-1-butene**
(69%)　　　　　　(31%)
(more stable)　　　(less stable)

2-Methyl-2-butene is a trisubstituted alkene (three methyl groups are attached to carbon atoms of the double bond), whereas 2-methyl-1-butene is only disubstituted. 2-Methyl-2-butene is the major product.

　　The reason for this behavior appears to be related to the double-bond character that develops in the transition state (cf. Section 6.18) for each reaction:

$$C_2H_5O^- + \ \overset{\overset{H}{\downarrow}}{—\overset{|}{\underset{|}{C}}}\overset{}{\underset{\underset{Br}{\uparrow}}{\overset{|}{C}}}— \longrightarrow \left[\begin{array}{c} \overset{\delta-}{C_2H_5O}\text{---}H \\ —\overset{|}{C}{=}\overset{|}{C}— \\ Br^{\delta-} \end{array}\right] \longrightarrow C_2H_5OH + \ \overset{\diagdown}{}C{=}C\overset{\diagup}{} + Br^-$$

Transition state for an E2 reaction
The carbon–carbon bond has some of the
character of a double bond

The transition state for the reaction leading to 2-methyl-2-butene (Fig. 7.8) resembles the product of the reaction: a trisubstituted alkene. The transition state for the reaction leading to 2-methyl-1-butene resembles its product: a disubstituted alkene. Because

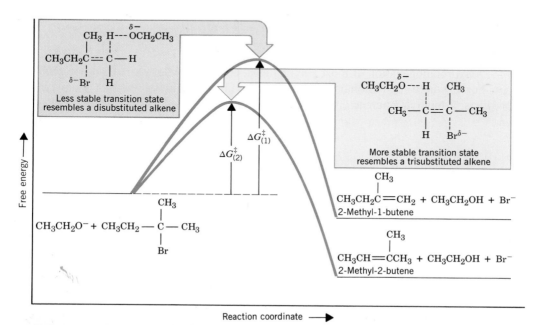

FIGURE 7.8 Reaction (2) leading to the more stable alkene occurs faster than reaction (1) leading to the less stable alkene; $\Delta G^{\ddagger}_{(2)}$ is less than $\Delta G^{\ddagger}_{(1)}$.

the transition state leading to 2-methyl-2-butene resembles a more stable alkene, this transition state is more stable. Because this transition state is more stable (occurs at lower free energy), the free energy of activation for this reaction is lower and 2-methyl-2-butene is formed faster. This explains why 2-methyl-2-butene is the major product. These reactions are known to be under kinetic control (Section 6.6).

Whenever an elimination occurs to give the most stable, most highly substituted alkene, chemists say that the elimination follows the **Zaitsev rule,** named for the nineteenth-century Russian chemist A. N. Zaitsev (1841–1910) who formulated it. (Zaitsev's name is also transliterated as Zaitzev, Saytzeff, Saytseff, or Saytzev.)

PROBLEM 7.10

List the alkenes that would be formed when each of the following alkenes is subjected to dehydrohalogenation with potassium ethoxide in ethanol and use Zaitsev's rule to predict the major product of each reaction: (a) 2-bromo-3-methylbutane and (b) 2-bromo-2,3-dimethylbutane.

7.12B An Exception to Zaitsev's Rule

Carrying out dehydrohalogenations with a base such as potassium *tert*-butoxide in *tert*-butyl alcohol favors the formation of **the less substituted alkene:**

2-Methyl-2-butene	2-Methyl-1-butene
(27.5%)	(72.5%)
(more substituted)	(less substituted)

The reasons for this behavior are complicated but seem to be related in part to the steric bulk of the base and to the fact that in *tert*-butyl alcohol the base is associated with solvent molecules and thus made even larger. The large *tert*-butoxide ion appears to have difficulty removing one of the internal (2°) hydrogen atoms because of greater crowding at that site in the transition state. It removes one of the more exposed (1°) hydrogen atoms of the methyl group instead. When an elimination yields the less substituted alkene, we say that it follows the **Hofmann rule** (see Section 20.13A).

7.12C The Stereochemistry of E2 Reactions: the Orientation of Groups in the Transition State

Considerable experimental evidence indicates that the five atoms involved in the transition state of an E2 reaction (including the base) must lie in the same plane. The requirement for coplanarity of the H—C—C—L unit arises from a need for proper overlap of orbitals in the developing π bond of the alkene that is being formed. There are two ways that this can happen:

Anti periplanar
transition state
(preferred)

Syn periplanar
transition state
(only with certain
rigid molecules)

Evidence also indicates that of these two arrangements for the transition state, the arrangement called the **anti periplanar** conformation is the preferred one. The **syn periplanar** transition state occurs only with rigid molecules that are unable to assume the anti arrangement. The reason: The antiperiplanar transition state is staggered (and therefore of lower energy), while the syn periplanar transition state is eclipsed.

PROBLEM 7.11

Consider a simple molecule such as ethyl bromide and show with Newman projection formulas how the anti periplanar transition state would be favored over the syn periplanar one.

Part of the evidence for the preferred anti periplanar arrangement of groups comes from experiments done with cyclic molecules. As examples, let us consider the different behavior shown in E2 reactions by two compounds containing cyclohexane rings that have the common names *neomenthyl chloride* and *menthyl chloride*.

Neomenthyl chloride

Menthyl chloride

The β hydrogen and the leaving group on a cyclohexane ring can assume an anti periplanar conformation **only when they are both axial:**

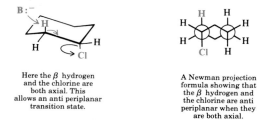

Here the β hydrogen
and the chlorine are
both axial. This
allows an anti periplanar
transition state.

A Newman projection
formula showing that
the β hydrogen and
the chlorine are anti
periplanar when they
are both axial.

Neither an axial–equatorial nor an equatorial–equatorial orientation of the groups allows the formation of an anti periplanar transition state.

In the more stable conformation of neomenthyl chloride (see following mechanism), the alkyl groups are both equatorial and the chlorine is axial. There are also axial hydrogen atoms on both C2 and C4. The base can attack either of these hydrogen atoms and achieve an anti periplanar transition state for an E2 reaction. Products corresponding to each of these transition states (2-menthene and 3-menthene) are formed rapidly. In accordance with Zaitsev's rule, 3-menthene (with the more highly substituted double bond) is the major product.

A MECHANISM FOR THE ELIMINATION

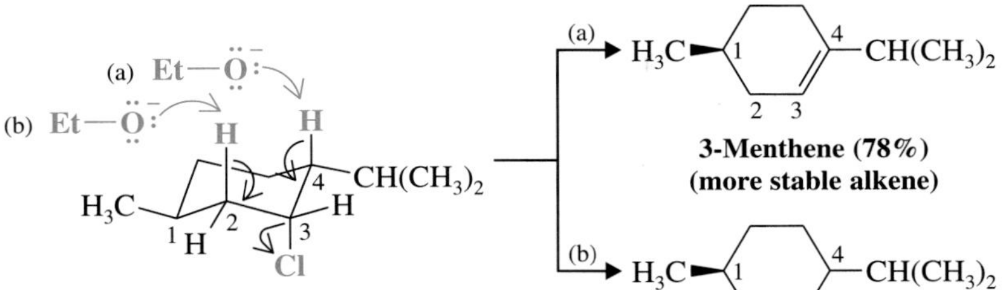

Neomenthyl chloride

Both red hydrogens are anti to
the chlorine in this the more stable
conformation. Elimination by path (a)
leads to 3-menthene; by path (b)
to 2-menthene.

3-Menthene (78%)
(more stable alkene)

2-Menthene (22%)
(less stable alkene)

On the other hand, the more stable conformation of menthyl chloride has all three groups (including the chlorine) equatorial. For the chlorine to become axial, menthyl chloride has to assume a conformation in which the large isopropyl group and the methyl group are also axial. This conformation is of much higher energy, and the free energy of activation for the reaction is large because it includes the energy necessary for the conformational change. Consequently, menthyl chloride undergoes an E2 reaction very slowly, and the product is entirely 2-menthene (contrary to Zaitsev's rule). This product (or any resulting from an elimination to yield the less substituted alkene) is sometimes called *the Hofmann product* (Section 20.13A).

A MECHANISM FOR THE ELIMINATION

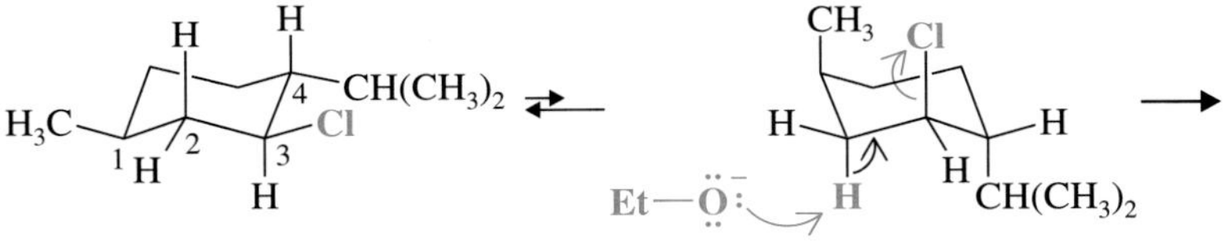

Menthyl chloride
(*more stable conformation*)

Elimination is not possible
for this conformation because
no hydrogen is anti to the
leaving group.

Menthyl chloride
(*less stable conformation*)

Elimination is possible from this
conformation because the red
hydrogen is anti to the chlorine.

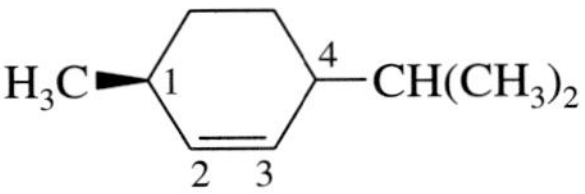

2-Menthene (100%)

PROBLEM 7.12 _____

When *cis*-1-bromo-4-*tert*-butylcyclohexane is treated with sodium ethoxide in ethanol, it reacts rapidly; the product is 4-*tert*-butylcyclohexene. Under the same conditions, *trans*-1-bromo-4-*tert*-butylcyclohexane reacts very slowly. Write conformational structures and explain the difference in reactivity of these cis–trans isomers.

PROBLEM 7.13 _____

(a) When *cis*-1-bromo-2-methylcyclohexane undergoes an E2 reaction, two products (cycloalkenes) are formed. What are these two cycloalkenes, and which would you expect to be the major product? Write conformational structures showing how each is formed.   (b) When *trans*-1-bromo-2-methylcyclohexane reacts in an E2 reaction, only one cycloalkene is formed. What is this product? Write conformational structures showing why it is the only product.

7.13 DEHYDRATION OF ALCOHOLS

Heating most alcohols with a strong acid causes them to lose a molecule of water (to **dehydrate**) and form an alkene:

$$-\overset{\displaystyle |}{\underset{\displaystyle H}{C}}-\overset{\displaystyle |}{\underset{\displaystyle OH}{C}}-\ \xrightarrow[\text{heat}]{\text{H}^+}\ \overset{\diagdown}{\diagup}C{=}C\overset{\diagup}{\diagdown}\ +H_2O$$

The reaction is an **elimination** and is favored at higher temperatures (Section 6.20). The most commonly used acids in the laboratory are Brønsted acids—proton donors such as sulfuric acid and phosphoric acid. Lewis acids such as alumina (Al_2O_3) are often used in industrial, gas-phase dehydrations.

Dehydration reactions of alcohols show several important characteristics, which we shall soon explain.

1. **The experimental conditions—temperature and acid concentration—that are required to bring about dehydration are closely related to the structure of the individual alcohol.** Alcohols in which the hydroxyl group is attached to a primary carbon (primary alcohols) are the most difficult to dehydrate. Dehydration of ethanol, for example, requires concentrated sulfuric acid and a temperature of 180°C.

$$H-\overset{\displaystyle H}{\underset{\displaystyle H}{C}}-\overset{\displaystyle H}{\underset{\displaystyle O-H}{C}}-H\ \xrightarrow[\substack{\text{H}_2\text{SO}_4\\180°C}]{\text{concd.}}\ H-\overset{\displaystyle H}{C}{=}\overset{\displaystyle H}{C}-H\ +\ H_2O$$

Ethanol
(a 1° alcohol)

Secondary alcohols usually dehydrate under milder conditions. Cyclohexanol, for example, dehydrates in 85% phosphoric acid at 165–170°C.

$$\text{Cyclohexanol} \xrightarrow[\text{165-170°C}]{\text{85\% H}_3\text{PO}_4} \text{Cyclohexene} + \text{H}_2\text{O}$$

Cyclohexanol   **Cyclohexene**
(80%)

Tertiary alcohols are usually so easily dehydrated that extremely mild conditions can be used. *tert*-Butyl alcohol, for example, dehydrates in 20% aqueous sulfuric acid at a temperature of 85°C.

$$\underset{\substack{| \\ \text{CH}_3}}{\overset{\substack{\text{CH}_3 \\ |}}{\text{CH}_3-\text{C}-\text{OH}}} \xrightarrow[\text{85°C}]{\text{20\% H}_2\text{SO}_4} \underset{\substack{\| \\ \text{CH}_2}}{\overset{\substack{\text{CH}_3 \\ |}}{\text{CH}_3-\text{C}}} + \text{H}_2\text{O}$$

tert-**Butyl**   **2-Methylpropene**
alcohol   **(84%)**

Thus, overall, the relative ease with which alcohols undergo dehydration is in the following order:

Ease of Dehydration

$$\underset{\substack{| \\ \text{R}}}{\overset{\substack{\text{R} \\ |}}{\text{R}-\text{C}-\text{OH}}} > \underset{\substack{| \\ \text{H}}}{\overset{\substack{\text{R} \\ |}}{\text{R}-\text{C}-\text{OH}}} > \underset{\substack{| \\ \text{H}}}{\overset{\substack{\text{H} \\ |}}{\text{R}-\text{C}-\text{OH}}}$$

3° Alcohol   **2° Alcohol**   **1° Alcohol**

This behavior, as we shall see in Section 7.14, is related to the stability of the carbocation formed in each reaction.

2. **Some primary and secondary alcohols also undergo rearrangements of their carbon skeleton during dehydration.** Such a rearrangement occurs in the dehydration of 3,3-dimethyl-2-butanol.

$$\underset{\substack{| \qquad | \\ \text{CH}_3 \ \ \text{OH}}}{\overset{\substack{\text{CH}_3 \\ |}}{\text{CH}_3-\text{C}-\!\!-\!\!\text{CH}-\text{CH}_3}} \xrightarrow[\text{80°C}]{\text{85\% H}_3\text{PO}_4} \overset{\substack{\text{CH}_3 \ \ \text{CH}_3 \\ | \qquad |}}{\text{CH}_3-\text{C}=\!\!=\text{C}-\text{CH}_3} + \overset{\substack{\text{CH}_3 \ \ \text{CH}_3 \\ | \qquad |}}{\text{CH}_2=\!\!=\text{C}-\!\!-\text{CHCH}_3}$$

3,3-Dimethyl-2-butanol   **2,3-Dimethyl-2-butene**   **2,3-Dimethyl-1-butene**
(80%)   **(20%)**

Notice that the carbon skeleton of the reactant is

$$\underset{\substack{| \\ \text{C}}}{\overset{\substack{\text{C} \\ |}}{\text{C}-\text{C}-\text{C}-\text{C}}} \text{ while that of the products is } \overset{\substack{\text{C} \ \ \text{C} \\ | \quad |}}{\text{C}-\text{C}-\text{C}-\text{C}}$$

We shall see in Section 7.15 that this reaction involves the migration of a methyl group from one carbon to the next.

7.13A Mechanism of Alcohol Dehydration: An E1 Reaction

Explanations for all of these observations can be based on a stepwise mechanism originally proposed by F. Whitmore (of the Pennsylvania State University). The mechanism is *an E1 reaction in which the substrate is a protonated alcohol (or an alkyloxonium ion*, see Section 6.5A). Consider the dehydration of *tert*-butyl alcohol as an example.

Step 1

$$CH_3-\underset{\underset{CH_3}{|}}{\overset{\overset{CH_3}{|}}{C}}-\ddot{\underset{\cdot\cdot}{O}}-H + H-\overset{+}{\underset{H}{\overset{H}{\ddot{O}}}} \rightleftharpoons CH_3-\underset{\underset{CH_3}{|}}{\overset{\overset{CH_3}{|}}{C}}-\overset{H}{\overset{|}{\overset{+}{O}}}-H + H:\ddot{\underset{\cdot\cdot}{O}}-H$$

**Protonated alcohol
or alkyloxonium ion**

In this step, an acid–base reaction, a proton is rapidly transferred from the acid to one of the unshared electron pairs of the alcohol. In dilute sulfuric acid the acid is a hydronium ion; in concentrated sulfuric acid the proton donor is sulfuric acid itself. This step is characteristic of all reactions of an alcohol with a strong acid.

The presence of the positive charge on the oxygen of the protonated alcohol weakens all bonds from oxygen including the carbon–oxygen bond, and in step 2 the carbon–oxygen bond breaks. The leaving group is a molecule of water:

Step 2

$$CH_3-\underset{\underset{CH_3}{|}}{\overset{\overset{CH_3}{|}}{C}}-\overset{H}{\overset{|}{\overset{+}{O}}}-H \rightleftharpoons CH_3-\underset{\underset{CH_3}{|}}{\overset{\overset{CH_3}{|}}{\overset{+}{C}}} + :\overset{H}{\underset{}{\overset{|}{O}}}-H$$

A carbocation

The carbon–oxygen bond breaks **heterolytically.** The bonding electrons depart with the water molecule and leave behind a carbocation. The carbocation is, of course, highly reactive because the central carbon atom has only six electrons in its valence level, not eight.

Finally, in step 3, the carbocation transfers a proton to a molecule of water. The result is the formation of a hydronium ion and an alkene.

Step 3

noble prize

$$\underset{CH_3-\underset{\underset{CH_3}{|}}{\overset{+}{C}}}{H-\overset{\overset{H}{|}}{\underset{}{C}}-H} + :\overset{H}{\underset{}{\overset{|}{O}}}-H \rightleftharpoons CH_3-\overset{CH_3}{\underset{\underset{CH_3}{|}}{\overset{||}{C}}} + H-\overset{H}{\overset{|}{\overset{+}{O}}}-H$$

2-Methylpropene

In step 3, also an acid–base reaction, any one of the nine protons available at the three methyl groups can be transferred to a molecule of water. The electron pair left behind when a proton is removed becomes the second bond of the double bond of the alkene. Notice that this step restores an octet of electrons to the central carbon atom.

PROBLEM 7.14

Dehydration of 2-propanol occurs in 75% H_2SO_4 at 100°C. (a) Using curved arrows write all steps in a mechanism for the dehydration. (b) Explain the essential role performed in alcohol dehydrations by the acid catalyst. *(Hint: Consider what would have to happen if no acid were present.)*

By itself, the Whitmore mechanism does not explain the observed order of reactivity of alcohols: **tertiary > secondary > primary.** Taken alone, it does not explain the formation of more than one product in the dehydration of certain alcohols nor the occurrence of a rearranged carbon skeleton in the dehydration of others. But when coupled with what is known about *the stability of carbocations,* the Whitmore mechanism *does* eventually account for all of these observations.

7.14 CARBOCATION STABILITY AND THE TRANSITION STATE

We saw in Section 6.13 that the order of stability of carbocations is tertiary > secondary > primary > methyl:

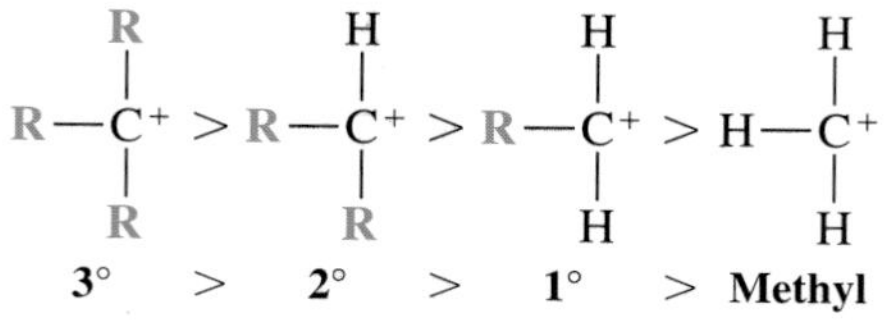

$$R-\overset{R}{\underset{R}{C^+}} > R-\overset{H}{\underset{R}{C^+}} > R-\overset{H}{\underset{H}{C^+}} > H-\overset{H}{\underset{H}{C^+}}$$

$$3° \quad > \quad 2° \quad > \quad 1° \quad > \quad \textbf{Methyl}$$

In the dehydration of alcohols (i.e., following steps 1–3 in the forward direction) the slowest step is step 2 because as we shall see, it is *a highly endergonic step* (Section 6.9): the formation of the carbocation from the protonated alcohol. The first and third steps are simple acid–base reactions. Proton-transfer reactions of this type occur very rapidly.

A MECHANISM FOR ACID-CATALYZED DEHYDRATION OF AN ALCOHOL

Step 1

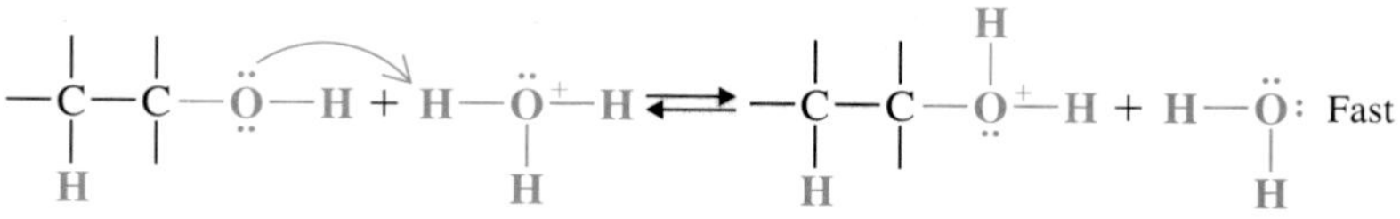

Alcohol   **Protonated alcohol**
The alcohol accepts a proton from the acid in a fast step.

Step 2

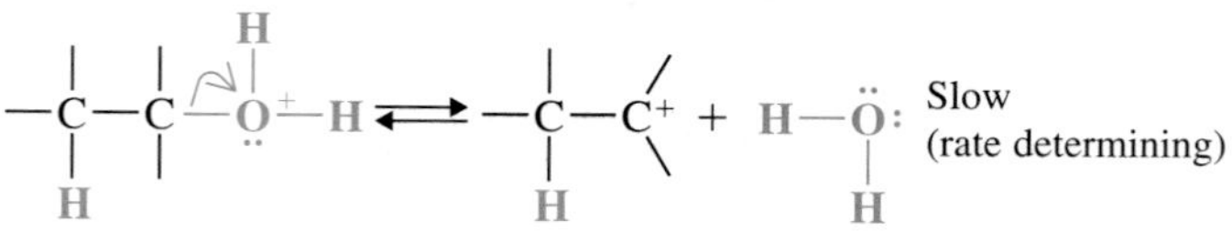

Slow
(rate determining)

The protonated alcohol loses a molecule of water to become a
carbocation. This step is slow and rate determining.

Step 3

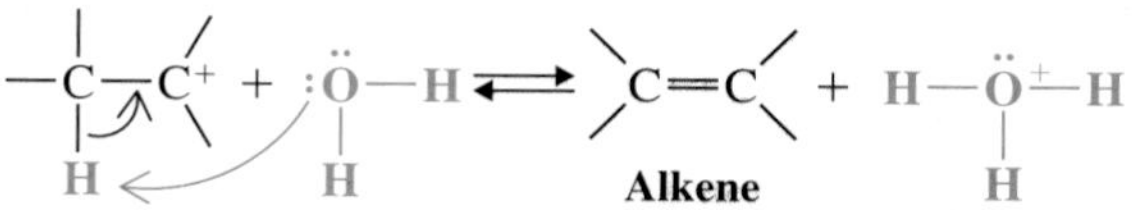

Alkene
The carbocation loses a proton to a base. This proton
transfer results in the formation of the alkene.

Because step 2 is, then, the rate-determining step, it is the step that determines the reactivity of alcohols toward dehydration. With this in mind, we can now understand why tertiary alcohols are the most easily dehydrated. The formation of a tertiary carbocation is easiest because the free energy of activation for step 2 of a reaction leading to a tertiary carbocation is lowest (see Fig. 7.9).

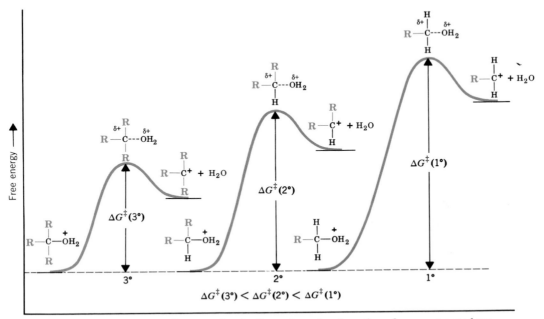

FIGURE 7.9 Free-energy diagrams for the formation of carbocations from protonated tertiary, secondary, and primary alcohols. The relative free energies of activation are tertiary < secondary < primary.

The reactions by which carbocations are formed from protonated alcohols are all highly *endergonic*. According to the Hammond–Leffler postulate (Section 6.15A), there should be a strong resemblance between the transition state and the product in each case. Of the three, *the transition state that leads to the tertiary carbocation is lowest in free energy because it resembles the most stable product.* By contrast, the transition state that leads to the primary carbocation occurs at highest free energy because it resembles the least stable product. In each instance, moreover, the same factor stabilizes the transition state that stabilizes the carbocation itself: **delocalization of the charge.** We can understand this if we examine the process by which the transition state is formed.

The oxygen atom of the protonated alcohol bears a full positive charge. As the transition state develops this oxygen atom begins to separate from the carbon atom to which it is attached. The carbon atom, because it is losing the electrons that bonded it to the oxygen atom, begins to develop a partial positive charge. This developing positive charge *is most effectively delocalized in the transition state leading to a tertiary carbocation because of the presence of three electron-releasing alkyl groups.* The positive charge is less effectively delocalized in the transition state leading to a secondary carbocation (*two* electron-releasing groups) and is least effectively delocalized in the transition state leading to a primary carbocation (*one* electron-releasing group).

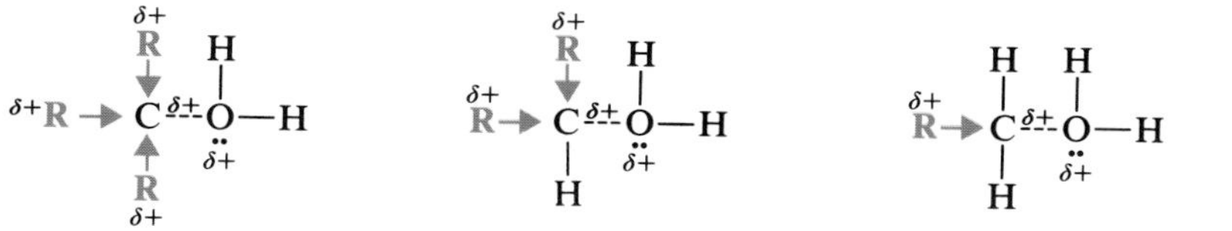

Transition state leading to 3° carbocation (most stable)   **Transition state leading to 2° carbocation**   **Transition state leading to 1° carbocation (least stable)**

7.15   CARBOCATION STABILITY AND THE OCCURRENCE OF MOLECULAR REARRANGEMENTS

With an understanding of carbocation stability and its effect on transition states behind us, we now proceed to explain the rearrangements of carbon skeletons that occur in some alcohol dehydrations. For example, let us consider again the rearrangement that occurs when 3,3-dimethyl-2-butanol is dehydrated.

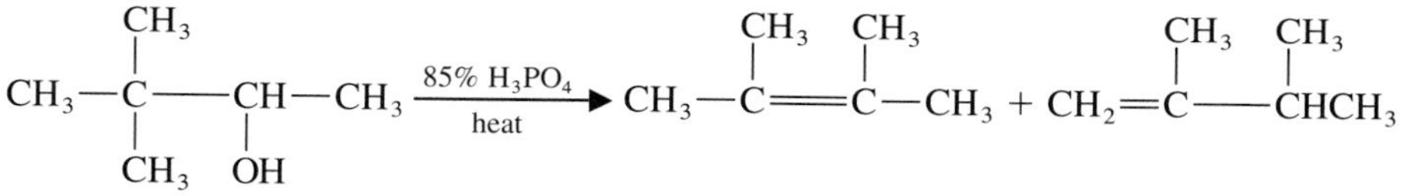

3,3-Dimethyl-2-butanol   **2,3-Dimethyl-2-butene (major product)**   **2,3-Dimethyl-1-butene (minor product)**

The first step of this dehydration is the formation of the protonated alcohol in the usual way:

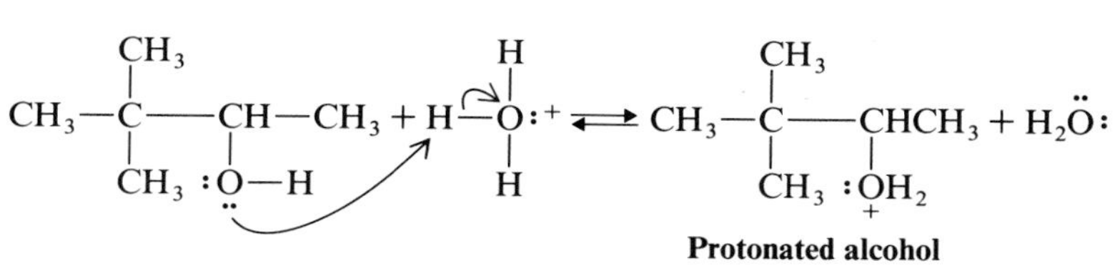

Protonated alcohol

In the second step the protonated alcohol loses water and a secondary carbocation forms:

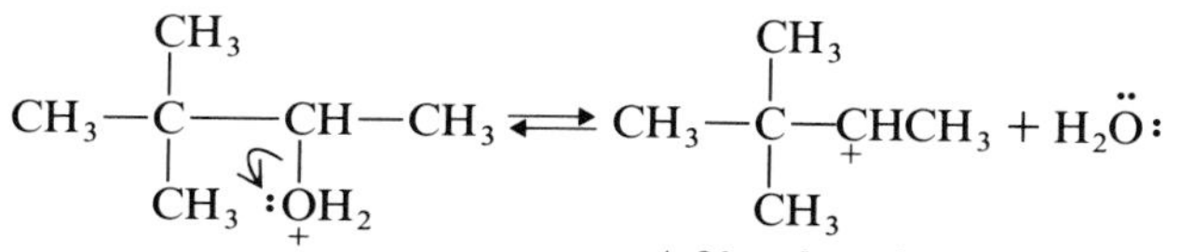

A 2° carbocation

Now the rearrangement occurs. *The less stable, secondary carbocation rearranges to a more stable tertiary carbocation.*

$$Step\ 3 \quad CH_3-\underset{\underset{CH_3}{|}}{\overset{\overset{CH_3}{|}}{C}}-\overset{+}{C}HCH_3 \longrightarrow \left[CH_3-\underset{\underset{CH_3}{|}}{\overset{\overset{CH_3}{|}}{\overset{\delta+}{C}}}\cdots\overset{\delta+}{C}HCH_3 \right] \longrightarrow CH_3-\underset{\underset{CH_3}{|}}{\overset{\overset{CH_3}{|}}{\overset{+}{C}}}-CH-CH_3$$

2° Carbocation   Transition state   3° Carbocation
(less stable)   (more stable)

The rearrangement occurs through the migration of an alkyl group (methyl) from the carbon atom adjacent to the one with the positive charge. The methyl group migrates **with its pair of electrons,** that is, as a methyl anion, $^-{:}CH_3$ (called a methanide ion). After the migration is complete, the carbon atom that the methyl anion left has become a carbocation, and the positive charge on the carbon atom to which it migrated has been neutralized. Because a group migrates from one carbon to the next, this kind of rearrangement is often called **a 1,2-shift.**

In the transition state the shifting methyl is partly bonded to both carbon atoms by the pair of electrons with which it migrates. It never leaves the carbon skeleton.

The final step of the reaction is the loss of a proton from the new carbocation and the formation of an alkene. This step, however, can occur in two ways.

$$Step\ 4 \quad H-\overset{(a)}{\overset{\frown}{CH_2}}-\overset{+}{\underset{\underset{CH_3}{|}}{C}}-\overset{\overset{H\ (b)}{|}}{\underset{\underset{CH_3}{|}}{C}}-CH_3$$

(a) → $CH_2{=}\underset{\underset{CH_3}{|}}{C}-\underset{\underset{CH_3}{|}}{C}HCH_3$ **Less stable alkene**

(minor product)

(b) → $CH_3-\underset{\underset{CH_3}{|}}{C}{=}\underset{\underset{CH_3}{|}}{C}-CH_3$ **More stable alkene**

(major product)

The more favored route is dictated by the type of alkene being formed. Path (b) leads to the highly stable tetrasubstituted alkene, and this is the path followed by most of the carbocations. Path (a), on the other hand, leads to a less stable, disubstituted alkene and produces the minor product of the reaction. *The formation of the more stable alkene is the general rule (Zaitsev's rule) in the acid-catalyzed dehydration reactions of alcohols.*

Studies of thousands of reactions involving carbocations show that rearrangements like those just described are general phenomena. *They occur almost invariably when the migration of an alkanide ion or hydride ion can lead to a more stable carbocation.* The following are examples:

$$CH_3-\underset{\underset{CH_3}{|}}{\overset{\overset{CH_3}{|}}{C}}-\overset{+}{C}H-CH_3 \xrightarrow[\text{migration}]{\text{methanide}} CH_3-\underset{\underset{CH_3}{|}}{\overset{\overset{CH_3}{|}}{\overset{+}{C}}}-CH-CH_3$$

2° Carbocation   3° Carbocation

$$CH_3-\underset{\underset{CH_3}{|}}{\overset{\overset{H}{|}}{C}}-\overset{+}{C}H-CH_3 \xrightarrow[\text{migration}]{\text{hydride}} CH_3-\underset{\underset{CH_3}{|}}{\overset{\overset{H}{|}}{\overset{+}{C}}}-CH-CH_3$$

2° Carbocation   3° Carbocation

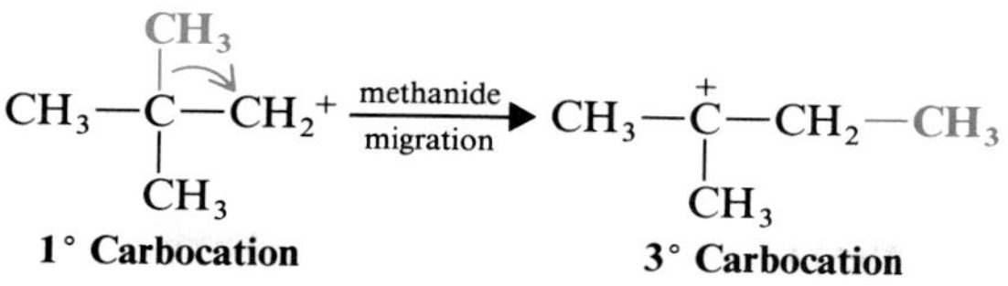

$$CH_3-\underset{\underset{CH_3}{|}}{\overset{\overset{CH_3}{|}}{C}}-CH_2^+ \xrightarrow[\text{migration}]{\text{methanide}} CH_3-\underset{\underset{CH_3}{|}}{\overset{+}{C}}-CH_2-CH_3$$

1° Carbocation   3° Carbocation

Rearrangements of carbocations can also lead to a change in ring size, as the following example shows:

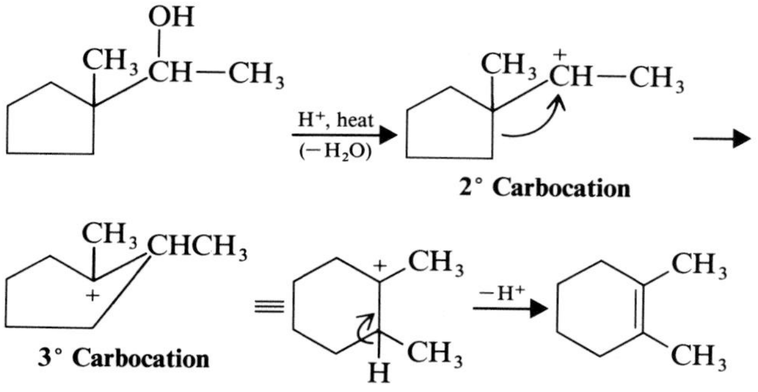

2° Carbocation

3° Carbocation

PROBLEM 7.15

Acid-catalyzed dehydration of neopentyl alcohol, $(CH_3)_3CCH_2OH$, yields 2-methyl-2-butene as the major product. Outline a mechanism showing all steps in its formation.

PROBLEM 7.16

Acid-catalyzed dehydration of either 2-methyl-1-butanol or 3-methyl-1-butanol gives 2-methyl-2-butene as the major product. Write plausible mechanisms that explain these results.

PROBLEM 7.17

When the compound called *isoborneol* is heated with 50% sulfuric acid, the product of the reaction is the compound called camphene and not bornylene as one might expect. Using models to assist you, write a step-by-step mechanism showing how camphene is formed.

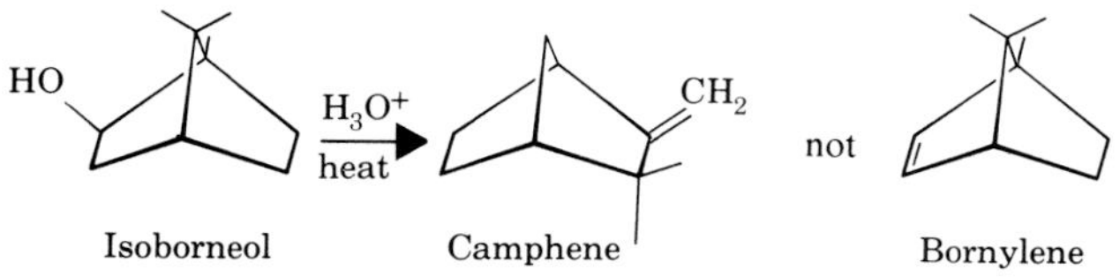

Isoborneol   Camphene   not   Bornylene

7.16 ALKENES BY DEBROMINATION OF VICINAL DIBROMIDES

Vicinal (or *vic*) dihalides are dihalo compounds in which the halogens are situated on adjacent carbon atoms. The name **geminal** (or *gem*) dihalide is used for those dihalides where both halogen atoms are attached to the same carbon atom.

$$
\begin{array}{cc}
\underset{\displaystyle X \quad X}{-\overset{\displaystyle |}{C}-\overset{\displaystyle |}{C}-} & \underset{\displaystyle X}{-\overset{\displaystyle X}{\overset{\displaystyle |}{C}}-\overset{\displaystyle |}{C}-} \\
\textbf{A } \textit{vic}\textbf{-dihalide} & \textbf{A } \textit{gem}\textbf{-dihalide}
\end{array}
$$

vic-Dibromides undergo **debromination** when they are treated with a solution of sodium iodide in acetone or with a mixture of zinc dust in acetic acid (or ethanol).

$$
-\overset{|}{\underset{\underset{\textstyle Br}{|}}{C}}-\overset{|}{\underset{\underset{\textstyle Br}{|}}{C}}- \ + \ 2NaI \xrightarrow{\text{acetone}} \ \diagdown C{=}C\diagup \ + \ I_2 \ + \ 2NaBr
$$

$$
-\overset{|}{\underset{\underset{\textstyle Br}{|}}{C}}-\overset{|}{\underset{\underset{\textstyle Br}{|}}{C}}- \ + \ Zn \xrightarrow[\substack{\text{or} \\ \text{CH}_3\text{CH}_2\text{OH}}]{\text{CH}_3\text{CO}_2\text{H}} \ \diagdown C{=}C\diagup \ + \ ZnBr_2
$$

Debromination by sodium iodide takes place by an E2 mechanism similar to that for dehydrohalogenation.

A MECHANISM FOR DEBROMINATION

Step 1

An iodide ion becomes bonded to a bromine atom in a step that is, in effect, an S$_N$2 attack on the bromine; removal of the bromine brings about an E2 elimination and the formation of a double bond.

Step 2

Here, an S$_N$2-type attack by an iodide ion on IBr leads to the formation of I$_2$ and a bromide ion.

Debromination by zinc takes place on the surface of the metal and the mechanism is uncertain. Other electropositive metals (e.g., Na, Ca, and Mg) also cause debromination of *vic*-dibromides.

vic-Dibromides are usually prepared by the addition of bromine to an alkene (cf. Section 8.6). Consequently, dehalogenation of a *vic*-dibromide is of little use as a general preparative reaction. Bromination followed by debromination is useful, however, in the purification of alkenes (see Problem 7.42) and in "protecting" the double bond. We shall see an example of this later.

7.17 SUMMARY OF METHODS FOR THE PREPARATION OF ALKENES

In this chapter we described four general methods for the preparation of alkenes.

1. **Dehydrohalogenation of alkyl halides (Section 7.12)**

General Reaction

$$-\overset{\displaystyle |}{\underset{\displaystyle H}{C}}-\overset{\displaystyle |}{\underset{\displaystyle X}{C}}-\ \xrightarrow[\text{heat} \atop (-\text{HX})]{\text{base}}\ \ \text{C}=\text{C}$$

Specific Examples

$$\text{CH}_3\text{CH}_2\underset{\underset{\displaystyle \text{Br}}{|}}{\text{CHCH}_3}\ \xrightarrow[\text{C}_2\text{H}_5\text{OH}]{\text{C}_2\text{H}_5\text{ONa}}\ \text{CH}_3\text{CH}=\text{CHCH}_3\ +\ \text{CH}_3\text{CH}_2\text{CH}=\text{CH}_2$$

$$\text{(cis and trans, 81\%)}\qquad\qquad\text{(19\%)}$$

$$\text{CH}_3\text{CH}_2\underset{\underset{\displaystyle \text{Br}}{|}}{\text{CHCH}_3}\ \xrightarrow[\text{(CH}_3)_3\text{COH}]{(\text{CH}_3)_3\text{COK} \atop 70°C}\ \text{CH}_3\text{CH}=\text{CHCH}_3\ +\ \text{CH}_3\text{CH}_2\text{CH}=\text{CH}_2$$

$$\qquad\qquad\quad\text{Disubstituted alkenes}\qquad\text{Monosubstituted alkene}$$
$$\qquad\qquad\quad\text{(cis and trans, 47\%)}\qquad\qquad\text{(53\%)}$$

2. **Dehydration of alcohols (Sections 7.13–7.15)**

General Reaction

$$-\overset{\displaystyle |}{\underset{\displaystyle H}{C}}-\overset{\displaystyle |}{\underset{\displaystyle OH}{C}}-\ \xrightarrow[\text{heat}]{\text{acid}}\ \ \text{C}=\text{C}\ +\ \text{H}_2\text{O}$$

Specific Examples

$$\text{CH}_3\text{CH}_2\text{OH}\ \xrightarrow[180°C]{\text{concd H}_2\text{SO}_4}\ \text{CH}_2=\text{CH}_2\ +\ \text{H}_2\text{O}$$

$$\text{CH}_3\underset{\underset{\displaystyle \text{CH}_3}{|}}{\overset{\overset{\displaystyle \text{CH}_3}{|}}{\text{C}}}-\text{OH}\ \xrightarrow[85°C]{20\% \text{ H}_2\text{SO}_4}\ \text{CH}_3\overset{\overset{\displaystyle \text{CH}_3}{|}}{\text{C}}=\text{CH}_2\ +\ \text{H}_2\text{O}$$

$$\text{(83\%)}$$

3. **Dehalogenation of *vic*-dibromides (Section 7.16)**

General Reaction

$$-\overset{\displaystyle |}{\underset{\displaystyle Br}{C}}-\overset{\displaystyle |}{\underset{\displaystyle Br}{C}}-\ \xrightarrow[\text{CH}_3\text{CO}_2\text{H}]{\text{Zn}}\ \ \text{C}=\text{C}\ +\ \text{ZnBr}_2$$

4. **Hydrogenation of alkynes (Section 7.7B)**

General Reaction

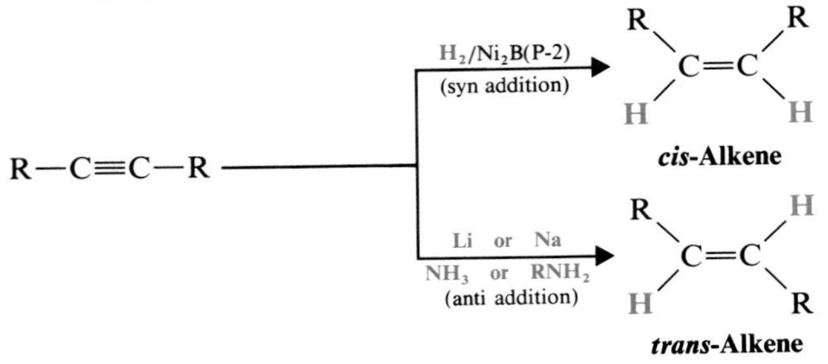

R—C≡C—R
$\xrightarrow[\text{(syn addition)}]{\text{H}_2/\text{Ni}_2\text{B(P-2)}}$
cis-Alkene

$\xrightarrow[\substack{\text{NH}_3 \quad \text{or} \quad \text{RNH}_2 \\ \text{(anti addition)}}]{\text{Li} \quad \text{or} \quad \text{Na}}$
trans-Alkene

In subsequent chapters we shall see a number of other methods for alkene synthesis.

7.18 SYNTHESIS OF ALKYNES BY ELIMINATION REACTIONS

Alkynes can also be synthesized from alkenes. In this method an alkene is first treated with bromine to form a *vic*-dibromo compound.

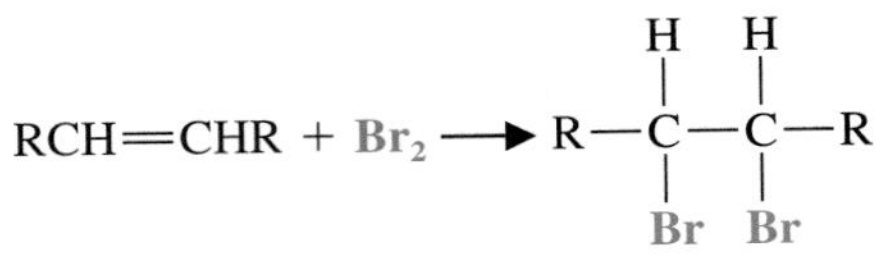

$$\text{RCH}=\text{CHR} + \text{Br}_2 \longrightarrow \underset{\substack{| \quad | \\ \text{Br} \quad \text{Br}}}{\overset{\substack{\text{H} \quad \text{H} \\ | \quad |}}{\text{R}-\text{C}-\text{C}-\text{R}}}$$

vic-**Dibromide**

Then the *vic*-dibromide is dehydrohalogenated through its reaction with a strong base. The dehydrohalogenation occurs in two steps. The first step yields a bromoalkene.

A MECHANISM FOR THE REACTION

REACTION:

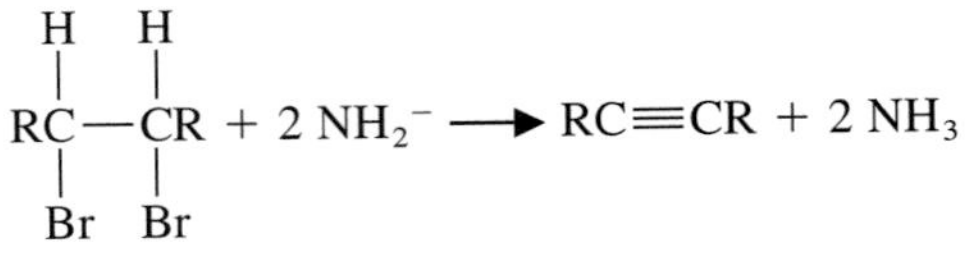

$$\underset{\substack{| \quad | \\ \text{Br} \quad \text{Br}}}{\overset{\substack{\text{H} \quad \text{H} \\ | \quad |}}{\text{RC}-\text{CR}}} + 2\,\text{NH}_2^- \longrightarrow \text{RC}\equiv\text{CR} + 2\,\text{NH}_3$$

MECHANISM:

Step 1

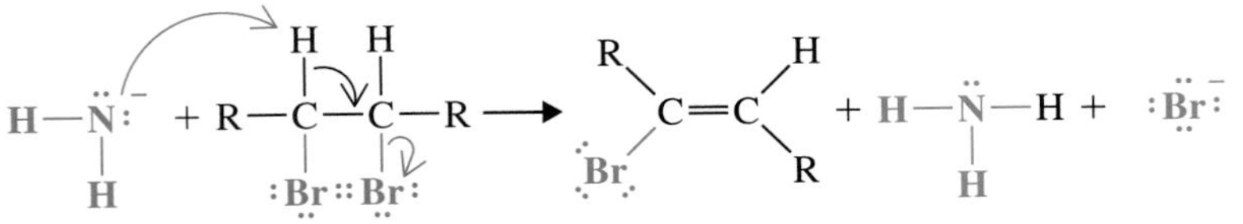

Amide ion   *vic*-Dibromide   Bromoalkene   Ammonia   Bromide ion

The strongly basic amide ion brings about an E2 reaction.

Step 2

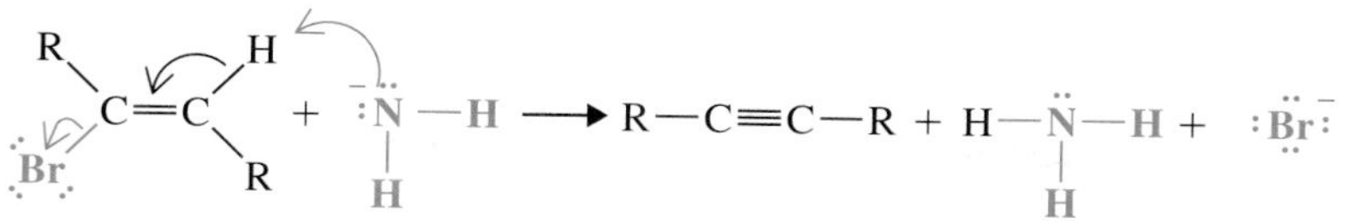

| Bromoalkene | Amide ion | Alkyne | Ammonia | Bromide ion |

A second E2 reaction
produces the alkyne.

Depending on the conditions, these two dehydrohalogenations may be carried out as separate reactions, or they may be carried out consecutively in a single mixture. The strong base, sodium amide, is capable of effecting both dehydrohalogenations in a single reaction mixture. (At least two molar equivalents of sodium amide per mole of the dihalide must be used, and if the product is a terminal alkyne, three molar equivalents must be used because the terminal alkyne reacts with sodium amide as it is formed in the mixture.) Dehydrohalogenations with sodium amide are usually carried out in liquid ammonia or in an inert medium such as mineral oil.

The following example illustrates this method.

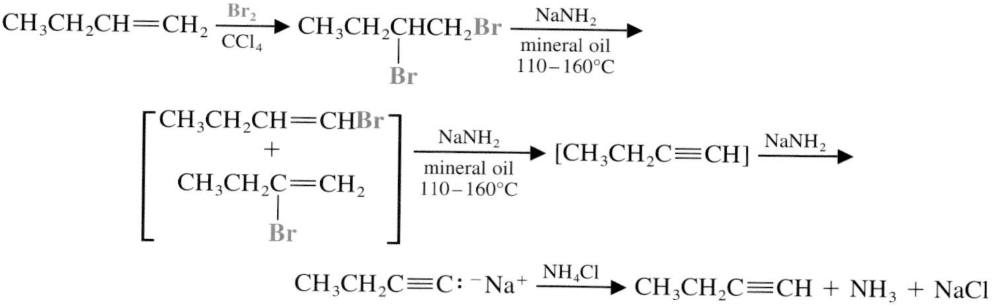

Ketones can be converted to *gem*-dichlorides through their reaction with phosphorus pentachloride, and these can also be used to synthesize alkynes.

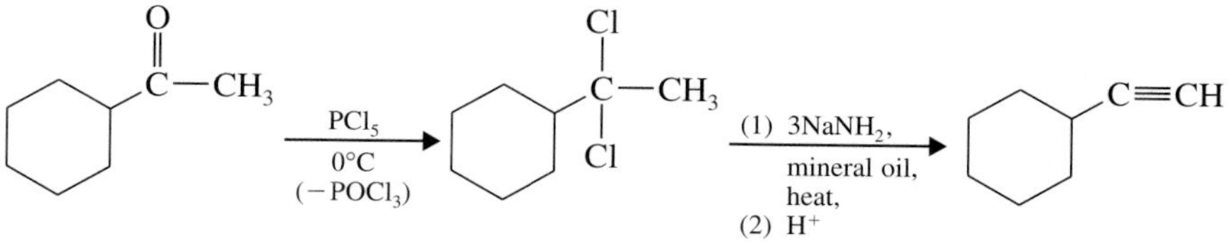

| Methyl cyclohexyl ketone | A *gem*-dichloride (70–80%) | Cyclohexylacetylene (46%) |

PROBLEM 7.18

Outline all steps in a synthesis of propyne from each of the following:

(a) CH_3COCH_3 (c) $CH_3CHBrCH_2Br$
(b) $CH_3CH_2CHBr_2$ (d) $CH_3CH=CH_2$

7.19 THE ACIDITY OF TERMINAL ALKYNES

The hydrogen atoms of ethyne are considerably more acidic than those of ethene or ethane (Section 3.7).

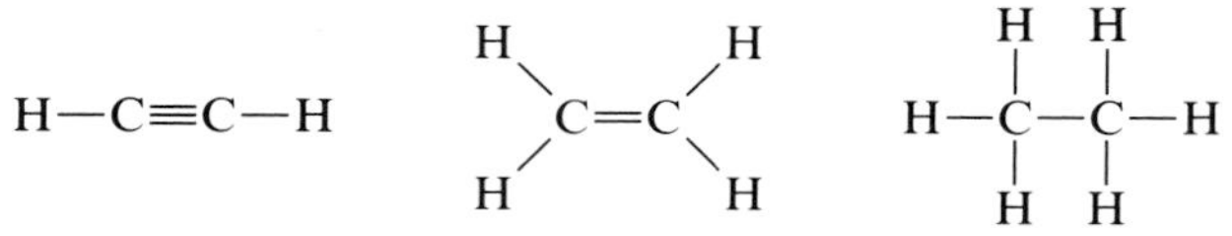

$$H—C\equiv C—H \qquad \underset{H}{\overset{H}{\diagup}}C=C\underset{H}{\overset{H}{\diagdown}} \qquad H—\overset{\overset{H}{|}}{\underset{\underset{H}{|}}{C}}—\overset{\overset{H}{|}}{\underset{\underset{H}{|}}{C}}—H$$

$$pK_a = 25 \qquad\qquad pK_a = 44 \qquad\qquad pK_a = 50$$

The order of basicities of the anions is opposite the relative acidity of the hydrocarbons. The ethanide ion is the most basic and the ethynide ion is the least basic.

Relative Basicity

$$CH_3CH_2:^- > CH_2\!=\!CH:^- > HC\equiv C:^-$$

What we have said about ethyne and ethynide ions is true of any terminal alkyne $(RC\equiv CH)$ and any alkynide ion $(RC\equiv C:^-)$. If we include other hydrogen compounds of the first-row elements of the periodic table, we can write the following orders of relative acidities and basicities:

Relative Acidity

$$H—\overset{..}{\underset{..}{O}}H > H—\overset{..}{\underset{..}{O}}R > H—C\equiv CR > H—\overset{..}{N}H_2 > H—CH\!=\!CH_2 > H—CH_2CH_3$$
$$pK_a \quad 15.7 \qquad 16\text{–}17 \qquad\quad 25 \qquad\quad 38 \qquad\qquad 44 \qquad\qquad 50$$

Relative Basicity

$$^-\!:\overset{..}{\underset{..}{O}}H <\ ^-\!:\overset{..}{\underset{..}{O}}R <\ ^-\!:C\equiv CR <\ ^-\!:\overset{..}{N}H_2 <\ ^-\!:CH\!=\!CH_2 <\ ^-\!:CH_2CH_3$$

We see from the order just given that while terminal alkynes are more acidic than ammonia, they are less acidic than alcohols and are less acidic than water.

The arguments just made apply only to acid–base reactions that take place in solution. In the gas phase, acidities and basicities are very much different. For example, in the gas phase the hydroxide ion is a stronger base than the acetylide ion. The explanation for this shows us again the important roles solvents play in reactions that involve ions (cf. Section 6.15). In solution, smaller ions (e.g., hydroxide ions) are more effectively solvated than larger ones (e.g., ethynide ions). Because they are more effectively solvated, smaller ions are more stable and are therefore less basic. In the gas phase, large ions are stabilized by polarization of their bonding electrons and the bigger a group is the more polarizable it will be. Thus in the gas phase larger ions are less basic.

PROBLEM 7.19 _____

Predict the products of the following acid–base reactions. If the equilibrium would not result in the formation of appreciable amounts of products, you should so indicate. In each case label the stronger acid, the stronger base, the weaker acid, and the weaker base.

(a) $CH_3CH{=}CH_2 + NaNH_2 \longrightarrow$

(b) $CH_3C{\equiv}CH + NaNH_2 \longrightarrow$

(c) $CH_3CH_2CH_3 + NaNH_2 \longrightarrow$

(d) $CH_3C{\equiv}C{:}^- + CH_3CH_2OH \longrightarrow$

(e) $CH_3C{\equiv}C{:}^- + NH_4Cl \longrightarrow$

7.20 REPLACEMENT OF THE ACETYLENIC HYDROGEN ATOM OF TERMINAL ALKYNES

Sodium ethynide and other sodium alkynides can be prepared by treating terminal alkynes with sodium amide in liquid ammonia.

$$H{-}C{\equiv}C{-}H + NaNH_2 \xrightarrow{\text{liq. } NH_3} H{-}C{\equiv}C{:}^- Na^+ + NH_3$$

$$CH_3C{\equiv}C{-}H + NaNH_2 \xrightarrow{\text{liq. } NH_3} CH_3C{\equiv}C{:}^- Na^+ + NH_3$$

These are acid–base reactions. The amide ion, by virtue of its being the anion of the very weak acid, ammonia ($pK_a = 38$), is able to remove the acetylenic protons of terminal alkynes ($pK_a = 25$). These reactions, for all practical purposes, go to completion.

Sodium alkynides are useful intermediates for the synthesis of other alkynes. These syntheses can be accomplished by treating the sodium alkynide with a primary alkyl halide.

$$R{-}C{\equiv}C{:}^- Na^+ + R'CH_2{-}Br \longrightarrow R{-}C{\equiv}C{-}CH_2R' + NaBr$$

| **Sodium alkynide** | **Primary alkyl halide** | **Mono- or disubstituted acetylene** |

(R or R′ or both may be hydrogen.)

The following examples illustrate this synthesis of alkynes:

$$HC{\equiv}C{:}^- Na^+ + CH_3{-}Br \xrightarrow[\text{5 h}]{\text{liq. } NH_3} H{-}C{\equiv}C{-}CH_3 + NaBr$$

Propyne (84%)

$$CH_3CH_2C{\equiv}C{:}^- Na^+ + CH_3CH_2{-}Br \xrightarrow[\text{6 h}]{\text{liq. } NH_3} CH_3CH_2C{\equiv}CCH_2CH_3 + NaBr$$

3-Hexyne (75%)

In all of these examples the alkynide ion acts as a nucleophile and displaces a halide ion from the primary alkyl halide. The result is **an S_N2 reaction** (Section 6.7).

$$RC{\equiv}C{:}^- \xrightarrow[\text{substitution } S_N2]{\text{nucleophilic}} RC{\equiv}C{-}CH_2R' + NaBr$$

| **Sodium alkynide** | **1° Alkyl halide** |

The unshared electron pair of the alkynide ion attacks the backside of the carbon atom that bears the halogen atom and forms a bond to it. The halogen atom departs as a halide ion.

This synthesis fails when secondary or tertiary halides are used because the alkynide ion acts as a base rather than as a nucleophile, and the major result is an **E2 elimination** (Section 6.18). The products of the elimination are an alkene and the alkyne from which the sodium alkynide was originally formed.

$$RC{\equiv}CH + R'CH = CHR'' + Br^-$$

2° Alkyl halide

PROBLEM 7.20

Your goal is to synthesize 3,3-dimethyl-2-pentyne. You have a choice of beginning with any of the following reagents.

$$CH_3C{\equiv}CH \qquad CH_3-\underset{\underset{CH_3}{|}}{\overset{\overset{CH_3}{|}}{C}}-Br \qquad CH_3-\underset{\underset{CH_3}{|}}{\overset{\overset{CH_3}{|}}{C}}-C{\equiv}CH \qquad CH_3I$$

Assume that you also have available sodium amide and liquid ammonia and outline the best synthesis of the required compound.

ADDITIONAL PROBLEMS

✗ 7.21   Each of the following names is incorrect. Give the correct name and explain your reasoning.

(a) *trans*-3-Pentene   (c) 2-Methylcyclohexene   (e) (*E*)-3-Chloro-2-butene
(b) 1,1-Dimethylethene   (d) 4-Methylcyclobutene   (f) 5,6-Dichlorocyclohexene

✗ 7.22   Write a structural formula for each of the following:

(a) 3-Methylcyclobutene   (g) (*Z*,4*R*)-4-Methyl-2-hexene
(b) 1-Methylcyclopentene   (h) (*E*,4*S*)-4-Chloro-2-pentene
(c) 2,3-Dimethyl-2-pentene   (i) (*Z*)-1-Cyclopropyl-1-pentene
(d) (*Z*)-3-Hexene   (j) 5-Cyclobutyl-1-pentene
(e) (*E*)-2-Pentene   (k) (*R*)-4-Chloro-2-pentyne
(f) 3,3,3-Tribromopropene   (l) (*E*)-3-Methyl-2-hexen-5-yne

✱ 7.23   Write three-dimensional formulas for, and give names using (*R–S*) and (*E–Z*) designations for the isomers of:

(a) 4-Bromo-2-hexene   (c) 2,4-Dichloro-2-pentene
(b) 3-Chloro-1,4-hexadiene   (d) 5-Bromo-3-chloro-4-hexen-1-yne

7.24   Give the IUPAC names for each of the following:

(a) 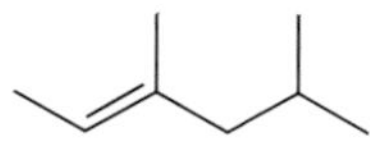 [structure]   (c) 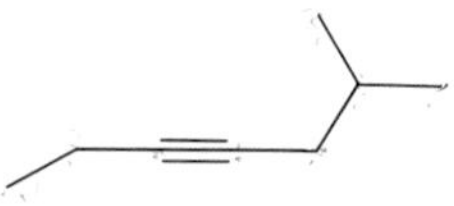 [structure]   (e) 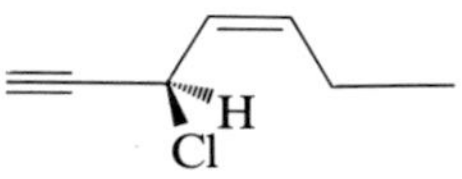 [structure with H, Cl]

(b) 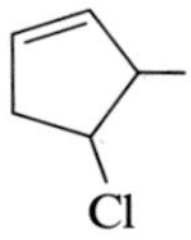 [structure with Cl]   (d) 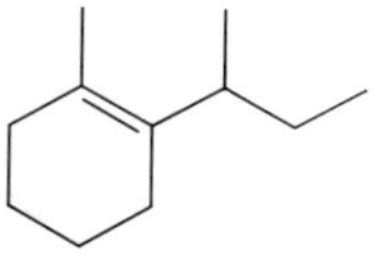 [structure]   (f) 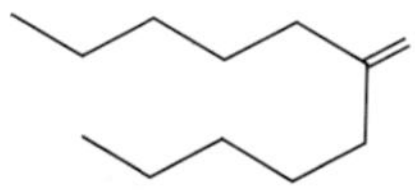 [structure]

7.25   Outline a synthesis of propene from each of the following:
(a) Propyl chloride   (d) Isopropyl alcohol
(b) Isopropyl chloride   (e) 1,2-Dibromopropane
(c) Propyl alcohol   (f) Propyne

7.26   Outline a synthesis of cyclopentene from each of the following:
(a) Bromocyclopentane
(b) 1,2-Dichlorocyclopentane
(c) Cyclopentanol [i.e., $(CH_2)_4CHOH$]

7.27   Starting with ethyne, outline syntheses of each of the following. You may use any other needed reagents, and you need not show the synthesis of compounds prepared in earlier parts of this problem.

(a) Propyne   (f) 1-Pentyne   (k) $CH_3CH_2C{\equiv}CD$
(b) 1-Butyne   (g) 2-Hexyne   (l) 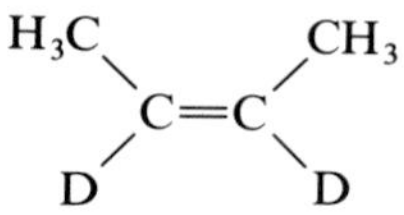 [structure H_3C, CH_3, $C{=}C$, D, D]
(c) 2-Butyne   (h) (Z)-2-Hexene
(d) cis-2-Butene   (i) (E)-2-Hexene
(e) trans-2-Butene   (j) 3-Hexyne

7.28   Starting with 1-methylcyclohexene and using any other needed reagents, outline a synthesis of the following deuterium-labeled compound.

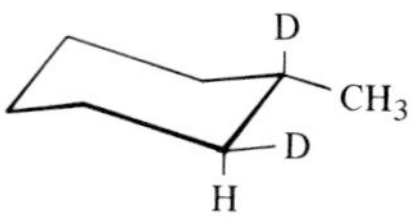

[structure with D, CH_3, D, H]

7.29   When trans-2-methylcyclohexanol (see following reaction) is subjected to acid-catalyzed dehydration, the major product is 1-methylcyclohexene:

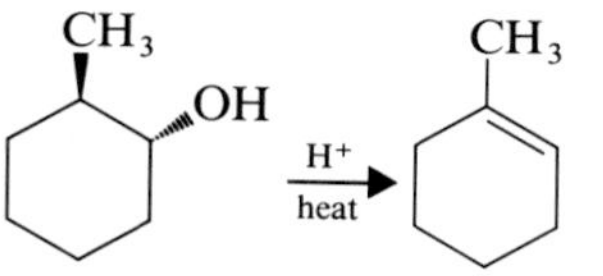

[structure CH_3, OH] $\xrightarrow[\text{heat}]{H^+}$ [structure CH_3]

However, when trans-1-bromo-2-methylcyclohexane is subjected to dehydrohalogenation, the major product is 3-methylcyclohexene:

Account for the different products of these two reactions.

7.30   Outline a synthesis of phenylethyne from each of the following:

(a) 1,1-Dibromo-1-phenylethane   (c) Phenylethene

(b) 1,1-Dibromo-2-phenylethane   (d) Acetophenone ($C_6H_5COCH_3$)

7.31   The heat of combustion of cyclobutane is 646.3 kcal mol^{-1}. What does this mean about the relative stability of cyclobutane when compared to the butene isomers (Section 7.9B)?

7.32   Match the following heats of combustion (806.7 kcal mol^{-1}, 805.2 kcal mol^{-1}, 804.2 kcal mol^{-1}, 803.4 kcal mol^{-1}, 801.8 kcal mol^{-1}) with the following alkenes: *cis*-2-pentene, *trans*-2-pentene, 2-methyl-2-butene, 1-pentene, 2-methyl-1-butene.

7.33   Without consulting tables, arrange the following compounds in order of decreasing acidity.

pentane   1-pentene   1-pentyne   1-pentanol

7.34   Write structural formulas for all the products that would be obtained when each of the following alkyl halides is heated with sodium ethoxide in ethanol. When more than one product results you should indicate which would be the major product, and which would be the minor products. You may neglect cis–trans isomerism of the products when answering this question.

(a) 2-Bromo-3-methylbutane   (d) 1-Bromo-1-methylcyclohexane

(b) 3-Bromo-2,2-dimethylbutane   (e) *cis*-1-Bromo-2-ethylcyclohexane

(c) 3-Bromo-3-methylpentane   (f) *trans*-1-Bromo-2-ethylcyclohexane

7.35   Write structural formulas for all the products that would be obtained when each of the following alkyl halides is heated with potassium *tert*-butoxide in *tert*-butyl alcohol. When more than one product results you should indicate which would be the major product, and which would be the minor products. You may neglect cis–trans isomerism of the products when answering this question.

(a) 2-Bromo-3-methylbutane   (d) 4-Bromo-2,2-dimethylpentane

(b) 4-Bromo-2,2-dimethylbutane   (e) 1-Bromo-1-methylcyclohexane

(c) 3-Bromo-3-methylpentane

7.36   Starting with an appropriate alkyl halide and base, outline syntheses that would yield each of the following alkenes as the major (or only) product.

(a) 1-Pentene   (d) 4-Methylcyclohexene

(b) 3-Methyl-1-butene   (e) 1-Methylcyclopentene

(c) 2,3-Dimethyl-1-butene

7.37   Arrange the following alcohols in order of their reactivity toward acid-catalyzed dehydration (with the most reactive first):

1-pentanol   2-methyl-2-butanol   3-methyl-2-butanol

7.38 Give the products that would be formed when each of the following alcohols is subjected to acid-catalyzed dehydration. If more than one product would be formed, designate the alkene that would be the major product (Neglect cis–trans isomerism.)

(a) 2-Methyl-2-propanol (d) 2,2-Dimethyl-1-propanol

(b) 3-Methyl-2-butanol (e) 1,4-Dimethylcyclohexanol

(c) 3-Methyl-3-pentanol

7.39 1-Bromobicyclo[2.2.1]heptane does not undergo elimination (below) when heated with base. Explain this failure to react.

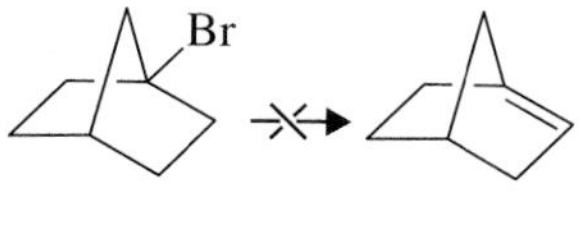

7.40 When the deuterium-labeled compound shown below is subjected to dehydrohalogenation using sodium ethoxide in ethanol, the only alkene product is 3-methylcyclohexene. (The product contains no deuterium.) Provide an explanation for this result.

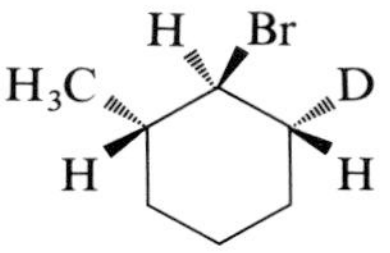

7.41 Provide mechanistic explanations for each of the following occurrences:

(a) Acid-catalyzed dehydration of 1-butanol produces *trans*-2-butene as the major product.

(b) Acid-catalyzed dehydration of 2,2-dimethylcyclohexanol produces 1,2-dimethylcyclohexene as the major product.

(c) Treating 3-iodo-2,2-dimethylbutane with silver nitrate in ethanol produces 2,3-dimethyl-2-butene as the major product.

(d) Dehydrohalogenation of (1S,2R)-1-bromo-1,2-diphenylpropane produces only (E)-1,2-diphenylpropene.

7.42 Cholesterol is an important steroid found in nearly all body tissues; it is also the major component of gallstones. Impure cholesterol can be obtained from gallstones by extracting them with an organic solvent. The crude cholesterol thus obtained can be purified by (a) treatment with Br_2 in $CHCl_3$, (b) careful crystallization of the product, and (c) treatment of the latter with zinc in ethanol. What reactions are involved in this procedure?

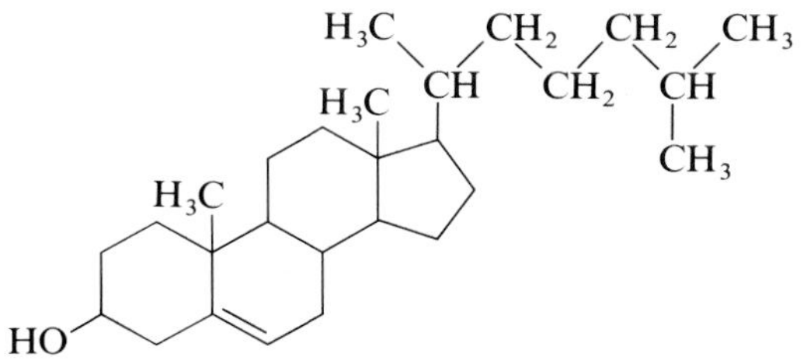

Cholesterol

7.43 Caryophyllene, a compound found in oil of cloves, has the molecular formula $C_{15}H_{24}$ and has no triple bonds. Reaction of caryophyllene with an excess of hydrogen in the presence of a platinum catalyst produces a compound with the formula $C_{15}H_{28}$. How many (a) double bonds, and (b) rings does a molecule of caryophyllene have?

7.44   Squalene, an important intermediate in the biosynthesis of steroids, has the molecular formula $C_{30}H_{50}$ and has no triple bonds.   (a) What is the index of hydrogen deficiency of squalene?   (b) Squalene undergoes catalytic hydrogenation to yield a compound with the molecular formula $C_{30}H_{62}$. How many double bonds does a molecule of squalene have?   (c) How many rings?

7.45   Consider the interconversion of *cis*-2-butene and *trans*-2-butene.   (a) What is the value of $\Delta H°$ for the reaction, *cis*-2-butene $\longrightarrow$ *trans*-2-butene?   (b) Assume $\Delta H° \approx \Delta G°$, what minimum value of $\Delta G^{\ddagger}$ would you expect for this reaction?   (c) Sketch a free-energy diagram for the reaction and label $\Delta G°$ and $\Delta G^{\ddagger}$.

7.46   Propose structures for compounds **E–H**.   (a) Compound **E** has the molecular formula C_5H_8 and is optically active. On catalytic hydrogenation **E** yields **F**. Compound **F** has the molecular formula C_5H_{10}, is optically inactive, and cannot be resolved into separate enantiomers.   (b) Compound **G** has the molecular formula C_6H_{10} and is optically active. Compound **G** contains no triple bonds. On catalytic hydrogenation **G** yields **H**. Compound **H** has the molecular formula C_6H_{14}, is optically inactive, and cannot be resolved into separate enantiomers.

7.47   Compounds **I** and **J** both have the molecular formula C_7H_{14}. Compounds **I** and **J** are both optically active and both rotate plane-polarized light in the same direction. On catalytic hydrogenation **I** and **J** yield the same compound **K**(C_7H_{16}). Compound **K** is optically active. Propose possible structures for **I**, **J**, and **K**.

7.48   Compounds **L** and **M** have the molecular formula C_7H_{14}. Compounds **L** and **M** are optically inactive, are nonresolvable, and are diastereomers of each other. Catalytic hydrogenation of either **L** or **M** yields **N**. Compound **N** is optically inactive but can be resolved into separate enantiomers. Propose possible structures for **L**, **M**, and **N**.

7.49   (a) Partial dehydrohalogenation of either (1R,2R)-1,2-dibromo-1,2-diphenylethane or (1S,2S)-1,2-dibromo-1,2-diphenylethane enantiomers (or a racemate of the two) produces (Z)-1-bromo-1,2-diphenylethene as the product, whereas   (b) partial dehydrohalogenation of (1R,2S)-1,2-dibromo-1,2-diphenylethane (the meso compound) gives only (E)-1-bromo-1,2-diphenylethene.   (c) Treating (1R,2S)-1,2-dibromo-1,2-diphenylethane with sodium iodide in acetone produces only (E)-1,2-diphenylethene. Explain these results.

CHAPTER **8**

Osmium tetroxide addition
to cyclopentene
(Section 8.9)

ALKENES AND ALKYNES II.
ADDITION REACTIONS

8.1 INTRODUCTION: ADDITIONS TO ALKENES

A characteristic reaction of compounds with a carbon–carbon double bond is an
addition—a reaction of the general type shown below.

$$\underset{/}{\overset{\backslash}{C}}=\underset{\backslash}{\overset{/}{C}} + A—B \xrightarrow{\text{addition}} A—\overset{|}{\underset{|}{C}}—\overset{|}{\underset{|}{C}}—B$$

We saw in Section 7.5 that alkenes undergo the addition of hydrogen. In this
chapter we shall study other examples of additions to the double bonds of alkenes. We
begin with the additions of hydrogen halides, sulfuric acid, water (in the presence of an
acid catalyst), and halogens.

H—X → H—C—C—X **Alkyl halide**
(Sections 8.2, 8.3,
and 9.9)

H—OSO₃H → H—C—C—OSO₃H **Alkyl hydrogen
sulfate**
(Section 8.4)

Alkene H—OH → H—C—C—OH **Alcohol**
(Section 8.5)

X—X → X—C—C—X **Dihaloalkane**
(Sections 8.6–8.8)

327

Two characteristics of the double bond help us understand why these addition reactions occur:

1. An addition reaction results in the conversion of one π bond (Section 2.4) and one σ bond into two σ bonds. The result of this change is usually energetically favorable. The heat evolved in making two σ bonds exceeds that needed to break one σ bond and one π bond (because π bonds are weaker), and, therefore, addition reactions are usually exothermic.

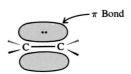

2. The electrons of the π bond are exposed. Because the π bond results from overlapping p orbitals, the π electrons lie above and below the plane of the double bond:

The π bond is particularly susceptible to electron-seeking reagents. Such reagents are said to be **electrophilic** (electron seeking) and are called **electrophiles.** Electrophiles include positive reagents such as protons (H^+), neutral reagents such as bromine (because it can be polarized so that one end is positive), and the Lewis acids BF_3 and $AlCl_3$. Metal ions that contain vacant orbitals—the silver ion (Ag^+), the mercuric ion (Hg^{2+}), and the platinum ion (Pt^{2+}), for example—also act as electrophiles.

Hydrogen halides, for example, react with alkenes by donating a proton to the π bond. The proton uses the two electrons of the π bond to form a σ bond to one of the carbon atoms. This leaves a vacant p orbital and a + charge on the other carbon. The overall result is the formation of a carbocation and a halide ion from the alkene and HX:

Being highly reactive, the carbocation then combines with the halide ion by accepting one of its electron pairs:

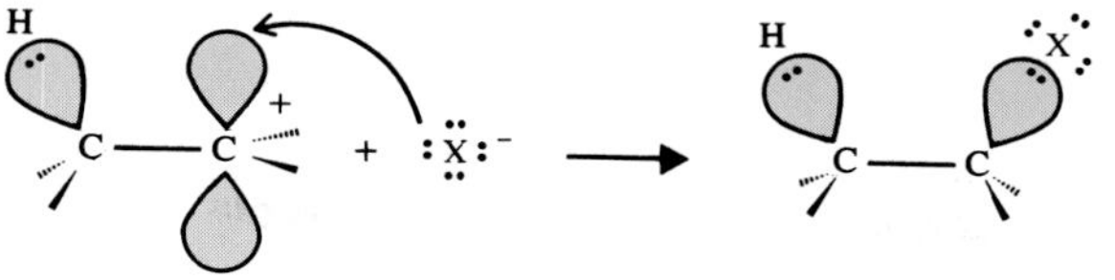

Electrophiles Are Lewis Acids They are molecules or ions that can accept an electron pair. Nucleophiles are molecules or ions that can furnish an electron pair (i.e., Lewis bases). Any reaction of an electrophile also involves a nucleophile. In the protonation of an alkene the electrophile is the proton donated by an acid; the nucleophile is the alkene.

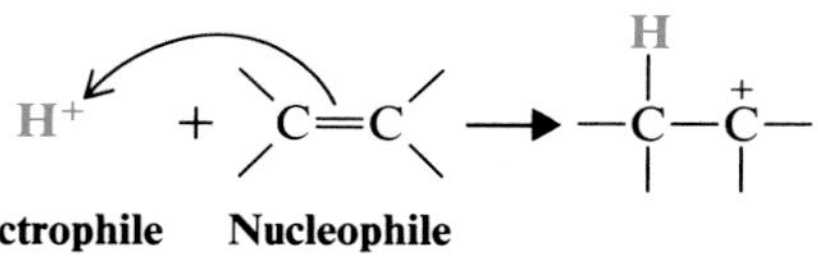

Electrophile   Nucleophile

In the next step, the reaction of the carbocation with a halide ion, the carbocation is the electrophile and the halide ion is the nucleophile.

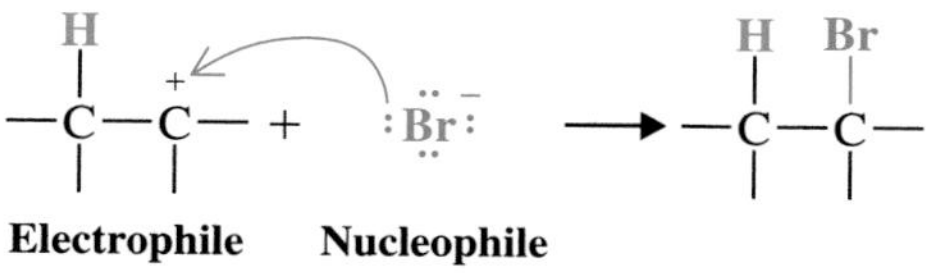

Electrophile   Nucleophile

8.2 ADDITION OF HYDROGEN HALIDES TO ALKENES: MARKOVNIKOV'S RULE

Hydrogen halides (HCl, HBr, and HI) add to the double bond of alkenes:

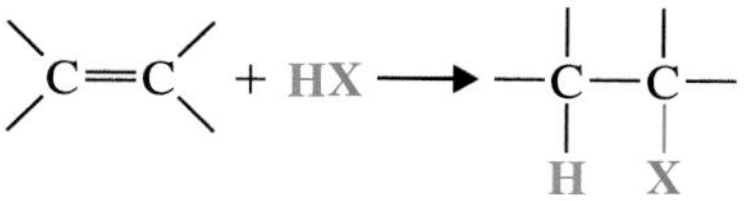

In the past these additions have been carried out by dissolving the hydrogen halide in a solvent, such as acetic acid or CH_2Cl_2, or by bubbling the gaseous hydrogen halide directly into the alkene and using the alkene itself as the solvent. The order of reactivity of the hydrogen halides is HI > HBr > HCl, and unless the alkene is highly substituted, HCl reacts so slowly that the reaction is not one that is useful as a preparative method. HBr adds readily, but as we will learn in Section 9.9, unless precautions are taken, the reaction may follow an alternate course. Fortunately, however, researchers have found recently that adding silica gel or alumina to the mixture of the alkene and HCl or HBr in CH_2Cl_2 increases the rate of addition dramatically and makes the reaction an easy one to carry out.*

The addition of HX to an unsymmetrical alkene could conceivably occur in two ways. In practice, however, one product usually predominates. The addition of HBr to

* See P. J. Kropp, K. A. Daus, S. D. Crawford, M. W. Tubergen, K. D. Kepler, S. L. Craig, and V. P. Wilson, *J. Am. Chem. Soc.,* **1990,** *112,* 7433–7434.

propene, for example, could conceivably lead to either 1-bromopropane or 2-bromo-propane. The main product, however, is 2-bromopropane.

$$CH_2{=}CHCH_3 + HBr \longrightarrow \underset{\underset{Br}{|}}{CH_3CHCH_3} \qquad (\textit{little } BrCH_2CH_2CH_3)$$

2-Bromopropane   **1-Bromopropane**

When 2-methylpropene reacts with HBr, the main product is *tert*-butyl bromide, not isobutyl bromide.

$$\underset{H_3C}{\overset{H_3C}{\diagdown}}C{=}CH_2 + HBr \longrightarrow CH_3{-}\underset{\underset{Br}{|}}{\overset{\overset{CH_3}{|}}{C}}{-}CH_3 \qquad \left(\textit{little } \quad CH_3{-}\underset{}{\overset{\overset{CH_3}{|}}{CH}}{-}CH_2{-}Br\right)$$

2-Methylpropene   ***tert*-Butyl bromide**   **Isobutyl bromide**
(isobutylene)

Consideration of many examples like this led the Russian chemist Vladimir Markovnikov in 1870 to formulate what is now known as **Markovnikov's rule.** One way to state this rule is to say that *in the addition of HX to an alkene, the hydrogen atom adds to the carbon atom of the double bond that already has the greater number of hydrogen atoms.** The addition of HBr to propene is an illustration.

Carbon atom with the greater number of hydrogen atoms

$$CH_2{=}CHCH_3 \longrightarrow \underset{\underset{H}{|}\quad\underset{Br}{|}}{CH_2{-}CHCH_3}$$

$$\overset{\uparrow}{\vdots}\quad\overset{\uparrow}{\vdots}$$

H   Br

Markovnikov addition product

Reactions that illustrate Markovnikov's rule are said to be *Markovnikov additions.*

A mechanism for addition of a hydrogen halide to an alkene involves the following two steps:

A MECHANISM FOR THE REACTION

Step 1
$$\overset{\diagdown}{\underset{\diagup}{C}}{=}\overset{\diagup}{\underset{\diagdown}{C}} + H{-}\ddot{\underset{\cdot\cdot}{X}}{:} \xrightarrow{\text{slow}} \underset{+}{{-}\overset{|}{C}{-}\overset{\overset{H}{|}}{C}{-}} + {:}\ddot{\underset{\cdot\cdot}{X}}{:}^{-}$$

The alkene accepts a proton from HX
to form a carbocation and a halide ion.

Step 2
$${:}\ddot{\underset{\cdot\cdot}{X}}{:}^{-} + \underset{+}{{-}\overset{\overset{H}{|}}{C}{-}\overset{\overset{H}{|}}{C}{-}} \xrightarrow{\text{fast}} {-}\overset{\overset{H}{|}}{C}{-}\underset{\underset{{:}\ddot{X}{:}}{|}}{\overset{\overset{H}{|}}{C}}{-}$$

The halide ion reacts with the carbocation by
donating an electron pair; the result is an alkyl halide.

---

* In his original publication, Markovnikov described the rule in terms of the point of attachment of the halogen atom, stating that "if an unsymmetrical alkene combines with a hydrogen halide, the halide ion adds to the carbon atom with the fewer hydrogen atoms."

The important step—because it is the **rate-determining step**—is step 1. In step 1 the alkene accepts a proton from the hydrogen halide and forms a carbocation. This step (Fig. 8.1) is highly endothermic and has a high free energy of activation. Consequently, it takes place slowly. In step 2 the highly reactive carbocation stabilizes itself by combining with a halide ion. This exothermic step has a very low free energy of activation and takes place very rapidly.

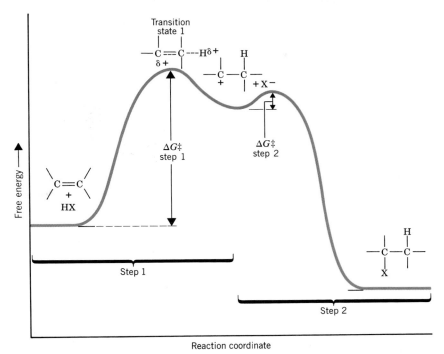

FIGURE 8.1 Free energy diagram for the addition of HX to an alkene. The free energy of activation for step 1 is much larger than for step 2.

8.2A Theoretical Explanation of Markovnikov's Rule

If the alkene that undergoes addition of a hydrogen halide is an unsymmetrical alkene such as propene, then step 1 could conceivably lead to two different carbocations:

$$CH_3CH = CH_2 + H^+ \longrightarrow CH_3\overset{H}{\underset{}{C}}H - \overset{+}{C}H_2$$

1° Carbocation
(less stable)

$$CH_3CH = CH_2 + H^+ \longrightarrow CH_3\overset{+}{C}H - CH_2 - H$$

2° Carbocation
(more stable)

These two carbocations are not of equal stability, however. The secondary carbocation is *more stable,* and it is the greater stability of the secondary carbocation that accounts for the correct prediction of the overall addition by Markovnikov's rule. In the addition of HBr to propene, for example, the reaction takes the following course:

$$CH_3CH{=}CH_2 \xrightarrow[slow]{H^+} \begin{cases} \xrightarrow{\times} CH_3CH_2\overset{+}{CH_2} \xrightarrow{Br^-} CH_3CH_2CH_2Br \\ \qquad\qquad 1^\circ \qquad\qquad\qquad \textbf{1-Bromopropane} \\ \qquad\qquad\qquad\qquad\qquad\quad (\textit{not}\textbf{ formed}) \\[2em] \longrightarrow CH_3\overset{+}{C}HCH_3 \xrightarrow[fast]{Br^-} CH_3CHCH_3 \\ \qquad\qquad\qquad\qquad\qquad\qquad\qquad | \\ \qquad\qquad\qquad\qquad\qquad\qquad\quad Br \\ \qquad\quad 2^\circ \qquad\qquad\qquad \textbf{2-Bromopropane} \\ \qquad\qquad\qquad\qquad\qquad (\textbf{actual product}) \end{cases}$$

|— Step 1 ——————————+—————————— Step 2 —|

The ultimate product of the reaction is 2-bromopropane because the more stable secondary carbocation is formed preferentially in the first step.

The more stable carbocation predominates because it is formed faster. We can understand why this is true if we examine the free energy diagrams in Fig. 8.2.

The reaction (Fig. 8.2) leading to the secondary carbocation (and ultimately to 2-bromopropane) has the lower free energy of activation. That is reasonable because its transition state resembles the more stable carbocation. The reaction leading to the primary carbocation (and ultimately to 1-bromopropane) has a higher free energy of

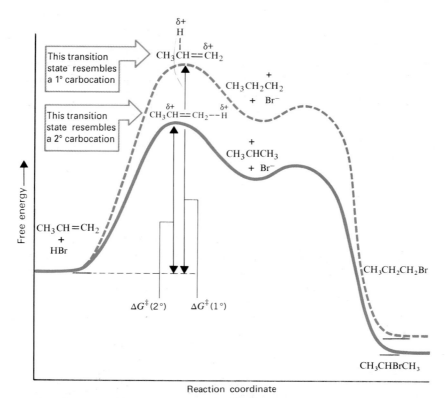

FIGURE 8.2 Free energy diagram for the addition of HBr to propene. $\Delta G^\ddagger$ (2°) is less than $\Delta G^\ddagger$ (1°).

activation because its transition state resembles a less stable primary carbocation. This second reaction is much slower and does not compete with the first reaction.

The reaction of HBr with 2-methylpropene produces only *tert*-butyl bromide, and for the same reason. Here, in the first step (i.e., the attachment of the proton) the choice is even more pronounced—between a tertiary carbocation and a primary carbocation.

Thus, isobutyl bromide is *not* obtained as a product of the reaction because its formation would require the formation of a primary carbocation. Such a reaction would have a much higher free energy of activation than that leading to a tertiary carbocation.

A MECHANISM FOR THE REACTION

This reaction takes place:

3° Carbocation
(more stable carbocation)   *tert*-Butyl bromide

This reaction does not occur appreciably:

1° Carbocation
(less stable carbocation)   Isobutyl bromide

Because carbocations are formed in the addition of HX to an alkene, rearrangements invariably occur when the carbocation initially formed can rearrange to a more stable one (see Problem 8.3).

8.2B Modern Statement of Markovnikov's Rule

With this understanding of the mechanism for the ionic addition of hydrogen halides to alkenes behind us, we are now in a position to give the following modern statement of Markovnikov's rule: *In the ionic addition of an unsymmetrical reagent to a double bond, the positive portion of the adding reagent attaches itself to a carbon atom of the double bond so as to yield the more stable carbocation as an intermediate.* Because this is the step that occurs first (before the addition of the negative portion of the adding reagent), it is the step that determines the overall orientation of the reaction.

Notice that this formulation of Markovnikov's rule allows us to predict the outcome of the addition of a reagent such as ICl. Because of the greater electronegativity of chlorine, the positive portion of this molecule is iodine. The addition of ICl to 2-methylpropene takes place in the following way and produces 2-chloro-1-iodo-2-methylpropane.

2-Methylpropene

2-Chloro-1-iodo-2-methylpropane

PROBLEM 8.1

Give the structure and name of the product that would be obtained from the ionic addition of IBr to propene.

PROBLEM 8.2

Outline mechanisms for the ionic additions   (a) of HBr to 2-methyl-1-butene, (b) of ICl to 2-methyl-2-butene, and   (c) of HI to 1-methylcyclopentene.

PROBLEM 8.3

Provide mechanistic explanations for the following observations: (a) The addition of hydrogen chloride to 3-methyl-1-butene produces two products: 2-chloro-3-methylbutane and 2-chloro-2-methylbutane.   (b) The addition of hydrogen chloride to 3,3-dimethyl-1-butene produces two products: 3-chloro-2,2-dimethylbutane and 2-chloro-2,3-dimethylbutane. (The first explanations for the course of both of these reactions were provided by Frank Whitmore, see page 309.)

8.2C Regioselective Reactions

Chemists describe reactions like the Markovnikov additions of hydrogen halides to alkenes as being *regioselective*. *Regio* comes from the Latin word *regionem* meaning direction. When a reaction *that can potentially yield two or more constitutional isomers actually produces only one* (or a predominance of one), the reaction is said to be *regioselective*. The addition of HX to an unsymmetrical alkene such as propene could conceivably yield two constitutional isomers, for example. However, as we have seen, the reaction yields only one, and therefore it is regioselective.

8.2D An Exception to Markovnikov's Rule

In Section 9.9 we shall study an exception to Markovnikov's rule. This exception concerns the addition of HBr to alkenes **when the addition is carried out in the presence of peroxides** (i.e., compounds with the general formula ROOR). When alkenes are treated with HBr in the presence of peroxides the addition occurs in an anti-Markovnikov manner in the sense that the hydrogen atom becomes attached to the carbon atom with the fewer hydrogen atoms. With propene, for example, the addition takes place as follows:

$$CH_3CH{=}CH_2 + HBr \xrightarrow{ROOR} CH_3CH_2CH_2Br$$

In Section 9.9 we shall find that this addition occurs by *a radical mechanism,* and not by the ionic mechanism given in Section 8.2. This anti-Markovnikov addition occurs *only when HBr is used in the presence of peroxides* and does not occur significantly with HF, HCl, and HI even when peroxides are present.

8.3 STEREOCHEMISTRY OF THE IONIC ADDITION TO AN ALKENE

Consider the following addition of HX to 1-butene and notice that the reaction leads to the formation of a product, 2-halobutane, that contains a stereocenter.

$$CH_3CH_2CH{=}CH_2 + HX \longrightarrow CH_3CH_2\overset{*}{C}HCH_3$$
$$\underset{X}{|}$$

The product, therefore, can exist as a pair of enantiomers. The question now arises as to how these enantiomers are formed. Is one enantiomer formed in greater amount than the other? The answer is *no*; the carbocation that is formed in the first step of the addition (see following scheme) is trigonal planar and *is achiral* (a model will show that it has a plane of symmetry). When the halide ion reacts with this achiral carbocation in the second step, *reaction is equally likely at either face.* The reactions leading to the two enantiomers occur at the same rate, and the enantiomers, therefore, are produced in equal amounts *as a racemic form.*

A MECHANISM FOR THE REACTION

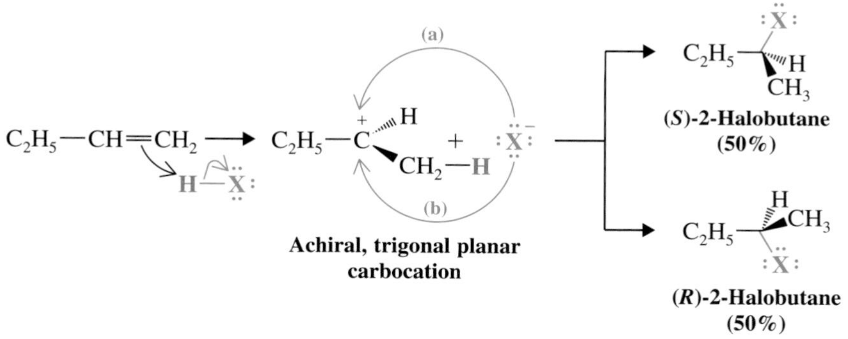

(a)

$C_2H_5{-}CH{=}CH_2 \longrightarrow C_2H_5{-}\overset{+}{C}\overset{H}{\underset{CH_2{-}H}{}} + \;:\!\ddot{X}\!:\!^-$

Achiral, trigonal planar carbocation

(b)

$C_2H_5{-}\overset{:\ddot{X}:}{\underset{CH_3}{\overset{|}{C}}}{}^{\prime\prime\prime\prime}H$

(S)-2-Halobutane (50%)

$C_2H_5{-}\overset{H}{\underset{:\ddot{X}:}{\overset{|}{C}}}CH_3$

(R)-2-Halobutane (50%)

1-Butene accepts a proton from HX to form an achiral carbocation.

The carbocation reacts with the halide ion at equal rates by path (a) or (b) to form the enantiomers as a racemate.

8.4 ADDITION OF SULFURIC ACID TO ALKENES

When alkenes are treated with **cold** concentrated sulfuric acid, *they dissolve* because they react by addition to form alkyl hydrogen sulfates. The mechanism is similar to that for the addition of HX. In the first step of this reaction the alkene accepts a proton from sulfuric acid to form a carbocation; in the second step the carbocation reacts with a hydrogen sulfate ion to form an alkyl hydrogen sulfate:

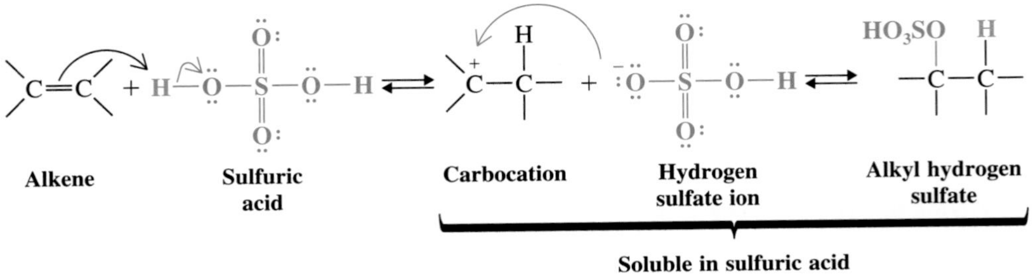

Alkene   Sulfuric acid   Carbocation   Hydrogen sulfate ion   Alkyl hydrogen sulfate

Soluble in sulfuric acid

The addition of sulfuric acid is also regioselective, and it follows Markovnikov's rule. Propene, for example, reacts to yield isopropyl hydrogen sulfate rather than propyl hydrogen sulfate.

3° Carbocation
(more stable carbocation)

Isopropyl hydrogen sulfate

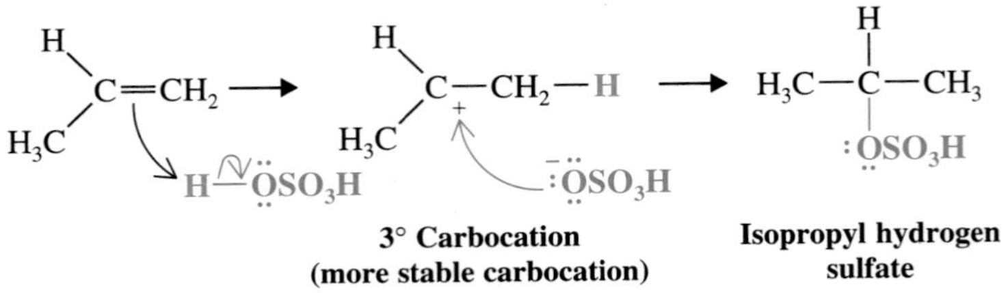

8.4A Alcohols from Alkyl Hydrogen Sulfates

Alkyl hydrogen sulfates can be easily hydrolyzed to alcohols by **heating** them with water. The overall result of the addition of sulfuric acid to an alkene followed by hydrolysis is the Markovnikov addition of H— and —OH.

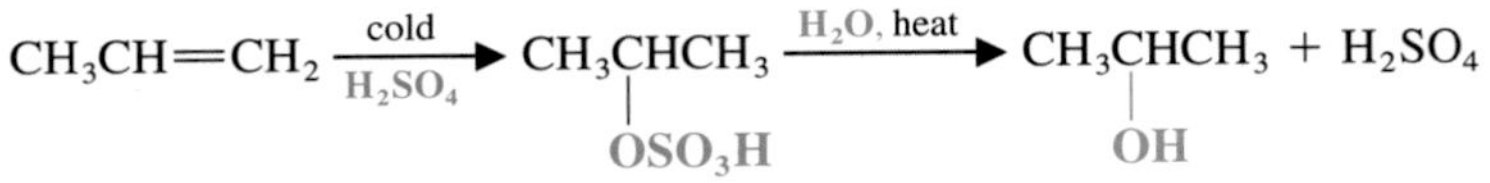

$$CH_3CH{=}CH_2 \xrightarrow[H_2SO_4]{cold} \underset{\overset{|}{OSO_3H}}{CH_3CHCH_3} \xrightarrow{H_2O, \text{ heat}} \underset{\overset{|}{OH}}{CH_3CHCH_3} + H_2SO_4$$

PROBLEM 8.4

In one industrial synthesis of ethanol, ethene is first dissolved in 95% sulfuric acid. In a second step water is added and the mixture is heated. Outline the reactions involved.

8.5 ADDITION OF WATER TO ALKENES: ACID-CATALYZED HYDRATION

The acid-catalyzed addition of water to the double bond of an alkene (hydration of an alkene) is a method for the preparation of low molecular weight alcohols that has its greatest utility in large-scale industrial processes. The acids most commonly used to catalyze the hydration of alkenes are dilute solutions of sulfuric acid and phosphoric acid. These reactions, too, are usually regioselective, and the addition of water to the double bond follows Markovnikov's rule. In general the reaction takes the form that follows:

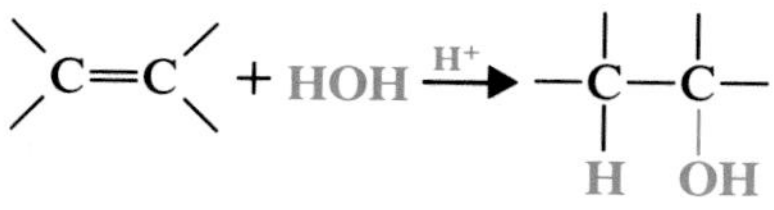

$$\ce{>C=C< + HOH ->[H+] -\underset{H}{C}-\underset{OH}{C}-}$$

An example is the hydration of 2-methylpropene.

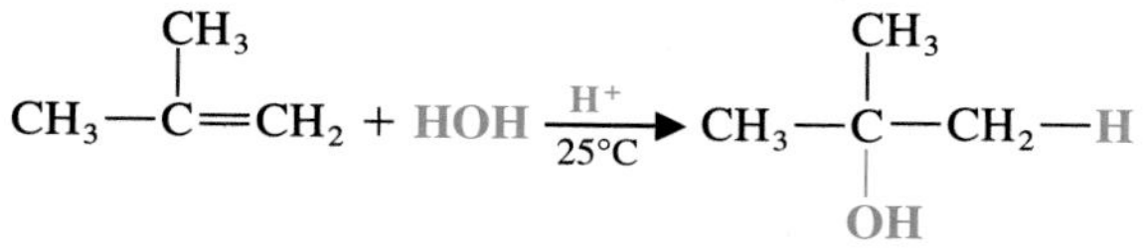

$$\underset{\substack{\text{2-Methylpropene}\\ \text{(isobutylene)}}}{CH_3-\overset{CH_3}{\underset{}{C}}=CH_2} + HOH \xrightarrow[25°C]{H^+} \underset{\textit{tert-Butyl alcohol}}{CH_3-\overset{CH_3}{\underset{OH}{C}}-CH_2-H}$$

Because the reactions follow Markovnikov's rule, acid-catalyzed hydrations of alkenes do not yield primary alcohols except in the special case of the hydration of ethene.

$$CH_2{=}CH_2 + HOH \xrightarrow[300°C]{H_3PO_4} CH_3CH_2OH$$

The mechanism for the hydration of an alkene is simply the reverse of the mechanism for the dehydration of an alcohol. We can illustrate this by giving the mechanism for the **hydration** of 2-methylpropene and by comparing it with the mechanism for the **dehydration** of *tert*-butyl alcohol given in Section 7.13.

A MECHANISM FOR HYDRATION OF AN ALKENE

Step 1

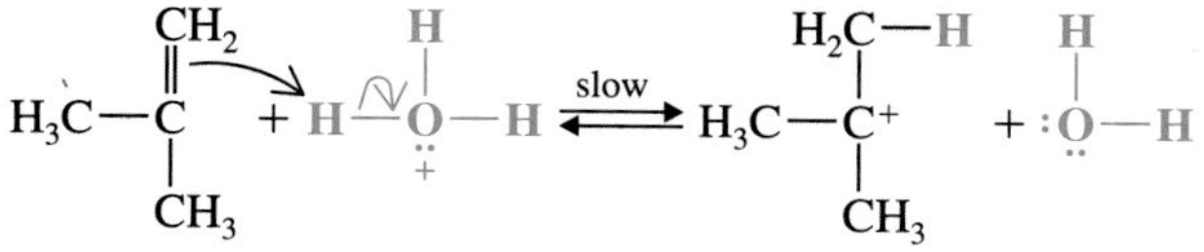

The alkene accepts a proton to form the more stable 3° carbocation.

Step 2

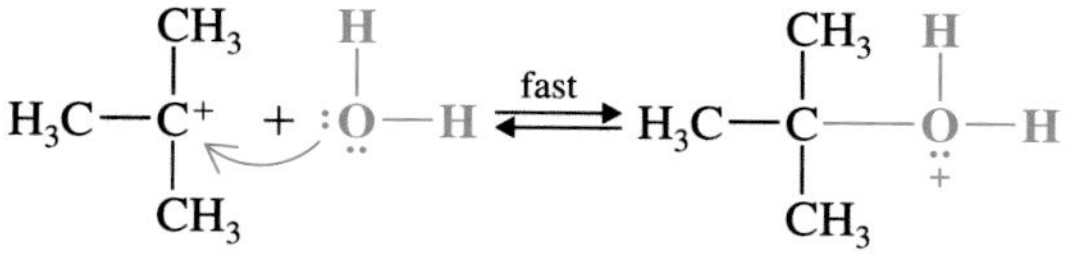

The carbocation reacts with a molecule of water to form a protonated alcohol.

Step 3

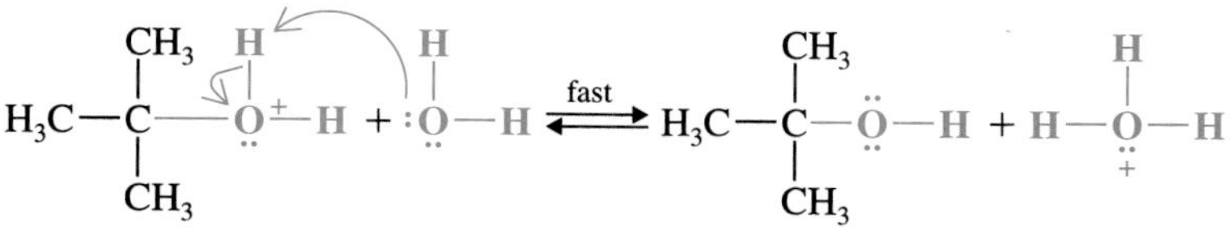

A transfer of a proton to a molecule of water leads to the product.

The rate-determining step in the *hydration* mechanism is step 1: the formation of the carbocation. It is this step, too, that accounts for the Markovnikov addition of water to the double bond. The reaction produces *tert*-butyl alcohol because step 1 leads to the formation of the more stable *tert*-butyl cation rather than the much less stable isobutyl cation:

$$
\begin{array}{l}
CH_3-\underset{\underset{CH_3}{|}}{\overset{\overset{CH_2}{\|}}{C}} \quad +H-\overset{\overset{H}{\frown}}{\underset{\underset{+}{\cdot\cdot}}{O}}-H \xrightleftharpoons[\text{slow}]{\text{very}} CH_3\underset{\underset{CH_3}{|}}{\overset{\overset{CH_2^+}{|}}{C}}-H + \overset{\overset{H}{|}}{:\underset{\cdot\cdot}{O}}-H
\end{array}
$$

For all practical purposes this reaction does not take place because it produces a 1° carbocation

The reactions whereby *alkenes are hydrated or alcohols are dehydrated* are reactions in which the ultimate product is governed by the position of an equilibrium. Therefore, in the *dehydration of an alcohol* it is best to use a concentrated acid so that the concentration of water is low. (The water can be removed as it is formed, and it helps to use a high temperature.) In the *hydration of an alkene* it is best to use dilute acid so that the concentration of water is high. (It also usually helps to use a lower temperature.)

PROBLEM 8.5

(a) Show all steps in the acid-catalyzed hydration of ethene to produce ethanol and give the general conditions that you would use to ensure a good yield of the product. (b) Give the general conditions that you would use to carry out the reverse reaction, the dehydration of ethanol to produce ethene. (c) What product would you expect to obtain from the acid-catalyzed hydration of propene? Explain your answer.

One complication associated with alkene hydrations is the occurrence of **rearrangements.** Because the reaction involves the formation of a carbocation in the first step, the carbocation formed initially invariably rearranges to a more stable one if such a rearrangement is possible. An illustration is the formation of 2,3-dimethyl-2-butanol as the major product when 3,3-dimethyl-1-butene is hydrated:

$$
CH_3-\underset{\underset{CH_3}{|}}{\overset{\overset{CH_3}{|}}{C}}-CH=CH_2 \xrightarrow[H_2O]{H_2SO_4} CH_3-\underset{\underset{CH_3}{|}}{\overset{\overset{OH}{|}}{C}}-\underset{\underset{CH_3}{|}}{CH}-CH_3
$$

3,3-Dimethyl-1-butene   2,3-Dimethyl-2-butanol
                                (major product)

PROBLEM 8.6

Outline all steps in a mechanism showing how 2,3-dimethyl-2-butanol is formed in the acid-catalyzed hydration of 3,3-dimethyl-1-butene.

PROBLEM 8.7

The following order of reactivity is observed when the following alkenes are subjected to acid-catalyzed hydration.

$$(CH_3)_2C=CH_2 > CH_3CH=CH_2 > CH_2=\overset{\diagdown}{C}H_2$$

Explain this order of reactivity.

PROBLEM 8.8 ⎯⎯⎯⎯⎯⎯⎯⎯⎯⎯⎯⎯⎯⎯⎯⎯⎯⎯⎯⎯

When 2-methylpropene (isobutylene) is dissolved in methanol containing a strong acid, a reaction takes place to produce *tert*-butyl methyl ether, $CH_3OC(CH_3)_3$. Write a mechanism that accounts for this.

⎯⎯⎯⎯⎯⎯⎯⎯⎯⎯⎯⎯⎯⎯⎯⎯⎯⎯⎯⎯⎯⎯⎯⎯⎯⎯⎯⎯⎯⎯⎯⎯⎯⎯⎯⎯⎯⎯

The occurrence of carbocation rearrangements limits the utility of alkene hydrations as a laboratory method for preparing alcohols. In Chapter 10 we shall study two very useful laboratory syntheses. One, called **oxymercuration–demercuration,** allows the Markovnikov addition of H— and —OH *without rearrangements.* Another, called **hydroboration–oxidation,** permits the *anti-Markovnikov* and *syn addition* of H— and —OH.

8.6 ADDITION OF BROMINE AND CHLORINE TO ALKENES

In the absence of light, *alkanes* do not react appreciably with chlorine or bromine at room temperature (Section 9.4). If we add an alkane to a solution of bromine in carbon tetrachloride, the red-brown color of the bromine persists in the solution as long as we keep the mixture away from sunlight and as long as the solution is not heated.

$$R-H \quad + \quad Br_2 \quad \xrightarrow[\text{in the dark, CCl}_4]{\text{room temperature}} \quad \text{no appreciable reaction}$$

Alkane   **Bromine**
(colorless) **(red brown)**

On the other hand, if we expose the reactants to sunlight, the bromine color will fade slowly. If we now place a small piece of moist blue litmus paper in the region above the liquid, the litmus paper will turn red because of the hydrogen bromide that evolves as the alkane and bromine react. (Hydrogen bromide is not very soluble in carbon tetrachloride.)

$$R-H \quad + \quad Br_2 \quad \xrightarrow[\text{sunlight, CCl}_4]{\text{room temperature}} \quad R-Br \quad + \quad HBr$$

Alkane   **Bromine**   **Alkyl halide**   **Hydrogen bromide**
(colorless) **(red brown)** **(colorless)**   **(detected by**
                                      moist blue litmus)

The behavior of **alkenes** toward bromine in carbon tetrachloride contrasts markedly with that of alkanes *and is a useful test for carbon–carbon multiple bonds.* Alkenes react rapidly with bromine at room temperature and in the *absence of light.* If we add bromine to an alkene, the red-brown color of the bromine disappears almost instantly as long as the alkene is present in excess. If we test the atmosphere above the solution with moist blue litmus paper, we shall find that no hydrogen bromide is present. The reaction is one of addition. (Alkynes, as we shall see in Section 8.12, also undergo addition of bromine.)

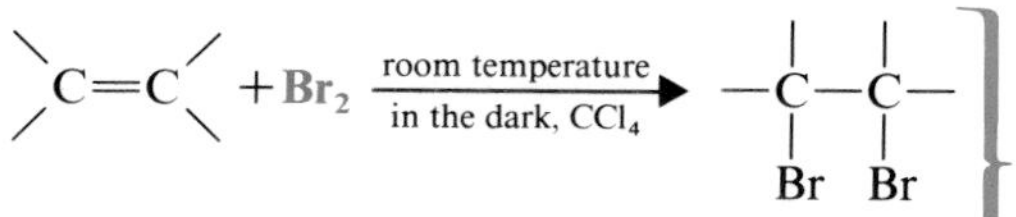

$$\text{C==C} + Br_2 \xrightarrow[\text{in the dark, CCl}_4]{\text{room temperature}} \begin{array}{c} | \quad | \\ -C-C- \\ | \quad | \\ Br \quad Br \end{array}$$

An alkene (colorless)	*vic*-Dibromide (colorless)

Rapid decolorization of Br$_2$/CCl$_4$ is a test for alkenes and alkynes

The addition reaction between alkenes and chlorine or bromine is a general one. The products are vicinal dihalides.

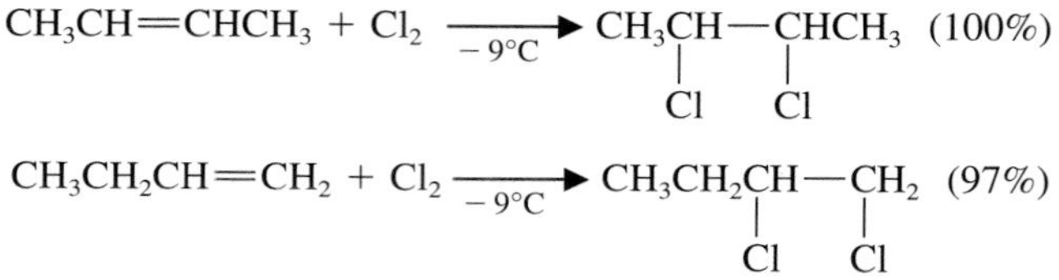

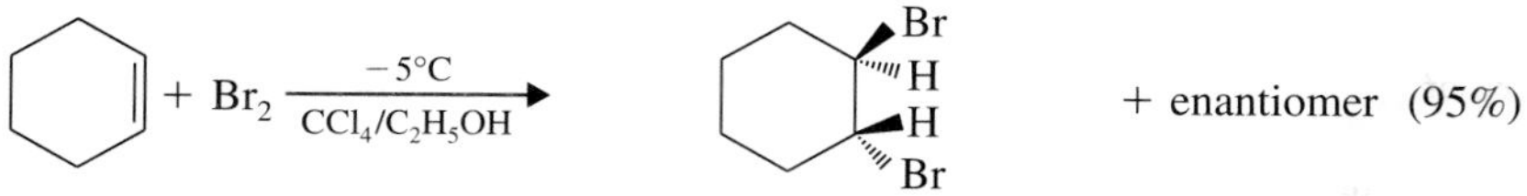

$$CH_3CH{=}CHCH_3 + Cl_2 \xrightarrow[-9°C]{} \begin{array}{c} CH_3CH{-}CHCH_3 \\ | \quad\quad | \\ Cl \quad\quad Cl \end{array} \ (100\%)$$

$$CH_3CH_2CH{=}CH_2 + Cl_2 \xrightarrow[-9°C]{} \begin{array}{c} CH_3CH_2CH{-}CH_2 \\ | \quad\quad | \\ Cl \quad\quad Cl \end{array} \ (97\%)$$

+ Br$_2$ $\xrightarrow[\text{CCl}_4/\text{C}_2\text{H}_5\text{OH}]{-5°C}$

+ enantiomer (95%)

trans-**1,2-Dibromocyclohexane**
(as a racemic form)

8.6A Mechanism of Halogen Addition

One mechanism that has been proposed for halogen addition is **an ionic mechanism.*** In the first step the exposed electrons of the π bond of the alkene attack the halogen in the following way:

A MECHANISM FOR BROMINE ADDITION

Step 1

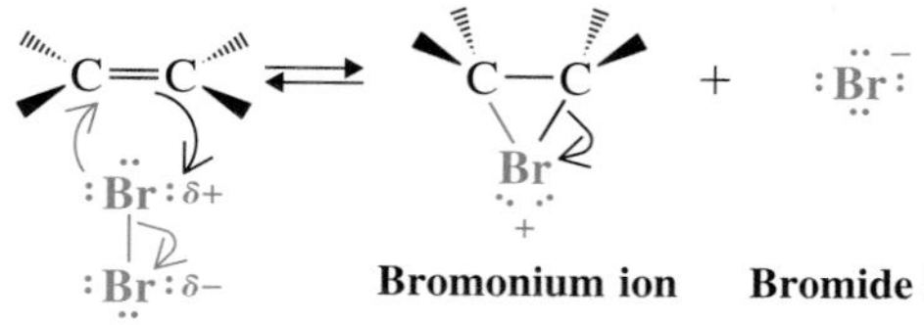

Bromonium ion   Bromide ion

A bromine molecule becomes polarized as it approaches the alkene. The polarized bromine molecule transfers a positive bromine atom (with six electrons in its valence shell) to the alkene resulting in the formation of a bromonium ion.

---

* There is evidence that in the absence of oxygen some reactions between alkenes and chlorine proceed through a radical mechanism. We shall not discuss this mechanism here, however.

Step 2

Bromonium ion   **Bromide ion**   *vic*-**Dibromide**

A bromide ion attacks one carbon (or the other) of the
bromonium ion in an S_N2 reaction, causing the
ring to open and resulting in the formation
of a *vic*-dibromide.

As the π electrons of the alkene approach the bromine molecule, the electrons of the
bromine–bromine bond drift in the direction of the bromine atom more distant from
the approaching alkene. The bromine molecule becomes *polarized* as a result. The
more distant bromine develops a partial negative charge; the nearer bromine becomes
partially positive. Polarization weakens the bromine–bromine bond, causing it to
break heterolytically. A bromide ion departs, and a *bromonium ion* forms. In the
bromonium ion a positively charged bromine atom is bonded to two carbon atoms by
two pairs of electrons: one pair from the π bond of the alkene, the other pair from the
bromine atom (one of its unshared pairs).

Although the three-membered ring of the bromonium ion is strained, the bromine
atom achieves stability by acquiring an octet of electrons.

In the second step, one of the bromide ions produced in step 1 attacks one of the
carbon atoms of the bromonium ion. The nucleophilic attack results in the formation of
a *vic*-dibromide by opening the three-membered ring.

This ring opening (see preceding scheme) is an S_N2 reaction. The bromide ion,
acting as a nucleophile, uses a pair of electrons to form a bond to one carbon atom of
the bromonium ion while the positive bromine of the bromonium ion acts as a leaving
group.

8.7 STEREOCHEMISTRY OF THE ADDITION OF HALOGENS TO ALKENES

The addition of bromine to cyclopentene provides additional evidence for bromonium
ion intermediates in bromine additions. When cyclopentene reacts with bromine in
carbon tetrachloride, **anti addition** occurs, and the products of the reaction are *trans*-
1,2-dibromocyclopentane enantiomers (as a racemate).

trans-1, 2-Dibromo-
cyclopentane

+ enantiomer

This anti addition of bromine to cyclopentene can be explained by a mechanism
that involves the formation of a bromonium ion in the first step. In the second step, a

bromide ion attacks a carbon atom of the ring from the side opposite that of the bromonium ion. The reaction is an S_N2 reaction. Nucleophilic attack by the bromide ion causes **inversion of the configuration of the carbon being attacked** (Section 6.10). This inversion of configuration at one carbon atom of the ring leads to the formation of one *trans*-1,2-dibromocyclopentane enantiomer. (The other enantiomer results from attack of the bromide ion at the other carbon of the bromonium ion.)

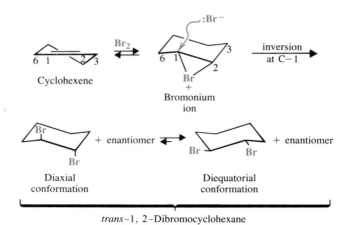

Bromonium   *trans*-**1,2-Dibromo-**
ion   **cyclopentane**

When cyclohexene undergoes addition of bromine, the product is a racemate of the *trans*-1,2-dibromocyclohexane enantiomers (Section 8.6). In this instance, too, *anti* addition results from the intermediate formation of a bromonium ion followed by S_N2 attack by a bromide ion. The reaction shown here illustrates the formation of one enantiomer. (The other enantiomer is formed when the bromide ion attacks the other carbon of the bromonium ion.)

Notice that the initial product of the reaction is the *diaxial conformer*. This rapidly converts into the diequatorial form, and when equilibrium is reached the diequatorial form predominates. We saw earlier (Section 7.12C) that when cyclohexane derivatives undergo elimination, the required conformation is the diaxial one. Here we find that when cyclohexene undergoes addition (the opposite of elimination), the initial product is also diaxial.

8.7A Stereospecific Reactions

The anti addition of a halogen to an alkene provides us with an example of what is called a **stereospecific reaction.**

A reaction is **stereospecific** when **a particular stereoisomeric form of the starting** *material reacts in such a way that it gives a specific stereoisomeric form of the product.* It does this because the reaction mechanism requires that the configurations of the atoms involved change in a characteristic way.

Consider the reactions of *cis-* and *trans*-2-butene with bromine shown below. When *trans*-2-butene adds bromine, the product is the meso compound, (2R,3S)-2,3-dibromobutane. When *cis*-2-butene adds bromine, the product is a *racemic form* of (2R,3R)-2,3-dibromobutane and (2S,3S)-2,3-dibromobutane.

Reaction 1

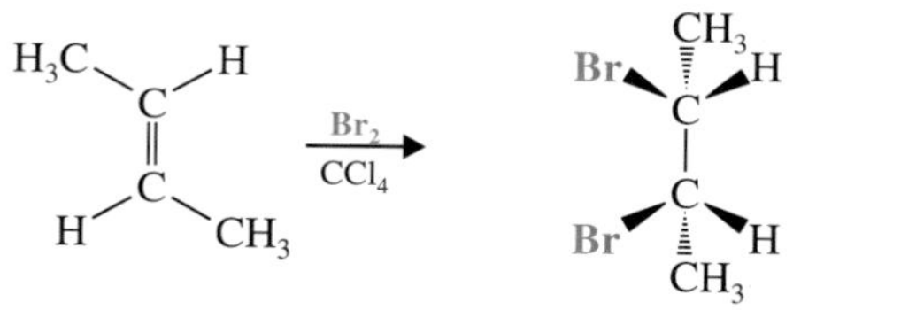

trans-**2-Butene**   **(2R,3S)-2,3-Dibromobutane**
                                  (a meso compound)

Reaction 2

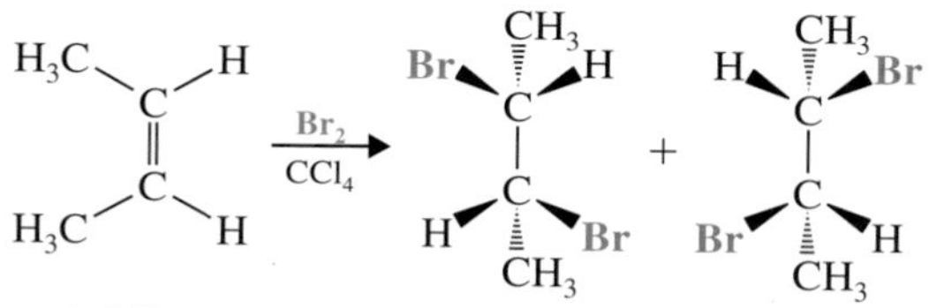

cis-**2-Butene**   **(2R,3R)**   **(2S,3S)**

The reactants *cis*-2-butene and *trans*-2-butene are stereoisomers; they are *dia-stereomers.* The product of reaction 1, (2R,3S)-2,3-dibromobutane, is a meso compound, and it is a stereoisomer of either of the products of reaction 2 (the enantiomeric 2,3-dibromobutanes). Thus, by definition, both reactions are stereospecific. One stereoisomeric form of the reactant (e.g., *trans*-2-butene) gives one product (the meso compound), whereas the other stereoisomeric form of the reactant (*cis*-2-butene) gives a stereoisomerically different product (the enantiomers).

We can better understand the results of these two reactions if we examine their mechanisms.

Figure 8.3 shows how *cis*-2-butene adds bromine to yield intermediate bromonium ions that are achiral. (The bromonium ion has a plane of symmetry. Can you find it?) These bromonium ions can then react with bromide ions by either path (a) or path (b). Reaction by path (a) yields one 2,3-dibromobutane enantiomer; reaction by path (b) yields the other enantiomer. Reaction occurs at the same rate by either path; therefore the two enantiomers are produced in equal amounts (as a racemic form).

Figure 8.4 shows how *trans*-2-butene reacts at the bottom face to yield an intermediate bromonium ion that is chiral. (Reaction at the other face would produce the enantiomeric bromonium ion.) Reaction of this chiral bromonium ion (or its enantiomer) with a bromide ion either by path (a) or by path (b) yields the same compound, the *meso*-2,3-dibromobutane.

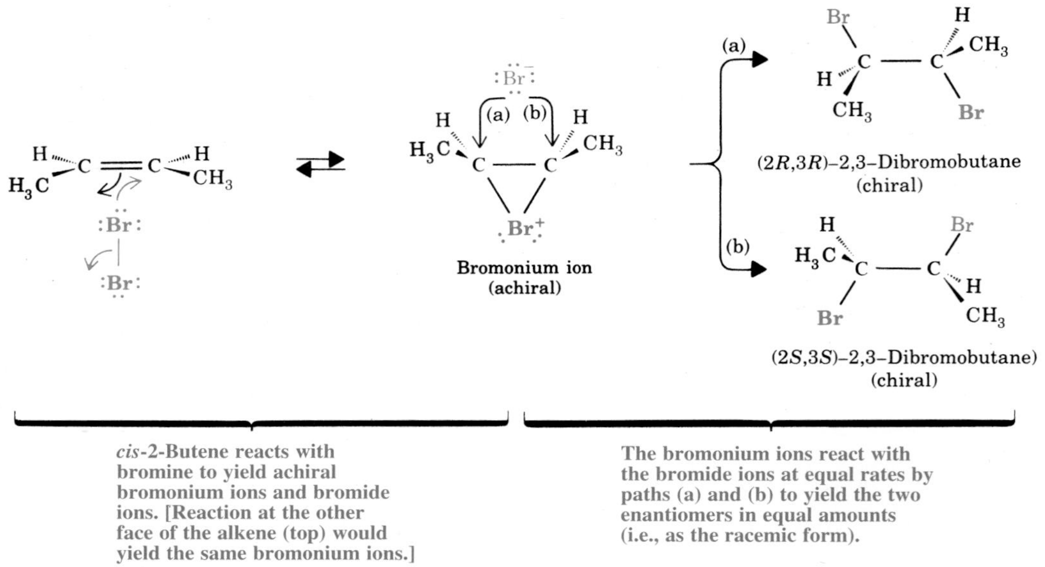
(a)
(b)
(2R,3R)-2,3-Dibromobutane
(chiral)
(2S,3S)-2,3-Dibromobutane)
(chiral)
Bromonium ion
(achiral)
cis-2-Butene reacts with bromine to yield achiral bromonium ions and bromide ions. [Reaction at the other face of the alkene (top) would yield the same bromonium ions.]
The bromonium ions react with the bromide ions at equal rates by paths (a) and (b) to yield the two enantiomers in equal amounts (i.e., as the racemic form).

FIGURE 8.3 A mechanism showing how *cis*-2-butene reacts with bromine to yield the enantiomeric 2,3-dibromobutanes.

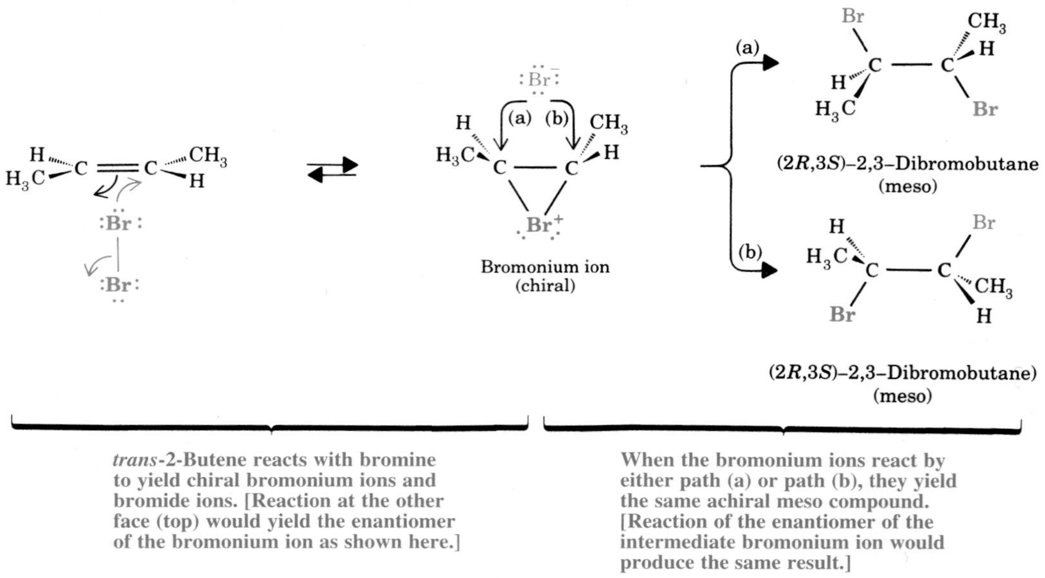
(a)
(b)
(2R,3S)-2,3-Dibromobutane
(meso)
(2R,3S)-2,3-Dibromobutane)
(meso)
Bromonium ion
(chiral)
trans-2-Butene reacts with bromine to yield chiral bromonium ions and bromide ions. [Reaction at the other face (top) would yield the enantiomer of the bromonium ion as shown here.]
When the bromonium ions react by either path (a) or path (b), they yield the same achiral meso compound. [Reaction of the enantiomer of the intermediate bromonium ion would produce the same result.]

FIGURE 8.4 A mechanism showing how *trans*-2-butene reacts with bromine to yield *meso*-2,3-dibromobutane.

PROBLEM 8.9

In Section 8.7 you studied a mechanism for the formation of one enantiomer of *trans*-1,2-dibromocyclopentane when bromine adds to cyclopentene. Now write a mechanism showing how the other enantiomer forms.

8.8 HALOHYDRIN FORMATION

If the halogenation of an alkene is carried out in aqueous solution (rather than in carbon tetrachloride), the major product of the overall reaction is not a *vic*-dihalide; instead, it is a **halo alcohol** called a **halohydrin.** In this case, molecules of the solvent become reactants, too.

X = Cl or Br   **Halohydrin**   *vic*-**Dihalide**

Halohydrin formation can be explained by the following mechanism:

A MECHANISM FOR HALOHYDRIN FORMATION

Step 1

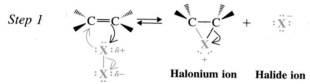

Halonium ion   **Halide ion**

This step is the same as for halogen addition to an alkene (See Section 8.6).

Steps 2 and 3

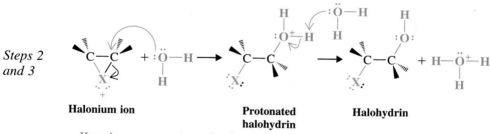

Halonium ion   **Protonated halohydrin**   **Halohydrin**

Here, however, a water molecule acts as the nucleophile and attacks a carbon of the ring, causing the formation of a protonated halohydrin.

The protonated halohydrin loses a proton (it is transferred to a molecule of water). This step produces the halohydrin and a hydronium ion.

The first step is the same as that for halogen addition. In the second step, however, the two mechanisms differ. In halohydrin formation, water acts as the nucleophile and attacks one carbon atom of the halonium ion. The three-membered ring opens, and a protonated halohydrin is produced. Loss of a proton then leads to the formation of the halohydrin itself.

Water, because of its unshared electron pairs, acts as a nucleophile in this and in many other reactions. In this instance water molecules far outnumber halide ions because water is the solvent for the reactants. This accounts for the halohydrin being the major product.

PROBLEM 8.10

Outline a mechanism that accounts for the formation of *trans*-2-bromocyclopentanol and its enantiomer, when cyclopentene is treated with an aqueous solution of bromine.

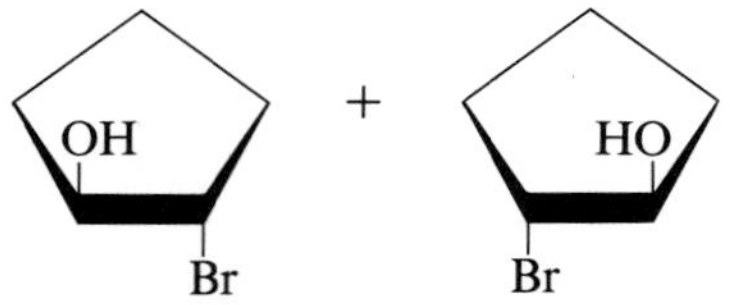

***trans*-2-Bromocyclopentanol enantiomers**

If the alkene is unsymmetrical, the halogen ends up on the carbon atom with the greater number of hydrogen atoms. Bonding in the intermediate bromonium ion (next page) is apparently *unsymmetrical.* The more highly substituted carbon atom bears the greater positive charge because it resembles the more stable carbocation. Consequently, water attacks this carbon atom preferentially. The greater positive charge on the tertiary carbon permits a pathway with a lower free energy of activation even though attack at the primary carbon atom is less hindered.

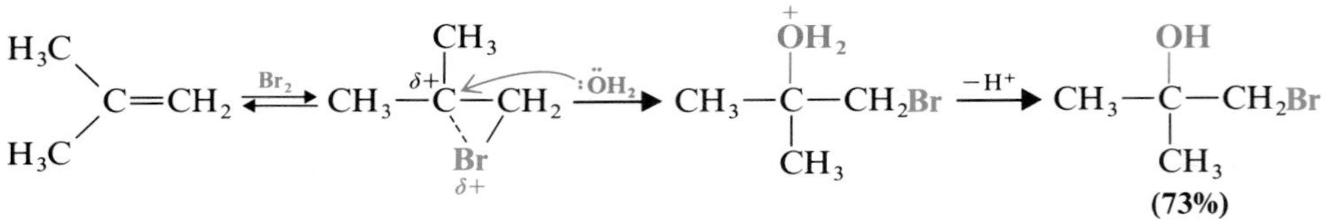

(73%)

PROBLEM 8.11

When ethene gas is passed into an aqueous solution containing bromine and sodium chloride, the products of the reaction are $BrCH_2CH_2Br$, $BrCH_2CH_2OH$, *and* $BrCH_2CH_2Cl$. Write mechanisms showing how each product is formed.

8.9 OXIDATIONS OF ALKENES: SYN HYDROXYLATION

Alkenes undergo a number of reactions in which the carbon–carbon double bond is oxidized. Potassium permanganate or osmium tetroxide, for example, can be used to oxidize alkenes to **1,2-diols** called **glycols.**

$$CH_2{=}CH_2 + KMnO_4 \xrightarrow[OH^-]{\text{cold}} \underset{\text{OH}}{\underset{|}{H_2C}}{-}\underset{\text{OH}}{\underset{|}{CH_2}}$$

Ethene → **1,2-Ethanediol (ethylene glycol)**

$$CH_3CH{=}CH_2 \xrightarrow[\text{(2) } Na_2SO_3/H_2O \text{ or } NaHSO_3/H_2O]{\text{(1) } OsO_4\text{, pyridine}} CH_3\underset{\text{OH}}{\underset{|}{CH}}{-}\underset{\text{OH}}{\underset{|}{CH_2}}$$

Propene → **1,2-Propanediol (propylene glycol)**

8.9A Syn Hydroxylation of Alkenes

The mechanisms for the formation of glycols by permanganate ion and osmium tetroxide oxidations first involve the formation of cyclic intermediates. Then in several steps cleavage at the oxygen–metal bond takes place (at the dashed lines in the following reactions) ultimately producing the glycol and MnO_2 or osmium metal.

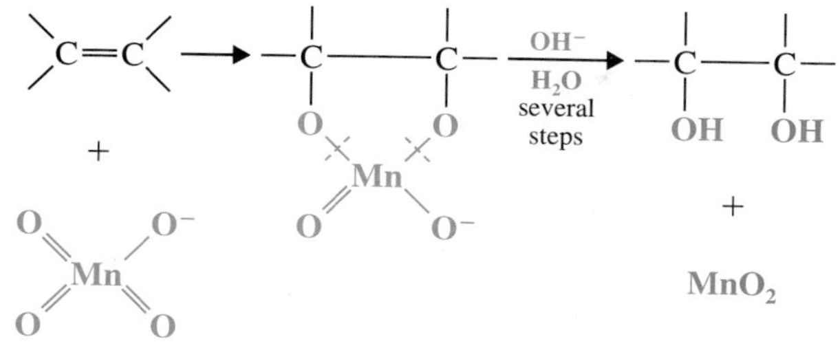

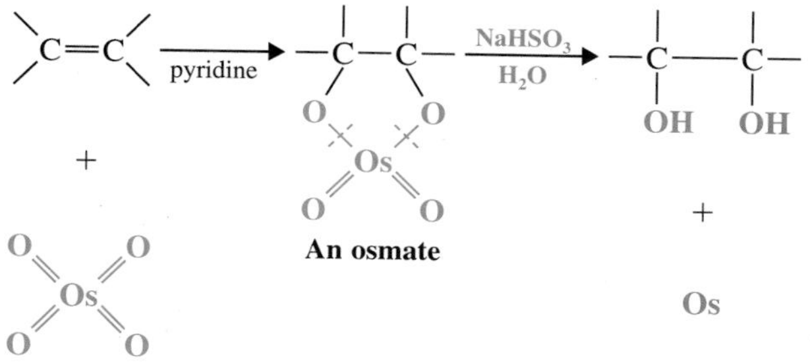

An osmate

The course of these reactions is **syn hydroxylation.** This can readily be seen when cyclopentene reacts with cold dilute potassium permanganate (in base) or with osmium tetroxide (followed by treatment with $NaHSO_3$ or Na_2SO_3). The product in either case is *cis*-1,2-cyclopentanediol. (*cis*-1,2-Cyclopentanediol is a meso compound.)

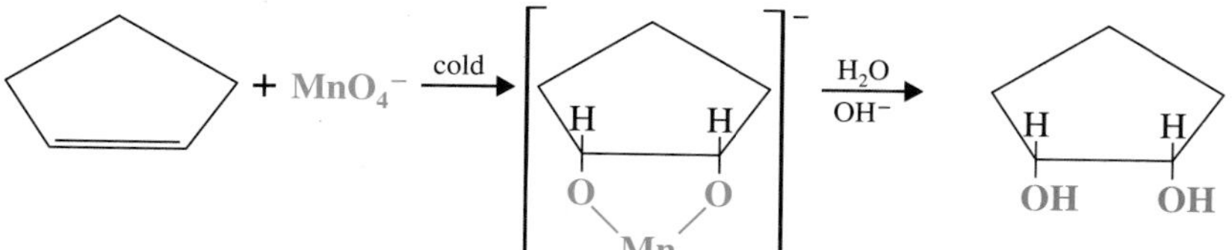

cis-**1,2-Cyclopentanediol**
(a meso compound)

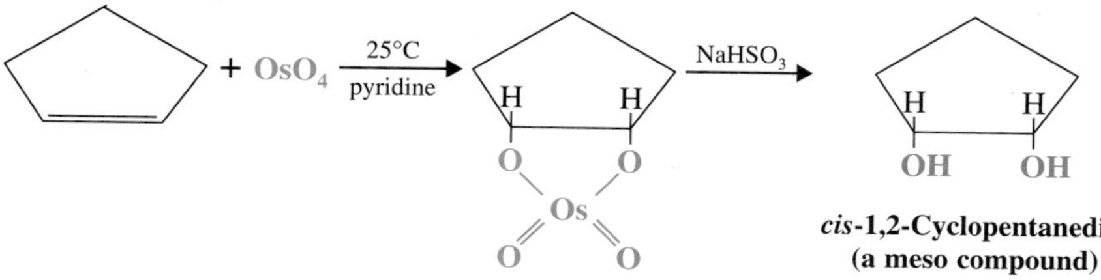

cis-**1,2-Cyclopentanediol**
(a meso compound)

Of the two reagents used for syn hydroxylation, osmium tetroxide gives the higher yields. However, osmium tetroxide is highly toxic and is very expensive. For this reason, methods have been developed that permit OsO_4 to be used catalytically in conjunction with a cooxidant. (We shall not discuss these methods further here, however.) Potassium permanganate is a very powerful oxidizing agent and, as we shall see in Section 8.10, *it is easily capable of causing further oxidation of the glycol.* Limiting the reaction to hydroxylation alone is often difficult but is usually attempted by using cold, dilute, and basic solutions of potassium permanganate. Even so, yields are sometimes very low.

PROBLEM 8.12 _____

Starting with an alkene, outline syntheses of each of the following:

(a)   (b)   (c)

(+ enantiomer)   (+ enantiomer)

PROBLEM 8.13 _____

Explain the following facts: (a) Treating (*Z*)-2-butene with OsO_4 in pyridine and then $NaHSO_3$ in water gives a diol that is optically inactive and cannot be resolved.   (b) Treating (*E*)-2-butene with OsO_4 and then $NaHSO_3$ gives a diol that is optically inactive but can be resolved into enantiomers.

8.10 OXIDATIVE CLEAVAGE OF ALKENES

Alkenes are oxidatively cleaved to salts of carboxylic acids *by hot permanganate solutions.* We can illustrate this reaction with the oxidative cleavage of either *cis-* or *trans*-2-butene to two molar equivalents of acetate ion. The intermediate in this reaction may be a glycol that is oxidized further with cleavage at the carbon–carbon bond.

$$CH_3CH{=}CHCH_3 \xrightarrow[\text{heat}]{\text{KMnO}_4,\ \text{OH}^-} 2\ CH_3C\overset{\text{O}}{\underset{\text{O}^-}{\diagdown}} \xrightarrow{\text{H}^+} 2\ CH_3C\overset{\text{O}}{\underset{\text{OH}}{\diagdown}}$$

(cis or trans)   **Acetate ion**   **Acetic acid**

Acidification of the mixture, after the oxidation is complete, produces 2 mol of acetic acid for each mole of 2-butene.

The terminal CH_2 group of a 1-alkene is completely oxidized to carbon dioxide and water by hot permanganate. A disubstituted carbon atom of a double bond becomes the $\diagup^{\diagdown}C{=}O$ group of a ketone (Section 2.13).

$$\underset{\text{CH}_3\text{CH}_2\text{C}{=}\text{CH}_2}{\overset{\text{CH}_3}{|}} \xrightarrow[\text{(2) H}^+]{\substack{\text{(1) KMnO}_4,\ \text{OH}^- \\ \text{heat}}} \underset{\text{CH}_3\text{CH}_2\text{C}{=}\text{O}}{\overset{\text{CH}_3}{|}} + \text{O}{=}\text{C}{=}\text{O} + \text{H}_2\text{O}$$

The oxidative cleavage of alkenes has frequently been used to prove the location of the double bond in an alkene chain or ring. The reasoning process requires us to think backward much as we do with retrosynthetic analysis. Here we are required to work backward from the products to the reactant that might have led to those products. We can see how this might be done with the following examples:

EXAMPLE A

An unknown alkene with the formula C_8H_{16} was found, on oxidation with hot basic permanganate, to yield a three-carbon carboxylic acid (propanoic acid) and a five-carbon carboxylic acid (pentanoic acid).

$$C_8H_{16} \xrightarrow[\text{(2) H}^+]{\text{(1) KMnO}_4,\ \text{H}_2\text{O,} \atop \text{OH}^-,\ \text{heat}} \underset{\textbf{Propanoic acid}}{CH_3CH_2\overset{\displaystyle O}{\overset{\|}{C}}-OH} + \underset{\textbf{Pentanoic acid}}{HO-\overset{\displaystyle O}{\overset{\|}{C}}CH_2CH_2CH_2CH_3}$$

Oxidative cleavage must have occurred as follows, and the unknown alkene must have been *cis*- or *trans*-3-octene.

Cleavage occurs here

$$\underset{\substack{\textbf{Unknown alkene} \\ \textbf{(either }\textit{cis-}\textbf{ or }\textit{trans-}\textbf{3-octene)}}}{CH_3CH_2CH\!=\!CHCH_2CH_2CH_2CH_3} \xrightarrow[\text{(2) H}^+]{\text{(1) KMnO}_4,\ \text{H}_2\text{O,} \atop \text{OH}^-,\ \text{heat}} CH_3CH_2\overset{\displaystyle O}{\overset{\|}{C}}-OH + HO-\overset{\displaystyle O}{\overset{\|}{C}}CH_2CH_2CH_2CH_3$$

EXAMPLE B

An unknown alkene with the formula C_7H_{12} undergoes oxidation by hot basic $KMnO_4$ to yield, after acidification, *only one product:*

$$C_7H_{12} \xrightarrow[\text{(2) H}^+]{\text{(1) KMnO}_4,\ \text{H}_2\text{O,} \atop \text{OH}^-,\ \text{heat}} CH_3\overset{\displaystyle O}{\overset{\|}{C}}CH_2CH_2CH_2CH_2\overset{\displaystyle O}{\overset{\|}{C}}-OH$$

Since the product contains the same number of carbon atoms as the reactant, the only reasonable explanation is that the reactant has a double bond contained in a ring. Oxidative cleavage of the double bond opens the ring.

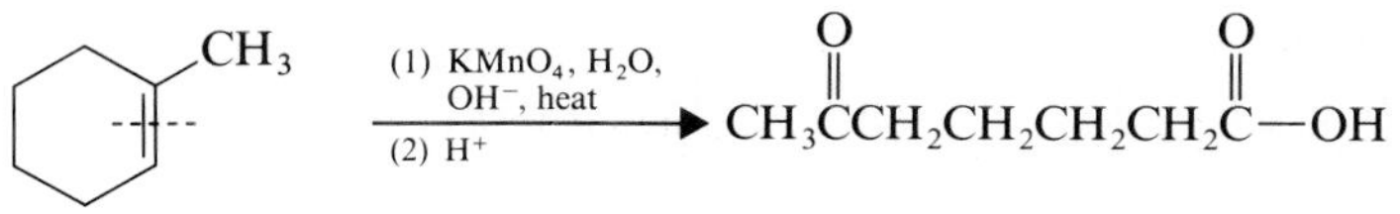

$$\underset{\substack{\textbf{Unknown alkene} \\ \textbf{(1-methylcyclohexene)}}}{}\ \xrightarrow[\text{(2) H}^+]{\text{(1) KMnO}_4,\ \text{H}_2\text{O,} \atop \text{OH}^-,\ \text{heat}}\ CH_3\overset{\displaystyle O}{\overset{\|}{C}}CH_2CH_2CH_2CH_2\overset{\displaystyle O}{\overset{\|}{C}}-OH$$

8.10A Ozonolysis of Alkenes

A more widely used method for locating the double bond of an alkene involves the use of ozone (O_3). Ozone reacts vigorously with alkenes to form unstable compounds called *initial ozonides,* which rearrange spontaneously (and often noisily) to form compounds known as **ozonides.**

A MECHANISM FOR OZONIDE FORMATION

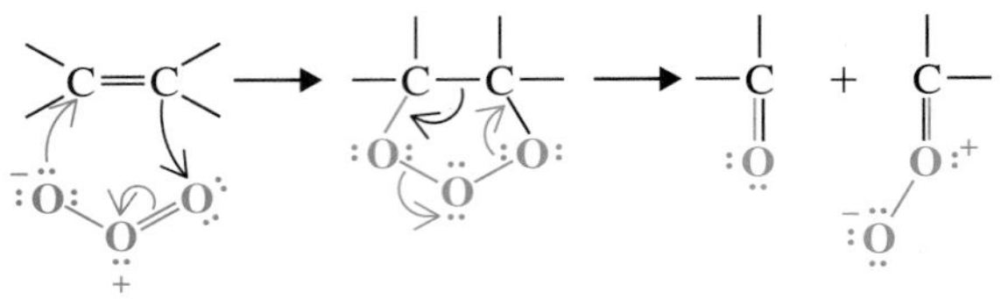

Initial ozonide

Ozone adds to the alkene   The initial ozonide fragments.
to form an initial ozonide.

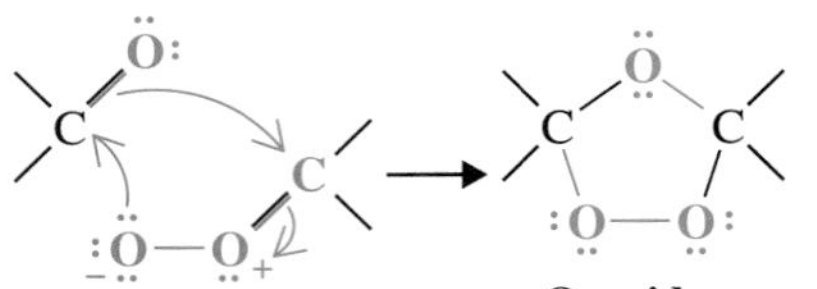

Ozonide
The fragments recombine to form the ozonide.

This rearrangement is thought to occur through dissociation of the initial ozonide into reactive fragments that recombine to yield the ozonide.

Ozonides, themselves, are very unstable compounds and low molecular weight ozonides often explode violently. Because of this property they are not usually isolated but are reduced directly by treatment with zinc and water. The reduction produces carbonyl compounds (either aldehydes or ketones, see Section 2.13) that can be safely isolated and identified.

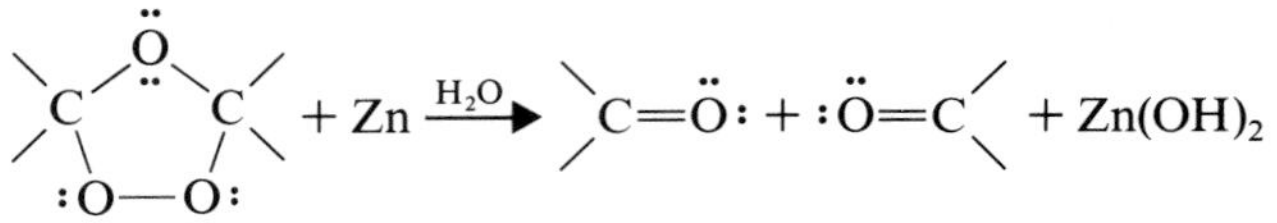

$$\overset{\ddots}{\underset{:O—O:}{C—C}} + Zn \xrightarrow{H_2O} \ \ \ C=\ddot{O}: + :\ddot{O}=C \ \ \ + Zn(OH)_2$$

Ozonide   **Aldehydes and/or
ketones**

The overall process of ozonolysis followed by reduction with zinc and water amounts to a disconnection of the carbon–carbon double bond in the following fashion.

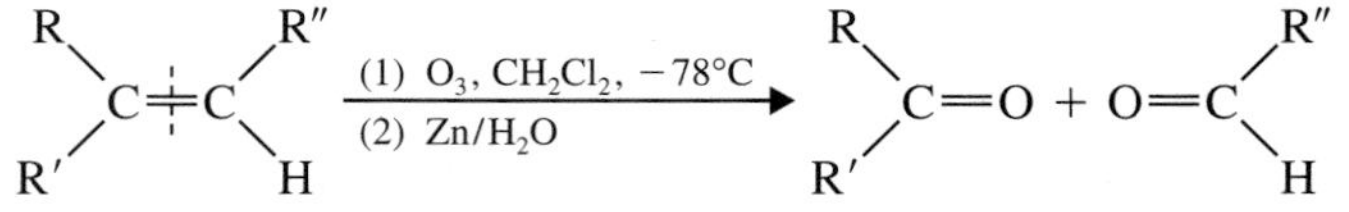

$$\underset{R'}{\overset{R}{>}}C \! \mid \! C\underset{H}{\overset{R''}{<}} \xrightarrow[\text{(2) Zn/H}_2\text{O}]{\text{(1) O}_3, \text{CH}_2\text{Cl}_2, -78°C} \underset{R'}{\overset{R}{>}}C=O + O=C\underset{H}{\overset{R''}{<}}$$

Notice that a —H attached to the double bond is not oxidized to —OH as it is with permanganate oxidations. Consider the following examples as illustrations of the overall process.

$$CH_3C{=}CHCH_3 \xrightarrow[\text{(2) Zn/H}_2\text{O}]{\text{(1) O}_3,\ \text{CH}_2\text{Cl}_2,\ -78°\text{C}} CH_3C{=}O + CH_3CH$$

2-Methyl-2-butene   Acetone   Acetaldehyde

$$CH_3CH{-}CH{=}CH_2 \xrightarrow[\text{(2) Zn/H}_2\text{O}]{\text{(1) O}_3,\ \text{CH}_2\text{Cl}_2,\ -78°\text{C}} CH_3CH{-}CH + HCH$$

3-Methyl-1-butene   Isobutyraldehyde   Formaldehyde

PROBLEM 8.14

Write the structures of the alkenes that would produce the following products when treated with ozone and then with zinc and water.

(a) CH_3COCH_3 and $CH_3CH(CH_3)CHO$

(b) CH_3CH_2CHO only (2 mol are produced from 1 mol of alkene)

(c) ⬠=O   and   HCHO

8.11 SUMMARY OF ADDITION REACTIONS OF ALKENES

The stereochemistry and regiospecificity (where appropriate) of the addition reactions of alkenes that we have studied thus far are summarized in Figure 8.5. We have used 1-methylcyclopentene as the starting alkene.

8.12 ADDITION OF BROMINE AND CHLORINE TO ALKYNES

Alkynes show the same kind of reactions toward chlorine and bromine that alkenes do: *They react by addition.* However, with alkynes the addition may occur once or twice, depending on the number of molar equivalents of halogen we employ.

$$-C{\equiv}C- \xrightarrow[\text{CCl}_4]{\text{Br}_2} \overset{\text{Br}}{\underset{\text{Br}}{\,}}C{=}C \xrightarrow[\text{CCl}_4]{\text{Br}_2} \overset{\text{Br}\ \ \text{Br}}{\underset{\text{Br}\ \ \text{Br}}{-C{-}C-}}$$

Dibromoalkene   Tetrabromoalkane

$$-C{\equiv}C- \xrightarrow[\text{CCl}_4]{\text{Cl}_2} \overset{\text{Cl}}{\underset{\text{Cl}}{\,}}C{=}C \xrightarrow[\text{CCl}_4]{\text{Cl}_2} \overset{\text{Cl}\ \ \text{Cl}}{\underset{\text{Cl}\ \ \text{Cl}}{-C{-}C-}}$$

Dichloroalkene   Tetrachloroalkane

It is usually possible to prepare a dihaloalkene by simply adding one molar equivalent of the halogen.

$$CH_3CH_2CH_2CH_2C{\equiv}CCH_2OH \xrightarrow[\substack{\text{CCl}_4 \\ 0°\text{C}}]{\text{Br}_2\ (1\ \text{mol})} CH_3CH_2CH_2CH_2CBr{=}CBrCH_2OH$$

(80%)

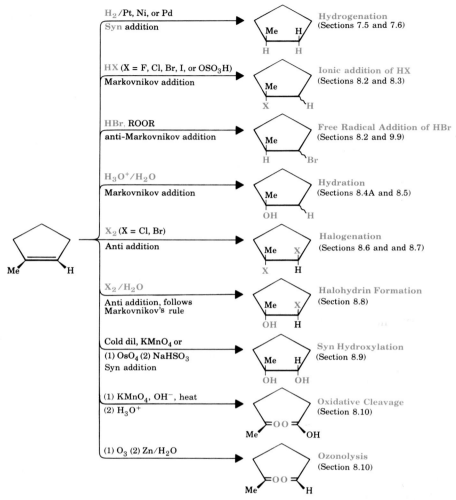

FIGURE 8.5 A summary of addition reactions of alkenes with 1-methylcyclopentene as the organic substrate. A bond designated ∿ means that the stereochemistry of the group is unspecified. For brevity we have shown the structure of only one enantiomer of the product, even though racemic forms would be produced in all instances in which the product is chiral.

Most additions of chlorine and bromine to alkynes are anti additions and yield *trans*-dihaloalkenes. Addition of bromine to acetylenedicarboxylic acid, for example, gives the trans isomer in 70% yield.

$$HO_2C-C\equiv C-CO_2H \xrightarrow{Br_2} \underset{\substack{\text{Acetylenedicarboxylic}\\ \text{acid}}}{} \underset{(70\%)}{\begin{array}{c} HO_2C \\ \diagdown \\ Br \end{array} C=C \begin{array}{c} Br \\ \diagup \\ CO_2H \end{array}}$$

Acetylenedicarboxylic acid (70%)

PROBLEM 8.15

Alkenes are more reactive than alkynes toward addition of electrophilic reagents (i.e., Br_2, Cl_2, or HCl). Yet when alkynes are treated with one molar equivalent of these same electrophilic reagents, it is easy to stop the addition at the "alkene stage." This appears to be a paradox and yet it is not. Explain.

8.13 ADDITION OF HYDROGEN HALIDES TO ALKYNES

Alkynes react with hydrogen chloride and hydrogen bromide to form haloalkenes or geminal dihalides depending, once again, on whether one or two molar equivalents of the hydrogen halide are used. **Both additions are regioselective and follow Markovnikov's rule:**

$$-C\equiv C- \xrightarrow{HX} \begin{array}{c} H \\ \diagdown \\ C=C \\ \diagup \\ X \end{array} \xrightarrow{HX} \begin{array}{c} H \quad X \\ | \quad | \\ -C-C- \\ | \quad | \\ H \quad X \end{array}$$

Haloalkene   ***gem*-Dihalide**

The hydrogen atom of the hydrogen halide becomes attached to the carbon atom that has the greater number of hydrogen atoms. 1-Hexyne, for example, reacts slowly with one molar equivalent of hydrogen bromide to yield 2-bromohexene and with two molar equivalents to yield 2,2-dibromohexane.

$$C_4H_9C\equiv CH \xrightarrow{HBr} \underset{\underset{Br}{|}}{C_4H_9-C=CH_2} \xrightarrow{HBr} \underset{\underset{Br}{|}}{\overset{\overset{Br}{|}}{C_4H_9-C-CH_3}}$$

2-Bromohexene   **2,2-Dibromohexane**

The addition of HBr to an alkyne can be facilitated by using acetyl bromide (CH$_3$COBr) and alumina instead of aqueous HBr. Acetyl bromide acts as an HBr precursor by reacting with the alumina to generate HBr. The alumina by its presence (Section 8.2) increases the rate of reaction. For example, 1-heptyne can be converted to 2-bromoheptene in good yield using this method.

$$C_5H_{11}C\equiv CH \xrightarrow[\substack{CH_3COBr/alumina \\ CH_2Cl_2}]{\text{"HBr"}} \underset{C_5H_{11}}{\overset{Br}{\diagdown}}C=CH_2$$

(82%)

Anti-Markovnikov addition of hydrogen bromide to alkynes occurs when peroxides are present in the reaction mixture. These reactions take place through a free radical mechanism (Section 9.9).

$$CH_3CH_2CH_2CH_2C\equiv CH \xrightarrow[peroxides]{HBr} CH_3CH_2CH_2CH_2CH=CHBr$$

(74%)

The addition of water to alkynes, a method for synthesizing ketones, is discussed in Section 16.5.

8.14 OXIDATIVE CLEAVAGE OF ALKYNES

Treating alkynes with ozone or with basic potassium permanganate leads to cleavage at the carbon–carbon triple bond. The products are carboxylic acids.

$$R-C\equiv C-R' \xrightarrow[\text{(2) } H_2O]{\text{(1) } O_3} RCO_2H + R'CO_2H$$

or

$$R-C\equiv C-R' \xrightarrow[\text{(2) } H^+]{\text{(1) } KMnO_4, OH^-} RCO_2H + R'CO_2H$$

PROBLEM 8.16

Three different alkynes, **A**, **B**, and **C**, have the molecular formula C_8H_{12}. Treating either compound **A** or **B** with excess hydrogen in the presence of a metal catalyst leads to the formation of octane. Similar treatment of compound **C** leads to a product with the formula C_8H_{16}. Treating alkyne **A** with ozone and then water furnishes a single product, $CH_3CH_2CH_2CO_2H$. Treating alkyne **C** with ozone and then water produces a single product, $HO_2C(CH_2)_6CO_2H$. Treating alkyne **B** with $Ag(NH_3)_2OH$ produces a silver-containing precipitate. Treating this precipitate with sodium cyanide converts it back to **B**. What are compounds **A**, **B**, and **C**?

8.15 SUMMARY OF ADDITION REACTIONS OF ALKYNES

Figure 8.6 summarizes the addition reactions of alkynes.

8.16 SYNTHETIC STRATEGIES REVISITED

In planning a synthesis we often have to consider four interrelated aspects:

1. Construction of the carbon skeleton
2. Functional group interconversions
3. Control of regiochemistry
4. Control of stereochemistry

You have had some experience in the first two aspects of synthetic strategy in earlier sections. In Section 4.16 you were introduced to the ideas of *retrosynthetic analysis* and how this kind of thinking could be applied to the construction of carbon skeletons of alkanes and cycloalkanes. In Section 6.16 you learned the meaning of a *functional group interconversion* and learned how nucleophilic substitution reactions could be used for this purpose. In other sections, perhaps without realizing it, you have begun adding to your basic store of methods for construction of carbon skeletons, and for making functional group interconversions. This is the time to begin keeping a notebook that lists all the reactions that you have learned, noting especially their applications to synthesis. This notebook will become your **Tool Kit for Organic Synthesis.**

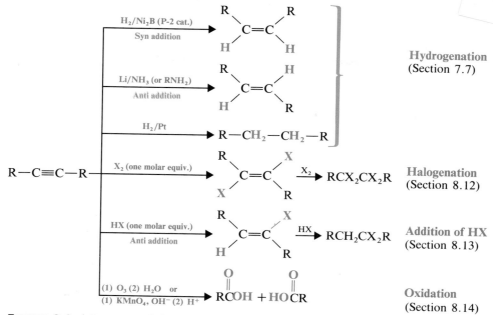

FIGURE 8.6 A summary of the addition reactions of alkynes.

Now is the time to look at some new examples and to see how we integrate all four aspects of synthesis into our planning.

Consider a problem in which we are asked to outline a synthesis of 2-bromobutane from compounds of two carbon atoms or fewer. This synthesis, as we shall see, involves construction of the carbon skeleton, functional group interconversion, and control of regiochemistry.

We begin by thinking backward. One way to make 2-bromobutane is by addition of hydrogen bromide to 1-butene. The regiochemistry of this functional group interconversion must be Markovnikov addition.

Analysis

$$CH_3CH_2CHCH_3 \Longrightarrow CH_3CH_2CH{=}CH_2 + H{-}Br \quad \substack{\textbf{Markovnikov} \\ \textbf{addition}}$$
$$\underset{Br}{|}$$

Synthesis

$$CH_3CH_2CH{=}CH_2 + HBr \xrightarrow[\text{peroxides}]{\text{no}} CH_3CH_2CHCH_3$$
$$\underset{Br}{|}$$

Remember: The open arrow is a symbol used to show a retrosynthetic process that relates the target molecule to its precursors.

Target molecule $\Longrightarrow$ precursors

Next we try to think of a way to synthesize 1-butene, keeping in mind that we have to construct the carbon skeleton from compounds with two carbon atoms or fewer. One retrosynthetic route might be

Analysis

$$CH_3CH_2CH{=}CH_2 \Longrightarrow CH_3CH_2C{\equiv}CH + H_2$$

$$CH_3CH_2C{\equiv}CH \Longrightarrow CH_3CH_2Br + NaC{\equiv}CH$$

$$NaC{\equiv}CH \Longrightarrow HC{\equiv}CH + NaNH_2$$

Synthesis

$$HC{\equiv}C{-}H + Na^+\ ^-NH_2 \xrightarrow[-33°C]{\text{liq. NH}_3} HC{\equiv}C{:}^-Na^+$$

$$CH_3CH_2{-}Br + Na^+\ ^-{:}C{\equiv}CH \xrightarrow[-33°C]{\text{liq. NH}_3} CH_3CH_2C{\equiv}CH$$

$$CH_3CH_2C{\equiv}CH + H_2 \xrightarrow{\text{Ni}_2B(P-2)} CH_3CH_2CH{=}CH_2$$

One approach to retrosynthetic analysis is to consider a retrosynthetic step as a "disconnection" of one of the bonds (Section 4.16).* For example, an important step in the synthesis that we have just given is the one in which a new carbon–carbon bond is formed. Retrosynthetically, it can be shown in the following way:

$$CH_3CH_2{-}C{\equiv}CH \Longrightarrow CH_3\overset{+}{C}H_2 + ^-{:}C{\equiv}CH$$

The fragments of this disconnection are an ethyl cation and an ethynide anion. These fragments are called **synthons.** Seeing these synthons may help us to reason as follows: "We could, in theory, synthesize a molecule of 1-butyne by combining an ethyl cation with an ethynide anion." We know, however, that bottles of carbocations and carbanions are not to be found on our laboratory shelves. What we need are the **synthetic equivalents** of these synthons. The synthetic equivalent of an ethynide ion is sodium ethynide, because sodium ethynide contains an ethynide ion (and a sodium cation). The synthetic equivalent of an ethyl cation is ethyl bromide. To understand how this is true, we reason as follows: If ethyl bromide were to react by an S_N1 reaction, it would produce an ethyl cation and a bromide ion. However, we know that being a primary halide, ethyl bromide is unlikely to react by an S_N1 reaction. Ethyl bromide, however, will react readily with a strong nucleophile such as sodium ethynide by an S_N2 reaction, and when it reacts, the product that is obtained is the same as the product that would have been obtained from the reaction of an ethyl cation with sodium ethynide. Thus, ethyl bromide, in this reaction, functions as the synthetic equivalent of an ethyl cation.

Consider another example, a synthesis that requires stereochemical control: the synthesis of the enantiomeric 2,3-butanediols, (2R,3R)-2,3-butanediol and (2S,3S)-2,3-butanediol, from compounds of two carbon atoms or fewer.

Here (recall Problem 8.13) we see that a possible final step to the enantiomers is syn hydroxylation of *trans*-2-butene.

---

* For an excellent detailed treatment of this approach you may want to read: Stuart Warren, *Organic Synthesis, The Disconnection Approach,* Wiley, New York, 1982, and Stuart Warren, *Workbook for Organic Synthesis, the Disconnection Approach,* Wiley, New York, 1982.

Analysis

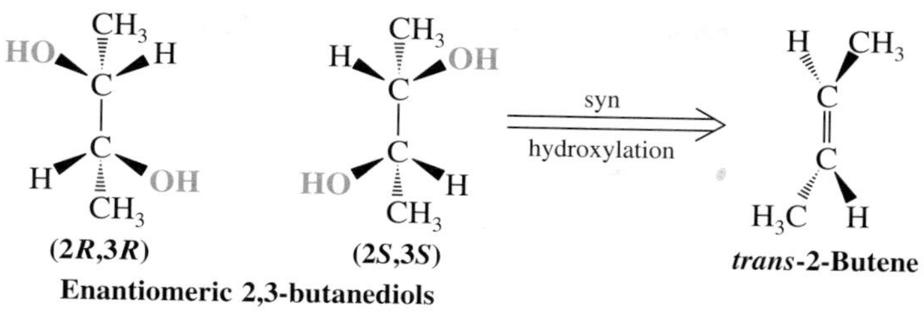

Enantiomeric 2,3-butanediols   *trans*-2-Butene

Synthesis

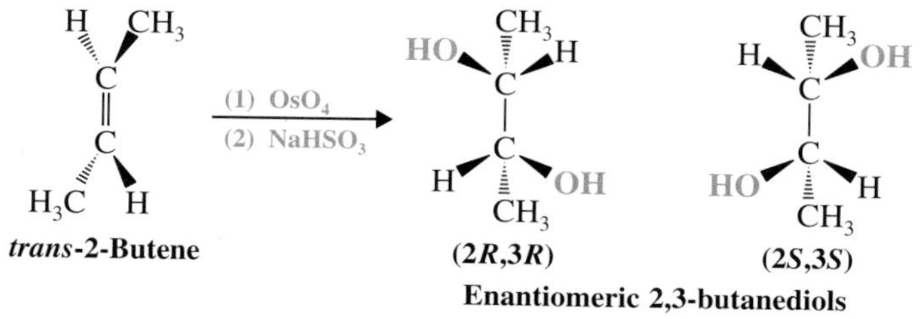

trans-2-Butene   (2R,3R)   (2S,3S)

Enantiomeric 2,3-butanediols

This reaction is stereospecific and produces the desired enantiomeric 2,3-butanediols as a racemic form.

Next, a synthesis of *trans*-2-butene can be accomplished by treating 2-butyne with lithium in liquid ammonia. This is a reaction that is also stereospecific and the addition of hydrogen is anti giving us the trans product that we need.

Analysis

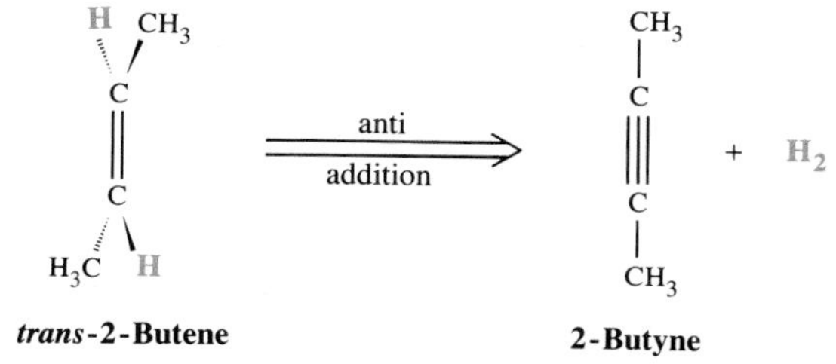

trans-2-Butene   2-Butyne

Synthesis

2-Butyne   *trans*-2-Butene

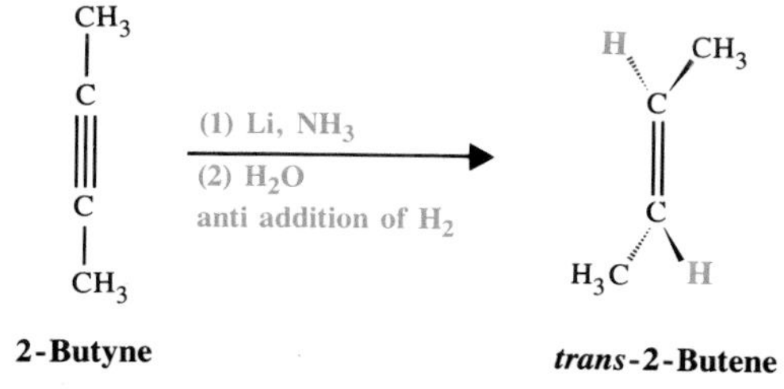

Finally, we can synthesize 2-butyne from propyne by first converting it to sodium propynide and then alkylating sodium propynide with methyl iodide:

Analysis

$$CH_3-C\equiv C\rlap{/}{-}CH_3 \Longrightarrow CH_3-C\equiv C\!:^-Na^+ + CH_3-I$$

$$CH_3-C\equiv C\!:^-Na^+ \Longrightarrow CH_3-C\equiv C-H + NaNH_2$$

Synthesis

$$CH_3-C\equiv C-H \xrightarrow[\text{(2) }CH_3I]{\text{(1) }NaNH_2/\text{liq. }NH_3} CH_3-C\equiv C-CH_3$$

Finally, we can synthesize propyne from ethyne:

Analysis

$$H-C\equiv C\rlap{/}{-}CH_3 \Longrightarrow H-C\equiv C\!:^-Na^+ + CH_3-I$$

Synthesis

$$H-C\equiv C-H \xrightarrow[\text{(2) }CH_3I]{\text{(1) }NaNH_2/\text{liq. }NH_3} CH_3-C\equiv C-H$$

SAMPLE PROBLEM

Illustrating a Stereospecific Multistep Synthesis

Starting with compounds of two carbon atoms or fewer, outline a stereospecific synthesis of *meso*-3,4-dibromohexane.

ANSWER:

We begin by working backward from the product. The addition of bromine to an alkene is stereospecifically anti. Therefore, adding bromine to *trans*-3-hexene will give *meso*-3,4-dibromohexane:

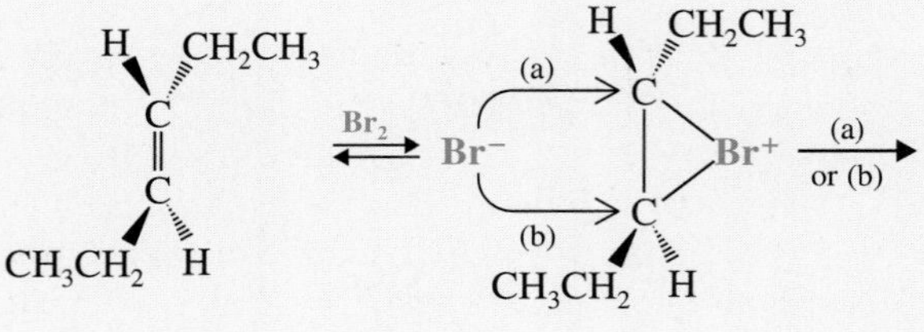

trans-3-Hexene

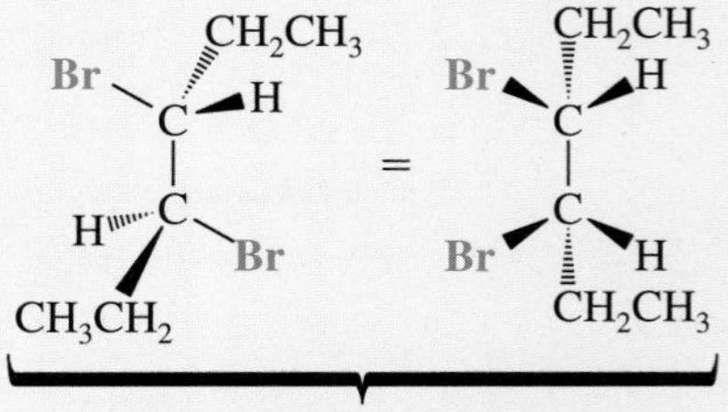

meso-3,4-Dibromohexane

We can make *trans*-3-hexene in a stereospecific way from 3-hexyne by reducing it with lithium in liquid ammonia (Section 7.7). Again the addition is anti.

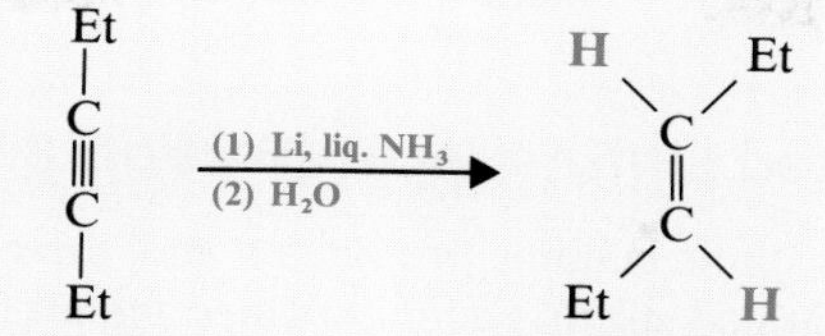

3-Hexyne **trans-3-Hexene**

3-Hexyne can be made from acetylene and ethyl bromide by successive alkylations using sodium amide as a base:

$$H{-}C{\equiv}C{-}H \xrightarrow[\text{(2) } CH_3CH_2Br]{\text{(1) } NaNH_2 \text{ liq. } NH_3}$$

$$CH_3CH_2C{\equiv}CH \xrightarrow[\text{(2) } CH_3CH_2Br]{\text{(1) } NaNH_2 \text{ liq. } NH_3} CH_3CH_2C{\equiv}CCH_2CH_3$$

PROBLEM 8.17

How would you modify the procedure given in the sample problem so as to synthesize a racemic form of (3*R*,4*R*)- and (3*S*,4*S*)-3,4-dibromohexane?

8.17 SIMPLE CHEMICAL TESTS FOR ALKANES, ALKENES, ALKYNES, ALKYL HALIDES, AND ALCOHOLS

Very often in the course of laboratory work we need to decide what functional groups are present in a compound that we have isolated. We may have isolated a compound from a synthesis, for example, and the presence of a particular functional group may tell us whether our synthesis has succeeded or failed. Or we may have isolated a compound from some natural material. Before we subject it to elaborate procedures for structure determination, it is often desirable to know something about the kind of compound we have.

Spectroscopic methods are available that will do all of these things for us, and we shall study these procedures in Chapter 13. Spectrometers are expensive instruments, however. It is helpful, therefore, to have simpler means to identify a particular functional group.

Very often this can be done by a simple chemical test. Such a test often consists of a single reagent that, when mixed with the compound in question, indicates the presence of a particular functional group. Not all reactions of a functional group serve as chemical tests, however. To be useful the reaction must proceed with a clear signal: a color change, the evolution of a gas, or the appearance of a precipitate.

8.17A Chemical Tests

A number of reagents that are used as tests for some of the functional groups that we have studied so far are summarized in the following sections. We are restricting our attention at this point to alkanes, alkenes, alkynes, alkyl halides, and alcohols.

8.17B Concentrated Sulfuric Acid (Section 8.4)

Alkenes, alkynes, and alcohols become protonated and, therefore, dissolve when they are added to cold concentrated sulfuric acid. Alkanes and alkyl halides are insoluble in cold concentrated sulfuric acid.

8.17C Bromine in Carbon Tetrachloride (Section 8.6)

Alkenes and alkynes both add bromine at room temperature and in the absence of light. Alkanes, alkyl halides, and alcohols do not react with bromine unless the reaction mixture is heated or exposed to strong irradiation. Thus, rapid decolorization of bromine in carbon tetrachloride at room temperature and in the absence of strong irradiation by light indicates the presence of a carbon–carbon double bond or a carbon–carbon triple bond.

8.17D Cold Dilute Potassium Permanganate (Section 8.9)

Alkenes and alkynes are oxidized by cold dilute solutions of potassium permanganate. If the alkene or alkyne is present in excess, the deep-purple color of the permanganate solution disappears and is replaced by the brown color of precipitated manganese dioxide.

Alkanes, alkyl halides, and pure alcohols do not react with cold dilute potassium permanganate. When these compounds are tested, the purple color is not discharged and a precipitate of manganese dioxide does not appear. (Impure alcohols often contain aldehydes and aldehydes give a positive test with cold dilute potassium permanganate.)

Cold dilute potassium permanganate is often called Baeyer's reagent.

8.17E Alcoholic Silver Nitrate

Alkyl and allylic halides (Section 12.2) react with silver ion to form a precipitate of silver halide. Ethanol is a convenient solvent because it dissolves silver nitrate and the alkyl halide. It does not dissolve the silver halide.

$$\underset{\substack{\textbf{Alkyl}\\ \textbf{halide}}}{R-X} + AgNO_3 \xrightarrow{\text{alcohol}} AgX{\downarrow} + R^+ \longrightarrow \text{other products}$$

$$\underset{\textbf{An allylic halide}}{R-CH=CHCH_2X} + AgNO_3 \xrightarrow{\text{alcohol}} AgX{\downarrow} + RCH=CHCH_2^+ \longrightarrow \text{other products}$$

Vinylic halides and phenyl halides (Section 6.16A) do not give a silver halide precipitate when treated with silver nitrate in alcohol because vinylic cations and phenyl cations are very unstable and, therefore, do not form readily.

8.17F Silver Nitrate in Ammonia

Silver nitrate reacts with aqueous ammonia to give a solution containing $Ag(NH_3)_2OH$. This reacts with terminal alkynes to form a precipitate of the silver alkynide (Section 7.21).

$$R—C\equiv CH + Ag(NH_3)_2{}^+ + OH^- \longrightarrow R—C\equiv CAg\downarrow + HOH + 2\,NH_3$$

Nonterminal alkynes do not give a precipitate. *Silver alkynides can be distinguished from silver halides on the basis of their solubility in nitric acid; silver alkynides dissolve, whereas silver halides do not.*

ADDITIONAL PROBLEMS

8.18   Write structural formulas for the products that form when 1-butene reacts with each of the following reagents:

(a) HI

(b) H_2, Pt

(c) Dilute H_2SO_4, warm

(d) Cold concentrated H_2SO_4

(e) Cold concentrated H_2SO_4, then H_2O and heat

(f) HBr in the presence of alumina

(g) Br_2 in CCl_4

(h) Br_2 in CCl_4, then KI in acetone

(i) Br_2 in H_2O

(j) HCl in the presence of alumina

(k) Cold dilute $KMnO_4$, OH^-

(l) O_3, then Zn, H_2O

(m) OsO_4, then $NaHSO_3/H_2O$

(n) $KMnO_4$, OH^-, heat, then H^+

8.19   Repeat Problem 8.18 using cyclohexene instead of 1-butene.

8.20   Give the structure of the products that you would expect from the reaction of 1-butyne with:

(a) One molar equivalent of Br_2

(b) One molar equivalent of HBr in the presence of alumina

(c) Two molar equivalents of HBr in the presence of alumina

(d) H_2 (in excess)/Pt

(e) H_2, Ni_2B(P-2)

(f) $NaNH_2$ in liq. NH_3, then CH_3I

(g) $NaNH_2$ in liq. NH_3, then $(CH_3)_3CBr$

(h) $Ag(NH_3)_2OH$

(i) $Cu(NH_3)_2OH$

8.21   Give the structure of the products you would expect from the reaction (if any) of 2-butyne with:

(a) One molar equivalent of HBr in the presence of alumina

(b) Two molar equivalents of HBr in the presence of alumina

(c) One molar equivalent of Br_2

(d) Two molar equivalents of Br_2

(e) H_2, Ni_2B(P-2)

(f) One molar equivalent of HCl in the presence of alumina

(g) Li/liq. NH_3 (h) H_2 (in excess)/Pt

(i) $Ag(NH_3)_2OH$ (l) O_3, H_2O

(j) Two molar equivalents of H_2, Pt   (m) $NaNH_2$, liq. NH_3

(k) $KMnO_4$, OH^-, then H^+

8.22 Show how 1-butyne could be synthesized from each of the following:

(a) 1-Butene   (c) 1-Chloro-1-butene   (e) Ethyne and ethyl bromide

(b) 1-Chlorobutane   (d) 1,1-Dichlorobutane

✻ **8.23** Starting with 2-methylpropene (isobutylene) and using any other needed reagents, outline a synthesis of each of the following:

(a) $(CH_3)_3COH$ (d) $(CH_3)_3CF$

(b) $(CH_3)_3CCl$ (e) $(CH_3)_2C(OH)CH_2Cl$

(c) $(CH_3)_3CBr$

8.24 Myrcene, a fragrant compound found in bayberry wax, has the formula $C_{10}H_{16}$ and is known not to contain any triple bonds. (a) What is the index of hydrogen deficiency of myrcene? When treated with excess hydrogen and a platinum catalyst, myrcene is converted to a compound (**A**) with the formula $C_{10}H_{22}$. (b) How many rings does myrcene contain? (c) How many double bonds? Compound **A** can be identified as 2,6-dimethyloctane. Ozonolysis of myrcene followed by treatment with zinc and water yields 2 mol of formaldehyde (HCHO), 1 mol of acetone (CH_3COCH_3), and a third compound (**B**) with the formula $C_5H_6O_3$. (d) What is the structure of myrcene? (e) Of compound **B**?

8.25 When propene is treated with hydrogen chloride in ethanol, one of the products of the reaction is ethyl isopropyl ether. Write a plausible mechanism that accounts for the formation of this product.

8.26 When in separate reactions, 2-methylpropene, propene, and ethene are allowed to react with HI under the same conditions (i.e., identical concentration and temperature), 2-methylpropene is found to react fastest and ethene slowest. Provide an explanation for these relative rates.

8.27 Farnesene (below) is a compound found in the waxy coating of apples. Give the structure and IUPAC name of the product formed when farnesene is allowed to react with excess hydrogen in the presence of a platinum catalyst.

Farnesene

8.28 Write structural formulas for the products that would be formed when geranial, a component of lemongrass oil, is treated with ozone and then with zinc and water.

Geranial

8.29   Limonene is a compound found in orange and lemon oil. When limonene is treated with excess hydrogen and a platinum catalyst, the product of the reaction is 1-isopropyl-4-methylcyclohexane. When limonene is treated with ozone and then with zinc and water the products of the reaction are HCHO and the following compound. Write a structural formula for limonene.

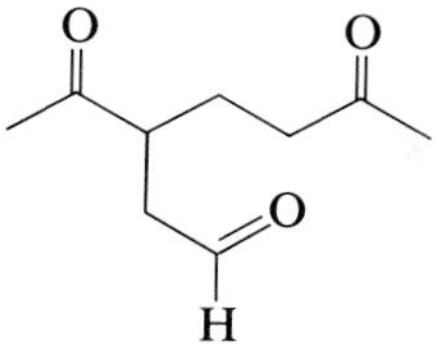

8.30   When 2,2-diphenyl-1-ethanol is treated with aqueous HI, the main product of the reaction is 1-iodo-1,1-diphenylethane. Propose a plausible mechanism that accounts for this product.

8.31   When a 3,3-dimethyl-2-butanol is treated with concentrated HI, a rearrangement takes place. Which alkyl iodide would you expect from the reaction? (Show the mechanism by which it is formed.)

8.32   Pheromones (Section 4.17) are substances secreted by animals (especially insects) that produce a specific behavioral response in other members of the same species. Pheromones are effective at very low concentrations and include sex attractants, warning substances, and "aggregation" compounds. The sex attractant pheromone of the codling moth has the molecular formula $C_{13}H_{24}O$. On catalytic hydrogenation this pheromone absorbs two molar equivalents of hydrogen and is converted to 3-ethyl-7-methyl-1-decanol. On treatment with ozone followed by treatment with zinc and water the pheromone produces: $CH_3CH_2CH_2COCH_3$, $CH_3CH_2COCH_2CH_2CHO$, and $OHCCH_2OH$.   (a) Neglecting the stereochemistry of the double bonds, write a general structure for this pheromone.   (b) The double bonds are known (on the basis of other evidence) to be (2Z,6E). Write a stereochemical formula for the codling moth sex attractant.

8.33   The sex attractant of the common house fly *(Musca domestica)* is a compound called muscalure. Muscalure is (Z)-$CH_3(CH_2)_{12}CH$=$CH(CH_2)_7CH_3$. Starting with ethyne and any other needed reagents outline a possible synthesis of muscalure.

8.34   Starting with ethyne and 1-bromopentane as your only organic reagents (except for solvents) and using any needed inorganic compounds outline a synthesis of the compound shown below.

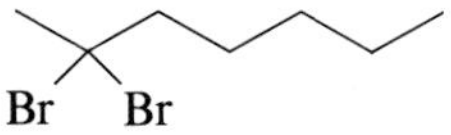

8.35   Shown below is the final step in a synthesis of the important perfume constituent, *cis*-jasmone. Which reagents would you choose to carry out this last step?

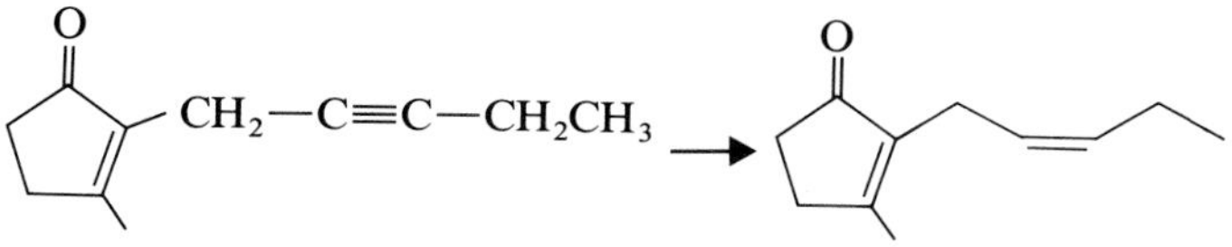

***cis*-Jasmone**

8.36   Write stereochemical formulas for all of the products that you would expect from each of the following reactions. (You may find models helpful.)

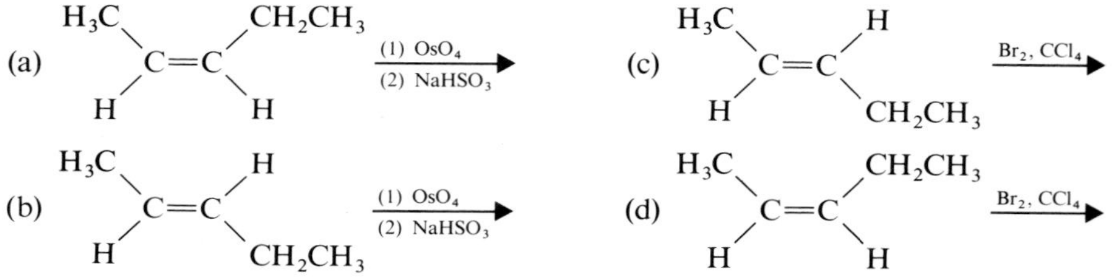

(a)
$$H_3C, H \quad C=C \quad CH_2CH_3, H \xrightarrow[(2)\ NaHSO_3]{(1)\ OsO_4}$$

(c)
$$H_3C, H \quad C=C \quad H, CH_2CH_3 \xrightarrow{Br_2,\ CCl_4}$$

(b)
$$H_3C, H \quad C=C \quad H, CH_2CH_3 \xrightarrow[(2)\ NaHSO_3]{(1)\ OsO_4}$$

(d)
$$H_3C, H \quad C=C \quad CH_2CH_3, H \xrightarrow{Br_2,\ CCl_4}$$

8.37   Give $(R-S)$ designations for each different compound given as an answer to Problem 8.36.

8.38   When cyclohexene is allowed to react with bromine in an aqueous solution of sodium chloride, the products are *trans*-1,2-dibromocyclohexane, *trans*-2-bromocyclohexanol, and *trans*-1-bromo-2-chlorocyclohexane. Write a plausible mechanism that explains the formation of this last product.

8.39   Describe with equations a simple chemical test that you could use to distinguish between the members of the following pairs of compounds. (In each case tell what the visible result would be.)

(a) Pentane and 1-pentyne
(b) Pentane and 1-pentene
(c) 1-Pentene and 1-pentyne
(d) Pentane and 1-bromopentane
(e) 2-Pentyne and 1-pentyne

(f) 1-Pentene and 1-pentanol
(g) Pentane and 1-pentanol
(h) 1-Bromo-2-pentene and 1-bromopentane
(i) 1-Pentanol and 2-penten-1-ol

8.40   The double bond of tetrachloroethene is undetectable in the bromine/carbon tetrachloride test for unsaturation. Give a plausible explanation for this behavior.

8.41   Three compounds **A**, **B**, and **C** all have the formula C_6H_{10}. All three compounds rapidly decolorize bromine in CCl_4; all three are soluble in cold concentrated sulfuric acid. Compound **A** gives a precipitate when treated with silver nitrate in ammonia, but compounds **B** and **C** do not. Compounds **A** and **B** both yield hexane when they are treated with excess hydrogen in the presence of a platinum catalyst. Under these conditions **C** absorbs only one molar equivalent of hydrogen and gives a product with the formula C_6H_{12}. When **A** is oxidized with hot basic potassium permanganate and the resulting solution acidified, the only organic product that can be isolated is $CH_3(CH_2)_3CO_2H$. Similar oxidation of **B** gives only $CH_3CH_2CO_2H$, and similar treatment of **C** gives only $HO_2C(CH_2)_4CO_2H$. What are structures for **A**, **B**, and **C**?

8.42   Ricinoleic acid, a compound that can be isolated from castor oil, has the structure $CH_3(CH_2)_5CHOHCH_2CH=CH(CH_2)_7CO_2H$.   (a) How many stereoisomers of this structure are possible?   (b) Write these structures.

8.43   There are two dicarboxylic acids with the general formula $HO_2CCH=CHCO_2H$. One dicarboxylic acid is called maleic acid; the other is called fumaric acid. In 1880, Kekulé found that on treatment with cold dilute $KMnO_4$, maleic acid yields *meso*-tartaric acid and that fumaric acid yields ($\pm$)-tartaric acid. Show how this information allows one to write stereochemical formulas for maleic acid and fumaric acid.

8.44   Use your answers to the preceding problem to predict the stereochemical outcome of the addition of bromine to maleic acid and to fumaric acid.   (a) Which dicarboxylic acid would add bromine to yield a meso compound?   (b) Which would yield a racemic form?

8.45   An optically active compound **A** (assume that it is dextrorotatory) has the molecular formula $C_7H_{11}Br$. **A** reacts with hydrogen bromide, in the absence of peroxides, to yield isomeric products, **B** and **C**, with molecular formula $C_7H_{12}Br_2$. Compound **B** is optically active; **C** is not. Treating **B** with 1 mol of potassium *tert*-butoxide yields (+)**A**. Treating **C** with 1 mol of potassium *tert*-butoxide yields (±)**A**. Treating **A** with potassium *tert*-butoxide yields **D** (C_7H_{10}). Subjecting 1 mol of **D** to ozonolysis followed by treatment with zinc and water yields 2 mol of formaldehyde and 1 mol of 1,3-cyclopentanedione.

1,3-Cyclopentanedione

Propose stereochemical formulas for **A**, **B**, **C**, and **D** and outline the reactions involved in these transformations.

8.46   A naturally occurring antibiotic called mycomycin has the structure shown here. Mycomycin is optically active. Explain this by writing structures for the enantiomeric forms of mycomycin.

$$HC\equiv C-C\equiv C-CH=C=CH-(CH=CH)_2CH_2CO_2H$$
Mycomycin

8.47   An optically active compound **D** has the molecular formula C_6H_{10}. The compound gives a precipitate when treated with a solution containing $Ag(NH_3)_2OH$. On catalytic hydrogenation **D** yields **E**(C_6H_{14}). Compound **E** is optically inactive and cannot be resolved. Propose structures for **D** and **E**.

CHAPTER 9

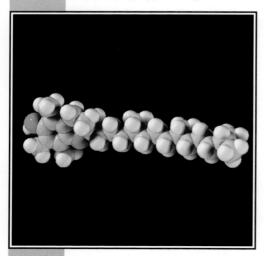

Vitamin E
(Section 9.11B)

RADICAL REACTIONS

9.1 INTRODUCTION

So far all of the reactions whose mechanisms we have studied have been **ionic reactions.** Ionic reactions are those in which covalent bonds break **heterolytically,** and in which ions are involved as reactants, intermediates, or products.

Another broad category of reactions have mechanisms that involve **homolysis** of covalent bonds with the production of intermediates possessing unpaired electrons called **radicals** (or **free radicals**).

$$A : B \xrightarrow{\text{homolysis}} A \cdot + \cdot B$$
Radicals

This simple example also illustrates another way curved arrows are used in writing reaction mechanisms. Here we show the movement of **a single electron** (not of an electron pair as we have done earlier) so we use **single-barbed arrows.** In this instance, each group, A and B, comes away with one of the electrons of the covalent bond that joined them.

Production of Radicals Energy must be supplied to cause homolysis of covalent bonds (Section 9.2) and this is usually done in two ways: by heating or by irradiation with light. For example, compounds with an oxygen–oxygen single bond, called **peroxides,** undergo homolysis readily when heated, because the oxygen–oxygen bond is weak. The products are two radicals, called alkoxyl radicals.*

---

* Notice that when we use curved arrows to show the movement of a single electron, we use a single-barbed curved arrow.

$$R-\overset{..}{\underset{..}{O}} : \overset{..}{\underset{..}{O}}-R \xrightarrow{\text{heat}} 2\ R-\overset{..}{\underset{..}{O}} \cdot$$

Dialkyl peroxide   **Alkoxyl radicals**

Halogen molecules (X_2) also contain a relatively weak bond. As we shall soon see, halogens undergo homolysis readily when heated, or when irradiated with light of a wavelength that can be absorbed by the halogen molecule.

$$:\overset{..}{\underset{..}{X}} : \overset{..}{\underset{..}{X}} : \xrightarrow[\text{heat or light}]{\text{homolysis}} 2\ :\overset{..}{\underset{..}{X}} \cdot$$

The products of this homolysis are halogen atoms, and because halogen atoms contain an unpaired electron, they are radicals.

Reactions of Radicals Almost all small radicals are short-lived, highly reactive species. When they collide with other molecules they tend to react in a way that leads to pairing of their unpaired electron. One way they can do this is by abstracting an atom from another molecule. For example, a halogen atom may abstract a hydrogen atom from an alkane. This hydrogen abstraction gives the halogen atom an electron (from the hydrogen atom) to pair with its unpaired electron. Notice, however, that the other product of this abstraction *is another radical,* in this case, an alkyl radical, R · .

General Reaction

$$:\overset{..}{\underset{..}{X}} \cdot + \ H : R \longrightarrow H : \overset{..}{\underset{..}{X}} : + \ R \cdot$$

          Alkane   **Alkyl**
                                     radical

Specific Example

$$:\overset{..}{\underset{..}{Cl}} \cdot + \ H : CH_3 \longrightarrow H : \overset{..}{\underset{..}{Cl}} : + \ CH_3 \cdot$$

          Methane   **Methyl**
                                     radical

This behavior is characteristic of radical reactions. Consider another example, one that shows another way that radicals can react: They can combine with a compound containing a multiple bond to produce a new, larger radical. (We shall study reactions of this type in Section 9.10.)

$$R \cdot \quad \overset{R}{\underset{|}{}}$$
$$-\overset{|}{\underset{|}{C}}=\overset{|}{\underset{|}{C}}- \longrightarrow -\overset{|}{\underset{|}{C}}-\overset{\cdot}{\underset{|}{C}}-$$

      Alkene   **New radical**

Radicals in Industry Radical reactions are important in many industrial processes. We shall learn later, for example (Section 9.10), how radical reactions are used to produce a whole class of useful "plastics" or *polymers* such as polyethylene, Teflon, polystyrene, and so on. Radical reactions are also central in the "cracking" process by which gasoline and other fuels are made from petroleum. And, the combustion process by which these fuels are converted to energy involves radical reactions (Section 9.11A).

Radicals in Biology and Medicine   Radical reactions are of vital importance in biology and medicine. Radical reactions are ubiquitous in living things, because radicals are produced in the normal course of metabolism. However, because radicals are highly reactive, they are also capable of randomly damaging all components of the body. Accordingly, they are believed to be important in the "aging process" in the sense that radicals are involved in the development of the chronic diseases that are life limiting. For example, there is steadily growing evidence that radical reactions are important in the development of cancers and in the development of atherosclerosis. Radicals in cigarette smoke have been implicated in inactivation of an antiprotease in the lungs, an inactivation that leads to the development of emphysema.

In Section 9.11C, we shall learn how radicals produced in the upper atmosphere from chlorofluorocarbons are causing the depletion of the ozone layer that protects living organisms from highly damaging ultraviolet (UV) radiation.

Before exploring radical chemistry further it will be useful to examine the energy changes that occur when covalent bonds break homolytically. We shall see that these changes provide us with information about an important aspect of radicals that will help us to understand their reactions: their relative stability.

9.2 HOMOLYTIC BOND DISSOCIATION ENERGIES

When atoms combine to form molecules, energy is released as covalent bonds form. The molecules of the products have lower enthalpy than the separate atoms. When hydrogen atoms combine to form hydrogen molecules, for example, the reaction is *exothermic;* it evolves 104 kcal of heat for every mole of hydrogen that is produced. Similarly, when chlorine atoms combine to form chlorine molecules, the reaction evolves 58 kcal mol^{-1} of chlorine produced.

$$H\cdot + H\cdot \longrightarrow H-H \qquad \Delta H° = -104 \text{ kcal mol}^{-1}$$
$$Cl\cdot + Cl\cdot \longrightarrow Cl-Cl \qquad \Delta H° = -58 \text{ kcal mol}^{-1}$$

Bond formation is an exothermic process

To break covalent bonds, energy must be supplied. Reactions in which only bond breaking occurs are always endothermic. The energy required to break the covalent bonds of hydrogen or chlorine homolytically is exactly equal to that evolved when the separate atoms combine to form molecules. In the bond cleavage reaction, however, $\Delta H°$ is positive.

$$H-H \longrightarrow H\cdot + H\cdot \qquad \Delta H° = +104 \text{ kcal mol}^{-1}$$
$$Cl-Cl \longrightarrow Cl\cdot + Cl\cdot \qquad \Delta H° = +58 \text{ kcal mol}^{-1}$$

The energies required to break covalent bonds homolytically have been determined experimentally for many types of covalent bonds. These energies are called **homolytic bond dissociation energies,** and they are usually abbreviated by the symbol $DH°$. The homolytic bond dissociation energies of hydrogen and chlorine, for example, might be written in the following way.

$$H-H \qquad\qquad Cl-Cl$$
$$(DH° = 104 \text{ kcal mol}^{-1}) \qquad (DH° = 58 \text{ kcal mol}^{-1})$$

The homolytic bond dissociation energies of a variety of covalent bonds are listed in Table 9.1.

9.2A Homolytic Bond Dissociation Energies and Heats of Reaction

Bond dissociation energies have, as we shall see, a variety of uses. They can be used, for example, to calculate the enthalpy change ($\Delta H°$) for a reaction. To make such a calculation (see following reaction) we must remember that for bond breaking $\Delta H°$ is positive and for bond formation $\Delta H°$ is negative. Let us consider, for example, the reaction of hydrogen and chlorine to produce 2 mol of hydrogen chloride. From Table 9.1 we get the following values of $DH°$.

$$H-H \quad + \quad Cl-Cl \longrightarrow \qquad 2\ H-Cl$$
$$(DH° = 104) \quad (DH° = 58) \qquad (DH° = 103) \times 2$$
$$+162 \text{ kcal mol}^{-1} \text{ is required} \qquad -206 \text{ kcal mol}^{-1} \text{ is evolved}$$
$$\text{for bond cleavage} \qquad\qquad \text{in bond formation}$$

Overall, the reaction is exothermic:

$$\Delta H° = (-206 \text{ kcal mol}^{-1} + 162 \text{ kcal mol}^{-1}) = -44 \text{ kcal mol}^{-1}$$

For the purpose of our calculation, we have assumed a particular pathway, that amounts to:

$$H-H \longrightarrow 2\ H\cdot$$

and

$$Cl-Cl \longrightarrow 2\ Cl\cdot$$

then

$$2\ H\cdot + 2\ Cl\cdot \longrightarrow 2\ H-Cl$$

This is not the way the reaction actually occurs. Nevertheless, the heat of reaction, $\Delta H°$, is a thermodynamic quantity that is dependent *only* on the initial and final states of the reacting molecules. $\Delta H°$ is independent of the path followed and, for this reason, our calculation is valid.

PROBLEM 9.1

Calculate the heat of reaction, $\Delta H°$, for the following reactions:

(a) $H_2 + F_2 \longrightarrow 2\ HF$

(b) $CH_4 + F_2 \longrightarrow CH_3F + HF$

(c) $CH_4 + Cl_2 \longrightarrow CH_3Cl + HCl$

(d) $CH_4 + Br_2 \longrightarrow CH_3Br + HBr$

(e) $CH_4 + I_2 \longrightarrow CH_3I + HI$

(f) $CH_3CH_3 + Cl_2 \longrightarrow CH_3CH_2Cl + HCl$

(g) $CH_3CH_2CH_3 + Cl_2 \longrightarrow CH_3CHClCH_3 + HCl$

(h) $(CH_3)_3CH + Cl_2 \longrightarrow (CH_3)_3CCl + HCl$

TABLE 9.1   Single-bond homolytic dissociation energies $DH°$ at 25°C

$$A:B \longrightarrow A\cdot + B\cdot$$

BOND BROKEN (shown in red)	kcal mol^{-1}	kJ mol^{-1}	BOND BROKEN (shown in red)	kcal mol^{-1}	kJ mol^{-1}
H—H	104	435	$(CH_3)_2CH$—H	94.5	395
D—D	106	444	$(CH_3)_2CH$—F	105	439
F—F	38	159	$(CH_3)_2CH$—Cl	81	339
Cl—Cl	58	243	$(CH_3)_2CH$—Br	68	285
Br—Br	46	192	$(CH_3)_2CH$—I	53	222
I—I	36	151	$(CH_3)_2CH$—OH	92	385
H—F	136	569	$(CH_3)_2CH$—OCH_3	80.5	337
H—Cl	103	431	$(CH_3)_2CHCH_2$—H	98	410
H—Br	87.5	366	$(CH_3)_3C$—H	91	381
H—I	71	297	$(CH_3)_3C$—Cl	78.5	328
CH_3—H	104	435	$(CH_3)_3C$—Br	63	264
CH_3—F	108	452	$(CH_3)_3C$—I	49.5	207
CH_3—Cl	83.5	349	$(CH_3)_3C$—OH	90.5	379
CH_3—Br	70	293	$(CH_3)_3C$—OCH_3	78	326
CH_3—I	56	234	$C_6H_5CH_2$—H	85	356
CH_3—OH	91.5	383	$CH_2{=}CHCH_2$—H	85	356
CH_3—OCH_3	80	335	$CH_2{=}CH$—H	108	452
CH_3CH_2—H	98	410	C_6H_5—H	110	460
CH_3CH_2—F	106	444	$HC{\equiv}C$—H	125	523
CH_3CH_2—Cl	81.5	341	CH_3—CH_3	88	368
CH_3CH_2—Br	69	289	CH_3CH_2—CH_3	85	356
CH_3CH_2—I	53.5	224	$CH_3CH_2CH_2$—CH_3	85	356
CH_3CH_2—OH	91.5	383	CH_3CH_2—CH_2CH_3	82	343
CH_3CH_2—OCH_3	80	335	$(CH_3)_2CH$—CH_3	84	351
			$(CH_3)_3C$—CH_3	80	335
$CH_3CH_2CH_2$—H	98	410	HO—H	119	498
$CH_3CH_2CH_2$—F	106	444	HOO—H	90	377
$CH_3CH_2CH_2$—Cl	81.5	341	HO—OH	51	213
$CH_3CH_2CH_2$—Br	69	289	CH_3CH_2O—OCH_3	44	184
$CH_3CH_2CH_2$—I	53.5	224	CH_3CH_2O—H	103	431
$CH_3CH_2CH_2$—OH	91.5	383	$CH_3\overset{\displaystyle O}{\overset{\|}{C}}$—H	87	364
$CH_3CH_2CH_2$—OCH_3	80	335			

9.2B Homolytic Bond Dissociation Energies and the Relative Stabilities of Radicals

Homolytic bond dissociation energies also provide us with a convenient way to estimate the relative stabilities of radicals. If we examine the data given in Table 9.1, we find the following values of $DH°$ for the primary and secondary C—H bonds of propane:

$$CH_3CH_2CH_2—H \qquad\qquad (CH_3)_2CH—H$$
$$(DH° = 98 \text{ kcal mol}^{-1}) \qquad (DH° = 94.5 \text{ kcal mol}^{-1})$$

This means that for the reaction in which the designated C—H bonds are broken homolytically, the values of $\Delta H°$ are those given here.

$$CH_3CH_2CH_2—H \longrightarrow CH_3CH_2CH_2\cdot + H\cdot \qquad \Delta H° = +98 \text{ kcal mol}^{-1}$$
Propyl radical
(a 1° radical)

$$CH_3\underset{\underset{H}{|}}{C}HCH_3 \longrightarrow CH_3\dot{C}HCH_3 + H\cdot \qquad \Delta H° = +94.5 \text{ kcal mol}^{-1}$$
Isopropyl radical
(a 2° radical)

These reactions resemble each other in two respects: They both begin with the same alkane (propane), and they both produce an alkyl radical and a hydrogen atom. They differ, however, in the amount of energy required and in the type of carbon radical being produced.* These two differences are related to each other.

More energy must be supplied to produce a primary alkyl radical (the propyl radical) from propane than is required to produce a secondary carbon radical (the isopropyl radical) from the same compound. This must mean that the primary radical has absorbed more energy and thus has greater *potential energy*. Because the relative stability of a chemical species is inversely related to its potential energy, the secondary radical must be the *more stable* radical (Fig. 9.1a). In fact, the secondary isopropyl radical is more stable than the primary propyl radical by 3.5 kcal mol^{-1}.

We can use the data in Table 9.1 to make a similar comparison of the *tert*-butyl radical (a 3° radical) and the isobutyl radical (a 1° radical) relative to isobutane.

$$CH_3—\underset{\underset{H}{|}}{\overset{\overset{CH_3}{|}}{C}}—CH_2—H \longrightarrow CH_3\overset{\overset{CH_3}{|}}{C}CH_3 + H\cdot \qquad \Delta H° = +91 \text{ kcal mol}^{-1}$$
***tert*-Butyl**
radical
(a 3° radical)

$$CH_3—\underset{\underset{H}{|}}{\overset{\overset{CH_3}{|}}{C}}—CH_2—H \longrightarrow CH_3\overset{\overset{CH_3}{|}}{C}HCH_2\cdot + H\cdot \qquad \Delta H° = +98 \text{ kcal mol}^{-1}$$
Isobutyl radical
(a 1° radical)

*Alkyl radicals are classified as being 1°, 2°, or 3° on the basis of the carbon atom that has the unpaired electron.

Here we find (Fig. 9.1*b*) that the difference in stability of the two radicals is even larger. The tertiary radical is more stable than the primary radical by 7 kcal mol^{-1}.

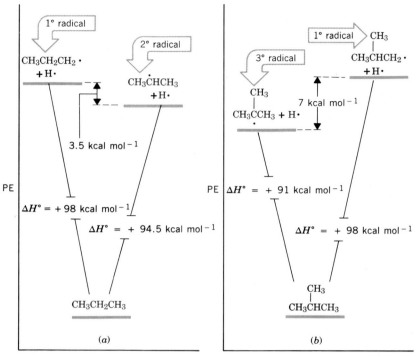

FIGURE 9.1 (*a*) A comparison of the potential energies of the propyl radical (+ H·) and the isopropyl radical (+ H·) relative to propane. The isopropyl radical—a 2° radical—is more stable than the 1° radical by 3.5 kcal mol^{-1}. (*b*) A comparison of the potential energies of the *tert*-butyl radical (+ H·) and the isobutyl radical (+ H ·) relative to isobutane. The 3° radical is more stable than the 1° radical by 7 kcal mol^{-1}.

The kind of pattern that we find in these examples is found with alkyl radicals generally; overall, their relative stabilities are the following:

Tertiary > Secondary > Primary > Methyl

$$
\begin{array}{cccc}
\text{C} & \text{C} & \text{H} & \text{H} \\
| & | & | & | \\
\text{C—C·} > & \text{C—C·} > & \text{C—C·} > & \text{H—C·} \\
| & | & | & | \\
\text{C} & \text{H} & \text{H} & \text{H}
\end{array}
$$

The order of stability of alkyl radicals is the same as for carbocations (Section 6.13B), and the reasons are similar. Although alkyl radicals are uncharged, the carbon that bears the odd electron is *electron deficient*. Therefore, electron-releasing alkyl groups attached to this carbon provide a stabilizing effect, and the more alkyl groups that are attached to this carbon the more stable the radical is.

PROBLEM 9.2

(a) Construct potential energy diagrams similar to those given in Fig. 9.1 showing the potential energy of a propyl radical and a hydrogen atom relative to propane and a diagram showing the potential energy of an ethyl radical and a hydrogen atom relative to ethane. Align these two diagrams side by side so that the relative potential energies of the two alkanes correspond. Now compare these two diagrams and explain how this comparison justifies the assertion that relative to the alkane from which it is formed the two 1° radicals have similar stabilities. (b) Make a new comparison by including a diagram showing the relative potential energy of a methyl radical and a hydrogen atom relative to methane. Align this diagram with the previous two so that the energy levels of the alkanes correspond and explain how this comparison justifies the assertion that, relative to the alkane from which it is formed, a methyl radical is less stable than a primary radical.

PROBLEM 9.3

List the following radicals in order of decreasing stability:

$$CH_3\cdot \qquad (CH_3)_2CHCH_2\cdot \qquad CH_3CH_2\overset{\overset{\displaystyle CH_3}{|}}{\underset{\underset{\displaystyle CH_3}{|}}{C}}\cdot \qquad CH_3CH_2\overset{\displaystyle CH\cdot}{\underset{\underset{\displaystyle CH_3}{|}}{}}$$

9.3 THE REACTIONS OF ALKANES WITH HALOGENS

Methane, ethane, and other alkanes react with the first three members of the halogen family: fluorine, chlorine, and bromine. Alkanes do not react appreciably with iodine. With methane the reaction produces a mixture of halomethanes and a hydrogen halide.

Methane   **Halogen**   **Halomethane**   **Dihalomethane**

Trihalomethane   **Tetrahalomethane**   **Hydrogen halide**

X = F, Cl, or Br

The reaction of an alkane with a halogen is a **substitution reaction,** called **halogenation.** The general reaction to produce a monohaloalkane can be written as follows:

$$R\!-\!H + X_2 \longrightarrow R\!-\!X + HX$$

In these reactions a halogen atom replaces one or more of the hydrogen atoms of the alkane.

9.3A Halogenation Reactions

One complicating characteristic of alkane halogenations is that multiple substitution reactions almost always occur. As we saw at the beginning of this section, the halogenation of methane produces a mixture of monohalomethane, dihalomethane, trihalomethane, and tetrahalomethane.

This happens because all hydrogen atoms attached to carbon are capable of reacting with fluorine, chlorine, or bromine.

Let us consider the reaction that takes place between chlorine and methane as an example. If we mix methane and chlorine (both substances are gases at room temperature) and then either heat the mixture or irradiate it with light, a reaction begins to occur vigorously. At the outset, the only compounds that are present in the mixture are chlorine and methane, and the only reaction that can take place is one that produces chloromethane and hydrogen chloride.

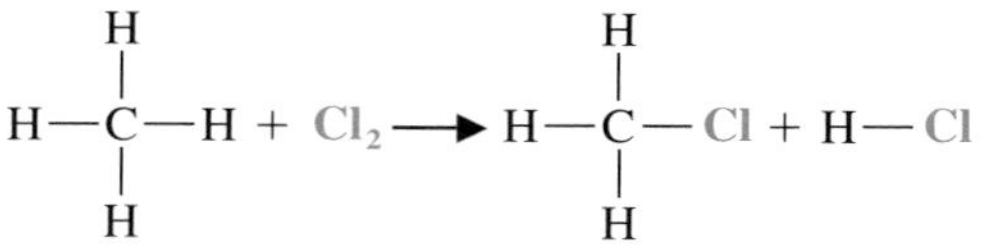

$$\underset{\substack{| \\ H}}{\overset{\substack{H \\ |}}{H-C-H}} + Cl_2 \longrightarrow \underset{\substack{| \\ H}}{\overset{\substack{H \\ |}}{H-C-Cl}} + H-Cl$$

As the reaction progresses, however, the concentration of chloromethane in the mixture increases, and a second substitution reaction begins to occur. Chloromethane reacts with chlorine to produce dichloromethane.

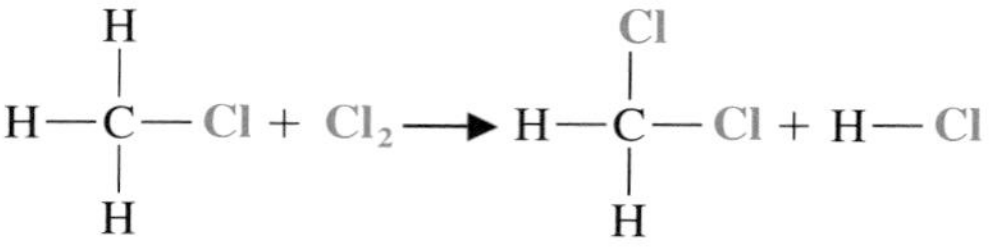

$$\underset{\substack{| \\ H}}{\overset{\substack{H \\ |}}{H-C-Cl}} + Cl_2 \longrightarrow \underset{\substack{| \\ H}}{\overset{\substack{Cl \\ |}}{H-C-Cl}} + H-Cl$$

The dichloromethane produced can then react to form trichloromethane, and trichloromethane, as it accumulates in the mixture, can react with chlorine to produce tetrachloromethane. Each time a substitution of —Cl for —H takes place a molecule of H—Cl is produced.

SAMPLE PROBLEM

If the goal of a synthesis is to prepare chloromethane (CH_3Cl), its formation can be maximized and the formation of CH_2Cl_2, $CHCl_3$, and CCl_4 minimized by using a large excess of methane in the reaction mixture. Explain why this is possible.

ANSWER:

The use of a large excess of methane maximizes the probability that chlorine will attack methane molecules because the concentration of methane in the mixture will always be relatively large. It also minimizes the probability that chlorine will attack molecules of CH_3Cl, CH_2Cl_2, and $CHCl_3$, because their concentrations will always be relatively small. After the reaction is over, the unreacted excess methane can be recovered and recycled.

Chlorination of most higher alkanes gives a mixture of isomeric monochloro products as well as more highly halogenated compounds. Chlorine is relatively **unselective**; it does not discriminate greatly among the different types of hydrogen atoms (primary, secondary, and tertiary) in an alkane. An example is the light-promoted chlorination of isobutane.

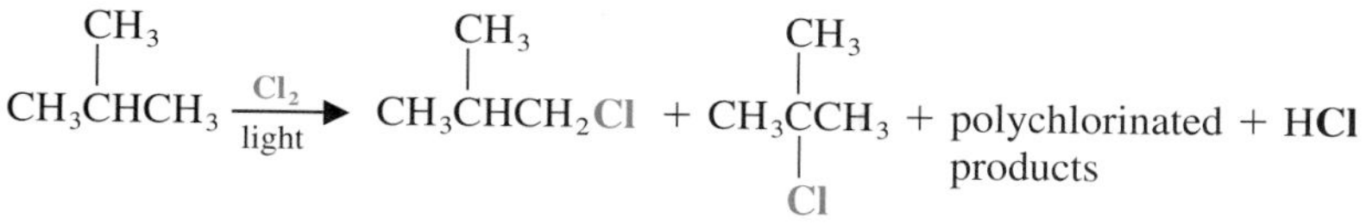

$$
\underset{\substack{\text{Isobutane}}}{CH_3CHCH_3}
\xrightarrow[\text{light}]{Cl_2}
\underset{\substack{\text{Isobutyl chloride}\\(48\%)}}{CH_3CHCH_2Cl}
+
\underset{\substack{\textit{tert}\text{-Butyl}\\\text{chloride}\\(29\%)}}{CH_3CCH_3}
+ \text{polychlorinated} + HCl
$$

with CH_3 groups shown above the carbons bearing the chain, CH_3 and Cl substituents on the central carbon, and (23%) above the products label.

Because alkane chlorinations usually yield a complex mixture of products, they are not generally useful as synthetic methods when our goal is the preparation of a specific alkyl chloride. An exception is the halogenation of an alkane (or cycloalkane) whose hydrogen atoms *are all equivalent.** Neopentane, for example, can form only one monohalogenation product, and the use of a large excess of neopentane minimizes polychlorination.

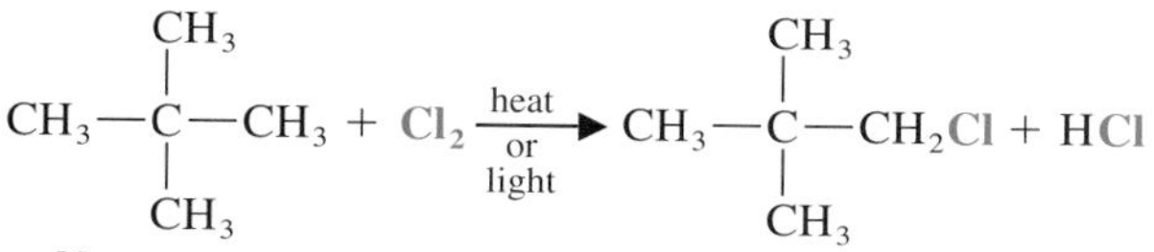

$$
\underset{\substack{\text{Neopentane}\\(\text{excess})}}{CH_3-\overset{\displaystyle CH_3}{\underset{\displaystyle CH_3}{C}}-CH_3}
+ Cl_2
\xrightarrow[\substack{\text{or}\\\text{light}}]{\text{heat}}
\underset{\substack{\text{Neopentyl chloride}}}{CH_3-\overset{\displaystyle CH_3}{\underset{\displaystyle CH_3}{C}}-CH_2Cl}
+ HCl
$$

Bromine is generally less reactive toward alkanes than chlorine, and bromine is *more selective* in the site of attack when it does react. We shall examine this topic further in Section 9.6A.

9.4 CHLORINATION OF METHANE: MECHANISM OF REACTION

The *halogenation* reactions of alkanes take place by a radical mechanism. Let us begin our study of them by examining a simple example of an alkane halogenation: the reaction of methane with chlorine that takes place in the gas phase.

Several important experimental observations can be made about this reaction:

$$
CH_4 + Cl_2 \longrightarrow CH_3Cl + HCl \,(+ \, CH_2Cl_2, \, CHCl_3, \text{ and } CCl_4)
$$

1. **The reaction is promoted by heat or light.** At room temperature methane and chlorine do not react at a perceptible rate as long as the mixture is kept away from light. Methane and chlorine do react, however, at room temperature if the gaseous reaction mixture is irradiated with UV light, and methane and chlorine do react in the dark, if the gaseous mixture is heated to temperatures greater than 100°C.

---

*Equivalent hydrogen atoms are defined as those which on replacement by some other group (e.g., chlorine) yield the same compound.

2. **The light-promoted reaction is highly efficient.** A relatively small number of light photons permits the formation of relatively large amounts of chlorinated product.

A mechanism that is consistent with these observations has several steps. The first step involves the fragmentation of a chlorine molecule, by heat or light, into two chlorine atoms. Chlorine is known, from other evidence, to undergo such reactions. It can be shown, moreover, that the frequency of light that promotes the chlorination of methane is a frequency that is absorbed by chlorine molecules and not by methane molecules.

The mechanism follows.*

A MECHANISM FOR THE REACTION

REACTION:

$$CH_4 + Cl_2 \xrightarrow[\text{or light}]{\text{heat}} CH_3Cl + HCl$$

MECHANISM:

Step 1

$$:\!\overset{..}{\underset{..}{Cl}}\!\!:\overset{..}{\underset{..}{Cl}}\!\!: \xrightarrow[\text{or light}]{\text{heat}} :\!\overset{..}{\underset{..}{Cl}}\!\cdot \; + \; \cdot\overset{..}{\underset{..}{Cl}}\!\!:$$

Under the influence of heat or light a molecule of chlorine dissociates; each atom takes one of the bonding electrons.

This step produces two highly reactive chlorine atoms.

Step 2

$$:\!\overset{..}{\underset{..}{Cl}}\!\cdot \; + \; H\!:\!\overset{\overset{\displaystyle H}{|}}{\underset{\underset{\displaystyle H}{|}}{C}}\!\!-\!H \longrightarrow :\!\overset{..}{\underset{..}{Cl}}\!\!:\!H \; + \; \cdot\overset{\overset{\displaystyle H}{|}}{\underset{\underset{\displaystyle H}{|}}{C}}\!\!-\!H$$

A chlorine atom abstracts a hydrogen atom from a methane molecule.

This step produces a molecule of hydrogen chloride and a methyl radical.

Step 3

$$H\!-\!\overset{\overset{\displaystyle H}{|}}{\underset{\underset{\displaystyle H}{|}}{C}}\!\cdot \; + \; :\!\overset{..}{\underset{..}{Cl}}\!\!:\overset{..}{\underset{..}{Cl}}\!\!: \longrightarrow H\!-\!\overset{\overset{\displaystyle H}{|}}{\underset{\underset{\displaystyle H}{|}}{C}}\!\!:\!\overset{..}{\underset{..}{Cl}}\!\!: \; + \; \cdot\overset{..}{\underset{..}{Cl}}\!\!:$$

A methyl radical abstracts a chlorine atom from a chlorine molecule.

This step produces a molecule of methyl chloride and a chlorine atom. The chlorine atom can now cause a repetition of step 2.

Remember: These conventions are used in illustrating reaction mechanisms in this text.

1. Arrows ⌒ or ⌒ always show the direction of movement of electrons.
2. Single-barbed arrows ⌒ show the attack (or movement) of an unpaired electron.
3. Double-barbed arrows ⌒ show the attack (or movement) of an electron pair.

In step 3 the highly reactive methyl radical reacts with a chlorine molecule by abstracting a chlorine atom. This results in the formation of a molecule of chloromethane (one of the ultimate products of the reaction) and a *chlorine atom.* This latter product is particularly significant, for the chlorine atom formed in step 3 can attack another methane molecule and cause a repetition of step 2. Then, step 3 is repeated, and so forth, for hundreds or thousands of times. (With each repetition of step 3 a molecule of chloromethane is produced.) This type of sequential, stepwise mechanism, in which each step generates the reactive intermediate that causes the next step to occur, is called a **chain reaction.**

Step 1 is called the **chain-initiating step.** In the chain-initiating step *radicals are created.* Steps 2 and 3 are called **chain-propagating steps.** In chain-propagating steps *one radical generates another.*

Chain Initiation

Step 1 $Cl_2 \xrightarrow[\text{light}]{\text{heat}} 2\ Cl \cdot$

Chain Propagation

Step 2 $CH_4 + Cl \cdot \longrightarrow H\!-\!Cl + CH_3 \cdot$

Step 3 $CH_3 \cdot + Cl_2 \cdot \longrightarrow CH_3Cl + Cl \cdot$

The chain nature of the reaction accounts for the observation that the light-promoted reaction is highly efficient. The presence of a relatively few atoms of chlorine at any given moment is all that is needed to cause the formation of many thousands of molecules of chloromethane.

What causes the chains to terminate? Why does one photon of light not promote the chlorination of all of the methane molecules present? We know that this does not happen because we find that at low temperatures, continuous irradiation is required or the reaction slows and stops. The answer to these questions is the existence of *chain-terminating steps:* steps that occur infrequently, but occur often enough to *use up one or both of the reactive intermediates.* The continuous replacement of intermediates used up by chain-terminating steps requires continuous irradiation. Plausible chain-terminating steps are:

Chain Termination

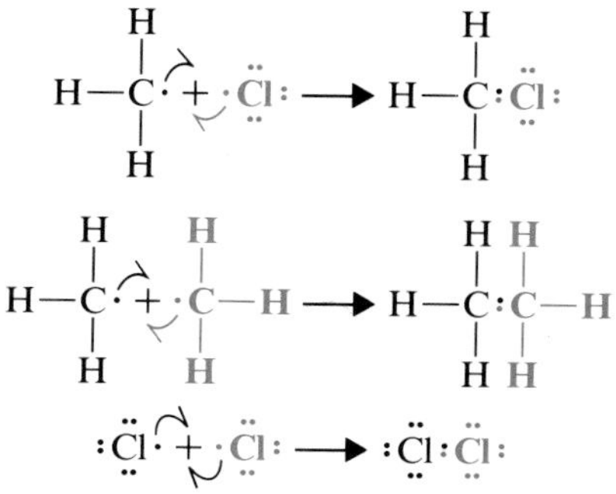

and

This last step probably occurs least frequently. The two chlorine atoms are highly energetic; as a result, the simple diatomic chlorine molecule that is formed must dissipate its excess energy rapidly by colliding with some other molecule or the walls of the container. Otherwise it simply flies apart again. By contrast, chloromethane and ethane, formed in the other two chain-terminating steps, can dissipate their excess energy through vibrations of their C—H bonds.

Our radical mechanism also explains how the reaction of methane with chlorine produces the more highly halogenated products, CH_2Cl_2, $CHCl_3$, and CCl_4 (as well as additional HCl). As the reaction progresses, chloromethane (CH_3Cl) accumulates in the mixture and its hydrogen atoms, too, are susceptible to abstraction by chlorine. Thus chloromethyl radicals are produced that lead to dichloromethane (CH_2Cl_2).

Step 2a
$$Cl\cdot + H\!:\!\overset{\displaystyle Cl}{\underset{\displaystyle H}{C}}\!\!-\!H \longrightarrow H\!:\!Cl + \cdot\overset{\displaystyle Cl}{\underset{\displaystyle H}{C}}\!\!-\!H$$

Step 3a
$$H\!-\!\overset{\displaystyle Cl}{\underset{\displaystyle H}{C}}\!\cdot + Cl\!:\!Cl \longrightarrow H\!-\!\overset{\displaystyle Cl}{\underset{\displaystyle H}{C}}\!:\!Cl + Cl\cdot$$

Then step 2a is repeated, then step 3a is repeated, and so on. Each repetition of step 2a yields a molecule of HCl, and each repetition of step 3a yields a molecule of CH_2Cl_2.

PROBLEM 9.4

Suggest a method for separating the mixture of CH_4, CH_3Cl, CH_2Cl_2, $CHCl_3$, and CCl_4 that is formed when methane is chlorinated. (You may want to consult a handbook.)

PROBLEM 9.5

When methane is chlorinated, among the products found are traces of chloro-ethane. How is it formed? Of what significance is its formation?

PROBLEM 9.6

If our goal is to synthesize CCl_4 in maximum yield, this can be accomplished by using a large excess of chlorine. Explain.

9.5 CHLORINATION OF METHANE: ENERGY CHANGES

We saw in Section 9.2A that we can calculate the overall heat of reaction from bond dissociation energies. We can also calculate the heat of reaction for each individual step of a mechanism.

Chain Initiation

Step 1 $Cl-Cl \longrightarrow 2\ Cl\cdot$ $\Delta H^\circ = +58$ kcal mol^{-1}
$(DH^\circ = 58)$

Chain Propagation

Step 2 CH_3—H + Cl· $\longrightarrow$ H—Cl + CH_3· $\Delta H° = +1$ kcal mol^{-1}
             $(DH° = 104)$ $(DH° = 103)$

Step 3 CH_3· + Cl—Cl $\longrightarrow$ CH_3—Cl + Cl· $\Delta H° = -25.5$ kcal mol^{-1}
             $(DH° = 58)$ $(DH° = 83.5)$

Chain Termination

CH_3· + Cl· $\longrightarrow$ CH_3—Cl $\Delta H° = -83.5$ kcal mol^{-1}
                       $(DH° = 83.5)$

CH_3· + ·CH_3 $\longrightarrow$ CH_3—CH_3 $\Delta H° = -88$ kcal mol^{-1}
                       $(DH° = 88)$

Cl· + Cl· $\longrightarrow$ Cl—Cl $\Delta H° = -58$ kcal mol^{-1}
                 $(DH° = 58)$

In the chain-initiating step only one bond is broken—the bond between two chlorine atoms—and no bonds are formed. The heat of reaction for this step is simply the bond dissociation energy for a chlorine molecule, and it is highly endothermic.

In the chain-terminating steps bonds are formed, but no bonds are broken. As a result, all of the chain-terminating steps are highly exothermic.

Each of the chain-propagating steps, on the other hand, requires the breaking of one bond and the formation of another. The value of $\Delta H°$ for each of these steps is the difference between the bond dissociation energy of the bond that is broken and the bond dissociation energy for the bond that is formed. The first chain-propagating step is slightly endothermic ($\Delta H° = +1$ kcal mol^{-1}), but the second is exothermic by a large amount ($\Delta H° = -25.5$ kcal mol^{-1}).

PROBLEM 9.7

Assuming the same mechanism applies, calculate $\Delta H°$ for the chain-initiating, chain-propagating, and chain-terminating steps involved in the fluorination of methane.

The addition of the chain-propagating steps yields the overall equation for the chlorination of methane:

Cl· + CH_3—H $\longrightarrow$ CH_3· + H—Cl $\Delta H° = +\ 1$ kcal mol^{-1}
CH_3· + Cl—Cl $\longrightarrow$ CH_3—Cl + Cl· $\Delta H° = -25.5$ kcal mol^{-1}
CH_3—H + Cl—Cl $\longrightarrow$ CH_3—Cl + H—Cl $\Delta H° = -24.5$ kcal mol^{-1}

and the addition of the values of $\Delta H°$ for the individual chain-propagating steps yields the overall value of $\Delta H°$ for the reaction.

PROBLEM 9.8

Show how you can use the chain-propagating steps (see Problem 9.7) to calculate the overall value of $\Delta H°$ for the fluorination of methane.

9.5A The Overall Free-Energy Change

For many reactions the entropy change is so small that the term $T\Delta S°$ in the expression

$$\Delta G° = \Delta H° - T\Delta S°$$

is almost zero, and $\Delta G°$ is approximately equal to $\Delta H°$. This happens when the reaction is one in which the relative order of reactants and products is about the same. Recall (Section 3.8) that entropy measures the relative disorder or randomness of a system. For a chemical system the relative disorder of the molecules can be related to the number of *degrees of freedom* available to the molecules and their constituent atoms. Degrees of freedom are associated with ways in which *movement or changes in relative position can occur*. Molecules have three sorts of degrees of freedom: translational degrees of freedom associated with movements of the whole molecule through space, rotational degrees of freedom associated with the tumbling motions of the molecule, and vibrational degrees of freedom associated with the stretching and bending motion of atoms about the bonds that connect them (Fig. 9.2). If the atoms of the products of a reaction have more degrees of freedom available than they did as reactants, the entropy change ($\Delta S°$) for the reaction will be positive. If, on the other hand, the atoms of the products are more constrained (have fewer degrees of freedom) than the reactants, a negative $\Delta S°$ results.

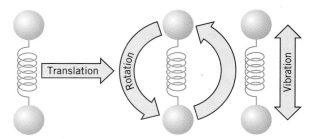

FIGURE 9.2 Translational, rotational, and vibrational degrees of freedom for a simple diatomic molecule.

Consider the reaction of methane with chlorine.

$$CH_4 + Cl_2 \longrightarrow CH_3Cl + HCl$$

Here, 2 mol of the products are formed from the same number of moles of the reactants. Thus the number of translational degrees of freedom available to products and reactants are the same. Furthermore, CH_3Cl is a tetrahedral molecule like CH_4, and HCl is a diatomic molecule like Cl_2. This means that vibrational and rotational degrees of freedom available to products and reactants should also be approximately the same. The actual entropy change for this reaction is quite small, $\Delta S° = +0.67$ cal K^{-1} mol^{-1}. Therefore, at room temperature (298 K) the $T\Delta S°$ term is 0.2 kcal mol^{-1}. The enthalpy change for the reaction and the free-energy change are almost equal, $\Delta H° = -24.5$ kcal mol^{-1}, and $\Delta G° = -24.7$ kcal mol^{-1}.

In situations like this one it is often convenient to make predictions about whether or not a reaction will proceed to completion on the basis of $\Delta H°$ rather than $\Delta G°$ since $\Delta H°$ values are readily obtained from bond dissociation energies.

9.5B Activation Energies

For many reactions that we shall study in which entropy changes are small, it is also often convenient to base our estimates of reaction rates on what are called simply **energies of activation, E_{act},** rather than on free energies of activation, $\Delta G^{\ddagger}$. Without going into detail, suffice it to say that these two quantities are closely related, and that **both measure the difference in energy between the reactants and the transition state.** Therefore, a low energy of activation means a reaction will take place rapidly; a high energy of activation means that a reaction will take place slowly.

Having seen earlier in this section how to calculate ΔH° for each step in the chlorination of methane, let us consider the energy of activation for each step. These values are as follows:

Chain Initiation

Step 1 $Cl_2 \longrightarrow 2\ Cl\cdot$ $E_{act} = +58$ kcal mol^{-1}

Chain Propagation

Step 2 $Cl\cdot\ + CH_4 \longrightarrow HCl + CH_3\cdot$ $E_{act} = +3.8$ kcal mol^{-1}

Step 3 $CH_3\cdot\ + Cl_2 \longrightarrow CH_3Cl + Cl\cdot$ $E_{act} = {\sim}2$ kcal mol^{-1}

How does one know what the energy of activation for a reaction will be? Could we, for example, have predicted from bond dissociation energies that the energy of activation for the reaction, $Cl\cdot\ + CH_4 \longrightarrow HCl + CH_3\cdot$, would be precisely 3.8 kcal mol^{-1}? The answer is *no*. The energy of activation must be determined from other experimental data. It cannot be directly measured—it is calculated. Certain principles can be established, however, that enable one to arrive at estimates of energies of activation:

1. **Any reaction in which *bonds are broken* will have an energy of activation greater than zero. This will be true even if a stronger bond is formed and the reaction is exothermic. The reason: Bond formation and bond breaking do not occur simultaneously in the transition state. Bond formation lags behind, and its energy is not all available for bond breaking.**

2. **Activation energies of *endothermic reactions that involve both bond formation and bond rupture* will be greater than the heat of reaction, ΔH°.** Two examples illustrate this principle: the first chain-propagating step in the chlorination of methane and the corresponding step in the bromination of methane:

$Cl\cdot\ +\ CH_3{-}H\ \longrightarrow\ H{-}Cl\ +\ CH_3\cdot$ $\Delta H^{\circ} = +1$ kcal mol^{-1}
    $(DH^{\circ} = 104)$ $(DH^{\circ} = 103)$ $E_{act} = +3.8$ kcal mol^{-1}

$Br\cdot\ +\ CH_3{-}H\ \longrightarrow\ H{-}Br\ +\ CH_3\cdot$ $\Delta H^{\circ} = +16.5$ kcal mol^{-1}
    $(DH^{\circ} = 104)$ $(DH^{\circ} = 87.5)$ $E_{act} = +18.6$ kcal mol^{-1}

In both of these reactions the energy released in bond formation is less than that required for bond rupture; both reactions are, therefore, endothermic. We can easily see why the energy of activation for each reaction is greater than the heat of reaction by looking at the potential energy diagrams in Fig. 9.3. In each case the path from reactants to products is from a lower energy plateau to a higher one. In each case the intervening energy hill is higher still, and since the energy of activation is the vertical (energy) distance between the plateau of reactants and the top of this hill the energy of activation exceeds the heat of reaction.

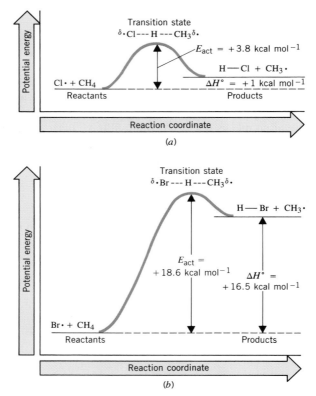

FIGURE 9.3 Potential energy diagrams for (*a*) the reaction of a chlorine atom with methane and (*b*) for the reaction of a bromine atom with methane.

3. **The energy of activation of a gas-phase reaction where bonds are broken homolytically but no bonds are formed is equal to $\Delta H°$.*** An example of this type of reaction is the chain-initiating step in the chlorination of methane: the dissociation of chlorine molecules into chlorine atoms.

$$Cl—Cl \longrightarrow 2\ Cl·\qquad \Delta H° = +58\ kcal\ mol^{-1}$$
$$(DH° = 58)\qquad\qquad E_{act} = +58\ kcal\ mol^{-1}$$

The potential energy diagram for this reaction is shown in Fig. 9.4.

4. **The energy of activation for a gas-phase reaction in which small radicals combine to form molecules is usually zero.** In reactions of this type the problem of nonsimultaneous bond formation and bond rupture does not exist; only one process occurs: that of bond formation. All of the chain-terminating steps in the chlorination of methane fall into this category. An example is the combination of two methyl radicals to form a molecule of ethane.

$$2\ CH_3· \longrightarrow CH_3—CH_3\qquad \Delta H° = -88\ kcal\ mol^{-1}$$
$$(DH° = 88)\qquad\qquad E_{act} = 0$$

Figure 9.5 illustrates the potential energy changes that occur in this reaction.

---

*This rule applies only to radical reactions taking place in the gas phase. It does not apply to reactions taking place in solution, especially where ions are involved, because solvation energies are also important.

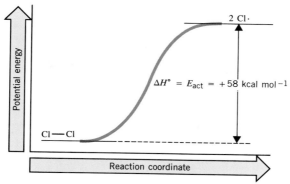

FIGURE 9.4 The potential energy diagram for the dissociation of a chlorine molecule into chlorine atoms.

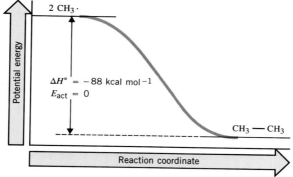

FIGURE 9.5 The potential energy diagram for the combination of two methyl radicals to form a molecule of ethane.

PROBLEM 9.9

When pentane is heated to a very high temperature, radical reactions take place that produce (among other products) methane, ethane, propane, and butane. This type of change is called **thermal cracking.** Among the reactions that take place are the following:

(1) $CH_3CH_2CH_2CH_2CH_3 \longrightarrow CH_3\cdot + CH_3CH_2CH_2CH_2\cdot$

(2) $CH_3CH_2CH_2CH_2CH_3 \longrightarrow CH_3CH_2\cdot + CH_3CH_2CH_2\cdot$

(3) $CH_3\cdot + CH_3\cdot \longrightarrow CH_3CH_3$

(4) $CH_3\cdot + CH_3CH_2CH_2CH_2CH_3 \longrightarrow CH_4 + CH_3CH_2CH_2CH_2CH_2\cdot$

(5) $CH_3\cdot + CH_3CH_2\cdot \longrightarrow CH_3CH_2CH_3$

(6) $CH_3CH_2\cdot + CH_3CH_2\cdot \longrightarrow CH_3CH_2CH_2CH_3$

(a) For which of these reactions would you expect E_{act} to equal zero? (b) To be greater than zero? (c) To equal $\Delta H°$?

PROBLEM 9.10

The energy of activation for the first chain-propagating step in the fluorination of methane (cf. Problem 9.7) is known to be $+1.2$ kcal mol^{-1}. The energy of activation for the second chain-propagating step is known to be very small. Assuming that it is $+1.0$ kcal mol^{-1} (a) and (b) Sketch potential energy diagrams for these two chain-propagating steps. (c) Sketch a potential energy diagram for the chain-initiating step in the fluorination of methane and (d) for the chain-terminating step that produces CH_3F. (e) Sketch a potential energy diagram for the following reaction:

$$CH_3 \cdot + HF \longrightarrow CH_4 + F \cdot$$

9.5C Reaction of Methane with Other Halogens

The *reactivity* of one substance toward another is measured by the *rate* at which the two substances react. A reagent that reacts very rapidly with a particular substance is said to be highly reactive toward that substance. One that reacts slowly or not at all under the same experimental conditions (e.g., concentration, pressure, and temperature) is said to have a low relative reactivity or to be unreactive. The reactions of the halogens (fluorine, chlorine, bromine, and iodine) with methane show a wide spread of relative reactivities. Fluorine is most reactive—so reactive, in fact, that without special precautions mixtures of fluorine and methane explode. Chlorine is the next most reactive. However, the chlorination of methane is easily controlled by the judicious control of heat and light. Bromine is much less reactive toward methane than chlorine, and iodine is so unreactive that for all practical purposes we can say that no reaction takes place at all.

If the mechanisms for fluorination, bromination, and iodination of methane are the same as for its chlorination, we can explain the wide variation in reactivity of the halogens by a careful examination of $\Delta H°$ and E_{act} for each step.

FLUORINATION

	$\Delta H°$ (kcal mol^{-1})	E_{act} (kcal mol^{-1})
Chain Initiation		
$F_2 \longrightarrow 2 F\cdot$	$+\ 38$	$+38$
Chain Propagation		
$F\cdot + CH_4 \longrightarrow HF + CH_3\cdot$	$-\ 32$	$+\ 1.2$
$CH_3\cdot + F_2 \longrightarrow CH_3F + F\cdot$	$-\ 70$	Small
Overall $\Delta H° = -102$		

The chain-initiating step in fluorination is highly endothermic and thus has a high energy of activation.

If we did not know otherwise, we might carelessly conclude from the energy of activation of the chain-initiating step alone that fluorine would be quite unreactive toward methane. (If we then proceeded to try the reaction, as a result of this careless assessment, the results would be literally disastrous.) We know, however, that the chain-initiating step occurs only infrequently relative to the chain-propagating steps. One initiating step is able to produce thousands of fluorination reactions. As a result, the high activation energy for this step is not an impediment to the reaction.

Chain-propagating steps, by contrast, cannot afford to have high energies of activation. If they do, the highly reactive intermediates are consumed by chain-terminating steps before the chains progress very far. Both of the chain-propagating steps in fluorination have very small energies of activation. This allows a relatively large fraction of energetically favorable collisions even at room temperature. Moreover, the overall heat of reaction, $\Delta H°$, is very large. This means that as the reaction occurs, a large quantity of heat is evolved. This heat may accumulate in the mixture faster than it dissipates to the surroundings, causing the temperature to rise and with it a rapid increase in the frequency of additional chain-initiating steps that would generate additional chains. These two factors, the low energy of activation for the chain-propagating steps, and the large overall heat of reaction, account for the high reactivity of fluorine toward methane.*

CHLORINATION

	$\Delta H°$ (kcal mol^{-1})	E_{act} (kcal mol^{-1})

Chain Initiation

$Cl_2 \longrightarrow 2\ Cl\cdot$	$+58$	$+58$

Chain Propagation

$Cl\cdot + CH_4 \longrightarrow HCl + CH_3\cdot$	$+\ 1$	$+\ 3.8$
$CH_3\cdot + Cl_2 \longrightarrow CH_3Cl + Cl\cdot$	-25.5	Small
Overall $\Delta H° = -24.5$		

The higher energy of activation of the first chain-propagating step (the hydrogen abstraction step) in chlorination of methane ($+3.8$ kcal mol^{-1}), versus the lower energy of activation ($+1.2$ kcal mol^{-1}) in fluorination, partly explains the lower reactivity of chlorine. The greater energy required to break the chlorine–chlorine bond in the initiating step ($+58$ kcal mol^{-1} for Cl_2 versus $+38$ kcal mol^{-1} for F_2) has some effect, too. However, the much greater overall heat of reaction in fluorination probably plays the greatest role in accounting for the much greater reactivity of fluorine.

BROMINATION

	$\Delta H°$ (kcal mol^{-1})	E_{act} (kcal mol^{-1})

Chain Initiation

$Br_2 \longrightarrow 2\ Br\cdot$	$+46$	$+46$

Chain Propagation

$Br\cdot + CH_4 \longrightarrow HBr + CH_3\cdot$	$+16.5$	$+18.6$
$CH_3\cdot + Br_2 \longrightarrow CH_3Br + Br\cdot$	-24	Small
Overall $\Delta H° = -\ 7.5$		

In contrast to chlorination, the hydrogen atom–abstraction step in bromination has a very high energy of activation ($E_{act} = 18.6$ kcal mol^{-1}). This means that only a very

---

*Fluorination reactions can be controlled. This is usually accomplished by diluting both the hydrocarbon and the fluorine with an inert gas such as helium before bringing them together. The reaction is also carried out in a reactor packed with copper shot. The copper, by absorbing the heat produced, moderates the reaction.

tiny fraction of all of the collisions between bromine atoms and methane molecules will be energetically effective even at a temperature of 300°C. Bromine, as a result, is much less reactive toward methane than chlorine, even though the net reaction is slightly exothermic.

IODINATION

	$\Delta H°$ (kcal mol^{-1})	E_{act} (kcal mol^{-1})
Chain Initiation		
$I_2 \longrightarrow 2 I\cdot$	$+36$	$+36$
Chain Propagation		
$I\cdot + CH_4 \longrightarrow HI + CH_3\cdot$	$+33$	$+33.5$
$CH_3\cdot + I_2 \longrightarrow CH_3I + I\cdot$	-20	Small
Overall $\Delta H° = +13$		

The thermodynamic quantities for iodination of methane make it clear that the chain-initiating step is not responsible for the observed order of reactivities: $F_2 > Cl_2 > Br_2 > I_2$. The iodine–iodine bond is even weaker than the fluorine–fluorine bond. On this basis alone, one would predict iodine to be the most reactive of the halogens. This clearly is not the case. Once again, it is the hydrogen-atom abstraction step that correlates with the experimentally determined order of reactivities. The energy of activation of this step in the iodine reaction (33.5 kcal mol^{-1}) is so large that only two collisions out of every 10^{12} have sufficient energy to produce reactions at 300°C. As a result, iodination is not a feasible reaction experimentally.

Before we leave this topic, one further point needs to be made. We have given explanations of the relative reactivities of the halogens toward methane that have been based on energy considerations alone. This has been possible *only because the reactions are quite similar and thus have similar entropy changes.* Had the reactions been of different types, this kind of analysis would not have been proper and might have given incorrect explanations.

9.6 HALOGENATION OF HIGHER ALKANES

Higher alkanes react with halogens by the same kind of chain mechanisms as those that we have just seen. Ethane, for example, reacts with chlorine to produce chloroethane (ethyl chloride). The mechanism is as follows:

Chain Initiation

Step 1 $\quad Cl_2 \xrightarrow[\text{or}\\ \text{heat}]{\text{light}} 2\,Cl\cdot$

Chain Propagation

Step 2 $\quad CH_3\overset{\frown}{C}H_2:H + \cdot Cl \longrightarrow CH_3CH_2\cdot + H:Cl$

Step 3 $\quad CH_3CH_2\cdot + Cl:Cl \longrightarrow CH_3CH_2:Cl + Cl\cdot$

Then steps 2, 3, 2, 3, and so on.

Chain Termination

$$CH_3CH_2\cdot + \cdot Cl \longrightarrow CH_3CH_2\!:\!Cl$$

$$CH_3CH_2\cdot + \cdot CH_2CH_3 \longrightarrow CH_3CH_2\!:\!CH_2CH_3$$

$$Cl\cdot + \cdot Cl \longrightarrow Cl\!:\!Cl$$

PROBLEM 9.11

The energy of activation for the hydrogen atom–abstraction step in the chlorination of ethane is 1.0 kcal mol^{-1}. (a) Use the homolytic bond dissociation energies in Table 9.1 to calculate $\Delta H°$ for this step. (b) Sketch a potential energy diagram for the hydrogen-atom abstraction step in the chlorination of ethane similar to that for the chlorination of methane shown in Fig. 9.3a. (c) When an equimolar mixture of methane and ethane is chlorinated, the reaction yields far more ethyl chloride than methyl chloride (~ 400 molecules of ethyl chloride for every molecule of methyl chloride). Explain this greater yield of ethyl chloride.

PROBLEM 9.12

When ethane is chlorinated, 1,1-dichloroethane and 1,2-dichloroethane, as well as more highly chlorinated ethanes are formed in the mixture (cf. Section 9.3A). Write chain mechanisms accounting for the formation of 1,1-dichloroethane and 1,2-dichloroethane.

Chlorination of most alkanes whose molecules contain more than two carbon atoms gives a mixture of isomeric monochloro products (as well as more highly chlorinated compounds). Several examples follow. The percentages given are based on the total amount of monochloro products formed in each reaction.

$$CH_3CH_2CH_3 \xrightarrow[\text{light, 25°C}]{Cl_2} CH_3CH_2CH_2Cl \ + \ CH_3\underset{\underset{Cl}{|}}{C}HCH_3$$

| Propane | Propyl chloride (45%) | Isopropyl chloride (55%) |

$$CH_3\underset{\underset{CH_3}{|}}{C}HCH_3 \xrightarrow[\substack{\text{light}\\25°C}]{Cl_2} CH_3\underset{\underset{CH_3}{|}}{C}HCH_2Cl \ + \ CH_3\underset{\underset{Cl}{|}}{\overset{\overset{CH_3}{|}}{C}}CH_3$$

| Isobutane | Isobutyl chloride (63%) | *tert*-Butyl chloride (37%) |

$$CH_3CCH_2CH_3 \xrightarrow[300°C]{Cl_2} ClCH_2CHCH_2CH_3 \ + \ CH_3CCH_2CH_3$$

(2-Methylbutane) with CH₃ and H substituents → 1-Chloro-2-methyl-butane (30%) + 2-Chloro-2-methyl-butane (22%)

2-Methylbutane

1-Chloro-2-methyl-butane
(30%)

2-Chloro-2-methyl-butane
(22%)

$$+ \ CH_3CHCHCH_3 \ + \ CH_3CHCH_2CH_2Cl$$

2-Chloro-3-methyl-butane
(33%)

1-Chloro-3-methyl-butane
(15%)

The ratios of products that we obtain from chlorination reactions of higher alkanes are not identical with what we would expect if all the hydrogen atoms of the alkane were equally reactive. We find that there is a correlation between reactivity of different hydrogen atoms and the type of hydrogen atom (1°, 2°, or 3°) being replaced. The tertiary hydrogen atoms of an alkane are most reactive, secondary hydrogen atoms are next most reactive, and primary hydrogen atoms are the least reactive (see Problem 9.13).

PROBLEM 9.13

(a) What percentages of propyl chloride and isopropyl chloride would you expect to obtain from the chlorination of propane if 1° and 2° hydrogen atoms were equally reactive?   (b) What percentages of isobutyl chloride and *tert*-butyl chloride would you expect from the chlorination of isobutane if the 1° and 3° hydrogen atoms were equally reactive?   (c) Compare these calculated answers with the results actually obtained (above) and justify the assertion that the order of reactivity of the hydrogen atoms is 3° > 2° > 1°.

We can account for the relative reactivities of the primary, secondary, and tertiary hydrogen atoms in a chlorination reaction on the basis of the homolytic bond dissociation energies we saw earlier (Table 9.1). Of the three types, breaking a tertiary C—H bond requires the least energy, and breaking a primary C—H bond requires the most. Since the step in which the C—H bond is broken (i.e., the hydrogen atom–abstraction step) determines the location or orientation of the chlorination, we would expect the E_{act} for abstracting a tertiary hydrogen atom to be least and the E_{act} for abstracting a primary hydrogen atom to be greatest. Thus tertiary hydrogen atoms should be most reactive, secondary hydrogen atoms should be the next most reactive, and primary hydrogen atoms should be the least reactive.

The differences in the rates with which primary, secondary, and tertiary hydrogen atoms are replaced by chlorine are not large, however. Chlorine, as a result, does not discriminate among the different types of hydrogen atoms in a way that makes chlorination of higher alkanes a generally useful laboratory synthesis. (Alkane chlorinations do find use in some industrial processes, especially in those instances where mixtures of alkyl chlorides can be used.)

PROBLEM 9.14 _____

Chlorination reactions of certain alkanes can be used for laboratory preparations. Examples are the preparation of cyclopropyl chloride from cyclopropane and cyclobutyl chloride from cyclobutane. What structural feature of these molecules makes this possible?

PROBLEM 9.15 _____

Each of the following alkanes reacts with chlorine to give a single monochloro substitution product. On the basis of this information, deduce the structure of each alkane.

(a) C_5H_{10} (b) C_8H_{18} (c) C_5H_{12}

9.6A Selectivity of Bromine

Bromine is less reactive toward alkanes in general than chlorine, but bromine is more *selective* in the site of attack when it does react. Bromine shows a much greater ability to discriminate among the different types of hydrogen atoms. The reaction of isobutane and bromine, for example, gives almost exclusive replacement of the tertiary hydrogen atom.

$$CH_3\!-\!\underset{\underset{H}{|}}{\overset{\overset{CH_3}{|}}{C}}\!-\!CH_3 \xrightarrow[\text{light, 127°C}]{Br_2} CH_3\!-\!\underset{\underset{Br}{|}}{\overset{\overset{CH_3}{|}}{C}}\!-\!CH_3 + CH_3\!-\!\underset{\underset{H}{|}}{\overset{\overset{CH_3}{|}}{C}}\!-\!CH_2Br$$

$$(>99\%) \qquad\qquad (\text{Trace})$$

A very different result is obtained when isobutane reacts with chlorine.

$$CH_3CHCH_3 \xrightarrow[25°C]{Cl_2,\ h\nu} CH_3\underset{\underset{Cl}{|}}{\overset{\overset{CH_3}{|}}{C}}CH_3 + CH_3\overset{\overset{CH_3}{|}}{C}HCH_2Cl$$

$$(37\%) \qquad\qquad (63\%)$$

Fluorine, being much more reactive than chlorine, *is even less selective than chlorine.* Because the energy of activation for the abstraction of any type of hydrogen by a fluorine atom is low, there is very little difference in the rate at which a 1°, 2°, or 3° hydrogen reacts with fluorine. Reactions of alkanes with fluorine give (almost) the distribution of products that we would expect if all of the hydrogens of the alkane were equally reactive.

PROBLEM 9.16 _____

Why is temperature an important variable to consider when using isomer distribution to evaluate the reactivities of the hydrogens of an alkane towards radical chlorination?

9.7 THE GEOMETRY OF ALKYL RADICALS

Experimental evidence indicates that the geometrical structure of most alkyl radicals is trigonal planar at the carbon having the unpaired electron. This structure can be accommodated by an sp^2-hybridized central carbon. In an alkyl radical, the p orbital contains the unpaired electron (Fig. 9.6).

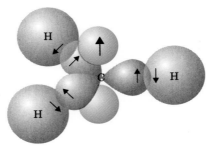

FIGURE 9.6 sp^2-Hybridized carbon atom at the center of a methyl radical showing the odd electron in one lobe of the half-filled p orbital. It could be shown in the other lobe.

9.8 REACTIONS THAT GENERATE TETRAHEDRAL STEREOCENTERS

When achiral molecules react to produce a compound with a single tetrahedral stereocenter, the product will be obtained as a racemic form. This will always be true in the absence of any chiral influence on the reaction such as an enzyme or the use of a chiral solvent.

Let us examine a reaction that illustrates this principle, the radical chlorination of pentane.

$$CH_3CH_2CH_2CH_2CH_3 \xrightarrow[\text{(achiral)}]{Cl_2} CH_3CH_2CH_2CH_2CH_2Cl$$

Pentane
(achiral)

1-Chloropentane
(achiral)

$$+ \; CH_3CH_2CH_2\overset{*}{C}HClCH_3 \; + \; CH_3CH_2CHClCH_2CH_3$$

(±)-2-Chloropentane
(a racemic form)

3-Chloropentane
(achiral)

The reaction will lead to the products shown here, as well as more highly chlorinated products. (We can use an excess of pentane to minimize multiple chlorinations.) Neither 1-chloropentane nor 3-chloropentane contains a stereocenter, but 2-chloropentane does, and it is *obtained as a racemic form*. If we examine the mechanism in Fig. 9.7 we shall see why.

Abstraction of a hydrogen atom from C2 produces a trigonal planar radical that is achiral. This radical then reacts with chlorine at either face [by path (a) or path (b)]. Because the radical is achiral the probability of reaction by either path is the same, therefore, the two enantiomers are produced in equal amounts.

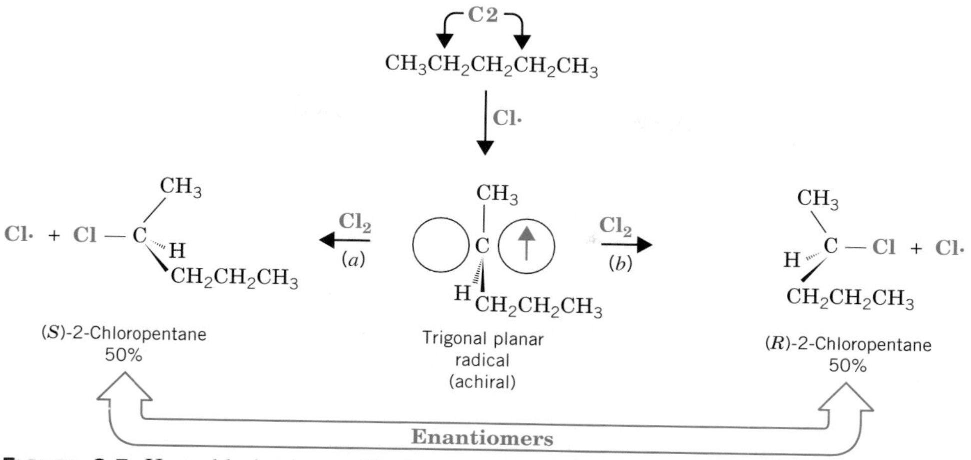

FIGURE 9.7 How chlorination at C2 of pentane yields a racemic form of 2-chloropentane.

9.8A Generation of a Second Stereocenter in a Radical Halogenation

Let us now examine what happens when a chiral molecule (containing one stereocenter) reacts so as to yield a product with a second stereocenter. As an example consider what happens when (S)-2-chloropentane undergoes chlorination at C3 (other products are formed, of course, by chlorination at other carbon atoms). The results of chlorination at C3 are shown in Fig. 9.8.

The products of the reactions are $(2S,3S)$-2,3-dichloropentane and $(2S,3R)$-2,3-dichloropentane. These two compounds are **diastereomers.** (They are stereoisomers but they are not mirror images of each other.) The two diastereomers are *not* produced in equal amounts. Because the intermediate radical itself is chiral, reactions at the two faces are not equally likely. The radical reacts with chlorine to a greater extent at one face than the other (although we cannot easily predict which). That is, the presence of a stereocenter in the radical (at C2) influences the reaction that introduces the new stereocenter (at C3).

Both of the 2,3-dichloropentane diastereomers are chiral and, therefore, each exhibits optical activity. Moreover, because the two compounds are *diastereomers,* they have different physical properties (e.g., different melting points and boiling points) and are separable by conventional means (by gas–liquid chromatography or by careful fractional distillation).

PROBLEM 9.17

Consider the chlorination of (S)-2-chloropentane at C4. (a) Write stereochemical structures for the products that would be obtained and give each its proper $(R-S)$ designation. (b) What is the stereoisomeric relationship between these products? (c) Are both products chiral? (d) Are both optically active? (e) Could the products be separated by conventional means? (f) What other

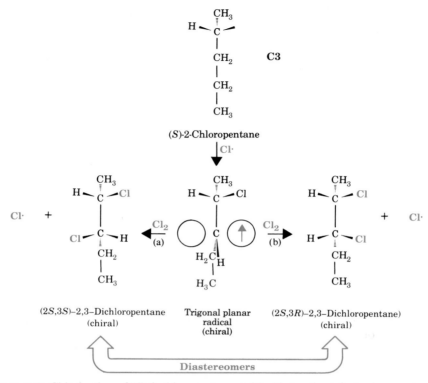

FIGURE 9.8 Chlorination of (*S*)-2-chloropentane at C3. Abstraction of a hydrogen atom from C3 produces a radical that is chiral (it contains a stereocenter at C2). This chiral radical can then react with chlorine at one face [path (a)] to produce (2*S*,3*S*)-2,3-dichloropentane and at the other face [path (b)] to yield (2*S*,3*R*)-2,3-dichloropentane. These two compounds are diastereomers and they are not produced in equal amounts. Each product is chiral and each alone would be optically active.

dichloropentanes would be obtained by chlorination of (*S*)-2-chloropentane? (g) Which of these are optically active?

PROBLEM 9.18

Consider the bromination of butane using sufficient bromine to cause dibromination. After the reaction is over, you isolate all of the dibromoisomers by gas–liquid chromatography or by fractional distillation.   (a) How many fractions would you obtain?   (b) What compound (or compounds) would the individual fractions contain?   (c) Which, if any, of the fractions would show optical activity?

PROBLEM 9.19

The chlorination of 2-methylbutane yields 1-chloro-2-methylbutane, 2-chloro-2-methylbutane, 2-chloro-3-methylbutane, and 1-chloro-3-methylbutane.   (a) Assuming that these compounds were separated after the reaction by fractional distillation, tell whether any fractions would show optical activity.   (b) Would any of these fractions be resolvable into enantiomers?

9.9 RADICAL ADDITION TO ALKENES: THE ANTI-MARKOVNIKOV ADDITION OF HYDROGEN BROMIDE

Before 1933, the orientation of the addition of hydrogen bromide to alkenes was the subject of much confusion. At times addition occurred in accordance with Markovnikov's rule; at other times it occurred in just the opposite manner. Many instances were reported where, under what seemed to be the same experimental conditions, Markovnikov additions were obtained in one laboratory and anti-Markovnikov additions in another. At times even the same chemist would obtain different results using the same conditions but on different occasions.

The mystery was solved in 1933 by the research of M. S. Kharasch and F. R. Mayo (of the University of Chicago). The explanatory factor turned out to be organic peroxides present in the alkenes—peroxides that were formed by the action of atmospheric oxygen on the alkenes (Section 9.11B). Kharasch and Mayo found that when alkenes that contained peroxides or hydroperoxides reacted with hydrogen bromide, anti-Markovnikov addition of hydrogen bromide occurred.

$$R—\overset{\cdot\cdot}{\underset{\cdot\cdot}{O}}—\overset{\cdot\cdot}{\underset{\cdot\cdot}{O}}—R \qquad\qquad R—\overset{\cdot\cdot}{\underset{\cdot\cdot}{O}}—\overset{\cdot\cdot}{\underset{\cdot\cdot}{O}}—H$$

An organic peroxide   **An organic hydroperoxide**

Under these conditions, for example, propene yields 1-bromopropane. In the absence of peroxides, or in the presence of compounds that would "trap" radicals, normal Markovnikov addition occurs.

$$CH_3CH{=}CH_2 + HBr \xrightarrow{ROOR} CH_3CH_2CH_2Br \qquad \text{Anti-Markovnikov addition}$$

$$CH_3CH{=}CH_2 + HBr \xrightarrow[\text{peroxides}]{\text{no}} \underset{\underset{\text{2-Bromopropane}}{Br}}{CH_3\underset{|}{CH}CH_3} \qquad \text{Markovnikov addition}$$

Hydrogen fluoride, hydrogen chloride, and hydrogen iodide *do not* **give anti-Markovnikov addition even when peroxides are present.**

According to Kharasch and Mayo, the mechanism for anti-Markovnikov addition of hydrogen bromide is a *radical chain reaction* initiated by peroxides:

A MECHANISM FOR ANTI-MARKOVNIKOV ADDITION

Chain Initiation

Step 1 $R—\overset{\cdot\cdot}{\underset{\cdot\cdot}{O}} \colon \overset{\cdot\cdot}{\underset{\cdot\cdot}{O}}—R \xrightarrow{\text{heat}} 2\,R—\overset{\cdot\cdot}{\underset{\cdot\cdot}{O}}\cdot$

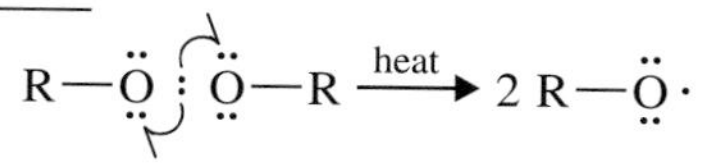

Heat brings about homolytic
cleavage of the weak
oxygen–oxygen bond.

Step 2 $R—\overset{\cdot\cdot}{\underset{\cdot\cdot}{O}}\cdot + H \colon \overset{\cdot\cdot}{\underset{\cdot\cdot}{Br}} \colon \longrightarrow R—\overset{\cdot\cdot}{\underset{\cdot\cdot}{O}} \colon H + \colon \overset{\cdot\cdot}{\underset{\cdot}{Br}}\cdot$

The alkoxy radical abstracts a
hydrogen atom from HBr, producing a
bromine atom.

Chain Propagation

Step 3 $: \ddot{Br} \cdot + H_2C = CH - CH_3 \longrightarrow : \ddot{Br} : CH_2 - \dot{C}H - CH_3$

2° Radical

A bromine atom adds to the double bond
to produce the more stable 2° radical.

Step 4 $: \ddot{Br} - CH_2 - \dot{C}H - CH_3 + H : \ddot{Br} : \longrightarrow : \ddot{Br} - CH_2 - \underset{H}{CH} - CH_3 + \cdot \ddot{Br} :$

1-Bromopropane

The 2° radical abstracts a hydrogen atom from HBr. This leads to the
product and regenerates a bromide atom. Then repetitions of steps 3 and 4
lead to a chain reaction.

Step 1 is the simple homolytic cleavage of the peroxide molecule to produce two alkoxyl radicals. The oxygen–oxygen bond of peroxides is weak and such reactions are known to occur readily.

$$R - \ddot{O} : \ddot{O} - R \longrightarrow 2 R - \ddot{O} \cdot \qquad \Delta H° \cong + 35 \text{ kcal mol}^{-1}$$

Peroxide   **Alkoxyl radical**

Step 2 of the mechanism, abstraction of a hydrogen atom by the radical, is exothermic and has a low energy of activation.

$$R - \ddot{O} \cdot + H : \ddot{Br} : \longrightarrow R - \ddot{O} : H + : \ddot{Br} \cdot \qquad \begin{array}{l} \Delta H° \cong - 23 \text{ kcal mol}^{-1} \\ E_{\text{act}} \text{ is low} \end{array}$$

Step 3 of the mechanism determines the final orientation of bromine in the product. It occurs as it does because a *more stable secondary radical* is produced and because *attack at the primary carbon atom is less hindered.* Had the bromine attacked propene at the secondary carbon atom, a less stable, primary radical would have been the result,

$$Br \cdot + CH_2 = CHCH_3 \xrightarrow{\times} \cdot \underset{\underset{Br}{|}}{CH_2}CHCH_3$$

**1° Radical
(less stable)**

and attack at the secondary carbon atom would have been more hindered.

Step 4 of the mechanism is simply the abstraction of a hydrogen atom from hydrogen bromide by the radical produced in step 3. This hydrogen-atom abstraction produces a bromine atom that can bring about step 3 again, then step 4 occurs again— a chain reaction.

We can now see the contrast between the two ways that HBr can add to an alkene.

In the absence of peroxides, the reagent that attacks the double bond is a proton. Because a proton is small, steric effects are unimportant. It attaches itself to a carbon atom in the way that yields the more stable carbocation. The result is Markovnikov addition.

Ionic Addition

$$CH_3CH = CH_2 \xrightarrow{H^+} CH_3\overset{+}{C}HCH_3 \xrightarrow{Br^-} CH_3\underset{\underset{Br}{|}}{C}HCH_3$$

More stable   **Markovnikov**
carbocation   **product**

In the presence of peroxides, the reagent that attacks the double bond is a larger bromine atom. It attaches itself to the less hindered carbon atom in the way that yields the more stable radical. The result is anti-Markovnikov addition.

Radical Addition

$$CH_3CH = CH_2 \xrightarrow{Br\cdot} CH_3\overset{\cdot}{C}HCH_2Br \xrightarrow{HBr} CH_3CH_2CH_2Br \ + \ Br\cdot$$

More stable   **Anti-Markovnikov**
radical   **product**

9.10 RADICAL POLYMERIZATION OF ALKENES: CHAIN-GROWTH POLYMERS

Polymers are substances that consist of very large molecules called **macromolecules** that are made up of many repeating subunits. The molecular subunits that are used to synthesize polymers are called **monomers,** and the reactions by which monomers are joined together are called **polymerizations.** Many polymerizations can be initiated by radicals.

Ethylene (ethene), for example, is the monomer that is used to synthesize the familiar polymer called *polyethylene.*

Monomeric units

$$m \ CH_2 = CH_2 \xrightarrow{\text{polymerization}} -CH_2CH_2 (\!\!-CH_2CH_2\!\!-)_n CH_2CH_2-$$

Ethylene   **Polyethylene**
monomer   *polymer*
(*m* and *n* are large numbers)

Because polymers such as polyethylene are made by addition reactions, they are often called **chain-growth polymers or addition polymers.** Let us now examine in some detail how polyethylene is made.

Ethylene polymerizes by a radical mechanism when it is heated at a pressure of 1000 atm with a small amount of an organic peroxide (called a diacylperoxide). The diacylperoxide dissociates to produce radicals, which in turn initiate chains.

Chain Initiation

Step 1 $R-\overset{\overset{O}{\|}}{C}-O\!:\!O-\overset{\overset{O}{\|}}{C}-R \longrightarrow 2\,R\!:\!\overset{\overset{O}{\|}}{C}-O\cdot \longrightarrow 2\,CO_2 + 2\,R\cdot$

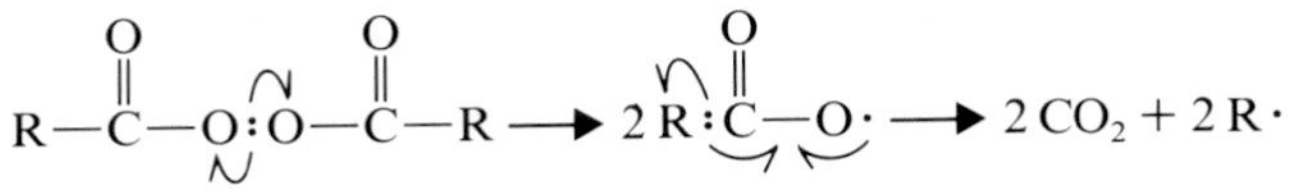

Diacylperoxide

Step 2 $R\cdot + CH_2{=}CH_2 \longrightarrow R\!:\!CH_2-CH_2\cdot$

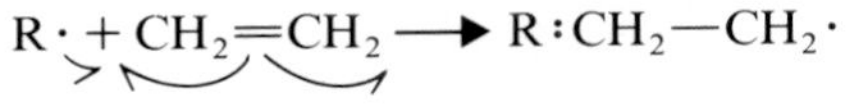

Chain Propagation

Step 3 $R-CH_2CH_2\cdot + n\,CH_2{=}CH_2 \longrightarrow R{+}CH_2CH_2{\rightarrow}_n CH_2CH_2\cdot$

Chains propagate by adding successive ethylene units, until their growth is stopped by combination or disproportionation.

Chain Termination

Step 4 $2\,R{+}CH_2CH_2{\rightarrow}_n CH_2CH_2\cdot$

$\xrightarrow{\text{combination}} [R{+}CH_2CH_2{\rightarrow}_n CH_2CH_2{+}_2$

$\xrightarrow{\text{disproportionation}} R{+}CH_2CH_2{\rightarrow}_n CH{=}CH_2 +$
$R{+}CH_2CH_2{\rightarrow}_n CH_2CH_3$

The radical at the end of the growing polymer chain can also abstract a hydrogen atom from itself by what is called "back biting." This leads to chain branching.

Chain Branching

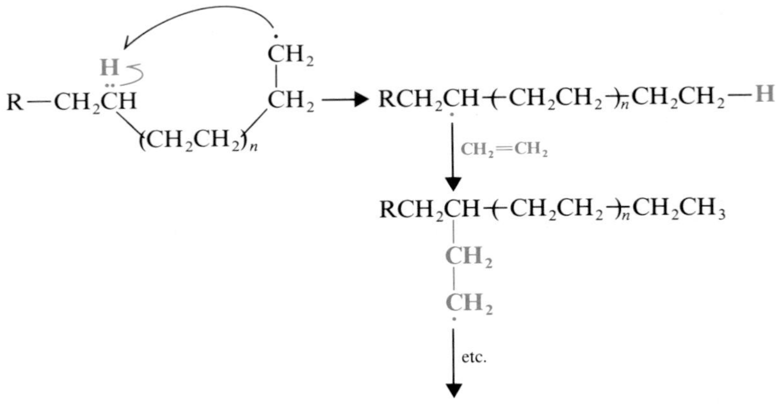

$R-CH_2\overset{..}{C}H\;\underset{(CH_2CH_2)_n}{\overset{H}{\diagdown}}\;\overset{\overset{\overset{\cdot}{CH_2}}{|}}{CH_2} \longrightarrow RCH_2\overset{\cdot}{C}H{+}CH_2CH_2{\rightarrow}_n CH_2CH_2{-}H$

$\downarrow CH_2{=}CH_2$

$RCH_2\underset{\overset{|}{CH_2}}{\overset{|}{C}H}{+}CH_2CH_2{\rightarrow}_n CH_2CH_3$

$\underset{\overset{|}{CH_2}}{\overset{|}{}}$

$\downarrow \text{etc.}$

The polyethylene produced by radical polymerization is not generally useful unless it has a molecular weight of nearly 1,000,000. Very high molecular weight polyethylene can be obtained by using a low concentration of the initiator. This initiates the growth of only a few chains and ensures that each chain will have a large excess of the monomer available. More initiator may be added as chains terminate during the polymerization and, in this way, new chains are begun.

Polyethylene has been produced commercially since 1943. It is used in manufacturing flexible bottles, films, sheets, and insulation for electric wires. Polyethylene produced by radical polymerization has a softening point of about 110°C.

Polyethylene can be produced in a different way using catalysts called **Ziegler–Natta catalysts** that are organometallic complexes of transition metals. In this process no radicals are produced, no back biting occurs, and, consequently, there is no chain branching. The polyethylene that is produced is of higher density, has a higher melting point, and has greater strength. (Ziegler–Natta catalysts are discussed in greater detail in Special Topic C.)

Another familiar polymer is *polystyrene*. The monomer used in making polystyrene is phenylethene, a compound commonly known as *styrene*.

Styrene   **Polystyrene**

Table 9.2 lists several other common addition polymers.

TABLE 9.2 Other common chain-growth polymers

MONOMER	POLYMER	NAMES
$CH_2{=}CHCH_3$	$\\{+CH_2{-}CH\\}_n$ with CH_3	Polypropylene
$CH_2{=}CHCl$	$\\{+CH_2{-}CH\\}_n$ with Cl	Poly(vinyl chloride), PVC
$CH_2{=}CHCN$	$\\{+CH_2{-}CH\\}_n$ with CN	Polyacrylonitrile, Orlon
$CF_2{=}CF_2$	$\\{+CF_2{-}CF_2\\}_n$	Polytetrafluoroethene, Teflon
$CH_2{=}CCO_2CH_3$ with CH_3	$\\{+CH_2{-}C\\}_n$ with CH_3 and CO_2CH_3	Poly(methyl methacrylate), Lucite, Plexiglas, Perspex

9.11 OTHER IMPORTANT RADICAL CHAIN REACTIONS

Radical chain mechanisms are important in understanding many other organic reactions. We shall see other examples in later chapters, but let us examine three here: the combustion of alkanes, autoxidation, and some reactions of chlorofluoromethanes that have threatened the protective layer of ozone in the stratosphere.

9.11A Combustion of Alkanes

When alkanes react with oxygen (e.g., in oil furnaces and in internal combustion engines) a complex series of reactions takes place, ultimately converting the alkane to carbon dioxide and water (Section 4.8A). Although our understanding of the detailed mechanism of combustion is incomplete, we do know that the important reactions occur by radical chain mechanisms with chain-initiating and chain-propagating steps such as the following reactions.

$$RH + O_2 \longrightarrow R\cdot + \cdot OOH \qquad \textbf{Initiating}$$

$$R\cdot + O_2 \longrightarrow R-OO\cdot$$
$$R-OO\cdot + R-H \longrightarrow R-OOH + R\cdot \Bigg\} \textbf{Propagating}$$

One product of the second step is R—OOH, called an alkyl hydroperoxide. The oxygen–oxygen bond of an alkyl hydroperoxide is quite weak, and it can break and produce radicals that can initiate other chains:

$$RO-OH \longrightarrow RO\cdot + HO\cdot$$

9.11B Autoxidation

Linoleic acid is an example of a *polyunsaturated fatty acid,* the kind of polyunsaturated acid that occurs as an ester in **polyunsaturated fats** (Section 7.5A and Chapter 23). By polyunsaturated, we mean that the compound contains two or more double bonds.

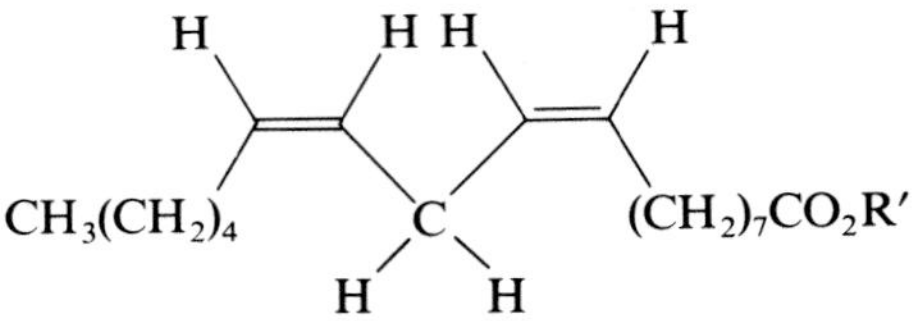

Linoleic acid
(as an ester)

Polyunsaturated fats occur widely in the fats and oils that are components of our diets. They are also widespread in the tissues of the body where they perform numerous vital functions.

The hydrogen atoms of the —CH$_2$— group located between the two double bonds of linoleic ester (Lin—H) are especially susceptible to abstraction by radicals (we shall see why in Chapter 12). Abstraction of one of these hydrogen atoms produces a new radical (Lin·) that can react with oxygen in a chain reaction that belongs to a general type of reaction called **autoxidation** (Fig. 9.9). The result of autoxidation is the formation of a hydroperoxide. Autoxidation is a process that occurs in many substances; for example, autoxidation is responsible for the development of the rancidity that occurs when fats and oils spoil and for the spontaneous combustion of oily rags left open to the air. Autoxidation also occurs in the body, and here it may cause irreversible damage. Autoxidation is inhibited when compounds are present that can rapidly "trap" peroxyl radicals by reacting with them to give stabilized radicals that do not continue the chain.

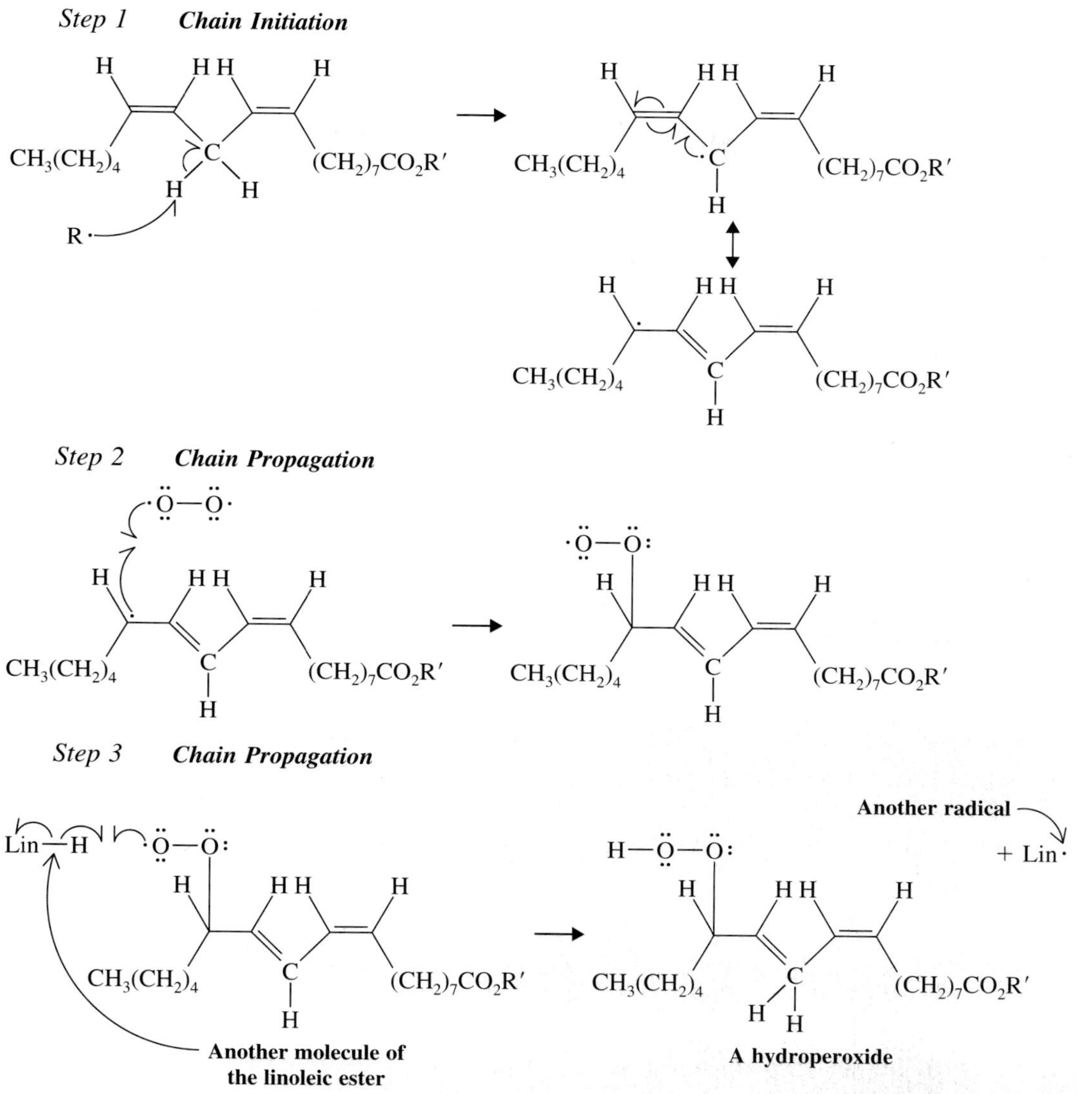

Step 1   **Chain Initiation**

Step 2   **Chain Propagation**

Step 3   **Chain Propagation**

**Another molecule of
the linoleic ester**

A hydroperoxide

FIGURE 9.9 Autoxidation of a linoleic acid ester. In step 1 the reaction is initiated by the attack of a radical on one of the hydrogen atoms of the —CH$_2$— group between the two double bonds; this hydrogen abstraction produces a radical that is a resonance hybrid. In step 2 this radical reacts with oxygen in the first of two chain-propagating steps to produce an oxygen-containing radical, which in step 3 can abstract a hydrogen from another molecule of the linoleic ester (Lin—H). The result of this second chain-propagating step is the formation of a hydroperoxide and a radical (Lin ·) that can bring about a repetition of step 2.

$$\text{Lin—O—O·} + \text{In—H} \longrightarrow \text{Lin—O—OH} + \text{In·}$$

Vitamin E (α-tocopherol) is capable of acting as a radical trap in this way, and one of the important roles that vitamin E plays in the body may be in inhibiting radical reactions that could cause cell damage. Such compounds as BHT are added to foods to prevent autoxidation. BHT is also known to trap radicals.

Vitamin E
(α-tocopherol)

BHT

9.11C Ozone Depletion and Chlorofluorocarbons (CFCs)

In the stratosphere at altitudes of about 25 km, very high energy (very short wave-length) UV light converts diatomic oxygen (O_2) into ozone (O_3). The reactions that take place may be represented as follows:

Step 1 $O_2 + h\nu \longrightarrow O + O$

Step 2 $O + O_2 + M \longrightarrow O_3 + M + heat$

where M is some other particle that can absorb some of the energy released in the second step.

The ozone produced in step 2 can also interact with high-energy UV light in the following way:

Step 3 $O_3 + h\nu \longrightarrow O_2 + O + heat$

The oxygen atom formed in step 3 can cause a repetition of step 2, and so forth. The net result of these steps is to convert highly energetic UV light into heat. This is important because the existence of this cycle shields the Earth from radiation that is destructive to living organisms. This shield makes life possible on the Earth's surface. Even a relatively small increase in high-energy UV radiation at the Earth's surface would cause a large increase in the incidence of skin cancers.

Production of chlorofluoromethanes (and of chlorofluoroethanes) called chloro-fluorocarbons (CFCs) or *freons* began in 1930. These compounds have been used as refrigerants, solvents, and propellants in aerosol cans. Typical freons are trichloro-fluoromethane, $CFCl_3$ (called Freon-11), and dichlorodifluoromethane, CF_2Cl_2 (called Freon-12).

By 1974 world freon production was about 2 billion pounds annually. Most freon, even that used in refrigeration, eventually makes its way into the atmosphere where it diffuses unchanged into the stratosphere. In June 1974 F. S. Rowland and M. J. Molina published an article indicating, for the first time, that in the stratosphere freon is able to initiate radical chain reactions that can upset the natural ozone balance. The reactions that take place are the following. (Freon-12 is used as an example.)

Chain Initiation

Step 1 $CF_2Cl_2 + h\nu \longrightarrow CF_2Cl\cdot + Cl\cdot$

Chain Propagation

Step 2 $Cl\cdot + O_3 \longrightarrow ClO\cdot + O_2$

Step 3 $ClO\cdot + O \longrightarrow O_2 + Cl\cdot$

In the chain-initiating step, UV light causes homolytic cleavage of one C—Cl bond of the freon. The chlorine atom thus produced is the real villain; it can set off a chain reaction that destroys thousands of molecules of ozone before it diffuses out of the stratosphere or reacts with some other substance.

In 1975 a study by the National Academy of Science supported the predictions of Rowland and Molina and since January 1978 the use of freons in aerosol cans in the United States has been banned.

In 1985 a hole was discovered in the ozone layer above Antarctica. Studies done since then strongly suggest that chlorine atom destruction of the ozone is a factor in the formation of the hole. This ozone hole has continued to grow in size and such a hole has also been discovered in the Arctic ozone layer. Should the ozone layer be depleted, more of the Sun's damaging rays would penetrate to the surface of the Earth.

Recognizing the global nature of the problem, in 1987, 24 nations signed the "Montreal Protocol," which among other things, requires these nations to reduce their consumption of chlorofluorocarbons by 50% before the turn of the century.

9.12 SOME IMPORTANT TERMS AND CONCEPTS

Radicals (or free radicals) are reactive intermediates that have an unpaired electron. The relative stability of alkyl radicals is as follows:

$$\underset{3°}{\overset{C}{\underset{C}{C-\overset{|}{\underset{|}{C}}\cdot}}} > \underset{2°}{\overset{C}{\underset{H}{C-\overset{|}{\underset{|}{C}}\cdot}}} > \underset{1°}{\overset{H}{\underset{H}{C-\overset{|}{\underset{|}{C}}\cdot}}} > \underset{Methyl}{\overset{H}{\underset{H}{H-\overset{|}{\underset{|}{C}}\cdot}}}$$

$3° \quad > \quad 2° \quad > \quad 1° \quad > \quad$ **Methyl**

Bond dissociation energy (abbreviated $DH°$) is the amount of energy required for homolysis of a covalent bond.

Halogenations of alkanes are substitution reactions in which a halogen replaces one (or more) of the alkane's hydrogen atoms.

$$RH + X_2 \longrightarrow RX + HX$$

The reactions occur by a radical mechanism.

Chain Initiation

Step 1 $X_2 \longrightarrow 2\,X\cdot$

Chain Propagation

Step 2   RH + X· $\longrightarrow$ R· + HX

Step 3   R· + X$_2$ $\longrightarrow$ RX + X·

Chain reactions are reactions whose mechanisms involve a series of steps with each step producing a reactive intermediate that causes the next step to occur. The halogenation of an alkane is a chain reaction.

ADDITIONAL PROBLEMS

9.20 The radical reaction of propane with chlorine yields (in addition to more highly halogenated compounds) 1-chloropropane and 2-chloropropane. Write chain-initiating and chain-propagating steps showing how each compound is formed.

9.21 In addition to more highly chlorinated products, chlorination of butane yields a mixture of compounds with the formula C$_4$H$_9$Cl. (a) Taking stereochemistry into account, how many different isomers with the formula C$_4$H$_9$Cl would you expect to be produced? (b) If the mixture of C$_4$H$_9$Cl isomers were subjected to fractional distillation how many fractions would you expect to obtain? (c) Which fractions would be optically inactive? (d) Which would you be able to resolve into enantiomers?

9.22 Chlorination of (*R*)-2-chlorobutane yields a mixture of isomers with the formula C$_4$H$_8$Cl$_2$. (a) How many different isomers would you expect to be produced? Write their structures. (b) If the mixture of C$_4$H$_8$Cl$_2$ isomers were subjected to fractional distillation how many fractions would you expect to obtain? (c) Which of these fractions would be optically active?

9.23 Peroxides are often used to initiate radical chain reactions such as alkane halogenations. (a) Examine the bond energies in Table 9.1 and give reasons that explain why peroxides are especially effective as radical initiators. (b) Illustrate your answer by outlining how di-*tert*-butyl peroxide, (CH$_3$)$_3$CO—OC(CH$_3$)$_3$, might initiate an alkane halogenation.

9.24 List in order of decreasing stability all of the radicals that can be obtained by abstraction of a hydrogen atom from 2-methylbutane.

9.25 Shown below is an alternative mechanism for the chlorination of methane.

(1) Cl$_2$ $\longrightarrow$ 2 Cl·

(2) Cl· + CH$_4$ $\longrightarrow$ CH$_3$Cl + H·

(3) H· + Cl$_2$ $\longrightarrow$ HCl + Cl·

Calculate $\Delta H°$ for each step of this mechanism and then explain whether or not this mechanism is likely to compete with the one discussed in Sections 9.4 and 9.5A.

9.26 Starting with the compound or compounds indicated in each part and using any other needed reagents, outline syntheses of each of the following compounds. (You need not repeat steps carried out in earlier parts of this problem.)

(a) Ethylcyclopentane from ethane and cyclopentane.

(b) 2,2-Dimethylbutane from methane and neopentane.

(c) Ethyl iodide from ethane.

(d) Diethyl ether from ethane.

(e) Cyclopentene from cyclopentane.

—(f) $(CH_3)_3CT$ from 2-methylpropane.

(g) Ethyl azide $(CH_3CH_2N_3)$ from ethane.

9.27 Consider the relative rates of chlorination at the various positions in 1-fluorobutane:

$$H_3C—CH_2—CH_2—CH_2—F$$
$$1.0 \quad\quad 3.7 \quad\quad 1.7 \quad\quad 0.9$$

Explain this order of reactivity.

Polypropylene (syndiotactic)

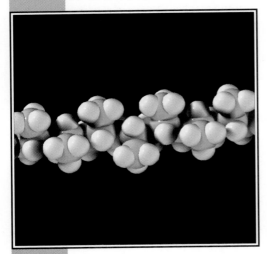

CHAIN-GROWTH POLYMERS

The names *Orlon, Plexiglas, Lucite, polyethylene,* and *Teflon* are now familiar to most of us. These "plastics" or polymers are used in the construction of many objects around us—from the clothing we wear to portions of the houses we live in. Yet all of these compounds were unknown 60 years ago. The development of the processes by which synthetic polymers are made, more than any other single factor, has been responsible for the remarkable growth of the chemical industry in this century.

At the same time, some scientists are now expressing concern about the reliance we have placed on these synthetic materials. Because they are the products of laboratory and industrial processes rather than processes that occur in nature, nature often has no way of disposing of many of them. Although progress has been made in the development of "biodegradable plastics" in recent years, many materials are still used that are not biodegradable. Although most of these objects are combustible, incineration is not always a feasible method of disposal because of attendant air pollution.

Not all polymers are synthetic. Many naturally occurring compounds are polymers as well. Silk and wool are polymers that we call proteins. The starches of our diet are polymers and so is the cellulose of cotton and wood.

Polymers are compounds that consist of very large molecules made up of many repeating subunits. The molecular subunits that are used to synthesize polymers are called *monomers,* and the reactions by which monomers are joined together are called polymerization reactions.

Propylene (propene), for example, can be polymerized to form *polypropylene.* This polymerization occurs by a chain reaction and, as a consequence, polymers such as polypropylene are called **chain-growth** or **addition polymers.**

$$CH_2\!=\!CH \xrightarrow{\ \text{polymerization}\ } -CH_2CH\!\left(\!CH_2CH\!\right)_{\!n}\!-CH_2CH-$$
$$\underset{CH_3}{|} \qquad\qquad \underset{CH_3}{|}\ \ \underset{CH_3}{|} \qquad \underset{CH_3}{|}$$

Propylene   **Polypropylene**

As we saw in Section 9.10, alkenes are convenient starting materials for the preparation of chain-growth polymers. The addition reactions occur through radical, cationic, or anionic mechanisms depending on how they are initiated. The following examples illustrate these mechanisms. All of these reactions are chain reactions.

Radical Polymerization

Cationic Polymerization

Anionic Polymerization

Radical polymerization of chloroethene (vinyl chloride) produces a polymer called poly(vinyl chloride) also known as **PVC**.

Vinyl chloride   **Poly(vinyl chloride)**
(PVC)

This reaction produces a polymer that has a molecular weight of about 1,500,000 and that is a hard, brittle, and rigid material. In this form it is often used to make pipes, rods, and compact discs. Poly(vinyl chloride) can be softened by mixing it with esters (called plasticizers). The softer material is used for making "vinyl leather," plastic raincoats, shower curtains, and garden hoses.

Exposure to vinyl chloride has been linked to the development of a rare cancer of the liver called angiocarcinoma. This link was first noted in 1974 and 1975 among workers in vinyl chloride factories. Since that time, standards have been set to limit workers' exposure to less than one part per million average over an 8-h day. The U.S. Food and Drug Administration (FDA) has banned the use of PVC in packages for food. [There is evidence that poly(vinyl chloride) contains traces of vinyl chloride.]

Acrylonitrile (CH_2=CHCN) polymerizes to form polyacrylonitrile or Orlon. The initiator for the polymerization is a mixture of ferrous sulfate and hydrogen peroxide. These two compounds react to produce hydroxyl radicals ($\cdot OH$), which act as chain initiators.

Acrylonitrile   **Polyacrylonitrile**
(Orlon)

Polyacrylonitrile decomposes before it melts, so melt spinning cannot be used for the production of fibers. Polyacrylonitrile, however, is soluble in N,N-dimethyl-formamide, and these solutions can be used to spin fibers. Fibers produced in this way are used in making carpets and clothing.

Teflon is made by polymerizing tetrafluoroethylene in aqueous suspension.

$$n\ CF_2{=}CF_2 \xrightarrow[\substack{H_2O_2 \\ H_2O}]{Fe^{2+}} {+}CF_2{-}CF_2{\rightarrow}_{\overline{n}}$$

The reaction is highly exothermic and water helps dissipate the heat that is produced. Teflon has a melting point (327°C) that is unusually high for an addition polymer. It is also highly resistant to chemical attack and has a low coefficient of friction. Because of these properties Teflon is used in greaseless bearings, in liners for pots and pans, and in many special situations that require a substance that is highly resistant to corrosive chemicals.

Vinyl alcohol is an unstable compound that rearranges spontaneously to acetalde-hyde (cf. Section 17.2). Consequently, the water-soluble polymer, poly(vinyl alco-

Vinyl alcohol   Acetaldehyde

hol), cannot be made directly. It can be made, however, by an indirect method that begins with the polymerization of vinyl acetate to poly(vinyl acetate). This is then hydrolyzed to poly(vinyl alcohol). Hydrolysis is rarely carried to completion, how-ever, because the presence of a few ester groups helps confer water solubility on the product. The ester groups apparently help keep the polymer chains apart and this permits hydration of the hydroxyl groups. Poly(vinyl alcohol) in which 10% of the ester groups remain dissolves readily in water. Poly(vinyl alcohol) is used to manu-facture water-soluble films and adhesives. Poly(vinyl acetate) is used as an emulsion in water-base paints.

Vinyl acetate   Poly(vinyl   Poly(vinyl
                       acetate)   alcohol)

A polymer with excellent optical properties can be made by the free radical polymerization of methyl methacrylate. Poly(methyl methacrylate) is marketed under the names Lucite, Plexiglas, and Perspex.

Methyl methacrylate   Poly(methyl
                             methacrylate)

A mixture of vinyl chloride and vinylidene chloride polymerizes to form what is known as a *copolymer*. The familiar *Saran Wrap* used in food packaging is made by polymerizing a mixture in which the vinylidene chloride predominates.

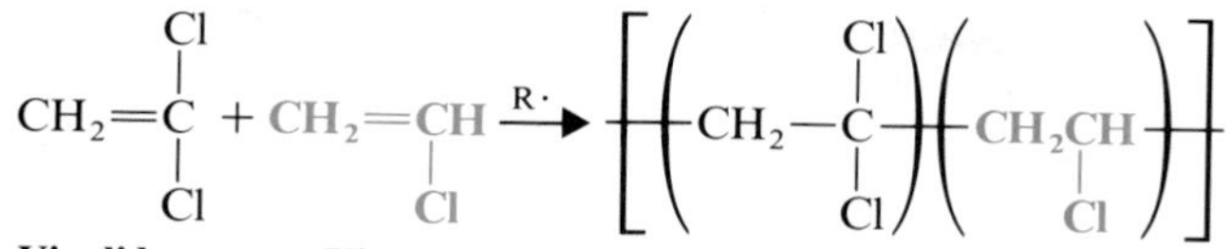

$$CH_2{=}\underset{\underset{Cl}{|}}{\overset{\overset{Cl}{|}}{C}} + CH_2{=}\underset{\underset{Cl}{|}}{CH} \xrightarrow{R\cdot} \left[\left(CH_2{-}\underset{\underset{Cl}{|}}{\overset{\overset{Cl}{|}}{C}}\right)\left(CH_2CH\underset{\underset{Cl}{|}}{}\right)\right]$$

Vinylidene chloride (excess)   **Vinyl chloride**   **Saran Wrap**

The subunits do not necessarily alternate regularly along the polymer chain.

PROBLEM C.1

Can you suggest an explanation that accounts for the fact that the radical polymerization of styrene ($C_6H_5CH{=}CH_2$) to produce polystyrene occurs in a head-to-tail fashion

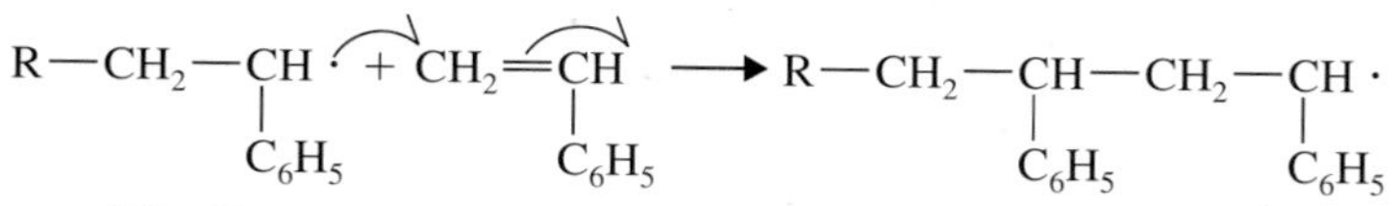

$$R{-}CH_2{-}\underset{\underset{C_6H_5}{|}}{CH} \cdot + CH_2{=}\underset{\underset{C_6H_5}{|}}{CH} \longrightarrow R{-}CH_2{-}\underset{\underset{C_6H_5}{|}}{CH}{-}CH_2{-}\underset{\underset{C_6H_5}{|}}{CH} \cdot$$

"Head"   "Tail"   **Polystyrene**

rather than the head-to-head manner shown here?

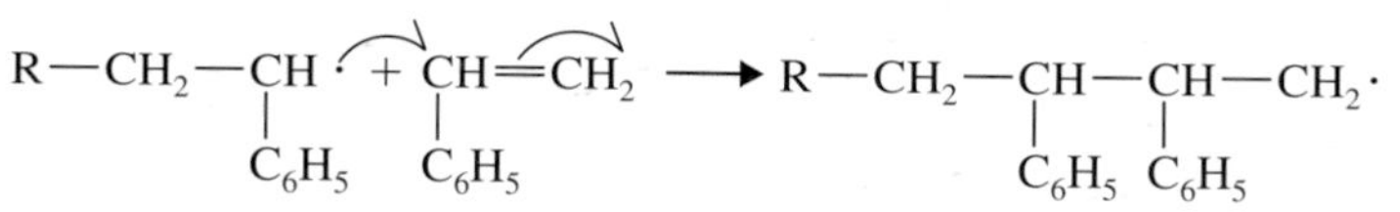

$$R{-}CH_2{-}\underset{\underset{C_6H_5}{|}}{CH} \cdot + \underset{\underset{C_6H_5}{|}}{CH}{=}CH_2 \longrightarrow R{-}CH_2{-}\underset{\underset{C_6H_5}{|}}{CH}{-}\underset{\underset{C_6H_5}{|}}{CH}{-}CH_2 \cdot$$

"Head"   "Head"

PROBLEM C.2

Outline a general method for the synthesis of each of the following polymers by radical polymerization. Show the monomers that you would use.

(a) $+CH_2{-}\underset{\underset{OCH_3}{|}}{CH}{-}CH_2{-}\underset{\underset{OCH_3}{|}}{CH}{-}CH_2{-}\underset{\underset{OCH_3}{|}}{CH}+_n$

(b) $+CH_2{-}CCl_2{-}CH_2{-}CCl_2{-}CH_2{-}CCl_2+_n$

Alkenes also polymerize when they are treated with strong acids. The growing chains in acid-catalyzed polymerizations are *cations* rather than radicals. The following reactions illustrate the cationic polymerization of isobutylene.

Step 1 $H_2O + BF_3 \rightleftharpoons H^+ + BF_3(OH)^-$

Step 2 $H^+ + CH_2{=}\underset{\underset{CH_3}{|}}{\overset{\overset{CH_3}{|}}{C}} \longrightarrow CH_3{-}\underset{\underset{CH_3}{|}}{\overset{\overset{CH_3}{|}}{C}}{}^+$

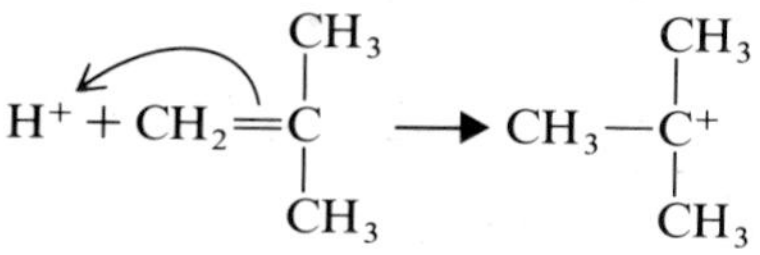

Step 3

$$CH_3-\overset{\overset{\displaystyle CH_3}{|}}{\underset{\underset{\displaystyle CH_3}{|}}{C}}{}^{+} + CH_2{=}\overset{\overset{\displaystyle CH_3}{|}}{\underset{\underset{\displaystyle CH_3}{|}}{C}} \longrightarrow CH_3-\overset{\overset{\displaystyle CH_3}{|}}{\underset{\underset{\displaystyle CH_3}{|}}{C}}-CH_2-\overset{\overset{\displaystyle CH_3}{|}}{\underset{\underset{\displaystyle CH_3}{|}}{C}}{}^{+}$$

Step 4

$$CH_3-\overset{\overset{\displaystyle CH_3}{|}}{\underset{\underset{\displaystyle CH_3}{|}}{C}}-CH_2-\overset{\overset{\displaystyle CH_3}{|}}{\underset{\underset{\displaystyle CH_3}{|}}{C}}{}^{+} \xrightarrow{\quad CH_2{=}\overset{CH_3}{\underset{CH_3}{C}} \quad} CH_3-\overset{\overset{\displaystyle CH_3}{|}}{\underset{\underset{\displaystyle CH_3}{|}}{C}}-CH_2-\overset{\overset{\displaystyle CH_3}{|}}{\underset{\underset{\displaystyle CH_3}{|}}{C}}-CH_2-\overset{\overset{\displaystyle CH_3}{|}}{\underset{\underset{\displaystyle CH_3}{|}}{C}}{}^{+} \xrightarrow{etc.}$$

The catalysts used for cationic polymerizations are usually Lewis acids that contain a small amount of water. The polymerization of isobutylene illustrates how the catalyst (BF_3 and H_2O) functions to produce growing cationic chains.

PROBLEM C.3

Alkenes such as ethylene, vinyl chloride, and acrylonitrile do not undergo cationic polymerization very readily. On the other hand, isobutylene undergoes cationic polymerization rapidly. Provide an explanation for this behavior.

Alkenes containing electron-withdrawing groups polymerize in the presence of strong bases. Acrylonitrile, for example, polymerizes when it is treated with sodium amide ($NaNH_2$) in liquid ammonia. The growing chains in this polymerization are anions.

$$H_2\ddot{N}{:}^{-} + CH_2{=}\underset{\underset{\displaystyle CN}{|}}{CH} \xrightarrow{NH_3} H_2N-CH_2-\underset{\underset{\displaystyle CN}{|}}{CH}{:}^{-}$$

$$H_2N-CH_2-\underset{\underset{\displaystyle CN}{|}}{CH}{:}^{-} \xrightarrow{CH_2{=}CHCN} H_2N-CH_2-\underset{\underset{\displaystyle CN}{|}}{CH}-CH_2-\underset{\underset{\displaystyle CN}{|}}{CH}{:}^{-} \xrightarrow{etc.}$$

Anionic polymerization of acrylonitrile is less important in commercial production than the free radical process we illustrated earlier.

PROBLEM C.4

The remarkable adhesive called ''super glue'' is a result of anionic polymerization. Super glue is a solution containing highly pure methyl α-cyanoacrylate:

$$CH_2{=}C\overset{\displaystyle CN}{\underset{\displaystyle CO_2CH_3}{\big<}}$$

Methyl α-cyanoacrylate

Methyl cyanoacrylate can be polymerized by anions such as hydroxide ion, but it is even polymerized by traces of water found on the surfaces of the two objects that are being glued together. (These two objects, unfortunately, have often been two fingers of the person doing the gluing.) Show how methyl α-cyanoacrylate would undergo anionic polymerization.

C.1 Stereochemistry of Chain-Growth Polymerization

Head-to-tail polymerization of propylene produces a polymer in which every other atom is a stereocenter. Many of the physical properties of the polypropylene produced in this way depend on the stereochemistry of these stereocenters.

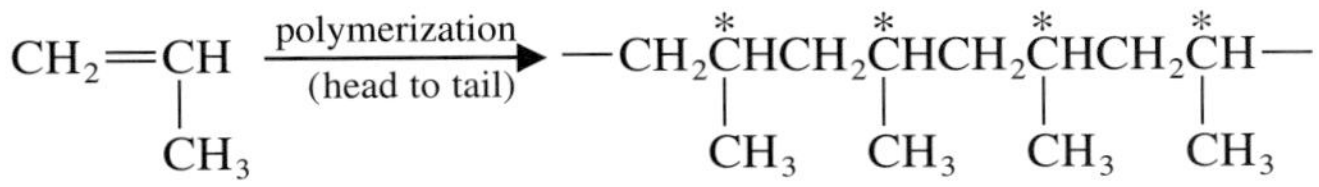

$$CH_2{=}CH \xrightarrow[\text{(head to tail)}]{\text{polymerization}} -CH_2\overset{*}{C}HCH_2\overset{*}{C}HCH_2\overset{*}{C}HCH_2\overset{*}{C}H-$$

with CH_3 on the $CH_2{=}CH$ and CH_3, CH_3, CH_3, CH_3 on the product chain.

There are three general arrangements of the methyl groups and hydrogen atoms along the chain. These arrangements are described as being *atactic, syndiotactic,* and *isotactic.*

If the stereochemistry at the stereocenters is random (Fig. C.1), the polymer is said to be atactic (*a,* without + Greek: *taktikos,* order).

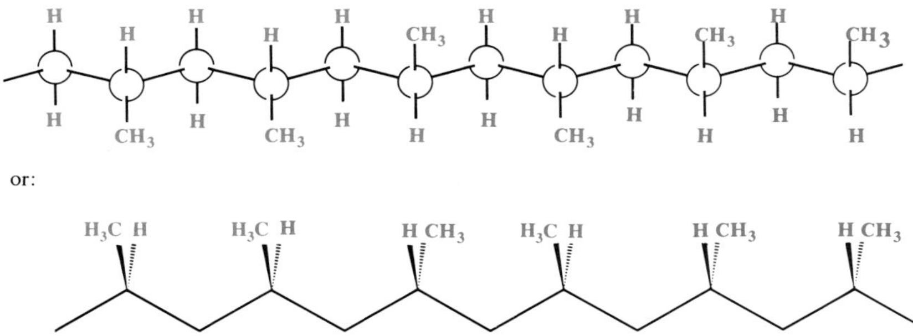

or:

FIGURE C.1 Atactic polypropylene. (In this illustration a ''stretched'' carbon chain is used for clarity.)

In atactic polypropylene the methyl groups are randomly disposed on either side of the stretched carbon chain. If we were to arbitrarily designate one end of the chain as having higher preference than the other, we could give $(R{-}S)$ designations (Section 5.6) to the stereocenters. In atactic polypropylene the sequence of $(R{-}S)$ designations along the chain is random.

Polypropylene produced by radical polymerization at high pressures is atactic. Because the polymer is atactic, it is noncrystalline, has a low softening point, and has poor mechanical properties.

A second possible arrangement of the groups along the carbon chain is that of *syndiotactic* polypropylene. In syndiotactic polypropylene the methyl groups alternate regularly from one side of the stretched chain to the other (Fig. C.2). If we were to arbitrarily designate one end of the chain of syndiotactic polypropylene as having higher preference, the configuration of the stereocenters would alternate, (R), (S), (R), (S), (R), (S), (R), (S), and so on.

The third possible arrangement of stereocenters is the *isotactic* arrangement shown in Fig. C.3. In the isotactic arrangement all of the methyl groups are on the same

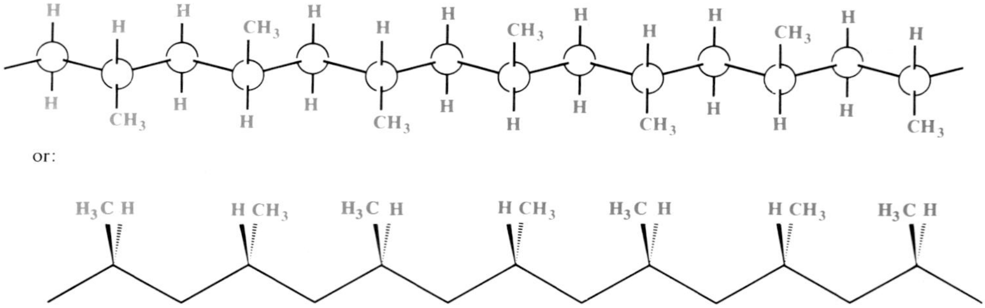

FIGURE C.2 Syndiotactic polypropylene.

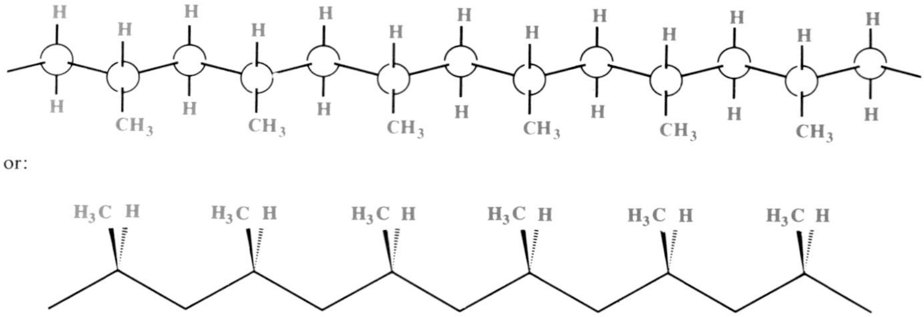

FIGURE C.3 Isotactic polypropylene.

side of the stretched chain. The configurations of the stereocenters are either all (*R*) or all (*S*) depending on which end of the chain is assigned higher preference.

The names isotactic and syndiotactic come from the Greek term *taktikos* (order) plus *iso* (same) and *syndyo* (two together).

Before 1953 isotactic and syndiotactic addition polymers were unknown. In that year, however, a German chemist, Karl Ziegler, and an Italian chemist, Giulio Natta, announced independently the discovery of catalysts that permit stereochemical control of polymerization reactions.* The Ziegler–Natta catalysts, as they are now called, are prepared from transition metal halides and a reducing agent. The catalysts most commonly used are prepared from titanium tetrachloride ($TiCl_4$) and a trialkylaluminum (R_3Al).

Ziegler–Natta catalysts are generally employed as suspended solids, and polymerization probably occurs at metal atoms on the surfaces of the particles. The mechanism for the polymerization is an ionic mechanism, but its details are not fully understood. There is evidence that polymerization occurs through an insertion of the alkene monomer between the metal and the growing polymer chain.

Both syndiotactic and isotactic polypropylene have been made using Ziegler–Natta catalysts. The polymerizations occur at much lower pressures and the polymers that are produced are much higher melting than atactic polypropylene. Isotactic polypropylene, for example, melts at 175°C. Isotactic and syndiotactic polymers are also much more crystalline than atactic polymers. The regular arrangement of groups along the chains allows them to fit together better in a crystal structure.

Atactic, syndiotactic, and isotactic forms of poly(methyl methacrylate) are known.

*Ziegler and Natta were awarded the Nobel Prize for their discoveries in 1963.

The atactic form is a noncrystalline glass. The crystalline syndiotactic and isotactic forms melt at 160 and 200°C, respectively.

PROBLEM C.5

(a) Write structural formulas for portions of the chain of the atactic, syndiotactic, and isotactic forms of polystyrene (see Problem C.1). (b) If solutions were made of each of these forms of polystyrene, which solutions would you expect to show optical activity?

Bicyclo[4.1.0]heptane

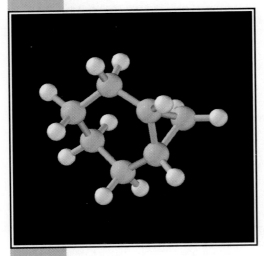

Divalent Carbon Compounds: Carbenes

In recent years considerable research has been devoted to investigating the structures and reactions of a group of compounds in which carbon forms only *two bonds*. These neutral divalent carbon compounds are called *carbenes*.

Most carbenes are highly unstable compounds that are capable of only fleeting existence. Soon after carbenes are formed they usually react with another molecule. The reactions of carbenes are especially interesting because, in many instances, the reactions show a remarkable degree of stereospecificity. The reactions of carbenes are also of great synthetic use in the preparation of compounds that have three-membered rings.

D.1 Structure of Methylene

The simplest carbene is the compound called methylene (CH_2). Methylene can be prepared by the decomposition of diazomethane* (CH_2N_2), a very poisonous yellow gas. This decomposition can be accomplished by heating diazomethane (thermolysis) or by irradiating it with light of a wavelength that it can absorb (photolysis).

$$:\overset{-}{C}H_2 \!-\! \overset{+}{N} \!\equiv\! N: \xrightarrow[\text{or light}]{\text{heat}} \ \ :CH_2 \ \ + :N \!\equiv\! N:$$

| **Diazomethane** | **Methylene** | **Nitrogen** |

* Diazomethane is a resonance hybrid of the three structures, **I**, **II**, and **III**, shown below.

$$:\overset{-}{C}H_2 \!-\! \overset{+}{N} \!\equiv\! N: \longleftrightarrow CH_2 \!=\! \overset{+}{N} \!=\! \overset{-}{\underset{..}{N}}: \longleftrightarrow :\overset{-}{C}H_2 \!-\! N \!=\! \overset{+}{\overset{..}{N}}:$$

$$\textbf{I} \qquad\qquad\qquad \textbf{II} \qquad\qquad\qquad \textbf{III}$$

We have chosen resonance structure **I** to illustrate the decomposition of diazomethane because with **I** it is readily apparent that heterolytic cleavage of the carbon–nitrogen bond results in the formation of methylene and molecular nitrogen.

D.2 Reactions of Methylene

Methylene reacts with alkenes by adding to the double bond to form cyclopropanes.

Alkene   Methylene   Cyclopropane

D.3 Reactions of Other Carbenes: Dihalocarbenes

Dihalocarbenes are also frequently employed in the synthesis of cyclopropane derivatives from alkenes. Most reactions of dihalocarbenes are stereospecific.

The addition of $:CX_2$ is stereospecific. If the R groups of the alkene are trans, they will be trans in the product

Dichlorocarbene can be synthesized by the *α elimination* of the elements of hydrogen chloride from chloroform. This reaction resembles the *β* elimination reactions by which alkenes are synthesized from alkyl halides (Section 6.17).

$$R-\ddot{\underset{\cdot\cdot}{O}}\!:^-K^+ + H\!:\!CCl_3 \rightleftharpoons R-\ddot{\underset{\cdot\cdot}{O}}\!:\!H + {}^-\!:\!CCl_3 + K^+ \xrightarrow{\text{slow}} \quad :CCl_2 \quad + \quad :\ddot{\underset{\cdot\cdot}{Cl}}\!:^-$$

Dichlorocarbene

Compounds *with a β hydrogen* react by *β* elimination preferentially. Compounds with no *β* hydrogen but with an *α* hydrogen (such as chloroform) react by *α* elimination.

    A variety of cyclopropane derivatives have been prepared by generating dichlorocarbene in the presence of alkenes. Cyclohexene, for example, reacts with dichlorocarbene generated by treating chloroform with potassium *tert*-butoxide to give a bicyclic product.

7,7-Dichlorobicyclo[4.1.0]heptane
(59%)

D.4 Carbenoids: The Simmons–Smith Cyclopropane Synthesis

A useful cyclopropane synthesis has been developed by H. E. Simmons and R. D. Smith of the du Pont Company. In this synthesis diiodomethane and a zinc–copper couple are stirred together with an alkene. The diiodomethane and zinc react to produce a carbenelike species called a *carbenoid*.

$$CH_2I_2 + Zn(Cu) \longrightarrow ICH_2ZnI$$
A carbenoid

The carbenoid then brings about the stereospecific addition of a CH_2 group directly to the double bond.

PROBLEM D.1

What products would you expect from each of the following reactions?

(a) *trans*-2-Butene $\xrightarrow[\text{CHCl}_3]{\text{KOC(CH}_3)_3}$

(b) Cyclopentene $\xrightarrow[\text{CHBr}_3]{\text{KOC(CH}_3)_3}$

(c) *cis*-2-Butene $\xrightarrow[\text{Diethyl ether}]{\text{CH}_2I_2/\text{Zn(Cu)}}$

PROBLEM D.2

Starting with cyclohexene and using any other needed reagents, outline a synthesis of 7,7-dibromobicyclo[4.1.0]heptane.

PROBLEM D.3

Treating cyclohexene with 1,1-diiodoethane and a zinc–copper couple leads to two isomeric products. What are their structures?

CHAPTER **10**

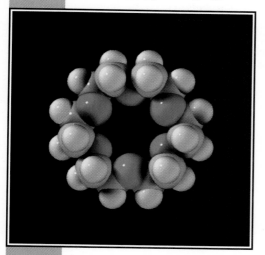

18-Crown-6
(Section 10.22A)

ALCOHOLS AND ETHERS

10.1 STRUCTURE AND NOMENCLATURE

Alcohols are compounds whose molecules have a hydroxyl group attached to a *saturated* carbon atom.* The saturated carbon atom may be that of a simple alkyl group:

$$CH_3OH \qquad CH_3CH_2OH \qquad CH_3\underset{\underset{\textstyle OH}{|}}{C}HCH_3 \qquad CH_3\underset{\underset{\textstyle OH}{|}}{\overset{\overset{\textstyle CH_3}{|}}{C}}CH_3$$

Methanol	Ethanol	2-Propanol	2-Methyl-2-propanol
(methyl alcohol)	(ethyl alcohol)	(isopropyl alcohol)	(*tert*-butyl alcohol)
	a 1° alcohol	*a 2° alcohol*	*a 3° alcohol*

The carbon atom may be a saturated carbon atom of an alkenyl or alkynyl group, or the carbon atom may be a saturated carbon atom that is attached to a benzene ring.

$$\langle\!\!\!\!\bigcirc\!\!\!\!\rangle\!\!-\!CH_2OH \qquad CH_2{=}CHCH_2OH \qquad H{-}C{\equiv}CCH_2OH$$

Benzyl alcohol
a benzylic alcohol

2-Propenol
(allyl alcohol)
an allylic alcohol

2-Propynol
(propargyl alcohol)

Compounds that have a hydroxyl group attached directly to a benzene ring are called *phenols*. (Phenols are discussed in detail in Chapter 21.)

---

* Compounds in which a hydroxyl group is attached to an unsaturated carbon atom of a double bond (i.e., C=C) are called enols, cf. Section 17.2.

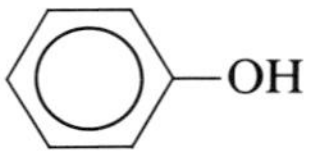

| Phenol | p-Methylphenol *a substituted phenol* | General formula for a phenol |

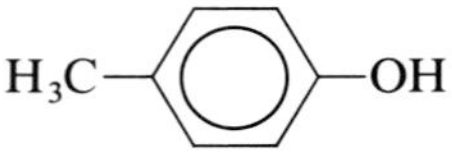

Ethers differ from alcohols in that the oxygen atom of an ether is bonded to two carbon atoms. The hydrocarbon groups may be alkyl, alkenyl, vinyl, alkynyl, or aryl. Several examples are shown here.

$$CH_3CH_2\!-\!O\!-\!CH_2CH_3 \qquad CH_2\!=\!CHCH_2\!-\!O\!-\!CH_3$$

Diethyl ether   **Allyl methyl ether**

$$CH_2\!=\!CH\!-\!O\!-\!CH\!=\!CH_2$$

Divinyl ether   **Methyl phenyl ether**

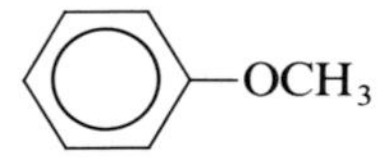

10.1A Nomenclature of Alcohols

We studied the IUPAC system of nomenclature for alcohols in Section 4.3F. As a review consider the following example.

SAMPLE PROBLEM

Give IUPAC substitutive names for the following alcohols:

(a) $CH_3CHCH_2CHCH_2OH$
         | |
       CH_3 CH_3

(b) $CH_3CHCH_2CHCH_3$
         | |
        OH C_6H_5

(c) $CH_3CHCH_2CH\!=\!CH_2$
         |
        OH

ANSWER:
The longest chain *to which the hydroxyl group is attached* gives us the *base name.* The ending is *-ol.* We then number *the longest chain from the end that gives the carbon bearing the hydroxyl group the lower number.* Thus, the names are

(a) $\overset{5}{C}H_3\overset{4}{C}H\overset{3}{C}H_2\overset{2}{C}H\overset{1}{C}H_2OH$
              | |
            CH_3 CH_3
2,4-Dimethyl-1-pentanol

(b) $\overset{1}{C}H_3\overset{2}{C}H\overset{3}{C}H_2\overset{4}{C}H\overset{5}{C}H_3$
              | |
             OH C_6H_5
4-Phenyl-2-pentanol

(c) $\overset{1}{C}H_3\overset{2}{C}H\overset{3}{C}H_2\overset{4}{C}H\!=\!\overset{5}{C}H_2$
              |
             OH
4-Penten-2-ol

The hydroxyl group [see example (c)] has precedence over double bonds and triple bonds in deciding which functional group to name as the suffix.

In common radicofunctional nomenclature (Section 2.10) alcohols are called **alkyl alcohols** such as methyl alcohol, ethyl alcohol, and so on.

PROBLEM 10.1

What is wrong with the use of such names as "isopropanol" and "*tert*-butanol" ?

10.1B Nomenclature of Ethers

Simple ethers are frequently given common radicofunctional names. One simply lists (in alphabetical order) both groups that are attached to the oxygen atom and adds the word *ether*.

$$CH_3OCH_2CH_3 \qquad CH_3CH_2OCH_2CH_3 \qquad C_6H_5OC\overset{\overset{\displaystyle CH_3}{|}}{\underset{\underset{\displaystyle CH_3}{|}}{}}CH_3$$

Ethyl methyl ether   **Diethyl ether**   *tert*-**Butyl phenyl ether**

IUPAC substitutive names should be used for complicated ethers, however, and for compounds with more than one ether linkage. In this IUPAC style, ethers are named as alkoxyalkanes, alkoxyalkenes, and alkoxyarenes. The RO— group is an **alkoxy** group.

$$CH_3\underset{\underset{\displaystyle OCH_3}{|}}{CH}CH_2CH_2CH_3 \qquad CH_3CH_2O-\!\!\!\bigcirc\!\!\!-CH_3 \qquad CH_3OCH_2CH_2OCH_3$$

2-Methoxypentane   **1-Ethoxy-4-methylbenzene**   **1,2-Dimethoxyethane**

Cyclic ethers can be named in several ways. One simple way is to use **replacement nomenclature,** in which we relate the cyclic ether to the corresponding hydrocarbon ring system and use the prefix **oxa-** to indicate that an oxygen atom replaces a CH_2 group. In another system, a cyclic three-membered ether is name **oxirane** and a four-membered ether is called **oxetane.** Several simple cyclic ethers also have common names; in the examples below, these common names are given in parentheses. Tetrahydrofuran (THF) and 1,4-dioxane are useful sovents.

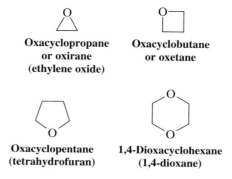

Oxacyclopropane   **Oxacyclobutane**
or oxirane   or oxetane
(ethylene oxide)

Oxacyclopentane   **1,4-Dioxacyclohexane**
(tetrahydrofuran)   (1,4-dioxane)

PROBLEM 10.2 _____

Give bond-line formulas and appropriate names for all of the alcohols and ethers with the formulas (a) C_3H_8O, and (b) $C_4H_{10}O$

10.2 PHYSICAL PROPERTIES OF ALCOHOLS AND ETHERS

The physical properties of a number of alcohols and ethers are given in Tables 10.1 and 10.2.

Ethers have boiling points that are roughly comparable with those of hydrocarbons of the same molecular weight. For example, the boiling point of diethyl ether (MW = 74) is 34.6°C; that of pentane (MW = 72) is 36°C. Alcohols, on the other hand, have much higher boiling points than comparable ethers or hydrocarbons. The boiling point of butyl alcohol (MW = 74) is 117.7°C. We learned the reason for this behavior in Section 2.16C; the molecules of alcohols can associate with each other through **hydrogen bonding,** whereas those of ethers and hydrocarbons cannot.

Hydrogen bonding between molecules of an alcohol

TABLE 10.1 Physical properties of ethers

NAME	FORMULA	mp (°C)	bp (°C)	DENSITY d_4^{20} (g mL^{-1})
Dimethyl ether	CH_3OCH_3	− 138	− 24.9	0.661
Ethyl methyl ether	$CH_3OCH_2CH_3$		10.8	0.697
Diethyl ether	$CH_3CH_2OCH_2CH_3$	− 116	34.6	0.714
Dipropyl ether	$(CH_3CH_2CH_2)_2O$	− 122	90.5	0.736
Diisopropyl ether	$(CH_3)_2CHOCH(CH_3)_2$	− 86	68	0.725
Dibutyl ether	$(CH_3CH_2CH_2CH_2)_2O$	− 97.9	141	0.769
1,2-Dimethoxyethane	$CH_3OCH_2CH_2OCH_3$	− 68	83	0.863
Tetrahydrofuran	$(CH_2)_4O$	− 108	65.4	0.888
1,4-Dioxane		11	101	1.033
Anisole (methoxybenzene)	—OCH$_3$	− 37.3	158.3	0.994

TABLE 10.2 Physical properties of alcohols

COMPOUND	NAME	mp (°C)	bp (°C) (1 atm)	DENSITY d_4^{20} (g mL^{-1})	WATER SOLUBILITY (g 100 mL^{-1} H$_2$O)
Monohydroxy Alcohols					
CH$_3$OH	Methanol	− 97	64.7	0.792	∞
CH$_3$CH$_2$OH	Ethanol	− 117	78.3	0.789	∞
CH$_3$CH$_2$CH$_2$OH	Propyl alcohol	− 126	97.2	0.804	∞
CH$_3$CH(OH)CH$_3$	Isopropyl alcohol	− 88	82.3	0.786	∞
CH$_3$CH$_2$CH$_2$CH$_2$OH	Butyl alcohol	− 90	117.7	0.810	8.3
CH$_3$CH(CH$_3$)CH$_2$OH	Isobutyl alcohol	− 108	108.0	0.802	10.0
CH$_3$CH$_2$CH(OH)CH$_3$	*sec*-Butyl alcohol	− 114	99.5	0.808	26.0
(CH$_3$)$_3$COH	*tert*-Butyl alcohol	25	82.5	0.789	∞
CH$_3$(CH$_2$)$_3$CH$_2$OH	Pentyl alcohol	− 78.5	138.0	0.817	2.4
CH$_3$(CH$_2$)$_4$CH$_2$OH	Hexyl alcohol	− 52	156.5	0.819	0.6
CH$_3$(CH$_2$)$_5$CH$_2$OH	Heptyl alcohol	− 34	176	0.822	0.2
CH$_3$(CH$_2$)$_6$CH$_2$OH	Octyl alcohol	− 15	195	0.825	0.05
CH$_3$(CH$_2$)$_7$CH$_2$OH	Nonyl alcohol	− 5.5	212	0.827	
CH$_3$(CH$_2$)$_8$CH$_2$OH	Decyl alcohol	6	228	0.829	
CH$_2$=CHCH$_2$OH	Allyl alcohol	− 129	97	0.855	∞
(CH$_2$)$_4$CHOH	Cyclopentanol	− 19	140	0.949	
(CH$_2$)$_5$CHOH	Cyclohexanol	24	161.5	0.962	3.6
C$_6$H$_5$CH$_2$OH	Benzyl alcohol	− 15	205	1.046	4
Diols and Triols					
CH$_2$OHCH$_2$OH	Ethylene glycol	− 12.6	197	1.113	∞
CH$_3$CHOHCH$_2$OH	Propylene glycol	− 59	187	1.040	∞
CH$_2$OHCH$_2$CH$_2$OH	Trimethylene glycol	− 30	215	1.060	∞
CH$_2$OHCHOHCH$_2$OH	Glycerol	18	290	1.261	∞

Ethers, however, *are* able to form hydrogen bonds with compounds such as water. Ethers, therefore, have solubilities in water that are similar to those of alcohols of the same molecular weight and that are very different from those of hydrocarbons.

Diethyl ether and 1-butanol, for example, have the same solubility in water, approximately 8 g per 100 mL at room temperature. Pentane, by contrast, is virtually insoluble in water.

Methanol, ethanol, both propyl alcohols, and *tert*-butyl alcohol are completely miscible with water (Table 10.2). The remaining butyl alcohols have solubilities in water between 8.3 and 26.0 g per 100 mL. The solubility of alcohols in water gradually decreases as the hydrocarbon portion of the molecule lengthens; long-chain alcohols are more "alkane-like" and are, therefore, less like water.

PROBLEM 10.3 _____

1,2-Propanediol and 1,3-propanediol (propylene glycol and trimethylene glycol, respectively; see Table 10.2) have higher boiling points than any of the butyl alcohols, even though all of the compounds have roughly the same molecular weight. How can you explain this observation?

10.3 IMPORTANT ALCOHOLS AND ETHERS

10.3A Methanol

At one time, most methanol was produced by the destructive distillation of wood (i.e., heating wood to a high temperature in the absence of air). It was because of this method of preparation that methanol came to be called "wood alcohol." Today, most methanol is prepared by the catalytic hydrogenation of carbon monoxide. This reaction takes place under high pressure and at a temperature of 300–400°C.

$$CO + 2\,H_2 \xrightarrow[\substack{200-300\ \text{atm} \\ ZnO-Cr_2O_3}]{300-400°C} CH_3OH$$

Methanol is highly toxic. Ingestion of even small quantities of methanol can cause blindness; large quantities cause death. Methanol poisoning can also occur by inhalation of the vapors or by prolonged exposure to the skin.

10.3B Ethanol

Ethanol can be made by the fermentation of sugars, and it is the alcohol of all alcoholic beverages. The synthesis of ethanol in the form of wine by the fermentation of the sugars of fruit juices was probably our first accomplishment in the field of organic synthesis. Sugars from a wide variety of sources can be used in the preparation of alcoholic beverages. Often, these sugars are from grains, and it is this derivation that accounts for ethanol having the synonym "grain alcohol."

Fermentation is usually carried out by adding yeast to a mixture of sugars and water. Yeast contains enzymes that promote a long series of reactions that ultimately convert a simple sugar ($C_6H_{12}O_6$) to ethanol and carbon dioxide.

$$C_6H_{12}O_6 \xrightarrow{\text{yeast}} 2\,CH_3CH_2OH + 2\,CO_2$$
$$(\sim 95\%\ \textbf{yield})$$

Fermentation alone does not produce beverages with an ethanol content greater than 12–15% because the enzymes of the yeast are deactivated at higher concentrations. To produce beverages of higher alcohol content the aqueous solution must be distilled. Brandy, whiskey, and vodka are produced in this way. The "proof" of an alcoholic beverage is simply twice the percentage of ethanol (by volume). One hundred proof whiskey is 50% ethanol. The flavors of the various distilled liquors result from other organic compounds that distill with the alcohol and water.

Distillation of a solution of ethanol and water does not yield ethanol more concentrated than 95%. A mixture of 95% ethanol and 5% water boils at a lower temperature (78.15°C) than either pure ethanol (bp, 78.3°C) or pure water (bp, 100°C). Such a mixture is an example of an **azeotrope**.* Pure ethanol can be prepared by adding benzene to the mixture of 95% ethanol and water and then distilling this solution. Benzene forms a different azeotrope with ethanol and water that is 7.5% water. This azeotrope boils at 64.9°C and allows removal of the water (along with some ethanol). Eventually pure ethanol distills over. Pure ethanol is called **absolute alcohol**.

Ethanol is quite cheap, but when it is used for beverages it is highly taxed. (The tax is greater than $20 per gallon in most states.) Federal law requires that some ethanol used for scientific and industrial purposes be adulterated or "denatured" to make it undrinkable. Various denaturants are used, including methanol.

Ethanol is an important industrial chemical. Most ethanol for industrial purposes is produced by the acid-catalyzed hydration of ethene.

$$CH_2{=}CH_2 + H_2O \xrightarrow[\text{acid}]{} CH_3CH_2OH$$

Ethanol is a *hypnotic* (sleep producer). It depresses activity in the upper brain even though it gives the illusion of being a stimulant. Ethanol is also toxic, but it is much less toxic than methanol. In rats the lethal dose of ethanol is 13.7 g kg^{-1} of body weight. Abuse of ethanol is a major drug problem in most countries.

10.3C Ethylene Glycol

Ethylene glycol ($HOCH_2CH_2OH$) has a low molecular weight and a high boiling point and is miscible with water. These properties make ethylene glycol an ideal automobile antifreeze. Much ethylene glycol is sold for this purpose under a variety of trade names. Unfortunately, ethylene glycol is toxic.

10.3D Diethyl Ether

Diethyl ether is a very low-boiling, highly flammable liquid. Care should always be taken when diethyl ether is used in the laboratory, because open flames or sparks from light switches can cause explosive combustion of mixtures of diethyl ether and air.

Most ethers react slowly with oxygen by a radical process called **autoxidation** (see Section 9.11B) to form hydroperoxides and peroxides.

Step 1 $\quad R\cdot + -\overset{|}{\underset{|}{C}}{-}OR' \longrightarrow R{:}H + -\overset{|}{\underset{|}{\overset{\cdot}{C}}}{-}OR'$

Step 2 $\quad -\overset{|}{\underset{|}{\overset{\cdot}{C}}}{-}OR' + O_2 \longrightarrow -\overset{\overset{OO\cdot}{|}}{\underset{|}{C}}{-}OR'$

Step 3a $\quad -\overset{\overset{OO\cdot}{|}}{\underset{|}{C}}{-}OR' + -\overset{|}{\underset{|}{C}}{-}OR' \longrightarrow -\overset{\overset{OOH}{|}}{\underset{|}{C}}{-}OR' + -\overset{|}{\underset{|}{\overset{\cdot}{C}}}{-}OR'$

A hydroperoxide

*Azeotropes can also have boiling points that are higher than that of either of the pure components.

or

$$
\textit{Step 3b} \quad -\overset{\overset{\displaystyle OO\cdot}{|}}{\underset{|}{C}}-OR' + -\overset{\cdot}{\underset{|}{C}}-OR' \longrightarrow R'O-\overset{|}{\underset{|}{C}}-OO-\overset{|}{\underset{|}{C}}-OR'
$$

A peroxide

These hydroperoxides and peroxides, which often accumulate in ethers that have been left standing for long periods of time in contact with air (the air in the top of the bottle is enough), are dangerously explosive. They often detonate without warning when ether solutions are distilled to near dryness. Since ethers are used frequently in extractions, one should take care to test for and decompose any peroxides present in the ether before a distillation is carried out. (Consult a laboratory manual for instructions.)

Diethyl ether was first employed as a surgical anesthetic by C. W. Long of Jefferson, Georgia, in 1842. Long's use of diethyl ether was not published, but shortly thereafter, diethyl ether was introduced into surgical use at the Massachusetts General Hospital in Boston by J. C. Warren.

The most popular modern anesthetic is halothane ($CF_3CHBrCl$). Unlike diethyl ether, halothane is not flammable.

10.4 SYNTHESIS OF ALCOHOLS FROM ALKENES

We have already studied one method for the synthesis of alcohols from alkenes: **acid-catalyzed hydration** (or by the addition of sulfuric acid followed by hydrolysis, which amounts to the same thing).

10.4A Hydration of Alkenes (Discussed in Section 8.5)

Alkenes add water in the presence of an acid catalyst. The addition follows Markovnikov's rule; thus, except for the hydration of ethylene, the reaction produces secondary and tertiary alcohols. The reaction is reversible and the mechanism for the hydration of an alkene is simply the reverse of that for the dehydration of an alcohol (Section 7.13).

$$
\underset{\textbf{Alkene}}{\overset{\diagdown}{\underset{\diagup}{C}}=\overset{\diagup}{\underset{\diagdown}{C}}}
\underset{-H^+}{\overset{+H^+}{\rightleftharpoons}}
-\overset{|}{\underset{\underset{H}{|}}{C}}-\overset{|}{\underset{+}{C}}-
\underset{-H_2O}{\overset{+H_2O}{\rightleftharpoons}}
-\overset{|}{\underset{\underset{H}{|}}{C}}-\overset{|}{\underset{+OH_2}{C}}-
\underset{+H^+}{\overset{-H^+}{\rightleftharpoons}}
\underset{\textbf{Alcohol}}{-\overset{|}{\underset{\underset{H}{|}}{C}}-\overset{|}{\underset{OH}{C}}-}
$$

Because rearrangements often occur, this method has limited usefulness as a laboratory method. [*Remember:* Rearrangements occur whenever a less stable carbocation can rearrange (by a hydride or alkanide shift) to a more stable one.]

PROBLEM 10.4 ──────────────────────────────

What products would you expect from acid-catalyzed hydration of each of the following alkenes?

(a) Ethene   (c) 2-Methylpropene

(b) Propene   (d) 2-Methyl-1-butene

PROBLEM 10.5

Treating 3,3-dimethyl-1-butene with dilute sulfuric acid is largely unsuccessful as a method for preparing 3,3-dimethyl-2-butanol because an isomeric compound is the major product. What is this isomeric compound and how is it formed?

In the sections that follow we shall study two new methods for synthesizing alcohols from alkenes. One of these methods, **oxymercuration–demercuration** (Section 10.5), complements acid-catalyzed hydration in that it gives us an additional method for **Markovnikov addition** of H— and —OH, **and one that is not complicated by rearrangements.** The other method, **hydroboration–oxidation** (Section 10.7), *gives us a method for the net anti-Markovnikov addition of H— and —OH.*

10.5 ALCOHOLS FROM ALKENES THROUGH OXYMERCURATION–DEMERCURATION

A useful laboratory procedure for synthesizing alcohols from alkenes is a two-step method called **oxymercuration–demercuration.**

Alkenes react with mercuric acetate* in a mixture of THF and water to produce (hydroxyalkyl)mercury compounds. These (hydroxyalkyl)mercury compounds can be reduced to alcohols with sodium borohydride:

Step 1: **Oxymercuration**

$$-\overset{|}{C}=\overset{|}{C}- + H_2O + Hg\left(\overset{O}{\overset{\|}{OCCH_3}}\right)_2 \xrightarrow{\text{THF}} -\overset{|}{\underset{HO}{C}}-\overset{|}{\underset{Hg-OCCH_3}{C}}- + CH_3\overset{O}{\overset{\|}{C}}OH$$

Step 2: **Demercuration**

$$-\overset{|}{\underset{HO}{C}}-\overset{|}{\underset{Hg-OCCH_3}{C}}- + OH^- + NaBH_4 \longrightarrow -\overset{|}{\underset{HO}{C}}-\overset{|}{\underset{H}{C}}- + Hg + CH_3\overset{O}{\overset{\|}{C}}O^-$$

In the first step, **oxymercuration,** water and mercuric acetate add to the double bond; in the second step, **demercuration,** sodium borohydride reduces the acetoxymercuri group and replaces it with hydrogen. (The acetate group is often abbreviated —OAc.)

Both steps can be carried out in the same vessel, and both reactions take place very rapidly at room temperature or below. The first step—oxymercuration—usually goes to completion within a period of 20 s–10 min. The second step—demercuration—normally requires less than an hour. The overall reaction gives alcohols in very high yields, usually greater than 90%.

Oxymercuration–demercuration is also highly regioselective. The net orientation of the addition of the elements of water, H— and —OH, *is in accordance with*

---

*Mercury compounds are extremely hazardous. Before you carry out a reaction involving mercury or its compounds you should familiarize yourself with current procedures for its use and disposal.

Markovnikov's rule. The H— becomes attached to the carbon atom of the double bond with the greater number of hydrogen atoms:

$$
R-\underset{\underset{HO-H}{+}}{\overset{\overset{H\ \ \ H}{|\ \ \ \ |}}{C}}=\overset{\overset{H}{|}}{C}-H \xrightarrow[\text{(2) NaBH}_4,\ \text{OH}^-]{\text{(1) Hg(OAc)}_2/\text{THF}-\text{H}_2\text{O}} R-\underset{\underset{HO\ \ \ H}{|\ \ \ \ |}}{\overset{\overset{H\ \ \ H}{|\ \ \ \ |}}{C}}-\overset{}{C}-H
$$

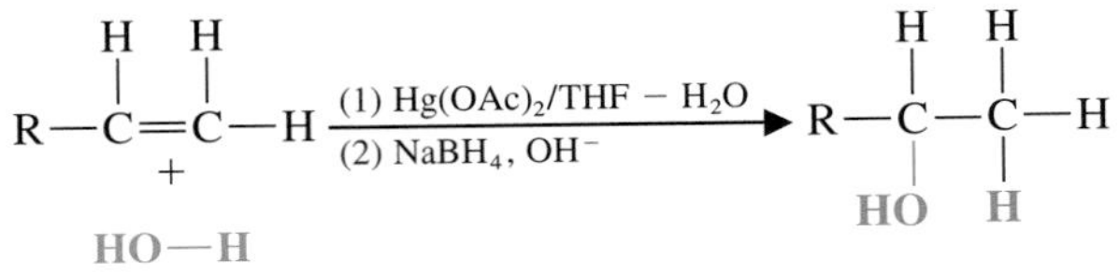

The following specific examples are illustrations:

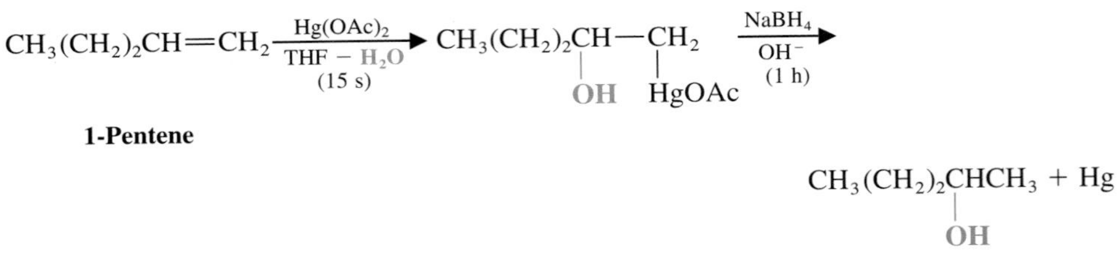

$$
CH_3(CH_2)_2CH{=}CH_2 \xrightarrow[\text{THF}-\text{H}_2\text{O}\ (15\ s)]{\text{Hg(OAc)}_2} CH_3(CH_2)_2\underset{\underset{OH\ \ \ HgOAc}{|\ \ \ \ \ \ \ |}}{CH}-CH_2 \xrightarrow[\substack{\text{OH}^- \\ (1\ h)}]{\text{NaBH}_4}
$$

1-Pentene

$$
CH_3(CH_2)_2\underset{\underset{OH}{|}}{CH}CH_3 + Hg
$$

2-Pentanol
(93%)

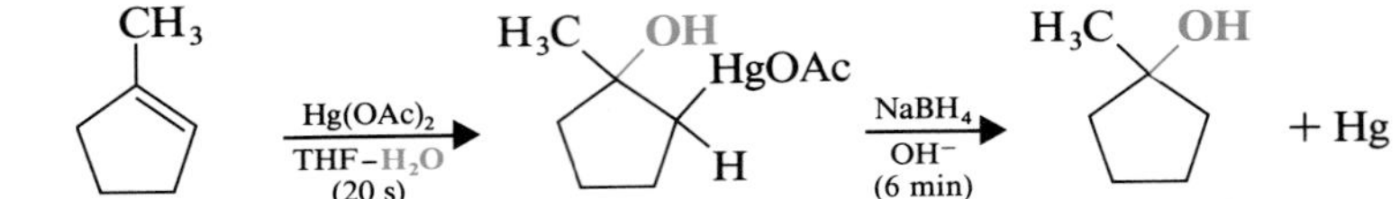

1-Methylcyclo-
pentene

1-Methylcyclo-
pentanol

Rearrangements of the carbon skeleton seldom occur in oxymercuration–demercuration. The following oxymercuration–demercuration of 3,3-dimethyl-1-butene is a striking example illustrating this feature.

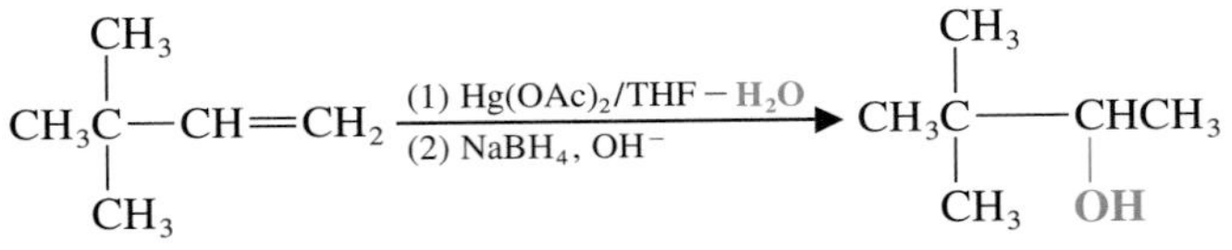

$$
\underset{\underset{CH_3}{|}}{\overset{\overset{CH_3}{|}}{CH_3C}}-CH{=}CH_2 \xrightarrow[\text{(2) NaBH}_4,\ \text{OH}^-]{\text{(1) Hg(OAc)}_2/\text{THF}-\text{H}_2\text{O}} \underset{\underset{CH_3\ \ \ OH}{|\ \ \ \ \ |}}{\overset{\overset{CH_3}{|}}{CH_3C}}-CHCH_3
$$

3,3-Dimethyl-1-
butene

3,3-Dimethyl-2-
butanol
(94%)

Analysis of the mixture of products by gas–liquid chromatography failed to reveal the presence of any 2,3-dimethyl-2-butanol. The acid-catalyzed hydration of 3,3-dimethyl-1-butene, by contrast, gives 2,3-dimethyl-2-butanol as the major product (Section 8.5).

A mechanism that accounts for the orientation of addition in the oxymercuration stage, and one that also explains the general lack of accompanying rearrangements, is shown below. Central to this mechanism is an electrophilic attack by the mercury species, $\overset{+}{\text{H}}\text{gOAc}$, at the less substituted carbon of the double bond (i.e., at the carbon atom that bears the greater number of hydrogen atoms). We illustrate the mechanism using 3,3-dimethyl-1-butene as the example:

A MECHANISM FOR OXYMERCURATION

Step 1 $Hg(OAc)_2 \rightleftharpoons \overset{+}{H}gOAc + OAc^-$

Mercuric acetate dissociates to form an H^+gOAc ion and an acetate ion.

Step 2 $H_3C-\underset{\underset{CH_3}{|}}{\overset{\overset{CH_3}{|}}{C}}-CH=CH_2 + \overset{+}{H}gOAc \longrightarrow H_3C-\underset{\underset{CH_3}{|}}{\overset{\overset{CH_3}{|}}{C}}-\overset{\delta+}{CH}-CH_2$ ⋯ HgOAc $_{\delta+}$

3,3-Dimethyl-1-butene   **Mercury-bridged carbocation**

The electrophilic $HgOAc^+$ ion accepts a pair of electrons from
the alkene to form a mercury-bridged carbocation. In this carbocation,
the positive charge is shared between the 2° carbon atom and
the mercury atom. The charge on the carbon atom is large
enough to account for the Markovnikov orientation of the addition,
but not large enough for a rearrangment to occur.

Step 3 $H_3C-\underset{\underset{CH_3}{|}}{\overset{\overset{CH_3}{|}}{C}}-\overset{\delta+}{CH}-\underset{\underset{HgOAc}{|}}{CH_2}$ $:\!\overset{..}{O}\!-H$ with H $\longrightarrow$ $H_3C-\underset{\underset{H_3C}{|}}{\overset{\overset{CH_3}{|}}{C}}-CH-\underset{\underset{HgOAc}{|}}{CH_2}$ with $H_3C\ \overset{+}{:}\!O\!-H$ and H

A water molecule attacks the carbon bearing the partial positive charge.

Step 4 $H_3C-\underset{\underset{H_3C}{|}}{\overset{\overset{H_3C\ \overset{+}{:}O-H}{|}}{C}}-CH-\underset{\underset{HgOAc}{|}}{CH_2}$ $:\!\overset{..}{O}\!-H$ with H $\longrightarrow$ $H_3C-\underset{\underset{H_3C}{|}}{\overset{\overset{H_3C\ :O:}{|}}{C}}-CH-\underset{\underset{HgOAc}{|}}{CH_2}$ $+\ H-\overset{..}{\underset{\underset{H}{|}}{O}}-H$

**(Hydroxyalkyl)mercury
compound**

An acid–base reaction transfers a proton to another water molecule (or to an
acetate ion). This step produces the (hydroxyalkyl)mercury compound.

Calculations indicate that mercury-bridged carbocations such as those formed in
this reaction retain much of the positive charge on the mercury moiety. Only a small
portion of the positive charge resides on the more substituted carbon atom. The charge
is large enough to account for the observed Markovnikov addition, but it is too small to
allow the usual rapid carbon-skeleton rearrangements that take place with more fully
developed carbocations.

The mechanism for the replacement of the acetoxymercuri group by hydrogen is
not well understood. Radicals are thought to be involved.

PROBLEM 10.6

Starting with an appropriate alkene, show all steps in the synthesis of each of the
following alcohols by oxymercuration–demercuration.

(a) *tert*-Butyl alcohol   (b) Isopropyl alcohol   (c) 2-Methyl-2-butanol

PROBLEM 10.7

When an alkene is treated with mercuric trifluoroacetate, $Hg(O_2CCF_3)_2$, in THF containing an alcohol, ROH, the product is an (alkoxyalkyl)mercury compound. Treating this product with $NaBH_4/OH^-$ results in the formation of an ether. The overall process is called *solvomercuration–demercuration.*

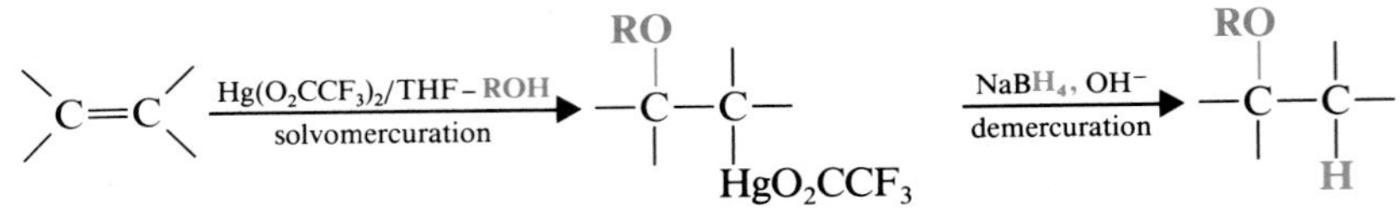

| Alkene | (Alkoxyalkyl)mercuric trifluoroacetate | Ether |

(a) Outline a likely mechanism for the solvomercuration step of this ether synthesis. (b) Show how you would use solvomercuration–demercuration to prepare *tert*-butyl methyl ether.

10.6 HYDROBORATION: SYNTHESIS OF ORGANOBORANES

The addition of a compound containing a hydrogen–boron bond, H—B (called a **boron hydride**), to an alkene is the starting point for a number of highly useful synthetic procedures. This addition, called **hydroboration,** was discovered by Herbert C. Brown* (of Purdue University). In its simplest terms, hydroboration can be represented as follows:

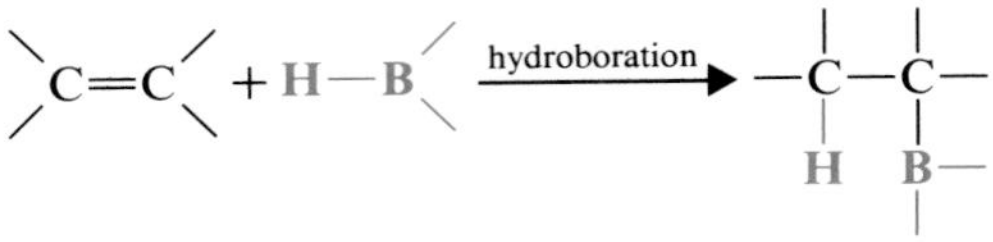

| Alkene | Boron hydride | Organoborane |

Hydroboration can be carried out by using the boron hydride (B_2H_6) called **diborane.** It is much more convenient, however, to use a solution of diborane in THF. When diborane dissolves in THF each B_2H_6 dissociates to produce two molecules of a complex between BH_3 (called **borane**) and THF:

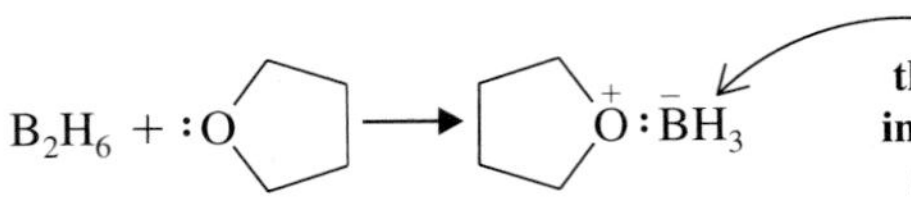

BH_3 is a Lewis acid (because the boron has only six electrons in its valence shell). It accepts an electron pair from the oxygen atom of THF.

| Diborane | THF | THF : BH₃ |

Solutions containing the $THF : BH_3$ complex can be obtained commercially. Hydroboration reactions are usually carried out in ethers: either in diethyl ether, $(C_2H_5)_2O$, or

*Brown's discovery of hydroboration led to his being a co-winner of the Nobel Prize for Chemistry in 1979.

in some higher molecular weight ether such as "diglyme" [($CH_3OCH_2CH_2$)$_2$O, *di*ethylene *gly*col *di*methyl ether].

> *Great care must be used in handling diborane and alkylboranes because they ignite spontaneously in air (with a green flame). The solution of THF : BH_3 is considerably less prone to spontaneous ignition but still must be used in an inert atmosphere and with care.*

10.6A Mechanism of Hydroboration

When a 1-alkene such as propene is treated with a solution containing the THF : BH_3 complex, the boron hydride adds successively to the double bonds of three molecules of the alkene to form a trialkylborane:

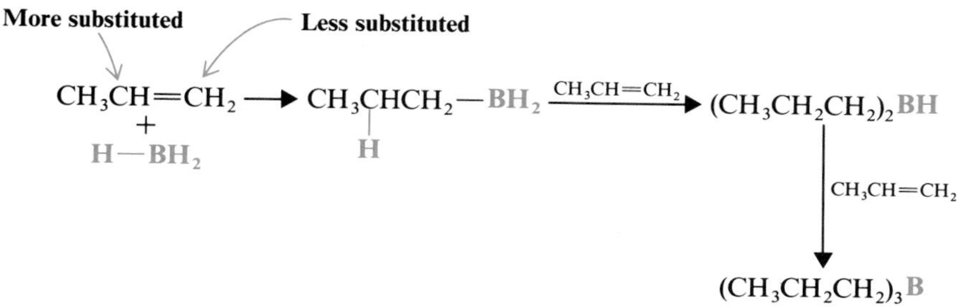

More substituted   Less substituted

$$CH_3CH{=}CH_2 \longrightarrow CH_3\underset{\underset{\displaystyle H}{|}}{CH}CH_2{-}BH_2 \xrightarrow{CH_3CH=CH_2} (CH_3CH_2CH_2)_2BH$$
$$+$$
$$H{-}BH_2$$

$\downarrow CH_3CH{=}CH_2$

$(CH_3CH_2CH_2)_3B$
Tripropylborane

In each addition step *the boron atom becomes attached to the less substituted carbon atom of the double bond,* and a hydrogen atom is transferred from the boron atom to the other carbon atom of the double bond. Thus, hydroboration is regioselective and it is **anti-Markovnikov** (the hydrogen atom becomes attached to the carbon atom with **fewer** hydrogen atoms).

Other examples that illustrate this tendency for the boron atom to become attached to the less substituted carbon atom are shown here. The percentages designate where the boron atom becomes attached.

Less substituted   Less substituted

$$\underset{\underset{1\%}{\nearrow}\quad\underset{99\%}{\nwarrow}}{CH_3CH_2\overset{\overset{\displaystyle CH_3}{|}}{C}{=}CH_2}\qquad\qquad\underset{\underset{2\%}{\nearrow}\quad\underset{98\%}{\nwarrow}}{CH_3\overset{\overset{\displaystyle CH_3}{|}}{C}{=}CHCH_3}$$

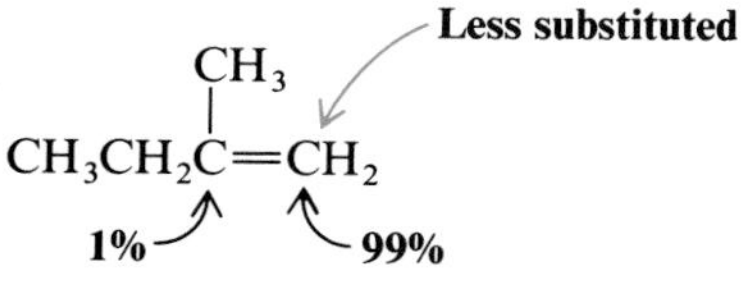

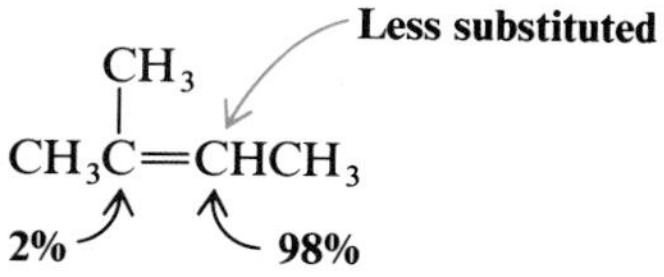

This observed attachment of boron to the less substituted carbon atom of the double bond seems to result in part from **steric factors**—the bulky boron-containing group can approach the less substituted carbon atom more easily.

A mechanism that has been proposed for the addition of BH_3 to the double bond begins with a donation of π electrons from the double bond to the vacant p orbital of BH_3 (see mechanism below). In the next step this complex becomes the addition product by passing through a four-center transition state in which the boron atom is partially bonded to the less substituted carbon atom of the double bond and one hydrogen atom is partially bonded to the other carbon atom. As this transition state is approached, electrons shift in the direction of the boron atom and away from the more substituted carbon atom of the double bond. This makes the more substituted carbon atom develop a partial positive charge *and because it bears an electron-releasing alkyl group, it is better able to accommodate this positive charge.* Thus, both *electronic* and *steric factors* account for the anti-Markovnikov orientation of the addition.

A MECHANISM FOR HYDROBORATION

π complex   **Four-center transition state**

Addition takes place through the intial formation of a π complex, which changes into a cyclic four-center transition state with the boron atom adding to the less hindered carbon atom. The dashed bonds in the transition state represent bonds that are partially formed or partially broken.

The transition state passes over to become an alkylborane. The other B—H bonds of the alkylborane can undergo similar additions, leading finally to a trialkylborane.

10.6B The Stereochemistry of Hydroboration

The transition state for hydroboration requires that *the boron atom and the hydrogen atom add to the same face of the double bond* (look at the previous mechanism again). The addition, therefore, is a **syn** addition:

We can see the results of a syn addition in the hydroboration of 1-methylcyclopentene.

Formation of the enantiomer, which is equally likely, results when the boron hydride adds to the top face of 1-methylcyclopentene.

PROBLEM 10.8

Starting with an appropriate alkene, show the synthesis of (a) tributylborane, (b) triisobutylborane, and (c) tri-*sec*-butylborane. (d) Show the stereochemistry involved in the hydroboration of 1-methylcyclohexene.

PROBLEM 10.9

Treating a hindered alkene such as 2-methyl-2-butene with THF : BH$_3$ leads to the formation of a dialkylborane instead of a trialkylborane. When 2 mol of 2-methyl-2-butene add to 1 mol of BH$_3$, the product formed has the nickname "disiamylborane." Write its structure. (The name "disiamyl" comes from "*disecondary-isoamyl*" a completely unsystematic and unacceptable name. The name "amyl" is an old common name for a five-carbon alkyl group.) Disiamylborane is a useful reagent in certain syntheses that require a sterically hindered borane.

10.7 ALCOHOLS FROM ALKENES THROUGH HYDROBORATION–OXIDATION

Addition of the elements of water to a double bond can also be achieved in the laboratory through the use of diborane or THF:BH$_3$. The addition of water is indirect and two reactions are involved. The first is the addition of borane to the double bond, **hydroboration**; the second is the **oxidation** and hydrolysis of the organoboron intermediate to an alcohol and boric acid. We can illustrate these steps with the hydroboration–oxidation of propene.

$$3\ CH_3CH{=}CH_2 \xrightarrow{\ THF\,:\,BH_3\ } (CH_3CH_2CH_2)_3B \xrightarrow{\ H_2O_2/OH^-\ } 3\ CH_3CH_2CH_2OH$$

Propene	Tripropylborane	Propyl alcohol
⟵ Hydroboration ⟶ ⟵		Oxidation ⟶

The alkylboranes produced in the hydroboration step usually are not isolated. They are oxidized and hydrolyzed to alcohols in the same reaction vessel by the addition of hydrogen peroxide in an aqueous base.

$$R_3B \xrightarrow[\text{NaOH, 25°C}]{H_2O_2} 3\ R{-}OH + Na_3BO_3$$
$$\text{oxidation}$$

The mechanism for the oxidation step begins with the addition of a hydroperoxide ion (HOO$^-$) to the electron-deficient boron atom.

A MECHANISM FOR THE OXIDATION

$$R-\overset{\overset{\displaystyle R}{|}}{\underset{\underset{\displaystyle R}{|}}{B}} + :\ddot{\overset{..}{O}}-\ddot{\overset{..}{O}}-H \longrightarrow \left[R-\overset{\overset{\displaystyle R}{|}}{\underset{\underset{\displaystyle R}{|}}{\overset{-}{B}}} - \ddot{O} \overset{\frown}{} \ddot{\overset{..}{O}}-H \right] \longrightarrow R-\overset{\overset{\displaystyle R}{|}}{B}-\ddot{\overset{..}{O}}-R + :\ddot{\overset{..}{O}}-H$$

Trialkyl-   Hydroperoxide   **Unstable intermediate**   **Borate ester**
borane   ion

The boron atom accepts an electron   An alkyl group migrates from boron to the
pair from the hydroperoxide ion to   adjacent oxygen atom as a hydroxide ion departs.
form an unstable intermediate.

The alkyl migration takes place with retention of configuration of the alkyl group.
Repetitions of these two steps occur until all of the alkyl groups have become attached
to oxygen atoms. The result is the formation of a trialkyl borate, an ester, $B(OR)_3$. This
ester then undergoes basic hydrolysis to produce three molecules of the alcohol and a
borate ion.

$$B(OR)_3 + 3\ OH^- \xrightarrow{\ H_2O\ } 3\ ROH + BO_3{}^{3-}$$

Because hydroboration reactions are regioselective, the net result of hydroboration–oxidation is an apparent **anti-Markovnikov addition of water.** As a consequence, *hydroboration–oxidation gives us a method for the preparation of alcohols
that cannot normally be obtained through the acid-catalyzed hydration of alkenes or
by oxymercuration–demercuration.* For example, acid-catalyzed hydration (or
oxymercuration–demercuration) of 1-hexene yields 2-hexanol:

$$CH_3CH_2CH_2CH_2CH{=}CH_2 \xrightarrow{\ H_3O^+,\ H_2O\ } CH_3CH_2CH_2CH_2\underset{\underset{\displaystyle OH}{|}}{C}HCH_3$$

1-Hexene   **2-Hexanol**

Hydroboration–oxidation, by contrast, yields 1-hexanol:

$$CH_3CH_2CH_2CH_2CH{=}CH_2 \xrightarrow[\ (2)\ H_2O_2,\ OH^-\ ]{\ (1)\ THF{:}BH_3\ } CH_3CH_2CH_2CH_2CH_2CH_2OH$$

1-Hexene   **1-Hexanol (90%)**

Other examples of hydroboration–oxidation are the following:

$$CH_3{-}\overset{\overset{\displaystyle CH_3}{|}}{C}{=}CHCH_3 \xrightarrow[\ (2)\ H_2O_2,\ OH^-\ ]{\ (1)\ THF{:}BH_3\ } CH_3{-}\overset{\overset{\displaystyle CH_3}{|}}{C}H{-}\underset{\underset{\displaystyle OH}{|}}{C}HCH_3$$

2-Methyl-2-butene   **3-Methyl-2-butanol (59%)**

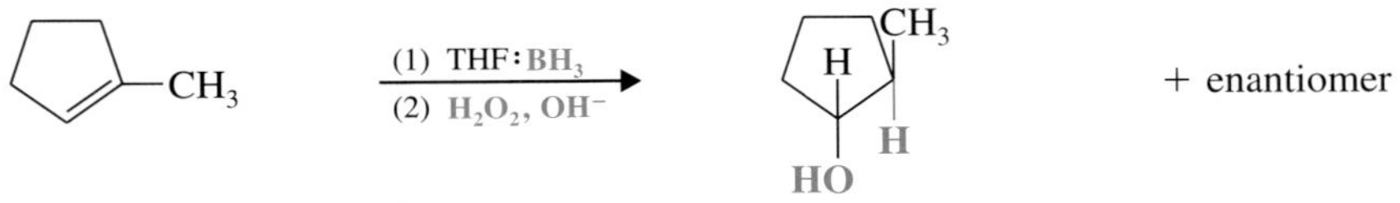

1-Methylcyclopentene   **trans-2-Methylcyclopentanol (86%)**

10.7A The Stereochemistry of the Oxidation of Organoboranes

Because the oxidation step in the hydroboration–oxidation synthesis of alcohols takes place with retention of configuration, *the hydroxyl group replaces the boron atom where it stands in the organoboron compound.* The net result of the two steps (hydroboration and oxidation) is the *syn addition* of —H and —OH. We can see this if we examine the hydroboration–oxidation of 1-methylcyclopentene (Fig. 10.1).

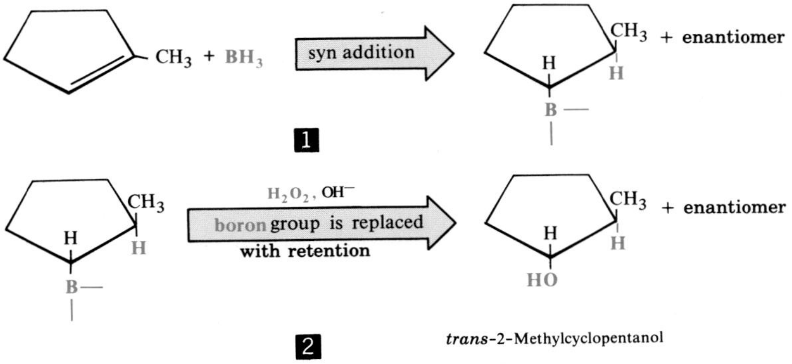

FIGURE 10.1 The hydroboration–oxidation of 1-methylcyclopentene. The first reaction is a syn addition of borane. (In this illustration we have shown the boron and hydrogen both entering from the bottom side of 1-methylcyclopentene. The reaction also takes place from the top side at an equal rate to produce the enantiomer.) In the second reaction the boron atom is replaced by a hydroxyl group with retention of configuration. The product is a trans compound (*trans*-2-methylcyclopentanol) and the overall result is the syn addition of —H and —OH.

PROBLEM 10.10

Starting with the appropriate alkene show how you could use hydroboration–oxidation to prepare each of the following alcohols.

(a) 1-Pentanol

(b) 2-Methyl-1-pentanol

(c) 3-Methyl-2-pentanol

(d) 2-Methyl-3-pentanol

(e) *trans*-2-Methylcyclobutanol

10.7B Protonolysis of Organoboranes

Heating an organoborane with acetic acid causes cleavage of the carbon–boron bond in the following way:

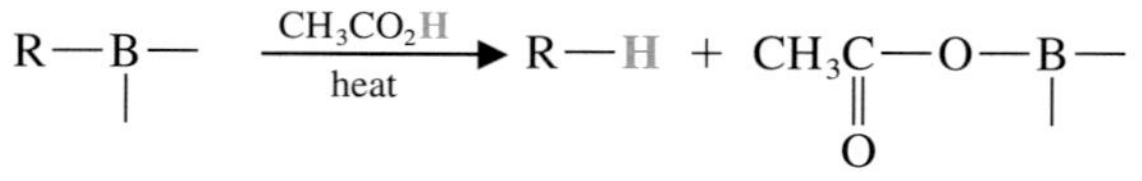

$$R-B- \xrightarrow[\text{heat}]{CH_3CO_2H} R-H + CH_3C-O-B-$$

Organoborane   **Alkane**

This reaction also takes place with retention of configuration; therefore, hydrogen replaces boron *where it stands* in the organoborane. The stereochemistry of this reaction, therefore, is like that of the oxidation of organoboranes, and it can be very useful in introducing deuterium or tritium in a specific way.

PROBLEM 10.11

Starting with any needed alkene (or cycloalkene), and assuming you have deuterioacetic acid (CH_3CO_2D) available, outline syntheses of the following deuterium-labeled compounds.

(a) $(CH_3)_2CHCH_2CH_2D$ (b) $(CH_3)_2CHCHDCH_3$ (c) 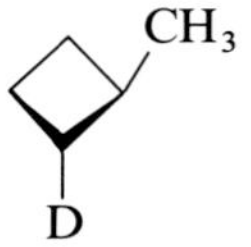 [structure] (+ enantiomer)

(d) Assuming you also have available THF : BD_3 and CH_3CO_2T, can you suggest a synthesis of the following?

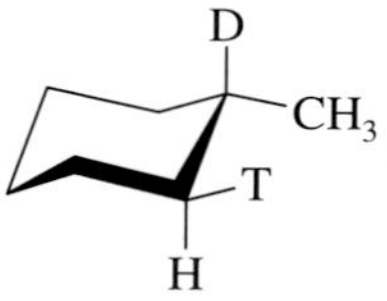

[structure] (+ enantiomer)

10.8 REACTIONS OF ALCOHOLS

Our understanding of the reactions of alcohols will be aided by an initial examination of the electron distribution in the alcohol functional group, and of how this distribution affects its reactivity. The oxygen atom of an alcohol polarizes both the C—O bond and the O—H bond of an alcohol:

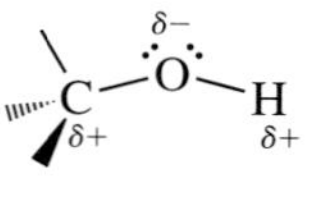

[structure]

Polarization of the O—H bond makes the hydrogen partially positive and explains why alcohols are weak acids (Section 10.9). Polarization of the C—O bond makes the carbon atom partially positive, and if it were not for the fact that OH^- is a strong base, and, therefore, a very poor leaving group, this carbon would be susceptible to nucleophilic attack.

The electron pairs on the oxygen atom make it both *basic* and *nucleophilic.* In the presence of strong acids, alcohols act as bases and accept protons in the following way:

$$\underset{\text{Alcohol}}{\overset{|}{\underset{|}{C}}-\ddot{O}-H} + \underset{\text{Strong acid}}{H-A} \rightleftharpoons \underset{\substack{\text{Protonated} \\ \text{alcohol}}}{-\overset{|}{\underset{|}{C}}-\overset{\overset{\displaystyle H}{|}}{\underset{\cdot\cdot}{O}}\!\!{}^{+}\!\!-H} + A^-$$

Protonation of the alcohol converts a poor leaving group (OH^-) into a good one (H_2O). It also makes the carbon atom even more positive (because $-OH_2{}^+$ is more electron withdrawing than $-OH$) and, therefore, even more susceptible to nucleophilic attack. Now S_N2 reactions become possible (Section 10.14).

$$Nu:^- + \underset{\substack{\text{Protonated} \\ \text{alcohol}}}{-\overset{|}{\underset{|}{C}}-\overset{\overset{\displaystyle H}{|}}{\underset{\cdot\cdot}{O}}\!\!{}^{+}\!\!-H} \xrightarrow{\ S_N2\ } Nu-\overset{|}{\underset{|}{C}}- + :\overset{\overset{\displaystyle H}{|}}{\underset{\cdot\cdot}{O}}-H$$

Because alcohols are nucleophiles they, too, can react with protonated alcohols. This, as we shall see in Section 10.16, is an important step in the synthesis of ethers.

$$R-\overset{\displaystyle \cdot\cdot}{\underset{\displaystyle H}{\ddot{O}}}: + \underset{\substack{\text{Protonated} \\ \text{ether}}}{-\overset{|}{\underset{|}{C}}-\overset{\overset{\displaystyle H}{|}}{\underset{\cdot\cdot}{O}}\!\!{}^{+}\!\!-H} \xrightarrow{\ S_N2\ } R-\overset{\overset{\displaystyle H}{|}}{\underset{\displaystyle H}{\ddot{O}}}\!\!{}^{+}\!\!-\overset{|}{\underset{|}{C}}- + :\overset{\overset{\displaystyle H}{|}}{\underset{\cdot\cdot}{O}}-H$$

At a high enough temperature, and in the absence of a good nucleophile, protonated alcohols are capable of undergoing E1 reactions. This is what happens in alcohol dehydrations (Section 7.13).

Alcohols also react with PBr_3 and $SOCl_2$ to yield alkyl bromides and alkyl chlorides. These reactions, as we shall see in Section 10.15, are initiated by the alcohol using its unshared electron pairs to act as a nucleophile.

10.9 ALCOHOLS AS ACIDS

As we might expect, alcohols have acidities similar to that of water. Methanol is a slightly stronger acid than water ($pK_a = 15.7$) but most alcohols are somewhat weaker acids. Values of pK_a for several alcohols are listed in Table 10.3.

TABLE 10.3 pK_a Values for some weak acids

ACID	pK_a
CH_3OH	15.5
H_2O	15.74
CH_3CH_2OH	15.9
$(CH_3)_3COH$	18.0

The reason sterically hindered alcohols such as *tert*-butyl alcohol are less acidic arises from solvation effects. With unhindered alcohols, water molecules are able to surround and solvate the negative oxygen of the alkoxide ion formed when an alcohol loses a proton. Solvation stabilizes the alkoxide ion and increases the acidity of the alcohol. (*Remember:* Any factor that stabilizes the conjugate base of an acid increases its acidity.)

$$R-\overset{..}{\underset{..}{O}}-H \quad :\overset{H}{\underset{..}{O}}-H \rightleftharpoons \quad R-\overset{..}{\underset{..}{O}}:^{-} \quad +H-\overset{H}{\underset{..}{O}}{^{\pm}}H$$

Alcohol   **Alkoxide ion**
(stabilized by solvation)

If the R— group of the alcohol is bulky, solvation of the alkoxide ion is hindered, and the alkoxide ion is not as effectively stabilized. The alcohol, consequently, is a weaker acid.

All alcohols, however, are much stronger acids than terminal alkynes, and are very much stronger acids than hydrogen, ammonia, and alkanes (see Table 3.1).

Relative Acidity

$$H_2O > ROH > RC{\equiv}CH > H_2 > NH_3 > RH$$

The conjugate base of an alcohol is an **alkoxide ion.** Sodium and potassium alkoxides can be prepared by treating alcohols with sodium or potassium metal or with the metal hydride (Section 6.17B). Because most alcohols are weaker acids than water, most alkoxide ions are stronger bases than the hydroxide ion.

Relative Basicity

$$R^- > NH_2^- > H^- > RC{\equiv}C^- > RO^- > OH^-$$

PROBLEM 10.12 _____

Write equations for the acid–base reactions that would occur (if any) if ethanol were added to solutions of each of the following compounds. In each reaction, label the stronger acid, the stronger base, and so forth. (a) Sodium amide, (b) sodium ethynide, and (c) ethyllithium.

Sodium and potassium alkoxides are often used as bases in organic syntheses (Section 6.17B). We use alkoxides, such as ethoxide and *tert*-butoxide, when we carry out reactions that require stronger bases than hydroxide ion but do not require exceptionally powerful bases, such as the amide ion or the anion of an alkane. We also use alkoxide ions when (for reasons of solubility) we need to carry out a reaction in an alcohol solvent rather than in water.

10.10 CONVERSION OF ALCOHOLS INTO MESYLATES AND TOSYLATES

Alcohols react with sulfonyl chlorides to form esters that are called **sulfonates.** Ethanol, for example, reacts with methanesulfonyl chloride to form ethyl methanesulfonate and with *p*-toluenesulfonyl chloride to form ethyl *p*-toluenesulfonate. These reactions involve cleavage of the O—H bond of the alcohol and not the C—O bond. [If the alcohol had been chiral no change of configuration would have occurred (see Section 10.11)].

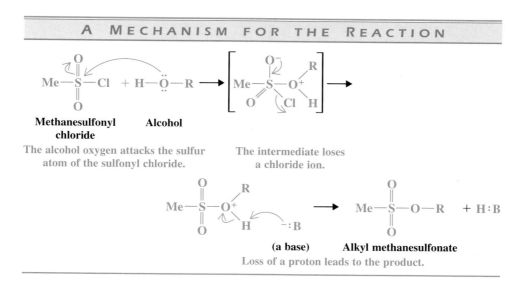

The mechanism that follows (using methanesulfonyl chloride as the example) accounts for the fact that the C—O bond of the alcohol does not break.

A MECHANISM FOR THE REACTION

Methanesulfonyl chloride   **Alcohol**

The alcohol oxygen attacks the sulfur atom of the sulfonyl chloride.

The intermediate loses a chloride ion.

(a base)   **Alkyl methanesulfonate**

Loss of a proton leads to the product.

PROBLEM 10.13

Suggest an experiment using an isotopically labeled alcohol that would prove that the formation of an alkyl sulfonate does not cause cleavage at the C—O bond of the alcohol.

Sulfonyl chlorides are usually prepared by treating sulfonic acids with phosphorus pentachloride. (We shall study syntheses of sulfonic acids in Chapter 15.)

p-Toluenesulfonic acid — **p-Toluenesulfonyl chloride (tosyl chloride)**

Methanesulfonyl chloride and *p*-toluenesulfonyl chloride are used so often that organic chemists have shortened their rather long names to "mesyl chloride" and "tosyl chloride," respectively. The methanesulfonyl group is often called a "mesyl" group and the *p*-toluenesulfonyl group is called a "tosyl" group. Methanesulfonates are known as "mesylates" and *p*-toluenesulfonates are known as "tosylates."

The mesyl group — The tosyl group

An alkyl mesylate — An alkyl tosylate

PROBLEM 10.14 ————————————————————

Starting with the appropriate sulfonic acid and PCl$_5$, or with the appropriate sulfonyl chloride, show how you would prepare each of the following alkyl sulfonates: (a) Methyl *p*-toluenesulfonate,   (b) isobutyl methanesulfonate, (c) *tert*-butyl methanesulfonate.

10.11   MESYLATES AND TOSYLATES IN S$_N$2 REACTIONS

Alkyl sulfonates are frequently used as substrates for nucleophilic substitution reactions because sulfonate ions are excellent leaving groups.

Alkyl sulfonate (tosylate, mesylate, etc.) — **Sulfonate ion (very weak base— a good leaving group)**

The trifluoromethanesulfonate ion (CF$_3$SO$_2$O$^-$) is one of the best of all known leaving groups. Alkyl trifluoromethanesulfonates—called *alkyl triflates*—react extremely rapidly in nucleophilic substitution reactions. The triflate ion is such a good leaving group that even vinylic triflates undergo S$_N$1 reactions and yield vinylic cations.

$$\diagdown C = C \diagup^{OSO_2CF_3} \xrightarrow{\text{solvolysis}} \diagdown C = C^{\pm} + {}^-OSO_2CF_3$$

Vinylic triflate   **Vinylic cation**   **Triflate ion**

Alkyl sulfonates give us an indirect method for carrying out nucleophilic substitution reactions on alcohols. We first convert the alcohol to an alkyl sulfonate and then we allow the sulfonate to react with a nucleophile. When the carbon atom bearing the —OH is a stereocenter, the first step—sulfonate formation—proceeds with **retention of configuration** because no bonds to the stereocenter are broken. Only the O—H bond breaks. The second step—if the reaction is S_N2—proceeds with *inversion of configuration*.

Step 1

$$H^{\text{\tiny IIIII}}\overset{R}{\underset{R'}{C}}{-}O{\vdots}H + Cl{\vdots}Ts \xrightarrow[-HCl]{\text{retention}} H^{\text{\tiny IIIII}}\overset{R}{\underset{R'}{C}}{-}O{-}Ts$$

Step 2

$$Nu:^- + H^{\text{\tiny IIIII}}\overset{R}{\underset{R'}{C}}{-}O{-}Ts \xrightarrow[S_N2]{\text{inversion}} Nu{-}\overset{R}{\underset{R'}{C}}{}^{\text{\tiny IIII}}H + {}^-O{-}Ts$$

Alkyl sulfonates (tosylates, etc.) undergo all the nucleophilic substitution reactions that alkyl halides do.

PROBLEM 10.15

Show the configurations of products formed when   (a) (*R*)-2-butanol is converted to a tosylate, and   (b) when this tosylate reacts with hydroxide ion by an S_N2 reaction.   (c) Converting *cis*-4-methylcyclohexanol to a tosylate and then allowing the tosylate to react with LiCl (in an appropriate solvent) yields *trans*-1-chloro-4-methylcyclohexane. Outline the stereochemistry of these steps.

10.12 ALKYL PHOSPHATES

Alcohols react with phosphoric acid to yield alkyl phosphates:

$$ROH + HO{-}\overset{\displaystyle O}{\overset{\|}{\underset{\underset{\displaystyle OH}{|}}{P}}}{-}OH \xrightarrow[(-H_2O)]{} RO{-}\overset{\displaystyle O}{\overset{\|}{\underset{\underset{\displaystyle OH}{|}}{P}}}{-}OH \xrightarrow[(-H_2O)]{ROH}$$

Phosphoric acid   **Alkyl dihydrogen phosphate**

$$RO{-}\overset{\displaystyle O}{\overset{\|}{\underset{\underset{\displaystyle OR}{|}}{P}}}{-}OH \xrightarrow[(-H_2O)]{ROH} RO{-}\overset{\displaystyle O}{\overset{\|}{\underset{\underset{\displaystyle OR}{|}}{P}}}{-}OR$$

Dialkyl hydrogen phosphate   **Trialkyl phosphate**

Esters of phosphoric acids are important in biochemical reactions. Especially important are triphosphate esters. Although hydrolysis of the ester group or of one of the anhydride linkages of an alkyl triphosphate is exothermic, these reactions occur very slowly in aqueous solutions. Near pH 7, these triphosphates exist as negatively charged ions and hence are much less susceptible to nucleophilic attack. Alkyl triphosphates are, consequently, relatively stable compounds in the aqueous medium of a living cell.

Enzymes, on the other hand, are able to catalyze reactions of these triphosphates in which the energy made available when their anhydride linkages break helps the cell make other chemical bonds. We have more to say about this in Chapter 22 when we discuss the important triphosphate called adenosine triphosphate (or ATP).

10.13 CONVERSION OF ALCOHOLS INTO ALKYL HALIDES

Alcohols react with a variety of reagents to yield alkyl halides. The most commonly used reagents are hydrogen halides (HCl, HBr, or HI), phosphorus tribromide (PBr_3), and thionyl chloride ($SOCl_2$). Examples of the use of these reagents are the following. All of these reactions result in cleavage at the C—O bond of the alcohol.

10.14 ALKYL HALIDES FROM THE REACTIONS OF ALCOHOLS WITH HYDROGEN HALIDES

When alcohols react with a hydrogen halide, a substitution takes place producing an alkyl halide and water:

$$R \text{--} OH + HX \longrightarrow R\text{--}X + H_2O$$

The order of reactivity of the hydrogen halides is HI > HBr > HCl (HF is generally unreactive), and the order of reactivity of alcohols is 3° > 2° > 1° < methyl.

The reaction is *acid catalyzed.* Alcohols react with the strongly acidic hydrogen halides, HCl, HBr, and HI, but they do not react with nonacidic NaCl, NaBr, or NaI. Primary and secondary alcohols can be converted to alkyl chlorides and bromides by allowing them to react with a mixture of a sodium halide and sulfuric acid.

$$ROH + NaX \xrightarrow{H_2SO_4} RX + NaHSO_4 + H_2O$$

10.14A Mechanisms of the Reactions of Alcohols with HX

Secondary, tertiary, allylic, and benzylic alcohols appear to react by a mechanism that involves the formation of a carbocation—one that we studied first in Section 3.12, and that now is recognizable *as an S_N1 reaction with the protonated alcohol acting as the substrate.* We again illustrate this mechanism with the reaction of *tert*-butyl alcohol and hydrochloric acid.

The first two steps are the same as in the mechanism for the dehydration of an alcohol (Section 7.13). The alcohol accepts a proton and then the protonated alcohol dissociates to form a carbocation and water.

In step 3 the mechanisms for the dehydration of an alcohol and the formation of an alkyl halide differ. In dehydration reactions the carbocation loses a proton in an E1 reaction to form an alkene. In the formation of an alkyl halide, the carbocation reacts with a nucleophile (a halide ion) in an S_N1 reaction.

How can we account for the different course of these two reactions?

When we dehydrate alcohols we usually carry out the reaction in concentrated sulfuric acid. The only nucleophiles present in this reaction mixture are water and hydrogen sulfate (HSO_4^-) ions. Both are poor nucleophiles and both are usually present in low concentrations. Under these conditions, the highly reactive carbocation stabilizes itself by losing a proton and becoming an alkene. The net result is *an E1 reaction.*

In the reverse reaction, that is, the hydration of an alkene (Section 8.5), the carbocation *does* react with a nucleophile. It reacts with water. Alkene hydrations are carried out in dilute sulfuric acid, where the water concentration is high. In some instances, too, carbocations may react with HSO_4^- ions or with sulfuric acid, itself. When they do they form alkyl hydrogen sulfates ($R-OSO_2OH$).

When we convert an alcohol to an alkyl halide, we carry out the reaction in the presence of acid and *in the presence of halide ions.* Halide ions are good nucleophiles (much stronger nucleophiles than water), and since they are present in high concentration, most of the carbocations stabilize themselves by accepting the electron pair of a halide ion. The overall result is an S_N1 reaction.

These two reactions, dehydration and the formation of an alkyl halide, also furnish us another example of the competition between nucleophilic substitution and elimination (cf. Section 6.20). Very often, in conversions of alcohols to alkyl halides, we find that the reaction is accompanied by the formation of some alkene (i.e., by elimination). The free energies of activation for these two reactions of carbocations are not very different from one another. Thus, not all of the carbocations react with nucleophiles; some stabilize themselves by losing protons.

Not all acid-catalyzed conversions of alcohols to alkyl halides proceed through the formation of carbocations. Primary alcohols and methanol apparently react through a mechanism that we recognize as an S_N2 type. In these reactions the function of the acid is to produce *a protonated alcohol.* The halide ion then displaces a molecule of water (a good leaving group) from carbon; this produces an alkyl halide.

(protonated 1°
or methanol)

(a good
leaving
group)

Although halide ions (particularly iodide and bromide ions) are strong nucleophiles, they are not strong enough to carry out substitution reactions with alcohols themselves. That is, reactions of the following type do not occur to any appreciable extent.

They do not occur because the leaving group would have to be a strongly basic hydroxide ion.

The reverse reaction, that is, the reaction of an alkyl halide with hydroxide ion, does occur and is a method for the synthesis of alcohols. We saw this reaction in Chapter 6.

We can see now why the reactions of alcohols with hydrogen halides are acid catalyzed. With tertiary and secondary alcohols, the function of the acid is to help produce a carbocation. With methanol and primary alcohols, the function of the acid is to produce a substrate in which the leaving group is a weakly basic water molecule rather than a strongly basic hydroxide ion.

As we might expect, many reactions of alcohols with hydrogen halides, particularly those in which carbocations are formed, *are accompanied by rearrangements.*

Because the chloride ion is a weaker nucleophile than bromide or iodide ions, hydrogen chloride does not react with primary or secondary alcohols unless zinc chloride or some similar Lewis acid is added to the reaction mixture as well. Zinc chloride, a good Lewis acid, forms a complex with the alcohol through association with an unshared pair of electrons on the oxygen atom. This provides a better leaving group for the reaction than H_2O.

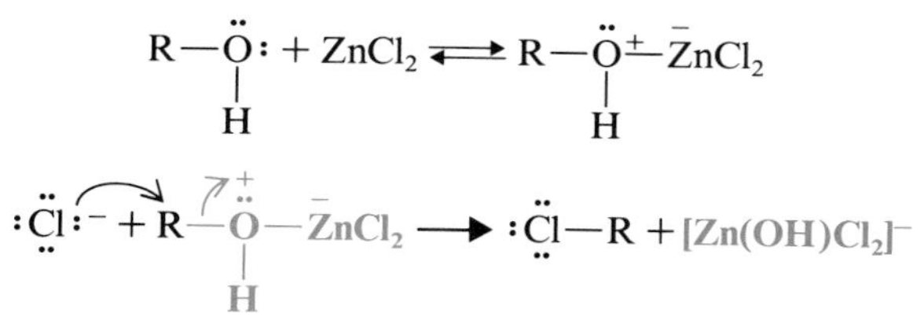

$$R-\overset{\cdot\cdot}{\underset{\underset{H}{|}}{O}}: + ZnCl_2 \rightleftharpoons R-\overset{\pm}{\underset{\underset{H}{|}}{O}}-\bar{Z}nCl_2$$

$$:\overset{\cdot\cdot}{\underset{\cdot\cdot}{Cl}}:^- + R-\overset{+}{\underset{\underset{H}{|}}{O}}-\bar{Z}nCl_2 \longrightarrow :\overset{\cdot\cdot}{\underset{\cdot\cdot}{Cl}}-R + [Zn(OH)Cl_2]^-$$

$$[Zn(OH)Cl_2]^- + H^+ \rightleftharpoons ZnCl_2 + H_2O$$

PROBLEM 10.16

(a) What factor explains the observation that tertiary alcohols react with HX faster than secondary alcohols?   (b) What factor explains the observation that methanol reacts with HX faster than a primary alcohol?

PROBLEM 10.17

Treating 3-methyl-2-butanol (see following reaction) with HBr yields 2-bromo-2-methylbutane as the sole product. Outline a mechanism for the reaction.

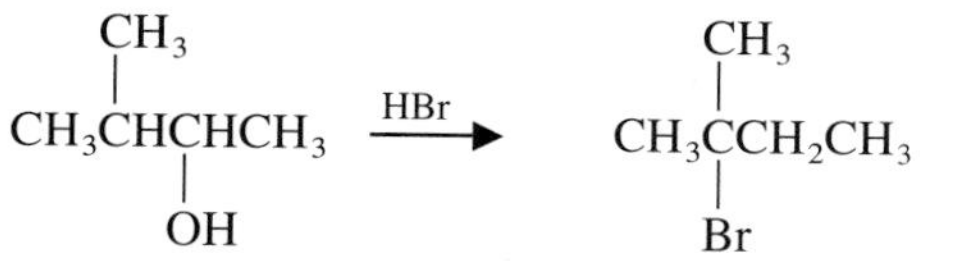

$$\underset{\substack{\textbf{3-Methyl-2-butanol}}}{\overset{\overset{\displaystyle CH_3}{|}}{\underset{\underset{\displaystyle OH}{|}}{CH_3CHCHCH_3}}} \xrightarrow{HBr} \underset{\substack{\textbf{2-Bromo-2-methylbutane}}}{\overset{\overset{\displaystyle CH_3}{|}}{\underset{\underset{\displaystyle Br}{|}}{CH_3CCH_2CH_3}}}$$

10.15 ALKYL HALIDES FROM THE REACTIONS OF ALCOHOLS WITH PBR₃ OR SOCL₂

Primary and secondary alcohols react with phosphorus tribromide to yield alkyl bromides.

$$3 \text{ R} \overset{|}{-} OH + PBr_3 \longrightarrow 3 \text{ R} - Br + H_3PO_3$$
$$(1° \text{ or } 2°)$$

Unlike the reaction of an alcohol with HBr, the reaction of an alcohol with PBr_3 does not involve the formation of a carbocation and *usually occurs without rearrangement* of the carbon skeleton (especially if the temperature is kept below 0°C). For this reason phosphorus tribromide is often preferred as a reagent for the transformation of an alcohol to the corresponding alkyl bromide.

The mechanism for the reaction involves the initial formation of a protonated alkyl dibromophosphite (see following reaction) by a nucleophilic displacement on phosphorus; the alcohol acts as the nucleophile:

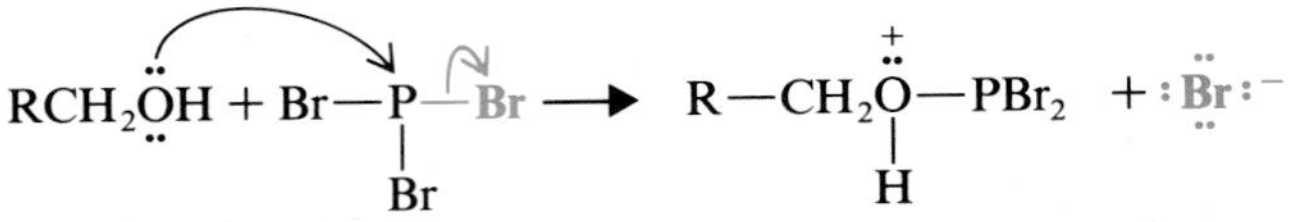

$$RCH_2\ddot{O}H + Br-\underset{\underset{Br}{|}}{P}-Br \longrightarrow R-CH_2\overset{+}{\underset{\underset{H}{|}}{\ddot{O}}}-PBr_2 \ +\ :\ddot{Br}:^-$$

**Protonated
alkyl dibromophosphite**

Then a bromide ion acts as a nucleophile and displaces $HOPBr_2$.

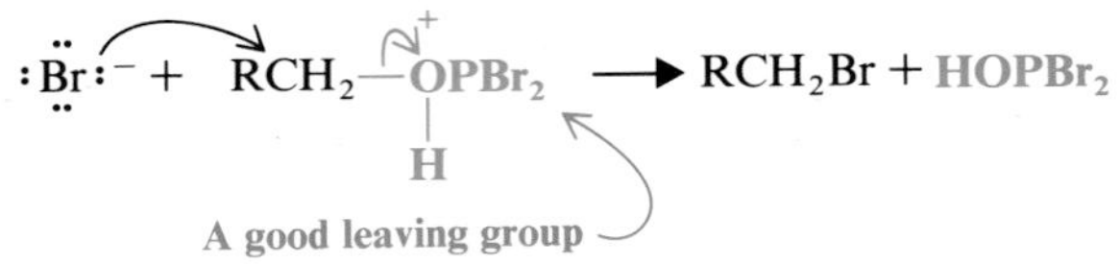

$$:\ddot{Br}:^- +\ RCH_2-\overset{+}{\underset{\underset{H}{|}}{O}}PBr_2 \longrightarrow RCH_2Br + HOPBr_2$$

A good leaving group

The $HOPBr_2$ can react with more alcohol so the net result is the conversion of 3 mol of alcohol to alkyl bromide by 1 mol of phosphorus tribromide.

Thionyl chloride ($SOCl_2$) converts primary and secondary alcohols to alkyl chlorides (usually without rearrangement).

$$R-OH\ +\ SOCl_2 \xrightarrow{\text{reflux}} R-Cl + SO_2 + HCl$$
(1° or 2°)

Often a tertiary amine is added to the mixture to promote the reaction by reacting with the HCl.

$$R_3N: + HCl \longrightarrow R_3NH^+ + Cl^-$$

The reaction mechanism involves initial formation of the alkyl chlorosulfite:

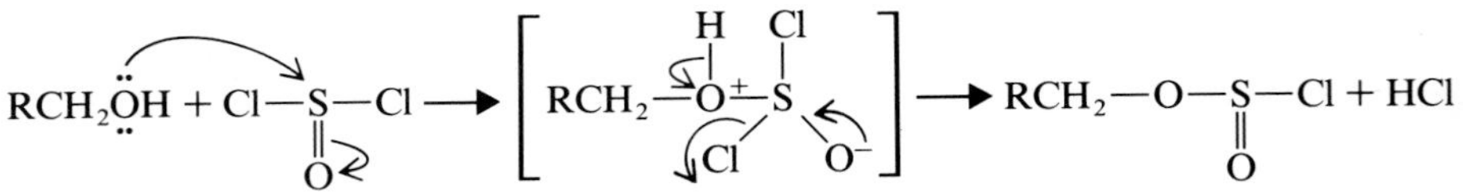

$$RCH_2\ddot{O}H + Cl-\underset{\underset{O}{\|}}{S}-Cl \longrightarrow \left[RCH_2-\overset{+}{\underset{\underset{Cl}{|}}{O}}\cdot\cdot\overset{\overset{H\ \ Cl}{|\ \ |}}{S}\cdot\cdot\underset{O^-}{} \right] \longrightarrow RCH_2-O-\underset{\underset{O}{\|}}{S}-Cl + HCl$$

**Alkyl
chlorosulfite**

Then a chloride ion (from $R_3N + HCl \longrightarrow R_3NH^+ + Cl^-$) can bring about an S_N2

displacement of a very good leaving group, $ClSO_2^-$, which, by decomposing (to the gas SO_2 and Cl^- ion), helps drive the reaction to completion.

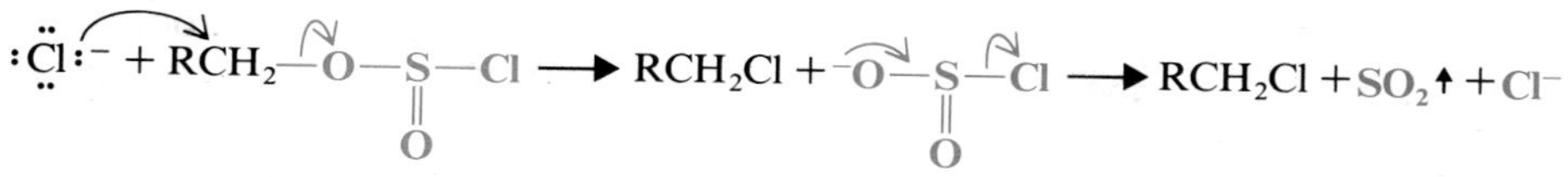

SAMPLE PROBLEM

Starting with alcohols, outline a synthesis of each of the following: (a) benzyl bromide, (b) cyclohexyl chloride, and (c) butyl bromide.

POSSIBLE ANSWERS:

(a) $C_6H_5CH_2OH \xrightarrow{PBr_3} C_6H_5CH_2Br$

(b) 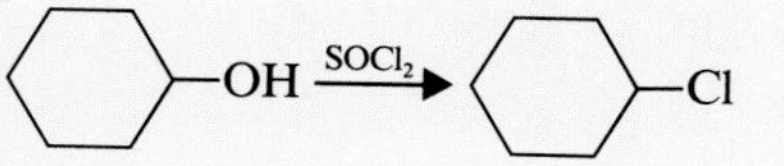

(c) $CH_3CH_2CH_2CH_2OH \xrightarrow{PBr_3} CH_3CH_2CH_2CH_2Br$

10.16 SYNTHESIS OF ETHERS

10.16A Ethers by Intermolecular Dehydration of Alcohols

Alcohols can dehydrate to form alkenes. We studied this in Sections 7.13–7.15. Primary alcohols can also dehydrate to form ethers.

$$R{-}OH + HO{-}R \xrightarrow[(-H_2O)]{H^+} R{-}O{-}R$$

Dehydration to an ether usually takes place at a lower temperature than dehydration to the alkene, and dehydration to the ether can be aided by distilling the ether as it is formed. Diethyl ether is made commercially by dehydration of ethanol. Diethyl ether is the predominant product at 140°C; ethene is the major product at 180°C:

$$CH_3CH_2OH \xrightarrow[180\ °C]{H_2SO_4} CH_2{=}CH_2 \quad \textbf{Ethene}$$

$$CH_3CH_2OH \xrightarrow[140\ °C]{H_2SO_4} CH_3CH_2OCH_2CH_3 \quad \textbf{Diethyl ether}$$

The formation of the ether occurs by an S_N2 mechanism with one molecule of the alcohol acting as the nucleophile and with another protonated molecule of the alcohol acting as the substrate (see Section 10.8).

A MECHANISM FOR THE REACTION

Step 1 $CH_3CH_2-\overset{..}{\underset{..}{O}}-H + H-OSO_3H \rightleftharpoons CH_3CH_2-\overset{H}{\underset{|}{\overset{+}{O}}}-H + {}^-OSO_3H$

This is an acid–base reaction in which the alcohol accepts a proton from the sulfuric acid.

Step 2 $CH_3CH_2-\overset{..}{\underset{..}{O}}-H + CH_3CH_2-\overset{H}{\underset{|}{\overset{+}{O}}}-H \rightleftharpoons CH_3CH_2-\overset{H}{\underset{|}{\overset{+}{O}}}-CH_2CH_3 + :\overset{H}{\underset{}{O}}-H$

Another molecule of the alcohol acts as a nucleophile and attacks the protonated alcohol in an S_N2 reaction.

Step 3 $CH_3CH_2-\overset{H}{\underset{|}{\overset{+}{O}}}-CH_2CH_3 + :\overset{H}{\underset{}{O}}-H \rightleftharpoons CH_3CH_2-\overset{..}{\underset{..}{O}}-CH_2CH_3 + H-\overset{H}{\underset{+}{O}}-H$

Another acid–base reaction converts the protonated ether to an ether by transferring a proton to a molecule of water (or to another molecule of the alcohol).

This method of preparing ethers is of limited usefulness, however. Attempts to synthesize ethers with secondary alkyl groups by intermolecular dehydration of secondary alcohols are usually unsuccessful because alkenes form too easily. Attempts to make ethers with tertiary alkyl groups lead exclusively to the alkenes. And, finally, this method is not useful for the preparation of unsymmetrical ethers from primary alcohols because the reaction leads to a mixture of products.

$$\underbrace{ROH + R'OH}_{1° \text{ alcohols}} \underset{H_2SO_4}{\rightleftharpoons} \begin{array}{c} ROR \\ + \\ ROR' \\ + \\ R'OR' \end{array} + H_2O$$

PROBLEM 10.18

An exception to what we have just said has to do with syntheses of unsymmetrical ethers in which one alkyl group is a *tert*-butyl group and the other group is primary. This synthesis can be accomplished by adding *tert*-butyl alcohol to a mixture of the primary alcohol and H_2SO_4 at room temperature. Give a likely mechanism for this reaction and explain why it is successful.

10.16B The Williamson Synthesis of Ethers

An important route to unsymmetrical ethers is a nucleophilic substitution reaction known as the Williamson synthesis. This synthesis consists of an S_N2 reaction of a sodium alkoxide with an alkyl halide, alkyl sulfonate, or alkyl sulfate:

A MECHANISM FOR THE WILLIAMSON ETHER SYNTHESIS

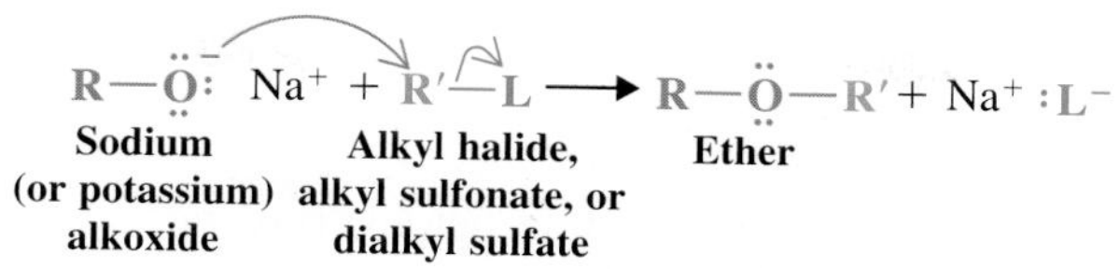

$$R-\overset{..}{\underset{..}{O}}: \ Na^+ \ + \ R'\overset{\frown}{\underset{}{-}}L \longrightarrow R-\overset{..}{\underset{..}{O}}-R' + Na^+ :L^-$$

Sodium   Alkyl halide,   Ether
(or potassium)   alkyl sulfonate, or
alkoxide   dialkyl sulfate

The alkoxide ion reacts with the substrate in an S_N2 reaction, with the resulting formation of an ether. The substrate must bear a good leaving group. Typical substrates are alkyl halides, alkyl sulfonates, and dialkyl sulfates, i.e.,

$$-L \ = \ -\overset{..}{\underset{..}{Br}}:, \ \ -\overset{..}{\underset{..}{I}}:, \ \ -OSO_2R'', \text{ or } \ -OSO_2OR''$$

The following reaction is a specific example of the Williamson synthesis. The sodium alkoxide can be prepared by allowing an alcohol to react with NaH.

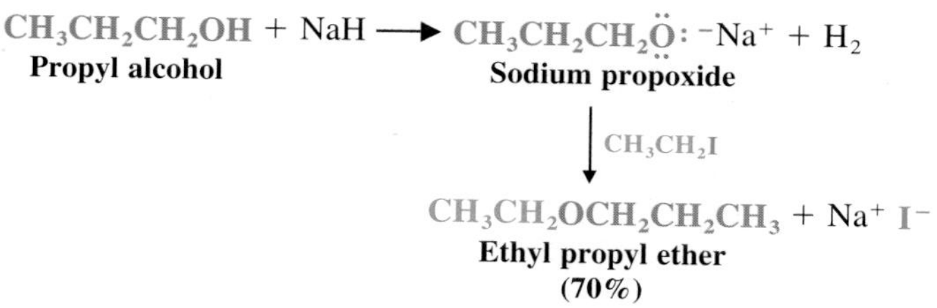

$$CH_3CH_2CH_2OH + NaH \longrightarrow CH_3CH_2CH_2\overset{..}{\underset{..}{O}}: \ ^-Na^+ + H_2$$

Propyl alcohol   Sodium propoxide

$$\downarrow CH_3CH_2I$$

$$CH_3CH_2OCH_2CH_2CH_3 + Na^+ \ I^-$$

Ethyl propyl ether
(70%)

The usual limitations of S_N2 reactions apply here. Best results are obtained when the alkyl halide, sulfonate, or sulfate is primary (or methyl). If the substrate is tertiary, elimination is the exclusive result. Substitution is also favored over elimination at lower temperatures.

PROBLEM 10.19

(a) Outline two methods for preparing isopropyl methyl ether by a Williamson synthesis. (b) One method gives a much better yield of the ether than the other. Explain which is the better method and why.

PROBLEM 10.20

The two syntheses of 2-ethoxy-1-phenylpropane shown here give products with opposite optical rotations.

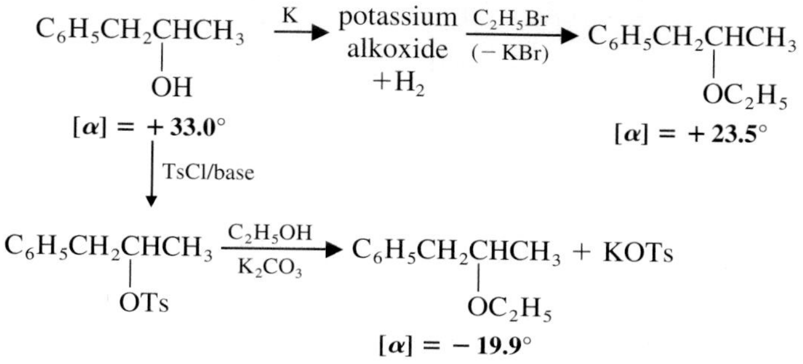

$$\underset{\underset{[\alpha] = +33.0°}{\overset{|}{OH}}}{C_6H_5CH_2CHCH_3} \xrightarrow[\text{alkoxide}]{\overset{K}{\underset{+H_2}{\text{potassium}}}} \xrightarrow[(-KBr)]{C_2H_5Br} \underset{\underset{[\alpha] = +23.5°}{\overset{|}{OC_2H_5}}}{C_6H_5CH_2CHCH_3}$$

$$\downarrow \text{TsCl/base}$$

$$\underset{\overset{|}{OTs}}{C_6H_5CH_2CHCH_3} \xrightarrow[K_2CO_3]{C_2H_5OH} \underset{\underset{[\alpha] = -19.9°}{\overset{|}{OC_2H_5}}}{C_6H_5CH_2CHCH_3} + KOTs$$

How can you explain this result?

PROBLEM 10.21

Write a mechanism that explains the formation of tetrahydrofuran (THF) from the reaction of 4-chloro-1-butanol and aqueous sodium hydroxide.

PROBLEM 10.22

Epoxides can be synthesized by treating halohydrins with aqueous base. For example, treating $ClCH_2CH_2OH$ with aqueous sodium hydroxide yields ethylene oxide. (a) Propose a mechanism for this reaction. (b) *trans*-2-Chlorocyclohexanol reacts readily with sodium hydroxide to yield cyclohexene oxide. *cis*-2-Chlorocyclohexanol does not undergo this reaction, however. How can you account for this difference?

10.16C *Tert*-Butyl Ethers by Alkylation of Alcohols. Protecting Groups

Primary alcohols can be converted to *tert*-butyl ethers by dissolving them in a strong acid such as sulfuric acid and then adding isobutylene to the mixture. (This procedure minimizes dimerization and polymerization of the isobutylene.)

$$RCH_2OH + CH_2{=}CCH_3 \;\xrightarrow{H_2SO_4}\; RCH_2O{-}CCH_3$$

with CH_3 groups shown:

$$CH_2{=}CCH_3 \text{ (with } CH_3 \text{ below)} \qquad RCH_2O{-}\underset{CH_3}{\overset{CH_3}{C}}CH_3 \quad \text{[} \textit{tert}\text{-Butyl protecting group]}$$

This method is often used to "protect" the hydroxyl group of a primary alcohol while another reaction is carried out on some other part of the molecule. The protecting *tert*-butyl group can be removed easily by treating the ether with dilute aqueous acid.

Suppose, for example, we wanted to prepare 4-pentyn-1-ol from 3-bromo-1-propanol and sodium acetylide. If we allow them to react directly, the strongly basic sodium acetylide will react first with the hydroxyl group.

$$HOCH_2CH_2CH_2Br + NaC{\equiv}CH \longrightarrow NaOCH_2CH_2CH_2Br + HC{\equiv}CH$$

3-Bromo-1-propanol

However, if we protect the —OH group first, the synthesis becomes feasible.

$$HOCH_2CH_2CH_2Br \xrightarrow[\text{(2) } CH_2=C(CH_3)_2]{\text{(1) } H_2SO_4} (CH_3)_3COCH_2CH_2CH_2Br \xrightarrow{NaC\equiv CH}$$

$$(CH_3)_3COCH_2CH_2CH_2C{\equiv}CH \xrightarrow{H_3O^+/H_2O} HOCH_2CH_2CH_2C{\equiv}CH + (CH_3)_3COH$$

4-Pentyn-1-ol

PROBLEM 10.23

(a) The mechanism for the formation of the *tert*-butyl ether from a primary alcohol and isobutylene is similar to that discussed in Problem 10.18. Propose such a mechanism.   (b) What factor makes it possible to remove the protecting *tert*-butyl group so easily? (Other ethers require much more forcing conditions for their cleavage, as we shall see in Section 10.17.)   (c) Propose a mechanism for the removal of the protecting *tert*-butyl group.

10.16D Trimethylsilyl Ethers. Silylation

A hydroxyl group can also be protected in neutral or basic solutions by converting it to a trimethylsilyl ether group, $-OSi(CH_3)_3$. This reaction, called **silylation,** is done by allowing the alcohol to react with chlorotrimethylsilane in the presence of a tertiary amine:

$$R-OH + \quad (CH_3)_3SiCl \quad \xrightarrow{(CH_3CH_2)_3N} R-O-Si(CH_3)_3$$
$$\textbf{Chlorotrimethylsilane}$$

This protecting group can also be removed with aqueous acid.

$$R-O-Si(CH_3)_3 \xrightarrow{H_3O^+/H_2O} R-OH + (CH_3)_3SiOH$$

Converting an alcohol to a trimethylsilyl ether also makes it much more volatile. (Why?) This increased volatility makes the alcohol (as a trimethylsilyl ether) much more amenable to analysis by gas–liquid chromatography.

10.17 REACTIONS OF ETHERS

Dialkyl ethers react with very few reagents other than acids. The only reactive sites that molecules of a dialkyl ether present to another reactive substance are the C—H bonds of the alkyl groups and the $-\overset{..}{O}-$ group of the ether linkage. Ethers resist attack by nucleophiles (why?) and by bases. This lack of reactivity, coupled with the ability of ethers to solvate cations (by donating an electron pair from their oxygen atom) makes ethers especially useful as solvents for many reactions.

Ethers are like alkanes in that they undergo halogenation reactions (Chapter 9), but these are of little synthetic importance.

The oxygen of the ether linkage makes ethers basic. Ethers can react with proton donors to form **oxonium salts.**

$$CH_3CH_2\overset{..}{\underset{..}{O}}CH_2CH_3 + HBr \rightleftharpoons CH_3CH_2-\overset{+}{\underset{\underset{H}{|}}{O}}-CH_2CH_3 \ Br^-$$

An oxonium salt

Heating dialkyl ethers with very strong acids (HI, HBr, and H_2SO_4) causes them to undergo reactions in which the carbon–oxygen bond breaks. Diethyl ether, for example, reacts with hot concentrated hydrobromic acid to give two molar equivalents of ethyl bromide.

$$CH_3CH_2OCH_2CH_3 + 2\ HBr \longrightarrow 2\ CH_3CH_2Br + H_2O \qquad \textbf{Cleavage of an ether}$$

The mechanism for this reaction begins with formation of an oxonium ion. Then an S_N2 reaction with a bromide ion acting as the nucleophile produces ethanol and ethyl bromide.

A MECHANISM FOR THE REACTION

$$CH_3CH_2\ddot{O}CH_2CH_3 + H\ddot{Br}: \rightleftharpoons CH_3CH_2\overset{+}{\underset{\underset{H}{|}}{\ddot{O}}}CH_2CH_3 + :\ddot{Br}:^- \longrightarrow$$

$$CH_3CH_2\underset{\underset{H}{|}}{\ddot{O}}: + CH_3CH_2Br$$

Ethanol   Ethyl bromide

In the next step the ethanol (just formed) reacts with HBr to form a second molar equivalent of ethyl bromide.

$$CH_3CH_2\ddot{O}H + H\ddot{Br}: \rightleftharpoons :\ddot{Br}:^- + CH_3CH_2\overset{+}{\underset{\underset{H}{|}}{\ddot{O}}}-H \longrightarrow CH_3CH_2-\ddot{Br}: + :\underset{\underset{H}{|}}{\ddot{O}}-H$$

PROBLEM 10.24

When an ether is treated with *cold* concentrated HI, cleavage occurs as follows:

$$R-O-R + HI \longrightarrow ROH + RI$$

When mixed ethers are used, the alcohol and alkyl iodide that form depend on the nature of the alkyl groups. Explain the following observations. (a) When (*R*)-2-methoxybutane reacts, the products are methyl iodide and (*R*)-2-butanol. (b) When *tert*-butyl methyl ether reacts, the products are methanol and *tert*-butyl iodide.

10.18 EPOXIDES

Epoxides are cyclic ethers with three-membered rings. In IUPAC nomenclature epoxides are called **oxiranes.** The simplest epoxide has the common name ethylene oxide.

$$\underset{\textbf{An epoxide}}{\overset{\displaystyle -\underset{|}{C}-\underset{|}{C}-}{\underset{\displaystyle\ddot{O}}{\diagdown\diagup}}} \qquad \underset{\substack{\textbf{IUPAC: Oxirane}\\\textbf{Common: Ethylene oxide}}}{\overset{\displaystyle H_2\overset{2}{C}-\overset{3}{C}H_2}{\underset{\displaystyle\ddot{O}}{\diagdown\underset{1}{\diagup}}}}$$

The most widely used method for synthesizing epoxides is the reaction of an alkene with an organic **peroxy acid** (sometimes called simply a **peracid**), a process that is called **epoxidation.**

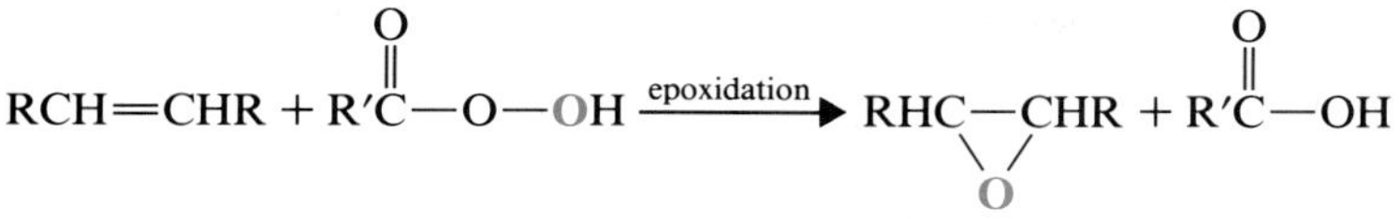

$$RCH=CHR + R'\overset{\overset{\displaystyle O}{\|}}{C}-O-OH \xrightarrow{\text{epoxidation}} RHC-CHR + R'\overset{\overset{\displaystyle O}{\|}}{C}-OH$$

An alkene   **A peroxy acid**   **An epoxide**
                                                      (or oxirane)

In this reaction the peroxy acid transfers an oxygen atom to the alkene. The following mechanism has been proposed.

A   MECHANISM   FOR   THE   REACTION

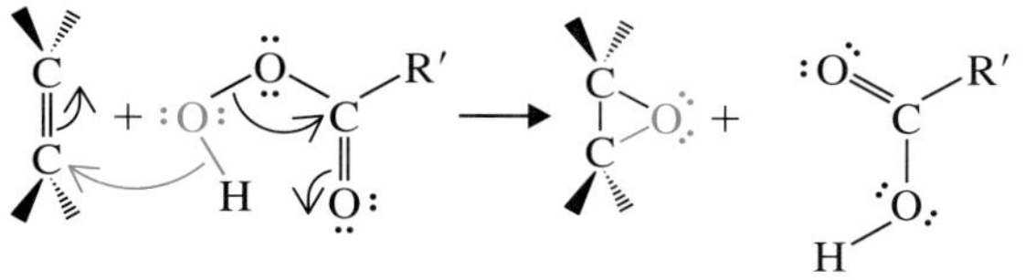

Alkene   Peroxyacid   Epoxide   Carboxylic acid

The peroxyacid transfers an oxygen atom to the alkene in a cyclic, single-step
mechanism. The result is the syn addition of the oxygen to the alkene, with
the formation of an epoxide and a carboxylic acid.

The addition of oxygen to the double bond in an epoxidation reaction is, of necessity, a **syn** addition. In order to form a three-membered ring, the oxygen atom must add to both carbon atoms of the double bond at the same face.

Some peroxy acids that have been used in the past for preparing epoxides are unstable and, therefore, are possibly unsafe. Now, because of its stability, the peroxy acid most often used is magnesium monoperoxyphthalate (MMPP).

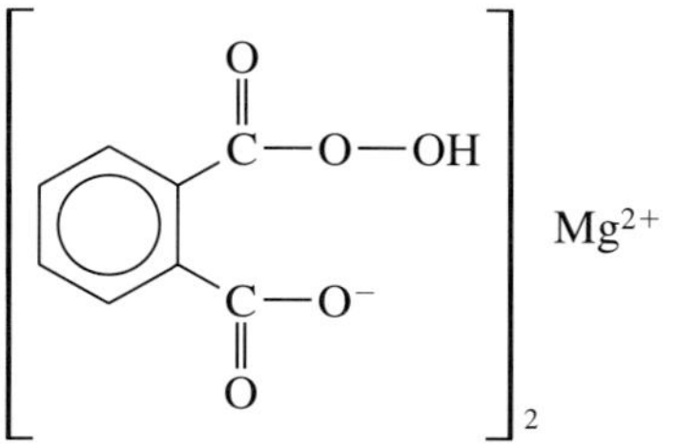

Magnesium monoperoxyphthalate
(MMPP)

Cyclohexene, for example, reacts with magnesium monoperoxyphthalate in ethanol to give 1,2-epoxycyclohexane.

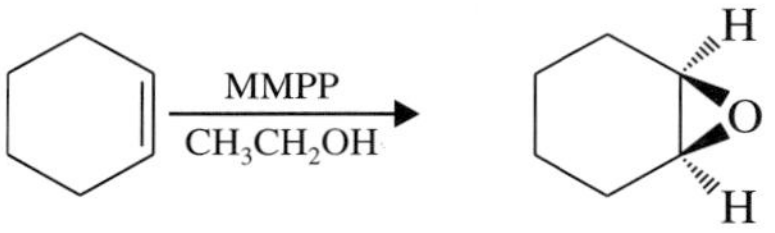

1,2-Epoxycyclohexane
(cyclohexene oxide)
(85%)

The reaction of alkenes with peroxy acids takes place in a stereospecific way: *cis*-2-Butene, for example, yields only *cis*-2,3-dimethyloxirane, and *trans*-2-butene yields only the *trans*-2,3-dimethyloxiranes.

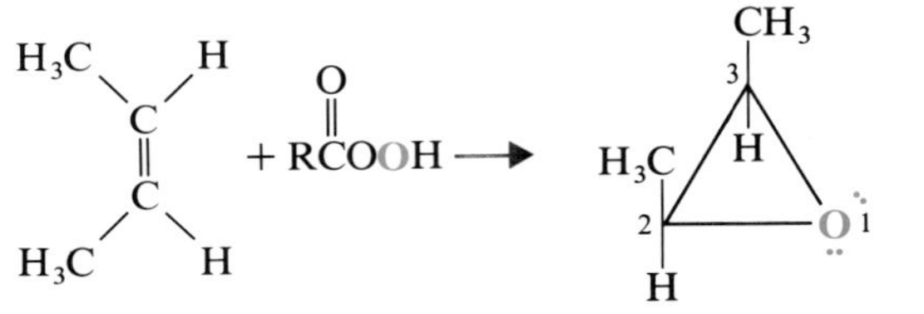

cis-2-Butene → **cis-2,3-Dimethyloxirane**
(a meso compound)

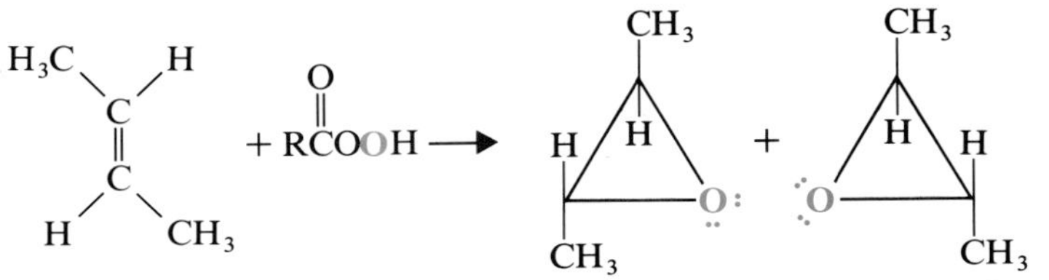

trans-2-Butene → **Enantiomeric *trans*-2,3-dimethyloxiranes**

10.19 REACTIONS OF EPOXIDES

The highly strained three-membered ring in molecules of epoxides makes them much more reactive toward nucleophilic substitution than other ethers.

Acid catalysis assists epoxide ring opening by providing a better leaving group (an alcohol) at the carbon atom undergoing nucleophilic attack. This catalysis is especially important if the nucleophile is a weak nucleophile such as water or an alcohol. An example is the acid-catalyzed hydrolysis of an epoxide:

A MECHANISM FOR ACID-CATALYZED RING OPENING

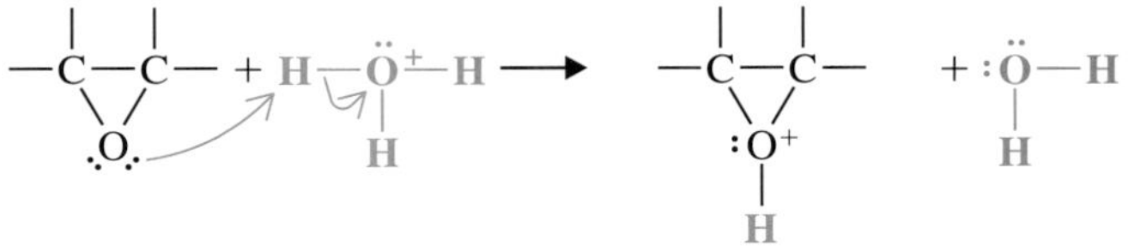

Epoxide   **Protonated epoxide**
The acid reacts with the epoxide to produce a protonated epoxide.

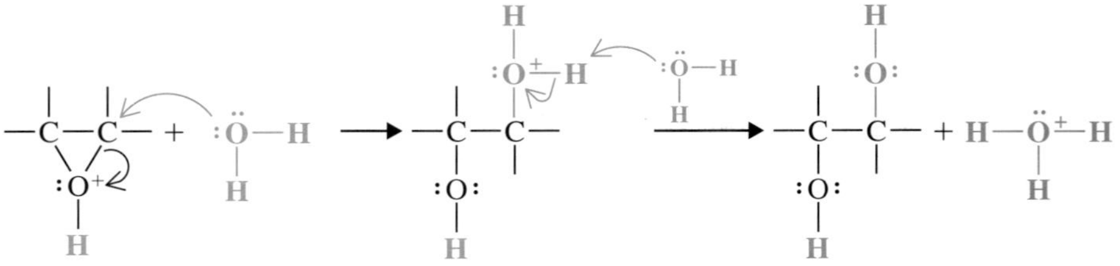

Protonated epoxide   **Weak nucleophile**   **Protonated glycol**   **Glycol**

The protonated epoxide reacts with the weak nucleophile (water) to form a protonated glycol, which then transfers a proton to a molecule of water to form the glycol and a hydronium ion.

Epoxides can also undergo base-catalyzed ring opening. Such reactions do not occur with other ethers, but they are possible with epoxides (because of ring strain), provided that the attacking nucleophile is also a strong base such as an alkoxide ion or hydroxide ion.

A MECHANISM FOR BASE-CATALYZED RING OPENING

Strong nucleophile   **Epoxide**   **An alkoxide ion**

$+ RO^{-}$

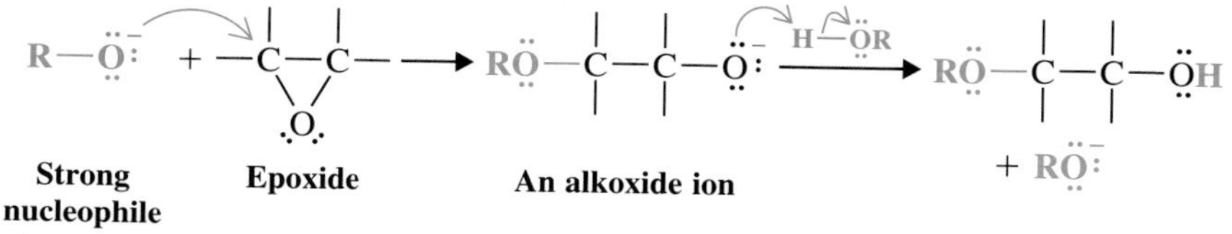

A strong nucleophile such as an alkoxide ion or a hydroxide ion is able to open the strained epoxide ring in a direct S_N2 reaction.

If the epoxide is unsymmetrical, in **base-catalyzed ring opening,** attack by the alkoxide ion occurs primarily *at the less substituted carbon atom.* For example, methyloxirane reacts with an alkoxide ion mainly at its primary carbon atom:

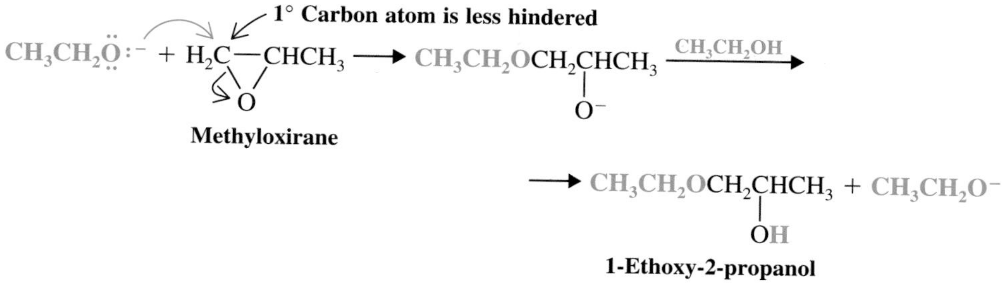

1° Carbon atom is less hindered

$CH_3CH_2\ddot{O}:^- + H_2C-CHCH_3 \longrightarrow CH_3CH_2OCH_2CHCH_3 \xrightarrow{CH_3CH_2OH}$

Methyloxirane

$\longrightarrow CH_3CH_2OCH_2CHCH_3 + CH_3CH_2O^-$
$\qquad\qquad\qquad\quad |$
$\qquad\qquad\qquad OH$

1-Ethoxy-2-propanol

This is just what we should expect: The reaction is, after all, an S_N2 reaction, and as we learned earlier (Section 6.15A), primary substrates react more rapidly in S_N2 reactions because they are less sterically hindered.

In the **acid-catalyzed ring opening** of an unsymmetrical epoxide the nucleophile attacks primarily *at the more substituted carbon atom.* For example,

$$CH_3OH + CH_3-\underset{\underset{O}{\diagup\diagdown}}{C}-CH_2 \xrightarrow{H^+} CH_3-\underset{\underset{OCH_3}{|}}{\overset{\overset{CH_3}{|}}{C}}-CH_2OH$$

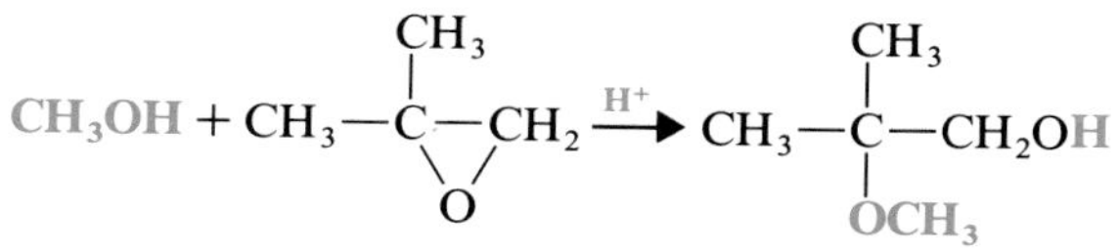

The reason: Bonding in the protonated epoxide (see following reaction) is unsymmetrical, with the more highly substituted carbon atom bearing a considerable positive charge; the reaction is S_N1 like. The nucleophile, therefore, attacks this carbon atom even though it is more highly substituted.

This carbon
resembles a
3° carbocation

$$CH_3\ddot{O}H + CH_3\overset{\delta+}{-}\overset{CH_3}{\underset{\underset{H}{O\,\delta+}}{C}}-CH_2 \longrightarrow CH_3\overset{CH_3}{\underset{\underset{H}{^+OCH_3}}{C}}-CH_2OH$$

Protonated
epoxide

The more highly substituted carbon atom bears a greater positive charge because it resembles a more stable tertiary carbocation. [Notice how this reaction (and its explanation) resembles that given for halohydrin formation from unsymmetrical alkenes in Section 8.8.]

PROBLEM 10.25

Propose structures for each of the following products:

(a) Oxirane $\xrightarrow[CH_3OH]{H^+}$ $C_3H_8O_2$ (an industrial solvent called Methyl Cellosolve)

(b) Oxirane $\xrightarrow[CH_3CH_2OH]{H^+}$ $C_4H_{10}O_2$ (Ethyl Cellosolve)

(c) Oxirane $\xrightarrow[H_2O]{KI}$ C_2H_4IO

(d) Oxirane $\xrightarrow{NH_3}$ C_2H_7NO

(e) Oxirane $\xrightarrow[CH_3OH]{CH_3ONa}$ $C_3H_8O_2$

PROBLEM 10.26

Treating 2,2-dimethyloxirane, $H_2C\!-\!C(CH_3)_2$, with sodium methoxide in methanol gives primarily 1-methoxy-2-methyl-2-propanol. What factor accounts for this result?

PROBLEM 10.27

When sodium ethoxide reacts with 1-(chloromethyl)oxirane, labeled with ^{14}C as shown by the asterisk in **I**, the major product is an epoxide bearing the label as in **II**. Provide an explanation for this reaction.

$$Cl\!-\!CH_2\!-\!\underset{\underset{O}{\diagdown\diagup}}{CH}\!-\!\overset{*}{C}H_2 \xrightarrow{NaOC_2H_5} \underset{\underset{O}{\diagdown\diagup}}{CH_2}\!-\!CH\!-\!\overset{*}{C}H_2\!-\!OC_2H_5$$

I   **II**
1-(Chloromethyl)oxirane
(epichlorohydrin)

10.19A Polyether Formation

Treating ethylene oxide with sodium methoxide (in the presence of a small amount of methanol) can result in the formation of a **polyether**:

$$H_3C-\ddot{O}\colon^- + H_2C-CH_2 \longrightarrow H_3C-\ddot{O}-CH_2-CH_2-\ddot{O}\colon^- + H_2C-CH_2 \longrightarrow$$

$$H_3C-\ddot{O}-CH_2-CH_2-\ddot{O}-CH_2-CH_2-\ddot{O}\colon^- \longrightarrow \text{etc.} \xrightarrow{\text{CH}_3\text{OH}}$$

$$H_3C-\ddot{O}(CH_2-CH_2-\ddot{O})_n CH_2-CH_2-\ddot{O}-H + H_3C-\ddot{O}\colon^-$$

Poly(ethylene glycol)
(a polyether)

This is an example of **anionic polymerization** (Special Topic C). The polymer chains continue to grow until methanol protonates the alkoxide group at the end of the growing chain. The average length of the growing chains and, therefore, the average molecular weight of the polymer can be controlled by the amount of methanol present. The physical properties of the polymer depend on its average molecular weight.

Polyethers have high water solubilities because of their ability to form multiple hydrogen bonds to water molecules. Marketed commercially as **carbowaxes** these polymers have a variety of uses, ranging from use in gas–liquid chromatography columns, to applications in cosmetics.

10.20 ANTI HYDROXYLATION OF ALKENES VIA EPOXIDES

Epoxidation of cyclopentene produces 1,2-epoxycyclopentane:

Cyclopentene   1.2–Epoxycyclopentane

Acid-catalyzed hydrolysis of 1,2-epoxycyclopentane yields a *trans*-diol, *trans*-1,2-cyclopentanediol. Water acting as a nucleophile attacks the protonated epoxide from the side opposite the epoxide group. The carbon atom being attacked undergoes an inversion of configuration. We show here only one carbon atom being attacked. Attack at the other carbon atom of this symmetrical system is equally likely and produces the enantiomeric form of *trans*-1,2-cyclopentanediol.

trans-**1,2-Cyclopentanediol**

Epoxidation followed by acid-catalyzed hydrolysis gives us, therefore, a method for *anti hydroxylation* of a double bond (as opposed to syn hydroxylation, Section 8.9). The stereochemistry of this technique parallels closely the stereochemistry of the bromination of cyclopentene given earlier (Section 8.7).

PROBLEM 10.28

Outline a mechanism similar to the one just given that shows how the enantiomeric form of *trans*-1,2-cyclopentanediol is produced.

SAMPLE PROBLEM

In Section 10.18 we showed the epoxidation of *cis*-2-butene to yield *cis*-2,3-dimethyloxirane and epoxidation of *trans*-2-butene to yield *trans*-2,3-dimethyloxirane.   (a) Now consider acid-catalyzed hydrolysis of these two epoxides and show what product or products would result from each.   (b) Are these reactions stereospecific?

ANSWER:

The meso compound, *cis*-2,3-dimethyloxirane (Fig. 10.2, p. 455), yields on hydrolysis (2*R*,3*R*)-2,3-butanediol and (2*S*,3*S*)-2,3-butanediol. These products are enantiomers. Since the attack by water at either carbon [path (a) or path (b)] occurs at the same rate, the product is obtained in a racemic form.

When either of the *trans*-2,3-dimethyloxirane enantiomers undergoes acid-catalyzed hydrolysis, the only product that is obtained is the meso compound, (2*R*,3*S*)-2,3-butanediol. The hydrolysis of one enantiomer is shown in Fig. 10.3, p. 456. (You might construct a similar diagram showing the hydrolysis of the other enantiomer to convince yourself that it, too, yields the same product.)

Since both steps in this method for the conversion of an alkene to a diol (glycol) are stereospecific (i.e., both the epoxidation step and the acid-catalyzed hydrolysis), the net result is a stereospecific anti hydroxylation of the double bond (Fig. 10.4, p. 457).

10.21   SUMMARY OF REACTIONS OF ALCOHOLS AND ETHERS

Most of the important reactions of alcohols and ethers that we have studied so far are summarized in Fig. 10.5, p. 457.

10.21A Alkenes in Synthesis

We have studied reactions in this chapter and in Chapter 8 that can be extremely useful in designing syntheses. For example, if we want to **hydrate a double bond in a Markovnikov orientation,** we have three methods for doing so: (1) *oxymercuration–*

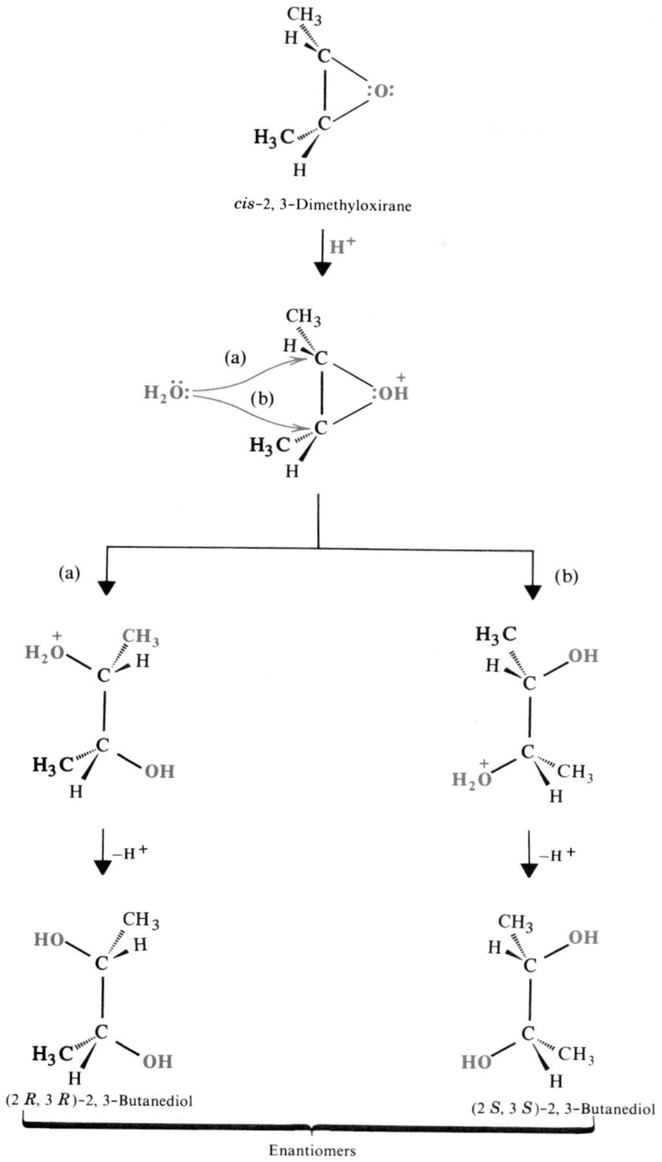

FIGURE 10.2 Acid-catalyzed hydrolysis of *cis*-2,3-dimethyloxirane yields (2R,3R)-2,3-butanediol by path (a) and (2S,3S)-2,3-butanediol by path (b). (Use models to convince yourself.)

demercuration (Section 10.5), (2) *acid-catalyzed hydration* (Section 8.5), and (3) *addition of sulfuric acid followed by hydrolysis* (Section 8.4). Of these methods oxymercuration–demercuration is the most useful in the laboratory because it is easy to carry out and because it *is not accompanied by rearrangements.*

If we want to **hydrate a double bond in an anti-Markovnikov orientation,** we can use *hydroboration–oxidation* (Section 10.7). With hydroboration–oxidation we can also achieve a *syn addition of the H— and —OH groups.* Remember, too, **the boron group of an organoborane can be replaced by hydrogen, deuterium, or**

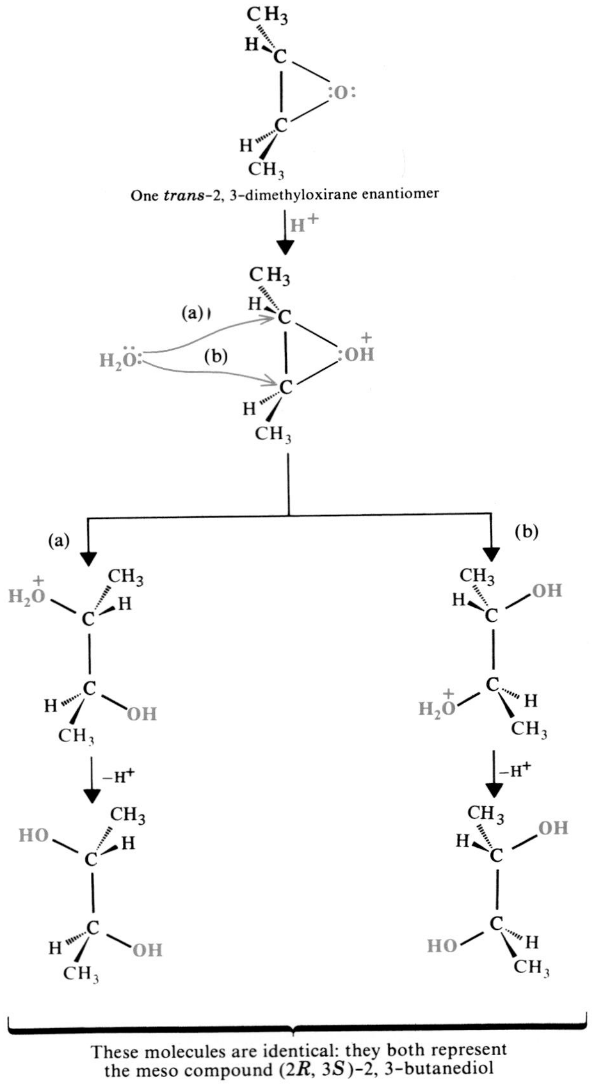

FIGURE 10.3 The acid-catalyzed hydrolysis of one *trans*-2,3-dimethyloxirane enantiomer produces the meso compound, (2R,3S)-2,3-butanediol, by path (a) or path (b). Hydrolysis of the other enantiomer (or the racemic modification) would yield the same product. (You should use models to convince yourself that the two structures given for the products above do represent the same compound.)

tritium (Section 10.7B), and that hydroboration, itself, involves a *syn addition of H—and —B.*

If we want to **add HX to a double bond in a Markovnikov sense** (Section 8.2), we treat the alkene with HF, HCl, HBr, or HI.

If we want to **add HBr in an anti-Markovnikov orientation** (Section 9.9), we treat the alkene with HBr *and a peroxide.* (The other hydrogen halides do not undergo anti-Markovnikov addition when peroxides are present.)

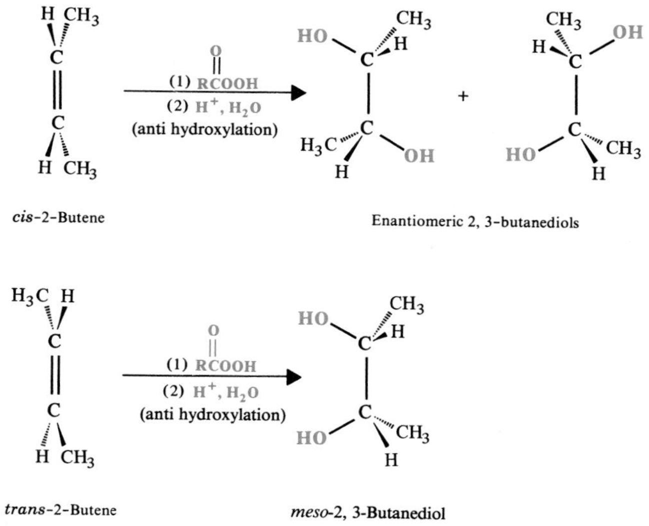

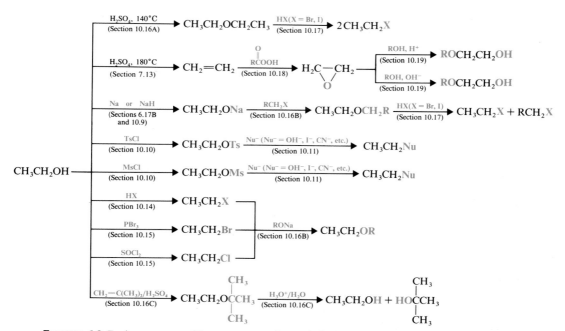

FIGURE 10.4 The overall result of epoxidation followed by acid-catalyzed hydrolysis is a stereospecific anti hydroxylation of the double bond. *cis*-2-Butene yields the enantiomeric 2,3-butanediols; *trans*-2-butene yields the meso compound.

FIGURE 10.5 A summary of important reactions of alcohols and ethers starting with ethanol.

We can **add bromine or chlorine to a double bond** (Section 8.6), and the addition is an *anti addition* (Section 8.7). We can also **add X— and —OH** to a double bond (i.e., synthesize a halohydrin) by carrying out the bromination or chlorination in water (Section 8.8). This addition, too, is an *anti addition.*

If we want to carry out a **syn hydroxylation of a double bond,** we can use either $KMnO_4$ in a cold, dilute, and basic solution or use OsO_4 followed by $NaHSO_3$ (Section 8.9). Of these two methods, the latter is preferable because of the tendency of $KMnO_4$ to overoxidize the alkene and cause cleavage at the double bond.

Anti hydroxylation of a double bond can be achieved by converting the alkene to an *epoxide* and then carrying out an acid-catalyzed hydrolysis (Section 10.20). Equations for most of these reactions are given in Figs. 8.5 and 10.6.

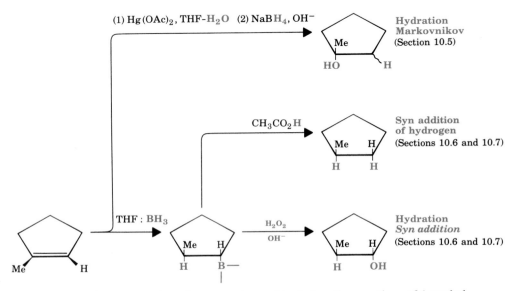

FIGURE 10.6 Oxymercuration–demercuration and hydroboration reactions of 1-methyl-cyclopentene. This figure supplements Fig. 8.5.

10.22 CROWN ETHERS: NUCLEOPHILIC SUBSTITUTION REACTIONS IN RELATIVELY NONPOLAR APROTIC SOLVENTS BY PHASE-TRANSFER CATALYSIS

When we studied the effect of the solvent on nucleophilic substitution reactions in Section 6.15C, we found that S_N2 reactions take place much more rapidly in polar aprotic solvents such as dimethyl sulfoxide and *N,N*-dimethylformamide. The reason: *In these polar aprotic solvents the nucleophile is only very slightly solvated and is, consequently, highly reactive.*

This increased reactivity of nucleophiles is a distinct advantage. Reactions that might have taken many hours or days are often over in a matter of minutes. There are, unfortunately, certain disadvantages that accompany the use of solvents such as DMSO and DMF. These solvents have very high boiling points and, as a result, they are often difficult to remove after the reaction is over. Purification of these solvents is also time consuming, and they are expensive. At high temperatures certain of these polar aprotic solvents decompose.

In some ways the ideal solvent for an S_N2 reaction would be a *nonpolar* aprotic solvent such as a hydrocarbon or relatively nonpolar chlorinated hydrocarbon. They have low boiling points, they are cheap, and they are relatively stable.

Until recently, aprotic solvents such as a hydrocarbon or chlorinated hydrocarbon were seldom used for nucleophilic substitution reactions because of their inability to dissolve ionic compounds. This situation has changed with the development of a procedure called **phase-transfer catalysis.**

With phase-transfer catalysis, we usually use two immiscible phases that are in contact—often an aqueous phase containing an ionic reactant and an organic phase (benzene, $CHCl_3$, etc.) containing the organic substrate. Normally the reaction of two substances in separate phases like this is inhibited because of the inability of the reagents to come together. Adding a phase-transfer catalyst solves this problem by transferring the ionic reactant into the organic phase. And again, because the reaction medium is aprotic, an S_N2 reaction occurs rapidly.

An example of phase-transfer catalysis is outlined in Fig. 10.7. The phase-transfer catalyst (Q^+X^-) is usually a quaternary ammonium halide ($R_4N^+X^-$) such as tetra-butylammonium halide, $(CH_3CH_2CH_2CH_2)_4N^+X^-$. The phase-transfer catalyst causes the transfer of the nucleophile (e.g., CN^-) as an ion pair $[Q^+CN^-]$ into the organic phase. This transfer apparently takes place because the cation (Q^+) of the ion pair, with its four alkyl groups, resembles a hydrocarbon in spite of its positive charge. It is said to be **lipophilic**—it prefers a nonpolar environment to an aqueous one. In the organic phase the nucleophile of the ion pair (CN^-) reacts with the organic substrate RX. The cation (Q^+) [and anion (X^-)] then migrate back into the aqueous phase to complete the cycle. This process continues until all of the nucleophile or the organic substrate has reacted.

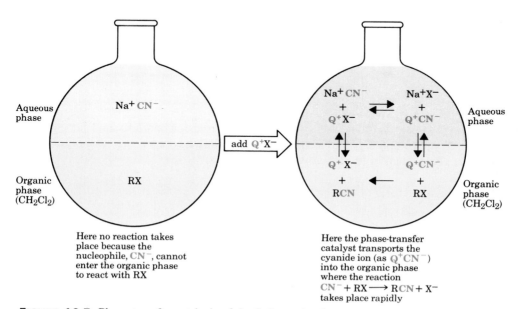

FIGURE 10.7 Phase-transfer catalysis of the S_N2 reaction between sodium cyanide and an alkyl halide.

An example of a nucleophilic substitution reaction carried out with phase-transfer catalysis is the reaction of 1-chlorooctane (in decane) and sodium cyanide (in water). The reaction (at 105°C) is complete in less than 2 h and gives a 95% yield of the substitution product.

$$CH_3(CH_2)_6CH_2Cl \text{ (in decane)} \xrightarrow[\text{aqueous NaCN, } 105°C]{R_4N^+Br^-} CH_3(CH_2)_6CH_2CN$$
$$(95\%)$$

Many other nucleophilic substitution reactions have been carried out in a similar way.

Phase-transfer catalysis, however, is not limited to nucleophilic substitutions. Many other types of reactions are also amenable to phase-transfer catalysis. Oxidations of alkenes dissolved in benzene can be accomplished in excellent yield using potassium permanganate (in water) when a quaternary ammonium salt is present:

$$CH_3(CH_2)_5CH{=}CH_2 \text{ (benzene)} \xrightarrow[\text{aqueous KMnO}_4, \ 35°C]{R_4N^+X^-} CH_3(CH_2)_5CO_2H$$
$$(99\%)$$

Potassium permanganate can also be transferred to benzene by quaternary ammonium salts for the purpose of chemical tests. The resulting "purple benzene" can be used as a test reagent for unsaturated compounds. As an unsaturated compound is added to the benzene solution of $KMnO_4$, the purple color disappears and the solution becomes brown (because of the presence of MnO_2), indicating a positive test (see Section 8.17D).

PROBLEM 10.29

Outline a scheme such as the one shown in Fig. 10.7 showing how the reaction of $CH_3(CH_2)_6CH_2Cl$ with cyanide ion (just shown) takes place by phase-transfer catalysis. Be sure to indicate which ions are present in the organic phase, which are in the aqueous phase, and which pass from one phase to the other.

10.22A Crown Ethers

Compounds called **crown ethers** are also phase-transfer catalysts and are able to transport ionic compounds into an organic phase. Crown ethers are cyclic polymers of ethylene glycol such as the 18-crown-6 that follows:

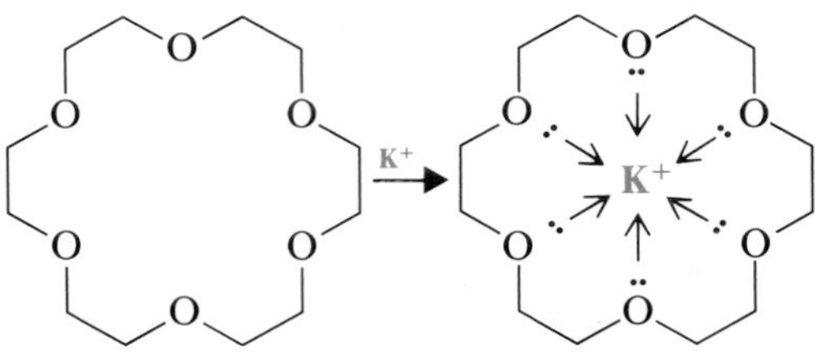

18-Crown-6

Crown ethers are named as *x*-crown-*y* where *x* is the total number of atoms in the ring and *y* is the number of oxygen atoms. The relationship between the crown ether and the ion that it transports is called a **host–guest** relationship. The crown ether acts as the **host,** and the coordinated cation is the **guest.***

When crown ethers coordinate with a metal cation, they thereby convert the metal ion into a species with a hydrocarbonlike exterior. The crown ether 18-crown-6, for example, coordinates very effectively with potassium ions because the cavity size is correct and because the six oxygen atoms are ideally situated to donate their electron pairs to the central ion.

Crown ethers render many salts soluble in nonpolar solvents. Salts such as KF, KCN, and CH_3CO_2K, for example, can be transferred into aprotic solvents by using catalytic amounts of 18-crown-6. In the organic phase the relatively unsolvated anions of these salts can carry out a nucleophilic substitution reaction on an organic substrate.

$$K^+CN^- + RCH_2X \xrightarrow[\text{benzene}]{\text{18-crown-6}} RCH_2CN + K^+X^-$$

$$C_6H_5CH_2Cl + K^+F^- \xrightarrow[\text{acetonitrile}]{\text{18-crown-6}} C_6H_5CH_2F + K^+Cl^-$$
(100%)

Crown ethers can also be used as phase-transfer catalysts for many other types of reactions. The following reaction is one example of the use of a crown ether in an oxidation.

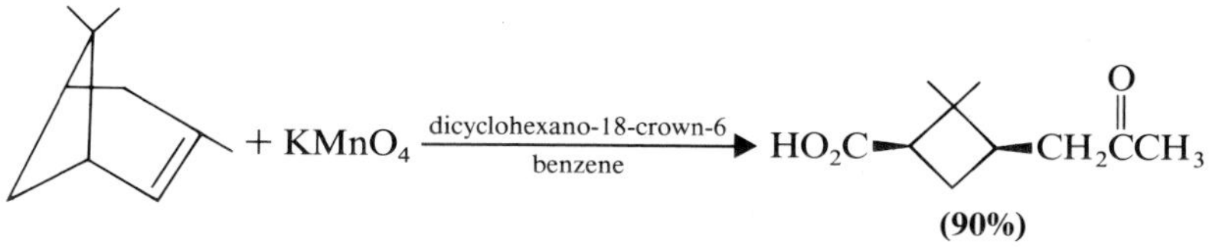

Dicyclohexano-18-crown-6 has the following structure:

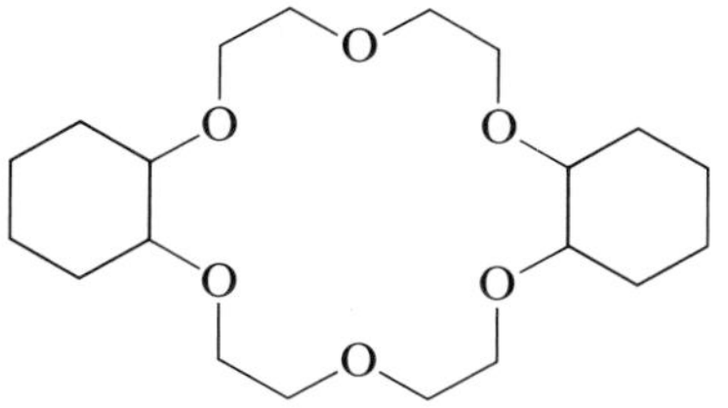

Dicyclohexano-18-crown-6

PROBLEM 10.30

Write structures for   (a) 15-crown-5 and   (b) 12-crown-4.

---

* The Nobel Prize for Chemistry in 1987 was awarded to Charles J. Pedersen (retired from the Du Pont company), Donald J. Cram (University of California, Los Angeles), and to Jean-Marie Lehn (Louis Pasteur University, Strasbourg, France) for their development of crown ethers and other molecules ''with structure specific interactions of high selectivity.'' Their contributions to our understanding of what is now called ''molecular recognition'' has implications for how enzymes recognize their substrates, how hormones cause their effects, how antibodies recognize antigens, how neurotransmitters propagate their signals, and many other aspects of biochemistry. Molecular recognition is one of the most exciting areas in chemical research today.

10.22B Transport Antibiotics and Crown Ethers

There are several antibiotics called ionophores, most notably *nonactin* and *valinomycin,* that coordinate with metal cations in a manner similar to that of crown ethers. Normally, cells must maintain a gradient between the concentrations of sodium and potassium ions inside and outside the cell wall. Potassium ions are "pumped" in; sodium ions are pumped out. The cell membrane, in its interior, is like a hydrocarbon, because it consists in this region primarily of the hydrocarbon portions of lipids (Chapter 23). The transport of hydrated sodium and potassium ions through the cell membrane is slow, and this transport requires an expenditure of energy by the cell. Nonactin, for example, upsets the concentration gradient of these ions by coordinating more strongly with potassium ions than with sodium ions. Because the potassium ions are bound in the interior of the nonactin, this host–guest complex becomes hydrocarbon-like on its surface and passes readily through the interior of the membrane. The cell membrane thereby becomes permeable to potassium ions, and the essential concentration gradient is destroyed.

Nonactin

ADDITIONAL PROBLEMS

✳10.31   Give an IUPAC substitutive name for each of the following alcohols:

(a) $(CH_3)_3CCH_2CH_2OH$

(b) CH_2=$CHCH_2\overset{\underset{\displaystyle |}{CH_3}}{CHOH}$

(c) $HOCH_2\overset{\underset{\displaystyle |}{CH_3}}{CHCH_2}CH_2OH$

(d) $C_6H_5CH_2CH_2OH$

(e)

(f)

✳10.32   Write structural formulas for each of the following:

(a) (Z)-2-Buten-1-ol

(b) (R)-1,2,4-Butanetriol

(c) (1R,2R)-1,2-Cyclopentanediol

(d) 1-Ethylcyclobutanol

(e) 2-Chloro-3-hexyn-1-ol

(f) Tetrahydrofuran

(g) 2-Ethoxypentane

(h) Ethyl phenyl ether

(i) Diisopropyl ether

(j) 2-Ethoxyethanol

10.33   Starting with each of the following, outline a practical synthesis of 1-butanol.

(a) 1-Butene   (b) 1-Chlorobutane   (c) 2-Chlorobutane   (d) 1-Butyne

10.34   Show how you might prepare 2-bromobutane from

(a) 2-Butanol, $CH_3CH_2CHOHCH_3$

(b) 1-Butanol, $CH_3CH_2CH_2CH_2OH$

(c) 1-Butene

(d) 1-Butyne

10.35   Show how you might carry out the following transformations:
(a) Cyclohexanol ⟶ chlorocyclohexane
(b) Cyclohexene ⟶ chlorocyclohexane
(c) 1-Methylcyclohexene ⟶ 1-bromo-1-methylcyclohexane
(d) 1-Methylcyclohexene ⟶ *trans*-2-methylcyclohexanol
(e) 1-Bromo-1-methylcyclohexane ⟶ cyclohexylmethanol

10.36   Give the structures and acceptable names for the compounds that would be formed when 1-butanol is treated with each of the following reagents:

(a) Sodium hydride
(b) Sodium hydride, then 1-bromopropane
(c) Methanesulfonyl chloride and base
(d) *p*-Toluenesulfonyl chloride
(e) Product of (c), then sodium methoxide

(f) Product of (d), then KI
(g) Phosphorus trichloride
(h) Thionyl chloride
(i) Sulfuric acid at 140°C
(j) Refluxing concentrated HBr

10.37   Give the structures and names for the compounds that would be formed when 2-butanol is treated with each of the reagents in Problem 10.36.

10.38   What compounds would you expect to be formed when each of the following ethers is refluxed with excess concentrated hydrobromic acid?

(a) Ethyl methyl ether   (b) *tert*-Butyl ethyl ether   (c) Tetrahydrofuran   (d) 1,4-Dioxane

10.39   Write a mechanism that accounts for the following reaction:

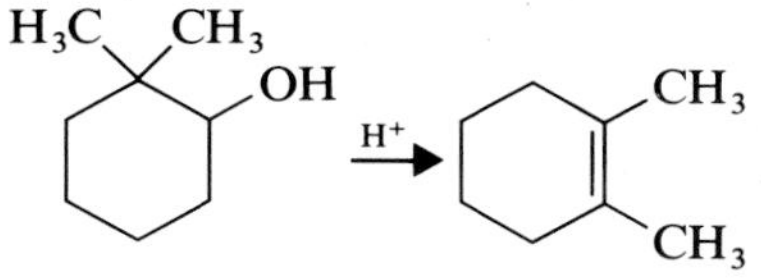

10.40   Show how you would utilize the hydroboration–oxidation procedure to prepare each of the following alcohols:

(a) 3,3-Dimethyl-1-butanol
(b) 1-Hexanol

(c) 2-Phenylethanol
(d) *trans*-2-Methylcyclopentanol

10.41   Write a three-dimensional formula for the product formed when 1-methylcyclohexene is treated with each of the following reagents. In each case, designate the location of deuterium or tritium atoms.

(a) (1) THF:BH₃ (2) CH₃CO₂T
(b) (1) THF : BD₃ (2) CH₃CO₂D

(c) (1) THF : BD₃ (2) NaOH, H₂O₂, H₂O

10.42   Starting with isobutane show how each of the following could be synthesized. (You need not repeat the synthesis of a compound prepared in an earlier part of this problem.)

(a) *tert*-Butyl bromide
(b) 2-Methylpropene
(c) Isobutyl bromide
(d) Isobutyl iodide
(e) Isobutyl alcohol (two ways)

(i) CH₃CHCH₂CN with CH₃ substituent

(j) CH₃CHCH₂SCH₃ (two ways) with CH₃ substituent

(f) *tert*-Butyl alcohol

(g) Isobutyl methyl ether

(h) $CH_3CHCH_2OCCH_3$ with CH$_3$ and O substituents

(k) $CH_3CCH_2CBr_3$ with CH$_3$ and Br substituents

10.43   Vicinal halo alcohols (halohydrins) can be synthesized by treating epoxides with HX. (a) Show how you would use this method to synthesize 2-chlorocyclopentanol from cyclopentene. (b) Would you expect the product to be *cis*-2-chlorocyclopentanol or *trans*-2-chlorocyclopentanol; that is, would you expect a net syn addition or a net anti addition of —Cl and —OH? Explain.

10.44   Outlined below is a synthesis of the gypsy moth sex attractant **E** (a type of pheromone, see Section 4.17). Give the structure of the gypsy moth sex attractant **E** and of the intermediates in the synthesis.

$$HC\equiv CNa \xrightarrow[\text{liq. NH}_3]{\text{1-bromo-5-methylhexane}} A\ (C_9H_{16}) \xrightarrow[\text{liq. NH}_3]{\text{NaNH}_2} B\ (C_9H_{15}Na)$$

$$\xrightarrow{\text{1-Bromodecane}} C\ (C_{19}H_{36}) \xrightarrow[\text{Ni}_2\text{B (P-2)}]{\text{H}_2} D\ (C_{19}H_{38}) \xrightarrow{C_6H_5CO_3H} E\ (C_{19}H_{38}O)$$

10.45   Starting with 2-methylpropene (isobutylene) and using any other needed reagents, outline a synthesis of each of the following:

(a) $(CH_3)_2CHCH_2OH$

(b) $(CH_3)_2CHCH_2T$

(c) $(CH_3)_2CDCH_2T$

(d) $(CH_3)_2CHCH_2OCH_2CH_3$

10.46   Show how you would use oxymercuration–demercuration to prepare each of the following alcohols from the appropriate alkene:

(a) 2-Pentanol

(b) 1-Cyclopentylethanol

(c) 3-Methyl-3-pentanol

(d) 1-Ethylcyclopentanol

10.47   Give stereochemical formulas for each product **A–L** and answer the questions given in parts (b) and (g).

(a) 1-Methylcyclobutene $\xrightarrow[\text{(2) H}_2\text{O}_2,\ \text{OH}^-]{\text{(1) THF:BH}_3} A\ (C_5H_{10}O) \xrightarrow[\text{OH}^-]{\text{TsCl}}$

$B\ (C_{12}H_{16}SO_3) \xrightarrow{\text{OH}^-} C\ (C_5H_{10}O)$

(b) What is the stereoisomeric relationship between **A** and **C**?

(c) $B\ (C_{12}H_{16}SO_3) \xrightarrow{I^-} D\ (C_5H_9I)$

(d) *trans*-4-Methylcyclohexanol $\xrightarrow[\text{OH}^-]{\text{MsCl}} E\ (C_8H_{16}SO_3) \xrightarrow{HC\equiv CNa}$

$F\ (C_9H_{14})$

(e) (R)-2-Butanol $\xrightarrow{\text{NaH}} [H\ (C_4H_9ONa)] \xrightarrow{CH_3I} J\ (C_5H_{12}O)$

(f) (R)-2-Butanol $\xrightarrow{\text{MsCl}} K\ (C_5H_{12}SO_3) \xrightarrow{CH_3ONa} L\ (C_5H_{12}O)$

(g) What is the stereoisomeric relationship between **J** and **L**?

10.48   When the 3-bromo-2-butanol with the stereochemical structure **A** is treated with concentrated HBr it yields *meso*-2,3-dibromobutane; a similar reaction of the 3-bromo-2-butanol **B** yields (±)-2,3-dibromobutane. This classic experiment performed in 1939 by S. Winstein and H. J. Lucas was the starting point for a series of investigations of what are called *neighboring group effects. Propose mechanisms that will account for the stereochemistry of these reactions.*

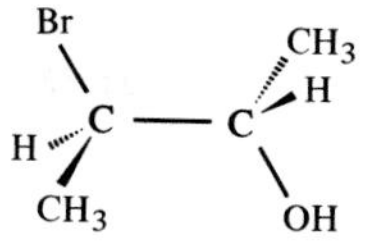

A

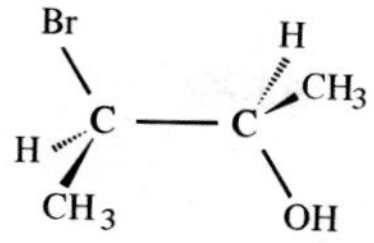

B

CHAPTER 11

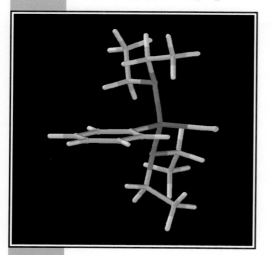

Phenylmagnesium bromide solvated by diethyl ether (Section 11.6)

ALCOHOLS FROM CARBONYL COMPOUNDS. OXIDATION–REDUCTION AND ORGANOMETALLIC COMPOUNDS

11.1 INTRODUCTION

Carbonyl compounds are a broad group of compounds that includes aldehydes, ketones, carboxylic acids, and esters.

The carbonyl group	An aldehyde	A ketone	A carboxylic acid	A carboxylate ester
$\underset{/}{\overset{\backslash}{}}C=\overset{..}{\underset{..}{O}}:$	$\underset{H}{\overset{R}{\diagdown}}C=\overset{..}{\underset{..}{O}}:$	$\underset{R'}{\overset{R}{\diagdown}}C=\overset{..}{\underset{..}{O}}:$	$\underset{HO}{\overset{R}{\diagdown}}C=\overset{..}{\underset{..}{O}}:$	$\underset{R'O}{\overset{R}{\diagdown}}C=\overset{..}{\underset{..}{O}}:$

Although we shall not study the chemistry of these compounds in detail until we reach Chapters 16–19, it is useful now to consider reactions by which these compounds are converted to alcohols. Before we do this, however, let us consider the structure of the carbonyl group and its relationship to the reactivity of carbonyl compounds.

11.1A Structure of the Carbonyl Group

The carbonyl carbon atom is sp^2 hybridized; thus it and the three groups attached to it lie in the same plane. The bond angles between the three attached atoms are what we would expect of a trigonal planar structure; they are approximately 120°.

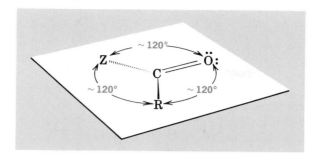

The carbon–oxygen double bond consists of a σ bond and a π bond.

The π bond is formed by overlap of the carbon p orbital with a p orbital from the oxygen atom.

The more electronegative oxygen atom strongly attracts the electrons of both the σ bond and the π bond causing the carbonyl group to be highly polarized; the carbon atom bears a substantial positive charge and the oxygen atom bears a substantial negative charge. Polarization of the π bond can be represented by the following resonance structures for the carbonyl group (see also Section 3.9).

Resonance structures for the carbonyl group   or   **Hybrid**

Evidence for the polarity of the carbon–oxygen bond can be found in the rather large dipole moments associated with carbonyl compounds.

Formaldehyde   **Acetone**
$\mu = 2.27$ D $\mu = 2.88$ D

11.1B Reaction of Carbonyl Compounds with Nucleophiles

From a synthetic point of view, one of the most important reactions of carbonyl compounds is one in which the carbonyl compound undergoes **nucleophilic addition.** The carbonyl group is susceptible to nucleophilic attack because, as we have just seen, the carbonyl carbon bears a partial positive charge. When the nucleophile adds to the carbonyl group it uses its electron pair to form a bond to the carbonyl carbon atom. The carbonyl carbon can accept this electron pair because one pair of electrons of the carbon–oxygen double bond can shift out to the oxygen.

$$Nu:^- \quad + \quad \overset{\delta+ \; \delta-}{\underset{}{\diagdown}C{=}\ddot{O}:} \longrightarrow Nu{-}\underset{|}{\overset{|}{C}}{-}\ddot{O}:^-$$

As the reaction takes place, the carbon atom undergoes a change in its geometry and its hybridization state. It goes from a trigonal planar geometry and sp^2 hybridization to a tetrahedral geometry and sp^3 hybridization.

Two important nucleophiles that add to carbonyl compounds are **hydride ions** from compounds such as $NaBH_4$ or $LiAlH_4$ (Section 11.3), and **carbanions** from compounds such as RLi or RMgX (Section 11.7C).

Another related set of reactions are reactions in which alcohols and carbonyl compounds are **oxidized** and **reduced** (Sections 11.2–11.4). For example, primary alcohols can be oxidized to aldehydes and aldehydes can be reduced to alcohols.

$$R{-}\underset{\underset{H}{|}}{\overset{\overset{H}{|}}{C}}{-}\ddot{O}{-}H \underset{\underset{[H]}{reduction}}{\overset{\overset{[O]}{oxidation}}{\rightleftarrows}} \underset{H}{\overset{R}{\diagdown}}C{=}\ddot{O}:$$

A primary   **An aldehyde**
alcohol

Let us begin by examining some general principles that apply to the oxidation and reduction of organic compounds.

11.2 OXIDATION–REDUCTION REACTIONS IN ORGANIC CHEMISTRY

Reduction of an organic molecule usually corresponds to increasing its hydrogen content or to decreasing its oxygen content. For example, converting a carboxylic acid to an aldehyde is a reduction because the oxygen content is decreased.

oxygen content decreases

$$R{-}\overset{\overset{O}{||}}{C}{-}OH \xrightarrow[reduction]{[H]} R{-}\overset{\overset{O}{||}}{C}{-}H$$

Carboxylic   **Aldehyde**
acid

Converting an aldehyde to an alcohol is also a reduction.

hydrogen content increases

$$R{-}\overset{\overset{O}{||}}{C}{-}H \xrightarrow[reduction]{[H]} RCH_2OH$$

Converting an alcohol to an alkane is also a reduction.

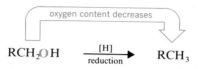

In these examples we have used the symbol [H] to indicate that a reduction of the organic compound has taken place. We do this when we want to write a general equation without specifying what the reducing agent is.

The opposite of reduction is **oxidation.** *Thus, increasing the oxygen content of an organic molecule or decreasing its hydrogen content is an oxidation* of the organic molecule. The reverse of each reaction that we have given is an oxidation of the organic molecule, and we can summarize these oxidation–reduction reactions as follows below. We use the symbol [O] to indicate in a general way that the organic molecule has been oxidized.

$$
RCH_3 \underset{[H]}{\overset{[O]}{\rightleftharpoons}} RCH_2OH \underset{[H]}{\overset{[O]}{\rightleftharpoons}} R\overset{\overset{O}{\parallel}}{C}H \underset{[H]}{\overset{[O]}{\rightleftharpoons}} R\overset{\overset{O}{\parallel}}{C}OH
$$

Lowest oxidation state   **Highest oxidation state**

Oxidation of an organic compound may be more broadly defined as a reaction that increases its content of any element more electronegative than carbon. For example, replacing hydrogen atoms by chlorine atoms is an oxidation:

$$
Ar-CH_3 \underset{[H]}{\overset{[O]}{\rightleftharpoons}} Ar-CH_2Cl \underset{[H]}{\overset{[O]}{\rightleftharpoons}} Ar-CHCl_2 \underset{[H]}{\overset{[O]}{\rightleftharpoons}} Ar-CCl_3
$$

Of course, when an organic compound is reduced, something else—**the reducing agent**—must be oxidized. And when an organic compound is oxidized, something else—**the oxidizing agent**—is reduced. These oxidizing and reducing agents are often inorganic compounds, and in the next two sections we shall see what some of them are.

PROBLEM 11.1

One method for assigning an oxidation state to a carbon atom of an organic compound is to base that assignment on the groups attached to the carbon; a bond to hydrogen (or anything less electronegative than carbon) makes it -1, a bond to oxygen, nitrogen, or halogen (or to anything more electronegative than carbon) makes it $+1$, and a bond to another carbon makes it 0. Thus the carbon of methane is assigned an oxidation state of -4, and that of carbon dioxide, $+4$. (a) Use this method to assign oxidation states to the carbon atoms of methanol (CH_3OH), formic acid $\left(H\overset{\overset{O}{\parallel}}{C}OH \right)$, and formaldehyde $\left(H\overset{\overset{O}{\parallel}}{C}H \right)$. (b) Arrange the compounds methane, carbon dioxide, methanol, formic acid, and formaldehyde in order of increasing oxidation state. (c) What change in oxidation state accompanies the reaction, methanol $\longrightarrow$ formaldehyde? (d) Is this an oxidation or a reduction? (e) When H_2CrO_4 acts as an oxidizing agent in this reaction, the chromium of H_2CrO_4 becomes Cr^{3+}. What change in oxidation state does chromium undergo?

PROBLEM 11.2

(a) Use the method described in the preceding problem to assign oxidation states to each carbon of ethanol and to each carbon of acetaldehyde. (b) What do these numbers reveal about the site of oxidation when ethanol is oxidized to acetaldehyde? (c) Repeat this procedure for the oxidation of acetaldehyde to acetic acid.

PROBLEM 11.3

(a) Although we have described the hydrogenation of an alkene as an addition reaction, organic chemists often refer to it as a ''reduction.'' Refer to the method described in Problem 11.1 and explain. (b) Make similar comments about the reaction:

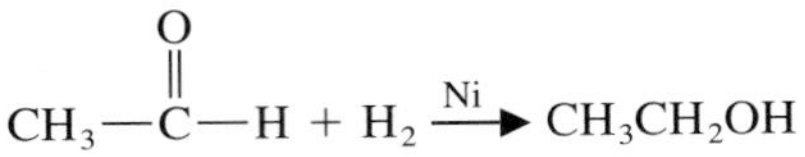

$$CH_3-\overset{\overset{\displaystyle O}{\|}}{C}-H + H_2 \xrightarrow{\text{Ni}} CH_3CH_2OH$$

11.2A Balancing Oxidation–Reduction Equations

A method for balancing organic oxidation–reduction reactions is described in the Study Guide that accompanies this text.

11.3 ALCOHOLS BY REDUCTION OF CARBONYL COMPOUNDS

Primary and secondary alcohols can be synthesized by the reduction of a variety of compounds that contain the carbonyl $\left(\overset{\diagdown}{\underset{\diagup}{C}}=O\right)$ group. Several general examples are shown here.

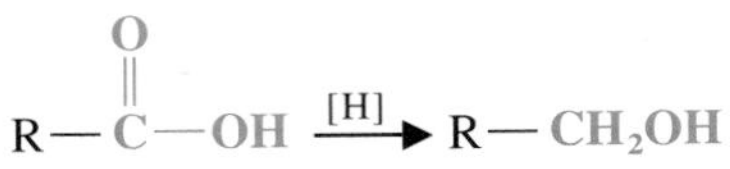

$$R-\overset{\overset{\displaystyle O}{\|}}{C}-OH \xrightarrow{\text{[H]}} R-CH_2OH$$

Carboxylic acid   **1° Alcohol**

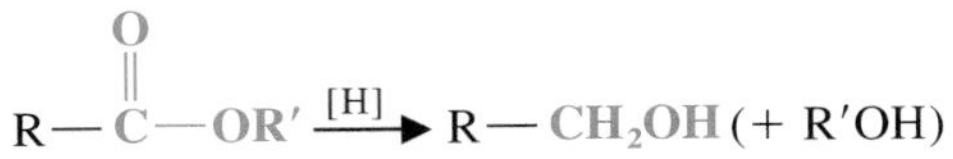

$$R-\overset{\overset{\displaystyle O}{\|}}{C}-OR' \xrightarrow{\text{[H]}} R-CH_2OH\,(+\,R'OH)$$

Ester   **1° Alcohol**

$$R-\overset{\overset{\displaystyle O}{\|}}{C}-H \xrightarrow{\text{[H]}} R-CH_2OH$$

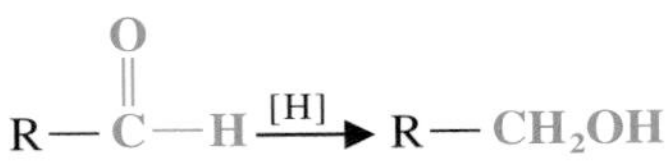

Aldehyde   **1° Alcohol**

$$R-\overset{\overset{\displaystyle O}{\|}}{C}-R' \xrightarrow{[H]} R-\underset{\underset{\displaystyle OH}{|}}{C}H-R'$$

Ketone   **2° Alcohol**

Reductions of carboxylic acids are the most difficult, but they can be accomplished with the powerful reducing agent **lithium aluminum hydride** ($LiAlH_4$, abb. LAH). It reduces acids to primary alcohols in excellent yields.

$$4\ RCO_2H + 3\ LiAlH_4 \xrightarrow{Et_2O} [(RCH_2O)_4Al]Li + 4\ H_2 + 2\ LiAlO_2$$

Lithium aluminum hydride

$$\xrightarrow[H_2O]{} 4\ RCH_2OH + Al(OH)_3 + LiOH$$

An example is the lithium aluminum hydride reduction of 2,2-dimethylpropanoic acid.

$$CH_3-\underset{\underset{\displaystyle CH_3}{|}}{\overset{\overset{\displaystyle CH_3}{|}}{C}}-CO_2H \xrightarrow[\text{(2) }H_2O]{\text{(1) }LiAlH_4/Et_2O} CH_3-\underset{\underset{\displaystyle CH_3}{|}}{\overset{\overset{\displaystyle CH_3}{|}}{C}}-CH_2OH$$

2,2-Dimethylpropanoic acid   **Neopentyl alcohol (92%)**

Esters can be reduced by high-pressure hydrogenation (a reaction preferred for industrial processes and often referred to as "hydrogenolysis" because a carbon–oxygen bond is cleaved in the process), or through the use of lithium aluminum hydride.

$$R-\overset{\overset{\displaystyle O}{\|}}{C}OR' + H_2 \xrightarrow[\underset{\text{5000 psi}}{175°C}]{CuO \cdot CuCr_2O_4} RCH_2OH + R'OH$$

$$R-\overset{\overset{\displaystyle O}{\|}}{C}OR' \xrightarrow[\text{(2) }H_2O]{\text{(1) }LiAlH_4/Et_2O} RCH_2OH + R'OH$$

The latter method is the one most commonly used now in small-scale laboratory synthesis.

Aldehydes and ketones can also be reduced to alcohols by hydrogen and a metal catalyst, by sodium in alcohol, and by lithium aluminum hydride. The reducing agent most often used, however, is sodium borohydride ($NaBH_4$).

$$4\ R\overset{\overset{\displaystyle O}{\|}}{C}H + NaBH_4 + 3\ H_2O \longrightarrow 4\ RCH_2OH + NaH_2BO_3$$

$$CH_3CH_2CH_2\overset{\overset{\displaystyle O}{\|}}{C}H \xrightarrow[H_2O]{NaBH_4} CH_3CH_2CH_2CH_2OH$$

Butanal   **1-Butanol (85%)**

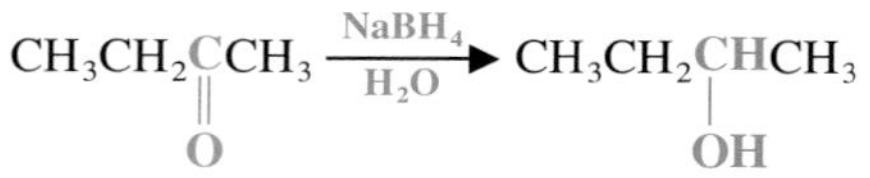

$$CH_3CH_2\underset{\underset{O}{\|}}{C}CH_3 \xrightarrow[H_2O]{NaBH_4} CH_3CH_2\underset{\underset{OH}{|}}{C}HCH_3$$

2-Butanone   **2-Butanol**
                                              (87%)

The key step in the reduction of a carbonyl compound by either lithium aluminum hydride or sodium borohydride is the transfer of a **hydride ion** from the metal to the carbonyl carbon. In this transfer the hydride ion acts as a *nucleophile*. The mechanism for the reduction of a ketone by sodium borohydride is illustrated here.

A MECHANISM FOR THE REDUCTION

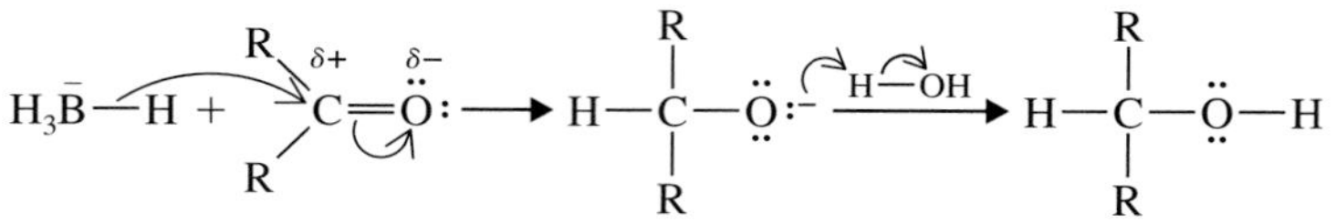

Hydride transfer   **Alkoxide ion**   **Alcohol**

These steps are repeated until all hydrogen atoms attached to boron have been transferred.

Sodium borohydride is a less powerful reducing agent than lithium aluminum hydride. Lithium aluminum hydride reduces acids, esters, aldehydes, and ketones; but sodium borohydride reduces only aldehydes and ketones.

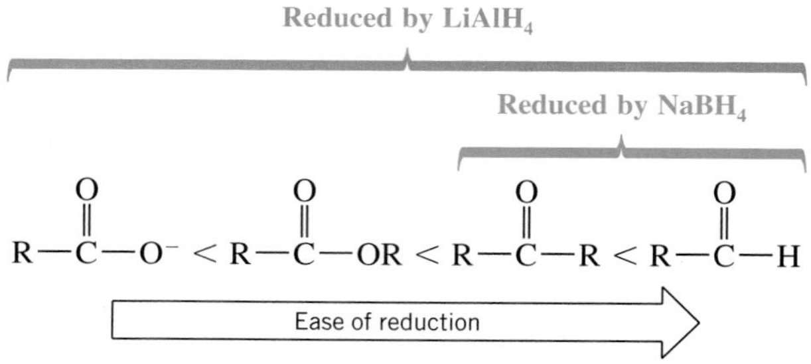

Lithium aluminum hydride reacts violently with water and therefore reductions with lithium aluminum hydride must be carried out in anhydrous solutions, usually in anhydrous ether. (Ethyl acetate is added cautiously after the reaction is over to decompose excess LiAlH$_4$, then water is added to decompose the aluminum complex.)*
Sodium borohydride reductions, by contrast, can be carried out in water or alcohol solutions.

---

*Unless special precautions are taken, lithium aluminum hydride reductions can be very dangerous. You should consult an appropriate laboratory manual before attempting such a reduction, and the reaction should be carried out on a small scale.

PROBLEM 11.4

Which reducing agent, $LiAlH_4$ or $NaBH_4$, would you use to carry out the following transformations?

(a)

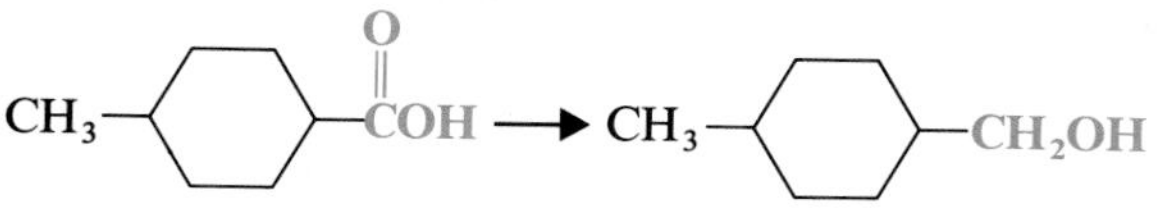

(b)

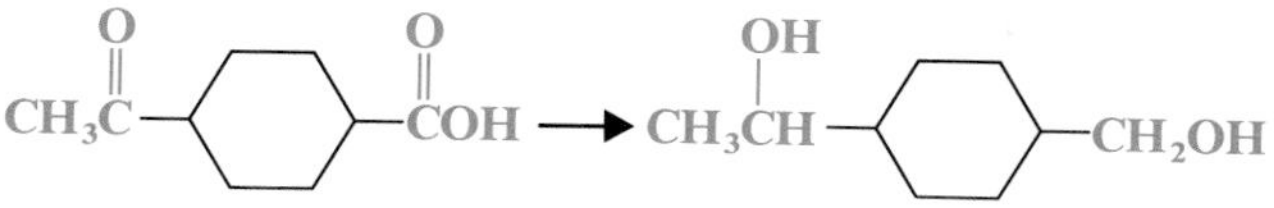

(c)

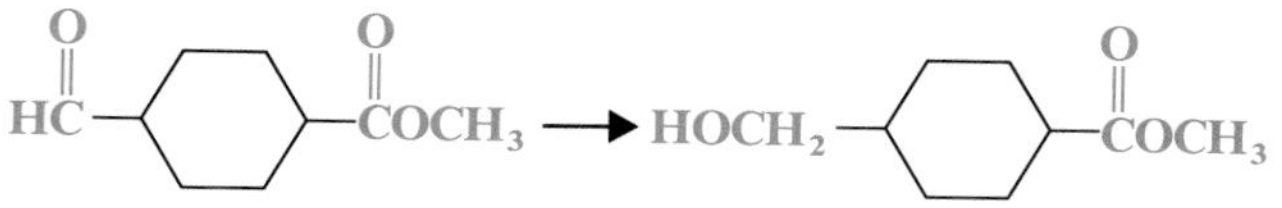

11.4 OXIDATION OF ALCOHOLS

11.4A Oxidation of 1° Alcohols to Aldehydes: $RCH_2OH \longrightarrow RCHO$

Primary alcohols can be oxidized to aldehydes and carboxylic acids.

$$R-CH_2OH \xrightarrow{[O]} R-\overset{O}{\overset{\|}{C}}-H \xrightarrow{[O]} R-\overset{O}{\overset{\|}{C}}-OH$$

1° Alcohol   **Aldehyde**   **Carboxylic acid**

The oxidation of aldehydes to carboxylic acids in aqueous solutions usually takes place with less powerful oxidizing agents than those required to oxidize primary alcohols to aldehydes; thus it is difficult to stop the oxidation at the aldehyde stage. One way of avoiding this problem is to remove the aldehyde as soon as it is formed. This can often be done because aldehydes have lower boiling points than alcohols (why?), and therefore aldehydes can be distilled from the reaction mixture as they are formed. An example is the synthesis of butanal from 1-butanol using a mixture of $K_2Cr_2O_7$ and sulfuric acid:

$$CH_3CH_2CH_2CH_2OH \xrightarrow[H_2SO_4]{K_2Cr_2O_7} CH_3CH_2CH_2\overset{O}{\overset{\|}{C}}H$$

1-Butanol   **Butanal**
bp, 117.7°C   **bp, 75.7°C**
                      (50%)

This procedure does not give good yields with aldehydes that boil above 100°C, however.

An industrial process for preparing low molecular weight aldehydes is **dehydrogenation** of a primary alcohol:

$$CH_3CH_2OH \xrightarrow[300°C]{Cu} CH_3\overset{\overset{\displaystyle O}{\|}}{C}H + H_2$$

[Notice that dehydrogenation of an organic compound corresponds to oxidation, whereas hydrogenation (cf. Problem 11.3) corresponds to reduction.]

In most laboratory preparations we must rely on special oxidizing agents to prepare aldehydes from primary alcohols. A variety of reagents is available and to discuss them all here is beyond our scope. An excellent reagent for this purpose is the compound formed when CrO_3 is dissolved in hydrochloric acid and then treated with pyridine.

$$CrO_3 + HCl + \underset{\substack{\textbf{Pyridine} \\ \textbf{(C}_5\textbf{H}_5\textbf{N)}}}{\bigcirc N:} \longrightarrow \underset{\substack{\textbf{Pyridinium chlorochromate} \\ \textbf{(PCC)}}}{\bigcirc \overset{+}{N}-H \quad Cr\,O_3Cl^-}$$

This compound, called **pyridinium chlorochromate** [abbreviated (PCC)], when dissolved in CH_2Cl_2, will oxidize a primary alcohol to an aldehyde and stop at that stage.

$$\underset{\substack{\textbf{2-Ethyl-2-methyl-1-} \\ \textbf{butanol}}}{(C_2H_5)_2\overset{\overset{\displaystyle CH_3}{|}}{C}-CH_2OH} + PCC \xrightarrow[25°C]{CH_2Cl_2} \underset{\textbf{2-Ethyl-2-methylbutanal}}{(C_2H_5)_2\overset{\overset{\displaystyle CH_3}{|}}{C}-\overset{\overset{\displaystyle O}{\|}}{C}H}$$

Pyridinium chlorochromate also does not attack double bonds.

One reason for the success of oxidation with pyridinium chlorochromate is that the oxidation can be carried out in a solvent such as CH_2Cl_2, in which PCC is soluble. Aldehydes themselves are not nearly as easily oxidized as are the *aldehyde hydrates, RCH(OH)$_2$*, that form (Section 16.7A) when aldehydes are dissolved in water, the usual medium for oxidation by chromium compounds.

$$RCHO + H_2O \rightleftharpoons RCH(OH)_2$$

We explain this further in Section 11.4D.

11.4B Oxidation of 1° Alcohols to Carboxylic Acids: $RCH_2OH \longrightarrow RCO_2H$

Primary alcohols can be oxidized to **carboxylic acids** by potassium permanganate. The reaction is usually carried out in basic aqueous solution from which MnO_2 precipitates as the oxidation takes place. After the oxidation is complete, filtration allows removal of the MnO_2 and acidification of the filtrate gives the carboxylic acid.

$$R-CH_2OH + KMnO_4 \xrightarrow[\substack{H_2O \\ \text{heat}}]{OH^-} RCO_2^-K^+ + MnO_2 \\ \qquad\qquad\qquad\qquad \downarrow{\scriptstyle H^+} \\ \qquad\qquad\qquad RCO_2H$$

11.4C Oxidation of 2° Alcohols to Ketones:

$$\underset{RCHR'}{\overset{OH}{|}} \longrightarrow \underset{RCR'}{\overset{O}{||}}$$

Secondary alcohols can be oxidized to ketones. The reaction usually stops at the ketone stage because further oxidation requires the breaking of a carbon–carbon bond.

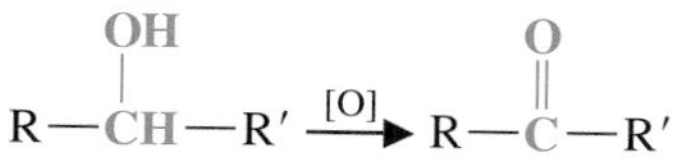

$$\underset{\text{2° Alcohol}}{R-\overset{\overset{OH}{|}}{C}H-R'} \xrightarrow{\text{[O]}} \underset{\text{Ketone}}{R-\overset{\overset{O}{||}}{C}-R'}$$

Various oxidizing agents based on chromium(VI) have been used to oxidize secondary alcohols to ketones. The most commonly used reagent is chromic acid (H_2CrO_4). Chromic acid is usually prepared by adding chromium(VI) oxide (CrO_3) or sodium dichromate $(Na_2Cr_2O_7)$ to aqueous sulfuric acid. Oxidations of secondary alcohols are generally carried out in acetone or acetic acid solutions. The balanced equation is shown here.

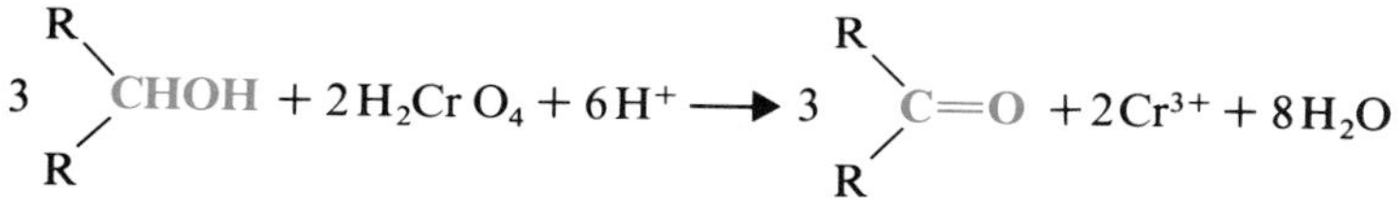

$$3\ \overset{R}{\underset{R}{\diagdown}}CHOH + 2H_2CrO_4 + 6H^+ \longrightarrow 3\ \overset{R}{\underset{R}{\diagdown}}C{=}O + 2Cr^{3+} + 8H_2O$$

As chromic acid oxidizes the alcohol to the ketone, chromium is reduced from the $+6$ oxidation state (H_2CrO_4) to the $+3$ oxidation state (Cr^{3+}).* Chromic acid oxidations of secondary alcohols generally give ketones in excellent yields if the temperature is controlled. A specific example is the oxidation of cyclooctanol to cyclooctanone.

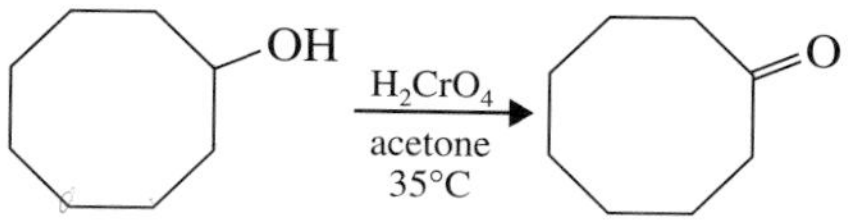

Cyclooctanol → Cyclooctanone (92–96%)

with H_2CrO_4, acetone, 35°C

The use of CrO_3 in aqueous acetone is usually called the **Jones oxidation** (or oxidation by the Jones reagent). This procedure rarely affects double bonds present in the molecule.

11.4D Mechanism of Chromate Oxidations

The mechanism of chromic acid oxidations of alcohols has been investigated thoroughly. It is interesting because it shows how changes in oxidation states occur in a reaction between an organic and an inorganic compound. The first step is the formation of a chromate ester of the alcohol. Here we show this step using a 2° alcohol.

---

*It is the color change that accompanies this change in oxidation state that allows chromic acid to be used as a test for primary and secondary alcohols (Section 11.4E).

A MECHANISM FOR FORMATION OF THE CHROMATE ESTER

Step 1

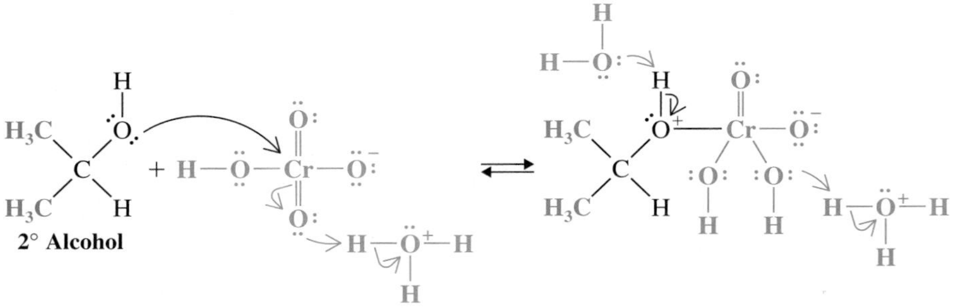

The alcohol donates an electron pair to the chromium atom, as an oxygen accepts a proton.

One oxygen loses a proton; another oxygen accepts a proton.

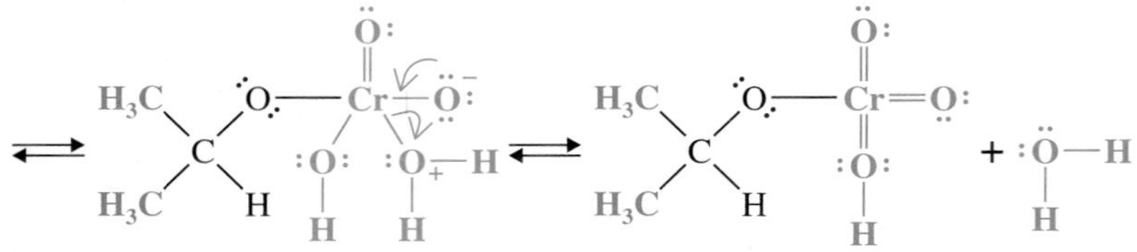

Chromate ester

A molecule of water departs as a leaving group as a chromium-oxygen double bond forms.

The chromate ester is unstable and is not isolated. It transfers a proton to a base (usually water) and simultaneously eliminates an $HCrO_3^-$ ion.

A MECHANISM FOR THE OXIDATION STEP

Step 2

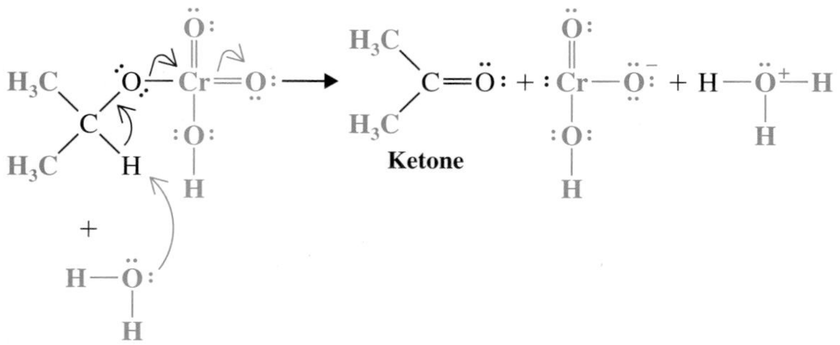

Ketone

The chromium atom departs with a pair of electrons that formerly belonged to the alcohol; the alcohol is thereby oxidized and the chromium reduced.

The overall result of the second step is the reduction of $HCrO_4^-$ to $HCrO_3^-$, a two-electron ($2\,e^-$) change in the oxidation state of chromium, from $Cr(VI)$ to $Cr(IV)$. At the same time the alcohol undergoes a $2\,e^-$ oxidation to the ketone.

The remaining steps of the mechanism are complicated and we need not give them in detail. Suffice it to say that further oxidations (and disproportionations) take place, ultimately converting $Cr(IV)$ compounds to Cr^{3+} ions.

The requirement for the formation of a chromate ester in step 1 of the mechanism helps us understand why 1° alcohols are easily oxidized beyond the aldehyde stage in aqueous solutions (and, therefore, why oxidation with PCC in CH_2Cl_2 stops at the aldehyde stage). The aldehyde initially formed from the 1° alcohol (produced by a mechanism similar to the one we have just given) reacts with water to form an aldehyde hydrate. The aldehyde hydrate can then react with $HCrO_4^-$ (and H^+) to form a chromate ester, and this can then be oxidized to the carboxylic acid. In the absence of water (i.e., using PCC in CH_2Cl_2), the aldehyde hydrate does not form; therefore, further oxidation does not take place.

Aldehyde hydrate

Carboxylic acid

The elimination that takes place in step 2 of the mechanism on page 476 helps us to understand why 3° alcohols do not generally react in chromate oxidations. While 3° alcohols have no difficulty in forming chromate esters, the ester that is formed does not bear a hydrogen that can be eliminated and, therefore, no oxidation takes place.

3° Alcohol

**Chromate ester
cannot undergo
elimination of H_2CrO_3**

11.4E A Chemical Test for Primary and Secondary Alcohols

The relative ease of oxidation of primary and secondary alcohols compared with the difficulty of oxidizing tertiary alcohols forms the basis for a convenient chemical test. Primary and secondary alcohols are rapidly oxidized by a solution of CrO_3 in aqueous sulfuric acid. Chromic oxide (CrO_3) dissolves in aqueous sulfuric acid to give a clear orange solution containing $Cr_2O_7{}^{2-}$ ions. A positive test is indicated when this clear orange solution becomes opaque and takes on a greenish cast within 2 s.*

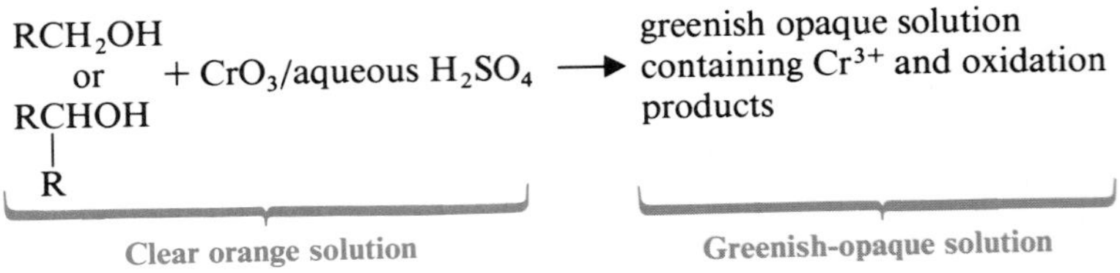

$$\begin{matrix} RCH_2OH \\ or \\ RCHOH \\ | \\ R \end{matrix} + CrO_3/aqueous\ H_2SO_4 \longrightarrow \begin{matrix} greenish\ opaque\ solution \\ containing\ Cr^{3+}\ and\ oxidation \\ products \end{matrix}$$

Clear orange solution   Greenish-opaque solution

Not only will this test distinguish primary and secondary alcohols from tertiary alcohols, it will distinguish primary and secondary alcohols from most other compounds except aldehydes.

PROBLEM 11.5 _____

Show how each of the following transformations could be accomplished.

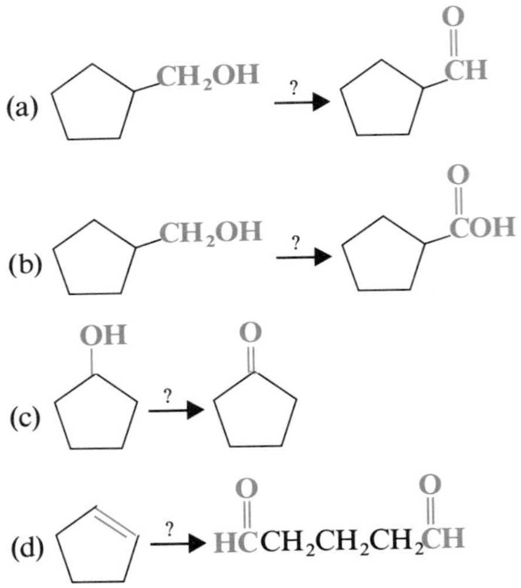

(a)

(b)

(c)

(d)

11.5 ORGANOMETALLIC COMPOUNDS

Compounds that contain carbon–metal bonds are called **organometallic compounds.** The natures of the carbon–metal bonds vary widely, ranging from bonds that are essentially ionic to those that are primarily covalent. Whereas the structure of the organic portion of the organometallic compound has some effect on the nature of the carbon–metal bond, the identity of the metal itself is of far greater importance. Carbon–sodium and carbon–potassium bonds are largely ionic in character; carbon–lead, carbon–tin, carbon–thallium, and carbon–mercury bonds are essentially covalent. Carbon–lithium and carbon–magnesium bonds lie between these extremes.

---

*This color change, associated with the reduction of $Cr_2O_7{}^{2-}$ to Cr^{3+}, forms the basis for "breathalyzer tubes" used to detect intoxicated motorists. In the breathalyzer the dichromate salt is coated on granules of silica gel.

$$\underset{\substack{\text{Primarily ionic} \\ (M = Na^+ \text{ or } K^+)}}{-\overset{|}{\underset{|}{C}}{:}^- M^+} \qquad \underset{(M = Mg \text{ or } Li)}{-\overset{|}{\underset{|}{C}} \overset{\delta- \quad \delta+}{:} M} \qquad \underset{\substack{\text{Primarily covalent} \\ (M = Pb, Sn, Hg, \text{ or } Tl)}}{-\overset{|}{\underset{|}{C}}-M}$$

The reactivity of organometallic compounds increases with the percent ionic character of the carbon–metal bond. Alkylsodium and alkylpotassium compounds are highly reactive and are among the most powerful of bases. They react explosively with water and burst into flame when exposed to air. Organomercury and organolead compounds are much less reactive; they are often volatile and are stable in air. They are all poisonous. They are generally soluble in nonpolar solvents. Tetraethyllead, for example, has been used as an "antiknock" compound in gasoline.

Organometallic compounds of lithium and magnesium are of great importance in organic synthesis. They are relatively stable in ether solutions, but their carbon–metal bonds have considerable ionic character. Because of this ionic nature, the carbon atom that is bonded to the metal atom of an organolithium or organomagnesium compound is a strong base and powerful nucleophile. We shall soon see reactions that illustrate both of these properties.

11.6 PREPARATION OF ORGANOLITHIUM AND ORGANOMAGNESIUM COMPOUNDS

11.6A Organolithium Compounds

Organolithium compounds are often prepared by the reduction of organic halides with lithium metal (see Section 3.13). These reductions are usually carried out in ether solvents, and since organolithium compounds are strong bases, care must be taken to exclude moisture (why?). The ethers most commonly used as solvents are diethyl ether and tetrahydrofuran. (Tetrahydrofuran is a cyclic ether.)

$$CH_3CH_2\overset{..}{\underset{..}{O}}CH_2CH_3$$

Diethyl ether
(Et₂O)

Tetrahydrofuran
(THF)

For example, butyl bromide reacts with lithium metal in diethyl ether to give a solution of butyllithium.

$$CH_3CH_2CH_2CH_2Br + 2\,Li \xrightarrow[\text{Et}_2O]{-10°C} CH_3CH_2CH_2CH_2Li + LiBr$$

Butyl bromide   **Butyllithium**
(80–90%)

Other organolithium compounds, such as methyllithium, ethyllithium, and phenyllithium, can be prepared in the same general way.

$$\underset{\textbf{(or Ar—X)}}{R-X} + 2\,Li \xrightarrow{\text{Et}_2O} \underset{\textbf{(or ArLi)}}{RLi} + LiX$$

The order of reactivity of halides is $RI > RBr > RCl$. (Alkyl and aryl fluorides are seldom used in the preparation of organolithium compounds.)

Most organolithium compounds slowly attack ethers by bringing about an elimination reaction.

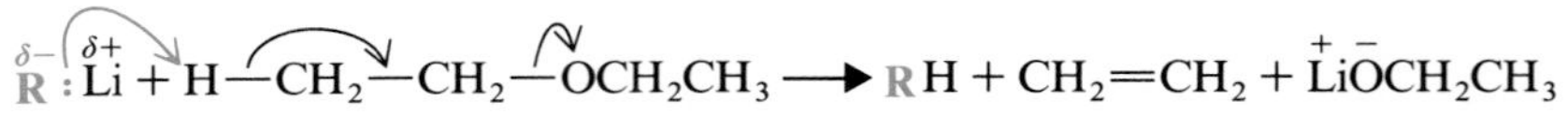

$$\overset{\delta-}{R} \overset{\delta+}{\underset{}{Li}} + H-CH_2-CH_2-OCH_2CH_3 \longrightarrow RH + CH_2{=}CH_2 + \overset{+}{Li}\overset{-}{O}CH_2CH_3$$

For this reason, ether solutions of organolithium reagents are not usually stored but are used immediately after preparation. Organolithium compounds are much more stable in hydrocarbon solvents. Several alkyl- and aryllithium reagents are commercially available in hexane, paraffin wax, or mineral oil.

11.6B Grignard Reagents

Organomagnesium halides were discovered by the French chemist Victor Grignard in 1900. Grignard received the Nobel Prize for his discovery in 1912, and organomagnesium halides are now called **Grignard reagents** in his honor. Grignard reagents have great use in organic synthesis.

Grignard reagents are usually prepared by the reaction of an organic halide and magnesium metal (turnings) in an ether solvent.

$$\left.\begin{array}{l} RX + Mg \xrightarrow{Et_2O} RMgX \\[6pt] ArX + Mg \xrightarrow{Et_2O} ArMgX \end{array}\right\} \begin{array}{l} \textbf{Grignard} \\ \textbf{reagents} \end{array}$$

The order of reactivity of halides with magnesium is also $RI > RBr > RCl$. Very few organomagnesium fluorides have been prepared. Aryl Grignard reagents are more easily prepared from aryl bromides and aryl iodides than from aryl chlorides, which react very sluggishly.

Grignard reagents are seldom isolated but are used for further reactions in ether solution. The ether solutions can be analyzed for the content of the Grignard reagent, however, and the yields of Grignard reagents are almost always very high (85–95%). Two examples are shown here.

$$CH_3I + Mg \xrightarrow[35°C]{Et_2O} CH_3MgI$$

**Methylmagnesium
iodide
(95%)**

$$C_6H_5Br + Mg \xrightarrow[35°C]{Et_2O} C_6H_5MgBr$$

**Phenylmagnesium
bromide
(95%)**

The actual structures of Grignard reagents are more complex than the general formula RMgX indicates. Experiments done with radioactive magnesium have established that, for most Grignard reagents, there is an equilibrium between an alkylmagnesium halide and a dialkylmagnesium.

$$2\,RMgX \rightleftharpoons R_2Mg + MgX_2$$

**Alkylmagnesium   Dialkylmagnesium
halide**

For convenience in this text, however, we shall write the formula for the Grignard reagent as though it were simply RMgX.

A Grignard reagent forms a complex with its ether solvent; the structure of the complex can be represented as follows:

$$
\begin{array}{c}
\text{R} \qquad \text{R} \\
\diagdown \; \ddot{\text{O}} \; \diagup \\
| \\
\text{R}-\text{Mg}-\text{X} \\
| \\
\diagup \; \ddot{\text{O}} \; \diagdown \\
\text{R} \qquad \text{R}
\end{array}
$$

Complex formation with molecules of ether is an important factor in the formation and stability of Grignard reagents. Organomagnesium compounds can be prepared in non-ethereal solvents, but the preparations are more difficult.

The mechanism by which Grignard reagents form is complicated and is still a matter of debate.* There seems to be general agreement that radicals are involved and that a mechanism similar to the following is likely.

$$\text{R}-\text{X} + \text{:Mg} \longrightarrow \text{R}\cdot + \cdot\text{MgX}$$

$$\text{R}\cdot + \cdot\text{MgX} \longrightarrow \text{RMgX}$$

11.7 REACTIONS OF ORGANOLITHIUM AND ORGANOMAGNESIUM COMPOUNDS

11.7A Reactions with Compounds Containing Acidic Hydrogen Atoms

Grignard reagents and organolithium compounds are very strong bases. They react with any compound that has a hydrogen attached to an electronegative atom such as oxygen, nitrogen, or sulfur. We can understand how these reactions occur if we represent the Grignard reagent and organolithium compounds in the following ways:

$$\overset{\delta-\;\;\delta+}{\text{R:MgX}} \quad \text{and} \quad \overset{\delta-\;\delta+}{\text{R:Li}}$$

When we do this, we can see that the reactions of Grignard reagents with water and alcohols are nothing more than acid–base reactions; they lead to the formation of the weaker conjugate acid and weaker conjugate base. The Grignard reagent behaves as if it contained the anion of an alkane, *as if it contained a carbanion.*

$$\overset{\delta-\;\;\delta+}{\text{R:MgX}} + \text{H:\ddot{O}H} \longrightarrow \text{R:H} + \quad \text{H\ddot{O}:}^- \quad + \text{Mg}^{2+} + \text{X}^-$$

| **Grignard reagent** (stronger base) | **Water** (stronger acid) | **Alkane** (weaker acid) | **Hydroxide ion** (weaker base) | | |

*Those interested may want to read the recent articles: John L. Garst and Brian L. Swift, *J. Am. Chem. Soc.,* **1989,** *111,* 241–250; H. M. Walborsky, *Acc. Chem Res.,* **1990,** *23,* 286–293; and John L. Garst, *Acc. Chem. Res.,* **1991,** *24,* 95–97.

$$R:MgX + H:\overset{..}{\underset{..}{O}}R \longrightarrow R:H + R\overset{..}{\underset{..}{O}}:^- + Mg^{2+} + X^-$$

| Grignard reagent (stronger base) | Alcohol (stronger acid) | Alkane (weaker acid) | Alkoxide ion (weaker base) |

PROBLEM 11.6

Write similar equations for the reactions that take place when phenyllithium is treated with (a) water and (b) ethanol. Designate the stronger and weaker acids and stronger and weaker bases.

PROBLEM 11.7

Assuming you have bromobenzene (C_6H_5Br), magnesium, dry ether, and deuterium oxide (D_2O) available, show how you might synthesize the following deuterium-labeled compound.

Grignard reagents and organolithium compounds abstract protons that are much less acidic than those of water and alcohols. They react with the terminal hydrogen atoms of 1-alkynes, for example, and this is a useful method for the preparation of alkynylmagnesium halides and alkynyllithiums. These reactions are also acid–base reactions.

$$RC\equiv CH + R':MgX \longrightarrow RC\equiv C:MgX + R':H$$

| Terminal alkyne (stronger acid) | Grignard reagent (stronger base) | Alkynylmagnesium halide (weaker base) | Alkane (weaker acid) |

$$R-C\equiv CH + R':Li \longrightarrow R-C\equiv C:Li + R':H$$

| Terminal alkyne (stronger acid) | Alkyllithium (stronger base) | Alkynyllithium (weaker base) | Alkane (weaker acid) |

The fact that these reactions go to completion is not surprising when we recall that alkanes have pK_a values ≈ 50, whereas those of terminal alkynes are ≈ 25 (Table 3.1).

Not only are Grignard reagents strong bases, they are also *powerful nucleophiles*. Reactions in which Grignard reagents act as nucleophiles are by far the most important. At this point, let us consider general examples that illustrate the ability of a Grignard reagent to act as a nucleophile by attacking saturated and unsaturated carbon atoms.

11.7B Reaction of Grignard Reagents with Oxiranes (Epoxides)

Grignard reagents carry out nucleophilic attack at a saturated carbon when they react with oxiranes. These reactions take the general form shown below and give us a convenient synthesis of primary alcohols.

The nucleophilic alkyl group of the Grignard reagent attacks the partially positive carbon of the oxirane ring. Because it is highly strained, the ring opens, and the reaction leads to the salt of a primary alcohol. Subsequent acidification produces the alcohol. (Compare this reaction with the base-catalyzed ring opening we studied in Section 10.19).

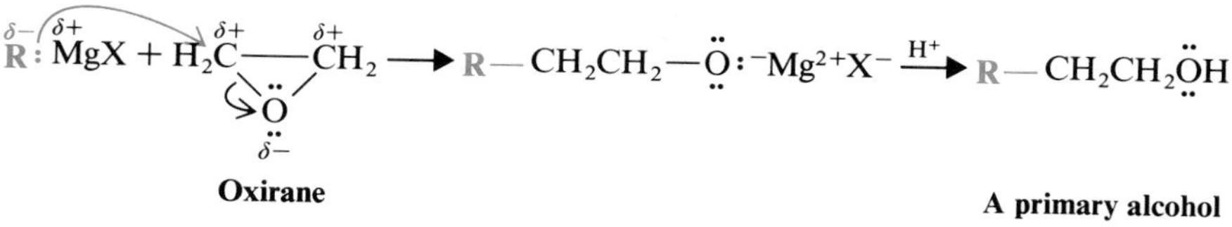

$$R\overset{\delta-}{:}\overset{\delta+}{MgX} + H_2\overset{\delta+}{C}\overset{\delta+}{-CH_2} \longrightarrow R-CH_2CH_2-\overset{..}{\underset{..}{O}}:^-Mg^{2+}X^- \overset{H^+}{\longrightarrow} R-CH_2CH_2\overset{..}{\underset{..}{O}}H$$

Oxirane   **A primary alcohol**

Specific Example

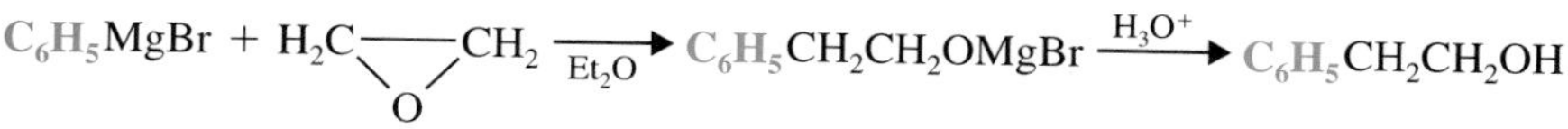

$$C_6H_5MgBr + H_2C\overset{}{\underset{O}{\diagup\!\diagdown}}CH_2 \xrightarrow{Et_2O} C_6H_5CH_2CH_2OMgBr \xrightarrow{H_3O^+} C_6H_5CH_2CH_2OH$$

Grignard reagents react primarily at the less-substituted ring carbon atom of substituted oxiranes.

Specific Example

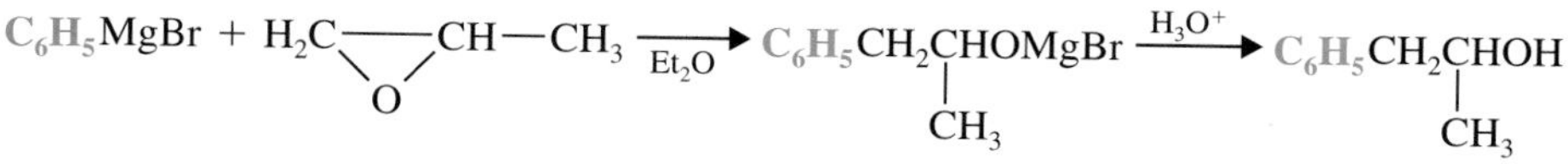

$$C_6H_5MgBr + H_2C\overset{}{\underset{O}{\diagup\!\diagdown}}CH-CH_3 \xrightarrow{Et_2O} C_6H_5CH_2\underset{CH_3}{CHO}MgBr \xrightarrow{H_3O^+} C_6H_5CH_2\underset{CH_3}{CHOH}$$

11.7C Reactions of Grignard Reagents with Carbonyl Compounds

From a synthetic point of view, the most important reactions of Grignard reagents and organolithium compounds are those in which these reagents act as nucleophiles and attack an unsaturated carbon—*especially the carbon of a carbonyl group.*

We saw in Section 11.1B that carbonyl compounds are highly susceptible to nucleophilic attack. Grignard reagents react with carbonyl compounds (aldehydes and ketones) in the following way:

A MECHANISM FOR THE GRIGNARD REACTION

REACTION:

$$RMgX + \overset{}{\underset{}{\diagdown}}C=O \xrightarrow[\text{(2) } H_3O^+ X^-]{\text{(1) ether*}} R-\overset{|}{\underset{|}{C}}-O-H + MgX_2$$

MECHANISM:

Step 1

$$\overset{\delta-}{R}\overset{\delta+}{:}MgX \quad + \quad \diagdown C = \overset{..}{\underset{..}{O}}: \longrightarrow \quad R - \overset{|}{\underset{|}{C}} - \overset{..}{\underset{..}{O}}:^- \ Mg^{2+} \ X^-$$

Grignard   Carbonyl   Halomagnesium alkoxide
reagent   compound

The strongly nucleophilic Grignard reagent uses its electron pair to form a bond to the carbon atom. One electron pair of the carbonyl group shifts out to the oxygen. This reaction is a nucleophilic addition to the carbonyl group and it results in the formation of an alkoxide ion associated with Mg^{2+} and X^-.

Step 2

$$R - \overset{|}{\underset{|}{C}} - \overset{..}{\underset{..}{O}}:^- \ Mg^{2+} \ X^- \ + \ H - \overset{..}{\underset{|}{O}}{}^+ - H \ + \ X^- \longrightarrow$$
$$\underset{\qquad H}{}$$

Halomagnesium alkoxide

$$R - \overset{|}{\underset{|}{C}} - \overset{..}{O} - H \ + \ :\overset{..}{\underset{|}{O}} - H \ + \ MgX_2$$
$$\underset{\qquad\qquad H}{}$$

Alcohol

In the second step, the addition of aqueous HX causes protonation of the alkoxide ion; this leads to the formation of the alcohol and MgX_2.

---

*By writing "(1) ether" over the arrow and "(2) H_3O^+ X^-" under the arrow we mean that in the first step the Grignard reagent and the carbonyl compound are allowed to react in an ether solvent. Then in a second step, after the reaction of the Grignard reagent and the carbonyl compound is over, we add aqueous acid (e.g., dilute HX) to convert the salt of the alcohol (ROMgX) to the alcohol itself. If the alcohol is tertiary, it will be susceptible to acid-catalyzed dehydration. In this case, a solution of NH_4Cl in water is often used because it is acidic enough to convert ROMgX to ROH without causing dehydration.

11.8 ALCOHOLS FROM GRIGNARD REAGENTS

Grignard additions to carbonyl compounds are especially useful because they can be used to prepare primary, secondary, or tertiary alcohols.

1. A Grignard reagent reacts with formaldehyde, to give a **primary alcohol.**

$$\overset{\delta-}{R}\overset{\delta+}{:}MgX \ + \ \underset{H}{\overset{H}{\diagup}}C = \overset{..}{\underset{..}{O}}: \ \longrightarrow \ R - \overset{H}{\underset{H}{\overset{|}{C}}} - \overset{..}{\underset{..}{O}}:MgX \ \xrightarrow{H_3O^+} \ R - \overset{H}{\underset{H}{\overset{|}{C}}} - \overset{..}{\underset{..}{O}}H$$

Formaldehyde   **1° Alcohol**

2. Grignard reagents react with higher aldehydes to give **secondary alcohols.**

$$\overset{\delta-}{R}\overset{\delta+}{:}MgX \ + \ \underset{H}{\overset{R'}{\diagup}}C = \overset{..}{\underset{..}{O}}: \ \longrightarrow \ R - \overset{R'}{\underset{H}{\overset{|}{C}}} - \overset{..}{\underset{..}{O}}:MgX \ \xrightarrow{H_3O^+} \ R - \overset{R'}{\underset{H}{\overset{|}{C}}} - \overset{..}{\underset{..}{O}}H$$

Higher   **2° Alcohol**
aldehyde

3. And Grignard reagents react with ketones to give **tertiary alcohols.**

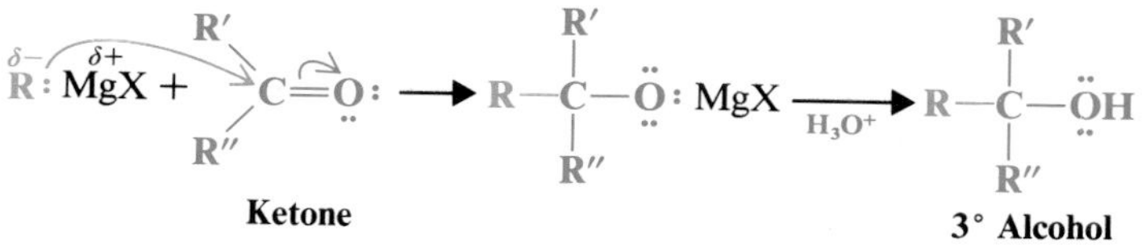

$$\overset{\delta-}{R}:\overset{\delta+}{MgX} + \underset{R''}{\overset{R'}{C}}=\ddot{O}: \longrightarrow R-\underset{R''}{\overset{R'}{\underset{|}{C}}}-\ddot{O}:MgX \xrightarrow{H_3O^+} R-\underset{R''}{\overset{R'}{\underset{|}{C}}}-\ddot{O}H$$

Ketone   3° Alcohol

Specific examples of these reactions are shown here.

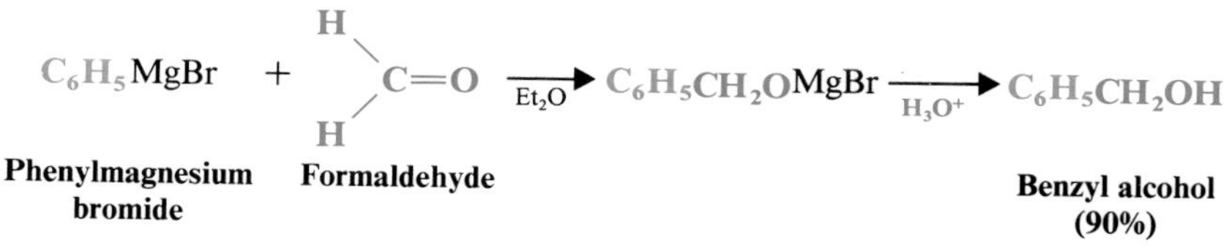

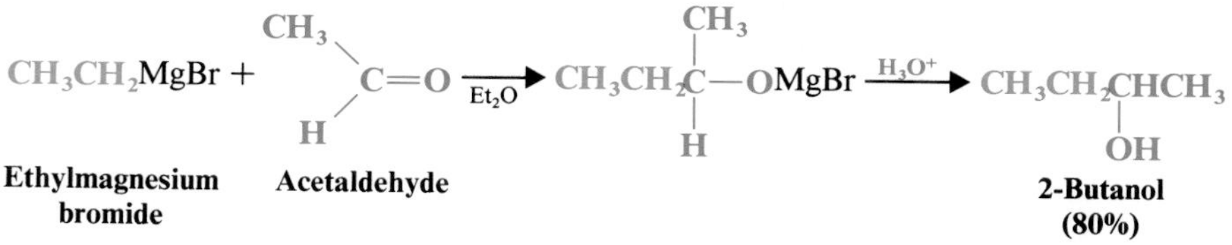

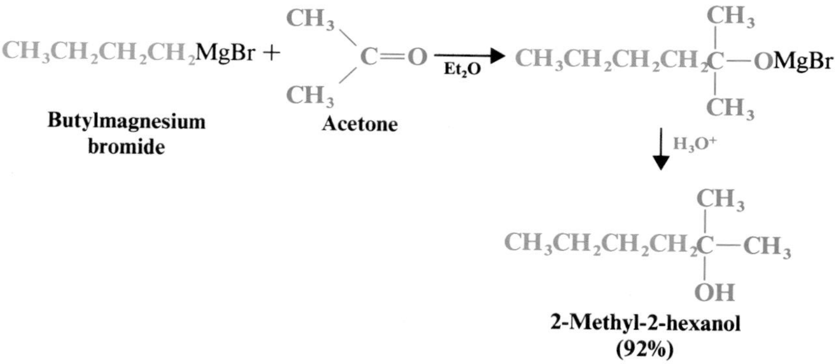

$$C_6H_5MgBr + \underset{H}{\overset{H}{C}}=O \xrightarrow{Et_2O} C_6H_5CH_2OMgBr \xrightarrow{H_3O^+} C_6H_5CH_2OH$$

Phenylmagnesium   **Formaldehyde**   **Benzyl alcohol**
bromide   **(90%)**

$$CH_3CH_2MgBr + \underset{H}{\overset{CH_3}{C}}=O \xrightarrow{Et_2O} CH_3CH_2\overset{CH_3}{\underset{H}{C}}-OMgBr \xrightarrow{H_3O^+} CH_3CH_2CHCH_3 \atop OH$$

Ethylmagnesium   **Acetaldehyde**   **2-Butanol**
bromide   **(80%)**

$$CH_3CH_2CH_2CH_2MgBr + \underset{CH_3}{\overset{CH_3}{C}}=O \xrightarrow{Et_2O} CH_3CH_2CH_2CH_2\overset{CH_3}{\underset{CH_3}{C}}-OMgBr$$

Butylmagnesium   **Acetone**
bromide

$$\downarrow H_3O^+$$

$$CH_3CH_2CH_2CH_2\overset{CH_3}{\underset{OH}{C}}-CH_3$$

2-Methyl-2-hexanol
(92%)

4. A Grignard reagent also adds to the carbonyl group of an ester. The initial product is unstable and it loses a magnesium alkoxide to form a ketone. Ketones are more reactive toward Grignard reagents than esters. Therefore, as soon as a molecule of the ketone is formed in the mixture, it reacts with a second molecule of the Grignard reagent. After hydrolysis, **the product is a tertiary alcohol with two identical alkyl groups,** groups that correspond to the alkyl portion of the Grignard reagent.

$$R:MgX + \underset{R''\ddot{O}}{\overset{R'}{C}}=\ddot{O}: \longrightarrow \left[R-\underset{:\ddot{O}-R''}{\overset{R'}{\underset{|}{C}}}-\ddot{O}-MgX \right] \xrightarrow[\text{spontaneously}]{-R''OMgX}$$

Ester   **Initial product**
                                                 (unstable)

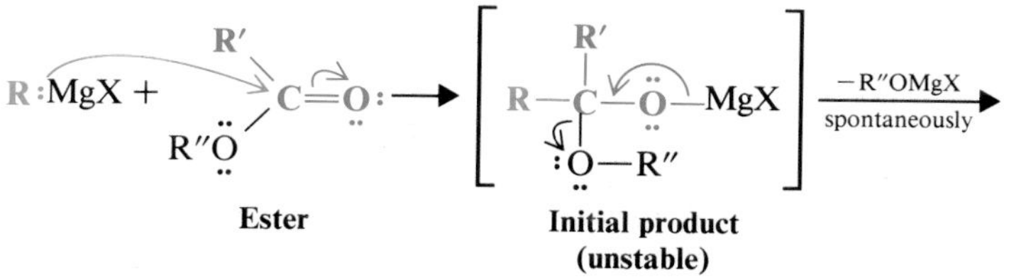

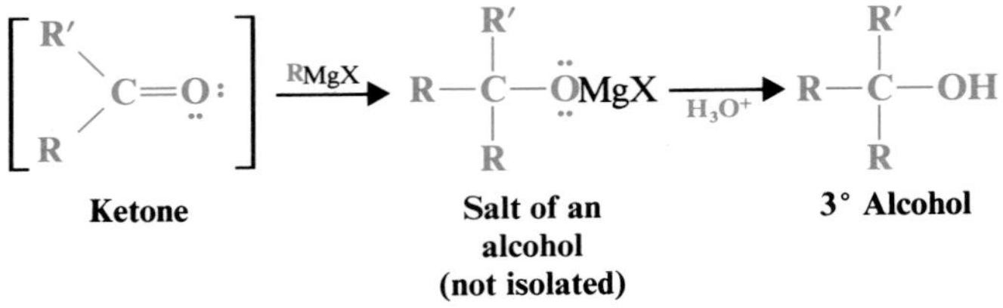

$$\left[\begin{array}{c} R' \\ \diagdown \\ \diagup \\ R \end{array} C=\ddot{O}: \right] \xrightarrow{RMgX} R-\overset{\displaystyle R'}{\underset{\displaystyle R}{\overset{|}{\underset{|}{C}}}}-\ddot{O}MgX \xrightarrow{H_3O^+} R-\overset{\displaystyle R'}{\underset{\displaystyle R}{\overset{|}{\underset{|}{C}}}}-OH$$

Ketone   **Salt of an**   **3° Alcohol**
alcohol
(not isolated)

A specific example of this reaction is shown here.

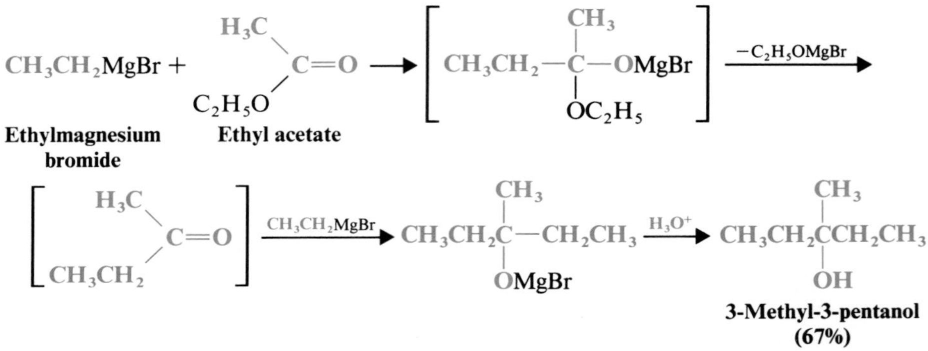

$$CH_3CH_2MgBr + \begin{array}{c} H_3C \\ \diagdown \\ \diagup \\ C_2H_5O \end{array} C=O \longrightarrow \left[CH_3CH_2-\overset{\displaystyle CH_3}{\underset{\displaystyle OC_2H_5}{\overset{|}{\underset{|}{C}}}}-OMgBr \right] \xrightarrow{-C_2H_5OMgBr}$$

Ethylmagnesium   **Ethyl acetate**
bromide

$$\left[\begin{array}{c} H_3C \\ \diagdown \\ \diagup \\ CH_3CH_2 \end{array} C=O \right] \xrightarrow{CH_3CH_2MgBr} CH_3CH_2\overset{\displaystyle CH_3}{\underset{\displaystyle OMgBr}{\overset{|}{\underset{|}{C}}}}-CH_2CH_3 \xrightarrow{H_3O^+} CH_3CH_2\overset{\displaystyle CH_3}{\underset{\displaystyle OH}{\overset{|}{\underset{|}{C}}}}CH_2CH_3$$

3-Methyl-3-pentanol
(67%)

PROBLEM 11.8

Phenylmagnesium bromide reacts with benzoyl chloride, $C_6H_5\overset{\displaystyle O}{\overset{\|}{C}}Cl$, to form tri-
phenylmethanol, $(C_6H_5)_3COH$. This reaction is typical of the reaction of Gri-
gnard reagents with acyl chlorides, and the mechanism is similar to that for the
reaction of a Grignard reagent with an ester just shown. Show the steps that lead
to the formation of triphenylmethanol.

11.8A Planning a Grignard Synthesis

By using Grignard synthesis skillfully we can synthesize almost any alcohol we wish.
In planning a Grignard synthesis we must simply choose the correct Grignard reagent
and the correct aldehyde, ketone, ester, or epoxide. We do this by examining the
alcohol we wish to prepare and by paying special attention to the groups attached to the
carbon atom bearing the —OH group. Many times there may be more than one way of
carrying out the synthesis. In these cases our final choice will probably be dictated by
the availability of starting compounds. Let us consider an example.

EXAMPLE

Suppose we want to prepare 3-phenyl-3-pentanol. We examine its structure and we see
that the groups attached to the carbon atom bearing the —OH are a *phenyl group*

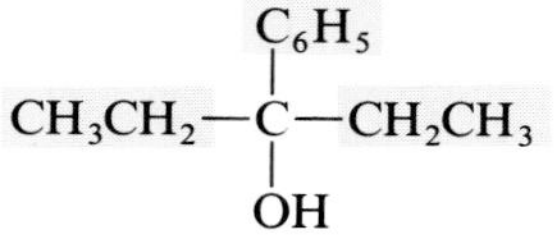

$$CH_3CH_2-\overset{\displaystyle C_6H_5}{\underset{\displaystyle OH}{\overset{|}{\underset{|}{C}}}}-CH_2CH_3$$

3-Phenyl-3-pentanol

and *two ethyl groups*. This means that we can synthesize this compound in different ways.

1. We can use a ketone with two ethyl groups (3-pentanone) and allow it to react with phenylmagnesium bromide:

Analysis

$$CH_3CH_2-\overset{\overset{\displaystyle C_6H_5}{|}}{\underset{\underset{\displaystyle OH}{|}}{C}}-CH_2CH_3 \Longrightarrow CH_3CH_2-\overset{\overset{\displaystyle }{}}{\underset{\underset{\displaystyle O}{\|}}{C}}-CH_2CH_3 + C_6H_5MgBr$$

Synthesis

$$C_6H_5MgBr \quad + \quad CH_3CH_2\underset{\underset{\displaystyle O}{\|}}{C}CH_2CH_3 \xrightarrow[\text{(2) } H_3O^+]{\text{(1) } Et_2O} CH_3CH_2-\overset{\overset{\displaystyle C_6H_5}{|}}{\underset{\underset{\displaystyle OH}{|}}{C}}-CH_2CH_3$$

| **Phenylmagnesium bromide** | **3-Pentanone** | **3-Phenyl-3-pentanol** |

2. We can use a ketone containing an ethyl group and a phenyl group (ethyl phenyl ketone) and allow it to react with ethylmagnesium bromide:

Analysis

$$CH_3CH_2-\overset{\overset{\displaystyle C_6H_5}{|}}{\underset{\underset{\displaystyle OH}{|}}{C}}\!\!\not{\;}\!CH_2CH_3 \Longrightarrow CH_3CH_2-\overset{\overset{\displaystyle C_6H_5}{|}}{\underset{\underset{\displaystyle O}{\|}}{C}} \quad + CH_3CH_2MgBr$$

Synthesis

$$CH_3CH_2MgBr + \underset{\displaystyle CH_3CH_2}{\overset{\displaystyle C_6H_5}{\diagdown}}C\!\!=\!\!O \xrightarrow[\text{(2) } H_3O^+]{\text{(1)}Et_2O} CH_3CH_2-\overset{\overset{\displaystyle C_6H_5}{|}}{\underset{\underset{\displaystyle OH}{|}}{C}}-CH_2CH_3$$

| **Ethylmagnesium bromide** | **Ethyl phenyl ketone** | **3-Phenyl-3-pentanol** |

3. Or we can use an ester of benzoic acid and allow it to react with two molar equivalents of ethylmagnesium bromide:

Analysis

$$CH_3CH_2\!\!\not{\;}\!\overset{\overset{\displaystyle C_6H_5}{|}}{\underset{\underset{\displaystyle OH}{|}}{C}}\!\!\not{\;}\!CH_2CH_3 \Longrightarrow \underset{O\diagup\diagdown OCH_3}{\overset{\overset{\displaystyle C_6H_5}{|}}{C}} + 2\,CH_3CH_2MgBr$$

Synthesis

$$2\;CH_3CH_2MgBr + C_6H_5\overset{\overset{\displaystyle O}{\|}}{C}OCH_3 \xrightarrow[\text{(2) } H_3O^+]{\text{(1) } Et_2O} CH_3CH_2-\overset{\overset{\displaystyle C_6H_5}{|}}{\underset{\underset{\displaystyle OH}{|}}{C}}-CH_2CH_3$$

| **Ethylmagnesium bromide** | **Methyl benzoate** | **3-Phenyl-3-pentanol** |

All of these methods will be likely to give us our desired compound in yields greater than 80%.

SAMPLE PROBLEM

Illustrating a Multistep Synthesis

Using an alcohol of no more than four carbon atoms as your only organic starting material, outline a synthesis of **A**.

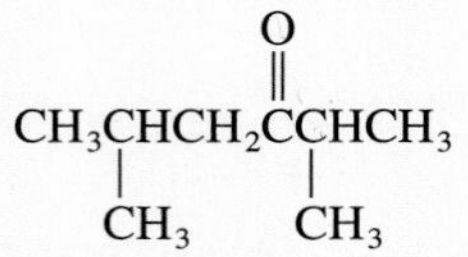

$$
\underset{A}{\underset{\overset{|}{CH_3} \quad \overset{|}{CH_3}}{CH_3CHCH_2\overset{\overset{O}{\|}}{C}CHCH_3}}
$$

ANSWER:

We can construct the carbon skeleton from two four-carbon atom compounds using a Grignard reaction. Then oxidation of the alcohol produced will yield the desired ketone.

Analysis

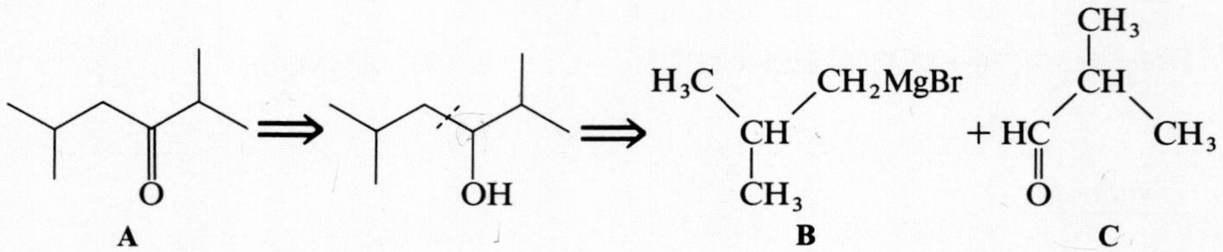

Synthesis

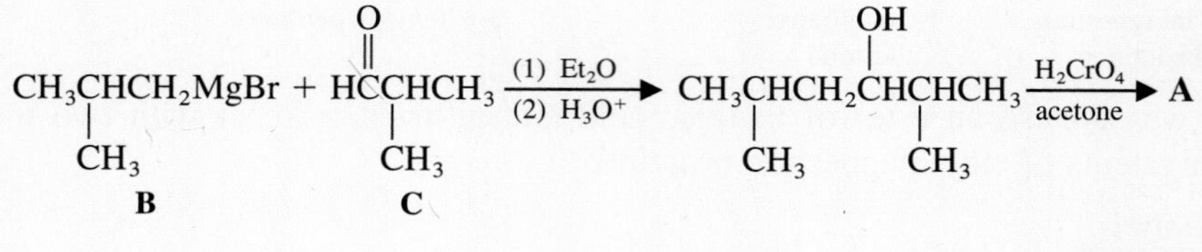

We can synthesize the Grignard reagent (**B**) and the aldehyde (**C**) from isobutyl alcohol.

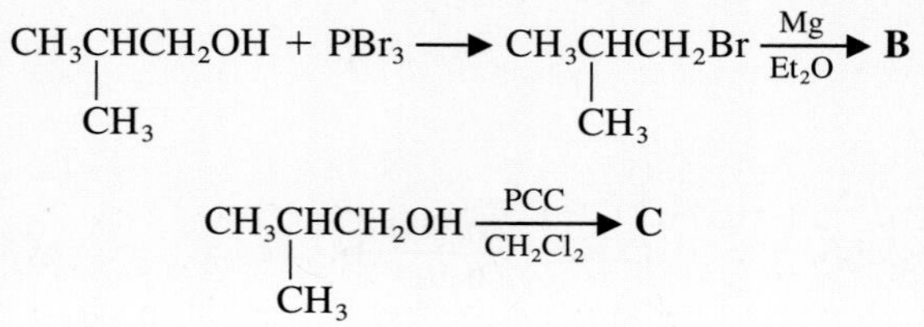

SAMPLE PROBLEM

Illustrating a Multistep Synthesis

Starting with bromobenzene and any other needed reagents, outline a synthesis of the following aldehyde.

ANSWER:
Working backwards, we remember that we can synthesize the aldehyde from the corresponding alcohol by oxidation with PCC (Section 11.4A). The alcohol can be made by treating phenylmagnesium bromide with oxirane. [Adding oxirane to a Grignard reagent is a very useful method for adding a —CH_2CH_2OH unit to an organic group (Section 11.7B).] Phenylmagnesium bromide can be made in the usual way, by treating bromobenzene with magnesium in an ether solvent.

Analysis

Synthesis

$$C_6H_5Br \xrightarrow[Et_2O]{Mg} C_6H_5MgBr \xrightarrow[(2)\ H_3O^+]{(1)\ \triangle O} C_6H_5CH_2CH_2OH \xrightarrow[CH_2Cl_2]{PCC} C_6H_5CH_2CHO$$

PROBLEM 11.9

Show how Grignard reactions could be used to synthesize each of the following compounds. (You must start with an organic halide and you may use any other compounds you need.)

(a) 2-Methyl-2-butanol (three ways)

(b) 3-Methyl-3-pentanol (three ways)

(c) 3-Ethyl-2-pentanol (two ways)

(d) 2-Phenyl-2-pentanol (three ways)

(e) Triphenylmethanol (two ways)

PROBLEM 11.10

Outline a synthesis of each of the following. Permitted starting materials are phenylmagnesium bromide, oxirane, formaldehyde, and alcohols or esters of four carbon atoms or fewer. You may use any inorganic reagents and oxidizing agents such as pyridinium chlorochromate (PCC).

(a) $C_6H_5CHCH_2CH_3$
           |
          OH

(b) C_6H_5CH (with O double bonded)

(c) $C_6H_5CCH_2CH_3$ (with OH above, C_6H_5 below)

(d) $C_6H_5CHCHCH_3$ (with OH above, CH_3 below)

11.8B Restrictions on the Use of Grignard Reagents

Although the Grignard synthesis is one of the most versatile of all general synthetic procedures, it is not without its limitations. Most of these limitations arise from the very feature of the Grignard reagent that makes it so useful, its *extraordinary reactivity as a nucleophile and a base.*

The Grignard reagent is a very powerful base; in effect it contains a carbanion. Thus, it is not possible to prepare a Grignard reagent from an organic group that contains an *acidic hydrogen;* and by an acidic hydrogen we mean any hydrogen more acidic than the hydrogen atoms of an alkane or alkene. We cannot, for example, prepare a Grignard reagent from a compound containing an —OH group, an —NH— group, an —SH group, a —CO₂H group, or an —SO₃H group. If we were to attempt to prepare a Grignard reagent from an organic halide containing any of these groups, the formation of the Grignard reagent would simply fail to take place. (Even if a Grignard reagent were to form, it would immediately react with the acidic group.)

Since Grignard reagents are powerful nucleophiles we cannot prepare a Grignard reagent from any organic halide that contains a carbonyl, epoxy, nitro, or cyano (—CN) group. If we were to attempt to carry out this kind of reaction, any Grignard reagent that formed would only react with the unreacted starting material.

$-OH, -NH_2, -NHR, -CO_2H, -SO_3H, -SH, -C\equiv C-H$

$-\overset{O}{\overset{\|}{C}}H, -\overset{O}{\overset{\|}{C}}R, -\overset{O}{\overset{\|}{C}}OR, -\overset{O}{\overset{\|}{C}}NH_2, -NO_2, -C\equiv N, -\overset{|}{C}\underset{O}{\overset{}{\diagdown\diagup}}\overset{|}{C}-$

> Grignard reagents containing these groups cannot be prepared.

This means that when we prepare Grignard reagents, we are effectively limited to alkyl halides or to analogous organic halides containing carbon–carbon double bonds, internal triple bonds, ether linkages, and —NR₂ groups.

Grignard reactions are so sensitive to acidic compounds that when we prepare a Grignard reagent we must take special care to exclude moisture from our apparatus, and we must use an anhydrous ether as our solvent.

As we saw earlier, acetylenic hydrogens are acidic enough to react with Grignard reagents. This is a limitation that we can use, however. We can make acetylenic Grignard reagents by allowing terminal alkynes to react with alkyl Grignard reagents (cf. Section 11.7A). We can then use these acetylenic Grignard reagents to carry out other syntheses. For example,

$$C_6H_5C\equiv CH + C_2H_5MgBr \longrightarrow C_6H_5C\equiv CMgBr + C_2H_6\uparrow$$

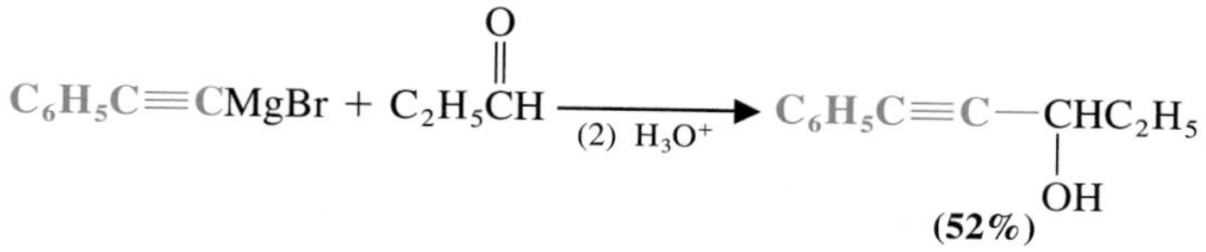

$$C_6H_5C\equiv CMgBr + C_2H_5\overset{\overset{\displaystyle O}{\|}}{C}H \xrightarrow[\text{(2) } H_3O^+]{} C_6H_5C\equiv C-\underset{\underset{\displaystyle OH}{|}}{C}HC_2H_5$$

(52%)

When we plan Grignard syntheses we must also take care not to plan a reaction in which a Grignard reagent is treated with an aldehyde, ketone, epoxide, or ester that contains an acidic group (other than when we deliberately let it react with a terminal alkyne). If we were to do this, the Grignard reagent would simply react as a base with the acidic hydrogen rather than reacting at the carbonyl or epoxide carbon as a nucleophile. If we were to treat 4-hydroxy-2-butanone with methylmagnesium bromide, for example, the following reaction would take place first,

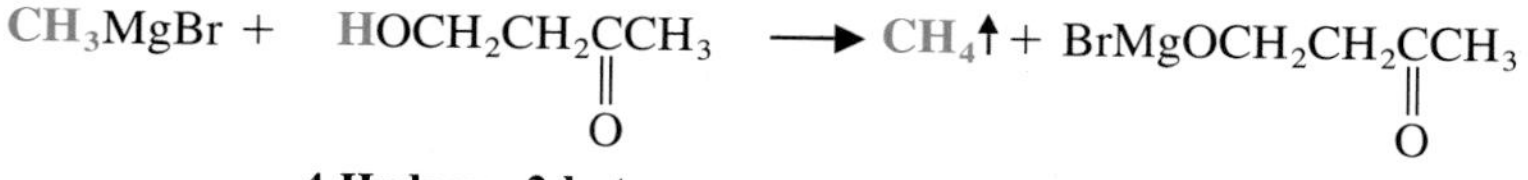

$$CH_3MgBr + \underset{\underset{\displaystyle\text{4-Hydroxy-2-butanone}}{}}{HOCH_2CH_2\overset{\overset{\displaystyle O}{\|}}{C}CH_3} \longrightarrow CH_4\uparrow + BrMgOCH_2CH_2\overset{\overset{\displaystyle O}{\|}}{C}CH_3$$

rather than

$$CH_3MgBr + HOCH_2CH_2\overset{\overset{\displaystyle O}{\|}}{C}CH_3 \xrightarrow{\;\;\times\;\;} HOCH_2CH_2\underset{\underset{\displaystyle OMgBr}{|}}{\overset{\overset{\displaystyle CH_3}{|}}{C}}-CH_3$$

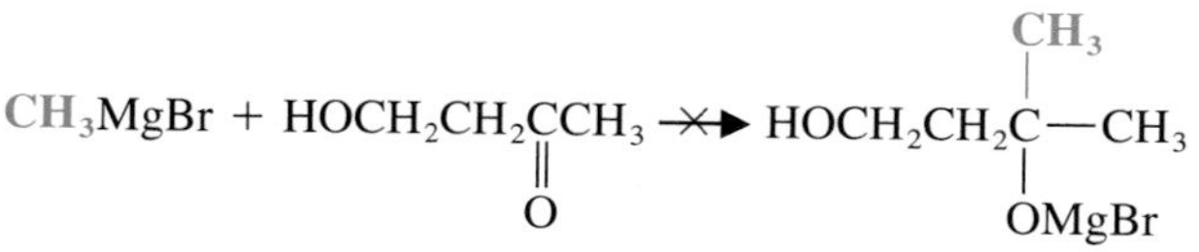

If we are prepared to waste one molar equivalent of the Grignard reagent, we can treat 4-hydroxy-2-butanone with two molar equivalents of the Grignard reagent and thereby get addition to the carbonyl group.

$$HOCH_2CH_2\overset{\overset{\displaystyle O}{\|}}{C}CH_3 \xrightarrow[-CH_4]{2\,CH_3MgBr} BrMgOCH_2CH_2\underset{\underset{\displaystyle OMgBr}{|}}{\overset{\overset{\displaystyle CH_3}{|}}{C}}CH_3 \xrightarrow{2\,H_3O^+} HOCH_2CH_2\underset{\underset{\displaystyle OH}{|}}{\overset{\overset{\displaystyle CH_3}{|}}{C}}CH_3$$

This technique is sometimes employed in small-scale reactions when the Grignard reagent is inexpensive and the other reagent is expensive.

11.8C The Use of Lithium Reagents

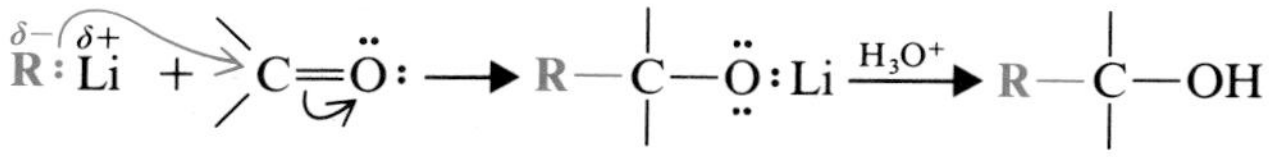

Organolithium reagents (RLi) react with carbonyl compounds in the same way as Grignard reagents and thus provide an alternative method for preparing alcohols.

$$\overset{\delta-}{R}\!:\!\overset{\delta+}{Li} + \underset{}{\overset{}{C}}=\ddot{O}: \longrightarrow R-\underset{|}{\overset{|}{C}}-\ddot{O}\!:\!Li \xrightarrow{H_3O^+} R-\underset{|}{\overset{|}{C}}-OH$$

Organo-   **Aldehyde**   **Lithium**   **Alcohol**
lithium   **or**   **alkoxide**
reagent   **ketone**

Organolithium reagents have the advantage of being somewhat more reactive than Grignard reagents.

11.8D The Use of Sodium Alkynides

Sodium alkynides also react with aldehydes and ketones to yield alcohols. An example is the following:

$$CH_3C\equiv CH \xrightarrow[-NH_3]{NaNH_2} CH_3C\equiv CNa$$

Then,

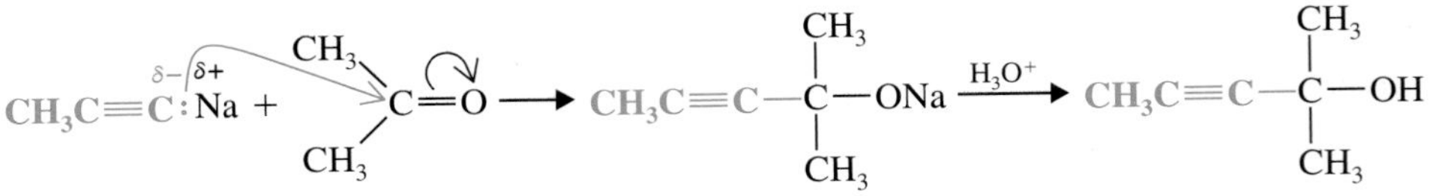

$$CH_3C\equiv C \overset{\delta-}{:} \overset{\delta+}{Na} + \underset{CH_3}{\overset{CH_3}{C}}C=O \longrightarrow CH_3C\equiv C-\underset{CH_3}{\overset{CH_3}{\underset{|}{\overset{|}{C}}}}-ONa \xrightarrow{H_3O^+} CH_3C\equiv C-\underset{CH_3}{\overset{CH_3}{\underset{|}{\overset{|}{C}}}}-OH$$

SAMPLE PROBLEM

Illustrating Multistep Syntheses

Starting with hydrocarbons, organic halides, alcohols, aldehydes, ketones, or esters containing six carbon atoms or fewer and using any other needed reagents, outline a synthesis of each of the following:

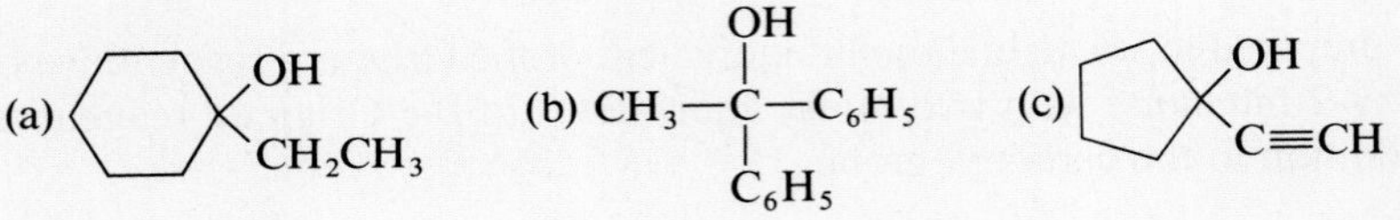

(a) OH, CH$_2$CH$_3$

(b) $CH_3-\underset{C_6H_5}{\overset{OH}{\underset{|}{\overset{|}{C}}}}-C_6H_5$

(c) OH, C≡CH

ANSWERS:

(a) $CH_3CH_2OH \xrightarrow{PBr_3} CH_3CH_2Br \xrightarrow[Et_2O]{Mg} CH_3CH_2MgBr \xrightarrow[(2)\ H_3O^+]{(1)\ \text{}} $

(b) $C_6H_5Br \xrightarrow[Et_2O]{Mg} C_6H_5MgBr \xrightarrow[(2)\ H_3O^+]{(1)\ CH_3COCH_3} CH_3-\underset{C_6H_5}{\overset{OH}{\underset{|}{\overset{|}{C}}}}-C_6H_5$

(c) $HC\equiv CH \xrightarrow{NaNH_2} HC\equiv CNa \xrightarrow[(2)\ H_3O^+]{(1)\ \text{}}$

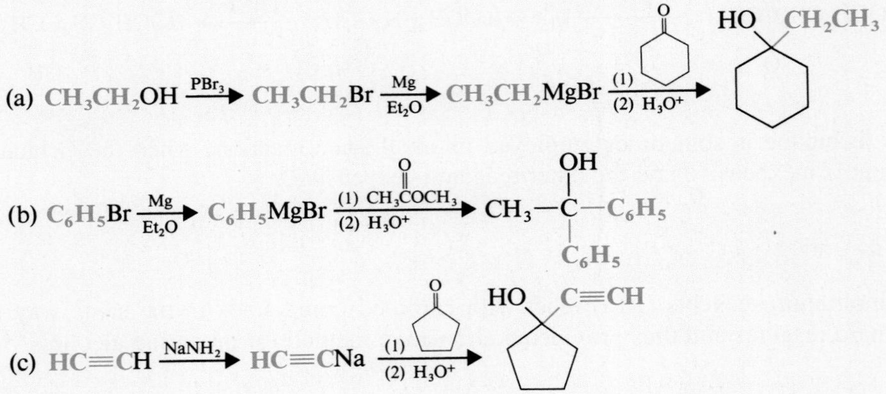

ADDITIONAL PROBLEMS

11.11   What product (or products) would be formed from the reaction of isobutyl bromide, $(CH_3)_2CHCH_2Br$, with each of the following reagents?

(a) OH⁻, H₂O

(b) CN⁻, ethanol

(c) (CH₃)₃CO⁻, (CH₃)₃COH

(d) CH₃O⁻, CH₃OH

(e) Li, Et₂O, then CH₃CCH₃, then H₃O⁺
   (with O double bond)

(f) Mg, Et₂O, then CH₃CH, then H₃O⁺
   (with O double bond)

(g) Mg, Et₂O, then CH₃COCH₃, then H₃O⁺
   (with O double bond)

(h) Mg, Et₂O, then H₂C—CH₂, then H₃O⁺
   (with O epoxide)

(i) Mg, Et₂O, then H—C—H, then H₃O⁺
   (with O double bond)

(j) Li, Et₂O, then CH₃OH

(k) Li, Et₂O, then CH₃C≡CH

11.12 What products would you expect from the reaction of ethylmagnesium bromide (CH₃CH₂MgBr) with each of the following reagents?

(a) H₂O

(b) D₂O

(c) C₆H₅CH, then H₃O⁺
   (with O double bond)

(d) C₆H₅CC₆H₅, then H₃O⁺
   (with O double bond)

(e) C₆H₅COCH₃, then H₃O⁺
   (with O double bond)

(f) C₆H₅CCH₃, then H₃O⁺
   (with O double bond)

(g) CH₃CH₂C≡CH, then CH₃CH, then H₃O⁺
   (with O double bond)

(h) Cyclopentadiene

11.13 What products would you expect from the reaction of propyllithium (CH₃CH₂CH₂Li) with each of the following reagents?

(a) (CH₃)₂CHCH, then H₃O⁺
   (with O double bond)

(b) (CH₃)₂CHCCH₃, then H₃O⁺
   (with O double bond)

(c) 1-Pentyne, then CH₃CCH₃, then H₃O⁺
   (with O double bond)

(d) Ethanol X

(e) CuI, then CH₂=CHCH₂Br

(f) CuI, then cyclopentyl bromide

(g) CuI, then (Z)-1-iodopropene

(h) CuI, then CH₃I

(i) CH₃CO₂D

11.14 Which oxidizing or reducing agent would you use to carry out the following transformations?

(a) CH₃COCH₂CH₂CO₂CH₃ ⟶ CH₃CHOHCH₂CH₂CH₂OH LiAlH₄

(b) CH₃COCH₂CH₂CO₂CH₃ ⟶ CH₃CHOHCH₂CH₂CO₂CH₃ NaBH₄

(c) HO₂CCH₂CH₂CH₂CO₂H ⟶ HOCH₂CH₂CH₂CH₂CH₂OH Li AlH₄

(d) HOCH₂CH₂CH₂CH₂CH₂OH ⟶ HO₂CCH₂CH₂CH₂CO₂H KMnO₄ Heat

(e) HOCH₂CH₂CH₂CH₂CH₂OH ⟶ OHCCH₂CH₂CH₂CHO PCC/CH₂Cl₂

11.15 Outline all steps in a synthesis that would transform isopropyl alcohol, CH₃CH(OH)CH₃, into each of the following:

(a) (CH₃)₂CHCH(OH)CH₃

(b) (CH₃)₂CHCH₂OH

(c) (CH₃)₂CHCH₂CH₂Cl

(d) (CH₃)₂CHCH(OH)CH(CH₃)₂

(e) CH₃CHDCH₃ — r D₂O H⁺

(f) cyclohexyl-CH(CH₃)₂

11.16   What organic products would you expect from each of the following reactions?

(a) Methyllithium + 1-butyne ⟶

(b) Product of (a) + cyclohexanone, then H_3O^+ ⟶

(c) Product of (b) + $Ni_2B(P-2)$ and H_2 ⟶

(d) Product of (b) + NaH, then $CH_3CH_2OSO_2CH_3$ ⟶

(e) $CH_3CH_2COCH_3$ + $NaBH_4$ ⟶

(f) Product of (e) + mesyl chloride $\xrightarrow{\text{base}}$

(g) Product of (f) + CH_3CO_2Na ⟶

(h) Product of (g) + $LiAlH_4$, then H_2O ⟶

11.17   Show how 1-pentanol could be transformed into each of the following compounds. (You may use any needed inorganic reagents and you need not show the synthesis of a particular compound more than once.)

(a) 1-Bromopentane

(b) 1-Pentene

(c) 2-Pentanol

(d) 2-Pentanone

(e) 2-Bromopentane

(f) 1-Hexanol

(g) 1-Heptanol

(h) Pentanal ($CH_3CH_2CH_2CH_2CHO$))

(i) 2-Pentanone ($CH_3COCH_2CH_2CH_3$)

(j) Pentanoic acid ($CH_3CH_2CH_2CH_2CO_2H$)

(k) Dipentyl ether (two ways)

(l) 1-Pentyne

(m) 2-Bromo-1-pentene

(n) Pentyllithium

(o) Decane

(p) 4-Methyl-4-nonanol

11.18   Show how each of the following transformations could be carried out.

(a) Phenylethyne ⟶ $C_6H_5C{\equiv}CC(OH)(CH_3)_2$

(b) $C_6H_5COCH_3$ ⟶ 1-phenylethanol

(c) Phenylethyne ⟶ phenylethene

(d) Phenylethene ⟶ 2-phenylethanol

(e) 2-Phenylethanol ⟶ 4-phenylbutanol

(f) 2-Phenylethanol ⟶ 1-methoxy-2-phenylethane

11.19   Assuming that you have available only alcohols or esters containing no more than four carbon atoms show how you might synthesize each of the following compounds. You must use a Grignard reagent at one step in the synthesis. If needed you may use oxirane and you may use bromobenzene, but you must show the synthesis of any other required organic compounds. Assume you have available any solvents, or any inorganic compounds, including oxidizing and reducing agents, that you require.

(a) $(CH_3)_2CHCOC_6H_5$

(b) 4-Ethyl-4-heptanol

(c) 1-Cyclobutyl-2-methyl-1-propanol

(d) $C_6H_5CH_2CHO$

(e) $(CH_3)_2CHCH_2CH_2CO_2H$

(f) 1-Propylcyclobutanol

(g) $CH_3CH_2CH_2COCH_2CH(CH_3)_2$

(h) 3-Phenyl-3-bromopentane

11.20   The alcohol shown below is used in making perfumes. Outline a synthesis of this alcohol from bromobenzene and 1-butene.

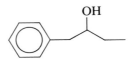

11.21   Show how a Grignard reagent might be used in the following synthesis.

11.22   Starting with compounds of four carbon atoms or fewer, outline a synthesis of racemic Meparfynol, a mild hypnotic (sleep-inducing compound).

$$CH_3-CH_2-\underset{\underset{OH}{|}}{\overset{\overset{CH_3}{|}}{C}}-C\equiv CH$$

Meparfynol

11.23   Although oxirane (oxacyclopropane) and oxetane (oxacyclobutane) react with Grignard reagents and organolithiums to form alcohols, tetrahydrofuran (oxacyclopentane) is so unreactive that it can be used as the solvent in which these organometallic compounds are prepared. Explain the difference in reactivity of these oxygen heterocycles.

11.24   Predict the products that would result from the reaction of a Grignard reagent with:

(a) Diethyl carbonate, $C_2H_5-O-\overset{\overset{O}{\|}}{C}-O-C_2H_5$

(b) Ethyl formate, $H-\overset{\overset{O}{\|}}{C}-O-C_2H_5$

CHAPTER 12

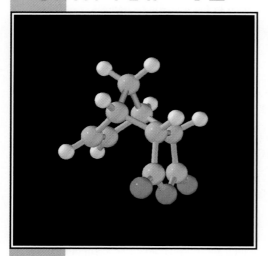

Diels Alder adduct of cyclopentadiene
and maleic anhydride
(Section 12.10)

CONJUGATED UNSATURATED SYSTEMS

12.1 INTRODUCTION

In our study of the reactions of alkenes in Chapter 8 we saw how important the π bond is in understanding the chemistry of unsaturated compounds. In this chapter we shall study a special group of unsaturated compounds and again we shall find that the π bond is the important part of the molecule. Here we shall examine *species that have a p orbital on an atom adjacent to a double bond.* The p orbital may be one that contains a single electron as in the allyl radical (CH_2=$CHCH_2\cdot$) (Section 12.2); it may be a vacant p orbital as in the allyl cation (CH_2=$CHCH_2^+$) (Section 12.4); or it may be the p orbital of another double bond as in 1,3-butadiene (CH_2=CH—CH=CH_2) (Section 12.7). We shall see that having a p orbital on an atom adjacent to a double bond allows the formation of an extended π bond—one that encompasses more than two nuclei.

Systems that have a p orbital on an atom adjacent to a double bond—molecules with delocalized π bonds—are called **conjugated unsaturated systems.** This general phenomenon is called **conjugation.** As we shall see, conjugation gives these systems special properties. We shall find, for example, that conjugated radicals, ions, or molecules are more stable than nonconjugated ones. We shall demonstrate this with the allyl radical, the allyl cation, and 1,3-butadiene. Conjugation also allows molecules to undergo unusual reactions, and we shall study these, too, including an important reaction for forming rings called the Diels–Alder reaction (Section 12.10).

12.2 ALLYLIC SUBSTITUTION AND THE ALLYL RADICAL

When propene reacts with bromine or chlorine at low temperatures, the reaction that takes place is the usual addition of halogen to the double bond.

$$CH_2{=}CH{-}CH_3 + X_2 \xrightarrow[\substack{CCl_4 \\ \text{(addition reaction)}}]{\text{low temperature}} \begin{array}{c} CH_2{-}CH{-}CH_3 \\ \mid \quad \mid \\ X \quad X \end{array}$$

However, when propene reacts with chlorine or bromine at very high temperatures or under conditions in which the concentration of the halogen is very small, the reaction that occurs is a **substitution.** These two examples illustrate how we can often change the course of an organic reaction simply by changing the conditions. (They also illustrate the need for specifying the conditions of a reaction carefully when we report experimental results.)

$$\underset{\textbf{Propene}}{CH_2{=}CH{-}CH_3} + X_2 \xrightarrow[\substack{\text{or} \\ \text{low conc of } X_2 \\ \text{(substitution reaction)}}]{\text{high temperature}} CH_2{=}CH{-}CH_2X + HX$$

In this substitution a halogen atom replaces one of the hydrogen atoms of the methyl group of propene. These hydrogen atoms are called the **allylic hydrogen atoms** and the substitution reaction is known as an **allylic substitution.**

Allylic hydrogen atoms

These are general terms as well. The hydrogen atoms of any saturated carbon atom adjacent to a double bond, that is,

are called *allylic* hydrogen atoms and any reaction in which an allylic hydrogen atom is replaced is called an *allylic substitution.*

12.2A Allylic Chlorination (High Temperature)

Propene undergoes allylic chlorination when propene and chlorine react in the gas phase at 400°C. This method for synthesizing allyl chloride is called the "Shell Process."

$$CH_2{=}CH{-}CH_3 + Cl_2 \xrightarrow[\text{gas phase}]{400°C} \underset{\substack{\textbf{3-Chloropropene} \\ \textbf{(allyl chloride)}}}{CH_2{=}CH{-}CH_2Cl} + HCl$$

The mechanism for allylic substitution is the same as the chain mechanism for alkane halogenations that we saw in Chapter 9. In the chain-initiating step, the chlorine molecule dissociates into chlorine atoms.

Chain-Initiating Step

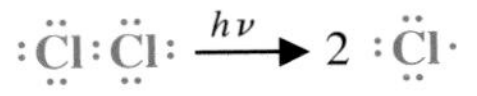

$$:\ddot{C}l:\ddot{C}l: \xrightarrow{h\nu} 2 :\ddot{C}l\cdot$$

In the first chain-propagating step the chlorine atom abstracts one of the allylic hydrogen atoms.

First Chain-Propagating Step

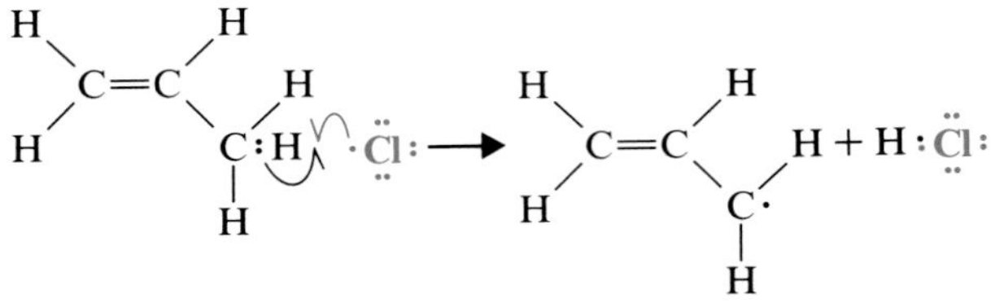

Allyl radical

The radical that is produced in this step is called an *allyl radical.**
In the second chain-propagating step the allyl radical reacts with a molecule of chlorine.

Second Chain-Propagating Step

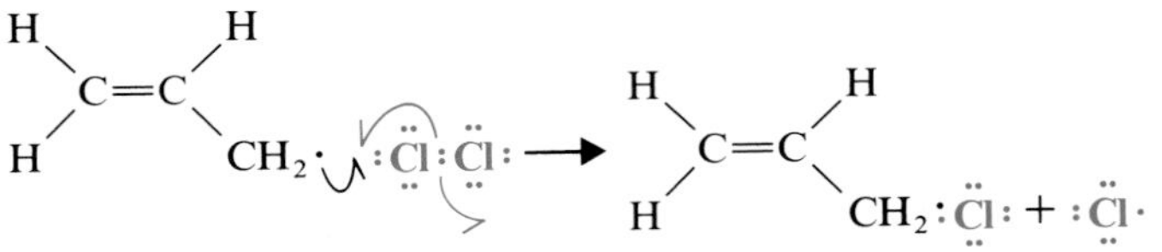

Allyl chloride

This step results in the formation of a molecule of allyl chloride and a chlorine atom. The chlorine atom then brings about a repetition of the first chain-propagating step. The chain reaction continues until the usual chain-terminating steps consume the radicals.

The reason for substitution at the allylic hydrogen atoms of propene will be more understandable if we examine the bond dissociation energy of an allylic carbon–hydrogen bond and compare it with the bond dissociation energies of other carbon–hydrogen bonds (cf. Table 9.1).

$$CH_2{=}CHCH_2{-}H \longrightarrow CH_2{=}CHCH_2\cdot + H\cdot \qquad DH° = 85 \text{ kcal mol}^{-1}$$

   Propene   **Allyl radical**

$$(CH_3)_3C{-}H \longrightarrow (CH_3)_3C\cdot + H\cdot \qquad DH° = 91 \text{ kcal mol}^{-1}$$

   Isobutane   **3° Radical**

* A radical of the general type $-\overset{|}{C}{=}\overset{|}{C}-\overset{|}{\underset{\cdot}{C}}-$ is called an *allylic* radical.

$$(CH_3)_2CH—H \longrightarrow (CH_3)_2CH\cdot + H\cdot \qquad DH° = 94.5 \text{ kcal mol}^{-1}$$

$\quad$ **Propane** $\qquad\qquad$ **2° Radical**

$$CH_3CH_2CH_2—H \longrightarrow CH_3CH_2CH_2\cdot + H\cdot \qquad DH° = 98 \text{ kcal mol}^{-1}$$

$\quad$ **Propane** $\qquad\qquad$ **1° Radical**

$$CH_2{=}CH—H \longrightarrow CH_2{=}CH\cdot + H\cdot \qquad DH° = 108 \text{ kcal mol}^{-1}$$

$\quad$ **Ethene** $\qquad$ **Vinyl radical**

We see that an allylic carbon–hydrogen bond of propene is broken with greater ease than even the tertiary carbon–hydrogen bond of isobutane and with far greater ease than a vinylic carbon–hydrogen bond.

$$CH_2{=}CH—CH_2{\overset{\frown}{}}H + \cdot\ddot{X}: \longrightarrow CH_2{=}CH—CH_2\cdot + HX \qquad E_{act} \text{ is low}$$

$\qquad\qquad\qquad\qquad$ **Allyl radical**

$$:\ddot{X}\cdot + H{\overset{\frown}{}}CH{=}CH—CH_3 \longrightarrow \cdot CH{=}CH—CH_3 + HX \qquad E_{act} \text{ is high}$$

$\qquad\qquad\qquad\qquad$ **Vinylic radical**

The ease with which an allylic carbon–hydrogen bond is broken means that relative to primary, secondary, tertiary, and vinylic free radicals the allyl radical is the *most stable* (Fig. 12.1).

$\qquad$ Relative stability $\quad$ allylic or allyl $> 3° > 2° > 1° >$ vinyl

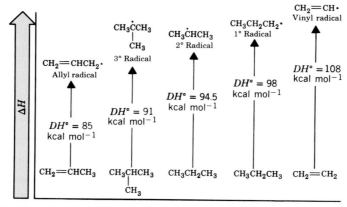

FIGURE 12.1 The relative stability of the allyl radical compared to 1°, 2°, 3°, and vinyl radicals. (The stabilities of the radicals are relative to the hydrocarbon from which each was formed, and the overall order of stability is allyl $> 3° > 2° > 1° >$ vinyl.)

12.2B Allylic Bromination With *N*-Bromosuccinimide (Low Concentration of Br₂)

Propene undergoes allylic bromination when it is treated with *N*-bromosuccinimide (NBS) in CCl_4 in the presence of peroxides or light.

$$CH_2{=}CH{-}CH_3 \;+\; \underset{\substack{\textbf{\textit{N}-Bromosuccinimide} \\ \textbf{(NBS)}}}{\overset{\displaystyle :O}{\bigg\langle}\;:N{-}Br} \;\xrightarrow[CCl_4]{\text{light or ROOR}}\; CH_2{=}CH{-}CH_2Br \;+\; \underset{\substack{\textbf{3-Bromopropene} \\ \textbf{(allyl bromide)}}}{CH_2{=}CH{-}CH_2Br}\;\; \underset{\textbf{Succinimide}}{\overset{\displaystyle :O}{\bigg\langle}\;:N{-}H}$$

The reaction is initiated by the formation of a small amount of Br· (possibly formed by dissociation of the N—Br bond of the NBS). The main propagation steps for this reaction are the same as for allylic chlorination (Section 12.2A).

$$CH_2{=}CH{-}CH_2{-}H + {\cdot}Br \longrightarrow CH_2{=}CH{-}CH_2{\cdot} + HBr$$

$$CH_2{=}CH{-}CH_2{\cdot} + Br{-}Br \longrightarrow CH_2{=}CH{-}CH_2Br + {\cdot}Br$$

N-Bromosuccinimide is nearly insoluble in CCl_4 and provides a constant but very low concentration of bromine in the reaction mixture. It does this by reacting very rapidly with the HBr formed in the substitution reaction. Each molecule of HBr is replaced by one molecule of Br_2.

$$\overset{\displaystyle :O}{\bigg\langle}\;:N{-}Br + HBr \longrightarrow \overset{\displaystyle :O}{\bigg\langle}\;:N{-}H + Br_2$$

Under these conditions, that is, *in a nonpolar solvent and with a very low concentration of bromine,* very little bromine adds to the double bond; it reacts by substitution and replaces an allylic hydrogen atom instead.

Optional Material

Why, we might ask, does a low concentration of bromine favor allylic substitution over addition? To understand this we must recall the mechanism for addition and notice that in the first step only one atom of the bromine molecule becomes attached to the alkene *in a reversible step.*

$$Br{-}Br + \overset{\displaystyle \diagdown\;\diagup}{\underset{\diagup\;\diagdown}{\overset{C}{\underset{C}{\|}}}} \rightleftharpoons Br\overset{+}{\diagdown}\!\underset{\diagdown C{-}}{\overset{{-}C}{\big\langle}} + Br^- \longrightarrow \underset{Br{-}C{-}}{\overset{-C{-}Br}{\big|}}$$

The other atom (now the bromide ion) becomes attached in the second step. Now, if the concentration of bromine is low, the equilibrium for the first step will lie far to the left. Moreover, even when the bromonium ion forms, the probability of its finding a bromide ion in its vicinity is also low. These two factors slow the addition so that allylic substitution competes successfully.

The use of a nonpolar solvent also slows addition. Since there are no polar molecules to solvate (and thus stabilize) the bromide ion formed in the first step, the bromide ion uses a bromine molecule as a substitute:

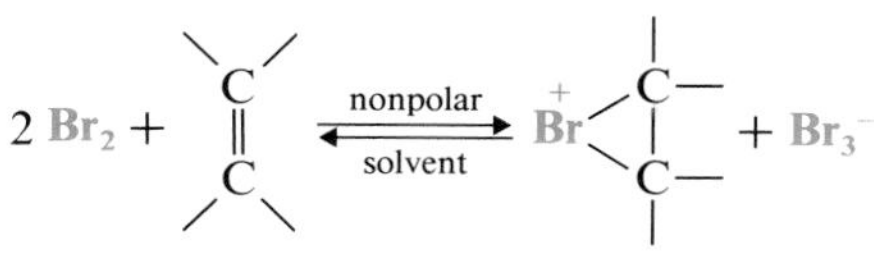

$$2 \ Br_2 + \quad \overset{\displaystyle \underset{\diagdown}{C} \diagup}{\underset{\diagup C \diagdown}{\parallel}} \quad \underset{\text{solvent}}{\overset{\text{nonpolar}}{\rightleftharpoons}} \quad \overset{+}{Br} \diagup \overset{\mid}{\underset{\mid}{\overset{C—}{\underset{C—}{\mid}}}} \ + \ Br_3^-$$

This means that in a nonpolar solvent the rate equation is second order with respect to bromine,

$$\text{rate} = k \left[\ \overset{\diagdown}{\underset{\diagup}{C}} = C \overset{\diagup}{\underset{\diagdown}{}} \ \right] [Br_2]^2$$

and that the low bromine concentration has an even more pronounced effect in slowing the rate of addition.

To understand why a high temperature favors allylic substitution over addition requires a consideration of the effect of entropy changes on equilibria (Section 3.8). The addition reaction, because it combines two molecules into one, has a substantial negative entropy change. At low temperatures, the $T\Delta S°$ term in $\Delta G° = \Delta H° - T\Delta S°$, is not large enough to offset the favorable $\Delta H°$ term. But as the temperature is increased, the $T\Delta S°$ term becomes more significant, $\Delta G°$ becomes more positive, and the equilibrium becomes more unfavorable.

12.3 THE STABILITY OF THE ALLYL RADICAL

An explanation of the stability of the allyl radical can be approached in two ways: in terms of molecular orbital theory and in terms of resonance theory (Section 1.8). As we shall see soon, both approaches give us equivalent descriptions of the allyl radical. The molecular orbital approach is easier to visualize, so we shall begin with it. (As preparation for this section, it would be a good idea to review the molecular orbital theory given in Sections 1.12 and 2.4.)

12.3A Molecular Orbital Description of the Allyl Radical

As an allylic hydrogen atom is abstracted from propene (see the following diagram), the sp^3-hybridized carbon atom of the methyl group changes its hybridization state to sp^2 (cf. Section 9.7). The p orbital of this new sp^2-hybridized carbon atom overlaps with the p orbital of the central carbon atom. Thus, in the allyl radical three p orbitals overlap to form a set of π molecular orbitals that encompass all three carbon atoms. The new p orbital of the allyl radical is said to be *conjugated* with those of the double bond and the allyl radical is said to be a *conjugated unsaturated system*.

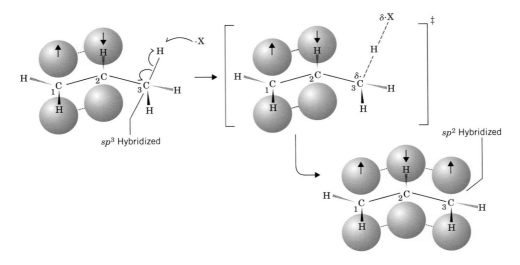

The unpaired electron of the allyl radical and the two electrons of the π bond are **delocalized** over all three carbon atoms. This delocalization of the unpaired electron accounts for the greater stability of the allyl radical when compared to primary, secondary, and tertiary radicals. Although some delocalization occurs in primary, secondary, and tertiary radicals, delocalization is not as effective because it occurs through σ bonds.

The diagram in Figure 12.2 illustrates how the three p orbitals of the allyl radical

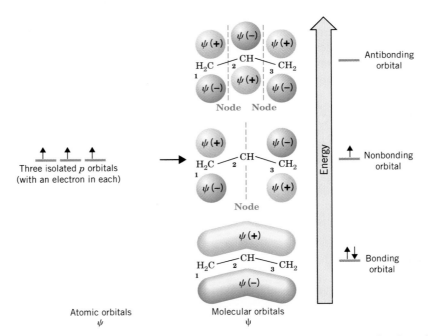

FIGURE 12.2 The combination of three atomic p orbitals to form three π molecular orbitals in the allyl radical. The bonding π molecular orbital is formed by the combination of the three p orbitals with lobes of the same sign overlapping above and below the plane of the atoms. The nonbonding π molecular orbital has a node at C2. The antibonding π molecular orbital has two nodes: between C1 and C2, and between C2 and C3.

combine to form three π molecular orbitals. (*Remember:* The number of molecular orbitals that results always equals the number of atomic orbitals that combine; cf. Section 1.12.) The bonding π molecular orbital is of lowest energy; it encompasses all three carbon atoms and is occupied by two spin-paired electrons. This bonding π orbital is the result of having *p* orbitals with lobes of the same sign overlap between adjacent carbon atoms. This type of overlap, as we recall, increases the π electron density in the regions between the atoms where it is needed for bonding. The non-bonding π orbital is occupied by one unpaired electron and it has a node at the central carbon atom. This node means that the unpaired electron is located in the vicinity of carbon atoms **1** and **3** only. The antibonding π molecular orbital results when orbital lobes of opposite sign overlap between adjacent carbon atoms: Such overlap means that in the antibonding π orbital there is a node between each pair of carbon atoms. This antibonding orbital of the allyl radical is of highest energy and is empty in the ground state of the radical.

We can illustrate the picture of the allyl radical given by molecular orbital theory in simpler terms with the following structure.

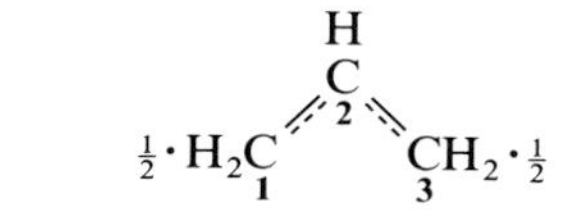

We indicate with dotted lines that both carbon–carbon bonds are partial double bonds. This accommodates one of the things that molecular orbital theory tells us: *that there is a π bond encompassing all three atoms.* We also place the symbol $\frac{1}{2}$· beside C1/C3 atoms. This denotes a second thing molecular orbital theory tells us: *that the unpaired electron spends its time in the vicinity of C1 and C3.* Finally, implicit in the molecular orbital picture of the allyl radical is this: The two ends of the allyl radical are *equivalent.* This aspect of the molecular orbital description is also implicit in the formula just given.

12.3B Resonance Description of the Allyl Radical

Earlier in this section we wrote the structure of the allyl radical as **A**.

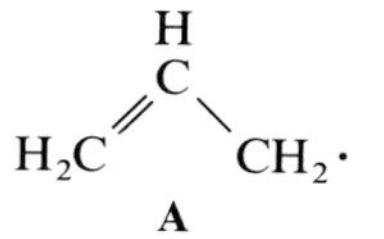

However, we might just as well have written the equivalent structure, **B**.

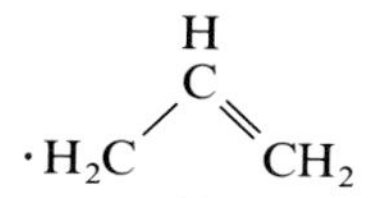

In writing structure **B** we do not mean to imply that we have simply taken structure **A** and turned it over. What we have done is moved the electrons in the following way:

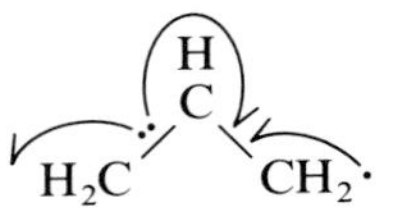

We have not moved the atomic nuclei themselves.

Resonance theory (Section 1.8) tells us that whenever we can write two structures for a chemical entity *that differ only in the positions of the electrons,* the entity cannot be represented by either structure alone but is a *hybrid* of both. We can represent the hybrid in two ways: We can write both structures **A** and **B** and connect them with a double-headed arrow, a special sign in resonance theory, which indicates they are resonance structures.

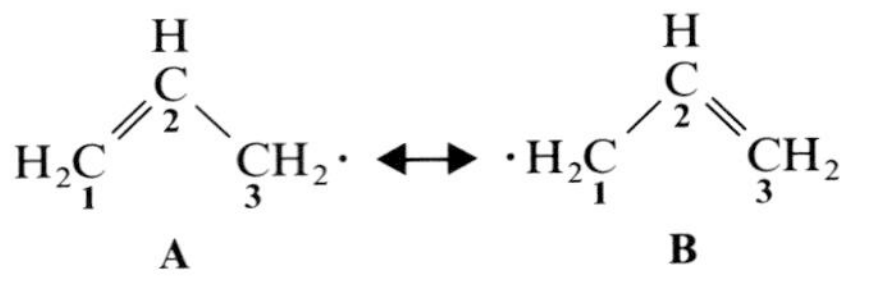

Or we can write a single structure, **C**, that blends the features of both resonance structures.

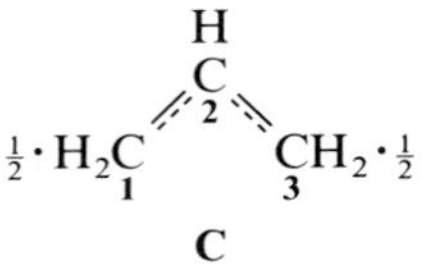

We see, then, that resonance theory gives us exactly the same picture of the allyl radical as we got from molecular orbital theory. Structure **C** describes the carbon–carbon bonds of the allyl radical as partial double bonds. The resonance structures **A** and **B** also tell us that the unpaired electron is associated only with the C1 and C3 atoms. We indicate this in structure **C** by placing a $\frac{1}{2}\cdot$ beside C1 and C3.* Because resonance structures **A** and **B** are equivalent, *C1 and C3 are also equivalent.*

Another rule in resonance theory is that *whenever equivalent resonance structures* can be written for a chemical species, *the chemical species is much more stable than either resonance structure (when taken alone) would indicate.* If we were to examine either **A** or **B** alone, we might decide that it resembled a primary radical. Thus, we might estimate the stability of the allyl radical as approximately that of a primary radical. In doing so, we would greatly underestimate the stability of the allyl radical. Resonance theory tells us, however, that since **A** and **B** are *equivalent resonance structures,* the allyl radical should be much more stable than either, that is, much more stable than a primary radical. This correlates with what experiments have shown to be true; the allyl radical is even more stable than a tertiary radical.

---

* A resonance structure such as the one shown below would indicate that an unpaired electron is associated with C2. This structure is not a proper resonance structure because resonance theory dictates that *all resonance structures must have the same number of unpaired electrons* (cf. Section 12.5).

$$\cdot CH_2 \!-\! \overset{\displaystyle\cdot}{C}H \!-\! CH_2 \cdot$$
(an incorrect resonance structure)

PROBLEM 12.1

(a) What product(s) would you expect to obtain if propene labeled with ^{14}C at C1 were subjected to allylic chlorination or bromination?   (b) Explain your answer.

$$^{14}CH_2{=}CHCH_3 + X_2 \xrightarrow[\substack{\text{or} \\ \text{low conc of } X_2}]{\text{high temperature}} ?$$

(c) If more than one product would be obtained, what relative proportions would you expect?

12.4 THE ALLYL CATION

Although we cannot go into the experimental evidence here, the allyl cation $(CH_2{=}CHCH_2{}^+)$ is an unusually stable carbocation. It is even more stable than a secondary carbocation and is almost as stable as a tertiary carbocation. In general terms, the relative order of stabilities of carbocations is that given here.

Relative Order of Carbocation Stability

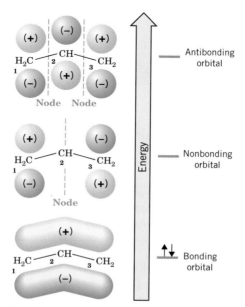

As we might expect, the unusual stability of the allyl cation and other allylic cations can also be accounted for in terms of molecular orbital or resonance theory.

The molecular orbital description of the allyl cation is shown in Fig. 12.3.

FIGURE 12.3 The π molecular orbitals of the allyl cation. The allyl cation, like the allyl radical (Fig. 12.2), is a conjugated unsaturated system.

The bonding π molecular orbital of the allyl cation, like that of the allyl radical (Fig. 12.2), contains two spin-paired electrons. The nonbonding π molecular orbital of the allyl cation, however, is empty. Since an allyl cation is what we would get if we removed an electron from an allyl radical, we can say, in effect, that we remove the electron from the nonbonding molecular orbital.

$$CH_2=CHCH_2\cdot \xrightarrow{-e^-} CH_2=CHCH_2^+$$

Removal of an electron from a nonbonding orbital (cf. Fig. 12.2) is known to require less energy than removal of an electron from a bonding orbital. In addition, the positive charge that forms on the allyl cation is *effectively delocalized* between C1 and C3. Thus, in molecular orbital theory these two factors, the ease of removal of a nonbonding electron and the delocalization of charge, account for the unusual stability of the allyl cation.

Resonance theory depicts the allyl cation as a hybrid of structures **D** and **E** represented here.

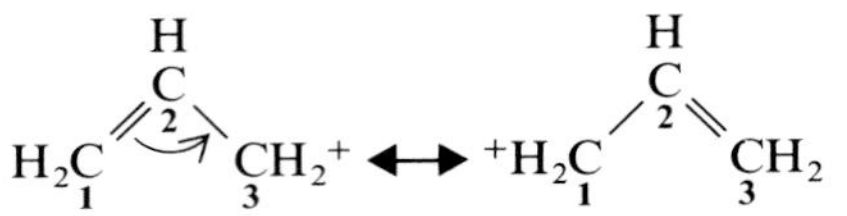

Because **D** and **E** are *equivalent* resonance structures, resonance theory predicts that the allyl cation should be unusually stable. Since the positive charge is located on C3 in **D** and on C1 in **E**, resonance theory also tells us that the positive charge should be delocalized over both carbon atoms. Carbon atom **2** carries none of the positive charge. The hybrid structure **F** (see following structure) includes charge and bond features of both **D** and **E**.

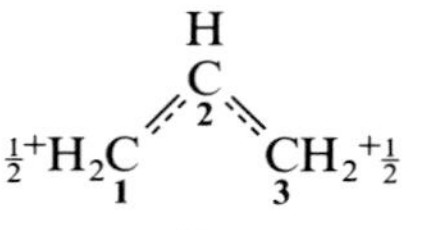

PROBLEM 12.2 _____

(a) Write structures corresponding to **D**, **E**, and **F** for the carbocation shown.

$$CH_3\!-\!\overset{+}{CH}\!-\!CH=CH_2$$

(b) This carbocation appears to be even more stable than a tertiary carbocation; how can you explain this?   (c) What product(s) would you expect to be formed if this carbocation reacted with a chloride ion?

12.5 SUMMARY OF RULES FOR RESONANCE

We have used resonance theory extensively in earlier sections of this chapter because we have been describing radicals and ions with delocalized electrons (and charges) in

π bonds. Resonance theory is especially useful with systems like this, and we shall use it again and again in the chapters that follow. We had an introduction to resonance theory in Section 1.8 and it should be helpful now to summarize the rules for writing resonance structures and for estimating the relative contribution a given structure will make to the overall hybrid.

12.5A Rules for Writing Resonance Structures

1. **Resonance structures exist only on paper.** They have no real existence of their own. Resonance structures are useful because they allow us to describe molecules, radicals, and ions for which a single Lewis structure is inadequate. We write two or more Lewis structures, calling them resonance structures or resonance contributors. We connect these structures by double-headed arrows ($\longleftrightarrow$), and we say that the real molecule, radical, or ion is a hybrid of all of them.

2. **In writing resonance structures we are only allowed to move electrons.** The positions of the nuclei of the atoms must remain the same in all of the structures. Structure **3** is not a resonance structure for the allylic cation, for example, because in order to form it we would have to move a hydrogen atom and this is not permitted.

$$CH_3\!-\!\overset{+}{C}H\!-\!CH\!=\!CH_2 \longleftrightarrow CH_3\!-\!CH\!=\!CH\!-\!\overset{+}{C}H_2 \qquad \overset{+}{C}H_2\!-\!CH_2\!-\!CH\!=\!CH_2$$

$$\underset{\textbf{1}}{} \qquad\qquad \underset{\textbf{2}}{} \qquad\qquad \underset{\textbf{3}}{}$$

These are resonance structures for the allylic cation formed when 1,3-butadiene accepts a proton	**This is not a proper resonance structure for the allylic cation because a hydrogen atom has been moved**

Generally speaking, when we move electrons we move only those of π bonds (as in the example above) and those of lone pairs.

3. **All of the structures must be proper Lewis structures.** We should not write structures in which carbon has five bonds, for example.

$$H\!-\!\overset{\overset{\displaystyle H}{|}}{\underset{\underset{\displaystyle H}{|}}{C}}\!=\!\overset{..}{\underset{..}{O}}{}^{\pm}\!-\!H$$

This is not a proper resonance structure for methanol because carbon has five bonds. Elements of the first major row of the periodic table cannot have more than eight electrons in their valence shell.

4. **All resonance structures must have the same number of unpaired electrons.** The following structure is not a resonance structure for the allyl radical because it contains three unpaired electrons and the allyl radical contains only one.

$$\cdot H_2C\overset{\displaystyle \overset{\displaystyle \cdot}{C}H}{\diagup\diagdown}CH_2\cdot \quad = \quad \uparrow H_2C\overset{\displaystyle \overset{\displaystyle \uparrow}{C}H}{\diagup\diagdown}CH_2\uparrow$$

This is not a proper resonance structure for the allyl radical because it does not contain the same number of unpaired electrons as $CH_2\!=\!CHCH_2\cdot$.

5. **All atoms that are a part of the delocalized system must lie in a plane or be nearly planar.** For example, 2,3-di-*tert*-butyl-1,3-butadiene behaves like a *nonconjugated* diene because the large *tert*-butyl groups twist the structure and prevent the double bonds from lying in the same plane. Because they are not in the same plane, the *p* orbitals at C2 and C3 do not overlap and delocalization (and therefore resonance) is prevented.

2,3-Di-*tert*-butyl-1,3-butadiene

6. **The energy of the actual molecule is lower than the energy that might be estimated for any contributing structure.** The actual allyl cation, for example, is more stable than either resonance structure **4** or **5** taken separately would indicate. Structures **4** and **5** resemble primary carbocations and yet the allyl cation is more stable (has lower energy) than a secondary carbocation. Chemists often call this kind of stabilization *resonance stabilization.*

$$CH_2\!=\!CH\!-\!\overset{+}{C}H_2 \longleftrightarrow \overset{+}{C}H_2\!-\!CH\!=\!CH_2$$
$$\textbf{4} \qquad\qquad\qquad \textbf{5}$$

In Chapter 14 we shall find that benzene is highly resonance-stabilized because it is a hybrid of the two equivalent forms that follow:

or

Resonance structures   **Representation**
for benzene   **of hybrid**

7. **Equivalent resonance structures make equal contributions to the hybrid, and a system described by them has a large resonance stabilization.** Structures **4** and **5** make equal contributions to the allylic cation because they are equivalent. They also make a large stabilizing contribution and account for allylic cations being unusually stable. The same can be said about the contributions made by the equivalent structures **A** and **B** (Section 12.3B) for the allyl radical and by the equivalent structures for benzene.

8. Structures that are not equivalent do not make equal contributions. Generally speaking, **the more stable a structure is (when taken by itself), the greater is its contribution to the hybrid.** For example, the following cation is a hybrid of structures **6** and **7**. Structure **6** makes a greater contribution than **7** because structure **6** is a more stable tertiary carbocation while structure **7** is a primary cation.

That **6** makes a larger contribution means that the partial positive charge on carbon *b* of the hybrid will be larger than the partial positive charge on carbon *d*. It also means that the bond between carbon atoms *c* and *d* will be more like a double bond than the bond between carbon atoms *b* and *c*.

12.5B Estimating the Relative Stability of Resonance Structures

The following rules will help us in making decisions about the relative stabilities of resonance structures.

   a. The more covalent bonds a structure has, the more stable it is. This is exactly what we would expect because we know that forming a covalent bond lowers the energy of atoms. This means that of the following structures for 1,3-butadiene, **8** is by far the most stable and makes by far the largest contribution because it contains one more bond. (It is also more stable for the reason given under rule **c.**)

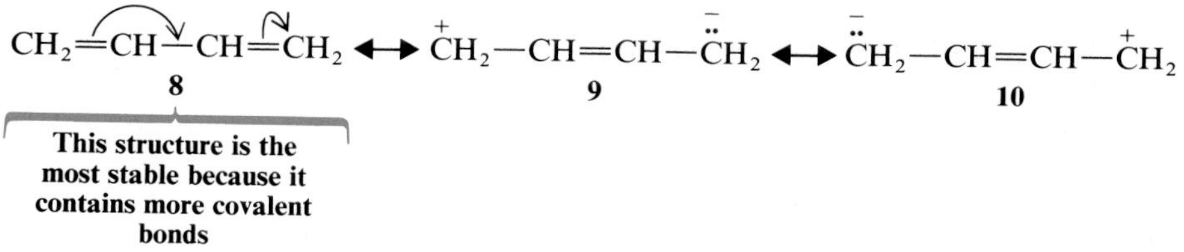

$$CH_2\!=\!CH\!-\!CH\!=\!CH_2 \longleftrightarrow \overset{+}{C}H_2\!-\!CH\!=\!CH\!-\!\overset{-}{C}H_2 \longleftrightarrow \overset{-}{C}H_2\!-\!CH\!=\!CH\!-\!\overset{+}{C}H_2$$

$$\qquad\quad \mathbf{8} \qquad\qquad\qquad\qquad \mathbf{9} \qquad\qquad\qquad\qquad \mathbf{10}$$

This structure is the most stable because it contains more covalent bonds

   b. Structures in which all of the atoms have a complete valence shell of electrons (i.e., the noble gas structure) **are especially stable and make large contributions to the hybrid.** Again, this is what we would expect from what we know about bonding. This means, for example, that **12** makes a larger stabilizing contribution to the cation below than **11** because all of the atoms of **12** have a complete valence shell. (Notice too that **12** has more covalent bonds than **11**, cf. rule **a.**)

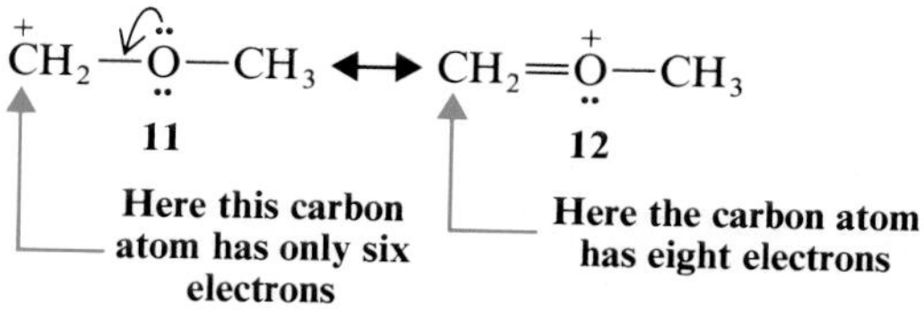

$$\overset{+}{C}H_2\!-\!\overset{..}{\underset{..}{O}}\!-\!CH_3 \longleftrightarrow CH_2\!=\!\overset{+}{\underset{..}{O}}\!-\!CH_3$$

$$\qquad\qquad \mathbf{11} \qquad\qquad\qquad\qquad \mathbf{12}$$

Here this carbon atom has only six electrons   Here the carbon atom has eight electrons

   c. Charge separation decreases stability. Separating opposite charges requires energy. Therefore, structures in which opposite charges are separated have greater energy (lower stability) than those that have no charge separation. This means that of the following two structures for vinyl chloride, structure **13** makes a larger contribution because it does not have separated charges. (This does not mean that structure **14** does not contribute to the hybrid, it just means that the contribution made by **14** is smaller.)

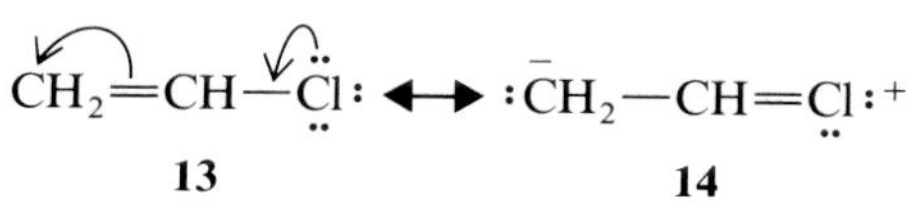

$$CH_2\!=\!CH\!-\!\overset{..}{\underset{..}{Cl}}: \longleftrightarrow :\overset{-}{C}H_2\!-\!CH\!=\!\overset{+}{\underset{..}{Cl}}:$$

$$\qquad\quad \mathbf{13} \qquad\qquad\qquad\qquad \mathbf{14}$$

PROBLEM 12.3

Give the important resonance structures for each of the following:

(a) $CH_2=\overset{\overset{\displaystyle CH_3}{|}}{C}-CH_2\cdot$

(b) $CH_2=CH-\overset{+}{CH}-CH=CH_2$

(c)

(d)

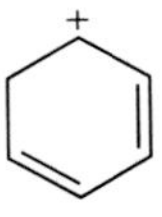

(e) $CH_3CH=CH-CH=\overset{+}{\underset{..}{O}}H$

(f) $CH_2=CH-Br$

(g) 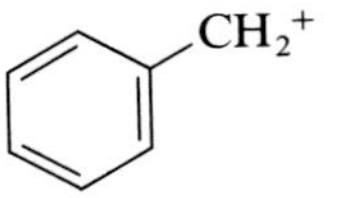 [benzene ring]—CH_2^+

(h) $^-:CH_2-\overset{\overset{\displaystyle O}{||}}{C}-CH_3$

(i) $CH_3-S-CH_2^+$

(j) CH_3-NO_2

PROBLEM 12.4

From each set of resonance structures that follow, designate the one that would contribute most to the hybrid and explain your choice.

(a) $CH_3CH_2\overset{\overset{\displaystyle CH_3}{|}}{C}=CH-CH_2^+ \longleftrightarrow CH_3CH_2\overset{\overset{\displaystyle CH_3}{|}}{\underset{+}{C}}-CH=CH_2$

(b) 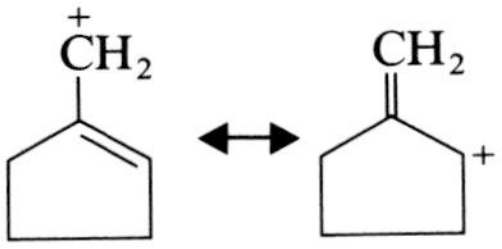 [cyclopentene ring with $\overset{+}{CH_2}$] $\longleftrightarrow$ [cyclopentane ring with CH_2 and +]

(c) $\overset{+}{CH_2}-\overset{..}{N}(CH_3)_2 \longleftrightarrow CH_2=\overset{+}{N}(CH_3)_2$

(d) $CH_3-\overset{\overset{\displaystyle O}{||}}{C}-O-H \longleftrightarrow CH_3-\overset{\overset{\displaystyle O^-}{|}}{C}=\overset{+}{O}-H$

(e) $\overset{.}{C}H_2CH=CHCH=CH_2 \longleftrightarrow CH_2=CH\overset{.}{C}HCH=CH_2 \longleftrightarrow$

$CH_2=CHCH=CH\overset{.}{C}H_2$

(f) $:NH_2-C\equiv N: \longleftrightarrow \overset{+}{N}H_2=C=\overset{..}{N}:^-$

PROBLEM 12.5

The following keto and enol forms differ in the positions for their electrons but they are not resonance structures. Explain why they are not.

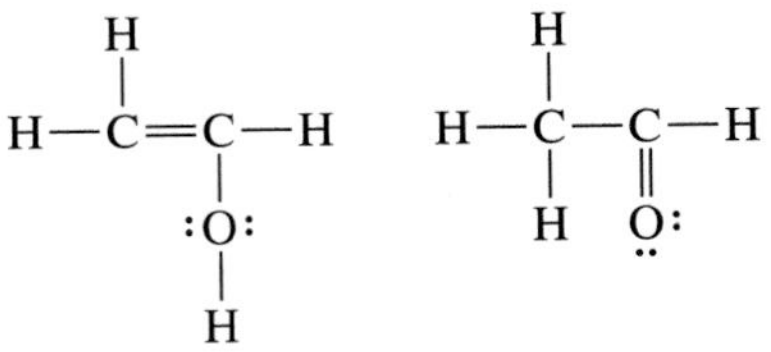

$$H-\overset{\overset{\displaystyle H}{|}}{C}=C-H \qquad H-\overset{\overset{\displaystyle H}{|}}{\underset{\underset{\displaystyle H}{|}}{C}}-\overset{\overset{\displaystyle }{}}{\underset{\underset{\displaystyle \overset{..}{O}:}{||}}{C}}-H$$

Enol form: $H-C=C-H$ with $:\overset{}{O}:$ and H below

Keto form: $H-\overset{H}{\underset{H}{C}}-\overset{O:}{C}-H$

Enol form   Keto form

12.6 ALKADIENES AND POLYUNSATURATED HYDROCARBONS

Many hydrocarbons are known whose molecules contain more than one double or triple bond. A hydrocarbon whose molecules contain two double bonds is called an **alkadiene;** one whose molecules contain three double bonds is called an **alkatriene,** and so on. Colloquially, these compounds are often referred to simply as "dienes" or "trienes." A hydrocarbon with two triple bonds is called an **alkadiyne,** and a hydrocarbon with a double and triple bond is called an **alkenyne.**

The following examples of polyunsaturated hydrocarbons illustrate how specific compounds are named.

$$\overset{1}{C}H_2=\overset{2}{C}=\overset{3}{C}H_2$$

1,2-Propadiene
(allene)

$$\overset{1}{C}H_2=\overset{2}{C}H-\overset{3}{C}H=\overset{4}{C}H_2$$

1,3-Butadiene

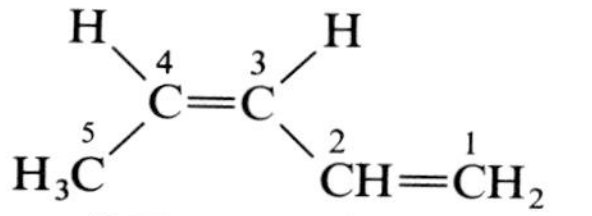

(3Z)-1,3-Pentadiene
(cis-1,3-pentadiene)

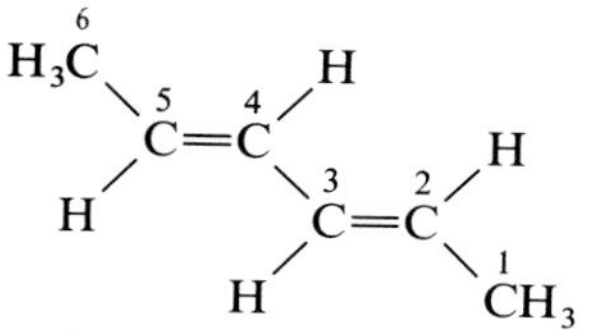

(2E,4E)-2,4-Hexadiene
(trans,trans-2,4-hexadiene)

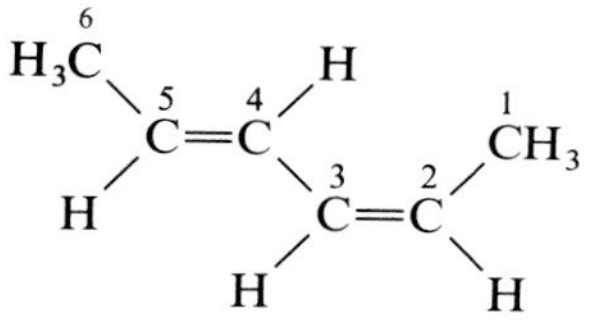

(2Z,4E)-2,4-Hexadiene
(cis,trans-2,4-hexadiene)

$$\overset{5}{H}C\equiv\overset{4}{C}-\overset{3}{C}H_2\overset{2}{C}H=\overset{1}{C}H_2$$

1-Penten-4-yne

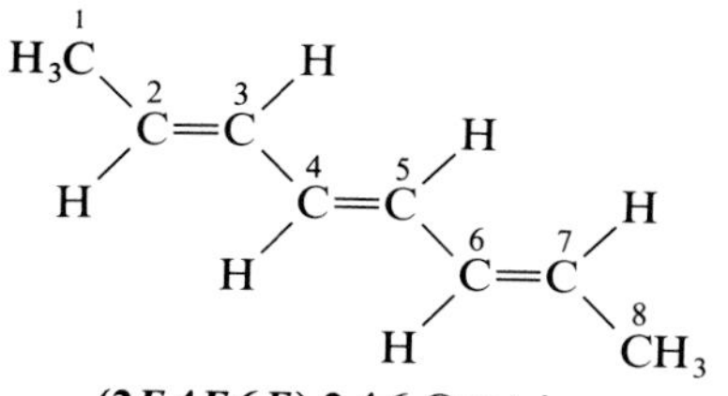

(2E,4E,6E)-2,4,6-Octatriene
(trans,trans,trans-2,4,6-octatriene)

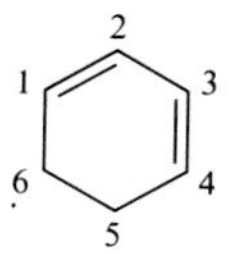

1,3-Cyclohexadiene 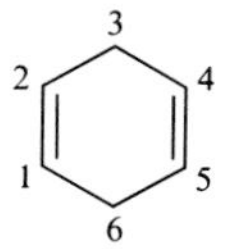   **1,4-Cyclohexadiene**

The multiple bonds of polyunsaturated compounds are classified as being **cumulated, conjugated,** or **isolated.** The double bonds of allene (1,2-propadiene) are said to be cumulated because one carbon (the central carbon) participates in two double bonds. Hydrocarbons whose molecules have cumulated double bonds are called **cumulenes.** The name **allene** (Section 5.17) is also used as a class name for molecules with two cumulated double bonds.

$$CH_2=C=CH_2 \qquad \overset{\diagdown}{\underset{\diagup}{}}C=C=C\overset{\diagup}{\underset{\diagdown}{}}$$

Allene   **A cumulated
diene**

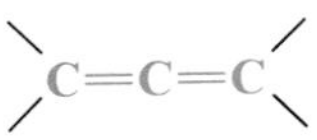

An example of a conjugated diene is 1,3-butadiene. In conjugated polyenes the double and single bonds *alternate* along the chain.

$$CH_2=CH-CH=CH_2 \qquad \overset{\diagdown}{\underset{\diagup}{}}C=C\overset{\diagup}{\underset{\diagdown}{C=C}}\overset{\diagup}{\underset{\diagdown}{}}$$

1,3-Butadiene   **A conjugated diene**

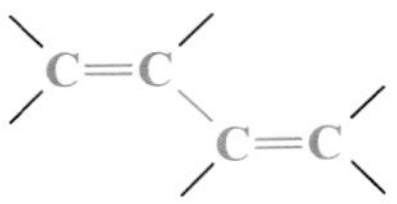

(2*E*,4*E*,6*E*)-2,4,6-Octatriene is an example of a conjugated alkatriene.

If one or more saturated carbon atoms intervene between the double bonds of an alkadiene, the double bonds are said to be *isolated.* An example of an isolated diene is 1,4-pentadiene.

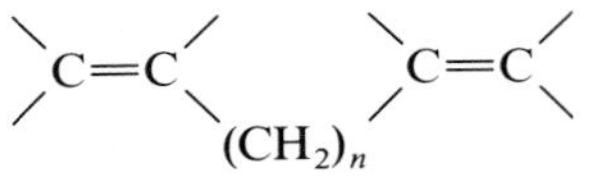

$$\overset{\diagdown}{\underset{\diagup}{}}C=C\overset{\diagup}{\underset{\diagdown}{}} \quad \overset{\diagdown}{\underset{\diagup}{}}C=C\overset{\diagup}{\underset{\diagdown}{}} \qquad CH_2=CH-CH_2-CH=CH_2$$

$$(CH_2)_n$$

An isolated diene   **1,4-Pentadiene**
(*n* ≠ 0)

PROBLEM 12.6

(a) Which other compounds in Section 12.6 are conjugated dienes? (b) Which other compound is an isolated diene? (c) Which compound is an isolated enyne?

In Chapter 5 we saw that appropriately substituted cumulated dienes (allenes) give rise to chiral molecules. Cumulated dienes have had some commercial importance and cumulated double bonds are occasionally found in naturally occurring molecules. In general, cumulated dienes are less stable than isolated dienes.

The double bonds of isolated dienes behave just as their name suggests—as isolated "enes." They undergo all of the reactions of alkenes, and except for the fact that they are capable of reacting twice, their behavior is not unusual. Conjugated dienes are far more interesting because we find that their double bonds interact with each other. This interaction leads to unexpected properties and reactions. We shall therefore consider the chemistry of conjugated dienes in detail.

12.7 1,3-BUTADIENE: ELECTRON DELOCALIZATION

12.7A Bond Lengths of 1,3-Butadiene

The carbon–carbon bond lengths of 1,3-butadiene have been determined and are shown here.

$$\overset{1}{CH_2}=\overset{2}{CH}-\overset{3}{CH}=\overset{4}{CH_2}$$

1.34 Å   1.47 Å   1.34 Å

TABLE 12.1 *Carbon–carbon single bond lengths and hybridization state*

COMPOUND	HYBRIDIZATION STATE	BOND LENGTH (Å)
$H_3C—CH_3$	sp^3-sp^3	1.54
CH_2=$CH—CH_3$	sp^2-sp^3	1.50
CH_2=$CH—CH$=CH_2	sp^2-sp^2	1.47
HC≡$C—CH_3$	sp-sp^3	1.46
HC≡$C—CH$=CH_2	sp-sp^2	1.43
HC≡$C—C$≡CH	sp-sp	1.37

The C1—C2 bond and the C3—C4 bond are (within experimental error) the same length as the carbon–carbon double bond of ethene. The central bond of 1,3-butadiene (1.47 Å), however, is considerably shorter than the single bond of ethane (1.54 Å).

This should not be surprising. All of the carbon atoms of 1,3-butadiene are sp^2 hybridized and, as a result, the central bond of butadiene results from overlapping sp^2 orbitals. And, as we know, a sigma bond that is sp^3-sp^3 is *longer*. There is, in fact, a steady decrease in bond length of carbon–carbon single bonds as the hybridization state of the bonded atoms changes from sp^3 to sp (Table 12.1).

12.7B Conformations of 1,3-Butadiene

There are two possible planar conformations of 1,3-butadiene: the s-cis and the s-trans conformations.

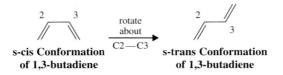

s-cis Conformation of 1,3-butadiene — rotate about C2—C3 → **s-trans Conformation of 1,3-butadiene**

These are not true cis and trans forms since the s-cis and s-trans conformations of 1,3-butadiene can be interconverted through rotation about the single bond (hence the prefix s). The s-trans conformation is the predominant one at room temperature.

12.7C Molecular Orbitals of 1,3-Butadiene

The central carbon atoms of 1,3-butadiene (Fig. 12.4) are close enough for overlap to occur between the p orbitals of C2 and C3. This overlap is not as great as that between

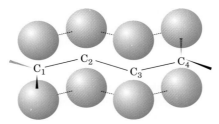

FIGURE 12.4 The p orbitals of 1,3-butadiene.

the orbitals of C1 and C2 (or those of C3 and C4). The C2–C3 orbital overlap, however, gives the central bond partial double bond character and allows the four π electrons of 1,3-butadiene to be delocalized over all four atoms.

Figure 12.5 shows how the four p orbitals of 1,3-butadiene combine to form a set of four π molecular orbitals.

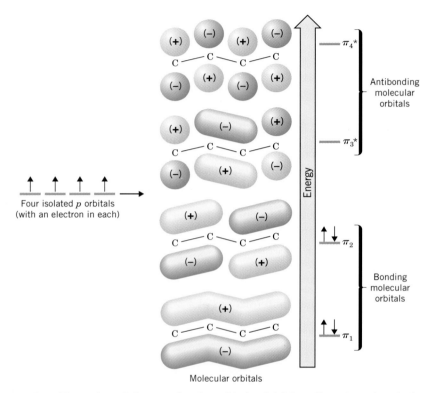

FIGURE 12.5 Formation of the π molecular orbitals of 1,3-butadiene from four isolated p orbitals.

Two of the π molecular orbitals of 1,3-butadiene are bonding molecular orbitals. In the ground state these orbitals hold the four π electrons with two spin-paired electrons in each. The other two π molecular orbitals are antibonding molecular orbitals. In the ground state these orbitals are unoccupied. An electron can be excited from the highest occupied molecular orbital (HOMO) to the lowest unoccupied molecular orbital (LUMO) when 1,3-butadiene absorbs light with a wavelength of 217 nm. (We shall study the absorption of light by unsaturated molecules in Chapter 13.)

The delocalized bonding that we have just described for 1,3-butadiene is characteristic of all conjugated polyenes.

12.8 THE STABILITY OF CONJUGATED DIENES

Conjugated alkadienes are thermodynamically more stable than isomeric isolated alkadienes. Two examples of this extra stability of conjugated dienes can be seen in an analysis of the heats of hydrogenation given in Table 12.2.

TABLE 12.2 Heats of hydrogenation of alkenes and alkadienes

COMPOUND	H_2 (mol)	$\Delta H°$ (kcal mol^{-1})	(kJ mol^{-1})
1-Butene	1	− 30.3	− 126.8
1-Pentene	1	− 30.1	− 125.9
trans-2-Pentene	1	− 27.6	− 115.5
1,3-Butadiene	2	− 57.1	− 238.9
trans-1,3-Pentadiene	2	− 54.1	− 226.4
1,4-Pentadiene	2	− 60.8	− 254.4
1,5-Hexadiene	2	− 60.5	− 253.1

In itself, 1,3-butadiene cannot be compared directly with an isolated diene of the same chain length. However, a comparison can be made between the heat of hydrogenation of 1,3-butadiene and that obtained when two molar equivalents of 1-butene are hydrogenated.

$$\Delta H° \text{ (kcal mol}^{-1})$$

$$2 \ CH_2{=}CHCH_2CH_3 + 2 \ H_2 \longrightarrow 2 \ CH_3CH_2CH_2CH_3 \ (2 \times -30.3) = -60.6$$
1-Butene

$$CH_2{=}CHCH{=}CH_2 + 2 \ H_2 \longrightarrow CH_3CH_2CH_2CH_3 \qquad = \underline{-57.1}$$
1,3-Butadiene \qquad\qquad\qquad\qquad\qquad Difference \qquad 3.5 kcal mol^{-1}

Because 1-butene has the same kind of monosubstituted double bond as either of those in 1,3-butadiene, we might expect hydrogenation of 1,3-butadiene to liberate the same amount of heat (-60.6 kcal mol^{-1}) as two molar equivalents of 1-butene. We find, however, that 1,3-butadiene liberates only 57.1 kcal mol^{-1}, 3.5 kcal mol^{-1} *less* than expected. We conclude, therefore, that conjugation imparts some extra stability to the conjugated system (Fig. 12.6).

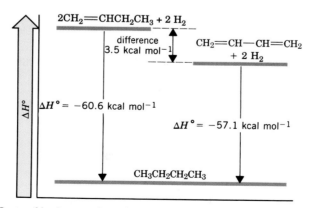

FIGURE 12.6 Heats of hydrogenation of 2 mol of 1-butene and 1 mol of 1,3-butadiene.

An assessment of the stabilization that conjugation provides *trans*-1,3-pentadiene can be made by comparing the heat of hydrogenation of *trans*-1,3-pentadiene to the sum of the heats of hydrogenation of 1-pentene and *trans*-2-pentene. This way we are comparing double bonds of comparable types.

$$CH_2{=}CHCH_2CH_2CH_3$$
1-Pentene

$\Delta H° = -30.1 \text{ kcal mol}^{-1}$

CH_3CH_2 ⟍   ⟋ H
        $C{=}C$
H ⟋   ⟍ CH_3

***trans*-2-Pentene**

$\Delta H° = -27.6 \text{ kcal mol}^{-1}$
$\overline{\text{Sum} = -57.7 \text{ kcal mol}^{-1}}$

$CH_2{=}CH$ ⟍   ⟋ H
         $C{=}C$
H ⟋   ⟍ CH_3

***trans*-1,3-Pentadiene**

$\Delta H° = -54.1 \text{ kcal mol}^{-1}$
$\overline{\text{Difference} = \quad 3.6 \text{ kcal mol}^{-1}}$

We see from these calculations that conjugation affords *trans*-1,3-pentadiene an extra stability of 3.6 kcal mol^{-1}, a value that is very close to the one we obtained for 1,3-butadiene (3.5 kcal mol^{-1}).

When calculations like these are carried out for other conjugated dienes, similar results are obtained; *conjugated dienes are found to be more stable than isolated dienes.* The question, then, is this: What is the source of the extra stability associated with conjugated dienes? There are two factors that contribute. The extra stability of conjugated dienes arises in part from the stronger central bond that they contain, and in part from the additional delocalization of the π electrons that occurs in conjugated dienes.

12.9 ELECTROPHILIC ATTACK ON CONJUGATED DIENES: 1,4 ADDITION

Not only are conjugated dienes somewhat more stable than nonconjugated dienes, they also display special behavior when they react with electrophilic reagents. For example, 1,3-butadiene reacts with one molar equivalent of hydrogen chloride to produce two products: 3-chloro-1-butene and 1-chloro-2-butene.

$$CH_2{=}CH{-}CH{=}CH_2 \xrightarrow[25°C]{HCl} CH_3{-}\underset{\underset{Cl}{|}}{CH}{-}CH{=}CH_2 + CH_3{-}CH{=}CH{-}CH_2Cl$$

1,3-Butadiene   **3-Chloro-1-butene**   **1-Chloro-2-butene**
                                    (78%)   **(22%)**

If only the first product (3-chloro-1-butene) were formed, we would not be particularly surprised. We would conclude that hydrogen chloride had added to one double bond of 1,3-butadiene in the usual way.

$$\overset{1}{C}H_2{=}\overset{2}{C}H{-}\overset{3}{C}H{=}\overset{4}{C}H_2 \xrightarrow{1,2 \text{ addition}} \underset{\underset{H}{|}}{C}H_2{-}\underset{\underset{Cl}{|}}{C}H{-}CH{=}CH_2$$
$$+$$
$$H{-}Cl$$

3-Chloro-1-butene

It is the second product, 1-chloro-2-butene, that is unusual. Its double bond is between the central atoms, and the elements of hydrogen chloride have added to the C1 and C4 atoms.

$$\overset{1}{CH_2}=\overset{2}{CH}-\overset{3}{CH}=\overset{4}{CH_2} \xrightarrow{\text{1,4 addition}} CH_2-CH=CH-CH_2$$
$$+ \qquad\qquad\qquad\qquad\quad | \qquad\qquad\qquad\quad |$$
$$H-Cl \qquad\qquad\qquad\qquad\quad H \qquad\qquad\qquad\quad Cl$$

1-Chloro-2-butene

This unusual behavior of 1,3-butadiene can be attributed directly to the stability and the delocalized nature of an allylic cation (Section 12.4). In order to see this, consider a mechanism for the addition of hydrogen chloride.

Step 1

$$H^+ + CH_2=CH-CH=CH_2 \longrightarrow CH_3-\underset{+}{CH}-CH=CH_2 \longleftrightarrow CH_3-CH=CH-\underset{+}{CH_2}$$

**An allylic cation
equivalent to**

$$CH_3-\underset{\delta+}{CH}\text{=}\text{=}\text{=}CH\text{=}\text{=}\text{=}\underset{\delta+}{CH_2}$$

Step 2 $$CH_3\underset{\delta+}{CH}\text{=}\text{=}\text{=}CH\text{=}\text{=}\text{=}\underset{\delta+}{CH_2} + \overset{..}{\underset{..}{Cl}}:^-$$

(a) → $CH_3CH-CH=CH_2$ **1,2 Addition**
         |
         Cl

(b) → $CH_3CH=CHCH_2Cl$ **1,4 Addition**

In step 1 a proton adds to one of the terminal carbon atoms of 1,3-butadiene to form, as usual, the more stable carbocation, in this case a resonance-stabilized allylic cation. Addition to one of the inner carbon atoms would have produced a much less stable primary cation, one that could not be stabilized by resonance.

$$CH_2=CH-CH=CH_2 \xmapsto{} {}^+CH_2-CH_2-CH=CH_2$$
$$\searrow H^+ \qquad\qquad\qquad \textbf{A 1}° \textbf{ carbocation}$$

In step 2 a chloride ion forms a bond to one of the carbon atoms of the allylic cation that bears a partial positive charge. Reaction at one carbon atom results in the 1,2-addition product; reaction at the other gives the 1,4-addition product.

PROBLEM 12.7 _____

(a) What products would you expect to obtain if hydrogen chloride were allowed to react with a 2,4-hexadiene, $CH_3CH=CHCH=CHCH_3$? (b) With 1,3-pentadiene, $CH_2=CHCH=CHCH_3$? (Neglect cis–trans isomerism.)

1,3-Butadiene shows 1,4-addition reactions with electrophilic reagents other than hydrogen chloride. Two examples are shown here, the addition of hydrogen bromide (in the absence of peroxides) and the addition of bromine.

$$CH_2=CHCH=CH_2 \xrightarrow[40°C]{HBr} CH_3CHBrCH=CH_2 + CH_3CH=CHCH_2Br$$
$$(20\%) \qquad\qquad (80\%)$$

$$CH_2=CHCH=CH_2 \xrightarrow[-15°C]{Br_2} CH_2BrCHBrCH=CH_2 + CH_2BrCH=CHCH_2Br$$
$$(54\%) \qquad\qquad (46\%)$$

Reactions of this type are quite general with other conjugated dienes. Conjugated trienes often show 1,6 addition. An example is the 1,6 addition of bromine to 1,3,5-cyclooctatriene:

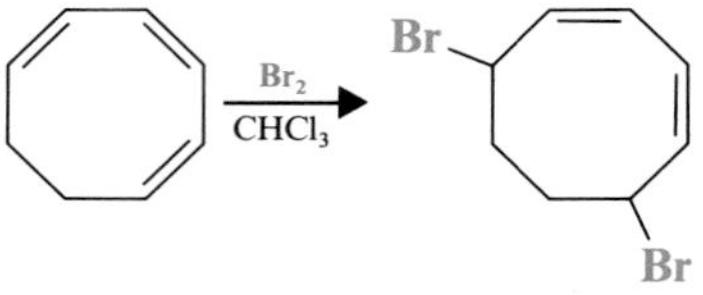

$$(>68\%)$$

12.9A Kinetic Control Versus Thermodynamic Control of a Chemical Reaction

The addition of hydrogen bromide to 1,3-butadiene is interesting in another respect. The relative amounts of 1,2- and 1,4-addition products that we obtain are dependent on the temperature at which we carry out the reaction.

When 1,3-butadiene and hydrogen bromide react at a low temperature ($-80°C$) in the absence of peroxides, the major reaction is 1,2 addition; we obtain about 80% of the 1,2 product and only about 20% of the 1,4 product. At a higher temperature ($40°C$) the result is reversed. The major reaction is 1,4 addition; we obtain about 80% of the 1,4 product and only about 20% of the 1,2 product.

When the mixture formed at the lower temperature is brought to the higher temperature, moreover, the relative amounts of the two products change. This new reaction mixture eventually contains the same proportion of products given by the reaction carried out at the higher temperature.

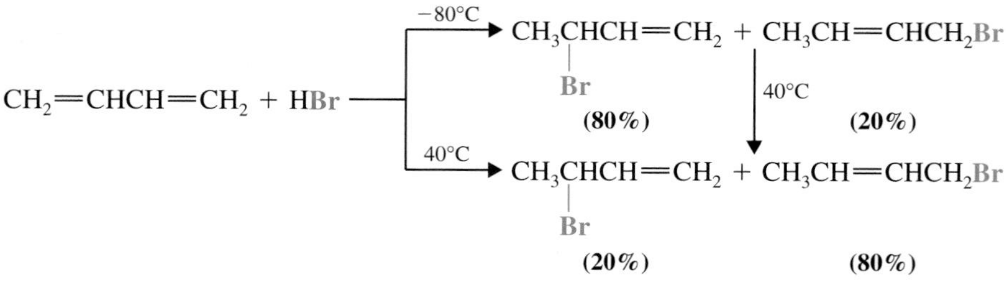

It can also be shown that at the higher temperature and in the presence of hydrogen bromide, the 1,2-addition product rearranges to the 1,4 product and that an equilibrium exists between them.

$$CH_3CHCH\!=\!CH_2 \xrightarrow{\text{40°C, HBr}} CH_3CH\!=\!CHCH_2Br$$

Br

1,2-Addition
product

1,4-Addition
product

Because this equilibrium favors the 1,4-addition product, *it must be more stable.*

The reactions of hydrogen bromide with 1,3-butadiene serve as a striking illustration of the way that the outcome of a chemical reaction can be determined, in one instance, by relative rates of competing reactions and, in another, by the relative stabilities of the final products. At the lower temperature, the relative amounts of the products of the addition are determined by the relative rates at which the two additions occur; 1,2 addition occurs faster so the 1,2-addition product is the major product. At the higher temperature, the relative amounts of the products are determined by the position of an equilibrium. The 1,4-addition product is the more stable, so it is the major product.

This behavior of 1,3-butadiene and hydrogen bromide can be more fully understood if we examine the diagram shown in Fig. 12.7.

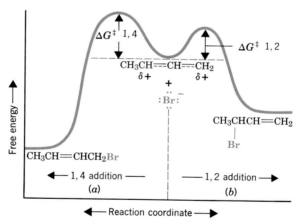

FIGURE 12.7 Free energy versus reaction coordinate diagram for the reactions of an allylic cation with a bromide ion. One reaction pathway (a) leads to the 1,4-addition product and the other (b) leads to the 1,2-addition product.

The step that determines the overall outcome of the reaction is the step in which the hybrid allylic cation combines with a bromide ion, that is,

$$CH_2\!=\!CH\!-\!CH\!=\!CH_2 \xrightarrow{H^+}$$

$$CH_3\!-\!\overset{\delta+}{CH}\!=\!=\!CH\!=\!=\!\overset{\delta+}{CH_2}$$

$$\xrightarrow{Br^-} CH_3\!-\!\underset{Br}{CH}\!-\!CH\!=\!CH_2$$

1,2 Product

$$\xrightarrow{Br^-} CH_3\!-\!CH\!=\!CH\!-\!CH_2Br$$

1,4 Product

{ This step determines the regioselectivity of the reaction

We see in Fig. 12.7 that, for this step, the free energy of activation leading to the 1,2-addition product is less than the free energy of activation leading to the 1,4-addition product, even though the 1,4 product is more stable. At low temperatures, a larger fraction of collisions between the intermediate ions has enough energy to cross the lower barrier (leading to the 1,2 product), and only a very small fraction of collisions has enough energy to cross the higher barrier (leading to the 1,4 product). In either case (and this is the *key point*), whichever barrier is crossed, product formation is *irreversible* because there is not enough energy available to lift either product out of its deep potential energy valley. Since 1,2 addition occurs faster, the 1,2 product predominates and the reaction is said to be under **kinetic control** or **rate control.**

At higher temperatures, the intermediate ions have sufficient energy to cross both barriers with relative ease. More importantly, however, *both reactions are reversible.* Sufficient energy is also available to take the products back over their energy barriers to the intermediate level of allylic cations and bromide ions. The 1,2 product is still formed faster, but being less stable than the 1,4 product, it also reverts to the allylic cation faster. Under these conditions, that is, at higher temperatures, the relative proportions of the products *do not reflect* the relative heights of the energy barriers leading from allylic cation to products. Instead, *they reflect the relative stabilities of the products themselves.* Since the 1,4 product is more stable, it is formed at the expense of the 1,2 product because the overall change from 1,2 product to 1,4 product is energetically favored. Such a reaction is said to be under **thermodynamic control** or **equilibrium control.**

Before we leave this subject one final point should be made. This example clearly demonstrates that predictions of relative reaction rates made on the basis of product stabilities alone can be wrong. This is not always the case, however. For many reactions in which a common intermediate leads to two or more products, the most stable product is formed fastest.

PROBLEM 12.8 _____

(a) Can you suggest a possible explanation for the fact that the 1,2-addition reaction of 1,3-butadiene and hydrogen bromide occurs faster than 1,4 addition? (*Hint:* Consider the relative contributions that the two forms $CH_3\overset{+}{C}HCH=CH_2$ and $CH_3CH=CH\overset{+}{C}H_2$ make to the resonance hybrid of the allylic cation.)   (b) How can you account for the fact that the 1,4-addition product is more stable?

12.10 THE DIELS–ALDER REACTION: A 1,4-CYCLOADDITION REACTION OF DIENES

In 1928 two German chemists, Otto Diels and Kurt Alder, developed a 1,4-cycloaddition reaction of dienes that has since come to bear their names. The reaction proved to be one of such great versatility and synthetic utility that Diels and Alder were awarded the Nobel Prize for Chemistry in 1950.

An example of the Diels–Alder reaction is the reaction that takes place when 1,3-butadiene and maleic anhydride are heated together at 100°C. The product is obtained in quantitative yield.

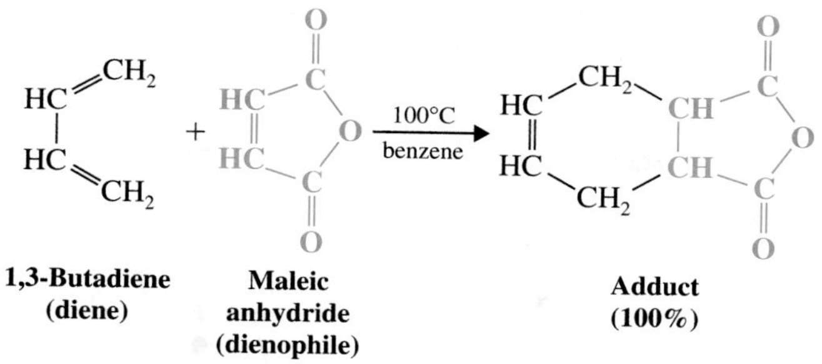

1,3-Butadiene   **Maleic**   **Adduct**
(diene)   **anhydride**   **(100%)**
              (dienophile)

or

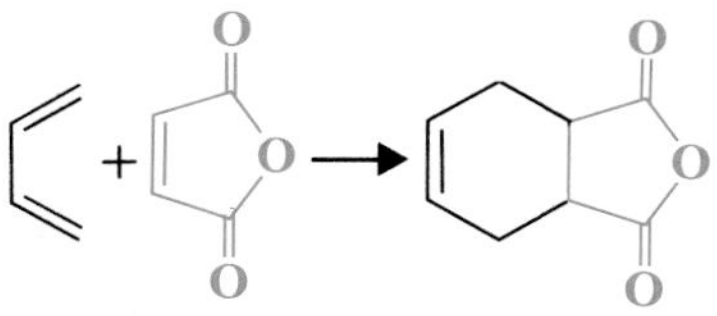

In general terms, the reaction is one between a conjugated **diene** (a 4π-electron system) and a compound containing a double bond (a 2π-electron system) called a **dienophile** (diene + Greek: *philein*, to love). The product of a Diels–Alder reaction is often called an **adduct.** In the Diels–Alder reaction, two new σ bonds are formed at the expense of two π bonds of the diene and dienophile. Since σ bonds are usually stronger than π bonds, formation of the adduct is usually favored energetically, *but most Diels–Alder reactions are reversible.*

We can account for all of the bond changes in a Diels–Alder reaction by using curved arrows in the following way:

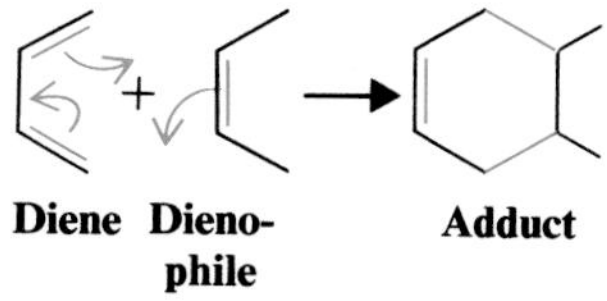

Diene Dieno-   **Adduct**
       phile

We do not intend to imply a mechanism by the use of these curved arrows; they are used only to account for the electrons.

The simplest example of a Diels–Alder reaction is the one that takes place between 1,3-butadiene and ethene. This reaction, however, takes place much more slowly than the reaction of butadiene with maleic anhydride and must also be carried out under pressure.

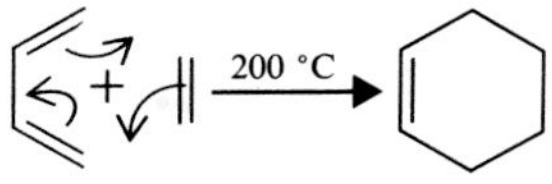

(20%)

Alder originally stated that the Diels–Alder reaction is favored by the presence of electron-withdrawing groups in the dienophile and by electron-releasing groups in the diene. Maleic anhydride, a very potent dienophile, has two electron-withdrawing carbonyl groups on carbon atoms adjacent to the double bond.

The helpful effect of electron-releasing groups in the diene can also be demonstrated; 2,3-dimethyl-1,3-butadiene, for example, is nearly five times as reactive in Diels–Alder reactions as is 1,3-butadiene. When 2,3-dimethyl-1,3-butadiene reacts with propenal (acrolein) at only 30°C, the adduct is obtained in quantitative yield.

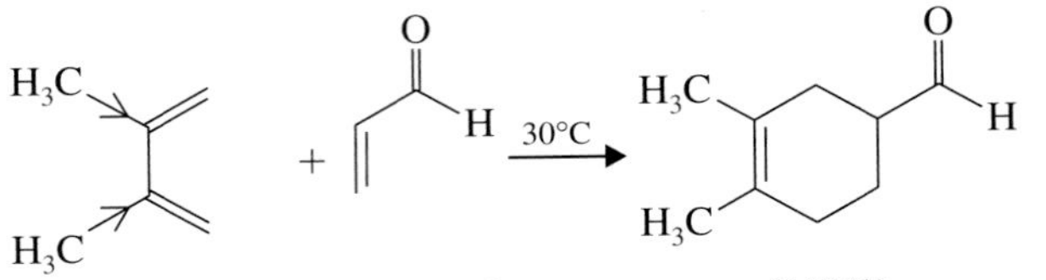

2,3-Dimethyl-1,3-   Propenal   (100%)
butadiene

Research (by C. K. Bradsher of Duke University) has shown that the locations of electron-withdrawing and electron-releasing groups in the dienophile and diene can be reversed without reducing the yields of the adducts. Dienes with electron-withdrawing groups have been found to react readily with dienophiles containing electron-releasing groups. Additional facts about the reaction are these.

1. **The Diels–Alder reaction is highly stereospecific: The reaction is a *syn* addition and the configuration of the dienophile is *retained* in the product.** Two examples that illustrate this aspect of the reaction are shown here.

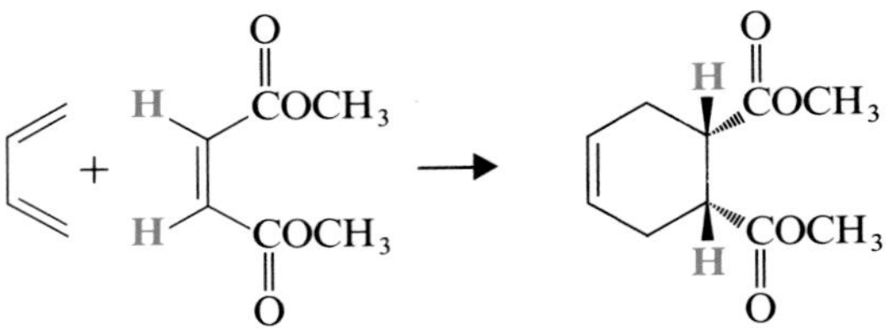

Dimethyl maleate   Dimethyl 4-cyclohexene-*cis*-
(a *cis*-dienophile)   1,2-dicarboxylate

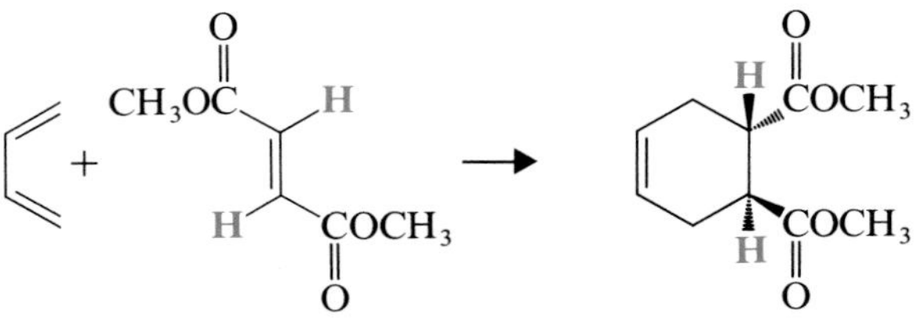

Dimethyl fumarate   Dimethyl 4-cyclohexene-*trans*-
(a *trans*-dienophile)   1,2-dicarboxylate

In the first example, a dienophile with cis ester groups reacts with 1,3-butadiene to give an adduct with cis ester groups. In the second example just the reverse is true. A *trans* dienophile gives a trans adduct.

2. **The diene, of necessity, reacts in the s-cis conformation rather than the s-trans.**

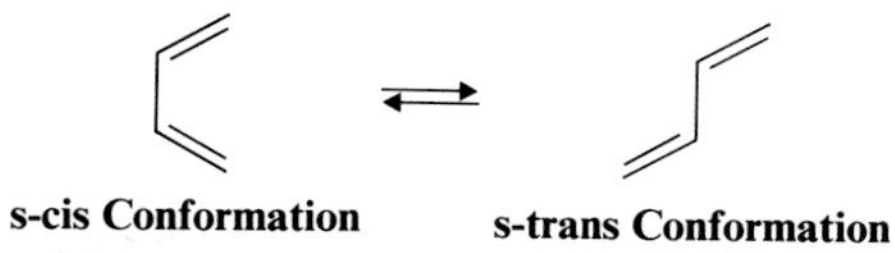

s-cis Conformation   s-trans Conformation

Reaction in the s-trans conformation would, if it occurred, produce a six-membered ring with a highly strained trans double bond. This course of the Diels–Alder reaction has never been observed.

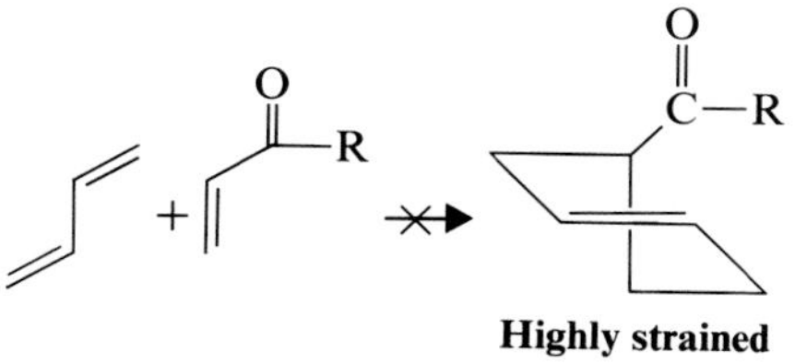

Highly strained

Cyclic dienes in which the double bonds are held in the s-cis configuration are usually highly reactive in the Diels–Alder reaction. Cyclopentadiene, for example, reacts with maleic anhydride at room temperature to give the following adduct in quantitative yield.

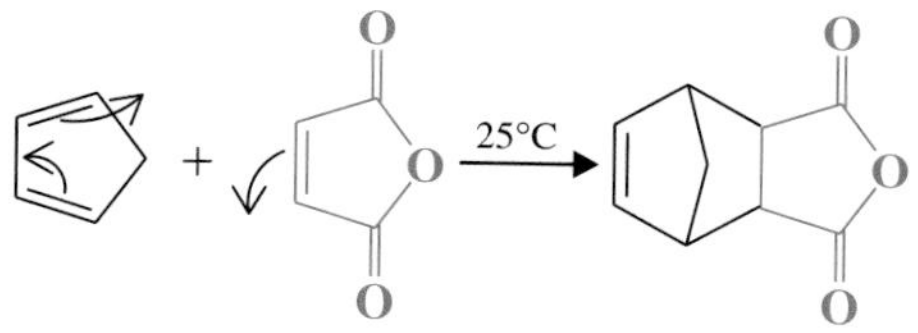

Cyclopentadiene is so reactive that on standing at room temperature it slowly undergoes a Diels–Alder reaction with itself.

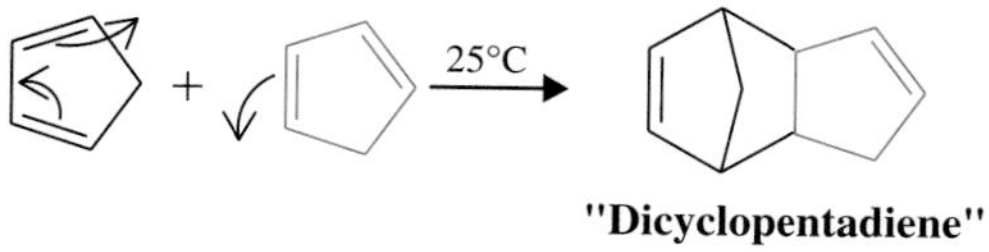

"Dicyclopentadiene"

The reaction is reversible, however. When "dicyclopentadiene" is distilled, it dissociates into two molar equivalents of cyclopentadiene.

The reactions of cyclopentadine illustrate a third stereochemical characteristic of the Diels–Alder reaction.

3. **The Diels–Alder reaction occurs primarily in an *endo* rather than an *exo* fashion when the reaction is kinetically controlled** (cf. Problem 12.25). Endo and exo are terms used to designate the stereochemistry of bridged rings such as bicyclo[2.2.1]heptane. The point of reference is the longest bridge. A group that is

anti to the longest bridge (the two-carbon bridge) is said to be exo; if it is on the same side, it is endo.*

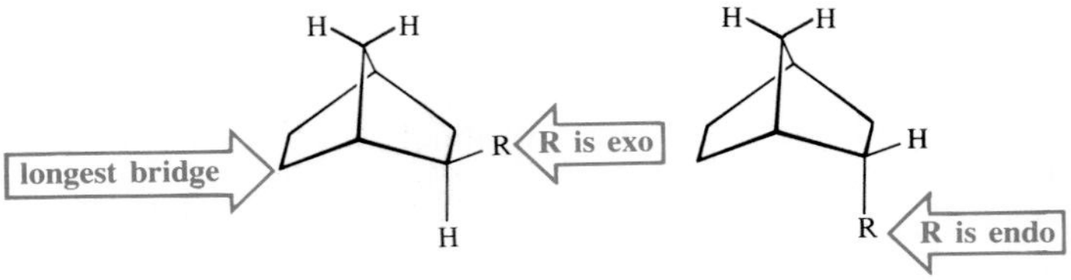

In the Diels–Alder reaction of cyclopentadiene with maleic anhydride the major product is the one in which the anhydride linkage, $-\overset{\displaystyle ||}{\underset{\displaystyle O}{C}}-O-\overset{\displaystyle ||}{\underset{\displaystyle O}{C}}-$, has assumed the endo configuration. See the following illustration. This favored endo stereochemistry seems to arise from favorable interactions between the π electrons of the developing double bond in the diene and the π electrons of unsaturated groups of the dienophile. In this example, the π electrons of the $-\overset{\displaystyle ||}{\underset{\displaystyle O}{C}}-O-\overset{\displaystyle ||}{\underset{\displaystyle O}{C}}-$ linkage of the anhydride interact with the π electrons of the developing double bond in cyclopentadiene.

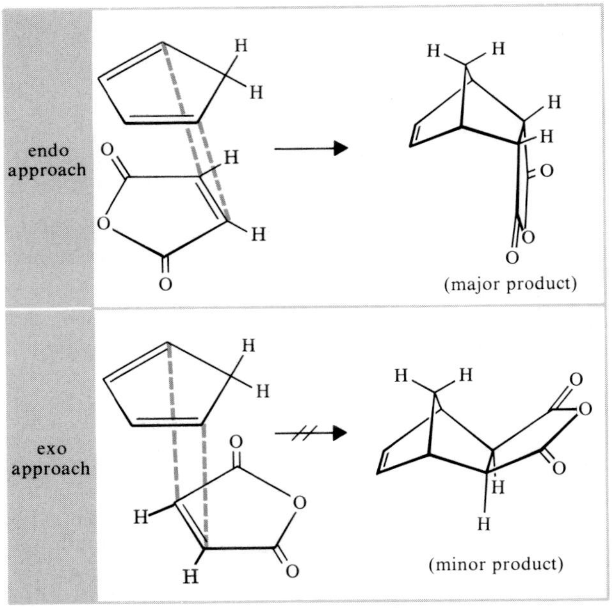

* In general, the exo substituent is always on the side anti to the *longer* bridge of a bicyclic structure (exo, outside; endo, inside). For example,

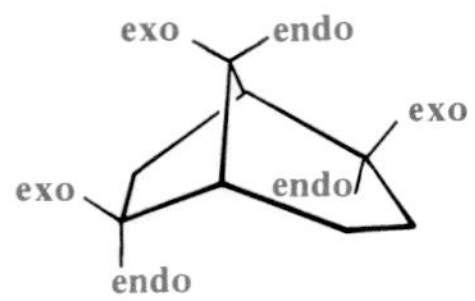

PROBLEM 12.9

The dimerization of cyclopentadiene also occurs in an endo way.   (a) Show how this happens.   (b) Which π electrons interact?   (c) What is the three-dimensional structure of the product?

PROBLEM 12.10

What products would you expect from the following reactions?

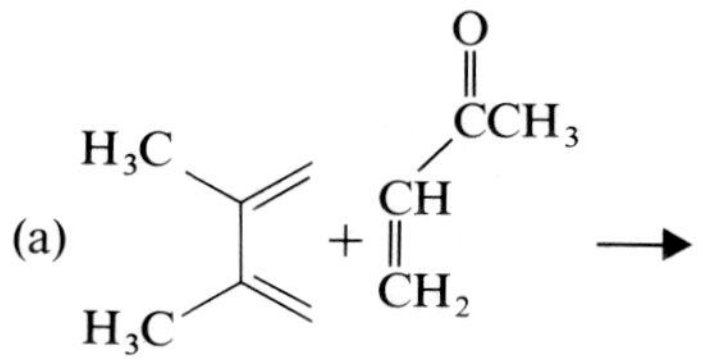

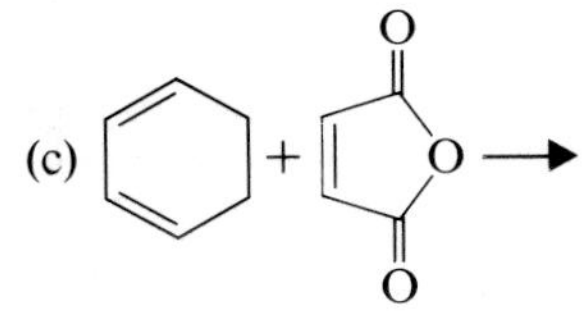

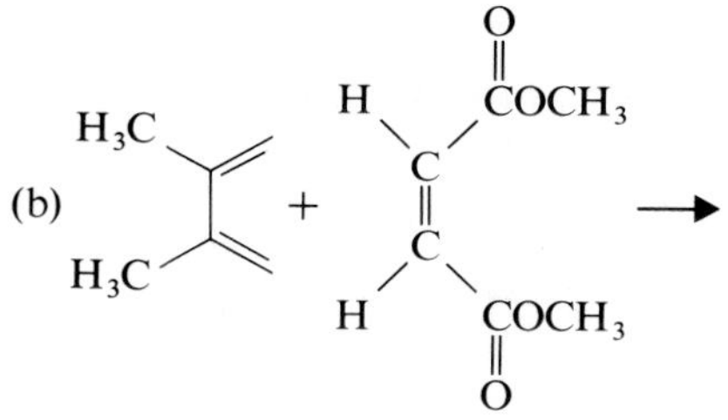

(a)

(c)

(b)

PROBLEM 12.11

Which diene and dienophile would you employ to synthesize the following compound?

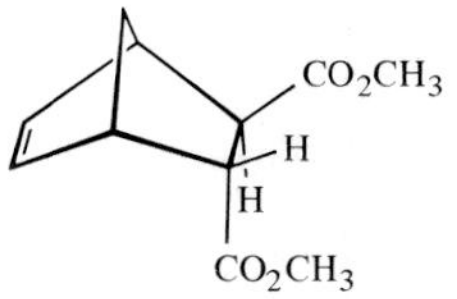

PROBLEM 12.12

Diels–Alder reactions also take place with triple-bonded (acetylenic) dienophiles. Which diene and which dienophile would you use to prepare:

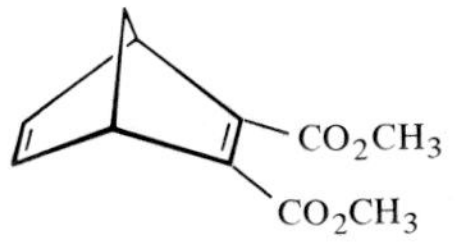

ADDITIONAL PROBLEMS

12.13 Outline a synthesis of 1,3-butadiene starting from

(a) 1,4-Dibromobutane

(b) $HOCH_2(CH_2)_2CH_2OH$

(c) CH_2=$CHCH_2CH_2OH$

(d) CH_2=$CHCH_2CH_2Cl$

(e) CH_2=$CHCHClCH_3$

(f) CH_2=$CHCH(OH)CH_3$

(g) HC≡CCH=CH_2

12.14 What product would you expect from the following reaction?

$$(CH_3)_2\underset{\underset{Cl}{|}}{C}-\underset{\underset{Cl}{|}}{C}(CH_3)_2 + 2\ KOH \xrightarrow[\text{heat}]{\text{ethanol}}$$

12.15 What products would you expect from the reaction of 1 mol of 1,3-butadiene and each of the following reagents? (If no reaction would occur, you should indicate that as well.)

(a) One mole of Cl_2

(b) Two moles of Cl_2

(c) Two moles of Br_2

(d) Two moles of H_2, Ni

(e) $Ag(NH_3)_2^+OH^-$

(f) One mole of Cl_2 in H_2O

(g) Hot $KMnO_4$

(h) H^+, H_2O

12.16 Show how you might carry out each of the following transformations. (In some transformations several steps may be necessary.)

(a) 1-Butene $\longrightarrow$ 1,3-butadiene

(b) 1-Pentene $\longrightarrow$ 1,3-pentadiene

(c) $CH_3CH_2CH_2CH_2OH \longrightarrow CH_2BrCH$=$CHCH_2Br$

(d) CH_3CH=$CHCH_3 \longrightarrow CH_3CH$=$CHCH_2Br$

(f)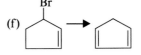

(e)

12.17 Conjugated dienes react with free radicals by both 1,2 and 1,4 addition. Account for this fact by using the peroxide-promoted addition of one molar equivalent of HBr to 1,3-butadiene as an illustration.

12.18 Outline a simple chemical test that would distinguish between the members of each of the following pairs of compounds.

(a) 1,3-Butadiene and 1-butyne

(b) 1,3-Butadiene and butane

(c) Butane and CH_2=$CHCH_2CH_2OH$

(d) 1,3-Butadiene and CH_2=$CHCH_2CH_2Br$

(e) CH_2BrCH=$CHCH_2Br$ and CH_3CBr=$CBrCH_3$

12.19 (a) The hydrogen atoms attached to C3 of 1,4-pentadiene are unusually susceptible to abstraction by radicals. How can you account for this? (b) Can you also provide an explanation for the fact that the protons attached to C3 of 1,4-pentadiene are more acidic than the methyl hydrogen atoms of propene?

12.20 When 2-methyl-1,3-butadiene (isoprene) undergoes a 1,4 addition of hydrogen chloride, the major product that is formed is 1-chloro-3-methyl-2-butene. Little or no 1-chloro-2-methyl-2-butene is formed. How can you explain this?

12.21   Which diene and dienophile would you employ in a synthesis of each of the following?

(a)

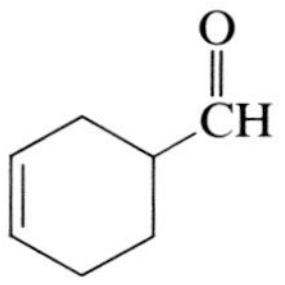

(b)

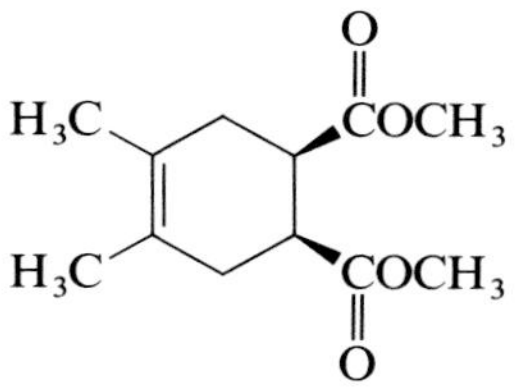

(c)

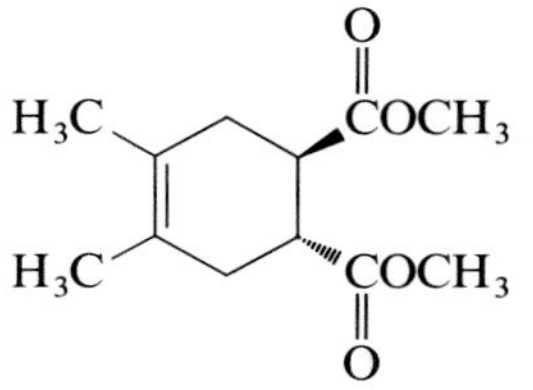

(d)

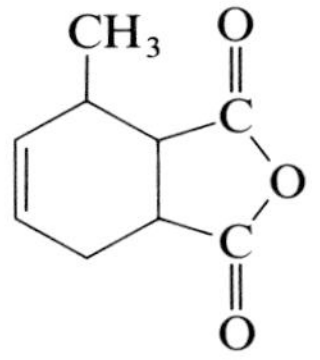

(e)

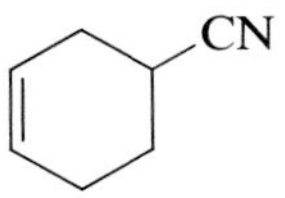

(f)

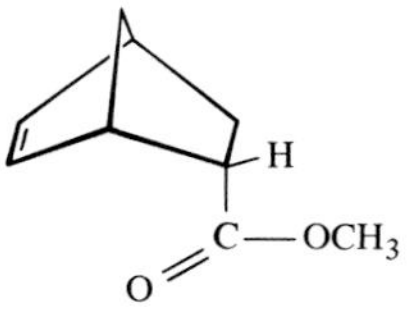

12.22   Account for the fact that neither of the following compounds undergoes a Diels–Alder reaction with maleic anhydride.

$$HC \equiv C - C \equiv CH \quad \text{or} \quad \bigcirc = CH_2$$

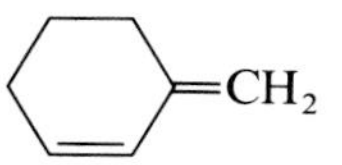

12.23   Acetylenic compounds may be used as dienophiles in the Diels–Alder reaction (cf. Problem 12.12). Write structures for the adducts that you expect from the reaction of 1,3-butadiene with:

(a) $CH_3OCC \equiv CCOCH_3$ (dimethyl acetylenedicarboxylate)

(b) $CF_3C \equiv CCF_3$ (hexafluoro-2-butyne)

12.24   Cyclopentadiene undergoes a Diels–Alder reaction with ethene at 160–180°C. Write the structure of the product of this reaction.

12.25   When furan and maleimide undergo a Diels–Alder reaction at 25°C, the major product is the endo adduct **G**. When the reaction is carried out at 90°C, however, the major product is the exo isomer **H**. The endo adduct isomerizes to the exo adduct when it is heated to 90°C. Propose an explanation that will account for these results.

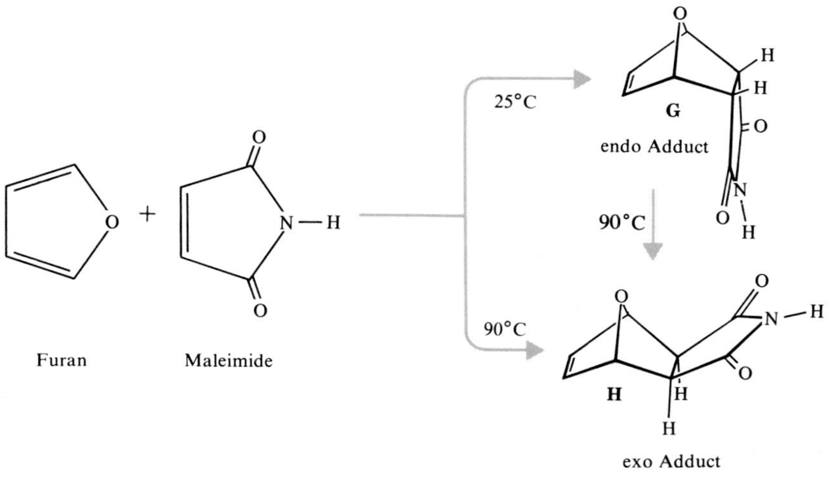

Furan   Maleimide

endo Adduct

exo Adduct

12.26   Two controversial "hard" insecticides are aldrin and dieldrin (see following diagram). [The Environmental Protection Agency (EPA) halted the use of these insecticides because of possible harmful side effects and because they are not biodegradable.] The commercial synthesis of aldrin begins with hexachlorocyclopentadiene and norbornadiene. Dieldrin is synthesized from aldrin. Show how these syntheses might be carried out.

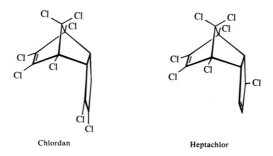

12.27   (a) Norbornadiene for the aldrin synthesis (Problem 12.26) can be prepared from cyclopentadiene and acetylene. Show the reaction involved. (b) It can also be prepared by allowing cyclopentadiene to react with vinyl chloride and treating the product with base. Outline this synthesis.

12.28   Two other hard insecticides (cf. Problem 12.26) are chlordan and heptachlor. Their commercial syntheses begin with cyclopentadiene and hexachlorocyclopentadiene. Show how these syntheses might be carried out.

Chlordan

Heptachlor

12.29   Isodrin, an isomer of aldrin, is obtained when cyclopentadiene reacts with the hexachloronorbornadiene, shown here. Propose a structure for isodrin.

+ ⟶ isodrin

12.30 When $CH_3CH=CHCH_2OH$ is treated with concentrated HCl, two products are produced, $CH_3CH=CHCH_2Cl$ and $CH_3CHClCH=CH_2$. Outline a mechanism that explains this.

12.31 When a solution of 1,3-butadiene in CH_3OH is treated with chlorine, the products are $ClCH_2CH=CHCH_2OCH_3$ (30%) and $ClCH_2CHCH=CH_2$ (70%). Write a mechanism that accounts for their formation.
$$\quad\quad\quad\quad\quad\quad\quad\quad\quad\quad\quad\quad\quad\quad\quad\quad\quad\quad\quad\quad | $$
$$\quad\quad\quad\quad\quad\quad\quad\quad\quad\quad\quad\quad\quad\quad\quad\quad\quad\quad OCH_3$$

12.32 Dehydrohalogenation of *vic*-dihalides (with the elimination of two molar equivalents of HX) normally leads to an alkyne rather than to a conjugated diene. However, when 1,2-dibromocyclohexane is dehydrohalogenated, 1,3-cyclohexadiene is produced in good yield. What factor accounts for this?

12.33 When 1-pentene reacts with *N*-bromosuccinimide, two products with the formula C_5H_9Br are obtained. What are these products and how are they formed?

12.34 Treating either 1-chloro-3-methyl-2-butene or 3-chloro-3-methyl-1-butene with Ag_2O in water gives (in addition to AgCl) the same mixture of alcohols: $(CH_3)_2C=CHCH_2OH$ (15%) and $(CH_3)_2CCH=CH_2$ (85%). (a) Write a mechanism that accounts for the formation
$$\quad\quad\quad\quad\quad\quad\quad\quad\quad | $$
$$\quad\quad\quad\quad\quad\quad\quad\quad\quad OH$$
of these products. (b) What might explain the relative proportions of the two alkenes that are formed?

12.35 The heat of hydrogenation of allene is 71.3 kcal mol^{-1}, whereas that of propyne is 69.3 kcal mol^{-1}. (a) Which compound is more stable? (b) Treating allene with a strong base causes it to isomerize to propyne. Explain.

12.36 Mixing furan (Problem 12.25) with maleic anhydride in diethyl ether yields a crystalline solid with a melting point of 125°C. When melting of this compound takes place, however, one can notice that the melt evolves a gas. If the melt is allowed to resolidify, one finds that it no longer melts at 125°C but instead it melts at 56°C. Consult an appropriate chemistry handbook and provide an explanation for what is taking place.

12.37 Give the structures of the products that would be formed when 1,3-butadiene reacts with each of the following:

(a) (E)-$CH_3CH=CHCO_2CH_3$ (c) (E)-$CH_3CH=CHCN$

(b) (Z)-$CH_3CH=CHCO_2CH_3$ (d) (Z)-$CH_3CH=CHCN$

12.38 Although both butyl bromide and 4-bromo-1-butene are primary halides, the latter undergoes elimination more rapidly. How can this behavior be explained?

12.39 Why does the molecule shown below, although a conjugated diene, fail to undergo a Diels–Alder reaction?

1. Provide a reasonable mechanism for the following reactions:

(a)

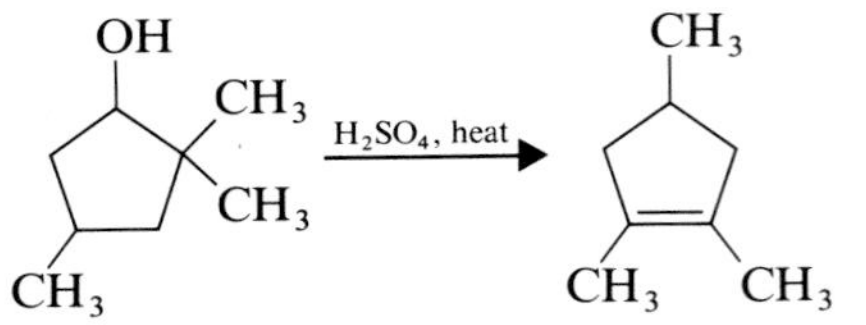
OH
CH3
CH3
CH3
H2SO4, heat
CH3
CH3
CH3

(b)

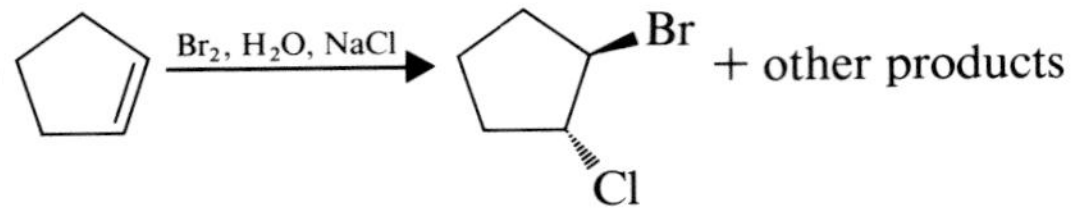
Br2, H2O, NaCl
Br
Cl
+ other products

(c) What other products would you expect from the reaction given in part (b)?

2. What are compounds **A–C?**

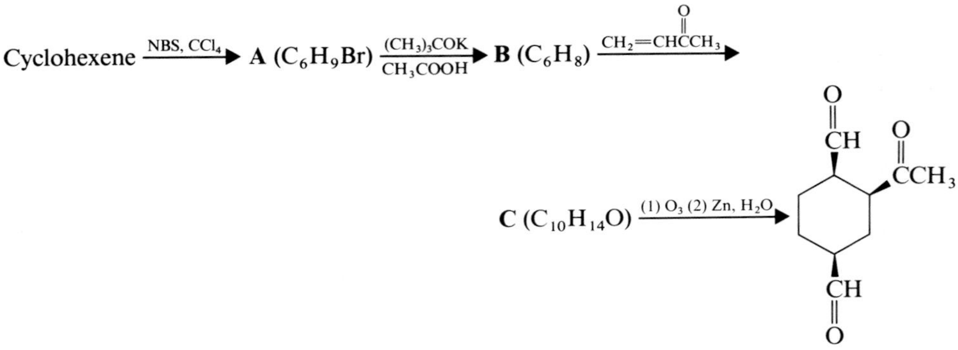
Cyclohexene
NBS, CCl4
A (C6H9Br)
(CH3)3COK
CH3COOH
B (C6H8)
CH2=CHCCH3
O
C (C10H14O)
(1) O3 (2) Zn, H2O
CH
O
CCH3
O
CH
O

3. Given the following data:

	$CH_2{=}CH_2$	CH_3CH_2Cl	$CH_2{=}CHCl$
C—Cl bond length		1.76 Å	1.69 Å
C=C bond length	1.34 Å		1.38 Å
C—C bond length		1.54 Å	
Dipole moment	0	2.05 D	1.44 D

Use resonance theory to explain each of the following: (a) The shorter C—Cl bond length in $CH_2{=}CHCl$ when compared to that in CH_3CH_2Cl. (b) The longer C=C bond in $CH_2{=}CHCl$ when compared to that in $CH_2{=}CH_2$. (c) The greater dipole moment of CH_3CH_2Cl when compared to $CH_2{=}CHCl$.

4. The following is a synthesis of "muscalure," the sex-attractant pheromone of the common house fly. Give the structure of each intermediate and of muscalure itself.

$$CH_3(CH_2)_{11}CH_2Br \xrightarrow{HC\equiv CNa} \textbf{A} (C_{15}H_{28}) \xrightarrow[\text{liq. } NH_3]{NaNH_2} \textbf{B} (C_{15}H_{27}Na) \xrightarrow{\text{1-bromooctane}}$$

$$\textbf{C} (C_{23}H_{44}) \xrightarrow{H_2, Ni_2B \ (P-2)} \text{muscalure} (C_{23}H_{46})$$

5. Write structures for the diastereomers of 2,3-diphenyl-2-butene and assign each diastereomer its (*E*) or (*Z*) designation. Hydrogenation of one of these diastereomers using a palladium catalyst produces a racemic form; similar treatment of the other produces a meso compound. On the basis of these experiments, tell which diastereomer is (*E*) and which is (*Z*).

6. A hydrocarbon (**A**) has the formula C_7H_{10}. On catalytic hydrogenation, **A** is converted to **B** (C_7H_{12}). On treatment with cold, dilute, and basic $KMnO_4$, **A** is converted to **C** ($C_7H_{12}O_2$). When heated with $KMnO_4$ in basic solution, followed by acidification, either **A** or **C** produces the meso form of 1,3-cyclopentanedicarboxylic acid (see the following structure). Give structural formulas for **A–C**.

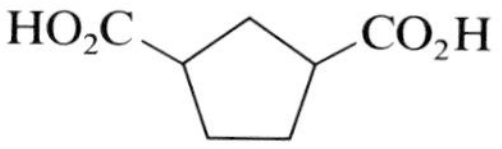

HO$_2$C   CO$_2$H

1,3-Cyclopentanedicarboxylic acid

7. Starting with propyne, and using any other required reagents, show how you would synthesize each of the following compounds. You need not repeat steps carried out in earlier parts of this problem.

(a) 2-Butyne

(b) *cis*-2-Butene

(c) *trans*-2-Butene

(d) 1-Butene

(e) 1,3-Butadiene

(f) 1-Bromobutane

(g) 2-Bromobutane (as a racemic form)

(h) (2*R*,3*S*)-2,3-Dibromobutane

(i) (2*R*,3*R*)- and (2*S*,3*S*)-2,3-Dibromobutane (as a racemic form)

(j) *meso*-2,3-Butanediol

(k) (*Z*)-2-Bromo-2-butene

8. Bromination of 2-methylbutane yields predominantly one product with the formula $C_5H_{11}Br$. What is this product? Show how you could use this compound to synthesize each of the following. (You need not repeat steps carried out in earlier parts.)

(a) 2-Methyl-2-butene

(b) 2-Methyl-2-butanol

(c) 3-Methyl-2-butanol

(d) 3-Methyl-1-butyne

(e) 1-Bromo-3-methylbutane

(f) 2-Chloro-3-methylbutane

(g) 2-Chloro-2-methylbutane

(h) 1-Iodo-3-methylbutane

(i) $CH_3\overset{O}{\overset{\|}{C}}CH_3$ and $CH_3\overset{O}{\overset{\|}{C}}H$

(j) $(CH_3)_2CH\overset{O}{\overset{\|}{C}}H$

9. An alkane (**A**) with the formula C_6H_{14} reacts with chlorine to yield three compounds with the formula $C_6H_{13}Cl$, **B**, **C**, and **D**. Of these only **C** and **D** undergo dehydrohalogenation with sodium ethoxide in ethanol to produce an alkene. Moreover, **C** and **D** yield the same alkene **E** (C_6H_{12}). Hydrogenation of **E** produces **A**. Treating **E** with HCl produces a compound (**F**) that is an isomer of **B**, **C**, and **D**. Treating **F** with Zn and acetic acid gives a compound (**G**) that is isomeric with **A**. Propose structures for **A–G**.

10. Compound **A** (C_4H_6) reacts with hydrogen and a platinum catalyst to yield butane. Compound **A** decolorizes Br_2 in CCl_4 and aqueous $KMnO_4$, but it does not react with

$Ag(NH_3)_2^+$. On treatment with hydrogen and Ni_2B (P-2 catalyst), **A** is converted to **B** (C_4H_8). When **B** is treated with OsO_4 and then with $NaHSO_3$, **B** is converted to **C** ($C_4H_{10}O_2$). Compound **C** cannot be resolved. Provide structures for **A–C**.

11. Dehalogenation of *meso*-2,3-dibromobutane occurs when it is treated with potassium iodide in ethanol. The product is *trans*-2-butene. Similar dehalogenation of either of the enantiomeric forms of 2,3-dibromobutane produces *cis*-2-butene. Give a mechanistic explanation of these results.

12. Dehydrohalogenation of *meso*-1,2-dibromo-1,2-diphenylethane by the action of sodium ethoxide in ethanol yields (*E*)-1-bromo-1,2-diphenylethene. Similar dehydrohalogenation of either of the enantiomeric forms of 1,2-dibromo-1,2-diphenylethane yields (*Z*)-1-bromo-1,2-diphenylethene. Provide an explanation for the results.

13. Give conformational structures for the major product formed when 1-*tert*-butylcyclohexene reacts with each of the following reagents. If the product would be obtained as a racemic form you should so indicate.

(a) Br_2, CCl_4

(b) OsO_4, then aqueous $NaHSO_3$

(c) $C_6H_5CO_3H$, then H_3O^+, H_2O

(d) THF : BH_3, then H_2O_2, OH^-

(e) $Hg(OAc)_2$ in THF–H_2O, then $NaBH_4$, OH^-

(f) Br_2, H_2O

(g) ICl

(h) O_3, then Zn, H_2O (conformational structure not required)

(i) D_2, Pt

(j) THF : BD_3, then CH_3CO_2T

14. Give structures for **A–C**.

$$\underset{\underset{Br}{|}}{\overset{\overset{CH_3}{|}}{CH_3\overset{|}{\underset{|}{C}}CH_2CH_2CH_3}} \xrightarrow{EtO^-/EtOH} \textbf{A }(C_6H_{12}) \text{ major product} \xrightarrow{THF\,:\,BH_3}$$

$$\textbf{B }(C_6H_{13})_2BH \xrightarrow{H_2O_2,\,OH^-} \textbf{C }(C_6H_{14}O)$$

15. (*R*)-3-Methyl-1-pentene is treated separately with the following reagents, and the products in each case are separated by fractional distillation. Write appropriate formulas for all of the components of each fraction, and tell whether each fraction would be optically active.

(a) Br_2, CCl_4 (d) THF : BH_3, then H_2O_2, OH^-

(b) H_2, Pt   (e) $Hg(OAc)_2$, THF–H_2O, then $NaBH_4$, OH^-

(c) OsO_4, then $NaHSO_3$ (f) Magnesium perphthalate, then H_3O^+, H_2O

16. Compound **A** ($C_8H_{15}Cl$) exists as a racemic form. Compound **A** does not decolorize either Br_2/CCl_4 or dilute aqueous $KMnO_4$. When **A** is treated with zinc and acetic acid and the mixture is separated by gas–liquid chromatography, two fractions **B** and **C** are obtained. The components of both fractions have the formula C_8H_{16}. Fraction **B** consists of a racemic form and can be resolved. Fraction **C** cannot be resolved. Treating **A** with sodium ethoxide in ethanol converts **A** into **D** (C_8H_{14}). Hydrogenation of **D** using a platinum catalyst yields **C**. Ozonolysis of **D** followed by treatment with zinc and water yields

$$\underset{\substack{\parallel \\ CH_3CCH_2CH_2CH_2CH_2CCH_3}}{O \qquad\qquad O}$$

Propose structures for **A**, **B**, **C**, and **D** including, where appropriate, their stereochemistry.

17. There are nine stereoisomers of 1,2,3,4,5,6-hexachlorocyclohexane. Seven of these isomers are meso compounds, and two are a pair of enantiomers. (a) Write structures for all of these stereoisomers, labeling meso forms and the pair of enantiomers. (b) One of these stereoisomers undergoes E2 reactions much more slowly than any of the others. Which isomer is this and why does it react so slowly in an E2 reaction?

18. In addition to more highly fluorinated products, fluorination of 2-methylbutane yields a mixture of compounds with the formula $C_5H_{11}F$. (a) How many different isomers with the formula $C_5H_{11}F$ would you expect to be produced, taking stereochemistry into account? (b) If the mixture of $C_5H_{11}F$ isomers were subjected to fractional distillation, how many fractions would you expect to obtain? (c) Which fractions would be optically inactive? (d) Which would you be able to resolve into enantiomers?

19. Fluorination of (R)-2-fluorobutane yields a mixture of isomers with the formula $C_4H_8F_2$. (a) How many different isomers would you expect to be produced? Write their structures. (b) If the mixture of $C_4H_8F_2$ isomers were subjected to fractional distillation, how many fractions would you expect to obtain? (c) Which of these fractions would be optically active?

20. There are two optically inactive (and nonresolvable) forms of 1,3-di-*sec*-butylcyclohexane. Write their structures.

21. When the following deuterium-labeled isomer undergoes elimination, the reaction yields *trans*-2-butene and *cis*-2-butene-2-*d* (as well as some 1-butene-3-*d*). The reaction does not yield *cis*-2-butene or *trans*-2-butene-2-*d*.

(+ $CH_3CHDCH{=}CH_2$)

but no

How can you explain these results?

CHAPTER 13

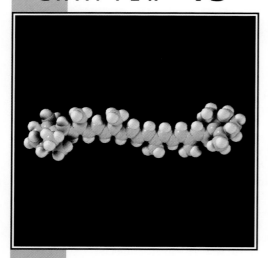

β-Carotene
(Section 13.2)

SPECTROSCOPIC METHODS OF STRUCTURE DETERMINATION

13.1 INTRODUCTION. THE ELECTROMAGNETIC SPECTRUM

The names of most forms of electromagnetic energy have become familiar terms. The *X-rays* used in medicine, the *light* that we see, the *ultraviolet* (UV) rays that produce sunburns, and the *radio* and *radar* waves used in communication are all different forms of the same phenomenon: electromagnetic radiation.

According to quantum mechanics, electromagnetic radiation has a dual and seemingly contradictory nature. Electromagnetic radiation has the properties of both a wave and a particle. Electromagnetic radiation can be described as a wave occurring simultaneously in electrical and magnetic fields. It can also be described as if it consisted of particles called quanta or photons. Different experiments disclose these two different aspects of electromagnetic radiation. They are not seen together in the same experiment.

A wave is usually described in terms of its **wavelength** (λ) or its **frequency** (ν). A simple wave is shown in Fig. 13.1. The distance between consecutive crests (or troughs) is the wavelength. The number of full cycles of the wave that pass a given point each second, as the wave moves through space, is called the *frequency,* and is measured in cycles per second or **hertz**.*

All electromagnetic radiation travels through a vacuum at the same velocity. This velocity (c), called the velocity of light, is 2.99792458×10^8 m s^{-1} and $c = \lambda\nu$. The wavelengths of electromagnetic radiation are expressed either in meters (m), milli-

---

* The term hertz (after the German physicist H. R. Hertz), abbreviated Hz, is now used in place of *cycles per second* (cps). Frequency of electromagnetic radiation is also sometimes expressed in *wavenumbers,* that is, the number of waves per centimeter.

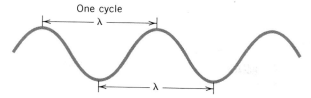

FIGURE 13.1 A simple wave and wavelength, λ.

meters (1 mm = 10^{-3} m), micrometers (1 μm = 10^{-6} m), or nanometers (1 nm = 10^{-9} m). [An older term for micrometer is *micron* (abbreviated μ) and an older term for nanometer is *millimicron.*]

The energy of a quantum of electromagnetic energy is directly related to its frequency.

$$E = h\nu$$

where h = Planck's constant, 6.63×10^{-34} J s,

and

$$\nu = \text{the frequency (Hz)}$$

This means that the higher the frequency of radiation the greater is its energy. X-Rays, for example, are much more energetic than rays of visible light. The frequencies of X-rays are on the order of 10^{19} Hz, while those of visible light are on the order of 10^{15} Hz.

Since $\nu = c/\lambda$, the energy of electromagnetic radiation is inversely proportional to its wavelength.

$$E = \frac{hc}{\lambda} \qquad (c = \text{the velocity of light})$$

Thus, per quantum, electromagnetic radiation of long wavelength has low energy, whereas that of short wavelength has high energy. X-Rays have wavelengths on the order of 0.1 nm, whereas visible light has wavelengths between 400 and 750 nm.*

It may be helpful to point out, too, that for visible light, wavelengths (and, thus, frequencies) are related to what we perceive as colors. The light that we call red light has a wavelength of approximately 750 nm. The light we call violet light has a wavelength of approximately 400 nm. All of the other colors of the visible spectrum (the rainbow) lie in between these wavelengths.

The different regions of the electromagnetic spectrum are shown in Fig. 13.2. Nearly every portion of the electromagnetic spectrum from the region of X-rays to that of microwaves and radio waves has been used in elucidating structures of atoms and molecules. Although techniques differ according to the portion of the electromagnetic

---

* A convenient formula that relates wavelength (in nm) to the energy of electromagnetic radiation is the following:

$$E \text{ (in kcal mol}^{-1}) = \frac{2.86 \times 10^{-10} \text{ kcal} \cdot \text{mol}^{-1}}{\text{wavelength in nanometers}}$$

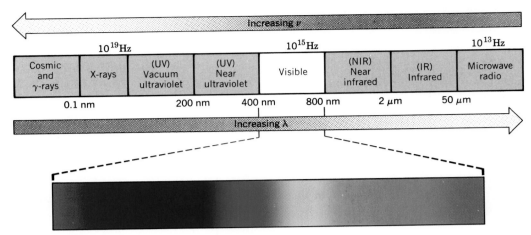

FIGURE 13.2 The electromagnetic spectrum.

spectrum in which we are working, there is a consistency and unity of basic principles. Later in this chapter we discuss the use that can be made of the infrared (IR) and radio regions when we take up IR spectroscopy and nuclear magnetic resonance (NMR) spectroscopy. At this point we direct our attention to electromagnetic radiation in the near UV and visible regions and see how it interacts with conjugated polyenes.

13.2 VISIBLE AND ULTRAVIOLET SPECTROSCOPY

When electromagnetic radiation in the UV and visible regions passes through a compound containing multiple bonds, a portion of the radiation is usually absorbed by the compound. Just how much of the radiation is absorbed depends on the wavelength of the radiation and the structure of the compound. The absorption of radiation is caused by the subtraction of energy from the radiation beam when electrons in orbitals of lower energy are excited into orbitals of higher energy.

Instruments called visible–UV spectrometers are used to measure the amount of light absorbed at each wavelength of the visible and UV regions. In these instruments a beam of light is split; one half of the beam (the sample beam) is directed through a transparent cell containing a solution of the compound being analyzed, and one half (the reference beam) is directed through an identical cell that does not contain the compound but contains the solvent. Solvents are chosen to be transparent in the region being analyzed. The instrument is designed so that it can make a comparison of the intensities of the two beams at each wavelength of the region. If the compound absorbs light at a particular wavelength, the intensity of the sample beam (I_S) will be less than that of the reference beam (I_R). The instrument indicates this by producing a graph—a plot of the wavelength of the entire region versus the absorbance (A) of light at each wavelength. [The absorbance at a particular wavelength is defined by the equation: $A_\lambda = \log (I_R/I_S)$.] Such a graph is called an **absorption spectrum.**

A typical UV absorption spectrum, that of 2,5-dimethyl-2,4-hexadiene, is shown in Fig. 13.3. It shows a broad absorption band in the region between 210 and 260 nm. The absorption is at a maximum at 242.5 nm. It is this wavelength that is usually reported in the chemical literature as λ_{max}.

In addition to reporting the wavelength of maximum absorption (λ_{max}), chemists

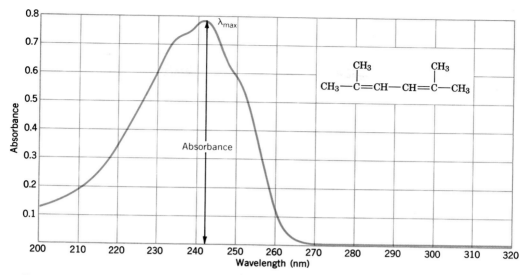

FIGURE 13.3 The UV absorption spectrum of 2,5-dimethyl-2,4-hexadiene in methanol at a concentration of $5.95 \times 10^{-5}\ M$ in a 1.00-cm cell. (Spectrum courtesy of Sadtler Research Laboratories, Inc., Philadelphia.)

often report another quantity that indicates the strength or the *intensity* of the absorption, called the **molar absorptivity, ε.***

The molar absorptivity is simply the proportionality constant that relates the observed absorbance (A) at a particular wavelength (λ) to the molar concentration (C) of the sample and the length (l) (in centimeters) of the path of the light beam through the sample cell.

$$A = \varepsilon \times C \times l \quad \text{or} \quad \varepsilon = \frac{A}{C \times l}$$

For 2,5-dimethyl-2,4-hexadiene dissolved in methanol the molar absorptivity at the wavelength of maximum absorbance (242.5 nm) is 13,100 $M^{-1}\ cm^{-1}$. In the chemical literature this would be reported as

2,5-Dimethyl-2,4-hexadiene, $\lambda_{max}^{methanol}$ 242.5 nm   ($\varepsilon = 13{,}100$)

As we noted earlier, when compounds absorb light in the UV and visible regions, electrons are excited from lower electronic energy levels to higher ones. For this reason, visible and UV spectra are often called **electronic spectra.** The absorption spectrum of 2,5-dimethyl-2,4-hexadiene is a typical electronic spectrum because the absorption band (or peak) is very broad. Most absorption bands in the visible and UV region are broad because each electronic energy level has associated with it vibrational and rotational levels. Thus, electron transitions may occur from any of several vibrational and rotational states of one electronic level to any of several vibrational and rotational states of a higher level.

Alkenes and nonconjugated dienes usually have absorption maxima below 200 nm. Ethene, for example, gives an absorption maximum at 171 nm; 1,4-penta-

* In older literature, the molar absorptivity (ε) is often referred to as the molar extinction coefficient.

diene gives an absorption maximum at 178 nm. These absorptions occur at wavelengths that are out of the range of operation of most visible–ultraviolet spectrometers because they occur where the oxygen in air also absorbs. Special air-free techniques must be employed in measuring them.

Compounds whose molecules contain *conjugated* multiple bonds have maxima at wavelengths longer than 200 nm. For example, 1,3-butadiene absorbs at 217 nm. This longer wavelength absorption by conjugated dienes is a direct consequence of conjugation.

We can understand how conjugation of multiple bonds brings about absorption of light at longer wavelengths if we examine Fig. 13.4.

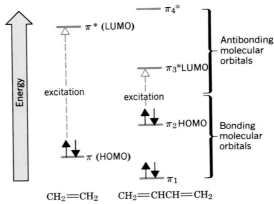

FIGURE 13.4 The relative energies of the π molecular orbitals of ethene and 1,3-butadiene (Section 13.2).

When a molecule absorbs light at its longest wavelength, an electron is excited from its highest occupied molecular orbital (HOMO) to the lowest unoccupied molecular orbital (LUMO). For most alkenes and alkadienes the HOMO is a bonding π orbital and the LUMO is an antibonding π^* orbital. The wavelength of the absorption maximum is determined by the difference in energy between these two levels. The energy gap between the HOMO and LUMO of ethene is greater than that between the corresponding orbitals of 1,3-butadiene. Thus, the $\pi \longrightarrow \pi^*$ electron excitation of ethene requires absorption of light of greater energy (shorter wavelength) than the corresponding $\pi_2 \longrightarrow \pi_3^*$ excitation in 1,3-butadiene. The energy difference between the HOMOs and the LUMOs of the two compounds is reflected in their absorption spectra. Ethene has its λ_{max} at 171 nm; 1,3-butadiene has a λ_{max} at 217 nm.

The narrower gap between the HOMO and the LUMO in 1,3-butadiene results from the conjugation of the double bonds. Molecular orbital calculations indicate that a much larger gap should occur in isolated alkadienes. This is borne out experimentally. Isolated alkadienes give absorption spectra similar to those of alkenes. Their λ_{max} are at shorter wavelengths, usually below 200 nm. As we mentioned, 1,4-pentadiene has its λ_{max} at 178 nm.

Conjugated alkatrienes absorb at longer wavelengths than conjugated alkadienes, and this too can be accounted for in molecular orbital calculations. The energy gap between the HOMO and the LUMO of an alkatriene is even smaller than that of an alkadiene. In fact, there is a general rule that states that *the greater the number of conjugated multiple bonds a compound contains, the longer will be the wavelength at which the compound absorbs light.*

Polyenes with eight or more conjugated double bonds absorb light in the visible region of the spectrum. For example, β-carotene, a precursor of vitamin A and a compound that imparts its orange color to carrots, has 11 conjugated double bonds; β-carotene has an absorption maximum at 497 nm, well into the visible region. Light of 497 nm has a blue-green color; this is the light that is absorbed by β-carotene. We perceive the complementary color of blue green, which is red orange.

β-Carotene

Lycopene, a compound partly responsible for the red color of tomatoes, also has 11 conjugated double bonds. Lycopene has an absorption maximum at 505 nm, and it absorbs there intensely. (Approximately 0.02 g of lycopene can be isolated from 1 kg of fresh, ripe tomatoes.)

Lycopene

Table 13.1 gives the values of λ_{max} for a number of unsaturated compounds.

Compounds with carbon–oxygen double bonds also absorb light in the UV region. Acetone, for example, has a broad absorption peak at 280 nm that corresponds to the excitation of an electron from one of the unshared pairs (a nonbonding or "n" electron) to the π^* orbital of the carbon–oxygen double bond:

Acetone
$\lambda_{max} = 280\,nm$
$\varepsilon_{max} = 15$

Compounds in which the carbon–oxygen double bond is conjugated with a carbon–carbon double bond have absorption maxima corresponding to $n \longrightarrow \pi^*$ excitations and $\pi \longrightarrow \pi^*$ excitations. The $n \longrightarrow \pi^*$ absorption maxima occur at longer wavelengths but are much weaker (i.e., have smaller molar absorbtivities).

$$CH_2{=}CH{-}\underset{\underset{CH_3}{|}}{C}{=}O$$

$n \longrightarrow \pi^* \; \lambda_{max} = 324$ nm, $\varepsilon_{max} = 24$
$\pi \longrightarrow \pi^* \; \lambda_{max} = 219$ nm, $\varepsilon_{max} = 3600$

TABLE 13.1 Long-wavelength absorption maxima of unsaturated hydrocarbons

COMPOUND	STRUCTURE	λ_{max} (nm)	ε_{max}
Ethene	$CH_2{=}CH_2$	171	15,530
trans-3-Hexene		184	10,000
Cyclohexene		182	7,600
1-Octene	$CH_3(CH_2)_5CH{=}CH_2$	177	12,600
1-Octyne	$CH_3(CH_2)_5C{\equiv}CH$	185	2,000
1,3-Butadiene	$CH_2{=}CHCH{=}CH_2$	217	21,000
cis-1,3-Pentadiene		223	22,600
trans-1,3-Pentadiene		223.5	23,000
1-Buten-3-yne	$CH_2{=}CHC{\equiv}CH$	228	7,800
1,4-Pentadiene	$CH_2{=}CHCH_2CH{=}CH_2$	178	17,000
1,3-Cyclopentadiene		239	3,400
1,3-Cyclohexadiene		256	8,000
trans-1,3,5-Hexatriene		274	50,000

PROBLEM 13.1

Two compounds, **A** and **B**, have the same molecular formula, C_6H_8. Both **A** and **B** decolorize bromine in carbon tetrachloride and both give positive tests with cold dilute potassium permanganate. Both **A** and **B** react with two molar equivalents of hydrogen in the presence of platinum to yield cyclohexane. Compound **A** shows an absorption maximum at 256 nm, whereas **B** shows no absorption maximum beyond 200 nm. What are the structures of **A** and **B**?

PROBLEM 13.2

Three compounds, **D**, **E**, and **F**, have the same molecular formula, C_5H_6. In the presence of a platinum catalyst, all three compounds absorb 3 molar equivalents of hydrogen and yield pentane. Compounds **E** and **F** give a precipitate when treated with ammoniacal silver nitrate; compound **D** gives no reaction. Compounds **D** and **E** show an absorption maximum near 230 nm. Compound **F** shows no absorption maximum beyond 200 nm. Propose structures for **D**, **E**, and **F**.

13.3 INFRARED SPECTROSCOPY

We saw in Section 13.2 that many organic compounds absorb radiation in the visible and UV regions of the electromagnetic spectrum. We also saw that when compounds absorb radiation of the visible and UV regions, electrons are excited from lower energy molecular orbitals to higher ones.

Organic compounds also absorb electromagnetic energy in the infrared (IR) region of the spectrum. Infrared radiation does not have sufficient energy to cause the excitation of electrons, but it does cause atoms and groups of atoms of organic compounds to vibrate faster and with increased amplitude about the covalent bonds that connect them. These vibrations are *quantized,* and as they occur, the compounds absorb IR energy in particular regions of the spectrum.

Infrared spectrometers operate in a manner similar to that of visible–UV spectrometers. A beam of IR radiation is passed through the sample and is constantly compared with a reference beam as the frequency of the incident radiation is varied. The spectrometer plots the results as a graph showing absorption versus frequency or wavelength.

The location of an IR absorption band (or peak) can be specified in **frequency-related units** by its **wavenumber** ($\bar{\nu}$) measured in reciprocal centimeters (cm^{-1}), or by its **wavelength** (λ) measured in micrometers (μm; old name micron, μ). The wavenumber is the number of cycles of the wave in each centimeter along the light beam, and the wavelength is the length of the wave, crest to crest.

$$\bar{\nu} = \frac{1}{\lambda} \text{ (with } \lambda \text{ in cm)} \quad \text{or} \quad \bar{\nu} = \frac{10,000}{\lambda} \text{ (with } \lambda \text{ in } \mu m)$$

In their vibrations covalent bonds behave as if they were tiny springs connecting the atoms. When the atoms vibrate they can do so only at certain frequencies, as if the bonds were "tuned." Because of this, covalently bonded atoms have only particular vibrational energy levels, i.e., the levels are quantized. The excitation of a molecule from one vibrational energy level to another occurs only when the compound absorbs IR radiation of a particular energy, meaning a particular wavelength or frequency (since $\Delta E = h\nu$).

Molecules can vibrate in a variety of ways. Two atoms joined by a covalent bond can undergo a stretching vibration where the atoms move back and forth as if joined by a spring.

A stretching vibration

Three atoms can also undergo a variety of stretching and bending vibrations:

Symmetric stretching

Asymmetric stretching

**An in-plane bending
vibration
(scissoring)**

**An out-of-plane bending
vibration
(twisting)**

The *frequency* of a given stretching vibration *in an IR spectrum* can be related to two factors. These are *the masses of the bonded atoms*—light atoms vibrate at higher frequencies than heavier ones—*and the relative stiffness of the bond.* Triple bonds are stiffer (and vibrate at higher frequencies) than double bonds and double bonds are stiffer (and vibrate at higher frequencies) than single bonds. We can see some of these effects in Table 13.2. Notice that stretching frequencies of groups involving hydrogen (a light atom) such as C—H, N—H, and O—H all occur at relatively high frequencies:

Group	Bond	Frequency Range (cm^{-1})
Alkyl	C—H	2853–2962
Alcohol	O—H	3590–3650
Amine	N—H	3300–3500

Notice too, that triple bonds vibrate at higher frequencies than double bonds:

Bond	Frequency Range (cm^{-1})
C≡C	2100–2260
C≡N	2220–2260
C=C	1620–1680
C=O	1630–1780

The IR spectra of even relatively simple compounds contain many absorption peaks.

Not all molecular vibrations result in the absorption of IR energy. ***In order for a vibration to occur with the absorption of IR energy, the dipole moment of the molecule must change as the vibration occurs.*** Thus when the four hydrogen atoms of methane vibrate symmetrically, methane does not absorb IR energy. Symmetrical vibrations of the carbon–carbon double and triple bonds of ethene and ethyne do not result in the absorption of IR radiation, either.

TABLE 13.2 Characteristic infrared absorptions of groups

GROUP		FREQUENCY RANGE (cm^{-1})	INTENSITY[a]
A. Alkyl			
C—H (stretching)		2853–2962	(m–s)
Isopropyl, —CH(CH$_3$)$_2$		1380–1385	(s)
	and	1365–1370	(s)
tert-Butyl, —C(CH$_3$)$_3$		1385–1395	(m)
	and	~1365	(s)
B. Alkenyl			
C—H (stretching)		3010–3095	(m)
C=C (stretching)		1620–1680	(v)
R—CH=CH$_2$		985–1000	(s)
	and	905–920	(s)
R$_2$C=CH$_2$	(out-of-plane C—H bendings)	880–900	(s)
cis-RCH=CHR		675–730	(s)
trans-RCH=CHR		960–975	(s)
C. Alkynyl			
≡C—H (stretching)		~3300	(s)
C≡C (stretching)		2100–2260	(v)
D. Aromatic			
Ar—H (stretching)		~3030	(v)
Aromatic substitution type (C—H out-of-plane bendings)			
Monosubstituted		690–710	(very s)
	and	730–770	(very s)
o Disubstituted		735–770	(s)
m Disubstituted		680–725	(s)
	and	750–810	(very s)
p Disubstituted		800–840	(very s)
E. Alcohols, Phenols, and Carboxylic Acids			
O—H (stretching)			
Alcohols, phenols (dilute solutions)		3590–3650	(sharp, v)
Alcohols, phenols (hydrogen bonded)		3200–3550	(broad, s)
Carboxylic acids (hydrogen bonded)		2500–3000	(broad, v)
F. Aldehydes, Ketones, Esters, and Carboxylic Acids			
C=O (stretching)		1630–1780	(s)
Aldehydes		1690–1740	(s)

TABLE 13.2 (continued)

GROUP	FREQUENCY RANGE (cm^{-1})	INTENSITY[a]
Ketones	1680–1750	(s)
Esters	1735–1750	(s)
Carboxylic acids	1710–1780	(s)
Amides	1630–1690	(s)
G. Amines		
N—H	3300–3500	(m)
H. Nitriles		
C≡N	2220–2260	(m)

[a] Abbreviations: s = strong, m = medium, w = weak, v = variable, ~ = approximately.

Vibrational absorption may occur outside the region measured by a particular IR spectrophotometer and vibrational absorptions may occur so closely together that peaks fall on top of peaks.

Other factors bring about even more absorption peaks. Overtones (harmonics) of fundamental absorption bands may be seen in IR spectra even though these overtones occur with greatly reduced intensity. Bands called combination bands and difference bands also appear in IR spectra.

Because IR spectra contain so many peaks, the possibility that two different compounds will have the same IR spectrum is exceedingly small. It is because of this that an IR spectrum has been called the "fingerprint" of a molecule. Thus, with organic compounds, if two pure samples give different IR spectra, one can be certain that they are different compounds. If they give the same IR spectrum then they are the same compound.

In the hands of one skilled in their interpretation, IR spectra contain a wealth of information about the structures of compounds. We show some of the information that can be gathered from the spectra of octane and methylbenzene (also called toluene) in Figs. 13.5 and 13.6. We have neither the time nor the space here to develop the skill that would lead to complete interpretations of IR spectra, but we can learn how to recognize the presence of absorption peaks in the IR spectrum that result from vibrations of characteristic functional groups in the compound. By doing only this, however, we shall be able to use the information we gather from IR spectra in a powerful way, particularly when we couple it with the information we gather from NMR spectra.

Let us now see how we can apply the data given in Table 13.2 to the interpretation of IR spectra.

13.3A Hydrocarbons

All hydrocarbons give absorption peaks in the 2800–3300-cm^{-1} region that are associated with carbon–hydrogen stretching vibrations. We can use these peaks in interpreting IR spectra because the exact location of the peak depends on the strength (and stiffness) of the C—H bond, which in turn depend on the hybridization state of the carbon that bears the hydrogen. We have already seen that C—H bonds involving

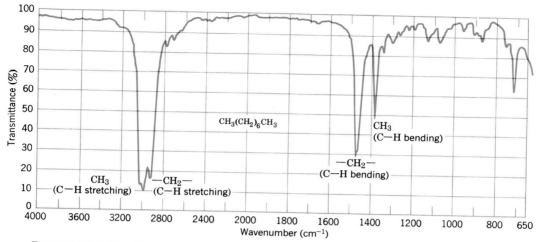

FIGURE 13.5 The IR spectrum of octane. (Notice that, in IR spectra, the peaks are usually measured in % transmittance. Thus, the peak at 2900 cm^{-1} has 10% transmittance (that is, an absorbance, A, of 0.95).

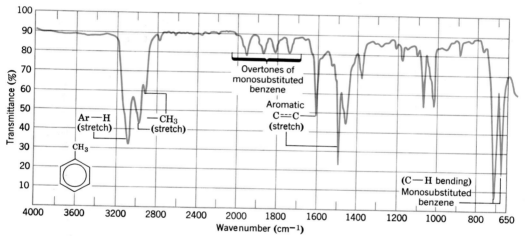

FIGURE 13.6 The IR spectrum of methylbenzene.

sp-hybridized carbon are strongest and those involving sp^3-hybridized carbon are weakest. The order of bond strength is

$$sp > sp^2 > sp^3$$

This too is the order of the bond stiffness.

The carbon–hydrogen stretching peaks of hydrogen atoms attached to sp-hybridized carbon atoms occur at highest frequencies, about 3300 cm^{-1}. Thus, $\equiv$C—H groups of terminal alkynes give peaks in this region. We can see the absorption of the acetylenic C—H bond of 1-hexyne at 3320 cm^{-1} in Fig. 13.7.

The carbon–hydrogen stretching peaks of hydrogen atoms attached to sp^2-hybridized carbon atoms occur in the 3000–3100-cm^{-1} region. Thus, alkenyl C—H bonds and the C—H groups of aromatic rings give absorption peaks in this region. We

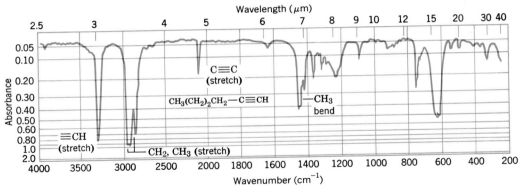

FIGURE 13.7 The IR spectrum of 1-hexyne. (Spectrum courtesy of Sadtler Research Laboratories, Inc., Philadelphia.)

can see the alkenyl C—H absorption peak of 3080 cm^{-1} in the spectrum of 1-hexene (Fig. 13.8) and we can see the C—H absorption of the aromatic hydrogen atoms at 3090 cm^{-1} in the spectrum of methylbenzene (Fig. 13.6).

The carbon–hydrogen stretching bands of hydrogen atoms attached to sp^3-hybridized carbon atoms occur at lowest frequencies, in the 2800–3000-cm^{-1} region. We can see methyl and methylene absorption peaks in the spectra of octane (Fig. 13.5), methylbenzene (Fig. 13.6), 1-hexyne (Fig. 13.7), and 1-hexene (Fig. 13.8).

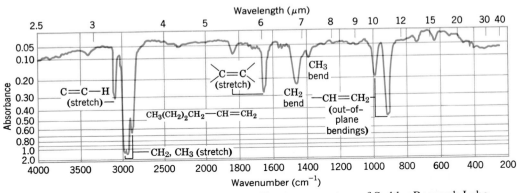

FIGURE 13.8 The IR spectrum of 1-hexene. (Spectrum courtesy of Sadtler Research Laboratories, Inc., Philadelphia.)

Hydrocarbons also give absorption peaks in their IR spectra that result from carbon–carbon bond stretchings. Carbon–carbon single bonds normally give rise to very weak peaks that are usually of little use in assigning structures. More useful peaks arise from multiple carbon–carbon bonds, however. Carbon–carbon double bonds give absorption peaks in the 1620–1680-cm^{-1} region and carbon–carbon triple bonds give absorption peaks between 2100 and 2260 cm^{-1}. These absorptions are not usually strong ones and they are not present at all if the double or triple bond is symmetrically substituted. (No dipole moment change will be associated with the vibration.) The stretchings of the carbon–carbon bonds of benzene rings usually give a set of characteristic sharp peaks in the 1450–1600-cm^{-1} region.

Absorptions arising from carbon–hydrogen bending vibrations of alkenes occur in the 600–1000-cm^{-1} region. The exact location of these peaks can often be used to determine the *substitution pattern of the double bond and its configuration.*

Monosubstituted alkenes give two strong peaks in the 905–920- and the 985–1000-cm^{-1} regions. Disubstituted alkenes of the type $R_2C{=}CH_2$ give a strong peak in the 880–900-cm^{-1} range. *cis*-Alkenes give an absorption peak in the 675–730-cm^{-1} region and *trans*-alkenes give a peak between 960 and 975 cm^{-1}. These ranges for the carbon–hydrogen bending vibrations can be used with fair reliability for alkenes that do not have an electron-releasing or electron-withdrawing substituent (other than an alkyl group) on one of the carbon atoms of the double bond. When electron-releasing or electron-withdrawing substituents are present on a double-bond carbon, the bending absorption peaks may be shifted out of the regions we have given.

13.3B Other Functional Groups

Infrared spectroscopy gives us an invaluable method for recognizing quickly and simply the presence of certain functional groups in a molecule. One important functional group that gives a prominent absorption peak in IR spectra is the **carbonyl group** $\diagdown\!\!\!\diagup\!\!C{=}O$. This group is present in aldehydes, ketones, esters, carboxylic acids, amides, and so forth. The carbon–oxygen double bond stretching frequency of all these groups gives a strong peak between 1630 and 1780 cm^{-1}. The exact location of the peak depends on whether it arises from an aldehyde, ketone, ester, and so forth. These locations are the following and we shall have more to say about carbonyl absorption peaks when we discuss these compounds in later chapters.

$$
\begin{array}{ccc}
\overset{\textstyle O}{\underset{\textstyle |}{\|}} & \overset{\textstyle O}{\|} & \overset{\textstyle O}{\|} \\
R{-}C{-}H & R{-}C{-}R & R{-}C{-}OR \\
\textbf{Aldehyde} & \textbf{Ketone} & \textbf{Ester} \\
\textbf{1690–1740 cm}^{-1} & \textbf{1680–1750 cm}^{-1} & \textbf{1735–1750 cm}^{-1}
\end{array}
$$

$$
\begin{array}{cc}
\overset{\textstyle O}{\|} & \overset{\textstyle O}{\|} \\
R{-}C{-}OH & R{-}C{-}NH_2 \\
\textbf{Carboxylic acid} & \textbf{Amide} \\
\textbf{1710–1780 cm}^{-1} & \textbf{1630–1690 cm}^{-1}
\end{array}
$$

The **hydroxyl groups** of alcohols and phenols are also easy to recognize in IR spectra by their O—H stretching absorptions. These bonds also give us direct evidence for hydrogen bonding. If an alcohol or phenol is present as a very dilute solution in CCl_4, O—H absorption occurs as a very sharp peak in the 3590–3650-cm^{-1} region. In very dilute solution or in the gas phase, formation of intermolecular hydrogen bonds does not take place because the molecules are too widely separated. The sharp peak in the 3590–3650-cm^{-1} region, therefore, is attributed to "free" (unassociated) hydroxyl groups. Increasing the concentration of the alcohol or phenol causes the sharp peak to be replaced by a broad band in the 3200–3550-cm^{-1} region. This absorption is attributed to OH groups that are associated through intermolecular hydrogen bonding.

Very dilute solutions of 1° and 2° **amines** also give sharp peaks in the 3300–3500-cm^{-1} region arising from free N—H stretching vibrations. Primary amines give two sharp peaks; secondary amines give only one. Tertiary amines, because they have no N—H bond, do not absorb in this region.

RNH_2	R_2NH
1° Amine	2° Amine
Two peaks in	One peak in
3300–3500-cm^{-1}	3300–3500-cm^{-1}
region	region

Hydrogen bonding causes these peaks to broaden. The NH groups of **amides** give similar absorption peaks.

13.4 NUCLEAR MAGNETIC RESONANCE SPECTROSCOPY

The nuclei of certain elements and isotopes behave as though they were magnets spinning about an axis. The nuclei of ordinary hydrogen (^1H) and carbon-13 (^{13}C) nuclei have this property. When one places a compound containing ^1H or ^{13}C atoms in a very strong magnetic field and simultaneously irradiates it with electromagnetic energy, the nuclei of the compound may absorb energy through a process called magnetic resonance.* This absorption of energy is quantized and produces a characteristic spectrum for the compound. Absorption of energy does not occur unless the strength of the magnetic field and the frequency of electromagnetic radiation are at specific values.

Instruments known as nuclear magnetic resonance (NMR) spectrometers allow chemists to measure the absorption of energy by ^1H or ^{13}C nuclei and by the nuclei of other elements that we shall discuss in Section 13.5. These instruments use very powerful magnets and irradiate the sample with electromagnetic radiation in the radio frequency (rf) region. Two types of NMR spectrometers, based on different designs, are now used by organic chemists: sweep [or continuous wave, (CW)] and Fourier transform (FT) spectrometers.

Sweep (CW) NMR Spectrometers

Nuclear magnetic resonance spectrometers can be designed so that they irradiate the compound with electromagnetic energy of a constant frequency while the magnetic field strength is varied or swept (Fig. 13.9). When the magnetic field reaches the

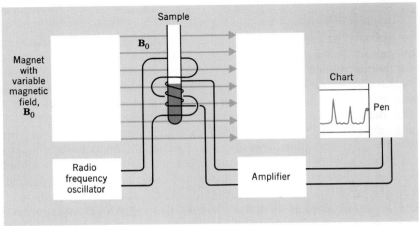

FIGURE 13.9 Diagram of a sweep (CW) NMR spectrometer.

*Magnetic resonance is an entirely different phenomenon from the resonance theory that we have discussed in earlier chapters.

correct strength, the nuclei absorb energy and resonance occurs. This absorption causes a tiny electrical current to flow in a receiver coil surrounding the sample. The instrument then amplifies this current and displays it as a signal (a peak or series of peaks) on a strip of chart paper calibrated in frequency units (Hz). The result is an NMR spectrum.

Fourier Transform (FT) NMR Spectrometers

Currently, the state of the art NMR instruments use superconducting magnets that have a much higher magnetic field strength than their predecessors. They also employ computers that allow signal averaging followed by a mathematical calculation known as a Fourier transform. In practice, the instrument accumulates a number of data scans and averages them so as to cancel out random electronic noise, thus enhancing the actual NMR signals. Fourier transform instruments (Fig. 13.10) achieve very high

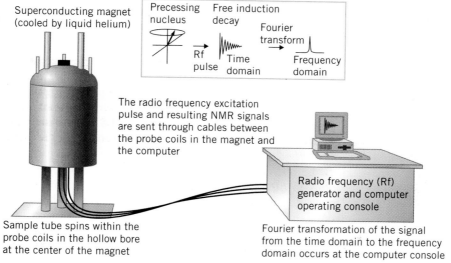

FIGURE 13.10 Diagram of a Fourier transform nuclear magnetic resonance spectrometer.

resolution and much greater sensitivity than CW NMR instruments. Instead of sweeping the magnetic field while irradiating the sample with electromagnetic energy in the rf region, as in the CW spectrometer, the FT instrument irradiates the sample with a short pulse of rf radiation (for $\sim 10^{-5}$ s). This rf pulse excites all the nuclei at once, as opposed to each nucleus being individually excited as with the sweep method. The data obtained from the pulse excitation method, however, are different from those obtained by the sweep method. One difference is that the sweep method takes 2–5 min to give a complete spectrum, whereas the pulse method can produce a spectrum in as little as 5 s. Another difference is that the sweep method provides a spectrum directly as a function of frequency (in Hz). With the pulse method, however, the data are collected as a function of time. After the pulse, a signal that contains information about all the peaks simultaneously is detected in the probe. This signal must then be transformed by computer into a function of frequency before the individual peaks can be identified.

To transform the signal from the time domain to the frequency domain, the computer carries out what is called a Fourier transformation (or FT). The mathematics of this process need not be of concern to us; the only important point is that the data are sampled and stored as discrete points—that is—they are *digitized.* In a pulsed NMR experiment, then, after excitation by an rf pulse the signal is detected as a voltage in the NMR probe. Upon amplification, the signal is converted to a series of data points and stored in the memory of the computer. This process is repeated until enough data are collected that signal averaging produces a strong signal. The acquired data are then transformed by the Fourier method to the frequency spectrum.

We begin our study of NMR spectroscopy with a brief examination of the main features of spectra arising from hydrogen nuclei. These spectra are often called *proton magnetic resonance (PMR) spectra* or **¹H NMR spectra.** We shall discuss NMR spectroscopy **in terms that apply to CW instruments** in which the magnetic field is being varied because by doing this we can make the explanations simpler. Equivalent but somewhat more complicated explanations can be given in terms that apply to FT instruments, in which the magnetic field is kept constant and the data are presented as though the frequency were varied.

The features we shall explore are chemical shift, peak area (integration), and signal splitting. After this brief overview we shall examine these and other aspects of NMR spectroscopy in more detail.

13.4A Chemical Shift. Peak Position in an NMR Spectrum

If hydrogen nuclei were stripped of their electrons and were completely isolated from other nuclei, all hydrogen nuclei (protons) would absorb energy at the same magnetic field strength for a given frequency of electromagnetic radiation. Fortunately this is not the case. In a given molecule some hydrogen nuclei are in regions of greater electron density than others, and as a result, the nuclei (protons) absorb energy at *slightly different* magnetic field strengths. The signals for these protons consequently occur at different positions in the NMR spectrum; they are said to have different **chemical shifts.** The actual field strength at which absorption occurs (the chemical shift) is highly dependent on the magnetic environment of each proton. This magnetic environment depends on two factors: on magnetic fields generated by circulation electrons, and on magnetic fields that result from other nearby protons (or other magnetic nuclei).

Figure 13.11 shows the ¹H NMR spectrum of 1,4-dimethylbenzene (a compound that is also called *p*-xylene).

Chemical shifts are measured along the bottom of the spectrum on a delta (δ) scale (in units of parts per million). We shall have more to say about these units later; for the moment, we need only point out that the externally applied magnetic field strength increases from left to right. A signal that occurs at a chemical shift of $\delta = 7$ occurs at a lower external magnetic field strength than one that occurs at $\delta = 2$. Signals on the left of the spectrum are said to occur **downfield** and those on the right are said to be **upfield.**

The spectrum in Fig. 13.11 shows a small signal $\delta = 0$. This signal arises from a compound that has been added to the sample to allow calibration of the chemical shift scale.

The first feature we want to notice is **the relation between the number of signals in the spectrum and the number of different types of hydrogen atoms in the compound.**

Because of its symmetry, 1,4-dimethylbenzene has only *two* different types of hydrogen atoms, and it gives only *two* signals in its NMR spectrum. The two types of

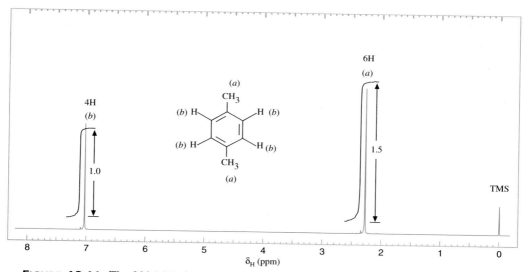

FIGURE 13.11 The 300 MHz ^{1}H NMR spectrum of 1,4-dimethylbenzene.

hydrogen atoms of 1,4-dimethylbenzene are the hydrogen atoms of the methyl groups and the hydrogen atoms of the benzene ring. The six methyl hydrogens of 1,4-dimethylbenzene *are all equivalent* and they are in a different environment from the four *equivalent* hydrogens of the ring. The six methyl hydrogens give rise to a signal that occurs at δ 2.30. The four hydrogen atoms of the benzene ring give a signal at δ 7.05.

13.4B Integration of Peak Areas. The Integral Curve

Next we want to examine the relative magnitude of the signals, for these are often helpful in assigning signals to particular groups of hydrogen atoms. What is important here is not necessarily the height of each peak, *but the area underneath it.* These areas, when accurately measured are in the same ratio as the number of hydrogen atoms causing each signal. We can see that the area under the signal for the methyl hydrogen atoms of 1,4-dimethylbenzene (6H) is larger than that for the phenyl hydrogen atoms (4H). Spectrometers measure these areas automatically and plot curves, called integral curves, over each signal. The heights of the integral curves are proportional to the areas under the signals. In this instance, the ratio of heights is 1.5 : 1 or 6 : 4.

13.4C Signal Splitting

A third feature of ^{1}H NMR spectra that provides us with information about the structure of a compound can be illustrated if we examine the spectrum for 1,1,2-trichloroethane (Fig. 13.12).

In Fig. 13.12 we have an example of **signal splitting.** Signal splitting is a phenomenon that arises from magnetic influences of hydrogens on atoms adjacent to those bearing the hydrogen atoms causing the signal being considered. The signal (*b*) from the two equivalent hydrogen atoms of the —CH$_2$Cl group is split into two peaks (a doublet) by the magnetic influence of the hydrogen of the —CHCl$_2$ group. Conversely, the signal (*a*) from the hydrogen of the —CHCl$_2$ group is split into three peaks (a triplet) by the magnetic influences of the two equivalent hydrogens of the —CH$_2$Cl group.

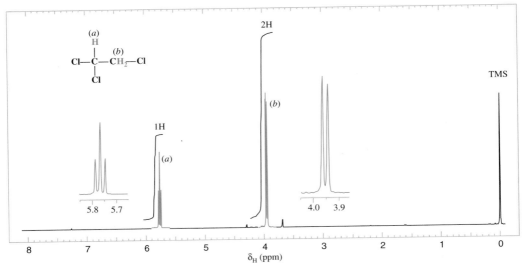

FIGURE 13.12 The 300 MHz proton ^1H NMR spectrum of 1,1,2-trichloroethane. Expansions of the signals are shown in the offset plots.

At this point signal splitting may seem like an unnecessary complication. As we gain experience in interpreting ^1H NMR spectra, we shall find that because signal splitting occurs in a predictable way, it often provides us with important information about the structure of the compound.

Now that we have had an introduction to the important features of ^1H NMR spectra, we are in a position to consider them in greater detail.

13.5 NUCLEAR SPIN: THE ORIGIN OF THE SIGNAL

We are already familiar with the concept of electron spin and with the fact that the spins of electrons confer on them the spin quantum states of $+\frac{1}{2}$ or $-\frac{1}{2}$. Electron spin is the basis for the Pauli exclusion principle (Section 1.11); it allows us to understand how two electrons with paired spins may occupy the same atomic or molecular orbital.

The nuclei of certain isotopes also spin and therefore these nuclei possess spin quantum numbers, I. The nucleus of ordinary hydrogen, ^1H (i.e., a proton), is like the electron; its spin quantum number I is $\frac{1}{2}$ and it can assume either of two spin states: $+\frac{1}{2}$ or $-\frac{1}{2}$. These correspond to the magnetic moments allowed for $I = \frac{1}{2}$, $m = +\frac{1}{2}$ or $-\frac{1}{2}$. Other nuclei with spin quantum numbers $I = \frac{1}{2}$ are ^{13}C, ^{19}F, and ^{31}P. Some nuclei, such as ^{12}C, ^{16}O, and ^{32}S, have no spin ($I = 0$) and these nuclei do not give an NMR spectrum. Other nuclei have spin quantum numbers greater than $\frac{1}{2}$. In our treatment here, however, we shall be concerned primarily with the spectra that arise from protons and from ^{13}C, both of which have $I = \frac{1}{2}$. We shall begin with proton spectra.

Since the proton is electrically charged, the spinning proton generates a tiny magnetic moment—one that coincides with the axis of spin (Fig. 13.13). This tiny magnetic moment confers on the spinning proton the properties of a tiny bar magnet.

In the absence of a magnetic field (Fig. 13.14*a*), the magnetic moments of the protons of a given sample are randomly oriented. When a compound containing hydrogen (and thus protons) is placed in an applied external magnetic field, however, the

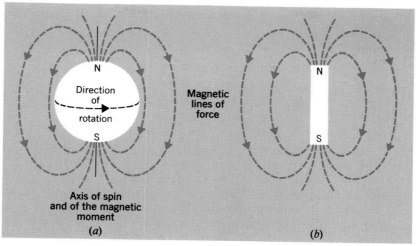

FIGURE 13.13 (*a*) The magnetic field associated with a spinning proton. (*b*) The spinning proton resembles a tiny bar magnet.

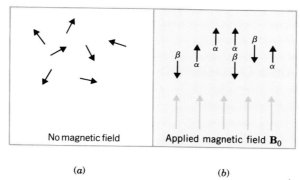

FIGURE 13.14 (*a*) In the absence of a magnetic field the magnetic moments of protons (represented by arrows) are randomly oriented. (*b*) When an external magnetic field ($\mathbf{B_0}$) is applied the protons orient themselves. Some are aligned with the applied field (α spin state) and some against it (β spin state).

protons may assume one of two possible orientations with respect to the external magnetic field. The magnetic moment of the proton may be aligned "with" the external field or "against" it (Fig. 13.14*b*). These alignments correspond to the two spin states mentioned earlier.

As we might expect, the two alignments of the proton in an external field are not of equal energy. When the proton is aligned with the magnetic field, its energy is lower than when it is aligned against the magnetic field.

Energy is required to "flip" the proton from its lower energy state (with the field) to its higher energy state (against the field). In an NMR spectrometer this energy is supplied by electromagnetic radiation in the rf region. When this energy absorption occurs, the nuclei are said to be *in resonance* with the electromagnetic radiation. The energy required is proportional to the strength of the magnetic field (Fig. 13.15). One can show by relatively simple calculations that, in a magnetic field of approximately 7.04 Tesla, for example, electromagnetic radiation of 300×10^6 cycles per second (cps)

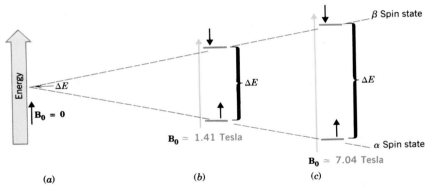

FIGURE 13.15 The energy difference between the two spin states of a proton depends on the strength of the applied external magnetic field, $\mathbf{B_0}$. (a) If there is no applied field ($\mathbf{B_0} = 0$), there is no energy difference between the two states. (b) If $\mathbf{B_0} \simeq 1.41$ Tesla, the energy difference corresponds to that of electromagnetic radiation of 60×10^6 Hz (60 MHz). (c) In a magnetic field of approximately 7.04 Tesla, the energy difference corresponds to electromagnetic radiation of 300×10^6 Hz (300 MHz). Instruments are available that operate at these and even higher frequencies (as high as 600 MHz).

(300 MHz) supplies the correct amount of energy for protons.* The proton NMR spectra given in this chapter are 300-MHz spectra.

13.6 SHIELDING AND DESHIELDING OF PROTONS

All protons do not absorb energy at the same external magnetic field strength. The two spectra that we examined earlier demonstrate this for us. The aromatic protons of 1,4-dimethylbenzene absorb at lower field strength (δ 7.05); the various alkyl protons of 1,4-dimethylbenzene and 1,1,2-trichloroethane all absorb at higher magnetic field strengths.

The general position of a signal in an NMR spectrum—that is, the strength of the magnetic field required to bring about absorption of energy—can be related to electron densities and electron circulations in the compounds. Under the influence of an external magnetic field the electrons move in certain preferred paths. Because they do, and because electrons are charged particles, they generate tiny magnetic fields.

We can see how this happens if we consider the electrons around the proton in a σ bond of a C—H group. In doing so, we shall oversimplify the situation by assuming that σ electrons move in generally circular paths. The magnetic field generated by these σ electrons is shown in Fig. 13.16.

The small magnetic field generated by the electrons is called **an induced field. *At the proton, the induced magnetic field opposes the external magnetic field.*** This means that the actual magnetic field sensed by the proton is slightly less than the external field. The electrons are said *to shield* the proton.

A proton strongly shielded by electrons does not, of course, absorb at the same external field strength as a proton that is less shielded by electrons. A shielded proton

---

* The relationship between the frequency of the radiation (ν) and the strength of the magnetic field ($\mathbf{B_0}$) is,

$$\nu = \frac{\gamma \mathbf{B_0}}{2\pi}$$

where γ is the magnetogyric (or gyromagnetic) ratio. For a proton, $\gamma = 26.753$ rad s^{-1} Tesla^{-1}.

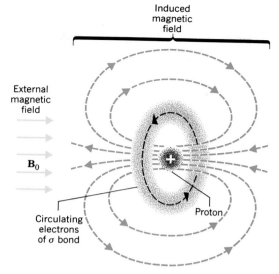

FIGURE 13.16 The circulations of the electrons of a C—H bond under the influence of an external magnetic field. The electron circulations generate a small magnetic field (an induced field) that shields the proton from the external field.

will absorb *at higher external field strengths* (or in an FT instrument at *higher frequencies*); the external field must be made larger by the spectrometer in order to compensate for the small induced field (Fig. 13.17).

The extent to which a proton is shielded by the circulation of σ electrons depends on the relative electron density around the proton. This electron density depends

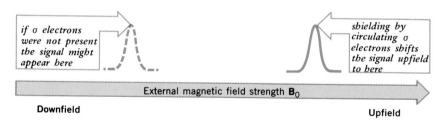

FIGURE 13.17 Shielding by σ electrons causes ^{1}H NMR absorption to be shifted to higher external magnetic field strengths.

largely on the presence or absence of electronegative groups. Electronegative groups withdraw electron density from the C—H bond, particularly if they are attached to the same carbon. We can see an example of this effect in the spectrum of 1,1,2-trichloroethane (Fig. 13.12). The proton of C1 absorbs at a lower magnetic field strength (δ 5.77) than the protons of C2 (δ 3.95). Carbon 1 bears two highly electronegative chloro groups, whereas C2 bears only one. The protons of C2, consequently, are more effectively shielded because the σ electron density around them is greater.

The circulations of delocalized π electrons generate magnetic fields that can either **shield or deshield** nearby protons. Whether shielding or deshielding occurs depends on the location of the proton in the *induced* field. The aromatic protons of benzene derivatives (Section 14.6C) are *deshielded* because their locations are such that the induced magnetic field reinforces the applied magnetic field.

Because of this deshielding effect, the absorption of energy by phenyl protons occurs downfield at relatively low magnetic field strength. The protons of benzene itself absorb at δ 7.27. The aromatic protons of 1,4-dimethylbenzene (Fig. 13.11) absorb at δ 7.05.

Magnetic fields created by circulating π electrons *shield* the protons of ethyne (and other terminal alkynes) causing them to absorb at higher magnetic field strengths than we might otherwise expect. If we were to consider *only* the relative electronegativities of carbon in its three hybridization states, we might expect the following order of protons attached to each type of carbon:

$$\text{(low field strength)}\qquad sp < sp^2 < sp^3 \qquad \text{(high field strength)}$$

In fact, protons of terminal alkynes absorb between δ 2.0 and δ 3.0 and the order is

$$\text{(low field strength)}\qquad sp^2 < sp < sp^3 \qquad \text{(high field strength)}$$

This upfield shift of the absorption of protons of terminal alkynes is a result of shielding produced by the circulating π electrons of the triple bond. The origin of this shielding is illustrated in Fig. 13.18.

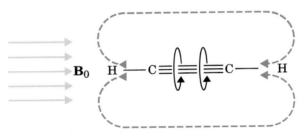

FIGURE 13.18 The shielding of protons of ethyne by π electron circulations. Shielding causes protons attached to the *sp*-carbons to absorb further upfield than vinylic protons.

13.7 THE CHEMICAL SHIFT

We see now that shielding and deshielding effects cause the absorptions of protons to be shifted from the position at which a bare proton would absorb (i.e., a proton stripped of its electrons). Since these shifts result from the circulation of electrons in *chemical* bonds, they are called **chemical shifts.**

Chemical shifts are measured with reference to the absorption of protons of reference compounds. A reference is used because it is impractical to measure the actual value of the magnetic field at which absorptions occur. The reference compound most often used is tetramethylsilane (TMS). A small amount of tetramethylsilane is usually added to the sample whose spectrum is being measured, and the signal from the 12 equivalent protons of TMS is used to establish the zero point on the delta (δ) scale.

$$Si(CH_3)_4$$
Tetramethylsilane
(TMS)

Tetramethylsilane was chosen as a reference compound for several reasons. It has 12 hydrogen atoms and, therefore, a very small amount of TMS gives a relatively large

TABLE 13.3 Approximate proton chemical shifts

TYPE OF PROTON	CHEMICAL SHIFT (δ, ppm)
1° Alkyl, RCH_3	0.8–1.0
2° Alkyl, RCH_2R	1.2–1.4
3° Alkyl, R_3CH	1.4–1.7
Allylic, $R_2C{=}\overset{\displaystyle }{\underset{\displaystyle R}{C}}{-}CH_3$	1.6–1.9
Ketone, $R\overset{\displaystyle }{\underset{\displaystyle \\ O}{C}}CH_3$	2.1–2.6
Benzylic, $ArCH_3$	2.2–2.5
Acetylenic, $RC{\equiv}CH$	2.5–3.1
Alkyl iodide, RCH_2I	3.1–3.3
Ether, $ROCH_2R$	3.3–3.9
Alcohol, $HOCH_2R$	3.3–4.0
Alkyl bromide, RCH_2Br	3.4–3.6
Alkyl chloride, RCH_2Cl	3.6–3.8
Vinylic, $R_2C{=}CH_2$	4.6–5.0
Vinylic, $R_2C{=}\overset{\displaystyle }{\underset{\displaystyle R}{C}}H$	5.2–5.7
Aromatic, ArH	6.0–9.5
Aldehyde, $R\overset{\displaystyle }{\underset{\displaystyle O}{C}}H$	9.5–9.6
Alcohol hydroxyl, ROH	0.5–6.0[a]
Amino, $R{-}NH_2$	1.0–5.0[a]
Phenolic, $ArOH$	4.5–7.7[a]
Carboxylic, $RC\overset{\displaystyle }{\underset{\displaystyle O}{O}}H$	10–13[a]

[a] The chemical shifts of these protons vary in different solvents and with temperature and concentration.

signal. Because the hydrogen atoms are all equivalent, they give a *single signal.* Since silicon is less electronegative than carbon, the protons of TMS are in regions of high electron density. They are, as a result, highly shielded, and the signal from TMS occurs in a region of the spectrum where few other hydrogen atoms absorb. Thus, their signal seldom interferes with the signals from other hydrogen atoms. Tetramethylsilane, like an alkane, is relatively inert. Finally, it is volatile; its boiling point is 27°C. After the spectrum has been determined, the TMS can be removed easily by evaporation.

Chemical shifts are measured in hertz (cps), as if the frequency of the electromagnetic radiation were being varied. In actuality, in a CW instrument, it is the magnetic field that is changed. But since the values of frequency and the strength of the magnetic field are mathematically proportional, frequency units (Hz) are appropriate ones.

The chemical shift of a proton, when expressed in hertz, is proportional to the strength of the external magnetic field. Since spectrometers with different magnetic field strengths are commonly used, it is desirable to express chemical shifts in a form that is independent of the strength of the external field. This can be done easily by dividing the chemical shift by the frequency of the spectrometer, with both numerator and denominator of the fraction expressed in frequency units (Hz). Since chemical shifts are always very small (typically < 5000 Hz) compared with the total field strength (commonly the equivalent of 60, 300, or 600 *million* Hz), it is convenient to express these fractions in units of *parts per million* (ppm). This is the origin of the delta scale for the expression of chemical shifts relative to TMS.

$$\delta = \frac{(\text{observed shift from TMS in hertz}) \times 10^6}{(\text{operating frequency of the instrument in hertz})}$$

For example, the chemical shift for benzene is 2181 Hz when the instrument is operating at 300 MHz. Therefore,

$$\delta = \frac{2181 \text{ Hz} \times 10^6}{300 \times 10^6 \text{ Hz}} = 7.27$$

Table 13.3 gives the *approximate* values of proton chemical shifts for some common hydrogen-containing groups.

13.8 CHEMICAL SHIFT EQUIVALENT AND NONEQUIVALENT PROTONS

Two or more protons that are in identical environments have the same chemical shift and, therefore, give only one ^{1}H NMR signal. How do we know when protons are in the same environment? For most compounds, protons that are in the same environment are also equivalent in chemical reactions. That is, **chemically equivalent** protons are **chemical shift equivalent** in ^{1}H NMR spectra.

13.8A Homotopic Hydrogen Atoms

A simple way to decide whether two or more protons in a given compound are chemical shift equivalent is to replace each hydrogen in turn by some other group. The group may be real or imaginary. If, in making these replacements, you get the same compound, then the hydrogens being replaced are said to be **chemically equivalent** or **homotopic,** and **homotopic atoms (or groups) are chemical shift equivalent.**

Consider 2-methylpropene as an example:

3-Chloro-2-methylpropene   1-Chloro-2-methylpropene

Here there are two sets of homotopic hydrogens. The six methyl hydrogens (*b*) form one set; replacing any one of them with chlorine, for example, leads to the same compound, 3-chloro-2-methylpropene. The two methylene hydrogens (*a*) form another set; replacing either of these leads to 1-chloro-2-methylpropene. 2-Methylpropene, therefore, gives two ^{1}H NMR signals.

PROBLEM 13.3 _____

Carry out a similar analysis for 1,4-dimethylbenzene (Section 13.4C) and show that it has two sets of chemical shift equivalent protons, explaining why its spectrum (Fig. 13.11) has only two signals.

PROBLEM 13.4 _____

How many signals would each compound give in its ^{1}H NMR spectrum?

(a) Ethane

(b) Propane

(c) *tert*-Butyl methyl ether

(d) 2,3-Dimethyl-2-butene

(e) (*Z*)-2-butene

(f) (*E*)-2-butene

13.8B Enantiotopic and Diastereotopic Hydrogen Atoms

If replacement of each of two hydrogen atoms by the same group yields compounds that are enantiomers, the two hydrogen atoms are said to be **enantiotopic.** *Enantiotopic hydrogen atoms have the same chemical shift and give only one ^{1}H NMR signal.**

Enantiomers

The two hydrogen atoms of the —CH$_2$Br group of ethyl bromide are enantiotopic. Ethyl bromide, then, gives two signals in its ^{1}H NMR spectrum. The three equivalent protons of the —CH$_3$ group give one signal; the two enantiotopic protons of the —CH$_2$Br group give the other signal. (The ^{1}H NMR spectrum of ethyl bromide as we shall see, actually consists of seven peaks (three in one signal, four in the other). This is a result of signal splitting, which is explained in Section 13.9.)

If replacement of each of two hydrogen atoms by a group, **Z,** gives compounds that are diastereomers, the two hydrogens are said to be **diastereotopic.** Except for accidental coincidence, *diastereotopic protons do not have the same chemical shift and give rise to different ^{1}H NMR signals.*

The two protons of the ═CH$_2$ group of chloroethene are diastereotopic.

---

*Enantiotopic hydrogen atoms may not have the same chemical shift if the compound is dissolved in a chiral solvent. However, most ^{1}H NMR spectra are determined using achiral solvents and in this situation enantiotopic protons have the same chemical shift.

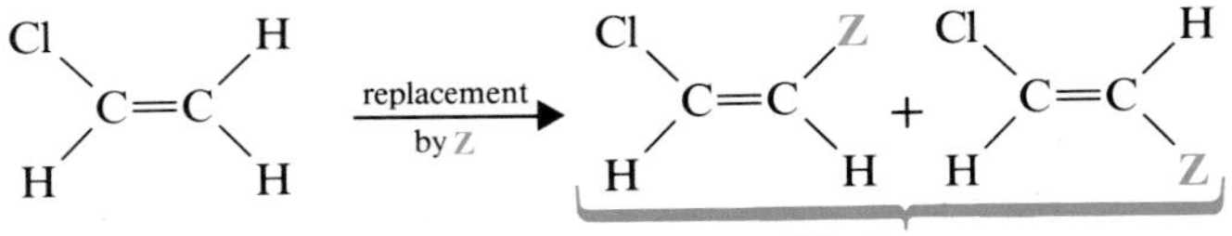

Diastereomers

Chloroethene, then, should give signals from three nonequivalent protons; one for the proton of the $ClCH=$ group, and one for each of the diastereotopic protons of the $=CH_2$ group.

The two methylene ($-CH_2-$) protons of *sec*-butyl alcohol are also diastereotopic. We can illustrate this with one enantiomer of *sec*-butyl alcohol in the following way:

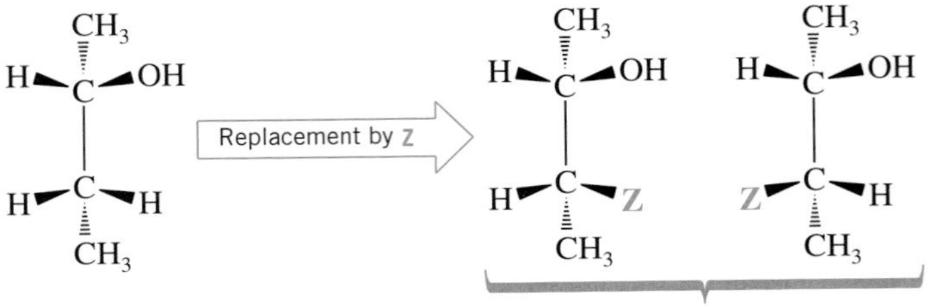

sec-**Butyl alcohol**
(one enantiomer)

Diastereomers

These two protons have different chemical shifts and give two signals in the ^{1}H NMR spectrum. The two signals may be close enough to overlap, however.

PROBLEM 13.5

(a) Show that replacing each of the two methylene protons of the other *sec*-butyl alcohol enantiomer by **Z** also leads to a pair of diastereomers. (b) How many chemically different kinds of protons are there in *sec*-butyl alcohol? (c) How many ^{1}H NMR signals would you expect to find in the spectrum of *sec*-butyl alcohol?

PROBLEM 13.6

How many ^{1}H NMR signals would you expect from each of the following compounds?

(a) $CH_3CH_2CH_2CH_3$ (f) 1,1-Dimethylcyclopropane

(b) CH_3CH_2OH (g) *trans*-1,2-Dimethylcyclopropane

(c) $CH_3CH=CH_2$ (h) *cis*-1,2-Dimethylcyclopropane

(d) *trans*-2-Butene   (i) 1-Pentene

(e) 1,2-Dibromopropane   (j) 1-Chloro-2-propanol

13.9 SIGNAL SPLITTING: SPIN–SPIN COUPLING

Signal splitting is caused by the effect of magnetic fields of protons on nearby atoms. We have seen an example of signal splitting in the spectrum of 1,1,2-trichloroethane

(Fig. 13.12). The signal from the two equivalent protons of the —CH$_2$Cl group of 1,1,2-trichloroethane is split into two peaks by the single proton of the CHCl$_2$— group. The signal from the proton of the CHCl$_2$— group is split into three peaks by the two protons of the —CH$_2$Cl group. This is further illustrated in Fig. 13.19.

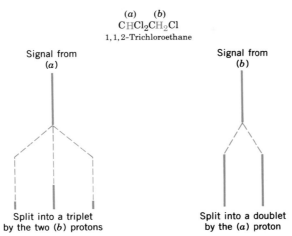

(a) (b)
CHCl$_2$CH$_2$Cl
1,1,2-Trichloroethane

Signal from (a)

Signal from (b)

Split into a triplet
by the two (b) protons

Split into a doublet
by the (a) proton

FIGURE 13.19 Signal splitting in 1,1,2-trichloroethane.

Signal splitting arises from a phenomenon known as **spin–spin coupling,** which we shall soon examine. Spin–spin coupling effects are transferred primarily through the bonding electrons and *are not usually observed if the coupled protons are separated by more than three σ bonds.* Thus, we observe signal splitting from the protons of *adjacent σ-bonded* atoms as in 1,1,2-trichloroethane (Fig. 13.12). However, we would not observe splitting of either signal of *tert*-butyl methyl ether (see following structure) because the protons labeled (b) are separated from those labeled (a) by more than three σ bonds. Both signals from *tert*-butyl methyl ether are singlets.

(a)
CH$_3$
(a) | (b)
CH$_3$—C—O—CH$_3$
          |
          CH$_3$
          (a)

tert-**Butyl methyl ether**
(no signal splitting)

Signal splitting is not observed for protons that are chemically equivalent (homotopic) or enantiotopic. That is, signal splittings do not occur between protons that have *exactly the same chemical shift.* Thus, we would not expect, and do not find, signal splitting in the signal from the six equivalent hydrogen atoms of ethane.

CH$_3$CH$_3$ (no signal splitting)

Nor do we find signal splitting occurring from enantiotopic protons of methoxy-acetonitrile (Fig. 13.20).

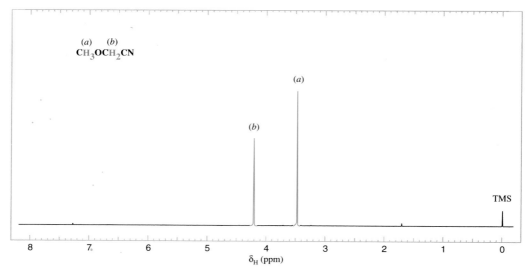

FIGURE 13.20 The 300 MHz ^{1}H NMR spectrum of methoxyacetonitrile. The signal of the enantiotopic protons (*b*) is not split.

There is a subtle distinction between *spin–spin coupling* and signal splitting. Spin–spin coupling often occurs between sets of protons that have the same chemical shift (and this coupling can be detected by methods that we shall not go into here). However, spin–spin coupling *leads to signal splitting only when the sets of protons have different chemical shifts.*

Let us now explain how signal splitting arises from coupled sets of protons that are not chemical shift equivalent.

We have seen that protons can be aligned in only two ways in an external magnetic field: with the field or against it. Therefore, the magnetic moment of a proton on an adjacent atom may affect the magnetic field at the proton whose signal we are observing in only one of two ways. The occurrence of these two slightly different effects causes the appearance of a smaller peak somewhat upfield (from where the signal may have occurred) and another peak somewhat downfield.

Figure 13.21 shows how two possible orientations of a neighboring proton, H_b, split the signal of the proton H_a. (H_b and H_a are not equivalent.)

The separation of these peaks in frequency units is called the **coupling constant** and is abbreviated J_{ab}. Coupling constants are generally reported in hertz. Because coupling is caused entirely by internal forces, the magnitudes of coupling constants *are not* dependent on the magnitude of the applied field. Coupling constants measured (in Hz) on an instrument operating at 60 MHz will be the same as those measured on an instrument operating at 300 MHz.

When we determine ^{1}H NMR spectra we are, of course, observing effects produced by billions of molecules. Since the difference in energy between the two possible orientations of the proton of H_b is very small, the two orientations will be present in roughly (but not exactly) equal numbers. The signal that we observe from H_a is, therefore, split into two peaks of roughly equal intensity, *a 1:1 doublet.*

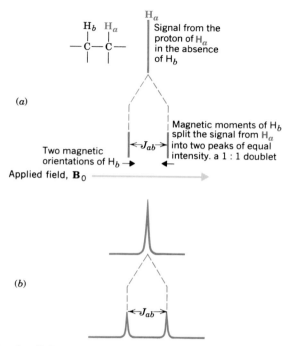

FIGURE 13.21 Signal splitting arising from spin–spin coupling with one nonequivalent proton of a neighboring hydrogen atom. A theoretical analysis is shown in (*a*) and the actual appearance of the spectrum in (*b*). The distance between the centers of the peaks of the doublet is called the coupling constant, J_{ab}. The term J_{ab} is expressed in hertz. The magnitudes of coupling constants are *not* dependent on the magnitude of the applied field and their values (in Hz) are the same, regardless of the operating frequency of the spectrometer.

PROBLEM 13.7

Sketch the ^{1}H NMR spectrum of $CHBr_2CHCl_2$. Which signal would you expect to occur at lower magnetic field strength; that of the proton of the $CHBr_2$— group or of the —$CHCl_2$ group? Why?

Two equivalent protons on an adjacent carbon atom (or on adjacent carbon atoms) split the signal from an absorbing proton into a 1:2:1 *triplet*. Figure 13.22 illustrates how this pattern occurs.

In compounds of either type shown in Figure 13.22, both protons may be aligned with the applied field. This orientation causes a peak to appear at a lower applied field strength than would occur in the absence of the two hydrogen atoms H_b. Conversely, both protons may be aligned against the applied field. This orientation of the protons of H_b causes a peak to appear at higher applied field strengths than would occur in their absence. Finally, there are two ways in which the two protons may be aligned in which one opposes the applied field and one reinforces it. These arrangements do not displace the signal. Since the probability of this last arrangement is twice that of either of the other two, the center peak of the triplet is twice as intense.

The proton of the $CHCl_2$— group of 1,1,2-trichloroethane is an example of a proton of the type having two equivalent protons on an adjacent carbon. The signal

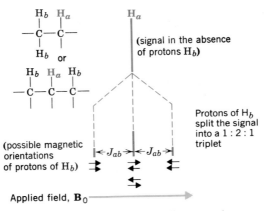

FIGURE 13.22 Two equivalent protons (H_b) on an adjacent carbon atom split the signal from H_a into a $1:2:1$ triplet.

from the $CHCl_2$— group (Fig. 13.12) appears as a $1:2:1$ triplet and, as we would expect, the signal of the —CH_2Cl group of 1,1,2-trichloroethane is split into a $1:1$ doublet by the proton of the $CHCl_2$— group.

The spectrum of 1,1,2,3,3-pentachloropropane (Fig. 13.23) is similar to that of 1,1,2-trichloroethane in that it also consists of a $1:2:1$ triplet and a $1:1$ doublet. The two hydrogen atoms H_b of 1,1,2,3,3-pentachloropropane are equivalent even though they are on separate carbon atoms.

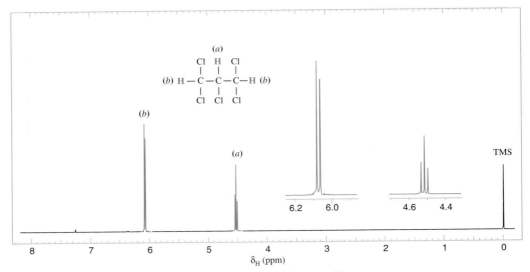

FIGURE 13.23 The ^{1}H NMR spectrum of 1,1,2,3,3-pentachloropropane.

PROBLEM 13.8

The relative positions of the doublet and triplet of 1,1,2-trichloroethane (Fig. 13.12) and 1,1,2,3,3-pentachloropropane (Fig. 13.23) are reversed. Explain this.

Three equivalent protons (H_b) on a neighboring carbon split the signal from the H_a proton into a 1:3:3:1 quartet. This pattern is shown in Fig. 13.24.

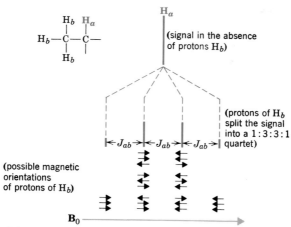

FIGURE 13.24 Three equivalent protons (H_b) on an adjacent carbon split the signal from H_a into a 1:3:3:1 quartet.

The signal from two equivalent protons of the —CH_2Br group of ethyl bromide (Fig. 13.25) appears as a 1:3:3:1 quartet because of this type of signal splitting. The three equivalent protons of the CH_3— group are split into a 1:2:1 triplet by the two protons of the —CH_2Br group.

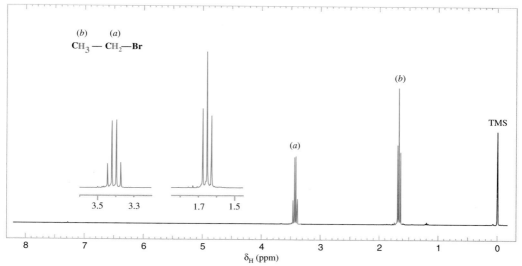

FIGURE 13.25 The 300 MHz ^{1}H NMR spectrum of ethyl bromide. Expansions of the signals are shown in the offset plots.

The kind of analysis that we have just given can be extended to compounds with even larger numbers of equivalent protons on adjacent atoms. These analyses show that *if there are n equivalent protons on adjacent atoms these will split a signal into n + 1 peaks.* (We may not always see all of these peaks in actual spectra, however, because some of them may be very small.)

PROBLEM 13.9

What kind of ^{1}H NMR spectrum would you expect the following compound to give?

$$(Cl_2CH)_3CH$$

Sketch the spectrum, showing the splitting patterns and relative position of each signal.

PROBLEM 13.10

Propose structures for each of the compounds whose spectra are shown in Fig. 13.26, and account for the splitting pattern of each signal.

The splitting patterns shown in Fig. 13.26 are fairly easy to recognize because in each compound there are only two sets of nonequivalent hydrogen atoms. One feature present in all spectra, however, helps us recognize splitting patterns in more complicated spectra: the **reciprocity of coupling constants.**

The separation of the peaks in hertz gives us the value of the coupling constants. Therefore, if we look for doublets, triplets, quartets, and so on, that have *the same coupling constants,* the chances are good that these multiplets are related to each other because they arise from reciprocal spin–spin couplings.

The two sets of protons of an ethyl group, for example, appear as a triplet and a quartet as long as the ethyl group is attached to an atom that does not bear any hydrogen atoms. The spacings of the peaks of the triplet and the quartet of an ethyl group will be the same because the coupling constants (J_{ab}) are the same (Fig. 13.27).

Other techniques, made possible by FT NMR methods, allow easy recognition of complex splitting relationships. One such technique is ^{1}H–^{1}H correlation spectroscopy, called COSY (see Special Topic E).

Proton NMR spectra have other features, however, that are not at all helpful when we try to determine the structure of a compound.

1. Signals may overlap. This happens when the chemical shifts of the signals are very nearly the same. In the 60 MHz spectrum of ethyl chloroacetate (Fig. 13.28a) we see that the singlet of the —CH_2Cl group falls directly on top of one of the outermost peaks of the ethyl quartet. Using NMR spectrometers with higher magnetic field strength (corresponding to ^{1}H resonance frequencies of 300, 500, or 600 MHz) often allows separation of signals that would overlap at lower magnetic field strengths (for an example, see Fig. 13.28b).

2. Spin–spin couplings between the protons of nonadjacent atoms may occur. This long-range coupling happens frequently when π-bonded atoms intervene between the atoms bearing the coupled protons.

3. The splitting patterns of aromatic groups are difficult to analyze. A monosubstituted benzene ring (a phenyl group) has three different kinds of protons.

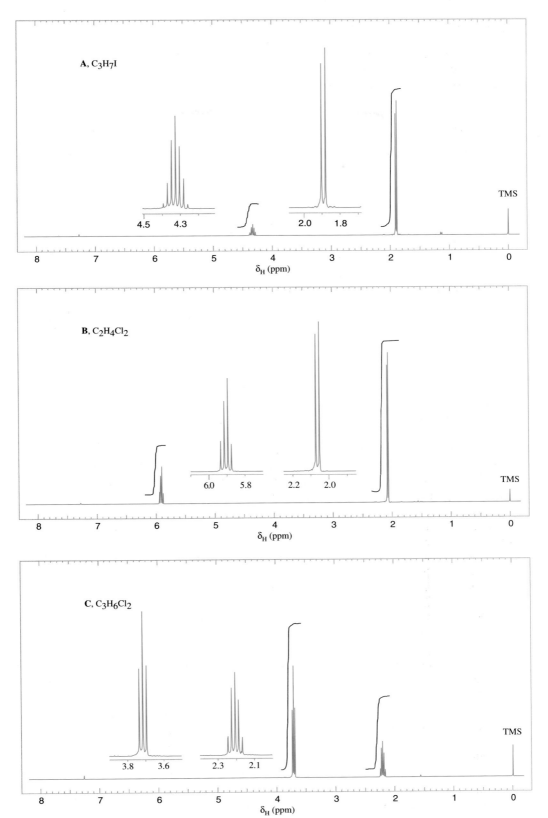

FIGURE 13.26 The 300 MHz ^{1}H NMR spectra Problem 13.10. Expansions are shown in the offset plots.

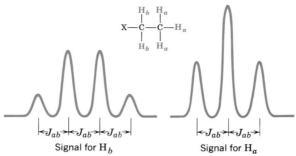

FIGURE 13.27 A theoretical splitting pattern for an ethyl group. For an example, see the spectrum of ethyl bromide (Fig. 13.25).

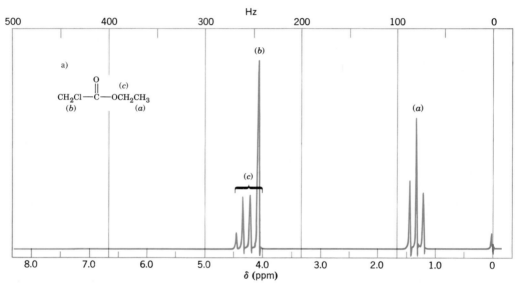

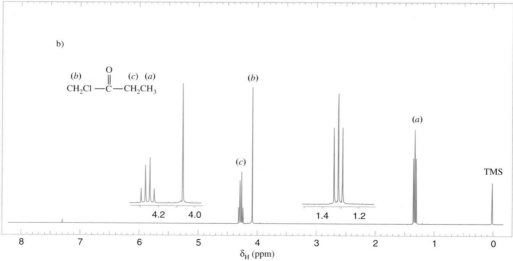

FIGURE 13.28 a) The 60 MHz ^{1}H NMR spectrum of ethyl chloroacetate. Note the overlapping signals at 4 ppm. b) The 300 MHz ^{1}H NMR spectrum of ethyl chloroacetate, showing resolution at higher magnetic field strength of the signals which overlapped at 60 MHz. Expansions of the signals are shown in the offset plots.

The chemical shifts of these protons may be so similar that the phenyl group gives a signal that resembles a singlet. Or the chemical shifts may be different and because of long-range couplings, the phenyl group signal appears as a very complicated multiplet.

In all of the ^{1}H NMR spectra that we have considered so far, we have restricted our attention to signal splittings arising from interactions of only two sets of equivalent protons on adjacent atoms. What kind of patterns should we expect from compounds in which more than two sets of equivalent protons are interacting? We cannot answer this question completely because of limitations of space, but we can give an example that illustrates the kind of analysis that is involved. Let us consider a 1-substituted propane.

$$\overset{(a)}{CH_3}-\overset{(b)}{CH_2}-\overset{(c)}{CH_2}-Z$$

Here, there are three sets of equivalent protons. We have no problem in deciding what kind of signal splitting to expect from the protons of the CH_3— group or the —CH_2Z group. The methyl group is spin–spin coupled only to the two protons of the central —CH_2— group. Therefore, the methyl group should appear as a triplet. The protons of the —CH_2Z group are similarly coupled only to the two protons of the central —CH_2— group. Thus, the protons of the —CH_2Z group should also appear as a triplet.

But what about the protons of the central —CH_2— group (b)? They are spin–spin coupled with the three protons at (a) and with two protons at (c). The protons at (a) and (c), moreover, are not equivalent. If the coupling constants J_{ab} and J_{bc} have quite different values, then the protons at (b) could be split into a quartet by the three protons (a) and each line of the quartet could be split into a triplet by the two protons (c) (Fig. 13.29).

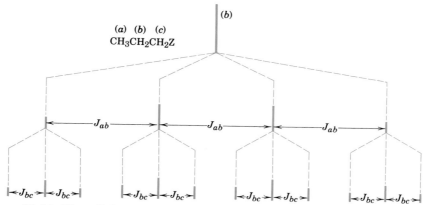

FIGURE 13.29 The splitting pattern that would occur for the (b) protons of $CH_3CH_2CH_2Z$ if J_{ab} is much larger than J_{bc}. Here $J_{ab} = 3J_{bc}$.

It is unlikely, however, that we would observe as many as 12 peaks in an actual spectrum because the coupling constants are such that peaks usually fall on top of peaks. The ^{1}H NMR spectrum of 1-nitropropane (Fig. 13.30) is typical of 1-substituted propane compounds. We see that the (b) protons are split into six major peaks, each of which shows a slight sign of further splitting.

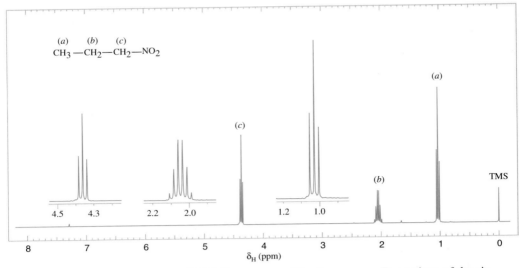

FIGURE 13.30 The 300 MHz ^{1}H NMR spectrum of 1-nitropropane. Expansions of the signals are shown in the offset plots.

PROBLEM 13.11

Carry out an analysis like that shown in Fig. 13.29 and show how many peaks the signal from (b) would be split into if $J_{ab} = 2J_{bc}$ and if $J_{ab} = J_{bc}$. (*Hint:* In both cases peaks will fall on top of peaks so that the total number of peaks in the signal is fewer than 12.)

The presentation we have given here applies only to what are called *first-order spectra.* In first-order spectra, the distance in hertz ($\Delta\nu$) that separates the coupled signals is very much larger than the coupling constant, J. That is, $\Delta\nu \gg J$. In *second-order spectra* (which we have not discussed) $\Delta\nu$ approaches J in magnitude and the situation becomes much more complex. The number of peaks increases and the intensities are not those that might be expected from first-order considerations.

13.10   PROTON NMR SPECTRA AND RATE PROCESSES

J. D. Roberts (of the California Institute of Technology), a pioneer in the application of NMR spectroscopy to problems of organic chemistry, has compared the NMR spectrometer to a camera with a relatively slow shutter speed. Just as a camera with a slow shutter speed blurs photographs of objects that are moving rapidly, the NMR spectrometer blurs its picture of molecular processes that are occurring rapidly.

What are some of the rapid processes that occur in organic molecules?

At temperatures near room temperature, groups connected by carbon–carbon single bonds rotate very rapidly. Because of this, when we determine spectra of compounds with single bonds, the spectra that we obtain often reflect the individual hydrogen atoms in their average environment—that is, in an environment that is an average of all the environments that the protons have as a result of the group rotations.

To see an example of this effect, let us consider the spectrum of ethyl bromide again. The most stable conformation of ethyl bromide is the one in which the groups are perfectly staggered. In this staggered conformation one hydrogen of the methyl group (in red in following figure) is in a different environment from that of the other two methyl hydrogen atoms. If the NMR spectrometer were to detect this particular conformation of ethyl bromide, it would show the proton of this hydrogen of the methyl group at *a different chemical shift*. We know, however, that in the spectrum of ethyl bromide (Fig. 13.25), the three protons of the methyl group give *a single signal* (a signal that is split into a triplet by spin–spin coupling with the two protons of the adjacent carbon).

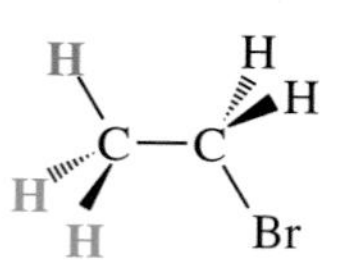

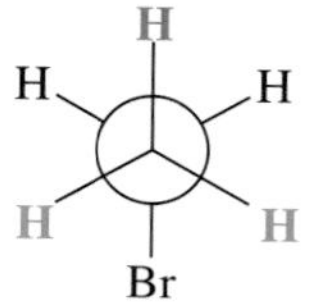

The methyl protons of ethyl bromide give a single signal because at room temperature the groups connected by the carbon–carbon single bond rotate approximately 1 million times each second. The "shutter speed" of the NMR spectrometer is too slow to "photograph" this rapid rotation; instead, it photographs the methyl hydrogen atoms in their average environments, and in this sense, it gives us a blurred picture of the methyl group.

Rotations about single bonds slow down as the temperature of the compound is lowered. Sometimes, this slowing of rotations allows us to "see" the different conformations of a molecule when we determine the spectrum at a sufficiently low temperature.

An example of this phenomenon, and one that also shows the usefulness of deuterium labeling, can be seen in the low-temperature ^{1}H NMR spectra of cyclohexane and of undecadeuteriocyclohexane. (These experiments originated with F. A. L. Anet of the University of California, Los Angeles, another pioneer in the applications of NMR spectroscopy to organic chemistry, especially to conformational analysis.)

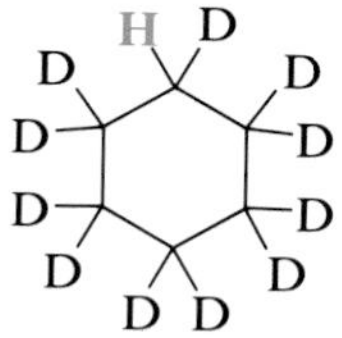

Undecadeuteriocyclohexane

At room temperature, ordinary cyclohexane gives one signal because interconversions between the various chair forms occur very rapidly. At low temperatures, however, ordinary cyclohexane gives a very complex ^{1}H NMR spectrum. At low temperatures interconversions are slow; the axial and equatorial protons have different chemical shifts, and complex spin–spin couplings occur.

At − 100°C, however, undecadeuteriocyclohexane gives only two signals of equal intensity. These signals correspond to the axial and equatorial hydrogen atoms of the following two chair conformations. Interconversions between these conformations occur at this low temperature, but they happen slowly enough for the NMR spectrometer to detect the individual conformations. [The nucleus of a deuterium atom (a deuteron) has a much smaller magnetic moment than a proton, and signals from deuteron absorption do not occur in ¹H NMR spectra.]

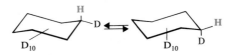

PROBLEM 13.12

What kind of ¹H NMR spectrum would you expect to obtain from undecadeuteriocyclohexane at room temperature?

Another example of a rapidly occurring process can be seen in ¹H NMR spectra of ethanol. The ¹H NMR spectrum of ordinary ethanol shows the hydroxyl proton as a singlet and the protons of the —CH₂— group as a quartet (Fig. 13.31). In ordinary ethanol we observe *no signal splitting arising from coupling between the hydroxyl proton and the protons of the —CH₂— group even though they are on adjacent atoms.*

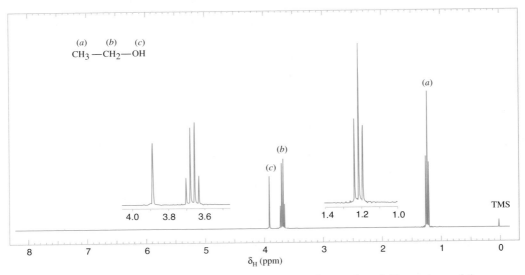

FIGURE 13.31 The 300 MHz ¹H NMR spectrum of ordinary ethanol. Expansions of the signals are shown in the offset plots.

If we were to examine a ¹H NMR spectrum of *very pure* ethanol, however, we would find that the signal from the hydroxyl proton was split into a triplet, and that the signal from the protons of the —CH₂— group was split into a multiplet of eight peaks.

Clearly, in very pure ethanol the spin of the proton of the hydroxyl group is coupled with the spins of the protons of the $-CH_2-$ groups.

Whether or not coupling occurs between the hydroxyl protons and the methylene protons depends on the length of time the proton spends on a particular ethanol molecule. Protons attached to electronegative atoms with lone pairs such as oxygen (or nitrogen) can undergo rapid **chemical exchange.** That is, they can be transferred rapidly from one molecule to another. The chemical exchange in very pure ethanol is slow and, as a consequence, we see the signal splitting of and by the hydroxyl proton in the spectrum. In ordinary ethanol, acidic and basic impurities catalyze the chemical exchange; the exchange occurs so rapidly that the hydroxyl proton gives an unsplit signal and that of the methylene protons is split only by coupling with the protons of the methyl group. We say, then, that rapid exchange causes **spin decoupling.**

Spin decoupling is found in the ^1H NMR spectra of alcohols, amines, and carboxylic acids and the signals of OH and NH protons are normally unsplit.

Protons that undergo rapid chemical exchange (i.e., those attached to oxygen or nitrogen) can be easily detected by placing the compound in D_2O. The protons are rapidly replaced by deuterons and the proton signal disappears from the spectrum.

13.11 CARBON-13 NMR SPECTROSCOPY

13.11A Interpretation of ^{13}C NMR Spectra

We begin our study of ^{13}C NMR spectroscopy with a brief examination of some special features of spectra arising from carbon-13 nuclei. These spectra are often called carbon magnetic resonance (CMR) spectra or ^{13}C NMR spectra. Although ^{13}C accounts for only 1.1% of naturally occurring carbon, the fact that ^{13}C can produce an NMR signal has profound importance for the analysis of organic compounds. The major isotope of carbon, on the other hand, carbon-12 (^{12}C), with natural abundance of about 99%, has no net magnetic spin and therefore cannot produce NMR signals. In some important ways ^{13}C spectra are usually less complex and easier to interpret than proton (^1H) NMR spectra.

13.11B One Peak for Each Type of Carbon Atom

One aspect of ^{13}C NMR spectra that greatly simplifies the interpretation process is that **each type of carbon atom in an ordinary organic molecule produces only one ^{13}C NMR peak.** There is no carbon–carbon coupling that causes splitting of signals into multiple peaks. Recall that in ^1H NMR spectra, hydrogen nuclei that are near each other (within a few bonds) couple with each other and cause the signal for each hydrogen to become a multiplet of peaks. This does not occur for adjacent carbons because only one carbon atom of every 100 carbon atoms is a carbon-13 nucleus (1.1% natural abundance). Therefore, the probability of there being two carbon-13 atoms adjacent to each other in a molecule is only about 1 in 10,000 (1.1% × 1.1%), essentially eliminating the possibility of two neighboring carbon atoms splitting each other's signal into a multiplet of peaks. The low natural abundance of ^{13}C nuclei and its inherently low sensitivity also have another effect: Carbon-13 NMR spectra can be obtained only on pulse FT NMR spectrometers, where signal averaging is possible.

Whereas carbon–carbon signal splitting does not occur in ^{13}C NMR spectra, hydrogen atoms attached to carbon can split ^{13}C NMR signals into multiple peaks.

However, it is possible to eliminate signal splitting by $^1\text{H}-^{13}\text{C}$ coupling by choosing instrumental parameters for the NMR spectrometer that decouple the proton–carbon interactions. A ^{13}C NMR spectrum that has the proton interactions eliminated is said to be **broad band (BB) proton decoupled,** since proton–carbon interactions of all types are decoupled from the carbon signals. Thus, in a typical broad band decoupled ^{13}C NMR spectrum, each type of carbon atom produces only one peak. Most ^{13}C NMR spectra are obtained in this simplified decoupled mode.

13.11C ^{13}C Chemical Shifts

As we found with ^1H spectra, the chemical shift of a given nucleus depends on the relative electron density around that atom. Decreased electron density around an atom **deshields** the atom from the magnetic field and causes its signal to occur further **downfield** (higher ppm, to the left) in the NMR spectrum. Relatively higher electron density around an atom **shields** the atom from the magnetic field and causes the signal to occur **upfield** (lower ppm, to the right) in the NMR spectrum. For example, carbon atoms that are attached only to other carbon and hydrogen atoms are relatively shielded from the magnetic field by the density of electrons around them, and, as a consequence, carbon atoms of this type produce peaks that are upfield in ^{13}C NMR spectra. On the other hand, carbon atoms bearing electronegative groups are deshielded from the magnetic field by the electron-withdrawing effects of these groups and, therefore, produce peaks that are downfield in the NMR spectrum. Electronegative groups such as halogens, hydroxyl groups, and other electron-withdrawing functional groups deshield the carbons to which they are attached, causing their ^{13}C NMR peaks to occur more downfield than those of unsubstituted carbon atoms. Reference tables of approximate chemical shift ranges for carbons bearing different substituents are available. Table 13.4 is an example. [The reference standard assigned as zero ppm in ^{13}C NMR spectra is also tetramethylsilane (TMS), $(\text{CH}_3)_4\text{Si.}$]

As a first example of the interpretation of ^{13}C NMR spectra, let us consider the ^{13}C spectrum of 1-chloro-2-propanol (Fig. 13.32a).

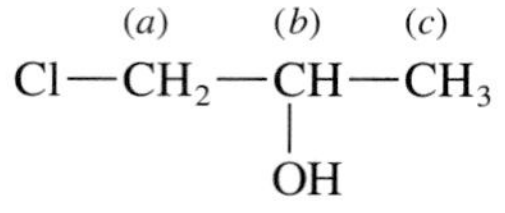

$$
\begin{array}{cccc}
(a) & (b) & (c) \\
\text{Cl}-\text{CH}_2-\text{CH}-\text{CH}_3 \\
& & | \\
& & \text{OH}
\end{array}
$$

1-Chloro-2-propanol

This compound contains three carbons and therefore produces only three peaks in its broad band decoupled ^{13}C NMR spectrum: approximately at δ 20, δ 51, and δ 67. Figure 13.32, also shows a close grouping of three peaks at δ 77. These peaks come from the deuteriochloroform (CDCl_3) used as a solvent for the sample. Many ^{13}C NMR spectra contain these peaks. Although not of concern to us here, the single carbon of deuteriochloroform is split into three peaks by an effect of the attached deuterium. The CDCl_3 peaks should be disregarded when interpreting ^{13}C spectra.

As we can see, the chemical shifts of the three peaks from 1-chloro-2-propanol are well separated from one another. This separation results from differences in shielding by circulating electrons in the local environment of each carbon. Remember: The lower the electron density in the vicinity of a given carbon, the less that carbon will be shielded, and the more downfield will be the signal for that carbon. The oxygen of the hydroxyl group is the most electronegative atom; it withdraws electrons most effectively. Therefore, the carbon bearing the —OH group is the most *deshielded* carbon,

TABLE 13.4 Approximate carbon-13 chemical shifts

TYPE OF CARBON ATOM	CHEMICAL SHIFT (δ, ppm)
1° Alkyl, RCH_3	0–40
2° Alkyl, RCH_2R	10–50
3° Alkyl, $RCHR_2$	15–50
Alkyl halide or amine, $-\overset{\vert}{\underset{\vert}{C}}-X$ $\left(X = Cl, Br, or \overset{\vert}{N}- \right)$	10–65
Alcohol or ether, $-\overset{\vert}{\underset{\vert}{C}}-O$	50–90
Alkyne, $-C\equiv$	60–90
Alkene, $\overset{\diagdown}{\underset{\diagup}{C}}=$	100–170
Aryl, ⬡$C-$	100–170
Nitriles, $-C\equiv N$	120–130
Amides, $-\overset{O}{\overset{\|}{C}}-\overset{\vert}{N}-$	150–180
Carboxylic acids, esters, $-\overset{O}{\overset{\|}{C}}-O$	160–185
Aldehydes, ketones, $-\overset{O}{\overset{\|}{C}}-$	182–215

and so this carbon gives the signal that is the most downfield, at δ 67. Chlorine is less electronegative than oxygen, causing the peak for the carbon to which it is attached to occur more upfield, at δ 51. The methyl group carbon has no electronegative groups directly attached to it, so it occurs the furthest upfield, at δ 20. Using tables of typical chemical shift values (such as Table 13.4) one can usually assign ^{13}C NMR signals to each carbon in a molecule, based on the groups attached to each carbon.

13.11D Off-Resonance Decoupled Spectra

At times, more information than a predicted chemical shift is needed to assign an NMR peak to a specific carbon atom of a molecule. Fortunately, NMR spectrometers can differentiate among carbon atoms on the basis of the number of hydrogen atoms that are attached to each carbon. Several methods to accomplish this are available. An early method is called **off-resonance decoupling.** In an off-resonance decoupled ^{13}C NMR spectrum, each carbon signal is split into a multiplet of peaks, depending on how many hydrogens are attached to that carbon. An $n + 1$ rule applies, where n is the number of hydrogens on the carbon in question. Thus, a carbon with no hydrogens produces a

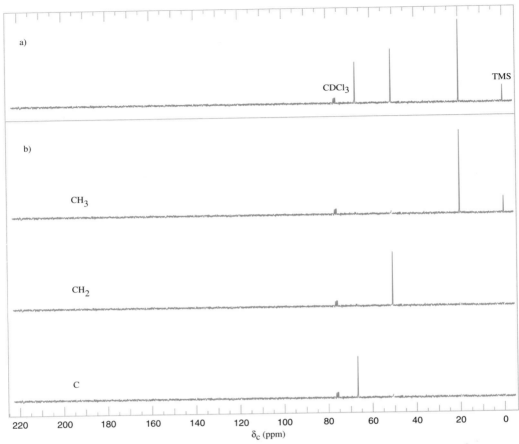

FIGURE 13.32 a) The broad band proton-decoupled ^{13}C NMR spectrum of 1-chloro-2-pro-panol. b) These three spectra show the DEPT ^{13}C NMR data from 1-chloro-2-propanol (See Section 13.11E). (This will be the only full display of a DEPT spectrum in the text. Other ^{13}C NMR figures will show the full broad band proton-decoupled spectrum but with information from the DEPT $^{13}CNMR$ spectra indicated near each peak as C, CH, CH_2, or CH_3.)

singlet ($n = 0$), a carbon with one hydrogen produces a doublet (two peaks), a carbon with two hydrogens produces a triplet (three peaks), and a methyl group carbon produces a quartet (four peaks). Interpretation of off-resonance decoupled ^{13}C spectra is often complicated by overlapping peaks from the multiplets.

13.11E DEPT ^{13}C Spectra

It is now possible to obtain ^{13}C NMR spectra that are much simpler to interpret with respect to carbon type. One such spectrum is called the **DEPT** ^{13}C spectrum. A DEPT (distortionless enhanced polarization transfer) spectrum actually consists of several spectra, with the final data presentation depicting one spectrum for each type of carbon atom. Thus, CH, CH_2, and CH_3 carbons are each printed out on separate spectra, together with a ^{13}C spectrum where all carbon types are shown (see Figure 13.32b). Each carbon type is thus identified unambiguously. **From this point forward in this text, rather than reproducing the entire family of four sorted DEPT spectra for** all subsequent spectra we will show the ^{13}C peaks labeled according to the information gained from the DEPT spectrum for the compound being considered.

As a second example of interpreting ^{13}C NMR spectra, let us look at the spectrum of methyl methacrylate (Fig 13.33). (This compound is the monomeric starting material for the commercial polymers Lucite and Plexiglas; c.f. Special Topic C.) The five carbons of methyl methacrylate represent carbon types from several chemical shift regions of ^{13}C spectra. Furthermore, because there is no symmetry in the structure of methyl methacrylate, all of its carbon atoms are chemically unique and so produce five distinct carbon NMR signals. Making use of our table of approximate ^{13}C chemical shifts (Table 13.4), we can readily deduce that the peak at δ 167.3 is due to the ester carbonyl carbon, the peak at δ 51.5 is for the methyl carbon attached to the ester oxygen, the peak at δ 18.3 is for the methyl attached to C2, and the peaks at δ 136.9 and δ 124.7 are for the alkene carbons. Additionally, employing the information from the DEPT ^{13}C spectrum we can unambiguously assign the alkene carbons. The DEPT spectrum tells us definitively that the peak at δ 124.7 has two attached hydrogens, and so it is due to C3, the terminal alkene carbon of methyl methacrylate. The alkene carbon with no attached hydrogens is then, of course, C2.

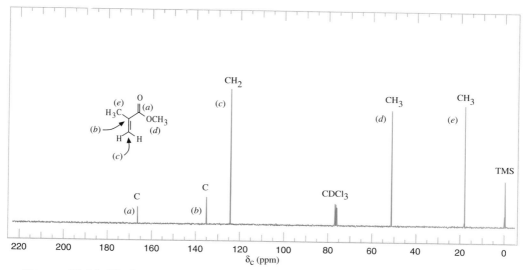

FIGURE 13.33 The broad band proton-decoupled ^{13}C NMR spectrum of methyl methacrylate.

PROBLEM 13.13

Compounds **A**, **B**, and **C** are isomers with the formula $C_5H_{11}Br$. Their broad band proton-decoupled ^{13}C NMR spectra are given in Fig. 13.34. Information from the DEPT ^{13}C NMR spectra is given near each peak. Give structures for **A**, **B**, and **C**.

13.12 MAGNETIC RESONANCE IMAGING IN MEDICINE

An important application of ^{1}H NMR spectroscopy in medicine today is a technique called **magnetic resonance imaging** or **MRI**. One great advantage of MRI is that,

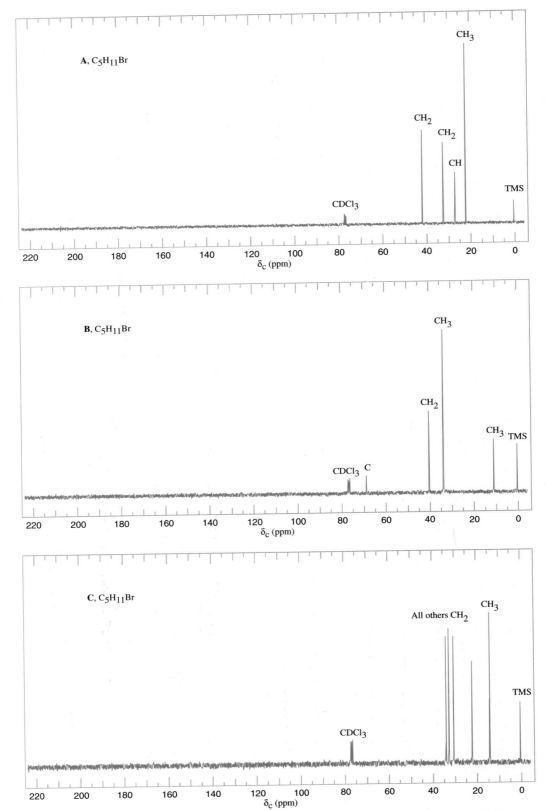

FIGURE 13.34A The broad band proton-decoupled ^{13}C NMR spectra of Compounds **A**, **B**, and **C**, Problem 13.13. Information from the DEPT ^{13}C NMR spectrum is given above the peaks.

unlike X-rays, it does not use dangerous ionizing radiation, and it does not require the injection of potentially harmful chemicals in order to produce contrasts in the image. In MRI, a portion of the patient's body is placed in a powerful magnetic field and irradiated with rf energy.

A typical MRI image is given in Fig. 13.35. The instruments used in producing images like this one use the pulse method (Section 13.4) to excite the protons in the tissue under observation and use a Fourier transformation to translate the information into an image. The brightness of various regions of the image are related to two things. The first factor is the number of protons in the tissue at that particular place. The second factor arises from what are called the **relaxation times** of the protons. When protons are excited to a higher energy state by the pulse of rf energy, they absorb energy. They must lose this energy to return to the lower energy spin state before they can be excited again by a second pulse. The process by which the nuclei lose this energy is called **relaxation,** and the time it takes to occur is the relaxation time. There are two basic modes of relaxation available to protons. In one, called *spin–lattice relaxation,* the extra energy is transferred to neighboring molecules in the surroundings (or lattice). The time required for this to happen is called T_1 and is characteristic of the time required for the spin system to return to thermal equilibrium with its surroundings. In solids, T_1 can be hours long. For protons in pure liquid water, T_1 is only a few seconds. In the other type of relaxation, called *spin–spin relaxation,* the extra energy is dissipated by being transferred to nuclei of nearby atoms. The time required for this is called T_2. In liquids the magnitude of T_2 is approximately equal to T_1. In solids, however, the T_1 is very much larger.

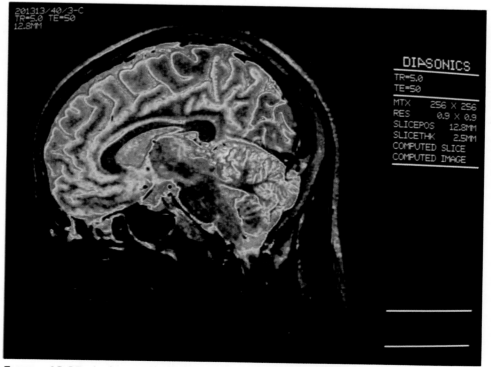

FIGURE 13.35   An image obtained by magnetic resonance imaging.

Various techniques based on the time between pulses of rf radiation have been developed to utilize the differences in relaxation times in order to produce contrasts between different regions in soft tissues. The soft tissue contrast is inherently higher than that produced with X-ray techniques. Magnetic resonance imaging is being used to great effect in locating tumors, lesions, and edemas. Improvements in this technique are occurring rapidly, and the method is not restricted to observation of proton signals.

One important area of medical research is based on the observation of signals from ^{31}P. Compounds that contain phosphorus as phosphate esters (Section 10.12) such as adenosine triphosphate (ATP) and adenosine diphosphate (ADP) are involved in most metabolic processes. By using techniques based on NMR, researchers now have a noninvasive way to follow cellular metabolism.

ADDITIONAL PROBLEMS

13.14   Listed here are ^1H NMR absorption peaks for several compounds. Propose a structure that is consistent with each set of data. (In some cases characteristic IR absorptions are given as well.)

(a) $C_4H_{10}O$

^1H NMR spectrum
singlet, δ 1.28 (9H)
singlet, δ 1.35 (1H)

(b) C_3H_7Br

^1H NMR spectrum
doublet, δ 1.71 (6H)
septet, δ 4.32 (1H)

(c) C_4H_8O

^1H NMR spectrum
triplet, δ 1.05 (3H)
singlet, δ 2.13 (3H)
quartet, δ 2.47 (2H)

IR spectrum
strong peak
near 1720 cm^{-1}

(d) C_7H_8O

^1H NMR spectrum
singlet, δ 2.43 (1H)
singlet, δ 4.58 (2H)
multiplet, δ 7.28 (5H)

IR spectrum
broad peak in
3200–3550-cm^{-1}
region

(e) C_4H_9Cl

^1H NMR spectrum
doublet, δ 1.04 (6H)
multiplet, δ 1.95 (1H)
doublet, δ 3.35 (2H)

(f) $C_{15}H_{14}O$

^1H NMR spectrum
singlet, δ 2.20 (3H)
singlet, δ 5.08 (1H)
multiplet, δ 7.25 (10H)

IR spectrum
strong peak
near 1720 cm^{-1}

(g) $C_4H_7BrO_2$

^1H NMR spectrum
triplet, δ 1.08 (3H)
multiplet, δ 2.07 (2H)
triplet, δ 4.23 (1H)
singlet, δ 10.97 (1H)

IR spectrum
broad peak in
2500–3000-cm^{-1}
region and a peak
at 1715 cm^{-1}

(h) C_8H_{10}

^1H NMR spectrum
triplet, δ 1.25 (3H)
quartet, δ 2.68 (2H)
multiplet, δ 7.23 (5H)

(i) $C_4H_8O_3$

^1H NMR spectrum
triplet, δ 1.27 (3H)
quartet, δ 3.66 (2H)
singlet, δ 4.13 (2H)
singlet, δ 10.95 (1H)

IR spectrum
broad peak in
2500–3000-cm^{-1}
region and a peak
at 1715 cm^{-1}

(j) $C_3H_7NO_2$ **^{1}H NMR spectrum**
doublet, δ 1.55 (6H)
septet, δ 4.67 (1H)

(k) $C_4H_{10}O_2$ **^{1}H NMR spectrum**
singlet, δ 3.25 (6H)
singlet, δ 3.45 (4H)

(l) $C_5H_{10}O$ **^{1}H NMR spectrum**   **IR spectrum**
doublet, δ 1.10 (6H)   strong peak
singlet, δ 2.10 (3H)   near 1720 cm^{-1}
septet, δ 2.50 (1H)

(m) C_8H_9Br **^{1}H NMR spectrum**
doublet, δ 2.0 (3H)
quartet, δ 5.15 (1H)
multiplet, δ 7.35 (5H)

13.15 The IR spectrum of compound **E** (C_8H_6) is shown in Fig. 13.36. Compound **E** decolorizes bromine in carbon tetrachloride and gives a precipitate when treated with ammonia-cal silver nitrate. What is the structure of **E**?

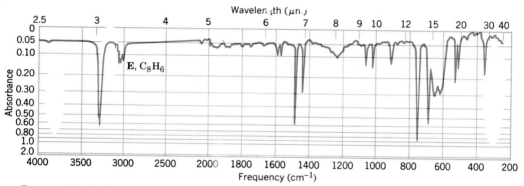

FIGURE 13.36 The IR spectrum of compound **E**, Problem 13.15. (Spectrum courtesy of Sadtler Research Laboratories, Inc., Philadelphia.)

13.16 Propose structures for the compounds **G** and **H** whose ^{1}H NMR spectra are shown in Figs. 13.37 and 13.38.

13.17 A two-carbon compound (**J**) contains only carbon, hydrogen, and chlorine. Its IR spectrum is relatively simple and shows the following absorption peaks: 3125 cm^{-1}(m), 1625 cm^{-1}(m), 1280 cm^{-1}(m), 820 cm^{-1}(s), 695 cm^{-1}(s). The ^{1}H NMR spectrum of **J** consists of a singlet at δ 6.3. Using Table 13.2, make as many assignments as you can and propose a structure for compound **J**.

13.18 When dissolved in $CDCl_3$, a compound (**K**) with molecular formula $C_4H_8O_2$ gives a ^{1}H NMR spectrum that consists of a doublet at δ 1.35, a singlet at δ 2.15, a broad singlet at δ 3.75 (1H), and a quartet at δ 4.25 (1H). When dissolved in D_2O, the compound gives a similar ^{1}H NMR spectrum, with the exception that the signal at δ 3.75 has disappeared. The IR spectrum of the compound shows a strong absorption peak near 1720 cm^{-1}. (a) Propose a structure for compound **K** and (b) explain why the NMR signal at δ 3.75 disappears when D_2O is used as the solvent.

13.19 Assume that in a certain NMR spectrum you find two peaks of roughly equal intensity. You are not certain whether these two peaks are *singlets* arising from uncoupled

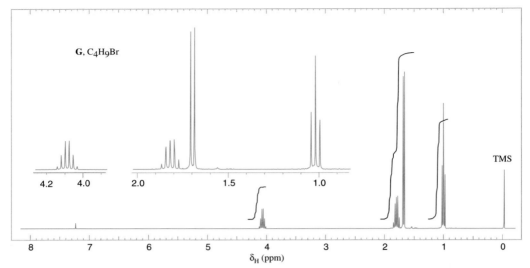

FIGURE 13.37 The 300 MHz ^1H NMR spectrum of compound **G**, Problem 13.16. Expansions of the signals are shown in the offset plots.

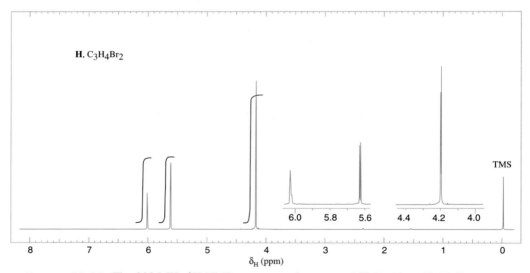

FIGURE 13.38 The 300 MHz ^1H NMR spectrum of compound **H**, Problem 13.16. Expansions of the signals are shown in the offset plots.

protons at different chemical shifts, or whether they are two peaks of a *doublet* that arises from protons coupling with a single adjacent proton. What simple experiment would you perform to distinguish between these two possibilities?

13.20 Compound **O** (C_6H_8) reacts with two molar equivalents of hydrogen in the presence of a catalyst to produce **P** (C_6H_{12}). The proton-decoupled ^{13}C spectrum of **O** consists of two singlets, one at δ 26.0 and one at δ 124.5. In the DEPT ^{13}C spectrum of **O** the signal at δ 26.0 appears as a CH_2 group and the one at δ 124.5 appears as a CH group. Propose structures for **O** and **P**.

13.21   Compound **Q** has the molecular formula C_7H_8. On catalytic hydrogenation **Q** is converted to **R** (C_7H_{12}). The broad band proton-decoupled ^{13}C spectrum of **Q** is given in Fig. 13.39. Propose structures for **Q** and **R**.

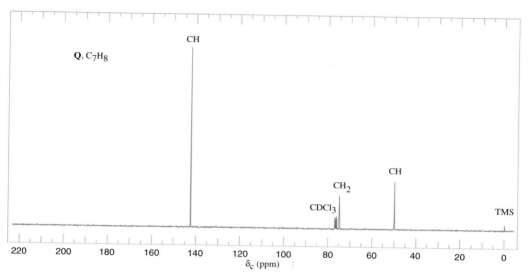

FIGURE 13.39 The broad band proton-decoupled ^{13}C NMR spectrum of Compound **Q**, Problem 13.21. Information from the DEPT ^{13}C NMR spectrum is given above each peak.

13.22   Compound **S** (C_8H_{16}) decolorizes a solution of bromine in carbon tetrachloride. The broad band proton-decoupled ^{13}C spectrum of **S** is given in Fig. 13.40. Propose a structure for **S**.

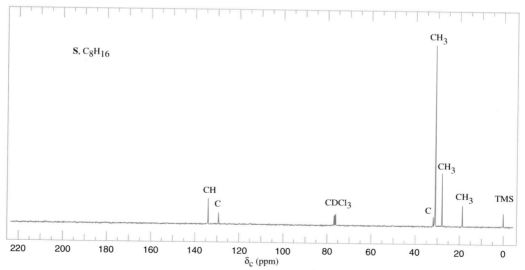

FIGURE 13.40 The broad band proton-decoupled ^{13}C NMR spectrum of Compound **S**, Problem 13.22. Information from the DEPT ^{13}C NMR spectrum is given above each peak.

13.23   Compound **T** (C_5H_8O) has a strong IR absorption band at 1745 cm^{-1}. The broad band proton-decoupled ^{13}C spectrum of **T** is given in Fig. 13.41. Propose a structure for **T**.

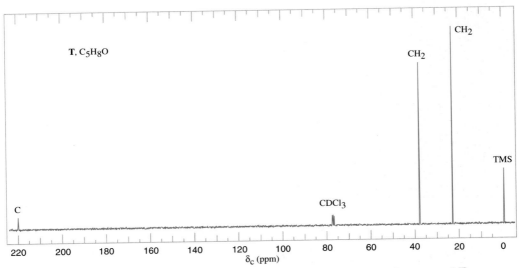

FIGURE 13.41 The broad band proton-decoupled ^{13}C NMR spectrum of compound **T**, Problem 13.23. Information from the DEPT ^{13}C NMR spectrum is given near each peak.

TWO-DIMENSIONAL (2D) NMR TECHNIQUES

Many NMR techniques are now available that greatly simplify the interpretation of NMR spectra. Chemists can now readily glean information about spin–spin coupling and the exact *connectivity* of atoms in molecules through techniques called **multidimensional NMR spectroscopy**. These techniques require an NMR spectrometer of the pulse (Fourier transform) type. The most common multidimensional techniques utilize two-dimensional NMR (2D NMR) and go by acronyms such as COSY, HETCOR, and a variety of others.* The two-dimensional sense of 2D NMR spectra does not refer to the way they appear on paper but instead reflects the fact that the data is accumulated using two radio frequency pulses with a varying time delay between them. Sophisticated application of other instrumental parameters is involved as well. Discussion of these parameters and the physics behind multidimensional NMR is beyond the scope of this discussion, however. The result is an NMR spectrum with the usual one-dimensional spectrum along the horizontal and vertical axes, and a set of correlation peaks that appear in the $x-y$ field of the graph. When 2D NMR is applied to ^{1}H NMR it is called **^{1}H–^{1}H correlation spectroscopy** (or **COSY** for short). COSY spectra are exceptionally useful for deducing proton–proton coupling relationships. 2D NMR spectra can also be obtained that indicate coupling between hydrogens and the *carbons* to which they are attached. In this case it is called **heteronuclear correlation spectroscopy (HETCOR, or C–H HETCOR).** When ambiguities are present in the one-dimensional ^{1}H and ^{13}C NMR spectra, a HETCOR spectrum can be very useful for assigning precisely which hydrogens and carbons are producing their respective peaks.

---

*Even three-dimensional techniques (and beyond) are possible, although computational requirements can limit their availability.

Figure E.1 shows the COSY spectrum for 1-chloro-2-propanol. In a COSY spectrum the ordinary one-dimensional ^1H spectrum is shown along both the horizontal and the vertical axes. Meanwhile, the $x-y$ field of a COSY spectrum is similar to a topographic map and can be thought of as looking down on the contour lines of a map of a mountain range. Along the diagonal of the COSY spectrum is a view that corresponds to looking down on the ordinary one-dimensional spectrum of 1-chloro-2-propanol as though each peak were a mountain. The one-dimensional counterpart of a given peak on the diagonal lies directly below that peak on each axis. The peaks on the diagonal provide no new information relative to that obtained from the one-dimensional spectrum along each axis. The important and new information from the COSY spectrum, however, comes from the correlation peaks ("mountains") that appear off the diagonal (called "cross peaks"). If one starts at a given cross peak and imagines two perpendicular lines (that is, parallel to each spectrum axis) leading back to the diagonal, the peaks intersected on the diagonal by these lines are coupled to each other. Hence, the peaks on the one-dimensional spectrum directly below the coupled diagonal peaks are coupled to each other. The cross peaks above the diagonal are mirror reflections of those below the diagonal, thus the information is redundant and only cross peaks on one side of the diagonal need be interpreted. The $x-y$ field cross peak correlations are the result of instrumental parameters used to obtain the COSY spectrum.

Let's trace the coupling relationships in 1-chloro-2-propanol made evident in its COSY spectrum (Fig. E.1). (Even though coupling relationships from the ordinary one-dimensional spectrum for 1-chloro-2-propanol are fairly readily interpreted, this compound makes a good beginning example for interpretation of COSY spectra.) First, one chooses a starting point in the COSY spectrum from which to begin tracing the coupling relationships. A peak whose assignment is relatively apparent in the one-dimensional spectrum is a good point of reference. For this compound, the doublet from the methyl groups at 1.2 ppm is quite obvious and readily assigned. If we find the peak on the diagonal that corresponds to the methyl doublet (labeled ^cH in Fig. E.1 and directly above the one-dimensional methyl doublet on both axes), an imaginary line can be drawn parallel to the vertical axis that intersects a correlation peak (labeled $^bH-^cH$) in the $x-y$ field off the diagonal. From here a perpendicular imaginary line can be drawn back to its intersection with the diagonal peaks. At its intersection we see that this diagonal peak is directly above the one-dimensional spectrum peak at δ 3.9. Thus, the methyl hydrogens at δ 1.2 are coupled to the hydrogen whose signal appears at δ 3.9. It is now clear that the peaks at δ 3.9 are due to the hydrogen on the alcohol carbon in 1-chloro-2-propanol (^bH on carbon 2). Returning to the peak on the diagonal above δ 3.9, we trace a line back parallel to the horizontal axis that intersects a pair of cross peaks between δ 3.4 and δ 3.5. Moving back up to the diagonal from each of these cross peaks ($^aH'-^bH$ and $^aH''-^bH$) indicates that the hydrogen whose signal appears at δ 3.9 is coupled to the hydrogens whose signals appear at δ 3.4 and δ 3.5. The hydrogens at δ 3.4 and δ 3.5 are therefore the two hydrogens on the carbon that bears the chlorine ($^aH'$ and $^aH''$). One can even see that $^aH'$ and $^aH''$ couple with each other by the cross peak they have in common between them right next to their diagonal peaks. Thus, from the COSY spectrum we can quickly see which hydrogens are coupled to each other. Furthermore, from the reference starting point, we can "walk around" a molecule, tracing the neighboring coupling relationships along the molecule's carbon skeleton as we go through the COSY.

In a HETCOR spectrum a ^{13}C spectrum is presented along one axis and a ^1H spectrum is shown along the other. Cross peaks relating the two types of spectra to each other are found in the $x-y$ field. Specifically, the cross peaks in a HETCOR

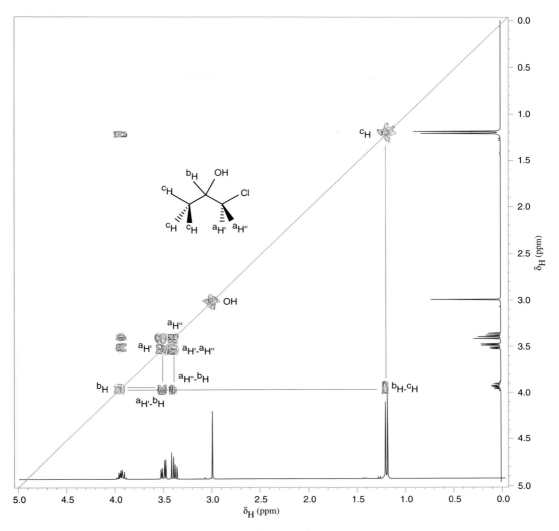

FIGURE E.1 COSY Spectrum of 1-chloro-2-propanol.

spectrum indicate which hydrogens are attached to which carbons in a molecule, or vice versa. These cross peak correlations are the result of instrumental parameters used to obtain the HETCOR spectrum. There is no diagonal spectrum in the $x-y$ field like that found in the COSY. If imaginary lines are drawn from a given cross peak in the $x-y$ field to each respective axis, the cross peak indicates that the hydrogen giving rise to the ^{1}H NMR signal on one axis is coupled (and attached) to the carbon that gives rise to the corresponding ^{13}C NMR signal on the other axis. Therefore, it is readily apparent which hydrogens are attached to which carbons.

Let's take a look at the HETCOR spectrum for 1-chloro-2-propanol (Fig. E.2). Having interpreted the COSY spectrum above, we know precisely which hydrogens of 1-chloro-2-propanol produce each signal in the ^{1}H spectrum. If an imaginary line is taken from the methyl doublet of the proton spectrum at 1.2 ppm (vertical axis) out to the correlation peak in the $x-y$ field and then dropped down to the ^{13}C spectrum axis (horizontal axis), it is apparent that the ^{13}C peak at 20 ppm is produced by the methyl

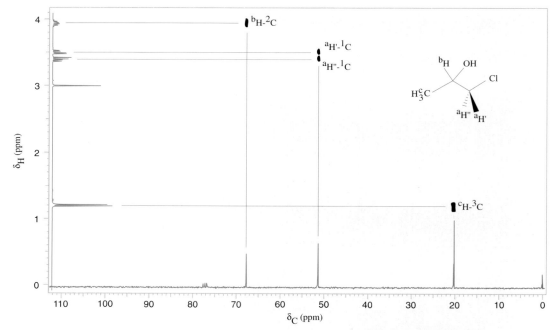

FIGURE E.2 ^{1}H-^{13}C HETCOR NMR spectrum of 1-chloro-2-propanol. The ^{1}H NMR spectrum is shown in blue and the ^{13}C NMR spectrum is shown in green. Correlations of the ^{1}H-^{13}C cross-peaks with the one-dimensional spectra are indicated by red lines.

carbon of 1-chloro-2-propanol (C3). Having assigned the ^{1}H NMR peak at 3.9 ppm to the hydrogen on the alcohol carbon of the molecule (C2), tracing out to the correlation peak and down to the ^{13}C spectrum indicates that the ^{13}C NMR signal at 67 ppm arises from the alcohol carbon (C2). Finally, from the ^{1}H NMR peaks at 3.4–3.5 for the two hydrogens on the carbon bearing the chlorine, our interpretation leads us out to the cross peak and down to the ^{13}C peak at 51 ppm (C1). Thus, by a combination of COSY and HETCOR spectra all ^{13}C and ^{1}H peaks can be unambiguously assigned to their respective carbon and hydrogen atoms in 1-chloro-2-propanol. (In this simple example using 1-chloro-2-propanol we could have arrived at complete assignment of these spectra without COSY and HETCOR data. For many compounds, however, the assignments are quite difficult to make without the aid of these 2D NMR techniques.)

Tropylium ion

MASS SPECTROMETRY

F.1 THE MASS SPECTROMETER

In a mass spectrometer (Fig. F.1) molecules in the gaseous state under low pressure are bombarded with a beam of high-energy electrons. The energy of the beam of electrons is usually 70 eV (electron volts) and one of the things this bombardment can do is dislodge one of the electrons of the molecule and produce a positively charged ion called *the molecular ion.*

$$M \quad + \quad e^- \quad \longrightarrow \quad M^{\ddagger} \quad + 2e^-$$

Molecule   **High-energy**   **Molecular**
                electron   **ion**

The molecular ion is not only a cation, but because it contains an odd number of electrons, it also is a free radical. Thus it belongs to a general group of ions called *radical cations.* If, for example, the molecule under bombardment is a molecule of ammonia, the following reaction will take place.

$$H\!:\!\overset{\cdot\cdot}{N}\!:\!H + e^- \longrightarrow \left[H\!:\!\overset{\cdot}{N}\!:\!H \right]^+ + 2e^-$$
$$\overset{\cdot\cdot}{H} \qquad\qquad\qquad \overset{\cdot\cdot}{H}$$

Molecular ion, M⁺
(a radical cation)

An electron beam with an energy of 70 eV ($\sim$ 1600 kcal mol^{-1}) not only dislodges electrons from molecules, producing molecular ions, it also imparts to the molecular ions considerable surplus energy. Not all molecular ions have the same amount of surplus energy, but for most, the surplus is far in excess of that required to break covalent bonds (50–100 kcal mol^{-1}). Soon after they are formed, therefore, most

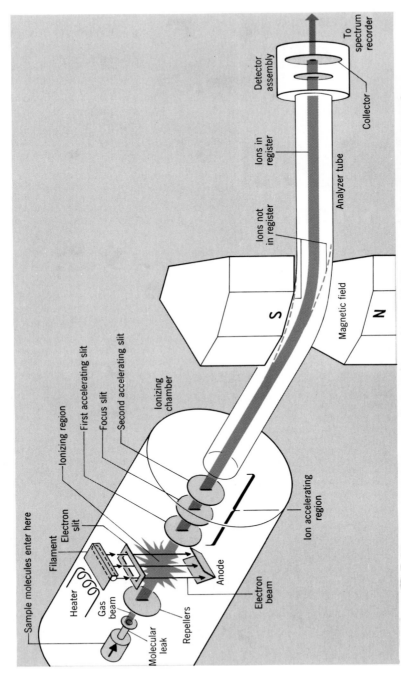

FIGURE F.1 Mass spectrometer. Schematic diagram of CEG model 21-103. The magnetic field that brings ions of vary charge (*m/z*) ratios into register is perpendicular to the page. (From John R. Holum, *Organic Chemistry: A Brief Course* New York, 1975. Used with permission.)

molecular ions literally fly apart—they undergo *fragmentation*. Fragmentation can take place in a variety of ways depending on the nature of the particular molecular ion, and as we shall see later, the way a molecular ion fragments can give us highly useful information about the structure of a complex molecule. Even with a relatively simple molecule such as ammonia, however, fragmentation can produce several new cations. The molecular ion can eject a hydrogen atom, for example, and produce the cation NH_2^+.

$$\overset{\cdot\,+}{H{:}N{:}H} \longrightarrow H{:}\overset{+}{\underset{\cdot\cdot}{N}}{:} + H\cdot$$
$$\underset{\cdot\cdot}{\overset{}{H}} \qquad\qquad \underset{\cdot\cdot}{\overset{}{H}}$$

This $\overset{+}{N}H_2$ cation can then lose a hydrogen atom to produce $\overset{+}{N}H\cdot$, which can lead, in turn, to $\overset{+}{N}$.

$$H{:}\overset{+}{\underset{\cdot\cdot}{N}}{:} \longrightarrow H{:}\overset{+}{\underset{\cdot}{N}}{:} + H\cdot$$
$$\underset{\cdot\cdot}{\overset{}{H}}$$

$$H{:}\overset{+}{\underset{\cdot}{N}}{:} \longrightarrow {:}\overset{+}{N}{:} + H\cdot$$

The mass spectrometer then *sorts* these cations on the basis of their mass/charge or m/z ratio. Since for all practical purposes the charge on all of the ions is $+1$, this amounts to sorting them on the basis of their mass. The conventional mass spectrometer does this by accelerating the ions through a series of slits and then it sends the ion beam into a curved tube (see Fig. F.1 again). This curved tube passes through a variable magnetic field and the magnetic field exerts an influence on the moving ions. Depending on its strength at a given moment, the magnetic field will cause ions with a particular m/z ratio to follow a curved path that exactly matches the curvature of the tube. These ions are said to be "in register." Because they are in register, these ions pass through another slit and impinge on an ion collector where the intensity of the ion beam is measured electronically. The intensity of the beam is simply a measure of the relative abundance of the ions with a particular m/z ratio. Some mass spectrometers are so sensitive that they can detect the arrival of a *single ion*.

The actual sorting of ions takes place in the magnetic field, and this sorting takes place because laws of physics govern the paths followed by charged particles when they move through magnetic fields. Generally speaking, a magnetic field such as this will cause ions moving through it to move in a path that represents part of a circle. The radius of curvature of this circular path is related to the m/z ratio of the ions, to the strength of the magnetic field (B_0, in Tesla), and to the accelerating voltage. If we keep the accelerating voltage constant and progressively increase the magnetic field, ions whose m/z ratios are progressively larger will travel in a circular path that exactly matches that of the curved tube. Hence, by steadily increasing B_0, ions with progressively increasing m/z ratios will be brought into register and so will be detected at the ion collector. Since, as we said earlier, the charge on nearly all of the ions is unity, this means that *ions of progressively increasing mass arrive at the collector and are detected*.

What we have described is called "magnetic focusing" (or "magnetic scanning"), and all of this is done automatically by the mass spectrometer. The spectrometer displays the results by plotting a series of peaks of varying intensity in which each peak corresponds to ions of a particular m/z ratio. This display (Fig. F.2) is one form of a *mass spectrum*.

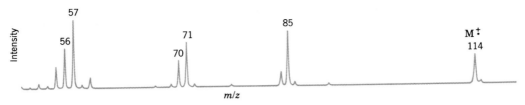

FIGURE F.2 A portion of the mass spectrum of octane.

Ion sorting can also be done with "electrical focusing." In this technique, the magnetic field is held constant and the accelerating voltage is varied. Both methods, of course, accomplish the same thing, and some high-resolution mass spectrometers employ both techniques.

To summarize: A mass spectrometer bombards organic molecules with a beam of high-energy electrons, causing them to ionize and fragment. It then separates the resulting mixture of ions on the basis of their m/z ratios and records the relative abundance of each ionic fragment. It displays this result as a plot of ion abundance versus m/z.

F.2 THE MASS SPECTRUM

Mass spectra are usually published as bar graphs or in tabular form, as illustrated in Fig. F.3 for the mass spectrum of ammonia. In either presentation, the most intense peak—called the *base peak*—is arbitrarily assigned an intensity of 100%. The intensities of all other peaks are given proportionate values, as percentages of the base peak.

The masses of the ions given in a mass spectrum are those that we would calculate

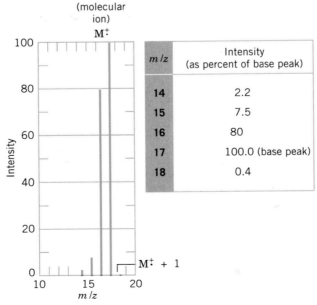

m/z	Intensity (as percent of base peak)
14	2.2
15	7.5
16	80
17	100.0 (base peak)
18	0.4

FIGURE F.3 The mass spectrum of NH_3 presented as a bar graph and in tabular form.

for the ion by assigning to the constituent atoms *masses rounded off to the nearest whole number.* For the commonly encountered atoms the nearest whole-number masses are

$$H = 1 \qquad O = 16$$
$$C = 12 \qquad F = 19$$
$$N = 14$$

In the mass spectrum of ammonia we see peaks at $m/z = 14, 15, 16,$ and 17. These correspond to the molecular ion and to the fragments we saw earlier.

$$NH_3 \xrightarrow{-e^-} [NH_3]^{\ddagger} \xrightarrow{-H\cdot} [NH_2]^+ \xrightarrow{-H\cdot} [NH]^{\ddagger} \xrightarrow{-H\cdot} [N]^+$$

$$m/z \quad = \quad \begin{array}{cccc} 17 & 16 & 15 & 14 \\ \text{(molecular} & & & \\ \text{ion)} & & & \end{array}$$

By convention we express,

$$\overset{\cdot\,+}{\underset{\underset{\displaystyle H}{}}{H\!:\!N\!:\!H}} \quad \text{as} \quad [NH_3]^{\ddagger}$$

$$\overset{+}{\underset{\underset{\displaystyle \ddot{H}}{}}{H\!:\!\ddot{N}\!:}} \quad \text{as} \quad [NH_2]^+$$

$$\overset{+}{H\!:\!\underset{\displaystyle \cdot}{N}\!:} \quad \text{as} \quad [NH]^{\ddagger}$$

and

$$\overset{+}{:\!N\!:} \quad \text{as} \quad [N]^+$$

In the case of ammonia, the base peak is the peak arising from the molecular ion. This is not always the case, however; in many of the spectra that we shall see later the base peak (the most intense peak) will be at an m/z value different from that of the molecular ion. This happens because in many instances the molecular ion fragments so rapidly that some other ion at a smaller m/z value produces the most intense peak. In a few cases the molecular ion peak is extremely small, and sometimes it is absent altogether.

One other feature in the spectrum of ammonia requires explanation: the small peak that occurs at m/z 18. In the bar graph we have labeled this peak $M^{\ddagger} + 1$ to indicate that it is one mass unit greater than the molecular ion. The $M^{\ddagger} + 1$ peak appears in the spectrum because most elements (e.g., nitrogen and hydrogen) have more than one naturally occurring isotope (Table F.1). Although most of the NH_3 molecules in a sample of ammonia are composed of $^{14}N^1H_3$, a small but detectable fraction of molecules is composed of $^{15}N^1H_3$. (A very tiny fraction of molecules is also composed of $^{14}N^1H_2{}^2H$.) These molecules ($^{15}N^1H_3$ or $^{14}N^1H_2{}^2H$) produce molecular ions at m/z 18, that is at $M^{\ddagger} + 1$.

The spectrum of ammonia begins to show us with a simple example how the masses (or m/z values) of individual ions can give us information about the composition of the ions and how this information can allow us to arrive at possible structures for a compound. Problems F.1 to F.3 will allow us further practice with this technique.

TABLE F.1   Principal stable isotopes of common elements[a]

ELEMENT	MOST COMMON ISOTOPE		NATURAL ABUNDANCE OF OTHER ISOTOPES (BASED ON 100 ATOMS OF MOST COMMON ISOTOPE)			
Carbon	^{12}C	100	^{13}C	1.11		
Hydrogen	^{1}H	100	^{2}H	0.016		
Nitrogen	^{14}N	100	^{15}N	0.38		
Oxygen	^{16}O	100	^{17}O	0.04	^{18}O	0.20
Fluorine	^{19}F	100				
Silicon	^{28}Si	100	^{29}Si	5.10	^{30}Si	3.35
Phosphorus	^{31}P	100				
Sulfur	^{32}S	100	^{33}S	0.78	^{34}S	4.40
Chlorine	^{35}Cl	100	^{37}Cl	32.5		
Bromine	^{79}Br	100	^{81}Br	98.0		
Iodine	^{127}I	100				

[a] Data obtained from R. M. Silverstein, G. C. Bassler, and T. C. Morrill, *Spectrometric Identification of Organic Compounds,* 5th ed., Wiley, New York, 1991, p. 9.

PROBLEM F.1

Propose a structure for the compound whose mass spectrum is given in Fig. F.4 and make reasonable assignments for each peak.

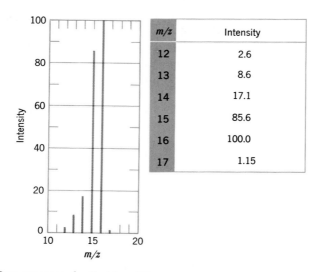

m/z	Intensity
12	2.6
13	8.6
14	17.1
15	85.6
16	100.0
17	1.15

FIGURE F.4 Mass spectrum for Problem F.1.

PROBLEM F.2

Propose a structure for the compound whose mass spectrum is given in Fig. F.5 and make reasonable assignments for each peak.

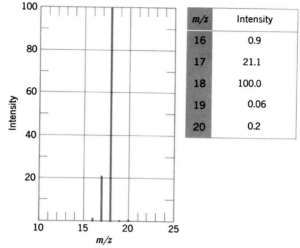

m/z	Intensity
16	0.9
17	21.1
18	100.0
19	0.06
20	0.2

FIGURE F.5 Mass spectrum for Problem F.2.

PROBLEM F.3

The compound whose mass spectrum is given in Fig. F.6 contains three elements, one of which is fluorine. Propose a structure for the compound and make reasonable assignments for each peak.

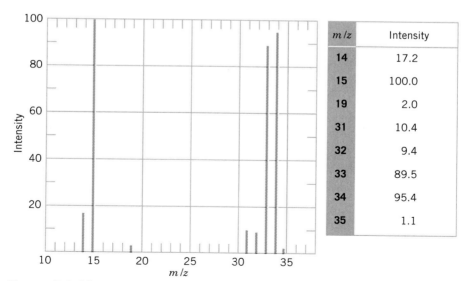

m/z	Intensity
14	17.2
15	100.0
19	2.0
31	10.4
32	9.4
33	89.5
34	95.4
35	1.1

FIGURE F.6 Mass spectrum for Problem F.3.

F.3 Determination of Molecular Formulas and Molecular Weights

F.3A The Molecular Ion and Isotopic Peaks

Look at Table F.1 for a moment. Notice that most of the common elements found in organic compounds have naturally occurring *heavier* isotopes. For three of the elements—carbon, hydrogen, and nitrogen—the principal heavier isotope is one mass unit greater than the most common isotope. The presence of these elements in a compound gives rise to a small isotopic peak one unit greater than the molecular ion—at $M^{\ddagger} + 1$. For four of the elements—oxygen, sulfur, chlorine, and bromine—the principal heavier isotope is two mass units greater than the most common isotope. The presence of these elements in a compound gives rise to an isotopic peak at $M^{\ddagger} + 2$.

$M^{\ddagger} + 1$ Elements: C, H, N

$M^{\ddagger} + 2$ Elements: O, S, Br, Cl

Isotopic   ks give ι one method for determining molecular formulas. To understand how th   can be done, let us begin by noticing that the isotope abundances in Table F.1 are ased on 100 atoms of the normal isotope. Now let us suppose, as ar example, that we have 100 molecules of methane (CH_4). On the average there will be 1.11 molecules that contain ^{13}C and 4×0.016 molecules that contain ^2H. Altogether then, these heavier isotopes should contribute an $M^{\ddagger} + 1$ peak whose intensity is about 1.17% of the intensity of the peak for the molecular ion.

$$1.11 + 4(0.016) \simeq 1.17\%$$

ιs correlates well with the ι   erved intensity of the $M^{\ddagger} + 1$ peak in the actual spe   m of methane given in Fig   4.
  : molecules with a modest   mber of atoms we can determine molecular formulas in the following way. If the   ‡ peak is not the base peak, the first thing we do with the mass spectrum of an un   own compound is to recalculate the intensities of the $M^{\ddagger} + 1$ and $M^{\ddagger} + 2$ peaks ιo express them as percentages of the intensity of the $M^{\ddagger}$ peak. Consider, for example, the mass spectrum given in Fig. F.7. The $M^{\ddagger}$ peak at $m/z = 72$ is not the base peak. Therefore, we need to recalculate the intensities of the pe   s in our spectrum at m/z 72, 73, and 74 as percentages of the peak at m/z 72. We do tl   by dividing each inte   ity by the intensity of the $M^{\ddagger}$ peak, which is 73%, and mul   lying by 100. These ι   ults are shown here and in the second column of Fig. F.7.

m/z	INTENSITY (% OF $M^{\ddagger}$)
72	١/73 × 100 = 100
73	.3/73 × 100 = 4.5
74	0.2/73 × 100 = 0.3

Then we use the following guides to determine the molecular formula.

m/z	Intensity (as percent of base peak)	m/z		Intensity (as percent of $M^{\ddagger}$)
27	59.0	72	$M^{\ddagger}$	100.0
28	15.0	73	$M^{\ddagger} + 1$	4.5
29	54.0	74	$M^{\ddagger} + 2$	0.3
39	23.0			
41	60.0			Recalculated to base on $M^{\ddagger}$
42	12.0			
43	79.0			
44	100.0 (base)			
72	73.0 $M^{\ddagger}$			
73	3.3			
74	0.2			

FIGURE F.7 Mass spectrum of an unknown compound.

1. **Is $M^{\ddagger}$ odd or even? According to the nitrogen rule, if it is even, then the compound must contain an even number of nitrogen atoms (zero is an even number).**

   For our unknown, $M^{\ddagger}$ is even. The compound must have an even number of nitrogen atoms.

2. **The relative abundance of the $M^{\ddagger} + 1$ peak indicates the number of carbon atoms. Number of C atoms = relative abundance of $(M^{\ddagger} + 1)/1.1$**

   For our unknown (Fig. F.7), number of C atoms $= \dfrac{4.5}{1.1} = 4$

   (This formula works because ^{13}C is the most important contributor to the $M^{\ddagger} + 1$ peak and the approximate natural abundance of ^{13}C is 1.1%.)

3. **The relative abundance of the $M^{\ddagger} + 2$ peak indicates the presence (or absence) of S (4.4%), Cl (33%), or Br (98%) (see Table F.1).**

   For our unknown, $M^{\ddagger} + 2 = 0.2\%$; thus we can assume that S, Cl, and Br are absent.

4. **The molecular formula can now be established by determining the number of hydrogen atoms and adding the appropriate number of oxygen atoms, if necessary.**

   For our unknown the $M^{\ddagger}$ peak at m/z 72 gives us the molecular weight. It also tells us (since it is even) that nitrogen is absent because C_4N_2 has a molecular weight (76) greater than that of our compound.

   For a molecule composed of C and H only:

   $$H = 72 - (4 \times 12) = 24$$

   but C_4H_{24} is impossible.

   For a molecule composed of C, H, and one O:

$$H = 72 - (4 \times 12) - 16 = 8$$

and thus our unknown has the molecular formula C_4H_8O.

PROBLEM F.4

(a) Write structural formulas for at least 14 stable compounds that have the formula C_4H_8O.   (b) The IR spectrum of the unknown compound shows a strong peak near 1730 cm^{-1}. Which structures now remain as possible formulas for the compound? (We continue with this compound in Problem F.14.)

PROBLEM F.5

Determine the molecular formula of the following compound.

m/z	INTENSITY (as % of base peak)
86 M⁺	10.00
87	0.56
88	0.04

PROBLEM F.6

(a) What approximate intensities would you expect for the M⁺ and M⁺ + 2 peaks of CH_3Cl?   (b) For the M⁺ and M⁺ + 2 peaks of CH_3Br?   (c) An organic compound gives an M⁺ peak at m/z 122 and a peak of nearly equal intensity at m/z 124. What is a likely molecular formula for the compound?

PROBLEM F.7

Use the mass spectral data given in Fig. F.8 to determine the molecular formula for the compound.

FIGURE F.8 Mass spectrum for Problem F.7.

m/z	INTENSITY (as % of base peak)
14	8.0
15	38.6
18	16.3
28	39.7
29	23.4
42	46.6
43	10.7
44	100.0 (base)
73	86.1 M⁺
74	3.2
75	0.2

PROBLEM F.8

(a) Determine the molecular formula of the compound whose mass spectrum is given in the following tabulation.   (b) The ^{1}H NMR spectrum of this compound consists only of a large doublet and a small septet. What is the structure of the compound?

m/z	INTENSITY (as % of base peak)
27	34
39	11
41	22
43	100 (base)
63	26
65	8
78	24 M⁺
79	0.8
80	8

As the number of atoms in a molecule increases, calculations like this become more and more complex and time consuming. Fortunately, however, these calculations can be done readily with computers, and tables are now available that give relative values for the $M^+ + 1$ and $M^+ + 2$ peaks for all combinations of common elements with molecular formulas up to mass 500. Part of the data obtained from one of these tables is given in Table F.2. Use Table F.2 to check the results of our example (Fig. F.7) and your answer to Problem F.7.

F.3B High-Resolution Mass Spectrometry

All of the spectra that we have described so far have been determined on what are called "low-resolution" mass spectrometers. These spectrometers, as we noted earlier, measure m/z values to the nearest whole-number mass unit. Some laboratories are equipped with this type of mass spectrometer.

Many laboratories, however, are equipped with the more expensive "high-resolution" mass spectrometers. These spectrometers can measure m/z values to three or four decimal places and thus they provide an extremely accurate method for determining molecular weights. And because molecular weights can be measured so accurately, these spectrometers also allow us to determine molecular formulas.

The determination of a molecular formula by an accurate measurement of a molecular weight is possible because the actual masses of atomic particles (nuclides) are not integers (see Table F.3). Consider, as examples, the three molecules O_2, N_2H_4, and CH_3OH. The actual atomic masses of the molecules are all different.

$$O_2 = 2(15.9949) = 31.9898$$

$$N_2H_4 = 2(14.0031) + 4(1.00783) = 32.0375$$

$$CH_4O = 12.0000 + 4(1.00783) + 15.9949 = 32.0262$$

TABLE F.2   Relative intensities of $M^{\ddagger} + 1$ and $M^{\ddagger} + 2$ peaks for various combinations of C, H, N, and O for masses 72 and 73

$M^{\ddagger}$	FORMULAS	PERCENTAGE OF $M^{\ddagger}$ INTENSITY		$M^{\ddagger}$	FORMULAS	PERCENTAGE OF $M^{\ddagger}$ INTENSITY	
		$M^{\ddagger} + 1$	$M^{\ddagger} + 2$			$M^{\ddagger} + 1$	$M^{\ddagger} + 2$
72	CH_2N_3O	2.30	0.22	73	CHN_2O_2	1.94	0.41
	CH_4N_4	2.67	0.03		CH_3N_3O	2.31	0.22
	$C_2H_2NO_2$	2.65	0.42		CH_5N_4	2.69	0.03
	$C_2H_4N_2O$	3.03	0.23		C_2HO_3	2.30	0.62
	$C_2H_6N_3$	3.40	0.04		$C_2H_3NO_2$	2.67	0.42
	$C_3H_4O_2$	3.38	0.44		$C_2H_5N_2O$	3.04	0.23
	C_3H_6NO	3.76	0.25		$C_2H_7N_3$	3.42	0.04
	$C_3H_8N_2$	4.13	0.07		$C_3H_5O_2$	3.40	0.44
	C_4H_8O	4.49	0.28		C_3H_7NO	3.77	0.25
	$C_4H_{10}N$	4.86	0.09		$C_3H_9N_2$	4.15	0.07
	C_5H_{12}	5.60	0.13		C_4H_9O	4.51	0.28
					$C_4H_{11}N$	4.88	0.10
					C_6H	6.50	0.18

Data from J. H. Beynon, *Mass Spectrometry and Its Application to Organic Chemistry,* Elsevier, Amsterdam, 1960.

High-resolution mass spectrometers are available that are capable of measuring mass with an accuracy of 1 part in 40,000 or better. Thus, such a spectrometer can easily distinguish among these three molecules and, in effect, tell us the molecular formula.

F.4   FRAGMENTATION

In most instances the molecular ion is a highly energetic species, and in the case of a complex molecule a great many things can happen to it. The molecular ion can break apart in a variety of ways and the fragments that are produced can then undergo further fragmentation, and so on. In a certain sense mass spectroscopy is a "brute force" technique. Striking an organic molecule with 70-eV electrons is a little like firing a howitzer at a house made of matchsticks. That fragmentation takes place in any sort of predictable way is truly remarkable—and yet it does. Many of the same factors that govern ordinary chemical reactions seem to apply to fragmentation processes, and many of the principles that we have learned about the relative stabilities of carbocations, radicals, and molecules will help us to make some sense out of what takes place. And as we learn something about what kind of fragmentations to expect, we shall be much better able to use mass spectra as aids in determining the structures of organic molecules.

We cannot, of course, in the limited space that we have here, look at these processes in great detail, but we can examine some of the more important ones.

TABLE F.3 Exact masses of nuclides

ISOTOPE	MASS	ISOTOPE	MASS
^1H	1.00783	^{19}F	18.9984
^2H	2.01410	^{32}S	31.9721
^{12}C	12.00000 (std)	^{33}S	32.9715
^{13}C	13.00336	^{34}S	33.9679
^{14}N	14.0031	^{35}Cl	34.9689
^{15}N	15.0001	^{37}Cl	36.9659
^{16}O	15.9949	^{79}Br	78.9183
^{17}O	16.9991	^{81}Br	80.9163
^{18}O	17.9992	^{127}I	126.9045

As we begin, keep two important principles in mind. (1) The reactions that take place in a mass spectrometer are usually *unimolecular*—that is, they involve only a *single* molecular fragment. This is true because the pressure in a mass spectrometer is kept so low ($\sim 10^{-6}$ torr) that reactions requiring bimolecular collisions usually do not occur. (2) The relative ion abundances, as measured by peak intensities, are extremely important. We shall see that the appearance of certain prominent peaks in the spectrum gives us important information about the structures of the fragments produced and about their original locations in the molecule.

F.4A Fragmentation by Cleavage at a Single Bond

One important type of fragmentation is the simple cleavage at a single bond. With a radical cation this cleavage can take place in at least two ways; each way produces a *cation* and a *radical*. Only the cations are detected in a mass spectrometer. (The radicals, because they are not charged, are not deflected by the magnetic field and, therefore, are not detected.) With the molecular ion obtained from propane, for example, two possible modes of cleavage are

$$CH_3CH_2 \overset{+}{\cdot} CH_3 \longrightarrow CH_3CH_2^+ + \cdot CH_3$$
$$\textit{m/z } \mathbf{29}$$

$$CH_3CH_2 \overset{+}{\cdot} CH_3 \longrightarrow CH_3CH_2 \cdot + {}^+CH_3$$
$$\textit{m/z } \mathbf{15}$$

These two modes of cleavage do not take place at equal rates, however. Although the relative abundance of cations produced by such a cleavage is influenced both by the stability of the carbocation and by the stability of the free radical, *the carbocation's stability is more important.** In the spectrum of propane the peak at *m/z* 29 ($CH_3CH_2^+$) is the most intense peak; the peak at *m/z* 15 (CH_3^+) has an intensity of only 5.6%. This reflects the greater stability of $CH_3CH_2^+$ when compared to CH_3^+.

---

*This can be demonstrated through thermochemical calculations that we cannot go into here. The interested student is referred to F. W. McLafferty, *Interpretation of Mass Spectra,* 2nd ed., Benjamin, Reading, MA, 1973, p. 41 and pp. 210–211.

F.4B Fragmentation Equations

Before we go further, we need to examine some of the conventions that are used in writing equations for fragmentation reactions. In the two equations for cleavage at the single bond of propane that we have just written, we have localized the odd electron and the charge on one of the carbon–carbon sigma bonds of the molecular ion. When we write structures this way, the choice of just where to localize the odd electron and the charge is sometimes arbitrary. When possible, however, we write the structure showing the molecular ion that would result from the removal of one of the most loosely held electrons of the original molecule. Just which electrons these are can usually be estimated from ionization potentials (Table F.4). [The ionization potential of a molecule is the amount of energy (in eV) required to remove an electron from the molecule.] As we might expect, ionization potentials indicate that the nonbonding electrons of nitrogen and oxygen and the π electrons of alkenes and aromatic molecules are held more loosely than the electrons of carbon–carbon and carbon–hydrogen sigma bonds. Therefore, the convention of localizing the odd electron and charge is especially applicable when the molecule contains an oxygen, nitrogen, a double bond, or an aromatic ring. If the molecule contains only carbon–carbon and carbon–hydrogen sigma bonds, and if it contains a great many of these, then the choice of where to localize the odd electron and the charge is so arbitrary as to be impractical. In these instances we usually resort to another convention. We write the formula for the radical cation in brackets and place the odd electron and charge outside. Using this convention we would write the two fragmentation reactions of propane in the following way:

$$[CH_3CH_2CH_3]^{+} \longrightarrow CH_3CH_2^+ + \cdot CH_3$$

$$[CH_3CH_2CH_3]^{+} \longrightarrow CH_3CH_2 \cdot + CH_3^+$$

PROBLEM F.9

The most intense peak in the mass spectrum of 2,2-dimethylbutane occurs at m/z 57. (a) What carbocation does this peak represent? (b) Using the convention that we have just described, write an equation that shows how this carbocation arises from the molecular ion.

TABLE F.4 Ionization potentials of selected molecules

COMPOUND	IONIZATION POTENTIAL (eV)
$CH_3(CH_2)_3NH_2$	8.7
C_6H_6	9.2
C_2H_4	10.5
CH_3OH	10.8
C_2H_6	11.5
CH_4	12.7

Figure F.9 shows us the kind of fragmentation a longer chain alkane can undergo. The example here is hexane and we see a reasonably abundant molecular ion at m/z 86 accompanied by a small $M^+ + 1$ peak. There is also a smaller peak at m/z 71 ($M^+ - 15$) corresponding to the loss of $\cdot CH_3$, and the base peak is at m/z 57 ($M^+ - 29$) corresponding to the loss of $\cdot CH_2CH_3$. The other prominent peaks are at m/z 43 ($M^+ - 43$) and m/z 29 ($M^+ - 57$), corresponding to the loss of $\cdot CH_2CH_2CH_3$ and $\cdot CH_2CH_2CH_2CH_3$, respectively. The important fragmentations are just the ones we would expect:

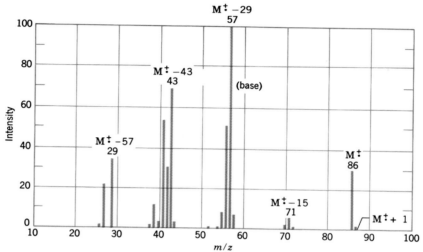

FIGURE F.9 Mass spectrum of hexane.

$$[CH_3CH_2CH_2CH_2CH_2CH_3]^{+\cdot} \longrightarrow$$

$$\rightarrow CH_3CH_2CH_2CH_2CH_2^+ + \cdot CH_3$$
$$\textit{m/z } \textbf{71}$$

$$\rightarrow CH_3CH_2CH_2CH_2^+ + \cdot CH_2CH_3$$
$$\textit{m/z } \textbf{57}$$

$$\rightarrow CH_3CH_2CH_2^+ + \cdot CH_2CH_2CH_3$$
$$\textit{m/z } \textbf{43}$$

$$\rightarrow CH_3CH_2^+ + \cdot CH_2CH_2CH_2CH_3$$
$$\textit{m/z } \textbf{29}$$

Chain branching increases the likelihood of cleavage at a branch point because a more stable carbocation can result. When we compare the mass spectrum of 2-methylbutane (Fig. F.10) with the spectrum of hexane, we see a much more intense peak at $M^+ - 15$. Loss of a methyl radical from the molecular ion of 2-methylbutane can give a secondary carbocation:

$$\begin{bmatrix} & CH_3 \\ & | \\ CH_3CHCH_2CH_3 \end{bmatrix}^{+\cdot} \longrightarrow CH_3\overset{+}{C}HCH_2CH_3 + \cdot CH_3$$
$$\begin{array}{cc} \textit{m/z } \textbf{72} & \qquad\qquad \textit{m/z } \textbf{57} \\ \textbf{M}^+ & \qquad\qquad \textbf{M}^+ - \textbf{15} \end{array}$$

whereas with hexane, loss of a methyl radical can yield only a primary carbocation.

With neopentane (Fig. F.11), this effect is even more dramatic. Loss of a methyl radical by the molecular ion produces a *tertiary* carbocation, and this reaction takes

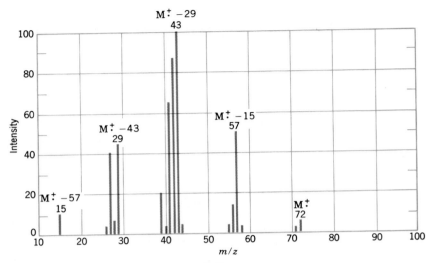

FIGURE F.10 The mass spectrum of 2-methylbutane.

place so readily that virtually none of the molecular ions survive long enough to be detected.

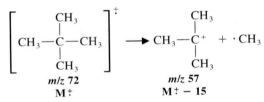

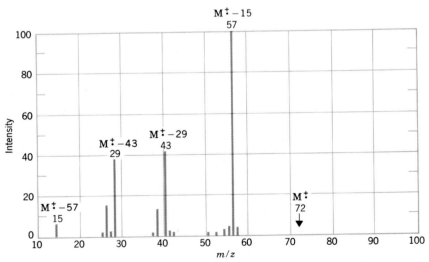

FIGURE F.11 Mass spectrum of neopentane.

PROBLEM F.10

In contrast to 2-methylbutane and neopentane, the mass spectrum of 3-methylpentane (not given) has a peak of very low intensity at $M^{+} - 15$. It has a peak of very high intensity at $M^{+} - 29$, however. Explain.

Carbocations stabilized by resonance are usually also prominent in mass spectra. Several ways that resonance-stabilized carbocations can be produced are outlined in the following list.

1. Alkenes frequently undergo fragmentations that yield allylic cations.

$$CH_2\overset{+\cdot}{=}CH\overset{}{-}CH_2\!:\!R \longrightarrow \overset{+}{C}H_2-CH=CH_2 + \cdot R$$
$$m/z\ \mathbf{41}$$

2. Carbon–carbon bonds next to an atom with an unshared electron pair usually break readily because the resulting carbocation is resonance stabilized.

$$R\overset{+\cdot}{-}Z\overset{}{-}CH_2\!:\!CH_3 \longrightarrow R-\overset{+}{Z}=CH_2 + \cdot CH_3$$

$$\updownarrow$$

$$R-\overset{..}{Z}-\overset{+}{C}H_2$$

Z = N, O, or S; R may also be H

3. Carbon–carbon bonds next to the carbonyl group of an aldehyde or ketone break readily because resonance-stabilized ions called acylium ions are produced.

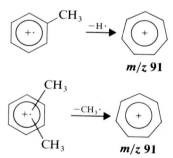

$$R'-\overset{+}{C}\equiv\overset{+}{O}\!: \longrightarrow R'-C\equiv\overset{+}{O}\!: + R\cdot$$

$$\updownarrow$$

$$R'-\overset{+}{C}=\underset{..}{O}\!:$$

Acylium ion

or

$$R-C\equiv\overset{+}{O}\!: + R'\cdot$$

$$\updownarrow$$

$$R-\overset{+}{C}=\overset{..}{O}\!:$$

Acylium ion

4. Alkyl-substituted benzenes undergo loss of a hydrogen atom or methyl group to yield the relatively stable tropylium ion (cf. Section 14.8). This fragmentation gives a prominent peak (sometimes the base peak) at m/z 91.

$$m/z\ \mathbf{91}$$

$$m/z\ \mathbf{91}$$

5. Monosubstituted benzenes also lose their substituent and yield a phenyl cation at m/z 77.

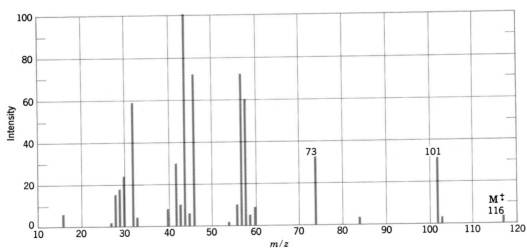

$$m/z\ 77$$

$$Y = halogen, -NO_2, -\overset{\overset{\displaystyle O}{\|}}{C}R, -R, \text{ and so on}$$

PROBLEM F.11

The mass spectrum of 4-methyl-1-hexene (not given) shows intense peaks at m/z 57 and m/z 41. What fragmentation reactions account for these peaks?

PROBLEM F.12

Explain the following observations that can be made about the mass spectra of alcohols: (a) The molecular ion peak of a primary or secondary alcohol is very small; with a tertiary alcohol it is usually undetectable. (b) Primary alcohols show a prominent peak at m/z 31. (c) Secondary alcohols usually give prominent peaks at m/z 45, 59, 73, and so on. (d) Tertiary alcohols have prominent peaks at m/z 59, 73, 87, and so on.

PROBLEM F.13

The mass spectra of butyl isopropyl ether and butyl propyl ether are given in Figs. F.12 and F.13. (a) Which spectrum represents which ether? (b) Explain your choice.

FIGURE F.12 Mass spectrum for Problem F.13.

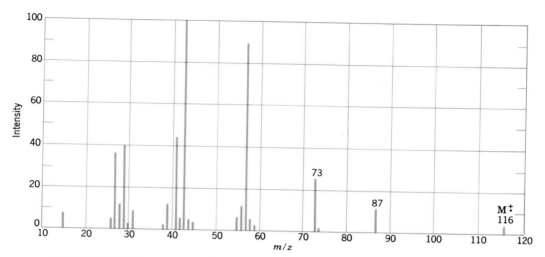

FIGURE F.13 Mass spectrum for Problem F.13.

F.4C Fragmentation by Cleavage of Two Bonds

Many peaks in mass spectra can be explained by fragmentation reactions that involve the breaking of two covalent bonds. When a radical–cation undergoes this type of fragmentation the products are *a new radical–cation* and *a neutral molecule*. Some important examples are the following:

1. Alcohols frequently show a prominent peak at $M^{\ddot{+}} - 18$. This corresponds to the loss of a molecule of water.

$$R-CH-CH_2 \longrightarrow R-CH\overset{\cdot+}{=}CH_2 + H-\overset{..}{\underset{..}{O}}-H$$
$$\mathbf{M^{\ddot{+}}} \qquad\qquad \mathbf{M^{\ddot{+}} - 18}$$

or

$$[R-CH_2-CH_2-OH]^{\ddot{+}} \longrightarrow [R-CH=CH_2]^{\ddot{+}} + H_2O$$
$$\mathbf{M^{\ddot{+}}} \qquad\qquad \mathbf{M^{\ddot{+}} - 18}$$

2. Cycloalkenes can undergo a retro-Diels–Alder reaction that produces an alkene and an alkadienyl radical cation.

$$\left[\bigcirc \right]^{\ddot{+}} \longrightarrow \left[\bigcirc \right]^{\ddot{+}} + \overset{CH_2}{\underset{CH_2}{\|}}$$

3. Carbonyl compounds with a hydrogen on their γ carbon undergo a fragmentation called the *McLafferty rearrangement.*
   Y may be R, H, OR, OH, and so on

$$\left[\begin{array}{c} \underset{H_2C}{\overset{O}{\underset{\|}{Y-C}}}\underset{\overset{|}{CH_2}}{\overset{H}{\underset{\searrow}{\overset{\nearrow}{\,}}}CHR} \end{array}\right]^{\ddagger} \longrightarrow \left[\begin{array}{c} \underset{Y-C}{\overset{O}{\underset{\|}{\,}}}\underset{CH_2}{\overset{H}{\underset{\diagdown}{\,}}} \end{array}\right]^{\ddagger} + RCH{=\!\!=}CH_2$$

Y may be R, H, OR, OH, and so on

In addition to these reactions, we frequently find peaks in mass spectra that result from the elimination of other small stable neutral molecules, for example, H_2, NH_3, CO, HCN, H_2S, alcohols, and alkenes.

F.5 Other Mass Spectrometric Techniques

A very powerful technique for analyzing mixtures uses a **gas chromatograph (GC)** combined with a mass spectrometer, called **gas chromatography/mass spectrometry (GC/MS).** The mixture is allowed to pass through the gas chromatograph first; this separates the components of the mixture. Then each separate component is allowed to pass into the mass spectrometer, where it is analyzed.

An even more powerful technique involves two coupled mass spectrometers, called **mass spectrometry-mass spectrometry (MS/MS).** The two mass spectrometers are linked in tandem. The first mass spectrometer generates (and separates) the molecular ions of a mixture and the second mass spectrometer analyzes individual molecular ions.

Mass Spectrometry of Biomolecules. Relatively recent advances have made mass spectrometry a tool of exceptional potential for analysis of large biomolecules. Although the electron impact mass spectrometry techniques described above are invaluable for analysis of volatile compounds with molecular weights up to about 1000 Daltons, it is now possible to use other ''soft'' ionization techniques for analysis of nonvolatile compounds such as proteins, nucleic acids, and other biologically relevant compounds with molecular weights up to and in excess of 100,000 Daltons. Techniques such as electrospray ionization (ESI), ion spray, matrix-assisted laser desorption ionization (MALDI), and fast atom bombardment (FAB) mass spectrometry have proven to be powerful tools for analysis of protein molecular weights, enzyme-substrate complexes, antibody-antigen binding, drug-receptor interactions, and DNA oligonucleotide sequence determination.

ADDITIONAL PROBLEMS

F.14 Reconsider Problem F.4 and the spectrum given in Fig. F.7. Important clues to the structure of this compound are the peaks at m/z 44 (the base peak) and m/z 29. Propose a structure for the compound and write fragmentation equations showing how these peaks arise.

F.15 The homologous series of primary amines, $CH_3(CH_2)_nNH_2$, from CH_3NH_2 to $CH_3(CH_2)_{13}NH_2$ all have their base (largest) peak at m/z 30. What ion does this peak represent and how is it formed?

F.16 The mass spectrum of compound **A** is given in Fig. F.14. The ^1H NMR spectrum of **A** consists of two singlets with area ratios of 9:2. The larger singlet is at δ 1.2, the smaller one at δ 1.3. Propose a structure for compound **A**.

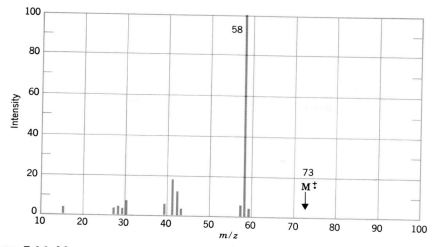

FIGURE F.14 Mass spectrum of compound **A** (Problem F.16).

F.17   The mass spectrum of compound **B** is given in Fig. F.15. The IR spectrum of **B** shows a broad peak between 3200 and 3550 cm^{-1}. The ^{1}H NMR spectrum of **B** shows the following peaks: a triplet at δ 0.9, a singlet at δ 1.1, and a quartet at δ 1.6. The area ratio of these peaks is 3 : 7 : 2, respectively. Propose a structure for **B**.

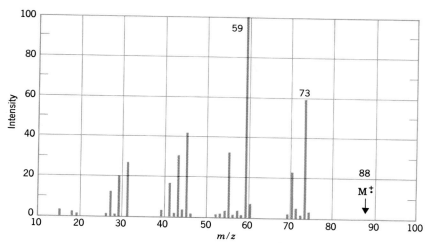

FIGURE F.15 Mass spectrum of compound **B** (Problem F.17).

SPECIAL TOPIC G

trans-Retinal

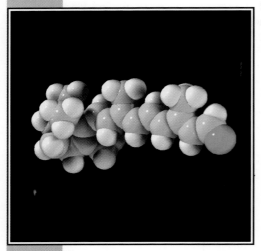

PHOTOCHEMISTRY OF VISION

The chemical changes that occur when light impinges on the retina of the eye involve several of the phenomena that we have studied in earlier chapters. Central to an understanding of the visual process at the molecular level are two phenomena in particular: the absorption of light by conjugated polyenes and the interconversion of cis–trans isomers.

The retina of the human eye contains two types of receptor cells. Because of their shapes, these cells have been named *rods* and *cones*. Rods are located primarily at the periphery of the retina and are responsible for vision in dim light. Rods, however, are color-blind and ''see'' only in shades of gray. Cones are found mainly in the center of the retina and are responsible for vision in bright light. Cones also possess the pigments that are responsible for color vision.

Some animals do not possess both rods and cones. The retinas of pigeons contain only cones. Thus, while pigeons have color vision, they see only in the bright light of day. The retinas of owls, on the other hand, have only rods; owls see very well in dim light but are color blind.

The chemical changes that occur in rods are much better understood than those in cones. For this reason we shall concern ourselves here with rod vision alone.

When light strikes rod cells, it is absorbed by a compound called rhodopsin. This initiates a series of chemical events that ultimately results in the transmission of a nerve impulse to the brain.

Our understanding of the chemical nature of rhodopsin and the conformational changes that occur when rhodopsin absorbs light has resulted largely from the research of George Wald and co-workers at Harvard University. Wald's research began in 1933 when he was a graduate student in Berlin; work with rhodopsin, however, began much earlier in other laboratories.

Rhodopsin was discovered in 1877 by the German physiologist Franz Boll. Boll noticed that the initial red-purple color of a pigment in the retina of frogs was "bleached" by the action of light. The bleaching process led first to a yellow retina and then to a colorless one. A year later, another German scientist, Willy Kuhne, isolated the red-purple pigment and named it, because of its color, *Sehpurpur* or "visual purple." The name visual purple is still commonly used for rhodopsin.

In 1952, Wald and one of his students, Ruth Hubbard, showed that the chromophore (light-absorbing group) of rhodopsin is the polyunsaturated aldehyde, 11-*cis*-retinal. Rhodopsin is the product of the reaction of this aldehyde and a protein called opsin (Fig. G.1). The reaction is between the aldehyde group of 11-*cis*-retinal and an amino group on the chain of the protein and involves the loss of a molecule of water. Other secondary interactions involving —SH groups of the protein probably also hold the *cis*-retinal in place. The site on the chain of the protein is one on which *cis*-retinal fits precisely.

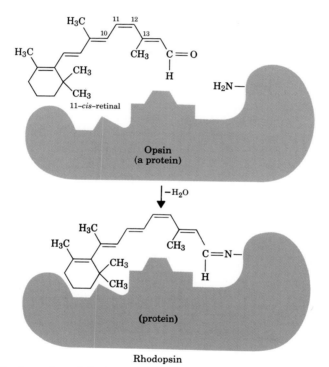

FIGURE G.1 The formation of rhodopsin from 11-*cis*-retinal and opsin.

The conjugated polyunsaturated chain of 11-*cis*-retinal gives rhodopsin the ability to absorb light over a broad region of the visible spectrum. Figure G.2 shows the absorption curve of rhodopsin in the visible region and compares it with the sensitivity curve for human rod vision. The fact that these two curves coincide provides strong evidence that rhodopsin is the light-sensitive material in rod vision.

When rhodopsin absorbs a photon of light, the 11-*cis*-retinal chromophore isomerizes to the all-trans form. The first photoproduct is an intermediate called bathorhodopsin, a compound that has about 35 kcal mol^{-1} more energy than rhodopsin. Bathorhodopsin then, through a series of steps, becomes metarhodopsin II (also all-trans). The high energy of the all-trans chromophore–protein combination causes it to change its shape. As a result, a cascade of enzymatic reactions culminates in the

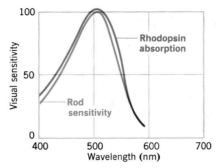

FIGURE G.2 A comparison of the visible absorption spectrum of rhodopsin and the sensitivity curve for rod vision. [Adapted from S. Hecht, S. Shlaer, and M. H. Pirenne, *J. Gen. Chem. Physiol.,* **25,** 819 (1942).]

transmission of a neural signal to the brain. The chromophore is ultimately hydrolyzed and expelled as all-*trans*-retinal. These steps are illustrated in Fig. G.3.

Rhodopsin has an absorption maximum at 498 nm. This gives rhodopsin its red-purple color. Together, all-*trans*-retinal and opsin have an absorbance maximum at 387 nm and, thus, are yellow. The light-initiated transformation of rhodopsin to all-*trans*-retinal and opsin corresponds to the initial bleaching that Boll observed in the retinas of frogs. Further bleaching to a colorless form occurs when all-*trans*-retinal is reduced enzymatically to all-*trans*-vitamin A. This reduction converts the aldehyde group of retinal to the primary alcohol function of vitamin A.

all-*trans*-Retinal

[H] | enzyme

all-*trans*-Vitamin A

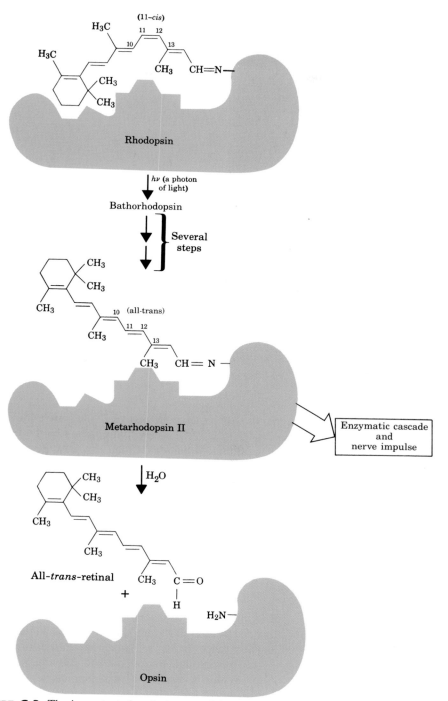

FIGURE G.3 The important chemical steps of the visual process. Absorption of a photon of light by the 11-*cis*-retinal portion of rhodopsin generates a nerve impulse as a result of an isomerization that leads, through a series of steps, to metarhodopsin II. Then hydrolysis of metarhodopsin II produces all-*trans*-retinal and opsin. This illustration greatly oversimplifies the shape of rhodopsin; the retinal portion is actually embedded in the center of a very complex protein structure. For a much more detailed representation of the structure of rhodopsin, and for a description of how a cascade of reactions results in a nerve signal, see L. Stryer, "The Molecules of Visual Excitation," *Scientific American,* **257,** 32 (1987).

CHAPTER 14

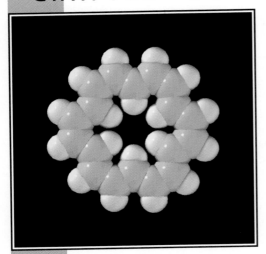

[18]Annulene
(Section 14.7A)

AROMATIC COMPOUNDS

14.1 INTRODUCTION

The study of the class of compounds that organic chemists call aromatic compounds (Section 2.7) began with the discovery in 1825 of a new hydrocarbon by the English chemist Michael Faraday (Royal Institution). Faraday called this new hydrocarbon "bicarburet of hydrogen"; we now call it benzene. Faraday isolated benzene from a compressed illuminating gas that had been made by pyrolyzing whale oil.

In 1834 the German chemist Eilhardt Mitscherlich (University of Berlin) synthesized benzene by heating benzoic acid with calcium oxide. Using vapor density measurements Mitscherlich further showed that benzene has the molecular formula C_6H_6.

$$C_6H_5CO_2H + CaO \xrightarrow{\text{heat}} C_6H_6 + CaCO_3$$

Benzoic acid **Benzene**

The molecular formula itself was surprising. Benzene has *only as many hydrogen atoms as it has carbon atoms*. Most compounds that were known then had a far greater proportion of hydrogen atoms, usually twice as many. Benzene with the formula of C_6H_6 (or C_nH_{2n-6}) should be a highly unsaturated compound, because it has an index of hydrogen deficiency equal to four. Within a very short time chemists began to find that benzene had unusual properties, and eventually they began to recognize that benzene was a member of a new class of organic compounds with unusual and interesting properties. As we shall see in Section 14.3 benzene does not show the behavior expected of a highly unsaturated compound.

During the latter part of the nineteenth century the Kekulé–Couper–Butlerov theory of valence was systematically applied to all known organic compounds. One result of this effort was the placing of organic compounds in either of two broad categories; compounds were classified as being either **aliphatic** or **aromatic.** To be classified as aliphatic meant then that the chemical behavior of a compound was "fatlike." (Now it means that the compound reacts like an alkane, an alkene, an alkyne, or one of their derivatives.) To be classified as aromatic meant then that the compound had a low hydrogen/carbon ratio and that it was "fragrant." Most of the early aromatic compounds were obtained from balsams, resins, or essential oils. Included among these were benzaldehyde (from oil of bitter almonds), benzoic acid and benzyl alcohol (from gum benzoin), and toluene (from tolu balsam).

Kekulé was the first to recognize that these early aromatic compounds all contain a six-carbon unit and that they retain this six-carbon unit through most chemical transformations and degradations. Benzene was eventually recognized as being the parent compound of this new series.

Since this new group of compounds proved to be distinctive in ways that are far more important than their odors, the term *aromatic* began to take on a purely chemical connotation. We shall see in this chapter that the meaning of aromatic has evolved as chemists have learned more about the reactions and properties of aromatic compounds.

14.2 NOMENCLATURE OF BENZENE DERIVATIVES

Two systems are used in naming monosubstituted benzenes. In certain compounds, *benzene* is the parent name and the substituent is simply indicated by a prefix. We have, for example,

Fluorobenzene   **Chlorobenzene**   **Bromobenzene**   **Nitrobenzene**

For other compounds, the substituent and the benzene ring taken together may form a new parent name. Methylbenzene is usually called *toluene,* hydroxybenzene is almost always called *phenol,* and aminobenzene is almost always called *aniline.* These and other examples are indicated here.

Toluene   **Phenol**   **Aniline**

Benzenesulfonic acid   **Benzoic acid**   **Acetophenone**   **Anisole**

When two substituents are present, their relative positions are indicated by the prefixes ***ortho, meta,*** and ***para*** (abbreviated ***o-, m-,*** and ***p-***) or by the use of numbers.* For the dibromobenzenes we have

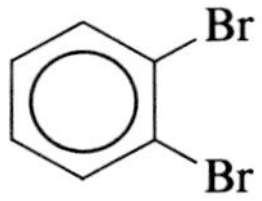

1,2-Dibromobenzene
(***o*-dibromobenzene**)
ortho

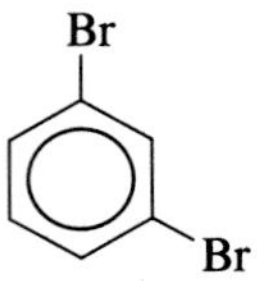

1,3-Dibromobenzene
(***m*-dibromobenzene**)
meta

1,4-Dibromobenzene
(***p*-dibromobenzene**)
para

and for the nitrobenzoic acids:

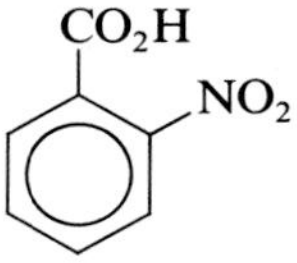

2-Nitrobenzoic acid
(***o*-nitrobenzoic acid**)

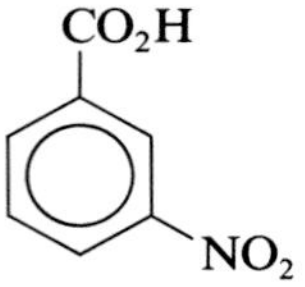

3-Nitrobenzoic acid
(***m*-nitrobenzoic acid**)

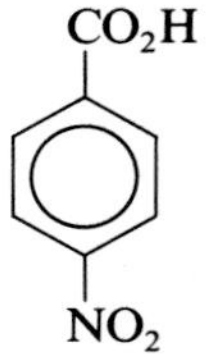

4-Nitrobenzoic acid
(***p*-nitrobenzoic acid**)

The dimethylbenzenes are often called *xylenes.*

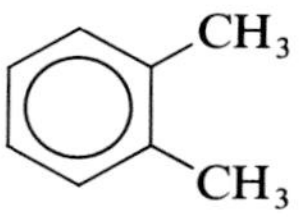

1,2-Dimethylbenzene
(***o*-xylene**)

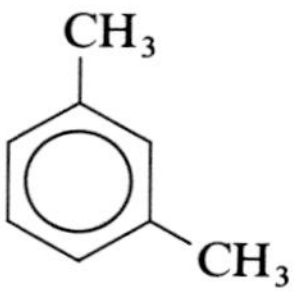

1,3-Dimethylbenzene
(***m*-xylene**)

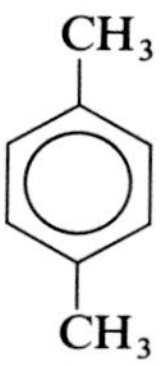

1,4-Dimethylbenzene
(***p*-xylene**)

If more than two groups are present on the benzene ring, their positions must be indicated by the use of *numbers.* As examples, consider the following two compounds.

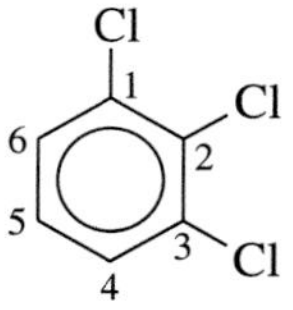

1,2,3-Trichlorobenzene

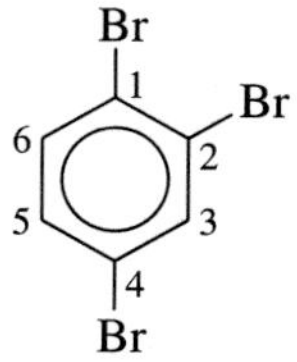

1,2,4-Tribromobenzene
(*not* **1,3,4-tribromobenzene**)

We notice, too, that the benzene ring is numbered so as to give ***the lowest possible numbers to the substituents.***

*Numbers can be used for two or more substituents, but ortho, meta, and para must never be used for more than two.

When more than two substituents are present and the substituents are different, they are listed in alphabetical order.

When a substituent is one that when taken together with the benzene ring gives a new base name, that substituent is assumed to be in position 1 and the new parent name is used:

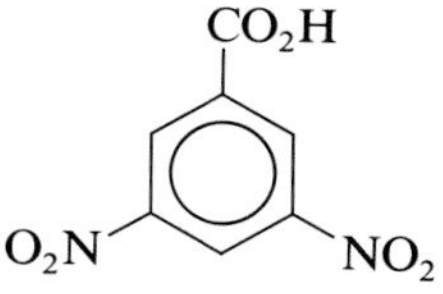

CO₂H
O₂N
NO₂

3,5-Dinitrobenzoic acid

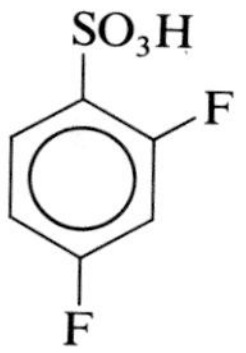

SO₃H
F
F

2,4-Difluorobenzenesulfonic acid

When the C_6H_5— group is named as a substituent, it is called a **phenyl** group. A hydrocarbon composed of one saturated chain and one benzene ring is usually named as a derivative of the larger structural unit. However, if the chain is unsaturated, the compound may be named as a derivative of that chain, regardless of ring size. The following are examples:

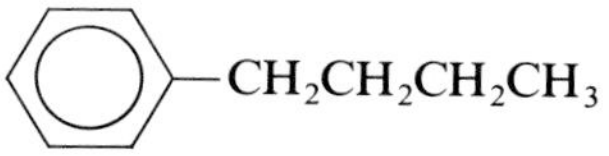

CH₂CH₂CH₂CH₃

Butylbenzene

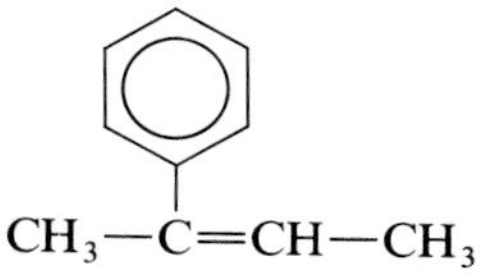

CH₃—C=CH—CH₃

2-Phenyl-2-butene

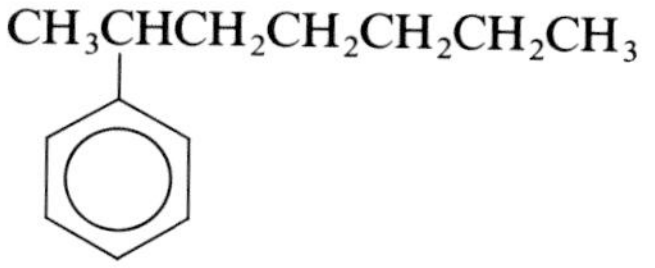

CH₃CHCH₂CH₂CH₂CH₂CH₃

2-Phenylheptane

The phenyl group is often abbreviated as C_6H_5—, Ph—, or ϕ—.

The name **benzyl** is an alternative name for the phenylmethyl group:

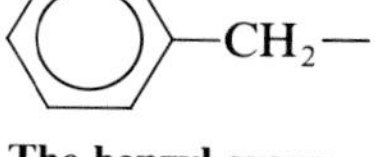

CH₂—

The benzyl group (the phenylmethyl group)

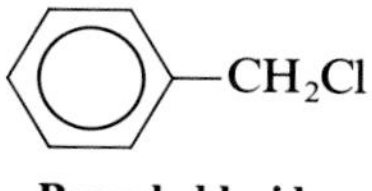

CH₂Cl

Benzyl chloride (phenylmethyl chloride)

14.3 REACTIONS OF BENZENE

In the mid-nineteenth century, benzene presented chemists with a real puzzle. They knew from its formula (Section 14.1) that benzene was highly unsaturated, and they expected it to react accordingly. They expected it to react like an alkene by decolorizing bromine in carbon tetrachloride by *adding bromine.* They expected that it would change the color of aqueous potassium permanganate by being *oxidized,* that it would *add hydrogen* rapidly in the presence of a metal catalyst, and that it would *add water* in the presence of strong acids.

Benzene does none of these. When benzene is treated with bromine in carbon tetrachloride in the dark or with aqueous potassium permanganate or with dilute acids, none of the expected reactions occurs. Benzene does add hydrogen in the presence of finely divided nickel, but only at high temperatures and under high pressures.

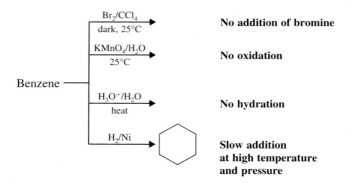

Benzene *does* react with bromine but only in the presence of a Lewis acid catalyst such as ferric bromide. Most surprisingly, however, it reacts not by addition but by *substitution*.

Substitution

$$C_6H_6 + Br_2 \xrightarrow{FeBr_3} C_6H_5Br + HBr \quad \text{Observed}$$

Addition

$$C_6H_6 + Br_2 \xarrow{} C_6H_6Br_2 + C_6H_6Br_4 + C_6H_6Br_6 \quad \text{Not observed}$$

When benzene reacts with bromine *only one monobromobenzene* is formed. That is, only one compound with the formula C_6H_5Br is found among the products. Similarly, when benzene is chlorinated *only one monochlorobenzene* results.

Two possible explanations can be given for these observations. The first is that only one of the six hydrogen atoms in benzene is reactive toward these reagents. The second is that all six hydrogen atoms in benzene are equivalent, and replacing any one of them with a substituent results in the same product. As we shall see, the second explanation is correct.

PROBLEM 14.1

Listed below are four compounds that have the molecular formula C_6H_6. Which of these compounds would yield only one monosubstitution product, if, for example, one hydrogen were replaced by bromine?

(a) $CH_3C \equiv C - C \equiv CCH_3$ (b)

(c)   (d)

14.4 THE KEKULÉ STRUCTURE FOR BENZENE

In 1865, August Kekulé, the originator of the structural theory (Section 1.3), proposed the first definite structure for benzene,* a structure that is still used today (although as we shall soon see, we give it a meaning different from the meaning Kekulé gave it). Kekulé suggested that the carbon atoms of benzene are in a ring, that they are bonded to each other by alternating single and double bonds, and that one hydrogen atom is attached to each carbon atom. This structure satisfied the requirements of the structural theory that carbon atoms form four bonds and that all the hydrogen atoms of benzene are equivalent.

The Kekulé formula for benzene

A problem soon arose with the Kekulé structure, however. The Kekulé structure predicts that there should be two different 1,2-dibromobenzenes. In one of these hypothetical compounds (below), the carbon atoms that bear the bromines are separated by a single bond, and in the other they are separated by a double bond. *Only one, 1,2-dibromobenzene, has ever been found, however.*

To accommodate this objection, Kekulé proposed that the two forms of benzene (and of benzene derivatives) are in a state of equilibrium, and that this equilibrium is so rapidly established that it prevents isolation of the separate compounds. Thus, the two 1,2-dibromobenzenes would also be rapidly equilibrated, and this would explain why chemists had not been able to isolate the two forms.

We now know that this proposal was incorrect and that *no such equilibrium exists.* Nonetheless, the Kekulé formulation of benzene's structure was an important step forward and, for very practical reasons, it is still used today. We understand its meaning differently, however.

The tendency of benzene to react by substitution rather than addition gave rise to another concept of aromaticity. For a compound to be called aromatic meant, experi-

---

* In 1861 the Austrian chemist Johann Josef Loschmidt represented the benzene ring with a circle, but he made no attempt to indicate how the carbon atoms were actually arranged in the ring.

mentally, that it gave substitution reactions rather than addition reactions even though it was highly unsaturated.

Before 1900, chemists assumed that the ring of alternating single and double bonds was the structural feature that gave rise to the aromatic properties. Since benzene and benzene derivatives (i.e., compounds with six-membered rings) were the only aromatic compounds known, chemists naturally sought other examples. The compound cyclooctatetraene seemed to be a likely candidate.

Cyclooctatetraene

In 1911, Richard Willstätter succeeded in synthesizing cyclooctatetraene. Willstätter found, however, that it is not at all like benzene. Cyclooctatetraene reacts with bromine by addition, it adds hydrogen readily, it is oxidized by solutions of potassium permanganate, and thus it is clearly *not aromatic*. While these findings must have been a keen disappointment to Willstätter, they were very significant for what they did not prove. Chemists, as a result, had to look deeper to discover the origin of benzene's aromaticity.

14.5 THE STABILITY OF BENZENE

We have seen that benzene shows unusual behavior by undergoing substitution reactions when, on the basis of its Kekulé structure, we should expect it to undergo addition. Benzene is unusual in another sense: It is *more stable* than the Kekulé structure suggests. To see how, consider the following thermochemical results.

Cyclohexene, a six-membered ring containing one double bond, can be hydrogenated easily to cyclohexane. When the $\Delta H°$ for this reaction is measured it is found to be -28.6 kcal mol^{-1}, very much like that of any similarly substituted alkene.

$$\text{Cyclohexene} + H_2 \xrightarrow{Pt} \text{Cyclohexane} \qquad \Delta H° = -28.6 \text{ kcal mol}^{-1}$$

We would expect that hydrogenation of 1,3-cyclohexadiene would liberate roughly twice as much heat and thus have a $\Delta H°$ equal to about -57.2 kcal mol^{-1}. When this experiment is done, the result is a $\Delta H° = -55.4$ kcal mol^{-1}. This result is quite close to what we calculated, and the difference can be explained by taking into account the fact that compounds containing conjugated double bonds are usually somewhat more stable than those that contain isolated double bonds (Section 12.8).

$$\text{1,3-Cyclohexadiene} + 2 H_2 \xrightarrow{Pt} \text{Cyclohexane}$$

Calculated
$\Delta H° = (2 \times -28.6) = -57.2 \text{ kcal mol}^{-1}$
Observed
$\Delta H° = -55.4 \text{ kcal mol}^{-1}$

If we extend this kind of thinking, and if benzene is simply 1,3,5-cyclohexatriene, we would predict benzene to liberate approximately 85.8 kcal mol^{-1} (3 × −28.6) when it is hydrogenated. When the experiment is actually done the result is surprisingly different. The reaction is exothermic, but only by 49.8 kcal mol^{-1}.

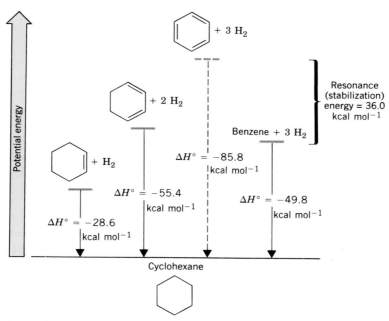

	Calculated
	$\Delta H° = (3 \times -28.6) = -85.8$ kcal mol^{-1}
	Observed $\quad \Delta H° = -49.8$ kcal mol^{-1}
	Difference $\quad = \quad 36.0$ kcal mol^{-1}

Benzene   Cyclohexane

When these results are represented as in Fig. 14.1, it becomes clear that benzene is much more stable than we calculated it to be. Indeed, it is more stable than the hypothetical 1,3,5-cyclohexatriene by 36 kcal mol^{-1}. This difference between the amount of heat actually released and that calculated on the basis of the Kekulé structure is now called the **resonance energy** of the compound.

FIGURE 14.1 Relative stabilities of cyclohexene, 1,3-cyclohexadiene, 1,3,5-cyclohexatriene (hypothetical), and benzene.

14.6 MODERN THEORIES OF THE STRUCTURE OF BENZENE

It was not until the development of quantum mechanics in the 1920s that the unusual behavior and stability of benzene began to be understood. Quantum mechanics, as we have seen, produced two ways of viewing bonds in molecules: resonance theory and molecular orbital theory. We now look at both of these as they apply to benzene.

14.6A The Resonance Explanation of the Structure of Benzene

A basic postulate of resonance theory (Sections 1.8 and 12.5) is that whenever two or more Lewis structures can be written for a molecule that *differ only in the positions of their electrons,* none of the structures will be in complete accord with the compound's chemical and physical properties. If we recognize this, we can now understand the true nature of the two Kekulé structures (**I** and **II**) for benzene. The two Kekulé structures differ only in the positions of their electrons. Structures **I** and **II**, then, do not represent two separate molecules in equilibrium as Kekulé had proposed. Instead, they are the closest we can get to a structure for benzene within the limitations of its molecular formula, the classical rules of valence, and the fact that the six hydrogen atoms are chemically equivalent. The problem with the Kekulé structures is that they are Lewis structures, and Lewis structures portray electrons in localized distributions. (With benzene, as we shall see, the electrons are delocalized.) Resonance theory, fortunately, does not stop with telling us when to expect this kind of trouble; it also gives us a way out. Resonance theory tells us to use structures **I** and **II** as resonance contributors to a picture of the real molecule of benzene. As such, **I** and **II** should be connected with a double-headed arrow and not with two separate ones (because we must reserve the symbol of two separate arrows for chemical equilibria). Resonance contributors, we emphasize again, are not in equilibrium. They are not structures of real molecules. They are the closest we can get if we are bound by simple rules of valence, but they are very useful in helping us visualize the actual molecule as a hybrid.

Look at the structures carefully. All of the single bonds in structure **I** are double bonds in structure **II**. If we blend **I** and **II**, that is, if we fashion a hybrid of them, then the carbon–carbon bonds in benzene are neither single bonds nor double bonds. Rather, they have a bond order between that of a single bond and that of a double bond. This is exactly what we find experimentally. Spectroscopic measurements show that the molecule of benzene is planar and that all of its carbon–carbon bonds are of equal length. Moreover, the carbon–carbon bond lengths in benzene (Fig. 14.2) are 1.39 Å, a value in between that for a carbon–carbon single bond between sp^2-hybridized atoms (1.47 Å) (cf. Table 12.1) and that for a carbon–carbon double bond (1.33 Å).

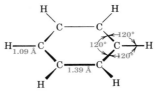

FIGURE 14.2 Bond lengths and angles in benzene. (Only the σ bonds are shown.)

The hybrid structure is represented by inscribing a circle in the hexagon, and it is this new formula (**III**) that is most often used for benzene today. There are times, however, when an accounting of the electrons must be made, and for these purposes we may use one or the other of the Kekulé structures. We do this simply because the electron count in a Kekulé structure is obvious, whereas the number of electrons represented by a circle or portion of a circle is ambiguous. With benzene the circle represents the six electrons that are delocalized about the six carbon atoms of the benzene ring. With other systems, however, a circle in a ring may represent numbers of delocalized electrons other than six.

III

PROBLEM 14.2

If benzene were 1,3,5-cyclohexatriene, the carbon–carbon bonds would be alternately long and short as indicated in the following structures. However, to consider the structures here as resonance contributors (or to connect them by a double-headed arrow) violates a basic principle of resonance theory. Explain.

Resonance theory (Section 12.5) also tells us that whenever equivalent resonance structures can be drawn for a molecule, the molecule (or hybrid) is much more stable than any of the resonance structures would be individually if they could exist. In this way resonance theory accounts for the much greater stability of benzene when compared to the hypothetical 1,3,5-cyclohexatriene. For this reason the extra stability associated with benzene is called its *resonance energy*.

14.6B The Molecular Orbital Explanation of the Structure of Benzene

The fact that the bond angles of the carbon atoms in the benzene ring are 120° strongly suggests that the carbon atoms are *sp²* *hybridized*. If we accept this suggestion and construct a planar six-membered ring from *sp²* carbon atoms as shown in Fig. 14.3, another picture of benzene begins to emerge. Because the carbon–carbon bond lengths are all 1.39 Å, the *p* orbitals are close enough to overlap effectively. The *p* orbitals overlap equally all around the ring.

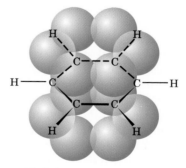

FIGURE 14.3 Overlapping *p* orbitals in benzene.

According to molecular orbital theory, the six overlapping p orbitals combine to form a set of six π molecular orbitals. Molecular orbital theory also allows us to calculate the relative energies of the π molecular orbitals. These calculations are beyond the scope of our discussion, but the energy levels are shown in Fig. 14.4.

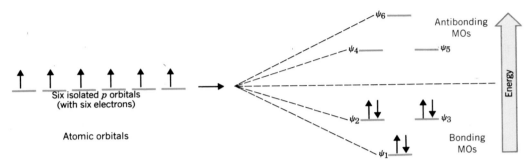

FIGURE 14.4 How six p atomic orbitals (one from each carbon of the benzene ring) combine to form six π molecular orbitals. Three of the molecular orbitals have energies lower than that of an isolated p orbital; these are the bonding molecular orbitals. Three of the molecular orbitals have energies higher than that of an isolated p orbital; these are the antibonding molecular orbitals. Orbitals ψ_2 and ψ_3 have the same energy and are said to be degenerate; the same is true of orbitals ψ_4 and ψ_5.

A molecular orbital, as we have seen, can accommodate two electrons if their spins are opposed. Thus, the electronic structure of the ground state of benzene is obtained by adding the six electrons to the π molecular orbitals, starting with those of lowest energy, as shown in Fig. 14.4. Notice that in benzene, all of the bonding orbitals are filled, all of the electrons have their spins paired, and there are no electrons in antibonding orbitals. Benzene is therefore said to have a *closed bonding shell* of delocalized π electrons. This closed bonding shell accounts, in part, for the stability of benzene.

14.7 HÜCKEL'S RULE: THE $(4n + 2)$ π ELECTRON RULE

In 1931 the German physicist Erich Hückel carried out a series of mathematical calculations based on the kind of theory that we have just described. Hückel concerned himself with compounds containing **one planar ring in which each atom has a p orbital** as in benzene. His calculations show that planar monocyclic rings containing $(4n + 2)$ π electrons where $n = 0, 1, 2, 3, \ldots$, and so on (i.e., rings containing 2, 6, 10, 14, . . . , etc., π electrons), have closed shells of delocalized electrons like benzene, and should have substantial resonance energies. In other words **planar monocyclic rings with 2, 6, 10, 14, . . . , delocalized electrons should be aromatic.**

Although Hückel's calculations are beyond our scope, we can get a picture of the relative energies of the π molecular orbitals of conjugated monocyclic systems in a relatively easy way. *We simply inscribe in a circle a regular polygon corresponding to the ring of the compound being considered so that one corner of the polygon is at the bottom.* **The points where the corners of the polygon touch the circle correspond to the energy levels of the π molecular orbitals of the system.** With benzene, for example, this method (Fig. 14.5) furnishes the same energy levels that we saw earlier in Fig. 14.4, energy levels that were based on quantum mechanical calculations.

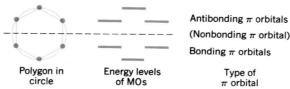

FIGURE 14.5 The polygon-and-circle method for deriving the relative energies of the π molecular orbitals of benzene. A horizontal line halfway up the circle divides the bonding orbitals from the antibonding orbitals. If an orbital falls on this line, it is a nonbonding orbital. This method was developed by C. A. Coulson (of Oxford University).

We can now understand why cyclooctatetraene is not aromatic. Cyclooctatetraene has a total of 8 π electrons. Eight is not a Hückel number; it is a *4n number,* not a *4n + 2 number.* Using the polygon-and-circle method (Fig. 14.6) we find that cyclooctatetraene, if it were planar, *would not* have a closed shell of π electrons like benzene; it would have an unpaired electron in each of two nonbonding orbitals. Molecules with unpaired electrons (radicals) are *not* unusually stable; they are typically highly reactive and unstable. A planar form of cyclooctatetraene, therefore, should not be at all like benzene and should not be aromatic.

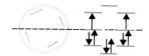

FIGURE 14.6 The π molecular orbitals that cyclooctatetraene would have if it were planar. Notice that, unlike benzene, this molecule is predicted to have two nonbonding orbitals and because it has eight π electrons it would have an unpaired electron in each of the two nonbonding orbitals. Such a system would not be expected to be aromatic.

Because cyclooctatetraene does not gain stability by becoming planar, it assumes the tub shape shown below. (In Section 14.7D we shall see that cyclooctatetraene would actually lose stability by becoming planar.)

The bonds of cyclooctatetraene are known to be alternately long and short; X-ray studies indicate that they are 1.48 and 1.34 Å, respectively.

14.7A The Annulenes

The name **annulene** has been proposed as a general name for monocyclic compounds that can be represented by structures having alternating single and double bonds. The ring size of an annulene is indicated by a number in brackets. Thus, benzene is [6]annulene and cyclooctatetraene is [8]annulene.* Hückel's rule predicts that annulenes will be aromatic, provided their molecules have (4n + 2) π electrons and have a planar carbon skeleton.

*These names are seldom used for benzene and cyclooctatetraene. They are often used, however, for conjugated rings of 10 or more carbon atoms.

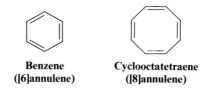

Benzene
([6]annulene)

Cyclooctatetraene
([8]annulene)

Before 1960 the only annulenes that were available to test Hückel's predictions were benzene and cyclooctatetraene. During the 1960s, and largely as a result of research by Franz Sondheimer, a number of large-ring annulenes were synthesized, and the predictions of Hückel's rule were verified.

Consider the [14], [16], [20], [22], and [24]annulenes as examples. Of these, *as Hückel's rule predicts,* the [14], [18], and [22]annulenes ($4n + 2$ when $n = 3, 4, 5,$ respectively) have been found to be aromatic. The [16]annulene and the [24]annulene are not aromatic. They are $4n$ compounds, not $4n + 2$ compounds.

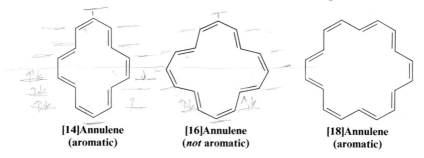

[14]Annulene
(aromatic)

[16]Annulene
(*not* aromatic)

[18]Annulene
(aromatic)

Examples of [10] and [12]annulenes have also been synthesized and none is aromatic. We would not expect [12]annulenes to be aromatic since they have 12 π electrons and do not obey Hückel's rule. The following [10]annulenes would be expected to be aromatic on the basis of electron count, but their rings are not planar.

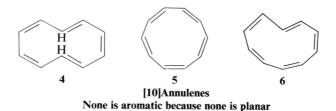

4

5

6

[10]Annulenes
None is aromatic because none is planar

The [10]annulene (**4**) has two trans double bonds. Its bond angles are approximately 120°; therefore, it has no appreciable angle strain. The carbon atoms of its ring, however, are prevented from becoming coplanar because the two hydrogen atoms in the center of the ring interfere with each other. Because the ring is not planar, the *p* orbitals of the carbon atoms are not parallel and, therefore, cannot overlap effectively around the ring to form the π molecular orbitals of an aromatic system.

The [10]annulene with all cis double bonds (**5**) would, if it were planar, have considerable angle strain because the internal bond angles would be 144°. Consequently, any stability this isomer gained by becoming planar in order to become aromatic would be more than offset by the destabilizing effect of the increased angle strain. A similar problem of a large angle strain associated with a planar form prevents molecules of the [10]annulene isomer with one trans double bond (**6**) from being aromatic.

After many unsuccessful attempts over many years, in 1965 [4]annulene (or cyclobutadiene) was synthesized by R. Pettit and his co-workers at the University of Texas, Austin. Cyclobutadiene is a $4n$ molecule not a $4n + 2$ molecule, and, as we would expect, it is a highly unstable compound and *it is not aromatic.*

**Cyclobutadiene
or [4]annulene
(*not* aromatic)**

PROBLEM 14.3

Use the polygon-and-circle method to outline the π molecular orbitals of cyclobutadiene and explain why, on this basis, you would not expect it to be aromatic.

14.7B NMR Spectroscopy. Evidence for Electron Delocalization in Aromatic Compounds

The ¹H NMR spectrum of benzene consists of a single unsplit signal at δ 7.27. That only a single unsplit signal is observed is further proof that all of the hydrogens of benzene are equivalent. That the signal occurs at such a low magnetic field strength, is, as we shall see, compelling evidence for the assertion that the π electrons of benzene are delocalized.

We learned in Section 13.6 that circulations of σ electrons of C—H bonds cause the protons of alkanes to be *shielded* from the applied magnetic field of an NMR spectrometer and, consequently, these protons absorb at higher magnetic field strengths. We shall now explain the low field strength of absorption of benzene on the basis of *deshielding caused by circulation of the π electrons of benzene,* and this explanation, as you will see, requires that the π electrons be delocalized.

When benzene molecules are placed in the powerful magnetic field of the NMR spectrometer, electrons circulate in the direction shown in Fig. 14.7; by doing so, they generate a **ring current.** (If you have studied physics you will understand why the electrons circulate in this way.) This π electron circulation creates an induced magnetic field that, *at the position of the protons of benzene, reinforces the applied magnetic field,* and this reinforcement causes the protons to be strongly *deshielded.* By "deshielded" we mean that the protons sense the sum of the two fields, and, therefore, the applied magnetic field strength does not have to be as high as might have been required in the absence of the induced field. This strong deshielding, which we attribute to a ring current created by the *delocalized* π electrons, explains why aromatic protons absorb at very low magnetic field strengths.

The deshielding of external aromatic protons that results from the ring current is one of the best pieces of physical evidence that we have for π electron delocalization in aromatic rings. In fact, low field strength proton absorption is often used as a criterion for aromaticity in newly synthesized conjugated cyclic compounds.

Not all aromatic protons absorb at low magnetic field strengths, however. The internal protons of large-ring aromatic compounds that have hydrogens in the center of the ring (in the π electron cavity) absorb at unusually high magnetic field strengths because they are highly shielded by the opposing induced magnetic field in the center of the ring (see Fig. 14.7). An example is [18]annulene (Fig. 14.8). The internal

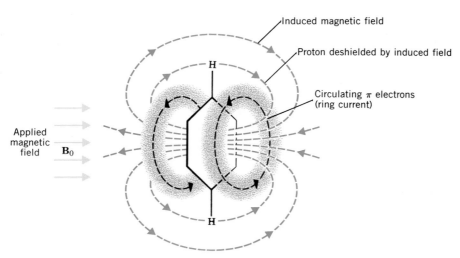

FIGURE 14.7 The induced magnetic field of the π electrons of benzene deshields the benzene protons. Deshielding occurs because at the location of the protons the induced field is in the same direction as the applied field.

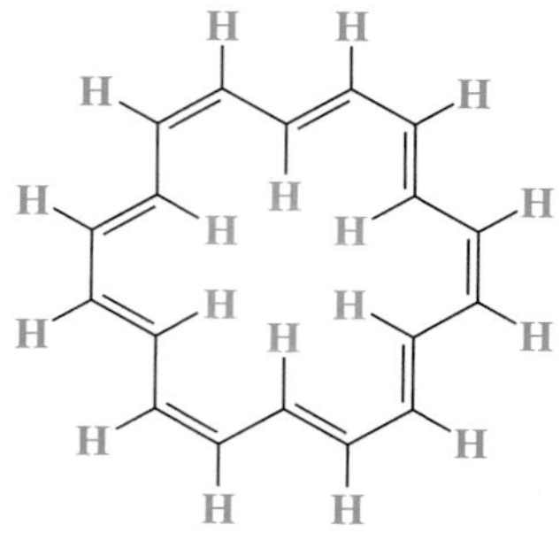

FIGURE 14.8 [18]Annulene. The internal protons (red) are highly shielded and absorb at δ -3.0 ppm. The external protons (blue) are highly deshielded and absorb at δ 9.3.

protons of [18]annulene absorb far upfield at $\delta -3.0$, above the signal for TMS; the external protons, on the other hand, absorb far downfield at δ 9.3. Considering that [18]annulene has $(4n + 2)$ π electrons, this evidence provides strong support for π electron delocalization as a criterion for aromaticity and for the predictive power of Hückel's rule.

14.7C Aromatic Ions

In addition to the neutral molecules that we have discussed so far, there are a number of monocyclic species that bear either a positive or a negative charge. Some of these ions show unexpected stabilities that suggest that they, too, are aromatic. Hückel's rule is helpful in accounting for the properties of these ions as well. We shall consider two examples: the cyclopentadienyl anion and the cycloheptatrienyl cation.

Cyclopentadiene is not aromatic; however, it is unusually acidic for a hydrocarbon. (The pK_a for cyclopentadiene is 16 and, by contrast, the pK_a for cycloheptatriene is 36.) Because of its acidity, cyclopentadiene can be converted to its anion by treatment with moderately strong bases. The cyclopentadienyl anion, moreover, is unusually stable and NMR spectroscopy shows that all five hydrogen atoms in the cyclo-pentadienyl anion are equivalent and absorb downfield.

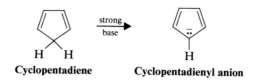

Cyclopentadiene   Cyclopentadienyl anion

The orbital structure of cyclopentadiene (Fig. 14.9) shows why cyclopentadiene, itself, is not aromatic. Not only does it not have the proper number of π electrons, but the π electrons cannot be delocalized about the entire ring because of the intervening sp^3-hybridized $-CH_2-$ group with no available p orbital.

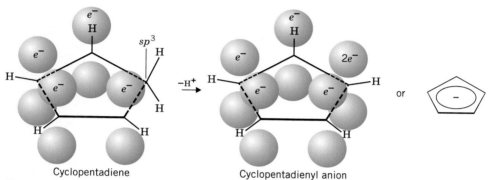

Cyclopentadiene   Cyclopentadienyl anion

FIGURE 14.9 The p orbitals of cyclopentadiene and of the cyclopentadienyl anion.

On the other hand, if the $-CH_2-$ carbon atom becomes sp^2 hybridized after it loses a proton (Fig. 14.9), the two electrons left behind can occupy the new p orbital that is produced. Moreover, this new p orbital can overlap with the p orbitals on either side of it and give rise to a ring with *six* delocalized π electrons. Because the electrons are delocalized, all of the hydrogen atoms are equivalent and this agrees with what NMR spectroscopy tells us.

Six is, of course, a Hückel number (4n + 2, where n = 1), and the cyclopenta-dienyl anion is, in fact, an **aromatic anion.** The unusual acidity of cyclopentadiene is a result of the unusual stability of its anion.

PROBLEM 14.4

(a) Outline the π molecular orbitals of the cyclopentadienyl system by inscribing a regular pentagon in a circle and explain on this basis why the cyclopentadienyl anion is aromatic.   (b) What electron distribution would you expect for the cyclopentadienyl cation?   (c) Would you expect it to be aromatic? Explain your answer.   (d) Would you expect the cyclopentadienyl cation to be aromatic on the basis of Hückel's rule?

Cycloheptatriene (Fig. 14.10) (a compound with the common name tropylidene) has six π electrons. However, the six π electrons of cycloheptatriene cannot be fully delocalized because of the presence of the —CH_2— group, a group that does not have an available p orbital (Fig. 14.10).

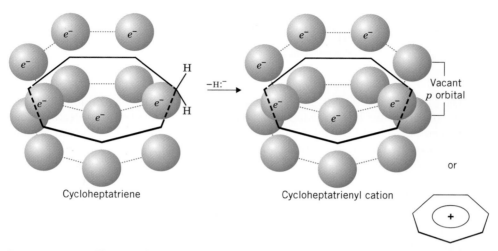

FIGURE 14.10 The p orbitals of cycloheptatriene and of the cycloheptatrienyl (tropylium) cation.

When cycloheptatriene is treated with a reagent that can abstract a hydride ion, it is converted to the cycloheptatrienyl (or tropylium) cation. The loss of a hydride ion from cycloheptatriene occurs with unexpected ease, and the cycloheptatrienyl cation is found to be unusually stable. The NMR spectrum of the cycloheptatrienyl cation indicates that all seven hydrogen atoms are equivalent. If we look closely at Fig. 14.10, we see how we can account for these observations.

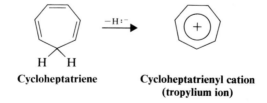

Cycloheptatriene   **Cycloheptatrienyl cation
(tropylium ion)**

As a hydride ion is removed from the —CH_2— group of cycloheptatriene, a vacant p orbital is created, and the carbon atom becomes sp^2 hybridized. The cation that results has seven overlapping p orbitals containing *six* delocalized π electrons. The cycloheptatrienyl cation is, therefore, an aromatic cation, and all of its hydrogen atoms should be equivalent; again this is exactly what we find experimentally.

PROBLEM 14.5

(a) Use the polygon-and-circle method to sketch the relative energies of the π molecular orbitals of the cycloheptatrienyl cation and explain why it is aromatic. (b) Would you expect the cycloheptatrienyl anion to be aromatic on the basis of the electron distribution in its π molecular orbitals? Explain. (c) Would you expect the cycloheptatrienyl anion to be aromatic on the basis of Hückel's rule?

PROBLEM 14.6

1,3,5-Cycloheptatriene (shown on the preceding page) is even less acidic than 1,3,5-heptatriene. Explain how this experimental observation might help to confirm your answer to parts (b) and (c) of the previous problem.

PROBLEM 14.7

When 1,3,5-cycloheptatriene reacts with one molar equivalent of bromine in CCl_4 at 0°C, it undergoes 1,6 addition.   (a) Write the structure of this product. (b) On heating, this 1,6-addition product loses HBr readily to form a compound with the molecular formula C_7H_7Br, called *tropylium bromide*. Tropylium bromide is insoluble in nonpolar solvents, but is soluble in water; it is unexpectedly high melting (mp = 203°C) and when treated with silver nitrate, an aqueous solution of tropylium bromide gives a precipitate of AgBr. What do these experimental results suggest about the bonding in tropylium bromide?

14.7D Aromatic, Antiaromatic, and Nonaromatic Compounds

What do we mean when we say that a compound is aromatic? We mean that its π electrons are *delocalized* over the entire ring and that it is *stabilized* by the π-electron delocalization.

As we have seen, the best way to determine whether or not the π electrons of a cyclic system are delocalized is through the use of NMR spectroscopy. It provides direct physical evidence of whether or not the π electrons are delocalized.

But what do we mean by saying that a compound is stabilized by π-electron delocalization? We have an idea of what this means from our comparison of the heat of hydrogenation of benzene and that calculated for the hypothetical 1,3,5-cyclohexatriene. We saw that benzene—in which the π electrons are delocalized—is much more stable than 1,3,5-cyclohexatriene (a model in which the π electrons are not delocalized). We call the energy difference between them the resonance energy (delocalization energy) or stabilization energy.

In order to make similar comparisons for other aromatic compounds we need to choose proper models. But what should these models be?

One proposal is that we should compare the π-electron energy of the cyclic system with that of the corresponding open-chain compound. This approach is particularly useful because it furnishes us with models not only for annulenes but for aromatic cations and anions as well. (Corrections need to be made, of course, when the cyclic system is strained.)

When we use this approach we take as our model a linear chain of sp^2-hybridized atoms that carries the same number of π electrons as our cyclic compound. Then we imagine ourselves removing two hydrogen atoms from the ends of this chain and joining the ends to form a ring. If the ring has *lower* π-electron energy than the open chain, then the ring is *aromatic*. If the ring and chain have *the same* π-electron energy, then the ring is *nonaromatic*. If the ring has *greater* π-electron energy than the open chain, then the ring is *antiaromatic*.

The actual calculations and experiments used in determining π-electron energies are beyond our scope, but we can study four examples that illustrate how this approach has been used.

Cyclobutadiene   For cyclobutadiene we consider the change in π-electron energy for the following *hypothetical* transformation.

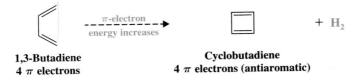

1,3-Butadiene
4 π electrons

Cyclobutadiene
4 π electrons (antiaromatic)

Calculations indicate and experiments appear to confirm that the π-electron energy of cyclobutadiene is higher than that of its open-chain counterpart. Thus cyclobutadiene is classified as antiaromatic.

Benzene   Here our comparison is based on the following hypothetical transformation.

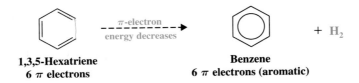

1,3,5-Hexatriene
6 π electrons

Benzene
6 π electrons (aromatic)

Calculations indicate and experiments confirm that benzene has a much lower π-electron energy than 1,3,5-hexatriene. Benzene is classified as being aromatic on the basis of this comparison as well.

Cyclopentadienyl Anion   Here we use a linear anion for our hypothetical transformation:

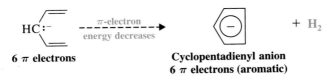

6 π electrons

Cyclopentadienyl anion
6 π electrons (aromatic)

Both calculations and experiments confirm that the cyclic anion has a lower π-electron energy than its open-chain counterpart. Therefore the cyclopentadienyl anion is classified as aromatic.

Cyclooctatetraene   For cyclooctatetraene we consider the following hypothetical transformation.

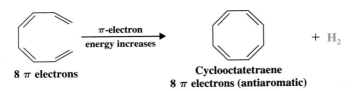

8 π electrons

Cyclooctatetraene
8 π electrons (antiaromatic)

Here calculations and experiments indicate that a planar cyclooctatetraene would have higher π-electron energy than the open-chain octatetraene. Therefore, a planar form of cyclooctatetraene would, if it existed, be *antiaromatic*. As we saw earlier, cyclooctatetraene is not planar and behaves like a simple cyclic polyene.

PROBLEM 14.8

Calculations indicate that the π-electron energy decreases for the hypothetical transformation from the allyl cation to the cyclopropenyl cation below.

$$CH_2{=}CH{-}CH_2{}^+ \longrightarrow \overset{+}{\triangle} + H_2$$

What does this indicate about the possible aromaticity of the cyclopropenyl cation? (See Problem 14.10 for more on this cation.)

PROBLEM 14.9

The cyclopentadienyl cation is apparently *antiaromatic*. Explain what this means in terms of the π-electron energies of a cyclic and an open-chain compound.

PROBLEM 14.10

In 1967 Ronald Breslow (of Columbia University) and his co-workers showed that adding $SbCl_5$ to a solution of 3-chlorocyclopropene in CH_2Cl_2 caused the precipitation of a white solid with the composition $C_3H_3{}^+SbCl_6{}^-$. The NMR spectroscopy of a solution of this salt showed that all of its hydrogen atoms were equivalent. What new aromatic ion had Breslow and his co-workers prepared?

14.8 OTHER AROMATIC COMPOUNDS

14.8A Benzenoid Aromatic Compounds

In addition to those that we have seen so far, there are many other examples of aromatic compounds. Representatives of one broad class of aromatic compounds, called **polycyclic benzenoid aromatic hydrocarbons,** are illustrated in Fig. 14.11.

All of these consist of molecules having two or more benzene rings *fused* together. A close look at one, naphthalene, will illustrate what we mean by this.

According to resonance theory, a molecule of naphthalene can be considered to be a hybrid of three Kekulé structures. One of these Kekulé structures is shown in Fig. 14.12. There are two carbon atoms in naphthalene (C4*a* and C8*a*) that are common to both rings. These two atoms are said to be at the points of *ring fusion*. They direct all of their bonds toward other carbon atoms and do not bear hydrogen atoms.

Naphthalene
$C_{10}H_8$

Anthracene
$C_{14}H_{10}$

Phenanthrene
$C_{14}H_{10}$

Pyrene
$C_{16}H_{10}$

Benzo[*a*]pyrene
$C_{20}H_{12}$

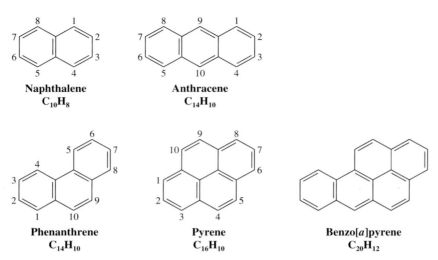

FIGURE 14.11 Benzenoid aromatic hydrocarbons.

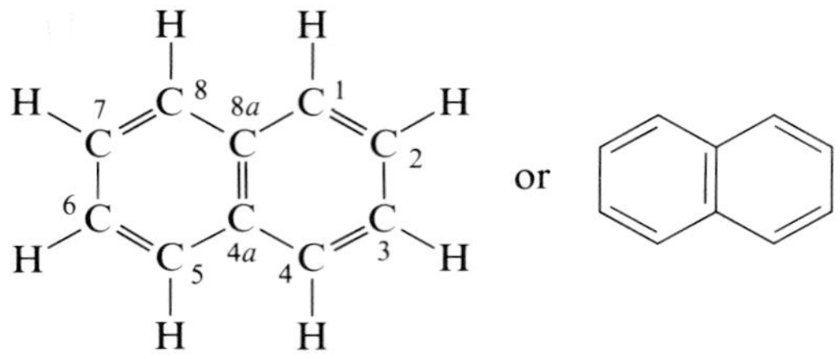

FIGURE 14.12 One Kekulé structure for naphthalene.

Molecular orbital calculations for naphthalene begin with the model shown in Fig. 14.13. The *p* orbitals overlap around the periphery of both rings and across the points of ring fusion.

When molecular orbital calculations are carried out for naphthalene using the model shown in Fig. 14.13, the results of the calculations correlate well with our experimental knowledge of naphthalene. The calculations indicate that delocalization of the 10 π electrons over the two rings produces a structure with considerably lower energy than that calculated for any individual Kekulé structure. Naphthalene, consequently, has a substantial resonance energy. Based on what we know about benzene, moreover, naphthalene's tendency to react by substitution rather than addition and to show other properties associated with aromatic compounds is understandable.

Anthracene and phenanthrene are isomers. In anthracene the three rings are fused in a linear way, and in phenanthrene they are fused so as to produce an angular molecule. Both of these molecules also show large resonance energies and chemical properties typical of aromatic compounds.

Pyrene is also aromatic. Pyrene itself has been known for a long time; a pyrene derivative, however, has been the object of research that shows another interesting application of Hückel's rule.

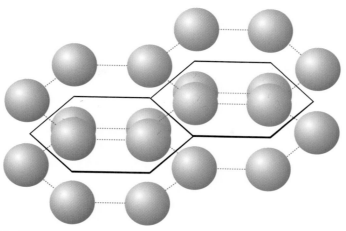

FIGURE 14.13 The *p* orbitals of naphthalene.

To understand this particular research we need to pay special attention to the Kekulé structure for pyrene (Fig. 14.14). The total number of π electrons in pyrene is 16 (8 double bonds = 16 π electrons). Sixteen is a non-Hückel number, but Hückel's

FIGURE 14.14 One Kekulé structure for pyrene. The internal double bond is enclosed in a dotted circle for emphasis.

rule is intended to be applied only to monocyclic compounds and pyrene is clearly tetracyclic. If we disregard the internal double bond of pyrene, however, and look only at the periphery, we see that the periphery is a planar ring with 14 π electrons. The periphery is, in fact, very much like that of [14]annulene. Fourteen *is* a Hückel number ($4n + 2$, where $n = 3$) and one might then predict that the periphery of pyrene would be aromatic by itself, in the absence of the internal double bond.

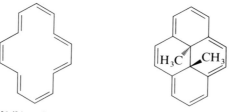

[14]Annulene   *trans*-15,16-Dimethyldihydropyrene

This prediction was confirmed when V. Boekelheide (University of Oregon) synthesized *trans*-15,16-dimethyldihydropyrene and showed that it is aromatic.

PROBLEM 14.11

In addition to a signal downfield, the ^{1}H NMR spectrum of 15, 16-dimethyldihydropyrene has a signal at $\delta - 4.2$. Account for the presence of this high field signal.

14.8B Nonbenzenoid Aromatic Compounds

Naphthalene, phenanthrene, and anthracene are examples of *benzenoid* aromatic compounds. On the other hand, the cyclopentadienyl anion, the cycloheptatrienyl cation, *trans*-15,16-dimethyldihydropyrene, and the aromatic annulenes (except for [6]annulene) are classified as **nonbenzenoid aromatic compounds.**

Another example of a *nonbenzenoid* aromatic hydrocarbon is the compound azulene. Azulene has a resonance energy of 49 kcal mol^{-1}.

Azulene

PROBLEM 14.12

Azulene has an appreciable dipole moment. Write resonance structures for azulene that explain this dipole moment and that help explain its aromaticity.

14.8C Fullerenes

In 1990 Wolfgang Krätschmer (Max Planck Institute, Heidelberg), Donald Huffman (University of Arizona), and their co-workers described the first practical synthesis of C_{60}, a molecule shaped like a soccer ball, and called buckminsterfullerene. Formed by the resistive heating of graphite in an inert atmosphere, C_{60} is a member of an exciting new group of aromatic compounds called fullerenes. Fullerenes are cagelike molecules with the geometry of a truncated icosahedron or geodesic dome, named after the architect Buckminster Fuller, renowned for his development of structures with geodesic domes. The structure of C_{60} and its existence had been established 5 years earlier, by H. W. Kroto (University of Sussex), R. E. Smalley and R. F. Curl (Rice University) and their co-workers. Kroto, Curl, and Smalley had found both C_{60} and C_{70} (Fig. 14.15) as highly stable components of a mixture of carbon clusters formed by laser-vaporizing graphite. Since 1990 chemists have synthesized many other higher and lower fullerenes and have begun exploring their interesting chemistry.

Like a geodesic dome, a fullerene is composed of a network of pentagons and hexagons. To close into a spheroid, a fullerene must have exactly 12 five-membered faces, but the number of six-membered faces can vary widely. The structure of C_{60} has 20 hexagonal faces; C_{70} has 25. Each carbon of a fullerene is sp^2 hybridized and forms σ bonds to three other carbon atoms. The remaining electron at each carbon is delocalized into a system of molecular orbitals that gives the whole molecule aromatic character.

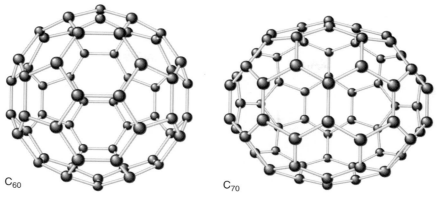

FIGURE 14.15 The structures of C_{60} and C_{70}. [Adapted from F. Diederich and R. L. Whetten, *Acc. Chem. Res.,* **1992,** *25,* 120.]

The chemistry of fullerenes is proving to be even more fascinating than their synthesis. Fullerenes have a high electron affinity and readily accept electrons from alkali metals to produce a new metallic phase—a "buckide" salt. One such salt, K_3C_{60}, is a stable metallic crystal consisting of a face-centered cubic structure of buckyballs with a potassium atom in between; it becomes a superconductor when cooled below 18 K. Fullerenes have even been synthesized that have metal atoms in the interior of the carbon atom cage.

14.9 HETEROCYCLIC AROMATIC COMPOUNDS

Almost all of the cyclic molecules that we have discussed so far have had rings composed solely of carbon atoms. However, in molecules of many cyclic compounds an element other than carbon is present in the ring. These compounds are called **heterocyclic compounds.** Heterocyclic molecules are quite commonly encountered in nature. For this reason, and because the structures of some of these molecules are closely related to the compounds that we discussed earlier, we shall now describe a few examples.

Heterocyclic compounds containing nitrogen, oxygen, or sulfur are by far the most common. Four important examples are given here in their Kekulé forms. *These four compounds are all aromatic.*

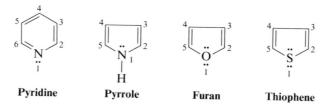

If we examine these structures, we shall see that pyridine is electronically related to benzene, and that pyrrole, furan, and thiophene are related to the cyclopentadienyl anion.

The nitrogen atoms in molecules of both pyridine and pyrrole are sp^2 hybridized. In pyridine (Fig. 14.16) the sp^2-hybridized nitrogen donates one electron to the π system. This electron, together with one from each of the five carbon atoms, gives

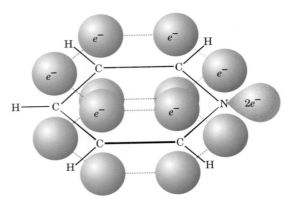

FIGURE 14.16 The orbital structure of pyridine.

pyridine a sextet of electrons like benzene. The two unshared electrons of the nitrogen of pyridine are in an sp^2 orbital that lies in the same plane as the atoms of the ring. This sp^2 orbital does not overlap with the p orbitals of the ring (it is, therefore, said to be *orthogonal* to the p orbitals). The unshared pair on nitrogen is not a part of the π system, and these electrons confer on pyridine the properties of a weak base.

In pyrrole (Fig. 14.17) the electrons are arranged differently. Because only four π electrons are contributed by the carbon atoms of the pyrrole ring, the sp^2-hybridized nitrogen must contribute two electrons to give an aromatic sextet. Because these electrons are a part of the aromatic sextet, they are not available for donation to a proton. Thus, in aqueous solution, pyrrole is not appreciably basic.

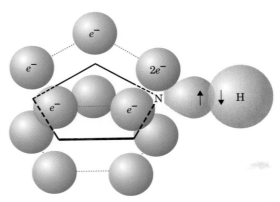

FIGURE 14.17 The orbital structure of pyrrole. (Compare with the orbital structure of the cyclopentadienyl anion in Fig. 14.9.)

Furan and thiophene are structurally quite similar to pyrrole. The oxygen atom in furan and the sulfur atom in thiophene are sp^2 hybridized. In both compounds the p orbital of the heteroatom donates two electrons to the π system. The oxygen and sulfur atoms of furan and thiophene carry an unshared pair of electrons in an sp^2 orbital (Fig. 14.18) that is orthogonal to the π system.

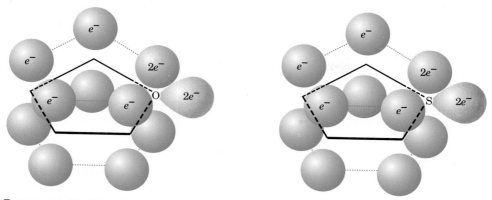

FIGURE 14.18 The orbital structures of furan and thiophene.

14.10 AROMATIC COMPOUNDS IN BIOCHEMISTRY

Compounds with aromatic rings occupy numerous and important positions in reactions that occur in living systems. It would be impossible to describe them all in this chapter. We shall, however, point out a few examples now and we shall see others later.

Two amino acids necessary for protein synthesis contain the benzene ring:

Phenylalanine

Tyrosine

A third aromatic amino acid, tryptophan, contains a benzene ring fused to a pyrrole ring. (This aromatic ring system is called an indole system, cf. Section 20.1B.)

Tryptophan

Indole

It appears that humans, because of the course of evolution, do not have the biochemical ability to synthesize the benzene ring. As a result, phenylalanine and tryptophan derivatives are essential in the human diet. Because tyrosine can be synthesized from phenylalanine in a reaction catalyzed by an enzyme known as *phenylalanine hydroxylase,* it is not essential in the diet as long as phenylalanine is present.

Heterocyclic aromatic compounds are also present in many biochemical systems. Derivatives of purine and pyrimidine are essential parts of DNA and RNA.

Purine

Pyrimidine

DNA is the molecule responsible for the storage of genetic information and RNA is prominently involved in the synthesis of enzymes and other proteins (Chapter 25).

PROBLEM 14.13

(a) The —SH group is sometimes called the *mercapto group*. 6-Mercaptopurine is used in the treatment of acute leukemia. Write its structure. (b) Allopurinol, a compound used to treat gout, is 6-hydroxypurine. Write its structure.

Both a pyridine derivative (nicotinamide) and a purine derivative (adenine) are present in one of the most important coenzymes (Section 24.9) in biological oxidations. This molecule, **nicotinamide adenine dinucleotide** (NAD$^+$, the oxidized form), is shown in Fig. 14.19. NADH is the reduced form.

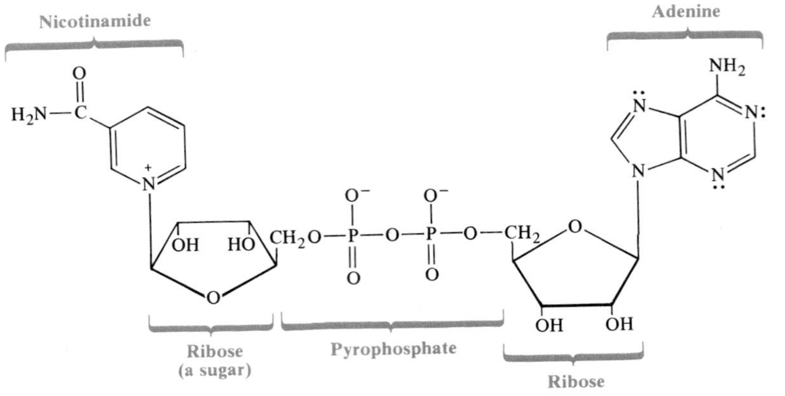

FIGURE 14.19 Nicotinamide adenine dinucleotide (NAD$^+$).

NAD$^+$, as part of a liver enzyme called alcohol dehydrogenase, is capable of oxidizing alcohols to aldehydes. While the overall change is quite complex, a look at one aspect of it will illustrate a *biological use* of the extra stability (resonance or delocalization energy) associated with an aromatic ring.

A simplified version of the oxidation of an alcohol to an aldehyde is illustrated here:

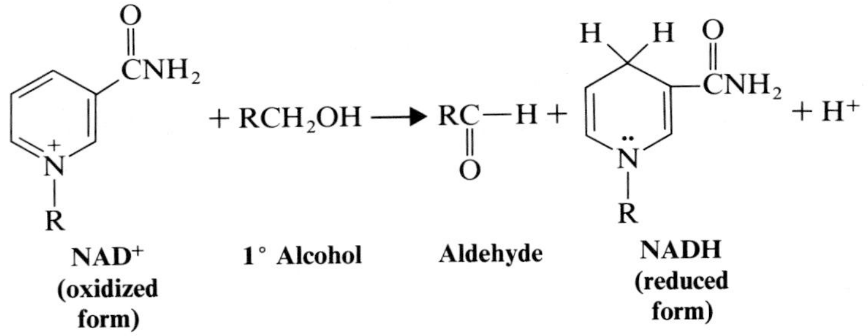

The *aromatic* pyridine ring (actually a *pyridinium* ring, because it is positively charged) in NAD$^+$ is converted to a *nonaromatic* ring in NADH. The extra stability of the pyridine ring is lost in this change and, as a result, the free energy of NADH is greater than that of NAD$^+$. The conversion of the alcohol to the aldehyde, however, occurs with a decrease in free energy. Because these reactions are coupled in biological systems (Fig. 14.20), a portion of the free energy contained in the alcohol becomes chemically contained in NADH. This stored energy in NADH is used to bring about other biochemical reactions that require energy and that are necessary to life.

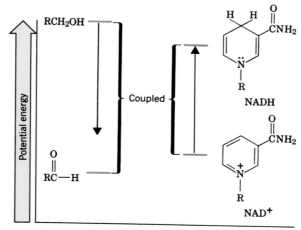

FIGURE 14.20 Potential energy diagram for the biologically coupled oxidation of an alcohol and reduction of nicotinamide adenine dinucleotide.

Although many aromatic compounds are essential to life, others are hazardous. Many are quite toxic and several benzenoid compounds, including benzene itself, are **carcinogenic.** Two other examples are benzo[a]pyrene and 7-methylbenz[a] anthracene.

Benzo[a]pyrene   **7-Methylbenz[a]anthracene**

The hydrocarbon benzo[a]pyrene has been found in cigarette smoke and in the exhaust from automobiles. It is also formed in the incomplete combustion of any fossil fuel. It is found on charcoal-broiled steaks and exudes from asphalt streets on a hot summer day. Benzo[a]pyrene is so carcinogenic that one can induce skin cancers in mice with almost total certainty simply by shaving an area of the body of the mouse and applying a coating of benzo[a]pyrene.

14.11 SPECTROSCOPY OF AROMATIC COMPOUNDS

14.11A ^{1}H NMR Spectra

As we learned in Chapter 13, the hydrogens of benzene derivatives absorb downfield in the region between δ 6.0 and δ 9.5. In Section 14.7B we found that absorption takes place far downfield because a ring current generated in the benzene ring creates a magnetic field, called "the induced field," which reinforces the applied magnetic field at the position of the protons of the ring. This reinforcement causes the protons of benzene to be highly deshielded.

We also learned in Section 14.7B that internal hydrogens of large-ring aromatic compounds such as [18]annulene, because of their position, are highly shielded by this induced field. They therefore absorb at unusually high field strength, often at negative delta values.

14.11B ^{13}C NMR Spectra

The carbon atoms of benzene rings generally absorb in the δ 100–170 region of ^{13}C NMR spectra. Figure 14.21 gives the broad band proton-decoupled ^{13}C NMR spectrum of 4-N,N-diethylaminobenzaldehyde and permits an exercise in making ^{13}C assignments of a compound with both aromatic and aliphatic carbon atoms.

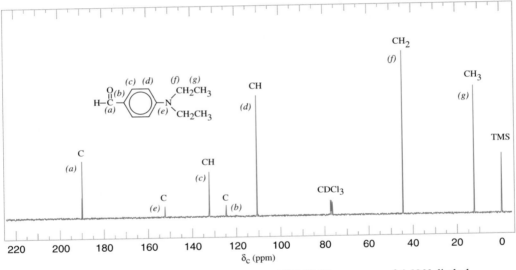

FIGURE 14.21 The broad band proton-decoupled ^{13}C NMR spectrum of 4-N,N-diethyl-aminobenzaldehyde. DEPT information and carbon assignments are shown by each peak.

The DEPT spectra (not given to save space) show that the signal at δ 45 arises from a CH_2 group and the one at δ 13 arises from a CH_3 group. This allows us to assign these two signals immediately to the two carbons of the equivalent ethyl groups.

The signals at δ 126 and δ 153 appear in the DEPT spectra as carbon atoms that do not bear hydrogen atoms and are assigned to carbons (b) and (e) (see Fig. 14.21). The greater electronegativity of nitrogen (when compared to carbon) causes the signal from (e) to be further downfield (at δ 153). The signal at δ 190 appears as a CH group in the DEPT spectra and arises from the carbon of the aldehyde group. Its chemical shift is the most downfield of all the peaks because of the great electronegativity of its oxygen and because the second resonance structure below makes the major contribution to the hybrid. Both factors cause the electron density at this carbon to be very low and, therefore, this carbon is strongly deshielded.

Resonance contributors for an aldehyde group

This leaves the signals at δ 112 and δ 133 and the two sets of carbon atoms of the benzene ring labeled (*c*) and (*d*) to be accounted for. Both signals are indicated as CH groups in the DEPT spectra. But which signal belongs to which set of carbon atoms? Here we find another interesting application of resonance theory.

If we write resonance structures **A–D** involving the unshared electron pair of the amino group, we see that contributions made by **B** and **D** increase the electron

density at the set of carbon atoms labeled (*d*). On the other hand, writing structures **E–H** involving the aldehyde group shows us that contributions made by **F** and **H**

decrease the electron density at the set of carbon atoms labeled (*c*). (Other resonance structures are possible but are not pertinent to the argument here.)

Increasing the electron density at a carbon should increase its shielding and should shift its signal upfield. Therefore, we assign the signal at δ 112 to the set of carbon atoms labeled (*d*). Conversely, decreasing the electron density at a carbon should shift its signal downfield, so we assign the signal at δ 133 to the set labeled (*c*).

Carbon-13 spectroscopy can be especially useful in recognizing a compound with a high degree of symmetry. The following sample problem illustrates one such application.

SAMPLE PROBLEM

The broad band proton-decoupled ^{13}C spectrum given in Fig. 14.22 is of a tribromobenzene ($C_6H_3Br_3$). Which tribromobenzene is it?

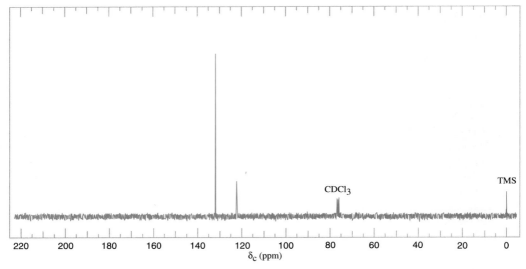

FIGURE 14.22 The broad band proton-decoupled ^{13}C NMR spectrum of a tribromobenzene.

ANSWER:
There are three possible tribromobenzenes:

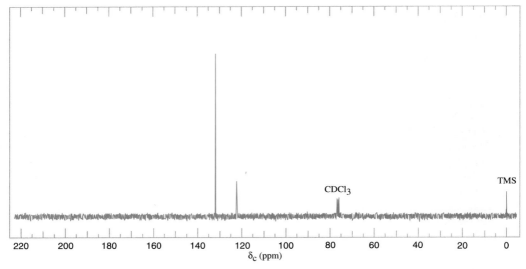

1,2,3-Tribromobenzene   1,2,4-Tribromobenzene   1,3,5-Tribromobenzene

Our spectrum (Fig. 14.22) consists of only two signals, indicating that only two different types of carbon atoms are present in the compound. Only 1,3,5-tribromobenzene has a degree of symmetry such that it would give only two signals, and, therefore, it is the correct answer. 1,2,3-Tribromobenzene would give four ^{13}C signals and 1,2,4-tribromobenzene would give six.

PROBLEM 14.14

Explain how ^{13}C spectroscopy could be used to distinguish the *ortho-, meta-,* and *para-*dibromobenzene isomers one from another.

14.11C Infrared Spectra of Substituted Benzenes

Benzene derivatives give characteristic C—H stretching peaks near 3030 cm^{-1} (Table 13.2). A complex series of motions of the benzene ring can give as many as four bands in the 1450–1600 cm^{-1} region; with two peaks near 1500 and 1600 cm^{-1} being stronger.

Absorption peaks in the 680–860 cm^{-1} region can often (but not always) be used to characterize the substitution patterns of benzene compounds (Table 14.1). **Monosubstituted benzenes** give two very strong peaks, between 690 and 710 cm^{-1} and between 730 and 770 cm^{-1}.

Ortho-disubstituted benzenes show a strong absorption peak between 735 and 770 cm^{-1} that arises from bending motions of the C—H bonds. **Meta-disubstituted benzenes** show two peaks: one strong peak between 680 and 725 cm^{-1} and one very strong peak between 750 and 810 cm^{-1}. **Para-disubstituted benzenes** give a single very strong absorption between 800 and 860 cm^{-1}.

PROBLEM 14.15

Four benzenoid compounds, all with the formula C_7H_7Br, gave the following IR peaks in the 680–840 cm^{-1} region.

TABLE 14.1   Infrared absorptions in the 680–860 cm^{-1} region[a]

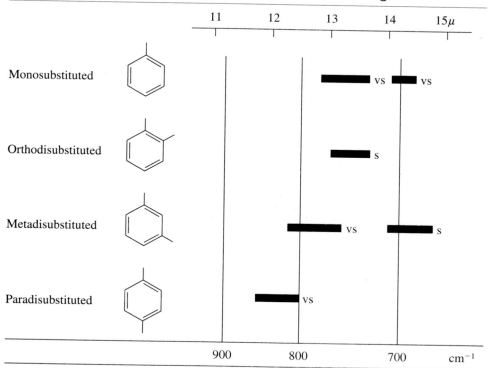

[a] s, strong; vs, very strong.

A, 740 cm^{-1} (strong)

C, 680 cm^{-1} (strong) and 760 cm^{-1} (very strong)

B, 800 cm^{-1} (very strong)

D, 693 cm^{-1} (very strong) and 765 cm^{-1} (very strong)

Propose structures for **A, B, C,** and **D.**

14.11D Visible–Ultraviolet Spectra of Aromatic Compounds

The conjugated π electrons of a benzene ring give characteristic ultraviolet absorptions that indicate the presence of a benzene ring in an unknown compound. One absorption band of moderate intensity occurs near 205 nm and another, less intense band appears in the 250–275 nm range.

14.12 A Summary of Important Terms and Concepts

An *aliphatic compound* is a compound such as an alkane, alkene, alkyne, cycloalkane, cycloalkene, or any of their derivatives.

An *aromatic compound* traditionally means one "having the chemistry typified by benzene." The molecules of aromatic compounds are cyclic, planar, and conjugated. They have a stability significantly greater than that of a hypothetical resonance structure (e.g., a Kekulé structure). Many aromatic compounds react with electrophilic reagents (Br_2, HNO_3, H_2SO_4) by substitution rather than addition, even though they are unsaturated (Chapter 15). A modern definition of any aromatic compound compares the energy of the π electrons of the cyclic conjugated molecule or ion with that of its open-chain counterpart. If on ring closure the π-electron energy *decreases,* the molecule is classified as being **aromatic,** if it *increases* the molecule is classified as being **antiaromatic,** and if it remains the same the molecule is classified as being **nonaromatic.**

The *resonance energy* of an aromatic compound is the difference in energy between the actual aromatic compound and that calculated for the most stable of the hypothetical resonance structures (e.g., a Kekulé structure). The resonance energy is also referred to as *stabilization energy* or *delocalization energy.*

Hückel's rule states that monocyclic compounds with planar conjugated rings having $(4n + 2)$ π electrons (i.e., with 2, 6, 10, 14, 18, or 22 π electrons) should be aromatic.

An *annulene* is a monocyclic compound that can be represented as a structure having alternating single and double bonds. For example, cyclobutadiene is [4]annulene and benzene is [6]annulene.

A *benzenoid* aromatic compound is one whose molecules contain benzene rings or fused benzene rings. Examples are benzene, naphthalene, anthracene, and phenanthrene.

A *nonbenzenoid* aromatic compound is one whose molecules contain a ring that is not six membered. Examples are [14]annulene, azulene, the cyclopentadienyl anion, and the cycloheptatrienyl cation.

A *heterocyclic* compound is one whose molecules have a ring containing an element other than carbon. Some heterocyclic compounds (e.g., pyridine, pyrrole, and thiophene) are aromatic.

ADDITIONAL PROBLEMS

✕ 14.16   Write structural formulas for each of the following:

(a) 3-Nitrobenzoic acid

(b) *p*-Bromotoluene

(c) *o*-Dibromobenzene

(d) *m*-Dinitrobenzene

(e) 3,5-Dinitrophenol

(f) *p*-Nitrobenzoic acid

(g) 3-Chloro-1-ethoxybenzene

(h) *p*-Chlorobenzenesulfonic acid

(i) Methyl *p*-toluenesulfonate

(j) Benzyl bromide

(k) *p*-Nitroaniline

(l) *o*-Xylene

(m) *tert*-Butylbenzene

(n) *p*-Cresol

(o) *p*-Bromoacetophenone

(p) 3-Phenylcyclohexanol

(q) 2-Methyl-3-phenyl-1-butanol

(r) *o*-Chloroanisole

✕ 14.17   Write structural formulas and give acceptable names for all representatives of the following:

(a) Tribromobenzenes

(b) Dichlorophenols

(c) Nitroanilines

(d) Methylbenzenesulfonic acids

(e) Isomers of C_6H_5—C_4H_9

14.18   Although Hückel's rule (Section 14.7) strictly applies only to monocyclic compounds, it does appear to have application to certain bicyclic compounds, provided the important resonance structures involve only the perimeter double bonds, as in naphthalene below.

Both naphthalene (Section 14.8A) and azulene (Section 14.8B) have 10 π electrons and are aromatic. Pentalene (below) is apparently antiaromatic and is unstable even at $-100°C$. Heptalene has been made but it adds bromine, reacts with acids, and its molecules are not planar. Is Hückel's rule applicable to these compounds, and, if so, explain their lack of aromaticity.

Pentalene   **Heptalene**

14.19   (a) In 1960 Thomas Katz (Columbia University) showed that cyclooctatetraene adds two electrons when treated with potassium metal and forms a stable, planar dianion, $C_8H_8^{2-}$ (as the dipotassium salt).

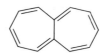

$$\xrightarrow[\text{THF}]{2\text{ K}} 2\text{ K}^+ C_8H_8^{2-}$$

Use the molecular orbital diagram given in Fig. 14.6 and explain this result.   (b) In 1964 Katz also showed that removing two protons from the compound below (using butyllithium as the base) leads to the formation of a stable dianion with the formula $C_8H_6^{2-}$ (as the dilithium salt).

$$+ 2 \text{ BuLi} \longrightarrow 2 \text{ Li}^+\text{C}_8\text{H}_6{}^{2-} + 2 \text{ BuH}$$

Consider your answer to Problem 14.18 and provide an explanation for what is happening here.

14.20   Although none of the [10]annulenes given in Section 14.7A is aromatic, the following 10 π-electron system is aromatic.

What factor makes this possible?

14.21   Cycloheptatrienone (**I**) is very stable. Cyclopentadienone (**II**) by contrast is quite unstable and rapidly undergoes a Diels–Alder reaction with itself.   (a) Propose an explanation for the different stabilities of these two compounds.   (b) Write the structure of the Diels–Alder adduct of cyclopentadienone.

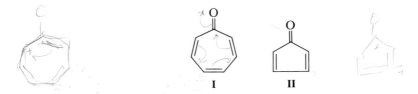

I   **II**

14.22   5-Chloro-1,3-cyclopentadiene (below) undergoes S_N1 solvolysis in the presence of silver ion extremely slowly even though the chlorine is doubly allylic and allylic halides normally ionize readily (Section 15.15). Provide an explanation for this behavior.

14.23   Explain the following: (a) The cycloheptatrienyl anion is antiaromatic, whereas the cyclononatetraenyl anion is planar (in spite of the angle strain involved) and appears to be aromatic.   (b) Although [16]annulene is not aromatic, it adds two electrons readily to form an aromatic dianion.

14.24   Furan possesses less aromatic character than benzene as measured by their resonance energies (23 kcal mol^{-1} for furan; 36 kcal mol^{-1} for benzene). What chemical evidence have we studied earlier that shows that furan is less aromatic than benzene.

14.25   Assign structures to each of the compounds, **A**, **B**, and **C** shown in Fig. 14.23.

14.26   The ^{1}H NMR spectrum of cyclooctatetraene consists of a single line located at δ 5.78. What does the location of this signal suggest about electron delocalization in cyclo-octatetraene?

14.27   Give a structure for compound **F** that is consistent with the ^{1}H NMR and IR spectra in Fig. 14.24.

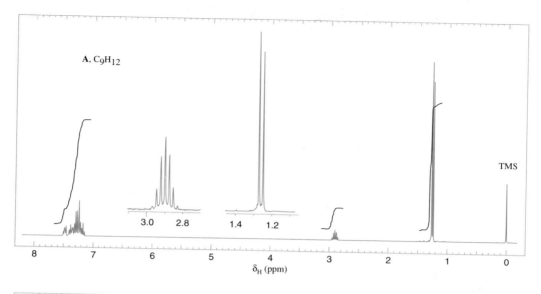

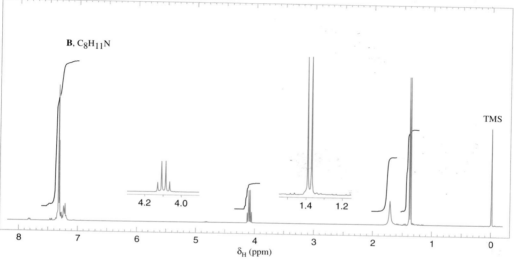

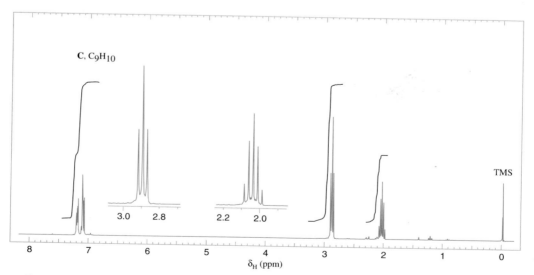

FIGURE 14.23 The 300 MHz ¹H NMR spectra for problem 14.25. Expansions of the signals are shown in the offset plots.

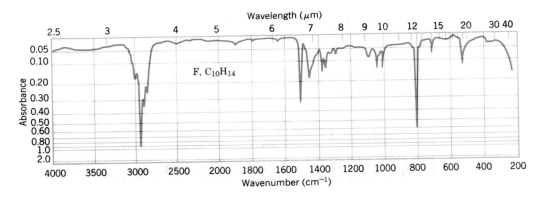

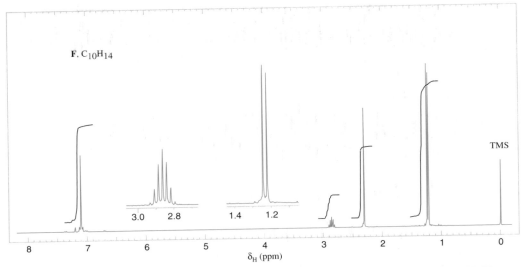

FIGURE 14.24 The 300 MHz ^{1}H NMR and IR spectra of Compound **F**, Problem 14.27. Expansions of the signals are shown in the offset plots.

14.28   A compound (**L**) with the molecular formula C_9H_{10} decolorizes bromine in carbon tetrachloride and gives an IR absorption spectrum that includes the following absorption peaks: 3035 cm^{-1}(m), 3020 cm^{-1}(m), 2925 cm^{-1}(m), 2853 cm^{-1}(w), 1640 cm^{-1}(m), 990 cm^{-1}(s), 915 cm^{-1}(s), 740 cm^{-1}(s), 695 cm^{-1}(s). The ^{1}H NMR spectrum of **L** consists of:

Doublet δ 3.1 (2H)	Multiplet δ 5.1	Multiplet δ 7.1 (5H)
Multiplet δ 4.8	Multiplet δ 5.8	

The UV spectrum shows a maximum at 255 nm. Propose a structure for compound **L** and make assignments for each of the IR peaks.

14.29   Compound **M** has the molecular formula C_9H_{12}. The ^{1}H NMR spectrum of **M** is given in Fig. 14.25 and the IR spectrum in Fig. 14.26. Propose a structure for **M**.

14.30   A compound (**N**) with the molecular formula $C_9H_{10}O$ gives a positive test with cold dilute aqueous potassium permanganate. The ^{1}H NMR spectrum of **N** is shown in Fig. 14.27 and the IR spectrum of **N** is shown in Fig. 14.28. Propose a structure for **N**.

14.31   The IR and ^{1}H NMR spectra for compound **X** (C_8H_{10}) are given in Fig. 14.29. Propose a structure for compound **X**.

14.32   The IR and ^{1}H NMR spectra of compound **Y** ($C_9H_{12}O$) are given in Fig. 14.30. Propose a structure for **Y**.

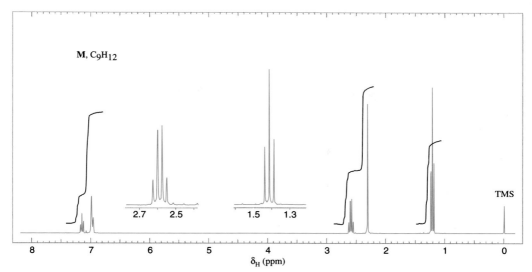

FIGURE 14.25 The 300 MHz ^1H NMR spectrum of compound **M**, Problem 14.29. Expansions of the signals are shown in the offset plots.

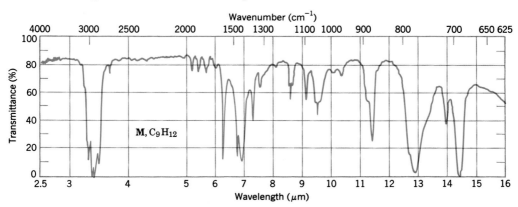

FIGURE 14.26 The IR spectrum of compound **M**, Problem 14.29. (Spectrum courtesy of Aldrich Chemical Co., Milwaukee, WI.)

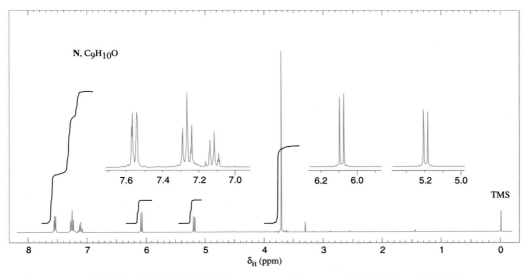

FIGURE 14.27 The 300 MHz ^1H NMR spectrum of compound **N**, Problem 14.30. Expansions of the signals are shown in the offset plots.

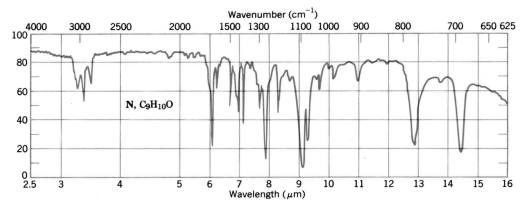

FIGURE 14.28 The IR spectrum of compound **N**, Problem 14.30. (Spectrum courtesy of Aldrich Chemical Co., Milwaukee, WI.)

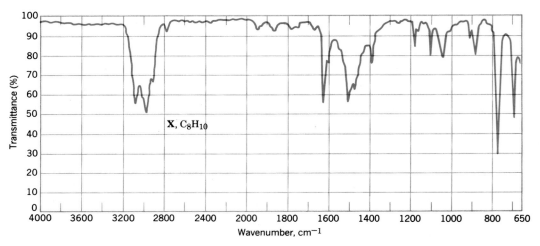

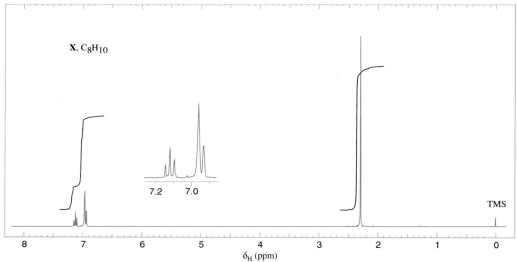

FIGURE 14.29 The IR and 300 MHz ^{1}H NMR spectra of compound **X**, Problem 14.31. Expansions of the signals are shown in the offset plots.

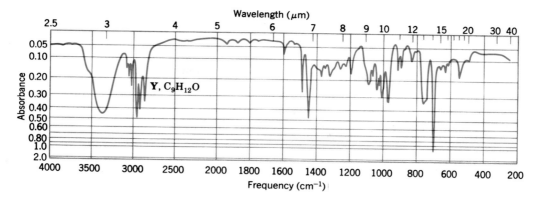

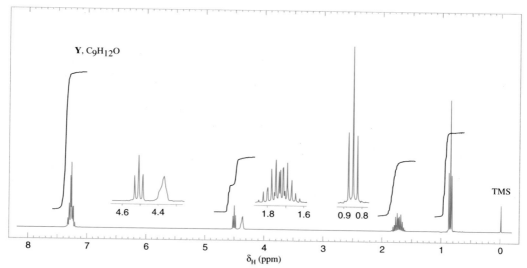

FIGURE 14.30 The IR and 300 MHz ^{1}H NMR spectra of compound **Y**, Problem 14.32. Expansions of the signals are shown in the offset plots.

14.33   (a) How many peaks would you expect to find in the ^{1}H NMR spectrum of caffeine?

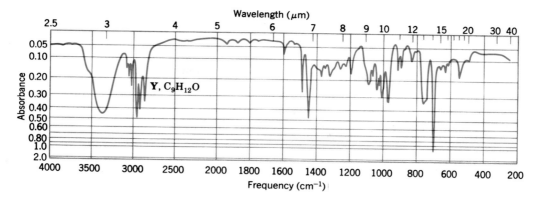

(b) What characteristic peaks would you expect to find in the IR spectrum of caffeine?

14.34   Given the following information predict the appearance of the ^{1}H NMR spectrum given by the vinyl hydrogen atoms of *p*-chlorostyrene.

Deshielding by the induced magnetic field of the ring is greatest at proton (c) (δ 6.7) and is least at proton (b) (δ 5.3). The chemical shift of (a) is about δ 5.7. The coupling constants have the following approximate magnitudes: $J_{ac} \simeq 18$ Hz, $J_{bc} \simeq 11$ Hz, and $J_{ab} \simeq 2$ Hz. (These coupling constants are typical of those given by vinylic systems: coupling constants for trans hydrogen atoms are larger than those for cis hydrogen atoms and coupling constants for geminal hydrogen atoms are very small.)

CHAPTER 15

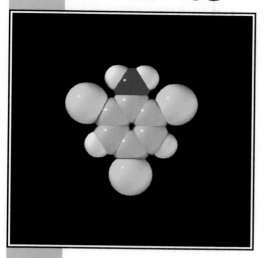

2,4,6 Tribromoaniline
(Section 15.10A)

REACTIONS OF AROMATIC COMPOUNDS

15.1 ELECTROPHILIC AROMATIC SUBSTITUTION REACTIONS

Aromatic hydrocarbons are known generally as **arenes.** An **aryl group** is one derived from an arene by removal of a hydrogen atom and its symbol is Ar— . Thus, arenes are designated ArH just as alkanes are designated RH.

The most characteristic reactions of benzenoid arenes are the substitution reactions that occur when they react with electrophilic reagents. These reactions are of the general type shown below.

$$ArH + E^+ \longrightarrow Ar-E + H^+$$

or $\bigcirc + E^+ \longrightarrow \bigcirc^{E} + H^+$

The electrophiles are either a positive ion (E^+) or some other electron-deficient species with a large partial positive charge. As we shall learn in Section 15.3, for example, benzene can be brominated when it reacts with bromine in the presence of $FeBr_3$. Bromine and $FeBr_3$ react to produce positive bromine ions, Br^+. These positive bromine ions act as electrophiles and attack the benzene ring, replacing one of the hydrogen atoms in a reaction that is called an **electrophilic aromatic substitution** (EAS).

Electrophilic aromatic substitutions allow the direct introduction of a wide variety of groups onto an aromatic ring and because of this they provide synthetic routes to many important compounds. The five electrophilic aromatic substitutions that we shall study in this chapter are outlined in Fig. 15.1. In Sections 15.3–15.7 we shall learn what the electrophile is in each instance.

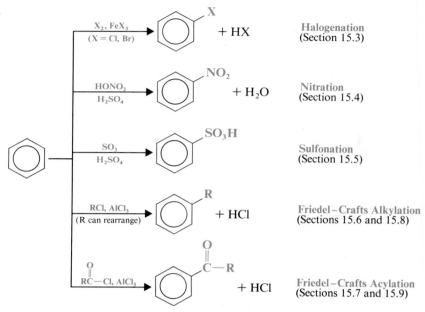

FIGURE 15.1 Electrophilic aromatic substitution reactions.

15.2 A GENERAL MECHANISM FOR ELECTROPHILIC AROMATIC SUBSTITUTION: ARENIUM IONS

Benzene is susceptible to electrophilic attack primarily because of its exposed π electrons. In this respect benzene resembles an alkene, for in the reaction of an alkene with an electrophile the site of attack is the exposed π bond.

We saw in Chapter 14, however, that benzene differs from an alkene in a very significant way. Benzene's closed shell of six π electrons gives it a special stability. So although benzene is susceptible to electrophilic attack, it undergoes *substitution reactions* rather than *addition reactions*. Substitution reactions allow the aromatic sextet of π electrons to be regenerated after attack by the electrophile has occurred. We can see how this happens if we examine a general mechanism for electrophilic aromatic substitution.

A considerable body of experimental evidence indicates that electrophiles attack the π system of benzene to form a **nonaromatic carbocation** known as an **arenium ion** (or sometimes as a σ *complex*). In showing this step it is convenient to use Kekulé structures, because these make it much easier to keep track of the π electrons.

Step 1

Arenium ion
(σ complex)

In step 1 the electrophile takes two electrons of the six-electron π system to form a σ bond to one carbon atom of the benzene ring. Formation of this bond interrupts the cyclic system of π electrons, because in the formation of the arenium ion the carbon that forms a bond to the electrophile becomes sp^3 hybridized and, therefore, no longer has an available p orbital. Now only five carbon atoms of the ring are still sp^2 hybridized and still have p orbitals. The four π electrons of the arenium ion are delocalized through these five p orbitals.

In step 2 the arenium ion loses a proton from the carbon atom that bears the electrophile. The two electrons that bonded this proton to carbon become a part of the π system. The carbon atom that bears the electrophile becomes sp^2 hybridized again, and a benzene derivative with six fully delocalized π electrons is formed. We can represent step 2 with any one of the resonance structures for the arenium ion.

Step 2

(The proton displaced is transferred to any of the bases present, for example, to the anion derived from the electrophile.)

PROBLEM 15.1

Show how loss of a proton can be represented using each of the three resonance structures for the arenium ion, and show how each representation leads to the formation of a benzene ring with three alternating double bonds (i.e., six fully delocalized π electrons).

Kekulé structures are more appropriate for writing mechanisms such as electrophilic aromatic substitution because they permit the use of resonance theory, which as we shall soon see, is invaluable as an aid to our understanding. If, for brevity, however, we wish to show the mechanism using the modern formula for benzene we can do it in the following way:

Step 1

Arenium ion

Step 2

There is firm experimental evidence that the arenium ion is a true *intermediate* in electrophilic substitution reactions. It is not a transition state. This means that in a free energy diagram (Fig. 15.2) the arenium ion lies in an energy valley between two transition states.

The free energy of activation, $\Delta G^{\ddagger}$, for the reaction leading from benzene and the electrophile, E^+, to the arenium ion has been shown to be much greater than the free energy of activation, $\Delta G^{\ddagger}$, leading from the arenium ion to the final product. This is consistent with what we would expect. The reaction leading from benzene and an

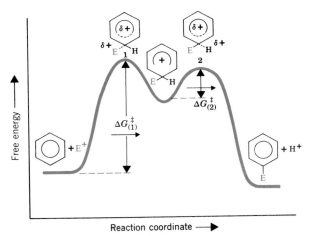

FIGURE 15.2 The free-energy diagram for an electrophilic aromatic substitution reaction. The arenium ion is a true intermediate lying between transition states **1** and **2**. In transition state **1** the bond between the electrophile and one carbon atom of the benzene ring is only partially formed. In transition state **2** the bond between the same benzene carbon atom and its hydrogen atom is partially broken.

electrophile to the arenium ion is highly endothermic, because the benzene ring loses its resonance energy. The reaction leading from the arenium ion to the substituted benzene, by contrast, is highly exothermic because in it the benzene ring regains its resonance energy.

Of the following two steps, step 1—the formation of the arenium ion—is usually the rate-determining step in electrophilic aromatic substitution.

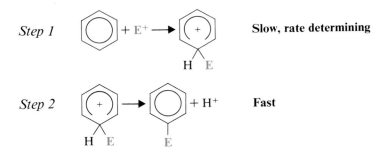

Step 2, the loss of a proton, occurs rapidly relative to step 1 and has no effect on the overall rate of reaction.

15.3 HALOGENATION OF BENZENE

Benzene does not react with bromine or chlorine unless a Lewis acid is present in the mixture. (As a consequence, benzene does not decolorize a solution of bromine in carbon tetrachloride.) When Lewis acids are present, however, benzene reacts readily with bromine or chlorine, and the reactions give bromobenzene and chlorobenzene in good yields.

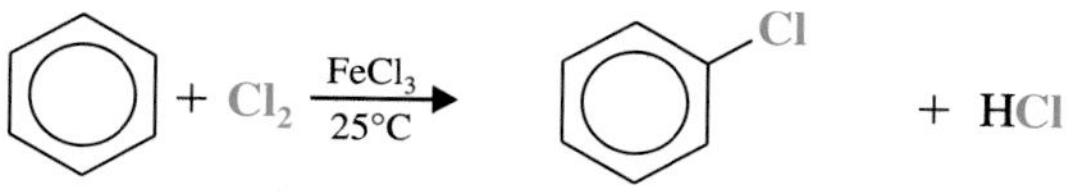

Chlorobenzene (90%)

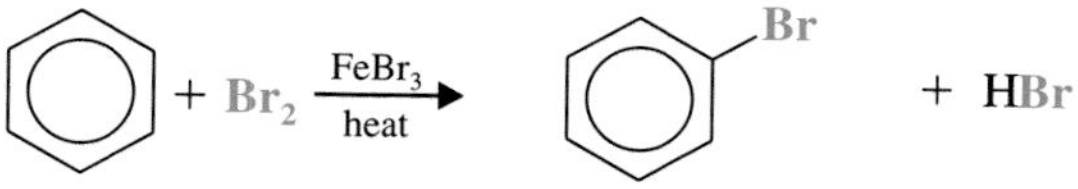

Bromobenzene (75%)

The Lewis acids most commonly used to effect chlorination and bromination reactions are $FeCl_3$, $FeBr_3$, and $AlCl_3$ all in the anhydrous form. Ferric chloride and ferric bromide are usually generated in the reaction mixture by adding iron to it. The iron then reacts with halogen to produce the ferric halide:

$$2 \, Fe + 3 \, X_2 \longrightarrow 2 \, FeX_3$$

The mechanism for aromatic bromination is as follows:

A MECHANISM FOR THE REACTION

Step 1

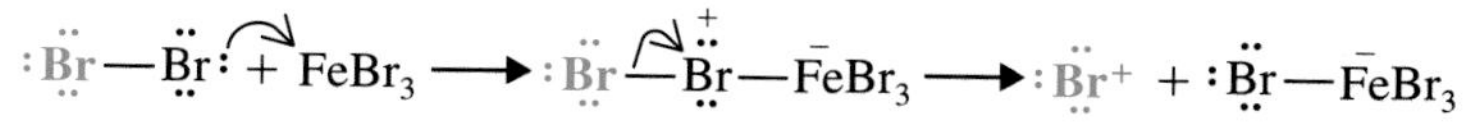

Bromine combines with $FeBr_3$ to form a complex that dissociates to form a positive bromine ion and $FeBr_4^-$.

Step 2

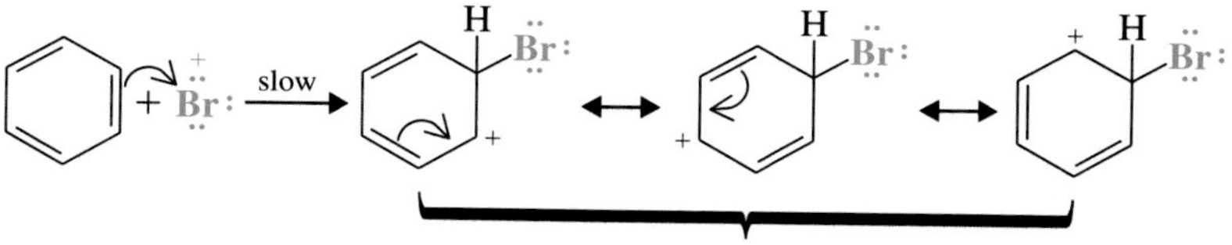

Arenium ion

The positive bromine ion attacks benzene to form an arenium ion.

Step 3

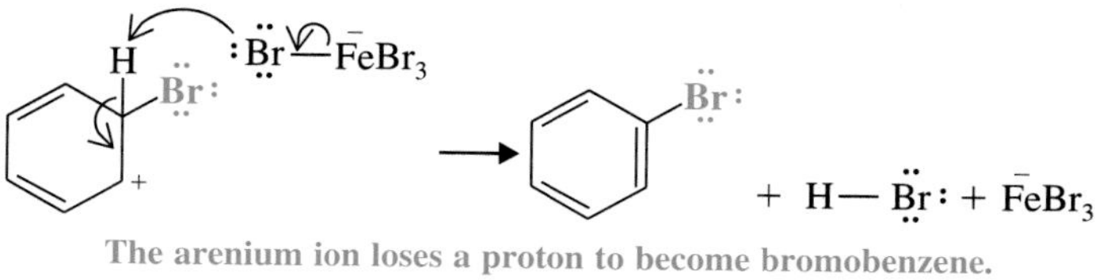

The arenium ion loses a proton to become bromobenzene.

The function of the Lewis acid can be seen in step 1. The ferric bromide reacts with bromine to produce a positive bromine ion, Br^+ (and $FeBr_4^-$). In step 2 this Br^+ ion attacks the benzene ring to produce an arenium ion. Then finally in step 3 the arenium ion transfers a proton to $FeBr_4^-$. This results in the formation of bromobenzene and hydrogen bromide, the products of the reaction. At the same time this step regenerates the catalyst, $FeBr_3$.

The mechanism of the chlorination of benzene in the presence of ferric chloride is analogous to the one for bromination. Ferric chloride serves the same purpose in aromatic chlorinations as ferric bromide does in aromatic brominations. It assists in the generation and transfer of a positive halogen ion.

Fluorine reacts so rapidly with benzene that aromatic fluorination requires special conditions and special types of apparatus. Even then, it is difficult to limit the reaction to monofluorination. Fluorobenzene can be made, however, by an indirect method that we shall see in Section 20.8D.

Iodine, on the other hand, is so unreactive that a special technique has to be used to effect direct iodination; the reaction has to be carried out in the presence of an oxidizing agent such as nitric acid:

$$\text{benzene} + I_2 \xrightarrow{\text{HNO}_3} \text{iodobenzene}$$

(86%)

15.4 NITRATION OF BENZENE

Benzene reacts slowly with hot concentrated nitric acid to yield nitrobenzene. The reaction is much faster if it is carried out by heating benzene with a mixture of concentrated nitric acid and concentrated sulfuric acid.

$$\text{benzene} + HNO_3 + H_2SO_4 \xrightarrow{50-55^\circ C} \text{nitrobenzene (NO}_2) + H_3O^+ + HSO_4^-$$

(85%)

Concentrated sulfuric acid increases the rate of the reaction by increasing the concentration of the electrophile, the nitronium ion (NO_2^+), as shown in the first two steps of the following mechanism.

A MECHANISM FOR THE NITRATION OF BENZENE

Step 1 $HO_3SO{-}H + H{-}\overset{..}{\underset{..}{O}}{-}\overset{+}{N}\overset{\overset{\overset{..}{O}:}{\parallel}}{\underset{\underset{..}{O}:^-}{}} \rightleftharpoons H{-}\overset{H}{\underset{}{\overset{+}{O}}}{-}\overset{+}{N}\overset{\overset{\overset{..}{O}:}{\parallel}}{\underset{\underset{..}{O}:^-}{}} + HSO_4^-$

(H₂SO₄)

In this step nitric acid accepts a proton from the stronger acid, sulfuric acid.

Step 2 $H{-}\overset{H}{\underset{..}{\overset{+}{O}}}{-}\overset{+}{N}\overset{\overset{..}{O}:}{\underset{..}{O}:^-} \rightleftharpoons H_2O + \overset{\overset{..}{O}:}{\underset{\underset{..}{O}:}{N^+}}$

Nitronium ion

Now that it is protonated, nitric acid can dissociate to form a nitronium ion.

Step 3

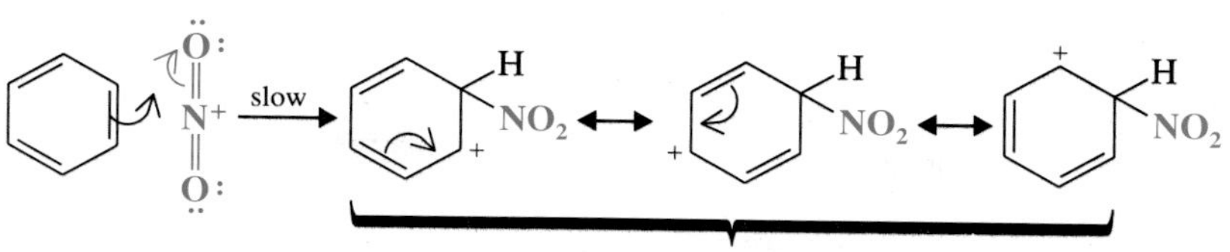

Arenium ion

The nitronium ion is the actual electrophile in nitration; it reacts with
benzene to form a resonance-stabilized arenium ion.

Step 4

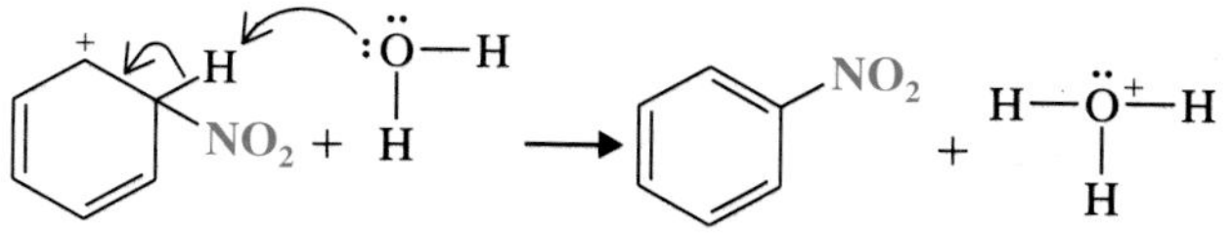

The arenium ion then transfers a proton to a base and becomes nitrobenzene.

PROBLEM 15.2

Given that the pK_a of H_2SO_4 is -9, and that of HNO_3 is -1.3, explain why
nitration occurs more rapidly in a mixture of concentrated nitric and sulfuric
acids, rather than in concentrated nitric acid alone.

15.5 SULFONATION OF BENZENE

Benzene reacts with fuming sulfuric acid at room temperature to produce benzenesulfonic acid. Fuming sulfuric acid is sulfuric acid that contains added sulfur trioxide
(SO_3). Sulfonation also takes place in concentrated sulfuric acid alone, but more
slowly.

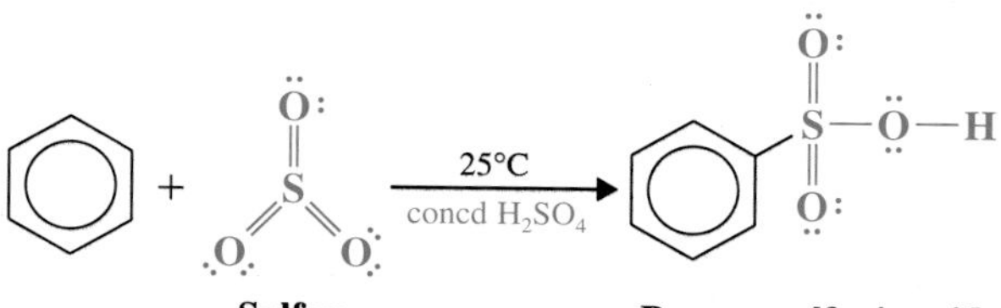

**Sulfur
trioxide**

Benzenesulfonic acid
(56%)

In either reaction the electrophile appears to be sulfur trioxide. In concentrated
sulfuric acid, sulfur trioxide is produced in an equilibrium in which H_2SO_4 acts as both
an acid and a base (see step 1 of the following mechanism).

A MECHANISM FOR THE SULFONATION OF BENZENE

Step 1 $2 H_2SO_4 \rightleftharpoons SO_3 + H_3O^+ + HSO_4^-$

This equilibrium produces SO_3 in
concentrated H_2SO_4.

Step 2

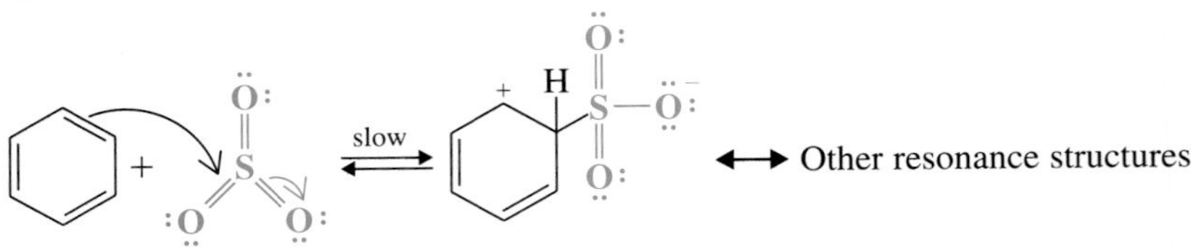

SO₃ is the actual electrophile that reacts with benzene to form an arenium ion.

Step 3   HSO₄⁻ +

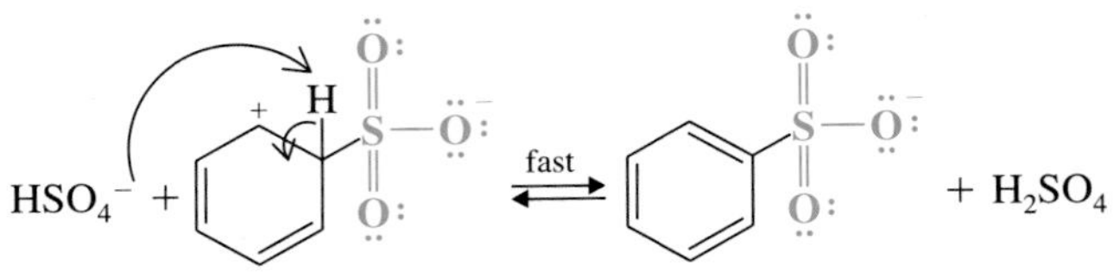

The arenium ion loses a proton to form the benzenesulfonate ion.

Step 4

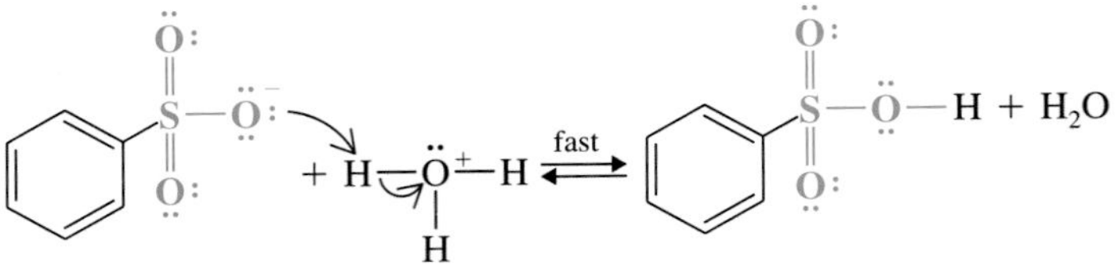

The benzenesulfonate ion accepts a proton to become benzenesulfonic acid.

All of the steps are equilibria, including step 1, in which sulfur trioxide is formed from sulfuric acid. This means that the overall reaction is an equilibrium as well. In concentrated sulfuric acid, the overall equilibrium is the sum of steps 1–4.

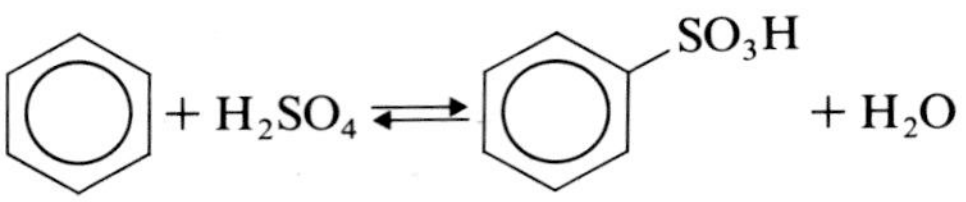

In fuming sulfuric acid, step 1 is unimportant because the dissolved sulfur trioxide reacts directly.

Because all of the steps are equilibria, the position of equilibrium can be influenced by the conditions we employ. If we want to sulfonate benzene we use concentrated sulfuric acid or—better yet—fuming sulfuric acid. Under these conditions the position of equilibrium lies appreciably to the right and we obtain benzenesulfonic acid in good yield.

On the other hand, we may want to remove a sulfonic acid group from a benzene ring. To do this we employ dilute sulfuric acid and usually pass steam through the mixture. Under these conditions—with a high concentration of water—the equilibrium lies appreciably to the left and desulfonation occurs. The equilibrium is shifted even further to the left with volatile aromatic compounds because the aromatic compound distills with the steam.

We shall see later that sulfonation and desulfonation reactions are often used in synthetic work. We may, for example, introduce a sulfonic acid group into a benzene ring to influence the course of some further reaction. Later, we may remove the sulfonic acid group by desulfonation.

15.6 FRIEDEL–CRAFTS ALKYLATION

In 1877 a French chemist, Charles Friedel, and his American collaborator, James M. Crafts, discovered new methods for the preparation of alkylbenzenes (ArR) and acylbenzenes (ArCOR). These reactions are now called the Friedel–Crafts alkylation and acylation reactions. We shall study the Friedel–Crafts alkylation reaction here and take up the Friedel–Crafts acylation reaction in Section 15.7.

A general equation for a Friedel–Crafts alkylation reaction is the following:

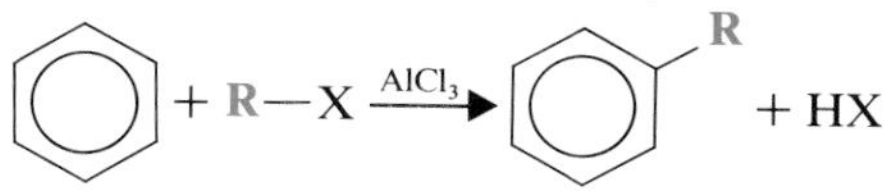

The mechanism for the reaction (shown in the following steps—with isopropyl chloride as R—X) starts with the formation of a carbocation (step 1). The carbocation then acts as an electrophile (step 2) and attacks the benzene ring to form an arenium ion. The arenium ion (step 3) then loses a proton to generate isopropylbenzene.

A MECHANISM FOR FRIEDEL–CRAFTS ALKYLATION

Step 1

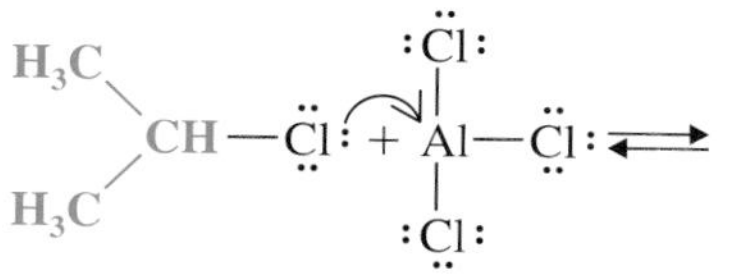

This is a Lewis acid–base reaction (see Section 3.1B)

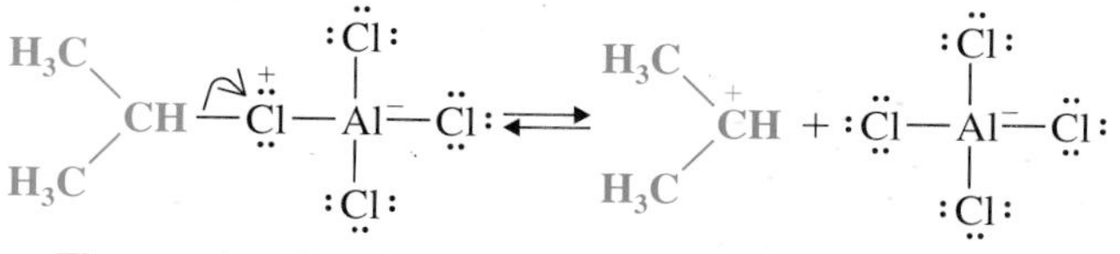

The complex dissociates to form a carbocation and $AlCl_4^-$.

Step 2

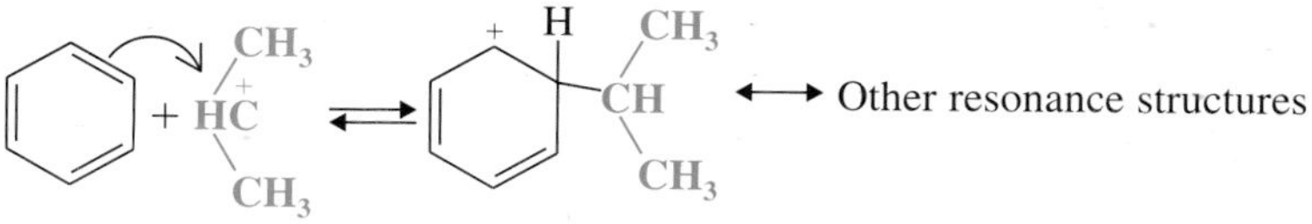

Other resonance structures

The carbocation, acting as an electrophile, reacts with benzene to produce an arenium ion.

Step 3

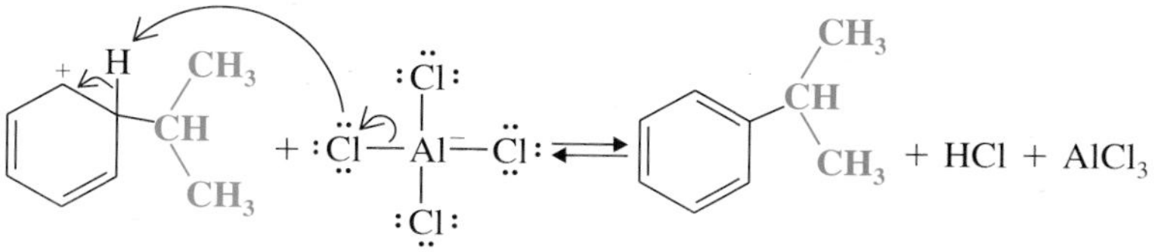

The arenium ion loses a proton to form isopropylbenzene. This step also regenerates the $AlCl_3$ and liberates HCl.

When R—X is a primary halide, a simple carbocation probably does not form. Instead, the aluminum chloride forms a complex with the alkyl halide and this complex acts as the electrophile. The complex is one in which the carbon–halogen bond is nearly broken— and one in which the carbon atom has a considerable positive charge:

$$\overset{\delta+}{RCH_2} \text{---} \overset{\delta-}{Cl} \text{:} AlCl_3$$

Even though this complex is not a simple carbocation, it acts as if it were and it transfers a positive alkyl group to the aromatic ring. As we shall see in Section 15.8, these complexes are so carbocationlike that they also undergo typical carbocation rearrangements.

Friedel–Crafts alkylations are not restricted to the use of alkyl halides and aluminum chloride. Many other pairs of reagents that form carbocations (or carbocationlike species) may be used as well. These possibilities include the use of a mixture of an alkene and an acid.

Propene

**Isopropylbenzene
(84%)**

Cyclohexene

**Cyclohexylbenzene
(62%)**

A mixture of an alcohol and an acid may also be used

Cyclohexanol

**Cyclohexylbenzene
(56%)**

There are several important limitations of the Friedel–Crafts reaction. These are discussed in Section 15.8.

PROBLEM 15.3

Outline all steps in a reasonable mechanism for the formation of isopropylbenzene from propene and benzene in liquid HF (just shown). Your mechanism must account for the product being isopropylbenzene, and not propylbenzene.

15.7 FRIEDEL–CRAFTS ACYLATION

The $\overset{\overset{\displaystyle O}{\|}}{RC}$— group is called an **acyl group,** and a reaction whereby an acyl group is introduced into a compound is called an **acylation** reaction. Two common acyl groups are the acetyl group and the benzoyl group.

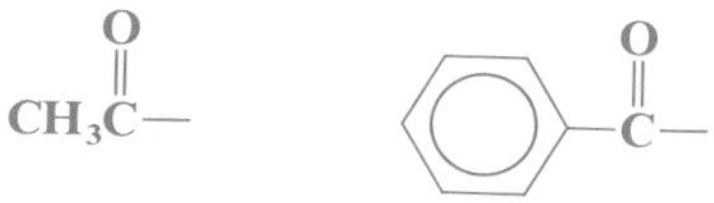

Acetyl group (ethanoyl group) **Benzoyl group**

The Friedel–Crafts acylation reaction is an effective means of introducing an acyl group into an aromatic ring. The reaction is often carried out by treating the aromatic compound with an acyl halide. Unless the aromatic compound is one that is highly reactive, the reaction requires the addition of at least one equivalent of a Lewis acid (such as $AlCl_3$) as well. The product of the reaction is an aryl ketone.

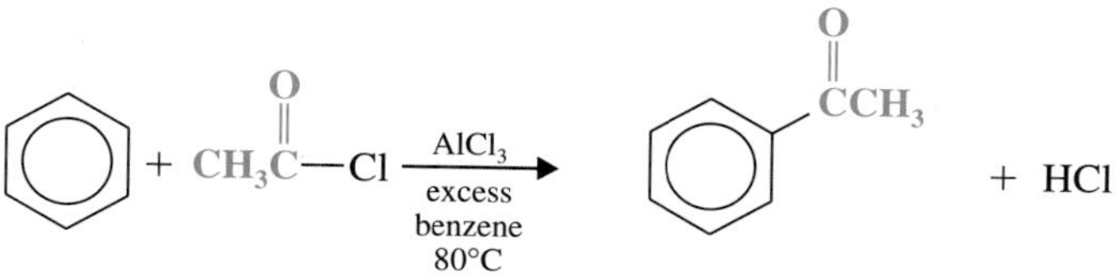

Acetyl chloride **Acetophenone (methyl phenyl ketone) (97%)**

Acyl chlorides, also called **acid chlorides,** are easily prepared by treating carboxylic acids with thionyl chloride $(SOCl_2)$ or phosphorus pentachloride (PCl_5).

$$CH_3\overset{O}{\overset{\|}{C}}OH + SOCl_2 \xrightarrow{80°C} CH_3\overset{O}{\overset{\|}{C}}Cl + SO_2 + HCl$$

Acetic acid **Thionyl chloride** **Acetyl chloride (80–90%)**

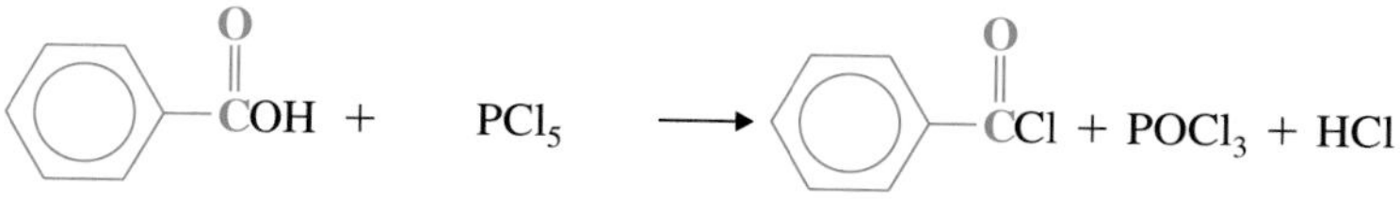

Benzoic acid **Phosphorus pentachloride** **Benzoyl chloride (90%)**

Friedel–Crafts acylations can also be carried out using carboxylic acid anhydrides. For example:

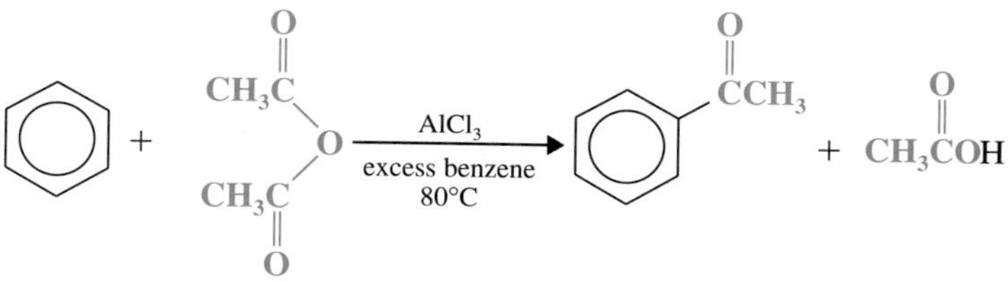

Acetic anhydride (a carboxylic acid anhydride) **Acetophenone (82–85%)**

In most Friedel–Crafts acylations the electrophile appears to be an **acylium ion** formed from an acyl halide in the following way:

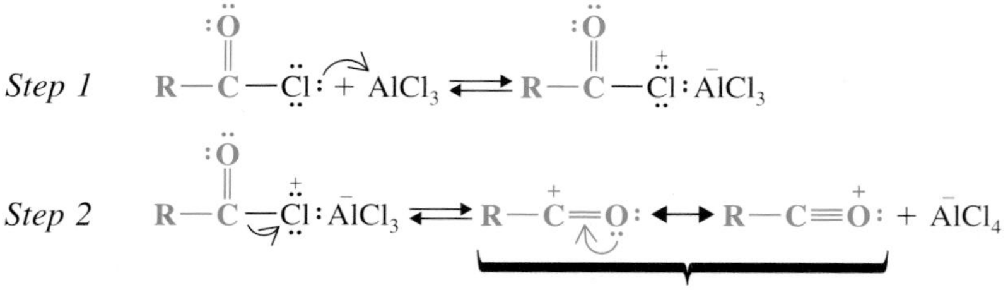

Step 1 $\overset{\displaystyle :\ddot{O}:}{\underset{\displaystyle \|}{R-C}}-\ddot{\underset{..}{C}l: + AlCl_3 \rightleftharpoons R-\overset{:\ddot{O}:}{\underset{\|}{C}}-\overset{+}{\ddot{\underset{..}{C}l}}:\bar{A}lCl_3$

Step 2 $R-\overset{:\ddot{O}:}{\underset{\|}{C}}-\overset{+}{\ddot{\underset{..}{C}l}}:\bar{A}lCl_3 \rightleftharpoons R-\overset{+}{C}=\overset{..}{\overset{..}{O}}: \longleftrightarrow R-C\equiv\overset{+}{\overset{..}{O}}: + \bar{A}lCl_4$

An acylium ion\
(**a resonance hybrid**)

PROBLEM 15.4

Show how an acylium ion could be formed from acetic anhydride in the presence of $AlCl_3$.

The remaining steps in the Friedel–Crafts acylation of benzene are the following:

A MECHANISM FOR FRIEDEL – CRAFTS\
ACYLATION

Step 3   benzene + $\overset{R}{\underset{\overset{\|}{\overset{..}{\overset{+}{O}}}}{C^+}}$ → arenium ion with **Arenium ion**

Step 4   arenium ion $+ AlCl_4^- \longrightarrow$ product $+ HCl + AlCl_3$

Step 5   ketone $: + AlCl_3 \rightleftharpoons$ complex $:AlCl_3$

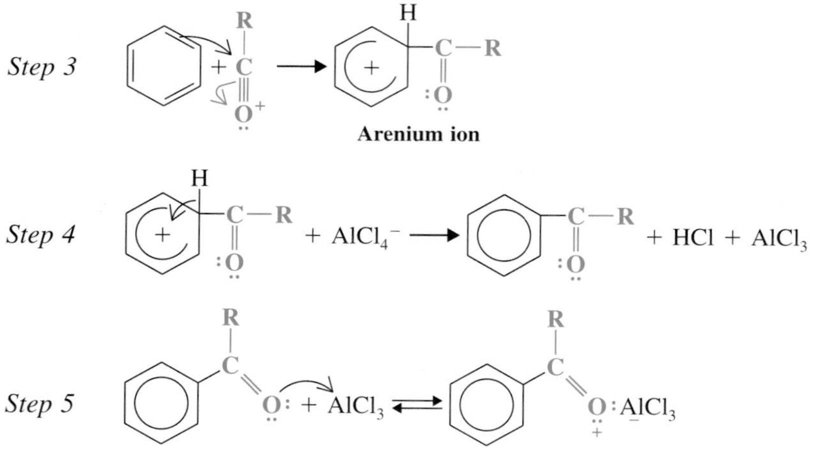

In the last step aluminum chloride (a Lewis acid) forms a complex with the ketone (a Lewis base). After the reaction is over, treating the complex with water liberates the ketone.

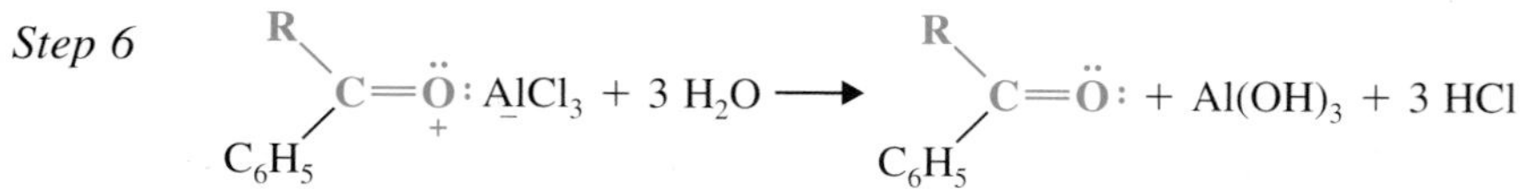

Step 6 $\overset{R}{\underset{C_6H_5}{>}}C=\overset{..}{\overset{..}{O}}:\overset{+}{\bar{A}}lCl_3 + 3\ H_2O \longrightarrow \overset{R}{\underset{C_6H_5}{>}}C=\overset{..}{\overset{..}{O}}: + Al(OH)_3 + 3\ HCl$

Several important synthetic applications of the Friedel–Crafts reaction are given in Section 15.9C.

15.8 LIMITATIONS OF FRIEDEL–CRAFTS REACTIONS

Several restrictions limit the usefulness of Friedel–Crafts reactions.

1. **When the carbocation formed from an alkyl halide, alkene, or alcohol can rearrange to a more stable carbocation, it usually does so and the major product obtained from the reaction is usually the one from the more stable carbocation.**

When benzene is alkylated with butyl bromide, for example, some of the developing butyl cations rearrange by a hydride shift—some developing 1° carbocations (see following reactions) become more stable 2° carbocations. Then benzene reacts with both kinds of carbocations to form both butylbenzene and *sec*-butylbenzene:

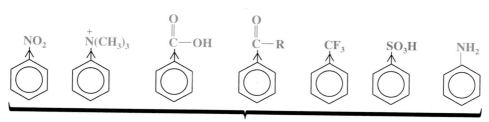

$$CH_3CH_2CH_2CH_2Br \xrightarrow{AlCl_3} CH_3CH_2\overset{\delta+}{\underset{\underset{H}{|}}{CH}}\overset{\delta-}{CH_2}\text{---}Br\overset{}{AlCl_3} \xrightarrow[(-BrAlCl_3^-)]{} CH_3CH_2\overset{+}{CH}CH_3$$

Butylbenzene
(32–36% of mixture)

sec-Butylbenzene
(64–68% of mixture)

2. **Friedel–Crafts reactions usually give poor yields when powerful electron-withdrawing groups** (Section 15.11) **are present on the aromatic ring** or when the ring bears an **—NH₂, —NHR, or —NR₂ group.** This applies to alkylations and acylations.

These usually give poor yields in Friedel-Crafts reactions.

We shall learn in Section 15.10 that groups present on an aromatic ring can have a large effect on the reactivity of the ring toward electrophilic aromatic substitution. Electron-withdrawing groups make the ring less reactive by making it electron deficient. Any substituent more electron withdrawing (or deactivating) than a halogen, that is, **any meta-directing group** (Section 15.11C), **makes an aromatic ring too electron deficient to undergo a Friedel–Crafts reaction.** The amino groups, $-NH_2$, $-NHR$, and $-NR_2$, are changed into powerful electron-withdrawing groups by the Lewis acids used to catalyze Friedel–Crafts reactions. For example:

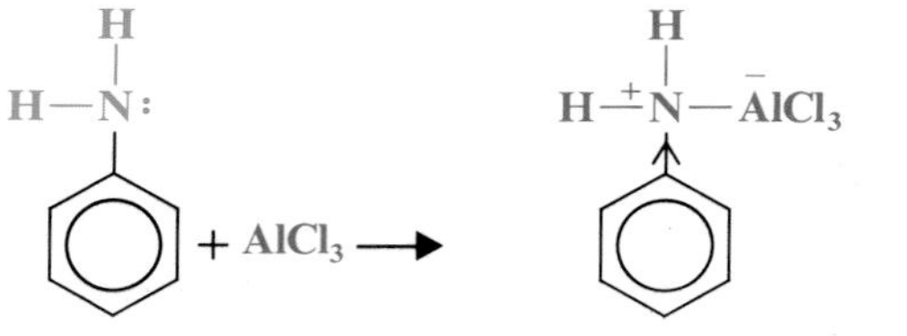

**Does not undergo
a Friedel–Crafts reaction**

3. **Aryl and vinylic halides cannot be used as the halide component because they do not form carbocations readily** (cf. Section 6.16).

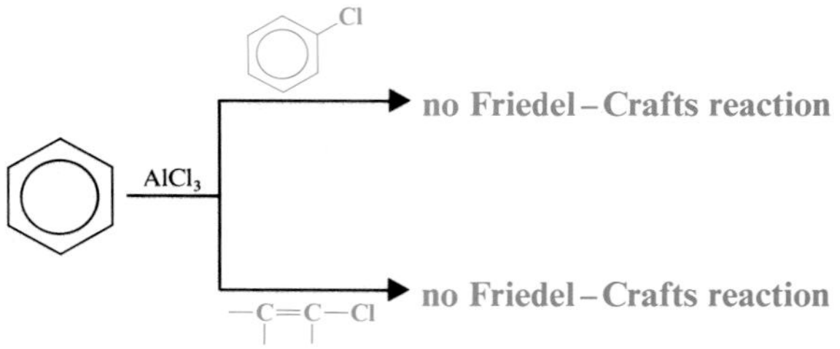

no Friedel–Crafts reaction

no Friedel–Crafts reaction

4. **Polyalkylations often occur.** Alkyl groups are electron-releasing groups, and once one is introduced into the benzene ring it activates the ring toward further substitution (cf. Section 15.10).

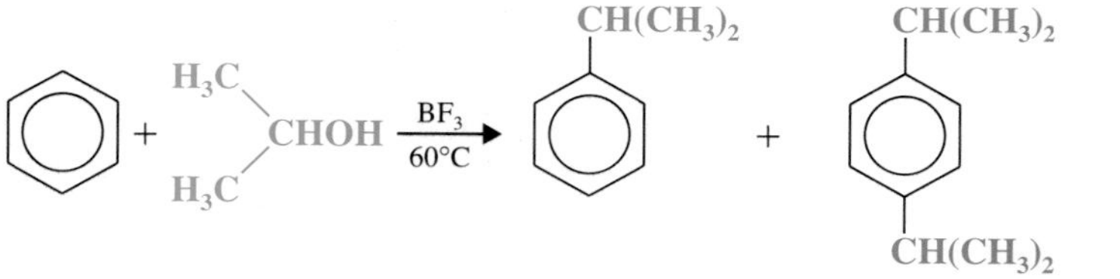

**Isopropyl-
benzene
(24%)**   **p-Diisopropylbenzene
(14%)**

Polyacylations are not a problem in Friedel–Crafts acylations, however. The acyl group (RCO—) by itself is an electron-withdrawing group, and when it forms a complex with $AlCl_3$ in the last step of the reaction (Section 15.7), it is made even more electron withdrawing. This strongly inhibits further substitution and makes monoacylation easy.

PROBLEM 15.5

When benzene reacts with neopentyl chloride, $(CH_3)_3CCH_2Cl$, in the presence of aluminum chloride, the major product is 2-methyl-2-phenylbutane, not neopentylbenzene. Explain this result.

PROBLEM 15.6

When benzene reacts with propyl alcohol in the presence of boron trifluoride, both propylbenzene and isopropylbenzene are obtained as products. Write a mechanism that accounts for this.

15.9 SYNTHETIC APPLICATIONS OF FRIEDEL–CRAFTS ACYLATIONS: THE CLEMMENSEN REDUCTION

Rearrangements of the carbon chain do not occur in Friedel–Crafts acylations. The acylium ion, because it is stabilized by resonance, is more stable than most other carbocations. Thus, there is no driving force for a rearrangement. Because rearrangements do not occur, Friedel–Crafts acylations often give us much better routes to unbranched alkylbenzenes than do Friedel–Crafts alkylations.

As an example, let us consider the problem of synthesizing propylbenzene. If we attempt this synthesis through a Friedel–Crafts alkylation, a rearrangement occurs and the major product is isopropylbenzene (see also Problem 15.6).

By contrast, the Friedel–Crafts acylation of benzene with propanoyl chloride produces a ketone with an unrearranged carbon chain in excellent yield.

This ketone can then be reduced to propylbenzene by several methods. One general method—called **the Clemmensen reduction**—consists of refluxing the ketone with hydrochloric acid containing amalgamated zinc. [*Caution:* As we shall discuss later (Section 19.5), zinc and hydrochloric acid will also reduce nitro groups to amino groups.]

When cyclic anhydrides are used as one component, the Friedel–Crafts acylation provides a means of adding a new ring to an aromatic compound. One illustration is shown here.

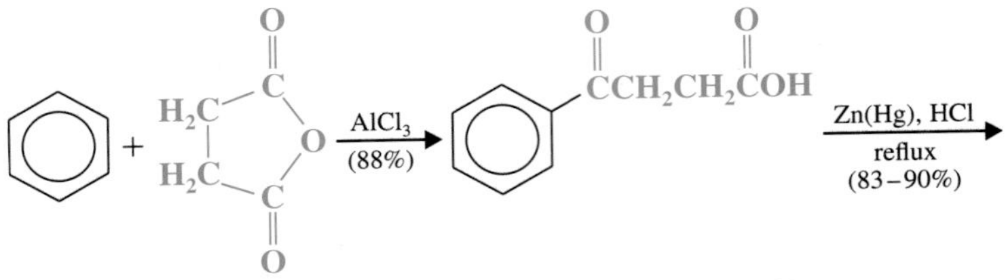

| Benzene (excess) | Succinic anhydride | | 3-Benzoylpropanoic acid | |

| 4-Phenylbutanoic acid | 4-Phenylbutanoyl chloride | α-Tetralone |

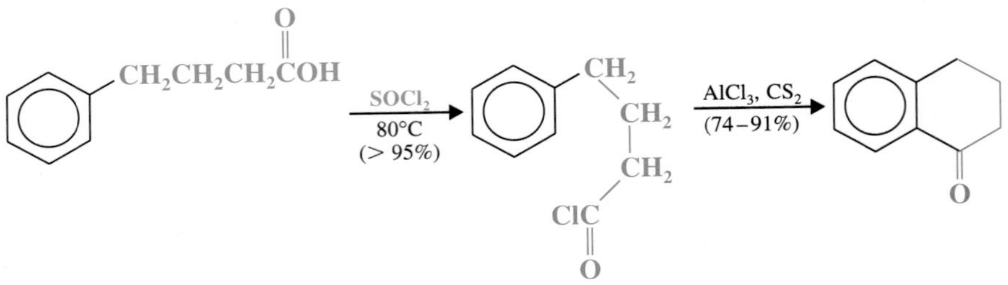

PROBLEM 15.7

Starting with benzene and the appropriate acyl chloride or acid anhydride, outline a synthesis of each of the following:

(a) Butylbenzene

(b) $(CH_3)_2CHCH_2CH_2C_6H_5$

(c) Benzophenone $(C_6H_5COC_6H_5)$

(d) 9,10-Dihydroanthracene

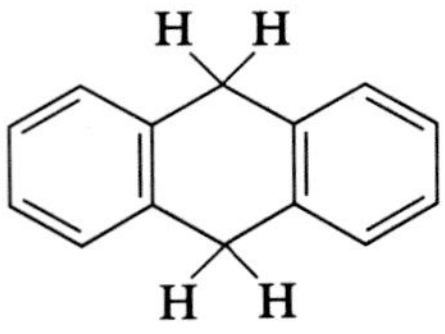

9,10-Dihydroanthracene

15.10 EFFECT OF SUBSTITUENTS ON REACTIVITY AND ORIENTATION

When substituted benzenes undergo electrophilic attack, groups already on the ring affect both the rate of the reaction and the site of attack. We say, therefore, that substituent groups affect both **reactivity** and **orientation** in electrophilic aromatic substitutions.

We can divide substituent groups into two classes according to their influence on the reactivity of the ring. Those that cause the ring to be more reactive than benzene itself we call **activating groups.** Those that cause the ring to be less reactive than benzene we call **deactivating groups.**

We also find that we can divide substituent groups into two classes according to the way they influence the orientation of attack by the incoming electrophile. Substituents in one class tend to bring about electrophilic substitution primarily at the positions *ortho* and *para* to themselves. We call these groups *ortho–para directors* because they tend to *direct* the incoming group into the ortho and para positions. Substituents in the second category tend to direct the incoming electrophile to the *meta* position. We call these groups *meta directors.*

Several examples will illustrate more clearly what we mean by these terms.

15.10A Activating Groups: Ortho–Para Directors

The methyl group is an **activating** group and **an ortho–para director.** Toluene reacts considerably faster than benzene in all electrophilic substitutions.

**More reactive than benzene
toward electrophilic substitution**

We observe the greater reactivity of toluene in several ways. We find, for example, that with toluene, milder conditions—lower temperatures and lower concentrations of the electrophile—can be used in electrophilic substitutions than with benzene. We also find that under the same conditions, toluene reacts faster than benzene. In nitration, for example, toluene reacts 25 times as fast as benzene.

We find, moreover, that when toluene undergoes electrophilic substitution, most of the substitution takes place at its ortho and para positions. When we nitrate toluene with nitric and sulfuric acids, we get mononitrotoluenes in the following relative proportions.

Of the mononitrotoluenes obtained from the reaction, 96% (59% + 37%) has the nitro group in an ortho or para position. Only 4% has the nitro group in a meta position.

PROBLEM 15.8

Explain how the percentages just given show that the methyl group exerts an ortho–para directive effect by considering the percentages that would be obtained if the methyl group had no effect on the orientation of the incoming electrophile.

Predominant substitution at the ortho and para positions of toluene is not restricted to nitration reactions. The same behavior is observed in halogenation, sulfonation, and so forth.

All alkyl groups are activating groups, and they are all also ortho–para directors. The methoxyl group, $CH_3O—$, and the acetamido group, $CH_3CONH—$, are strong activating groups and both are ortho–para directors.

The hydroxyl group and the amino group are very powerful activating groups and are also powerful ortho–para directors. Phenol and aniline react with bromine in water (no catalyst is required) to produce products in which both of the ortho positions and the para position are substituted. These tribromo products are obtained in nearly quantitative yield.

2,4,6-Tribromophenol
(~100%)

2,4,6-Tribromoaniline
(~100%)

15.10B Deactivating Groups: Meta Directors

The nitro group is a very strong **deactivating group.** Nitrobenzene undergoes nitration at a rate only 10^{-4} times that of benzene. The nitro group is a meta director. When nitrobenzene is nitrated with nitric and sulfuric acids, 93% of the substitution occurs at the meta position.

| (6%) | (1%) | (93%) |

The carboxyl group ($—CO_2H$), the sulfo group ($—SO_3H$), and the trifluoromethyl group ($—CF_3$) are also deactivating groups; they are also meta directors.

15.10C Halo Substituents: Deactivating Ortho–Para Directors

The chloro and bromo groups are weak deactivating groups. Chlorobenzene and bromobenzene undergo nitration at rates that are, respectively, 33 and 30 times slower

TABLE 15.1   Electrophilic substitutions of chlorobenzene

REACTION	ORTHO PRODUCT (%)	PARA PRODUCT (%)	TOTAL ORTHO AND PARA (%)	META PRODUCT (%)
Chlorination	39	55	94	6
Bromination	11	87	98	2
Nitration	30	70	100	
Sulfonation		100	100	

than for benzene. The chloro and bromo groups are ortho–para directors, however. The relative percentages of monosubstituted products that are obtained when chlorobenzene is chlorinated, brominated, nitrated, and sulfonated are shown in Table 15.1.

Similar results are obtained from electrophilic substitutions of bromobenzene.

15.10D Classification of Substituents

Studies like the ones that we have presented in this section have been done for a number of other substituted benzenes. The effects of these substituents on reactivity and orientation are included in Table 15.2.

TABLE 15.2   Effect of substituents on electrophilic aromatic substitution

ORTHO–PARA DIRECTORS	META DIRECTORS
Strongly Activating	**Moderately Deactivating**
$-\ddot{N}H_2, -\ddot{N}HR, -\ddot{N}R_2$	$-C\equiv N$
$-\ddot{O}H, -\ddot{O}:^-$	$-SO_3H$
	$-CO_2H, -CO_2R$
Moderately Activating	$-CHO, -COR$
$-\ddot{N}HCOCH_3, -\ddot{N}HCOR$	**Strongly Deactivating**
$-\ddot{O}CH_3, -\ddot{O}R$	$-NO_2$
Weakly Activating	$-NR_3^+$
$-CH_3, -C_2H_5, -R$	$-CF_3, -CCl_3$
$-C_6H_5$	
Weakly Deactivating	
$-\ddot{\underset{..}{F}}:, -\ddot{\underset{..}{C}l}:, -\ddot{\underset{..}{B}r}:, -\ddot{\underset{..}{I}}:$	

PROBLEM 15.9 _____

Use Table 15.2 to predict the major products formed when:

(a) Toluene is sulfonated.

(b) Benzoic acid is nitrated.

(c) Nitrobenzene is brominated.

(d) Phenol is subjected to Friedel–Crafts acetylation.

If the major products would be a mixture of ortho and para isomers you should so state.

15.11 THEORY OF SUBSTITUENT EFFECTS ON ELECTROPHILIC AROMATIC SUBSTITUTION

15.11A Reactivity: The Effect of Electron-Releasing and Electron-Withdrawing Groups

We have now seen that certain groups *activate* the benzene ring toward electrophilic substitution, whereas other groups *deactivate* the ring. When we say that a group activates the ring, what we mean, of course, is that the group increases the relative rate of the reaction. We mean that an aromatic compound with an activating group reacts faster in electrophilic substitutions than benzene. When we say that a group deactivates the ring, we mean that an aromatic compound with a deactivating group reacts slower than benzene.

We have also seen that we can account for relative reaction rates by examining the transition state for the rate-determining steps. We know that any factor that increases the energy of the transition state relative to that of the reactants decreases the relative rate of the reaction. It does this because it increases the free energy of activation of the reaction. In the same way, any factor that decreases the energy of the transition state relative to that of the reactants lowers the free energy of activation and increases the relative rate of the reaction.

The rate-determining step in electrophilic substitutions of substituted benzenes is the step that results in the formation of the arenium ion. We can write the formula for a substituted benzene in a generalized way if we use the letter **S** to represent any ring substituent including hydrogen. (If **S** is hydrogen the compound is benzene itself.) We can also write the structure for the arenium ion in the way shown here. By this formula we mean that **S** can be in any position—ortho, meta, or para—relative to the electrophile, **E**. Using these conventions, then, we are able to write the rate-determining step for electrophilic aromatic substitution in the following general way.

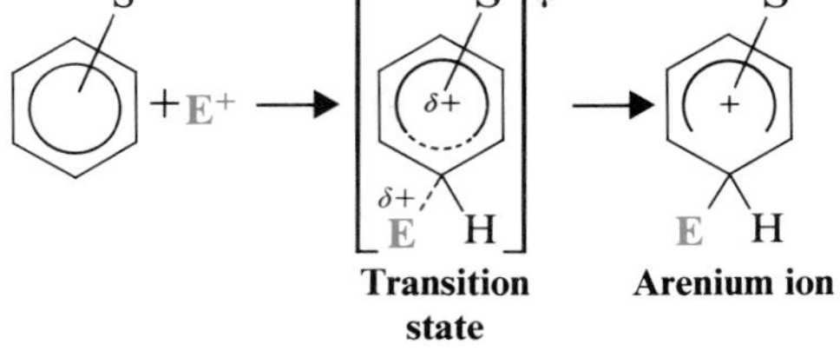

Transition state   **Arenium ion**

When we examine this step for a large number of reactions, we find that the relative rates of the reactions depend on whether **S withdraws** or **releases** electrons. If

S is an electron-releasing group (relative to hydrogen), the reaction occurs faster than the corresponding reaction of benzene. If S is an electron-withdrawing group, the reaction is slower than that of benzene.

It appears, then, that the substituent (S) must affect the stability of the transition state relative to that of the reactants. Electron-releasing groups apparently make the transition state more stable, whereas electron-withdrawing groups make it less stable. That this is so is entirely reasonable, because the transition state resembles the arenium ion, and the arenium ion is a delocalized *carbocation.*

This effect illustrates another application of the Hammond–Leffler postulate (Section 6.15A). The arenium ion is a high-energy intermediate, and the step that leads to it is a *highly endothermic step.* Thus, according to the Hammond–Leffler postulate there should be a strong resemblance between the arenium ion itself and the transition state leading to it.

Since the arenium ion is positively charged, we would expect an electron-releasing group to stabilize it *and the transition state leading to the arenium ion,* for the transition state is a developing delocalized carbocation. We can make the same kind of arguments about the effect of electron-withdrawing groups. An electron-withdrawing group should make the arenium ion *less stable* and in a corresponding way it should make the transition state leading to the arenium ion *less stable.*

Figure 15.3 shows how the electron-withdrawing and electron-releasing abilities of substituents affect the relative free energies of activation of electrophilic aromatic substitution reactions.

15.11B Inductive and Resonance Effects: Theory of Orientation

We can account for the electron-withdrawing and electron-releasing properties of groups on the basis of two factors: *inductive effects* and *resonance effects.* We shall also see that these two factors determine orientation in aromatic substitution reactions.

The **inductive effect** of a substituent S arises from the electrostatic interaction of the polarized S to ring bond with the developing positive charge in the ring as it is attacked by an electrophile. If, for example, S is a more electronegative atom (or group) than carbon, then the ring will be at the positive end of the dipole:

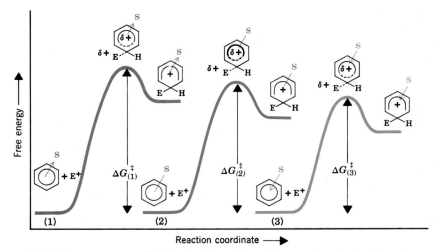

FIGURE 15.3 Free-energy profiles for the formation of the arenium ion in three electrophilic aromatic substitution reactions. In (1), **S** is an electron-withdrawing group. In (2), **S** = H. In (3), **S** is an electron-releasing group. $\Delta G_{(1)}^{\ddagger} > \Delta G_{(2)}^{\ddagger} > \Delta G_{(3)}^{\ddagger}$.

$$\overset{\delta-\ \ \delta+}{S\!\!\leftarrow}\!\!\bigcirc \qquad \text{(e.g., S = F, Cl, or Br)}$$

Attack by an electrophile will be retarded because this will lead to an additional full positive charge on the ring. The halogens are all more electronegative than carbon and exert an electron-withdrawing inductive effect. Other groups have an electron-withdrawing inductive effect because the atom directly attached to the ring bears a full or partial positive charge. Examples are the following:

$$\overset{+}{\rightarrow}\!NR_3 \qquad (R = \text{alkyl or H}) \qquad \overset{X^{\delta-}}{\underset{X^{\delta-}}{\rightarrow}}\!\!\overset{\uparrow}{\underset{\downarrow}{C}}{}^{\delta+}\!\!\rightarrow\!X^{\delta-} \qquad \rightarrow\!\overset{O}{\underset{O^-}{N^+}} \qquad \overset{O^-}{\underset{O}{\rightarrow}\!S^\pm}\!\!-OH$$

$$\overset{\ddots}{\underset{\overset{\ }{\rightarrow}}{\overset{O:}{C}}}\!\!-G \longleftrightarrow \overset{:\overset{\ddots}{O}:^-}{\underset{\ }{\rightarrow\!\overset{+}{C}}}\!\!-G \qquad (G = \text{H, R, OH, or OR})$$

**Electron-withdrawing groups with a full or partial
charge on the atom attached to the ring**

The **resonance effect** of a substituent **S** refers to the possibility that the presence of **S** may increase or decrease the resonance stabilization of the intermediate arenium ion. The **S** substituent may, for example, cause one of the three contributors to the resonance hybrid for the arenium ion to be better or worse than the case when **S** is hydrogen. Moreover, when **S** is an atom bearing one or more nonbonding electron pairs, it may lend extra stability to the arenium ion by providing a *fourth* resonance contributor in which the positive charge resides on **S**.

This electron-donating resonance effect applies with decreasing strength in the following order:

Most electron donating → $-\ddot{N}H_2$, $-\ddot{N}R_2$ > $-\ddot{O}H$, $-\ddot{O}R$ > $-\ddot{X}:$ ← Least electron donating

This is also the order of the activating ability of these groups. Amino groups are highly activating, hydroxyl and alkoxyl groups are somewhat less activating, and halogen substituents are weakly deactivating. When X = F, this order can be related to the electronegativity of the atoms with the nonbonding pair. The more electronegative the atom is the less able it is to accept the positive charge (fluorine is the most electronegative, nitrogen the least). When X = Cl, Br, or I, the relatively poor electron-donating ability of the halogens by resonance is understandable on a different basis. These atoms (Cl, Br, and I) are all larger than carbon and, therefore, the orbitals that contain the nonbonding pairs are further from the nucleus and do not overlap well with the 2*p* orbital of carbon. (This is a general phenomenon: Resonance effects are not transmitted well between atoms of different rows in the periodic table.)

15.11C Meta-Directing Groups

All meta-directing groups have either a partial positive charge or a full positive charge on the atom directly attached to the ring. As a typical example let us consider the trifluoromethyl group.

The trifluoromethyl group, because of the three highly electronegative fluorine atoms, is strongly electron withdrawing. It is a strong deactivating group and a powerful meta director in electrophilic aromatic substitution reactions. We can account for both of these characteristics of the trifluoromethyl group in the following way.

The trifluoromethyl group affects reactivity by causing the transition state leading to the arenium ion to be highly unstable. It does this by withdrawing electrons from the developing carbocation thus increasing the positive charge on the ring.

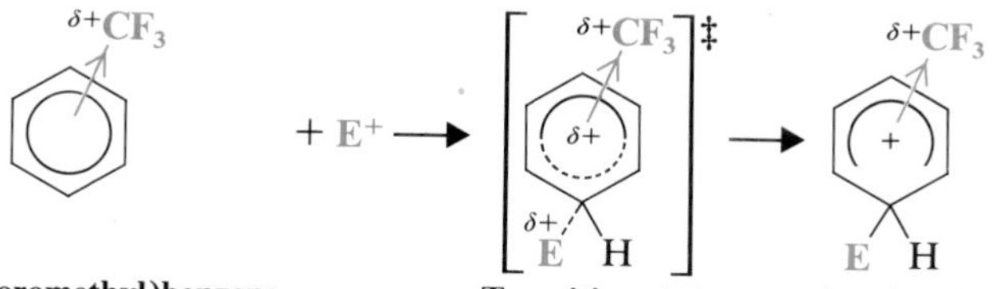

(Trifluoromethyl)benzene   Transition state   Arenium ion

We can understand how the trifluoromethyl group affects *orientation* in electrophilic aromatic substitution if we examine the resonance structures for the arenium ion that would be formed when an electrophile attacks the ortho, meta, and para positions of (trifluoromethyl)benzene.

Ortho attack

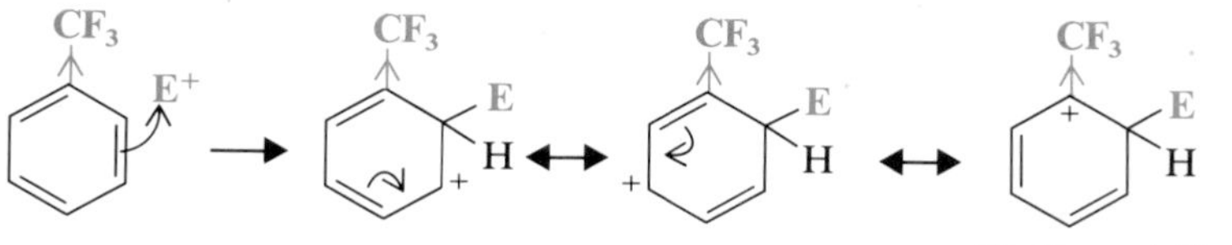

Highly unstable contributor

Meta attack

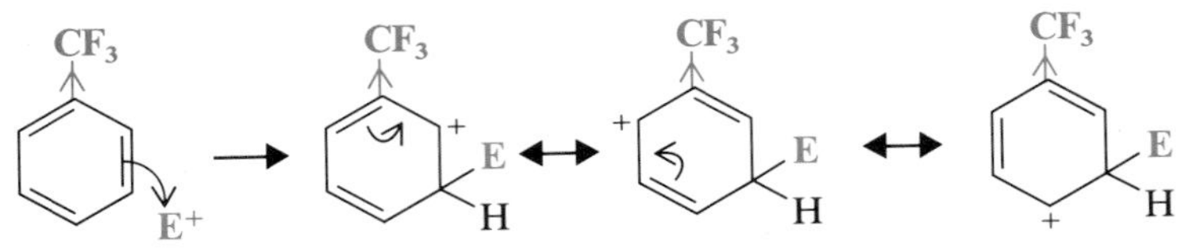

Para attack

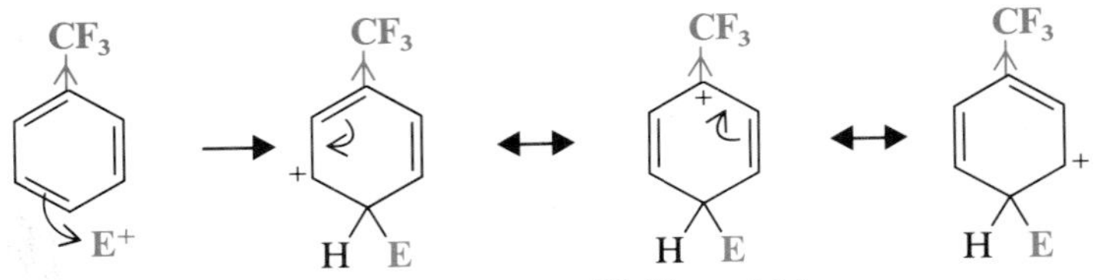

Highly unstable contributor

We see in the resonance structures for the arenium ion arising from ortho and para attack that *one contributing structure is highly unstable relative to all the others because the positive charge is located on the ring carbon that bears the electron-withdrawing group*. We see *no* such highly unstable resonance structure in the arenium ion arising from meta attack. This means that the arenium ion formed by meta attack should be the most stable of the three. By the usual reasoning we would also expect the transition state leading to the meta-substituted arenium ion to be the most stable and, therefore, that meta attack would be favored. This is exactly what we find experimentally. The trifluoromethyl group is a powerful meta director.

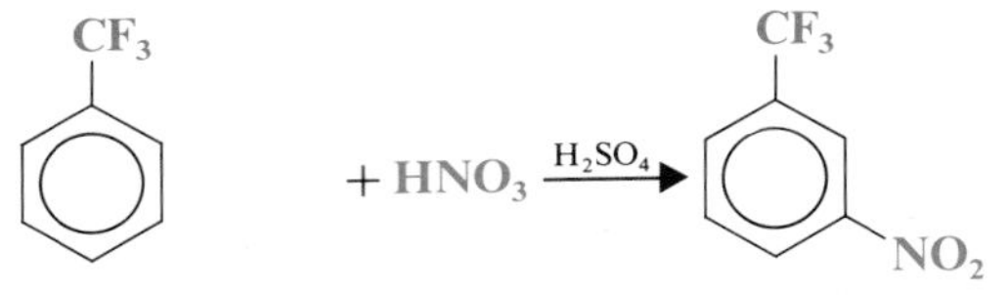

(Trifluoromethyl)benzene   **(∼100%)**

Bear in mind, however, that meta substitution is favored only in the sense that *it is the least unfavorable of three unfavorable pathways*. The free energy of activation for substitution at the meta position of (trifluoromethyl)benzene is less than that for attack at an ortho or para position, but it is still far greater than that for an attack on benzene. Substitution occurs at the meta position of (trifluoromethyl)benzene faster than substitution takes place at the ortho and para positions, but it occurs much more slowly than it does with benzene.

The nitro group, the carboxyl group, and other meta-directing groups are all powerful electron-withdrawing groups and all act in a similar way.

15.11D Ortho–Para-Directing Groups

Except for the alkyl and phenyl substituents, all of the ortho–para-directing groups in Table 15.2 are of the following general type:

Nonbonding electron pair

as in

Aniline   **Phenol**   **Chlorobenzene**

All of these ortho–para directors have at least one pair of nonbonding electrons on the atom adjacent to the benzene ring.

This structural feature—an unshared electron pair on the atom adjacent to the ring—determines the orientation and influences reactivity in electrophilic substitution reactions.

The *directive effect* of these groups with an unshared pair is predominantly caused by an electron-releasing resonance effect. The resonance effect, moreover, operates primarily in the arenium ion and, consequently, in the transition state leading to it.

Except for the halogens, the primary effect on reactivity of these groups is also caused by an electron-releasing resonance effect. And, again, this effect operates primarily in the transition state leading to the arenium ion.

In order to understand these resonance effects let us begin by recalling the effect of the amino group on electrophilic aromatic substitution reactions. The amino group is not only a powerful activating group, it is also a powerful ortho–para director. We saw earlier (Section 15.10A) that aniline reacts with bromine in aqueous solution at room temperature and in the absence of a catalyst to yield a product in which both ortho positions and the para position are substituted.

The inductive effect of the amino group makes it slightly electron withdrawing. Nitrogen, as we know, is more electronegative than carbon. The difference between the electronegativities of nitrogen and carbon in aniline is not large, however, because the carbon of the benzene ring is sp^2 hybridized and so it is somewhat more electronegative than it would be if it were sp^3 hybridized.

The resonance effect of the amino group is far more important than its inductive effect in electrophilic aromatic substitution, and this resonance effect makes the amino group electron releasing. We can understand this effect if we write the resonance structures for the arenium ions that would arise from ortho, meta, and para attack on aniline.

Ortho attack

Relatively stable contributor

Meta attack

Para attack

Relatively stable
contributor

We see that four reasonable resonance structures can be written for the arenium ions resulting from ortho and para attack, whereas only three can be written for the arenium ion that results from meta attack. This, in itself, suggests that the ortho- and para-substituted arenium ions should be more stable. Of greater importance, however, are the relatively stable structures that contribute to the hybrid for the ortho- and para-substituted arenium ions. In these structures, nonbonding pairs of electrons from nitrogen form an extra bond to the carbon of the ring. This extra bond—and the fact that every atom in each of these structures has a complete outer octet of electrons—makes these structures the most stable of all of the contributors. Because these structures are unusually stable, they make a large—*and stabilizing*—contribution to the hybrid. This means, of course, that the ortho- and para-substituted arenium ions themselves are considerably more stable than the arenium ion that results from the meta attack. The transition states leading to the ortho- and para-substituted arenium ions occur at unusually low potential energies. As a result, electrophiles react at the ortho and para positions very rapidly.

PROBLEM 15.10

Use resonance theory to explain why the hydroxyl group of phenol is an activating group and an ortho–para director. Illustrate your explanation by showing the arenium ion formed when phenol reacts with a Br^+ ion at the ortho, meta, and para position.

PROBLEM 15.11

Phenol reacts with acetic anydride, in the presence of sodium acetate, to produce the ester, phenyl acetate.

Phenol   **Phenyl acetate**

The CH_3COO— group of phenyl acetate, like the —OH group of phenol (Problem 15.10), is an ortho–para director. (a) What structural feature of the CH_3COO— group explains this? (b) Phenyl acetate, although undergoing reaction at the *o* and *p* positions, is less reactive toward electrophilic aromatic substitution than phenol. Use resonance theory to explain why this is so. (c) Aniline is often so highly reactive toward electrophilic substitution that undesirable reactions take place (see Section 15.14). One way to avoid these undesirable reactions is to convert aniline to acetanilide (below), by treating aniline with acetyl chloride or acetic anhydride.

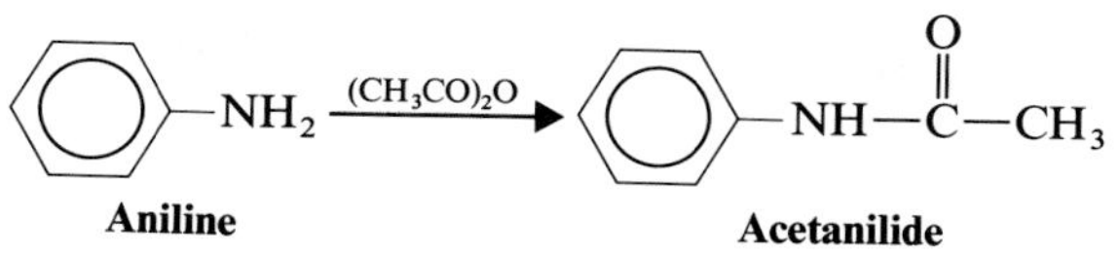

What kind of directive effect would you expect the acetamido group (CH_3CONH—) to have? (d) Explain why it is much less activating than the amino group, —NH_2.

The directive and reactivity effects of halo substituents may, at first, seem to be contradictory. *The halo groups are the only ortho–para directors* (in Table 15.2) *that are deactivating groups.* All other deactivating groups are meta directors. We can readily account for the behavior of halo substituents, however, if we assume that their electron-withdrawing inductive effect influences reactivity and their electron-donating resonance effect governs orientation.

Let us apply these assumptions specifically to chlorobenzene. The chlorine atom is highly electronegative. Thus, we would expect a chlorine atom to withdraw electrons from the benzene ring and thereby deactivate it.

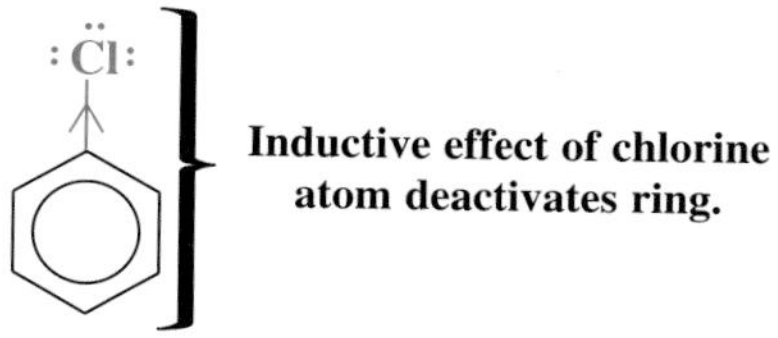

Inductive effect of chlorine atom deactivates ring.

On the other hand, when electrophilic attack does take place, the chlorine atom stabilizes the arenium ions resulting from ortho and para attack relative to that from meta attack. The chlorine atom does this in the same way as amino groups and hydroxyl groups do—*by donating an unshared pair of electrons.* These electrons give rise to relatively stable resonance structures contributing to the hybrids for the ortho- and para-substituted arenium ions (Section 15.11D).

Ortho attack

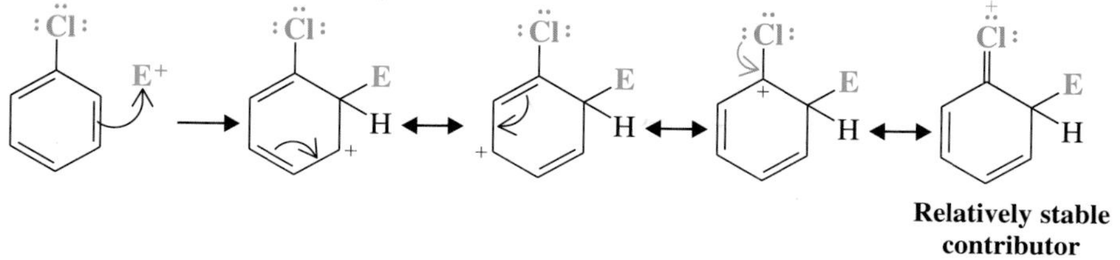

Relatively stable contributor

Meta attack

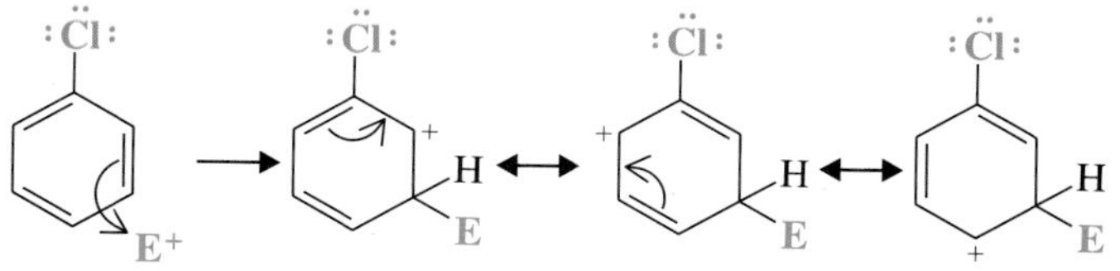

Para attack

**Relatively stable
contributor**

What we have said about chlorobenzene is, of course, true of bromobenzene.

We can summarize the inductive and resonance effects of halo substituents in the following way. Through their electron-withdrawing inductive effect, halo groups make the ring more positive than that of benzene. This causes the free energy of activation for any electrophilic aromatic substitution reaction to be greater than that for benzene, and, therefore, halo groups are deactivating. Through their electron-donating resonance effect, however, halo substituents cause the free energies of activation leading to ortho and para substitution to be lower than the free energy of activation leading to meta substitution. This makes halo substituents ortho–para directors.

You may have noticed an apparent contradiction between the rationale offered for the unusual effects of the halogens and that offered earlier for amino or hydroxyl groups. That is, oxygen is *more* electronegative than chlorine or bromine (and especially iodine). Yet, the hydroxyl group is an activating group, whereas halogens are deactivating groups. An explanation for this can be obtained if we consider the relative stabilizing contributions made to the transition state leading to the arenium ion by resonance structures involving a group $-\ddot{Z}$ ($-\ddot{Z} = -\ddot{N}H_2, -\ddot{O}-H, -\ddot{F}:, -\ddot{C}l:, -\ddot{B}r:, -\ddot{I}:$) that is directly attached to the benzene ring in which Z donates an electron pair. If $-\ddot{Z}$ is $-\ddot{O}H$ or $-\ddot{N}H_2$, these resonance structures arise because of the overlap of a $2p$ orbital of carbon with that of oxygen or nitrogen. Such overlap is favorable because the atoms are almost the same size. With chlorine, however, donation of an electron pair to the benzene ring requires overlap of a carbon $2p$ orbital with a chlorine $3p$ orbital. Such overlap is less effective; the chlorine atom is much larger and its $3p$ orbital is much further from its nucleus. With bromine and iodine, overlap is even less effective. Justification for this explanation can be found in the observation that fluorobenzene ($Z = -\ddot{F}:$) is the most reactive halobenzene in spite of the high electronegativity of fluorine and the fact that $-\ddot{F}:$ is the most powerful ortho–para director of the halogens. With fluorine, donation of an electron pair arises from overlap of a $2p$ orbital of fluorine with a $2p$ orbital of carbon (as with $-\ddot{N}H_2$ and $-\ddot{O}H$). This overlap is effective because the atoms $=\overset{|}{C}-$ and $-\ddot{F}:$ are of the same relative size.

PROBLEM 15.12

Chloroethene adds hydrogen chloride more slowly than ethene and the product is 1,1-dichloroethane. How can you explain this using resonance and inductive effects?

$$Cl-CH=CH_2 \xrightarrow{\text{HCl}} Cl-\underset{\underset{Cl}{|}}{C}H-\underset{\underset{H}{|}}{C}H_2$$

15.11E Ortho–Para Direction and Reactivity of Alkylbenzenes

Alkyl groups are better electron-releasing groups than hydrogen. Because of this they can activate a benzene ring toward electrophilic substitution by stabilizing the transition state leading to the arenium ion:

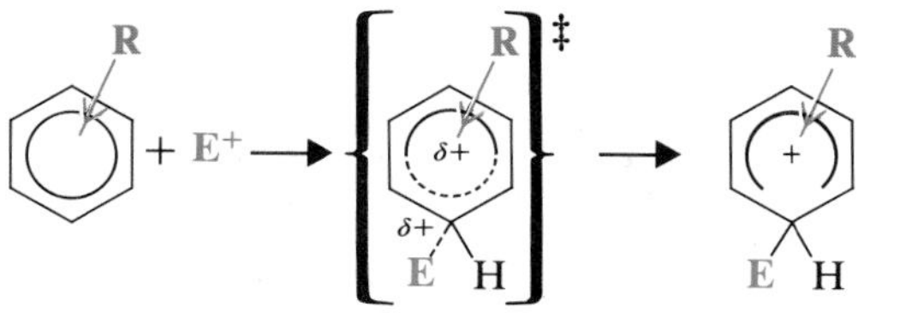

Transition state   **Arenium ion**
is stabilized   **is stabilized**

For an alkylbenzene the free energy of activation of the step leading to the arenium ion (just shown) is lower than that for benzene, and alkylbenzenes react faster.

Alkyl groups are ortho–para directors. We can also account for this property of alkyl groups on the basis of their ability to release electrons—an effect that is particularly important when the alkyl group is attached directly to a carbon that bears a positive charge. (Recall the ability of alkyl groups to stabilize carbocations that we discussed in Section 6.13 and in Fig. 6.10.)

If, for example, we write resonance structures for the arenium ions formed when toluene undergoes electrophilic substitution, we get the following results:

Ortho attack

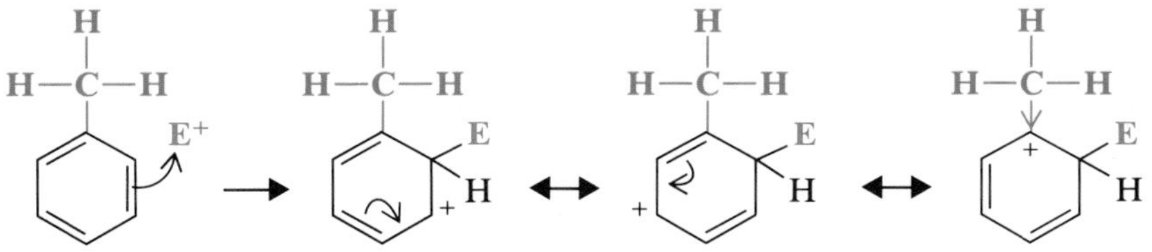

Relatively stable contributor

Meta attack

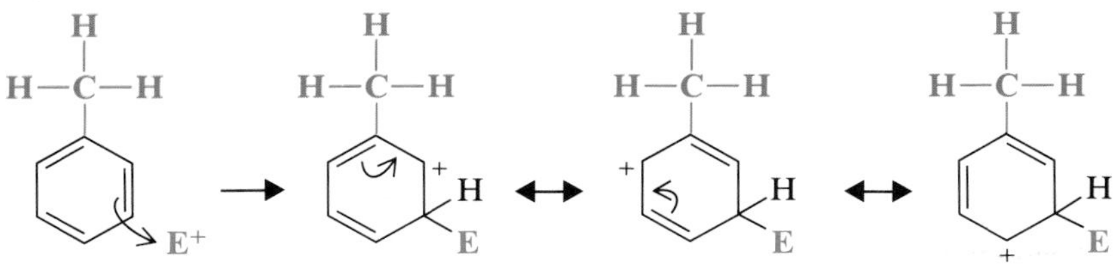

Para attack

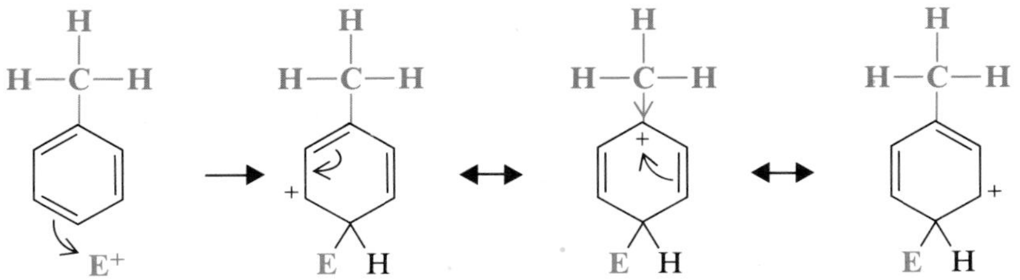

Relatively stable contributor

In ortho attack and para attack we find that we can write resonance structures in which the methyl group is directly attached to a positively charged carbon of the ring. These structures are more *stable* relative to any of the others because in them the stabilizing influence of the methyl group (by electron release) is most effective. These structures, therefore, make a large (stabilizing) contribution to the overall hybrid for ortho- and para-substituted arenium ions. No such relatively stable structure contributes to the hybrid for the meta-substituted arenium ion, and as a result it is less stable than the ortho- or para-substituted arenium ion. Since the ortho- and para-substituted arenium ions are more stable, the transition states leading to them occur at lower energy and ortho and para substitution take place most rapidly.

PROBLEM 15.13

Write resonance structures for the ortho and para arenium ions formed when ethylbenzene reacts with a Br^+ ion (as formed from $Br_2/FeBr_3$).

PROBLEM 15.14

When biphenyl (C_6H_5—C_6H_5) undergoes nitration, it reacts more rapidly than benzene, and the major products are 1-nitro-2-phenylbenzene and 1-nitro-4-phenylbenzene. Explain these results.

15.11F Summary of Substituent Effects on Orientation and Reactivity

We can summarize the effects that groups have on orientation and reactivity in the following way.

Full or partial (+) charge on directly attached atoms	At least one nonbonding pair on directly attached atom		Alkyl or aryl
	Halogen	$—\ddot{N}H_2, —\ddot{O}H$, etc.	
←meta directing→	←————ortho–para directing————→		
←————deactivating————→	←————activating————→		

15.12 REACTIONS OF THE SIDE CHAIN OF ALKYLBENZENES

Hydrocarbons that consist of both aliphatic and aromatic groups are also known as **arenes.** Toluene, ethylbenzene, and isopropylbenzene are **alkylbenzenes.**

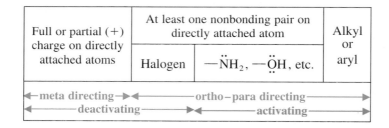

| Methylbenzene (toluene) | Ethylbenzene | Isopropyl-benzene (cumene) | Phenylethene (styrene or vinylbenzene) |

Phenylethene, usually called styrene, is an example of an **alkenylbenzene.** The aliphatic portion of these compounds is commonly called the **side chain.**

Styrene is one of the most important industrial chemicals—more than 6 billion lb is produced each year. The starting material for the commercial synthesis of styrene is ethylbenzene, produced by Friedel–Crafts alkylation of benzene:

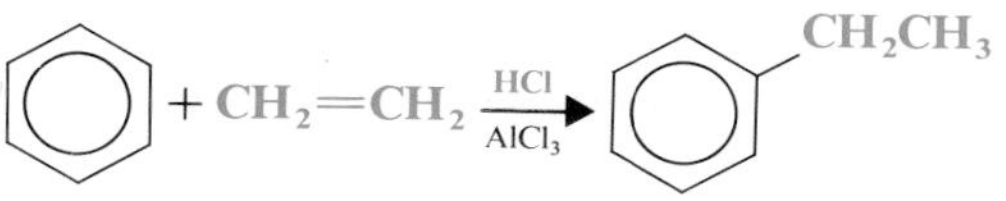

+ CH2=CH2
HCl
AlCl3
CH2CH3

Ethylbenzene

Ethylbenzene is then dehydrogenated in the presence of a catalyst (zinc oxide or chromium oxide) to produce styrene.

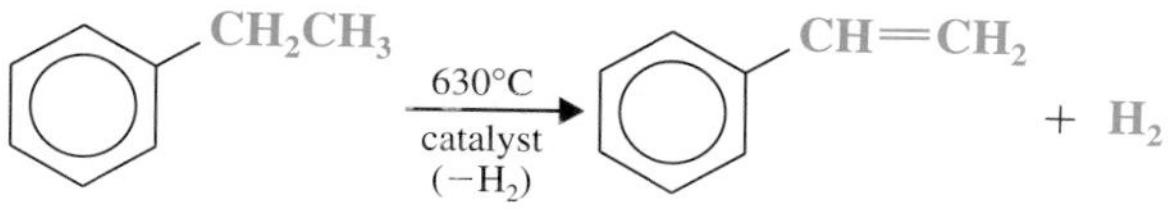

CH2CH3
630°C
catalyst
(−H2)
CH=CH2
+ H2

Styrene
(90–92% yield)

Most styrene is polymerized (Special Topic C) to the familiar plastic, polystyrene.

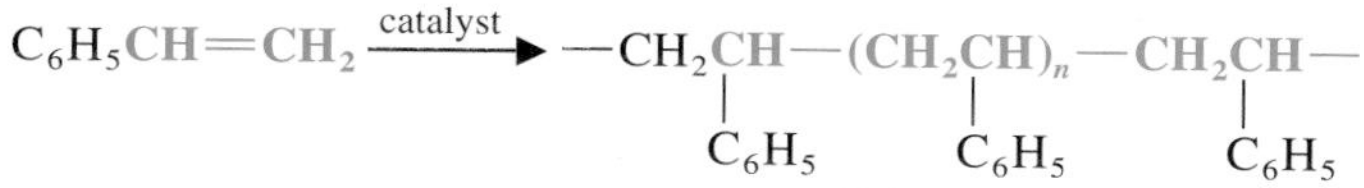

C6H5CH=CH2
catalyst
—CH2CH—(CH2CH)n—CH2CH—
C6H5
C6H5
C6H5

Polystyrene

15.12A Benzylic Radicals and Cations

Removal of a hydrogen atom from the methyl group of methylbenzene (toluene) produces a radical called the **benzyl radical:**

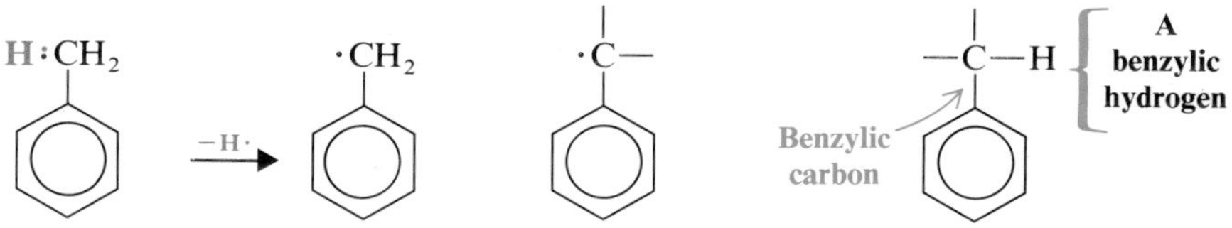

H:CH2
−H·
·CH2
·C—
—C—H
A benzylic hydrogen
Benzylic carbon

Methylbenzene (toluene) **The benzyl radical** **A benzylic radical**

The name benzyl radical is used as a specific name for the radical produced in this reaction. The general name **benzylic radical** applies to all radicals that have an unpaired electron on the side chain carbon atom that is directly attached to the benzene ring. The hydrogen atoms of the carbon atom directly attached to the benzene ring are called **benzylic hydrogen atoms.**

Removal of an electron from a benzylic radical produces a **benzylic cation:**

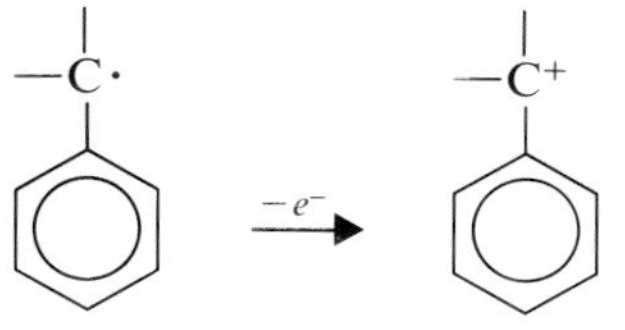

—C·
−e−
—C+

Benzylic radical **Benzylic cation**

Benzylic radicals and benzylic cations are *conjugated unsaturated systems* and *both are unusually stable.* They have approximately the same stabilities as allylic radicals and cations. This exceptional stability of benzylic radicals and cations can be explained by resonance theory. In the case of each entity, resonance structures can be written that place either the unpaired electron (in the case of the radical) or the positive charge (in the case of the cation) on an ortho or para carbon of the ring (see the following structures). Thus resonance delocalizes the unpaired electron or the charge, and this delocalization causes the radical or cation to be highly stabilized.

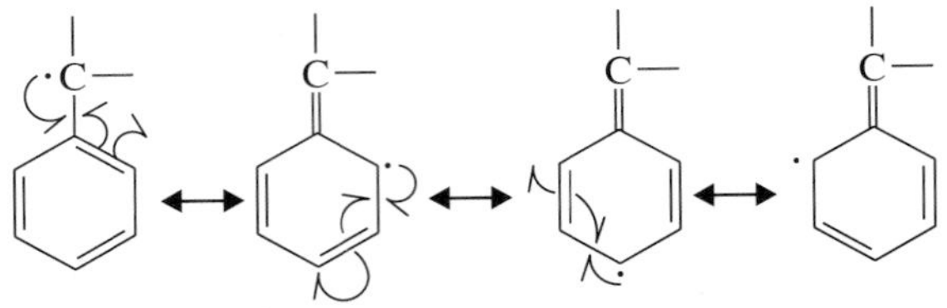

**Benzylic radicals are
stabilized by resonance**

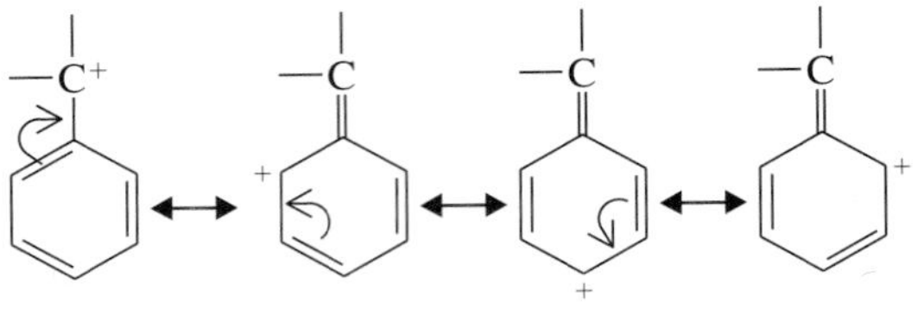

**Benzylic cations are
stabilized by resonance**

15.12B Halogenation of the Side Chain. Benzylic Radicals

We have seen that bromine and chlorine replace hydrogen atoms on the ring of toluene when the reaction takes place in the presence of a Lewis acid. In ring halogenations the electrophiles are *positive* chlorine or bromine ions or they are Lewis acid complexes that have positive halogens. These positive electrophiles attack the π electrons of the benzene ring and aromatic substitution takes place.

Chlorine and bromine can also be made to replace hydrogens of the methyl group of toluene. Side-chain, or *benzylic*, halogenation takes place when the reaction is carried out *in the absence of Lewis acids* and **under conditions that favor the formation of radicals.** When toluene reacts with *N*-bromosuccinimide (NBS) in the presence of light, for example, the major product is benzyl bromide. *N*-Bromosuccinimide furnishes a low concentration of Br_2 and the reaction is analogous to that for allylic bromination that we studied in Section 12.2.

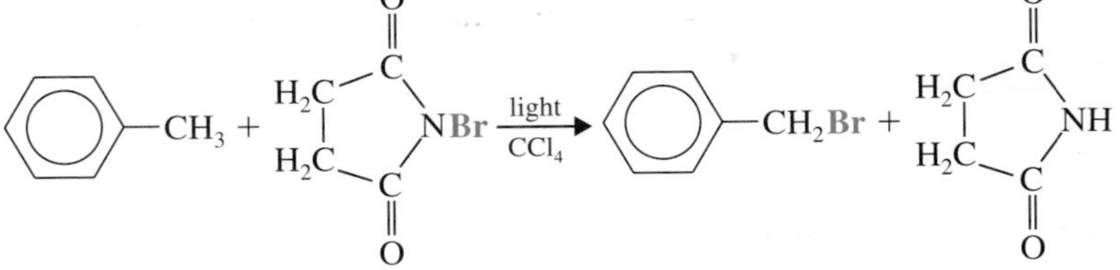

NBS

**Benzyl bromide
(α-bromotoluene)
(64%)**

Side-chain chlorination of toluene also takes place in the gas phase at 400–600°C or in the presence of UV light. When an excess of chlorine is used, multiple chlorinations of the side chain occur.

| Benzyl chloride | (Dichloromethyl)- benzene | (Trichloromethyl)- benzene |

These halogenations take place through the same radical mechanisms we saw for alkanes in Section 9.4. The halogens dissociate to produce halogen atoms and then the halogen atoms initiate chains by abstracting hydrogens of the methyl group.

A MECHANISM FOR BENZYLIC HALOGENATION

Chain Initiation

Step 1 $X_2 \xrightarrow[\text{or light}]{\substack{\text{peroxides,} \\ \text{heat,}}} 2\,X\cdot$

Chain Propagation

Step 2 $C_6H_5CH_3 + X\cdot \longrightarrow C_6H_5CH_2\cdot + HX$
Benzyl radical

Step 3 $C_6H_5CH_2\cdot + X_2 \longrightarrow C_6H_5CH_2X + X\cdot$
Benzyl radical   Benzyl halide

Abstraction of a hydrogen from the methyl group of toluene produces **a benzyl radical.** The benzyl radical then reacts with a halogen molecule to produce a benzyl halide and a halogen atom. The halogen atom then brings about a repetition of step 2, then step 3 occurs again, and so on.

Benzylic halogenations are similar to allylic halogenations (Section 12.2) in that they involve the formation of *unusually stable radicals* (Section 15.12A). Benzylic and allylic radicals are even more stable than tertiary radicals.

The greater stability of benzylic radicals accounts for the fact that when ethylbenzene is halogenated the major product is the 1-halo-1-phenylethane. The benzylic radical is formed much faster than the 1° radical:

PROBLEM 15.15

When propylbenzene reacts with chlorine in the presence of UV radiation, the major product is 1-chloro-1-phenylpropane. Both 2-chloro-1-phenylpropane and 3-chloro-1-phenylpropane are minor products. Write the structure of the radical leading to each product and account for the fact that 1-chloro-1-phenylpropane is the major product.

SAMPLE PROBLEM

Illustrating a Multistep Synthesis

Starting with ethylbenzene, outline a synthesis of phenylacetylene $(C_6H_5C\equiv CH)$.

ANSWER:
Working backward, that is, using *retrosynthetic analysis,* we find that we could make phenylacetylene by dehydrohalogenating either of the following compounds using sodium amide in mineral oil (Section 7.18).

$$C_6H_5CBr_2CH_3 \xrightarrow[\text{(2) H}^+]{\text{(1) NaNH}_2, \text{ mineral oil, heat}} C_6H_5C\equiv CH$$

$$C_6H_5CHBrCH_2Br \xrightarrow[\text{(2) H}^+]{\text{(1) NaNH}_2, \text{ mineral oil, heat}} C_6H_5C\equiv CH$$

We could make the first compound from ethylbenzene by allowing it to react with 2 mol of NBS.

$$C_6H_5CH_2CH_3 \xrightarrow[\text{CCL}_4]{\text{NBS, light}} C_6H_5CBr_2CH_3$$

We could make the second compound by adding bromine to styrene, and we could make styrene from ethylbenzene as follows:

$$C_6H_5CH_2CH_3 \xrightarrow[\text{CCL}_4]{\text{NBS, light}} C_6H_5CHBrCH_3 \xrightarrow{\text{KOH, heat}}$$

$$C_6H_5CH=CH_2 \xrightarrow{\text{Br}_2, \text{ CCl}_4} C_6H_5CHBrCH_2Br$$

PROBLEM 15.16

Starting with phenylacetylene ($C_6H_5C{\equiv}CH$), outline a synthesis of   (a) 1-phenylpropyne,   (b) 1-phenyl-1-butyne,   (c) (*Z*)-1-phenylpropene,   and (d) (*E*)-1-phenylpropene.

15.13 ALKENYLBENZENES

15.13A Stability of Conjugated Alkenylbenzenes

Alkenylbenzenes that have their side-chain double bond conjugated with the benzene ring are more stable than those that do not.

Conjugated system   **more stable than**   **Nonconjugated system**

Part of the evidence for this comes from acid-catalyzed alcohol dehydrations, which are known to yield the most stable alkene (Section 7.15). For example, dehydration of an alcohol such as the one that follows yields exclusively the conjugated system.

Because conjugation always lowers the energy of an unsaturated system by allowing the π electrons to be delocalized, this behavior is just what we would expect.

15.13B Additions to the Double Bond of Alkenylbenzenes

In the presence of peroxides, hydrogen bromide adds to the double bond of 1-phenylpropene to give 2-bromo-1-phenylpropane as the major product.

1-Phenylpropene   **2-Bromo-1-phenylpropane**

In the absence of peroxides, HBr adds in just the opposite way.

1-Phenylpropene   **1-Bromo-1-phenylpropane**

The addition of hydrogen bromide to 1-phenylpropene proceeds through a benzylic radical in the presence of peroxides, and through a benzylic cation in their absence (cf. Problem 15.17 and Section 9.9).

PROBLEM 15.17

Write mechanisms for the reactions whereby HBr adds to 1-phenylpropene (a) in the presence of peroxides and (b) in the absence of peroxides. In each case account for the regiochemistry of the addition (i.e., explain why the major product is 2-bromo-1-phenylpropane when peroxides are present, and why it is 1-bromo-1-phenylpropane when peroxides are absent.

PROBLEM 15.18

(a) What would you expect to be the major product when 1-phenylpropene reacts with HCl?   (b) When it is subjected to oxymercuration–demercuration?

15.13C Oxidation of the Side Chain

Strong oxidizing agents oxidize toluene to benzoic acid. The oxidation can be carried out by the action of hot alkaline potassium permanganate. This method gives benzoic acid in almost quantitative yield.

Benzoic acid
(~100%)

An important characteristic of side-chain oxidations is that oxidation takes place initially at the benzylic carbon; **alkylbenzenes with alkyl groups longer than methyl are ultimately degraded to benzoic acids.**

An alkylbenzene   **Benzoic acid**

Side-chain oxidations are similar to benzylic halogenations, because in the first step the oxidizing agent abstracts a benzylic hydrogen. Once oxidation is begun at the benzylic carbon, it continues at that site. Ultimately, the oxidizing agent oxidizes the benzylic carbon to a carboxyl group and, in the process, it cleaves off the remaining carbon atoms of the side chain. (*tert*-Butylbenzene is resistant to side-chain oxidation. Why?)

Side-chain oxidation is not restricted to alkyl groups. **Alkenyl, alkynyl, and acyl groups are oxidized by hot alkaline potassium permanganate in the same way.**

15.13D Oxidation of the Benzene Ring

The benzene ring of an alkylbenzene can be converted to a carboxyl group by ozonolysis, followed by treatment with hydrogen peroxide:

$$R-C_6H_5 \xrightarrow[\text{(2) } H_2O_2]{\text{(1) } O_3,\ CH_3CO_2H} R-\overset{\displaystyle O}{\overset{\|}{C}}OH$$

15.14 SYNTHETIC APPLICATIONS

The substitution reactions of aromatic rings and the reactions of the side chains of alkyl- and alkenylbenzenes, when taken together, offer us a powerful set of reactions for organic synthesis. By using these reactions skillfully, we shall be able to synthesize a large number of benzene derivatives.

Part of the skill in planning a synthesis is in deciding the order in which reactions should be carried out. Let us suppose, for example, that we want to synthesize *o*-bromonitrobenzene. We can see very quickly that we should introduce the bromine into the ring first because it is an ortho–para director.

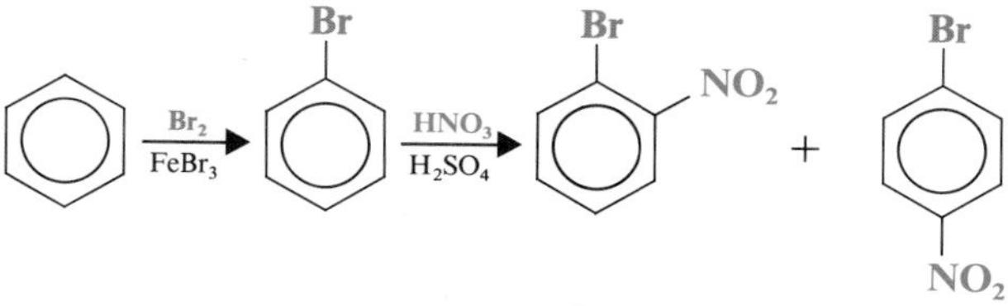

o-Bromonitro-
benzene

p-Bromonitro-
benzene

The ortho and para compounds that we get as products can be separated by various methods. However, had we introduced the nitro group first, we would have obtained *m*-bromonitrobenzene as the major product.

Other examples in which choosing the proper order for the reactions is important are the syntheses of the *ortho-, meta-,* and *para*-nitrobenzoic acids. We can synthesize the *ortho-* and *para*-nitrobenzoic acids from toluene by nitrating it, separating the *ortho-* and *para*-nitrotoluenes, and then oxidizing the methyl groups to carboxyl groups.

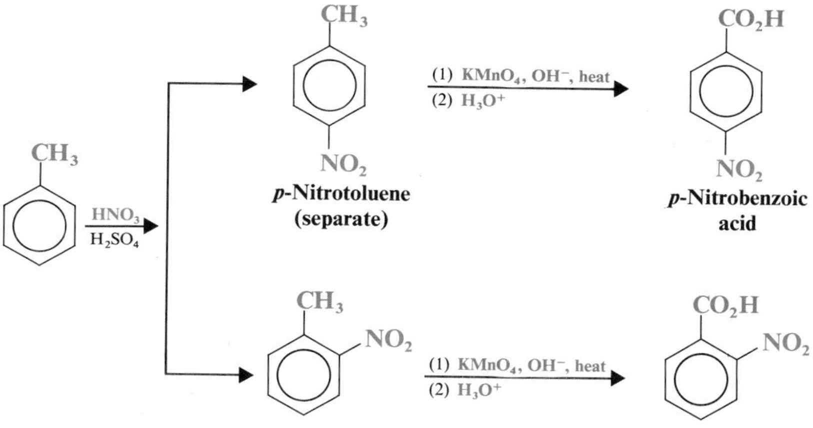

p-Nitrotoluene
(separate)

p-Nitrobenzoic
acid

o-Nitrotoluene

o-Nitrobenzoic acid

We can synthesize *m*-nitrobenzoic acid by reversing the order of the reactions.

Benzoic acid

m-Nitrobenzoic acid

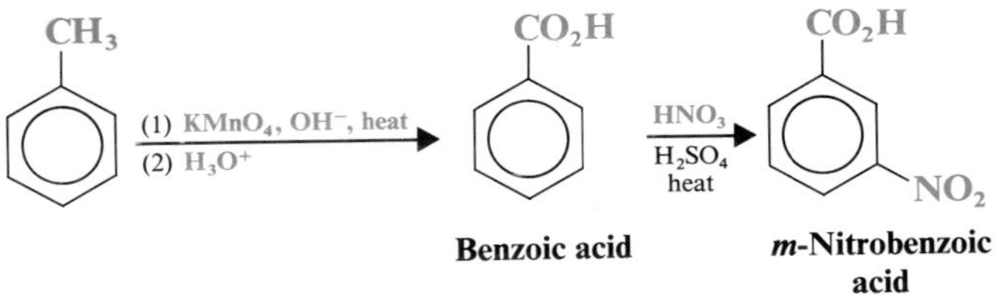

SAMPLE PROBLEM

Starting with toluene, outline a synthesis of (a) 1-bromo-2-(trichloromethyl)-benzene, (b) 1-bromo-3-(trichloromethyl)benzene, and (c) 1-bromo-4-(trichloromethyl)benzene.

ANSWER:

Compounds (a) and (c) can be obtained by ring bromination of toluene followed by chlorination of the side chain using three molar equivalents of chlorine:

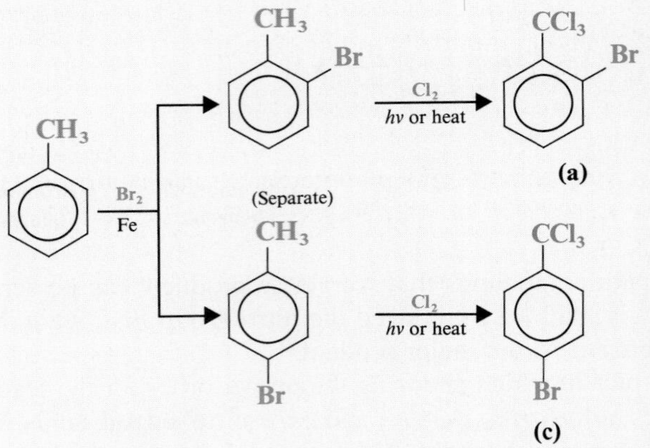

To make compound (b) we reverse the order of the reactions. By converting the side chain to a —CCl_3 group first, we create a meta director, which causes the bromine to enter the desired position.

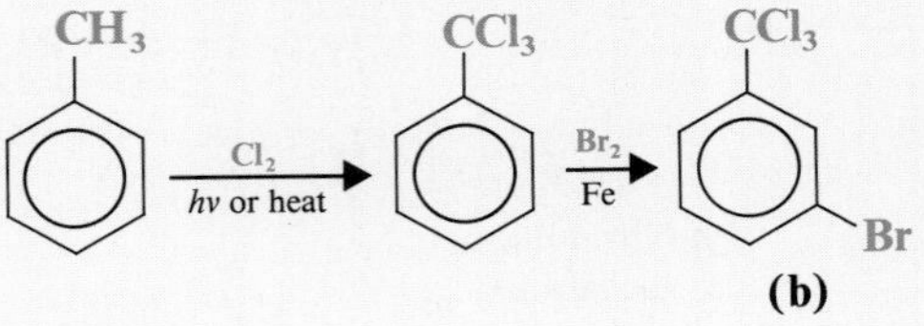

PROBLEM 15.19

Suppose you needed to synthesize *m*-chloroethylbenzene from benzene.

You could begin by chlorinating benzene and then follow with a Friedel–Crafts alkylation using CH_3CH_2Cl and $AlCl_3$, or you could begin with a Friedel–Crafts alkylation followed by chlorination. Neither method will give the desired product, however. (a) Why not? There is a three-step method that will work if the steps are done in the right order. (b) What is this method?

Very powerful activating groups such as amino groups and hydroxyl groups cause the benzene ring to be so reactive that undesirable reactions may take place. Some reagents used for electrophilic substitution reactions, such as nitric acid, are also strong *oxidizing agents.* (Both electrophiles and oxidizing agents seek electrons.) Thus, amino groups and hydroxyl groups not only activate the ring toward electrophilic substitution, they also activate it toward oxidation. Nitration of aniline, for example, results in considerable destruction of the benzene ring because it is oxidized by the nitric acid. Direct nitration of aniline, consequently, is not a satisfactory method for the preparation of *o*- and *p*-nitroaniline.

Treating aniline with acetyl chloride, CH_3COCl, or acetic anhydride, $(CH_3CO)_2O$, converts aniline to acetanilide. The amino group is converted to an acetamido group ($—NHCOCH_3$), a group that is only moderately activating and one that does not make the ring highly susceptible to oxidation (see Problem 15.11). With acetanilide, direct nitration becomes possible.

Nitration of acetanilide gives *p*-nitroacetanilide in excellent yield with only a trace of the ortho isomer. Acidic hydrolysis of *p*-nitroacetanilide (Section 18.8F) removes the acetyl group and gives *p*-nitroaniline, also in good yield.

Suppose, however, that we need *o*-nitroaniline. The synthesis that we just outlined would obviously not be a satisfactory method, for only a trace of *o*-nitroacetanilide is obtained in the nitration reaction. (The acetamido group is purely a para director in many reactions. Bromination of acetanilide, for example, gives *p*-bromoacetanilide almost exclusively.)

We can synthesize *o*-nitroacetanilide, however, through the reactions that follow:

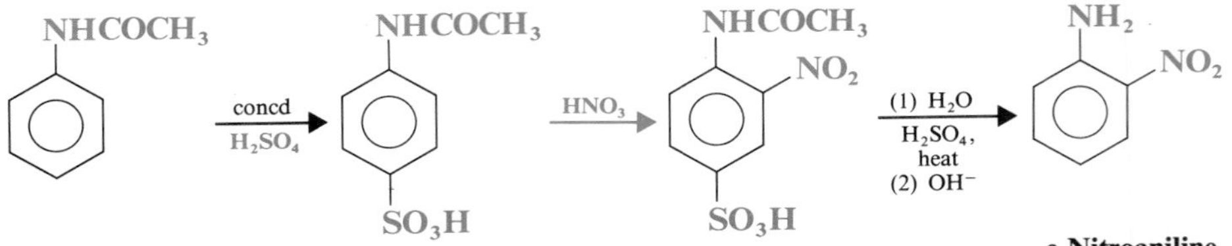

Acetanilide **o-Nitroaniline (56%)**

Here we see how a sulfonic acid group can be used as a "blocking group." We can remove the sulfonic acid group by desulfonation at a later stage. In this example, the reagent used for desulfonation (dilute H_2SO_4) also conveniently removes the acetyl group that we employed to "protect" the benzene ring from oxidation by nitric acid.

15.14A Orientation in Disubstituted Benzenes

When two groups are present on a benzene ring, ***the more powerful activating group*** (Table 15.2) ***generally determines the outcome of the reaction.*** Let us consider, as an example, the orientation of electrophilic substitution of *p*-methylacetanilide. The acetamido group is a much stronger activating group than the methyl group. The following example shows that the acetamido group determines the outcome of the reaction. Substitution occurs primarily at the position ortho to the acetamido group.

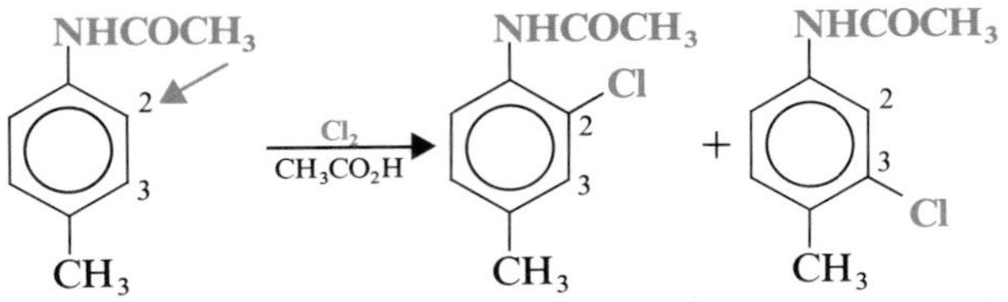

(major product) **(minor product)**

Because all ortho–para- directing groups are more activating than meta directors, ***the ortho–para director determines the orientation of the incoming group.***

Steric effects are also important in aromatic substitutions. ***Substitution does not occur to an appreciable extent between meta substituents if another position is open.*** A good example of this effect can be seen in the nitration of *m*-bromochlorobenzene.

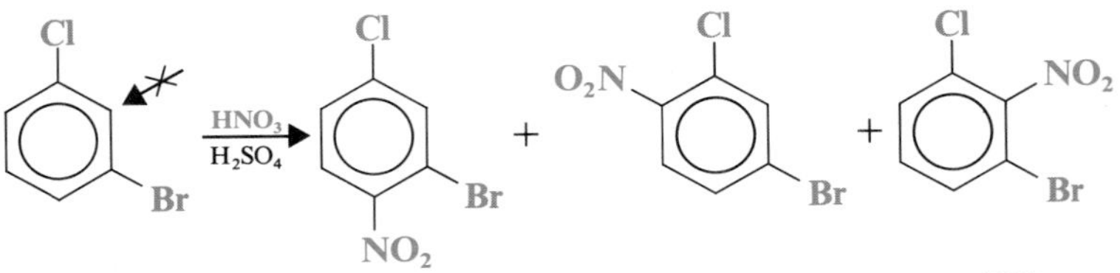

(62%) **(37%)** **(1%)**

Only 1% of the mononitro product has the nitro group between the bromine and chlorine.

PROBLEM 15.20

Predict the major product (or products) that would be obtained when each of the following compounds is nitrated.

(a) OH / CF$_3$ substituted benzene ring

(b) CN / SO$_3$H substituted benzene ring

(c) OCH$_3$ / NO$_2$ substituted benzene ring

15.15 ALLYLIC AND BENZYLIC HALIDES IN NUCLEOPHILIC SUBSTITUTION REACTIONS

Allylic and benzylic halides can be classified in the same way that we have classified other organic halides:

$$-\overset{|}{C}=\overset{|}{C}-CH_2X \qquad -\overset{|}{C}=\overset{|}{C}-\overset{\overset{R}{|}}{C}HX \qquad -\overset{|}{C}=\overset{|}{C}-\overset{\overset{R}{|}}{\underset{\underset{R'}{|}}{C}}X$$

1° Allylic   **2° Allylic**   **3° Allylic**

$$ArCH_2X \qquad Ar\overset{\overset{R}{|}}{C}HX \qquad Ar\overset{\overset{R}{|}}{\underset{\underset{R'}{|}}{C}}X$$

1° Benzylic   **2° Benzylic**   **3° Benzylic**

All of these compounds undergo nucleophilic substitution reactions. As with other tertiary halides (Section 6.15A), the steric hindrance associated with having three bulky groups on the carbon bearing the halogen prevents tertiary allylic and tertiary benzylic halides from reacting by an S_N2 mechanism. They react with nucleophiles only by an S_N1 mechanism.

Primary and secondary allylic and benzylic halides can react either by an S_N2 mechanism or by an S_N1 mechanism in ordinary nonacidic solvents. We would expect these halides to react by an S_N2 mechanism because they are structurally similar to primary and secondary alkyl halides. (Having only one or two groups attached to the carbon bearing the halogen does not prevent S_N2 attack.) But primary and secondary allylic and benzylic halides can also react by an S_N1 mechanism because they can form relatively stable carbocations and in this regard they differ from primary and secondary alkyl halides.*

---

*There is some dispute as to whether 2° alkyl halides react by an S_N1 mechanism to any appreciable extent in ordinary nonacidic solvents such as mixtures of water and alcohol or acetone, but it is clear that react by an S_N2 mechanism is, for all practical purposes, the more important pathway.

Overall we can summarize the effect of structure on the reactivity of alkyl, allylic, and benzylic halides in the ways shown in Table 15.3.

SAMPLE PROBLEM

When either enantiomer of 3-chloro-1-butene [(*R*) or (*S*)] is subjected to hydrolysis, the products of the reaction are optically inactive. Explain these results.

ANSWER:

The solvolysis reaction is S_N1. The intermediate allylic cation is achiral and therefore reacts with water to give the enantiomeric 3-buten-2-ols in equal amounts and to give some of the achiral 2-buten-1-ol.

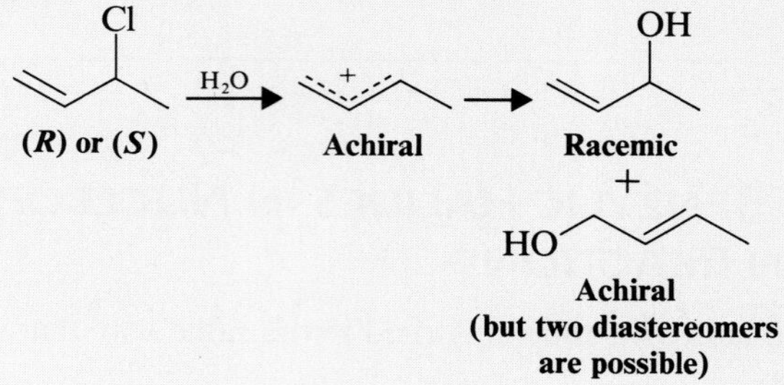

PROBLEM 15.21

Account for the following observations: (a) When 1-chloro-2-butene is allowed to react with a relatively concentrated solution of sodium ethoxide in ethanol, the reaction rate depends on the concentration of the allylic halide and on the con-

TABLE 15.3 A summary of alkyl, allylic, and benzylic halides in S_N reactions

These halides give mainly S_N2 reactions	These halides give mainly S_N1 reactions

CH_3—X   R—CH_2—X   R—CH—X
                                       |
                                       R′	

These halides may give either S_N1 or S_N2 reactions	R'—C—X (with R above, R″ below)

Ar—CH_2—X   Ar—CH—X
                              |
                              R
	Ar—C—X (with R above, R′ below)

—C=CCH₂—X   —C=CCH—X
                                    |
                                    R
	—C=C—C—X (with R above, R′ below)

centration of ethoxide ion. The product of the reaction is almost exclusively $CH_3CH=CHCH_2OCH_2CH_3$. (b) When 1-chloro-2-butene is allowed to react with very dilute solutions of sodium ethoxide in ethanol (or with ethanol alone), the reaction rate is independent of the concentration of ethoxide ion; it depends only on the concentration of the allylic halide. Under these conditions the reaction produces a mixture of $CH_3CH=CHCH_2OCH_2CH_3$ and $CH_3\underset{\underset{\displaystyle OCH_2CH_3}{|}}{C}HCH=CH_2$. (c) In the presence of traces of water 1-chloro-2-bu-
tene is slowly converted to a mixture of 1-chloro-2-butene and 3-chloro-1-bu-
tene.

PROBLEM 15.22

1-Chloro-3-methyl-2-butene undergoes hydrolysis in a mixture of water and dioxane at a rate that is more than a thousand times that of 1-chloro-2-butene. (a) What factor accounts for the difference in reactivity? (b) What products would you expect to obtain? [Dioxane is a cyclic ether (below) that is miscible with water in all proportions and is a useful cosolvent for conducting reactions like these. Dioxane is carcinogenic (i.e., cancer causing), however, and like most ethers, it tends to form peroxides.]

Dioxane

PROBLEM 15.23

Primary halides of the type $ROCH_2X$ apparently undergo S_N1 type reactions, whereas most primary halides do not. Can you propose a resonance explanation for the ability of halides of the type $ROCH_2X$ to undergo S_N1 reactions?

PROBLEM 15.24

The following chlorides undergo solvolysis in ethanol at the relative rates given in parentheses. How can you explain these results?

$C_6H_5CH_2Cl$	$C_6H_5\underset{\underset{\displaystyle Cl}{|}}{C}HCH_3$	$(C_6H_5)_2CHCl$	$(C_6H_5)_3CCl$
(0.08)	**(1)**	**(300)**	**(3 × 10⁶)**

15.16 REDUCTION OF AROMATIC COMPOUNDS: THE BIRCH REDUCTION

Hydrogenation of benzene under pressure using a metal catalyst such as nickel results in the addition of three molar equivalents of hydrogen and the formation of cyclohexane (Section 14.3). The intermediate cyclohexadienes and cyclohexene cannot be isolated because these undergo catalytic hydrogenation faster than benzene does.

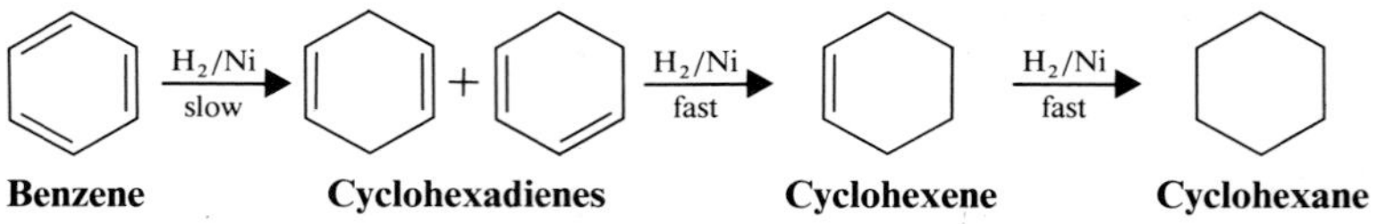

Benzene   **Cyclohexadienes**   **Cyclohexene**   **Cyclohexane**

15.16A The Birch Reduction

Benzene can be reduced to 1,4-cyclohexadiene by treating it with an alkali metal (sodium, lithium, or potassium) in a mixture of liquid ammonia and an alcohol.

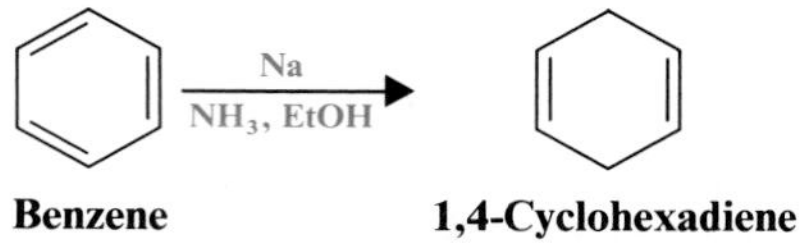

Benzene   **1,4-Cyclohexadiene**

This is another dissolving metal reduction and the mechanism for it resembles the mechanism for the reduction of alkynes that we studied in Section 7.7B. A sequence of electron transfers from the alkali metal and proton transfers from the alcohol takes place. (See the following reaction sequence.) The first electron transfer produces a delocalized benzene radical anion. Protonation produces a cyclohexadienyl radical (also a delocalized species). Transfer of another electron leads to the formation of a delocalized cyclohexadienyl anion and protonation of this produces the 1,4-cyclohexadiene.

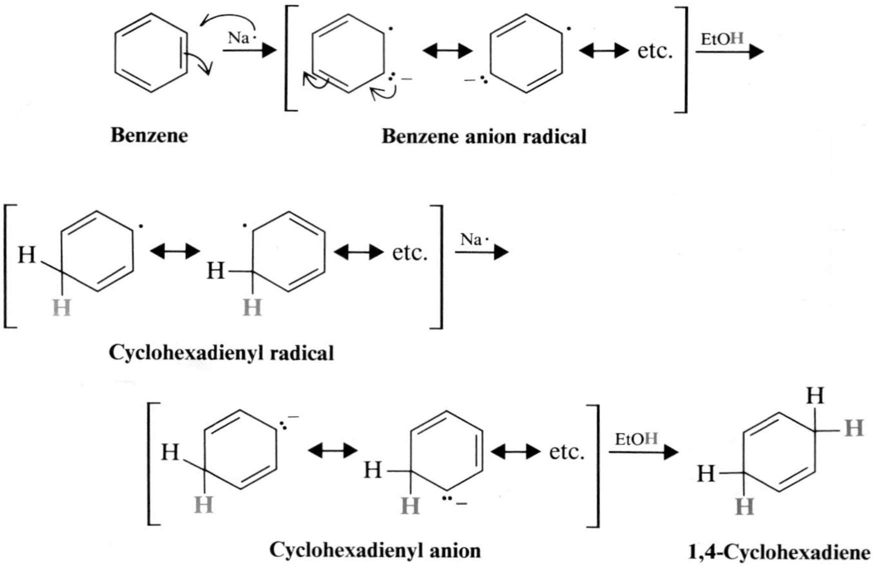

Benzene   **Benzene anion radical**

Cyclohexadienyl radical

Cyclohexadienyl anion   **1,4-Cyclohexadiene**

Formation of a 1,4-cyclohexadiene in a reaction of this type is quite general, but the reason for its formation in preference to the more stable conjugated 1,3-cyclohexadiene is not understood.

Dissolving metal reductions of this type were developed by the Australian chemist A. J. Birch and have come to be known as **Birch reductions.**

Substituent groups on the benzene ring influence the course of the reaction. Birch reduction of methoxybenzene (anisole) leads to the formation of 1-methoxy-1,4-cyclohexadiene, a compound that can be hydrolyzed by dilute acid to 2-cyclohexenone. This method provides a useful synthesis of 2-cyclohexenones.

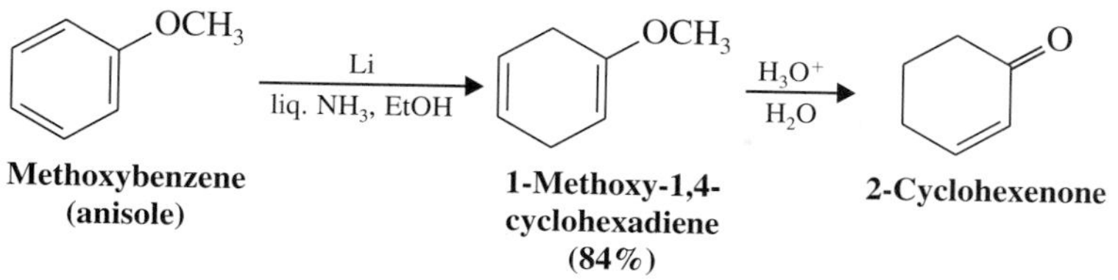

| Methoxybenzene (anisole) | 1-Methoxy-1,4-cyclohexadiene (84%) | 2-Cyclohexenone |

PROBLEM 15.25

Birch reduction of toluene leads to a product with the molecular formula C_7H_{10}. On ozonolysis followed by reduction with zinc and water, the product is transformed into CH_3COCH_2CHO and $OHCCH_2CHO$. What is the structure of the Birch reduction product?

ADDITIONAL PROBLEMS

15.26 Give the major product (or products) that would be obtained when each of the following compounds is subjected to ring chlorination with Cl_2 and $FeCl_3$.

(a) Ethylbenzene

(b) Anisole ($C_6H_5OCH_3$)

(c) Fluorobenzene

(d) Benzoic acid

(e) Nitrobenzene

(f) Chlorobenzene

(g) Biphenyl (C_6H_5—C_6H_5)

(h) Ethyl phenyl ether

15.27 Predict the major product (or products) formed when each of the following compounds is subjected to ring nitration.

(a) Acetanilide ($C_6H_5NHCOCH_3$)

(b) Phenyl acetate ($CH_3CO_2C_6H_5$)

(c) 4-Chlorobenzoic acid

(d) 3-Chlorobenzoic acid

(e) $C_6H_5COC_6H_5$

15.28 Give the structures of the major products of the following reactions:

(a) Styrene + HCl $\longrightarrow$

(b) 2-Bromo-1-phenylpropane + $C_2H_5ONa \longrightarrow$

(c) $C_6H_5CH_2CHOHCH_2CH_3 \xrightarrow{\text{H}^+, \text{ heat}}$

(d) Product of (c) + HBr $\xrightarrow{\text{peroxides}}$

(e) Product of (c) + $H_2O \xrightarrow[\text{heat}]{\text{H}^+}$

(f) Product of (c) + H_2 (1 molar equivalent) $\xrightarrow[25°C]{\text{Pt}}$

(g) Product of (f) $\xrightarrow{\substack{(1)\ \text{KMnO}_4,\ \text{OH}^-,\ \text{heat} \\ (2)\ \text{H}_3\text{O}+}}$

15.29   Starting with benzene, outline a synthesis of each of the following:

(a) Isopropylbenzene

(b) *tert*-Butylbenzene

(c) Propylbenzene

(d) Butylbenzene

(e) 1-*tert*-Butyl-4-chlorobenzene

(f) 1-Phenylcyclopentene

(g) *trans*-2-Phenylcyclopentanol

(h) *m*-Dinitrobenzene

(i) *m*-Bromonitrobenzene

(j) *p*-Bromonitrobenzene

(k) *p*-Chlorobenzenesulfonic acid

(l) *o*-Chloronitrobenzene

(m) *m*-Nitrobenzenesulfonic acid

15.30   Starting with styrene, outline a synthesis of each of the following:

(a) $C_6H_5CHClCH_2Cl$

(b) $C_6H_5CH_2CH_3$

(c) $C_6H_5CHOHCH_2OH$

(d) $C_6H_5CO_2H$

(e) $C_6H_5CHOHCH_3$

(f) $C_6H_5CHBrCH_3$

(g) $C_6H_5CH_2CH_2OH$

(h) $C_6H_5CH_2CH_2D$

(i) $C_6H_5CH_2CH_2Br$

(j) $C_6H_5CH_2CH_2I$

(k) $C_6H_5CH_2CH_2CN$

(l) $C_6H_5CHDCH_2D$

(m) Cyclohexylbenzene

(n) $C_6H_5CH_2CH_2OCH_3$

15.31   Starting with toluene, outline a synthesis of each of the following:

(a) *m*-Chlorobenzoic acid

(b) *p*-Acetyltoluene

(c) 2-Bromo-4-nitrotoluene

(d) *p*-Bromobenzoic acid

(e) 1-Chloro-3-(trichloromethyl)benzene

(f) *p*-Isopropyltoluene

(g) 1-Cyclohexyl-4-methylbenzene

(h) 2,4,6-Trinitrotoluene (TNT)

(i) 4-Chloro-2-nitrobenzoic acid

(j) 1-Butyl-4-methylbenzene

15.32   Starting with aniline, outline a synthesis of each of the following:

(a) *p*-Bromoaniline

(b) *o*-Bromoaniline

(c) 2-Bromo-4-nitroaniline

(d) 4-Bromo-2-nitroaniline

(e) 2,4,6-Tribromoaniline

15.33   Both of the following syntheses will fail. Explain what is wrong with each one.

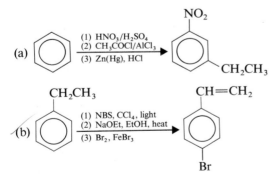

15.34   One ring of phenyl benzoate undergoes electrophilic aromatic substitution much more readily than the other.   (a) Which one is it?   (b) Explain your answer.

Phenyl benzoate

15.35 What product (or products) would you expect to obtain when the following compounds undergo ring bromination with Br_2 and $FeBr_3$?

(a) [structure: phenyl–CH_2–C(=O)–phenyl]

(c) [structure: phenyl–C(=O)–C(=O)–phenyl]

(b) [structure: phenyl–NH–C(=O)–phenyl]

15.36 Many polycyclic aromatic compounds have been synthesized by a cyclization reaction known as the **Bradsher reaction** or **aromatic cyclodehydration.** This method can be illustrated by the following synthesis of 9-methylphenanthrene.

[reaction scheme: biphenyl derivative with CH_2 and $C(=O)-CH_3$ groups, reacting with HBr / acetic acid / heat to give 9-methylphenanthrene with CH_3 group]

9-Methylphenanthrene

An arenium ion is an intermediate in this reaction, and the last step involves the dehydration of an alcohol. Propose a plausible mechanism for this example of the Bradsher reaction.

15.37 Propose structures for compounds **G–I.**

[reaction scheme: resorcinol (benzene with two OH groups) → concd H_2SO_4, 60-65°C → **G** $(C_6H_6S_2O_8)$ → concd HNO_3, concd H_2SO_4 → **H** $(C_6H_5NS_2O_{10})$ → H_3O^+, H_2O, heat → **I** $(C_6H_5NO_4)$]

15.38 2,6-Dichlorophenol has been isolated from the females of two species of ticks (*Amblyomma americanum* and *A. maculatum*), where it apparently serves as a sex attractant. Each female tick yields about 5 ng of 2,6-dichlorophenol. Assume that you need larger quantities than this, and outline a synthesis of 2,6-dichlorophenol from phenol. (*Hint:* When phenol is sulfonated at 100°C, the product is chiefly *p*-hydroxybenzenesulfonic acid.)

15.39 The addition of a hydrogen halide (hydrogen bromide or hydrogen chloride) to 1-phenyl-1,3-butadiene produces (only) 1-phenyl-3-halo-1-butene. (a) Write a mechanism that accounts for the formation of this product. (b) Is this 1,4 addition or 1,2 addition to the butadiene system? (c) Is the product of the reaction consistent with the formation of the most stable intermediate carbocation? (d) Does the reaction appear to be under kinetic control or equilibrium control? Explain.

15.40 2-Methylnaphthalene can be synthesized from *p*-toluene through the following sequence of reactions. Write the structure of each intermediate.

Toluene + succinic anhydride $\xrightarrow{\text{AlCl}_3}$ **A** $\underset{(C_{11}H_{12}O_3)}{}$ $\xrightarrow[\text{HCl}]{\text{Zn(Hg)}}$ **B** $\underset{(C_{11}H_{12}O_2)}{}$

$\xrightarrow{\text{SOCl}_2}$ **C** $\underset{(C_{11}H_{13}ClO)}{}$ $\xrightarrow{\text{AlCl}_3}$ **D** $\underset{(C_{11}H_{12}O)}{}$ $\xrightarrow{\text{NaBH}_4}$ **E** $\underset{(C_{11}H_{14}O)}{}$

$\xrightarrow[\text{heat}]{\text{H}^+}$ **F** $\underset{(C_{11}H_{12})}{}$ $\xrightarrow[\text{CCl}_4,\text{ light}]{\text{NBS}}$ **G** $\underset{(C_{11}H_{11}Br)}{}$ $\xrightarrow[\substack{\text{EtOH} \\ \text{heat}}]{\text{NaOEt}}$ 2-Methylnaphthalene

15.41   Ring nitration of a dimethylbenzene (a xylene) results in the formation of only one nitrodimethylbenzene. What is the structure of the dimethylbenzene?

15.42   Write mechanisms that account for the products of the following reactions:

(a) 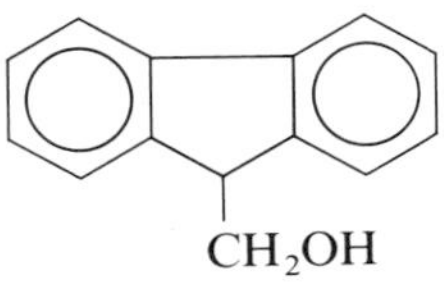 $\xrightarrow[(-\text{H}_2\text{O})]{\text{H}^+}$ phenanthrene   (b) $2\ \text{CH}_3-\overset{\displaystyle C_6H_5}{\underset{\displaystyle C_6H_5}{C}}=CH_2 \xrightarrow{\text{H}^+}$

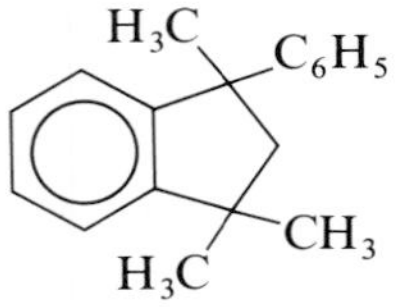

15.43   Show how you might synthesize each of the following starting with α-tetralone (Section 15.9).

(a) 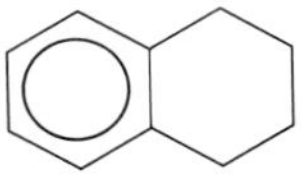   (b) 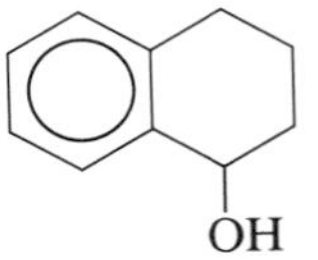   (c) 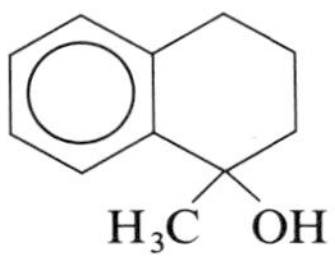   (d)

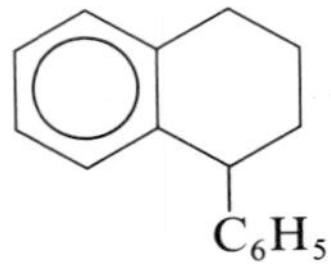

15.44   The compound phenylbenzene ($C_6H_5-C_6H_5$) is called *biphenyl* and the rings are numbered in the following manner.

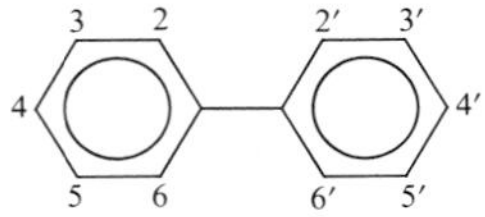

Use models to answer the following questions about substituted biphenyls.   (a) When certain large groups occupy three or four of the *ortho* positions (e.g., 2, 6, 2′, and 6′), the substituted biphenyl may exist in enantiomeric forms. An example of a biphenyl that exists in enantiomeric forms is the compound in which the following substituents are present. 2-NO_2, 6-CO_2H, 2′-NO_2, 6′-CO_2H. What factors account for this?   (b) Would you expect a biphenyl with 2-Br, 6-CO_2H, 2′-CO_2H, 6′-H to exist in enantiomeric forms?   (c) The biphenyl with 2-NO_2, 6-NO_2, 2′-CO_2H, 6′-Br cannot be resolved into enantiomeric forms. Explain.

15.45   Give structures (including stereochemistry where appropriate) for compounds **A–G**.

(a) Benzene + $CH_3CH_2\overset{\displaystyle O}{\overset{\displaystyle \|}{C}}Cl \xrightarrow{\text{AlCl}_3}$ **A** $\xrightarrow[0°C]{\text{PCl}_5}$ **B** $(C_9H_{10}Cl_2)$ $\xrightarrow[\substack{\text{mineral oil,} \\ \text{heat}}]{2\ \text{NaNH}_2}$

C (C_9H_8) $\xrightarrow{\text{H}_2,\ \text{Ni}_2\text{B (P-2)}}$ **D** (C_9H_{10})

Hint: The ^{1}H NMR spectrum of compound **C** consists of a multiplet at δ 7.20 (5H) and a singlet at δ 2.0 (3H).

(b) **C** $\xrightarrow[(2)\ \text{H}_2\text{O}]{(1)\ \text{Li, liq. NH}_3}$ **E** (C_9H_{10})

(c) **D** $\xrightarrow[2-5°C]{\text{Br}_2,\ \text{CCl}_4}$ **F** + enantiomer (major products)

(d) **E** $\xrightarrow[2-5°C]{\text{Br}_2,\ \text{CCl}_4}$ **G** + enantiomer (major products)

15.46 Treating cyclohexene with acetyl chloride and AlCl$_3$ leads to the formation of a product with the molecular formula C$_8$H$_{13}$ClO. Treating this product with a base leads to the formation of 1-acetylcyclohexene. Propose mechanisms for both steps of this sequence of reactions.

15.47 The *tert*-butyl group can be used as a blocking group in certain syntheses of aromatic compounds. (a) How would you introduce a *tert*-butyl group and (b) how would you remove it? (c) What advantage might a *tert*-butyl group have over a —SO$_3$H group as a blocking group.

15.48 When toluene is sulfonated (concentrated H$_2$SO$_4$) at room temperature, predominant (about 95% of the total) ortho and para substitution occurs. If elevated temperatures (150–200°C) and longer reaction times are employed, meta (chiefly) and para substitution account for some 95% of the products. Account for these differences. (Hint: *m*-Toluenesulfonic acid is the most stable isomer.)

15.49 A C—D bond is harder to break than a C—H bond and, consequently, reactions in which C—D bonds are broken proceed more slowly than reactions in which C—H bonds are broken. What mechanistic information comes from the observation that perdeuterated benzene, C$_6$D$_6$, is nitrated at the same rate as normal benzene, C$_6$H$_6$?

15.50 Show how you might synthesize each of the following compounds starting with either benzyl bromide or allyl bromide.

(a) C$_6$H$_5$CH$_2$CN

(b) C$_6$H$_5$CH$_2$OCH$_3$

(c) C$_6$H$_5$CH$_2$O$_2$CCH$_3$

(d) C$_6$H$_5$CH$_2$I

(e) CH$_2$=CHCH$_2$N$_3$

(f) CH$_2$=CHCH$_2$OCH$_2$CH(CH$_3$)$_2$

15.51 Provide structures for compounds **A, B,** and **C.**

Benzene $\xrightarrow[\text{liq. NH}_3,\ \text{EtOH}]{\text{Na}}$ **A** (C$_6$H$_8$) $\xrightarrow[\text{CCl}_4]{\text{NBS}}$ **B** (C$_6$H$_7$Br) $\xrightarrow{(\text{CH}_3)_2\text{CuLi}}$ **C** (C$_7$H$_{10}$)

15.52 Heating 1,1,1-triphenylmethanol with ethanol containing a trace of a strong acid causes the formation of 1-ethoxy-1,1,1-triphenylmethane. Write a plausible mechanism that accounts for the formation of this product.

15.53 (a) Which of the following halides would you expect to be most reactive in an S$_N$2 reaction? (b) In an S$_N$1 reaction? Explain your answers.

CH$_3$CH$_2$CH=CHCH$_2$Br   CH$_3$CH=CHCHBrCH$_3$ CH$_2$=CHCBr(CH$_3$)$_2$

15.54 Acetanilide was subjected to the following sequence of reactions: (1) conc. H$_2$SO$_4$; (2) HNO$_3$, heat; (3) H$_2$O, H$_2$SO$_4$, heat, then OH$^-$. The ^{13}C NMR spectrum of the final product gives 6 signals. Write the structure of the final product.

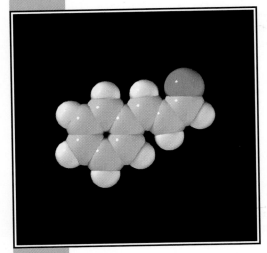

CHAPTER 16

Cinnamaldehyde
(Section 16.3)

ALDEHYDES AND KETONES I. NUCLEOPHILIC ADDITIONS TO THE CARBONYL GROUP

16.1 INTRODUCTION

Except for formaldehyde, the simplest aldehyde, all aldehydes have a carbonyl group,

$$\overset{O}{\overset{\|}{-C-}}$$

, bonded on one side to a carbon, and on the other side to a hydrogen. In ketones, the carbonyl group is situated between two carbon atoms.

$$\overset{O}{\overset{\|}{H-C-H}}$$ **Formaldehyde**

$$\overset{O}{\overset{\|}{R-C-H}}$$ **General formula for an aldehyde**

$$\overset{O}{\overset{\|}{R-C-R'}}$$ **General formula for a ketone**

Although earlier chapters have given us some insight into the chemistry of carbonyl compounds, we shall now consider their chemistry in detail. The reason: The chemistry of the carbonyl group is central to the chemistry of most of the chapters that follow.

In this chapter our attention is focused on the preparation of aldehydes and ketones, on their physical properties, and especially on *the nucleophilic addition reactions that take place at their carbonyl groups*. In Chapter 17 we shall study the chemistry of aldehydes and ketones *that results from the acidity of the hydrogen atoms on the carbon atoms adjacent to their carbonyl groups*.

16.2 NOMENCLATURE OF ALDEHYDES AND KETONES

In the IUPAC system aliphatic aldehydes are named *substitutively* by replacing the final **-e** of the name of the corresponding alkane with **-al**. Since the aldehyde group must be at the end of the chain of carbon atoms, there is no need to indicate its position. When other substituents are present, however, the carbonyl group carbon is assigned position 1. Many aldehydes also have common names; these are given here in parentheses. These common names are derived from the common names for the corresponding carboxylic acids (Section 18.2A) and some of them are retained by the IUPAC as acceptable names.

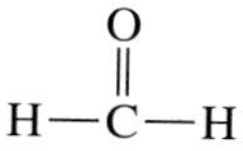

$$\underset{\substack{\textbf{Methanal} \\ \textbf{(formaldehyde)}}}{\text{H}-\overset{\displaystyle O}{\overset{\|}{\text{C}}}-\text{H}} \qquad \underset{\substack{\textbf{Ethanal} \\ \textbf{(acetaldehyde)}}}{\text{CH}_3\overset{\displaystyle O}{\overset{\|}{\text{C}}}-\text{H}} \qquad \underset{\substack{\textbf{Propanal} \\ \textbf{(propionaldehyde)}}}{\text{CH}_3\text{CH}_2\overset{\displaystyle O}{\overset{\|}{\text{C}}}-\text{H}}$$

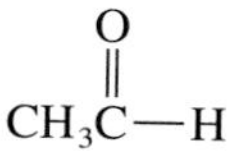

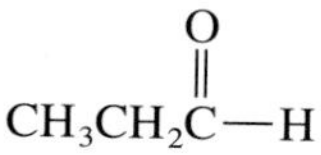

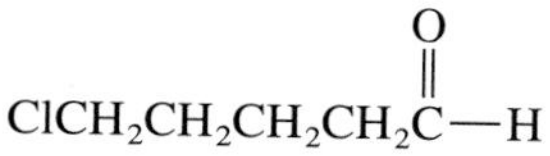

$$\underset{\textbf{5-Chloropentanal}}{\text{ClCH}_2\text{CH}_2\text{CH}_2\text{CH}_2\overset{\displaystyle O}{\overset{\|}{\text{C}}}-\text{H}} \qquad \underset{\substack{\textbf{Phenylethanal} \\ \textbf{(phenylacetaldehyde)}}}{\text{C}_6\text{H}_5\text{CH}_2\overset{\displaystyle O}{\overset{\|}{\text{C}}}-\text{H}}$$

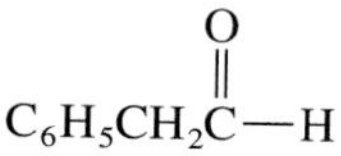

Aldehydes in which the —CHO group is attached to a ring system are named substitutively by adding the suffix *carbaldehyde*. Several examples follow:

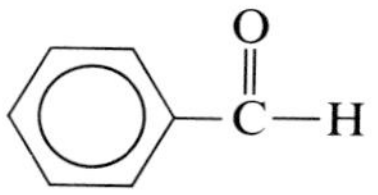

Benzenecarbaldehyde   **Cyclohexanecarbaldehyde** 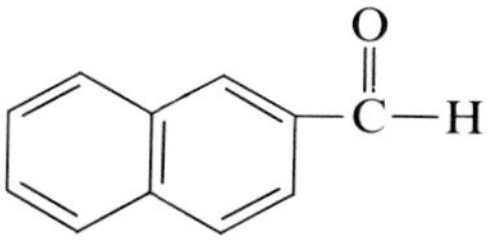   **2-Naphthalenecarbaldehyde**
(benzaldehyde)

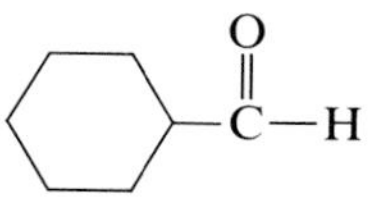

The common name *benzaldehyde* is far more frequently used than benzenecarbaldehyde for C_6H_5CHO, and it is the name we shall use in this text.

Aliphatic ketones are named substitutively by replacing the final **-e** of the name of the corresponding alkane with **-one**. The chain is then numbered in the way that gives the carbonyl carbon atom the lower possible number, and this number is used to designate its position.

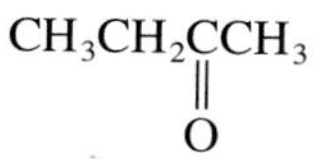

$$\underset{\substack{\textbf{Butanone} \\ \textbf{(ethyl methyl ketone)}}}{\text{CH}_3\text{CH}_2\underset{\displaystyle O}{\overset{\|}{\text{C}}}\text{CH}_3} \qquad \underset{\substack{\textbf{2-Pentanone} \\ \textbf{(methyl propyl ketone)}}}{\text{CH}_3\overset{\displaystyle O}{\overset{\|}{\text{C}}}\text{CH}_2\text{CH}_2\text{CH}_3} \qquad \underset{\substack{\textbf{4-Penten-2-one} \\ \textbf{(}not\textbf{ 1-penten-4-one)} \\ \textbf{(allylmethyl ketone)}}}{\text{CH}_3\overset{\displaystyle O}{\overset{\|}{\text{C}}}\text{CH}_2\text{CH}=\text{CH}_2}$$

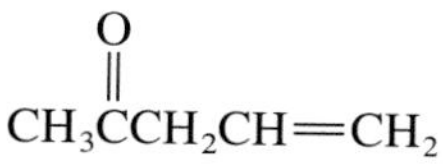

Common radicofunctional names for ketones (in parentheses above) are obtained simply by separately naming the two groups attached to the carbonyl group and adding the word **ketone** as a separate word.

Some ketones have common names that are retained in the IUPAC system.

$$CH_3\overset{\overset{\displaystyle O}{\|}}{C}CH_3$$

Acetone
(propanone or
dimethyl ketone)

Acetophenone
(1-phenylethanone or
methyl phenyl ketone)

Benzophenone
(diphenylmethanone or
diphenyl ketone)

When it is necessary to name the $-\overset{\overset{\displaystyle O}{\|}}{C}H$ group as a prefix, it is the **methanoyl** or

formyl group. The $CH_3\overset{\overset{\displaystyle O}{\|}}{C}$—group is called the **ethanoyl** or **acetyl group.** When

$R\overset{\overset{\displaystyle O}{\|}}{C}$— groups are named as substituents, they are called **alkanoyl** or **acyl groups.**

2-Methanoylbenzoic acid
(*o*-formylbenzoic acid)

4-Ethanoylbenzenesulfonic acid
(*p*-acetylbenzenesulfonic acid)

PROBLEM 16.1

(a) Give IUPAC substitutive names for the seven isomeric aldehydes and ketones with the formula $C_5H_{10}O$. (b) Give structures and names (common or IUPAC substitutive names) for all the aldehydes and ketones that contain a benzene ring and have the formula C_8H_8O.

16.3 PHYSICAL PROPERTIES

The carbonyl group is a polar group; therefore aldehydes and ketones have higher boiling points than hydrocarbons of the same molecular weight. However, since aldehydes and ketones cannot have strong hydrogen bonds *between their molecules*, they have lower boiling points than the corresponding alcohols.

$CH_3CH_2CH_2CH_3$
Butane
bp, −0.5°C

$CH_3CH_2\overset{\overset{\displaystyle O}{\|}}{C}H$
Propanal
bp, 49°C

$CH_3\overset{\overset{\displaystyle O}{\|}}{C}CH_3$
Acetone
bp, 56.1°C

$CH_3CH_2CH_2OH$
1-Propanol
bp, 97.2°C

PROBLEM 16.2

Which compound in each of the following pairs listed has the higher boiling point? (Answer this problem without consulting tables.)

(a) Pentanal or 1-pentanol

(d) Acetophenone or 2-phenylethanol

(b) 2-Pentanone or 2-pentanol

(e) Benzaldehyde or benzyl alcohol

(c) Pentane or pentanal

The carbonyl oxygen atom allows molecules of aldehydes and ketones to form strong hydrogen bonds to molecules of water. As a result, low molecular weight aldehydes and ketones show appreciable solubilities in water. Acetone and acetaldehyde are soluble in water in all proportions.

Table 16.1 lists the physical properties of a number of common aldehydes and ketones.

Some aromatic aldehydes obtained from natural sources have very pleasant fragrances. Some of these are the following:

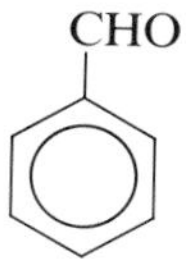

Benzaldehyde
(from bitter almonds)

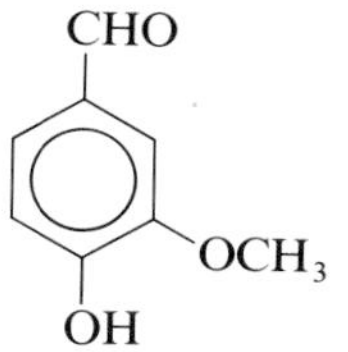

Vanillin
(from vanilla beans)

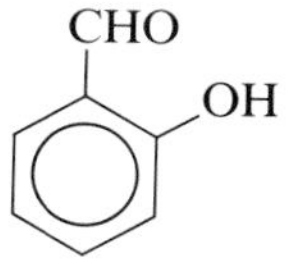

Salicylaldehyde
(from meadowsweet)

TABLE 16.1 Physical properties of aldehydes and ketones

FORMULA	NAME	mp (°C)	bp (°C)	SOLUBILITY IN WATER
HCHO	Formaldehyde	−92	−21	Very soluble
CH_3CHO	Acetaldehyde	−125	21	∞
CH_3CH_2CHO	Propanal	−81	49	Very soluble
$CH_3(CH_2)_2CHO$	Butanal	−99	76	Soluble
$CH_3(CH_2)_3CHO$	Pentanal	−91.5	102	Sl. soluble
$CH_3(CH_2)_4CHO$	Hexanal	−51	131	Sl. soluble
C_6H_5CHO	Benzaldehyde	−26	178	Sl. soluble
$C_6H_5CH_2CHO$	Phenylacetaldehyde	33	193	Sl. soluble
CH_3COCH_3	Acetone	−95	56.1	∞
$CH_3COCH_2CH_3$	Butanone	−86	79.6	Very soluble
$CH_3COCH_2CH_2CH_3$	2-Pentanone	−78	102	Soluble
$CH_3CH_2COCH_2CH_3$	3-Pentanone	−39	102	Soluble
$C_6H_5COCH_3$	Acetophenone	21	202	Insoluble
$C_6H_5COC_6H_5$	Benzophenone	48	306	Insoluble

Cinnamaldehyde
(from cinnamon)

Piperonal
(made from safrole;
odor of heliotrope)

16.4 SYNTHESIS OF ALDEHYDES

16.4A Aldehydes by Oxidation of 1° Alcohols

We learned in Section 11.4A that the oxidation state of aldehydes lies between that of 1° alcohols and carboxylic acids and that aldehydes can be prepared from 1° alcohols by oxidation with pyridinium chlorochromate (PCC):

$$R-CH_2OH \underset{[H]}{\overset{[O]}{\rightleftarrows}} R-\overset{O}{\overset{\|}{C}}-H \underset{[H]}{\overset{[O]}{\rightleftarrows}} R-\overset{O}{\overset{\|}{C}}-OH$$

1° Alcohol **Aldehyde** **Carboxylic acid**

$$R-CH_2OH \xrightarrow[CH_2Cl_2]{C_5H_5NH^+CrO_3Cl^- \ (PCC)} R-\overset{O}{\overset{\|}{C}}-H$$

1° Alcohol **Aldehyde**

An example of this synthesis of aldehydes is the oxidation of 1-heptanol to heptanal:

$$CH_3(CH_2)_5CH_2OH \xrightarrow[CH_2Cl_2]{C_5H_5NH^+CrO_3Cl^- \ (PCC)} CH_3(CH_2)_5CHO$$

1-Heptanol **Heptanal**
(93%)

16.4B Aldehydes by Reduction of Acyl Chlorides, Esters, and Nitriles

Theoretically, it ought to be possible to prepare aldehydes by reduction of carboxylic acids. In practice, this is not possible, because the reagent normally used to reduce a carboxylic acid directly is lithium aluminum hydride ($LiAlH_4$ or LAH) and when any carboxylic acid is treated with LAH, it is reduced all the way to the 1° alcohol. This happens because LAH is a very powerful reducing agent and aldehydes are very easily reduced. Any aldehyde that might be formed in the reaction mixture is immediately reduced by the LAH to the 1° alcohol. (It does not help to use a stoichiometric amount of LAH, because as soon as the first few molecules of aldehyde are formed in the mixture, there will still be much unreacted LAH present and it will reduce the aldehyde.)

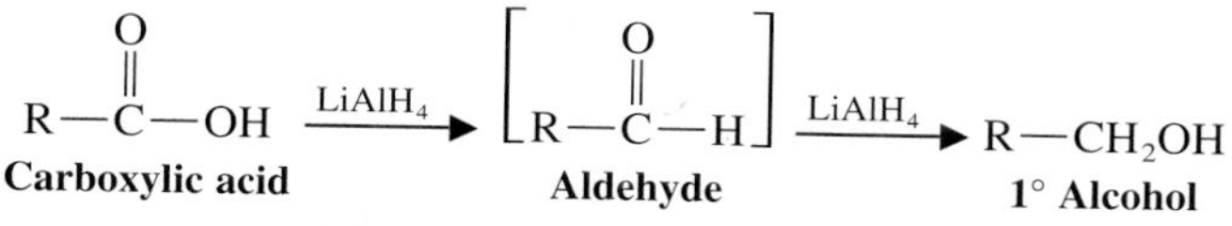

$$R-\overset{\displaystyle O}{\overset{\|}{C}}-OH \xrightarrow{\text{LiAlH}_4} \left[R-\overset{\displaystyle O}{\overset{\|}{C}}-H\right] \xrightarrow{\text{LiAlH}_4} R-CH_2OH$$

Carboxylic acid   **Aldehyde**   **1° Alcohol**

The secret to success here is not to use a carboxylic acid itself, but to use a derivative of a carboxylic acid that is more easily reduced, and to use an aluminum hydride derivative that is less reactive than LAH. We shall study derivatives of carboxylic acids in detail in Chapter 18, but suffice it to say here that acyl chlorides ($RCOCl$), esters (RCO_2R'), and nitriles (RCN) are all easily prepared from carboxylic acids, and they all are more easily reduced. (Acyl chlorides, esters, and nitriles all also have the same oxidation state as carboxylic acids. Convince yourself of this by applying the principles that you learned in Problem 11.1.) Two derivatives of aluminum hydride that are less reactive than LAH (in part because they are much more sterically hindered and, therefore, have difficulty in transferring hydride ions) are lithium tri-*tert*-butoxyaluminum hydride and diisobutylaluminum hydride (DIBAL-H).

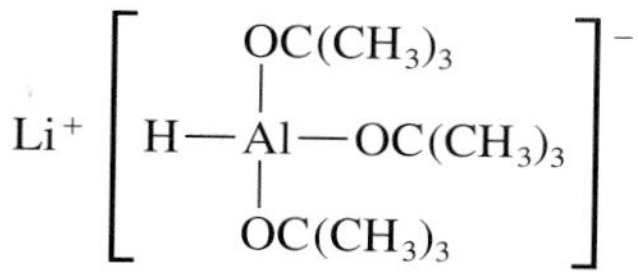

$$Li^+ \left[H-\overset{\displaystyle OC(CH_3)_3}{\underset{\displaystyle OC(CH_3)_3}{\overset{\displaystyle |}{\underset{\displaystyle |}{Al}}}}-OC(CH_3)_3 \right]^-$$

Lithium tri-*tert*-butoxy-aluminum hydride

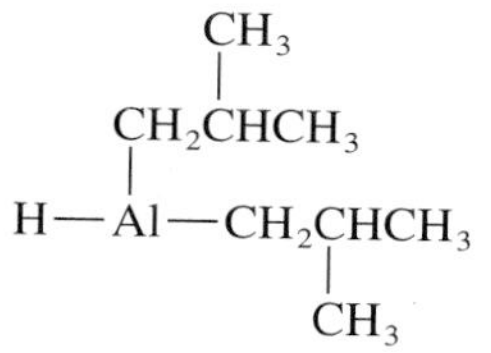

$$H-\overset{\displaystyle CH_2CHCH_3}{\underset{\displaystyle CH_3}{\overset{\displaystyle |}{\underset{\displaystyle |}{Al}}}}-\overset{\displaystyle CH_3}{\overset{\displaystyle |}{\underset{\displaystyle |}{CH_2CHCH_3}}}$$

Diisobutylaluminum hydride (abbreviated *i*-Bu$_2$AlH or DIBAL-H)

The following scheme summarizes how these reagents are used to synthesize aldehydes from acid derivatives.

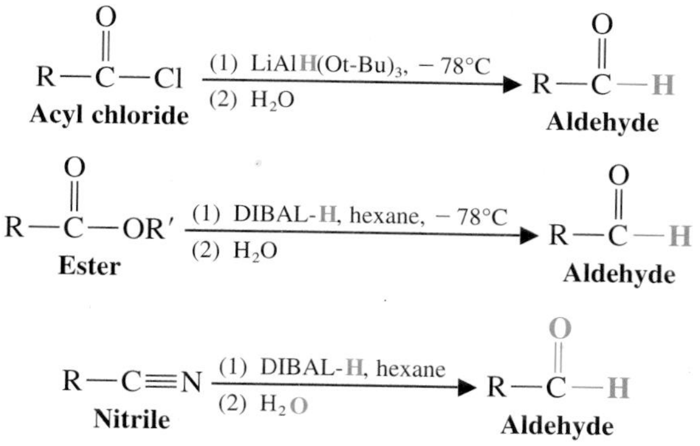

$$R-\overset{\displaystyle O}{\overset{\|}{C}}-Cl \xrightarrow[\text{(2) H}_2\text{O}]{\text{(1) LiAlH(O}t\text{-Bu)}_3, -78°C} R-\overset{\displaystyle O}{\overset{\|}{C}}-H$$

Acyl chloride   **Aldehyde**

$$R-\overset{\displaystyle O}{\overset{\|}{C}}-OR' \xrightarrow[\text{(2) H}_2\text{O}]{\text{(1) DIBAL-H, hexane}, -78°C} R-\overset{\displaystyle O}{\overset{\|}{C}}-H$$

Ester   **Aldehyde**

$$R-C\equiv N \xrightarrow[\text{(2) H}_2\text{O}]{\text{(1) DIBAL-H, hexane}} R-\overset{\displaystyle O}{\overset{\|}{C}}-H$$

Nitrile   **Aldehyde**

We now examine each of these aldehyde syntheses in more detail.

Aldehydes from Acyl Chlorides: $RCOCl \longrightarrow RCHO$ Acyl chlorides can be reduced to aldehydes by treating them with lithium tri-*tert*-butoxyaluminum hydride, $LiAlH[OC(CH_3)_3]_3$, at $-78°C$. (Carboxylic acids can be converted to acyl chlorides by using $SOCl_2$, Section 15.7.)

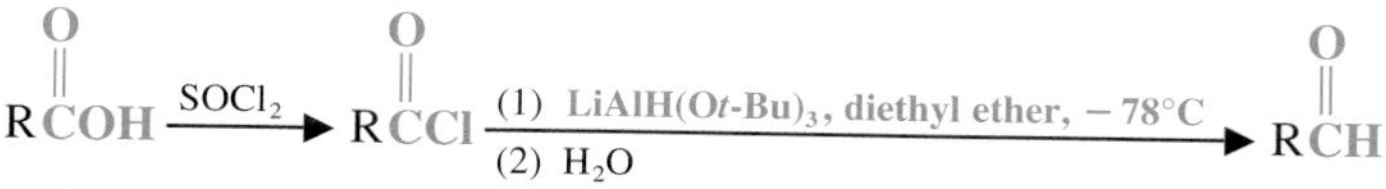

$$R\overset{\displaystyle O}{\overset{\|}{C}}OH \xrightarrow{\text{SOCl}_2} R\overset{\displaystyle O}{\overset{\|}{C}}Cl \xrightarrow[\text{(2) H}_2\text{O}]{\text{(1) LiAlH(O}t\text{-Bu)}_3, \text{ diethyl ether}, -78°C} R\overset{\displaystyle O}{\overset{\|}{C}}H$$

The following is a specific example:

Mechanistically, the reduction is brought about by the transfer of a hydride ion from the aluminum atom to the carbonyl carbon of the acyl chloride (cf. Section 11.3). Subsequent hydrolysis frees the aldehyde.

A MECHANISM FOR THE REDUCTION OF AN ACYL CHLORIDE

The aluminum atom accepts an electron pair from the carbonyl carbon atom in a Lewis acid–base reaction.

Transfer of a hydride ion to the carbonyl carbon brings about its reduction.

This intermediate loses a chloride ion as an electron pair from the oxygen assists.

The addition of water causes hydrolysis of this aluminum complex to take place, producing the aldehyde. (Several steps are involved.)

Aldehydes from Esters and Nitriles: $RCO_2R' \longrightarrow RCHO$ and $RC\equiv N \longrightarrow$ RCHO.   Both esters and nitriles can be reduced to aldehydes by use of DIBAL-H. Carefully controlled amounts of the reagent must be used to avoid overreduction and the ester reduction must be carried out at low temperatures. Both reductions result in the formation of a relatively stable intermediate by the addition of a hydride ion to the carbonyl carbon of the ester or to the carbon of the $-C\equiv N$ group of the nitrile. Hydrolysis of the intermediate liberates the aldehyde. Schematically, the reactions can be viewed this way:

A MECHANISM FOR THE REDUCTION OF AN ESTER

The aluminum atom accepts an electron pair from the carbonyl carbon atom in a Lewis acid–base reaction.

Transfer of a hydride ion to the carbonyl carbon brings about its reduction.

This intermediate loses an alkoxide ion as an electron pair from the oxygen assists.

The addition of water causes hydrolysis of this aluminum complex to take place, producing the aldehyde. (Several steps are involved.)

A MECHANISM FOR THE REDUCTION OF A NITRILE

The aluminum atom accepts an electron pair from the nitrile in a Lewis acid–base reaction.

Transfer of a hydride ion to the nitrile carbon brings about its reduction.

The addition of water causes hydrolysis of this aluminum complex to take place, producing the aldehyde. (Several steps are involved.)

The following specific examples illustrate these syntheses.

$$CH_3(CH_2)_{10}COEt \xrightarrow[\text{hexane, } -78°C]{(i\text{-Bu})_2AlH} CH_3(CH_2)_{10}\overset{\overset{OAl(i\text{-Bu})_2}{|}}{C}H \xrightarrow{H_2O} CH_3(CH_2)_{10}CH$$

$$\underset{\text{OEt}}{}$$

(88%)

$$\text{CH}_3\text{CH}=\text{CHCH}_2\text{CH}_2\text{CH}_2\text{C}\equiv\text{N} \xrightarrow[\text{hexane}]{(i\text{-Bu})_2\text{AlH}} \overset{\overset{\displaystyle \text{NAl}(i\text{-Bu})_2}{\|}}{\text{CH}_3\text{CH}=\text{CHCH}_2\text{CH}_2\text{CH}_2\text{CH}}$$

$$\xrightarrow{\text{H}_2\text{O}} \overset{\overset{\displaystyle O}{\|}}{\text{CH}_3\text{CH}=\text{CHCH}_2\text{CH}_2\text{CH}_2\text{CH}}$$

PROBLEM 16.3

Show how you would synthesize propanal from each of the following: (a) 1-propanol and (b) propanoic acid ($\text{CH}_3\text{CH}_2\text{CO}_2\text{H}$).

16.5 SYNTHESIS OF KETONES

16.5A Ketones from Alkenes, Arenes, and 2° Alcohols

We have seen three laboratory methods for the preparation of ketones in earlier chapters.

1. **Ketones (and Aldehydes) by Ozonolysis of Alkenes** (discussed in Section 8.10A).

Ketone   Aldehyde

2. **Ketones from Arenes by Friedel–Crafts Acylations** (discussed in Section 15.7).

$$\text{ArH} + \text{R}-\overset{\overset{\displaystyle O}{\|}}{\text{C}}-\text{Cl} \xrightarrow{\text{AlCl}_3} \text{Ar}-\overset{\overset{\displaystyle O}{\|}}{\text{C}}-\text{R} + \text{HCl}$$

**An alkyl aryl
ketone**

or

$$\text{ArH} + \text{Ar}-\overset{\overset{\displaystyle O}{\|}}{\text{C}}-\text{Cl} \xrightarrow{\text{AlCl}_3} \text{Ar}-\overset{\overset{\displaystyle O}{\|}}{\text{C}}-\text{Ar} + \text{HCl}$$

A diaryl ketone

3. **Ketones from Secondary Alcohols by Oxidation** (discussed in Section 11.4).

$$\text{R}-\overset{\overset{\displaystyle \text{OH}}{|}}{\text{CH}}-\text{R}' \xrightarrow{\text{H}_2\text{CrO}_4} \text{R}-\overset{\overset{\displaystyle O}{\|}}{\text{C}}-\text{R}'$$

A useful method for preparing ketones is based on the hydration of alkynes.

16.5B Ketones from Alkynes

Alkynes add water readily when the reaction is catalyzed by strong acids and mercuric (Hg^{2+}) ions. Aqueous solutions of sulfuric acid and mercuric sulfate are often used for this purpose. The vinylic alcohol that is initially produced is usually unstable, and it rearranges rapidly to a ketone [or in the case of ethyne, to ethanal]. The rearrangement involves the loss of a proton from the hydroxyl group, the addition of a proton to the vicinal carbon atom, and the relocation of the double bond.

$$-C \equiv C- \ + \ H-OH \ \xrightarrow[\text{H}_2\text{SO}_4]{\text{HgSO}_4} \ \left[\underset{\substack{\text{A vinylic} \\ \text{alcohol} \\ \text{(unstable)}}}{-\overset{\text{H}}{\underset{\text{OH}}{C}}=C-} \right] \longrightarrow \underset{\text{Ketone}}{-\overset{\text{H}}{\underset{\text{H}}{C}}-\overset{}{\underset{\text{O}}{C}}-}$$

This kind of rearrangement, known as a **tautomerization,** is acid catalyzed and occurs in the following way:

$$H-\overset{+}{\underset{\cdots}{O}}\overset{\text{H}}{{\longleftarrow}} H + \underset{\substack{\text{Vinylic} \\ \text{alcohol}}}{-\overset{\text{H}}{C}=\overset{}{\underset{\overset{\text{H}}{\underset{\cdots}{O}}-H}{C}}-} \longrightarrow -\overset{\text{H}}{\underset{\text{H}}{C}}-\overset{}{\underset{\overset{+}{\underset{\cdots}{O}}-H}{C}}- \xrightarrow{\text{H}_2\overset{\cdots}{\underset{\cdots}{O}}:} \underset{\text{Ketone}}{-\overset{\text{H}}{\underset{\text{H}}{C}}-\overset{}{\underset{\overset{\cdots}{\underset{\cdots}{O}}:}{C}}-} + H_3O^+$$

The vinylic alcohol accepts a proton at one carbon atom of the double bond to yield a cationic intermediate, which then loses a proton from the oxygen atom to produce a ketone.

Vinylic alcohols are often called **enols** (*-en*, the ending for alkenes, plus *-ol*, the ending for alcohols). The product of the rearrangement is usually a ketone, and these rearrangements are known as **keto–enol tautomerizations.**

$$\underset{\substack{\text{Enol form}}}{-\overset{}{C}=\overset{}{\underset{\overset{\cdots}{\underset{\cdots}{O}}-H}{C}}-} \ \underset{}{\overset{\text{H}^+}{\rightleftharpoons}} \ \underset{\substack{\text{Keto form}}}{-\overset{}{\underset{\text{H}}{C}}-\overset{}{\underset{\overset{\cdots}{\underset{\cdots}{O}}:}{C}}-}$$

We examine this phenomenon in greater detail in Section 17.2.

Only when ethyne itself undergoes addition of water is the product an aldehyde.

$$\underset{\text{Ethyne}}{H-C \equiv C-H} + H_2O \ \xrightarrow[\text{H}_2\text{SO}_4]{\text{HgSO}_4} \ \left[\underset{H}{\overset{H}{C}}=\underset{OH}{\overset{H}{C}} \right] \longrightarrow \underset{\substack{\text{Ethanal} \\ \text{(acetaldehyde)}}}{H-\overset{\text{H}}{\underset{\text{H}}{C}}-\overset{}{\underset{\text{O}}{C}}-H}$$

This method has been important in the commercial production of ethanal.

The addition of water to alkynes also follows Markovnikov's rule—the hydrogen atom becomes attached to the carbon atom with the greater number of hydrogen atoms. Therefore, when higher terminal alkynes are hydrated, ketones, rather than aldehydes, are the products.

$$R-C\equiv C-H + H_2O \xrightarrow{Hg^{2+}}{H^+} \left[\begin{matrix} H \\ | \\ RC=C-H \\ | \\ OH \end{matrix} \right] \longrightarrow \begin{matrix} H \quad H \\ | \quad\; | \\ R-C-C-H \\ \| \quad | \\ O \quad H \end{matrix}$$

A ketone

Two examples of this ketone synthesis are listed here.

$$CH_3C\equiv CH + H_2O \xrightarrow{Hg^{2+}}{H^+} \left[\begin{matrix} CH_3-C=CH_2 \\ | \\ OH \end{matrix} \right] \longrightarrow \begin{matrix} CH_3-C-CH_3 \\ \| \\ O \end{matrix}$$

Acetone

$$CH_3CH_2CH_2CH_2C\equiv CH + H_2O \xrightarrow{HgSO_4}{H_2SO_4} \begin{matrix} CH_3CH_2CH_2CH_2CCH_3 \\ \| \\ O \end{matrix}$$

(80%)

Two other laboratory methods for the preparation of ketones are based on the use of organometallic compounds.

16.5C Ketones from Lithium Dialkylcuprates

When an ether solution of a lithium dialkylcuprate is treated with an acyl chloride at −78°C, the product is a ketone. This ketone synthesis is a variation of the Corey–Posner, Whitesides–House alkane synthesis (Section 4.15C).

General Reaction

$$R_2CuLi \quad + \quad \begin{matrix} O \\ \| \\ R'-C-Cl \end{matrix} \longrightarrow \begin{matrix} O \\ \| \\ R'-C-R \end{matrix}$$

Lithium   **Acyl**   **Ketone**
dialkylcuprate   **chloride**

Specific Example

$$\left(\text{cyclohexyl}\right)\begin{matrix} O \\ \| \\ C-Cl \end{matrix} + (CH_3)_2CuLi \xrightarrow{-78°C}{diethyl\ ether} \left(\text{cyclohexyl}\right)\begin{matrix} O \\ \| \\ C-CH_3 \end{matrix}$$

(81%)

16.5D Ketones from Nitriles

Treating a nitrile (R—C≡N) with either a Grignard reagent or an organolithium reagent followed by hydrolysis yields a ketone.

General Reactions

$$R-C\equiv N + R'-MgX \longrightarrow \begin{matrix} N^-MgX^+ \\ \| \\ R-C-R' \end{matrix} \xrightarrow{H_3O^+} \begin{matrix} O \\ \| \\ R-C-R' \end{matrix} + NH_4^+ + Mg^{2+} + X^-$$

$$R-C\equiv N + R'-Li \longrightarrow R-\overset{\overset{\displaystyle N^-Li^+}{\|}}{C}-R' \xrightarrow{H_3O^+} R-\overset{\overset{\displaystyle O}{\|}}{C}-R' + NH_4^+ + Li^+$$

Specific Examples

$$C_6H_5-C\equiv N + CH_3CH_2CH_2CH_2Li \xrightarrow[\text{(2) } H_3O^+]{\text{(1) diethyl ether}} C_6H_5-\overset{\overset{\displaystyle O}{\|}}{C}-CH_2CH_2CH_2CH_3$$

$$\underset{\overset{\displaystyle |}{CH_3}}{CH_3CH}-C\equiv N + C_6H_5MgBr \xrightarrow[\text{(2) } H_3O^+]{\text{(1) diethyl ether}} \underset{\overset{\displaystyle |}{CH_3}}{CH_3CH}-\overset{\overset{\displaystyle O}{\|}}{C}-C_6H_5$$

Even though a nitrile has a triple bond, addition of the Grignard or lithium reagent takes place only once. The reason: If addition took place twice, this would place a double negative charge on the nitrogen:

$$R-C\equiv N: \xrightarrow{R'-Li} R-\overset{\overset{\displaystyle N^-Li^+}{\|}}{C}-R' \overset{R'-Li}{\underset{\times}{\longrightarrow}} R-\underset{\overset{\displaystyle |}{R'}}{\overset{\overset{\displaystyle N^{2-} 2\ Li^+}{|}}{C}}-R'$$

(does not form)

SAMPLE PROBLEM

Illustrating a Multistep Synthesis

With 1-butanol as your only organic starting compound, outline a synthesis of 5-nonanone.

ANSWER:

5-Nonanone can be synthesized by adding butylmagnesium bromide to the following nitrile.

Analysis:

$$CH_3CH_2CH_2CH_2-\overset{\overset{\displaystyle O}{\|}}{C}\!\!\mid\!\!CH_2CH_2CH_2CH_3 \Longrightarrow CH_3CH_2CH_2CH_2-\overset{\delta+}{C}\!\!\overset{\delta-}{\equiv}\!\!N:$$
$$+ \overset{\delta+}{Br}Mg-\overset{\delta-}{CH_2}CH_2CH_2CH_3$$

Synthesis

$$CH_3CH_2CH_2CH_2C\equiv N + CH_3CH_2CH_2CH_2MgBr \xrightarrow[\text{(2) } H_3O^+]{}$$
$$\overset{\overset{\displaystyle O}{\|}}{CH_3(CH_2)_3C(CH_2)_3CH_3}$$
$$\textbf{5-Nonanone}$$

The nitrile can be synthes   n butyl bromide and sodium cyanide in an S_N2 reaction.

$$CH_3CH_2CH_2CH_2Br + N \longrightarrow CH_3CH_2CH_2CH_2C\equiv N + NaBr$$

Butyl bromide also can b   to prepare the Grignard reagent.

$$CH_3CH_2CH_2CH_2Br + Mg \xrightarrow[\text{diethyl ether}]{} CH_3CH_2CH_2CH_2MgBr$$

And, finally, butyl bromide can be prepared from 1-butanol.

$$CH_3CH_2CH_2CH_2OH \xrightarrow{PBr_3} CH_3CH_2CH_2CH_2Br$$

PROBLEM 16.4

Which reagents would you use to carry out each of the following reactions?

(a) Benzene $\longrightarrow$ bromobenzene $\longrightarrow$ phenylmagnesium bromide $\longrightarrow$ benzyl alcohol $\longrightarrow$ benzaldehyde

(b) Toluene $\longrightarrow$ benzoic acid $\longrightarrow$ benzoyl chloride $\longrightarrow$ benzaldehyde

(c) Ethyl bromide $\longrightarrow$ 1-butyne $\longrightarrow$ 2-butanone

(d) 2-Butyne $\longrightarrow$ 2-butanone

(e) 1-Phenylethanol $\longrightarrow$ acetophenone

(f) Benzene $\longrightarrow$ acetophenone

(g) Benzoyl chloride $\longrightarrow$ acetophenone

(h) Benzoic acid $\longrightarrow$ acetophenone

(i) Benzyl bromide $\longrightarrow$ $C_6H_5CH_2CN$ $\longrightarrow$ 1-phenyl-2-butanone

(j) $C_6H_5CH_2CN$ $\longrightarrow$ 2-phenylethanal

(k) $CH_3(CH_2)_4CO_2CH_3$ $\longrightarrow$ hexanal

16.6 NUCLEOPHILIC ADDITION TO THE CARBON–OXYGEN DOUBLE BOND

The most characteristic reaction of aldehydes and ketones is *nucleophilic addition* to the carbon–oxygen double bond.

$$\underset{H}{\overset{R}{\diagdown}}C=O + H-Nu \rightleftarrows R-\underset{H}{\overset{Nu}{\underset{|}{\overset{|}{C}}}}-OH$$

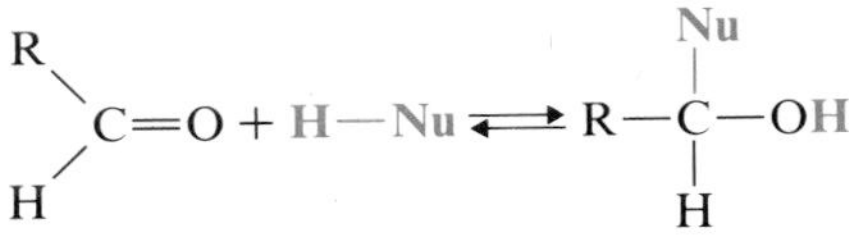

Specific Examples

$$H_3C \diagdown \atop H \diagup C=O \; + \; H-OCH_2CH_3 \; \rightleftharpoons \; CH_3 \overset{\displaystyle OCH_2CH_3}{\underset{\displaystyle |}{}} -OH$$

(a hemiacetal; see
Section 16.7)

$$H_3C \diagdown \atop H_3C \diagup C=O \; + \; H-CN \; \rightleftharpoons \; CH_3 \overset{\displaystyle CN}{\underset{\displaystyle CH_3}{- \underset{|}{\overset{|}{C}} -}} OH$$

(a cyanohydrin; see
Section 16.9)

Aldehydes and ketones are especially susceptible to nucleophilic addition because of the structural features that we discussed in Section 11.1 and which are shown below.

$$\overset{R'}{\underset{R}{}} \diagup C \overset{\delta+ \quad \delta-}{=} \ddot{O}:$$

**Aldehyde or ketone
(R or R′ may be H)**

The trigonal planar arrangement of groups around the carbonyl carbon atom means that the carbonyl carbon atom is relatively open to attack from above or below. The positive charge on the carbonyl carbon atom means that it is especially susceptible to attack by a nucleophile. The negative charge on the carbonyl oxygen atom means that nucleophilic addition is susceptible to acid catalysis. We can visualize nucleophilic addition to the carbon–oxygen double bond occurring in either of two general ways:

1. When the reagent is a strong nucleophile (Nu), addition usually takes place in the following way, converting the trigonal planar aldehyde or ketone into a tetrahedral product:

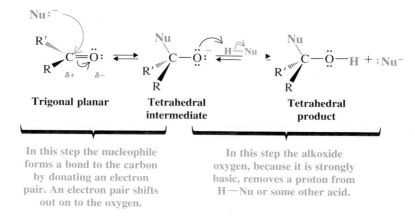

Trigonal planar   **Tetrahedral
intermediate**   **Tetrahedral
product**

In this step the nucleophile
forms a bond to the carbon
by donating an electron
pair. An electron pair shifts
out on to the oxygen.

In this step the alkoxide
oxygen, because it is strongly
basic, removes a proton from
H—Nu or some other acid.

In this type of addition the nucleophile uses its electron pair to form a bond to the carbonyl carbon atom. As this happens the electron pair of the carbon–oxygen π bond shifts out to the carbonyl oxygen atom and the hybridization state of the carbon changes from sp^2 to sp^3. *The important aspect of this step is the ability of the carbonyl oxygen atom to accommodate the electron pair of the carbon–oxygen double bond.*

In the second step the oxygen atom accepts a proton. This happens because the oxygen atom is now much more basic; it carries a full negative charge, and it is an alkoxide ion.

2. A second general mechanism that operates in nucleophilic additions to carbon–oxygen double bonds is an acid-catalyzed mechanism:

(or a Lewis acid)

In this step an electron pair of the carbonyl oxygen accepts a proton from the acid (or associates with a Lewis acid), producing an oxonium ion. Because one resonance structure of the oxonium ion has a full positive charge on the carbon, the carbon is more susceptible to nucleophilic attack.

In the first of these two steps, the oxonium ion accepts the electron pair of the nucleophile. In the second step, a base removes a proton from the positively charged atom, regenerating the acid.

This mechanism operates when carbonyl compounds are treated with *strong acids* in the presence of *weak nucleophiles.* In the first step the acid donates a proton to an electron pair of the carbonyl oxygen atom. The resulting protonated carbonyl compound, an **oxonium ion,*** is highly reactive toward nucleophilic attack at the carbonyl carbon atom because the carbonyl carbon atom carries more positive charge than it does in the unprotonated compound.

16.6A Reversibility of Nucleophilic Additions to the Carbon–Oxygen Double Bond

Many nucleophilic additions to carbon–oxygen double bonds are reversible; the overall results of these reactions depend, therefore, on the position of an equilibrium. This behavior contrasts markedly with most electrophilic additions to carbon–carbon double bonds and with nucleophilic substitutions at saturated carbon atoms. The latter reactions are essentially irreversible, and overall results are a function of relative reaction rates.

---

* Any compound containing a positively charged oxygen atom that forms three covalent bonds is an *oxonium ion.*

16.6B Relative Reactivity: Aldehydes Versus Ketones

In general, **aldehydes are more reactive in nucleophilic substitutions than are ketones.** Both steric and electronic factors favor aldehydes: With one group being the small hydrogen atom, the central carbon of the tetrahedral product formed from an aldehyde is less crowded and the product is more stable. Formation of the product, therefore, is favored at equilibrium. With ketones, the two alkyl substituents at the carbonyl carbon cause greater steric crowding in the tetrahedral product and make it less stable. Therefore, a smaller concentration of the product is present at equilibrium.

Because alkyl groups are electron releasing, **aldehydes are more reactive on electronic grounds, as well.** Aldehydes have only one electron-releasing group to partially neutralize, and thereby stabilize, the positive charge at their carbonyl carbon atom. Ketones have two electron-releasing groups and are stabilized more. Greater stabilization of the ketone (the reactant) relative to its product means that the equilibrium constant for the formation of the tetrahedral product from a ketone is smaller and the reaction is less favorable.

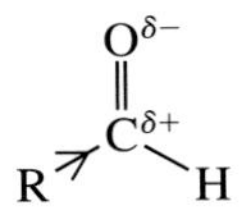

Aldehyde	**Ketone**
Carbonyl carbon is more positive	Carbonyl carbon is less positive

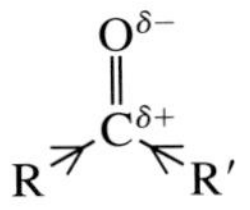

In this regard, too, electron-withdrawing substituents (e.g., $-CF_3$ or $-CCl_3$ groups) cause the carbonyl carbon to be more positive (the starting compound becomes less stable) and cause the addition reaction to be more favorable.

16.6C Subsequent Reactions of Addition Products

Nucleophilic addition to a carbon–oxygen double bond may lead to a product that is stable under the reaction conditions that we employ. If this is the case we are then able to isolate products with the following general structure:

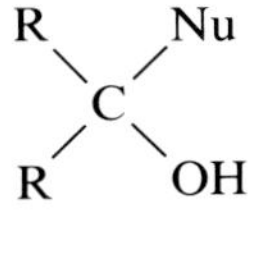

In other reactions the product formed initially may be unstable and may spontaneously undergo subsequent reactions. Even if the initial addition product is stable, however, we may deliberately bring about a subsequent reaction by changing the reaction conditions. When we begin our study of specific reactions, we shall see that one common subsequent reaction is an *elimination reaction,* especially *dehydration.*

PROBLEM 16.5 ————————————————————————————

The reaction of an aldehyde or ketone with a Grignard reagent (Section 11.8) is a nucleophilic addition to the carbon–oxygen double bond.   (a) What is the nucleophile?   (b) The magnesium portion of the Grignard reagent plays an important part in this reaction. What is its function? (c) What product is formed initially? (d) What product forms when water is added?

PROBLEM 16.6

The reactions of aldehydes and ketones with $LiAlH_4$ and $NaBH_4$ (Section 11.3) are nucleophilic additions to the carbonyl group. What is the nucleophile in these reactions?

16.7 THE ADDITION OF WATER AND ALCOHOLS: HYDRATES AND ACETALS

16.7A Aldehyde Hydrates: *Gem*-Diols

Dissolving an aldehyde such as acetaldehyde in water causes the establishment of an equilibrium between the aldehyde and its **hydrate.** This hydrate is in actuality a 1,1-diol, called a *gem*-diol.

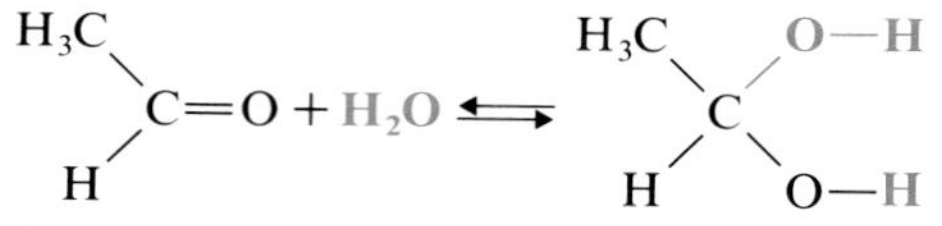
H3C
C=O + H2O
H
H3C
O—H
C
H
O—H

Acetaldehyde **Hydrate (a *gem*-diol)**

The *gem*-diol results from a nucleophilic addition of water to the carbonyl group of the aldehyde:

A MECHANISM FOR HYDRATE FORMATION

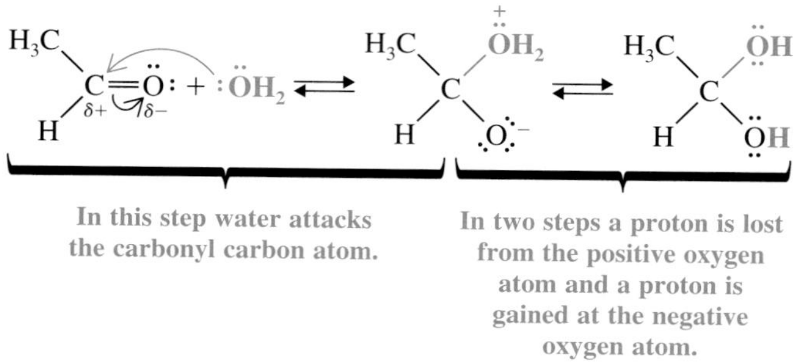
H3C
C=O: + :OH2
δ+ δ−
H
H3C
OH2
C
H
O:−
H3C
OH
C
H
OH
In this step water attacks the carbonyl carbon atom.
In two steps a proton is lost from the positive oxygen atom and a proton is gained at the negative oxygen atom.

PROBLEM 16.7

Dissolving formaldehyde in water leads to a solution containing primarily the *gem*-diol, $CH_2(OH)_2$. Show the steps in its formation from formaldehyde.

The addition of water is subject to catalysis by both acids and bases. That is, addition takes place much more rapidly in the presence of small amounts of acids or bases than it does in pure water.

The mechanism for the **base-catalyzed reaction** is as follows:

A MECHANISM FOR THE BASE-CATALYZED REACTION

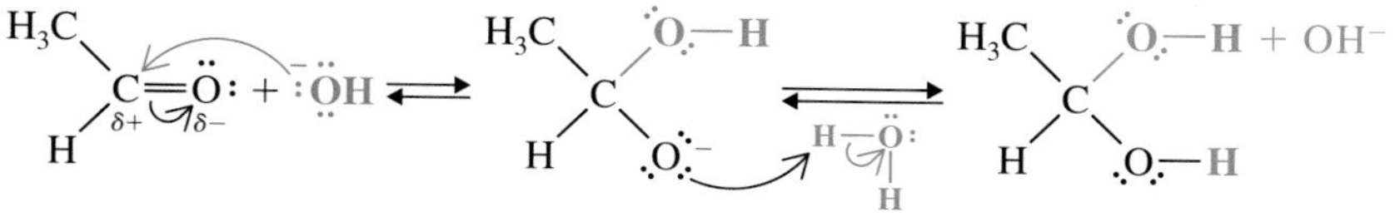

A hydroxide ion acting as a nucleophile attacks the carbonyl carbon atom. An electron pair shifts onto the oxygen atom, producing an alkoxide ion.

The alkoxide ion abstracts a proton from a water molecule to produce the *gem*-diol and regenerate a hydroxide ion (the catalyst).

The important factor here in increasing the rate is the greater nucleophilicity of the hydroxide ion when compared to water.

The **acid-catalyzed mechanism** involves an initial rapid protonation of the carbonyl oxygen atom:

A MECHANISM FOR THE ACID-CATALYZED REACTION

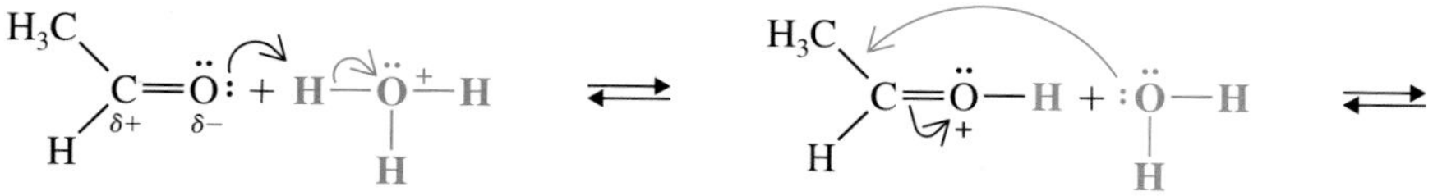

Protonation of the carbonyl oxygen atom makes the carbonyl carbon atom more susceptible to nucleophilic attack.

A water molecule adds to the carbonyl carbon leading to a protonated *gem*-diol.

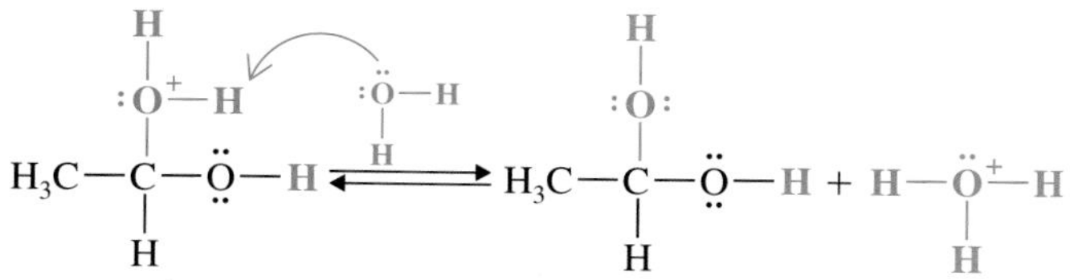

The transfer of a proton to a molecule of water leads to the *gem*-diol and regenerates the catalyst.

Protonation makes the carbonyl carbon atom more susceptible to attack by water, and here this factor is the key to the rate acceleration.

PROBLEM 16.8

When acetone is dissolved in water containing ^{18}O instead of ordinary ^{16}O (i.e., $H_2{}^{18}O$ instead of $H_2{}^{16}O$), the acetone soon begins to acquire ^{18}O and becomes

$$CH_3\overset{\overset{\displaystyle ^{18}O}{\|}}{C}CH_3.$$ The formation of this oxygen-labeled acetone is catalyzed by traces of strong acids and by strong bases (e.g., OH^-). Show the steps that explain both the acid-catalyzed reaction and the base-catalyzed reaction.

The equilibrium for the addition of water to most ketones is unfavorable, whereas some aldehydes (e.g., formaldehyde) exist primarily as the *gem*-diol in aqueous solution.

It is not possible to isolate most *gem*-diols from the aqueous solutions in which they are formed. Evaporation of the water, for example, simply displaces the overall equilibrium to the left and the *gem*-diol (or hydrate) reverts to the carbonyl compound.

$$\underset{\underset{H}{\overset{R}{|}}}{\underset{OH}{\overset{OH}{|}}}C \xrightarrow{\text{distillation}} \underset{H}{\overset{R}{\diagdown}}C{=}O + H_2O$$

Compounds with strong electron-withdrawing groups attached to the carbonyl group can form stable *gem*-diols. An example is the compound called chloral hydrate.

$$\underset{\underset{H}{}}{\overset{CCl_3}{\diagdown}}C\overset{OH}{\diagup}OH$$

Chloral hydrate

16.7B Hemiacetals

Dissolving an aldehyde or ketone in an alcohol causes the slow establishment of an equilibrium between these two compounds and a new compound called a **hemiacetal:**

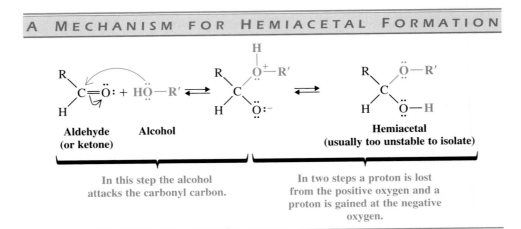

A MECHANISM FOR HEMIACETAL FORMATION

Aldehyde (or ketone)   Alcohol   Hemiacetal (usually too unstable to isolate)

In this step the alcohol attacks the carbonyl carbon.   In two steps a proton is lost from the positive oxygen and a proton is gained at the negative oxygen.

The essential structural features of a hemiacetal are an —OH and an —OR group attached to the same carbon atom.

Most open-chain hemiacetals are not sufficiently stable to allow their isolation. Cyclic hemiacetals with five- or six-membered rings, however, are usually much more stable:

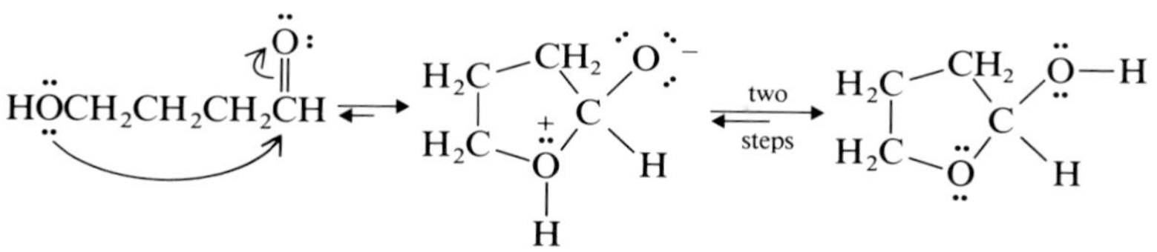

Most simple sugars (Chapter 22) exist primarily in a cyclic hemiacetal form. Glucose is an example:

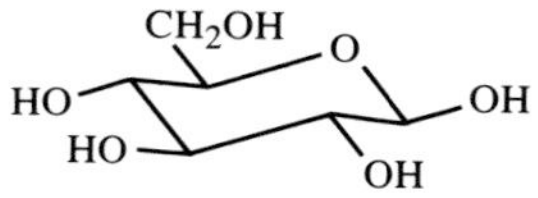

(+)-Glucose
(a cyclic hemiacetal)

Ketones undergo similar reactions when they are dissolved in an alcohol. The products (also unstable in open-chain compounds) are sometimes called **hemiketals,** but this usage is no longer recommended by the IUPAC. The preferred usage is to call them hemiacetals, as well.

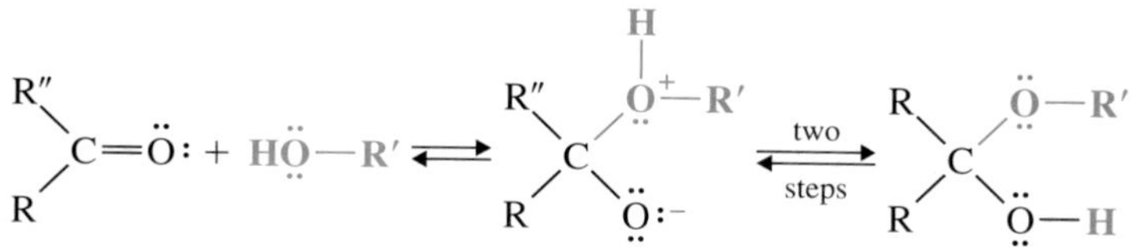

Ketone   **Hemiacetal**

The formation of hemiacetals is also catalyzed by acids and bases. The mechanisms are similar to those for the formation of aldehyde hydrates that we studied in Section 16.7A.

Acid-Catalyzed Hemiacetal Formation

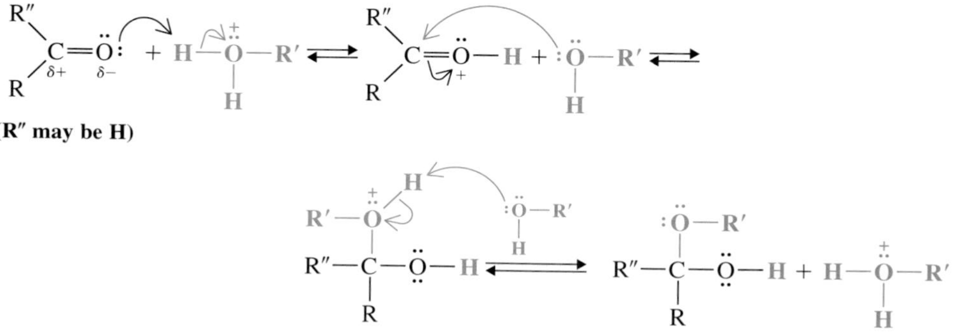

(R″ may be H)

___

Base-Catalyzed Hemiacetal Formation

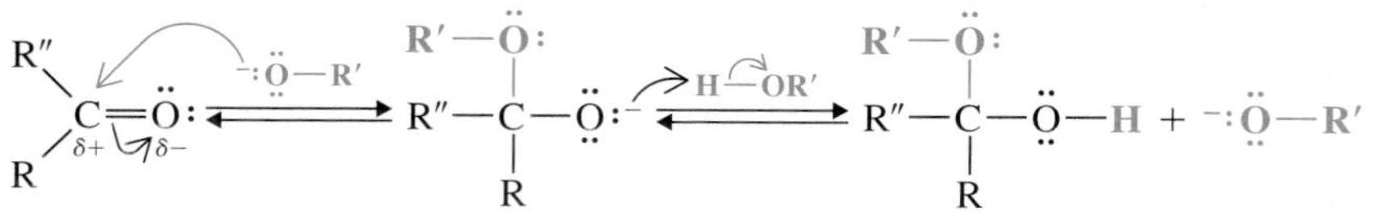

$$R''\overset{\displaystyle\diagup}{\underset{\displaystyle R}{\diagdown}}C{=}\ddot{O}\!: \;\; \overset{:\ddot{O}-R'}{} \;\rightleftharpoons\; R''{-}\underset{\displaystyle R}{\overset{\displaystyle R'-\ddot{O}:}{C}}{-}\ddot{O}\!:^- \;\overset{H{-}OR'}{\rightleftharpoons}\; R''{-}\underset{\displaystyle R}{\overset{\displaystyle R'-\ddot{O}:}{C}}{-}\ddot{O}{-}H \;+\; {}^-\!:\!\ddot{O}{-}R'$$

(**R″ may be H**)

16.7C Acetals

If we take an alcohol solution of an aldehyde (or ketone) and pass into it a small amount of gaseous HCl the hemiacetal forms, and then a second reaction takes place. The hemiacetal reacts with a second molar equivalent of the alcohol to produce an **acetal (sometimes called a ketal)**. An acetal has two —OR groups attached to the same carbon atom.

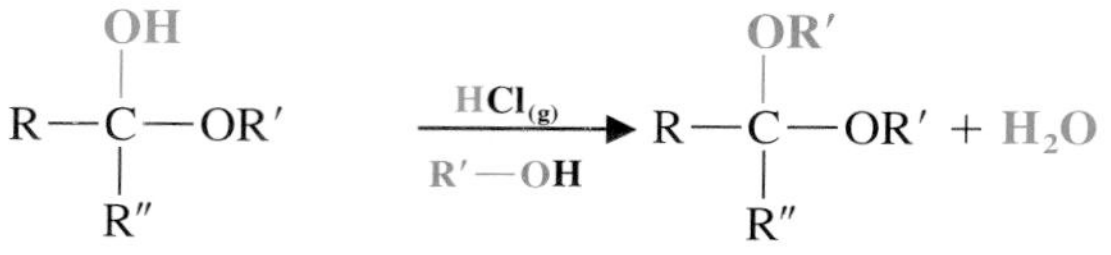

$$R{-}\underset{\displaystyle R''}{\overset{\displaystyle OH}{C}}{-}OR' \;\;\xrightarrow[R'-OH]{HCl_{(g)}}\;\; R{-}\underset{\displaystyle R''}{\overset{\displaystyle OR'}{C}}{-}OR' \;+\; H_2O$$

Hemiacetal (R″ may be H)   **An acetal (R″ may be H)**

PROBLEM 16.9

Shown below is the structural formula for sucrose (table sugar). Sucrose has two acetal groupings. Identify these.

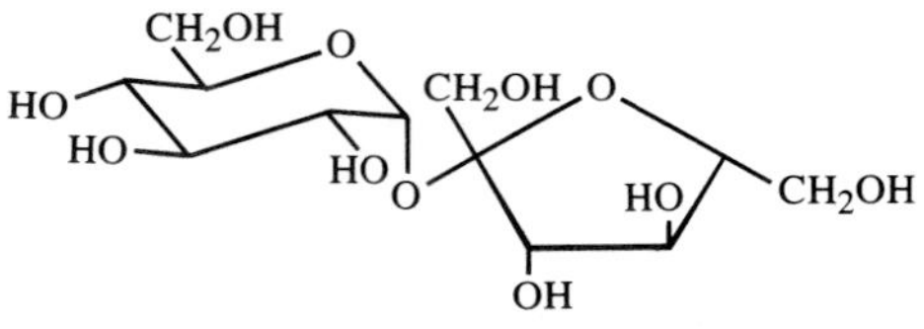

Sucrose

___

The mechanism for acetal formation involves acid-catalyzed formation of the hemiacetal, then an acid-catalyzed elimination of water, followed by a second *addition* of the alcohol and loss of a proton.

___

A   M E C H A N I S M   F O R   A C E T A L   F O R M A T I O N

$$\underset{\displaystyle R}{\overset{\displaystyle H}{\diagdown}}C{=}\ddot{O}\!: \;\; H{-}\underset{\displaystyle H}{\overset{\displaystyle +}{\ddot{O}}}{-}R' \;\rightleftharpoons\; \underset{\displaystyle R}{\overset{\displaystyle H}{\diagdown}}C{=}\overset{+}{\ddot{O}}{-}H \;+\; :\underset{\displaystyle H}{\ddot{O}}{-}R' \;\rightleftharpoons$$

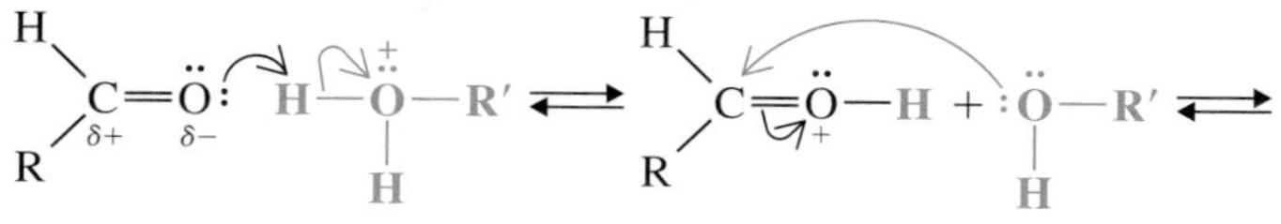

Acid-catalyzed formation of a hemiacetal

Acid-catalyzed elimination of water

Reaction with a second molecule of the alcohol

PROBLEM 16.10

Write out all the steps (as just shown) for the formation of an acetal from benzaldehyde and methanol in the presence of an acid catalyst.

All steps in the formation of an acetal from an aldehyde are reversible. If we dissolve an aldehyde in a large excess of an anhydrous alcohol and add a small amount of an anhydrous acid (e.g., gaseous HCl or conc'd H_2SO_4), the equilibrium will strongly favor the formation of an acetal. After the equilibrium is established, we can isolate the acetal by neutralizing the acid and evaporating the excess alcohol.

If we then place the acetal in water and add a small amount of acid, all of the steps reverse. Under these conditions (an excess of water), the equilibrium favors the formation of the aldehyde. The acetal undergoes *hydrolysis*.

Acetal   **Aldehyde**

Acetal formation is not favored when ketones are treated with simple alcohols and gaseous HCl. Cyclic acetal formation *is* favored, however, when a ketone is treated with an excess of a 1,2-diol and a trace of acid.

$$\underset{\substack{\text{Ketone}}}{\overset{R'}{\underset{R}{>}}C=O} + \underset{\substack{\text{(excess)}}}{\overset{HOCH_2}{\underset{HOCH_2}{|}}} \underset{\overleftarrow{}}{\overset{H^+}{\rightleftharpoons}} \underset{\substack{\text{Cyclic acetal}}}{\overset{R}{\underset{R}{>}}C\overset{O-CH_2}{\underset{O-CH_2}{|}}} + H_2O$$

This reaction, too, can be reversed by treating the acetal with aqueous acid.

$$\overset{R}{\underset{R}{>}}C\overset{O-CH_2}{\underset{O-CH_2}{|}} + H_2O \overset{H^+}{\rightleftharpoons} \overset{R}{\underset{R}{>}}C=O + \overset{CH_2OH}{\underset{CH_2OH}{|}}$$

PROBLEM 16.11

Outline all steps in the mechanism for the formation of a cyclic acetal from acetone and ethylene glycol in the presence of gaseous HCl.

16.7D Acetals as Protecting Groups

Although acetals are hydrolyzed to aldehydes and ketones in aqueous acid, **they are stable in basic solutions.**

$$\overset{R}{\underset{H}{>}}C\overset{OR'}{\underset{OR'}{<}} + H_2O \overset{OH^-}{\longrightarrow} \text{no reaction}$$

$$\overset{R}{\underset{R}{>}}C\overset{O-CH_2}{\underset{O-CH_2}{|}} + H_2O \overset{OH^-}{\longrightarrow} \text{no reaction}$$

Because of this property, acetals give us a convenient method for **protecting aldehyde and ketone groups from undesired reactions in basic solutions.** (Acetals are really *gem*-diethers and, like ethers, are relatively unreactive toward bases.) We can convert an aldehyde or ketone to an acetal, carry out a reaction on some other part of the molecule, and then hydrolyze the acetal with aqueous acid.

As an example, let us consider the problem of converting

$$\underset{\textbf{A}}{O\!\!=\!\!\overset{\overset{\displaystyle O}{\|}}{\diagdown}\!\!COC_2H_5} \quad \text{to} \quad \underset{\textbf{B}}{O\!\!=\!\!\diagdown\!\!CH_2OH}$$

Keto groups are more easily reduced than ester groups. Any reducing agent (e.g., $LiAlH_4$ or H_2/Ni) that can reduce the ester group of **A** reduces the keto group as well. But if we "protect" the keto group by converting it to a cyclic acetal (the ester group does not react), we can reduce the ester group in basic solution without affecting the cyclic acetal. After we finish the ester reduction, we can hydrolyze the cyclic acetal and obtain our desired product, **B.**

PROBLEM 16.12

What product would be obtained if **A** were treated with lithium aluminum hydride without first converting it to a cyclic acetal?

PROBLEM 16.13

(a) Show how you might use a cyclic acetal in carrying out the following transformation:

(b) Why would a direct addition of methylmagnesium bromide to **A** fail to give **C**?

PROBLEM 16.14

Dihydropyran reacts readily with an alcohol in the presence of a trace of anhydrous HCl or H_2SO_4 to form a tetrahydropyranyl ether.

Dihydropyran   **Tetrahydropyranyl ether**

(a) Write a plausible mechanism for this reaction.   (b) Tetrahydropyranyl ethers are stable in aqueous base but hydrolyze rapidly in aqueous acid to yield the original alcohol and another compound. Explain. (What is the other compound?)   (c) The tetrahydropyranyl group can be used as a protecting group for alcohols and phenols. Show how you might use it in a synthesis of 5-methyl-1,5-hexanediol starting with 4-chloro-1-butanol.

$$\overset{\overset{\textstyle O}{\|}}{16.7E \text{ Thioacetals: } RCR'} \longrightarrow RCH_2R'$$

Aldehydes and ketones react with thiols to form *thioacetals*.

$$\underset{H}{\overset{R}{\diagdown}}C=O + 2\,CH_3CH_2SH \xrightarrow{H^+} \underset{H}{\overset{R}{\diagdown}}C\overset{S-CH_2CH_3}{\underset{S-CH_2CH_3}{\diagup}} + H_2O$$

Thioacetal

$$\underset{R'}{\overset{R}{\diagdown}}C=O + HSCH_2CH_2SH \xrightarrow{BF_3} \underset{R'}{\overset{R}{\diagdown}}C\overset{S-CH_2}{\underset{S-CH_2}{\diagup}}\Big| + H_2O$$

Cyclic thioacetal

Thioacetals are important in organic synthesis because they react with Raney nickel to yield hydrocarbons.* These reactions (i.e., thioacetal formation and subsequent "desulfurization") give us an additional method for converting carbonyl groups

$$\underset{R'}{\overset{R}{\diagdown}}C\overset{S-CH_2}{\underset{S-CH_2}{\diagup}}\Big| \xrightarrow[\text{(H}_2)]{\text{Raney Ni}} \underset{R'}{\overset{R}{\diagdown}}CH_2 + H-CH_2CH_2-H + NiS$$

of aldehydes and ketones to $-CH_2-$ groups. The other method we have studied is the **Clemmensen reduction** (Section 15.9). In Section 16.8C, we shall see how this can also be accomplished with the **Wolff–Kishner reduction.**

PROBLEM 16.15

Show how you might use thioacetal formation and Raney nickel desulfurization to convert: (a) cyclohexanone to cyclohexane and (b) benzaldehyde to toluene.

16.8 THE ADDITION OF DERIVATIVES OF AMMONIA

Aldehydes or ketones react with primary amines (RNH_2) to form compounds with a carbon–nitrogen double bond called **imines** ($RCH=NR$ or $R_2C=NR$). The reaction is acid catalyzed.

$$\overset{\diagdown}{\underset{\diagup}{}}C=O + H_2N-R \underset{}{\overset{H_3O^+}{\rightleftharpoons}} \overset{\diagdown}{\underset{\diagup}{}}C=N-R + H_2O$$

Aldehyde   **1°Amine**   **Imine**
or ketone

Imine formation is slow at very low and at very high pH and generally takes place fastest between pH 4 and pH 5. We can understand why an acid catalyst is necessary if we consider the mechanism that has been proposed for imine formation. The important step is the step in which the protonated amino alcohol loses a molecule of water to become an iminium ion. By protonating the alcohol group, the acid converts a poor leaving group (an $-OH$ group) into a good one (an $-OH_2{}^+$ group).

*Raney nickel is a special nickel catalyst that contains adsorbed hydrogen.

A MECHANISM FOR IMINE FORMATION

Aldehyde **1°Amine**
or ketone

The amine adds to the carbonyl group
to form a dipolar tetrahedral intermediate.

Dipolar **Aminoalcohol**
intermediate

A proton transfer from nitrogen
to oxygen produces an aminoalcohol.

Protonated
aminoalcohol

Protonation of the oxygen
produces a good leaving group.
Loss of a molecule of water
yields an iminium ion.

Iminium ion **Imine**

Transfer of a proton to water
produces the imine and regenerates
the catalytic hydronium ion.

The reaction proceeds more slowly if the hydronium ion concentration is too high, because protonation of the amine itself takes place to a considerable extent; this has the effect of decreasing the concentration of the nucleophile needed in the first step. If the concentration of the hydronium ion is too low, the reaction becomes slower because the concentration of the protonated aminoalcohol becomes lower. A pH between pH 4 and pH 5 is an effective compromise.

Imine formation occurs in many biochemical reactions because enzymes often use an —NH_2 group to react with an aldehyde or ketone. Formation of an imine linkage is important in one step of the reactions that take place during the visual process (see Special Topic G).

Imines are also formed as intermediates in a useful laboratory synthesis of amines that we will study in Section 20.5.

Imines from compounds such as hydroxylamine (NH_2OH), hydrazine (NH_2NH_2), and semicarbazide ($NH_2NHCONH_2$) find use as derivatives of aldehydes and ketones (Section 16.8A). Table 16.2 lists examples of these compounds. The mechanisms by which these derivatives form are similar to the mechanism for the formation of an imine from a primary amine that we have just studied.

16.8A 2,4-Dinitrophenylhydrazones, Semicarbazones, and Oximes

The products of the reactions of aldehydes and ketones with 2,4-dinitrophenylhydrazine, semicarbazide, and hydroxylamine are often used to identify unknown aldehydes and ketones. These compounds, that is, 2,4-dinitrophenylhydrazones, semicarbazones, and oximes, are usually relatively insoluble solids that have sharp characteristic melting points. Table 16.3 gives representative examples from the very extensive tables of these derivatives.

TABLE 16.2 Reactions of aldehydes and ketones with derivatives of ammonia

1. Reaction with Hydroxylamine

General Reaction

$$\text{C=O} + \text{H}_2\text{N—OH} \longrightarrow \text{C=N—OH} + \text{H}_2\text{O}$$

Aldehyde or ketone   Hydroxylamine   An oxime

Specific Example

$$\begin{matrix}\text{H}_3\text{C} \\ \quad \text{C=O} + \text{H}_2\text{NOH} \\ \text{H}\end{matrix} \longrightarrow \begin{matrix}\text{H}_3\text{C} \\ \quad \text{C=NOH} + \text{H}_2\text{O} \\ \text{H}\end{matrix}$$

Acetaldehyde   Acetaldoxime

2. Reactions with Hydrazine, Phenylhydrazine, and 2,4-Dinitrophenylhydrazine

General Reactions

Aldehyde or ketone

$$\text{C=O} + \text{H}_2\text{NNH}_2 \longrightarrow \text{C=NNH}_2 + \text{H}_2\text{O}$$

Hydrazine   A hydrazone

$$\text{C=O} + \text{H}_2\text{NNHC}_6\text{H}_5 \longrightarrow \text{C=NNHC}_6\text{H}_5 + \text{H}_2\text{O}$$

Phenylhydrazine   A phenylhydrazone

$$\text{C=O} + \text{H}_2\text{NNH}-\!\!\bigcirc\!\!-\text{NO}_2 \longrightarrow \text{C=NNH}-\!\!\bigcirc\!\!-\text{NO}_2 + \text{H}_2\text{O}$$

       NO$_2$ NO$_2$

2,4-Dinitrophenylhydrazine   A 2,4-Dinitrophenylhydrazone

Specific Example

$$\begin{matrix}\text{C}_6\text{H}_5 \\ \quad \text{C=O} + \text{H}_2\text{NNHC}_6\text{H}_5 \\ \text{H}_3\text{C}\end{matrix} \xrightarrow[\text{CH}_3\text{CO}_2\text{H}]{\text{H}_3\text{O}^+} \begin{matrix}\text{C}_6\text{H}_5 \\ \quad \text{C=NNHC}_6\text{H}_5 \quad + \text{H}_2\text{O} \\ \text{H}_3\text{C}\end{matrix}$$

Acetophenone   Acetophenone phenylhydrazone

3. Reaction with Semicarbazide

General Reaction

$$\begin{matrix} & & \text{O} & & & & \text{O} \\ & & \| & & & & \| \\ \text{C=O} + \text{H}_2\text{NNHCNH}_2 & \longrightarrow & \text{C=NNHCNH}_2 + \text{H}_2\text{O}\end{matrix}$$

Aldehyde   Semicarbazide   A semicarbazone
or ketone

TABLE 16.2   (continued)

Specific Example

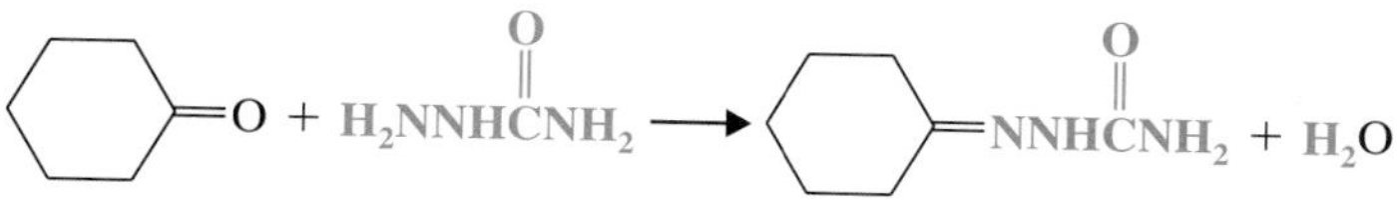

Cyclohexanone
semicarbazone

16.8B Hydrazones: The Wolff–Kishner Reduction

Hydrazones are occasionally used to identify aldehydes and ketones. But, unlike 2,4-dinitrophenylhydrazones, simple hydrazones often have low melting points. Hydrazones, however, are the basis for a useful method to reduce carbonyl groups of aldehydes and ketones to —CH$_2$— groups, called the **Wolff–Kishner reduction:**

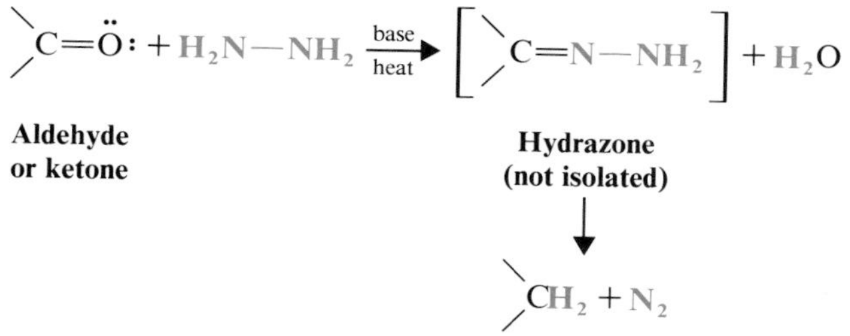

**Aldehyde
or ketone**

**Hydrazone
(not isolated)**

$$\diagdown CH_2 + N_2$$

Specific Example

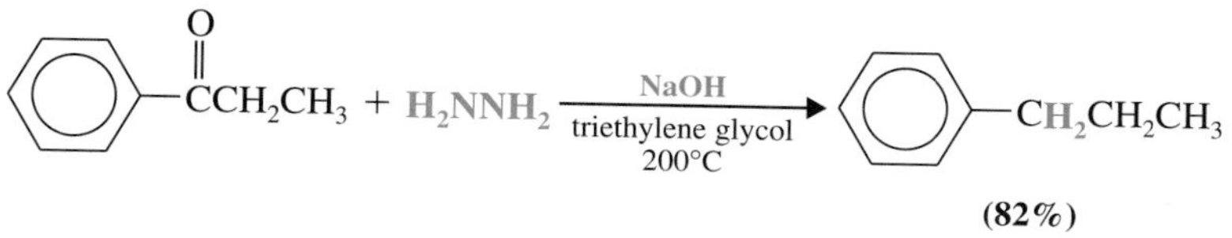

(82%)

TABLE 16.3   Derivatives of aldehydes and ketone

ALDEHYDE OR KETONE	mp (°C) OF 2,4-DINITRO-PHENYLHYDRAZONE	mp (°C) OF SEMICARBAZONE	mp (°C) OF OXIME
Acetaldehyde	168.5	162	46.5
Acetone	128	187 dec	61
Benzaldehyde	237	222	35
2-Methylbenzaldehyde	195	208	49
3-Methylbenzaldehyde	211	204	60
4-Methylbenzaldehyde	233	234	79
Phenylacetaldehyde	121	156	103

The Wolff–Kishner reduction can be accomplished at much lower temperatures if dimethyl sulfoxide is used as the solvent.

The Wolff–Kishner reduction complements the Clemmensen reduction (Section 15.9) and the reduction of thioacetals (Section 16.7E), because all three reactions convert $\diagdown$C=O groups into —CH$_2$— groups. The Clemmensen reduction takes place in strongly acidic media and can be used for those compounds that are sensitive to base. The Wolff–Kishner reduction takes place in strongly basic solutions and can be used for those compounds that are sensitive to acid. The reduction of thioacetals takes place in neutral solution and can be used for compounds that are sensitive to both acids and bases.

The mechanism of the Wolff–Kishner reduction is as follows. The first step is the formation of the hydrazone. Then the strong base brings about an isomerization of the hydrazone to a derivative with the structure $\diagdown$CH—N=NH. This derivative then undergoes the base-catalyzed elimination of a molecule of nitrogen. The loss of the especially stable molecule of nitrogen provides the driving force for the reaction.

A MECHANISM FOR THE WOLFF-KISHNER REDUCTION

1. —C=O + H$_2$N̈—N̈H$_2$ $\rightleftharpoons$ —C=N̈—N̈H$_2$ + H$_2$O

2. —C=N̈—N̈H$_2$ $\underset{\text{H}_2\text{O}}{\overset{\text{OH}^-}{\rightleftharpoons}}$ $\left[\text{—C=N̈—N̈H} \leftrightarrow \text{—C̈—N̈=NH} \right]$ $\underset{\text{OH}^-}{\overset{\text{H}_2\text{O}}{\rightleftharpoons}}$

—C—N̈=N̈—H $\underset{\text{H}_2\text{O}}{\overset{\text{OH}^-}{\rightleftharpoons}}$ —C—N̈=N̈: $\overset{-\text{N}_2}{\longrightarrow}$ —C:$^-$ $\overset{\text{H}_2\text{O}}{\longrightarrow}$ —C—H

16.9 THE ADDITION OF HYDROGEN CYANIDE

Hydrogen cyanide adds to the carbonyl groups of aldehydes and most ketones to form compounds called **cyanohydrins.** Ketones in which the carbonyl group is highly hindered do not undergo this reaction.

$$\underset{\text{O}}{\overset{\text{O}}{\parallel}}$$

RCH + HCN $\rightleftharpoons$ R—C(OH)(H)—CN

R—C(=O)—R' + HCN $\rightleftharpoons$ R—C(OH)(R')—CN

Cyanohydrins

The addition of hydrogen cyanide itself takes place very slowly because HCN is a poor nucleophile. The addition of potassium cyanide, or any base that can generate cyanide ions from the weak acid HCN, causes a dramatic increase in the rate of reaction. This effect was discovered in 1903 by the British chemist Arthur Lapworth, and in his studies of the addition of HCN, Lapworth became one of the originators of the mechanistic view of organic chemistry. Lapworth assumed that the addition was ionic in nature (a remarkable insight considering that Lewis and Kössel's theories of bonding were some 13 years in the future). He proposed "that the formation of cyanohydrins is to be represented as a comparatively slow union of the negative cyanide ion with carbonyl, followed by almost instantaneous combination of the complex with hydrogen." *

A MECHANISM FOR CYANOHYDRIN FORMATION

The cyanide ion, being a stronger nucleophile, is able to attack the carbonyl carbon atom much more rapidly than HCN itself and this is the source of its catalytic effect. Once the addition of cyanide ion has taken place, the strongly basic alkoxide oxygen atom of the intermediate removes a proton from any available acid. If this acid is HCN, this step regenerates the cyanide ion.

Bases stronger than cyanide ion catalyze the reaction by converting HCN ($pK_a \approx 9$) to cyanide ion in an acid–base reaction. The cyanide ions, thus formed, can go on to attack the carbonyl group.

Liquid hydrogen cyanide can be used for this reaction (HCN is a gas at room temperature), but since HCN is very toxic and volatile, it is safer to generate it in the reaction mixture. This can be done by mixing the aldehyde or ketone with aqueous sodium cyanide and then slowly adding sulfuric acid to the mixture. *Even with this procedure, however, great care must be taken and the reaction must be carried out in a very efficient fume hood.*

Cyanohydrins are useful intermediates in organic synthesis. Depending on the conditions used, acidic hydrolysis converts cyanohydrins to α-hydroxy acids or to α,β-unsaturated acids. (The mechanism for this hydrolysis is discussed in Section 18.8H.)

---

*A. Lapworth, *J. Chem. Soc.*, *83*, 995, **1903**. For a fine review of Lapworth's work see M. Saltzman, *J. Chem. Ed.*, *49*, 750, **1972**.

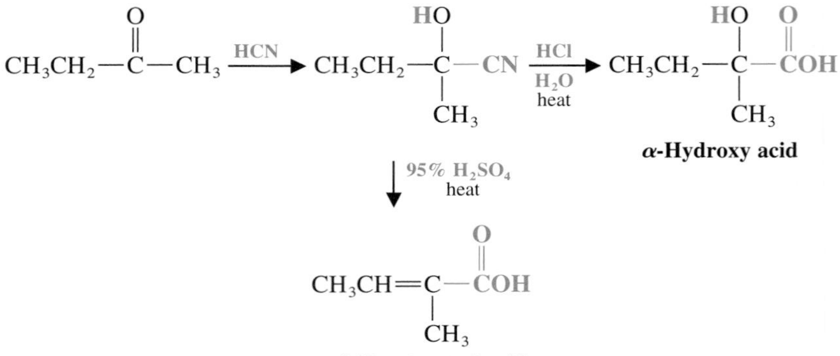

$$CH_3CH_2-\overset{\overset{\displaystyle O}{\|}}{C}-CH_3 \xrightarrow{HCN} CH_3CH_2-\overset{\overset{\displaystyle HO}{|}}{\underset{\underset{\displaystyle CH_3}{|}}{C}}-CN \xrightarrow[\substack{H_2O \\ heat}]{HCl} CH_3CH_2-\overset{\overset{\displaystyle HO}{|}}{\underset{\underset{\displaystyle CH_3}{|}}{C}}-\overset{\overset{\displaystyle O}{\|}}{C}OH$$

α-Hydroxy acid

$$\downarrow \substack{95\% \ H_2SO_4 \\ heat}$$

$$CH_3CH=\overset{\overset{\displaystyle O}{\|}}{\underset{\underset{\displaystyle CH_3}{|}}{C}}-\overset{\displaystyle O}{C}OH$$

α,β-Unsaturated acid

Reducing a cyanohydrin with lithium aluminum hydride gives a β-aminoalcohol:

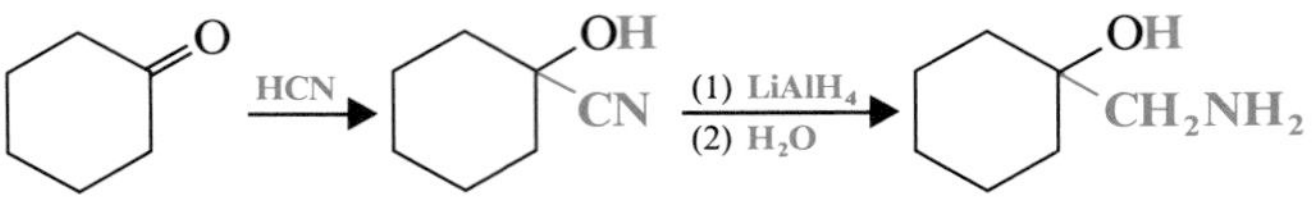

PROBLEM 16.16

(a) Show how you might prepare lactic acid ($CH_3CHOHCO_2H$) from acetaldehyde through a cyanohydrin intermediate. (b) What stereoisomeric form of lactic acid would you expect to obtain?

16.10 THE ADDITION OF YLIDES: THE WITTIG REACTION

Aldehydes and ketones react with phosphorus ylides to yield *alkenes* and triphenylphosphine oxide. (An ylide is a neutral molecule having a negative carbon adjacent to a positive heteroatom.) Phosphorus ylides are also called phosphoranes.

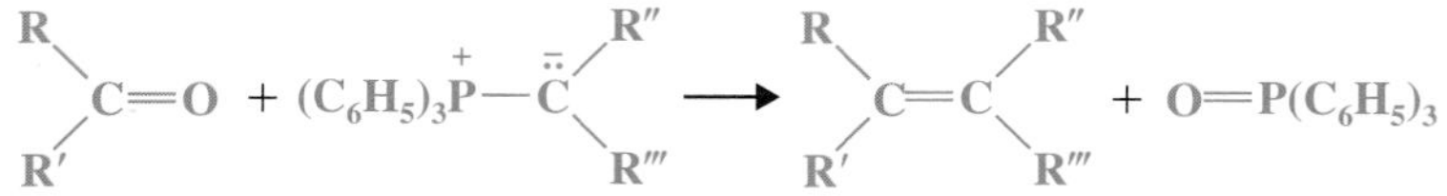

$$\underset{R'}{\overset{R}{\diagdown}}C=O + (C_6H_5)_3\overset{+}{P}-\overset{\overset{\displaystyle ..}{}}{\underset{R'''}{\overset{R''}{\diagup}}}\overset{..}{C} \longrightarrow \underset{R'}{\overset{R}{\diagdown}}C=C\underset{R'''}{\overset{R''}{\diagup}} + O=P(C_6H_5)_3$$

| Aldehyde or ketone | Phosphorus ylide (or phosphorane) | Alkene | Triphenyl-phosphine oxide |

This reaction, known as the **Wittig reaction,*** has proved to be a valuable method for synthesizing alkenes. The Wittig reaction is applicable to a wide variety of compounds, and although a mixture of (E) and (Z) isomers may result, the Wittig reaction gives a great advantage over most other alkene syntheses in that *no ambiguity exists as to the location of the double bond in the product.*

---

*Discovered in 1954 by Georg Wittig, then at the University of Tübingen. Wittig was a cowinner of the Nobel prize for chemistry in 1979.

Phosphorus ylides are easily prepared from triphenylphosphine and alkyl halides. Their preparation involves two reactions:

General Reaction

Reaction 1 $(C_6H_5)_3P:$ + $\underset{R'''}{\overset{R''}{>}}CH-X$ ⟶ $(C_6H_5)_3\overset{+}{P}-\underset{R'''}{\overset{R''}{C}H}$ X^-

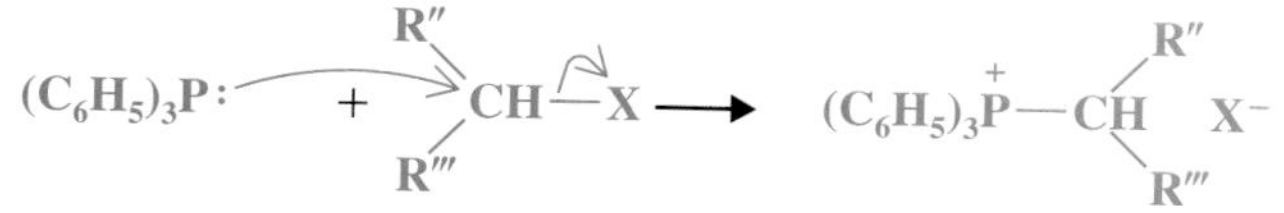

Triphenylphosphine   **An Alkyltriphenylphosphonium halide**

Reaction 2 $(C_6H_5)_3\overset{+}{P}-\underset{R'''}{\overset{R''}{C}}H$ $:\bar{B}$ ⟶ $(C_6H_5)_3\overset{+}{P}-\underset{R'''}{\overset{R''}{C}}:^-$ + H:B

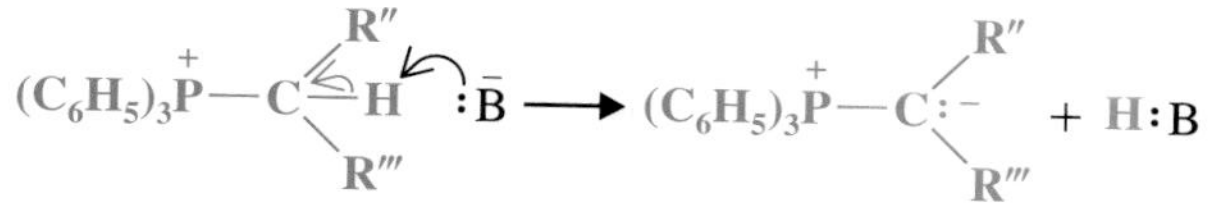

A Phosphorus ylide

Specific Example

Reaction 1 $(C_6H_5)_3P: + CH_3Br \xrightarrow{C_6H_6}$ $(C_6H_5)_3\overset{+}{P}-CH_3$ Br^-

Methyltriphenylphosphonium bromide (89%)

Reaction 2 $(C_6H_5)_3\overset{+}{P}-\underset{Br^-}{CH_3} + C_6H_5Li$ ⟶ $(C_6H_5)_3\overset{+}{P}-CH_2:^- + C_6H_6$ + LiBr

The first reaction is a nucleophilic substitution reaction. Triphenylphosphine is an excellent nucleophile and a weak base. It reacts readily with 1° and 2° alkyl halides by an S_N2 mechanism to displace a halide ion from the alkyl halide to give an alkyltriphenylphosphonium salt. The second reaction is an acid–base reaction. A strong base (usually an alkyllithium or phenyllithium) removes a proton from the carbon that is attached to phosphorus to give the ylide.

Phosphorus ylides can be represented as a hybrid of the two resonance structures shown here. Quantum mechanical calculations indicate that the contribution made by the first structure is relatively unimportant.

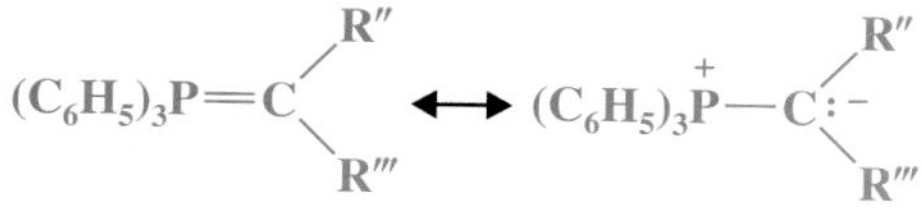

$(C_6H_5)_3P{=}\underset{R'''}{\overset{R''}{C}}$ ⟷ $(C_6H_5)_3\overset{+}{P}-\underset{R'''}{\overset{R''}{C}}:^-$

The mechanism of the Wittig reaction has been the subject of considerable study. An early mechanistic proposal suggested that the ylide, acting as a carbanion, attacks the carbonyl carbon of the aldehyde or ketone to form an unstable intermediate with separated charges called a **betaine.** In the next step, the betaine is envisioned as becoming an unstable four-membered cyclic system called an **oxaphosphatane,** which then spontaneously loses triphenylphosphine oxide to become an alkene. More recently, studies by E. Vedejs (of the University of Wisconsin) and others suggest that the betaine is not an intermediate and that the oxaphosphatane is formed directly by a cycloaddition reaction. The driving force for the Wittig reaction is the formation of the very strong ($DH° = 130$ kcal mol^{-1}) phosphorus–oxygen bond in triphenylphosphine oxide.

A MECHANISM FOR THE WITTIG REACTION

Aldehyde   **Ylide**   **Betaine**   **Oxaphosphatane**
or ketone   **(may not be formed)**

$$\underset{R}{\overset{R'}{C}}=\underset{R}{\overset{R''}{C}} \quad + \quad :O{=}P(C_6H_5)_3$$

Alkene   **Triphenylphosphine**
(+ diastereomer)   **oxide**

Specific Example

Methylenecyclohexane
(86% from cyclohexanone
and methyltriphenylphosphonium
bromide)

The elimination of triphenylphosphine oxide from the betaine if, indeed, it forms, may occur in two separate steps as we have just shown, or both steps may occur simultaneously.

While Wittig syntheses may appear to be complicated, in actual practice they are easy to carry out. Most of the steps can be carried out in the same reaction vessel, and the entire synthesis can be accomplished in a matter of hours.

The overall result of a Wittig synthesis is

Planning a Wittig synthesis begins with recognizing in the desired alkene what can be the aldehyde or ketone component and what can be the halide component. Any or all of the R groups may be hydrogen. The halide component must be a primary, secondary, or methyl halide.

SAMPLE PROBLEM

Outline a Wittig synthesis of 2-methyl-1-phenyl-1-propene.

ANSWER:

We examine the structure of the compound, paying attention to the groups on each side of the double bond.

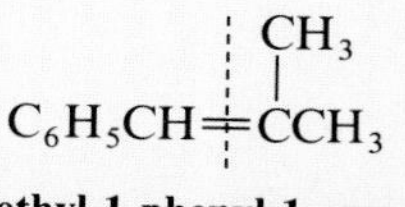

$$C_6H_5CH{=}CCH_3$$
$$\overset{\displaystyle CH_3}{|}$$

2-Methyl-1-phenyl-1-propene

Two general approaches to the synthesis are possible:

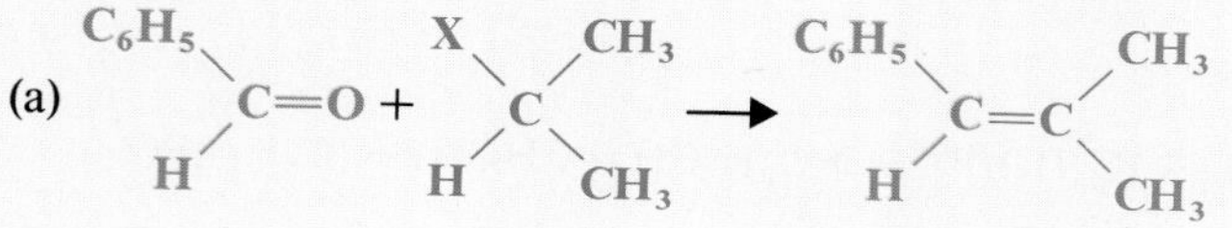

(a)

$$\underset{H}{\overset{C_6H_5}{>}}C{=}O + \underset{H}{\overset{X}{>}}C\underset{CH_3}{\overset{CH_3}{<}} \longrightarrow \underset{H}{\overset{C_6H_5}{>}}C{=}C\underset{CH_3}{\overset{CH_3}{<}}$$

or

(b)

$$\underset{H}{\overset{C_6H_5}{>}}C\underset{X}{\overset{H}{<}} + O{=}C\underset{CH_3}{\overset{CH_3}{<}} \longrightarrow \underset{H}{\overset{C_6H_5}{>}}C{=}C\underset{CH_3}{\overset{CH_3}{<}}$$

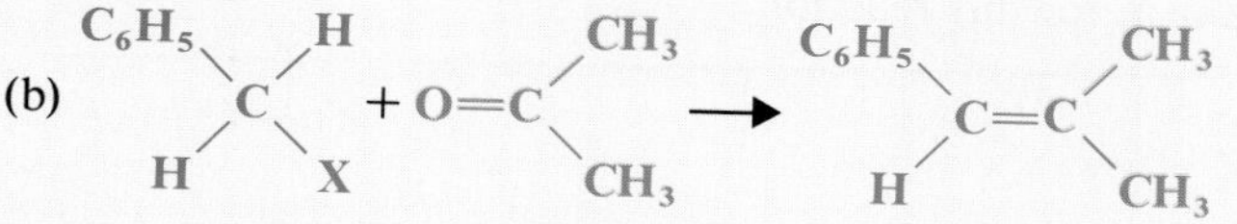

In (a) we first make the ylide from a 2-halopropane and then allow it to react with benzaldehyde.

(a) $(CH_3)_2CHBr + (C_6H_5)_3P \longrightarrow (CH_3)_2CH-\overset{+}{P}(C_6H_5)_3 \ Br^- \xrightarrow{RLi}$

$(CH_3)_2\overset{..}{C}-\overset{+}{P}(C_6H_5)_3 \xrightarrow{C_6H_5CHO} (CH_3)_2C{=}CHC_6H_5 + (C_6H_5)_3P{=}O$

In (b) we would make the ylide from a benzyl halide and allow it to react with acetone.

(b) $C_6H_5CH_2Br + (C_6H_5)_3P \longrightarrow C_6H_5CH_2-\overset{+}{P}(C_6H_5)_3 \ Br^- \xrightarrow{RLi}$

$C_6H_5\overset{..}{C}H-\overset{+}{P}(C_6H_5)_3 \xrightarrow{(CH_3)_2C{=}O} C_6H_5CH{=}C(CH_3)_2 + (C_6H_5)_3P{=}O$

PROBLEM 16.17

In addition to triphenylphosphine, assume that you have available as starting materials any necessary aldehydes, ketones, and organic halides. Show how you might synthesize each of the following alkenes using the Wittig reaction:

(a) $C_6H_5\underset{\underset{CH_3}{|}}{C}{=}CH_2$

(d)

(b) $C_6H_5\underset{\underset{CH_3}{|}}{C}{=}CHCH_3$

(c) $\underset{H_3C}{\overset{H_3C}{\diagdown}}C{=}CH_2$

(e) $CH_3CH_2CH{=}\underset{\underset{CH_3}{|}}{C}CH_2CH_3$

(f) $C_6H_5CH{=}CHCH{=}CH_2$

(g) $C_6H_5CH{=}CHC_6H_5$

PROBLEM 16.18

Triphenylphosphine can be used to convert epoxides to alkenes, for example,

$$C_6H_5\overset{\overset{\ddot{O}}{\diagup\!\!\diagdown}}{CH}{-}CHCH_3 + (C_6H_5)_3P\!: \longrightarrow C_6H_5CH{=}CHCH_3 + (C_6H_5)_3PO$$

Propose a likely mechanism for this reaction.

16.11   THE ADDITION OF ORGANOMETALLIC REAGENTS: THE REFORMATSKY REACTION

In Section 11.8 we studied the addition of Grignard reagents, organolithium compounds, and sodium alkynides to aldehydes and ketones. These reactions, as we saw then, can be used to produce a wide variety of alcohols:

$$\overset{\delta-}{R}\!:\!\overset{\delta+}{MgX} + \underset{\diagup}{\overset{\diagdown}{}}C{=}O \longrightarrow R{-}\overset{|}{\underset{|}{C}}{-}OMgX \xrightarrow{H_3O^+} R{-}\overset{|}{\underset{|}{C}}{-}OH$$

$$\overset{\delta-}{R}\!:\!\overset{\delta+}{Li} + \underset{\diagup}{\overset{\diagdown}{}}C{=}O \longrightarrow R{-}\overset{|}{\underset{|}{C}}{-}OLi \xrightarrow{H_3O^+} R{-}\overset{|}{\underset{|}{C}}{-}OH$$

$$RC{\equiv}\overset{\delta-}{C}\!:\!\overset{\delta+}{Na} + \underset{\diagup}{\overset{\diagdown}{}}C{=}O \longrightarrow RC{\equiv}C{-}\overset{|}{\underset{|}{C}}{-}ONa \xrightarrow{H_3O^+} RC{\equiv}C{-}\overset{|}{\underset{|}{C}}{-}OH$$

We shall now examine a similar reaction that involves the addition of an organozinc reagent to the carbonyl group of an aldehyde or ketone. This reaction, called the *Reformatsky reaction*, extends the carbon skeleton of an aldehyde or ketone and yields *β*-hydroxy esters. It involves treating an aldehyde or ketone with an *α*-bromo ester in the presence of zinc metal; the solvent most often used is benzene. The initial product is a zinc alkoxide, which must be hydrolyzed to yield the *β*-hydroxy ester.

The intermediate in the reaction appears to be an organozinc reagent that adds to the carbonyl group in a manner analogous to that of a Grignard reagent.

A MECHANISM FOR THE REFORMATSKY REACTION

Because the organozinc reagent is less reactive than a Grignard reagent, it does not add to the ester group. The β-hydroxy esters produced in the Reformatsky reaction are easily dehydrated to α,β-unsaturated esters, because dehydration yields a system in which the carbon–carbon double bond is conjugated with the carbon–oxygen double bond of the ester.

β-Hydroxy ester → α,β-Unsaturated ester

Examples of the Reformatsky reaction are the following:

$$CH_3CH_2CH_2\overset{O}{\overset{\|}{C}}H + BrCH_2CO_2Et \xrightarrow[(2)\ H_3O^+]{(1)\ Zn} CH_3CH_2CH_2\overset{OH}{\overset{|}{C}}HCH_2CO_2Et$$

where Et = CH_3CH_2—

$$CH_3\overset{O}{\overset{\|}{C}}H + Br\overset{CH_3}{\underset{CH_3}{\overset{|}{\underset{|}{C}}}}CO_2Et \xrightarrow[(2)\ H_3O^+]{(1)\ Zn} CH_3\overset{OH}{\overset{|}{C}}H\overset{CH_3}{\underset{CH_3}{\overset{|}{\underset{|}{C}}}}CO_2Et$$

$$\underset{\substack{\| \\ C_6H_5CH}}{O} + \underset{\substack{| \\ CH_3}}{Br-CH-CO_2Et} \xrightarrow[\text{(2) } H_3O^+]{\text{(1) Zn}} \underset{\substack{OH \quad CH_3 \\ | \qquad |}}{C_6H_5CH-CH-CO_2Et}$$

PROBLEM 16.19

Show how you would use a Reformatsky reaction in the synthesis of each of the following compounds. (Additional steps may be necessary in some instances.)

(a) $(CH_3)_2\overset{\overset{\displaystyle OH}{|}}{C}CH_2CO_2CH_2CH_3$ (c) $CH_3CH_2CH_2CH_2CO_2CH_2CH_3$

(b) [cyclohexane ring]—$\underset{\substack{| \\ CH_3}}{\overset{\substack{OH \\ |}}{C}HCO_2CH_2CH_3}$

16.12 OXIDATION OF ALDEHYDES AND KETONES

Aldehydes are much more easily oxidized than ketones. Aldehydes are readily oxidized by strong oxidizing agents such as potassium permanganate, and they are also oxidized by such mild oxidizing agents as silver oxide.

$$\underset{\substack{\| \\ RCH}}{O} \xrightarrow{KMnO_4,\ OH^-} \underset{\substack{\| \\ RCO^-}}{O} \xrightarrow{H_3O^+} \underset{\substack{\| \\ RCOH}}{O}$$

$$\underset{\substack{\| \\ RCH}}{O} \xrightarrow{Ag_2O,\ OH^-} \underset{\substack{\| \\ RCO^-}}{O} \xrightarrow{H_3O^+} \underset{\substack{\| \\ RCOH}}{O}$$

Notice that in these oxidations aldehydes lose the hydrogen that is attached to the carbonyl carbon atom. Because ketones lack this hydrogen, they are more resistant to oxidation.

16.12A The Baeyer–Villiger Oxidation of Aldehydes and Ketones

Both aldehydes and ketones are oxidized by peroxy acids. This reaction, called the *Baeyer–Villiger oxidation,* is especially useful with ketones, because it converts them to carboxylic esters. For example, treating acetophenone with a peroxy acid converts it to the ester, phenyl acetate.

$$\underset{\substack{\| \\ C_6H_5-C-CH_3 \\ \textbf{Acetophenone}}}{O} \xrightarrow{RCOOH} \underset{\substack{\| \\ C_6H_5-O-C-CH_3 \\ \textbf{Phenyl acetate}}}{O}$$

The mechanism proposed for this reaction involves the following steps:

A MECHANISM FOR THE BAEYER-VILLIGER OXIDATION

$$CH_3-\overset{\overset{O}{\|}}{\underset{C_6H_5}{C}} + :O-O-\overset{\overset{O}{\|}}{C}-R \underset{}{\overset{(1)}{\rightleftharpoons}} CH_3-\overset{\overset{O-H}{|}}{\underset{C_6H_5}{C}}-O-O-\overset{\overset{O}{\|}}{C}-R \overset{(2)\ H^+}{\underset{}{\rightleftharpoons}}$$

$$CH_3-\overset{\overset{O-H}{|}}{\underset{C_6H_5}{C}}-O-O-\overset{\overset{\overset{+}{O}-H}{\|}}{C}-R \overset{-R\overset{O}{\overset{\|}{C}}-OH}{\underset{(3)}{\longrightarrow}} CH_3-\overset{\overset{O-H}{|}}{\underset{C_6H_5}{\overset{+}{C}}}-O \overset{-H^+}{\longrightarrow} CH_3\overset{\overset{O}{\|}}{C}-O-C_6H_5 + H^+$$

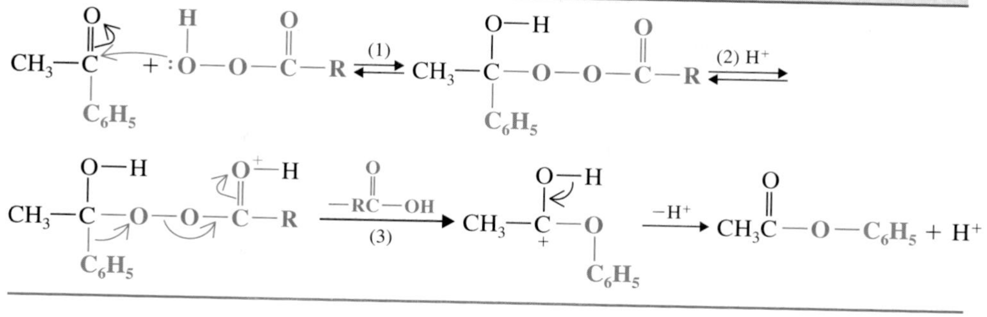

In step (1), the peroxy acid adds to the carbonyl group of the ketone. At this point there are several equilibria involving the attachment of a proton to one of the oxygen atoms of this addition product. When a proton attaches itself to one of the oxygen atoms of the carboxylic acid portion, it makes this part a good leaving group, which in step (3) departs. Simultaneously with the departure of RCO_2H, the phenyl group migrates (as an anion) to the electron-deficient oxygen that is being created. After that, the loss of a proton produces the ester.

Step 3 and the following reaction show that a phenyl group has a greater tendency to migrate than a methyl group; otherwise the product would have been $C_6H_5COOCH_3$ and not $CH_3COOC_6H_5$. This tendency of a group to migrate is called its **migratory aptitude.** Studies of the Baeyer–Villiger oxidation and other reactions have shown that the migratory aptitude of groups is H > phenyl > 3° alkyl > 2° alkyl > 1° alkyl > methyl. In all cases, this order is for groups migrating with their electron pairs, that is, as anions.

PROBLEM 16.20

When benzaldehyde reacts with a peroxy acid, the product is benzoic acid. The mechanism for this reaction is analogous to the one just given for the oxidation of acetophenone, and the outcome illustrates the greater migratory aptitude of a hydrogen atom compared to phenyl. Outline all the steps involved.

PROBLEM 16.21

Give the structure of the product that would result from a Baeyer–Villiger oxidation of cyclopentanone.

PROBLEM 16.22

What would be the major product formed in the Baeyer–Villiger oxidation of 3-methyl-2-butanone?

16.13 CHEMICAL ANALYSIS FOR ALDEHYDES AND KETONES

16.13A Derivatives of Aldehydes and Ketones

Aldehydes and ketones can be differentiated from noncarbonyl compounds through their reactions with derivatives of ammonia (Section 16.8). Semicarbazide, 2,4-dinitrophenylhydrazine, and hydroxylamine react with aldehydes and ketones to form precipitates. Semicarbazones and oximes are usually colorless, whereas 2,4-dinitrophenylhydrazones are usually orange. The melting points of these derivatives can also be used in identifying specific aldehydes and ketones.

16.13B Tollens' Test (Silver Mirror Test)

The ease with which aldehydes undergo oxidation provides a useful test that differentiates aldehydes from most ketones. Mixing aqueous silver nitrate with aqueous ammonia produces a solution known as Tollens' reagent. The reagent contains the diamminosilver(I) ion, $Ag(NH_3)_2^+$. Although this ion is a very weak oxidizing agent, it oxidizes aldehydes to carboxylate ions. As it does this, silver is reduced from the $+1$ oxidation state [of $Ag(NH_3)_2^+$] to metallic silver. If the rate of reaction is slow and the walls of the vessel are clean, metallic silver deposits on the walls of the test tube as a mirror; if not, it deposits as a gray to black precipitate. Tollens' reagent gives a negative result with all ketones except α-hydroxy ketones.

16.14 SPECTROSCOPIC PROPERTIES OF ALDEHYDES AND KETONES

16.14A IR Spectra of Aldehydes and Ketones

Carbonyl groups of aldehydes and ketones give rise to very strong $C{=}O$ stretching absorption bands in the $1665-1780$ cm^{-1} region. The exact location of the absorption (Table 16.4) depends on the structure of the aldehyde or ketone and is one of the most useful and characteristic absorptions in the IR spectrum. Saturated acyclic aldehydes typically absorb near 1730 cm^{-1}; similar ketones absorb near 1715 cm^{-1}.

Conjugation of the carbonyl group with a double bond or a benzene ring shifts the $C{=}O$ *absorption to lower frequencies by about 40 cm⁻¹.* This shift to lower frequencies occurs because the double bond of a conjugated compound has more single bond

TABLE 16.4 Carbonyl stretching bands of aldehydes and ketones

	C=O STRETCHING FREQUENCIES		
COMPOUND	RANGE (cm^{-1})	COMPOUND	RANGE (cm^{-1})
R—CHO	1720–1740	RCOR	1705–1720
Ar—CHO	1695–1715	ArCOR	1680–1700
—C=C—CHO	1680–1690	—C=C—COR	1665–1680
		Cyclohexanone	1715
		Cyclopentanone	1751
		Cyclobutanone	1785

character (see the resonance structures below), and single bonds are easier to stretch than double bonds.

The location of the carbonyl absorption of cyclic ketones depends on the size of the ring (compare the cyclic compounds in Table 16.4). *As the ring grows smaller the C=O stretching peak is shifted to higher frequencies.*

Vibrations of the C—H bond of the CHO group of aldehydes also give two weak bands in the 2700–2775 and 2820–2900 cm^{-1} regions that are easily identified.

Figure 16.1 shows the IR spectrum of phenylethanal.

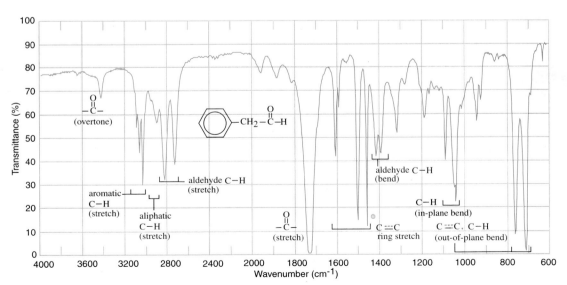

FIGURE 16.1 The infrared spectrum of phenylethanal.

16.14B NMR Spectra of Aldehydes and Ketones

^{13}C NMR Spectra

The carbonyl carbon atom of an aldehyde or ketone gives characteristic NMR signals in the δ 180–215 region of ^{13}C spectra. Since almost no other signals occur in this region, *the presence of a signal near in this region (near δ 200) strongly suggests the presence of a carbonyl group.*

^{1}H NMR Spectra

The aldehydic proton gives a signal far downfield in ^{1}H NMR spectra in a region (δ 9–10) where almost no other protons absorb; therefore, it is easily identified. The aldehydic proton of an aliphatic aldehyde shows spin–spin coupling with protons on the adjacent α carbon, and the splitting pattern reveals the degree of substitution of the α carbon. For example, in acetaldehyde (CH_3CHO) the aldehydic proton signal is split into a quartet by the three methyl protons, and the methyl proton is split into a doublet by the aldehydic proton. The coupling constant is about 3 Hz.

Protons on the α carbon are deshielded by the carbonyl group and their signals generally appear in the δ 2.0–2.3 region. Methyl ketones show a characteristic (3H) singlet near δ 2.1.

Figures 16.2 and 16.3 show annotated ^{1}H and ^{13}C spectra of phenylethanal.

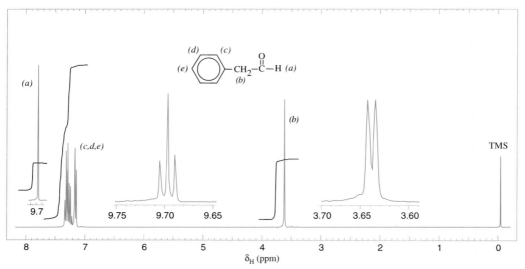

FIGURE 16.2 The 300 MHz ^{1}H NMR spectrum of phenylethanal. The small coupling between the aldehyde and methylene protons (2.6 Hz) is shown in the expanded offset plots.

16.14C UV Spectra

The carbonyl groups of saturated aldehydes and ketones give a weak absorption band in the UV region between 270 and 300 nm. This band is shifted to longer wavelengths (300–350 nm) when the carbonyl group is conjugated with a double bond.

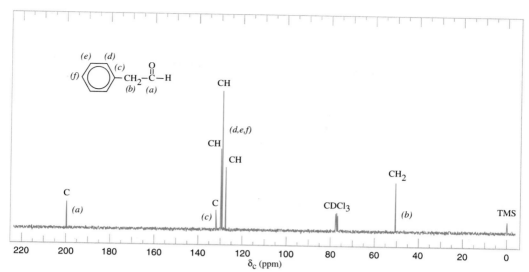

FIGURE 16.3 The broad band proton-decoupled ^{13}C NMR spectrum of phenylethanal. DEPT ^{13}C NMR information and carbon assignments are shown near each peak.

16.15 SUMMARY OF THE ADDITION REACTIONS OF ALDEHYDES AND KETONES

Table 16.5 summarizes the nucleophilic addition reactions of aldehydes and ketones occurring at the carbonyl carbon atom that we have studied so far. In Chapter 17 we shall see other examples.

TABLE 16.5 *Nucleophilic addition reactions of aldehydes and ketones*

1. Addition of Organometallic Compounds

General Reaction

$$\overset{\delta-}{R}:\overset{\delta+}{M} + \overset{}{\underset{\curvearrowleft}{C}}=\overset{\cdot\cdot}{\underset{\curvearrowleft}{O}}: \longrightarrow R-\overset{|}{\underset{|}{C}}-O^-M^+ \xrightarrow{H^+} R-\overset{|}{\underset{|}{C}}-O-H$$

Specific Example Using a Grignard Reagent (Section 11.7)

$$CH_3CH_2MgBr + CH_3\overset{O}{\overset{\|}{C}}-H \xrightarrow[(2)\ H^+]{(1)\ diethyl\ ether} CH_3CH_2\overset{OH}{\overset{|}{C}HCH_3}$$

(67%)

Specific Example Using the Reformatsky Reaction (Section 16.11)

$$CH_3\overset{}{\underset{\overset{|}{CH_3}}{C}}HCH_2\overset{O}{\overset{\|}{C}}H + Br-\overset{CH_3}{\underset{\overset{|}{CH_3}}{C}}-CO_2CH_2CH_3 \xrightarrow[(2)\ H_3O^+]{(1)\ Zn} CH_3\overset{}{\underset{\overset{|}{CH_3}}{C}}HCH_2\overset{HO}{\overset{}{C}}H\overset{CH_3}{\underset{\overset{|}{CH_3}}{C}}-CO_2CH_2CH_3$$

(65%)

(continued)

TABLE 16.5 (continued)

2. Addition of Hydride Ion

General Reaction

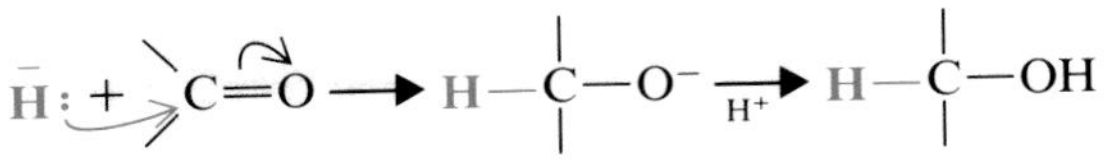

$$\bar{H}: + \underset{}{\overset{}{{>}}}C{=}O \longrightarrow H-\underset{|}{\overset{|}{C}}-O^- \xrightarrow[H^+]{} H-\underset{|}{\overset{|}{C}}-OH$$

Specific Examples Using Metal Hydrides (Section 11.3)

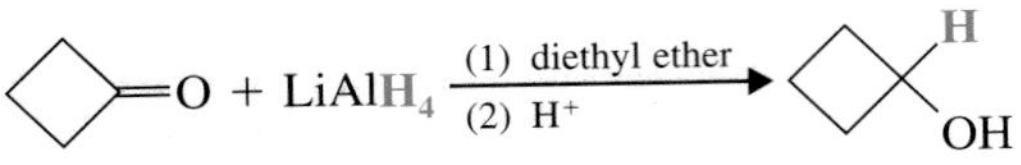

$$\square{=}O + LiAlH_4 \xrightarrow[\text{(2) H}^+]{\text{(1) diethyl ether}} \square\underset{OH}{\overset{H}{<}}$$

(90%)

$$\underset{}{\overset{O}{\underset{\|}{}}}CH_3CCH_2CH_2CH_3 + NaBH_4 \xrightarrow[OH^-]{CH_3OH} CH_3\underset{}{\overset{OH}{\underset{|}{C}H}}CH_2CH_2CH_3$$

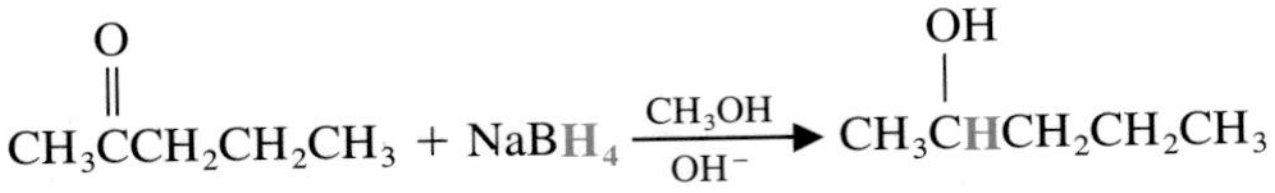

(100%)

3. Addition of Hydrogen Cyanide (Section 16.9)

General Reaction

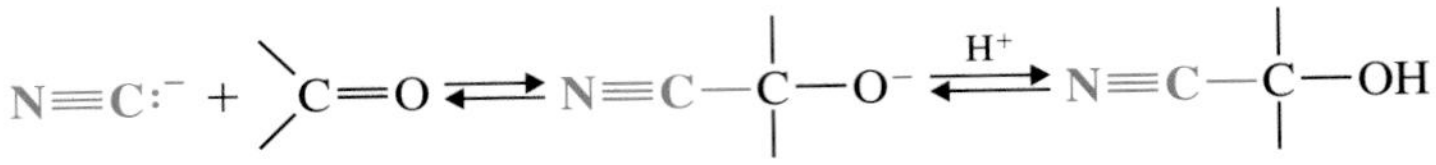

$$N{\equiv}C{:}^- + \underset{}{\overset{}{{>}}}C{=}O \rightleftharpoons N{\equiv}C-\underset{|}{\overset{|}{C}}-O^- \overset{H^+}{\rightleftharpoons} N{\equiv}C-\underset{|}{\overset{|}{C}}-OH$$

Specific Example

$$\underset{H_3C}{\overset{H_3C}{{>}}}C{=}O \xrightarrow[H^+]{NaCN} \underset{H_3C}{\overset{H_3C}{>}}\underset{CN}{\overset{OH}{C}}$$

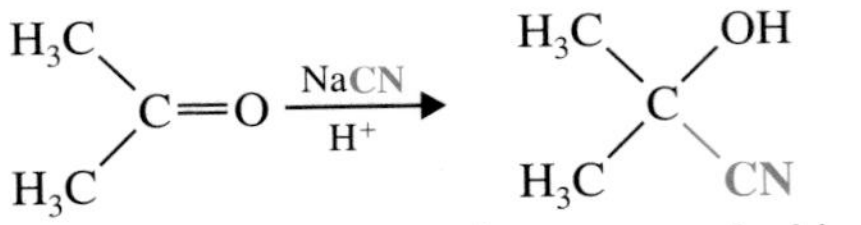

Acetone cyanohydrin
(78%)

4. Addition of Ylides (Section 16.10)

The Wittig Reaction

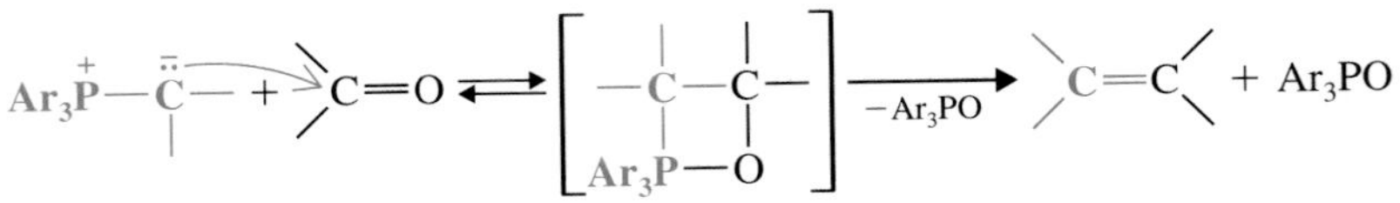

$$Ar_3\overset{+}{P}-\overset{..}{\underset{|}{C}}- + \underset{}{\overset{}{{>}}}C{=}O \rightleftharpoons \left[-\underset{Ar_3P-O}{\overset{|}{C}}-\underset{|}{\overset{|}{C}}- \right] \xrightarrow{-Ar_3PO} \underset{}{\overset{}{{>}}}C{=}C\underset{}{\overset{}{{<}}} + Ar_3PO$$

5. Addition of Alcohols (Section 16.7)

General Reaction

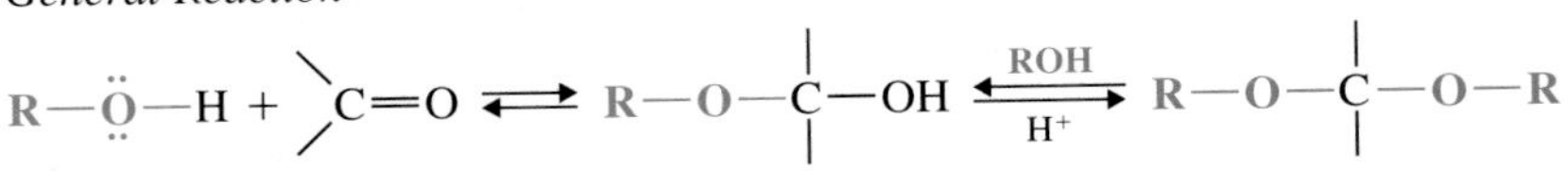

$$R-\overset{..}{\underset{..}{O}}-H + \underset{}{\overset{}{{>}}}C{=}O \rightleftharpoons R-O-\underset{|}{\overset{|}{C}}-OH \underset{H^+}{\overset{ROH}{\rightleftharpoons}} R-O-\underset{|}{\overset{|}{C}}-O-R$$

Hemiacetal   Acetal

(continued)

TABLE 16.5 (continued)

Specific Example

$$C_2H_5OH + CH_3CH \rightleftharpoons C_2H_5O-\underset{\underset{H}{|}}{\overset{\overset{CH_3}{|}}{C}}-OH \xrightarrow[H^+]{C_2H_5OH} C_2H_5O-\underset{\underset{H}{|}}{\overset{\overset{CH_3}{|}}{C}}-OC_2H_5$$

6. Addition of Derivatives of Ammonia (Section 16.8)

General Reaction

$$-\overset{|}{\underset{H}{N}}-H + \overset{\diagdown}{\diagup}C=O \rightleftharpoons \left[-\overset{|}{\underset{H}{N}}-\overset{|}{\underset{|}{C}}-OH \right] \xrightarrow[-H_2O]{} -\overset{..}{N}=C\overset{\diagup}{\diagdown}$$

Specific Examples

$$CH_3\overset{\overset{O}{||}}{C}H + NH_2OH \longrightarrow CH_3CH{=}NOH$$
Acetaldoxime

ADDITIONAL PROBLEMS

16.23 Give a structural formula and another acceptable name for each of the following compounds:

(a) Formaldehyde

(b) Acetaldehyde

(c) Phenylacetaldehyde

(d) Acetone

(e) Ethyl methyl ketone

(f) Acetophenone

(g) Benzophenone

(h) Salicylaldehyde

(i) Vanillin

(j) Diethyl ketone

(k) Ethyl isopropyl ketone

(l) Diisopropyl ketone

(m) Dibutyl ketone

(n) Dipropyl ketone

(o) Cinnamaldehyde

16.24 Write structural formulas for the products formed when propanal reacts with each of the following reagents:

(a) $NaBH_4$ in aqueous NaOH

(b) C_6H_5MgBr, then H_2O

(c) $LiAlH_4$, then H_2O

(d) Ag_2O, OH^-

(e) $(C_6H_5)_3P{=}CH_2$

(f) H_2 and Pt

(g) $HOCH_2CH_2OH$ and H^+

(h) $CH_3\overset{..}{C}H-\overset{+}{P}(C_6H_5)_3$

(i) (1) $BrCH_2CO_2C_2H_5$, Zn; (2) H_3O^+

(j) $Ag(NH_3)_2{}^+$

(k) Hydroxylamine

(l) Semicarbazide

(m) Phenylhydrazine

(n) Cold dilute $KMnO_4$

(o) $HSCH_2CH_2SH$, H^+

(p) $HSCH_2CH_2SH$, H^+, then Raney nickel

16.25 Give structural formulas for the products formed (if any) from the reaction of acetone with each reagent in Problem 16.24.

16.26   What products would be obtained from each of the following reactions of aceto-phenone?

(a) Acetophenone + HNO$_3$ $\xrightarrow[H_2SO_4]{}$

(d) Acetophenone + NaBH$_4$ $\xrightarrow[OH^-]{H_2O}$

(b) Acetophenone + C$_6$H$_5$NHNH$_2$ $\longrightarrow$

(e) Acetophenone + C$_6$H$_5$MgBr $\xrightarrow[(2)\ H_2O]{}$

(c) Acetophenone + $^-$:CH$_2$—$\overset{+}{P}$(C$_6$H$_5$)$_3$ $\longrightarrow$

16.27   (a) Give three methods for synthesizing phenyl propyl ketone from benzene and any other needed reagents.   (b) Give three methods for transforming phenyl propyl ketone into butylbenzene.

16.28   Show how you would convert benzaldehyde into each of the following. You may use any other needed reagents, and more than one step may be required.

(a) Benzyl alcohol   (g) 3-Methyl-1-phenyl-1-butanol   (m) C$_6$H$_5$CH(OH)CN

(b) Benzoic acid   (h) Benzyl bromide   (n) C$_6$H$_5$CH=NOH

(c) Benzoyl chloride   (i) Toluene   (o) C$_6$H$_5$CH=NNHC$_6$H$_5$

(d) Benzophenone   (j) C$_6$H$_5$CH(OCH$_3$)$_2$ (p) C$_6$H$_5$CH=NNHCONH$_2$

(e) Acetophenone   (k) C$_6$H$_5$CH^{18}O   (q) C$_6$H$_5$CH=CHCH=CH$_2$

(f) 1-Phenylethanol   (l) C$_6$H$_5$CHDOH   (r) C$_6$H$_5$CH(OH)SO$_3$Na

16.29   Show how ethyl phenyl ketone (C$_6$H$_5$COCH$_2$CH$_3$) could be synthesized from each of the following:

(a) Benzene   (c) Benzonitrile, C$_6$H$_5$CN

(b) Benzoyl chloride   (d) Benzaldehyde

16.30   Show how benzaldehyde could be synthesized from each of the following:

(a) Benzyl alcohol   (c) Phenylethyne   (e) C$_6$H$_5$CO$_2$CH$_3$

(b) Benzoic acid   (d) Phenylethene (styrene)   (f) C$_6$H$_5$C≡N

16.31   Give structures for compounds **A–E.**

Cyclohexanol $\xrightarrow[\text{acetone}]{H_2CrO_4}$ **A** (C$_6$H$_{10}$O) $\xrightarrow[(2)\ H_3O^+]{(1)\ CH_3MgI}$ **B** (C$_7$H$_{14}$O) $\xrightarrow[\text{heat}]{H^+}$

C (C$_7$H$_{12}$) $\xrightarrow[(2)\ Zn,\ H_2O]{(1)\ O_3}$ **D** (C$_7$H$_{12}$O$_2$) $\xrightarrow[(2)\ H_3O^+]{(1)\ Ag_2O,\ OH^-}$ **E** (C$_7$H$_{12}$O$_3$)

16.32   The following reaction sequence shows how the carbon chain of an aldehyde may be lengthened by two carbon atoms. What are the intermediates **K–M?**

Ethanal $\xrightarrow[(2)\ H^+]{(1)\ BrCH_2CO_2Et,\ Zn}$ **K** (C$_6$H$_{12}$O$_3$) $\xrightarrow{H^+,\ heat}$ **L** (C$_6$H$_{10}$O$_2$) $\xrightarrow{H_2,\ Pt}$

M (C$_6$H$_{12}$O$_2$) $\xrightarrow[(2)\ H_2O]{(1)\ DIBAL-H}$ butanal

HINT: The ^{13}C NMR spectrum of **L** consists of signals at δ 166.7, δ 144.5, δ 122.8, δ 60.2, δ 17.9, and δ 14.3.

16.33   Warming piperonal (Section 16.3) with dilute aqueous HCl converts it to a compound with the formula C$_7$H$_6$O$_3$. What is this compound and what type of reaction is involved?

16.34   Starting with benzyl bromide, show how you would synthesize each of the following:

(a) $C_6H_5CH_2CHOHCH_3$ (c) $C_6H_5CH=CH—CH=CHC_6H_5$

(b) $C_6H_5CH_2CH_2CHO$ (d) $C_6H_5CH_2COCH_2CH_3$

16.35   Compounds **A** and **D** do not give positive Tollens' tests; however, compound **C** does. Give structures for **A–D**.

4-Bromobutanal $\xrightarrow{\text{HOCH}_2\text{CH}_2\text{OH, H}^+}$ **A** $(C_6H_{11}O_2Br)$ $\xrightarrow{\text{Mg, diethyl ether}}$

[**B** $(C_6H_{11}MgO_2Br)$] $\xrightarrow[\text{(2) H}_3\text{O}^+,\ \text{H}_2\text{O}]{\text{(1) CH}_3\text{CHO}}$ **C** $(C_6H_{12}O_2)$ $\xrightarrow[\text{H}^+]{\text{CH}_3\text{OH}}$ **D** $(C_7H_{14}O_2)$

16.36   Provide the missing reagents and intermediate in the following synthesis:

$HO-$⬡$-CH_2OH \xrightarrow{\text{(a)}} CH_3O-$⬡$-CH_2OH \xrightarrow{\text{(b)}} ? \xrightarrow{\text{(c)}}$

CH_3O-⬡$-\overset{\overset{\text{OH}}{|}}{C}\text{H}\overset{|}{\underset{|}{\underset{\text{CH}_3}{C}}}\text{HCO}_2\text{Et} \xrightarrow{\text{(d)}} CH_3O-$⬡$-\overset{\overset{\text{OH}}{|}}{C}\text{H}\overset{|}{\underset{\text{CH}_3}{C}}\text{HCH}_2\text{OH}$

16.37   Outlined here is a synthesis of glyceraldehyde (Section 5.14A). What are the intermediates **A–C**, and what stereoisomeric form of glyceraldehyde would you expect to obtain?

$CH_2=CHCH_2OH \xrightarrow[\text{CH}_2\text{Cl}_2]{\text{PCC}}$ **A** $(C_3H_4O) \xrightarrow{\text{CH}_3\text{OH, H}^+}$

B $(C_5H_{10}O_2) \xrightarrow[\text{cold, dilute}]{\text{KMnO}_4,\ \text{OH}^-}$ **C** $(C_5H_{12}O_4) \xrightarrow[\text{H}_2\text{O}]{\text{H}_3\text{O}^+}$ glyceraldehyde

16.38   Consider the reduction of (*R*)-3-phenyl-2-pentanone by sodium borohydride. After the reduction is complete, the mixture is separated by gas–liquid chromatography into two fractions. These fractions contain isomeric compounds, and each isomer is optically active. What are these two isomers, and what is the stereoisomeric relationship between them?

16.39   The structure of the sex pheromone (attractant) of the female tsetse fly has been confirmed by the following synthesis. Compound **C** appears to be identical to the natural pheromone in all respects (including the response of the male tsetse fly). Provide structures for **A, B,** and **C.**

$BrCH_2(CH_2)_7CH_2Br \xrightarrow[\text{(2) 2 RLi}]{\text{(1) 2 (C}_6\text{H}_5)_3\text{P}}$ **A** $(C_{45}H_{46}P_2) \xrightarrow{2\ \text{CH}_3(\text{CH}_2)_{11}\overset{\overset{\text{O}}{\|}}{\text{C}}\text{CH}_3}$

B $(C_{37}H_{72}) \xrightarrow{\text{H}_2,\ \text{Pt}}$ **C** $(C_{37}H_{76})$

16.40   Outline simple chemical tests that would distinguish between each of the following:

(a) Benzaldehyde and benzyl alcohol

(b) Hexanal and 2-hexanone

(c) 2-Hexanone and hexane

(d) 2-Hexanol and 2-hexanone

(e) $C_6H_5CH{=}CHCOC_6H_5$ and $C_6H_5COC_6H_5$

(f) Pentanal and diethyl ether

(g) $\underset{\text{O}}{\overset{\text{O}}{\|}}\,\, \underset{\text{O}}{\overset{\text{O}}{\|}}$ $CH_3CCH_2CCH_3$ and $CH_3C{=}CHCCH_3$ (with OH and O labels)

(h) 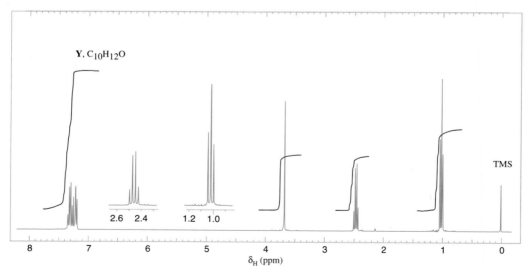 and

16.41 Compounds **W** and **X** are isomers; they have the molecular formula C_9H_8O. The IR spectrum of each compound shows a strong absorption band near 1715 cm^{-1}. Oxidation of either compound with hot, basic potassium permanganate followed by acidification yields phthalic acid. The ^{1}H NMR spectrum of **W** shows a multiplet at δ 7.3 and a singlet at δ 3.4. The ^{1}H NMR spectrum of **X** shows a multiplet at δ 7.5, a triplet at δ 3.1, and a triplet at δ 2.5. Propose structures for **W** and **X**.

Phthalic acid

16.42 Compounds **Y** and **Z** are isomers with the molecular formula $C_{10}H_{12}O$. The IR spectrum of each compound shows a strong absorption band near 1710 cm^{-1}. The ^{1}H NMR spectra of **Y** and **Z** are given in Figs. 16.4 and 16.5. Propose structures for **Y** and **Z**.

FIGURE 16.4 The 300 MHz ^{1}H NMR spectrum of compound **Y**, problem 16.42. Expansions of the signals are shown in the offset plots.

16.43 Compound **A** ($C_9H_{18}O$) forms a phenylhydrazone, but gives a negative Tollens' test. The IR spectrum of **A** has a strong band near 1710 cm^{-1}. The broad band proton-decoupled ^{13}C NMR spectrum of **A** is given in Fig. 16.6. Propose a structure for **A**.

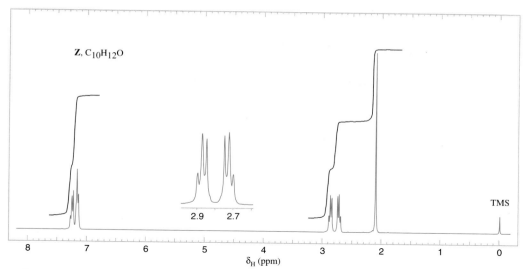

FIGURE 16.5 The 300 MHz ^{1}H NMR spectrum of compound **Z**, problem 16.42. Expansions of the signals are shown in the offset plots.

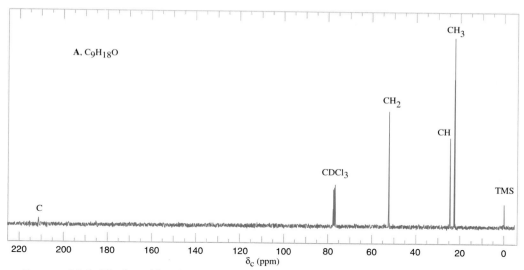

FIGURE 16.6 The broad band proton-decoupled ^{13}C NMR spectrum of Compound **A**, Problem 16.43. Information from the DEPT ^{13}C NMR spectrum is given above the peaks.

16.44   Compound **B** ($C_8H_{12}O_2$) shows a strong carbonyl absorption in its IR spectrum. The broad band proton-decoupled ^{13}C NMR spectrum of **B** is given in Fig. 16.7. Propose a structure for **B**.

16.45   When semicarbazide ($H_2NNHCONH_2$) reacts with a ketone (or an aldehyde) to form a semicarbazone (Section 16.8) only one nitrogen atom of semicarbazide acts as a nucleophile and attacks the carbonyl carbon atom of the ketone. The product of the reaction, consequently, is $R_2C=NNHCONH_2$ rather than $R_2C=NCONHNH_2$. What factor accounts for the fact that two nitrogen atoms of semicarbazide are relatively nonnucleophilic?

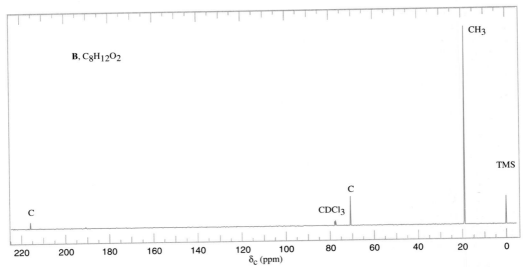

FIGURE 16.7 The broad band proton-decoupled ^{13}C NMR spectrum of Compound **B**, Problem 16.44. Information from the DEPT ^{13}C NMR spectrum is given above the peaks

16.46   Dutch elm disease is caused by a fungus transmitted to elm trees by the elm bark beetle. The female beetle, when she has located an attractive elm tree, releases several pheromones, including multistriatin, below. These pheromones attract male beetles which bring with them the deadly fungus.

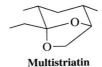

Multistriatin

Treating multistriatin with dilute aqueous acid at room temperature leads to the formation of a product, $C_{10}H_{18}O_3$, which shows a strong infrared peak near 1715 cm^{-1}. Propose a structure for this product.

16.47   An industrial synthesis of benzaldehyde makes use of toluene and molecular chlorine as starting materials to produce $C_6H_5CHCl_2$. This compound is then converted to benzaldehyde. Suggest what steps are involved in the process.

16.48   An optically active compound, $C_6H_{12}O$, gives a positive test with 2,4-dinitrophenylhydrazine but a negative test with Tollens' reagent. What is the structure of the compound?

16.49   In the case of aldehydes and unsymmetrical ketones, two isomeric oximes are possible. What is the origin of this isomerism?

CHAPTER 17

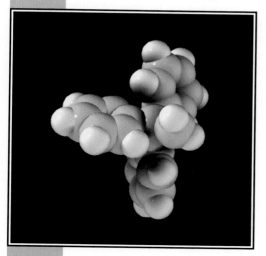

A Wittig reagent
(Problem 17.36)

ALDEHYDES AND KETONES II.
ALDOL REACTIONS

17.1 THE ACIDITY OF THE α HYDROGENS OF CARBONYL COMPOUNDS: ENOLATE IONS

In Chapter 16, we found that one important characteristic of aldehydes and ketones is their ability to undergo nucleophilic addition at their carbonyl groups.

$$\diagdown C{=}O + H{-}Nu \longrightarrow \diagup C \diagdown^{OH}_{Nu}$$

Nucleophilic addition

A second important characteristic of carbonyl compounds is an unusual acidity of hydrogen atoms on carbon atoms adjacent to the carbonyl group. (These hydrogen atoms are usually called the **α hydrogens,** and the carbon to which they are attached is called the **α carbon.**)

$$R{-}\overset{\overset{\displaystyle \ddot{O}}{\|}}{C}{-}\overset{\overset{\alpha}{|}}{\underset{|}{C}}{-}\overset{\overset{\beta}{|}}{\underset{|}{C}}{-}$$
$$\phantom{R{-}C{-}}HH$$

α Hydrogens are unusually acidic ($pK_a = 19-20$).

β Hydrogens are not acidic ($pK_a = 40-50$).

753

When we say that the α hydrogens are acidic, *we mean that they are unusually acidic for hydrogen atoms attached to carbon.* The pK_a values for the α hydrogens of most simple aldehydes or ketones are of the order of $19-20$ $(K_a = 10^{-19}-10^{-20})$. This means that they are more acidic than hydrogen atoms of ethyne, $pK_a = 25$ $(K_a = 10^{-25})$, and are far more acidic than the hydrogens of ethene $(pK_a = 44)$ or of ethane $(pK_a = 50)$.

The reasons for the unusual acidity of the α hydrogens of carbonyl compounds are straightforward: The carbonyl group is strongly electron withdrawing (Section 3.9), and when a carbonyl compound loses an α proton, the anion that is produced is *stabilized by resonance. The negative charge of the anion is delocalized.*

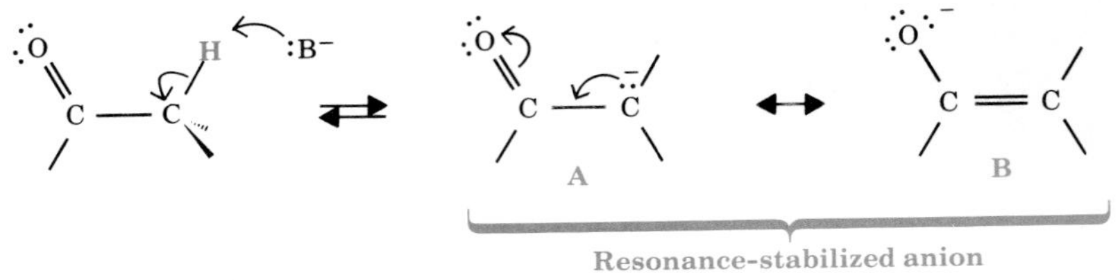

Resonance-stabilized anion

We see from this reaction that two resonance structures, **A** and **B**, can be written for the anion. In structure **A** the negative charge is on carbon and in structure **B** the negative charge is on oxygen. Both structures contribute to the hybrid. Although structure **A** is favored by the strength of its carbon–oxygen π bond relative to the weaker carbon–carbon π bond of **B**, structure **B** makes a greater contribution to the hybrid because oxygen, being highly electronegative, is better able to accommodate the negative charge. We can depict the hybrid in the following way:

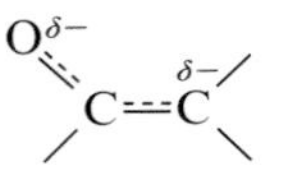

When this resonance-stabilized anion accepts a proton, it can do so in either of two ways: It can accept the proton at carbon to form the original carbonyl compound in what is called the **keto form,** or it may accept the proton at oxygen to form an **enol.**

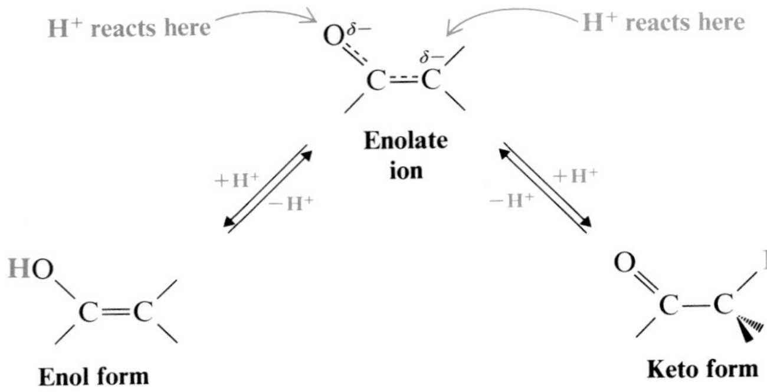

Both of these reactions are reversible. Because of its relation to the enol, the resonance-stabilized anion is called an **enolate ion.**

17.2 KETO AND ENOL TAUTOMERS

The keto and enol forms of carbonyl compounds are constitutional isomers, but of a special type. Because they are easily interconverted in the presence of traces of acids and bases, chemists use a special term to describe this type of constitutional isomerism. Interconvertible keto and enol forms are said to be **tautomers,** and their interconversion is called **tautomerization.**

Under most circumstances, we encounter keto–enol tautomers in a state of equilibrium. (The surfaces of ordinary laboratory glassware are able to catalyze the interconversion and establish the equilibrium.) For simple monocarbonyl compounds such as acetone and acetaldehyde, the amount of the enol form present at equilibrium is *very small.* In acetone it is much less than 1%; in acetaldehyde the enol concentration is too small to be detected. The greater stability of the following keto forms of monocarbonyl compounds can be related to the greater strength of the carbon–oxygen π bond compared to the carbon–carbon π bond ($\sim$ 87 kcal mol^{-1} vs $\sim$ 60 kcal mol^{-1}).

In compounds whose molecules have two carbonyl groups separated by one —CH$_2$— group (called β-dicarbonyl compounds), the amount of enol present at equilibrium is far higher. For example, 2,4-pentanedione exists in the enol form to an extent of 76%.

The greater stability of the enol form of β-dicarbonyl compounds can be attributed to stability gained through resonance stabilization of the conjugated double bonds and (in a cyclic form) through hydrogen bonding.

Resonance stabilization of the enol form

PROBLEM 17.1

For all practical purposes, the compound 2,4-cyclohexadien-1-one exists totally in its enol form. Write the structure of 2,4-cyclohexadien-1-one and of its enol form. What special factor accounts for the stability of the enol form?

17.3 REACTIONS VIA ENOLS AND ENOLATE IONS

17.3A Racemization

When a solution of (+)-*sec*-butyl phenyl ketone (see following reaction) in aqueous ethanol is treated with either acids or bases, the solution gradually loses its optical activity. After a time, isolation of the ketone shows that it has been racemized.

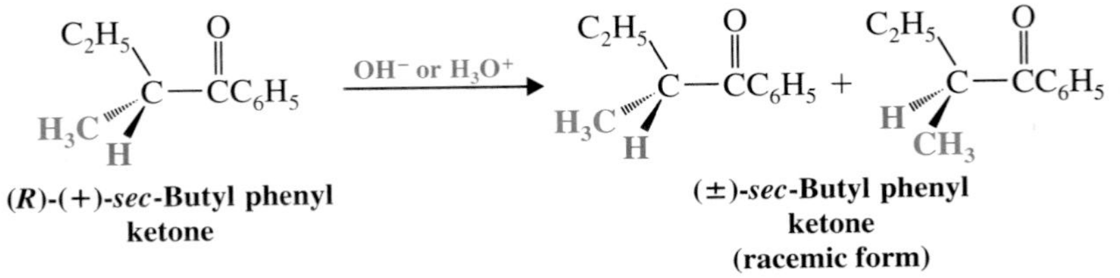

(R)-(+)-*sec*-Butyl phenyl ketone

$\xrightarrow{\text{OH}^- \text{ or } H_3O^+}$

(±)-*sec*-Butyl phenyl ketone (racemic form)

Racemization takes place in the presence of acids or bases because the ketone slowly but reversibly changes to its enol *and the enol is achiral.* When the enol reverts to the keto form, it produces equal amounts of the two enantiomers.

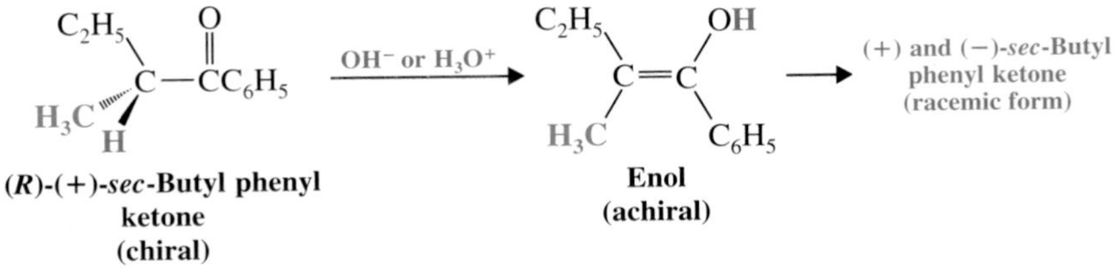

(R)-(+)-*sec*-Butyl phenyl ketone (chiral)

$\xrightarrow{\text{OH}^- \text{ or } H_3O^+}$

Enol (achiral)

(+) and (−)-*sec*-Butyl phenyl ketone (racemic form)

Base catalyzes the formation of an enol through the intermediate formation of an enolate ion:

A MECHANISM FOR BASE-CATALYZED ENOLIZATION

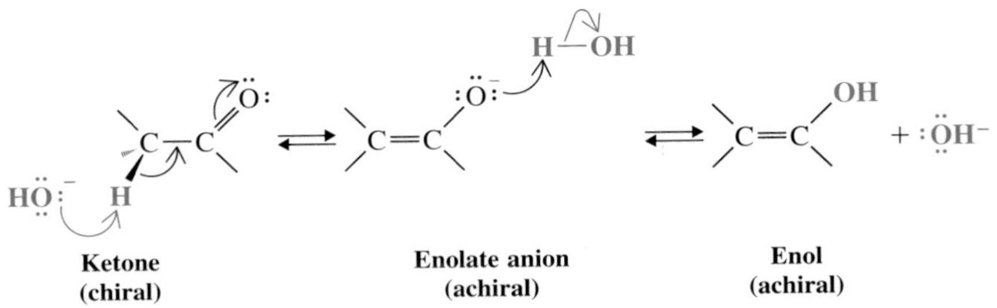

Ketone (chiral)

Enolate anion (achiral)

Enol (achiral)

Acid can catalyze enolization in the following way:

A MECHANISM FOR ACID-CATALYZED ENOLIZATION

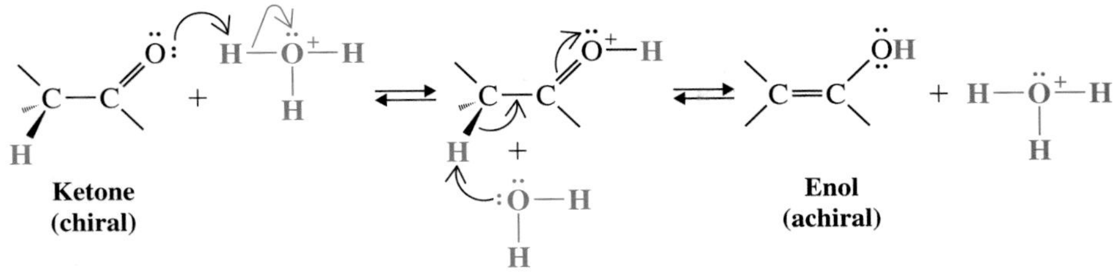

Ketone (chiral)   Enol (achiral)

PROBLEM 17.2

Would you expect optically active ketones such as the following to undergo acid- or base-catalyzed racemization? Explain your answer.

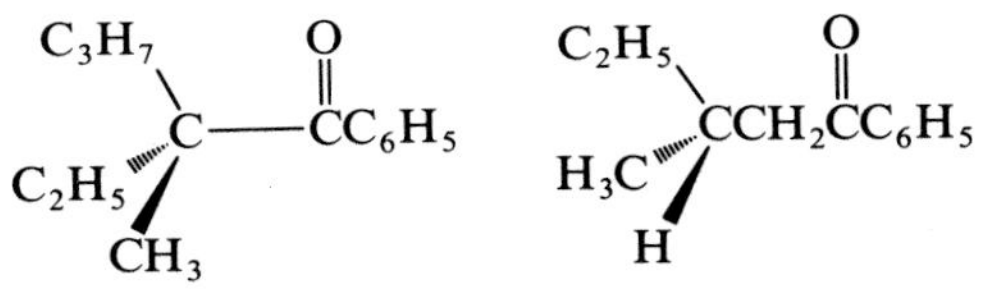

PROBLEM 17.3

When *sec*-butyl phenyl ketone is treated with either OD^- or D_3O^+ in the presence of D_2O, the ketone undergoes hydrogen–deuterium exchange and produces:

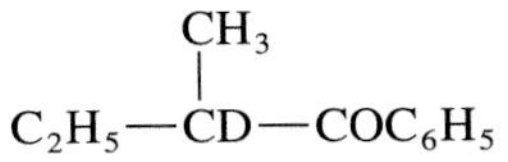

$$\underset{\text{CH}_3}{\overset{\text{CH}_3}{\underset{|}{}}}$$
$$C_2H_5-CD-COC_6H_5$$

Write mechanisms that account for this behavior.

17.3B Halogenation of Ketones

Ketones that have an α hydrogen react readily with halogens by substitution. The rates of these halogenation reactions *increase when acids or bases are added and substitution takes place almost exclusively at the α carbon:*

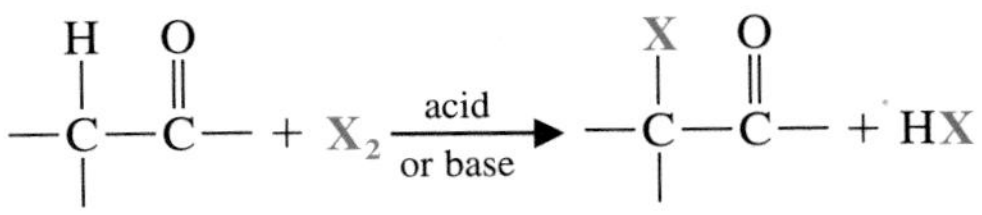

This behavior of ketones can be accounted for in terms of two related properties that we have already encountered: the acidity of the α hydrogens of ketones and the tendency of ketones to form enols.

Base-Promoted Halogenation

In the presence of bases, halogenation takes place through the slow formation of an enolate ion or an enol, followed by a rapid reaction of the enolate ion or enol with halogen.

A MECHANISM FOR BASE-PROMOTED HALOGENATION

1. B:⁻ + —C—C + (slow) ⇌ B:H + C==C (Enolate ion) ⇌ (fast) C=C—OH + B:⁻ (Enol)

2a. X—X + —C—C ⟶ (fast) —C—C + X⁻

Enolate ion

or

2b. X—X + C=C···O—H-X ⇌ (fast) —C—C + HX + X⁻

Enol

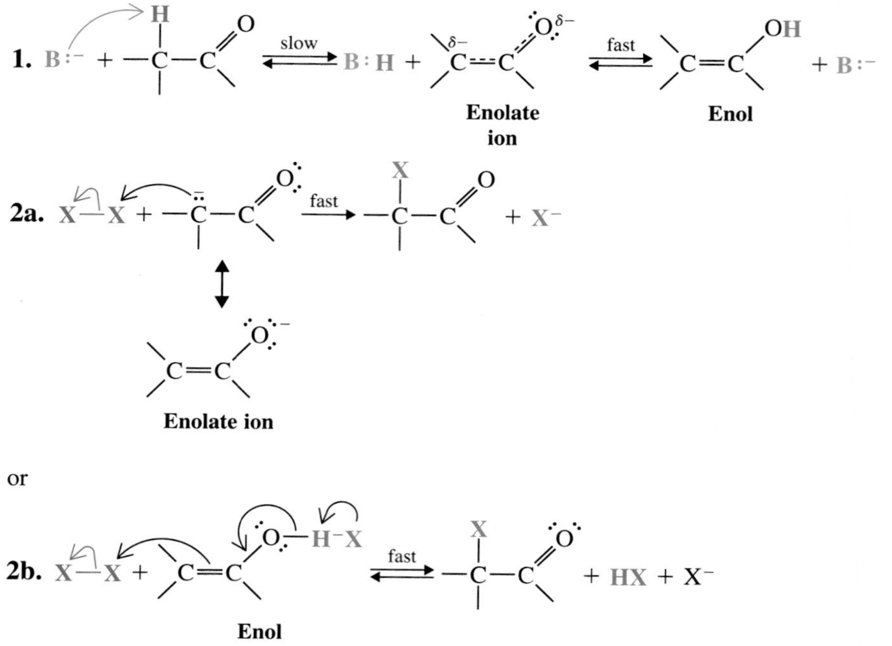

As we shall see in Section 17.4 multiple halogenations can occur.

Acid-Catalyzed Halogenation

In the presence of acids, halogenation takes place through slow formation of an enol followed by rapid reaction of the enol with the halogen.

A MECHANISM FOR ACID-CATALYZED HALOGENATION

1. —C—C + H:B ⇌ (fast) —C—C ⟶ (slow) C=C—OH + H:B (Enol)

2. X—X + C=C···O—H-X ⇌ (fast) —C—C + HX

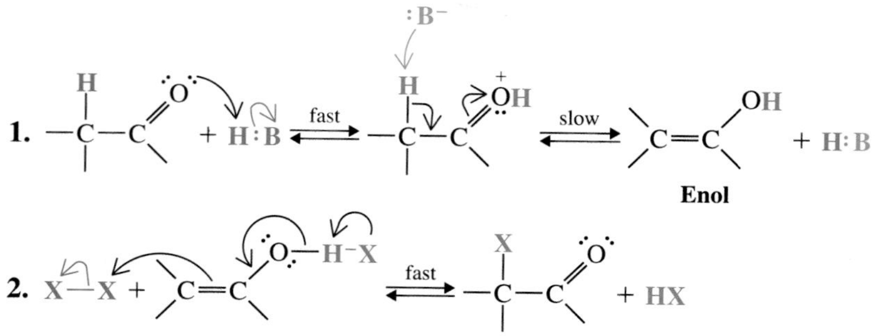

Part of the evidence that supports these mechanisms comes from studies of reaction kinetics. Both base-promoted and acid-catalyzed halogenations of ketones *show initial rates that are independent of the halogen concentration.* The mechanisms that we have written are in accord with this observation: In both instances the slow step of the mechanism occurs before the intervention of the halogen. (The initial rates are also independent of the nature of the halogen; see Problem 17.5.)

PROBLEM 17.4

Why do we say that the first reaction is "base promoted" rather than "base catalyzed?"

PROBLEM 17.5

Additional evidence for the halogenation mechanisms that we just presented comes from the following facts: (a) Optically active *sec*-butyl phenyl ketone undergoes acid-catalyzed racemization at a rate exactly equivalent to the rate at which it undergoes acid-catalyzed halogenation. (b) *sec*-Butyl phenyl ketone undergoes acid-catalyzed iodination at the same rate that it undergoes acid-catalyzed bromination. (c) *sec*-Butyl phenyl ketone undergoes base-catalyzed hydrogen–deuterium exchange at the same rate that it undergoes base-promoted halogenation. Explain how each of these observations supports the mechanisms that we have presented.

17.4 THE HALOFORM REACTION

When methyl ketones react with halogens in the presence of base (cf. Section 17.3), multiple halogenations always occur at the carbon of the methyl group. Multiple

halogenations occur because introduction of the first halogen (owing to its electronegativity) makes the remaining α hydrogens on the methyl carbon more acidic.

A MECHANISM FOR THE HALOFORM
REACTION: HALOGENATION STEP

Enolate ion

$$C_6H_5-\overset{\overset{\displaystyle O}{\|}}{C}-\overset{\overset{\displaystyle X}{|}}{\underset{\underset{\displaystyle H}{|}}{C}}-H + :\bar{B} \;\rightleftharpoons\; C_6H_5-\overset{\overset{\displaystyle O}{\|}}{C}-\overset{\overset{\displaystyle X}{|}}{\underset{\underset{\displaystyle H}{|}}{C}}:^- \overset{X-X}{\longrightarrow} C_6H_5-\overset{\overset{\displaystyle O}{\|}}{C}-\overset{\overset{\displaystyle X}{|}}{\underset{\underset{\displaystyle H}{|}}{C}}-X + X^-$$

Acidity is
increased by
electron-withdrawing
halogen

$\downarrow$:$\bar{B}$ then X_2

$$C_6H_5-\overset{\overset{\displaystyle O}{\|}}{C}-\overset{\overset{\displaystyle X}{|}}{\underset{\underset{\displaystyle X}{|}}{C}}-X$$

When methyl ketones react with halogens in aqueous sodium hydroxide (i.e., in *hypohalite solutions**), an additional reaction takes place. Hydroxide ion attacks the carbonyl carbon atom of the trihalo ketone and causes a cleavage at the carbon–carbon bond between the carbonyl group and the trihalomethyl group, a moderately good leaving group. This cleavage ultimately produces a carboxylate ion and a *haloform* (i.e., either $CHCl_3$, $CHBr_3$, or CHI_3). The initial step is a nucleophilic attack by hydroxide ion on the carbonyl carbon atom. In the next step carbon–carbon bond cleavage occurs and the trihalomethyl anion ($:CX_3^-$) departs. This is one of the rare instances in which a carbanion acts as a leaving group. This step can occur because the trihalomethyl anion is unusually stable; its negative charge is dispersed by the three electronegative halogen atoms. In the last step, a proton transfer takes place between the carboxylic acid and the trihalomethyl anion.

A MECHANISM FOR THE HALOFORM REACTION: CLEAVAGE STEP

$$C_6H_5-\overset{\overset{\displaystyle :\ddot{O}}{\|}}{C}-\overset{\overset{\displaystyle X}{|}}{\underset{\underset{\displaystyle X}{|}}{C}}-X + ^-:\ddot{O}H \;\rightleftharpoons\; C_6H_5-\overset{\overset{\displaystyle ^-:\ddot{O}:}{|}}{\underset{\underset{\displaystyle :OH}{|}}{C}}-\overset{\overset{\displaystyle X}{|}}{\underset{\underset{\displaystyle X}{|}}{C}}-X$$

$$C_6H_5-\overset{\overset{\displaystyle ^-:\ddot{O}:}{|}}{\underset{\underset{\displaystyle :OH}{|}}{C}}-\overset{\overset{\displaystyle X}{|}}{\underset{\underset{\displaystyle X}{|}}{C}}-X \;\rightleftharpoons\; C_6H_5-\overset{\overset{\displaystyle :\ddot{O}}{\|}}{C}\underset{.\ddot{O}-H}{\diagdown} \quad + \; ^-:\overset{\overset{\displaystyle X}{|}}{\underset{\underset{\displaystyle X}{|}}{C}}-X$$

$\downarrow\uparrow$

$$C_6H_5-\overset{\overset{\displaystyle :\ddot{O}}{\|}}{C}\underset{\ddot{O}:^-}{\diagdown} \quad + \; H-\overset{\overset{\displaystyle X}{|}}{\underset{\underset{\displaystyle X}{|}}{C}}-X$$

Carboxylate   Haloform
ion

*Dissolving a halogen in aqueous sodium hydroxide produces a solution containing sodium hypohalite (NaOX) because of the following equilibrium:

$$X_2 + 2\,NaOH \;\rightleftharpoons\; NaOX + NaX + H_2O$$

The haloform reaction is of synthetic utility as a means of converting methyl ketones to carboxylic acids. When the haloform reaction is used in synthesis, chlorine and bromine are most commonly used as the halogen component. Chloroform ($CHCl_3$) and bromoform ($CHBr_3$) are both liquids which are immiscible with water and are easily separated from the aqueous solution containing the carboxylate ion.

When water is chlorinated to purify it for public consumption, chloroform is produced from organic impurities in the water via the haloform reaction. (Many of these organic impurities are naturally occurring, such as humic substances.) The presence of chloroform in public water is of concern for water treatment plants and environmental officers, because chloroform is carcinogenic. Thus, the technology that solves one problem creates another. It is worthwhile recalling, however, that before chlorination of water was introduced, thousands of people died in epidemics of diseases, such as cholera and dysentery.

17.4A The Iodoform Test

The haloform reaction using iodine and aqueous sodium hydroxide is called the *iodoform test*. The iodoform test was once frequently used in structure determinations (before the advent of NMR spectral analysis) because it allows identification of the following two groups:

$$-\underset{\underset{O}{\|}}{C}-CH_3 \quad \text{and} \quad -\underset{\underset{OH}{|}}{CH}-CH_3$$

Compounds containing either of these groups react with iodine in sodium hydroxide to give a bright yellow precipitate of *iodoform* (CHI_3, mp 119°C). Compounds containing the $-CHOHCH_3$ group give a positive iodoform test because they are first oxidized to methyl ketones:

$$-\underset{\underset{OH}{|}}{CH}CH_3 + I_2 + 2\,OH^- \longrightarrow -\underset{\underset{O}{\|}}{C}CH_3 + 2\,I^- + 2\,H_2O$$

Methyl ketones then react with iodine and hydroxide ion to produce iodoform:

$$-\underset{\underset{O}{\|}}{C}-CH_3 + 3\,I_2 + 3\,OH^- \longrightarrow -\underset{\underset{O}{\|}}{C}-CI_3 + 3\,I^- + 3\,H_2O$$

$$-\underset{\underset{O}{\|}}{C}-CI_3 + OH^- \longrightarrow -\underset{\underset{O}{\|}}{C}-O^- + \quad CHI_3\downarrow$$

Yellow precipitate

The group to which the $-COCH_3$ or $-CHOHCH_3$ function is attached can be aryl, alkyl, or hydrogen. Thus, even ethanol and acetaldehyde give positive iodoform tests.

PROBLEM 17.6 _____

Which of the following compounds would give a positive iodoform test?

(a) Acetone (e) 3-Pentanone (i) Methyl 2-naphthyl ketone
(b) Acetophenone (f) 1-Phenylethanol (j) 3-Pentanol
(c) Pentanal (g) 2-Phenylethanol
(d) 2-Pentanone (h) 2-Butanol

17.5 THE ALDOL REACTION: THE ADDITION OF ENOLATE IONS TO ALDEHYDES AND KETONES

When acetaldehyde reacts with dilute sodium hydroxide at room temperature (or below), a dimerization takes place producing 3-hydroxybutanal. Since 3-hydroxy-

$$2\ CH_3CH\ \xrightarrow[5°C]{10\%\ NaOH,\ H_2O}\ CH_3CHCH_2CH$$

3-Hydroxybutanal
"aldol"
(50%)

butanal is both an **alde**hyde and an alcoh**ol**, it has been given the common name **"aldol,"** and reactions of this general type have come to be known as **aldol additions** (or **aldol reactions**).

The mechanism for the aldol addition illustrates two important characteristics of carbonyl compounds: the acidity of their α hydrogens and the tendency of their carbonyl groups to undergo nucleophilic addition.

A MECHANISM FOR THE ALDOL ADDITION

Step 1

$$H-\ddot{\underset{..}{O}}:^- + H-CH_2-\overset{\overset{\ddot{O}:}{\|}}{C}-H \rightleftharpoons \left[:CH_2-\overset{\overset{\ddot{O}:}{\|}}{C}-H \longleftrightarrow CH_2=\overset{\overset{:\ddot{O}:^-}{|}}{C}-H \right] + H-\ddot{\underset{H}{O}}:$$

Enolate ion

In this step the base (a hydroxide ion) removes a proton from the α carbon of one molecule of acetaldehyde to give a resonance-stabilized enolate ion.

Step 2 $$CH_3-\overset{\overset{\ddot{O}:}{\|}}{C}-H + \ ^-:CH_2-\overset{\overset{\ddot{O}:}{\|}}{C}-H \rightleftharpoons CH_3-\overset{\overset{:\ddot{O}:^-}{|}}{CH}-CH_2-\overset{\overset{\ddot{O}:}{\|}}{C}-H$$

An alkoxide ion

$$\updownarrow$$

$$CH_2=\overset{\overset{:\ddot{O}:^-}{|}}{C}-H$$

The enolate ion then acts as a nucleophile—as a carbanion—and attacks the carbonyl carbon of a second molecule of acetaldehyde, producing an alkoxide ion.

Step 3

$$CH_3-CH-CH_2-C-H + H-\ddot{O}-H \longrightarrow CH_3-CH-CH_2-C-H + \ ^-\!\!:\!\ddot{O}-H$$

Stronger base   **Aldol**   **Weaker base**

The alkoxide ion now removes a proton from a molecule of water to form the aldol.

17.5A Dehydration of Addition Product

If the basic mixture containing the aldol (in the previous example) is heated, dehydration takes place and 2-butenal (crotonaldehyde) is formed. Dehydration occurs readily because of the acidity of the remaining α hydrogens (even though the leaving group is a hydroxide ion) *and because the product is stabilized by having conjugated double bonds.*

A MECHANISM FOR THE DEHYDRATION

These hydrogens are acidic.   Here the double bonds are conjugated.

$$CH_3-CH-CH-C-H \longrightarrow CH_3-CH=CH-C-H + H-\ddot{O}: + H-\ddot{O}:^-$$

**2-Butenal
(crotonaldehyde)**

$$H-\ddot{O}:^-$$

In some aldol reactions, dehydration occurs so readily that we cannot isolate the product in the aldol form; we obtain the derived *enal* instead. An **aldol condensation** occurs instead of an aldol *addition*. A condensation reaction is one in which molecules are joined through the intermolecular elimination of a small molecule such as water or an alcohol.

	Addition product	**Condensation product**

$$2\ RCH_2CH \xrightarrow{\text{base}} \left[RCH_2CHCHCH \right] \xrightarrow{-H_2O} RCH_2CH=C-CH$$

Not isolated   **An enal**
(an α,β unsaturated aldehyde)

17.5B Synthetic Applications

The aldol reaction is a general reaction of aldehydes that possess an α hydrogen. Propanal, for example, reacts with aqueous sodium hydroxide to give 3-hydroxy-2-methylpentanal.

$$2 \ CH_3CH_2\overset{\overset{\displaystyle O}{\|}}{C}H \quad \xrightarrow[0-10°C]{OH^-} \quad CH_3CH_2\overset{\overset{\displaystyle OH}{|}}{C}H\overset{|}{\underset{CH_3}{C}}H\overset{\overset{\displaystyle O}{\|}}{C}H$$

Propanal   **3-Hydroxy-2-methylpentanal**
                                        (55–60%)

PROBLEM 17.7

(a) Show all steps in the aldol addition that occur when propanal

$\left(CH_3CH_2\overset{\overset{\displaystyle O}{\|}}{C}H \right)$ is treated with base.   (b) How can you account for the fact

that the product of the aldol addition is $CH_3CH_2\overset{\overset{\displaystyle OH}{|}}{C}H\overset{|}{\underset{CH_3}{C}}H\overset{\overset{\displaystyle O}{\|}}{C}H$ and not

$CH_3CH_2\overset{\overset{\displaystyle OH}{|}}{C}HCH_2CH_2\overset{\overset{\displaystyle O}{\|}}{C}H$?   (c) What product would be formed if the reaction mixture were heated?

The aldol reaction is important in organic synthesis because it gives us a method for linking two smaller molecules by introducing a carbon–carbon bond between them. Because aldol products contain two functional groups, —OH and —CHO, we can use them to carry out a number of subsequent reactions. Examples are the following:

$$2 \ RCH_2\overset{\overset{\displaystyle O}{\|}}{C}H \xrightarrow[H_2O]{OH^-} RCH_2\overset{\overset{\displaystyle OH}{|}}{C}H\overset{|}{\underset{R}{C}}H\overset{\overset{\displaystyle O}{\|}}{C}H \xrightarrow{NaBH_4} RCH_2\overset{\overset{\displaystyle OH}{|}}{C}H\overset{|}{\underset{R}{C}}HCH_2OH$$

Aldehyde   **An aldol**   **A 1,3-diol**

$$\downarrow H^+ \ -H_2O$$

$$RCH_2CH_2\overset{|}{\underset{R}{C}}HCH_2OH \xleftarrow[\substack{high \\ pressure}]{H_2/Ni} RCH_2CH{=}\overset{|}{\underset{R}{C}}\overset{\overset{\displaystyle O}{\|}}{C}H \xrightarrow{LiAlH_4^*} RCH_2CH{=}\overset{|}{\underset{R}{C}}CH_2OH$$

A saturated alcohol   **An α, β-unsaturated**   **An allylic alcohol**
                                  aldehyde

$$\downarrow H_2, \ Pd{—}C$$

$$RCH_2CH_2\overset{|}{\underset{R}{C}}H\overset{\overset{\displaystyle O}{\|}}{C}H$$

An aldehyde

*LiAlH$_4$ reduces the carbonyl group of α,β-unsaturated aldehydes and ketones cleanly. NaBH$_4$ often reduces the carbon–carbon double bond as well.

PROBLEM 17.8 _____

One industrial process for the synthesis of 1-butanol begins with acetaldehyde. Show how this synthesis might be carried out.

PROBLEM 17.9 _____

Show how each of the following products could be synthesized from butanal:
(a) 2-Ethyl-3-hydroxyhexanal
(b) 2-Ethyl-2-hexen-1-ol
(c) 2-Ethyl-1-hexanol
(d) 2-Ethyl-1,3-hexanediol (the insect repellent ''6–12'')

Ketones also undergo base-catalyzed aldol additions, but for them the equilibrium is unfavorable. This complication can be overcome, however, by carrying out the reaction in a special apparatus that allows the product to be removed from contact with the base as it is formed. This removal of product displaces the equilibrium to the right and permits successful aldol additions with many ketones. Acetone, for example, reacts as follows:

$$2 \ CH_3\overset{\overset{\displaystyle O}{\|}}{C}CH_3 \underset{}{\overset{OH^-}{\rightleftharpoons}} CH_3\overset{\overset{\displaystyle OH}{|}}{\underset{\underset{\displaystyle CH_3}{|}}{C}}CH_2\overset{\overset{\displaystyle O}{\|}}{C}CH_3$$

(80%)

17.5C The Reversibility of Aldol Additions

The aldol addition is reversible. If, for example, the aldol addition product obtained from acetone (see Problem 17.10) is heated with a strong base, it reverts to an equilibrium mixture that consists largely (~ 95%) of acetone. This type of reaction is called a *retro-aldol* reaction.

$$CH_3\overset{\overset{\displaystyle OH}{|}}{\underset{\underset{\displaystyle CH_3}{|}}{C}}-CH_2\overset{\overset{\displaystyle O}{\|}}{C}CH_3 \underset{H_2O}{\overset{OH^-}{\rightleftharpoons}} CH_3\overset{\overset{\displaystyle O^-}{|}}{\underset{\underset{\displaystyle CH_3}{|}}{C}}CH_2\overset{\overset{\displaystyle O}{\|}}{C}CH_3 \rightleftharpoons CH_3\overset{\overset{\displaystyle O}{\|}}{\underset{\underset{\displaystyle CH_3}{|}}{C}} + \ ^-\!:CH_2\overset{\overset{\displaystyle O}{\|}}{C}CH_3 \underset{OH^-}{\overset{H_2O}{\rightleftharpoons}} 2 \ CH_3\overset{\overset{\displaystyle O}{\|}}{C}CH_3$$

(5%)   (95%)

17.6 CROSSED ALDOL REACTIONS

An aldol reaction that starts with two different carbonyl compounds is called a **crossed aldol reaction.** Crossed aldol reactions using aqueous sodium hydroxide solutions are of little synthetic importance if both reactants have α hydrogens, because these reactions give a complex mixture of products. If, for example, we were to carry out a crossed aldol addition using acetaldehyde and propanal, we would obtain at least four products.

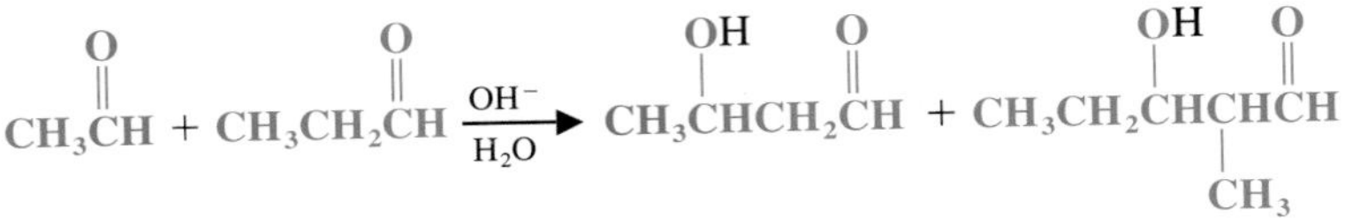

$$CH_3CH + CH_3CH_2CH \xrightarrow[H_2O]{OH^-} CH_3CHCH_2CH + CH_3CH_2CHCHCH$$

3-Hydroxybutanal
(from 2 molecules
of acetaldehyde)

3-Hydroxy-2-
methylpentanal
(from 2 molecules
of propanal)

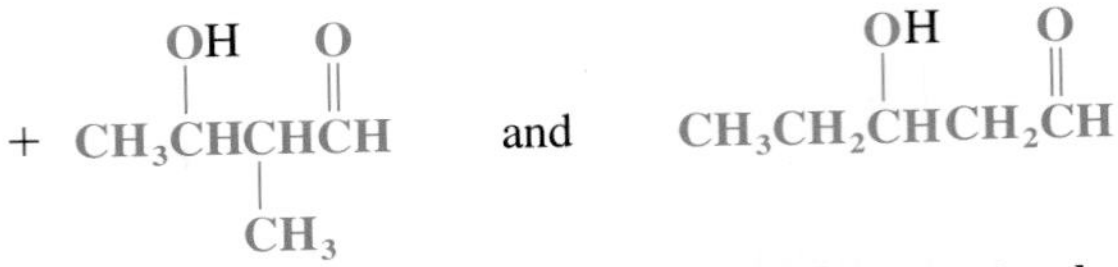

$$+ \ CH_3CHCHCH \quad \text{and} \quad CH_3CH_2CHCH_2CH$$

3-Hydroxy-2-methylbutanal   **3-Hydroxypentanal**
(from 1 molecule of acetaldehyde and 1 molecule of propanal)

SAMPLE PROBLEM

Show how each of the four products just given is formed in the crossed aldol addition between acetaldehyde and propanal.

ANSWER:

In the basic aqueous solution, four organic entities will initially be present: molecules of acetaldehyde, molecules of propanal, enolate ions derived from acetaldehyde, and enolate ions derived from propanal.

We have already seen (Section 17.5) how a molecule of acetaldehyde can react with its enolate ion to form 3-hydroxybutanal (aldol).

Reaction 1

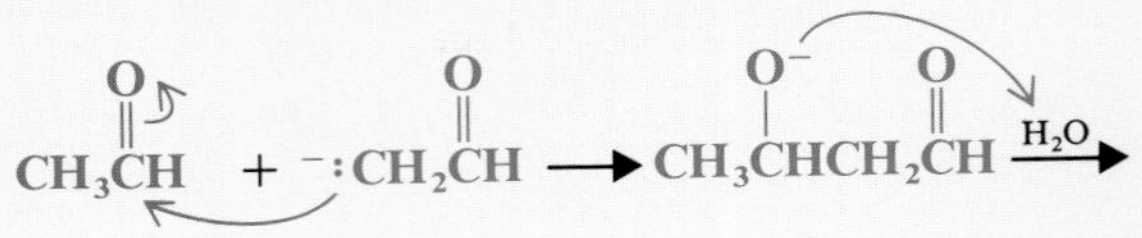

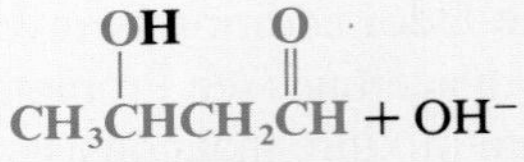

$$CH_3CH \ + \ ^-\!:CH_2CH \longrightarrow CH_3CHCH_2CH \xrightarrow{H_2O}$$

$$CH_3CHCH_2CH + OH^-$$

3-Hydroxy-
butanal

We have also seen (Problem 17.7) how propanal can react with its enolate ion to form 3-hydroxy-2-methylpentanal.

Reaction 2

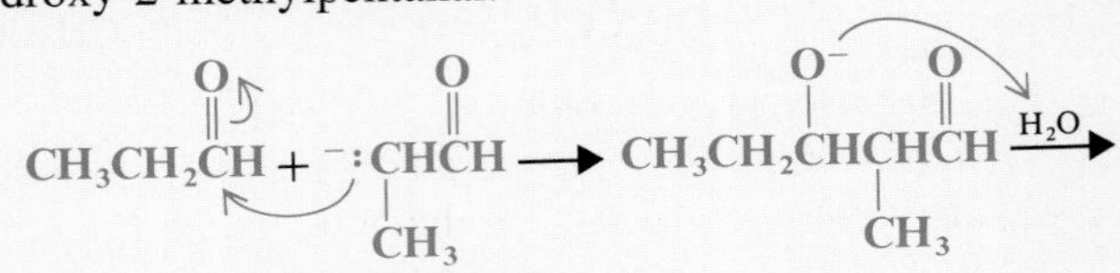

$$CH_3CH_2CH + \ ^-\!:CHCH \longrightarrow CH_3CH_2CHCHCH \xrightarrow{H_2O}$$
$$\qquad\qquad\quad CH_3 \qquad\qquad\qquad\quad CH_3$$

Propanal   **Enolate**
of
propanal

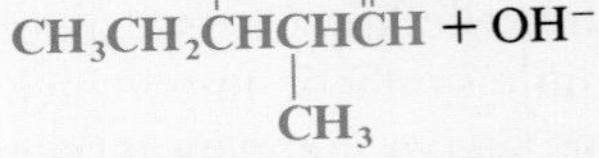

$$CH_3CH_2CHCHCH + OH^-$$
$$\qquad\qquad CH_3$$

3-Hydroxy-2-
methylpentanal

Acetaldehyde can also react with the enolate of propanal. This reaction leads to the third product, 3-hydroxy-2-methylbutanal.

Reaction 3

$$CH_3\overset{\overset{O}{\|}}{CH} \ + \ \ ^-:\overset{}{\underset{\underset{CH_3}{|}}{CH}}\overset{\overset{O}{\|}}{CH} \longrightarrow CH_3\overset{\overset{O^-}{|}}{CH}\underset{\underset{CH_3}{|}}{CH}\overset{\overset{O}{\|}}{CH} \xrightarrow{H_2O}$$

Acetaldehyde	**Enolate of propanal**

$$CH_3\overset{\overset{OH}{|}}{CH}\underset{\underset{CH_3}{|}}{CH}\overset{\overset{O}{\|}}{CH} + OH^-$$

3-Hydroxy-2-methylbutanal

And finally, propanal can react with the enolate of acetaldehyde. This reaction accounts for the fourth product.

Reaction 4

$$CH_3CH_2\overset{\overset{O}{\|}}{CH} + \ ^-:CH_2\overset{\overset{O}{\|}}{CH} \longrightarrow CH_3CH_2\overset{\overset{O^-}{|}}{CH}CH_2\overset{\overset{O}{\|}}{CH} \xrightarrow{H_2O}$$

Propanal	**Enolate of acetaldehyde**

$$CH_3CH_2\overset{\overset{OH}{|}}{CH}CH_2\overset{\overset{O}{\|}}{CH} + OH^-$$

3-Hydroxypentanal

17.6A Practical Crossed Aldol Reactions

Crossed aldol reactions are practical, using bases such as NaOH, when one reactant does not have an α hydrogen and so cannot undergo self-condensation because it cannot form an enolate ion. We can avoid other side reactions by placing this component in base and then slowly adding the reactant with an α hydrogen to the mixture. Under these conditions the concentration of the reactant with an α hydrogen is always low and much of the reactant is present as an enolate ion. The main reaction that takes place is one between this enolate ion and the component that has no α hydrogen. The examples listed in Table 17.1 illustrate this technique.

As the examples in Table 17.1 also show, the crossed aldol reaction is often accompanied by dehydration. Whether or not dehydration occurs can, at times, be determined by our choice of reaction conditions, but *dehydration is especially easy when it leads to an extended conjugated system.*

PROBLEM 17.10

Outlined below is a synthesis of a compound used in perfumes, called lily aldehyde. Provide all of the missing structures.

TABLE 17.1   Crossed aldol reactions

THIS REACTANT WITH NO α HYDROGEN IS PLACED IN BASE	THIS REACTANT WITH AN α HYDROGEN IS ADDED SLOWLY		PRODUCT
$\underset{\text{Benzaldehyde}}{\overset{\displaystyle O}{\underset{\|}{C_6H_5CH}}}$ +	$\underset{\text{Propanal}}{\overset{\displaystyle O}{\underset{\|}{CH_3CH_2CH}}}$	$\xrightarrow[10°C]{OH^-}$	$\overset{\displaystyle CH_3\ \ O}{C_6H_5CH=C\overset{\|}{\quad}CH}$ 2-Methyl-3-phenyl-2-propenal (α-methylcinnamaldehyde) (68%)
$\underset{\text{Benzaldehyde}}{\overset{\displaystyle O}{\underset{\|}{C_6H_5CH}}}$ +	$\underset{\text{Phenylacetaldehyde}}{\overset{\displaystyle O}{\underset{\|}{C_6H_5CH_2CH}}}$	$\xrightarrow[20°C]{OH^-}$	$\underset{\underset{C_6H_5}{\|}}{C_6H_5CH=CCH}$ 2,3-Diphenyl-2-propenal
$\underset{\text{Formaldehyde}}{\overset{\displaystyle O}{\underset{\|}{HCH}}}$ +	$\underset{\text{2-Methylpropanal}}{\overset{\displaystyle CH_3\ \ O}{CH_3CH-CH}}$	$\xrightarrow[40°C]{\text{dil. } Na_2CO_3}$	$\underset{\underset{CH_2OH}{\|}}{\overset{\displaystyle CH_3\ \ O}{CH_3-C\overset{\|}{\quad}CH}}$ 3-Hydroxy-2,2-dimethylpropanal (>64%)

p-tert-Butylbenzyl alcohol $\xrightarrow[CH_2Cl_2]{PCC}$ $C_{11}H_{14}O$ $\xrightarrow{\text{Propanal}}_{OH^-}$

$C_{14}H_{18}O$ $\xrightarrow{H_2,\ Pd—C}$ Lily aldehyde ($C_{14}H_{20}O$)

PROBLEM 17.11

Show how you could use a crossed aldol reaction to synthesize cinnamaldehyde ($C_6H_5CH=CHCHO$).

PROBLEM 17.12

When excess formaldehyde in basic solution is treated with acetaldehyde, the following reaction takes place:

$$3\ \overset{\displaystyle O}{\underset{\|}{HCH}} + \overset{\displaystyle O}{\underset{\|}{CH_3CH}} \xrightarrow[40°C]{\text{dil. } Na_2CO_3} \underset{\underset{CH_2OH}{\|}}{\overset{\displaystyle CH_2OH}{HOCH_2-C-CHO}}$$

(82%)

Write a mechanism that accounts for the formation of the product.

17.6B Claisen–Schmidt Reactions

When ketones are used as one component, the crossed aldol reactions are called **Claisen–Schmidt reactions,** after the German chemists, J. G. Schmidt (who discovered the reaction in 1880) and Ludwig Claisen (who developed it between 1881 and 1889). These reactions are practical when bases such as sodium hydroxide are used because under these conditions ketones do not self-condense appreciably. (The equilibrium is unfavorable; cf. Section 17.5C.)

Two examples of Claisen–Schmidt reactions are the following:

$$C_6H_5\overset{\overset{\displaystyle O}{\|}}{C}H + CH_3\overset{\overset{\displaystyle O}{\|}}{C}CH_3 \xrightarrow[100°C]{OH^-} C_6H_5CH=CH\overset{\overset{\displaystyle O}{\|}}{C}CH_3$$

4-Phenyl-3-buten-2-one
(benzalacetone)
(70%)

$$C_6H_5\overset{\overset{\displaystyle O}{\|}}{C}H + CH_3\overset{\overset{\displaystyle O}{\|}}{C}C_6H_5 \xrightarrow[20°C]{OH^-} C_6H_5CH=CH\overset{\overset{\displaystyle O}{\|}}{C}C_6H_5$$

1,3-Diphenyl-2-propen-1-one
(benzalacetophenone)
(85%)

A MECHANISM FOR THE CLAISEN–SCHMIDT REACTION

Step 1

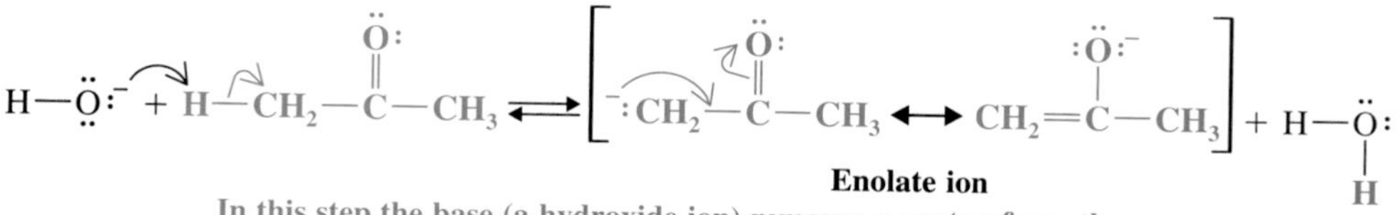

In this step the base (a hydroxide ion) removes a proton from the α carbon of one molecule of the ketone to give a resonance-stabilized enolate ion.

Step 2

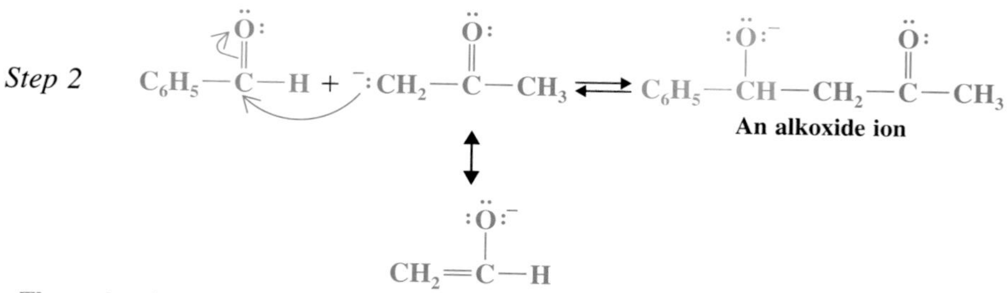

An alkoxide ion

The enolate ion then acts as a nucleophile—as a carbanion—and attacks the carbonyl carbon of the molecule of aldehyde, producing an alkoxide ion.

Step 3

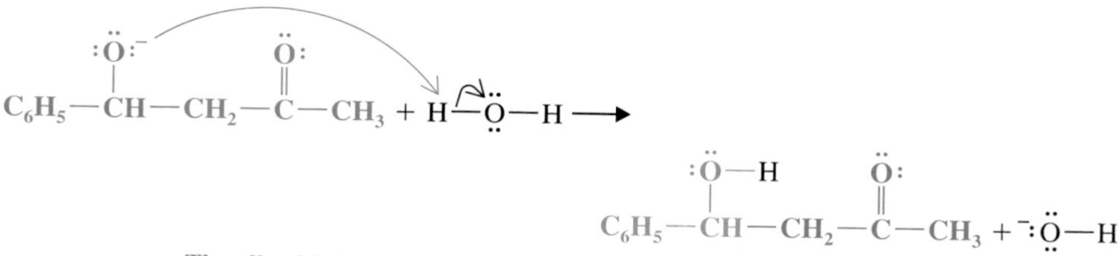

The alkoxide ion now removes a proton from a molecule of water.

Step 4

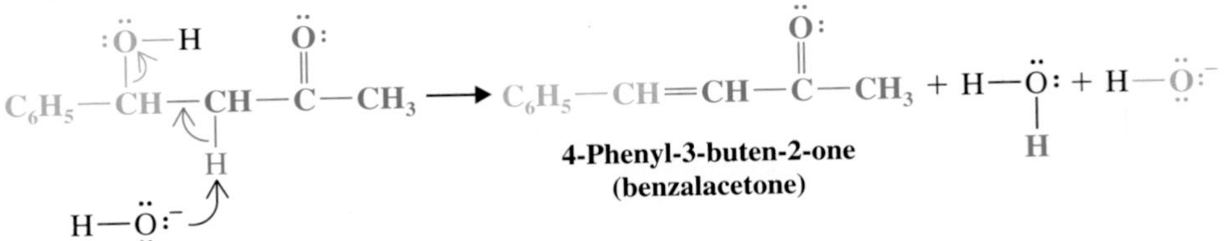

4-Phenyl-3-buten-2-one
(benzalacetone)

Dehydration produces the conjugated product.

In the Claisen–Schmidt reactions given above dehydration occurs readily because the double bond that forms is conjugated both with the carbonyl group and with the benzene ring. The conjugated system is thereby extended.

An important step in a commercial synthesis of vitamin A makes use of a Claisen–Schmidt reaction between geranial and acetone:

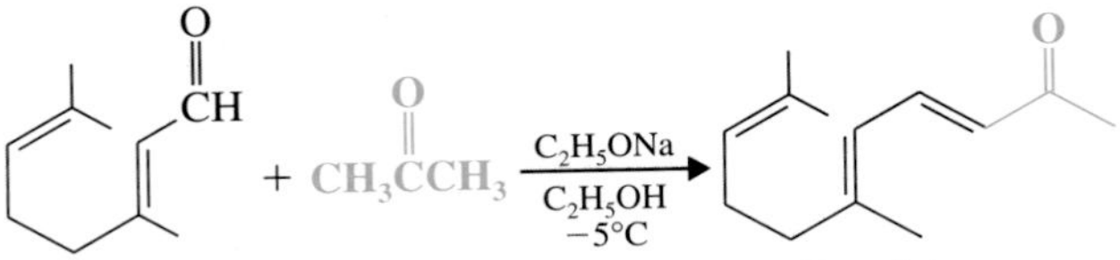

Geranial   **Pseudoionone**
                                          (49%)

Geranial is a naturally occurring aldehyde that can be obtained from lemongrass oil. Its α hydrogen is *vinylic* and, therefore, not appreciably acidic. Notice, in this reaction, too, dehydration occurs readily because dehydration extends the conjugated system.

PROBLEM 17.13

When pseudoionone is treated with BF_3 in acetic acid, ring closure takes place and α- and β-ionone are produced. This is the next step in the vitamin A synthesis.

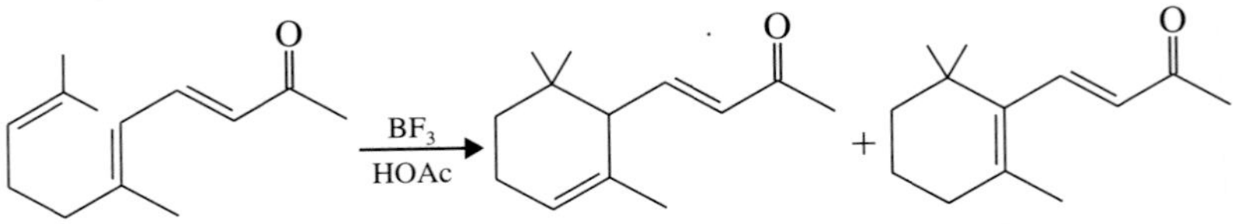

Pseudoionone   **α-Ionone**   **β-Ionone**

(a) Write mechanisms that explain the formation of α- and β-ionone.   (b) β-Ionone is the major product. How can you explain this?   (c) Which ionone would you expect to absorb at longer wavelengths in the visible–UV region? Why?

17.6C Condensations with Nitroalkanes

The α hydrogens of nitroalkanes are appreciably acidic ($pK_a = 10$), much more acidic than those of aldehydes and ketones. The acidity of these hydrogen atoms, like the α hydrogens of aldehydes and ketones, can be explained by the powerful electron-with-drawing effect of the nitro group and by resonance stabilization of the anion that is produced.

$$R{-}CH_2{-}\overset{+}{N}\overset{O}{\diagdown}_{O^-} + ^-{:}B \longrightarrow R{-}\overset{\cdot\cdot}{\overset{}{CH}}{-}\overset{+}{N}\overset{O}{\diagdown}_{O^-} \longleftrightarrow R{-}CH{=}\overset{+}{N}\overset{O^-}{\diagdown}_{O^-} + H{:}B$$

Resonance-stabilized anion

Nitroalkanes that have α hydrogens undergo base-catalyzed condensations with alde-hydes and ketones that resemble aldol condensations. An example is the condensation of benzaldehyde with nitromethane.

$$C_6H_5\overset{O}{\overset{\|}{C}}H + CH_3NO_2 \xrightarrow{\text{OH}^-} C_6H_5CH{=}CHNO_2$$

This condensation is especially useful because the nitro group of the product can be easily reduced to an amino group. One technique that brings about this transformation uses hydrogen and a nickel catalyst. This combination not only reduces the nitro group but also reduces the double bond:

$$C_6H_5CH{=}CHNO_2 \xrightarrow{\text{H}_2,\ \text{Ni}} C_6H_5CH_2CH_2NH_2$$

PROBLEM 17.14

Assuming that you have available the required aldehydes, ketones, and nitro-alkanes, show how you would synthesize each of the following:

(a) $C_6H_5CH{=}\underset{\underset{CH_3}{|}}{C}NO_2$ (b) $HOCH_2CH_2NO_2$

17.6D Condensations with Nitriles

The α hydrogens of nitriles are also appreciably acidic, but less so than those of aldehydes and ketones. The acidity constant for acetonitrile (CH_3CN) is about 10^{-25} ($pK_a \sim 25$). Other nitriles with α hydrogens show comparable acidities, and conse-quently these nitriles undergo condensations of the aldol type. An example is the condensation of benzaldehyde with phenylacetonitrile.

$$C_6H_5\overset{O}{\overset{\|}{C}}H + C_6H_5CH_2CN \xrightarrow[\text{EtOH}]{\text{EtO}^-} C_6H_5CH{=}\underset{\underset{C_6H_5}{|}}{C}{-}CN$$

PROBLEM 17.15 _____

(a) Write resonance structures for the anion of acetonitrile that account for its being much more acidic than ethane. (b) Give a step-by-step mechanism for the condensation of benzaldehyde with acetonitrile.

17.7 CYCLIZATIONS VIA ALDOL CONDENSATIONS

The aldol condensation also offers a convenient way to synthesize molecules with five- and six-membered rings (and sometimes even larger rings). This can be done by an intramolecular aldol condensation using a dialdehyde, a keto aldehyde, or a diketone as the substrate. For example, the following keto aldehyde cyclizes to yield 1-cyclopentenyl methyl ketone.

$$CH_3CCH_2CH_2CH_2CH_2CH \xrightarrow{OH^-}$$

(73%)

This reaction almost certainly involves the formation of at least three different enolates. However, it is the enolate from the ketone side of the molecule that adds to the aldehyde group leading to the product.

A MECHANISM FOR THE ALDOL CYCLIZATION

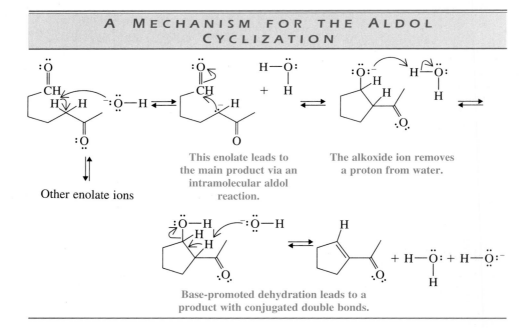

This enolate leads to the main product via an intramolecular aldol reaction.

The alkoxide ion removes a proton from water.

Other enolate ions

Base-promoted dehydration leads to a product with conjugated double bonds.

The reason the aldehyde group undergoes addition preferentially may arise from the greater reactivity of aldehydes toward nucleophilic addition generally. The carbonyl carbon atom of a ketone is less positive (and therefore less reactive toward a nucleophile) because it bears two electron-releasing alkyl groups; it is also more sterically hindered.

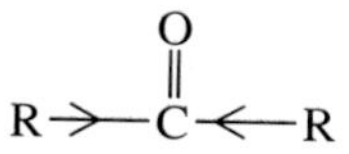

Ketones are less reactive toward nucleophiles.

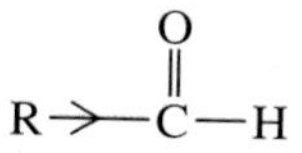

Aldehydes are more reactive toward nucleophiles.

In reactions of this type, five-membered rings form far more readily than seven-membered rings.

PROBLEM 17.16

Assuming that dehydration occurs in all instances, write the structures of the two other products that might have resulted from the aldol cyclization just given. (One of these products will have a five-membered ring and the other will have a seven-membered ring.)

PROBLEM 17.17

What starting compound would you use in an aldol cyclization to prepare each of the following?

(a) 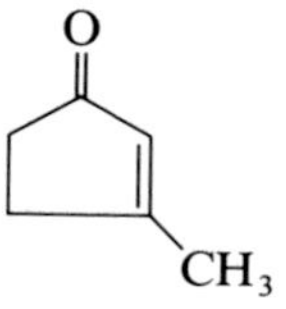

(b) 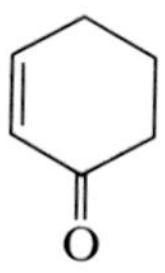

(c) 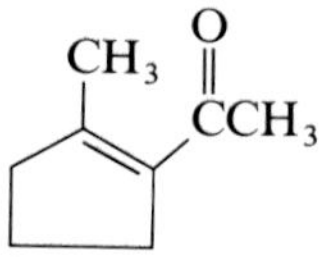

PROBLEM 17.18

What experimental conditions would favor the cyclization process in the intramolecular aldol reaction, over intermolecular condensation?

17.8 ACID-CATALYZED ALDOL CONDENSATIONS

Aldol condensations can also be brought about with acid catalysis. Treating acetone with hydrogen chloride, for example, leads to the formation of 4-methyl-3-penten-2-one, the aldol condensation product. In general, acid-catalyzed aldol reactions lead to dehydration of the initially formed aldol addition product.

A MECHANISM FOR THE ACID-CATALYZED ALDOL REACTION

REACTION:

$$2\,H_3C-\overset{O}{\overset{\|}{C}}-CH_3 \xrightarrow{HCl} H_3C-\overset{O}{\overset{\|}{C}}-CH=\overset{CH_3}{\overset{|}{C}}-CH_3 + H_2O$$

4-Methyl-3-penten-2-one

MECHANISM:

$$H_3C-\overset{\overset{\ddot{O}:}{\|}}{C}-CH_3 + H-\ddot{\underset{..}{C}}l: \rightleftharpoons H_3C-\overset{\overset{+\ddot{O}-H}{\|}}{C}-CH_2-H + :\ddot{\underset{..}{C}}l^- \rightleftharpoons$$

$$H_3C-\overset{\overset{:\ddot{O}-H}{|}}{C}=CH_2 + H-\ddot{\underset{..}{C}}l:$$

The mechanism begins with the acid-catalyzed formation of the enol.

$$H_3C-\overset{\overset{:\ddot{O}-H}{|}}{C}=CH_2 + \overset{CH_3}{\underset{\underset{CH_3}{|}}{C}}=\overset{+}{\underset{..}{O}}-H \rightleftharpoons H_3C-\overset{\overset{+\ddot{O}-H}{\|}}{C}-CH_2-\overset{\overset{CH_3}{|}}{\underset{\underset{CH_3}{|}}{C}}-\ddot{\underset{..}{O}}-H$$

Then the enol adds to the protonated carbonyl group of another molecule of acetone.

$$H_3C-\overset{\overset{+\ddot{O}-H}{\|}}{C}-CH_2-\overset{\overset{CH_3}{|}}{\underset{\underset{CH_3}{|}}{C}}-\ddot{\underset{..}{O}}-H \rightleftharpoons H_3C-\overset{\overset{\ddot{O}:}{\|}}{C}-\underset{\underset{H}{|}}{CH}-\overset{\overset{CH_3}{|}}{\underset{\underset{H}{|}}{C}}-\overset{+}{\underset{..}{O}}-H \rightleftharpoons$$

$$:\ddot{\underset{..}{C}}l:^-$$

$$H_3C-\overset{\overset{\ddot{O}:}{\|}}{C}-CH=\overset{\overset{CH_3}{|}}{C}-CH_3$$

$$+ H-\ddot{\underset{..}{C}}l:$$

Finally, acid-catalyzed dehydration occurs leading to the product.

$$+ :\overset{\ddot{O}-H}{\underset{H}{|}}$$

PROBLEM 17.19

The acid-catalyzed aldol condensation of acetone (just shown) also produces some 2,6-dimethyl-2,5-heptadien-4-one. Give a mechanism that explains the formation of this product.

PROBLEM 17.20

Heating acetone with sulfuric acid leads to the formation of mesitylene (1,3,5-trimethylbenzene). Propose a mechanism for this reaction.

17.9 ADDITIONS TO α,β-UNSATURATED ALDEHYDES AND KETONES

When α,β-unsaturated aldehydes and ketones react with nucleophilic reagents, they may do so in two ways. They may react by a *simple addition,* that is, one in which the nucleophile adds across the double bond of the carbonyl group; or they may react by a *conjugate addition.* These two processes resemble the 1,2- and the 1,4-addition reactions of conjugated dienes (Section 12.9).

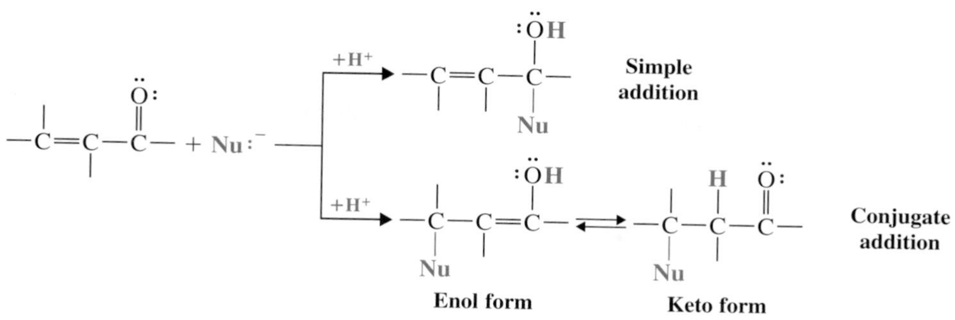

$$-\overset{|}{C}=\overset{|}{C}-\overset{\overset{\ddot{O}:}{\|}}{C}- \; + \; Nu:^-$$

Simple addition (via +H⁺):

$$-\overset{|}{C}=\overset{|}{C}-\overset{\overset{:\ddot{O}H}{|}}{\underset{\underset{Nu}{|}}{C}}-$$

Conjugate addition (via +H⁺):

$$-\overset{\overset{:\ddot{O}H}{|}}{\underset{\underset{Nu}{|}}{C}}-\overset{|}{C}=\overset{|}{C}- \;\rightleftharpoons\; -\overset{\overset{H}{|}}{\underset{\underset{Nu}{|}}{C}}-\overset{|}{C}-\overset{\overset{\ddot{O}:}{\|}}{C}-$$

Enol form   Keto form

In many instances both modes of addition occur in the same mixture. As an example, let us consider the Grignard reaction shown here.

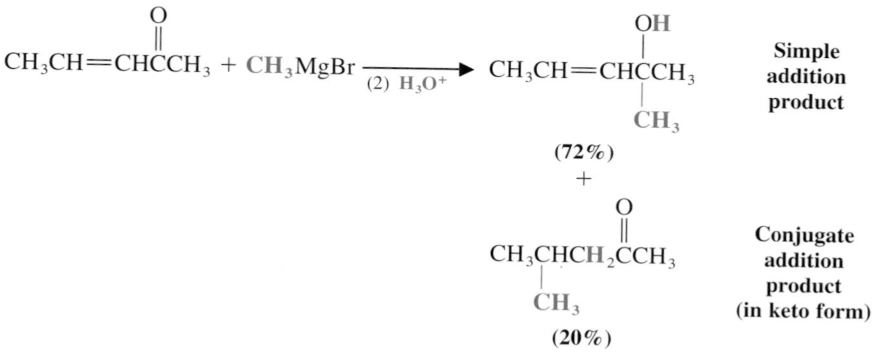

$$CH_3CH{=}CHCCH_3 \; + \; CH_3MgBr \xrightarrow{(2)\ H_3O^+}$$

$$\overset{\overset{O}{\|}}{}$$

Simple addition product

$$CH_3CH{=}CH\overset{\overset{OH}{|}}{\underset{\underset{CH_3}{|}}{C}}CH_3$$

(72%)

+

Conjugate addition product (in keto form)

$$CH_3\overset{\overset{}{\underset{\underset{CH_3}{|}}{C}}}HCH_2\overset{\overset{O}{\|}}{C}CH_3$$

(20%)

In this example we see that simple addition is favored, and this is generally the case with strong nucleophiles. Conjugate addition is favored when weaker nucleophiles are employed.

If we examine the resonance structures that contribute to the overall hybrid for an α,β-unsaturated aldehyde or ketone (see structures **A–C**), we shall be in a better position to understand these reactions.

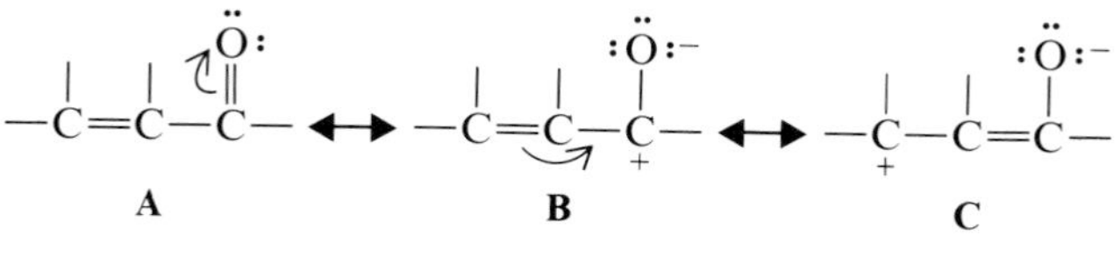

$$-\overset{|}{C}=\overset{|}{C}-\overset{\overset{\ddot{O}:}{\|}}{C}- \;\longleftrightarrow\; -\overset{|}{C}=\overset{|}{\underset{+}{C}}-\overset{\overset{:\ddot{O}:^-}{|}}{C}- \;\longleftrightarrow\; -\overset{|}{\underset{+}{C}}-\overset{|}{C}=\overset{\overset{:\ddot{O}:^-}{|}}{C}-$$

A   B   C

Although structures **B** and **C** involve separated charges, they make a significant contribution to the hybrid because, in each, the negative charge is carried by electronegative oxygen. Structures **B** and **C** also indicate that *both the carbonyl carbon and the β carbon should bear a partial positive charge.* They indicate that we should represent the hybrid in the following way:

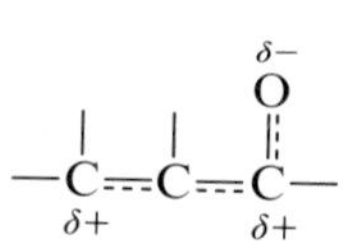

$$-\underset{\delta+}{\overset{|}{C}}{=\!=}\underset{}{\overset{|}{C}}{=\!=}\underset{\delta+}{\overset{\overset{\overset{\delta-}{O}}{\|}}{C}}-$$

This structure tells us that we should expect a nucleophilic reagent to attack either the carbonyl carbon or the β carbon.

Almost every nucleophilic reagent that adds at the carbonyl carbon of a simple aldehyde or ketone is capable of adding at the β carbon of an α,β-unsaturated carbonyl compound. In many instances when weaker nucleophiles are used, conjugate addition is the major reaction path. Consider the following addition of hydrogen cyanide.

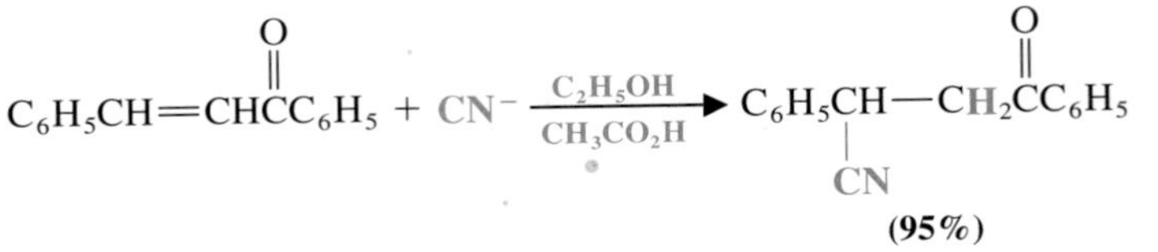
C6H5CH=CHCC6H5 + CN−
C2H5OH
CH3CO2H
C6H5CH—CH2CC6H5
CN
(95%)

A MECHANISM FOR THE CONJUGATE ADDITION OF HCN

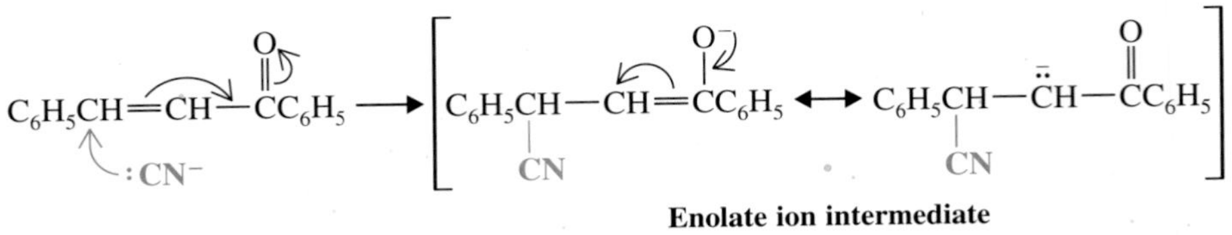
C6H5CH=CH—CC6H5
:CN−
C6H5CH—CH=CC6H5
C6H5CH—CH—CC6H5
CN
Enolate ion intermediate

Then, the enolate intermediate accepts a proton in either of two ways:

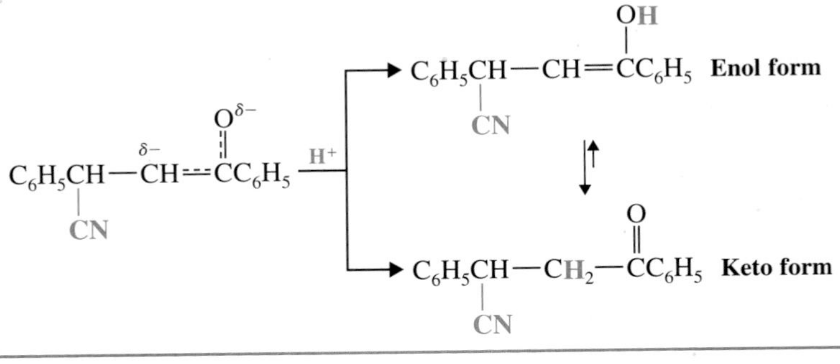
C6H5CH—CH=CC6H5
CN
H+
C6H5CH—CH=CC6H5 Enol form
C6H5CH—CH2—CC6H5 Keto form

Another example of this type of addition is the following:

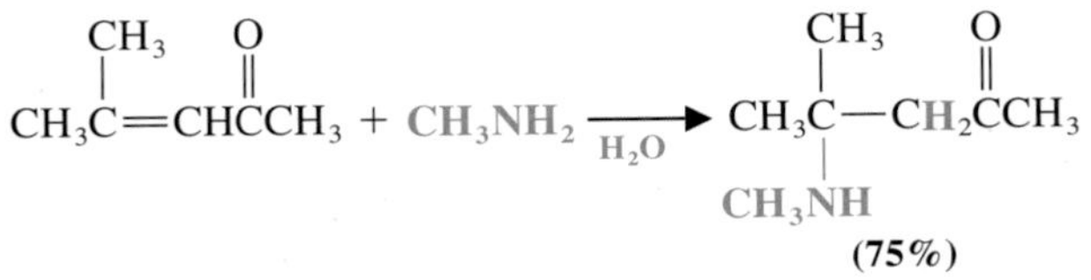
CH3C=CHCCH3 + CH3NH2
H2O
CH3C—CH2CCH3
CH3NH
(75%)

A MECHANISM FOR THE CONJUGATE ADDITION OF AN AMINE

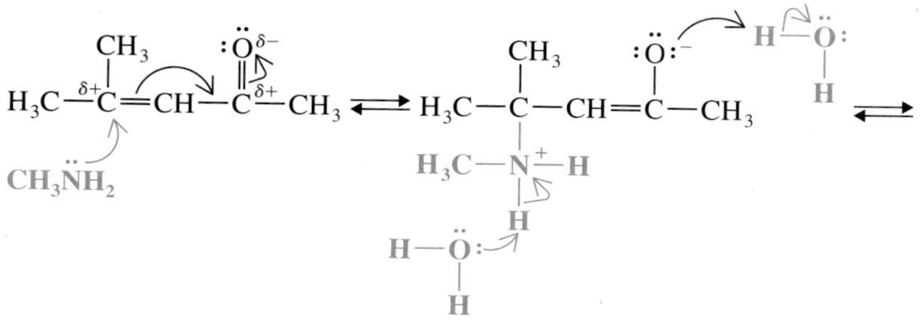

The nucleophile attacks the partially positive β carbon.

A proton is lost from the nitrogen atom and a proton is gained at the oxygen.

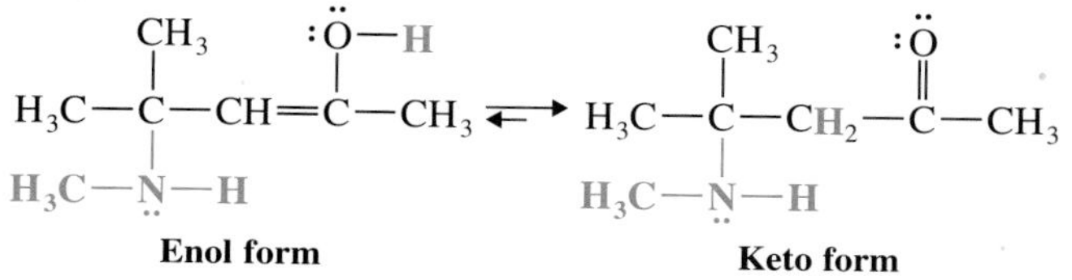

Enol form

Keto form

17.9A Conjugate Addition of Organocopper Reagents

Organocopper reagents, either RCu or R_2CuLi, add to α,β-unsaturated carbonyl compounds and **they add almost exclusively in the conjugate manner.**

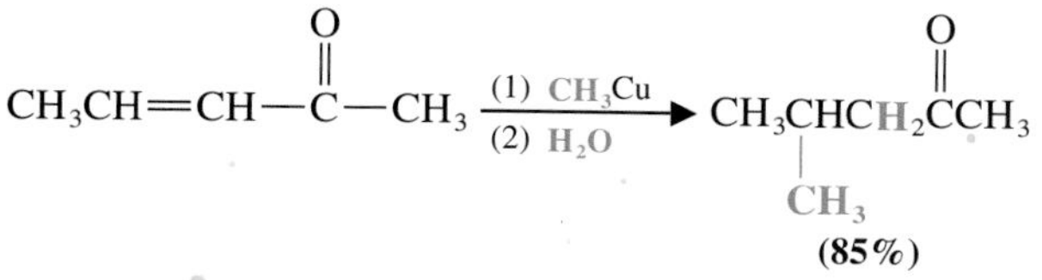

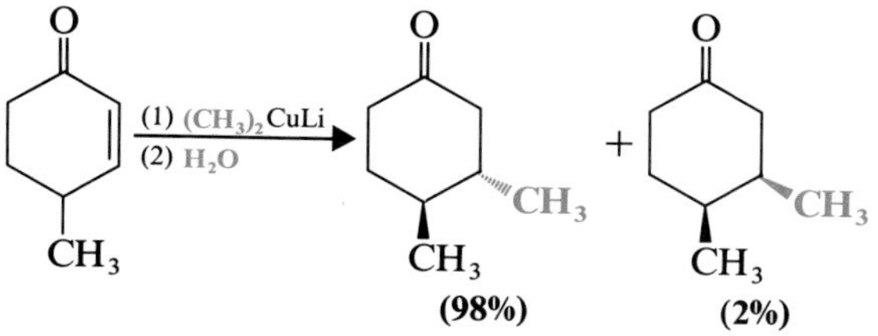

$$CH_3CH=CH-\overset{O}{\underset{}{C}}-CH_3 \xrightarrow[\text{(2) H}_2\text{O}]{\text{(1) CH}_3\text{Cu}} CH_3CHCH_2CCH_3$$

(85%)

$$\xrightarrow[\text{(2) H}_2\text{O}]{\text{(1) (CH}_3)_2\text{CuLi}}$$

(98%)   (2%)

With an alkyl-substituted cyclic α,β-unsaturated ketone, as the example just cited shows, lithium dialkylcuprates add predominantly in the less hindered way to give the product with the alkyl groups trans to each other.

17.9B Michael Additions

Conjugate additions of enolate ions to α,β-unsaturated carbonyl compounds are known generally as Michael additions (after their discovery, in 1887, by Arthur Michael, of Tufts University and later of Harvard University). An example is the addition of cyclohexanone to $C_6H_5CH{=}CHCOC_6H_5$:

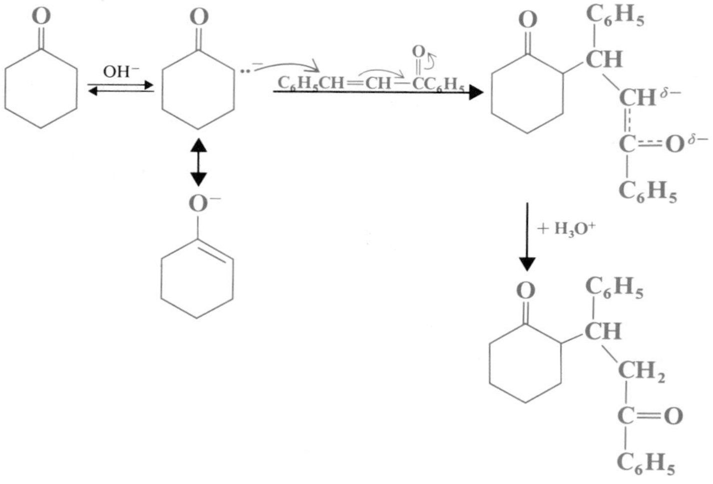

The sequence that follows illustrates how a conjugate aldol addition (Michael addition) followed by a simple aldol condensation may be used to build one ring onto another. This procedure is known as the *Robinson annulation* (ring forming) reaction (after the English chemist, Sir Robert Robinson, who won the Nobel Prize for chemistry in 1947 for his research on naturally occurring compounds).

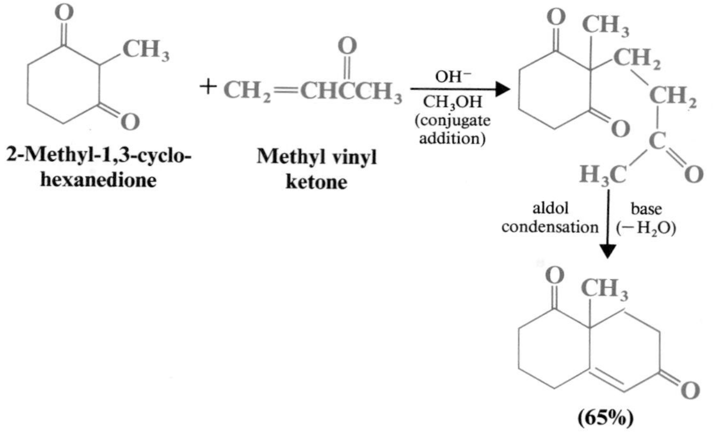

2-Methyl-1,3-cyclo-hexanedione

Methyl vinyl ketone

(65%)

PROBLEM 17.21

(a) Propose step-by-step mechanisms for both transformations of the Robinson annulation sequence just shown.   (b) Would you expect 2-methyl-1,3-cyclo-hexanedione to be more or less acidic than cyclohexanone? Explain your answer.

PROBLEM 17.22

What product would you expect to obtain from the base-catalyzed Michael reaction   (a) of 1,3-diphenyl-2-propen-1-one (Section 17.6B) and acetophenone?

(b) of 1,3-diphenyl-2-propen-1-one and cyclopentadiene? Show all steps in each mechanism.

PROBLEM 17.23

When acrolein reacts with hydrazine, the product is a dihydropyrazole:

$$CH_2=CHCHO + H_2N-NH_2 \longrightarrow$$

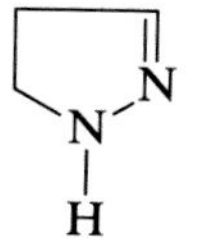

| Acrolein | Hydrazine | A dihydropyrazole |

Suggest a mechanism that explains this reaction.

We shall study further examples of the Michael addition in Chapter 19.

ADDITIONAL PROBLEMS

17.24 Give structural formulas for the products of the reaction (if one occurs) when propanal is treated with each of the following reagents:

(a) OH^-, H_2O

(b) C_6H_5CHO, OH^-

(c) HCN

(d) $NaBH_4$

(e) $HOCH_2CH_2OH$, H^+

(f) Ag_2O, OH^-, then H^+

(g) CH_3MgI, then H^+

(h) $Ag(NH_3)_2{}^+OH^-$, then H^+

(i) NH_2OH

(j) $C_6H_5\overset{-}{C}H-\overset{+}{P}(C_6H_5)_3$

(k) C_6H_5Li, then H^+

(l) $HC\equiv CNa$, then H^+

(m) $HSCH_2CH_2SH$, H^+, then Raney Ni, H_2

(n) $CH_3CH_2CHBrCO_2Et$ and Zn, then H^+

17.25 Give structural formulas for the products of the reaction (if one occurs) when acetone is treated with each reagent of the preceding problem.

17.26 What products would form when 4-methylbenzaldehyde reacts with each of the following?

(a) CH_3CHO, OH^-

(b) $CH_3C\equiv CNa$, then H^+

(c) CH_3CH_2MgBr, then H^+

(d) Cold dilute $KMnO_4$, OH^-, then H^+

(e) Hot $KMnO_4$, OH^-, then H^+

(f) $^-:CH_2-\overset{+}{P}(C_6H_5)_3$

(g) $CH_3COC_6H_5$, OH^-

(h) $BrCH_2CO_2Et$ and Zn, then H^+

17.27 Show how each of the following transformations could be accomplished. You may use any other required reagents.

(a) $CH_3COC(CH_3)_3 \longrightarrow C_6H_5CH=CHCOC(CH_3)_3$

(b) $C_6H_5CHO \longrightarrow C_6H_5CH=$

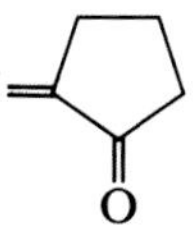

(c) $C_6H_5CHO \longrightarrow C_6H_5CH_2\underset{\underset{CH_3}{|}}{C}HNH_2$

(d) $CH_3\overset{\overset{O}{\|}}{C}(CH_2)_4\overset{\overset{O}{\|}}{C}CH_3 \longrightarrow$

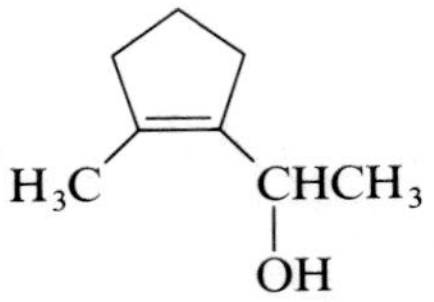

(e) $CH_3CN \longrightarrow CH_3O-$$-CH=CHCN$

(f) $CH_3CH_2CH_2CH_2\overset{\overset{O}{\|}}{C}H \longrightarrow CH_3(CH_2)_3CH=\underset{\underset{CH_2OH}{|}}{C}(CH_2)_2CH_3$

(g)

17.28 The following reaction illustrates the Robinson annulation reaction (Section 17.9B). Give mechanisms for the steps that occur.

$$C_6H_5COCH_2CH_3 + CH_2=\underset{\underset{CH_3}{|}}{\overset{\overset{O}{\|}}{C}}CCH_3 \xrightarrow{\text{base}}$$

17.29 Write structural formulas for **A**, **B**, and **C**.

$$HC\equiv CH \xrightarrow[\substack{(2)\ CH_3COCH_3 \\ (3)\ NH_4Cl/H_2O}]{(1)\ NaNH_2,\ \text{liq. } NH_3} \textbf{A } (C_5H_8O) \xrightarrow[H_2O]{Hg^{2+},\ H_3O^+} \textbf{B } (C_5H_{10}O_2) \xrightarrow{C_6H_5CHO,\ OH^-} \textbf{C } (C_{12}H_{14}O_2)$$

17.30 The hydrogen atoms of the γ carbon of crotonaldehyde are appreciably acidic $(pK_a \sim 20)$.

(a) Write resonance structures that will explain this fact.

$$\overset{\gamma}{C}H_3\overset{\beta}{C}H=\overset{\alpha}{C}HCHO$$
Crotonaldehyde

(b) Write a mechanism that accounts for the following reaction:

$$C_6H_5CH=CHCHO + CH_3CH=CHCHO \xrightarrow[C_2H_5OH]{\text{base}} C_6H_5(CH=CH)_3CHO$$
$$\textbf{(87\%)}$$

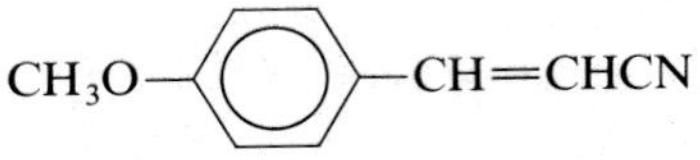

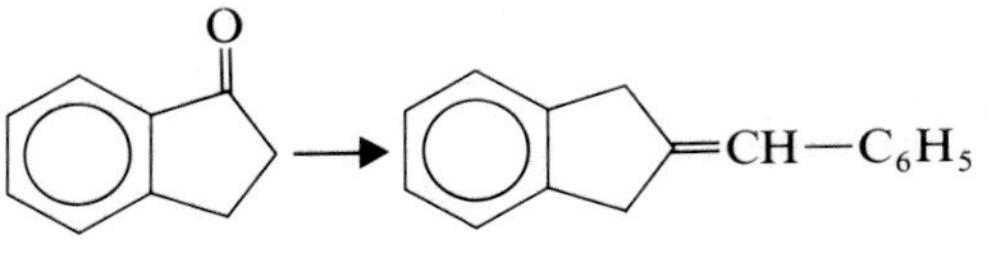

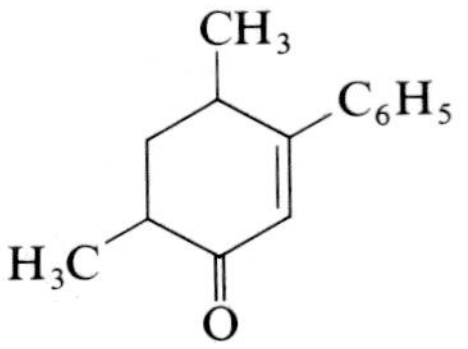

17.31 What reagents would you use to bring about each step of the following syntheses?

(a)

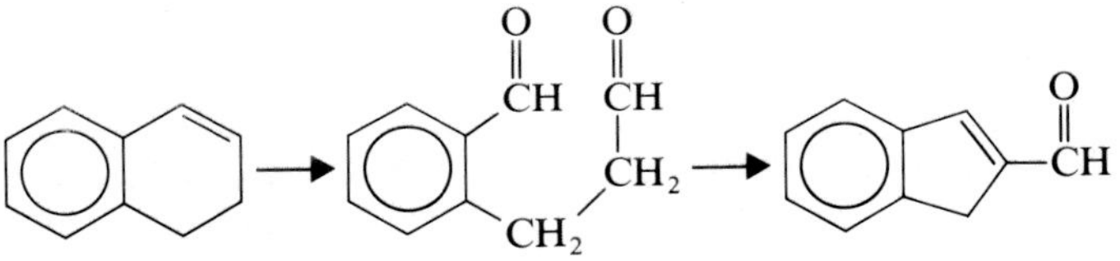

(b)

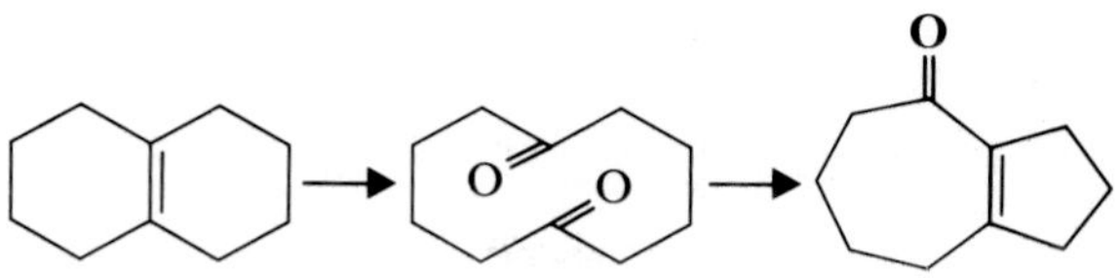

(c)

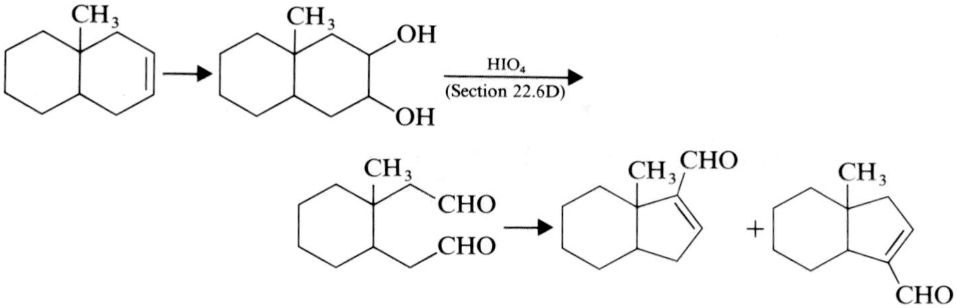

$$\xrightarrow[\text{(Section 22.6D)}]{\text{HIO}_4}$$

(d)

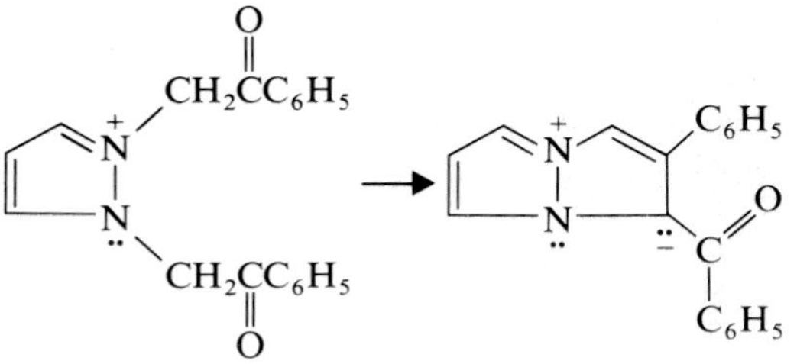

17.32   (a) Infrared spectroscopy gives an easy method for deciding whether the product obtained from the addition of a Grignard reagent to an α,β-unsaturated ketone is the simple addition product or the conjugate addition product. Explain. (What peak or peaks would you look for?)   (b) How might you follow the rate of the following reaction using UV spectroscopy?

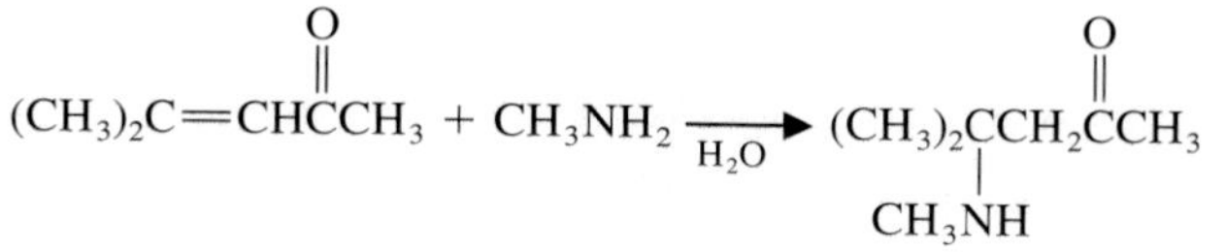

$$(CH_3)_2C{=}CHCCH_3 + CH_3NH_2 \xrightarrow[H_2O]{} (CH_3)_2CCH_2CCH_3$$
$$\underset{CH_3NH}{}$$

17.33   (a) A compound **U** ($C_9H_{10}O$) gives a negative iodoform test. The IR spectrum of **U** shows a strong absorption peak at 1690 cm^{-1}. The ^{1}H NMR spectrum of **U** gives the following:

Triplet	δ 1.2 (3H)
Quartet	δ 3.0 (2H)
Multiplet	δ 7.7 (5H)

What is the structure of **U**?

(b) A compound **V** is an isomer of **U**. Compound **V** gives a positive iodoform test; its IR spectrum shows a strong peak at 1705 cm^{-1}. The ^{1}H NMR spectrum of **V** gives the following:

Singlet	δ 2.0 (3H)
Singlet	δ 3.5 (2H)
Multiplet	δ 7.1 (5H)

What is the structure of **V**?

17.34   Compound **A** has the molecular formula $C_6H_{12}O_3$ and shows a strong IR absorption peak at 1710 cm^{-1}. When treated with iodine in aqueous sodium hydroxide, **A** gives a yellow precipitate. When **A** is treated with Tollens' reagent, no reaction occurs; however, if **A** is treated first with water containing a drop of sulfuric acid and then treated with Tollens' reagent, a silver mirror forms in the test tube. Compound **A** shows the following ^{1}H NMR spectrum.

Singlet	δ 2.1
Doublet	δ 2.6
Singlet	δ 3.2 (6H)
Triplet	δ 4.7

Write a structure for **A**.

17.35 Treating a solution of *cis*-1-decalone with base causes an isomerization to take place. When the system reaches equilibrium, the solution is found to contain about 95% *trans*-1-decalone and about 5% *cis*-1-decalone. Explain this isomerization.

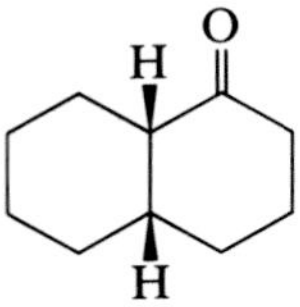

cis-**1-Decalone**

17.36 The Wittig reaction (Section 16.10) can be used in the synthesis of aldehydes, for example,

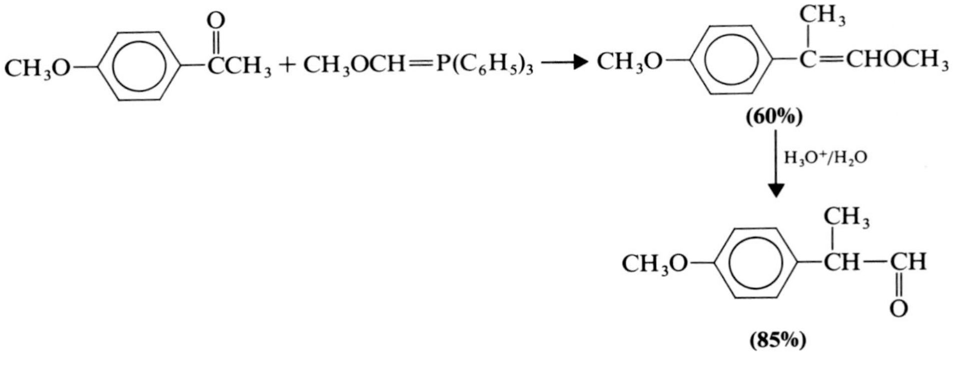

(a) How would you prepare $CH_3OCH\!=\!P(C_6H_5)_3$? (b) Show with a mechanism how the second reaction produces an aldehyde. (c) How would you use this method to prepare

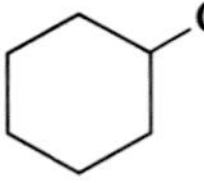

CHO from cyclohexanone?

17.37 Aldehydes that have no α hydrogen undergo an intermolecular oxidation–reduction called the **Cannizzaro reaction** when they are treated with concentrated base. An example is the following reaction of benzaldehyde.

$$2\ C_6H_5\!-\!CHO \xrightarrow[\text{H}_2\text{O}]{\text{OH}^-} C_6H_5\!-\!CH_2OH + C_6H_5\!-\!CO_2^-$$

(a) When the reaction is carried out in D_2O, the benzyl alcohol that is isolated contains no deuterium bound to carbon. It is $C_6H_5CH_2OD$. What does this suggest about the mechanism for the reaction? (b) When $(CH_3)_2CHCHO$ and $Ba(OH)_2/H_2O$ are heated in a sealed tube the reaction produces only $(CH_3)_2CHCH_2OH$ and $[(CH_3)_2CHCO_2]_2Ba$. Provide an explanation for the formation of these products instead of those expected from an aldol reaction.

17.38 When the aldol reaction of acetaldehyde is carried out in D_2O, no deuterium is found in the methyl group of unreacted aldehyde. However, in the aldol reaction of acetone, deuterium is incorporated in the methyl group of the unreacted acetone. Explain this difference in behavior.

17.39 Shown below is a synthesis of the elm bark beetle pheromone, multistriatin (see Problem 16.46). Give structures for compounds **A**, **B**, **C**, and **D**.

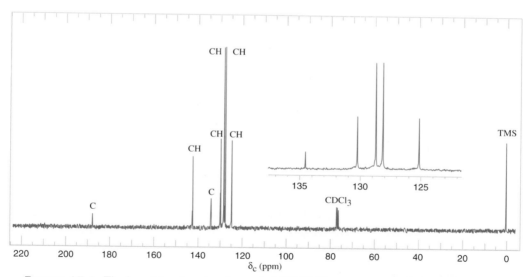

Multistriatin

17.40 Allowing acetone to react with 2 molar equivalents of benzaldehyde in the presence of KOH in ethanol leads to the formation of compound **X**. The ^{13}C NMR spectrum of **X** is given in Figure 17.1. Propose a structure for compound **X**.

FIGURE 17.1 The broad band proton-decoupled ^{13}C NMR spectrum of compound **X**, Problem 17.40. Information from the DEPT ^{13}C NMR spectrum is given above the peaks.

Chlorotrimethylsilane

LITHIUM ENOLATES IN ORGANIC SYNTHESIS

H.1 Enolate Ions

Enolate ions are formed when a carbonyl compound with an α hydrogen is treated with a base (Section 17.1). The extent to which the enolate ion forms will depend on the strength of the base used. If the base employed is a weaker base than the enolate ion, then the equilibrium lies to the left. This is the case, for example, when a ketone is treated with an aqueous solution containing sodium hydroxide.

$$CH_3-\overset{\overset{\displaystyle O}{\|}}{C}-CH_3 + Na^+OH^- \rightleftarrows CH_3-\overset{\overset{\displaystyle O^{\delta-}Na^+}{\|}}{C}\text{===}CH_2^{\delta-} + H_2O$$

Weaker acid   **Weaker base**   **Stronger base**   **Stronger acid**
($pK_a = 20$)   ($pK_a = 16$)

On the other hand, if a very strong base is employed, the equilibrium lies far to the right. One very useful strong base for converting ketones to enolates is lithium diisopropylamide, $(i\text{-}C_3H_7)_2N^-Li^+$.

$$CH_3-\overset{\overset{\displaystyle O}{\|}}{C}-CH_3 + (i\text{-}C_3H_7)_2N^-Li^+ \longrightarrow CH_3-\overset{\overset{\displaystyle O^{\delta-}Li^+}{\|}}{C}\text{===}CH_2^{\delta-} + (i\text{-}C_3H_7)_2NH$$

Stronger acid   **Stronger base**   **Weaker base**   **Weaker acid**
($pK_a = 20$)   ($pK_a = 38$)

Lithium diisopropylamide (abbreviated **LDA**) can be prepared by dissolving diisopropylamine in a solvent such as diethyl ether or THF, and treating it with an alkyllithium.

$$(i\text{-}C_3H_7)_2NH + CH_3^-Li^+ \xrightarrow{\text{diethyl ether}} (i\text{-}C_3H_7)_2N^-Li^+ + CH_4\uparrow$$

Stronger acid	**Stronger base**		**Weaker base**	**Weaker acid**
($pK_a = 38$)				($pK_a = 50$)

H.1A Enolate Ions as Ambident Nucleophiles

Because enolate ions have a partial negative charge on an oxygen atom they can react in nucleophilic substitution reactions as if they were **alkoxide ions.** Because they have a partial negative charge on a carbon atom they can also react as **carbanions.** Nucleophiles like this, *those that are capable of reacting at two sites,* are called **ambident nucleophiles.**

This site reacts as an alkoxide ion

$$CH_3-\overset{\overset{\displaystyle O^{\delta-}}{\|}}{C}\overset{\delta-}{=\!=\!=}CH_2$$

This site reacts as a carbanion

Just how an enolate ion reacts depends, in part, on the substrate with which it reacts. *One substrate that tends to react almost exclusively at the oxygen atom of an enolate is chlorotrimethylsilane,* $(CH_3)_3SiCl.$

$$CH_3\overset{\overset{\displaystyle O^{\delta-}}{\|}}{C}\overset{\delta-}{=\!=\!=}CH_2 + (CH_3)_3Si-Cl \xrightarrow{\text{THF}} CH_3\overset{\overset{\displaystyle OSi(CH_3)_3}{|}}{C}=CH_2 + Cl^-$$

Enol trimethylsilyl ether (85%)

This reaction, called **silylation** (cf. Section 10.16D), is a nucleophilic substitution at the silicon atom by the oxygen atom of the enolate, and it takes place as it does because the oxygen–silicon bond that forms in the enol trimethylsilyl ether is very strong (much stronger than a carbon–silicon bond). This factor makes formation of the enol trimethylsilyl ether highly exothermic, and, consequently, the free energy of activation for reaction at the oxygen atom is lower than that for reaction at the α carbon.

Enolate ions display their carbanionic character when they react with alkyl halides. In these reactions the major product is usually the one in which alkylation occurs at the carbon atom, called *C*-alkylation. (Alkylation at the oxygen atom is called *O*-alkylation.)

$$R-\overset{\overset{\displaystyle O^{\delta-}}{\|}}{C}\overset{\delta-}{=\!=\!=}CHR' + R''CH_2-X \longrightarrow R-\overset{\overset{\displaystyle O}{\|}}{C}-\overset{\overset{\displaystyle CH_2R''}{|}}{C}HR'$$

C-Alkylated product
(major product)

Alkylation reactions like these have an important limitation. Because the reactions are S_N2 reactions and because enolate ions are strong bases, *successful alkylations occur only when primary alkyl, primary benzylic, and primary allylic halides are used.* With secondary and tertiary halides, elimination becomes the main course of the reaction.

PROBLEM H.1

Write structures for the *C*-alkylated and *O*-alkylated products that form when the enolate derived from cyclohexanone reacts with benzyl bromide.

H.1B Regioselective Formation of Enolate Ions

An unsymmetrical ketone such as 2-methylcyclohexanone can form two possible enolates. Just which enolate is formed predominantly depends on the base used and on the conditions employed. The enolate *with the more highly substituted double bond is the thermodynamically more stable enolate* in the same way that an al-kene with the more highly substituted double bond is the more stable alkene (Section 7.9). This enolate, called the **thermodynamic enolate,** is formed predom-inantly under conditions that permit the establishment of an equilibrium. This will generally be the case if the enolate is produced using a relatively weak base in a protic solvent.

2-Methylcyclohexanone   Thermodynamic
                                      enolate

This enolate is more stable because the double bond is more highly substituted. It is the predominant enolate present at equilibrium

On the other hand, *the enolate with the less substituted double bond is us-ually formed faster,* because removal of the hydrogen necessary to produce this enolate is less sterically hindered. This enolate, called the **kinetic enolate,** is formed predominantly when the reaction is kinetically controlled (or rate con-trolled).

The kinetically favored enolate can be formed cleanly through the use of lith-ium diisopropylamide (LDA). This strong, sterically hindered base rapidly re-moves the proton from the less substituted α carbon of the ketone. The following example, using 2-methylcyclohexanone, is an illustration. The solvent for the reaction is 1,2-dimethoxyethane ($CH_3OCH_2CH_2OCH_3$), abbreviated **DME.** The LDA removes the hydrogen from the $-CH_2-$ α carbon more rapidly be-cause it is less hindered (and because there are twice as many hydrogens there to re-act).

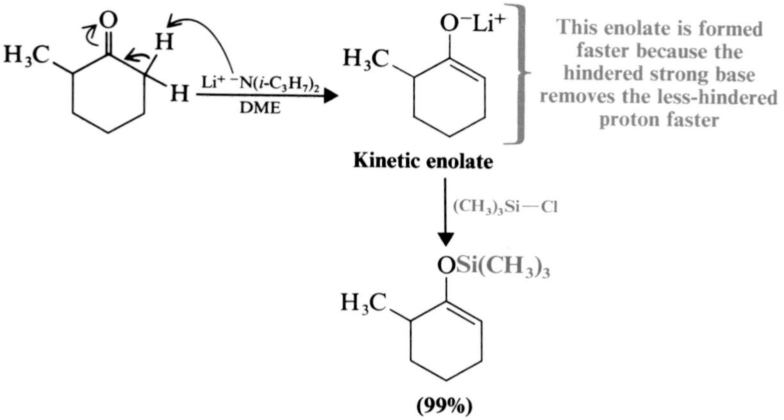

This enolate is formed faster because the hindered strong base removes the less-hindered proton faster

Kinetic enolate

(99%)

The example just given also shows how the enolate ion can be "trapped" by converting it to the enol trimethylsilyl ether. This procedure is especially useful because the enol trimethylsilyl ether can be purified, if necessary, and then converted back to an enolate. One way of achieving this conversion is by treating the enol trimethylsilyl ether with a solution containing fluoride ions.

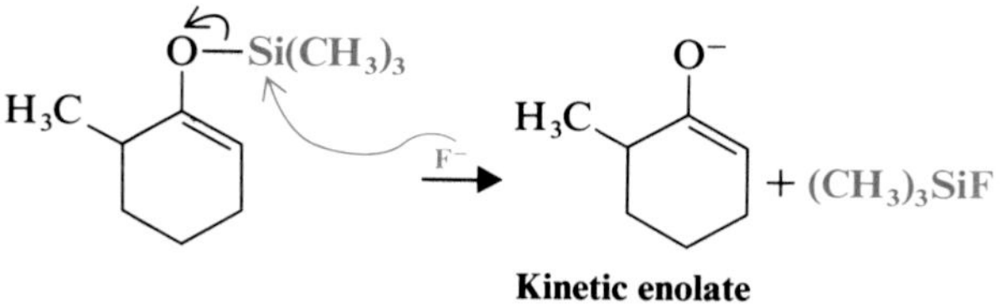

Kinetic enolate

This reaction is a nucleophilic substitution at the silicon atom brought about by a fluoride ion. Fluoride ions have an extremely high affinity for silicon atoms because Si—F bonds are very strong.

Another way to convert an enol trimethylsilyl ether back to an enolate is to treat it with methyllithium.

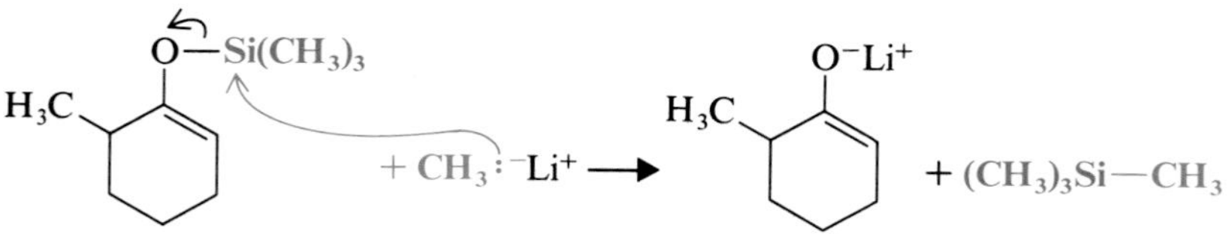

H.2 Direct Alkylation of Ketones via Lithium Enolates

The formation of lithium enolates using lithium diisopropylamide furnishes a useful way of alkylating ketones in a regioselective way. For example, the lithium enolate formed from 2-methylcyclohexanone (Section H.1B) can be methylated and benzylated by allowing it to react with methyl iodide and benzyl bromide, respectively.

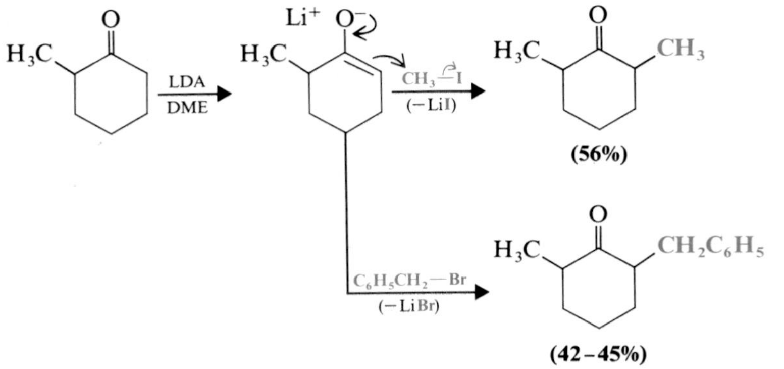

H.3 Lithium Enolates in Directed Aldol Reactions

One of the most effective and versatile ways to bring about a crossed aldol reaction is to use a lithium enolate obtained from a ketone as one component and an aldehyde or ketone as the other. An example of what is called a **directed aldol reaction** is shown in Fig. H.1.

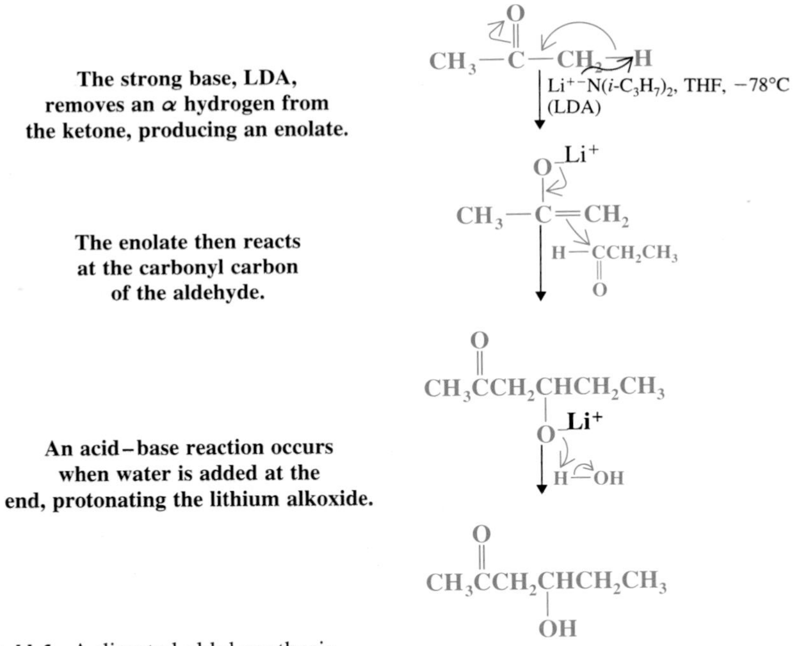

The strong base, LDA, removes an α hydrogen from the ketone, producing an enolate.

The enolate then reacts at the carbonyl carbon of the aldehyde.

An acid–base reaction occurs when water is added at the end, protonating the lithium alkoxide.

FIGURE H.1 A directed aldol synthesis.

Regioselectivity can be achieved when unsymmetrical ketones are used in directed aldol reactions by generating the kinetic enolate using lithium diisopropylamide. This ensures production of the enolate in which the proton has been removed from the less substituted α carbon. The following is an example:

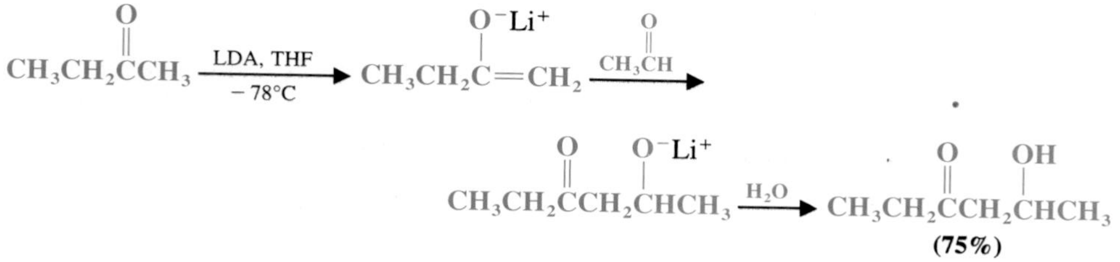

$$CH_3CH_2CCH_3 \xrightarrow[-78°C]{LDA, THF} CH_3CH_2C=CH_2 \xrightarrow{CH_3CH}$$

$$CH_3CH_2CCH_2CHCH_3 \xrightarrow{H_2O} CH_3CH_2CCH_2CHCH_3$$
$$\text{(75\%)}$$

If the aldol (Claisen–Schmidt) reaction had been carried out in the classical way (Section 17.6B) using hydroxide ion as the base then at least two products would have been formed in significant amounts. Both the kinetic and thermodynamic enolates would have been formed from the ketone and each of these would have added to the carbonyl carbon of the aldehyde:

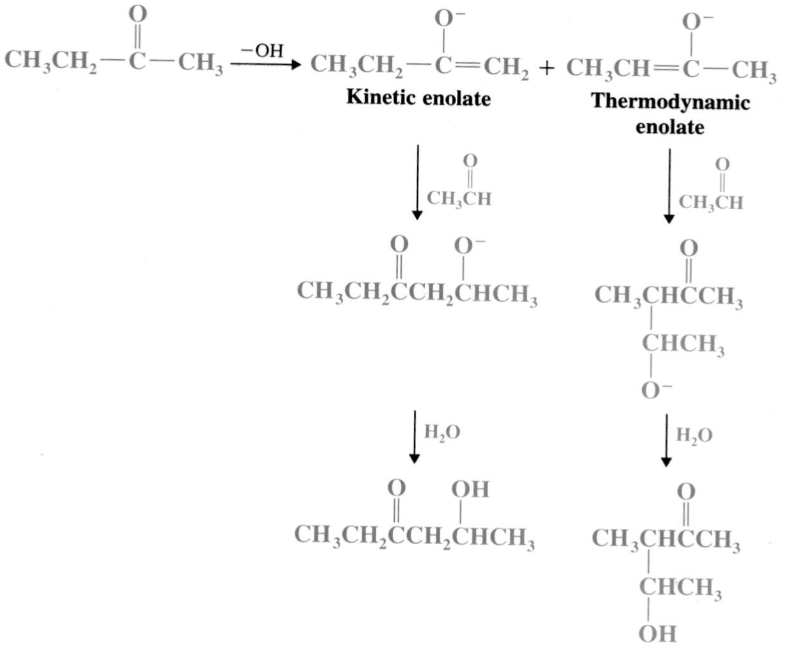

$$CH_3CH_2-C-CH_3 \xrightarrow{-OH} CH_3CH_2-C=CH_2 + CH_3CH=C-CH_3$$
Kinetic enolate   **Thermodynamic enolate**

$$\downarrow CH_3CH \qquad\qquad \downarrow CH_3CH$$

$$CH_3CH_2CCH_2CHCH_3 \qquad CH_3CHCCH_3$$
$$\qquad\qquad\qquad\qquad CHCH_3$$
$$\qquad\qquad\qquad\qquad O^-$$

$$\downarrow H_2O \qquad\qquad \downarrow H_2O$$

$$CH_3CH_2CCH_2CHCH_3 \qquad CH_3CHCCH_3$$
$$\qquad\qquad\qquad\qquad CHCH_3$$
$$\qquad\qquad\qquad\qquad OH$$

PROBLEM H.2

Starting with ketones and aldehydes of your choice outline a directed aldol synthesis of each of the following:

(a)

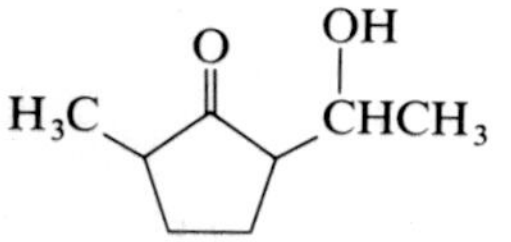

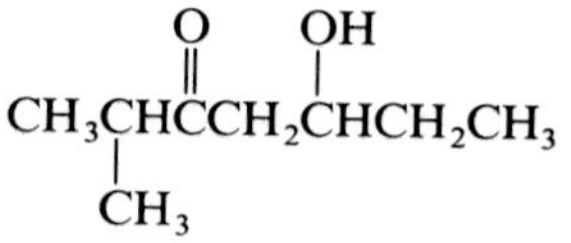

(b) 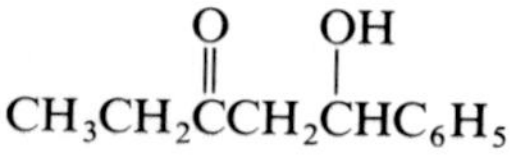 $CH_3CH_2CCH_2CHC_6H_5$

(c) $CH_3CHCCH_2CHCH_2CH_3$
$\qquad CH_3$

(d)  $CH_3CH=CHCCH_2CHCH_3$

PROBLEM H.3

The compounds called α-bisabolanone and ocimenone have both been synthesized by directed aldol syntheses. In both syntheses one starting compound was $(CH_3)_2C{=}CHCOCH_3$. Choose other appropriate starting compounds and outline syntheses of (a) α-bisabolanone and (b) ocimenone.

α-Bisabolanone   Ocimenone

PROBLEM H.4

Treating the enol trimethylsilyl ether derived from cyclohexanone with benzaldehyde and tetrabutylammonium fluoride, $(C_4H_9)_4N^+F^-$ (abbreviated TBAF), gave the following product. Outline the steps that occur in this reaction.

H.4 α-Selenation: A Synthesis of α,β-Unsaturated Carbonyl Compounds

Lithium enolates react with benzeneselenenyl bromide (C_6H_5SeBr) (or with C_6H_5SeCl) to yield products containing a $C_6H_5Se{-}$ group at the α position.

Treating the α-benzeneselenenyl ketone with hydrogen peroxide at room temperature converts it to an α,β-unsaturated ketone.

$$C_6H_5 \overset{\overset{\displaystyle O}{\|}}{C} \underset{\underset{\displaystyle SeC_6H_5}{|}}{CH} CH_3 \xrightarrow[25°C]{H_2O_2} C_6H_5 \overset{\overset{\displaystyle O}{\|}}{C} CH=CH_2 + C_6H_5SeOH$$

(84–89% overall
from the ketone)

These are very mild conditions for the introduction of a double bond (room temperature and a neutral solution), and this is one reason why this method is a valuable one.

Mechanistically, two steps are involved in the conversion of the α-benzeneselenenyl ketone to the α,β-unsaturated ketone. The first step is an oxidation brought about by the H_2O_2. The second step is a spontaneous intramolecular elimination in which the negatively charged oxygen atom attached to the selenium atom acts as a base.

When we study the Cope elimination in Section 20.13B, we shall find another example of this kind of intramolecular elimination.

PROBLEM H.5

Starting with 2-methylcyclohexanone, show how you would use α-selenation in a synthesis of the following compound:

CHAPTER **18**

Vitamin C
(Section 18.7C)

CARBOXYLIC ACIDS AND THEIR DERIVATIVES. NUCLEOPHILIC SUBSTITUTION AT THE ACYL CARBON

18.1 INTRODUCTION

The carboxyl group, $-\overset{\overset{\text{O}}{\|}}{\text{C}}\text{OH}$ (abbreviated $-CO_2H$ or $-COOH$), is one of the most widely occurring functional groups in chemistry and biochemistry. Not only are carboxylic acids themselves important, but the carboxyl group is the parent group of a large family of related compounds called **acyl compounds** or **carboxylic acid derivatives** shown in Table 18.1.

18.2 NOMENCLATURE AND PHYSICAL PROPERTIES

18.2A Carboxylic Acids

IUPAC systematic or substitutive names for carboxylic acids are obtained by dropping the final -e of the name of the alkane corresponding to the longest chain in the acid and by adding -oic acid. The carboxyl carbon atom is assigned number 1. The examples listed here illustrate how this is done.

TABLE 18.1   Carboxylic acid derivatives

STRUCTURE	NAME	STRUCTURE	NAME
$R-\overset{\overset{\displaystyle O}{\|\|}}{C}-Cl$	Acyl (or acid) chloride	$R-\overset{\overset{\displaystyle O}{\|\|}}{C}-NH_2$	
$R-\overset{\overset{\displaystyle O}{\|\|}}{C}-O-\overset{\overset{\displaystyle O}{\|\|}}{C}-R$	Acid anhydride	$R-\overset{\overset{\displaystyle O}{\|\|}}{C}-NHR'$	Amide
$R-\overset{\overset{\displaystyle O}{\|\|}}{C}-O-R'$	Ester	$R-\overset{\overset{\displaystyle O}{\|\|}}{C}-NR'R''$	
$R-C\equiv N$	Nitrile		

$$\overset{6}{CH_3}\overset{5}{CH_2}\overset{4}{CH}\overset{3}{CH_2}\overset{2}{CH_2}\overset{\overset{\displaystyle O}{\|\|}}{C}OH$$
$$|$$
$$CH_3$$

4-Methylhexanoic acid

$$\overset{6}{CH_3}\overset{5}{CH}=\overset{4}{CH}\overset{3}{CH_2}\overset{2}{CH_2}\overset{\overset{\displaystyle O}{\|\|}}{C}OH$$

4-Hexenoic acid

Many carboxylic acids have common names that are derived from Latin or Greek words that indicate one of their natural sources (Table 18.2). Methanoic acid is called formic acid (from the Latin, *formica,* or ant). Ethanoic acid is called acetic acid (from the Latin, *acetum,* or vinegar). Butanoic acid is one compound responsible for the odor of rancid butter, so its common name is butyric acid (from the Latin, *butyrum,* or butter). Pentanoic acid, as a result of its occurrence in valerian, a perennial herb, is named valeric acid. Hexanoic acid is one compound associated with the odor of goats, hence its common name, caproic acid (from the Latin *caper,* or goat). Octadecanoic acid takes its common name, stearic acid, from the Greek word *stear,* for tallow.

Most of these common names have been with us for a long time and some are likely to remain in common usage for even longer, so it is helpful to be familiar with them. In this text we shall always refer to methanoic acid and ethanoic acid as formic acid and acetic acid. However, in almost all other instances we shall use IUPAC systematic or substitutive names.

Carboxylic acids are polar substances. Their molecules can form strong hydrogen bonds with each other and with water. As a result, carboxylic acids generally have high boiling points, and low molecular weight carboxylic acids show appreciable solubility in water. The first four carboxylic acids (Table 18.2) are miscible with water in all proportions. As the length of the carbon chain increases, water solubility declines.

18.2B Carboxylic Salts

Salts of carboxylic acids are named as *-ates;* in both common and systematic names, *-ate* replaces *-ic acid.* Thus, CH_3CO_2Na is sodium acetate or sodium ethanoate.

Table 18.2   Carboxylic acids

STRUCTURE	SYSTEMATIC NAME	COMMON NAME	mp (°C)	bp (°C)	WATER SOLUBILITY (g 100 mL^{-1} H$_2$O) 25°C	pK_a
HCO$_2$H	Methanoic acid	Formic acid	8	100.5	∞	3.75
CH$_3$CO$_2$H	Ethanoic acid	Acetic acid	16.6	118	∞	4.76
CH$_3$CH$_2$CO$_2$H	Propanoic acid	Propionic acid	−21	141	∞	4.87
CH$_3$(CH$_2$)$_2$CO$_2$H	Butanoic acid	Butyric acid	−6	164	∞	4.81
CH$_3$(CH$_2$)$_3$CO$_2$H	Pentanoic acid	Valeric acid	−34	187	4.97	4.82
CH$_3$(CH$_2$)$_4$CO$_2$H	Hexanoic acid	Caproic acid	−3	205	1.08	4.84
CH$_3$(CH$_2$)$_6$CO$_2$H	Octanoic acid	Caprylic acid	16	239	0.07	4.89
CH$_3$(CH$_2$)$_8$CO$_2$H	Decanoic acid	Capric acid	31	269	0.015	4.84
CH$_3$(CH$_2$)$_{10}$CO$_2$H	Dodecanoic acid	Lauric acid	44	179^{18}	0.006	5.30
CH$_3$(CH$_2$)$_{12}$CO$_2$H	Tetradecanoic acid	Myristic acid	59	200^{20}	0.002	
CH$_3$(CH$_2$)$_{14}$CO$_2$H	Hexadecanoic acid	Palmitic acid	63	219^{17}	0.0007	6.46
CH$_3$(CH$_2$)$_{16}$CO$_2$H	Octadecanoic acid	Stearic acid	70	383	0.0003	
CH$_2$ClCO$_2$H	Chloroethanoic acid	Chloroacetic acid	63	189	Very soluble	2.86
CHCl$_2$CO$_2$H	Dichloroethanoic acid	Dichloroacetic acid	10.8	192	Very soluble	1.48
CCl$_3$CO$_2$H	Trichloroethanoic acid	Trichloroacetic acid	56.3	198	Very soluble	0.70
CH$_3$CHClCO$_2$H	2-Chloropropanoic acid	α-Chloropropionic acid		186	Soluble	2.83
CH$_2$ClCH$_2$CO$_2$H	3-Chloropropanoic acid	β-Chloropropionic acid	61	204	Soluble	3.98
C$_6$H$_5$CO$_2$H	Benzoic acid	Benzoic acid	122	250	0.34	4.19
p-CH$_3$C$_6$H$_4$CO$_2$H	4-Methylbenzoic acid	p-Toluic acid	180	275	0.03	4.36
p-ClC$_6$H$_4$CO$_2$H	4-Chlorobenzoic acid	p-Chlorobenzoic acid	242		0.009	3.98
p-NO$_2$C$_6$H$_4$CO$_2$H	4-Nitrobenzoic acid	p-Nitrobenzoic acid	242		0.03	3.41
(CO$_2$H structure)	1-Naphthoic acid	α-Naphthoic acid	160	300	Insoluble	3.70
(CO$_2$H structure)	2-Naphthoic acid	β-Naphthoic acid	185	>300	Insoluble	4.17

Sodium and potassium salts of most carboxylic acids are readily soluble in water. This is true even of the long-chain carboxylic acids. Sodium or potassium salts of long-chain carboxylic acids are the major ingredients of soap (cf. Section 23.2C).

PROBLEM 18.1

Give an IUPAC systematic name for each of the following:

(a) CH$_3$CH$_2$CHCO$_2$H
          |
          CH$_3$

(b) CH$_3$CH=CHCH$_2$CO$_2$H

(c) BrCH$_2$CH$_2$CH$_2$CO$_2$Na

(d) C$_6$H$_5$CH$_2$CH$_2$CH$_2$CH$_2$CO$_2$H

(e) CH$_3$CH=CCH$_2$CO$_2$H
              |
              CH$_3$

PROBLEM 18.2

Experiments show that the molecular weight of acetic acid in the vapor state (just above its boiling point) is approximately 120. Explain the discrepancy between this experimental value and the true value of approximately 60.

18.2C Acidity of Carboxylic Acids

Most unsubstituted carboxylic acids have K_a values in the range of 10^{-4}–10^{-5} ($pK_a = 4$–5) as seen in Table 18.2. The pK_a of water is about 16 and the apparent pK_a of H_2CO_3 is about 7. These relative acidities mean that carboxylic acids react readily with aqueous solutions of sodium hydroxide and sodium bicarbonate to form soluble sodium salts. We can use solubility tests, therefore, to distinguish water-insoluble carboxylic acids from water-insoluble phenols (Chapter 21) and alcohols. Water-insoluble carboxylic acids dissolve in either aqueous sodium hydroxide or aqueous sodium bicarbonate:

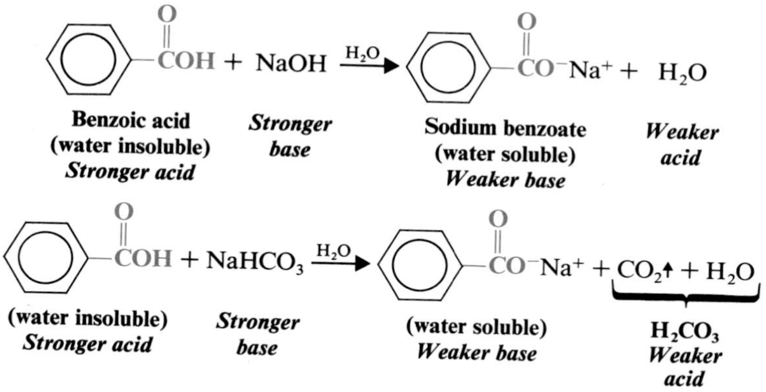

Water-insoluble phenols (Section 21.5) dissolve in aqueous sodium hydroxide but (except for some nitrophenols) do not dissolve in aqueous sodium bicarbonate. Water-insoluble alcohols do not dissolve in either aqueous sodium hydroxide or sodium bicarbonate.

We see in Table 18.2 that carboxylic acids having electron-withdrawing groups are stronger than unsubstituted acids. The chloroacetic acids, for example, show the following order of acidities:

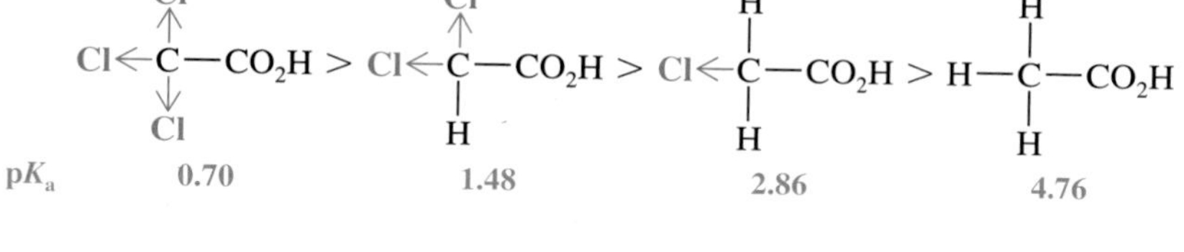

As we saw in Sections 3.5B and 3.7 this acid-strengthening effect of electron-withdrawing groups arises from a combination of inductive effects and entropy effects. Since inductive effects are not transmitted very effectively through covalent bonds, the acid-strengthening effect decreases as distance between the electron-withdrawing group and the carboxyl group increases. Of the chlorobutanoic acids that follow, the strongest acid is 2-chlorobutanoic acid:

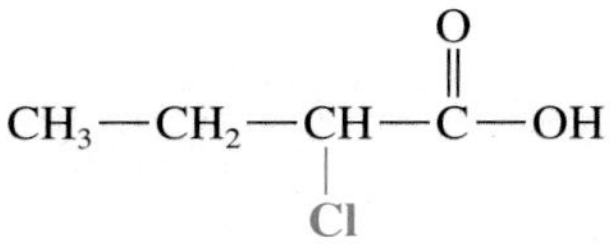

| 2-Chlorobutanoic acid ($pK_a = 2.85$) | 3-Chlorobutanoic acid ($pK_a = 4.05$) | 4-Chlorobutanoic acid ($pK_a = 4.50$) |

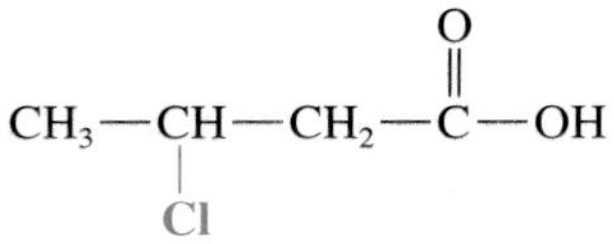

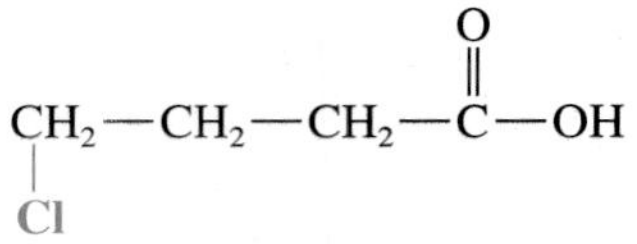

PROBLEM 18.3

Which acid of each pair shown here would you expect to be stronger?

(a) CH_3CO_2H or CH_2FCO_2H

(b) CH_2FCO_2H or CH_2ClCO_2H

(c) CH_2ClCO_2H or CH_2BrCO_2H

(d) $CH_2FCH_2CH_2CO_2H$ or $CH_3CHFCH_2CO_2H$

(e) $CH_3CH_2CHFCO_2H$ or $CH_3CHFCH_2CO_2H$

(f)

(g)

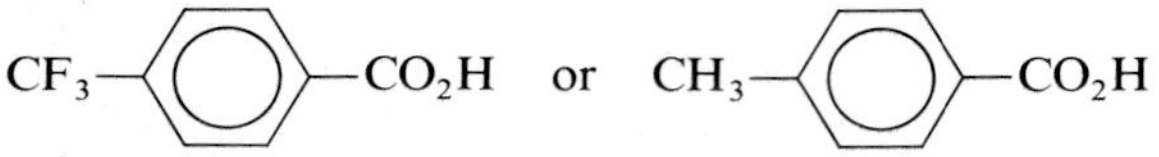

18.2D Dicarboxylic Acids

Dicarboxylic acids are named as **alkanedioic acids** in the IUPAC systematic or substitutive system. Most simple dicarboxylic acids have common names (Table 18.3), and these are the names that we shall use.

PROBLEM 18.4

Suggest explanations for the following facts: (a) pK_1 for all of the dicarboxylic acids in Table 18.3 are smaller than the pK_a for monocarboxylic acids with the same number of carbon atoms.   (b) The difference between pK_1 and pK_2 for dicarboxylic acids of type $HO_2C(CH_2)_nCO_2H$ decreases as n increases.

8.2E Esters

The names of esters are derived from the names of the alcohol (with the ending **-yl**) and the acid (with the ending **-ate** or **-oate**). The portion of the name derived from the alcohol comes first.

TABLE 18.3 Dicarboxylic acids

STRUCTURE	COMMON NAME	mp (°C)	pK_a (at 25°C) pK_1	pK_2
$HO_2C—CO_2H$	Oxalic acid	189 dec	1.2	4.2
$HO_2CCH_2CO_2H$	Malonic acid	136	2.9	5.7
$HO_2C(CH_2)_2CO_2H$	Succinic acid	187	4.2	5.6
$HO_2C(CH_2)_3CO_2H$	Glutaric acid	98	4.3	5.4
$HO_2C(CH_2)_4CO_2H$	Adipic acid	153	4.4	5.6
$cis\text{-}HO_2C—CH{=}CH—CO_2H$	Maleic acid	131	1.9	6.1
$trans\text{-}HO_2C—CH{=}CH—CO_2H$	Fumaric acid	287	3.0	4.4
Phthalic acid (structure)	Phthalic acid	206–208 dec	2.9	5.4
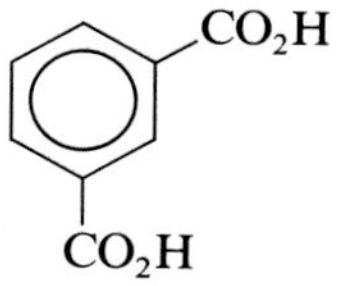 Isophthalic acid (structure)	Isophthalic acid	345–348	3.5	4.6
Terephthalic acid (structure)	Terephthalic acid	Sublimes	3.5	4.8

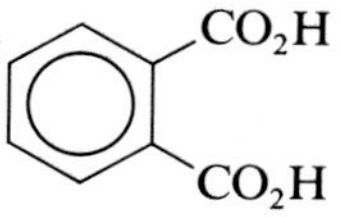

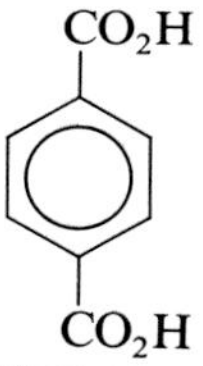

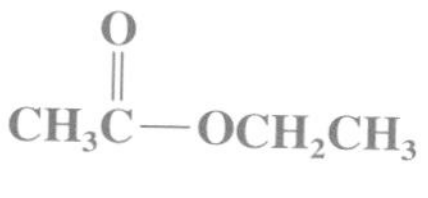

$$CH_3\overset{\overset{\displaystyle O}{\|}}{C}—OCH_2CH_3$$

**Ethyl acetate or
ethyl ethanoate**

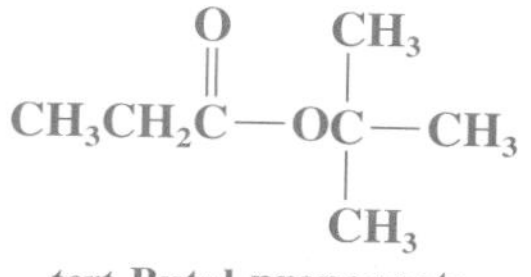

$$CH_3CH_2\overset{\overset{\displaystyle O}{\|}}{C}—O\overset{\overset{\displaystyle CH_3}{|}}{\underset{\underset{\displaystyle CH_3}{|}}{C}}—CH_3$$

tert-**Butyl propanoate**

$$CH_3\overset{\overset{\displaystyle O}{\|}}{C}OCH{=}CH_2$$

**Vinyl acetate or
ethenyl ethanoate**

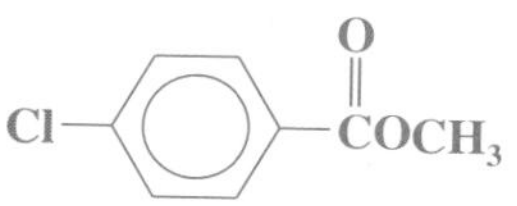

$$Cl—\langle\rangle—\overset{\overset{\displaystyle O}{\|}}{C}OCH_3$$

Methyl *p*-chlorobenzoate

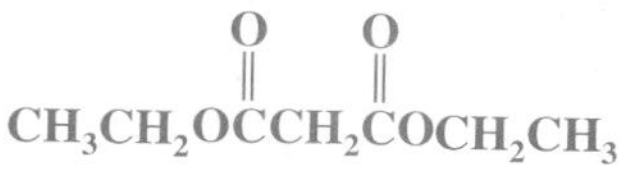

$$CH_3CH_2O\overset{\overset{\displaystyle O}{\|}}{C}CH_2\overset{\overset{\displaystyle O}{\|}}{C}OCH_2CH_3$$

Diethyl malonate

Esters are polar compounds but lacking a hydrogen attached to oxygen, their molecules cannot form strong hydrogen bonds to each other. As a result, esters have boiling points that are lower than those of acids and alcohols of comparable molecular weight. The boiling points (Table 18.4) of esters are about the same as those of comparable aldehydes and ketones.

TABLE 18.4   Carboxylic esters

NAME	STRUCTURE	mp (°C)	bp (°C)	SOLUBILITY IN WATER (g 100 mL^{-1} at 20°C)
Methyl formate	HCO_2CH_3	− 99	31.5	Very soluble
Ethyl formate	$HCO_2CH_2CH_3$	− 79	54	Soluble
Methyl acetate	$CH_3CO_2CH_3$	− 99	57	24.4
Ethyl acetate	$CH_3CO_2CH_2CH_3$	− 82	77	7.39 (25°C)
Propyl acetate	$CH_3CO_2CH_2CH_2CH_3$	− 93	102	1.89
Butyl acetate	$CH_3CO_2CH_2(CH_2)_2CH_3$	− 74	125	1.0 (22°C)
Ethyl propanoate	$CH_3CH_2CO_2CH_2CH_3$	− 73	99	1.75
Ethyl butanoate	$CH_3(CH_2)_2CO_2CH_2CH_3$	− 93	120	0.51
Ethyl pentanoate	$CH_3(CH_2)_3CO_2CH_2CH_3$	− 91	145	0.22
Ethyl hexanoate	$CH_3(CH_2)_4CO_2CH_2CH_3$	− 68	168	0.063
Methyl benzoate	$C_6H_5CO_2CH_3$	− 12	199	0.15
Ethyl benzoate	$C_6H_5CO_2CH_2CH_3$	− 35	213	0.08
Phenyl acetate	$CH_3CO_2C_6H_5$		196	V. sl. soluble
Methyl salicylate	$o\text{-}HOC_6H_4CO_2CH_3$	− 9	223	0.74 (30°C)

Unlike the low molecular weight acids, esters usually have pleasant odors, some resembling those of fruits, and these are used in the manufacture of synthetic flavors:

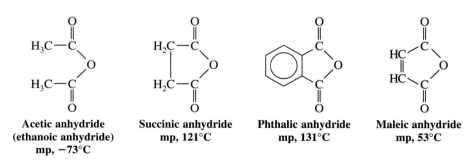

Isopentyl acetate
(used in synthetic banana flavor)

Isopentyl pentanoate
(used in synthetic apple flavor)

18.2F Carboxylic Anhydrides

Most anhydrides are named by dropping the word **acid** from the name of the carboxylic acid and then adding the word **anhydride.**

Acetic anhydride
(ethanoic anhydride)
mp, −73°C

Succinic anhydride
mp, 121°C

Phthalic anhydride
mp, 131°C

Maleic anhydride
mp, 53°C

18.2G Acyl Chlorides

Acyl chlorides are also called **acid chlorides.** They are named by dropping **-ic acid** from the name of the acid and then adding **-yl chloride.** Examples are

$$CH_3\overset{\displaystyle O}{\overset{\|}{C}}-Cl$$

Acetyl chloride
(ethanoyl chloride)
mp, − 112°C; bp, 51°C

$$CH_3CH_2\overset{\displaystyle O}{\overset{\|}{C}}-Cl$$

Propanoyl chloride
mp, − 94°C; bp, 80°C

$$C_6H_5\overset{\displaystyle O}{\overset{\|}{C}}-Cl$$

Benzoyl chloride
mp, − 1°C; bp, 197°C

Acyl chlorides and carboxylic anhydrides have boiling points in the same range as esters of comparable molecular weight.

18.2H Amides

Amides that have no substituent on nitrogen are named by dropping **-ic acid** from the common name of the acid (or *-oic acid* from the substitutive name) and then adding **-amide.** Alkyl groups on the nitrogen atom of amides are named as substituents and the named substituent is prefaced by *N-*, or *N,N-*. Examples are

$$CH_3\overset{\displaystyle O}{\overset{\|}{C}}-NH_2$$

Acetamide
(ethanamide)
mp, 82°C; bp, 221°C

$$CH_3\overset{\displaystyle O}{\overset{\|}{C}}-N\overset{\displaystyle CH_3}{\underset{\displaystyle CH_3}{}}$$

***N,N*-Dimethylacetamide**
mp,− 20°C; bp, 166°C

$$CH_3\overset{\displaystyle O}{\overset{\|}{C}}-NHC_2H_5$$

***N*-Ethylacetamide**
bp, 205°C

$$CH_3\overset{\displaystyle O}{\overset{\|}{C}}-N\overset{\displaystyle C_6H_5}{\underset{\displaystyle CH_2CH_2CH_3}{}}$$

***N*-Phenyl-*N*-propylacetamide**
mp, 49°C; bp, 266°C at 712 torr

Benzamide
mp, 130°C; bp, 290°C

Molecules of amides with one (or no) substituent on nitrogen are able to form strong hydrogen bonds to each other and, consequently, such amides have high melting points and boiling points. Molecules of *N,N*-disubstituted amides cannot form strong hydrogen bonds to each other; they have lower melting points and boiling points.

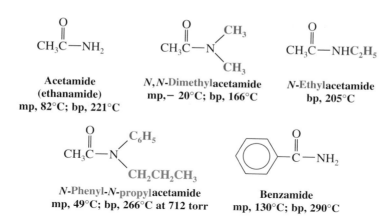

Hydrogen bonding
between molecules of an amide

18.2I Nitriles

In IUPAC substitutive nomenclature, acyclic nitriles are named by adding the suffix *nitrile* to the name of the corresponding hydrocarbon. The carbon atom of the —C≡N group is assigned number 1. The name acetonitrile is an acceptable common name for CH_3CN and acrylonitrile is an acceptable common name for CH_2=CHCN.

$$\overset{2}{C}H_3 — \overset{1}{C}≡N:$$
Ethanenitrile
(acetonitrile)

$$\overset{4}{C}H_3\overset{3}{C}H_2\overset{2}{C}H_2 — \overset{1}{C}≡N:$$
Butanenitrile

$$\overset{3}{C}H_2 = \overset{2}{C}H — \overset{1}{C}≡N:$$
Propenenitrile
(acrylonitrile)

$$\overset{5}{C}H_2 = \overset{4}{C}H\overset{3}{C}H_2\overset{2}{C}H_2 — \overset{1}{C}≡N:$$
4-Pentenenitrile

Cyclic nitriles are named by adding the suffix *carbonitrile* to the name of the ring system to which the —CN group is attached. Benzonitrile is an acceptable common name for C_6H_5CN.

—C≡N:
Benzenecarbonitrile
(benzonitrile)

—C≡N:
Cyclohexanecarbonitrile

PROBLEM 18.5

Write structural formulas for the following:

(a) Methyl propanoate

(b) Ethyl *p*-nitrobenzoate

(c) Dimethyl malonate

(d) *N*,*N*-Dimethylbenzamide

(e) Pentanenitrile

(f) Dimethyl phthalate

(g) Dipropyl maleate

(h) *N*,*N*-Dimethylformamide

(i) 2-Bromopropanoyl bromide

(j) Diethyl succinate

18.2J Spectroscopic Properties of Acyl Compounds

IR Spectra   Infrared spectroscopy is of considerable importance in identifying carboxylic acids and their derivatives. The C=O stretching band is one of the most prominent in their IR spectra since it is always a strong band. The C=O stretching band occurs at different frequencies for acids, esters, and amides, and its precise location is often helpful in structure determination. Table 18.5 gives the location of this band for most acyl compounds. Notice that conjugation shifts the location of the C=O absorption to lower frequencies.

The hydroxyl groups of carboxylic acids also give rise to a broad peak in the 2500–3100-cm^{-1} region arising from O—H stretching vibrations. The N—H stretching vibrations of amides absorb between 3140 and 3500 cm^{-1}.

TABLE 18.5 *Carbonyl stretching absorptions of acyl compounds*

TYPE OF COMPOUND	FREQUENCY RANGE (cm^{-1})
Carboxylic Acids	
R—CO$_2$H	1700–1725
—C=C—CO$_2$H	1690–1715
ArCO$_2$H	1680–1700
Acid Anhydrides	
R—C(O)—O—C(O)—R	1800–1850 and 1740–1790
Ar—C(O)—O—C(O)—Ar	1780–1860 and 1730–1780
Acyl Chlorides	
R—CCl(O) and Ar—CCl(O)	1780–1850
Esters	
R—C(O)—OR′	1735–1750
Ar—C(O)—OR′	1715–1730
Amides	
RCNH$_2$(O), RCNHR(O), and RCNR$_2$(O)	1630–1690
Carboxylate Ions	
RCO$_2^-$	1550–1630
Nitriles (C≡N stretch)	
R—C≡N	Near 2250

Figure 18.1 shows an annotated spectrum of propanoic acid. Nitriles show an intense and characteristic infrared absorption band near 2250 cm^{-1} that arises from stretching of the carbon–nitrogen triple bond.

^{1}H NMR Spectra   The acidic protons of carboxylic acids are highly deshielded and absorb far downfield in the δ 10–12 region. The protons of the α carbon of carboxylic

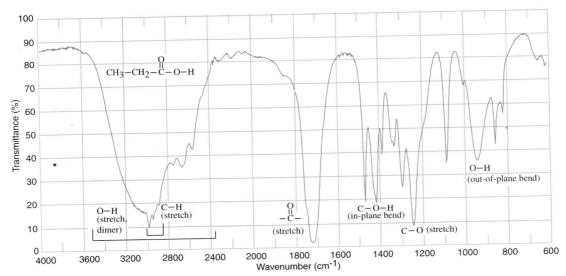

FIGURE 18.1 The infrared spectrum of propanoic acid.

acids absorb in the δ 2.0–2.5 region. Figure 18.2 gives an annotated ^1H NMR spectrum of an ester, methyl propanoate and shows the normal splitting pattern (quartet and triplet) of an ethyl group and, as we would expect, it shows an unsplit methyl group.

^{13}C NMR Spectra   The carbonyl carbon of carboxylic acids and their derivatives occur far downfield in the δ 166–178 region (see the following examples). The nitrile carbon is not shifted so far downfield and absorbs in the δ 115–120 region. Notice, however, that the downfield shift for these derivatives is not as much as for aldehydes and ketones (δ 182–215).

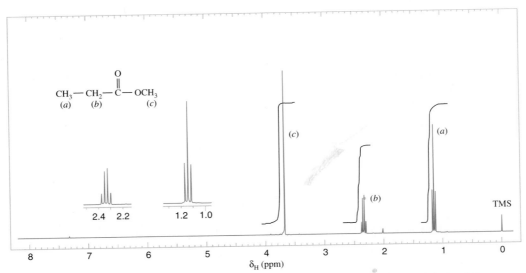

FIGURE 18.2 The 300 MHz ^1H NMR spectrum of methyl propanoate. Expansions of the signals are shown in the offset plots.

δ 177.2 δ 170.7 δ 170.3 δ 172.6 δ 117.4

^{13}C NMR chemical shifts for the carbonyl or nitrite carbon atom

The carbon atoms of the alkyl groups of carboxylic acids and their derivatives have ^{13}C chemical shifts much further upfield. The chemical shifts for each carbon of pentanoic acid are as follows:

δ 13.5   22.0   27.0   34.1   179.7

^{13}C NMR chemical shifts

18.3 PREPARATION OF CARBOXYLIC ACIDS

Most of the methods for the preparation of carboxylic acids have been presented previously.

1. **By oxidation of alkenes.** We learned in Section 8.10 that alkenes can be oxidized to carboxylic acids with hot alkaline $KMnO_4$.

$$RCH=CHR' \xrightarrow[\substack{\text{heat} \\ \text{(2) } H_3O^+}]{\text{(1) } KMnO_4, OH^-} RCO_2H + R'CO_2H$$

Alternatively, ozonides (Section 8.10A) can be subjected to an oxidative workup that yields carboxylic acids.

$$RCH=CHR' \xrightarrow[\text{(2) } H_2O_2]{\text{(1) } O_3} RCO_2H + R'CO_2H$$

2. **By oxidation of aldehydes and primary alcohols.** Aldehydes can be oxidized to carboxylic acids with mild oxidizing agents such as $Ag(NH_3)_2{}^+OH^-$ (Section 16.12). Primary alcohols can be oxidized with $KMnO_4$.

$$R-CHO \xrightarrow[\text{(2) } H_3O^+]{\text{(1) } Ag_2O \text{ or } Ag(NH_3)_2{}^+OH^-} RCO_2H$$

$$RCH_2OH \xrightarrow[\substack{\text{heat} \\ \text{(2) } H_3O^+}]{\text{(1) } KMnO_4, OH^-} RCO_2H$$

3. **By oxidation of alkylbenzenes.** Primary and secondary alkyl groups (but not 3° groups) directly attached to a benzene ring are oxidized by $KMnO_4$ to a $-CO_2H$ group (Section 15.12).

4. **By oxidation of methyl ketones.** Methyl ketones can be converted to carboxylic acids via the haloform reaction (Section 17.4).

$$Ar-\overset{\overset{\displaystyle O}{\|}}{C}-CH_3 \xrightarrow[\text{(2) } H_3O^+]{\text{(1) } X_2/NaOH} Ar-\overset{\overset{\displaystyle O}{\|}}{C}OH + CHX_3$$

5. **By hydrolysis of cyanohydrins and other nitriles.** We saw, in Section 16.9A, that aldehydes and ketones can be converted to cyanohydrins, and that these can be hydrolyzed to α-hydroxy acids. In the hydrolysis the $-CN$ group is converted to a $-CO_2H$ group. The mechanism of hydrolysis is discussed in Section 18.8H.

$$\underset{R'}{\overset{R}{>}}C=O + HCN \rightleftharpoons \underset{R'}{\overset{R}{>}}\underset{CN}{\overset{OH}{C}} \xrightarrow[H_2O]{H^+} R-\underset{\underset{R'}{|}}{\overset{\overset{OH}{|}}{C}}-CO_2H$$

Nitriles can also be prepared by nucleophilic substitution reactions of alkyl halides with sodium cyanide. Hydrolysis of the nitrile yields a carboxylic acid *with a chain one carbon atom longer* than the original alkyl halide.

**General Reaction**

$$R-CH_2X + CN^- \longrightarrow RCH_2CN \xrightarrow[\substack{H_2O \\ heat}]{H^+} RCH_2CO_2H + NH_4^+$$
$$\xrightarrow[\substack{H_2O \\ heat}]{OH^-} RCH_2CO_2^- + NH_3$$

**Specific Examples**

$$HOCH_2CH_2Cl \xrightarrow[(80\%)]{NaCN} HOCH_2CH_2CN \xrightarrow[\substack{\text{(2) } H_3O^+ \\ (75-80\%)}]{\text{(1) } OH^-, H_2O} HOCH_2CH_2CO_2H$$

3-Hydroxy-
propanenitrile   **3-Hydroxypropanoic**
                              acid

$$BrCH_2CH_2CH_2Br \xrightarrow[(77-86\%)]{NaCN} NCCH_2CH_2CH_2CN \xrightarrow[(83-85\%)]{H_3O^+} HO_2CCH_2CH_2CH_2CO_2H$$

Pentanedinitrile   **Glutaric acid**

This synthetic method is generally limited to the use of *primary alkyl* halides. The cyanide ion is a relatively strong base, and the use of a secondary or tertiary alkyl halide leads primarily to an alkene (through elimination) rather than to a nitrile (through substitution). Aryl halides (except for those with ortho and para nitro groups) do not react with sodium cyanide.

6. **By carbonation of Grignard reagents.** Grignard reagents react with carbon dioxide to yield magnesium carboxylates. Acidification produces carboxylic acids.

$$R-X + Mg \xrightarrow[\text{diethyl ether}]{} RMgX \xrightarrow{CO_2} RCO_2MgX \xrightarrow{H_3O^+} RCO_2H$$

or

$$Ar-Br + Mg \xrightarrow[\text{diethyl ether}]{} ArMgBr \xrightarrow{CO_2} ArCO_2MgBr \xrightarrow{H_3O^+} ArCO_2H$$

This synthesis of carboxylic acids is applicable to primary, secondary, tertiary, allyl, benzyl, and aryl halides, provided they have no groups incompatible with a Grignard reaction (cf. Section 11.8B).

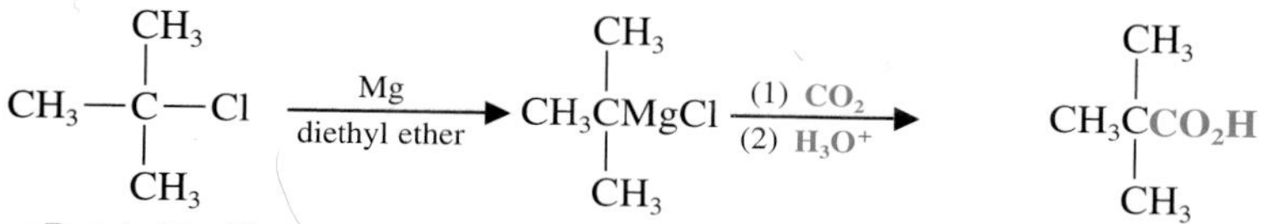

tert-Butyl chloride

2,2-Dimethylpropanoic acid
(79–80% overall)

CH$_3$CH$_2$CH$_2$CH$_2$Cl $\xrightarrow[\text{diethyl ether}]{\text{Mg}}$ CH$_3$CH$_2$CH$_2$CH$_2$MgCl $\xrightarrow[\text{(2) H}_3\text{O}^+]{\text{(1) CO}_2}$ CH$_3$CH$_2$CH$_2$CH$_2$CO$_2$H

Butyl chloride

Pentanoic acid
(80% overall)

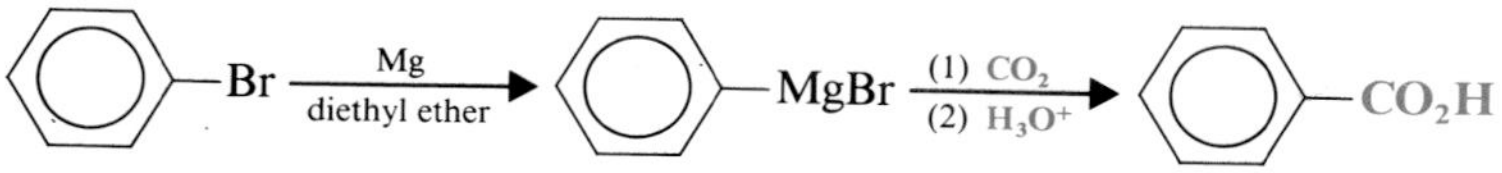

Benzoic acid
(85%)

PROBLEM 18.6

Show how each of the following compounds could be converted to benzoic acid.
(a) Ethylbenzene   (c) Acetophenone   (e) Benzyl alcohol
(b) Bromobenzene   (d) Phenylethene (styrene)   (f) Benzaldehyde

PROBLEM 18.7

Show how you would prepare each of the following carboxylic acids through a Grignard synthesis.
(a) Phenylacetic acid   (c) 3-Butenoic acid   (e) Hexanoic acid
(b) 2,2-Dimethylpentanoic acid (d) 4-Methylbenzoic acid

PROBLEM 18.8

(a) Which of the carboxylic acids in Problem 18.7 could be prepared by a nitrile synthesis as well?   (b) Which synthesis, Grignard or nitrile, would you choose to prepare HOCH$_2$CH$_2$CH$_2$CH$_2$CO$_2$H from HOCH$_2$CH$_2$CH$_2$CH$_2$Br? Why?

18.4 NUCLEOPHILIC SUBSTITUTIONS AT THE ACYL CARBON

In our study of carbonyl compounds in Chapter 17, we saw that a characteristic reaction of aldehydes and ketones is one of *nucleophilic addition* to the carbon–oxygen double bond.

Nucleophilic addition

As we study carboxylic acids and their derivatives in this chapter we shall find that their reactions are characterized by **nucleophilic substitutions** that take place at their acyl (carbonyl) carbon atoms. We shall encounter many reactions of the following general type.

Nucleophilic addition   Elimination

Nucleophilic substitution

Many reactions of this type occur in living organisms and biochemists call them *acyl transfer* reactions.

Although the final results obtained from the reactions of acyl compounds with nucleophiles (substitutions) differ from those obtained from aldehydes and ketones (additions), the two reactions have one characteristic in common. *The initial step in both reactions involves nucleophilic addition at the carbonyl carbon atom.* With both groups of compounds, this initial attack is facilitated by the same factors: the relative steric openness of the carbonyl carbon atom and the ability of the carbonyl oxygen atom to accommodate an electron pair of the carbon–oxygen double bond.

It is after the initial nucleophilic attack has taken place that the two reactions differ. The tetrahedral intermediate formed from an aldehyde or ketone usually accepts a proton to form a stable addition product. By contrast, the intermediate formed from an acyl compound usually *eliminates* a leaving group; this **elimination** leads to regeneration of the carbon–oxygen double bond and to a *substitution product*. The overall process in the case of **acyl substitution** occurs, therefore, by a **nucleophilic addition–elimination** mechanism.

Acyl compounds react as they do because they all have good, or reasonably good, leaving groups (or they can be protonated to have good leaving groups) attached to the carbonyl carbon atom. An acyl chloride, for example, generally reacts by losing *a chloride ion*—a very weak base, and thus, a very good leaving group.

EXAMPLE

The reaction of an acyl chloride with water.

An acid anhydride generally reacts by losing *a carboxylate ion* or a molecule of a *carboxylic acid*—both are weak bases and good leaving groups.

As we shall see later, esters generally undergo nucleophilic substitution by losing a molecule of an *alcohol,* acids react by losing a molecule of *water,* and amides react by losing a molecule of *ammonia* or of an *amine.* All of the molecules lost in these reactions are weak bases and are reasonably good leaving groups.

For an aldehyde or ketone to react by substitution, the tetrahedral intermediate would have to eject a hydride ion (H :⁻) or an alkanide ion (R :⁻). Both are *very powerful bases* and both are, therefore, *very poor leaving groups.*

[The haloform reaction (Section 17.4) is one of the rare instances in which an alkanide ion can act as a leaving group, but then only because the leaving group is a weakly basic trichloromethyl anion.]

18.4A Relative Reactivity of Acyl Compounds

Of the acid derivatives that we shall study in this chapter, acyl chlorides are the most reactive toward nucleophilic substitution, and amides are the least reactive. In general, the overall order of reactivity is

| Acyl chloride | Acid anhydride | Ester | Amide |

The general order of reactivity of acid derivatives can be explained by taking into account the basicity of the leaving groups. When acyl chlorides react, the leaving group is a *chloride ion.* When acid anhydrides react, the leaving group is a carboxylic acid or a carboxylate ion. When esters react, the leaving group is an alcohol, and when amides react, the leaving group is an amine (or ammonia). Of all of these bases, chloride ions are the *weakest bases* and acyl chlorides are the *most reactive* acyl compounds. Amines (or ammonia) are the *strongest bases* and amides are the *least reactive* acyl compounds.

18.4B Synthesis of Acid Derivatives

As we begin now to explore the syntheses of carboxylic acid derivatives we shall find that in many instances one acid derivative can be synthesized through a nucleophilic substitution reaction of another. The order of reactivities that we have presented gives us a clue as to which syntheses are practical and which are not. In general, *less reactive acyl compounds can be synthesized from more reactive ones, but the reverse is usually difficult and, when possible, requires special conditions.*

18.5 ACYL CHLORIDES

18.5A Synthesis of Acyl Chlorides

Since acyl chlorides are the most reactive of the acid derivatives, we must use special reagents to prepare them. We use other acid chlorides, *the acid chlorides of inorganic acids:* we use **PCl₅** (an acid chloride of phosphoric acid), **PCl₃** (an acid chloride of phosphorous acid), and **SOCl₂** (an acid chloride of sulfurous acid).

All of these reagents react with carboxylic acids to give acyl chlorides in good yield.

General Reactions

$$\underset{\text{Thionyl chloride}}{\overset{O}{\underset{\|}{R\text{C}OH}} + SOCl_2} \longrightarrow R\overset{O}{\underset{\|}{-\text{C}}}-Cl + SO_2 + HCl$$

$$\underset{\substack{\text{Phosphorus}\\\text{trichloride}}}{3\;\overset{O}{\underset{\|}{R\text{C}OH}} + PCl_3} \longrightarrow 3\;\overset{O}{\underset{\|}{R\text{C}Cl}} + H_3PO_3$$

$$\underset{\substack{\text{Phosphorus}\\\text{pentachloride}}}{\overset{O}{\underset{\|}{R\text{C}OH}} + PCl_5} \longrightarrow \overset{O}{\underset{\|}{R\text{C}Cl}} + POCl_3 + HCl$$

These reactions all involve nucleophilic substitutions by chloride ion on a highly reactive intermediate: a protonated acyl chlorosulfite, a protonated acyl chlorophosphite, or a protonated acyl chlorophosphate. Thionyl chloride, for example, reacts with a carboxylic acid in the following way:

A MECHANISM FOR A SYNTHESIS OF ACYL CHLORIDES

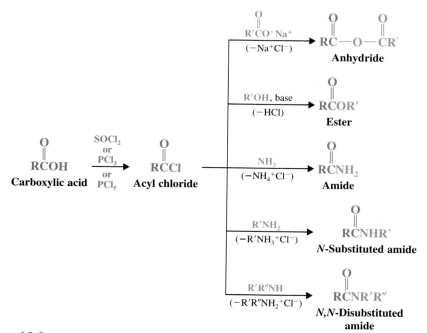

Protonated acyl chlorosulfite

HCl + SO₂

18.5B Reactions of Acyl Chlorides

Because acyl chlorides are the most reactive of the acyl derivatives, they are easily converted to less reactive ones. Many times, therefore, the best synthetic route to an anhydride, an ester, or an amide will involve an initial synthesis of the acyl chloride from the acid, and then conversion of the acyl chloride to the desired acid derivative. The scheme given in Fig. 18.3 illustrates how this can be done. We will examine these reactions in detail in Sections 18.6–18.8.

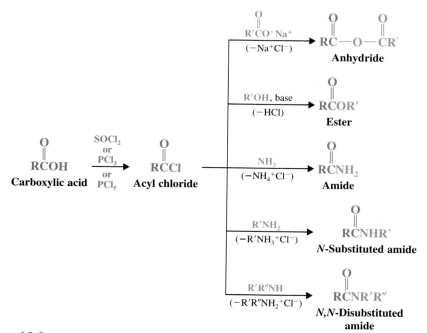

FIGURE 18.3

Acyl chlorides also react with water and (even more rapidly) with aqueous base:

$$
\underset{\substack{\text{Acyl chloride}}}{\overset{\displaystyle O}{\underset{\displaystyle \|}{R-C-Cl}}}
\begin{array}{l}
\xrightarrow{\;H_2O\;}\;\; \overset{\displaystyle O}{\underset{\displaystyle \|}{R-C-OH}} \\[3em]
\xrightarrow{\;OH^-/H_2O\;}\;\; \overset{\displaystyle O}{\underset{\displaystyle \|}{R-C-O^-}} + Cl^-
\end{array}
$$

18.6 CARBOXYLIC ACID ANHYDRIDES

18.6A Synthesis of Carboxylic Acid Anhydrides

Carboxylic acids react with acyl chlorides in the presence of pyridine to give carboxylic acid anhydrides.

$$
R-\overset{O}{\overset{\|}{C}}-OH + R'-\overset{O}{\overset{\|}{C}}-Cl + \underset{N}{\bigcirc} \longrightarrow R-\overset{O}{\overset{\|}{C}}-O-\overset{O}{\overset{\|}{C}}-R' + \underset{\underset{H}{\overset{+}{N}}}{\bigcirc} Cl^-
$$

This method is frequently used in the laboratory for the preparation of anhydrides. The method is quite general and can be used to prepare mixed anhydrides ($R \neq R'$) or simple anhydrides ($R = R'$).

Sodium salts of carboxylic acids also react with acyl chlorides to give anhydrides:

$$
R-\overset{O}{\overset{\|}{C}}-O^-Na^+ + R'-\overset{O}{\overset{\|}{C}}-Cl \longrightarrow R-\overset{O}{\overset{\|}{C}}-O-\overset{O}{\overset{\|}{C}}-R' + Na^+Cl^-
$$

In this reaction a carboxylate ion acts as a nucleophile and brings about a nucleophilic substitution reaction at the acyl carbon of the acyl chloride.

Cyclic anhydrides can sometimes be prepared by simply heating the appropriate dicarboxylic acid. This method succeeds, however, only when anhydride formation leads to a five- or six-membered ring.

$$
\underset{\substack{\text{Succinic} \\ \text{acid}}}{
\begin{array}{l}
H_2C\!-\!\overset{O}{\overset{\|}{C}}\!-\!OH \\
| \\
H_2C\!-\!\underset{O}{\underset{\|}{C}}\!-\!OH
\end{array}}
\xrightarrow{\;300^\circ C\;}
\underset{\substack{\text{Succinic} \\ \text{anhydride}}}{
\begin{array}{l}
H_2C\!-\!\overset{O}{\overset{\|}{C}} \\
| \qquad\quad O \\
H_2C\!-\!\underset{O}{\underset{\|}{C}}
\end{array}}
+ H_2O
$$

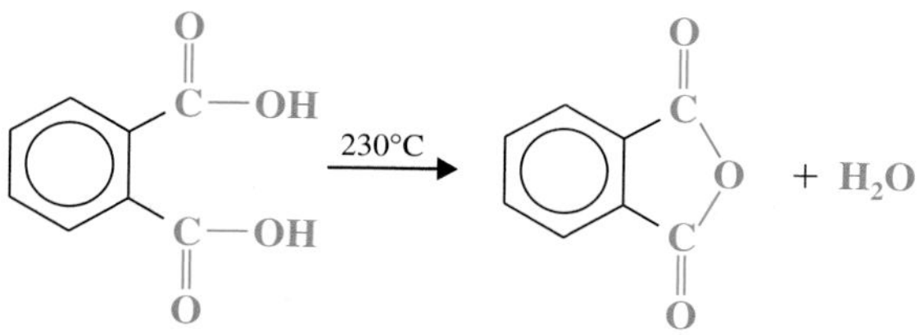

Phthalic acid   **Phthalic anhydride**
(~100%)

PROBLEM 18.9

When maleic acid is heated to 200°C, it loses water and becomes maleic anhydride. Fumaric acid, a diastereomer of maleic acid, requires a much higher temperature before it dehydrates; when it does, it also yields maleic anhydride. Provide an explanation for these observations.

18.6B Reactions of Carboxylic Acid Anhydrides

Because carboxylic acid anhydrides are highly reactive they can be used to prepare esters and amides (Fig. 18.4). We shall study these reactions in detail in Sections 18.7 and 18.8.

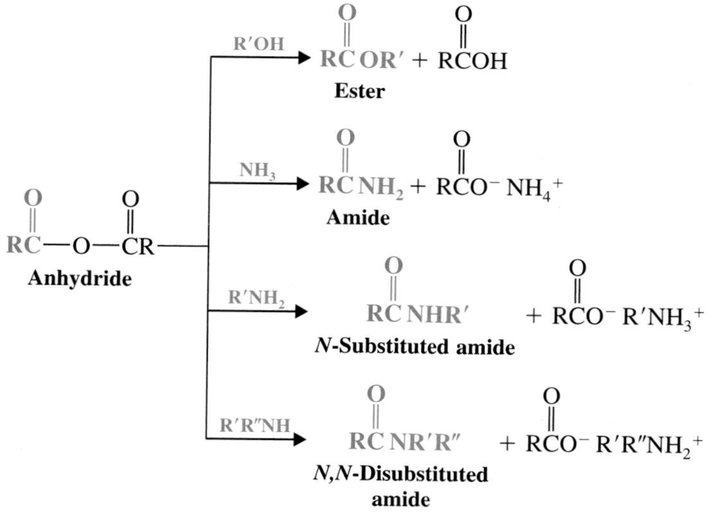

FIGURE 18.4

Carboxylic acid anhydrides also undergo hydrolysis:

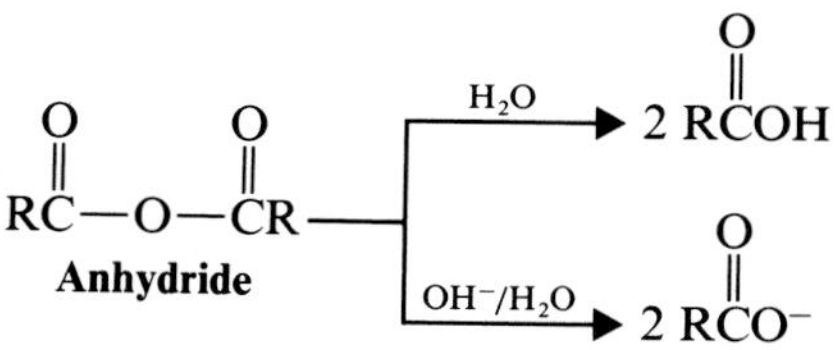

18.7 ESTERS

18.7A Synthesis of Esters: Esterification

Carboxylic acids react with alcohols to form esters through a condensation reaction known as **esterification:**

General Reaction

$$R-\overset{\overset{\text{O}}{\|}}{C}-OH + R'-OH \xrightleftharpoons{H^+} R-\overset{\overset{\text{O}}{\|}}{C}-OR' + H_2O$$

Specific Examples

$$CH_3\overset{\overset{\text{O}}{\|}}{C}OH + CH_3CH_2OH \xrightleftharpoons{H^+} CH_3\overset{\overset{\text{O}}{\|}}{C}OCH_2CH_3 + H_2O$$

Acetic acid   Ethanol   Ethyl acetate

$$C_6H_5\overset{\overset{\text{O}}{\|}}{C}OH + CH_3OH \xrightleftharpoons{H^+} C_6H_5\overset{\overset{\text{O}}{\|}}{C}OCH_3 + H_2O$$

Benzoic acid   Methanol   Methyl benzoate

Esterification reactions are acid catalyzed. They proceed very slowly in the absence of strong acids, but they reach equilibrium within a matter of a few hours when an acid and an alcohol are refluxed with a small amount of concentrated sulfuric acid or hydrogen chloride. Since the position of equilibrium controls the amount of the ester formed, the use of an excess of either the carboxylic acid or the alcohol increases the yield based on the limiting reagent. Just which component we choose to use in excess will depend on its availability and cost. The yield of an esterification reaction can also be increased by removing water from the reaction mixture as it is formed.

When benzoic acid reacts with methanol that has been labeled with ^{18}O, the labeled oxygen appears in the ester. This result reveals just which bonds break in the esterification.

$$C_6H_5\overset{\overset{\text{O}}{\|}}{C}\!-\!OH + CH_3\!-\!^{18}O\!-\!H \xrightleftharpoons{H^+} C_6H_5\overset{\overset{\text{O}}{\|}}{C}\!-\!^{18}OCH_3 + H_2O$$

The results of the labeling experiment and the fact that esterifications are acid catalyzed are both consistent with the mechanism that follows. This mechanism is typical of acid-catalyzed nucleophilic substitution reactions at acyl carbon atoms.

A Mechanism for Acid-Catalyzed Esterification

The carboxylic acid accepts a proton from the strong acid catalyst.

The alcohol attacks the protonated carbonyl group to give a tetrahedral intermediate.

A proton is lost at one oxygen atom and gained at another.

Loss of a molecule of water gives a protonated ester.

Transfer of a proton to a base leads to the ester.

If we follow the forward reactions in this mechanism, we have the mechanism for the *acid-catalyzed esterification of an acid*. If, however, we follow the reverse reactions, we have the mechanism for the *acid-catalyzed hydrolysis of an ester*:

Acid-Catalyzed Ester Hydrolysis

$$\underset{O}{\overset{O}{R-C-OR'}} + H_2O \overset{H_3O^+}{\rightleftharpoons} \underset{O}{\overset{O}{R-C-OH}} + R'-OH$$

Which result we obtain will depend on the conditions we choose. If we want to esterify an acid, we use an excess of the alcohol and, if possible, remove the water as it is formed. If we want to hydrolyze an ester, we use a large excess of water; that is, we reflux the ester with dilute aqueous HCl or dilute aqueous H_2SO_4.

Problem 18.10

Where would you expect to find the labeled oxygen if you carried out an acid-catalyzed hydrolysis of methyl benzoate in ^{18}O-labeled water?

Steric factors strongly affect the rates of acid-catalyzed hydrolyses of esters. Large groups near the reaction site, whether in the alcohol component or the acid component, slow both reactions markedly. Tertiary alcohols, for example, react so slowly in acid-catalyzed esterifications that they usually undergo elimination instead. However, they can be converted to esters safely through the use of acyl chlorides and anhydrides in the ways that follow.

Esters from Acyl Chlorides.
Esters can also be synthesized by the reaction of acyl chlorides with alcohols. Since acyl chlorides are much more reactive toward nucleophilic substitution than carboxylic acids, the reaction of an acyl chloride and an alcohol occurs rapidly and does not require an acid catalyst. Pyridine is often added to the reaction mixture to react with the HCl that forms. (Pyridine may also react with the acyl chloride to form an acylpyridinium ion, an intermediate that is even more reactive toward the nucleophile than the acyl chloride is.)

General Reaction

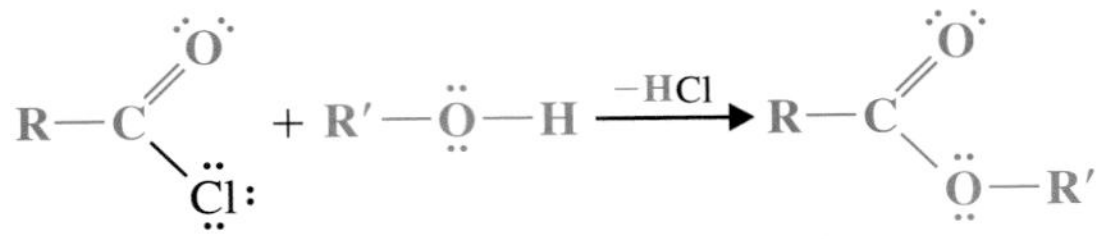

Specific Example

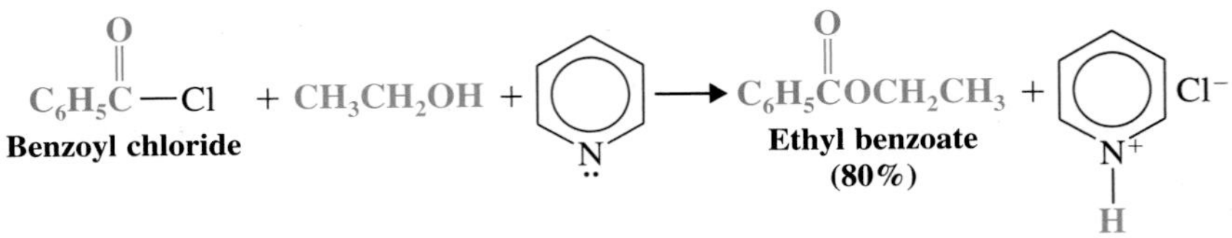

Esters from Carboxylic Acid Anhydrides.
Carboxylic acid anhydrides also react with alcohols to form esters in the absence of an acid catalyst.

General Reaction

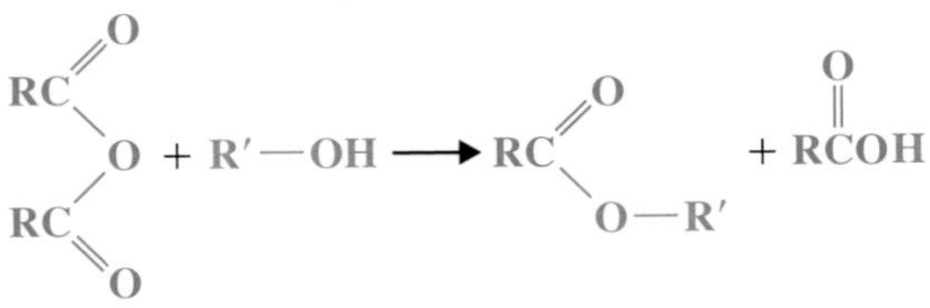

Specific Example

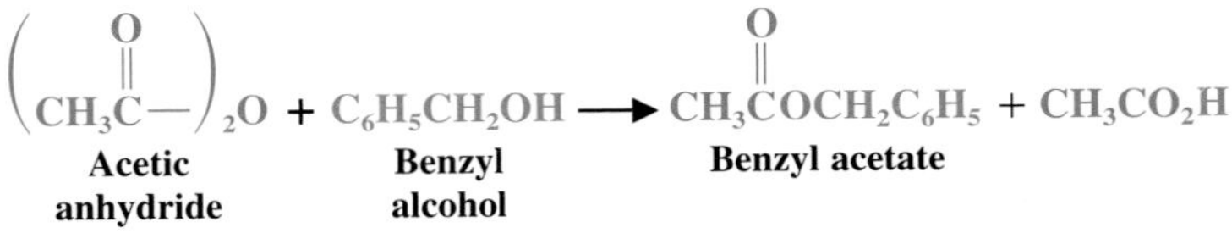

The reaction of an alcohol with an anhydride or an acyl chloride is often the best method for preparing an ester.

Cyclic anhydrides react with one molar equivalent of an alcohol to form compounds that are *both esters and acids*.

Phthalic anhydride + *sec*-**Butyl alcohol** $\xrightarrow{110°C}$ *sec*-**Butyl hydrogen phthalate** (97%)

PROBLEM 18.11

Esters can also be synthesized by *transesterification:*

$$\underset{\substack{\text{High-boiling}\\\text{ester}}}{R-\overset{\overset{\displaystyle O}{\|}}{C}-OR'} + \underset{\substack{\text{High-boiling}\\\text{alcohol}}}{R''-OH} \underset{}{\overset{H^+,\ heat}{\rightleftharpoons}} \underset{\substack{\text{Higher boiling}\\\text{ester}}}{R\overset{\overset{\displaystyle O}{\|}}{C}-OR''} + \underset{\substack{\text{Lower boiling}\\\text{alcohol}}}{R'-OH}$$

In this procedure we shift the equilibrium to the right by allowing the low-boiling alcohol to distill from the reaction mixture. The mechanism for transesterification is similar to that for an acid-catalyzed esterification (or an acid-catalyzed ester hydrolysis). Write a mechanism for the following transesterification.

$$\underset{\substack{\text{Methyl acrylate}}}{CH_2=CH\overset{\overset{\displaystyle O}{\|}}{C}OCH_3} + \underset{\substack{\text{Butyl alcohol}}}{CH_3CH_2CH_2CH_2OH} \overset{H^+}{\rightleftharpoons}$$

$$\underset{\substack{\text{Butyl acrylate}\\(94\%)}}{CH_2=CH\overset{\overset{\displaystyle O}{\|}}{C}OCH_2CH_2CH_2CH_3} + \underset{\substack{\text{Methanol}}}{CH_3OH}$$

18.7B Base-Promoted Hydrolysis of Esters: Saponification

Esters not only undergo acid hydrolysis, they also undergo *base-promoted hydrolysis.* Base-promoted hydrolysis is sometimes called *saponification* (from the Latin *sapo* for soap (see Section 23.2C). Refluxing an ester with aqueous sodium hydroxide, for example, produces an alcohol and the sodium salt of the acid:

$$\underset{\substack{\text{Ester}}}{R\overset{\overset{\displaystyle O}{\|}}{C}-OR'} + NaOH \xrightarrow{H_2O} \underset{\substack{\text{Sodium carboxylate}}}{R\overset{\overset{\displaystyle O}{\|}}{C}-O^-Na^+} + \underset{\substack{\text{Alcohol}}}{R'OH}$$

The carboxylate ion is very unreactive toward nucleophilic substitution because it is negatively charged. Base-promoted hydrolysis of an ester, as a result, is an essentially irreversible reaction.

The accepted mechanism for the base-promoted hydrolysis of an ester also involves a nucleophilic substitution at the acyl carbon:

A MECHANISM FOR BASE-PROMOTED HYDROLYSIS OF AN ESTER

A hydroxide ion attacks the carbonyl carbon atom.

The tetrahedral intermediate expels an alkoxide ion.

Transfer of a proton leads to the products of the reaction.

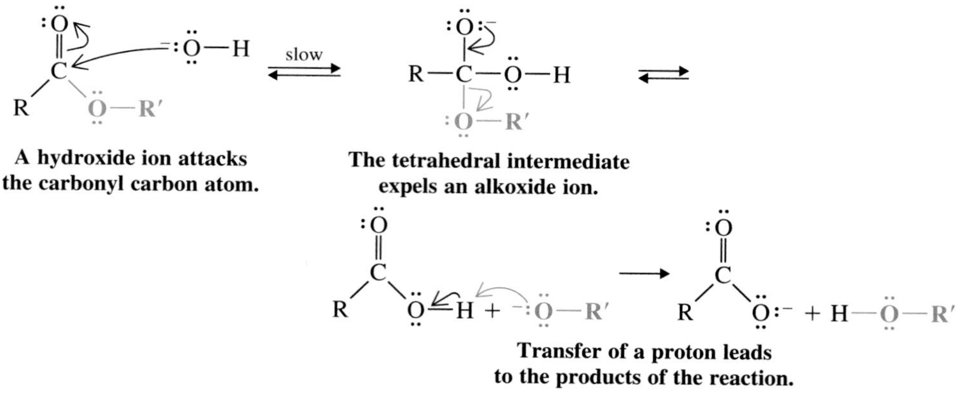

Evidence for this mechanism comes from studies done with isotopically labeled esters. When ethyl propanoate labeled with ^{18}O in the ether-type oxygen of the ester (below) is subjected to hydrolysis with aqueous NaOH, all of the ^{18}O shows up in the ethanol that is produced. None of the ^{18}O appears in the propanoate ion.

$$CH_3CH_2-\overset{O}{\overset{\|}{C}}-^{18}O-CH_2CH_3 + NaOH \xrightarrow{H_2O} CH_3CH_2-\overset{O}{\overset{\|}{C}}-O^-Na^+ + H-^{18}O-CH_2CH_3$$

This labeling result is completely consistent with the mechanism given above (outline the steps for yourself and follow the labeled oxygen through to the products). If the hydroxide ion had attacked the alkyl carbon instead of the acyl carbon, the alcohol obtained would not have been labeled. Attack at the alkyl carbon is almost never observed. (For one exception see Problem 18.13.)

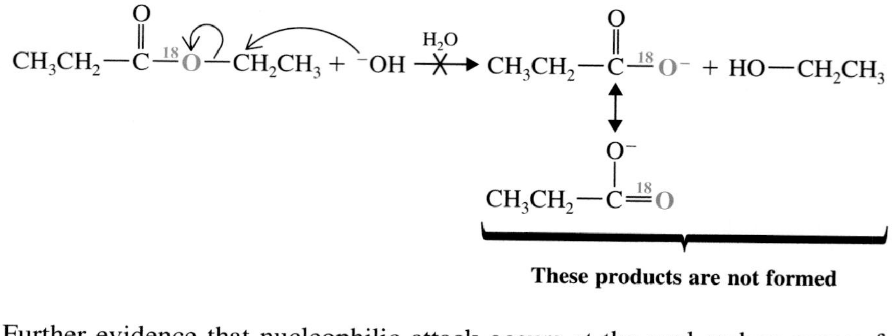

$$CH_3CH_2-\overset{O}{\overset{\|}{C}}-^{18}O-CH_2CH_3 + {}^-OH \xrightarrow[X]{H_2O} CH_3CH_2-\overset{O}{\overset{\|}{C}}-^{18}O^- + HO-CH_2CH_3$$

$$CH_3CH_2-\overset{O^-}{\overset{|}{C}}=^{18}O$$

These products are not formed

Further evidence that nucleophilic attack occurs at the acyl carbon comes from studies in which esters of chiral alcohols were subjected to base-promoted hydrolysis. Reaction by path A (at the acyl carbon) should lead to retention of configuration in the alcohol. Reaction by path B (at the alkyl carbon) should lead to an inversion of configuration of the alcohol. ***Inversion of configuration is almost never observed.*** In almost every instance basic hydrolysis of a carboxylic ester of a chiral alcohol proceeds with ***retention of configuration.***

Path A: Nucleophilic Substitution at the Acyl Carbon

This mechanism is preferred with alkyl carboxylates.

Path B: Nucleophilic Substitution at the Alkyl Carbon

This mechanism is seldom observed.

Although nucleophilic attack at the alkyl carbon seldom occurs with esters of carboxylic acids, it is the preferred mode of attack with esters of sulfonic acids (Section 10.11).

An alkyl sulfonate

This mechanism is preferred with alkyl sulfonates.

PROBLEM 18.12

(a) Write stereochemical formulas for compounds **A**–**F**.

1. *cis*-3-Methylcyclopentanol + $C_6H_5SO_2Cl \longrightarrow$

$$A \xrightarrow[\text{heat}]{OH^-} B + C_6H_5SO_3^-$$

2. *cis*-3-Methylcyclopentanol + $C_6H_5\overset{\displaystyle O}{\overset{\displaystyle \|}{C}}-Cl \longrightarrow$

$$C \xrightarrow[\text{reflux}]{OH^-} D + C_6H_5CO_2^-$$

3. (R)-2-Bromooctane + $CH_3CO_2^-Na^+ \longrightarrow$ **E** + NaBr

$\downarrow$ OH$^-$, H$_2$O (reflux)

F

4. (R)-2-Bromooctane + OH$^-$ $\xrightarrow{\text{acetone}}$ **F** + Br$^-$

(b) Which of the last two methods, **(3)** or **(4)**, would you expect to give a higher yield of **F**? Why?

PROBLEM 18.13

Base-promoted hydrolysis of methyl mesitoate occurs through an attack on the alcohol carbon instead of the acyl carbon.

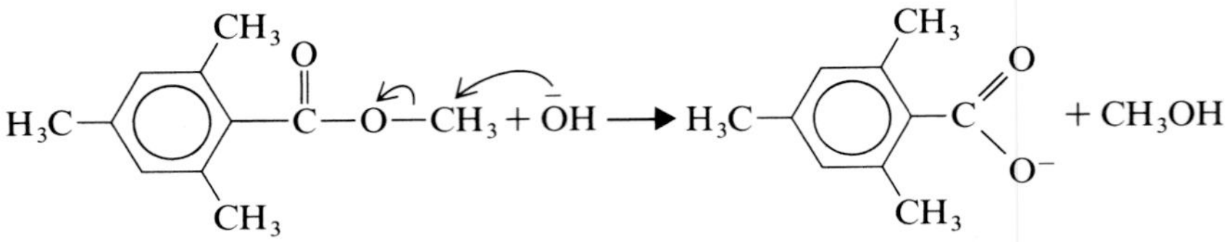

Methyl mesitoate

(a) Can you suggest a reason that accounts for this unusual behavior?
(b) Suggest an experiment with labeled compounds that would confirm this mode of attack.

18.7C Lactones

Carboxylic acids whose molecules have a hydroxyl group on a γ or δ carbon undergo an intramolecular esterification to give cyclic esters known as γ- or δ-*lactones*. The reaction is acid catalyzed:

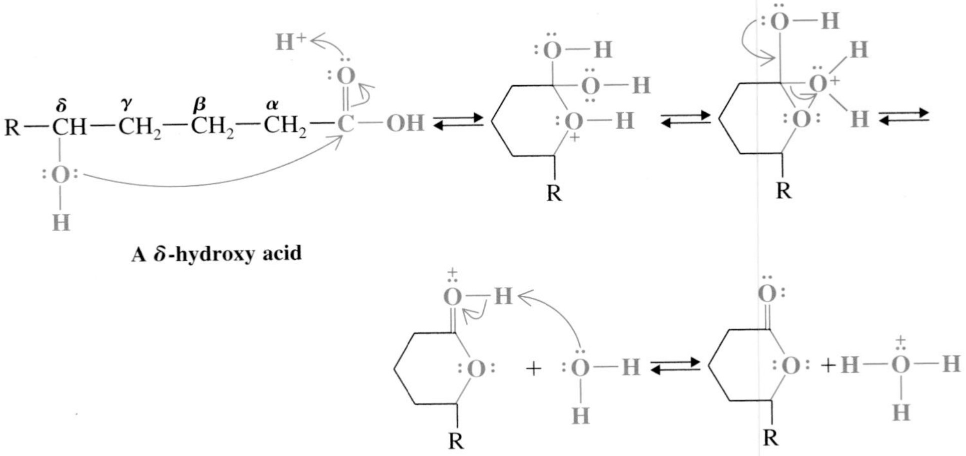

A δ-hydroxy acid

A δ-lactone

Lactones are hydrolyzed by aqueous base just as other esters are. Acidification of the sodium salt, however, may lead spontaneously back to the γ- or δ-lactone, particularly if excess acid is used.

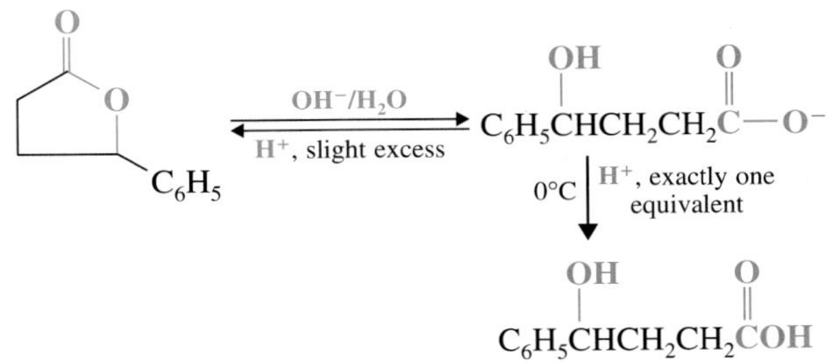

Many lactones occur in nature. Vitamin C (below), for example, is a γ-lactone. Some antibiotics, such as erythromycin, are lactones with very large rings, but most naturally occurring lactones are γ- or δ-lactones; that is, most contain five- or six-membered rings.

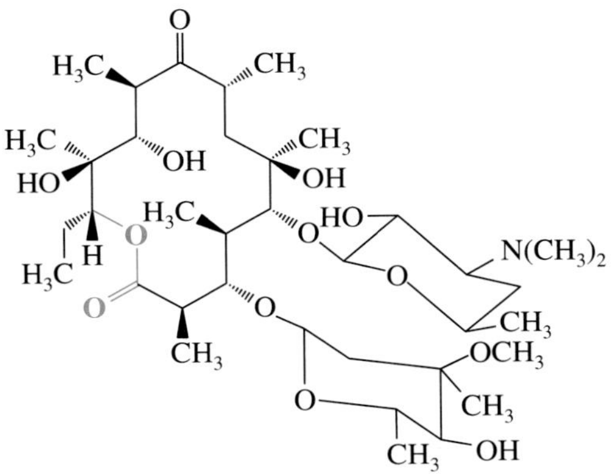

Erythromycin A

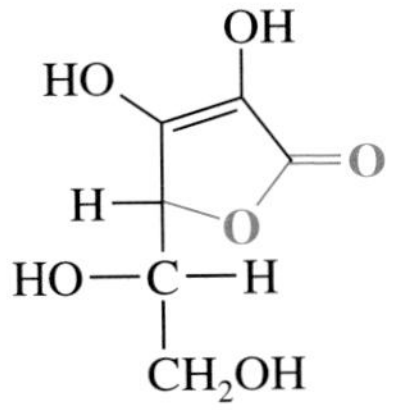

**Vitamin C
(ascorbic acid)**

β-Lactones (lactones with four-membered rings) have been detected as intermediates in some reactions and several have been isolated. They are highly reactive, however. If one attempts to prepare a β-lactone from a β-hydroxy acid, β elimination usually occurs instead:

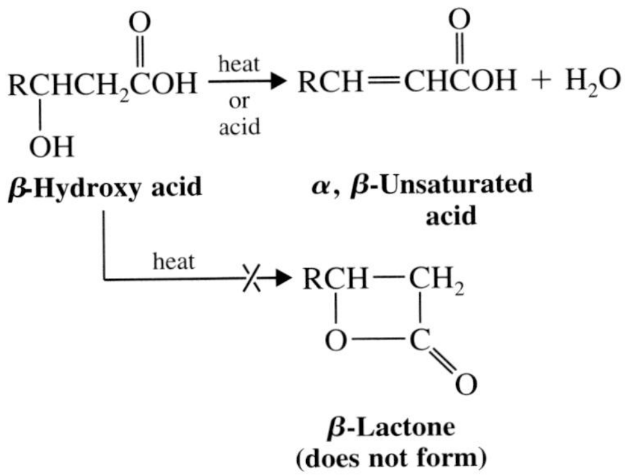

β-Hydroxy acid

**α, β-Unsaturated
acid**

**β-Lactone
(does not form)**

When α-hydroxy acids are heated, they form cyclic diesters called *lactides*.

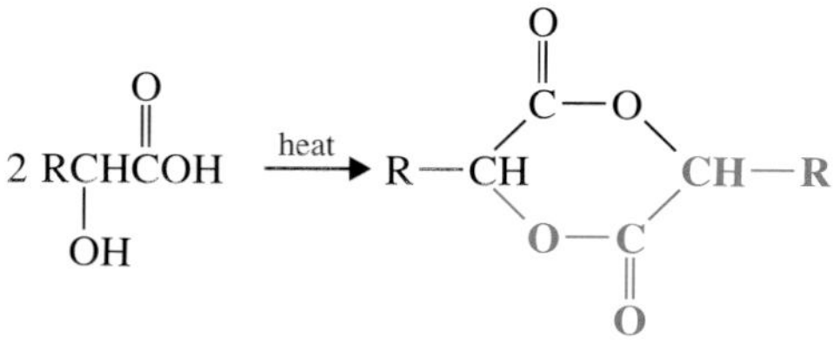

$$2 \text{ RCHCOH} \xrightarrow{\text{heat}}$$

α-Hydroxy acid   **A lactide**

α-Lactones occur as intermediates in some reactions (cf. Special Topic B).

18.8 AMIDES

18.8A Synthesis of Amides

Amides can be prepared in a variety of ways, starting with acyl chlorides, acid anhydrides, esters, carboxylic acids, and carboxylic salts. All of these methods involve nucleophilic substitution reactions by ammonia or an amine at an acyl carbon. As we might expect, acid chlorides are the most reactive and carboxylate ions are the least.

18.8B Amides from Acyl Chlorides

Primary amines, secondary amines, and ammonia all react rapidly with acid chlorides to form amides. An excess of ammonia or amine is used to neutralize the HCl that would be formed otherwise.

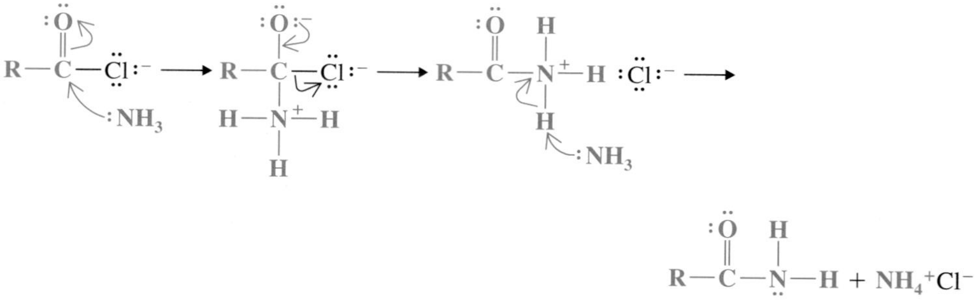

An amide

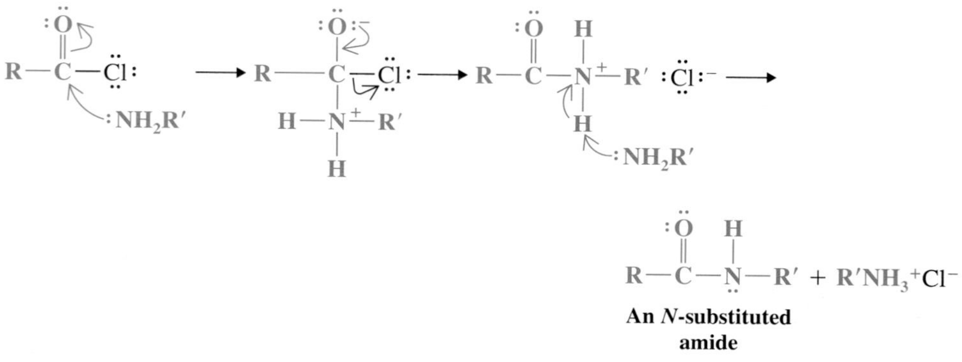

An *N*-substituted amide

An *N,N*-disubstituted
amide

Since acyl chlorides are easily prepared from carboxylic acids this is one of the most widely used laboratory methods for the synthesis of amides. The reaction between the acyl chloride and the amine (or ammonia) usually takes place at room temperature (or below) and produces the amide in high yield.

Acyl chlorides also react with tertiary amines by a nucleophilic substitution reaction. The acylammonium ion that forms, however, is not stable in the presence of water or any hydroxylic solvent.

Acylpyridinium ions are probably involved as intermediates in those reactions of acyl chlorides that are carried out in the presence of pyridine.

18.8C Amides from Carboxylic Anhydrides

Acid anhydrides react with ammonia and with primary and secondary amines and form amides through reactions that are analogous to those of acyl chlorides.

Cyclic anhydrides react with ammonia or an amine in the same general way as acyclic anhydrides; however, the reaction produces a product that is both an amide and an ammonium salt. Acidifying the ammonium salt gives a compound that is both an amide and an acid:

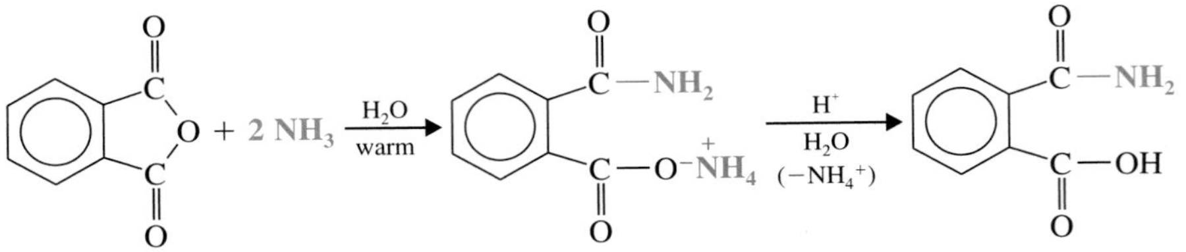

Phthalic anhydride

Ammonium phthalamate (94%)

Phthalamic acid (81%)

Heating the amide acid causes dehydration to occur and gives an *imide*. Imides contain the linkage

$$-\overset{\overset{\displaystyle O}{\|}}{C}-NH-\overset{\overset{\displaystyle O}{\|}}{C}-.$$

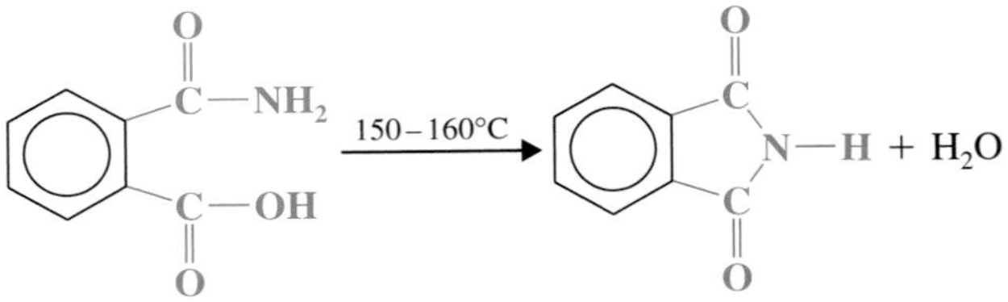

Phthalamic acid

Phthalimide (~100%)

18.8D Amides from Esters

Esters undergo nucleophilic substitution at their acyl carbon atoms when they are treated with ammonia (called *ammonolysis*) or with primary and secondary amines. These reactions take place more slowly than those of acyl chlorides and anhydrides, but they are synthetically useful.

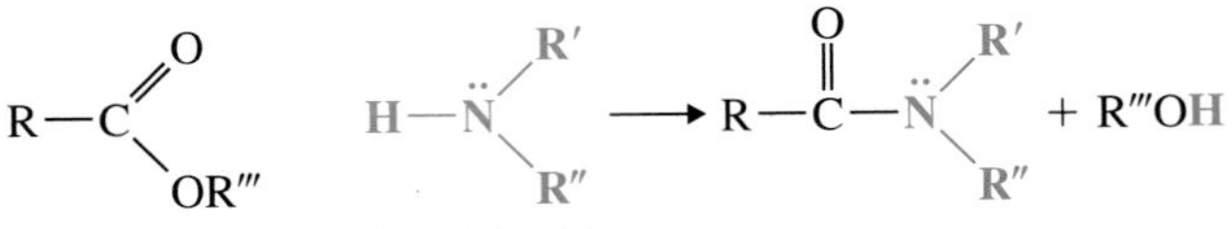

R′ and/or R″ may be H

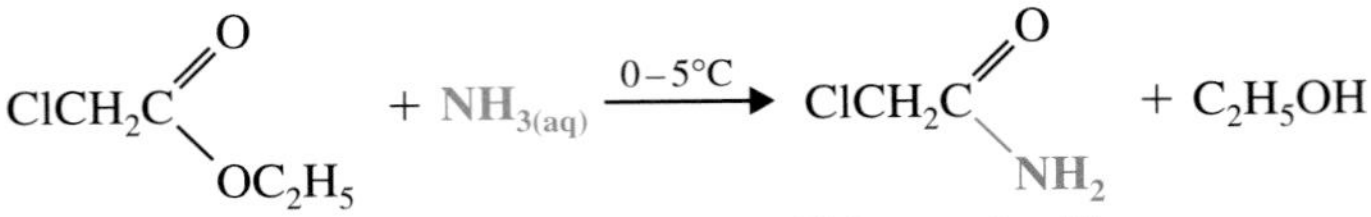

Ethyl chloroacetate

Chloroacetamide (62–87%)

18.8E Amides from Carboxylic Acids and Ammonium Carboxylates

Carboxylic acids react with aqueous ammonia to form ammonium salts.

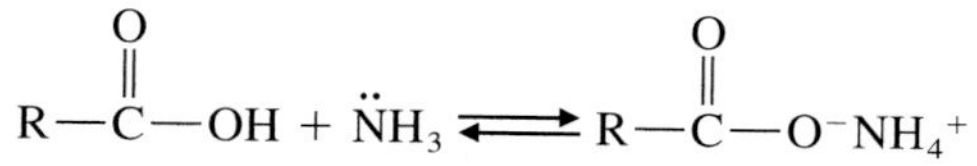

$$R-\overset{\displaystyle O}{\overset{\|}{C}}-OH + \overset{\cdot\cdot}{N}H_3 \rightleftharpoons R-\overset{\displaystyle O}{\overset{\|}{C}}-O^-NH_4{}^+$$

An ammonium carboxylate

Because of the low reactivity of the carboxylate ion toward nucleophilic substitution, further reaction does not usually take place in aqueous solution. However, if we evaporate the water and subsequently heat the dry salt, dehydration produces an amide.

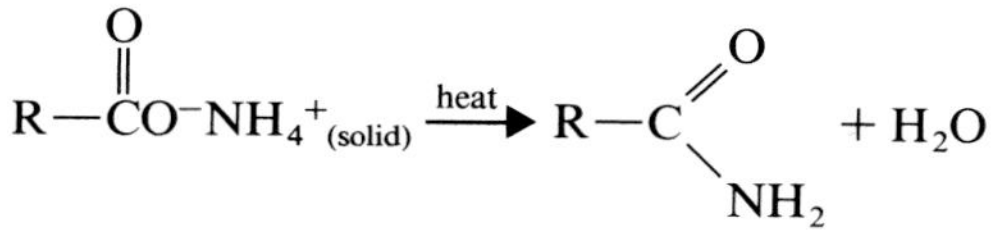

$$R-\overset{\displaystyle O}{\overset{\|}{C}}O^-NH_4{}^+{}_{(solid)} \xrightarrow{\text{heat}} R-C\overset{\displaystyle O}{\underset{NH_2}{\diagup}} + H_2O$$

This is generally a poor method for preparing amides. A much better method is to convert the acid to an acyl chloride and then treat the acyl chloride with ammonia or an amine (Section 18.8B).

Amides are of great importance in biochemistry. The linkages that join individual amino acids together to form proteins are primarily amide linkages. As a consequence, much research has been done to find new and mild ways for amide synthesis. One especially useful reagent is the compound dicyclohexylcarbodiimide (DCC), $C_6H_{11}-N=C=N-C_6H_{11}$. Dicyclohexylcarbodiimide promotes amide formation by reacting with the carboxyl group of an acid and activating it toward nucleophilic substitution.

A MECHANISM FOR DCC-PROMOTED AMIDE SYNTHESIS

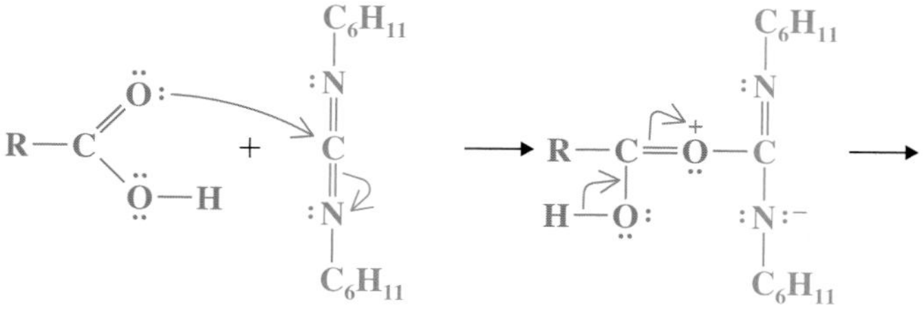

Dicyclohexyl-
carbodiimide
(DCC)

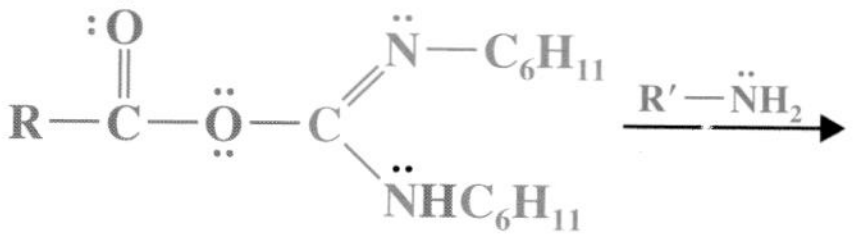

Reactive
intermediate

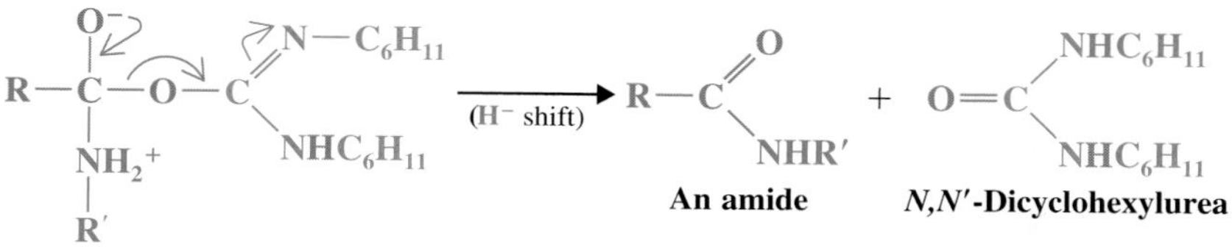

An amide   N,N'-Dicyclohexylurea

The intermediate in this synthesis does not need to be isolated, and both steps take place at room temperature. Amides are produced in very high yield. In Chapter 24 we shall see how dicyclohexylcarbodiimide can be used in an automated synthesis of proteins.

18.8F Hydrolysis of Amides

Amides undergo hydrolysis when they are heated with aqueous acid or aqueous base.

Acidic Hydrolysis

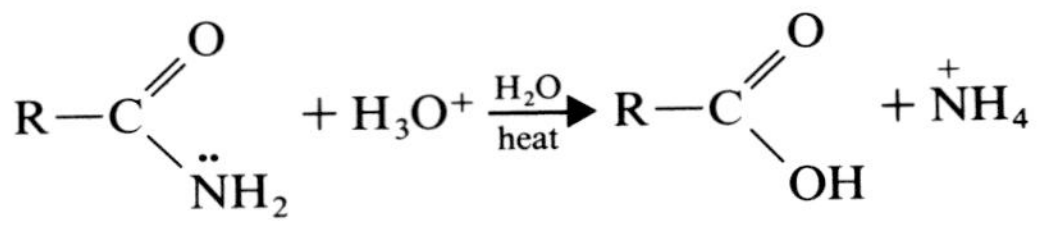

$$R-C\overset{O}{\underset{NH_2}{}} + H_3O^+ \xrightarrow[\text{heat}]{H_2O} R-C\overset{O}{\underset{OH}{}} + \overset{+}{NH_4}$$

Basic Hydrolysis

$$R-C\overset{O}{\underset{NH_2}{}} + Na^+OH^- \xrightarrow[\text{heat}]{H_2O} R-C\overset{O}{\underset{O^-Na^+}{}} + \ddot{N}H_3$$

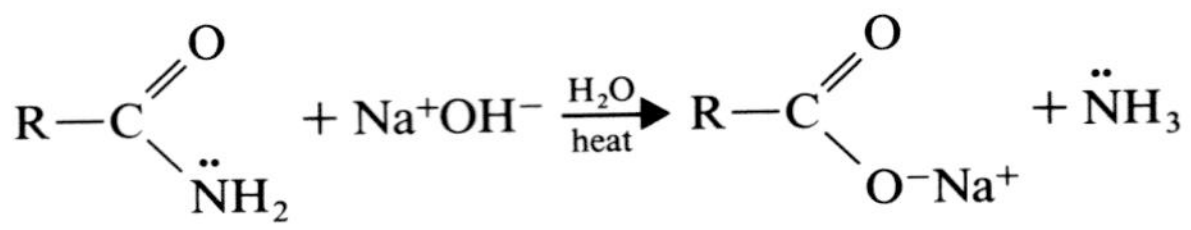

N-Substituted amides and *N,N*-disubstituted amides also undergo hydrolysis in aqueous acid or base. Amide hydrolysis by either method takes place more slowly than the corresponding hydrolysis of an ester. Thus, amide hydrolyses generally require more forcing conditions.

The mechanism for acid hydrolysis of an amide is similar to that given in Section 18.7A for the acid hydrolysis of an ester. Water acts as a nucleophile and attacks the protonated amide. The leaving group in the acidic hydrolysis of an amide is ammonia (or an amine).

A MECHANISM FOR ACIDIC HYDROLYSIS OF AN AMIDE

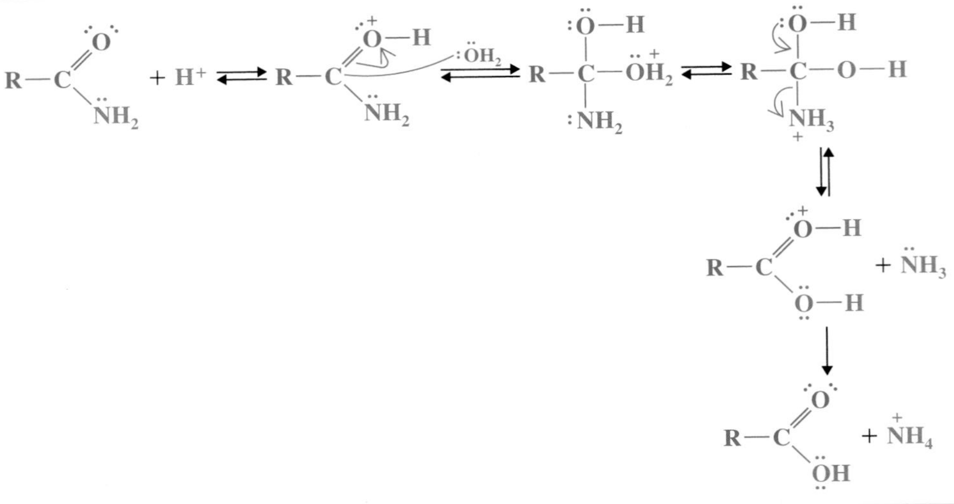

There is evidence that in basic hydrolyses of amides, hydroxide ions act both as nucleophiles and as bases. In the first step (in the following reaction) a hydroxide ion attacks the acyl carbon of the amide. In the second step, a hydroxide ion removes a proton to give a dianion. In the final step, the dianion loses a molecule of ammonia (or an amine); this step is synchronized with a proton transfer from water.

A MECHANISM FOR BASIC HYDROLYSIS OF AN AMIDE

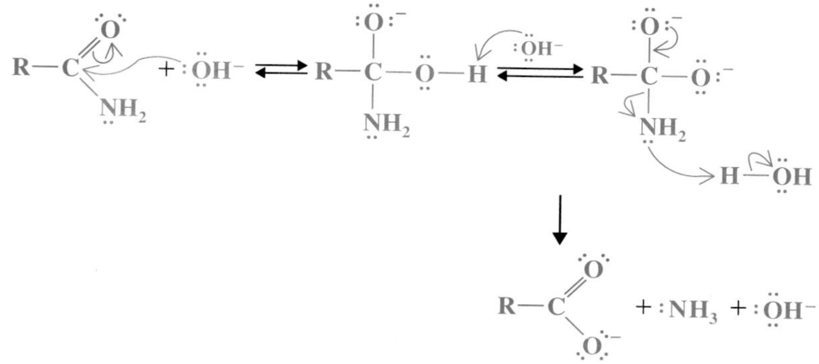

PROBLEM 18.14

What products would you obtain from acidic and basic hydrolysis of each of the following amides?

(a) *N,N*-Diethylbenzamide

(b)

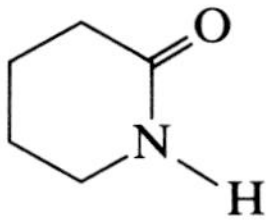

(c) $HO_2CCH-NHC-CHNH_2$ (a dipeptide)
with $\overset{O}{\overset{\|}{C}}$ above, CH_3 below first CH, CH_2 and C_6H_5 below second CH

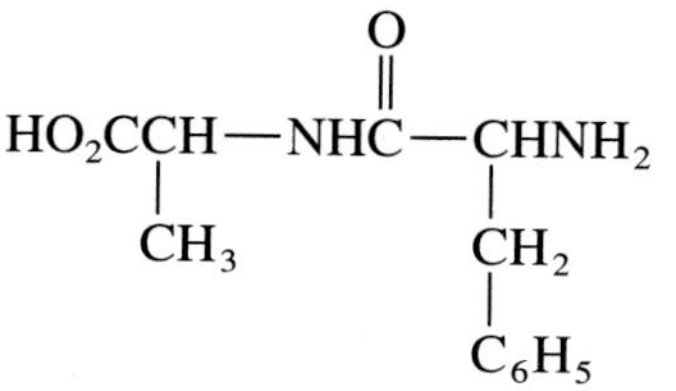

18.8G Nitriles from the Dehydration of Amides

Amides react with P_4O_{10} (a compound that is often called phosphorus pentoxide and written P_2O_5) or with boiling acetic anhydride to form nitriles.

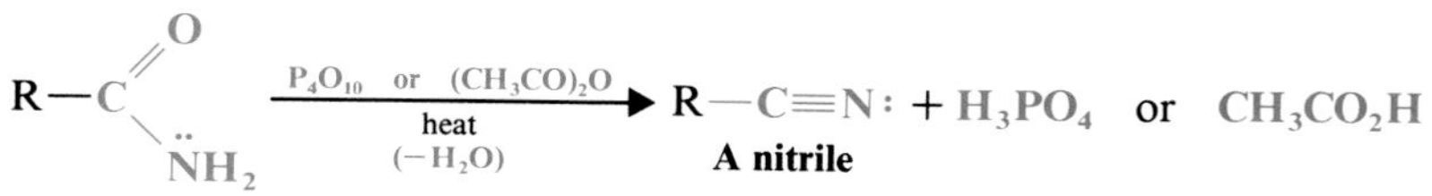

$$R-C\overset{O}{\underset{NH_2}{\diagdown}} \xrightarrow[\substack{\text{heat} \\ (-H_2O)}]{P_4O_{10} \text{ or } (CH_3CO)_2O} R-C\equiv N: + H_3PO_4 \text{ or } CH_3CO_2H$$

A nitrile

This is a useful synthetic method for preparing nitriles that are not available by nucleophilic substitution reactions between alkyl halides and cyanide ion.

PROBLEM 18.15

(a) Show all steps in the synthesis of $(CH_3)_3CCN$ from $(CH_3)_3CCO_2H$.
(b) What product would you expect to obtain if you attempted to synthesize $(CH_3)_3CCN$ using the following method?

$$(CH_3)_3C—Br + CN^- \longrightarrow$$

18.8H Hydrolysis of Nitriles

Although nitriles do not contain a carbonyl group, they are usually considered to be derivatives of carboxylic acids because complete hydrolysis of a nitrile produces a carboxylic acid or a carboxylate ion (Sections 16.9A and 18.3).

$$R—C\equiv N \begin{cases} \xrightarrow{H_3O^+,\ H_2O,\ heat} RCO_2H \\ \xrightarrow{OH^-,\ H_2O,\ heat} RCO_2^- \end{cases}$$

The mechanisms for these hydrolyses are related to those for the acidic and basic hydrolyses of amides. In **acidic hydrolysis** of a nitrile the first step is protonation of the nitrogen atom. This protonation (in the following sequence) polarizes the nitrile group and makes the carbon atom more susceptible to nucleophilic attack by the weak nucleophile, water. The loss of a proton from the oxygen atom then produces a tautomeric form of an amide. Gain of a proton at the nitrogen atom gives a **protonated amide,** and from this point on the steps are the same as those given for the acidic hydrolysis of an amide in Section 18.8F. (In concentrated H_2SO_4 the reaction stops at the protonated amide and this is a useful way of making amides from nitriles.)

A MECHANISM FOR ACIDIC HYDROLYSIS OF A NITRILE

In **basic hydrolysis,** a hydroxide ion attacks the nitrile carbon atom and subsequent protonation leads to the amide tautomer. Further attack by hydroxide ion leads to hydrolysis in a manner analogous to that for the basic hydrolysis of an amide (Section 18.8F). (Under the appropriate conditions, amides can be isolated when nitriles are hydrolyzed.)

A MECHANISM FOR BASIC HYDROLYSIS OF A NITRILE

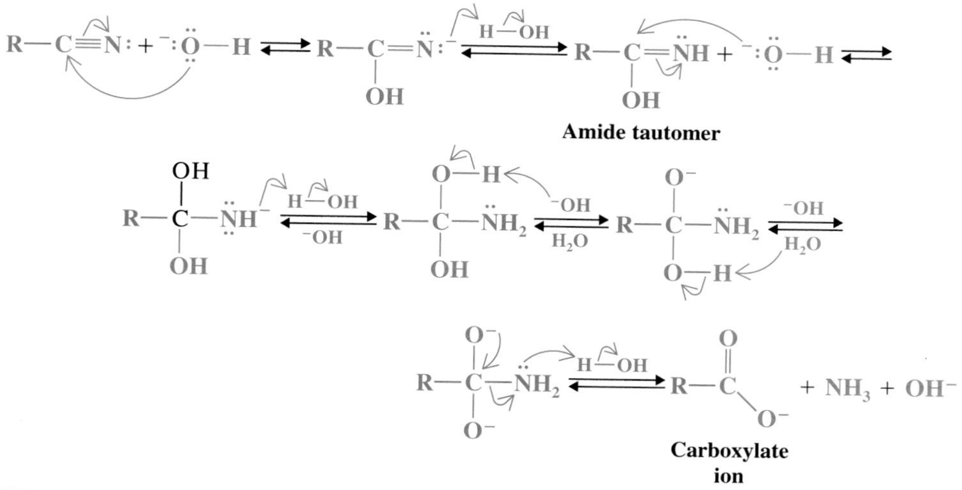

Amide tautomer

Carboxylate ion

18.8I Lactams

Cyclic amides are called lactams. The size of the lactam ring is designated by Greek letters in a way that is analogous to lactone nomenclature.

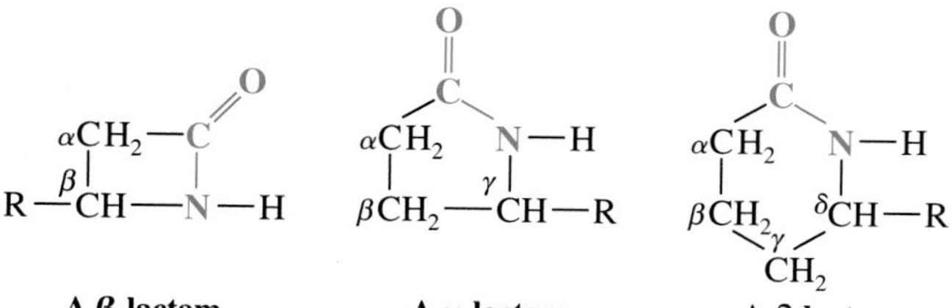

A β-lactam   **A γ-lactam**   **A δ-lactam**

γ-Lactams and δ-lactams often form spontaneously from γ- and δ-amino acids. β-Lactams, however, are highly reactive; their strained four-membered rings open easily in the presence of nucleophilic reagents. The penicillin antibiotics (see following structures) contain a β-lactam ring.

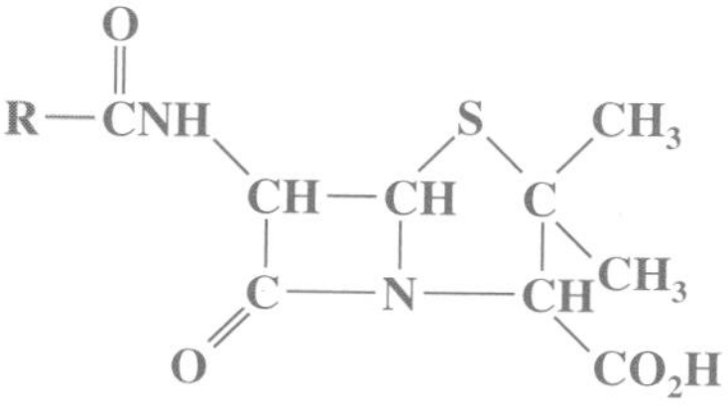

$R = C_6H_5CH_2-$ (penicillin G)

$R = C_6H_5CH-$ (ampicillin)
$\qquad\quad |$
$\qquad\;\; NH_2$

$R = C_6H_5OCH_2-$ (penicillin V)

The penicillins apparently act by interfering with the synthesis of the bacterial cell walls. It is thought that they do this by reacting with an amino group of an essential enzyme of the cell wall biosynthetic pathway. This reaction involves ring opening of the β-lactam and acylation of the enzyme, inactivating it.

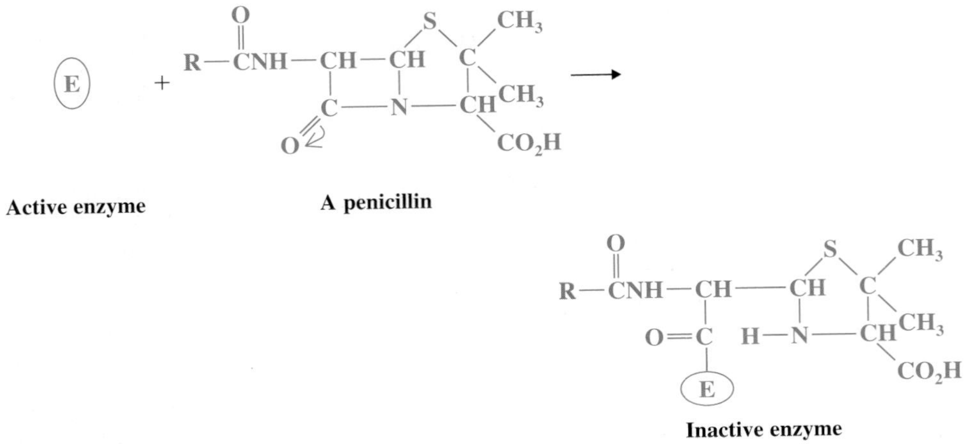

Active enzyme   A penicillin

Inactive enzyme

18.9   α-HALO ACIDS: THE HELL–VOLHARD–ZELINSKI REACTION

Aliphatic carboxylic acids react with bromine or chlorine in the presence of phosphorus (or a phosphorus halide) to give α-halo acids through a reaction known as the Hell–Volhard–Zelinski (or HVZ) reaction.

General Reaction

$$RCH_2CO_2H \xrightarrow[\text{(2) } H_2O]{\text{(1) } X_2,\, P} RCHCO_2H$$
$$\overset{|}{X}$$

α-Halo acid

Specific Example

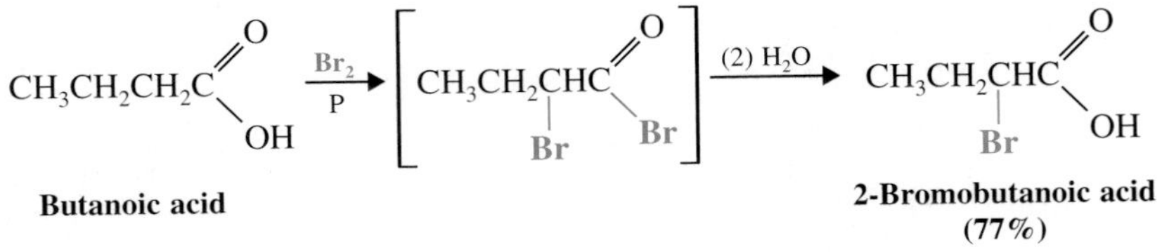

Butanoic acid   2-Bromobutanoic acid
                                              (77%)

Halogenation occurs specifically at the α carbon. If more than one molar equivalent of bromine or chlorine is used in the reaction, the products obtained are α,α-dihalo acids or α,α,α-trihalo acids.

   The mechanism for the Hell–Volhard–Zelinski reaction is outlined on the next page. The key step involves the formation of an enol from an acyl halide. (Carboxylic acids do not form enols readily.) Enol formation accounts for specific halogenation at the α position.

A MECHANISM FOR THE HVZ REACTION

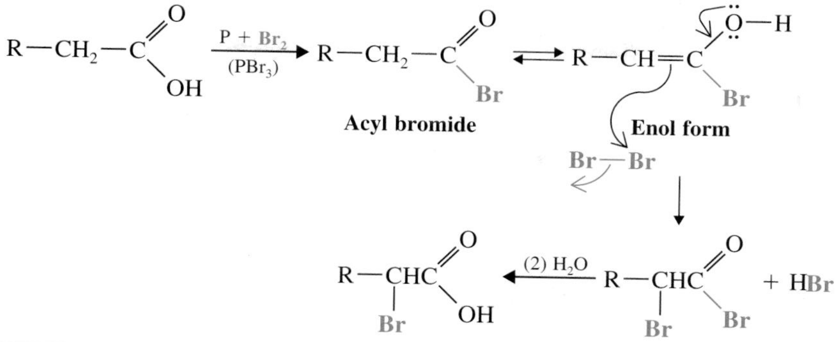

Acyl bromide   **Enol form**

A more versatile method for α-halogenation has been developed by David N. Harpp (of McGill University). Acyl chlorides, formed *in situ* by the reaction of the carboxylic acid with SOCl₂, are treated with the appropriate *N*-halosuccinimide and a trace of HX to produce α-chloro and α-bromo acyl chlorides.

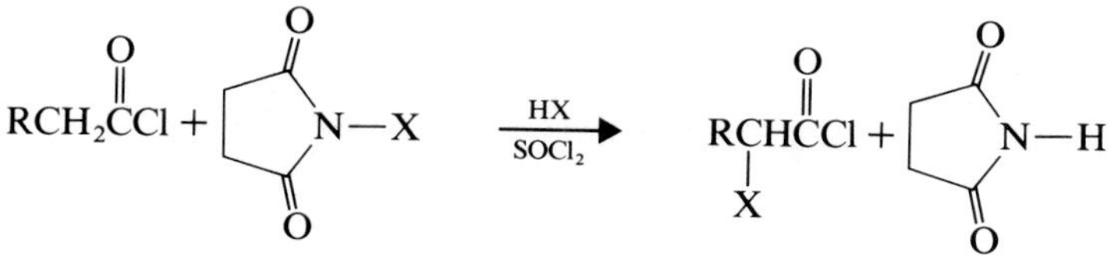

(X = Cl or Br)

α-Iodo acyl chlorides can be obtained by using molecular iodine in a similar reaction.

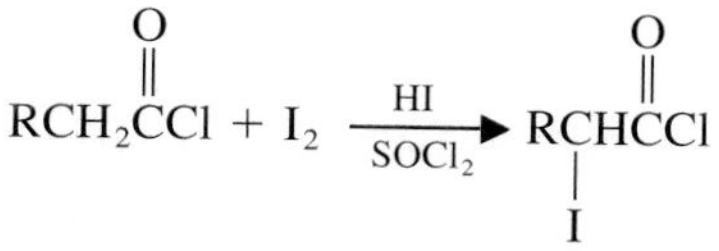

α-Halo acids are important synthetic intermediates because they are capable of reacting with a variety of nucleophiles:

Conversion to α-Hydroxy Acids

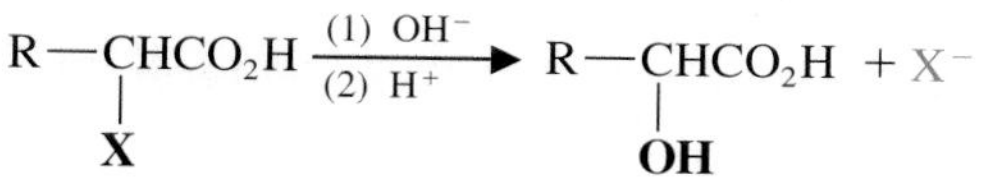

α-Halo acid   **α-Hydroxy acid**

Specific Example

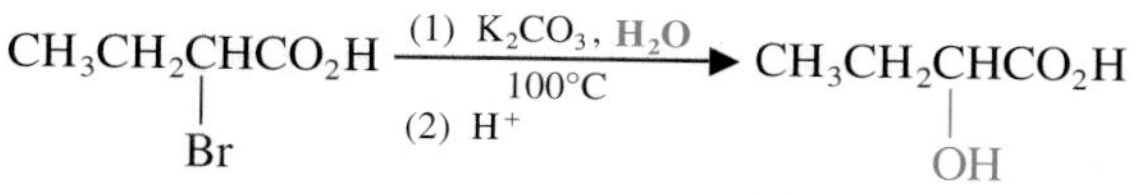

2-Hydroxybutanoic acid
(69%)

Conversion to α-Amino Acids

$$R-\underset{\underset{X}{|}}{C}HCO_2H + 2\ NH_3 \longrightarrow R\underset{\underset{NH_3^+}{|}}{C}HCO_2^- + NH_4X$$

α-Halo	α-Amino
acid	acid

Specific Example

$$\underset{\underset{Br}{|}}{C}H_2CO_2H + 2\ NH_3 \longrightarrow \underset{\underset{NH_3^+}{|}}{C}H_2CO_2^- + NH_4Br$$

Aminoacetic acid
(glycine)
(60–64%)

18.10 DERIVATIVES OF CARBONIC ACID

Carbonic acid $\left(\underset{}{\overset{O}{\underset{\|}{HOCOH}}}\right)$ is an unstable compound that decomposes spontaneously (to produce carbon dioxide and water) and, therefore, cannot be isolated. However, many acyl chlorides, esters, and amides that are derived from carbonic acid (on paper, not in the laboratory) are stable compounds that have important applications.

Carbonyl dichloride (ClCOCl), a highly toxic compound that is also called *phosgene,* can be thought of as the diacyl chloride of carbonic acid. Carbonyl dichloride reacts with two molar equivalents of an alcohol to yield a **dialkyl carbonate.**

$$\overset{O}{\underset{\|}{Cl-C-Cl}} + 2\ CH_3CH_2OH \longrightarrow CH_3CH_2O\overset{O}{\underset{\|}{C}}OCH_2CH_3 + 2\ HCl$$

Carbonyl   **Diethyl carbonate**
dichloride

A tertiary amine is usually added to the reaction to neutralize the hydrogen chloride that is produced.

Carbonyl dichloride reacts with ammonia to yield **urea** (Section 1.2A).

$$\overset{O}{\underset{\|}{Cl-C-Cl}} + 4\ NH_3 \longrightarrow H_2N\overset{O}{\underset{\|}{C}}NH_2 + 2\ NH_4Cl$$

Urea

Urea is the end product of the metabolism of nitrogen-containing compounds in most mammals and is excreted in the urine.

18.10A Alkyl Chloroformates and Carbamates (Urethanes)

Treating carbonyl dichloride with one molar equivalent of an alcohol leads to the formation of an alkyl chloroformate:

$$\text{ROH} + \text{Cl}-\overset{\overset{\displaystyle O}{\|}}{C}-\text{Cl} \longrightarrow \text{RO}-\overset{\overset{\displaystyle O}{\|}}{C}-\text{Cl} + \text{HCl}$$

Alkyl chloroformate

Specific Example

$$\text{C}_6\text{H}_5\text{CH}_2\text{OH} + \text{Cl}-\overset{\overset{\displaystyle O}{\|}}{C}-\text{Cl} \longrightarrow \text{C}_6\text{H}_5\text{CH}_2\text{O}-\overset{\overset{\displaystyle O}{\|}}{C}-\text{Cl} + \text{HCl}$$

Benzyl chloroformate

Alkyl chloroformates react with ammonia or amines to yield compounds called *carbamates* or *urethanes:*

$$\text{RO}-\overset{\overset{\displaystyle O}{\|}}{C}-\text{Cl} + \text{R}'\text{NH}_2 \xrightarrow{\text{OH}^-} \text{RO}-\overset{\overset{\displaystyle O}{\|}}{C}-\text{NHR}'$$

A carbamate (or urethane)

Benzyl chloroformate is used to install on an amino group a protecting group called the benzyloxycarbonyl group. We shall see in Section 24.7 how use is made of this protecting group in the synthesis of peptides and proteins. One advantage of the benzyloxycarbonyl group is that it can be removed under mild conditions. Treating the benzyloxycarbonyl derivative with hydrogen and a catalyst or with cold HBr in acetic acid removes the protecting group.

$$\text{R}-\text{NH}_2 + \text{C}_6\text{H}_5\text{CH}_2\text{O}\overset{\overset{\displaystyle O}{\|}}{C}\text{Cl} \xrightarrow{\text{OH}^-} \text{R}-\text{NH}-\overset{\overset{\displaystyle O}{\|}}{C}\text{OCH}_2\text{C}_6\text{H}_5 \left.\right\} \begin{array}{l}\textbf{Protected} \\ \textbf{amine}\end{array}$$

$$\text{R}-\text{NH}-\overset{\overset{\displaystyle O}{\|}}{C}\text{OCH}_2\text{C}_6\text{H}_5 \begin{cases} \xrightarrow{\text{H}_2,\ \text{Pd}} \text{R}-\text{NH}_2 + \text{CO}_2 + \text{C}_6\text{H}_5\text{CH}_3 \\ \\ \xrightarrow{\text{HBr, CH}_3\text{CO}_2\text{H}} \text{R}-\overset{+}{\text{NH}}_3 + \text{CO}_2 + \text{C}_6\text{H}_5\text{CH}_2\text{Br} \end{cases}$$

Carbamates can also be synthesized by allowing an alcohol to react with an isocyanate, $\text{R}-\text{N}{=}\text{C}{=}\text{O}$. (Carbamates tend to be nicely crystalline solids and are useful derivatives for identifying alcohols.) The reaction is an example of nucleophilic addition to the acyl carbon.

$$ROH + C_6H_5—N=C=O \longrightarrow RO\overset{\overset{\displaystyle O}{\|}}{C}—NH—C_6H_5$$

**Phenyl
isocyanate**

The insecticide called *Sevin* is a carbamate made by allowing 1-naphthol to react with methyl isocyanate.

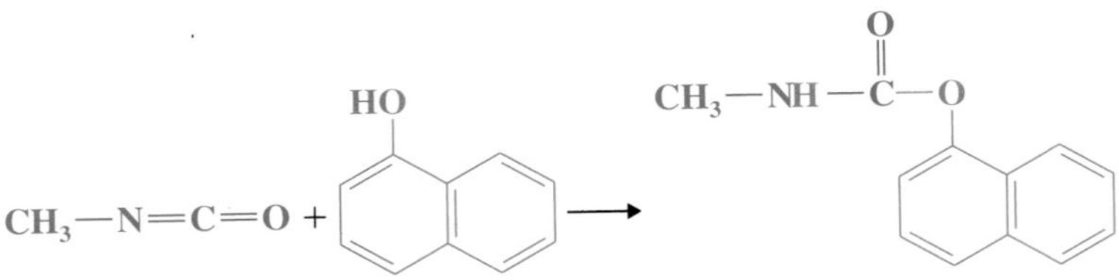

$$CH_3—N=C=O \;+ \qquad\qquad \longrightarrow$$

HO

CH$_3$—NH—$\overset{\overset{\displaystyle O}{\|}}{C}$—O

Methyl isocyanate   **1-Naphthol**   **Sevin**

A tragic accident that occurred at Bhopal, India, in 1984, was caused by leakage of methyl isocyanate from a manufacturing plant. Methyl isocyanate is a highly toxic gas and more than 1800 people living near the plant lost their lives.

PROBLEM 18.16

Write structures for the products of the following reactions:

(a) $C_6H_5CH_2OH + C_6H_5N=C=O \longrightarrow$

(b) $ClCOCl$ + excess $CH_3NH_2 \longrightarrow$

(c) Glycine $(H_3\overset{+}{N}CH_2CO_2^-) + C_6H_5CH_2OCOCl \xrightarrow{OH^-}$

(d) Product of (c) $+ H_2$, Pd $\longrightarrow$

(e) Product of (c) + cold HBr, $CH_3CO_2H \longrightarrow$

(f) Urea $+ OH^-$, H_2O, heat

Although alkyl chloroformates $\left(RO\overset{\overset{\displaystyle O}{\|}}{C}Cl\right)$, dialkyl carbonates $\left(RO\overset{\overset{\displaystyle O}{\|}}{C}OR\right)$, and carbamates $\left(RO\overset{\overset{\displaystyle O}{\|}}{C}NH_2, RO\overset{\overset{\displaystyle O}{\|}}{C}NHR, \text{ etc.}\right)$ are stable, chloroformic acid $\left(HO\overset{\overset{\displaystyle O}{\|}}{C}Cl\right)$, alkyl hydrogen carbonates $\left(RO\overset{\overset{\displaystyle O}{\|}}{C}OH\right)$, and carbamic acid $\left(HO\overset{\overset{\displaystyle O}{\|}}{C}NH_2\right)$ are not. These latter compounds decompose spontaneously to liberate carbon dioxide.

$$HO—\overset{\overset{\displaystyle O}{\|}}{C}—Cl \longrightarrow HCl + CO_2$$

Unstable

$$RO-\overset{\overset{\displaystyle O}{\|}}{C}-OH \longrightarrow ROH + CO_2$$
Unstable

$$HO-\overset{\overset{\displaystyle O}{\|}}{C}-NH_2 \longrightarrow NH_3 + CO_2$$
Unstable

This instability is a characteristic that these compounds share with their functional parent, carbonic acid.

$$HO-\overset{\overset{\displaystyle O}{\|}}{C}-OH \longrightarrow H_2O + CO_2$$
Unstable

18.11 DECARBOXYLATION OF CARBOXYLIC ACIDS

The reaction whereby a carboxylic acid loses CO_2 is called a *decarboxylation*.

$$R-\overset{\overset{\displaystyle O}{\|}}{C}-OH \xrightarrow{\text{decarboxylation}} R-H + CO_2$$

Although the unusual stability of carbon dioxide means that decarboxylation of most acids is exothermic, in practice the reaction is not always easy to carry out because the reaction is very slow. Special groups usually have to be present in the molecule for decarboxylation to be rapid enough to be synthetically useful.

Acids whose molecules have a carbonyl group one carbon removed from the carboxylic acid group, **called β-keto acids,** decarboxylate readily when they are heated to $100-150°C$. (Some β-keto acids even decarboxylate slowly at room temperature.)

$$R\overset{\overset{\displaystyle O}{\|}}{C}CH_2\overset{\overset{\displaystyle O}{\|}}{C}OH \xrightarrow{100-150°C} R\overset{\overset{\displaystyle O}{\|}}{C}CH_3 + CO_2$$
A β-keto acid

There are two reasons for this ease of decarboxylation:

1. When the carboxylate anion decarboxylates, it forms a resonance-stabilized enolate anion:

$$R-\overset{\overset{\displaystyle \ddot{O}:}{\|}}{C}-CH_2-\overset{\overset{\displaystyle \ddot{O}:}{\|}}{C}-\ddot{O}:^- \xrightarrow{-CO_2} R-\overset{\overset{\displaystyle \ddot{O}:}{\|}}{C}-\overset{-}{\ddot{C}H_2} \xrightarrow{H^+} R-\overset{\overset{\displaystyle \ddot{O}:}{\|}}{C}-CH_3$$

Acylacetate ion

$$R-\overset{\overset{\displaystyle :\ddot{O}:^-}{|}}{C}=CH_2$$

Resonance-stabilized anion

This anion is much more stable than the anion $RCH_2:^-$ that would be produced by decarboxylation of an ordinary carboxylic acid anion.

2. When the acid itself decarboxylates, it can do so through a six-membered cyclic transition state:

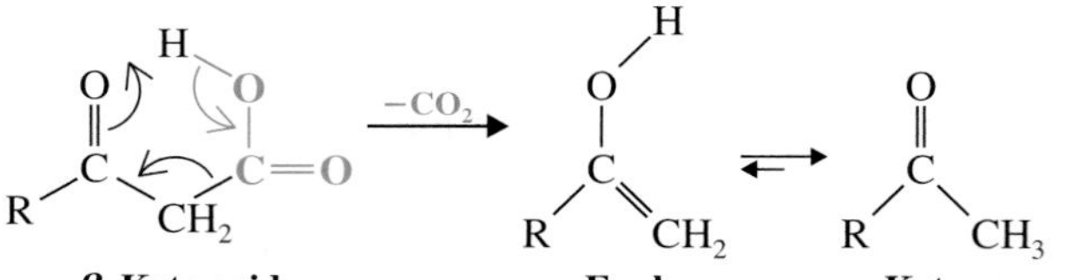

β-Keto acid **Enol** **Ketone**

This reaction produces an enol directly and avoids an anionic intermediate. The enol then tautomerizes to a methyl ketone.

Malonic acids also decarboxylate readily for similar reasons.

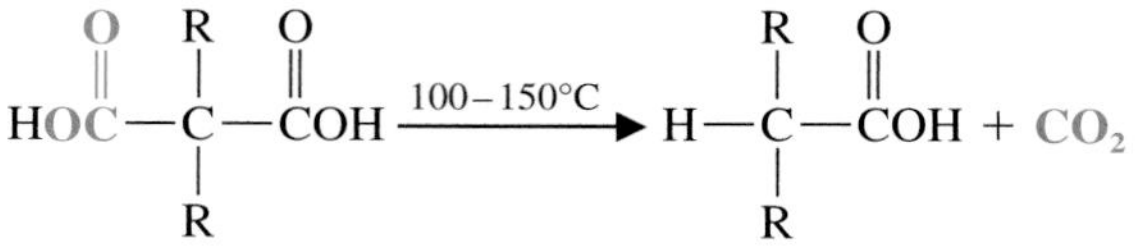

A malonic acid

Notice that malonic acids undergo decarboxylation so readily that they do not form cyclic anhydrides (Section 18.6).

We shall see in Chapter 19 how decarboxylations of β-keto acids and malonic acids are synthetically useful.

Aromatic carboxylic acids can be decarboxylated by converting them to a cuprous salt and then heating the salt with quinoline in an inert atmosphere in the presence of cuprous oxide.

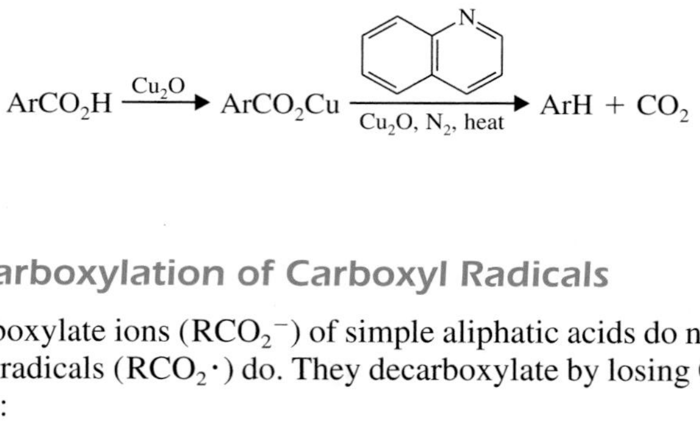

18.11A Decarboxylation of Carboxyl Radicals

Although the carboxylate ions (RCO_2^-) of simple aliphatic acids do not decarboxylate readily, carboxyl radicals ($RCO_2\cdot$) do. They decarboxylate by losing CO_2 and producing alkyl radicals:

$$RCO_2\cdot \longrightarrow R\cdot + CO_2$$

PROBLEM 18.17

Using decarboxylation reactions, outline a synthesis of each of the following from appropriate starting materials.

(a) 2-Hexanone
(b) 2-Methylbutanoic acid
(c) Cyclohexanone
(d) Pentanoic acid

PROBLEM 18.18 _____

Diacyl peroxides $\left(\begin{array}{cc} O & O \\ \parallel & \parallel \\ RC-O-O-CR \end{array} \right)$ decompose readily when heated. (a) What factor accounts for this instability? (b) The decomposition of a diacyl peroxide produces CO_2. How is it formed? (c) Diacyl peroxides are often used to initiate free radical reactions, for example, the polymerization of an alkene:

$$n\ CH_2{=}CH_2 \xrightarrow[\substack{(-\,CO_2)}]{\substack{O \quad\quad O \\ \parallel \quad\quad \parallel \\ RC-O-O-CR}} R{-}(CH_2CH_2{-})_{\!n}H$$

Show the steps involved.

_____

18.12 CHEMICAL TESTS FOR ACYL COMPOUNDS

Carboxylic acids are weak acids and their acidity helps us to detect them. Aqueous solutions of water-soluble carboxylic acids give an acid test with blue litmus paper. Water-insoluble carboxylic acids dissolve in aqueous sodium hydroxide and aqueous sodium bicarbonate (cf. Section 18.2C). The latter reagent helps us distinguish carboxylic acids from most phenols. Except for the di- and trinitrophenols, phenols do not dissolve in aqueous sodium bicarbonate. When carboxylic acids dissolve in aqueous sodium bicarbonate they also cause the evolution of carbon dioxide.

Acyl chlorides hydrolyze in water and the resulting chloride ion gives a precipitate when treated with aqueous silver nitrate. Acid anhydrides dissolve when heated briefly with aqueous sodium hydroxide.

Esters and amides hydrolyze slowly when they are refluxed with sodium hydroxide. An ester produces a carboxylate ion and an alcohol; an amide produces a carboxylate ion and an amine or ammonia. The hydrolysis products, the acid and the alcohol or amine, can be isolated and identified. Since base-promoted hydrolysis of an unsubstituted amide produces ammonia, this ammonia can often be detected by holding moist red litmus paper in the vapors above the reaction mixture.

Amides can be distinguished from amines with dilute HCl. Most amines dissolve in dilute HCl, whereas most amides do not (cf. Problem 18.37).

18.13 SUMMARY OF THE REACTIONS OF CARBOXYLIC ACIDS AND THEIR DERIVATIVES

The reactions of carboxylic acids and their derivatives are summarized here.

18.13A Reactions of Carboxylic Acids

1. As acids (discussed in Sections 3.9, and 18.2C).

$$RCO_2H + NaOH \longrightarrow RCO_2{}^-Na^+ + H_2O$$

$$RCO_2H + NaHCO_3 \longrightarrow RCO_2{}^-Na^+ + H_2O + CO_2$$

2. Reduction (discussed in Section 11.3).

$$RCO_2H + LiAlH_4 \xrightarrow[(2)\ H_2O]{(1)\ \text{diethyl ether}} RCH_2OH$$

3. Conversion to acyl chlorides (discussed in Section 18.5).

$$RCO_2H + SOCl_2 \longrightarrow RCOCl + SO_2 + HCl$$

$$3\ RCO_2H + PCl_3 \longrightarrow 3\ RCOCl + H_3PO_3$$

$$RCO_2H + PCl_5 \longrightarrow RCOCl + POCl_3 + HCl$$

4. Conversion to acid anhydrides (discussed in Section 18.6).

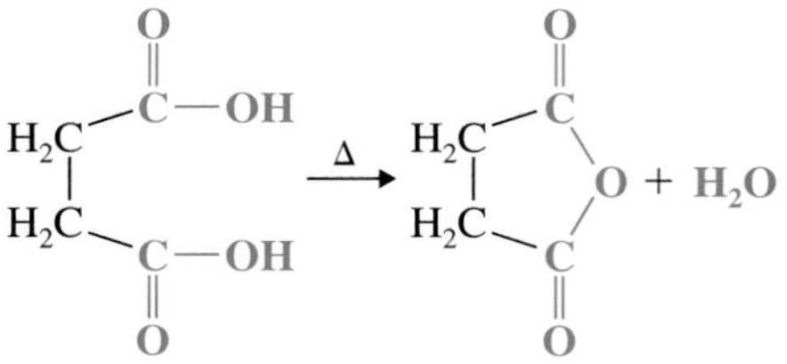

5. Conversion to esters (discussed in Section 18.7).

$$R{-}\overset{O}{\overset{\|}{C}}OH + R'{-}OH \overset{H^+}{\rightleftharpoons} R{-}\overset{O}{\overset{\|}{C}}{-}OR' + H_2O$$

6. Conversion to lactones (discussed in Section 18.7).

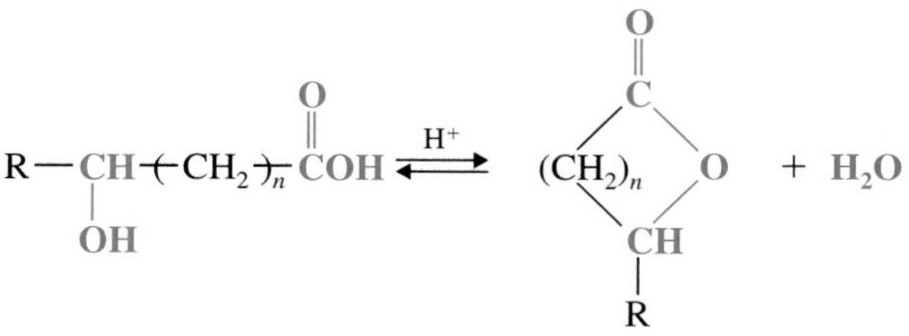

n = 2, a γ-lactone
n = 3, a δ-lactone

7. Conversion to amides and imides (discussed in Section 18.8).

$$R{-}\overset{O}{\overset{\|}{C}}{-}OH + NH_3 \rightleftharpoons R{-}\overset{O}{\overset{\|}{C}}{-}O^-NH_4^+ \xrightarrow{\text{heat}} R{-}\overset{O}{\overset{\|}{C}}{-}NH_2 + H_2O$$

An amide

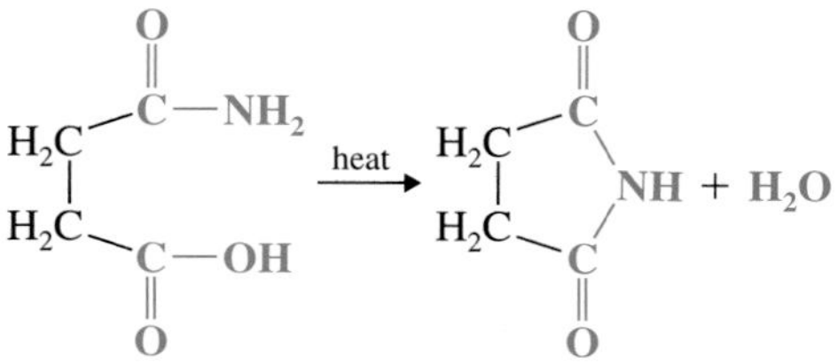

A cyclic imide

8. Conversion to lactams (discussed in Section 18.8).

$$\underset{\text{NH}_2}{\text{R}-\overset{|}{\text{CH}}(\text{CH}_2)_n}\overset{O}{\overset{||}{\text{C}}}\text{OH} \xrightarrow{\text{heat}} \underset{R}{\underset{|}{\underset{\diagdown}{\text{C}}}}\underset{(\text{CH}_2)_n}{\overset{\text{H}}{\overset{|}{\text{N}}}}\text{C}{=}\text{O} + \text{H}_2\text{O}$$

$$n = 2, \text{ a } \gamma\text{-lactam}$$
$$n = 3, \text{ a } \delta\text{-lactam}$$

9. α-Halogenation (discussed in Section 18.9).

$$\text{R}-\text{CH}_2\text{CO}_2\text{H} + \text{X}_2 \xrightarrow{\text{P}} \text{R}-\underset{\overset{|}{\text{X}}}{\overset{|}{\text{CH}}}\text{CO}_2\text{H} + \text{HX}$$

$$\text{X}_2 = \text{Cl}_2 \quad \text{or} \quad \text{Br}_2$$

10. Decarboxylation (discussed in Section 18.11).

$$\overset{O}{\overset{||}{\text{R}\text{C}}}\text{CH}_2\overset{O}{\overset{||}{\text{C}}}\text{OH} \xrightarrow{\text{heat}} \overset{O}{\overset{||}{\text{R}\text{C}}}\text{CH}_3 + \text{CO}_2$$

$$\text{HO}\overset{O}{\overset{||}{\text{C}}}\text{CH}_2\overset{O}{\overset{||}{\text{C}}}\text{OH} \xrightarrow{\text{heat}} \text{CH}_3\overset{O}{\overset{||}{\text{C}}}\text{OH} + \text{CO}_2$$

18.13B Reactions of Acyl Chlorides

1. Conversion to acids (discussed in Section 18.4).

$$\text{R}-\overset{O}{\overset{||}{\text{C}}}-\text{Cl} + \text{H}_2\text{O} \longrightarrow \text{R}-\overset{O}{\overset{||}{\text{C}}}-\text{OH} + \text{HCl}$$

2. Conversion to anhydrides (discussed in Section 18.6).

$$\text{R}-\overset{O}{\overset{||}{\text{C}}}-\text{Cl} + \text{R}-\overset{O}{\overset{||}{\text{C}}}-\text{O}^- \longrightarrow \text{R}-\overset{O}{\overset{||}{\text{C}}}-\text{O}-\overset{O}{\overset{||}{\text{C}}}-\text{R} + \text{Cl}^-$$

3. Conversion to esters (discussed in Section 18.7).

$$\text{R}-\overset{O}{\overset{||}{\text{C}}}-\text{Cl} + \text{R}'-\text{OH} \xrightarrow{\text{pyridine}} \text{R}-\overset{O}{\overset{||}{\text{C}}}-\text{OR}' + \text{Cl}^- + \text{pyr-H}^+$$

4. Conversion to amides (discussed in Section 18.8).

$$\text{R}-\overset{O}{\overset{||}{\text{C}}}-\text{Cl} + \text{NH}_3(\text{excess}) \longrightarrow \text{R}-\overset{O}{\overset{||}{\text{C}}}-\text{NH}_2 + \text{NH}_4\text{Cl}$$

$$\text{R}-\overset{O}{\overset{||}{\text{C}}}-\text{Cl} + \text{R}'\text{NH}_2(\text{excess}) \longrightarrow \text{R}-\overset{O}{\overset{||}{\text{C}}}-\text{NHR}' + \text{R}'\text{NH}_3\text{Cl}$$

$$\text{R}-\overset{O}{\overset{||}{\text{C}}}-\text{Cl} + \text{R}'_2\text{NH}(\text{excess}) \longrightarrow \text{R}-\overset{O}{\overset{||}{\text{C}}}-\text{NR}'_2 + \text{R}'_2\text{NH}_2\text{Cl}$$

5. Conversion to ketones

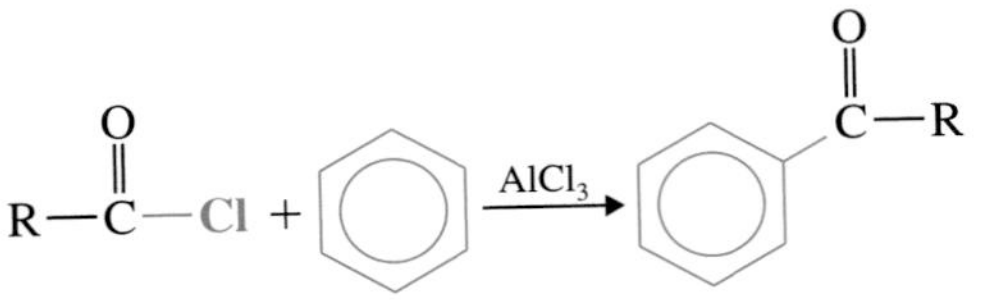

(discussed in Sections 15.7–15.9)

(discussed in Sections 16.5)

6. Conversion to aldehydes (discussed in Section 16.4).

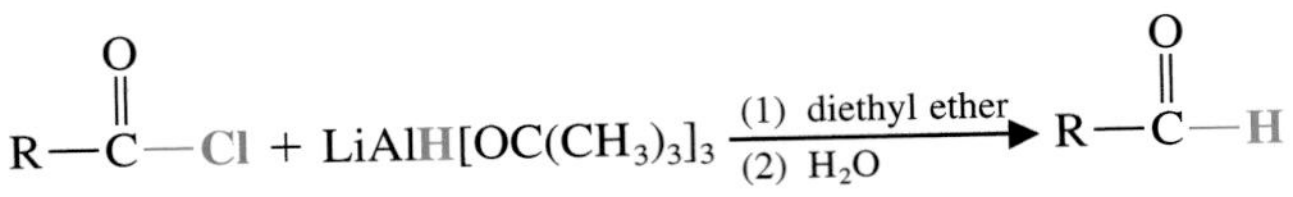

18.13C Reactions of Acid Anhydrides

1. Conversion to acids (cf. Section 18.12).

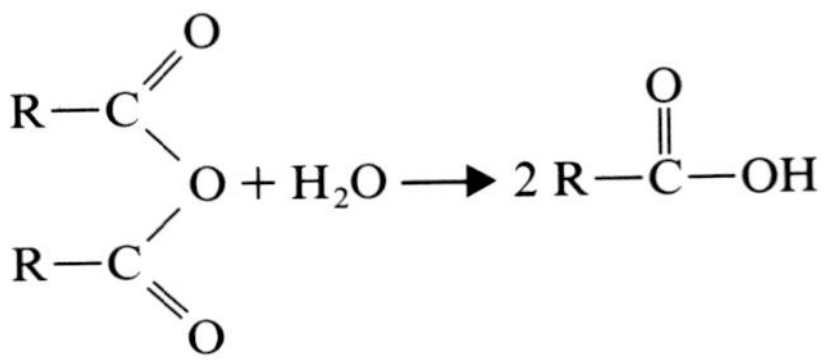

2. Conversion to esters (discussed in Sections 18.4 and 18.7).

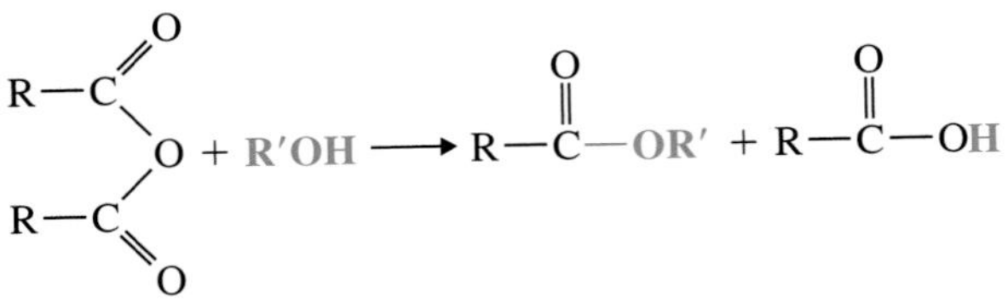

3. Conversion to amides and imides (discussed in Section 18.8).

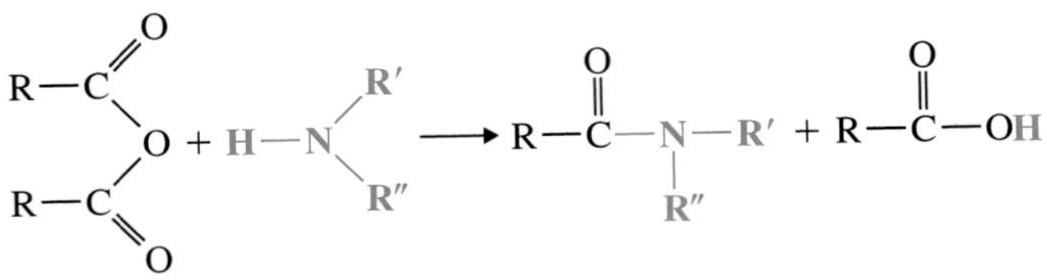

R′ and/or R″ may be H

or

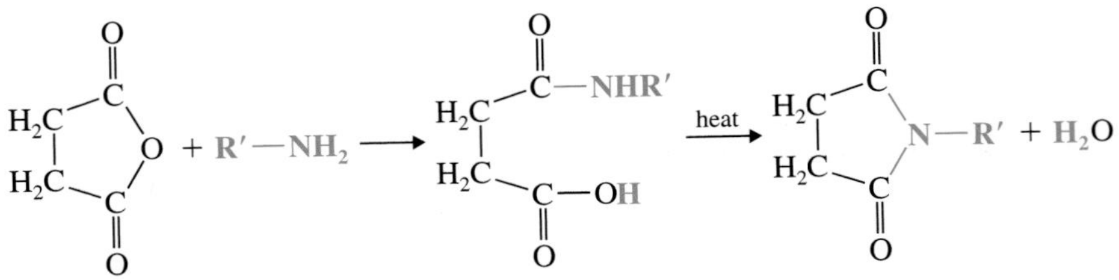

R′ may be H

4. Conversion to ketones (discussed in Sections 15.7–15.9).

$$R{-}\underset{O}{\overset{O}{C}}{-}O{-}\underset{O}{\overset{O}{C}}{-}R \ + \ \bigcirc \ \xrightarrow{\text{AlCl}_3} \ \bigcirc{-}\underset{O}{\overset{O}{C}}{-}R \ + \ R{-}\underset{O}{\overset{O}{C}}{-}OH$$

18.13D Reactions of Esters

1. Hydrolysis (discussed in Section 18.7).

$$R{-}\overset{O}{\overset{\|}{C}}{-}O{-}R' + H_2O \underset{}{\overset{H^+}{\rightleftharpoons}} R{-}\overset{O}{\overset{\|}{C}}{-}OH + R'{-}OH$$

$$R{-}\overset{O}{\overset{\|}{C}}{-}O{-}R' + OH^- \xrightarrow[H_2O]{} R{-}\overset{O}{\overset{\|}{C}}{-}O^- + R'{-}OH$$

2. Conversion to other esters: transesterification (discussed in Problem 18.11).

$$R{-}\overset{O}{\overset{\|}{C}}{-}O{-}R' + R''{-}OH \underset{}{\overset{H^+}{\rightleftharpoons}} R{-}\overset{O}{\overset{\|}{C}}{-}O{-}R'' + R'{-}OH$$

3. Conversion to amides (discussed in Section 18.8).

$$R{-}\overset{O}{\overset{\|}{C}}{-}OR' + HN\overset{R''}{\underset{R'''}{\diagup}} \ \longrightarrow \ R{-}\overset{O}{\overset{\|}{C}}{-}N\overset{R''}{\underset{R'''}{\diagup}} \ + \ R'{-}OH$$

R″ and/or R‴ may be H

4. Reaction with Grignard reagents (discussed in Section 11.8).

$$R{-}\overset{O}{\overset{\|}{C}}{-}OR' + 2\,R''MgX \xrightarrow{\text{diethyl ether}} R{-}\underset{R''}{\overset{OMgX}{\underset{|}{\overset{|}{C}}}}{-}R'' + R'OMgX$$

$$\downarrow H^+$$

$$R{-}\underset{R''}{\overset{OH}{\underset{|}{\overset{|}{C}}}}{-}R''$$

5. Reduction (discussed in Section 11.3).

$$R{-}\overset{O}{\overset{\|}{C}}{-}O{-}R' + H_2 \xrightarrow{\text{Ni}} R{-}CH_2OH + R'{-}OH$$

$$R-\overset{\overset{\displaystyle O}{\|}}{C}-O-R' + LiAlH_4 \xrightarrow[\text{(2)}\quad H_2O]{\text{(1)}\quad \text{diethyl ether}} R-CH_2OH + R'-OH$$

$$R-\overset{\overset{\displaystyle O}{\|}}{C}-O-R' + Na \xrightarrow[\text{(2)}\quad H^+]{\text{(1)}\quad C_2H_5OH} R-CH_2OH + R'-OH$$

18.13E Reactions of Amides

1. Hydrolysis (discussed in Section 18.8).

$$R-\overset{\overset{\displaystyle O}{\|}}{C}-\underset{\underset{\displaystyle R''}{|}}{N}R' + H_3O^+ \xrightarrow{H_2O} R-\overset{\overset{\displaystyle O}{\|}}{C}-OH + R'-\overset{+}{\underset{\underset{\displaystyle R''}{|}}{N}}H_2$$

$$R-\overset{\overset{\displaystyle O}{\|}}{C}-\underset{\underset{\displaystyle R''}{|}}{N}R' + OH^- \xrightarrow{H_2O} R\overset{\overset{\displaystyle O}{\|}}{C}-O^- + R'-\underset{\underset{\displaystyle R''}{|}}{N}H$$

R, R', and/or R″ may be H

2. Conversion to nitriles: dehydration (discussed in Section 18.8).

$$R-\overset{\overset{\displaystyle O}{\|}}{C}NH_2 \xrightarrow[\text{heat}]{P_4O_{10}} R-C\equiv N$$
$$(-H_2O)$$

3. Conversion to imides (discussed in Section 18.8).

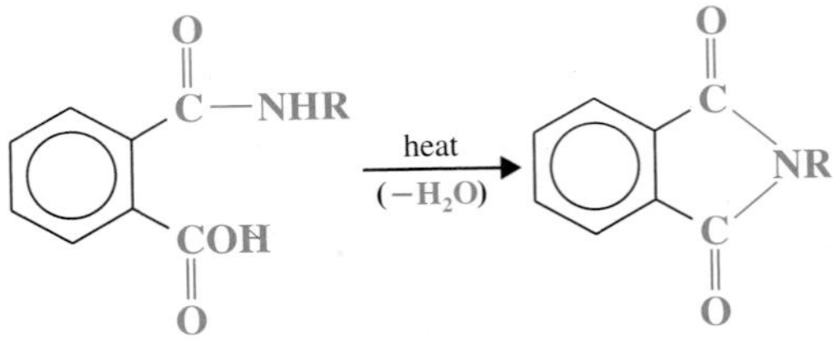

ADDITIONAL PROBLEMS

✲18.19   Write a structural formula for each of the following compounds:

(a) Hexanoic acid

(b) Hexanamide

(c) *N*-Ethylhexanamide

(d) *N,N*-Diethylhexanamide

(e) 3-Hexenoic acid

(f) 2-Methyl-4-hexenoic acid

(g) Hexanedioic acid

(h) Phthalic acid

(i) Isophthalic acid

(j) Terephthalic acid

(k) Diethyl oxalate

(l) Diethyl adipate

(m) Isobutyl propanoate

(n) 2-Naphthoic acid

(o) Maleic acid

(p) 2-Hydroxybutanedioic acid (malic acid)

(q) Fumaric acid

(r) Succinic acid

(s) Succinimide

(t) Malonic acid

(u) Diethyl malonate

*18.20   Give an IUPAC systematic or common name for each of the following compounds:

(a) $C_6H_5CO_2H$

(b) C_6H_5COCl

(c) $C_6H_5CONH_2$

(d) $(C_6H_5CO)_2O$

(e) $C_6H_5CO_2CH_2C_6H_5$

(f) $C_6H_5CO_2C_6H_5$

(g) $CH_3CO_2CH(CH_3)_2$

(h) $CH_3CON(CH_3)_2$

(i) CH_3CN

*18.21   Show how p-chlorotoluene could be converted to each of the following.

(a) p-Chlorobenzoic acid

(b) p-Chlorophenylacetic acid

(c) $p\text{-}ClC_6H_4CH(OH)CO_2H$

(d) $p\text{-}ClC_6H_4CH{=}CHCO_2H$

*18.22   Outline each of the following syntheses.

(a)   Malonic acid from acetic acid

(b)   Succinic acid from 1,4-butanediol

(c)   Adipic acid from cyclohexanol

*18.23   Show how pentanoic acid can be prepared from each of the following:

(a) 1-Pentanol

(b) 1-Bromobutane (two ways)

(c) 5-Decene

(d) Pentanal

18.24   What major organic product would you expect to obtain when acetyl chloride reacts with each of the following?

(a) H_2O

(b) $AgNO_3/H_2O$

(c) $CH_3(CH_2)_2CH_2OH$ and pyridine

(d) NH_3 (excess)

(e) $C_6H_5CH_3$ and $AlCl_3$

(f) $LiAlH[OC(CH_3)_3]_3$

(g) $(CH_3)_2CuLi$

(h) $NaOH/H_2O$

* (i) CH_3NH_2 (excess)

(j) $C_6H_5NH_2$ (excess)

(k) $(CH_3)_2NH$ (excess)

(l) CH_3CH_2OH and pyridine

(m) $CH_3CO_2^-Na^+$

(n) CH_3CO_2H and pyridine

(o) Phenol and pyridine

(p) NBS, HBr, and $SOCl_2$

18.25   What major organic product would you expect to obtain when acetic anhydride reacts with each of the following?

(a) NH_3 (excess)

(b) H_2O

(c) $CH_3CH_2CH_2OH$

(d) $C_6H_6 + AlCl_3$

(e) $CH_3CH_2NH_2$ (excess)

(f) $(CH_3CH_2)_2NH$ (excess)

18.26   What major organic product would you expect to obtain when succinic anhydride reacts with each of the reagents given in Problem 18.25?

18.27   Starting with benzene and succinic anhydride, and using any other needed reagents, outline a synthesis of 1-phenylnapthalene.

18.28 Starting with either *cis*- or *trans*-$HO_2C-CH=CH-CO_2H$ (i.e., either maleic or fumaric acid), and using any other needed compounds, outline syntheses of each of the following:

(a) (b) (c)

18.29 What products would you expect to obtain when ethyl propanoate reacts with each of the following?

(a) H_3O^+, H_2O (c) 1-Octanol, HCl   (e) $LiAlH_4$, then H_2O

(b) OH^-, H_2O (d) CH_3NH_2 (f) C_6H_5MgBr, then H_2O

18.30 What products would you expect to obtain when propanamide reacts with each of the following?

(a) H_3O^+, H_2O (b) OH^-, H_2O (c) P_4O_{10} and heat

18.31 Outline a simple chemical test that would serve to distinguish between

(a) Benzoic acid and methyl benzoate   (e) Ethyl benzoate and benzamide

(b) Benzoic acid and benzoyl chloride   (f) Benzoic acid and cinnamic acid

(c) Benzoic acid and benzamide   (g) Ethyl benzoate and benzoyl chloride

(d) Benzoic acid and 4-methylphenol   (h) 2-Chlorobutanoic acid and butanoic acid

18.32 What products would you expect to obtain when each of the following compounds is heated?

(a) 4-Hydroxybutanoic acid

(f)

(b) 3-Hydroxybutanoic acid

(c) 2-Hydroxybutanoic acid

(d) Glutaric acid

(e) $CH_3CHCH_2CH_2CH_2CO^-$
   |
   NH_3^+

18.33 Give stereochemical formulas for compounds **A–Q**.

(a) (R)-(−)-2-Butanol $\xrightarrow[\text{pyridine}]{\text{TsCl}}$ **A** $\xrightarrow{CN^-}$ **B** (C_5H_9N) $\xrightarrow[\text{H}_2\text{O}]{\text{H}_2\text{SO}_4}$

(+)-**C** $(C_5H_{10}O_2)\xrightarrow[\text{(2) H}_2\text{O}]{\text{(1) LiAlH}_4}$ (−)-**D** $(C_5H_{12}O)$

(b) (R)-(−)-2-Butanol $\xrightarrow[\text{pyridine}]{\text{PBr}_3}$ **E** (C_4H_9Br) $\xrightarrow{CN^-}$ **F** (C_5H_9N) $\xrightarrow[\text{H}_2\text{O}]{\text{H}_2\text{SO}_4}$

(−)-**C** $(C_5H_{10}O_2)$ $\xrightarrow[\text{(2) H}_2\text{O}]{\text{(1) LiAlH}_4}$ (+)-**D** $(C_5H_{12}O)$

(c) $A \xrightarrow{CH_3CO_2^-} G (C_6H_{12}O_2) \xrightarrow{OH^-} (+)\text{-}H (C_4H_{10}O) + CH_3CO_2^-$

(d) $(-)\text{-}D \xrightarrow{PBr_3} J (C_5H_{11}Br) \xrightarrow[\text{diethyl ether}]{Mg} K (C_5H_{11}MgBr) \xrightarrow[(2) \ H^+]{(1) \ CO_2} L (C_6H_{12}O_2)$

(e) $(R)\text{-}(+)\text{-Glyceraldehyde} \xrightarrow{HCN} M (C_4H_7NO_3) + N (C_4H_7NO_3)$

$$\underbrace{}$$

**Diastereomers, separated
by fractional crystallization**

(f) $M \xrightarrow[H_2O]{H_2SO_4} P (C_4H_8O_5) \xrightarrow[HNO_3]{[O]} meso\text{-tartaric acid}$

(g) $N \xrightarrow[H_2O]{H_2SO_4} Q (C_4H_8O_5) \xrightarrow[HNO_3]{[O]} (-)\text{-tartaric acid}$

18.34   (a) $(\pm)$-Pantetheine and $(\pm)$-pantothenic acid, important intermediates in the synthesis of coenzyme A, were prepared by the following route. Give structures for compounds **A–D**.

$$\underset{\displaystyle \overset{\displaystyle CH_3}{|}}{CH_3CHCHO} + \overset{O}{\overset{\|}{HCH}} \xrightarrow[H_2O]{K_2CO_3} A (C_5H_{10}O_2) \xrightarrow{HCN}$$

$$(\pm)\text{-}B (C_6H_{11}NO_2) \xrightarrow{H_3O^+} [(\pm)\text{-}C (C_6H_{12}O_4)] \xrightarrow{-H_2O}$$

$$(\pm)\text{-}D (C_6H_{10}O_3) \xrightarrow{H_3\overset{+}{N}CH_2CH_2CO^-} (CH_3)_2\overset{\displaystyle CH_2OH}{\underset{\displaystyle OH}{C}}\!\!-\!\!\overset{O}{\underset{}{CHC}}\!\!-\!\!\overset{O}{\overset{\|}{NHCH_2CH_2COH}}$$

γ-lactone from A   **$(\pm)$-Pantothenic acid**

$$\downarrow \quad H_2NCH_2CH_2\overset{O}{\overset{\|}{C}}NHCH_2CH_2SH$$

$$(CH_3)_2\overset{\displaystyle CH_2OH}{\underset{\displaystyle OH}{C}}\!\!-\!\!CH\!\!-\!\!\overset{O}{\overset{\|}{C}}\!\!-\!\!NHCH_2CH_2\overset{O}{\overset{\|}{C}}NHCH_2CH_2SH$$

$(\pm)$-Pantetheine

(b) The γ-lactone, $(\pm)$ **D**, can be resolved. If the $(-)$-γ-lactone is used in the last step, the pantetheine that is obtained is identical with that obtained naturally. The $(-)$-γ-lactone has the (R) configuration. What is the stereochemistry of naturally occurring pantetheine?   (c) What products would you expect to obtain when $(\pm)$-pantetheine is heated with aqueous sodium hydroxide?

18.35   The IR and ^1H NMR spectra of phenacetin ($C_{10}H_{13}NO_2$) are given in Fig. 18.5. Phenacetin is an analgesic and antipyretic compound and was the P of A-P-C tablets (**a**spirin–**p**henacetin–**c**affeine). (Because of its toxicity phenacetin is no longer used medically.) When phenacetin is heated with aqueous sodium hydroxide, it yields phenetidine ($C_8H_{11}NO$) and sodium acetate. Propose structures for phenacetin and phenetidine.

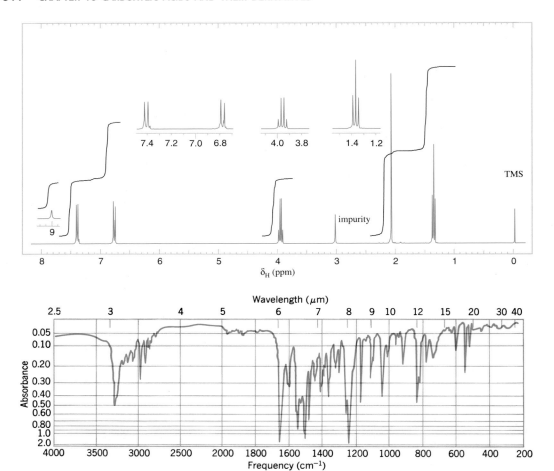

FIGURE 18.5 The 300 MHz ^{1}H NMR and IR spectra of phenacetin. Expansions of the ^{1}H NMR signals are shown in the offset plots. (Infrared spectrum, courtesy Sadtler Research Laboratories, Philadelphia.)

✕ 18.36 Given here are the ^{1}H NMR spectra and carbonyl absorption peaks of five acyl compounds. Propose structures for each.

(a) $C_8H_{14}O_4$

^{1}H NMR spectrum
Triplet δ 1.2 (6H)
Singlet δ 2.5 (4H)
Quartet δ 4.1 (4H)

IR spectrum
1740 cm^{-1}

(b) $C_{11}H_{14}O_2$

^{1}H NMR spectrum
Doublet δ 1.0 (6H)
Multiplet δ 2.1 (1H)
Doublet δ 4.1 (2H)
Multiplet δ 7.8 (5H)

IR spectrum
1720 cm^{-1}

(c) $C_{10}H_{12}O_2$

^{1}H NMR spectrum
Triplet δ 1.2 (3H)
Singlet δ 3.5 (2H)
Quartet δ 4.1 (2H)
Multiplet δ 7.3 (5H)

IR spectrum
1740 cm^{-1}

(d) $C_2H_2Cl_2O_2$ ¹H NMR spectrum

Singlet δ 6.0

Singlet δ 11.70

IR spectrum

Broad peak 2500–2700 cm⁻¹

1705 cm⁻¹

(e) $C_4H_7ClO_2$ ¹H NMR spectrum

Triplet δ 1.3

Singlet δ 4.0

Quartet δ 4.2

IR spectrum

1745 cm⁻¹

18.37 The active ingredient of the insect repellent "Off" is N,N-diethyl-m-toluamide, m-CH₃C₆H₄CON(CH₂CH₃)₂. Outline a synthesis of this compound starting with m-toluic acid.

18.38 Amides are weaker bases than corresponding amines. For example, most water-insoluble amines (RNH₂) will dissolve in dilute aqueous acids (e.g., aqueous HCl, H₂SO₄, etc.) by forming water-soluble alkylammonium salts (RNH₃⁺X⁻). Corresponding amides (RCONH₂) *do not dissolve in dilute aqueous acids,* however. Propose an explanation for the much lower basicity of amides when compared to amines.

18.39 While amides are much less basic than amines, they are much stronger acids. Amides have pK_a values in the range 14–16, whereas for amines, pK_a = 33–35. (a) What factor accounts for the much greater acidity of amides? (b) *Imides,* that is, compounds with the structure (RC)₂NH, with the carbonyl O above, are even stronger acids than amides. For imides, pK_a = 9–10 and, as a consequence, water-insoluble imides dissolve in aqueous NaOH by forming soluble sodium salts. What extra factor accounts for the greater acidity of imides?

18.40 Compound **X** ($C_7H_{12}O_4$) is insoluble in aqueous sodium bicarbonate. The IR spectrum of **X** has a strong absorption peak near 1740 cm⁻¹, and its broad band proton-decoupled ¹³C spectrum is given in Fig. 18.6. Propose a structure for **X**.

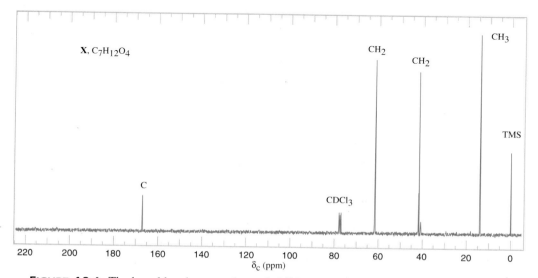

FIGURE 18.6 The broad band proton-decoupled ¹³C NMR spectrum of compound **X**, Problem 18.40. Information from the DEPT ¹³C NMR spectrum is given above each peak.

18.41   Alkylthiyl acetates $\left(\begin{array}{c} O \\ \parallel \\ CH_3CSCH_2CH_2R \end{array}\right)$ can be prepared by a peroxide-initiated

reaction between thiolacetic acid $\left(\begin{array}{c} O \\ \parallel \\ CH_3CSH \end{array}\right)$ and an alkene (CH_2=CHR).   (a) Outline a reasonable mechanism for this reaction.   (b) Show how you might use this reaction in a synthesis of 3-methyl-2-butanethiol from 2-methyl-2-butene.

18.42   On heating, *cis*-4-hydroxycyclohexanecarboxylic acid forms a lactone but *trans*-4-hydroxycyclohexanecarboxylic acid does not. Explain.

18.43   (R)-(+)-Glyceraldehyde can be transformed into (+)-malic acid by the following synthetic route. Give stereochemical structures for the products of each step.

(R)-(+)-Glyceraldehyde $\xrightarrow[\text{oxidation}]{Br_2, H_2O}$ (−)-glyceric acid $\xrightarrow{PBr_3}$

(−)-3-bromo-2-hydroxypropanoic acid $\xrightarrow{NaCN}$ $C_4H_5NO_3$ $\xrightarrow[\text{heat}]{H_3O^+}$ (+)-malic acid

18.44   (R)-(+)-Glyceraldehyde can also be transformed into (−)-malic acid. This synthesis begins with the conversion of (R)-(+)-glyceraldehyde into (−)-tartaric acid, as shown in Problem 18.33, parts (e) and (g). Then (−)-tartaric acid is allowed to react with phosphorus tribromide in order to replace one alcoholic —OH group with —Br. This step takes place with inversion of configuration at the carbon that undergoes attack. Treating the product of this reaction with zinc and acid produces (−)-malic acid.   (a) Outline all steps in this synthesis by writing stereochemical structures for each intermediate.   (b) The step in which (−)-tartaric acid is treated with phosphorus tribromide produces only one stereoisomer, even though there are two replaceable —OH groups. How is this possible?   (c) Suppose that the step in which (−)-tartaric acid is treated with phosphorus tribromide had taken place with "mixed" stereochemistry, that is, with both inversion and retention at the carbon under attack. How many stereoisomers would have been produced?   (d) What difference would this have made to the overall outcome of the synthesis?

✳ 18.45   Cantharidin is a powerful vesicant that can be isolated from dried beetles (*Cantharis vesicatoria* or "Spanish fly"). Outlined here is the stereospecific synthesis of cantharidin reported by Gilbert Stork of Columbia University in 1953. Supply the missing reagents (a)−(n).

HIO₄
(cf. Section 22.6D)

(i)

(j)

(k)

(l)

Heat
(−CH₃CO₂H)

(m)

(n)

Cantharidin

18.46 Examine the structure of cantharidin (Problem 18.45) carefully and (a) suggest a possible two-step synthesis of cantharidin starting with furan (Section 13.12).   (b) F. von Bruchhausen and H. W. Bersch at the University of Münster attempted this two-step synthesis in 1928, only a few months after Diels and Alder published their first paper describing their new diene addition, and found that the expected addition failed to take place. von Bruchhausen and Bersch also found that although cantharidin is stable at relatively high temperatures, heating cantharidin with a palladium catalyst causes cantharidin to decompose. They identified furan and dimethylmaleic anhydride among the decomposition products. What has happened in the decomposition and what does this suggest about why the first step of their attempted synthesis failed?

18.47 Compound **Y** ($C_8H_4O_3$) dissolves slowly when warmed with aqueous sodium bicarbonate. The IR spectrum of **Y** has strong peaks at 1779 and at 1854 cm^{-1}. The broad band proton-decoupled ^{13}C spectrum of **Y** is given in Fig. 18.7 (p. 848). Acidification of the bicarbonate solution of **Y** gave compound **Z**. The proton-decoupled ^{13}C NMR spectrum of **Z** showed four signals. When **Y** was warmed in ethanol, a compound **AA** was produced. The ^{13}C NMR spectrum of **AA** displayed 10 signals. Propose structures for **Y**, **Z**, and **AA**.

18.48 Ketene, $H_2C{=}C{=}O$, an important industrial chemical, can be prepared by dehydration of acetic acid at a high temperature or by the pyrolysis of acetone. Predict the products that would be formed when ketene reacts with   (a) ethanol,   (b) acetic acid,   (c) ethylamine. (Hint: Markovnikov addition occurs.)

18.49 The two unsymmetrical anhydrides react with ethylamine as follows:

$$HCOCCH_3 + CH_3CH_2NH_2 \longrightarrow CH_3CH_2NHCHO + CH_3CO_2^- \ CH_3CH_2NH_3^+$$

$$CF_3COCCH_3 + CH_3CH_2NH_2 \longrightarrow CH_3CH_2NHCCH_3 + CF_3CO_2^- \ CH_3CH_2NH_3^+$$

Explain the factors that might account for the formation of the products in each reaction.

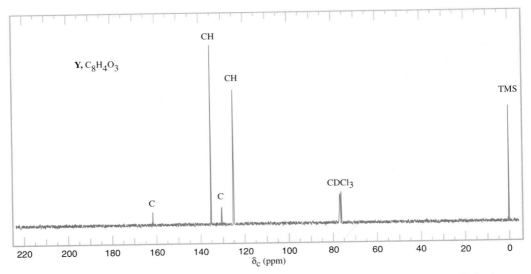

FIGURE 18.7 The broad band proton-decoupled ^{13}C NMR spectrum of compound **Y**, Problem 18.47. Information from the DEPT ^{13}C NMR spectrum is given above each peak.

18.50   Starting with 1-naphthol, suggest an alternative synthesis of the insecticide Sevin to the one given in Section 18.10A.

18.51   Suggest a synthesis of ibuprofen (Section 5.10) from benzene, employing **chloromethylation** as one step. Chloromethylation is a special case of the Friedel–Crafts reaction in which a mixture of HCHO and HCl, in the presence of $ZnCl_2$, introduces a $-CH_2Cl$ group into an aromatic ring.

18.52   An alternative synthesis of ibuprofen is given below. Supply the structural formulas for compounds **A–D**.

18.53   As a method for the synthesis of cinnamaldehyde (3-phenyl-2-propenal) a chemist treated 3-phenyl-2-propen-1-ol with $K_2Cr_2O_7$ in sulfuric acid. The product obtained from the reaction gave a signal at δ 164.5 in its ^{13}C NMR spectrum. Alternatively, when the chemist treated 3-phenyl-2-propen-1-ol with PCC in CH_2Cl_2, the ^{13}C NMR spectrum of the product displayed a signal at δ 193.8. (All other signals in the spectra of both compounds appeared at similar chemical shifts.)   (a) Which reaction produced cinnamaldehyde?   (b) What was the other product?

Nylon-6,6

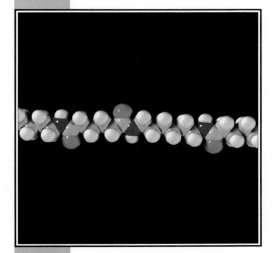

STEP-GROWTH POLYMERS

We saw, in Special Topic C, that large molecules with many repeating subunits—called *polymers*—can be prepared by addition reactions of alkenes. These polymers, we noted, are called *chain-growth polymers* or *addition polymers*.

Another broad group of polymers has been called *condensation polymers*, but are now more often called *step-growth polymers*. These polymers, as their older name suggests, are prepared by condensation reactions—reactions in which monomeric subunits are joined through intermolecular eliminations of small molecules such as water or alcohols. Among the most important condensation polymers are *polyamides, polyesters, polyurethanes,* and *formaldehyde resins.*

I.1 Polyamides

Silk and wool are two naturally occurring polymers that humans have used for centuries to fabricate articles of clothing. They are examples of a family of compounds that are called *proteins*—a group of compounds that we shall discuss in detail in Chapter 24. At this point we need only to notice (below) that the repeating subunits of proteins are derived from α-amino acids and that these subunits are joined by amide linkages. Proteins, therefore, are polyamides.

$$H_2N-CH-\overset{\overset{\displaystyle O}{\|}}{C}-OH$$

$$|$$
$$R$$

An α-amino acid

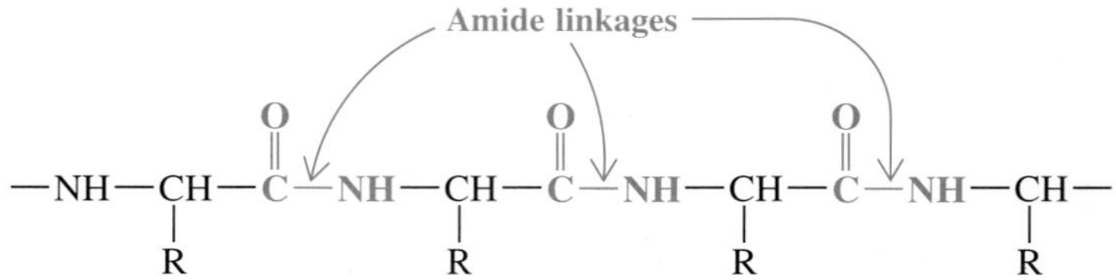

**A portion of polyamide chain as
it might occur in a protein**

The search for a synthetic material with properties similar to those of silk led to the discovery of a family of synthetic polyamides called nylons.

One of the most important nylons, called *nylon 6,6,* can be prepared from the six-carbon dicarboxylic acid, adipic acid, and the six-carbon diamine, hexamethyl-enediamine. In the commercial process these two compounds are allowed to react in equimolar proportions in order to produce a 1 : 1 salt,

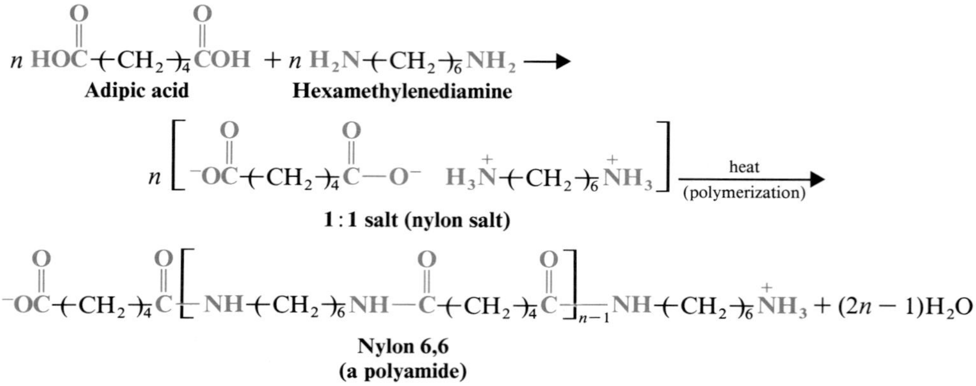

1 : 1 salt (nylon salt)

**Nylon 6,6
(a polyamide)**

Then, heating the 1 : 1 salt (nylon salt) to a temperature of 270°C at a pressure of 250 psi (pounds per square inch) causes a polymerization to take place. Water molecules are lost as condensation reactions occur between $-\overset{O}{\overset{\|}{C}}-O^-$ and $-NH_3^+$ groups of the salt to give the polyamide.

The nylon 6,6 produced in this way has a molecular weight of about 10,000, has a melting point of about 250°C, and when molten can be spun into fibers from a melt. The fibers are then stretched to about four times their original length. This orients the linear polyamide molecules so that they are parallel to the fiber axis and allows hydrogen bonds to form between $-NH-$ and $C=O$ groups on adjacent chains. Called "cold drawing," stretching greatly increases the fibers' strength.

Another type of nylon, nylon 6, can be prepared by a ring-opening polymerization of ε-caprolactam:

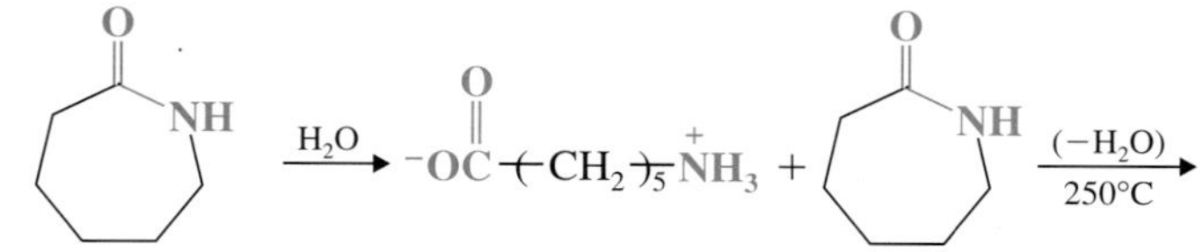

**ε-Caprolactam
(a cyclic amide)**

$$-NH-\left[\overset{\overset{\displaystyle O}{\|}}{C}(CH_2)_5NH-\overset{\overset{\displaystyle O}{\|}}{C}+CH_2)_5NH\right]_n\overset{\overset{\displaystyle O}{\|}}{C}-$$

Nylon 6

In this process ε-caprolactam is allowed to react with water, converting some of it to ε-aminocaproic acid. Then, heating this mixture at 250°C drives off water as ε-caprolactam and ε-aminocaproic acid react to produce the polyamide. Nylon 6 can also be converted into fibers by melt spinning.

PROBLEM I.1

The raw materials for the production of nylon 6,6 can be obtained in several ways, as indicated below. Give equations for each synthesis of adipic acid and of hexamethylenediamine.

(a) Cyclohexanone $\xrightarrow{[O]}$ adipic acid

(b) Adipic acid $\xrightarrow{\text{2 NH}_3}$ a salt $\xrightarrow{\text{heat}}$ $C_6H_{12}N_2O_2 \xrightarrow[\text{catalyst}]{350°C}$

$C_6H_8N_2 \xrightarrow[\text{catalyst}]{\text{4 H}_2}$ hexamethylenediamine

(c) 1,3-Butadiene $\xrightarrow{\text{Cl}_2}$ $C_4H_6Cl_2 \xrightarrow{\text{2 NaCN}} C_6H_6N_2 \xrightarrow[\text{Ni}]{\text{H}_2}$

$C_6H_8N_2 \xrightarrow[\text{catalyst}]{\text{4 H}_2}$ hexamethylenediamine

(d) Tetrahydrofuran $\xrightarrow{\text{2 HCl}}$ $C_4H_8Cl_2 \xrightarrow{\text{2 NaCN}}$

$C_6H_8N_2 \xrightarrow[\text{catalyst}]{\text{4 H}_2}$ hexamethylenediamine

I.2 Polyesters

One of the most important polyesters is poly(ethylene terephthalate), a polymer that is marketed under the names *Dacron, Terylene,* and *Mylar.*

$$-OCH_2CH_2-O\left[\overset{\overset{\displaystyle O}{\|}}{C}-\bigcirc-\overset{\overset{\displaystyle O}{\|}}{C}-OCH_2CH_2-O\right]_n\overset{\overset{\displaystyle O}{\|}}{C}-\bigcirc-\overset{\overset{\displaystyle O}{\|}}{C}-$$

Poly(ethylene terephthalate)
(Dacron, Terylene, or Mylar)

One can obtain poly(ethylene terephthalate) by a direct acid-catalyzed esterification of ethylene glycol and terephthalic acid.

$$HO-CH_2CH_2-OH + HO-\overset{\overset{\displaystyle O}{\|}}{C}-\bigcirc-\overset{\overset{\displaystyle O}{\|}}{C}-OH \overset{\text{H}^+}{\underset{\text{heat}}{\Big\downarrow}}$$

Ethylene glycol   **Terephthalic acid**

Poly(ethylene terephthalate) + H_2O

Another method for synthesizing poly(ethylene terephthalate) is based on trans-esterification reactions—reactions in which one ester is converted into another. One commercial synthesis utilizes two transesterifications. In the first, dimethyl terephthalate and excess ethylene glycol are heated to 200°C in the presence of a basic catalyst. Distillation of the mixture results in the loss of methanol (bp, 64.7°C) and the formation of a new ester, one formed from 2 mol of ethylene glycol and 1 mol of terephthalic acid. When this new ester is heated to a higher temperature (~280°C), ethylene glycol (bp, 198°C) distills and polymerization (the second transesterification) takes place.

Dimethyl terephthalate   **Ethylene glycol**

Poly(ethylene terephthalate)

The poly(ethylene terephthalate) thus produced melts at about 270°C. It can be melt-spun into fibers to produce Dacron or Terylene; it can also be made into a film, in which form it is marketed as Mylar.

PROBLEM I.2

Transesterifications are catalyzed by either acids or bases. Using the transesterification reaction that takes place when dimethyl terephthalate is heated with ethylene glycol as an example, outline reasonable mechanisms for   (a) the base-catalyzed reaction and   (b) the acid-catalyzed reaction.

PROBLEM I.3

Kodel is another polyester that enjoys wide commercial use.

Kodel

Kodel is also produced by a transesterification.   (a) What methyl ester and what alcohol are required for the synthesis of Kodel?   (b) The alcohol can be prepared from dimethyl terephthalate. How might this be done?

PROBLEM I.4

Heating phthalic anhydride and glycerol together yields a polyester called a glyptal resin. A glyptal resin is especially rigid because the polymer chains are "cross-linked." Write a portion of the structure of a glyptal resin and show how cross-linking occurs.

PROBLEM I.5

Lexan, a high molecular weight "polycarbonate," is manufactured by mixing bisphenol A with phosgene in the presence of pyridine. Suggest a structure for Lexan.

$$HO—\underset{\textstyle\text{Bisphenol A}}{\bigcirc—\overset{\displaystyle CH_3}{\underset{\displaystyle CH_3}{C}}—\bigcirc}—OH \qquad \underset{\textstyle\text{Phosgene}}{Cl—\overset{\displaystyle O}{C}—Cl}$$

PROBLEM I.6

The familiar "epoxy resins" or "epoxy glues" usually consist of two components that are sometimes labeled "resin" and "hardener." The resin is manufactured by allowing bisphenol A (Problem I.5) to react with an excess of epichlorohydrin, $H_2C—CHCH_2Cl$, in the presence of a base until a low molec-
         \/
          O
ular weight polymer is obtained.   (a) What is a likely structure for this polymer and   (b) what is the purpose of using an excess of epichlorohydrin? The hardener is usually an amine such as $H_2NCH_2CH_2NHCH_2CH_2NH_2$.   (c) What reaction takes place when the resin and hardener are mixed?

I.3 Polyurethanes

A *urethane* is the product formed when an alcohol reacts with an isocyanate:

$$R—OH + O{=}C{=}N—R' \longrightarrow R—O—\overset{\displaystyle O}{\overset{\displaystyle \|}{C}}—\overset{\displaystyle H}{\overset{\displaystyle |}{N}}—R'$$

Alcohol   Isocyanate   A urethane
                                               (a carbamate)

The reaction takes place in the following way:

$$R—\overset{..}{O}H + \underset{\underset{R'}{\overset{|}{N}}}{\overset{\displaystyle O}{\overset{\displaystyle \|}{C}}} \longrightarrow R—\overset{+}{O}—\underset{\overset{|}{H}}{C{=}N—R'} \longrightarrow R—O—\overset{\displaystyle O}{\overset{\displaystyle \|}{C}}—\overset{\displaystyle H}{\overset{\displaystyle |}{N}}—R'$$

A urethane is also called a *carbamate* because formally it is an ester of an alcohol (ROH) and a carbamic acid ($R'NHCO_2H$).

Polyurethanes are usually made by allowing a *diol* to react with a *diisocyanate*. The diol is typically a polyester with —CH$_2$OH end groups. The diisocyanate is usually toluene 2,4-diisocyanate.*

HOCH$_2$—polymer—CH$_2$OH

O=C=N
N=C=O
CH$_3$

Toluene 2,4-diisocyanate

$$\left[\begin{array}{c} NH-\overset{\overset{\displaystyle O}{\|}}{C}-OCH_2-polymer-CH_2O-\overset{\overset{\displaystyle O}{\|}}{C}-NH \\ CH_3 \end{array} \right]_n$$

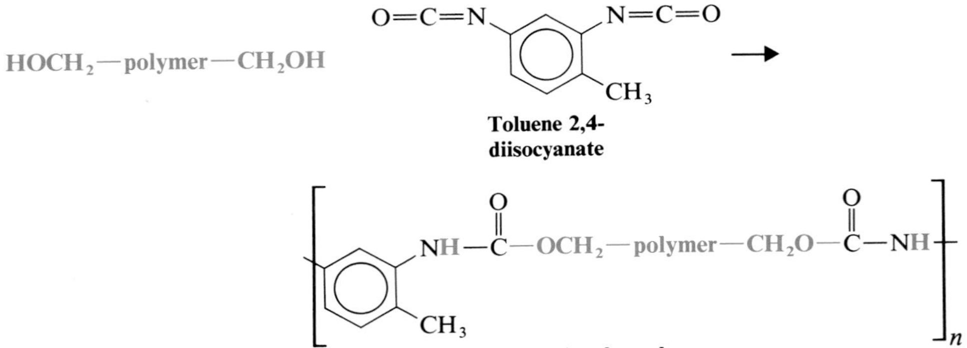

A polyurethane

PROBLEM I.7

A typical polyurethane can be made in the following way. Adipic acid is polymerized with an excess of ethylene glycol. The resulting polyester is then treated with toluene 2,4-diisocyanate. (a) Write the structure of the polyurethane. (b) Why is an excess of ethylene glycol used in making the polyester?

Polyurethane foams, as used in pillows and paddings, are made by adding small amounts of water to the reaction mixture during the polymerization with the diisocyanate. Some of the isocyanate groups react with water to produce carbon dioxide, and this gas acts as the foaming agent.

$$R-N=C=O + H_2O \longrightarrow R-NH_2 + CO_2\uparrow$$

I.4 Phenol–Formaldehyde Polymers

One of the first synthetic polymers to be produced was a polymer (or resin) known as *Bakelite*. Bakelite is made by a condensation reaction between phenol and formaldehyde; the reaction can be catalyzed by either acids or bases. The base-catalyzed reaction probably takes place in the general way shown here. Reaction can take place at the ortho and para positions of phenol.

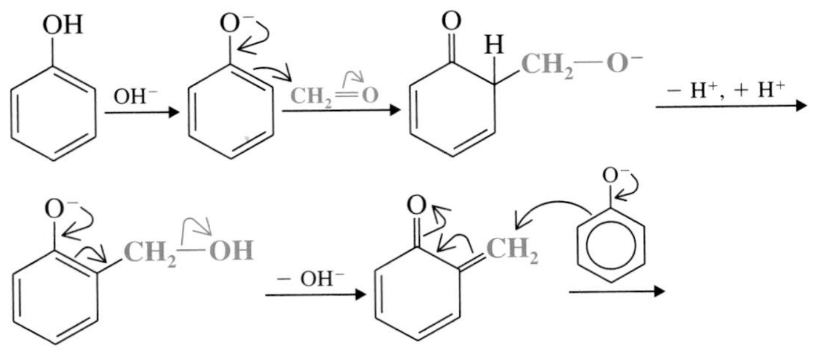

* Toluene 2,4-diisocyanate is a hazardous chemical that has caused acute respiratory problems among workers synthesizing polyurethanes.

Bakelite

Generally, the polymerization is carried out in two stages. The first polymerization produces a low molecular weight fusible (meltable) polymer called a *resole*. The resole can be molded to the desired shape, and then further polymerization produces a very high molecular weight polymer, which, because it is highly cross-linked, is infusible.

PROBLEM I.8

Using a para-substituted phenol such as *p*-cresol yields a phenol–formaldehyde polymer that is *thermoplastic* rather than *thermosetting*. That is, the polymer remains fusible; it does *not* become impossible to melt. What accounts for this?

PROBLEM I.9

Outline a general mechanism for acid-catalyzed polymerization of phenol and formaldehyde.

I.5 Cascade Polymers

One exciting development in polymer chemistry in the last 10 years has been the synthesis of high molecular weight, symmetrical, highly branched, polyfunctional molecules called **cascade polymers.** George R. Newkome (of the University of South

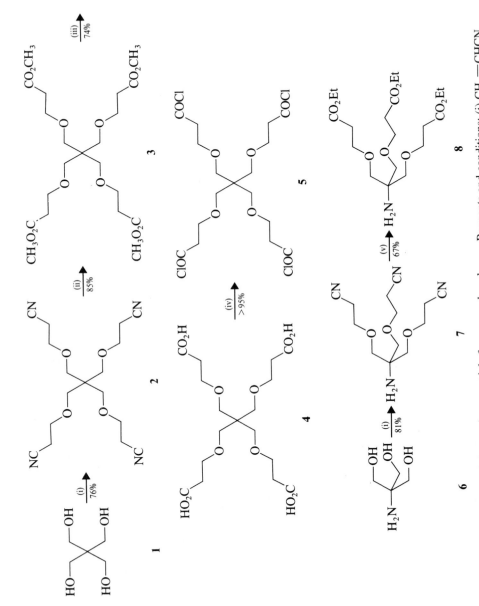

FIGURE I.1 Synthesis of the starting materials for a cascade polymer. Reagents and conditions: (i) CH$_2$=CHCN, KOH, p-dioxane, 25°C, 24 h; (ii) MeOH, dry HCl, reflux, 2 h; (iii) 3 N NaOH, 70°C, 24 h; (iv) SOCl$_2$, CH$_2$Cl$_2$, reflux, 1 h; (v) EtOH, dry HCl, reflux, 3 h. (Adapted from George R. Newkome and Xiaofeng Lin, *Macromolecules, 1991, 24,* 1443–1444.)

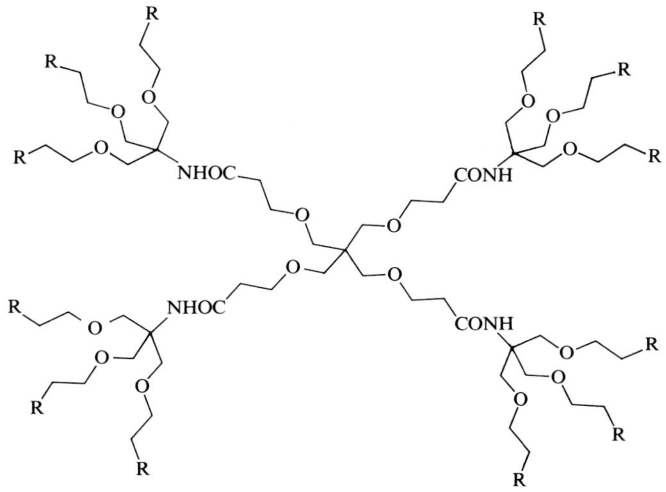

9 R = $CO_2CH_2CH_3$
10 R = COOH

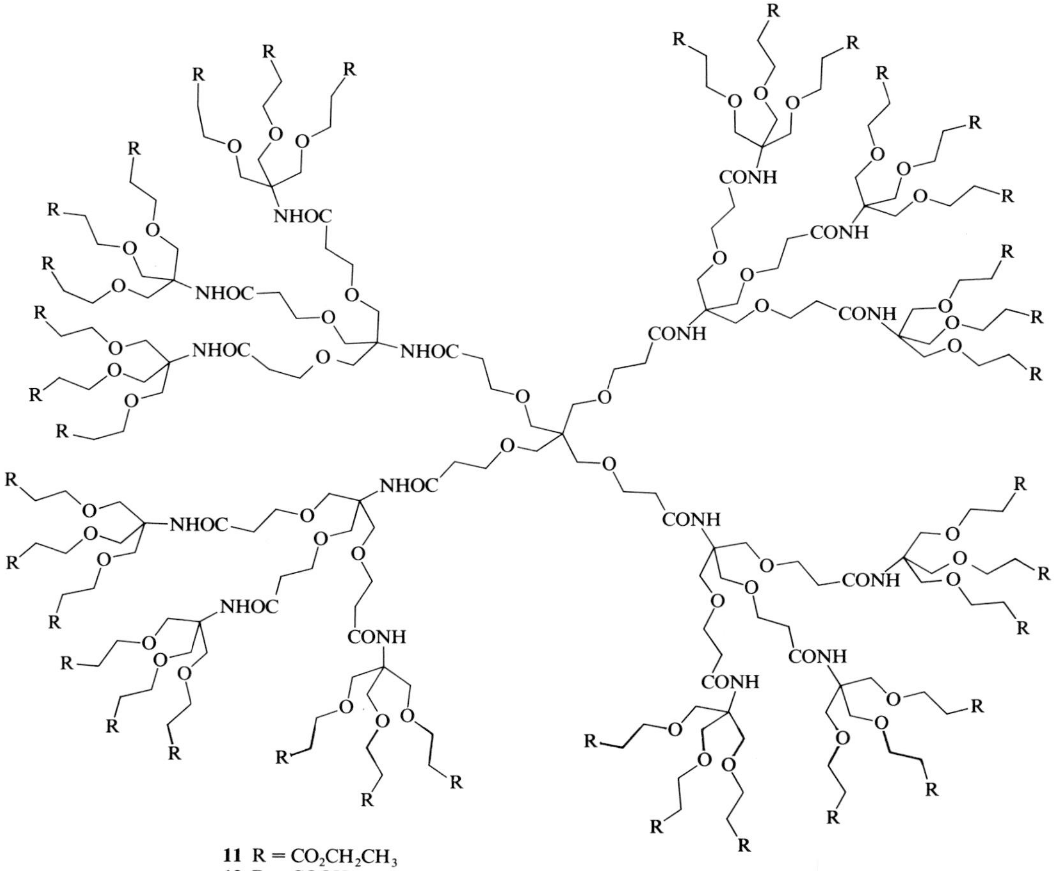

11 R = $CO_2CH_2CH_3$
12 R = COOH
13 R = $CONHC(CH_2OCH_2CH_2CO_2CH_2CH_3)_3$
14 R = $CONHC(CH_2OCH_2CH_2CO_2H)_3$
15 R = $CONHC[CH_2OCH_2CH_2CONHC(CH_2OCH_2CH_2CO_2CH_2CH_3)_3]_3$

FIGURE I.2 Cascade polymers. (Adapted from George R. Newkome and Xiaofeng Lin, *Macromolecules, 1991, 24,* 1443–1444.)

Florida) and Donald A. Tomalia (of the Michigan Molecular Institute) have been pioneers in this area of research.

All of the polymers that we have considered so far are inevitably nonhomogeneous. Although they consist of molecules with common repetitive monomeric units, the molecules of the material obtained from the polymerization reactions vary widely in molecular weight (and, therefore, in size). Cascade polymers, by contrast, can be synthesized in ways that yield polymers consisting of molecules ·of uniform molecular weight and size.

Syntheses of cascade polymers begin with a core building block that can lead to branching in one, two, three, or even four directions. Starting with this core molecule, through repetitive reactions, layers (called **cascade spheres**) are added. Each new sphere increases (usually by three times) the number of branch points from which the next sphere can be constructed. Because of this multiplying effect, very large molecules can be built up very quickly.

Figures I.1 and I.2 show how a four directional cascade molecule has been constructed. All of the reactions are closely related to ones that we have studied already. The starting material for construction of the core molecule is a branched tetraol, **1**. In the first step (i), **1** is allowed to react with propenenitrile (CH_2=CHCN) in a conjugate addition called *cyanoethylation* to produce **2**. Treating **2** with methanol and acid [step (ii)] converts the cyano groups to methyl carboxylate groups. (Instead of hydrolyzing the cyano groups to carboxylic acids and then esterifying them, this process accomplishes the same result in one step.) In step (iii), the ester groups are hydrolyzed, and in step (iv), the carboxyl groups are converted to acyl chlorides. Compound **5** is the core building block.

The synthesis of the compound used in constructing the next cascade sphere is shown in the sequence **6** → **7** → **8** (cyanoethylation followed by esterification). Treating the core compound, **5**, with an excess of the amine, **8**, produced compound **9** with 12 surface ester groups (called, for convenience, the [12]-ester). The key to this step is the formation of amide linkages between **5** and four molecules of **8**. The [12-ester], **9**, was hydrolyzed to the [12]-acid, **10**. Treating **10** with **8** using dicyclohexylcarbodiimide (Section 18.8E) to promote amide formation led to the [36]-ester, **11**. The [36]-ester, **11**, was then hydrolyzed to the [36]-acid **12** which, in turn, was allowed to react with **8** to produce the next cascade molecule, a [108]-ester, **13**.

Repeating these steps one more time produced the [324]-ester, **15**, a compound with a molecular weight of 60,604! At each step the cascade molecules were isolated, purified, and identified. Because the yields for each step are 40–60%, and because the starting materials are inexpensive, this method offers a reasonable route to large homogeneous spherical polymers.

CHAPTER 19

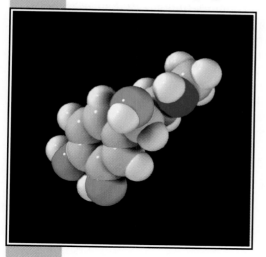

Phenobarbital
(Section 19.12)

SYNTHESIS AND REACTIONS OF β-DICARBONYL COMPOUNDS: MORE CHEMISTRY OF ENOLATE IONS

19.1 INTRODUCTION

Compounds having two carbonyl groups separated by an intervening carbon are called β-dicarbonyl compounds, and these compounds are highly versatile reagents for organic synthesis. In this chapter we shall explore some of the methods for preparing β-dicarbonyl compounds and some of their important reactions.

The β-dicarbonyl system   A β-keto ester   A malonic ester

$$\underset{\beta \quad \alpha}{\overset{O \quad O}{-\!C\!-\!C\!-\!C\!-}}$$ $$\underset{\beta \; \alpha}{\overset{O \quad O}{RCCH_2COR'}}$$ $$\underset{\beta \; \alpha}{\overset{O \quad O}{ROCCH_2COR}}$$

The β-dicarbonyl system   A β-keto ester   A malonic ester
                    (Section 19.2)   (Section 19.4)

Central to the chemistry of β-dicarbonyl compounds is the acidity of protons located on the carbon between two carbonyl groups. The pK_a for such a proton is in the range 9–11.

$$\underset{H}{\overset{O \quad O}{-\!C\!-\!C\!-\!C\!-}} \quad pK_a = 9\text{--}11$$

Early in this chapter we shall see how the acidity of these protons allows the synthesis of β-dicarbonyl compounds through reactions that are called *Claisen syntheses* (Section 19.2). Later in the chapter we shall study the *acetoacetic ester synthesis* (Section 19.3) and the *malonic ester synthesis* (Section 19.4), in which the acidity of these hydrogen atoms forms the basis for the synthesis of substituted acetones and substituted acetic acids. The acidity of the hydrogen atoms of a carbon located between two carbonyl groups allows easy conversion of the compound to an enolate ion, and these enolate ions can be alkylated and acylated. Similar chemistry underlies syntheses using a variety of other useful reactions (Section 19.5), including the Knoevenagel condensation (Section 19.8).

One other feature that appears again and again in the syntheses that we study here is the decarboxylation of a β-keto acid:

$$-\overset{\overset{\displaystyle O}{\|}}{C}-\overset{|}{\underset{|}{C}}-\overset{\overset{\displaystyle O}{\|}}{C}OH \xrightarrow{\text{heat}} -\overset{\overset{\displaystyle O}{\|}}{C}-\overset{|}{\underset{|}{C}}-H + CO_2$$

We learned in Section 18.11 that these decarboxylations occur at relatively low temperatures, and it is this ease of decarboxylation that makes many of the syntheses in this chapter such useful ones.

19.2 THE CLAISEN CONDENSATION: THE SYNTHESIS OF β-KETO ESTERS

When ethyl acetate reacts with sodium ethoxide, it undergoes *a condensation reaction.* After acidification, the product is a β-keto ester, ethyl acetoacetate (commonly called *acetoacetic ester*).

$$2\ CH_3\overset{\overset{\displaystyle O}{\|}}{C}OC_2H_5 \xrightarrow{\text{NaOC}_2\text{H}_5} \left[CH_3\overset{\overset{\displaystyle O}{\|}}{C}\overset{\overset{\displaystyle O}{\|}}{\underset{\underset{\displaystyle Na^+}{\cdot\cdot}}{C}}HCOC_2H_5 \right] + C_2H_5OH$$

Sodioacetoacetic ester   (removed by distillation)

$$\downarrow \text{HCl}$$

$$CH_3\overset{\overset{\displaystyle O}{\|}}{C}CH_2\overset{\overset{\displaystyle O}{\|}}{C}OC_2H_5$$

Ethyl acetoacetate (acetoacetic ester) (75–76%)

Condensations of this type occur with many other esters and are known generally as *Claisen condensations*. Like the aldol condensation (Section 17.15), Claisen condensations involve the α carbon of one molecule and the carbonyl group of another.

Ethyl pentanoate, for example, reacts with sodium ethoxide to give the β-keto ester that follows:

$$2\ CH_3CH_2CH_2CH_2COC_2H_5 \xrightarrow{\text{NaOCH}_2\text{CH}_3} \left[\begin{array}{c} Na^+ \\ CH_3CH_2CH_2CH_2C-\overset{..}{C}-COC_2H_5 \\ | \\ CH_2 \\ | \\ CH_2 \\ | \\ CH_3 \end{array} \right] + C_2H_5OH$$

Ethyl pentanoate

$$\downarrow CH_3CO_2H$$

$$CH_3CH_2CH_2CH_2\overset{O}{\overset{||}{C}}-CH-\overset{O}{\overset{||}{C}}OC_2H_5$$
$$| \\ CH_2 \\ | \\ CH_2 \\ | \\ CH_3$$

(77%)

If we look closely at these examples, we can see that, overall, both reactions involve a condensation in which one ester loses an α hydrogen and the other loses an ethoxide ion; that is,

$$R-CH_2\overset{O}{\overset{||}{C}}\underset{}{\boxed{OC_2H_5}} + H\underset{}{\boxed{}}CH\overset{O}{\overset{||}{C}}-OC_2H_5 \xrightarrow[\text{(2) H}^+]{\text{(1) NaOC}_2\text{H}_5}$$
$$| \\ R$$

(R may also be H)

$$R-CH_2\overset{O}{\overset{||}{C}}-CHCOC_2H_5 + C_2H_5OH$$
$$| \\ R$$

A β-keto ester

We can understand how this happens if we examine the reaction mechanism.

The first step of a Claisen condensation resembles that of an aldol addition. Ethoxide ion abstracts an α proton from the ester. Although the α protons of an ester are not as acidic as those of aldehydes and ketones, the enolate anion that forms is stabilized by resonance in a similar way.

Step 1 $R\overset{\bullet}{C}H-\overset{O}{\overset{||}{C}}OC_2H_5 + :\overset{..}{\overset{..}{O}}C_2H_5 \rightleftharpoons R\overset{..}{\overset{\overset{\displaystyle\overset{..}{O}:}{|}}{C}H-COC_2H_5} + C_2H_5OH$
$|$
H

$$\updownarrow$$

$$:\overset{..}{\overset{..}{O}}:^-$$
$$|$$
$$RCH=COC_2H_5$$

In the second step the enolate anion attacks the carbonyl carbon atom of a second molecule of the ester. It is at this point that the Claisen condensation and the aldol addition *differ,* and they differ in an understandable way. In the aldol reaction nucleophilic attack leads to *addition;* in the Claisen condensation it leads to *substitution.*

Step 2

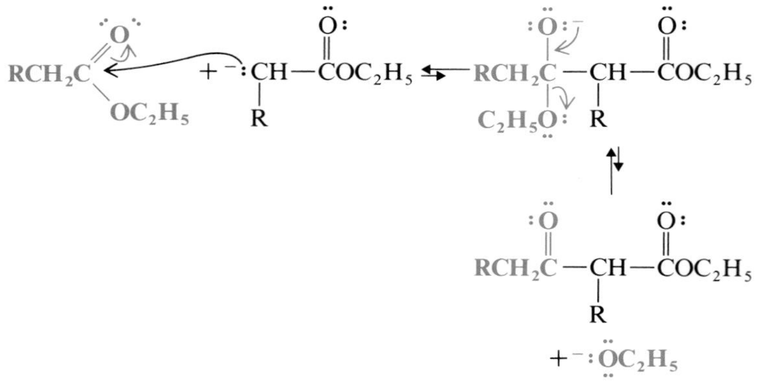

$$RCH_2C \overset{\ddots{O}}{\underset{OC_2H_5}{\big\langle}} + ^-\!:CH-\overset{\ddot{O}:}{\overset{\|}{C}}OC_2H_5 \; \rightleftharpoons \; RCH_2C-CH-\overset{\ddot{O}:}{\overset{\|}{C}}OC_2H_5$$

with R on the CH group, and $C_2H_5\ddot{O}:$ and R shown below; leading to:

$$RCH_2\overset{:\ddot{O}}{\overset{\|}{C}}-CH-\overset{\ddot{O}:}{\overset{\|}{C}}OC_2H_5$$
$$\underset{R}{|}$$
$$+\;^-\!:\ddot{O}C_2H_5$$

Although the products of this second step are a β-keto ester and ethoxide ion, all of the equilibria up to this point have been unfavorable. Very little product would be formed if this were the last step in the reaction.

The final step of a Claisen condensation is an acid–base reaction that takes place between ethoxide ion and the β-keto ester. *The position of equilibrium for this step is favorable,* and we can make it even more favorable by distilling ethanol from the reaction mixture as it forms.

Step 3

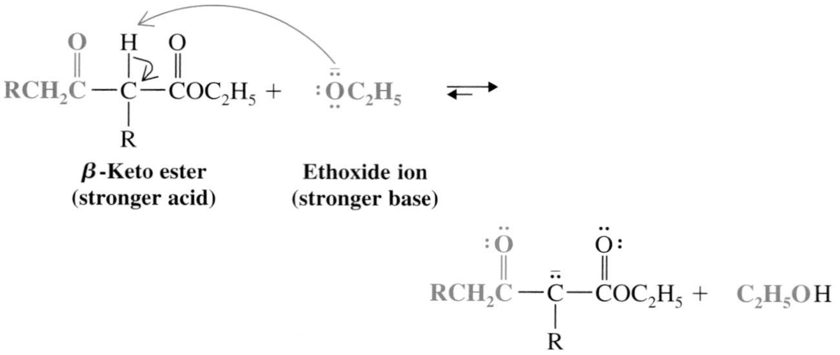

$$RCH_2\overset{O}{\overset{\|}{C}}-\overset{H}{\underset{R}{\overset{|}{C}}}-\overset{O}{\overset{\|}{C}}OC_2H_5 \; + \; {}^-\!:\ddot{O}C_2H_5 \; \rightleftharpoons$$

β-Keto ester   **Ethoxide ion**
(stronger acid)   **(stronger base)**

$$RCH_2\overset{:\ddot{O}}{\overset{\|}{C}}-\overset{..}{\underset{R}{\overset{|}{C}}}-\overset{\ddot{O}:}{\overset{\|}{C}}OC_2H_5 \; + \; C_2H_5OH$$

β-Keto ester anion   **Ethanol**
(weaker base)   **(weaker acid)**

β-Keto esters are stronger acids than ethanol. They react with ethoxide ion almost quantitatively to produce ethanol and anions of β-keto esters. (It is this reaction that pulls the equilibrium to the right.) β-Keto esters are much more acidic than ordinary esters, because their enolate anions are more stabilized by resonance: Their negative charge is delocalized into two carbonyl groups:

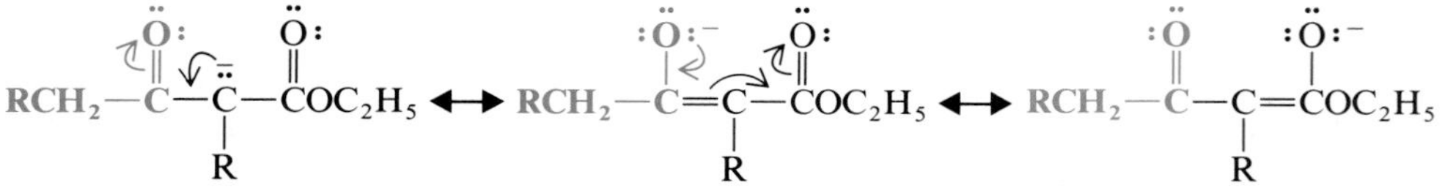

$$RCH_2-\overset{\ddot{O}:}{\overset{\|}{C}}-\overset{..}{\underset{R}{\overset{|}{C}}}-\overset{\ddot{O}:}{\overset{\|}{C}}OC_2H_5 \longleftrightarrow RCH_2-\overset{:\ddot{O}:^-}{\overset{\|}{C}}=\overset{\ddot{O}:}{\underset{R}{\overset{\|}{C}}}-\overset{}{C}OC_2H_5 \longleftrightarrow RCH_2-\overset{:\ddot{O}}{\overset{\|}{C}}-\overset{}{\underset{R}{\overset{|}{C}}}=\overset{:\ddot{O}:^-}{C}OC_2H_5$$

$$\overset{\delta-}{O} \qquad \overset{\delta-}{O}$$
$$RCH_2-\overset{\|}{C}\overset{\delta-}{=\!=\!=}\overset{\|}{C}-\overset{\|}{C}OC_2H_5$$
$$\underset{R}{\mid}$$

Resonance hybrid

After steps 1–3 of a Claisen condensation have taken place, we add an acid to the reaction mixture. This addition brings about a rapid protonation of the anion and produces the β-keto ester as an equilibrium mixture of its keto and enol forms.

Step 4 $RCH_2-\overset{\delta-}{\overset{O}{\|}}C \overset{\delta-}{=\!=\!=} \overset{\delta-}{\overset{O}{\|}}C-COC_2H_5 \xrightarrow[\text{(rapid)}]{H^+} RCH_2-\overset{O}{\overset{\|}{C}}-\overset{\mid}{C}H-\overset{O}{\overset{\|}{C}}OC_2H_5$

Keto form

$$RCH_2-\overset{OH}{\underset{}{C}}=\overset{}{C}-\overset{O}{\overset{\|}{C}}OC_2H_5$$
$$\underset{R}{\mid}$$

Enol form

Esters that have only one α hydrogen do not undergo the usual Claisen condensation. An example of an ester that does not react in a normal Claisen condensation is ethyl 2-methylpropanoate.

Only one α hydrogen

$$\underset{\underset{CH_3}{\mid}}{CH_3CHCOCH_2CH_3}$$

Does not undergo a Claisen condensation.

Ethyl 2-methylpropanoate

Inspection of the mechanism just given will make clear why this is so. An ester with only one α hydrogen will not have an acidic hydrogen when step 3 is reached, and step 3 provides the favorable equilibrium that ensures the success of the reaction. (In Section 19.2A we shall see how esters with only one α hydrogen can be converted to a β-keto ester through the use of very strong bases.)

PROBLEM 19.1

(a) Write a mechanism for all steps of the Claisen condensation that takes place when ethyl propanoate reacts with ethoxide ion. (b) What products form when the reaction mixture is acidified?

When diethyl hexanedioate is heated with sodium ethoxide, subsequent acidification of the reaction mixture gives ethyl 2-oxocyclopentanecarboxylate.

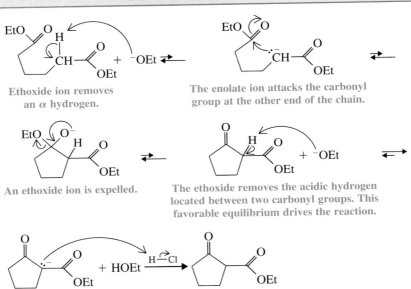

**Diethyl hexanedioate
(diethyl adipate)**

**Ethyl 2-oxocyclopentane-
carboxylate
(74–81%)**

This reaction, called the *Dieckmann condensation,* is an intramolecular Claisen condensation. The α carbon atom and the ester group for the condensation come from the same molecule. In general, the Dieckmann condensation is useful only for the preparation of five- and six-membered rings.

A MECHANISM FOR THE DIECKMANN CONDENSATION

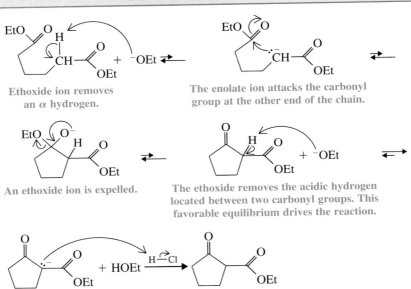

Ethoxide ion removes an α hydrogen.

The enolate ion attacks the carbonyl group at the other end of the chain.

An ethoxide ion is expelled.

The ethoxide removes the acidic hydrogen located between two carbonyl groups. This favorable equilibrium drives the reaction.

Addition of HCl rapidly protonates the anion, giving the final product.

PROBLEM 19.2

(a) What product would you expect from a Dieckmann condensation of diethyl heptanedioate (diethyl pimelate)?   (b) Can you account for the fact that diethyl pentanedioate (diethyl glutarate) does not undergo a Dieckmann condensation?

19.2A Crossed Claisen Condensations

Crossed Claisen condensations (like crossed aldol condensations) are possible **when one ester component has no α hydrogens** and is, therefore, unable to form an enolate ion and undergo self-condensation. Ethyl benzoate, for example, condenses with ethyl acetate to give ethyl benzoylacetate.

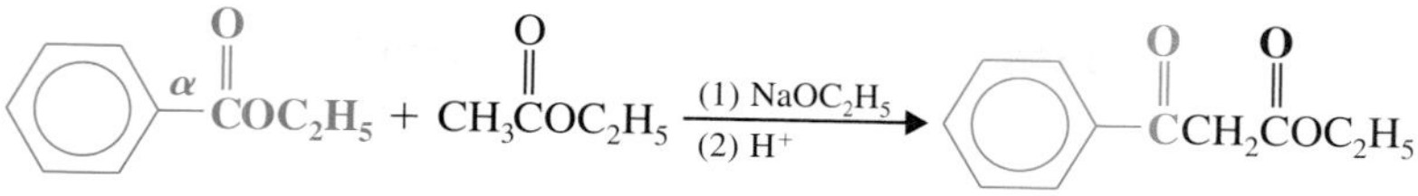

Ethyl benzoate
(no α hydrogen)

Ethyl benzoylacetate
(60%)

Ethyl phenylacetate condenses with diethyl carbonate to give diethyl phenylmalonate.

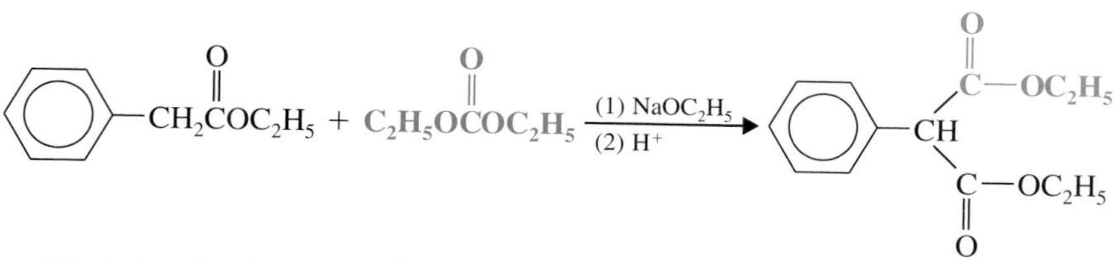

Ethyl phenylacetate   **Diethyl carbonate**
                               (no α carbon)

Diethyl phenylmalonate
(65%)

PROBLEM 19.3

Write mechanisms that account for the products that are formed in the two crossed Claisen condensations just illustrated.

PROBLEM 19.4

What products would you expect to obtain from each of the following crossed Claisen condensations?

(a) Ethyl propanoate + diethyl oxalate $\xrightarrow[\text{(2) H}^+]{\text{(1) NaOCH}_2\text{CH}_3}$

(b) Ethyl acetate + ethyl formate $\xrightarrow[\text{(2) H}^+]{\text{(1) NaOCH}_2\text{CH}_3}$

As we learned earlier in this section, esters that have only one α hydrogen cannot be converted to β-keto esters by sodium ethoxide. However, they can be converted to β-keto esters by reactions that use very strong bases. The strong base converts the ester to its enolate anion in nearly quantitative yield. This allows us to *acylate* the enolate anion by treating it with an acyl chloride or an ester. An example of this technique that makes use of the very powerful base sodium triphenylmethanide is shown next.

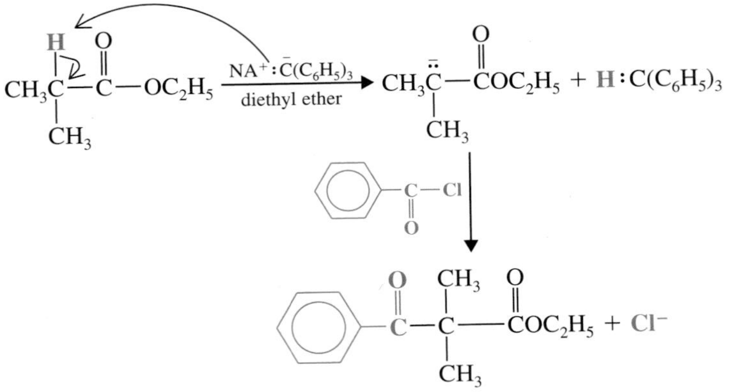

Ethyl 2,2-dimethyl-3-oxo-3-phenylpropanoate

19.2B Acylation of Other Carbanions

Enolate anions derived from ketones also react with esters in nucleophilic substitution reactions that resemble Claisen condensations. In the following first example, although two anions are possible from the reaction of the ketone with sodium amide, the major product is derived from the primary carbanion. The primary α hydrogens are more acidic than the secondary α hydrogens.

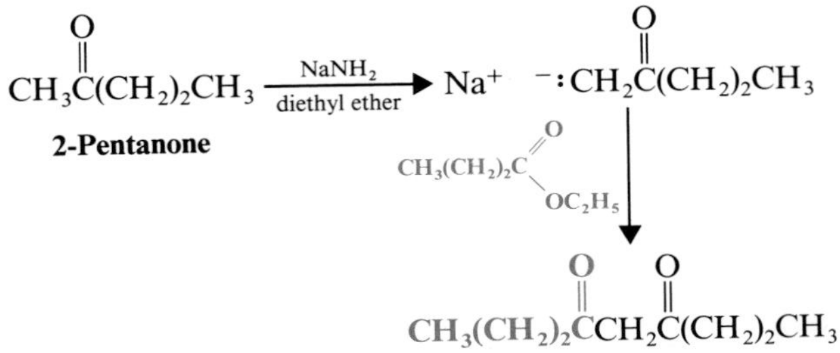

2-Pentanone

4,6-Nonanedione
(76%)

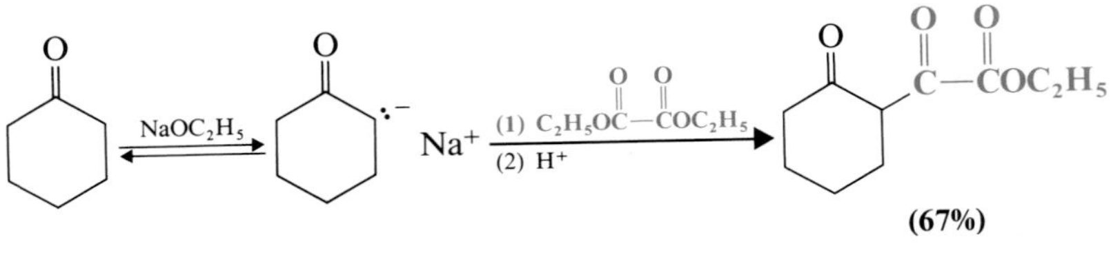

(67%)

PROBLEM 19.5

Show how you might synthesize each of the following compounds using, as your starting materials, esters, ketones, acyl halides, and so on.

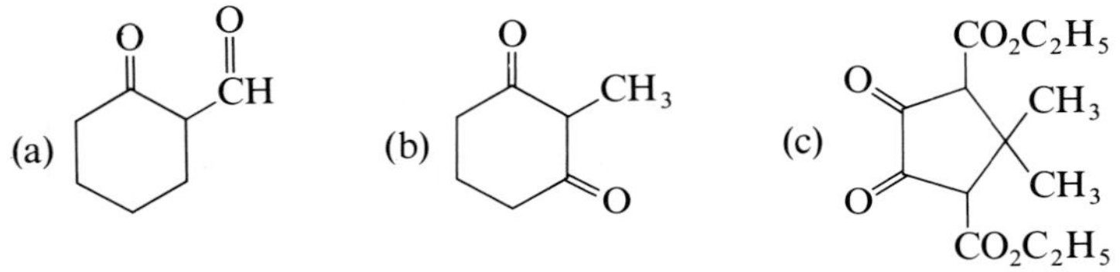

(a)   (b)   (c)

PROBLEM 19.6

Keto esters are capable of undergoing cyclization reactions similar to the Dieckmann condensation. Write a mechanism that accounts for the product formed in the following reaction:

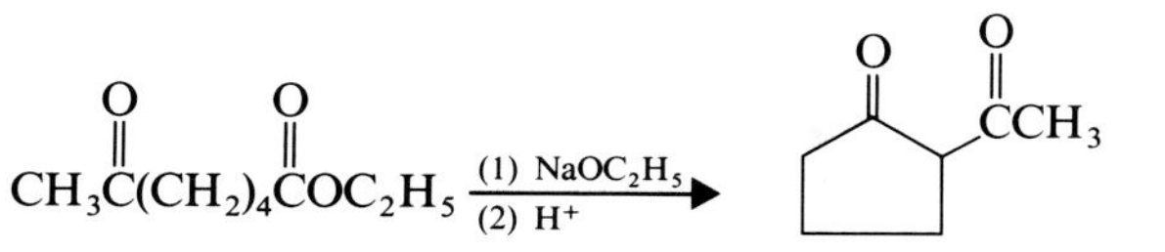

$$CH_3\overset{\overset{\displaystyle O}{\|}}{C}(CH_2)_4\overset{\overset{\displaystyle O}{\|}}{C}OC_2H_5 \xrightarrow[\text{(2) H}^+]{\text{(1) NaOC}_2\text{H}_5}$$

2-Acetylcyclopentanone

19.3 THE ACETOACETIC ESTER SYNTHESIS: SYNTHESIS OF SUBSTITUTED ACETONES

Acetoacetic esters are useful reagents for the preparation of methyl ketones of the types shown here:

$$CH_3-\overset{\overset{\displaystyle O}{\|}}{C}-CH_2-R \quad \text{or} \quad CH_3-\overset{\overset{\displaystyle O}{\|}}{C}-\overset{\overset{\displaystyle }{|}}{\underset{\underset{\displaystyle R}{|}}{C}H}-R$$

| Monosubstituted acetone | Disubstituted acetone |

Acetoacetic ester acts as the *synthetic equivalent* (Section 8.18) of the following three-carbon fragment:

$$CH_3-\overset{\overset{\displaystyle O}{\|}}{C}-\overset{\overset{\displaystyle }{|}}{C}H-$$

Two factors make such syntheses practical: (1) The methylene protons of β-keto esters are appreciably acidic and (2) β-keto acids decarboxylate readily (cf. Section 18.11).

As we have seen (Section 19.2) the methylene protons of acetoacetic ester are more acidic than the —OH proton of ethanol because they are located between two carbonyl groups and yield a highly stabilized enolate anion. This acidity means that we can convert acetoacetic ester to an enolate anion using sodium ethoxide as a base. We can then carry out an alkylation reaction by treating the enolate anion with an alkyl halide.

Since the alkylation (see following reaction) is an S_N2 reaction, best yields are obtained from the use of primary alkyl halides (including primary allylic and benzylic halides) or methyl halides. Secondary halides give lower yields, and tertiary halides give only elimination.

$$CH_3\overset{\displaystyle \overset{..}{\underset{..}{O}}:}{\overset{\|}{C}}-CH_2-\overset{\displaystyle \overset{..}{\underset{..}{O}}:}{\overset{\|}{C}}OC_2H_5 + C_2H_5O^-Na^+ \rightleftharpoons CH_3\overset{\displaystyle \overset{..}{\underset{..}{O}}:}{\overset{\|}{C}}-\overset{\overset{\displaystyle Na^+}{\underset{..}{\overline{\underline{C}}}H}}{}-\overset{\displaystyle \overset{..}{\underset{..}{O}}:}{\overset{\|}{C}}-OC_2H_5 + C_2H_5OH$$

Acetoacetic ester   **Sodium**   **Sodioacetoacetic**
                              ethoxide   **ester**

$$\Big\downarrow R-X$$

$$CH_3\overset{\displaystyle \overset{..}{\underset{..}{O}}:}{\overset{\|}{C}}-\underset{\underset{\displaystyle R}{|}}{CH}-\overset{\displaystyle \overset{..}{\underset{..}{O}}:}{\overset{\|}{C}}-OC_2H_5 + NaX$$

Monoalkylacetoacetic ester

The monoalkylacetoacetic ester still has one appreciably acidic hydrogen and, if we desire, we can carry out a second alkylation. Because the monoalkylacetoacetic ester is somewhat less acidic than acetoacetic ester itself (why?), it is usually helpful to use a base stronger than ethoxide ion.

$$CH_3\overset{\displaystyle O}{\overset{\|}{C}}-\underset{\underset{\displaystyle R}{|}}{CH}-\overset{\displaystyle O}{\overset{\|}{C}}-OC_2H_5 + (CH_3)_3CO^-K^+ \rightleftharpoons CH_3\overset{\displaystyle O}{\overset{\|}{C}}-\overset{\overset{\displaystyle K^+}{\underset{\underset{\displaystyle R}{|}}{\overline{\underline{C}}}}}{}-\overset{\displaystyle O}{\overset{\|}{C}}OC_2H_5 + (CH_3)_3COH$$

Monoalkylacetoacetic   **Potassium *tert-***
ester   **butoxide**

$$\Big\downarrow R'-X$$

$$CH_3\overset{\displaystyle O}{\overset{\|}{C}}-\overset{\overset{\displaystyle R'}{\underset{\underset{\displaystyle R}{|}}{|}}{C}}{}-\overset{\displaystyle O}{\overset{\|}{C}}-OC_2H_5 + KX$$

Dialkylacetoacetic ester

If our goal is the preparation of a monosubstituted acetone, we carry out only one alkylation reaction. We then hydrolyze the monoalkylacetoacetic ester using dilute sodium or potassium hydroxide. Subsequent acidification of the mixture gives an alkylacetoacetic acid, and heating this β-keto acid to 100°C brings about decarboxylation (Section 18.11).

$$CH_3\overset{\displaystyle O}{\overset{\|}{C}}-\underset{\underset{\displaystyle R}{|}}{CH}-\overset{\displaystyle O}{\overset{\|}{C}}OC_2H_5 \xrightarrow[\text{heat}]{\text{dil. NaOH}} CH_3\overset{\displaystyle O}{\overset{\|}{C}}-\underset{\underset{\displaystyle R}{|}}{CH}-\overset{\displaystyle O}{\overset{\|}{C}}-O^-Na^+$$

Basic hydrolysis of the ester group

$$\xrightarrow{H_3O^+} CH_3\overset{\displaystyle O}{\overset{\|}{C}}-\underset{\underset{\displaystyle R}{|}}{CH}-\overset{\displaystyle O}{\overset{\|}{C}}-OH \xrightarrow[100°C]{\text{heat}} CH_3\overset{\displaystyle O}{\overset{\|}{C}}-CH-R + CO_2$$

Acidification   Decarboxylation of the β-keto acid

A specific example is the following synthesis of 2-heptanone:

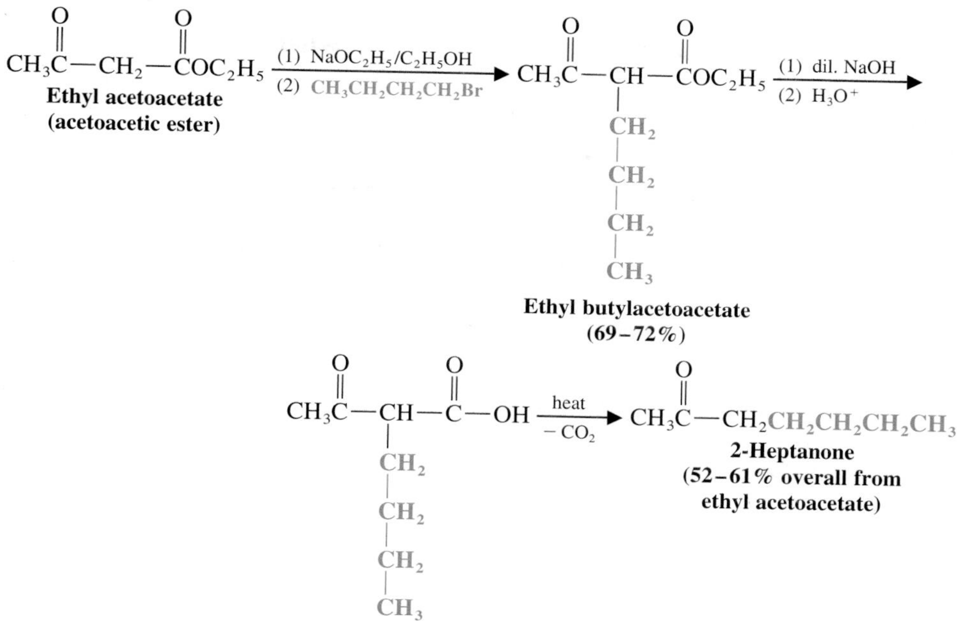

Ethyl acetoacetate
(acetoacetic ester)

$\xrightarrow[\text{(2) } CH_3CH_2CH_2CH_2Br]{\text{(1) } NaOC_2H_5/C_2H_5OH}$

Ethyl butylacetoacetate
(69–72%)

$\xrightarrow[\text{(2) } H_3O^+]{\text{(1) dil. NaOH}}$

$\xrightarrow[-CO_2]{\text{heat}}$

2-Heptanone
(52–61% overall from
ethyl acetoacetate)

If our goal is the preparation of a disubstituted acetone, we carry out two successive alkylations, we hydrolyze the dialkylacetoacetic ester that is produced, and then we decarboxylate the dialkylacetoacetic acid. An example of this procedure is the synthesis of 3-butyl-2-heptanone.

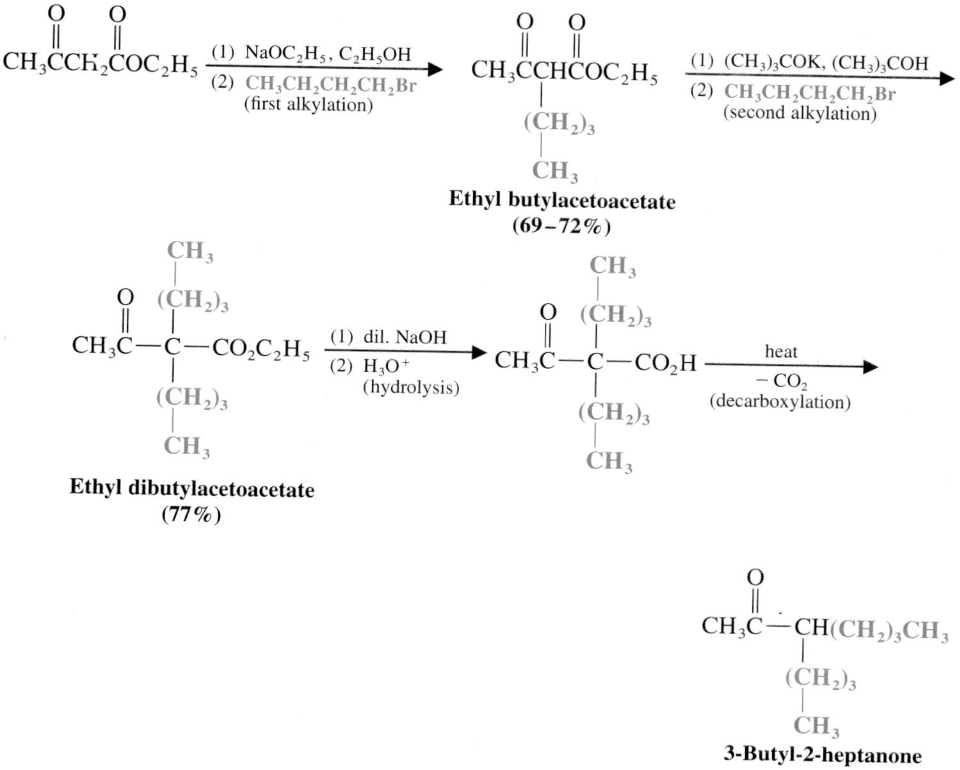

$\xrightarrow[\substack{\text{(2) } CH_3CH_2CH_2CH_2Br \\ \text{(first alkylation)}}]{\text{(1) } NaOC_2H_5, C_2H_5OH}$

$\xrightarrow[\substack{\text{(2) } CH_3CH_2CH_2CH_2Br \\ \text{(second alkylation)}}]{\text{(1) } (CH_3)_3COK, (CH_3)_3COH}$

Ethyl butylacetoacetate
(69–72%)

$\xrightarrow[\substack{\text{(2) } H_3O^+ \\ \text{(hydrolysis)}}]{\text{(1) dil. NaOH}}$

$\xrightarrow[\substack{-CO_2 \\ \text{(decarboxylation)}}]{\text{heat}}$

Ethyl dibutylacetoacetate
(77%)

3-Butyl-2-heptanone

Although both alkylations in the example just given were carried out with the same alkyl halide, we could have used different alkyl halides if our synthesis had required it.

PROBLEM 19.7

Occasional side products of alkylations of sodioacetoacetic esters are compounds with the following general structure:

$$\overset{\displaystyle\overset{\displaystyle..}{\overset{..}{RO:}}}{\underset{}{}} \qquad \overset{\displaystyle\overset{..}{\overset{..}{O:}}}{\underset{}{}}$$

$$CH_3C\!\!=\!\!CHCOC_2H_5$$

Explain how these are formed.

PROBLEM 19.8

Show how you would use the acetoacetic ester synthesis to prepare each of the following:

(a) 2-pentanone, (b) 3-propyl-2-hexanone, and (c) 4-phenyl-2-butanone.

PROBLEM 19.9

The acetoacetic ester synthesis generally gives best yields when primary halides are used in the alkylation step. Secondary halides give low yields, and tertiary halides give practically no alkylation product at all. (a) Explain. (b) What products would you expect from the reaction of sodioacetoacetic ester and *tert*-butyl bromide? (c) Bromobenzene cannot be used as an arylating agent in an acetoacetic ester synthesis in the manner we have just described. Why not?

PROBLEM 19.10

Since the products obtained from Claisen condensations are β-keto esters, subsequent hydrolysis and decarboxylation of these products gives a general method for the synthesis of ketones. Show how you would employ this technique in a synthesis of 4-heptanone.

The acetoacetic ester synthesis can also be carried out using halo esters and halo ketones. The use of an α-halo ester provides a convenient synthesis of γ-keto acids:

$$CH_3C-CH_2CH_2-C-OH$$
4-Oxopentanoic acid

PROBLEM 19.11

In the synthesis of the keto acid just given, the dicarboxylic acid decarboxylates in a specific way; it gives

$$CH_3CCH_2CH_2COH \quad \text{rather than} \quad CH_3CCHCOH$$
$$CH_3$$

Explain.

The use of an α-halo ketone in an acetoacetic ester synthesis provides a general method for preparing γ-diketones:

$$CH_3C-CH_2CH_2-C-R$$
A γ-diketone

PROBLEM 19.12

How would you use the acetoacetic ester synthesis to prepare the following?

Anions obtained from acetoacetic esters undergo acylation when they are treated with acyl chlorides or acid anhydrides. Because both of these acylating agents react with alcohols, acylation reactions cannot be carried out in ethanol and must be carried out in aprotic solvents such as DMF or DMSO (Section 6.15C). (If the reaction were to be carried out in ethanol, using sodium ethoxide, for example, then the acyl chloride would be rapidly converted to an ethyl ester and the ethoxide ion would be neutralized.) Sodium hydride can be used to generate the enolate anion in an aprotic solvent.

$$
\underset{\substack{O \\ \parallel}}{CH_3-C}-CH_2-\underset{\substack{O \\ \parallel}}{C}-OC_2H_5 \xrightarrow[\substack{\text{aprotic solvent} \\ (-H_2)}]{Na^+H^-}
$$

$$
\underset{\substack{Na^+}}{\underset{\substack{O \\ \parallel}}{CH_3-C}-\overset{..}{\underset{\substack{}}{C}}H-\underset{\substack{O \\ \parallel}}{C}-OC_2H_5} \xrightarrow[\substack{(-NaCl)}]{RCCl} \underset{\substack{C=O \\ | \\ R}}{\underset{\substack{O \\ \parallel}}{CH_3-C}-CH-\underset{\substack{O \\ \parallel}}{C}-OC_2H_5} \xrightarrow[\substack{(2)\ H_3O^+}]{(1)\ \text{dil. NaOH}}
$$

$$
\underset{\substack{C=O \\ | \\ R}}{\underset{\substack{O \\ \parallel}}{CH_3-C}-CH-\underset{\substack{O \\ \parallel}}{C}-OH} \xrightarrow[\substack{-CO_2}]{heat} \underset{\textbf{A β-diketone}}{\underset{\substack{O \\ \parallel}}{CH_3-C}-CH_2-\underset{\substack{O \\ \parallel}}{C}-R}
$$

Acylations of acetoacetic esters followed by hydrolysis and decarboxylation give us a method for preparing β-diketones.

PROBLEM 19.13

How would you use the acetoacetic ester synthesis to prepare the following?

$$
\text{C}_6\text{H}_5 - \underset{\substack{O \\ \parallel}}{C}CH_2\underset{\substack{O \\ \parallel}}{C}CH_3
$$

Acetoacetic ester cannot be phenylated in a manner analogous to the alkylation reactions we have studied because bromobenzene is not susceptible to S_N2 reactions [Section 6.16A and Problem 19.9(c)]. However, if acetoacetic ester is treated with bromobenzene and *two molar equivalents of sodium amide*, then phenylation does occur *by a benzyne mechanism* (Section 21.11). The overall reaction is as follows:

$$
\underset{\substack{O \quad O \\ \parallel \quad \parallel}}{CH_3CCH_2COC_2H_5} + C_6H_5Br + 2\ NaNH_2 \xrightarrow{\text{liq. NH}_3} \underset{\substack{| \\ C_6H_5}}{\underset{\substack{O \quad O \\ \parallel \quad \parallel}}{CH_3CCHCOC_2H_5}}
$$

Malonic esters $\left(\begin{array}{cc} O & O \\ \| & \| \\ ROCCH_2COR \end{array}\right)$ can be phenylated in an analogous way.

PROBLEM 19.14

(a) Outline a step-by-step mechanism for the phenylation of acetoacetic ester by bromobenzene and two molar equivalents of sodium amide. (Why are two molar equivalents of $NaNH_2$ necessary?)   (b) What product would be obtained by hydrolysis and decarboxylation of the phenylated acetoacetic ester?   (c) How would you prepare phenylacetic acid from malonic ester?

One further variation of the acetoacetic ester synthesis involves the conversion of an acetoacetic ester to a resonance-stabilized *dianion* by using a very strong base such as potassium amide in liquid ammonia.

etc.

When this dianion is treated with 1 mol of a primary (or methyl) halide, it undergoes alkylation at its terminal carbon rather than at its interior one. This orientation of the alkylation reaction apparently results from the greater basicity (and thus nucleophilicity) of the terminal carbanion. This carbanion is more basic because it is stabilized by only one adjacent carbonyl group. After monoalkylation has taken place, the anion that remains can be protonated by adding ammonium chloride.

PROBLEM 19.15 ───────────────────────────────

Show how you could use ethyl acetoacetate in a synthesis of

$$C_6H_5CH_2CH_2\overset{\overset{\displaystyle O}{\|}}{C}CH_2\overset{\overset{\displaystyle O}{\|}}{C}OC_2H_5$$

19.4   THE MALONIC ESTER SYNTHESIS: SYNTHESIS OF SUBSTITUTED ACETIC ACIDS

A useful counterpart of the acetoacetic ester synthesis—one that allows the synthesis of *mono- and disubstituted acetic acids*—is called the *malonic ester synthesis.*

The malonic ester synthesis resembles the acetoacetic ester synthesis in several respects.

1. Diethyl malonate (malonic ester), the starting compound, forms a relatively stable enolate ion:

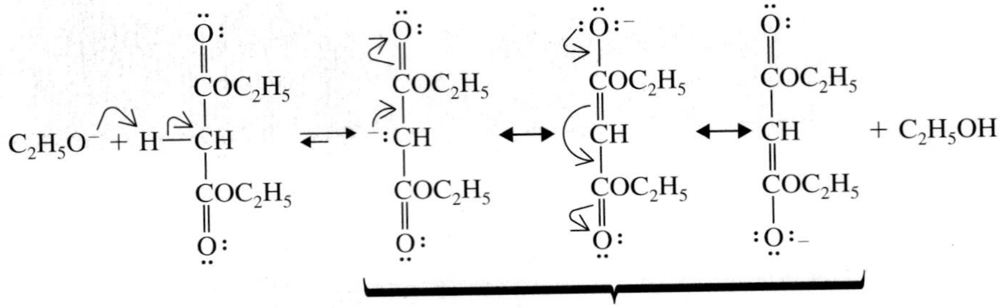

Resonance-stabilized anion

2. This enolate ion can be alkylated,

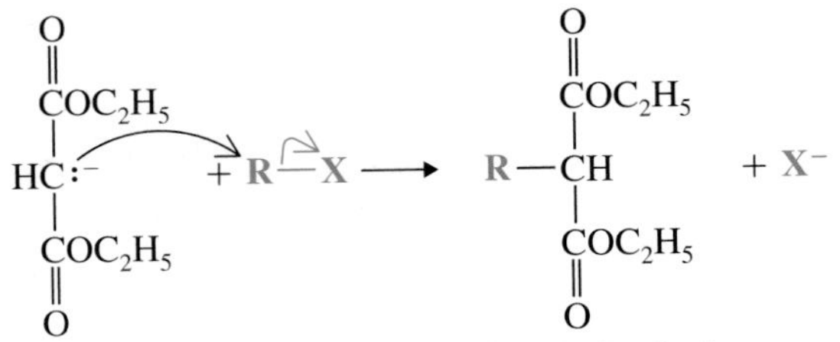

| Enolate ion | Monoalkylmalonic ester |

and the product can be alkylated again if our synthesis requires it:

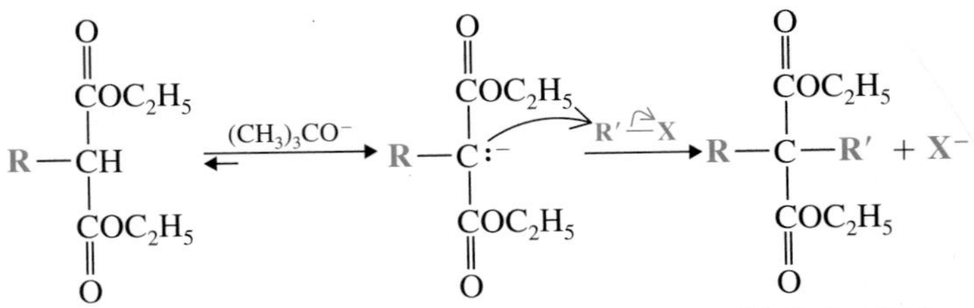

Dialkylmalonic ester

3. The mono- or dialkylmalonic ester can then be hydrolyzed to a mono- or di-alkylmalonic acid, and substituted malonic acids decarboxylate readily. Decarboxylation gives a mono- or disubstituted acetic acid.

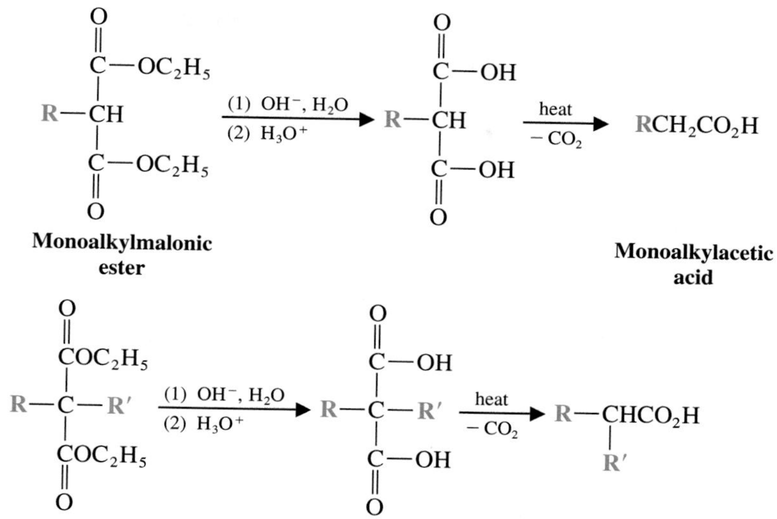

Monoalkylmalonic
ester

Monoalkylacetic
acid

Dialkylmalonic
ester

Dialkylacetic
acid

Two specific examples of the malonic ester synthesis are the syntheses of hexanoic acid and 2-ethylpentanoic acid that follow.

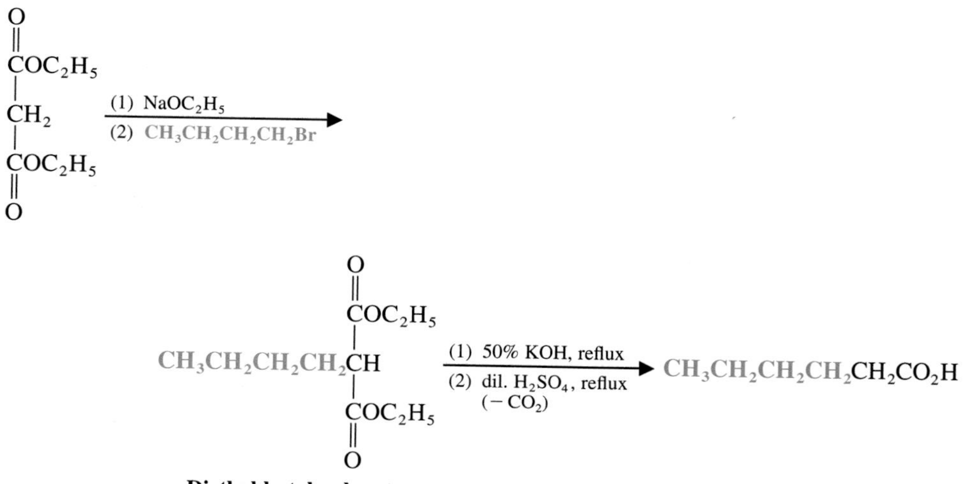

Diethyl butylmalonate
(80–90%)

Hexanoic acid
(75%)

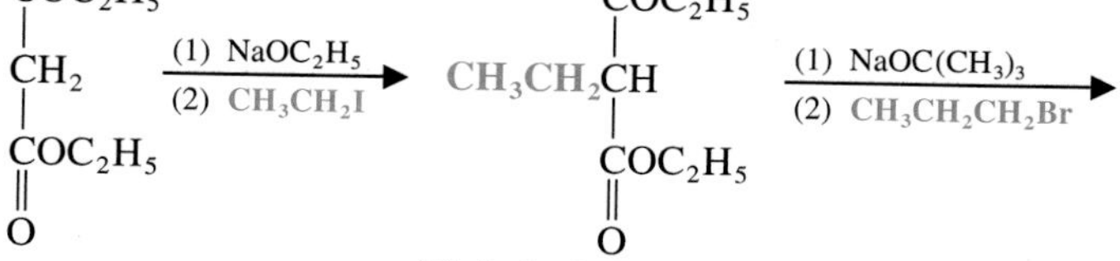

Diethyl ethylmalonate

Diethyl ethylpropylmalonate   **Ethylpropylmalonic acid**

$$CH_3CH_2CH_2\underset{\underset{CH_3}{\overset{|}{CH_2}}}{\overset{|}{CH}}CO_2H$$

2-Ethylpentanoic acid

PROBLEM 19.16

Outline all steps in a malonic ester synthesis of each of the following:

(a) pentanoic acid,   (b) 2-methylpentanoic acid, and   (c) 4-methylpentanoic acid.

Two variations of the malonic ester synthesis make use of dihaloalkanes. In the first of these, two molar equivalents of sodiomalonic ester are allowed to react with a dihaloalkane. Two consecutive alkylations occur giving a tetraester; hydrolysis and decarboxylation of the tetraester yield a dicarboxylic acid. An example is the synthesis of glutaric acid:

$$\underset{\substack{\textbf{Glutaric acid} \\ \textbf{(80\% from tetraester)}}}{HOCCH_2CH_2CH_2COH} + 2\ CO_2 + 4\ C_2H_5OH$$

In a second variation, one molar equivalent of sodiomalonic ester is allowed to react with one molar equivalent of a dihaloalkane. This reaction gives a halo-alkylmalonic ester, which when treated with sodium ethoxide, undergoes an internal alkylation reaction. This method has been used to prepare three-, four-, five-, and six-membered rings. An example is the synthesis of cyclobutanecarboxylic acid.

Cyclobutanecarboxylic acid

19.5 FURTHER REACTIONS OF ACTIVE HYDROGEN COMPOUNDS

Because of the acidity of their methylene hydrogens, malonic esters, acetoacetic esters, and similar compounds are often called *active hydrogen compounds* or *active methylene compounds*. Generally speaking, active hydrogen compounds have two electron-withdrawing groups attached to the same carbon atom:

$$Z-CH_2-Z'$$

Active hydrogen compound
(Z and Z' are electron-withdrawing groups)

The electron-withdrawing groups can be a variety of substituents including:

For example, ethyl cyanoacetate reacts with base to yield a resonance-stabilized anion:

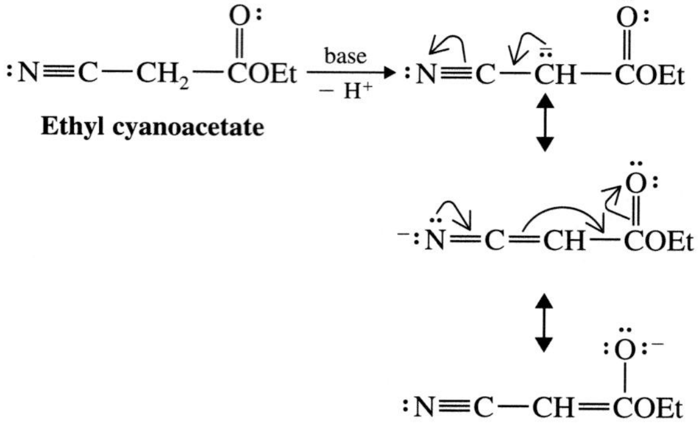

Ethyl cyanoacetate anions also undergo alkylations. They can be dialkylated with isopropyl iodide, for example.

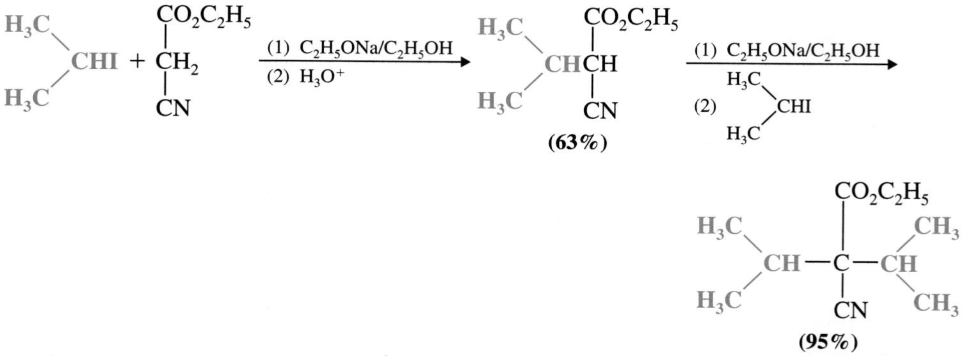

Another way of preparing ketones is to use a β-keto sulfoxide as an active hydrogen compound:

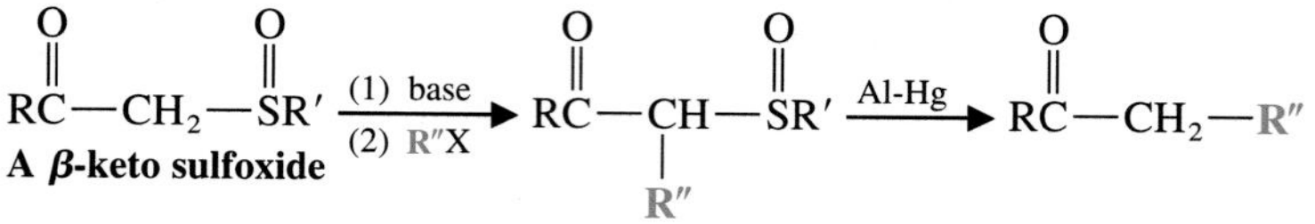

The β-keto sulfoxide is first converted to an anion and then the anion is alkylated. Treating the product of these steps with aluminum amalgam (Al-Hg) causes cleavage at the carbon–sulfur bond and gives the ketone in high yield.

PROBLEM 19.17

The antiepileptic drug valproic acid is 2-propylpentanoic acid (administered as the sodium salt). One commercial synthesis of valproic acid begins with ethyl cyanoacetate. The penultimate step of this synthesis involves a decarboxylation, and the last step involves hydrolysis of a nitrile. Outline this synthesis.

19.6 DIRECT ALKYLATION OF ESTERS AND NITRILES

We have seen in Sections 19.3–19.5 that it is easy to alkylate β-keto esters and other active hydrogen compounds. The hydrogens situated on the carbon atom between the two electron-withdrawing groups are unusually acidic and are easily removed by bases such as ethoxide ion. It is also possible, however, to alkylate esters and nitriles that do not have a β-keto group. To do this we must use a stronger base, one that will convert the ester or nitrile into its enolate anion rapidly so that all of the ester or nitrile is converted to its enolate before it can undergo Claisen condensation. We must also use a base that is sufficiently bulky not to react at the carbonyl carbon of the ester or at the carbon of the nitrile group. Such a base is lithium diisopropylamide (LDA).

Lithium diisopropylamide is a very strong base because it is the conjugate base of the very weak acid, diisopropylamine ($pK_a = 38$). Lithium diisopropylamide is prepared by treating diisopropylamine with methyllithium. Solvents commonly used for reactions in which LDA is the base are ethers such as tetrahydrofuran (THF) and 1,2-dimethoxyethane (DME). (The use of LDA in other syntheses is described in Special Topic H.)

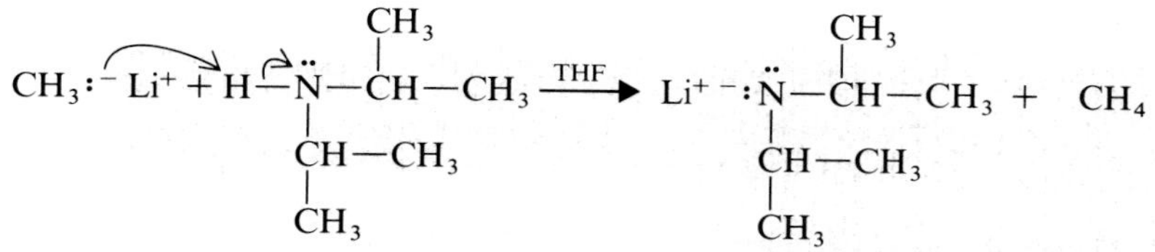

$pK_a = 38$ **Lithium diisopropylamide [abbreviated LDA or $(i\text{-}C_3H_7)_2N^-Li^+$]** **$pK_a = 50$**

Examples of the direct alkylation of esters are shown below. In the second example the ester is a lactone (Section 18.7C).

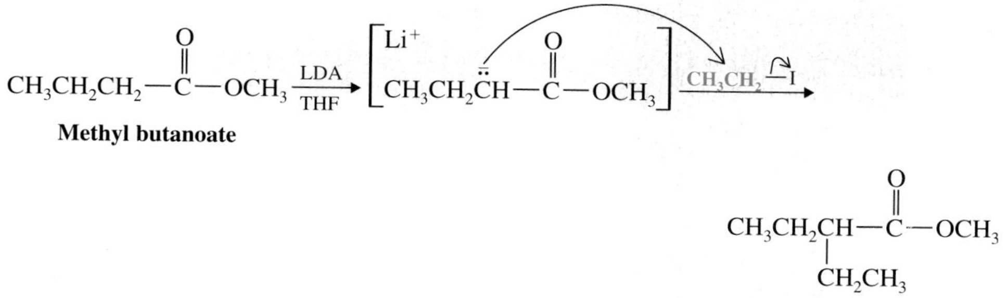

Methyl butanoate

Methyl 2-ethylbutanoate (96%)

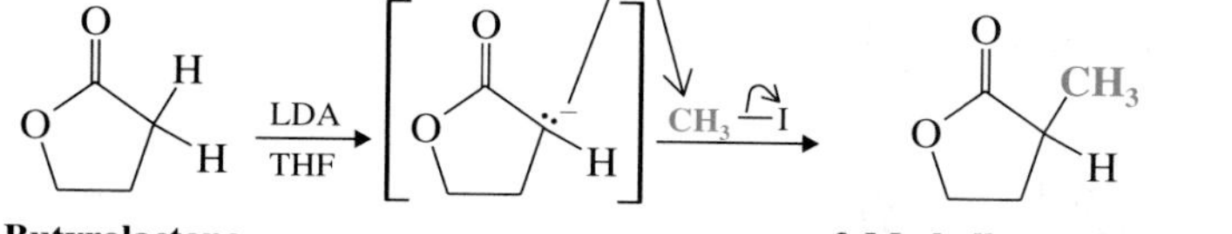

Butyrolactone **2-Methylbutyrolactone (88%)**

19.7 ALKYLATION OF 1,3-DITHIANES

Two sulfur atoms attached to the same carbon of 1,3-dithiane cause the hydrogen atoms of that carbon to be more acidic ($pK_a = 32$) than those of most alkyl carbon atoms.

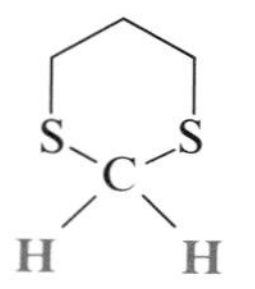

1,3-Dithiane
$pK_a = 32$

Sulfur atoms, because they are easily polarized, can aid in stabilizing the negative charge of the anion. Strong bases such as butyllithium are usually used to convert a dithiane to its anion.

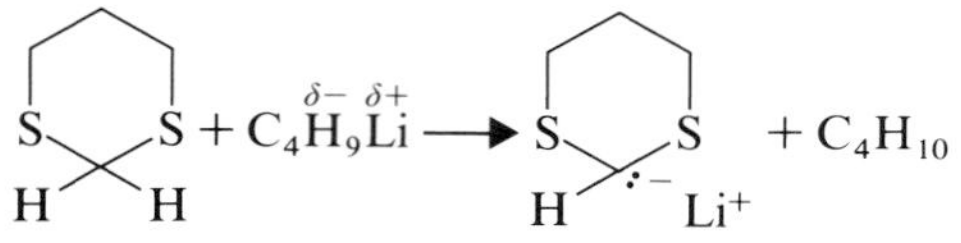

1,3-Dithianes are thioacetals (cf. Section 16.7E); they can be prepared by treating an aldehyde with 1,3-propanedithiol in the presence of a trace of acid.

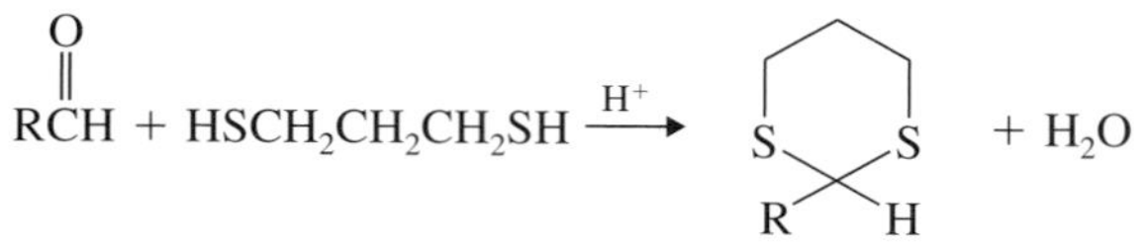

A 1,3-dithiane

Alkylating the 1,3-dithiane with a primary halide and then hydrolyzing the product (a thioketal) is a method for converting an aldehyde to a ketone. Hydrolysis is usually carried out by using $HgCl_2$ either in methanol or in aqueous acetonitrile, CH_3CN.

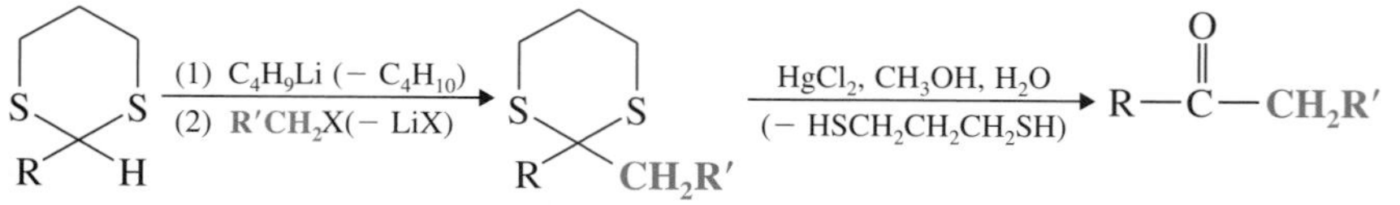

Thioacetal   **Ketone**

Notice that in these 1,3-dithiane syntheses the usual mode of reaction of an aldehyde is reversed. Normally the carbonyl carbon atom of an aldehyde is partially positive; it is electrophilic and, consequently, it reacts with nucleophiles. When the aldehyde is converted to a 1,3-dithiane and treated with butyllithium, this same carbon atom becomes negatively charged and reacts with electrophiles. This reversal of polarity of the carbonyl carbon atom is called **umpolung** (German for **polarity reversal**).

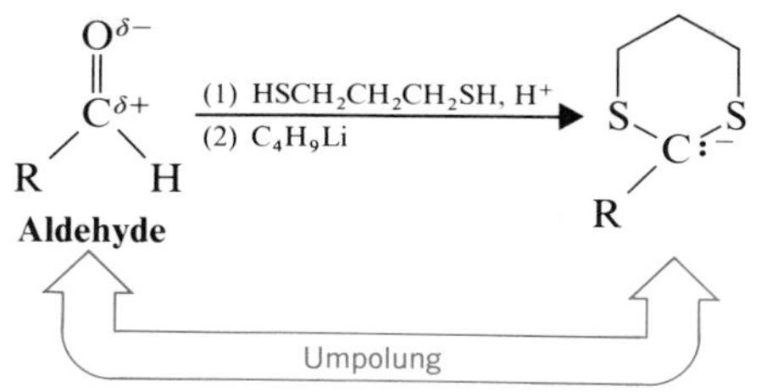

The synthetic use of 1,3-dithianes was developed by E. J. Corey (Section 4.15) and D. Seebach and is often called the *Corey–Seebach* method.

PROBLEM 19.18

(a) Which aldehyde would you use to prepare 1,3-dithiane itself? (b) How would you synthesize $C_6H_5CH_2CHO$ using a 1,3-dithiane as an intermediate? (c) How would you convert benzaldehyde to acetophenone?

PROBLEM 19.19

The Corey–Seebach method can also be used to synthesize molecules with the structure RCH_2CH_2R'. How might this be done?

PROBLEM 19.20

(a) The Corey–Seebach method has been used to prepare the following highly strained molecule called a metaparacyclophane. What are the structures of the intermediates **A–D**?

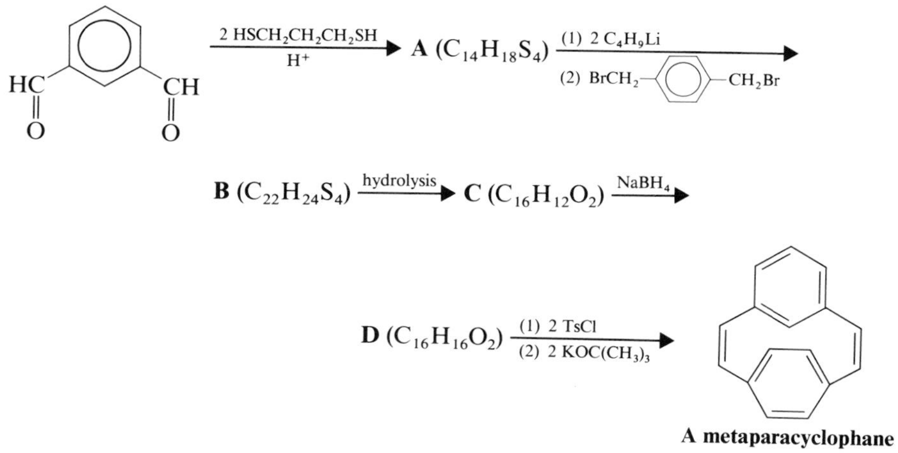

A metaparacyclophane

(b) What compound would be obtained by treating **B** with excess Raney Ni?

19.8 THE KNOEVENAGEL CONDENSATION

Active hydrogen compounds condense with aldehydes and ketones. Known as Knoevenagel condensations, these aldol-like condensations are catalyzed by weak bases. An example is the following:

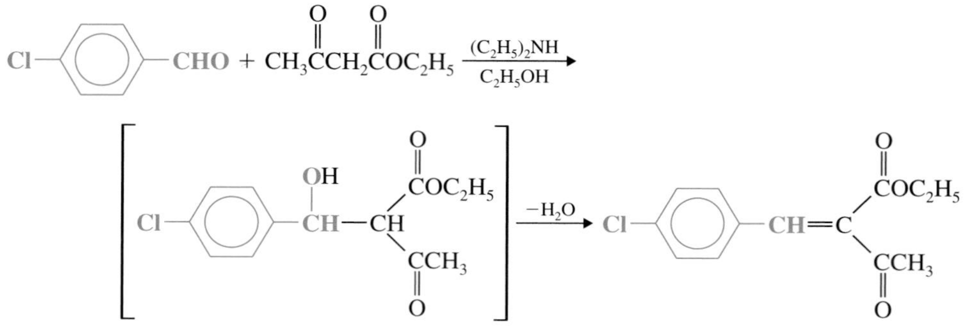

(86%)

19.9 MICHAEL ADDITIONS

Active hydrogen compounds also undergo conjugate additions to α,β-unsaturated carbonyl compounds. These reactions are known as Michael additions, a reaction that we studied in Section 17.9C. An example of the Michael addition of an active hydrogen compound is the following:

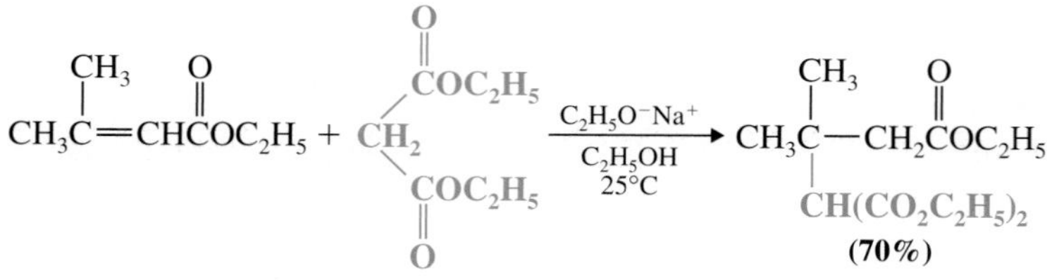

(70%)

The mechanism for this reaction begins with formation of an anion from the active hydrogen compound,

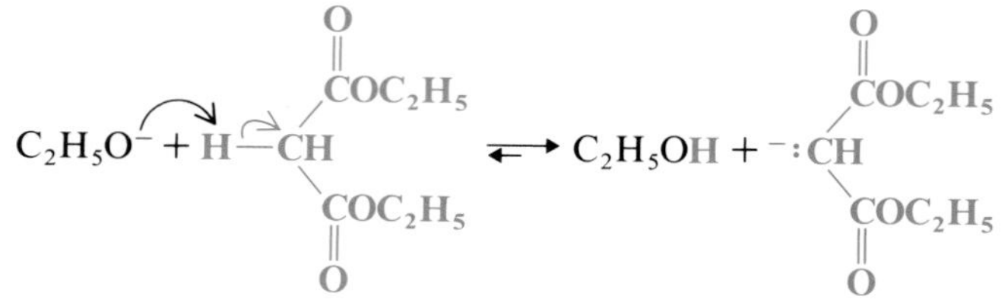

then conjugate addition of the anion to the α,β-unsaturated ester (step 2) is followed by the acceptance of a proton (step 3). Weak nucleophiles such as enolates tend to undergo conjugate addition (Section 17.9).

Step 2

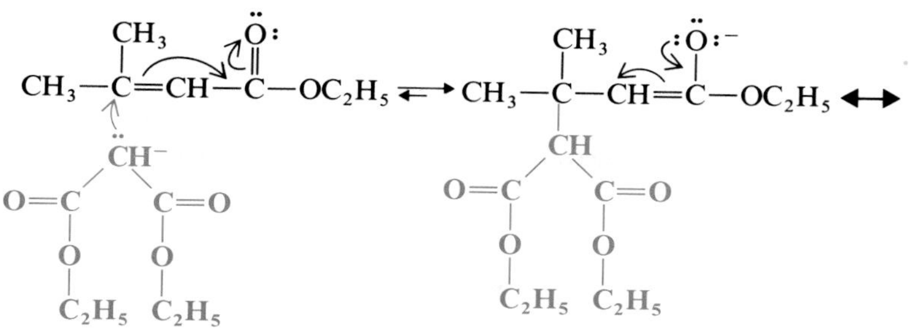

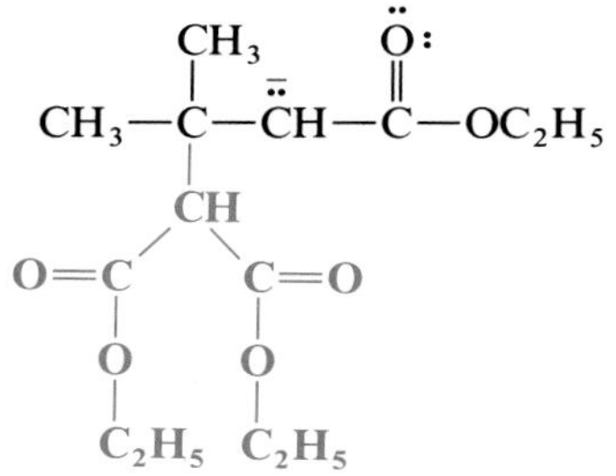

$$CH_3-\underset{\overset{|}{\underset{\overset{|}{CH^-}}{}}}{\overset{CH_3}{\underset{}{C}}}=CH-\overset{\overset{\ddot{\overset{..}{O}}:}{\|}}{C}-OC_2H_5 \rightleftharpoons CH_3-\overset{CH_3}{\underset{}{C}}-CH=\overset{:\overset{..}{O}:^-}{C}-OC_2H_5 \longleftrightarrow$$

with structures $O=C$ and $C=O$, O and O, C_2H_5 and C_2H_5 below.

$$CH_3-\overset{CH_3}{\underset{\overset{|}{CH}}{C}}-\overset{..}{\overset{..}{C}H}-\overset{\overset{\ddot{O}:}{\|}}{C}-OC_2H_5$$

with structures $O=C$ and $C=O$, O and O, C_2H_5 and C_2H_5 below.

Step 3

$$CH_3-\overset{CH_3}{\underset{\overset{|}{CH}}{C}}-\overset{..}{\overset{..}{C}H}-\overset{\overset{\ddot{O}:}{\|}}{C}-OC_2H_5 \xrightarrow{+H^+} CH_3-\overset{CH_3}{\underset{\overset{|}{CH}}{C}}-CH_2-\overset{\overset{\ddot{O}:}{\|}}{C}-OC_2H_5$$

with structures $O=C$ and $C=O$, O and O, C_2H_5 and C_2H_5 below both.

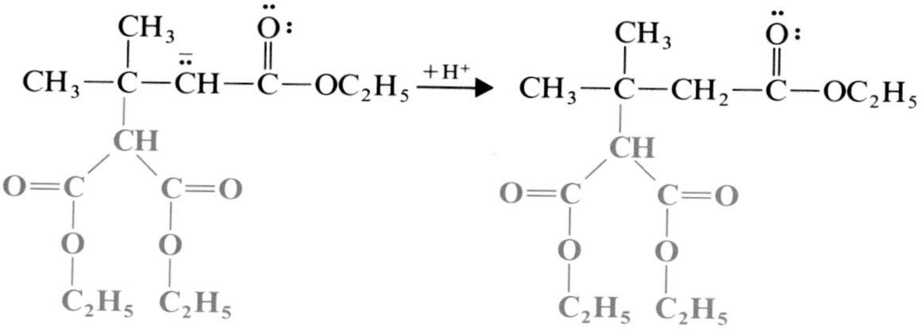

PROBLEM 19.21

How would you prepare $HO\overset{\overset{O}{\|}}{C}CH_2\overset{\overset{CH_3}{|}}{\underset{\overset{|}{CH_3}}{C}}CH_2\overset{\overset{O}{\|}}{C}OH$ from the product of the Michael addition given previously?

Michael additions take place with a variety of other reagents; these include acetylenic esters and α,β-unsaturated nitriles:

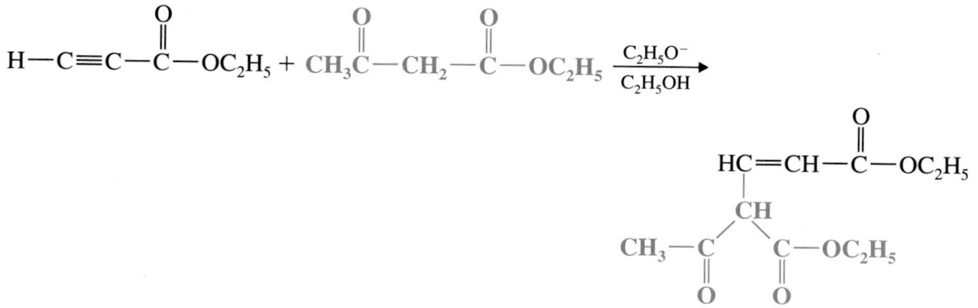

$$H-C\equiv C-\overset{\overset{O}{\|}}{C}-OC_2H_5 + CH_3\overset{\overset{O}{\|}}{C}-CH_2-\overset{\overset{O}{\|}}{C}-OC_2H_5 \xrightarrow[C_2H_5OH]{C_2H_5O^-}$$

$$HC=CH-\overset{\overset{O}{\|}}{C}-OC_2H_5$$
$$\underset{CH_3-\overset{}{\underset{\overset{\|}{O}}{C}} \qquad \overset{}{\underset{\overset{\|}{O}}{C}}-OC_2H_5}{CH}$$

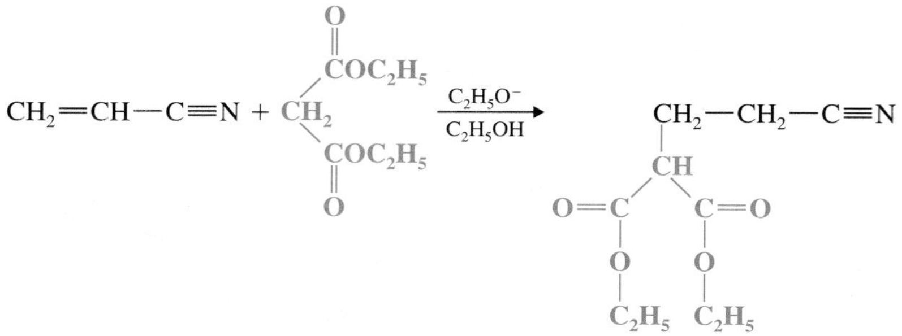

$$CH_2=CH-C\equiv N + \underset{\underset{\displaystyle \underset{O}{\overset{\|}{C}}OC_2H_5}{|}}{\overset{\overset{\displaystyle \overset{O}{\overset{\|}{C}}OC_2H_5}{|}}{CH_2}} \xrightarrow[\text{C}_2\text{H}_5\text{OH}]{\text{C}_2\text{H}_5\text{O}^-} \underset{\underset{C_2H_5 \quad C_2H_5}{|\qquad |}}{\underset{O \qquad O}{\underset{\|\qquad \|}{\underset{O=C \qquad C=O}{\underset{\diagdown \quad \diagup}{CH}}}}}\overset{\displaystyle CH_2-CH_2-C\equiv N}{|}$$

19.10 THE MANNICH REACTION

Compounds capable of forming an enol react with formaldehyde and a primary or secondary amine to yield compounds called Mannich bases. The following reaction of acetone, formaldehyde, and diethylamine is an example.

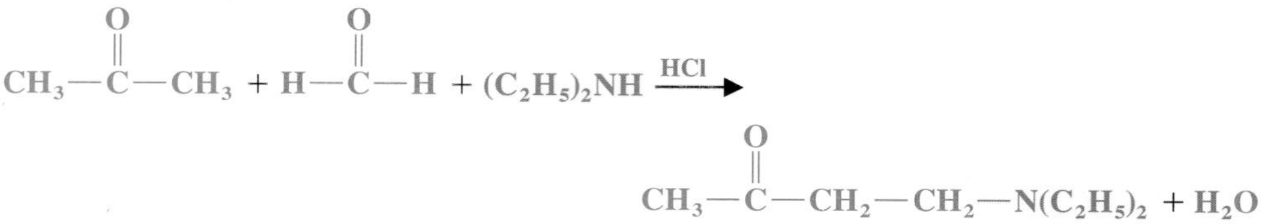

$$CH_3-\overset{\overset{\displaystyle O}{\|}}{C}-CH_3 + H-\overset{\overset{\displaystyle O}{\|}}{C}-H + (C_2H_5)_2NH \xrightarrow{\text{HCl}}$$

$$CH_3-\overset{\overset{\displaystyle O}{\|}}{C}-CH_2-CH_2-N(C_2H_5)_2 + H_2O$$
A Mannich base

The Mannich reaction apparently proceeds through a variety of mechanisms depending on the reactants and the conditions that are employed. One mechanism that appears to operate in neutral or acidic media involves (step 1) initial reaction of the secondary amine with formaldehyde to yield an iminium ion and (step 2) subsequent reaction of the iminium ion with the enol form of the active hydrogen compound.

Step 1 $$R_2\overset{..}{N}H + \underset{H}{\overset{H}{>}}C=\overset{..}{O}: \rightleftharpoons R_2\overset{..}{N}-\overset{\overset{\displaystyle H}{|}}{\underset{\underset{\displaystyle H}{|}}{C}}-\overset{..}{O}-H \xrightarrow{\text{H}^+}$$

$$R_2\overset{..}{N}-\overset{\overset{\displaystyle H}{|}}{\underset{\underset{\displaystyle H}{|}}{C}}-\overset{\overset{\displaystyle H}{|}}{\underset{\underset{\displaystyle ..}{|}}{O}}{}^+-H \xrightarrow{-\text{H}_2\text{O}} R_2\overset{+}{N}=CH_2$$

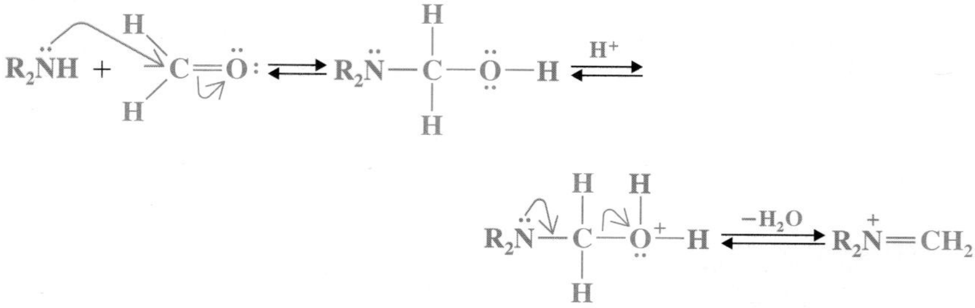

Iminium ion

Step 2 $$CH_3-\overset{\overset{\displaystyle O}{\|}}{C}-CH_3 \underset{\text{H}^+}{\overset{}{\rightleftharpoons}} CH_3-\overset{\overset{\displaystyle O-H}{|}}{C}=CH_2 \qquad \textbf{Enol}$$

$$\overset{+}{CH_2}=\overset{..}{N}R_2 \qquad \textbf{Iminium ion}$$

$$CH_3-\overset{\overset{\displaystyle O}{\|}}{C}-CH_2-CH_2-\overset{..}{N}R_2 + H^+$$
Mannich base

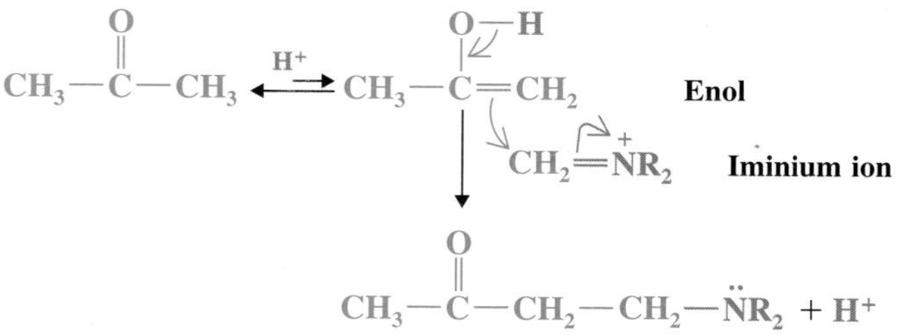

PROBLEM 19.22

Outline reasonable mechanisms that account for the products of the following Mannich reactions:

(a)

$$\text{cyclohexanone} + CH_2O + (CH_3)_2NH \longrightarrow \text{2-((dimethylamino)methyl)cyclohexanone}$$

(b)

$$\text{Ph—CCH}_3 + CH_2O + \underset{H}{\text{pyrrolidine}} \longrightarrow \text{Ph—CCH}_2CH_2\text{—N(pyrrolidine)}$$

(c)

$$\text{4-methylphenol} + 2\ CH_2O + 2\ (CH_3)_2NH \longrightarrow \text{2,6-bis((dimethylamino)methyl)-4-methylphenol}$$

19.11 SYNTHESIS OF ENAMINES: STORK ENAMINE REACTIONS

Aldehydes and ketones react with secondary amines to form compounds called *enamines*. The general reaction for enamine formation can be written as follows:

$$\underset{\substack{\text{Aldehyde} \\ \text{or ketone}}}{\overset{\ddot{O}}{\underset{H}{\text{C}}}} + \underset{\substack{2°\text{Amine}}}{\text{HN—R}} \rightleftharpoons \underset{\substack{H}}{-\overset{OH}{\underset{}{C}}-\overset{}{\underset{}{C}}-\overset{R}{\underset{R}{N}}} \rightleftharpoons \underset{\text{Enamine}}{\overset{R}{\underset{R}{N}}\overset{}{C}=C} + H_2O$$

Since enamine formation requires the loss of a molecule of water, enamine preparations are usually carried out in a way that allows water to be removed as an azeotrope or by a drying agent. This removal of water drives the reversible reaction to completion. Enamine formation is also catalyzed by the presence of a trace of an acid. The secondary amines most commonly used to prepare enamines are cyclic amines such as pyrrolidine, piperidine, and morpholine.

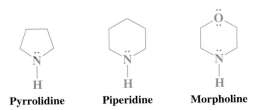

Pyrrolidine   **Piperidine**   **Morpholine**

Cyclohexanone, for example, reacts with pyrrolidine in the following way:

N-(1-Cyclohexenyl)pyrrolidine
(an enamine)

Enamines are good nucleophiles, and an examination of the resonance structures that follow will show us that we should expect enamines to have both a nucleophilic nitrogen and a *nucleophilic carbon.*

| **Contribution to the hybrid made by this structure confers nucleophilicity on nitrogen.** | **Contribution to the hybrid made by this structure confers nucleophilicity on carbon and decreases nucleophilicity of nitrogen.** |

The nucleophilicity of the carbon of enamines makes them particularly useful reagents in organic synthesis because they can be **acylated, alkylated,** and used in **Michael additions.** Development of these techniques originated with the work of Gilbert Stork of Columbia University and in his honor they have come to be known as **Stork enamine reactions.**

When an enamine reacts with an acyl halide or an acid anhydride, the product is the *C*-acylated compound. The iminium ion that forms hydrolyzes when water is added and the overall reaction provides a synthesis of β-diketones.

Iminium salt

2-Acetylcyclohexanone
(a β-diketone)

Although *N*-acylation may occur in this synthesis, the *N*-acyl product is unstable and can act as an acylating agent itself.

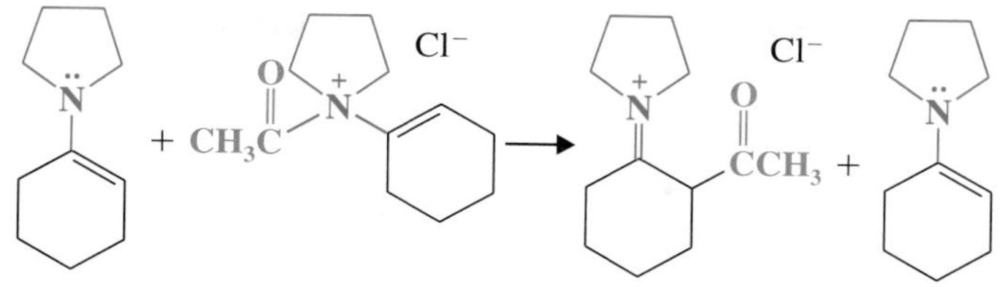

| Enamine | N-Acylated enamine | C-Acylated iminium salt | Enamine |

As a consequence, the yields of *C*-acylated products are generally high.

Enamines can be alkylated as well as acylated. Although alkylation may lead to the formation of a considerable amount of *N*-alkylated product, heating the *N*-alkylated product often converts it to a *C*-alkyl compound. This rearrangement is particularly favored when the alkyl halide is an allylic halide, benzylic halide, or α-haloacetic ester.

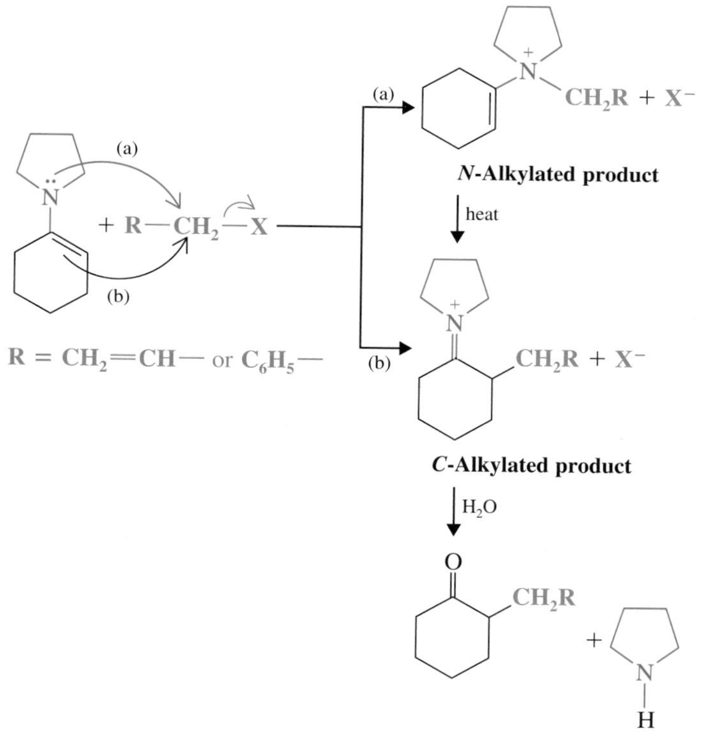

N-Alkylated product

C-Alkylated product

$R = CH_2$=CH— or C_6H_5—

Enamine alkylations are S_N2 reactions; therefore, when we choose our alkylating agents, we are usually restricted to the use of methyl, primary, allylic, and benzylic halides. α-Halo esters can also be used as the alkylating agents, and this reaction provides a convenient synthesis of γ-keto esters:

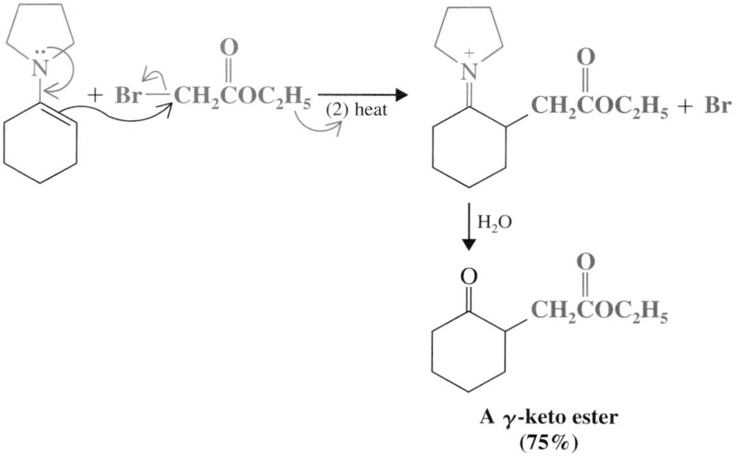

**A γ-keto ester
(75%)**

PROBLEM 19.23

Show how you could employ enamines in syntheses of the following compounds:

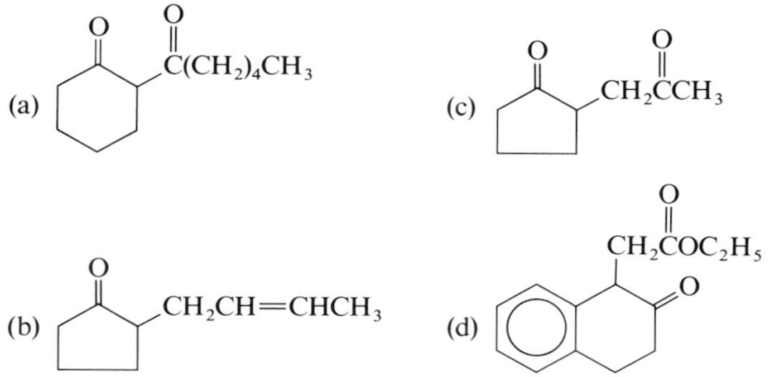

(a)

(c)

(b)

(d)

An especially interesting set of enamine alkylations is shown in the following reactions (developed by J. K. Whitesell of the University of Texas at Austin). The enamine (prepared from a single enantiomer of the secondary amine) is chiral. Alkylation from the bottom of the enamine is severely hindered by the methyl group. (Notice that this hindrance exists even if rotation of the groups takes place about the bond connecting the two rings.) Consequently, alkylation takes place much more rapidly from the top side. This reaction yields (after hydrolysis) 2-substituted cyclohexanones consisting almost entirely of a single enantiomer.

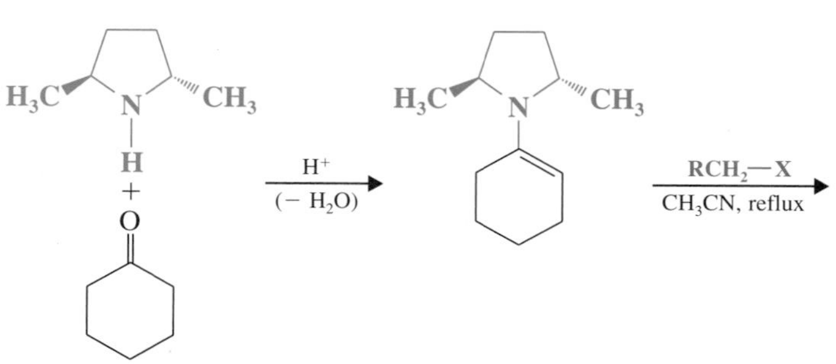

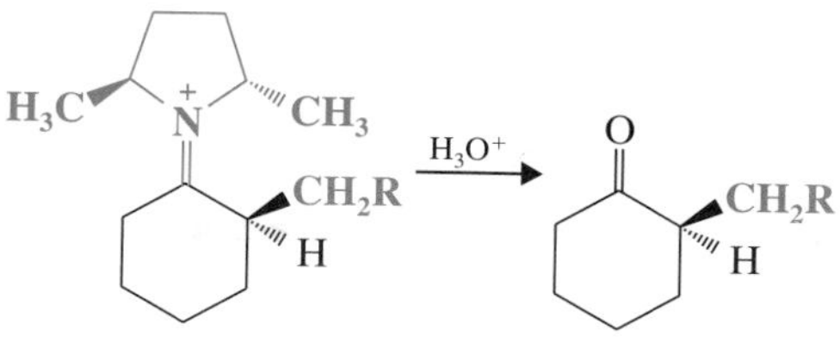

R Group	Chemical Yield (%)	Enantiomeric Excess (%)
H—	50	83
CH₃CH₂—	57	93
CH₂=CH—	80	82

Enamines can also be used in Michael additions. An example is the following:

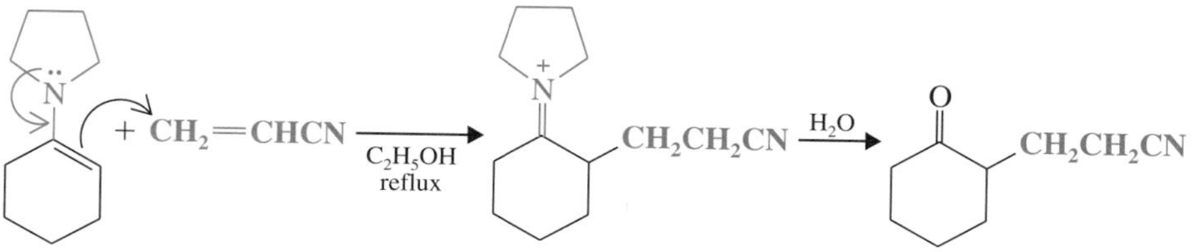

19.12 BARBITURATES

In the presence of sodium ethoxide, diethyl malonate reacts with urea to yield a compound called barbituric acid.

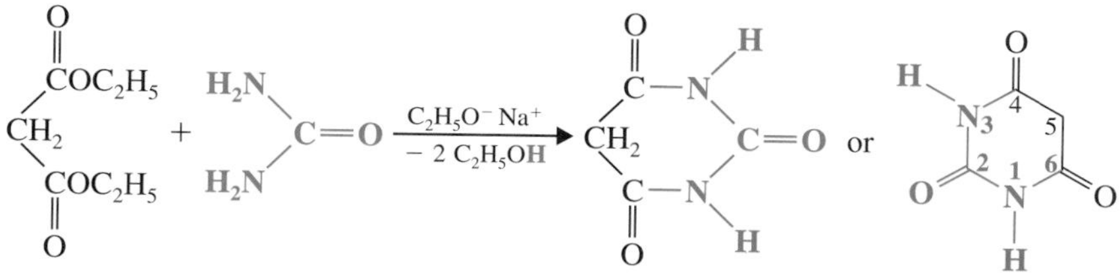

Barbituric acid

Barbituric acid is a pyrimidine derivative (cf. Section 20.1), and it exists in several tautomeric forms including one with an aromatic ring.

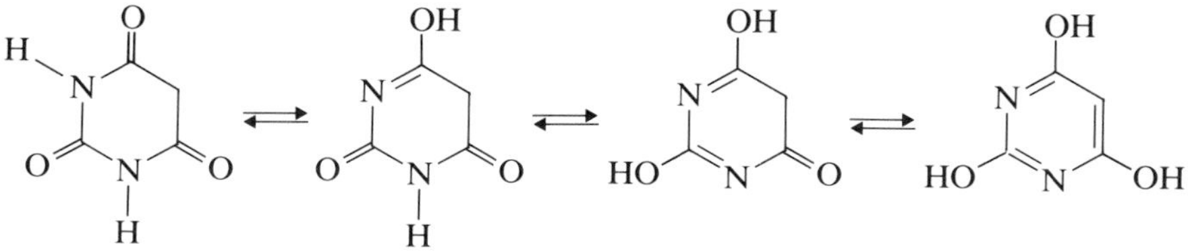

As its name suggests barbituric acid is a moderately strong acid, stronger even than acetic acid. Its anion is highly resonance stabilized.

Derivatives of barbituric acid are *barbiturates*. Barbiturates have been used in medicine as soporifics (sleep producers) since 1903. One of the earliest barbiturates introduced into medical use is the compound veronal (5,5-diethylbarbituric acid).

Veronal is usually used as its sodium salt. Other barbiturates are seconal and pheno-barbital.

Veronal	Seconal	Phenobarbital
(5,5-diethylbarbituric acid)	**[5-allyl-5-(1-methylbutyl)**	**(5-ethyl-5-phenylbarbituric**
	barbituric acid)]	**acid)**

Although barbiturates are very effective soporifics, their use is also hazardous. They are addictive, and overdosage, often with fatal results, is common.

PROBLEM 19.24

Outlined here is a synthesis of phenobarbital.

(a) What are compounds **A–F**?   (b) Propose an alternative synthesis of **E** from diethyl malonate.

$$C_6H_5—CH_3 \xrightarrow[CCl_4]{NBS} A\ (C_7H_7Br) \xrightarrow[(2)\ CO_2,\ then\ H^+]{(1)\ Mg,\ Et_2O} B\ (C_8H_8O_2) \xrightarrow{SOCl_2}$$

$$C\ (C_8H_7ClO) \xrightarrow{EtOH} D\ (C_{10}H_{12}O_2) \xrightarrow[NaOEt]{EtOCOEt} E\ (C_{13}H_{16}O_4) \xrightarrow[CH_3CH_2Br]{KOC(CH_3)_3}$$

$$F\ (C_{15}H_{20}O_4) \xrightarrow{H_2NCNH_2,\ NaOEt} phenobarbital$$

PROBLEM 19.25

Starting with diethyl malonate, urea, and any other required reagents, outline a synthesis of veronal and seconal.

19.13 SUMMARY OF IMPORTANT REACTIONS

1. Claisen condensation (Section 19.2)

$$2\ R—CH_2—\overset{O}{\overset{\|}{C}}—OEt \xrightarrow[(2)\ H^+]{(1)\ NaOEt} R—CH_2—\overset{O}{\overset{\|}{C}}—\underset{\underset{R}{|}}{CH}—\overset{O}{\overset{\|}{C}}—OEt$$

2. Crossed Claisen condensation (Section 19.2A)

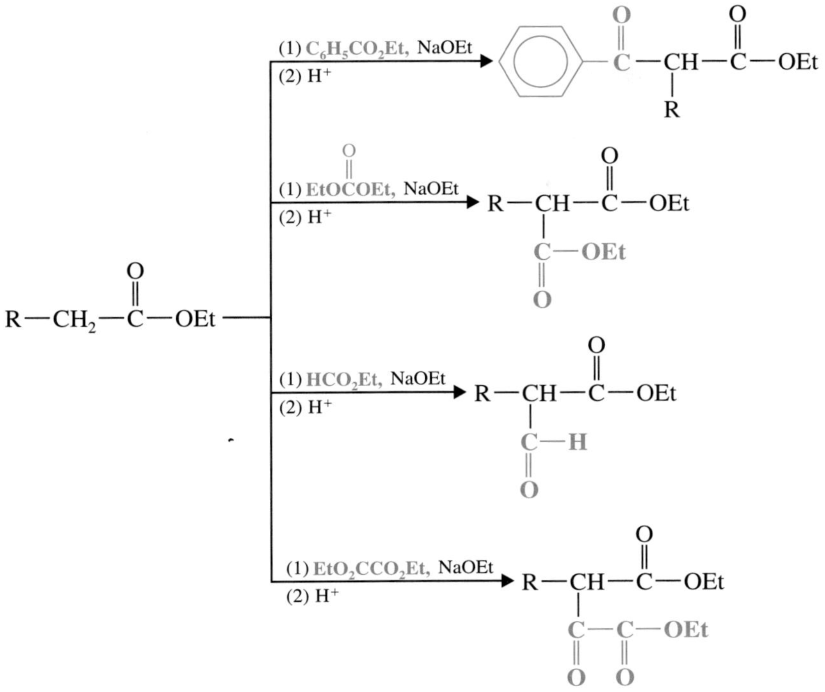

$$\text{R—CH}_2\text{—}\overset{\displaystyle O}{\overset{\|}{C}}\text{—OEt}$$

$\xrightarrow[\text{(2) H}^+]{\text{(1) C}_6\text{H}_5\text{CO}_2\text{Et, NaOEt}}$ C$_6$H$_5$—$\overset{\displaystyle O}{\overset{\|}{C}}$—CH—$\overset{\displaystyle O}{\overset{\|}{C}}$—OEt with R below the CH

$\xrightarrow[\text{(2) H}^+]{\text{(1) EtO}\overset{O}{\overset{\|}{C}}\text{OEt, NaOEt}}$ R—CH—$\overset{\displaystyle O}{\overset{\|}{C}}$—OEt with C—OEt (C=O) below

$\xrightarrow[\text{(2) H}^+]{\text{(1) HCO}_2\text{Et, NaOEt}}$ R—CH—$\overset{\displaystyle O}{\overset{\|}{C}}$—OEt with C—H (C=O) below

$\xrightarrow[\text{(2) H}^+]{\text{(1) EtO}_2\text{CCO}_2\text{Et, NaOEt}}$ R—CH—$\overset{\displaystyle O}{\overset{\|}{C}}$—OEt with C—C—OEt (both C=O) below

3. Acetoacetic ester synthesis (Section 19.3)

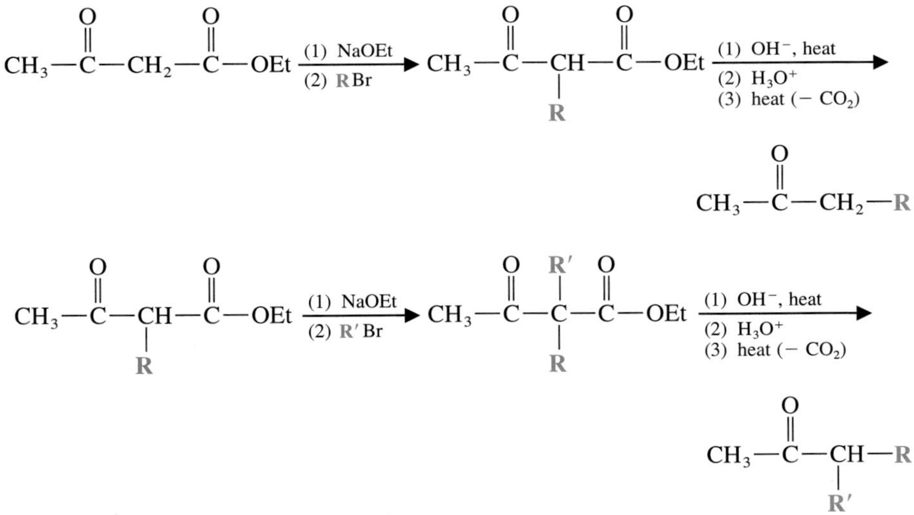

$$\text{CH}_3\text{—}\overset{\displaystyle O}{\overset{\|}{C}}\text{—CH}_2\text{—}\overset{\displaystyle O}{\overset{\|}{C}}\text{—OEt} \xrightarrow[\text{(2) R Br}]{\text{(1) NaOEt}} \text{CH}_3\text{—}\overset{\displaystyle O}{\overset{\|}{C}}\text{—CH—}\overset{\displaystyle O}{\overset{\|}{C}}\text{—OEt} \xrightarrow[\substack{\text{(2) H}_3\text{O}^+\\ \text{(3) heat }(-\text{CO}_2)}]{\text{(1) OH}^-\text{, heat}}$$

(with R below CH)

$$\text{CH}_3\text{—}\overset{\displaystyle O}{\overset{\|}{C}}\text{—CH}_2\text{—R}$$

$$\text{CH}_3\text{—}\overset{\displaystyle O}{\overset{\|}{C}}\text{—CH—}\overset{\displaystyle O}{\overset{\|}{C}}\text{—OEt} \xrightarrow[\text{(2) R' Br}]{\text{(1) NaOEt}} \text{CH}_3\text{—}\overset{\displaystyle O}{\overset{\|}{C}}\text{—}\overset{\displaystyle R'}{\overset{|}{C}}\text{—}\overset{\displaystyle O}{\overset{\|}{C}}\text{—OEt} \xrightarrow[\substack{\text{(2) H}_3\text{O}^+\\ \text{(3) heat }(-\text{CO}_2)}]{\text{(1) OH}^-\text{, heat}}$$

(with R below CH in reactant; R below C in product)

$$\text{CH}_3\text{—}\overset{\displaystyle O}{\overset{\|}{C}}\text{—CH—R}$$
(with R' below CH)

4. Malonic ester synthesis (Section 19.4)

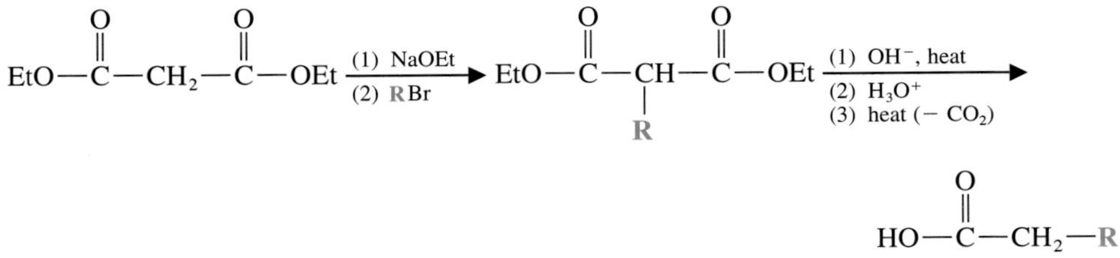

$$\text{EtO—}\overset{\displaystyle O}{\overset{\|}{C}}\text{—CH}_2\text{—}\overset{\displaystyle O}{\overset{\|}{C}}\text{—OEt} \xrightarrow[\text{(2) R Br}]{\text{(1) NaOEt}} \text{EtO—}\overset{\displaystyle O}{\overset{\|}{C}}\text{—CH—}\overset{\displaystyle O}{\overset{\|}{C}}\text{—OEt} \xrightarrow[\substack{\text{(2) H}_3\text{O}^+\\ \text{(3) heat }(-\text{CO}_2)}]{\text{(1) OH}^-\text{, heat}}$$

(with R below CH)

$$\text{HO—}\overset{\displaystyle O}{\overset{\|}{C}}\text{—CH}_2\text{—R}$$

$$\underset{\substack{|\\R}}{EtO-\overset{O}{\overset{||}{C}}-CH-\overset{O}{\overset{||}{C}}-OEt} \xrightarrow[\text{(2) } R'Br]{\text{(1) NaOEt}} \underset{\substack{|\\R}}{EtO-\overset{O}{\overset{||}{C}}-\overset{R'}{\underset{}{C}}-\overset{O}{\overset{||}{C}}-OEt} \xrightarrow[\substack{\text{(2) } H_3O^+ \\ \text{(3) heat } (-CO_2)}]{\text{(1) OH}^-,\text{ heat}}$$

$$\underset{\substack{|\\R'}}{HO-\overset{O}{\overset{||}{C}}-CH-R}$$

5. Direct alkylation of esters (Section 19.6)

$$R-CH_2-\overset{O}{\overset{||}{C}}-OEt \xrightarrow[\text{THF}]{\text{LDA}} \overset{Li^+}{R-\overset{..}{C}H-\overset{O}{\overset{||}{C}}-OEt} \xrightarrow{R'CH_2\frown Br} \underset{\substack{|\\CH_2\\|\\R'}}{R-CH-\overset{O}{\overset{||}{C}}-OEt}$$

6. Alkylation of dithianes (Section 19.7)

$$R-\overset{O}{\overset{||}{C}}-H \xrightarrow[H^+]{HSCH_2CH_2CH_2SH} \underset{R\quad H}{\overset{S\quad S}{\bigcirc}} \xrightarrow[\text{(2) } R'CH_2X]{\text{(1) BuLi}} \underset{R\quad CH_2R'}{\overset{S\quad S}{\bigcirc}} \xrightarrow[H_2O]{HgCl_2,\ CH_3OH} R-\overset{O}{\overset{||}{C}}-CH_2R'$$

7. Knoevenagel condensation (Section 19.8)

$$\bigcirc\!\!-CHO + \underset{\substack{|\\CO_2R}}{\overset{CO_2R}{\overset{|}{CH_2}}} \xrightarrow[(-H_2O)]{\text{base}} \bigcirc\!\!-CH=C\overset{CO_2R}{\underset{CO_2R}{}}$$

8. Michael addition (Section 19.9)

$$\underset{\substack{|\\R'}}{R-\overset{R'}{\underset{}{C}}=CH-\overset{O}{\overset{||}{C}}-R''} + \underset{\substack{|\\CO_2R}}{\overset{CO_2R}{\overset{|}{CH_2}}} \xrightarrow{\text{base}} \underset{\substack{CH-CO_2R\\|\\CO_2R}}{R-\overset{R'}{\underset{|}{C}}-CH_2-\overset{O}{\overset{||}{C}}-R''}$$

α,β-Unsaturated   **Or other active**
carbonyl compound   **methylene compound**

9. Mannich reaction (Section 19.10)

$$R-\overset{O}{\overset{||}{C}}-CH_3 + H-\overset{O}{\overset{||}{C}}-H + H-N\overset{R'}{\underset{R''}{}} \longrightarrow R-\overset{O}{\overset{||}{C}}-CH_2-CH_2-N\overset{R'}{\underset{R''}{}}$$

10. Stork enamine reaction (Section 19.11)

$$RCH_2-\overset{O}{\overset{||}{C}}-CH_2R + R_2'NH \longrightarrow RCH_2-\overset{NR_2'}{\underset{}{C}}=CHR \xrightarrow[\substack{\text{(2) heat}\\\text{(3) } H_2O}]{\text{(1) } R''CH_2Br} \underset{\substack{|\\CH_2R''}}{RCH_2-\overset{O}{\overset{||}{C}}-CHR}$$

Enamine

ADDITIONAL PROBLEMS

19.26 Show all steps in the following syntheses. You may use any other needed reagents but should begin with the compound given. You need not repeat steps carried out in earlier parts of this exercise.

(a) $CH_3CH_2CH_2\overset{O}{\underset{\|}{C}}OC_2H_5 \longrightarrow CH_3CH_2CH_2\overset{O}{\underset{\|}{C}}\overset{}{C}H\overset{O}{\underset{\|}{C}}OC_2H_5$ with CH_2CH_3 branch

$$CH_3CH_2CH_2\overset{O}{\underset{\|}{C}}OC_2H_5 \longrightarrow CH_3CH_2CH_2\overset{O}{\underset{\|}{C}}\overset{|}{C}H\overset{O}{\underset{\|}{C}}OC_2H_5 \quad \begin{array}{c} CH_2 \\ | \\ CH_3 \end{array}$$

(b) $CH_3CH_2CH_2\overset{O}{\underset{\|}{C}}OC_2H_5 \longrightarrow CH_3CH_2CH_2\overset{O}{\underset{\|}{C}}CH_2CH_2CH_3$

(c) $C_6H_5CH_2\overset{O}{\underset{\|}{C}}OC_2H_5 \longrightarrow C_6H_5\overset{CH_3}{\underset{|}{C}}HCO_2H$

(d) $CH_3CH_2CH_2\overset{O}{\underset{\|}{C}}OC_2H_5 \longrightarrow CH_3CH_2\overset{O}{\underset{\|}{C}}H\overset{O}{\underset{\|}{C}}OC_2H_5$ with $\overset{O}{\underset{\|}{C}}\!-\!\overset{O}{\underset{\|}{C}}OC_2H_5$

(e) $CH_3CH_2CH_2\overset{O}{\underset{\|}{C}}OC_2H_5 \longrightarrow CH_3CH_2CH_2\overset{O}{\underset{\|}{C}}\!-\!\overset{O}{\underset{\|}{C}}OC_2H_5$

(f) $C_6H_5CH_2\overset{O}{\underset{\|}{C}}OC_2H_5 \longrightarrow C_6H_5\overset{|}{C}H\overset{O}{\underset{\|}{C}}OC_2H_5$ with $\overset{CH}{\underset{\|}{O}}$ branch

(g) cyclopentanone $\longrightarrow$ 2-acetylcyclopentanone ($\overset{O}{\underset{\|}{C}}CH_3$)

(h) cyclopentanone $\longrightarrow$ 2-methyl-2-acetylcyclopentanone (CH_3, $\overset{O}{\underset{\|}{C}}CH_3$)

(i) 2-(ethoxycarbonyl)cyclohexanone ($\overset{O}{\underset{\|}{C}}OC_2H_5$) $\longrightarrow$ 2-ethylcyclohexanone (CH_2CH_3)

19.27 Outline syntheses of each of the following from acetoacetic ester and any other required reagents.

(a) *tert*-Butyl methyl ketone
(b) 2-Hexanone
(c) 2,5-Hexanedione
(d) 4-Hydroxypentanoic acid
(e) 2-Ethyl-1,3-butanediol
(f) 1-Phenyl-1,3-butanediol

19.28   Outline syntheses of each of the following from diethyl malonate and any other required reagents.

(a) 2-Methylbutanoic acid   (d) $HOCH_2CH_2CH_2CH_2OH$

(b) 4-Methyl-1-pentanol

(c) $CH_3CH_2CHCH_2OH$
$\quad\quad\quad |$
$\quad\quad\quad CH_2OH$

19.29   The synthesis of cyclobutanecarboxylic acid given in Section 19.4 was first carried out by William Perkin, Jr., in 1883, and it represented one of the first syntheses of an organic compound with a ring smaller than six carbon atoms. (There was a general feeling at the time that such compounds would be too unstable to exist.) Earlier in 1883, Perkin reported what he mistakenly believed to be a cyclobutane derivative obtained from the reaction of acetoacetic ester and 1,3-dibromopropane. The reaction that Perkin had expected to take place was the following:

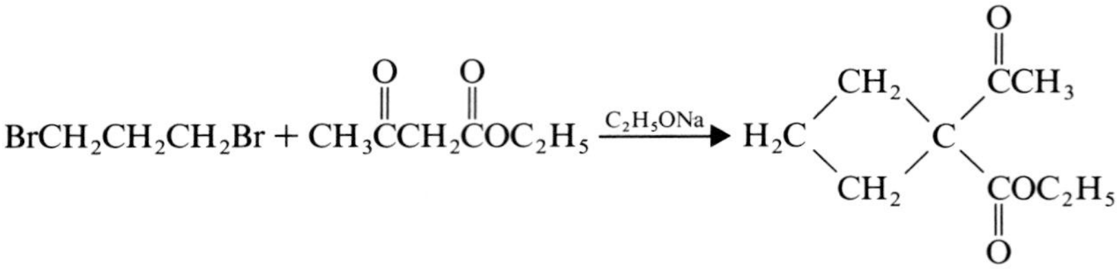

The molecular formula for his product agreed with the formulation given in the preceding reaction, and alkaline hydrolysis and acidification gave a nicely crystalline acid (also having the expected molecular formula). The acid, however, was quite stable to heat and resisted decarboxylation. Perkin later found that both the ester and the acid contained six-membered rings (five carbon atoms and one oxygen atom). Recall the charge distribution in the enolate ion obtained from acetoacetic ester and propose structures for Perkin's ester and acid.

19.30   (a) In 1884 Perkin achieved a successful synthesis of cyclopropanecarboxylic acid from sodiomalonic ester and 1,2-dibromoethane. Outline the reactions involved in this synthesis.   (b) In 1885 Perkin synthesized five-membered carbocyclic compounds **D** and **E** in the following way:

$$2\ Na^+ : \overset{-}{C}H(CO_2C_2H_5)_2 + BrCH_2CH_2CH_2Br \longrightarrow A\ (C_{17}H_{28}O_8) \xrightarrow{2\ C_2H_5O^-Na^+} \xrightarrow{Br_2}$$

$$B\ (C_{17}H_{26}O_8) \xrightarrow[\text{(2) } H_3O^+]{\text{(1) } OH^-/H_2O} C\ (C_9H_{10}O_8) \xrightarrow{\text{heat}} D\ (C_7H_{10}O_4) + E\ (C_7H_{10}O_4)$$

D and **E** are diastereomers; **D** can be resolved into enantiomeric forms while **E** cannot. What are the structures of **A–E**?   (c) Ten years later Perkin was able to synthesize 1,4-dibromobutane; he later used this compound and diethyl malonate to prepare cyclopentanecarboxylic acid. Show the reactions involved.

19.31   Write mechanisms that account for the products of the following reactions:

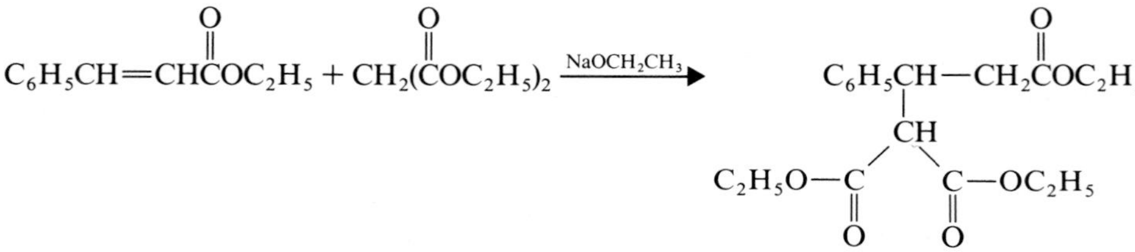

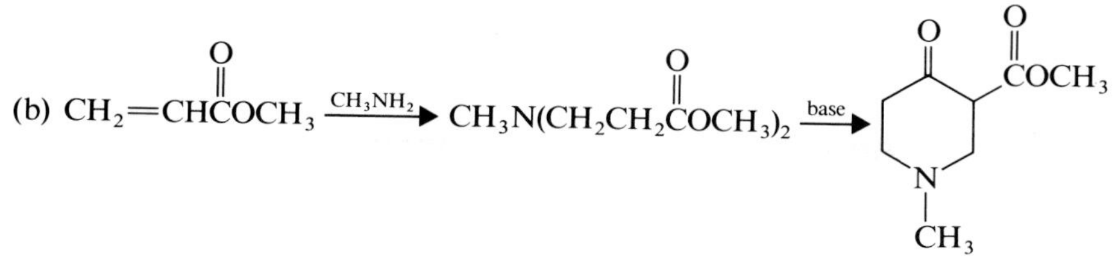

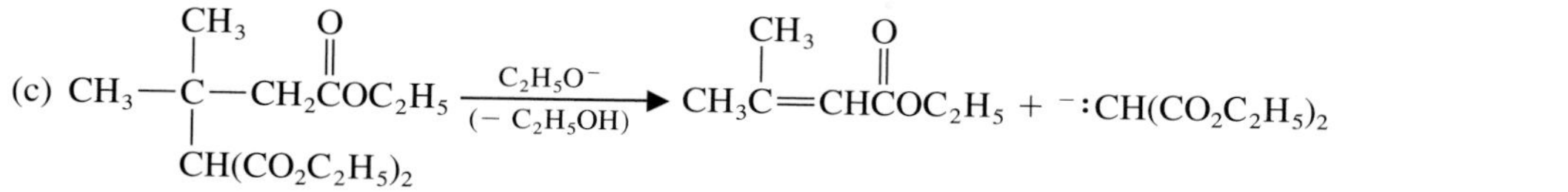

(b) $CH_2{=}CHCOCH_3$ $\xrightarrow{CH_3NH_2}$ $CH_3N(CH_2CH_2COCH_3)_2$ $\xrightarrow{base}$

(c)

$CH_3{-}\underset{\underset{CH(CO_2C_2H_5)_2}{|}}{\overset{\overset{CH_3}{|}}{C}}{-}CH_2\overset{\overset{O}{\|}}{C}OC_2H_5$ $\xrightarrow[(-\,C_2H_5OH)]{C_2H_5O^-}$ $CH_3\overset{\overset{CH_3}{|}}{C}{=}CH\overset{\overset{O}{\|}}{C}OC_2H_5 + {}^-{:}CH(CO_2C_2H_5)_2$

19.32 Knoevenagel condensations in which the active hydrogen compound is a β-keto ester or a β-diketone often yield products that result from one molecule of aldehyde or ketone and two molecules of the active methylene component. For example,

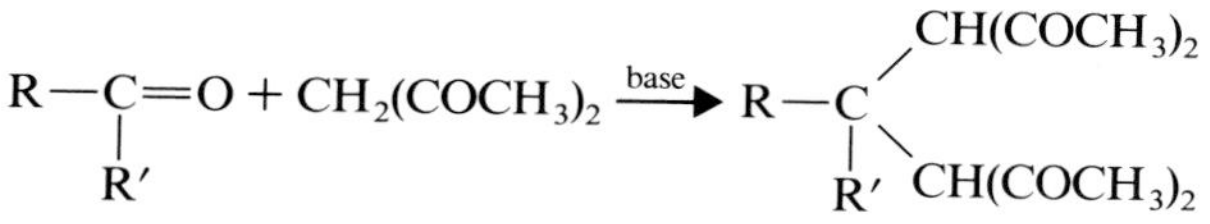

$$R{-}\underset{\underset{R'}{|}}{C}{=}O + CH_2(COCH_3)_2 \xrightarrow{base} R{-}\underset{\underset{R'}{|}}{C}\overset{CH(COCH_3)_2}{\diagup}_{\diagdown CH(COCH_3)_2}$$

Suggest a reasonable mechanism that accounts for the formation of these products.

19.33 Thymine is one of the heterocyclic bases found in DNA. Starting with ethyl propanoate and using any other needed reagents, show how you might synthesize thymine.

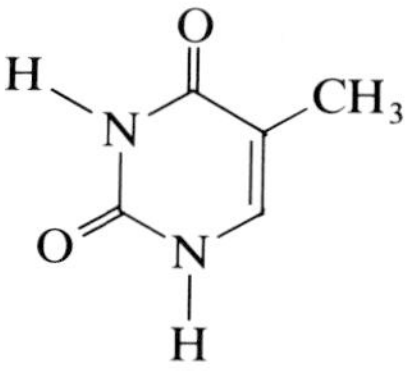

Thymine

19.34 The mandibular glands of queen bees secrete a fluid that contains a remarkable compound known as "queen substance." When even an exceedingly small amount of the queen substance is transferred to worker bees, it inhibits the development of their ovaries and prevents the workers from bearing new queens. Queen substance, a monocarboxylic acid with the molecular formula $C_{10}H_{16}O_3$, has been synthesized by the following route:

Cycloheptanone $\xrightarrow[(2)\ H_3O^+]{(1)\ CH_3MgI}$ $A\ (C_8H_{16}O)$ $\xrightarrow{H^+,\ heat}$ $B\ (C_8H_{14})$ $\xrightarrow[(2)\ Zn,\ H_2O]{(1)\ O_3}$

$C\ (C_8H_{14}O_2)$ $\xrightarrow[pyridine]{CH_2(CO_2H)_2}$ queen substance $(C_{10}H_{16}O_3)$

On catalytic hydrogenation queen substance yields compound **D**, which, on treatment with iodine in sodium hydroxide and subsequent acidification, yields a dicarboxylic acid **E**; that is,

Queen substance $\xrightarrow[Pd]{H_2}$ $D\ (C_{10}H_{18}O_3)$ $\xrightarrow[(2)\ H_3O^+]{(1)\ I_2\ in\ aq\ NaOH}$ $E\ (C_9H_{16}O_4)$

Provide structures for the queen substance and compounds **A–E**.

19.35   Linalool, a fragrant compound that can be isolated from a variety of plants, is 3,7-di-methyl-1,6-octadien-3-ol. Linalool is used in making perfumes and it can be synthesized in the following way:

$$CH_2=\overset{\underset{\textstyle CH_3}{|}}{C}-CH=CH_2 + HBr \longrightarrow F\ (C_5H_9Br) \xrightarrow[\text{ester}]{\text{sodioacetoacetic}}$$

$$G\ (C_{11}H_{18}O_3) \xrightarrow[\text{(2) } H_3O^+,\text{ (3) heat}]{\text{(1) dil. NaOH}} H\ (C_8H_{14}O) \xrightarrow[\text{(2) } H_3O^+]{\text{(1) LiC}\equiv\text{CH}}$$

$$I\ (C_{10}H_{16}O) \xrightarrow[\substack{\text{Lindlar's}\\\text{catalyst}}]{H_2} \text{linalool}$$

Outline the reactions involved. (*Hint:* Compound **F** is the more stable isomer capable of being produced in the first step.)

19.36   Compound **J**, a compound with two four-membered rings, has been synthesized by the following route. Outline the steps that are involved.

$$NaCH(CO_2C_2H_5)_2 + BrCH_2CH_2CH_2Br \longrightarrow (C_{10}H_{17}BrO_4) \xrightarrow{NaOC_2H_5}$$

$$C_{10}H_{16}O_4 \xrightarrow[\text{(2) } H_2O]{\text{(1) LiAlH}_4} C_6H_{12}O_2 \xrightarrow{HBr} C_6H_{10}Br_2 \xrightarrow[\text{2 NaOC}_2H_5]{CH_2(CO_2C_2H_5)_2}$$

$$C_{13}H_{20}O_4 \xrightarrow[\text{(2) } H_3O^+]{\text{(1) OH}^-,\ H_2O} C_9H_{12}O_4 \xrightarrow{heat} J\ (C_8H_{12}O_2) + CO_2$$

19.37   When an aldehyde or a ketone is condensed with ethyl α-chloroacetate in the presence of sodium ethoxide, the product is an α,β-epoxy ester called a *glycidic ester.* The synthesis is called the Darzens condensation.

$$R-\overset{\underset{\textstyle R'}{|}}{C}=O + ClCH_2CO_2C_2H_5 \xrightarrow{C_2H_5ONa} R-\overset{\underset{\textstyle }{|}}{\underset{O}{C}}\overset{R'}{\diagdown\diagup}CHCO_2C_2H_5 + NaCl + C_2H_5OH$$

A glycidic ester

(a) Outline a reasonable mechanism for the Darzens condensation.   (b) Hydrolysis of the epoxy ester leads to an epoxy acid that, on heating with pyridine, furnishes an aldehyde.

$$R-\overset{\underset{\textstyle }{|}}{\underset{O}{C}}\overset{R'}{\diagdown\diagup}CHCO_2H \xrightarrow[\text{heat}]{C_5H_5N} R-\overset{R'}{\underset{|}{C}H}-\overset{O}{\overset{||}{C}}H + CO_2$$

What is happening here?   (c) Starting with β-ionone (Problem 17.13), show how you might synthesize the following aldehyde. (This aldehyde is an intermediate in an industrial synthesis of vitamin A.)

19.38   The *Perkin condensation* is an aldol-type condensation in which an aromatic aldehyde (ArCHO) reacts with a carboxylic acid anhydride $(RCH_2CO)_2O$, to give an α,β-unsaturated acid $(ArCH=CRCO_2H)$. The catalyst that is usually employed is the potassium salt of the carboxylic acid (RCH_2CO_2K).   (a) Outline the Perkin condensation that takes place

when benzaldehyde reacts with propanoic anhydride in the presence of potassium propanoate.
(b) How would you use a Perkin condensation to prepare *p*-chlorocinnamic acid,
p-ClC$_6$H$_4$CH=CHCO$_2$H?

19.39   (+)-Fenchone is a terpenoid that can be isolated from fennel oil. (±)-Fenchone has
been synthesized through the following route. Supply the missing intermediates and reagents.

19.40   Outline a synthesis of the analgesic Darvon (below) starting with ethyl phenyl
ketone.

Darvon

19.41   Explain the variation in enol content that is observed for solutions of acetylacetone
(2,4-pentanedione) in the several solvents indicated.

solvent	% enol
H$_2$O	15
CH$_3$CN	58
C$_6$H$_{14}$	92
gas phase	92

19.42   When a Dieckmann condensation is attempted with diethyl succinate, the product
obtained has the molecular formula C$_{12}$H$_{16}$O$_6$. What is the structure of this compound?

19.43   Ethyl crotonate, $CH_3CH{=}CHCOOC_2H_5$, reacts with diethyl oxalate, $C_2H_5OOC{-}COOC_2H_5$, to form a Claisen-type condensation product,

$$\underset{\displaystyle C_2H_5OC}{\overset{\displaystyle O}{\|}} {-} \underset{\displaystyle CCH_2CH}{\overset{\displaystyle O}{\|}} {=} CH \underset{\displaystyle COC_2H_5}{\overset{\displaystyle O}{\|}}$$

Account for the formation of this compound.

19.44   Show how this diketone could be prepared by a condensation reaction:

CHAPTER 20

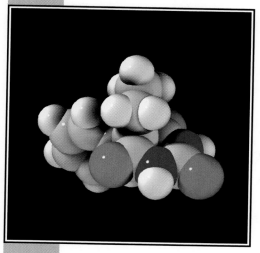

Adrenalin
(Section 20.4)

AMINES

20.1 NOMENCLATURE

In common nomenclature most primary amines are named as *alkylamines*. In systematic nomenclature (in parentheses below) they are named by adding the suffix *-amine* to the name of the chain or ring system to which the NH_2 group is attached with elision of the final *-e*. Amines are classified as being primary (1°), secondary (2°), or tertiary (3°) on the basis of the number of organic groups attached to the nitrogen (Section 2.12).

Primary Amines

CH_3NH_2
Methylamine
(methanamine)

$CH_3CH_2NH_2$
Ethylamine
(ethanamine)

$CH_3CHCH_2NH_2$
|
CH_3
Isobutylamine
(2-methyl-1-propanamine)

—NH_2
Cyclohexylamine
(cyclohexanamine)

Most secondary and tertiary amines are named in the same general way. In common nomenclature we either designate the organic groups individually if they are different, or use the prefixes *di-* or *tri-* if they are the same. In systematic nomenclature we use the locant *N* to designate substituents attached to a nitrogen atom.

Secondary Amines

$CH_3NHCH_2CH_3$
Ethylmethylamine
(*N*-methylethanamine)

$(CH_3CH_2)_2NH$
Diethylamine
(*N*-ethylethanamine)

Tertiary Amines

$(CH_3CH_2)_3N$
Triethylamine
(*N,N*-diethylethanamine)

$$CH_2CH_3$$
$$|$$
$$CH_3NCH_2CH_2CH_3$$
Ethylmethylpropylamine
(*N*-ethyl-*N*-methyl-1-propanamine)

In the IUPAC system, the substituent —NH$_2$ is called the *amino* group. We often use this system for naming amines containing an OH group or a CO$_2$H group.

$H_2NCH_2CH_2OH$
2-Aminoethanol

$$O$$
$$||$$
$$H_2NCH_2CH_2COH$$
3-Aminopropanoic acid

20.1A Arylamines

Three common arylamines have the following names:

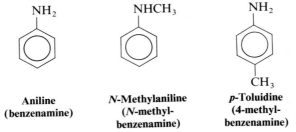

Aniline
(benzenamine)

N-Methylaniline
(*N*-methyl-benzenamine)

p-Toluidine
(4-methyl-benzenamine)

20.1B Heterocyclic Amines

The important *heterocyclic* amines all have common names. In systematic replacement nomenclature the prefixes *aza-*, *diaza-*, and *triaza-* are used to indicate that nitrogen atoms have replaced carbon atoms in the corresponding hydrocarbon.

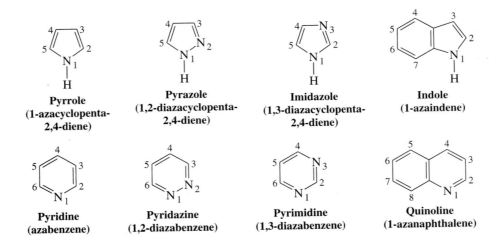

Pyrrole
(1-azacyclopenta-2,4-diene)

Pyrazole
(1,2-diazacyclopenta-2,4-diene)

Imidazole
(1,3-diazacyclopenta-2,4-diene)

Indole
(1-azaindene)

Pyridine
(azabenzene)

Pyridazine
(1,2-diazabenzene)

Pyrimidine
(1,3-diazabenzene)

Quinoline
(1-azanaphthalene)

Piperidine
(azacyclohexane)

Pyrrolidine
(azacyclopentane)

Thiazole
(1-thia-3-
azacyclopenta-
2,4-diene)

Purine

20.2 PHYSICAL PROPERTIES AND STRUCTURE OF AMINES

20.2A Physical Properties

Amines are moderately polar substances; they have boiling points that are higher than those of alkanes but generally lower than those of alcohols of comparable molecular weight. Molecules of primary and secondary amines can form strong hydrogen bonds to each other and to water. Molecules of tertiary amines cannot form hydrogen bonds to each other, but they can form hydrogen bonds to molecules of water or other hydroxylic solvents. As a result, tertiary amines generally boil at lower temperatures than primary and secondary amines of comparable molecular weight, but all low molecular weight amines are very water soluble.

Table 20.1 lists the physical properties of some common amines.

20.2B Structure of Amines

The nitrogen atom of most amines is like that of ammonia; it is approximately sp^3 hybridized. The three alkyl groups (or hydrogen atoms) occupy corners of a tetrahedron; the sp^3 orbital containing the unshared electron pair is directed toward the other corner. We describe the geometry of the amine by the location of the atoms as being **trigonal pyramidal** (Section 1.17). However, if we were to consider the unshared electron pair as being a group we would describe the amine as being tetrahedral.

Structure of an amine

The bond angles are what one would expect of a tetrahedral structure; they are very close to 109.5°. The bond angles for trimethylamine, for example, are 108°.

If the alkyl groups of a tertiary amine are all different the amine will be chiral. There will be two enantiomeric forms of the tertiary amine and, theoretically, we ought to be able to resolve (separate) these enantiomers. In practice, however, resolution is usually impossible because the enantiomers interconvert rapidly.

TABLE 20.1 Physical properties of amines

NAME	STRUCTURE	mp (°C)	bp (°C)	WATER SOLUBILITY (25°C) (g 100 mL^{-1})	pK_a (aminium ion)
Primary Amines					
Methylamine	CH_3NH_2	-94	-6	Very soluble	10.64
Ethylamine	$CH_3CH_2NH_2$	-81	17	Very soluble	10.75
Propylamine	$CH_3CH_2CH_2NH_2$	-83	49	Very soluble	10.67
Isopropylamine	$(CH_3)_2CHNH_2$	-101	33	Very soluble	10.73
Butylamine	$CH_3(CH_2)_2CH_2NH_2$	-51	78	Very soluble	10.61
Isobutylamine	$(CH_3)_2CHCH_2NH_2$	-86	68	Very soluble	10.49
sec-Butylamine	$CH_3CH_2CH(CH_3)NH_2$	-104	63	Very soluble	10.56
tert-Butylamine	$(CH_3)_3CNH_2$	-68	45	Very soluble	10.45
Cyclohexylamine	Cyclo-$C_6H_{11}NH_2$	-18	134	Sl. soluble	10.64
Benzylamine	$C_6H_5CH_2NH_2$	10	185	Sl. soluble	9.30
Aniline	$C_6H_5NH_2$	-6	184	3.7	4.58
p-Toluidine	p-$CH_3C_6H_4NH_2$	44	200	Sl. soluble	5.08
p-Anisidine	p-$CH_3OC_6H_4NH_2$	57	244	V. sl. soluble	5.30
p-Chloroaniline	p-$ClC_6H_4NH_2$	73	232	Insoluble	4.00
p-Nitroaniline	p-$NO_2C_6H_4NH_2$	148	332	Insoluble	1.00
Secondary Amines					
Dimethylamine	$(CH_3)_2NH$	-92	7	Very soluble	10.72
Diethylamine	$(CH_3CH_2)_2NH$	-48	56	Very soluble	10.98
Dipropylamine	$(CH_3CH_2CH_2)_2NH$	-40	110	Very soluble	10.98
N-Methylaniline	$C_6H_5NHCH_3$	-57	196	Sl. soluble	4.70
Diphenylamine	$(C_6H_5)_2NH$	53	302	Insoluble	0.80
Tertiary Amines					
Trimethylamine	$(CH_3)_3N$	-117	2.9	Very soluble	9.70
Triethylamine	$(CH_3CH_2)_3N$	-115	90	14	10.76
Tripropylamine	$(CH_3CH_2CH_2)_3N$	-93	156	Sl. soluble	10.64
N,N-Dimethylaniline	$C_6H_5N(CH_3)_2$	3	194	Sl. soluble	5.06

Interconversion of amine enantiomers

This interconversion occurs through what is called a **pyramidal or nitrogen inversion.** The barrier to the interconversion is about 6 kcal mol^{-1} for most simple amines. In the transition state for the inversion, the nitrogen atom becomes sp^2 hybridized with the unshared electron pair occupying a p orbital.

Ammonium salts cannot undergo inversion because they do not have an unshared pair. Therefore, those quaternary ammonium salts with four different groups are chiral and can be resolved into separate (relatively stable) enantiomers.

Quaternary ammonium salts such as these can be resolved.

20.3 BASICITY OF AMINES: AMINE SALTS

Amines are relatively weak bases. They are stronger bases than water but are far weaker bases than hydroxide ions, alkoxide ions, and carbanions.

A convenient way to compare the base strength of amines is to compare the acidity constants (or pK_a values) of their conjugate acids, the alkylaminium ions (Section 3.5C). The expression for this acidity constant is as follows:

$$\overset{+}{R}NH_3 + H_2O \rightleftharpoons RNH_2 + H_3O^+$$

$$K_a = \frac{[RNH_2]\,[H_3O^+]}{[RNH_3^+]} \qquad pK_a = -\log K_a$$

If the amine is *strongly basic,* the aminium ion will hold the proton tightly and, consequently, will not be very acidic (it will have a large pK_a). On the other hand if the amine is *weakly basic,* the aminium ion will not hold the proton tightly and will be much more acidic (it will have a small pK_a).

When we examine the basicities of the amines given in Table 20.1, we see that most primary aliphatic amines (e.g., methylamine and ethylamine) are somewhat stronger bases than ammonia:

conj. acid pK_a	$\ddot{N}H_3$	$CH_3\ddot{N}H_2$	$CH_3CH_2\ddot{N}H_2$	$CH_3CH_2CH_2\ddot{N}H_2$
	9.26	10.64	10.75	10.67

We can account for this on the basis of the electron-releasing ability of an alkyl group. An alkyl group releases electrons, and it *stabilizes* the alkylaminium ion that results from the acid–base reaction *by dispersing its positive charge.* It stabilizes the alkylaminium ion to a greater extent than it stabilizes the amine.

By releasing electrons, R⟶
stabilizes the alkylaminium ion
through dispersal of charge.

This explanation is supported by measurements showing that in the *gas phase* the basicities of the following amines increase with increasing methyl substitution:

$$(CH_3)_3N > (CH_3)_2NH > CH_3NH_2 > NH_3$$

This is not the order of basicity of these amines in aqueous solution, however. In aqueous solution (Table 20.1) the order is

$$(CH_3)_2NH > CH_3NH_2 > (CH_3)_3N > NH_3$$

The reason for this apparent anomaly is now known. In aqueous solution the aminium ions formed from secondary and primary amines are stabilized by solvation through hydrogen bonding much more effectively than are the aminium ions formed from tertiary amines. The aminium ion formed from a tertiary amine, $(CH_3)_3NH^+$, has only one hydrogen to use in hydrogen bonding to water molecules, whereas the aminium ions from secondary and primary amines have two and three hydrogens, respectively. Poorer solvation of the aminium ion formed from a tertiary amine more than counteracts the electron-releasing effect of the three methyl groups and makes the tertiary amine less basic than primary and secondary amines in aqueous solution. However, the electron-releasing effect does make the tertiary amine more basic than ammonia.

20.3A Basicity of Arylamines

When we examine the pK_a values of the aromatic amines (e.g., aniline and *p*-toluidine) in Table 20.1, we see that they are much weaker bases than the corresponding nonaromatic amine, cyclohexylamine.

conj. acid pK_a	Cyclo-$C_6H_{11}NH_2$	$C_6H_5NH_2$	*p*-$CH_3C_6H_4NH_2$
	10.64	4.58	5.08

We can account for this effect, in part, on the basis of resonance contributions to the overall hybrid of an arylamine. For aniline, the following contributors are important.

Structures **1** and **2** are the Kekulé structures that contribute to any benzene derivative.

Structures **3–5**, however, *delocalize* the unshared electron pair of the nitrogen over the ortho and para positions of the ring. This delocalization of the electron pair makes it less available to a proton, and *delocalization of the electron pair stabilizes aniline.* When aniline accepts a proton it becomes an anilinium ion.

$$C_6H_5\ddot{N}H_2 + H_2O \rightleftharpoons C_6H_5\overset{+}{N}H_3 + \overset{-}{O}H$$

**Anilinium
ion**

Since the electron pair of the nitrogen atom accepts the proton, we are able to write only *two* resonance structures for the anilinium ion — the two Kekulé structures:

Structures corresponding to **3–5** are not possible for the anilinium ion and, consequently, although resonance does stabilize the anilinium ion considerably, it does not stabilize the anilinium ion to as great an extent as it does aniline itself. This greater stabilization of the reactant (aniline) when compared to that of the product (anilinium ion) means that $\Delta H°$ for the reaction,

$$\text{Aniline} + H_2O \longrightarrow \text{anilinium ion} + OH^- \qquad (\text{Fig. 20.1})$$

will be a larger positive quantity than that for the reaction,

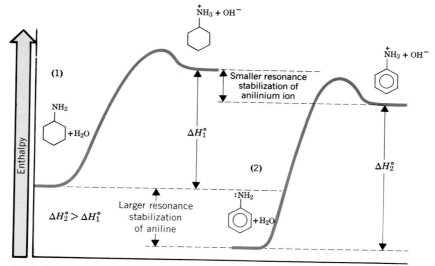

FIGURE 20.1 Enthalpy diagram (1) for the reaction of cyclohexylamine with H_2O, and (2) for the reaction of aniline with H_2O. (The curves are aligned for comparison only and are not to scale.)

Cyclohexylamine + H_2O ⟶ cyclohexylaminium ion + OH^-

Aniline, as a result, is the weaker base.

Another important effect in explaining the lower basicity of aromatic amines is **the electron-withdrawing effect of a phenyl group.** Because the carbon atoms of a phenyl group are sp^2 hybridized, they are more electronegative (and therefore more electron withdrawing) than the sp^3-hybridized carbon atoms of alkyl groups. We shall discuss this effect further in Section 21.5A.

20.3B Amines Versus Amides

Although amides are superficially similar to amines, they are far less basic (even less basic than arylamines). The pK_a of the conjugate acid of a typical amide is about 0.

This lower basicity of amides when compared to amines can also be understood in terms of resonance and inductive effects. An amide is stabilized by resonance involving the nonbonding pair of electrons on the nitrogen atom. However, an amide protonated on its nitrogen atom lacks this type of resonance stabilization. This is shown in the following resonance structures.

Amide

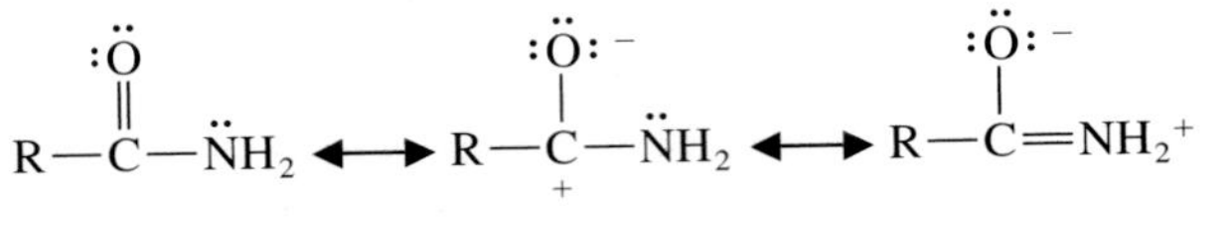

Larger resonance stabilization

N-Protonated Amide

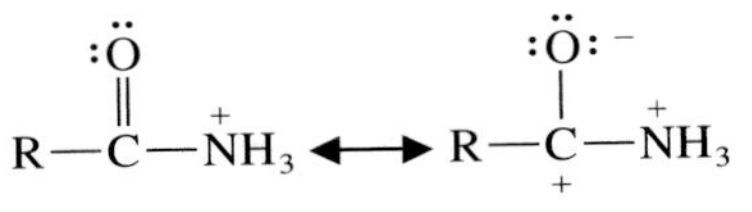

Smaller resonance stabilization

However, a more important factor accounting for amides being weaker bases than amines is the powerful electron-withdrawing effect of the carbonyl group of the amide. This means that the equilibrium for the following reaction,

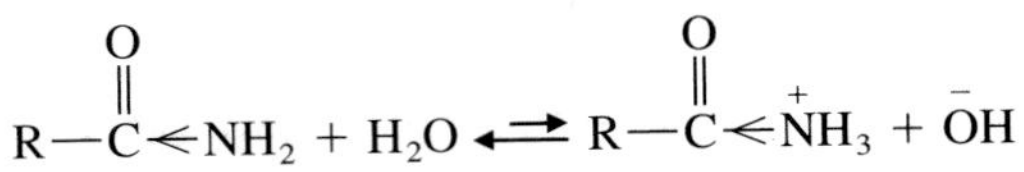

lies more to the left than that for the reaction,

$$R{\to}\ddot{N}H_2 + H_2O \rightleftharpoons R{\to}\overset{+}{N}H_3 + \overset{-}{O}H$$

and explains why the amide is the weaker base.

The nitrogen atoms of amides are so weakly basic that when an amide accepts a proton, it does so on its oxygen atom instead. Protonation on the oxygen atom occurs even though oxygen atoms (because of their greater electronegativity) are typically less basic than nitrogen atoms. Notice, however, that if an amide accepts a proton on its oxygen atom, resonance stabilization involving the nonbonding electron pair of the nitrogen atom is possible.

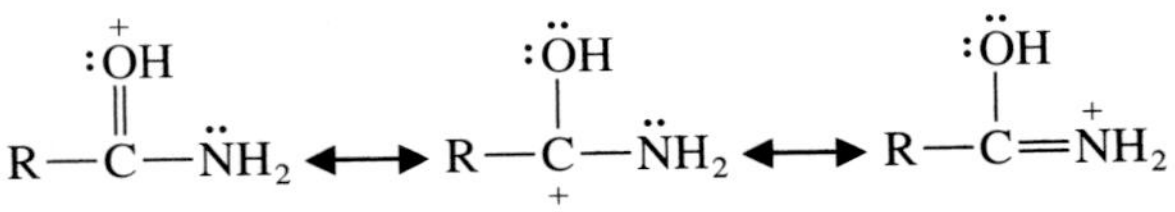

20.3C Aminium Salts and Quaternary Ammonium Salts

When primary, secondary, and tertiary amines act as bases and react with acids, they form compounds called **aminium salts.** In an aminium salt the positively charged nitrogen atom is attached to at least one hydrogen atom.

$$CH_3CH_2\overset{..}{N}H_2 + HCl \xrightarrow{H_2O} CH_3CH_2\overset{+}{N}H_3 \quad Cl^-$$

Ethylaminium chloride
(an amine salt)

$$(CH_3CH_2)_2\overset{..}{N}H + HBr \xrightarrow{H_2O} (CH_3CH_2)_2\overset{+}{N}H_2 \quad Br^-$$

Diethylaminium bromide

$$(CH_3CH_2)_3\overset{..}{N} + HI \xrightarrow{H_2O} (CH_3CH_2)_3\overset{+}{N}H \quad I^-$$

Triethylaminium iodide

When the central nitrogen atom of a compound is positively charged *but is not attached to a hydrogen atom* the compound is called a **quaternary ammonium salt.** For example,

$$CH_3CH_2-\overset{\overset{\displaystyle CH_2CH_3}{|}}{\underset{\underset{\displaystyle CH_2CH_3}{|}}{N}}\overset{+}{-}CH_2CH_3 \quad Br^-$$

Tetraethylammonium bromide
(a quaternary ammonium salt)

Quaternary ammonium halides—because they do not have an unshared electron pair on the nitrogen atom—cannot act as bases.

$$(CH_3CH_2)_4\overset{+}{N} \quad Br^-$$

Tetraethylammonium bromide
(does not undergo reaction with acid)

Quaternary ammonium *hydroxides,* however, are strong bases. As solids, or in solution, they consist *entirely* of quaternary ammonium cations (R_4N^+) and hydroxide ions (OH^-); they are, therefore, strong bases—as strong as sodium or potassium hydroxide. Quaternary ammonium hydroxides react with acids to form quaternary ammonium salts:

$$(CH_3)_4\overset{+}{N} \quad OH^- + HCl \longrightarrow (CH_3)_4\overset{+}{N} \quad Cl^- + H_2O$$

20.3D Solubility of Amines in Aqueous Acids

Almost all alkylaminium chlorides, bromides, iodides, and sulfates are soluble in water. Thus, primary, secondary, or tertiary amines that are not soluble in water do dissolve in dilute aqueous HCl, HBr, HI, or H_2SO_4. Solubility in dilute acid provides a convenient chemical method for distinguishing amines from nonbasic compounds that are insoluble in water. Solubility in dilute acid also gives us a useful method for separating amines from nonbasic compounds that are insoluble in water. The amine can be extracted into aqueous acid (dil. HCl) and then recovered by making the aqueous solution basic and extracting the amine into ether or CH_2Cl_2.

Because amides are far less basic than amines, water-insoluble amides *do not dissolve* in dilute aqueous HCl, HBr, HI, or H_2SO_4.

Water-insoluble amide
(not soluble in aqueous acids)

PROBLEM 20.1

Outline a procedure for separating hexylamine from cyclohexane using dilute HCl, aqueous NaOH, and diethyl ether.

PROBLEM 20.2

Outline a procedure for separating a mixture of benzoic acid, *p*-cresol, aniline, and benzene using acids, bases, and organic solvents.

20.3E Amines as Resolving Agents

Enantiomerically pure amines are often used to resolve racemic forms of acidic compounds. We can illustrate the principles involved in this procedure by showing how a racemic form of an organic acid might be resolved (separated) into its enantiomers with the single enantiomer of an amine (Fig. 20.2) used as a resolving agent.

In this procedure the single enantiomer of an amine (*R*)-1-phenylethylamine is added to a solution of the racemic form of an acid. The salts that form are not enantiomers. They are diastereomers. (The stereocenters of the acid portion of the salts are enantiomerically related to each other, but the stereocenters of the amine portion are not.) The diastereomers have different solubilities and can be separated by careful crystallization. The separated salts are then acidified with hydrochloric acid and the enantiomeric acids are obtained from the separate solutions. The amine remains in solution as its hydrochloride salt.

Single enantiomers that are employed as resolving agents are often readily available from natural sources. Because most of the chiral organic molecules that occur in living organisms are synthesized by enzymatically catalyzed reactions, most of them occur as single enantiomers. Naturally occurring optically active amines such as (−)-quinine (Section 20.4), (−)-strychnine, and (−)-brucine are often employed as resolving agents for racemic acids. Acids such as (+)- or (−)-tartaric acid (Section 5.14) are often used for resolving racemic bases.

(−)-Strychnine

(−)-Brucine

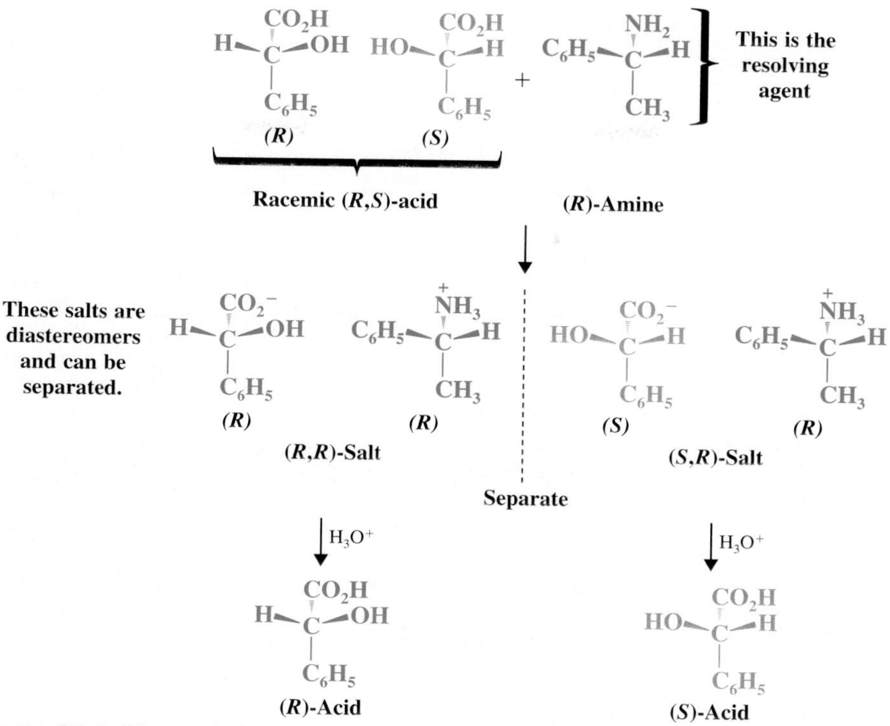

FIGURE 20.2 The resolution of the racemic form of an organic acid by the use of an optically active amine. Acidification of the separated diastereomeric salts causes the enantiomeric acids to precipitate (assuming they are insoluble in water) and leaves the resolving agent in solution as its conjugate acid.

One of the newest techniques for resolving racemates is based on high-performance liquid chromatography (HPLC) using a **chiral stationary phase** (CSP). This technique, developed by William H. Pirkle of the University of Illinois, has been used to resolve many racemic amines, alcohols, amino acids, and related compounds. We do not have the space here to discuss this technique in detail,* but suffice it to say that a solution of the racemate is passed through a column (called a **Pirkle column**) containing small silica microporous beads. Chemically attached to the surface of the beads is a chiral group such as the one that follows:

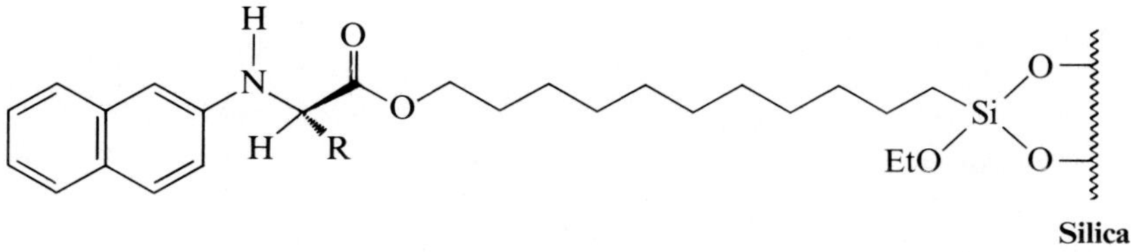

Silica

The compound to be resolved is first converted to a derivative containing a 3,5-dinitrophenyl group. An amine, for example, is converted to a 3,5-dinitrobenzamide:

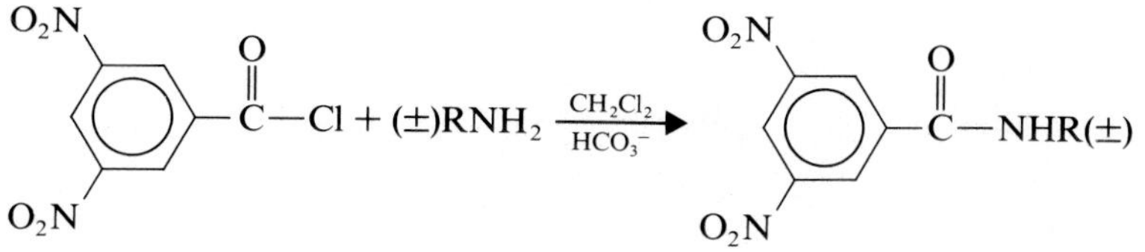

---

* You may want to consult a laboratory manual or read the following article: W. H. Pirkle, T. C. Pochapsky, G. S. Mahler, D. E. Corey, D. S. Reno, and D. M. Alessi, *J. Org. Chem., 51,* 4991, 1986.

An alcohol is converted to a carbamate (Section 18.10A) through a variation of the Curtius rearrangement (Section 20.5E):

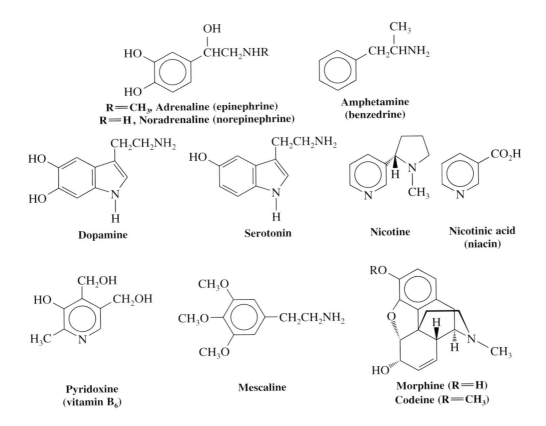

The stationary phase, because it is chiral, binds one enantiomer much more tightly than the other. This binding increases the retention time of that enantiomer and permits separation. The binding comes partially from hydrogen-bonding interactions between the derivative and the CSP, but highly important is a $\pi-\pi$ interaction between the electron-deficient 3,5-dinitrophenyl ring of the derivative and the electron-rich naphthalene ring of the CSP.

20.4 SOME BIOLOGICALLY IMPORTANT AMINES

A large number of medically and biologically important compounds are amines. Listed here are some important examples.

R=CH₃, Adrenaline (epinephrine)
R=H, Noradrenaline (norepinephrine)

Amphetamine (benzedrine)

Dopamine

Serotonin

Nicotine

Nicotinic acid (niacin)

Pyridoxine (vitamin B₆)

Mescaline

Morphine (R=H)
Codeine (R=CH₃)

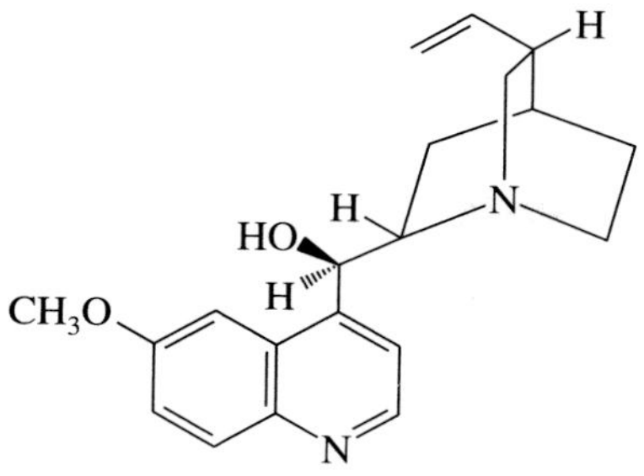

(−)-Quinine

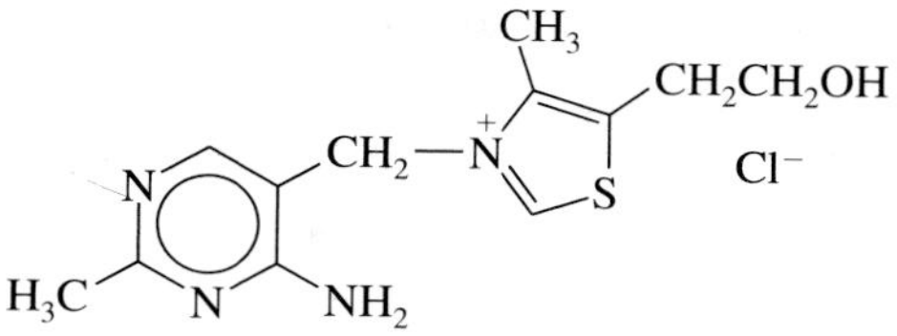

Thiamine chloride
(vitamin B$_1$)

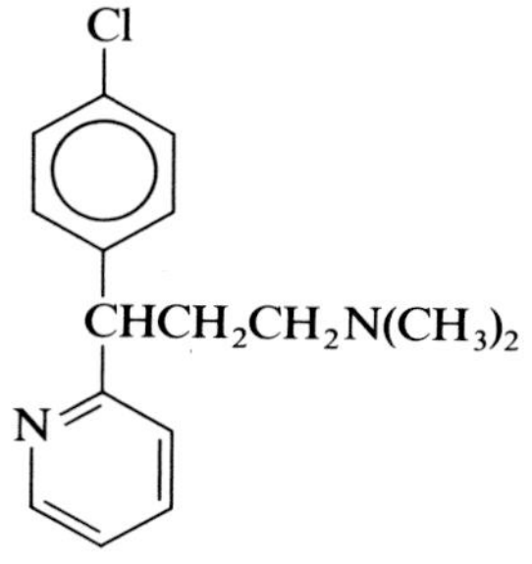

Chlorpheniramine

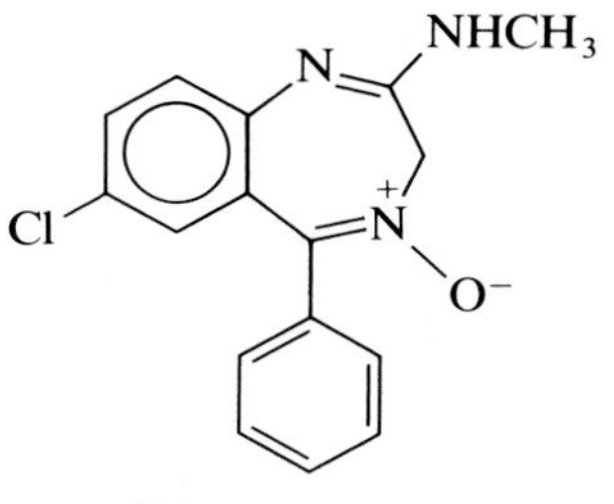

Chlorodiazepoxide
(librium)

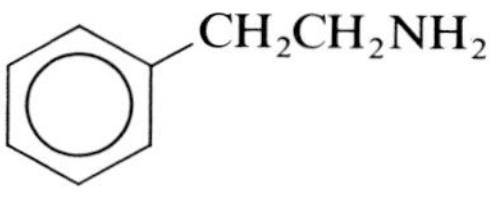

2-Phenylethylamine

2-Phenylethylamines Many phenylethylamines compounds have powerful physiological and psychological effects. Adrenaline and noradrenaline are two hormones secreted in the medulla of the adrenal gland. Released into the bloodstream when an animal senses danger, adrenaline causes an increase in blood pressure, a strengthening of the heart rate, and a widening of the passages of the lungs. All of these effects prepare the animal to fight or flee. Noradrenaline also causes an increase in blood pressure, and it is involved in the transmission of impulses from the end of one nerve fiber to the next. Dopamine and serotonin are important neurotransmitters in the brain. Abnormalities in the level of dopamine in the brain are associated with many psychiatric disorders including Parkinson's disease. Dopamine plays a pivotal role in the regulation and control of movement, motivation, and cognition. Serotonin is a compound of particular interest because it appears to be important in maintaining stable mental processes. It has been suggested that the mental disorder schizophrenia may be connected with abnormalities in the metabolism of serotonin.

Amphetamine (a powerful stimulant) and mescaline (a hallucinogen) have structures similar to those of serotonin, adrenaline, and noradrenaline. They are all derivatives of 2-phenylethylamine (see structure given). (In serotonin the nitrogen is connected to the benzene ring to create a five-membered ring.) The structural similarities of these compounds must be related to their physiological and psychological effects because many other compounds with similar properties are also derivatives of 2-phenylethylamine. Examples (not shown) are *N*-methylamphetamine and LSD. Even morphine and codeine, two powerful analgesics, have a 2-phenylethylamine system as a part of their structures. (Morphine and codeine are examples of compounds called alkaloids. Try to locate the 2-phenylethylamine system in their structures now, however.)

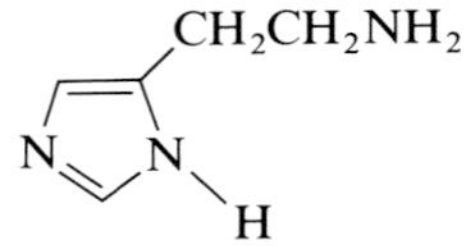

Histamine

Vitamins and Antihistamines A number of amines are vitamins. These include nicotinic acid and nicotinamide (the antipellagra factors), pyridoxine (vitamin B_6), and thiamine chloride (vitamin B_1). Nicotine is a toxic alkaloid found in tobacco that makes smoking habit forming. Histamine, another toxic amine, is found bound to proteins in nearly all tissues of the body. Release of free histamine causes the symptoms associated with allergic reactions and the common cold. Chlorpheniramine, an ''antihistamine,'' is an ingredient of many over-the-counter cold remedies.

Tranquilizers Chlordiazepoxide, an interesting compound with a seven-membered ring, is one of the most widely prescribed tranquilizers. (Chlordiazepoxide also contains a positively charged nitrogen, present as an *N*-oxide.)

Neurotransmitters Nerve cells interact with other nerve cells or with muscles at junctions, or gaps, called **synapses.** Nerve impulses are carried across the synaptic gap by chemical compounds, called *neurotransmitters.* Acetylcholine (see the following reaction) is an important neurotransmitter at neuromuscular synapses called *cholinergic synapses.* Acetylcholine contains a quaternary ammonium group. Being small and ionic, acetylcholine is highly soluble in water and highly diffusible, qualities that suit its role as a neurotransmitter. Acetylcholine molecules are released by the presynaptic membrane in the neuron in packets of about 10^4 molecules. The packet of molecules then diffuses across the synaptic gap.

$$\underset{\textbf{Acetylcholine}}{CH_3\overset{O}{\overset{\|}{C}}OCH_2CH_2\overset{+}{N}(CH_3)_3} + H_2O \xrightleftharpoons{\text{acetylcholinesterase}} CH_3CO_2H + \underset{\textbf{Choline}}{HOCH_2CH_2\overset{+}{N}(CH_3)_3}$$

Having carried a nerve impulse across the synapse to the muscle where it triggers an electrical response, the acetylcholine molecules must be hydrolyzed (to choline) within a few milliseconds to allow the arrival of the next impulse. This hydrolysis is catalyzed by an enzyme of almost perfect efficiency called *acetylcholinesterase.*

The acetylcholine receptor on the postsynaptic membrane of muscle is the target for some of the most deadly neurotoxins. Among them are **tubocurarine,** a component of the Amazonian arrow poison **curare.** Notice that tubocurarine has a quaternary ammonium group (actually two), and in this respect resembles acetylcholine.

Tubocurarine

The quaternary ammonium groups are apparently important in binding both acetylcholine, and substances like tubocurarine, to the receptor. By binding to the muscle receptor, tubocurarine blocks the transfer of the nerve impulse (via acetycholine) and causes paralysis, including paralysis of the muscles involved in respiration. Tubocurarine, in tiny doses, is used in medicine as a muscle relaxant during surgery. So, too, is the compound, succinylcholine bromide.

Succinylcholine bromide

20.5 PREPARATION OF AMINES

20.5A Through Nucleophilic Substitution Reactions

Salts of primary amines can be prepared from ammonia and alkyl halides by nucleophilic substitution reactions. Subsequent treatment of the resulting aminium salts with base gives primary amines.

$$\ddot{N}H_3 + R\overset{\frown}{-}X \longrightarrow R-\overset{+}{N}H_3 X^- \xrightarrow{\ OH^-\ } RNH_2$$

This method is of very limited synthetic application because multiple alkylations occur. When ethyl bromide reacts with ammonia, for example, the ethylaminium bromide that is produced initially can react with ammonia to liberate ethylamine. Ethylamine can then compete with ammonia and react with ethyl bromide to give diethylaminium bromide. Repetitions of acid–base and alkylation reactions ultimately produce some tertiary amines and even some quaternary ammonium salts if the alkyl halide is present in excess.

A MECHANISM FOR THE
ALKYLATION OF NH₃

$$\ddot{N}H_3 + CH_3CH_2\overset{\frown}{-}Br \longrightarrow CH_3CH_2-\overset{+}{N}H_3 + Br^-$$

$$CH_3CH_2-\overset{H}{\underset{H}{\overset{|}{N}}}\overset{+}{-}H + \ddot{:}NH_3 \longrightarrow CH_3CH_2\ddot{N}H_2 + \overset{+}{N}H_4$$

$$CH_3CH_2\ddot{N}H_2 + CH_3CH_2\overset{\frown}{-}Br \longrightarrow (CH_3CH_2)_2\overset{+}{N}H_2 + Br^-, \text{ etc.}$$

Multiple alkylations can be minimized by using a large excess of ammonia. (Why?) An example of this technique can be seen in the synthesis of alanine from 2-bromopropanoic acid:

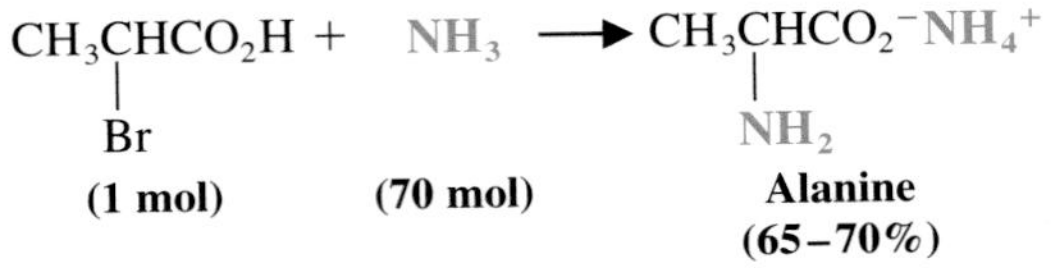

$$CH_3CHCO_2H + NH_3 \longrightarrow CH_3CHCO_2^- NH_4^+$$

Br		NH₂
(1 mol)	(70 mol)	Alanine
		(65–70%)

A much better method for preparing a primary amine from an alkyl halide is first to convert the alkyl halide to an alkyl azide ($R-N_3$) by a nucleophilic substitution reaction:

$$R-X + N_3^- \xrightarrow[(-X^-)]{S_N2} R-N=\overset{+}{N}=\overset{-}{N} \xrightarrow[\text{or}]{\text{Na/alcohol}} RNH_2$$
$$\text{LiAlH}_4$$

Azide   **Alkyl**
ion   **azide**

Then the alkyl azide can be reduced to a primary amine with sodium and alcohol or with lithium aluminum hydride. A word of *caution:* Alkyl azides are explosive and low molecular weight alkyl azides should not be isolated but should be kept in solution.

Potassium phthalimide (see following reaction) can also be used to prepare primary amines by a method known as the *Gabriel synthesis*. This synthesis also avoids the complications of multiple alkylations that occur when alkyl halides are treated with ammonia:

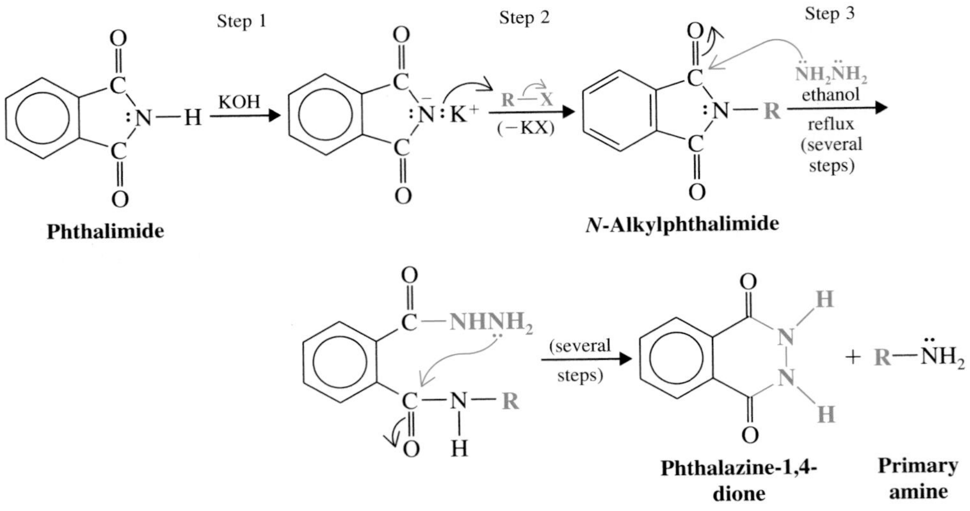

Phthalimide   **N-Alkylphthalimide**

Phthalazine-1,4-   **Primary**
dione   **amine**

Phthalimide is quite acidic ($pK_a = 9$); it can be converted to potassium phthalimide by potassium hydroxide (step 1). The phthalimide anion is a strong nucleophile and (in step 2) it reacts with an alkyl halide to give an *N*-alkylphthalimide. At this point, the *N*-alkylphthalimide can be hydrolyzed with aqueous acid or base, but the hydrolysis is often difficult. It is often more convenient to treat the *N*-alkylphthalimide with hydrazine (NH_2NH_2) in refluxing ethanol (step 3) to give a primary amine and phthalazine-1,4-dione.

Syntheses of amines using the Gabriel synthesis are, as we might expect, restricted to the use of methyl, primary, and secondary alkyl halides. The use of tertiary halides leads almost exclusively to eliminations.

PROBLEM 20.3

(a) Write resonance structures for the phthalimide anion that account for the acidity of phthalimide.   (b) Would you expect phthalimide to be more or less acidic than benzamide? Why?   (c) After step (2) of our reaction, several steps have been omitted. Propose reasonable mechanisms for these steps.

PROBLEM 20.4

Outline a preparation of benzylamine using the Gabriel synthesis.

Multiple alkylations are not a problem when tertiary amines are alkylated with methyl or primary halides. Reactions such as the following take place in good yield.

$$R_3N: + RCH_2\!-\!Br \longrightarrow R_3\overset{+}{N}\!-\!CH_2R \; Br^-$$

20.5B Preparation of Amines Through Reduction of Nitro Compounds

The most widely used method for preparing aromatic amines involves nitration of the ring and subsequent reduction of the nitro group to an amino group.

$$Ar\!-\!H \xrightarrow[H_2SO_4]{HNO_3} Ar\!-\!NO_2 \xrightarrow{[H]} Ar\!-\!NH_2$$

We studied ring nitration in Chapter 15 and saw there that it is applicable to a wide variety of aromatic compounds. Reduction of the nitro group can also be carried out in a number of ways. The most frequently used methods employ catalytic hydrogenation, or treatment of the nitro compound with acid and iron. Zinc, or tin, or a metal salt such as $SnCl_2$ can also be used. Overall, this is a $6e^-$ reduction.

$$Ar\!-\!NO_2 \xrightarrow[\text{or} \quad (1) \; Fe, \; HCl \quad (2) \; OH^-]{H_2, \; cat.} Ar\!-\!NH_2$$

Specific Example

(97%)

Selective reduction of one nitro group of a dinitro compound can often be achieved through the use of hydrogen sulfide in aqueous (or alcoholic) ammonia:

m-Dinitrobenzene   **m-Nitroaniline**
                                          (70–80%)

When this method is used, the amount of the hydrogen sulfide must be carefully measured because the use of an excess may result in the reduction of more than one nitro group.

It is not always possible to predict just which nitro group will be reduced, however. Treating 2,4-dinitrotoluene with hydrogen sulfide and ammonia results in reduction of the 4-nitro group:

On the other hand, monoreduction of 2,4-dinitroaniline causes reduction of the 2-nitro group:

(52–58%)

20.5C Preparation of Amines Through Reductive Amination

Aldehydes and ketones can be converted to amines through catalytic or chemical reduction in the presence of ammonia or an amine. Primary, secondary, and tertiary amines can be prepared this way:

This process, called *reductive amination* of the aldehyde or ketone (or *reductive alkyl-*

ation of the amine), appears to proceed through the following general mechanism (illustrated with a 1° amine):

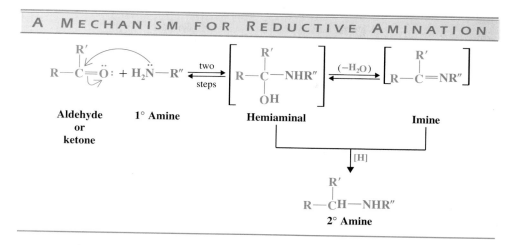

A MECHANISM FOR REDUCTIVE AMINATION

Aldehyde or ketone 1° Amine Hemiaminal Imine

R—CH—NHR″
R′
2° Amine

When ammonia or a primary amine is used, there are two possible pathways to the product—via an amino alcohol that is similar to a hemiacetal and is called a *hemiaminal,* or via an imine. When secondary amines are used, an imine cannot form and, therefore, the pathway is through the hemiaminal or through an iminium ion.

$$R—C=\overset{+}{N}—R''$$

with R′ on the carbon and R‴ on the nitrogen

Iminium ion

The reducing agents employed include hydrogen and a catalyst (such as nickel), or NaBH₃CN or LiBH₃CN (sodium or lithium cyanoborohydride). The latter two reducing agents are similar to NaBH₄ and are especially effective in reductive aminations. Three specific examples of reductive amination follow:

Benzaldehyde $\xrightarrow[\text{40–70°C}]{\substack{\text{NH}_3, \text{H}_2, \text{Ni} \\ \text{90 atm}}}$ **Benzylamine (89%)**

Benzaldehyde $\xrightarrow[\text{LiBH}_3\text{CN}]{\text{CH}_3\text{CH}_2\text{NH}_2}$ **N-Benzylethanamine (89%)**

Cyclohexanone $\xrightarrow[\text{NaBH}_3\text{CN}]{(\text{CH}_3)_2\text{NH}}$ **N,N-Dimethylcyclohexanamine (52–54%)**

PROBLEM 20.5

Show how you might prepare each of the following amines through reductive amination:

(a) $CH_3(CH_2)_3CH_2NH_2$

(c) $CH_3(CH_2)_4CH_2NHC_6H_5$

(b) $C_6H_5CH_2CHCH_3$
$\quad\quad\quad\quad\;\;|$
$\quad\quad\quad\quad NH_2$
(amphetamine)

(d) $C_6H_5CH_2N(CH_3)_2$

PROBLEM 20.6

Reductive amination of a ketone is almost always a better method for the prepa-

$$\overset{\displaystyle R'}{\underset{\displaystyle |}{}}$$

ration of amines of the type $RCHNH_2$ than treatment of an alkyl halide with ammonia. Why would this be true?

20.5D Preparation of Amines Through Reduction of Amides, Oximes, and Nitriles

Amides, oximes, and nitriles can be reduced to amines. Reduction of a nitrile or an oxime yields a primary amine; reduction of an amide can yield a primary, secondary, or tertiary amine.

$$R-C\equiv N \xrightarrow{\text{[H]}} RCH_2NH_2$$

Nitrile 1° Amine

{ Nitriles can be prepared from alkyl halides and CN^- (Section 18.3) or from aldehydes and ketones as cyanohydrins (Section 16.9A)

$$RCH=NOH \xrightarrow{\text{[H]}} RCH_2NH_2$$

Oxime 1° Amine

{ Oximes can be prepared from aldehydes and ketones (Section 16.8A)

$$\overset{\displaystyle O}{\overset{\displaystyle \|}{R-C-\underset{\underset{\displaystyle R''}{|}}{N}-R'}} \xrightarrow{\text{[H]}} \underset{\underset{\displaystyle R''}{|}}{RCH_2N}-R'$$

Amide 3° Amine

{ Amides can be prepared from acid chlorides, acid anhydrides, and esters (Section 18.8)

(In the last example, if R' and R″ = H, the product is a 1° amine; if R′ = H, the product is a 2° amine.)

All of these reductions can be carried out with hydrogen and a catalyst or with $LiAlH_4$. Oximes are also conveniently reduced with sodium in alcohol—a safer method than the use of $LiAlH_4$ (Section 11.3).

Specific examples follow:

$$=N-OH \xrightarrow{\text{Na, } C_2H_5OH} -NH_2$$

(50–60%)

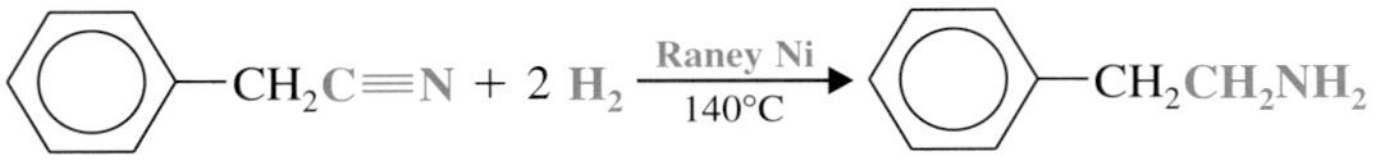

2-Phenylethanenitrile
(phenylacetonitrile)

2-Phenylethanamine
(71%)

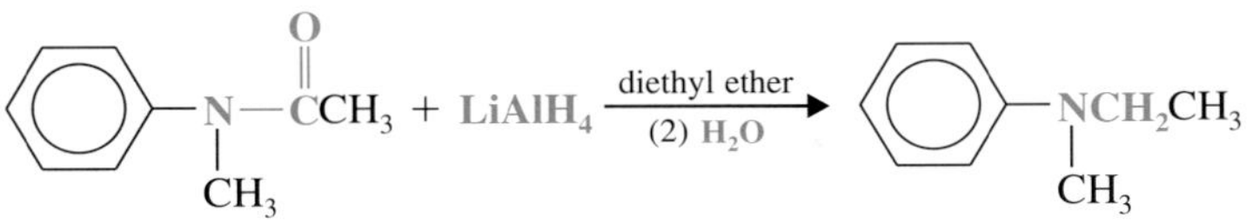

N-Methylacetanilide

N-Ethyl-N-methylaniline

Reduction of an amide is the last step in a useful procedure for **monoalkylation of an amine.** The process begins with *acylation* of the amine using an acyl chloride or acid anhydride; then the amide is reduced with lithium aluminum hydride. For example,

Benzylamine

Benzylethylamine

PROBLEM 20.7

Show how you might utilize the reduction of an amide, oxime, or nitrile to carry out each of the following transformations:

(a) Benzoic acid ⟶ benzylethylamine

(b) 1-Bromopentane ⟶ hexylamine

(c) Propanoic acid ⟶ tripropylamine

(d) 2-Butanone ⟶ *sec*-butylamine

20.5E Preparation of Amines Through the Hofmann and Curtius Rearrangements

Amides with no substituent on the nitrogen react with solutions of bromine or chlorine in sodium hydroxide to yield amines through a reaction known as the *Hofmann rearrangement* or *Hofmann degradation:*

$$R—\overset{\overset{\displaystyle O}{\|}}{C}—NH_2 + Br_2 + 4\ NaOH \xrightarrow{H_2O} RNH_2 + 2\ NaBr + Na_2CO_3 + 2\ H_2O$$

From this equation we can see that the carbonyl carbon atom of the amide is lost (as CO_3^{2-}) and that the R group of the amide becomes attached to the nitrogen of the amine. Primary amines made this way are not contaminated by 2° or 3° amines.

The mechanism for this interesting reaction is shown in the following scheme. In the first two steps the amide undergoes a base-promoted bromination, in a manner analogous to the base-promoted halogenation of a ketone that we studied in Section 17.3B. (The electron-withdrawing acyl group of the amide makes the amido hydrogens much more acidic than those of an amine.) The N-bromo amide then reacts with hydroxide ion to produce an anion, which spontaneously rearranges with the loss of a bromide ion to produce an isocyanate (Section 18.10A). In the rearrangement the R— group migrates with its electrons from the acyl carbon to the nitrogen atom at the same time the bromide ion departs. The isocyanate that forms in the mixture is quickly hydrolyzed by the aqueous base to a carbamate ion, which undergoes spontaneous decarboxylation resulting in the formation of the amine.

A MECHANISM FOR THE HOFMANN REARRANGEMENT

Base-promoted *N*-bromination of the amide.

Base removes a proton from the nitrogen to give a bromo amide anion.

The R— group migrates to the nitrogen as a bromide ion departs. This produces an isocyanate.

The isocyanate undergoes hydrolysis and decarboxylation to produce the amine.

An examination of the first two steps of this mechanism shows that, initially, two hydrogen atoms must be present on the nitrogen of the amide for the reaction to occur. Consequently, the Hofmann rearrangement is limited to amides of the type $RCONH_2$.

Studies of Hofmann rearrangement of optically active amides in which the stereocenter is directly attached to the carbonyl group have shown that these reactions occur with *retention of configuration*. Thus, the R group migrates to nitrogen with its electrons, *but without inversion*.

The Curtius rearrangement is a rearrangement that occurs with acyl azides. It resembles the Hofmann rearrangement in that an R— group migrates from the acyl carbon to the nitrogen atom as the leaving group departs. In this instance the leaving group is N_2 (the best of all possible leaving groups since it is highly stable, virtually nonbasic, and being a gas removes itself from the medium). Acyl azides are easily prepared by allowing acyl chlorides to react with sodium azide. Heating the acyl azide brings about the rearrangement; afterwards, adding water causes hydrolysis and decarboxylation of the isocyanate.

$$
R-\overset{\overset{\displaystyle ..}{\overset{\displaystyle O}{\|}}}{C}-\ddot{\underset{..}{Cl}}: \xrightarrow[(-\,NaCl)]{NaN_3} R-\overset{\overset{\displaystyle ..}{\overset{\displaystyle O}{\|}}}{C}-\ddot{N}-\overset{+}{N}\equiv N: \xrightarrow[-N_2]{heat} R-\ddot{N}=C=\ddot{O}: \xrightarrow{H_2O} R-\ddot{N}H_2 + CO_2
$$

Acyl chloride   **Acyl azide**   **Isocyanate**   **Amine**

PROBLEM 20.8

Using a different method for each part, but taking care in each case to select a *good* method, show how each of the following transformations might be accomplished:

(a) $CH_3O-\langle\bigcirc\rangle \longrightarrow CH_3O-\langle\bigcirc\rangle-NH_2$

(b) $CH_3O-\langle\bigcirc\rangle \longrightarrow CH_3O-\langle\bigcirc\rangle-\underset{\underset{\displaystyle NH_2}{|}}{CHCH_3}$

(c) $\langle\bigcirc\rangle-CH_3 \longrightarrow \langle\bigcirc\rangle-CH_2\overset{+}{N}(CH_3)_3\ \overset{-}{Cl}$

(d) $NO_2-\langle\bigcirc\rangle-CH_3 \longrightarrow NO_2-\langle\bigcirc\rangle-NH_2$

(e) $\langle\bigcirc\rangle-CH_3 \longrightarrow \langle\bigcirc\rangle-CH_2CH_2NH_2$

20.6 REACTIONS OF AMINES

We have encountered a number of important reactions of amines in earlier sections of this book. In Section 20.3 we saw reactions in which primary, secondary, and tertiary amines act *as bases*. In Section 20.5 we saw their reactions as *nucleophiles in alkylation reactions,* and in Chapter 18 as *nucleophiles in acylation reactions.* In Chapter 15 we saw that an amino group on an aromatic ring acts as a powerful *activating group* and as an *ortho–para director.*

The structural feature of amines that underlies all of these reactions and that forms a basis for our understanding of most of the chemistry of amines is the ability of nitrogen to share an electron pair:

$$
\overset{\diagdown}{\underset{\diagup}{N}}:+\ H^+ \rightleftharpoons \overset{\diagdown}{\underset{\diagup}{\overset{+}{N}}}-H
$$

An amine acting as a base

$$\overset{|}{-}\overset{}{N}: + R-CH_2\overset{\curvearrowright}{-}Br \longrightarrow \overset{|}{-}\overset{+}{N}-CH_2R + Br^-$$

An amine acting as a nucleophile in an alkylation reaction

$$\overset{|}{-}\overset{}{\underset{H}{N}}: + R-\overset{O}{\overset{||}{C}}-Cl \longrightarrow \overset{|}{-}\overset{+}{\underset{H}{N}}-\overset{O\overset{-}{|}}{\underset{Cl}{\overset{|}{C}}}-R \xrightarrow[(-HCl)]{} \overset{|}{-}\overset{\cdot\cdot}{N}-\overset{O}{\overset{||}{C}}-R$$

An amine acting as a nucleophile in an acylation reaction

In the preceding examples the amine acts as a nucleophile by donating its electron pair to an electrophilic reagent. In the following example, resonance contributions involving the nitrogen electron pair make *carbon* atoms nucleophilic.

The amino group acting as an activating group and as an ortho–para director in electrophilic aromatic substitution

PROBLEM 20.9

Review the chemistry of amines given in earlier sections and provide a specific example for each of the previously illustrated reactions.

20.6A Oxidation of Amines

Primary and secondary aliphatic amines are subject to oxidation, although in most instances useful products are not obtained. Complicated side reactions often occur, causing the formation of complex mixtures.

Tertiary amines can be oxidized cleanly to tertiary amine oxides. This transformation can be brought about by using hydrogen peroxide or a peroxy acid.

$$R_3N: \xrightarrow{H_2O_2 \quad or \quad RCOOH} R_3\overset{+}{N}-\overset{\cdot\cdot}{\underset{\cdot\cdot}{O}}:^-$$

A tertiary amine oxide

Tertiary amine oxides undergo a useful elimination reaction as discussed in Section 20.13B.

Arylamines are very easily oxidized by a variety of reagents, including the oxygen in air. Oxidation is not confined to the amino group but also occurs in the ring. (The amino group through its electron-donating ability makes the ring electron rich and hence especially susceptible to oxidation.) The oxidation of other functional groups on an aromatic ring cannot usually be accomplished when an amino group is present on the ring, because oxidation of the ring takes place first.

20.7 REACTIONS OF AMINES WITH NITROUS ACID

Nitrous acid (HONO) is a weak, unstable acid. It is always prepared *in situ,* usually by treating sodium nitrite ($NaNO_2$) with an aqueous solution of a strong acid:

$$HCl_{(aq)} + NaNO_{2(aq)} \longrightarrow HONO_{(aq)} + NaCl_{(aq)}$$

$$H_2SO_4 + 2\,NaNO_{2(aq)} \longrightarrow 2\,HONO_{(aq)} + Na_2SO_{4(aq)}$$

Nitrous acid reacts with all classes of amines. The products that we obtain from these reactions depend on whether the amine is primary, secondary, or tertiary and whether the amine is aliphatic or aromatic.

20.7A Reactions of Primary Aliphatic Amines with Nitrous Acid

Primary aliphatic amines react with nitrous acid through a reaction called *diazotization* to yield highly unstable aliphatic *diazonium salts.* Even at low temperatures, *aliphatic diazonium salts* decompose spontaneously by losing nitrogen to form carbocations. The carbocations go on to produce mixtures of alkenes, alcohols, and alkyl halides by elimination of H^+, reaction with H_2O, and reaction with X^-.

General Reaction

$$R-NH_2 + NaNO_2 + 2\,HX \xrightarrow[\text{H}_2\text{O}]{\text{(HONO)}} \left[R-\overset{+}{N}\equiv N: \quad X^- \right] + NaX + 2\,H_2O$$

1° Aliphatic amine

Aliphatic diazonium salt (highly unstable)

$$\big\downarrow\ -N_2\ (\text{i.e., } :N\equiv N:)$$

$$R^+ + X^-$$

$$\big\downarrow$$

Alkenes, alcohols, alkyl halides

Diazotizations of primary aliphatic amines are of little synthetic importance because they yield such a complex mixture of products. Diazotizations of primary aliphatic amines are used in some analytical procedures, however, because the evolution of nitrogen is quantitative. They can also be used to generate and thus study the behavior of carbocations in water, acetic acid, and other solvents.

20.7B Reactions of Primary Arylamines with Nitrous Acid

The most important reaction of amines with nitrous acid, by far, is the reaction of primary arylamines. We shall see why in Section 20.8. Primary arylamines react with nitrous acid to give arenediazonium salts. Whereas arenediazonium salts are unstable, they are far more stable than aliphatic diazonium salts; they do not decompose at an appreciable rate in solution when the temperature of the reaction mixture is kept below 5°C.

$$Ar-NH_2 \quad + NaNO_2 + 2\ HX \longrightarrow Ar-\overset{+}{N}\equiv N\!:\ \overset{-}{X} + NaX + 2\ H_2O$$

Primary arylamine

**Arenediazonium
salt
(stable if kept
below 5°C)**

Diazotization of a primary amine takes place through a series of steps. In the presence of strong acid, nitrous acid dissociates to produce ^+NO ions. These ions then react with the nitrogen of the amine to form an unstable *N*-nitrosoammonium ion as an intermediate. This intermediate then loses a proton to form an *N*-nitrosoamine, which, in turn, tautomerizes to a diazohydroxide in a reaction that is similar to keto–enol tautomerization. Then, in the presence of acid, the diazohydroxide loses water to form the diazonium ion.

A MECHANISM FOR DIAZOTIZATION

$$HO\ddot{N}O + H_3O^+ \rightleftharpoons H_2\overset{+}{O}\!-\!\ddot{N}O + H_2O \rightleftharpoons 2\ H_2O + \overset{+}{\underset{..}{N}}\!=\!O$$

$$Ar-\underset{\underset{H}{|}}{\overset{\overset{H}{|}}{N}}\!:\ +\ ^+\ddot{N}\!=\!O \longrightarrow Ar-\underset{\underset{H}{|}}{\overset{\overset{H}{|}}{\overset{+}{N}}}\!-\!\ddot{N}\!=\!O \xrightarrow{-\,H^+} Ar-\underset{\underset{H}{|}}{\ddot{N}}\!-\!\ddot{N}\!=\!O \underset{+\,H^+}{\overset{-\,H^+}{\longrightarrow}}$$

**1° Aryl or
alkyl
amine**   **N-Nitroso-
ammonium
ion**   **N-Nitrosoamine**

$$Ar-\ddot{N}\!=\!\ddot{N}\!-\!O^- \underset{-\,H^+}{\overset{+\,H^+}{\rightleftharpoons}} Ar-\ddot{N}\!=\!\ddot{N}\!-\!OH \underset{-\,H^+}{\overset{+\,H^+}{\rightleftharpoons}} Ar-\ddot{N}\!=\!\ddot{N}\!-\!\overset{+}{O}H_2 \rightleftharpoons$$

Diazohydroxide

$$Ar-\overset{+}{N}\!\equiv\!N\!:\ \longleftrightarrow Ar-\ddot{N}\!=\!\overset{+}{N} + H_2O$$

Diazonium ion

Diazotization reactions of primary arylamines are of considerable synthetic importance because the diazonium group, $-\overset{+}{N}\equiv N\!:$, can be replaced by a variety of other functional groups. We shall examine these reactions in Section 20.8.

20.7C Reactions of Secondary Amines with Nitrous Acid

Secondary amines—both aryl and alkyl—react with nitrous acid to yield *N*-nitrosoamines. *N*-Nitrosoamines usually separate from the reaction mixture as oily yellow liquids.

Specific Examples

$$(CH_3)_2\overset{..}{N}H \;+\; HCl + NaNO_2 \xrightarrow[H_2O]{(HONO)} (CH_3)_2\overset{..}{N}-\overset{..}{N}=O$$

Dimethylamine

N-Nitrosodimethylamine
(a yellow oil)

N-Methylaniline $+ HCl + NaNO_2 \xrightarrow[H_2O]{(HONO)}$ N-Nitroso-N-methyl-aniline

N-Methylaniline

N-Nitroso-N-methyl-aniline
(87–93%)
(a yellow oil)

N-Nitrosoamines are very powerful carcinogens which scientists fear may be present in many foods, especially in cooked meats that have been cured with sodium nitrite. Sodium nitrite is added to many meats (e.g., bacon, ham, frankfurters, sausages, and corned beef) to inhibit the growth of *Clostridium botulinum* (the bacterium that produces botulinus toxin) and to keep red meats from turning brown. (Food poisoning by botulinus toxin is often fatal.) In the presence of acid or under the influence of heat, sodium nitrite reacts with amines always present in the meat to produce *N*-nitrosoamines. Cooked bacon, for example, has been shown to contain *N*-nitrosodimethylamine and *N*-nitrosopyrrolidine. There is also concern that nitrites from food may produce nitrosoamines when they react with amines in the presence of the acid found in the stomach. In 1976, the FDA reduced the permissible amount of nitrite allowed in cured meats from 200 parts per million (ppm) to 50–125 ppm. Nitrites (and nitrates that can be converted to nitrites by bacteria) also occur naturally in many foods. Cigarette smoke is known to contain *N*-nitrosodimethylamine. Someone smoking a pack of cigarettes a day inhales about 0.8 μg of *N*-nitrosodimethylamine and even more has been shown to be present in the side-stream smoke.

20.7D Reactions of Tertiary Amines with Nitrous Acid

When a tertiary aliphatic amine is mixed with nitrous acid, an equilibrium is established among the tertiary amine, its salt, and an *N*-nitrosoammonium compound.

$$2\,R_3N: \quad + HX + NaNO_2 \rightleftharpoons R_3\overset{+}{N}H\,\overset{-}{X} \;+\; R_3\overset{+}{N}-N=O\;X^-$$

Tertiary aliphatic
amine

Amine salt

N-Nitrosoammonium
compound

While *N*-nitrosoammonium compounds are stable at low temperatures, at higher temperatures and in aqueous acid they decompose to produce aldehydes or ketones. These reactions are of little synthetic importance, however.

Tertiary arylamines react with nitrous acid to form *C*-nitroso aromatic compounds. Nitrosation takes place almost exclusively at the para position if it is open and, if not, at the ortho position. The reaction (see Problem 20.10) is another example of electrophilic aromatic substitution.

Specific Example

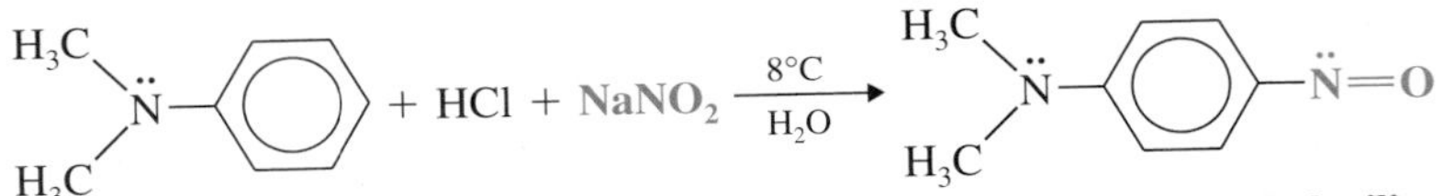

p-Nitroso-N,N-dimethylaniline
(80–90%)

PROBLEM 20.10 _____

Para-nitrosation of N,N-dimethylaniline (C-nitrosation) is believed to take place through an electrophilic attack of $\overset{+}{N}O$ ions. (a) Show how $\overset{+}{N}O$ ions might be formed in an aqueous solution of $NaNO_2$ and HCl. (b) Write a mechanism for p-nitrosation of N,N-dimethylaniline. (c) Tertiary aromatic amines and phenols undergo C-nitrosation reactions, whereas most other benzene derivatives do not. How can you account for this difference?

20.8 REPLACEMENT REACTIONS OF ARENEDIAZONIUM SALTS

Diazonium salts are highly useful intermediates in the synthesis of aromatic compounds, because the diazonium group can be replaced by any one of a number of other atoms or groups, including —F, —Cl, —Br, —I, —CN, —OH, and —H.

   Diazonium salts are almost always prepared by diazotizing primary aromatic amines. Primary arylamines can be synthesized through reduction of nitro compounds that are readily available through direct nitration reactions.

20.8A Syntheses Using Diazonium Salts

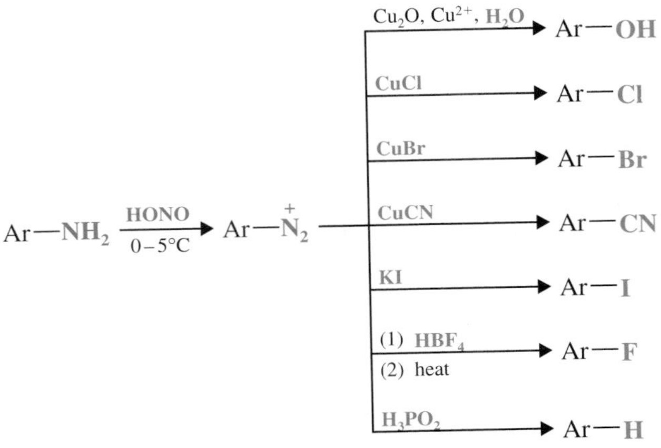

Most arenediazonium salts are unstable at temperatures above 5–10°C, and many explode when dry. Fortunately, however, most of the replacement reactions of diazonium salts do not require their isolation. We simply add another reagent (CuCl, CuBr,

KI, etc.) to the mixture, gently warm the solution, and the replacement (accompanied by the evolution of nitrogen) takes place.

Only in the replacement of the diazonium group by —F need we isolate a diazonium salt. We do this by adding HBF_4 to the mixture, causing the sparingly soluble and reasonably stable arenediazonium fluoborate, $ArN_2^+ BF_4^-$, to precipitate.

20.8B The Sandmeyer Reaction: Replacement of the Diazonium Group by —Cl, —Br, or —CN

Arenediazonium salts react with cuprous chloride, cuprous bromide, and cuprous cyanide to give products in which the diazonium group has been replaced by —Cl, —Br, and —CN, respectively. These reactions are known generally as *Sandmeyer reactions.* Several specific examples follow. The mechanisms of these replacement reactions are not fully understood; the reactions appear to be radical in nature, not ionic.

o-Toluidine → HCl, NaNO₂ / H₂O / (0–5°C) → CuCl / 15–60°C → *o*-Chlorotoluene (74–79% overall) + N₂

m-Chloroaniline → HBr, NaNO₂ / H₂O / (0–10°C) → CuBr / 100°C → *m*-Bromochlorobenzene (70% overall) + N₂

o-Nitroaniline → HCl, NaNO₂ / H₂O / (room temp.) → CuCN / 90–100°C → *o*-Nitrobenzonitrile (65% overall) + N₂

20.8C Replacement by —I

Arenediazonium salts react with potassium iodide to give products in which the diazonium group has been replaced by —I. An example is the synthesis of *p*-iodonitrobenzene:

p-Nitroaniline → H₂SO₄, NaNO₂ / H₂O / 0–5°C → KI → *p*-Iodonitrobenzene (81% overall) + N₂

20.8D Replacement by —F

The diazonium group can be replaced by fluorine by treating the diazonium salt with fluoroboric acid (HBF$_4$). The diazonium fluoroborate that precipitates is isolated, dried, and heated until decomposition occurs. An aryl fluoride is produced.

m-Toluidine   *m*-Toluenediazonium fluoborate (79%)   *m*-Fluorotoluene (69%)

(1) HONO, H$^+$
(2) HBF$_4$

heat

+ N$_2$ + BF$_3$

20.8E Replacement by —OH

The diazonium group can be replaced by a hydroxyl group by adding cuprous oxide to a dilute solution of the diazonium salt containing a large excess of cupric nitrate.

p-Toluenediazonium hydrogen sulfate   *p*-Cresol (93%)

This variation of the Sandmeyer reaction (developed by Theodore Cohen of the University of Pittsburgh) is a much simpler and safer procedure than an older method for phenol preparation, which required heating the diazonium salt with concentrated aqueous acid.

PROBLEM 20.11

In the preceding examples of diazonium reactions, we have illustrated syntheses beginning with the compounds (a)–(e) here. Show how you might prepare each of the following compounds from benzene.

(a) *m*-Nitroaniline   (c) *m*-Bromoaniline   (e) *p*-Nitroaniline

(b) *m*-Chloroaniline   (d) *o*-Nitroaniline

20.8F Replacement by —H: Deamination by Diazotization

Arenediazonium salts react with hypophosphorous acid (H$_3$PO$_2$) to yield products in which the diazonium group has been replaced by —H.

Since we usually begin a synthesis using diazonium salts by nitrating an aromatic compound, that is, replacing —H by —NO$_2$ and then by —NH$_2$, it may seem strange that we would ever want to replace a diazonium group by —H. However, replacement of the diazonium group by —H can be a useful reaction. We can introduce an amino group into an aromatic ring to influence the orientation of a subsequent reaction. Later we can remove the amino group (i.e., carry out a *deamination*) by diazotizing it and treating the diazonium salt with H$_3$PO$_2$.

We can see an example of the usefulness of a deamination reaction in the following synthesis of *m*-bromotoluene. We cannot prepare *m*-bromotoluene by

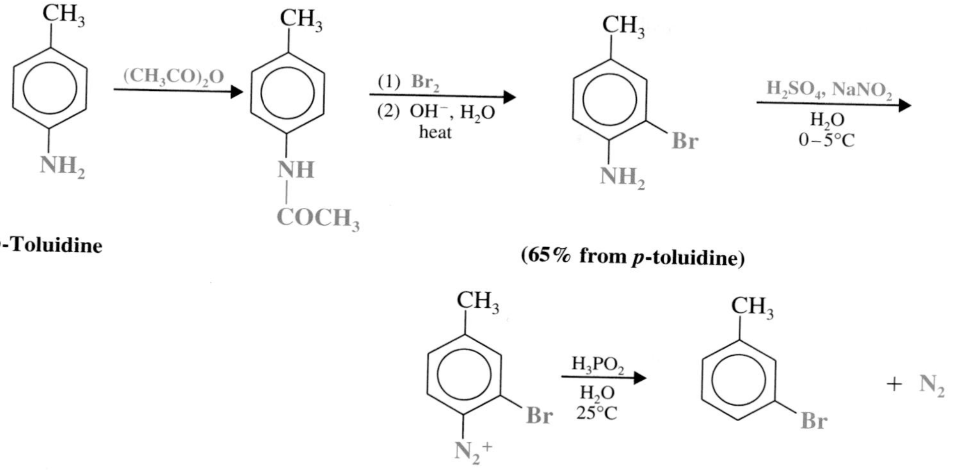

p-Toluidine

(65% from *p*-toluidine)

m-Bromotoluene
(85% from 2-bromo-4-methylaniline)

direct bromination of toluene or by a Friedel–Crafts alkylation of bromobenzene because both reactions give *o*- and *p*-bromotoluene. (Both CH₃— and Br— are ortho–para directors.) However, if we begin with *p*-toluidine (prepared by nitrating toluene, separating the para isomer, and reducing the nitro group), we can carry out the sequence of reactions shown and obtain *m*-bromotoluene in good yield. The first step, synthesis of the *N*-acetyl derivative of *p*-toluidine, is done to reduce the activating effect of the amino group. (Otherwise both ortho positions would be brominated.) Later, the acetyl group is removed by hydrolysis.

PROBLEM 20.12

Suggest how you might modify the preceding synthesis in order to prepare 3,5-dibromotoluene.

PROBLEM 20.13

(a) In Section 20.8D we showed a synthesis of *m*-fluorotoluene starting with *m*-toluidine. How would you prepare *m*-toluidine from toluene?   (b) How would you prepare *m*-chlorotoluene?   (c) *m*-Bromotoluene?   (d) *m*-Iodotoluene?   (e) *m*-Tolunitrile (m-CH₃C₆H₄CN)?   (f) *m*-Toluic acid?

PROBLEM 20.14

Starting with *p*-nitroaniline [Problem 20.11(e)] show how you might synthesize 1,2,3-tribromobenzene.

20.9 COUPLING REACTIONS OF ARENEDIAZONIUM SALTS

Arenediazonium ions are weak electrophiles; they react with highly reactive aromatic compounds—with phenols and tertiary arylamines—to yield *azo* compounds. This electrophilic aromatic substitution is often called a *diazo coupling reaction.*

General Reaction

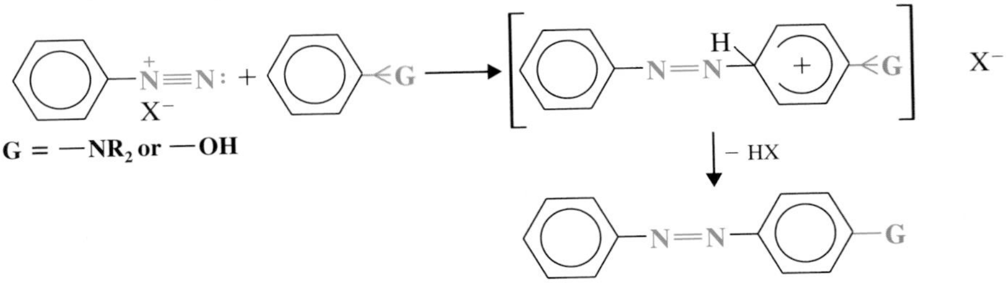

$$G = -NR_2 \text{ or } -OH$$

An azo compound

Specific Examples

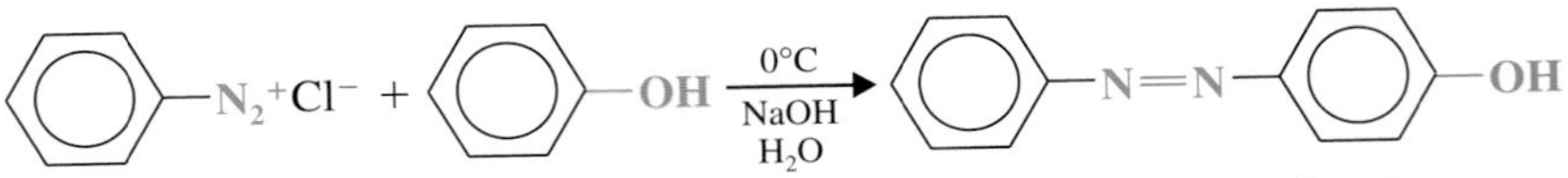

| Benzenediazonium chloride | Phenol | | p-(Phenylazo)phenol (orange solid) |

| Benzenediazonium chloride | N,N-Dimethylaniline | | N,N-Dimethyl-p-(phenylazo)aniline (yellow solid) |

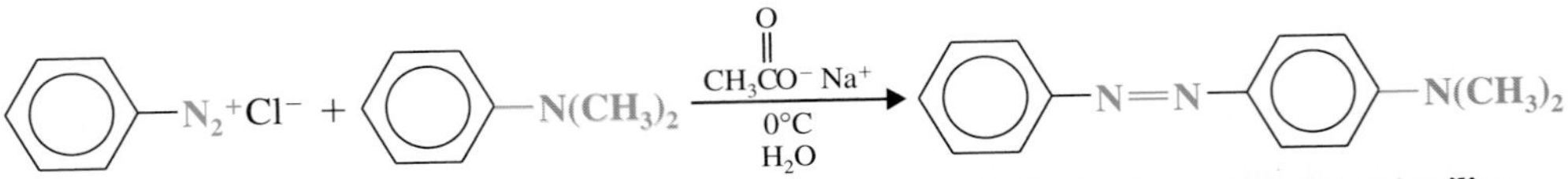

Couplings between arenediazonium cations and phenols take place most rapidly in *slightly* alkaline solution. Under these conditions an appreciable amount of the phenol is present as a phenoxide ion, ArO^-, and phenoxide ions are even more reactive toward electrophilic substitution than are phenols themselves. (Why?) If the solution is too alkaline (pH > 10), however, the arenediazonium salt itself reacts with hydroxide ion to form a relatively unreactive diazohydroxide or diazotate ion:

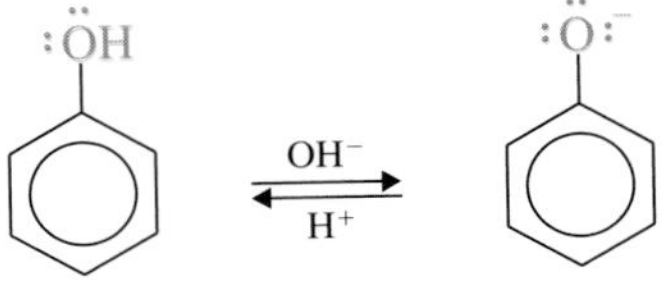

Phenol (couples slowly)   Phenoxide ion (couples rapidly)

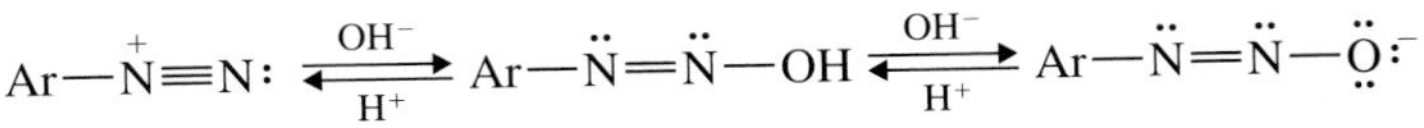

| Arenediazonium ion (couples) | Diazohydroxide (does not couple) | Diazotate ion (does not couple) |

Couplings between arenediazonium cations and amines take place most rapidly in slightly acidic solutions (pH 5–7). Under these conditions the concentration of the arenediazonium cation is at a maximum; at the same time an excessive amount of the amine has not been converted to an unreactive aminium salt:

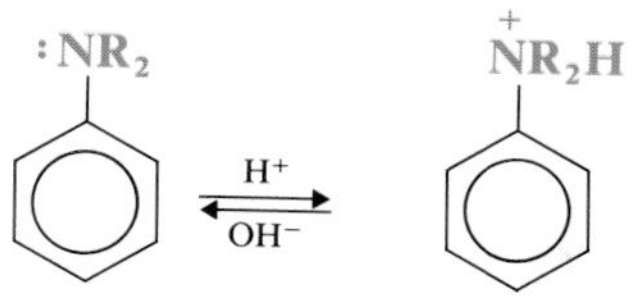

Amine
(couples)

Aminium salt
(does not couple)

If the pH of the solution is lower than 5, the rate of amine coupling is low.

With phenols and aniline derivatives, coupling takes place almost exclusively at the para position if it is open. If it is not, coupling takes place at the ortho position.

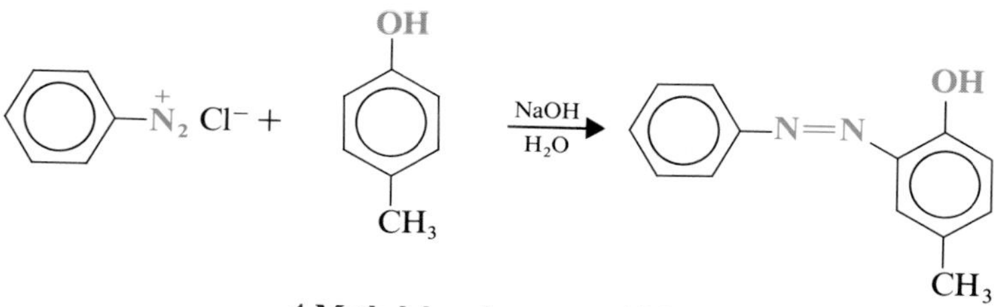

4-Methylphenol
(p-cresol)

4-Methyl-2-(phenylazo)phenol

Azo compounds are usually intensely colored because the azo (diazenediyl) linkage —N=N— brings the two aromatic rings into conjugation. This gives an extended system of delocalized π electrons and allows absorption of light in the visible region. Azo compounds, because of their intense colors, and because they can be synthesized from relatively inexpensive compounds, are used extensively as *dyes*.

Azo dyes almost always contain one or more —SO_3^- Na^+ groups to confer water solubility on the dye and assist in binding the dye to the surfaces of polar fibers (wool, cotton, or nylon). Many dyes are made by coupling reactions of naphthylamines and naphthols.

Orange II, a dye introduced in 1876, is made from 2-naphthol.

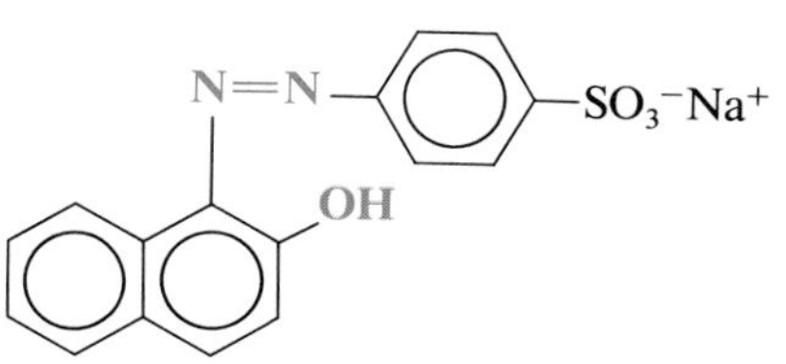

Orange II

PROBLEM 20.15

Outline a synthesis of Orange II from 2-naphthol and *p*-aminobenzenesulfonic acid.

PROBLEM 20.16

Butter Yellow is a dye once used to color margarine. It has since been shown to be carcinogenic and its use in food is no longer permitted. Outline a synthesis of Butter Yellow from benzene and N,N-dimethylaniline.

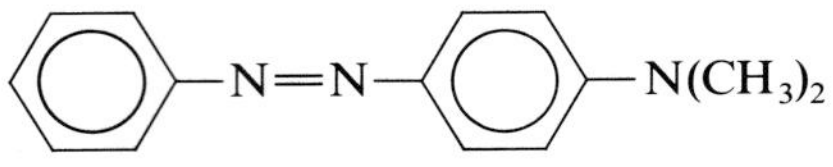

$$\text{C}_6\text{H}_5-\text{N}=\text{N}-\text{C}_6\text{H}_4-\text{N(CH}_3)_2$$

Butter Yellow

PROBLEM 20.17

Azo compounds can be reduced to amines by a variety of reagents including stannous chloride ($SnCl_2$).

$$\text{Ar}-\text{N}=\text{N}-\text{Ar}' \xrightarrow{\text{SnCl}_2} \text{ArNH}_2 + \text{Ar}'\text{NH}_2$$

This reduction can be useful in synthesis as the following example shows:

4-Ethoxyaniline $\xrightarrow[\text{(2) phenol, OH}^-]{\text{(1) HONO, H}_3\text{O}^+}$ **A** ($C_{14}H_{14}N_2O_2$) $\xrightarrow{\text{NaOH, CH}_3\text{CH}_2\text{Br}}$

B ($C_{16}H_{18}N_2O_2$) $\xrightarrow{\text{SnCl}_2}$ two molar equivalents of

C ($C_8H_{11}NO$) $\xrightarrow{\text{acetic anhydride}}$ phenacetin ($C_{10}H_{13}NO_2$)

Give a structure for phenacetin and for the intermediates **A**, **B**, **C**. (Phenacetin, formerly used as an analgesic, is also the subject of Problem 18.35.)

20.10   REACTIONS OF AMINES WITH SULFONYL CHLORIDES

Primary and secondary amines react with sulfonyl chlorides to form *sulfonamides*.

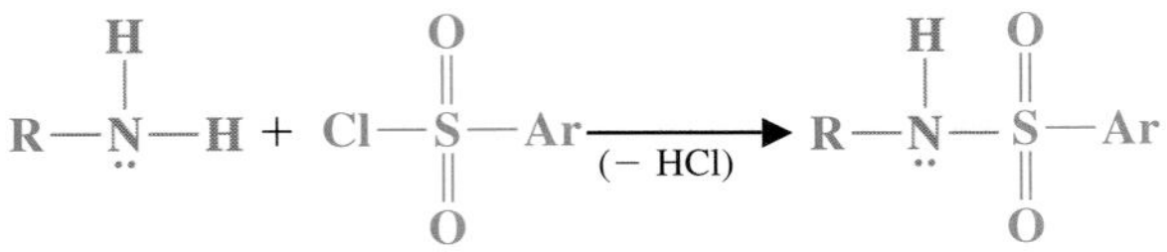

| 1° Amine | Sulfonyl chloride | | N-Substituted sulfonamide |

| 2° Amine | | | N,N-Disubstituted sulfonamide |

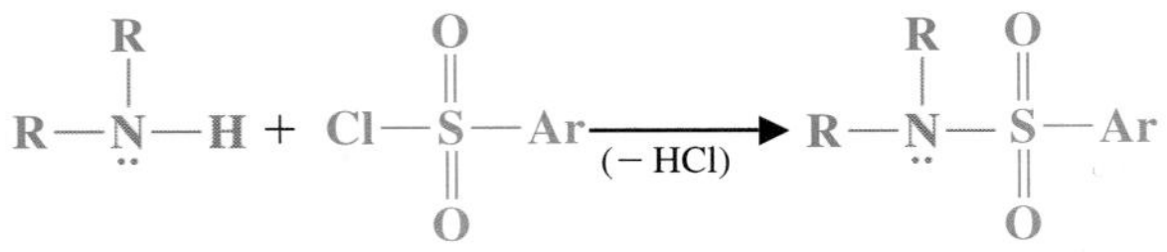

When heated with aqueous acid, sulfonamides are hydrolyzed to amines:

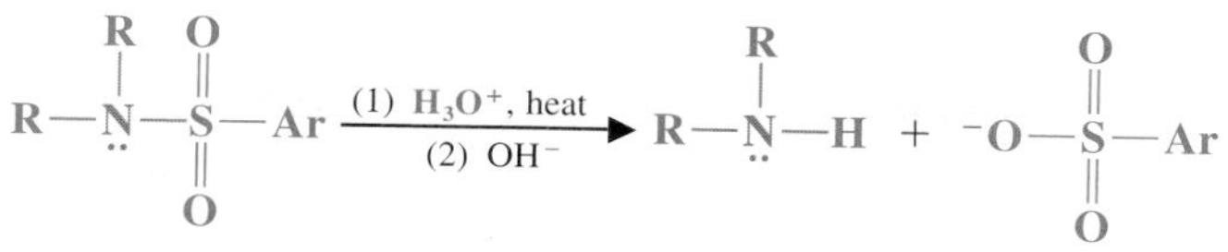

$$
\underset{\overset{\underset{..}{\big|}}{R}}{R-N-\underset{\overset{\|}{O}}{\overset{\|}{S}}-Ar} \xrightarrow[\text{(2) OH}^-]{\text{(1) H}_3\text{O}^+, \text{heat}} R-\underset{..}{\underset{\big|}{N}}-H \; + \; {}^-O-\underset{\overset{\|}{O}}{\overset{\|}{S}}-Ar
$$

This hydrolysis is much slower, however, than hydrolysis of carboxamides.

20.10A The Hinsberg Test

Sulfonamide formation is the basis for a chemical test, called the Hinsberg test, that can be used to demonstrate whether an amine is primary, secondary, or tertiary. A Hinsberg test involves two steps. First, a mixture containing a small amount of the amine and benzenesulfonyl chloride is shaken with *excess* potassium hydroxide. Next, after allowing time for a reaction to take place, the mixture is acidified. Each type of amine—primary, secondary, or tertiary—gives a different set of *visible* results after each of these two stages of the test.

Primary amines react with benzenesulfonyl chloride to form *N*-substituted benzenesulfonamides. These, in turn, undergo acid–base reactions with the excess potassium hydroxide to form water-soluble potassium salts. (These reactions take place because the hydrogen attached to nitrogen is made acidic by the strongly electron-withdrawing —SO$_2$— group.) At this stage our test tube contains a clear solution. Acidification of this solution will, in the next stage, cause the water-insoluble *N*-substituted sulfonamide to precipitate.

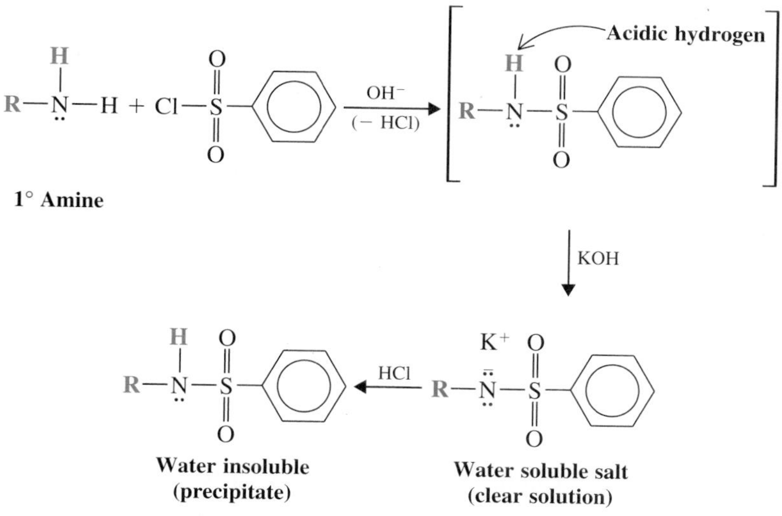

Secondary amines react with benzenesulfonyl chloride in aqueous potassium hydroxide to form insoluble *N*,*N*-disubstituted sulfonamides that precipitate after the first stage. *N*,*N*-Disubstituted sulfonamides do not dissolve in aqueous potassium hydroxide because they do not have an acidic hydrogen. Acidification of the mixture obtained from a secondary amine produces no visible result—the nonbasic *N*,*N*-disubstituted sulfonamide remains as a precipitate and no new precipitate forms.

**Water insoluble
(precipitate)**

If the amine is a tertiary amine and if it is water insoluble, no apparent change will take place in the mixture as we shake it with benzenesulfonyl chloride and aqueous KOH. When we acidify the mixture, the tertiary amine dissolves because it forms a water-soluble salt.

PROBLEM 20.18

An amine **A** has the molecular formula C_7H_9N. Compound **A** reacts with benzenesulfonyl chloride in aqueous potassium hydroxide to give a clear solution; acidification of the solution gives a precipitate. When **A** is treated with $NaNO_2$ and HCl at 0–5°C, and then with 2-naphthol, an intensely colored compound is formed. Compound **A** gives a single strong IR absorption peak 815 cm^{-1}. What is the structure of **A**?

PROBLEM 20.19

Sulfonamides of primary amines are often used to synthesize *pure* secondary amines. Suggest how this synthesis is carried out.

20.11 THE SULFA DRUGS: SULFANILAMIDE

20.11A Chemotherapy

Chemotherapy is defined as the use of chemical agents selectively to destroy infectious organisms without simultaneously destroying the host. Although it may be difficult to believe in this age of "wonder drugs," chemotherapy is a relatively modern phenomenon. Before 1900 only three specific chemical remedies were known: mercury (for syphilis—but often with disastrous results), cinchona bark (for malaria), and ipecacuanha (for dysentery).

Modern chemotherapy began with the work of Paul Ehrlich early in this century —particularly with his discovery in 1907 of the curative properties of a dye called Trypan Red I when used against experimental trypanosomiasis and with his discovery in 1909 of Salvarsan as a remedy for syphilis (Special Topic N). Ehrlich invented the term "chemotherapy," and in his research sought what he called "magic bullets," that is, chemicals that would be toxic to infectious microorganisms but harmless to humans.*

As a medical student, Ehrlich had been impressed with the ability of certain dyes to stain tissues selectively. Working on the idea that "staining" was a result of a chemical reaction between the tissue and the dye, Ehrlich sought dyes with selective affinities for microorganisms. He hoped that in this way he might find a dye that could be modified so as to render it specifically lethal to microorganisms.

*P. Ehrlich was awarded the Nobel Prize for medicine in 1908.

20.11B Sulfa Drugs

Between 1909 and 1935, tens of thousands of chemicals, including many dyes, were tested by Ehrlich and others in a search for such "magic bullets." Very few compounds, however, were found to have any promising effect. Then, in 1935, an amazing event happened. The daughter of Gerhard Domagk, a doctor employed by a German dye manufacturer, contracted a streptococcal infection from a pin prick. As his daughter neared death, Domagk decided to give her an oral dose of a dye called Prontosil. Prontosil had been developed at Domagk's firm (I. G. Farbenindustrie) and tests with mice had shown that Prontosil inhibited the growth of streptococci. Within a short time the little girl recovered. Domagk's gamble not only saved his daughter's life, but it also initiated a new and spectacularly productive phase in modern chemotherapy.*

A year later, in 1936, Ernest Fourneau of the Pasteur Institute in Paris demonstrated that Prontosil breaks down in the human body to produce sulfanilamide, and that sulfanilamide is the actual active agent against streptococci.

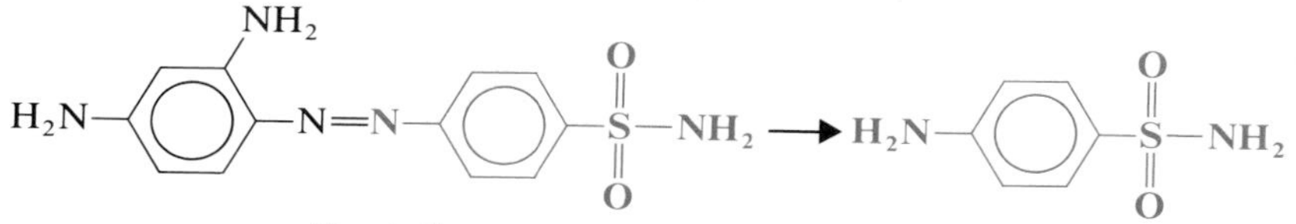

Prontosil → **Sulfanilamide**

Fourneau's announcement of this result set in motion a search for other chemicals (related to sulfanilamide) that might have even better chemotherapeutic effects. Literally thousands of chemical variations were played on the sulfanilamide theme; the structure of sulfanilamide was varied in almost every imaginable way. The best therapeutic results were obtained from compounds in which one hydrogen of the —SO_2NH_2 group was replaced by some other group, usually a heterocyclic ring (shown in blue in the following structures). Among the most successful variations were the following compounds. Sulfanilamide itself is too toxic for general use.

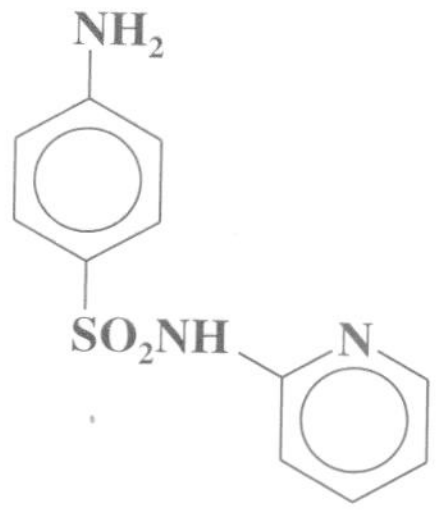

Sulfapyridine 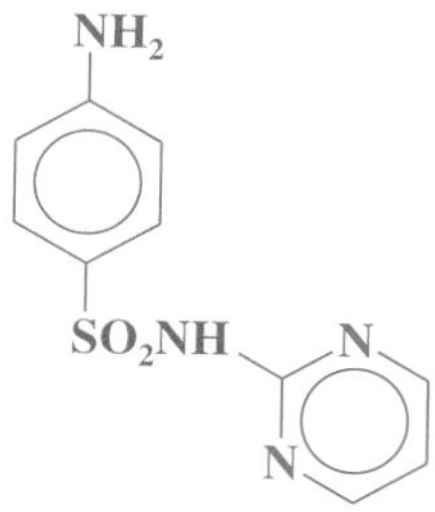   **Sulfadiazine** 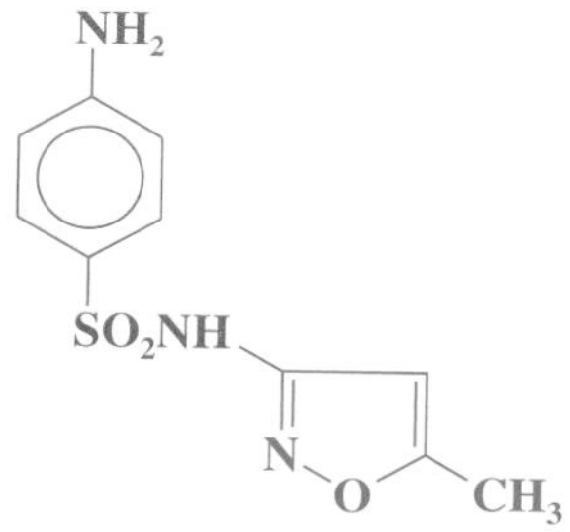   **Sulfamethoxazole**

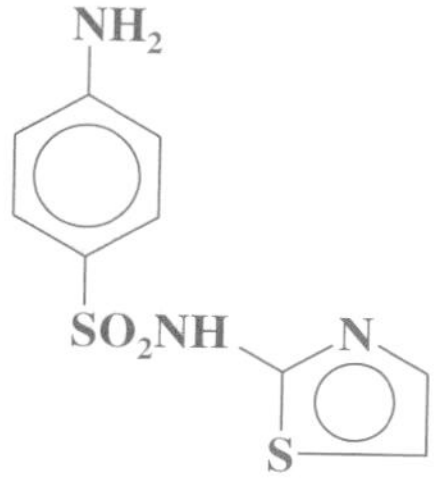

Sulfathiazole 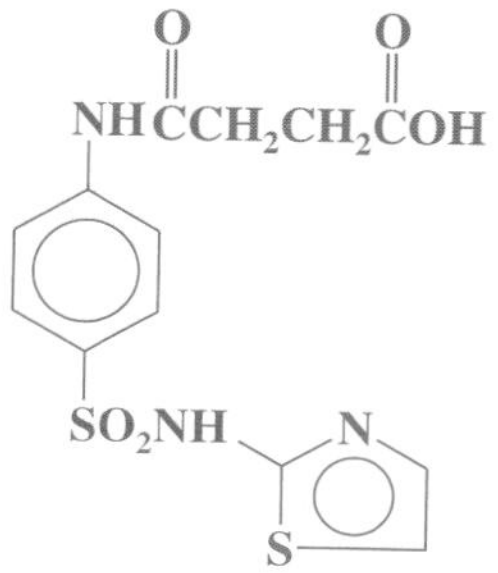   **Succinylsulfathiazole** 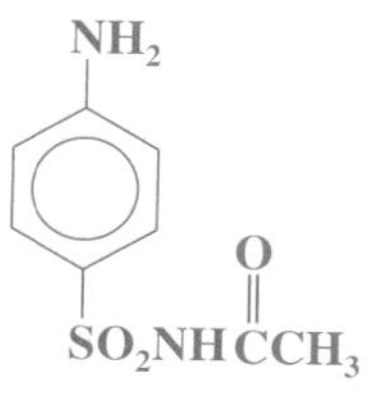   **Sulfacetamide**

*G. Domagk was awarded the Nobel Prize for medicine in 1939 but was unable to accept it until 1947.

Sulfapyridine was shown to be effective against pneumonia in 1938. (Before that time pneumonia epidemics had brought death to tens of thousands.) Sulfacetamide was first used successfully in treating urinary tract infections in 1941. Succinoylsulfathiazole and the related compound phthalylsulfathiazole were used as chemotherapeutic agents against infections of the gastrointestinal tract beginning in 1942. (Both compounds are slowly hydrolyzed internally to sulfathiazole.) Sulfathiazole saved the lives of countless wounded soldiers during World War II.

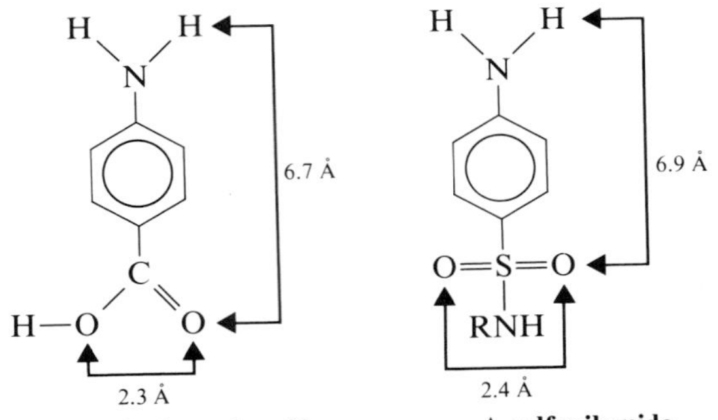

FIGURE 20.3 The structural similarity of *p*-aminobenzoic acid and a sulfanilamide. (From A. Korolkovas, *Essentials of Molecular Pharmacology*, Wiley, New York, 1970, p. 105. Used with permission.)

In 1940 a discovery by D. D. Woods laid the groundwork for our understanding of how the sulfa drugs work. Woods observed that the inhibition of growth of certain microorganisms by sulfanilamide is competitively overcome by *p*-aminobenzoic acid. Woods noticed the structural similarity between the two compounds (Fig. 20.3) and reasoned that the two compounds compete with each other in some essential metabolic process.

20.11C Essential Nutrients and Antimetabolites

All higher animals and many microorganisms lack the biochemical ability to synthesize certain essential organic compounds. These essential nutrients include vitamins, certain amino acids, unsaturated carboxylic acids, purines, and pyrimidines. The aromatic amine *p*-aminobenzoic acid is an essential nutrient for those bacteria that are sensitive to sulfanilamide therapy. Enzymes within these bacteria use *p*-aminobenzoic acid to synthesize another essential compound called *folic acid*.

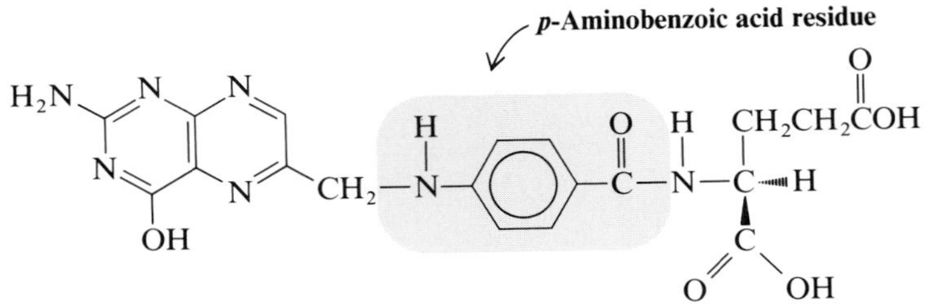

Folic acid

Chemicals that inhibit the growth of microbes are called *antimetabolites*. The sulfanilamides are antimetabolites for those bacteria that require *p*-aminobenzoic acid.

The sulfanilamides apparently inhibit those enzymatic steps of the bacteria that are involved in the synthesis of folic acid. The bacterial enzymes are apparently unable to distinguish between a molecule of a sulfanilamide and a molecule of *p*-aminobenzoic acid; thus, sulfanilamide ''inhibits'' the bacterial enzyme. Because the microorganism is unable to synthesize enough folic acid when sulfanilamide is present, it dies. Humans are unaffected by sulfanilamide therapy because we derive our folic acid from dietary sources (folic acid is a vitamin) and do not synthesize it from *p*-aminobenzoic acid.

The discovery of the mode of action of the sulfanilamides has led to the discovery of many new and effective antimetabolites. One example is *methotrexate,* a derivative of folic acid that has been used successfully in treating certain carcinomas:

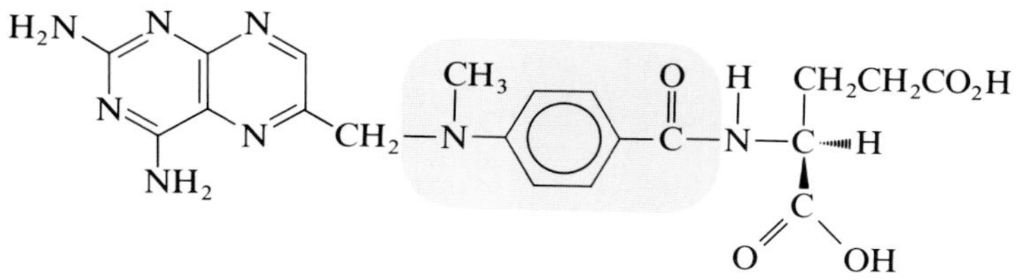

Methotrexate

Methotrexate, by virtue of its resemblance to folic acid, can enter into some of the same reactions as folic acid, but it cannot serve the same function, particularly in important reactions involved in cell division. Although methotrexate is toxic to all dividing cells, those cells that divide most rapidly—*cancer cells*—are most vulnerable to its effect.

20.11D Synthesis of Sulfa Drugs

Sulfanilamides can be synthesized from aniline through the following sequence of reactions.

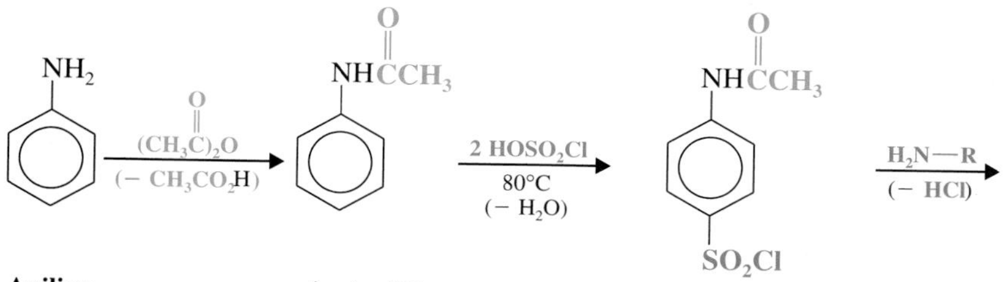

Aniline
1

Acetanilide
2

***p*-Acetamidobenzene-sulfonyl chloride**
3

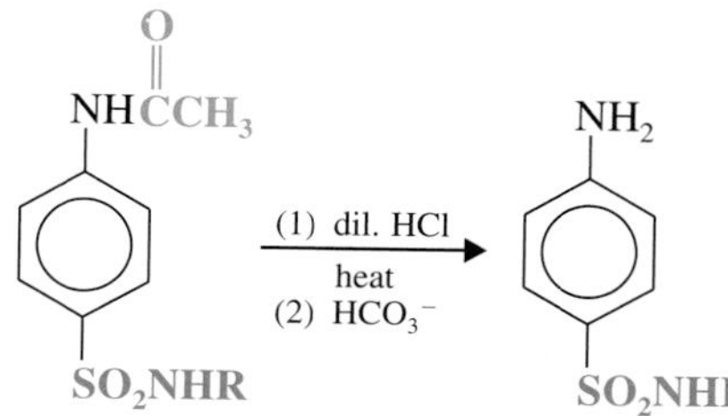

4

A sulfanilamide
5

Acetylation of aniline produces acetanilide, **2**, and protects the amino group from the reagent to be used next. Treatment of **2** with chlorosulfonic acid brings about an electrophilic aromatic substitution reaction and yields *p*-acetamidobenzenesulfonyl chloride, **3**. Addition of ammonia or a primary amine gives the diamide, **4** (an amide of both a carboxylic acid and a sulfonic acid). Finally, refluxing **4** with dilute hydrochloric acid selectively hydrolyzes the carboxamide linkage and produces sulfanilamide. (Hydrolysis of carboxamides is much more rapid than that of sulfonamides.)

PROBLEM 20.20

(a) Starting with aniline and assuming that you have 2-aminothiazole available, show how you would synthesize sulfathiazole.   (b) How would you convert sulfathiazole to succinylsulfathiazole?

2-Aminothiazole

20.12 ANALYSIS OF AMINES

20.12A Chemical Analysis

Amines are characterized by their basicity and, thus, by their ability to dissolve in dilute aqueous acid (Section 20.3A). Primary, secondary, and tertiary amines can be distinguished from each other on the basis of the Hinsberg test (Section 20.10A). Primary aromatic amines are often detected through diazonium salt formation and subsequent coupling with 2-naphthol to form a brightly colored azo dye (Section 20.9).

20.12B Spectroscopic Analysis

Infrared Spectra   Primary and secondary amines are characterized by IR absorption bands in the 3300–3555 cm^{-1} region that arise from N—H stretching vibrations. Primary amines give two bands in this region; secondary amines generally give only one. Tertiary amines, because they have no N—H group, do not absorb in this region. Absorption bands arising from C—N stretching vibrations of aliphatic amines occur in the 1020–1220 cm^{-1} region but are usually weak and difficult to identify. Aromatic amines generally give a strong C—N stretching band in the 1250–1360 cm^{-1} region. Figure 20.4 shows an annotated IR spectrum of 4-methylaniline.

^{1}H NMR Spectra   Primary and secondary amines show N—H proton signals in the region δ 1–5. These signals are usually broad and their exact position depends on the nature of the solvent, the purity of the sample, the concentration, and the temperature. Because of proton exchange, N—H protons are not usually coupled to protons on adjacent carbons. As such, they are difficult to identify and are best detected by proton counting or by adding a small amount of D$_2$O to the sample. Exchange of N—D for the N—H protons takes place and the N—H signal disappears from the spectrum.

Protons on the α carbon of an aliphatic amine are deshielded by the electron-withdrawing effect of the nitrogen and absorb typically in the δ 2.2–2.9 region; protons on the β carbon are not deshielded as much and absorb in the range δ 1.0–1.7.

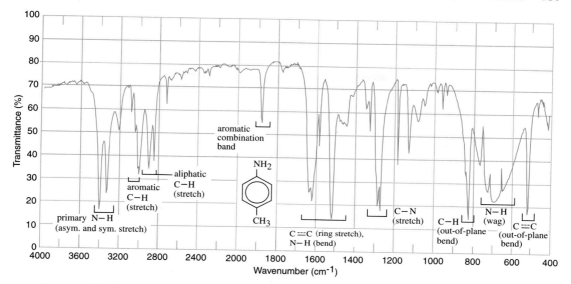

FIGURE 20.4 Annotated IR spectrum of 4-methylaniline.

Figure 20.5 shows an annotated ¹H NMR spectrum of diisopropylamine.

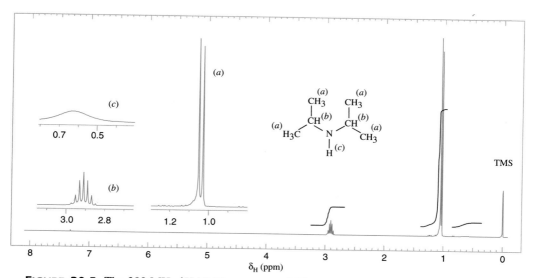

FIGURE 20.5 The 300 MHz ¹H NMR spectrum of diisopropylamine. Note the integral for the broad NH peak at approximately 0.7 ppm. Vertical expansions are not too scale.

¹³C NMR Spectra   The α carbon of an aliphatic amine experiences deshielding by the electronegative nitrogen and its absorption is shifted downfield. The shift is not as great as for the α carbon of an alcohol, however, because nitrogen is less electronegative than oxygen. The downfield shift is even less for the β carbon, and so on down the chain, as the chemical shifts of the carbons of pentylamine show:

$$H_3C-CH_2-CH_2-CH_2-CH_2-NH_2$$
$$\delta \quad 14.3 \qquad 23.0 \qquad 29.7 \qquad 34.0 \qquad 42.5$$
¹³C NMR chemical shifts

20.13 ELIMINATIONS INVOLVING AMMONIUM COMPOUNDS

20.13A The Hofmann Elimination

All of the eliminations that we have described so far have involved electrically neutral substrates. However, eliminations are known in which the substrate bears a positive charge. One of the most important of these is the E2-type elimination that takes place when a quaternary ammonium hydroxide is heated. The products are an alkene, water, and tertiary amine.

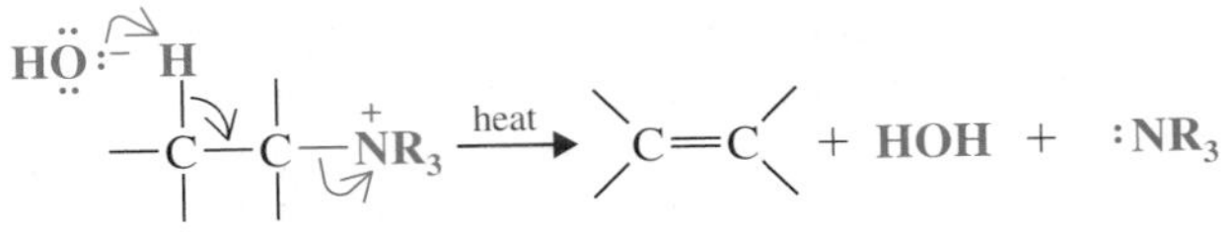

$$\text{HO}:\overset{\frown}{}\text{H}$$

$$-\overset{|}{\underset{|}{C}}\!-\!\overset{|}{\underset{|}{C}}\!-\!\overset{+}{N}R_3 \xrightarrow{\text{heat}} \;\; \overset{\backslash}{}C\!=\!C\overset{/}{} + \text{HOH} + \;:NR_3$$

A quaternary
ammonium hydroxide ⟶ an alkene + water + **a tertiary amine**

This reaction was discovered in 1851 by August W. von Hofmann and has since come to bear his name.

Quaternary ammonium hydroxides can be prepared from quaternary ammonium halides in aqueous solution through the use of silver oxide or an ion exchange resin.

$$2\ RCH_2CH_2\overset{+}{N}(CH_3)_3\ X^- + Ag_2O + H_2O \longrightarrow 2\ RCH_2CH_2\overset{+}{N}(CH_3)_3\ OH^- + 2\ AgX\!\!\downarrow$$

A quaternary ammonium halide   **A quaternary ammonium hydroxide**

Silver halide precipitates from the solution and can be removed by filtration. The quaternary ammonium hydroxide can then be obtained by evaporation of the water.

Although most eliminations involving neutral substrates tend to follow the *Zaitsev rule* (Section 7.12A), eliminations with charged substrates tend to follow what is called the *Hofmann rule* and *yield mainly the least substituted alkene.* We can see an example of this behavior if we compare the following reactions.

$$C_2H_5O^-\,Na^+ + CH_3CH_2\underset{\underset{Br}{|}}{C}HCH_3 \xrightarrow[25°C]{C_2H_5OH}$$

$$CH_3CH\!=\!CHCH_3 + CH_3CH_2CH\!=\!CH_2 + NaBr + C_2H_5OH$$
$$\text{(75\%)} \qquad\qquad \text{(25\%)}$$

$$CH_3CH_2\underset{\underset{\overset{+}{N}(CH_3)_3}{|}}{C}HCH_3\ OH^- \xrightarrow[150°C]{}$$

$$CH_3CH\!=\!CHCH_3 + CH_3CH_2CH\!=\!CH_2 + (CH_3)_3N: + H_2O$$
$$\text{(5\%)} \qquad\qquad \text{(95\%)}$$

$$CH_3CH_2\underset{\underset{\overset{+}{S}(CH_3)_2}{|}}{C}HCH_3\ \overset{-}{O}C_2H_5 \longrightarrow$$

$$CH_3CH\!=\!CHCH_3 + CH_3CH_2CH\!=\!CH_2 + (CH_3)_2S + C_2H_5OH$$
$$\text{(26\%)} \qquad\qquad \text{(74\%)}$$

The precise mechanistic reasons for these differences are complex and are not yet fully understood. One possible explanation is that the transition states of elimination reactions with charged substrates have considerable carbanion character. Therefore, these transition states show little resemblance to the final alkene product and are not stabilized appreciably by a developing double bond.

Carbanion-like transition state
(gives Hofmann orientation)

Alkene-like transition state
(gives Zaitsev orientation)

With a charged substrate, the base attacks the most acidic hydrogen instead. A primary hydrogen atom is more acidic because its carbon atom bears only one electron-releasing group.

20.13B The Cope Elimination

Tertiary amine oxides undergo the elimination of a dialkylhydroxylamine when they are heated. This reaction is called the Cope elimination.

$$RCH_2CH_2\overset{+}{N}-CH_3 \xrightarrow{150°C} RCH=CH_2 \quad + \quad :N-CH_3$$

A tertiary amine oxide   **An alkene**   *N,N*-**Dimethylhydroxylamine**

The Cope elimination is a syn elimination and proceeds through a cyclic transition state:

$$R-CH-CH_2 \quad CH_3 \longrightarrow R-CH=CH_2 + H-\overset{..}{O}$$

Tertiary amine oxides are easily prepared by treating tertiary amines with hydrogen peroxide (Section 20.6A).

The Cope elimination is useful synthetically. Consider the following synthesis of methylenecyclohexane.

$$\xrightarrow{160°C} \quad =CH_2 + (CH_3)_2\overset{..}{N}OH$$

(98%)

20.14 SUMMARY OF PREPARATIONS AND REACTIONS OF AMINES

20.14A Preparation of Amines

1. Gabriel synthesis (discussed in Section 20.5A).

2. By reductions of alkyl azides (discussed in Section 20.5A).

$$R-Br \xrightarrow[\text{ethanol}]{NaN_3} R-\overset{+}{N}=\overset{}{N}=\overset{-}{N} \xrightarrow[\substack{\text{or} \\ LiAlH_4}]{Na/alcohol} R-NH_2$$

3. By amination of alkyl halides (discussed in Section 20.5A).

$$R-Br + NH_3 \longrightarrow RNH_3{}^+Br^- + R_2NH_2{}^+Br^- + R_3N^+Br^- + R_4N^+Br^-$$
$$\downarrow OH^-$$
$$RNH_2 + R_2NH + R_3N + R_4N^+OH^-$$

(**R** is a 1° alkyl group)

4. By reduction of nitroarenes (discussed in Section 20.5B).

$$Ar-NO_2 \xrightarrow[\substack{\text{or} \\ (1)\ Fe/HCl\ (2)\ NaOH}]{H_2,\ cat} Ar-NH_2$$

5. By reductive amination (discussed in Section 20.5C).

6. By reduction of amides, nitriles, and oximes (discussed in Section 20.5D).

$$R-\overset{\overset{\displaystyle O}{\|}}{C}-\underset{\underset{\displaystyle H}{|}}{N}-H \xrightarrow[\text{(2) } H_2O]{\text{(1) } LiAlH_4,\text{ diethyl ether}} R-CH_2-\underset{\underset{\displaystyle H}{|}}{N}-H \qquad \textbf{1° Amine}$$

$$R-\overset{\overset{\displaystyle O}{\|}}{C}-\underset{\underset{\displaystyle H}{|}}{N}-R' \xrightarrow[\text{(2) } H_2O]{\text{(1) } LiAlH_4,\text{ diethyl ether}} R-CH_2-\underset{\underset{\displaystyle H}{|}}{N}-R' \qquad \textbf{2° Amine}$$

$$R-\overset{\overset{\displaystyle O}{\|}}{C}-\underset{\underset{\displaystyle R''}{|}}{N}-R' \xrightarrow[\text{(2) } H_2O]{\text{(1) } LiAlH_4,\text{ diethyl ether}} R-CH_2-\underset{\underset{\displaystyle R''}{|}}{N}-R' \qquad \textbf{3° Amine}$$

$$R-C{\equiv}N \xrightarrow[\text{(2) } H_2O]{\text{(1) } LiAlH_4,\text{ diethyl ether}} R-CH_2-\underset{\underset{\displaystyle H}{|}}{N}-H \qquad \textbf{1° Amine}$$

$$R-\overset{\overset{\displaystyle N-OH}{\|}}{C}-R' \xrightarrow{\text{Na/ethanol}} R-\overset{\overset{\displaystyle NH_2}{|}}{CH}-R' \qquad \textbf{1° Amine}$$

7. Through the Hofmann and Curtius rearrangements (discussed in Section 20.5E).

Hofmann Rearrangement

$$R-\overset{\overset{\displaystyle O}{\|}}{C}-\underset{\underset{\displaystyle H}{|}}{N}-H \xrightarrow{Br_2,\ OH^-} R-NH_2 + CO_3{}^{2-}$$

Curtius Rearrangement

$$R-\overset{\overset{\displaystyle O}{\|}}{C}-Cl \xrightarrow[(-NaCl)]{NaN_3} R-\overset{\overset{\displaystyle O}{\|}}{C}-N_3 \xrightarrow[-N_2]{\text{heat}} R-N{=}C{=}O \xrightarrow{H_2O} R-NH_2 + CO_2$$

20.14B Reactions of Amines

1. As bases (discussed in Section 20.3).

$$R-CH_2-\underset{\underset{\displaystyle R''}{|}}{\overset{..}{N}}-R' + H-A \longrightarrow R-CH_2-\underset{\underset{\displaystyle R''}{|}}{\overset{\overset{\displaystyle H}{|}}{N^{\pm}}}-R'\ A^-$$

(R, R′, or R″ may be H or Ar)

2. Diazotization of 1° arylamines and replacement of the diazonium group (discussed in Sections 20.8 and 20.9).

$$\text{Ar}-\text{NH}_2 \xrightarrow[0-5°]{\text{HONO}} \text{Ar}-\overset{+}{\text{N}}_2$$

Reagent	Product
$\text{Cu}_2\text{O, Cu}^{2+}, \text{H}_2\text{O}$	$\text{Ar}-\text{OH}$
CuCl	$\text{Ar}-\text{Cl}$
CuBr	$\text{Ar}-\text{Br}$
CuCN	$\text{Ar}-\text{CN}$
KI	$\text{Ar}-\text{I}$
(1) HBF_4 (2) heat	$\text{Ar}-\text{F}$
H_3PO_2	$\text{Ar}-\text{H}$

$$\overset{+}{\text{N}}_2 \xrightarrow[G = NR_2 \text{ or OH}]{} -\text{N}=\text{N}- \leftarrow G$$

3. Conversion to sulfonamides (discussed in Section 20.10).

$$\text{R}-\overset{\underset{\displaystyle H}{|}}{\text{N}}-\text{H} \xrightarrow[\text{(2) HCl}]{\text{(1) ArSO}_2\text{Cl, OH}^-} \text{R}-\overset{\underset{\displaystyle H}{|}}{\text{N}}-\overset{\overset{\displaystyle O}{\parallel}}{\underset{\underset{\displaystyle O}{\parallel}}{\text{S}}}-\text{Ar}$$

$$\text{R}-\overset{\underset{\displaystyle R'}{|}}{\text{N}}-\text{H} \xrightarrow{\text{ArSO}_2\text{Cl, OH}^-} \text{R}-\overset{\underset{\displaystyle R'}{|}}{\text{N}}-\overset{\overset{\displaystyle O}{\parallel}}{\underset{\underset{\displaystyle O}{\parallel}}{\text{S}}}-\text{Ar}$$

4. Conversion to amides (discussed in Section 18.8).

$$\text{R}-\overset{\underset{\displaystyle H}{|}}{\text{N}}-\text{H} \xrightarrow[\text{base}]{\overset{\displaystyle O}{\overset{\parallel}{\text{R''C}}}-\text{Cl}} \text{R}-\overset{\underset{\displaystyle H}{|}}{\text{N}}-\overset{\overset{\displaystyle O}{\parallel}}{\text{C}}-\text{R''} + \text{Cl}^-$$

$$\text{R}-\overset{\underset{\displaystyle H}{|}}{\text{N}}-\text{H} \xrightarrow{(\text{R''C}\overset{O}{\parallel})_2\text{O}} \text{R}-\overset{\underset{\displaystyle H}{|}}{\text{N}}-\overset{\overset{\displaystyle O}{\parallel}}{\text{C}}-\text{R''} + \text{R''}-\overset{\overset{\displaystyle O}{\parallel}}{\text{C}}-\text{OH}$$

$$\text{R}-\overset{\underset{\displaystyle R'}{|}}{\text{N}}-\text{H} \xrightarrow[\text{base}]{\overset{\displaystyle O}{\overset{\parallel}{\text{R''C}}}-\text{Cl}} \text{R}-\overset{\underset{\displaystyle R'}{|}}{\text{N}}-\overset{\overset{\displaystyle O}{\parallel}}{\text{C}}-\text{R''} + \text{Cl}^-$$

5. Hofmann and Cope eliminations (discussed in Section 20.13).

Hofmann Elimination

$$-\overset{\underset{\displaystyle |}{|}}{\text{C}}-\overset{\underset{\displaystyle |}{\overset{\displaystyle H}{|}}}{\text{C}}-\overset{+}{\text{N}}\text{R}_3\text{OH}^- \xrightarrow{\text{heat}} \text{C}=\text{C} + \text{H}_2\text{O} + \text{NR}_3$$

Cope Elimination

$$-\overset{\overset{\displaystyle H}{|}}{\underset{|}{C}}-\overset{\overset{\displaystyle {}^+N(CH_3)_2 \atop O^-}{|}}{\underset{|}{C}}- \quad \xrightarrow[\text{(syn elimination)}]{\text{heat}} \quad \overset{}{}C=C\overset{}{} + (CH_3)_2NOH$$

ADDITIONAL PROBLEMS

20.21 Write structural formulas for each of the following compounds:

(a) Benzylmethylamine
(b) Triisopropylamine
(c) *N*-Ethyl-*N*-methylaniline
(d) *m*-Toluidine
(e) 2-Methylpyrrole
(f) *N*-Ethylpiperidine
(g) *N*-Ethylpyridinium bromide
(h) 3-Pyridinecarboxylic acid
(i) Indole
(j) Acetanilide

(k) Dimethylaminium chloride
(l) 2-Methylimidazole
(m) 3-Amino-1-propanol
(n) Tetrapropylammonium chloride
(o) Pyrrolidine
(p) *N,N*-Dimethyl-*p*-toluidine
(q) 4-Methoxyaniline
(r) Tetramethylammonium hydroxide
(s) *p*-Aminobenzoic acid
(t) *N*-Methylaniline

20.22 Give common or systematic names for each of the following compounds:

(a) $CH_3CH_2CH_2NH_2$
(b) $C_6H_5NHCH_3$
(c) $(CH_3)_2CH\overset{+}{N}(CH_3)_3\ I^-$
(d) *o*-$CH_3C_6H_4NH_2$
(e) *o*-$CH_3OC_6H_4NH_2$
(f)

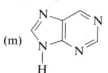

(g)

(h) $C_6H_5CH_2NH_3{}^+Cl^-$
(i) $C_6H_5N(CH_2CH_2CH_3)_2$
(j) $C_6H_5SO_2NH_2$
(k) $CH_3NH_3{}^+CH_3CO_2{}^-$
(l) $HOCH_2CH_2CH_2NH_2$

(m)

(n)

20.23 Show how you might prepare benzylamine from each of the following compounds:

(a) Benzonitrile
(b) Benzamide
(c) Benzyl bromide (two ways)

(d) Benzyl tosylate
(e) Benzaldehyde
(f) Phenylnitromethane

(g) Phenylacetamide

20.24 Show how you might prepare aniline from each of the following compounds:

(a) Benzene   (b) Bromobenzene   (c) Benzamide

20.25 Show how you might synthesize each of the following compounds from butyl alcohol:

(a) Butylamine (free of 2° and 3° amines)
(b) Pentylamine

(c) Propylamine
(d) Butylmethylamine

20.26   Show how you might convert aniline into each of the following compounds. (You need not repeat steps carried out in earlier parts of this problem.)

(a) Acetanilide
(b) *N*-Phenylphthalimide
(c) *p*-Nitroaniline
(d) Sulfanilamide
(e) *N,N*-Dimethylaniline
(f) Fluorobenzene
(g) Chlorobenzene
(h) Bromobenzene

(i) Iodobenzene
(j) Benzonitrile
(k) Benzoic acid
(l) Phenol
(m) Benzene
(n) *p*-(Phenylazo)phenol
(o) *N,N*-Dimethyl-*p*-(phenylazo)aniline

20.27   What products would you expect to be formed when each of the following amines reacts with aqueous sodium nitrite and hydrochloric acid?

(a) Propylamine
(b) Dipropylamine

(c) *N*-Propylaniline
(d) *N,N*-Dipropylaniline

(e) *p*-Propylaniline

20.28   (a) What products would you expect to be formed when each of the amines in the preceding problem reacts with benzenesulfonyl chloride and excess aqueous potassium hydroxide?   (b) What would you observe in each reaction?   (c) What would you observe when the resulting solution or mixture is acidified?

20.29   (a) What product would you expect to obtain from the reaction of piperidine with aqueous sodium nitrite and hydrochloric acid?   (b) From the reaction of piperidine and benzenesulfonyl chloride in excess aqueous potassium hydroxide?

20.30   Give structures for the products of each of the following reactions:

(a) Ethylamine + benzoyl chloride $\longrightarrow$
(b) Methylamine + acetic anhydride $\longrightarrow$
(c) Methylamine + succinic anhydride $\longrightarrow$
(d) Product of (c) $\xrightarrow{\text{heat}}$
(e) Pyrrolidine + phthalic anhydride $\longrightarrow$
(f) Pyrrole + acetic anhydride $\longrightarrow$
(g) Aniline + propanoyl chloride $\longrightarrow$
(h) Tetraethylammonium hydroxide $\xrightarrow{\text{heat}}$
(i) *m*-Dinitrobenzene + H_2S $\xrightarrow[\text{C}_2\text{H}_5\text{OH}]{\text{NH}_3}$
(j) *p*-Toluidine + Br_2(excess) $\xrightarrow[\text{H}_2\text{O}]{}$

20.31   Starting with benzene or toluene, outline a synthesis of each of the following compounds using diazonium salts as intermediates. (You need not repeat syntheses carried out in earlier parts of this problem.)

(a) *o*-Cresol

(b) *m*-Cresol

(c) *p*-Cresol

(d) *m*-Dichlorobenzene

(e) *m*-$C_6H_4(CN)_2$

(f) *m*-Iodophenol

(g) *m*-Bromobenzonitrile

(h) 1,3-Dibromo-5-nitrobenzene

(i) 3,5-Dibromoaniline

(j) 3,4,5-Tribromophenol

(k) 3,4,5-Tribromobenzonitrile

(l) 2,6-Dibromobenzoic acid

(m) 1,3-Dibromo-2-iodobenzene

(n) 4-Bromo-2-nitrotoluene

(o) 4-Methyl-3-nitrophenol

(p) CH_3—⟨ring⟩—Br, CN

(q) CH_3—⟨ring⟩—N=N—⟨ring⟩—OH

(r) CH_3—⟨ring⟩—N=N—⟨ring with OH and CH₃⟩

20.32   Write equations for simple chemical tests that would distinguish between

(a) Benzylamine and benzamide

(b) Allylamine and propylamine

(c) *p*-Toluidine and *N*-methylaniline

(d) Cyclohexylamine and piperidine

(e) Pyridine and benzene

(f) Cyclohexylamine and aniline

(g) Triethylamine and diethylamine

(h) Tripropylaminium chloride and tetrapropylammonium chloride

(i) Tetrapropylammonium chloride and tetrapropylammonium hydroxide

20.33   Describe with equations how you might separate a mixture of aniline, *p*-cresol, benzoic acid, and toluene using ordinary laboratory reagents.

20.34   Show how you might synthesize β-aminopropionic acid ($H_3\overset{+}{N}CH_2CH_2CO_2^-$) from succinic anhydride. (β-Aminopropionic acid is used in the synthesis of pantothenic acid; cf. Problem 18.34.)

20.35   Show how you might synthesize each of the following from the compounds indicated and any other needed reagents.

(a) $(CH_3)_3\overset{+}{N}(CH_2)_{10}\overset{+}{N}(CH_3)_3$ $2Br^-$ from 1,10-decanediol.

(b) Succinylcholine bromide from succinic acid, 2-bromoethanol, and trimethylamine (Section 20.4)

20.36   A commercial synthesis of folic acid consists of heating the following three compounds with aqueous sodium bicarbonate. Propose reasonable mechanisms for the reactions that lead to folic acid.

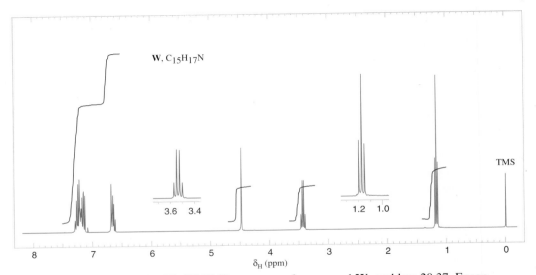

20.37 When compound **W** ($C_{15}H_{17}N$) is treated with benzenesulfonyl chloride and aqueous potassium hydroxide, no apparent change occurs. Acidification of this mixture gives a clear solution. The ^1H NMR spectrum of **W** is shown in Fig. 20.6. Propose a structure for **W**.

FIGURE 20.6 The 300 MHz ^1H NMR spectrum of compound **W**, problem 20.37. Expansions of the signals are shown in the offset plots.

20.38 Propose structures for compounds **X**, **Y**, and **Z**.

$$\mathbf{X}(C_7H_7Br) \xrightarrow{\text{NaCN}} \mathbf{Y}(C_8H_7N) \xrightarrow{\text{LiAlH}_4} \mathbf{Z}(C_8H_{11}N)$$

The ^1H NMR spectrum of **X** gives two signals, a multiplet at δ 7.3 (5H) and a singlet at δ 4.25 (2H); the 680–840-cm^{-1} region of the IR spectrum of **X** shows peaks at 690 and 770 cm^{-1}. The ^1H NMR spectrum of **Y** is similar to that of **X**: multiplet δ 7.3 (5H), singlet δ 3.7 (2H). The ^1H NMR and IR spectra of **Z** are shown in Fig. 20.7.

20.39 Using reactions that we have studied in this chapter, propose a mechanism that accounts for the following reaction:

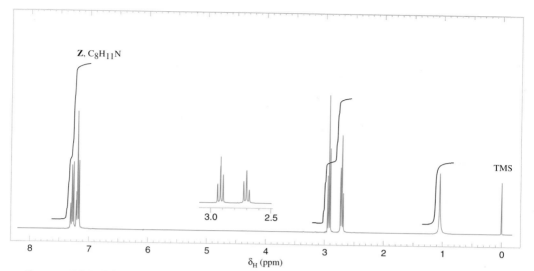

FIGURE 20.7 The 300 MHz ^{1}H NMR spectrum of compound **Z**, problem 20.38. Expansions of the signals are shown in the offset plots.

20.40   Give structures for compounds **R–W**:

$$N\text{-Methylpiperidine} + CH_3I \longrightarrow R\ (C_7H_{16}NI) \xrightarrow[\text{H}_2\text{O}]{\text{Ag}_2\text{O}}$$

$$S\ (C_7H_{17}NO) \xrightarrow[(-\text{H}_2\text{O})]{\text{heat}} T\ (C_7H_{15}N) \xrightarrow{CH_3I} U\ (C_8H_{18}NI) \xrightarrow[\text{H}_2\text{O}]{\text{Ag}_2\text{O}}$$

$$V\ (C_8H_{19}NO) \xrightarrow{\text{heat}} W\ (C_5H_8) + H_2O + (CH_3)_3N$$

20.41   Compound **A** ($C_{10}H_{15}N$) is soluble in dilute HCl. The IR absorption spectrum shows two bands in the 3300–3500-cm^{-1} region. The broad band proton-decoupled ^{13}C spectrum of **A** is given in Fig. 20.8. Propose a structure for **A**.

20.42   Compound **B**, an isomer of **A** (Problem 20.41), is also soluble in dilute HCl. The IR spectrum of **B** shows no bands in the 3300–3500-cm^{-1} region. The broad band proton-decoupled ^{13}C spectrum of **B** is given in Fig. 20.8. Propose a structure for **B**.

20.43   Compound **C** ($C_9H_{11}NO$) gives a positive Tollens' test and is soluble in dilute HCl. The IR spectrum of **C** shows a strong band near 1695 cm^{-1} but shows no bands in the 3300–3500-cm^{-1} region. The broad band proton-decoupled ^{13}C NMR spectrum of **C** is shown in Fig. 20.8. Propose a structure for **C**.

20.44   Outline a synthesis of acetylcholine iodide using as organic starting materials: dimethylamine, oxirane, and acetyl choloride.

$$H_3C-\overset{\overset{\displaystyle CH_3}{|}}{\underset{\underset{\displaystyle CH_3}{|}}{N^+}}-CH_2-CH_2-O-\overset{\overset{\displaystyle O}{||}}{C}-CH_3$$

$$I^-$$

Acetylcholine iodide

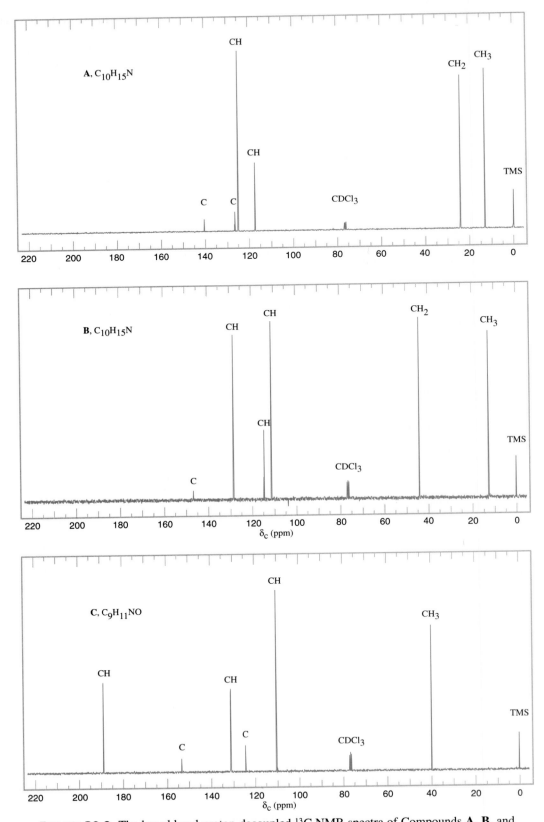

FIGURE 20.8 The broad band proton-decoupled ^{13}C NMR spectra of Compounds **A**, **B**, and **C**, Problems 20.41–20.43 Information from the DEPT ^{13}C NMR spectrum is given above each peak.

20.45   Ethanolamine, HOCH$_2$CH$_2$NH$_2$, and diethanolamine, (HOCH$_2$CH$_2$)$_2$NH, are used commercially to form emulsifying agents and to absorb acidic gases. Propose syntheses of these two compounds.

20.46   Diethylpropion (see following structure) is a compound used in the treatment of anorexia. Propose a synthesis of diethylpropion starting with benzene and using any other needed reagents.

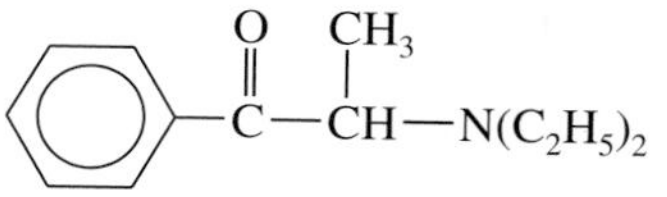

Diethylpropion

20.47   Suggest an experiment to test the proposition that the Hofmann reaction is an intramolecular rearrangement, i.e., one in which the migrating R group never fully separates from the amide molecule.

20.48   Using as starting materials propanoic acid, aniline, and 2-naphthol, propose a synthesis of naproanilide, a herbicide used in rice paddies in the Orient.

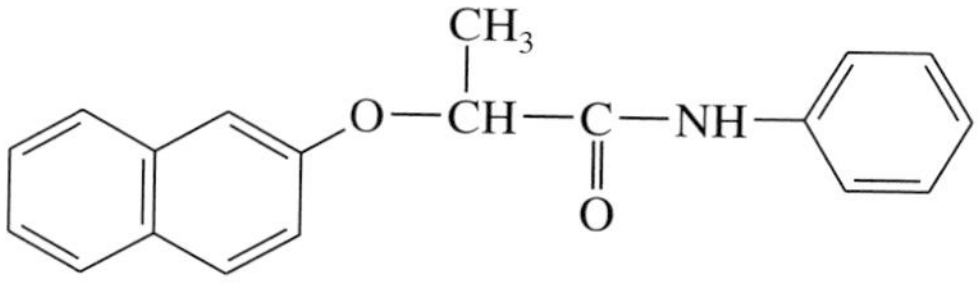

Naproanilide

Quinoline

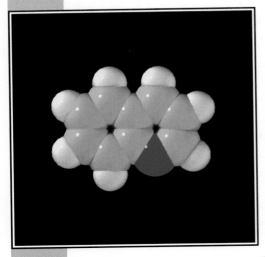

REACTIONS AND SYNTHESIS OF HETEROCYCLIC AMINES

Heterocyclic amines undergo many reactions that are similar to those of the amines that we have studied in earlier chapters.

J.1 Heterocyclic Amines as Bases

Nonaromatic heterocyclic amines have basicities that are approximately the same as those of acyclic amines.

Piperidine
$pK_b = 2.80$

Pyrrolidine
$pK_b = 2.89$

Diethylamine
$pK_b = 3.02$

In aqueous solution, aromatic heterocyclic amines such as pyridine, pyrimidine, and pyrrole are much weaker bases than nonaromatic amines or ammonia. (In the gas phase, however, pyridine and pyrrole are more basic than ammonia, indicating that solvation has a very important effect on their relative basicities, cf. Section 20.3.)

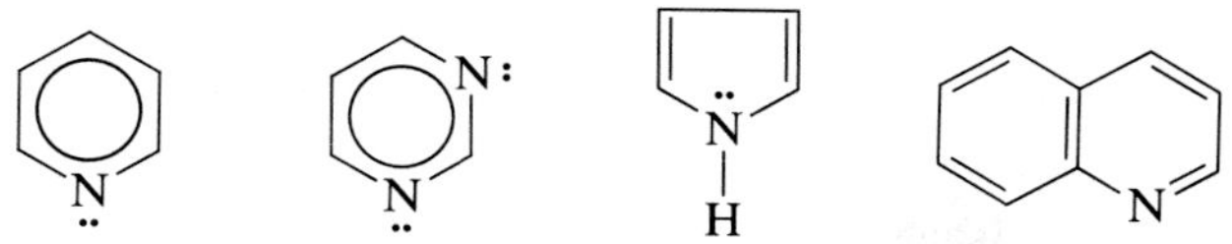

Pyridine
$pK_b = 8.77$

Pyrimidine
$pK_b = 11.30$

Pyrrole
$pK_b = 13.60$

Quinoline
$pK_b = 9.5$

J.2 Heterocyclic Amines as Nucleophiles in Alkylation and Acylation Reactions

Most heterocyclic amines undergo alkylation and acylation reactions in much the same way as acyclic amines.

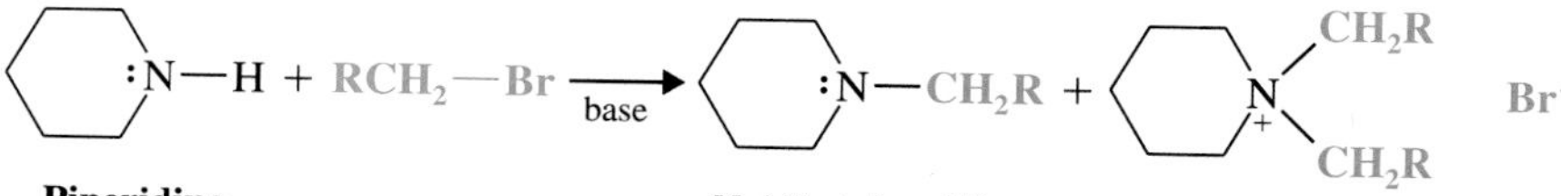

Piperidine

N-Alkylpiperidine

N,N-Dialkylpiper-
idinium bromide

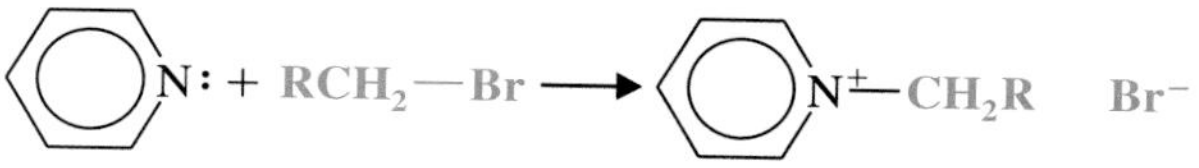

Pyridine

N-Alkylpyridinium
bromide

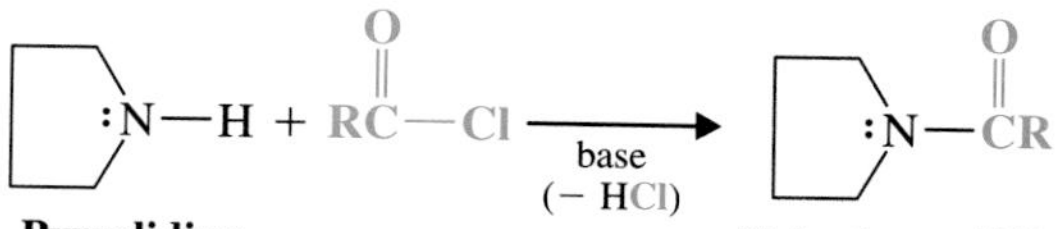

Pyrrolidine

N-Acylpyrrolidine
(an amide)

PROBLEM J.1

What products would you expect to obtain from the following reactions?

(a) Piperidine + acetic anhydride $\longrightarrow$

(b) Pyridine + methyl iodide $\longrightarrow$

(c) Pyrrolidine + phthalic anhydride $\longrightarrow$

(d) Pyrrolidine + (excess) methyl iodide $\xrightarrow{\text{(base)}}$

(e) Product of (d) + Ag_2O, H_2O, then heat $\longrightarrow$

J.3 Electrophilic Substitution Reactions of Aromatic Heterocyclic Amines

Pyrrole is highly reactive toward electrophilic substitution and it takes place primarily at position 2.

General Reaction

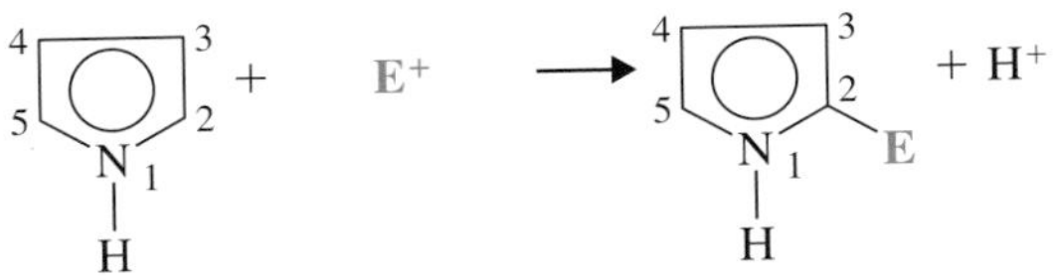
4
3
5
2
N
1
H
+
E+
4
3
5
2
N
1
E
H
+ H+

Pyrrole **Electrophile** **2-Substituted pyrrole**

Specific Example

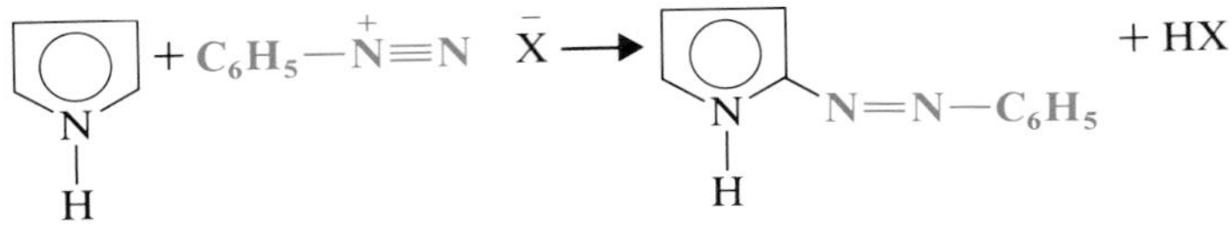
N
H
+ C6H5—N≡N X
N
N=N—C6H5
H
+ HX

We can understand why electrophilic substitution at the 2 position is preferred if we examine the resonance structures that follow:

Substitution at the 2 Position of Pyrrole

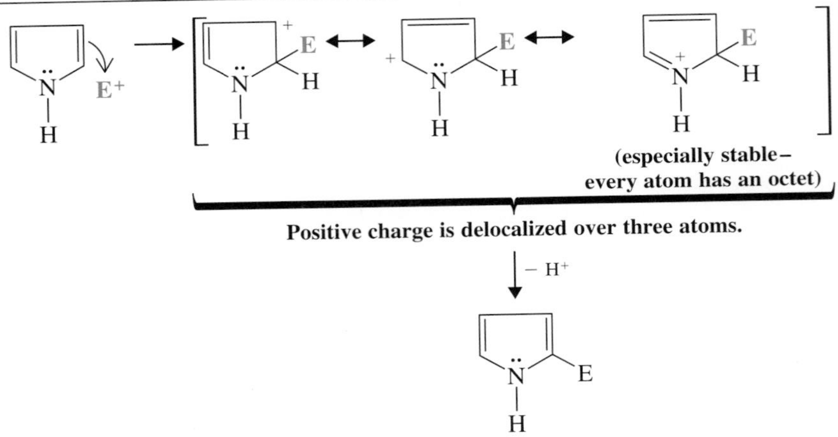
E+
E
H
E
H
E
H
(especially stable–
every atom has an octet)
Positive charge is delocalized over three atoms.
– H+
E

Substitution at the 3 Position of Pyrrole

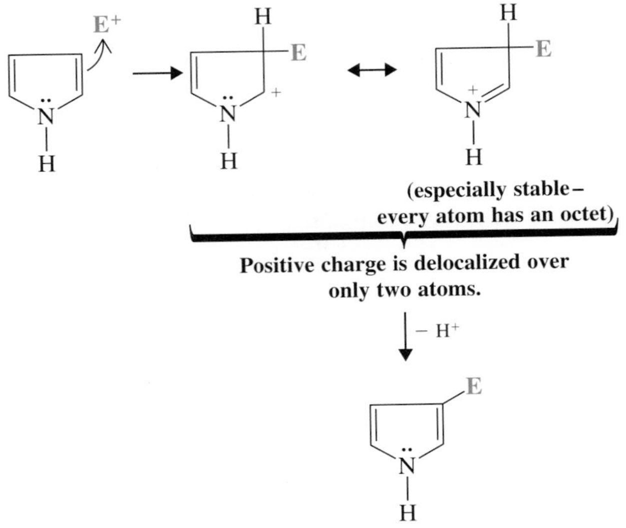
E+
H
E
H
E
(especially stable–
every atom has an octet)
Positive charge is delocalized over
only two atoms.
– H+
E

We see that whereas a relatively stable structure contributes to the hybrid for both intermediates, the intermediate arising from attack at the 2 position is stabilized by one additional resonance structure, and the positive charge is delocalized over three atoms rather than two. This means that this intermediate is more stable, and that attack at the 2 position has a lower free energy of activation.

Pyridine is much less reactive toward electrophilic substitution than benzene. Pyridine does not undergo Friedel–Crafts acylation or alkylation; it does not couple with diazonium compounds. Bromination of pyridine can be accomplished, but only in the vapor phase at 200°C, where a free radical mechanism may operate. Nitration and sulfonation also require forcing conditions. Electrophilic substitution, when it occurs, nearly always takes place at the 3 position.

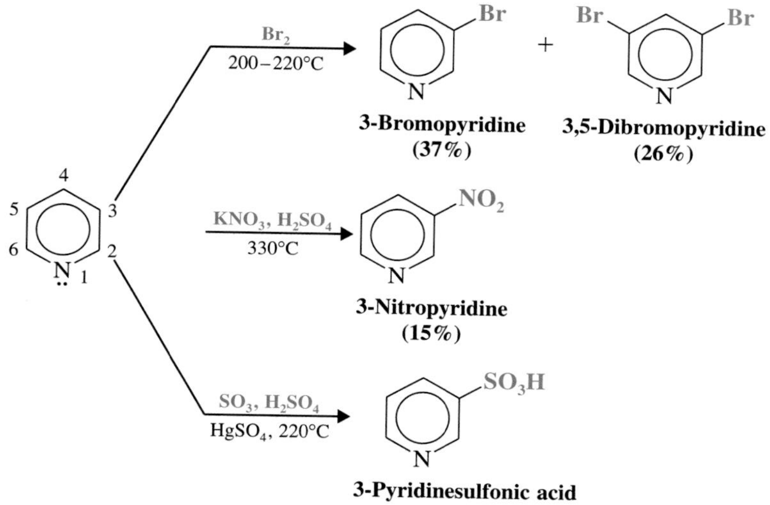

3-Bromopyridine
(37%)

3,5-Dibromopyridine
(26%)

3-Nitropyridine
(15%)

3-Pyridinesulfonic acid

We can, in part, attribute the lower reactivity of pyridine (when compared to benzene) to the greater electronegativity of nitrogen (when compared to carbon). Nitrogen, being more electronegative, is less able to accommodate the electron deficiency that characterizes the transition state leading to the positively charged ion (similar to an arenium ion) in electrophilic substitution.

Pyridine

Transition state is of higher energy because of greater electronegativity of nitrogen.

Similar to (but less stable than) an arenium ion.

Benzene

Transition state is of lower energy because of lower electronegativity of carbon.

Arenium ion

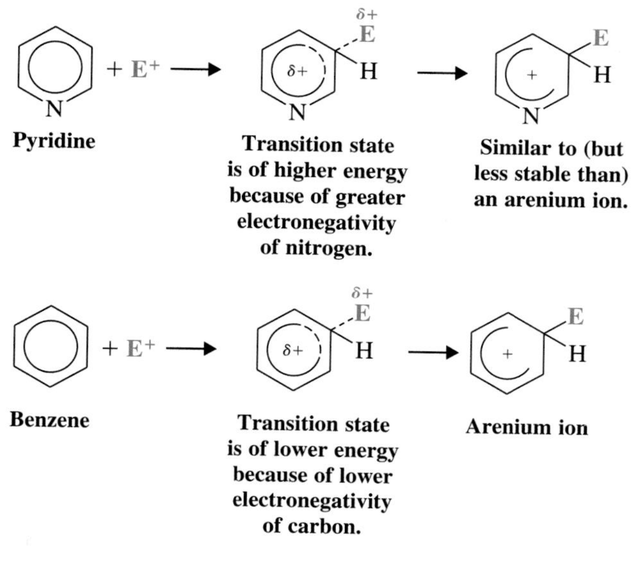

The low reactivity of pyridine toward electrophilic substitution may arise mainly from the fact that pyridine is converted initially to a pyridinium ion by a proton or other electrophile.

Pyridinium ion
(highly unreactive because
of positive charge)

Electrophilic attack at the 4 position (or the 2 position) is unfavorable because an especially unstable resonance structure contributes to the intermediate hybrid.

Especially unstable because
nitrogen has a sextet
and two positive charges

Similar resonance structures can be written for attack at the 2 position.

No especially unstable *or stable* structure contributes to the hybrid arising from attack at the 3 position; as a result, attack at the 3 position is preferred but occurs slowly.

No especially unstable or stable
structure contributes to the hybrid

Pyrimidine is even less reactive toward electrophilic substitution than pyridine. (Why?) When electrophilic substitution takes place, it occurs at the 5 position.

Pyrimidine

Imidazole is much more susceptible to electrophilic substitution than pyridine or pyrimidine but is less reactive than pyrrole. Imidazoles with 1 substituents undergo electrophilic substitution at the 4 position.

1-Methyl-4-nitroimidazole

Imidazole, itself, undergoes electrophilic substitution in a similar fashion. Tautomerism, however, makes the 4 and 5 positions equivalent.

4-(5)-Bromoimidazole

PROBLEM J.2

Both pyrrole and imidazole are weak acids; they react with strong bases to form anions:

Pyrrole anion   **Imidazole anion**

(a) These anions resemble a carbocyclic anion that we have studied before. What is it?   (b) Write resonance structures that account for the stabilities of pyrrole and imidazole anions.

J.4 Nucleophilic Substitutions of Pyridine

In its reactions, the pyridine ring resembles a benzene ring with a strong electron-withdrawing group; pyridine is relatively unreactive toward electrophilic substitution but appreciably reactive toward nucleophilic substitution.

In the previous section we compared the reactivity of pyridine and benzene toward electrophilic substitution and there we attributed pyridine's lower reactivity to the greater electronegativity of its ring nitrogen. Because nitrogen is more electronegative than carbon, it is less able to accommodate the electron deficiency in the transition state of the rate-determining step in electrophilic aromatic substitution. On the other hand, nitrogen's greater electronegativity makes it *more* able to accommodate the excess *negative* charge that an aromatic ring must accept in *nucleophilic substitution.*

Pyridine reacts with sodium amide, for example, to form 2-aminopyridine. In this remarkable reaction (called the Chichibabin reaction), amide ion (NH_2^-) displaces a hydride ion (H^-).

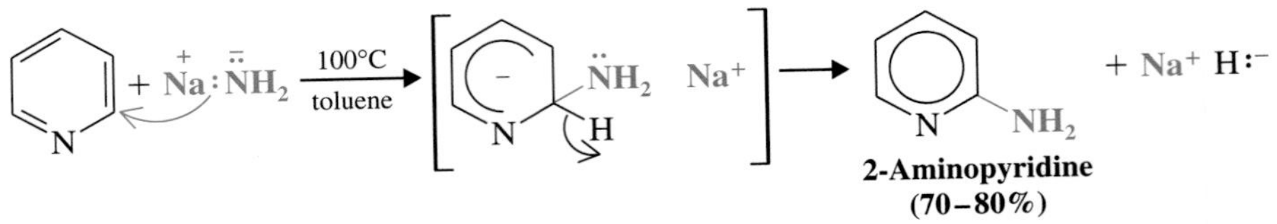

2-Aminopyridine
(70–80%)

If we examine the resonance structures that contribute to the intermediate in this reaction, we shall be able to see how the ring nitrogen atom accommodates the negative charge:

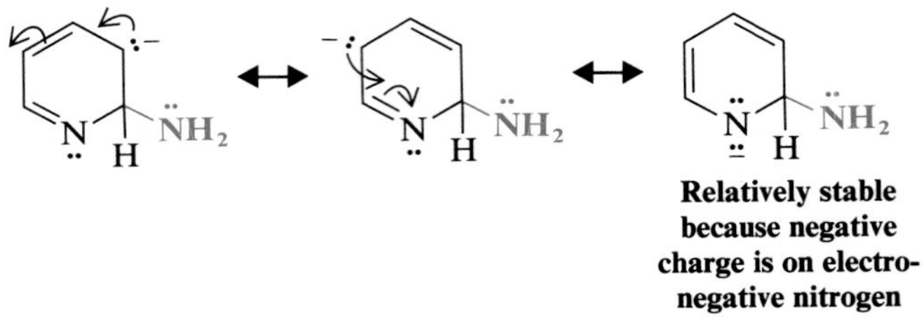

Relatively stable
because negative
charge is on electro-
negative nitrogen

In the next step the intermediate loses a hydride ion and becomes 2-amino-pyridine.*

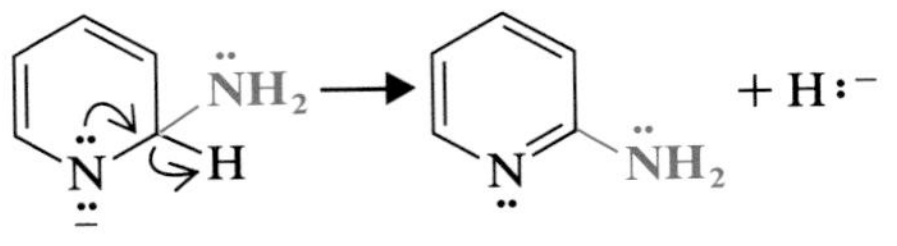

Pyridine undergoes similar nucleophilic substitution reactions with phenyllithium, butyllithium, and potassium hydroxide.

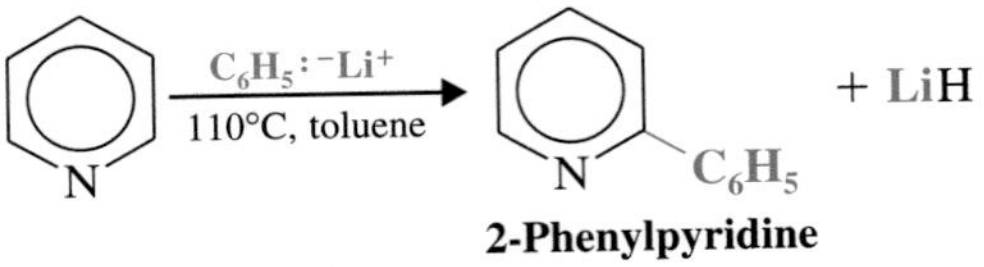

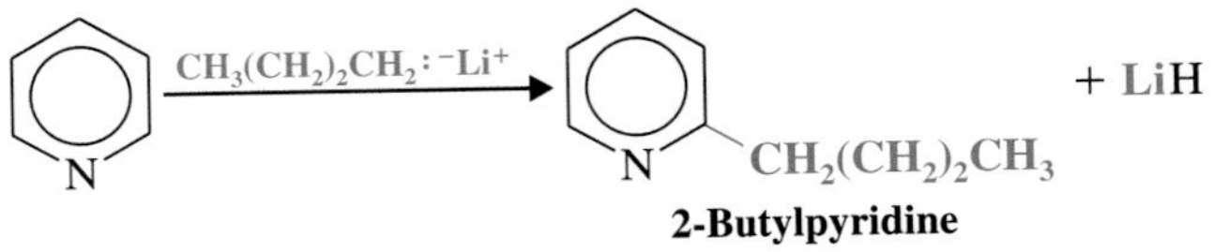

2-Phenylpyridine

2-Butylpyridine

---

* In practice, a subsequent reaction occurs; 2-aminopyridine reacts with the sodium hydride to produce a sodio derivative, and this helps shift the equilibrium to the right.

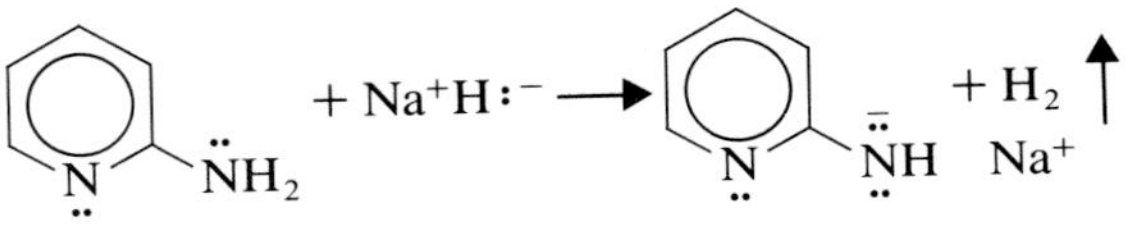

When the reaction is over, the addition of cold water to the reaction mixture converts the sodio derivative to 2-aminopyridine.

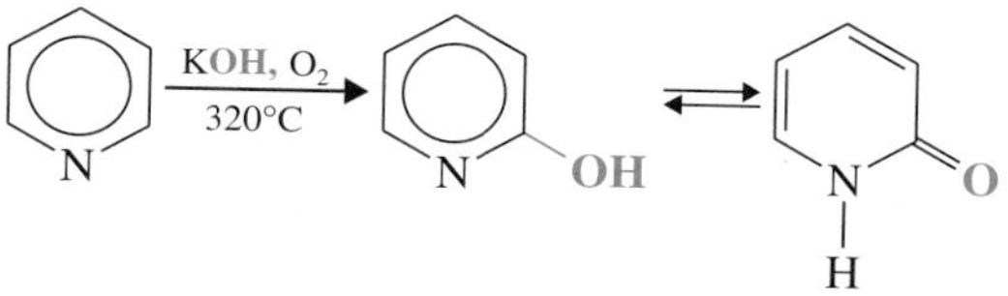

2-Pyridinol **2-Pyridone**
(50%)

2-Chloropyridine reacts with sodium methoxide to yield 2-methoxypyridine:

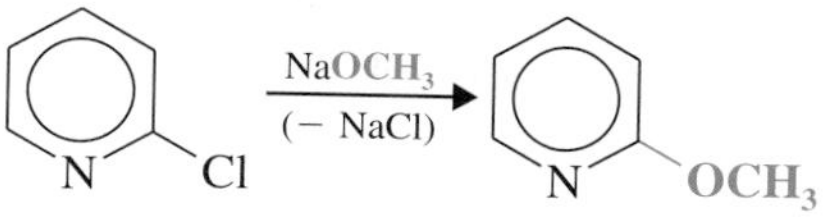

PROBLEM J.3

An alternative mechanism to the one given for the amination of pyridine in Section J.4, involves a "pyridyne" intermediate, that is,

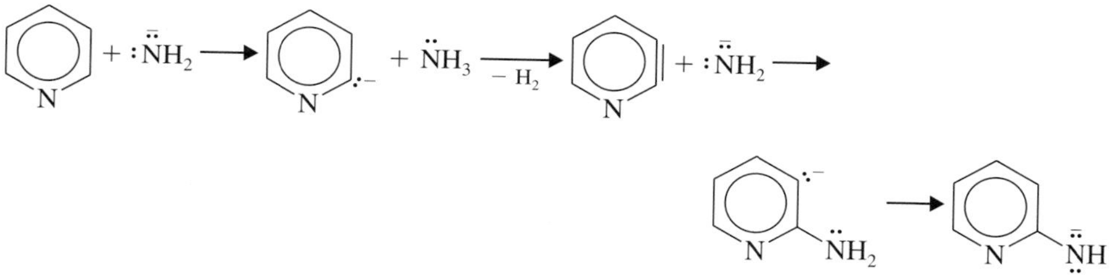

This mechanism was disallowed on the basis of an experiment in which 3-deuteriopyridine was allowed to react with sodium amide. Consider the fate of deuterium in both mechanisms and explain.

PROBLEM J.4

2-Halopyridines undergo nucleophilic substitution much more readily than pyridine itself. What factor accounts for this?

J.5 Nucleophilic Additions to Pyridinium Ions

Pyridinium ions are especially susceptible to nucleophilic attack at the 2 or 4 position because of the contributions of the resonance forms shown here.

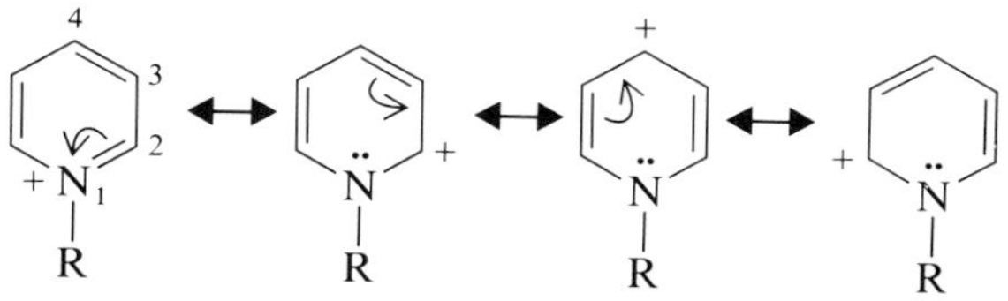

N-Alkylpyridinium halides, for example, react with hydroxide ions primarily at position 2; this causes the formation of an addition product called a *pseudo base*.

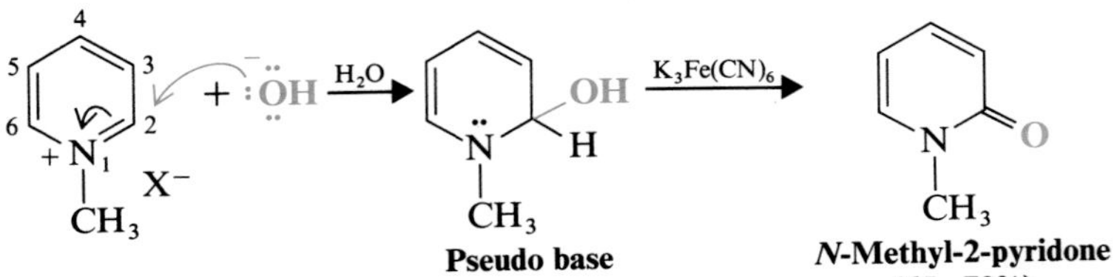

Pseudo base   **N-Methyl-2-pyridone**
(65–70%)

Oxidation of the pseudo base with potassium ferricyanide (see previous reaction) produces an *N*-alkylpyridone.

Nucleophilic additions to pyridinium ions, especially the addition of *hydride ions,* have been of considerable interest to chemists because these reactions resemble the biological reduction of the important coenzyme, nicotinamide adenine dinucleotide (NAD$^+$) (Section 14.10).

A number of model reactions have been carried out in connection with these studies. Treating an *N*-alkylpyridinium ion with sodium borohydride, for example, brings about hydride addition, but addition occurs at position 2 and is usually accompanied by over-reduction:

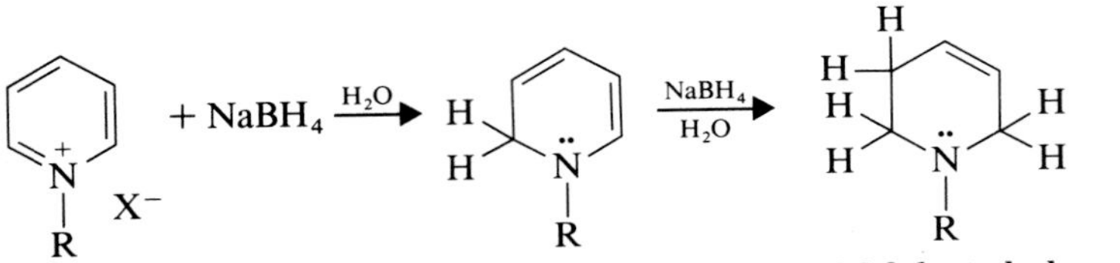

N-Alkyl-   A 1,2-dihydro-   A 1,2,3,6-tetrahydro-
pyridinium   pyridine   pyridine
halide

Treating a pyridinium ion with basic sodium dithionite ($Na_2S_2O_4$), however, brings about specific addition to position 4:

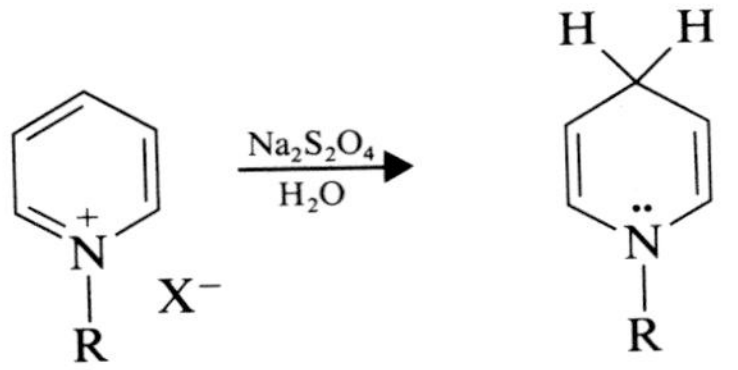

A 1,4-dihydropyridine

Sodium dithionite in aqueous base also reduces NAD$^+$ to NADH. The NADH formed by dithionite reduction has been shown to be biologically active and can be oxidized to NAD$^+$ with potassium ferricyanide.

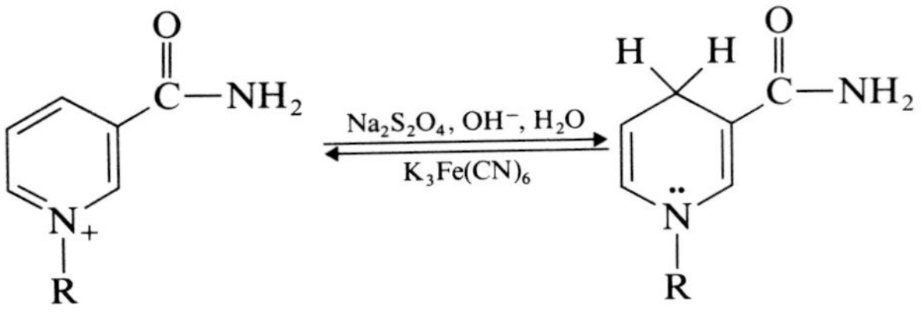

NAD$^+$ **NADH**
(see Section 13.12 for
the structure of R)

J.6 Synthesis of Heterocyclic Amines

The most general and widely used method for synthesizing pyrroles is to condense an α-amino ketone or α-amino-β-keto ester with a ketone or keto ester. This reaction, called the Knorr synthesis, is catalyzed by acids or bases. Two examples are shown here.

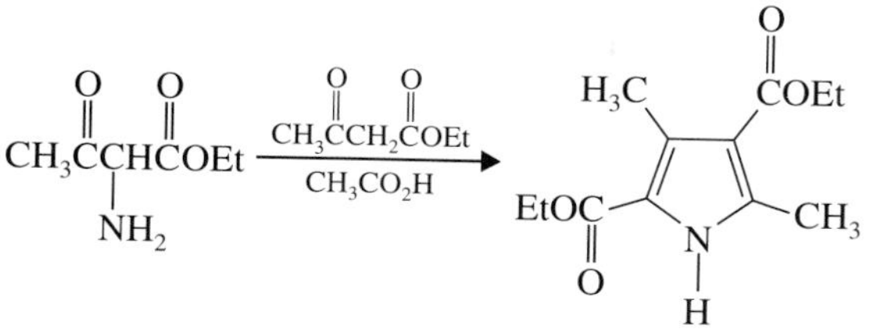

(57–64%)

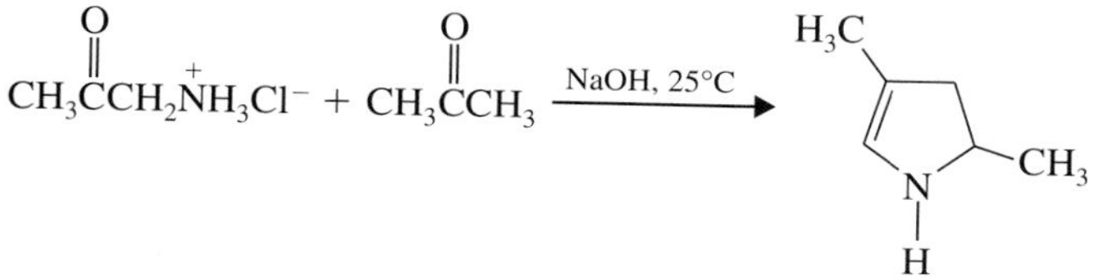

PROBLEM J.5

Propose reasonable mechanisms for the two syntheses of substituted pyrroles just given.

Pyridine and many of its derivatives can be isolated from coal tar. Many pyridine derivatives are prepared from these coal-tar derivatives through substitution reactions. The most general overall pyridine synthesis is one called the Hantzsch synthesis. In this method a β-keto ester, an aldehyde, and ammonia are allowed to condense to produce a dihydropyridine; oxidation of the dihydropyridine yields the substituted pyridine. An example is the following:

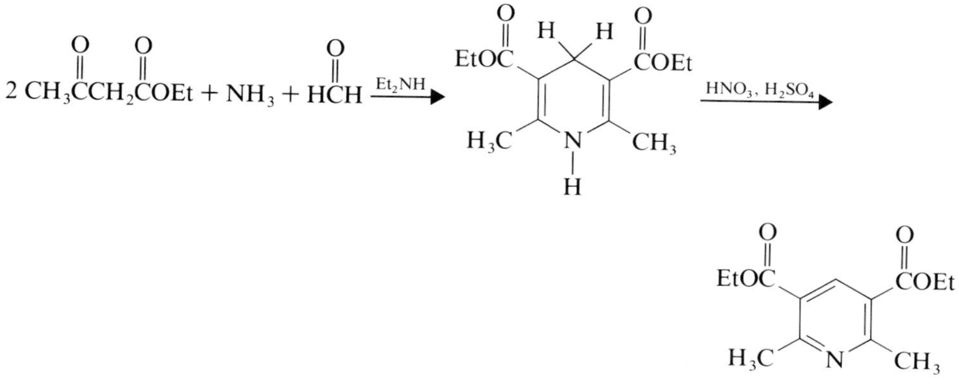

(58–65%)

The most general quinoline synthesis is the Skraup synthesis. In this method, aniline is heated with glycerol in the presence of sulfuric acid and an oxidizing agent. Various oxidizing agents have been used, including nitrobenzene and air.

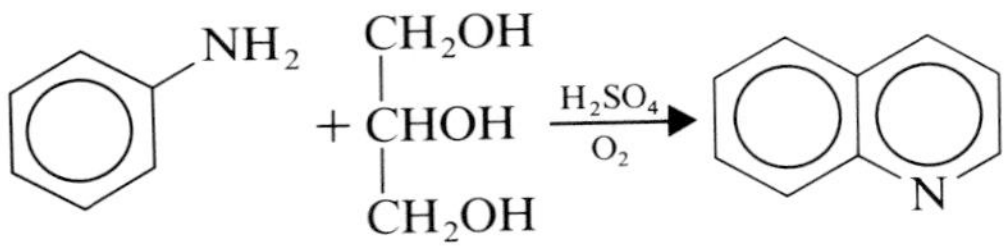

The mechanism for this reaction consists of the following steps:

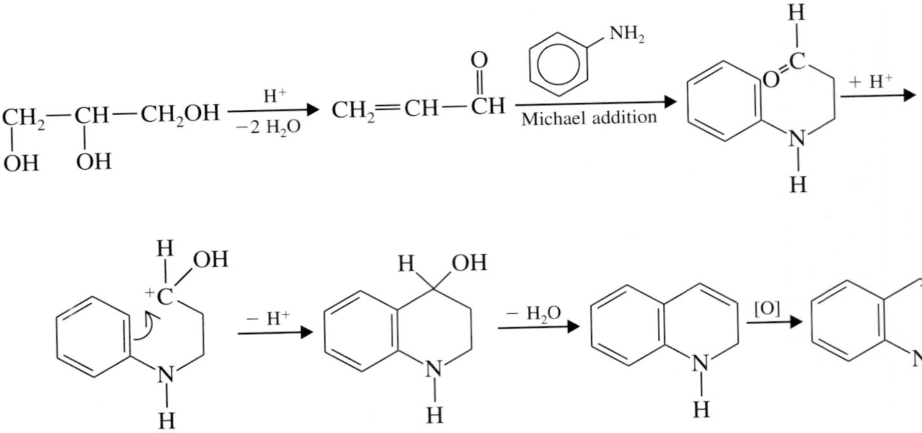

In the first step glycerol dehydrates in the presence of the acid to produce propenal (acrolein). Then a Michael addition of aniline to the propenal is followed by an acid-catalyzed cyclization to yield dihydroquinoline. Finally, oxidation of the dihydroquinoline produces quinoline.

PROBLEM J.6

Give structures for compounds **A–H**.

(a) 2,5-Hexanedione + $(NH_4)_2CO_3$ $\xrightarrow{100°C}$ **A** (C_6H_9N)
A pyrrole

(b) $CH_3\overset{O}{\overset{\|}{C}}CH_2NH_2$ + acetone $\xrightarrow{base}$ **B** (C_6H_9N)
An isomer of A

(c) CH_3NHNH_2 + $(CH_3O)_2CHCH_2CH(OCH_3)_2$ $\xrightarrow[H_2O]{H^+}$ **C** $(C_4H_6N_2)$
A pyrazole

(d) 2,5-Hexanedione + hydrazine $\xrightarrow{heat}$ **D** $(C_6H_{10}N_2)$ $\xrightarrow{O_2}$ **E** $(C_6H_8N_2)$
A dihydropyridazine   **A pyridazine**

(e) Aniline + $CH_2{=}CHCCH_3$ (with O) $\xrightarrow[FeCl_3]{ZnCl_2}$ **F** $(C_{10}H_9N)$
A quinoline

(f) 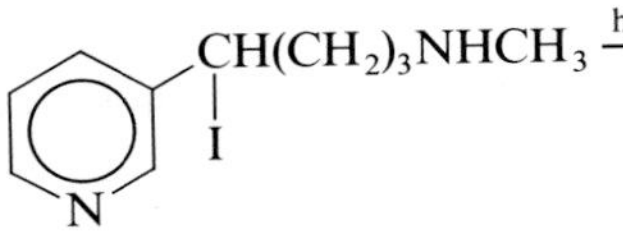 $\xrightarrow{heat}$ $\xrightarrow{OH^-}$ **G** $(C_{10}H_{14}N_2)$, $\xrightarrow[(2)\ H^+]{(1)\ KMnO_4,\ OH^-}$
Nicotine

H $(C_6H_5NO_2)$
Nicotinic acid

CHAPTER 21

Estradiol
(Section 21.2)

PHENOLS AND ARYL HALIDES: NUCLEOPHILIC AROMATIC SUBSTITUTION

21.1 STRUCTURE AND NOMENCLATURE OF PHENOLS

Compounds that have a hydroxyl group directly attached to a benzene ring are called **phenols.** Thus, **phenol** is the specific name for hydroxybenzene and it is the general name for the family of compounds derived from hydroxybenzene:

Phenol

4-Methylphenol (a phenol)

Compounds that have a hydroxyl group attached to a polycyclic benzenoid ring are chemically similar to phenols, but they are called **naphthols** and **phenanthrols,** for example.

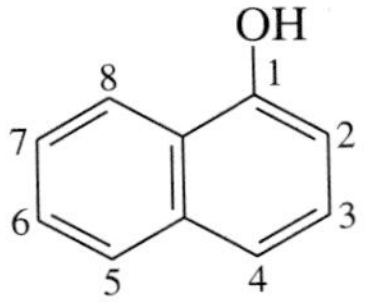

1-Naphthol
(α-naphthol)

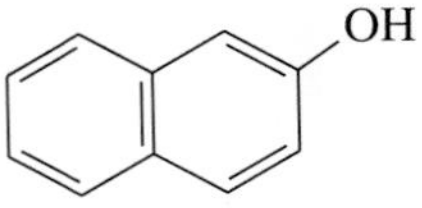

2-Naphthol
(β-naphthol)

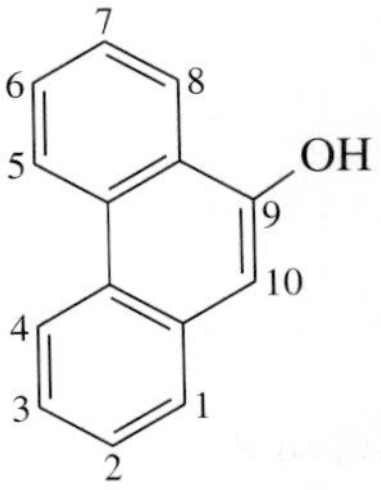

9-Phenanthrol

21.1A Nomenclature of Phenols

We studied the nomenclature of some of the phenols in Chapter 14. In many compounds *phenol* is the base name:

4-Chlorophenol
(*p*-chlorophenol)

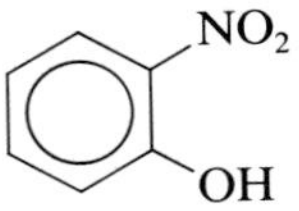

2-Nitrophenol
(*o*-nitrophenol)

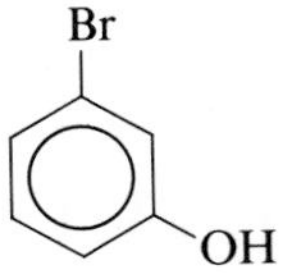

3-Bromophenol
(*m*-bromophenol)

The methylphenols are commonly called *cresols:*

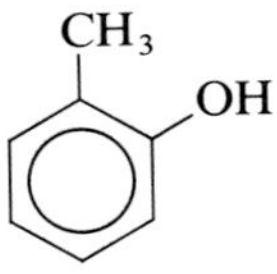

2-Methylphenol
(*o*-cresol)

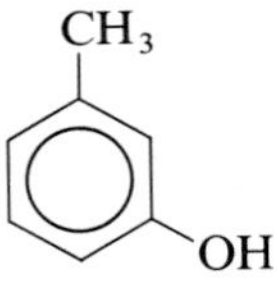

3-Methylphenol
(*m*-cresol)

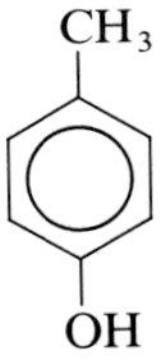

4-Methylphenol
(*p*-cresol)

The benzenediols also have common names:

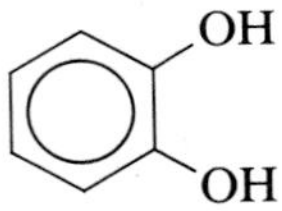

1,2-Benzenediol
(catechol)

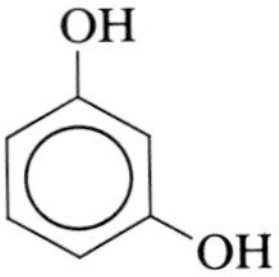

1,3-Benzenediol
(resorcinol)

1,4-Benzenediol
(hydroquinone)

21.2 NATURALLY OCCURRING PHENOLS

Phenols and related compounds occur widely in nature. Tyrosine is an amino acid that occurs in proteins. Methyl salicylate is found in oil of wintergreen, eugenol is found in oil of cloves, and thymol is found in thyme.

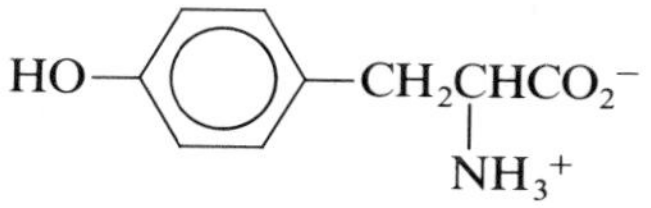

Tyrosine

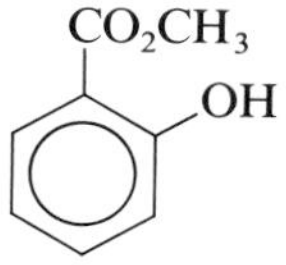

Methyl salicylate
(oil of wintergreen)

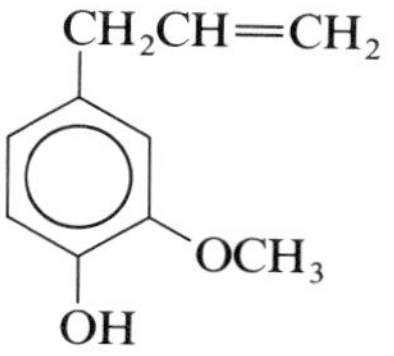

Eugenol
(oil of cloves)

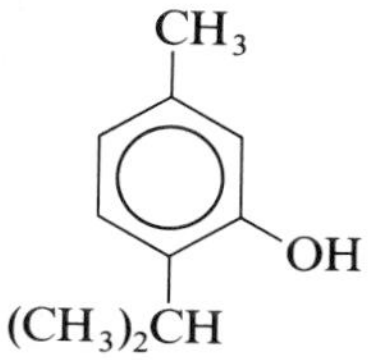

Thymol
(thyme)

The urushiols are blistering agents (vesicants) found in poison ivy.

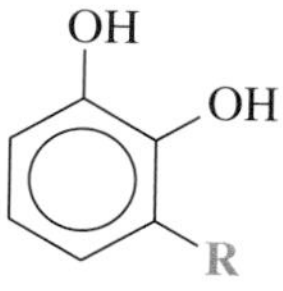

Urushiols

$R = -(CH_2)_{14}CH_3,$

$-(CH_2)_7CH=CH(CH_2)_5CH_3,$ or

$-(CH_2)_7CH=CHCH_2CH=CH(CH_2)_2CH_3$

   Estradiol is a female sex hormone and the tetracyclines are important anti-biotics.

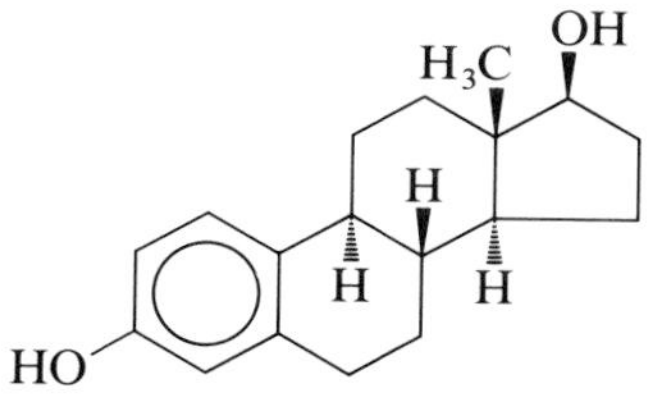

Estradiol

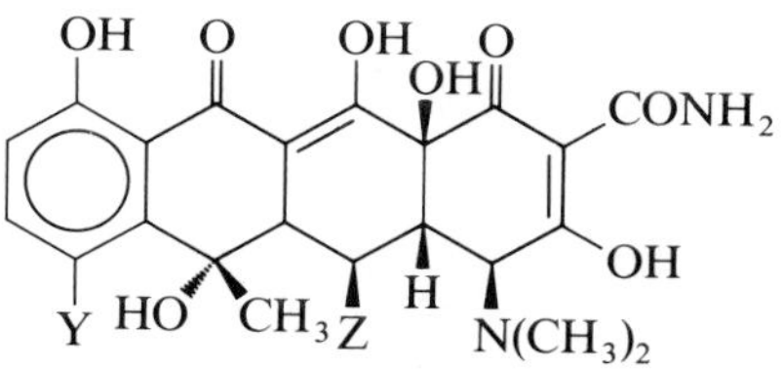

Tetracyclines
(Y = Cl, Z = H; Aureomycin)
(Y = H, Z = OH; Terramycin)

21.3 PHYSICAL PROPERTIES OF PHENOLS

The presence of hydroxyl groups in the molecules of phenols means that phenols are like alcohols (Section 10.2) in being able to form strong intermolecular hydrogen bonds. This hydrogen bonding causes phenols to be associated and, therefore, to have higher boiling points than hydrocarbons of the same molecular weight. For example,

TABLE 21.1   Physical properties of phenols

NAME	FORMULA	mp (°C)	bp (°C)	WATER SOLUBILITY (g 100 mL^{-1} of H$_2$O)
Phenol	C_6H_5OH	43	182	9.3
2-Methylphenol	$o\text{-}CH_3C_6H_4OH$	30	191	2.5
3-Methylphenol	$m\text{-}CH_3C_6H_4OH$	11	201	2.6
4-Methylphenol	$p\text{-}CH_3C_6H_4OH$	35.5	201	2.3
2-Chlorophenol	$o\text{-}ClC_6H_4OH$	8	176	2.8
3-Chlorophenol	$m\text{-}ClC_6H_4OH$	33	214	2.6
4-Chlorophenol	$p\text{-}ClC_6H_4OH$	43	220	2.7
2-Nitrophenol	$o\text{-}O_2NC_6H_4OH$	45	217	0.2
3-Nitrophenol	$m\text{-}O_2NC_6H_4OH$	96		1.4
4-Nitrophenol	$p\text{-}O_2NC_6H_4OH$	114		1.7
2,4-Dinitrophenol	(structure)	113		0.6
2,4,6-Trinitrophenol (picric acid)	(structure)	122		1.4

phenol (bp, 182°C) has a boiling point more than 70°C higher than toluene (bp, 110.6°C), even though the two compounds have almost the same molecular weight.

The ability to form strong hydrogen bonds to molecules of water confers on phenols a modest solubility in water. Table 21.1 lists the physical properties of a number of common phenols.

21.4 SYNTHESIS OF PHENOLS

21.4A Laboratory Synthesis

The most important laboratory synthesis of phenols is by hydrolysis of arenediazonium salts (Section 20.8E). This method is highly versatile, and the conditions required for the diazotization step and the hydrolysis step are mild. This means that other groups present on the ring are unlikely to be affected.

General Reaction

$$Ar-NH_2 \xrightarrow{HONO} Ar-\overset{+}{N_2} \xrightarrow[Cu^{2+},\ H_2O]{Cu_2O} Ar-OH$$

Specific Example

2-Bromo-4-methylphenol
(80-92%)

21.4B Industrial Syntheses

Phenol is a highly important industrial chemical; it serves as the raw material for a large number of commercial products ranging from aspirin to a variety of plastics. Worldwide production of phenol is more than 3 million tons per year! Several methods have been used to synthesize phenol commercially.

1. **Hydrolysis of Chlorobenzene (Dow process).** In this process chlorobenzene is heated at 350°C (under high pressure) with aqueous sodium hydroxide. The reaction produces sodium phenoxide which, on acidification, yields phenol. The mechanism for the reaction probably involves the formation of benzyne (Section 21.11B).

2. **Alkali Fusion of Sodium Benzenesulfonate.** This, the first commercial process for synthesizing phenol, was developed in Germany in 1890. Sodium benzenesulfonate is melted (fused) with sodium hydroxide (at 350°C) to produce sodium phenoxide. Acidification then yields phenol.

This procedure can also be used in the laboratory and works quite well for the preparation of *p*-cresol, as the following example shows. However, the conditions required to bring about the reaction are so vigorous that this method cannot be used for the preparation of many phenols.

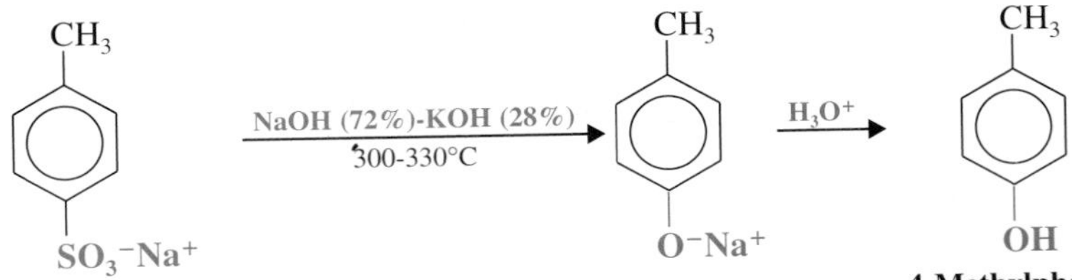

Sodium
p-toluenesulfonate → 4-Methylphenol
(63–70% overall)

3. **From Cumene Hydroperoxide.** This process illustrates industrial chemistry at its best. Overall, it is a method for converting two relatively inexpensive organic compounds—benzene and propene—into two more valuable ones—phenol and acetone. The only other substance consumed in the process is oxygen from air. Most of the worldwide production of phenol is now based on this method. The synthesis begins with the Friedel–Crafts alkylation of benzene with propene to produce cumene (isopropylbenzene).

Reaction 1

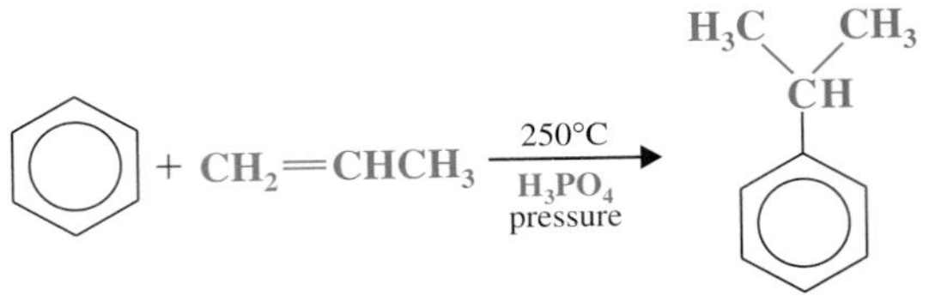

Cumene

Then cumene is oxidized to cumene hydroperoxide:

Reaction 2

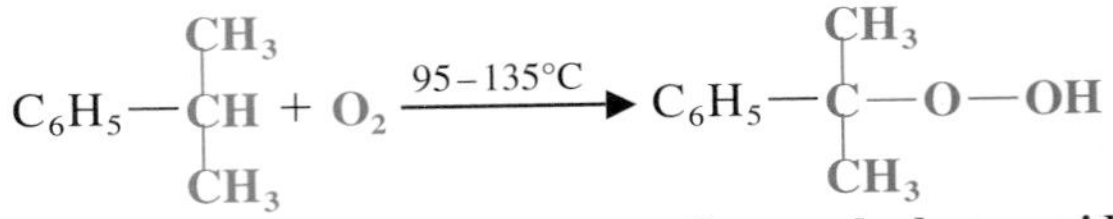

Cumene hydroperoxide

Finally, when treated with 10% sulfuric acid, cumene hydroperoxide undergoes a hydrolytic rearrangement that yields phenol and acetone:

Reaction 3

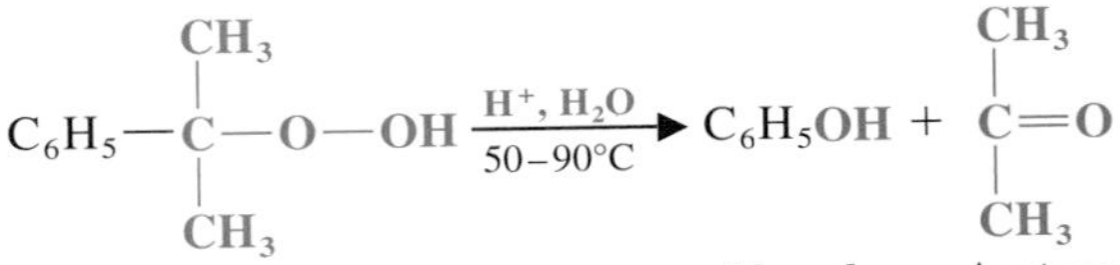

Phenol   Acetone

The mechanism of each of these reactions requires some comment. The first is a familiar one. The isopropyl cation generated by the reaction of propene with the acid (H_3PO_4) alkylates benzene in a typical electrophilic aromatic substitution:

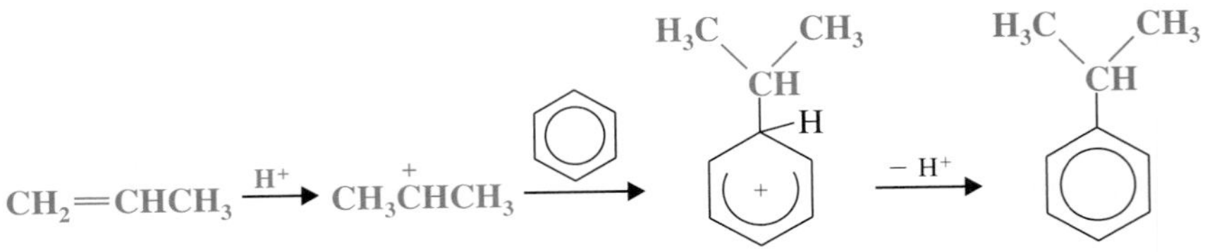

The second reaction is a radical chain reaction. A radical initiator abstracts the benzylic hydrogen atom of cumene producing a 3° benzylic radical. Then a chain reaction with oxygen produces cumene hydroperoxide:

Chain Initiation

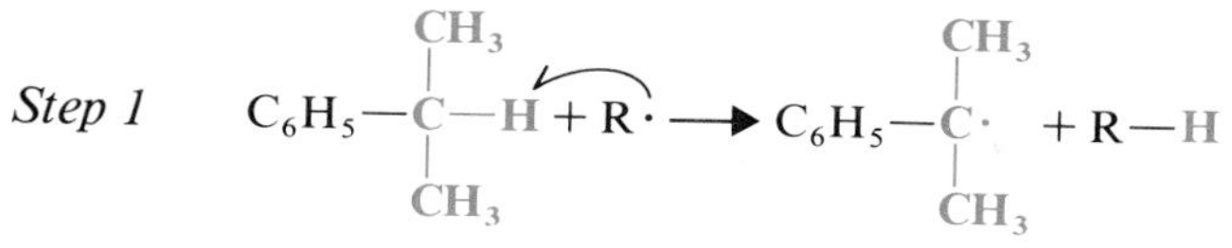

Step 1

$$C_6H_5-\underset{\underset{CH_3}{|}}{\overset{\overset{CH_3}{|}}{C}}-H + R\cdot \longrightarrow C_6H_5-\underset{\underset{CH_3}{|}}{\overset{\overset{CH_3}{|}}{C}}\cdot + R-H$$

Chain Propagation

Step 2

$$C_6H_5-\underset{\underset{CH_3}{|}}{\overset{\overset{CH_3}{|}}{C}}\cdot + O_2 \longrightarrow C_6H_5-\underset{\underset{CH_3}{|}}{\overset{\overset{CH_3}{|}}{C}}-O-O\cdot$$

Step 3

$$C_6H_5-\underset{\underset{CH_3}{|}}{\overset{\overset{CH_3}{|}}{C}}-O-O\cdot + H-\underset{\underset{CH_3}{|}}{\overset{\overset{CH_3}{|}}{C}}-C_6H_5 \longrightarrow$$

$$C_6H_5-\underset{\underset{CH_3}{|}}{\overset{\overset{CH_3}{|}}{C}}-O-O-H + C_6H_5-\underset{\underset{CH_3}{|}}{\overset{\overset{CH_3}{|}}{C}}\cdot$$

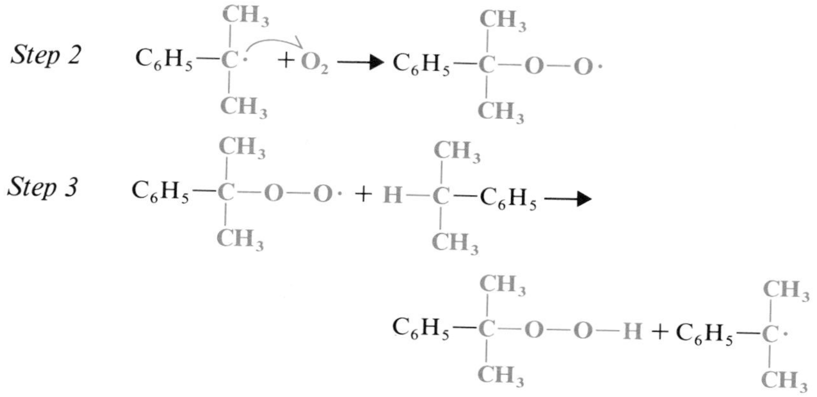

Then, step 2, step 3, step 2, step 3, etc.

The third reaction—the hydrolytic rearrangement—resembles the carbocation rearrangements that we have studied before. In this instance, however, the rearrangement involves the migration of a phenyl group to *a cationic oxygen atom.* Phenyl groups have a much greater tendency to migrate to a cationic center than do methyl groups (see Section 16.12A). The following equations show all the steps of the mechanism.

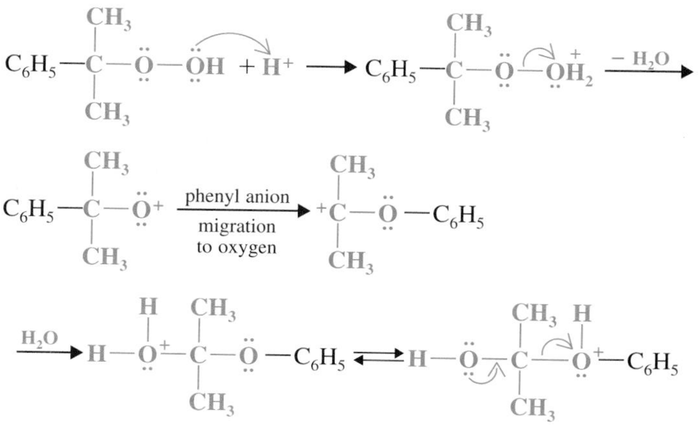

$$C_6H_5-\underset{\underset{CH_3}{|}}{\overset{\overset{CH_3}{|}}{C}}-\ddot{\underset{..}{O}}-\ddot{\underset{..}{O}}H + H^+ \longrightarrow C_6H_5-\underset{\underset{CH_3}{|}}{\overset{\overset{CH_3}{|}}{C}}-\ddot{\underset{..}{O}}-\overset{+}{O}H_2 \xrightarrow{-H_2O}$$

$$C_6H_5-\underset{\underset{CH_3}{|}}{\overset{\overset{CH_3}{|}}{C}}-\overset{+}{\underset{..}{O}} \xrightarrow[\substack{\text{migration}\\\text{to oxygen}}]{\text{phenyl anion}} \overset{+}{C}-\ddot{\underset{..}{O}}-C_6H_5$$

$$\xrightarrow{H_2O} H-\overset{+}{\underset{..}{O}}-\underset{\underset{CH_3}{|}}{\overset{\overset{CH_3}{|}}{C}}-\ddot{\underset{..}{O}}-C_6H_5 \rightleftharpoons H-\ddot{\underset{..}{O}}-\underset{\underset{CH_3}{|}}{\overset{\overset{CH_3\ H}{|}}{C}}-\overset{+}{O}-C_6H_5$$

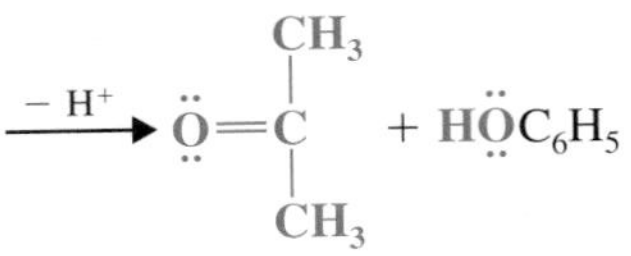

$$\xrightarrow{-H^+} \ddot{O}{=}\underset{\underset{CH_3}{|}}{\overset{\overset{CH_3}{|}}{C}} + H\ddot{O}C_6H_5$$

Acetone   **Phenol**

The second and third steps of the mechanism may actually take place at the same time; that is, the loss of H_2O and the migration of $C_6H_5{-}$ may be concerted.

21.5 REACTIONS OF PHENOLS AS ACIDS

21.5A Strength of Phenols as Acids

Although phenols are structurally similar to alcohols, they are much stronger acids. The pK_a values of most alcohols are of the order of 18. However, as we see in Table 21.2, the pK_a values of phenols are smaller than 11.

Let us compare two *superficially* similar compounds, cyclohexanol and phenol.

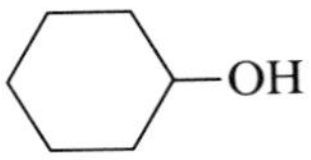

        —OH   —OH

Cyclohexanol   **Phenol**
$pK_a = 18$ $pK_a = 9.89$

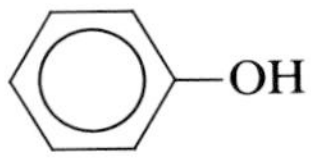

TABLE 21.2 The acidity constants of phenols

NAME	pK_a (in H_2O at 25°C)
Phenol	9.89
2-Methylphenol	10.20
3-Methylphenol	10.01
4-Methylphenol	10.17
2-Chlorophenol	8.11
3-Chlorophenol	8.80
4-Chlorophenol	9.20
2-Nitrophenol	7.17
3-Nitrophenol	8.28
4-Nitrophenol	7.15
2,4-Dinitrophenol	3.96
2,4,6-Trinitrophenol (picric acid)	0.38
1-Naphthol	9.31
2-Naphthol	9.55

Although phenol is a weak acid when compared with a carboxylic acid such as acetic acid ($pK_a = 4.76$), phenol is a much stronger acid than cyclohexanol (by a factor of 8 pK_a units).

Recent experimental and theoretical results have shown that the greater acidity of phenol owes itself primarily to an electrical charge distribution in phenol that causes the —OH oxygen to be more positive; therefore, the proton is held less strongly. In effect, the benzene ring of phenol acts as if it were an electron-withdrawing group when compared with the cyclohexane ring of cyclohexanol.

We can understand this effect by noting that the carbon atom that bears the hydroxyl group in phenol is sp^2 hybridized, whereas, in cyclohexane, it is sp^3 hybridized. Because of their greater *s* character, sp^2-hybridized carbon atoms are more electronegative than sp^3-hybridized carbon atoms (Section 6.18).

Another factor influencing the electron distribution may be the contributions to the overall resonance hybrid of phenol made by structures **2–4**. Notice that the effect of these structures is to withdraw electrons from the hydroxyl group and to make the oxygen positive.*

Resonance structures for phenol

1a   **1b**   **2**   **3**   **4**

PROBLEM 21.1

If we examine Table 21.2 we find that the methylphenols (cresols) are less acidic than phenol itself. For example,

Phenol
$pK_a = 9.89$

4-Methylphenol
$pK_a = 10.17$

This behavior is characteristic of phenols bearing electron-releasing groups. Provide an explanation.

---

* An alternative explanation for the greater acidity of phenol relative to cyclohexanol can be based on similar resonance structures for the phenoxide ion. Unlike the structures for phenol, **2–4**, resonance structures for the phenoxide ion do not involve charge separation. According to resonance theory, such structures should stabilize the phenoxide ion more than stuctures **2–4** stabilize phenol. (No resonance structures can be written for cyclohexanol or its anion, of course.) Greater stabilization of the phenoxide ion (the conjugate base) than of phenol (the acid) has an acid strengthening effect. Those who may be interested in pursuing this subject further should see the following articles: M. R. F. Siggel and T. D. Thomas, *J. Am. Chem. Soc.* **1986**, *108*, 4360–4362, and M. R. F. Siggel, A. R. Streitwieser, and T. D. Thomas, *J. Am. Chem. Soc.* **1988**, *110*, 8022–8028.

PROBLEM 21.2

If we examine Table 21.2 we see that phenols having electron-withdrawing groups (Cl— or O_2N—) attached to the benzene ring are more acidic than phenol itself. Account for this trend on the basis of resonance and inductive effects. Your answer should also explain the large acid strengthening effect of nitro groups, an effect that makes 2,4,6-trinitrophenol (also called *picric acid*) so exceptionally acidic ($pK_a = 0.38$) that it is more acidic than acetic acid ($pK_a = 4.76$).

21.5B Distinguishing and Separating Phenols from Alcohols and Carboxylic Acids

Because phenols are more acidic than water, the following reaction goes essentially to completion and produces water-soluble sodium phenoxide.

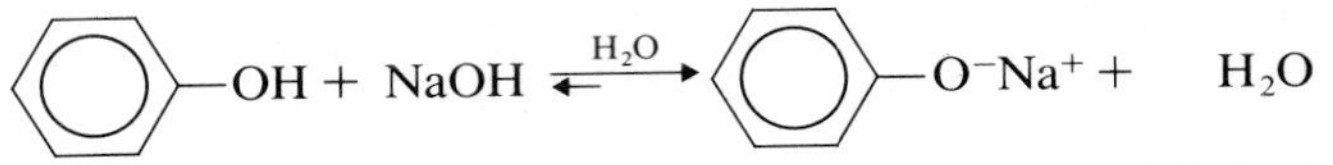

$$\text{OH} + \text{NaOH} \underset{H_2O}{\rightleftharpoons} \text{O}^-\text{Na}^+ + H_2O$$

| Stronger acid $pK_a \approx 10$ (slightly soluble) | Stronger base | | Weaker base (soluble) | Weaker acid $pK_a \approx 16$ |

The corresponding reaction of 1-hexanol with aqueous sodium hydroxide does not occur to a significant extent because 1-hexanol is a weaker acid than water.

$$CH_3(CH_2)_4CH_2OH + NaOH \underset{H_2O}{\rightleftharpoons} CH_3(CH_2)_4CH_2O^-Na^+ + H_2O$$

| Weaker acid $pK_a \approx 18$ (very slightly soluble) | Weaker base | | Stronger base (soluble) | Stronger acid $pK_a \approx 16$ |

The fact that phenols dissolve in aqueous sodium hydroxide, whereas most alcohols with six carbon atoms or more do not, gives us a convenient means for distinguishing and separating phenols from most alcohols. (Alcohols with five carbon atoms or fewer are quite soluble in water—some are infinitely so—and so they dissolve in aqueous sodium hydroxide even though they are not converted to sodium alkoxides in appreciable amounts.)

Most phenols, however, are not soluble in aqueous sodium bicarbonate ($NaHCO_3$), but carboxylic acids are soluble. Thus, aqueous $NaHCO_3$ provides a method for distinguishing and separating most phenols from carboxylic acids.

PROBLEM 21.3

Your laboratory instructor gives you a mixture of 4-methylphenol, benzoic acid, and toluene. Assume that you have available common laboratory acids, bases, and solvents and explain how you would proceed to separate this mixture by making use of solubility differences of its components.

21.6 OTHER REACTIONS OF THE O—H GROUP OF PHENOLS

Phenols react with carboxylic acid anhydrides and acid chlorides to form esters. These reactions are quite similar to those of alcohols (Section 18.7).

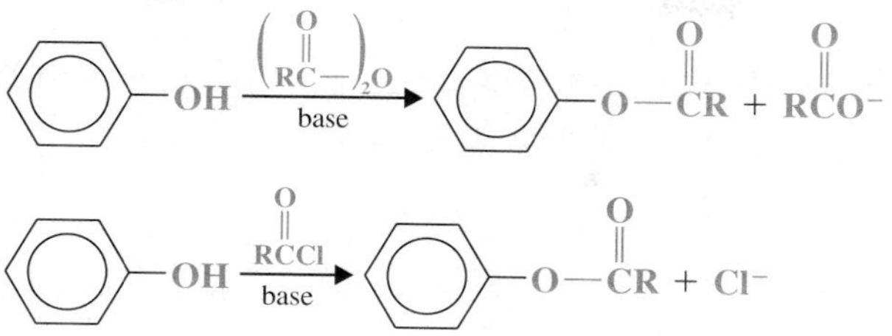

21.6A Phenols in the Williamson Synthesis

Phenols can be converted to ethers through the Williamson synthesis (Section 10.16B). Because phenols are more acidic than alcohols, they can be converted to sodium phenoxides through the use of sodium hydroxide (rather than metallic sodium, the reagent used to convert alcohols to alkoxide ions).

General Reaction

$$ArOH \xrightarrow{\text{NaOH}} ArO^-Na^+ \xrightarrow[\substack{(X = Cl, Br, I, \\ OSO_2OR \text{ or} \\ OSO_2R')}]{R-X} ArOR + NaX$$

Specific Examples

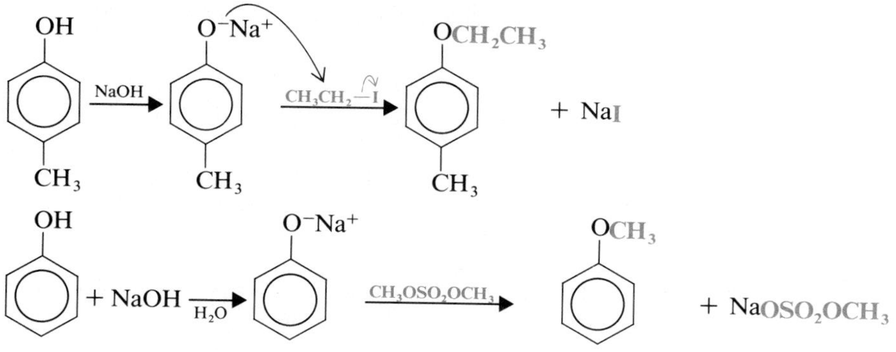

Anisole
(methoxybenzene)

21.7 CLEAVAGE OF ALKYL ARYL ETHERS

We learned in Section 10.17 that when dialkyl ethers are heated with excess concentrated HBr or HI, the ethers are cleaved and alkyl halides are produced from both alkyl groups.

$$R-O-R' \xrightarrow[\text{heat}]{\text{concd HX}} R-X + R'-X + H_2O$$

When alkyl aryl ethers react with strong acids such as HI and HBr, the reaction produces an alkyl halide and a phenol. The phenol does not react further to produce an aryl halide because the carbon–oxygen bond is very strong (cf. Problem 21.1) and because phenyl cations do not form readily.

General Reaction

$$Ar-O-R \xrightarrow[\text{heat}]{\text{concd HX}} Ar-OH + R-X$$

Specific Example

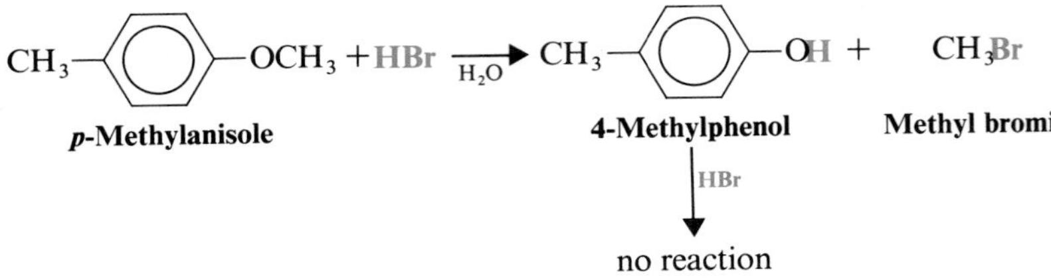

$$CH_3-\text{⟨⟩}-OCH_3 + HBr \xrightarrow{H_2O} CH_3-\text{⟨⟩}-OH + CH_3Br$$

p-Methylanisole   4-Methylphenol   Methyl bromide

$$\downarrow \text{HBr}$$

no reaction

21.8 REACTIONS OF THE BENZENE RING OF PHENOLS

Bromination   The hydroxyl group is a powerful activating group—and an ortho–para director—in electrophilic substitutions. Phenol itself reacts with bromine in aqueous solution to yield 2,4,6-tribromophenol in nearly quantitative yield. Note that a Lewis acid is not required for the bromination of this highly activated ring.

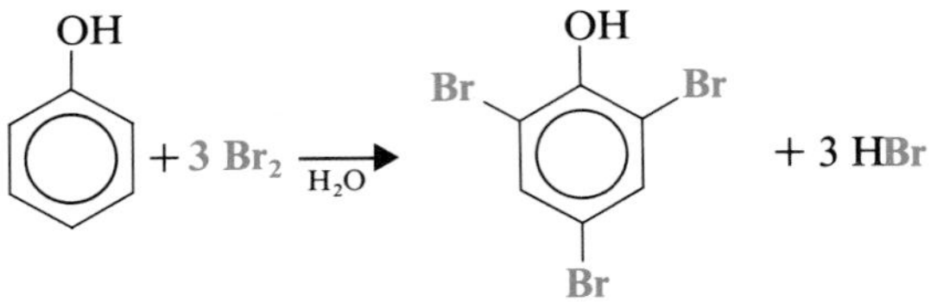

$$\text{⟨⟩}-OH + 3 Br_2 \xrightarrow{H_2O} \text{⟨⟩} + 3 HBr$$

2,4,6-Tribromophenol
(~100%)

Monobromination of phenol can be achieved by carrying out the reaction in carbon disulfide at a low temperature, conditions that reduce the electrophilic reactivity of bromine. The major product is the para isomer.

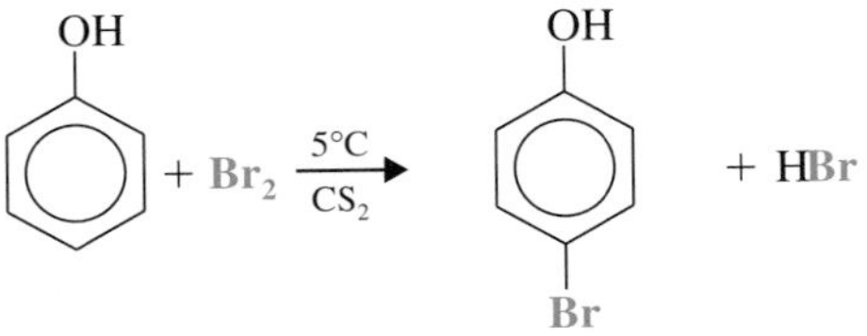

$$\text{⟨⟩}-OH + Br_2 \xrightarrow[\text{CS}_2]{5°C} \text{⟨⟩} + HBr$$

p-Bromophenol
(80–84%)

Nitration   Phenol reacts with dilute nitric acid to yield a mixture of *o*- and *p*-nitro-phenol.

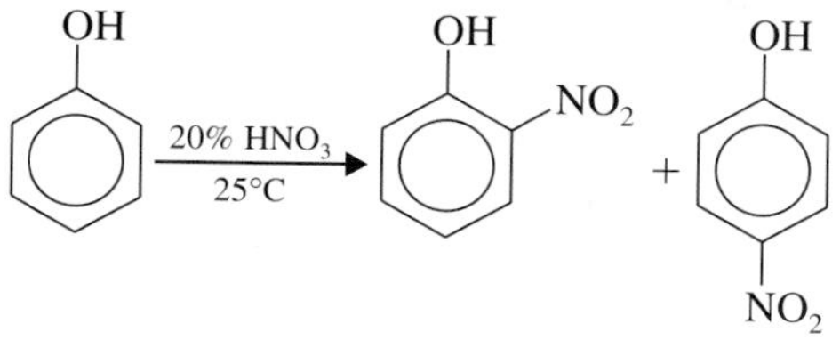

(30–40%)   (15%)

Although the yield is relatively low (because of oxidation of the ring), the ortho and para isomers can be separated by steam distillation. *o*-Nitrophenol is the more volatile isomer because its hydrogen bonding (see following structures) is *intramolecular*. *p*-Nitrophenol is less volatile because intermolecular hydrogen bonding causes association among its molecules. Thus, *o*-nitrophenol passes over with the steam, and *p*-nitrophenol remains in the distillation flask.

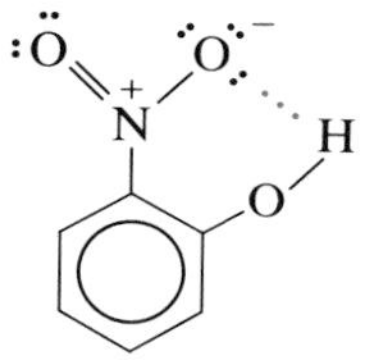

o-Nitrophenol
(more volatile because of
intramolecular hydrogen bonding)

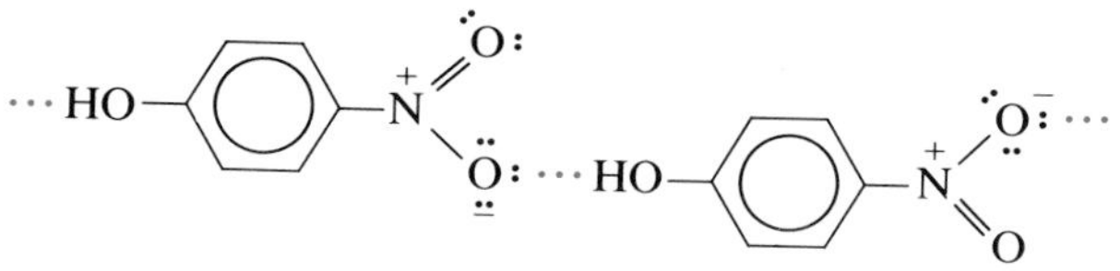

p-Nitrophenol
(less volatile because of

Sulfonation   Phenol reacts with concentrated sulfuric acid to yield mainly the ortho-sulfonated product if the reaction is carried out at 25°C and mainly the para-sulfonated product at 100°C. This is another example of equilibrium versus rate control of a reaction.

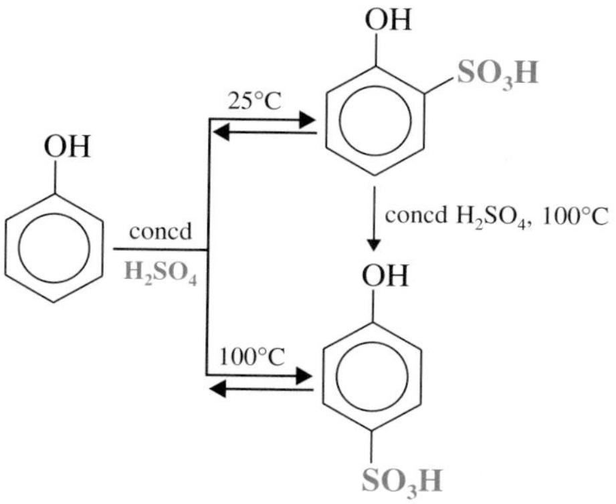

Major product, rate control

Major product, equilibrium control

PROBLEM 21.4

(a) Which sulfonic acid (see previous reactions) is more stable?   (b) For which sulfonation (ortho or para) is the free energy of activation lower?

Kolbe Reaction   The phenoxide ion is even more susceptible to electrophilic aromatic substitution than phenol itself. (Why?) Use is made of the high reactivity of the phenoxide ring in a reaction called the *Kolbe reaction.* In the Kolbe reaction carbon dioxide acts as the electrophile.

A   MECHANISM   FOR   THE   KOLBE   REACTION

Sodium salicylate   **Salicylic acid**

The reaction is usually carried out by allowing sodium phenoxide to absorb carbon dioxide and then heating the product to 125°C under a pressure of several atmospheres of carbon dioxide. The unstable intermediate undergoes a proton shift (a keto–enol tautomerization; see Section 17.2) that leads to sodium salicylate. Subsequent acidification of the mixture produces *salicylic acid.*

Reaction of salicylic acid with acetic anhydride yields the widely used pain reliever—*aspirin.*

| Salicylic
acid | Acetic
anhydride | Acetylsalicylic acid
(aspirin) | |

21.9 THE CLAISEN REARRANGEMENT

Heating allyl phenyl ether to 200°C effects an intramolecular reaction called a **Claisen rearrangement.** The product of the rearrangement is *o*-allylphenol:

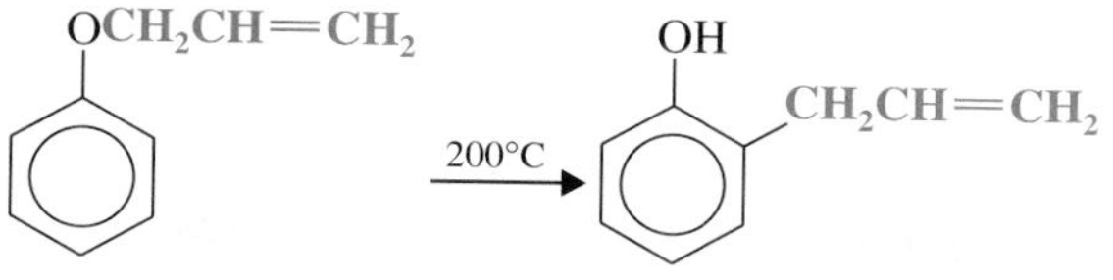

Allyl phenyl ether → **o-Allylphenol**

The reaction takes place through a **concerted rearrangement** in which the bond between C3 of the allyl group and the ortho position of the benzene ring forms at the same time that the carbon–oxygen bond of the allyl phenyl ether breaks. The product of this rearrangement is an unstable intermediate that, like the unstable intermediate in the Kolbe reaction (Section 21.8) undergoes a proton shift (a keto–enol tautomerization, see Section 17.2) that leads to the *o*-allylphenol.

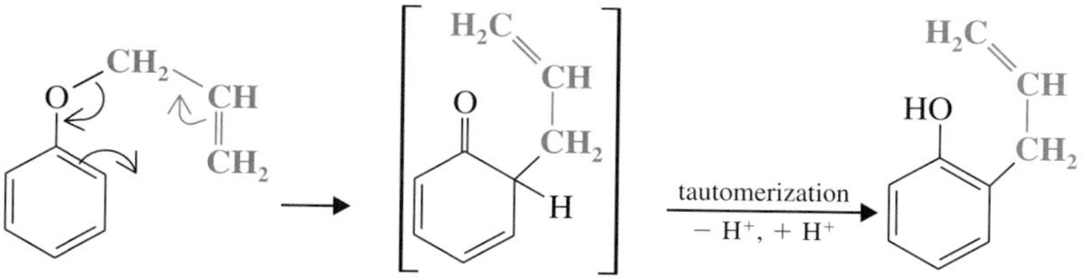

Unstable intermediate

That only C3 of the allyl group becomes bonded to the benzene ring was demonstrated by carrying out the rearrangement with allyl phenyl ether containing ^{14}C at C3. All of the product of this reaction had the labeled carbon atom bonded to the ring.

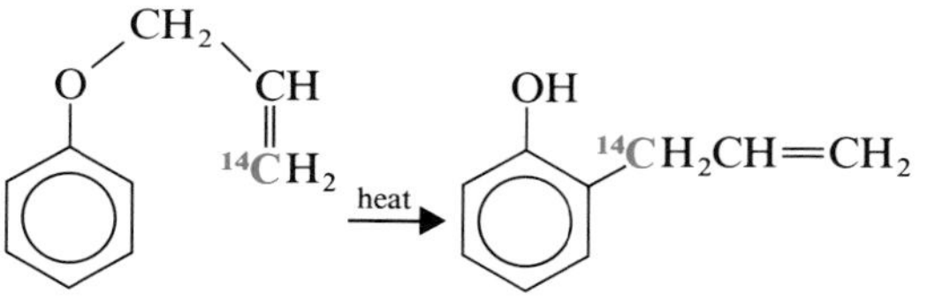

Only product

PROBLEM 21.5

The labeling experiment just described eliminates from consideration a mechanism in which the allyl phenyl ether dissociates to produce an allyl cation (Section 12.4) and a phenoxide ion, which then subsequently undergo a Friedel–Crafts alkylation (Section 15.6) to produce the *o*-allylphenol. Explain how this alternative mechanism can be discounted by showing the product (or products) that would result from it.

PROBLEM 21.6

Show how you would synthesize allyl phenyl ether through a Williamson synthesis (Section 21.6A) starting with phenol and allyl bromide.

A Claisen rearrangement also takes place when allyl vinyl ethers are heated. For example:

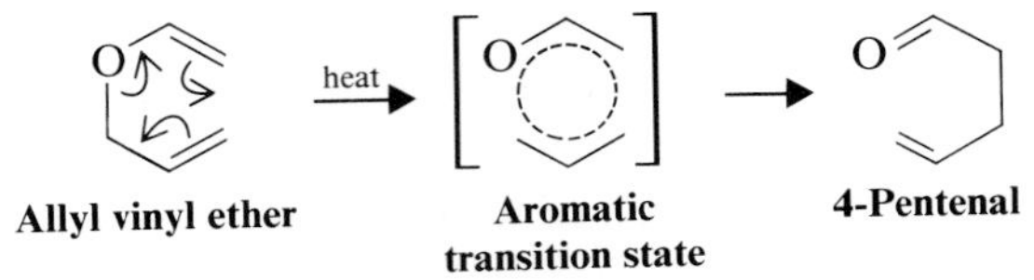

Allyl vinyl ether   **Aromatic**   **4-Pentenal**
                               transition state

The transition state for the Claisen rearrangement involves a cycle of six orbitals and six electrons. Having six electrons suggests that the transition state has aromatic character (Section 14.7). Other reactions of this general type are known and they are called **pericyclic reactions.**

Another similar pericyclic reaction is the **Cope rearrangement** shown here.

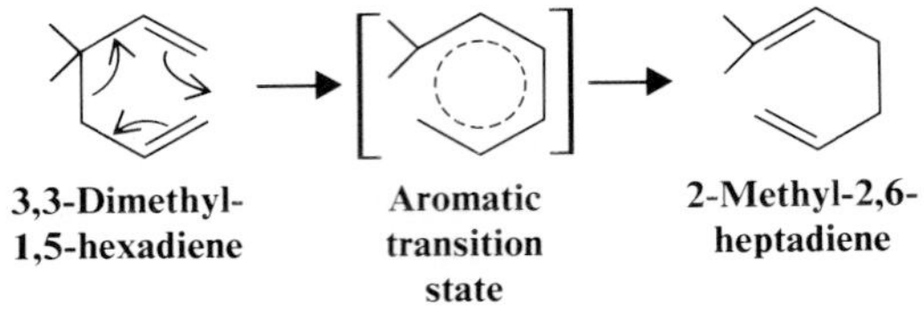

3,3-Dimethyl-   **Aromatic**   **2-Methyl-2,6-**
1,5-hexadiene   **transition**   **heptadiene**
                        state

The Diels–Alder reaction (Section 12.10) is also a pericyclic reaction. The transition state for the Diels–Alder reaction also involves six orbitals and six electrons.

Aromatic
transition
state

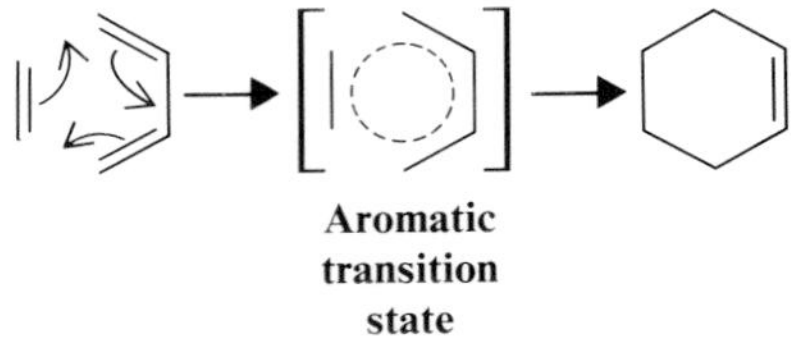

We shall discuss the mechanism of the Diels–Alder reaction further in Special Topic O.

21.10 QUINONES

Oxidation of hydroquinone (1,4-benzenediol) produces a compound known as *p*-benzoquinone. The oxidation can be brought about by mild oxidizing agents, and, overall, the oxidation amounts to the removal of a pair of electrons ($2e^-$) and two protons from hydroquinone. (Another way of visualizing the oxidation is as the loss of a hydrogen molecule, H : H, making it a dehydrogenation.)

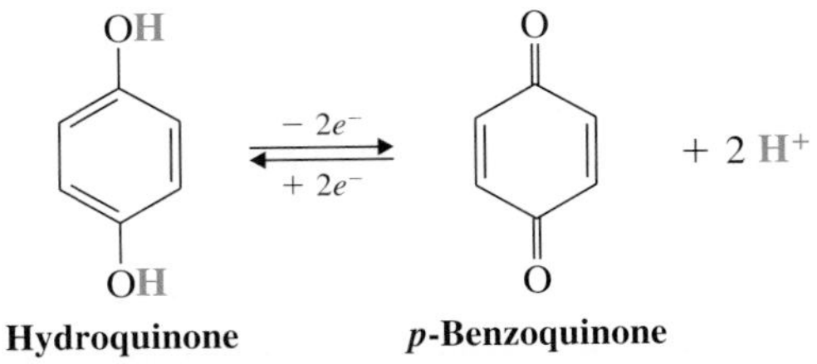

Hydroquinone   ***p*-Benzoquinone**

This reaction is reversible; *p*-benzoquinone is easily reduced by mild reducing agents to hydroquinone.

Nature makes much use of this type of reversible oxidation–reduction to transport a pair of electrons from one substance to another in enzyme-catalyzed reactions. Important compounds in this respect are the compounds called **ubiquinones** (from *ubiquitous* + quinone—these quinones are found everywhere in biological systems). Ubiquinones are also called coenzymes Q (CoQ).

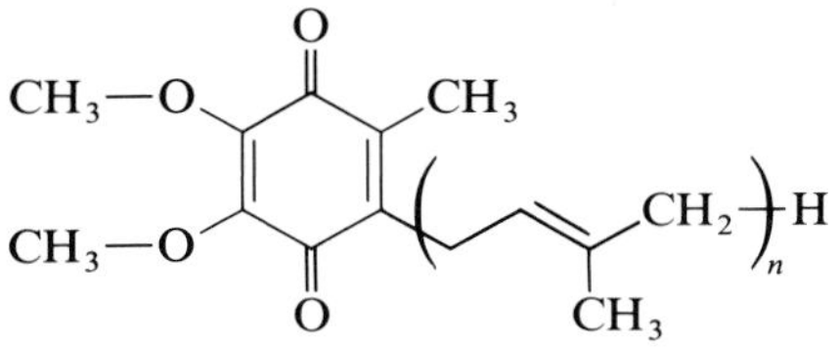

Ubiquinones ($n = 6 - 10$)
(coenzymes Q)

Vitamin K_1, the important dietary factor that is instrumental in maintaining the coagulant properties of blood, contains a 1,4-naphthoquinone structure.

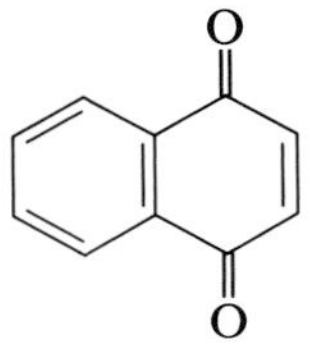

1,4-Naphthoquinone 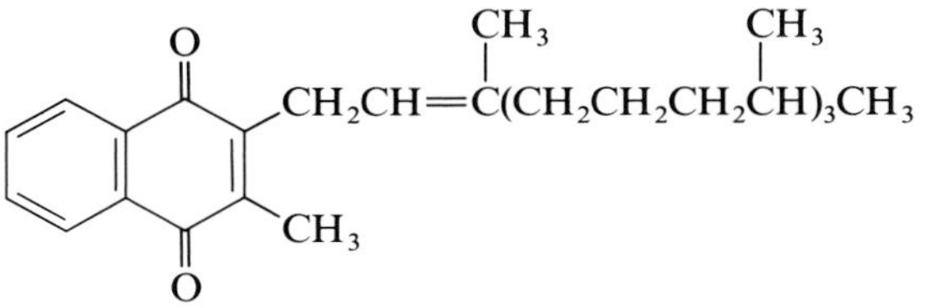   **Vitamin K_1**

PROBLEM 21.7

p-Benzoquinone and 1,4-naphthoquinone act as dienophiles in Diels–Alder reactions. Give the structures of the products of the following reactions: (a) *p*-Benzoquinone + butadiene, (b) 1,4-Naphthoquinone + butadiene, and (c) *p*-Benzoquinone + 1,3-cyclopentadiene.

PROBLEM 21.8

Outline a possible synthesis of the following compound.

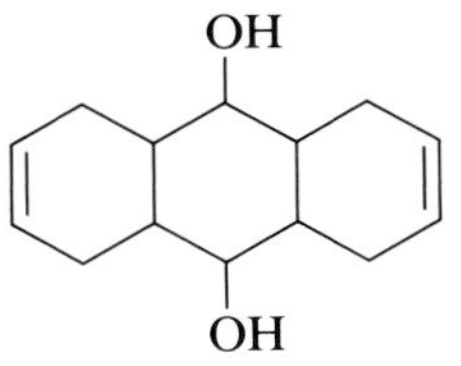

21.11 ARYL HALIDES AND NUCLEOPHILIC AROMATIC SUBSTITUTION

Simple aryl halides are like vinylic halides (Section 6.16A) in that they are relatively unreactive toward nucleophilic substitution under conditions that give facile nucleophilic substitution with alkyl halides. Chlorobenzene, for example, can be boiled with sodium hydroxide for days without producing a detectable amount of phenol (or sodium phenoxide). Similarly, when vinyl chloride is heated with sodium hydroxide, no substitution occurs:

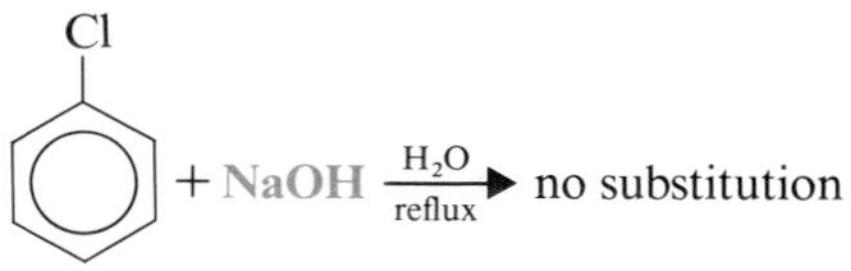

$$CH_2 = CHCl + NaOH \xrightarrow[\text{reflux}]{H_2O} \text{no substitution}$$

Aryl halides and vinylic halides do not give a positive test (a silver halide precipitate) when treated with alcoholic silver nitrate (Section 8.17).

We can understand this lack of reactivity on the basis of several factors. The benzene ring of an aryl halide prevents backside attack in an S_N2 reaction:

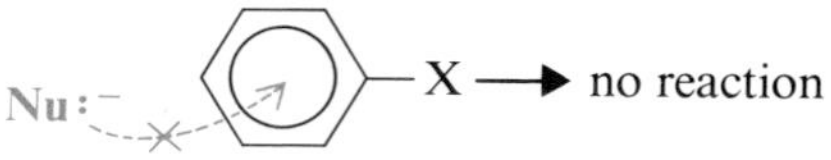

Phenyl cations are very unstable; thus S_N1 reactions do not occur. The carbon–halogen bonds of aryl (and vinylic) halides are shorter and stronger than those of alkyl, allylic, and benzylic halides. Stronger carbon–halogen bonds mean that bond breaking by either an S_N1 or S_N2 mechanism will require more energy.

Two effects make the carbon–halogen bonds of aryl and vinylic halides shorter and stronger. (1) The carbon of either type of halide is sp^2 hybridized and therefore the electrons of the carbon orbital are closer to the nucleus than those of an sp^3-hybridized carbon. (2) Resonance of the type shown here strengthens the carbon–halogen bond by giving it *double-bond character.*

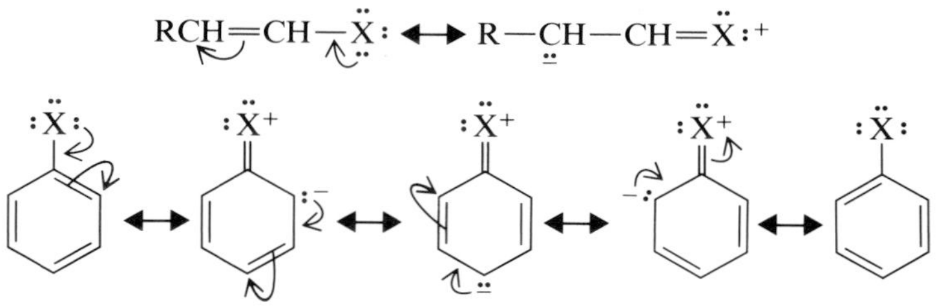

Having said all this, we shall find in the next two subsections that *aryl halides can be remarkably reactive toward nucleophiles* if they bear certain substituents or when we allow them to react under the proper conditions.

21.11A Nucleophilic Aromatic Substitution by Addition–Elimination: The $S_N Ar$ Mechanism

Nucleophilic substitution reactions of aryl halides *do* occur readily when an electronic factor makes the aryl bonded to the halogen carbon susceptible to nucleophilic attack. *Nucleophilic substitution can occur when strong electron-withdrawing groups are ortho or para to the halogen atom:*

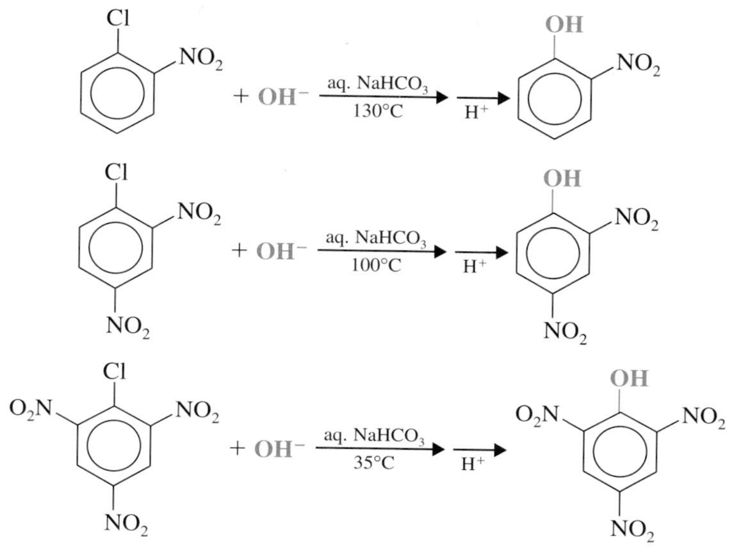

We also see in these examples that the temperature required to bring about the reaction is related to the number of ortho or para nitro groups. Of the three compounds, *o*-nitrochlorobenzene requires the highest temperature (*p*-nitrochlorobenzene reacts at 130°C as well) and 2,4,6-trinitrochlorobenzene requires the lowest temperature.

A meta-nitro group does not produce a similar activating effect. For example, *m*-nitrochlorobenzene gives no corresponding reaction.

The mechanism that operates in these reactions is an *addition–elimination* mechanism involving the formation of a *carbanion* with delocalized electrons called a **Meisenheimer complex** after the German chemist, Jacob Meisenheimer, who proposed its correct structure. In the following first step addition of a hydroxide ion to *p*-nitrochlorobenzene, for example, produces the carbanion; then elimination of a chloride ion yields the substitution product as the aromaticity of the ring is recovered. This mechanism is called the $S_N Ar$ mechanism.

THE $S_N AR$ MECHANISM

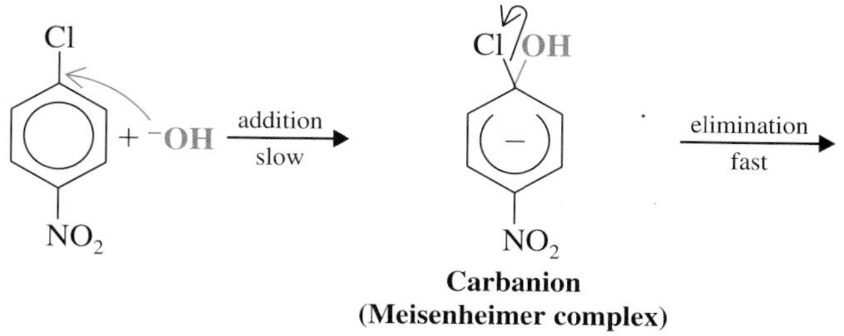

Carbanion
(Meisenheimer complex)

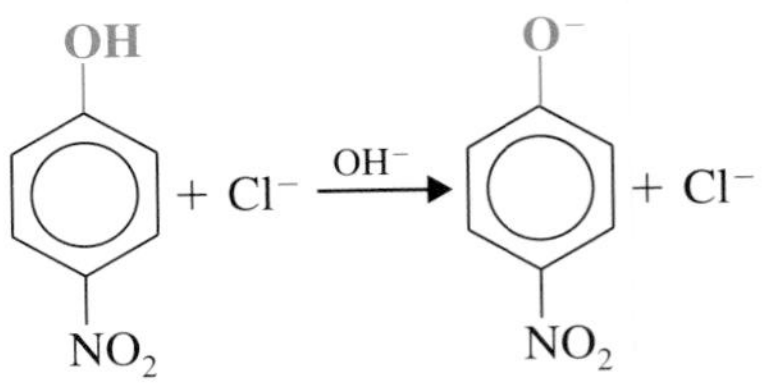

The carbanion is stabilized by *electron-withdrawing groups* in the positions ortho and para to the halogen atom. If we examine the following resonance structures, we can see how.

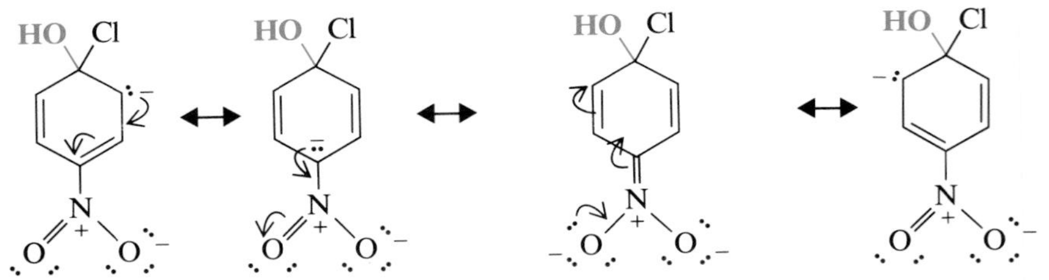

Especially stable
(negative charges
are both on oxygen atoms)

PROBLEM 21.9

1-Fluoro-2,4-dinitrobenzene is highly reactive toward nucleophilic substitution through an $S_N Ar$ mechanism. (In Section 24.5A we shall see how this reagent can be used in determining the structures of proteins.) What product would be formed when 1-fluoro-2,4-dinitrobenzene reacts with each of the following reagents.

(a) CH_3CH_2ONa (c) $C_6H_5NH_2$

(b) NH_3 (d) CH_3CH_2SNa

21.11B Nucleophilic Aromatic Substitution Through an Elimination–Addition Mechanism: Benzyne

Although aryl halides such as chlorobenzene and bromobenzene do not react with most nucleophiles under ordinary circumstances, they do react under highly forcing conditions. Chlorobenzene can be converted to phenol by heating it with aqueous sodium hydroxide in a pressurized reactor at 350°C (Section 21.4).

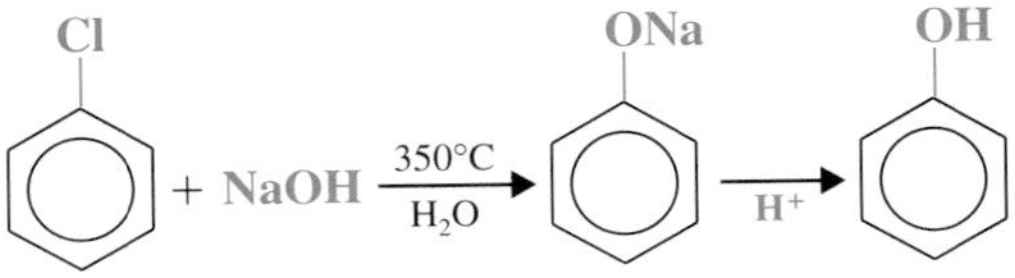

Phenol

Bromobenzene reacts with the very powerful base, $\overset{-}{N}H_2$, in liquid ammonia:

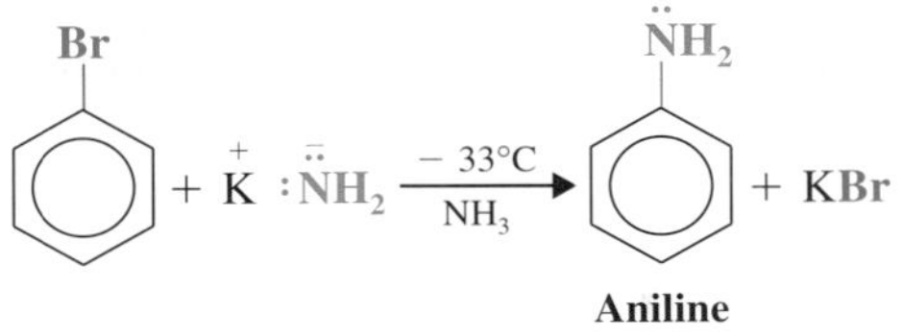

Aniline

These reactions take place through an **elimination–addition mechanism** that involves the formation of an interesting intermediate called *benzyne (or dehydrobenzene)*. We can illustrate this mechanism with the reaction of bromobenzene and amide ion.

In the first step (see following mechanism), the amide ion initiates an elimination by abstracting one of the ortho protons because they are the most acidic. The negative charge that develops on the ortho carbon is stabilized by the inductive effect of the bromine. The anion then loses a bromide ion. This elimination produces the highly unstable, and thus highly reactive, **benzyne.** Benzyne then reacts with any available nucleophile (in this case, an amide ion) by a two-step addition reaction to produce aniline.

THE ELIMINATION-ADDITION MECHANISM

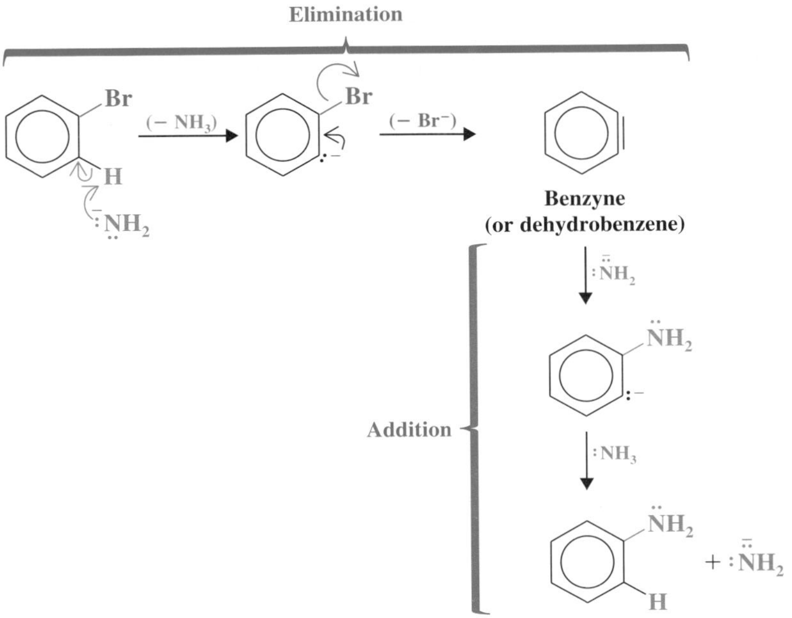

The nature of benzyne itself will become clearer if we examine the following orbital diagram.

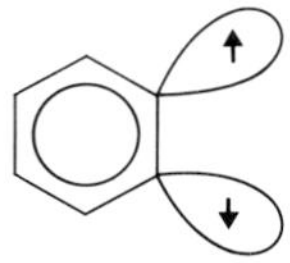

Benzyne

The extra bond in benzyne results from the overlap of sp^2 orbitals on adjacent carbon atoms of the ring. The axes of these sp^2 orbitals lie in the same plane as that of the ring, and consequently they are perpendicular to, and do not overlap with, the π orbitals of the aromatic system. They do not appreciably disturb the aromatic system and they do not make an appreciable resonance contribution to it. The extra bond is weak. Even though the ring hexagon is probably somewhat distorted in order to bring the sp^2 orbitals closer together, overlap between them is not large. Benzyne, as a result, is highly unstable and highly reactive. It has never been isolated.

What, then, is some of the evidence for the existence of benzyne and for an elimination–addition mechanism in some nucleophilic aromatic substitutions?

The first piece of clearcut evidence was an experiment done by J. D. Roberts (Section 13.10) in 1953 — one that marked the beginning of benzyne chemistry. Roberts showed that when ^{14}C-labeled (C*) bromobenzene is treated with amide ion in liquid ammonia, the aniline that is produced has the label equally divided between the 1 and 2 positions. This result is consistent with the following elimination–addition mechanism but is, of course, not at all consistent with a direct displacement or with an addition–elimination mechanism. (Why?)

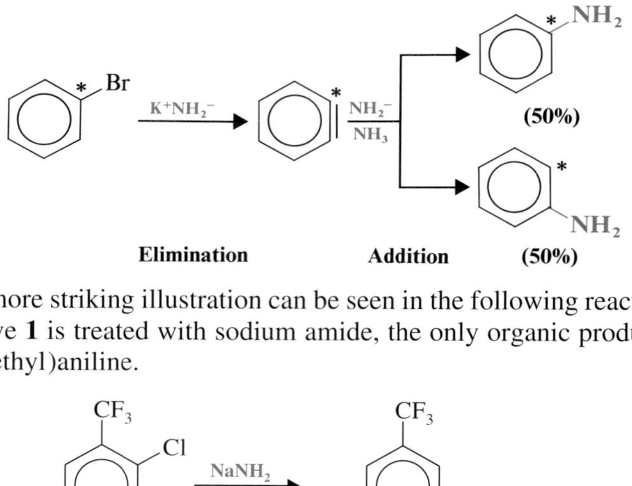

Elimination   Addition   (50%)

An even more striking illustration can be seen in the following reaction. When the ortho derivative **1** is treated with sodium amide, the only organic product obtained is *m*-(trifluoromethyl)aniline.

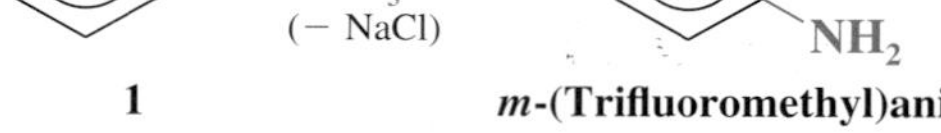

1   *m*-(Trifluoromethyl)aniline

This result can also be explained by an elimination–addition mechanism. The first step produces the benzyne **2**:

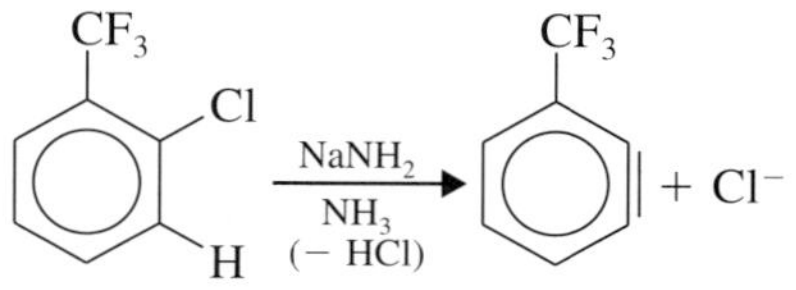

1   **2**

This benzyne then adds an amide ion in the way that produces the more stable carbanion **3** rather than the less stable carbanion **4**.

4
Less stable carbanion

3
More stable carbanion (negative charge is closer to the electronegative trifluoromethyl group)

Carbanion **3** then accepts a proton from ammonia to form *m*-(trifluoromethyl)aniline.

Carbanion **3** is more stable than **4** because the carbon atom bearing the negative charge is closer to the highly electronegative trifluoromethyl group. The trifluoromethyl group stabilizes the negative charge through its inductive effect. (Resonance effects are not important here because the sp^2 orbital that contains the electron pair does not overlap with the π orbitals of the aromatic system.)

Benzyne intermediates have been "trapped" through the use of Diels–Alder reactions. When benzyne is generated in the presence of the diene *furan*, the product is a Diels–Alder adduct.

Benzyne (generated by an elimination reaction)   **Furan**   **Diels–Alder adduct**

PROBLEM 21.10

When *o*-chlorotoluene is subjected to the conditions used in the Dow process (i.e., aqueous NaOH at 350°C at high pressure) the products of the reaction are *o*-cresol and *m*-cresol. What does this result suggest about the mechanism of the Dow process?

PROBLEM 21.11

When 2-bromo-1,3-dimethylbenzene is treated with sodium amide in liquid ammonia, no substitution takes place. This result can be interpreted as providing evidence for the elimination–addition mechanism. Explain how this interpretation can be given.

21.12 SPECTROSCOPIC ANALYSIS OF PHENOLS AND ARYL HALIDES

Infrared Spectra Phenols show a characteristic absorption band (usually broad) arising from O—H stretching in the 3400–3600 cm^{-1} region. Phenols and aryl halides also show the characteristic absorptions that arise from their benzene rings (see Section 14.11C)

¹H NMR Spectra The hydroxylic proton of a phenol is more deshielded than that of an alcohol. The exact position of the O—H signal depends on the extent of hydrogen bonding, and on whether the hydrogen bonding is *intermolecular* or *intramolecular*. The extent of intermolecular hydrogen bonding depends on the concentration of the phenol, and this strongly affects the position of the O—H signal. In phenol, itself, for example, the position of the O—H signal varies from δ 2.55 for pure phenol to δ 5.63 at 1% concentration in CCl$_4$. Phenols with strong intramolecular hydrogen bonding, such as salicylaldehyde, show O—H signals between δ 0.5 and δ 1.0, and the position of the signal varies only slightly with concentration. As with other protons that undergo exchange (Section 13.10), the identity of the O—H proton of a phenol can be determined by adding D$_2$O to the sample. The O—H proton undergoes rapid exchange with deuterium and the proton signal disappears. The aromatic protons of phenols and aryl halides give signals in the δ 7–9 region.

¹³C NMR Spectra The carbon atoms of the aromatic ring of phenols and aryl halides appear in the region δ 135–170.

21.13 SUMMARY OF IMPORTANT REACTIONS

21.13A Synthesis of Phenols (Section 21.4)

1. Via arenediazonium salts

$$Ar—NH_2 \xrightarrow[0-5°C]{HONO} Ar—\overset{+}{N}_2 \xrightarrow[Cu^{2+},H_2O]{Cu_2O} Ar—OH$$

2. Dow process

3. From sodium benzenesulfonates

4. Via cumene hydroperoxide

21.13B Reactions of Phenols

1. As acids (Section 21.5A)

2. Williamson synthesis (Section 21.6A)

3. Acylation (Section 18.7)

4. Electrophilic aromatic substitution (Section 21.8)

21.13C Claisen Rearrangement (Section 21.9)

21.13D Synthesis of Aryl Halides

1. By electrophilic aromatic substitution (Section 15.3)

2. Via arenediazonium salts (Section 20.8)

21.13E Reactions of Aryl Halides

1. Electrophilic aromatic substitution (Sections 15.3–15.7)

2. Nucleophilic aromatic substitution (Section 21.11)

ADDITIONAL PROBLEMS

21.12 What products would be obtained from each of the following acid–base reactions?
(a) Sodium ethoxide in ethanol + phenol ⟶
(b) Phenol + aqueous sodium hydroxide ⟶
(c) Sodium phenoxide + aqueous hydrochloric acid ⟶
(d) Sodium phenoxide + H_2O + CO_2 ⟶

21.13 Complete the following equations:

(a) Phenol + Br_2 $\xrightarrow{\text{5°C, CS}_2}$

(b) Phenol + concd H_2SO_4 $\xrightarrow{\text{25°C}}$

(c) Phenol + concd H_2SO_4 $\xrightarrow{\text{100°C}}$

(d) CH_3—⟨○⟩—OH + p-toluenesulfonyl chloride $\xrightarrow[\text{OH}^-]{}$

(e) Phenol + Br_2 $\xrightarrow{\text{H}_2\text{O}}$

(f) Phenol + ⟶

(g) p-Cresol + Br_2 $\xrightarrow{\text{H}_2\text{O}}$

(h) Phenol + $C_6H_5\overset{\overset{\displaystyle O}{\|}}{C}Cl$ $\xrightarrow[\text{base}]{}$

(i) Phenol + $\left(C_6H_5\overset{\overset{\displaystyle O}{\|}}{C}-\right)_2O$ $\xrightarrow[\text{base}]{}$

(j) Phenol + NaOH $\longrightarrow$

(k) Product of (j) + $CH_3OSO_2OCH_3$ $\longrightarrow$

(l) Product of (j) + CH_3I $\longrightarrow$

(m) Product of (j) + $C_6H_5CH_2Cl$ $\longrightarrow$

21.14 Describe a simple chemical test that could be used to distinguish between each of the following pairs of compounds.

(a) 4-Chlorophenol and 4-chloro-1-methylbenzene

(b) 4-Methylphenol and 4-methylbenzoic acid

(c) Vinyl phenyl ether and ethyl phenyl ether

(d) 4-Methylphenol and 2,4,6-trinitrophenol

(e) Ethyl phenyl ether and 4-ethylphenol

21.15 When *m*-chlorotoluene is treated with sodium amide in liquid ammonia, the products of the reaction are *o*-, *m*-, and *p*-toluidine (i.e., o-$CH_3C_6H_4NH_2$, m-$CH_3C_6H_4NH_2$, and p-$CH_3C_6H_4NH_2$). Propose plausible mechanisms that account for the formation of each product.

21.16 Without consulting tables, select the stronger acid from each of the following pairs.

(a) 4-Methylphenol and 4-fluorophenol

(b) 4-Methylphenol and 4-nitrophenol

(c) 4-Nitrophenol and 3-nitrophenol

(d) 4-Methylphenol and benzyl alcohol

(e) 4-Fluorophenol and 4-bromophenol

21.17 Phenols are often effective antioxidants (see Problem 21.20 and Section 9.11) because they are said to "trap" radicals. The trapping occurs when phenols react with highly reactive radicals to produce less reactive (more stable) phenolic radicals. (a) Show how phenol itself might react with an alkoxy radical ($RO\cdot$) in a hydrogen abstraction reaction involving the phenolic —OH. (b) Write resonance structures for the resulting radical that account for its being relatively unreactive.

21.18 The first synthesis of a crown ether (Section 10.22A) by C. J. Pedersen (of the Du Pont Company) involved treating 1,2-benzenediol with di(2-chloroethyl) ether, $(ClCH_2CH_2)_2O$, in the presence of NaOH. The product was a compound called dibenzo-18-crown-6. Give the structure of dibenzo-18-crown-6 and provide a plausible mechanism for its formation.

21.19 A compound **X** ($C_{10}H_{14}O$) dissolves in aqueous sodium hydroxide but is insoluble in aqueous sodium bicarbonate. Compound **X** reacts with bromine in water to yield a dibromo derivative, $C_{10}H_{12}Br_2O$. The 3000–4000-cm^{-1} region of the IR spectrum of **X** shows a broad peak centered at 3250 cm⁻¹; the 680–840-cm⁻¹ region shows a strong peak at 830 cm⁻¹. The ¹H NMR spectrum of **X** gives the following:

Singlet	δ 1.3 (9H)
Singlet	δ 4.9 (1H)
Multiplet	δ 7.0 (4H)

What is the structure of **X**?

21.20   The widely used antioxidant and food preservative called **BHA** (**b**utylated **h**ydroxy **a**nisole) is actually a mixture of 2-*tert*-butyl-4-methoxyphenol and 3-*tert*-butyl-4-methoxy-phenol. **BHA** is synthesized from *p*-methoxyphenol and 2-methylpropene.   (a) Suggest how this is done.   (b) Another widely used antioxidant is **BHT** (**b**utylated **h**ydroxy **t**oluene). **BHT** is actually 2,6-di-*tert*-butyl-4-methylphenol, and the raw materials used in its production are *p*-cresol and 2-methylpropene. What reaction is used here?

21.21   The herbicide **2,4-D** can be synthesized from phenol and chloroacetic acid. Outline the steps involved.

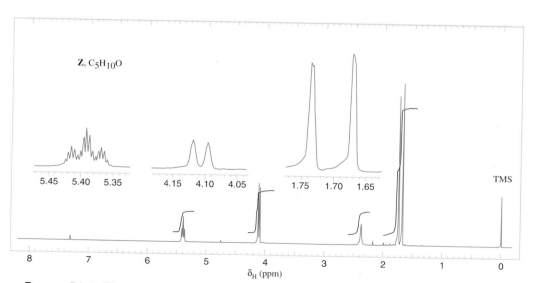

2,4-D
2,4-Dichlorophenoxyacetic acid

ClCH₂CO₂H

Chloroacetic acid

21.22   Compound **Z** (C₅H₁₀O) decolorizes bromine in carbon tetrachloride. The IR spec-trum of **Z** shows a broad peak in the 3200–3600-cm⁻¹ region. The 300 MHz ¹H NMR spectrum of **Z** is given in Fig. 21.1. Propose a structure for **Z**.

FIGURE 21.1   The 300 MHz ¹H NMR spectrum of compound **Z** (Problem 21.22). Expan-sions of the signals are shown in the offset plots.

21.23   A synthesis of the β-receptor blocker called toliprolol begins with a reaction between 3-methylphenol and epichlorohydrin. The synthesis is outlined below. Give the structures of the intermediates and toliprolol.

$$3\text{-Methylphenol} + H_2C\!-\!\!CH\!-\!CH_2Cl \longrightarrow C_{10}H_{13}O_2Cl \xrightarrow{OH^-}$$

$$\overset{O}{\diagdown\diagup}$$

Epichlorohydrin

$$C_{10}H_{12}O_2 \xrightarrow{(CH_3)_2CHNH_2} \textit{Toliprolol, } C_{13}H_{21}NO_2$$

21.24   Explain how it is possible for 2,2'-dihydroxy-1,1'-binaphthyl (below) to exist in enantiomeric forms.

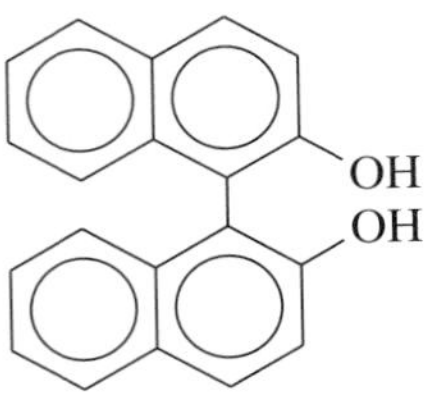

21.25   *p*-Chloronitrobenzene was allowed to react with sodium 2,6-di-*tert*-butylphenoxide with the intention of preparing the diphenyl ether **1**. The product was not **1**, but rather was an isomer of **1** that still possessed a phenolic hydroxyl group.

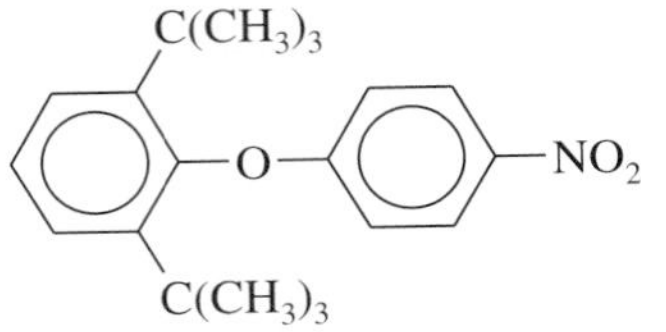

1

What was this product, and how can one account for its formation?

21.26   Account for the fact that the Dow process for the production of phenol produces both diphenyl ether **(1)** and 4-hydroxybiphenyl **(2)** as byproducts.

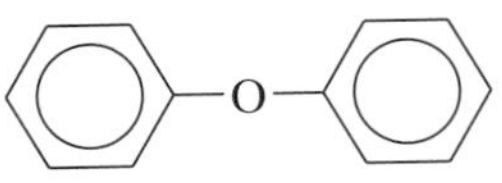

1 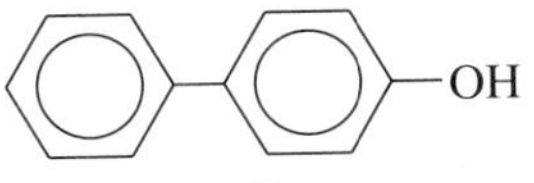   2

21.27   Commonly, the leaving group in S_NAr reactions is halide ion. Explain the success of the following reaction where hydride is the leaving group.

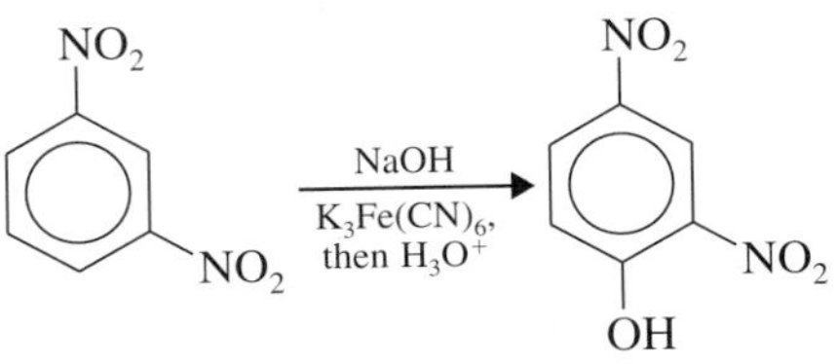

1. Arrange the compounds of each of the following series in order of increasing acidity.

(a) CH_3CH_2OH $CH_3\overset{\overset{O}{\|}}{C}OH$ $CH_3O\overset{\overset{O}{\|}}{C}CH_2\overset{\overset{O}{\|}}{C}OCH_3$ $CH_3\overset{\overset{O}{\|}}{C}CH_3$

(b) ⬡—OH   ⬡—OH   ⬡—C≡CH   ⬡—$\overset{\overset{O}{\|}}{C}OH$

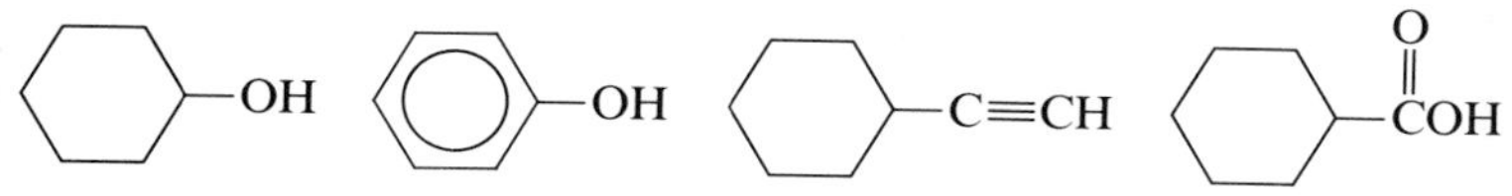

(c) $(CH_3)_3\overset{+}{N}$—⬡—$\overset{\overset{O}{\|}}{C}OH$ $(CH_3)_3C$—⬡—$\overset{\overset{O}{\|}}{C}OH$ ⬡—$\overset{\overset{O}{\|}}{C}OH$

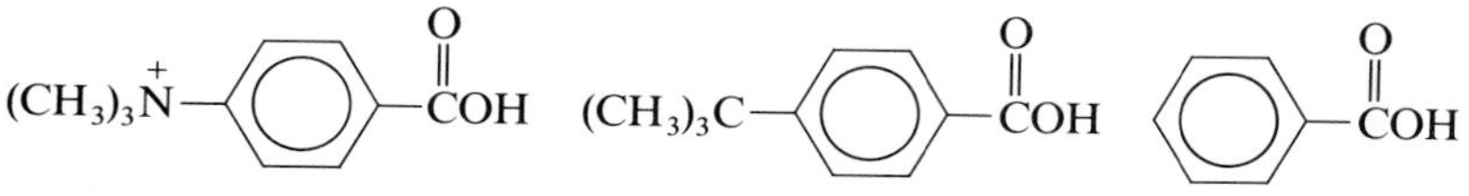

(d) $CH_3CCl_2\overset{\overset{O}{\|}}{C}OH$ $CH_3CH_2\overset{\overset{O}{\|}}{C}OH$ $CH_3CHCl\overset{\overset{O}{\|}}{C}OH$

(e) ⬡—$\overset{\overset{O}{\|}}{C}NH_2$ ⬡(—$\overset{\overset{O}{\|}}{C}$—NH—$\overset{\underset{O}{\|}}{C}$—)   ⬡—$NH_2$

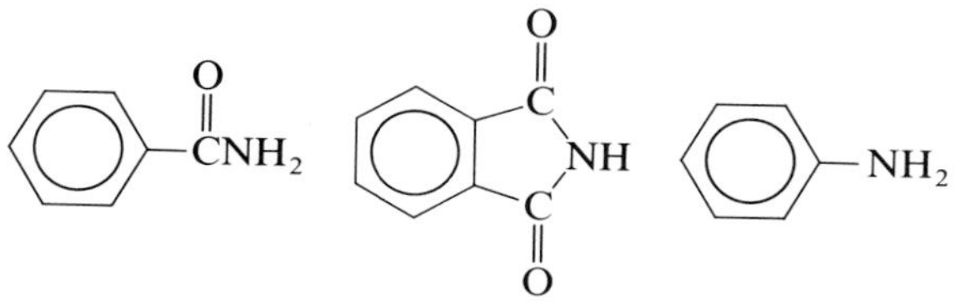

2. Arrange the compounds of each of the following series in order of increasing basicity.

(a) $CH_3\overset{\overset{O}{\|}}{C}NH_2$ $CH_3CH_2NH_2$ NH_3

(b) ⬡—NH_2 ⬡—NH_2 CH_3—⬡—NH_2

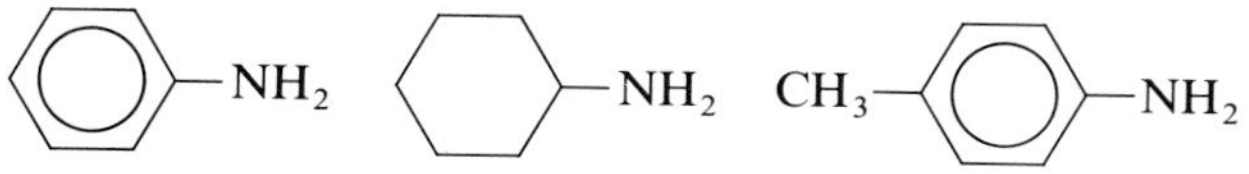

(c) O_2N—⬡—NH_2 CH_3—⬡—NH_2 ⬡—NH_2

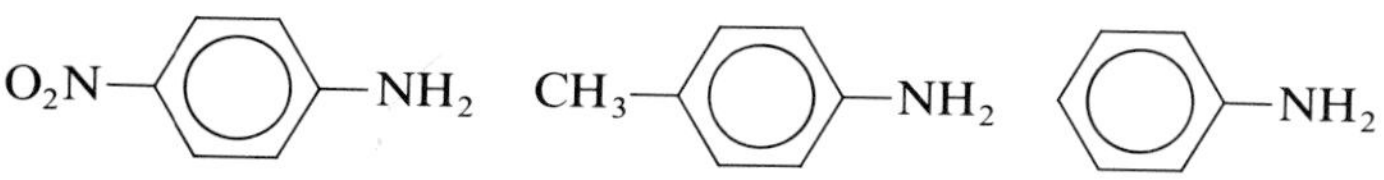

(d) $CH_3CH_2CH_3$ CH_3NHCH_3 CH_3OCH_3

3. Starting with 1-butanol and using any other required reagents, outline a synthesis of each of the following compounds. You need not repeat steps carried out in earlier parts of this problem.

(a) Butyl bromide
(b) Butylamine
(c) Pentylamine
(d) Butanoic acid
(e) Pentanoic acid
(f) Butanoyl chloride
(g) Butanamide
(h) Butyl butanoate
(i) Propylamine
(j) Butylbenzene
(k) Butanoic anhydride
(l) Hexanoic acid

4. Starting with benzene, toluene, or aniline and any other required reagents, outline a synthesis of each of the following.

(a) 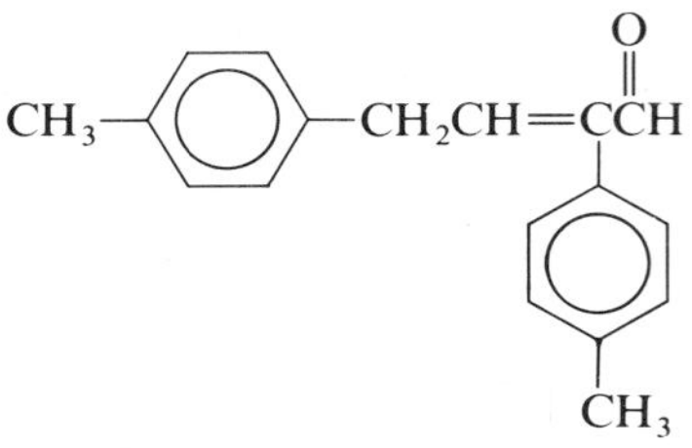 CH_3—⟨⟩—CH_2CH=$\overset{\overset{O}{\parallel}}{C}CH$

with a CH_3-substituted benzene ring below

(d) 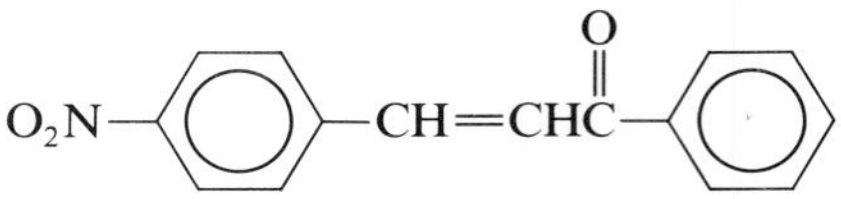 O_2N—⟨⟩—CH=$\overset{\overset{O}{\parallel}}{CH}C$—⟨⟩

(e) C_6H_5CH=$\underset{\underset{CO_2H}{\mid}}{C}CH_3$

(b) 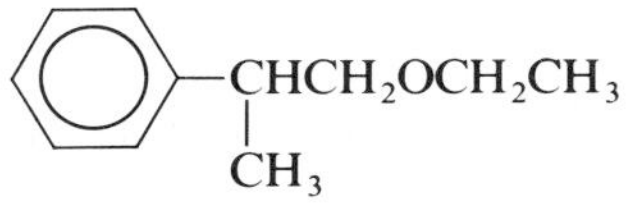 ⟨⟩—$\underset{\underset{CH_3}{\mid}}{C}HCH_2OCH_2CH_3$

(c) Cl—⟨⟩ with Br at top and Br at bottom

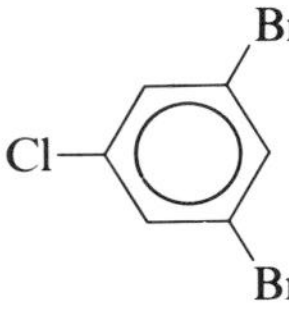

5. Give stereochemical structures for compounds **A–D.**

2-Methyl-1,3-butadiene + diethyl fumarate $\longrightarrow$ **A** $(C_{13}H_{20}O_4)$ $\xrightarrow{\text{(1) LiAlH}_4\text{, (2) H}_2O}$

B $(C_9H_{16}O_2)$ $\xrightarrow{\text{PBr}_3}$ **C** $(C_9H_{14}Br_2)$ $\xrightarrow{\text{Zn, H}^+}$ **D** (C_9H_{16})

6. A Grignard reagent that is a key intermediate in an industrial synthesis of vitamin A (Section 17.6B) can be prepared in the following way:

HC≡CLi + CH_2=$CH\overset{\overset{O}{\parallel}}{C}CH_3$ $\xrightarrow[\text{(2) NH}_4^+]{\text{(1) liq. NH}_3}$ **A** (C_6H_8O) $\xrightarrow{H^+}$

B $HOCH_2CH$=$\underset{\underset{\mid}{}}{\overset{\overset{CH_3}{\mid}}{C}}$—$C$≡$CH$ $\xrightarrow{2\ C_2H_5MgBr}$ **C** $(C_6H_6Mg_2Br_2O)$

(a) What are the structures of compounds **A** and **C**?

(b) The acid-catalyzed rearrangement of **A–B** takes place very readily. What two factors account for this?

7. The remaining steps in the industrial synthesis of vitamin A (as an acetate) are as follows: The Grignard reagent **C** from Problem 6 is allowed to react with the aldehyde shown here.

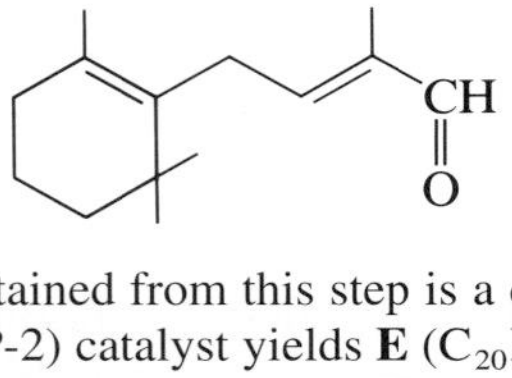

structure of an aldehyde: a cyclohexene ring with a methyl group bearing a chain —CH₂CH=C(CH₃)—CH=O

After acidification, the product obtained from this step is a diol **D**. Selective hydrogenation of the triple bond of **D** using Ni_2B (P-2) catalyst yields **E** $(C_{20}H_{32}O_2)$. Treating **E** with one molar equivalent of acetic anhydride yields a monoacetate (**F**) and dehydration of **F** yields vitamin A acetate. What are the structures of **D–F**?

8.   Heating acetone with an excess of phenol in the presence of hydrogen chloride is the basis for an industrial process used in the manufacture of a compound called "bisphenol A." (Bisphenol A is used in the manufacture of epoxy resins and a polymer called Lexan.) Bisphenol A has the molecular formula $C_{15}H_{16}O_2$ and the reactions involved in its formation are similar to those involved in the synthesis of DDT (Special Topic N). Write out these reactions and give the structure of bisphenol A.

9.   Outlined here is a synthesis of the local anesthetic *procaine*. Provide structures for procaine and the intermediates **A–C.**

p-Nitrotoluene $\xrightarrow[\text{(2) } H_3O^+]{\text{(1) } KMnO_4, OH^-, \text{heat}}$ **A** $(C_7H_5NO_4)$ $\xrightarrow{SOCl_2}$

B $(C_7H_4ClNO_3)$ $\xrightarrow{HOCH_2CH_2N(C_2H_5)_2}$ **C** $(C_{13}H_{18}N_2O_4)$ $\xrightarrow{H_2, \text{ cat.}}$ procaine $(C_{13}H_{20}N_2O_2)$

10.   The sedative–hypnotic *ethinamate* can be synthesized by the following route. Provide structures for ethinamate and the intermediates **A** and **B**.

Cyclohexanone $\xrightarrow{\text{(1) } HC\equiv CNa, \text{ (2) } H^+}$ **A** $(C_8H_{12}O)$ $\xrightarrow{ClCOCl}$

B $(C_9H_{11}ClO_2)$ $\xrightarrow{NH_3}$ ethinamate $(C_9H_{13}NO_2)$

11.   The prototype of the antihistamines, *diphenhydramine* (also called Benadryl), can be synthesized by the following sequence of reactions.   (a) Give structures for diphenhydramine and for the intermediates **A** and **B**.   (b) Comment on a possible mechanism for the last step of the synthesis.

Benzaldehyde $\xrightarrow{\text{(1) } C_6H_5MgBr, \text{ (2) } H_3O^+}$ **A** $(C_{13}H_{12}O)$ $\xrightarrow{PBr_3}$

B $(C_{13}H_{11}Br)$ $\xrightarrow{(CH_3)_2NCH_2CH_2OH}$ diphenhydramine $(C_{17}H_{21}NO)$

12.   Show how you would modify the synthesis given in the previous problem to synthesize the following drugs.

(a) Br—⟨C₆H₄⟩—CHOCH₂CH₂N(CH₃)₂ with C₆H₅   **Bromodiphenhydramine (an antihistamine)**

(b) [o-CH₃-C₆H₄]—CHOCH₂CH₂N(CH₃)₂ with C₆H₅   **Orphenadrine (an antispasmodic, used in controlling Parkinson's disease)**

13.   Outlined here is a synthesis of 2-methyl-3-oxocyclopentanecarboxylic acid. Give the structure of each intermediate.

$CH_3CHCO_2C_2H_5$ (with Br substituent) $\xrightarrow{CH_2(CO_2C_2H_5)_2, \text{ EtO}^-}$ **A** $(C_{12}H_{20}O_6)$ $\xrightarrow{CH_2=CHCN, \text{ EtO}^-}$

B $(C_{15}H_{23}NO_6)$ $\xrightarrow{EtOH, H^+}$ **C** $(C_{17}H_{28}O_8)$ $\xrightarrow{EtO^-}$ **D** $(C_{15}H_{22}O_7)$ $\xrightarrow[\text{(2) } H_3O^+, \text{ (3) heat}]{\text{(1) } OH^-, H_2O, \text{heat}}$

14. Give structures for compounds **A–D**. Compound **D** decolorizes bromine in carbon tetrachloride and gives a strong IR absorption band near 1720 cm⁻ .

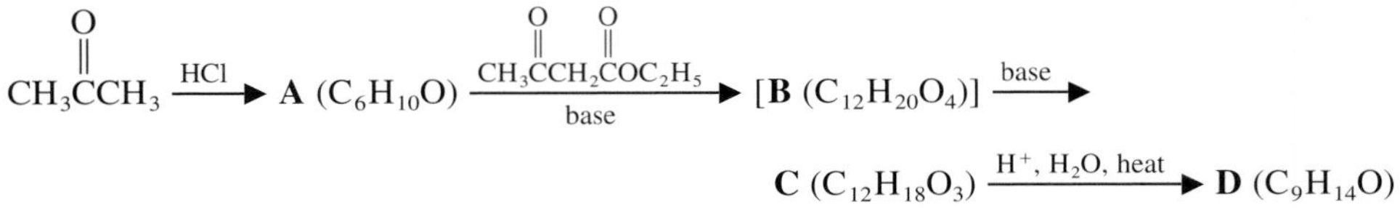

$$CH_3\overset{\overset{\displaystyle O}{\|}}{C}CH_3 \xrightarrow{HCl} \textbf{A} (C_6H_{10}O) \xrightarrow[\text{base}]{CH_3\overset{\overset{\displaystyle O}{\|}}{C}CH_2\overset{\overset{\displaystyle O}{\|}}{C}OC_2H_5} [\textbf{B} (C_{12}H_{20}O_4)] \xrightarrow{\text{base}}$$

$$\textbf{C} (C_{12}H_{18}O_3) \xrightarrow{H^+, H_2O, \text{ heat}} \textbf{D} (C_9H_{14}O)$$

15. A synthesis of the broad-spectrum antibiotic *chloramphenicol* is shown here. In the last step basic hydrolysis selectively hydrolyzes ester linkages in the presence of an amide group. What are the intermediates **A–E**?

$$\text{Benzaldehyde } + \text{ HOCH}_2CH_2NO_2 \xrightarrow{EtO^-} \textbf{A} (C_9H_{11}NO_4) \xrightarrow{H_2, \text{ cat.}}$$

$$\textbf{B} (C_9H_{13}NO_2) \xrightarrow{Cl_2CHCOCl} \textbf{C} (C_{11}H_{13}Cl_2NO_3) \xrightarrow{\text{excess } (CH_3CO)_2O}$$

$$\textbf{D} (C_{15}H_{17}Cl_2NO_5) \xrightarrow{HNO_3, H_2SO_4} \textbf{E} (C_{15}H_{16}Cl_2N_2O_7) \xrightarrow{OH^-, H_2O}$$

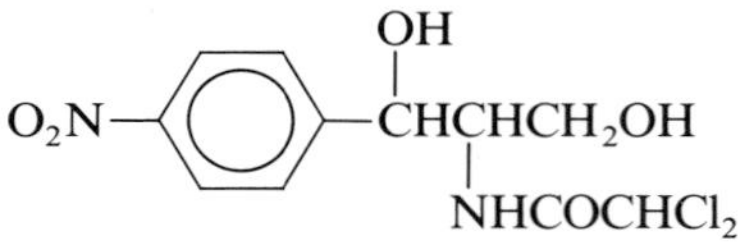

$$O_2N-\overset{OH}{\underset{NHCOCHCl_2}{\underset{|}{\overset{|}{C}H}CHCH_2OH}}$$

Chloramphenicol

16. The tranquilizing drug *meprobamate* (Equanil or Miltown) can be synthesized from 2-methylpentanal as follows. Give structures for meprobamate and for the intermediates **A–C**.

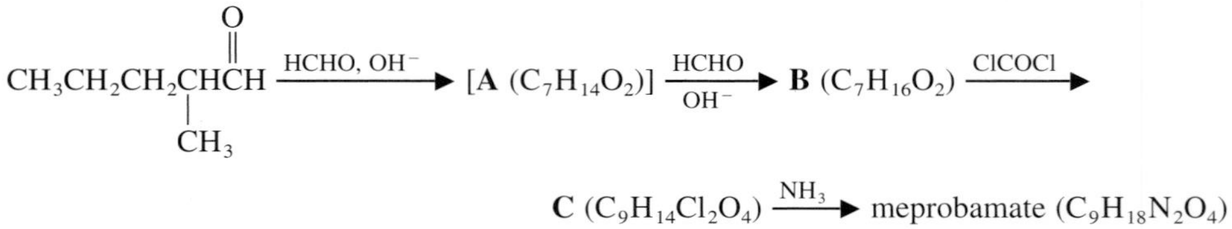

$$CH_3CH_2CH_2\underset{\underset{CH_3}{|}}{C}H\overset{\overset{\displaystyle O}{\|}}{C}H \xrightarrow{HCHO, OH^-} [\textbf{A} (C_7H_{14}O_2)] \xrightarrow[OH^-]{HCHO} \textbf{B} (C_7H_{16}O_2) \xrightarrow{ClCOCl}$$

$$\textbf{C} (C_9H_{14}Cl_2O_4) \xrightarrow{NH_3} \text{meprobamate } (C_9H_{18}N_2O_4)$$

17. What are compounds **A–C**? Compound **C** is useful as an insect repellent.

$$\text{Succinic anhydride} \xrightarrow{CH_3CH_2CH_2OH} \textbf{A} (C_7H_{12}O_4) \xrightarrow{SOCl_2}$$

$$\textbf{B} (C_7H_{11}ClO_3) \xrightarrow{(CH_3CH_2)_2NH} \textbf{C} (C_{11}H_{21}NO_3)$$

18. Outlined here is the synthesis of a central nervous system stimulant called *fencamfamine*. Provide structural formulas for each intermediate and for fencamfamine itself.

$$\text{1,3-Cyclopentadiene} + (E)\text{-}C_6H_5CH{=}CHNO_2 \longrightarrow \textbf{A} (C_{13}H_{13}NO_2) \xrightarrow{H_2, Pt}$$

$$\textbf{B} (C_{13}H_{17}N) \xrightarrow{CH_3CHO} [\textbf{C} (C_{15}H_{19}N)] \xrightarrow{H_2, Ni} \text{fencamfamine } (C_{15}H_{21}N)$$

19.   What are compounds **A** and **B**? Compound **B** has a strong IR absorption band in the 1650–1730-cm^{-1} region and a broad strong band in the 3200–3550-cm^{-1} region.

$$1\text{-Methylcyclohexene} \xrightarrow[\text{(2) NaHSO}_3]{\text{(1) OsO}_4} \mathbf{A} \ (C_7H_{14}O_2) \xrightarrow[\text{CH}_3\text{CO}_2\text{H}]{\text{CrO}_3} \mathbf{B} \ (C_7H_{12}O_2)$$

20.   Starting with phenol, outline a stereoselective synthesis of methyl *trans*-4-isopropyl-cyclohexanecarboxylate, that is,

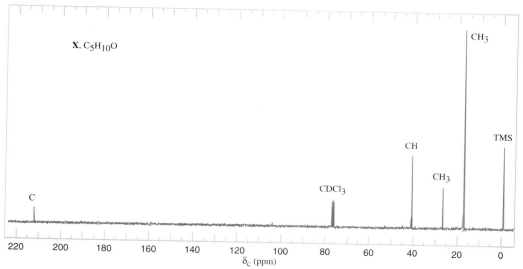

21.   Compound **X** (C$_5$H$_{10}$O) gives a positive iodoform test and shows a strong IR absorption band near 1710 cm^{-1}. The broad band proton-decoupled ^{13}C NMR spectrum of **X** is shown in Fig. 1. Propose a structure for **X**.

X, C$_5$H$_{10}$O

CH$_3$

TMS

CH

CH$_3$

CDCl$_3$

C

| 220 | 200 | 180 | 160 | 140 | 120 | 100 | 80 | 60 | 40 | 20 | 0 |

δ_C (ppm)

FIGURE 1 The broad band proton-decoupled ^{13}C NMR spectra of Compound **X** (Problem 21). Information from the DEPT ^{13}C NMR spectrum is given near each peak.

22.   Compound **Y** (C$_6$H$_{14}$O) gives a green opaque solution when treated with CrO$_3$ in aqueous H$_2$SO$_4$. Compound **Y** gives a negative iodoform test. The broad band proton-decoupled ^{13}C NMR spectrum of **Y** is given in Fig. 2. Propose a structure for **Y**.

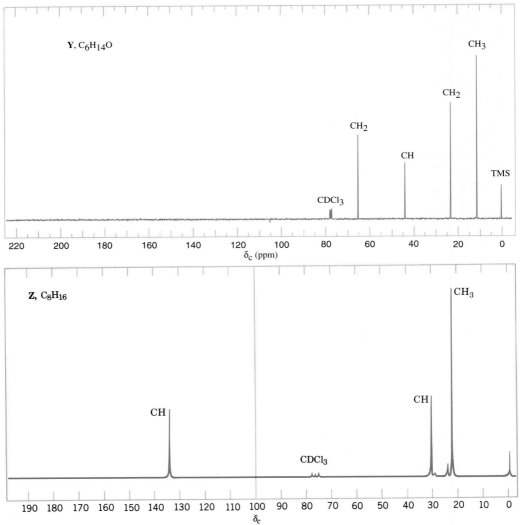

FIGURE 2 The broad band proton-decoupled ^{13}C NMR spectra of Compounds **Y** and **Z** (Problem 22). Information from the DEPT ^{13}C NMR spectrum is given above the peaks.

23.   Compound **Z** (C_8H_{16}) decolorizes bromine in carbon tetrachloride and is the more stable of a pair of stereoisomers. Ozonolysis of **Z** gives a single product. The broad band proton-decoupled ^{13}C NMR spectrum of **Z** is given in Fig. 2. Propose a structure for **Z**.

SPECIAL TOPIC K

Cocaine

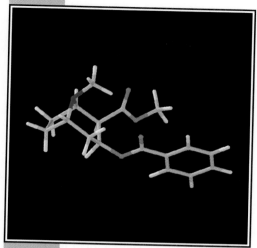

ALKALOIDS

Extracting the bark, roots, leaves, berries, and fruits of plants often yields nitrogen-containing bases called *alkaloids*. The name alkaloid comes from the fact that these substances are "alkali-like"; that is, since alkaloids are amines they often react with acids to yield soluble salts. The nitrogen atoms of most alkaloids are present in heterocyclic rings. In a few instances, however, nitrogen may be present as a primary amine or as a quaternary ammonium group.

When administered to animals most alkaloids produce striking physiological effects, which effects *vary greatly* from alkaloid to alkaloid. Some alkaloids stimulate the central nervous system, others cause paralysis; some alkaloids elevate blood pressure, others lower it. Certain alkaloids act as pain relievers; others act as tranquilizers; still others act against infectious microorganisms. Most alkaloids are toxic when their dosage is large enough, and with some this dosage is very small. In spite of this, many alkaloids find use in medicine.

Systematic names are seldom used for alkaloids, and their common names have a variety of origins. In many instances the common name reflects the botanical source of the compound. The alkaloid strychnine, for example, comes from the seeds of the *Strychnos* plant. In other instances the names are more whimsical: The name of the opium alkaloid morphine comes from Morpheus, the ancient Greek god of dreams; the name of the tobacco alkaloid nicotine comes from Nicot, an early French ambassador who sent tobacco seeds to France. The one characteristic that alkaloid names have in common is the ending *-ine*, reflecting the fact that they are all amines.

Alkaloids have been of interest to chemists for nearly two centuries, and in that time thousands of alkaloids have been isolated. Most of these have had their structures determined through the application of chemical and physical methods, and in many instances these structures have been confirmed by independent synthesis. A complete account of the chemistry of the alkaloids would (and does) occupy volumes; here we have space to consider only a few representative examples.

K.1 Alkaloids Containing a Pyridine or Reduced Pyridine Ring

The predominant alkaloid of the tobacco plant is nicotine:

Nicotine   Nicotinic acid

In very small doses nicotine acts as a stimulant, but in larger doses it causes depression, nausea, and vomiting. In still larger doses it is a violent poison. Nicotine salts are used as insecticides.

Oxidation of nicotine by concentrated nitric acid produces pyridine-3-carboxylic acid—a compound that is called *nicotinic acid*. Whereas the consumption of nicotine is of no benefit to humans, nicotinic acid is a vitamin; it is incorporated into the important coenzyme, nicotinamide adenine dinucleotide, commonly referred to as NAD^+ (oxidized form).

PROBLEM K.1

Nicotine has been synthesized by the following route. All of the steps involve reactions that we have seen before. Suggest reagents that could be used for each.

A number of alkaloids contain a piperidine ring. These include coniine (from the poison hemlock, *Conium maculatum,* a member of the carrot family, Umbelliferae), atropine (from *Atropa belladonna* and other genera of the plant family, Solanaceae), and cocaine (from *Erythroxylon coca*).

Coniine
[(+)-2-propylpiperidine)]

Atropine

Cocaine

Coniine is toxic; its ingestion may cause weakness, drowsiness, nausea, labored respiration, paralysis, and death. Coniine was one toxic substance of the "hemlock" used in the execution of Socrates (other poisons may have been included as well).

In small doses cocaine decreases fatigue, increases mental activity, and gives a general feeling of wellbeing. Prolonged use of cocaine, however, leads to physical addiction and to periods of deep depression. Cocaine is also a local anesthetic and, for a time, it was used medically in that capacity. When its tendency to cause addiction was recognized, efforts were made to develop other local anesthetics. This led, in 1905, to the synthesis of Novocain, a compound also called Procaine, that has some of the same structural features as cocaine (e.g., its benzoic ester and tertiary amine groups).

Novocain
(Procaine)

Atropine is an intense poison. In dilute solutions (0.5–1.0%) it is used to dilate the pupil of the eye in ophthalmic examinations. Compounds related to atropine are contained in the 12-h continuous-release capsules used to relieve symptoms of the common cold.

PROBLEM K.2

The principal alkaloid of *Atropa belladonna* is the optically active alkaloid *hyoscyamine*. During its isolation hyoscyamine is often racemized by bases to optically inactive atropine.   (a) What stereocenter is likely to be involved in the racemization?   (b) In hyoscyamine this stereocenter has the (S) configuration. Write a three-dimensional structure for hyoscyamine.

PROBLEM K.3

Hydrolysis of atropine gives tropine and ($\pm$)-tropic acid. (a) What are their structures? (b) Even though tropine has a stereocenter, it is optically inactive. Explain. (c) An isomeric form of tropine called ψ-tropine has also been prepared by heating tropine with base. ψ-Tropine is also optically inactive. What is its structure?

PROBLEM K.4

In 1891 G. Merling transformed tropine (cf. Problem K.3) into 1,3,5-cycloheptatriene (tropylidene) through the following sequence of reactions.

$$\text{Tropine } (C_8H_{15}NO) \xrightarrow{-H_2O} C_8H_{13}N \xrightarrow{CH_3I} C_9H_{16}NI \xrightarrow[\text{(2) heat}]{\text{(1) Ag}_2O/H_2O}$$

$$C_9H_{15}N \xrightarrow{CH_3I} C_{10}H_{18}NI \xrightarrow[\text{(2) heat}]{\text{(1) Ag}_2O/H_2O}$$

$$1,3,5\text{-cycloheptatriene} + (CH_3)_3N + H_2O$$

Write out all of the reactions that take place.

PROBLEM K.5

Many alkaloids appear to be synthesized in plants by reactions that resemble the Mannich reaction (Section 19.10). Recognition of this (by R. Robinson in 1917) led to a synthesis of tropinone that takes place under "physiological conditions," that is, at room temperature and at pH values near neutrality. This synthesis is shown here. Propose reasonable mechanisms that account for the overall course of the reaction.

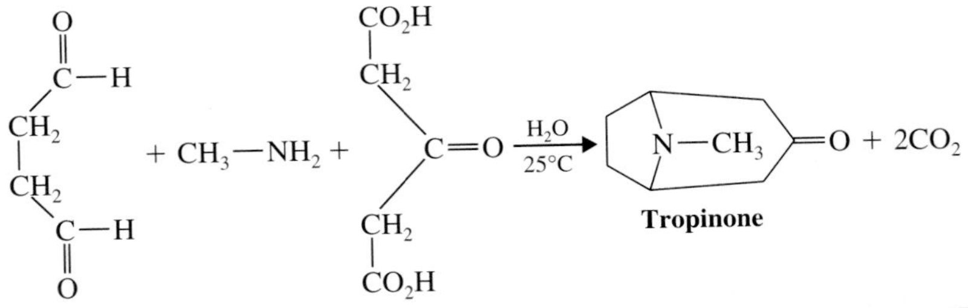

K.2 Alkaloids Containing an Isoquinoline or Reduced Isoquinoline Ring

Papaverine, morphine, and codeine are all alkaloids obtained from the opium poppy, *Papaver somniferum*.

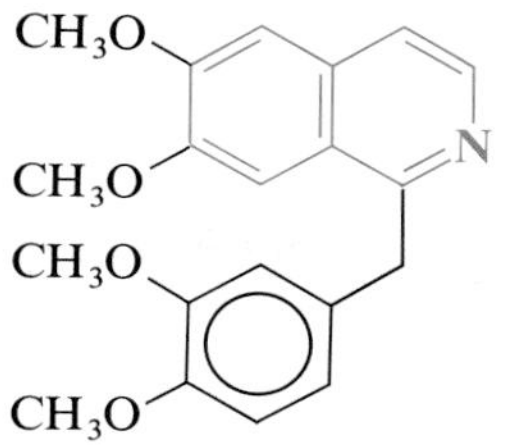

Papaverine

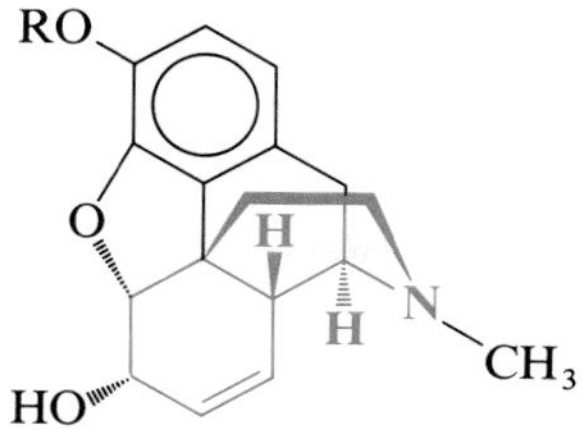

Morphine (R = H)
Codeine (R = CH₃)

Papaverine has an isoquinoline ring; in morphine and codeine the isoquinoline ring is partially hydrogenated (reduced).

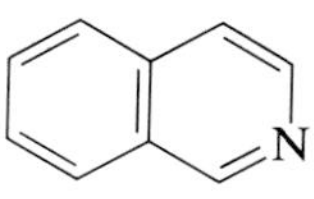

Isoquinoline

Opium has been used since earliest recorded history. Morphine was first isolated from opium in 1803, and its isolation represented one of the first instances of the purification of the active principle of a drug. One hundred and twenty years were to pass, however, before the complicated structure of morphine was deduced, and its final confirmation through independent synthesis (by Marshall Gates of the University of Rochester) did not take place until 1952.

Morphine is one of the most potent analgesics known, and it is still used extensively in medicine to relieve pain, especially "deep" pain. Its greatest drawbacks, however, are its tendencies to lead to addiction and to depress respiration. These disadvantages have brought about a search for morphine-like compounds that do not have these disadvantages. One of the newest candidates is the compound pentazocine. Pentazocine is a highly effective analgesic and is nonaddictive; unfortunately however, like morphine, it depresses respiration.

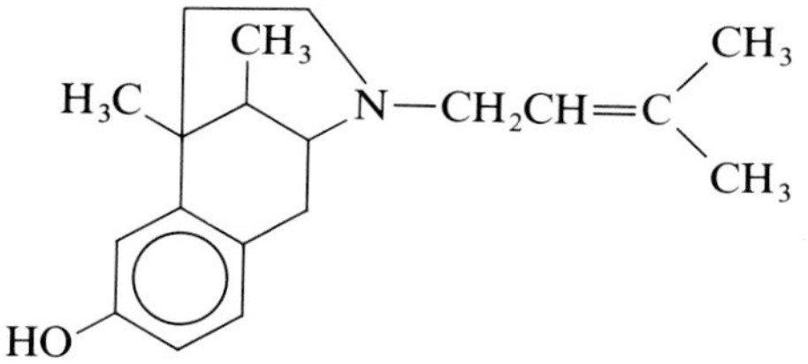

Pentazocine

PROBLEM K.6

Papaverine has been synthesized by the following route:

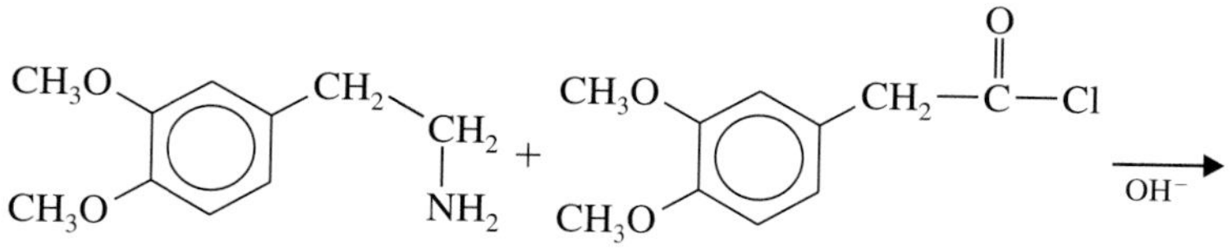

$$C_{20}H_{25}NO_5 \xrightarrow[\substack{\text{heat} \\ (-\ H_2O)}]{P_4O_{10}} \text{dihydropapaverine} \xrightarrow[\substack{\text{heat} \\ (-\ H_2)}]{\text{Pd}} \text{papaverine}$$

Outline the reactions involved.

PROBLEM K.7

One of the important steps in the synthesis of morphine involved the following transformation:

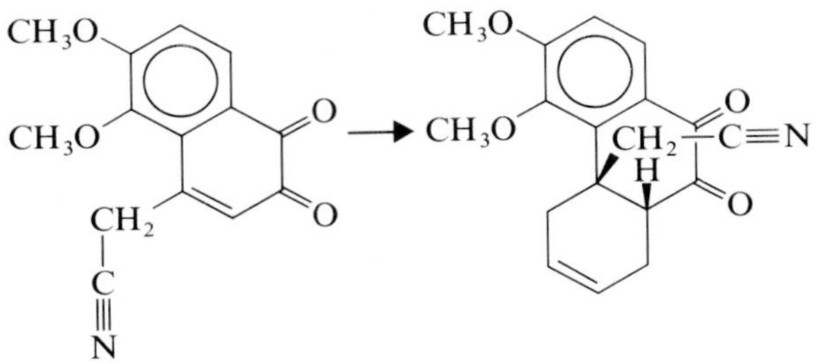

Suggest how this step was accomplished.

PROBLEM K.8

When morphine reacts with 2 mol of acetic anhydride, it is transformed into the highly addictive narcotic, heroin. What is the structure of heroin?

K.3 Alkaloids Containing Indole or Reduced Indole Rings

A large number of alkaloids are derivatives of an indole ring system. These range from the relatively simple *gramine* to the highly complicated structures of *strychnine* and *reserpine.*

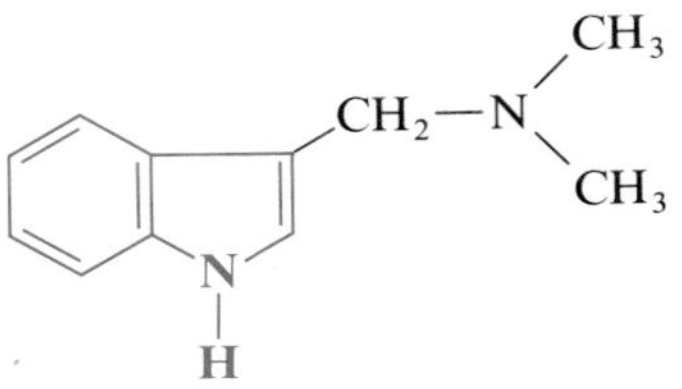

Gramine

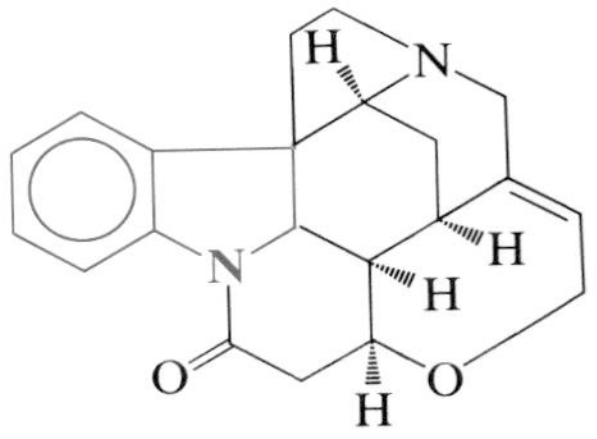

Strychnine

Reserpine

Gramine can be obtained from chlorophyll-deficient mutants of barley. Strych-nine, a very bitter and highly poisonous compound, comes from the seeds of *Strychnos nux-vomica.* Strychnine is a central nervous system stimulant and has been used medically (in low dosage) to counteract poisoning by central nervous system depressants. Reserpine can be obtained from the Indian snakeroot *Rauwolfia serpen-tina,* a plant that has been used in native medicine for centuries. Reserpine is used in modern medicine as a tranquilizer and as an agent to lower blood pressure.

PROBLEM K.9

Gramine has been synthesized by heating a mixture of indole, formaldehyde, and dimethylamine. (a) What general reaction is involved here? (b) Outline a rea-sonable mechanism for the gramine synthesis.

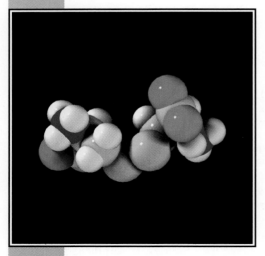

Cystine

THIOLS, THIOETHERS, AND THIOPHENOLS

Sulfur is directly below oxygen in Group **VI** of the periodic table and, as we might expect, there are sulfur counterparts of the oxygen compounds that we studied in Chapters 10 and 21 such as thiols, thioethers, and thiophenols.

These and other important examples of organosulfur compounds are the following:

R—SH	R—S—R′	ArSH	R—S—S—R′	$R—\overset{\overset{\displaystyle R'}{\vert}}{S}{\pm}R''$
Thiols	**Thioethers**	**Thiophenols**	**Disulfides**	**Trialkylsulfonium ions**

$R—\overset{\overset{\displaystyle O}{\Vert}}{S}—R'$	$R—\overset{\overset{\displaystyle O}{\Vert}}{\underset{\underset{\displaystyle O}{\Vert}}{S}}—R'$	$R—\overset{\overset{\displaystyle S}{\Vert}}{C}—R'$	$R—\overset{\overset{\displaystyle O}{\Vert}}{S}—OH$	$R—\overset{\overset{\displaystyle O}{\Vert}}{\underset{\underset{\displaystyle O}{\Vert}}{S}}—OH$
Sulfoxides	**Sulfones**	**Thioketones**	**Sulfinic acids**	**Sulfonic acids**

The sulfur counterpart of an alcohol is called a *thiol* or a *mercaptan.* The name mercaptan comes from the Latin, *mercurium captans,* meaning "capturing mercury." Mercaptans react with mercuric ions and the ions of other heavy metals to form precipitates. The compound CH_2CHCH_2OH, known as British Anti-Lewisite (BAL),

$$\underset{\underset{\displaystyle SH}{\vert}}{CH_2}\underset{\underset{\displaystyle SH}{\vert}}{CH}CH_2OH$$

was developed as an antidote for poisonous arsenic compounds used as war gases. British Anti-Lewisite is also an effective antidote for mercury poisoning.

Several simple thiols are shown below.

CH_3CH_2SH
Ethanethiol
(ethyl mercaptan)

$CH_3CH_2CH_2SH$
1-Propanethiol
(propyl mercaptan)

$$\overset{\displaystyle CH_3}{\underset{\displaystyle}{|}}$$
$CH_3CHCH_2CH_2SH$
3-Methyl-1-butanethiol
(isopentyl mercaptan)

$CH_2=CHCH_2SH$
2-Propene-1-thiol
(allyl mercaptan)

Compounds of sulfur, in general, and the low molecular weight thiols, in particular, are noted for their disagreeable odors. Anyone who has passed anywhere near a general chemistry laboratory when hydrogen sulfide (H_2S) was being used has noticed the strong odor of that substance—the odor of rotten eggs. Another sulfur compound, 3-methyl-1-butanethiol, is one unpleasant constituent of the liquid that skunks use as a defensive weapon. 1-Propanethiol evolves from freshly chopped onions, and allyl mercaptan is one of the compounds responsible for the odor and flavor of garlic.

Aside from their odors, analogous sulfur and oxygen compounds show other chemical differences. These arise largely from the following features of sulfur compounds.

1. The sulfur atom is larger and more polarizable than the oxygen atom. As a result, sulfur compounds are more powerful nucleophiles and compounds containing —SH groups are stronger acids than their oxygen analogs. The ethanethiolate ion ($CH_3CH_2\ddot{\underset{..}{S}}:^-$), for example, is a much stronger nucleophile when it reacts at carbon atoms than is the ethoxide ion ($CH_3CH_2\ddot{\underset{..}{O}}:^-$). On the other hand, since ethanol is a weaker acid than ethanethiol, the ethoxide ion is the stronger of the two conjugate bases.

2. The bond dissociation energy of the S—H bond of thiols (~ 80 kcal mol^{-1}) is much less than that of the O—H bond of alcohols (~ 100 kcal mol^{-1}). The weakness of the S—H bond allows thiols to undergo an oxidative coupling reaction when they react with mild oxidizing agents; the product is a disulfide:

$$2\ RS-H + H_2O_2 \longrightarrow RS-SR + 2\ H_2O$$
$$\text{A thiol} \qquad\qquad\qquad \text{A disulfide}$$

Alcohols do not undergo an analogous reaction. When alcohols are treated with oxidizing agents, oxidation takes place at the weaker C—H (~ 85 kcal mol^{-1}) bond rather than at the strong O—H bond.

3. Because sulfur atoms are easily polarized they can stabilize a negative charge on an adjacent atom. This means that hydrogen atoms on carbon atoms that are adjacent to an alkylthio group are more acidic than those adjacent to an alkoxy group. Thioanisole, for example, reacts with butyllithium in the following way:

$\langle\!\!\!\bigcirc\!\!\!\rangle$—$SCH_3 + C_4H_9:^-\ Li^+ \longrightarrow \langle\!\!\!\bigcirc\!\!\!\rangle$—$SCH_2:^-\ Li^+ + C_4H_{10}$

Thioanisole

Anisole ($CH_3OC_6H_5$) does not undergo an analogous reaction.

The $\diagdown\!\!S{=}O$ group of sulfoxides and the positive sulfur of sulfonium ions are even more effective in delocalizing negative charge on an adjacent atom:

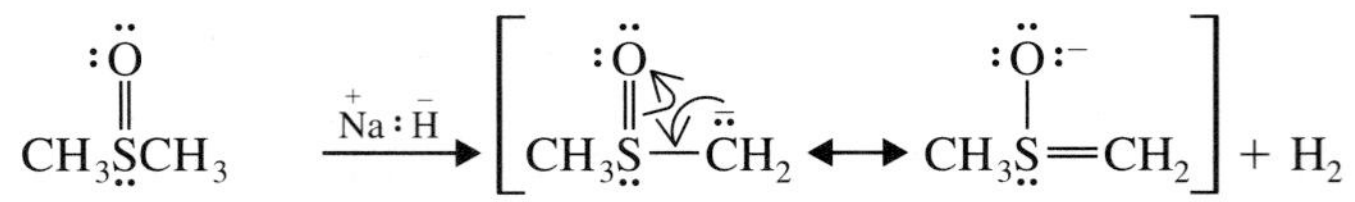

$$
\begin{array}{c}
\overset{:\ddot{O}:}{\underset{}{\overset{\|}{CH_3\ddot{S}CH_3}}} \xrightarrow{\overset{+}{Na}:\bar{H}} \left[\overset{:\ddot{O}:}{\underset{}{\overset{\|}{CH_3\ddot{S}\!-\!\ddot{C}H_2}}} \longleftrightarrow \overset{:\ddot{O}:^-}{\underset{}{\overset{|}{CH_3\ddot{S}{=}CH_2}}} \right] + H_2
\end{array}
$$

Dimethyl sulfoxide

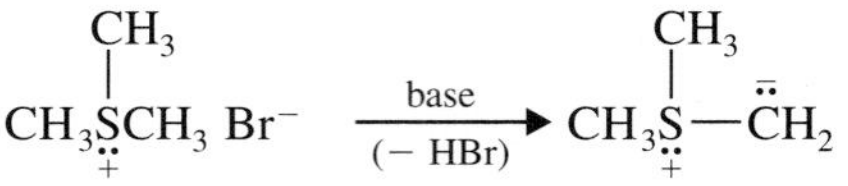

$$
\underset{\overset{+}{CH_3\ddot{S}CH_3}\ Br^-}{\overset{\overset{CH_3}{|}}{}} \xrightarrow[(-HBr)]{\text{base}} \underset{\overset{+}{CH_3\ddot{S}{-}\ddot{C}H_2}}{\overset{\overset{CH_3}{|}}{}}
$$

**Trimethylsulfonium   An ylide
bromide**

The anions formed in the reactions just given are of synthetic use. They can be used to synthesize epoxides, for example (cf. Section L.3).

L.1 Preparation of Thiols

Alkyl bromides and iodides react with potassium hydrogen sulfide to form thiols. (Potassium hydrogen sulfide can be generated by passing gaseous H_2S into an alcoholic solution of potassium hydroxide.)

$$
R{-}Br + KOH + \underset{\textbf{(excess)}}{H_2S} \xrightarrow[\text{heat}]{C_2H_5OH} R{-}SH + KBr + H_2O
$$

The thiol that forms is sufficiently acidic to form a thiolate ion in the presence of potassium hydroxide. Thus, if excess H_2S is not employed in the reaction, the major product of the reaction will be a thioether. The thioether results from the following reactions:

$$
R{-}SH + KOH \longrightarrow R{-}\ddot{S}:^- K^+ + H_2O
$$

$$
R{-}\ddot{S}:^- K^+ + R{-}\ddot{B}r: \longrightarrow R{-}\ddot{S}{-}R + KBr
$$

Thioether

Alkyl halides also react with thiourea to form (stable) *S*-alkylisothiouronium salts. These can be used to prepare thiols.

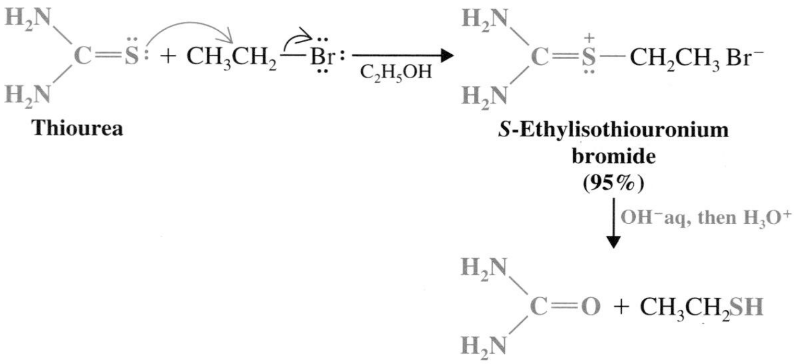

$$
\underset{\underset{H_2N}{H_2N}\diagdown}{C}{=}\ddot{S}: + CH_3CH_2{-}\ddot{B}r: \xrightarrow{C_2H_5OH} \underset{\underset{H_2N}{H_2N}\diagdown}{C}{=}\overset{+}{\ddot{S}}{-}CH_2CH_3\ Br^-
$$

Thiourea   ***S*-Ethylisothiouronium
bromide
(95%)**

$$
\Big\downarrow \text{OH}^-\text{aq, then }H_3O^+
$$

$$
\underset{\underset{H_2N}{H_2N}\diagdown}{C}{=}O + CH_3CH_2SH
$$

**Urea   Ethanethiol
(90%)**

L.2 Physical Properties of Thiols

Thiols form very weak hydrogen bonds; their hydrogen bonds are not nearly as strong as those of alcohols. Because of this, low molecular weight thiols have lower boiling points than corresponding alcohols. Ethanethiol, for example, boils more than 40°C lower than ethanol (37 vs 78°C). The relative weakness of hydrogen bonds between molecules of thiols is also evident when we compare the boiling points of ethanethiol and its isomer dimethyl sulfide:

$$CH_3CH_2SH \qquad CH_3SCH_3$$
$$\textbf{bp, 37°C} \qquad \textbf{bp, 38°C}$$

Physical properties of several thiols are given in Table L.1.

L.3 The Addition of Sulfur Ylides to Aldehydes and Ketones

Sulfur ylides also react as nucleophiles at the carbonyl carbon of aldehydes and ketones. The betaine that forms usually decomposes to an *epoxide* rather than to an alkene.

Trimethylsulfonium iodide

$$\left[(CH_3)_2\overset{+}{S}-\overset{..}{\overset{-}{C}}H_2 \longleftrightarrow (CH_3)_2S=CH_2 \right]$$

Resonance-stabilized sulfur ylide

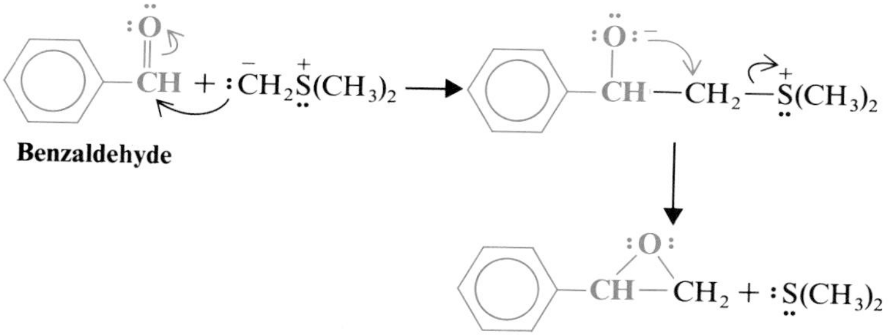

Benzaldehyde

(75%)

TABLE L.1 *Physical properties of thiols*

COMPOUND	STRUCTURE	mp (°C)	bp (°C)
Methanethiol	CH_3SH	−123	6
Ethanethiol	CH_3CH_2SH	−144	37
1-Propanethiol	$CH_3CH_2CH_2SH$	−113	67
2-Propanethiol	$(CH_3)_2CHSH$	−131	58
1-Butanethiol	$CH_3(CH_2)_2CH_2SH$	−116	98

PROBLEM L.1

Show how you might use a sulfur ylide to prepare

(a) 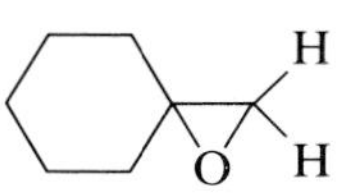 [structure: cyclohexane fused to epoxide with CH bearing H, H]

(b) 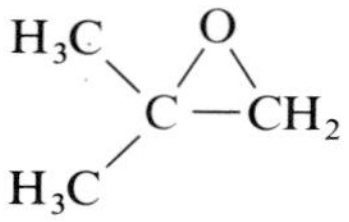 [structure: H_3C — C(—CH_3)(=O epoxide)—CH_2]

L.4 Thiols and Disulfides in Biochemistry

Thiols and disulfides are important compounds in living cells, and in many biochemical oxidation–reduction reactions they are interconverted.

$$2\ RSH \underset{[H]}{\overset{[O]}{\rightleftharpoons}} R\!-\!S\!-\!S\!-\!R$$

Lipoic acid, for example, an important cofactor in biological oxidations, undergoes this oxidation–reduction reaction.

[structure: S—S ring with —$(CH_2)_4CO_2H$] $\underset{[O]}{\overset{[H]}{\rightleftharpoons}}$ [structure: SH SH (H H / S S) with —$(CH_2)_4CO_2H$]

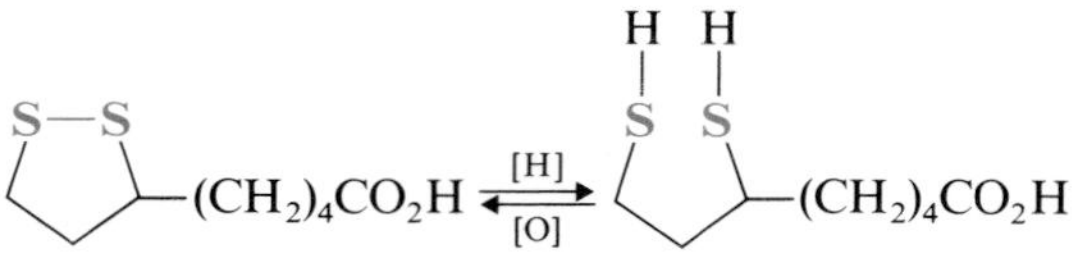

Lipoic acid   **Dihydrolipoic acid**

The amino acids *cysteine* and *cystine* are interconverted in a similar way:

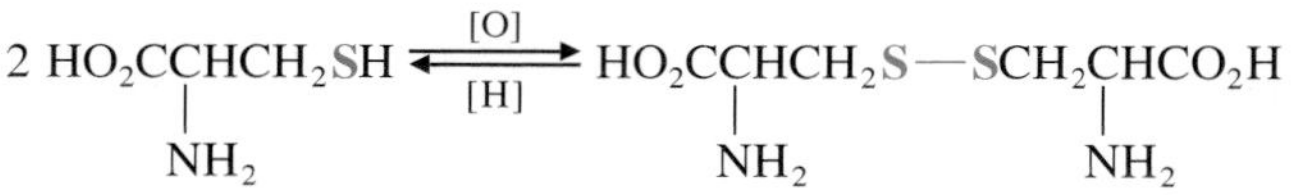

$$2\ HO_2CCHCH_2SH \underset{[H]}{\overset{[O]}{\rightleftharpoons}} HO_2CCHCH_2S\!-\!SCH_2CHCO_2H$$

with NH_2 groups below; **Cysteine**   **Cystine**

As we shall see in Chapter 24, the disulfide linkages of cystine units are important in determining the overall shapes of protein molecules.

PROBLEM L.2

Give structures for the products of the following reactions:

(a) Benzyl bromide + thiourea $\longrightarrow$

(b) Product of (a) + OH⁻/H_2O, then H⁺ $\longrightarrow$

(c) Product of (b) + H_2O_2 $\longrightarrow$

(d) Product of (b) + NaOH $\longrightarrow$

(e) Product of (d) + benzyl bromide $\longrightarrow$

L.2 Physical Properties of Thiols

Thiols form very weak hydrogen bonds; their hydrogen bonds are not nearly as strong as those of alcohols. Because of this, low molecular weight thiols have lower boiling points than corresponding alcohols. Ethanethiol, for example, boils more than 40°C lower than ethanol (37 vs 78°C). The relative weakness of hydrogen bonds between molecules of thiols is also evident when we compare the boiling points of ethanethiol and its isomer dimethyl sulfide:

CH_3CH_2SH CH_3SCH_3
bp, 37°C   **bp, 38°C**

Physical properties of several thiols are given in Table L.1.

L.3 The Addition of Sulfur Ylides to Aldehydes and Ketones

Sulfur ylides also react as nucleophiles at the carbonyl carbon of aldehydes and ketones. The betaine that forms usually decomposes to an *epoxide* rather than to an alkene.

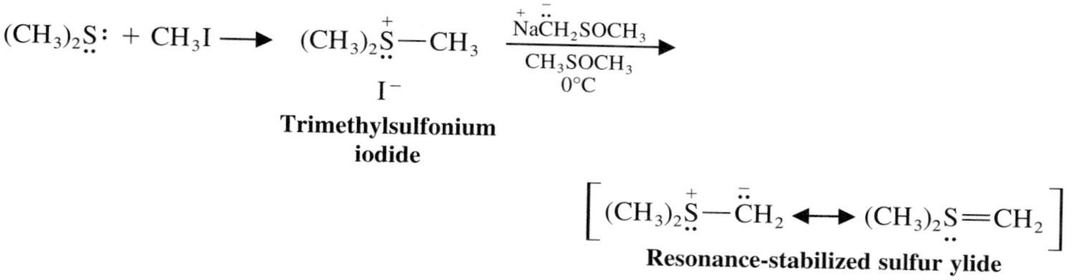

$$(CH_3)_2\overset{..}{\underset{..}{S}}: + CH_3I \longrightarrow (CH_3)_2\overset{+}{\underset{..}{S}}{-}CH_3 \quad \underset{\underset{0°C}{CH_3SOCH_3}}{\xrightarrow{Na\overset{..}{\overset{-}{C}}H_2SOCH_3}}$$

$$\underset{\textbf{iodide}}{\underset{\textbf{Trimethylsulfonium}}{I^-}}$$

$$\left[(CH_3)_2\overset{+}{\underset{..}{S}}{-}\overset{..}{\overset{-}{C}}H_2 \longleftrightarrow (CH_3)_2\underset{..}{S}{=}CH_2 \right]$$

Resonance-stabilized sulfur ylide

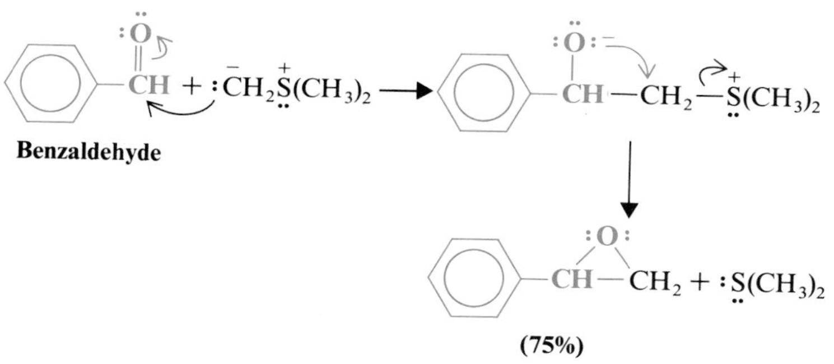

Benzaldehyde

(75%)

TABLE L.1 *Physical properties of thiols*

COMPOUND	STRUCTURE	mp (°C)	bp (°C)
Methanethiol	CH_3SH	−123	6
Ethanethiol	CH_3CH_2SH	−144	37
1-Propanethiol	$CH_3CH_2CH_2SH$	−113	67
2-Propanethiol	$(CH_3)_2CHSH$	−131	58
1-Butanethiol	$CH_3(CH_2)_2CH_2SH$	−116	98

PROBLEM L.1

Show how you might use a sulfur ylide to prepare

(a) 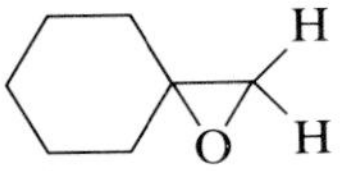   (b)

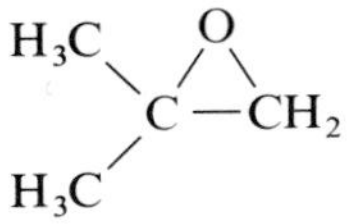

L.4 Thiols and Disulfides in Biochemistry

Thiols and disulfides are important compounds in living cells, and in many biochemical oxidation–reduction reactions they are interconverted.

$$2 \text{ RSH} \underset{[H]}{\overset{[O]}{\rightleftarrows}} R-S-S-R$$

Lipoic acid, for example, an important cofactor in biological oxidations, undergoes this oxidation–reduction reaction.

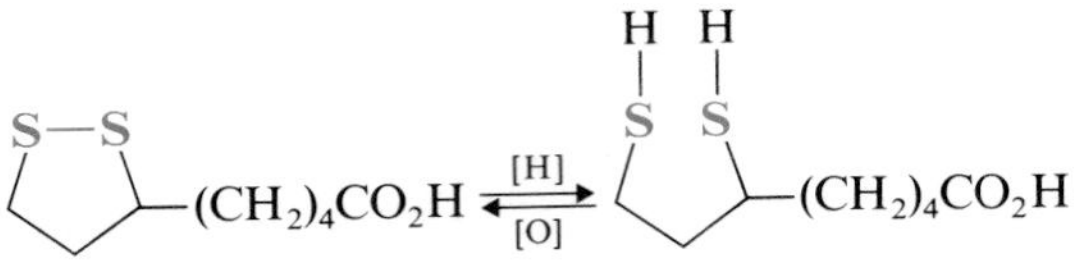

Lipoic acid   **Dihydrolipoic acid**

The amino acids *cysteine* and *cystine* are interconverted in a similar way:

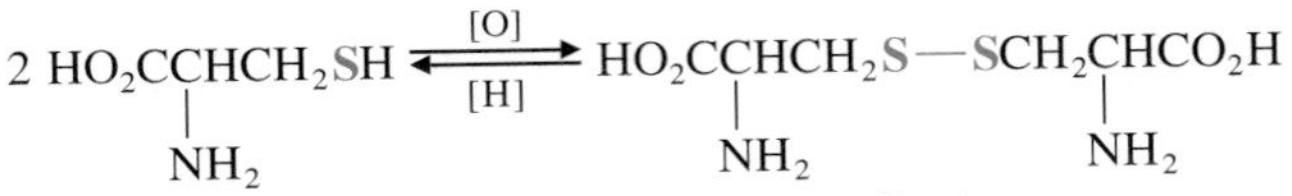

$$2 \text{ HO}_2\text{CCHCH}_2\text{SH} \underset{[H]}{\overset{[O]}{\rightleftarrows}} \text{HO}_2\text{CCHCH}_2\text{S}-\text{SCH}_2\text{CHCO}_2\text{H}$$

NH₂	NH₂	NH₂

Cysteine   **Cystine**

As we shall see in Chapter 24, the disulfide linkages of cystine units are important in determining the overall shapes of protein molecules.

PROBLEM L.2

Give structures for the products of the following reactions:

(a) Benzyl bromide + thiourea ⟶
(b) Product of (a) + OH^-/H_2O, then H^+ ⟶
(c) Product of (b) + H_2O_2 ⟶
(d) Product of (b) + NaOH ⟶
(e) Product of (d) + benzyl bromide ⟶

PROBLEM L.3

Allyl disulfide, CH_2=$CHCH_2S$—SCH_2CH=CH_2, is another important component of oil of garlic. Suggest a synthesis of allyl disulfide starting with allyl bromide.

PROBLEM L.4

Starting with allyl alcohol, outline a synthesis of BAL, $CH_2SHCHSHCH_2OH$.

PROBLEM L.5

A synthesis of lipoic acid (see structure just given) is outlined here. Supply the missing reagents and intermediates.

$$Cl-\overset{\overset{O}{\|}}{C}(CH_2)_4CO_2C_2H_5 \xrightarrow[AlCl_3]{CH_2=CH_2} (a)\ C_{10}H_{17}ClO_3 \xrightarrow{NaBH_4}$$

$$\underset{\underset{OH}{|}}{ClCH_2CH_2CH}(CH_2)_4CO_2C_2H_5 \xrightarrow{(b)} \underset{\underset{Cl}{|}}{ClCH_2CH_2CH}(CH_2)_4CO_2C_2H_5 \xrightarrow[(d)]{(c)}$$

$$\underset{\underset{SCH_2C_6H_5}{|}}{C_6H_5CH_2SCH_2CH_2CH}(CH_2)_4CO_2H \xrightarrow[(2)\ H^+]{(1)\ Na,\ liq.\ NH_3}$$

$$(e)\ C_8H_{16}S_2O_2 \xrightarrow{O_2} lipoic\ acid$$

PROBLEM L.6

One chemical-warfare agent used in World War I is a powerful vesicant called "mustard gas." (The name comes from its mustardlike odor; mustard gas, however, is not a gas but a high-boiling liquid that was dispersed as a mist of tiny droplets.) Mustard gas can be synthesized from oxirane in the following manner. Outline the reactions involved.

$$2\ H_2C-CH_2 + H_2S \longrightarrow C_4H_{10}SO_2 \xrightarrow[ZnCl_2]{HCl} C_4H_8SCl_2$$
$$\underset{O}{\diagdown\diagup}$$
$$\text{"Mustard gas"}$$

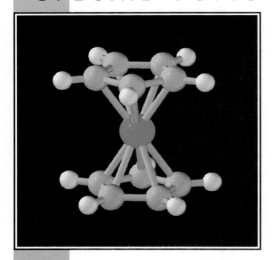

TRANSITION METAL ORGANOMETALLIC COMPOUNDS

M.1 Introduction

One of the most active areas of chemical research in recent years has involved studying compounds in which a bond exists between the carbon atom of an organic group and a transition metal. This field, which combines aspects of organic chemistry and inorganic chemistry, has led to many important applications in organic synthesis. Many of these transition metal organic compounds act as catalysts of extraordinary selectivity.

The transition metals are defined as those elements that have partly filled *d* (or *f*) shells, either in the elemental state or in their important compounds. The transition metals that are of most concern to organic chemists are those shown in the green and yellow portion of the periodic table given in Fig. M.1. Transition metals react with a variety of molecules or groups, called *ligands,* to form *transition metal complexes.* In forming a complex, the ligands donate electrons to vacant orbitals of the metal. The bonds between the ligand and the metal range from very weak to very strong. The bonds are covalent but often have considerable polar character.

Transition metal complexes can assume a variety of geometries depending on the metal and on the number of ligands around it. Rhodium can form complexes with four ligands, for example, that are *square planar.* On the other hand, rhodium can form complexes with five or six ligands that are trigonal bipyramidal or octahedral. These typical shapes are shown on the following page with the letter L used to indicate a ligand.

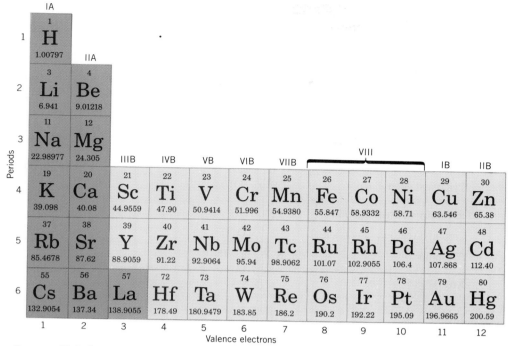

FIGURE M.1 Important transition elements are shown in the green and yellow portion of the periodic table. Given across the bottom is the total number of valence electrons (s and d) of each element.

M.2 Electron Counting. The 18-Electron Rule

Transition metals are like the elements that we have studied earlier in that they are most stable when they have the electronic configuration of a noble gas. In addition to s and p orbitals, transition metals have five d orbitals (which can hold a total of 10 electrons). Therefore, the noble gas configuration for a transition metal is *18 electrons,* not 8 as with carbon, nitrogen, oxygen, and so on. When the metal of a transition metal complex has 18 valence electrons, it is said to be *coordinatively saturated.**

To determine the valence electron count of a transition metal in a complex, we take the total number of valence electrons of the metal in the elemental state (see Fig. M.1) and subtract from this number the oxidation state of the metal in the com-

---

*We do not usually show the unshared electron pairs of a metal complex in our structures, because to do so would make the structure unnecessarily complicated.

TABLE M.1 Common ligands in transition metal complexes[a]

LIGAND	COUNT AS	NUMBER OF ELECTRONS DONATED
Negatively charged ligands		
H	H:$^-$	2
R	R:$^-$	2
X	X:$^-$	2
Allyl	⌒⌒ (−)	4
Cyclopentadienyl, Cp	(−) pentagon	6
Electrically neutral ligands		
Carbonyl (carbon monoxide)	CO	2
Phosphine	R$_3$P or Ph$_3$P	2
Alkene	$C{=}C$	2
Diene	∥ ⌢ ∖	4
Benzene	⬡	6

[a]Adapted from J. Schwartz and J. A. Labinger, *J. Chem. Ed., 57,* 170, 1980.

plex. This gives us what is called the *d* electron count, d^n. The oxidation state of the metal is the charge that would be left on the metal if all the ligands (Table M.1) were removed.

$$d^n = \begin{array}{l}\text{total number of valence electrons}\\ \text{of the elemental metal}\end{array} - \begin{array}{l}\text{oxidation state of}\\ \text{the metal in the complex}\end{array}$$

Then to get the total valence electron count of the metal *in the complex,* we add to d^n the number of electrons donated by all of the ligands. Table M.1 gives the number of electrons donated by several of the most common ligands.

$$\begin{array}{l}\text{total number of valence electrons}\\ \text{of the metal in the complex}\end{array} = d^n + \begin{array}{l}\text{electrons donated}\\ \text{by ligands}\end{array}$$

Let us now work out the valence electron count of two examples.

EXAMPLE A

Consider iron pentacarbonyl, Fe(CO)$_5$, a toxic liquid that forms when finely divided iron reacts with carbon monoxide.

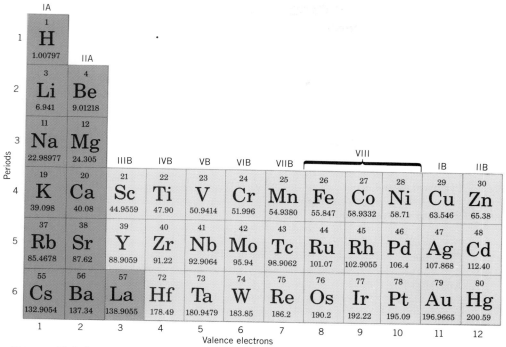

FIGURE M.1 Important transition elements are shown in the green and yellow portion of the periodic table. Given across the bottom is the total number of valence electrons (*s* and *d*) of each element.

Square planar
rhodium complex

Trigonal bipyramidal
rhodium complex

Octahedral
rhodium complex

M.2 Electron Counting. The 18-Electron Rule

Transition metals are like the elements that we have studied earlier in that they are most stable when they have the electronic configuration of a noble gas. In addition to *s* and *p* orbitals, transition metals have five *d* orbitals (which can hold a total of 10 electrons). Therefore, the noble gas configuration for a transition metal is *18 electrons,* not 8 as with carbon, nitrogen, oxygen, and so on. When the metal of a transition metal complex has 18 valence electrons, it is said to be *coordinatively saturated.**

To determine the valence electron count of a transition metal in a complex, we take the total number of valence electrons of the metal in the elemental state (see Fig. M.1) and subtract from this number the oxidation state of the metal in the com-

---

*We do not usually show the unshared electron pairs of a metal complex in our structures, because to do so would make the structure unnecessarily complicated.

TABLE M.1   Common ligands in transition metal complexes[a]

LIGAND	COUNT AS	NUMBER OF ELECTRONS DONATED
Negatively charged ligands		
H	H:⁻	2
R	R:⁻	2
X	X:⁻	2
Allyl		4
Cyclopentadienyl, Cp		6
Electrically neutral ligands		
Carbonyl (carbon monoxide)	CO	2
Phosphine	R₃P or Ph₃P	2
Alkene	>C=C<	2
Diene		4
Benzene		6

[a]Adapted from J. Schwartz and J. A. Labinger, *J. Chem. Ed., 57,* 170, 1980.

plex. This gives us what is called the *d* electron count, d^n. The oxidation state of the metal is the charge that would be left on the metal if all the ligands (Table M.1) were removed.

$$d^n = \frac{\text{total number of valence electrons}}{\text{of the elemental metal}} - \frac{\text{oxidation state of}}{\text{the metal in the complex}}$$

Then to get the total valence electron count of the metal *in the complex,* we add to d^n the number of electrons donated by all of the ligands. Table M.1 gives the number of electrons donated by several of the most common ligands.

$$\frac{\text{total number of valence electrons}}{\text{of the metal in the complex}} = d^n + \frac{\text{electrons donated}}{\text{by ligands}}$$

Let us now work out the valence electron count of two examples.

EXAMPLE A

Consider iron pentacarbonyl, $Fe(CO)_5$, a toxic liquid that forms when finely divided iron reacts with carbon monoxide.

$$Fe + 5\ CO \longrightarrow Fe(CO)_5 \qquad \text{or}$$

Iron pentacarbonyl

From Fig. M.1 we find that an iron atom in the elemental state has 8 valence electrons. We arrive at the oxidation state of iron in iron pentacarbonyl by noting that the charge on the complex as a whole is zero (it is not an ion), and that the charge on each CO ligand is also zero. Therefore, the iron is in the zero oxidation state.

Using these numbers, we can now calculate d^n and, from it, the total number of valence electrons of the iron in the complex.

$$d^n = 8 - 0 = 8$$

$$\text{total number of valence electrons} = d^n + 5(CO) = 8 + 5(2) = 18$$

We find that the iron of $Fe(CO)_5$ has 18 valence electrons and is, therefore, coordinatively saturated.

EXAMPLE B

Consider the rhodium complex $Rh[(C_6H_5)_3P]_3H_2Cl$, a complex that, as we shall see later, is an intermediate in certain alkene hydrogenations.

$L = Ph_3P$ [i.e., $(C_6H_5)_3P$]

The oxidation state of rhodium in the complex is $+3$. (The two hydrogen atoms and the chlorine are each counted as -1, and the charge on each of the triphenylphosphine ligands is zero. Removing all the ligands would leave a Rh^{3+} ion.) From Fig. M.1 we find that in the elemental state, rhodium has 9 valence electrons. We can now calculate d^n for the rhodium of the complex.

$$d^n = 9 - 3 = 6$$

Each of the six ligands of the complex donates two electrons to the rhodium in the complex, and therefore, the total number of valence electrons of the rhodium is 18. The rhodium of $Rh[(C_6H_5)_3P]_3H_2Cl$ is coordinatively saturated.

$$\text{total number of valence electrons of rhodium} = d^n + 6(2) = 6 + 12 = 18$$

M.3 Metallocenes: Organometallic Sandwich Compounds

Cyclopentadiene reacts with phenylmagnesium bromide to give the Grignard reagent of cyclopentadiene. This reaction is not unusual for it is simply another acid–base reaction like those we saw earlier. The methylene hydrogen atoms of cyclopentadiene are much more acidic than the hydrogen atoms of benzene and, therefore, the reaction goes to completion. (The methylene hydrogen atoms of cyclopentadiene are acidic relative to ordinary methylene hydrogen atoms because the cyclopentadienyl anion is aromatic; cf. Section 14.7B.)

$$\text{Cyclopentadiene} + C_6H_5MgBr \xrightarrow{\text{diethyl ether}} \text{Cyclopentadienylmagnesium bromide} + C_6H_6$$

Cyclopentadiene   Phenylmagnesium bromide   Cyclopenta-dienylmagnesium bromide   Benzene

When the Grignard reagent of cyclopentadiene is treated with ferrous chloride, a reaction takes place that produces a product called *ferrocene*.

$$2\ \overset{2+}{\underset{}{\bigcirc}}^{-}\ MgBr + FeCl_2 \longrightarrow (C_5H_5)_2Fe + 2\ MgBrCl$$

Ferrocene
(71% overall yield
from cyclopentadiene)

Ferrocene is an orange solid with a melting point of 174°C. It is a highly stable compound; ferrocene can be sublimed at 100°C and is not damaged when heated to 400°C.

Many studies, including X-ray analysis, show that ferrocene is a compound in which the iron(II) ion is located between two cyclopentadienyl rings.

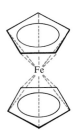

The carbon–carbon bond distances are all 1.40 Å and the carbon–iron bond distances are all 2.04 Å. Because of their structures, molecules such as ferrocene have been called "sandwich" compounds.

The carbon–iron bonding in ferrocene results from overlap between the inner lobes of the *p* orbitals of the cyclopentadienyl anions and 3*d* orbitals of the iron atom. Studies have shown, moreover, that this bonding is such that the rings of ferrocene are capable of essentially free rotation about an axis that passes through the iron atom and that is perpendicular to the rings.

The iron of ferrocene has 18 valence electrons and is, therefore, coordinatively saturated. We calculate this number as follows:

Iron has 8 valence electrons in the elemental state and its oxidation state in ferrocene is +2. Therefore, $d^n = 6$.

$$d^n = 8 - 2 = 6$$

Each cyclopentadienyl (Cp) ligand of ferrocene donates 6 electrons to the iron. Therefore, for the iron, the valence electron count is 18.

$$\text{total number of valence electrons} = d^n + 2(Cp) = 6 + 2(6) = 18$$

Ferrocene is an *aromatic compound.* It undergoes a number of electrophilic aromatic substitutions, including sulfonation and Friedel–Crafts acylation.

The discovery of ferrocene (in 1951) was followed by the preparation of a number of similar aromatic compounds. These compounds, as a class, are called *metallocenes.* * Metallocenes with five-, six-, seven-, and even eight-membered rings have been synthesized from metals as diverse as zirconium, manganese, cobalt, nickel, chromium, and uranium.

''Half-sandwich'' compounds have been prepared through the use of metal carbonyls. Several are shown here.

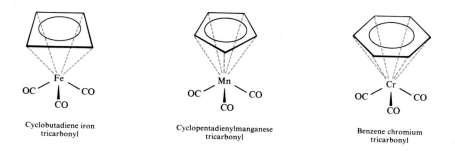

Cyclobutadiene iron tricarbonyl

Cyclopentadienylmanganese tricarbonyl

Benzene chromium tricarbonyl

Although cyclobutadiene itself is *not* stable, the cyclobutadiene iron tricarbonyl is.

PROBLEM M.1

The metal of each of the previously given half-sandwich compounds is coordinatively saturated. Show that this is true by working out the valence electron count for the metal in each complex.

M.4 Reactions of Transition Metal Complexes

Much of the chemistry of organic transition metal compounds becomes more understandable if we are able to follow the mechanisms of the reactions that occur. These mechanisms, in most cases, amount to nothing more than a sequence of reactions, each of which represents *a fundamental reaction type that is characteristic of a transition metal complex.* Let us examine three of the fundamental reaction types now. In each instance we shall use steps that occur when an alkene is hydrogenated using a catalyst called Wilkinson's catalyst. Later (in Section M.5) we shall examine the entire hydrogenation mechanism.

* Ernst O. Fischer (of the Technical University, Munich) and Geoffrey Wilkinson (of Imperial College, London) received the Nobel Prize in 1973 for their pioneering work (performed independently) on the chemistry of organometallic sandwich compounds—or metallocenes.

1. **Ligand Dissociation–Association (Ligand Exchange).** A transition metal complex can lose a ligand (by dissociation) and combine with another ligand (by association). In the process it undergoes *ligand exchange*. For example, the rhodium complex that we encountered in Example B can react with an alkene (in this example, with ethene) as follows:

$$\text{H}-\overset{\overset{\text{H}}{|}}{\underset{\underset{\text{Cl}}{|}}{\text{Rh}}}-\text{L} + \text{CH}_2{=}\text{CH}_2 \rightleftharpoons \text{H}-\overset{\overset{\text{H}}{|}}{\underset{\underset{\text{Cl}}{|}}{\text{Rh}}} \leftarrow \overset{\text{CH}_2}{\underset{\text{CH}_2}{\|}} + \text{L}$$

$$\text{L} = \text{Ph}_3 \text{ [i.e., } (\text{C}_6\text{H}_5)_3\text{P]}$$

Two steps are actually involved. In the first step, one of the triphenylphosphine ligands dissociates. This leads to a complex in which the rhodium has only 16 electrons and is, therefore, coordinatively *unsaturated.*

$$\text{H}-\overset{\overset{\text{H}}{|}}{\underset{\underset{\text{Cl}}{|}}{\text{Rh}}}-\text{L} \rightleftharpoons \text{H}-\overset{\overset{\text{H}}{|}}{\underset{\underset{\text{Cl}}{|}}{\text{Rh}}} + \text{L}$$

(18 electrons)   **(16 electrons)**

$$\text{L} = \text{Ph}_3\text{P}$$

In the second step, the rhodium associates with the alkene to become coordinatively saturated again.

$$\text{H}-\overset{\overset{\text{H}}{|}}{\underset{\underset{\text{Cl}}{|}}{\text{Rh}}} + \text{CH}_2{=}\text{CH}_2 \rightleftharpoons \text{H}-\overset{\overset{\text{H}}{|}}{\underset{\underset{\text{Cl}}{|}}{\text{Rh}}} \leftarrow \overset{\text{CH}_2}{\underset{\text{CH}_2}{\|}}$$

(16 electrons)   **(18 electrons)**

The complex between the rhodium and the alkene is called a π *complex.* In it, two electrons are donated by the alkene to the rhodium. Alkenes are often called π donors to distinguish them from σ donors such as $\text{Ph}_3\text{P}\!:\,$, Cl^-, and so on.

    In a π complex such as the one just given, there is also a donation of electrons from a populated d orbital of the metal back to the vacant π^* orbital of the alkene. This kind of donation is called ''back-bonding.''

2. **Insertion–Deinsertion.** An unsaturated ligand such as an alkene can undergo *insertion* into a bond between the metal of a complex and a hydrogen or a carbon. These reactions are reversible, and the reverse reaction is called *deinsertion.*

    The following is an example of insertion–deinsertion.

$$\text{H}-\overset{\overset{\text{H}}{|}}{\underset{\underset{\text{Cl}}{|}}{\text{Rh}}} \leftarrow \overset{\text{CH}_2}{\underset{\text{CH}_2}{\|}} \underset{\text{deinsertion}}{\overset{\text{insertion}}{\rightleftharpoons}} \text{H}-\overset{\overset{\overset{\overset{\text{CH}_3}{|}}{\text{CH}_2}}{|}}{\underset{\underset{\text{Cl}}{|}}{\text{Rh}}}$$

(18 electrons)   **(16 electrons)**

In this process, a π bond (between the rhodium and the alkene) and a σ bond (between the rhodium and the hydrogen) are exchanged for a new σ bond (between rhodium and carbon). The valence electron count of the rhodium decreases from 18 to 16.

This insertion–deinsertion occurs in a stereospecific way, as a *syn addition* of the M—H unit to the alkene.

$$\text{C=C} + \text{M—H} \rightleftharpoons \text{C—C}$$

3. Oxidative Addition–Reductive Elimination. Coordinatively unsaturated metal complexes can undergo oxidative addition of a variety of substrates in the following way.*

$$\ddot{M} + A—B \xrightarrow{\text{oxidative addition}} —M—B$$

The substrates, A—B, can be H—H, H—X, R—X, RCO—H, RCO—X, and a number of other compounds.

In this type of oxidative addition, the metal of the complex undergoes an increase in the number of its valence electrons *and in its oxidation state*. Consider, as an example, the oxidative addition of hydrogen to the rhodium complex that follows (L = Ph_3P).

$$\text{L—Rh—Cl (with L's)} + H—H \underset{\text{reductive elimination}}{\overset{\text{oxidative addition}}{\rightleftharpoons}} \text{H—Rh—H}$$

(16 electrons, Rh is in +1 oxidation state)   **(18 electrons, Rh is in +3 oxidation state)**

Reductive elimination is the reverse of oxidative addition. With this background, we are now in a position to examine a few interesting applications of transition metal complexes in organic synthesis.

M.5 Homogeneous Hydrogenation

Until now, all of the hydrogenations that we have examined have been heterogeneous processes. Two phases have been involved in the reaction: the solid phase of the catalyst (Pt, Pd, Ni, etc.), containing the adsorbed hydrogen, and the liquid phase of the solution, containing the unsaturated compound. In homogeneous hydrogenation using a transition metal complex such as $Rh[(C_6H_5)_3P]_3Cl$ (called Wilkinson's catalyst), hydrogenation takes place *in a single phase*—in solution.

When Wilkinson's catalyst is used to carry out the hydrogenation of an alkene, the following steps take place (L = Ph_3P).

Step 1 $\text{L—Rh—Cl (with L's)} + H—H \longrightarrow \text{L—Rh—H (with L's, H)}$ **Oxidative addition**

(16 electrons, Rh^I)   **(18 electrons, Rh^{III})**

*Coordinatively saturated complexes also undergo oxidative addition.

Step 2

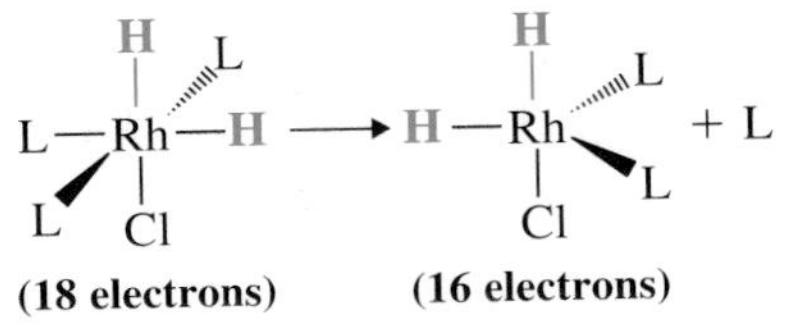

$$H-Rh-H \longrightarrow H-Rh + L$$

Ligand dissociation

(18 electrons) (16 electrons)

Step 3

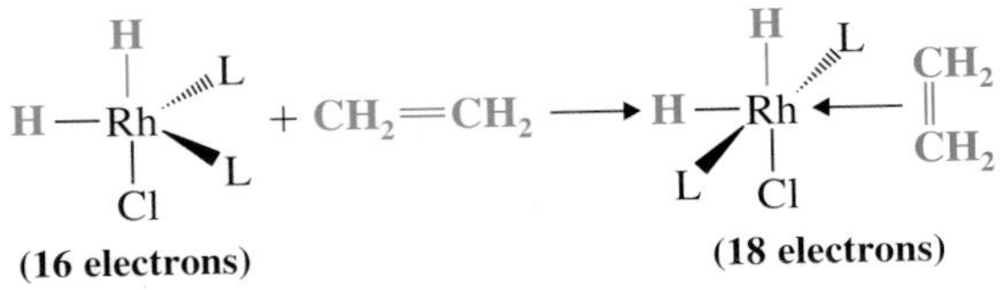

$$H-Rh + CH_2=CH_2 \longrightarrow H-Rh \leftarrow \begin{matrix} CH_2 \\ \parallel \\ CH_2 \end{matrix}$$

Ligand association

(16 electrons) (18 electrons)

Step 4

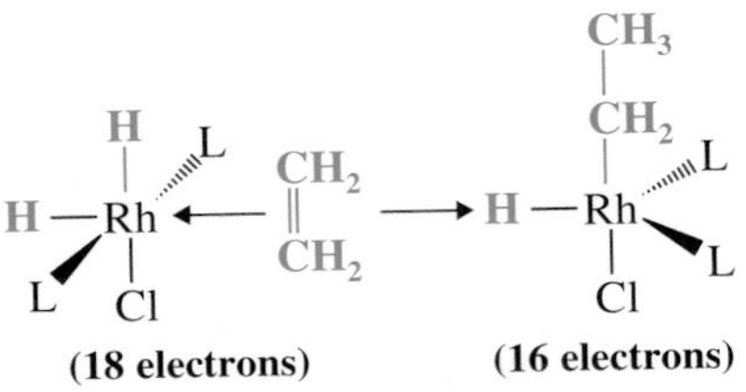

$$H-Rh \leftarrow \begin{matrix} CH_2 \\ \parallel \\ CH_2 \end{matrix} \longrightarrow H-Rh$$

Insertion

(18 electrons) (16 electrons)

Step 5

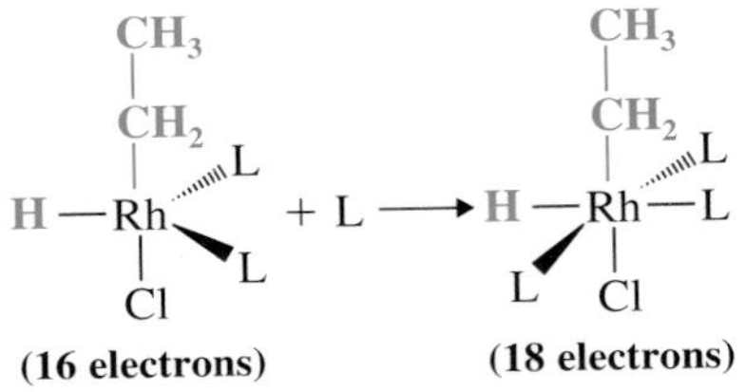

$$H-Rh + L \longrightarrow H-Rh-L$$

Ligand association

(16 electrons) (18 electrons)

Step 6

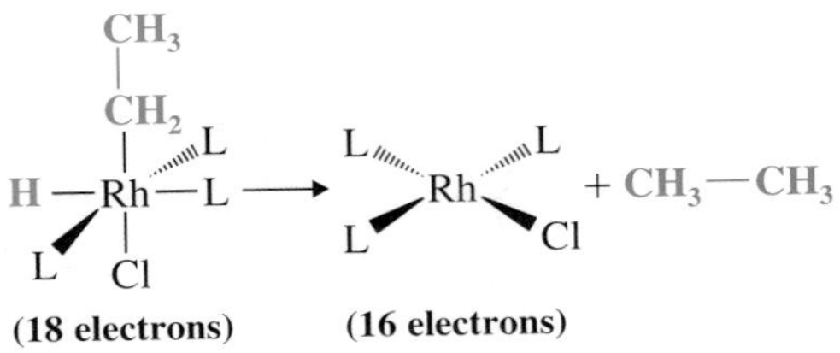

$$H-Rh-L \longrightarrow Rh + CH_3-CH_3$$

Reductive elimination

(18 electrons) (16 electrons)

Then steps 1, 2, 3, 4, 5, 6, and so on.

Step 6 regenerates the catalyst, which can then cause hydrogenation of another molecule of the alkene.

Because the insertion step 4 and the reductive elimination step 6 are stereospecific, the net result of hydrogenation using Wilkinson's catalyst is a *syn addition* of hydrogen to the alkene. The following example (with D_2 in place of H_2) illustrates this aspect.

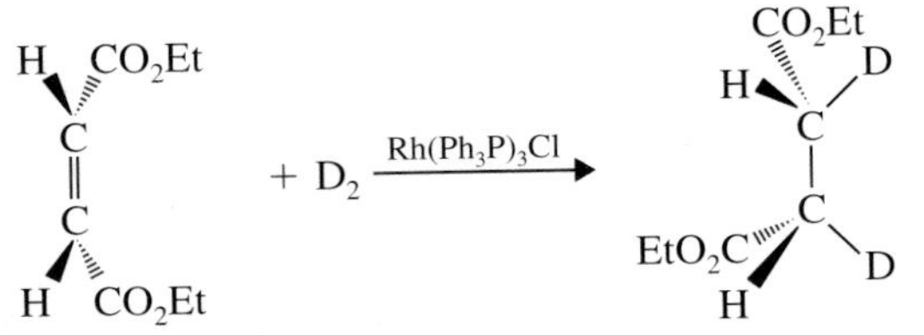

$$\begin{matrix} H & CO_2Et \\ & C \\ & \parallel \\ & C \\ H & CO_2Et \end{matrix} + D_2 \xrightarrow{Rh(Ph_3P)_3Cl} \begin{matrix} CO_2Et \\ H-C-D \\ EtO_2C-C-D \\ H \end{matrix}$$

A *cis*-alkene
(diethyl maleate)

A meso compound

PROBLEM M.2

What product (or products) would be formed if the *trans* alkene corresponding to the *cis* alkene (see previous reaction) had been hydrogenated with D_2 and Wilkinson's catalyst?

M.6 Carbon–Carbon Bond Forming Reactions Using Rhodium Complexes

Rhodium complexes have also been used to synthesize compounds in which the formation of a carbon–carbon bond is required. An example is the synthesis that follows:

$$(Ph_3P)_3RhCl + CH_3Li \xrightarrow[\text{exchange}]{\text{ligand}} (Ph_3P)_3RhCH_3 + LiCl$$

The first step, *a ligand exchange,* occurs by a combination of ligand association–dissociation steps and incorporates the methyl group into the coordination sphere of the rhodium. The next step, *an oxidative addition,* incorporates the phenyl group into the rhodium coordination sphere. Then, in the last step, *a reductive elimination* joins the methyl group and the benzene ring to form toluene.

PROBLEM M.3

Give the total valence electron count for rhodium in each complex in the synthesis outlined previously.

Another example is the following ketone synthesis.

$$(Ph_3P)_2Rh(CO)Cl + CH_3Li \xrightarrow{(a)} (Ph_3P)_2Rh(CO)(CH_3) + LiCl$$
$$\textbf{1} \qquad\qquad\qquad\qquad\qquad \textbf{2}$$

PROBLEM M.4

Give the valence electron count and the oxidation state of rhodium in the complexes labeled **1, 2, 3**; then describe each step (a), (b), and (c) as to its fundamental type (oxidative addition, ligand exchange, etc.).

Still another carbon–carbon bond-forming reaction (below) illustrates the stereospecificity of these reactions.

PROBLEM M.5

Give, in detail, a possible mechanism for the synthesis just outlined, describing each step according to its fundamental type.

PROBLEM M.6

The actual mechanism of the Corey–Posner, Whitesides–House synthesis is not known with certainty. One possible mechanism involves the oxidative addition of R′—X or Ar—X to R$_2$CuLi followed by a reductive elimination to generate R—R′ or R—Ar. Outline the steps in this mechanism using $(CH_3)_2CuLi$ and C_6H_5I.

M.7 Vitamin B$_{12}$: A Transition Metal Biomolecule

The discovery (in 1926) that pernicious anemia can be overcome by the ingestion of large amounts of liver led ultimately to the isolation (in 1948) of the curative factor, called vitamin B$_{12}$. The complete three-dimensional structure of vitamin B$_{12}$ (Fig. M.2) was elucidated in 1956 through the X-ray studies of Dorothy Hodgkin (Nobel Prize, 1964), and in 1972 the synthesis of this complicated molecule was announced by R. B. Woodward (Harvard University) and A. Eschenmoser (Swiss Federal Institute of Technology). The synthesis took 11 years and involved more than 90 separate reactions. One hundred co-workers took part in the project.

Vitamin B$_{12}$ is the only known biomolecule that possesses a carbon–metal bond. In the stable commercial form of the vitamin, a cyano group is bonded to the cobalt, and the cobalt is in the +3 oxidation state. The core of the vitamin B$_{12}$ molecule is a

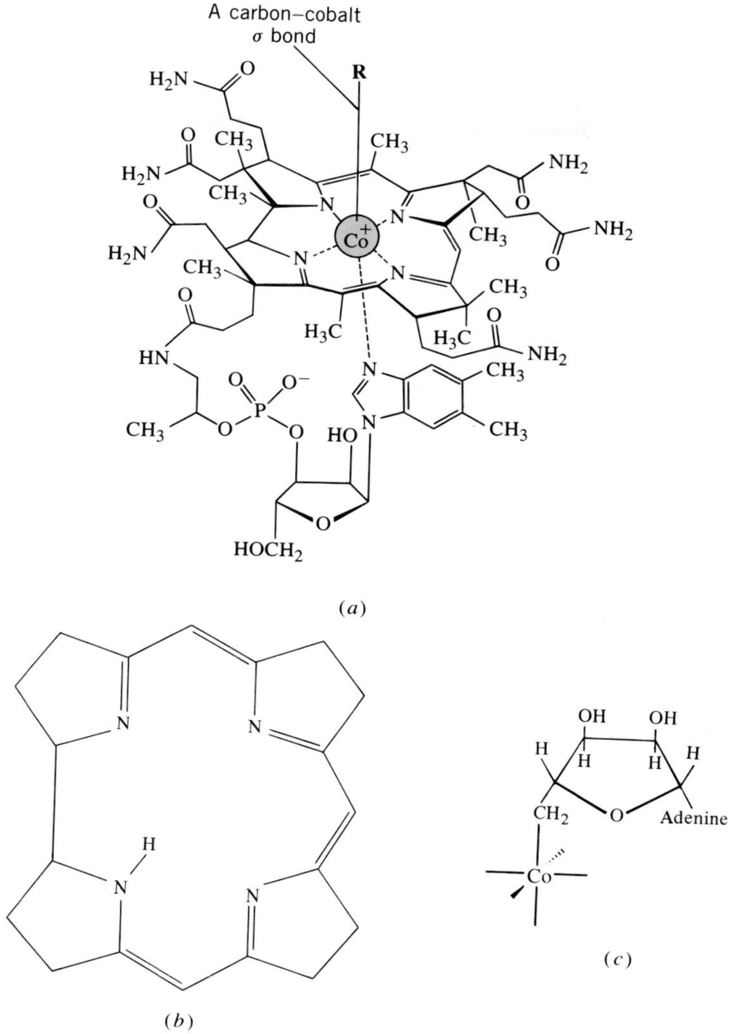

A carbon–cobalt
σ bond

(a)

(b)

(c)

FIGURE M.2 (a) The structure of vitamin B_{12}. In the commercial form of the vitamin (cyanocobalamin) R = CN. (b) The corrin ring system. (c) In the biologically active form of the vitamin (5′-deoxyadenosylcobalamin) the 5′-carbon atom of 5′-deoxyadenosine is coordinated to the cobalt atom. For the structure of adenine see Section 25.2.

corrin ring with various attached side groups. The corrin ring consists of four pyrrole subunits, the nitrogen of each of which is coordinated to the central cobalt. The sixth ligand (below the corrin ring in Fig. M.2) is a nitrogen of a heterocyclic molecule called 5,6-dimethylbenzimidazole.

The cobalt of vitamin B_{12} can be reduced to a $+ 2$ or a $+ 1$ oxidation state. When the cobalt is in the $+ 1$ oxidation state, vitamin B_{12} (called B_{12s}) becomes one of the most powerful nucleophiles known, being more nucleophilic than methanol by a factor of 10^{14}.

Acting as a nucleophile, vitamin B_{12s} reacts with adenosine triphosphate (Fig. 22.2) to yield the biologically active form of the vitamin (Fig. M.2c).

Organic Halides and Organometallic Compounds in the Environment

N.1 Organic Halides as Insecticides

Since the discovery of the insecticidal properties of DDT in 1942, vast quantities of chlorinated hydrocarbons have been sprayed over the surface of the earth in an effort to destroy insects. These efforts initially met with incredible success in ridding large areas of the earth of disease-carrying insects, particularly those of typhus and malaria. As time has passed, however, we have begun to understand that this prodigious use of chlorinated hydrocarbons has not been without harmful—indeed tragic—side effects. Chlorinated hydrocarbons are usually highly stable compounds and are only slowly destroyed by natural processes in the environment. As a result, many chloroorganic insecticides remain in the environment for years. These persistent pesticides are called "hard" pesticides.

Chlorohydrocarbons are also fat soluble and tend to accumulate in the fatty tissues of most animals. The food chain that runs from plankton to small fish to larger fish to birds and to larger animals, including humans, tends to magnify the concentrations of chloroorganic compounds at each step.

The chlorohydrocarbon DDT is prepared from inexpensive starting materials, chlorobenzene and trichloroacetaldehyde. The reaction is catalyzed by acid.

$$2\ Cl\text{—}\bigcirc + H\overset{O}{\overset{\|}{C}}CCl_3 \xrightarrow{H_2SO_4} Cl\text{—}\bigcirc\text{—}\underset{\underset{CCl_3}{|}}{CH}\text{—}\bigcirc\text{—}Cl$$

DDT
[1,1,1-Trichloro-2,2-
bis(*p*-chlorophenyl)ethane]

In nature the principal decomposition product of DDT is DDE.

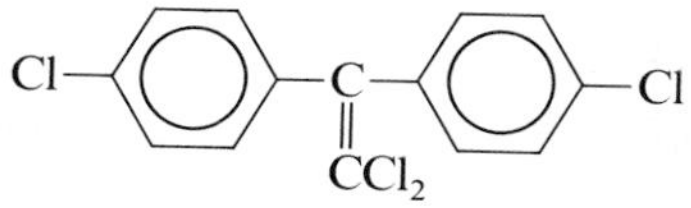

DDE
[1,1-Dichloro-2,2-
bis(*p*-chlorophenyl)ethene]

Estimates indicate that nearly 1 billion lb of DDE were spread throughout the world ecosystem. One pronounced environmental effect of DDE has been in its action on eggshell formation of many birds. DDE inhibits the enzyme *carbonic anhydrase* that controls the calcium supply for shell formation. As a consequence, the shells are often very fragile and do not survive to the time of hatching. During the late 1940s the populations of eagles, falcons, and hawks dropped dramatically. There can be little doubt that DDE was primarily responsible.

DDE also accumulates in the fatty tissues of humans. Although humans appear to have a short-range tolerance to moderate DDE levels, the long-range effects are far from certain.

Other hard insecticides are aldrin, dieldrin, and chlordan. Aldrin can be manufactured through the Diels–Alder reaction of hexachlorocyclopentadiene and norbornadiene.

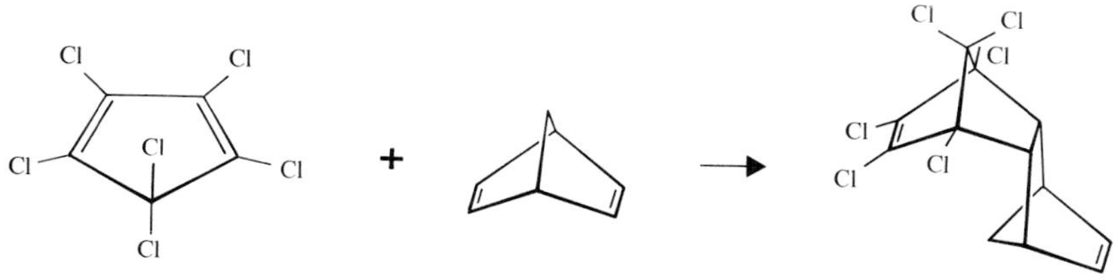

Hexachloro-
cyclopentadiene

Norbornadiene

Aldrin

Chlordan can be made by adding chlorine to the unsubstituted double bond of the Diels–Alder adduct obtained from hexachlorocyclopentadiene and cyclopentadiene. Dieldrin can be made by converting an aldrin double bond to an epoxide. (This reaction also takes place in nature.)

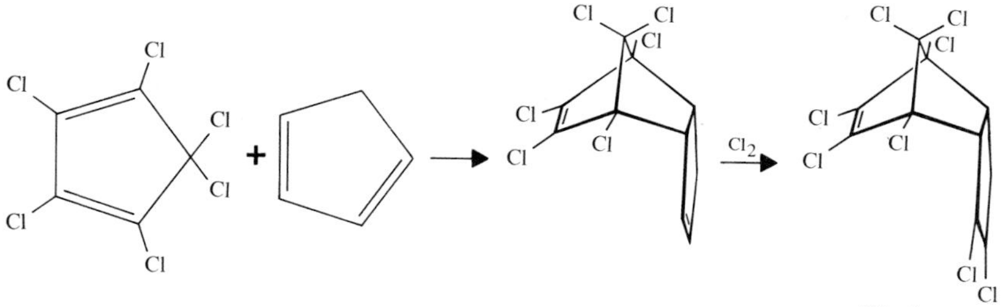

Chlordan

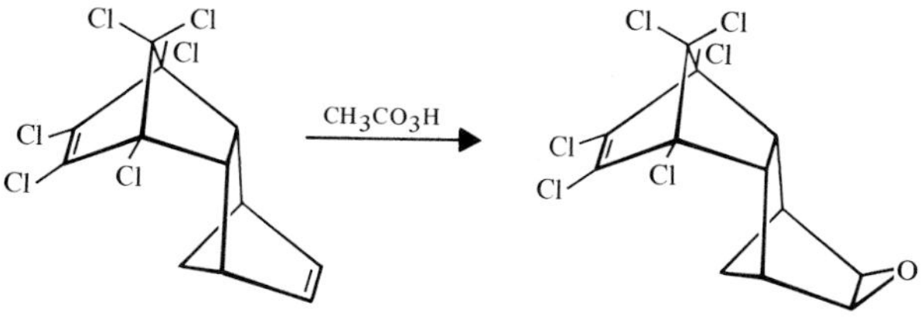

Aldrin

CH_3CO_3H

Dieldrin

During the 1970s the Environmental Protection Agency (EPA) banned the use of DDT, aldrin, dieldrin, and chlordan because of known or suspected hazards to human life. All of the compounds are suspected of causing cancers.

PROBLEM N.1

The mechanism for the formation of DDT from chlorobenzene and trichloro-acetaldehyde in sulfuric acid involves two electrophilic aromatic substitution reactions. In the first electrophilic substitution reaction, the electrophile is pro-tonated trichloroacetaldehyde. In the second, the electrophile is a carbocation. Propose a mechanism for the formation of DDT.

PROBLEM N.2

What kind of reaction is involved in the conversion of DDT to DDE?

Mirex, kepone, and lindane are also hard insecticides whose use has been banned.

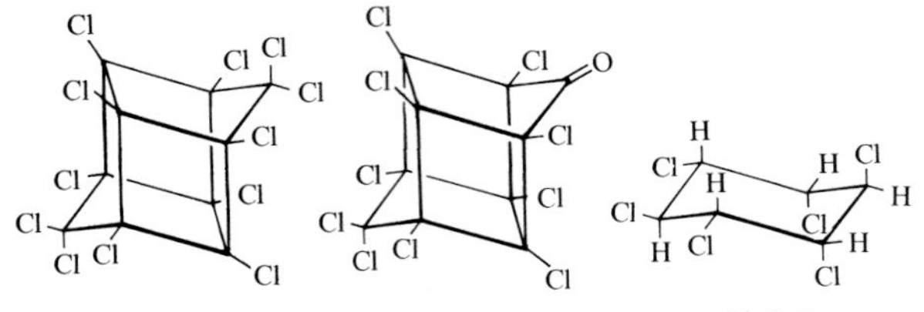

Mirex   Kepone   Lindane

N.2 Organic Halides as Herbicides

Other chlorinated organic compounds have been used extensively as herbicides. The following two examples are 2,4-D and 2,4,5-T.

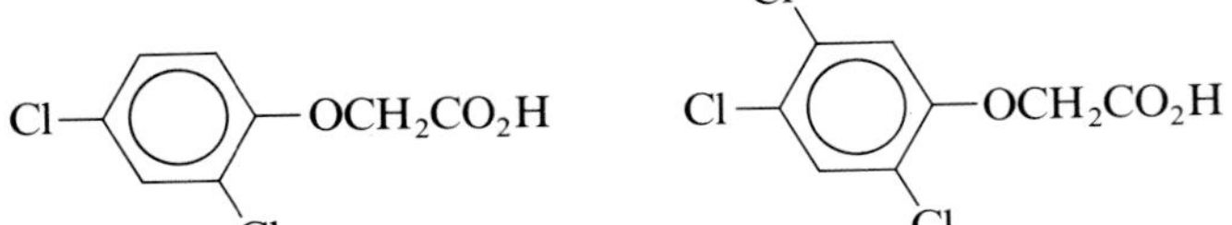

2,4-D
(2,4-Dichlorophenoxy-
acetic acid)

2,4,5-T
(2,4,5-Trichlorophenoxy-
acetic acid)

Enormous quantities of these two compounds were used as defoliants in the jun-gles of Indochina during the Vietnam war. Some samples of 2,4,5-T have been shown to be a teratogen (a fetus-deforming agent). This teratogenic effect was the result of an impurity present in commercial 2,4,5-T, the compound 2,3,7,8-tetrachlorodibenzo-dioxin. 2,3,7,8-Tetrachlorodibenzodioxin is also highly toxic; it is more toxic, for example, than cyanide ion, strychnine, and the nerve gases.

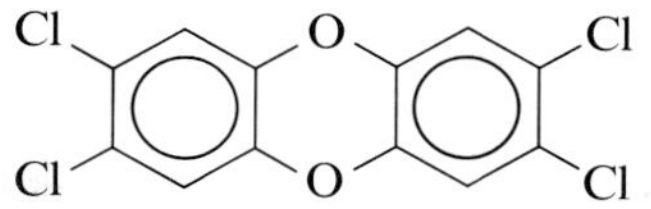

2,3,7,8-Tetrachlorodibenzodioxin
(also called TCDD)

This dioxin is also highly stable; it persists in the environment and because of its fat solubility can be passed up the food chain. In sublethal amounts it can cause a disfiguring skin disease called chloracne.

In July 1976 an explosion at a chemical plant in Seveso, Italy, caused the release of between 22 and 132 lb of this dioxin into the atmosphere. The plant was engaged in the manufacture of 2,4,5-trichlorophenol (used in making 2,4,5-T) using the following method:

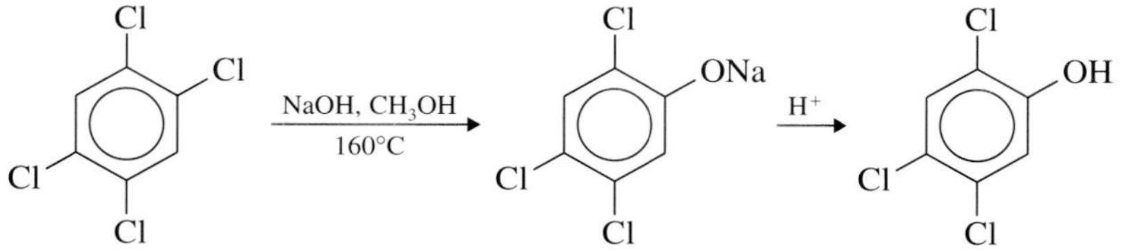

| **1,2,4,5-** | **Sodium 2,4,5-** | **2,4,5-** |
| **Tetrachlorobenzene** | **trichlorophenoxide** | **Trichlorophenol** |

The temperature of the first reaction must be very carefully controlled; if it is not, this dioxin forms in the reaction mixture:

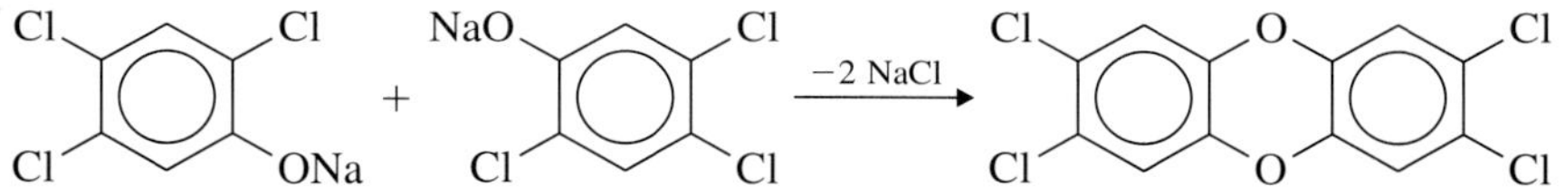

Apparently at the Italian factory, the temperature got out of control, causing the pressure to build up. Eventually, a valve opened and released a cloud of trichlorophenol and the dioxin into the atmosphere. Many wild and domestic animals were killed and many people, especially children, were afflicted with severe skin rashes.

PROBLEM N.3

(a) Assume that the ortho and para chlorine atoms provide enough activation by electron withdrawal for nucleophilic substitution to occur by an addition–elimination pathway and outline a possible mechanism for the conversion of 1,2,4,5-tetrachlorobenzene to sodium 2,4,5-trichlorophenoxide. (b) Do the same for the conversion of 2,4,5-trichlorophenoxide to the dioxin of Section N.2.

PROBLEM N.4

2,4,5-T is made by allowing sodium 2,4,5-trichlorophenoxide to react with sodium chloroacetate ($ClCH_2COONa$). (This produces the sodium salt of 2,4,5-T that, on acidification, gives 2,4,5-T itself.) What kind of mechanism accounts for the reaction of sodium 2,4,5-trichlorophenoxide with $ClCH_2COONa$? Write the equation.

N.3 Germicides

2,4,5-Trichlorophenol is also used in the manufacture of hexachlorophene, a germicide once widely used in soaps, shampoos, deodorants, mouthwashes, aftershave lotions, and other over-the-counter products.

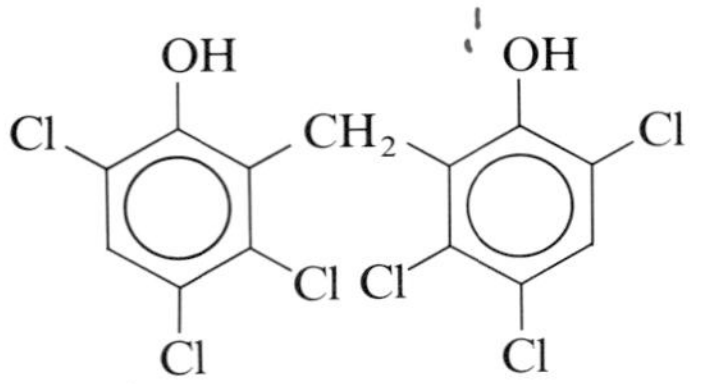

Hexachlorophene

Hexachlorophene is absorbed intact through the skin and tests with experimental animals have shown that it causes brain damage. Since 1972, the use of hexachlorophene in cleansers and cosmetics sold over the counter has been banned by the Food and Drug Administration.

N.4 Polychlorinated Biphenyls (PCBs)

Mixtures of polychlorinated biphenyls have been produced and used commercially since 1929. In these mixtures, biphenyls with chlorine atoms at any of the numbered positions (see following structure) may be present. In all, there are 210 possible compounds. A typical commercial mixture may contain as many as 50 different PCBs. Mixtures are usually classified on the basis of their chlorine content, and most industrial mixtures contain from 40 to 60% chlorine.

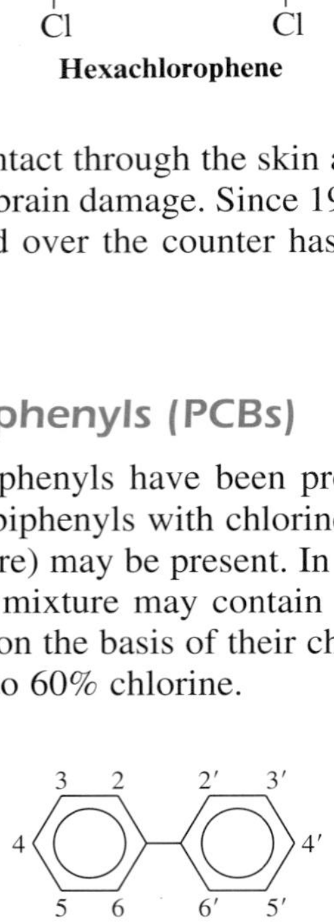

Biphenyl

Polychlorinated biphenyls have had a multitude of uses: as heat-exchange agents in transformers; in capacitors, thermostats, and hydraulic systems; as plasticizers in polystyrene coffee cups, frozen food bags, bread wrappers, and plastic liners for baby bottles. They have been used in printing inks, in carbonless carbon paper, and as waxes for making molds for metal castings. Between 1929 and 1972, about 500,000 metric tons of PCBs were manufactured.

Although they were never intended for release into the environment, PCBs have become, perhaps more than any other chemical, the most widespread pollutant. They have been found in rainwater, in many species of fish, birds, and other animals (including polar bears) all over the globe, and in human tissue.

Polychlorinated biphenyls are highly persistent and, being fat soluble, tend to accumulate in the food chain. Fish that feed in PCB-contaminated waters, for example, have PCB levels 1000–100,000 times the level of the surrounding water and this amount is further magnified in birds that feed on the fish. The toxicity of PCBs depends on the composition of the individual mixture. The largest incident of human poisoning by PCBs occurred in Japan in 1968 when about 1000 people ingested a cooking oil accidentally contaminated with PCBs.

As late as 1975, industrial concerns were legally discharging PCBs into the Hudson River. In 1977, the EPA banned the direct discharge into waterways and, since 1979, their manufacture, processing, and distribution have been prohibited.

N.5 Polybromobiphenyls (PBBs)

Polybromobiphenyls are bromine analogs of PCBs that have been used as flame retardants. In 1973, in Michigan, a mistake at a chemical company led to PBBs being mixed into animal feeds that were sold to farmers. Before the mistake was recognized, PBBs had affected thousands of dairy cattle, hogs, chickens, and sheep, necessitating their destruction.

N.6 Organometallic Compounds

With few exceptions, organometallic compounds are toxic. This toxicity varies greatly depending on the nature of the organometallic compound and the identity of the metal. Organic compounds of arsenic, antimony, lead, thallium, and mercury are toxic because the metal ions, themselves, are toxic. Certain organic derivatives of silicon are toxic even though silicon and most of its inorganic compounds are nontoxic.

Early in this century the recognition of the biocidal effects of organoarsenic compounds led Paul Ehrlich to his pioneering work in chemotherapy. Ehrlich sought compounds (which he called ''magic bullets'') that would show greater toxicity toward disease-causing microorganisms than they would toward their hosts. Ehrlich's research led to the development of Salvarsan and Neosalvarsan, two organoarsenic compounds that were used successfully in the treatment of diseases caused by spirochetes (e.g., syphilis) and trypanosomes (e.g., sleeping sickness). Salvarsan and Neosalvarsan are no longer used in the treatment of these diseases; they have been displaced by safer and more effective antibiotics. Ehrlich's research, however, initiated the field of chemotherapy (cf. Section 20.11).

Many microorganisms actually synthesize organometallic compounds, and this discovery has an alarming ecological aspect. Mercury metal is toxic, but mercury metal is also unreactive. In the past, untold tons of mercury metal present in industrial wastes have been disposed of by simply dumping such wastes into lakes and streams. Since mercury is toxic, many bacteria protect themselves from its effect by converting mercury metal to methylmercury ions (CH_3Hg^+) and to gaseous dimethylmercury $(CH_3)_2Hg$. These organic mercury compounds are passed up the food chain (with modification) through fish to humans, where methylmercury ions act as a deadly nerve poison. Between 1953 and 1964, 116 people in Minamata, Japan, were poisoned by eating fish containing methylmercury compounds. Arsenic is also methylated by organisms to the poisonous dimethylarsine, $(CH_3)_2AsH$.

Ironically, chlorinated hydrocarbons appear to inhibit the biological reactions that bring about mercury methylation. Lakes polluted with organochlorine pesticides show significantly lower mercury methylation. Whereas this particular interaction of two pollutants may, in a certain sense, be beneficial, it is also instructive of the complexity of the environmental problems that we face.

Tetraethyllead and other alkyllead compounds have been used as antiknock agents in gasoline since 1923. Although this use has now been phased out, more than 1 trillion lb of lead have been introduced into the atmosphere. In the northern hemisphere, gasoline burning alone has spread about 10 mg of lead on each square meter of the Earth's surface. In highly industrialized areas the amount of lead per square meter is probably several hundred times higher. Because of the well-known toxicity of lead, these facts are of great concern.

cis-Tetramethylcyclobutene

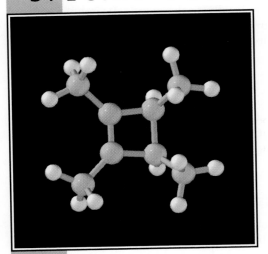

ELECTROCYCLIC AND CYCLOADDITION REACTIONS

O.1 Introduction

In recent years, chemists have found that there are many reactions in which certain symmetry characteristics of molecular orbitals control the overall course of the reaction. These reactions are often called *pericyclic reactions* because they take place through cyclic transition states. Now that we have a background knowledge of molecular orbital theory—especially as it applies to conjugated polyenes (dienes, trienes, etc.)—we are in a position to examine some of the intriguing aspects of these reactions. We shall look in detail at two basic types: *electrocyclic reactions* and *cycloaddition reactions.*

O.2 Electrocyclic Reactions

A number of reactions, like the one shown here, transform a conjugated polyene into a cyclic compound.

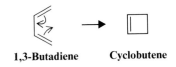

1,3-Butadiene   **Cyclobutene**

In many other reactions, the ring of a cyclic compound opens and a conjugated polyene forms.

Cyclobutene   1,3-Butadiene

Reactions of either type are called *electrocyclic reactions*.

In electrocyclic reactions, σ and π bonds are interconverted. In our first example, one π bond of 1,3-butadiene becomes a σ bond in cyclobutene. In our second example, the reverse is true; a σ bond of cyclobutene becomes a π bond in 1,3-butadiene.

Electrocyclic reactions have several characteristic features:

1. They require only heat or light for initiation.
2. Their mechanisms do not involve radical or ionic intermediates.
3. Bonds are made and broken in *a single concerted step involving a cyclic transition state*.
4. The reactions are *highly stereospecific*.

The examples that follow demonstrate this last characteristic of electrocyclic reactions.

trans,trans-2,4-Hexadiene   *cis*-3,4-Dimethylcyclobutene

trans,cis,trans-2,4,6-Octatriene   *cis*-5,6-Dimethyl-1,3-cyclohexadiene

In each of these examples, a single stereoisomeric form of the reactant yields a single stereoisomeric form of the product. The concerted photochemical cyclization of *trans,trans*-2,4-hexadiene, for example, yields only *cis*-3,4-dimethylcyclobutene; it does not yield *trans*-3,4-dimethylcyclobutene.

trans,trans-2,4-Hexadiene   *trans*-3,4-Dimethylcyclobutene   (not formed)

The other two concerted reactions are characterized by the same stereospecificity.

The electrocyclic reactions that we shall study here and the concerted cyclo-addition reactions that we shall study in the next section were poorly understood by chemists before 1960. In the years that followed, several scientists, most notably K. Fukui in Japan, H. C. Longuet-Higgins in England, and R. B. Woodward and R. Hoffmann in the United States provided us with a basis for understanding how these reactions occur and why they take place with such remarkable stereospecificity.*

All of these men worked from molecular orbital theory. In 1965, Woodward and Hoffmann formulated their theoretical insights into a set of rules that not only enabled chemists to understand reactions that were already known but that correctly predicted the outcome of many reactions that had not been attempted.

The Woodward–Hoffmann rules are formulated for concerted reactions only. Concerted reactions are reactions in which bonds are broken and formed simulta-neously and, thus, no intermediates occur. The Woodward–Hoffmann rules are based on this hypothesis: *In concerted reactions molecular orbitals of the reactant are con-tinuously converted into molecular orbitals of the product.* This conversion of molec-ular orbitals is not a random one, however. Molecular orbitals have symmetry charac-teristics. Because they do, restrictions exist on which molecular orbitals of the reactant may be transformed into particular molecular orbitals of the product.

According to Woodward and Hoffmann, certain reaction paths are said to be *symmetry allowed,* whereas others are said to be *symmetry forbidden.* To say that a particular path is symmetry forbidden does not necessarily mean, however, that the reaction will not occur. It simply means that if the reaction were to occur through a symmetry-forbidden path, the concerted reaction would have a much higher free en-ergy of activation. The reaction may occur, but it will probably do so in a different way: through another path that is symmetry allowed or through a nonconcerted path.

A complete analysis of electrocyclic reactions using the Woodward–Hoffmann rules requires a correlation of symmetry characteristics of *all* of the molecular orbitals of the reactants and product. Such analyses are beyond the scope of our discussion here. We shall find, however, that a simplified approach can be undertaken, one that is easy to visualize and, at the same time, is accurate in most instances. In this simplified approach to electrocyclic reactions we focus our attention only on the *highest occupied molecular orbital (HOMO) of the conjugated polyene.* This approach is based on a method developed by Fukui called the *frontier orbital method.*

O.2A Electrocyclic Reactions of $4n$ π-Electron Systems

Let us begin with an analysis of the thermal interconversion of *cis*-3,4-dimethyl-cyclobutene and *cis,trans*-2,4-hexadiene shown here.

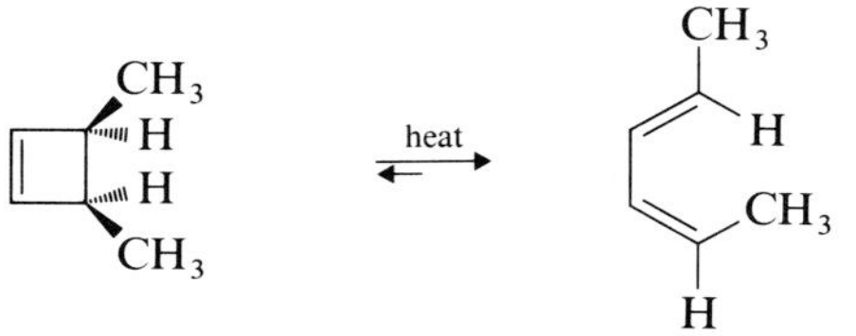

cis-3,4-Dimethylcyclobutene   *cis,trans*-2,4-Hexadiene

Electrocyclic reactions are reversible, and so the path for the forward reaction is the same as that for the reverse reaction. In this example it is easier to see what happens

---

* Hoffmann and Fukui were awarded the Nobel Prize in 1981 for this work.

to the orbitals if we follow the *cyclization* reaction, *cis,trans*-2,4-hexadiene ⟶ *cis*-3,4-dimethylcyclobutene.

In this cyclization one π bond of the hexadiene is transformed into a σ bond of the cyclobutene. But which π bond? And, how does the conversion occur?

Let us begin by examining the π molecular orbitals of 2,4-hexadiene and, in particular, let us look at *the HOMO of the ground state* (Fig. O.1).

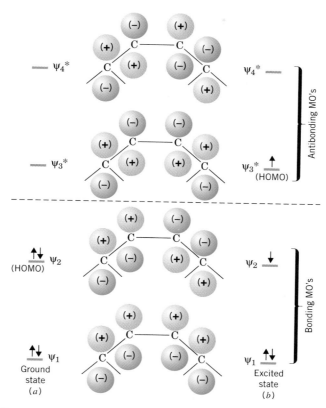

FIGURE O.1 The π molecular orbitals of a 2,4-hexadiene. (*a*) The electron distribution of the ground state. (*b*) The electronic distribution of the first excited state. (The first excited state is formed when the molecule absorbs a photon of light of the proper wavelength.) Notice that the orbitals of a 2,4-hexadiene are like those of 1,3-butadiene shown in Fig. 12.5.

The cyclization that we are concerned with now, *cis,trans*-2,4-hexadiene ⇌ *cis*-3,4-dimethylcyclobutene, requires heat alone. We conclude, therefore, that excited states of the hexadiene are not involved, for these would require the absorption of light. If we focus our attention on Ψ_2—the HOMO of the ground state—we can see how the *p* orbitals at C2 and C5 can be transformed into a σ bond in the cyclobutene.

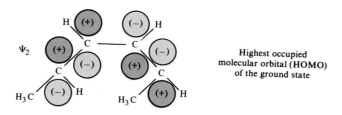

Highest occupied
molecular orbital (**HOMO**)
of the ground state

A bonding σ molecular orbital between C2 and C5 is formed when the *p* orbitals *rotate in the same direction* (both clockwise, as shown, or both counterclockwise, which leads to an equivalent result). The term *conrotatory* is used to describe this type of motion of the two *p* orbitals relative to each other.

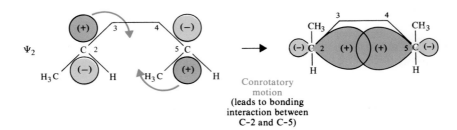

Conrotatory motion allows *p*-orbital lobes of the *same phase sign* to overlap. It also places the two methyl groups on the same side of the molecule in the product, that is, in the cis configuration.*

The pathway with conrotatory motion of the methyl groups is consistent with what we know from experiments to be true: The *thermal reaction* results in the interconversion of *cis*-3,4-dimethylcyclobutene and *cis,trans*-2,4-hexadiene.

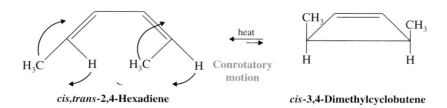

cis,trans-**2,4-Hexadiene**   *cis*-**3,4-Dimethylcyclobutene**

We can now examine another 2,4-hexadiene ⇌ 3,4-dimethylcyclobutene interconversion: one that takes place under the influence of light. This reaction is shown here.

---

* Notice that if conrotatory motion occurs in the opposite (counterclockwise) direction, lobes of the same phase sign still overlap, and the methyl groups are still cis.

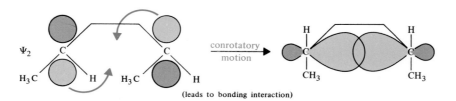

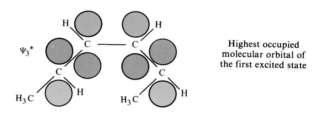

trans,trans-2,4-Hexadiene   Disrotatory motion   *cis*-3,4-Dimethylcyclobutene

In the photochemical reaction, *cis*-3,4-dimethylcyclobutene and *trans,trans*-2,4-hexa-diene are interconverted. The photochemical interconversion occurs with the methyl groups rotating in *opposite directions,* that is, with the methyl groups undergoing *disrotatory motion.*

    The photochemical reaction can also be understood by considering orbitals of the 2,4-hexadiene. In this reaction, however—since the absorption of light is in-volved—we want to look at the first *excited state* of the hexadiene. We want to examine Ψ_3^*, because in the first excited state Ψ_3^* is the *highest occupied molecular orbital.*

Ψ_3^*

Highest occupied molecular orbital of the first excited state

We find that disrotatory motion of the orbitals at C2 and C5 of Ψ_3^* allows lobes of the same sign to overlap and form a bonding sigma molecular orbital between them. Disrotatory motion of the orbitals, of course, also requires disrotatory motion of the methyl groups and, once again, this is consistent with what we find experimentally. The *photochemical reaction* results in the interconversion of *cis*-3,4-dimethylcyclobu-tene and *trans,trans*-2,4-hexadiene.

Ψ_3^*

Disrotatory motion (leads to bonding interaction between C-2 and C-5)

trans,trans-2,4-Hexadiene   *cis*-3,4-Dimethylcyclobutene

    Since both of the interconversions that we have presented so far involve *cis*-3,4-dimethylcyclobutene, we can summarize them in the following way:

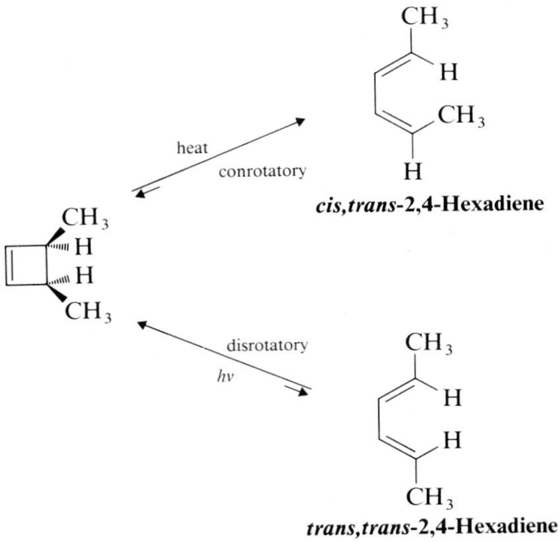

We see that these two interconversions occur with precisely opposite stereochemistry. We also see that the stereochemistry of the interconversions depends on whether the reaction is brought about by the application of heat or light.

The first Woodward–Hoffmann rule can be stated as follows:

1. **A thermal electrocyclic reaction involving $4n$ π electrons (where $n = 1, 2, 3, \ldots$) proceeds with conrotatory motion; the photochemical reaction proceeds with disrotatory motion.**

Both of the interconversions that we have studied involve systems of 4 π electrons and both follow this rule. Many other $4n$ π-electron systems have been studied since Woodward and Hoffmann stated their rule. Virtually all have been found to follow it.

PROBLEM O.1

What product would you expect from a concerted photochemical cyclization of *cis,trans*-2,4-hexadiene?

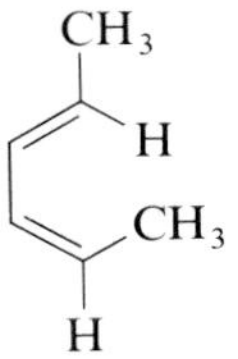

cis,trans-2,4-Hexadiene

PROBLEM O.2

(a) Show the orbitals involved in the following thermal electrocyclic reaction.

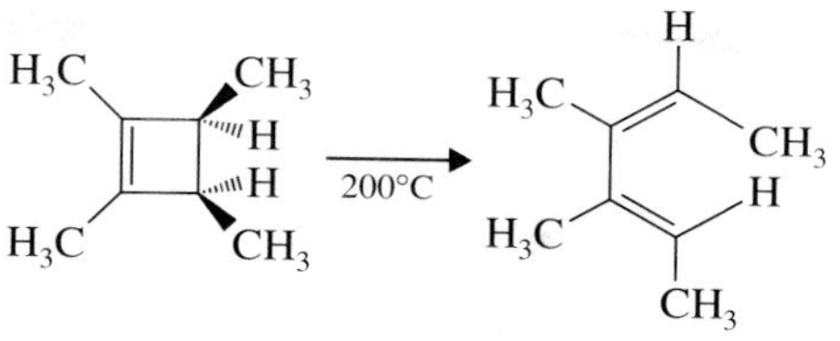

(b) Do the groups rotate in a conrotatory or disrotatory manner?

PROBLEM O.3

Can you suggest a method for carrying out a stereospecific conversion of trans,trans-2,4-hexadiene into cis,trans-2,4-hexadiene?

PROBLEM O.4

The following 2,4,6,8-decatetraenes undergo ring closure to dimethylcyclo-octatrienes when heated or irradiated. What product would you expect from each reaction?

(a) 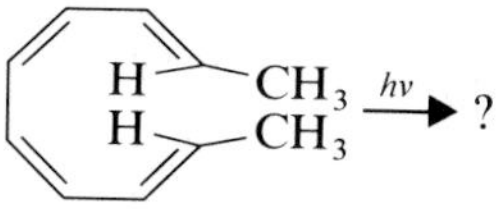   (b)

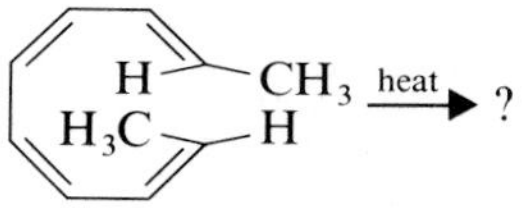

PROBLEM O.5

(a) For each of the following reactions, state whether conrotatory or disrotatory motion of the groups is involved and   (b) state whether you would expect the reaction to occur under the influence of heat or of light.

(a)

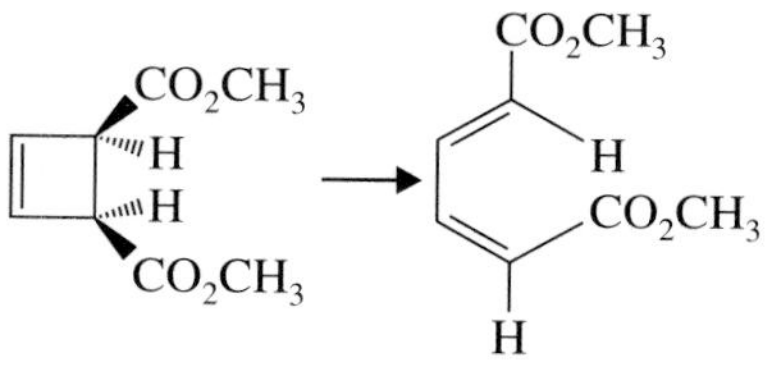

(b)

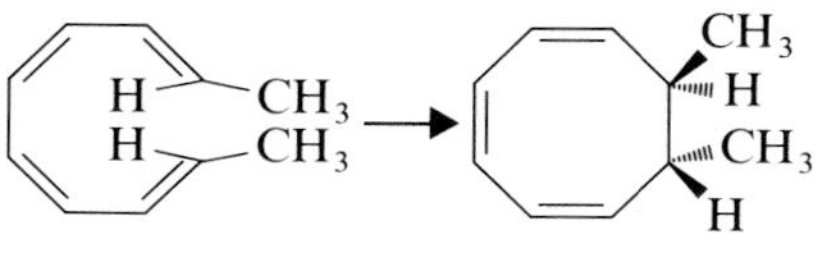

(c) 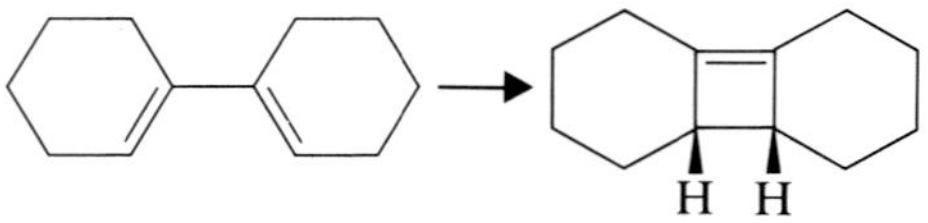

O.2B Electrocyclic Reactions of (4*n* + 2) π-Electron Systems

The second Woodward–Hoffmann rule for electrocyclic reactions is stated as follows:

2. **A thermal electrocyclic reaction involving (4*n* + 2) π electrons (where *n* = 0, 1, 2, . . .) proceeds with disrotatory motion; the photochemical reaction proceeds with conrotatory motion.**

According to this rule, the direction of rotation of the thermal and photochemical reactions of (4*n* + 2) π-electron systems is the opposite of that for corresponding 4*n* systems. Thus, we can summarize both systems in the way shown in Table O.1.

The interconversions of *trans*-5,6-dimethyl-1,3-cyclohexadiene and the two different 2,4,6-octatrienes that follow illustrate thermal and photochemical interconversions of 6 π-electron systems (4*n* + 2, where *n* = 1).

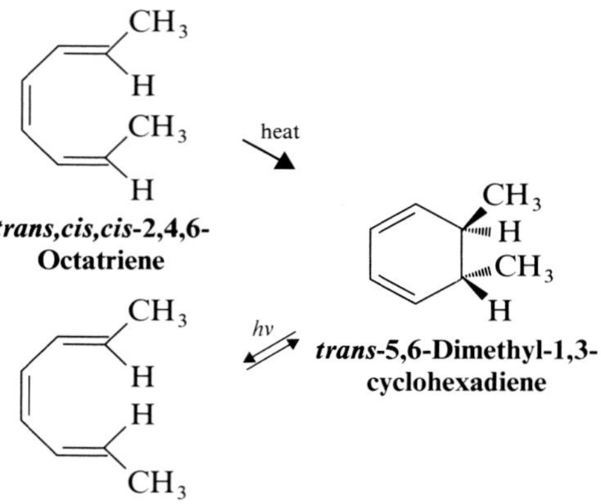

***trans,cis,cis*-2,4,6-Octatriene**

***trans*-5,6-Dimethyl-1,3-cyclohexadiene**

***trans,cis,trans*-2,4,6-Octatriene**

In the following thermal reaction, the methyl groups rotate in a disrotatory fashion.

TABLE O.1 Woodward–Hoffmann rules for electrocyclic reactions

NUMBER OF ELECTRONS	MOTION	RULE
4*n*	Conrotatory	Thermally allowed, photochemically forbidden
4*n*	Disrotatory	Photochemically allowed, thermally forbidden
4*n* + 2	Disrotatory	Thermally allowed, photochemically forbidden
4*n* + 2	Conrotatory	Photochemically allowed, thermally forbidden

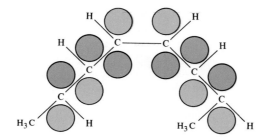

In the photochemical reaction, the groups rotate in a conrotatory way.

We can understand how these reactions occur if we examine the π molecular orbitals shown in Fig. O.2. Once again, we want to pay attention to the highest occupied molecular orbitals. For the thermal reaction of a 2,4,6-octatriene, the highest occupied orbital is Ψ_3 because the molecule reacts in its ground state.

Ψ_3 of *trans, cis, cis,–2, 4, 6-Octatriene*

We see in the following figure that disrotatory motion of orbitals at C2 and C7 of Ψ_3 allows the formation of a bonding sigma molecular orbital between them. Disrotatory motion of the orbitals, of course, also requires disrotatory motion of the groups attached to C2 and C7. And, disrotatory motion of the groups is what we observe in the thermal reaction: *trans,cis,cis-2,4,6-octatriene* ⟶ *trans-5,6-dimethyl-1,3-cyclohexadiene*.

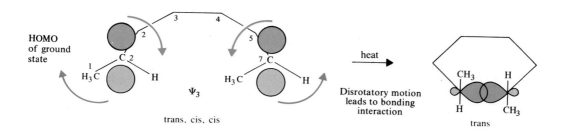

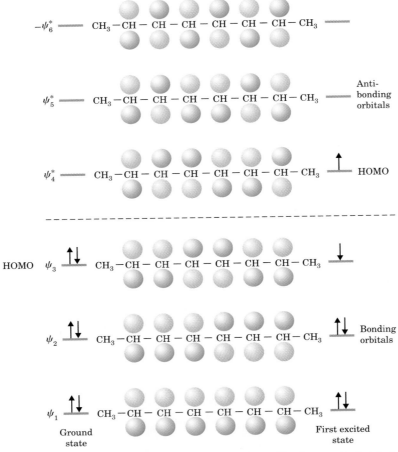

FIGURE O.2 The π molecular orbitals of a 2,4,6-octatriene. The first excited state is formed when the molecule absorbs light of the proper wavelength. (These molecular orbitals are obtained using procedures that are beyond the scope of our discussions.)

When we consider the photochemical reaction, *trans,cis,trans*-2,4,6-octatriene $\rightleftharpoons$ *trans*-5,6-dimethyl-1,3-cyclohexadiene, we want to focus our attention on Ψ_4^*. In the photochemical reaction, light causes the promotion of an electron from Ψ_3 to Ψ_4^*, and thus Ψ_4^* becomes the HOMO. We also want to look at the symmetry of the orbitals at C2 and C7 of Ψ_4^*, for these are the orbitals that form a σ bond. In the interconversion shown here, conrotatory motion of the orbitals allows lobes of the same sign to overlap. Thus, we can understand why conrotatory motion of the groups is what we observe in the photochemical reaction.

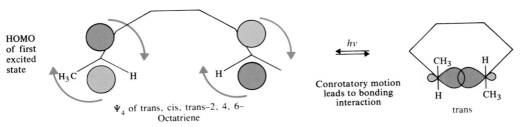

PROBLEM O.6

Give the stereochemistry of the product that you would expect from each of the following electrocyclic reactions.

(a) 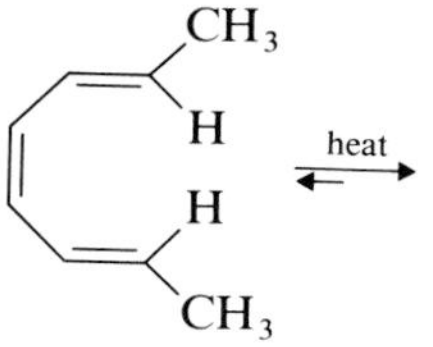 ... heat ...   (b) 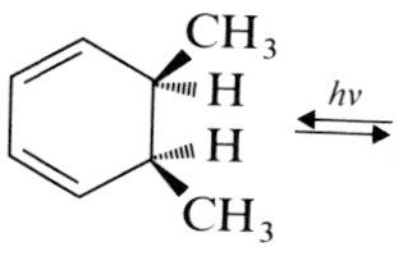 ... hv ...

PROBLEM O.7

Can you suggest a stereospecific method for converting *trans*-5,6-dimethyl-1,3-cyclohexadiene into *cis*-5,6-dimethyl-1,3-cyclohexadiene?

PROBLEM O.8

When compound **A** is heated, compound **B** can be isolated from the reaction mixture. A sequence of two electrocyclic reactions occurs; the first involves a 4 π-electron system, the second involves a 6 π-electron system. Outline both electrocyclic reactions and give the structure of the intermediate that intervenes.

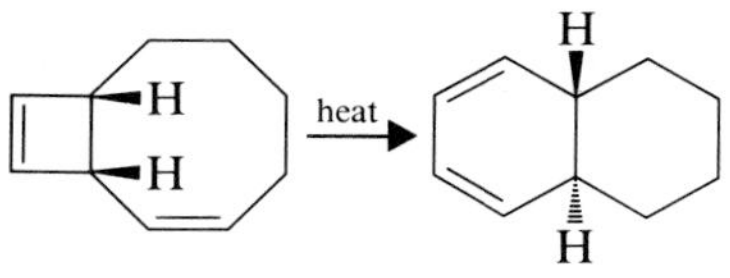

A   B

O.3 Cycloaddition Reactions

There are a number of reactions of alkenes and polyenes in which two molecules react to form a cyclic product. These reactions, called *cycloaddition* reactions, are shown next.

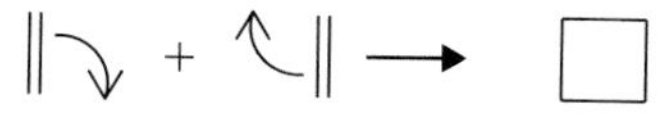

Alkene   Alkene   Cyclobutane   A [2 + 2] cycloaddition

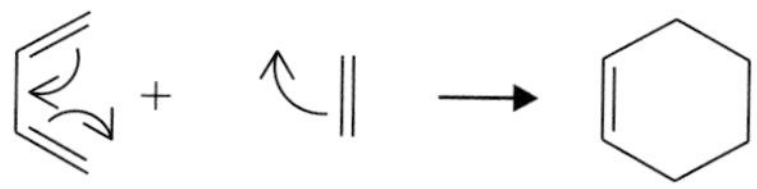

Diene   Alkene   Cyclohexene   A [4 + 2] cycloaddition
           (dienophile)   (adduct)

Chemists classify cycloaddition reactions on the basis of the number of π electrons involved in each component. The reaction of two alkenes to form a cyclobutane is a [2 + 2] cycloaddition; the reaction of a diene and an alkene to form a cyclohexene is called a [4 + 2] cycloaddition. We are already familiar with the [4 + 2] cycloaddition, because it is the Diels–Alder reaction that we studied in Section 12.10.

Cycloaddition reactions resemble electrocyclic reactions in the following important ways:

1. Sigma and pi bonds are interconverted.
2. Cycloaddition reactions require only heat or light for initiation.
3. Radicals and ionic intermediates are not involved in the mechanisms for cycloadditions.
4. Bonds are made and broken in a single concerted step involving a cyclic transition state.
5. Cycloaddition reactions are highly stereospecific.

As we might expect, concerted cycloaddition reactions resemble electrocyclic reactions in still another important way: The symmetry elements of the interacting molecular orbitals allow us to account for their stereochemistry. The symmetry elements of the interacting molecular orbitals also allow us to account for two other observations that have been made about cycloaddition reactions:

1. *Photochemical [2 + 2] cycloaddition reactions occur readily, whereas thermal [2 + 2] cycloadditions take place only under extreme conditions.* When thermal [2 + 2] cycloadditions do take place, they occur through radical (or ionic) mechanisms, not through a concerted process.
2. *Thermal [4 + 2] cycloaddition reactions occur readily and photochemical [4 + 2] cycloadditions are difficult.*

O.3A [2 + 2] Cycloadditions

Let us begin with an analysis of the [2 + 2] cycloaddition of two ethylene molecules to form a molecule of cyclobutane.

$$2 \underset{CH_2}{\overset{CH_2}{\parallel}} \longrightarrow \begin{array}{c} H_2C-CH_2 \\ | \qquad | \\ H_2C-CH_2 \end{array}$$

In this reaction we see that two π bonds are converted into two σ bonds. But how does this conversion take place? One way of answering this question is by examining the frontier orbitals of the reactants. The frontier orbitals are the HOMO of one reactant and the LUMO of the other.

We can see how frontier orbital interactions come into play if we examine the possibility of a *concerted thermal* conversion of two ethene molecules into cyclobutane.

Thermal reactions involve molecules reacting in their ground states. The following is the orbital diagram for ethene in its ground state.

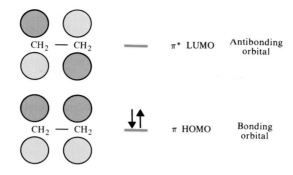

π^* LUMO   Antibonding orbital

π HOMO   Bonding orbital

The ground state of ethene

The HOMO of ethene in its ground state is the π orbital. Since this orbital contains two electrons, it interacts with an *unoccupied* molecular orbital of another ethene molecule. The LUMO of the ground state of ethene is, of course, π^*.

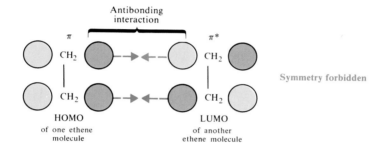

Antibonding interaction

π π^*

Symmetry forbidden

HOMO
of one ethene molecule

LUMO
of another ethene molecule

We see from the previous diagram, however, that overlapping the π orbital of one ethene molecule with the π^* orbital of another does not lead to bonding between both sets of carbon atoms because orbitals of opposite signs overlap between the top pair of carbon atoms. This reaction is said to be *symmetry forbidden*. What does this mean? It means that a thermal (or ground state) cycloaddition of ethene would be unlikely to occur in a concerted process. This is exactly what we find experimentally; thermal cycloadditions of ethene, when they occur, take place through nonconcerted, radical mechanisms.

What, then, can we decide about the other possibility—a photochemical [2 + 2] cycloaddition? If an ethene molecule absorbs a photon of light of the proper wavelength, an electron is promoted from π to π^*. In this excited state the HOMO of an ethene molecule is π^*. The following diagram shows how the HOMO of an excited state ethene molecule interacts with the LUMO of a ground state ethene molecule.

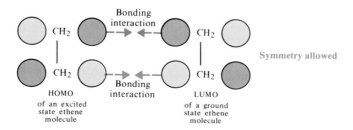

Bonding interaction

Symmetry allowed

Bonding interaction

HOMO
of an excited state ethene molecule

LUMO
of a ground state ethene molecule

Here we find that bonding interactions occur between both CH_2 groups, that is, lobes of the same sign overlap between both sets of carbon atoms. Complete correlation diagrams also show that the photochemical reaction is *symmetry allowed* and should occur readily through a concerted process. This, moreover, is what we observe experimentally: Ethene reacts readily in a *photochemical* cycloaddition.

The analysis that we have given for the [2 + 2] ethene cycloaddition can be made for any alkene [2 + 2] cycloaddition because the symmetry elements of the π and π^* orbitals of all alkenes are the same.

PROBLEM O.9

What products would you expect from the following concerted cycloaddition reactions? (Give stereochemical formulas.)

(a) *cis*-2-Butene $\xrightarrow{hv}$

(b) *trans*-2-Butene $\xrightarrow{hv}$

PROBLEM O.10

Show what happens in the following reaction:

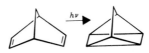

O.3B [4 + 2] Cycloadditions

Concerted [4 + 2] cycloadditions—Diels–Alder reactions—are *thermal reactions.* Considerations of orbital interactions allow us to account for this fact as well. To see how, let us consider the diagrams shown in Fig. O.3.

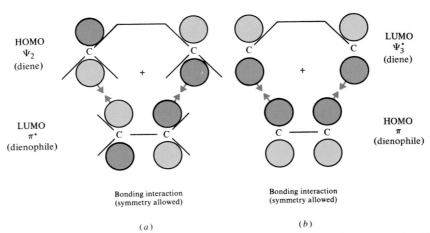

FIGURE O.3 Two symmetry-allowed interactions for a thermal [4 + 2] cycloaddition. (*a*) Bonding interaction between the HOMO of a diene and the LUMO of a dienophile. (*b*) Bonding interaction between the LUMO of the diene and the HOMO of the dienophile.

Both modes of orbital overlap shown in Fig. O.3 lead to bonding interactions and both involve *ground states* of the reactants. The ground state of a diene has two electrons in Ψ_2 (its HOMO). The overlap shown in part (*a*) allows these two electrons to flow into the LUMO, π^*, of the dienophile. The overlap shown in part (*b*) allows two electrons to flow from the HOMO of the dienophile, π, into the LUMO of the diene, Ψ_3^*. This thermal reaction is said to be symmetry allowed.

In Section 12.10 we saw that the Diels–Alder reaction proceeds with retention of configuration of the dienophile. Because the Diels–Alder reaction is usually con-certed, it also proceeds with retention of configuration of the diene.

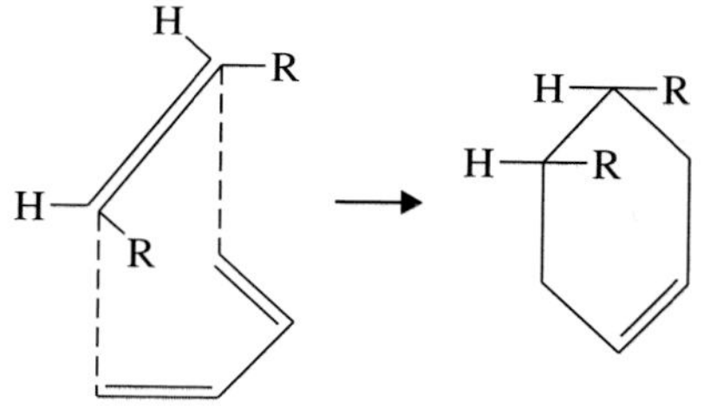

Retention of
configuration of
the dienophile

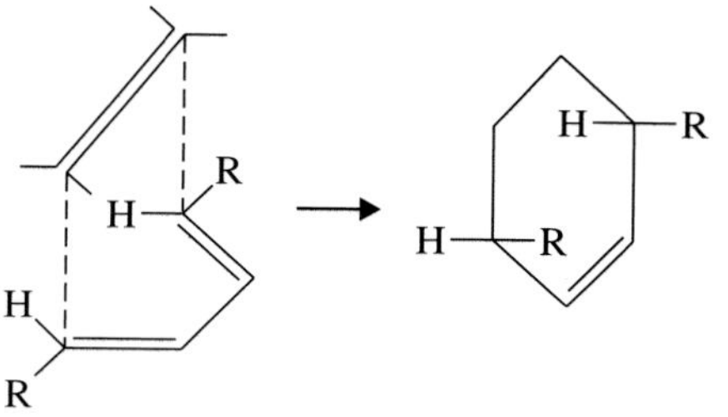

Retention of
configuration of
the diene

PROBLEM O.11

What products would you expect from the following reaction?

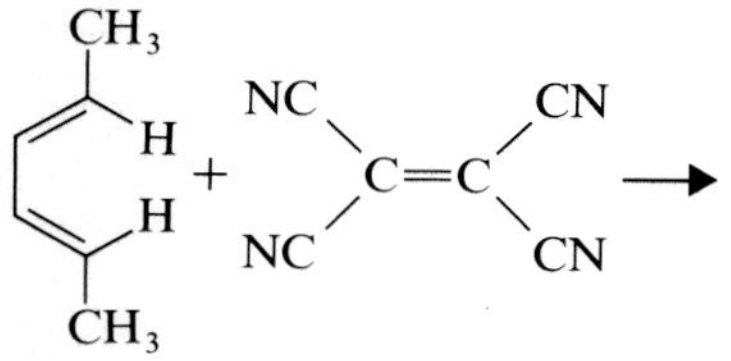

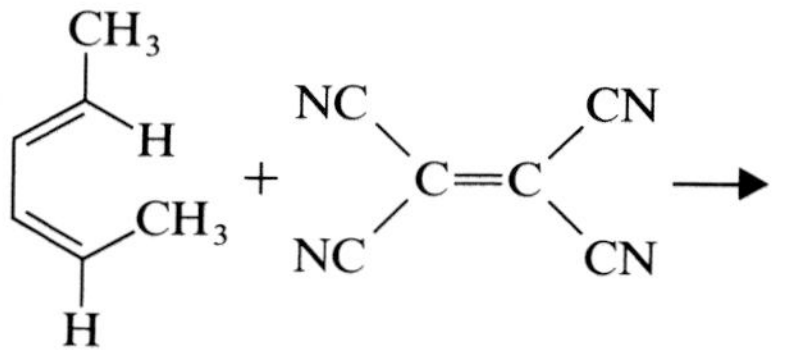

CHAPTER 22

Sucrose
(Section 22.12A)

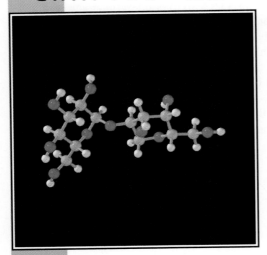

CARBOHYDRATES

22.1 INTRODUCTION

22.1A Classification of Carbohydrates

The group of compounds known as carbohydrates received their general name because of early observations that they often have the formula $C_x(H_2O)_y$—that is, they appear to be "hydrates of carbon." Simple carbohydrates are also known as sugars or saccharides (Latin *saccharum,* Greek, *sakcharon,* sugar) and the ending of the names of most sugars is -*ose.* Thus, we have such names as *sucrose* for ordinary table sugar, *glucose* for the principal sugar in blood, *fructose* for a sugar in fruits and honey, and *maltose* for malt sugar.

Carbohydrates are usually defined as *polyhydroxy aldehydes and ketones or substances that hydrolyze to yield polyhydroxy aldehydes and ketones.* Although this definition draws attention to the important functional groups of carbohydrates, it is not entirely satisfactory. We shall later find that because carbohydrates contain $\diagdown \!\! \underset{\diagup}{C} \!\!=\!\! O$

groups and —OH groups, they exist, primarily, as *hemiacetals* or *acetals* (Section 16.7).

The simplest carbohydrates, those that cannot be hydrolyzed into simpler carbohydrates, are called *monosaccharides.* On a molecular basis, carbohydrates that undergo hydrolysis to produce only two molecules of monosaccharide are called *disaccharides;* those that yield three molecules of monosaccharide are called *trisaccharides;* and so on. (Carbohydrates that hydrolyze to yield 2–10 molecules of monosaccharide are sometimes called *oligosaccharides.*) Carbohydrates that yield a large number of molecules of monosaccharide (> 10) are known as *polysaccharides.*

Maltose and sucrose are examples of disaccharides. On hydrolysis, 1 mol of maltose yields 2 mol of the monosaccharide glucose; sucrose undergoes hydrolysis to yield 1 mol of glucose and 1 mol of the monosaccharide fructose. Starch and cellulose are examples of polysaccharides; both are glucose polymers. Hydrolysis of either yields a large number of glucose units.

$$1 \text{ mol of maltose} \xrightarrow[\text{H}_3\text{O}^+]{\text{H}_2\text{O}} 2 \text{ mol of glucose}$$

A disaccharide   **A monosaccharide**

$$1 \text{ mol of sucrose} \xrightarrow[\text{H}_3\text{O}^+]{\text{H}_2\text{O}} 1 \text{ mol of glucose} + 1 \text{ mol of fructose}$$

A disaccharide   **Monosaccharides**

$$\begin{array}{c} 1 \text{ mol of starch} \\ \text{or} \\ 1 \text{ mol of cellulose} \end{array} \xrightarrow[\text{H}_3\text{O}^+]{\text{H}_2\text{O}} \text{many moles of glucose}$$

Polysaccharides   **Monosaccharide**

Carbohydrates are the most abundant organic constituents of plants. They not only serve as an important source of chemical energy for living organisms (sugars and starches are important in this respect), but also in plants and in some animals they serve as important constituents of supporting tissues (this is the primary function of the cellulose found in wood, cotton, and flax, for example).

We encounter carbohydrates at almost every turn of our daily lives. The paper on which this book is printed is largely cellulose; so too, is the cotton of our clothes and the wood of our houses. The flour from which we make bread is mainly starch, and starch is also a major constituent of many other foodstuffs, such as potatoes, rice, beans, corn, and peas.

22.1B Photosynthesis and Carbohydrate Metabolism

Carbohydrates are synthesized in green plants by *photosynthesis*—a process that uses solar energy to reduce, or "fix," carbon dioxide. The overall equation for photosynthesis can be written as follows:

$$x\,CO_2 + y\,H_2O + \text{solar energy} \longrightarrow \underset{\textbf{Carbohydrate}}{C_x(H_2O)_y} + x\,O_2$$

Many individual enzyme-catalyzed reactions take place in the general photosynthetic process and not all are fully understood. We know, however, that photosynthesis begins with the absorption of light by the important green pigment of plants, chlorophyll (Fig. 22.1). The green color of chlorophyll and, therefore, its ability to absorb sunlight in the visible region are due primarily to its extended conjugated system. As photons of sunlight are trapped by chlorophyll, energy becomes available to the plant in a chemical form that can be used to carry out the reactions that reduce carbon dioxide to carbohydrates and oxidize water to oxygen.

Carbohydrates act as a major chemical repository for solar energy. Their energy is released when animals or plants metabolize carbohydrates to carbon dioxide and water.

$$C_x(H_2O)_y + x\,O_2 \longrightarrow x\,CO_2 + y\,H_2O + \text{energy}$$

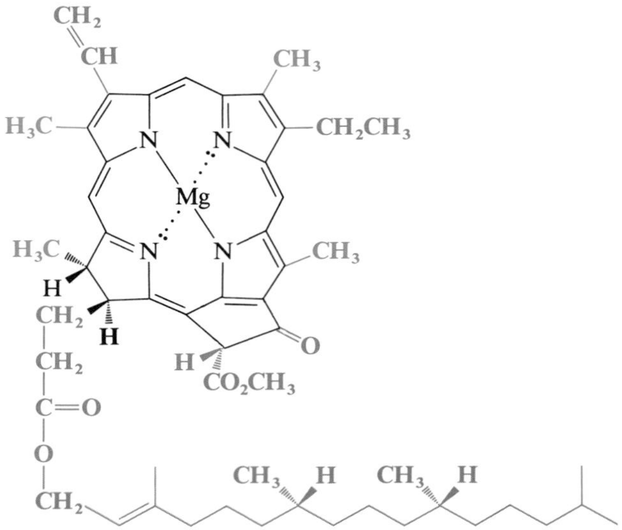

FIGURE 22.1 Chlorophyll *a*. [The structure of chlorophyll *a* was established largely through the work of H. Fischer (Munich), R. Willstätter (Munich), and J. B. Conant (Harvard). A synthesis of chlorophyll *a* from simple organic compounds was achieved by R. B. Woodward (Harvard) in 1960, who won the Nobel Prize in 1965 for his outstanding contributions to synthetic organic chemistry.]

The metabolism of carbohydrates also takes place through a series of enzyme-catalyzed reactions in which each energy-yielding step is an oxidation (or the consequence of an oxidation).

Although some of the energy released in the oxidation of carbohydrates is inevitably converted to heat, much of it is conserved in a new chemical form through reactions that are coupled to the synthesis of adenosine triphosphate (ATP) from adenosine diphosphate (ADP) and inorganic phosphate (P_i) (Fig. 22.2). The phosphoric anhydride bond that forms between the terminal phosphate group of ADP and the phosphate ion becomes another repository of chemical energy. Plants and animals can use the conserved energy of ATP (or very similar substances) to carry out all of their energy-requiring processes: the contraction of a muscle, the synthesis of a macromolecule, and so on. When the energy in ATP is used, a coupled reaction takes place in which ATP is hydrolyzed:

$$\text{ATP} + H_2O \longrightarrow \text{ADP} + P_i + \text{energy}$$

or a new anhydride linkage is created:

$$\underset{\displaystyle R-\overset{\displaystyle \overset{O}{\|}}{C}-OH}{} + \text{ATP} \longrightarrow \underset{\displaystyle R-\overset{\displaystyle \overset{O}{\|}}{C}-O-\overset{\displaystyle \overset{O}{\|}}{\underset{\displaystyle \underset{O^-}{|}}{P}}-O^-}{} + \text{ADP}$$

Acyl phosphate

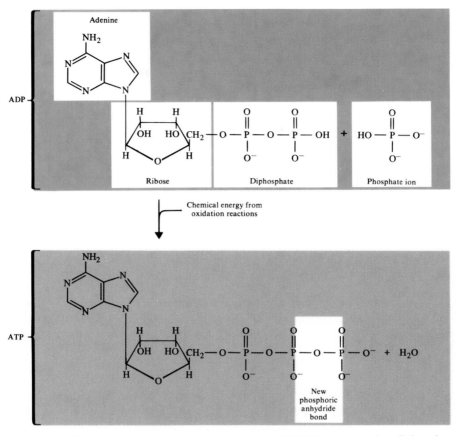

FIGURE 22.2 The synthesis of adenosine triphosphate (ATP) from adenosine diphosphate (ADP), and hydrogen phosphate ion. This reaction takes place in all living organisms, and adenosine triphosphate is the major compound into which the chemical energy released by biological oxidations is transformed.

22.2 MONOSACCHARIDES

22.2A Classification of Monosaccharides

Monosaccharides are classified according to (1) the number of carbon atoms present in the molecule and (2) whether they contain an aldehyde or keto group. Thus, a mono-saccharide containing three carbon atoms is called a *triose;* one containing four carbon atoms is called a *tetrose;* one containing five carbon atoms is a *pentose;* and one containing six carbon atoms is a *hexose.* A monosaccharide containing an aldehyde group is called an *aldose;* one containing a keto group is called a *ketose.* These two classifications are frequently combined. A C_4 aldose, for example, is called an *aldo-tetrose;* a C_5 ketose is called a *ketopentose.*

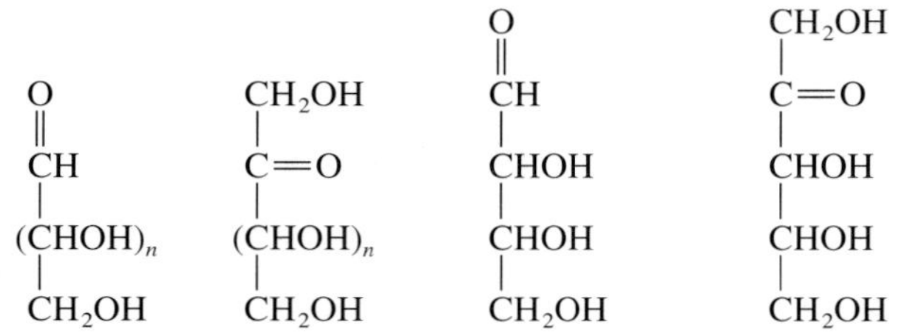

$$
\begin{array}{llll}
\begin{array}{c}
\text{O} \\
\parallel \\
\text{CH} \\
\mid \\
\text{(CHOH)}_n \\
\mid \\
\text{CH}_2\text{OH} \\
\textbf{An aldose}
\end{array}
&
\begin{array}{c}
\text{CH}_2\text{OH} \\
\mid \\
\text{C}{=}\text{O} \\
\mid \\
\text{(CHOH)}_n \\
\mid \\
\text{CH}_2\text{OH} \\
\textbf{A ketose}
\end{array}
&
\begin{array}{c}
\text{O} \\
\parallel \\
\text{CH} \\
\mid \\
\text{CHOH} \\
\mid \\
\text{CHOH} \\
\mid \\
\text{CH}_2\text{OH} \\
\textbf{An aldotetrose}
\end{array}
&
\begin{array}{c}
\text{CH}_2\text{OH} \\
\mid \\
\text{C}{=}\text{O} \\
\mid \\
\text{CHOH} \\
\mid \\
\text{CHOH} \\
\mid \\
\text{CH}_2\text{OH} \\
\textbf{A ketopentose}
\end{array}
\end{array}
$$

PROBLEM 22.1

How many stereocenters are contained in the (a) aldotetrose and (b) ketopentose just given? (c) How many stereoisomers would you expect from each general structure?

22.2B D and L Designations of Monosaccharides

The simplest monosaccharides are the compounds glyceraldehyde and dihydroxyacetone (see following structures). Of these two compounds, only glyceraldehyde contains a stereocenter.

$$
\begin{array}{cc}
\begin{array}{c}
\text{CHO} \\
\mid \\
{*}\text{CHOH} \\
\mid \\
\text{CH}_2\text{OH} \\
\textbf{Glyceraldehyde} \\
\textbf{(an aldotriose)}
\end{array}
&
\begin{array}{c}
\text{CH}_2\text{OH} \\
\mid \\
\text{C}{=}\text{O} \\
\mid \\
\text{CH}_2\text{OH} \\
\textbf{Dihydroxyacetone} \\
\textbf{(a ketotriose)}
\end{array}
\end{array}
$$

Glyceraldehyde exists, therefore, in two enantiomeric forms that are known to have the absolute configurations shown here.

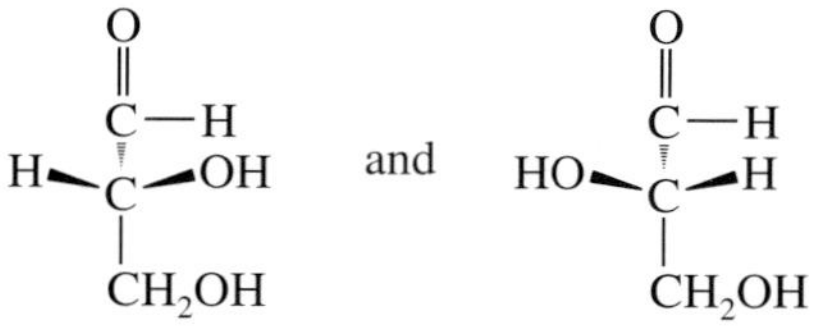

$$
\begin{array}{ccc}
\begin{array}{c}
\text{O} \\
\parallel \\
\text{C}-\text{H} \\
\text{H}{-}\text{C}{-}\text{OH} \\
\mid \\
\text{CH}_2\text{OH} \\
\textbf{(+)-Glyceraldehyde}
\end{array}
& \text{and} &
\begin{array}{c}
\text{O} \\
\parallel \\
\text{C}-\text{H} \\
\text{HO}{-}\text{C}{-}\text{H} \\
\mid \\
\text{CH}_2\text{OH} \\
\textbf{(−)-Glyceraldehyde}
\end{array}
\end{array}
$$

We saw in Section 5.6 that, according to the Cahn–Ingold–Prelog convention, (+)-glyceraldehyde should be designated (R)-(+)-glyceraldehyde and (−)-glyceraldehyde should be designated (S)-(−)-glyceraldehyde.

Early in this century, before the absolute configurations of any organic compounds were known, another system of stereochemical designations was introduced. According to this system (first suggested by M. A. Rosanoff of New York University in 1906), (+)-glyceraldehyde is designated D-(+)-glyceraldehyde and (−)-glyceraldehyde is designated L-(−)-glyceraldehyde. These two compounds, moreover, serve as configurational standards for all monosaccharides. A monosaccharide *whose highest numbered stereocenter* (the penultimate carbon) has the same configuration as D-(+)-glyceraldehyde is designated as a D sugar; one whose highest numbered stereocenter has the same configuration as L-glyceraldehyde is designated as an L sugar. By convention, acyclic forms of monosaccharides are drawn vertically with the aldehyde or keto group at or nearest the top. When drawn in this way, D-sugars have the —OH on their penultimate carbon on the right.

A D-aldopentose   A L-ketohexose

D and L designations are like (R) and (S) designations in that they are not necessarily related to the optical rotations of the sugars to which they are applied. Thus, one may encounter other sugars that are D-(+)- or D-(−)- and that are L-(+)- or L-(−)- .

The D–L system of stereochemical designations is thoroughly entrenched in the literature of carbohydrate chemistry, and even though it has the disadvantage of specifying the configuration of only one stereocenter—that of the highest numbered stereocenter—we shall employ the D–L system in our designations of carbohydrates.

PROBLEM 22.2

Write three-dimensional formulas for each aldotetrose and ketopentose isomer in Problem 22.1 and designate each as a D or L sugar.

22.2C Structural Formulas for Monosaccharides

Later in this chapter we shall see how the great carbohydrate chemist Emil Fischer* was able to establish the stereochemical configuration of the aldohexose D-(+)-glucose, the most abundant monosaccharide. In the meantime, however, we can use D-(+)-glucose as an example illustrating the various ways of representing the structures of monosaccharides.

Fischer represented the structure of D-(+)-glucose with the cross formulation (1) in Fig. 22.3. This type of formulation is now called a Fischer projection formula (Section 5.12) and is still useful for carbohydrates. In Fischer projection formulas, by convention, *horizontal lines project out toward the reader and vertical lines project behind the plane of the page. When we use Fischer projection formulas, however, we must not* (in our mind's eye) *remove them from the plane of the page in order to test*

---

*Emil Fischer (1852–1919) was professor of organic chemistry at the University of Berlin. In addition to monumental work in the field of carbohydrate chemistry, where Fischer and his co-workers established the configuration of most of the monosaccharides, Fischer also made important contributions to studies of amino acids, proteins, purines, indoles, and stereochemistry generally. As a graduate student, Fischer discovered phenylhydrazine, a reagent that was highly important in his later work with carbohydrates. Fischer was the second recipient (in 1902) of the Nobel Prize for Chemistry.

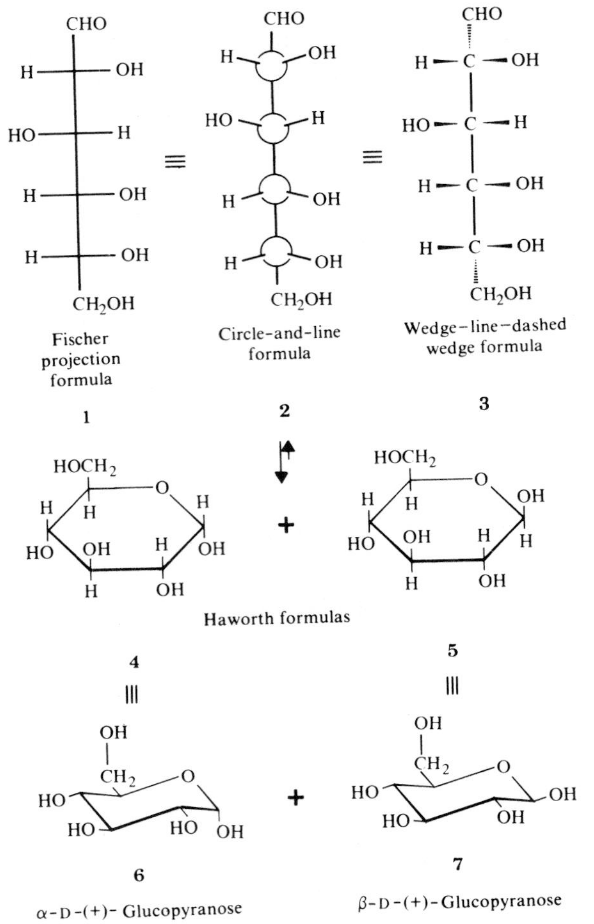

CHO

H——OH

HO——H

H——OH

H——OH

CH₂OH

Fischer
projection
formula

1

$\equiv$

CHO

H——OH

HO——H

H——OH

H——OH

CH₂OH

Circle-and-line
formula

2

$\equiv$

CHO

H—C—OH

HO—C—H

H—C—OH

H—C—OH

CH₂OH

Wedge–line–dashed
wedge formula

3

HOCH₂

H H   H
                O
HO OH   H OH
       H   OH

4

$+$

HOCH₂

H H   OH
                O
HO OH   H H
       H   OH

5

Haworth formulas

OH

CH₂
                O
HO
   HO   HO OH

6

α-D-(+)-Glucopyranose

$+$

OH

CH₂
                O
HO   OH
   HO   OH

7

β-D-(+)-Glucopyranose

FIGURE 22.3 (*a*) **1–3** are formulas used for the open-chain structure of D-(+)-glucose. (*b*) **4–7** are formulas used for the two cyclic hemiacetal forms of D-(+)-glucose.

their superposability and we must not rotate them by 90°. In terms of more familiar formulations, the Fischer projection formula translates into formulas **2** and **3**.*

In IUPAC nomenclature and with the Cahn–Ingold–Prelog system of stereochemical designations, the open-chain form of D-(+)-glucose is (2*R*, 3*S*, 4*R*, 5*R*)-2,3,4,5,6-pentahydroxyhexanal.

Although many of the properties of D-(+)-glucose can be explained in terms of an open-chain structure (**1**, **2**, or **3**), a considerable body of evidence indicates that the open-chain structure exists, primarily, in equilibrium with two cyclic forms. These can

*The meaning of formulas **1**, **2**, and **3** can be seen best through the use of molecular models: We first construct a chain of six carbon atoms with the —CHO group at the top and a —CH₂OH group at the bottom. We then bring the —CH₂OH group up behind the chain until it almost touches the —CHO group. Holding this model so that the —CHO and —CH₂OH groups are directed generally away from us, we then begin placing —H and —OH groups on each of the four remaining carbon atoms. The —OH group of C2 is placed on the right; that of C3 on the left; and those of C4 and C5 on the right.

be represented by structures **4** and **5** or **6** and **7**. The cyclic forms of D-(+)-glucose are *hemiacetals* formed by an intramolecular reaction of the —OH group at C5 with the aldehyde group (Fig. 22.4). Cyclization creates a new stereocenter at C1 and this stereocenter explains how two cyclic forms are possible. These two cyclic forms are *diastereomers* that differ only in the configuration of C1. In carbohydrate chemistry diastereomers of this type are called *anomers,* and the hemiacetal carbon atom is called the *anomeric carbon atom.*

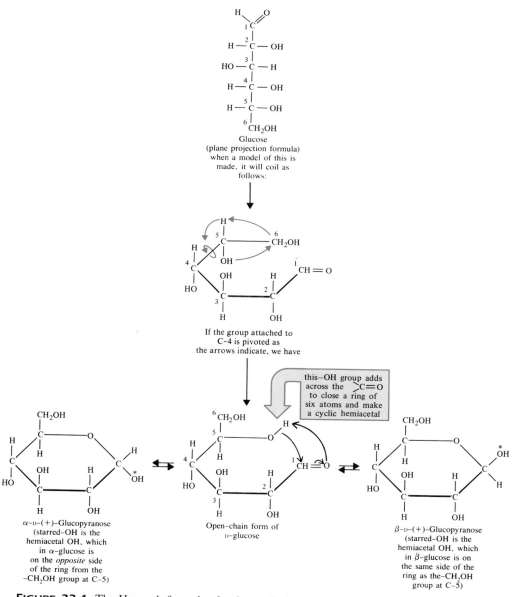

FIGURE 22.4 The Haworth formulas for the cyclic hemiacetal forms of D-(+)-glucose and their relation to the open-chain polyhydroxy aldehyde structure. [From John R. Holum, *Organic Chemistry: A Brief Course,* Wiley, New York, 1975, p. 332. Used by permission.]

Structures **4** and **5** for the glucose anomers are called Haworth formulas* and, although they do not give an accurate picture of the shape of the six-membered ring, they have many practical uses. Figure 22.4 demonstrates how the representation of each stereocenter of the open-chain form can be correlated with its representation in the Haworth formula.

Each glucose anomer is designated as an α *anomer* or a β *anomer* depending on the location of the —OH group of C1. When we draw the cyclic forms of a D sugar in the orientation shown in Figs. 22.3 or 22.4, the α anomer has the —OH trans to the CH₂OH group and the β anomer has the —OH cis to the CH₂OH group.

Studies of the structures of the cyclic hemiacetal forms of D-(+)-glucose using X-ray analysis have demonstrated that the actual conformations of the rings are the chair forms represented by conformational formulas **6** and **7** in Fig. 22.3. This shape is exactly what we would expect from our studies of the conformations of cyclohexane (Chapter 4) and, it is especially interesting to notice that in the β anomer of D-glucose, all of the large substituents, —OH or —CH₂OH, are equatorial. In the α anomer, the only bulky axial substituent is the —OH at C1.

It is convenient at times to represent the cyclic structures of a monosaccharide without specifying whether the configuration of the anomeric carbon atom is α or β. When we do this, we shall use formulas such as the following:

Not all carbohydrates exist in equilibrium with six-membered hemiacetal rings; in several instances the ring is five membered. (Even glucose exists, to a small extent, in equilibrium with five-membered hemiacetal rings.) Because of this variation, a system of nomenclature has been introduced to allow designation of the ring size. If the monosaccharide ring is six membered, the compound is called a *pyranose;* if the ring is five membered, the compound is designated as a *furanose.*† Thus, the full name of compound **4** (or **6**) is α-D-(+)-glucopyranose, while that of **5** (or **7**) is β-D-(+)-glucopyranose.

---

*After the English chemist W. N. Haworth (University of Birmingham) who, in 1926, along with E. L. Hirst, demonstrated that the cyclic form of glucose acetals consists of a six-membered ring. Haworth received the Nobel Prize for his work in carbohydrate chemistry in 1937. For an excellent discussion of Haworth formulas and their relation to open-chain forms see the following article: D. M. S. Wheeler, M. M. Wheeler, and T. S. Wheeler, "The Conversion of Open Chain Structures of Monosaccharides into the Corresponding Haworth Formulas," *J. Chem. Ed. 59,* 969, 1982.

†These names come from the names of the oxygen heterocycles *pyran* and *furan* + *ose.*

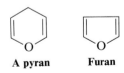

A pyran   **Furan**

22.3 MUTAROTATION

Part of the evidence for the cyclic hemiacetal structure for D-(+)-glucose comes from experiments in which both α and β forms have been isolated. Ordinary D-(+)-glucose has a melting point of 146°C. However, when D-(+)-glucose is crystallized by evaporating an aqueous solution kept above 98°C, a second form of D-(+)-glucose with a melting point of 150°C can be obtained. When the optical rotations of these two forms are measured, they are found to be significantly different, but when an aqueous solution of either form is allowed to stand, its rotation changes. The specific rotation of one form decreases and the rotation of the other increases, *until both solutions show the same value.* A solution of ordinary D-(+)-glucose (mp, 146°C) has an initial specific rotation of +112° but, ultimately, the specific rotation of this solution falls to +52.7°. A solution of the second form of D-(+)-glucose (mp, 150°C) has an initial specific rotation of +18.7°; but slowly, the specific rotation of this solution rises to +52.7°. This change in rotation toward an equilibrium value is called *mutarotation.*

The explanation for this mutarotation lies in the existence of an equilibrium between the open-chain form of D-(+)-glucose and the α and β forms of the cyclic hemiacetals.

α-D-(+)-Glucopyranose
(mp, 146°C $[\alpha]_D^{25} = +112°$)

Open-chain
form of
D-(+)-glucose

β-D-(+)-Glucopyranose
(mp, 150°C $[\alpha]_D^{25} = +18.7°$)

X-Ray analysis has confirmed that ordinary D-(+)-glucose has the α configuration at the anomeric carbon atom and that the higher melting form has the β configuration.

The concentration of open-chain D-(+)-glucose in solution at equilibrium is very small. Solutions of D-(+)-glucose give no observable UV or IR absorption band for a carbonyl group, and solutions of D-(+)-glucose give a negative test with Schiff's reagent—a special reagent that requires a relatively high concentration of a free aldehyde group (rather than a hemiacetal) in order to give a positive test.

Assuming that the concentration of the open-chain form is negligible, one can, by use of the specific rotations in the preceding figures, calculate the percentages of the α and β anomers present at equilibrium. These percentages, 36% α anomer and 64% β anomer, are in accord with a greater stability for β-D-(+)-glucopyranose. This preference is what we might expect on the basis of its having only equatorial groups.

α-D-(+)-Glucopyranose
(36% at equilibrium)

β-D-(+)-Glucopyranose
(64% at equilibrium)

The β anomer of a pyranose is not always the more stable, however. With D-mannose, the equilibrium favors the α anomer and this result is called an *anomeric effect*.

α-D-Mannopyranose
(69% at equilibrium)

β-D-Mannopyranose
(31% at equilibrium)

We shall not discuss anomeric effects further except to say that they arise from conformational aspects of the interactions of two electronegative oxygen atoms. An anomeric effect will frequently cause an electronegative substituent, such as a hydroxyl or alkoxy group, to prefer the axial orientation.

22.4 GLYCOSIDE FORMATION

When a small amount of gaseous hydrogen chloride is passed into a solution of D-(+)-glucose in methanol, a reaction takes place that results in the formation of anomeric methyl *acetals*.

D−(+)− Glucose

Methyl α−D−glucopyranoside
(mp, 165°C $[\alpha]_D^{25}$ = +158°)

Methyl β−D−glucopyranoside
(mp, 107°C $[\alpha]_D^{25}$ = −33°)

Carbohydrate acetals, generally, are called *glycosides* (see the following mechanism) and an acetal of glucose is called a *glucoside*. (Acetals of mannose are *mannosides*, acetals of fructose are *fructosides*, and so on.) The methyl D-glucosides have been shown to have six-membered rings (Section 22.2C) so they are properly named methyl α-D-glucopyranoside and methyl β-D-glucopyranoside.

The mechanism for the formation of the methyl glucosides (starting arbitrarily with β-D-glucopyranose) is as follows:

A MECHANISM FOR THE FORMATION OF A GLYCOSIDE

You should review the mechanism for acetal formation given in Section 16.7C and compare it with the steps given here. Notice, again, the important role played by the electron pair of the adjacent oxygen atom in stabilizing the carbocation that forms in the second step.

The mixed acetal of an aldose (or ketose) is called a **glycoside.** Because glycosides are acetals, they are stable in basic aqueous solutions. In acidic solutions, however, glycosides (again because they are acetals, Section 16.7C) undergo hydrolysis to produce a sugar and an alcohol. The alcohol obtained from a glycoside is known as an **aglycone.**

Glycoside   **Sugar**   **Aglycone**
(stable in basic solutions)

For example, when an aqueous solution of methyl β-D-glucopyranoside is made acidic, the glycoside undergoes hydrolysis to produce D-glucose as a mixture of the two pyranose forms (in equilibrium with a small amount of the open-chain form).

Methyl β-D-glucopyranoside

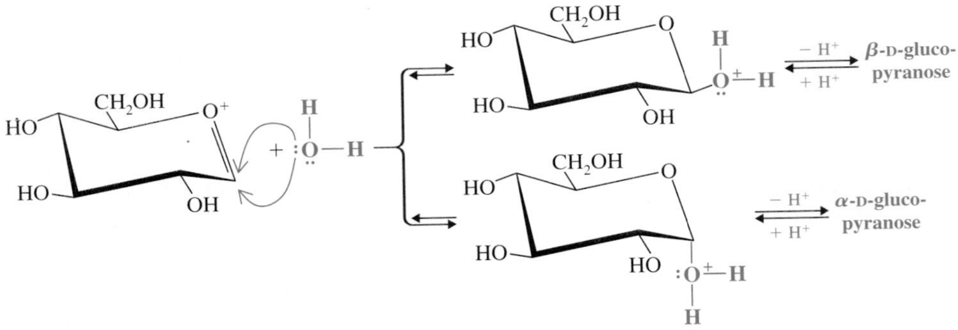

Glycosides may be as simple as the methyl glucosides that we have just studied, or they may be considerably more complex. Many naturally occurring compounds are glycosides. An example is *salicin,* a compound found in the bark of willow trees. The moiety bound to the sugar through a glycosidic linkage is called an **aglycone.**

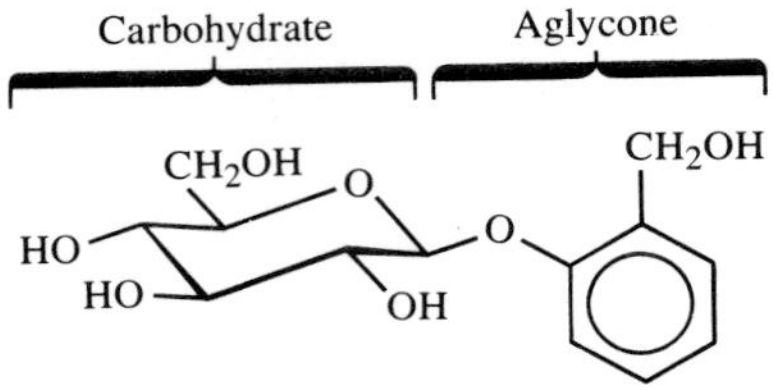

Salicin

As early as the time of the ancient Greeks, preparations made from willow bark were used in relieving pain. Eventually, chemists isolated salicin from other plant materials and were able to show that it was responsible for the analgesic effect of the willow bark preparations. Salicin can be converted to salicylic acid which, in turn, can be converted into the most widely used modern analgesic, *aspirin* (Section 21.8).

PROBLEM 22.3

(a) What products would be formed if salicin were treated with dilute aqueous HCl?   (b) Outline a mechanism for the reactions involved in their formation.

PROBLEM 22.4

How would you convert D-glucose to a mixture of ethyl α-D-glucopyranoside and ethyl β-D-glucopyranoside. Show all steps in the mechanism for their formation.

PROBLEM 22.5

In neutral or basic aqueous solutions, glycosides do not show mutarotation. However, if the solutions are made acidic, glycosides do show mutarotation. Explain why this occurs.

22.5 OTHER REACTIONS OF MONOSACCHARIDES

22.5A Enolization, Tautomerization, and Isomerization

Dissolving monosaccharides in aqueous base causes them to undergo enolizations and a series of keto–enol tautomerizations that lead to isomerizations. For example, if a solution of D-glucose containing calcium hydroxide is allowed to stand for several days, several products can be isolated, including D-fructose and D-mannose (Fig. 22.5). This type of reaction is called the **Lobry de Bruyn–Alberda van Ekenstein transformation** after the two Dutch chemists who discovered it in 1895.

When carrying out reactions with monosaccharides it is usually important to prevent these isomerizations and thereby to preserve the stereochemistry at all of the stereocenters. One way to do this is to convert the monosaccharide to the methyl glycoside first. We can then safely carry out reactions in basic media because the aldehyde group has been converted to an acetal and acetals are stable in aqueous base.

22.5B Formation of Ethers

A methyl glucoside, for example, can be converted to the pentamethyl derivative by treating it with excess dimethyl sulfate in aqueous sodium hydroxide. This reaction is just a multiple Williamson synthesis (Section 10.16). The hydroxyl groups of mono-

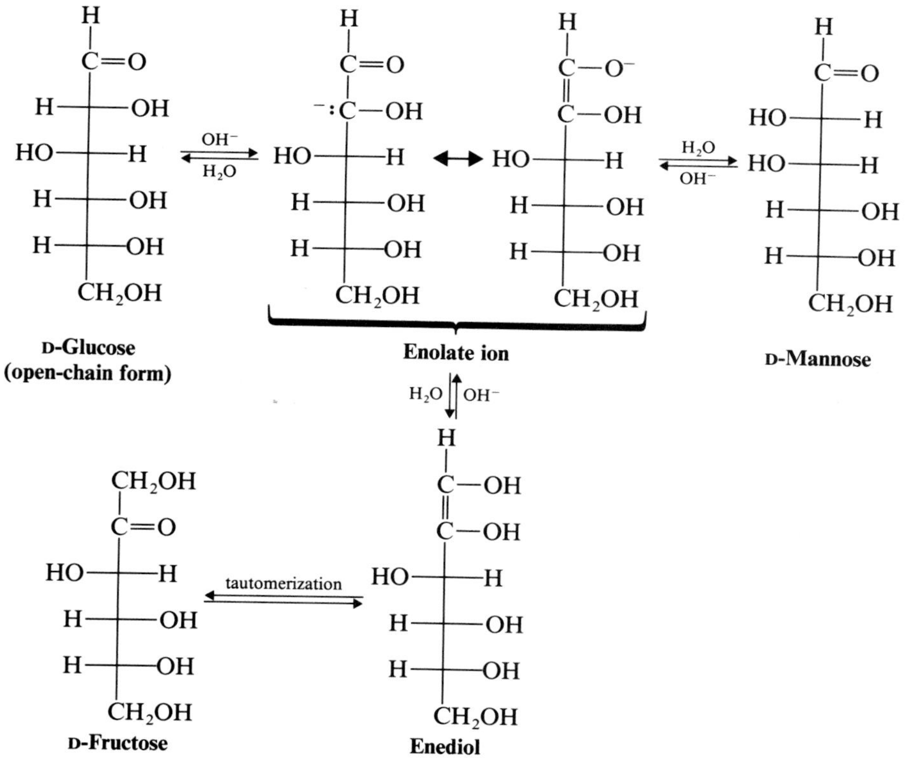

FIGURE 22.5 Monosaccharides undergo isomerizations via enolate ions and enediols when placed in aqueous base. Here we show how D-glucose isomerizes to D-mannose and to D-fructose.

saccharides are more acidic than those of ordinary alcohols because the monosaccharide contains so many electronegative oxygen atoms, all of which exert electron-withdrawing inductive effects on nearby hydroxyl groups. In aqueous NaOH, the hydroxyl groups are converted to alkoxide ions and each of these, in turn, reacts with dimethyl sulfate in an S_N2 reaction to yield a methyl ether. The process is called *exhaustive methylation.*

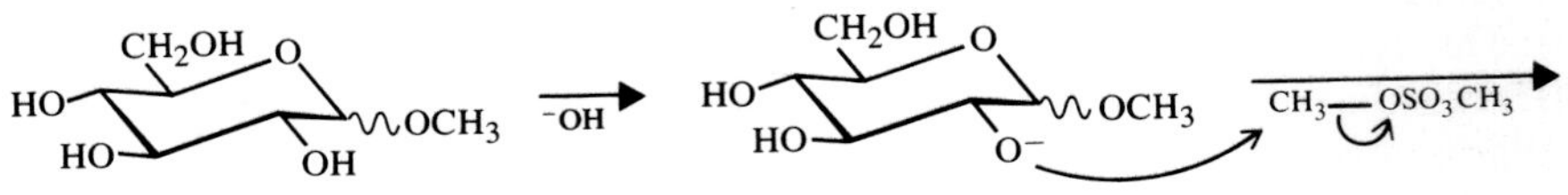

Methyl glucoside

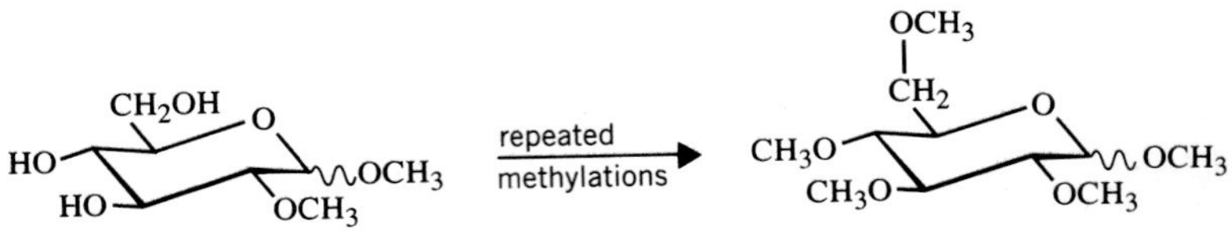

Pentamethyl derivative

The methoxy groups at C2, C3, C4, and C6 atoms of the pentamethyl derivative are ordinary ether groups. These groups, consequently, are stable in dilute aqueous acid. (To cleave ethers requires heating with concentrated HBr or HI, Section 10.17.) The methoxy group at C1, however, is different from the others because it is part of an acetal linkage (it is glycosidic). Therefore, treating the pentamethyl derivative with dilute aqueous acid causes hydrolysis of this glycosidic methoxy group and produces 2,3,4,6-tetra-*O*-methyl-D-glucose. (The *O* in this name means that the methyl groups are attached to oxygen atoms.)

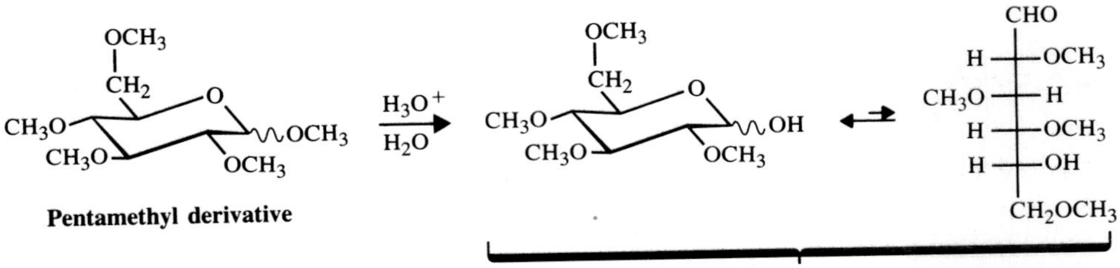

Pentamethyl derivative

2, 3, 4, 6-Tetra-*O*-methyl-D-glucose

Notice in the open-chain form that the oxygen at C5 does not bear a methyl group because it was originally a part of the cyclic hemiacetal linkage of D-glucose.

22.5C Conversion to Esters

Treating a monosaccharide with excess acetic anhydride and a weak base (such as pyridine or sodium acetate) converts all of the hydroxyl groups, including the anomeric hydroxyl, to ester groups. If the reaction is carried out at a low temperature (e.g., 0°C), the reaction occurs stereospecifically; the α anomer gives the α-acetate and the β-anomer gives the β-acetate.

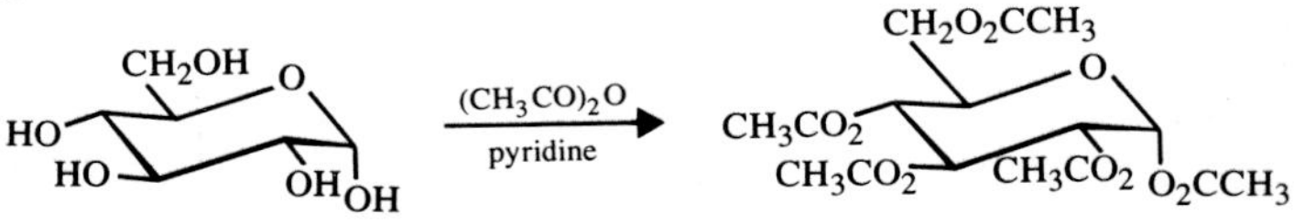

22.5D Conversion to Cyclic Acetals

In Section 16.7C we learned that aldehydes and ketones react with open-chain 1,2-diols to produce cyclic acetals.

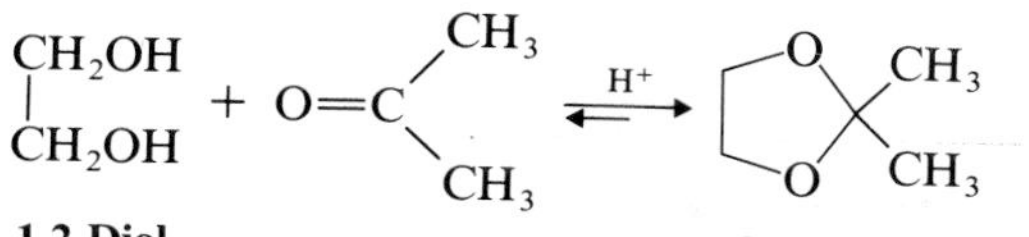

1,2-Diol   **Cyclic ketal**

If the 1,2-diol is attached to a ring, as in a monosaccharide, **formation of the cyclic acetal occurs only when the vicinal hydroxyl groups are cis to each other.** For example, α-D-galactopyranose reacts with acetone in the following way:

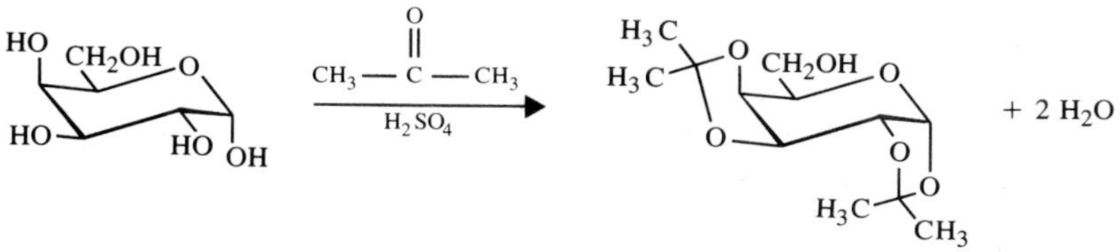

The formation of cyclic acetals can be used to protect certain hydroxyl groups of a sugar while reactions are carried out on other parts of the molecule. We shall see examples of this in Problems 22.19, 22.41, and 22.42 and in Chapter 25. The acetals formed from acetone are called **acetonides.**

22.6 OXIDATION REACTIONS OF MONOSACCHARIDES

A number of oxidizing agents are used to identify functional groups of carbohydrates, in elucidating their structures, and for syntheses. The most important are (1) Benedict's or Tollens' reagents, (2) bromine water, (3) nitric acid, and (4) periodic acid. Each of these reagents produces a different and usually specific effect when it is allowed to react with a monosaccharide. We should now examine what these effects are.

22.6A Benedict's or Tollens' Reagents: Reducing Sugars

Benedict's reagent (an alkaline solution containing a cupric citrate complex ion) and Tollens' solution [$Ag(NH_3)_2OH$] oxidize and thus give positive tests with *aldoses and ketoses.* The tests are positive even though aldoses and ketoses exist primarily as cyclic hemiacetals.

We studied the use of Tollens' silver mirror test in Section 16.13. Benedict's solution (and the related Fehling's solution that contains a cupric tartrate complex ion) give brick-red precipitates of Cu_2O when they oxidize an aldose. (In alkaline solution ketoses are converted to aldoses (Section 22.5A), which are then oxidized by the cupric complexes.) Since the solutions of cupric tartrates and citrates are blue, the appearance of a brick-red precipitate is a vivid and unmistakable indication of a positive test.

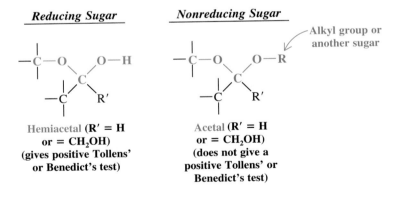

O
‖
CH
|
(CHOH)$_n$
|
CH$_2$OH
Aldose

Cu^{2+} (complex) +   or $\longrightarrow$ Cu$_2$O ↓ + oxidation products
Benedict's   (**brick-red**
solution   **reduction**
(**blue**)   **product**)

CH$_2$OH
|
C=O
|
(CHOH)$_2$
|
CH$_2$OH
Ketose

Sugars that give positive tests with Tollens' or Benedict's solutions are known as *reducing sugars,* and all carbohydrates that contain a *hemiacetal group* give positive tests. In aqueous solution these hemiacetals exist in equilibrium with relatively small, but not insignificant, concentrations of noncyclic aldehydes or α-hydroxy ketones. It is the latter two that undergo the oxidation, perturbing the equilibrium to produce more aldehyde or α-hydroxy ketone, which then undergoes oxidation until one reactant is exhausted.

Carbohydrates that contain only acetal groups do not give positive tests with Benedict's or Tollens' solution, and they are called *nonreducing sugars.* Acetals do not exist in equilibrium with aldehydes or α-hydroxy ketones in the basic aqueous media of the test reagents.

Reducing Sugar   *Nonreducing Sugar*

Hemiacetal (**R′ = H**   Acetal (**R′ = H**
or **= CH$_2$OH**)   or **= CH$_2$OH**)
(**gives positive Tollens'**   (**does not give a**
or Benedict's test)   **positive Tollens' or**
                            Benedict's test)

PROBLEM 22.6

How might you distinguish between α-D-glucopyranose (i.e., D-glucose) and methyl α-D-glucopyranoside?

Although Benedict's and Tollens' reagents have some use as diagnostic tools [Benedict's solution can be used in quantitative determinations of reducing sugars (reported as glucose) in blood or urine], neither of these reagents is useful as a preparative reagent in carbohydrate oxidations. Oxidations with both reagents take place in alkaline solution, *and in alkaline solutions sugars undergo a complex series of reactions that lead to isomerizations* (Section 22.5A).

22.6B Bromine Water: The Synthesis of Aldonic Acids

Monosaccharides do not undergo isomerization and fragmentation reactions in mildly acidic solution. Thus, a useful oxidizing reagent for preparative purposes is bromine in water (pH 6.0). Bromine water is a general reagent that selectively oxidizes the —CHO group to a —CO_2H group. It converts an aldose to an *aldonic acid:*

$$
\begin{array}{ccc}
\text{CHO} & & \text{CO}_2\text{H} \\
| & \xrightarrow[\text{H}_2\text{O}]{\text{Br}_2} & | \\
(\text{CHOH})_n & & (\text{CHOH})_n \\
| & & | \\
\text{CH}_2\text{OH} & & \text{CH}_2\text{OH} \\
\textbf{Aldose} & & \textbf{Aldonic acid}
\end{array}
$$

Experiments with aldopyranoses have shown that the actual course of the reaction is somewhat more complex than we have indicated above. Bromine water specifically oxidizes the *β* anomer, and the initial product that forms is a *δ-aldonolactone.* This compound may then hydrolyze to an aldonic acid, and the aldonic acid may undergo a subsequent ring closure to form a *γ-aldonolactone.*

β-D-Glucopyranose   D-Glucono-δ-lactone   D-Gluconic acid   D-Glucono-γ-lactone

22.6C Nitric Acid Oxidation: Aldaric Acids

Dilute nitric acid—a stronger oxidizing agent than bromine water—oxidizes both the —CHO group and the terminal —CH_2OH group of an aldose to —CO_2H groups. These dicarboxylic acids are known as *aldaric acids.*

$$
\begin{array}{ccc}
\text{CHO} & & \text{CO}_2\text{H} \\
| & \xrightarrow{\text{HNO}_3} & | \\
(\text{CHOH})_n & & (\text{CHOH})_n \\
| & & | \\
\text{CH}_2\text{OH} & & \text{CO}_2\text{H} \\
\textbf{Aldose} & & \textbf{Aldaric acid}
\end{array}
$$

It is not known whether a lactone is an intermediate in the oxidation of an aldose to an aldaric acid; however, aldaric acids form γ- and δ-lactones readily.

$$
\begin{array}{ccc}
\underset{\text{Aldaric acid}}{\begin{array}{c} \overset{O}{\underset{\parallel}{C}}\!\!-\!OH \\ CHOH \\ CHOH \\ CHOH \\ CHOH \\ \underset{\parallel}{\overset{}{C}}\!\!-\!OH \\ O \end{array}}
& \xrightarrow{-\,H_2O} &
\begin{array}{c}
\overset{O}{\underset{\parallel}{C}} \\ CHOH \\ CHOH \\ CH \\ CHOH \\ C\!-\!OH \\ O
\end{array}
\quad \text{or} \quad
\begin{array}{c}
\overset{O}{\underset{\parallel}{C}}\!\!-\!OH \\ CHOH \\ CH \\ CHOH \\ CHOH \\ C \\ O
\end{array}
\end{array}
$$

Aldaric acid
(from an aldohexose)   **γ-Lactones of an aldaric acid**

Corners such as this do not represent a CH_2 group

The aldaric acid obtained from D-glucose is called D-glucaric acid.*

D-Glucose ⇌

$$
\begin{array}{c}
\overset{O}{\underset{\parallel}{C}}H \\
H\!-\!\!-\!OH \\
HO\!-\!\!-\!H \\
H\!-\!\!-\!OH \\
H\!-\!\!-\!OH \\
CH_2OH
\end{array}
$$

$\xrightarrow{HNO_3}$

$$
\begin{array}{c}
\overset{O}{\underset{\parallel}{C}}\!-\!OH \\
H\!-\!\!-\!OH \\
HO\!-\!\!-\!H \\
H\!-\!\!-\!OH \\
H\!-\!\!-\!OH \\
\overset{}{\underset{\parallel}{C}}\!-\!OH \\
O
\end{array}
$$

D-Glucaric acid

PROBLEM 22.7

(a) Would you expect D-glucaric acid to be optically active?

(b) Write the open-chain structure for the aldaric acid (mannaric acid) that would be obtained by nitric acid oxidation of D-mannose.

(c) Would you expect it to be optically active?

(d) What aldaric acid would you expect to obtain from D-erythrose?

$$
\begin{array}{c}
CHO \\
H\!-\!\!-\!OH \\
H\!-\!\!-\!OH \\
CH_2OH
\end{array}
$$

D-Erythrose

(e) Would it show optical activity?

(f) D-Threose, a diastereomer of D-erythrose, yields an optically active aldaric acid when it is subjected to nitric acid oxidation. Write Fischer projection formulas for D-threose and its nitric acid oxidation product.

(g) What are the names of the aldaric acids obtained from D-erythrose and D-threose? (See Section 5.14A.)

*Older terms for an aldaric acid are a *glycaric* acid or a *saccharic* acid.

PROBLEM 22.8

D-Glucaric acid undergoes lactonization to yield two different γ-lactones. What are their structures?

22.6D Periodate Oxidations: Oxidative Cleavage of Polyhydroxy Compounds

Compounds that have hydroxyl groups on adjacent atoms undergo oxidative cleavage when they are treated with aqueous periodic acid (HIO_4). The reaction breaks carbon–carbon bonds and produces carbonyl compounds (aldehydes, ketones, or acids). The stoichiometry of the reaction is

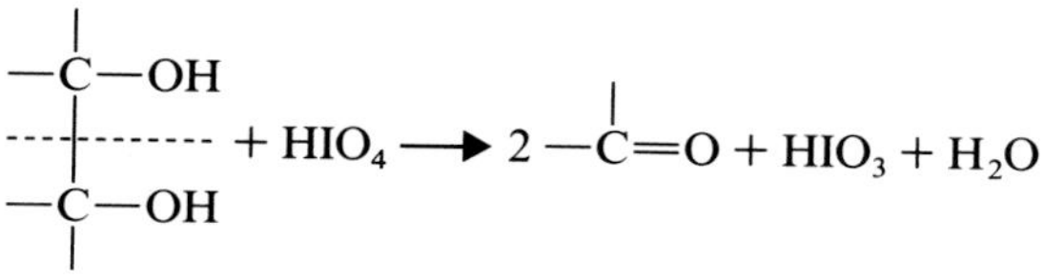

$$\begin{array}{c} -\overset{|}{C}-OH \\ \text{-------} \\ -\overset{|}{C}-OH \end{array} + HIO_4 \longrightarrow 2 -\overset{|}{C}=O + HIO_3 + H_2O$$

Since the reaction usually takes place in quantitative yield, valuable information can often be gained by measuring the number of molar equivalents of periodic acid that is consumed in the reaction as well as by identifying the carbonyl products.*

Periodate oxidations are thought to take place through a cyclic intermediate:

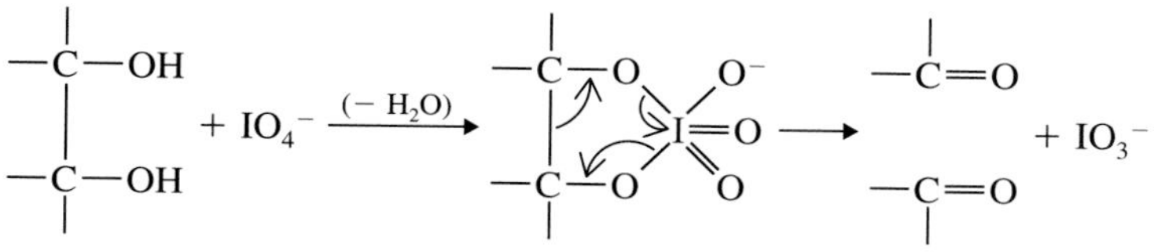

Before we discuss the use of periodic acid in carbohydrate chemistry, we should illustrate the course of the reaction with several simple examples. Notice in these periodate oxidations that *for every C—C bond broken, a C—O bond is formed at each carbon.*

1. When three or more —CHOH groups are contiguous, the internal ones are obtained as *formic acid.* Periodate oxidation of glycerol, for example, gives two molar equivalents of formaldehyde and one molar equivalent of formic acid.

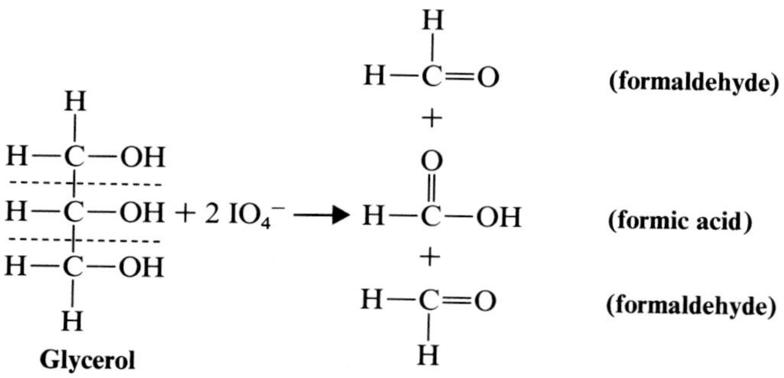

---

* The reagent lead tetraacetate, $Pb(O_2CCH_3)_4$, brings about cleavage reactions similar to those of periodic acid. The two reagents are complementary; periodic acid works well in aqueous solutions and lead tetraacetate gives good results in organic solvents.

2. Oxidative cleavage also takes place when an —OH group is adjacent to the carbonyl group of an aldehyde or ketone (but not that of an acid or an ester). Glyceraldehyde yields two molar equivalents of formic acid and one molar equivalent of formaldehyde, while dihydroxyacetone gives two molar equivalents of formaldehyde and one molar equivalent of carbon dioxide.

$$
\begin{array}{ll}
& \text{O} \\
& \parallel \\
& \text{H--C--OH} \quad \text{(formic acid)} \\
& + \\
\text{O} & \text{O} \\
\parallel & \parallel \\
\text{C--H} & \text{H--C--OH} \quad \text{(formic acid)} \\
\text{H--C--OH} + 2\ \text{IO}_4^- \longrightarrow & + \\
\text{H--C--OH} & \text{H--C=O} \quad \text{(formaldehyde)} \\
\text{H} & \text{H} \\
\textbf{Glyceraldehyde} &
\end{array}
$$

$$
\begin{array}{ll}
& \text{H} \\
& \mid \\
\text{H} & \text{H--C=O} \quad \text{(formaldehyde)} \\
\mid & + \\
\text{H--C--OH} & \\
\text{C=O} \quad + 2\ \text{IO}_4^- \longrightarrow & \text{O=C=O} \quad \text{(carbon dioxide)} \\
\text{H--C--OH} & + \\
\mid & \text{H--C=O} \quad \text{(formaldehyde)} \\
\text{H} & \text{H} \\
\textbf{Dihydroxyacetone} &
\end{array}
$$

3. Periodic acid does not cleave compounds in which the hydroxyl groups are separated by an intervening —CH₂— group, nor those in which a hydroxyl group is adjacent to an ether or acetal function.

$$
\begin{array}{lll}
\text{CH}_2\text{OH} & & \text{CH}_2\text{OCH}_3 \\
\mid & & \mid \\
\text{CH}_2 \quad + \text{IO}_4^- \longrightarrow \text{ no cleavage} & & \text{CHOH} \quad + \text{IO}_4^- \longrightarrow \text{ no cleavage} \\
\mid & & \mid \\
\text{CH}_2\text{OH} & & \text{CH}_2\text{R}
\end{array}
$$

PROBLEM 22.9

What products would you expect to be formed when each of the following compounds is treated with an appropriate amount of periodic acid? How many molar equivalents of HIO_4 would be consumed in each case?

(a) 2,3-Butanediol

(b) 1,2,3-Butanetriol

(c) $CH_2OHCHOHCH(OCH_3)_2$

(d) $CH_2OHCHOHCOCH_3$

(e) $CH_3COCHOHCOCH_3$

(f) *cis*-1,2-Cyclopentanediol

(g) $CH_3\overset{\displaystyle CH_3}{\underset{\displaystyle HO \quad OH}{\text{C}-\text{CH}_2}}$

(h) D-Erythrose

PROBLEM 22.10

Show how periodic acid could be used to distinguish between an aldohexose and a ketohexose. What products would you obtain from each, and how many molar equivalents of HIO_4 would be consumed?

22.7 REDUCTION OF MONOSACCHARIDES: ALDITOLS

Aldoses (and ketoses) can be reduced with sodium borohydride to compounds called *alditols*.

$$
\begin{array}{c}
\text{CHO} \\
| \\
\text{(CHOH)}_n \\
| \\
\text{CH}_2\text{OH} \\
\textbf{Aldose}
\end{array}
\xrightarrow[\substack{\text{or} \\ H_2,\, Pt}]{\text{NaBH}_4}
\begin{array}{c}
\text{CH}_2\text{OH} \\
| \\
\text{(CHOH)}_n \\
| \\
\text{CH}_2\text{OH} \\
\textbf{Alditol}
\end{array}
$$

Reduction of D-glucose, for example, yields D-glucitol.

D-Glucitol
(or D-sorbitol)

PROBLEM 22.11

(a) Would you expect D-glucitol to be optically active?   (b) Write Fischer projection formulas for all of the D-aldohexoses that would yield *optically inactive* alditols.

22.8 REACTIONS OF MONOSACCHARIDES WITH PHENYLHYDRAZINE: OSAZONES

The aldehyde group of an aldose reacts with such carbonyl reagents as hydroxylamine and phenylhydrazine (Section 16.8). With hydroxylamine, the product is the expected oxime. With enough phenylhydrazine, however, three molar equivalents of phenylhydrazine are consumed and a second phenylhydrazone group is introduced at C2. The product is called a *phenylosazone*. Phenylosazones crystallize readily (unlike sugars) and are useful derivatives for identifying sugars.

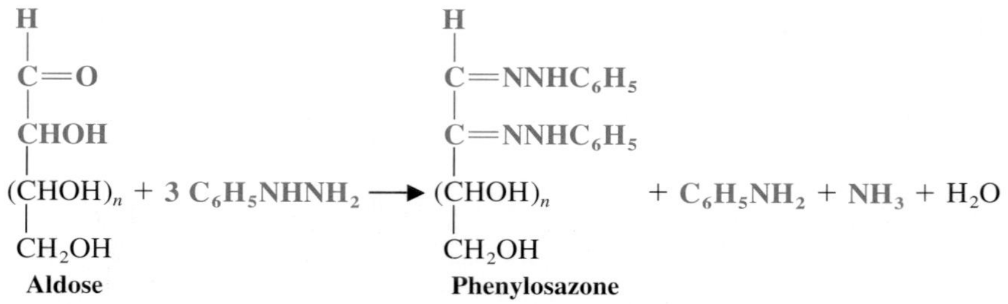

$$
\begin{array}{c}
\text{H} \\
| \\
\text{C}{=}\text{O} \\
| \\
\text{CHOH} \\
| \\
(\text{CHOH})_n \\
| \\
\text{CH}_2\text{OH} \\
\textbf{Aldose}
\end{array}
+ 3\ \text{C}_6\text{H}_5\text{NHNH}_2 \longrightarrow
\begin{array}{c}
\text{H} \\
| \\
\text{C}{=}\text{NNHC}_6\text{H}_5 \\
| \\
\text{C}{=}\text{NNHC}_6\text{H}_5 \\
| \\
(\text{CHOH})_n \\
| \\
\text{CH}_2\text{OH} \\
\textbf{Phenylosazone}
\end{array}
+ \text{C}_6\text{H}_5\text{NH}_2 + \text{NH}_3 + \text{H}_2\text{O}
$$

The mechanism for osazone formation probably depends on a series of reactions in which $\text{C}{=}\text{N}-$ behaves very much like $\text{C}{=}\text{O}$ in giving a nitrogen version of an enol.

A MECHANISM FOR PHENYLOSAZONE FORMATION

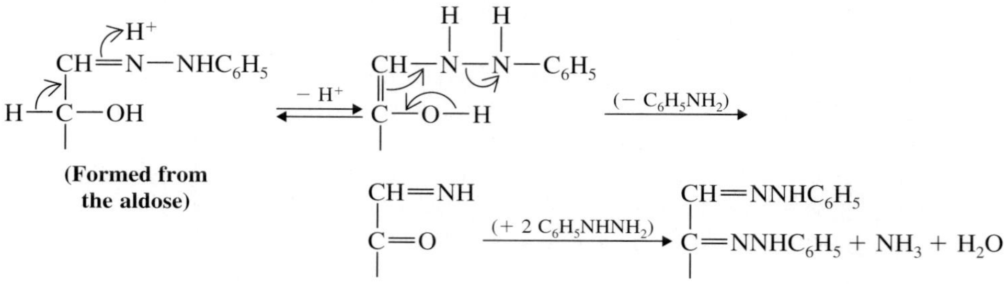

$$
\begin{array}{c}
\text{CH}{=}\text{N}-\text{NHC}_6\text{H}_5 \\
| \\
\text{H}-\text{C}-\text{OH} \\
| \\
\textbf{(Formed from} \\
\textbf{the aldose)}
\end{array}
\xrightleftharpoons[\;-\,\text{H}^+\;]{}
\begin{array}{c}
\text{H}\quad \text{H} \\
|\quad | \\
\text{CH}-\text{N}-\text{N}-\text{C}_6\text{H}_5 \\
\| \\
\text{C}-\text{O}-\text{H} \\
|
\end{array}
\xrightarrow{(-\;\text{C}_6\text{H}_5\text{NH}_2)}
$$

$$
\begin{array}{c}
\text{CH}{=}\text{NH} \\
| \\
\text{C}{=}\text{O} \\
|
\end{array}
\xrightarrow{(+\;2\;\text{C}_6\text{H}_5\text{NHNH}_2)}
\begin{array}{c}
\text{CH}{=}\text{NNHC}_6\text{H}_5 \\
| \\
\text{C}{=}\text{NNHC}_6\text{H}_5 \\
|
\end{array}
+ \text{NH}_3 + \text{H}_2\text{O}
$$

Osazone formation results in a loss of the stereocenter at C2 but does not affect other stereocenters; D-glucose and D-mannose, for example, yield the same phenylosazone:

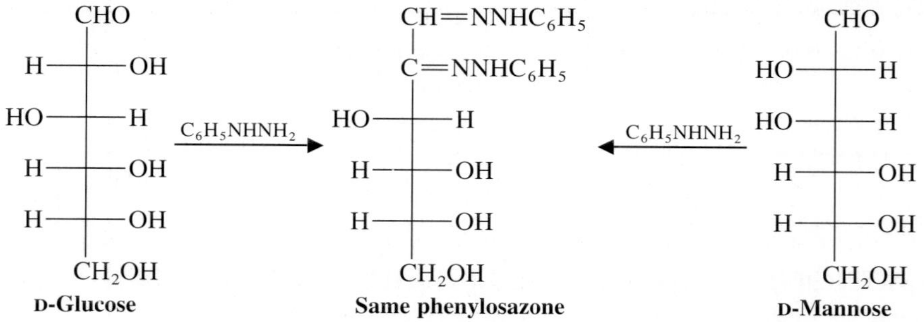

D-Glucose $\xrightarrow{\text{C}_6\text{H}_5\text{NHNH}_2}$ Same phenylosazone $\xleftarrow{\text{C}_6\text{H}_5\text{NHNH}_2}$ D-Mannose

This experiment, first done by Emil Fischer, establishes that D-glucose and D-mannose have the same configurations about C3, C4, and C5. Diastereomeric aldoses (such as D-glucose and D-mannose) that differ only in configuration at C2 are called *epimers*.*

---

*The term *epimer* has taken on a broader meaning and is now often applied to any pair of diastereomers that differ only in the configuration at a single atom.

PROBLEM 22.12

Although D-fructose is not an epimer of D-glucose or D-mannose (D-fructose is a ketohexose), all three yield the same phenylosazone. (a) Using Fischer projection formulas, write an equation for the reaction of fructose with phenyl-hydrazine. (b) What information about the stereochemistry of D-fructose does this experiment yield?

22.9 SYNTHESIS AND DEGRADATION OF MONOSACCHARIDES

22.9A Kiliani–Fischer Synthesis

In 1885, Heinrich Kiliani (Freiburg, Germany) discovered that an aldose can be converted to the epimeric aldonic acids having one additional carbon through the addition of hydrogen cyanide and subsequent hydrolysis of the epimeric cyanohydrins. Fischer later extended this method by showing that aldonolactones obtained from the aldonic acids can be reduced to aldoses. Today, this method for lengthening the carbon chain of an aldose is called the Kiliani–Fischer synthesis.

We can illustrate the Kiliani–Fischer synthesis with the synthesis of D-threose and D-erythrose (aldotetroses) from D-glyceraldehyde (an aldotriose) in Fig. 22.6.

Addition of hydrogen cyanide to glyceraldehyde produces two epimeric cyanohydrins because the reaction creates a new stereocenter. The cyanohydrins can be separated easily (since they are diastereomers) and each can be converted to an aldose through hydrolysis, acidification, lactonization, and reduction with Na-Hg at pH 3–5. One cyanohydrin ultimately yields D-(−)-erythrose and the other yields D-(−)-threose.

We can be sure that the aldotetroses that we obtain from this Kiliani–Fischer synthesis are both D sugars because the starting compound is D-glyceraldehyde and its stereocenter is unaffected by the synthesis. On the basis of the Kiliani–Fischer synthesis we cannot know just which aldotetrose has both —OH groups on the right and which has the top —OH on the left in the Fischer projection formulas. However, if we oxidize both aldotetroses to aldaric acids, one [D-(−)-erythrose] will yield an *optically inactive* (meso) product while the other [D-(−)-threose] will yield a product that is *optically active* (cf. Problem 22.7).

PROBLEM 22.13

(a) What are the structures of L-(+)-threose and L-(+)-erythrose? (b) What aldotriose would you use to prepare them in a Kiliani–Fischer synthesis?

PROBLEM 22.14

(a) Outline a Kiliani–Fischer synthesis of epimeric aldopentoses starting with D-(−)-erythrose (use Fischer projection formulas). (b) The two epimeric aldopentoses that one obtains are D-(−)-arabinose and D-(−)-ribose. Nitric acid oxidation of D-(−)-ribose yields an optically inactive aldaric acid, whereas similar oxidation of D-(−)-arabinose yields an optically active product. On the basis of this information alone, which Fischer projection formula represents D-(−)-arabinose and which represents D-(−)-ribose?

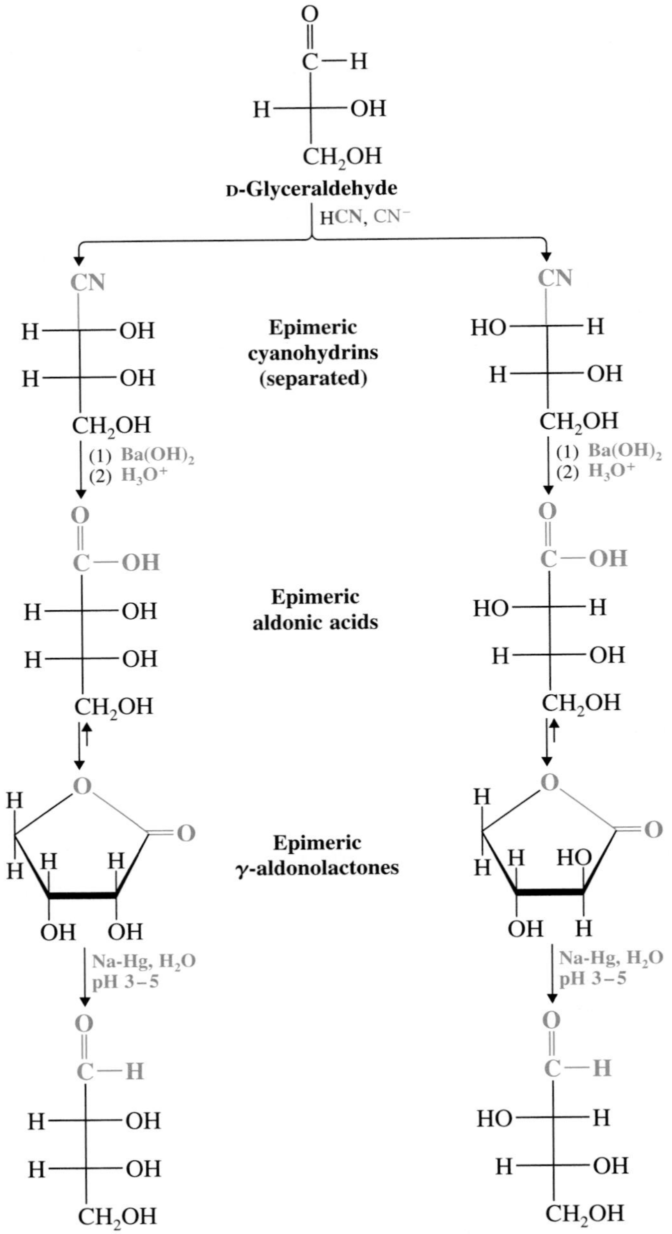

FIGURE 22.6 A Kiliani–Fischer synthesis of D-(−)-erythrose and D-(−)-threose from D-glyceraldehyde.

PROBLEM 22.15

Subjecting D-(−)-threose to a Kiliani–Fischer synthesis yields two other epimeric aldopentoses, D-(+)-xylose and D-(−)-lyxose. D-(+)-Xylose can be oxidized (with nitric acid) to an optically inactive aldaric acid, while similar oxidation of D-(−)-lyxose gives an optically active product. What are the structures of D-(+)-xylose and D-(−)-lyxose?

PROBLEM 22.16

There are eight aldopentoses. In Problems 22.14 and 22.15 you have arrived at the structures of four. What are the names and structures of the four that remain?

22.9B The Ruff Degradation

Just as the Kiliani–Fischer synthesis can be used to lengthen the chain of an aldose by one carbon atom, the Ruff degradation* can be used to shorten the chain by a similar unit. The Ruff degradation involves (1) oxidation of the aldose to an aldonic acid using bromine water and (2) oxidative decarboxylation of the aldonic acid to the next lower aldose using hydrogen peroxide and ferric sulfate. D-(−)-Ribose, for example, can be degraded to D-(−)-erythrose:

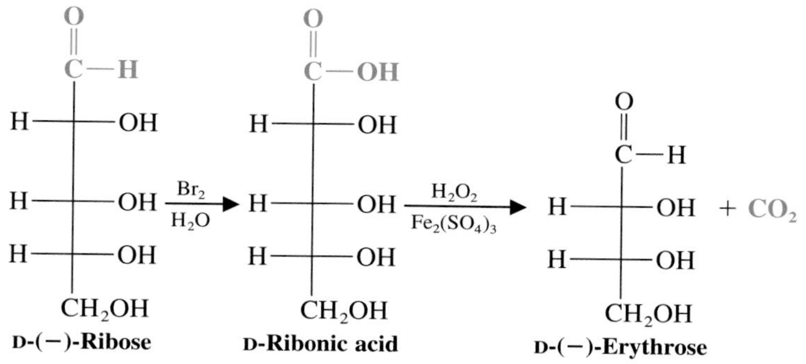

PROBLEM 22.17

The aldohexose D-(+)-galactose can be obtained by hydrolysis of *lactose,* a disaccharide found in milk. When D-(+)-galactose is treated with nitric acid, it yields an optically inactive aldaric acid. When D-(+)-galactose is subjected to a Ruff degradation, it yields D-(−)-lyxose (cf. Problem 22.15). Using only these data, write the Fischer projection formula for D-(+)-galactose.

22.10 THE D FAMILY OF ALDOSES

The Ruff degradation and the Kiliani–Fischer synthesis allow us to place all of the aldoses into families or "family trees" based on their relation to D- or L-glyceraldehyde. Such a tree is constructed in Fig. 22.7 and includes the structures of the D-aldohexoses, **1–8**.

Most, but not all, of the naturally occurring aldoses belong to the D family with D-(+)-glucose being by far the most common. D-(+)-Galactose can be obtained from milk sugar (lactose), but L-(−)-galactose occurs in a polysaccharide obtained from the vineyard snail, *Helix pomatia.* L-(+)-Arabinose is found widely, but D-(−)-arabinose is scarce, being found only in certain bacteria and sponges. Threose, lyxose, gulose, and allose do not occur naturally, but one or both forms (D or L) of each have been synthesized.

*Developed by Otto Ruff, a German chemist, 1871–1939.

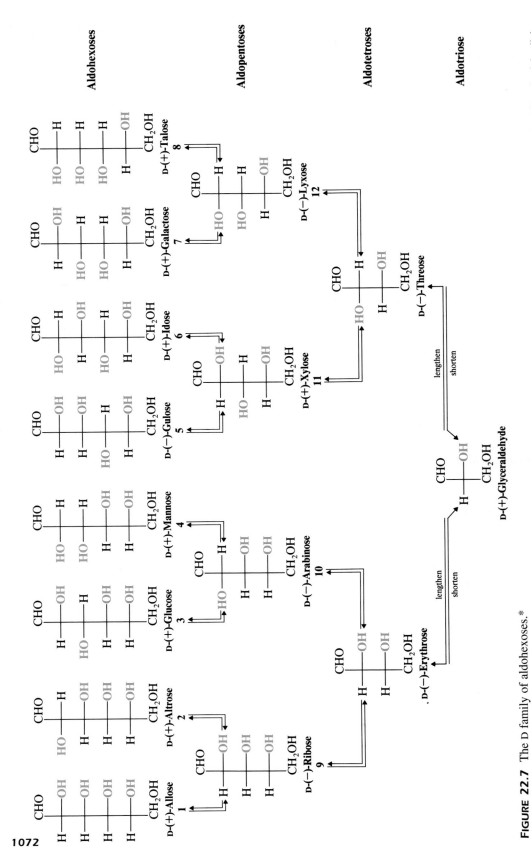

FIGURE 22.7 The D family of aldohexoses.*

* A useful mnemonic for the D-aldohexoses: All altruists gladly make gum in gallon tanks. Write the names in a line and above each write CH₂OH. Then, for C5 write OH to the right all the way across. For C4 write OH to the right four times, then four to the left; for C3, write OH twice to the right, twice to the left, and repeat; C2, alternate OH and H to the right. (From L. F. Fieser and Mary Fieser, *Organic Chemistry*, Reinhold, New York, 1956, p. 359.)

22.11 FISCHER'S PROOF OF THE CONFIGURATION OF D-(+)-GLUCOSE

Emil Fischer began his work on the stereochemistry of (+)-glucose in 1888, only 12 years after van't Hoff and Le Bel had made their proposal concerning the tetrahedral structure of carbon. Only a small body of data was available to Fischer at the beginning: Only a few monosaccharides were known, including (+)-glucose, (+)-arabinose, and (+)-mannose. [(+)-Mannose had just been synthesized by Fischer.] The sugars (+)-glucose and (+)-mannose were known to be aldohexoses; (+)-arabinose was known to be an aldopentose.

Since an aldohexose has four stereocenters, 2^4 (or 16) stereoisomers are possible —*one of which is* (+)-*glucose.* Fischer arbitrarily decided to limit his attention to the eight structures with the D configuration given in Fig. 22.7 (structures **1–8**). Fischer realized that he would be unable to differentiate between enantiomeric configurations because methods for determining the absolute configuration of organic compounds had not been developed. It was not until 1951, when Bijvoet (Section 5.14A) determined the absolute configuration of L-(+)-tartaric acid [and, hence, D-(+)-glyceraldehyde] that Fischer's arbitrary assignment of (+)-glucose to the family we call the D family was known to be correct.

Fischer's assignment of structure **3** to (+)-glucose was based on the following reasoning:

1. Nitric acid oxidation of (+)-glucose gives an optically active aldaric acid. This eliminates structures **1** and **7** from consideration because both compounds would yield *meso*-aldaric acids.

2. *Degradation of* (+)-*glucose gives* (−)-*arabinose, and nitric acid oxidation of* (−)-*arabinose gives an optically active aldaric acid.* This means that (−)-arabinose cannot have configurations **9** or **11** and must have either structure **10** or **12**. It also establishes that (+)-glucose cannot have configuration **2**, **5**, or **6**. This leaves structures **3**, **4**, and **8** as possibilities for (+)-glucose.

3. A Kiliani–Fischer synthesis beginning with (−)-arabinose gives (+)-glucose and (+)-mannose; nitric acid oxidation of (+)-mannose gives an optically active aldaric acid. This, together with the fact that (+)-glucose yields a different but also optically active aldaric acid, establishes structure **10** as the structure of (−)-arabinose and eliminates structure **8** as a possible structure for (+)-glucose. Had (−)-arabinose been represented by structure **12**, a Kiliani–Fischer synthesis would have given the two aldohexoses, **7** and **8**, one of which (**7**) would yield an optically inactive aldaric acid on nitric acid oxidation.

4. Two structures now remain, **3** and **4**; one structure represents (+)-glucose and one represents (+)-mannose. Fischer realized that (+)-glucose and (+)-mannose were epimeric (at C2), but a decision as to which compound was represented by which structure was most difficult.

5. Fischer had already developed a method for effectively *interchanging the two end groups* (CHO and CH_2OH) *of an aldose chain.* And, with brilliant logic, Fischer realized that if (+)-glucose had structure **4**, an interchange of end groups *would yield the same aldohexose:*

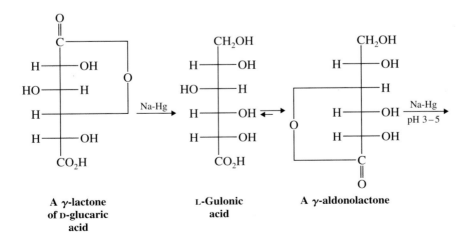

On the other hand, if (+)-glucose has structure **3**, *an end-group interchange will yield a different aldohexose,* **13**:

This new aldohexose, if it were formed, would be an L-sugar and it would be the mirror reflection of D-gulose. Thus its name would be L-gulose.

Fischer carried out the end-group interchange starting with (+)-glucose and *the product was the new aldohexose* **13**. This outcome proved that (+)-glucose has structure **3**. It also established **4** as the structure for (+)-mannose, and it proved the structure of L-(+)-gulose as **13**.

The procedure Fischer used for interchanging the ends of the (+)-glucose chain began with one of the γ-lactones of D-glucaric acid (cf. Problem 22.8) and was carried out as follows:

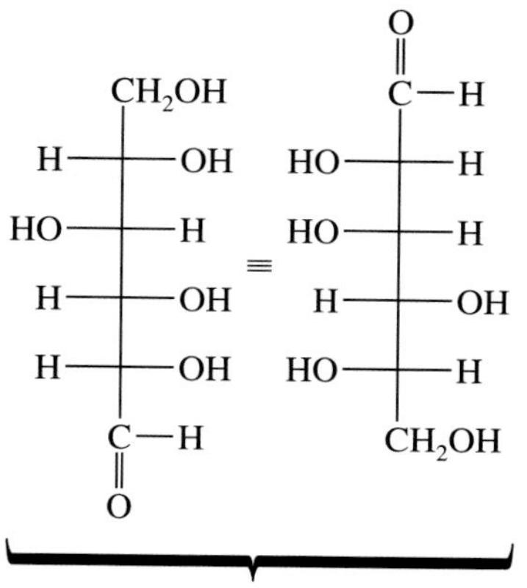

L-(+)-Gulose
13

Notice in this synthesis that the second reduction with Na-Hg is carried out at pH 3–5. Under these conditions, reduction of the lactone yields an aldehyde and not a primary alcohol.

PROBLEM 22.18

Fischer actually had to subject both γ-lactones of D-glucaric acid (Problem 22.8) to the procedure just outlined. What product does the other γ-lactone yield?

22.12 DISACCHARIDES

22.12A Sucrose

Ordinary table sugar is a disaccharide called *sucrose*. Sucrose, the most widely occurring disaccharide, is found in all photosynthetic plants and is obtained commercially from sugar cane or sugar beets. Sucrose has the structure shown in Fig. 22.8.

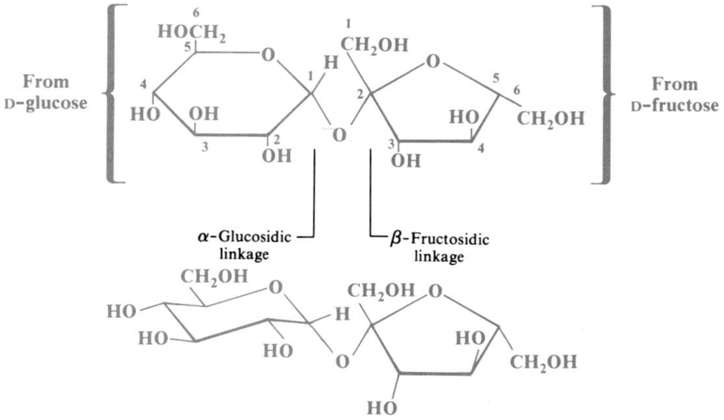

FIGURE 22.8 Two representations of the formula for (+)-sucrose (α-D-glucopyranosyl β-D-fructofuranoside).

The structure of sucrose is based on the following evidence:

1. Sucrose has the molecular formula $C_{12}H_{22}O_{11}$.
2. Acid-catalyzed hydrolysis of 1 mol of sucrose yields 1 mol of D-glucose and 1 mol of D-fructose.

Fructose
(as a β – furanose)

3. Sucrose is a nonreducing sugar; it gives negative tests with Benedict's and Tollens' solutions. Sucrose does not form an osazone and does not undergo mutarotation. These facts mean that neither the glucose nor the fructose portion of sucrose has a hemiacetal group. Thus, the two hexoses must have a glycoside linkage that involves C1 of glucose and C2 of fructose, for only in this way will both carbonyl groups be present as full acetals (i.e., as glycosides).
4. The stereochemistry of the glycoside linkages can be inferred from experiments done with enzymes. Sucrose is hydrolyzed by an *α-glucosidase* obtained from yeast but not by β-glucosidases. This hydrolysis indicates *an α configuration at the glucoside portion.* Sucrose is also hydrolyzed by *sucrase,* an enzyme known to hydrolyze β-fructofuranosides but not α-fructofuranosides. This hydrolysis indicates a *β configuration at the fructoside portion.*
5. Methylation of sucrose gives an octamethyl derivative that, on hydrolysis, gives 2,3,4,6-tetra-*O*-methyl-D-glucose and 1,3,4,6-tetra-*O*-methyl-D-fructose. The identities of these two products demonstrate that the glucose portion is a *pyranoside* and that the fructose portion is a *furanoside.*

The structure of sucrose has been confirmed by X-ray analysis and by an unambiguous synthesis.

22.12B Maltose

When starch (Section 22.13A) is hydrolyzed by the enzyme *diastase,* one product is a disaccharide known as *maltose* (Fig. 22.9).

1. When 1 mol of maltose is subjected to acid-catalyzed hydrolysis, it yields 2 mol of D-(+)-glucose.
2. Unlike sucrose, *maltose is a reducing sugar;* it gives positive tests with Fehling's, Benedict's, and Tollens' solutions. Maltose also reacts with phenylhydrazine to form a monophenylosazone (i.e., it incorporates two molecules of phenylhydrazine).
3. Maltose exists in two anomeric forms; α-(+)-maltose, $[\alpha]_D^{25} = +168°$, and β-(+)-maltose, $[\alpha]_D^{25} = +112°$. The maltose anomers undergo mutarotation to yield an equilibrium mixture, $[\alpha]_D^{25} = +136°$.

Facts **2** and **3** demonstrate that one of the glucose residues of maltose is present in a hemiacetal form; the other, therefore, must be present as a glucoside. The configuration of this glucosidic linkage can be inferred as α, because maltose is hydrolyzed by α-glucosidases and not by β-glucosidases.

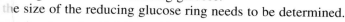

FIGURE 22.9 Two representations of the structure of the β anomer of (+)-maltose, 4-O-(α-D-glucopyranosyl)-β-D-glucopyranose.

4. Maltose reacts with bromine water to form a monocarboxylic acid, maltonic acid (Fig. 22.10a). This fact, too, is consistent with the presence of only one hemiacetal group.

5. Methylation of maltonic acid followed by hydrolysis gives 2,3,4,6-tetra-O-methyl-D-glucose and 2,3,5,6-tetra-O-methyl-D-gluconic acid. That the first product has a free —OH at C5 indicates that the nonreducing glucose portion is present as a pyranoside; that the second product, 2,3,5,6-tetra-O-methyl-D-gluconic acid, has a free —OH at C4 indicates that this position was involved in a glycosidic linkage with the nonreducing glucose.

the size of the reducing glucose ring needs to be determined.

of maltose itself, followed by hydrolysis (Fig. 22.10b), gives 2,3,4,6-hyl-D-glucose and 2,3,6-tri-O-methyl-D-glucose. The free —OH at atter product indicates that it must have been involved in the oxide ring ne reducing glucose is present as a *pyranose*.

Cellobiose

Partial hydrolysis of cellulose (Section 22.13C) gives the disaccharide cellobiose $(C_{12}H_{22}O_{11})$ (Fig. 22.11). Cellobiose resembles maltose in every respect except one: the configuration of its glycosidic linkage.

Cellobiose, like maltose, is a reducing sugar that, on acid-catalyzed hydrolysis, yields two molar equivalents of D-glucose. Cellobiose also undergoes mutarotation and forms a monophenylosazone. Methylation studies show that C1 of one glucose unit is connected in glycosidic linkage with C4 of the other and that both rings are six membered. Unlike maltose, however, cellobiose is hydrolyzed by β-*glucosidases* and not by α-glucosidases: This indicates that the glycosidic linkage in cellobiose is β (Fig. 22.11).

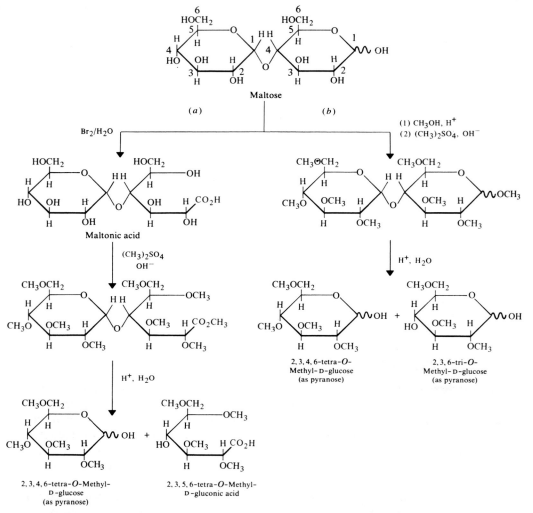

FIGURE 22.10 (*a*) Oxidation of maltose to maltonic acid followed by methylation and hydrolysis. (*b*) Methylation and subsequent hydrolysis of maltose itself.

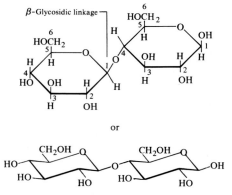

FIGURE 22.11 Two representations of the *β* anomer of cellobiose, 4-*O*-(*β*-D-glucopyranosyl)-*β*-D-glucopyranose.

22.12D Lactose

Lactose (Fig. 22.12) is a disaccharide present in the milk of humans, cows, and almost all other mammals. Lactose is a reducing sugar that hydrolyzes to yield D-glucose and D-galactose; the glycosidic linkage is β.

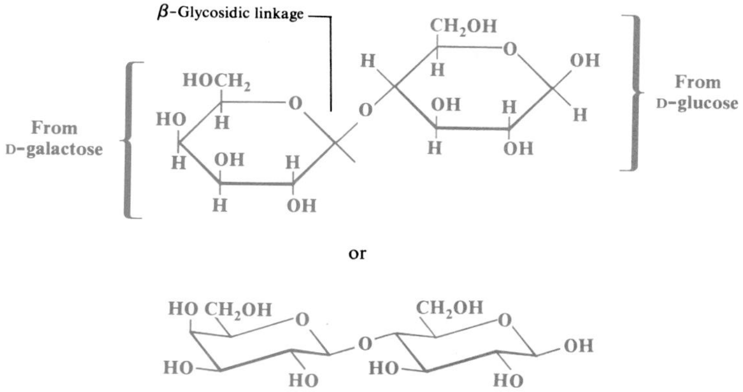

FIGURE 22.12 Two representations of the β anomer of lactose, 4-*O*-(β-D-galactopyranosyl)-β-D-glucopyranose.

22.13 POLYSACCHARIDES

Polysaccharides, also known as **glycans,** consist of monosaccharides joined together by glycosidic linkages. Polysaccharides that are polymers of a single monosaccharide are called **homopolysaccharides;** those made up of more than one type of monosaccharide are called **heteropolysaccharides.** Homopolysaccharides are also classified on the basis of their monosaccharide units. A homopolysaccharide consisting of glucose monomeric units is called a **glucan;** one consisting of galactose units is a **galactan,** and so on.

Three important polysaccharides, all of which are glucans, are starch, glycogen, and cellulose. Starch is the principal food reserve of plants; glycogen functions as a carbohydrate reserve for animals; and cellulose serves as structural material in plants. As we examine the structures of these three polysaccharides, we shall be able to see how each is especially suited for its function.

22.13A Starch

Starch occurs as microscopic granules in the roots, tubers, and seeds of plants. Corn, potatoes, wheat, and rice are important commercial sources of starch. Heating starch with water causes the granules to swell and produce a colloidal suspension from which two major components can be isolated. One fraction is called *amylose* and the other *amylopectin.* Most starches yield 10–20% amylose and 80–90% amylopectin.

Physical measurements show that amylose typically consists of more than 1000 D-glucopyranoside units *connected in α linkages* between C1 of one unit and C4 of the next (Fig. 22.13). Thus, in the ring size of its glucose units and in the configuration of the glycosidic linkages between them, amylose resembles maltose.

Chains of D-glucose units with α-glycosidic linkages such as those of amylose tend to assume a helical arrangement (Fig. 22.14). This arrangement results in a com-

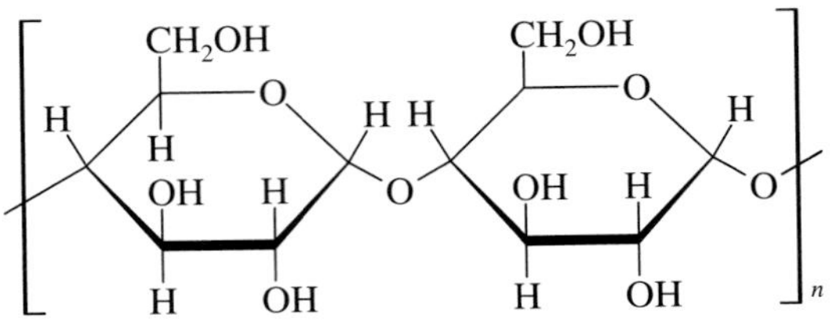

n > 500

FIGURE 22.13 Partial structure of amylose, an unbranched polymer of D-glucose connected in α,1 : 4-glycosidic linkages.

FIGURE 22.14 Amylose. The α,1 : 4 linkages cause it to assume the shape of a left-handed helix. (Figure copyrighted © by Irving Geis. From D. Voet and J. G. Voet, *Biochemistry,* 2nd ed., Wiley, New York, 1995, p. 262. Used with permission.)

pact shape for the amylose molecule even though its molecular weight is quite large (150,000–600,000).

Amylopectin has a structure similar to that of amylose (i.e., α,1 : 4 links), with the exception that in amylopectin the chains are branched. Branching takes place between C6 of one glucose unit and C1 of another and occurs at intervals of 20–25 glucose units (Fig. 22.15). Physical measurements indicate that amylopectin has a molecular weight of 1–6 million; thus amylopectin consists of hundreds of interconnecting chains of 20–25 glucose units each.

22.13B Glycogen

Glycogen has a structure very much like that of amylopectin; however, in glycogen the chains are much more highly branched. Methylation and hydrolysis of glycogen indi-

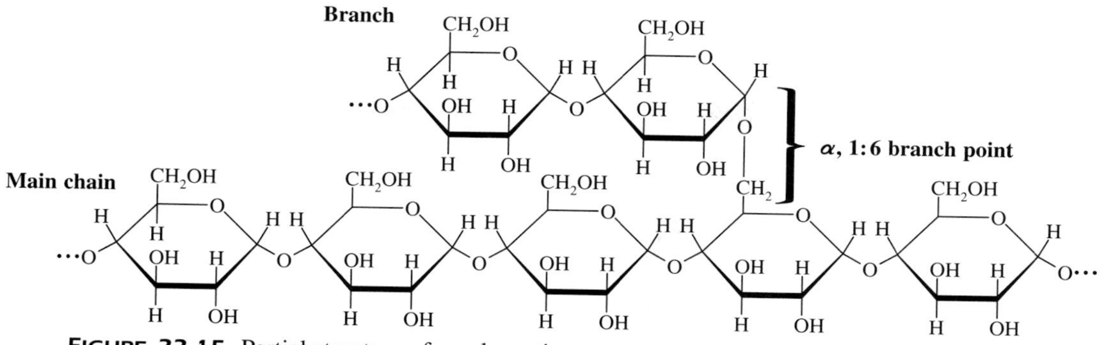

FIGURE 22.15 Partial structure of amylopectin.

cates that there is one end group for every 10–12 glucose units; branches may occur as often as every 6 units. Glycogen has a very high molecular weight. Studies of glycogens isolated under conditions that minimize the likelihood of hydrolysis indicate molecular weights as high as 100 million.

The size and structure of glycogen beautifully suit its function as reserve carbohydrate for animals. First, its size makes it too large to diffuse across cell membranes; thus, glycogen remains inside the cell where it is needed as an energy source. Second, because glycogen incorporates tens of thousands of glucose units in a single molecule, it solves an important osmotic problem for the cell. Were so many glucose units present in the cell as individual molecules, the osmotic pressure within the cell would be enormous—so large that the cell membrane would almost certainly break.* Finally, the localization of glucose units within a large, highly branched structure simplifies one of the cell's logistical problems: that of having a ready source of glucose when cellular glucose concentrations are low and of being able to store glucose rapidly when cellular glucose concentrations are high. There are enzymes within the cell that catalyze the reactions by which glucose units are detached from (or attached to) glycogen. These enzymes operate at end groups by hydrolyzing (or forming) $\alpha,1:4$ glycosidic linkages. Because glycogen is so highly branched, a very large number of end groups is available at which these enzymes can operate. At the same time the overall concentration of glycogen (in moles per liter) is quite low because of its enormous molecular weight.

Amylopectin presumably serves a similar function in plants. The fact that amylopectin is less highly branched than glycogen is, however, not a serious disadvantage. Plants have a much lower metabolic rate than animals—and plants, of course, do not require sudden bursts of energy.

Animals store energy as fats (triacylglycerols) as well as glycogen. Fats, because they are more highly reduced, are capable of furnishing much more energy. The metabolism of a typical fatty acid, for example, liberates more than twice as much energy per carbon as glucose or glycogen. Why then, we might ask, has Nature developed two different repositories? Glucose (from glycogen) is readily available and is highly water soluble.† Glucose, as a result, diffuses rapidly through the aqueous medium of the cell and serves as an ideal source of "ready energy." Long-chain fatty

*The phenomenon of osmotic pressure occurs whenever two solutions of different concentrations are separated by a membrane that allows penetration (by osmosis) of the solvent but not of the solute. The osmotic pressure (π) on one side of the membrane is related to the number of moles of solute particles (n), the volume of the solution (V), and the gas constant times the absolute temperature (RT): $\pi V = nRT$.

†Glucose is actually liberated as glucose-6-phosphate (G6P), which is also water soluble.

acids, by contrast, are almost insoluble in water and their concentration inside the cell could never be very high. They would be a poor source of energy if the cell were in an energy pinch. On the other hand, fatty acids (as triacylglycerols) because of their caloric richness are an excellent energy repository for long-term energy storage.

22.13C Cellulose

When we examine the structure of cellulose, we find another example of a polysaccharide in which nature has arranged monomeric glucose units in a manner that suits its function. Cellulose contains D-glucopyranoside units linked in 1 : 4 fashion in very long unbranched chains. Unlike starch and glycogen, however, the linkages in cellulose are *β-glycosidic linkages* (Fig. 22.16). This configuration of the anomeric carbon atoms of cellulose makes cellulose chains essentially linear; they do not tend to coil into helical structures as do glucose polymers when linked in an $\alpha,1:4$ manner.

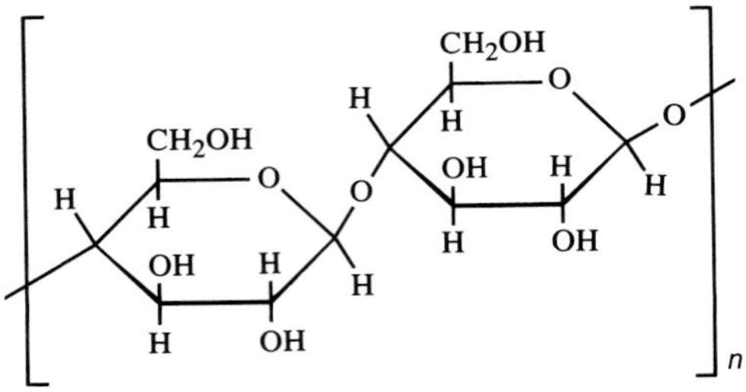

FIGURE 22.16 A portion of a cellulose chain. The glycosidic linkages are $\beta,1:4$.

The linear arrangement of β-linked glucose units in cellulose presents a uniform distribution of —OH groups on the outside of each chain. When two or more cellulose chains make contact, the hydroxyl groups are ideally situated to "zip" the chains together by forming hydrogen bonds (Fig. 22.17). Zipping many cellulose chains together in this way gives a highly insoluble, rigid, and fibrous polymer that is ideal as cell-wall material for plants.

This special property of cellulose chains, we should emphasize, is not just a result of $\beta,1:4$ glycosidic linkages; it is also a consequence of the precise stereochemistry of D-glucose at each stereocenter. Were D-galactose or D-allose units linked in a similar fashion, they almost certainly would not give rise to a polymer with properties like cellulose. Thus, we get another glimpse of why D-glucose occupies such a special position in the chemistry of plants and animals. Not only is it the most stable aldohexose (because it can exist in a chair conformation that allows all of its bulky groups to occupy equatorial positions), but its special stereochemistry also allows it to form helical structures when α linked as in starches, and rigid linear structures when β linked as in cellulose.

Another interesting and important fact about cellulose: The digestive enzymes of humans cannot attack its $\beta,1:4$ linkages. Hence, cellulose cannot serve as a food source for humans, as can starch. Cows and termites, however, can use cellulose (of grass and wood) as a food source because symbiotic bacteria in their digestive systems furnish β-glucosidases.

Perhaps we should ask ourselves one other question: Why has Nature "chosen" D-(+)-glucose for its special role rather than L-(−)-glucose, its mirror image? Here an answer cannot be given with any certainty. The selection of D-(+)-glucose may simply

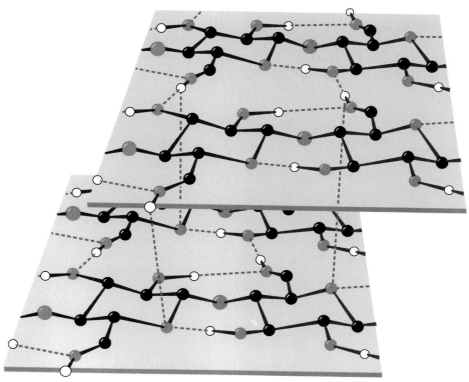

FIGURE 22.17 A proposed structure for cellulose. A fiber of cellulose may consist of about 40 parallel strands of glucose molecules linked in a $\beta,1:4$ fashion. Each glucose unit in a chain is turned over with respect to the preceding glucose unit, and is held in this position by hydrogen bonds (dashed lines) between the chains. The glucan chains line up laterally to form sheets and these sheets stack vertically so that they are staggered by one half of a glucose unit. (Hydrogen atoms that do not participate in hydrogen bonding have been omitted for clarity.) (From D. Voet and J. G. Voet, *Biochemistry,* 2nd ed., Wiley, New York, 1995, p. 261. Used with permission.)

have been a random event early in the course of the evolution of enzyme catalysts. Once this selection was made, however, the stereogenicity of the active sites of the enzymes involved would retain a bias toward D-(+)-glucose and against L-(−)-glucose (because of the improper fit of the latter). Once introduced, this bias would be perpetuated and extended to other catalysts.

Finally, when we speak of Nature selecting or choosing a particular molecule for a given function, we do not mean to imply that evolution operates on a molecular level. Evolution, of course, takes place at the level of organism populations, and molecules are selected only in the sense that their use gives the organism an increased likelihood of surviving and procreating.

22.13D Cellulose Derivatives

A number of derivatives of cellulose are used commercially. Most of these are compounds in which two or three of the free hydroxyl groups of each glucose unit have been converted to an ester or an ether. This conversion substantially alters the physical properties of the material, making it more soluble in organic solvents and allowing it to be made into fibers and films. Treating cellulose with acetic anhydride produces the triacetate known as "Arnel" or "acetate," used widely in the textile industry. Cellulose trinitrate, also called "gun cotton" or nitrocellulose, is used in explosives.

Rayon is made by treating cellulose (from cotton or wood pulp) with carbon disulfide in a basic solution. This reaction converts cellulose to a soluble xanthate:

$$\text{Cellulose}-\text{OH} + \text{CS}_2 \xrightarrow{\text{NaOH}} \underset{\textbf{Cellulose xanthate}}{\text{cellulose}-\text{O}-\overset{\displaystyle \text{S}}{\underset{}{\overset{\|}{\text{C}}}}-\text{S}^-\text{Na}^+}$$

The solution of cellulose xanthate is then passed through a small orifice or slit into an acidic solution. This operation regenerates the —OH groups of cellulose, causing it to precipitate as a fiber or a sheet.

$$\text{Cellulose}-\text{O}-\overset{\displaystyle \text{S}}{\overset{\|}{\text{C}}}-\text{S}^-\text{Na}^+ \xrightarrow{\text{H}_3\text{O}^+} \underset{\textbf{Rayon or Cellophane}}{\text{cellulose}-\text{OH}}$$

The fibers are *Rayon;* the sheets, after softening with glycerol, are *Cellophane.*

22.14 OTHER BIOLOGICALLY IMPORTANT SUGARS

Monosaccharide derivatives in which the —CH$_2$OH group at C6 has been specifically oxidized to a carboxyl group are called **uronic acids.** Their names are based on the monosaccharide from which they are derived. For example, specific oxidation of C6 of glucose to a carboxyl group converts *glucose* to **glucuronic acid.** In the same way, specific oxidation of C6 of *galactose* would yield **galacturonic acid.**

D-Glucuronic acid   D-Galacturonic acid

PROBLEM 22.19

Direct oxidation of an aldose affects the aldehyde group first, converting it to a carboxylic acid (Section 22.6B), and most oxidizing agents that will attack 1° alcohol groups will also attack 2° alcohol groups. Clearly, then, a laboratory synthesis of a uronic acid from an aldose requires protecting these groups from oxidation. Keeping this in mind, suggest a method for carrying out a specific oxidation that would convert D-galactose to D-galacturonic acid. (*Hint:* See Section 22.5C.)

Monosaccharides in which an —OH group has been replaced by —H are known as **deoxy sugars.** The most important deoxy sugar, because it occurs in DNA, is **deoxyribose.** Other deoxy sugars that occur widely in polysaccharides are L-rhamnose and L-fucose.

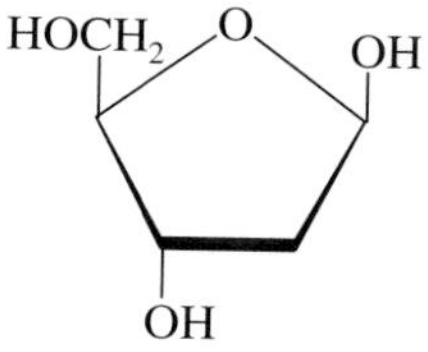

β-2-Deoxy-D-ribose

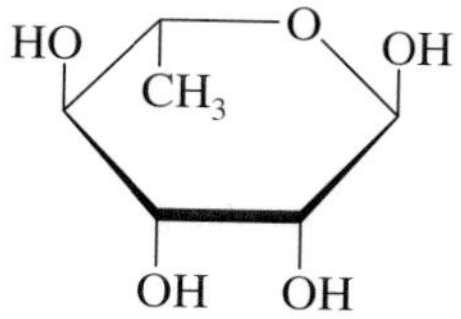

α-L-Rhamnose
(6-deoxy-L-mannose)

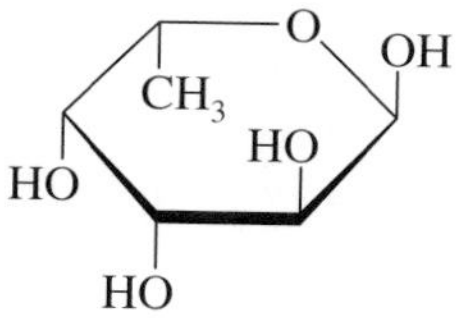

α-L-Fucose
(6-deoxy-L-galactose)

22.15 SUGARS THAT CONTAIN NITROGEN

22.15A Glycosylamines

A sugar in which an amino group replaces the anomeric —OH is called a glycosylamine. Examples are β-D-glucopyranosylamine and adenosine.

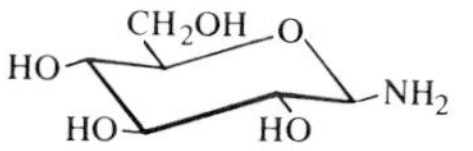

β–D–Glucopyranosyl amine

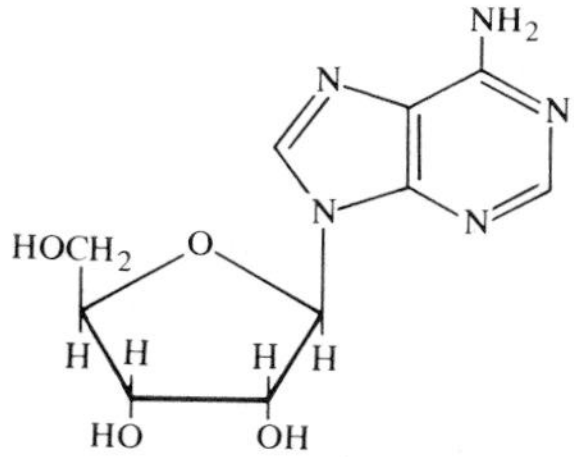

Adenosine

Adenosine is an example of a glycosylamine that is also called a **nucleoside.** Nucleosides are glycosylamines in which the amino component is a pyrimidine or a purine (Section 20.1B) and in which the sugar component is either D-ribose or 2-deoxy-D-ribose (i.e., D-ribose minus the oxygen at the 2 position). Nucleosides are the important components of RNA (ribonucleic acid) and DNA (deoxyribonucleic acid). We shall describe their properties in detail in Section 25.2.

22.15B Amino Sugars

A sugar in which an amino group replaces a nonanomeric —OH group is called an **amino sugar.** An example is **D-glucosamine.** In many instances the amino group is acetylated as in **N-acetyl-D-glucosamine. N-Acetylmuramic acid** is an important component of bacterial cell walls (Section 24.10).

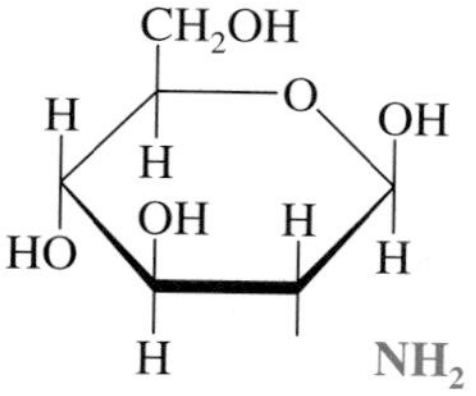

β-D-Glucosamine

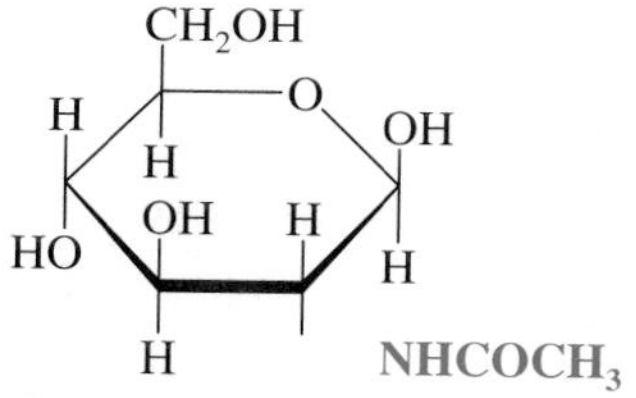

β-N-Acetyl-D-glucosamine
(NAG)

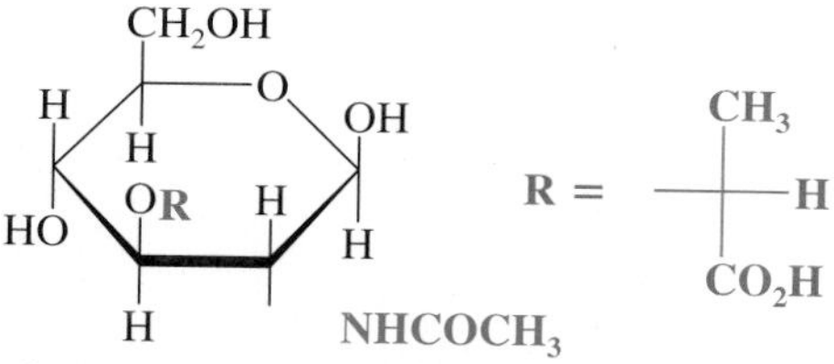

β-N-Acetylmuramic acid
(NAM)

$R = $

D-Glucosamine can be obtained by hydrolysis of **chitin,** a polysaccharide found in the shells of lobsters and crabs and in the external skeletons of insects and spiders. The amino group of D-glucosamine as it occurs in chitin, however, is acetylated; thus, the repeating unit is actually *N*-acetylglucosamine (Fig. 22.18). The glycosidic linkages in chitin are β,1 : 4. X-ray analysis indicates that the structure of chitin is similar to that of cellulose.

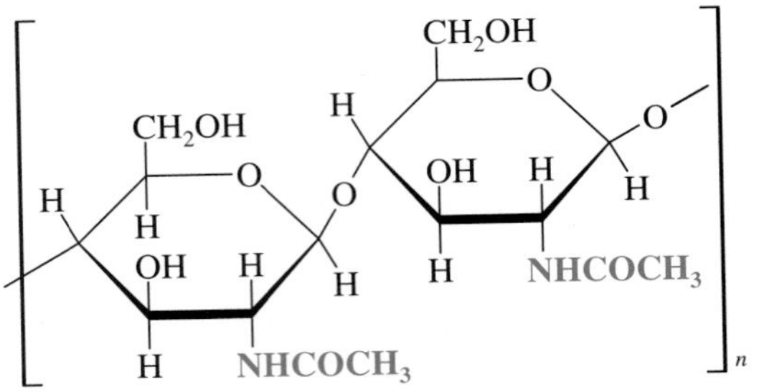

FIGURE 22.18 A partial structure of chitin. The repeating units are *N*-acetylglucosamines linked β,1 : 4.

D-Glucosamine can also be isolated from **heparin,** a sulfated polysaccharide that consists predominately of alternating units of D-glucuronate-2-sulfate and *N*-sulfo-D-glucosamine-6-sulfate (Fig. 22.19). Heparin occurs in intracellular granules of mast cells that line arterial walls, where, when released through injury, it inhibits the clotting of blood. Its purpose seems to be to prevent runaway clot formation. Heparin is widely used in medicine to prevent blood clotting in postsurgical patients.

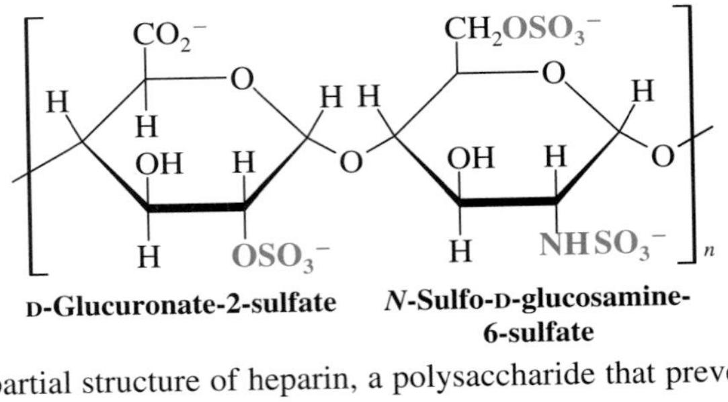

D-Glucuronate-2-sulfate   **N-Sulfo-D-glucosamine-6-sulfate**

FIGURE 22.19 A partial structure of heparin, a polysaccharide that prevents blood clotting.

22.16 GLYCOLIPIDS AND GLYCOPROTEINS OF THE CELL SURFACE

Before 1960, it was thought that the biology of carbohydrates was rather uninteresting; that, in addition to being a kind of inert filler in cells, carbohydrates served only as an energy source and, in plants, as structural materials. Research of the last 30 years has shown, however, that carbohydrates joined through glycosidic linkages to lipids (Chapter 23) and to proteins (Chapter 24), called **glycolipids** and **glycoproteins,** respectively, have functions that span the entire spectrum of activities in the cell. Indeed, most proteins are glycoproteins, and the carbohydrate content of glycoproteins can vary from less than 1% to greater than 90%.

Glycolipids and glycoproteins on the cell surface (Section 23.6A) are now known to be the agents by which cells interact with other cells and with invading bacteria and viruses. The human blood groups offer an example of how carbohydrates, in the form of glycolipids and glycoproteins, act as biochemical markers. The A, B, and O blood types are determined, respectively, by the A, B, and H determinants on the blood cell surface. (The odd naming of the type O determinant came about for complicated historical reasons.) Type AB blood cells have both A and B determinants. These determinants are the carbohydrate portions of the A, B, and H **antigens.**

Antigens are characteristic chemical groups that cause the production of **antibodies** when injected into an animal. Each antibody can bind at least two of its corresponding antigen molecules, causing them to become linked. Linking of red blood cells causes them to agglutinate (clump together). In a transfusion this agglutination can lead to a fatal blockage of the blood vessels.

Individuals with type A antigens on their blood cells carry anti-B antibodies in their serum; those with type B antigens on their blood cells carry anti-A antibodies in their serum. Individuals with type AB cells have both A and B antigens, but have neither anti-A nor anti-B antibodies. Type O individuals have neither A nor B antigens on their blood cells, but have both anti-A and anti-B antibodies.

The A, B, and H antigens differ only in the monosaccharide units at their nonreducing ends. The type H antigen (Fig. 22.20) is the precursor oligosaccharide of the type A and B antigens. Individuals with blood type A have an enzyme that specifically adds an *N*-acetylgalactosamine unit to the 3-OH group of the terminal galactose unit of the H antigen. Individuals with blood type B have an enzyme that specifically adds galactose instead. In individuals with type O blood, the enzyme is inactive.

Antigen–antibody interactions like those that determine blood types are the basis of the immune system. These interactions almost always involve the chemical recognition of a glycolipid or glycoprotein in the antigen by a glycolipid or glycoprotein of the antibody.

22.17 CARBOHYDRATE ANTIBIOTICS

One of the important discoveries in carbohydrate chemistry was the isolation (in 1944) of the carbohydrate antibiotic called *streptomycin.* Streptomycin is made up of the following three subunits:

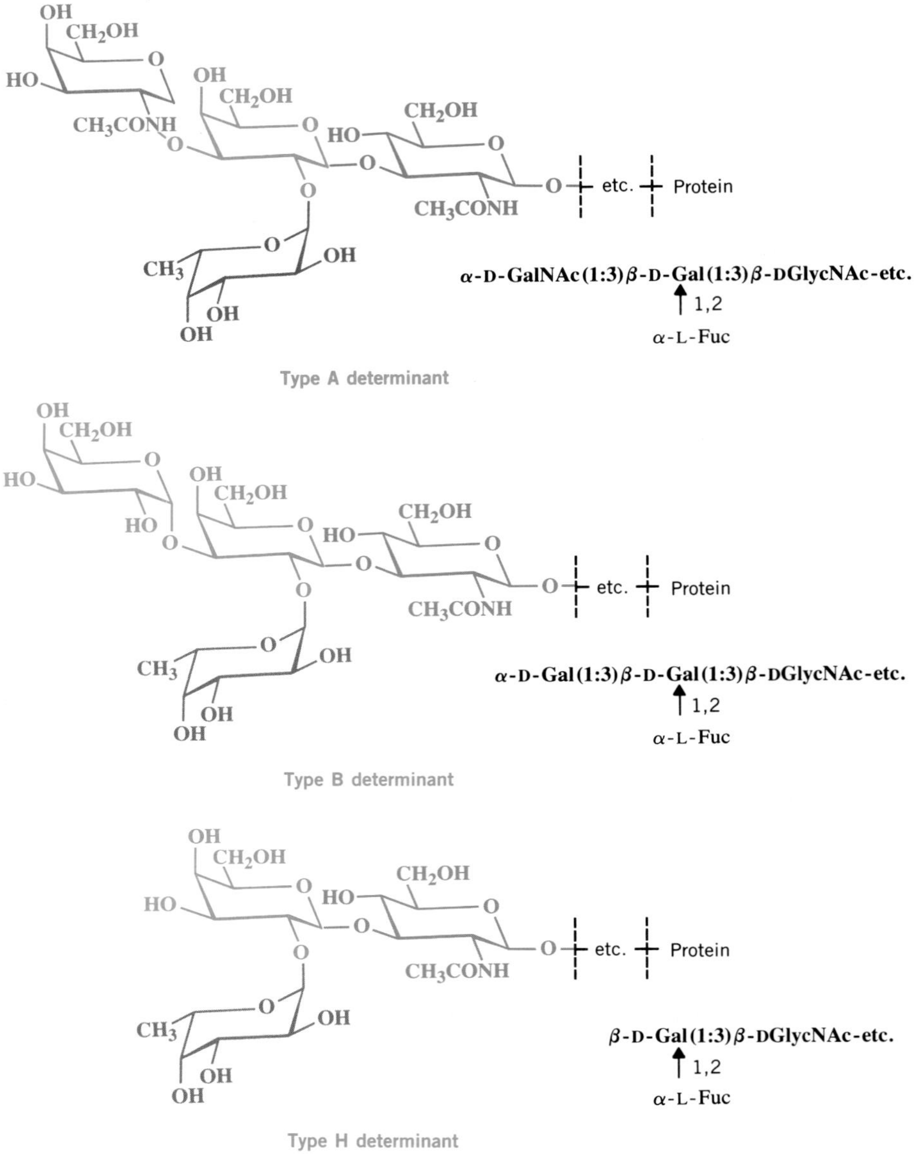

α-D-GalNAc(1:3)β-D-Gal(1:3)β-DGlycNAc-etc.
↑ 1,2
α-L-Fuc

Type A determinant

α-D-Gal(1:3)β-D-Gal(1:3)β-DGlycNAc-etc.
↑ 1,2
α-L-Fuc

Type B determinant

β-D-Gal(1:3)β-DGlycNAc-etc.
↑ 1,2
α-L-Fuc

Type H determinant

FIGURE 22.20 The terminal monosaccharides of the antigenic determinants for types A, B, and O blood. The type H determinant is present in individuals with blood type O and is the precursor of the type A and B determinants. These oligosaccharide antigens are attached to carrier lipid or protein molecules that are anchored in the red blood cell membrane (see Fig. 23.8, for a depiction of a cell membrane.) Ac = acetyl, Gal = D-galactose, GalNAc = *N*-acetylgalactosamine, GlycNAc = *N*-acetylglucosamine, Fuc = Fucose.

All three components are unusual: The amino sugar is based on L-glucose; streptose is a branched-chain monosaccharide; and streptidine is not a sugar at all, but a cyclohexane derivative called an amino cyclitol.

Other members of this family are antibiotics called kanamycins, neomycins, and gentamicins (not shown). All are based on an amino cyclitol linked to one or more amino sugars. The glycosidic linkage is nearly always α. These antibiotics are especially useful against bacteria that are resistant to penicillins.

ADDITIONAL PROBLEMS

22.20 Give appropriate structural formulas to illustrate each of the following:

(a) An aldopentose	(g) An aldonolactone	(m) Epimers
(b) A ketohexose	(h) A pyranose	(n) Anomers
(c) An L-monosaccharide	(i) A furanose	(o) A phenylosazone
(d) A glycoside	(j) A reducing sugar	(p) A disaccharide
(e) An aldonic acid	(k) A pyranoside	(q) A polysaccharide
(f) An aldaric acid	(l) A furanoside	(r) A nonreducing sugar

22.21 Draw conformational formulas for each of the following: (a) α-D-allopyranose, (b) methyl β-D-allopyranoside, and (c) methyl 2,3,4,6-tetra-O-methyl-β-D-allopyranoside.

22.22 Draw structures for furanose and pyranose forms of D-ribose. Show how you could use periodate oxidation to distinguish between a methyl ribofuranoside and a methyl ribopyranoside.

22.23 One reference book lists D-mannose as being dextrorotatory; another lists it as being levorotatory. Both references are correct. Explain.

22.24 The starting material for a commercial synthesis of vitamin C is L-sorbose (see following reaction); it can be synthesized from D-glucose through the following reaction sequence:

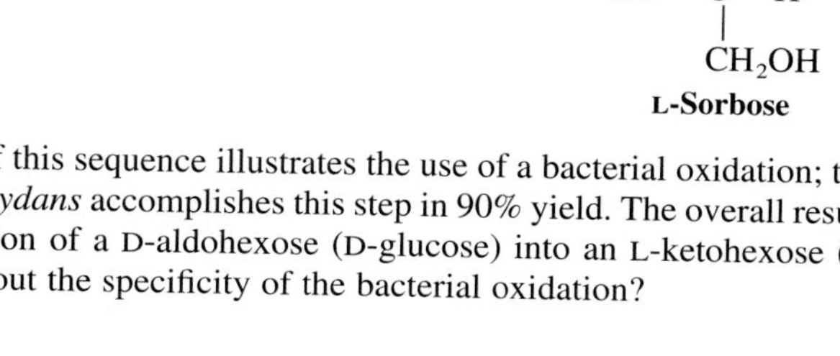

$$\text{D-Glucose} \xrightarrow[\text{Ni}]{\text{H}_2} \text{D-glucitol} \xrightarrow[\substack{\textit{Acetobacter} \\ \textit{suboxydans}}]{\text{O}_2}$$

$$\begin{array}{c} \text{CH}_2\text{OH} \\ | \\ \text{C}=\text{O} \\ | \\ \text{HO}-\text{C}-\text{H} \\ | \\ \text{H}-\text{C}-\text{OH} \\ | \\ \text{HO}-\text{C}-\text{H} \\ | \\ \text{CH}_2\text{OH} \end{array}$$

L-Sorbose

The second step of this sequence illustrates the use of a bacterial oxidation; the microorganism *Acetobacter suboxydans* accomplishes this step in 90% yield. The overall result of the synthesis is the transformation of a D-aldohexose (D-glucose) into an L-ketohexose (L-sorbose). What does this mean about the specificity of the bacterial oxidation?

22.25 What two aldoses would yield the same phenylosazone as L-sorbose (Problem 22.24)?

22.26 In addition to fructose (Problem 22.12) and sorbose (Problem 22.24) there are two other 2-ketohexoses, *psicose* and *tagatose*. D-Psicose yields the same phenylosazone as D-allose (or D-altrose); D-tagatose yields the same osazone as D-galactose (or D-talose). What are the structures of D-psicose and D-tagatose?

22.27 **A**, **B**, and **C** are three aldohexoses. Compounds **A** and **B** yield the same optically active alditol when they are reduced with hydrogen and a catalyst; **A** and **B** yield different phenylosazones when treated with phenylhydrazine; **B** and **C** give the same phenylosazone but different alditols. Assuming that all are D-sugars, give names and structures for **A**, **B**, and **C**.

22.28 D-Xylitol is a sweetener that is used in sugarless chewing gum. Starting with an appropriate monosaccharide, outline a possible synthesis of D-xylitol.

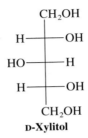

D-Xylitol

22.29 Although monosaccharides undergo complex isomerizations in base (cf. Section 22.5), aldonic acids are epimerized specifically at C2 when they are heated with pyridine. Show how you could make use of this reaction in a synthesis of D-mannose from D-glucose.

22.30 The most stable conformation of most aldopyranoses is one in which the largest group, the —CH$_2$OH group, is equatorial. However, D-idopyranose exists primarily in a conformation with an axial —CH$_2$OH group. Write formulas for the two chair conformations of α-D-idopyranose (one with the —CH$_2$OH group axial and one with the —CH$_2$OH group equatorial) and provide an explanation.

22.31 (a) Heating D-altrose with dilute acid produces a nonreducing *anhydro sugar* (C$_6$H$_{10}$O$_5$). Methylation of the anhydro sugar followed by acid hydrolysis yields 2,3,4-tri-*O*-methyl-D-altrose. The formation of the anhydro sugar takes place through a chair conformation of β-D-altropyranose in which the —CH$_2$OH group is axial. What is the structure of the anhydro sugar and how is it formed? (b) D-Glucose also forms an anhydro sugar but the conditions required are much more drastic than for the corresponding reaction of D-altrose. Explain.

22.32 Show how the following experimental evidence can be used to deduce the structure of lactose (Section 22.12D).

1. Acid hydrolysis of lactose (C$_{12}$H$_{22}$O$_{11}$) gives equimolar quantities of D-glucose and D-galactose. Lactose undergoes a similar hydrolysis in the presence of a *β-galactosidase*.
2. Lactose is a reducing sugar and forms a phenylosazone; it also undergoes mutarotation.
3. Oxidation of lactose with bromine water followed by hydrolysis with dilute acid gives D-galactose and D-gluconic acid.
4. Bromine water oxidation of lactose followed by methylation and hydrolysis gives 2,3,6-tri-*O*-methylgluconolactone and 2,3,4,6-tetra-*O*-methyl-D-galactose.
5. Methylation and hydrolysis of lactose gives 2,3,6-tri-*O*-methyl-D-glucose and 2,3,4,6-tetra-*O*-methyl-D-galactose.

22.33 Deduce the structure of the disaccharide *melibiose* from the following data:

1. Melibiose is a reducing sugar that undergoes mutarotation and forms a phenylosazone.
2. Hydrolysis of melibiose with acid or with an *α-galactosidase* gives D-galactose and D-glucose.
3. Bromine water oxidation of melibiose gives *melibionic acid*. Hydrolysis of melibionic acid gives D-galactose and D-gluconic acid. Methylation of melibionic acid followed by

hydrolysis gives 2,3,4,6-tetra-*O*-methyl-D-galactose and 2,3,4,5-tetra-*O*-methyl-D-gluconic acid.

4. Methylation and hydrolysis of melibiose gives 2,3,4,6-tetra-*O*-methyl-D-galactose and 2,3,4-tri-*O*-methyl-D-glucose.

22.34 Trehalose is a disaccharide that can be obtained from yeasts, fungi, sea urchins, algae, and insects. Deduce the structure of trehalose from the following information:

1. Acid hydrolysis of trehalose yields only D-glucose.

2. Trehalose is hydrolyzed by α-glucosidases but not by β-glucosidases.

3. Trehalose is a nonreducing sugar; it does not mutarotate, form a phenylosazone, or react with bromine water.

4. Methylation of trehalose followed by hydrolysis yields two molar equivalents of 2,3,4,6-tetra-*O*-methyl-D-glucose.

22.35 Outline chemical tests that will distinguish between each of the following:

(a) D-Glucose and D-glucitol

(b) D-Glucitol and D-glucaric acid

(c) D-Glucose and D-fructose

(d) D-Glucose and D-galactose

(e) Sucrose and maltose

(f) Maltose and maltonic acid

(g) Methyl β-D-glucopyranoside and 2,3,4,6-tetra-*O*-methyl-β-D-glucopyranose

(h) Methyl α-D-ribofuranoside **(I)** and methyl 2-deoxy-α-D-ribofuranoside **(II)**

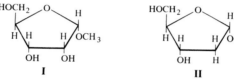

22.36 A group of oligosaccharides called *Schardinger dextrins* can be isolated from *Bacillus macerans* when the bacillus is grown on a medium rich in amylose. These oligosaccharides are all *nonreducing*. A typical Schardinger dextrin undergoes hydrolysis when treated with an acid or an α-glucosidase to yield six, seven, or eight molecules of D-glucose. Complete methylation of a Schardinger dextrin followed by acid hydrolysis yields only 2,3,6-tri-*O*-methyl-D-glucose. Propose a general structure for a Schardinger dextrin.

22.37 *Isomaltose* is a disaccharide that can be obtained by enzymatic hydrolysis of amylopectin. Deduce the structure of isomaltose from the following data:

1. Hydrolysis of 1 mol of isomaltose by acid or by an α-glucosidase gives 2 mol of D-glucose.

2. Isomaltose is a reducing sugar.

3. Isomaltose is oxidized by bromine water to isomaltonic acid. Methylation of isomaltonic acid and subsequent hydrolysis yields 2,3,4,6-tetra-*O*-methyl-D-glucose and 2,3,4,5-tetra-*O*-methyl-D-gluconic acid.

4. Methylation of isomaltose itself followed by hydrolysis gives 2,3,4,6-tetra-*O*-methyl-D-glucose and 2,3,4-tri-*O*-methyl-D-glucose.

22.38 *Stachyose* occurs in the roots of several species of plants. Deduce the structure of stachyose from the following data:

1. Acidic hydrolysis of 1 mol of stachyose yields 2 mol of D-galactose, 1 mol of D-glucose, and 1 mol of D-fructose.

2. Stachyose is a nonreducing sugar.

3. Treating stachyose with an α-galactosidase produces a mixture containing D-galactose, sucrose, and a nonreducing trisaccharide called *raffinose*.
Acidic hydrolysis of raffinose gives D-glucose, D-fructose, and D-galactose. Treating raffinose with an α-galactosidase yields D-galactose and sucrose. Treating raffinose with

invertase (an enzyme that hydrolyzes sucrose) yields fructose and *melibiose* (cf. Problem 22.33).

5. Methylation of stachyose followed by hydrolysis yields 2,3,4,6-tetra-*O*-methyl-D-galactose, 2,3,4-tri-*O*-methyl-D-galactose, 2,3,4-tri-*O*-methyl-D-glucose, and 1,3,4,6-tetra-*O*-methyl-D-fructose.

22.39 *Arbutin,* a compound that can be isolated from the leaves of barberry, cranberry, and pear trees, has the molecular formula $C_{12}H_{16}O_7$. When arbutin is treated with aqueous acid or with a β-glucosidase, the reaction produces D-glucose and a compound **X** with the molecular formula $C_6H_6O_2$. The ^{1}H NMR spectrum of compound **X** consists of two singlets, one at δ 6.8 (4H) and one at δ 7.9 (2H). Methylation of arbutin followed by acidic hydrolysis yields 2,3,4,6-tetra-*O*-methyl-D-glucose and a compound **Y** ($C_7H_8O_2$). Compound **Y** is soluble in dilute aqueous NaOH but is insoluble in aqueous $NaHCO_3$. The ^{1}H NMR spectrum of **Y** shows a singlet at δ 3.9 (3H), a singlet at δ 4.8 (1H), and a multiplet (that resembles a singlet) at δ 6.8 (4H). Treating compound **Y** with aqueous NaOH and $(CH_3)_2SO_4$ produces compound **Z** ($C_8H_{10}O_2$). The ^{1}H NMR spectrum of **Z** consists of two singlets, one at δ 3.75 (6H) and one at δ 6.8 (4H). Propose structures for arbutin and for compounds **X**, **Y**, and **Z**.

22.40 When subjected to a Ruff degradation a D-aldopentose, **A**, is converted to an aldotetrose, **B**. When reduced with sodium borohydride, the aldotetrose **B** forms an optically active alditol. The ^{13}C NMR spectrum of this alditol displays only two signals. The alditol obtained by direct reduction of **A** with sodium borohydride is not optically active. When **A** is used as the starting material for a Kiliani-Fischer synthesis, two diastereomeric aldohexoses, **C** and **D**, are produced. On treatment with sodium borohydride, **C** leads to an alditol **E**, and **D** leads to **F**. The ^{13}C NMR spectrum of **E** consists of three signals, that of **F** consists of six. Propose structures for **A-F**.

22.41 Figure 22.21 shows the ^{13}C NMR spectrum for the product of the reaction of D-(+)-mannose with acetone containing a trace of acid. This compound is a mannofuranose with some hydroxyl groups protected as acetone acetals (as acetonides). Use the ^{13}C NMR spectrum to determine how many acetonide groups are present in the compound.

22.42 D-(+)-Mannose can be reduced with sodium borohydride to form D-mannitol. When D-mannitol is dissolved in acetone containing a trace amount of acid and the product of this reaction subsequently oxidized with $NaIO_4$, a compound whose ^{13}C NMR spectrum consists of six signals is produced. One of these signals is near δ 200. What is the structure of this compound?

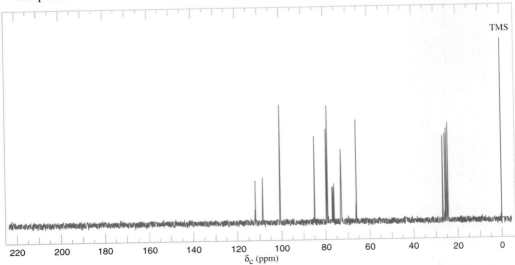

FIGURE 22.21 The broad-band proton-decoupled ^{13}C NMR spectrum for the reaction product in Problem 22.41. Information from the DEPT ^{13}C NMR spectrum is given above the peaks.

CHAPTER 23

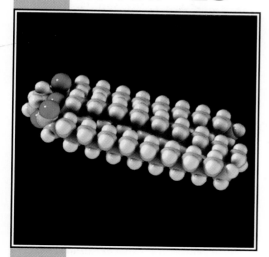

Saturated triacylglycerol
(Section 23.2)

LIPIDS

23.1 INTRODUCTION

Lipids are compounds of biological origin that dissolve in nonpolar solvents, such as chloroform or diethyl ether. The name lipid comes from the Greek word *lipos,* for fat. Unlike carbohydrates and proteins, which are defined in terms of their structures, lipids are defined by the physical operation that we use to isolate them. Not surprisingly, then, lipids include a variety of structural types. Examples are the following:

A fat or oil
(a triacylglycerol)

Menthol
(a terpene)

Vitamin A
(a terpenoid)

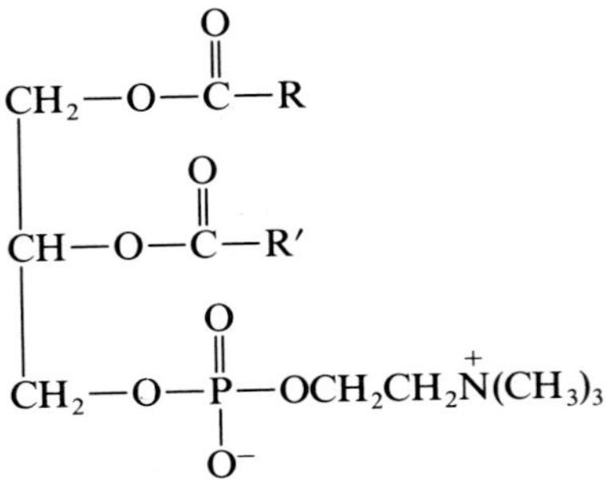

A lecithin
(a phosphatide)

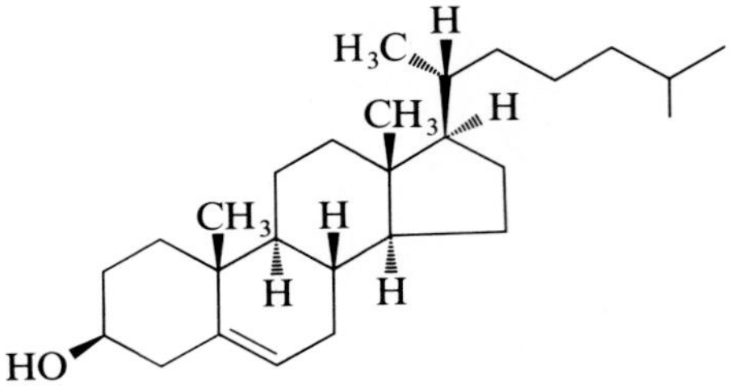

Cholesterol
(a steroid)

23.2 FATTY ACIDS AND TRIACYLGLYCEROLS

Only a small portion of the total lipid fraction obtained by extraction with a nonpolar solvent consists of long-chain carboxylic acids. Most of the carboxylic acids of biological origin are found as *esters of glycerol,* that is, as **triacylglycerols** (Fig. 23.1).*

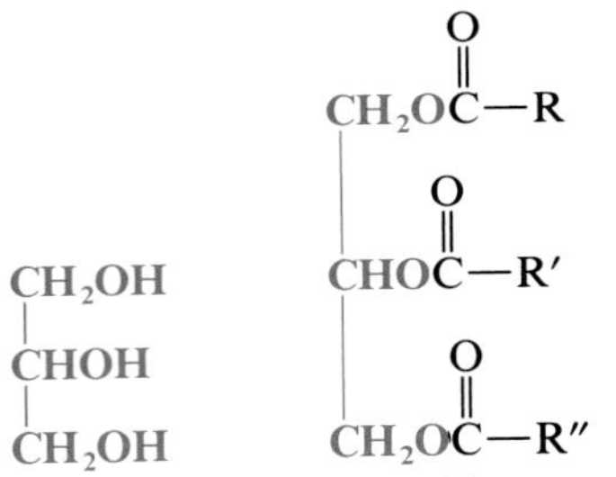

FIGURE 23.1 (*a*) Glycerol. (*b*) A triacylglycerol. The groups R, R′, and R″ are usually long-chain alkyl groups. R, R′, and R″ may also contain one or more carbon–carbon double bonds. In a triacylglycerol R, R′, and R″ may all be different.

Triacylglycerols are the oils of plants or the fats of animal origin. They include such common substances as peanut oil, soybean oil, corn oil, sunflower oil, butter, lard, and tallow. Triacylglycerols that are liquids at room temperature are generally called **oils;** those that are solids are called *fats.* Triacylglycerols can be **simple triacylglycerols** in which all three acyl groups are the same. More commonly, however, the triacylglycerol is a **mixed triacylglycerol** in which the acyl groups are different.

---

* In the older literature triacylglycerols were referred to as triglycerides, or simply as glycerides. In IUPAC nomenclature, because they are esters of glycerol, they should be named as glyceryl trialkanoates, glyceryl trialkenoates, and so on.

Hydrolysis of a fat or oil produces a mixture of fatty acids:

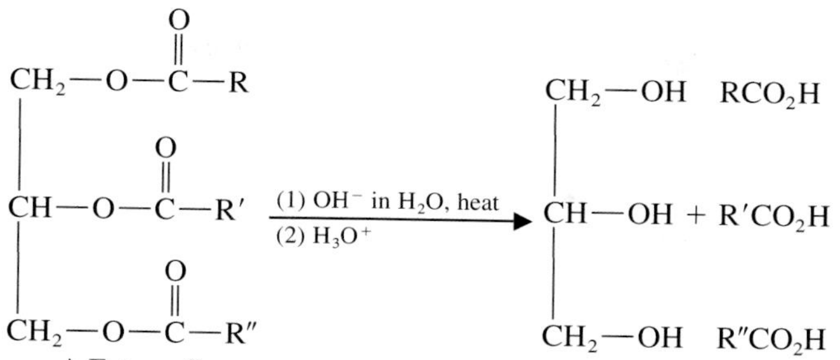

A Fat or oil → **Glycerol** + **Fatty acids**

(1) OH^- in H_2O, heat
(2) H_3O^+

TABLE 23.1 Common fatty acids

	mp (°C)
Saturated Carboxylic Acids	
$CH_3(CH_2)_{12}CO_2H$ **Myristic acid** **(tetradecanoic acid)**	54
$CH_3(CH_2)_{14}CO_2H$ **Palmitic acid** **(hexadecanoic acid)**	63
$CH_3(CH_2)_{16}CO_2H$ **Stearic acid** **(octadecanoic acid)**	70
Unsaturated Carboxylic Acids	
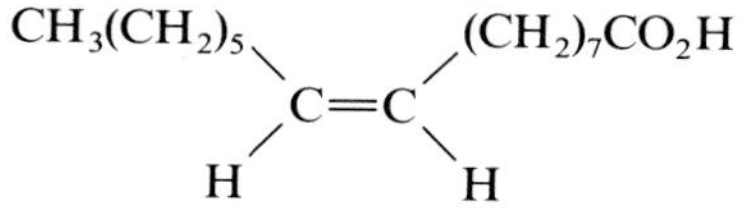 **Palmitoleic acid** **(cis-9-hexadecenoic acid)**	32
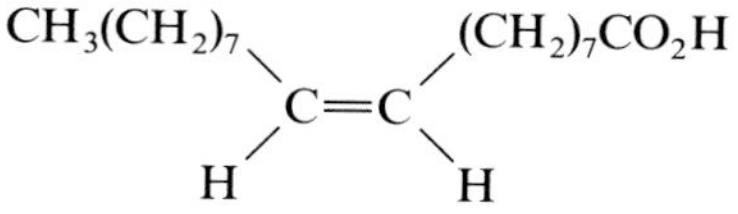 **Oleic acid** **(cis-9-octadecenoic acid)**	4
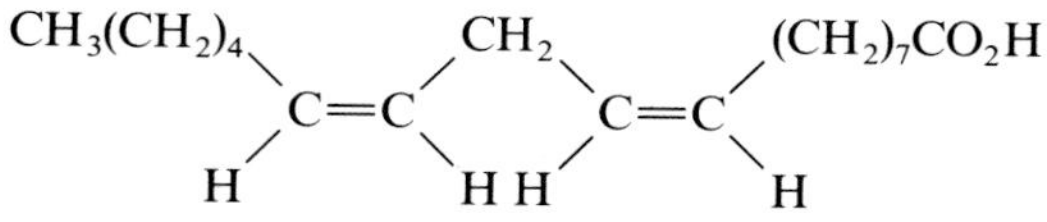 **Linoleic acid** **(cis,cis-9,12-octadecadienoic acid)**	−5
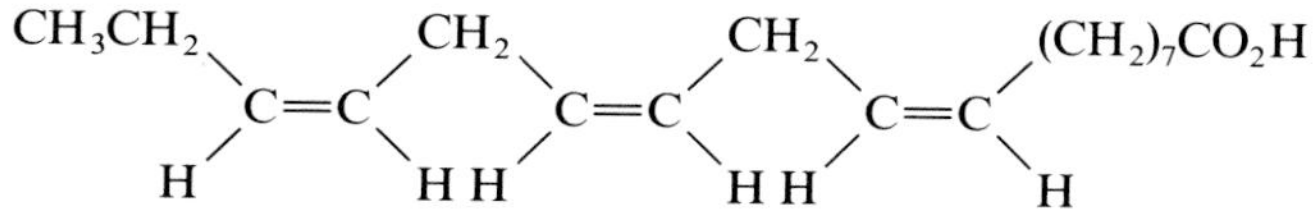 **Linolenic acid** **(cis,cis,cis-9,12,15-octadecatrienoic acid)**	−11

TABLE 23.2 Fatty acid composition obtained by hydrolysis of common fats and oils[a]

| FAT OR OIL | AVERAGE COMPOSITION OF FATTY ACIDS (mol %) | | | | | | | | | | | |
| | SATURATED | | | | | | | | UNSATURATED | | | |
	C_4 BUTYRIC ACID	C_6 CAPROIC ACID	C_8 CAPRYLIC ACID	C_{10} CAPRIC ACID	C_{12} LAURIC ACID	C_{14} MYRISTIC ACID	C_{16} PALMITIC ACID	C_{18} STEARIC ACID	C_{16} PALMIT-OLEIC ACID	C_{18} OLEIC ACID	C_{18} LINOLEIC ACID	C_{18} LINOLENIC ACID
Animal Fats												
Butter	3–4	1–2	0–1	2–3	2–5	8–15	25–29	9–12	4–6	18–33	2–4	0–1
Lard						1–2	25–30	12–18	4–6	48–60	6–12	0–1
Beef tallow						2–5	24–34	15–30		35–45	1–3	
Vegetable Oils												
Olive						0–1	5–15	1–4		67–84	8–12	
Peanut							7–12	2–6		30–60	20–38	
Corn						1–2	7–11	3–4	1–2	25–35	50–60	
Cottonseed						1–2	18–25	1–2	1–3	17–38	45–55	
Soybean						1–2	6–10	2–4		20–30	50–58	5–10
Linseed							4–7	2–4		14–30	14–25	45–60
Coconut		0–1	5–7	7–9	40–50	15–20	9–12	2–4	0–1	6–9	0–1	
Marine Oils												
Cod liver						5–7	8–10	0–1	18–22	27–33	27–32	

[a] Data adapted from John R. Holum, *Organic and Biological Chemistry*, Wiley, New York, 1978, p. 220, and from *Biology Data Book*, Philip L. Altman and Dorothy S. Ditmer, Eds., Federation of American Societies for Experimental Biology, Washington, DC, 1964.

Most natural fatty acids have **unbranched chains** and, because they are synthesized from two-carbon units, **they have an even number of carbon atoms.** Table 23.1 lists some of the most common fatty acids and Table 23.2 gives the fatty acid composition of a number of common fats and oils. Notice that in the unsaturated fatty acids in Table 23.1 **the double bonds are all cis.** Many naturally occurring fatty acids contain two or three double bonds. The fats or oils that these come from are called **polyunsaturated fats or oils.** The first double bond of an unsaturated fatty acid commonly occurs between C9 and C10, the remaining double bonds tend to begin with C12 and C15 (as in linoleic acid and linolenic acid). The double bonds, therefore, *are not conjugated.* Triple bonds rarely occur in fatty acids.

The carbon chains of saturated fatty acids can adopt many conformations but tend to be fully extended because this minimizes steric repulsions between neighboring methylene groups. Saturated fatty acids pack efficiently into crystals, and because van der Waals attractions are large, they have relatively high melting points. The melting points increase with increasing molecular weight. The cis configuration of the double bond of an unsaturated fatty acid puts a rigid bend in the carbon chain that interferes with crystal packing, causing reduced van der Waals attractions between molecules. Unsaturated fatty acids, consequently, have lower melting points.

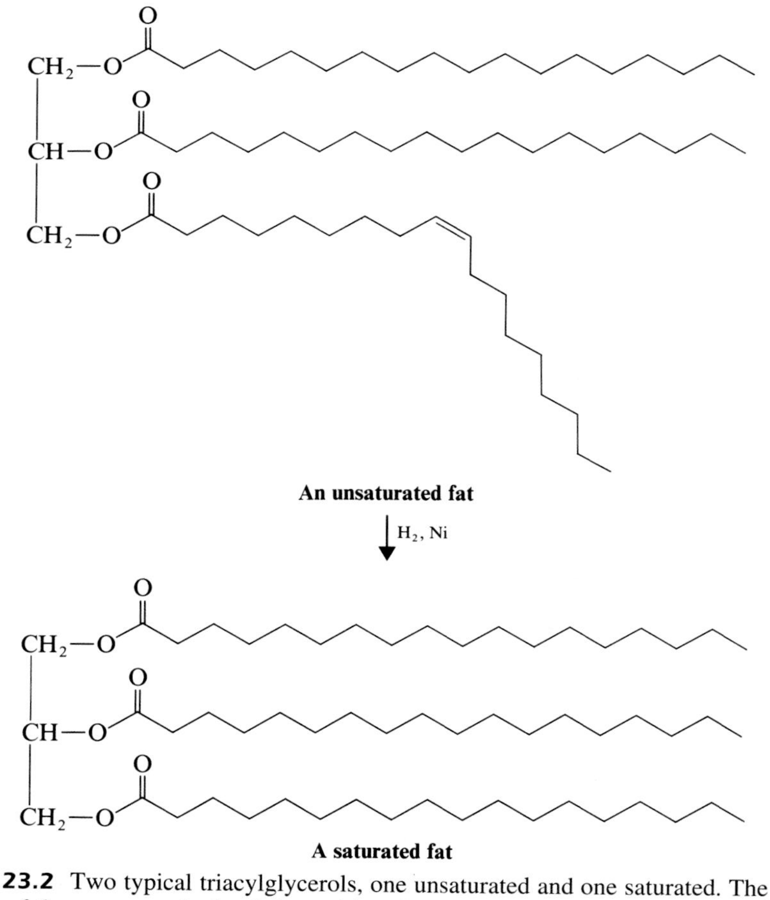

An unsaturated fat

H$_2$, Ni

A saturated fat

FIGURE 23.2 Two typical triacylglycerols, one unsaturated and one saturated. The cis double bond of the unsaturated triacylglycerol interferes with efficient crystal packing and causes it to have a lower melting point. Hydrogenation of the double bond causes an unsaturated triacylglycerol to become saturated.

What we have just said about the fatty acids applies to the triacylglycerols as well. Triacylglycerols made up of largely saturated fatty acids have high melting points and are solids at room temperature. They are what we call *fats*. Triacylglycerols with a high proportion of unsaturated and polyunsaturated fatty acids have lower melting points. They are *oils*. Figure 23.2 shows how the introduction of a single cis double bond affects the shape of a triacylglycerol and how catalytic hydrogenation can be used to convert an unsaturated triacylglycerol into a saturated one.

23.2A Hydrogenation of Triacylglycerols

Solid commercial cooking fats are manufactured by partial hydrogenation of vegetable oils. The result is the familiar "partially hydrogenated fat" present in so many pre-pared foods. Complete hydrogenation of the oil is avoided because a completely saturated triacylglycerol is very hard and brittle. Typically, the vegetable oil is hydrogenated until a semisolid of appealing consistency is obtained. One commercial advantage of partial hydrogenation is to give the fat a longer shelf life. Polyunsaturated oils tend to react by autoxidation (Section 9.11B), causing them to become rancid. One problem with partial hydrogenation, however, is that the catalyst isomerizes some of the unreacted double bonds from the natural cis arrangement to the unnatural trans arrangement, and there is accumulating evidence that "trans" fats are associated with an increased risk of cardiovascular disease.

23.2B Biological Functions of Triacylglycerols

The primary function of triacylglycerols in animals is as an energy reserve. When triacylglycerols are converted to carbon dioxide and water by biochemical reactions (i.e., when triacylglycerols are *metabolized*), they yield more than twice as many kilocalories per gram as do carbohydrates or proteins. This is largely because of the high proportion of carbon–hydrogen bonds per molecule.

In animals, specialized cells called **adipocytes** (fat cells) synthesize and store triacylglycerols. The tissue containing these cells, adipose tissue, is most abundant in the abdominal cavity and in the subcutaneous layer. Men have a fat content of about 21%, women about 26%. This fat content is sufficient to enable us to survive starvation for 2–3 months. By contrast, glycogen, our carbohydrate reserve, can provide only one day's energy need.

All of the saturated triacylglycerols of the body, and some of the unsaturated ones, can be synthesized from carbohydrates and proteins. Certain polyunsaturated fatty acids, however, are essential in the diets of higher animals.

The amount of fat in the diet, especially the proportion of saturated fat, has been a health concern for many years. There is compelling evidence that too much saturated fat in the diet is a factor in the development of heart disease and cancer.

23.2C Saponification of Triacylglycerols

Alkaline hydrolysis (i.e., saponification) of triacylglycerols produces glycerol and a mixture of salts of long-chain carboxylic acids:

$$
\begin{array}{ccc}
\underset{\underset{}{\overset{\overset{\displaystyle O}{\parallel}}{}}}{CH_2OCR} & & \underset{}{\overset{}{CH_2OH}} \quad \underset{}{\overset{\overset{\displaystyle O}{\parallel}}{RCO^-}} \quad Na^+ \\[2ex]
\mid & & \mid \\[1ex]
\underset{\underset{}{\overset{\overset{\displaystyle O}{\parallel}}{}}}{CHOCR'} + 3\,NaOH \xrightarrow{\;H_2O\;} & & \underset{}{\overset{}{CHOH}} + \quad \underset{}{\overset{\overset{\displaystyle O}{\parallel}}{R'CO^-}} \quad Na^+ \\[2ex]
\mid & & \mid \\[1ex]
\underset{}{\overset{\overset{\displaystyle O}{\parallel}}{CH_2OCR''}} & & \underset{}{\overset{}{CH_2OH}} \quad \underset{}{\overset{\overset{\displaystyle O}{\parallel}}{R''CO^-}} \quad Na^+ \\[1ex]
& & \textbf{Glycerol} \quad \textbf{Sodium carboxylates} \\
& & \textbf{``soap''}
\end{array}
$$

These salts of long-chain carboxylic acids are **soaps,** and this saponification reaction is the way most soaps are manufactured. Fats and oils are boiled in aqueous sodium hydroxide until hydrolysis is complete. Adding sodium chloride to the mixture then causes the soap to precipitate. (After the soap has been separated, glycerol can be isolated from the aqueous phase by distillation.) Crude soaps are usually purified by several reprecipitations. Perfumes can be added if a toilet soap is the desired product. Sand, sodium carbonate, and other fillers can be added to make a scouring soap, and air can be blown into the molten soap if the manufacturer wants to market a soap that floats.

The sodium salts of long-chain carboxylic acids (soaps) are almost completely miscible with water. However, they do not dissolve as we might expect, that is, as individual ions. Except in very dilute solutions, soaps exist as **micelles** (Fig. 23.3). Soap micelles are usually spherical clusters of carboxylate ions that are dispersed throughout the aqueous phase. The carboxylate ions are packed together with their negatively charged (and thus, *polar*) carboxylate groups at the surface and with their nonpolar hydrocarbon chains on the interior. The sodium ions are scattered throughout the aqueous phase as individual solvated ions.

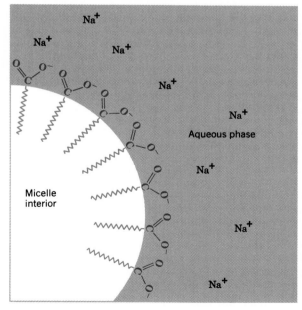

FIGURE 23.3 A portion of a soap micelle showing its interface with the polar dispersing medium.

Micelle formation accounts for the fact that soaps dissolve in water. The nonpolar (and thus **hydrophobic**) alkyl chains of the soap remain in a nonpolar environment—in the interior of the micelle. The polar (and therefore **hydrophilic**) carboxylate groups are exposed to a polar environment—that of the aqueous phase. Because the surfaces of the micelles are negatively charged, individual micelles repel each other and remain dispersed throughout the aqueous phase.

Soaps serve their function as "dirt removers" in a similar way. Most dirt particles (e.g., on the skin) become surrounded by a layer of an oil or fat. Water molecules alone are unable to disperse these greasy globules because they are unable to penetrate the oily layer and separate the individual particles from each other or from the surface to which they are stuck. Soap solutions, however, *are* able to separate the individual particles because their hydrocarbon chains can "dissolve" in the oily layer (Fig. 23.4). As this happens, each individual particle develops an outer layer of carboxylate ions and presents the aqueous phase with a much more compatible exterior—a polar surface. The individual globules now repel each other and thus become dispersed throughout the aqueous phase. Shortly thereafter, they make their way down the drain.

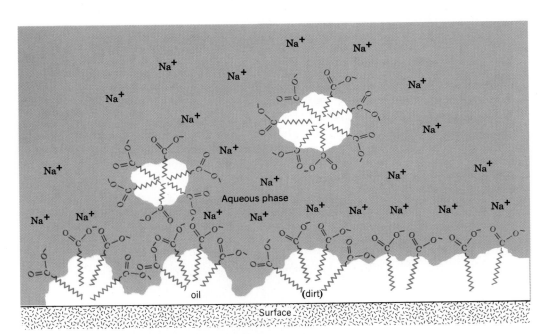

FIGURE 23.4 Dispersal of oil-coated dirt particles by a soap.

Synthetic detergents (Fig. 23.5) function in the same way as soaps; they have long nonpolar alkane chains with polar groups at the end. The polar groups of most synthetic detergents are sodium sulfonates or sodium sulfates. (At one time, extensive use was made of synthetic detergents with highly branched alkyl groups. These detergents proved to be nonbiodegradable, and their use was discontinued.)

Synthetic detergents offer an advantage over soaps; they function well in "hard" water, that is, water containing Ca^{2+}, Fe^{2+}, Fe^{3+}, and Mg^{2+} ions. Calcium, iron, and magnesium salts of alkanesulfonates and alkyl hydrogen sulfates are largely water soluble, and thus synthetic detergents remain in solution. Soaps, by contrast, form precipitates—the ring around the bathtub—when they are used in hard water.

$$CH_3(CH_2)_nCH_2SO_2O^- \ Na^+$$
Sodium alkanesulfonates

$$CH_3(CH_2)_nCH_2OSO_2O^- \ Na^+$$
Sodium alkyl sulfates

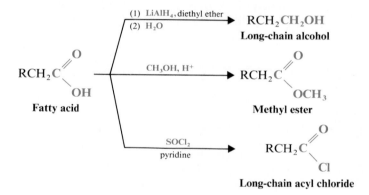

Sodium alkylbenzenesulfonates

FIGURE 23.5 Typical synthetic detergents ($n = 10$).

23.2D Reactions of the Carboxyl Group of Fatty Acids

Fatty acids, as we might expect, undergo reactions typical of carboxylic acids. They react with $LiAlH_4$ to form alcohols, with alcohols and mineral acid to form esters, and with thionyl chloride to form acyl chlorides:

23.2E Reactions of the Alkyl Chain of Saturated Fatty Acids

Fatty acids are like other carboxylic acids in that they undergo specific α-halogenation when they are treated with bromine or chlorine in the presence of phosphorus. This is the familiar Hell–Volhard–Zelinski reaction.

23.2F Reactions of the Alkenyl Chain of Unsaturated Fatty Acids

The double bonds of the carbon chains of fatty acids undergo characteristic alkene addition reactions:

$$CH_3(CH_2)_nCH=CH(CH_2)_mCO_2H$$

$$\xrightarrow[\text{Ni}]{\text{H}_2,} CH_3(CH_2)_nCH_2-CH_2(CH_2)_mCO_2H$$

$$\xrightarrow[\text{CCl}_4]{\text{Br}_2,} CH_3(CH_2)_nCHBrCHBr(CH_2)_mCO_2H$$

$$\xrightarrow[\text{(2) NaHSO}_3]{\text{(1) OsO}_4} CH_3(CH_2)_n\underset{\underset{OH}{|}}{CH}-\underset{\underset{OH}{|}}{CH}(CH_2)_mCO_2H$$

$$\xrightarrow{\text{HBr}} CH_3(CH_2)_nCH_2CHBr(CH_2)_mCO_2H$$
$$+$$
$$CH_3(CH_2)_nCHBrCH_2(CH_2)_mCO_2H$$

PROBLEM 23.1

(a) How many stereoisomers are possible for 9,10-dibromohexadecanoic acid?
(b) The addition of bromine to palmitoleic acid yields primarily one set of enantiomers, (±)-*threo*-9,10-dibromohexadecanoic acid. The addition of bromine is an anti addition to the double bond (i.e., it apparently takes place through a bromonium ion intermediate). Taking into account the cis stereochemistry of the double bond of palmitoleic acid and the stereochemistry of the bromine addition, write three-dimensional structures for the (±)-*threo*-9,10-dibromohexadecanoic acids.

23.3 TERPENES AND TERPENOIDS

People have isolated organic compounds from plants since antiquity. By gently heating or by steam distilling certain plant materials, one can obtain mixtures of odoriferous compounds known as *essential oils*. These compounds have had a variety of uses, particularly in early medicine and in the making of perfumes.

As the science of organic chemistry developed, chemists separated the various components of these mixtures and determined their molecular formulas and, later, their structural formulas. Even today these natural products offer challenging problems for chemists interested in structure determination and synthesis. Research in this area has also given us important information about the ways the plants themselves synthesize these compounds (see Special Topic P).

Hydrocarbons known generally as **terpenes**, and oxygen-containing compounds called **terpenoids** are the most important constituents of essential oils. Most terpenes have skeletons of 10, 15, 20, or 30 carbon atoms and are classified in the following way:

NUMBER OF CARBON ATOMS	CLASS
10	Monoterpenes
15	Sesquiterpenes
20	Diterpenes
30	Triterpenes

One can view terpenes as being built up from two or more C$_5$ units known as *isoprene units.* Isoprene is 2-methyl-1,3-butadiene. Isoprene and the isoprene unit can be represented in various ways.

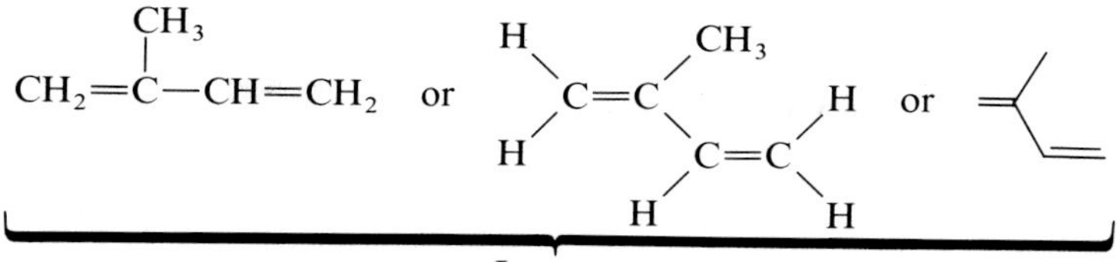

$$CH_2{=}C{-}CH{=}CH_2 \quad \text{or} \quad$$

Isoprene

$$C{-}C{-}C{-}C \quad \text{or}$$

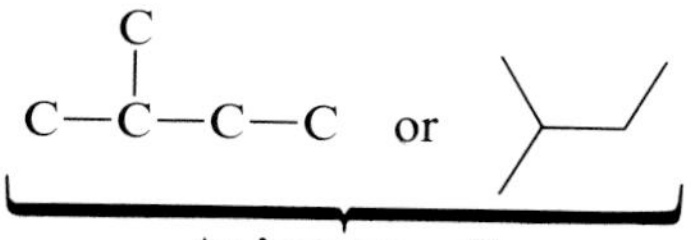

An isoprene unit

We now know that plants do not synthesize terpenes from isoprene (see Special Topic P). However, recognition of the isoprene unit as a component of the structure of terpenes has been a great aid in elucidating their structures. We can see how, if we examine the following structures.

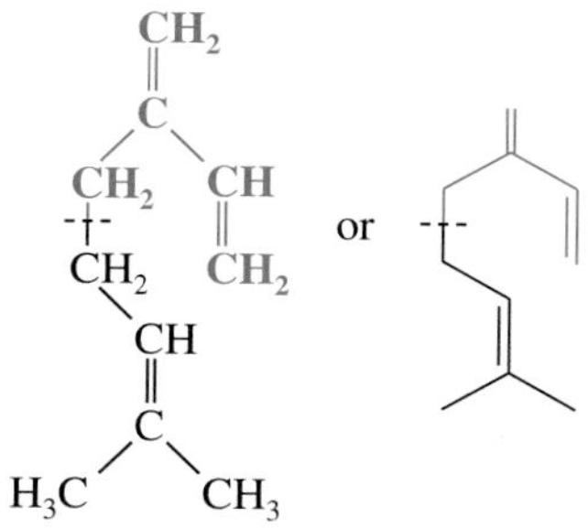

Myrcene
(isolated from bay oil)

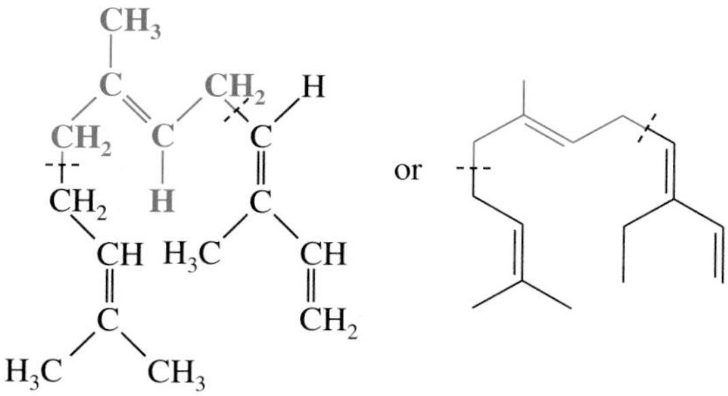

α-Farnesene
(from natural coating of apples)

Using dashed lines to separate isoprene units, we can see that the monoterpene (myrcene) has two isoprene units; and that the sesquiterpene (α-farnesene) has three. In both compounds the isoprene units are linked head to tail.

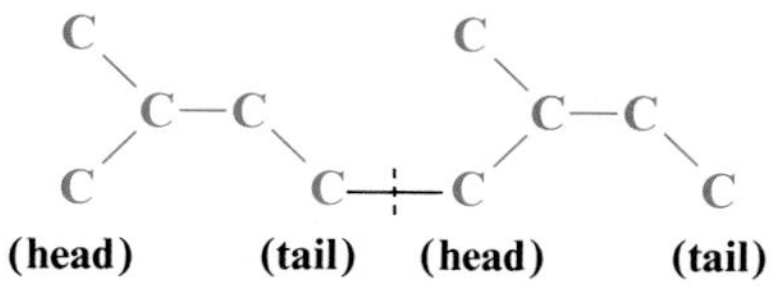

(head)   (tail)   (head)   (tail)

Many terpenes also have isoprene units linked in rings, and others (terpenoids) contain oxygen.

Limonene
(from oil of lemon or orange)

β-Pinene
(from oil of turpentine)

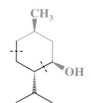

Geraniol
(from roses and other flowers)

Menthol
(from peppermint)

PROBLEM 23.2

(a) Show the isoprene units in each of the following terpenes. (b) Classify each as a monoterpene, sesquiterpene, diterpene, and so on.

Zingiberene
(from oil of ginger)

β-Selinene
(from oil of celery)

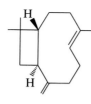

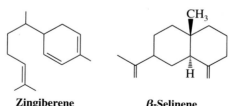

Caryophyllene
(from oil of cloves)

Squalene
(from shark liver oil)

PROBLEM 23.3

What products would you expect to obtain if each of the following terpenes were subjected to ozonization and subsequent treatment with zinc and water?

(a) Myrcene   (c) α-Farnesene   (e) Squalene

(b) Limonene   (d) Geraniol

PROBLEM 23.4

Give structural formulas for the products that you would expect from the following reactions:

(a) β-Pinene + hot KMnO₄ ⟶

(b) Zingiberene + H₂ $\xrightarrow{\text{Pt}}$

(c) Caryophyllene + HCl ⟶

(d) β-Selinene + 2 THF : BH₃ $\xrightarrow{\text{(2) H}_2\text{O}_2,\text{ OH}^-}$

PROBLEM 23.5

What simple chemical test could you use to distinguish between geraniol and menthol?

The carotenes are tetraterpenes. They can be thought of as two diterpenes linked in tail-to-tail fashion.

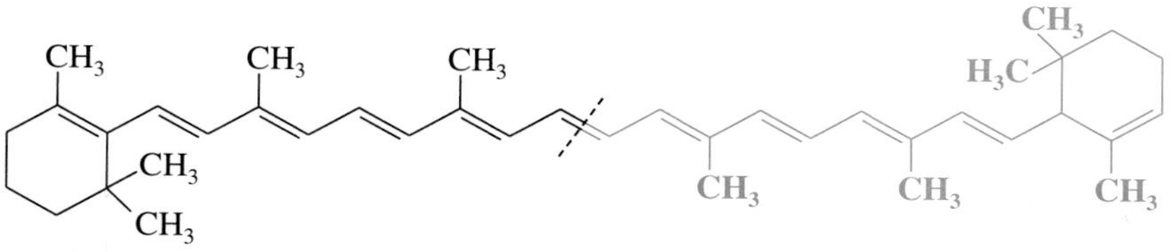

α-Carotene

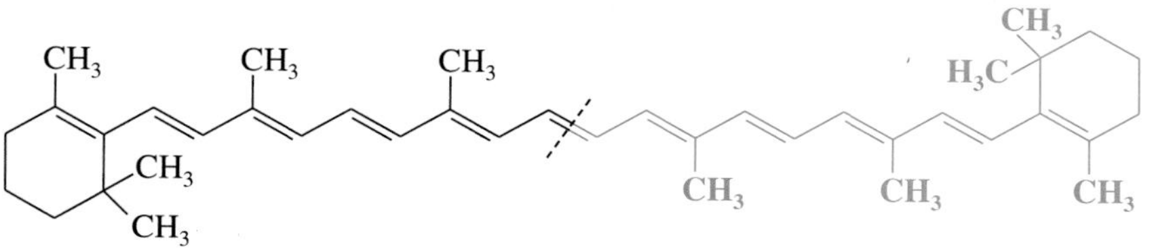

β-Carotene

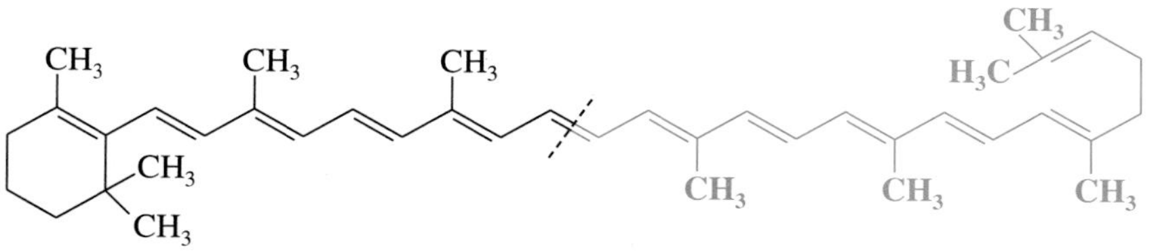

γ-Carotene

The carotenes are present in almost all green plants. All three carotenes serve as precursors for vitamin A, for they all can be converted to vitamin A by enzymes in the liver.

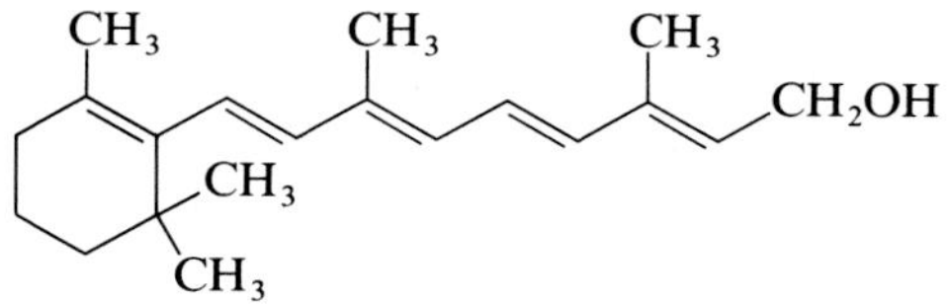

Vitamin A

In this conversion, one molecule of β-carotene yields two of vitamin A: α- and γ-carotene give only one. Vitamin A is important not only in vision but in many other ways as well. For example, young animals whose diets are deficient in vitamin A fail to grow.

23.3A Natural Rubber

Natural rubber can be viewed as a 1,4-addition polymer of isoprene. In fact, pyrolysis degrades natural rubber to isoprene. Pyrolysis (Greek: *pyros,* a fire, + *lysis*) is the heating of something in the absence of air until it decomposes. The isoprene units of natural rubber are all linked in a head-to-tail fashion and all of the double bonds are cis.

Natural rubber
(*cis*-1,4-polyisoprene)

Ziegler–Natta catalysts make it possible to polymerize isoprene and obtain a synthetic product that is identical with the rubber obtained from natural sources.

Pure natural rubber is soft and tacky. To be useful, natural rubber has to be *vulcanized.* In vulcanization, natural rubber is heated with sulfur. A reaction takes place that produces cross-links between the *cis*-polyisoprene chains and makes the rubber much harder. Sulfur reacts both at the double bonds and at allylic hydrogen atoms.

Vulcanized rubber

23.4 STEROIDS

The lipid fractions obtained from plants and animals contain another important group of compounds known as **steroids.** Steroids are important "biological regulators" that nearly always show dramatic physiological effects when they are administered to living organisms. Among these important compounds are male and female sex hormones, adrenocortical hormones, D vitamins, the bile acids, and certain cardiac poisons.

23.4A Structure and Systematic Nomenclature of Steroids

Steroids are derivatives of the following perhydrocyclopentanophenanthrene ring system.

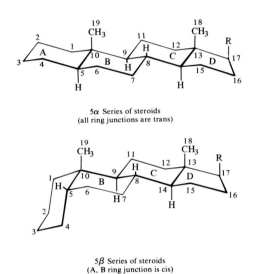

The carbon atoms of this ring system are numbered as shown. The four rings are designated with letters.

In most steroids the **B,C** and **C,D** ring junctions are trans. The **A,B** ring junction, however, may be either cis or trans and this possibility gives rise to two general groups of steroids having the three-dimensional structures shown in Fig. 23.6.

FIGURE 23.6 The basic ring systems of the 5α and 5β series of steroids.

The methyl groups that are attached at points of ring junction (i.e., those numbered 18 and 19) are called **angular methyl groups** and they serve as important reference points for stereochemical designations. The angular methyl groups protrude above the general plane of the ring system when it is written in the manner shown in Fig. 23.6. By convention, other groups that lie on the same general side of the molecule as the angular methyl groups (i.e., on the top side) are designated **β substituents** (these are written with a solid wedge). Groups that lie generally on the bottom (i.e., are trans to the angular methyl groups) are designated **α substituents** (these are written with a dashed wedge). When α and β designations are applied to the hydrogen atom at position 5, the ring system in which the **A,B** ring junction is trans becomes the 5α series; and the ring system in which the **A,B** ring junction is cis becomes the 5β series.

PROBLEM 23.6

Draw the two basic ring systems given in Fig. 23.6 for the 5α and 5β series showing all hydrogen atoms of the cyclohexane rings. Label each hydrogen atom as to whether it is axial or equatorial.

In systematic nomenclature the nature of the R group at position 17 determines (primarily) the base name of an individual steroid. These names are derived from the steroid hydrocarbon names given in Table 23.3.

The following two examples illustrate the way these base names are used.

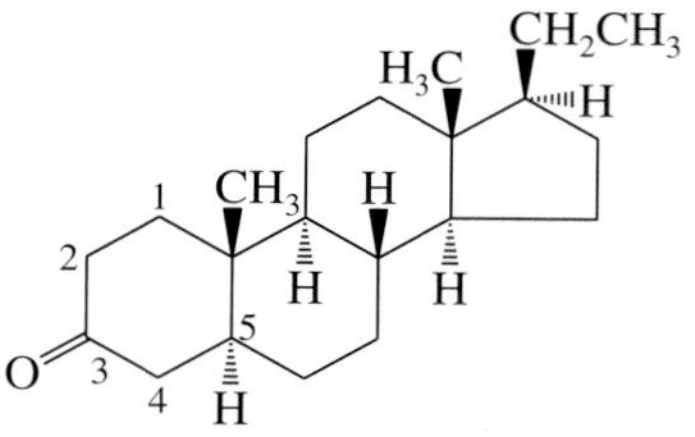

5α-Pregnan-3-one

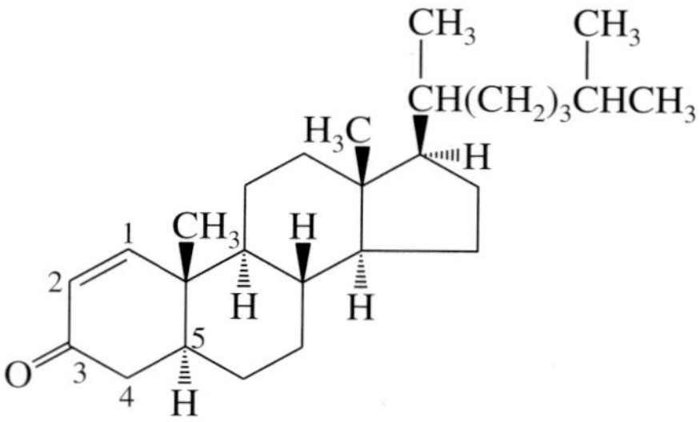

5α-Cholest-1-en-3-one

TABLE 23.3 Names of steroid hydrocarbons

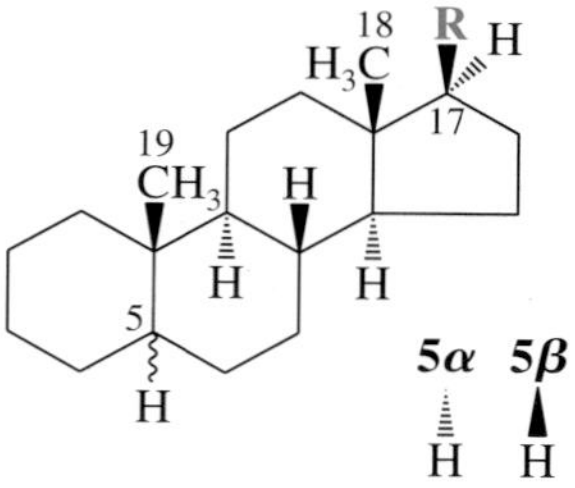

R	NAME
—H	Androstane
—H (with —H also replacing $\overset{19}{—CH_3}$)	Estrane
$\overset{20}{—CH_2}\overset{21}{CH_3}$	Pregnane
$\overset{20}{—CH}\overset{22}{CH_2}\overset{23}{CH_2}\overset{24}{CH_3}$ $\underset{21}{\overset{\vert}{CH_3}}$	Cholane
$\overset{20}{—CH}\overset{22}{CH_2}\overset{23}{CH_2}\overset{24}{CH_2}\overset{25}{CH}\overset{26}{CH_3}$ $\underset{21}{\overset{\vert}{CH_3}}$ $\underset{27}{\overset{\vert}{CH_3}}$	Cholestane

We shall see that many steroids also have common names and that the names of the steroid hydrocarbons given in Table 23.3 are derived from these common names.

PROBLEM 23.7

(a) Androsterone, a secondary male sex hormone, has the systematic name 3α-hydroxy-5α-androstan-17-one. Give a three-dimensional formula for androsterone. (b) Norethynodrel, a synthetic steroid that has been widely used in oral contraceptives, has the systematic name 17α-ethynyl-17β-hydroxy-5(10)-estren-3-one. Give a three-dimensional formula for norethynodrel.

23.4B Cholesterol

Cholesterol, one of the most widely occurring steroids, can be isolated by extraction of nearly all animal tissues. Human gallstones are a particularly rich source.

Cholesterol was first isolated in 1770. In the 1920s, two German chemists, Adolf Windaus (University of Göttingen) and Heinrich Wieland (University of Munich), were responsible for outlining a structure for cholesterol; they received Nobel Prizes for their work in 1927 and 1928.*

Part of the difficulty in assigning an absolute structure to cholesterol is that cholesterol contains *eight* tetrahedral stereocenters. This feature means that 2^8 or 256 possible stereoisomeric forms of the basic structure are possible, *only one of which is cholesterol.*

5-Cholesten-3β-ol
(absolute configuration of cholesterol)

PROBLEM 23.8

Designate with asterisks the eight stereocenters of cholesterol.

*The original cholesterol structure proposed by Windaus and Wieland was incorrect. This became evident in 1932 as a result of X-ray diffraction studies done by the British physicist J. D. Bernal. By the end of 1932, however, English scientists, and Wieland himself, using Bernal's results, were able to outline the correct structure of cholesterol.

Cholesterol occurs widely in the human body, but not all of the biological functions of cholesterol are yet known. Cholesterol is known to serve as an intermediate in the biosynthesis of all of the steroids of the body. Cholesterol, therefore, is essential to life. We do not need to have cholesterol in our diet, however, because our body can synthesize all we need. When we ingest cholesterol, our body synthesizes less than if we ate none at all, but the total cholesterol is more than if we ate none at all. Far more cholesterol is present in the body than is necessary for steroid biosynthesis. High levels of blood cholesterol have been implicated in the development of arteriosclerosis (hardening of the arteries) and in heart attacks that occur when cholesterol-containing plaques block arteries of the heart. Considerable research is being carried out in the area of cholesterol metabolism with the hope of finding ways of minimizing cholesterol levels through the use of dietary adjustments or drugs.

23.4C Sex Hormones

The sex hormones can be classified into three major groups: (1) the female sex hormones, or **estrogens,** (2) the male sex hormones, or **androgens,** and (3) the pregnancy hormones, or **progestins.**

The first sex hormone to be isolated was an estrogen, *estrone.* Working independently, Adolf Butenandt (in Germany at the University of Göttingen) and Edward Doisy (in the United States at St. Louis University) isolated estrone from the urine of pregnant women. They published their discoveries in 1929. Later, Doisy was able to isolate the much more potent estrogen, *estradiol.* In this research Doisy had to extract *4 tons* of sow ovaries in order to obtain just 12 mg of estradiol. Estradiol, it turns out, is the true female sex hormone, and estrone is a metabolized form of estradiol that is excreted.

Estrone
[3-hydroxy-1,3,5(10)-
estratrien-17-one]

Estradiol
[1,3,5(10)-estra-
triene-3,17 β-diol]

Estradiol is secreted by the ovaries and promotes the development of the secondary female characteristics that appear at the onset of puberty. Estrogens also stimulate the development of the mammary glands during pregnancy and induce estrus (heat) in animals.

In 1931, Butenandt and Kurt Tscherning isolated the first androgen, *androsterone.* They were able to obtain 15 mg of this hormone by extracting approximately 15,000 L of male urine. Soon afterwards (in 1935), Ernest Laqueur (in Holland) isolated another male sex hormone, *testosterone,* from bull testes. It soon became clear that testosterone is the true male sex hormone and that androsterone is a metabolized form of testosterone that is excreted in the urine.

Androsterone
(3 α-hydroxy-5 α-androstan-17-one)

Testosterone
(17 β-hydroxy-4-androsten-3-one)

Testosterone, secreted by the testes, is the hormone that promotes the development of secondary male characteristics: the growth of facial and body hair; the deepening of the voice; muscular development; and the maturation of the male sex organs.

Testosterone and estradiol, then, are the chemical compounds from which "maleness" and "femaleness" are derived. It is especially interesting to examine their structural formulas and see how very slightly these two compounds differ. Testosterone has an angular methyl group at the **A,B** ring junction that is missing in estradiol. Ring **A** of estradiol is a benzene ring and, as a result, estradiol is a phenol. Ring **A** of testosterone contains an α,β-unsaturated keto group.

PROBLEM 23.9

The estrogens (estrone and estradiol) are easily separated from the androgens (androsterone and testosterone) on the basis of one of their chemical properties. What is the property and how could such a separation be accomplished?

Progesterone
(4-pregnene-3,20-dione)

Progesterone is the most important *progestin* (pregnancy hormone). After ovulation occurs, the remnant of the ruptured ovarian follicle (called the *corpus luteum*) begins to secrete progesterone. This hormone prepares the lining of the uterus for implantation of the fertilized ovum, and continued progesterone secretion is necessary for the completion of pregnancy. (Progesterone is secreted by the placenta after secretion by the *corpus luteum* declines.)

Progesterone *also suppresses ovulation,* and it is the chemical agent that apparently accounts for the fact that pregnant women do not conceive again while pregnant. It was this observation that led to the search for synthetic progestins that could be used as oral contraceptives. (Progesterone itself requires very large doses to be effective in suppressing ovulation when taken orally because it is degraded in the intestinal tract.) A number of such compounds have been developed and are now widely used. In addition to norethynodrel (cf. Problem 23.7), another widely used synthetic progestin is its double-bond isomer, *norethindrone*.

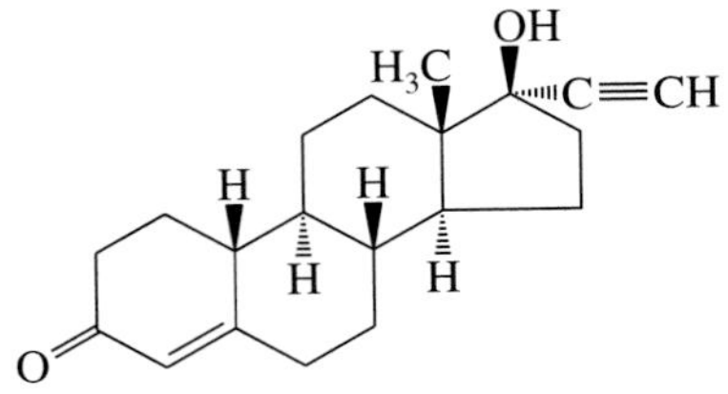

Norethindrone
(17α-ethynyl-17β-hydroxy-4-estren-3-one)

Synthetic estrogens have also been developed and these are often used in oral contraceptives in combination with synthetic progestins. A very potent synthetic estrogen is the compound called *ethynylestradiol* or *novestrol*.

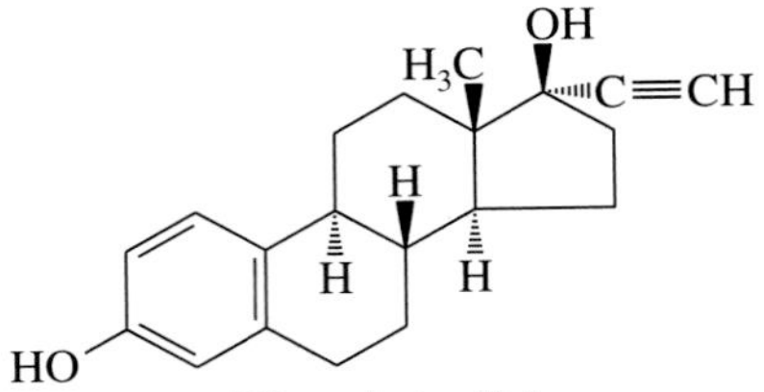

Ethynylestradiol
[17α-ethynyl-1,3,5(10)-estratriene-3,17β-diol]

23.4D Adrenocortical Hormones

At least 28 different hormones have been isolated from the adrenal cortex, part of the adrenal glands that sit on top of the kidneys. Included in this group are the following two steroids:

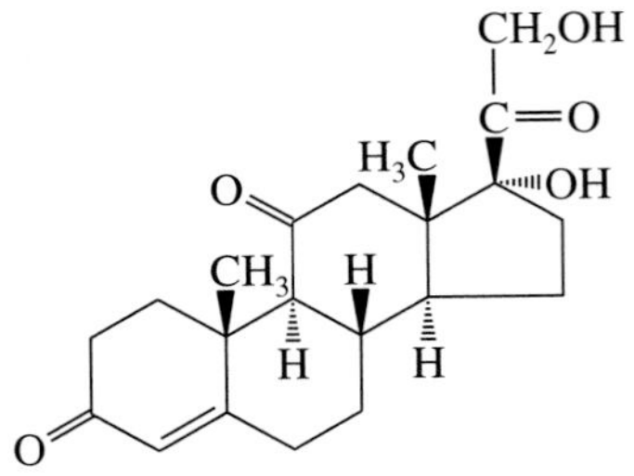

Cortisone
(17α,21-dihydroxy-4-pregnene-3,11,20-trione)

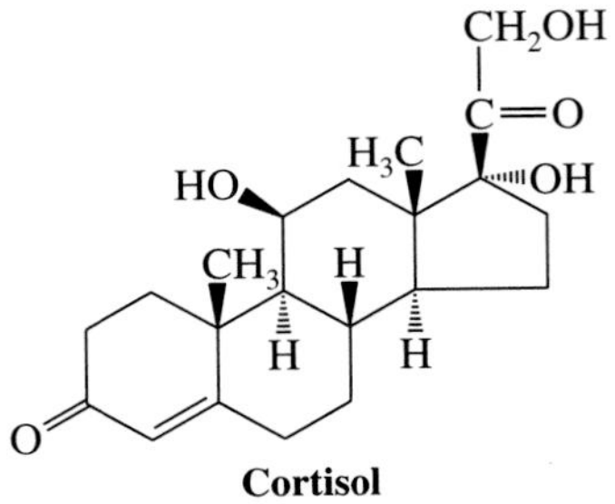

Cortisol
(11β,17α,21-trihydroxy-4-pregnene-3,20-dione)

Most of the adrenocortical steroids have an oxygen function at position 11 (a keto group in cortisone, for example, and a β-hydroxyl in cortisol). Cortisol is the major hormone synthesized by the human adrenal cortex.

The adrenocortical steroids are apparently involved in the regulation of a large number of biological activities, including carbohydrate, protein, and lipid metabolism; water and electrolyte balance and reactions to allergic and inflammatory phenomena. Recognition, in 1949, of the antiinflammatory effect of cortisone and its usefulness in the treatment of rheumatoid arthritis led to extensive research in this area. Many 11-oxygenated steroids are now used in the treatment of a variety of disorders ranging from Addison's disease, to asthma, and to skin inflammations.

23.4E D Vitamins

The demonstration, in 1919, that sunlight helped cure rickets—a childhood disease characterized by poor bone growth—began a long search for a chemical explanation. Soon it was discovered that irradiation of certain foodstuffs increased their antirachitic properties and, in 1930, the search led to a steroid that can be isolated from yeast, called *ergosterol*. Irradiation of ergosterol was found to produce a highly active mate-rial. In 1932, Windaus (Section 23.4B) and his co-workers in Germany demonstrated that this highly active substance was vitamin D_2. The photochemical reaction that takes place is one in which the dienoid ring **B** of ergosterol opens to produce a conjugated triene:

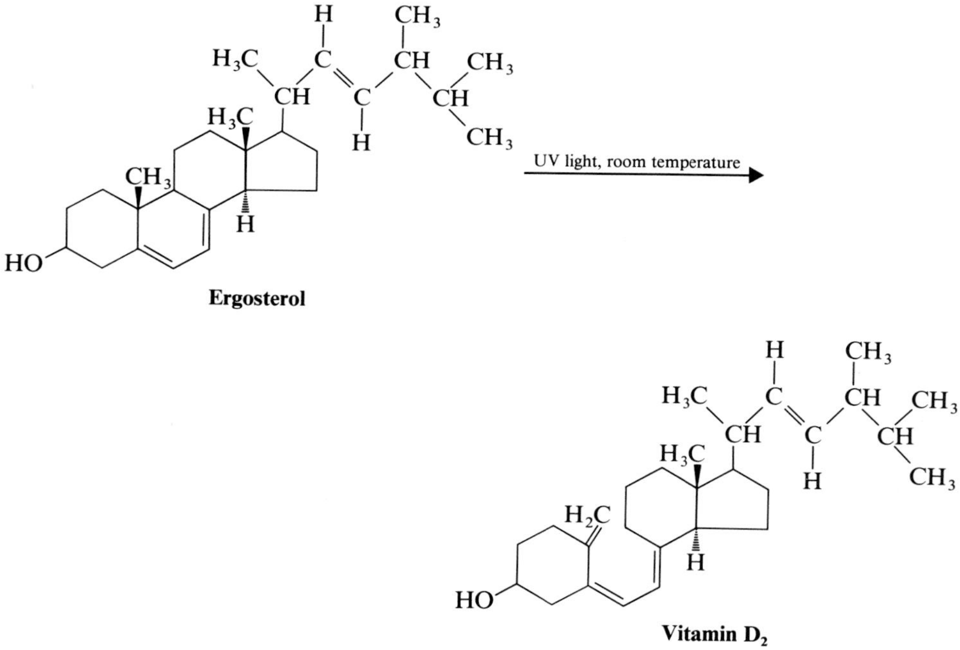

Ergosterol

UV light, room temperature

Vitamin D_2

23.4F Other Steroids

The structures, sources, and physiological properties of a number of other important steroids are given in Table 23.4.

23.4G Reactions of Steroids

Steroids undergo all of the reactions that we might expect of molecules containing double bonds, hydroxyl groups, keto groups, and so on. While the stereochemistry of steroid reactions is often quite complex, it is often strongly influenced by the steric

TABLE 23.4 *Other important steroids*

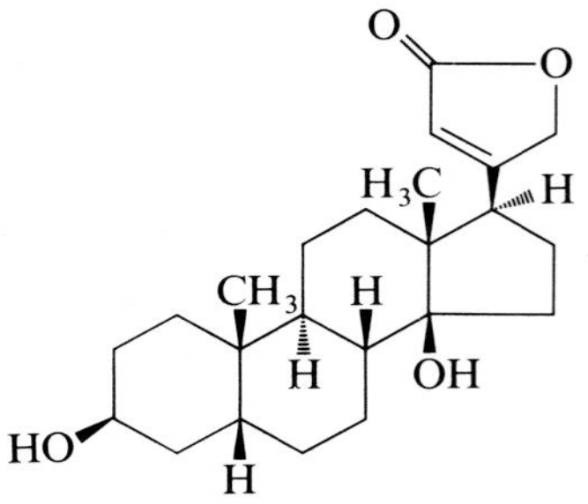

Digitoxigenin

Digitoxigenin is a cardiac agly-cone that can be isolated by hydrolysis of digitalis, a phar-maceutical that has been used in treating heart disease since 1785. In digitalis, sugar mole-cules are joined in acetal link-ages to the 3-OH group of the steroid. In small doses digitalis strengthens the heart muscle; in larger doses it is a powerful heart poison. The aglycone has only about one-fortieth the ac-tivity of digitalis.

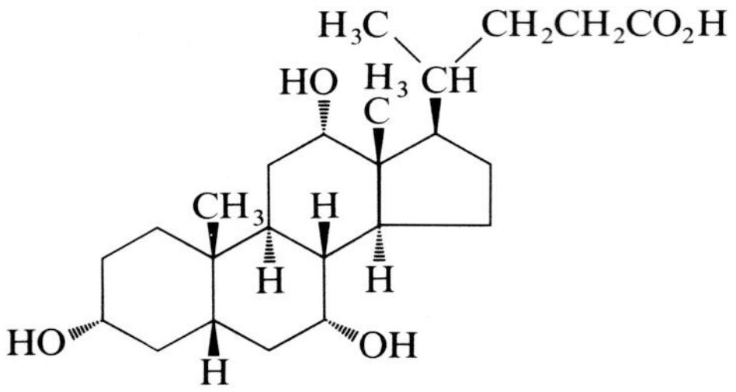

Cholic acid

Cholic acid is the most abun-dant acid obtained from the hydrolysis of human or ox bile. Bile is produced by the liver and stored in the gall-bladder. When secreted into the small intestine, bile emul-sifies lipids by acting as a soap. This action aids in the digestive process.

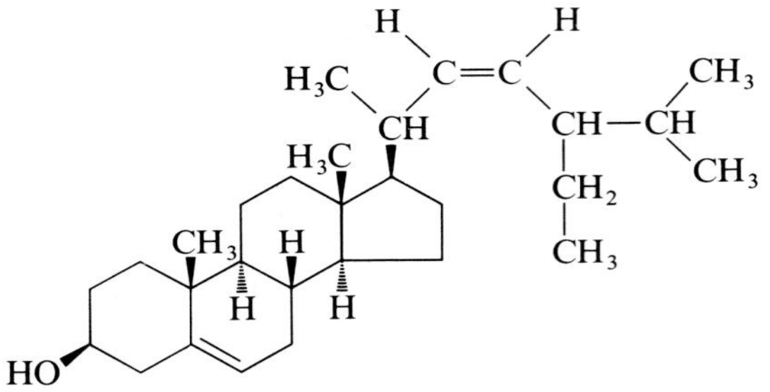

Stigmasterol

Stigmasterol is a widely occur-ring plant steroid that is ob-tained commercially from soy-bean oil.

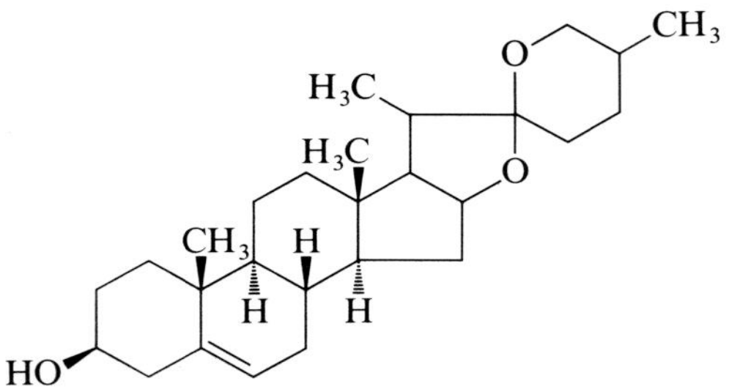

Diosgenin

Diosgenin is obtained from a Mexican vine, *cabeza de negro,* genus *Dioscorea.* It is used as the starting material for a commercial synthesis of cortisone and sex hormones.

hindrance presented at the β face of the molecule by the angular methyl groups. Many reagents react preferentially at the relatively unhindered α face, especially when the reaction takes place at a functional group very near an angular methyl group and when the attacking reagent is bulky. Examples that illustrate this tendency are shown in the following reactions.

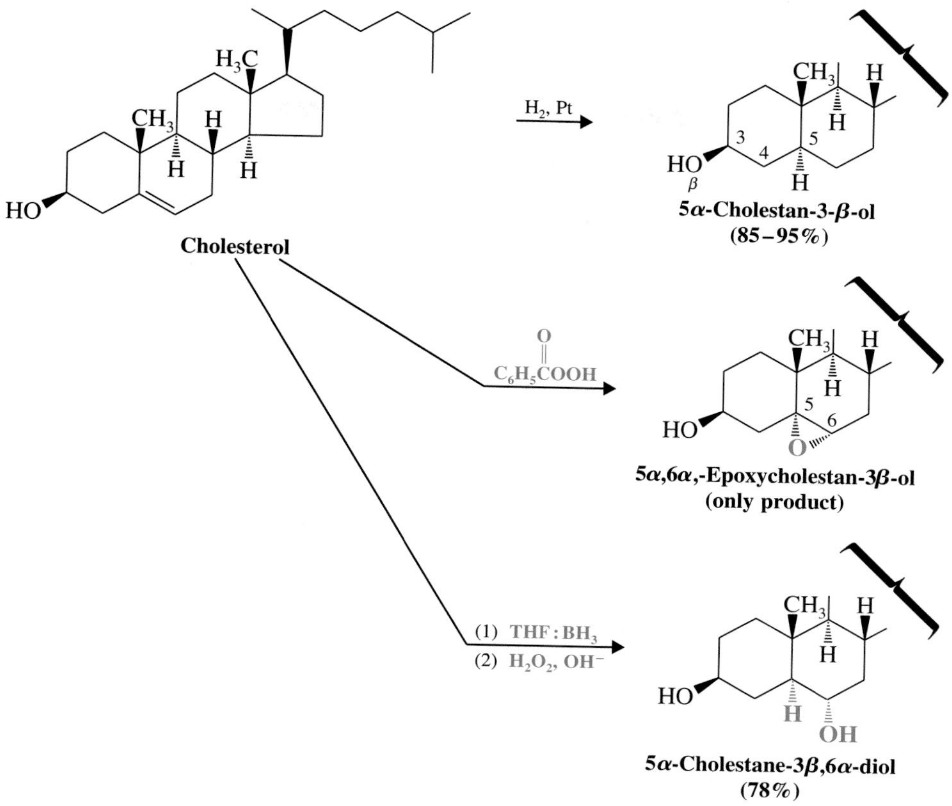

Cholesterol

H_2, Pt

5α-Cholestan-3-β-ol
(85–95%)

C_6H_5COOH

5α,6α,-Epoxycholestan-3β-ol
(only product)

(1) THF : BH$_3$
(2) H$_2$O$_2$, OH$^-$

5α-Cholestane-3β,6α-diol
(78%)

When the epoxide ring of 5α,6α-epoxycholestan-3β-ol (see following reaction) is opened, attack by chloride ion must occur from the β face, but it takes place at the more open 6 position. Notice that the 5- and 6- substituents in the product are *diaxial* (Section 8.7).

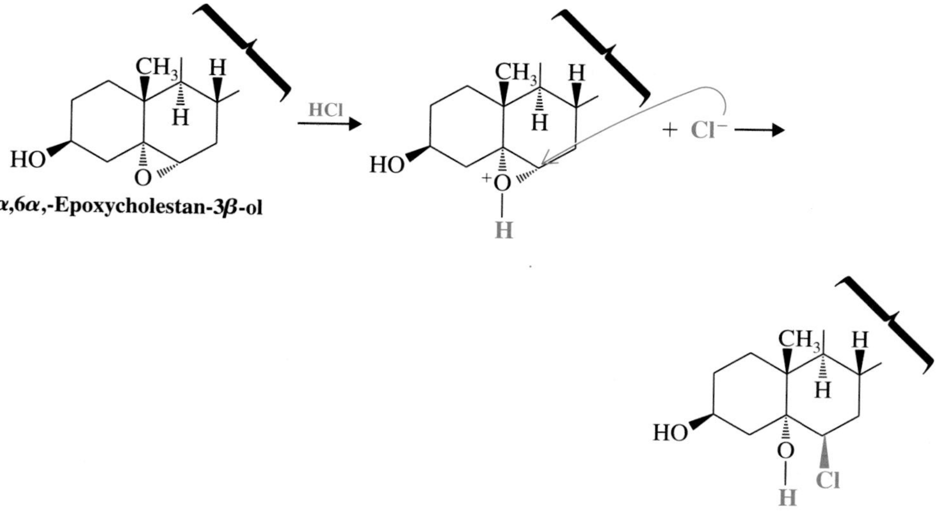

5α,6α,-Epoxycholestan-3β-ol

HCl

+ Cl$^-$ ⟶

PROBLEM 23.10

Show how you might convert cholesterol into each of the following compounds:

(a) 5α,6β-Dibromocholestan-3β-ol   (d) 6α-Deuterio-5α-cholestan-3β-ol

(b) Cholestane-3β,5α,6β-triol   (e) 6β-Bromocholestane-3β,5α-diol

(c) 5α-Cholestan-3-one

The relative openness of equatorial groups (when compared to axial groups) also influences the stereochemical course of steroid reactions. When 5α-cholestane-3β,7α-diol (see following reaction) is treated with excess ethyl chloroformate (C_2H_5OCOCl), only the equatorial 3β-hydroxyl becomes esterified. The axial 7α-hydroxyl is unaffected by the reaction.

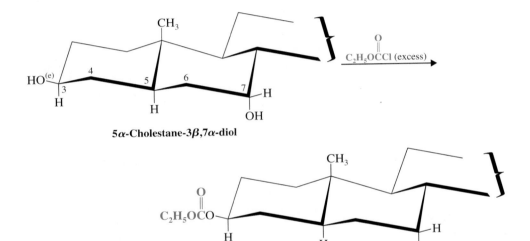

5α-Cholestane-3β,7α-diol

(only product)

By contrast, treating 5α-cholestane-3β,7β-diol with excess ethyl chloroformate esterifies both hydroxyl groups. In this instance both groups are equatorial.

5α–Cholestane–3β, 7β–diol

23.5 PROSTAGLANDINS

One very active area of current research is concerned with a group of lipids called prostaglandins.* Prostaglandins are C_{20}-carboxylic acids that contain a five-membered ring, at least one double bond, and several oxygen-containing functional groups. Two of the most biologically active prostaglandins are prostaglandin E_2 and prostaglandin $F_{1\alpha}$.

Prostaglandin E_2†
(PGE_2*)

Prostaglandin $F_{1\alpha}$
($PGF_{1\alpha}$)

Prostaglandins of the E type have a carbonyl group at C9 and a hydroxyl group at C11; those of the F type have hydroxyl groups at both positions. Prostaglandins of the 2 series have a double bond between C5 and C6; in the 1 series this bond is a single bond.

First isolated from seminal fluid, prostaglandins have since been found in almost all animal tissues. The amounts vary from tissue to tissue but are almost always very small. Most prostaglandins have powerful physiological activity, however, and this activity covers a broad spectrum of effects. Prostaglandins are known to affect heart rate, blood pressure, blood clotting, conception, fertility, and allergic responses.

The recent finding that prostaglandins can prevent formation of blood clots has great clinical significance, because heart attacks and strokes often result from the formation of abnormal clots in blood vessels. An understanding of how prostaglandins affect the formation of clots may lead to the development of drugs to prevent heart attacks and strokes.

The biosynthesis of prostaglandins of the 2 series begins with a C_{20} polyenoic acid, arachidonic acid. (Synthesis of prostaglandins of the 1 series begins with a fatty acid with one fewer double bond.) The first step requires two molecules of oxygen and is catalyzed by an enzyme called *cyclooxygenase*.

*The 1982 Nobel Prize in physiology or medicine was awarded to S. K. Bergström and B. I. Samuelson (of the Karolinska Institute, Stockholm, Sweden) and to J. R. Vane (of the Wellcome Foundation, Bechenham, England) for their work on prostaglandins.

†These names are abbreviated designations used by workers in the field; systematic names are seldom used for prostaglandins.

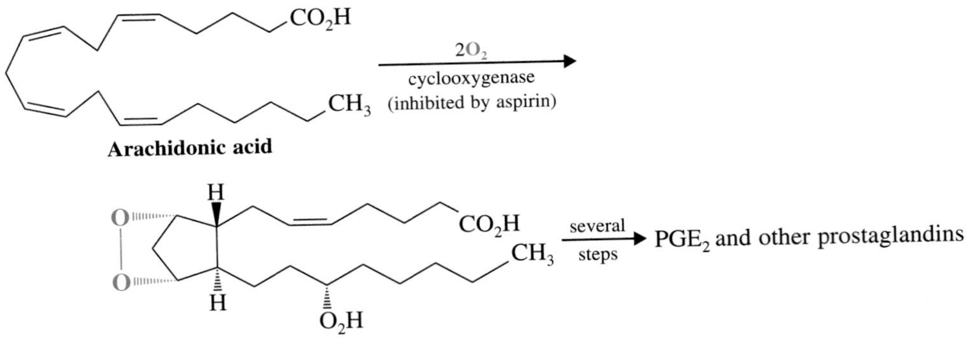

Arachidonic acid

$2O_2$
cyclooxygenase
(inhibited by aspirin)

PGG₂
(a cyclic endoperoxide)

PGE₂ and other prostaglandins

The involvement of prostaglandins in allergic and inflammation responses has also been of special interest. Some prostaglandins induce inflammation; others relieve it. The most widely used antiinflammatory drug is ordinary aspirin (cf. Section 21.8). Aspirin blocks the synthesis of prostaglandins from arachidonic acid, apparently by acetylating the enzyme cyclooxygenase, thus rendering it inactive (see the previous reaction). This reaction may represent the origin of aspirin's antiinflammatory properties. Another prostaglandin (PGE₁) is a potent fever-inducing agent (pyrogen), and aspirin's ability to reduce fever may also arise from its inhibition of prostaglandin synthesis.

23.6 PHOSPHOLIPIDS AND CELL MEMBRANES

Another large class of lipids are those called *phospholipids*. Most phospholipids are structurally derived from a glycerol derivative known as a *phosphatidic acid*. In a phosphatidic acid, two hydroxyl groups of glycerol are joined in ester linkages to fatty acids and one terminal hydroxyl group is joined in an ester linkage to *phosphoric acid*.

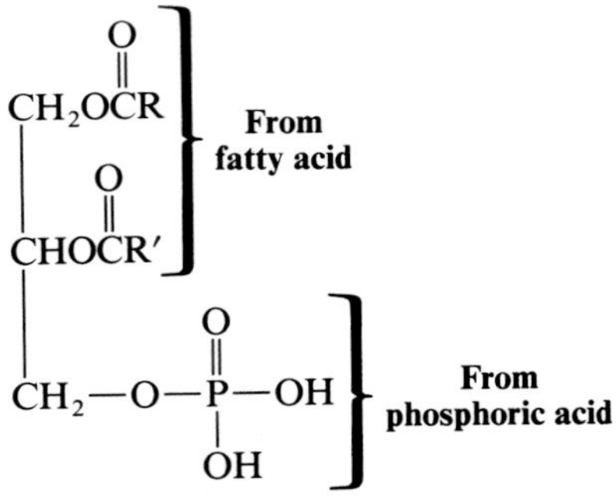

A phosphatidic acid
(a diacylglyceryl phosphate)

23.6A Phosphatides

In *phosphatides*, the phosphate group of a phosphatidic acid is bound through another phosphate ester linkage to one of the following nitrogen-containing compounds.

$$\overset{+}{\text{HOCH}_2\text{CH}_2\text{N}(\text{CH}_3)_3} \quad \text{HO}^- \qquad \text{HOCH}_2\text{CH}_2\text{NH}_2 \qquad \text{HOCH}_2-\overset{\overset{+}{\text{NH}_3}}{\underset{\text{H}}{\text{C}}}\text{---CO}_2^-$$

Choline   **2-Aminoethanol**   L-**Serine**
                                 (ethanolamine)

    The most important phosphatides are the **lecithins,** the **cephalins, phosphatidyl serines,** and the **plasmalogens** (a phosphatidyl derivative). Their general structures are shown in Table 23.5.

    Phosphatides resemble soaps and detergents in that they are molecules having both polar and nonpolar groups (Fig. 23.7*a*). Like soaps and detergents, too, phos-

TABLE 23.5 *Phosphatides*

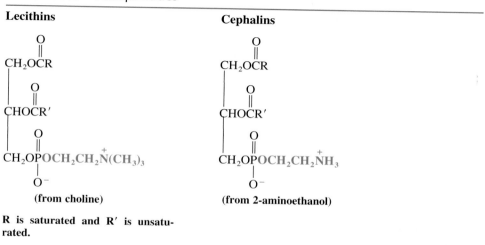

Lecithins

CH₂OCR (with O double bond)
CHOCR′ (with O double bond)
CH₂OPOCH₂CH₂N⁺(CH₃)₃ with O⁻
(from choline)

R is saturated and R′ is unsaturated.

Cephalins

CH₂OCR (with O double bond)
CHOCR′ (with O double bond)
CH₂OPOCH₂CH₂N⁺H₃ with O⁻
(from 2-aminoethanol)

Phosphatidyl Serines

CH₂OCR (with O double bond)
CHOCR′ (with O double bond)
CH₂OPOCH₂CHN⁺H₃ with O⁻ and CO₂⁻
(from L-serine)

R is saturated and R′ is unsaturated.

Plasmalogens

CH₂OR
CHOCR′ (with O double bond)
CH₂OPOCH₂CH₂N⁺H₃ with O⁻
(from 2-aminoethanol)

or
OCH₂CH₂N⁺(CH₃)₃
(from choline)

R is —CH=CH(CH₂)ₙCH₃ **(this linkage is that of an α,β-unsaturated ether).**
R′ is that of an unsaturated fatty acid.

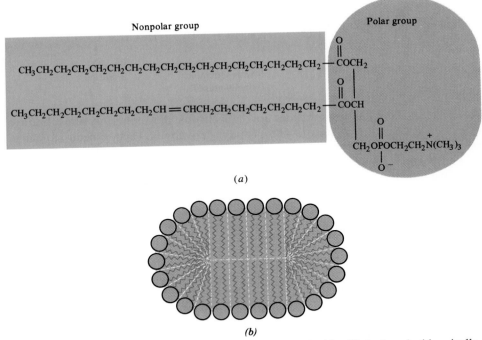

FIGURE 23.7 (a) Polar and nonpolar sections of a phosphatide. (b) A phosphatide micelle or lipid bilayer.

phatides "dissolve" in aqueous media by forming micelles. There is evidence that in biological systems the preferred micelles consist of three-dimensional arrays of "stacked" bimolecular micelles (Fig. 23.7) that are better described as **lipid bilayers.**

The hydrophilic and hydrophobic portions of phosphatides make them perfectly suited for one of their most important biological functions: They form a portion of a structural unit that creates an interface between an organic and an aqueous environment. This structure (Fig. 23.8) is found in cell walls and membranes where phosphatides are often found associated with proteins and glycolipids (Section 23.6B).

PROBLEM 23.11

Under suitable conditions all of the ester (and ether) linkages of a phosphatide can be hydrolyzed. What organic compounds would you expect to obtain from the complete hydrolysis of (a) a lecithin, (b) a cephalin, (c) a choline-based plasmalogen? [*Note:* Pay particular attention to the fate of the α,β-unsaturated ether in part (c).]

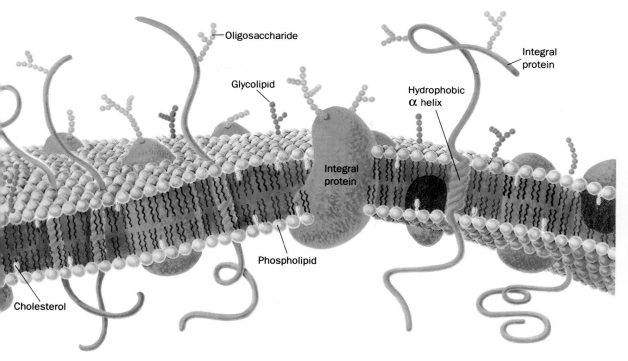

FIGURE 23.8 A schematic diagram of a plasma membrane. Integral proteins (*orange*) are embedded in a bilayer composed of phospholipids (*blue spheres with two wiggly tails;* shown, for clarity, in much greater proportion than they have in biological membranes) and cholesterol (*yellow*). The carbohydrate components of glycoproteins (*yellow beaded chains*) and glycolipids (*green beaded chains*) occur only on the external face of the membrane. (From D. Voet and J. G. Voet, *Biochemistry,* Wiley, New York, 1990, p. 286. Used with permission.)

23.6B Derivatives of Sphingosine

Another important group of lipids is derived from **sphingosine;** the derivatives are called **sphingolipids.** Two sphingolipids, a typical *sphingomyelin* and a typical *cerebroside,* are shown in Fig. 23.9.

On hydrolysis, sphingomyelins yield sphingosine, choline, phosphoric acid, and a C_{24} fatty acid called lignoceric acid. In a sphingomyelin this last component is bound to the —NH_2 group of sphingosine. The sphingolipids do not yield glycerol when they are hydrolyzed.

The cerebroside shown in Fig. 23.9 is an example of a **glycolipid.** Glycolipids have a polar group that is contributed by a *carbohydrate.* They do not yield phosphoric acid or choline when they are hydrolyzed.

The sphingolipids, together with proteins and polysaccharides, make up **myelin,** the protective coating that encloses nerve fibers or **axons.** The axons of nerve cells carry electrical nerve impulses. Myelin has been described as having a function relative to the axon similar to that of the insulation on an ordinary electric wire.

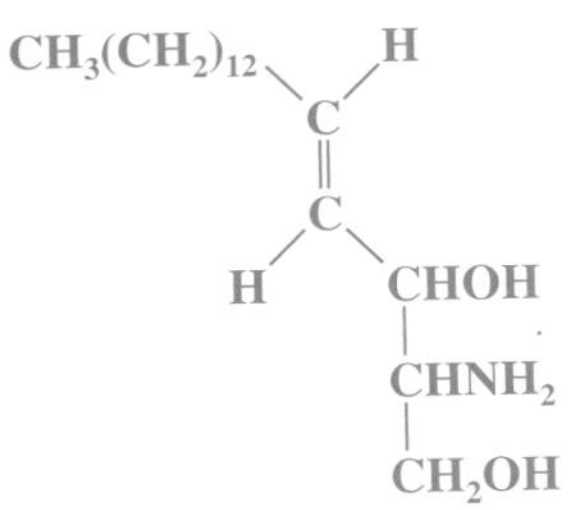

Sphingosine

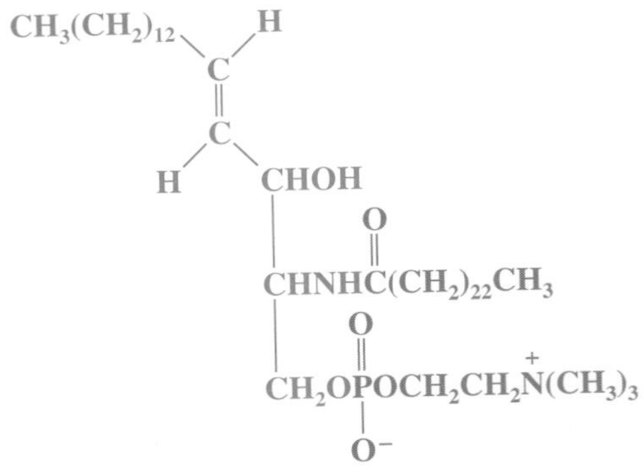

**Sphingomyelin
(a sphingolipid)**

From a
carbohydrate,
D-galactose

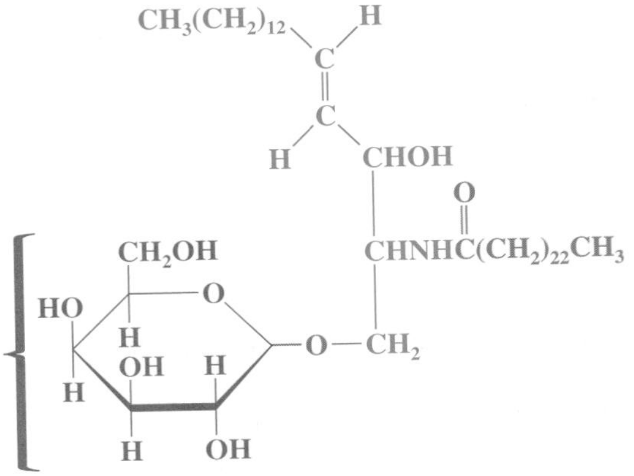

Cerebroside

FIGURE 23.9 A sphingosine and two sphingolipids.

23.7 WAXES

Most waxes are esters of long-chain fatty acids and long-chain alcohols. Waxes are
found as protective coatings on the skin, fur, or feathers of animals, and on the leaves
and fruits of plants. Several esters isolated from waxes are the following:

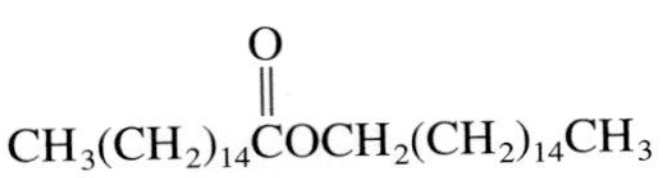

$$CH_3(CH_2)_{14}COCH_2(CH_2)_{14}CH_3$$
**Cetyl palmitate
(from spermaceti)**

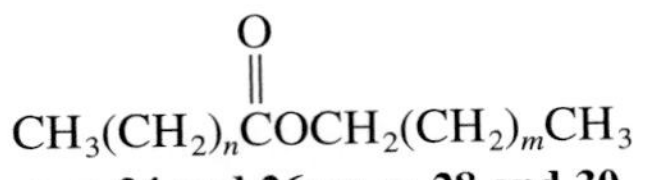

$$CH_3(CH_2)_nCOCH_2(CH_2)_mCH_3$$
$n = 24$ and 26; $m = 28$ and 30
(from beeswax)

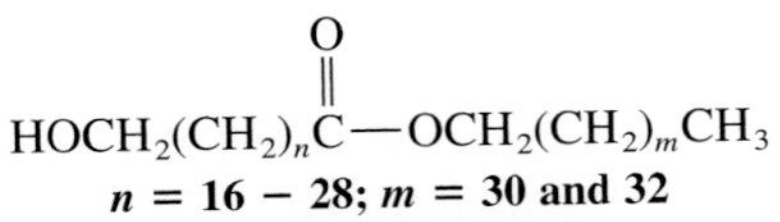

$$HOCH_2(CH_2)_nC-OCH_2(CH_2)_mCH_3$$
$n = 16 - 28$; $m = 30$ and 32
(from carnauba wax)

ADDITIONAL PROBLEMS

23.12   How would you convert stearic acid, $CH_3(CH_2)_{16}CO_2H$, into each of the following?

(a) Ethyl stearate, $CH_3(CH_2)_{16}CO_2C_2H_5$ (two ways)

(b) *tert*-Butyl stearate, $CH_3(CH_2)_{16}CO_2C(CH_3)_3$

(c) Stearamide, $CH_3(CH_2)_{16}CONH_2$

(d) *N,N*-Dimethylstearamide, $CH_3(CH_2)_{16}CON(CH_3)_2$

(e) Octadecylamine, $CH_3(CH_2)_{16}CH_2NH_2$

(f) Heptadecylamine, $CH_3(CH_2)_{15}CH_2NH_2$

(g) Octadecanal, $CH_3(CH_2)_{16}CHO$

(h) Octadecyl stearate, $CH_3(CH_2)_{16}\overset{\overset{\displaystyle O}{\|}}{C}OCH_2(CH_2)_{16}CH_3$

(i) 1-Octadecanol, $CH_3(CH_2)_{16}CH_2OH$ (two ways)

(j) 2-Nonadecanone, $CH_3(CH_2)_{16}\overset{\overset{\displaystyle O}{\|}}{C}CH_3$

(k) 1-Bromooctadecane, $CH_3(CH_2)_{16}CH_2Br$

(l) Nonadecanoic acid, $CH_3(CH_2)_{16}CH_2CO_2H$

23.13 How would you transform myristic acid into each of the following?

(a) $CH_3(CH_2)_{11}\underset{\underset{\displaystyle Br}{|}}{C}HCO_2H$

(b) $CH_3(CH_2)_{11}\underset{\underset{\displaystyle OH}{|}}{C}HCO_2H$

(c) $CH_3(CH_2)_{11}\underset{\underset{\displaystyle CN}{|}}{C}HCO_2H$

(d) $CH_3(CH_2)_{11}\underset{\underset{\displaystyle NH_3^+}{|}}{C}HCO_2^-$

23.14 Using palmitoleic acid as an example and neglecting stereochemistry, illustrate each of the following reactions of the double bond.

(a) Addition of bromine

(b) Addition of hydrogen

(c) Hydroxylation

(d) Addition of HCl

23.15 When oleic acid is heated to 180–200°C (in the presence of a small amount of selenium), an equilibrium is established between oleic acid (33%) and an isomeric compound called elaidic acid (67%). Suggest a possible structure for elaidic acid.

23.16 Gadoleic acid ($C_{20}H_{38}O_2$), a fatty acid that can be isolated from codliver oil, can be cleaved by hydroxylation and subsequent treatment with periodic acid to $CH_3(CH_2)_9CHO$ and $OHC(CH_2)_7CO_2H$. (a) What two stereoisomeric structures are possible for gadoleic acid? (b) What spectroscopic technique would make possible a decision as to the actual structure of gadoleic acid? (c) What peaks would you look for?

✶ 23.17 When limonene (Section 23.3) is heated strongly, it yields 2 mol of isoprene. What kind of reaction is involved here?

23.18 α-Phellandrene and β-phellandrene are isomeric compounds that are minor constituents of spearmint oil; they have the molecular formula $C_{10}H_{16}$. Each compound has a UV absorption maximum in the 230–270-nm range. On catalytic hydrogenation, each compound yields 1-isopropyl-4-methylcyclohexane. On vigorous oxidation with potassium permanganate, α-phellandrene yields $CH_3\overset{\overset{\displaystyle O}{\|}}{C}CO_2H$ and $CH_3\underset{\underset{\displaystyle CH_3}{|}}{C}HCH(CO_2H)CH_2CO_2H$. A similar oxidation of β-phellandrene yields $CH_3CHCH(CO_2H)CH_2CH_2CCO_2H$ as the only isolable

with the $\overset{O}{\overset{\|}{C}}$ shown above the sixth carbon and CH_3 below the second carbon:

$$CH_3CHCH(CO_2H)CH_2CH_2\overset{\overset{\displaystyle O}{\displaystyle \|}}{C}CO_2H$$
$$\overset{|}{CH_3}$$

product. Propose structures for α- and β-phellandrene.

23.19   Vaccenic acid, a constitutional isomer of oleic acid, has been synthesized through the following reaction sequence:

1-Octyne + NaNH$_2$ $\xrightarrow[\text{NH}_3]{\text{liq.}}$ **A** ($C_8H_{13}Na$) $\xrightarrow{\text{ICH}_2(\text{CH}_2)_7\text{CH}_2\text{Cl}}$

B ($C_{17}H_{31}Cl$) $\xrightarrow{\text{NaCN}}$ **C** ($C_{18}H_{31}N$) $\xrightarrow{\text{KOH, H}_2\text{O}}$ **D** ($C_{18}H_{31}O_2K$) $\xrightarrow{\text{H}_3\text{O}^+}$

E ($C_{18}H_{32}O_2$) $\xrightarrow[\text{BaSO}_4]{\text{H}_2, \text{Pd}}$ vaccenic acid ($C_{18}H_{34}O_2$)

Propose a structure for vaccenic acid and for the intermediates **A**–**E**.

23.20 ω-Fluorooleic acid can be isolated from a shrub, *Dechapetalum toxicarium*, that grows in Africa. The compound is highly toxic to warm-blooded animals; it has found use as an arrow poison in tribal warfare, in poisoning enemy water supplies, and by witch doctors "for terrorizing the native population." Powdered fruit of the plant has been used as a rat poison, hence ω-fluorooleic acid has the common name "ratsbane." A synthesis of ω-fluorooleic acid is outlined here. Give structures for compounds **F**–**I**.

1-Bromo-8-fluorooctane + sodium acetylide $\longrightarrow$ **F** ($C_{10}H_{17}F$) $\xrightarrow[\text{(2) I(CH}_2)_7\text{Cl}]{\text{(1) NaNH}_2}$

G ($C_{17}H_{30}FCl$) $\xrightarrow{\text{NaCN}}$ **H** ($C_{18}H_{30}NF$) $\xrightarrow[\text{(2) H}^+]{\text{(1) KOH}}$ **I** ($C_{18}H_{31}O_2F$) $\xrightarrow[\text{Ni}_2\text{B (P-2)}]{\text{H}_2}$

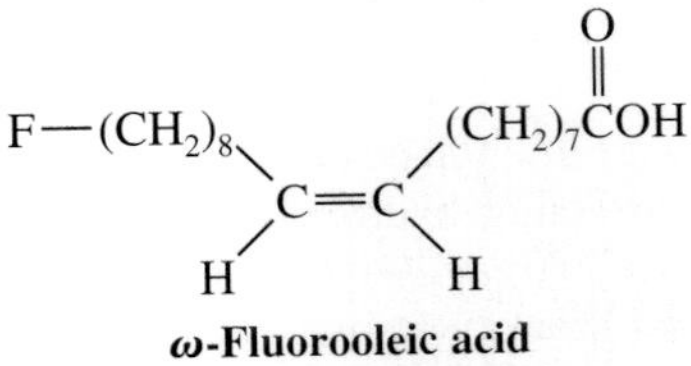

$$F-(CH_2)_8 \overset{\displaystyle }{\underset{\displaystyle H}{\diagdown}} C = C \overset{\displaystyle (CH_2)_7\overset{\overset{\displaystyle O}{\displaystyle \|}}{C}OH}{\underset{\displaystyle H}{\diagup}}$$

ω-Fluorooleic acid
(46% yield, overall)

23.21   Give formulas and names for compounds **A** and **B**.

$$5\alpha\text{-Cholest-2-ene} \xrightarrow{\text{C}_6\text{H}_5\overset{\overset{\displaystyle O}{\displaystyle \|}}{\text{C}}\text{OOH}} \textbf{A} \text{ (an epoxide)} \xrightarrow{\text{HBr}} \textbf{B}$$

(*Hint:* **B** is not the most stable stereoisomer.)

23.22   One of the first laboratory syntheses of cholesterol was achieved by R. B. Woodward and his students at Harvard University in 1951. Many of the steps of this synthesis are outlined here. Supply the missing reagents (a)–(w).

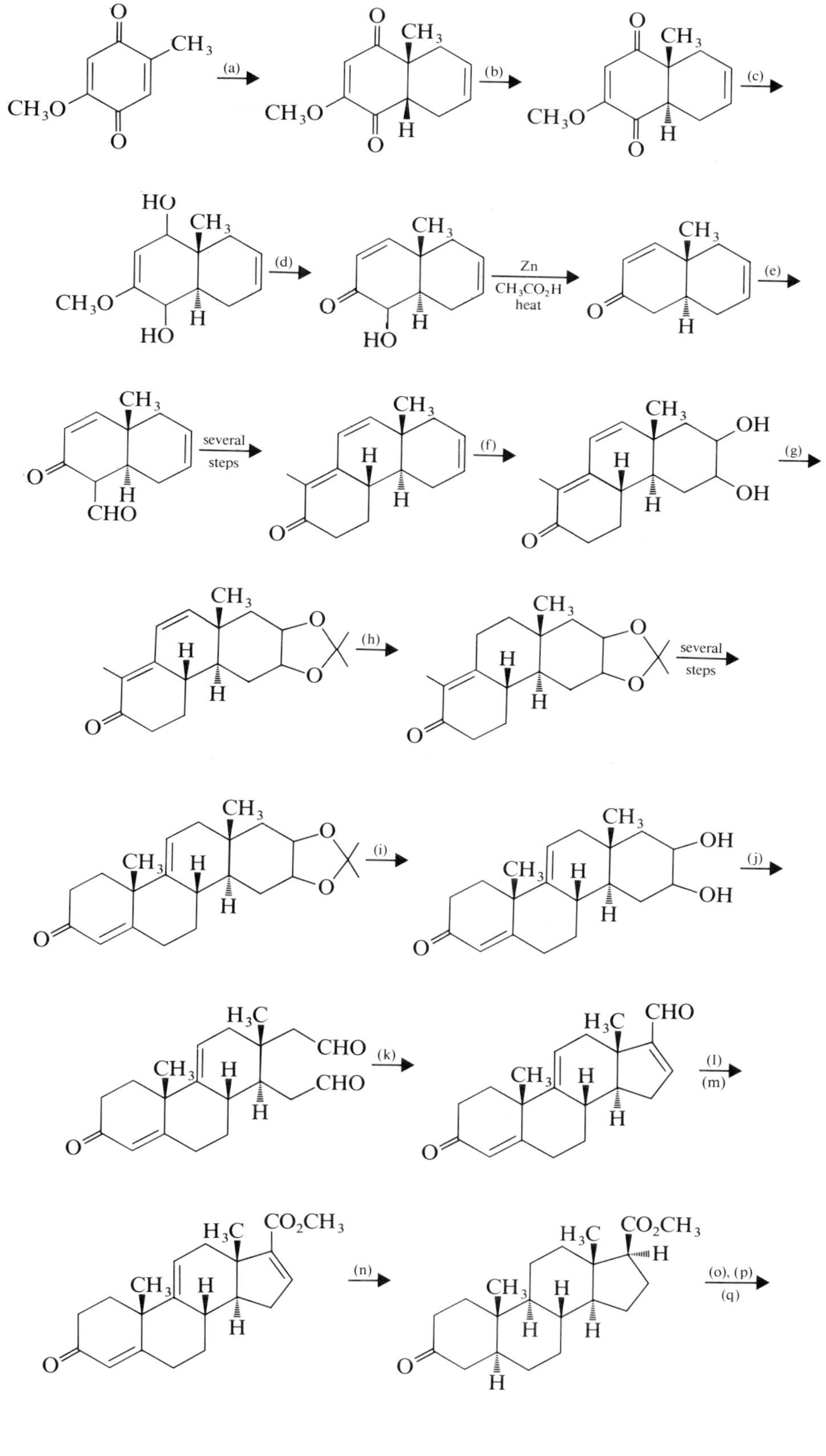
(a)
(b)
(c)
(d)
Zn
CH_3CO_2H
heat
(e)
several
steps
(f)
(g)
(h)
several
steps
(i)
(j)
(k)
(l)
(m)
(n)
(o), (p)
(q)

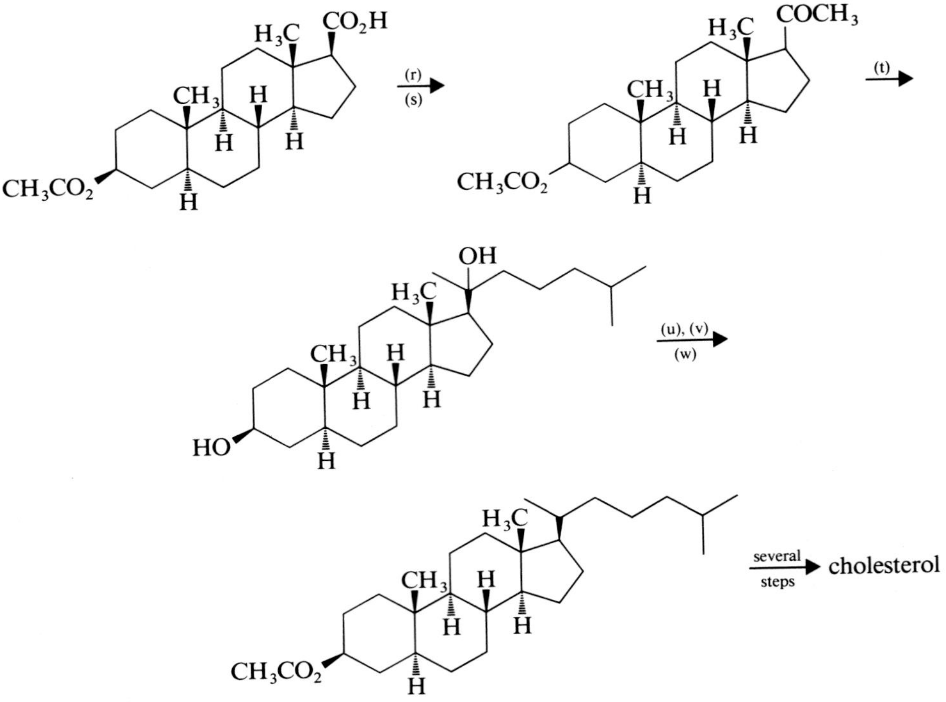

23.23   The initial steps of a laboratory synthesis of several prostaglandins reported by E. J. Corey (Section 4.15C) and his co-workers in 1968 are outlined here. Supply each of the missing reagents.

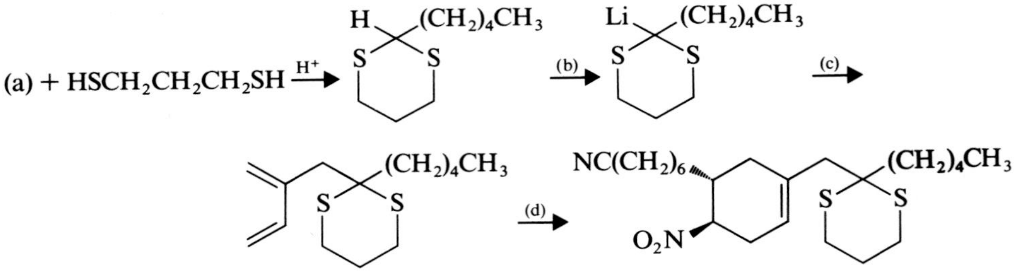

(e) The initial step in another prostaglandin synthesis is shown in the following reaction. What kind of reaction—and catalyst—is needed here?

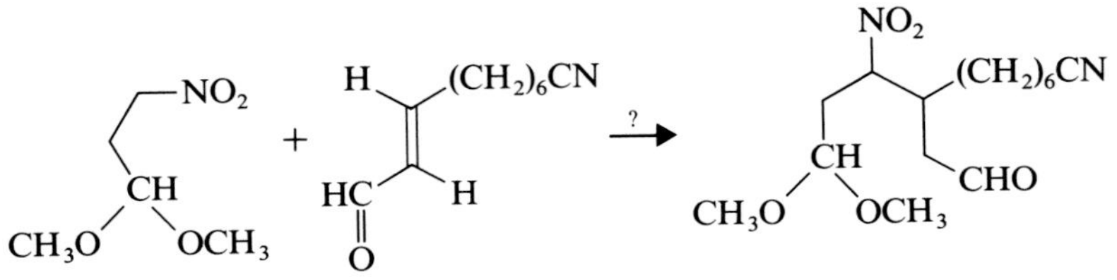

23.24   A useful synthesis of sesquiterpene ketones, called *cyperones,* was accomplished through a modification of the following Robinson annulation procedure (Section 17.9C)

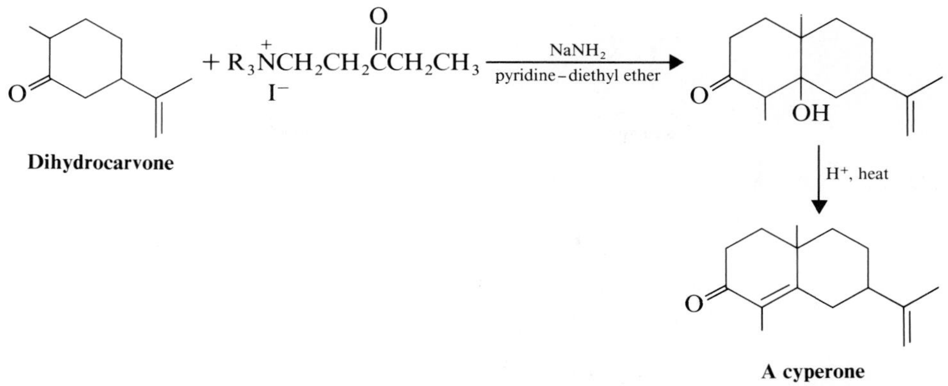

Dihydrocarvone

A cyperone

Write a mechanism that accounts for each step of this synthesis.

3-Methyl-3-butenyl pyrophosphate

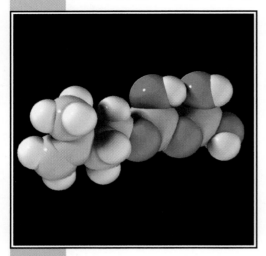

THIOL ESTERS AND LIPID BIOSYNTHESIS

P.1 Thiol Esters

Thiol esters can be prepared by reactions of a thiol with an acyl chloride.

$$R-C\overset{O}{\underset{Cl}{\diagup}} + R'-SH \longrightarrow R-C\overset{O}{\underset{S-R'}{\diagup}} + HCl$$

Thiol ester

$$CH_3C\overset{O}{\underset{Cl}{\diagup}} + CH_3SH \xrightarrow{\text{pyridine}} CH_3C\overset{O}{\underset{SCH_3}{\diagup}} + \underset{\underset{H}{\overset{N_+}{\bigcirc}}}{} \quad Cl^-$$

Although thiol esters are not often used in laboratory syntheses, they are of great importance in syntheses that occur within living cells. One of the important thiol esters in biochemistry is "acetyl-coenzyme A."

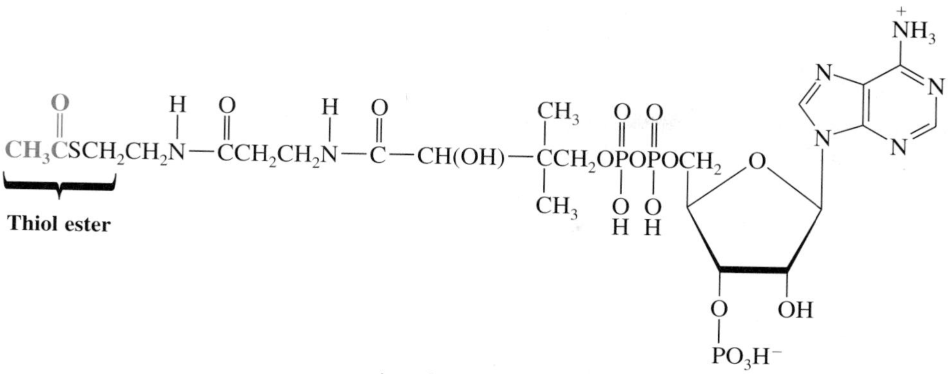

Acetyl coenzyme A

The important part of this rather complicated structure is the thiol ester at the beginning of the chain; because of this, acetyl-coenzyme A is usually abbreviated as follows:

$$\text{CoA}-\text{SCCH}_3 \quad (\text{C}=\text{O})$$

and coenzyme A, itself, is abbreviated:

$$\text{CoA}-\text{SH}$$

In certain biochemical reactions, an *acyl* coenzyme A operates as an *acylating agent;* it transfers an acyl group to another nucleophile in a reaction that involves a nucleophilic attack at the acyl carbon of the thiol ester. For example:

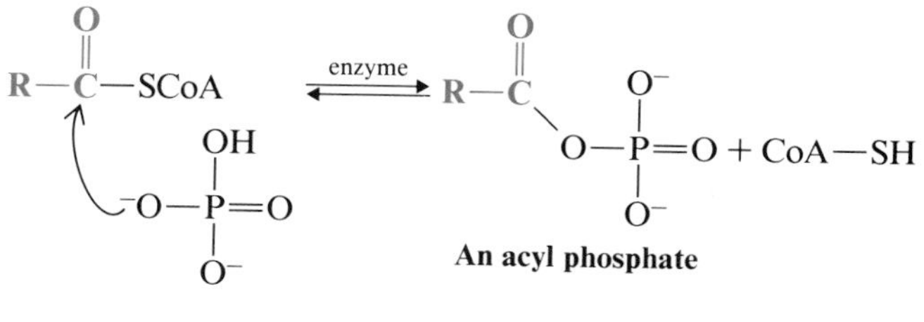

An acyl phosphate

This reaction is catalyzed by the enzyme *phosphotransacetylase.*

The α hydrogens of the acetyl group of acetyl coenzyme A are appreciably acidic. Acetyl-coenzyme A, as a result, also functions as an *alkylating agent.* Acetyl-coenzyme A, for example, reacts with oxaloacetate ion to form citrate ion in a reaction that resembles an aldol addition.

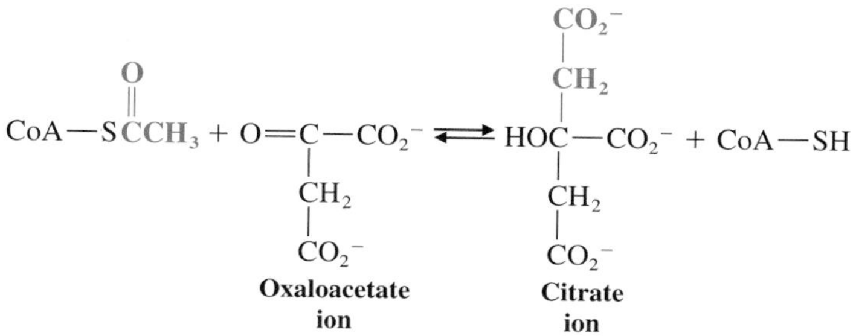

Oxaloacetate
ion

Citrate
ion

One might well ask, "Why has nature made such prominent use of thiol esters?" Or, "In contrast to ordinary esters, what advantages do thiol esters offer the cell?" In answering these questions we can consider three factors:

1. Resonance contributions of type (b) in the following reaction stabilize an ordinary ester and make the carbonyl group less susceptible to nucleophilic attack.

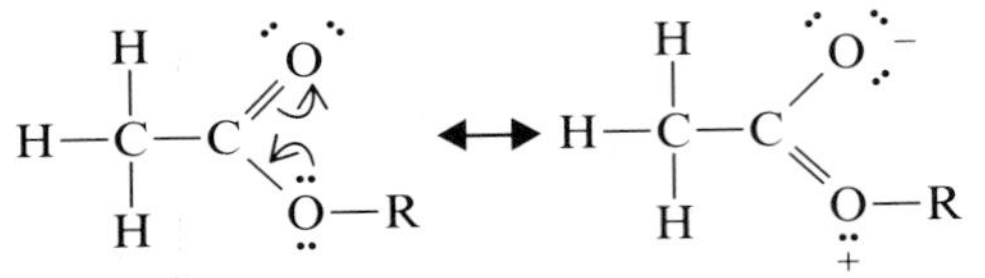

(a)   (b)
**This structure makes
an important contribution**

By contrast, thiol esters are not as effectively stabilized by a similar resonance contribution because structure (d) among the following ones requires overlap between the 3p orbital of sulfur and a 2p orbital of carbon. Since this overlap is not large, resonance stabilization by (d) is not as effective. Structure (e) does, however, make an important contribution—one that makes the carbonyl group more susceptible to nucleophilic attack.

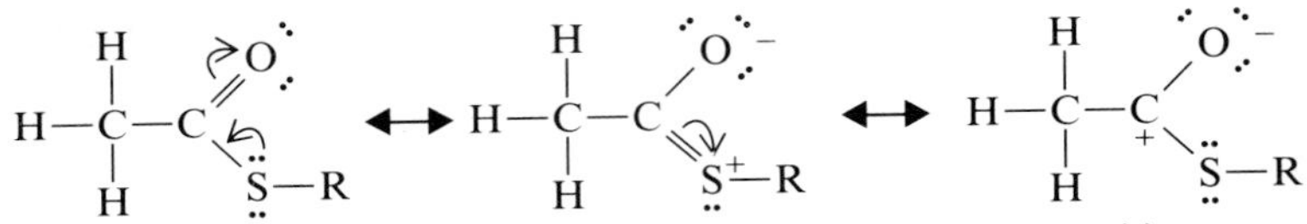

(c)   (d)
**This structure is
not an important
contributor**

(e)
**This structure
makes the carbonyl
carbon atom susceptible
to nucleophilic attack**

2. A resonance contribution from the similar structure (g) makes the α hydrogens of thiol esters more acidic than those of ordinary esters.

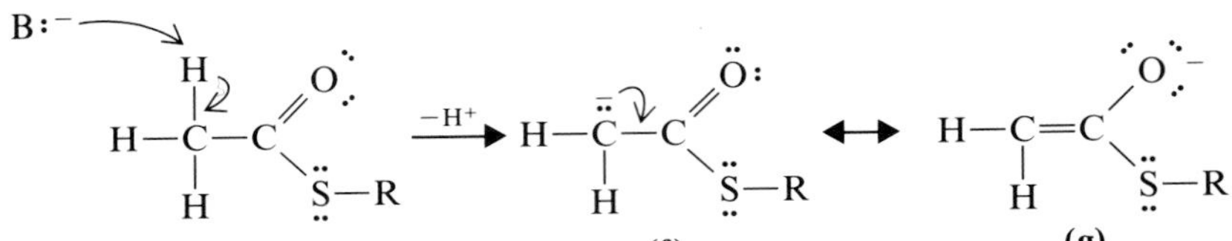

(f)

(g)
**This structure's
contribution stabilizes
the anion of a thiol ester**

3. The carbon–sulfur bond of a thiol ester is weaker than the carbon–oxygen bond of an ordinary ester; $^-$SR is a better leaving group than $^-$OR.

Factors **1** and **3** make thiol esters effective *acylating agents;* factor **2** makes them effective *alkylating* agents. Therefore, we should not be surprised when we encounter reactions similar to the following one:

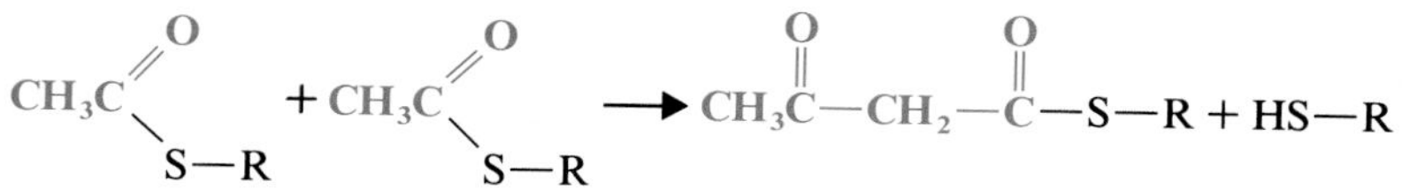

In this reaction 1 mol of a thiol ester acts as an acylating agent and the other acts as an alkylating agent (cf. Section P.2).

P.2 Biosynthesis of Fatty Acids

The fact that most naturally occurring fatty acids are made up of an even number of carbon atoms suggests that they are assembled from two-carbon units. The idea that these might be acetate ($CH_3CO_2^-$) units was put forth as early as 1893. Many years later, when radioactively labeled compounds became available, it became possible to test and confirm this hypothesis.

When an animal is fed acetic acid labeled with carbon-14 at the carboxyl group, the fatty acids that the animal synthesizes contain the label at alternate carbon atoms beginning with the carboxyl carbon:

$$CH_3\overset{*}{C}O_2H \qquad CH_3\overset{*}{C}H_2CH_2\overset{*}{C}H_2CH_2\overset{*}{C}H_2CH_2\overset{*}{C}H_2CH_2\overset{*}{C}H_2CH_2\overset{*}{C}H_2CH_2\overset{*}{C}H_2CH_2\overset{*}{C}O_2H$$

Feeding carboxyl-labeled acetic acid (**C* =** ^{14}**C**) · · · **yields palmitic acid labeled at these positions**

Conversely, feeding acetic acid labeled at the methyl carbon yields a fatty acid labeled at the other set of alternate carbon atoms:

$$\overset{*}{C}H_3CO_2H \qquad \overset{*}{C}H_3CH_2\overset{*}{C}H_2CH_2\overset{*}{C}H_2CH_2\overset{*}{C}H_2CH_2\overset{*}{C}H_2CH_2\overset{*}{C}H_2CH_2\overset{*}{C}H_2CH_2\overset{*}{C}H_2CO_2H$$

Feeding methyl-labeled acetic acid · · · **yields palmitic acid labeled at these positions.**

The biosynthesis of fatty acids is now known to begin with acetyl-coenzyme A:

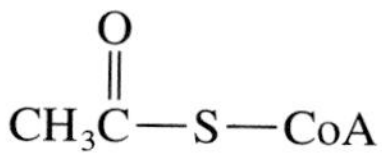

The acetyl portion of acetyl-coenzyme A can be synthesized in the cell from acetic acid; it can also be synthesized from carbohydrates, proteins, and other fats.

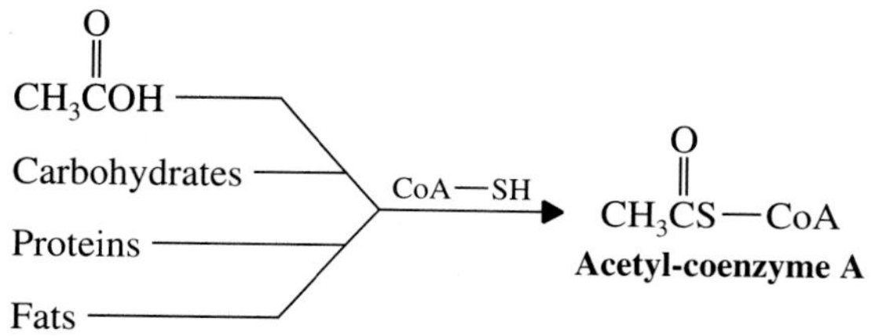

Although the methyl group of acetyl-coenzyme A is already activated toward condensation reactions by virtue of its being a part of a thiol ester (Section P.1), nature activates it again by converting it to *malonyl-coenzyme A.*

$$\underset{\textbf{Acetyl-CoA}}{CH_3\overset{\overset{O}{\|}}{C}S-CoA} + CO_2 \xrightleftharpoons{\text{acetyl-CoA carboxylase*}} \underset{\textbf{Malonyl-CoA}}{HO\overset{\overset{O}{\|}}{C}CH_2\overset{\overset{O}{\|}}{C}S-CoA}$$

The next steps in fatty acid synthesis involve the transfer of acyl groups of malonyl-CoA and acetyl-coenzyme A to the thiol group of a coenzyme called *acyl carrier protein* or ACP—SH.

$$\underset{\textbf{Malonyl-CoA}}{HO\overset{\overset{O}{\|}}{C}CH_2\overset{\overset{O}{\|}}{C}S-CoA} + ACP-SH \rightleftharpoons \underset{\textbf{Malonyl-S-ACP}}{HO\overset{\overset{O}{\|}}{C}CH_2\overset{\overset{O}{\|}}{C}S-ACP} + CoA-SH$$

$$\underset{\textbf{Acetyl-CoA}}{CH_3\overset{\overset{O}{\|}}{C}S-CoA} + ACP-SH \rightleftharpoons \underset{\textbf{Acetyl-S-ACP}}{CH_3\overset{\overset{O}{\|}}{C}S-ACP} + CoA-SH$$

Acetyl-S-ACP and malonyl-S-ACP then condense with each other to form acetoacetyl-S-ACP:

$$\underset{\textbf{Acetyl-S-ACP}}{CH_3\overset{\overset{O}{\|}}{C}S-ACP} + \underset{\textbf{Malonyl-S-ACP}}{HO\overset{\overset{O}{\|}}{C}CH_2\overset{\overset{O}{\|}}{C}S-ACP} \rightleftharpoons \underset{\textbf{Acetoacetyl-S-ACP}}{CH_3\overset{\overset{O}{\|}}{C}CH_2\overset{\overset{O}{\|}}{C}S-ACP} + CO_2 + ACP-SH$$

The molecule of CO_2 that is lost in this reaction is the same molecule that was incorporated into malonyl-CoA in the acetyl-CoA carboxylase reaction.

This remarkable reaction bears a strong resemblance to the malonic ester syntheses that we saw earlier (Section 19.4) and it deserves special comment. One can imagine, for example, a more economical synthesis of acetoacetyl-S-ACP, that is, a simple condensation between 2 mol of acetyl-S-ACP.

$$CH_3\overset{\overset{O}{\|}}{C}S-ACP + CH_3\overset{\overset{O}{\|}}{C}S-ACP \rightleftharpoons CH_3\overset{\overset{O}{\|}}{C}CH_2\overset{\overset{O}{\|}}{C}S-ACP + ACP-SH$$

Studies of this last reaction, however, have revealed that it is highly *endothermic* and that the position of equilibrium lies very far to the left. By contrast, the condensation of acetyl-S-ACP and malonyl-S-ACP is highly exothermic, and the position of equilibrium lies far to the right. The favorable thermodynamics of the condensation utilizing malonyl-S-ACP comes about because *the reaction also produces a highly stable substance: carbon dioxide.* Thus, decarboxylation of the malonyl group provides the condensation with thermodynamic assistance.

---

*This step also requires 1 mol of adenosine triphosphate (Section 22.1B) and an enzyme that transfers the carbon dioxide.

The next three steps in fatty acid synthesis transform the acetoacetyl group of acetoacetyl-S-ACP into a butyryl (butanoyl) group. These steps involve (1) reduction of the keto group (utilizing NADPH* as the reducing agent), (2) dehydration of an alcohol, and (3) reduction of a double bond (again utilizing NADPH).

Reduction of the Keto Group

$$CH_3CCH_2CS-ACP + NADPH + H^+ \rightleftharpoons CH_3CHCH_2CS-ACP + NADP^+$$

Acetoacetyl-S-ACP   **β-Hydroxybutyryl-S-ACP**

Dehydration of the Alcohol

$$CH_3CHCH_2CS-ACP \rightleftharpoons CH_3CH=CHCS-ACP + H_2O$$

β-Hydroxybutyryl-S-ACP   **Crotonyl-S-ACP**

Reduction of the Double Bond

$$CH_3CH=CHCS-ACP + NADPH + H^+ \rightleftharpoons CH_3CH_2CH_2CS-ACP + NADP^+$$

Crotonyl-S-ACP   **Butyryl-S-ACP**

These steps complete one cycle of the overall fatty acid synthesis. Their net result is the conversion of two acetate units into the four-carbon butyrate unit of butyryl-S-ACP. (This conversion requires, of course, the crucial intervention of a molecule of carbon dioxide.) At this point, another cycle begins and the chain is lengthened by two more carbon atoms.

Subsequent turns of the cycle continue to lengthen the chain by two-carbon units until a long-chain fatty acid is produced. The overall equation for the synthesis of palmitic acid, for example, can be written as follows:

$$CH_3CS-CoA + 7 HOCCH_2CS-CoA + 14 NADPH + 14 H^+ \longrightarrow$$
$$CH_3(CH_2)_{14}CO_2H + 7 CO_2 + 8 CoA-SH + 14 NADP^+ + 6 H_2O$$

One of the most remarkable aspects of fatty acid synthesis is that the entire cycle appears to be carried out by a complex of enzymes that are clustered into a single unit. The molecular weight of this cluster of proteins, called *fatty acid synthetase,* has been estimated as 2,300,000.† The synthesis begins with a single molecule of acetyl-S-ACP serving as a primer. Then, in the synthesis of palmitic acid, for example, successive condensations of seven molecules of malonyl-S-ACP occur with each condensation followed by reduction, dehydration, and reduction. All of these steps, which result in the synthesis of a C_{16} chain, take place before the fatty acid is released from the enzyme cluster.

---

* NADPH is *nicotinamide adenine dinucleotide phosphate (reduced form),* a coenzyme that is very similar in structure and function to NADH, Section 14.10.

† As isolated from yeast cells. Fatty acid synthetases from different sources have different molecular weights; that from pigeon liver, for example, has a molecular weight of 450,000.

The acyl carrier protein has been isolated and purified; its molecular weight is approximately 10,000. The protein contains a chain of groups called a *phosphopantetheine group* that is identical to that of coenzyme A (Section P.1). In ACP this chain is attached to a protein (rather than to an adenosine phosphate as it is in coenzyme A):

The length of the phosphopantetheine group is 20.2 Å, and it has been postulated that it acts as a "swinging arm" in transferring the growing acyl chain from one enzyme of the cluster to the next (Fig. P.1).

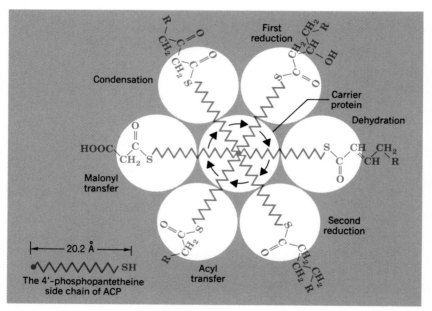

FIGURE P.1 The phosphopantetheine group as a swinging arm in the fatty acid synthetase complex. (Adapted from A. L. Lehninger, *Biochemistry,* Worth, New York, 1970, p. 519. Used with permission.)

P.3 Biosynthesis of Isoprenoids

The basic building block for the synthesis of terpenes and terpenoids (Section 23.3) is 3-methyl-3-butenyl pyrophosphate. The five carbon atoms of this compound are the source of the "isoprene units" of all "isoprenoids." The pyrophosphate group is a group that nature relies upon for a vast number of chemical processes. In the reactions that we shall soon see, the pyrophosphate group functions as a natural "leaving group."

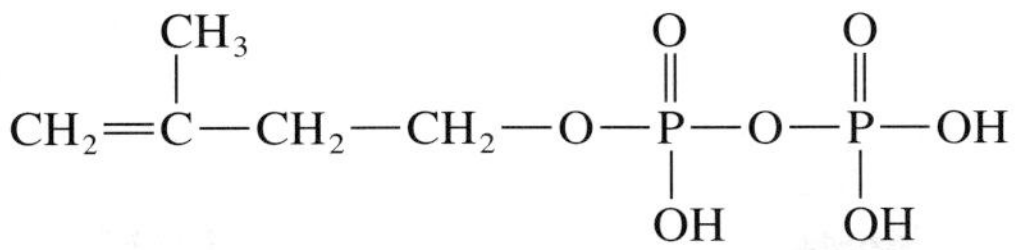

3-Methyl-3-butenyl pyrophosphate

3-Methyl-3-butenyl pyrophosphate is isomerized by an enzyme to 3-methyl-2-butenyl pyrophosphate. The isomerization establishes an equilibrium that makes both compounds available to the cell.

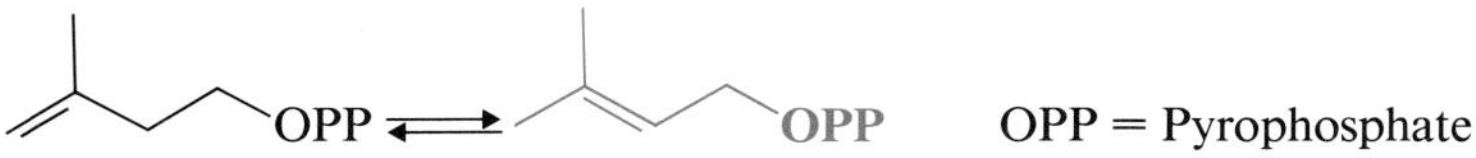

| **3-Methyl-3-butenyl pyrophosphate** | **3-Methyl-2-butenyl pyrophosphate** | OPP = Pyrophosphate |

These two C_5 compounds condense with each other in another enzymatic reaction to yield the C_{10} compound, geranyl pyrophosphate. The first step involves the formation of an allylic cation.

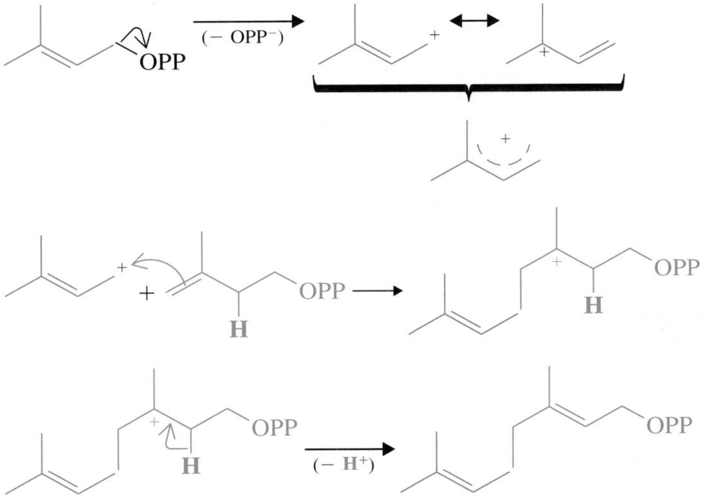

Geranyl pyrophosphate

Geranyl pyrophosphate is the precursor of the monoterpenes; hydrolysis of geranyl pyrophosphate, for example, yields geraniol.

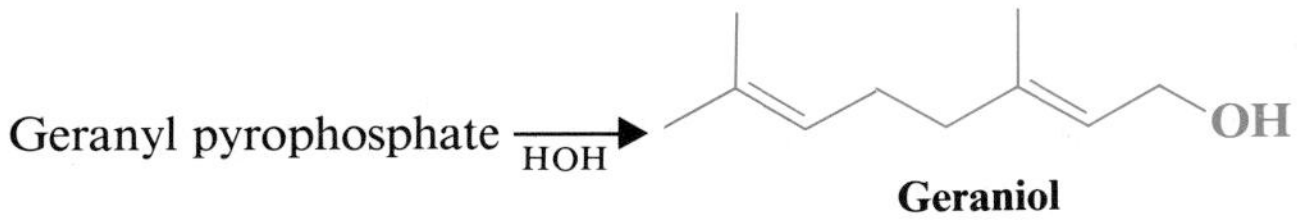

Geranyl pyrophosphate $\xrightarrow{\text{HOH}}$

Geraniol

Geranyl pyrophosphate can also condense with 3-methyl-3-butenyl pyrophosphate to form the C_{15} precursor for sesquiterpenes, farnesyl pyrophosphate.

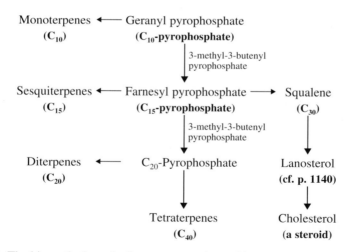

Geranyl pyrophosphate

Farnesyl pyrophosphate

other sesquiterpenes

Farnesol

Farnesol has been isolated from ambrette oil. It has the odor of lily of the valley. Farnesol also functions as a hormone in certain insects and initiates the change from caterpillar to pupa to moth.

Similar condensation reactions yield the precursors for all of the other terpenes (Fig. P.2). In addition, a tail-to-tail reductive coupling of two molecules of farnesyl pyrophosphate produces squalene, the precursor for the important group of isoprenoids known as *steroids* (cf. Sections 23.4 and P.4).

Monoterpenes ⟵ Geranyl pyrophosphate
(C_{10}) $(C_{10}$-**pyrophosphate**)

3-methyl-3-butenyl pyrophosphate

Sesquiterpenes ⟵ Farnesyl pyrophosphate ⟶ Squalene
(C_{15}) $(C_{15}$-**pyrophosphate**) (C_{30})

3-methyl-3-butenyl pyrophosphate

Diterpenes ⟵ C_{20}-Pyrophosphate   Lanosterol
(C_{20}) (cf. p. 1140)

Tetraterpenes   Cholesterol
(C_{40}) (a steroid)

FIGURE P.2 The biosynthetic paths for terpenes and steroids.

PROBLEM P.1

When farnesol is treated with sulfuric acid, it is converted to bisabolene. Outline a possible mechanism for this reaction.

Farnesol $\xrightarrow{\text{H}_2\text{SO}_4}$

Bisabolene

P.4 Biosynthesis Of Steroids

We saw in the previous section that the C_5 compound, 3-methyl-3-butenyl pyrophosphate, is the actual "isoprene unit" that nature uses in constructing terpenoids and carotenoids. We can now extend that biosynthetic pathway in two directions. We can show how 3-methyl-3-butenyl pyrophosphate (like the fatty acids) is ultimately derived from acetate units, and how cholesterol, the precursor of most of the important steroids, is synthesized from 3-methyl-3-butenyl pyrophosphate.

In the 1940s, Konrad Bloch of Harvard University used labeling experiments to demonstrate that all of the carbon atoms of cholesterol can be derived from acetic acid. Using *methyl-labeled* acetic acid, for example, Bloch found the following label distribution in the cholesterol that was synthesized.

$$\overset{*}{\text{C}}\text{H}_3\text{CO}_2\text{H} \longrightarrow$$

Bloch also found that feeding *carboxyl-labeled* acetic acid led to incorporation of the label into all of the other carbon atoms of cholesterol (the unstarred carbon atoms of the formula just given).

Subsequent research by a number of investigators has shown that 3-methyl-3-butenyl pyrophosphate is synthesized from acetate units through the following sequence of reactions:

$$\underset{\substack{\textbf{(C}_2\textbf{)} \\ \textbf{Acetyl-CoA}}}{\text{CH}_3\overset{\text{O}}{\overset{\|}{\text{C}}}\text{S}-\text{CoA}} + \underset{\substack{\textbf{(C}_4\textbf{)} \\ \textbf{Acetoacetyl-CoA}}}{\text{CH}_3\overset{\text{O}}{\overset{\|}{\text{C}}}\text{CH}_2\overset{\text{O}}{\overset{\|}{\text{C}}}\text{S}-\text{CoA}} \xrightarrow{\text{CoA}-\text{SH}} \underset{\substack{\textbf{(C}_6\textbf{)} \\ \boldsymbol{\beta}\textbf{-Hydroxy-}\boldsymbol{\beta}\textbf{-methylglutaryl-CoA}}}{\text{HO}\overset{\text{O}}{\overset{\|}{\text{C}}}\text{CH}_2 \underset{\substack{\text{H}_3\text{C} \quad \text{OH}}}{\overset{\displaystyle \text{C}}{}} \text{CH}_2\overset{\text{O}}{\overset{\|}{\text{C}}}\text{S}-\text{CoA}}$$

$$\text{— 2 NADPH + 2 H}^+$$
$$\text{2 NADP}^+ + \text{CoA}-\text{SH}$$

$$\underset{\substack{(\mathbf{C_6}) \\ \textbf{Mevalonic acid}}}{\overset{\displaystyle \underset{\text{H}_3\text{C}}{\overset{\displaystyle \text{HOCCH}_2}{\underset{\displaystyle}{\overset{\parallel}{\text{O}}}}}\overset{\displaystyle}{\underset{\displaystyle \text{OH}}{\text{C}}}\overset{\displaystyle \text{CH}_2\text{CH}_2\text{OH}}{}}{}}$$

Mevalonic acid (C$_6$)

3 ATP → 3 ADP   **Three successive steps**

3-Phospho-5-pyrophosphomevalonic acid (C$_6$)

$$\text{H}-\text{O}-\overset{\text{O}}{\overset{\parallel}{\text{C}}}-\text{CH}_2-\underset{\substack{| \\ \text{OPO}_3\text{H}^-}}{\overset{\substack{\text{CH}_3 \\ |}}{\text{C}}}-\text{CH}_2\text{CH}_2\text{O}-\overset{\text{O}}{\overset{\parallel}{\text{P}}}-\text{O}-\overset{\text{O}}{\overset{\parallel}{\text{P}}}-\text{O}^-$$

$\longrightarrow$ CO$_2$ + H$_2$PO$_3^-$

3-Methyl-3-butenyl pyrophosphate (C$_5$)

$$\underset{\text{H}_3\text{C}}{\overset{\text{H}_2\text{C}}{>}}\!\!\!C-\text{CH}_2\text{CH}_2-\text{O}-\overset{\text{O}}{\overset{\parallel}{\text{P}}}-\text{O}-\overset{\text{O}}{\overset{\parallel}{\text{P}}}-\text{O}^-$$

The first step of this synthetic pathway is straightforward. Acetyl-CoA (from 1 mol of acetate) and acetoacetyl-CoA (from 2 mol of acetate) condense to form the C$_6$ compound, β-hydroxy-β-methylglutaryl-CoA. This step is followed by an enzymatic reduction of the thiol ester group of β-hydroxy-β-methylglutaryl-CoA to the primary alcohol of mevalonic acid. The enzyme that catalyzes this step is called HMG-CoA reductase (HMG = β-hydroxy-β-methylglutaryl) and this step is the rate-limiting step in cholesterol biosynthesis. The key to finding this pathway was the discovery that mevalonic acid was an intermediate and that this C$_6$ compound could be transformed into the five-carbon 3-methyl-3-butenyl pyrophosphate by successive phosphorylations and decarboxylation.

Then 3-methyl-3-butenyl pyrophosphate isomerizes to produce an equilibrium mixture that contains 3-methyl-2-butenyl pyrophosphate, and these two compounds condense to form geranyl pyrophosphate, a C$_{10}$ compound. Geranyl pyrophosphate subsequently condenses with another mol of 3-methyl-3-butenyl pyrophosphate to form farnesyl pyrophosphate, a C$_{15}$ compound. (Geranyl pyrophosphate and farnesyl pyrophosphate are the precursors of the mono- and sesquiterpenes, cf. Section P.3.)

OP$_2$O$_6^{3-}$ $\rightleftharpoons$ OP$_2$O$_6^{3-}$

$\longrightarrow$ H$_2$P$_2$O$_7^{2-}$

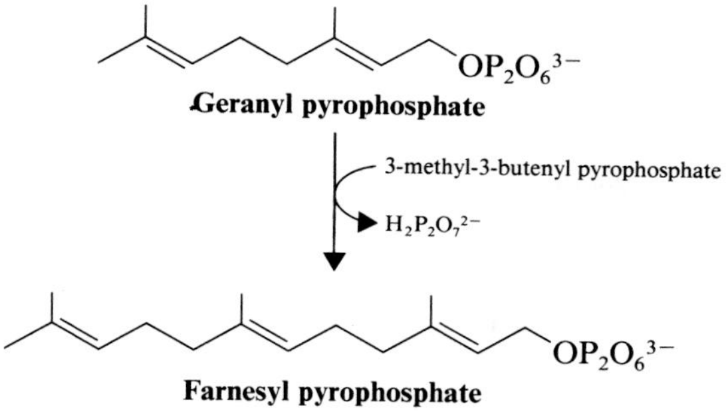

Geranyl pyrophosphate

3-methyl-3-butenyl pyrophosphate

$H_2P_2O_7^{2-}$

Farnesyl pyrophosphate

Two molecules of farnesyl pyrophosphate then undergo a reductive condensation to produce squalene.

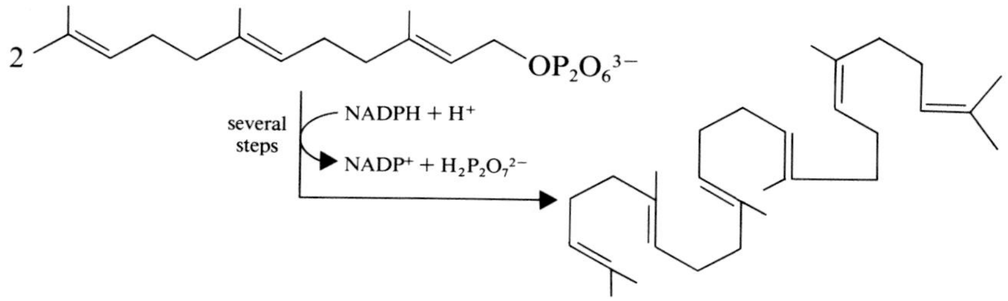

several steps

NADPH + H⁺

NADP⁺ + $H_2P_2O_7^{2-}$

Squalene

Squalene is the direct precursor of cholesterol. Oxidation of squalene yields squalene 2,3-epoxide, which undergoes a remarkable series of ring closures accompanied by concerted methyl and hydride migrations to yield lanosterol. Lanosterol is then converted to cholesterol through a series of enzyme-catalyzed reactions.

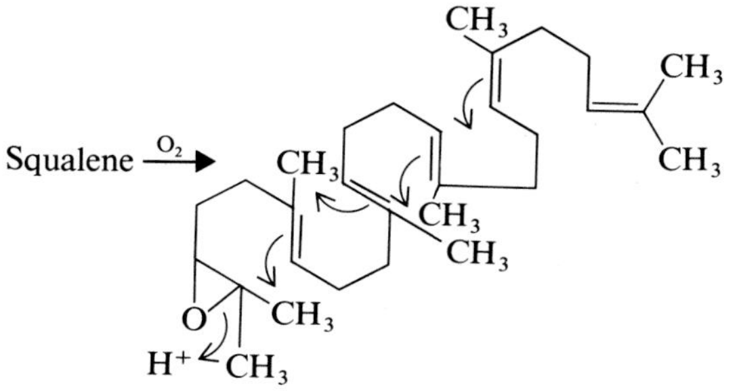

Squalene $\xrightarrow{O_2}$

Squalene 2,3-epoxide

H⁺

rearrangements

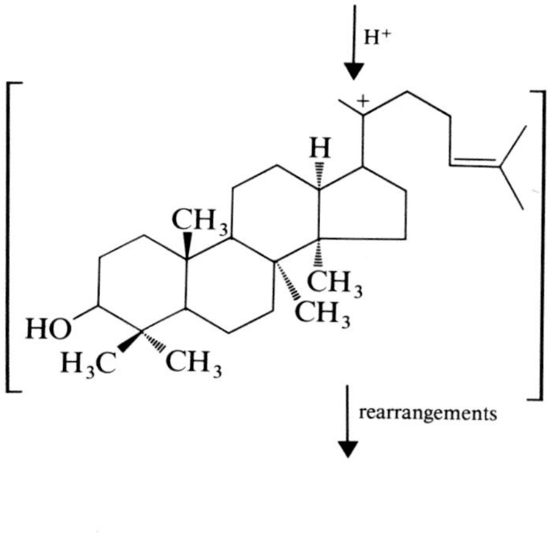

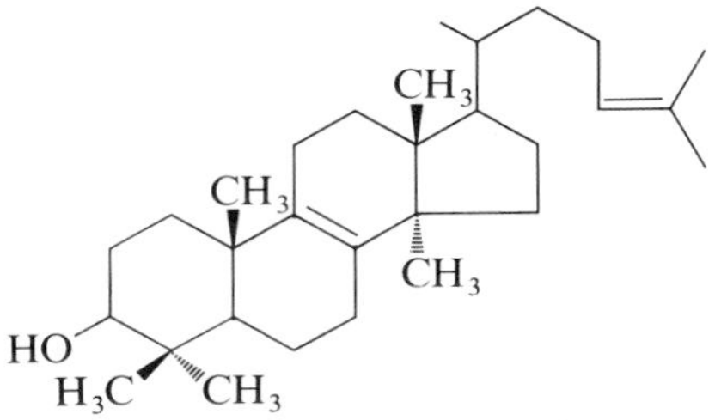

Lanosterol

several steps

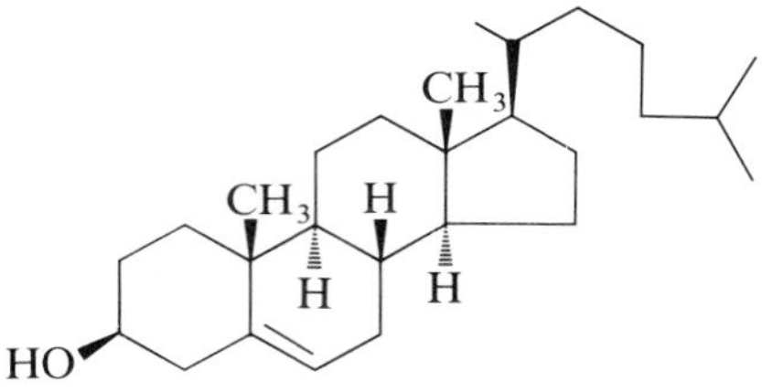

Cholesterol

P.5 Cholesterol and Heart Disease

Because cholesterol is the precursor of steroid hormones and is a vital constituent of cell membranes, it is essential to life. On the other hand, deposition of cholesterol in arteries is a cause of heart disease and atherosclerosis, two leading causes of death in humans. For an organism to remain healthy, there has to be an intricate balance between the biosynthesis of cholesterol and its utilization, so that arterial deposition is kept at a minimum.

For some individuals with high blood levels of cholesterol, the remedy is as simple as following a diet low in cholesterol and in fat. For those who suffer from the genetic disease **familial hypercholesterolemia** (FH), other means of blood cholesterol reduction are required. One remedy involves using the drug *lovastatin* (also called *mevinolin*).

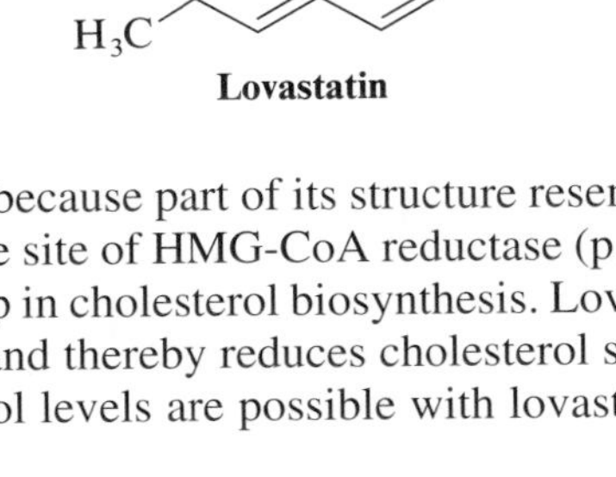

Mevalonate ion

Lovastatin

Lovastatin, because part of its structure resembles mevalonate ion, can apparently bind at the active site of HMG-CoA reductase (p. 1138), the enzyme that catalyzes the rate-limiting step in cholesterol biosynthesis. Lovastatin acts as a competitive inhibitor of this enzyme and thereby reduces cholesterol synthesis. Reductions of up to 30% in serum cholesterol levels are possible with lovastatin therapy.

Cholesterol synthesized in the liver is either converted to bile acids that are used in digestion or it is esterified for transport by the blood. Cholesterol is transported in the blood and taken up in cells, in the form of lipoprotein complexes named on the basis of their density. **Low density lipoproteins (LDL)** transport cholesterol from the liver to peripheral tissues. **High density lipoproteins (HDL)** transports cholesterol back to the liver, where surplus cholesterol is disposed of by the liver as bile acids. High density lipoproteins have come to be called ''good cholesterol'' because high levels of HDL may reduce cholesterol deposits in arteries. Because high levels of LDL are associated with the arterial deposition of cholesterol that causes cardiovascular disease, it has come to be called ''bad cholesterol.''

Bile acids that flow from the liver to the intestines, however, are efficiently recycled to the liver. Recognition of this has led to another method of cholesterol reduction, the ingestion of resins that bind bile acids and thereby prevent their reabsorption in the intestines.

CHAPTER 24

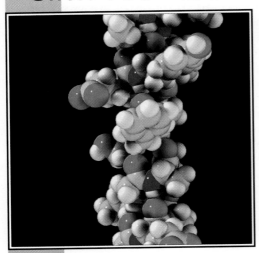

Protein alpha helix
(Section 24.8)

AMINO ACIDS AND PROTEINS

24.1 INTRODUCTION

The three groups of biological polymers are polysaccharides, proteins, and nucleic acids. We studied polysaccharides in Chapter 22 and saw that they function primarily as energy reserves, as biochemical labels on cell surfaces, and, in plants, as structural materials. When we study nucleic acids in Chapter 25 we shall find that they serve two major purposes: storage and transmission of information. Of the three groups of biopolymers, proteins have the most diverse functions. As enzymes and hormones, proteins catalyze and regulate the reactions that occur in the body; as muscles and tendons they provide the body with the means for movement; as skin and hair they give it an outer covering; as hemoglobins they transfer all-important oxygen to its most remote corners; as antibodies they provide it with a means of protection against disease; and in combination with other substances in bone they provide it with structural support.

Given such diversity of functions, we should not be surprised to find that proteins come in all sizes and shapes. By the standard of most of the molecules we have studied, even small proteins have very high molecular weights. Lysozyme, an enzyme, is a relatively small protein and yet its molecular weight is 14,600. The molecular weights of most proteins are much larger. Their shapes cover a range from the globular proteins such as lysozyme and hemoglobin to the helical coils of α-keratin (hair, nails, and wool) and the pleated sheets of silk fibroin.

And yet, in spite of such diversity of size, shape, and function, all proteins have common features that allow us to deduce their structures and understand their properties. Later in this chapter we shall see how this is done.

Proteins are **polyamides** and their monomeric units are comprised of about 20 different α-amino acids:

An α-amino acid   A portion of a protein molecule

Cells use the different α-amino acids to synthesize proteins. The exact sequence of the different α-amino acids along the protein chain is called the **primary structure** of the protein. This primary structure, as its name suggests, is of fundamental importance. For the protein to carry out its particular function, the primary structure must be correct. We shall see later that when the primary structure is correct, the polyamide chain folds in certain particular ways to give it the shape it needs for its particular task. This folding of the polyamide chain gives rise to higher levels of complexity called the **secondary** and **tertiary structures** of the protein.

Hydrolysis of proteins with acid or base yields a mixture of the different amino acids. Although hydrolysis of naturally occurring proteins may yield as many as 22 different amino acids, the amino acids have an important structural feature in common: With the exception of glycine (whose molecules are achiral), almost all naturally occurring amino acids have the L configuration at the α carbon.* That is, they have the same relative configuration as L-glyceraldehyde:

An L-α-amino acid   L-Glyceraldehyde
[usually an (S)-α-amino acid]   [(S)-glyceraldehyde]   Glycine

24.2 AMINO ACIDS

24.2A Structures and Names

The 22 α-amino acids that can be obtained from proteins can be subdivided into three different groups on the basis of the structures of their side chains, R. These are given in Table 24.1.

Only 20 of the 22 α-amino acids in Table 24.1 are actually used by cells when they synthesize proteins. Two amino acids are synthesized after the polyamide chain is intact. Hydroxyproline (present mainly in collagen) is synthesized from proline, and cystine (present in most proteins) is synthesized from cysteine.

This conversion of cysteine to cystine requires additional comment. The —SH group of cysteine makes cysteine a *thiol*. One property of thiols is that they can be converted to disulfides by mild oxidizing agents. This conversion, moreover, can be reversed by mild reducing agents.

*Some D-amino acids have been obtained from the material comprising the cell walls of bacteria and by hydrolysis of certain antibiotics.

TABLE 24.1 L-Amino acids found in proteins

STRUCTURE OF R	NAME[a]	ABBRE-VIATIONS	pK_{a_1} $\alpha\text{-CO}_2\text{H}$	pK_{a_2} $\alpha\text{-NH}_3{}^+$	pK_{a_3} R GROUP	pI
Neutral Amino Acids						
—H	Glycine	G or Gly	2.3	9.6		6.0
—CH$_3$	Alanine	A or Ala	2.3	9.7		6.0
—CH(CH$_3$)$_2$	Valine[e]	V or Val	2.3	9.6		6.0
—CH$_2$CH(CH$_3$)$_2$	Leucine[e]	L or Leu	2.4	9.6		6.0
—CHCH$_2$CH$_3$ CH$_3$	Isoleucine[e]	I or Ile	2.4	9.7		6.1
—CH$_2$—C$_6$H$_5$	Phenylalanine[e]	F or Phe	1.8	9.1		5.5
—CH$_2$CONH$_2$	Asparagine	N or Asn	2.0	8.8		5.4
—CH$_2$CH$_2$CONH$_2$	Glutamine	Q or Gln	2.2	9.1		5.7
—CH$_2$-(indole)	Tryptophan[e]	W or Trp	2.4	9.4		5.9
HOC—CH—CH$_2$ / HN CH$_2$ / CH$_2$ (complete structure)	Proline	P or Pro	2.0	10.6		6.3
—CH$_2$OH	Serine	S or Ser	2.2	9.2		5.7
—CHOH CH$_3$	Threonine[e]	T or Thr	2.6	10.4		6.5
—CH$_2$—C$_6$H$_4$—OH	Tyrosine	Y or Tyr	2.2	9.1	10.1	5.7
HOC—CH—CH$_2$ / HN CH / CH$_2$ OH (complete structure)	Hydroxyproline	Hyp	1.9	9.7		6.3

TABLE 24.1 (continued)

$$H_2N \overset{\displaystyle CO_2H}{\underset{\displaystyle R}{\rule[-0.4em]{0.03em}{1.4em}\!-\!H}} \quad \text{or} \quad R \overset{\displaystyle H}{\underset{\displaystyle NH_2}{\!\!\!\diagdown C\!-\!CO_2H}}$$

STRUCTURE OF R	NAME[a]	ABBRE-VIATIONS	pK_{a_1} $\alpha\text{-}CO_2H$	pK_{a_2} $\alpha\text{-}NH_3^+$	pK_{a_3} R GROUP	pI
$-CH_2SH$	Cysteine	C or Cys	1.7	10.8	8.3	5.0
$-CH_2-S$ $\mid$ $-CH_2-S$	Cystine	Cys-Cys	1.6 2.3	7.9 9.9		5.1
$-CH_2CH_2SCH_3$	Methionine[e]	M or Met	2.3	9.2		5.8
R Contains an Acidic (Carboxyl) Group						
$-CH_2CO_2H$	Aspartic acid	D or Asp	2.1	9.8	3.9	3.0
$-CH_2CH_2CO_2H$	Glutamic acid	E or Glu	2.2	9.7	4.3	3.2
R Contains a Basic Group						
$-CH_2CH_2CH_2CH_2NH_2$	Lysine[e]	K or Lys	2.2	9.0	10.5[b]	9.8
$-CH_2CH_2CH_2NH-\overset{\displaystyle NH}{\overset{\displaystyle \|}{C}}-NH_2$	Arginine	R or Arg	2.2	9.0	12.5[b]	10.8
$-CH_2\!-\!\underset{\underset{H}{\mid}}{\overset{N}{\diagup\!\!\diagdown}}\!\!N$	Histidine	H or His	1.8	9.2	6.0[b]	7.6

[a] e = essential amino acids.
[b] pK_a is of protonated amine of R group.

$$2\,R-S-H \underset{[H]}{\overset{[O]}{\rightleftharpoons}} R-S-S-R$$

Thiol **Disulfide**

Disulfide linkage

$$2\,\underset{\underset{\textbf{Cysteine}}{NH_2}}{HO_2CCHCH_2SH} \underset{[H]}{\overset{[O]}{\rightleftharpoons}} \underset{\underset{\textbf{Cystine}}{NH_2}}{HO_2CCHCH_2S}-\underset{NH_2}{SCH_2CHCO_2H}$$

We shall see later how the disulfide linkage between cysteine units in a protein chain contributes to the overall structure and shape of the protein.

24.2B Essential Amino Acids

Amino acids can be synthesized by all living organisms, plants and animals. Many higher animals, however, are deficient in their ability to synthesize all of the amino acids they need for their proteins. Thus, these higher animals require certain amino acids as a part of their diet. For adult humans there are eight essential amino acids; these are designated with the superscript e in Table 24.1.

24.2C Amino Acids as Dipolar Ions

Amino acids contain both a basic group ($-NH_2$) and an acidic group ($-CO_2H$). In the dry solid state, amino acids exist as **dipolar ions,** a form in which the carboxyl group is present as a carboxylate ion, $-CO_2^-$, and the amino group is present as an aminium group, $-NH_3^+$. (Dipolar ions are also called **zwitterions.**) In aqueous solution, an equilibrium exists between the dipolar ion and the anionic and cationic forms of an amino acid.

$$\overset{+}{H_3}NCHCO_2H \underset{+H^+}{\overset{-H^+}{\rightleftarrows}} \overset{+}{H_3}NCHCO_2^- \underset{+H^+}{\overset{-H^+}{\rightleftarrows}} H_2NCHCO_2^-$$

R	R	R
Cationic form	**Dipolar ion**	**Anionic form**
(predominant in		(predominant in
strongly acidic		strongly basic
solutions, e.g.,		solutions, e.g.,
at pH 0)		at pH 14)

The predominant form of the amino acid present in a solution depends on the pH of the solution and on the nature of the amino acid. In strongly acidic solutions all amino acids are present primarily as cations; in strongly basic solutions they are present as anions. At some intermediate pH, called the *isoelectric point* (pI), the concentration of the dipolar ion is at its maximum and the concentrations of the anions and cations are equal. Each amino acid has a particular isoelectric point. These are given in Table 24.1.

Let us consider first an amino acid with a side chain that contains neither acidic nor basic groups—an amino acid, for example, such as alanine.

If alanine is present in a strongly acidic solution (e.g., at pH 0), it is present mainly in the following cationic form. The pK_a for the carboxyl group of cationic form is 2.3. This is considerably smaller than the pK_a of a corresponding carboxylic acid (e.g., propanoic acid) and indicates that the cationic form of alanine is the stronger acid. But we should expect it to be. After all, it is a positively charged species and therefore should lose a proton more readily.

$$\begin{array}{cc} CH_3CHCO_2H & CH_3CH_2CO_2H \\ | & \\ \overset{}{\underset{+}{NH_3}} & \end{array}$$

Cationic form of alanine   **Propanoic acid**
p$K_{a_1} = 2.3$ p$K_a = 4.89$

The dipolar ion form of an amino acid is also a potential acid because the $-NH_3^+$ group can donate a proton. The pK_a of the dipolar ion form of alanine is 9.7.

$$\begin{array}{c} CH_3CHCO_2^- \\ | \\ \underset{+}{NH_3} \end{array}$$

p$K_{a_2} = 9.7$

The isoelectric point (pI) of an amino acid such as alanine is the average of pK_{a_1} and pK_{a_2}.

$$pI = \frac{2.3 + 9.7}{2} = 6.0 \quad \text{(isoelectric point of alanine)}$$

What does this mean about the behavior of alanine as the pH of a strongly acidic solution containing it is gradually raised by adding a base (i.e., OH⁻)? At first (pH 0) (Fig. 24.1), the predominant form will be the cationic form. But then, as the acidity reaches pH 2.3 (the pK_a of the cationic form, pK_{a_1}), one half of the cationic form will be converted to the dipolar ion.* As the pH increases further—from pH 2.3 to pH 9.7—the predominant form will be the dipolar ion. At pH 6.0, the pH equals pI and the concentration of the dipolar ion is at its maximum.

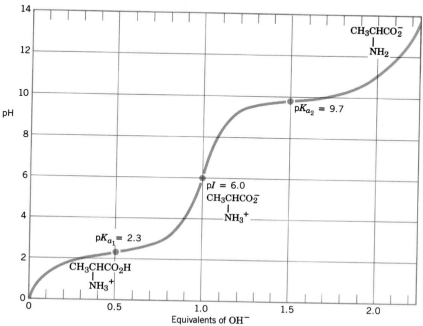

FIGURE 24.1 A titration curve for CH_3CHCO_2H.

When the pH rises to pH 9.7 (the pK_a of the dipolar ion), the dipolar ion is half-converted to the anionic form. Then, as the pH approaches pH 14, the anionic form becomes the predominant form present in the solution.

If the side chain of an amino acid contains an extra acidic or basic group, then the

*It is easy to show that for an acid:

$$pK_a = pH + \log \frac{[\text{acid}]}{[\text{conjugate base}]}$$

When the acid is half-neutralized, [acid] = [conjugate base] and $\log \dfrac{[\text{acid}]}{[\text{conjugate base}]} = 0$; thus pH = pK_a.

equilibria are more complex. Consider lysine, for example, an amino acid that has an extra —NH_2 group on its ε carbon. In strongly acidic solution, lysine is present as a dication because both amino groups are protonated. The first proton to be lost as the pH is raised is a proton of the carboxyl group ($pK_a = 2.2$), the next is from the α-aminium group ($pK_a = 9.0$), and the last is from the ε-aminium group.

$$\overset{+}{H_3N}(CH_2)_4CHCO_2H \underset{H^+}{\overset{OH^-}{\rightleftharpoons}} \overset{+}{H_3N}(CH_2)_4CHCO_2^- \underset{H^+}{\overset{OH^-}{\rightleftharpoons}}$$

NH_3	NH_3
$+$	$+$
Dicationic form of	**Monocationic**
lysine	**form**
($pK_{a_1} = 2.2$)	**($pK_{a_2} = 9.0$)**

$$\overset{+}{H_3N}(CH_2)_4CHCO_2^- \underset{H^+}{\overset{OH^-}{\rightleftharpoons}} H_2N(CH_2)_4CHCO_2^-$$

NH_2	NH_2
Dipolar ion	**Anionic form**
($pK_{a_3} = 10.5$)	

The isoelectric point of lysine is the average of pK_{a_2} (the monocation) and pK_{a_3} (the dipolar ion).

$$pI = \frac{9.0 + 10.5}{2} = 9.8 \qquad \text{(isoelectric point of lysine)}$$

PROBLEM 24.1

What form of glutamic acid would you expect to predominate in: (a) strongly acid solution? (b) strongly basic solution? (c) at its isoelectric point (pI 3.2)? (d) The isoelectric point of glutamine (pI 5.7) is considerably higher than that of glutamic acid. Explain.

PROBLEM 24.2

$$NH$$
$$\|$$

The guanidino group —NH—C—NH_2 of arginine is one of the most strongly basic of all organic groups. Explain.

24.3 LABORATORY SYNTHESIS OF α-AMINO ACIDS

A variety of methods has been developed for the laboratory synthesis of α-amino acids. We shall describe here three general methods, all of which are based on reactions we have seen before.

24.3A Direct Ammonolysis of an α-Halo Acid

$$R—CH_2CO_2H \xrightarrow[\text{(2) } H_2O]{\text{(1) } X_2, P_4} RCHCO_2H \xrightarrow{NH_3 \text{ (excess)}} R—CHCO_2^-$$

X	NH_3
	$+$

This method is probably used least often because yields tend to be poor. We saw an example of this method in Section 18.9.

24.3B From Potassium Phthalimide

This method is a modification of the Gabriel synthesis of amines (Section 20.5A). The yields are usually high and the products are easily purified.

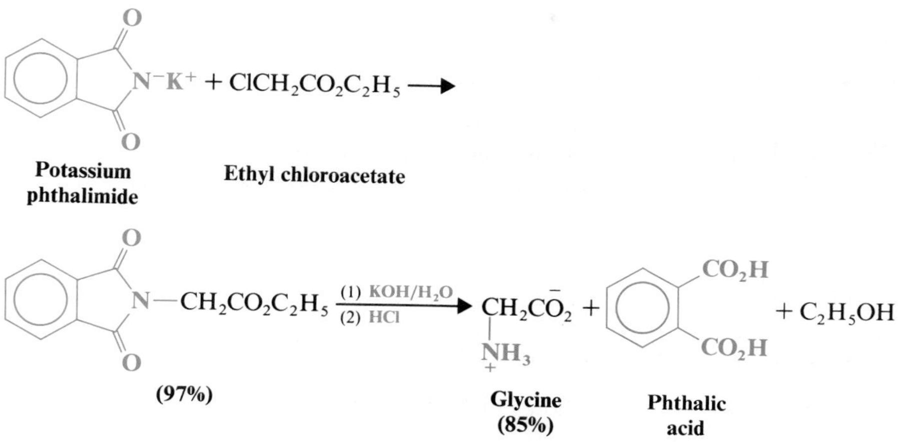

Potassium phthalimide   **Ethyl chloroacetate**

(97%)   **Glycine (85%)**   **Phthalic acid**

A variation of this procedure uses potassium phthalimide and diethyl α-bromomalonate to prepare an *imido* malonic ester. This method is illustrated with a synthesis of methionine.

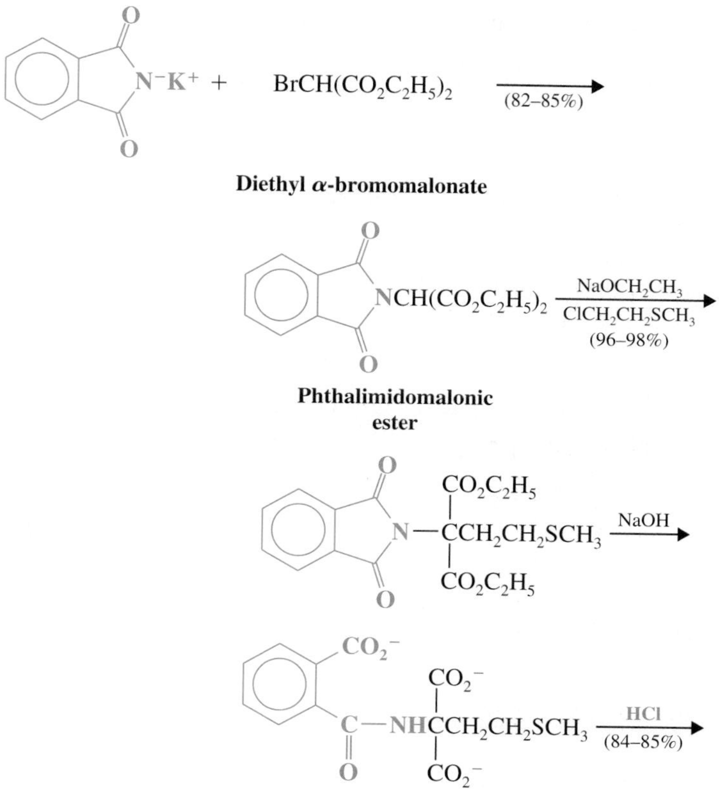

Diethyl α-bromomalonate

Phthalimidomalonic ester

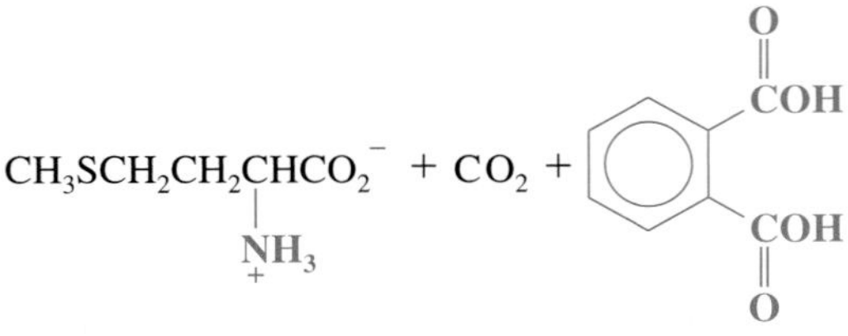

$$CH_3SCH_2CH_2CHCO_2^- + CO_2 + \underset{\substack{\displaystyle \Vert \\ O}}{}$$

$$\underset{NH_3^+}{|}$$

DL-Methionine

PROBLEM 24.3

Starting with diethyl α-bromomalonate and potassium phthalimide and using any other necessary reagents show how you might synthesize: (a) DL-leucine, (b) DL-alanine, and (c) DL-phenylalanine.

24.3C The Strecker Synthesis

Treating an aldehyde with ammonia and hydrogen cyanide produces an α-amino nitrile. Hydrolysis of the nitrile group (Section 18.3) of the α-amino nitrile converts the latter to an α-amino acid. This synthesis is called the Strecker synthesis.

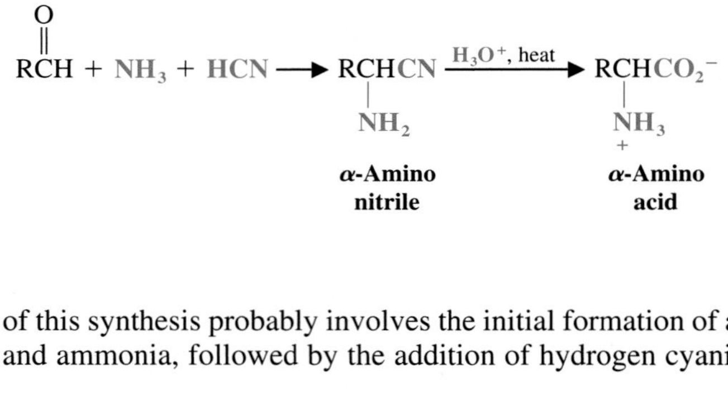

$$\underset{\substack{\displaystyle \Vert \\ O}}{RCH} + NH_3 + HCN \longrightarrow \underset{\substack{| \\ NH_2}}{RCHCN} \xrightarrow{H_3O^+,\ heat} \underset{\substack{| \\ NH_3 \\ +}}{RCHCO_2^-}$$

α**-Amino** $\qquad\qquad$ α**-Amino**
nitrile $\qquad\qquad\qquad$ **acid**

The first step of this synthesis probably involves the initial formation of an imine from the aldehyde and ammonia, followed by the addition of hydrogen cyanide.

A MECHANISM FOR THE
STRECKER SYNTHESIS

$$\underset{\substack{\displaystyle \Vert \\ O}}{RCH} + :NH_3 \rightleftharpoons \underset{\substack{| \\ O^-}}{RCH\overset{+}{N}H_3} \rightleftharpoons \underset{\substack{| \\ OH}}{RCHNH_2} \underset{-\ H_2O}{\rightleftharpoons}$$

$$RCH{=}NH \underset{CN^-}{\rightleftharpoons} \underset{\substack{| \\ CN}}{RCH{-}NH^-} \underset{H^+}{\rightleftharpoons} \underset{\substack{| \\ CN}}{RCH{-}NH_2}$$

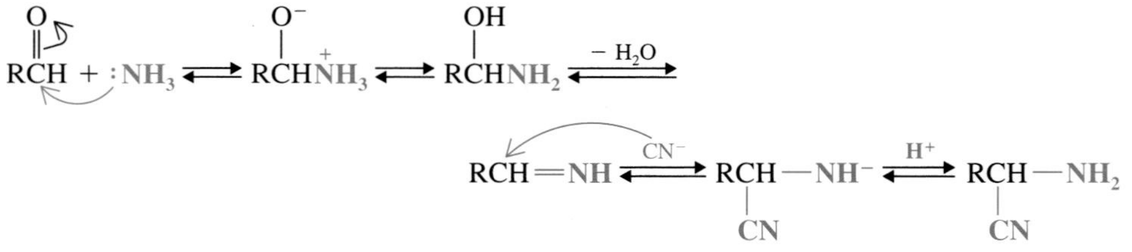

Imine $\qquad\qquad\qquad\qquad\qquad$ α**-aminonitrile**

SAMPLE PROBLEM

Outline a Strecker synthesis of DL-tyrosine.

ANSWER:

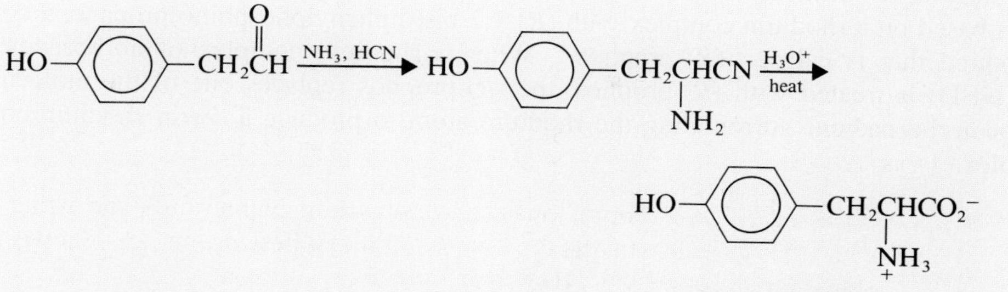

DL-Tyrosine

PROBLEM 24.4

(a) Outline a Strecker synthesis of DL-phenylalanine. (b) DL-Methionine can also be synthesized by a Strecker synthesis. The required starting aldehyde can be prepared from acrolein ($CH_2{=}CHCHO$) and methanethiol (CH_3SH). Outline all steps in this synthesis of DL-methionine.

24.3D Resolution of DL-Amino Acids

With the exception of glycine, which has no stereocenter, the amino acids that are produced by the methods we have outlined are all produced as racemic forms. To obtain the naturally occurring L-amino acid we must, of course, resolve the racemic form. This can be done in a variety of ways including the methods outlined in Section 20.3.

One especially interesting method for resolving amino acids is based on the use of enzymes called *deacylases.* These enzymes catalyze the hydrolysis of *N-acylamino acids* in living organisms. Since the active site of the enzyme is chiral, it hydrolyzes only *N*-acylamino acids of the L configuration. When it is exposed to a racemic modification of *N*-acylamino acids, only the derivative of the L-amino acid is affected and the products, as a result, are separated easily.

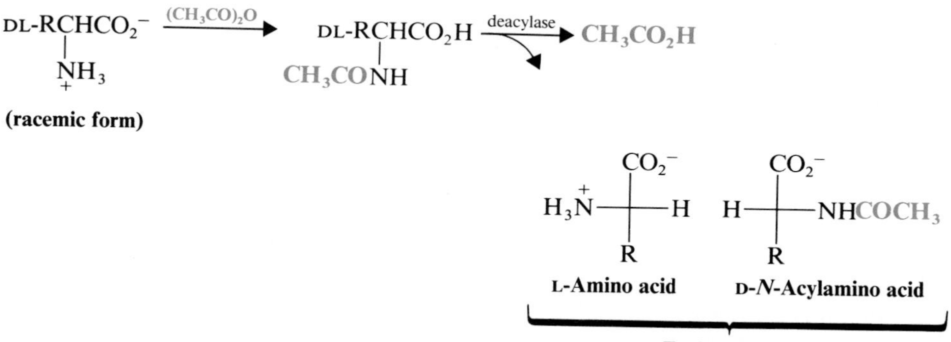

(racemic form)

L-Amino acid **D-*N*-Acylamino acid**

Easily separated

24.3E Stereoselective Syntheses of Amino Acids

The ideal synthesis of an amino acid, of course, would be one that would produce only the naturally occurring L-amino acid. This ideal has now been realized through the use of chiral hydrogenation catalysts derived from transition metals. A variety of catalysts has been used. One developed by B. Bosnich (of the University of Toronto) is based on a rhodium complex with (R)-1,2-bis(diphenylphosphino)propane, a compound that is called "(R)-prophos." When a rhodium complex of norbornadiene (NBD) is treated with (R)-prophos, the (R)-prophos replaces one of the molecules of norbornadiene surrounding the rhodium atom to produce a *chiral* rhodium complex.

$$
\begin{array}{c}
\text{H}_3\text{C} \\
\text{H} \diagdown \\
\text{C}-\text{CH}_2 \\
(\text{C}_6\text{H}_5)_2\text{P} \text{P}(\text{C}_6\text{H}_5)_2
\end{array}
$$

(R)-Prophos

$$[\text{Rh(NBD)}_2]\text{ClO}_4 + (R)\text{-prophos} \longrightarrow [\text{Rh}((R)\text{-prophos})(\text{NBD})]\text{ClO}_4 + \text{NBD}$$

Chiral rhodium complex

Treating this rhodium complex with hydrogen in a solvent such as ethanol yields a solution containing an active *chiral* hydrogenation catalyst, which probably has the composition $[\text{Rh}((R)\text{-prophos})(\text{H})_2(\text{EtOH})_2]^+$.

When 2-acetylaminopropenoic acid is added to this solution and hydrogenation is carried out, the product of the reaction is the *N*-acetyl derivative of L-alanine in 90% enantiomeric excess. Hydrolysis of the *N*-acetyl group yields L-alanine. Because the hydrogenation catalyst is chiral, it transfers its hydrogen atoms in a stereoselective way. This type of reaction is often called an **asymmetric synthesis** or **enantioselective synthesis** (Section 5.8C).

$$
\begin{array}{c}
\text{CH}_2{=}\text{C}-\text{CO}_2\text{H} \\
| \\
\text{NHCOCH}_3
\end{array}
\xrightarrow[\text{H}_2]{[\text{Rh}((R)\text{-prophos}) (\text{H})_2(\text{solvent})_2]^+}
\begin{array}{c}
\text{H} \\
\text{H}_3\text{C} \diagdown \\
\text{C}-\text{CO}_2\text{H} \\
\text{NHCOCH}_3
\end{array}
$$

**2-Acetylaminopropenoic
acid** — **N-Acetyl-L-alanine**

$$
\xrightarrow{\text{(1) OH}^-,\ \text{H}_2\text{O, heat (2) H}_3\text{O}^+}
\begin{array}{c}
\text{H} \\
\text{H}_3\text{C} \diagdown \\
\text{C}-\text{CO}_2^- \\
\overset{+}{\text{N}}\text{H}_3
\end{array}
$$

L-Alanine

This same procedure has been used to synthesize several other L-amino acids from 2-acetylaminopropenoic acids that have substituents at the 3 position. Use of the (R)-prophos catalyst in hydrogenation of the (Z)-isomer yields the L-amino acid with an enantiomeric excess of 87–93%.

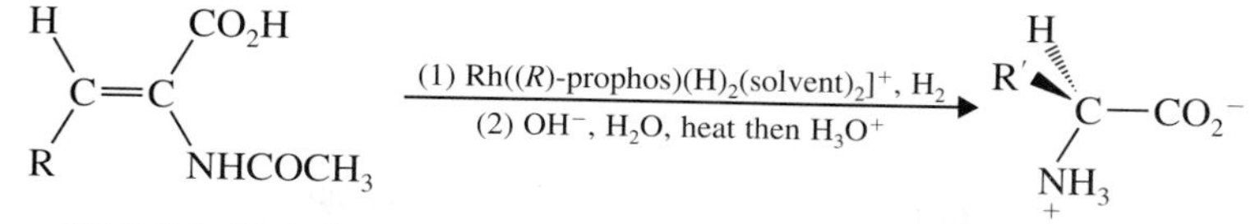

**(Z)-3-Substituted
2-acetylaminopropenoic
acid**

L-Amino acid

24.4 ANALYSIS OF POLYPEPTIDES AND PROTEINS

Enzymes can cause α-amino acids to polymerize through the elimination of water:

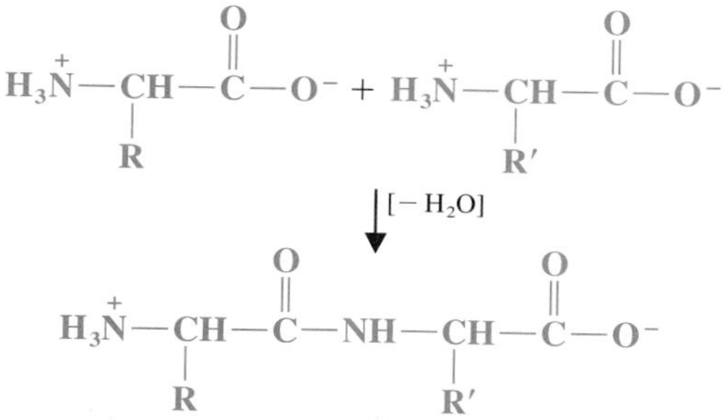

A dipeptide

The —CO—NH—(amide) linkage that forms between the amino acids is known as a **peptide bond** or **peptide linkage.** Amino acids, when joined in this way (as opposed to being free), are called **amino acid residues.** The polymers that contain 2, 3, a few (3–10), or many amino acid residues are called **dipeptides, tripeptides, oligopeptides,** and **polypeptides,** respectively. **Proteins** are molecules that contain one or more polypeptide chains.

Polypeptides are **linear polymers.** One end of a polypeptide chain terminates in an amino acid residue that has a free —NH$_3^+$ group; the other terminates in an amino acid residue with a free —CO$_2^-$ group. These two groups are called the **N-terminal** and the **C-terminal residues,** respectively.

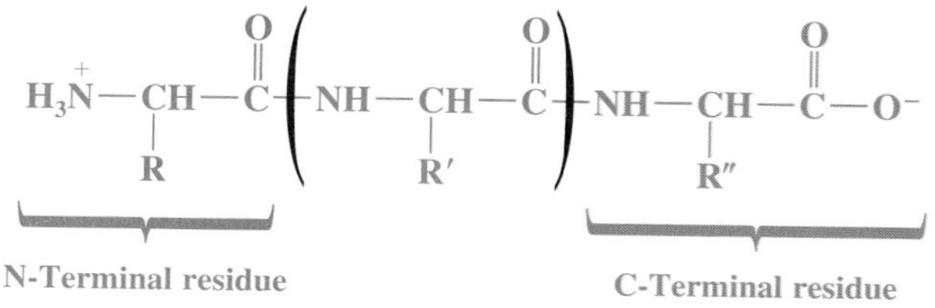

N-Terminal residue   C-Terminal residue

By convention, we write peptide and protein structures with the N-terminal amino acid on the left and the C-terminal residue on the right.

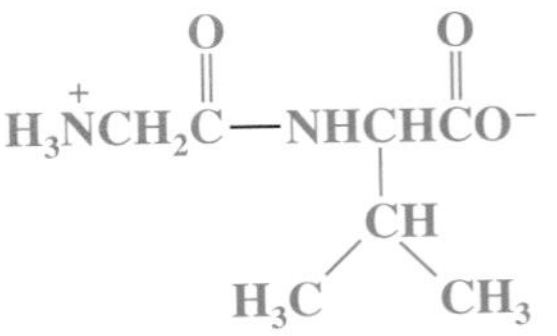

Glycylvaline
(Gly · Val)

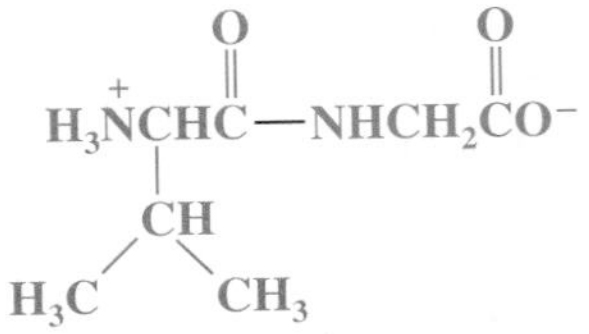

Valylglycine
(Val · Gly)

The tripeptide glycylvalylphenylalanine can be represented in the following way:

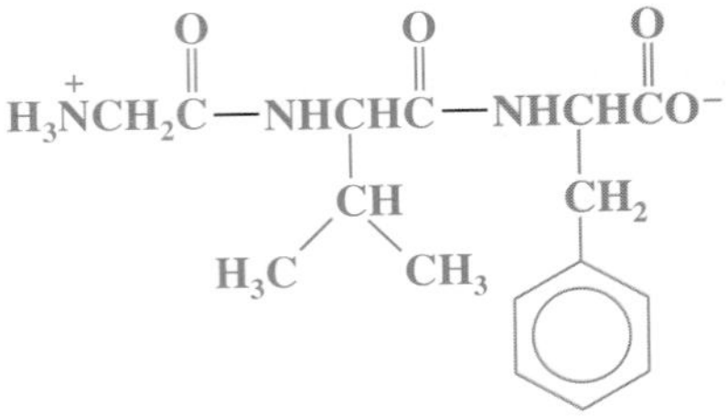

Glycylvalylphenylalanine
(Gly · Val · Phe)

When a protein or polypeptide is refluxed with 6*M* hydrochloric acid for 24 h, hydrolysis of all of the amide linkages usually takes place, and this produces a mixture of amino acids. One of the first tasks that we face when we attempt to determine the structure of a polypeptide or protein is the separation and identification of the individual amino acids in such a mixture. Since as many as 22 different amino acids may be present, this becomes a formidable task if we are restricted to conventional methods.

Fortunately, techniques have been developed, based on the principle of elution chromatography, that simplify this problem immensely and even allow its solution to be automated. Automatic amino acid analyzers were developed at the Rockefeller Institute in 1950 and have since become commercially available. They are based on the use of insoluble polymers containing sulfonate groups, called *cation-exchange resins* (Fig. 24.2).

If an acidic solution containing a mixture of amino acids is passed through a column packed with a cation-exchange resin, the amino acids will be adsorbed by the

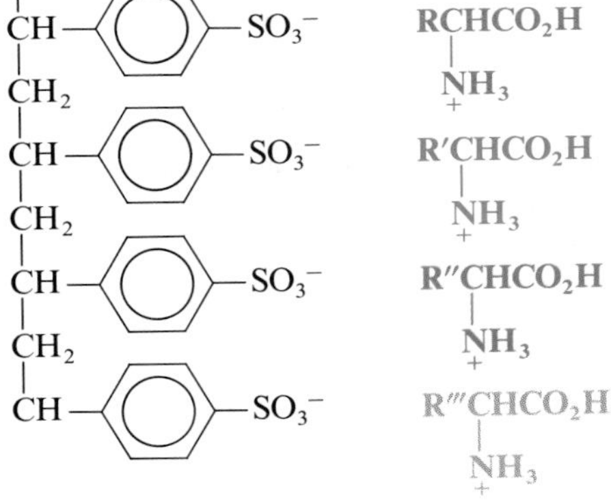

FIGURE 24.2 A section of a cation-exchange resin with adsorbed amino acids.

resin because of attractive forces between the negatively charged sulfonate groups and the positively charged amino acids. The strength of the adsorption varies with the basicity of the individual amino acids; those that are most basic are held most strongly. If the column is then washed with a buffered solution at a given pH, the individual amino acids move down the column at different rates and ultimately become separated. At the end of the column the eluate is allowed to mix with **ninhydrin,** a reagent that reacts with most amino acids to give a derivative with an intense purple color (λ_{max} 570 nm). The amino acid analyzer is designed so that it can measure the absorbance of the eluate (at 570 nm) continuously and record this absorbance as a function of the volume of the effluent.

A typical graph obtained from an automatic amino acid analyzer is shown in Fig. 24.3. When the procedure is standardized, the positions of the peaks are characteristic of the individual amino acids and the areas under the peaks correspond to their relative amounts.

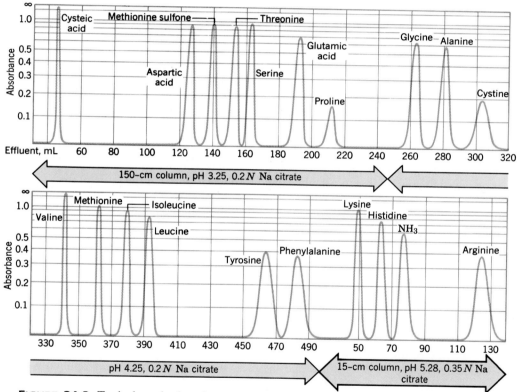

FIGURE 24.3 Typical result given by an automatic amino acid analyzer. [Adapted with permission from D. H. Spackman, W. H. Stein, and S. Moore, *Anal. Chem., 30,* 1190, **1958.** Copyright © by the American Chemical Society.]

Ninhydrin is the hydrate of indane-1,2,3-trione. With the exception of proline and hydroxyproline, all of the α-amino acids found in proteins react with ninhydrin to give the same intensely colored purple anion (λ_{max} 570 nm). We shall not go into the mechanism here, but notice that the only portion of the anion that is derived from the α-amino acid is the nitrogen.

Indane-1,2,3-trione   **Ninhydrin**

Purple anion

Proline and hydroxyproline do not react with ninhydrin in the same way because their α-amino groups are part of a five-membered ring.

24.5 AMINO ACID SEQUENCE OF POLYPEPTIDES AND PROTEINS

Once we have determined the amino acid composition of a protein or a polypeptide, we should then determine its molecular weight. Various methods are available for doing this, including chemical methods, ultracentrifugation, light scattering, osmotic pressure, and X-ray diffraction. With the molecular weight and amino acid composition we shall now be able to calculate the *molecular formula* of the protein; that is, we shall know how many of each type of amino acid are present as amino acid residues in each protein molecule. Unfortunately, however, we have only begun our task of determining its structure. The next step is a formidable one, indeed. We must determine the order in which the amino acids are connected; that is, we must determine the ***covalent structure (or primary structure) of the polypeptide.***

A simple tripeptide composed of 3 different amino acids can have 6 different amino acid sequences; a tetrapeptide composed of 4 different amino acids can have as many as 24. For a protein composed of 20 different amino acids in a single chain of 100 residues, there are $20^{100} = 1.27 \times 10^{130}$ possible polypeptides, a number much greater than the number of atoms estimated to be in the universe (9×10^{78}).

In spite of this, a number of methods have been developed that allow the amino acid sequences to be determined and these, as we shall see, have been applied with amazing success. In our discussion here we shall limit our attention to two methods that illustrate how sequence determinations can be done: **terminal residue analysis** and **partial hydrolysis.** In Section 25.6A we shall study an easier method.

24.5A Terminal Residue Analysis

One very useful method for determining the N-terminal amino acid residue, called the **Sanger method,** is based on the use of 2,4-dinitrofluorobenzene (DNFB).* When a

*This method was introduced by Frederick Sanger of Cambridge University in 1945. Sanger made extensive use of this procedure in his determination of the amino acid sequence of insulin and won the Nobel Prize for chemistry for the work in 1958.

polypeptide is treated with DNFB in mildly basic solution, a nucleophilic aromatic substitution reaction takes place involving the free amino group of the N-terminal residue. Subsequent hydrolysis of the polypeptide gives a mixture of amino acids in which the N-terminal amino acid bears a label, *the 2,4-dinitrophenyl group*. As a result, after separating this amino acid from the mixture, we can identify it.

$$O_2N-\text{⟨benzene⟩}-F \quad + \quad H_2\ddot{N}CHCO-NHCHCO\text{∿etc.} \xrightarrow[(-HF)]{HCO_3^-}$$

with substituents NO_2, R, R'

2,4-Dinitrofluorobenzene
(DNFB)

Polypeptide

$$O_2N-\text{⟨benzene⟩}-NHCHCO-NHCHCO\text{∿etc.} \xrightarrow{H_3O^+}$$

with substituents NO_2, R, R'

Labeled polypeptide

$$O_2N-\text{⟨benzene⟩}-NHCHCO_2H \quad + \quad \overset{+}{H_3}NCHCO_2^-$$

with substituents NO_2, R, R'

Labeled N-terminal amino acid   **Mixture of amino acids**

Separate and identify

PROBLEM 24.5

The electron-withdrawing property of the 2,4-dinitrophenyl group makes separation of the labeled amino acid very easy. Suggest how this is done.

Of course, 2,4-dinitrofluorobenzene will react with any free amino group present in a polypeptide, including the ε-amino group of lysine. But only the N-terminal amino acid residue will bear the label at the α-amino group.

A second method of N-terminal analysis is the *Edman degradation* (developed by Pehr Edman of the University of Lund, Sweden). This method offers an advantage over the Sanger method in that it removes the N-terminal residue and leaves the remainder of the peptide chain intact. The Edman degradation is based on a labeling reaction between the N-terminal amino group and phenyl isothiocyanate, $C_6H_5N=C=S$.

$$\text{⟨benzene⟩}-N=C=S \quad + \quad H_2\ddot{N}CHCO-NHCHCO\text{∿etc.} \xrightarrow{OH^-,\ pH\ 9}$$

with substituents R, R'

$$\text{⟨benzene⟩}-NH-\overset{\overset{S}{\|}}{C}-NHCHCO-NHCHCO\text{∿etc.} \xrightarrow{H^+}$$

with substituents R, R'

Labeled polypeptide

Unstable Intermediate

+

$$H_3\overset{+}{N}CHCO\sim\sim$$
$$\quad\quad |$$
$$\quad\quad R'$$

**Polypeptide with one
less amino acid residue**

Phenylthiohydantoin

When the labeled polypeptide is treated with acid, the N-terminal amino acid residue splits off as an unstable intermediate that undergoes rearrangement to a phenylthiohydantoin. This last product can be identified by comparison with phenylthiohydantoins prepared from standard amino acids.

The polypeptide that remains after the first Edman degradation can be submitted to another degradation to identify the next amino acid in the sequence, and this process has even been automated. Unfortunately, Edman degradations cannot be repeated indefinitely. As residues are successively removed, amino acids formed by hydrolysis during the acid treatment accumulate in the reaction mixture and interfere with the procedure. The Edman degradation, however, has been automated into what is called a **sequenator.** Each amino acid is automatically detected as it is removed. This technique has been successfully applied to polypeptides with as many as 60 amino acid residues.

C-Terminal residues can be identified through the use of digestive enzymes called *carboxypeptidases.* These enzymes specifically catalyze the hydrolysis of the amide bond of the amino acid residue containing a free —CO_2H group, liberating it as a free amino acid. A carboxypeptidase, however, will continue to attack the polypeptide chain that remains, successively lopping off C-terminal residues. As a consequence, it is necessary to follow the amino acids released as a function of time. The procedure can be applied to only a limited amino acid sequence for, at best, after a time the situation becomes too confused to sort out.

PROBLEM 24.6

(a) Write a reaction showing how 2,4-dinitrofluorobenzene could be used to identify the N-terminal amino acid of Val·Ala·Gly.   (b) What products would you expect (after hydrolysis) when Val·Lys·Gly is treated with 2,4-dinitrofluorobenzene?

PROBLEM 24.7

Write the reactions involved in a sequential Edman degradation of Met·Ile·Arg.

24.5B Partial Hydrolysis

Sequential analysis using the Edman degradation or carboxypeptidase becomes impractical with proteins or polypeptides of appreciable size. Fortunately, however, we can resort to another technique, that of **partial hydrolysis.** Using dilute acids or enzymes, we attempt to break the polypeptide chain into small fragments, ones that we can identify using DNFB or the Edman degradation. Then we examine the structures of these smaller fragments looking for points of overlap and attempt to piece together the amino acid sequence of the original polypeptide.

Consider a simple example: We are given a pentapeptide known to contain valine (two residues), leucine (one residue), histidine (one residue), and phenylalanine (one residue). With this information we can write the ''molecular formula'' of the protein in the following way, using commas to indicate that the sequence is unknown.

$$\text{Val}_2, \text{Leu}, \text{His}, \text{Phe}$$

Then, let us assume that by using DNFB and carboxypeptidase we discover that valine and leucine are the N-terminal and C-terminal residues, respectively. So far we know the following:

$$\text{Val (Val, His, Phe) Leu}$$

But the sequence of the three nonterminal amino acids is still unknown.

We then subject the pentapeptide to partial acid hydrolysis and obtain the following dipeptides. (We also get individual amino acids and larger pieces, i.e., tripeptides and tetrapeptides.)

$$\text{Val} \cdot \text{His} + \text{His} \cdot \text{Val} + \text{Val} \cdot \text{Phe} + \text{Phe} \cdot \text{Leu}$$

The points of overlap of the dipeptides (i.e., His, Val, and Phe) tell us that the original pentapeptide must have been the following:

$$\text{Val} \cdot \text{His} \cdot \text{Val} \cdot \text{Phe} \cdot \text{Leu}$$

Two enzymes are also frequently used to cleave certain peptide bonds in a large protein. *Trypsin* preferentially catalyzes the hydrolysis of peptide bonds in which the carboxyl group is a part of a lysine or arginine residue. *Chymotrypsin* preferentially catalyzes the hydrolysis of peptide bonds at the carboxyl groups of phenylalanine, tyrosine, and tryptophan. It also attacks the peptide bonds at the carboxyl groups of leucine, methionine, asparagine, and glutamine. Treating a large protein with trypsin or chymotrypsin will break it into smaller pieces. Then, each smaller piece can be subjected to an Edman degradation or to labeling followed by partial hydrolysis.

PROBLEM 24.8

Glutathione is a tripeptide found in most living cells. Partial acid-catalyzed hydrolysis of glutathione yields two dipeptides, Cys·Gly, and one composed of Glu and Cys. When this second dipeptide was treated with DNFB, acid hydrolysis gave *N*-labeled Glu.   (a) Based on this information alone, what structures are possible for glutathione? (b) Synthetic experiments have shown that the second dipeptide has the following structure:

$$\overset{+}{H_3N}CHCH_2CH_2CONHCHCO_2^-$$
$$|\qquad\qquad\qquad\qquad |$$
$$CO_2^-\qquad\qquad\quad CH_2SH$$

What is the structure of glutathione?

PROBLEM 24.9

Give the amino acid sequence of the following polypeptides using only the data given by partial acidic hydrolysis.

(a) Ser, Hyp, Pro, Thr $\xrightarrow[H_2O]{H^+}$ Ser · Thr + Thr · Hyp + Pro · Ser

(b) Ala, Arg, Cys, Val, Leu $\xrightarrow[H_2O]{H^+}$ Ala · Cys · + Cys · Arg + Arg · Val + Leu · Ala

24.6 PRIMARY STRUCTURES OF POLYPEPTIDES AND PROTEINS

The covalent structure of a protein or polypeptide is called its *primary structure* (Fig. 24.4). By using the techniques we described in the previous sections, chemists have had remarkable success in determining the primary structures of polypeptides and proteins. The compounds described in the following pages are important examples.

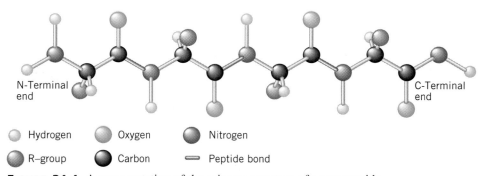

N-Terminal end   C-Terminal end

○ Hydrogen   ○ Oxygen   ● Nitrogen
● R–group   ● Carbon   ⬭ Peptide bond

FIGURE 24.4 A representation of the primary structure of a tetrapeptide.

24.6A Oxytocin and Vasopressin

Oxytocin and vasopressin (Fig. 24.5) are two rather small polypeptides with strikingly similar structures (where oxytocin has leucine, vasopressin has arginine and where oxytocin has isoleucine, vasopressin has phenylalanine). In spite of the similarity of their amino acid sequences, these two polypeptides have quite different physiological

Oxytocin

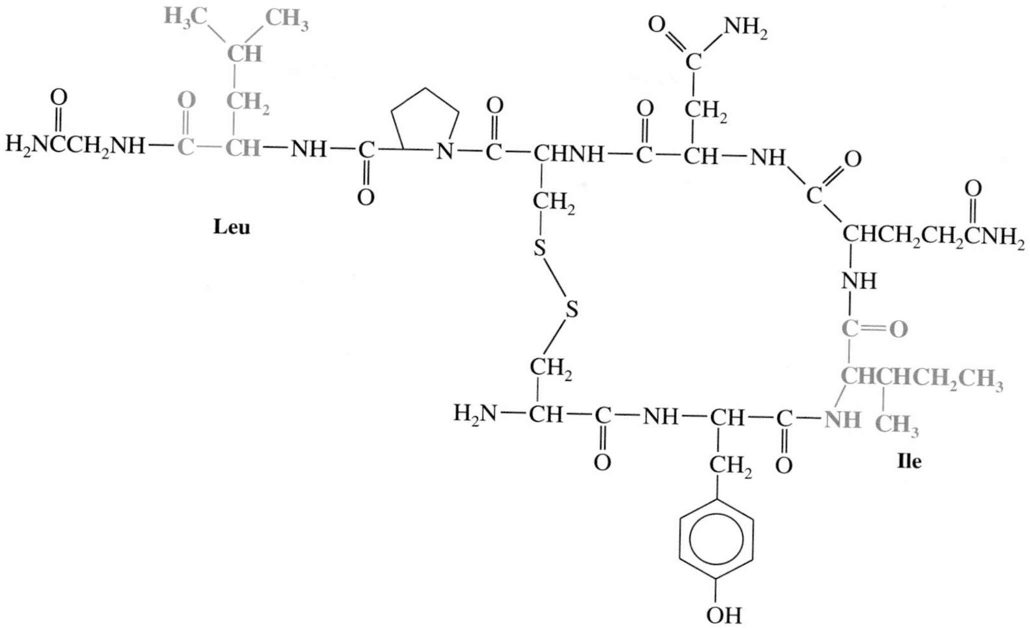

Vasopressin

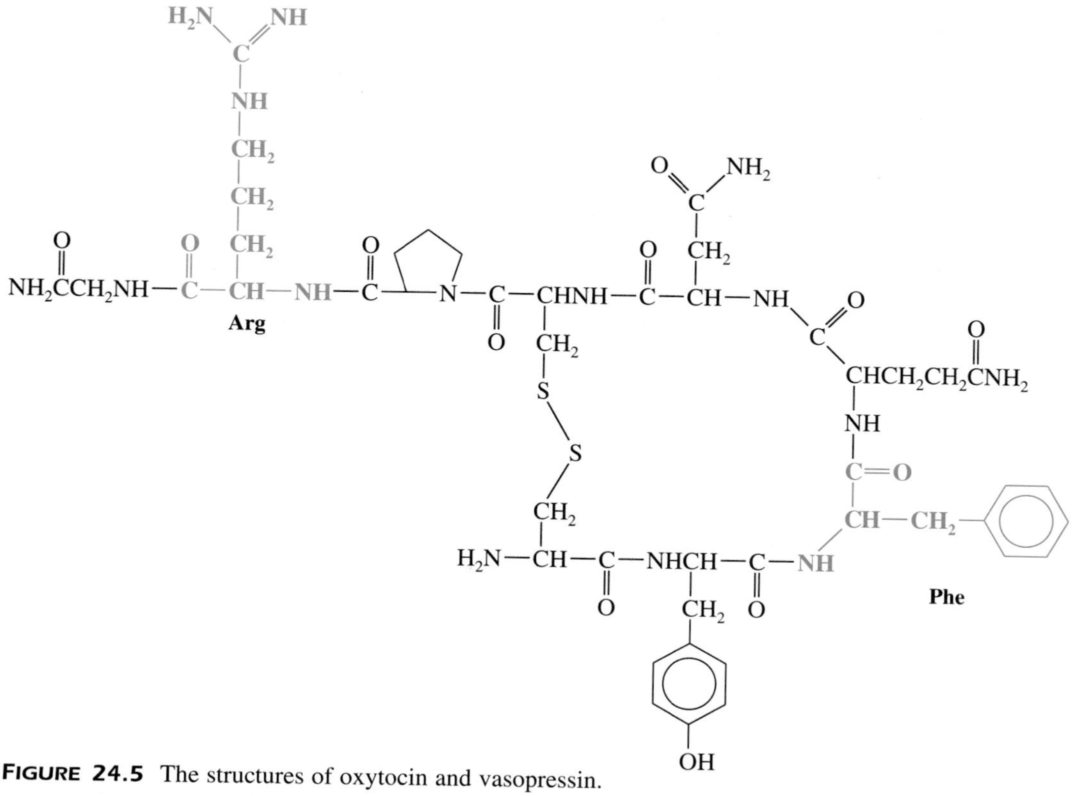

FIGURE 24.5 The structures of oxytocin and vasopressin.

effects. Oxytocin occurs only in the female of a species and stimulates uterine contractions during childbirth. Vasopressin occurs in males and females; it causes contraction of peripheral blood vessels and an increase in blood pressure. Its major function, however, is as an *antidiuretic;* physiologists often refer to vasopressin as an *antidiuretic hormone.*

   The structures of oxytocin and vasopressin also illustrate the importance of the disulfide linkage between cysteine residues (Section 24.2A) in the overall primary structure of a polypeptide. In these two molecules this disulfide linkage leads to a cyclic structure.*

PROBLEM 24.10

Treating oxytocin with certain reducing agents (e.g., sodium in liquid ammonia) brings about a single chemical change that can be reversed by air oxidation. What chemical changes are involved?

24.6B Insulin

Insulin, a hormone secreted by the pancreas, regulates glucose metabolism. Insulin deficiency in humans is the major problem in diabetes mellitus.

   The amino acid sequence of bovine insulin (Fig. 24.6) was determined by Sanger in 1953 after 10 years of work. Bovine insulin has a total of 51 amino acid residues in two polypeptide chains, called the A and B chains. These chains are joined by two

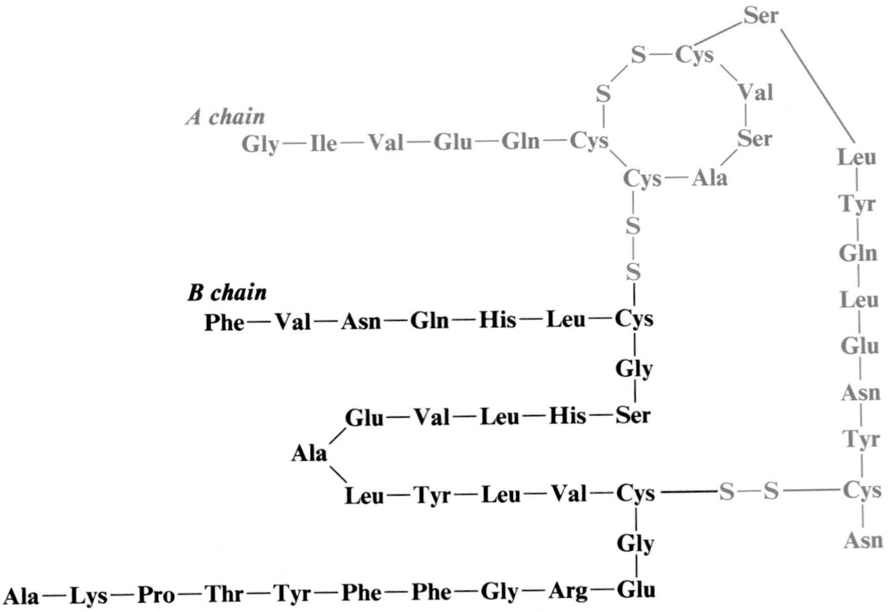

FIGURE 24.6 The amino acid sequence of bovine insulin.

---

*Vincent du Vigneaud of Cornell Medical College synthesized oxytocin and vasopressin in 1953; he received the Nobel Prize in 1955.

disulfide linkages. The A chain contains an additional disulfide linkage between cysteine residues at positions 6 and 11.

Human insulin differs from bovine insulin at only three amino acid residues: Threonine replaces alanine once in the A chain (residue 8) and once in the B chain (residue 30), and isoleucine replaces valine once in the A chain (residue 10). Insulins from most mammals have similar structures.

24.6C Other Polypeptides and Proteins

Successful sequential analyses have now been achieved with hundreds of other polypeptides and proteins including the following:

1. **Bovine ribonuclease.** This enzyme, which catalyzes the hydrolysis of ribonucleic acid (Chapter 25), has a single chain of 124 amino acid residues and four intrachain disulfide linkages.

2. **Human hemoglobin.** There are four peptide chains in this important oxygen-carrying protein. Two identical α chains have 141 residues each, and two identical β chains have 146 residues. The genetically based disease sickle-cell anemia results from a single amino acid error in the β chain. In normal hemoglobin, position 6 has a glutamic acid residue, whereas in sickle-cell hemoglobin position 6 is occupied by valine.

   Red blood cells (erythrocytes) containing hemoglobin with this amino acid residue error tend to become crescent shaped (''sickle'') when the partial pressure of oxygen is low, as it is in venous blood. These distorted cells are more difficult for the heart to pump through small capillaries. They may even block capillaries by clumping together; at other times the red cells may even split open. Children who inherit this genetic trait from both parents suffer from a severe form of the disease and usually do not live past the age of two. Children who inherit the disease from only one parent generally have a much milder form. Sickle-cell anemia arose among the populations of central and western Africa where, ironically, it may have had a beneficial effect. People with a mild form of the disease are far less susceptible to malaria than those with normal hemoglobin. Malaria, a disease caused by an infectious microorganism, is especially prevalent in central and western Africa. Mutational changes such as those that give rise to sickle-cell anemia are very common. Approximately 150 different types of mutant hemoglobin have been detected in humans; fortunately, most are harmless.

3. **Bovine trypsinogen and chymotrypsinogen.** These two enzyme precursors have single chains of 229 and 245 residues, respectively.

4. **Gamma globulin.** This immunoprotein has a total of 1320 amino acid residues in four chains. Two chains have 214 residues each; the other two have 446.

5. **p53, An anticancer protein.** The protein called p53 (the p stands for protein), consisting of 393 amino acid residues, has a variety of cellular functions, but the most important ones involve controlling the steps that lead to cell growth. It acts as a **tumor suppressor** by halting abnormal growth in normal cells, and by doing so it prevents cancer. Discovered in 1979, p53 was originally thought to be a protein synthesized by an oncogene (a gene that causes cancer). More recent research, however, has shown that the form of p53 originally thought to have this cancer-causing property was a mutant form of the normal protein. The unmutated (or *wild type*) of p53 is now called the ''guardian of the genome'' because it apparently coordinates a complex set of responses to changes in DNA that could otherwise lead to cancer. When p53 becomes mutated, it no longer provides the cell with its

cancer-preventing role; it apparently does the opposite, by acting to increase abnormal growth.

Of the 6.5 million people diagnosed with cancer each year, more than half have a mutant form of p53 in their cancers. Different forms of cancer have been shown to result from different mutations in the protein, and the list of cancer types associated with mutant p53 includes cancers of most of the body parts: breast, brain, bladder, cervix, colon, liver, lung, ovary, pancreas, prostate, skin, stomach, and so on.

Interest in p53 among scientists has become intense. In 1993 there were almost 1000 publications describing some aspect of research on p53, and in 1993, the magazine *Science* designated p53 "molecule of the year."*

24.7 POLYPEPTIDE AND PROTEIN SYNTHESIS

We saw in Chapter 18 that the synthesis of an amide linkage is a relatively simple one. We must first "activate" the carboxyl group of an acid by converting it to an anhydride or acid chloride and then allow it to react with an amine:

$$
\underset{\textbf{Anhydride}}{R-\overset{\overset{\textstyle O}{\|}}{C}-O-\overset{\overset{\textstyle O}{\|}}{C}-R} + \underset{\textbf{Amine}}{R'-NH_2} \longrightarrow \underset{\textbf{Amide}}{R-\overset{\overset{\textstyle O}{\|}}{C}-NHR'} + R-CO_2H
$$

The problem becomes somewhat more complicated, however, when both the acid group and the amino group are present in the same molecule, as they are in an amino acid and, especially, when our goal is the synthesis of a naturally occurring polyamide where the sequence of different amino acids is all important. Let us consider, as an example, the synthesis of the simple dipeptide alanylglycine, Ala·Gly. We might first activate the carboxyl group of alanine by converting it to an acid chloride, and then we might allow it to react with glycine. Unfortunately, however, we cannot prevent alanyl chloride from reacting with itself. So our reaction would yield not only Ala·Gly but also Ala·Ala. It could also lead to Ala·Ala·Ala and Ala·Ala·Gly, and so on. The yield of our desired product would be low, and we would also have a difficult problem separating the dipeptides, tripeptides, and so on.

$$
\underset{\substack{\overset{+}{N}H_3 \\ \textbf{Ala}}}{CH_3CHCO^-} \xrightarrow[\text{(2) } H_3\overset{+}{N}CH_2CO_2^-]{\text{(1) SOCl}_2} \underset{\substack{\overset{+}{N}H_3 \\ \textbf{Ala·Gly}}}{CH_3CHCNHCH_2CO_2^-} + \underset{\substack{\overset{+}{N}H_3 \quad CH_3 \\ \textbf{Ala·Ala}}}{CH_3CHCNHCHCO_2^-}
$$

$$
+ \underset{\substack{\overset{+}{N}H_3 \quad CH_3 \quad CH_3 \\ \textbf{Ala·Ala·Ala}}}{CH_3CHCNHCHCNHCHCO_2^-} + \underset{\substack{\overset{+}{N}H_3 \quad CH_3 \\ \textbf{Ala·Ala·Gly}}}{CH_3CHCNHCHCNHCH_2CO_2^-}
$$

where Gly corresponds to $H_3\overset{+}{N}CH_2CO_2^-$

*See *Science* 262, pp. 1953, 1958–1959, and 1980–1981, 1993.

24.7A Protecting Groups

The solution of this problem is to "protect" the amino group of the first amino acid before we activate it and allow it to react with the second. By protecting the amino group we mean that we must convert it to some other group of low nucleophilicity—*one that will not react with a reactive acyl derivative.* The protecting group must be carefully chosen because after we have synthesized the amide linkage between the first amino acid and the second, we will want to be able to remove the protecting group without disturbing the new amide bond.

A number of reagents have been developed to meet these requirements. Two that are often used are *benzyl chloroformate* and di-*tert*-butyl carbonate.

$$C_6H_5CH_2-O-\overset{\overset{\displaystyle O}{\|}}{C}-Cl \qquad\qquad (CH_3)_3C-O-\overset{\overset{\displaystyle O}{\|}}{C}-O-C(CH_3)_3$$

Benzyl chloroformate   **Di-*tert*-butyl carbonate**

Both reagents react with the following amino groups to form derivatives that are unreactive toward further acylation. Both derivatives, however, are of a type that allow removal of the protecting group under conditions that do not affect peptide bonds. The benzyloxycarbonyl group (abbreviated Z-) can be removed by catalytic hydrogenation or by treating the derivative with cold HBr in acetic acid. The *tert*-butyloxycarbonyl group (abbreviated Boc-) can be removed through treatment with HCl or CF_3CO_2H in acetic acid.

Benzyloxycarbonyl Group

tert-Butyloxycarbonyl Group

The easy removal of both groups (Z- and Boc-) in acidic media results from the exceptional stability of the carbocations that are formed initially. The benzyloxycarbonyl group gives a *benzyl cation;* the *tert*-butyloxycarbonyl group yields, initially, a *tert*-butyl cation.

Removal of the benzyloxycarbonyl group with hydrogen and a catalyst depends on the fact that benzyl–oxygen bonds are weak and are subject to hydrogenolysis at low temperatures.

$$C_6H_5CH_2-\overset{\overset{O}{\|}}{O}CR \xrightarrow[25°C]{H_2,\ Pd} C_6H_5CH_3 + HO\overset{\overset{O}{\|}}{C}R$$

A benzyl ester

24.7B Activation of the Carboxyl Group

Perhaps the most obvious way to activate a carboxyl group is to convert it to an acyl chloride. This method was used in early peptide syntheses, but acyl chlorides are actually more reactive than necessary. As a result, their use leads to complicating side reactions. A much better method is to convert the carboxyl group of the "protected" amino acid to a mixed anhydride using ethyl chloroformate, $Cl-\overset{\overset{O}{\|}}{C}-OC_2H_5$.

$$Z-NHCH\overset{\overset{O}{\|}}{C}-OH \underset{\underset{R}{|}}{} \xrightarrow[(2)\ ClCO_2C_2H_5]{(1)\ (C_2H_5)_3N} Z-NHCH-\overset{\overset{O}{\|}}{C}-O-\overset{\overset{O}{\|}}{C}-OC_2H_5 \underset{\underset{R}{|}}{}$$

"Mixed anhydride"

The mixed anhydride can then be used to acylate another amino acid and form a peptide linkage.

$$Z-NHCH\overset{\overset{O}{\|}}{C}-O-\overset{\overset{O}{\|}}{C}OC_2H_5 \underset{\underset{R}{|}}{} \xrightarrow{\overset{\overset{\overset{+}{H_3N}-CHCO_2^-}{|}}{R'}}$$

$$Z-NHCH\overset{\overset{O}{\|}}{C}-NHCHCO_2H + CO_2 + C_2H_5OH \underset{\underset{R}{|}\ \ \ \ \underset{R'}{|}}{}$$

Dicyclohexylcarbodiimide (Section 18.8E) can also be used to activate the carboxyl group of an amino acid. In Section 24.7D we shall see how it is used in an automated peptide synthesis.

24.7C Peptide Synthesis

Let us examine now how we might use these reagents in the preparation of the simple dipeptide, Ala·Leu. The principles involved here can, of course, be extended to the synthesis of much longer polypeptide chains.

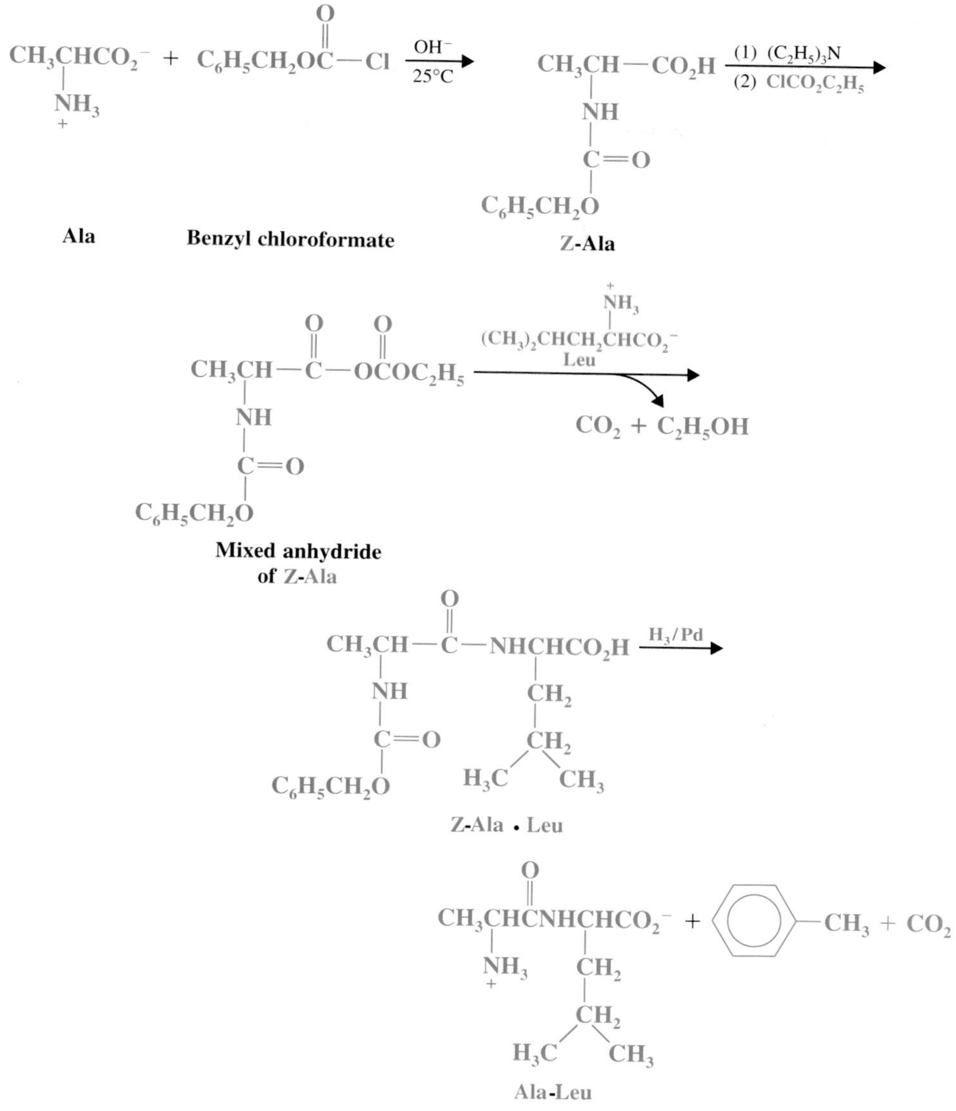
OH⁻
25°C
(1) (C2H5)3N
(2) ClCO2C2H5
Ala
Benzyl chloroformate
Z-Ala
Leu
CO2 + C2H5OH
Mixed anhydride
of Z-Ala
H2/Pd
Z-Ala • Leu
Ala-Leu

PROBLEM 24.11

Show all steps in the synthesis of Gly · Val · Ala using the *tert*-butyloxycarbonyl (Boc-) group as a protecting group.

PROBLEM 24.12

The synthesis of a polypeptide containing lysine requires the protection of both amino groups. (a) Show how you might do this in a synthesis of Lys · Ile using the benzyloxycarbonyl group as a protecting group. (b) The benzyloxycarbonyl group can also be used to protect the guanidino group, $-\mathrm{NH}\overset{\mathrm{NH}}{\overset{\|}{\mathrm{C}}}-\mathrm{NH_2}$, of arginine. Show a synthesis of Arg · Ala.

PROBLEM 24.13

The terminal carboxyl groups of glutamic acid and aspartic acid are often protected through their conversion to benzyl esters. What mild method could be used for removal of this protecting group?

24.7D Automated Peptide Synthesis

Although the methods that we have described thus far have been used to synthesize a number of polypeptides, including ones as large as insulin, they are extremely time consuming. One must isolate and purify the product at almost every stage. Thus, a real advance in peptide synthesis came with the development by R. B. Merrifield (at Rockefeller University) of a procedure for automating peptide synthesis. Merrifield received the Nobel Prize in chemistry in 1984 for this work.

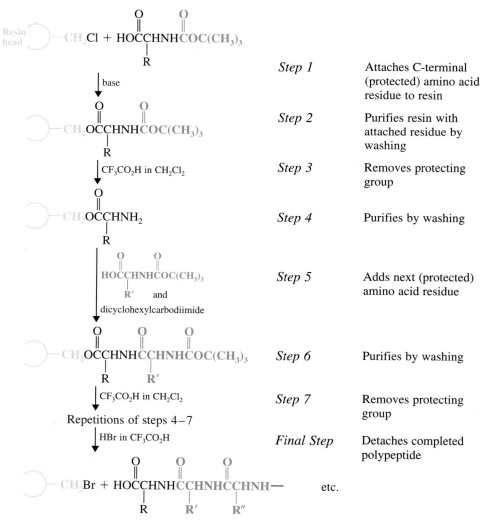

Step 1 Attaches C-terminal (protected) amino acid residue to resin

Step 2 Purifies resin with attached residue by washing

Step 3 Removes protecting group

Step 4 Purifies by washing

Step 5 Adds next (protected) amino acid residue

Step 6 Purifies by washing

Step 7 Removes protecting group

Final Step Detaches completed polypeptide

FIGURE 24.7 The Merrifield method for automated protein synthesis.

The Merrifield method is based on the use of a polystyrene resin similar to the one we saw in Fig. 24.2, *but one that contains* —CH₂Cl *groups* instead of sulfonic acid groups. This resin is used in the form of small beads and is insoluble in most solvents.

The first step in automated peptide synthesis (Fig. 24.7) involves a reaction that attaches the first protected amino acid residue to the resin beads. After this step is complete, the protecting group is removed and the next amino acid (also protected) is condensed with the first using dicyclohexylcarbodiimide (Section 18.8E) to activate its carboxyl group. Then, removal of the protecting group of the second residue readies the resin-dipeptide for the next step.

The great advantage of this procedure is that purification of the resin with its attached polypeptide can be carried out at each stage by simply washing the resin with an appropriate solvent. Impurities, because they are not attached to the insoluble resin, are simply carried away by the solvent. In the automated procedure each cycle of the ''protein-making machine'' requires only 4 h and attaches one new amino acid residue.*

The Merrifield technique has been applied successfully to the synthesis of ribonuclease, a protein with 124 amino acid residues. The synthesis involved 369 chemical reactions and 11,931 automated steps—all were carried out without isolating an intermediate. The synthetic ribonuclease not only had the same physical characteristics as the natural enzyme, it possessed the identical biological activity as well. The overall yield was 17%, which means that the average yield of each individual step was greater than 99%.

PROBLEM 24.14

The resin for the Merrifield procedure is prepared by treating polystyrene,

$$\left(-CH_2CH- \atop \underset{C_6H_5}{|} \right)_n,$$

with CH_3OCH_2Cl and a Lewis acid catalyst.   (a) What reaction is involved?   (b) After purification, the complete polypeptide or protein can be detached from the resin by treating it with HBr in trifluoroacetic acid under conditions mild enough not to affect the amide linkages. What structural feature of the resin makes this possible?

PROBLEM 24.15

Outline the steps in the synthesis of Lys·Phe·Ala using the Merrifield procedure.

24.8 SECONDARY AND TERTIARY STRUCTURES OF PROTEINS

We have seen how amide and disulfide linkages constitute the covalent or *primary structure* of proteins. Of equal importance in understanding how proteins function is knowledge of the way in which the peptide chains are arranged in three dimensions. Involved here are the secondary and tertiary structures of proteins.

---

*Protein synthesis in the body catalyzed by enzymes and directed by DNA and RNA takes only 1 min to add 150 amino acids in a specific sequence (cf. Section 25.5).

24.8A Secondary Structure

The **secondary structure** of a protein is defined by the local conformation of its polypeptide backbone. These local conformations have come to be specified in terms of regular folding patterns called *helices, pleated sheets,* and *turns.* The major experimental techniques that have been used in elucidating the secondary structures of proteins are X-ray analysis and NMR (including 2D NMR).

When X-rays pass through a crystalline substance they produce diffraction patterns. Analysis of these patterns indicates a regular repetition of particular structural units with certain specific distances between them, called **repeat distances.** X-Ray analyses have revealed that the polypeptide chain of a natural protein can interact with itself in two major ways: through formation of a **β-pleated sheet** and an **α helix.***

To understand how these interactions occur let us look first at what X-ray analysis has revealed about the geometry at the peptide bond itself. Peptide bonds tend to assume a geometry such that six atoms of the amide linkage are coplanar (Fig. 24.8). The carbon–nitrogen bond of the amide linkage is unusually short, indicating that resonance contributions of the type shown here are important.

$$
\overset{..}{\underset{/}{N}}-C\diagup\overset{O}{\diagdown} \quad \longleftrightarrow \quad \overset{+}{\underset{/}{N}}=C\diagup\overset{O^-}{\diagdown}
$$

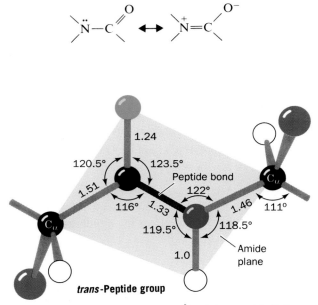

trans-Peptide group

FIGURE 24.8 The geometry and bond lengths (in Å) of the peptide linkage. The six enclosed atoms tend to be coplanar and assume a "transoid" arrangement. (From D. Voet and J. G. Voet, *Biochemistry,* 2nd ed., Wiley, New York, 1995. Used with permission.)

The carbon–nitrogen bond, consequently, has considerable double-bond character (~40%) and rotations of groups about this bond are severely hindered.

---

*Two American scientists, Linus Pauling and Robert B. Corey, were pioneers in the X-ray analysis of proteins. Beginning in 1939, Pauling and Corey initiated a long series of studies of the conformations of peptide chains. At first, they used crystals of single amino acids, then dipeptides and tripeptides, and so on. Moving on to larger and larger molecules and using the precisely constructed molecular models, they were able to understand the secondary structures of proteins for the first time.

Rotations of groups attached to the amide nitrogen and the carbonyl carbon are relatively free, however, and these rotations allow peptide chains to form different conformations.

The transoid arrangement of groups around the relatively rigid amide bond would cause the R groups to alternate from side to side of a single fully extended peptide chain:

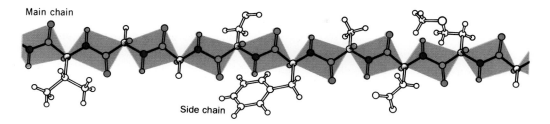

Calculations show that such a polypeptide chain would have a repeat distance (i.e., distance between alternating units) of 7.2 Å.

Fully extended polypeptide chains could conceivably form a flat-sheet structure, with each alternating amino acid in each chain forming two hydrogen bonds with an amino acid in the adjacent chain:

Hypothetical flat-sheet structure

This structure does not exist in naturally occurring proteins because of the crowding that would exist between R groups. If such a structure did exist, it would have the same repeat distance as the fully extended peptide chain, that is, 7.2 Å.

Slight rotations of bonds, however, can transform a flat-sheet structure into what is called the **β-pleated sheet** or **β configuration** (Fig. 24.9). The pleated-sheet structure gives small and medium-sized R groups room enough to avoid van der Waals repulsions and is the predominant structure of silk fibroin (48% glycine and 38% serine and alanine residues). The pleated-sheet structure has a slightly shorter repeat distance, 7.0 Å, than the flat sheet.

Of far more importance in naturally occurring proteins is the secondary structure called the α helix (Fig. 24.10). This structure is a right-handed helix with 3.6 amino acid residues per turn. Each amide group in the chain has a hydrogen bond to an amide group at a distance of three amino acid residues in either direction, and the R groups all extend away from the axis of the helix. The repeat distance of the α helix is 1.5 Å.

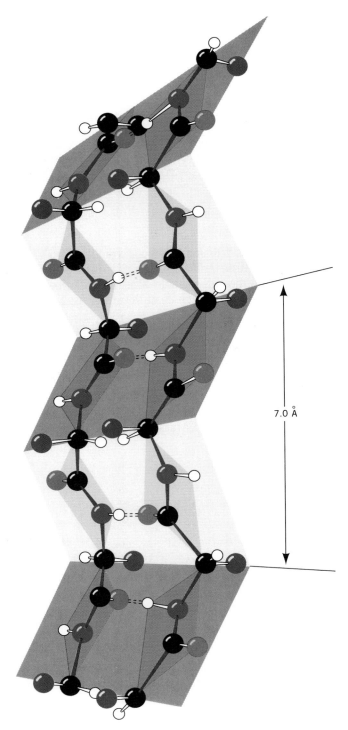

7.0 Å

FIGURE 24.9 The β-pleated sheet or β configuration of a protein. (Figure copyrighted © by Irving Geis. From D. Voet and J. G. Voet, *Biochemistry,* 2nd ed., Wiley, New York, 1995. Used with permission.)

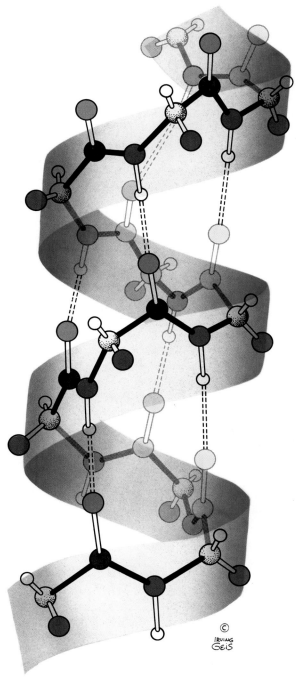

FIGURE 24.10 A representation of the α-helical structure of a polypeptide. Hydrogen bonds are denoted by dotted lines. (Figure copyrighted © by Irving Geis. From D. Voet and J. G. Voet, *Biochemistry,* 2nd ed., Wiley, New York, 1995. Used with permission.)

The α-helical structure is found in many proteins; it is the predominant structure of the polypeptide chains of fibrous proteins such as *myosin,* the protein of muscle, and of *α-keratin,* the protein of hair, unstretched wool, and nails.

Helices and pleated sheets account for only about one half of the structure of the average globular protein. The remaining polypeptide segments have what is called a **coil** or **loop conformation.** These nonrepetitive structures are not random, they are just more difficult to describe. Globular proteins also have stretches, called **reverse turns** or **β bends,** where the polypeptide chain abruptly changes direction. These often connect successive strands of β sheets and almost always occur at the surface of proteins.

Figure 24.11 is a schematic representation of an enzyme, human carbonic anhydrase, showing how segments of α helix and β sheets intervene between reverse turns and nonrepetitive structures.

The locations of the side chains of amino acids of globular proteins are usually those that we would expect from their polarities.

1. Residues with **nonpolar, hydrophobic, side chains,** such as *valine, leucine, iso-leucine, methionine, and phenylalanine* are almost always found in the interior of the protein, out of contact with the aqueous solvent. (These hydrophobic inter-

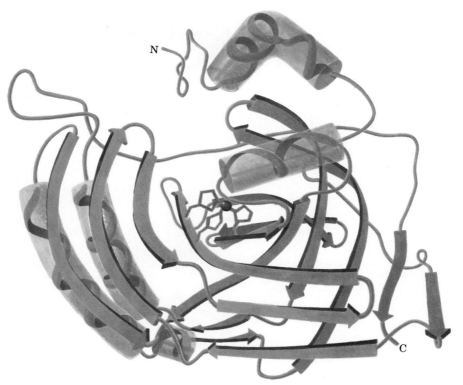

FIGURE 24.11 A schematic representation of the enzyme, human carbonic anhydrase. Alpha helices are represented as cylinders and strands of β-pleated sheets are drawn with an arrow pointing toward the C-terminus of the polypeptide. The side chains of three histidine residues (shown in blue) coordinate with a zinc atom (gray). Notice that the C-terminus is tucked through a loop of the polypeptide chain, making carbonic anhydrase a rare example of a native protein in which the polypeptide chain forms a knot. (From D. Voet and J. G. Voet, *Biochemistry,* 2nd ed., Wiley, New York, 1995.)

actions are largely responsible for the tertiary structure of proteins that we discuss in Section 24.8B.)

2. Side chains of **polar residues with + or − charges,** such as *arginine, lysine, aspartic acid, and glutamic acid,* are usually on the surface of the protein in contact with the aqueous solvent.

3. **Uncharged polar side chains,** such as those of *serine, threonine, asparagine, glutamine, tyrosine, and tryptophan,* are most often found on the surface, but some of these are found in the interior as well. When they are found in the interior, they are virtually all hydrogen bonded to other similar residues. Hydrogen bonding apparently helps neutralize the polarity of these groups.

Certain peptide chains assume what is called a **random coil arrangement,** a structure that is flexible, changing, and statistically random. Synthetic polylysine, for example, exists as a random coil and does not normally form an α helix. At pH 7, the ε-amino groups of the lysine residues are positively charged and, as a result, repulsive forces between them are so large that they overcome any stabilization that would be gained through hydrogen bond formation of an α helix. At pH 12, however, the ε-amino groups are uncharged and polylysine spontaneously forms an α helix.

The presence of proline or hydroxyproline residues in polypeptide chains produces another striking effect: Because the nitrogen atoms of these amino acids are part of five-membered rings, the groups attached by the nitrogen−α carbon bond cannot rotate enough to allow an α-helical structure. Wherever proline or hydroxyproline occur in a peptide chain, their presence causes a kink or bend and interrupts the α helix.

24.8B Tertiary Structure

The tertiary structure of a protein is its three-dimensional shape that arises from further foldings of its polypeptide chains, foldings superimposed on the coils of the α helixes. These foldings do not occur randomly: Under the proper environmental conditions they occur in one particular way—a way that is characteristic of a particular protein and one that is often highly important to its function.

Various forces are involved in stabilizing tertiary structures, including the disulfide bonds of the primary structure. One characteristic of most proteins is that the folding takes place in such a way as to expose the maximum number of polar (hydrophilic) groups to the aqueous environment and enclose a maximum number of nonpolar (hydrophobic) groups within its interior.

The soluble globular proteins tend to be much more highly folded than fibrous proteins. However, fibrous proteins also have a tertiary structure; the α-helical strands of α-keratin, for example, are wound together into a "super helix." This super helix has a repeat distance of 5.1-Å units, indicating that the super helix makes one complete turn for each 35 turns of the α helix. The tertiary structure does not end here, however. Even the super helices can be wound together to give a ropelike structure of seven strands.

Myoglobin (Fig. 24.12) and hemoglobin (Section 24.12) were the first proteins (in 1957 and 1959) to be subjected to a completely successful X-ray analysis. This work was accomplished by J. C. Kendrew and Max Perutz at Cambridge University in England. (They received the Nobel Prize in 1962.) Since then many other proteins including lysozyme, ribonuclease, and α-chymotrypsin have yielded to complete structural analysis.

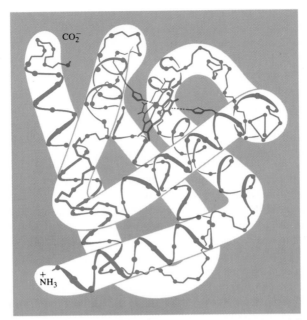

FIGURE 24.12 The three-dimensional structure of myoglobin. (Adapted from R. E. Dickerson, *The Proteins II,* H. Neurath, Ed., Academic Press, New York, 1964. p. 634. Used with permission.)

24.9   INTRODUCTION TO ENZYMES

All of the reactions that occur in living cells are mediated by remarkable biological catalysts called enzymes. Enzymes have the ability to bring about vast increases in the rates of reactions; in most instances, the rates of enzyme-catalyzed reactions are faster than those of uncatalyzed reactions by factors of $10^6 - 10^{12}$. For living organisms, rate enhancements of this magnitude are important because they permit reactions to take place at reasonable rates, even under the mild conditions that exist in living cells (i.e., approximately neutral pH and a temperature of about 35°C.)

Enzymes also show remarkable **specificity** for their reactants (called **substrates**) and for their products. This specificity is far greater than that shown by most chemical catalysts. In the enzymatic synthesis of proteins, for example (through reactions that take place on ribosomes, Section 25.5D), polypeptides consisting of well over 1000 amino acid residues are synthesized virtually without error. It was Emil Fischer's discovery, in 1894, of the ability of enzymes to distinguish between α- and β- glycosidic linkages (Section 22.12) that led him to formulate his **lock-and-key hypothesis** for enzyme specificity. According to this hypothesis, the specificity of an enzyme (the lock) and its substrate (the key) comes from their geometrically complementary shapes.

The enzyme and the substrate combine to form an **enzyme–substrate complex.** Formation of the complex often induces a conformational change in the enzyme that allows it to bind the substrate more effectively. This is called an **induced fit.** Binding the substrate also often causes certain of its bonds to become strained, and therefore more easily broken. The product of the reaction usually has a different shape from the substrate, and this altered shape, or in some instances, the intervention of another molecule, causes the complex to dissociate. The enzyme can then accept another molecule of the substrate and the whole process is repeated.

Enzyme + substrate $\rightleftharpoons$ enzyme–substrate $\longrightarrow$ enzyme + product
complex

Almost all enzymes are proteins.* The substrate is bound to the protein, and the reaction takes place, at what is called the **active site.** The noncovalent forces that bind the substrate to the active site are the same forces that account for the conformations of the proteins themselves: van der Waals forces, electrostatic forces, hydrogen bonding, and hydrophobic interactions. The amino acids located in the active site are arranged so that they can interact specifically with the substrate.

Reactions catalyzed by enzymes are completely **stereospecific,** and this specificity comes from the way enzymes bind their substrates. An α-glucosidase will only bind the α form of a glucoside, not the β form. Enzymes that metabolize sugars bind only D-sugars; enzymes that synthesize proteins bind only L-amino acids; and so on.

Although enzymes are absolutely stereospecific, they often vary considerably in what is called their **geometric specificity.** By geometric specificity we mean a specificity that is related to the identities of the chemical groups of the substrates. Some enzymes will accept only one compound as its substrate. Others, however, will accept a range of compounds with similar groups. Carboxypeptidase A, for example, will hydrolyze the C-terminal peptide from all polypeptides as long as the penultimate residue is not arginine, lysine, or proline, and as long as the next preceding residue is not proline. Chymotrypsin, a digestive enzyme that catalyzes the hydrolysis of peptide bonds, will also catalyze the hydrolysis of esters.

$$\underset{\textbf{Peptide}}{R-\overset{\overset{\displaystyle O}{\|}}{C}-NH-R'} + H_2O \xrightarrow{\text{chymotrypsin}} R-\overset{\overset{\displaystyle O}{\|}}{C}-O^- + H_3\overset{+}{N}-R'$$

$$\underset{\textbf{Ester}}{R-\overset{\overset{\displaystyle O}{\|}}{C}-O-R'} + H_2O \xrightarrow{\text{chymotrypsin}} R-\overset{\overset{\displaystyle O}{\|}}{C}-OH + HO-R'$$

A compound that can alter the activity of an enzyme is called an **inhibitor.** A compound that competes directly with the substrate for the active site is known as a **competitive inhibitor.** We learned in Section 20.11, for example, that sulfanilamide is a competitive inhibitor for a bacterial enzyme that incorporates p-aminobenzoic acid into folic acid.

Some enzymes require the presence of a **cofactor.** The cofactor may be a metal ion as, for example, the zinc atom of human carbonic anhydrase (Fig. 24.11). Others may require the presence of an organic molecule, such as NAD^+ (Section 14.10), called a **coenzyme.** Coenzymes become chemically changed in the course of the enzymatic reaction. NAD^+ becomes converted to **NADH.** In some enzymes the cofactor is permanently bound to the enzyme, in which case it is called a **prosthetic group.**

---

* We now know that certain RNA molecules can also act as enzymes. The 1989 Nobel Prize for Chemistry went to Sidney Altman (of Yale University) and to Thomas R. Cech (of the University of Colorado, Boulder) for this discovery.

Many of the water-soluble vitamins are the precursors of coenzymes. Niacin (nicotinic acid) is a precursor of NAD^+, for example. Pantothenic acid is a precursor of coenzyme A.

Niacin   **Pantothenic Acid**

24.10   LYSOZYME: MODE OF ACTION OF AN ENZYME

Lysozyme is made up of 129 amino acid residues (Fig. 24.13). Three short segments of the chain between residues 5 and 15, 24 and 34, and 88 and 96 have the structure of an α helix; the residues between 41 and 45 and 50 and 54 form pleated sheets, and a hairpin turn occurs at residues 46–49. The remaining polypeptide segments of lysozyme have a coil or loop conformation.

The discovery of lysozyme is an interesting story in itself:

> One day in 1922 Alexander Fleming was suffering from a cold. This is not unusual in London, but Fleming was a most unusual man and he took advantage of the cold in a characteristic way. He allowed a few drops of his nasal mucus to fall on a culture of bacteria he was working with and then put the plate to one side to see what would happen. Imagine his excitement when he discovered some time later that the bacteria near the mucus had dissolved away. For a while he thought his ambition of finding a universal antibiotic had been realized. In a burst of activity he quickly established that the antibacterial action of the mucus was due to the presence of an enzyme; he called this substance lysozyme because of its capacity to lyse, or dissolve, the bacterial cells. Lysozyme was soon discovered in many tissues and secretions of the human body, in plants, and most plentifully of all in the white of an egg. Unfortunately Fleming found that it is not effective against the most harmful bacteria. He had to wait 7 years before a strangely similar experiment revealed the existence of a genuinely effective antibiotic: penicillin.

This story was related by Professor David C. Phillips of Oxford University who many years later used X-ray analysis to discover the three-dimensional structure of lysozyme.*

Phillips' X-ray diffraction studies of lysozyme are especially interesting because they have also revealed important information about how this enzyme acts on its substrate. Lysozyme's substrate is a polysaccharide of amino sugars that makes up part of the bacterial cell wall. An oligosaccharide that has the same general structure as the cell wall polysaccharide is shown in Fig. 24.14.

By using oligosaccharides (made up of N-acetylglucosamine units only) on which lysozyme acts very slowly, Phillips and his co-workers were able to discover how the substrate fits into the enzyme's active site. This site is a deep cleft in the lysozyme structure (Fig. 24.15a). The oligosaccharide is held in this cleft by hydrogen bonds, and, as the enzyme binds the substrate, two important changes take place: The cleft in

---

*Quotation from David C. Phillips, *The Three-Dimensional Structure of an Enzyme Molecule.* Copyright © 1966 by Scientific American, Inc. All rights reserved.

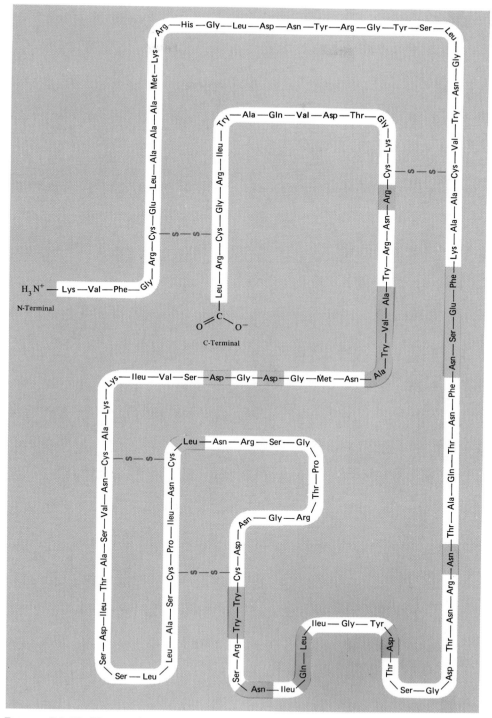

FIGURE 24.13 The covalent structure of lysozyme. The amino acids that line the active site of lysozyme are shown in color.

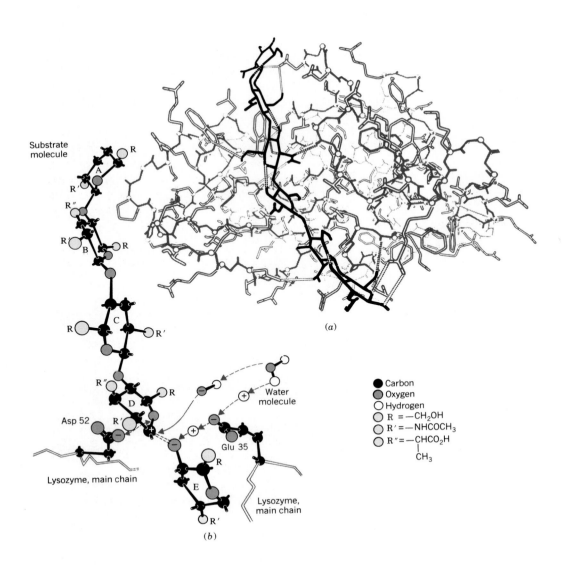

$R_1 = -CH_2OH$ $R_2 = NHCCH_3$ (with $\overset{O}{\overset{\|}{}}$) $R_3 = -CHCOH$ (with $\overset{O}{\overset{\|}{}}$ and CH_3)

FIGURE 24.14 A hexasaccharide that has the same general structure as the cell wall polysaccharide on which lysozyme acts. Two different amino sugars are present: rings **A**, **C**, and **E** are derived from a monosaccharide called *N*-acetylglucosamine; rings **B**, **D**, and **F** are derived from a monosaccharide called *N*-acetylmuramic acid. When lysozyme acts on this oligosaccharide, hydrolysis takes place and results in cleavage at the glycosidic linkage between rings **D** and **E**.

the enzyme closes slightly and ring **D** of the oligosaccharide is "flattened" out of its stable chair conformation. This flattening causes atoms 1, 2, 5, and 6 of ring **D** to become coplanar; it also distorts ring **D** in such a way as to make the glycosidic linkage between it and ring **E** more susceptible to hydrolysis.*

Hydrolysis of the glycosidic linkage probably takes the course illustrated in Fig. 24.15*b*. The carboxyl group of glutamic acid (residue number 35) donates a proton to the oxygen between rings **D** and **E**. Protonation leads to cleavage at the glycosidic link and to the formation of a carbocation at C1 of ring **D**. This carbocation is stabilized by the negatively charged carboxylate group of aspartic acid (residue number 52), which lies in close proximity. A water molecule diffuses in and supplies an OH$^-$ ion to the carbocation and a proton to replace that lost by glutamic acid.

When the polysaccharide is a part of a bacterial cell wall, lysozyme probably first attaches itself to the cell wall by hydrogen bonds. After hydrolysis has taken place, lysozyme falls away, leaving behind a bacterium with a punctured cell wall.

24.11   SERINE PROTEASES

Chymotrypsin, trypsin, and elastin are digestive enzymes secreted by the pancreas into the small intestines to catalyze the hydrolysis of peptide bonds. These enzymes are all called **serine proteases** because the mechanism for their proteolytic activity (one that they have in common) involves a particular serine residue that is essential for their enzymatic activity. As another example of how enzymes work, we shall examine the mechanism of action of chymotrypsin.

Chymotrypsin is formed from a precursor molecule called chymotrypsinogen, which has 245 amino acid residues. Cleavage of two dipeptide units of chymotrypsinogen produces chymotrypsin. Chymotrypsin folds in a way that brings together histidine at position 57, aspartic acid at position 102, and serine at position 195. Together, these residues constitute what is called the **catalytic triad** of the active site (Fig. 24.16). Near the active site is a hydrophobic binding site, a slitlike pocket that preferentially accommodates the nonpolar side chains of Phe, Tyr, and Trp.

---

*R. H. Lemieux and G. Huber, while with the National Research Council of Canada, showed that when an aldohexose is converted to a carbocation the ring of the carbocation assumes just this flattened conformation.

FIGURE 24.15 (facing page) (*a*) This drawing shows the backbones of the lysozyme–substrate complex. The substrate (in this drawing a hexasaccharide) fits into a cleft in the lysozyme structure and is held in place by hydrogen bonds. As lysozyme binds the oligosaccharide, the cleft in its structure closes slightly. (Adapted with permission from *Atlas of Protein Sequence and Structure,* 1969, Margaret O. Dayoff, Ed. National Biomedical Research Foundation, Washington, DC, 1969. The drawing was made by Irving Geis, based on his perspective painting of the molecule, which appeared in *Scientific American,* November 1966. The painting was made of an actual model assembled at the Royal Institution, London, by D. C. Phillips and his colleagues, based on their X-ray crystallography results.) (*b*) A possible mechanism for lysozyme action. This drawing shows an expanded portion of part (*a*) and illustrates how hydrolysis of the acetal linkage between rings **D** and **E** of the substrate may occur. Glutamic acid (residue 35) donates a proton to the intervening oxygen atom. This causes the formation of a carbocation that is stabilized by the carboxylate ion of aspartic acid (residue 52). A water molecule supplies an OH$^-$ to the carbocation and H$^+$ to glutamic acid. (Adapted with permission from *The Three-Dimensional Structures of an Enzyme Molecule,* by David C. Phillips, Copyright © Nov. 1966 by Scientific American, Inc. All rights reserved.)

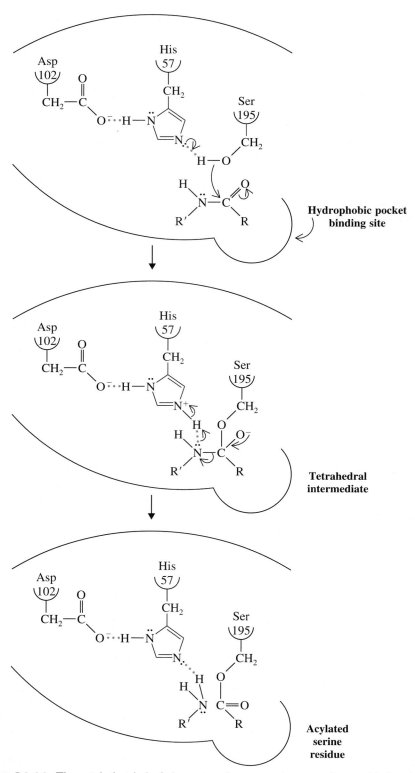

FIGURE 24.16 The catalytic triad of chymotrypsin causes cleavage of a peptide bond by acylation of the serine residue 195 of chymotrypsin. Near the active site is a hydrophobic binding site that accommodates nonpolar side chains of the protein.

After chymotrypsin has bound its protein substrate, the serine residue at position 195 is ideally situated to attack the acyl carbon of the peptide bond. This serine residue is made more nucleophilic by transferring its proton to the imidazole nitrogen of the histidine residue at position 57. The imidazolium ion that is formed is stabilized by the polarizing effect of the carboxylate ion of the aspartic acid residue at position 102. (Neutron diffraction studies, which show the positions of hydrogen atoms, confirm that the carboxylate ion remains as a carboxylate ion throughout and does not actually accept a proton from the imidazole.) Nucleophilic attack by the serine leads to an acylated serine through a tetrahedral intermediate. The new N-terminal end of the cleaved polypeptide chain diffuses away and is replaced by a water molecule.

Regeneration of the active site of chymotrypsin is shown in Fig. 24.17. In this process water acts as the nucleophile, and, in a series of steps analogous to those in Fig. 24.16, hydrolyzes the acyl–serine bond. The enzyme is now ready to repeat the whole process.

There is much evidence for this mechanism that, for reasons of space, we shall have to ignore. One bit of evidence deserves mention, however. There are compounds such as **diisopropylphosphofluoridate (DIPF)** that irreversibly inhibit serine proteases. It has been shown that they do this by reacting only with Ser 195.

$$\text{Ser 195} \rightarrow \text{CH}_2\text{OH} + \quad
\begin{array}{c}
\text{CH(CH}_3)_2 \\
| \\
\text{O} \\
| \\
\text{F} - \text{P} = \text{O} \\
| \\
\text{O} \\
| \\
\text{CH(CH}_3)_2
\end{array}
\quad \longrightarrow \quad
\text{Ser 195} \rightarrow \text{CH}_2 - \text{O} -
\begin{array}{c}
\text{CH(CH}_3)_2 \\
| \\
\text{O} \\
| \\
\text{P} = \text{O} \\
| \\
\text{O} \\
| \\
\text{CH(CH}_3)_2
\end{array}$$

**Diisopropylphospho-
fluoridate (DIPF)**   **DIP-Enzyme**

Recognition of the inactivating effect of DIPF came about as a result of the discovery that DIPF and related compounds are powerful **nerve poisons.** (They are the ''nerve gases'' of military use, even though they are liquids dispersed as fine droplets, and not gases.) Diisopropylphosphofluoridate inactivates **acetylcholinesterase** (Section 20.14) by reacting with it in the same way that it does with chymotrypsin. Acetylcholinesterase is a **serine esterase** rather than a serine protease.

24.12 HEMOGLOBIN: A CONJUGATED PROTEIN

Some proteins, called **conjugated proteins,** contain as a part of their structure a nonprotein group called a **prosthetic group.** An example is the oxygen-carrying protein, hemoglobin. Each of the four polypeptide chains of hemoglobin is bound to a prosthetic group called *heme* (Fig. 24.18). The four polypeptide chains of hemoglobin are wound in such a way as to give hemoglobin a roughly spherical shape (Fig. 24.19). Moreover, each heme group lies in a crevice with the hydrophobic vinyl groups of its porphyrin structure surrounded by hydrophobic side chains of amino residues. The two propanoate side chains of heme lie near positively charged amino groups of lysine and arginine residues.

The iron of the heme group is in the 2 + (ferrous) oxidation state and it forms a coordinate bond to a nitrogen of the imidazole group of histidine of the polypeptide chain. This leaves one valence of the ferrous ion free to combine with oxygen as follows:

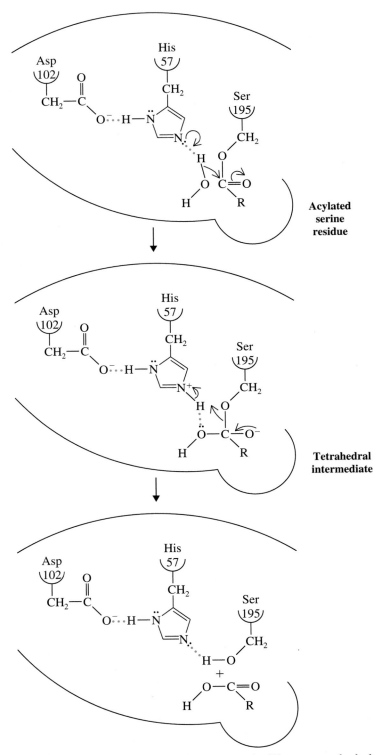

FIGURE 24.17 Regeneration of the active site of chymotrypsin. Water causes hydrolysis of the acyl–serine bond.

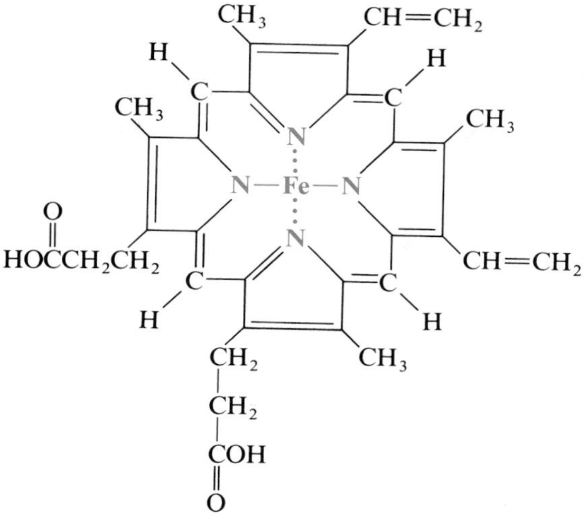

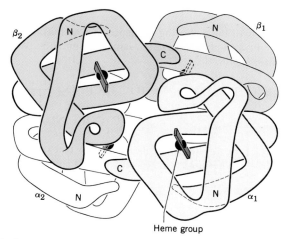

FIGURE 24.18 The structure of heme, the prosthetic group of hemoglobin. Heme has a structure similar to that of chlorophyll (Fig. 22.1) in that each is derived from the heterocyclic ring, porphyrin. The iron of heme is in the ferrous (2 +) oxidation state.

FIGURE 24.19 The hemoglobin molecule.

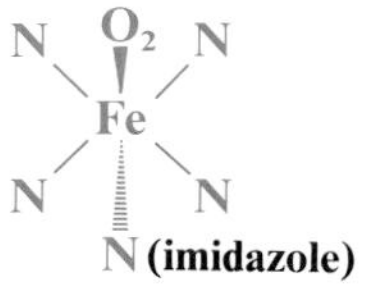

A portion of oxygenated
hemoglobin

The fact that the ferrous ion of the heme group combines with oxygen is not particularly remarkable; many similar compounds do the same thing. What is remarkable about hemoglobin is that when the heme combines with oxygen the ferrous ion does not become readily oxidized to the ferric state. Studies with model heme compounds in water, for example, show that they undergo a rapid combination with oxygen but they also undergo a rapid oxidation of the iron from Fe^{2+} to Fe^{3+}. When these same compounds are embedded in the hydrophobic environment of a polystyrene resin, however, the iron is easily oxygenated and deoxygenated and this occurs *with no change in oxidation state of iron*. In this respect, it is especially interesting to note that X-ray studies of hemoglobin have revealed that the polypeptide chains provide each heme group with a similar hydrophobic environment.

ADDITIONAL PROBLEMS

24.16 (a) Which amino acids in Table 24.1 have more than one stereocenter? (b) Write Fischer projection formulas for the isomers of each of these amino acids that would have the L configuration at the α carbon. (c) What kind of isomers have you drawn in each case?

24.17 (a) Which amino acid in Table 24.1 could react with nitrous acid (i.e., a solution of $NaNO_2$ and HCl) to yield lactic acid? (b) All of the amino acids in Table 24.1 liberate nitrogen when they are treated with nitrous acid except two; which are these? (c) What product would you expect to obtain from treating tyrosine with excess bromine water? (d) What product would you expect to be formed in the reaction of phenylalanine with ethanol in the presence of hydrogen chloride? (e) What product would you expect from the reaction of alanine and benzoyl chloride in aqueous base?

24.18 (a) On the basis of the following sequence of reactions, Emil Fischer was able to show that $(-)$-serine and L-$(+)$-alanine have the same configuration. Write Fischer projection formulas for the intermediates **A–C**.

$$(-)\text{-Serine} \xrightarrow[\text{CH}_3\text{OH}]{\text{HCl}} \textbf{A} \ (C_4H_{10}ClNO_3) \xrightarrow{\text{PCl}_5} \textbf{B} \ (C_4H_9Cl_2NO_2) \xrightarrow[\text{(2) OH}^-]{\text{(1) H}_3\text{O}^+,\ \text{H}_2\text{O, heat}}$$

$$\textbf{C} \ (C_3H_6ClNO_2) \xrightarrow[\text{dil. H}_3\text{O}^+]{\text{Na-Hg}} \text{L-}(+)\text{-alanine}$$

(b) The configuration of L-$(-)$-cysteine can be related to that of L-$(-)$-serine through the following reactions. Write Fischer projection formulas for **D** and **E**.

$$\textbf{B} \text{ [from part (a)]} \xrightarrow{\text{OH}^-} \textbf{D} \ (C_4H_8ClNO_2) \xrightarrow{\text{NaSH}}$$

$$\textbf{E} \ (C_4H_9NO_2S) \xrightarrow[\text{(2) OH}^-]{\text{(1) H}_3\text{O}^+,\ \text{H}_2\text{O, heat}} \text{L-}(+)\text{-cysteine}$$

(c) The configuration of L-$(-)$-asparagine can be related to that of L-$(-)$-serine in the following way. What is the structure of **F**?

$$\text{L-}(-)\text{-Asparagine} \xrightarrow[\substack{\text{Hofmann} \\ \text{rearrangement}}]{\text{NaOBr/OH}^-} \textbf{F} (C_3H_7N_2O_2)$$

$$\text{C [from part (a)]} \xrightarrow{\text{NH}_3} \uparrow$$

24.19 (a) DL-Glutamic acid has been synthesized from diethyl acetamidomalonate in the following way: Outline the reactions involved.

$$\underset{\substack{\textbf{Diethyl acetamido-} \\ \textbf{malonate}}}{CH_3\overset{\overset{\text{O}}{\|}}{C}NHCH(CO_2C_2H_5)_2} + CH_2{=}CH{-}C{\equiv}N \xrightarrow[\substack{\text{C}_2\text{H}_5\text{OH} \\ \text{(95\% yield)}}]{\text{NaOC}_2\text{H}_5}$$

$$\textbf{G} \ (C_{12}H_{18}N_2O_5) \xrightarrow[\substack{\text{reflux 6 h} \\ \text{(66\% yield)}}]{\text{concd HCl}} \text{DL-glutamic acid}$$

(b) Compound **G** has also been used to prepare the amino acid DL-ornithine through the following route. Outline the reaction involved here.

$$\textbf{G } (C_{12}H_{18}N_2O_5) \xrightarrow[\substack{68°C, 1000\ psi \\ (90\%\ yield)}]{H_2,\ Ni} \textbf{H } (C_{10}H_{16}N_2O_4,\ a\ \delta\text{-lactam}) \xrightarrow[\substack{reflux\ 4\ h \\ (97\%\ yield)}]{concd\ HCl}$$

$$\text{DL-ornithine hydrochloride } (C_5H_{13}ClN_2O_2)$$

(L-Ornithine is a naturally occurring amino acid but does not occur in proteins. In one metabolic pathway L-ornithine serves as a precursor for L-arginine.)

24.20   Bradykinin is a nonapeptide released by blood plasma globulins in response to a wasp sting. It is a very potent pain-causing agent. Its molecular formula is Arg_2, Gly, Phe_2, Pro_3, Ser. The use of 2,4-dinitrofluorobenzene and carboxypeptidase show that both terminal residues are arginine. Partial acid hydrolysis of bradykinin gives the following di- and tripeptides:

$$Phe \cdot Ser + Pro \cdot Gly \cdot Phe + Pro \cdot Pro + Ser \cdot Pro \cdot Phe + Phe \cdot Arg + Arg \cdot Pro$$

What is the amino acid sequence of bradykinin?

24.21   Complete hydrolysis of a heptapeptide showed that it had the following molecular formula:

$$Ala_2,\ Glu,\ Leu,\ Lys,\ Phe,\ Val$$

Deduce the amino acid sequence of this heptapeptide from the following data.

1.   Treatment of the heptapeptide with 2,4-dinitrofluorobenzene followed by incomplete hydrolysis gave, among other products: valine labeled at the α-amino group, lysine labeled at the ε-amino group, and a dipeptide, DNP—Val·Leu (DNP = 2,4- dinitrophenyl-).
2.   Hydrolysis of the heptapeptide with carboxypeptidase gives an initial high concentration of alanine, followed by a rising concentration of glutamic acid.
3.   Partial enzymatic hydrolysis of the heptapeptide gave a dipeptide (**A**) and a tripeptide (**B**).
   a.   Treatment of **A** with 2,4-dinitrofluorobenzene followed by hydrolysis gave DNP-labeled leucine and lysine labeled only at the ε-amino group.
   b.   Complete hydrolysis of **B** gave phenylalanine, glutamic acid, and alanine. When **B** was allowed to react with carboxypeptidase, the solution showed an initial high concentration of glutamic acid. Treatment of **B** with 2,4-dinitrofluorobenzene followed by hydrolysis gave labeled phenylalanine.

24.22   Synthetic polyglutamic acid exists as an α helix in solution at pH 2–3. When the pH of such a solution is gradually raised through the addition of base, a dramatic change in optical rotation takes place at pH 5. This change has been associated with the unfolding of the α helix and the formation of a random coil. What structural feature of polyglutamic acid, and what chemical change, can you suggest as an explanation of this transformation?

24.23   Part of the evidence for restricted rotation about the carbon–nitrogen bond in a peptide linkage (see Section 24.8A) comes from ^1H NMR studies done with simple amides. For example, at room temperature and with the instrument operating at 60 MHz, the ^1H NMR spectrum of N,N-dimethylformamide, $(CH_3)_2NCHO$, shows a doublet at δ 2.80 (3H), a doublet at δ 2.95 (3H), and a multiplet at δ 8.05 (1H). When the spectrum is determined at lower magnetic field strength (i.e., with the instrument operating at 30 MHz), the doublets are found to have shifted so that the distance (in hertz) that separates one doublet from the other is smaller. When the temperature at which the spectrum is determined is raised, the doublets persist until a temperature of 111°C is reached; then the doublets coalesce to become a single signal. Explain in detail how these observations are consistent with the existence of a relatively large barrier to rotation about the carbon–nitrogen bond of DMF.

CHAPTER 25

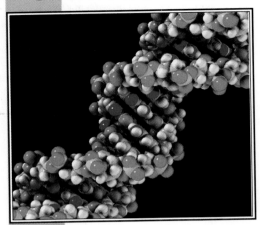

DNA
(Section 25.4)

NUCLEIC ACIDS AND PROTEIN SYNTHESIS

. . . I cannot help wondering whether some day an enthusiastic scientist will christen his newborn twins Adenine and Thymine.

F. H. C. Crick*

25.1 INTRODUCTION

The **nucleic acids,** deoxyribonucleic acid (DNA) and ribonucleic acid (RNA), are, respectively, the molecules that preserve hereditary information and that transcribe and translate it in a way that allows the synthesis of all the varied proteins of the cell. These biological polymers are sometimes found associated with proteins and in this form they are known as **nucleoproteins.**

Much of our knowledge of how genetic information is preserved, how it is passed on to succeeding generations of the organism, and how it is transformed into the working parts of the cell has come from the study of nucleic acids. For these reasons, we shall focus our attention on the structures and properties of nucleic acids and of their components, **nucleotides** and **nucleosides.**

25.2 NUCLEOTIDES AND NUCLEOSIDES

Mild degradations of nucleic acids yield their monomeric units, compounds that are called **nucleotides.** A general formula for a nucleotide and the specific structure of one, called adenylic acid, are shown in Fig. 25.1.

*Who along with J. D. Watson and Maurice Wilkins shared the Nobel Prize in 1962 for their proposal of (and evidence for) the double helix structure of DNA. (Taken from F. H. C. Crick, ''The Structure of the Hereditary Material,'' *Sci. Am., 191*, 20, 54–61, 1954.)

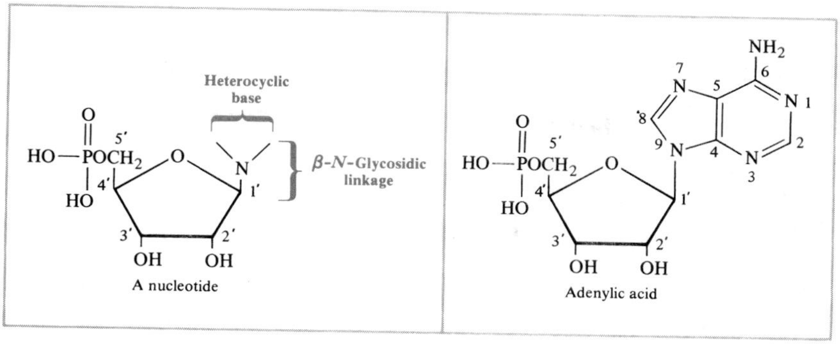

(a)   (b)

FIGURE 25.1 (a) General structure of a nucleotide obtained from RNA. The heterocyclic base is a purine or pyrimidine. In nucleotides obtained from DNA, the sugar component is 2'-deoxyribose; that is, the —OH at position 2' is replaced by —H. The phosphate group of the nucleotide is shown attached to the C5'; it may instead be attached to the C3' atom. The heterocyclic base is always attached through a β-N-glycosidic linkage at C1'. (b) Adenylic acid, a typical nucleotide.

Complete hydrolysis of a nucleotide furnishes:

1. A heterocyclic base, either a purine or pyrimidine.
2. A five-carbon monosaccharide, either D-ribose or 2-deoxy-D-ribose.
3. A phosphate ion.

The central portion of the nucleotide is the monosaccharide and it is always present as a five-membered ring, that is, as a furanoside. The heterocyclic base of a nucleotide is attached through an N-glycosidic linkage to C1' of the ribose or deoxyribose unit and this linkage is always β. The phosphate group of a nucleotide is present as a phosphate ester and it may be attached at C5' or C3'. (In nucleotides, the carbon atoms of the monosaccharide portion are designated with primed numbers, i.e., 1', 2', 3', etc.)

Removal of the phosphate group of a nucleotide converts it to a compound known as a **nucleoside** (Section 22.15A). The nucleosides that can be obtained from DNA all contain 2-deoxy-D-ribose as their sugar component and one of four heterocyclic bases, either adenine, guanine, cytosine, or thymine:

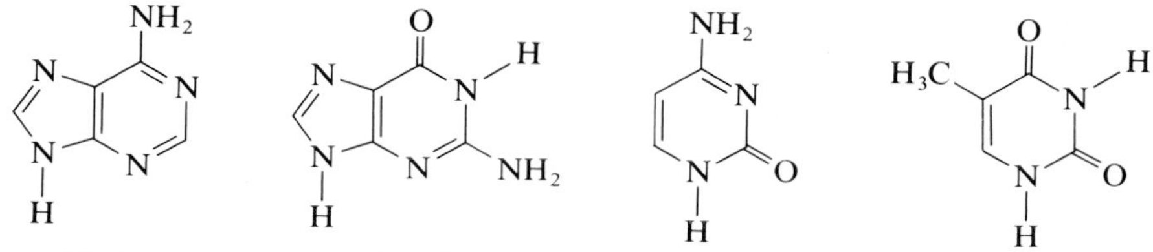

| Adenine (A) | Guanine (G) | Cytosine (C) | Thymine (T) |

◄————— Purines —————►   ◄————— Pyrimidines —————►

The nucleosides obtained from RNA contain D-ribose as their sugar component and either adenine, guanine, cytosine, or uracil as their heterocyclic base.*

---

*Notice that in an RNA nucleoside (or nucleotide), uracil replaces thymine. (Some nucleosides obtained from specialized forms of RNA may also contain other, but similar, purines and pyrimidines.)

Uracil
(a pyrimidine)

The heterocyclic bases obtained from nucleosides are capable of existing in more than one tautomeric form. The forms that we have shown are the predominant forms that the bases assume when they are present in nucleic acids.

PROBLEM 25.1

Write the structures of other tautomeric forms of adenine, guanine, cytosine, thymine, and uracil.

The names and structures of the nucleosides found in DNA are shown in Fig. 25.2; those found in RNA are given in Fig. 25.3.

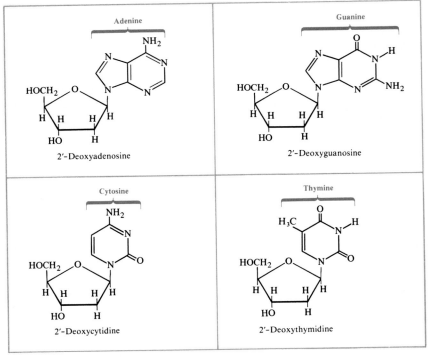

FIGURE 25.2 Nucleosides that can be obtained from DNA.

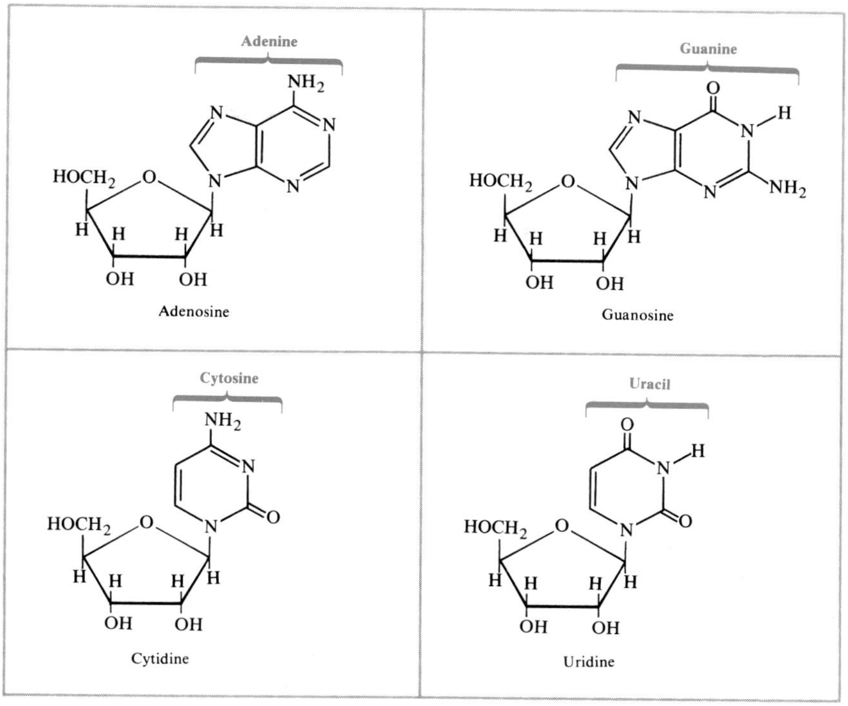

FIGURE 25.3 Nucleosides that can be obtained from RNA.

PROBLEM 25.2

The nucleosides shown in Figs. 25.2 and 25.3 are stable in dilute base. In dilute acid, however, they undergo rapid hydrolysis yielding a sugar (deoxyribose or ribose) and a heterocyclic base.   (a) What structural feature of the nucleoside accounts for this behavior?   (b) Propose a reasonable mechanism for the hydrolysis.

Nucleotides are named in several ways. Adenylic acid (Fig. 25.1), for example, is sometimes called 5′-adenylic acid in order to designate the position of the phosphate group; it is also called adenosine 5′-phosphate, or simply adenosine monophosphate (AMP). Uridylic acid is called 5′-uridylic acid, uridine 5′-phosphate, or uridine monophosphate (UMP), and so on.

Nucleosides and nucleotides are found in places other than as part of the structure of DNA and RNA. We have seen, for example, that adenosine units are part of the structures of two important coenzymes, NADH and coenzyme A. The 5′-triphosphate of adenosine is, of course, the important energy source, ATP (Section 22.1B). The compound called 3′,5′-cyclic adenylic acid (or cyclic AMP) (Fig. 25.4) is an important regulator of hormone activity. Cells synthesize this compound from ATP through the action of an enzyme, *adenyl cyclase.* In the laboratory, 3′,5′-cyclic adenylic

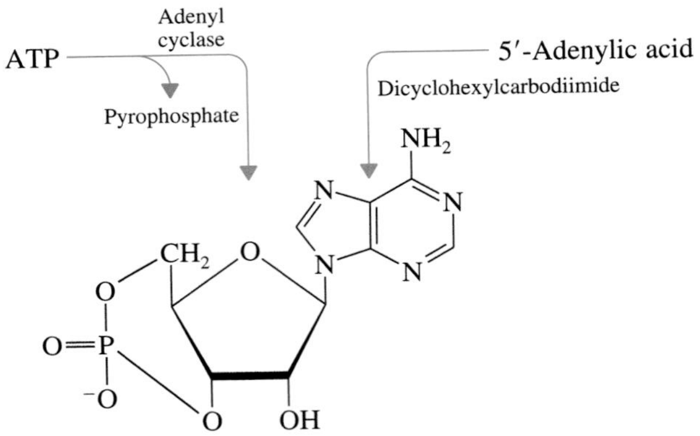

FIGURE 25.4 3′,5′-Cyclic adenylic acid and its biosynthesis and laboratory synthesis.

acid can be prepared through dehydration of 5′-adenylic acid with dicyclohexyl-carbodiimide.

PROBLEM 25.3 ————————————————————————————————

When 3′,5′-cyclic adenylic acid is treated with aqueous sodium hydroxide, the major product that is obtained is 3′-adenylic acid (adenosine 3′-phosphate) rather than 5′-adenylic acid. Suggest an explanation that accounts for the course of this reaction.

25.3 LABORATORY SYNTHESIS OF NUCLEOSIDES AND NUCLEOTIDES

A variety of methods has been developed for the synthesis of nucleosides. One technique uses reactions that assemble the nucleoside from suitably activated and protected ribose derivatives and heterocyclic bases. An example is the following synthesis of adenosine:

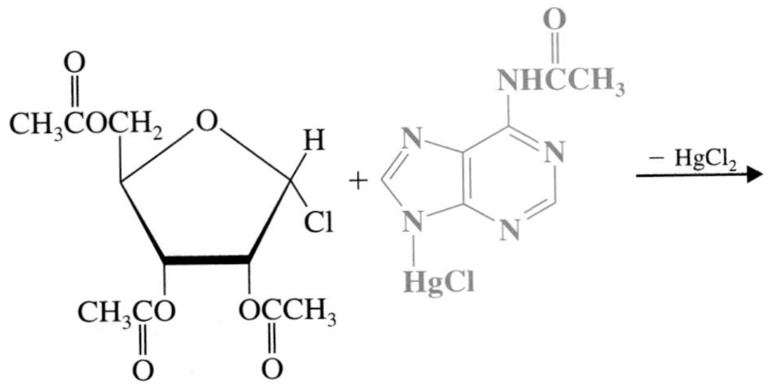

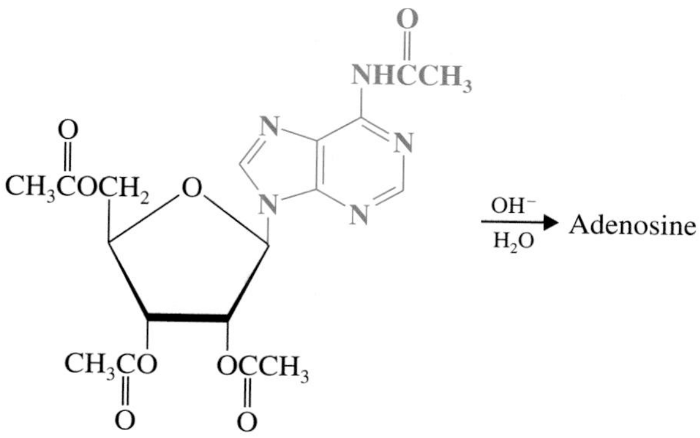

Another technique involves formation of the heterocyclic base on a protected ribosylamine derivative:

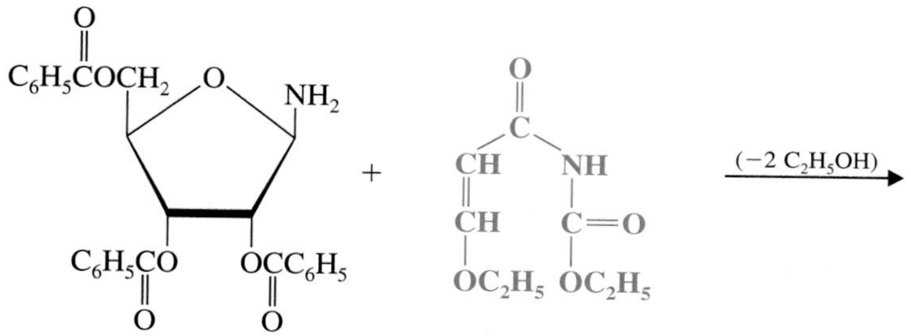

2,3,5-tri-O-Benzoyl-β-D-ribofuranosylamine

β-Ethoxy-N-ethoxycarbonylacrylamide

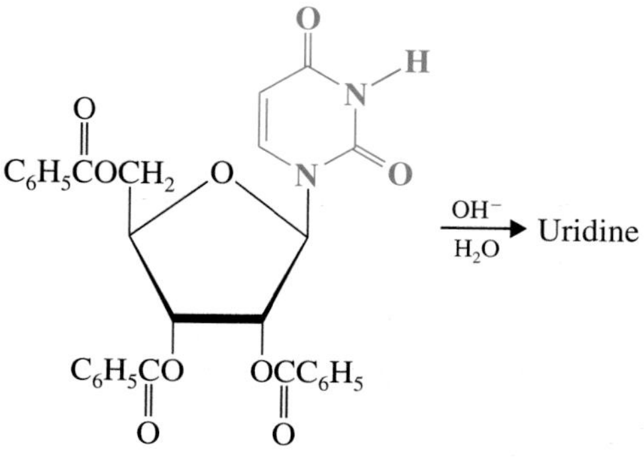

PROBLEM 25.4

Basing your answer on reactions that you have seen before, propose a likely mechanism for the condensation reaction in the first step of the preceding uridine synthesis.

Still a third technique involves the synthesis of a nucleoside with a substituent in the heterocyclic ring that can be replaced with other groups. This method has been used extensively to synthesize unusual nucleosides that do not necessarily occur naturally. The following example makes use of a 6-chloropurine derivative obtained from the appropriate ribofuranosyl chloride and chloromercuripurine.

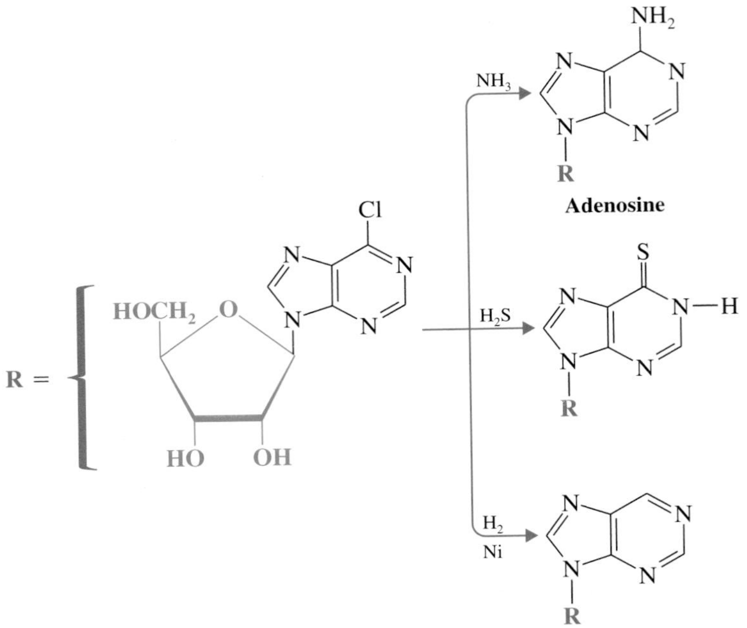

Numerous phosphorylating agents have been used to convert nucleosides to nucleotides. One of the most useful is dibenzyl phosphochloridate.

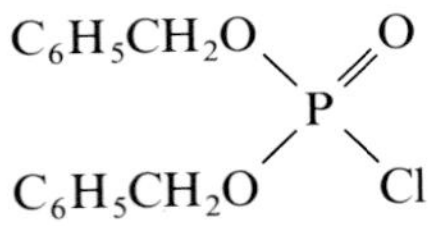

Dibenzyl phosphochloridate

Specific phosphorylation of the 5'-OH can be achieved if the 2'- and 3'-OH groups of the nucleoside are protected by an isopropylidene group (see following figure).

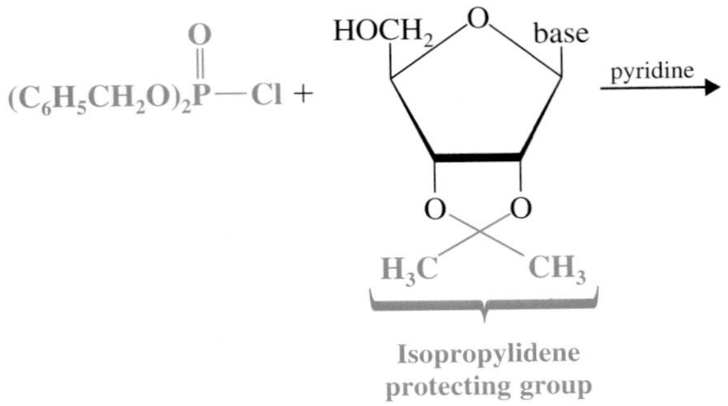

Isopropylidene
protecting group

Mild acid-catalyzed hydrolysis removes the isopropylidene group, and hydrogenolysis cleaves the benzyl phosphate bonds.

PROBLEM 25.5

(a) What kind of linkage is involved in the isopropylidene-protected nucleoside and why is it susceptible to mild acid-catalyzed hydrolysis?   (b) How might such a protecting group be installed?

25.3A Medical Applications

In the early 1950s, Gertrude Elion and George Hitchings (of the Wellcome Research Laboratories) discovered that 6-mercaptopurine had antitumor and antileukemic properties. This discovery led to the development of other purine derivatives and related compounds, including nucleosides, of considerable medical importance.* Three examples are the following.

6-Mercaptopurine   Allopurinol   Acyclovir

6-Mercaptopurine is used in combination with other chemotherapeutic agents to treat acute leukemia in children, and almost 80% of the children treated are now cured. Allopurinol, another purine derivative, is a standard therapy for the treatment of gout. Acyclovir, a nucleoside that lacks two carbon atoms of its ribose ring, is highly effective in treating diseases caused by certain herpes viruses, including *herpes simplex* type 1 (fever blisters), type 2 (genital herpes), and varicella-zoster (shingles).

---

* Elion and Hitchings shared the 1988 Nobel Prize for their work in the development of chemotherapeutic agents derived from purines.

25.4 DEOXYRIBONUCLEIC ACID: DNA

25.4A Primary Structure

Nucleotides bear the same relation to a nucleic acid that amino acids do to a protein; they are its monomeric units. The connecting links in proteins are amide groups; in nucleic acids they are phosphate ester linkages. Phosphate esters link the 3'-OH of one ribose (or deoxyribose) with the 5'-OH of another. This makes the nucleic acid a long unbranched chain with a "backbone" of sugar and phosphate units with heterocyclic bases protruding from the chain at regular intervals (Fig. 25.5). We would indicate the direction of the bases in Fig. 25.5 in the following way:

$$5' \longleftarrow A-T-G-C \longrightarrow 3'$$

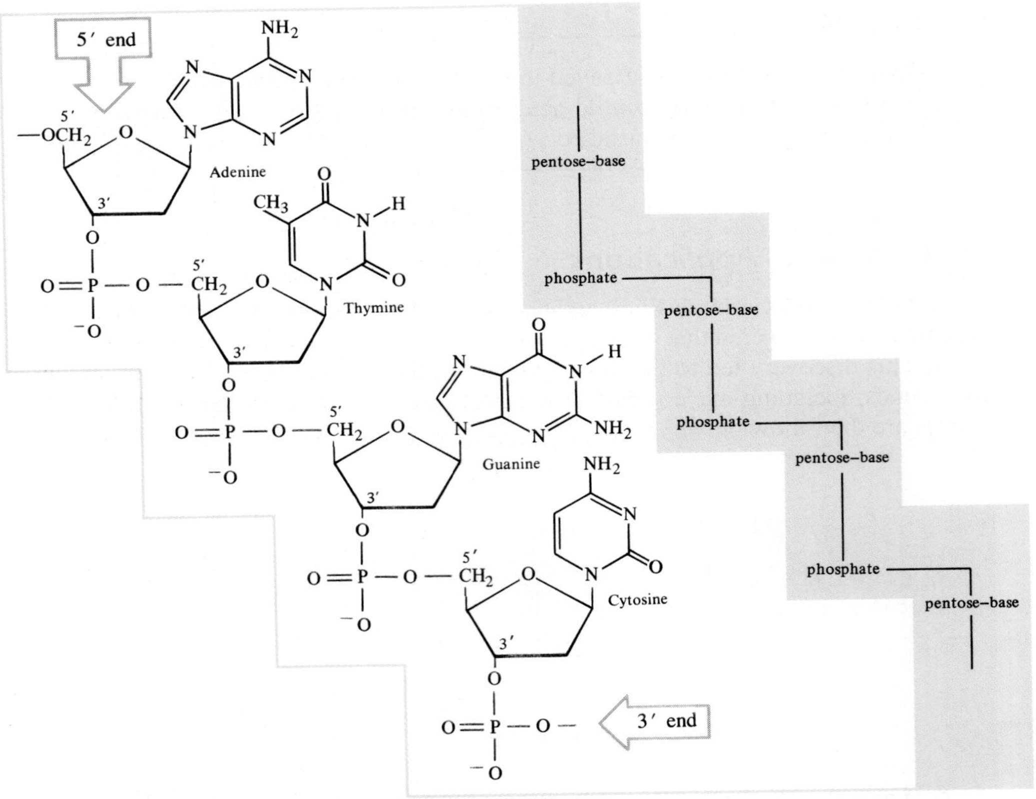

FIGURE 25.5 Hypothetical segment of a single DNA chain showing how phosphate ester groups link the 3'- and 5'-OH groups of deoxyribose units. RNA has a similar structure with two exceptions: A hydroxyl replaces a hydrogen atom at the 2' position of each ribose unit and uracil replaces thymine.

It is, as we shall see, the **base sequence** along the chain of DNA that contains the encoded genetic information. The sequence of bases can be determined through techniques based on selective enzymatic hydrolyses. The actual base sequences have been worked out for a number of nucleic acids (Section 25.6).

25.4B Secondary Structure

It was the now-classic proposal of Watson and Crick (made in 1953 and verified shortly thereafter through the X-ray analysis by Wilkins) that gave a model for the secondary structure of DNA. The secondary structure of DNA is especially important because it enables us to understand how the genetic information is preserved, how it can be passed on during the process of cell division, and how it can be transcribed to provide a template for protein synthesis.

Of prime importance to Watson and Crick's proposal was an earlier observation (late 1940s) by E. Chargaff that certain regularities can be seen in the percentages of heterocyclic bases obtained from the DNA of a variety of species. Table 25.1 gives results that are typical of those that can be obtained.

Chargaff pointed out that for all species examined:

1. The total mole percentage of purines is approximately equal to that of the pyrimidines, that is, $(\%G + \%A)/(\%C + \%T) \simeq 1$.
2. The mole percentage of adenine is nearly equal to that of thymine (i.e., $\%A/\%T \simeq 1$) and the mole percentage of guanine is nearly equal to that of cytosine (i.e., $\%G/\%C \simeq 1$).

Chargaff also noted that the ratio that varies from species to species is the ratio $(\%A + \%T)/(\%G + \%C)$. He noted, moreover, that whereas this ratio is characteristic of the DNA of a given species, it is the same for DNA obtained from different tissues of the same animal, and does not vary appreciably with the age or conditions of growth of individual organisms within the same species.

Watson and Crick also had X-ray data that gave them the bond lengths and angles of the purines and pyrimidines of model compounds. In addition, they had data from Wilkins that indicated an unusually long repeat distance, 34 Å, in natural DNA.

Reasoning from these data, Watson and Crick proposed a double helix as a model for the secondary structure of DNA. According to this model, two nucleic acid chains are held together by hydrogen bonds between base pairs on opposite strands. This

TABLE 25.1   DNA composition of various species

SPECIES	BASE PROPORTIONS (mol %)							
	G	A	C	T	$\dfrac{G + A}{C + T}$	$\dfrac{A + T}{G + C}$	$\dfrac{A}{T}$	$\dfrac{G}{C}$
Sarcina lutea	37.1	13.4	37.1	12.4	1.02	0.35	1.08	1.00
Escherichia coli K12	24.9	26.0	25.2	23.9	1.08	1.00	1.09	0.99
Wheat germ	22.7	27.3	22.8[a]	27.1	1.00	1.19	1.01	1.00
Bovine thymus	21.5	28.2	22.5[a]	27.8	0.96	1.27	1.01	0.96
Staphylococcus aureus	21.0	30.8	19.0	29.2	1.11	1.50	1.05	1.11
Human thymus	19.9	30.9	19.8	29.4	1.01	1.52	1.05	1.01
Human liver	19.5	30.3	19.9	30.3	0.98	1.54	1.00	0.98

[a] Cytosine + methylcytosine.

From *Principles of Biochemistry* by A. White, P. Handler, and E. L. Smith. Copyright © 1964 by McGraw–Hill, Inc. Used with permission of McGraw–Hill Book Company, New York.

double chain is wound into a helix with both chains sharing the same axis. The base pairs are on the inside of the helix and the sugar–phosphate backbone on the outside (Fig. 25.6). The pitch of the helix is such that 10 successive nucleotide pairs give rise to one complete turn in 34 Å (the repeat distance). The exterior width of the spiral is about 20 Å and the internal distance between 1′-positions of ribose units on opposite chains is about 11 Å.

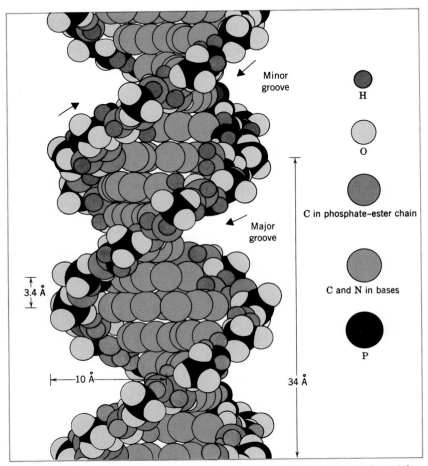

FIGURE 25.6 A molecular model of a portion of the DNA double helix. (Adapted from *Chemistry and Biochemistry: A Comprehensive Introduction* by A. L. Neal. Copyright © 1971 by McGraw–Hill Inc. Used with permission of McGraw–Hill Book Company, New York.)

Using molecular scale models, Watson and Crick observed that the internal distance of the double helix is such that it allows only a purine–pyrimidine type of hydrogen bonding between base pairs. Purine–purine base pairs do not occur because they would be too large to fit, and pyrimidine–pyrimidine base pairs do not occur because they would be too far apart to form effective hydrogen bonds.

Watson and Crick went one crucial step further in their proposal. Assuming that the oxygen-containing heterocyclic bases existed in keto forms, they argued that base pairing through hydrogen bonds can occur in only a specific way:

Adenine pairs with thymine	and	Guanine pairs with cytosine

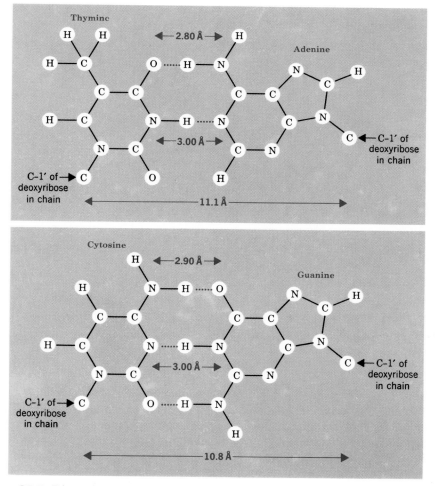

The bond lengths of these base pairs are shown in Fig. 25.7.

Specific base pairing of this kind is consistent with Chargaff's finding that the %A/%T $\simeq$ 1 and that the %G/%C $\simeq$ 1.

FIGURE 25.7 Dimensions of thymine–adenine and cytosine–guanine base pairs. The dimensions are such that they allow the formation of strong hydrogen bonds and also allow the base pairs to fit inside the two phosphate–ribose chains of the double helix. [Adapted from L. Pauling and R. B. Corey, *Arch. Biochem. Biophys., 65,* 164, 1956.]

Specific base pairing also means that the two chains of DNA are complementary. Wherever adenine appears in one chain, thymine must appear opposite it in the other; wherever cytosine appears in one chain, guanine must appear in the other (Fig. 25.8).

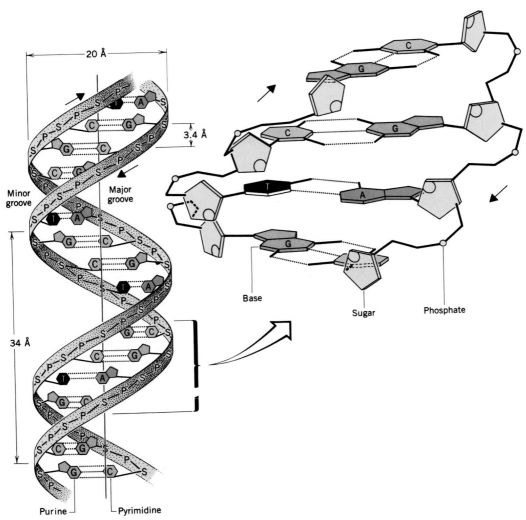

FIGURE 25.8 Diagram of the DNA double helix showing complementary base pairing. The arrows indicate the $3' \rightarrow 5'$ direction.

Notice that while the sugar–phosphate backbone of DNA is completely regular, the sequence of heterocyclic base pairs along the backbone can assume many different permutations. This is important because it is the precise sequence of base pairs that carries the genetic information. Notice, too, that one chain of the double strand is the complement of the other. If one knows the sequence of bases along one chain, one can write down the sequence along the other, because A always pairs with T and G always pairs with C. It is this complementarity of the two strands that explains how a DNA molecule replicates itself at the time of cell division and thereby passes on the genetic information to each of the two daughter cells.

25.4C Replication of DNA

Just prior to cell division the double strand of DNA begins to unwind. Complementary strands are formed along each chain (Fig. 25.9). Each chain acts, in effect, as a template, for the formation of its complement. When unwinding and duplication is complete, there are two identical DNA molecules where only one had existed before. These two molecules can then be passed on, one to each daughter cell.

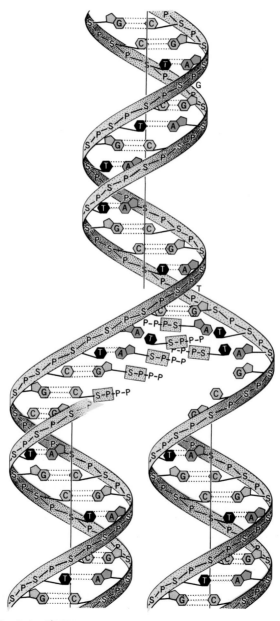

FIGURE 25.9 Replication of DNA. The double strand unwinds from one end and complementary strands are formed along each chain.

PROBLEM 25.6

(a) There are approximately 6 billion base pairs in the DNA of a single human cell. Assuming that this DNA exists as a double helix, calculate the length of all the DNA contained in a human cell.   (b) The weight of DNA in a single human cell is 6×10^{-12} g. Assuming that the earth's population is about 3 billion, we can conclude that all of the genetic information that gave rise to all human beings now alive was once contained in the DNA of a corresponding number of fertilized ova. What is the total weight of this DNA? (The volume that this DNA would occupy is approximately that of a raindrop, yet, if the individual molecules were laid end to end, they would stretch to the moon and back almost eight times.)

PROBLEM 25.7

(a) The most stable tautomeric form of guanine is the lactam form. This is the form normally present in DNA and, as we have seen, it pairs specifically with cytosine. If guanine tautomerizes to the abnormal lactim form, it pairs with thymine instead. Write structural formulas showing the hydrogen bonds in this abnormal base pair.

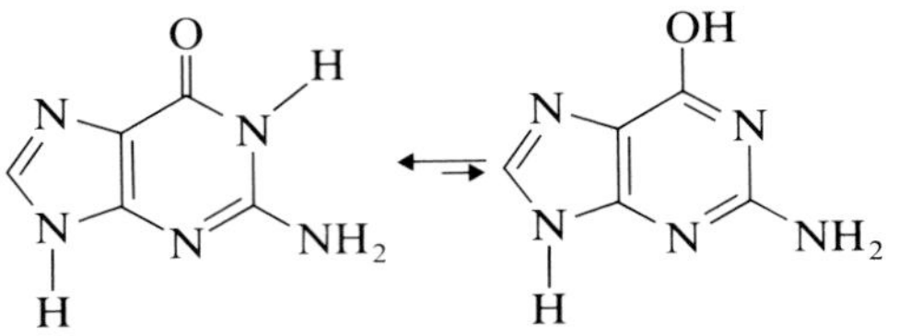

Lactam form
of guanine

Lactim form
of guanine

(b) Improper base pairings that result from tautomerizations occurring during the process of DNA replication have been suggested as a source of spontaneous mutations. We saw in part (a) that if a tautomerization of guanine occurred at the proper moment, it could lead to the introduction of thymine (instead of cytosine) into its complementary DNA chain. What error would this new DNA chain introduce into *its* complementary strand during the next replication even if no further tautomerizations take place?

PROBLEM 25.8

Mutations can also be caused chemically, and nitrous acid is one of the most potent chemical **mutagens.** One explanation that has been suggested for the mutagenic effect of nitrous acid is the deamination reactions that it produces with purines and pyrimidines bearing amino groups. When, for example, an adenine-containing nucleotide is treated with nitrous acid, it is converted to a hypoxanthine derivative:

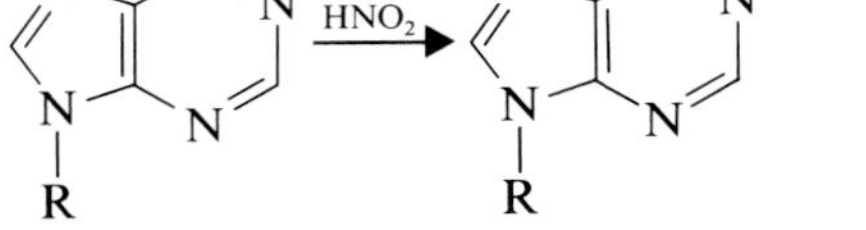

Adenine
nucleotide

Hypoxanthine
nucleotide

(a) Basing your answer on reactions you have seen before, what are likely intermediates in the adenine ⟶ hypoxanthine interconversion? (b) Adenine normally pairs with thymine in DNA, but hypoxanthine pairs with cytosine. Show the hydrogen bonds of a hypoxanthine–cytosine base pair. (c) Show what errors an adenine ⟶ hypoxanthine interconversion would generate in DNA through two replications.

25.5 RNA AND PROTEIN SYNTHESIS

Soon after the Watson–Crick hypothesis was published, scientists began to extend it to yield what Crick called "the central dogma of molecular genetics." This dogma stated that genetic information flows from:

$$DNA \longrightarrow RNA \longrightarrow proteins*$$

The synthesis of protein is, of course, all important to a cell's function because proteins (as enzymes) catalyze its reactions. Even the very primitive cells of bacteria require as many as 3000 different enzymes. This means that the DNA molecules of these cells must contain a corresponding number of genes to direct the synthesis of these proteins. A gene is that segment of the DNA molecule that contains the information necessary to direct the synthesis of one protein (or one polypeptide).

DNA is found primarily in the nucleus of the cell. Protein synthesis takes place primarily in that part of the cell called the *cytoplasm*. Protein synthesis requires that two major processes take place; the first takes place in the cell nucleus, the second in the cytoplasm. The first is **transcription,** a process in which the genetic message is transcribed onto a form of RNA called messenger RNA (mRNA). The second process involves two other forms of RNA, called ribosomal RNA (rRNA) and transfer RNA (tRNA).

25.5A Messenger RNA Synthesis—Transcription

Protein synthesis begins in the cell nucleus with the synthesis of mRNA. Part of the DNA double helix unwinds sufficiently to expose on a single chain a portion corresponding to at least one gene. Ribonucleotides, present in the cell nucleus, assemble along the exposed DNA chain pairing with the bases of DNA. The pairing patterns are the same as those in DNA with the exception that in RNA uracil replaces thymine. The ribonucleotide units of mRNA are joined into a chain by an enzyme called *RNA polymerase*. This process is illustrated in Fig. 25.10.

PROBLEM 25.9

Write structural formulas showing how the keto form of uracil (Section 25.2) in mRNA can pair with adenine in DNA through hydrogen bond formation.

After mRNA has been synthesized in the cell nucleus, it migrates into the cytoplasm where, as we shall see, it acts as a template for protein synthesis.

---

* There are viruses, called retroviruses, in which information flows from RNA to DNA. The virus that causes AIDS is a retrovirus.

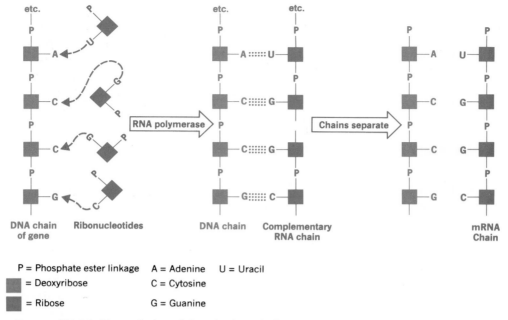

P = Phosphate ester linkage   A = Adenine   U = Uracil
■ = Deoxyribose   C = Cytosine
■ = Ribose   G = Guanine

FIGURE 25.10 Transcription of the genetic code from DNA to mRNA.

25.5B Ribosomes—rRNA

Scattered throughout the cytoplasm of most cells are small bodies called ribosomes. Ribosomes of *Escherichia coli (E. coli),* for example, are about 180 Å in diameter and are composed of approximately 60% RNA (ribosomal RNA) and 40% protein. They apparently exist as two associated subunits called the 50S and 30S subunits (Fig. 25.11); together they form a 70S ribosome.* Although the ribosomes are at the site of protein synthesis, rRNA itself does not direct protein synthesis. Instead, a number of ribosomes become attached to a chain of mRNA and form what is called a **polysome.** It is along the polysome—with mRNA acting as the template—that protein synthesis takes place. One of the functions of rRNA is to bind the ribosome to the mRNA chain.

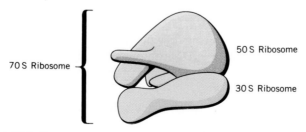

FIGURE 25.11 A 70S ribosome showing the two subunits.

*S stands for svedberg unit; it is used in describing the behavior of proteins in an ultracentrifuge.

25.5C Transfer RNA

Transfer RNA has a very low molecular weight when compared to those of mRNA or rRNA. Transfer RNA, consequently, is much more soluble than mRNA or rRNA and is sometimes referred to as soluble RNA. The function of tRNA is to transport amino acids to specific areas of the mRNA of the polysome. There are, therefore, many forms of tRNA, more than one for each of the 20 amino acids that are incorporated into proteins, including the redundancies in the genetic code (see Table 25.2).*

TABLE 25.2   The messenger RNA genetic code

AMINO ACID	BASE SEQUENCE $5' \rightarrow 3'$	AMINO ACID	BASE SEQUENCE $5' \rightarrow 3'$	AMINO ACID	BASE SEQUENCE $5' \rightarrow 3'$
Ala	GCA	His	CAC	Ser	AGC
	GCC		CAU		AGU
	GCG				UCA
	GCU	Ile	AUA		UCG
			AUC		UCC
Arg	AGA		AUU		UCU
	AGG				
	CGA	Leu	CUA	Thr	ACA
	CGC		CUC		ACC
	CGG		CUG		ACG
	CGU		CUU		ACU
			UUA		
Asn	AAC		UUG	Trp	UGG
	AAU			Tyr	UGG
		Lys	AAA		UAC
Asp	GAC		AAG		UAU
	GAU			Val	GUA
					GUG
Cys	UGC	Met	AUG		GUC
	UGU				GUU
		Phe	UUU		
Gln	CAA		UUC	Chain initiation	
	CAG	Pro	CCA		
			CCC	Met$_f$	AUG
Glu	GAA		CCG		
	GAG		CCU		
Gly	GGA				
	GGC	Chain termination			UAA
	GGG				UAG
	GGU				UGA

*Although proteins are composed of 22 different amino acids, protein synthesis requires only 20. Proline is converted to hydroxyproline and cysteine is converted to cystine after synthesis of the polypeptide chain has taken place.

The structures of most tRNAs have been determined. They are composed of a relatively small number of nucleotide units (70–90 units) folded into several loops or arms through base pairing along the chain (Fig. 25.12). One arm always terminates in the sequence cytosine–cytosine–adenine. It is to this arm that a specific amino acid becomes attached *through an ester* linkage to the 3′-OH of the terminal adenosine. This attachment reaction is catalyzed by an enzyme that is specific for the tRNA and for the amino acid. The specificity may grow out of the enzyme's ability to recognize base sequences along other arms of the tRNA.

At the loop of still another arm is a specific sequence of bases, called the **antico-don.** The anticodon is highly important because it allows the tRNA to bind with a specific site—called the **codon**—of mRNA. The order in which amino acids are

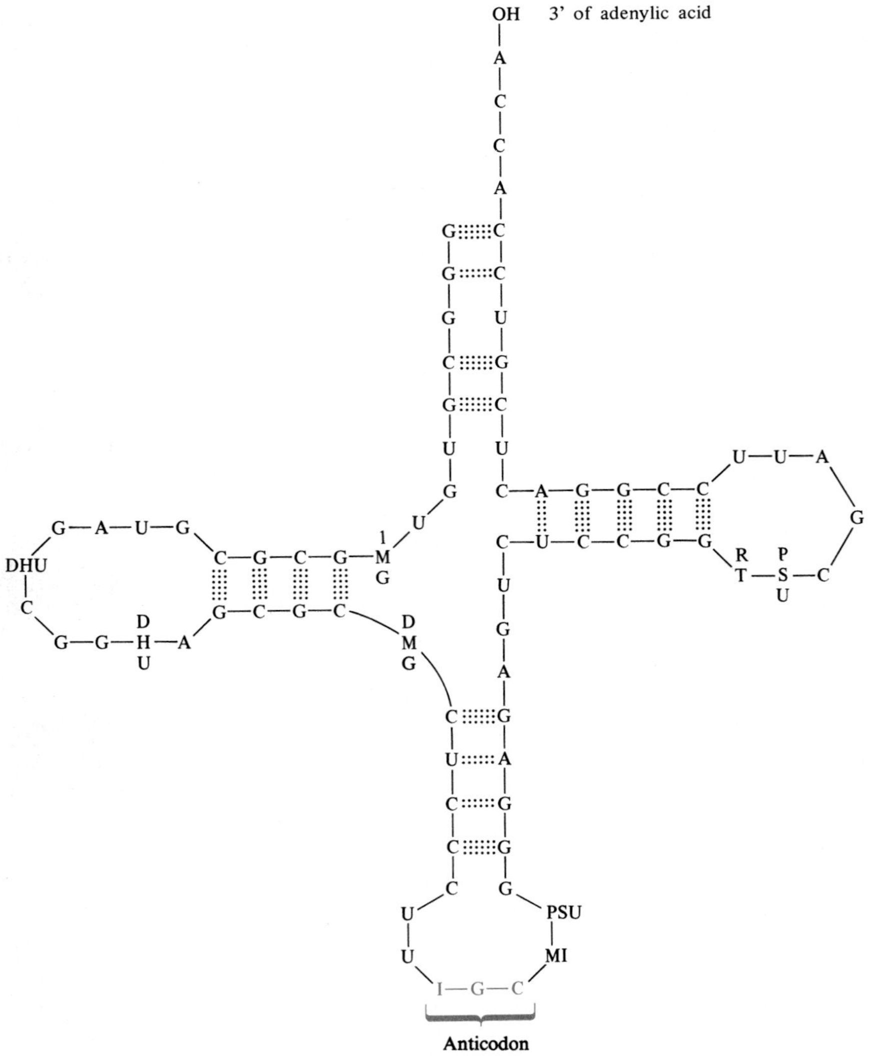

FIGURE 25.12 Structure of a tRNA isolated from yeast that has the specific function of transferring alanine residues. Transfer RNAs often contain unusual nucleosides. PSU = pseudouridine, RT = ribothymidine, MI = 1-methylinosine, I = inosine, DMG = N^2-methylguanosine, DHU = 4,5-dihydrouridine, 1 MG = 1-methylguanosine.

brought by their tRNA units to the mRNA strand is determined by the sequence of codons. This sequence, therefore, constitutes a genetic message. Individual units of that message (the individual words, each corresponding to an amino acid) are triplets of nucleotides.

25.5D The Genetic Code

Which triplet on mRNA corresponds to which amino acid is called the genetic code (see Table 25.2). The code must be in the form of three bases, not one or two, because there are 20 different amino acids used in protein synthesis but there are only four different bases in mRNA. If only two bases were used, there would be only 4^2 or 16 possible combinations, a number too small to accommodate all of the possible amino acids. However, with a three-base code, 4^3 or 64 different sequences are possible. This is far more than are needed, and it allows for multiple ways of specifying an amino acid. It also allows for sequences that punctuate protein synthesis, sequences that say, in effect, "start here" and "end here."

Both methionine (Met) and N-formylmethionine (Met$_f$) have the same mRNA code (AUG); however, N-formylmethionine is carried by a different tRNA from that which carries methionine. N-Formylmethionine appears to be the first amino acid incorporated into the chain of proteins in bacteria, and the tRNA that carries Met$_f$ appears to be the punctuation mark that says "start here." Before the polypeptide synthesis is complete, N-formylmethionine is removed from the protein chain by an enzymatic hydrolysis.

$$CH_3SCH_2CH_2CHCO_2H$$
$$|$$
$$NH$$
$$|$$
$$C{=}O$$
$$|$$
$$H$$

N-Formylmethionine

We are now in a position to see how the synthesis of a hypothetical polypeptide might take place. Let us imagine that a long strand of mRNA is in the cytoplasm of a cell and that it is in contact with ribosomes. Also in the cytoplasm are the 20 different amino acids, each acylated to its own specific tRNA.

As shown in Fig. 25.13, a tRNA bearing Met$_f$ uses its anticodon to associate with the proper codon (AUG) on that portion of mRNA that is in contact with a ribosome. The next triplet of bases on this particular mRNA chain is AAA; this is the codon that specifies lysine. A lysyl-tRNA with the matching anticodon UUU attaches itself to this site. The two amino acids, Met$_f$ and Lys, are now in the proper position for an enzyme to join them in peptide linkage. After this happens, the ribosome moves down the chain so that it is in contact with the next codon. This one, GUA, specifies valine. A tRNA bearing valine (and with the proper anticodon) binds itself to this site. Another enzymatic reaction takes place attaching valine to the polypeptide chain. Then the whole process repeats itself again and again. The ribosome moves along the mRNA chain, other tRNAs move up with their amino acids, new peptide bonds are formed, and the polypeptide chain grows. At some point an enzymatic reaction removes Met$_f$ from the beginning of the chain. Finally, when the chain is the proper length the ribosome reaches a punctuation mark, UAA, saying "stop here." The ribosome separates from the mRNA chain and so, too, does the protein.

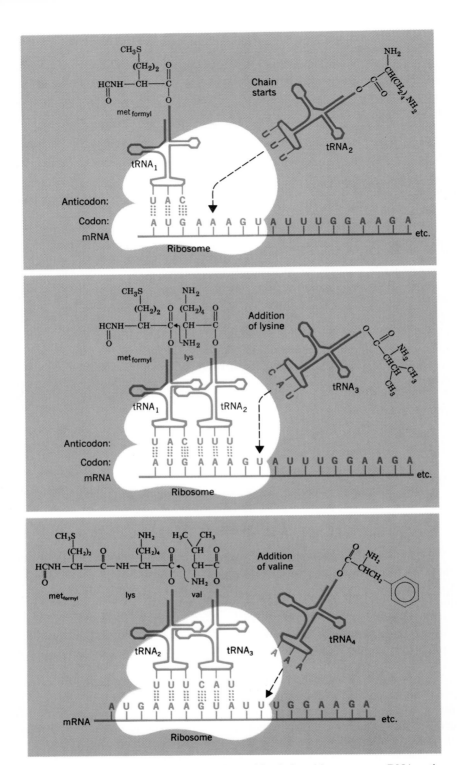

FIGURE 25.13 Step-by-step growth of a polypeptide chain with messenger RNA acting as a template. Transfer RNAs carry amino acid residues to the site of mRNA that is in contact with a ribosome. Codon–anticodon pairing occurs between mRNA and RNA at the ribosomal surface. An enzymatic reaction joins the amino acid residues through an amide linkage. After the first amide bond is formed the ribosome moves to the next codon on mRNA. A new tRNA arrives, pairs, and transfers its amino acid residue to the growing peptide chain, and so on.

Even before the polypeptide chain is fully grown, it begins to form its own specific secondary and tertiary structure (Fig. 25.14). This happens because its primary structure is correct — its amino acids are ordered in just the right way. Hydrogen bonds form, giving rise to specific segments of α helix, pleated sheet, and coil or loop. Then the whole chain folds and bends; enzymes install disulfide linkages, so that when the chain is fully grown, the whole protein has just the shape it needs to do its job.

If this protein happens to be lysozyme, it has a deep cleft, or jaw, where a specific polysaccharide fits. And if it is lysozyme, and a certain bacterium wanders by, that jaw begins to work; it bites its first polysaccharide in half.

In the meantime, other ribosomes nearer the beginning of the mRNA chain are already moving along, each one synthesizing another molecule of the polypeptide. The time required to synthesize a protein depends, of course, on the number of amino acid residues it contains, but indications are that each ribosome can cause 150 peptide bonds to be formed each minute. Thus, a protein, such as lysozyme, with 129 amino acid residues requires less than a minute for its synthesis. However, if four ribosomes are working their way along a single mRNA chain, the polysome can produce a lysozyme molecule every 13 seconds.

But why, we might ask, is all this protein synthesis necessary — particularly in a fully grown organism? The answer is that proteins are not permanent; they are not synthesized once and then left intact in the cell for the lifetime of the organism. They are synthesized when and where they are needed. Then they are taken apart, back to amino acids; enzymes disassemble enzymes. Some amino acids are metabolized for energy; others — new ones — come in from the food that is eaten and the whole process begins again.

PROBLEM 25.10

A segment of DNA has the following sequence of bases:

$$\ldots\; A\; C\; C\; C\; C\; A\; A\; A\; A\; T\; G\; T\; C\; G\; \ldots$$

(a) What sequence of bases would appear in mRNA transcribed from this segment? (b) Assume that the first base in this mRNA is the beginning of a codon. What order of amino acids would be translated into a polypeptide synthesized along this segment? (c) Give anticodons for each tRNA associated with the translation in part (b).

PROBLEM 25.11

(a) Using the first codon given for each amino acid in Table 25.2, write the base sequence of mRNA that would translate the synthesis of the following pentapeptide:

$$\text{Arg} \cdot \text{Ile} \cdot \text{Cys} \cdot \text{Tyr} \cdot \text{Val}$$

(b) What base sequence in DNA would transcribe a synthesis of the mRNA?
(c) What anticodons would appear in the tRNAs involved in the pentapeptide synthesis?

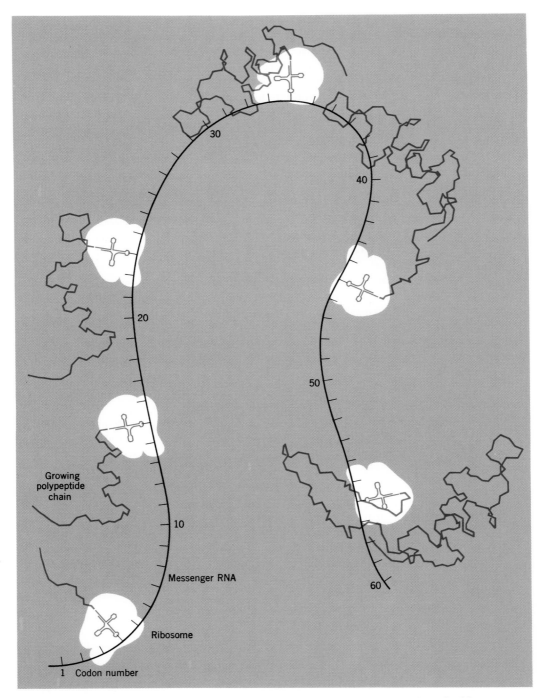

FIGURE 25.14 The folding of a protein molecule as it is synthesized. [Adapted with permission from D. C. Phillips, "The Three-Dimensional Structure of an Enzyme Molecule." *Bio-Organic Chemistry,* M. Calvin and M. J. Jorgenson, Eds., W. H. Freeman and Co., San Francisco, 1968, p. 62. Copyright © 1966 Scientific American, Inc. All rights reserved.]

PROBLEM 25.12 _____

Explain how an error of a single base in each strand of DNA could bring about the amino acid residue error that causes sickle-cell anemia (Section 24.6C).

_____

25.6 DETERMINING THE BASE SEQUENCE OF DNA

The basic strategy used to sequence DNA resembles the methods used to sequence proteins (Section 24.5). Because molecules of DNA are so large, it is first necessary to cleave them into smaller, manageable fragments. These fragments are sequenced individually, and then by identifying points of overlap, the fragments are ordered so as to reveal the nucleotide sequence of the original nucleic acid.

The first part of the process is accomplished by using enzymes called **restriction endonucleases.** These enzymes cleave double-stranded DNA at specific base sequences. Several hundred restriction endonucleases are now known. One, for example, called *Alu*I, cleaves the sequence AGCT between G and C. Another, called *Eco*RI, cleaves GAATTC between G and A. Most of the sites recognized by restriction enzymes have sequences of base pairs with the same order in both strands when read from the 5′ direction to the 3′ direction. For example:

$$5' \longleftarrow G-A-A-T-T-C \longrightarrow 3'$$
$$3' \longleftarrow C-T-T-A-A-G \longrightarrow 5'$$

Such sequences are known as **palindromes.** (Palindromes are words or sentences that read the same forward or backward. Examples are ''radar,'' or ''Madam, I'm Adam.'')

Sequencing of the fragments (called restriction fragments) can be done chemically (a method described below) or with the aid of enzymes. The first chemical method was introduced in 1977 by Allan M. Maxam and Walter Gilbert of Harvard University; the first enzymatic method was introduced in the same year by Frederick Sanger.*

25.6A Chemical Sequencing

The double-stranded restriction fragment to be sequenced is first enzymatically tagged at the 5′ end with a phosphate group containing radioactive phosphate. After this, the strands are separated and isolated. Next, the labeled, single-stranded fragment is treated with reagents that attack specific bases, modifying them in a way that allows cleavage of the chain next to the specific bases. For example, if we had a chain like the following (reading from 5′ → 3′, left to right),

$$^{32}P-GCAATCACGTC$$

treating the fragment with hydrazine (NH_2NH_2) in 1.5M NaCl, will (in a way that we cannot go into here) attack cytosine residues so that subsequent treatment with piperidine (Section 20.1B) will cause cleavage at the 5′ side of C residues. This will produce the following set of 5′-labeled fragments.

_____

* Gilbert and Sanger shared the Nobel Prize in 1980 with Paul Berg for their work on nucleic acids. Sanger (Section 24.5), who pioneered the sequencing of proteins, had won an earlier Nobel Prize in 1958 for the determination of the structure of insulin.

$$^{32}\text{P—GCAATCACGT}$$
$$^{32}\text{P—GCAATCA}$$
$$^{32}\text{P—GCAAT}$$
$$^{32}\text{P—G}$$

These fragments can then be separated by a technique called **gel electrophoresis** (Fig. 25.15). A sample containing a mixture of the fragments is placed at one end of the thin strip of a gel made up of a polyacrylamide [$-\!\!\!(\text{CH}_2\text{CHCONH}_2)_n\!\!-$]. The gel is designed to separate the radiolabeled fragments when a voltage difference is applied across the ends of the gel. The fragments move through the gel at different rates depending on the number of negatively charged phosphate groups that they contain and on their size. The smaller fragments move faster. After separation, the gel is placed in contact with a photographic plate. Radiation from a fragment containing a radioactive 5'-phosphate causes a dark spot to appear on the plate opposite where the fragment is located in the gel. The exposed plate is called an **autoradiograph** and this technique is called **autoradiography.** Unlabeled fragments from the middle of the chain are present, but these do not show up on the plate and are therefore ignored.

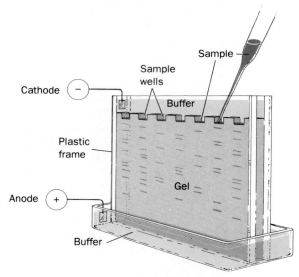

FIGURE 25.15 An apparatus for gel electrophoresis. Samples are applied in the slots at the top of the gel. Application of a voltage difference causes the samples to move. The samples move in parallel lanes. (From D. Voet and J. G. Voet, *Biochemistry,* 2nd ed., Wiley, New York, 1995. Used with permission.)

The DNA to be sequenced may be cleaved next to specific pairs by subjecting it in separate aliquots to four different treatments. In addition to cleavage next to **C only,** mentioned earlier, there are reagents that cleave on the 5' side of **G only,** on the 5' side of **A and G,** and on the 5' side of **C and T.** After cleavage, the separate aliquots are subjected to simultaneous electrophoresis in four parallel tracks of the gel. After autoradiography, results such as those in Fig. 25.16 allow reading of the DNA sequence directly from the gel.

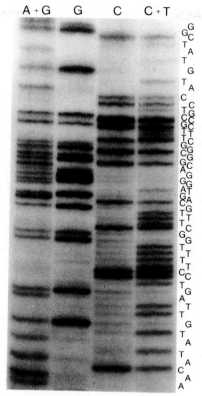

FIGURE 25.16 An autoradiograph of a sequencing gel containing fragments of a DNA segment that was subjected to chemical sequencing. The DNA was labeled with ^{32}P at its 5′ end. The deduced sequence of the DNA fragment is written beside the gel. Since the shorter fragments are at the bottom of the gel, the 5′ ⟶ 3′ direction in the sequence corresponds to the upward direction in the gel. (Courtesy of David Dressler, Harvard University Medical School. From D. Voet and J. G. Voet, *Biochemistry,* 2nd ed., Wiley, New York, 1995. Used with permission.)

Starting from the bottom: There is a dark spot in the A + G track but none in the G track. This indicates that the smallest labeled fragment was A. The same pattern occurs at the second level, indicating another A. The third level up is dark in the C track and light in the C + T track, indicating C as the third base in the sequence. The fourth is A again, and so on. The process is so regular that computerized devices are now used to aid in reading gels.

Developments in DNA sequencing have been so rapid that DNA sequencing of the gene corresponding to a protein is now the easier method for determining a protein's amino acid sequence. (Since the genetic code is known, we can deduce the amino acid sequence of the protein from the base sequence of the DNA that codes for the protein.) A recent high point in DNA sequencing was the determination of the entire 172,282 base pair sequence of Epstein–Barr virus (human herpes virus). Plans are now being made to sequence the 2.9 billion base pairs of the 100,000 genes that constitute the human genome. This task is so large, however, that even if a sequencing rate of 1 million base pairs a day can be achieved, it will take almost 10 years to complete.

25.7   LABORATORY SYNTHESIS OF OLIGONUCLEOTIDES

A gene is a blueprint for a protein that is encoded in a particular sequence of base pairs of DNA. What do the individual genes do, and how do they do it? These are the kinds of questions being addressed by biologists and biochemists today. In doing so, they are using a completely new approach to studying genetics, called **reverse genetics.** The traditional approach to genetics involves randomly altering or deleting genes by inducing mutations in an organism, and then observing the effects in its progeny. With higher organisms, such as vertebrates, there are serious disadvantages to this approach. The generations are inconveniently long, the number of offspring are few, and the most interesting mutations are usually lethal, making them difficult to propagate and study.

The approach of reverse genetics is to start with a cloned gene and to manipulate it in order to find out how it functions. One approach to this manipulation is to synthesize strands of DNA (oligonucleotides with about 15 bases) that are complementary to particular portions of the gene. These synthetic oligonucleotides, called **antisense nucleotides,** are capable of binding with what is called the **sense** sequence of the DNA. In doing so, they can alter the activity of the gene, or even turn it off entirely. For example, if the sense portion of DNA in a gene read:

$$A-G-A-C-C-G-T-G-G$$

The antisense oligonucleotide would read:

$$T-C-T-G-G-C-A-C-C$$

The ability to deactivate specific genes in this way holds out great medical promise. Many viruses and bacteria, during their life cycles, use a method like this to regulate some of their own genes. The hope, therefore, is to synthesize antisense oligonucleotides that will seek out and destroy viruses in a person's cells by binding with crucial sequences of the viral DNA or RNA. Synthesis of such oligonucleotides is an active area of research today and is directed at many viral diseases, including AIDS.

Current methods for oligonucleotide synthesis are similar to those used to synthesize proteins, including the use of automated solid phase techniques (Section 24.7D). A suitably protected nucleotide is attached to a solid phase called a "controlled pore glass" or CPG (Fig. 25.17) through a linkage that can ultimately be cleaved. The next protected nucleotide in the form of a **phosphoramidite** is added, and coupling is brought about by a coupling agent, usually 1,2,3,4-tetrazole. The phosphite triester that results from the coupling is oxidized to phosphate triester with iodine, producing a chain that has been lengthened by one nucleotide. The **dimethoxytrityl (DMTr)** group used to protect the 5′ end of the added nucleotide is removed by treatment with acid and the steps **coupling, oxidation, detritylation** are repeated. (All the steps are carried out in nonaqueous solvents.) With automatic synthesizers the process can be repeated at least 50 times and the time for a complete cycle is 40 min or less. After the desired oligonucleotide has been synthesized, it is released from the solid support and the various protecting groups, including those on the bases, are removed.

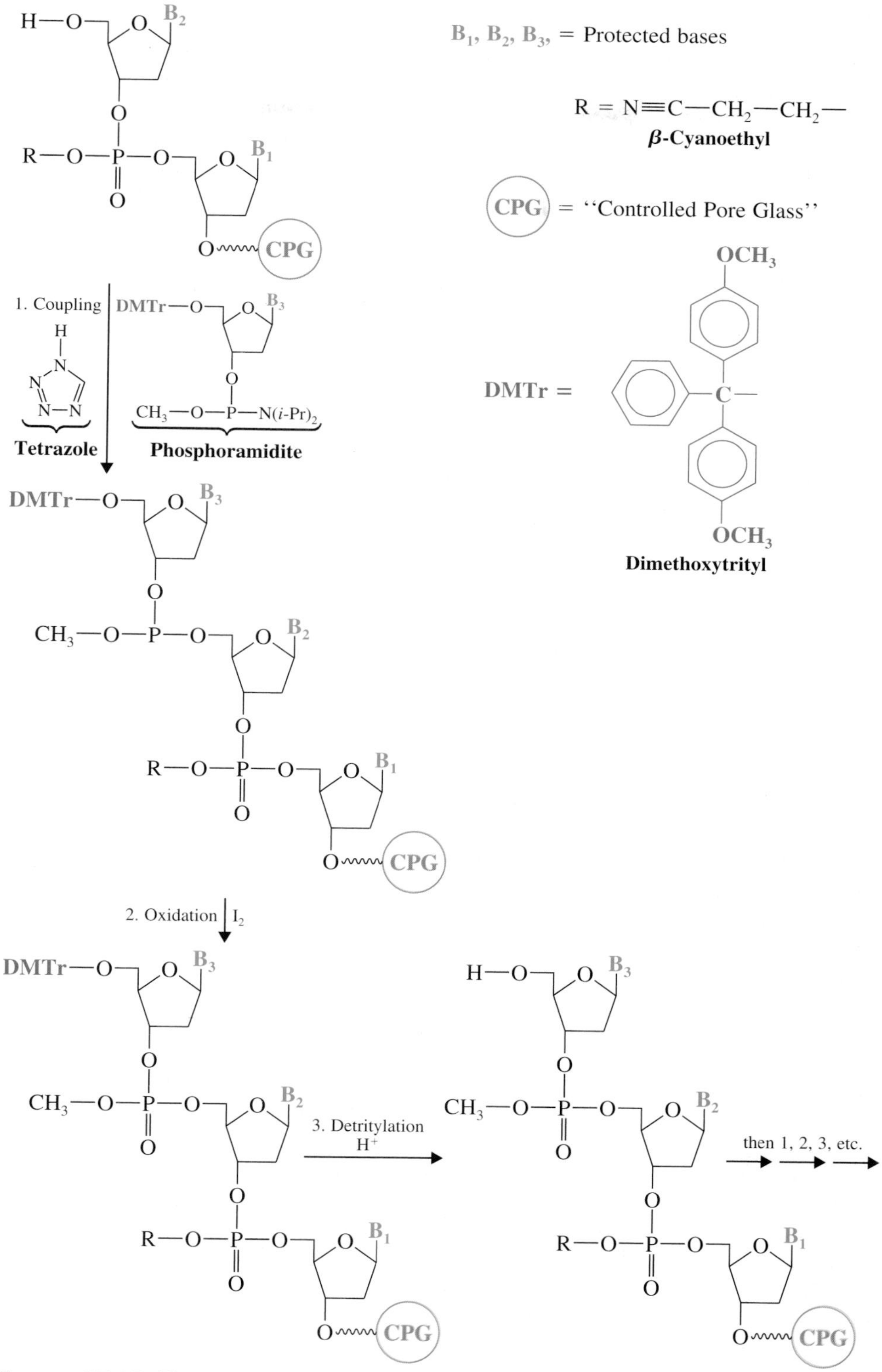

B1, B2, B3, = Protected bases
R = N≡C—CH2—CH2—
β-Cyanoethyl
CPG = "Controlled Pore Glass"
DMTr =
Dimethoxytrityl
1. Coupling
Tetrazole
Phosphoramidite
CH3—O—P—N(i-Pr)2
2. Oxidation I2
3. Detritylation H+
then 1, 2, 3, etc.

FIGURE 25.17 The steps involved in automated synthesis of oligonucleotides using the phosphoramidite coupling method.

25.8 THE POLYMERASE CHAIN REACTION

The polymerase chain reaction (PCR) is an extraordinarily simple and effective method for amplifying DNA sequences. Beginning with a single molecule of DNA, the polymerase chain reaction can generate 100 billion copies in a single afternoon. The reaction is easy to carry out; it requires only a few reagents, a test tube, and a source of heat.

The PCR has already had a major effect on molecular biology. It is being used in medicine to diagnose infectious and genetic diseases. One of the original aims in developing the PCR was to use it in increasing the speed and effectiveness of prenatal diagnosis of sickle-cell anemia (Section 24.6C). It is now being applied to the prenatal diagnosis of a number of other genetic diseases, including muscular dystrophy and cystic fibrosis. Among infectious diseases, the PCR has been used to detect cytomegalovirus and the viruses that cause AIDS, certain cervical carcinomas, hepatitis, measles, and Epstein–Barr disease.

The PCR has been used in forensic sciences, in human genetics, and in evolutionary biology. The DNA sample that is copied may have come from a drop of blood or semen, or from a hair left at the scene of a crime. It may even have come from the brain of a mummy or from a 40,000-year-old wooly mammoth.

The PCR was invented by Kary B. Mullis* and developed by him and his coworkers at Cetus Corporation. It makes use of the enzyme DNA polymerase, discovered in 1955 by Arthur Kornberg and his associates at Stanford University. In living cells, DNA polymerases help repair and replicate DNA. The PCR makes use of a particular property of DNA polymerases: Their ability to attach additional nucleotides to a short oligonucleotide "primer" when the primer is bound to a complementary strand of DNA called a template. The nucleotides are attached at the 3' end of the primer and the nucleotide that the polymerase attaches will be the one that is complementary to the base in the adjacent position on the template strand. If the adjacent template nucleotide is G, the polymerase adds C to the primer; if the adjacent template nucleotide is A, then the polymerase adds T, and so on. Polymerase repeats this process again and again as long as the requisite nucleotides (as triphosphates) are present in the solution, until it reaches the 5' end of the template.

Figure 25.18 shows one cycle of the PCR in the way that it is usually carried out. Knowledge of the nucleotide sequence of the target of the PCR is not necessary. However, one must know the sequence of a small stretch on each side of the target in order to synthesize two single-stranded oligonucleotides (of ~ 20 nucleotides) that will act as the primers. The primers have to have nucleotide sequences that are complementary to the flanking sequences on each strand of DNA.

At the outset of the PCR, the double-stranded (duplex) DNA is heated to separate its strands. The primers (one for each strand) are added and annealed to their respective flanking sequences. DNA polymerase and nucleotide triphosphates are then added and the polymerase causes each primer to become extended across the target sequence of each strand. If the extension of a given primer is long enough, it will include the sequence complementary to the other primer. Consequently, each new extension product can, after the strands are separated, act as the template for another cycle.

Each cycle doubles the amount of target DNA (Fig. 25.19). This means that the amount of DNA increases exponentially. After n cycles, the amount of DNA will have increased 2^n times. After 10 cycles, there is roughly 1000 times as much DNA; after 20 cycles, roughly 1 million times as much. Application of the PCR is extremely rapid and has been automated; 25 cycles can be carried out in 1 hour.

* Mullis was awarded the Nobel Prize for Chemistry for this work in 1993.

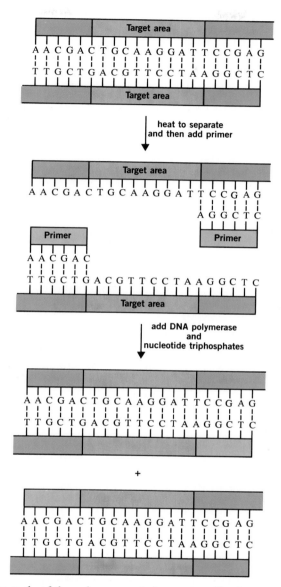

FIGURE 25.18 One cycle of the polymerase chain reaction. Heating separates the strands of DNA of the target to give two single-stranded templates. Primers, designed to complement the nucleotide sequences flanking the targets, anneal to each strand. DNA polymerase, in the presence of nucleotide triphosphates, catalyzes the synthesis of two pieces of DNA, each identical to the original target DNA.

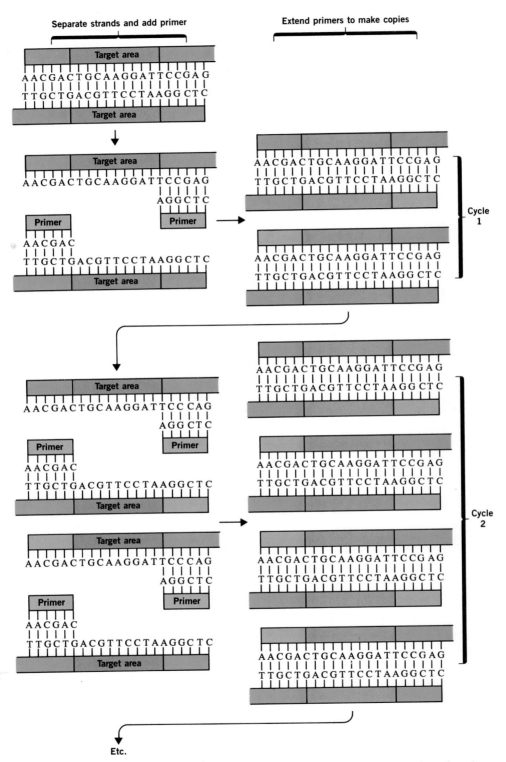

FIGURE 25.19 Each cycle of the polymerase chain reaction doubles the number of copies of the target area.

ANSWERS TO SELECTED PROBLEMS

CHAPTER 1

1.8 (a), (c), (f), (g) are tetrahedral; (e) is trigonal planar; (b) is linear; (d) is angular; (h) is trigonal pyramidal.

1.11 VSEPR theory predicts a trigonal structure for BF_3; this causes the bond moments to cancel.

1.13 The fact that SO_2 has a dipole moment indicates that the molecule is angular, not linear.

1.18 (a) and (d); (b) and (e); and (c) and (f).

1.26 (a), (g), (i), (l), represent different compounds that are not isomeric; (c–e), (h), (j), (m), (n), (o) represent the same compound; (b), (f), (k), (p) represent constitutional isomers.

1.33 (a) The structures differ in the positions of the nuclei.

1.35 (a) A negative charge; (b) a negative charge; (c) trigonal pyramidal.

CHAPTER 2

2.1 cis–trans Isomers are not possible for (a) and (c).

2.5 (c) Propyl bromide; (d) isopropyl fluoride; (e) phenyl iodide.

2.8 (a) $CH_3CH_2OCH_2CH_3$; (b) $CH_3CH_2OCH_2CH_2CH_3$; (e) diisopropyl ether.

2.12 (a) $CH_3CH_2CH_2CH_2OH$ would boil higher because its molecules can form hydrogen bonds to each other; (c) $HOCH_2CH_2CH_2OH$ would boil higher because it can form more hydrogen bonds.

2.14 (a) Ketone; (c) alcohol; (e) alcohol.

2.15 (a) 3 Alkene, and a 2° alcohol; (c) phenyl and 1° amine; (e) phenyl, ester and 3° amine; (g) alkene and 2 ester groups.

2.21 (f) $CH_3CH_2CH_2CH_2CH_2Br$; $(CH_3)_2CHCH_2CH_2Br$; $CH_3CH_2CH(CH_3)CH_2Br$; $(CH_3)_3CCH_2Br$.

2.24 Ester

CHAPTER 3

3.2 (a), (c), (d), and (f) are Lewis bases; (b) and (e) are Lewis acids.

3.4 (a) $[H_3O^+] = [HCO_2^-] = 0.013\ M$; (b) Ionization = 1.3%.

3.5 (a) $pK_a = 7$; (b) $pK_a = -0.7$; (c) Because the acid with a $pK_a = 5$ has a larger K_a, it is the stronger acid.

3.7 The pK_a of the methylaminium ion is equal to 10.6 (Section 3.3C). Because the pK_a of the anilinium ion is equal to 4.6, the anilinium ion is a stronger acid than the methylaminium ion, and aniline ($C_6H_5NH_2$) is a weaker base than methylamine (CH_3NH_2).

3.11 (a) $CHCl_2CO_2H$ would be the stronger acid because the electron-withdrawing inductive effect of two chlorine atoms would make its hydroxyl proton more positive. (c) CH_2FCO_2H would be the stronger acid because a fluorine atom is more electronegative than a bromine atom and would be more electron withdrawing.

3.23 (a) $pK_a = 3.752$; (b) $K_a = 10^{-13}$.

CHAPTER 4

4.5 (a) 1-(1-Methylethyl)-2-(1,1-dimethylethyl)cyclopentane or 1-*tert*-butyl-2-isopropylcyclopentane; (c) butylcyclohexane; (e) 2-chlorocyclopentanol.

4.6 (a) 2-Chlorobicyclo[1.1.0]butane; (c) bicyclo[2.1.1]hexane; (e) 2-methylbicyclo[2.2.2]octane.

4.18 (a) 3,3,4-Trimethylhexane;
(c) 3,5,7-trimethylnonane;
(e) 2-bromobicyclo[3.3.1]nonane;
(g) cyclobutylcyclopentane.

4.21 The alkane is 2-methylpentane,
$(CH_3)_2CHCH_2CH_2CH_3$.

4.23 The alkane is 2,3-dimethylbutane.

4.25 $(CH_3)_3CCH_3$ is the most stable isomer.

4.32 (a) Pentane would boil higher
because its chain is unbranched.
(c) 2-Chloropropane because it is more
polar and has a higher molecular weight.
(e) CH_3COCH_3 because its molecules are
more polar.

4.34 (a) The trans isomer is more stable
because both methyl groups can be
equatorial. (c) The trans isomer is more
stable because both methyl groups can be
equatorial.

4.39 (a) From Table 4.10 we find that this is
trans-1,2-dichlorocyclohexane. (b) The
chlorines must have added from opposite
sides of the double bond.

CHAPTER 5

5.1 (a) Achiral; (c) chiral; (e) chiral;
(i) chiral.

5.2 (a) Yes; (c) no.

5.3 (a) They are the same. (b) They are
enantiomers.

5.7 The following possess a plane of
symmetry and are, therefore, achiral:
screwdriver, baseball bat, hammer.

5.11 (a) $-Cl > -SH > -OH > -H$;
(c) $-OH > -CHO > -CH_3 > -H$;
(e) $-OCH_3 > -N(CH_3)_2 > -CH_3 >$
$-H$

5.13 (a) Enantiomers; (c) enantiomers.

5.17 (a) Diastereomers; (c) no; (e) no.

5.19 (a) Represents **A**; (b) represents **C**;
(c) represents **B**.

5.21 **B** (2S,3S)-2,3-dibromobutane;
C (2R,3S)-2,3-dibromobutane.

5.35 (a) Same; (c) diastereomers; (e) same;
(g) diastereomers; (i) same;
(k) diastereomers; (m) diastereomers;
(o) diastereomers; (q) same.

CHAPTER 6

6.3 (a) The reaction is S_N2 and, therefore,
occurs with inversion of configuration.
Consequently, the configuration of (+)-2-
chlorobutane is opposite [i.e., (*S*)] to that
of (−)-2-butanol [i.e., (*R*)]. (b) The
configuration of (−)-2-iodobutane is (*R*).

6.7 Protic solvents are formic acid,
formamide, ammonia, and ethylene glycol.
The others are aprotic.

6.9 (a) CH_3O^-; (c) $(CH_3)_3P$.

6.14 (a) 1-Bromopropane would react more
rapidly, because, being a primary halide, it
is less hindered. (c) 1-Chlorobutane, because
the carbon bearing the leaving group is less
hindered than in 1-chloro-2-methylpropane.
(e) 1-Chlorohexane because it is a primary
halide. Phenyl halides are unreactive in S_N2
reactions.

6.15 (a) Reaction (1) because ethoxide
ion is a stronger nucleophile than
ethanol; (c) reaction (2) because
triphenylphosphine, $(C_6H_5)_3P$, is a stronger
nucleophile than triphenylamine.
(Phosphorus atoms are larger than nitrogen
atoms.)

6.16 (a) Reaction (2) because bromide ion is
a better leaving group than chloride ion; (c)
reaction (2) because the concentration of the
substrate is twice that of reaction (1).

6.19 The best yield is obtained by using the
secondary halide, 1-bromo-1-phenylethane,
because the desired reaction is E2. Using the
primary halide will result in substantial S_N2
reaction as well, producing the alcohol
instead of the desired alkene.

6.29 (a) You should use a strong base, such
as RO^-, at a higher temperature to bring
about an E2 reaction. (b) Here we want an
S_N1 reaction. We use ethanol as the solvent
and as the nucleophile, and we carry out the
reaction at a low temperature so that
elimination is minimized.

CHAPTER 7

7.1 (a) *trans*-3-Heptene; (c) 4-ethyl-2-methyl-1-hexene

7.2 (a)

(c)

(e)

(g)

(i)

7.4

(*R*)-3-Methyl-
1-pentyne

(*S*)-3-Methyl-
1-pentyne

7.8 (a) 2,3-Dimethyl-2-butene would be the more stable because the double bond is tetrasubstituted. (c) *cis*-3-Hexene would be more stable because its double bond is disubstituted.

7.10 (a)

(trisubstituted,
more stable)

(monosubstituted
less stable)

Major product   **Minor product**

7.21 (a) We designate the position of the double bond by using the *lower* of the two numbers of the doubly bonded carbon atoms, and the chain is numbered from the end nearer the double bond. The correct name is *trans*-2-pentene. (c) We use the lower number of the two doubly bonded carbon atoms to designate the position of the double bond. The correct name is 1-methylcyclohexene.

7.22 (a)

(c)

(e)

(g)

7.24 (a) (*E*)-3,5-Dimethyl-2-hexene;
(c) 6-methyl-3-hexyne; (e) (*Z*,5*R*)-5-chloro-3-heptene-6-yne.

7.40 Only the deuterium atom can assume the antiperiplanar orientation necessary for an E2 reaction to occur.

CHAPTER 8

8.1 2-Bromo-1-iodopropane.

8.7 The order reflects the relative ease with which these alkenes accept a proton and form a carbocation. 2-Methylpropene reacts fastest because it leads to a 3° cation; ethene reacts slowest because it leads to a 1° cation.

8.13 (a) *syn*-Hydroxylation at either face of the (*Z*)-isomer leads to the meso compound (2*R*,3*S*)-2,3-butanediol. (b) *syn*-Hydroxylation at one face of the (*E*)-isomer leads to (2*R*,3*R*)-2,3-butanediol; at the other face, which is equally likely, it leads to the (2*S*,3*S*)-enantiomer.

8.18 (a) $CH_3CH_2CHICH_3$;
(c) $CH_3CH_2CH_2CH_3$;
(e) $CH_3CH_2CH(OH)CH_3$;
(h) $CH_3CH_2CH=CH_2$;
(j) $CH_3CH_2CHClCH_3$;
(l) $CH_3CH_2CHO + HCHO$.

8.20 (a) $CH_3CH_2CBr=CHBr$;
(c) $CH_3CH_2CBr_2CH_3$; (e) $CH_3CH_2CH=CH_2$.

8.23

(a)
$$CH_3\underset{\underset{CH_3}{|}}{C}{=}CH_2 \xrightarrow{H_3O^+, H_2O} CH_3\underset{\underset{OH}{|}}{\overset{\overset{CH_3}{|}}{C}}CH_3$$

(c)
$$CH_3\underset{\underset{CH_3}{|}}{C}{=}CH_2 \xrightarrow[\text{(no peroxides)}]{HBr} CH_3\underset{\underset{Br}{|}}{\overset{\overset{CH_3}{|}}{C}}CH_3$$

(d)
$$CH_3\underset{\underset{CH_3}{|}}{C}{=}CH_2 \xrightarrow{HF} CH_3\underset{\underset{F}{|}}{\overset{\overset{CH_3}{|}}{C}}CH_3$$

8.47

$$HC{\equiv}C\underset{\underset{CH_2CH_3}{|}}{\overset{\overset{CH_3}{|}}{C}}H \xrightarrow[\text{Pt}]{H_2} CH_3CH_2\overset{\overset{CH_3}{|}}{C}HCH_2CH_3$$

D   **E**

CHAPTER 9

9.1 (a) $\Delta H° = -130$ kcal mol^{-1};
(c) $\Delta H° = -24.5$ kcal mol^{-1}; (e) $\Delta H° = +13$ kcal mol^{-1}; (g) $\Delta H° = -31.5$ kcal mol^{-1}.

9.5 A small amount of ethane is formed by the combination of two methyl radicals; the ethane then reacts with chlorine to form chloroethane.

9.15 (a) Cyclopentane; (c) 2,2-dimethylpropane.

9.17

(a)
$$\underset{CH_3}{\overset{\overset{CH_3}{\vdots}}{\underset{\underset{\underset{\underset{CH_3}{|}}{CH_2}}{|}}{\underset{\underset{CH_2}{|}}{C}}}}\quad H\diagup\diagdown Cl \xrightarrow[\text{light}]{Cl_2}$$

$$\underset{CH_3}{\overset{\overset{CH_3}{\vdots}}{\underset{\underset{\underset{\underset{CH_3}{|}}{\underset{Cl\diagup\diagdown H}{C}}}{|}}{\underset{\underset{CH_2}{|}}{C}}}}\quad H\diagup\diagdown Cl \quad + \quad \underset{CH_3}{\overset{\overset{CH_3}{\vdots}}{\underset{\underset{\underset{\underset{CH_3}{|}}{\underset{H\diagup\diagdown Cl}{C}}}{|}}{\underset{\underset{CH_2}{|}}{C}}}}\quad H\diagup\diagdown Cl$$

(2S,4S)-2,4-Dichloro-   **(2R,4S)-2,4-Dichloro-**
pentane   **pentane**

(c) No, (2*R*,4*S*)-2,4-dichloropentane is achiral because it is a meso compound. (It has a plane of symmetry passing through C3.) (e) Yes, by fractional distillation or by gas–liquid chromatography. (Diastereomers have different physical properties. Therefore, the two isomers would have different vapor pressures.)

9.18 (a) Seven fractions; (c) none of the fractions would show optical activity.

9.19 (a) No, the only fractions that would contain chiral molecules (as enantiomers) would be those containing 1-chloro-2-methylbutane and the one containing 2-chloro-3-methylbutane. These fractions would not show optical activity, however, because they would contain racemic forms of the enantiomers. (b) Yes, the fraction containing 1-chloro-2-methylbutane and the one containing 2-chloro-3-methylbutane.

9.24 $CH_3\underset{\underset{CH_3}{|}}{\overset{\cdot}{C}}CH_2CH_3 > CH_3\overset{\cdot}{\underset{\underset{CH_3}{|}}{C}}HCHCH_3$
$\qquad\quad$ **(3°)** $\qquad\qquad\qquad$ **(2°)**

$> CH_3\underset{\underset{CH_3}{|}}{C}HCH_2CH_2\cdot \sim CH_3\overset{\cdot}{C}HCH_2\underset{\underset{}{|}}{\overset{\overset{CH_2\cdot}{|}}{}}CH_3$
$\qquad\qquad$ **(1°)** $\qquad\qquad\qquad$ **(1°)**

CHAPTER 10

10.3 The presence of two —OH groups in the glycols allows their molecules to form more hydrogen bonds.

10.4 (a) CH_3CH_2OH; (c) $(CH_3)_3COH$.

10.13 Use an alcohol containing labeled oxygen. If all of the labeled oxygen appears in the sulfonate ester, then it can be concluded that the alcohol C—O bond does not break during the reaction.

10.31 (a) 3,3-Dimethyl-1-butanol; (c) 2-methyl-1,4-butanediol; (e) 1-methyl-2-cyclopenten-1-ol.

10.32 (a)

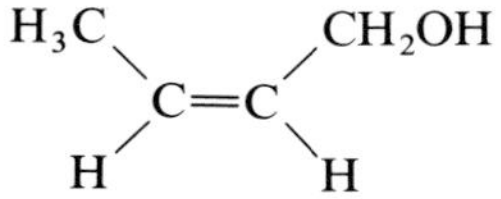

(c)

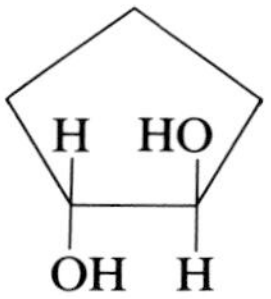

(e) $CH_3CH_2C{\equiv}CCHCH_2OH$
         |
         Cl

(g) $CH_3CHCH_2CH_2CH_3$
        |
        OCH_2CH_3

(i) $CH_3CH{-}O{-}CHCH_3$
        | |
        CH_3 CH_3

10.38 (a) $CH_3Br + CH_3CH_2Br$; (c) $Br{-}CH_2CH_2CH_2CH_2{-}Br$

10.41 (a)

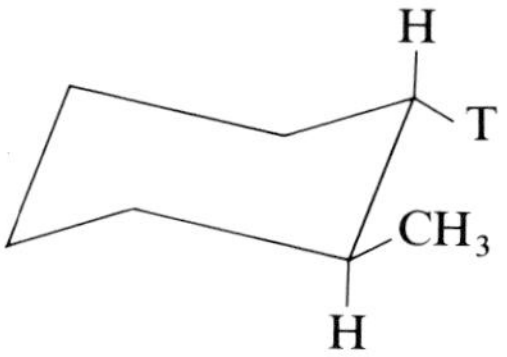

(c)

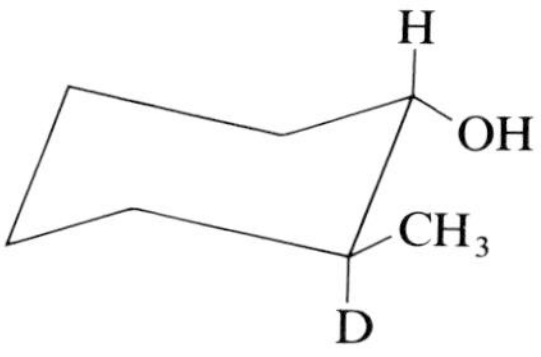

CHAPTER 11

11.4 (a) $LiAlH_4$; (c) $NaBH_4$

11.5

(a)

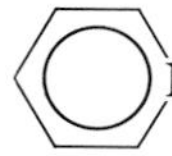

$NH^+CrO_3Cl^-$(PCC)/CH_2Cl_2

(c) H_2CrO_4/acetone

11.10 (a) $CH_3CH_2CH_2OH \xrightarrow[CH_2Cl_2]{PCC}$

11.12 (a) CH_3CH_3; (b) CH_3CH_2D;

(c) $C_6H_5CHCH_2CH_3$
         |
         OH

(g) $CH_3CH_3 + CH_3CH_2C{\equiv}C{-}CHCH_3$
                                          |
                                          OH

(h) $CH_3CH_3 +$

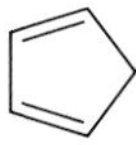

—MgBr

11.13 (a) $(CH_3)_2CHCHCH_2CH_2CH_3$
                           |
                           OH

(b) $(CH_3)_2CHCCH_2CH_2CH_3$
                 |
                 CH_3 (OH above)

(e) $CH_3CH_2CH_2CH_2CH{=}CH_2$

(f) $CH_3CH_2CH_2{-}$

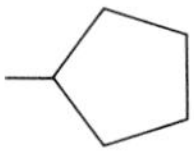

11.21

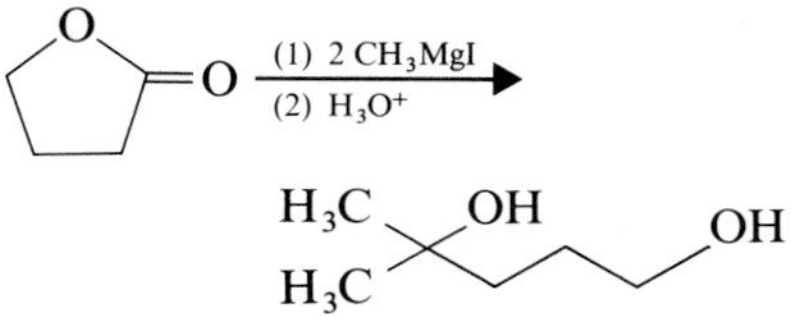

$\xrightarrow[(2)\ H_3O^+]{(1)\ 2\ CH_3MgI}$

CHAPTER 12

12.1 (a) $^{14}CH_2=CH-CH_2-X +$ $X-^{14}CH_2-CH=CH_2$; (c) in equal amounts.

12.6 (b) 1,4-Cyclohexadiene is an isolated diene.

12.13 (a) 1,4-Dibromobutane + $(CH_3)_3COK$, and heat; (g) $HC\equiv CCH=CH_2 + H_2$, Ni_2B (P-2).

12.16 (a) 1-Butene + N-bromosuccinimide, then $(CH_3)_3COK$ and heat; (e) cyclopentane + Br_2, $h\nu$, then $(CH_3)_3COK$ and heat, then N-bromosuccinimide.

12.18 (a) $Ag(NH_3)_2OH$; (c) H_2SO_4; (e) $AgNO_3$ in alcohol.

12.25 This is another example of rate versus equilibrium control of a reaction. The endo adduct, **G**, is formed faster, and at the lower temperature it is the major product. The exo adduct, **H**, is more stable, and at the higher temperature it is the major product.

CHAPTER 13

13.4 (a) One; (b) two; (c) two; (d) one; (e) two; (f) two.

13.9 A doublet (3H) downfield; a quartet (1H) upfield.

13.10 (a) CH_3CHICH_3; (b) CH_3CHCl_2; (c) $CH_2ClCH_2CH_2Cl$

13.15 Phenylacetylene.

13.16 **G**, $CH_3CH_2CHBrCH_3$
H, $CH_2=CBrCH_2Br$

13.21 **Q** is bicyclo[2.2.1]hepta-2,5-diene.

CHAPTER 14

14.1 Compounds (a) and (b).

14.8 It suggests that the cyclopropenyl cation should be aromatic.

14.10 The cyclopropenyl cation.

14.15 **A**, o-Bromotoluene; **B**, p-bromotoluene; **C**, m-bromotoluene; **D**, benzyl bromide.

14.18 Hückel's rule should apply to both pentalene and heptalene. Pentalene's antiaromaticity can be attributed to its having 8 π electrons. Heptalene's lack of aromaticity can be attributed to its having 12 π electrons. Neither 8 nor 12 is a Hückel number.

14.20 The bridging $-CH_2-$ group causes the 10 π electron ring system (below) to become planar. This allows the ring to become aromatic.

14.23 (a) The cycloheptatrienyl anion has 8 π electrons, and does not obey Hückel's rule; the cyclononatetraenyl anion with 10 π electrons obeys Hückel's rule.

14.25 (a) $C_6H_5CH(CH_3)_2$;
(b) $C_6H_5CH(NH_2)CH_3$;

(c)

14.27

CHAPTER 15

15.8 If the methyl group had no directive effect on the incoming electrophile, we would expect to obtain the products in purely statistical amounts. Since there are two ortho hydrogen atoms, two meta hydrogen atoms, and one para hydrogen, we would expect to get 40% ortho (2/5), 40% meta (2/5), and 20% para (1/5). Thus, we would expect that only 60% of the mixture of mononitrotoluenes would have the nitro group in the ortho or para position. And, we would expect to obtain 40% of meta-nitrotoluene. In actuality, we get 96% of combined o- and p-nitrotoluene and only 4% m-nitrotoluene. This result shows the ortho–para directive effect of the methyl group.

15.11 (b) Structures such as the following compete with the benzene ring for the oxygen electrons, making them less available to the benzene ring.

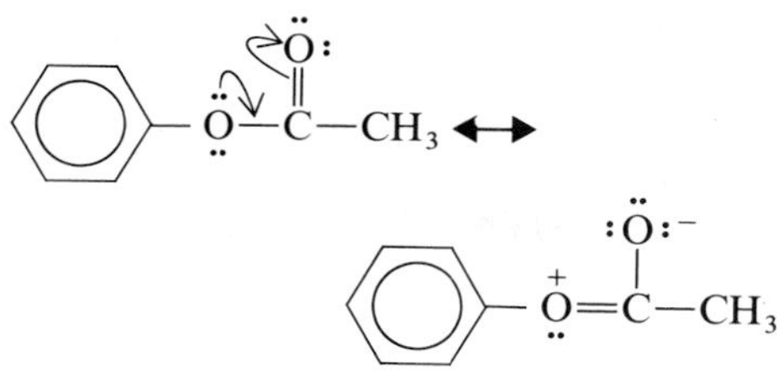

(d) Structures such as the following compete with the benzene ring for the nitrogen electrons, making them less available to the benzene ring.

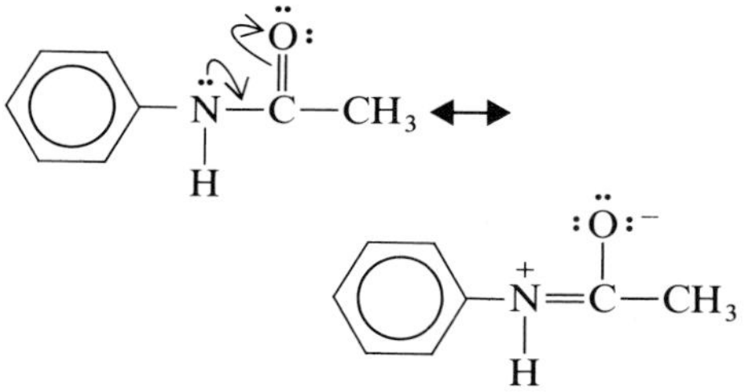

15.28 (a)

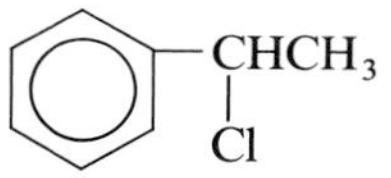

(c)

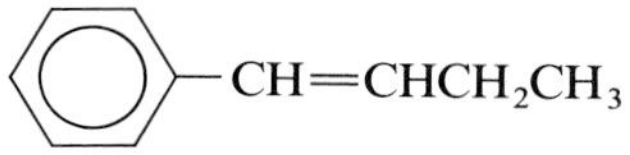

(e)

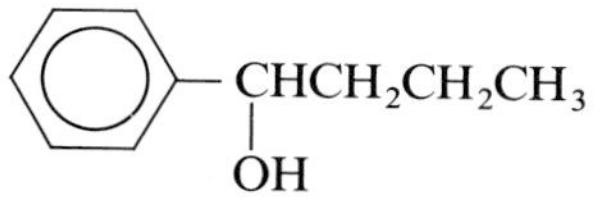

(g)

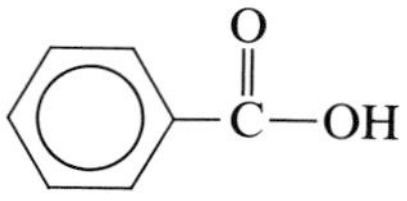

15.35 (a)

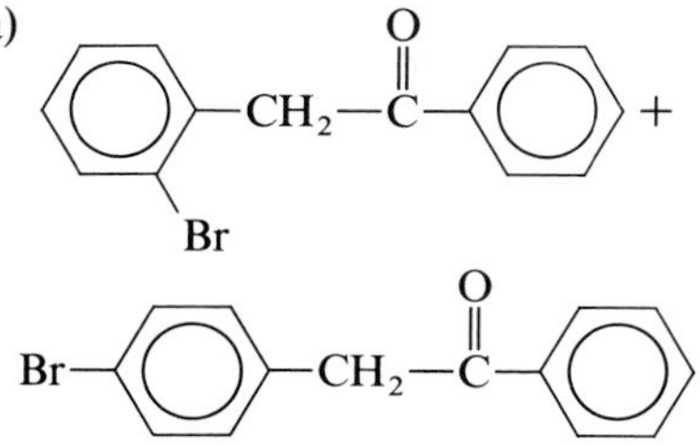

(c)

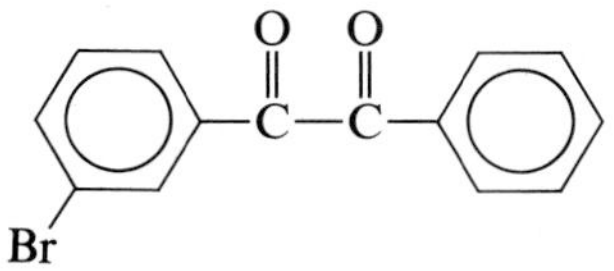

CHAPTER 16 _____

16.2 (a) 1-Pentanol; (c) pentanal; (e) benzyl alcohol.

16.6 A hydride ion.

16.17 (b) $CH_3CH_2Br + (C_6H_5)_3P$, then strong base, then $C_6H_5COCH_3$;
(d) $CH_3I + (C_6H_5)_3P$, then strong base, then cyclopentanone;
(f) $CH_2=CHCH_2Br + (C_6H_5)_3P$, then strong base, then C_6H_5CHO.

16.24 (a) $CH_3CH_2CH_2OH$
(c) $CH_3CH_2CH_2OH$
(h) $CH_3CH_2CH=CHCH_3$
(j) $CH_3CH_2CO_2^-NH_4^+ + Ag\downarrow$
(l) $CH_3CH_2CH=NNHCONH_2$
(n) $CH_3CH_2CO_2H$

16.35 (a) Tollens' reagent; (e) Br_2 in CCl_4; (f) Tollens' reagent; (h) Tollens' reagent.

16.41

X   is

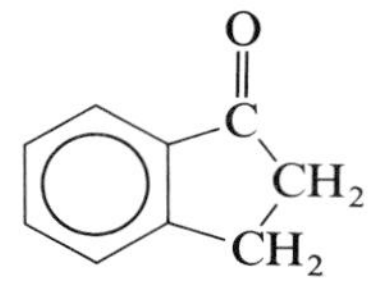

16.42 **Y** is 1-phenyl-2-butanone; **Z** is 4-phenyl-2-butanone.

CHAPTER 17 _____

17.1 The enol form is phenol. It is especially stable because it is aromatic.

17.4 Base is consumed as the reaction takes place. A catalyst, by definition, is not consumed.

17.6 (a), (b), (d), (f), (h), (i).

17.11 $C_6H_5CHO + OH^- \xrightarrow[heat]{CH_3CHO}$

$$C_6H_5CH=CHCHO$$

17.14 (b) 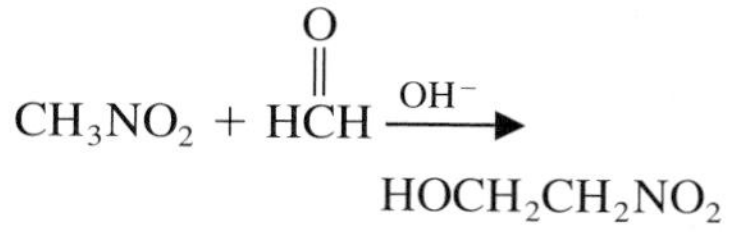 $CH_3NO_2 + H\overset{O}{\underset{\parallel}{C}}H \xrightarrow{OH^-}$

$$HOCH_2CH_2NO_2$$

17.24 (a) $CH_3CH_2CH(OH)\underset{\underset{CH_3}{|}}{C}HCHO$

(b) $C_6H_5CH=\underset{\underset{CH_3}{|}}{C}CHO$

(k) $CH_3CH_2CH(OH)C_6H_5$
(l) $CH_3CH_2CH(OH)C\equiv CH$

17.29 B is

$$CH_3\overset{\overset{\displaystyle O}{\|}}{C}-\overset{\overset{\displaystyle CH_3}{|}}{\underset{\underset{\displaystyle OH}{|}}{C}}-CH_3$$

CHAPTER 18

18.3 (a) CH_2FCO_2H; (c) CH_2ClCO_2H; (e) $CH_3CH_2CHClCO_2H$;

(g) CF_3—⟨○⟩—CO_2H

18.6 (a) $C_6H_5CH_2Br$ + Mg + diethyl ether, then CO_2, then H_3O^+; (c) CH_2=$CHCH_2Br$ + Mg + diethyl ether, then CO_2, then H_3O^+.

18.7 (a), (c), and (e).

18.10 In the carboxyl group of benzoic acid.

18.15 (a) $(CH_3)_3CCO_2H$ + $SOCl_2$, then NH_3, then P_4O_{10}, heat; (b) CH_2=$\overset{\underset{\displaystyle CH_3}{|}}{C}$—$CH_3$

18.24 (a) CH_3CO_2H
(c) $CH_3CO_2CH_2(CH_2)_2CH_3$
(e) p-$CH_3COC_6H_4CH_3$ + o-$CH_3COC_6H_4CH_3$
(g) CH_3COCH_3
(i) $CH_3CONHCH_3$
(k) $CH_3CON(CH_3)_2$
(m) $(CH_3CO)_2O$
(o) $CH_3CO_2C_6H_5$

18.31 (a) $NaHCO_3$ in H_2O; (c) $NaHCO_3$ in H_2O; (e) OH^- in H_2O, heat, detect NH_3 with litmus paper; (g) $AgNO_3$ in alcohol.

18.36 (a) Diethyl succinate; (c) ethyl phenylacetate; (e) ethyl chloroacetate.

18.40 X is diethyl malonate.

CHAPTER 19

19.4 (a) $CH_3\overset{\underset{\displaystyle CO_2C_2H_5}{|}}{CH}COCO_2C_2H_5$

(b) $H\overset{\overset{\displaystyle O}{\|}}{C}CH_2CO_2C_2H_5$

19.7 O-alkylation that results from the oxygen of the enolate ion acting as a nucleophile.

19.9 (a) Reactivity is the same as with any S_N2 reaction. With primary halides substitution is highly favored, with secondary halides elimination competes with substitution, and with tertiary halides elimination is the exclusive course of the reaction. (b) Acetoacetic ester and 2-methylpropene. (c) Bromobenzene is unreactive toward nucleophilic substitution.

19.30 (b) **D** is racemic *trans*-1,2-cyclopentanedicarboxylic acid, **E** is *cis*-1,2-cyclopentanedicarboxylic acid, a meso compound.

19.39 (a) CH_2=$C(CH_3)CO_2CH_3$
(b) $KMnO_4$, OH^-; H_3O^+
(c) CH_3OH, H^+
(d) CH_3ONa, then H^+
(e) and (f)

and

(g) OH^-, H_2O, then H_3O^+
(h) heat ($-CO_2$)
(i) CH_3OH, H^+
(j) Zn, $BrCH_2CO_2CH_3$, diethyl ether, then H_3O^+
(k)

(l) H_2, Pt
(m) CH_3ONa, then H^+
(n) 2 $NaNH_2$ + 2 CH_3I

CHAPTER 20

20.5 (a) $CH_3(CH_2)_3CHO$ + NH_3 $\xrightarrow{H_2, Ni}$ $CH_3(CH_2)_3CH_2NH_2$

(c) $CH_3(CH_2)_4CHO$ + $C_6H_5NH_2$ $\xrightarrow{H_2, Ni}$ $CH_3(CH_2)_4CH_2NHC_6H_5$

20.6 The reaction of a secondary halide with ammonia is almost always accompanied by some elimination.

20.8 (a) Methoxybenzene + HNO_3 + H_2SO_4, then Fe + HCl; (b) Methoxybenzene + CH_3COCl + $AlCl_3$, then NH_3 + H_2 + Ni; (c) toluene + Cl_2 and light, then $(CH_3)_3N$; (d) p-nitrotoluene + $KMnO_4$ + OH^-, then H_3O^+, then $SOCl_2$ followed by NH_3, then NaOBr (Br_2 in NaOH); (e) toluene + N-bromosuccinimide in CCl_4, then KCN, then $LiAlH_4$.

20.14 p-Nitroaniline + Br_2 + Fe, followed by $H_2SO_4/NaNO_2$ followed by CuBr, then Fe/HCl, then $H_2SO_4/NaNO_2$ followed by H_3PO_2.

20.37 **W** is N-benzyl-N-ethylaniline.

CHAPTER 21

21.4 (a) The para-sulfonated phenol. (b) For ortho sulfonation.

21.9 (a)

OCH₂CH₃ ... NO₂ ... NO₂   (b)   NH₂ ... NO₂ ... NO₂

21.10 That o-chlorotoluene leads to the formation of two products (o-cresol and m-cresol), when submitted to the conditions used in the Dow process, suggests that an elimination–addition mechanism takes place.

21.11 2-Bromo-1,3-dimethylbenzene, because it has no α-hydrogen atom, cannot undergo an elimination. Its lack of reactivity toward sodium amide in liquid ammonia suggests that those compounds (e.g., bromobenzene) that do react, react by a mechanism that begins with an elimination.

21.14 (a) 4-Chlorophenol will dissolve in aqueous NaOH; 4-chloro-1-methylbenzene will not. (c) Vinyl phenyl ether will react with bromine in carbon tetrachloride by addition (thus decolorizing the solution); ethyl phenyl ether will not. (e) 4-Ethylphenol will dissolve in aqueous NaOH; ethyl phenyl ether will not.

21.16 (a) 4-Fluorobenzene because a fluorine substituent is more electron withdrawing than a methyl group. (e) 4-Fluorophenol because fluorine is more electronegative than bromine.

CHAPTER 22

22.1 (a) Two; (b) two; (c) four.

22.5 Acid catalyzes hydrolysis of the glycosidic (acetal) group.

22.9 (a) 2 CH_3CHO, one molar equivalent HIO_4
(b) HCHO + HCO_2H + CH_3CHO, two molar equivalents HIO_4
(c) HCHO + $OHCCH(OCH_3)_2$, one molar equivalent HIO_4
(d) HCHO + HCO_2H + CH_3CO_2H, two molar equivalents HIO_4
(e) 2 CH_3CO_2H + HCO_2H, two molar equivalents, HIO_4

22.18 D-(+)-Glucose.

22.23 One anomeric form of D-mannose is dextrorotatory ($[\alpha]_D = +29.3°$), the other is levorotatory ($[\alpha]_D = -17.0°$).

22.24 The microorganism selectively oxidizes the —CHOH group of D-glucitol that corresponds to C5 of D-glucose.

22.27 **A** is D-altrose; **B** is D-talose, **C** is D-galactose.

CHAPTER 23

23.5 Br_2 in CCl_4 would be decolorized by geraniol but not by menthol.

23.12 (a) C_2H_5OH, H^+, heat or $SOCl_2$, then C_2H_5OH
(d) $SOCl_2$, then $(CH_3)_2NH$
(g) $SOCl_2$, then $LiAlH[OCC(CH_3)_3]$
(j) $SOCl_2$, then $(CH_3)_2CuLi$

23.15 Elaidic acid is trans-9-octadecenoic acid.

23.19 **A** is $CH_3(CH_2)_5C \equiv CNa$
B is $CH_3(CH_2)_5C \equiv CCH_2(CH_2)_7CH_2Cl$
C is $CH_3(CH_2)_5C \equiv CCH_2(CH_2)_7CH_2CN$
E is $CH_3(CH_2)_5C \equiv CCH_2(CH_2)_7CH_2CO_2H$
Vaccenic acid is

$$CH_3(CH_2)_5 \diagdown \diagup (CH_2)_9CO_2H$$
$$C=C$$
$$H H$$

23.20 **F** is $FCH_2(CH_2)_6CH_2C \equiv CH$
G is $FCH_2(CH_2)_6CH_2C \equiv C(CH_2)_7Cl$
H is $FCH_2(CH_2)_6CH_2C \equiv C(CH_2)_7CH_2CN$
I is $FCH_2(CH_2)_7C \equiv C(CH_2)_7CO_2H$

CHAPTER 24

24.5 The labeled amino acid no longer has a basic —NH_2 group; it is, therefore, insoluble in aqueous acid.

24.8 Glutathione is

$$\overset{+}{H_3}NCHCH_2CH_2CONHCHCONHCH_2CO_2H$$
$$||$$
$$CO_2^{-}CH_2SH$$

24.20
   Arg·Pro·Pro·Gly·Phe·Ser·Pro·Phe·Arg

24.21 Val·Leu·Lys·Phe·Ala·Glu·Ala

CHAPTER 25

25.2 (a) The nucleosides have an *N*-glycosidic linkage that (like an *O*-glycosidic linkage) is rapidly hydrolyzed by aqueous acid, but one that is stable in aqueous base.

25.3 The reaction appears to take place through an S_N2 mechanism. Attack occurs preferentially at the primary 5' carbon atom rather than at the secondary 3' carbon atom.

25.5 (a) The isopropylidene group is a cyclic acetal. (b) By treating the nucleoside with acetone and a trace of acid.

25.7 (b) Thymine would pair with adenine, and, therefore, adenine would be introduced into the complementary strand where guanine should occur.

25.12 A change from C-T-T to C-A-T, or a change from C-T-C to C-A-C.

GLOSSARY

Absolute configuration (Section 5.14A): The actual arrangement of groups in a molecule. The absolute configuration of a molecule can be determined by X-ray analysis or by relating the configuration of a molecule, using reactions of known stereochemistry, to another molecule whose absolute configuration is known.

Absorption spectrum (Section 13.2): A plot of the wavelength (λ) of a region of the spectrum versus the absorbance (A) at each wavelength. The absorbance at a particular wavelength (A_λ) is defined by the equation $A_\lambda = \log (I_R/I_S)$, where I_R is the intensity of the reference beam and I_S is the intensity of the sample beam.

Acetal (Section 16.7C): A functional group, consisting of a carbon bonded to alkoxy groups [i.e., $RCH(OR')_2$ or $R_2C(OR')_2$], derived by adding 2 molar equivalents of an alcohol to an aldehyde or ketone.

Acetylene (Sections 7.1 and 7.3A): A common name for ethyne. Also used as a general name for alkynes.

Acetylenic hydrogen atom (Section 7.3): A hydrogen atom attached to a carbon atom that is bonded to another carbon atom by a triple bond.

Achiral molecule (Section 5.2): A molecule that is superposable on its mirror image. Achiral molecules lack handedness and are incapable of existing as a pair of enantiomers.

Acid strength (Section 3.5): The strength of an acid is related to its acidity constant, K_a or to its pK_a. The larger the value of its K_a or the smaller the value of its pK_a, the stronger is the acid.

Acidity constant (Section 3.5A): An equilibrium constant related to the strength of an acid. For the reaction,

$$HA + H_2O \rightleftharpoons H_3O^+ + A^-$$

$$K_a = \frac{[H_3O^+][A^-]}{[HA]}$$

Activating group (Section 15.10): A group that when present on a benzene ring causes the ring to be more reactive in electrophilic substitution than benzene itself.

Activation Energy, E_{act}; see **Energy of activation.**

Acyl group (Section 15.7): The general name for groups with the structure RCO— or ArCO—.

Acyl halide (Section 15.7): Also called an *acid halide*. A general name for compounds with the structure RCOX or ArCOX.

Acylation (Section 15.8): The introduction of an acyl group into a molecule.

Acylium ion (Section 15.7): The resonance-stabilized cation:

$$R-\overset{+}{C}=\overset{\cdot\cdot}{\underset{\cdot\cdot}{O}}: \longleftrightarrow R-C\equiv\overset{+}{\underset{\cdot\cdot}{O}}:$$

Addition polymer (Section 9.10): A polymer that results from a stepwise addition of monomers to a chain (usually through a chain reaction) with no loss of other atoms or molecules in the process.

Addition reaction (Section 8.1): A reaction that results in an increase in the number of groups attached to a pair of atoms joined by a double or triple bond. An addition reaction is the opposite of an elimination reaction.

Aglycone (Section 22.4): The alcohol obtained by hydrolysis of a glycoside.

Aldaric acid (Section 22.6C): An α,ω-dicarboxylic acid that results from oxidation of the aldehyde group and the terminal 1° alcohol group of an aldose.

Alditol (Section 22.7): The alcohol that results from the reduction of the aldehyde or keto group of an aldose or ketose.

Aldonic acid (Section 22.6B): A monocarboxylic acid that results from oxidation of the aldehyde group of an aldose.

Aliphatic compound (Section 14.1): A nonaromatic compound such as an alkane, cycloalkane, alkene, or alkyne.

Alkaloid (Special Topic K): A naturally occurring basic compound that contains an amino group. Most alkaloids have profound physiological effects.

Alkylation (Section 15.7): The introduction of an alkyl group into a molecule.

Allyl group (Section 7.2): The $CH_2\!=\!CHCH_2-$ group.

Allylic substituent (Section 12.2): Refers to a substituent on a carbon atom adjacent to a carbon–carbon double bond.

Ambident nucleophile (Section H.1A): Nucleophiles that are capable of reacting at two different nucleophilic sites.

Angle strain (Section 4.9): The increased potential energy of a molecule (usually a cyclic one) caused by deformation of a bond angle away from its lowest energy value.

Annulene (Section 14.7A): Monocyclic hydrocarbons that can be represented by structures having alternating single and double bonds. The ring size of an annulene is represented by a number in brackets, e.g., benzene is [6]annulene and cyclooctatetraene is [8]annulene.

Anomers (Section 22.2C): A term used in carbohydrate chemistry. Anomers are diastereomers that differ only in configuration at the acetal or hemiacetal carbon of a sugar in its cyclic form.

Anti addition (Section 7.6A): An addition that places the parts of the adding reagent on opposite faces of the reactant.

Anti conformation (Section 4.7): A gauche conformation of butane, for example, has the methyl groups at an angle of 180° to each other:

CH$_3$

**Anti
conformation
of butane**

Antiaromatic compound (Section 14.7C): A cyclic conjugated system whose π electron energy is greater than that of the corresponding open-chain compound.

Antibonding molecular orbital (antibonding MO) (Section 1.12): A molecular orbital whose energy is higher than that of the isolated atomic orbitals from which it is constructed. Electrons in an antibonding molecular orbital destabilize the bond between the atoms that the orbital encompasses.

Anticodon (Section 25.5D): A sequence of three bases on transfer RNA (tRNA) that associates with a codon of messenger RNA (mRNA).

Aprotic solvent (Section 6.15C): A solvent whose molecules do not have a hydrogen atom attached to a strongly electronegative element (such as oxygen). For most purposes, this means that an aprotic solvent is one whose molecules lack an —OH group.

Arene (Section 15.1): A general name for an aromatic hydrocarbon.

Aromatic compound (Sections 14.1–14.7 and 14.11): A cyclic conjugated unsaturated molecule or ion that is stabilized by π electron delocalization. Aromatic compounds are characterized by having large resonance energies, by reacting by substitution rather than addition, and by deshielding of protons exterior to the ring in their ^{1}H NMR spectra caused by the presence of an induced ring current.

Aryl group (Section 15.1): The general name for a group obtained (on paper) by the removal of a hydrogen from a ring position of an aromatic hydrocarbon. Abbreviated Ar—.

Aryl halide (Section 6.1): An organic halide in which the halogen atom is attached to an aromatic ring, such as a benzene ring.

Atactic polymer (Special Topic C): A polymer in which the configuration at the stereocenters along the chain is random.

Atomic orbital (AO) (Section 1.11): A volume of space about the nucleus of an atom where there is a high probability of finding an electron. An atomic orbital can be described mathematically by its **wave function**. Atomic orbitals have characteristic quantum numbers; the *principle quantum number, n*, is related to the energy of the electron in an atomic orbital and can have the values 1, 2, 3, ... The *azimuthal quantum number, l*, determines the angular momentum of the electron that results from its motion around the nucleus, and can have the values 0, 1, 2, . . . , $(n-1)$. The *mag-*

netic quantum number, m, determines the orientation in space of the angular momentum and can have values from $+l$ to $-l$. The *spin quantum number, s*, measures the intrinsic angular momentum of an electron and can have the values of $+\frac{1}{2}$ and $-\frac{1}{2}$ only.

Aufbau principle (Section 1.11): A principle that guides us in assigning electrons to orbitals of an atom or molecule in its lowest energy state or **ground state**. The aufbau principle states that electrons are added so that orbitals of lowest energy are filled first.

Autoxidation (Section 9.11B): The reaction of an organic compound with oxygen to form a hydroperoxide.

Axial bond (Section 4.11): The six bonds of a cyclohexane ring (below) that are perpendicular to the general plane of the ring, and that alternate up

and down around the ring.

Base peak (Section F.2): The most intense peak in a mass spectrum.

Base strength (Section 3.5): The strength of a base is inversely related to the strength of its conjugate acid; the weaker the conjugate acid, the stronger is the base. In other words, if the conjugate acid has a large pK_a, the base will be strong.

Benzenoid aromatic compound (Section 14.8A): An aromatic compound whose molecules have one or more benzene rings.

Benzyl group (Section 15.15): The $C_6H_5CH_2$— group.

Benzylic substituent (Section 13.10): Refers to a substituent on a carbon atom adjacent to a benzene ring.

Benzyne (Section 21.11B): An unstable, highly reactive intermediate consisting of a benzene ring with an additional bond resulting from sideways overlap of sp^2 orbitals on adjacent atoms of the ring.

Betaine (Section 16.10): An electrically neutral molecule that has nonadjacent cationic and anionic sites and that does not possess a hydrogen atom bonded to the cationic site.

Bimolecular reaction (Section 6.7): A reaction whose rate-determining step involves two initially separate species.

Boat conformation (Section 4.10): A conformation of cyclohexane that resembles a boat and that has eclipsed bonds along its two sides:

It is of higher energy than the chair conformation.

Bond angle (Section 2.6A): The angle between two bonds originating at the same atom.

Bond dissociation energy, see **Homolytic bond dissociation energy.**

Bond length (Section 2.6A): The equilibrium distance between two bonded atoms or groups.

Bond-line formula (Section 1.20D): A formula that shows the carbon skeleton of a molecule with lines. The number of hydrogen atoms necessary to fulfill each carbon's valence is assumed to be present but not written in. Other atoms (e.g., O, Cl, N) are written in.

Bonding molecular orbital (bonding MO) (Section 1.12): The energy of a bonding molecular orbital is lower than the energy of the isolated atomic orbitals from which it arises. When electrons occupy a bonding molecular orbital they help hold together the atoms that the molecular orbital encompasses.

Bromination (Section 8.6): A reaction in which bromine atoms are introduced into a molecule.

Bromohydrin (Section 8.8): A bromo alcohol.

Bromonium ion (Section 8.6A): An ion containing a positive bromine atom bonded to two carbon atoms.

Brønsted–Lowry theory of acids and bases (Section 3.2A): An acid is a substance that can donate (or lose) a proton; a base is a substance that can accept (or remove) a proton. The **conjugate acid** of a base is the molecule or ion that forms when a base accepts a proton. The **conjugate base** of an acid is the molecule or ion that forms when an acid loses its proton.

Carbanion (Section 3.3): A chemical species in which a carbon atom bears a formal negative charge.

Carbene (Special Topic D): An uncharged species in which a carbon atom is divalent. The species, $:CH_2$, called methylene, is a carbene.

Carbenoid (Special Topic D): A carbene-like species. A species such as the reagent formed when diiodomethane reacts with a zinc–copper couple. This reagent, called the Simmons–Smith reagent, reacts with alkenes to add methylene to the double bond in a stereospecific way.

Carbocation (Section 3.3): A chemical species in which a trivalent carbon atom bears a formal positive charge.

Carbohydrate (Section 22.1A): A group of naturally occurring compounds that are usually defined as polyhydroxyaldehydes or polyhydroxyketones, or as substances that undergo hydrolysis to yield such compounds. In actuality, the aldehyde and ketone groups of carbohydrates are often present as hemiacetals and acetals. The name comes from the fact that many carbohydrates possess the empirical formula $C_x(H_2O)_y$.

Carbonyl group (Section 16.1): A functional group consisting of a carbon atom doubly bonded to an oxygen atom. The carbonyl group is found in aldehydes, ketones, esters, anhydrides, amides, acyl halides, and so on. Collectively these compounds are referred to as carbonyl compounds.

Cascade polymer (Section I.5): A polymer produced from a multifunctional central core by progressively adding layers of repeating units.

CFC (see **Freon**): A chlorofluorocarbon.

Chain reaction (Section 9.4): A reaction that proceeds by a sequential, stepwise mechanism, in which each step generates the reactive intermediate that causes the next step to occur. Chain reactions have *chain initiating steps, chain propagating steps,* and *chain terminating steps.*

Chair conformation (Section 4.10): The all-staggered conformation of cyclohexane that has no angle strain or torsional strain and is, therefore, the lowest energy conformation:

Chemical shift, δ (Section 13.7): The position in an NMR spectrum, relative to a reference compound, at which a nucleus absorbs. The reference compound most often used is tetramethylsilane (TMS), and its absorption point is arbitrarily designated zero. The chemical shift of a given nucleus is proportional to the strength of the magnetic field of the spectrometer. The chemical shift in delta units, δ, is determined by dividing the observed shift from TMS in hertz multiplied by 10^6 by the operating frequency of the spectrometer in hertz.

Chiral molecule (Section 5.2): A molecule that is not superposable on its mirror image. Chiral molecules have handedness and are capable of existing as a pair of enantiomers.

Chirality (Section 5.2): The property of having handedness.

Chlorination (Section 9.5): A reaction in which chlorine atoms are introduced into a molecule.

Chlorohydrin (Section 8.8): A chloro alcohol.

Cis–trans isomers (Sections 2.14B and 4.12): Diastereomers that differ in their stereochemistry at adjacent atoms of a double bond or on different atoms of a ring.

Codon (Section 25.5D): A sequence of three bases on messenger RNA (mRNA) that contains the genetic information for one amino acid. The codon associates, by hydrogen bonding, with an anticodon of a transfer RNA (tRNA) that carries the particular amino acid for protein synthesis on the ribosome.

Condensation polymer (Special Topic I): A polymer produced when bifunctional monomers (or potentially bifunctional monomers) react with each other through the intermolecular elimination of water or an alcohol. Polyesters, polyamides, and polyurethanes are all condensation polymers.

Condensation reaction (Section 17.5A): A reaction in which molecules become joined through the intermolecular elimination of water or an alcohol.

Configuration (Section 5.6): The particular arrangement of atoms (or groups) in space that is characteristic of a given stereoisomer.

Conformation (Section 4.5): A particular temporary orientation of a molecule that results from rotations about its single bonds.

Conformational analysis (Section 4.6): An analysis of the energy changes that a molecule undergoes as its groups undergo rotation (sometimes only partial) about the single bonds that join them.

Conformer (Section 4.6): A particular staggered conformation of a molecule.

Conjugate acid (Section 3.2A): The molecule or ion that forms when a base accepts a proton.

Conjugate addition (Section 17.9): A form of nucleophilic addition to an α,β-unsaturated carbonyl compound in which the nucleophile adds to the β-carbon.

Conjugate base (Section 3.2A): The molecule or ion that forms when an acid loses its proton.

Conjugated system (Section 12.1): Molecules or ions that have an extended π system. A conjugated system has a p orbital on an atom adjacent to a multiple bond; the p orbital may be that of another multiple bond or that of a radical, carbocation, or carbanion.

Connectivity (Section 1.3): The sequence, or order, in which the atoms of a molecule are attached to each other.

Constitutional isomers (Section 1.3A): Compounds that have the same molecular formula but that differ in their connectivity (i.e., molecules that have the same molecular formula but have their atoms connected in different ways).

Copolymer (Special Topic C): A polymer synthesized by polymerizing two monomers.

Coupling constant, J_{ab} (Section 13.9): The separation in frequency units (hertz) of the peaks of a multiplet caused by spin–spin coupling between atoms a and b.

Covalent bond (Section 1.4B): The type of bond that results when atoms share electrons.

Cracking (Section 4.1C): A process used in the petroleum industry for breaking down the molecules of larger alkanes into smaller ones. Cracking may be accomplished with heat (thermal cracking), or with a catalyst (catalytic cracking).

Crown ether (Section 10.22A): Cyclic polyethers that have the ability to form complexes with metal ions. Crown ethers are named as x-crown-y where x is the total number of atoms in the ring and y is the number of oxygen atoms in the ring.

Cyanohydrin (Section 16.9A): A functional group consisting of a carbon atom bonded to a cyano group and to a hydroxyl group, i.e., RHC(OH)(CN) or R_2C(OH)(CN), derived by adding HCN to an aldehyde or ketone.

Cycloaddition (Section 12.10): A reaction, like the Diels–Alder reaction, in which two connected groups add to the end of a π system to generate a new ring.

D and L designations (Section 22.2B): A method for designating the configuration of monosaccarides and other similar compounds in which the reference compound is (+)- or (−)-glyceraldehyde. According to this system, (+)-glyceraldehyde is designated D-(+)-glyceraldehyde and (−)-glyceraldehyde is designated L-(−)-glyceraldehyde. Therefore, a monosaccharide whose highest numbered stereocenter has the same general configuration as D-(+)-glyceraldehyde is designated a D-sugar; one whose highest numbered stereocenter has the same general configuration as L-(+)-glyceraldehyde is designated an L-sugar.

Deactivating group (Section 15.10): A group that when present on a benzene ring causes the ring to be less reactive in electrophilic substitution than benzene itself.

Debromination (Section 7.16): The elimination of two atoms of bromine from a *vic*-dibromide, or, more generally, the loss of bromine from a molecule.

Debye unit (Section 1.18): The unit in which dipole moments are stated. One debye, D, equals 1×10^{-18} esu cm.

Decarboxylation (Section 18.11): A reaction whereby a carboxylic acid loses CO_2.

Degenerate orbitals (Section 1.11): Orbitals of equal energy. For example, the three $2p$ orbitals are degenerate.

Dehydration reaction (Section 7.13): An elimination that involves the loss of a molecule of water from the substrate.

Dehydrohalogenation (Section 6.17): An elimination reaction that results in the loss of HX from adjacent carbons of the substrate and the formation of a π bond.

Delocalization (Section 6.13B): The dispersal of electrons (or of electrical charge). Delocalization of charge always stabilizes a system.

Dextrarotatory (Section 5.7B): A compound that rotates plane-polarized light clockwise.

Diastereomers (Section 5.1): Stereoisomers that are not mirror images of each other.

Diastereotopic hydrogens (or ligands) (Section 13.8A): If replacement of each of two hydrogens (or ligands) by the same group yields compounds that are diastereomers, the two hydrogen atoms (or ligands) are said to be diastereotopic.

Dielectric constant (Section 6.15D): A measure of a solvent's ability to insulate opposite charges from each other. The dielectric constant of a solvent roughly measures its polarity. Solvents with high dielectric constants are better solvents for ions than are solvents with low dielectric constants.

Dienophile (Section 12.10): The diene-seeking component of a Diels–Alder reaction.

Dipolar ion (Section 24.2C): The charge-separated form of an amino acid that results from the transfer of a proton from a carboxyl group to a basic group.

Dipole moment, μ (Section 1.18): A physical property associated with a polar molecule that can be measured experimentally. It is defined as the product of the charge in electrostatic units (esu) and the distance that separates them in centimeters: $\mu = e \times d$.

Dipole–dipole interaction (Section 2.16B): An interaction between molecules having permanent dipole moments.

Disaccharide (Section 22.1A): A carbohydrate that, on a molecular basis, undergoes hydrolytic cleavage to yield two molecules of a monosaccharide.

Distortionless enhanced polarization transfer (DEPT) spectra (Section 14.11E): A series of ^{13}C NMR spectra in which the signal for each type of carbon, C, CH, CH_2, and CH_3, is printed out separately. Data from DEPT spectra help identify the different types of carbon atoms in a ^{13}C NMR spectrum.

E–Z system (Section 7.2A): A system for designating the stereochemistry of alkene diastereomers based on the priorities of groups in the Cahn–Ingold–Prelog convention.

E1 reaction (Section 6.19): A unimolecular elimination in which, in a slow, rate-determining step, a leaving group departs from the substrate to form a carbocation. The carbocation then in a fast step loses a proton with the resulting formation of a π bond.

E2 reaction (Section 6.18): A bimolecular 1,2 elimination in which, in a single step, a base removes a proton and a leaving group departs from the substrate, resulting in the formation of a π bond.

Eclipsed conformation (Section 4.6): A temporary orientation of groups around two atoms joined by a single bond such that the groups directly oppose each other.

An eclipsed conformation

Electromagnetic spectrum (Section 13.1): The full range of energies propagated by wave fluctuations in electromagnetic field.

Electronegativity (Section 1.4A): A measure of the ability of an atom to attract electrons it is sharing with another and thereby polarize the bond.

Electrophile (Section 3.3 and 8.1): A Lewis acid, an electron-pair acceptor, an electron seeking reagent.

Electrophoresis (Section 25.6A): A technique for separating charged molecules based on their different mobilities in an electric field.

Elimination (Section 6.17): A reaction that results in the loss of two groups from the substrate and the formation of a π bond. The most common elimination is a 1,2 elimination or β elimination, in which the two groups are lost from adjacent atoms.

Empirical formula (Section 1.2B): A formula that expresses the relative proportions of atoms in a molecule as smallest whole numbers.

Enantiomeric excess or **enantiomeric purity** (Section 5.8B): A percentage calculated for a mixture of enantiomers by dividing the moles of one enantiomer minus the moles of the other enantiomer by the moles of both enantiomers and multiplying by 100. The enantiomeric excess equals the percentage optical purity.

Enantiomers (Section 5.1): Stereoisomers that are mirror images of each other.

Enantiotopic hydrogens (or **ligands**) (Section 13.8A): If replacement of each of two hydrogens (or ligands) by the same group yields compounds that are enantiomers, the two hydrogen atoms (or ligands) are said to be enantiotopic.

Endergonic reaction (Section 6.9): A reaction that proceeds with a positive free energy change.

Endothermic reaction (Section 1.9A): A reaction that absorbs heat. For an endothermic reaction $\Delta H°$ is positive.

Energy (Section 1.9): Energy is the capacity to do work.

Energy of activation, E_{act} (Section 9.5B): A measure of the difference in potential energy between the reactants and the transition state of a reaction. It is related to, but not the same as, the free energy of activation, $\Delta G^{\ddagger}$.

Enolate ion (Section 17.1): The delocalized anion formed when an enol loses its hydroxylic proton or when the aldehyde or ketone that is in equilibrium with the enol loses an α proton.

Enthalpy change (Sections 1.9A and 3.8): Also called the heat of reaction. The *standard enthalpy change*, $\Delta H°$, is the change in enthalpy after a system in its standard state has undergone a transformation to another system, also in its standard state. For a reaction, $\Delta H°$ is a measure of the difference in the total bond energy of the reactants and products. It is one way of expressing the change in potential energy of molecules as they undergo reaction. The enthalpy change is related to the **free-energy change**, $\Delta G°$, and to the **entropy change**, $\Delta S°$, through the expression:

$$\Delta H° = \Delta G° + T\Delta S°$$

Entropy change (Section 3.8): The standard entropy change, $\Delta S°$, is the change in entropy between two systems in their standard states. Entropy changes have to do with changes in the relative order of a system. The more random a system is, the greater is its entropy. When a system becomes more disorderly its entropy change is positive.

Epoxide (Section 10.18): An oxirane. A three-membered ring containing one oxygen atom.

Equatorial bond (Section 4.11): The six bonds of a cyclohexane ring that lie generally around the "equator" of the molecule:

Equilibrium constant (Section 3.5A): A constant that expresses the position of an equilibrium. The equilibrium constant is calculated by multiplying the molar concentrations of the products together and then dividing this number by the number obtained by multiplying together the molar concentrations of the reactants.

Equilibrium control, see **Thermodynamic control.**

Essential oil (Section 23.3): A volatile odoriferous compound obtained by steam distillation of plant material.

Exergonic reaction (Section 6.9): A reaction that proceeds with a negative free-energy change.

Exothermic reaction (Section 1.9A): A reaction that evolves heat. For an exothermic reaction, $\Delta H°$ is negative.

Fat (Section 23.2): A triacylglycerol. The triester of glycerol with carboxylic acids.

Fatty acid (Section 23.2): A long-chained carboxylic acid (usually with an even number of carbon atoms) that is isolated by the hydrolysis of a fat.

Fischer projection formula (Sections 5.12 and 22.2C): A two-dimensional formula for representing the configuration of a chiral molecule. By convention, Fischer projection formulas are written with the main carbon chain extending from top to bottom with all groups eclipsed. Vertical lines represent bonds that project behind the plane of the page (or that lie in it). Horizontal lines represent bonds that project out of the plane of the page.

Fischer   Wedge-dashed
projection   wedge formula

Fluorination (Section 9.3): A reaction in which fluorine atoms are introduced into a molecule.

Formal charge (Section 1.7): The difference between the number of electrons assigned to an atom in a molecule and the number of electrons it has in its outer shell in its elemental state. Formal charge can be calculated using the formula: $F = Z - S/2 - U$, where F is the formal charge, Z is the group number of the atom (i.e, the number of electrons the atom has in its outer shell in its elemental state), S equals the number of electrons the atom is sharing with another atom, and U is the number of unshared electrons the atom possesses.

Free energy of activation, $\Delta G^{\ddagger}$, (Section 6.9): The difference in free energy between the transition state and the reactants.

Free-energy change (Section 3.8): The *standard free-energy change,* $\Delta G°$, is the change in free energy between two systems in their standard states. At constant temperature, $\Delta G° = \Delta H° - T\Delta S° = -RT \ln K_{eq}$, where $\Delta H°$ is the standard enthalpy change, $\Delta S°$ is the standard entropy change, and K_{eq} is the equilibrium constant. A negative value of $\Delta G°$ for a reaction means that the formation of products is favored when the reaction reaches equilibrium.

Freon (Section 9.11C): A chlorofluorocarbon or CFC.

Frequency (abbreviated ν) (Section 13.1): The number of full cycles of a wave that pass a given point in each second.

Functional group (Section 2.8): The particular group of atoms in a molecule that primarily determines how the molecule reacts.

Functional group interconversion (Section 6.16): A process that converts one functional group into another.

Furanose (Section 22.1C): A sugar in which the cyclic acetal or hemiacetal ring is five membered.

Gauche conformation (Section 4.7): A gauche conformation of butane, for example, has the methyl groups at an angle of 60° to each other:

A gauche
conformation
of butane

Geminal (gem) substituents (Section 7.16): Substituents that are on the same atom.

Glycol (Section 4.3F): A diol.

Glycoside (Section 22.4): A cyclic mixed acetal of a sugar with an alcohol.

Grignard reagent (Section 11.6B): An organomagnesium halide, usually written RMgX.

Ground state (Section 1.11): The lowest electronic energy state of an atom or molecule.

Halogenation (Section 9.3): A reaction in which halogen atoms are introduced into a molecule.

Halohydrin (Section 8.8): A halo alcohol.

Halonium ion (Section 8.6A): An ion containing a positive halogen atom bonded to two carbon atoms.

Hammond–Leffler postulate (Section 6.15A): A postulate stating that the structure and geometry of the transition state of a given step will show a greater resemblance to the reactants or products of that step depending on which is closer to the transition state in energy. This means that the transition state of an endothermic step will resemble the products of that step more than the reactants, whereas the transition state of an exothermic step will resemble the reactants of that step more than the products.

Heat of combustion (Section 4.8): The standard enthalpy change for the complete combustion of 1 mol of a compound.

Heat of hydrogenation (Section 7.9A): The standard enthalpy change that accompanies the hydrogenation of 1 mol of a compound to form a particular product.

Heisenberg uncertainty principle (Section 1.12): A fundamental principle that states that both the position and momentum of an electron (or of any object) cannot be exactly measured simultaneously.

Hemiacetal (Section 16.7B): A functional group, consisting of a carbon atom bonded to an alkoxy group and to a hydroxyl group, [i.e., $RCH(OH)(OR')$ or $R_2C(OH)(OR')$]. Hemiacetals are synthesized by adding one molar equivalent of an alcohol to an aldehyde or a ketone.

Hemiketal (Section 16.7B): Also called a hemiacetal. A functional group, consisting of a carbon atom bonded to an alkoxy group and to a hydroxyl group, [i.e., $R_2C(OH)(OR')$]. Hemiketals are synthesized by adding one molar equivalent of an alcohol to a ketone.

Hertz (abbreviated Hz) (Section 13.1): Now used instead of the equivalent cycles per second as a measure of the frequency of a wave.

Heterocyclic compound (Section 14.9): A compound whose molecules have a ring containing an element other than carbon.

Heterolysis (Section 3.1B): The cleavage of a covalent bond so that one fragment departs with both of the electrons of the covalent bond that joined them. Heterolysis of a bond normally produces positive and negative ions.

Hofmann rule (Section 7.12B): When an elimination yields the alkene with the less substituted double bond, it is said to follow the Hofmann rule.

HOMO (Section 13.2): The highest occupied molecular orbital.

Homologous series (Section 4.5): A series of compounds in which each member differs from the next member by a constant unit.

Homolysis (Section 3.1B): The cleavage of a covalent bond so that each fragment departs with one of the electrons of the covalent bond that joined them.

Homolytic bond dissociation energy, $DH°$, (Section 9.2): The enthalpy change that accompanies the homolytic cleavage of a covalent bond.

Hückel's rule (Section 14.7): A rule stating that planar monocyclic rings with $(4n + 2)$ delocalized π electrons (i.e., with 2, 6, 10, 14, . . . , delocalized π electrons) will be aromatic.

Hund's rule (Section 1.11): A rule used in applying the **aufbau principle**. When orbitals are of equal energy (i.e., when they are **degenerate**), electrons are added to each orbital with their spins unpaired, until each degenerate orbital contains one electron. Then electrons are added to the orbitals so that the spins are paired.

Hybridization of atomic orbitals (Section 1.13): A mathematical (and theoretical) mixing of two or more atomic orbitals to give the same number of new orbitals, called *hybrid orbitals*, each of which has some of the character of the original atomic orbitals.

Hydration (Section 8.5): The addition of water to a molecule, such as the addition of water to an alkene to form an alcohol.

Hydroboration (Section 10.6): The addition of a boron hydride (either BH_3 or an alkylborane) to a multiple bond.

Hydrogen bond (Section 2.16C): A strong dipole–dipole interaction ($1-9$ kcal mol^{-1}) that occurs between hydrogen atoms bonded to small strongly electronegative atoms (O, N, or F) and the nonbonding electron pairs on other such electronegative atoms.

Hydrogenation (Section 4.15A): The addition of hydrogen to a molecule, usually to a multiple bond in the molecule.

Hydrogenation (Section 7.5): A reaction in which hydrogen adds to a double bond. Hydrogenation is often accomplished through the use of a metal catalyst such as platinum, palladium, rhodium, or ruthenium.

Hydrophilic group (Section 2.16E): A polar group that seeks an aqueous environment.

Hydrophobic group (also called a **lipophilic group**) (Sections 2.16E and 10.22): A nonpolar group that avoids an aqueous surrounding and seeks a nonpolar environment.

Hydroxylation (Sections 8.9 and 10.20): The addition of hydroxyl groups to each carbon or atom of a double bond.

Index of hydrogen deficiency (Section 7.8): The index of hydrogen deficiency (or IHD) equals the number of pairs of hydrogen atoms that must be subtracted from the molecular formula of the corresponding alkane to give the molecular formula of the compound under consideration.

Inductive effect (Sections 3.7B and 15.11B): An intrinsic electron-attracting or -releasing effect that results from a nearby dipole in the molecule and that is transmitted through space and through the bonds of a molecule.

Infrared (IR) spectroscopy (Section 13.3): A type of optical spectroscopy that measures the absorption of infrared radiation. Infrared spectroscopy provides structural information about functional groups present in the compound being analyzed.

Intermediate (Sections 3.1A, 6.11, and 6.12): A transient species that exists between reactants and products in a state corresponding to a local energy minimum on a potential energy diagram.

Iodination (Section 9.3): A reaction in which iodine atoms are introduced into a molecule.

Ion (Section 6.3A): A chemical species that bears an electrical charge.

Ion–dipole interaction (Section 2.16E): The interaction of an ion with a permanent dipole. Such interactions (resulting in solvation) occur between ions and the molecules of polar solvents.

Ionic bond (Section 1.4A): A bond formed by the transfer of electrons from one atom to another resulting in the creation of oppositely charged ions.

Ionic reaction (Section 3.1B): A reaction involving ions as reactants, intermediates, or products. Ionic reactions occur through the heterolysis of covalent bonds.

Isoelectric point (Section 24.2C): The pH at which the number of positive and negative charges on an amino acid or protein are equal.

Isomers (Sections 1.3A and 5.1): Different molecules that have the same molecular formula.

Isoprene unit (Section 23.3): A name for the structural unit:

found in all terpenes.

Isotactic polymer (Special Topic C): A polymer in which the configuration at each stereocenter along the chain is the same.

Kekulé structure (Sections 1.3 and 14.4): A structure in which lines are used to represent bonds. The Kekulé structure for benzene is a hexagon of carbon atoms with alternating single and double bonds around the ring, and with one hydrogen atom attached to each carbon.

Ketal (Section 16.7C): Properly called an acetal. A functional group, consisting of a carbon bonded to alkoxy groups, [i.e., $R_2C(OR')_2$], derived by adding two molar equivalents of an alcohol to an aldehyde or ketone.

Kinetic control (Sections 6.6 and 12.9A): A principle stating that when the ratio of products of a reaction is determined by relative rates of reaction, the most abundant product will be the one that is formed fastest.

Kinetic energy (Section 1.9): Energy that results from the motion of an object. Kinetic energy $(KE) = \frac{1}{2} mv^2$, where m is the mass of the object, and v is its velocity.

Kinetics (Section 6.6): A term that refers to rates of reactions.

Lactam (Section 18.8I): A cyclic amide.

Lactone (Section 18.7C): A cyclic ester.

Leaving group (or **nucleofuge**) (Section 6.5): The substituent with an unshared electron pair that departs from the substrate in a nucleophilic substitution reaction.

Leveling effect of a solvent (Section 3.13): An effect that restricts the use of certain solvents with strong acids and bases. In principle, no acid stronger than the conjugate acid of a particular solvent can exist to an appreciable extent in that solvent, and no base stronger than the conjugate base of the solvent can exist to an appreciable extent in that solvent.

Levorotatory (Section 5.7B): A compound that rotates plane-polarized light in a counterclockwise direction.

Lewis structure (or **electron-dot structure**) (Section 1.5): A representation of a molecule showing electron pairs as a pair of dots or as a dash.

Lewis theory of acids and bases (Section 3.2B): An acid is an electron pair donor, and a base is an electron pair acceptor.

Lipid (Section 23.1): A substance of biological origin that is soluble in nonpolar solvents. Lipids include fatty acids, triacylglycerols (fats and oils), steroids, prostaglandins, terpenes and terpenoids, and waxes.

Lipophilic group (or **hydrophopic group**) (Sections 2.16E and 10.22): A nonpolar group that avoids an aqueous surrounding and seeks a nonpolar environment.

LUMO (Section 13.2): The lowest unoccupied molecular orbital.

Macromolecule (Section 9.10): A very large molecule.

Magnetic resonance imaging (Section 13.12): A technique based on NMR spectroscopy that is used in medicine.

Markovnikov's rule (Section 8.2): A rule for predicting the regiochemistry of electrophilic additions to alkenes and alkynes that can be stated in various ways. As originally stated (in 1870) by Vladimir Markovnikov, the rule provides that "if an unsymmetrical alkene combines with a hydrogen halide, the halide ion adds to the carbon with the fewer hydrogen atoms." More commonly the rule has been stated in reverse: that in the addition of HX to an alkene or alkyne the hydrogen atom adds to the carbon atom that already has the greater number of hydrogen atoms. A modern expression of Markovnikov's rule is: *In the ionic addition of an unsymmetrical reagent to a multiple bond, the positive portion of the reagent (the electrophile) attaches itself to a carbon atom of the reagent in the way that leads to the formation of the more stable intermediate carbocation.*

Mass spectrometry (Special Topic F): A technique, useful in structure elucidation, that is based on generating ions of a molecule in a magnetic field, then instrumentally determining the mass/charge ratio and relative amounts of the ions that result.

Mechanism, see **Reaction mechanism**.

Meso compound (Section 5.10A): An optically inactive compound whose molecules are achiral even though they contain tetrahedral atoms with four different attached groups.

Mesylate (Section 10.10): A methanesulfonate ester.

Methylene (Special Topic D): The carbene with the formula CH_2.

Methylene group (Section 6.2): The $-CH_2-$ group.

Micelle (Section 23.2C): A spherical cluster of ions in aqueous solution (such as those from a soap) in which the nonpolar groups are in the interior and the ionic (or polar) groups are at the surface.

Molar absorptivity (abbreviated ε) (Section 14.2): A proportionality constant that relates the observed absorbance (A) at a particular wavelength (λ) to the molar concentration of the sample (C) and the length (l) (in centimeters) of the path of the light beam through the sample cell:

$$\varepsilon = A/C \times l$$

Molecular formula (Section 1.2B): A formula that gives the total number of each kind of atom in a molecule. The molecular formula is a whole number multiple of the empirical formula. For example the molecular formula for benzene is C_6H_6; the empirical formula is CH.

Molecular ion (Section F.1): The cation produced in a mass spectrometer when one electron is dislodged from the parent molecule.

Molecular orbital (MO) (Section 1.12): Orbitals that encompass more than one atom of a molecule. When atomic orbitals combine to form molecular orbitals, the number of molecular orbitals that results always equals the number of atomic orbitals that combine.

Molecularity (Section 6.8): The number of species involved in a single step of a reaction (usually the rate determining step).

Monomer (Section 9.10): The simple starting compound from which a polymer is made. For example, the polymer polyethylene is made from the monomer, ethylene.

Monosaccharide (Section 22.1A): The simplest carbohydrate, one that does not undergo hydrolytic cleavage to a simpler carbohydrate.

Mutarotation (Section 22.3): The spontaneous change that takes place in the optical rotation of α and β anomers of a sugar when they are dissolved in water. The optical rotations of the sugars change until they reach the same value.

Neighboring group participation (Special Topic B): The effect on the course or rate of a reaction brought about by another group near the functional group undergoing reaction.

Newman projection formula (Section 4.6): A means of representing the spatial relationships of groups attached to two atoms of a molecule. In writing a Newman projection formula we imagine ourselves viewing the molecule from one end directly along the bond axis joining the two atoms. Bonds that are attached to the front atom are shown as radiating from the center of a circle; those attached to the rear atom are shown as radiating from the edge of the circle:

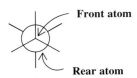

Nitrogen rule (Section F.3): A rule that states that if the mass of the molecular ion in a mass spectrum is an even number, the parent compound contains an even number of nitrogen atoms, and conversely.

Node (Section 1.10): A place where a wave function (Ψ) is equal to zero. The greater the number of nodes in an orbital, the greater is the energy of the orbital.

Nonbenzenoid aromatic compound (Section 14.8B): An aromatic compound, such as azulene, that does not contain benzene rings.

Nuclear magnetic resonance (NMR) spectroscopy (Section 13.4): A spectroscopic method for measuring the absorption of radiofrequency radiation by certain nuclei when the nuclei are in a strong magnetic field. The most important NMR spectra for organic chemists are ^{1}H NMR spectra and ^{13}C NMR spectra. These two types of spectra provide structural information about the carbon framework of the molecule, and about the number and environment of hydrogen atoms attached to each carbon atom.

Nucleic acids (Sections 25.1 and 25.2): Biological polymers of nucleotides. DNA and RNA are, respectively, nucleic acids that preserve and transcribe hereditary information within cells.

Nucleophile (Section 3.3): A Lewis base, an electron pair donor that seeks a positive center in a molecule.

Nucleophilic substitution reaction (Section 6.3): A reaction initiated by a nucleophile (a species with an unshared electron pair) in which the nucleophile reacts with a substrate to replace a substituent (called the leaving group), that departs with an unshared electron pair.

Nucleophilicity (Section 6.15B): The relative reactivity of a nucleophile in an S_N2 reaction as measured by relative rates of reaction.

Nucleoside (Section 25.2): A five-carbon monosaccharide bonded at the 1′ position to a purine or pyrimidine.

Nucleotide (Section 25.2): A five-carbon monosaccharide bonded at the 1′ position to a purine or pyrimidine and at the 3′ or 5′ position to a phosphate group.

Olefin (Section 7.1): An old name for an alkene.

Optical purity (Section 5.8B): A percentage calculated for a mixture of enantiomers by dividing the observed specific rotation for the mixture by the specific rotation of the pure enantiomer and multiplying by 100. The optical purity equals the enantiomeric purity or enantiomeric excess.

Optically active compound (Section 5.7): A compound that rotates the plane of polarization of plane polarized light.

Orbital (Section 1.11): A volume of space in which there is a high probability of finding an electron. Orbitals are described mathematically by wave functions, and each orbital has a characteristic energy. An orbital can hold two electrons when their spins are paired.

Organometallic compound (Section 11.5): A compound that contains a carbon–metal bond.

Oxidation (Section 11.2): A reaction that increases the oxidation state of atoms in a molecule or ion. For an organic substrate, oxidation usually involves increasing its oxygen content or decreasing its hydrogen content. Oxidation also accompanies any reaction in which a less electronegative substituent is replaced by a more electronegative one.

Oxonium ion (Section 3.11): A chemical species with an oxygen atom that bears a formal positive charge.

Oxymercuration (Section 10.5): The addition of —OH and —HgO_2CR to a multiple bond.

Ozonolysis (Section 8.10A): The cleavage of a multiple bond that makes use of the reagent O_3, called ozone. The reaction leads to the formation of a cyclic compound called an *ozonide*, which is then reduced to carbonyl compounds by treatment with zinc and water.

***p* orbitals** (Section 1.11): A set of three degenerate (equal energy) atomic orbitals shaped like two tangent spheres with a nodal plane at the nucleus. For *p* orbitals, the principal quantum number, n (see **atomic orbital**), is 2; the azimuthal quantum number, l, $= 1$; and the magnetic quantum numbers, m, are $+1$, 0, or -1.

Paraffin (Section 4.14): An old name for an alkane.

Pauli exclusion principle (Section 1.11): A principle that states that no two electrons of an atom or molecule may have the same set of four quantum numbers. It means that only two electrons can occupy the same orbital, and then only when their spin quantum numbers are opposite. When this is true, we say that the spins of the electrons are paired.

Periplanar (Section 7.12C): A conformation in which vicinal groups lie in the same plane.

Peroxide (Section 9.1): A compound with an oxygen–oxygen single bond.

Peroxyacid (Section 10.18): An acid with the general formula RCO_3H, containing an oxygen–oxygen single bond.

Phase sign (Section 1.10): Signs, either $+$ or $-$, that are characteristic of all equations that describe the amplitudes of waves.

Phase-transfer catalysts (Section 10.22): A reagent that transports an ion from an aqueous phase into a nonpolar phase where reaction takes place more rapidly. Tetra-alkyl-ammonium ions and crown ethers are phase-transfer catalysts.

Phospholipid (Section 23.6): Compounds that are structurally derived from *phosphatidic acids*. Phosphatidic acids are derivatives of glycerol in which two hydroxyl groups are joined to fatty acids, and one terminal hydroxyl group is joined in an ester linkage to phosphoric acid. In a phospholipid the phosphate group of the phosphatidic acid is joined in ester linkage to a nitrogen-containing compound such as choline, 2-aminoethanol, or L-serine.

Pi (π) bond (Section 2.4): A bond formed when electrons occupy a bonding π molecular orbital (i.e., the lower energy molecular orbital that results from overlap of parallel p orbitals on adjacent atoms).

Pi (π) molecular orbital (Section 2.4): A molecular orbital formed when parallel p orbitals on adjacent atoms overlap. Pi orbitals may be *bonding* (p lobes of the same phase sign overlap) or *antibonding* (p orbitals of opposite phase sign overlap).

pK_a (Section 3.5): The pK_a is the negative logarithm of the acidity constant, K_a.

Plane of symmetry (Section 5.5): An imaginary plane that bisects a molecule in a way such that the two halves of the molecule are mirror images of each other. Any molecule with a plane of symmetry will be achiral.

Plane-polarized light (Section 5.7A): Ordinary light in which the oscillations of the electrical field occur only in one plane.

Polar covalent bond (Section 1.18): A covalent bond in which the electrons are not equally shared because of differing electronegativities of the bonded atoms.

Polar molecule (Section 1.19): A molecule with a dipole moment.

Polarimeter (Section 5.7B): A device used for measuring optical activity.

Polarizability (Section 6.15C): The susceptibility of the electron cloud of an uncharged molecule to distortion by the influence of an electric charge.

Polymer (Section 9.10): A large molecule made up of many repeating subunits. For example, the polymer polyethylene is made up of the repeating subunit $-(CH_2CH_2)_2-$.

Polysaccharide (Section 22.1A): A carbohydrate that, on a molecular basis, undergoes hydrolytic cleavage to yield many molecules of a monosaccharide.

Potential energy (Section 1.9): Potential energy is stored energy; it exists when attractive or repulsive forces exist between objects.

Potential energy diagram (Section 6.9): A plot of potential energy changes that take place during a reaction versus the reaction coordinate. It displays potential energy changes as a function of changes in bond orders and distances as reactants proceed through the transition state to become products.

Primary carbon (Section 2.10): A carbon atom that has only one other carbon atom attached to it.

Primary structure (Section 24.5): The covalent structure of a polypeptide or protein. This structure is determined, in large part, by determining the sequence of amino acids in the protein.

Protecting group (Section 16.7D): A group that is introduced into a molecule to protect a sensitive group from reaction while a reaction is carried out at some other location in the molecule. Later, the protecting group is removed.

Protein (Section 24.1): A large biological polymer of α-amino acids joined by amide linkages.

Protic solvent (Sections 3.10 and 6.15C): A solvent whose molecules have a hydrogen atom attached to a strongly electronegative element such as oxygen or nitrogen. Molecules of a protic solvent can therefore form hydrogen bonds to unshared electron pairs of oxygen or nitrogen atoms of solute molecules or ions, thereby stabilizing them. Water, methanol, ethanol, formic acid, and acetic acid are typical protic solvents.

Proton decoupling (Section 13.11): An electronic technique used in ^{13}C NMR spectroscopy that allows decoupling of spin–spin interactions between ^{13}C nuclei and ^1H nuclei. In spectra obtained in this mode of operation all carbon resonances appear as singlets.

Proton off-resonance decoupling (Section 13.11D): An electronic technique used in ^{13}C NMR spectroscopy that allows one-bond couplings between ^{13}C nuclei and ^1H nuclei. In spectra obtained in this mode of operation, CH_3 groups appear as quartets, CH_2 groups appear as triplets, CH groups appear as doublets, and carbon atoms with no attached hydrogen atoms appear as singlets.

Psi function (Ψ function or wave function) (Section 1.10): A mathematical expression derived from *quantum mechanics* corresponding to an energy state for an electron. The square of the Ψ function, Ψ^2, gives the probability of finding the electron in a particular location in space.

Pyranose (Section 22.1C): A sugar in which the cyclic acetal or hemiacetal ring is six membered.

R (Section 2.8A): A symbol used to designate an alkyl group. Oftentimes it is taken to symbolize any organic group.

(R–S) System (Section 5.6): A method for designating the configuration of tetrahedral stereocenters.

Racemic form (racemate or **racemic mixture)** (Section 5.8A): An equimolar mixture of enantiomers. A racemic form is optically inactive.

Racemization (Section 6.14A): A reaction that transforms an optically active compound into a racemic form is said to proceed with racemization. Racemization takes place whenever a reaction causes chiral molecules to be converted to an achiral intermediate.

Radical (or free radical) (Section 3.1B): An uncharged chemical species that contains an unpaired electron.

Radical reaction (Section 9.1): A reaction involving radicals. Homolysis of covalent bonds occurs in radical reactions.

Radicofunctional nomenclature (Section 4.3E): A system for naming compounds that uses two or more words to describe the compound. The final word corresponds to the functional group present, the preceding words, usually listed in alphabetical order, describe the remainder of the molecule. Examples are: methyl alcohol, methyl ethyl ether, and ethyl bromide.

Rate control, see **Kinetic control.**

Rate-determining step (Section 6.11A): If a reaction takes place in a series of steps, and if the first step is intrinsically slower than all of the others, then the rate of the overall reaction will be the same as (will be determined by) the rate of this slow step.

Reaction coordinate (Section 6.9): The abscissa in a potential energy diagram that represents the progress of the reaction. It represents the changes in bond orders and bond distances that must take place as reactants are converted to products.

Reaction mechanism (Section 3.1): A step-by-step description of the events that are postulated to take place at the molecular level as reactants are converted to products. A mechanism will include a description of all intermediates and transition states. Any mechanism proposed for a reaction must be consistent with all experimental data obtained for the reaction.

Rearrangement (or 1,2 shift) (Section 7.15): A reaction that results in a product with a different carbon skeleton from the reactant. The type of rearrangement called a 1,2 shift involves the migration of an organic group (with its electrons) from one atom to the atom next to it.

Reducing sugar (Section 22.6A): Sugars that reduce Tollens' or Benedict's reagents. All sugars that contain hemiacetal or hemiketal groups (and therefore are in equilibrium with aldehydes or α-hydroxyketones) are reducing sugars. Sugars in which only acetal or ketal groups are present are nonreducing sugars.

Reduction (Section 11.2): A reaction that lowers the oxidation state of atoms in a molecule or ion. Reduction of an organic compound usually involves increasing its hydrogen content or decreasing its oxygen content. Reduction also accompanies any reaction that results in replacement of a more electronegative substituent by a less electronegative one.

Regioselective reaction (Section 8.2C): A reaction that yields only one (or a predominance of one) constitutional isomer as the product when two or more constitutional isomers are possible products.

Relative configuration (Section 5.14A): The relationship between the configurations of two chiral molecules. Molecules are said to have the same relative configuration when similar or identical groups in each occupy the same position in space. The configurations of molecules can be related to each other through reactions of known stereo-chemistry, for example, through reactions that cause no bonds to a stereocenter to be broken.

Resolution (Sections 5.15 and 19.3C): The process by which the enantiomers of a racemic form are separated.

Resonance effect (Sections 3.9 and 15.11B): An effect by which a substituent exerts either an electron-releasing or -withdrawing effect through the pi system of the molecule.

Resonance energy (Section 14.5): An energy of stabilization that represents the difference in energy between the actual compound and that calculated for a single resonance structure. The resonance energy arises from delocalization of electrons in a conjugated system.

Resonance structures (or **resonance contributors**) (Sections 1.8 and 12.5): Lewis structures that differ from one another only in the position of their electrons. A single resonance structure will not adequately represent a molecule. The molecule is better represented as a *hybrid* of all of the resonance structures.

Retrosynthetic analysis (Section 4.16): A method for planning syntheses that involves reasoning backward from the target molecule through various levels of precursors and thus finally to the starting materials.

Ring flip (Sections 4.10 and 4.11): The change in a cyclohexane ring (resulting from partial bond rotations) that converts one ring conformation to another. A ring flip converts any equatorial substitutent to an axial substituent and vice versa.

Ring strain (Section 4.8): The increased potential energy of the cyclic form of a molecule (usually measured by heats of combustion) when compared to its acyclic form.

s **orbital** (Section 1.11): A spherical atomic orbital. For *s* orbitals the azimuthal quantum number $l = 0$ (see atomic orbital).

Saponification (Section 18.7B): Base-promoted hydrolysis of an ester.

Saturated compound (Section 2.2): A compound that does not contain any multiple bonds.

Secondary carbon (Section 2.10): A carbon atom that has two other carbon atoms attached to it.

Secondary structure (Section 24.8): The local conformation of a polypeptide backbone. These local conformations are specified in terms of regular folding patterns such as pleated sheets, α-helixes, and turns.

Shielding and deshielding (Section 13.6): Effects observed in NMR spectra caused by the circulation of sigma and pi electrons within the molecule. Shielding causes signals to appear at higher magnetic fields (upfield), deshielding causes signals to appear at lower magnetic fields (downfield).

Sigma (σ) bond (Section 1.13): A single bond. A bond formed when electrons occupy the bonding σ orbital formed by the end-on overlap of atomic orbitals (or hybrid orbitals) on adjacent atoms. In a sigma bond the electron density has circular symmetry when viewed along the bond axis.

Sigma (σ) orbital (Section 1.13): An orbital formed by end-on overlap of orbitals (or lobes of orbitals) on adjacent atoms. Sigma orbitals may be *bonding* (orbitals or lobes of the same phase sign overlap) or *antibonding* (orbitals or lobes of opposite phase sign overlap).

S_N1 reaction (Sections 6.10 and 6.15): Literally, substitution nucleophilic unimolecular. A multistep nucleophilic substitution in which the leaving group departs in a unimolecular step before the attack of the nucleophile. The rate equation is first order in substrate but zero order in the attacking nucleophile.

S_N2 reaction (Sections 6.7 and 6.8): Literally, substitution nucleophilic bimolecular. A bimolecular nucleophilic substitution reaction that takes place in a single step in which a nucleophile attacks a carbon bearing a leaving group from the backside, causing an inversion of configuration at this carbon and displacement of the leaving group.

Solvent effect (Section 6.15D): An effect on relative rates of reaction caused by the solvent. For example, the use of a polar solvent will increase the rate of reaction of an alkyl halide in an S_N1 reaction.

Solvolysis (Section 6.14B): Literally, cleavage by the solvent. A nucleophilic substitution reaction in which the nucleophile is a molecule of the solvent.

sp **orbital** (Section 1.15): A hybrid orbital that is derived by mathematically combining one *s* atomic orbital and one *p* atomic orbital. Two *sp* hybrid orbitals are obtained by this process, and they are oriented in opposite directions with an angle of 180° between them.

*sp*² **orbital** (Section 1.14): A hybrid orbital that is derived by mathematically combining one *s* atomic orbital and two *p* atomic orbitals. Three *sp*² hybrid orbitals are obtained by this process, and they are directed toward the corners of a equilateral triangle with angles of 120° between them.

*sp*³ **orbital** (Section 1.13): A hybrid orbital that is derived by mathematically combining one *s* atomic orbital and three *p* atomic orbitals. Four *sp*³ hybrid orbitals are obtained by this process, and they are directed toward the corners of a regular tetrahedron with angles of 109.5° between them.

Specific rotation (Section 5.6C): A physical constant calculated from the observed rotation of a compound using the following equation:

$$[\alpha]_D = \frac{\alpha}{c \times l}$$

where α_D is the observed rotation using the D line of a sodium lamp, c is the concentration of the solution in grams per milliliter or the density of a neat liquid in g mL⁻¹, and *l* is the length of the tube in decimeters.

Spin decoupling (Section 13.10): An effect that causes spin–spin splitting not to be observed in NMR spectra.

Spin–spin splitting (Section 13.9): An effect observed in NMR spectra. Spin–spin splittings result in a signal appearing as a multiplet (i.e., doublet, triplet, quartet, etc.) and are caused by magnetic couplings of the nucleus being observed with nuclei of nearby atoms.

Staggered conformation (Section 4.6): A temporary orientation of groups around two atoms joined by a single bond such that the bonds of the back atom exactly bisect the angles formed by the bonds of the front atom in a Newman projection formula:

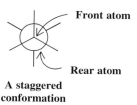

A staggered
conformation

Stereocenter (Section 5.2): An atom bearing groups of such nature that an interchange of any two groups will produce a stereoisomer.

Stereochemistry (Section 5.4): Chemical studies that take into account the spatial aspects of molecules.

Stereoisomers (Sections 2.4B, 5.1, and 5.2): Compounds with the same molecular formula that differ *only* in the arrangement of their atoms in space. Stereoisomers have the same connectivity and, therefore, are not constitutional isomers. Stereoisomers are classified further as being either enantiomers or diastereomers.

Stereospecific reaction (Section 8.7A): A reaction in which a particular stereoisomeric form of the reactant reacts in such a way that it leads to a specific stereoisomeric form of the product.

Steric effect (Section 6.15A): An effect on relative reaction rates caused by the space-filling properties of those parts of a molecule attached at or near the reacting site.

Steric hindrance (Section 6.15A): An effect on relative reaction rates caused when the spatial arrangement of atoms or groups at or near the reacting site hinders or retards a reaction.

Steroid (Section 23.4): Steroids are lipids that are derived from the following perhydrocyclopentanophenanthrene ring system:

Structural formula (Sections 1.2B and 1.20): A formula that shows how the atoms of a molecule are attached to each other.

Substituent effect (Sections 3.7B and 15.10): An effect on the rate of reaction (or on the equilibrium constant) caused by the replacement of a hydrogen atom by another atom or group. Substituent effects include those effects caused by the size of the atom or group, called steric effects, and those effects caused by the ability of the group to release or withdraw electrons, called electronic effects. Electronic effects are further classified as being inductive effects or resonance effects.

Substitution reaction (Sections 6.3 and 9.3): A reaction in which one group replaces another in a molecule.

Substitutive nomenclature (Section 4.3F): A system for naming compounds in which each atom or group, called a substituent, is cited as a prefix or suffix to a parent compound. In the IUPAC system only one group may be cited as a suffix. Locants (usually numbers) are used to tell where the group occurs.

Substrate (Section 6.3): The molecule or ion that undergoes reaction.

Sugar (Section 21.1A): A carbohydrate.

Syn addition (Section 7.6A): An addition that places both parts of the adding reagent on the same face of the reactant.

Syndiotactic polymer (Special Topic C): A polymer in which the configuration at the stereocenters along the chain alternate regularly: (R), (S), (R), (S), etc.

Synthon (Section 8.16): The fragments that result (on paper) from the disconnection of a bond. The actual reagent that will, in a synthetic step, provide the synthon is called the *synthetic equivalent.*

Tautomers (Section 17.2): Constitutional isomers that are easily interconverted. Keto and enol tautomers, for example, are rapidly interconverted in the presence of acids and bases.

Terpene (Section 23.3): Terpenes are lipids that have a structure that can be derived on paper by linking isoprene units.

Tertiary carbon (Section 2.10): A carbon atom that has three other carbon atoms attached to it.

Tertiary structure (Section 24.8B): The three-dimensional shape of a protein that arises from folding of its polypeptide chains superimposed on its α helixes and pleated sheets.

Thermodynamic control (Sections 6.6 and 12.9A): A principle stating that the ratio of products of a reaction that reaches equilibrium is determined by the relative stabilities of the products (as measured by their standard free energies, $\Delta G°$). The most abundant product will be the one that is the most stable.

Torsional barrier (Section 4.6): The barrier to rotation of groups joined by a single bond caused by repulsions between the aligned electron pairs in the eclipsed form.

Torsional strain (Section 4.9): The strain associated with an eclipsed conformation of a molecule; it is caused by repulsions between the aligned electron pairs of the eclipsed bonds.

Tosylate (Section 10.10): A *p*-toluenesulfonate ester.

Transition state (Sections 6.8 and 6.9): A state on a potential energy diagram corresponding to an energy maximum (i.e., characterized by having higher potential energy than immediately adjacent states). The term transition state is also used to refer to the species that occurs at this state of maximum potential energy; another term used for this species is *the activated complex.*

Unimolecular reaction (Section 6.11): A reaction whose rate-determining step involves only one species.

Unsaturated compound (Section 2.2): A compound that contains multiple bonds.

van der Waals force (or **London force**) (Section 2.16D and 4.7): Weak forces that act between nonpolar molecules or between parts of the same molecule. Bringing two groups (or molecules) together first results in an attractive force between them because a temporary unsymmetrical distribution of electrons in one group induces an opposite polarity in the other. When groups are brought closer than their *van der Waals* *radii,* the force between them becomes repulsive because their electron clouds begin to interpenetrate each other.

Vicinal (*vic*) substituents (Section 7.16): Substituents that are on adjacent atoms.

Vinyl group (Section 8.2): The $CH_2\!\!=\!\!CH\!\!-\!\!$ group.

Vinylic halide (Section 6.1): An organic halide in which the halogen atom is attached to a carbon atom of a double bond.

Vinylic substituent (Section 6.1): Refers to a substituent on a carbon atom that participates in a carbon–carbon double bond.

Visible–ultraviolet (visible–UV) spectroscopy (Section 13.2): A type of optical spectroscopy that measures the absorption of light in the visible and ultraviolet regions of the spectrum. Visible–UV spectra primarily provide structural information about the kind and extent of conjugation of multiple bonds in the compound being analyzed.

Wave function (or Ψ **function**) (Section 1.10): A mathematical expression derived from *quantum mechanics* corresponding to an energy state for an electron, i.e., for an orbital. The square of the Ψ function, Ψ^2, gives the probability of finding the electron in a particular place in space.

Wavelength (abbreviated λ) (Section 13.1): The distance between consecutive crests (or troughs) of a wave.

Wavenumber (Section 13.1): A way to express the frequency of a wave. The wavenumber is the number of waves per centimeter, expressed as cm^{-1}.

Ylide (Section 16.10): An electrically neutral molecule that has a negative carbon with an unshared electron pair adjacent to a positive heteroatom.

Zaitsev's rule (Section 7.12A): A rule stating that an elimination will give as the major product the most stable alkene (i.e., the alkene with the most highly substituted double bond).

Zwitterion (see **dipolar ion**): Another name for a dipolar ion.

INDEX

Periodic Table of the Elements

Periods

Group

Key

atomic mass →	**12.011**
electronegativity →	**2.5**
	[He]$2s^2 2p^2$ ← electronic configuration
symbol →	**C** atomic number
name →	**Carbon** 6

I A

									VII

1

1.0079
2.2
1s
H 1
Hydrogen

II A

2

6.941	9.01218
1.0	1.5
[He]$2s$	[He]$2s^2$
Li 3	**Be** 4
Lithium	Beryllium

3

22.98977	24.305
1.0	1.2
[Ne]$3s$	[Ne]$3s^2$
Na 11	**Mg** 12
Sodium	Magnesium

III B	IV B	V B	VI B	VII B			

4

39.0983	40.08	44.9559	47.88	50.9415	51.996	54.9380	55.847	58.9332
0.9	1.0	1.2	1.3	1.5	1.6	1.6	1.6	1.7
[Ar]$4s$	[Ar]$4s^2$	[Ar]$3d4s^2$	[Ar]$3d^24s^2$	[Ar]$3d^34s^2$	[Ar]$3d^54s$	[Ar]$3d^54s^2$	[Ar]$3d^64s^2$	[Ar]$3d^74s^2$
K 19	**Ca** 20	**Sc** 21	**Ti** 22	**V** 23	**Cr** 24	**Mn** 25	**Fe** 26	**Co** 27
Potassium	Calcium	Scandium	Titanium	Vanadium	Chromium	Manganese	Iron	Cobalt

5

85.4678	87.62	88.9059	91.22	92.9064	95.94	98.906	101.07	102.9055
0.9	1.0	1.1	1.2	1.2	1.3	1.4	1.4	1.5
[Kr]$5s$	[Kr]$5s^2$	[Kr]$4d5s^2$	[Kr]$4d^25s^2$	[Kr]$4d^45s$	[Kr]$4d^55s$	[Kr]$4d^65s$	[Kr]$4d^75s$	[Kr]$4d^85s$
Rb 37	**Sr** 38	**Y** 39	**Zr** 40	**Nb** 41	**Mo** 42	**Tc** 43	**Ru** 44	**Rh** 45
Rubidium	Strontium	Yttrium	Zirconium	Niobium	Molybdenum	Technetium	Ruthenium	Rhodium

6

132.9054	137.33	138.9055	178.49	180.9479	183.85	186.207	190.2	192.22
0.9	1.0	1.1	1.2	1.3	1.4	1.5	1.5	1.6
[Xe]$6s$	[Xe]$6s^2$	[Xe]$5d6s^2$	[Xe]$4f^{14}5d^26s^2$	[Xe]$4f^{14}5d^36s^2$	[Xe]$4f^{14}5d^46s^2$	[Xe]$4f^{14}5d^56s^2$	[Xe]$4f^{14}5d^66s^2$	[Xe]$4f^{14}5d^76s^2$
Cs 55	**Ba** 56	***La** 57	**Hf** 72	**Ta** 73	**W** 74	**Re** 75	**Os** 76	**Ir** 77
Cesium	Barium	Lanthanum	Hafnium	Tantalum	Tungsten	Rhenium	Osmium	Iridium

7

(223)	226.0254	227.0278	(261)	(262)	(263)
0.9	1.0	1.0			
[Rn]$7s$	[Rn]$7s^2$	[Rn]$6d7s^2$	[Rn]$5f^{14}6d^27s^2$	[Rn]$5f^{14}6d^37s^2$	[Rn]$5f^{14}6d^47s^2$
Fr 87	**Ra** 88	**†Ac** 89	**Unq** 104	**Unp** 105	**Unh** 106
Francium	Radium	Actinium	Unnilquadium	Unnilpentium	Unnilhexium

* Lanthanides

6

140.12	140.9077	144.24	145	150.4	151.96	157.25
1.1	1.1	1.1	1.1	1.1	1.0	1.1
[Xe]$4f^26s^2$	[Xe]$4f^36s^2$	[Xe]$4f^46s^2$	[Xe]$4f^56s^2$	[Xe]$4f^66s^2$	[Xe]$4f^76s^2$	[Xe]$4f^75d6s^2$
Ce 58	**Pr** 59	**Nd** 60	**Pm** 61	**Sm** 62	**Eu** 63	**Gd** 64
Cerium	Praseodymium	Neodymium	Promethium	Samarium	Europium	Gadolinium

...nides

7

232.0381	231.0359	238.029	237.0482	(244)	(243)	(247)
1.1	1.1	1.2	1.2	1.2	1.2	≈1.2
[Rn]$6d^27s^2$	[Rn]$5f^26d7s^2$	[Rn]$5f^36d7s^2$	[Rn]$5f^46d7s^2$	[Rn]$5f^67s^2$	[Rn]$5f^77s^2$	[Rn]$5f^76d7s^2$
Th 90	**Pa** 91	**U** 92	**Np** 93	**Pu** 94	**Am** 95	**Cm** 96
Thorium	Protactinium	Uranium	Neptunium	Plutonium	Americium	Curium

Approximate proton chemical shifts

TYPE OF PROTON	CHEMICAL SHIFT, δ (ppm)
1° Alkyl, RCH_3	0.8–1.0
2° Alkyl, RCH_2R	1.2–1.4
3° Alkyl, R_3CH	1.4–1.7
Allylic, $R_2C=C-CH_3$ R	1.6–1.9
Benzylic, $ArCH_3$	2.2–2.5
Alkyl chloride, RCH_2Cl	3.6–3.8
Alkyl bromide, RCH_2Br	3.4–3.6
Alkyl iodide, RCH_2I	3.1–3.3
Ether, $ROCH_2R$	3.3–3.9
Alcohol, $HOCH_2R$	3.3–4.0

Ketone, $RCCH_3$
         ‖
         O
	2.1–2.6

Aldehyde, RCH
            ‖
            O
	9.5–9.6

Vinylic, $R_2C=CH_2$	4.6–5.0
Vinylic, $R_2C=CH$ R	5.2–5.7
Aromatic, ArH	6.0–9.5
Acetylenic, $RC≡CH$	2.5–3.1
Alcohol hydroxyl, ROH	0.5–6.0[a]

Carboxylic, $RCOH$
           ‖
           O
	10–13[a]

Phenolic, $ArOH$	4.5–7.7[a]
Amino, $R-NH_2$	1.0–5.0[a]

[a]The chemical shifts of these protons vary in different solvents and with temperature and concentration.

Approximate carbon-13 chemical shifts

TYPE OF CARBON ATOM	CHEMICAL SHIFT, δ (ppm)
1° Alkyl, RCH_3	0–40
2° Alkyl, RCH_2R	10–50
3° Alkyl, $RCHR_2$	15–50
Alkyl halide or amine, $-\overset{\|}{\underset{\|}{C}}-X \left(X=Cl, Br, \text{ or } N- \right)$	10–65
Alcohol or ether, $-\overset{\|}{\underset{\|}{C}}-O$	50–90
Alkyne, $-C≡$	60–90
Alkene, $>C=$	100–170
Aryl,	100–170
Nitriles, $-C≡N$	120–130
Amides, $-\overset{O}{\overset{‖}{C}}-N-$	150–180
Carboxylic acids, esters, $-\overset{O}{\overset{‖}{C}}-O$	160–185
Aldehydes, ketones, $-\overset{O}{\overset{‖}{C}}-$	182–215